**Cotton tree** (*Ceiba pentandra* (Linn.) Gaertn., Bombacaceae) in the centre of Freetown, Sierra Leone. [Photo: T. L. Green, January, 1979]

A Revision of
Dalziel's

# THE USEFUL PLANTS OF WEST TROPICAL AFRICA

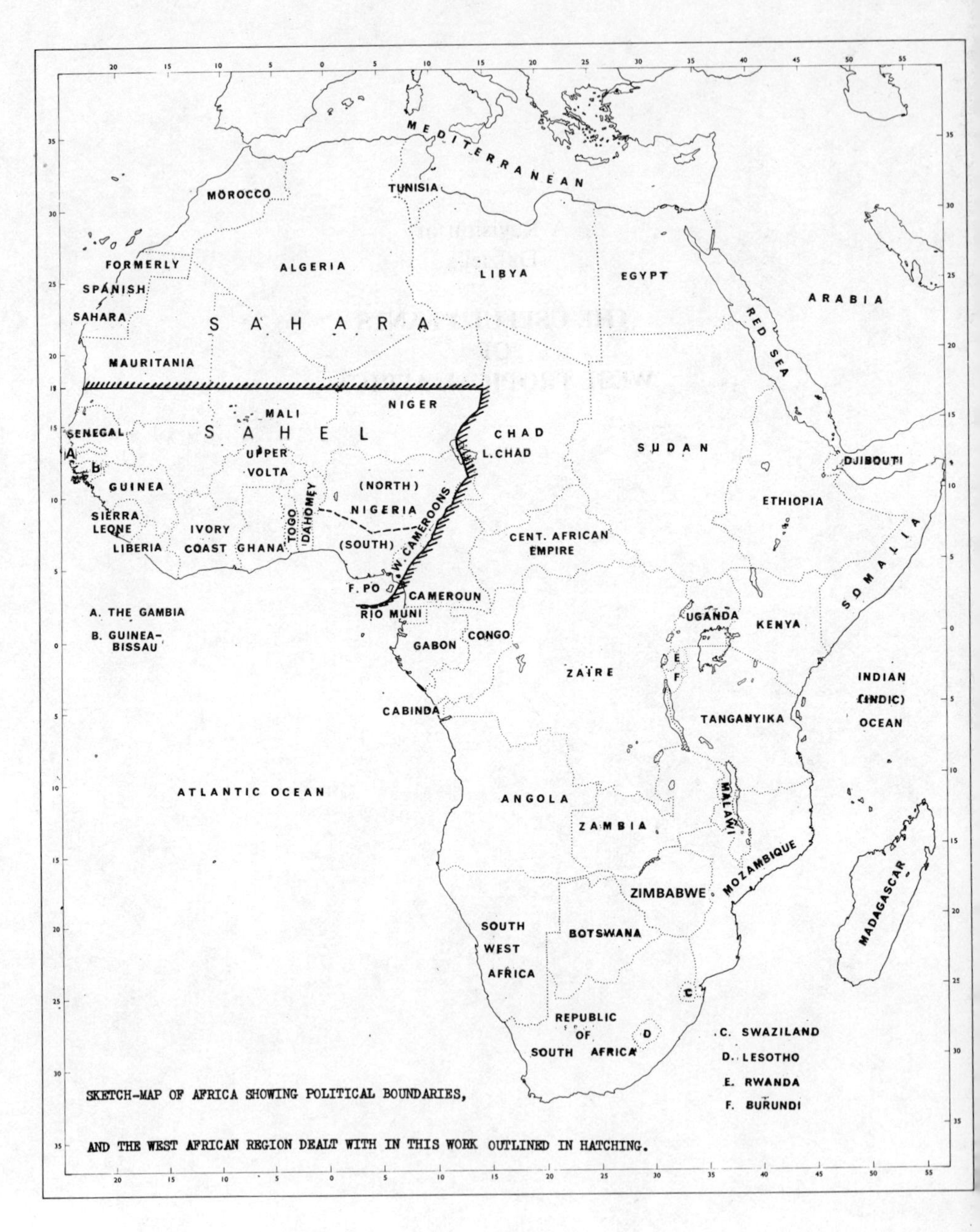


Sketch-map of Africa showing political boundaries, and the West African region dealt with in this work outlined in hatching.

# THE USEFUL PLANTS OF WEST TROPICAL AFRICA

Edition 2

H. M. BURKILL

VOL. 1
Families A–D

ROYAL BOTANIC GARDENS
KEW
1985

First Edition (1937) by J. M. Dalziel

This revision is a Supplement to the
Second Edition (1954–1972) of
*The Flora of West Tropical Africa*
edited by R. W. J. Keay and F. N. Hepper
and was compiled by
H. M. Burkill, O.B.E., M.A., F.L.S., M.I.Biol.
lately of H.M. Overseas Civil Service


First published 1985


ISBN 0 947643 01 X

Printed in Great Britain by
The Whitefriars Press Limited,
Medway Wharf Road,
Tonbridge, Kent TN9 1QR
for the Publishers
Royal Botanic Gardens, Kew,
Richmond, Surrey TW8 3AE

**The Bible,** Genesis 1

11. And God said, Let the earth bring forth grass, the herb yielding seed, and the fruit tree yielding fruit after his kind, whose seed is in itself, upon the earth: and it was so.
12. And the earth brought forth grass, and herb yielding seed after his kind, and the tree yielding fruit, whose seed was in itself, after his kind: and God saw that it was good.

**The Koran,** Surah 13

3. And He it is who hath outstretched the earth, and placed on it the firm mountains, and rivers; and of every fruit He hath placed on it two kinds: He causeth the night to enshroud the day. Verily in this are signs for those who reflect.
4. And on the earth hard by each other are its various portions: gardens of grapes and corn, and palm trees single or clustered. Though watered by the same water, yet some make we more excellent as food than other: Verily in all this are signs for those who understand.

## TABLE OF CONTENTS

# Preface

## "THE USEFUL PLANTS OF WEST TROPICAL AFRICA"

The first edition of "The Useful Plants of West Tropical Africa" by the late Dr. J. Hutchinson and the late Dr. J. M. Dalziel was published in 1937, with a preface by Sir Arthur Hill, the then Director of the Royal Botanic Gardens, Kew. As one of his successors in the post I am happy to be able to fulfil a similar function for the second edition.

The first edition, although a substandial volume in its own right, was modestly described as an "Appendix" to the Flora of West Tropical Africa, which had been completed the previous year. The preparation and completion of the second edition of the Flora of West Tropical Africa during the years 1954–1972 naturally aroused the feeling that there would be a good case for a parallel revised edition to "The Useful Plants of West Tropical Africa".

During the period since 1936 great changes have taken place in our knowledge of the economic uses and vernacular names of the plants of the Region. In the first place the number of plant species known from the area of the Flora has increased very greatly from 5,609 in the first edition to, 7,349 in the second. In the second place it was clear that the body of knowledge about the economic uses and vernacular names had itself increased proportionately. Some preliminary work in assessing and digesting this additional knowledge was carried out by Mr. T. A. Russell, B.Sc., A.R.C.S., A.I.C.T.A. while he was Principal Scientific Officer in the then Department of Economic Botany at Kew. However, time did not permit him to make more than the slightest impression on the formidable task of up-dating. It was thus fortunate that Mr. Humphrey Burkill, O.B.E., after his retirement from the Directorship of the Botanic Gardens at Singapore, accepted the task and, in spite of many difficulties and delays, has persevered indomitably.

The increase in the size of the work is to be gauged by the fact that the first edition was a single volume of 612 pages, while the second will be at least three times as large. The present is but the first of four projected volumes.

The importance and urgency of putting on record the economic uses of the plants of West Tropical Africa has increased rather than decreased over the years. Countries now independent and no longer colonies often have a greater need than previously to develop the potential of their economically important plants in the interests of prosperity. Plants of minor commercial significance against a world background may be locally of major importance in the areas where they are exploited. The drier parts of western Africa have shown themselves especially vulnerable to extremes of climate—the recent years of drought have left a bitter memory. The basic needs of food for man, forage for animals, timber for building and firewood for heating and cooking may become of greater immediate importance even than cash crops.

Knowledge of the uses of plants, though sometimes jealously guarded by their owners, formerly seemed reasonably secure from generation to generation. The changes imposed by modern life on social structures and attitudes now seem too often to cause the loss or rejection of such local knowledge. This may occur quite rapidly and result in the loss of information and skill often carefully and painfully acquired by processes of trial and error carried out over the centuries. There is thus an urgency to put this knowledge on record before it may be too late.

Many bodies and people have helped in the long process of compilation and in many ways. It is a pleasure to put on record, on behalf of the Royal Botanic Gardens, Kew, my gratitude and appreciation to them.

Royal Botanic Gardens, Kew, E. A. BELL
*December, 1984.*

# Introduction

This work, *A Revision of Dalziel's Useful Plants of West Tropical Africa* is, as was the original work, being prepared as a supplement to the *Flora of West Tropical Africa*, the former (1937) to the first edition of the Flora (1927–36), the present one to the second edition (1954–72). Dalziel's work dealt with a few species over 2,000 and the text was confined to one volume. The present work follows the format established in the first edition, and attention has been paid mainly to information published since 1937, the date of the first edition. The revised edition is expected to deal with over 4,600 species, and early in the preparation it was decided on the basis of the considerably greater number of species and the enlarged amount of information that four volumes would be necessary. Though this introduction does refer to the whole work, it is written for the first volume before volumes 2, 3 and 4 have advanced beyond an early stage.

**Taxonomic treatment.** In general the treatment of species is *sensu FWTA ed. 2*. However, subsequent to the beginning of the publication of *FWTA, ed. 2* much taxonomic work has been done on the flora of other parts of Africa and of the tropics generally, and revisions have often brought the status of the West African flora into the context of a much wider taxonomic treatment. Where such revisions have introduced changes which have been considered acceptable these have been followed. Thus, for example, Agavaceae (*sensu FWTA ed. 2*) now excludes *Cordyline, Dracaena* and *Sanseveria,* transferred to Liliaceae; Ficoidaceae and Molluginaceae of *FWTA, ed. 2,* appear under Aizoaceae; Rosaceae is split into Rosaceae (*sensu stricto*) and Chrysobalanaceae; Caesalpiniaceae, Mimosaceae and Papilionaceae will appear as subfamilies of Leguminosae, etc. Changes within families and within genera, and also within species, have also resulted from taxonomic revisions principally from floras for East Tropical Africa, Gabon, Cameroun, Zaïre, Central Africa, Angola, etc. For example, of the 26 genera of Cucurbitaceae listed in *FWTA ed. 2,* six now have to be known by other names, and of their 64 species listed, 38 have new binomials. Not necessarily all such changes affect the present work since only "useful" plants are included in it, but where there has been a change of name to any of them, the Flora name (and also that of *UPWTA ed. 1*) is given as synonymy beneath the presently accepted name. Some 15 per cent of the approximate 1,500 species in Volume 1 of this work have names other than in the *Flora*. The order of presentation is alphabetically: (1) by families, (2) by genera within families, and (3) species within genera. The present volume, No. 1, consists of the plants of Families A to D. A provisional distribution for subsequent volumes is: Volume 2, Families E to L, approximate estimate 1470 species; Volume 3, Families M to S, 1100 species; and Volume 4, Families T to Z, 600 species, plus Cryptogams of numbers not yet quantified, and Indexes.

**What is "useful"?** Semantically it is an anthropinism, i.e. a consideration of things in their relation to man, and in this work the term *useful* is intended to reflect some measure of the ethnological status of plants. It is interpreted in its widest sense; that is, usefulness is both positive and negative in value. If a plant is positively useful, it has some benign attributes: it can be eaten, turned into healing medicine, made into something to man's advantage. This is all well and good. There is no argument as to its utility. On the other hand some plants have malign attributes: they can cause distress and death. Such may be an ambivalent negative usefulness—one man's disadvantage and another's advantage, as for example in arrow-poisons, ordeal-poisons, etc., and by extension of the sophistry there are plants so dangerous to humans as to be studiously shunned, avoided at all costs and to be used for nothing at all. Such are at the limit of negative usefulness, and the fact of their utter malignity merits record and gains them also a place in this work. Between the two extremes of good and bad all degrees occur. It is plants at the neutral point that have been missed out for such appear to have neither positive nor negative usefulness, or it is that such has yet to be disclosed. Somewhere along the scale lie plants that enter religion and folk-lore, and have superstitious, fetishistic, magical and plain mumbo-jumbo uses. These too are included for such uses imply usefulness towards man's well-being in a spiritual sense, or by way of the charlatan's stock-in-trade demonstrate gullibility of the credulous.

**Geographical area.** The area of *FWTA ed. 2* is that part of Western Africa lying south of latitude 18°N from the Atlantic seaboard to the eastern limit of Niger and Nigeria through Lake Chad, along the eastern side of the former mandated territory of British Cameroons to the Bight of Benin and including Fernando Po. This is also the relevant area of the present work and is referred to as "the Region". The southernmost part of the ex-British Cameroons is now a part of the Cameroun Republic, and is specified in this work as West (or W) Cameroons. The balance of the Cameroun Republic is distinguished as East (or E) Cameroun and lies outside the Region. The accompanying map shows the regional political divisions.

The plants of the Region have been looked at in a wider context than regional. Uses in other areas have been recorded as, though there may be no known usage in the Region, use elsewhere indicates a potential. Further investigation may be called for to find out whether that potential exists in the Region, for it may be suppressed or enhanced by climate, edaphic and varietal factors, as for example, the production of fibre, narcotics and seed oil from *Cannabis sativa* (Cannabaceae). Many similar variations in potential will be found in the text.

Political territorial names have been followed whenever practical. The use of the name Benin for the territory formerly known as Dahomey has not been accepted in order to avoid confusion with the large province of Benin in SE Nigeria. In E Africa Tanzania has been identified in the text as Tanganyika when referring to the mainland, and Zanzibar (and Pemba) when the offshore islands are meant.

**Languages, orthography and vernacular names.** The presentation of vernacular names has proved very troublesome. The supply of names is legion. Acceptance of their reliability is often an act of good faith. Omission could, however, create an unnecessary gap in the Region's ethnobotany, for they may provide information on usage, origin, migration and so forth. For a number of species only a vernacular name is given without any detail of usage. Such plants are entered on the premise that the acquisition of a local name must imply that they have come to man's notice for some reason as yet undocumented, which may in time become recorded.

Two major difficulties have been met, if not wholly surmounted. Early collectors recorded the names of languages, often uncritically, often seemingly whimsically by nicknames, or even derogatory epithets used by neighbouring tribes, so that it has been necessary with the benefit of current linguistic research to sort out and to reduce the prolific assortment of language names to those currently accepted and so to render them by related groupings. To my aid have come Professor Kay Williamson of the School of Humanities, University of Port Harcourt, Dr Roger Blench, Research Graduate of Cambridge University and Mr Denis Winston of the School of Oriental and African Studies of the University of London.

The other difficulty has been in spelling of plant vernacular names. Most

African languages had till relatively recently no written form. A few under Islamic influence, e.g. Hausa, had a literature in Arabic script. Thus with modern education based on the Roman alphabet, the transcription of vernacular names has in recent times been virtually from scratch in all languages, and as between the work of English, French and Portuguese expatriates there has been the innate variance of sound values of their own native tongues. To this also collectors meeting but not recognising dialectal variation of emphasis on pronunciation, or failing to appreciate that a word spoken by an informant from an empty mouth sounds appreciably different from a mouth filled with a kola quid, have recorded the same word in many forms of transcription, as the lists in this book will show. This confusion will continue till languages acquire a stabilised form of writing. This is a task for the linguists and the educationalists, and is a process now being tackled.

For the purpose of this work the standard Roman alphabet is used to which certain extra cyphers have been added for particularly African sound values. The whole is shown in the following alpha-sequence:

| | | |
|---|---|---|
| A | a | |
| B | b | |
| 'B | ɓ | a glottalised *b* |
| Ḅ | ḅ | an implosive *b*—for Niger Delta languages |
| C | c | |
| D | d | |
| Đ | ɖ | a post-alveolar *d* |
| 'D | ɗ | a glottalised *d* |
| Ḍ | ḍ | an implosive *d*—for Niger Delta languages |
| E | e | |
| Ɛ | ɛ | an open *e* |
| Ə | ə | a central vowel—as at the end of English sod*a* |
| F | f | |
| Ƒ | ƒ | a bilabial *f* |
| G | g | |
| Ɣ | ɣ | a voiced velar fricative—*gh* |
| H | h | |
| I | i | |
| J | j | |
| K | k | |
| Ƙ | ƙ | an ejective *k* |
| L | l | |
| M | m | |
| N | n | |
| Ŋ | ŋ | *ng* as in English si*ng* |
| Ɲ | ɲ | *ny* |
| Ɔ | ɔ | a short vowel as in English n*o*t |
| Ọ | ọ | as the preceding in Nigerian languages |
| P | p | |
| Q | q | |
| R | r | |
| S | s | |
| Σ | ʃ | *sh* |
| T | t | |
| U | u | |
| V | v | |
| Ʋ | ʋ | a bilabial *v* |
| W | w | |
| X | x | phonetically as in Scottish lo*ch* |
| Y | y | |

Ƴ ƴ or 'y—a glottalised *y*
Z z
Ʒ ʒ a sibilant *zh*

Diacritical accents:

´ over a vowel: the standard acute accent—a rising tone in many languages
´ over d, k and t in Senegalese languages, hardens and shortens the letter
` over a vowel: the standard grave accent—a falling tone in many languages
̏ over a vowel: a double grave accent making a deeper falling tone, e.g. when following a syllable already under a falling tone
˘ over a vowel—shortens it
¯ over a vowel—lengthens it, but now commonly superseded by a doubling of the vowel to show length
under a vowel: low-level pitch
ˈ over a vowel: a mid-tone mark
ˆ over a vowel: the standard circumflex accent—a rising-falling tone mark in many languages
ˇ over a vowel: a falling-rising tone mark in Nigerian languages
ˇ over a consonant: in general producing a hardened aspiration, a fricative or sibilant
· over n: ng as in si*ng*
. under a vowel: makes it more open
. under a consonant in Isekiri and Yoruba indicates a sibilant sound, e.g. ṣ=sh
' apostrophe: in Hausa and Fula, a glottalisation of the following consonant, or when between two vowels, a glottal stop
¨ the German Umlaut: ä=ae; ö=ue
¨ diaeresis over one of two adjacent vowels—vowels sounded separately
~ Spanish tilde—a standard nasalisation. In Yoruba a tone mark or contraction of two syllables, but now outdated by the use of two vowels each with its own tone mark
, below a letter=the Continental cedilla

**Acknowledgements.** This work was financed from November 1969 to December 1977 by the Ministry of Overseas Development of the United Kingdom Government under grant R 2689. From April to December 1978, and again from January 1981 onwards, the Bentham-Moxon Trust has supported part-time clerical assistance, and the Research Awards Committee of the Leverhulme Trust Fund granted financial support over 1979–80 for the same purpose. These fundings have made possible the preparation of Volume 1 and the collection of notes towards succeeding volumes. Most of the work has been done at the Royal Botanic Gardens, Kew. I am indebted to the authorities of these bodies for the support and amenities given to me. I have to thank Sir George Taylor, who as Director of the Royal Botanic Gardens in 1969, put the work in my way, to Professor J. P. M. Brenan, Director, 1976–1981 and Mr P. S. Green, Deputy Director and Keeper of the Herbarium, 1976–82, to Professor E. A. Bell, thc present Director and Mr G. Ll. Lucas, O.B.E., the present Acting-Keeper of the Herbarium, who constantly showed their personal interest in the work, and to many members of the Kew Herbarium staff who have helped in various ways. I wish to express my gratitude to my wife who helped me to glean information from collectors' notes on sheets in the herbaria at Kew, Edinburgh, Ibadan, Ife, Legon, Kumasi, Fourah Bay, Njala and Dakar; and to the late Mrs Irene Scott, my clerical assistant from January 1973 to January 1984. Outside of Kew I have come into contact with numerous people in the course of this work and have received much kindly help. I must acknowledge in particular frequent assistance from Dr Joyce Lowe of Ibadan University and the late Mr John Hall, at one time of the University of Ghana. Information has been screened from many publications, and the sources are cited. In particular I acknowledge the doyens of West African phytochemistry and ethnobotany: Professor J. Kerharo, Professor of Pharmacognosy of the Faculty of Medicine of the University of Dakar, who showed my wife and myself much kindness and help during our visit to Dakar in 1973, and Dr J. G. Adam of the Natural History Museum, Paris, both of whose works have been freely drawn upon. The preparation of lists of vernacular plant names and the unstinting help (and encouragement when the going seemed particularly hopeless) from Professor K. Williamson, Dr R. Blench and Mr D. Winston have already been referred to in the preceding section. They themselves have checked some of the language lists, and through them or directly I have received help in checking languages of national territories or of certain specific languages as acknowledged herewith: THE GAMBIA—Mr J. Rooke, lately Department of Co-operatives; Mr R. J. McEwan, Director of Forestry; Mr G. Hallam, VSO, Department of Forests; Mr Sidia Jatta, School of Oriental and African Studies, London; Mr Amadou, Mr A. Dampha, Mr Y. Darbo and Mr S. Mauku of the Forest Department (Manding-Maninka); Mr O. Colley of the Forest Department, Mr Abba Sonko and Mr Lamin Sadhiou (Diola); Mr M. S. N'jie, Deputy Principal, Gambia College (Wolof): GUINEA BISSAU—Dr J. Paiva, University of Coimbra (Crioulo, Manding-Mandinka): SIERRA LEONE—Mr F. C. Deighton, lately Plant Pathologist, Department of Agriculture; Dr P. J. C. Harris, Mr. B. Turay, Dr B. Tegler and Mr J. L. Boboh of the University College of Njala; Professor G. Innes, School of Oriental and African studies (Mende): GHANA—Mr A. A. Enti (Akan languages, Ga, Grusi-Kasena, Nankanni): NIGERIA—Mr S. M. Adewole, School of Humanities, University of Port Harcourt (Yoruba); Mr G. B. Affia, Director of the Library, University of Port Harcourt (Anaang, Efik); Mr G. U. Akpakpan, Technician, University of Port Harcourt (Ibibio); Dr M. A. Amayo, Department of English, University of Port Harcourt (Edo); Dr R. Blench (Hausa, Igala and related languages, Kanuri, Nupe); Dr E. N. Emenanjo and students, Ikoku College of Education, Owerri (Igbo); Mr

Jackson Emwiogbon, Forest Research Institute, Ibadan (Edo); Mr N. Faraclas (Urhoho); Dr M. McIntosh, Afrikaanse Taalkunde, Leiden University (Fula, Hausa); Dr B. N. Okigbo, International Institute of Tropical Agriculture, Ibadan (Igbo); Mr J. Opayemi, School of Biological Sciences, University of Port Harcourt (Yoruba); Mr R. E. Sharland (Hausa); Mr S. A. Tilley (Tiv); Mr A. O. Timitimi (Ijo); Professor D. E. Vermeer, Department of Geography and Anthropology, Louisiana State University (Tiv); Professor K. Williamson (Anaang, Dera, Efik, Ekpeye, Gure-Kuhaga, Hausa, Hwana, Ibibio, Icen, Igbo, Ijo, Jera, Jukun, Karama); Mr F. D. D. Winston (Efik).

Funds for publication have come from depositions of money by the four West African Commonwealth Territories (Nigeria, Ghana, Sierra Leone and The Gambia) with the Crown Agents in London soon after the end of World War II in anticipation of this work being undertaken. Inflation made this sum quite inadequate even to publish Volume 1, but to it the Nigerian Government through the Forestry Institute of Nigeria, Ibadan, have made in 1978 a handsome addition making publication possible.

The Herbarium, Royal Botanic Gardens, Kew, H. M. BURKILL
*September, 1984.*

# ACANTHACEAE

**Acanthus guineensis** Heine & P. Taylor

FWTA, ed. 2, 2: 410.
West African: **SIERRA LEONE** TEMNE *a*-futɔ (NWT) **GHANA** AKAN-TWI nsu bibire *so called because water where it grows is seen to become black* (Plumtre)

An erect almost unbranched prickly semi-woody herb to nearly 2 m high, of the high-forest from Guinea to Ghana.

The plant is very similar in appearance to *A. montanus*. It is ornamental in flower.

In Sierra Leone the boiled fruits are used to loosen phlegm in children's coughs (1).

Reference:
1. Deighton 381, K.

**Acanthus montanus** (Nees) T. Anders.

FWTA, ed. 2, 2: 410. UPWTA, ed. 1, 449.
English: false thistle (Dalziel).
West African: **SIERRA LEONE** MENDE kpete-pela (FCD) kpete-wela (FCD) **NIGERIA** EDO àgámòbọ̀ (Dennett) IGBO (Asaba) ágàméèbù (NWT) IGBO (Awka) ágàméèbù (KW) IGBO (Onitsha) ágàméèbù, ágà: *needle* IGBO (Owerri) ágàméèbù (JMD) IJO-IZON (Kolokuma) ẹ̀dụ̀lẹ̀ ìmémèin = *leopard's claws* (KW) YORUBA ahọn ẹ̀kùn = *leopard's tongue* (auctt.) ahọ́n ẹ̀kùn dúdú = *leopard's black tongue* (Verger) èèkún-arúgbo = *old man's knee* (JMD)

A prickly semi-woody herb to nearly 2 m high, of the high-forest in Dahomey to W Cameroons and Fernando Po, and to the Congo Basin and Angola.

The plant in flower is ornamental with white or pinkish-white flowers green-veined, in a terminal spike 20–30 cm long.

The leaves are used [? in poultice] in S Nigeria for a sort of boil on the fingers called, *mleloṅko* (Igbo — Asaba) (6). Similarly in Congo the central portion of twigs or the leaves are applied as a hot poultice to maturate abcesses (2). The plant has various medicinal uses in Nigeria, chiefly as a cough-medicine, for women and children (3). Use against cough is recorded in Gabon (9) and Congo (2) either as a leaf-infusion or cooked with vegetable, and in Cameroun for cough (5) and chest-complaints (4). A decoction of leafy-twigs is taken in Congo as a purgative (2). A leaf-macerate is given to children in Gabon as an emetic and the fresh young growths are taken for heart-troubles (8, 9). The young shoots cooked with groundnuts or the kernel-butter of *Irvingia gabonensis* Baill. (Ixonanthaceae) are taken to settle upset-tummy and to counteract 'morning-sickness' in pregnant women (8, 9). The Ijo of S Nigeria take the pounded leaves cooked with pepper and salt to eat with fish for rheumatism: men also chew the stems (10). Diuretic action is claimed in Congo where the plant is pounded up with a stem of *Costus* and a young pineapple fruit and then soaked in palm-wine: this is held to be a good remedy for urethral discharge (2). A shoot-macerate enters into a Gabon treatment for syphilis (9), and the leaf-spines are used to make scarifications in treating an area of rheumatic pain which precedes yaws (8, 9).

Some alkaloids have been detected in the leaves of Nigerian material (1).

The Yoruba of Nigeria invoke the plant to rescue people attacked by witches (7). In Congo a few drops of sap from crushed leaves put on the eyebrows are believed to protect one from devils, and along with other plants a mixture is made for use in ceremonies of purification and exorcism (2).

References:

1. Adegoke & al. 1968. 2. Bouquet, 1969: 48. 3. Dalziel, 1937. 4. Endengle 2127, K. 5. Leeuwenberg 7342, K. 6. Thomas, N. W. 1665 (Nig. Ser.), K. 7. Verger, 1967: No. 24. 8. Walker, 1952: 183. 9. Walker & Sillans, 1961: 39. 10. Williamson, K. 32, UCI.

**Adhatoda buchholzii** (Lindau) S. Moore

FWTA, ed. 2, 2: 422. UPWTA, ed. 1, 449.

A semi-woody scandent herbaceous shrub to 1 m high, of the rain-forest zone in S Nigeria, and extending to Gabon.

In S Nigeria the leaves are pounded to a paste for external application in rheumatism (1). It is also used in Nigeria as a fish-poison, the whole plant being crushed and thrown into streams and rivers (3).

A related Indian species, *A. vasica* Nees, is well-known in traditional Indian Ayurvedic medicine for treatment of bronchitis, asthma, fever, jaundice and consumption. It is official in the Indian Pharmacopoea as an expectorant in respiratory affections and as a mild bronchial antiseptic. The plant contains a little essential oil and the alkaloid *vasicine,* up to 0.4% in the leaves and 0.35%in the roots, which produced prolonged bronchodilation. The plant has shown some antibiotic action against avian, bovine and human strains of *Mycobacterium tuberculosis,* but its efficacy in treatment of tuberculosis has been doubted. The plant can be used as a substitute for atropine (2, 4, 5). West African *Adhatoda spp.* merit examination.

References:

1. Ainslie, 1937: sp. no.14. 2. Bouquet & Debray, 1974: 11, under *A. vasica* Nees. 3. Dalziel, 1937. 4. Manjunath [Ed.], 1948: 31–32. 5. Oliver, 1960: 17, 42, 44.

**Adhatoda camerunensis** Heine

FWTA, ed. 2, 2: 422.

A shrub to 2 m high, of lowland rain-forest in W Cameroons and E Cameroun.

In E Cameroun the plant is used against cough (1).

Reference:
1. Leeuwenberg 7344, K.

**Adhatoda robusta** C. B. Cl.

FWTA, ed. 2, 2: 422.

A shrub or tree to 5 m high, in the undergrowth of primary and secondary rain-forest, from Ivory Coast to W Cameroons and Fernando Po, and extending to Gabon.

In S Nigeria the leaves are pounded to a paste for external application in rheumatism (1).

Reference:
1. Ainslie, 1937: sp. no. 14.

**Adhatoda tristis** Nees

FWTA, ed. 2, 2: 423.

A herbaceous undershrub to about 1 m high, of the rain-forest in S Nigeria and Fernando Po, and in Cameroun.

In S Nigeria the leaves are pounded to a paste for external application in rheumatism (1).

Reference:

1. Ainslie, 1937: sp. no. 14.

**Asystasia calycina** Benth.

FWTA, ed. 2, 2: 413. UPWTA, ed. 1, 449.
West African: **SIERRA LEONE** MANDING-MANINKA (Koranko) laŋbaŋgunɛ (NWT) MENDE legbule (*def.*-i) (FCD; JMD) TEMNE *a*-puth (NWT) *a*-tila (NWT) **LIBERIA** MANO namu (Har.) **IVORY COAST** BAULE borokwé (B&D) pinbéigna (B&D) 'NEKEDIE' borowka titi (B&D) naia naia roba (B&D) nayu anayu roba (B&D) **NIGERIA** IGBO (Agukwu) nci ogi, nci-oji (NWT)

A herb, erect or straggling to 60 cm high, recorded from Guinea to S Nigeria, and occurring also in E Cameroun.

The leaves are eaten as greens in Gabon (4). The plant is used for headache in S Nigeria (3), and in Sierra Leone the leaves are beaten, lime-juice is added, and the whole warmed for application to large craw-craw sores and children's yaws (2). In Ivory Coast a decoction of the plant is also used for yaws, and the twigs are said by the Shien to be aphrodisiac (1).

References:

1. Bouquet & Debray, 1974: 11. 2. Deighton 2342, K. 3. Thomas, N. W. 1689 (Nig. Ser.), K. 4. Walker & Sillans, 1961: 39–40.

**Asystasia gangetica** (Linn.) T. Anders.

FWTA, ed. 2, 2: 413. UPWTA, ed. 1, 449.
West African: **SENEGAL** DIOLA (Oussouye) hédul (K&A) DIOLA-FLUP hédul (K&A) WOLOF omigalin (JLT; K&A) umegarin (JLT; K&A) **THE GAMBIA** MANDING-MANDINKA namala = *slippery vine* (Fox; DF) **SIERRA LEONE** KISSI kɔlɔ (FCD) LOKO kɛrɛge (NWT) pɛrɛge (NWT) MANDING-MANINKA (Koranko) tɛgɔlo (NWT; JMD) MENDE legbule (*def.*-i) (FCD) lelegbule (FCD) TEMNE *a*-puth (NWT; FCD) **LIBERIA** MANO namu (JMD) **GHANA** AKAN-TWI fiangoro (Ll.W) ɔbɔfo fan = *hunters spinach* (FRI) DAGBANI n'chendua (Ll.W) **NIGERIA** EDO èbóghògíró (Ross) òvbiàkpẹ̀ (Kennedy) EFIK ọmiọ̀n̈ (AJC; Lowe) IBIBIO m̀mèmé (Okpon; Kaufman) IGBO ékèrè (AJC) IGBO (Agukwu) ùlì ókō (NWT) IGBO (Owerri) ìkéré (Lowe) IJO-IZON (Kolokuma) bèrìbà ịgbá = *plantain creeper* (KW) ọ̀kụ́rọ́, ọ̀bìrì òwéí (KW) YORUBA lobiri (Millson)

An annual or perennial semi-woody herb to 1 m high, flowers multi-coloured, distributed throughout the Region, and occurring widespread through tropical Africa and the tropics. The binomial *A. gangetica* covers a very complex, polymorphic aggregate, and it is possible that the West African material may be separable. The group is considered here as a whole.

The leaves are edible though consumption appears to be occasional rather than general (4; Nigeria, 10; Kenya, 6, 8; Philippines 11). The Ghanaian Twi name meaning 'hunter's spinach' suggests dietary improvisation while on the chase. In Kenya the plant is browsed by ungulates (8) and in Travancore, India, it is cut as cattle-fodder (9). In The Gambia the plant is used as a soap-substitute. It froths in water (5) suggesting the presence of saponins. In Ghana-/Togo region the plant is boiled and the infusion mixed with peppers is used for syringing [sic: ? vaginal douche] during the later months of pregnancy to ease childbirth pains, and the infusion is also drunk for the same purpose: such use is widespread (18). In Congo leaf-sap is placed on the stomach of women in childbirth to facilitate labour (2). In S Nigeria Igbo use the plant for bad feet (? sores) (15) and for sores (16), and it is applied by the Efik to the wound of the lobe of the ear after piercing (Carpenter fide 4). Sap of the leaf is put up the nostrils in Congo for nose-bleeding, and is used as an embrocation for stiffneck and for enlarged spleen in infants; the pulped leaf is used as a suppository for piles (2).

The plant roots are powdered and given in Tanganyika to a Sukuma child with a swollen tummy before suckling (13) and the Chagga rub a mixture of powdered root and the sap of a young banana into scarifications for enlarged spleen in a new-born baby (Bally fide 17). The plant appears to have some analgesic property: the powdered roots are a general remedy for stomach-pains in Tanganyika (14); in Congo a leaf-decoction is taken for fever-aches, epilepsy, stomach-pains, heart-pains and urethral discharge (2); and in India the sap is given to children with swellings and rheumatism (9) and as a vermifuge (3). The root is used as an emetic in treating snake-bite in Casamance (Senegal) (7), and the leaves (without explained function) are considered antidotal in snake-bite in Tanganyika (1). In Gabon the plant is applied as an anti-pruritic (12).

Leaf and flower are reported negative in tests for haemolysis, flavonols, alkaloids and tannins, and strongly positive for sterols, while other tests have shown a trace of alkaloids, and an abundance of potassium (11, 17).

In NE Ghana the plant is given as a juju to babies to make them fearless (19). In Congo it is considered a fetish protective for children especially against illness (2).

References:

1. Bally, 1937. 2. Bouquet, 1969: 48. 3. Chopra, 1933. 4. Dalziel, 1937. 5. Fox 135, K. 6. Jeffrey K.33, K. 7. Kerharo & Adam, 1963, a. 8. Magogo & Glover 423, K. 9. Manjunath [Ed.], 1948: 134. 10. Okpon s.n. (UIH. 3455), UCI. 11. Quisumbing, 1951: 884. 12. Sillans, 1953, a: 364. 13. Tanner 1375, K. 14. Tanner 5895, K. 15. Thomas, N. W. 1613 (Nig. Ser.), K. 16. Thomas, N. W. 1645 (Nig. Ser.), K. 17. Watt & Brejer-Brandwijk, 1962: 1–2. 18. Williams, L1. 50, K. 19. Williams, L1. 168, K.

## **Asystasia scandens** (Lindley) Hook.

FWTA, ed. 2, 2: 412.

English: climbing asystasia (Hooker, *Bot. Mag.*).
West African: **SIERRA LEONE** MENDE legbule (after L-P)

A scendent undershrub of the forest in Guinea to Liberia.

The flowers are cream-coloured, tinged purplish or bluish, 5 cm long, in terminal racemes, decorative and making the plant worthy of cultivation.

## **Barleria brownii** S. Moore

FWTA, ed. 2, 2: 426.

A straggling shrub or liane to 3 m high, recorded only in S Nigeria in the Region, but occurring across the Congo basin to Uganda and Angola.

In Gabon the young leaves are eaten (1).

Reference:

1. Sillans, 1953, b: 83 as *B. talbotii* S. Moore.

## **Barleria cristata** Linn.

FWTA, ed. 2, 2: 421. UPWTA, ed. 1.

English: blue bell (Howes); Philippine violet (U.S.A., Bates).

A bush to about 1.30 m high, native of India, Burma and China, and dispersed by man to many warm countries.

It is cultivated here and there in West Africa as an ornamental. The flowers are blue or white. It stands clipping and can be trained into an attractive and tidy low hedge (2). Propagation by cuttings in the rainy season is easy.

Leaves and roots are used for cough and inflammation in India (3), and root, leaf and twigs in an oil-base have been used for abdominal tumours (1). The seeds are supposed to be antidotal in snake-bite (4).

References:

1. Hartwell, 1967. 2. Howes, 1946. 3. Manjunath [Ed.], 1948: 158. 4. Quisumbing, 1951: 884–5.

**Barleria eranthemoides** R. Br.

FWTA, ed. 2, 2: 421.

A spiny shrub to 75 cm high of savanna in N Nigeria and W Cameroons, and extending across Africa to NE and E Africa and Mozambique.

In Kenya it provides some grazing for goats (1). The flowers are nectariferous, and are often sucked by Ethiopian children to get the honey (2).

References:

1. Glover & Samuel 3263, K. 2. Turton 13, K.

**Barleria oenotheroides** Dum. Cours.

FWTA, ed. 2, 2: 420.

English: yellow barleria (Hooker, *Bot. Mag.).*

An erect woody shrub with terminal spikes of many densely-packed yellow flowers, from Mali to S Nigeria.

The plant in flower is decorative and is sometimes cultivated as an ornamental.

**Barleria opaca** (Vahl) Nees

FWTA, ed. 2, 2: 421. UPWTA, ed. 1, 450.

West African: **GHANA** ADANGME mu (FRI) AKAN-KWAWU efanba = *child's vegetable, from* efaŋ : *spinach ;* ba : *child* (FRI) TWI ɛfaŋba = *child's vegetable, from* ɛfaŋ : *spinach ;* ba : *child* (FRI) GBE-VHE (Awlan) ala fango (Krause) **NIGERIA** YORUBA arenikosun (N&E)

A herb or straggling undershrub, occurring in Ivory Coast to W Cameroons, and in E Cameroun and Gabon.

The leaves are eaten as a vegetable in parts of the Region (2; Ghana, 3, 4), and in Gabon (5). The Twi name meaning 'children's spinach' no doubt refers to its consumption as such, but the plant is also used by the Twi and Krobo as a children's medicine for piles, the leaves being boiled in water and the child squatting in the warm liquor (3, 4). Leaf-analysis showed carbohydrates 47%, protein 19%, ash 17% with high mineral content especially calcium, aluminium, and iron (2). In Nigeria the whole plant is reported used in treatment for jaundice, rheumatism and paralysis, and the leaf-sap for catarrh (1).

References:

1. Ainslie, 1937: sp. no. 47. 2. Busson, 1965: 392–4, with leaf-analysis. 3. Irvine 1701, K. 4. Irvine 1961: 743. 5. Sillans, 1953, b: 5: 83.

**Barleria prionitis** Linn.

FWTA, ed. 2, 2: 421.

A stiff bushy plant to 1.70 m high, spiny, native of tropical E Africa and Asia and dispersed by man to other hot countries.

The plant is adapted to high rainfall areas. It has attractive yellow or light buff-coloured flowers, and is cultivated here and there in the West African region as an ornamental. In Asia is often grown as a hedge and with clipping it strengthens and by reason of its spines becomes impenetrable (1, 3, 4).

Leaf-sap is slightly acid and bitter. It is given with honey or sugar to children in India for catarrhal affections accompanied by fever and much phlegm. In the Philippines a bath using a decoction of leaves and leafy twigs is administered for febrile catarrh (5). Other Indian uses are for aphthae, intermittent fever, paralysis, rheumatism, liver diseases, jaundice and dropsy (6), whooping-cough, urinary troubles, bleeding gums, earache, and cracking and laceration of the feet in the rainy season (5). A root-decoction is taken as a mouth-wash in E Africa to relieve toothache (6).

Tests for the plant's antimalarial activity have proved negative on avian malaria (2).

The plant is found to be rich in potassium and this is said to contribute to its diuretic action. Important organic principles appear to be absent.

References:

1. Burkill, 1935: 303. 2. Claude & al., 1947. 3. Howes, 1946. 4. Manjunath [Ed.], 1948: 158. 5. Quisumbing, 1951: 885–6. 6. Watt & Breyer-Brandwijk, 1964: 2.

**Blepharis linariifolia** Pers.

FWTA, ed. 2, 2: 410. UPWTA, ed. 1, 450.

West African: **SENEGAL** MANDING-BAMBARA kuru ulu (JB) **MALI** DOGON nãman (Dieterlen) nãman doñolo ya (Dieterlen) MANDING-BAMBARA nãman (Dieterlen) nanam doñolo ya (Dieterlen) nyame danga (Dieterlen) **UPPER VOLTA** MOORE wiwintin wilinwitin (K&B) **NIGERIA** HAUSA daúdàr-maáguzaáwaa *protection for boys who have been circumcised* (RES; ZOG) fàskàrà toòyí = *resistant to burning* (JMD; ZOG) gagi, gigi (ZOG) gandarin damo (AST) HAUSA (West) gigi

A low-growing wiry herb with prickly bracts, from the mainly drier areas of the Region in Mauritania to N Nigeria and W Cameroons, and into Sudan, and NW India.

It is grazed by all stock in Senegal and is particularly relished by cattle (2). In NW Senegal in the area of Richard Toll it occurs in the irrigation bunds and provides good fodder for sheep in the dry season (1). In the Chad area cattle take it but little (2). Nomads in the Sahara collect the seeds for food (A. Chevalier fide 4).

In Upper Volta (5) and in Ivory Coast (3) a tisane of the whole plant is taken for syphilis.

The spiny bracts are used in N Nigeria to trace patterns on earthenware before firing (4).

References:

1. Adam, 1960: 366. 2. Adam, 1966, a. 3. Bouquet & Debray, 1974: 11. 4. Dalziel, 1937. 5. Kerharo & Bouquet, 1950: 229.

**Blepharis maderaspatensis** (Linn.) Heyne

FWTA, ed. 2, 2: 410.

West African: **MALI** DOGON énye taràà (Dieterlen; C-G) MANDING-BAMBARA eñe tara (Dieterlen) **NIGERIA** YORUBA abèròdéfẹ̀ (RJN)

A procumbent shrubby herb with spiny spikes, weed of roadsides, waste-places and forest-margins in the savanna zone, from Senegal to S Nigeria, and widespread in tropical Africa generally.

The plant is not grazed by cattle in Senegal (1). In Tanganyika Sukuma burn the plant to ash which is powdered and mixed with oil for rubbing onto swollen legs after they have been washed in warm water (3). Unspecified medicinal use in Kenya is recorded (2).

References:

1. Adam, 1966, a. 2. Mwangangi 293, K. 3. Tanner 1382, K.

## **Brillantaisia** P. Beauv.

FWTA, ed. 2, 2: 405. UPWTA, ed. 1, 450.

A genus of semi-woody herbaceous plants, occurring in damp places in forest regions. All species are very ornamental with striking flowers, generally blue or purple.

## **Brillantaisia lamium** (Nees) Benth.

FWTA, ed. 2, 2: 406. UPWTA, ed. 1, 450.
West African: **SIERRA LEONE** MANDING-MANINKA (Koranko) filamεse (NWT) MENDE kɔli-la (JMD) puta-puta-la (JMD) **GHANA** AKAN-ASANTE a-guare-(a)nsra (FRI) FANTE a-guare-(a)nsra (FRI) TWI *a*-guare-ansra (FRI) nsu-twẽ = *to draw water* (FRI) **NIGERIA** EDO ebohohede (JMD) IGBO ékèkè (JMD) IGBO (Awka) polo-polo (JMD) IGBO (Ogwashi) abwoluku (JMD) YORUBA akowaleji (RJN) ekun (JRA)

*NOTE: These names may refer to several spp.*

A perennial coarse herb to 1.5 m tall, of damp shaded localities from Guinea to W Cameroons, and extending across the Congo basin to E Africa.

In Sierra Leone the plant is boiled to make a vapour-bath for someone with fever to induce perspiration (2). The leaves are pounded in Nigeria to a paste in their own sap and this is applied externally for yaws (1). The leaves are used as a polishing material (2).

Other uses under *B. nitens* probably also apply.

References:

1. Ainslie, 1937: sp. no. 59. 2. Dalziel, 1937.

## **Brillantaisia nitens** Lindau

FWTA, ed. 2, 2: 406–7. UPWTA, ed. 1, 406.
West African: **GUINEA** KISSI pedjindo (FB) LOMA boloboloye (FB) MANDING-MANINKA bolobolo (FB) **SIERRA LEONE** MENDE daŋinyε-fuε̃ *from* daŋinyε, *a peculiar itch on the instep of the foot* (FCD) kɔli-la (FCD) putaputa-la (auctt.) **GHANA** AKAN-ASANTE a-guare-(a)nsra (FRI) FANTE a-guare-(a)nsra (FRI) TWI *a*-guare-ansra (FRI) nsu-twẽ = *to draw water* (FRI) **NIGERIA** EDO ebohohede (Dennett) IGBO ékèrè (JMD) IGBO (Awka) polo-polo (JMD) IGBO (Ogwashi) abwoluku (NWT; JMD)

*NOTE: These names may refer to several spp.*

A perennial herb to 3 m high, of damp sites in the forest zone, and up to montane elevations, from Guinea to W Cameroons, and into E Cameroun.

In Ghana the leaves are used to wash with, and as they contain an oil it is said that there is no need to use pomade afterwards, hence the Akan name which means 'if one washes [with it] one does not anoint [the body with pomade]' (3): cf. *Eremomastax speciosa* (Acanthaceae). Sap is squeezed in Ghana into the ears after bathing to relieve the symptoms due to water in them: from this use arises the Twi name meaning 'to draw water' (3). The fresh leaves passed through a flame are placed as a poultice in Sierra Leone on the feet for soreness, and specifically for a peculiar itch known as *danginye* occurring in the instep. One of the Mende names is derived from this application (2).

The root taken in soup is a woman's medicine in S Nigeria given during pregnancy (4) and for pains in pregnancy (5).

The defoliated plant is reduced to ash in Guinea to produce a cooking salt (1).

Other uses under *B. lamium* probably also apply.

References:

1. Busson, 1965: 392. 2. Deighton 3571, K. 3. Irvine, 1930: 64–65. 4. Thomas, N. W. 2034 (Nig. Ser.), K. 5. Thomas, N. W. 2285 (Nig. Ser.), K.

**Brillantaisia owariensis** P. Beauv.

FWTA, ed. 2, 2: 406.
West African: **NIGERIA** IGBO n̠rí átụ̄ = *bushcow food* (anon.)IGBO (Awka) polo polo (NWT) **IJO**-IZON (Kolokuma) bị̀làbẹ̀rí = *elephant's ears* (KW) IZON (Oporoma) bị́lábẹ̀rì = *elephant's ears* (KW)

A robust shrubby herb to 4 m, high, confined to Nigeria and W Cameroons.

**Brillantaisia patula** T. Anders.

FWTA, ed. 2, 2: 406. UPWTA, ed. 1, 450.
West African: **NIGERIA** YORUBA ọwọ́ (JMD)

A robust shrubby plant to 3 m high, from Togo to W Cameroons and across the Congo basin to Uganda and Angola.

The plant is a familiar medicine in S Nigeria taken by a bride or a barren wife to ensure conception: it is sometimes grown in local gardens (3). In Gabon the leaves are used for yaws and rheumatism, and a decoction is taken to ease childbirth and for menstrual pains and stomachache (4, 5). In Congo they are given to women with stomach-trouble, are taken for chest-conditions and infantile spleen affections, and are eaten as a vegatable for anaemia and malnutrition (2). Traces of alkaloid have been detected in the leaves of Nigerian plants (1). Leaf-sap is instilled into eyes and nose, and is given by draught in Congo as a seditive in epilepsy and insanity, and the sap alone, or mixed with that of other drug leaves, or diluted with palm-wine is taken for tachycardia (2). In cases of infection of circumcision wounds in Gabon the leaves lightly passed through a fire are used as dressing with snail shells ground to a powder (4, 5).

The plant is held in Congo to have magical power to exorcise evil spirits, and when treatment for epilepsy or insanity is being carried out (see above), the patient is submitted to lotions and hot vapour-baths made from the roots (2). In Gabon the plant is held to have the power to adjudicate in quarrels (5).

References:

1. Adegoke & al., 1968. 2. Bouquet, 1969: 48–49. 3. Dalziel, 1937. 4. Walker, 1952: 183. 5. Walker & Sillans, 1961: 40.

**Brillantaisia sp. indet.**

West African: **NIGERIA** EFIK ádàn umọ̀n (RFGA; Lowe)

An unidentified species of the above Efik name is used in an unspecified manner in medicine.

**Crossandra flava** Hook.

FWTA, ed. 2, 2: 408–9. UPWTA, ed. 1, 450, under *C. guineensis* Nees.
English: yellow-flowered Crossandra (*Bot. Mag.*).

A semi-woody herbaceous plant to only about 20 cm high, from Sierra Leone to S Nigeria.
It has bright yellow flowers in a compact spike and is suitable as an ornamental bedding plant.

**Crossandra nilotica** Oliv., **ssp. massaica** (Mildbr.) Napper

FWTA, ed. 2, 2: 409, as *Crossandra massaica* Mildbr.

A straggling shrub, in rocky hill woodland, recorded only from Ghana in the Region, but distributed in Congo, Kenya and Tanganyika.
Masai of Kenya say that the plant is poisonous (4), but it is grazed by goats and 'dikdik' (2). It is commonly used as a snake-antivenom. Persons bitten by a cobra may chew the leaf and apply the quid to the bite and wherever the poison may be (1). Cattle-folk similarly apply the chewed leaf and saliva to a bite on their cattle (3), or onto cattle that have been sprayed by the venom of the spitting cobra (2).

References:

1. Glower & al. 57, K. 2. Glover & al. 468, K. 3. Glover & al. 901, K. 4. Glover & al. 1819, K.

**Dicliptera elliotii** C. B. Cl.

FWTA, ed. 2, 2: 425–6.
West African: SIERRA LEONE MENDE kɔtu-gbɛ *probably general for several small herbs* (FCD)

A weak decumbent herb, to 30 cm long or more, rooting at the nodes, of the forest zone from Guinea to Nigeria, and dispersed over central Africa.
In Sierra Leone the leaves are ground up and rubbed on boils and other skin-affections (1).

Reference:

1. Deighton 1806, K.

**Dicliptera laxata** C. B. Cl.

FWTA, ed. 2, 2: 426.

A scrambling shrub of montane situations in W Cameroons and Fernando Po, and spread across the Congo basin and in E Africa.
The plant is used in Tanganyika as a remedy for general debility (2). A macerate in hot water is reported to result in a blood-like liquid ('à l'aspect du sang') (1).

Reference:

1. Dschang 1735, K. 2. Watt & Breyer-Brandwijk, 1962: 3.

ACANTHACEAE

**Dicliptera verticillata** (Forssk.) C. Christens.

FWTA, ed. 2, 2: 425.
West African: THE GAMBIA MANDING-MANDINKA timintiminjoŋ (*def.*-o) = *sour-sour* (Fox, DF)

A weak, decumbent perennial herb, woody towards the base with several-angled stems reaching to about 1 m long, throughout the Region, but seemingly not common, and occurring in Cape Verde Island and to the Congo, Angola and E Africa, and into Arabia.

The plant has unspecified medicinal use in The Gambia (2). In Congo sap expressed from the leaves after heating over a fire is taken to relieve whooping-cough (1). Dried powdered leaves are mixed with castor oil in Tanganyika and applied to scabies (3).

References:
1. Bouquet, 1969, 49. 2. Fox 69, K. 3. Haerdi, 1964: 182, as *D. umbellata* (Vahl) Nees.

**Dischistocalyx** T. Anders.

FWTA, ed. 2, 2: 397.

The two herbaceous undershrub species, *D. thunbergiiflorus* (T. Anders.) Benth. and *D. obanensis* S. Moore, recorded from S Nigeria and W Cameroons, have no reported usage, but the related *D. walkeri* R. Benoist is used in Gabon as a fish-poison in association with the bark of *Mallotus subulatus* Müll.-Arg. (Euphorbiaceae) (1, 2). The two Nigerian species merit examination.

References:
1. Walker s.n. Dec. 1938, K. 2. Walker & Sillans, 1961: 4.

**Dyschoriste pedicellata** C. B. Cl.

FWTA, ed. 2, 2: 404.
West African: THE GAMBIA FULA-PULAAR boru (Williams)

A small shrub recorded from The Gambia, Mali and N Nigeria, and also in Ubangi-Chari.

A leaf-infusion is given to children in The Gambia as a febrifuge (1).

Reference:
1. Williams, F. N., 1907: 370.

**Dyschoriste perrottetii** Nees) O. Ktze

FWTA, ed. 2, 2: 404. UPWTA, ed. 1, 450.
West African: NIGERIA FULA-FULFULDE (Nigeria) momodil (JMD) teɓete *from* teɓa: *to pluck* (JMD) HAUSA bidi diyan (ZOG) fid-dà-hákuúkúwá = *to remove spicules of grass or chaff* (JMD; ZOG)

A semi-woody herb or undershrub of savanna, widespread from Senegal to W Cameroons and to NE Africa.

The seed has a mucilaginous coat. It is used in N Nigeria by placing one in the eye to which foreign particles adhere and can then be removed. The Hausa name arises from this. The Fula name implies the same.

Reference:
Dalziel, 1937.

**Dyschoriste radicans** Nees

FWTA, ed. 2, 2: 404.

A spreading, creeping, matting semi-woody herb of grass savanna, and open spaces in woodland savanna, widely dispersed in NE, E, Central and South-central Africa. It has not been recorded in W Africa but *FWTA, l.c.,* states it may have been overlooked.

In Kenya it is grazed by all stock (1). The flowers are much frequented by bees (2). Stem and leaves are chopped up, powdered and mixed with a little water and taken for diarrhoea (3). In Tanganyika it is prepared with *Alysicarpus wallichii* (Leguminosae — Papilionoiidae) for bathing swollen limbs (4).

References:

1. Glover & al. 343, 1552, 2231, K. 2. Glover & al. 721, 835, K. 3. Mwangangi 522, K. 4. Watt & Breyer-Brandwijk, 1962: 3.

**Elytraria acaulis** (Linn. f.) Lindau, **var. lyrata** (Vahl) Brem.

FWTA, ed. 2, 2: 418, as *E. marginata* Vahl, in minor part.

A stemless annual herb recorded locally in Ghana and Nigeria; also in Congo, Angola, coastal tropical E Africa, SE India and Ceylon. Var. *acaulis* is widespread in India and S Tropical Africa. This plant has previously been confused with the larger perennial *E. marginata,* and possibly some of the attributes of the latter apply.

**Elytraria marginata** Vahl

FWTA, ed. 2, 2: 418, in major part. UPWTA, ed. 1, 450 as *E. acaulis* Lindau.
West African: **IVORY COAST** AKAN-ASANTE atiamassa (B&D) BAULE blatiki (B&D) GAGU bidabo (K&B) KRU-GUERE (Chiehn) dodowoa (B&D) kpadubu walé (K&B) kpakguéï (B&D) tiébé zago (B&D) zaméréni (K&B) GUERE (Wobe) nohotu (A&AA) 'KRU' nawésu (K&B) notu (K&B) **GHANA** AKAN-ASANTE otiemasa *the 'small otiemasa'* (FRI) **NIGERIA** IGBO (Umuahia) achulu nta (AJC) IJO-IZON (Kolokuma) kẹnï bụ̄ọ̀-tịẹ́ = *standing on one leg* (KW) YORUBA ẹṣọ̀ (JMD)

A perennial herb to 25 cm high, occurring from Guinea to W Cameroons and Fernando Po, and Cameroun, Congo and E Africa.

In SW Nigeria the plant is often grown in gardens and it is given in infusion to children for cough (4). It is used by the Edo as a medicine for stomachache (5), and the Yoruba use the sap squeezed from the leaves to treat open wounds (3). At Asaba, Igbo are reported using it, perhaps superstitiously, in medicine against poisoning caused by bats (8). The Ijo grind the leaves with *kument fuu* and *egbe pulou* (unidentified items) and lick the paste to cure chest-pains; they also claim that the plant is a cure for male impotence, the treatment being somewhat confused: the leaves are picked and made into a pepper soup with a fresh *ikpoki* (unidentified) without removing the spikes; the sap is then placed on [sic] a stick and the patient eats it standing on a plank (11). Perhaps this performance has some connexion with the Ijo name meaning 'standing on one leg'.

In Ghana the plant leaves are ground and boiled and the liquor is drunk for gonorrhoea (6). Throughout Ivory Coast the leaf-sap is administered as an enema and as a draught to women for sterility or haemorrhage, during pregnancy, and to prevent miscarriage; the plant is also used for haemoptysis and nausea (2,7). In Gabon the leaf-sap is said to be good for gummy eyes, and the leaves are given to a child to eat for easing heart-palpitations (9, 10). Leaf-sap is also taken in Ivory Coast to stop palpitations (7).

Some alkaloid is reported present in the leaves of Nigerian material (1). The plant is a fetish of prostitution and for warding off evil spirits in Gabon (10).

References:

1. Adegoke & al., 1968. 2. Bouquet & Debray, 1974: 11. 3. Chizea FHI.23982, K. 4. Dalziel, 1937. 5. Idahosa FHI.23880, K. 6. Irvine 444, K. 7. Kerharo & Bouquet, 1950: 229–30, as *E. acaulis* Lindau. 8. Thomas, N. W. 1647 (Nig. Ser.), K. 9. Walker, 1952: 184, as *E. acaulis.* 10. Walker & Sillans, 1961: 41, as *E. acaulis* Lind. 11. Williamson, K. 56, UCI.

**Eremomastax speciosa** (Hochst.) Cufod.

FWTA, ed. 2, 2: 397, as *E. polysperma* (Benth.) Dandy. UPWTA, ed. 1, 452, as *Paulowilhelmia polysperma* Benth.

West African: **SIERRA LEONE** MENDE wote (FCD) **IVORY COAST** KRU-GUERE (Chiehn) karagbéi (B&D) wotiaï (B&D) KWENI guégambo (B&D) **GHANA** AKAN-AKUAPEM adubiri (Easmon) ASANTE a-guare-(a)nsra, guare: *to wash the body;* ansra: *pomade* (FRI) FANTE a-guare-(a)nsra, guare: *to wash the body;* ansra: *pomade* (FRI) TWI a-guare-(a)nsra, guare: *to wash the body;* ansra: *pomade* (FRI) kwasi pɛtɛprɛ, pɛtɛprɛ *being onomatopoeic for the fruit capsule bursting* (FRI) **NIGERIA** IGBO (Ogwashi) ákwụ́kwọ̀ ijiji uku, ákwụ́kwọ̀: *leaf* (NWT) IGBO (Ukwuani) àgbọ̀lọ̀ úkwú (NWT) YORUBA aṣọyun (Millson)

A stout erect much-branched herb, of the forest zone, weed on cacao farms, dispersed from Guinea to W Cameroons, and widespread in tropical Africa.

The leaves are eaten in Zäire (5), but alkaloid is reported in Nigerian material (1). The plant also contains a saponin (3), and leaves are used in Ghana for washing the body, as the Asanti/Fante/Twi names imply. After such washing, there is said to be no need to use pomade (6, 7). In Ivory Coast (2, 8) and in Nigeria (9) pulped leaves are applied to guineaworm sores to kill the parasite, and in the former country the leaves are occasionally used for headache and as a poison-antidote (2). The plant has been used in Ghana by the Akuapem people as a fish-poison (Easmon fide 7). Crushed leaves are applied in Sierra Leone to *porro* cicatrices (reason unexplained) (4).

References:

1. Adegoke & al., 1968. 2. Bouquet & Debray, 1974: 11. 3. Dalziel, 1937. 4. Deighton 3919, K. 5. Evrard 3326, K. 6. Irvine 532, K. 7. Irvine, 1930: 325, as *Paulowilhelmia polysperma* Benth. 8. Kerharo & Bouquet, 1950: 230, as *P. polysperma* Benth. 9. Oliver, 1960: 33, as *P. polysperma.*

**Fittonia verschaffeltii** (Lem.) Coem., **var. argyroneura** (Coem.) Nichols

FWTA, ed. 2, 2: 417 as *F. argyroneura* Coem.

English: typical form — mosaic plant, nerve plant; var. *argyroneura* — silver-net plant (U.S.A., Bates).

A creeping rooting herb, of Colombia and Peru, of a few forms cultivated for their ornamental value. Var. *verschaffeltii* has dark green leaves with rosy-red reticulated venation. Var. *argyroneura* has light green leaves with white reticulated venation and is under cultivation in Ghana and S Nigeria. The pale yellow flowers are borne in a terminal spike.

**Graptophyllum pictum** (Linn.) Griffith

FWTA, ed. 2, 2: 423.

English: caricature plant, Joseph's coat.

A shrub to 2.50 m high, probably native of New Guinea, but widely dispersed by man to many warm countries, and commonly found in gardens in the W African region.

The leaves are elliptic to 15 cm long, purplish or green, attractively variegated with creamy white, yellow or crimson blotches. The flowers are in a spike, purple or crimson. The plant is easily propagated by cuttings and it is grown for its decorative value principally as a foliage plant, though it also makes an attractive hedge.

The plant is used in India and SE Asia for cuts and on certain skin-complaints. An infusion is taken for constipation, and the sap is dripped into the ears for earache. The leaves are emolient and resolvent and are applied to swellings and ulcers. In parts of India the plant is reportedly used in the same way as the sovereign drug-plant *Adhatoda vasica* (see under *A. buchholzii* (Lindau) S. Moore, Acanthaceae.) A trace of a non-toxic alkaloid is reported in the leaves, and when the plant is subjected to steam-distillation there is a smell suggestive of *coumarin* (1, 2).

References:

1. Burkill, 1935: 1110. 2. Sastri [Ed.] 1956: 259.

**Hygrophila auriculata** (Schumach.) Heine

FWTA, ed. 2, 2: 395. UPWTA, ed. 1, 451, as *H. spinosa* T. Anders.

West African: **SENEGAL** BANYUN balâ balâgan (K&A) DIOLA édalora (JB; K&A) MANDING-BAMBARA kélé béto kala (JB; K&A) 'SOCE' kositâbo (K&A) kostâbo (K&A) SERER dabasaw (JB; K&A) god a kos (JB; K&A) ndokarok (JB) orokarok (K&A) rokãndok (JLT; JB) rokarot (K&A) SERER-NON (Nyominka) orokarok (K&A) rokarot (K&A) WOLOF séber bụki = *hyena's drum* (JB; K&A) **THE GAMBIA** DIOLA (Fogny) edalora = *white leaf* (DF) MANDING-MANDINKA fastambo (Fox) **GUINEA-BISSAU** DIOLA-FLUP elolaè (JDES) FULA-PULAAR (Guinea-Bissau) lobóte (JDES; EPdS) PEPEL bechete (JDES; EPdS) **GHANA** AKAN-FANTE atwẽ (Deakin) GBE-VHE eyitrɔ (FRI) VHE (Awlan) ayetrɔ (FRI) **NIGERIA** HAUSA dàyiń giíwaá, danyin giwa = *elephant's Centaurea* (JMD; ZOG) ƙayar giíwaá = *elephant's thorn* (JMD) káyàr raàkúmií = *camel's thorn* (JMD; ZOG) saàre gwiáwà = *prick the knee* (JMD; ZOG) zaazar-giíwaá = *pubic hair of the elephant* (RES; ZOG) zakankau *the ash from which salt is prepared* (Robinson; RES) HAUSA (West) káyàr-giíwaá (ZOG) YORUBA mafọwọ kauruọnu (Macgregor)

A stout annual herb to 1.5 m high or more, quadrangular stem, leaves whorled with strong spines, of wet places, often brackish, littoral, estuarine, commonly from Senegal to W Cameroons and widespread in the Old World topics.

The plant is used by the Pedi of southern Africa as a vegetable (17). It is grazed by cattle in Senegal (1), and a little by goats in Sudan (13), but its pungent spines limit its value, and in parts of Malawi it is considered a serious weed of pasture (16). The whole plant is recognized in Senegal to have diuretic properties, and the roots in particular are used for blennorrhoea, hydropsy and anuria (8). The whole plant, or its ashes, and the roots are similarly used in Sudan as a cooling medicine and diuretic in cases of hepatic obstruction, dropsy, rheumatism, etc., and the seeds as a demulcent and diuretic (2). *H. salicifolia* Nees in India is official as a diuretic made effective by the presence of a large amount of mucilage with potassium salts in it (8, 12, 14). A vegetable salt is commonly prepared from the ashes of *H. auriculata* and in parts of Ubangui (3, 15); in Sudan (11) it is specially grown for the purpose. The plant is also used in Senegal for catarrh (7) and with another plant identified only as *kisâdor* (Banyun) to induce menstruation, particularly in young mothers on weaning a baby (6, 8). A remarkable treatment with good result of a sting-ray's sting on the sole of the foot by a decoction of the plant by alternation of bathing and steam fumigation is recorded (7, 8). In Ghana people with fever are bathed in water in which leaves have been boiled (4, 5). The plant is used in Lagos for stomachache *(iku)* (9) and elsewhere (3). Leaves, stems and flowers are burnt over a naked flame in Kenya and the ash is taken in tea for stomach-ache (10). The plant is occasionally used in N Nigeria for craw-craw (3).

A number of alkaloids is recorded present in Indian plant material, also mineral salts, fixed oil and mucilage. The seeds contain a semi-drying oil, sugars and enzymes. The essential oil from the roots has been found to possess bacteristatic action against both Gram +ve and Gram −ve organisms. (8, 17).

References:

1. Adam, 1966, a: as *H. longifolia.* 2. Broun & Massey, 1929: 338, as *H. spinosa* T. Anders. 3. Dalziel, 1937. 4. Irvine 682, K. 5. Irvine, 1930: 233, as *H. spinosa* T. And. 6. Kerharo & Adam, 1963, a: as *H. longifolia* (Linn.) Kurz. 7. Kerharo & Adam, 1964, c: 308–9. 8. Kerharo & Adam, 1974: 112–4, with phytochemistry and pharmacology. 9. Macgregor 64, K. 10. Mwangangi 283, K. 11. Myers 7839, K. 12. Oliver, 1960: 7, 67, as *H. spinosa.* 13. Peers s.n., 27/8/1953, K. 14. Sastri [Ed.], 1959: 148. 16. Tisserant 1883, K. 16. Van Rensburg 2092, K. 17. Watt & Breyer-Brandwijk, 1962: 1, as *Asteracanthus longifolius* Nees.

**Hygrophila barbata** (Nees) T. Anders.

FWTA, ed. 2, 2: 396.

An erect much-branched annual herb of marsh and weed of rice-paddy, in Senegal to Liberia.

**Hygrophila odora** (Nees) T. Anders.

FWTA, ed. 2, 2: 396.
West African: **SIERRA LEONE** MENDE kpawu (*def.* kpawii) *a group term* (NWT; FCD)

An aromatic semi-woody herb to nearly 1 m high, recorded from Senegal to Liberia.
The flowers have a musk-like scent.

**Hygrophila senegalensis** (Nees) T. Anders.

FWTA, ed. 2, 2: 396.
West African: **SENEGAL** MANDING-BAMBARA kuru ulu (JB; K&A) 'SOCE' buben (K&A) SERER maro (JB; K&A) mbubén (JB; K&A)

An erect herb to 1 m high, of swamp and wet places, and in grassy openings of savanna-forest from Senegal to N Nigeria.
The plant is grazed by stock except horses in Senegal (1). Soce medicinemen use the pounded seeds in the eyes for eye-complaints (2, 3).
The purplish-blue flowers are attractive.

References:

1. Adam, 1966, a. 2. Kerharo & Adam, 1964, c: 309. 3. Kerharo & Adam, 1974; 114.

**Hypoestes aristata** (Vahl) Soland.

FWTA, ed. 2, 2: 431.

An erect herb to 1 m high, of the forest from sea level to montane elevations in N and S Nigeria, W Cameroons and Fernando Po.
In Tanganyika leaf-sap and a root-decoction are taken in draught with the leaf-sap of *Ampelopteris prolifera* (Retz). Copel. (Thelypteridaceae) for meningitis and encephalitis (1).

Reference:

1. Haerdi, 1964: 182.

**Hypoestes cancellata** Nees

FWTA, ed. 2, 2: 430.
West African: **SENEGAL** BASARI ỹélixɛ̀r (Ferry)

A herb reaching to 1 m high recorded from Senegal to W Cameroons.

**Hypoestes rosea** P. Beauv.

FWTA, ed. 2, 2: 431.
West African: **NIGERIA** IGBO (Owerri) ikéré (AJC)

An erect herb reaching 1 m high of the forest in S Nigeria and W Cameroons.

**Hypoestes verticillaris** (Linn. f.) Soland.

FWTA, ed. 2, 2: 431. UPWTA, ed. 1, 451.
West African: **SENEGAL** BALANTA nguan (A. Chev.) MANDING-'SOCE' kénéfetéu (A. Chev.) WOLOF bumbop (A. Chev.) lugusque (A. Chev.) **GHANA** AKAN-ASANTE kwaaduko (FRI) TWI kwaduko (JMD) kwasi pɛtɛprɛ (FRI) **NIGERIA** IGBO (Umuahia) ékèrè, ékérè (KW) IGBO ('Ibugo') ákwụ́kwọ́ ìjìjí (NWT)

A polymorphic, tufted herb up to about 1 m high, in open places of wooded savanna, and secondary and deciduous forest, occurring throughout the Region, and widespread elsewhere in tropical Africa, and extending into Arabia.

The plant is sometimes eaten as a vegetable in Tanganyika (17, 19). It is grazed by all stock in the Narok district of Kenya (3–8), though elsewhere it has been recorded as being refused (24). The giraffe in Kenya will browse it (15). A decoction of the plant is used by the Swahili in Tanganyika as a remedy for chest-complaints (25) and for influenza (14). The root is chewed by the Masai and Kipsigis of Kenya as a sovereign cough-cure (3, 6, 7, 9–11) and for sore-throat (9, 12). In S Nigeria plant-sap is used to treat sores (21), and the leaves pounded to a paste are applied to swellings by the Zigua of Tanganyika (20), where also the whole plant is soaked in water which is then used to bathe a child with fever (18). The plant is burnt to an ash in the Mt. Elgon area of Kenya to prepare a form of cooking soda (16).

The flowers are much visited by honey-bees in Kenya (3–5, 10, 24).

In S Nigeria the plant is used to rub on a cross-bow to clean it (23). It is used to polish floors (1, 2), and is said to drive away flies (22).

The root is used in Sudan as the source of a dye in making mats (13).

References:

1. Carpenter 309, UCI. 2. Dalziel, 1937. 3. Glover & al. 673, 957, 1361, 1777, K. 4. Glover & al. 795, K. 5. Glover & al. 1165, K. 6. Glover & al. 1459, K. 7. Glover & al. 2082, K. 8. Glover & al. 2130B, K. 9. Glover & al. 2285, K. 10. Glover & al. 2401, K. 11. Glover & al. 2437, K. 12. Glover & al. 2585, K. 13. Greenway, 1941. 14. Koritschoner 1478, K. 15. Nesbit Evans 36, K. 16. Samuel 8, K. 17. Semsei FH.327, 2631, K. 18. Tanner 2001, K. 19. Tanner 2017, K. 20. Tanner 2935, K. 21. Thomas, N. W. 1842 (Nig. Ser.), K. 22. Thomas, N. W. 1988 (Nig. Ser.), K. 23. Thomas N. W. 2144 (Nig. Ser.), K. 24. Watt 1319, K. 25. Watt & Breyer-Brandwijk, 1962: 3.

**Justicia betonica** Linn.

FWTA, ed. 2, 2: 427.

An erect semi-woody herb or shrub to about 1 m high, of waste land and rocky places, recorded only from Mali, and doubtfully from Sierra Leone, and occurring commonly in E Africa and in the Asian tropics.

The plant is grazed by stock in Tanganyika (4), and cattle-folk in Uganda give a soup of the boiled leaves to cows-in-milk to drink as a galactogogue (5). The Sukuma of Tanganyika, prepare an ointment from the plant-ash in butter to apply to areas of scaly skin (7). An infusion of the green leaves is drunk in Kenya for snake bite (2). A poultice of the leaves is applied to boils in Ceylon (6), and to swellings in Malaya (1). A leaf-preparation is taken in Ceylon for diarrhoea (6).

The flowers are very attractive to bees in Kenya (3).

References:

1. Burkill, 1935: 1274. 2. Corral H.1, K. 3. Glover & Samuel 3095, K. 4. Hornby 92, K. 5. Howells 2, K. 6. Sastri [Ed.], 1959: 312. 7. Tanner 1282, K.

**Justicia extensa** T. Anders.

FWTA, ed. 2, 2: 428.
West African: **NIGERIA** YORUBA ahọ̀n ẹ̀kùn (JRA)

A herbaceous shrub to 3 m high, of the forest, occurring from Guinea to W Cameroons, and extending to the Congo basin, Angola and Mozambique.

The leaves when pounded and the sap squeezed out yield a bitter astrigent, and when eaten are said in Nigeria to cure diarrhoea. A leaf-infusion is drunk as a blood-purifier to cure boils and ulcers, and by children in smallpox cases, etc. A leaf-decoction is drunk for chronic rheumatism and arthritis (1). Anodynal action is also sought in the application of pulped leaves in frictions for fever stiffness and pains in Ivory Coast, where the pulped leaves are also rubbed onto babies for convulsions; it also has a good reputation as a haemostatic for wounds, vaginal bleeding and haemoptysis (3). Leaf-sap is used in Congo to kill filaria in the eyes and the plant is used as a fish-poison for which it is cultivated (2). In the Pescadore Islands off Taiwan, a related species, *J. hayatai* Yamamoto, has for centuries been used as a fish-poison. *Justicidin A* and *B* have been isolated and shown to have fish-killing properties equal to rotenone (4).

A root-decoction is said in Nigeria to possess anti-spasmodic properties and is prescribed for asthma and ague. The root, as also the leaf, is bruised and mixed with lime-juice for application to ring-worm and cutaneous affections. It is also prepared as an electuary taken at a dosage of one teaspoonful twice daily (1).

The flowers are pounded up and eaten in Nigeria to cure dysentery (1).

Traces of alkaloid have been detected in the leaves and bark (3).

References:

1. Ainslie, 1937: sp. no. 197. 2. Bouquet 1969: 49. 3. Bouquet & Debray, 1974: 11–12. 4. Munakata & al., 1965.

**Justicia flava** (Forssk.) Vahl

FWTA, ed. 2, 2: 427–8. UPWTA, ed. 1, 451.
West African: **GUINEA** KISSI yogbo (FB) LOMA budébeli (FB) MANDING-MANINKA (Koranko) negbé-negbé (FB) **IVORY COAST** AKAN-ASANTE aféma (K&B) assiadiomuro (B&D) **GHANA** AKAN-AKYEM ntumunum (auctt.) ASANTE afema (auctt.) ntumunum (auctt.) FANTE ntumunum (auctt.) TWI ntumunum (auctt.) NZEMA atinyɛkɛ (JMD) mubulighra (FRI)

An erect or straggling herb to about 1 m high, occurring throughout the Region, and in NE and E Africa.

It is semi-cultivated in parts of Guinea for eating as a vegetable (3), cf. *J. insularis*. Cattle and other domestic stock browse it in NE Africa (Sudan, 16:

Somalia, 15) and E Africa (Uganda, 5; Kenya, 8-good grazing, 9), but in Tanganyika the leaves are recorded as being emetic (14).

The plant has a reputation in Ivory Coast as a haemostatic, and preparations are used on cuts and for menorrhagia and blood in the sputum (1), and the whole crushed plant with vegetable ash, seed of maleguetta (*Aframomum melegueta* K. Schum. and perhaps other *A. spp.*, Zingiberaceae) and pimento (*Capsicum spp.*, Solenaceae) is administered by enema for painful menses, or this mixture with citron juice (*Citrus medica* Linn., Rutaceae) added is given to induce menstruation (13). The pulped leaves are rubbed on babies in Ivory Coast for convulsions and feverish aches and pains (1). In Ghana the plant is used in cases of fever (17), yaws and diarrhoea in children, internally or sometimes by external rubbing (12). Leaf-sap is taken by draught in Tanganyika for hookworm, and for hydrocele with also bathing of the affected parts (11).

In Kenya the leaves are burnt to an ash to produce a vegetable salt (2), and the roots are chewed, followed by a drink of water, by the Masai for diarrhoea (10) and by Masai and Kipsigis for coughs (8). The root is very bitter.

The inflorescence is said in Ghana to be a cure for dysentery (12). The flowers are meliferous and are visited in Kenya by bees (8) and other insects (7; Sudan, 6).

A form of this species, or a new species of very close affinity, occurs on coastal sand-dunes and on sandy river-banks in Kenya and contributes to the sand-binding vegetation.

The plant has religious use in Ghana being a ceremonial ingredient in the brass pan when making a new Asante shrine (4).

References:

1. Bouquet & Debray, 1974: 11. 2. Broadhurst-Hill 274, K. 3. Busson, 1965: 394. 4. Dalziel, 1937. 5. Dyson Hudson 461, K. 6. Gillett 3926, K. 7. Gillett 12876, K. 8. Glover & al. 392, K. 9. Glover & al. 2589, K. 10. Glover & Samual 3262, K. 11. Haerdi, 1964: 183. 12. Irvine, 1930: 248. 13. Kerharo & Bouquet, 1950: 230. 14. Koritschoner 1022, K. 15. Newbould 806, K. 16. Pritchard 80, K. 17. Sampeney 711, K.

## **Justicia glabra** Koenig

FWTA, ed. 2, 2: 428. UPTWA, ed. 1.

An erect or straggling herb to nearly 2 m high, recorded in Ivory Coast, N Nigeria and W Cameroons, and occurring elsewhere throughout tropical Africa and in India.

The leaves are pounded with butter by the Sukuma of Tanganyika who apply the preparation while hot to boils; this is followed by a plaster of the leaves (1).

Reference:

1. Tanner 1096, K.

## **Justicia laxa** T. Anders.

FWTA, ed. 2, 2: 428. UPWTA, ed. 1, 451
West African: **IVORY COAST** KRU-GUERE (Chiehn) gabué (B&D)

Herbaceous shrub to about 2 m high, of the forest in Ivory Coast, S Nigeria and Fernando Po.

The plant is said to be a fish-poison in S Nigeria (2). Traces of alkaloid have been reported in Ivorean material (1).

References:

1. Bouquet & Debray, 1974: 12. 2. Dalziel, 1937.

**Justicia schimperi** (Hochst.) Dandy **J. striata** (Klotzsch) Bullock complex

FWTA, ed. 2, 2: 427. UPWTA, ed. 1, 451.
West African: **GUINEA** KISSI bendu (JB) k-pendo (JB) LOMA pélévélé (JB) MANDING-MANINKA (Koranko) negbe (JB) ningbe (JB) **SIERRA LEONE** BULOM (Sherbro) bitagbɔ-lɛ (FCD) KISSI yɔkpɔ (FCD) KONO yɔgbɔɛ (FCD) yɔkpɔɛ (FCD) LIMBA kɔrɔ-kɔrɔ (FCD) MENDE tɛvɛ (FCD) tɛyɛ (FCD) MENDE (East) yɔkpɔ (FCD) SUSU-DYALONKE sɔyɔ-khudika-na (FCD) TEMNE *a*-kaikur (NWT) TEMNE (Kunike) *an*-lami (Glanville; FCD) TEMNE (Port Loko) *an*-kɔkɔrɔ (Glanville; FCD) TEMNE (Sanda) *an*-kɔkɔrɔ (Glanville; FCD) TEMNE (Yoni) *an*-lami (Glanville; FCD) **GHANA** AKAN-ASANTE asiapiriwa (CV; FRI) ɛfan *a loose term* (CV; FRI) GBE-VHE (Kpando) tasrɔɛ̃ (Kasamany) **NIGERIA** CHAMBA dang-dang (FNH) HAUSA budidiyo (AST) KANURI shigel kirgegu (Musa Daggash) YORUBA obe dudu (N&E)

Straggling, variable herbs to near 1 m high, recorded throughout the Region, and tropical Africa.

This group is one of the most appreciated of the sweet green vegetables. It is often eaten with the leaves of the sweet potato (*Ipomoea batatas* (Linn.) Lam., Convolvulaceae) rendering the preparation less gluey (4), and equally it is used to make vegetable soup or eaten cooked as spinach. In W Cameroon it is added to groundnut soup (7, 8). In Sierra Leone (6), Ghana (10) and in W Cameroons (3) the plant is specially cultivated, a practice which, of course, permits of ennoblement. Doubtless it is grown elsewhere, and in West Africa and Central Africa there have been official recommendations for its cultivation (5). It is recorded that the name of the Fante people (Akan: *fan* greens, *di*, to eat) was given to them as a nickname of ridicule by their former enemies, the Asante, when they were driven south out of their home-lands, and before they could establish new farms they had to subsist on this plant (10). The Akan word *fan* or *ɛfan* applies commonly to pot-herbs used as spinach. Cattle browse the plant in Senegal, but equally there are observations to the contrary (1). In Togoland, NE Ghana, an extract of the boiled leaves is given to babies to loosen their bowels, and the leaves are applied to wounds to promote healing (9). In Congo the leaves mixed with oil and salt are eaten for cardiac troubles (2), and in Gabon the boiled up leaves serve as a soap-substitute (11).

References:

1. Adam, 1966, a. 2. Bouquet, 1969: 49. 3. Brant 897, K. 4. Busson, 1965: 394. 5. Dalziel, 1937. 6. Gledhill SL.2434, K. 7. Hepper 1277, K. 8. Hepper, 1965: 443. 9. Kasamany 137, K. 10. Vigne 1012, K. 11. Walker & Sillans, 1961: 42.

**Justicia tenella** (Nees) T. Anders.

FWTA, ed. 2, 2: 428.
West African: **SIERRA LEONE** TEMNE *a*-sira-*a*-ro-kant (NWT)

A slender herb to 30 cm high, widespread throughout the Region and tropical Africa.

**Justicia spp.**

West African: **THE GAMBIA** FULA-PULAAR (The Gambia) tepilibo (DRR) MANDING-MANDINKA ninsi kumbaliŋ (*def.*-o) = *cow's knee* (auctt.)

A herbaceous plant of the names above occurs in The Gambia forming a dense ground cover in damp places, and is of similar habit to *Sphaeranthus senegalensis* DC. (Compositae) with which the names are shared (2).

Alkaloids *justicin A* and *B* have been isolated from certain non-regional *Justica spp.* which have strong piscicidal properties (1).

References:

1. Bouquet & Debray, 1974: 11. 2. Rosevear, 1961.

**Lankesteria brevior** C. B. Cl.

FWTA, ed. 2, 2: 407. UPWTA, ed. 1, 451.
West African: **SIERRA LEONE** MENDE gegba (SKS) kafa-hina (*def.*-hinei) (FCD) TEMNE *a*-kan-*a*-kei (Dawe) **GHANA** AKAN-ASANTE otiemasa *the 'large otiemasa'*, (cf. *Elytraria marginata*, the 'small *o*.') (FRI)

A shrub to 1.30 m high, of forest undergrowth, from Sierra Leone to Ghana, and in W Cameroons.
The leaves are ground and boiled and the liquor is drunk for gonorrhoea in Ghana (2, 3). Alkaloids have been recorded present: abundant in the roots, moderate amount in the twig-bark, and a trace in the leaves (1).
The plant is a common weed on cacao farms in Ghana (3).

References:
1. Bouquet & Debray, 1974: 12. 2. Irvine 545, K. 3. Irvine, 1930: 258–9.

**Lankesteria elegans** (P. Beauv.) T. Anders.

FWTA, ed. 2, 2: 407. UPWTA, ed. 1, 451.
West African: **NIGERIA** EDO àròbàọwẹ̀è (Kennedy) IGBO (Owerri) osisi-izi = *fly plant*, (JMD; Lowe) IGBO (Umuahia) osisi-izi = *fly plant*, (JMD; Lowe) YORUBA abilokun (JMD) abisuru (JRA) otta, otta ifo (Rowland) ọwọ̀ *misapplied, properly Brillantasia patula T. Anders* (Millson)

A shrub to 1.30 m high, of forest undergrowth from Sierra Leone to W Cameroons and across Africa to Sudan and Uganda.
A decoction of the pounded leaves is taken in Nigeria for gonorrhoea (1) and other undefined conditions (2).

References:
1. Ainslie, 1937: sp. no. 203. 2. Dalziel, 1937.

**Lankesteria hispida** (Willd.) T. Anders.

FWTA, ed. 2, 2: 408.

An undershrub to 60 cm high, of forested areas from Guinea to Ivory Coast.
The bush is quite decorative with lemon-yellow or pure white tubular flowers to 3 cm long in dense heads.

**Lankesteria thyrsoidea** S. Moore

FWTA, ed. 2, 2: 407.
West African: **NIGERIA** YORUBA owuyẹ (Macgregor)

A slight undershrub to 30 cm high of the forest in S Nigeria and W Cameroons.

**Lepidagathis alopecuroides** (Vahl) R. Br.

FWTA, ed. 2, 2: 414.

A decumbent herb from a woody base, widely dispersed throughout the Region, and widespread in the Old and New World tropics.

ACANTHACEAE

No usage is recorded in the Region. The whole plant has been screened for anti-malarial activity and no action on avian malaria was detected (1). Mild insecticidal activity is reported (2). In Dominica, Caribs make an infusion which is given to calm frightened children (3, 4).

References:

1. Claude & al. 1947. 2. Heal & al. 1950: 101. 3. Uphof, 1968: 307. 4. Usher, 1974: 350.

## **Lepidagathis collina** (Endl.) Milne-Redhead

FWTA, ed. 2, 2: 416.
West African: **GUINEA-BISSAU** BALANTA massinquessára (JDES)

A coarse woody herb to 60 cm high, of grassy savanna, from Senegal to W Cameroons, and across equatorial Africa to Ethiopia and to Uganda.

## **Lepidagathis heudelotiana** Nees

FWTA, ed. 2, 2: 416. UPWTA, ed. 1, 452.
West African: **THE GAMBIA** MANDING-MANDINKA wolokuŋ (*def*.-o) = *francolin's head* (Hayes) **GUINEA** FULA-PULAAR (Guinea) towatowei (Adames) MANDING-MANINKA bene-fi (CHOP)

A wiry woody undershrub occurring from Senegal to Ghana.

The plant grows in clumps with velvety hairy leaves and whitish-hairy flower spikes, and is somewhat ornamental. In Guinea the seeds are roasted and ground to make a decoction used as an eye-wash, and the stems and roots are similarly used as a purgative and depurative (1).

Reference:

1. Pobéguin, 1912: 44, as *L. sp.*

## **Lepidagathis serica** Benoist

FWTA, ed. 2, 2: 416.
West African: **SENEGAL** BASARI a-ỹingeny (Ferry) BEDIK ɓoyóm (Ferry)

A woody undershrub from S Senegal, Guinea and Mali.

## **Mimulopsis solmsii** Schweinf.

FWTA, ed. 2, 2: 403. UPWTA, ed. 1, 452, as *M. violacea* Lindau.
West African: **WEST CAMEROONS** KPE majamanjumbe (Waldau)

A herb or undershrub to 4 m high, of montane forest in W Cameroons and Fernando Po.

A decoction of the leaves is taken in the Cameroon Mountain area for gonorrhoea (Santesson fide 1).

Reference:

1. Dalziel, 1937.

**Monechma ciliatum** (Jacq.) Milne-Redhead

FWTA, ed. 2, 2: 429. UPWTA, ed. 1, 452.
West African: SENEGAL MANDING-BAMBARA nâbla (JB ex K&A) WOLOF ngatum béy (Merlier ex K&A) GHANA DAGBANI hunglade (Ll.W) NIGER HAUSA fisawa (JMD) kumudoi (Bartha) NIGERIA HAUSA alkama gidan tururuwa (RES) damfarƙami *loosely applied to weeds of fallows, etc.* (JMD; Lely) fidda hakukuwa (RES) fisawa (JMD; ZOG) kan-bunsuruu = *ram's head* (JMD; ZOG) wútsíyar-kadangaree = *lizard's tail* (JMD; ZOG)

An annual herb to about 1 m high of savanna and disturbed waste lands, throughout the Region, and occurring widely in tropical Africa.

The plant is not grazed in Senegal (1), and, indeed, it is often considered by herdsmen and cultivators to be toxic to stock (4).

In Nigeria the dried leaves are powdered and burnt as an inhalation for headcolds (2). In Sudan the dried seeds are used as a scent-preparation (3).

References:

1. Adam, 1966, a. 2. Dalziel, 1937. 3. Harrison 162, K. 4. Kerharo & Adam, 1974: 114–5.

**Monechma depauperatum** (T. Anders.) C. B. Cl.

FWTA, ed. 2, 2: 429.
West African: IVORY COAST MANDING-MANINKA sanuguélé (B&D)

An erect, variable shrub to over 1 m high, of wet grassland, occurring throughout the Region, and in Cameroun and Angola.

The plant is sometimes prescribed in the savanna areas of Ivory Coast for refactory and very painful headaches (1).

Reference:

1. Bouquet & Debray, 1974: 11.

**Nelsonia canescens** (Lam.) Spreng.

FWTA, ed. 2, 2; 418 (in part). UPWTA, ed. 1, 452 (as *N. campestris* R. Br., in part).
West African: SENEGAL BANYUN ɗapoli (K&A) BASARI ɛ-nám (K&A; Ferry) BEDIK ga-nány (Ferry) DIOLA ékolinkol (JB) DIOLA (Fogny) ékolingéol (K&A) MANDING-BAMBARA konikadlo (JB, ex K&A) MANDINKA ńokoto (K&A) MANDYAK bukobatané (K&A) SERER mbodahâ (JB; K&A) THE GAMBIA DIOLA (Fogny) ékolingéol = *charcoal* (DF) ekolinkol = *run from side to side growth* (DF) MANDING-MANDINKA farokonoɲaama (*def.*-farakonoɲpaa-moo) = *rice-farm grass* (Hayes) ninsikumbaliŋ (*def.*-o) = *cow's knee* (Hayes) GUINEA FULA-PULAAR (Guinea) landan-niari (A. Chev.) UPPER VOLTA HAUSA tsamysar kassa (K&B) NIGERIA FULA-FULFULDE (Nigeria) baali; manda mbaala = *sheep's salt* (auctt.) HAUSA dambun makiyaayaa = *food of pasturage* (JMD; RES) tsaamiyar kasa = *tamarind of the earth* (auctt.) tsaamiyar maharbaa = *tamarind of the hunter* (auctt.) tsaamiyar makiyaayaa = *tamarind of pasturage* (auctt.) HAUSA (West) damdun makiyaya (ZOG) TIV ítúmbā bùá (DA)

A softly pubescent decumbent herb of shaded locations, often stream-beds of the savanna and open country throughout the Region, and generally wide-spread in the tropics. Recent work (10) has shown that *N. canescens* as under-stood in *FWTA*, ed. 2, has included a second species, *N. smithii* Orsted, which is a plant of the damper forest zone. The information given here probably applies to both species equally.

The leaves are acid to taste (3), and in Tanganyika a droplet exudation on the leaf-surface is reported to be also distinctly acid-tasting (9). The plant grows abundantly in damp situations on Gambian rice-farms where the Man-dinka call it 'rice-farm grass.' Nigerian Fula refer to it as 'sheep's salt', and the plant is fed as fodder to goats and to sheep (3). In Guinea also it is known as a salt-substitute (2).

The sap of the leaves is applied topically to guineaworm sores in Ivory Coast-Upper Volta to kill the causative parasite (1, 7). Nupe of Northern Nigeria prepare a brew of the plant for treating smallpox (8), and in Tanganyika the sap is taken for diarrhoea and the root in decoction for schistosomiasis (6).

References:

1. Bouquet & Debray, 1974: 11. 2. Chevalier, 1920: 491–2 (as *N. campestris* R. Br.). 3. Dalziel 363, K. 4. Dalziel, 1937. 5. Deshaprabhu [Ed.], 1966: 7. 6. Haerdi 184. 7. Haerdi & Bouquet, 1950: 230. 8. Medler 21, IFE. 9. Michelmore 1143, 1408, K. 10. Morton, J. K., 1979.

**Nelsonia smithii** Orsted

FWTA, ed. 2, 418 (as *N. canescens* (Lam.) Spreng., in part).
UPWTA, ed. 1, 452 (as *N. campestris* R. Br., in part).
West African: **GHANA** AKAN-ASANTE asresidie (FRI)

A softly pubescent herb of damp situations in the forest zone of the Region from Guinea to West Cameroons, and across Africa to Sudan, Tanganyika and Angola. This plant has in the past been identified conspecifically as *N. canescens* (3) with which usages are probably identical.

In Ghana the leaf-sap is recorded as used as eye-drops for administration in cases of fever (1, 2).

References:

1. Irvine 475, K. 2. Irvine, 1930: 303–4 (as *N. campestris* R. Br.). 3. Morton, J. K., 1979.

**Odontonema cuspidatum** (Nees) O. Kuntze

FWTA, ed. 2, 2: 423.

A shrub to 2 m high, native of Central America and introduced by man to other warm countries. It is reported cultivated as an ornamental in S Nigeria. The flowers are in an erect spike with crimson corollas, 2.5 cm long.

**Pachystachys coccinea** (Aubl.) Nees

English: Cardinal's guard (U.S.A., Bates)

A shrub to 2 m high, native of the West Indies and northern S America and introduced to the Region as an ornamental.

A plant has a showy inflorescence with the flower corolla scarlet and about 5 cm long.

**Peristrophe bicalyculata** (Retz.) Nees

FWTA, ed. 2, 2: 424. UPWTA, ed. 1, 452.
West African: **SENEGAL** MANDING-BAMBARA bara kala (JB, ex K&A) baré (JB, ex K&A) SERER buben (K&A) mut (JB ex K&A) WOLOF môto (K&A) nopo sâdar = *swollen ear* (K&A) **NIGERIA** FULA-FULFULDE (Nigeria) fureya pucci = *fureya (meaning not known) of horses* (MM) HAUSA tubanin dawaki = *flour of the horse* (auctt.)

A perennial herb, stems several-angled, erect to 1.50 m high, occurring in the Sahel part of the Region from Mauritania to Niger and N Nigeria, and distributed throughout tropical Africa and to India, Burma and Thailand.

The plant is relished for grazing by cattle in Senegal (1) and provides fodder in many other countries (Lake Chad, 1; S Rhodesia, 4; Zambia, cattle and game, 7). In India it is cut for horesfeed, and it is ploughed in as green manure (5). There is record of the plant being used in Sudan (2) and India (3) as a remedy for snake-bite. In S Rhodesia a vegetable salt is extracted from it (4).

An unnamed alkaloid is reported present in the leaf and stem (8). A yellowish-brown essential oil can be extracted by steam-distillation and this shows tuberculostatic activity *in vitro,* and inhibits the growth of various strains of *Mycobacterium tuberculosis* (5).

The seeds carry a certain amount of mucilage which on drying can be stretched out into a fine thread. This is used in Senegal for fishing out foreign bodies from eyes and ears (6).

References:

1. Adam, 1966, a. 2. Broun & Massey, 1929: 348. 3. Dalziel, 1937. 4. Davies 2045, K. 5. Deshaprabhu, 1966: 313. 6. Kerharo & Adam, 1974: 115–6. 7. Verboorn 345, K. 8. Willaman & Li, 1970.

**Phaulopsis barteri** (T. Anders.) Lindau

FWTA, ed. 2, 2: 390–400
West African: **SIERRA LEONE** MANDING-MANDINKA pɛrɛtɛtɛ (FCD) **IVORY COAST** KRU-GUERE (Chiehn) ligrotiti (K&B)

An undershrub to 1.30 cm high, of woodland savanna from Guinea-Bissau and Mali to W Cameroons, and into the Congo basin.

The plant has similar uses in Ivory Coast as has *P. falcisepala* for sores, nausea, stomachache, aphrodisia and pains (1). Pimentos (*Capsicum spp.,* Solenaceae) may be added to the leaf-decoction used as an antiseptic in cleansing sores (2).

Like the other *Phaulopsis spp.* the fruit capsules burst with a cracking sound when they were wetted. The Mandinka name in Sierra Leone is onomatopoeic.

References:

1. Bouquet & Debray, 1974: 11. 2. Kerharo & Bouquet 1950: 230, as *Phaylopsis barteri* T. Anders.

**Phaulopsis falcisepala** C. B. Cl.

FWTA, ed. 2, 2: 399. UPWTA, ed. 1, 452, as *Phaylopsis falcisepala* C. B. Cl.
West African: **SIERRA LEONE** LOKO nyaebwe (JMD) nyaebwere (JMD) pɛgɛre (NWT) tɛgɛre (NWT) LOKO (Magbile) kɛgɛre (NWT) MENDE nyaebui (JMD) TEMNE *a*-punt-*a*-ro-kant (NWT) **IVORY COAST** MANDING-MANINKA tudu (B&D) **GHANA** AKAN-ASANTE kwaaduko (FRI) **NIGERIA** YORUBA apa ọgbẹ́ = *to heal wound* (Millen) ata-igbó (Clarke) ata(k)para (Millson)

A herb or undershrub to 1.30 m high, erect or decumbent of the forest zone from Senegal to S Nigeria and Fernando Po.

The plant dried and powdered is applied to staunch wounds in S Nigeria as is indicated by one of the Yoruba names (2), and the fresh sap is applied to small sores (3). The plant in decoction is taken as a laxative (2). In Ivory Coast the plant is used to treat sores caused by skin-parasites, e.g. mange, ringworm and fungal infections; the sap is given as a draught to stop nausea and stomachache, and as an aphrodisiac, and is sometimes put into baths or steam-baths to treat fever stiffness and rheumatic pains (1).

The Yoruba name, *atakpara,* means 'I fire with a sound like "akpara"', and is an onomatopoeic simile of the dried fruit-capsule exploding when placed in water (2).

References:

1. Bouquet & Debray, 1974: 11. 2. Millson, 1891: 217. 3. Thomas, N. W. 1842 (Nig. Ser.), K.

**Phaulopsis imbricata** (Forssk.) Sweet

FWTA, ed. 2, 2: 399. UPWTA, ed. 1, 452, as *Phaylopsis parviflora* Willd.

English: papleaf (The Gambia, Williams).
West African: **SENEGAL** DIOLA (Fogny) nirico — *salt of the cow* (Etesse) **THE GAMBIA** MANDING-MANDINKA fereteto = *sparks when exposed to fire* (Fox; DF) **SIERRA LEONE** LIMBA fureta (NWT) **IVORY COAST** GAGU bagogwé (B&D) KRU-GUERE (Chiehn) kukwékpo (B&D) vaka (B&D) waka (B&D) KWENI vaka (B&D)

*NOTE: Vernacular names under P. falcisepala may also apply.*

A herb or shrub to 60 cm high, distributed from Senegal to S Nigeria and in NE, E and South-central Africa.

The plant is grazed by all stock in the Narok area of Kenya (3), and the young leaves are eaten as a vegetable in Tanganyika (5). In view of the Diola name meaning 'the salt of the cow', it seems probable that cattle in Casamance graze it and benefit from some alleged mineral intake.

The leaves are used in The Gambia in hot fomentations called 'ague cake' over the spleen (7) [? fever treatment]. In Ivory Coast it has the same usages as has *P. falcisepala* for sores, nausea, stomachache, aphrodisia and pains (1). In Nigeria there is probably a similar identity of uses. The Sukuma of Tanganyika apply the powdered root to sores on the legs after washing in warm water (6). Also in Tanganyika, plant-ash in oil is rubbed into scarifications on the back for rheumatism, and on the temples for headache as an analgesic, and the leaf-sap is taken for diarrhoea (4).

The flowers are frequented by bees in Kenya (3).

Like the fruit-capsules of *P. falcisepala,* those of this species burst when placed in water, an early rains dispersal mechanism. This makes a plaything for small boys in The Gambia (2). The Mandinka name is onomatopoeic.

References:

1. Bouquet & Debray, 1974: 11. 2. Fox 70, K. 3. Glover & al. 793, K. 4. Haerdi, 1964: 184, as *Phaylopsis longifolia* Thoms., but probably correctly *P. imbricata* — cf. Haerdi 316/OH, 316/OB, K. 5. Semsei F. H.2561, K. 6. Tanner 1554, K. 7. Williams, F. N., 1907: 370, as *Phaylopsis parviflora* Willd.

**Pseuderanthemum atropurpureum** L. H. Bailey

English: purple false eranthemum (U.S.A., Bates).

A shrub to 1.30 m high, native of Polynesia and brought into cultivation in several countries as an ornamental. It is found in gardens in Ghana (1). The leaves are 15 cm long or so, purple, sometimes green, variously marked with white or yellow along the veins. Flowers are borne in a loose spike: corolla 2.5 cm across, white with purple markings.

Reference:

1. Irvine, 1961: 746.

**Pseuderanthemum dispersum** Milne-Redhead

FWTA, ed. 2, 2: 421. UPWTA, ed. 1, 453, as *P. hypocrateriforme* Radlk.

A herb or shrub to 80 cm high, of S Nigeria and W Cameroons; also in E Cameroun to Sudan.

In Sudan the root ground to a paste is given in Sorghum beer for stomachache (1).

References:

1. Broun & Massey, 1929: 344, as *Eranthemum hypocrateriforme* Roem. & Schult.

**Pseuderanthemum ludovicianum** (Büttner) Lindau

FWTA, ed. 2, 2: 421.

A weak scrambling shrub to 4 m high, of undergrowth of evergreen and deciduous forests, from Liberia to W Cameroons and Fernando Po, and also in E Cameroun, Congo basin, NE and E Africa, and Angola.

The plant has white, pale blue or purplish flowers with a spotted tube, 3.25 cm long and decorative.

The whole plant is cooked in Congo as a vegetable and eaten with meat or fish to prevent enlargement of the spleen, and to render women fecund. A leaf-decoction is drunk for pains in the sides and for menstural troubles,and also it is used as a wash for a newly-born baby (1).

Reference:

1. Bouquet, 1969: 50.

**Pseuderanthemum tunicatum** (Afzel.) Milne-Redhead

FWTA, ed. 2, 2: 421–2. UPTWA, ed. 1, 453, as *P. nigritianum* Radlk.
West African: **SIERRA LEONE** BULOM nolli (Afzelius) MENDE jɛbi (NWT) kɔkawuli (NWT) TEMNE *a*-foɛn (NWT) *a*-mɛs (Afzelius) **NIGERIA** IGBO (Obompa) ùchíchì-nà-èfífiè = *night and afternoon* (NWT; KW)

A woody herb or undershrub to 85 cm high, of the evergreen forest, on rocks and near streams, occurring commonly from Ghana to W Cameroons and Fernando Po, and widespread throughout tropical Africa.

The plant is eaten in Gabon (4). A decoction of the plant is drunk in Congo as a strengthening tonic, to restore appetite and to combat fatigue (1). Fresh leaves or a decoction of the whole plant is recorded as having been used in Sierra Leone as a fomentation for orchitis (probably venereal) (Afzelius fide 2). It is used in Gabon for yaws (7, 8). The root is used in Tanganyika for chest-complaints (5), and abdominal pain (6), and a root-decoction with leaf-sap is drunk for pneumonia (3). The plant is given to cattle in Tanganyika as a binding stomachic (4).

References:

1. Bouquet, 1969: 50. 2. Dalziel, 1937. 3. Haerdi, 1964: 184. 4. Irvine, 1961: 744, 746. 5. Koritschoner 1266, 1279, K. 6. Koritschoner 1321, K. 7. Walker, 1952: 184, as *Eranthemum nigritanum* T. Anders. 8. Walker & Sillans, 1961: 42, as *E. nigritanum* T. Anders.

**Rhinacanthus virens** (Nees) Milne-Redhead

FWTA, ed. 2, 2: 425. UPWTA, ed. 1, 453, as *R. communis* Nees.
West African: **SIERRA LEONE** LOKO ndɔwɔ (NWT) MENDE kpondo (GFSE) **GHANA** AKAN-ASANTE kwaaduko (FRI) TWI kwaduko (JMD) **NIGERIA** IGBO (Onitcha) áfịfịá àchàlà (KW) IGBO (Umuahia) mbe (Ariwaodo)

A semi-woody weak herb to 1 m high of forest under-growth, from Guinea to W Cameroons, and dispersed to Congo.

The plant is used medicinally in Sierra Leone for indigestion (Scott Elliot fide 2). The roots and leaves are used in Nigeria for indigestion, ringworm and other skin-diseases: antiparasitic *oxymethylanthro-quinones* are reported present (3). Leaves and stems are used in Gabon as an emetic (4). The Igbo at Umuahia use the plant as a fish-poison (1).

References:

1. Ariwaodo ARS.1130, K. 2. Dalziel, 1937. 3. Oliver, 1960: 36. as *R. communis*. 4. Walker & Sillans, 1961: 43, as *R. communis* Nees.

**Ruellia praetermissa** Schweinf.

FWTA, ed. 2, 2: 396–7.

A semi-woody herb or shrub to 30–40 cm high, in undergrowth of secondary forest and in grass savanna of Senegal to Dahomey, and in NE and E Africa.

The usages recorded for *R. patula* Jacq. as follow probably apply. In Kenya this latter plant is eaten as a vegetable (2), though in the Morogoro District of Tanganyika an edible and an inedible form are recognized (6). In Kenya (4) and in Sudan (1) goats, sheep, etc. relish the foliage. The Sukuma of Tanganyika apply the roots and leaves pounded up with ghee and made hot to swellings and to boils (5). The leaves are said to have (unspecified) medicinal use in Ghana (3).

References — as *R. patula* Jacq.:

1. Andrews A.2711, K. 2. Graham 1745, K. 3. Irvine, 1961: 746. 4. Mwangangi 1347, K. 5. Tanner 1298, K. 6. Wallace 557, K.

**Ruellia primuloides** (T. Anders.). H. Heine

FWTA, ed. 2, 2: 397–8, as *Endosiphon primuloides* T. Anders.
West African: **GHANA** AKAN-ASANTE ɔbɔfofan (FRI)

A herb or semi-woody-undershrub to 1 m high, of the forest zone from Sierra Leone to S Nigeria and Fernando Po, and extending to Gabon.

**Ruellia tuberosa** Linn.

English: meadow weed, menow weed (U.S.A., Bates).

An erect semi-woody perennial herb with an elongated fleshy tuberous root, native of the Caribbean region, introduced to many tropical and subtropical countries, and now recorded as naturalized in Ghana.

The blue-violet or violet-mauve flowers are attractive and for these the plant is grown as an ornamental. Two white-flowered forms are known and are in cultivation.

A decoction of leaves is said to relieve chronic bronchitis. The plant has emetic properties and is a substitute for ipecacuanha. It is also used to treat stones in the bladder (1). Tests for antimalarial activity in avian malaria have proved negative (2), and insecticidal action is reported to be very slight (3).

References:

1. Chadha, 1972: 90. 2. Claude & al. 1947. 3. Heal & al., 1950: 101.

**Rungia eriostachya** Hua

FWTA, ed. 2, 2: 429–30.
West African: **SENEGAL** BASARI doxonyá (Ferry) KONYAGI yunul, wa-yunul (Ferry)

An undershrub of scrub in S Senegal, Guinea-Bissau and Guinea.

**Rungia grandis** T. Anders.

FWTA, ed. 2, 2: 430, in part. UPWTA, ed. 1, 453, in part.
West African: **NIGERIA** YORUBA aládé oko = *prince of the field* (Millson)

A shrub to 3 m high, of the forest from Dahomey to W Cameroons and extending to the Congo basin and Uganda.

In Gabon the leaves are used as anthelmintic and the bark as an antidysenteric (2). A fair quantity of alkaloid has been detected present in the leaves of Nigerian material (1).

References:
1. Adegoke & al. 1968. 2. Walker & Sillans, 1961: 43.

**Rungia guineensis** Heine

FWTA, ed. 2, 2: 430, as *R. grandis* sensu Heine, pp. UPWTA, ed. 1, 453, as *R. grandis* sensu Dalziel, pp.
West African: **SIERRA LEONE** MENDE begafe (Macdonald)

An undershrub of the forest occurring from Guinea Bissau to Ghana.

**Ruspolia hypocrateriformis** (Vahl) Milne-Redhead

FWTA, ed. 2, 2: 431–2.
West African: **GHANA** GA wɔnane (FRI)

A straggling shrub to over 1 m high of the savanna, and secondary and deciduous forest areas from Senegal to W Cameroons, and dispersed to Uganda, Kenya and Transvaal.

The plant is distinctly ornamental. The flowers are borne in showy terminal inflorescences with coral-red tubular flowers 3.75 cm long. A cultivar grown in America is predominantly yellow with lobes crimson on the outer side (1). The plant makes a good hedge.

Reference:
1. Bates, 1976: 989.

**Satanocrater berhautii** Benoist

FWTA, ed. 2, 2: 403.
French: coupe de Satan (Berhaut)

Herb with attractive mauve flowers to 6.25 cm long, recorded from Guinea, and perhaps present also in Senegal.

**Sclerochiton vogelii** T. Anders.

FWTA, ed. 2, 2: 408.

A slender scandent shrub to 2 m high, recorded from Guinea to Ghana, and in the Congo basin, Angola and E Africa.

The plant occurs in Ghana on sand-dune formations behind beaches and may perhaps contribute to sandbinding.

Traces of alkaloids have been detected in the leaves and bark of an unspecified *Sclerochiton* (perhaps this species) in Congo, and at 0.1–0.3% concentration in the roots (1).

Reference:
1. Bouquet, 1972: 12.

**Stenandriopsis buntingii** (S. Moore) Heine

FWTA, ed. 2, 2: 409, as *Crossandra buntingii* S. Moore
West African: **SIERRA LEONE** MENDE kɔmafai (FCD) kɔmafali (FCD)

A small herb of damp sites in the forest of Sierra Leone and Ivory Coast. In Sierra Leone the plant is rubbed on babies suffering from fits (1).

Reference:
1. Deighton 3810, K.

**Stenandriopsis guineensis** (Nees) Benoist

FWTA, ed. 2, 2: 409, as *Crossandra guineensis* Nees. UPWTA, ed. 1, 450, as *C. guineensis* Nees.
West African: **GHANA** GURMA (Mango-Jendi) wobogu (Mellin)

A low herbaceous plant from a woody base with stem to about 15 cm high and spike a further 8–12 cm, of the forest floor in Sierra Leone to W Cameroons and Fernando Po, and across the Congo basin to Uganda.

The whole plant is very decorative and is suitable as a garden ornamental. The leaves are dark green above with golden reticulated nerves, and reddish beneath; the flowers are pale lilac borne in a slender compact spike (2). The dried powdered leaves are used medicinally in Togo: a spoonful morning and evening boiled along with cereal pap is a remedy for diarrhoea. Analysis has shown only mucilage and some sugar and tanin (Thoms fide 1).

References:
1. Dalziel, 1937. 2. Hooker, 1878: 104: t. 6346.

**Thomandersia anachoreata** Heine

FWTA, ed. 2, 2: 413, as *T. laurifolia* (T. Anders.) Baill., quoad Liberia and Ivory Coast.

A shrub to 2 m high, of river-banks and forest-margins of the primary forest in Liberia and Ivory Coast.

No usage is recorded for the Region but the medicinal uses recorded under *T. hensii* may apply.

**Thomandersia hensii** De Wild. & Th. Dur.

FWTA, ed. 2, 2: 413, as *T. laurifolia* (T. Anders.) Baill., in part.

A shrub or small tree 1–4 (–15) m high, of the primary forest of S Nigeria and W Cameroons, and extending to Zäire.

The wood is very hard and is difficult to cut with a matchet (4). It is used in hut-construction to make poles, stakes and screens (3).

A leaf-decoction is used in the Cameroun—Zäire area for diarrhoea treatment and a leaf-infusion for colic. The leaf, like the bark, is very bitter. Sap is irritant to the skin and is used in a general way on skin-troubles — furuncles, abscess, yaws sores, eczema, eruptions of chickenpox and smallpox, syphilitic sores and venereal ulcers, leprosy, etc. If the sores are open powdered dried leaves are put on after washing with a decoction of the whole plant. A tisane of roots or leafy twigs with the sap expressed from the softer parts is taken for infections of the urino-genital system (e.g. leucorrhoea, vaginitis, urethral discharge), for intestinal parasites, and as a tonic in cases of debility and fatigue. Pulped roots are used in frictions for oedemas and rheumatism, and sap from them is sometimes used in eyes and ears for inflammation. (2, 3).

The flowers are very nectariferous, and are reported to be frequented in W Cameroons by ants seeking the honey (1).

Old herbarium material retains a strong smell of *coumarin* (3). The phytochemistry and pharmacognosy merit investigation.

The tree has fetish attributes. Hunters in Zäire place a branch in their huts to ensure success in the hunt, and a tree is commonly planted before a village headman's house (3). The tree is planted in Congo by a hut-entrance to keep away demons, and the sap as a drink will exorcise evil spirits and heal illness caused by sorcery (2).

These particulars may refer also to other *Thomandersia spp.*

References:

1. Bates 324, K. 2. Bouquet, 1969: 51 under *T. longifolia* De Wild. 3. Heine, 1966: 228. 4. Obiorah FHI.20312, K.

**Thomandersia laurifolia** (T. Anders. ex Benth.) Baill.

FWTA, ed. 2.2: 413, in part.

A bush 1–2 m high, recorded only from W Cameroons in the Region, but occurring from E Cameroun to Zäire.

No usage is recorded in W Cameroons, but in the rest of its range it appears to have the same uses as *T. hensii* (1).

Reference:

1. Heine, 1966: 219.

**Thunbergia alata** Boj. ex Sims

FWTA, ed. 2, 2: 400.

English: black-eyed Susan.

A herbaceous twiner, native of E and S Africa, dispersed by man to most tropical and subtropical countries for its decorative value.

In Kenya it is readily grazed by all stock (4) and the leaves are eaten as a vegetable (5). In India (7) and Malaya (1) the leaves are applied to the head for headache. Unnamed alkaloid is reported present (1, 2, 7). In Tanganyika leaf-sap is dripped into the eyes for inflammation (probably conjunctivitis), and together with that of *Hyptis pectinata* (Linn.) Poit (Labiatae) is drunk for internal piles or for early rectal cancer (6).

The pale to dark yellow flowers with a red or dark purple throat are decorative and are the cause of the English name. Cultivars with white or orange-yellow flowers are known. The fruit resembles a diminuitive gourd and is used as such in play and as a top by the children in the Narok area of Kenya (3).

References:

1. Burkill, 1935: 2158. 2. Chadha 1976, a: 233–4. 3. Glover & al. 356, K. 4. Glover & al. 490, 2188, 2532, 2602, K. 5. Graham 1868, K. 6. Haerdi, 1964: 185. 7. Willaman & Li, 1970.

**Thunbergia chrysops** Hook.

FWTA, ed. 2, 2: 402. UPWTA, ed. 1, 453.

West African: **SIERRA LEONE** LOKO tɛbaga (NWT) tɛbaiya (NWT) MENDE gomakpo (FCD) kpɔkpo (*def.*-i) *a general term for Ipomoea spp. and similar plants* (JMD; FCD) ndondokɔ (*def.*-i) (FCD) **IVORY COAST** KWENI diatakpakpa (B&D)

A slender climber of the wooded zone from Guinea to W Cameroons.

A plant has a showy violet-purple flower 7.5–10 cm long with a yellow throat. It is often cultivated as an ornamental.

Leaf-sap is rubbed into cuts in Sierra Leone to promote healing (2). In Ivory Coast an aqueous macerate is sometimes used to relieve stomachache and cough in children, and is taken for smallpox (1).

References:
1. Bouquet & Debray, 1974: 11. 2. Deighton 2169, K.

**Thunbergia cynanchifolia** Benth.

FWTA, ed. 2, 2: 402.
West African: **SIERRA LEONE** SUSU-DYALONKE kosona (Haswell) **IVORY COAST** AKAN-ASANTE filidota (K&B) BAULE fita fita nzali (K&B)

A slender climber of the forest zone from Guinea to W Cameroons, and also in E Cameroun.

Sap from the leaves is mixed with palm-wine in Ivory Coast and taken for colic, and an aqueous macerate is given to children for cough and is taken for smallpox (1, 2).

References:
1. Bouquet & Debray, 1974, 11. 2. Kerharo & Bouquet, 1950: 230–1.

**Thunbergia erecta** (Benth.) T. Anders.

FWTA, ed. 2, 2: 402.
English: bush clock vine, king's mantle (U.S.A., Bates).

A shrub to 5 m or more, erect or scandent, with pale or dark violet-purple flowers 5.0–7.5 cm long with white tube, of the forest zone from Guinea-Bissau to W Cameroon.

The plant is often cultivated as an ornamental. It stands clipping and makes an excellent hedge. Two forms are known in Ghana: one with small leaves and violet flowers setting viable seed, and the other with larger leaves, pale blue flowers and seedless. The latter is the more drought-resistant. (1).

Reference:
1. Irvine, 1961: 748.

**Thunbergia grandiflora** (Roxb.) Roxb.

FWTA, ed. 2, 2: 402.
English: blue trumpet vine, clock vine, Bengal clock vine (U.S.A., Bates).

A woody climber, native of Assam, India, and introduced to most humid tropical and subtropical countries. It has become naturalized in the Calabar area of Nigeria.

The leaves are eaten as a vegetable in Assam, and are fed to tame rabbits. The light blue flowers are attractive and the plant can be trained over a frame to make a screen, or up over the canopy of trees. The leaves are sometimes taken in India for stomach-complaints and are used as poultices, and usage found also in Malaya. The leaves are said to be rich in potassium salts. The nectar is high in sucrose with also some glucose and fructose. (1, 2).

References:
1. Burkill, 1935: 2158. 2. Chadha 1976, a: 234.

**Thunbergia laevis** Nees

FWTA, ed. 2, 2: 400.

A slender climber, white tubular flowers 3.25 cm long, native of tropical Asia, and introduced as an ornamental to many tropical countries. It is present in Sierra Leone and Ghana as a garden-plant.

**Thunbergia laurifolia** Lindl.

FWTA, ed. 2, 2: 402.

A soft-wooded climber, attaining the top canopies of trees, native of the rain-forest of the Andamans, and widely introduced by man into tropical countries as an ornamental for the sake of its lilac-coloured flowers.

In Malaya (1) and India (2) there are medicinal application; the leaves are used for poulticing cuts and boils, and the leaf-sap is taken for menorrhagia and instilled into the ears for deafness.

The plant is of very vigorous growth and may smother and kill trees over which it grows. It can be trained with good horticultural effect over trellises.

References:
1. Burkill, 1936: 2158. 2. Chadha 1976, a: 234.

**Thunbergia vogeliana** Benth.

FWTA, ed. 2, 2: 402. UPWTA, ed. 1, 453.
West African: NIGERIA EDO ohohiro (Farquhar)

A shrub to 5 m or more with pale or dark violet-purple flowers 5.0–7.5 cm long with white tube, from Ghana to W Cameroons and Fernando Po.

The plant is sometimes cultivated as an ornamental.

**Whitefieldia colorata** C. B. Cl.

FWTA, ed. 2, 2: 399.

An undershrub to about 3 m high, of the forest zone of Sierra Leone to Ivory Coast. Ornamental with bright red or reddish-purple flowers 2.5–4.0 cm long, with bracts turning red.

**Whitfieldia elongata** (P. Beauv.) De Wild. & Th. Dur.

FWTA, ed. 2, 2: 398. UPWTA, ed. 1, 453, as *W. longifolia* T. Anders.

A shrub, erect or semi-scandent, to 3–5 m long, of the rain-forest in S Nigeria, W Cameroons and Fernando Po, extending to Sudan, Tanganyika and Angola.

In the Igbo area of S Nigeria the stems are used to make spinning spindles (4), and when the plant comes into fruit, it is said to be time to clear the bush for planting crops (5).

The leaves pulped up after being passed through a fire are used as an embrocation in bronchitis in Congo, and the same preparation is eaten as a vegetable by women to aid conception (1). A leaf-decoction in palm wine is drunk in Congo for stomach-complaints and food-poisoning (1). The leaves are used in Tanganyika as an antidote for headache (2). Unspecified medicinal use is made of the plant in Gabon (6). The leaves yield a black dye use in Gabon (6) and in Zaïre (3).

References:

1. Bouquet, 1969: 51–52. 2. Gane 68, K. 3. Greenway, 1941: as *W. longifolia* A. And. 4. Thomas, N. W. 2003 (Nig. Ser.), K. 5. Thomas, N. W. 2015 (Nig. Ser.), K(Arch.) 6. Walker & Sillans, 1961: 43, as *W. longifolia* T. And.

### **Whitfieldia lateritia** Hook.

FWTA, ed. 2, 2: 398–9.

An undershrub to about 3 m high, of the forest zone in Guinea to Ivory Coast. Distinctly ornamental with red or salmon-coloured flowers to 2.5–3.75 cm long in terminal racemes.

### **Acanthaceae indet.**

UPWTA, ed. 1, 453.
West African: **NIGERIA** HAUSA bi ta swaiswai, bi ta zaizai *seeds imported from Air, Niger, for sale in Sokoto market* (JMD)

The seeds of an unidentified plant are recorded as being sold in Sokoto market under the above names and were said to have been brought from the Aïr region of Niger by Berber Tuaregs. The seeds were used by youths as a love-charm to secure affection, the underlying idea being urgent persistance in following the object of desire. The words refer to a kind of ant which goes in pairs, the one following close on the other; syn. *bi ta ɗanɗan,* Bargery, (1).

Reference:
1. Dalziel, 1937.

## AGAVACEAE

### **Agave sisalana** Perrine

FWTA, ed. 2, 3: 154.
English: sisal; sisal hemp.
French: sisal
Portuguese: agave.
West African: **SENEGAL** DIOLA (Fogny) busata (K&A) SERER yos (JB; K&A) **IVORY COAST** KWENI bwee yúrú (Grégoire) **NIGERIA** KANURI gòdéngəlái (C&H)

A robust acaulous perennial plant with a rosette of thick fleshy spiny leaves to 2 m long, native of Mexico to Central America, and dispersed by man to many countries.

The plant is the source of the important fibre known as sisal. The name is taken from a small port of that name in Yukatan, Mexico, from where the first exports are thought to have been made. The crop is obtained from the leaves which are retted to obtain the fibre which now accounts for about 70% of the world's hard fibres. Tanaganyika is the prime producing country, with Brazil second. Small plantations have been established in the Region but cultivation has not assumed much importance.

The plant is frequently planted as a barrier plant. The thorny leaf-edges make in a closely planted line an impenetrable barrier in 2–3 years (4, 6). In this respect, however, the fiercer the thorns the better and some other species are more suitable. On the other hand, for harvesting for fibre extraction, the fewer the thorns the easier the leaves are to handle. *A. sisalana* is held to be a natural hybrid of, perhaps, *A. fourcroydes* Lem. in which the thorns are reduced, and breeding and selection work for fibre production is in part to further their elimination.

The genus consists of some 300 species. Most yield a fibre and several others have been distributed by man to countries beyond the New World. *A. fourcroydes* is the source of *henequen* fibre and is coarser and less attractive than sisal. It finds particular use in agriculture as binder-twine. *A. cantala* Roxb., is the *cantala* or *maguey* of Mexico. The fibre is finer, whiter and more flexible than other agave fibres. *A. americana* Linn. is the Century Plant of temperate countries, the name being conjured up on the belief that the plant flowers only when 100 years old, after which it dies. Interpretation of this species has been confused. Fibre plantations of agave under this name have been established in Mali (10), and it is said to be an important sand-binder near Timbuktoo (8). In India it is grown as a fence along railway tracks (4). It thrives in dry sandy conditions.

The national drink of Mexico is *pulque.* If the flowering stem is cut off, a considerable quantity of sap is exspissated and this on fermentation is pulque. It has been attributed to *A. americana,* but other more succulent species are the usual sources in Mexico. Alcoholic content of fresh pulque is 4–8%. It is distilled to produce a stronger drink called *mescal.* Other liquors prepared from *Agave spp.* are *aquamiel* and *tequile.*

The bagasse after retting the leaves of *A. sisalana* contains as much as 11% sugars. Fermentation for the production of alcohol should be feasible but an economic process does not appear to have been worked out. Four steroidal substances have been isolated from sisal waste of which *hecogenin (mecogenin)* is considered the most important. This has been found in the leaves of several other *A. spp.,* and is a possible source for production of cortisone, but cultivars of *A. sisalana* selected for sisal show reduced hecogenin and other strains have been selected the leaves of which can be processed without dilution by water (3, 5, 7, 11). Sisal waste also yields a very hard wax derived from the leaf-cuticle (11). Pectins and pectates are also present.

References:

1. Burkill, 1935: 65–70. 2. Goulding, 1937: 42. 3. Hardman, 1969: 208. 4. Howes, 1946. 5. Kerharo & Adam, 1974: 116–7, with phytochemistry and pharmacology. 6. Morton, 1961: 22–23. 7. Oliver, 1960: 18, 44. 8. Portères, s.d. 9. Purseglove, 1972: 1: 8–29. 10. Roberty, 1955: 13. 11. Watt & Breyer-Brandwijk, 1962: 19.

**Furcraea foetida** (Linn.) Harv.

FWTA, ed. 2, 3: 154, as *F. gigantea* Vent.

English: Mauritius hemp.

French: fourcroya.

A herbaceous perennial with a rosette of prickly and pointed leaves to 1–2 m long by 12–20 cm wide, acaulescent or occasionally with a trunk, native of tropical S America and dispersed by man to many tropical countries.

Man's first interest in the plant was as an ornamental, especially races with yellow leaf-margins. In many countries it has thrived and become naturalized. It is to be found around villages in W Africa. It is grown as a barrier plant, the fiercely hooked and pointed leaves creating an obstacle no large animal, nor man can easily penetrate. It is used in India and Ceylon along railway tracks as a live fence (3). The plant in Mauritius has prospered exeedingly filling hedges between sugar-cane plantings, and in rocky terrain unfit for cultivation. Its fibre was found to be good so that now Mauritius is a leading producer of Furcrea fibre which is known commercially as Mauritius hemp. The fibre is longer, finer and softer than that of sisal but it is not so strong. It is usable for making cloth, twine and rope, but it fails as a marine cordage (2). In Mauritius it is woven into material for sugar-bags (4, 5).

The leaves contain an irritant substance in the sap. Plantation workers require protective gloves. Saponin is present which causes foaming. Hecogenin and a derivative have been reported (6).

The genus contains some 20 species and a number of others also yield a fibre which is extracted in S America. *F. selloa* Koch. is introduced and is more or less naturalized in W Africa. It yields a fibre identical with Mauritius hemp.

References:

1. Burkill, 1935: 1039–41. 2. Goulding, 1937: 47–48. 3. Howes, 1946. 4. Kirby, 1963. 5. Purseglove, 1972: 1: 29–31. 6. Watt & Breyer-Brandwijk, 1962: 32.

## AÏZOACEAE

**Aizoon canariensis** Linn.

UPWTA, ed. 1, 30.

West African: **MAURITANIA** ARABIC (Hassaniya) lkumsa (AN) **MALI** TAMACHEK ehafief (AM) iheiffif (B&T)

A diffuse or prostrate herb, partly woody, branches widely spreading, occurring in Mauritania and along the Sahara border of the Region to NE Africa.

In Cape Verde Islands the plant is mashed up into a paste with shrimps and put into holes in rocks along the littoral as fish-poison (1). In Mauritania the plant is used in treatment of warts (2).

References:

1. Howes, 1930: 134. 2. Naegelé, 1958, b: 891–2.

**Gisekia pharnaceoides** Linn.

FWTA, ed. 2, 1: 134, 759. UPWTA, ed. 1, 30.

West African: **SENEGAL** MANDING-BAMBARA kona basi (A. Chev.; JB) **MALI** MANDING-BAMBARA melanguer (A. Chev.) **NIGER** FULA-FULFULDE (Niger) takka cijla (PF) **NIGERIA** FULA-FULFULDE (Nigeria) nalle waynaaɓe = *herdsmen's henna* (JMD; MM) takka cijla (MM) HAUSA dandami (JMD; ZOG) geron tsuntsaye = *bird's millet* (JMD) lallen shaamuwaa = *stork's henna* (JMD; ZOG) HAUSA (West) lenje (JMD; ZOG) nadedel (BM)

A glabrous fleshy annual herb, very variable in size with branches from 1–2 cm to 70 cm long, occurring widespread in the northerly and dryer part of Region from Senegal to N Nigeria, and common in tropical and southern Africa, Madagascar and Asia. The plant is subject to fungal attack by *Exobasidium gisekiae* Allesch. which renders it pink or wine-red in colour, a situation accounting for the Fula name of 'herdsmen's henna' and the Hausa name 'stork's henna.'

The plant is occasionally eaten: as an emergency food in West Africa (3) and India (7), and elsewhere as an occasional vegetable (Kenya, 7; Tanganyika, 15; Somalia, 4). In Tibesti of northern Chad it is subjected to prolonged cooking, with hashed-up meat till nearly dry, when it is eaten as a condiment (10). It is eaten as a condiment in Zaïre (6). In Lake Province of Tanganyika the whole plant is eaten as a general strength restorative (14). Cattle and goats graze it in Senegal but in Chad all stock refuse it (1). In Ghana goats will take it and sometimes the plant is collected for fodder (7). The fruit is reputed to be poisonous (15), and it is perhaps the stage of growth that determines whether or not stock will graze it.

The plant is sold in medicine markets to the north of the Region as a purgative (A. Chevalier fide 3). In E Africa (2) and in South Africa, Tanganyika and Madagascar the plant is taken for diarrhoea (15). In India, Indonesia, South Africa and Madagascar it is used as a taenicide (15). It is used in the West African region to rub on swellings in the same fashion as *Portulaca* (Portulacaceae) is used (3). In Lake Province of Tanganyika the stem pounded with butter is placed on aching muscles (13). In Sokoto it is reported as used on areas of pain, probably rheumatic (9). Pounded with other herbs and native natron it is made into poultices for sores in cattle in N Nigeria (3). In Tanganyika the green leaves are cooked and eaten for asthma (12); in Kenya the roots are made into a chest medicine (8), and Swahili of East Africa make the whole plant into a remedy against miscarriage (15). The sap is used on warts in India (5).

Tannins are present in the plant (11, 15), and tannin-like principles α- and β-*gisekia* are in the seeds and these are probably anthelmintic (15).

References:

1. Adam, 1966, a. 2. Bally, 1937. 3. Dalziel, 1937. 4. Fawcett 27, K. 5. Hartwell, 1970. 6. Hauman, 1951, b: 102–3. 7. Irvine, s.d. 8. Jeffery K.465, K. 9. Moiser s.n., 8/3/1922, K. 10. Monod, 1950: 65. 11. Oliver, 1960: 27. 12. Tanner 3357, K. 13. Tanner 4041, K. 14. Tanner 4072, K. 15. Watt & Breyer-Brandwijk, 1962: 833.

**Glinus lotoides** Linn.

FWTA, ed. 2, 1: 135, 759. UPWTA, ed. 1, 30 (as *G. lotoides* Loefl.).
West African: **NIGERIA** FULA-FULFULDE (Nigeria) rubbundehi = *it grows in cow-pats* (J&D)

A spreading or prostrate annual herb, very variable, stems to 50 cm long, of damp sandy places, occurring across the northern part of the Region from Senegal to Niger and N Nigeria, and widely distributed in tropics and sub-tropics generally.

The plant is recorded as being the first colonizer of pools on their drying out in the Jebel Marra area of Sudan and then to become the dominant herb (8). At places in Sudan it is eaten as a vegetable (2). In Senegal cattle will not graze it, but it is taken by goats and pigs (1), and occasionally by goats in Kenya (5). The flowers are much visited by honey-bees in Tanganyika (3).

The small black or brown seeds are numerous and are eaten in *cus-cus* in Tibesti (4).

The plant has unspecified medicinal uses in Guinea (6). In India the herb is given as a purgative for abdominal illnesses, and is also prescribed for diarrhoea. It is used for itches and skin-diseases (7).

References:

1. Adam, 1966, a. 2. Andrews A. 686, K. 3. Burtt 5232, K. 4. Monod, 1950: 71. 5. Mwangangi & Gwynne 1235, K. 6. Pobéguin 505, K. 7. Quisumbing, 1951: 278, (as *Mollugo lotoides* (Linn.) C. B. Clarke, with references). 8. Wickens 1133, K.

**Glinus oppositifolius** (Linn.) A. DC

FWTA, ed. 2, 1: 135.

A slender spreading or ascending annual herb, variable, stems to 40 cm high, of damp sandy sites, occurring across the Region from Senegal to S Nigeria, and widely distributed in the tropics and sub-tropics generally.

No usage is recorded in the Region. In Zaïre animals will graze it for lack of better forage (5). In Kenya the leaves are used for headache (3), and for stomach complaints (2). In Tanganyika the fresh leaves pounded into a paste with butter are used for swollen fingers and toes (4), and in Zaïre chopped up leaves serve as a vulnerary (1).

In the Philippines the whole plant, less the roots, is eaten as a vegetable for which proper cooking is necessary to disperse the great bitterness. The plant is reported to be exceptionally rich in iron and to be a good source of calcium. In India it is considered to be stomachic, aperient and antiseptic, and to be good for treating dermal infections (6).

References:

1. Hauman, 1951, b: 108–9. 2. Jeffrey K.466, K. 3. Jeffery K.515, K. 4. Tanner 1100, K. 5. Troupin 282, K. 6. Quisumbing, 1951: 279 (as *Mollugo oppositifolius* Linn., with references).

**Limeum diffusum** (Gay) Schinz

FWTA, ed. 2, 1: 134. UPWTA, ed. 1, 30 (as *L. linifolium* Fenzl).
West African: **MALI** SONGHAI maïtara (JMD)

A diffuse glabrous annual occurring in the northern part of the Region from Mauritania to Niger.

In Senegal it is said to provide grazing for cattle, sheep, goats and donkeys (1).

Reference:

1. Adam, 1966, a.

**Limeum indicum** Stocks

FWTA, ed. 2, 1: 134.

A diffuse to sub-prostrate herb of desert regions in Mauritania, Mali and Niger, and across the Sahara and N Africa to India.

In Tibesti, northern Chad, the seeds are collected for food after the rains, and the plant is used to treat burns (1).

Reference:
1. Monod, 1950: 63, 72.

**Limeum pterocarpum** (Gay) Heimerl

FWTA, ed. 2. 1: 134.

A diffusely erect herb, 30–85 cm high, a weed of fields and in waste places of the northern part of the Region from Mauritania to N Nigeria, and across Africa to Sudan, and into Southern Africa.

The plant is not grazed by stock in Senegal (1).

Reference:

1. Adam, 1966, a.

**Limeum viscosum** (Gay) Fenzl

FWTA, ed. 2. 1: 134.

An annual to 30 cm high or more recorded from the northern part of the Region from Mauritania to N Nigeria, and also in the Sahara and Sudan.

The plant is grazed by all stock in Senegal and is especially relished by cattle (1). In S Africa it has been suspect of being poisonous to cattle, but cattle appear to take it freely with no ill effect (3). The seed is considered edible in Morocco (2).

References:

1. Adam, 1966, a. 2. Monteil, 1953: 54. 3. Watt & Breyer-Brandwijk, 1962: 833.

**Mollugo cerviana** (Linn.) Seringe

FWTA, ed. 2, 1: 135.

A small erect annual herb to 10 cm high, usually less, of sandy sites, widespread in the Region from Senegal to N Nigeria, and in N Africa, Europe, Asia and Australia.

Cattle are said not to graze it in Senegal but sheep will take it (1). In Somaliland lambs graze it (3). In Ethiopia it is favoured by young locust hoppers (2), and in Malawi it is reported to be stored underground (for food?) by mole-crickets (4).

In Tanganyika it is collected in large quantities, dried, powdered and burnt to put in incisions for pleurisy, and the leaf is chewed for cough (6). Chewing the leaves is said also to reduce hang-over (5).

References:

1. Adam, 1966, a. 2. Ellis 194, K. 3. Fawcett 71, K. 4. Lawrence 32, K. 5. Tanner 790, K. 6. Tanner 4045, K.

**Mollugo nudicaulis** Lam.

FWTA, ed. 2, 1: 134. UPWTA, ed. 1, 30.

English: daisy-leaved chickweed (Dalziel).

West African: **NIGERIA** FULA-FULFULDE (Nigeria) buubiho = *fly plant* (J&D; MM) GOEMAI narba (JMD) HAUSA ankwai narba (ZOG)

An annual herb with a rosette of prostrate leaves and a flowering stem to 25 cm high, occurring in grassy savanna throughout the Region from Senegal to S Nigeria, and widespread in tropical Africa and Asia.

The plant is very bitter. It is said to be edible in Zaïre (7). Cattle will not graze it in Senegal, but it is taken by sheep (1).

The Ankwe of the Benue region use it as a vermifuge after maceration in water for a night and the addition of a little lime-juice (5). The Fula of N Nigeria take small fragments of the plant in fresh milk for hepatitis (8). A decoction of the plant, preferably taken cold, is said to be effective in relieving coughs (Porter fide 5), and in Madagascar it is commonly sold in markets as an antitussive (6). The sap expressed from plants softened over a fire is used in Congo in nasal instillations for nose-bleeding (2).

Six saponosides which account for the bitterness have been extracted from the plant. Sugars, sterols, flavones and other substances have been found present (4). A trace of alkaloids and saponins are also reported (3).

References:

1. Adam, 1966, a. 2. Bouquet, 1969: 167. 3. Bouquet, 1972: 35. 4. Bouquet & Debray, 1974: 123. 5. Dalziel, 1937. 6. Debray & al., 1971: 15, 51. 7. Hauman, 1951, b: 110–1. 8. Jackson, 1973.

**Mollugo pentaphylla** Linn.

FWTA, ed. 2, 1: 134.

A slender herb, spreading leaves and ascending, of sandy places, introduced and recorded only in W Cameroons, but native of Asia, occurring widespread in India, Japan, China and Malesia.

No usage is recorded in the Region. In India it is valued as a bitter vegetable. Saponins and a high concentration of saltpetre have been reported present. In Malesia it is used for poulticing sore legs and is taken for sprue. It is used as a mild aperient in India.

References:

1. Burkill, 1935: 1482. 2. Quisumbing, 1951: 279–80.

**Sesuvium portulacastrum** (Linn.) Linn.

FWTA, ed. 2. 1: 135. UPWTA, ed. 1. 30.

English: seaside purslane (Dalziel; Irvine).

French: pourpier de mer (Berhaut); pourpier maritime (Busson).

West African: **MAURITANIA** ARABIC (Hassaniya) heruy (AN) **SENEGAL** BANYUN ďalémukor (K&A) MANDING-MANDINKA sâbańańi (K&A) 'SOCE' sâbańańi (K&A) MANINKA sâbana (K&A) SERER dalôga (JB ex K&A) roltakar (K&A) sâbana (K&A) samba gnagna (JLT) sambaniani (K&A) SERER-NON (Nyominka) isâbańańa (K&A) sâbana (K&A) WOLOF horomdâta (K&A) kamina (JB, ex K&A) **GUINEA-BISSAU** BALANTA bóssaha (JDES; EPdS) burunquè (JDES) PEPEL ombuntutchus (EPdS) **SIERRA LEONE** TEMNE *a*-ret (Glanville) **GHANA** AKAN-FANTE mbrenkyi (FRI) GA eŋmebi (FRI) gbekebii amada = *children's plantain* (FRI) kleŋme? (FRI) GBE-VHE solĩ (FRI)

A fleshy creeping herb of the foreshore down to the upper littoral, a halophyte of foreshore sand-dunes and edges of salt-lagoons, occurring throughout the West African coastal margin, and on all tropical shores.

The plant is eaten in all parts of the West African coast. Thorough washing is necessary to remove excess of salt and perhaps better still boiling in two or three changes of water. The leaves have the acidulous flavour of sorrel. They are said to be antiscorbutic. Sheep and goats graze it (2), and it provides a good salty pasturage for camels in Mauritania (5). It is a favourite food for crabs (2), and as such might serve as bait in crab-traps. It is sometimes burnt in Ghana to smoke local 'herrings' (2).

The plant is used on the Senegal coast as a haemostatic and a decoction of it is considered to be the best known antidote for stings of venomous fish, prolonged external application being required (3, 4). The presence of active principles has not been detected. Indifferent insecticidal results have been reported (1).

References:

1. Heal & al., 1950: 101, 144. 2. Irvine, 1930: 283. 3. Kerharo & Adam, 1964, c: 321. 4. Kerharo & Adam, 1974: 482. 5. Naegele, 1958, b: 891.

**Tetragona tetragonioides** (Pallas) O. Ktze

English: New Zealand spinach.

French: tetragone; épinard de la Nouvelle-Zélande.

A herbaceous plant with decumbent trailing stems, fleshy and softly woody at the base, fleshy leaves, a seashore plant of the Pacific basin from Japan and eastern Australia to non-tropical S America, and for long cultivated in Europe as an edible plant under the name of *New Zealand spinach*. It has more recently been introduced to West Africa and is grown as a domesticated plant for use as a vegetable. It stands tropical temperatures well and shows an ability, in Senegal at least, to become naturalized, its seeding cycle fitting the alternations of wet and dry seasons (1). Propagation by cuttings is easy. There are cultivars with larger leaves.

The vegetative parts are rich in saponin which is apparently of low toxicity and this is lost on boiling for consumption. Alkaloid is said to be present. In Brazil it is used as an antiscorbutic and as a remedy for pulmonary and intestinal affections (2, 3).

References:

1. Adam, 1961: 13. 2. Quisumbing, 1951: 280 (as *Tetragona expansa* Murr., with references). 3. Watt & Breyer-Brandwijk, 1962: 12 (as *T. expansa* Murr., with references).

**Trianthema portulacastrum** Linn.

FWTA, ed. 2, 1: 136. UPWTA, ed. 1, 31.

English: horse purslane (Irvine)

French: pourpier courant (Berhaut).

West African: **SENEGAL** SERER nakamaɗoɗe (JB) nɗisan (JB) ñokèt (JB) **THE GAMBIA** MANDING-MANDINKA ninsi woroso (*def.*-o) = *cow-hoof* (Fox;DF) sádjambó = *snake leaf* (JMD) **MALI** SONGHAI malinguer (A. Chev.) **GHANA** AKAN-TWI adwera-akoa, adwera: *Portulaca oleracea;* akoa: *interior;* i.e., *the lesser adwera* (Enti) GA tʃalai (FRI; JMD) **NIGERIA** ARABIC rabah (JMD) ARABIC-SHUWA tamalege (JMD) CHAMBA bin-y-wa-só (FNH) HAUSA daábùrín sáaniyaá = *cow's gum, or palate* (JMD; ZOG) dánkálìn yáàraa = *children's sweet potato* (JMD; ZOG) deylu (JRA) gádón maciìjií (ZOG) gwàndàr kúnneé (JMD; ZOG) halshen rogo, hálshèn túnkìyaá = *sheep's tongue* (JMD; ZOG) roógòn yaáraá = *children's cassava* (JMD; ZOG) HAUSA (West) gadon machiji = *snake's bed* (JMD) KANURI kənáskì (JMD; C&H) YORUBA oluwóṉjẹja (Egunjobi)

A rather fleshy, prostrate herb, with stems to half metre in length, a weed of waste places and cultivated land, occurring throughout the Region from Senegal to S Nigeria, and widespread in tropical Africa, Asia and America.

The plant has anthropogenic tendencies and is often found around inhabited places. The foliage is eaten as a cooked vegetable in parts of Ghana, e.g. the Ga of the Accra Plain (5, 6). In the Vogel Peak area of W Cameroons, the Chamba at Gurun cook the foliage into soup with ground nuts (7, 8). It is reported that the whole plant is boiled and eaten as a vegetable in Tanganyika (3). It is eaten in India and the Philippines and is said to be a good source of calcium, iron and phosphorus (11). There is record of its consumption in the Kwae Noi River area of Thailand (9), a practice which parallels its use as a supplement to subsistence diet of prisoners-of-war in Japanese hands in 1942–45 employed on the Burma-Thailand railway construction along the course of the Kwae Noi River. In India, however, the plant is generally considered as a famine food, and sometimes to have poisonous effects causing diarrhoea and paralysis (14). Especially the older leaves are dangerous to eat (2). Horses, donkeys, cattle, sheep and goats refuse to graze it, unless it is very young, in Senegal (1). Camels are said to take it in Bornu (6).

The fleshy nature of the leaves lends to their use as a wound-dressing or poultice. The leaves are used in a rheumatism treatment in N Nigeria (6), and old leaves for gonorrhorea (2). The boiled leaves ground and mixed with cacao-butter are recorded as used for indurations of the liver (4). Leaf-sap is taken in India as a diuretic, and has been found to contain an alkaloid with properties similar to those of *punarnavine* from *Boerhavia diffusa* Linn. (Nyctaginaceae) (10).

The powdered root has cathartic and stomachic effects (2), and a decoction is taken by draught in Gabon for venereal discharge (12, 13). In Asian medicine it is given as an emmenagogue, and in larger dosage as an abortifacient, for ammenorhoea, jaundice, strangury, dropsy and asthma. A glycoside is reported present (11).

References:

1. Adam, 1966, a. 2. Ainslie, 1937: sp. no. 340. 3. [illegible], K. 4. Hartwell, 1967. 5. Irvine, 1930: 418. 6. Irvine, s.d. 7. Hepper 1296, K. 8. Hepper, 1965: 461. 9. Kwae Noi River Basin Expedition, 1946, no. 309, K. 10. Oliver, 1960: 10, 89. 11. Quisumbing, 1951: 281 (with references). 12. Walker, 1953, a: 38. 13. Walker & Sillans, 1961: 183. 14. Watt, G., 1889–93: 6, 4: 77–78 (as *T. monogyna* Linn., and under *T. pentandra* Linn.).

**Trianthema triquetra** Rottl., ssp. **triquetra.** var. **triquetra.**

FWTA, ed. 2, 1: 136, as *T. sedifolia* Visiani.

West African: **MAURITANIA** ARABIC (Hassaniya) mhalluch (AN)

A prostrate fleshy herb recorded from Mauritania to Niger and Chad, and into Asia and Australia.

The plant provides grazing for sheep and goats in Mauritania. Other stock will not take it (1).

Reference:

1. Naegelé, 1958, b: 891.

**Zaleya pentandra** (Linn.) Jeffrey

FWTA, ed. 2, 1: 136, as *Trianthema pentandra* Linn. UPWTA, ed. 1, 30, as *T. pentandra* Linn.

English: horse purslane (Dalziel).

West African: **MAURITANIA** ARABIC (Hassaniya) igarufal (AN) **THE GAMBIA** MANDING-MANDINKA saajamba (*def.* saajamboo) = *snake leaf* (auctt.) **MALI** SONGHAI malinguer (A. Chev.) **GHANA** AKAN-TWI adwera-akoa, adwera: *Portulaca oleracea;* akoa: *interior;* i.e., *the lesser adwera* GA tʃalai (JMD) **NIGERIA** ARABIC rabah (JMD) ARABIC-SHUWA tamalege (JMD) HAUSA daábùrín saániyaá = *cow's gum, or palate* (JMD; ZOG) dánkálìn-yáàráa = *children's sweet potato* (JMD; ZOG) gádón maciijií (ZOG) gwándàr kúnneé (JMD; ZOG) halshen rogo, hálshèn túnkìyaá = *sheep's tongue* (JMD; ZOG) hana taƙama = *prevents swagger* (JMD) roógòn-yaàraá = *children's cassava* (JMD; ZOG) HAUSA (West) gadon machiji = *snake's bed* (JMD) KANURI kənáskì (JMD)

A semi-succulent, prostrate herb, with stout stem, occurring in the northern, drier parts of the Region from Mauritania to Niger and N Nigeria, and into E Africa and Arabia.

The plant forms a close cover over waste ground and is supposed to harbour snakes which may account for the Hausa name, *hana takama,* 'prevents swagger' (3). It is myrmecochoric, the grain being collected by ants. It is grazed by all stock in Mauritania (6), and by cattle in Senegal (1). In contrast it is recorded that in the Tana River region of Kenya it is one of the few herbs shunned by cattle (5). In India the plant is considered a dangerous poison capable of causing diarrhoea, paralysis and death by acute nephritis (8), and is eaten only as a famine-food (9), though it is eaten as a vegetable by Arabs and Indians in Somalia (4). It has medicinal use in Sudan as a stomachic for man and cattle and as a gonorrhoea cure (2). For this latter purpose the plant is dried and powdered and taken with millet beer. This sets up acute inflammation of the urino-genitary tract resulting in haematuria, vomiting and bloody stools (3), and a 'purging' of the infection. This treatment may have homeopathic connotations.

At Timbuktu in Mali the plant has been burnt to extract potash and to furnish a vegetable salt (A. Chevalier fide 3). Saponins are present (7).

References:

1. Adam, 1966, a: 2. Broun, s.n., 26/9/1905, K. 3. Dalziel, 1937. 4. Gillett 4128, K. 5. Lucas 42, K. 6. Naegelé, 1958, b: 891. 7. Oliver, 1960: 39. 8. Watt, G. 1889–93: 6, 1: 311 (as *Trianthema pentandra* Linn.). 9. Watt, G. 1889–93, 6, 4: 77–78. (as *T. pentandra* Linn.).

# ALANGIACEAE

**Alangium chinense** (Lour.) Harms

FWTA, ed. 2, 1: 749.

A tree to about 23 m high of montane forest in W Cameroons and Fernando Po, and widely dispersed across Africa and in the countries of the Indic Basin to Malesia. The generic name comes from the S Indian Tamil *alangi.*

The tree is fast growing and is a pioneer species of partly cleared lowland and upland forest with a life span said to be only 25 years. The wood is white, soft and even-grained, and is used in Asian countries for building purposes, furniture and firewood (2).

The alkaloid *alangine* which is hypotensive and acts on the parasympathetic nervous system, and 20 other alkaloids including *cephaline, emetine* and *psychotrine* have been identified in the related species, *A. lamarckii* Thw. (1). This species merits examination as a source of ipecacuanha alkaloids.

References:

1. Albright & al., 1965. 2. Burkill, 1935: 80, as *A. begoniifolium* Baill.

# ALISMATACEAE

**Burnatia enneandra** M. Micheli

FWTA, ed. 2, 3: 14.

An aquatic herb to about 60 cm high from a rhizomatous rootstock and/or with corms to 1 cm diameter, of the margins of muddy pools, from Senegal to Nigeria, and widely dispersed over much of tropical Africa.

In Uganda the tuber is eaten in time of dearth (1).

Reference:

1. Thomas, A. S. 3558, K.

**Caldesia reniformis** (D. Don) Mackino

FWTA, ed. 2, 3: 11.
West African: **NIGERIA** HAUSA zakaran bado = *lily of the cock* (AST; RES)

An aquatic plant with floating leaves and erect emergent flowering stem, of swamps and shallow water, in Dahomey and N Nigeria, and across northern Africa and Asia to China and Australia.

No usage is recorded for the Region. In Burma the succulent stems and leaves are cooked and eaten (1). The leaves of a related species, *A. parnossifolia* (Bassi) Parl., are added to sweet potatoes in New Guinea which are being fed to pigs; this is said to make the pigs grow quickly (2).

References:

1. Robertson 336, K. 2. Vink 16512, K.

**Limnophyton angolense** Buchen.

FWTA, ed. 2, 3: 11. UPWTA, ed. 1, 464, as *L. obtusifolium* Sensu Dalziel.
West African: **SIERRA LEONE** TEMNE *a*-mɛs-*a*-ro-kant (NWT) **LIBERIA** MANO bolo (Har.)

An aquatic herb attaining 1.30 m high, recorded from Guinea to Nigeria, and also in E Africa and Zambia.

The plant is commonly used in Liberia reduced to ash to obtain a vegetable salt taken with food or as an ingredient of medicines (2). In Tanganyika the Sukuma rub the ash into small cuts around any sort of painful aching area of the body (3). The petioles and stem-bases serve as floats for fishing nets in Uganda (1).

References:

1. Drummond & Hemsley 4650, K. 2. Harkey s.n., 10/5/1932, K. 3. Tanner 1117, K.

**Limnophyton obtusifolium** (Linn.) Miq.

FWTA, ed. 2, 3: 11.
West African: **SENEGAL** SERER ñãmb gôr (JB) **SIERRA LEONE** KISSI niŋgikɔma (FCD) waye-pio (FCD) KONO tagbo-mɛsɛ (FCD) MENDE nja-gboji (FCD) **NIGERIA** HAUSA gautan kada = *tomato of the crocodile* (Lely; RES)

An erect tufted herb, of muddy pond edges, distributed across the Region from Senegal to N Nigeria, and widespread elsewhere in tropical Africa, and in India, Ceylon and Malesia.

The tuber is recognized in Kenya as being non-edible (1).

Reference:

1. Graham 2164, K.

**Sagittaria guayanensis Kunth,** ssp. **lappula** (D. Don) Bogin

FWTA, ed. 2, 3: 11.
English: arrowhead, swamp potato (the genus generally, Bates).
West African: **SENEGAL** FULA-PULAAR (Senegal) tamé (K&A) **THE GAMBIA** MANDING-MANDINKA kunto (Frith) **NIGERIA** IJO-IZON (Egbema) okuába (Tiemo)

An aquatic annual herb with floating leaves, small rootstock, of swamps and muddy pools, across the northern part of the Region from Senegal to N Nigeria, and widely distributed in NE Africa, Madagascar and tropical Asia and America.

The Fula of Senegal use the leaves in drinks and in lotions and baths to treat nutritional deficiency conditions (2), anaemia and rickets (3, 4). The tubers when present are edible. A related species, *S. sinensis* Sims, is grown in China for raising tubers as pigfood.

The plant is fairly common over the Indian subcontinent occurring in fresh water reservoirs and in marshes. It becomes troublesome where fish-farming is carried on. It occurs in rice padi fallows and is often ploughed in as a green manure (1).

References:

1. Chadha [Ed.], 1972: 168. 2. Kerharo & Adam, 1963, a: as *Lophotocarpus guyanensis* Dur. & Sch. 3. Kerharo & Adam, 1964, b: 557, as *Lophotocarpus guyanensis* Dur. & Sch. 4. Kerharo & Adam, 1974: 119–9 as *L. guayanensis* Dur. & Sch.

## AMARANTHACEAE

**Achyranthes aspera** Linn.

FWTA, ed. 2, 1: 152; FWTA, ed. 2, 1: 760, as *A. argentea* Lam. = *A. aspera* var *sicula* Linn. UPWTA, ed. 1, 34.
English: prickly chaff flower; snake's tail (Kenya, E. Polhill).

French: (Berhaut)—queue de rat; queue de rat annuelle (as *A. argentea* Lam.).

West African: **SENEGAL** MANDING-BAMBARA sien doro nani (JB, ex K&A) SERER mboratad (JB, ex K&A) WOLOF ndatukan bugor (JGA) nobisindaq = *ear of the lizard* (JB; K&A) **IVORY COAST** AKAN-ASANTE assiadiomo (B&D) **NIGER** FULA-FULFULDE (Niger) kebbe jawle (PF) mbagga (PF) **NIGERIA** EFIK ọ́kpọ̀kọ̀ údọ̀k ḿbieèt (KW) FULA-FULFULDE (Nigeria) kebbe jawle = *Guinea-fowl's bur-grass* (MM) kure pallaandi = *lizard's arrows* (JMD) mbagga (MM) GWARI weknawuhi (Edgar) HAUSA hákoórín máciìjií = *snake's tooth* (auctt.) ƙaimin kadangaree = *lizard's spur* (JMD; ZOG) HAUSA (West) kiban ƙadangaru (JMD; ZOG) IGBO (Agukwu) n̄lí átụ́lụ̄ = *sheep's food* (NWT) IGBO ('Ibugo') ọ́dụ̀dụ́ ǹgwèlè = *lizard's tail* (KW) YORUBA aboro (auctt.) èèmá àgbò (Abraham) èèmá-àgbò (N&E)

A much-branched erect or subscandant herb, to 2 m high, very variable and adaptable to many sorts of location, from shaded damp and swampy sites to savanna conditions and to high altitudes, occurring in the Region from Senegal to W Cameroon, and widespread elsewhere in Africa and throughout the hotter parts of the world.

There is no record of the leaf being an item of human consumption in West Africa. The leaves however are not poisonous and are eaten when young in Tanganyika (3, Quinn fide 25). Their consumption in the Moluccas is also recorded (8). Cattle in Senegal and the Lake Chad area browse it a little (1), and in Cameroun (7). It is grazed by all stock in Kenya (5, 11, 17) though the only reference to cattle on Tanganyikan collections in *Herb. K* states specifically that the plant is not grazed (14).

In Nigeria the leaf-infusion is used as a tonic and an astringent (2) and an expectorant and diuretic (18). In Senegal its application is also as an expectorant and diuretic, and also an inhalation for rhino-pharyngeal and pulmonary troubles (15). Asantes in Ivory Coast use the leaves to make suppositories for haemorrhoids (6) which are similarly used in India (26). The leaves are considered healing on new circumcision wounds in Gabon (25) and to maturate boils and abscesses in E Africa (4, 26). In Zaïre boiled leaves are used on itch and to cure headache (13). Leaves in fumigation are used on abscesses and for advanced yaws in Gabon, while smoke from burning leaves is considered helpful in the extraction of splinters from the feet (25). Leaves are eaten with *odika* butter, *Irvingia gabonensis* Baill. (Ixonanthaceae), to treat diarrhoea in Gabon (25). The plant is put into a medicine for constipation in children in Tanganyika (10) and dysentery and bowel-complaints are treated by plant-sap in the Philippines (19). Leaf-sap is dripped into eyes in Tanganyika (12) and in the Philippines leaf-juice is used to clear corneal opacity (19) and dim vision in Java (8). The plant is used in medicine to treat colds in Tanganyika (10), and in southern Africa inhalation of steam from a boiling plant infusion and use in hot baths is recorded for acute colds. This promotes sweating (26).

The plant is an ingredient of snuff to the Sukuma of Tanganyika (20) and to people in Botswana (26).

The plant, and especially the root, appears to have anodynal properties. A root-decoction is drunk and the plant-ash is rubbed into scarifications on painful areas for rheumatism, and root-scrapings are similarly applied for pneumonia in Tanganyika (12) and the root provides a remedy for stitch (4) and stomach pain (24). Alternatively the whole plant may be used for generalized pain in the chest and abdomen (23). A root-decoction is drunk for menstrual pain (21). The root-macerate is recorded as relieving scorpion-stings in India, acting promptly, leaving a mere numbness of the part (9) but this is not corroborated in *The Wealth of India* (16). Similar action is, however, reported from Ethiopia and on snake-bite in India, and of the seed on snake-bite in Malesia (26). The flower-spike when crushed and applied is said to cure venomous bites and stings in India (16).

Though the plant has such a multiplicity of medicinal uses, these in part may have arrived through magical attributes. In Sanskritic medicine it was a sorcerer's plant for driving away demons, removing enchantments, curses and all illnesses. In Tanganyika the Zigua add the plant to medicine to remove spirit possession (22). The plant may have some benefit as a diuretic, especially in renal dropsies, and as an external cicitrisant (8, 16). The seed of Indian material has been found to contain 2% saponins and a trace of alkaloid in the plant. A precis of chemistry and pharmacology may be found at (15).

The whole plant is rich in minerals, especially potash (19). The plant has been used in India as a source of potash, and in the Shari-Chad area the Abok people reduce the plant to ash which is used as a cooking-salt (Chevalier fide 9, 13). The ash is used in Ethiopia to treat scabies, and mixed with honey it is a cough-cure (26). The plant has some manurial potential and might be valuable as a green manure.

References:

1. Adam, 1966, a. 2. Ainslie, 1937: sp. no. 9. 3. Akeroyd 15, K. 4. Bally, 1937. 5. Bogdan AB 4459, K. 6. Bouquet & Debray, 1974: 13. 7. Breteler 604, K. 8. Burkill, 1935: 32–33. 9. Dalziel, 1937. 10. F. D. Tanganyika A. 19/49, K. 11. Glover & al., 351, 399, 1902, 2139, 2179, 2376, 2427, 2431, K. 12. Haerdi, 1964: 160. 13. Hauman, 1951, a: 53–55. 14. Hornby 8, K. 15 Kerharo & Adam, 1974: 119–20. 16. Manjunath, 1948: 24. 17. Mathew 6583, K. 18. Oliver, 1960: 17. 19. Quisumbing, 1951: 263–4, with mineral analysis. 20. Tanner 732, K. 21. Tanner 2693, K. 22. Tanner 2972, K. 23. Tanner 4225, K. 24. Tanner 4253, K. 25. Walker & Sillans, 1961: 47. 26. Watt & Breyer-Brandwijk, 1962: 13.

**Aerva javanica** (Burm. f.) Juss.

FWTA, ed. 2, 1: 149. UPWTA, ed. 1, 35, as *A. tomentosa* Forssk.

West African: **MAURITANIA** ARABIC (Hassaniya) moghamlé (AN) tamiye (AN) tuhmeya (AN) **SENEGAL** VULGAR vèhlay (JB) **MALI** SONGHAI bogon koreil (A. Chev.) bomon korré (A. Chev.) TAMACHEK makargis, makersas (A. Chev.) makhmila (RM) tamakerkait, tamakerzist, timekerkest (RM) **GHANA** AKAN-ASANTE bameha (FRI) **NIGER** SONGHAI fèeji-máaní (*pl.* -ó) = *sheep's fat* (D&C) **NIGERIA** ARABIC sheiba (JMD) ARABIC-SHUWA diginshayib (Musa Daggash) HAUSA ahaji (ZOG) alhaji *from its appearance in allusion to the white habit of Mecca pilgrims* (JMD) furfuraa ta gyaatumii (ZOG) HAUSA (East) furfura ta jatumi = *old man's hoary locks* (JMD)

An erect or suberect herb with perennial woody rootstock, to about 1 m high, with woolly white tomentum especially on the flower-buds, a weed of waste places in arid localities. Recorded from Mauritania across the northern part of the Region to Niger and N Nigeria, and across Africa into Arabia, India and Madagascar.

The whole plant is said to be edible (4) but it can only be of minor importance, for there is no record in *Herb. K* of its consumption as a vegetable in Africa. It provides grazing for stock in Mauritania (12), Central Sahara (9), Senegal (1), for sheep in Somalia (8), and in Kenya for stock (11) and game (10). Records of its acceptance by stock are however not unanimous. Cattle at Chad are said not to take it (1), nor camels at Hoggar (9), and in Somalia stock and camels find it distasteful and take it only in default of better browsing (6). In Somalia it is recorded as being a contact poison (2)—? urticant.

The plant has veterinary medical use. In Bornu it is made into a medicine for horses and camels as a purge and emetic, and given internally for snake-bite. In Hausa the powdered plant is applied topically on ulcers in domestic animals (5). In Kenya the Masai grind up the flowers into a paste for cattle with 'East Coast' fever (7).

The root is used in Sudan as a chewstick (3).

The most widespread use of the plant throughout E Africa, Arabia and India is of the woolly inflorescences for stuffing pillows, cushions and mattresses, a practice seemingly not recorded in West Africa. In some parts of West Africa the dried inflorescences may be used as tinder (5). This striking part of the plant evokes a Hausa name meaning *old man's hoary locks,* and another, *Alhaji,* likens the white inflorescence to the white clothing of the muslim Hadj pilgrim.

References:

1. Adam, 1966, a. 2. Andy Y92, K. 3. Broun 1586, K. 4. Busson, 1965: 133. 5. Dalziel, 1937. 6. Gillett 4649, 4650; Peck 6, K. 7. Glover & al. 2986, K. 8. Gooding 54, K. 9. Maire. 1933: 86–87. 10. Mwangangi 1526, K. 11. Mwangangi & Gwynne 1144, K. 12. Naegelé, 1958: 876.

**Aerva lanata** (Linn.) Juss.

FWTA, ed. 2, 1: 149. UPWTA, ed. 1, 34.

West African: **IVORY COAST** ABURE n-tanfa (B&D) AKYE munongbe (A&AA) BAULE akopinolé (B&D) KRU-GUERE ula (K&B) GUERE (Chiehn) ura oré, wore oré (K&B) wulo wulé (B&D) KWENI fianépumandia (B&D) **GHANA** AKAN-ASANTE bameha (FRI) **NIGERIA** IGBO (Obompa) òkpùn̩zú nọ́nụ̄ = *something that carries native chalk in the mouth* (NWT; KW) YORUBA ewé ajé = *money-leaf* (Dennett) ewé owó = *money-leaf* (Dawodu)

A straggling, prostrate or succulent herb, to about 1 m long from a perennial woody root-stock. Occupying damper sites than *A. javanica* (Burm. f.) Juss., and recorded from Sierra Leone to S Nigeria, and across to E Africa, and into Asia to India, Ceylon and Malesia.

The whole plant, especially the leaves, is edible (9). The leaves are put into soup in S Nigeria (5, 17) and in Sierra Leone, or eaten as a spinach (5). They are eaten in Zaïre (8) and in Tanganyika (15) and Kenya (13) as a vegetable. The plant provides grazing for stock in Zaïre (8) and in Kenya for stock and game (6), and in Sudan for goats only (3). In the Kwale District of Kenya chickens are reported to eat the plant (13).

A leaf-decoction is prepared as a gargle in Ivory Coast (2) and by the Igbo of S Nigeria (17) for treating sore-throat. The Akye of Ivory Coast take by draught a decoction of the whole plant and use the lees in frictions for oedemas (2), and in Ivory Coast also the plant has a reputation for relieving stomach-pains in women and of preventing miscarriage (4, 10), and is used in various complex treatments against guinea-worm (10). The leaf-sap is also used for eye-complaints (4). In Gabon an infusion is given to cure diarrhoea (18). The leaves are used in Tanganyika in an unspecified manner at childbirth (11), and on sores (7). The root is used in a snake-bite treatment (12). For pains in the lower part of the back leaves and flowers are reduced to ash which is rubbed into cuts on the back (16). In the Kwale District of Kenya, a leaf-decoction is used to wash babies that have become unconscious during an attack of malaria or of some other disease, and at the same time smoke from the burning plant is inhaled (13).

The plant is said to be diuretic and demulcent. Its diuretic action is said in the Philippines to be very effective in the treatment of urethral discharges and gonorrhoea (14). Its use in India as a diuretic is also recorded and to be of value in cases of lithiasis and as an anthelmintic (14).

A trace of alkaloid has been detected in Nigerian material (1).

The plant has fetish attributes. In Ivory Coast it is deemed to confer protection against evil spirits (10). In Zaïre it is a good-luck talisman for hunters, and in Ruanda it safeguards the well-being of widows (8).

References:

1. Adegoke & al.: 1968. 2. Adjanohoun & Aké Assi, 1972: 11–12, with leaf analysis. 3. Andrews 3467, K. 4. Bouquet & Debray, 1974: 13. 5. Busson, 1965: 133–4, with leaf analysis. 6. Glover & al. 3016, K. 7. Haarer 75B, K. 8. Hauman, 1951, a: 58. 9. Irvine, 1930: 10. 10. Kerharo & Bouquet, 1950: 34. 11. Koritschoner 773, K. 12. Koritschoner 1024, K. 13. Magogo & Glover 924, K. 14. Quisumbing, 1951: 265, with references. 15. Tanner 2066, K. 16. Tanner 2785, K. 17. Thomas, N. W. 2224 (Nig. Ser.), K. 18. Walker & Sillans, 1961: 47.

**Alternanthera bettzickiana** (Regel) Voss

FWTA, ed. 2, 1: 153, as of (Regel) Nicholson.

A much-branched dwarf herb attaining 50 cm height of diverse coloured leaves, mainly reds and purples, said to be native of tropical America but now widespread pantropically with origins blurred by naturalization. It may be a cultigen of *A. ficoidea* (Linn.) R. Br. (2).

The plant is present in West Africa as an ornamental edging plant. Its spread in Sierra Leone is said to be associated with the Krio population (1).

References:

1. Deighton 1858, 1976, K. 2. Townsend, 1974: 42–43.

**Alternanthera maritima** (Mart.) St.-Hil.

FWTA, ed. 2, 1: 154.

West African: **SENEGAL** SERER fèmbar no mag (JB) **SIERRA LEONE** TEMNE *a*-lil (NWT) **IVORY COAST** KRU-GUERE kwagnan (K&B) KWENI sadré bohué (B&D)

A prostrate creeping herb with stems to 10 m long rooting at the nodes, of coastal sands down to high water mark, recorded from the West African coast from Senegal to S Nigeria, and on to Angola, and also on the Atlantic Coast of S America.

The plant is a strong creeper and sand-binder. The leaves are eaten by various coastal peoples in West Africa (4, 5). The plant has a relatively high ash content (2).

In Ivory Coast the plant is used to strengthen debilitated children (1), and pulped up it is used in frictions for oedema of the body (3).

References:

1. Bouquet & Debray, 1974: 13. 2. Busson, 1965: 134, with leaf analysis. 3. Kerharo & Bouquet, 1950: 35. 4. Schnell, 1953, a. 5. Walker & Sillans, 1961: 48.

**Alternanthera nodiflora** R. Br.

FWTA, ed. 2, 1: 154. UPWTA, ed. 1, 35.

West African: **THE GAMBIA** MANDING-MANDINKA kunindingturo = *little bird's comb* (Fox; DF) **GUINEA** MANDING-MANINKA missinikumbré (Brossart, A. Chev.) **IVORY COAST** ANYI zufien (K&B) **GHANA** AKAN-TWI abanase-abanase (JMD) **NIGERIA** HAUSA buúzún maúràyaá = *cob antelope's mat* (JMD) mài-kái-dúbuú = *thousand-headed* (JMD; ZOG) HAUSA (West) gambarin marmoru, gambari: *shirt*, (BM; RES) YORUBA ewé owó (Kennedy)

*NOTE: Some vernacular names of A. sessilis may also refer.*

A decumbent herb with woody rootstock, of open spaces, often beside water, recorded from the northern part of the Region from Senegal to N Nigeria, and occurring widespread elsewhere in tropical Africa and in India, SE Asia and Australia.

The plant is eaten in Tanganyika as a vegetable (3). In N Nigeria it is considered to be a famine-food (8). It is grazed by all stock in Senegal (1) and elsewhere (6).

Sap from the leaves is sniffed up the nose in Ghana for neuralgia, the ground-up leaves mixed with palm oil are taken for colic, and the whole plant is pounded and used for headaches and vertigo (6). The plant also has uses for poulticing boils, etc. (5). In Ivory Coast pulped-up roots or leaves are used in enemas and frictions for feverish pains (4, 7).

A trace of alkaloid is reported present in the leaves of the Nigerian material (2).

References:

1. Adam, 1966, a. 2. Adegoke & al., 1968. 3. Archbold 115, K. 4. Bouquet & Debray, 1974: 13. 5. Dalziel, 1937. 6. Irvine, s.d. 7. Kerharo & Bouquet, 1950: 34. 8. Moiser s.n., 6/3/1922, K.

**Alternanthera pungens** H. B. & K.

FWTA, ed. 2, 1: 154, as *A. repens* (Linn.) Link. UPWTA, ed. 1, 35, as *A. repens* O. Ktze.

English: khaki bur, khaki weed (S Africa, Dalziel).

West African: **SENEGAL** MANDING-BAMBARA sièn goni (JB) **GUINEA** MANDING-MANINKA bonfu (Brossart, A. Chev.) **UPPER VOLTA** MANDING-DYULA monigbe (K&B) 'SENUFO' kamélé (K&B) sabara (K&B) **IVORY COAST** AKAN-ASANTE abéné mulo (K&B) abéné muro (K&B) BRONG nakuru (K&B) BAULE diangérofia (B&D) koto barani (B&D) kuto blamien (B&D) KRU-GUERE (Chiehn) sabré bué (K&B) MANDING-DYULA monigbe (K&B) 'SENUFO' kamélé (K&B) sabara (K&B) **GHANA** ADANGME-KROBO klãgbã (FRI) AKAN-AKYEM nkasa-nkasa = *thorn-thorn* (FRI) ASANTE abencburo (Rattray; FRI) nkasɛe-nkasɛe = *thorns* (Enti) FANTE mpatsiwansɔe, mpatsiwa: *a small fish*; nsɔe: *thorns, perhaps likening the spiky fruit to herring bones* (FRI) TWI abirimuro (auctt.) asamaŋkama (FRI) kronpe (FRI) nkasɛe-nkasɛe = *thorn-thorn* (Enti) nsɔe-nsɔe, nsɔe: *thorns* (FRI; JMD) sraha nsɔe (JMD) GA awusa-ŋmei = *Hausa thorns* (FRI) GBE-VHE awusagbe = *Hausa herb* (JMD) awusatsoe = *Hausa thorns* (JMD) mitsimitsi (FRI) VHE (Pecí) sɛnutsoe (FRI) yegbeetsoe, tsoe: *thorn* (FRI) **NIGERIA** YORUBA dágunró (DF; JMD)

A prostrate herb creeping, variable, abundant in dry waste places throughout the West African region and elsewhere in Africa, SW Asia and central America.

The plant is a pest wherever it grows. It is said to be difficult to eradicate and it has become a noxious weed in many countries. The fruits with long straight spines are a nuisance on bush footpaths to wayfarers with bare feet and cause intense irritation.

In Asante medicine in Ghana the plant is used to stop abdominal pains in women, as an abortifacient and a galactogogue (5, 6). In Ivory Coast it is given to promote delivery in childbirth (3). The plant has applications for constipation with griping, and as an enema for diarrhoea (3, 5). In Ivory Coast it is commonly used as a vermifuge for which an aqueous decoction is taken in the morning on an empty stomach. This produces no griping nor diarrhoea, and the worm is expelled without difficulty at the second bowel-motion (8). Plant pulp is also given to children to soothe diarrhoea (8). In Ghana the leaves are a remedy for dysentery (7).

Sap obtained by squeezing leaves passed over a fire is used in Ghana to sniff up the nose for neuralgia (Bunting fide 6). In Ivory Coast plant-pulp is made into a gargle for sore-throat (8) and is used in a bath or vapour-bath for fevers and localized oedemas (3), and in Guinea (A. Chevalier fide 5) and Gabon (4) furnishes a cooling lotion for the head which is applied in the case of fever. A decoction of the whole plant is used for venereal disease treatment in Congo (2).

A trace of alkaloid has been detected in Nigerian material (1).

References (as *A. repens* of O. Kuntze or (Linn.) Link):

1. Adegoke, 1968. 2. Bouquet, 1969: 52. 3. Bouquet & Debray, 1974: 13–14. 4. Cavaco, 1963: 13–14. 5. Dalziel, 1937. 6. Irvine, 1930: 20–21. 7. Irvine, s.d. 8. Kerharo & Bouquet, 1950: 34.

**Alternanthera sessilis** (Linn.) DC.

FWTA, ed. 2, 1: 154, as of (Linn.) R. Br.; 1: 760, as of (Linn.) Sweet. UPWTA, ed. 1, 35, as *A. sessilis* R. Br.

West African: **SENEGAL** FULA-TUKULOR lebleban (K&A) **GUINEA** KISSI mého (JB) LOMA mili (JB) MANDING-MANINKA missinikumbré (auctt.) **SIERRA LEONE** MENDE ndatawulo (FCD) **NIGERIA** HAUSA buúzún maáràyaá = *cob antelope's mat* (JMD; ZOG) mài-kái-dúbuú = *thousand-headed* (JMD; ZOG) IJO-IZON (Kolokuma) tịn diri = *calling (out) medicine* (KW) YORUBA ewé owó (Kennedy) ṣà wẹwẹ (auctt.)

A common prostrate creeping herb, annual or perennial, very variable according to location, usually of dry situations but also of swampy localities in savanna country, occurring throughout the Region, and generally widespread in warmer parts of the world.

The plant is eaten in Guinea in default of rice and the leaves are said to be satiating (4). In S Nigeria it is eaten in soup (13), in Zaïre (7) and in Tanganyika (3) as a vegetable, in Zambia as a relish (12), and in Madagascar it is commonly taken as a pot-herb (5). In Asia it is eaten as a vegetable and in India and Ceylon as a galactogogue and cholagogue (14).

It is grazed by sheep in Senegal (1) and other stock evidently shun it, though in Kenya it is said to be taken by all animals, rabbits choosing only the inflorescences (10).

The plant has use for simple stomach-disorders (2, 5) and for diarrhoea and dysentery (11). In Sierra Leone a hot concoction of the plant is applied to maturate boils and to treat other diseased parts of the body (6). In Nigeria the whole plant pounded up is used for headaches and vertigo, and sap from the leaves is sniffed up the nose for neuralgia (2). The Ijo vernacular name meaning *calling out medicine* arises from the use of the plant to draw out spines or any other object from the body, the result being achieved after a few days application of a specially prepared paste (15). The Ijo use the plant also to cure hernia (15). In Senegal the leafy twigs ground to a powder are used externally as part of a treatment for snake-bite (8, 9). It also features on a snake-bite remedy in India (14) but studies have shown neither preventive, antidotal nor therapeutic property (9).

A decoction of the whole plant has veterinary use in Kenya to treat goats and camels suffering from several diseases which remain not clearly understood (10).

References:

1. Adam, 1966, a. 2. Ainslie, 1937: 23. 3. Archbold 306, K. 4. Busson, 1961: 136, with leaf-analysis. 5. Dalziel, 1937. 6. Deighton 2077, K. 7. Hauman, 1951, a: 74. 8. Kerharo & Adam, 1964, b: 405. 9. Kerharo & Adam, 1974: 121–2, with references. 10. Mwangangi 1374, K. 11. Oliver, 1960: 18. 12. Rabson Phiri 17, 17A, K. 13. Thomas, N. W. 1950 (Nig. ser.), K. 14. Watt & Breyer-Brandwijk, 1962: 13–14, with references. 15. Williamson 3A (UIH 13805), UCI.

## **Amaranthus dubius** Mart.

FWTA, ed. 2, 1: 148.

West African: **SIERRA LEONE** KRIO red green (FCD) MENDE hɔndi (FCD) TEMNE *a*-bonthila (FCD) **NIGERIA** EFIK okpu-utere (N&E) YORUBA olorungbin (CWvE) tẹ̀tẹ̀ *the Dahomey form* (N&E; CWvE)

A simple, annual herb of clearings, disturbed land, around habitations and occasionally in the savanna, recorded from Sierra Leone and S Nigeria and occurring no doubt much more widely in the Region. It is widespread elsewhere in tropical Africa and in America.

In Sierra Leone it appears to be confused with *A. hybridus* forma and similarly has green and red forms. It is a bit coarser and Krio and Temme informants say it is not cultivated. Mendes say it is cultivated (2). It does however yield a spinach and is so used. It is commonly on sale in SW Nigerian markets, some material of which is attributed to a 'Dahomey form' (3). In Zaïre it is cultivated for the leaves, especially in time of dearth (4).

In Tanganyika it has been reported used as a relish (5), and the whole plant in a medicine for stomach-pains (6).

Natural hybridizing between this species and *A. spinosus* is reported to occur in the neighbourhood of Ibadan, Nigeria (1).

References:

1. Clifford, 1958. 2. Deighton, 1887, K. 3. Epenhuijsen 8, 11, K. 4. Hauman, 1951, a: 30–31. 5. Hornsby 135, K. 6. Tanner 4223, K.

**Amaranthus graecizans** Linn.

FWTA, ed. 2, 1: 148.

English: wild amaranth.

West African: **MAURITANIA** ARABIC (Hassaniya) bodbada (AN) bözbazö (AN) bözebazö (AN) sag el mohor (AN) **MALI** FULA-PULAAR (Mali) nafa-nafa (A. Chev.) SONGHAI walia bocio (A. Chev.) TAMACHEK tazelanghatai (A. Chev.) walia bocio (A. Chev.) **GHANA** MOORE ziliva (FRI) **NIGER** SONGHAI húbéy (*pl.* -à) (D&C) **NIGERIA** HAUSA námíjìn-gaásàyaá = *male Gynandropsis* (JMD; ZOG) námíjìn-zaákií-bánzaá *appl. loosely to several weeds* (JMD; ZOG) zaákia-bánzaá (Lely)

A herb, somewhat woody, very variable and of several recognizable taxa, either subspecies or varieties according to authority. Widely dispersed in the drier northern part of the Region, and also elsewhere, and in Europe and Asia.

The plant has edible leaves and is used as a pot-herb and vegetable, and also as a fodder for cattle. In Mauritania the seed is baked into thin cakes (1).

Reference:

1. Naegelé, 1958: 877.

**Amaranthus hybridus** Linn., ssp. **incurvatus** (Timeroy) Brenan

FWTA, ed. 2, 1: 148, as *A. hybridus* Linn., ssp. *cruentus* (Linn.) Thell. UPWTA, ed. 1, 35–36, as *A. caudatus* Linn.

English: spinach; African spinach; Inca wheat (S America); bush greens (Sierra Leone, Deighton); love-lies-ableeding (forms with red or purple leaves and inflorescences).

French: amarante du Soudan; épinard du Soudan (Busson): queue de renard (Walker).

West African: **SENEGAL** MANDING-BAMBARA bura bura ba (JB) SERER mbuma (JB) WOLOF mbum i dyombor (JMD) mbum i kör = *spinach of the house* (JGA) **GUINEA-BISSAU** CRIOULO bredo femea (JDES) **GUINEA** KISSI fondulo *green form* (FB) fondulo saman *purple form* (FB) fondulo sankura *bullous-leafed form* (FB) LOMA sedi boegui *purple form* (FB) sedi oregui *green form* (FB) sedi pelevegui *bullous-leafed form* (FB) MANDING-MANDINKA bóró MANINKA boron (JMD) moron (JMD) **SIERRA LEONE** BULOM (Sherbro) grins-ɛ (FCD) FULA-PULAAR (Sierra Leone) boroboro (FCD) GOLA mando (FCD) n-dolo (FCD) KISSI fondolo *red form* (JMD; FCD) fondolo-bɛndoŋ (FCD) KONO bɔbɔɛ (FCD) bɔbɔ-wa (FCD) KRIO grins = *greens* (FCD) LIMBA bɔrebɔre-ba (FCD) LIMBA (Tonko) gbuhaŋga (FCD) LOKO hɔndi (FCD) MANDING-MANDINKA bɔrəbɔrɛ (FCD) bɔrɔ̃ (FCD) bɔrɔ̃-mbɛdɛ (FCD) MANINKA (Koranko) bɔrəbɔrɛ (FCD) bɔrɔ̃ (FCD) bɔrɔ̃-m-bɛdɛ (FCD) MENDE hɔndi (FCD) hɔndi-gbogboli *red form* (JMD) hɔndi-gowuli = *white hondi, i.e., green form* (JMD) hɔndi-guli *green form* (FCD) SUSU-DYALONKE yambɛli-na (FCD) TEMNE *a*-bonthila (auctt.) *a*-senduku *cultivated forms* (FCD) VAI lolo (FCD) **MALI** MANDING-BAMBARA boron (JMD) boron gbé *green form* (FB) boron ulé *purple form* (FB) moron (JMD) moron dyé *green form* (FB) **GHANA** ADANGME muɔtsu *red forms* (FRI) sɔbu *green forms* (FRI) ADANGME-KROBO sɔbuɛ (FRI) AKAN-ASANTE ɔtofammẽn *red form*, ben *or* men: *red* (FRI) TWI ɛfan, fan *green forms, and incl. various pot herbs* (FRI; JMD) GBE-VHE (Pecí) madzẽ *red form* (FRI) **NIGERIA** BEROM enɛ̀p (LB) EDO èbáfọ̃ (JMD) EFIK ìnyàñ áfiá (Lowe) FULA-FULFULDE (Nigeria) haako ndiyam = *water leaf* (JMD; J&D) HAUSA áláyyafóo, àláyyàfuú (auctt.) háko-ndiyam (ZOG) námíjìn gaásàyaá = *male Gynandropsis; applies to Amaranthus genus generally* (JMD) námíjìn zaákií bánzaá *applies to several domestic weeds* (JMD) IGBO ìnìnè *applies to Amaranthus genus generally* (JMD) ìnìnè olū, olū: *farm* (NWT; JMD) TIV kashia (JMD) URHOBO ọtẹtẹ (JMD) YORUBA ẹ̀fọ́o tẹ̀tẹ̀, èfó: *vegetable* (JMD) tẹ̀tẹ̀ = *trample, trample* (Verger) tẹ̀tẹ̀ ńlá, nla: *big* (JMD) tẹ̀tẹ̀ òyìbó = *white man's tete* (JMD) tẹ̀tẹ̀ pòpó = *Dahomey tete* (JMD) YORUBA (Lokoja) ẹfo (Elliott)

A robust annual (? perennial) herb, but dispersed by man to all warm countries and said to have reached West Africa from Asia, and is recorded throughout the Region and in all parts of tropical Africa. It is naturalized, or subspontaneous on disturbed land of clearings, roadsides, a weed of cultivation, etc., but rare in savanna. It is also cultivated (2, 4, 6, 11–13).

The plant is eaten and produces the best of the amaranth spinaches. It is very variable and basically four forms have been recognized (17):

1. Green: tall, yielding abundantly and of the best quality. This is the most often cultivated.
2. Purple: leaves larger and coarser, spinach inferior and but little cultivated.
3. Rosy: intermediate to 1 and 2.
4. Bullous-leafed: much relished but with smaller leaves than the green form, and cultivated here and there.

It is rich in dietetic requirements (4, 15, 16). Its dietetic value is obliquely recognized by the Yorubas in an incantation to obtain money. Their name *tẹ̀tẹ̀* meaning 'trample, trample', refers to an Odu incantation to the inability of other plants of the field to prevent it from 'trampling' over them (19). In Ghana the Fante tribe are said to be the 'spinach-eaters', their name being derived from (ɛ)*fan*, spinach and *di*, to eat because of their partiality to it (10).

The plant is diuretic. Mineral content is high. Potassium oxalate has been recorded (21). The plant provides fodder for stock but samples have been examined in Nigeria with hydrocyanic acid present in sufficient quantity to constitute a danger to pasture (16). A form of soda is obtained from the ash of the whole plant in Tanganyika which is used in the preparation of tobacco (5).

The seed is known as *Inca wheat* in South America and sometimes is grown there replacing cereals. It is also widely grown in northern India as a grain crop (3). It is said to be grown in Sierra Leone for the seed (6), in Zaïre (9), in Ethiopia (18), and in other parts of Africa (3) and the Middle East (21). The seeds are also prepared in Zaïre into an alcoholic drink (9).

In Senegal the roots are boiled with honey as a laxative for infants (Sébire fide 6), and in Ghana the plant is macerated in water which is used to wash persons afflicted with *hurae* (Akan-Twi), a disease causing violent pains in the limbs (14). In Ethiopia it has use as a tapeworm-expellant (7). Ash from the burnt stem is used as a wound-dressing in Sudan (1). The leaf has been used as an infusion for the relief of pulmonary conditions and to prepare an application for scrophulous sores in India (21). In Gabon heated leaves have been used on tumours (20).

The coloured forms have horticultural and decorative use, and the redder ones especially evoke the name *love-lies-ableeding*. A red dye can be obtained from them (8).

References:

1. Andrews 204, K. 2. Berhaut, 1967: 334. 3. Burkill, 1935: 126–7, as *A. caudatus*. 4. Busson, 1965: 137, 142–3, with leaf and seed analyses. 5. Culwick 4, K. 6. Dalziel, 1937. 7. Getahun, 1975, as *A. caudatus*. 8. Greenway, 1941, as *A. caudatus*. 9. Hauman, 1951, a: 27–29. 10. Irvine, 1930: 22–23, as *A. caudatus*. 11. Irvine, 1948: 258, as *A. caudatus*. 12. Irvine, 1952, b, as *A. caudatus*. 13. Irvine, 1956: 40. 14. Irvine, s.d. 15. Oke, 1965, as *A. caudatus*. 16. Okiy, 1960: 121, as *A. caudatus*. 17. Portères, 1951, a: t. 2, p. 71. 18. Turton 61, K. 19. Verger, 1967: No. 176, as *A. caudatus*. 20. Walker, 1952: 184, as *A. oleraceus*. 21. Watt & Breyer-Brandwijk, 1962: p. 14, as *A. caudatus*.

**Amaranthus lividus** Linn.

FWTA, ed. 2, 1: 148. UPWTA, ed. 1, 35, as *A. blitum* Linn. and *A. oleraceus* Linn.

English: wild amaranth, green amaranth.

West African: **SENEGAL** WOLOF mbum bu digèn = *female spinach* (JGA) **GUINEA-BISSAU** CRIOULO bredo femea (EPdS) **GUINEA** KISSI fondulo (FB) LOMA pélé uluzedigpoie (FB) MANDING-MANINKA boromonema (FB) **GHANA** AKAN-ASANTE efai (Rattray) **NIGERIA** EFIK ìnyàñ (RFGA) IGBO ('Ibugo') ìnìnè ḿmē (NWT)

A glabrous herb, annual, to about 30 cm high, very polymorphic, widespread in West Africa from Senegal to S Nigeria, and elsewhere in tropical Africa, and in Asia and America, and into southern Europe.

The plant provides a palatable spinach (2). As long as 400 years ago it was cultivated in southern Europe, and has been considerably ameliorated. It has been cultivated in W Africa in the past but is not generally so now. The plant is a weed of cultivation and waste places. It is common around habitations and is collected from spontaneous generations for consumption. It is eaten raw or cooked everywhere (2, 3, 4), or made into sauces and side-dishes which can stand on their own without necessarily adding palatizers (8).

In India var. *oleracea* Duthie is cultivated for the seed which is used like a cereal (5). An intoxicating drink made in SW Africa from the seed of *A. blitum* (1) is probably of this species.

In S Nigeria it is recorded that Igbos use the plant in medicine for lung-trouble (6) and that to 'cool' a new house it is either laid under the first course of the house (7a) or sprinkled over the ground before building (7b).

References:

1. Burkill, 1935: 125. 2. Busson, 1965: 137, 142–4, with leaf and seed analyses. 3. Dalziel, 1937. 4. Hauman, 1951, a: 33–34. 5. Manjunath, 1948: 66, as *A. blitum*. 6. Thomas, N. W., 1779 (Nig. Ser.), K. 7a. Thomas, N. W., 2016 (Nig. Ser.), K. 7b. Thomas, N. W., 2016 (Nig. Ser.), K (Arch.). 8. Williamson, J., 1955: 17.

**Amaranthus spinosus** Linn.

FWTA, ed. 2, 1: 148. UPWTA, ed. 1, 36.

English: spiny amaranth; prickly amaranth.

French: amaranthe épineuse; épinard piquant.

West African: **SENEGAL** MANDING-BAMBARA ngoroba blé (JB; K&A) SERER dahdir gor (JB; K&A) WOLOF mbum bu gor = *male spinach, i.e. with spines* (auctt.) mbum i gor = *male mbum* (K&A) **GUINEA-BISSAU** CRIOULO bredo (JDES) **GUINEA** KISSI fondolo-ungu (FB) KPELLE g-bo (FB) LOMA pélé uluzédikolé (FB) MANDING-MANINKA bulunonima (FB) **SIERRA LEONE** BULOM (Sherbro) bɛntisɔk-lɛ (FCD) KONO sɔndi (FCD) MENDE hɔndi (FCD) ta-hɔndi *wild 'spp.'* (FCD; JMD) TEMNE *a*-bonthila (Glanville; Melville) *a*-niŋkuna = *cow-dung* (JMD; FCD) *a*-senduku (FCD) **MALI** DOGON gɔ́dɔ gɔ́dɔ (C-G) **IVORY COAST** KRU-GREBO (Nyabo) bedié (K&B) GUERE nionzan (K&B) GUERE (Chiehn) békiblé (K&B) **GHANA** ADANGME anago mio = *Nigerian's thorn, or Yoruba thorn, supporting the idea that the plant was brought from Lagos by soldiers in the Asante war, 1873–74* (FRI) AKAN-AKYEM nantwi nkase = *cow's thorn* (FRI) ASANTE asantewa (FRI) TWI asibe (FRI) esibe (FRI) nantwi bin = *cow dung, from* nantwi: *cow;* bin: *dung, suggesting dispersal by cattle* (FRI) sraha nkaseɛ (FRI) sraha nsoe (FRI) GA slaha-ŋmei = *Nigerian's thorn* (Bunting; KD) GBE-VHE matoŋui = *you cannot touch it* (*me*) *with your mouth* (FRI) sɛnutsõe (Bunting) VHE (Pecí) amma (Easmon) awusagbe, awusatso = *Hausa thorns* (FRI) sɛnutsõe (Bunting) HAUSA rukuɓu (Williams) **NIGERIA** FULA-FULFULDE (Nigeria) haako ndiyam = *water-leaf/plant* (J&D; MM) HAUSA cilli (RES) námijìn gaásàyaá (ZOG) námíjìn-zaákií-bánzaá (JMD; ZOG) IGBO (Awka) ìnine-ógwū (NWT; KW) IGBO (Obu) ìnìnè ógwū (auctt.) IGBO (Umuahia) nnuno uku (AJC) IJO èré ínínáịn = *female ininain* (KW) IJO-IZON (Kolokuma) ìnìnáịn (KW) YORUBA ẹfọ tẹ̀tẹ̀ (IFE) tẹ̀tẹ̀ ẹ̀gún, tẹ̀tẹ̀ eléègún = *spiny tete* (JMD; Egunjobi) **WEST CAMEROONS** DUALA èwɔ̀lɛ̀ a mùnàŋgà (Ithmann)

A robust cosmopolitan herb, to 1 m high, of waste places, clearings, weed of cultivation, anthropogenic and not generally occurring in virgin bush, recorded throughout the West African region, and in all warm countries.

The leaves are edible and are of a good nutritional value (4). They are eaten in all countries, but the plant is seldom cultivated for this purpose. Its thorniness presents some obstacle to its preparation as a vegetable, hence the Ewe name for it in Ghana. Stock graze it (1, 5, 12), and in Ghana the Twi name *nantwi bin* (7) and in Sierra Leone the Temne name *ka-ninkuna* (5) both mean 'cow dung' and suggest a belief of the plant's dispersal by means of cattle. However its thorniness may cause injury to the mouths of grazing animals (6), and cattle may shun it if more palatable fodder is available with a result that ungrazed *A. spinosus* may become a dominant plant in grazing land. Nevertheless, in Indo-China and India it is fed to cows-in-milk for it is held to increase the milk yield (3).

The plant is of Asian origin and there are other vernaculars which also indicate an idea, in Ghana at least, that it has an exotic origin. Dangme, Krepi and Ga names all mean *Nigerian, Yoruba* or *Hausa thorn,* and it is held that the plant came with Nigerian soldiers to Ghana in the Ashanti War of 1873–74 (7).

The plant is diuretic. Burnt to ash it yields an alimentary salt (4), and diuresis is attributed to the potassium salts present, principally in the roots (3). In Asia the plant is regarded as a specific for conditions symptomized by discharge, e.g. gonorrhoea, menorrhagia, milk-yield, colic, etc. (3, 10, 12). In Nyasaland (13) and in southern Africa (12) the plant-ash is used as snuff alone or with tobacco.

The leaf is emollient. In Ghana the plant is used in enema prescriptions for internal troubles, and it is used for treating piles (7). The plant-ash in solution is used to wash sores and sometimes leprosy sores in Ivory Coast (2, 9) and the plant-sap is made into an eye-wash for ophthalmia and especially to treat convulsions in children, and attacks of epilepsy and insanity (2), and in Gabon the leaves are used for ear-troubles. In Senegal the plant is used by midwives in caring for the comfort of newly-delivered young women and for stomach-upsets of babies (8). Similarly in Gabon the leaves are prepared into a ceremonial bath for cleansing new mothers approximately one month after delivery (11), and the plant is attributed by some Gabonese races with other magical properties. In S Nigeria, the Ijo eat the pounded fruit with a dish of plantain and fish specially prepared to prevent habitual miscarriage (14).

Hydrogen cyanide has been reported in Asian material (10). Ivorean plant material is rich in saponins (2). A fatty oil is present in the seeds (9).

References:

1. Adam, 1966, a. 2. Bouquet & Debray, 1974: 14. 3. Burkill, 1935: 127–8, with references. 4. Busson. 1965: 137, 142–3, with leaf analysis. 5. Dalziel, 1937. 6. Egunjobi, 1969. 7. Irvine, 1930, 23–24. 8. Kerharo & Adam, 1974: 122–3, with references. 9. Kerharo & Bouquet, 1950: 34. 10. Quisumbing, 1951: 267, 1019, 1030. 11. Walker & Sillans, 1961: 49. 12. Watt & Breyer-Brandwijk, 1962: 16. 13. Williamson, J. 1955: 17. 14. Williamson, K. s.d. no. 60 (8A).

## **Amaranthus tricolor** Linn.

FWTA, ed. 2, 1: 148.

English: red spinach (Deighton).

West African: **NIGERIA** TIV aleifu buter = *white man's aleifu* (Vermeer)

An introduced cultivated herbaceous spinach, widespread in Africa, Asia and Europe, with red and green forms, but so far only recorded from Sierra Leone and the Benue Plateau of Nigeria. Its exotic origin is reflected in the Tiv name meaning *white man's aleifu* (spinach) (1). Plant material grown in Nyasaland is reported to be similar to *A. lividus* in yield and palatability (2).

References:

1. Vermeer 27, UCI. 2. Williamson, J., 1956: 17.

## **Amaranthus viridis** Linn.

FWTA, ed. 2, 1: 148. UPWTA, ed. 1, 36.

English: green amaranth; local tete (SW Nigeria, Epenhuijsen).

French: épinard vert (Berhaut); épinard du Congo (Zaïre Hauman).

West African: **SENEGAL** SERER dahdir (JB) WOLOF mbum bu digèn = *female spinach, i.e. spineless* (JMD) mbum bu levet *wild* (*tasteless*) *spinach* (AS) mbum i kör = *spinach of the house* (JB) **THE GAMBIA** DIOLA (Fogny) ébayibadjay = *scar* (DF) **GUINEA-BISSAU** CRIOULO bredo (EPdS) FULA-PULAAR (Guinea-Bissau) bórborò-dórò (EPdS) bóròbóró-déo (JDES) **GUINEA**

MANDING-MANINKA boromonema (JMD) **SIERRA LEONE** BULOM (Sherbro) bɛntisɔk-lɛ (FCD) KONO sɔndi (FCD) MENDE hɔndi (FCD) ta-hɔndi TEMNE (Kunike) *a*-niŋkuna = *cow dung* (FCD) TEMNE (Port Loko) *an*-senduku (FCD) TEMNE (Sanda) *an*-senduku (FCD) TEMNE (Yoni) *a*-niŋkuna = *cow dung* (FCD) **MALI** FULA-PULAAR (Mali) nafa-nafa (A. Chev.) SONGHAI walia bocio TAMACHEK tazelanghatai (A. Chev.) **IVORY COAST** AKAN-ASANTE famufuéué (B&D) fankokoré (B&D) **GHANA** AKAN-ASANTE ɛfan (Enti) DAGBANI boboroa (Mellin/Thoms) **TOGO** GURMA (Mango) boboroa (Mellin/Thoms) **NIGERIA** FULA-FULFULDE (Nigeria) haako-ndiyam = *water-leaf/plant* (JMD; MM) rukuɓuho (JMD) HAUSA malan-kotshi (ZOG) námíjìn-gaásàyaá = *male Gynandropsis* (JMD; ZOG) námíjìn-zaákií-bánzaá *loosely applied to several weeds* (JMD; ZOG) rùb-dà-túkúnyá = *is cooked a lot* (RES; ZOG) rukuɓu (ZOG) taatsunyaa (ZOG) zaákií-bánzaá zarangade HAUSA (East) lábàshií (JMD; ZOG) HAUSA (West) malan kochi (JMD) IGBO ìnìnè ójīī (BNO) memomene (BNO) IGBO (Agukwu) ìnìnè ófīá óló (NWT) ìnìnèḿbèkwú (NWT) ìnìnèḿmē (NWT) IJO-IZON (Kolokuma) ìnìnàịn (KW) IZON (Oporoma) níníyà (KW) KANURI lekir (JMD) YORUBA tẹ̀tẹ̀ (Egunjobi) tẹ̀tẹ̀ atẹle danji (JMD) tẹ̀tẹ̀ àtètèdáyé (auctt.) tẹ̀tẹ̀ kékeré = *little tete* (JMD)

An annual herb to nearly 1 m high, pan-tropical and present throughout the West African region on wasteland, around habitation and as a weed of cultivation. Not normally cultivated, but occasionally so in the Region (5, 6, 10, 11), and in Zaïre where it goes under the name of *Congo spinach* (9).

It yields a palatable spinach of good nutritional value (5), the crop being taken normally before coming into flower (6). It provides a good cattle fodder valuable because it is without spines (1, 6, 7).

The leaf is diuretic and emollient and is used in poultices, on inflammations, boils, abscesses, etc. (6), topically to gonorrhoea and orchitis, and in enemas and suppository for piles (2). An infusion of the whole plant is used in Nigeria to purify the blood and as a cooling medicine of value in eye-complaints (2). The leaf-sap is used in Ivory Coast as an eye-wash for ophthalmias and for treating convulsions in children, attacks of epilepsy and fits of insanity (4). The leaf-sap is also said in Congo to be vermifugal and to be effective against filaria in the eye-lid or eye-ball, to be emmenagogue and to relieve tummy and heart-troubles (3). Leaves are given to persons suffering from fever in parts of the Region (6); and in Congo a decoction of the whole plant is used as a bath for feverish children as well as to wash new-born babies (3). The powdered dried leaves are applied to the pustules of yaws and smallpox in Congo (3), and the powdered charred plant is put into sores in Ivory Coast (4). The pounded root is given for dysentry in Nigeria (2).

As with other amaranths, the ash is rich in soda, and in the Region the plant is burnt for making soap (6). The powdered leaf contains tannin and reducing sugar and resin. No alkaloid has been detected (Thoms fide 6).

The root system is reported to be extensive and deep (7). The plant might be good as a green manure.

References:

1. Adam, 1966, a. 2. Ainslie, 1937: sp. no. 24. 3. Bouquet, 1969: 52. 4. Bouquet & Debray, 1974: 14. 5. Busson, 1965: 142–3, with leaf analysis. 6. Dalziel, 1937. 7. Egunjobi, 1969. 8. Epenhuijsen 4, K. 9. Hauman, 1951, a: 32–33, as *A. gracilis*. 10. Irvine, s.d. 11. Walker & Sillans, 1961: 49.

**Celosia argentea** Linn.

FWTA, ed. 2, 1: 146. UPWTA, ed. 1, 36.

English: quail grass (Irvine); Lagos spinach (Ghana, Irvine 4802, K); lizard bean (Lagos, Nigeria, because the plant is often frequented by lizards, Onwunyi 1092, UCI).

French: célosie argentée.

West African: **SENEGAL** MANDING-BAMBARA ngo ban ku (JB) **SIERRA LEONE** KRIO shɔkɔtɔ-yɔkɔtɔ (FCD) MENDE kikpɔi (FCD) kipɔi (FCD) kpange (FCD) yɔgɔtɔ (FCD) SUSU-

DYALONKE teŋge-na (FCD) TEMNE (Kunike) *an*-lami (FCD) TEMNE (Port Loko) *an*-kɔkɔrɔ (FCD) TEMNE (Sanda) *an*-kɔkɔrɔ (FCD) TEMNE (Yoni) *an*-lami (FCD) **LIBERIA** MANO bŏh (JMD) **GHANA** ADANGME-KROBO tɔbɔ (FRI) AKAN-FANTE nkyewɔduɛ (FRI) MANO ʃokotɔ (FRI) **NIGERIA** ARABIC sheiba (JMD) EDO àbórrà (Dennett; Singha) HAUSA farar áláyyafóo = *white alayafu* (JMD; Singha) riímií (ZOG) IGBO eriemiọnụ (BNO) TIV ígyár´ (Vermeer) YORUBA ṣọkọ (CWvE) sọkọ yọ̀kọ̀tọ̀ = *make husband fat* (Verger; Dawodu) sọkọ yọ̀kọ̀tọ̀ pupa, pupa: *red* (CWvE) ṣọkọtọ *a contracted form* (JMD) YORUBA (Ilorin) osun (Lamb)

An annual herb of great variability and of many forms, sometimes attaining 2 m height, with flower-spike usually silvery-white, sometimes pink or even red, elongated and pointed. It seems probable that the plant is of Asian origin for it has been identified in the Materia Medica of the ancient Chinese (2). It has become dispersed throughout the Asian and African tropics and is present in W Africa from Senegal to W Cameroons as a weed of cultivation and in open waste places of the forests zone, flowering throughout the year. Teratological cultivars occur as forma *cristata* (Linn.) Schinz with showy and broadly fasciated inflorescences which if left to run wild tend to revert to type.

The plant makes and excellent pot-herb and a good, if slightly bitter, spinach relatively rich in protein, and vitamins (4, 17, 18). The dietetic value was well proven by prisoners-of-war in Japanese hands in Thailand, 1942–45, who ate it as spinach with good results against beri-beri and pellegra (7). Though its occurrence in W Africa is widespread and its use as a pot-herb is common, there are indications that it is of fairly recent introduction. In Accra, Ghana, its English name of *Lagos spinach* is applied because it is eaten by Nigerian expatriates implying its exotic origin (14). In N Nigeria, the Chamba of the Vogel Peak area neither eat it nor have a name for it (13). In the Gurun area of Adamawa it has no use by the local people (12). It appears to be particularly relished by the Yorubas of SW Nigeria. It is commonly sold in southwestern markets (8), but to them also it has the hallmark of being a novelty. A Yoruba Odu incantation for curing cough runs:

> 'The vegetable called sọkọyọkọtọ has recently arrived. If you see it on the market, buy it to bring home. I shall have a big belly.'

To emphasize the esteem in which the plant is held by the Yorubas, the name *sọkọyọkọtọ* means *make husband fat* (20). Under this name use of the plant has travelled to Sierra Leone with slight variation to *shọkọtọyọkọtọ* in Krio.

The plant is often cultivated for the production of spinach, (5, 6, 16, 21) and ornamental varieties are grown for decoration which appear to have the same value for consumption (15). In The Gambia the leaves are boiled and eaten with rice to give it an acid flavour (22). The plant provides good fodder for stock. In Zaïre a fibre is prepared from the stems (11).

The leaves are without medicinal use in West Africa. They are probably slightly diuretic. They are used in poultices in China on infected sores, wounds and skin eruptions and in India mixed with honey on inflamed areas and painful afflictions such as buboes, abcesses, etc. The whole plant is used as an antidote for snake-bite and the root as a specific for colic, gonorrhoea and eczema (19). In Kenya the water in which leaves, flowers and stems have been boiled is used as a body-wash for convalescents by the Masai (10). In Ethiopian folk-medicine the seeds are used for diarrhoea and the flowers for dysentry and muscular troubles (9), and in SE Asia the flowers are considered medicinal for conditions whose symptoms include discharge of blood, e.g., dysentry, haemophythysis and menstruation (3, 19). In this connexion it is perhaps the red-flowered forms that apply on the Theory of Signatures. The seeds contain an oil. They are used for diarrhoea in Zaïre (11), and in Asia they are considered antiscorbutic and anthelminthic (6).

In parts of Zaïre the plant enters into a form of witchcraft to ensure good crops where it is growing (11). An interesting parallel to this is recorded from New Britain in Polynesia where it is planted in taro plots: The taro 'smells it' and produces large tubers (1). In Central Sumatra it enters into most occult ceremonies to propitiate the protective spirits (3).

References:

1. Blackwood 326, 327, K. 2. Bretschneider, 1895: 158. 3. Burkill, 1935: 506 with references. 4. Busson, 1965: 144–7, with leaf analysis. 5. Cavaco, 1963: 24. 6. Dalziel, 1937. 7. Den Hoed & Kostermans, Kwai Noi River Basin Exped. 1946, no. 700, K. 8. Epenhuijsen, 1, 18, K. 9. Getahun, 1975. 10. Glover & al. 2963, K. 11. Hauman, 1951, a: 16. 12. Hepper 1265, K. 13. Hepper, 1965: 443. 14. Irvine 4802, K. 15. Irvine 1952, b. 16. Irvine, 1956: 39. 17. Oke, 1965. 18. Okiy, 1960: 121. 19. Quisumbing, 1951: 269–72, incl. *C. cristata* Linn., with references. 20. Verger, 1967: 173. 21. Walker & Sillans, 1961: 49. 22. Williams, F. N. 1907: 94.

## **Celosia globosa** Schinz

FWTA, ed. 2, 1: 147.

A straggling herb to about 70 cm length, of forest undergrowth, clearings and in cultivated land, recorded in S Nigeria and W Cameroun, and in central Africa from Cameroun to Zaïre and Uganda.

The leaves are eaten as a vegetable in Zaïre (1).

Reference:

1. Hauman, 1951, a: 23–24.

## **Celosia isertii** C. C. Townsend

FWTA, ed. 2, 1: 147, as *C. laxa* sensu FWTA, ed. 2, non Schum. & Thonn. UPWTA, ed. 1, 36, as *C. laxa* sensu Dalziel.

West African: **THE GAMBIA** MANDING-MANDINKA furayɲamo = *death grass* (Hayes) **GUINEA-BISSAU** FULA-PULAAR (Guinea-Bissau) bóròbórò-déo (JDES) **SIERRA LEONE** MENDE gimbui (FCD) tɛgɔ (*def.*-i) (FCD; D G Thomas) TEMNE *a*-kɔkɔrɔ-*a*-ro-pet (FCD) **NIGERIA** FULA-FULFULDE (Adamawa) jarfundu (JMD) FULFULDE (Nigeria) nanafo (JMD) HAUSA bokan gida (JMD) nànnáfaá (JMD; ZOG) nànnàhoó (JMD; ZOG) YORUBA àjẹ fáwo, àjẹ fọ́wo = *eat and break plate, from* jẹ: *to eat*; fọ́: *to break*; áwo: *a plate,* meaning *something good to eat* (JMD) ajítàn *from* jí: *to awake*: tàn: *to shine, meaning unknown* (Millson)

A straggling herb, sometimes woody at the base, becoming somewhat bushy or sub-lianeous to 3 m tall, of secondary forest, stream-banks, damp sites, clearings and rarely in savanna. Recorded from Senegal to S Nigeria and Fernando Po, and in Cameroun across central Africa to Tanganyika, Zambia and Angola. While the species is held to be a good one, it is given in all literature as *C. laxa* Schum. & Thonn., the type of which is *C. trigyna* Linn. (9).

The plant is often eaten as a vegetable or prepared in soups and sauces in West Africa (2, 4, 7, 8), in Gabon (3, 10) and in Zaïre (6). In Sierra Leone the leaves are boiled and applied hot for rheumatism (5). Traces of flavones have been reported in the entire plant from the Congo area (1).

References:

1. Bouquet, 1972: 12 (as *C. laxa*). 2. Busson, 1965: 146–7, with leaf analysis (as *C. laxa*). 3. Cavaco, 1963: 26–27 (as *C. laxa*). 4. Dalziel, 1937. 5. Deighton 421, K. 6. Hauman, 1951, a: 20–21 (as *C. laxa*). 7. Irvine, 1956: 39 (as *C. laxa*). 8. Oke, 1965 (as *C. laxa*). 9. Townsend, 1975: 57–59, tab. 3736. 10. Walker & Sillans, 1961: 50 (as *C. laxa*).

## **Celosia leptostachya** Benth.

FWTA, ed. 2, 1: 147.

West African: **SIERRA LEONE** KRIO it-don-brok-plet = *eat-done-break-plate* (FCD) **NIGERIA** HAUSA farin-áláyyafóo (JRA) sokoyokotaw *from Yoruba* (JRA)

A straggly annual, reaching 70 cm tall, of undergrowth and roadsides, recorded from Sierra Leone, to W Cameroon and Fernando Po, also from Cameroun to Zaïre

The leaves are eaten as a vegetable in Zaïre (2). In Nigeria the fruits and seeds are pounded up for topical application in cases of ophthalmia (1).

In the case of the Krio vernacular in Sierra Leone, it is of interest to note it has the exact rendering of the Yoruba vernaculars in Nigeria for *C. trigyna* Linn. (see there).

References:

1. Ainslie, 1937: sp. no. 84. 2. Hauman, 1951, a: 22.

### **Celosia pseudovirgata** Schinz

FWTA, ed. 2, 1: 147, as *C. bonnivairii* sensu FWTA ed. 2, non Schinz.

A glabrous herb attaining 1 m high, of the forest, roadsides and cultivated land. Recorded in S Nigeria and W Cameroon, and also ocurring in Zaïre.

No usage is recorded in the Region. In Zaïre it is taken as a vegetable, and used (? bait) in fishing (1).

Reference:

1. Hauman, 1951, a: 21–22 (as *C. bonnivairii).*

### **Celosia trigyna** Linn.

FWTA, ed. 2, 1: 146. UPWTA, ed. 1, 36–37.

West African: **SENEGAL** SERER o légir (JB) o rigil (JB) WOLOF putur u mbam (JB) warko (JGA) **THE GAMBIA** MANDING-MANDINKA furayɲamo = *death grass* (Hayes) **GUINEA-BISSAU** FULA-PULAAR (Guinea-Bissau) bórbórò-déo (DS) **SIERRA LEONE** KRIO ajɛ-fawo (FCD) MENDE tegɔ (*def.*-i) (JMD) TEMNE *a*-kɔkɔrɔ-*a*-ro-pet (FCD) **UPPER VOLTA** HAUSA nanafo (K&B) **IVORY COAST** ABURE kalanbabié (B&D) kanébaguyé (B&D) KRU-BETE banjaraburu (AJML) **GHANA** GRUSI tono (FRI) yilina (Darko; FRI) **NIGER** FULA-FULFULDE (Niger) kure pallaandi (PF) **NIGERIA** FULA-FULFULDE (Nigeria) jarfundu (MM) kure pallaandi = *lizard's arrows* (MM) nanafo (JMD) HAUSA balansana (RES) nannaho (JMD; RES) HAUSA (East) jarfundu (JMD) HAUSA (West) bokan gida (JMD) nanafa (BM) nanafo (BM) nannafa (JMD) nanwafa (JMD) IGBO (Awka) èlí-émì-ọ́nụ̄ = *eating and pulling a face* (KW) YORUBA ajẹ́fọ́wo *see under C. isertii* (JMD) ajewafọ (N&E; CWvE) ajítàn, jí: *to awake;* tàn: *to shine* (Millson)

An annual herb of great variability, decumbent and sprawling or laxly erect and scandant to over 1 m long; a weed of cultivation and waste places in the savanna and forest zones throughout West Africa, and widespread in eastern and central Africa, Madagascar and Arabia.

Though the leaves are bitter they are eaten as a vegetable in many parts and may even be grown for this, but more especially the plant is a famine-food. The leaves are used in soups, sauces and seasonings, and they are sometimes sold in West African markets (9). Consumption 'in error' (whether in excess or at the wrong stage of growth is not stated) is reported to cause vertigo and delirium (12). All stock is recorded as grazing it in Senegal (1), though in Sudan there are opposing statements on its acceptability (2).

The leaves have a widespread reputation as an anthelmintic, particularly for children and to be especially effective against tapeworm (4, 7–9, 12, 20–23). In Ivory Coast the sap is administered to infants for this (7, 15). The plant is reported to contain *kosotoxin* which accounts for its anthelmintic action (7, 22). Its use requires caution and a common method is to take it in a draught of water in which unboiled husked grain has been soaked (9).

In Sierra Leone the leaf is eaten raw for heart complaints (11). The leaf is applied to pustular skin eruptions in N Nigeria (17), to sores in S Nigeria (19), and to sores and boils in Ghana (13), and to skin-diseases generally, and sometimes by warm poultice to rheumatic conditions (9). In Congo the pulped leaves are applied over scarifications to relieve costal pains (5), and in Ghana in cataplasm for chest troubles, and by mouth for stomach, liver and urethral disorders (14). The plant is said to be diuretic and haemostatic (7) and to be used in West Africa to hasten child-birth (22). In Ivory Coast it is known as 'medicine for wives' (16), perhaps, as in Congo, to treat ovarian trouble (5), and in Ethiopia for excessive menstruation (22). The leaves and flowers are used in Ethiopia in enemas and to treat diarrhoea (22), and in enemas in Ivory Coast (16), and the powdered leaves are taken in Sudan for diarrhoea (3) and in Tanganyika for sharp stomach-pains (18).

Leaf-sap is prepared as a collyrium in several areas for treating ophthalmias (7, 12, 14). In northern Ghana the pounded seed is also used in lotion (10).

A dye reported as red (Zaïre, 12) or black (Tanganyika, 23) is obtained from the plant which in Tanganyika is used to dye leather.

The seed contains a fatty oil and a quantity of potassium nitrate (15). The whole plant is rich in saponins (7) and a trace of flavones have been recorded (6).

The Mandinka name in The Gambia meaning *death grass* refers perhaps either to the use of the plant in death ceremonies, or on account of its smell (Hayes fide 9). The Yoruba name, *eat and break plate,* raises the conjecture that this is another example common throughout anthropology of breaking eating utensils after use described in detail by Sir James George Fraser in his classic study of religion and magic, *The Golden Bough*. Vernaculars in Zaïre (12) especially corroborate the diversity of usages this plant has been put to that the plant merits careful screening.

References:

1. Adam, 1966, a. 2. Andrews 32, 3605, K. 3. Andrews 122, K. 4. Bally, 1937. 5. Bouquet, 1969: 52. 6. Bouquet, 1972: 12. 7. Bouquet & Debray, 1974: 14. 8. Cavaco, 1963: 25–26. 9. Dalziel, 1937. 10. Darko 449, K. 11. Deighton 879, K. 12. Hauman, 1951, a: 17–18. 13. Irvine 523, K. 14. Irvine, s.d. 15. Kerharo & Bouquet, 1950: 35, with references. 16. Leeuwenberg 4522, K. 17. Parsons L.52, K. 18. Tanner 5041, K. 19. Thomas, N. W., 1622 (Nig. Ser.), K. 20. Walker, 1952: 184. 21. Walker & Sillans, 1961: 50. 22. Watt & Breyer-Brandwijk, 1962: 17–18, with references. 23. Williams, F. N., 1907: 94.

**Centrostachys aquatica** (R. Br.) Wall.

FWTA, ed. 2, 1: 153.

A stout aquatic herb with stem attaining several metres in length and rooting at nodes forming meadows in marshy places and in rivers. Recorded from Senegal and N Nigeria, but widespread in tropical Africa and Asia.

No use is recorded for this species in the Region. In Tanganyika the pounded roots are eaten with porridge, mornings only, by pregnant Sukuma women in treatment for syphilis (1).

Reference:

1. Tanner 1533, K.

**Cyathula achyranthoides** (H. B. & K.) Moq.

FWTA, ed. 2, 1: 151.
West African: **IVORY COAST** AKAN-ASANTE akuba (B&D)

A robust (? perennial) herb, woody, to 2 m high, an under-shrub of forest boundaries, swampy and waste places, and into plantations, recorded from Mali to S Nigeria and Fernando Po, and in Central Africa and tropical America.

The plant is used in Ivory Coast for antiseptic and analgesic purposes. Leaf-sap is applied to sores and chancres, and used as ear-drops for otitis and for headache. The sap is drunk to stop diarrhoea, nausea and bloody vomit (2). In Zaïre the leaves are used on skin-diseases and to heal circumcision wounds (3). Leprous and eczematous parts are treated with a leaf-decoction in Congo, first as a wash, then compounded with an ointment to rub on (1). It is also recognized as a cicitrisant for wounds and an anti-diarrhoetic. The sap is rubbed on areas with lumber pain, and to relieve headache; for pains in the ribs, bronchitis and chest affections, it is cooked with *Aframomum maleguetta* (Zingiberaceae), a part is eaten and a part rubbed on the chest. Another formula is used to mend fractured bones. A macerate is also drunk to relieve urethral discharges (1).

In Zaïre a vegetable-salt used in cooking is prepared from the plant (3).

References:

1. Bouquet, 1969: 53. 2. Bouquet & Debray, 1974: 14. 3. Hauman, 1951, a: 66.

## **Cyathula cylindrica** Moq. var. **mannii** (Bak.) Suesseng.

FWTA, ed. 2, 1: 151.

A lianiform herb to 3 m long, recorded from montane localities in W Cameroon and Fernando Po. The species occurs across Central Africa to Natal and into Madagascar.

No use is recorded in the Region. In Kenya it is grazed by all stock. The burred seedheads catching on the coats of cattle evoke the Masai name *the plant sticking to cow's tails* (1).

In Tanganyika the roots are recognized to be toxic (5) and are used to treat leprosy (3). The leaves are used in the treatment of *safura* (Swahili: 1, a swollen or dropsical condition; 2, ankylostomiasis—*Standard Swahili–English Dictionary)* (4).

A soap is made from the roots on Lesotho (2).

References:

1. Glover & al., 1437, 2265, 2510, K. 2. Guillarmod, 1971: 421. 3. Koritschoner 796, K. 4. Koritschoner 861, K. 5. Koritschoner 925, K.

## **Cyathula prostrata** (Linn.) Blume

FWTA, ed. 2, 1: 149, incl. *Digera alternifolia* sensu FWTA, ed. 2, non (Linn.) Asch., p. 1: 148 and *C. pedicellata* C. B. Clarke (= *C. prostrata* var. *pedicellata* (C. B. Clarke) Cavaco). UPWTA, ed. 1, 37.

French: cyathule couchée (Berhaut).

West African: **THE GAMBIA** MANDING-MANDINKA ŋaniŋjoŋ (*def*.-o) (auctt.) **SIERRA LEONE** KISSI dɔkpɔ̃ (FCD) MENDE nana (*def*. nanei) *a general name incl. other plants with prickly fruits* (FCD; DS) **LIBERIA** KRU-GUERE (Krahn) nah (JMD) BASA nah (JMD) **IVORY COAST** VULGAR naly (A&AA) ABURE ahué (B&D) AKYE n-kpè (A&AA) DAN blé (K&B) GAGU puin puin (K&B) KRU-BETE kukwo (K&B) GREBO (Nyabo) kumâ (K&B) GUERE déanoi (K&B) dianucha (A&AA) kuin (K&B) kukuan (K&B) kuma (K&B) GUERE (Chiehn) kubué (B&D) kukwé (B&D) KWENI babaléguésorué (B&D) péoléfu (B&D) KYAMA tutubi (B&D) 'KRU' kbulélé boté (K&B) manié (K&B) 'NEKEDIE' dogbo (B&D) **GHANA** AKAN-ASANTE mpupuã (FRI, Enti) FANTE akukuaba (auctt.) TWI apupua (Williams; FRI) **NIGERIA** FULA-FULFULDE (Nigeria) kebbe doombi = *rat's burgrass; from* kebbe: *Cenchrus sp., or bur-grass*; dombi: *rat* (JMD) kebbe jaule = *guinea fowl's bur-grass;* jaule: *Guinea fowl* (JMD) HAUSA dánká dáfi *a general name for plants with burred fruits* (JMD; ZOG) kàràngíyár beéraá = *mouse's burgrass* (JMD; ZOG) kàràngíyár kuúsuú = *mouse's burgrass* (JMD; ZOG) maɗaɗɗafin kuúsuù = *mouse's bur* (JMD; ZOG) marin kuusu (JMD) tsattsarar ɓera = *rat's basket trap* (JMD; ZOG) tsattsarar kura (ZOG) IGBO (Owerri) àgbịrịgbá (AJC) ajutọ *appl. to plants with burred fruit* (AJC) IJO-IZON (Kolokuma) òbórìkọ̀rìghá = *doesn't catch goat; meaning unknown since the fruits are burred and adhesive* (KW) YORUBA ṣawere pèpè (Millson) **WEST CAMEROONS** KPE krokos (Waldau)

*NOTE: The vernacular names of this sp. are more or less interchangeable with those of Pupalia lappacea.*

A straggling to more or less erect annual herb, up to 1 m long, young foliage often coloured red, with burred and adhesive fruits, a weed of cultivated land, of waste places and forest margins. Recorded from The Gambia to West Cameroon and Fernando Po in the savanna and forest zones, and widespread elsewhere in tropical Africa, and in Asia, Australia and tropical America.

The leaves are eaten as a vegetable in Gabon (4) and in Zaïre (6).

The plant has analgesic and antiseptic properties. The 'Kru' people of Guinea use the ash of the burnt plant with water to smear on the body for craw-craw, scabies, etc. (Afzelius fide 5) and in Nigeria the leaves are applied to craw-craw (5). The sap is applied to sores and chancres, and used as ear-drops for otitis and headache in Ivory Coast (3), or made into an ointment and applied to the head for headache (7). The seeds, lightly roasted and ground to a powder, may be mixed with pimento and natron in oil and used for earache, or the leaves powdered taken as a snuff for headache (7). In Cameroon the plant enters into prescriptions for articular rheumatism and for dysentery (Santesson fide 5), for dysentery in Ivory Coast (7) and for colic (1), and to check diarrhoea, nausea and bloody vomit (3). In Ghana the burned seeds enter into a popular remedy for dysentery (9). The plant is used in Congo for its anti-diarrhoetic properties where its action on dermal complaints is also used in treatment of leprosy and eczema to wash the areas first, and then to apply an ointment (2).

Leafy twigs, inflorescences and seeds pulped into a paste, and with or without clay are used on sores, burns and fractures in Ivory Coast (7), and as a haemostatic and cicitrisant (1). Fractures are treated in Congo with a preparation using the plant (2) which also heals wounds. The plant in mixture with *Synedrella nodiflora* (Compositae), *Aframomum maleguetta* (Zingiberaceae) and clay is used in Ivory Coast for heart-trouble (7), and a prescription with *A. maleguetta* is rubbed onto the chest in Congo for bronchial affections (2). The Ijo of S Nigeria use the fruits in a prescription to prevent miscarriage, and for general debility for which the powdered leaves in a mixture with other drugs may also be applied to the affected parts over incisions (10). In Gabon the plant has uses in treating eye-troubles, wounds and urethral discharges (8) and in Zaïre (6) and in Congo (2) for the last named.

References:

1. Adjanohoun & Aké Assi, 1972: 13. 2. Bouquet, 1969: 53. 3. Bouquet & Debray, 1974: 14. 4. Cavaco, 1963: 34. 5. Dalziel, 1937. 6. Hauman, 1951, a: 64. 7. Kerharo & Bouquet, 1950: 35. 8. Walker & Sillans, 1961: 50. 9. Williams 22, K. 10. Williamson, K. 38A, UCI.

**Cyathula uncinulata** (Schrad.) Schinz

FWTA, ed. 2, 1: 151.

A sprawling, rampant or lianous herb attaining 5 m length, in montane localities of W Cameroon, and in east, central and south tropical Africa and Madagascar.

No use of the plant is recorded for the Region.

The leaves are used in the form of a plaster over wounds in Zaïre (3). In Lesotho the plant is used as an emetic (2), a root-decoction is taken for urethral 'stricture', and the root-ash is used to make soap (6). In Madagascar the plant is considered to be anti-syphilitic (6), and to be good in decoction with other drug plants for abdominal and liver complaints and to be helpful at childbirth (1). In Tanganyika the root is used by barren women to induce pregnancy (4), and preparations of the plant are used to treat ankylostoma infected eyes (5).

AMARANTHACEAE

References:

1. Debray & al., 1971: 5. 2. Guillarmod, 1971: 421. 3. Hauman, 1951: 68. 4. Luckman 3, K. 5. Tanner 5915, K. 6. Watt & Breyer-Brandwijk, 1962: p. 18.

**Gomphrena celosioides** Mart.

FWTA, ed. 2, 1: 143, 760.

A sprawling or decumbent herb, native of Brazil, Paraguay, Uraguay and Argentina, and spreading throughout the Old World tropics (4). It has probably reached E Africa from Asia some 50–60 years ago and is now relatively common there. Its presence in Ghana and Nigeria is recently recorded.

In Zambia it has been reported 'good for yarding' (3). In Nigeria it has become a common weed on mown grass (2) and in Ghana a weed of lawns, common on roadsides and waste places (1). By its prostrate habit it might be grown to produce a cover where the establishment of grass is difficult or too slow, and to combat erosion.

References:

1. Adams s.n., 8/4/1955, K. 2. Hambler 309, K. 3. Rabson Phiri 90, K. 4. Sandwith, 1946.

**Gomphrena globosa** Linn.

FWTA, ed. 2, 1: 153. UPWTA, ed. 1, 37.

English: bachelor's button; globe amaranth.

French: immortelle à bouton (Walker).

West African: **SIERRA LEONE** KRIO bachilɔs-botin = *batchelor's button* (FCD) **GHANA** ADANGME adinamɔmɔ (FRI) **TOGO** TEM (Tshaudjo) atshuro-tshatsho (JMD) **NIGERIA** HAUSA (West) kandiirii (Bargery)

An annual herb with globose red or purple flower-heads, erect to nearly 1 m high. A native of S America and introduced to many tropical countries as an ornamental. In West Africa it has become subspontaneous in places.

While the plant is commonly cultivated in gardens for decoration, it is sometimes grown for medicinal and fetish purposes. In Gabon the leaves are triturated with leaves of *Eclipta alba* Hassk. (Compositae) and copper filings are added for application to gangrenous wounds, and to some tribes it is a protective fetish (4). In Brazil and the West Indies the plant is used as a cough-cure (5). In Malesia it has been used as a vegetable (2).

The plant has given positive tests for triterpenoids, and negative for alkaloids, bitters, volatile oils, hydrogencyanide and saponin (5). In Rhodesia (1) and in E Africa (3) a red dye is extracted from the plant or flowers.

References:

1. Allnutt 1, K. 2. Burkill, 1935: 1097. 3. Greenway, 1941. 4. Walker & Sillans, 1961: 50. 5. Watt & Breyer-Brandwijk, 1962: 18.

**Nothosaerva brachiata** (Linn.) Wight

FWTA, ed. 2, 1: 149.

West African: **SENEGAL** FULA-PULAAR (Senegal) ba nâ, banana (K&A)

An annual herb, prostrate to erect to about 60 cm high, appearing in the wet season of the northern part of the Region in Senegal and N Nigeria, and in Chad, Sudan, Kenya, Angola, Mascarenes and India.

The plant does not appear to provide any grazing for stock in West Africa (1) but it is reported grazed by goats in Northern Kenya (4).

In Senegal the nomadic Fula consider it a magical plant and use it in various medico-magical formulae for both stock and man (2, 3).

References:

1. Adam, 1966, a. 2. Kerharo & Adam, 1964, b: 562. 3. Kerharo & Adam, 1974: 123. 4. Mathew 6276, K.

**Pandiaka heudelotii** (Moq.) Hook. f.

FWTA, ed. 2, 1: 151. UPWTA, ed. 1, 37.

West African: **GHANA** DAGBANI barimbini (FRI) powanyakruga zuperi (FRI) **NIGERIA** HAUSA maasun kadangaree = *lizard's spears* (JMD; ZOG) HAUSA (West) damfarƙami (JMD) IGBO (Obu) ílé ágwọ̄ = *snake's tongue* (NWT; KW) YORUBA (Ilorin) arẹsẹ̀kosùn (JMD)

An annual herb to about 1 m high from a lignified base, of savanna and often a weed of cultivated land of dry sandy areas, occurring from Senegal to S Nigeria and across central Africa to Sudan.

The plant is not grazed by stock (1).

The leaves of material collected in Guinea are said to contain a colouring matter (? colour) (3). Tests for the presence of alkaloid in Nigerian material have been negative (2).

References:

1. Adam, 1966, a. 2. Adegoke & al., 1968. 3. Chevalier, 1920: 530 (as *Achyranthus heudelotii* Moq.).

**Pandiaka involucrata** (Moq.) Hook. f.

FWTA, ed. 2, 1: 151. UPWTA, ed. 1, 37.

West African: **SENEGAL** SERER ñohombil (JB); *perhaps P. heudelotii* (Moq.) Hook.f. **NIGERIA** HAUSA maasun kadangaree = *lizard's spears* (auctt.) HAUSA (West) damfarƙami (JMD) YORUBA abadan awurebe (EWF) YORUBA (Ilorin) arẹsẹ̀kosùn (JMD)

An erect herb to about 1.20 m high, of savanna and wooded savanna especially on dry hills and rocky sites from scattered places between Senegal and Nigeria, and in Cameroun, Chad and Central African Republic.

The plant is not grazed by stock (1).

An infusion is drunk by Hausa women after childbirth (2).

References:

1. Adam, 1966, a. 2. Moiser 117, K.

**Philoxerus vermicularis** (Linn.) P. Beauv.

FWTA, ed. 2, 1: 153. UPWTA, ed. 1, 37.

French: amarante bord de mer (Berhaut).

West African: **SENEGAL** MANDING-MANINKA (The Gambia) é nal (JB) SERER foranoh (JB) fuhohol (JB) SERER-NON (Nyominka) hogho (A. Chev.) **THE GAMBIA** DIOLA timbindingo (DRR) DIOLA (Fogny) butim biringabu, *from the way of branching* (DF) enial = *slippery* (DF) MANDING-MANDINKA konyamo = *monkey grass* (Fox; DF) sindiŋ (*def.*-o) (DRR) MANINKA kuuɲambi (*def.* kuuɲamboo) (Sidia Tatta) **GUINEA-BISSAU** MANDING-MANDINKA konyamo sinding **SIERRA LEONE** TEMNE *a*-bantan (Glanville) *a*-kɔrɔ (NWT) **GHANA** GBE-VHE koklotade = *fowl's pepper* (FRI) VHE (Awlan) dzogbe soĩĩ = *Sesuvium* (*soli*) *of the field* (FRI)

A fleshy herb with red creeping stems, rooting at nodes and attaining over 1 m in length with shortly erect nodal branches bearing terminal and lateral inflorescences. Common on all sandy beach-heads and in dry mangrove from Senegal to W Cameroon, and on to Angola, and on the Atlantic Coast of the American Continent.

The Diola of Casamance eat the leaves as a spinach (3). In Ghana poultry will eat it, hence the Ewe name meaning *fowl's pepper* (2). By its creeping habit it acts as a good sand-binder, a property recorded from all parts of West African Region Coast line. In The Gambia it is said to cover large areas of mangrove land in the last stages of land-building (4). In the brackish tidal rice-lands of the Scarcies Rivers in Sierra Leone it is a common weed and may become troublesome (1, 5,).

The leaves have religious use in Ghana for preparing water for ceremonial bathing (2).

References:

1. Glanville 257, K. 2. Irvine, 1930: 333. 3. Portères, s.d. 4. Rosevear, 1961. 5. Small 119, K.

**Pupalia lappacea** (Linn.) Juss.

FWTA, ed. 2, 1: 151, 760. UPWTA, ed. 1, 37–38.

West African: **SENEGAL** FULA-PULAAR (Senegal) ńagobéré (K&A) MANDING-BAMBARA norna ba (JB, ex K&A) SERER ńadâg (JB, ex K&A) ñèpñèp (JB) WOLOF ñapèntan = *that which sticks* (auctt.) **IVORY COAST** BAULE apopo amli (K&B) apopo aubri (K&B) MANDING-DYULA nro-nrobaha (K&B) **GHANA** AKAN-AKUAPEM apupũa, mpupuã (FRI; Enti) ASANTE akyere-nkura = *catcher of mice, referring to the use of the fruit for this purpose* (FRI) apɔsɔmpɔ (FRI) mpupuã (Enti) FANTE akukuaba (FRI) ekukuaba (FRI) TWI akyere-nkura = *catcher of mice* (FRI) apupuã (Enti) mpupua (FRI) DAGBANI koruntiya GA memleŋte (FRI) GBE-VHE mitsimitsi GUANG-NCHUMBULU daboyeju NZEMA amyole (?) **NIGER** FULA-FULFULDE (Niger) kebbe jodde (PF) SONGHAI dènjìrìm-denjí (*pl.* -ó) (D&C) **NIGERIA** EFIK ìkọ̀ñ (RFGA; Lowe) FULA-FULFULDE (Nigeria) kebbe jodde (MM) nyakkabre (MM) HAUSA dánká dáfi (AST) maɗaɗɗafin kuúsuù = *mouse's bur* (auctt.) marin kuúsuù = *mouse's shackles* (JMD) HAUSA (East) kaimin kadangari (JMD) IGBO agbiriba (JMD) IGBO (Owerri) ósè, ósò: *general name for 'pepper'* IGBO (Uzuakoli) ñtụ̀rụ̀ḿ ōkụ̄kọ̀, ọ̀kụ̄kọ̀: *fowl* (FRI; KW) YORUBA ẽmọ́ àgbò = *ram's bur* (JMD; IFE)

A robust, pubescent and much branched herb, woody at base, to 1 m high, flower-head a bur with recurved spines, of savanna and woodland localities and forest pathsides, from Senegal to S Nigeria, but less common in the western part of the Region. Distribution is widespread elsewhere in tropical Africa and in Asia to Polynesia.

The plant is not normally eaten by man. It is recorded as causing colic in Somalia 'if eaten prior to watering' (15) which suggests that it can be eaten after proper precaution has been taken. In Ethiopia (5) and in Somalia it is readily grazed by sheep and goats (3) and all stock (16), but it is untouched in Senegal (1).

The plant has a number of medicinal uses. The leaves are put into soups for coughs in Ghana (8, 9) and the Igbos of Aroka, S Nigeria, select the purple coloured-leaves to pound with palm-oil and salt as a cough-remedy (9). In Ivory Coast a tisane of leaves is taken for coughs (4). To treat boils, the leaves are mixed with palm-oil and applied in Ghana (8, 9) and the Sukuma of Tanganyika boil and pound the leaves with butter to put on them (17). Leaves are also used in topical applications to cuts, or used as an enema or febrifuge (6, 8, 9). In Ivory Coast a decoction is taken in draught and applied in frictions for oedema of the legs and is used in various remedies for dysenteriform diarrhoea (4, 12).

The burred flower-head is an ingredient of a rat-poison in Nigeria (6, 18), but in what way it is used is not described. A Somali story runs that a squirrel whose tail had become stuck full of seeds of this plant presented its rear end to a snake about to attack it. The snake struck the squirrel's tail and was killed by the seeds (16). The action may perhaps be mechanical in that the reflexed hooks on the spines of the seed-coat get fixed in the mouth and throat causing choking. At least the hooks show great resilience. Fantes of Ghana are said to make a football with a mass of the adhering seed-heads covered with a piece of cloth (8, 9). In Uganda a mass of seed-heads is used as a beer-pot cover (7), and as a symbolism in human emotions they lead to the plant being used in charms, mainly love charms. Because of the stiffness of the spines, the echinate fruits are sold in SW Nigerian markets for scarifying the skin prior to blood-letting or the application of ointment (6).

Crushed seeds are considered to be in Ivory Coast a good remedy for infected sores and phagodenic ulcers (4).

The root is beaten up in water which is drunk for sore-throat in Uganda, some of the liquor being left to dry from which the residue is made into an ointment with fat for external application (7). In Tanganyika the root has uses in treatment for snake-bite (13) and for syphilis (14).

Plant-ash is said to be used in Sudan mixed with water and drunk for flatulence, and applied to leprosy sores after they have been made to bleed (6).

A fair amount of saponins has been detected in the whole plant (4), and in the seed of Nigerian material a trace of alkaloid (2).

The plant is reported used in a fish-lure in the Senegal River, not as a fish-poison but in a manner not clearly defined (10, 11).

References:

1. Adam, 1966, a. 2. Adegoke & al., 1968. 3. Andrews 3564, K. 4. Bouquet & Debray, 1974: 14. 5. Carr 918, K. 6. Dalziel, 1937. 7. Dyson-Hudson 443, K. 8. Irvine, 1930: 359–60. 9. Irvine, s.d. 10. Kerharo & Adam, 1964, b: 571. 11. Kerharo & Adam, 1974: 123–4. 12. Kerharo & Bouquet, 1950: 35. 13. Koritschoner 1014, K. 14. Koritschoner 1019, K. 15. Peek 3B, K. 16. Peek 240, K. 17. Tanner 1272, K. 18. Thornewill 141, K.

### **Sericostachys scandens** Gilg & Lopr.

FWTA, ed. 2, 1: 151.

A slender liane reaching the canopy of forest areas in S Nigeria, W Cameroon and Fernando Po; also occurring in Cameroun, Zaïre and Uganda.

No use is recorded of the plant in the West African region. In Zaïre the leaves are considered edible and vulnerary. The bark is used in antivenereal medicine (1).

The liane when in fruit is strikingly decorative.

Reference:

1. Hauman, 1951, a: 72.

## AMARYLLIDACEAE

### **Crinum** Linn.

FWTA, ed. 2, 3: 134. UPWTA, ed. 1, 486.

English: general — Crinum lily; some spp. — Cape Coast lily; unspecified — Hanga or Mabanta lily (Sierra Leone, etymology unknown, Dalziel).

French: general — lis de brousse.

West African: **SENEGAL** BASARI a-támb ɛɓ ɛɗén (Ferry) MANDING-BAMBARA baga *= poison; a general term* (K&A) **IVORY COAST** AKAN-ASANTE niassatindé (B&D) KRU-GUERE (Chiehn) bobèbé (B&D) KWENI yupoké (B&D) 'NEKEDIE' kokoboba (B&D) okoyabo (B&D) **NIGERIA** FULA-FULFULDE (Nigeria) gaadal (J&D) HAUSA gadali *a sacred bulb, perhaps after Fula 'gaadal', a medicinal plant* (Meek) JUKUN abi *a sacred bulb* (Meek) KANURI lùwàsàrngàshó (C&H)

All species are of ornamental value and can be grown from dormant bulbs gathered during the dry season. The larger ones are commonly planted as fetish plants near shrines, and the bulbs of some are added to arrow-poison preparations. The Fula name *gaadal* imples a medicinal plant, and on the Benue Plateau a lily known as *gadali* is sacred and is used in ceremonial ritual. The Jukun *abi* is also sacred and both are used as charms by fishermen (4).

Several species have been recorded as toxic (2). In Ivory Coast the Baule consider one species, of imprecise determination, but slightly toxic and suitable to use in decoction for taking by mouth to relieve distension of the stomach, caused, perhaps, by criminal poisoning. Violent purging, followed sometimes by vomiting, results. Anyi use a preparation of this plant as an eye-wash in treatment for jaundice (1).

Several *Crinum spp.* have shown some activity against avian malaria, but the West African species have not been reported on (3).

References:

1. Bouquet & Debra, 1974: 15. 2. Chevalier, 1950. 3. Claude & al. 1947. 4. Dalziel, 1937.

## **Crinum distichum** Herb.

FWTA, ed. 2, 3: 136.
West African: **SENEGAL** MANDING-MANINKA barhô firila (K&A) **THE GAMBIA** MANDING-MANDINKA ba (Fox)

A bulbous plant with distichous leaves, and flowers white with a medium pink stripe, one or two borne on a scape about 30 cm high, of seasonally flooded places in savanna from Senegal to N Nigeria, and extending to Sudan.

It is reported to be toxic (1).

Reference:

1. Kerharo & Adam, 1974: 125, as *C. pauciflorum* Bak. (*C. distichum* Herb.).

## **Crinum glaucum** A. Chev.

FWTA, ed. 2, 3: 136.

A bulbous plant with thick, stiff erect glaucous leaves, flowers pure white borne on a scape reaching 60 cm high, of wooded savanna in damp sites in Guinea-Bissau to N Nigeria.

The bulbs are unusually large attaining 2.5 kg in weight. Alkaloids *lycorine ambelline* and two others unidentified have been reported present (1).

Reference:

1. Powell & Taylor, 1967.

## **Crinum jagus** (Thomps.) Dandy

FWTA, ed. 2, 3: 136. UPWTA, ed. 1, 486, as *C. giganteum* Andr.
English: the forest crinum (Morton).
West African: **SIERRA LEONE** KONO takbo (FCD) MENDE nengbe, nja-nengbe *general for water plants* (auctt.) nengbe-wawa (*def.*-i) (FCD; JMD) pupenda (*def.* pupendei) (FCD) TEMNE *a*-rotə-*a*-kɛrɛ (NWT) *a*-yaba-*a*-kɛrɛ (NWT) **IVORY COAST** KYAMA anbolo (B&D) **GHANA** AKAN-TWI sukooko (Plumtre; FRI) **NIGERIA** EFIK ékóp-éyèn = (?) *child's navel* (Winston; KW) FULA-FULFULDE (Nigeria) gaadal (*pl.* gaade) (JMD) IGBO édè-chúkwú = *God's cocoyam* (JMD) IGBO (Agolo) òlòdì (NWT) IGBO (Owerri) édẹ ọ̀bàsị̀ (AJC; JMD) IGBO (Umuahia) òlòdì, òlòdù (JMD) ozu (AJC) YORUBA ẹdẹ́suku *a hairstyle* (Verger)

A bulbous plant with thin, wide-spreading, rich green leaves, and flowers white tinged green borne on a scape to 60 cm high, in swamp forest sites from Sierra Leone to S Nigeria.

The bulb is somewhat elongated and may reach 2 kg in weight. The fresh bulbs are reported to have an action similar to digitalis and toxicity, especially of the dried bulb, is low (1, 4). *Lycorine,* and traces of a number of other alkaloids are reported present, and to be in higher amount in November after the end of the rainy season than in April (5).

The mashed-up bulb is used in Igboland as a beauty preparation (2), and the Yoruba name is given to a plaited hairstyle, *Suku* style which features in an invocation 'to send Sìgìdì to kill one's enemy' (6). The same word appears in the Twi name *sukõko* for a plant the bulb of which is used in the Aowin district of Ghana to prepare a yellow dye for dyeing cloth (3).

References:

1. Bouquet & Debray, 1974: 15. 2. Carpenter 21, UCI. 3. Irvine, 1930: 134, as *C. sp.* (Irvine 52). 4. Oliver, 1960: 23, 58. 5. Powell & Taylor, 1967. 6. Verger, 1967: 83.

**Crinum lane-poolei** Hutch.

FWTA, ed. 2, 3: 134.

West African: **SIERRA LEONE** KRIO jaiant-lili = *giant lily* (FCD)

A bulbous plant with numerous white flowers, recorded from Sierra Leone, probably exotic of unknown origin.

It is grown in Sierra Leone as an ornamental (1).

Reference:
1. Deighton, 1984, 6082, K.

**Crinum natans** Bak.

FWTA, ed. 2, 3: 134.

English: the floating crinum (Morton).

West African: **SIERRA LEONE** KISSI pe-folã (FCD) KONO sasa (FCD) sasã (FCD) takbo (FCD, ex JMD) MENDE nengbe, nja-nengbe *general for water plants* (JMD) nengbe-wawa (*def.*-i) (JMD) pupenda (Dawe; FCD) **GHANA** AKAN-TWI sukooko (FRI) **NIGERIA** EFIK ékóp-éyèn (KW) FULA-FULFULDE (Nigeria) gaadal (*pl.* gaade) (JMD) IGBO édè-chúkwú = *God's cocoyam, for C. spp. generally* (JMD) édē-ọ̀bàsị̀ = *God's cocoyam* (JMD) IGBO (Umuahia) òlòdì, òlòdù (JMD) ozu (AJC)

An aquatic plant of clear running water with small bulb anchored in firm substrate, leaves submerged and floating, flower-scape erect above water-level, widespread from Guinea to W Cameroons and Fernando Po, and on to Gabon.

The plant is ornamental. It is used in S Nigeria in a superstitious way to ward off ill-health in new-born infants (2).

The plant is a part of the hippopotamus's diet in Gabon (4). Assay of West African material has shown no alkaloid (3), or only a feeble toxicity (1).

References:

1. Bouquet & Debray, 1974: 15. 2. Gregory 1, K. 3. Powell & Taylor, 1967. 4. Walker & Sillans, 1961: 51.

**Crinum purpurascens** Herb.

FWTA, ed. 2, 3: 134.

English: starry crinum (Morton)

West African: **THE GAMBIA** MANDING-MANDINKA ba (Fox) **NIGERIA** IJO-IZON (Kolokuma) òlíélíé-gbàmú-tùlá (KW)

A bulbous plant with white star-like flowers of high rainfall forest areas, from The Gambia to S Nigeria and on to the Congo region and Angola.

The plant is very decorative.

Throughout the Congo the bulb of this, and of other *Crinum spp.* is used for emeto-purgative purposes, often being grown specially for medicinal use; a decoction or macerate is taken by draught for swollen stomach, food-poisoning, broncho-pneumonia, ovarian troubles and hernia; sap is embrocated onto children for enlarged spleen; leaves cooked with palm kernels are eaten as an aphrodisiac and are said to be efficacious against snake-bite (1).

Reference:

1. Bouquet, 1969: 53.

**Crinum zeylanicum** (Linn.) Linn.

FWTA, ed. 2, 3: 134, as *C. ornatum* Herb. UPWTA, ed. 1, 486–7; as *C. yuccaeflorum* Salisb. and *C. sanderianum* Baker.

English: the beautiful crinum (Morton).

West African: **SENEGAL** DIOLA étiram (JB; K&A) MANDING-BAMBARA baga (JB; K&A) 'SOCE' baa (K&A) SERER lar (JB; K&A) rar (JLT) SERER-NON (Nyominka) ilar (K&A) WOLOF dalkané (auctt.) table (DF) **THE GAMBIA** DIOLA (Fogny) etiram (DF) **GUINEA** MANDING-MANINKA baga (JMD) bakha (JMD) **SIERRA LEONE** TEMNE *an*-yaba *after* yaba: *onion* (Afzelius) **MALI** MANDING-BAMBARA baga (A. Chev.) **GHANA** ADANGME ngalenge (FRI) AKAN-ASANTE sukookoo (Enti) FANTE nsukookoo (Hall) GA dwɔ ɛmine brɔfo ngme (FRI) **NIGER** FULA-FULFULDE (Niger) gawri waaliya (PF) **NIGERIA** BEROM jwɛ̀ɛ̀ bót (LB) FULA-FULFULDE (Nigeria) albacce buru = *hyena's onions; incl. other bulbs* (JMD; MM) albacce dawaaɗi = *dog's onions; incl. other bulbs* (JMD; MM) gaadal (*pl.* gaade) (JMD) gaadal faaɓru (*pl.* gaade paaɓi) = *frog's gaadal* (MM) gawri waaliya (MM) gurguli *probably this sp.* (Taylor) GOEMAI ndan-murenang (JMD) HAUSA albásar kwaaɗíi = *frog's onion; commonly used throughout N. Nigeria* (JMD; LB) NUPE dinkòrò'bá *husband of the lily* (JMD) YORUBA ìṣumẹri *from* ìṣu: *yam* (Dawodu)

A bulbous plant with leaves reaching 75 cm long by 6 cm wide, and scape 60–100 cm high bearing 4–6 white flowers with a broad purple band along the centre in a large umbel to 15 cm across, of slightly damp sites in savanna. Common from Senegal to W Cameroons, and widespread throughout much of tropical Africa.

The plant is distinctly ornamental and is frequently cultivated. It has been much used in horticulture for hybridizing with other species.

The plant is not eaten by stock (1). In Senegal, Mali and Guinea has a common name meaning 'poison' and is recognized as causing diarrhoea that is difficult to control (1, 8). Thus in Senegal it is not put into medicines taken by draught, but it is used externally by placing layers of the bulb over areas of skin-trouble and on injuries and refractory ulcers (5, 6). Analogous use is recorded in parts of Nigeria: the sliced bulb is heated to cause exudation of sap which is mixed with copper filings for application to wounds (3). Also the ash of the incinerated bulb is mixed with a leaf-gall found on *Terminalia avicennioides* (Combretaceae) to produce an ointment said to be effective on swollen joints. The bulb is also used as a pomade and cosmetic (3). It is rubefacient (7) and in Sierra Leone the leaves are said to have been used in cold infusion as a stimulant for bathing young children suffering from general debility, rickets, etc. The bulb is used in decoction in Ghana as a vermifuge. In SW Nigeria it is often traded in for its medicinal properties, probably with the bulbs of other species (4).

The fresh fruit is pounded with red natron in N Nigeria for application to guinea-worm blister, and in a superstitious way the crushed fruits are rubbed on the feet of farmers to prevent injury by the hoe. The dried seeds are an ingredient of some fish-poison mixtures (3).

The alkaloid *lycorine* has been recorded in a number of other *Crinum spp.* Its presence in this species is probable. Lycorine has properties like *yohimbine* of importance in veterinary medicine and as an aphrodisiac. *Tyramine* is reported present (2, 6, 7).

References:

1. Adam, 1966, a: 518, as *C. sanderianum*. 2. Bouquet & Debray, 1974: 15 as *C. yuccaeflorum*. 3. Dalziel, 1937. 4. Dawodu 43, K. 5. Kerharo & Adam, 1964, c: 301, as *C. sanderianum* Bak. 6. Kerharo & Adam, 1974: 125–6, with pharmacology, as *C. sanderianum* Bak. *(C. ornatum* (Ait.) Bury). 7. Oliver, 1964: 23, 58. 8. Walker, 1952: 185 as *C. sanderianum* Bak. 9. Walker & Sillans, 1961: 51, as *C. sanderianum* Bak.

**Eucharis grandiflora** Planch. & Linden.

FWTA, ed. 2, 3: 131.

English: Amazon lily, eucharis lily, madonna lily.

A bulbous plant, leaves to 30 cm long, scape to 60 cm bearing an umbel of 3–6 fragrant flowers, of the Colombian and Peruvian Andes, and taken into cultivation by man throughout the world as a very attractive ornamental.

The alkaloids, *haemanthamine, lycorine* and *tazettine,* have been reported present in the plant (2); also the phenolic amine *synephrine* (1).

References:

1. Wheaton & Stewart, 1970. 2. Willaman & Li, 1970.

**Hippeastrum puniceum** (Lam.) Voss

FWTA, ed. 2, 3: 131 as *H. equestre* (Ait.) Herb.

English: amaryllis, Barbados lily (U.S.A.) harmattan lily, cacao lily (Irvine).

A bulbous plant, leaves to 60 cm long, scape to 60 cm long bearing 2 or 3 unscented flowers 10–12 cm long, rosy coloured with cream centre, of tropical America, but taken by man into cultivation in all countries as an ornamental, and naturalized in West Africa.

The plant is commonly grown in cemeteries in the Region, and its persistence in the bush may mark the site of abandoned villages (3). In West Africa it rarely sets seed and propagation is by bulbs. In Congo a decoction of the bulb is drunk for tachycardia, and is taken as a love-philtre to aid seduction (1).

Several other *H. spp.* are recorded as showing considerable activity against avian malaria (2). The property of W African *H. puniceum* should be examined.

References:

1. Bouquet, 1969: 54, as *H. equestre* Herb. 2. Claude & al. 1947: as *H. spp.* 3. Morton J. K., 1961: 27–28, as *H. equestre* Herb.

**Hymenocallis littoralis** Salisb.

FWTA, ed. 2, 3: 131. UPWTA, ed. 1, 487.

English: spider lily.

West African: **SIERRA LEONE** MENDE nengbe, nja-nengbe *a general name for water plants* (JMD) nengbe-wawa (*def.*-i) (JMD) TEMNE *a*-rotə-*a*-kɛrɛ (JMD) *an*-yaba *after* yaba: *onion* (Afzelius)

*NOTE: Names for Crinum in general apply.*

A tough vigorous bulbous plant, leaves to nearly 1 m long by 6.5 cm broad, scape to 60 cm long bearing about 4–7 white flowers, originally of tropical America, but widely dispersed by man in cultivation, and naturalized in W Africa.

The plant is commonly grown as an ornamental. Propagation is by division of the bulbous rootstock. In Gabon it is used in line-planting as a fire-break (3).

Other *H. spp.* are probably under cultivation in the Region. The name *H. senegambica* for a species recorded in W Africa (2) is based on a mixture of *H. littoralis* and *H. pedalis* Herb. The material of several species, not of W African origin, has shown marked activity against avian malaria (1).

References:

1. Claude & al., 1947. 2. Deighton 2766, K. 3. Walker & Sillans, 1961: 52.

**Pancratium tenuifolium** Hochst.

FWTA, ed. 2, 3: 136, as *P. hirtum* A. Chev.

A bulbous plant with leaves to 30 cm long, and bearing a white night-flowering scented flower, plants growing solitarily but flowering synchronously over an area, from Guinea-Bissau to S Nigeria and into equatorial Africa.

In N Nigeria the appearance of flowering is considered a harbinger of the first rains. There can be widespread overnight gregarious flowering after the first heavy rainfall (2). At Accra in Ghana flowering of this species and of *P. trianthum* has been observed to follow on the third day after heavy rain, or if the rain is before midday on the second day (1).

References:

1. Holdsworth, 1961. 2. Parsons L. 112, K.

**Pancratium trianthum** Herb.

FWTA, ed. 2, 3: 136. UPWTA, ed. 1, 487.

English: pancratium lily.

West African: **SENEGAL** FULA-PULAAR (Senegal) nagada (K&A) MANDING-MANINKA bahô (K&A) WOLOF tondut (JB; K&A) **GUINEA** MANDING-MANINKA boa (Brossart) **MALI** SONGHAI kor el basal *incl. liliaceous bulbs* (A. Chev.) **UPPER VOLTA** 'SENUFO' monsavara (K&B) **IVORY COAST** MANDING-MANINKA boa (Curasson) SENUFO-DYIMINI babuni (K&B) 'SENUFO' monsavara (K&B) **NIGERIA** HAUSA hátsín mánoòmaá = *farmers' corn* (JMD; ZOG) YORUBA àlùbọ́sà erin = *elephant's onion* (Dennett; JMD) ìṣumeri (JMD)

A bulbous plant, gregarious, leaves to 30 cm long by 1 cm across, flowers white, campanulate, to 12 cm long by 10 cm across, borne on a stalk to 7–10 cm long, appearing synchronously, opening at dusk and fading by the following morning, just before the rainy season sets in, in savanna woodland in damp seasonally marshy sites throughout the region and widespread in Africa.

The plant is not grazed by stock in Senegal (1), nor in Sudan (2). The leaves are, however, held to be a fattening food taken pounded up with cereals by women in N Nigeria (4). The plant is said to be poisonous in Guinea (Brossart fide 4). The bulb is recorded as being edible in Western Sahara (7) and a decoction is sometimes taken in Ivory Coast-Upper Volta for cough (6), though in Senegal it is recognized as being strongly toxic (5) and appears to have no medicinal use there. The seeds are poisonous (7).

The phytochemistry of a number of other *P. spp.* has been reported upon, several alkaloids being present, the most frequent being *lycorine*. There was a marked variation between species, and between the parts of the plant. *P. maritimum* from Egypt has been reported to have a moderate antimalarial activity (3). Examination of the phytochemistry and pharmacology of *P. trianthum* should be carried out (5).

The plant is valuable as an ornamental and has been widely taken into horticulture. It has fetish attributes. In Ivory Coast-Upper Volta a macerate of the entire plant used as a face-wash is considered protection against all illnesses (6). In the Shari-Chad area it is planted at village shrines (4).

References:

1. Adam, 1966, a. 2. Andrew A. 726, K. 3. Claude & al. 1947. 4. Dalziel, 1937. 5. Kerharo & Adam, 1974: 128. 6. Kerharo & Bouquet, 1950: 248. 7. Monteil, 1953: 42–43.

**Scadoxus cinnabarinus** (Decne) Friis & Nordal

FWTA, ed. 2, 3: 132, as *Haemanthus cinnabarinus* Decaisne, in major part. UPWTA, ed. 1, 487, as *H. cinnabarinus* Decaisne.

English: blood flower; fire-ball lily; forest fire-ball lily (Morton).

West African: **GUINEA** MANDING-MANINKA ko-ba (JMD) **SIERRA LEONE** TEMNE *a*-rotə-*a*-kɛrɛ (JMD) **GHANA** AKAN-ASANTE nkwantabisa = *ask at the crossroads* (*significance unknown*) (FRI) TWI nkwantabisa = *ask at the crossroads* (*significance unknown*) (FRI) GA ngbɔbo *a children's game to find the flower* **NIGERIA** YORUBA aréyínkosùn = *looking like teeth rubbed with camwood* (Millson) ịṣumeri (Millson) totó odò, totó: *Marantochloa;* odò: *brook* (JMD)

A rhizomatous herb with scape rising to 60 cm, inflorescence spherical, pink or red, of dense forest areas in Ivory Coast to W Cameroons and Fernando Po, and extending over much of tropical Africa.

The plant is ornamental. It is planted in villages, probably for fetish purposes on account of the bright flowers.

References:

1. Dalziel, 1937. 2. Kerharo & Bouquet, 1950: 247.

**Scadoxus multiflorus** (Martyn) Raf.

FWTA, ed. 2, 3: 132 as *Haemanthus multiflorus* Martyn, incl. *H. longitubus* C. H. Wright, *H. mannii* Bak., and *H. rupestris* Bak. UPWTA, ed. 1, 487, as *H. multiflorus* Martyn and *H. rupestris* Baker.

English: blood flower; fire-ball lily; common fire-ball (Morton); small fire-ball lily *(H. rupestris,* Morton).

French: boule de feu.

West African: **SENEGAL** BASARI a-tyàndíw̃ù lokwę́ta (Ferry) DIOLA enkem (JB, ex K&A) SERER gafaa toh (JB, ex K&A) mbah atun (JB, ex K&A) WOLOF bâtu bahoń (JB, ex K&A) baxa (K&A) táy táy (JB, ex K&A) **THE GAMBIA** MANDING-MANDINKA bàa (auctt.) **GUINEA-BISSAU** BALANTA móutulo (EPdS; JDES) MANDING-MANDINKA bahá (JDES) **GUINEA** MANDING-MANINKA ko-ba (JMD) **UPPER VOLTA** 'SENUFO' guopatréké (K&B) **IVORY COAST** 'SENUFO' guopatréké (K&B) **GHANA** AKAN-TWI nkwanta-bisa = *ask at the crossroads* (*significance unknown*) (FRI) GA nghɔbo *a child's game to find the flower* (FRI) **NIGERIA** HAUSA albásar kwaadíí (JMD) gaatarin kureegee = *ground squirrel's hatchet; from* gataari: *hatchet;* kureegee *jerboa* (Hepburn) YORUBA aréyínkosùn = *looking like teeth rubbed with camwood* (Millson) ișumeri *the bulb* (Millson) sandan maym (AST) totó odò, totó: *Marantochloa;* odò: *brook* (Millson)

A fleshy herbaceous plant with a large bulb bearing leaves to 25 cm long by 8 cm wide which appear during the rainy season. The inflorescence a globular head of red flowers to 15 cm in diameter borne on a spotted scape 20–40 cm long which is produced in the dry season while the bulb is still leafless; of savanna woodland throughout the Region, and widespread in tropical Africa.

The flowers are distinctly decorative and the plant is often grown as an ornamental, and also for superstitious purposes. In The Gambia it is planted as a juju to protect farms from thieving, and in N Nigeria the Fula plant it to mark boundaries in farm plots (1).

The bulb is generally considered to be toxic (3). They have been reported used in Guinea as a fish-poison (1), and in Nigeria as a fish-poison, ointment for treating ulcers and as a stimulant in debility (6). The crushed bulb is applied to serious wounds in Gabon (8). In Ivory Coast the crushed plant is applied in topical frictions to weakly children (4). In Tanganyika the dried bulb is rubbed over scarifications on the breasts to promote milk-flow (2), and the sap is said to raise swellings on the lips and tongue, and to hasten calving when rubbed on to the groin of a cow (9). In Senegal ascites is treated with a mixture of the bulb with roots of *Acacia sieberana* (Leguminosae: Mimosoideae) and *Piliostigma reticulatum* (Leguminosae: Caesalpiniodeae) (3). A decoction of thc bulb is taken in Tanganyika for hookworm (2), and the sap is instilled into the ear for earache (7).

Though porcupine are said to eat the bulb in Kenya (5), the bulb is reported poisonous to pigs (1). An amorphous alkaloid, *haemanthine,* which is toxic to animals has been reported present (6). A number of other alkaloids has also been detected (3, 9, 10).

References:

1. Dalziel, 1937. 2. Haerdi, 1964: 201, as *Haemanthus multiflorus* Martyn. 3. Kerharo & Adam, 1974: 126–7, as *H. multiflorus* Martyn. 4. Kerharo & Bouquet, 1950: 247, as *H. rupestris* Bak. 5. Magago & Glover 367, K. 6. Oliver, 1960: 28, 66, as *H. multiflorus*. 7. Tanner 3195, K. 8. Walker, 1952: 185 as *H. multiflorus* Martyn. 9. Watt & Breyer-Brandwijk, 1962: 35, as *H. multiflorus* Martyn. 10. Willaman & Li, 1970: as *H. multiflorus* Martyn.

## Scadoxus sp. indet.

West African: **IVORY COAST** ABURE bag(b)a (B&D) bara (B&D) kotodié (B&D) **NIGERIA** IJO-IZON (Kolokuma) bòù yàbásị = *bush onion* (KW) òlíélíégbàmútùlá (KW) òtóró, òwéí ótóró = *male otoro* (KW)

A herbaceous bulbous plant of the above Ijo names is recorded used at Kaiama in S Nigeria as a purgative: the inner part of the bulb is cooked and eaten with food (1).

Reference:

1. Williamson, K. 68, UCI.

## Zephyranthes Herb.

FWTA, ed. 2, 3: 13.

English: zephyranthes, zephyr flower, rain flower, thunder flower (India).

A genus of some 40 species of small bulbous plants of tropical and subtropical America. Several species have been dispersed by man under cultivation to warm countries, and the following are to be met with in the W African region:

1. *Z. candida* (Lindl.) Herb. with white flowers sometimes tinged rose outside, of the La Plate region of S America. The rush-like leaf has been reported in S Africa to have hypoglycaemic properties and to give some relief in diabetes (2, 4).
2. *Z. citrina* Bak., with bright yellow flowers reaching to 5 cm long, from S America.
3. *Z. grandiflora* Lindl., with leaves to 30 cm long and large rose to pink flowers to 10 cm across, from S Mexico to Guatemala. This plant has been confused with *Z. rosea* (see below) which is appreciably smaller.
4. *Z. rosea* Lindl. Similar to but smaller than *Z. grandiflora,* from Cuba.
5. *Z. tubispatha* Herb. with narrow linear leaves and white flowers, from W Indies.

A number of alkaloids has been reported from all these species (2, 4, 5). *Lycorine* is common to all. This is relatively non-toxic but causes vomiting and diarrhoea and fatality may follow due to general collapse.

The English names arise from the habit of the gregarious flowering following heavy rain. A study made in Ghana showed that *Z. citrina* flowered after ± 5 days and *Z. tubispatha* ± 4 days (3). In Singapore they are recorded as flowering concurrently after rain, while *Z. rosea* followed one day later (1), but the lapse of time after the stimulatory rain-storm is not stated.

References:

1. Burkill, 1935: 2292. 2. Chadha, 1976, b: 83–84. 3. Holdsworth, 1961. 4. Watt & Breyer-Brandwijk, 1962: 42. 5. Willaman & Li, 1970.

## ANACARDIACEAE

**Anacardium occidentale** Linn.

FWTA, ed. 2, 1: 727. UPWTA, ed. 1, 336–7.

English: cashew.

French: acajou; cajou; pomme d'acajou, or pomme cajou (the swollen peduncle); noix d'acajou, or noix de cajou (the nut).

Portuguese: cajueiro (the tree); cajú; castanha de cajú (the nut).

West African: SENEGAL BASARI a-nyɔ̀gúr and aɓemɛn (Ferry) BEDIK yálágèrè (Ferry) CRIOULO kadu (K&A) DIOLA bululumay (JB, ex K&A) DIOLA (Brin/Seleki) bukayu (K&A) DIOLA-FLUP kubisa (K&A) MANDING-BAMBARA finzâ (A. Chev.; K&A) 'SOCE' darkassu (K&A) SERER daf du rubab (JB, ex K&A) darkasé, darkasu (K&A) SERER-NON (Nyominka) darkassu (K&A) WOLOF darkasé (K&A) darkasu = *kasu, i.e. cashew* (JB, ex K&A) **THE GAMBIA** MANDING-MANDINKA kasuu (*def.* kasuo) (auctt.) (DRR) kasuwo (*def.*) WOLOF dakasso *a corruption of cashew* (DRR) **GUINEA-BISSAU** CRIOULO cadjú (JDES) FULA-PULAAR (Guinea-Bissau) ialaguei (JDES) **SIERRA LEONE** KRIO kushu (FCD) MENDE kundi (JMD) TEMNE *an*-lil-*a*-potho *cf. an*-lil = *Uapaca guineensis* (FCD) **MALI** MANDING-BAMBARA finzâ (A. Chev.) **GHANA** ADANGME atʃia (FRI) AKAN-TWI atẽã (FRI) GA atĩà (KD) GBE-VHE yevutsa = *European's akee* (*Blighia*) (FRI) VHE (Awlan) yevutsa = *European's akee* (*Blighia*) (FRI) **TOGO** GBE-VHE atsia (Volkens) **NIGERIA** HAUSA fisa *the tree and fruit; properly Blighia* (JMD) jambe *the nuts* (JMD; ZOG) kànjuú (auctt.) kashu (MM) KANURI kanju (JRA) YORUBA kajú (auctt.) kantonọyọ (JMD)

A tree to about 10 m high with a crooked trunk and rather sprawling crown, of tropical America, now dispersed throughout the tropics and found in West Africa naturalized in the bush, favouring sandy localities mainly in the coastal region, but occurring also inland.

The tree is native of the area Mexico to Peru and Brazil and the Caribbean. Portuguese adventurers dispersed it in the Seventeenth Century to Africa, India and the Far East adopting into their language a Brazilian vernacular name *caju,* under various forms of which the plant is now known in all European languages.

This root appears also in West African languages, but not universally, a number of races applying the same names as given to *Blighia sapida* (Sapindaceae), the well known Akee apple.

The relatively low stature of the crooked bole bearing a crown that often comes down near to the ground gives a screen. The tree grows well in waste sandy places, and has been satisfactorily used for reclamation of sand-dunes near the sea (8; Senegal, 3, 14, 16; The Gambia, 21). It is drought-resistant. The wood is reddish brown, moderately hard and termite-resistant; it is seldom in large pieces but is used in boat-building, for boxes, chests, mortars, house and fence-posts and firewood (8, 13). Since the fruit is the most valued product of the tree, the timber is not much used. Two varieties of the tree are recognized in Ghana by the Ga people: *meididgi-atiã* and *blɔfo-atiã* (13), but the distinction between them is not explained. The Portuguese are said to have brought 2 races into Malaya which are distinguishable on the colour of the swollen peduncle (5).

The bark has astringent properties. It contains 9–21% tannin and is used in tanning in Senegal (Sébire fide 8) and in Gabon (22). Bark and leaf infusions are used to relieve toothache and sore gums and are taken internally for dysenteric conditions (8; 13; Senegal, 15, 16). The bark is used in Lagos for a disease known in Yoruba as *èfu,* symptomized by a white tender tongue, chiefly in children ('thrush'), and for another disease known as *kolobo* or *ishanu* (Ijebu), a more serious condition with a black tongue (8). In Congo a bark-infusion is taken for urethral discharge, and with *Manilkara obovata* (Sapotaceae) a decoction is used to treat women's stomach pains (4). The bark is an ingredient in the Congo region of arrow-poison. Oral administration of a tincture or extract of bark lowers blood-sugar level within 15 minutes of ingestion and continues for some hours (16, 23). Extracts have been shown ineffective in antibiotic tests (23) and insecticidally (11).

Young leaves are eaten in SE Asia (5). Mature leaves are rich in tannin (23%) and are used for their astringent properties as is the bark. The Aku (Yoruba) people of Sierra Leone use the young leaves for dysentry, diarrhoea, haemorrhoids, etc. (8). In Tanganyika the sap may be expressed and taken for diarrhoea (9). A yellow dye is extracted from the leaves in Senegal and a leaf-decoction is used for dyeing fishing nets (Sébire fide 13). A trace of alkaloid has been reported present in the leaves.

The bark contains a gum known as cashew gum, bright yellow to dark brown. It is a mixture of true gum and bassorin and is but partially soluble in water. It is said to be obnoxious to insects and to be used in S America as a book-binding gum in the place of acacia gum (8, 23). Sap from the bark is vesicant, said to be due to the presence of *cardol* which is found in the fruit. The sap has physiological action in the treatment of leprosy. It has practical application in the treatment of wood against decay and termites (16).

The roots, with bark removed, are boiled in Ghana and the liquid, known as *odido* in Ga, is used as a foot-bath for yaws on the soles (13).

The flowers yield a nectar for honey-production (13).

Though morphologically inaccurate, the 'fruit', so-called, is the swollen, pear-shaped pedicel, likened to an apple in the French name *pomme d'acajou* (8), though an alternative explanation is that it has the smell of apple (3). The nut is borne at the distal end. Before becoming ripe the fruit is highly astringent. When ripe, a slight astringency remains, but it is edible, and the juice is pleasantly thirst-quenching. It is rich in sugars and vitamins, especially vitamin C. The fruit can be made into jam and preserves. In Senegal it is sold in markets for food and medicine, sundried or reduced to a syrup endowed with the attributes of a general panacea, an aphrodisiac, stimulant, strengthener and elixir of longevity (14). The Ga of Ghana make an intoxicating drink from it (13) as do other races elsewhere in the world (23). The pulp is a potential source of alcohol, and has been used to make vinegar (5, 17). The distilled spirit is said to be diuretic and rubefacient (5, 20, 23).

The nut is contained in a tough leathery shell of two layers enclosed in a soft honeycombed mesocarp filled with a caustic oil known as cashew nut shell oil that amounts to 15–20% of the fresh shell in West African material. This is somewhat lower than in Indian material (25–30%) from which is the world's principle source of supply. Selection of planting material should give a worth-while reward in West Africa. This oil has considerable industrial application in the manufacture of brake-linings, industrial belting and clutches, for reinforcing synthetic rubbers, in laminating and impregnating materials to confer oil and acid resistance, etc., etc. The oil is dark coloured, viscous, poisonous and strongly vesicant due to the presence of two phenolic substances, cardol, about 10%, and anacardic acid, about 90%. The toughness of the shell is such that it cannot be cracked without heat-treatment to make it brittle, but heating drives off the shell oil. Since both the oil and the kernels are valuable products, somewhat sophisticated machinery is necessary to carry out the heating process in an inert atmosphere to prevent loss of the oil (2, 7, 10, 18). If roasting is carried out simply to drive off the shell oil to obtain the kernels after cracking, this operation in a confined space is dangerous since contamination by the oil vapour on the body can cause swelling, rubefaciation, vesication and dermatitis. The fresh husk is likewise able to provide these symptoms (23).

The shell oil has local uses for tattooing, to remove warts, to put in a carious tooth, etc. It stains linen indelibly. In India it is painted on to boats and on to woodwork, books, etc. as protection against insects (8). It is used in Gabon on refractory leprosy and ulcers (22) and in Senegal it is compounded into a number of ointments for treating leprosy (14, 16).

The kernels are extracted from the husk after heating and cracking. Heating is necessary for their treatment to make them edible for they too contain some of the vesicant substances in the raw state. The treated kernels are considered a delicacy. There is a considerable import into western countries for use of the kernels in confectionary and as a dessert nut. They are of high food value. Oil content is 40–57%; protein 20%. The oil is a clear yellow, suitable for edible purposes, and of Ivory Coast material is recorded as being about 62% oleic acid, 21% linoleic acid, 9% palmitic acid, 8% stearic acid (6). Domestic consumption of the kernels in West Africa appears to be very limited.

References:

1. Adegoke & al., 1968. 2. Anon, 1946, a. 3. Aubréville, 1959: 191. 4. Bouquet, 1969: 54. 5. Burkill, 1935: 143–6. 6. Busson, 1905: 339–40, 346, with kernel analysis. 7. Cornelius, 1966: with chemistry. 8. Dalziel, 1937. 9. Haerdi, 1964: 125. 10. Hall & Banks, 1965. 11. Heal & al., 1950: 102. 12. Irvine, 1930: 27–28. 13. Irvine, 1961: 552–3, with analyses of fruits and seeds. 14. Kerharo & Adam, 1962. 15. Kerharo & Adam, 1964, c: 292. 16. Kerharo & Adam, 1974: 129–33, with phytochemistry and pharmacognosy. 17. Manjunath, [Ed] 1948: 70–74. 18. Morton, J. F. 1961: (a comprehensive account). 19. Purseglove, 1968: 19–23. 20. Quisumbing, 1951: 535–8. 21. Rosevear, 1961. 22. Walker & Sillans, 1961: 56–57. 23. Watt & Breyer-Brandwijk, 1962: 43–44.

**Antrocaryon klaineanum** Pierre

FWTA, ed. 2, 1: 728. UPWTA, ed. 1, 327.

English: 'white mahogany' (the timber, Cameroun, Dalziel).

A tree to 30 m high of the dense humid forest, widespread but rare in S Nigeria and Fernando Po, and extending to Zaïre.

The trunk is straight, cylindrical, and to over 3 m in girth. The wood is heavy, reddish-white, not unlike mahogany but paler. It is easy to work and is valuable for carpentry and construction work, but is liable to attack by borers (2, 4).

The bark contains a small quantity of glutinous oleo-resin. The powdered bark is used in Gabon for liver complaints (3, 4).

The fruit is a drupe resembling a small apple. The flesh is edible with an acid taste (1, 4).

The nut contains five seeds which are oily and edible, but the nut-shell is thick and the difficulty of getting to the seeds is an obstacle to their use.

ANACARDIACEAE

References:

1. Busson, 1965: 340. 2. Dalziel, 1937. 3. Walker, 1952: 186. 4. Walker & Sillans, 1961: 57.

**Antrocaryon micraster** A. Chev. & Guillaum.

FWTA, ed. 2, 1: 728. UPWTA. ed. 1, 337, incl. *A. polyneurum* Mildbr. ined.
Trade: akoua (the timber, from Akye, Ivory Coast, Dalziel).
West African: **SIERRA LEONE** MENDE gbɔdua (*def.* gbɔduɛi) (S&F; FCD) **IVORY COAST** ABE akua (auctt.) AKYE akmé (A. Chev.) akua (Aub.; FB) haddo (Aub.) ANYI ékio (Aub.; FB) éküé (Aub.) DAN dehadidié (Aub.) KRU-GUERE (Wobe) nitué (Aub.) KYAMA akorabahia (Aub.) **GHANA** AKAN-ASANTE aprokuma (CJT) aprukuma (auctt.) TWI aprokuma (DF) apurukuma (FRI) WASA kuminimba (CJT) ANYI etwe (FRI) etwo (FRI) ANYI-SEHWI etwi (CJT) **NIGERIA** BOKYI ojifo (Catterall; KO&S) EDO ekhuen (Ross; KO&S) ùgbézàrò (Kennedy) ISEKIRI egin egbo (KO&S) YORUBA ifá òkété (KO&S)

A large tree to 50 m high by 3·5 m girth, of the high closed and semi-deciduous forest, occurring from Sierra Leone to S Nigeria, and extending from Cameroun across Africa to Uganda.

The bole is straight, cylindrical and scarcely buttressed. The sap-wood is soft and yellowish-white, heart-wood reddish-brown (2, 5, 6), but some material is described as white and undifferentiated (Sierra Leone, 9; Ghana, 8). It is not durable and is normally without commercial use though it has been traded in Ivory Coast under the name *akoua* (2). The wood sinks when fresh. It is sawn locally in Sierra Leone (3) and in Ghana (5) for planking and is carpentered into furniture.

The bark is thick and rough. Slash is red streaked with white, fragrant and exudes a yellow sticky substance which sets to a golden-brown translucent gum (8). The bark is said to be put into soup in Nigeria (7).

The fruits, sometimes called 'plums' for their similarity to the temperate fruit, are slightly dorsi-ventrally flattened, 3·7 cm long by 5 cm diameter. They have a strong mango-like smell, and the flesh is eaten in various parts, but not ubiquitously. Sometimes the flesh is fermented to an alcoholic drink (5). The fruit are sought after by wild pig in Uganda (4). In Sierra Leone the fruit is taken medicinally for cough and heart-ache [sic! presumably pains] (9), and for stomach-ache (Edwardson fide 5).

The stone contains 3–5 cavities housing the seeds which are rich in oil and are edible. Oil content of the kernels is over 70%, and is composed of linoleic acid 40%, oleic acid 31%, stearic acid 18%, and others (1). The hard endocarp is difficult to open to get at the seeds, but will burst if put in hot cinders (5).

References:

1. Busson, 1965: 341, 346, with kernel analysis. 2. Dalziel, 1937. 3. Deighton 3089, K. 4. Eggeling & Dale, 1952: 5. 5. Irvine, 1961: 553. 6. Keay & al., 1964: 324. 7. Ross 196, K. 8. Taylor, 1960: 81. 9. Savill & Fox, 1967, 36.

**Fegimanra afzelii** Engl.

FWTA, ed. 2, 1: 728. UPWTA, ed. 1, 338.
West African: **SIERRA LEONE** TEMNE *an*-peri *? from English: pear* (AHU; FCD)

A shrub, or small tree, of sandstone and rocky places in Guinea and Sierra Leone.

The seed is edible (1, 2).

References:

1. Busson, 1965: 342. 2. Dalziel, 1937.

**Haematostaphis barteri** Hook. f.

FWTA, ed. 2, 1: 733. UPWTA, ed. 1, 338.

English: blood plum (of Nupe).

West African: **UPPER VOLTA** MOORE subtuluga (FB) **GHANA** DAGAARI zimbringa (AEK; FRI) MOORE subtulinyanga *the female* (AEK) subtuluga *the male* (AEK) WALA jumbrinza (?) (FRI) sumbringa (AEK) **NIGERIA** FULA-FULFULDE (Nigeria) tursuhi, tursahi (*pl.* tursuuje) (auctt.) HAUSA ɗan danya (MM) ján danyaá (ZOG) tsaamiyar Lamaruudu = *tamarind of the giant* (ZOG) tsada (MM) NUPE jinjèrè'gyà (auctt.)

A tree to 8 m high by 65 cm girth, of rocky situations of the dry savanna from Upper Volta to Nigeria, and in Cameroun and Sudan.

The bark contains a clear gum. A bark-infusion is taken by Fula of N Nigeria for *sawoora* (hepatitis) (4), and at Nassarawa a decoction along with natron (*kanwa*, 3) is taken for sleeping sickness, in copious draughts which cause vomiting and diarrhoea lasting several days (2).

The fruit is a red-purple drupe, nearly 2·5 cm long like the temperate plum. The pulp is thin, edible with an acid but resinous taste. The kernels are somewhat oily, and are also edible (1, 2).

References:

1. Busson, 1965: 341. 2. Dalziel, 1937. 3. Irvine, 1961: 555. 4. Jackson, 1973.

**Lannea acida** A. Rich.

FWTA, ed. 2, 1: 732–3. UPWTA, ed. 1, 338.

West African: **SENEGAL** BASARI a-ngəlia-tyán ɛ-téd (Ferry) BEDIK gi-peureùɓ (K&A; Ferry) DIOLA bubuka (K&A) bufira (auctt.) bukinebélége ditopédamen (JB) mutopédamen (K&A) DIOLA (Fogny) dégérélên (K&A) FULA-PULAAR (Senegal) bembey (K&A) tiñoli (K&A) tuko (K&A) TUKULOR tchingauli (FB) KONYAGI a-tyékẹnú (Ferry) MANDING-BAMBARA bembé (auctt.) mpéku ba (JB) pékuni (Aub. ex K&A) MANDINKA bembe (*def.* bemboo) (auctt.) bembéñaña (K&A) 'SOCE' bembéñaña (K&A) bembo (K&A) SERER duguy (K&A) n-duguñ (K&A) ndugutj (auctt.) SONINKE-SARAKOLE čiñoli (K&A) siñoli (K&A) WOLOF son (auctt.) **THE GAMBIA** DIOLA (Fogny) boukinebelege, ebelege: *home* (DF) bufira = *lime* (DF) dégérélên *alluding to seed dispersal* (DF) mutopedamen = *goat fat* (DF) FULA-PULAAR (The Gambia) chukon-kyodi (DRR; DAP) pegu (?) (DAP) sibigwi (?) (DAP) tingoli (JMD) MANDING-MANDINKA bembefiŋ (*def.*-o) = *black bembo* (auctt.) bembo-keo, bembo-muso = *male, female bembo respectively on variability* (DF) santaŋ maloro = *young santang: Daniellia oliveri* (DRR; DA) WOLOF son (JMD; DAP) **GUINEA-BISSAU** CRIOULO mantede (JDES) FULA-PULAAR (Guinea-Bissau) bembedje (*pl.*) (JDES) bembem-hei (JDES) tchingauli (Aub.) tchingole (JDES) MANDING-MANDINKA bembô (JDES; Fernandes) bembô (*def.*) bembô-fingo PEPEL betôlôdje (JDES) **GUINEA** FULA-PULAAR (Guinea) tiuko (Aub.) tuko (Bouronville) MANDING-MANINKA bembénugu (auctt.) **MALI** DOGON sá (C-G) FULA-PULAAR (Mali) tingoli (GR) MANDING-BAMBARA bembé (auctt.) bembé fing = *black bembé* (A. Chev.) MANINKA pékuni = *little pékou* (Aub.) **UPPER VOLTA** DAGAARI siribu (JMD, ex K&B) sisubu (JMD, ex K&B) MANDING-BAMBARA bembé (Bégué, ex K&B) pekuni (Aub., ex K&B) MOORE sambagha (K&B) santuluga (K&B) 'SENUFO' véké (Aub., ex K&B) **IVORY COAST** AKYE ébruhé (JMD; K&B) ébruké (JMD; K&B) BAULE kondro (auctt.) MANDING-MANINKA béssomo (B&D) borosomo (B&D) 'SENUFO' véké (auctt.) **GHANA** AKAN-ASANTE biribiriwa (?) (FRI) BRONG ama (BD&H; Brown, ex FRI) kuntunkori (BD&H; FRI) BAULE kondro (FRI) DAGAARI ansigne (AEK; FRI) sisibi (AEK; FRI) sisibu (AEK; FRI) sissulu (AEK; FRI) DAGBANI sinsopiegu (FRI) GUANG-KRACHI kujin (auctt.) NCHUMBULU kamma (FRI) MOORE samtuluga (AEK; FRI) NANKANNI kyekyɛbtuliga (Gaisie) sinsobaga (FRI) **TOGO** BASSARI tchintchérékutun (Aub.) GBE-VHE ekualokpoe (Volkens) GURMA (Mango) tyetyebu (Volkens) KABRE kudupu (Gaisser; FB) TEM kela (FB) kélao (Aub.) TEM (Tshaudjo) kela(-e) (Volkens) YORUBA-IFE OF TOGO asogedaka (Volkens) **DAHOMEY** YORUBA-NAGO akuhu (Aub.) **NIGERIA** FULA-FULFULDE (Nigeria) faruhi (*pl.* faruuje) (JMD) paruhi HAUSA faàrú (JMD; KO&S) faàrún mútaàneé, mútaáneé: *of men* (JMD) furu (ZOG) TIV hil gbur (DA) hil nimbiligh (DA) nimbiligh (DA) YORUBA ekika (JRA)

A tree to 10 m high, and bole 2–3 m in girth, of savanna, particularly in rocky situations and requiring a precipitation of 635 mm (7), occurring throughout the area from Senegal to Nigeria.

The thick fissured bark helps the tree resist bush-fires. In Ivory Coast-Upper Volta when the bush is being cleared for farming it is usually not cut down (13). It is considered a useful tree and is particularly esteemed in the Senegalese pharmacopoea (10, 12).

The wood is whitish and soft. It is considered hard enough to make small stools, planks and other utensils. It is flexible and used for bows (4).

The bark is fibrous, and a cordage can be made from it but quality is poor (13). It also yields a gum which is edible (5–7, 14). In Senegal bark is used internally for beriberi, schistosomiasis, and haemorrhoids, and externally for eye-troubles, and with other drug-plants for dysentery and sterility (10, 11). A bark-infusion is taken in N Nigeria for stomach-troubles (4). Vapour from a bark-decoction is inhaled into the mouth for dental caries and buccal infections in the Casamance of Senegal (9, 12). Root-bark is considered good for skin-infections. In Ivory Coast-Upper Volta it is used in baths and lotions for blotches, herpes, etc. (2, 13), and a similar preparation is taken internally after 4–5 hours fermentation for gonorrhoea. The powdered root with salt is made into a tampon for application to the scrotum in treating orchites in Casamance (8, 12). The Moore of Upper Volta give a root-bark decoction as an enema to rachitic children, but recognize that it is too dangerous for small babies (13). The bark has such a wide medicinal appeal that it is commonly sold in markets in Dakar as a 'women's' medicine to ensure easy childbirth, and with other plants to counter sterility (12). Powdered bark is recognized as a treatment for *danévéle,* a symptom of beriberi (12). Similarly the bark of *L. antiscorbutica* Engl. has been reported used in Angola for scorbutic ulcers of the mouth and other symptoms of scurvy (4). Sap obtained from pounding the bark is given in Ivory Coast to epileptics and to persons subject to giddiness and fainting (2).

The young leaves are eaten in West Africa. Leaf analysis of material from Ivory Coast showed: carbohydrates 67%, protein 18%, minerals 5%, etc. (3, 12). Cattle in Senegal browse the foliage (1). The leaves are astringent and are used in Senegal for toothache (11).

The trees are dioecious. The male inflorescences are scented. The fruits which hang in clusters are red to purple-black with a bloom. The pulp is edible with a slightly acid and resinous taste. In some parts the fruits are fermented to an alcoholic drink. The fruit can be dried for storage (4). The nut is thin-shelled, and the kernel within oily. The oil is used for making soap or for toilet purposes (4). The kernel is purgative (7).

References:

1. Adam, 1966, a. 2. Bouquet & Debray, 1974: 16. 3. Busson, 1965: 341, with leaf analysis. 4. Dalziel, 1937. 5. Irvine, 1952, a: 31. 6. Irvine, 1952, b. 7. Irvine, 1961: 557. 8. Kerharo & Adam, 1962. 9. Kerharo & Adam, 1963, a. 10. Kerharo & Adam, 1964, b: 554–5. 11. Kerharo & Adam, 1964, c: 312. 12. Kerharo & Adam, 1974: 135–6. 13. Kerharo & Bouquet, 1950: 166–7. 14. Roberty, 1953: 451.

## **Lannea barteri** (Oliv.) Engl.

FWTA, ed. 2, 1: 732, as *L. kerstingii* Engl. & K. Krause. UPWTA, ed. 1, 339, in part.

West African: **GUINEA** FULA-PULAAR (Guinea) tiuko (Aub.) **SIERRA LEONE** SUSU-DYALONKE liberi-na (FCD) **MALI** MANDING-BAMBARA bembe (A. Chev.) **UPPER VOLTA** DAGAARI sisibigolo (K&B) sussuguté (K&B) HAUSA namisinfaru (K&B) MOORE sabagha (K&B) sambituliga (Bégué; Aub.) **IVORY COAST** BAULE kondro (K&B) MANDING-MANINKA bembé (Aub.) bimbé, dinbé (B&D) peku (Aub.) **GHANA** AKAN-ASANTE kuntunkuni (BD&H; FRI) BRONG kuntunkurfi (FRI) BIMOBA sinsabega (FRI; JMD) DAGAARI sinsabga (JMD) sinsagbyetiliga (JMD) DAGBANI sinsabega (FRI) sinsagbyetiliga (FRI) GBE-VHE (Awlan) kɔbɛwu (FRI) GUANG-NCHUMBULU kutchuichivi (FRI) sinsabga (FRI) sinsagbyetiliga (FRI) MAMPRULI sisibega (FRI) SISAALA kanchimbelli (AEK) **TOGO** GURMA (Mango) benature (Volkens) TEM (Tshaudjo) patandĕu (Volkens) tingbatau (Volkens) YORUBA-IFE OF TOGO aku (Volkens) **DAHOMEY** BASSARI mon (Aub.) **NIGERIA** FULA-FULFULDE (Adamawa) nebbamhi = *butter-*

*oil tree* (JMD) FULFULDE (Nigeria) faruhi (*pl.* faruuje) (JMD) paruhi (JMD) soriyehi (*pl.* soriyeeje) (Taylor; MM) HAUSA bàbbán báraá *the seedlings* (Bargery) baƙin kukkuki = *black kukkuki* (*Sterculia setigera Del.*) (JMD) faàrú (auctt.) faàrún bírii = *monkey faru* (JMD) faàrún doóyaà (auctt.) HAUSA (West) tudi (JMD; ZOG) KANURI dalubu (JMD) dama (JMD) TIV hil gbur (DA) hil nimbiligh (DA) nimbiligh (DA)

A tree to 15 m high with a twisting trunk to 1 m girth, of savanna woodland, from Guinea and Mali to N Nigeria, and to Zaïre and Uganda. It is very similar to *L. schimperi* (Hochst.) Engl. and records are possibly confused, as is also the status of *L. barteri* (Oliv.) Engl. (1897) vis à vis *L. kerstingii* Engl. & Krause (1911).

The wood is soft, dirty white, coarse-grained and without worth (3, 5).

The bark is fibrous and is used to produce a cordage. It contains a gummy secretion which becomes white and friable in the air (7). The bark-slash, however, from a deep red turns brown on exposure (4). The bark is used in Ghana to prepare a reddish-brown dye known in Akan dialects as *kunkunkuri* (5) used for dyeing funerary cloths. The dye is also prepared in Ivory Coast-Upper Volta (6), and in those territories the bark is used externally for ulcers and sores, leprosy maculae, and internally after decoction for pains of gastric origin (6), diarrhoeas, oedemas, paralysis, epilepsy and madness (1). The Igbo of Nigeria boil the bark and drink the liquid as a stomachic, and with the addition of the bark of *ukuku* (Yoruba, sp. indet.) is taken as a vermifuge (5).

The root is ground and wrapped in certain leaves of undisclosed species and is applied as a poultice to wounds (5).

The flowers attract bees (2) and may be useful for producing honey.

References:

1. Bouquet & Debray, 1974: 16. 2. Deighton 5355, K. 3. Eggeling & Dale, 1952: 8–10, as *L. kerstingii* Engl. & K. Krause. 4. Enti FH.7505, K. 5. Irvine, 1961: 558, as *L. kerstingii* Engl. & K. Krause. 6. Kerharo & Bouquet, 1950: 397. 7. Walker & Sillans, 1961: 57.

## **Lannea egregia** Engl. & K. Krause

FWTA, ed. 2, 1: 733.
West African: **GUINEA** FULA-PULAAR (Guinea) tiuko (Aub.) **IVORY COAST** MOORE sambituliga (Aub.) **DAHOMEY** BASSARI mon (Aub.) **NIGERIA** YORUBA ekudan (KO&S)

A tree to 13 m high of the savanna in Guinea to Nigeria.

In Guinea, Ivory Coast and Dahomy, this species shares the same vernaculars as *L. barteri* (Oliv.) Engl. (1), and doubtless has the same usages.

Reference:

1. Aubréville, 1950: 402.

## **Lannea fruticosa** (Hochst.) Engl.

FWTA, ed. 2, 1: 732. UPWTA, ed. 1, 339.
West African: **NIGER** FULA-FULFULDE (Niger) pãru kanuri (Aub.) **NIGERIA** ARABIC leyun (JMD) HAUSA paàruú (ZOG)

A small tree to about 8 m high of the Sahel in Niger and N Nigeria, and occuring also in Ethiopia and Uganda.

The bark exudes a gum which is soluble (1). It is very attractive to ants (2).

In Ethiopia the tuberous root is eaten (3).

References:

1. Dalziel, 1937. 2. Styles 280, K. 3. Turton 113, K.

**Lannea humilis** (Oliv.) Engl.

FWTA, ed. 2, 1: 732.
West African: **SENEGAL** FULA-TUKULOR béluki (Aub.) SERER arđ a koy (JB) ngonar đid (JB) ngonaro (JB) WOLOF habugan (JB) **NIGERIA** HAUSA paàruú (ZOG) KANURI kərwúlú (C&H)

A shrub not more than 3 m, or occasionally becoming a tree to 5 m tall, of the dry savanna region in Senegal, Niger and N Nigeria, and also in eastern Africa from Sudan to Zambia.

No usage is recorded in the Region. The wood is said to be good for carpentry in Sudan (2), and the bark to yield a fibre for rope-making (1, 4). The bark-slash exudes a colourless sticky liquid (5). The roots are eaten in Uganda either raw, or cooked like a potato (3) and in Tanganyika they are powdered and taken to stop nausea, or with milk added as a general tonic in weakness (6).

References:

1. Andrews A520, K. 2. Cooke 57, K. 3. Dyson-Hudson 217, K. 4. Jackson 1289, K. 5. Kiley 5051, K. 6. Koritschoner 1732, K.

**Lannea microcarpa** Engl. & K. Krause

FWTA, ed. 2, 1: 733.
West African: **SENEGAL** BASARI a-ngeul a-níyàn (Ferry) a-ngli (K&A) BEDIK gi-tyèn (K&A; Ferry) FULA-PULAAR (Senegal) tukoneudu (K&A) KONYAGI a-ngeula (Ferry) MANDING-BAMBARA péhuni (Aub., ex K&A) MANINKA bembé (K&A) **GUINEA** FULA-PULAAR (Guinea) karfaillé (K & A) pébouillé (K&A) tuko (Bouronville) **MALI** DOGON kuré (Aub.) MANDING-MANINKA bembé (Aub.) pékuba = *big péku* (Aub.) **UPPER VOLTA** 'SENUFO' veké (Aub.) **IVORY COAST** BAULE kondro (Aub.) MOORE sambiga (Aub.) 'SENUFO' veké (Aub.) **GHANA** SISAALA chichuɛŋ (Blass) **TOGO** GBE-VHE niadukuko (Aub.) MOBA tchientchabu (Aub.) SOMBA mucéhuda (Aub.) **DAHOMEY** BATONNUN bahéma (Aub.) hioronedu (Aub.) GBE-FON zuzui kanté (Aub.) VHE niadukuko (Aub.) SOMBA mucéhuda (Aub.) YORUBA-NAGO oku (Aub.) **NIGER** FULA-FULFULDE (Niger) faruhi (Aub.) paruhi (PF) GURMA limantiabulu (Aub.) HAUSA farom mutane (Aub.) malga *correctly Cassia sieberana DC.* (*Leguminosae: Caesalpinioideae*) (Aub.) SONGHAI-ZARMA tamarza (Aub.) **NIGERIA** FULA-FULFULDE (Nigeria) paruhi (PF) HAUSA faàrú (JMD) faàrún mútaàneé = *men's faru* (Lely; JMD) paàruú (ZOG)

A tree to 16 m tall by 2 m girth of the soudanian and northern guinean savanna across the Region from Gambia to Niger and N Nigeria. In places it may be abundant (3). It seems to prefer friable deep soil and is often found on cultivated land (1).

The sapwood is white, light and works easily, but deteriorates quickly (2).

The bark is fibrous, and is used to make cordage (1, 2). It has a sweet smell and contains a gum which is soluble in water and is edible (2).

The leaves are edible (2). They have medicinal use as those of *L. acida* but are considered to be less active (4).

The fruits have an edible pulp said in some parts to be preferred to that of *L. acida* (1). They are eaten raw or dried, and a fermented drink is made from them (2).

References:

1. Aubréville, 1950: 394. 2. Irvine, 1961: 538. 3. Keay & al., 1964: 314–6. 4. Kerharo & Adam, 1974: 136.

**Lannea nigritana** (Sc. Elliot) Keay

FWTA, ed. 2, 1: 733. UPWTA, ed. 1, 339, as *L. afzelii* Engl.
French: loloti des savanes (from Abe: *loloti* = *L. welwitschii* of the savanna, Ivory Coast, Aubréville).

West African: **SENEGAL** DIOLA beigérelên (K&A) budigi élen (K&A) bumakurin (Aub.; K&A) fupéléyen (K&A) KONYAGI âgel (K&A) **THE GAMBIA** DIOLA (Fogny) fupeleyen = *bitch's vagina* (DF) **GUINEA-BISSAU** CRIOULO mantede (JDES) FULA-PULAAR (Guinea-Bissau) bembedje (*pl.*) (JDES) bembem-hei (JDES) tchingole (Fernandes; JDES) MANDING-MANDINKA bembô (JDES; Fernandes) betôlôdje (JDES) **GUINEA** SUSU lokure (Aub.) **SIERRA LEONE** BULOM (Kim) lep-pogɛ (FCD) LOKO g-bagin-kamia (S&F) g-baŋgiŋ-kamia (FCD) MENDE gboji-hina = *male Spondias* (FCD; S&F) kanamamusu (*def.*-i) (S&F) kwamamusu (FCD; S&F) MENDE-KPA kwamamusu (*def.*-i) (FCD) SUSU-DYALONKE liberi-na (FCD) TEMNE *an*-thaŋth-*an*-ləp (FCD; S&F) **UPPER VOLTA** MANDING-DYULA bimbé (K&B) **IVORY COAST** ABE kinon (A&AA) ADYUKRU erègb atch (A&AA) AKYE atchon (A&AA) kinan (Aub.; K&B) kino (Aub.; K&B) MANDING-DYULA bimbé (K&B) MANINKA bembé (auctt.) 'SENUFO' kunam (Aub.) kunangué (Aub.) **GHANA** ADANGME aadɔŋtʃo (FRI) DAGBANI sinsabgbetiliga (Osnan, ex FRI) GA ádɔ̀dɔ́ŋ (KD) GUANG-GONJA kàpííbí (Rytz) **NIGERIA** YORUBA ekika (JRA)

A tree to 15 m high of the drier parts of secondary growth of the savanna forest zone from Senegal to Dahomey (var. *nigritana)* and from Ivory Coast to S Nigeria (var. *pubescens* Keay).

Ash from the burnt wood is used in Sierra Leone to make soap (10).

The bark has uses for treating dermal infections. In Ivory Coast the pulped bark is put on to areas of skin-trouble, and into sores (3, 9). In Sierra Leone a variety is recognized which 'does not have knobs on the bark.' This is said to have a finer scent than other forms and women collect the frass made by borers in the stem, dry it out, grind it and mix it with water. This is used as a scented cosmetic to rub on the body, and also to rub on to areas of craw-craw (4). In Casamance powdered bark is considered a sovereign remedy for stubborn sores (8) and the inner bark layer is used in massage to restore muscular tone (7). A bark-decoction is given in draught in Casamance for intestinal pains and dysentery (6, 8), a decoction of the roots and bark in butter and in draught for chest-troubles and stiffness, and the powdered root with leaves of *Strychnos sp.* (Loganiaceae) for abscesses and inflammations (8). In Ivory Coast a decoction of leaves and twig-bark is administered as an enema for abdominal complaints (9), and sap from the pounded bark is given to epileptics and persons subject to giddiness and fainting (3).

The tree is often planted in Akye villages in Ivory Coast for the fruit which is edible (2). The fruit are also eaten in Guinea Bissau (5) and in Sierra Leone (10). The seeds are purgative (1).

References:

1. Ainslie, 1937: sp. no. 204. 2. Aubréville, 1959: 2: 200. 3. Bouquet & Debray, 1974: 16. 4. Deighton 2787, K. 5. Esp. Santo 1519, K. 6. Kerharo & Adam, 1962. 7. Kerharo & Adam, 1963, a. 8. Kerharo & Adam, 1974: 136–7. 9. Kerharo & Bouquet, 1950: 167. 10. Savill & Fox, 1967: 37.

**Lannea schimperi** (Hochst.) Engl.

FWTA, ed. 2, 1: 732.

West African: **NIGERIA** HAUSA bàbbán báraà (ZOG) faàrú (H&J) faàrún bírii (ZOG) faàrún doóyaà (auctt.) paàruú (ZOG) NUPE ègbanci (Yates)

A tree to 8 m tall with short bole and twisted branches, of the savanna, recorded from N Nigeria only in the Region, but occurring widely elsewhere in tropical Africa.

The wood is soft, whitish and very light. It is said to be of no value (6).

The bark is used to make a string in N Nigeria (12), and rope and cordage in Tanganyika (11) and Sudan (1, 4, 7). It yields a gum of no recorded usage (5, 11), but an enema of powdered bark in water is used in Tanganyika (10). The bark (?) is put into a snake-bite treatment in Tanganyika (9).

The root has astringent properties, and a decoction is used in Tanganyika as a mouth-wash for toothache (8).

The fruit is about 1 cm in diameter with a pulp which is reported eaten in Sudan, where it is much relished (3), Uganda (2) and Tanganyika (10). In the last-named territory the fruit is used in treatment of ankylostomiasis (10), the effective part being perhaps the kernel which is bitter.

The flowers are much visited in E Africa by bees in search of nectar. The plant may have some value as a bee-plant.

References:

1. Andrews A. 559, K. 2. Brasnett 110, K. 3. Cooke 110, K. 4. Cooke 216, K. 5. De Wilde 10466, K. 6. Eggeling & Dale, 1952: 10. 7. Gazal 916, K. 8. Koritschoner 1530, 1708, K. 9. Tanner 5047, K. 10. Tanner 6013, K. 11. Watt & Breyer-Brandwijk, 1962: 46. 12. Yates 61, K.

## **Lannea velutina** A. Rich.

FWTA, ed. 2, 1: 732. UPWTA, ed. 1, 339.

West African: **SENEGAL** BANYUN bâgok (K&A) BASARI a-ngeul anyígẁr (Ferry) BEDIK ga-pɔ́r-fɔr, ga-tyáp-tyápɛŋ (Ferry) DIOLA bubuka (auctt.) mutopédamen (JB, ex K&A) DIOLA (Fogny) butope édiamen (K&A) FULA-PULAAR (Senegal) tiṅolipoley (K&A) ťukoṅabé (Aub., ex K&A) KONYAGI a-pẹryok (Ferry) MANDING-BAMBARA bakorompéku (auctt.) MANDINKA bâbadembéké (K&A) bembemusu (*def.* b. musoo) (auctt.) MANINKA bembégaga (Aub., ex K&A) SERER ndabarndoki (JB, ex K&A) ndôg (Aub.; K&A) SERER-NON tâgba (AS, ex K&A; Aub.) tâgba (auctt.) WOLOF ndogot (auctt.) son a béy (JB, ex K&A) **THE GAMBIA** DIOLA (Fogny) boutepe édiamen = *goat's boutepe* (DF) bubuka = *a producer* (DF) mutopedamen = *goat fat* (DF) FULA-PULAAR (The Gambia) chukoniade (DRR; DAP) tingoli (DAP) MANDING-MANDINKA bembe (*def.* bemboo) (auctt.) WOLOF n-dogot (DAP) son (DAP) **GUINEA-BISSAU** CRIOULO mantede (auctt.) FULA-PULAAR (Guinea-Bissau) bembedje (auctt.) bembem-hei (JDES) tchingole (JDES) tchucó (Fernandes) MANDING-MANDINKA bembô (auctt.) PEPEL betôlôdje (JDES) **GUINEA** FULA-PULAAR (Guinea) tiuko niabé (Aub.) **MALI** MANDING-BAMBARA bakorompéku (Aub.) MANINKA bembé (JMD) bembé gua gua (Aub.) **UPPER VOLTA** DAGAARI sisibigolo (K&B) sussuguté (K&B) HAUSA namisinfaru (K&B) MOORE sabagha (K&B) sambituliga (Aub.; K&B) **IVORY COAST** BAULE kondro (K&B) MANDING-MANINKA bembé (K&B) peku (K&B) **GHANA** DAGBANI *p*-pegu (Osnan, ex FRI) sinsa(-b) (FRI) SISAALA chichɔgbiŋk pumuŋ (Blass) kanchimbelli (FRI)

A shrub or tree 15 m high, with bole to 50 cm diameter, of the wooded savanna from Senegal to Ghana.

The wood is used to make small stools, and is flexible enough for bows, and has other uses as for *L. acida* A. Rich.

The bark yields, as do other *Lannea spp.,* a popular red-brown dye (3). The bark and roots have medicinal use in Senegal. In macerate they are used to bathe rachitic children, and by baths and flapping of the muscles for adults with general pain (4, 6). A decoction of powdered root is considered good for diarrhoea (5, 6). In Ivory Coast the bark is used for diarrhoea, oedema, paralysis, epilepsy and insanity (2, 7). The bark is used in Ghana externally on wounds, ulcers and leprous spots (3).

The foliage is sometimes browsed by cattle in Senegal (1). The fruit is edible.

References:

1. Adam, 1966, a. 2. Bouquet & Debray, 1974: 16. 3. Irvine, 1961: 559. 4. Kerharo & Adam, 1962. 5. Kerharo & Adam, 1964, b: 555. 6. Kerharo & Adam, 1974: 137. 7. Kerharo & Bouquet, 1950: 167.

## **Lannea welwitschii** (Hiern) Engl.

FWTA, ed. 2, 1: 732. UPWTA, ed. 1, 339, as *L. acidissima* A. Chev.

West African: **IVORY COAST** VULGAR loloti ABE loloti (auctt.) ngdongoloti (auctt.) AKAN-FANTE kakoro (A. Chev.; Aub.) AKYE n-nu (A. Chev.) nu (A. Chev.) tchiko (K&B; Aub.) tchiwo (A&AA) ANYI baiséguma (Aub.; K&B) baopiré (Aub.; K&B) boré poré (A. Chev.; K&B) BAULE trongba (A. Chev.) GAGU tobero (K&B) KRU-GUERE (Chiehn) tétégné (K&B) KULANGO (Bondoukou) duko (A. Chev.; K&B) durgo (A. Chev.; K&B) duruku (A. Chev.; K&B) KYAMA adubruhia (A. Chev.) atukruhia (A. Chev.) dugbruhia (Aub., ex K&B) NZEMA kakoro (A. Chev.)

**GHANA** VULGAR kumenini (DF) AKAN-ASANTE kum-anini (Enti) kumenini (FRI) kum-onini (auctt.) kuntunkuni (FRI) FANTE kakoro (Enti) TWI aberewa nyansiŋ = *old woman's faggots* (FRI) kum-anini (auctt.) okum-nini (Enti) WASA kumenini (FRI) ANYI-SEHWI bopire (auctt.) NZEMA abalapuli (auctt.) **NIGERIA** EDO ẹ̀wínwán = *a tick: from the likeness of the fruits* (auctt.) YORUBA ekika (JRA; KO&S) ekika-ajá (Gilman)

A tree to 30 m high with a straight cylindrical bole to 2.70 m girth, of deciduous and secondary forests from Ivory Coast to W Cameroons and extending to Uganda and Angola.

The wood is white to pale pinkish, light and pliable. It has a resinous smell, and burns well, hence the Ghanaian Twi name (3, 7–9, 13). It is not durable, and logs left lying in the forest are readily attacked by borers (13, 14). It is nevertheless currently recommended in Ghana for general joinery and furniture (2).

The wood and bark contain an abundant colourless gummy sap (15), which is sometimes resinous (12). The bark is used to prepare a saffron dye used in Ghana for funerary cloths known as *kuntunkuri* (Asante) (14). In Congo the pulped-up bark is applied as a wet dressing on oedema of the legs, and diluted with palm-wine it is taken for epilepsy (4). A decoction is taken by women for menstrual trouble and sterility; it is also taken to treat dysenteries, diarrhoeas, urethral discharge and haemorrhoids (4). In Nigeria the stem-bark is used to treat skin-infections, but examination for anti-pathogenic activity has shown neither bacteristatic nor fungistatic action (11). Bark-extracts have been found to contain no alkaloid, but have some saponins and tannins (5).

A root-decoction is expectorant or emetic, and is taken in Congo for cough and pulmonary congestion, as a counter-poison, and as a mouth-wash for gingivitis and mouth-infections (4).

To some tribes in Ivory Coast the tree is fetish, the leaf-sap being given to the sick to hasten recovery (10).

The fruit is about 6–7mm long, blackish and flattened. Its resemblance to a tick is reflected in the Edo (Nigeria) name. It is viscous and resinous, and smells of turpentine. The taste is acidulous. The pulp is eaten in Zaire (Staner fide 8). Birds eat the fruit and thereby disperse the seeds (14). The seeds are used in Nigeria as a purgative (1).

References:

1. Ainslie, 1937: sp. no. 204. 2. Anon, For. Prod. Res. Inst. Ghana, s.d. 3. Aubréville, 1959: 2, 202. 4. Bouquet, 1969: 54–55. 5. Bouquet, 1972: 12. 6. Burtt Davy & Hoyle, 1937: 3. 7. Eggeling & Dale, 1952: 11. 8. Irvine, 1961: 559–60. 9. Keay & al., 1964: 313. 10. Kerharo & Bouquet, 1950: 168. 11. Malcolm & Sofowora, 1969: 512–7. 12. Portères, s.d. 13. Schnell, 1950, b: 259. 14. Taylor, 1960: 83. 15. Walker & Sillans, 1961: 57–58.

**Lannea spp.**

FWTA, ed. 2, 1: 731.

Wood of *Lannea spp.* in Nigeria has been reported suitable for making the inner and outer cases of match-boxes (1). A number of species occur widely in the state.

Reference:

1. Anon 1965, b.

**Mangifera indica** Linn.

FWTA, ed. 2, 1: 727. UPWTA, ed. 1, 340.

English: mango.

French: manguier.

Portuguese: mangueiro.

West African: **SENEGAL** BASARI a-mangɔ́ (K&A; Ferry) ɔ-wud ɔr a-ɓẹ mɛn (Ferry) BEDIK ga-mángɔ̀ (Ferry) FULA-PULAAR (Senegal) mãgo (K&A) KONYAGI a-ngud, vya-ngud (Ferry)

MANDING-BAMBARA imâguru (K&A) mâguru (JB; K&A) mankuru (JMD) MANDINKA tubabuduto, tubaabuduuta (*def.* tubaabuduutoo) = *whiteman's duto* (K&A) SERER imâguru (K&A) mâguru (JB; K&A) SERER-NON (Nyominka) imâguru (K&A) mâguru (K&A) WOLOF mangaro (FB) mãngaru (JB) **THE GAMBIA** MANDING-MANDINKA duutoo (auctt.) tubaabuduuta (*def.* t. duutoo) = *whiteman's duto* (auctt.) WOLOF mangaro (JMD) **GUINEA-BISSAU** CRIOULO mango (JDES) mangueira (JDES; EPdS) FULA-PULAAR (Guinea-Bissau) mango-sane (JDES) **GUINEA** MANDING-MANINKA bodo porto (FB) **SIERRA LEONE** BULOM (Sherbro) maŋgo-ɛ (JMD) FULA-PULAAR (Sierra Leone) maŋgo (FCD) GOLA pulɔŋ (FCD) KISSI maŋguiyo (FCD) KONO maŋgo (FCD) maŋko (FCD) KRIO mangoi (FCD) maŋgro (FCD) LIMBA (Tonko) maŋgo (FCD) LOKO maŋgoro (FCD) MENDE mango (*def.*-i) (JMD; FCD) TEMNE *a*-maŋko, *a*-maŋkoro (FCD) VAI plɔŋ (FCD) **MALI** DOGON mȁgoro (C-G) MANDING-BAMBARA mankuru (JB) MANINKA bodo porto, bodo: *Detarium;* porto: *Portuguese* (JMD) **GHANA** ADANGME maŋo (FRI) AKAN-ASANTE amano *the fruit* (Enti) mãngo (FRI) *a*-mango-dua *the tree* (Enti) TWI mãngo (auctt.) mano (FRI) mano-dua *the tree* (FRI) WASA amango (Enti) GA maŋgo (FRI) GBE-VHE (Awlan) amago (FRI) GUANG-GONJA mángo (Rytz) **NIGER** SONGHAI máŋgù (*pl.* -à) (D&C) **NIGERIA** BEROM máŋgɔ̀rɔ̀ (LB) EDO ógúi, ógúiébō *properly Irvingia gabonensis* (JMD) EFIK mañgoro *from English* (JMD) ùyó ḿbàkárá = *wild mango* (*Irvingia gabonensis*) *of the Europeans* (KW) FULA-FULFULDE (Nigeria) mangoro *the fruit* (J&D) mangorohi *the tree* (MM) HAUSA mángwàrò (ZOG) máŋgwaro *the tree* (MM; LB) IBIBIO màñkòrò (JMD; Kaufman) IDOMA umangohi (Odoh) IGALA mangolo (Odoh) IGBO àgbọ̀nọ̀, úgirī *properly Irvingia gabonensis* máŋgòlò (JMD) IGBO (Owerri) ọkpọkpa bẹkẹ KANURI áuré *a hybrid ?* (C&H) mángùlò (C&H) TIV ìcàmégh´ (JMD) YORUBA máńgòrò (JMD) òro òyìnbó (IFE) **WEST CAMEROONS** DUALA jaŋgòlò (Ithmann) NGEMBA máŋgòrə̀ (Chumbow)

A tree reaching 30 m tall with pronounced straight bole carrying a heavy crown, native of India, and now cultivated for its fruits in many tropical and subtropical parts of the world where the climate is right: a mean shade temperature of 26°C, with minimum above freezing, and rainfall of 750–2500 mm on a markedly seasonal distribution, rainfall during flowering and fruiting being deleterious.

The date and method of arrival of the mango in the Region is not well documented. Its exotic origin is clearly recognized in the adoption in almost every West African language of variants of the name *mango,* itself of South Indian origin *(maankai, máa,* the plant: [–n–]: *kai,* a fruit, Tamil), or failing this in a few instances of a loan from a known plant, e.g. in Mali Maninka *bo'do porto* or Portuguese (i.e. foreign) bo'do *(Detarium senegalense),* and in S Nigeria of the names for *Irvingia gabonensis*. The fruits of these two plants bear some superficial resemblance to the mango fruit. The early Portuguese voyagers in India picked up the word as *mangas* from which it has entered European languages. The plant must have been dispersed in the Indic basin at an early time, and certainly into E Africa under the influence of Arab traders. It does not appear to have spread to West Africa by the land connexion, but to have reached there very much later from the New World whither it has been taken in the Eighteenth Century. It was not till the following century that dispersal in the West African region took place (7, 9, 17, 19).

The tree is planted throughout the West African region. It is well acclimatized so that in places it has become subspontaneous (e.g. Ivory Coast, 15). It makes an excellent shade and avenue tree if grown from seed. The dense canopy persists in the dry season and shades out undergrowth so that an avenue planting becomes a good fire-break (3). Once established the trees are drought-resistant.

The wood is light grey or greenish and somewhat fibrous. It is of value for hut-posts, but seems seldom to be used (11, 32). In Tanganyika the wood is used for boat-building and door and window-frames (24) and in Malawi for building and for firewood (25). Mango wood smoke is reported to cause dermatitis (2, 11). In India the timber is valued for quite a number of indoor and outdoor purposes, though durability is limited (21). It produces high quality charcoal.

The bark has astringent properties. A macerate is widely taken for diarrhoea and dysentery (Senegal, 12–14; Ivory Coast, 5, 15; Nigeria, 18, 22; Congo, 4). In Ivory Coast a decoction is used as a wash for the head in migraine (1), and is widely used as a mouth-wash for the relief of toothache, sore gums, sore-throat, etc. in Nigeria (9, 22), and in Congo (4). This is taken in Ivory Coast as a diuretic against urethral discharge (15). Young callus bark is taken from the trunk to make an enema for piles in Ghana and is also given rectally as a stimulant in wasting disease (11). Sap from the bark is used in the Region (9, 22) and in Gabon (23) as an anti-syphilitic. An irritant oil extracted from this sap has been used in the treatment of syphilis (24). The sap in Gabon (23) and in Senegal (14) is considered saponific. The bark also yields an oleo-resin which in India is sometimes sold as 'gum arabic.' It too is considered anti-syphilitic and good for dermal infections (21). Examination of a bark extract for activity against avian malaria showed no action (8). A yellow dye can be obtained from the bark (10, 21).

The leaves are astringent like the bark, and are widely used for dysentery and diarrhoea, and as a mouth-wash-gargle for toothache, sore-throat, etc. A leaf-decoction is used in Gabon for asthma and bronchitis (23). In Ivory Coast a decoction is considered febrifugal (5), and also in Nigeria (2). The leaf is used in Nigeria to treat skin-diseases, and laboratory tests confirm an action of *Sarcina lutea* and *Staphylococcus aureus,* but with no action on Gram -ve organisms, nor on fungi (16). A resin is said to be the active substance (24). The bark, stem and fruits also exhibit bacteristatic action (24). A decoction of leaves along with those of other plants is taken in Ivory Coast by draught for giddiness (1). Young tender leaves are eaten as a vegetable in Indonesia and the Philippines (20, 24). They are a good source of ascorbic acid. In time of dearth the mature leaves have been fed to cattle in India. These leaves contain the glycoside *mangiferine* which with prolonged feeding causes death (21). The leaves are however very palatable and young trees growing where there are cattle need to be protected. At one time, now forbidden by law, the production of *peori* dye (Indian yellow) was a commercial enterprise in India by its separation from the urine of cattle fed on mature leaves (6, 10, 21, 24).

The fruit is by far the most important part of the plant. The fruit of wild trees is liable to taste strongly of turpentine and to be fibrous. Propagation of unselected seedlings is too chancy to be worthwhile. Very considerable selection work has been carried out in India and more recently elsewhere to produce fruit of high quality for eating. There appears to be ample scope for selection work to be tried out in West Africa. Multiplication of desirable clones is done by grafting. The fruit is a rich source of nutrients and minerals, and the chemistry has been extensively explored (7, 20, 21, 24). Even though the fruit is so highly esteemed, excessive consumption may cause renal inflammation. Besides consumption as a dessert fruit, unripe fruit can be turned into chutney. A few varieties may have been selected in India for this, but in the main the fruit of seedling trees with tart flavour is used (21). Sweet varieties can be made into jam and preserves. The halved fruit can also be preserved by sun-drying.

The kernel is starchy and can be eaten roasted, or dried and pickled (9). It is considered anthelmintic in Nigeria (22) and in Gabon (23) where it is also taken against diarrhoea.

References:

1. Adjanohoun & Aké Assi, 1972: 20. 2. Ainslie, 1937: sp. no. 221. 3. Aubréville, 1950: 393. 4. Bouquet, 1969: 55. 5. Bouquet & Debray, 1974: 16. 6. Burkill, 1935: 1402–6. 7. Busson, 1965: 344–5, with analysis of fruit pulp. 8. Claude & al., 1947: 145–74. 9. Dalziel, 1937. 10. Greenway, 1941; 222–45. 11. Irvine, 1961: 560–1, with leaf analysis. 12. Kerharo & Adam, 1964, b: 557. 13. Kerharo & Adam, 1964, c: 314. 14. Kerharo & Adam, 1974: 137–40, with phytochemistry and pharmacology. 15. Kerharo & Bouquet, 1950: 168. 16. Malcolm & Sofowora, 1969: 512–7. 17. Mauny, 1953: 709–10. 18. Oliver, 1960: 30. 19. Purseglove, 1968: 24–32. 20. Quisumbing, 1951: 538–41. 21. Sastri, 1962: 6, L–M: 265–85, with extensive information. 22. Singha, 1965. 23. Walker & Sillans, 1961: 58. 24. Watt & Breyer-Brandwijk, 1962: 46–48. 25. Williamson, J. 1956: 78.

## Ozoroa insignis Del.

FWTA, ed. 2, 1: 739, as *Heeria insignis* (Del.) O. Ktze. UPWTA, ed. 1, 338, as *H. insignis* O. Ktze.

**West African:** **SENEGAL** BASARI a-nuyɔmbɔni (K&A; Ferry) BEDIK ga-kuramb (Ferry) FULA-PULAAR (Senegal) gurugâhi (Aub.) kélel, kelel ɗéri, kélelel ɗéri, kéleli = *little kéleli* (K&A) tak ara kuléhi (Aub.) TUKULOR wasswassur (JMD) KONYAGI a-ngòy (Ferry) MANDING-BAMBARA doliségi (JB, ex K&A) MANDINKA kalakati (*def.* kalakatoo) = *bow tree* kalakato dimbadâbo (K&A) MANINKA kalakari (Aub.) 'SOCE' kalakato (K&A) kalakatô (K&A) kalakato dimbadâbo (K&A) kalakoto (K&A) SERER gaygésan (K&A) ngégésan (JB) sané sané (K&A) WOLOF vosvosor (JB) waswasor, waswasür (K&A) **THE GAMBIA** FULA-PULAAR (The Gambia) gyolo-kidime (DRR; DAP) MANDING-MANDINKA koroŋkondo (auctt.) **GUINEA-BISSAU** MANDING-MANDINKA kórón kondo **GUINEA** MANDING-MANINKA kalakari (Aub.) **MALI** BOBO vélooélé (Aub.) DOGON sene pelye (Aub.) FULA-PULAAR (Mali) gurugâhi (Aub.) takara kuléhi (Aub.) SONGHAI tâssa **UPPER VOLTA** MOORE niinoré (Aub.) 'SENUFO' tifahama (Aub.) **IVORY COAST** 'SENUFO' tifahama (Aub.) **GHANA** MOORE neabnɔya (FRI) SISAALA nasia (AEK; FRI) **NIGER** FULA-FULFULDE (Niger) gurugâhi (Aub.) takara kuléhi (Aub.) HAUSA kasheshi (Aub.) **NIGERIA** HAUSA cirini (RES) háwaàyén zaákii = *lion's tears* (JMD; ZOG) kàsheé-shí (ZOG) ƙashéshé (auctt.) keékásheéshè (JMD; ZOG)

A shrub or tree to 6.5 m high, of the soudanian savanna from Senegal to Niger and Nigeria, and across Africa to Ethiopia, Zaire and E Africa. The plant is sensitive to fires, and consequently it is often only in coppice growths to 1–2 m high (8, 13).

The wood is dark red (4, 6), hard and heavy (20). It is used in Zaire for cabinetry (19). In The Gambia it is considered a good firewood (16) and widely charcoal made from it is valued (Ghana, 8; Zaire, 19; Angola, 5; Sudan, for smelting iron, 2). The stems are commonly used in N Nigeria as sticks to apply cosmetics to the face. The name *hawayen zaki,* Hausa, meaning literally 'lion's tears', and applied to a certain form of decorative face mask, is used loosely and includes other species similarly used (5).

The sap is rich in a white resin. In southern Africa an infusion of bark or leaf is used as a remedy for diarrhoea, and a bark-decoction as a purgative (20). In Senegal a bark-macerate is considered antienteralgic and cholagogic (12, 13). In Tanganyika the bark is used to treat pink-eye (3).

The roots have a high reputation in Senegal as a vermifuge and are commonly sold in markets for this purpose, though some medicinemen prefer to use the leaves with *Stereospermum kunthianum* (Bignoniaceae). Roots are also considered good for treating intestinal pain and dysentery, as a galactogogue and for severe migraine (13). In East Africa the roots enter into treatment for stomach troubles (Kenya, 9; Tanganyika, for diarrhoea with or without blood, 17), and for low backache (Tanganyika, 18). The possibility of edaphic ecotypes is reflected in the Fula (Senegal) name *kélelеldéri, déri* implying the form from land which is not flooded; this is preferred for its laxative properties over *kéleli* (11). In Tanganyika also the root is eaten raw or the root-bark is powdered to make an infusion drunk like tea as an antimalarial, and the root cooked in rice is considered aphrodisiac (7).

The leaves like the roots have purgative action. In Nigeria crushed leaves boiled in milk are taken as a vermifuge (5, 15). In Senegal a macerate of pulped leaves is taken as an antidysenteric and to promote milk-flow (12). Poultices of pulped leaves are used for swellings on the feet (10, 13). In Sudan a paste of leaves and bark is applied to skin-diseases (21). The foliage is grazed by cattle in Senegal (1), but in Sudan it is said to be fatal to donkeys (21).

The flowers are much visited by bees (14).

References:

1. Adam, 1966, a: as *H. insignis*. 2. Andrews A. 1818, K. 3. Bally, 1937: as *H. reticulata* (Bak. f.) Engl. 4. Dale & Greenway, 1961: 21, as *H. reticulata* (Bak. f.) Engl. 5. Dalziel, 1937. 6. Eggeling & Dale, 1952: 5–7, as *H. reticulata* (Bak. f.) Engl. 7. Haerdi, 1964: 125, as *H. reticulata* (Bak. f.) Engl. 8. Irvine, 1961: 555, as *H. insignis* (Del.) O. Ktze. 9. Jeffery K. 469, K. 10. Kerharo & Adam, 1964, a: 420–1, as *H. insignis* O. Ktze. 11. Kerharo & Adam, 1964, b: 548, as *H. insignis*

(Del.) O. Ktze. 12. Kerharo & Adam, 1964, c: 308, as *H. insignis* O. Ktze. 13. Kerharo & Adam, 1974: 133, 135, as *H. insignis* O. Ktze. 14. Miszewski 50, K. 15. Oliver, 1960: 28, 66, as *H. insignis*. 16. Rosevear, 1961, as *H. insignis*. 17. Tanner 474, K. 18. Tanner 4427, K. 19. Van der Veken, 1960: 11–14, as *H. insignis* (Del.) O. Ktze. 20. Watt & Breyer-Brandwijk, 1962: 45, as *H. insignis* O. Ktze. 21. Wickens 1371, K.

## **Ozoroa pulcherrima** (Schweinf.) R. & A. Fernandes

FWTA, ed. 2, 1: 739, as *Heeria pulcherrima* (Schweinf.) O. Ktze. UPWTA, ed. 1, 338, as *H. pulcherrima* O. Ktze.

English: lady's tears (McIntosh).

West African: **MALI** MANDING-MANINKA bediko (A. Chev.) **NIGERIA** HAUSA háwaàyén zaákii (MM) hawaye (Abrahams) shiwaákár-ján-gárgárii, shiwaáká: *bitter-leaved Vernonia amygdalina;* ján gárgárií: *implying a red clay soil useless for farming* (Bargery)

A shrub, often stunted by bush fires, with softly hairy stems, of the wooded savanna from Guinea to N Nigeria, and across Africa to Sudan.

The split stems are used in E. Africa for raising fire by friction (Trotha fide 1).

The bark is powdered and added to milk or is taken as a cold infusion by the Masai to relieve abdominal pain and diarrhoea in both man and in young cattle (2). The root is used in West Africa in either hot or cold infusions for gastro-intestinal trouble (1).

References:

1. Dalziel, 1937. 2. Watt & Breyer-Brandwijk, 1962: 45, as *H. pulcherrima* O. Ktze.

## **Pseudospondias microcarpa** (A. Rich.) Engl.

FWTA, ed. 2, 1: 729, 731. UPWTA, ed. 1, 340.

West African: **SENEGAL** BANYUN torugal (K&A) DIOLA budek (Aub. ex K&A) burombon (JB; K&A) kununu (JB; K&A) FULA-PULAAR (Senegal) dologa (Aub. ex K&A) MANDING-BAMBARA koni mpéku (JB; K&A) MANINKA doréké (Aub. ex K&A) **THE GAMBIA** DIOLA (Fogny) burombon = *a boil* (DF) **GUINEA-BISSAU** FULA-PULAAR (Guinea-Bissau) cadjôdjáe (JDES) gobi (Fernandes) MANDING-MANDINKA bembô (JDES) PEPEL utime (JDES; Fernandes) **GUINEA** FULA-PULAAR (Guinea) n-dŏloga (auctt.) **SIERRA LEONE** KONO duekɛ (FCD; S&F) LIMBA kuro (FCD; KM) LOKO n-duɛge(-ŋ) (NWT) MANDING-MANDINKA kɔdolɔka, kɔ: *water* (FCD) MANINKA (Koranko) bɔgɔ (NWT) doleke (S&F) MANINKA dɔwɔ (*def.* dɔwei), dɔ: *water;* wei: *bears* (auctt.) SUSU-DYALONKE gɛrɛ-na (FCD) TEMNE *an*-dibia (FCD; S&F) **LIBERIA** KRU-BASA pohn (C&R) **IVORY COAST** ABE blékuré (auctt.) AKYE bé (A. Chev., ex K&B) DAN sritié (Aub., ex K&B) FULA-FULFULDE (Ivory Coast) dologa (Aub.) KRU-GUERE purié (K&B) GUERE (Wobe) tidé (Aub.; K&B) MANDING-MANINKA doréké (Aub.; K&B) **GHANA** VULGAR akatawani (DF) AKAN-AKYEM *a*-katawani = *close your eyes* (Enti) osunyane (FRI) ASANTE akatani (auctt.) akatawani = *close your eyes* (BD&H; FRI) KWAWU *a*-katawani *close your eyes* (FRI) TWI *a*-katani (Enti) *a*-katáwani = *close your eyes* (FRI) sunyãn *not commonly used* (FRI, Enti) tamia (FRI) WASA *a*-katani, *a*-katawani (Enti) *a*-katawani = *close your eyes* (Enti) **TOGO** YORUBA-IFE OF TOGO onyangba (Volkens) **NIGERIA** HAUSA rimin kuroni (Yates) YORUBA ekika-ajá (JMD) okika-ajá (JMD; KO&S)

A tree to 20 m high with a short crooked and fluted bole, of deciduous fringing and secondary forests of the guinean zone, in damp sites and often on stream-banks, occurring throughout the Region, and extending to NE and E Africa.

The tree's preference for proximity to water is reflected in the Sierra Leone Mandinka name *kɔdolɔka, kɔ* meaning water. The tree is suspected of having a narcotic effect on those who sit or sleep under it: hence the Ghanaian names meaning 'close your eyes' (10).

The wood is soft to moderately hard and greyish or greenish (6), liable to warp (1), perishable and subject to borer-attack (7, 9). The wood is sometimes used in Liberia for poles and planks (6).

The bark contains a little reddish resin (1), which is used in Liberia to treat jaundice and other diseases affecting the eyes (6). The bark is also used in Ivory Coast for its purgative and diuretic properties in the treatment of jaundice (4, 13) and of cough (4). A bark-decoction is used in Gabon for toothache and for diuretic effect in treating urethral discharge (15, 16a). It is also held to be good for ulcers on the soles of the feet when compounded with the heartwood of *Pterocarpus soyauxii* (Leguminosae: Papilionoiideae), the feet having been previously washed in the sap of *Aframomum giganteum* (Oliv. & Hanb.) K. Schum. (Zingiberaceae), the false maleguetta of the central African region (16a). In Congo the powdered bark is eaten for cough, febrile lumbago, pains in the ribs and asthenia; a bark-decoction is taken for stomach complaints, jaundice conditions and gonococcal complications, and the bark is put into vapour baths and the lees used in frictions for persistent rheumatic pain (2). The plant (? bark) is used in Casamance as a wash for scrophulous infants (11, 12). In Gabon the plant is considered an aphrodisiac (16a), and in Congo sap from a piece of bark cooked in a leaf is given to a girl suffering infatuation because she is under sorcery to make her vomit up the fetish (2).

Tannin is recorded in the bark and leaves (3, 4).

The fruits, nearly 2.5 cm long, red or bluish black when ripe, are resinous. The juice stains the fingers. They are eaten in various parts of Africa. Birds eat them and are responsible for dispersal (14). In Gabon they are used as fish-bait (16b). The seeds are used as beads in Ghana (5, 10). They are eaten in Casamance (1).

References:

1. Aubréville, 1959: 2: 204. 2. Bouquet, 1969: 55. 3. Bouquet, 1972: 12. 4. Bouquet & Debray, 1974: 17. 5. Burtt Davy & Hoyle, 1937: 3. 6. Cooper & Record, 1931: 95–96, with timber characters. 7. Dale & Greenway, 1961: 26, 28. 8. Dalziel, 1937. 9. Eggeling & Dale, 1952: 12. 10. Irvine, 1961: 561. 11. Kerharo & Adam, 1963, b. 12. Kerharo & Adam, 1974: 140–1. 13. Kerharo & Bouquet, 1950: 168. 14. Savill & Fox, 1967, 37–38. 15. Walker, 1952: 186. as *P. longifolia* Engl. 16a. Walker & Sillans, 1961: 59–60, as *P. longifolia* Engl. 16b. Walker & Sillans, 1961: 60, as *P. microcarpa* (A. Rich.) Engl.

**Rhus longipes** Engl.

FWTA, ed. 2, 1: 739

West African: **SENEGAL** WOLOF taa (JB; JGA) **GUINEA-BISSAU** FULA-PULAAR (Guinea) uaga guitẹl (Aub.)

A bush or small tree widely dispersed throughout the savanna regions of tropical Africa, recorded only from Senegal, Guinea, Sierra Leone and Nigeria in the Region, but doubtless more widespread.

No usage is reported for West Africa. In Tanganyika a root-decoction is drunk for malaria, and a root-decoction with leaf-sap as a laxative and abortifacient (1).

The fruit are probably edible.

Reference:

1. Haerdi, 1964: 126.

**Rhus natalensis** Bernh.

FWTA, ed. 2, 1: 739.

A shrub to about 3 m high, of the savanna from Guinea to N Nigeria, and widely dispersed elsewhere in tropical Africa.

In E Africa the plant attains a greater stature to 6.5 m tall, and in Kenya the poles are utilized for hut-building (4). In Somalia the twigs serve as toothbrushes (3).

The bark of E African material yields 15.2% tanning material (8). An infusion is drunk as tea in Kenya (5). The root in decoction of water or milk is considered in Kenya to be good for stomach complaints (6) and is taken in E Africa for gonorrhoea and influenza, and to treat wounds (1, 8). The root is also given to a woman suffering from repeated abortions or still-births, and to children with fits (8). The plant (part not stated) is also used as a taenicide (8).

The leaf is also used to treat gonorrhoea, and leaf-sap is applied to weeping dermatoses, and to furuncles (8). The sap of a number of *Rhus spp.* is highly irritant and vesicant (7), and perhaps applies to the sap of this species. Nevertheless, the leaves appear to be readily browsed by stock.

The globose fruits, about 6 mm in diameter, have an edible pulp which is very commonly eaten in Uganda and Kenya. The seed is also edible, but is eaten only perhaps in time of dearth, and has the name 'Food of Famine' in Uganda (2).

References:

1. Bally, 1937. 2. Dyson-Hudson 412, K. 3. Gillett 4411, K. 4. Glover & al. 541, K. 5. Glover & al. 949, 2113, K. 6. Vesey-Fitzgerald 126, K. 7. Watt J. M. 1967. 8. Watt & Breyer-Brandwijk, 1962: 50.

**Rhus tripartita** (Ucria) Grande

FWTA, ed. 2, 1: 739.

West African: **MALI** ARABIC jedari (RM) ARABIC ('Maure') jédari, djédari (Aub.) TAMACHEK tahonak (Aub.)

A low spiny bush of the Sahel and desert region in Mali and Niger, and across northern Africa to Palestine.

The plant is similar to *R. oxyacantha* Cav., the fruits of which are eaten and whose bark is used for tanning by the Touareg of Central Sahara (1). The two species bear the same Arabic name, and it is probable that *R. tripartita* has the same uses.

Reference:

1. Maire, 1933: 149.

**Schinus molle** Linn.

FWTA, ed. 2, 1: 272.

English: Brazilian pepper tree, California pepper tree (Usher); American mastic (the gum, Usher).

French: faux poivrier; faux poivier d'Amerique; résine de mollé (the gum).

A shrub, native of the Andes of S America, and introduced and cultivated here and there in the West African region, principally as an ornamental for its graceful foliage and showy long clusters of small red fruits. It is well adapted to dry sandy soils and is drought-resistant (1).

The wood is said to be immune to termite-attack, but has little use except as a fuel (6). A strong-smelling gum is obtained from the trunk and leaves. It is composed of 40% gum, 60% resin and traces of volatile oil and a bitter principle. The leaf and green berry also yield a volatile oil which has been used instead of cubebs for treating gonorrhoea. The oil varies in composition according to the country of origin. It gives an intense phellandrene-reaction and contains *phellandrene, carvacrol* and *pinene,* and is probably poisonous. The gum is a masticatory (4, 5).

The bark is said to yield 23% tannin (3). It is used in S America as a tan-bark, and to make a purgative for man and animals. The bark and leaves are also used on swellings and sores (2).

The berry is hot to the mouth and in America is used to make a drink. The alkaloid *piperine* is present. They have been used as a condiment and as an adulterant for true pepper. Eaten in quantity they set up irritation of the intestine, especially in children (2, 4, 5, 6).

References:

1. Aubréville, 1959: 2: 191. 2. Burkill, 1935: 1974. 3. Greenway, 1941. 4. Uphof, 1968: 479. 5. Usher, 1974: 529. 6. Watt & Breyer Brandwijk, 1962: 51.

**Schinus terebinthifolius** Raddi

FWTA, ed. 2, 1: 727.

English: Brazilian pepper.

A shrub, native of South America, introduced and cultivated here and there in the West African region.

The bark is resinous and aromatic and is the source of a resin known in S America as 'balsamo de misiones' (2, 3). The bark has been used internally as a stimulant, tonic and astringent, and externally for rheumatism, gout and syphilis (4).

The leaves and fruit are used as a lotion to bathe wounds and sores (4), and in Brazil they have been applied to tumours on the feet (1). The leaves are said to be tonic (2, 3). *Tyramine* is reported present in them (5).

The fruit are sometimes eaten, as are those of *S. molle,* but with severer effect. A volatile oil is suspected to be the active principle (4).

Tannin is present in the plant (4).

References:

1. Hartwell, 1967. 2. Uphof, 1968: 474. 3. Usher, 1974: 529. 4. Watt & Breyer-Brandwijk, 1962: 51. 5. Wheaton & Stewart, 1970.

**Sclerocarya birrea** (A. Rich.) Hochst.

FWTA, ed. 2, 1: 729. UPWTA, ed. 1, 340–1.

West African: **MAURITANIA** ARABIC ('Maure') dambu (Aub.) **SENEGAL** BASARI a-ngúɗy (Ferry) a-nguit (K&A) a-nguk (K&A) BEDIK gi-kúɗy (Ferry) DIOLA findibasu (JB, ex K&A) FULA-PULAAR (Senegal) béri (K&A) éri (auctt.) hédéhi (Aub.) hédi (Aub.) kédé (Aub.) KONYAGI a-tẹma (Ferry) MANDING-BAMBARA mguna (JB, ex K&A) MANDINKA kutaŋ (*def.* kuntaŋo) (Sidia Jatta) MANINKA konnan (Aub.) kuntan (Aub.) kuntango (Aub.) 'SOCE' kutan dao (K&A) kuten dao (K&A) SERER ari (K&A) aritj (auctt.) SERER-NON sugu (auctt.) sungul (auctt.) NON (Nyominka) arid (K&A) arik (K&A) indarid (K&A) SONINKE-SARAKOLE nôné (K&A) WOLOF bér (auctt.) bièt (auctt.) bir (auctt.) bör (auctt.) **THE GAMBIA** VULGAR dib (Ozanne) DIOLA (Fogny) findibasu = *grassy* (DF) FULA-PULAAR (The Gambia) eri (DAP) kundingho *from Maninka* (DRR) MANDING-MANDINKA kuntaŋ-jawo (DAP; DRR) WOLOF birr *hence the specific name* (DAP) **GUINEA-BISSAU** FULA-PULAAR (Guinea-Bissau) éri (JDES; EPdS) **GUINEA** MANDING-MANINKA kunan (Aub.) kuntan (Aub.) kuntango (Aub.) **MALI** DOGON bíí, bibíí (auctt.) FULA-PULAAR (Mali) éri (Aub.; GR) hédéhi (Aub.) hédi (Aub.) kédé (Aub.) MANDING-BAMBARA kunan (A. Chev.) kuntan (Aub.) kuntango (Aub.; FB) kutan (FB) mguna (GR) MANINKA kunan (Aub.) kuntan (Aub.) kuntango (Aub.) SONGHAI dineygna (A. Chev.) TAMACHEK tuila (Aub.; Rodd) **UPPER VOLTA** MOORE nobéga (Aub.; FB) SENUFO-TUSIA kegue (*pl.* kel) (G&H) **GHANA** DAGBANI mu-mugga (Osnan) MOORE nobiga (FRI) NANKANNI nanogba (Lynn) SISAALA burunogo (AEK) **NIGER** FULA-FULFULDE (Niger) éri (Aub.) hédéhi (Aub.) hédi (Aub.) kédé (Aub.) GURMA bunamabu (Aub.) HAUSA dama (Aub.) KANURI kəmáà (Aub.) SONGHAI díinéy (*pl.* -à) (auctt.) dinégna (Aub.; FB) lúuley (*pl.* -á) (D&C) **NIGERIA** ARABIC homeid (JMD) ARABIC-SHUWA homeid (JMD) FULA-FULFULDE (Nigeria) eedere (*pl.* eede, eedi), edere (*pl.* eedee, eedi) *the fruit, but perhaps Spondias sp.* (auctt.) hedi (J&D) heri *the tree* (JMD; KO&S) HAUSA danya *the tree* (auctt.) danyaá (ZOG) ludu, lule, nunu *the fruit* (JMD) HAUSA (West) huli (JMD; ZOG) KANURI kə̀máà (C&H)

A tree to about 13 m high, the short bole sometimes reaching over 2.5 m in girth, of the drier Sahel savanna from Senegal to Niger and N Nigeria, and across Africa to Ethiopia and Uganda.

The tree is often planted around villages in E Africa. It is easily propagated either from seed or cuttings. A borer which attacks the tree is eaten by some people (6).

The wood is soft, coarse and greyish dirty white. It is durable when well-seasoned. When big enough it is used for mortars and strong black-stained bowls called in Hausa *akushi, bukuru,* or specially *baƙi ɗan danya.* Of the woods used to make these bowls, this species is considered the best (6). In Jebel Marra wooden platters are made of it (8), and in Ethiopia milking vessels and axe-handles (16). Ash from the burnt wood, along with that from other trees is used for dehairing goat-skins before tanning (6).

The bark yields a strong fibre (9) and when injured exudes a nearly colourless gum which becomes brittle and friable on drying. The gum, dissolved in water and mixed with soot, is used to make ink (6, 9). A bark-infusion is used by the Fula of N Nigeria for stomach-pains and constipation (10) and mixed with natron is used by the Hausa for dysentery (6). In Jebel Marra an infusion is taken to ease labour-pains (8). Analgesic action is also sought by the Basari of Senegal who chew the bark to compact carious cavities in the teeth to quell toothache (11, 14). Bark, especially of the roots, but also of the trunk, is prescribed as a remedy for snake-bites by the Fula of Senegal. Pounded to a paste it must be rubbed on till a swelling is raised, then a bark-decoction is drunk and a dressing applied over the area. The leaves may also be used (1, 12). The Fula of Senegal also consider the bark to furnish a good anti-inflammation preparation for external use, and with butter added is applied to the forehead for headache, and to the eyes for blepharitis. They take a decoction internally as a purge (12, 14). Soce of Senegal use a macerate of twig-bark for snake-bite (13). With other drug-plants it is used in Senegal in leprosy and syphilis treatments (14). A bark-decoction is used by the Nankani of Ghana for skin-eruptions (9). In veterinary medicine the Fula give a decoction to their stock to increase appetite (12).

The root is pounded up with water by the Sukuma of Tanganyika and the water is drunk for schistosomiasis, and is used for washing scabies (15).

The foliage is grazed by cattle and camels in Senegal (2), and by camels in Jebel Marra (8). The higher branches may be cut for fodder in time of drought (6).

The fruit is pale yellow, obovoid, nearly 3.7 cm long with a leathery rind like a mango and a similar fibrous soft pulp covering the stone. They are eaten in many parts of tropical Africa and are commonly sold in West African markets. When fully ripe they have an acid but pleasant taste. Travellers out in the bush find them satisfying to suck for thirst-quenching. They are said to be laxative. The expressed juice (*mbajalle,* Fula, N Nigeria) makes an agreeable drink, and in many areas is fermented into an alcoholic beverage (4, 6; Senegal, 3, 13; Upper Volta, 7). The juice may also be boiled down to a thick black consistency used for sweetening guinea-corn gruel (6). It is believed in Jebel Marra that the fruit when eaten attracts scorpions (8) to the eater.

The nut contains two or three seeds with oily and edible kernels (Hausa, *kuɓe,* or *ku'e*) (6). The oil consists of 64% oleic acid, 17% myristic acid and small quantities of several others (5).

References:

1. Abdoul Oumar, 1962: 14, as *Poupartia birrea* (Hochst.) Aubr. 2. Adam, 1966, a: 517. 3. Anon., 1946, b: 151. 4. Aubréville, 1950: 405–7. 5. Busson, 1965: 345–6, with kernel analysis. 6. Dalziel, 1937. 7. Guilhem & Hébert, 1965: as *Poupartia birrea* (Hochst.) Aubr. 8. Hunting Technical Surveys, 1968. 9. Irvine, 1961: 563. 10. Jackson, 1973. 11. Kerharo & Adam, 1964, a: 424–5. 12. Kerharo & Adam, 1964, b: 573. 13. Kerharo & Adam, 1964: 319. 14. Kerharo & Adam, 1974: 141–2. 15. Tanner 1266, K. 16. Turton 98, K.

**Sorindeia collina** Keay

FWTA, ed. 2, 1: 737.
West African: **SIERRA LEONE** BULOM (Sherbro) tɔnt-lɛ (S&F) KISSI yɛŋgulɛ (S&F) KONO kambane (S&F) MENDE kafa (*def.* kafei) *a general term for certain shrubs* (FCD; S&F) TEMNE *an*-thaŋka (S&F)

A tree to about 10 m high of montane forest in Sierra Leone, with several recorded vernacular names, but apparently not utilized in any way.

**Sorindeia grandifolia** Engl.

FWTA, ed. 2, 1: 738.
West African: **NIGERIA** EDO èhę̀góġó (KO&S)

A tree to 16 m high, often on stream-banks of the forest zone of S Nigeria and W Cameroons, and in E Cameroun and Sao Tomé.

The bark-slash exudes a little milky juice (1).

Reference:

1. Keay & al., 1964: 319–20.

**Sorindeia juglandifolia** (A. Rich.) Planch.

FWTA, ed. 2, 1: 737. UPWTA, ed. 1, 341.
English: 'damson' (the fruit, Sierra Leone).
West African: **SENEGAL** BASARI a-mbírír (Ferry) DIOLA fukot (JB, ex K&A) usińgelimit (Aub., ex K&A) DIOLA ('Kwaatay') daguru muda (K&A) MANDYAK kuridêdê (Aub.; K&A) **THE GAMBIA** DIOLA (Fogny) fukot (DF) **GUINEA-BISSAU** FULA-PULAAR (Guinea-Bissau) sandje-bombo (Fernandes) sandji-bombro (JDES) MANDYAK lagari (Fernandes; JDES) PEPEL n'tata (untata) (auctt.) timbá (D'O) **GUINEA** SUSU kusi bumba (Aub.) **SIERRA LEONE** BULOM (Sherbro) tɔnt-lɛ (FCD) KISSI yɛŋgulɛ̃ (FCD) KONO kambane (FCD) LOKO majɔnjɔ (NWT) MANDING-MANDINKA banuforikili (NWT) masɔ̃sɔ̃ (FCD) MENDE bali-kafa (FCD) kafa (*def.* kafei) *a group term* (NWT; FCD) kafa-wawa (FCD) MENDE-KPA bai-kafa (FCD) UP MENDE ndɔgbɔ-jɛlɛ kafa SUSU-DYALONKE kude-kade-na (FCD) TEMNE *an*-thaŋka (NWT; FCD) *an*-thaŋka bombar (Oldfield)

A shrub or small tree to 6 m high occurring on the edge of dry deciduous forest, regrowth in humid forest and in the galleried soudanian forest of Senegal to Dahomey and also in Ubangi-Shari, Angola and Zambia.

The stems are used as chew-sticks, and the leaves have been used in Sierra Leone for bilious complaints, and in decoction as a gargle for sores in the mouth in children (1). In Senegal pulped leaves are applied to sores and ulcers, and a leaf-decoction is used in the Casamance as a laxative and diuretic (4).

The fruit is yellow ripening purple and is edible. It is sold in Freetown markets under the name 'damson' because of its similarity to the temperate fruit. The pulp is a bit astringent (1–3).

References:

1. Dalziel, 1937. 2. Irvine, 1952, a: 36. 3. Irvine, 1961: 563. 4. Kerharo & Adam, 1974: 142.

**Sorindeia mildbraedii** Engl. & v. Brehm.

FWTA, ed. 2, 1: 737.
West African: **SIERRA LEONE** MENDE kafa (*def.* kafei) (S&F) TEMNE *an*-thaŋka (S&F)

A tree about 12 m high of the understorey of the high forest, recorded from Sierra Leone (1) and Nigeria and W Cameroons, and extending into E Cameroun.

Reference:

1. Savill & Fox, 1967, 36.

**Sorindeia warneckei** Engl.

FWTA, ed. 2, 1: 738. UPWTA, ed. 1, 341.
West African: **GHANA** AKAN-TWI osee-efu (Williams) GBE-VHE (Awlan) akpɔkpoe (Akpabla) **NIGERIA** IGBO (Umu-Ahia) obunnu (Ariwaodo) YORUBA bùjé wẹ̀ẹrẹ = *small bùjé* (*Randia maculata*) (JMD) gboyìn-gboyìn (Millson)

A scandant shrub or small tree, of forest understorey or in secondary jungle from Sierra Leone to S Nigeria.

The plant is added to medicinal baths in Accra (Williams fide 4). Sap is used to make blue colour in tattooing in Ghana (2, 4). The fruit are yellow ripening black, and the pulp is sweet and edible (1, 3, 4).

References:

1. Dalziel, 1937. 2. Irvine 275, K. 3. Irvine 1956, K. 4. Irvine, 1961: 565.

**Spondias cytherea** Sonner

FWTA, ed. 2, 1: 728.
English: Otaheite-apple; English plum (Sierra Leone).
French: pomme cythère; prune cythère (Aubréville).
West African: **GUINEA-BISSAU** CRIOULO cajamanga (JDES) **SIERRA LEONE** KRIO iŋglish-plɔm = *English plum* (FCD)

A tree of the eastern Pacific, now widely dispersed through the tropics. It is recorded present in Sierra Leone where it is commonly cultivated (3), in Ivory Coast (1) and on the Benue Plateau of Nigeria where missionaries in Tivland have established it (4).

The tree is grown for its fruit, edible and reaching 5–7 cm long by 3.5–5.0 cm in diameter. They are eaten cooked and are said to have a slight flavour of turpentine (5). They are an item of market-produce in Freetown, Sierra Leone. Temnes recognize two varieties of tree: one with rough corky bark, the other smaller and smooth-barked known as *an-thank-an-lup,* whose fruit are not so sweet (3). There is no doubt scope for selection.

The young shoots are eaten as a potherb in Gabon, and a leaf-infusion is used for sore-throat (5). In SE Asia the young leaves are eaten, and they are cooked with tough meat to make it tender (2).

The wood is very brittle, and the bark contains a gum (2).

References:

1. Aubréville, 1959: 2, 191, as *S. dulcis* Forst. f. 2. Burkill, 1935: 2067. 3. Deighton 2770, K. 4. Vermeer 14, UCI. 5. Walker & Sillans, 1961: 61, as *S. dulcis* Forst.

**Spondias mombin** Linn.

FWTA, ed. 2, 1: 728. UPWTA, ed. 1, 341, as *S. monbin* Linn.
English: hog plum; Ashanti plum (Ghana, Irvine); (yellow) Spanish plum (West Indies, Dalziel); mombin or monbin.

French: mombin, or monbin; the fruit — prune mombin or p. monbin; prune icaque; prune myrobalan (West Indies).

West African: **SENEGAL** BALANTA m-sal (K&A) BANYUN kifosin (K&A) BASARI a-nyɔ́ká (K&A; Ferry) BEDIK ga-tyálẹká (K&A; Ferry) DIOLA bu lila (JB; K&A) buɗoy (aucct.) bueta (K&A) buétia (auctt.) FULA-PULAAR (Senegal) talé (K&A) tall (K&A) KONYAGI a-tyálax (Ferry) MANDING-BAMBARA nemkôô (K&A) ningô (K&A) ninkom (K&A) MANDINKA ninkoŋ (*def.*-o) MANINKA minegon (Aub.) nemkôô (K&A) nineko (Aub.) ningô (K&A) ninkom (K&A) 'SOCE' bumkumo (K&A) kumkumo (K&A) MANDYAK pilmé (K&A) MANKANYA umpélo (K&A) SERER yoga (auctt.) SERER-NON sul (Aub.; AS, ex K&A) sul (auctt.) sun (auctt.) WOLOF ninkom (auctt.) sob (auctt.) **THE GAMBIA** DIOLA (Fogny) ku lila = *grows when planted* (DF) FULA-PULAAR (The Gambia) chali (DAP) MANDING-MANDINKA ninkoŋ (*def.*-o) supa-jambo = *leaf for soup* (auctt.) WOLOF ninkom (JMD) **GUINEA-BISSAU** BALANTA sale (JDES) same (JDES; EPdS) BIDYOGO negae (JDES) ogáe (JDES) CRIOULO mandiple (auctt.) FULA-PULAAR (Guinea-Bissau) tchálè (EPdS; JDES) MANDING-MANDINKA nincom (*def.*-ô) (JDES; EPdS) MANDYAK pilme (JDES; EPdS) MANKANYA n'pela (JDES) umpela (JDES) PEPEL mupila (GES) n'pilo (umpilo) (JDES) **GUINEA** FULA-PULAAR (Guinea) talé (CHOP) tialé (auctt.) KONO mungo (RS) MANDING-MANINKA minegon (Aub.) nineko (Aub.) ninkom (CHOB) **SIERRA LEONE** BULOM (Kim) lep (FCD) BULOM (Sherbro) le-lɛ (FCD; S&F) FULA-PULAAR (Sierra Leone) chalɛ (FCD) GOLA g-bedi (FCD) g-beri (FCD) g-besi (FCD) KISSI lewo (FCD; S&F) KONO k-paŋgin(-ɛ) (FCD; S&F) KRIO fiks-plɔm = *fits plum* (FCD; S&F) hel-faiya-plɔm = *hell fire plum* (FCD; S&F) mɔki-plɔm = *monkey plum* (FCD; S&F) LIMBA (Tonko) kutoro (FCD) LOKO k-paki (FCD; S&F) MANDING-MANDINKA niŋkɔ̃ (FCD) MENDE gboji (*def.*-i) (auctt.) SUSU-DYALONKE lɔkhdɛ-na (FCD) lokhodi-na (FCD) TEMNE *an*-lɘp (auctt.) *an*-lup (FCD) VAI g-bosi (FCD) **LIBERIA** MANO buna (Har.) **MALI** DOGON énye vɛvɛy = *chicken crisis; the fruit said to poisonous to chickens* (C-G) MANDING-BAMBARA maincon (A. Chev.) minkon (Aub.) ningo (A. Chev.) nongo (A. Chev.) MANINKA minegon (Aub.) nineko (Aub.) **UPPER VOLTA** FULA-FULFULDE (Upper Volta) tiali porto (K&B) MANDING-BAMBARA monkon (Bégué) naingro (A. Chev., ex JMD) ningo (A. Chev., ex K&B) DYULA miningone (K&B) 'SENUFO' tana (K&B) tanma (K&B) **IVORY COAST** ABE n'gba (A&AA) ngua (auctt.) ABURE haperrie (A. Chev.; K&B) poposané (B&D) AKAN-ASANTE tuané (B&D) AKYE n-gba (A&AA) n-gba baté (A&AA) n-gua (auctt.) ANYI n(i)truma (auctt.) torima (K&B) BAULE iruma (Croix) truma (auctt.) FULA-FULFULDE (Ivory Coast) tiali porto (K&B) GAGU tola (B&D) KRU-BETE titi (K&B) GUERE (Chiehn) ténékwé (B&D) tété (K&B; B&D) KULANGO (Bondoukou) naingro (A. Chev.) KWENI muenla (B&D) uena (B&D) uinda (B&D) KYAMA aubé (Aub., ex K&B) MANDING-DYULA miningone (K&B) MANINKA naingro (K&B) nineko (K&B) ningon (K&B) SENUFO-BAMANA mune(i)ko (Aub.; RS) TAGWANA hénin (K&B) **GHANA** VULGAR ataba (DF) atoa (DF) ADANGME aadɔntʃo (FRI) adɔdɔŋtʃo (FRI) adombaatʃo (FRI) akɔle (FRI) AKAN-AKUAPEM ntowãã (Enti) ASANTE atõaa, atowãã (FRI Enti) atõwãã (Enti) ntowãã (Enti) ASANTE (Denkyera) tõaa (Enti) FANTE ?ataawa (FRI) atõaba (FRI; Enti) TWI atõaa (FRI) atõwãã (Enti) WASA atõaa (Enti) ANYI troma (FRI) ANYI-AOWIN troma (FRI) GA ádɔ̀dɔ́ŋ (KD) akɔle (FRI) GBE-VHE akukɔti (FRI) VHE (Awlan) akukan (BD&H) akukɔ (FRI) VHE (Pecí) akukan (BD&H) akukɔ (FRI) KONKOMBA enayele (Volkens; Gaisser) munayem (Volkens; Gaisser) nayile (Volkens; Gaisser) MOORE subtuluga NZEMA tuane (FRI) **TOGO** BASSARI ayě (Volkens; Gaisser) KABRE kinyělo (Volkens; Gaisser) TEM (Tshaudjo) kinyělu (Volkens; Gaisser) YORUBA-IFE OF TOGO agliko (Volkens; Gaisser) akiko (Volkens; Gaisser) **NIGERIA** BOKYI kechibo (auctt.) EDO óghèèghè *the fruit* (JMD) ókhíkhàn (auctt.) EFIK ǹsúkàkàrà (auctt.) EJAGHAM ekpi (KO&S) ENGENNI ugriya (JMD) FULA-FULFULDE (Nigeria) caɓɓulli Makka = *wild olive* (*Ximenia americana, Olacaceae*) *of Mecca* (MM) HAUSA tsaádàr Lamarudu = *plum of the giant; referring to Lamarudu, a mythical giant* (JMD) tsaádàr másàr = *plum of Egypt; from* tsaádà: *Ximenia;* másàr: *Egypt* (auctt.) HAUSA (East) iyawe (JRA) kafili (JMD) opon (JRA) IDOMA ichinkla (Odoh) IGALA ichikala (Odoh) IGBO íjíkàrà (BNO) íshíkèrè (Singha; KO&S) ísíkàlà (KO&S) IGBO (Bende) íshíkèrè (JMD) IJO-IZON (Kolokuma) ìgìnàịn, ìgìnèín (KW) IZON (Mein) àgìnèén (KW) IZON (Oporoma) ìgìnèìn (KW) ISEKIRI akikan (JMD) NUPE jinjèrè *the fruit* (Banfield) jinjèrèci *the tree* (Banfield; KO&S) TIV kakka (JMD; KO&S) konkuagh *the fruit* (JMD) URHOBO óghìghẹ̀n (auctt.) YORUBA akika, okika *the tree* (auctt.) ìyeyè *the fruit* (JMD; Ross) **WEST CAMEROONS** BAFOK masimaken (DA) DUALA mugaŋgà (DA) KOOSI mesa (DA) KPE maso (DA) KUNDU ndoa (DA) LONG masimaken (DA) LUNDU ndoa (DA) MBONGE ndoa (DA) TANGA ndoa (DA) WOVEA maso (DA)

A tree to 20 m tall, widespread and common in farmland and secondary vegetation of the soudanian and guinean savanna regions. Whether or not it is indigenous or introduced to West Africa is open to conjecture. The species is certainly native to the Caribbean and tropical America from whence it has been dispersed under Spanish influence to the Philippines and thence into Asia-proper.

The plant has anthropogenic tendencies. It is naturalized around villages in West Africa. It grows very easily from stakes to make live fences and enclosures (11; Guinea-Bissau, 13; Sierra Leone, 25; Ivory Coast-Upper Volta, 5, 6, 22; Ghana, 9, 16; N Nigeria, 10). The stakes are also planted for yam poles by the Igbo in S Nigeria (17) and for pepper in Gabon (28). Basari votaries of the Ichak fetish in eastern Senegal consider the plant a fetish-protector of millet (21). Poles are used in hut-frames (11, 13), and are carpentered into small articles such as cattle-yokes in Guinea-Bissau (13), and by the Yoruba of S Nigeria into axe and hoe-handles, for which the roots may also be used (10). The wood is whitish-grcy, soft and liable to termite-attack. The wood-ash is used in Sierra Leone in making soap and is an ingredient of a snuff (25), and in Guinea in indigo dyeing (Portères fide 16).

The bark is conspicuously thick, fissured and hard, characters which confer a protection of the tree against savanna fires. The Ijo of S Nigeria use the thick bark to cut stamps and seals (29). The bark contains some tannin (8, 11) and has been used in the West Indies and Guyana (16) and in E Africa (14) for tanning. A bark-slash exudes a clear sticky gum (25). It has, along with the leaves and fruit, medicinal uses. A decoction of the bark is taken for severe cough with inflammatory symptoms, acting by causing relief through vomiting (11, 26). In Guinea dry powdered bark is applied as a wound-dressing in circumcision (Afzelius fide 11). The bark is used in Ivory Coast to treat sores, to facilitate parturition, and sometimes as an anthelmintic, and plant extracts are in general use for stomach-ache, diarrhoea, cough, sore-throat, bronchitis, nausea and as a poison-antidote (8). In Ivory Coast-Upper Volta the bark is applied topically to areas of leprosy (22). In the Delta region of Nigeria, Ijo use the bark to treat fungal infections of the feet and an extract is used to ward off encroachment by driver ants (29). A trace of alkaloid is reported present in the plant (? bark) (1).

A root-macerate is used in Casamance (Senegal) for colic with pain (21). In Guinea-Bissau a root-infusion is considered to be an important tonic able to prolong the life by two or three days of those about to die (13). In Ivory Coast-Upper Volta root-preparations are said to be febrifugal and are sometimes prescribed with other drug plants in decoction or as a pulp taken in draught, lotion or baths for this effect (22).

The fresh leaves are purgative in effect. In infusion they are a common remedy for cough, and laxative given in fever with constipation (11). The sap from young fresh leaves is given to children for stomach-troubles in Guinea (23). A decoction is sometimes taken for gonorrhoea and as an aphrodisiac in Nigeria (3). The leaves crushed with citron are considered in Ivory Coast to be a sovereign remedy for helminthiasis in children (22). A leaf-decoction is taken in Senegal for dysentery (20) and by the Edo of SE Nigeria for dysentery and other intestinal disorders (27). The leaves are used in Nigeria on malignant tumours (3) and it is noted that the bark in water is administered in Cuba for uterine cancer (15). In Ivory Coast the leaves are part of a complex prescription taken by mouth for curvature of the spine caused by Pott's disease (2).

The young leaves in infusion are taken internally and applied topically as a lotion by women in confinement (11, 25), and in Senegal Mandinka women with a history of miscarriage take a preparation of young leaves (21). In Ivory Coast young leaves softened over a fire and rolled in the hands with some salt are given as a mouth instillation (? inhalation) to prevent miscarriage (2).

The leaves are held to have anodynal, healing and haemostatic properties. The Baule of Ivory Coast prepare a gargle of the leaves with those of *Alchornea cordifolia* (Euphorbiaceae) and a citron (22). A decoction of pounded leaves is used in Guinea (23) and in Senegal (21) as an eye-lotion. In Congo it is taken as a mouth-wash for toothache (7). In Ivory Coast-Upper Volta the leaves are applied to new sores to prevent infection, and in association with other plants to areas affected by leprosy (22). The Fula of Senegal use a leaf-cum-bark preparation to stem post-partum haemorrhage (19), and in Sierra Leone young leaves, or older pulped leaves, are warmed and tied over the vagina in case of tearing in childbirth (12).

The fruit is like the temperate plum, 3.7 cm long, ovoid, 1-seeded, yellow-skinned when ripe. The flesh has a sharp acidulous taste and is edible. The fruit is commonly sold in West African markets. Eaten in excess the fruit is said to cause 'dysentery.' It is perhaps to cause fear from over-indulgence that Ghanaian children are told that snakes will bite them if they eat the fruit (16). Nutrient value is not particularly high (16, 24). The fruit can be stewed and a refreshing drink can be made from them. The juice is said to be febrifugal and diuretic. In the soudanian region of Senegal the fruits are fermented into a kind of beer (4).

Reports that the seed kernel is edible (10, 16) need testing. In the Philippines the seed is said to be poisonous (24).

References:

1. Adegoke & al., 1968. 2. Adjanohoun & Aké Assi, 1972: 21. 3. Ainslie, 1937: sp. no. 322. 4. Anon, 1946, b: 151. 5. Aubréville, 1950: 403. 6. Aubréville, 1959: 2: 206. 7. Bouquet, 1969: 55. 8. Bouquet & Debray, 1974: 17. 9. Burtt Davy & Hoyle, 1937: 4. 10. Dalziel 912, K. 11. Dalziel, 1937. 12. Deighton 1133, K. 13. Gomes e Sousa, 1930: 73. 14. Greenway, 1941. 15. Hartwell, 1967. 16. Irvine, 1961: 565–6, with chemical analysis of the fruit. 17. Irvine, s.d. 18. Kerharo & Adam, 1963, b. 19. Kerharo & Adam, 1964, c: 577. 20. Kerharo & Adam, 1964: 322. 21. Kerharo & Adam, 1974: 143. 22. Kerharo & Bouquet, 1950: 169. 23. Pobéguin, 1912: 14, as *S. lutea*. 24. Quisumbing, 1951: 543–4, with chemical analysis, as *S. purpurea* Linn. 25. Savill & Fox, 1967, 38–39. 26. Singha, 1965. 27. Vermeer 71, UCI. 28. Walker & Sillans, 1961: 61. 29. Williamson, K. 44, UCI.

**Spondias purpurea** Linn.

FWTA, ed. 2, 1: 728.

English: Gambia plum.

French: mombin rouge; prune rouge.

West African: **SENEGAL** WOLOF sob tubab = *whiteman's 'sob'* (auctt.) **GUINEA-BISSAU** CRIOULO mandiple de Sera Leôa (JDES) **SIERRA LEONE** KRIO gambe-plɔm = *Gambia plum* (FCD)

A tree introduced to West Africa from tropical America for its edible fruit. The fruit are red, acquiring the epithet 'red' in French. The tree is grown in Sierra Leone and the fruit is sold in Freetown market.

**Trichoscypha acuminata** Engl.

FWTA, ed. 2, 1: 735.

A tree to 20 m tall with a straight bole, but up to 1 m girth, with a few branches, of the humid forest in S Nigeria and W Cameroons, and extending to Zaïre.

No usage is recorded in the Region.

In Gabon the bark of this species, and of other *Trichoscypha spp.* of close affinity, is used as a remedy against constipation in infants (3). In Congo a bark-decoction is used to wash small-pox pustules and to bathe rheumatics. It is given to women to treat sterility and dysmenorrhoea, and for haemorrhage during pregnancy. A decoction is also used in steam baths and in frictions of the lees for bronchial affections, headache, feverish stiffness, pains in the sides or stomach, as a vermifuge and an aphrodisiac (1).

A trace of saponin is reported present in the bark, and tannin in the bark and roots (2).

The fruit-pulp is edible and has a vinous taste (3). The fruit-juice is pleasant to drink. In Congo the fruit is given to convalescents and to anaemic persons as a tonic (1).

References:

1. Bouquet, 1969: 56. 2. Bouquet, 1972: 12. 3. Walker & Sillans, 1961: 61.

**Trichoscypha arborea** (A. Chev.) A. Chev.

FWTA, ed. 2, 1: 736. UPWTA, ed. 1, 342, incl. *T. ferruginea* Engl.
West African: **SIERRA LEONE** MENDE kpoma-luwa (*def.*-luwei) (S&F; SKS) TEMNE *an*-thaŋka (FCD; S&F) **LIBERIA** KRU-BASA blimah (auctt.) gbeh (C) **IVORY COAST** ABE dao (auctt.) AKYE n-dabo (A. Chev.) ANYI alakui (Aub., ex K&B) DAN wuinto (K&B) KYAMA allahia (Aub.; K&B) 'KRU' nenanan (K&B) **GHANA** VULGAR anaku (DF) AKAN-WASA anaku (auctt.) ANYI alakui (FRI)

A tree to 30 m high with a straight bole to over 1 m girth, of the evergreen forest from Sierra Leone to W Cameroon.

The wood is greenish or pinkish, fairly hard, tough and durable (6). It is used for hut-construction and in Liberia for canoes and planks. It is also carved into fetish masks, etc. (7).

The bark contains a resin (1, 3) which is used in Ivory Coast by the Guere to prevent miscarriage (10). The plant is also used in Ivory Coast for amenorrhoea and dysentery (2).

The fruit is red when ripe with a sweetish edible pulp that is a bit fibrous around the seed. They are relished by both man (3, 4, 6, 8) and monkeys (5).

References:

1. Aubréville, 1959: 2, 196. 2. Bouquet & Debray, 1974: 17. 3. Burtt Davy & Hoyle, 1937: 4. 4. Cooper 65, K. 5. Cooper 230, K. 6. Cooper & Record, 1931: 96–97, with timber characteristics. 7. Dalziel, 1937. 8. Deighton 3616, K. 9. Irvine, 1961: 566. 10. Kerharo & Bouquet, 1950: 169.

**Trichoschypha baldwinii** Keay

FWTA, ed. 2, 1: 736.
West African: **SIERRA LEONE** MENDE kafa (*def.* kafei) *a group term* (SKS)

A tree to 6.5 m high of the closed-forest from Sierra Leone to Ghana.

**Trichoscypha beguei** Aubrév. & Pellegr.

FWTA, ed. 2, 1: 736. UPWTA, ed. 1, 342.
West African: **IVORY COAST** ABE daokro (JMD)

A small tree of the forest understorey occurring in Liberia and Ivory Coast.

In Ivory Coast it is confused with *T. yapoensis* Aubrév. & Pellegr. (1), and is likely to be used for similar purposes.

Reference:

1. Aubréville, 1959: 2, 198.

**Trichoscypha cavalliensis** Aubrév. & Pellegr.

FWTA, ed. 2, 1: 736.

A tree to 20 m tall, of the closed-forest of Liberia to Ghana.

The wood is used in Ivory Coast for rice-mortars because of its durability (1).

Reference:

1. Cooper 107, K.

**Trichoscypha chevalieri** Aubrév. & Pellegr.

FWTA, ed. 2, 1: 736.
West African: **IVORY COAST** KYAMA anonkoya (B&D)

A tree to 6.5 m high, of the closed-forest of Ivory Coast and Ghana.

The plant, part not stated, is used by medicinemen in Ivory Coast for costal pains and stiff-neck (1).

The fruit pulp is sweet when ripe and is sometimes eaten (2).

References:

1. Bouquet & Debray, 1971: 17. 2. Irvine, 1961: 567.

**Trichoscypha longifolia** (Hook. f.) Engl.

FWTA, ed. 2, 1: 736. UPWTA, ed. 1, 341.
West African: **GUINEA** FULA-PULAAR (Guinea) dologa-chango (JMD) **SIERRA LEONE** KONO k-pomaluwei (Fox) MENDE kpoma-luwa (*def.*-luwei) *a group term* (Sawyerr; Fox) **LIBERIA** KRU-BASA blimah(-pu) (C&R)

A tree to 16 m high, clean bole, no buttresses, of the forest in Sierra Leone and Liberia.

The wood is strong and tough. It is sawn for planks and timbers in Liberia.

The bark exudes a clear, sticky, pungent resin which becomes black on exposure. This can stain hands and clothing and is difficult to remove even by petrol solvent. The bark itself when boiled turns black and the liquor is used in Liberia as an antiseptic wash for sores and wounds. Leaves are applied to ulcers.

The fruit yields a similar watery resinous sap that blackens and stains. They are in grape-like clusters, each about 2.5 cm long. The seed kernels are oily and edible.

References:

1. Cooper 435, K. 2. Cooper & Record, 1931: 96, with timber characteristics.

**Trichoscypha oba** Aubrév. & Pellegr.

FWTA, ed. 2, 1: 736. UPWTA, ed. 1, 342.
West African: **IVORY COAST** VULGAR bim (JMD) ABE gogo mango (JMD; Aub.) oba-oba (JMD; Aub.) ringala (A. Chev.) ADYUKRU manoskpoel aatjhe (JMD; Aub.)

A shrub or small tree of the closed-forest in Ivory Coast and Nigeria.

The wood is quite durable, and the fruits are probably edible. (1).

Reference:

1. Irvine, 1961: 567.

**Trichoscypha patens** (Oliv.) Engl.

FWTA, ed. 2, 1: 736.
West African: **IVORY COAST** KRU-GUERE (Chiehn) mukopé (K&B)

A tree to 10 m or more high, of the humid rain-forest of Nigeria and W Cameroons, and extending to Zaïre.

A plant under this name is said to find medicinal use in Ivory Coast for febrile pains (1); a bark-decoction is used to bathe a patient ill without diagnosable symptoms, bark being taken from the east and west sides of the tree (relic of sun-worship?); and the leaf-sap is an eye installation for treating jaundice (2).

References:

1. Bouquet & Debray, 1974: 17. 2. Kerharo & Bouquet, 1950: 169.

**Trichoscypha preussii** Engl.

FWTA, ed. 2, 1: 736.

A tree to 10 m high, of the understorey of the closed-forest of S Nigeria and W Cameroons and into E Cameroun.

The bark contains a gummy white latex (1).

Reference:

1. Keay & al., 1964: 323.

**Trichoscypha smeathmannii** Keay

FWTA, ed. 2, 1: 736.

A bush or small tree to 6.5 m high of the guinean woodland of Guinea, Sierra Leone and Liberia.

The flowers are fragrant and may be visited by bees.

**Trichoscypha yapoensis** Aubrév. & Pellegr.

FWTA, ed. 2, 1: 736. UPWTA, ed. 1, 342.

West African: **LIBERIA** KRU-BASA blimah (C&R) bo-in-dah (C&R) gbeh (C&R) punh (C&R) **IVORY COAST** ABE daokro (JMD; Aub.)

A forest tree to 25 m high occurring in Liberia and Ivory Coast.

The wood is strong, tough and durable, very similar to that of *T. arborea* and locally not distinguished apart (5). It is used in Liberia for canoes and planks (3), and is carved into fetish masks. In this connexion it is known as the 'male' tree in distinction from *Guarea thompsonii* Sprague & Hutch. which is the 'female' tree (4).

The bark exudes a dirty orange-yellow resin with an aromatic smell (1, 5).

The fruit is edible (2), and is also taken by birds (4).

References:

1. Aubréville, 1959: 2, 198. 2. Cooper 66, K. 3. Cooper 114, K. 4. Cooper 226, K. 5. Cooper & Record, 1931: 97.

## ANCISTROCLADACEAE

**Ancistrocladus abbreviatus** Airy Shaw

FWTA, ed. 2, 1: 234.

West African: **SIERRA LEONE** KISSI pikiũ (FCD) LIMBA ka-thak-ka-thoni (FCD) MENDE nja-nɛngu (FCD)

A massive liane of swamp-forests, especially on river-banks, recorded from Sierra Leone to S Nigeria.

ANCISTROCLADACEAE

**Ancistrocladus barteri** Sc. Elliot

FWTA, ed. 2, 1: 234.
West African: **SIERRA LEONE** MENDE jɛnigbo (FCD) jɛnigbulo (FCD)

A forest liane recorded from Sierra Leone and Liberia.
The roots are boiled and the decoction is drunk as a laxative (1).

Reference:

1. Deighton 3813, K.

# ANNONACEAE

**Annona arenaria** Thonn.

FWTA, ed. 2, 1: 52.
West African: **SIERRA LEONE** MANDING-MANDINKA ninkɔ (Boboh) sunsuŋ (FCD) SUSU-DYALONKE sumu-na (FCD) TEMNE *a*-mɔmina (NWT; FCD) **UPPER VOLTA** DAGAARI bar oda (K&B) GURMA lubualansanlu (K&B) KIRMA tobéré (K&B) MANDING-BAMBARA dan (K&B) danguan (K&B) mandé sunsun (K&B) DYULA lommo (K&B) n-dara (K&B) n-dara lommo (K&B) numbulombo (K&B) MOORE karoduidiga (K&B) NYENYEGE bondolo (K&B) SONGHAI-ZARMA mufania (K&B) 'SENUFO' damugurumo (K&B) damurana (K&B) namurga (K&B) **IVORY COAST** AKAN-BRONG boboma (K&B) tien (K&B) BAULE amlon (K&B) amron (K&B) KPALAGHA gala (K&B) n'beni (K&B) KULANGO kuggo (K&B) kumo (K&B) soropodio (K&B) KWENI bré (K&B) MANDING-DYULA lommo (K&B) n-dara (K&B) n-dara lommo (K&B) numbulombo (K&B) MANINKA karamoko sunzun (A&AA) sunsu (K&B; B&D) sunsun sunzu (K&B) SENUFO-TAGWANA m'ru (K&B) muro (K&B) DYIMINI gbanessan (K&B) 'SENUFO' damugurumo (K&B) damurana (K&B) namurga (K&B) **GHANA** AKAN-ASANTE aboboma (FRI) TWI aboboma (FRI)

A savanna tree to 8 m high occurring from Guinea to the Ivory Coast, and in Cameroun, Zaïre and Angola. This species is confused with *A. senegalensis* Pers. so that vernaculars and uses overlap. In *Flore du Gabon* (5) it is recognized as a subspecies of *A. senegalensis:* ssp. *oulotricha* Le Thomas.

The plant finds use, especially in the Ivory Coast, as an emeto-purgative and diuretic in treatment of dropsy, oedemas, stomach-aches, poisoning, female sterility and leprosy, as an anthelmintic, to treat sores, rheumatism, febrile pains and refractory headache (2). The Tagwana of the Ivory Coast consider it a sure cure for rickets in children: the Kweni use it for epilepsy and scorpion stings. The Brong consider it anti-abortive and the Gouins use it in treating leprosy and skin-affections. All parts of the plant are equally used in these treatments as also for diarrhoeas, dysenteries, cough and fever (4). In Ghana it has been used to treat epilepsy (7), a use repeated in Congo where the sap expressed from crushed young leaves is taken along with other preparations (1). In Congo it is also prescribed in cases of difficult breathing, asthma and bronchitis for its expectorant and emetic properties, and for illnesses connected with blennorrhoea (1). Here also the sap of the roots is considered haemostatic and cicitrisant on wounds, and is given in draught to women with painful and irregular menses (1).

Traces of alkaloids have been detected in the plant which have been shown to be slightly toxic to lower invertebrates and by injection into a dog to cause temporary hypertension (4). The Ghanaian preparations used to treat epilepsy may yield a curare product (7).

The fruit is edible. The seeds are eaten in Sierra Leone (6) and contain an edible oil (3).

On the Ivory Coast the plant enters into a fetish to ensure catching plenty of game, and also to make a sort of dope to improve the hunting ability of dogs (4). In Congo it is held to have magical properties in sorcery to keep away evil spirits, to advance a case before a tribunal, and to confer success 'with the girls' (1).

References:

1. Bouquet, 1969: 57. 2. Bouquet & Debray, 1974: 18. 3. Irvine, 1961: 1. 4. Kerharo & Bouquet, 1950: 19–20. 5. Le Thomas, 1969: 322 (as *A. senegalensis* Persoon ssp. *oulotricha* Le Thomas). 6. Thomas, N. W., 92, K. 7. Watt, J. M., 1967.

**Annona cherimola** Mill.

FWTA, ed. 2, 1: 52.
English: cherimola.
French: cherimoyer.

A shrub to 5 m high, a native of the Andes and spread by the Indians of S America to the Caribbean area whence it has been taken by Europeans to various parts of the tropics. It has been introduced to West Africa with but little success. It requires the coolth of the altitude of its native home and might with careful husbandry be successfully established at higher elevations. It hybridizes with the other *Annona* species and a hybrid with *A. reticulata* called *atemoya* has received some attention.
The seeds can be used as an insecticide.

Reference:

Burkill, 1935: 166.

**Annona chrysophylla** Boj.

FWTA, ed. 2, 1: 52.
English: wild custard apple, wild soursop.

A small tree reaching 6 m high, essentially a plant of equatorial east and south tropical Africa, but entering the Region on the Bauchi Plateau in Nigeria at the extreme NW limit of its range.
The wood is greenish grey, soft (2, 3) and with no recorded use, except that in Zanzibar the poles are used for hut-building and the stems as hoe-handles (8). There are taboos on it and as firewood it may be used only by medicine men.
The bark yields a yellow or brown dye in East Africa (4) and an ink (7). The bark is considered emetic and has medical and criminal uses in Zanzibar (7). It also furnishes an inferior twine, and on pounding with water a dressing for women's hair (8). In Tanganyika chewed bark and leaves are placed over snake-bite after the wound has been enlarged. The bark is also placed on buboes. Sap is used by the Lobedu to make arrow-poison adhere.
A decoction of roots with those of the E African species *Combretum microlepidotum* (Combretaceae) is taken in Tanganyika as an expectorant, and roots and leaves are pounded together into a paste which is put on pusy swellings as an emollient (5). Boiled with sour oranges the concoction is given for stomach-ache in Zanzibar, and roots grated to a pulp are used as poultices (8). The root is said to make people forget and some races use a preparation to wean a baby from the breast (6, 7). The root also has had criminal use (1, 7). The plant is reported to contain a glycoside, *anonacein,* and the bark two alkaloids, one being *anonaine,* and a cyanogenetic resin, the glycoside and the resin not being

destroyed by boiling, but further evidence of the presence of these substances is necessary (6, 7). Charcoal made from the roots, which for magical purposes must be made by burning the root on a hoe, is rubbed by the Lobedu around the eyes to relieve twitching (7).

The leaf is astringent and its sap is used in Tanganyika to treat diarrhoea (5).

The fruit is sub-globose, resembling somewhat that of *A. squamosa*. The fruit pulp is edible, yellow to orange, and less attractive.

References:

1. Bally, 1937. 2. Dale & Greenway, 1961: 32–34. 3. Eggeling & Dale, 1952: 16–18. 4. Greenway, 1941. 5. Haerdi, 1964: 37. 6. Watt, J. M., 1967. 7. Watt & Breyer-Brandwijck, 1962: 56–58. 8. Williams, R. O., 1949: 124.

**Annona glabra** Linn.

FWTA, ed. 2, 1: 52, 757. UPWTA, ed. 1, 2, as *A. palustris* Linn.

English: alligator apple, serpent apple, monkey apple, marsh corkwood.

French: cachiman-cochon (3), pomme channelle de mer (5).

West African: **SENEGAL** DIOLA bu ruruf (JB) WOLOF digor yéner (AS; JGA) dugur, dugor mèr (auctt.) **THE GAMBIA** DIOLA (Fogny) bululuk = *it cried* (DF) **GUINEA-BISSAU** MANDING-MANDINKA suncum (*def.*-ô) (JDES) **GHANA** AKAN-FANTE adadima (FRI) adasima (FRI) **NIGERIA** YORUBA àfe = *a fishing float* (JMD)

A shrub or small tree attaining 10 m high of the mangrove and coastal marshes, throughout the region from Senegal to West Cameroon and on to Zaïre. It is widespread in the Caribbean.

The wood is soft. It has been used as a cork-substitute in the West Indies, a use also found in the Gran Bassa area of Liberia (8), but its stopping capacity in bottles is poor (2). The root-wood is used by fishermen as floats in West Africa. It is slightly lighter than cork but is more permeable to water and loses some of its bouyancy after immersion. The wood makes a pulp suitable only for coarse brown paper (4).

Extracts from the bark and the wood have yielded a substance called *liriodenine* which inhibits cancer growths. Further examination of its effect on human cancers is merited (5). Nigerian material examined has however been found lacking alkaloids (1). A root-decoction is used as a poison-antedote in Gabon (7).

The tree carries a dense foliage which seems to attract precipitation of dew in the dry season in Nigeria, keeping the ground beneath damp and producing a rich compost of leaf-litter. The tree is considered to be a valuable soil-improver (8). It has been used as a grafting stock (4) — presumably for other *Annona spp.*, but this is not stated.

The leaves are used in Nigeria as cough-medicine (6). In Gabon they are given in time of fever as a nourishing food and febrifuge, and mixed with citron leaves they provide a soothing lotion. Flowers and leafy shoots are also made into a cough-mixture (7).

The fruit is edible, though it is not regarded everywhere as palatable. It is relished in Gabon and is eaten either raw or cooked, and the unripe fruit is cooked as a febrifuge, or dried and powdered as treatment for dysentery (7).

The seeds are emetic (7).

References:

1. Adegoke & al., 1968. 2. Anon., 1904 as *A. palustris* Linn.. 3. Berhaut, 1967: 256. 4. Dalziel, 1937. 5. Le Thomas, 1969: 319–22. 6. Oliver, 1960: 19. 7. Walker & Sillans, 1961: 62–63. 8. Unwin, 1920: 262 as *A. palustris* Linn.

**Annona glauca** Schum. & Thonn.

FWTA, ed. 2, 1: 52. UPWTA, ed. 1, 1.
West African: **SENEGAL** MANDING-BAMBARA dangan (A. Chev.) mandé sunsun (A. Chev.) sunsun (A. Chev.) tangasu (A. Chev.) tongasu (A. Chev.) WOLOF digor yénèr (Aub.) dugur, dugor mèr *the names are said to be onomatopoeic (FD), but in what way is not clear* (auctt.) **GHANA** DAGBANI mampihege (FRI) GUANG-NCHUMBULU bileka (FRI)

A shrub attaining 1.50 m high, found in Senegal and Ghana, represented by two varieties, var. *glauca* and var. *minor* Robyns & Ghesq., which is confined to Senegal only.

It is abundant in parts of the Northern Territories of Ghana. 'The roots spread out about a foot under the ground and then in all directions roughly parallel to its surface. From these other shrubs spring, and the country was thickly studded with them' (2). The plant may be useful in re-covering dry areas (2).

In Senegal the roots are held to be diuretic, and are used in treating blennorrhoea (4).

The powdered seeds are used in parts of the tropics as an insecticide and fish-poison (3).

The fruits contain a clear sweet pulp (1). Roots are put as girdles onto young children in Senegal as a fetish-protection against illness (4).

References:

1. Berhaut, 1967. 2. Dalziel, 1937. 3. Irvine, 1961:3. 4. Kerharox & Adam, 1974: 144 as *A. glauca* Schum. & Thonn. var. *glauca*.

**Annona muricata** Linn.

FWTA, ed. 2, 1: 52. UPWTA, ed. 1, 1–2.

English: soursop.

French: corossolier, grand corossolier, corossol épineux, cachimantier.

West African: **SENEGAL** FULA-PULAAR (Senegal) dukumé porto (K&A) SERER niom (K&A) **GUINEA-BISSAU** CRIOULO pinha (JDES) FULA-PULAAR (Guinea) dukume porto (JMD) **SIERRA LEONE** BULOM (Sherbro) sipisap-lɛ (FCD) KRIO sawa sap = *sour-sop* (FCD; Boboh) MENDE supusapui *from Krio* (Boboh) **IVORY COAST** BAULE amlon (A&AA) sunzun (A&AA) **GHANA** ADANGME alukuntum (FRI) AKAN-AKYEM abrɔfontunkum *from* abrɔfo: *European;* ntunkum: *sourly sweet, like newly-made palm-wine* (FRI) aprɛ *from apple* (Enti) FANTE apere (FRI) TWI aprɛ, aperɛ *i.e., apple* (FRI, Enti) dɛboo (Fri) duantũnkũm, dua: *a tree;* tunkum: *sourly sweet* (FRI) nkraŋmrobe, nkra: *Accra;* abrobɛ: *a pineapple, i.e. Accra people's pineapple* (FRI) GA alugùtuŋ̀gũ (FRI; KD) nkraŋmrobɛ = *pineapple of the Accra people* (FRI) GBE-VHE apre *i.e., apple* (FRI) nyaŋkiɛ (FRI) vo (FRI) voti (FRI) votsi (FRI) VHE (Peci) yevu-nyiklɛ (FRI) **TOGO** TEM (Tshaudjo) alola (Gaisser)

A small tree attaining a height of about 8 m. A native of tropical America, but now widespread in the tropics, it is thought to have reached Africa (Angola) by 1686 (11).

The trunk and timber do not appear to have any particular uses. Extracts of the wood have shown no activity against avian malaria (5). The bark contains no or only a trace of tannin but hydrogen cyanide is present in large quantity (14, 15), though less in the root and leaf and but a trace in the fruit. A toxic and a non-toxic alkaloid have been isolated from the bark, the former causing convulsions in a frog. Both belong to the aporphine group. *Muricine* and *muricinine* of the aporphine group have been identified. Work on bark alkaloids showed a powerful stimulation on respiration in the rabbit with scarcely any alteration of blood-pressure and heart-action (15).

The bark and roots are in general used for dysenteries and as vermifuges (15), and the leaves for dysenteries and fevers (6). A decoction of roots is a poison-antedote in the Ivory Coast (1). A decoction of leaves is used both as a lotion and internally as a soothing and sudorific medicine for fever especially in children (1, 3, 6, 9, 12, 13).

The fresh leaves are pounded and applied to wounds (13). In the Casamance they are a dressing for the circumcision wound (9).

The plant shows parasiticidal properties. The leaf is pungent when crushed. In The Gambia (6) and in Sierra Leone (2) they are used to get rid of bedbugs — unfortunately temporarily precluding the use of the bed because of the smell. The leaves are held to be vermifugal, anthelmintic and antiphogistic (15). In Malaya a poultice is used on skin-affections and in Indonesia the juice of young leaves is applied to pustules to make them burst (4). The leaves are antispasmodic (12, 14).

By far the most important part of the plant is the fruit for which it has been so widely cultivated. The fruit-pulp is soft with an agreeably sour flavour. It is usually eaten raw but unfortunately contains a quantity of fibre so that it may be more acceptable after some preparation, i.e. either as juice, ice-cream, jellies but not jams. In the West Indies a fermented drink like cider is made from it. The fruit contains over 11% sugar, mostly glucose and fructose (4). It is deficient in calcium, phosphorus and vitamin A but very rich in vitamins B and C (14). The unripe fruit is astringent and is used for lack of tone in the intestines and for scurvy (15). In the Ivory Coast unripe fruits are dried and powdered for use against dysentery (1). In Surinam the ripe fruit serves as a bait in fish-traps (15).

The flowers and buds are used in Gabon (10) and the Ivory Coast (1) for coughs. Both flower and fruit pod have been used for catarrh (15).

The seeds are emetic and have been used as a fish-poison (8, 15), and in Madagascar (7) for cough. They contain up to 45% of a yellow non-drying oil which is an irritant poison, lethal to the head louse, *Pediculus capitis,* and is applied to the head in India and Mexico (8). The oil has a disagreeable taste and a poor digestibility which had led to the rejection of the seeds as an article of food (15).

The tree is normally grown from seed but cuttings and graftings are easy to establish.

References:

1. Adjanohoun & Aké Assi, 1972: 23. 2. Boboh, 1974. 3. Bouquet, 1969: 57. 4. Burkill, 1935: 166–7. 5. Claude & al., 1974. 6. Dalziel, 1937. 7. Debray & al., 1971: 52. 8. Irvine, 1961: 3–4. 9. Kerharo & Adam, 1974: 144–5 (with many references). 10. Le Thomas, 1969: 324–5. 11. Mauny, 1953: 696. 12. Oliver, 1960: 19, 46. 13. Pobeǵuin, 1912: 16. 14. Quisumbing, 1951: 306–7, 1030 (with many references). 15. Watt & Breyer-Brandwijk, 1962: 58–59 (with many references).

**Annona reticulata** Linn.

FWTA, ed. 2, 1: 52. UPWTA, ed. 1, 2.

English: custard apple, bullock's heart.

French: cachiman, cachimantier, coeur de boeuf, corossolier sauvage.

A tree reaching 8–10 m in height of Caribbean and Central American origin, now pan-tropical. It is cultivated in West Africa and may even have become naturalized in places. The lack of African vernacular names is an indication of its relative indigenous unimportance.

All parts of the tree contain hydrogen cyanide especially the bark (5, 6). The bark is strongly astringent and contains tannins and a small percentage of *anonaine,* an alkaloid of the aporphine group (5). The root is also reported to contain a toxic alkaloid and *liridenine* has been identified (7). Another alkaloid, *reticuline,* has also been found in the plant (7).

The bark has been used in Nigeria as an astringent in treating dysentery (4). Malays and Chinese in Asia use this astringency for tonic effect (1).

The sap from cut branches is acrid and irritant (6). The leaves are anthelmintic and antiphlogestic (6). In the West Indies they are used for tanning and to obtain a black dye (1). The leaves are strong-smelling and are used to maturate abcesses (1), and their use on cancerous tumours is recorded (2). A decoction is vermifugal, and the juice will kill lice (1). Various parts are insecticidal. Use of the fruit is recorded in West Africa, and the seed in India and the Philippines for *Pediculus capitis*. Tests have shown the root to be less potent than *Derris elliptica* (Leguminosae: Papilionoiideae). The insecticidal action is due to the presence of a glyceride or glycerides of a hydroxylated unsaturated acid or acids of high molecular weight (3, 6).

The tree is mainly cultivated for the heart-shaped fruit which may reach 10–15 cm in length by 10 cm in diameter, and contains a cream-coloured sweet juicy pulp. It is rich in vitamin C. It is not so highly esteemed as other *Annona spp.* in those parts of the world where the genus is better represented than it is in West Africa. It is nevertheless widely consumed. As a squash the fruit furnishes a refreshing drink which is commonly used in the Caribbean in treatment of dysentery (5), a use also found in West Africa (6). The unripe fruit and dried fruit can also be used as an astringent for diarrhoea, dysentery and intestinal worms (5).

The seed contains 42% of a fixed oil. The seed kernel is considered highly poisonous and contains tannins (6). A powerful astringent can be extracted which is of use in treating diarrhoea and dysentery (1).

Propagation is usually by seed, but grafting and budding have been successfully applied.

References:

1. Burkill, 1935: 167–9. 2. Hartwell, 1967. 3. Kerharo & Adam, 1974: 146–7. 4. Oliver, 1960: 19. 5. Quisumbing, 1951: 309–10, 1030. 6. Watt & Breyer-Brandwijk, 1962: 59. 7. Willaman & Li, 1970.

**Annona senegalensis** Pers.

FWTA, ed. 2, 1: 52. UPWTA, ed. 1, 2–3.

English: wild custard apple.

French: pomme channelle du Sénégal; annone.

West African: **SENEGAL** BALANTA boré (K&A) sungh kuo (K&A) BANYUN kétatone (K&A) BASARI a-nɔ́kɔtɔ̀ (Ferry) a-nyɔ́gúr (K&A; Ferry) BEDIK ga-nyàmbẹtɛ́r (Ferry) gi-tumkuŋ (K&A; Ferry) DIOLA (Brin/Seleki) butor (K&A) DIOLA (Fogny) fulolok (K&A) fulölok (K&A) DIOLA ('Kwaatay') huloluk (K&A) DIOLA-FLUP bleuleuf (K&A) bleulof (K&A) blölöf (K&A)FULA-PULAAR (Senegal) dugumé (K&A) dukumé (K&A) dukumi (K&A) KONYAGI a-mbɔɓy (Ferry) MANDING-BAMBARA dâka (A. Chev.; K&A) dânha (K&A) mâdé susu (K&A) mandé sunsun (A. Chev.) ndânga (JB) sunsun (A. Chev.) tangasu (A. Chev.) tongasu (A. Chev.) MANDINKA sunkuŋ (*def.*-o) 'SOCE' sukum (K&A) sunékuu (K&A) sunku (K&A) sunkuno (K&A) MANDYAK benempé (K&A) MANKANYA bam (K&A) SERER dôg (K&A) m-dôb, n-dôb (K&A) ndong (auctt.) WOLOF digor (JMD; K&A) dogut (JGA) dugor (auctt.) dugor ianuri (auctt.) ndôg (JGA) **THE GAMBIA** FULA-PULAAR (The Gambia) dukumi (DRR; DAP) MANDING-MANDINKA sunkuŋ (*def.*-o) (auctt.) WOLOF diorgud (JMD) dorgot (JMD) jorgut (DAP) **GUINEA-BISSAU** BALANTA bórè (JDES) BIDYOGO bole (JDES) CRIOULO mambumba (JDES) FULA-PULAAR (Guinea-Bissau) ducume (JDES) MANDING-MANDINKA sucum (*def.*-ô) (JDES) sunkung (*def.*-ô) MANDYAK benempe(-le) (JDES) benotàro (JDES) PEPEL sampane (JDES) **GUINEA** FULA-PULAAR (Guinea) dukume (Bouronville; JMD) SUSU *mété* (CHOP; FB) sugni (Farmar; FB) **SIERRA LEONE** MANDING-MANDINKA walisa (FCD) SUSU-DYALONKE kɔrɛtɛ-na (FCD) TEMNE *a*-mɔmina (JMD) **MALI** DOGON kunu (Rogeon) MANDING-BAMBARA dangan (FB) mandi (A. Chev.) sunsun (FB) tangasu (FB) MANINKA (Wasulunka) nugu-méné (JMD) **UPPER VOLTA** DAGAARI bar oda (K&B) GURMA lubualansanlu (K&B) KIRMA tobéré (K&B) MANDING-BAMBARA dan (K&B) danguan (K&B) mandé sunsun (K&B) DYULA lommo (K&B) n-dara (K&B) n-dara lommo (K&B) numbulombo (K&B) MOORE karoduidiga (K&B) NYENYEGE bondolo (K&B) SONGHAI-ZARMA mufania (K&B) 'SENUFO' damugurumo (K&B) damurana (K&B) namurga (K&B) **IVORY COAST** AKAN-BRONG boboma (K&B) tien (K&B) BAULE amlon (K&B; A&AA) amron (K&B; B&D) KPALAGHA gala (K&B) n-beni (K&B) KULANGO kuggo (K&B) kumo (K&B) soropodio (K&B) KWENI bré (K&B) LOBI kontakpé (A&AA) MANDING-DYULA lommo (K&B) n-dara (K&B) n-dara lommo (K&B) numbulombo

(K&B) MANINKA karamoko sunzun (A&AA) sunsu (K&B) sunsun sunzu (K&B) SENUFO-TAGWANA m'ru (K&B) muro (K&B) DYIMINI gbanessan (K&B) 'SENUFO' damugurumo (K&B) damurana (K&B) namurga (K&B) **GHANA** AKAN-BRONG aboboma (FRI) abɔma (FRI) TWI aboboma (BD&H) saa-borɔfere (FRI) BAULE amomon (FRI) DAGAARI batanga (AEK; FRI) bulɔmbuxo (FRI) bulumbogo (AEK; FRI) ulungkwana *from* kwana: *leprosy* (FRI) GBE-VHE anyiklɛ (FRI) GUANG-GONJA púrù púshè (Rytz) NANKANNI bangoora (Lynn) SISAALA baawuluŋbiiŋ (AEK; Blass) WALA batani (AEK) **TOGO** GBE-VHE anigli (E&D; FB) GURMA (Mango - Jendi) suku (Mellin) **NIGERIA** ARABIC-SHUWA umm boro (JMD) BEROM ewura (LB) FULA-FULFULDE (Nigeria) boili (MM) cika kondoyi (MM) dukkuhi (*pl.* dukkuuje) (MM) dukkuhi ladde (*pl.* dukkuuje ladde), ladde: *bush vegetation* (J&D; MM) dukkumbe ladde (JMD; MM) GOEMAI ououd (JMD) GWARI kmijirihi (JMD) HAUSA gwándàr daájìi, daájìi: *bush vegetation* (auctt.) gwándar jééji (LB) gwangwale *the flowers* (JMD; ZOG) ján-bàroódó (ZOG) kaabuse ladde = *ladde of the bush; from Fula* (MM) HAUSA (West) tàllâfà màraàyú (ZOG) IDOMA uwu (Odoh) IGALA ukpokpo (Odoh) IGBO uburu-ọcha (JMD; KO&S) KANURI ngónówù (JMD; C&H) TIV anyam hul (JMD; KO&S) hul (JMD) YEKHEE oguoto (Kennedy) YORUBA àbo (KO&S) YORUBA (Ilorin) arere (JMD) **WEST CAMEROONS** 'BAMILEKE' falõ (Johnstone)

A small tree or shrub of the savanna of up to 6 m high, from Senegal to Nigeria, in the Cape Verde Islands and across Africa to Sudan. It is represented by two varieties throughout the Region, var. *senegalensis* and var. *deltoidea* Robyns & Ghesq., the latter seemingly the more abundant. Both varieties are considered together here.

The tree is fire-resisting, sprouts strongly from the stump and sends up root-suckers (17) (cf. *A. glauca* Schum. & Thonn.). The stems are occasionally used in hut-building and for hoe-handles (6). In Sudan the wood is used as firesticks (8). The wood ashes are used in snuff-making (8) and the soluble portions in making soap (8). The wood is mixed with tobacco in a chewing-cud for the potassium it contains and is much esteemed (3). The outer bark supplies a fibre in The Gambia (16).

In medicine the bark of the trunk, branches and roots is extensively used. Bark as well as the root is administered to horses as a vermifuge, and with the powdered leaves and mixed with the latices of *Calotropis procera* (Asclepiadaceae) and *Euphorbia balsamifera* (Euphorbiaceae) are applied as a dressing for the condition known as 'yaws' in horses (epizootic lymphangitis) (6). Stem- and root-barks are used for diarrhoea and gastro-intestinal troubles (13–15).

The plant is used (part unspecified) in Senegal to relieve chest-complaints, and cough and diarrhoea in infants (10), and Fula peoples after a delicate preparation obtain a cough-medicine from the young branches (13). Root-bark boiled with natron is used by Hausas for intestinal troubles in a treatment that is without purging (6). Bark is sold in Senegal markets and is chewed for stomach-ache, and with other barks makes an infusion for the treatment of skin-eruptions (6). A preparation is used in lotion, and is taken in draught for leprosy in Mali (5). This causes vomiting and diarrhoea. A bark infusion forms a mouth-wash to relieve toothache, and in the Zambesi region root-bark is used as an antidote to snake-bite (6). Leafy twigs and bark are used by Fula of Senegal against sterility. They also consider the bark to be a galactogogue for both humans and animals (13). Bark from which all the old corky material has been removed is taken in the Casamance of Senegal as a macerate for diarrhoea and passing of blood or mucus; a bark-macerate is also used to treat leprosy and sterility. Such a preparation has a marked diuretic effect (11).

On the Ivory Coast the Lobis take by mouth an extract of young twigs for dysentery, or take by mouth or by enema root-bark powdered with a pimento and suspended in water (2).

The roots are sold in Hausa markets in N Nigeria and are used in the treatment of venereal diseases, the patient feeding for five or six days on a pap made by boiling the root with guinea-corn meal and native natron (6). In the Fula areas of Senegal the root is considered to be diuretic and is used to treat blennorrhoea, mumps, etc. (13). The root is said to be used as an insecticide, and to be vermifugal (6, 15) and good for internal complaints and babesiosis (piroplasmosis) (3) in horses for which the finely ground roots after sun-drying are given mixed with millet or added to drinking water. Earlier analyses (6) of the powdered root showed no active ingredient but some mucilage and a little tannin, but later work (15) on the anthelmintic action on horses has shown the presence of a wax, glucides, glycosides, proteins and amino-acids. The wax is separable into two fractions, one hard and one soft; the latter may have larvicidal properties.

Pulverized roots are applied to sores (6) and in Senegal roots are used as a remedy for dysentery and diarrhoea (6) and for enteralgias with constipation and lack of urine (14). In the Casamance medicine-men use the roots on a sort of anthropomorphic simile to treat ailments of the feet and legs for which they administer an aqueous macerate in draught and as a friction. For kidney and joint pains of the upper parts of the body, leaves often with leaves of other plants, are added to the roots to make a paste which is applied topically (11).

In N Nigeria the roots are an ingredient of a Fula arrow-poison (9). These people and the Hausa use the roots (? and stems) to cure a disease known as *nyau ladde* ('disease', or 'sickness of the bush', Fula) symptomized by a swelling which may turn into a boil (18). The plant has many other uses so that the Fula respect it: the dry stems are not burnt as such action would show ungratefulness and unfaithfulness to nature. There is a strong belief that should the plant be burnt in a house there will be an outbreak of *nyau ladde* (18).

Leaves are considered to be a good strengthening food for humans and for horses. The translation of a Hausa name is 'sustenance of orphans' (7). The Maninka of Ivory Coast use a leaf-decoction in baths or by mouth as a powerful tonic (2). Leaves are readily grazed by horses, donkeys and cattle. In Senegal they are fed to horses after illness (1). Their beneficial tonic effect is recognized in N Nigeria where they are commonly sold in Hausa markets for feeding to horses in bolus form with natron and bran, to which the sludge of spent indigo may be added, or mixed with cereal cakes. Leaves are also a remedy for worms in horses, and for mucal diarrhoea, being given daily with large draughts of water (6). Dry powdered leaves are a medicine for human diarrhoea in Senegal (14). The Basari of Senegal use a water decoction of equal parts of leaves of this plant with those of *Icacina senegalensis* (Icacinaceae) in washes and massages for illnesses 'that come with the wind in the dry season' (12), i.e., pneumo-bronchial affections arising during the Harmattan. In N Nigeria an eye-lotion is made from a leaf-infusion (6).

The flowers are used to flavour food (6). The fruit, yellow when ripe and about 5 cm long, is edible and of a pleasant flavour (6). It has a sweet clear yellow jelly (4). Consumption of a large amount of the fruit is believed to have a favourable influence in the treatment of guinea-worm (6).

Seeds, bark and leaves have been found to contain *anonaine,* a minor alkaloid of the aporphine group, which on injection into dogs produced hypotension (15).

The plant is in general considered a panacea for most illnesses acquiring thereby not a little medico-magical attribute. In Kontagora (N Nigeria) the root along with that of *Ximenia americana* (Olacaceae) has been found as an ingredient of a treatment for sleeping-sickness. The plant, however, does not seem to be a specific, but is held in superstitious esteem by Muhammedans and is used as a charm (6). In eastern Senegal the plant has a general reputation with sorcerers for casting spells (14). Fula medicine-men follow a complicated prescription in treating diarrhoea and dysentery in which bark is taken only from the left fork of a V-shaped branch on the east side of a tree, i.e. the side facing the rising sun (13).

References:

1. Adam, 1966, a. 2. Adjanohoun & Aké Assi, 1972: 24. 3. Aubréville, 1950: 43. 4. Berhaut, 1967: 243. 5. Chevalier, 1937, b: 169. 6. Dalziel, 1937. 7. Irvine, 1952, b. 8. Irvine, 1961: 4–5. 9. Jackson, 1973. 10. Kerharo, 1967. 11. Kerharo & Adam, 1962. 12. Kerharo & Adam, 1964, a: 406. 13. Kerharo & Adam, 1964, b: 406–7. 14. Kerharo & Adam, 1964, c: 293. 15. Oliver, 1960: 19, 46. 16. Rosevear, 1961. 17. Unwin, 1920: 262. 18. McIntosh, 26/1/79: in litt.

**Annona squamosa** Linn.

FWTA, ed. 2, 1: 52. UPWTA, ed. 1, 3.

English: sweet sop, (scaly) custard apple, sugar apple, sweet apre (Ghana, from Twi, Irvine).

French: pomme-channelle.

Portuguese: fruta pinha.

West African: SIERRA LEONE KRIO switi-sap = *sweet sop* (FCD) TEMNE *a*-mɔmina (JMD) GHANA ADANGME hãbue = *charcoal pot, from the appearance of its dried fruit* (FRI) AKAN-ASANTE aprɛ (Enti) FANTE aprɛ *i.e., 'apple'* (FRI) TWI aprɛ, aperɛ *i.e., apple* (FRI) borɔfo nyankõma = *European nyamkoma* (nyankoma = *Myrianthus*) (FRI) DAGBANI bulumbogo (FRI) GA aprɛ (FRI) kplɛ́mi (KD) ŋawie (FRI) GBE-VHE apre *i.e., apple* (FRI) yevunniklẽ (FRI)

A small tree reaching 6 m high, a native of the West Indies and now pan-tropical. It does best in areas with an alternation of wet and dry seasons and prefers a dry situation. It has been introduced to the West African region as a cultivate and in places it has become naturalized.

The leaves, bark and roots contain hydrogen cyanide with traces in the wood and seeds. The root is a drastic purge, but contrarily the plant can be used to stop diarrhoea when administered as an enema of the unripe fruit (1, 6). Tannin has been recorded in the roots and absent from other parts (5). Scrapings of root-bark may be used on the gums for toothache (1). An alkaloid in the roots is recorded (1) and *anonaine* in the bark (6). Various parts of the plant have insecticidal, vermicidal and piscicidal properties. The leaf is used in The Gambia as an insecticide and to prevent bed-bugs (2). Powdered seeds are used in Gabon to kill fleas (7), and are used in the Philippines, China and elsewhere as a parasiticide and insecticide especially against body and head-lice. A high potency is ascribed to the presence of a glyceride or glycerides of a hydroxylated unsaturated acid or acids of high molecular weight, but this is said to be less than that of DDT (8). The leaf is used in India as a fish-poison and vermicide (8). The seed and green fruit are regarded as vermicidal, and the latter is insecticidal (8). The root is used in tropical Africa as an insecticide (8).

The leaves are used in various preparations for treating cancerous tumours in many parts of the world (4, 7). They are applied as a poultice to children with dyspepsia (6), and are boiled and applied to abscesses in India and Java, on insect-bites in India and skin-complaints in Java (1). A decoction is taken in Tanganyika for diabetes (3). An unidentified alkaloid is recorded in the leaf (9) and also a pleasant-smelling green oil is found in small quantity (5, 8).

The fruit is subspherical to about 10 cm diameter, contains a number of black seeds each surrounded by a white, sweet, juicy flesh which is edible raw and is the most delectable of the *Annona* fruits. But consumption must be delayed till the fruit is completely ripe for its astringency lingers till the very last moment, a condition indicated by the disappearance of the very dark yellow droplets of latex from the fruit shell. The ripe pulp can be successfully made up into an ice-cream. In the West Indies the fermented fruit is used to make a kind of cider (1). The flesh of the fruit contains up to 10% sugar, mostly glucose with some fructose (1). The unripe fruit is astringent like the bark and leaves and has vermicidal and insecticidal properties. It is given in treatment for diarrhoea, dysentery and atonal dyspepsia (6).

The seeds contain highly toxic substances. There is a 45% content of a yellow, non-drying oil which may be suitable for soap-manufacture (8). The seeds also contain 0.56% of a neutral resin. Powdered seeds are used in certain countries on the hair to kill lice, but this operation requires care as the resin, which contains an acrid principle, causes great pain if in contact with the eyeball (1, 8). The powdered seed can also be used as an abortifacient by application to the os uteri (8).

There are conflicting reports on antibiotic activity of the plant, some showing a positive action on *Staphylococcus aureus* and *Escherichia coli* and others a negative (5).

The plant is an important host for the lac insect.

References:

1. Burkill, 1935: 168–9. 2. Dalziel, 1937. 3. Haerdi, 1964: 37. 4. Hartwell, 1967. 5. Kerharo & Adam, 1974: 148–9. 6. Quisumbing, 1951: 310–11, 1031. 7. Walker & Sillans, 1961: 63–64. 8. Watt & Breyer-Brandwijk, 1962: 60. 9. Willaman & Li, 1970.

**Anonidium mannii** (Oliv.) Engl. & Diels

FWTA, ed. 2, 1: 51, incl. *A. friesianum* Exell. UPWTA, ed. 1, 3.

West African: **GHANA** AKAN-ASANTE asumpa (auctt.) **NIGERIA** BOKYI keche buchu (Catterall; KO&S) EDO ọghẹ́dẹ́gbó (auctt.) IGBO imido (DRR; KO&S) YORUBA ewûrò-igbó (KO&S)

A lower and middle storey tree of the lowland dense rain-forest, to 20 m high by 2 m girth, occurring from Ghana to West Cameroons and to Gabon and Zaïre.

The bark is used as an aqueous decoction in Congo (1) for gastro-intestinal affections, dysenteriform diarrhoeas and ovarian troubles, and sometimes for coughs; bark scrapings passed over a fire, wrapped in a Marantacea leaf are applied as a poultice for feverish pains, oedemas, rheumatism, or put in a little boiling water the vapour may be applied; powdered bark is put on sores, to maturate buboes, on snake-bite, and is given to epileptics to eat or to ill persons suffering from giddiness. In Gabon a macerate of bark-raspings is given as an enema for colic (3).

A trace of alkaloids is reported present in the leaves, bark and roots (2).

The fruit is well-fleshed, is edible and has a sweet-sour taste (3).

The tree is ascribed in Congo with magical properties to ward off evil spirits and ghosts, quell nightmares, and to protect a house and its occupants: all that is necessary is to sprinkle a bark-decoction about the house, to put some leaves in the roof or a small piece of trunk under the threshold. Washing in the decoction will confer individual protection (1).

References:

1. Bouquet, 1969: 57. 2. Bouquet, 1972: 13. 3. Walker & Sillans, 1961: 64.

**Artabotrys hispidus** Sprague & Hutch.

FWTA, ed. 2, 1: 41.

West African: **SIERRA LEONE** MENDE mɔigbama

An uncommon species recorded only from Sierra Leone and closely related to, and perhaps not distinct from, *A. velutinus, A. dahomensis* and *A. stenopetalus*.

**Artabotrys insignis** Engl. & Diels

FWTA, ed. 2, 1: 40.

West African: **SIERRA LEONE** MENDE mɔigbama (NWT) **GHANA** AKAN-ASANTE akoo-ano (FRI)

A liane of deciduous forest, recorded from Sierra Leone to Ghana, and in Cameroun, Gabon and Zaïre.

**Artabotrys stenopetalus** Engl. & Diels

FWTA, ed. 2, 1: 41.
West African: **GHANA** ADANGME nyotʃo (A.S.Thomas) AKAN-ASANTE akoo-ano (FRI)

A liane of the evergreen, deciduous and secondary forest areas in Ghana to S Nigeria, and in Cameroun, Gabon and to Zaïre.

No use is recorded for the Region. In Zaïre, the leaves are eaten for treatment of enlargement of the spleen, the twigs enter into a prescription to promote conception and the sap is considered to be aphrodisiac (1).

The flowers are strongly scented — cf. *A. odoratissimus* R. Br. a species of the Gabon which is cultivated there for its flowers (2).

References:

1. Bouquet, 1969: 58. 2. Walker & Sillans, 1961: 64.

**Artabotrys thomsonii** Oliv.

FWTA, ed. 2, 1: 40.

A scandant liane attaining 30 m length. Recorded only from West Cameroons in the Region but extending into Cameroun, Gabon and Zaïre.

No use is recorded for the area. In Zaïre it is known for its capacity to yield water (2). In Congo the sap is held to be an aphrodisiac, and the stem enters a prescription to promote conception. The leaves are eaten in a treatment for enlarged spleen (1).

References:

1. Bouquet, 1969: 58. 2. Boutique, 1951: 314, 316.

**Artabotrys velutinus** Sc. Elliot

FWTA, ed. 2, 1: 40. UPWTA, ed. 1, 3.
West African: **GUINEA** FULA-PULAAR (Guinea) boilé (JMD) **SIERRA LEONE** MANDING-MANDINKA jiri-madaŋba (FCD) SUSU-DYALONKE wudi-madaŋba-na (FCD) **GHANA** AKAN-ASANTE akoo-ano (FRI)

A scandant shrub recorded from Guinea to northern Nigeria.

A fairly strong presence of alkaloids has been recorded in the leaves (1).

Reference:

1. Bouquet & Debray, 1974: 19.

**Cananga odorata** (Lam.) Hook. f.

FWTA, ed. 2, 1: 35.
English: ylang-ylang (trade): perfume tree (Liberia, Baldwin Jr. 6145, K).
French: ylang-ylang (trade).

A medium to large-sized tree, native of southeast Asia, Tenasserim to Australia, widely introduced to the Region and to many tropical countries for cultivation for its perfumed flowers. The generic name comes from various Malesian vernaculars. The trade name 'ylang-ylang' occurs in some Philippine languages and is said to mean *something which flutters,* from the flag-like petals of the flower (1).

The timber is white to grey and not durable, and is used in SE Asia for small household objects. It has a resonance making it suitable for tom-toms. Trees have been planted in Ghana as a village shade tree (2).

The bark yields a fibre made into coarse ropes in Celebes. The leaves are aromatic and are rubbed on the skin for itch in Malaya and the bark against scurf in Java (1).

Wherever the tree is grown the flowers are valued by women for preparing a scent. In several countries commercial production of oil distilled from the flowers has been undertaken. Two fractions are normally obtained, a light one known as Ylang-ylang oil which is the more valuable, and a heavy fraction called Cananga oil. Climate may possibly have some influence on the value of the oils obtainable. West African material needs investigation.

References:

1. Burkill, 1935: 422–4, as *Canangium odoratum* Baill. (giving several references). 2. Enti, 588, K.

**Cleistopholis patens** (Benth.) Engl. & Diels

FWTA, ed. 2, 1: 38, 757. UWTA, ed. 1, 3, incl. *C. klanieana* Pierre.

English: 'salt-and-oil' tree (from Ghanaian names alluding to the taste of the bark: Burtt-Davy; Irvine; Taylor).

Trade: otu (Sierra Leone, Savill & Fox, 1967).

West African: **SIERRA LEONE** KISSI siopiando (S&F) KONO fubame (S&F) MANDING-MANINKA (Koranko) karakil-kɛnɛ (S&F) MENDE mɔigbama, mɔigbwama (*def.* mɔigbamei) *to this and several other Annonaceae* (auctt.) TEMNE *am*-bobəl (FCD; S&F) *am*-bobɔl (FCD; S&F) *am*-bók *not to be confused with am-*bòk: *snake* (auctt.) **LIBERIA** KRU-BASA nee-wahn-johr = *tree which gushes water* (C&R) MENDE mɔigbwama (C&R) **UPPER VOLTA** MANDING-DYULA koroguio iri (K&B) **IVORY COAST** ABE sobu (Aub.; K&B) ABURE owua (B&D) AKAN-ASANTE autié (B&D) ebitié (B&D) eutié (B&D) FANTE bofu (A. Chev.) AKYE botopuo (Aub.; K&B) potopu (A. Chev.) potopuhun (A&AA) BAULE kulima dèma (B&D) sonbuin (B&D) KRU-GUERE (Wobe) pahuéko (A&AA) KWENI bokubué (Aub.) KYAMA aguto (Aub.; K&B) héré (B&D) MANDING-DYULA koroguie iri (K&B) SENUFO-TAGWANA lossion (K&B) **GHANA** AKAN-AKYEM fifiriwa (FRI) ASANTE afirifiriwa (FRI) ngo ne nkyene = *oil and salt* (FRI) nkyene ne ngo = *salt and oil* (FRI) TWI ngo-ne-nkyene (DF) nwo-ne-nkyene = *oil and salt* (CJT) WASA ngo-na-nkyene (Enti) obitie *not commonly used* (FRI; CJT) wisa ne nkyene = *pepper and salt* (FRI) ANYI-AOWIN ewutie (FRI) fotie (FRI; CJT) SEHWI aninkyeri (CJT) ewutie (FRI; CJT) NZEMA aheri (CJT) ebitie (FRI) NZEMA-SEHWI fotie (FRI; CJT) **TOGO** TEM (Tshaudjo) baledia (Volkens) buledia (Volkens) welengele (Volkens) YORUBA-IFE OF TOGO aru (Volkens) muso (Volkens) **NIGERIA** EDO ótù (JMD; KO&S) IGBO ójō (KW) IJO-IZON (Kolokuma) pàá (Kennedy; KW) IZON (Oporoma) párá (KW) YORUBA apakọ́ (auctt.) YORUBA (Ijebu) ọkẹ (JMD)

A tree reaching to 30 m high, occurring from Sierra Leone eastwards into Uganda and Zaïre; sun-loving, fast-growing, common in disturbed forest areas, rapidly colonizing abandoned areas.

The bole is slender, cylindrical and straight, and the timber is straight-grained, light-coloured, soft and light, a bit woolly-textured. It is not durable. It can be easily cut and finished smoothly. Both sap- and heart-wood have a sheen. The wood generally has properties similar to those of balsa (*Ochroma sp.* Malvaceae) and it should be suitable for the same purposes (6, 12, 14).

In Liberia tree trunks are used to float heavy timber (6). In Ghana it has been used as floats, to make drums, canoes and roof-beams (9), and is said to be suitable for joinery (3, 9). In Nigeria it is used to make canoes and has been called 'canoe wood' (7), and it is also so used in Gabon (16). At the present time there appears to be no commercial use for the timber. It may be of value as plywood (9). Its capacity for very rapid growth — to 13 m height by 70 cm girth in 7 years recorded in Sierra Leone (12) — might lend it to exploitation for pulp or cellulose production.

The leaves are chewed with kola, and a leaf-infusion is taken in draught for fever in Sierra Leone (13) and in Gabon (16). In Ghana, a leaf-infusion, sometimes with lemon grass, papaya or other plants, has been used for infective hepatitis (9), and merits fuller investigation. Leaves have been used as a vermifuge in many territories (6, 7, 12), and as a vermifuge and febrifuge in Gabon (16). In the Upper Ivory Coast a leaf-decoction is given in draughts and washes for sleeping sickness (10).

The bark is very fibrous and peels readily from the wood. The inner-bark yields a lace-like material which is remarkably strong, and is made into matting onto which cacao beans are spread for drying (8, 9). Cordage and matting are made from the bark throughout the tree's distribution. In Sierra Leone strips of the bark are used for the brow-bands and shoulder-straps used by natives when carrying palm hampers (11). Basket straps are also made in Congo (1). Some tribes in southern Nigeria (7) and in Gabon (16) use the bark to make hut-walls and partitions. In Ghana the bark was formerly used to make sandals (8, 9).

The slash emits a strong spicy scent. The bark and sap-wood are turgid with sap. The bark is said to yield water when cut (5) and the Bassa (Liberia) name means 'tree which gushes water' in allusion to this (6). The sap is reddish and looks like palm-oil and has a salty taste, hence the Akan (Ghana) names (9).

Sap from the pounded bark, or a bark-decoction is taken in draught for tuberculosis and for simple bronchial affections in Congo and pulp from crushed bark, with pieces of stem of *Costus,* is applied as a poultice on whitlows and oedemas (1). The bark is steeped in cold water and taken as a purgative in Sierra Leone (12) and to sooth colic in Gabon (16) and Congo (1). In the Ivory Coast a bark-decoction enters into a general treatment in draught, washes, vapour-baths and topical friction using the lees for hunchbacks and also for rachitic children, while bark-sap is a nasal instillation for headache (2).

The widespread use of the bark and leaves in local medicine merits investigation for the presence of active substances. However, bark tested for action on avian malaria showed no activity (4).

The seeds are used as beads (7).

The plant has magical applications. It features in an Yoruba *Odu* incantation for curing coughs, hence the Yoruba name *apakó* meaning 'killing cough' (15). In Congo the tree is planted in compounds to ward off thunder-bolts (1).

References:

1. Bouquet, 1969: 58. 2. Bouquet & Debray, 1974: 18. 3. Boutique, 1951: 302–4. 4. Claude & al., 1947: 74. 5. Cooper 430, K. 6. Cooper & Record, 1931: 13–14. 7. Dalziel, 1937. 8. Irvine, 1930: 108. 9. Irvine, 1961: 6–8. 10. Kerharo & Bouquet, 1950: 20. 11. Lane-Poole 206, SL. 12. Savill & Fox, 1967: 41. 13. Scott-Elliot 5698, K. 14. Taylor, 1960: 86. 15. Verger, 1967: sp. no. 36. 16. Walker & Sillans, 1961: 65.

**Cleistopholis staudtii** Engl. & Diels

FWTA, ed. 2, 1: 39. UPWTA, ed. 1, 3–4.

A medium-sized tree of S Nigeria and W Cameroons extending into E Cameroun.

The bark is used in E Cameroun for hut-walls, and for the support bands of carrying frames.

**Dennettia tripetala** Bak. f.

FWTA, ed. 2, 1: 51. UPWTA, ed. 1, 4.

English: pepper fruit (Nigeria, Okiy).

West African: **IVORY COAST** KRU-GUERE (Wobe) kloadjémiéhuon (A&AA) **NIGERIA** EDO ákò *incl. several Annonaceae* (auctt.) IBIBIO ǹkàríkâ, ǹkàríkà (auctt.) IGBO m̲mị̀mị̀ = *pepper fruit* (KO&S) YORUBA átá = *pepperfruit* igberi (AHU; KO&S)

A tree of the rain-forest and occasionally in the savanna, to 18 m high by 60 cm girth, of limited distribution in Ivory Coast, S Nigeria and W Cameroons.

The wood is white, soft and not durable, and is susceptible to termites.

Bark is fibrous and strongly scented (1).

The young leaves are chewed on account of their pungent spicy taste.

The fruits, green at first then turning red, ripen in April and May have a peppery spicy taste and are chewed for this property (1). The fruit is held to be a good source of vitamin (2).

References:

1. Keay & al., 1960: 46. 2. Okiy, 1960: 121.

**Enantia chlorantha** Oliv.

FWTA, ed. 2, 1: 51. UPWTA, ed. 1, 4 as *E. chlorantha* Oliv. in part; and *E. polycarpa* Engl. & Diels in part.

English: African yellow wood.

French: moambe jaune (trade name, from a Gabonese vernacular).

West African: **NIGERIA** BOKYI kakerim (Catterall; KO&S) EDO èrhẹ̀nbàvbógò (auctt.) YORUBA osopupa (KO&S) YORUBA (Ikale) osomolu (KO&S) **WEST CAMEROONS** DUALA bòmùke (Ithmann) njié (JMD) KPE woyoyo *implies 'bitterness'* (JMD) KUNDU bololo *implies 'bitterness'* (JMD)

An understorey tree of the high rain-forest attaining 30 m height by 70 cm girth with a long clear bole. Distribution within the Region is limited to S Nigeria, W Cameroons and Fernando Po, but extends to Gabon, Angola and Zaïre.

The wood is yellow, uniform throughout, fairly fine-grained, splitting easily and rather soft, turning brown after long exposure, taking a smooth polish. It is used for house-building, furniture and general joinery (3); wooden shovels have been made of it in Benin (7) and in Gabon, paddles, hut-poles and planks (5).

The outer bark is blackish, but the inner bark is bright yellow and is the source of a yellow dye used to dye cotton and other fibres. Pieces of bark are used in the Cameroons (3) and in Gabon (6) for sides, partitions and doors of huts. The fibrous bark is the source for a matting in Gabon (6) and for caps in S Nigeria (7).

The bark enters into numerous medical applications. It is used as a powder on sores in Nigeria (3) and Congo (2), and as scrapings on ulcers in Gabon (6). Bark-extracts are widely used as an antipyretic in treating fevers (4), sometimes mixed with other vegetable medicines (3), and on ulcers and as a haemostatic on wounds (4). In Congo a bark-decoction is taken for tuberculosis and bloody vomit, in baths for fatigue, and by mouth and vapour baths for rheumatism, intercostal pain and to promote conception (2). Bark-sap is taken in decoction in Gabon for excess of bile and by injection for diarrhoea (6). The bark may offer medical use as a uterus stimulant and antibiotic (4).

The bark is intensely bitter. Investigation on its effective curative action on sores, the very strong presence of alkaloids, and the presence of a substance of empirical formula $C_{20}H_{25}O_8N$, m.p. 196° has been reported (1). *Berberine* is the principal alkaloid present (3, 4, 5) to which the medical properties must be mainly due. It is also present in the wood conferring to it some resistance to insect attack.

In Congo hunters will wash in a bark-macerate before setting forth on a hunt to ensure meeting game and shooting straight (2).

References:

1. Adegoke & al., 1968. 2. Bouquet, 1969: 58–59. 3. Dalziel, 1937. 4. Oliver, 1960: 26. 5. Robyns & Ghesquière, 1933: 308–10. 6. Walker & Sillans, 1961: 65. 7. Unwin, 1920: 259–60, 427.

**Enantia polycarpa** (DC.) Engl. ex Diels

FWTA, ed. 2, 1: 51. UPWTA, ed. 1, 4, as *E. polycarpa* Engl. ex Diels in part, and *E. chlorantha* Oliv. in part.

English: yellow-wood (Chipp); African yellow-wood (Cooper & Record); African whitewood (Dalziel — in error for *Triplochiton scleroxylon* K. Schum., Sterculiaceae); Abeokuta bark, Kandia or Canta bark (both after places in S Nigeria, Dalziel).

West African: **SIERRA LEONE** KONO gbende-kaima (S&F) MENDE bɛlo-hina (*def.*-hinei) (FCD; S&F) mbela-wulo (*def.*-wuli) (DS) **LIBERIA** KRU-BASA sohn = *hand; from the several-stalked fruits* (C&R) songh (C&R) MENDE kpaini (C&R) **IVORY COAST** ABE baué (Aub.; K&B) g-bawé, m-bawé (auctt.) AKAN-ASANTE silikokoré (B&D) AKYE esuro (A. Chev.) ts(a)in (auctt.) ANYI chibo (A. Chev.) chibo okéré (Aub.) essulo (Aub.) KRU-GUERE suin (K&B) GUERE (Chiehn) bi (K&B) sahui (K&B) GUERE (Wobe) sô (A&AA) KYAMA atinhia (Aub.; K&B) **GHANA** VULGAR dua-sika (DF) AKAN-ASANTE dua sika = *tree of gold, from the colour of the bark-slash* (Enti) dua sika ɔkoduben = *imitation oduben* (*Harungana*) (FRI) TWI dua sika = *tree of gold* (JMD) dubima (CV; FRI) WASA dubima (DF; FRI) dubuma (auctt.) ANYI esulo (FRI) kyibo okere (FRI) **NIGERIA** EDO ọviẹn-èrhẹ̀nbàvbógò (Amayo) YORUBA gbeyido (JHH; AHU) osopupa (KO&S) yaru *probably this sp.* (AHU)

An understorey tree of the high rain-forest to 20 m high by 1 m girth, distributed from Sierra Leone to W Cameroons. In the Region the genus is represented by this species and *E. chlorantha* only. They are closely alike in characters and uses, and there has been confusion over identity. The only overlap of their respective distributions is in SE Nigeria and W Cameroons. References therefore to *E. chlorantha* in Liberia (6) and in Ghana (3) must be in error for *E. polycarpa*.

The wood is yellow and very soft. The colour fades to brown on exposure (5). The wood is sought after for hut-construction in the Ivory Coast and is valued because of its resistance to insect-attack and because it can be easily split by axe into planks (1). It is extensively used in Ghana for building huts (4, 12), though curiously in the Dukwai River area of Liberia, there is said to be no use for the wood (5). In Ghana it is used to make beads (3).

The wood, and more especially the bark, yield a yellow dye which is used for dyeing cloth (1), and leather and mats (7). The dye of a bark-extract is self-mordanting on skins to a light yellow shade (9). The Abés of the Ivory Coast, however, use it to dye cotton and raphia a madder-colour (10), a change of colour which may be akin to the fading of the wood from yellow to brown on exposure.

The bark is bitter, and a decoction is said to have an antiseptic action for which it is used in the Ivory Coast to wash sores and on ulcers and leprosy blotches (9) and also to treat ophthalmias (2). It is used for sores and skin-infections in Liberia (6). In Sierra Leone a bark-decoction is used on ulcers and taken for jaundice ('yellow fever') (11).

The Guéré of the Ivory Coast often incorporate a bark-extract into arrow-poison (9).

The bark of the aerial parts and of the roots have been shown to have a strong presence of alkaloids (2) and *berberine* has been isolated.

References:

1. Aubréville, 1959: 1, 124. 2. Bouquet & Debray, 1974: 19. 3. Burtt-Davy & Hoyle, 1937: 5 as *E. chlorantha* Oliv. and *E. polycarpea* Engl. & Diels. 4. Chipp 233, K. 5. Cooper 146, K. 6. Cooper & Record, 1931: 14 as *E. chlorantha* Oliv. 7. Dalziel, 1937. 8. Irvine, 1961: 8–9. 9. Kerharo & Bouquet, 1950: 20. 10. Robyns & Ghesquière, 1933: 311–12. 11. Savill & Fox, 1967: 40. 12. Vigne 908, K.

**Friesodielsia gracilis** (Hook. f.) van Steenis

FWTA, ed 2, 1: 45, as *Oxymitra gracilis* (Hook. f.) Sprague & Hutch.

West African: **SIERRA LEONE** MENDE bɔlɛ (NWT) TEMNE *an*-sali (NWT)

A small liane recorded from Sierra Leone to S Nigeria.

**Greenwayodendron oliveri** (Engl.) Verdc.

FWTA, ed. 2, 1: 43, 757, as *Polyalthia oliveri* Engl. UPWTA, ed. 1, 6–7, as *P. oliveri* Engl.

West African: **SIERRA LEONE** LOKO beyiŋgi (NWT) MENDE fakaiwulo (*def.*-wuli) (S&F) gbelo, kpelo (*def.*-i) (auctt.) gbelo-hina (*def.*-hinei) (SKS) **LIBERIA** KRU-BASA pe-ohn (C&R) **IVORY COAST** ABE baué fu (Aub.) KRU-GUERE puluhé (Aub.) KYAMA dambrohia (Aub.) mfainhiabro (Aub.) **GHANA** VULGAR duabiri (DF) dubiri (DF) AKAN-ASANTE duabiri = *black tree; from the colour of the twigs* (auctt.)

A small tree to about 13 m high with a clean straight bole, of the high rain-forest from Sierra Leone to Eastern Nigeria. *FWTA*, ed. 2, refers to an extension of this species into Gabon, but according to the *Flore du Gabon* (7) this species is not represented there.

The stems are used for the framework of native houses in Liberia (4). The timber is suitable in short lengths for purposes requiring strength and resilience. It is attractively figured when quarter-sawn (5).

The bark is fibrous and scented (1). The Mendes of Sierra Leone use a decoction as a cure for blackwater fever (3), and at Joru the herbalists prescribe it as a wash and to take by mouth for 'yellow fever' (6).

Stem-bark has shown a fairly strong presence of alkaloids (2).

References:

1. Aubréville, 1959: 146 as *Polyalthia oliveri* Engl.. 2. Bouquet & Debray, 1971: 20 as *P. oliveri*, Engl. 3. Burbridge 431, K. 4. Cooper 62, K. 5. Cooper & Record, 1931: 16 with timber characters, as *P. oliveri* Engl. 6. Herbalists, Joru Village, Sierra Leone, Nov. 1974. 7. Le Thomas, 1969: 204 as *P. oliveri* Engl.

**Greenwayodendron suaveolens** (Engl. & Diels) Verdc., **var. gabonica** (Le Thomas) Verdc.

FWTA, ed. 2, 1: 43, as *Polyalthia suaveolens* Engl. & Diels. UPWTA, ed. 1, 7, as *P. suaveolens* Engl. & Diels.

West African: **NIGERIA** BOKYI nchua (KO&S) EDO ẹ̀wáé (auctt.) ISEKIRI eleku (Kennedy; KO&S) okeren (Kennedy) URHOBO atorewa (JMD) osharo (KO&S) YORUBA agudugbu (KO&S) awuje (JRA)

A tree of the high-forest reaching to about 20 m height in S Nigeria to where it is limited within the Region, but attaining appreciably more in the eastern part of its range to Uganda and Angola.

The bole is straight and cylindrical; sap-wood is yellowish-white, darkening on exposure, and the heart-wood is brown. The timber is hard and resistant to borers and termites. It is valued in Gabon (6) for joinery, furniture-making, pit-props and house-construction, and in the Congo, where it is sometimes planted (5). It is used in S Nigeria for posts, rafters and planks, hence the Edo name, *ẹ̀wáé*, meaning 'house-post' (7). It has been considered suitable for cooperage (5), and in Gabon it is used for the shafts of fishing and hunting-spears (8).

The bark is scented and fibrous, the smell being likened to that of balsam (4, 8). It is considered purgative and is used in Congo either as an aqueous extract or as a powder for constipation and hernia, and also to facilitate childbirth and to make sterile women fertile (2). In Gabon it is used for stomach-aches (6).

The roots have a high reputation in Congo as an aphrodisiac and vermifuge, especially those of young plants not over 2–3 cm in diameter. A length of root heated at one end will exude a froth at the other end and this is used in Congo on inflamed oedemas and buboes (2). Alkaloids are reported to be abundant in the bark and roots, and saponins in the latter (3).

Pulped leaves or bark mixed with the seeds of *Aframomum meleguetta* (Zingiberaceae) and palm-oil are applied to areas of feverish and rheumatic pain in Congo, while juice from such a preparation is used as a nasal instillation for headache (7). In Nigeria the leaf has been recorded being taken internally for menorrhagia (1).

In Gabon leaves boiled up with those of other plants are used in ritual baths to cleanse maledictions brought by witch-doctors against crop-thieves (8).

The fruit is edible.

References — all as *Polyalthia suaveolens* Engl. & Diels:

1. Ainslie, 1937: sp. no. 286. 2. Bouquet, 1969: 60. 3. Bouquet, 1972: 13. 4. Boutique, 1951: 339–40. 5. Eggeling & Dale, 1952: 20–21. 6. Le Thomas, 1969: 202–6. 7. Unwin 413, fide Dalziel, 1937. 8. Walker & Sillans, 1961: 70.

**Hexalobous crispiflorus** A. Rich.

FWTA, ed. 2, 1: 47–48, 757. UPWTA, ed. 1, 4.

West African: **GUINEA** FULA-PULAAR (Guinea) boilé bonno (JMD) **SIERRA LEONE** MENDE mɔigbama *appl. to several Annonaceae* (JMD; Boboh) nja-lahɛ (*def.*-lahɛi) (FCD; S&F) SUSU-DYALONKE kodoyagi-na (FCD) **IVORY COAST** AKYE chienlébé (Aub.) siélébé (Aub.) **GHANA** AKAN-ASANTE agrada *referring to the wood splitting easily* (Enti) endwa (BD&H) etwa prada = *it splits easily* (FRI) **DAHOMEY** YORUBA-NAGO akpodo (Aub.) **NIGERIA** IGBO oji ogoda (Kennedy; KO&S); cf. ọ́jị̀: *Chlorophora excelsa* (*Moraceae*)*: iroko* YORUBA apárá *joke* (Verger; JMD) lapawe (KO&S)

A tree of the dense and fringing forest, attaining up to 30 m height by 1.70 m girth, occurring from Guinea-Bissau to S Nigeria, and across Africa to Sudan and southwards to Angola.

The wood is white or pinkish, open-grained, not very dense and is easy to work. It has an attractive appearance and though it splits easily (hence the Asante name) it is used in joinery. It is also used for house-beams (8). In Dahomey the Nago people use the wood to make knife-handles (1) and in Gabon it is made into paddles and tiles (12), while in Zaïre the flutings of the base of the trunk are used to make gun-butts (3).

The outer bark is dry and stringy and cracks longitudinally. While young it is stripped off for fibre in Sierra Leone (7). The inner bark is soft and yellowish, rapidly turning to orange on exposure (9). It is slightly scented (2). It is used as a masticatory with kola in Sierra Leone (7). A decoction is used as a vapour-bath for feverish children, and in draught and baths for skin-troubles in the Ivory Coast (6). In Gabon a bark-macerate is taken against venereal disease (12, 13). Freshly pulped bark is applied as a wet dressing on wounds, buboes and furuncles in Congo (4).

The bark, and roots, are reported to contain some saponin, but no alkaloid (5).

The fruit is edible, and is said to be stored by drying in Calabar, S Nigeria (10).

The plant enters into an Yoruba Odu incantation against eye disease under the name *àpárá,* meaning 'a joke' (11).

References:

1. Aubréville, 1950: 38. 2. Aubréville, 1959: 129–30. 3. Boutique, 1951: 372. 4. Bouquet, 1969: 59. 5. Bouquet, 1972: 13. 6. Bouquet & Debray, 1974: 18. 7. Deighton 3140, K. 8. Irvine, 1961: 9–11. 9. Le Thomas, 1969: 82. 10. Ujor, E. FHI 27994, K. 11. Verger, 1967: 37. 12. Walker, 1952: 186. 13. Walker & Sillans, 1961: 66.

**Hexalobus monopetalus** (A. Rich.) Engl. & Diels var. **monopetalus.**

FWTA, ed. 2, 1: 48. UPWTA, ed. 1, 5.

var. **parviflorus** Bak. f.

FWTA, ed. 2, 1: 48, 757. UPWTA, ed. 1, 5, as *H. glabrescens* Hutch. & Dalz.

West African: **SENEGAL** BASARI a-pɛ́xì (K&A; Ferry) BEDIK gi-tyeúmbẹreudy (Ferry) DIOLA é kobolora (JB) DIOLA (Fogny) kayes sako (K&A) FULA-PULAAR (Senegal) boilé (K&A) koilé (K&A) KONYAGI a-pín (Ferry) MANDING-BAMBARA fuganâ (JB; K&A) MANDINKA kundé (K&A) SERER mbélam (JB) sékör (AS, ex K&A) SERER-NON yugutj (AS, ex K&A) WOLOF hasao sâtèr (JGA) **THE GAMBIA** DIOLA (Fogny) e kobolora = *tortoise* (DF) FULA-PULAAR (The Gambia) daafi (DRR; DAP) kelli (DAP) mbo-lemboke (DRR) MANDING-MANDINKA kisiro (Fox) suma (DRR) sume (DAP) WOLOF sume (DAP) **GUINEA-BISSAU** FULA-PULAAR (Guinea-Bissau) bácurè (JDES) boile (JDES) MANDING-MANINKA fuagnan (Aub.; FB) fugnangnu (Aub.; FB) kundié (Aub.; FB) **MALI** DOGON gɔ̀nyɔ (C-G) FULA-PULAAR (Mali) kelli (A. Chev.) kelli danei (A. Chev.) MANDING-BAMBARA fukagnan (A. Chev.; FB) 'SENUFO' so messin (Aub.) **UPPER VOLTA** 'SENUFO' so messin (Aub.) **IVORY COAST** MANDING-MANINKA yakbassa (B&D) 'SENUFO' so messin (Aub.) **TOGO** BATONNUN muésué (Aub.) GBE-VHE gnirtii (Aub.) TEM tchabola-buanda (FB) TEM (Tshaudjo) tshabola buanda (Volkens) YORUBA-IFE OF TOGO tumbalaka (Volkens) **DAHOMEY** BATONNUN muésué (Aub.) GBE-VHE gnirtii (Aub.) **NIGERIA** EDO òhúnẹ̀gbó, ẹ̀gbo: *bush* (JMD; KO&S) FULA-FULFULDE (Adamawa) boile (JMD) FULFULDE (Nigeria) boili (MM) kelli (KO&S)

A small tree of the savanna woodland, 8–10 m high, occurring from Senegal to N Nigeria, and into Sudan and East and South Tropical Africa. The two varieties are separated on leaf-form but as regards their uses no distinction is made here between them.

The wood is reddish, strong, tough and very durable. It is valued in many areas for poles in hut-building, and the wood for chair-legs (10), and tool-handles (1, 4).

The bark yields a fibre which is commonly used for cordage in The Gambia (5, 9) and in West and East Sudan (3) and for fishing lines in Uganda (4). Bark-preparations enter into the Senegalese pharmacopoea for treatment of colic, and cause neither constipation nor diarrhoea (6, 7), while, however, a root-decoction is recommended as a laxative (7).

A decoction of leaves is used in Senegal as a bechic, expectorant and promoter of bronchial secretions (6, 7), and in Nigeria (8) and Guinea (3) the whole plant is used as an expectorant and in treatment of colds and chest-complaints, and in Guinea for diarrhoea and as a horse medicine (3).

The red fruit is edible and has a pleasantly acidulous taste. The seeds are eaten in S Rhodesia (10).

References:

1. Aubréville, 1950: 38. 2. Close, N. C. 4767, K. 3. Dalziel, 1937. 4. Eggeling & Dale, 1952: 18–19. 5. Fox, R. H. 37, K. 6. Kerharo & Adam, 1964, b: 548–9 . 7. Kerharo & Adam, 1974: 149–50. 8. Oliver, 1960: 28. 9. Rosevear, 1961. 10. Turner, L. 81, 94, K.

## **Isolona campanulata** Engl. & Diels

FWTA, ed. 2, 1: 53, 757. UPWTA, ed. 1, 5.

West African: **SIERRA LEONE** MENDE gbelo, kpelo (*def.*-i) (FCD; S&F) **UPPER VOLTA** KRU-GUERE (Chiehn) lalé (K&B) **IVORY COAST** ABE oroviti (A. Chev.) DAN bleu (K&B) KRU-GUERE pué (K&B) GUERE (Chiehn) tébékalé (K&B) 'KRU' daïri (K&B) pu daïri (K&B) **NIGERIA** EDO àghàkẹ̀zẹ̀ = *xylopia of the waterside* (JMD; KO&S)

A small tree to 17 m high by 60 cm girth, occurring from Sierra Leone to S Nigeria and in Cameroun and Gabon.

Powdered bark is added to food as an aphrodisiac by Kru and Guere medicine-men of the Ivory Coast, and a decoction of bark from the twigs is taken in draught for bronchial affections and as a febrifuge. The plant is also used in treatment for schistosomiasis (3). Use for coughs, fever and schistosomiasis is also recorded in Nigeria (4). A preparation is used in the Ivory Coast to promote female fertility and is given throughout a pregnancy, and also for haematuria and skin-trouble (1). In Sierra Leone a root-decoction is taken for rheumatism (2).

References:

1. Bouquet & Debray, 1974: 19. 2. Deighton 3676, K. 3. Kerharo & Bouquet, 1950: 20. 4. Oliver, 1960: 28.

**Isolona cooperi** Hutch. & Dalz.

FWTA, ed. 2, 1: 53. UPWTA, ed. 1, 5.
West African: **LIBERIA** KRU-BASA koo-gbeh (C&R) ku-gbe (C&R) **IVORY COAST** AKYE m-bokto (A&AA) **GHANA** NZEMA nzotala *this sp?* (FRI)

A small tree or shrub to 2 m high of the evergreen forest, occurring in Liberia, Ivory Coast and Ghana.

In the Ivory Coast Akye and Abe prepare a paste by trituration of leaves in a little water as a reconstituant massage for the body (1).

In Liberia the bark is used as a charm against witchcraft by chiefs and important men. Bark-ashes are mixed with palm-oil into a paste, and in a strange country, or in time of danger of eating poisoned or bewitched food, one licks the paste and the poison or spell is cast out (2, 3).

In Ghana an unnamed species close to *I. cooperi* yields a fibre used for rope (4, 5).

References:

1. Adjanohoun & Aké Assi, 1972: 28. 2. Cooper 417, K. 3. Cooper & Record, 1931: 15. 4. Irvine 2194, K. 5. Irvine, 1961: 12.

**Isolona deightonii** Keay

FWTA, ed. 2, 1: 53.
West African: **SIERRA LEONE** MENDE kpɛndɛ-golo (*def.*-golei) (FCD)

An understorey forest tree to 5 m high, recorded only from Sierra Leone. The fruit is edible (1).

Reference:

1. Deighton, 3072, K.

**Isolona pleurocarpa** Diels ssp. **nigerica** Keay

FWTA, ed. 2, 1: 53.
West African: **NIGERIA** EDO àghàkẹ́zẹ̀ (KO&S)

An understorey tree to 20 m high of the lowland rain-forest of Benin and Cameroons.

**Isolona thonneri** (De Wild. & Th. Dur.) Engl. & Diels

FWTA, ed. 2, 1: 53.

A tree of the evergreen forest, to 11 m high, recorded only from S Nigeria in the Region, but extending into Zaïre.

No use is recorded of this species within the Region, but in Zaïre the wood is said to be very hard and used in construction work. It possesses a resonance for which it is used to make xylophones (1).

Reference:

1. Boutique, 1951: 262–3.

**Mischogyne elliotianum** (Engl. & Diels) R. E. Fries

FWTA, ed. 2, 1: 47, as *Uvariastrum elliotianum* (Engl. & Diels) Sprague & Hutch. UPWTA, ed. 1, 8, as *U. elliotianum* Sprague & Hutch.

A small tree of the high-forest undergrowth, seldom attaining 10 m height, occurring from Sierra Leone to S Nigeria, and eastwards to Zaïre. Besides var. *elliotianum,* shrub or bush forms vars. *sericeum* and *glabrum* occur at the western end of its range.

Variety *sericeum* is recorded as used in the Njala area of Sierra Leone: 'Old leaves lose their hairs and become hard and are used for cleaning caps and other cloth articles' (1).

A Sierra Leone collection, *niyagaya,* Mende, in *Herb. K.* as *Uvariastrum sp. indet.* is perhaps this species, var. *elliotianum.* The plant is said to produce a 'good pole' (2).

References:

1. Deighton 3015, K. 2. King 186, K.

**Monanthotaxis barteri** (Baill.) Verdc.

FWTA, ed. 2, 1: 48, 50, as *Enneastemon barteri* (Baill.) Keay.
West African: GUINEA-BISSAU BIDYOGO úrei (JDES)

A shrub, 2 m high, recorded from Guinea-Bissau to S Nigeria.

**Monanthotaxis diclina** (Sprague) Verdc.

FWTA, ed. 2, 1: 44, 757, as *Popowia diclina* Sprague.
West African: LIBERIA KRU-BASA ne-bor-vah (C&R)

A shrubby liane of the dense rain-forest, occurring in Liberia within the Region and extending eastwards into Cameroun, Gabon and Congo.

The leaves have a pleasant scent and are pounded and mixed with clay to make a cosmetic paste which women in Liberia apply to their skin (2, 3).

Alkaloids are reported present (+ + Dragendorf test) in the bark and roots, and also some tannin (1).

References:

1. Bouquet, 1972: 13 as *Popowia declina* Sprague. 2. Cooper 199, K. 3. Cooper & Record, 1931: 17 as *Popowia ferruginea* Engl. & Diels.

**Monanthotaxis foliosa** (Engl. & Diels) Verdc.

FWTA, ed. 2, 1: 50, as *Enneastemon foliosa* (Engl. & Diels) Robyns & Ghesq.
West African: GHANA ANYI-SEHWI ntetekɔn (FRI)

A shrub or woody climber to 2.75 m high of secondary forest, recorded from Ghana to W Cameroon, and in E Cameroun.

**Monanthotaxis laurentii** (De Wild.) Verdc.

FWTA, ed. 2, 1: 44, as *Popowia congensis* (Engl. & Diels) Engl. & Diels.

A scandant shrub to about 2 m high of the dense and galleried forest occurring from Sierra Leone to S Nigeria within the Region and eastwards to Ubangi and Zaïre.

Water in which the leaves have been boiled is taken internally for treatment of fever in S Nigeria (1).
The fruit is edible.

References:

1. Unwin 49, K.

**Monanthotaxis stenosepala** (Engl. & Diels) Verdc.

FWTA, ed. 2, 1: 44, as *Popowia stenosepala* Engl. & Diels.
West African: **SIERRA LEONE** TEMNE *an*-seŋkowe (FCD)

A shrub recorded only in northern Sierra Leone.

**Monanthotaxis vogelii** (Hook. f.) Verdc.

FWTA, ed. 2, 1: 48, as *Enneastemon vogelii* (Hook. f.) Keay.

A scrambling shrub or small tree distributed from Ghana to Nigeria.
The roots are ground up for treatment of stomach-ache in Ghana (1, 2).
The fruit is edible (1, 2).

References:

1. Irvine 2887, K. 2. Irvine, 1961: 9 as *Enneastemon vogelii* (Hook. f.) Keay.

**Monanthotaxis whytei** (Stapf) Verdc.

FWTA, ed. 2, 1: 44, as *Popowia whytei* Stapf.
West African: **IVORY COAST** ABURE bagba *probably this sp.* (B&D)

A shrub or scandant found only in Liberia and Ivory Coast.
Its use has been recorded in the Ivory Coast for female sterility and oedemas (1).

Reference:

1. Bouquet & Debray, 1974: 19 as *Popowia whytei* Stapf.

**Monodora** Dunal

FWTA, ed. 2, 1: 53–54.

A genus of shrubs or trees, occasionally lianescent of which the seeds generally are aromatic.

**Monodora brevipes** Benth.

FWTA, ed. 2, 1: 54. UPWTA, ed. 1, 5.
English: yellow-flowered (or flowering) nutmeg.
West African: **SIERRA LEONE** MENDE kpɔ (*def.*-ei) *perhaps refers to M. tenuifolia and M. crispata also* (FCD) nja-mɔigbama (*def.*-bamei) *perhaps refers to M. tenuifolia and M. crispata also* (FCD; S&F) sagbe *perhaps refers to M. tenuifolia and M. crispata also* (FCD) **LIBERIA** KRU-BASA creba cribah (C&R) kray-bu (C&R) **GHANA** AKAN-WASA abotokuradua (FRI) **NIGERIA** EDO úkpósà (auctt.) YORUBA ause (Millen) làkọ̀ṣẹ (JMD)

A shrub or tree to 12 m high of mixed deciduous forest, from Guinea to W Cameroon and Fernando Po, and also in E Cameroun.

The wood (5) appears to have no recorded usage. The bark is used to prepare a purgative liquor in Liberia (3). A trace of alkaloids has been reported, but none in the roots (1) which are used to make a liquor taken in draught for relieving urethral stricture caused by venereal disease (3).

The fruit, more or less spherical to 7.5 cm diameter, contains a number of seeds which are powdered and used as a condiment (2, 6) or dried and used to flavour soup (4), or mixed with 'country' pepper (? *Piper guinense* Schum. & Thonn.) and used for spice (3). The seeds enter into medicine often mixed with fruits of *Xylopia aethiopica* (Annonaceae) (6).

References:

1. Adegoke & al., 1967. 2. Busson, 1965: 181–2. 3. Cooper 83, K. 4. Cooper 441, K. 5. Cooper & Record, 1931: 15 (with timber characters). 6. Dalziel, 1937.

**Monodora crispata** Engl. & Diels

FWTA, ed. 2, 1: 54.

West African: **SIERRA LEONE** MENDE nja-mɔigbama (*def.*-bamei) (S&F) TEMNE *an*-sali (NWT)

A scrambling shrub or small tree to 10 m high of the evergreen forest or in secondary jungle occurring from Sierra Leone to S Nigeria, and in Cameroun and Gabon.

The tree has decorative flowers and is worthy of cultivation as an ornamental.

The wood is used in construction work in Gabon (2). The leaves have been found to contain a strong presence of alkaloids (1). The seeds are aromatic and are used in Gabon as a condiment for food-flavouring (2).

References:

1. Bouquet & Debray, 1974: 19. 2. Walker & Sillans, 1961: 68.

**Monodora myristica** (Gaertn.) Dunal

FWTA, ed. 2, 1: 54. UPWTA, ed. 1, 5.

English: African nutmeg, false nutmeg, calabash nutmeg, Calabar nutmeg (Unwin).

French: fausse noix muscade, muscadier de calabash.

West African: **GUINEA-BISSAU** BALANTA sambè (JDES) BIAFADA durétche (JDES) FULA-PULAAR (Guinea-Bissau) guélè (JDES) quélè-naí (JDES) MANDING-MANDINKA djambadim (*def.*-ô) (JDES) **SIERRA LEONE** MENDE gboite (L-P) gɔmbewulo (NWT) **LIBERIA** KRU-BASA kray-bu (C&A) MENDE gboite (C&R) **IVORY COAST** ABE mué (auctt.) AKYE m-bo (auctt.) m-kbo (A. Chev.) m-kpo (A&AA) ANYI effuin (auctt.) efom'ba (A. Chev.) efuéro (A. Chev.) BAULE fuin (B&D) KYAMA aduanéhia (Aub.) afruenba (B&D) annéhia (Aub.) hané (A. Chev.) **GHANA** AKAN-ASANTE abotokuradua (FRI) awerewa (Enti) dubidi (FRI) kotokorowa (auctt.) o-wedgeɛ-aba, moto kura dua (Enti) o-wediɛ-aba (Enti) wadiɛ-aba (Enti) wediɛba (Enti) FANTE ayerɛw-amba = *fire seeds; from the pungent taste* (auctt.) TWI awere-aba (DF) awerewa (FRI) motokuradua (Enti) o-wediɛ-aba (Enti) owerewa (Enti) wediɛba (Enti) WASA awerewa (Enti) awiara (FRI) awire-aba (Enti) owedia-aba (Enti) o-wediɛ-aba (Enti) wediɛba (Enti) ANYI efua (FRI) ANYI-AOWIN ɛfõɔbaa (FRI) SEHWI ɛfõɔbaa (FRI) efua (FRI) GA maalai (FRI) GBE-VHE ayiku(-i) = *kidney; referring to the shape of the seed* (FRI) yikwi (FRI) NZEMA avonoba (CV; FRI) ayẽneba (JMD) **NIGERIA** EDO èbènọ̀yọ̀ọ̀bá (auctt.) HAUSA gujiya dan miya *the seed; = ground nut used for soup* (JMD; ZOG) IGBO éghùrù, érùrù (BNO) éhùrù (KO&S; Singha) IJO (Nembe) òkògòló (KW) YORUBA abo làkọ̀ṣẹ = *female lakoshe* (JMD; KO&S) ariwo = *noise* (Verger) aríwó *the seed* (JMD) efure *cf. M. tenuifolia Benth* (Lowe) làkọ̀ṣẹ (Punch; JRA)

A tree of the evergreen and deciduous forest, to 35 m high by 2 m in girth, distributed from Liberia to W Cameroons, and on to Uganda and Angola.

The flowers are conspicuous, attractive and scented.

The bole is usually clear. The wood is white or greyish, hard, somewhat tough and does not split well (9). It is easy to work and is suitable for carpentry and turnery (6) and walking sticks (4, 7).

The bark is used in the Ivory Coast to treat haemorrhoids, stomach-ache and febrile pains, and mixed with that of *M. tenuifolia* a collyrium is prepared for use in various eye-troubles (3). The preparation of a collyrium for treating filaria in the eye is recorded from Congo where the bark is also used in a vapour-bath as a defatigant and to relieve febrile lumbago, and the juice expressed from it to paint over itch (2).

The fruit, subspherical, may attain 20 cm length by 15 cm diameter, and lends its pot shape to the English and French names, calabash nutmeg. The seeds are embedded in a white sweet-smelling pulp and are the most economically important part of the tree. These, as are the seeds of the other *Monodora spp.,* are sold all over the West African region. They are aromatic and are used after grinding to a powder as a condiment in food providing a flavour resembling that of nutmeg. They are used as an aromatic and stimulating addition to medicines and to snuff (6). Ground to a powder they may be taken as a stimulant or stomachic or to relieve constipation; the powder may be sprinkled on sores, especially those caused by the guinea-worm, or the powder fried and made up into an oily pomade may be applied (6, 8, 12). Dusting or application of the pomade is used to disinfest from fleas (11, 12) and lice (7). The seeds chewed up are applied to the forehead for headache (6), and for migraine in Gabon (11, 12), and ground up for headaches, rhino-pharyngitis, or loss of voice, to apply on sores, or eaten as an anti-emetic aperative and tonic in Congo (2).

The seeds contain 5–9% of a colourless essential oil consisting largely of terpenes and with a pleasant taste and smell (8) and about 35–36% of a reddish-brown fixed oil which is mainly *linoleic acid,* 46.9%, and *oleic acid,* 35% (5). An alkaloid is also present (1) which is identified as crystalline *anonaceine* (8). The volatile oil is of inferior value for perfuming soap, nor is the fixed oil of economic value. The residual cake might serve as manure, but is unsuitable as a feeding-stuff.

The seeds are made into necklaces worn by women for the scent (8, 12). They are also regarded as a lure for hunting manatees in Gabon (12). They are endowed with magical attributes for which they are valued in many medical preparations. The seeds under the name of *ariwo* enter into an Yoruba Odu incantation against disease, the word meaning 'noise' and inferring the happiness of having noise about the house (10).

References:

1. Adegoke & al., 1968. 2. Bouquet, 1969: 59. 3. Bouquet & Debray, 1974: 19. 4. Boutique, 1951: 268–9. 5. Busson, 1965: 181–3, with oil analysis. 6. Dalziel, 1937. 7. Irvine, 1961: 13. 8. Oliver, 1960: sp. no. 231. 9. Unwin, 1920: 265 as *m. myristica* var. *grandifolia.* 10. Verger, 1967: 46. 11. Walker, 1953, a: 19. 12. Walker & Sillans, 1961: 68–69.

**Monodora tenuifolia** Benth.

FWTA, ed. 2, 1: 54, 757. UPWTA, ed. 1, 6.

English: African nutmeg (Unwin).

West African: **GUINEA-BISSAU** VULGAR molhanei (DF) BALANTA setane (JDES) FULA-PULAAR (Guinea) boilé bonno (JMD) kelli nayé (JMD) PULAAR (Guinea-Bissau) bólhaneí *from* bolha, *a leaf;* nei, *a cow* (D'O) guélè (JDES) kélè (EPdS) quélè-nai (JDES) **SIERRA LEONE** MENDE mɔigbama (*def.*-bamei) (JMD; Burbridge) nja-mɔigbama (S&F) **IVORY COAST** VULGAR piti mué (DF) ABE oroviti (A. Chev.; A&AA) piti mué (A. Chev.) GAGU gewora (B&D) KRU-GUERE (Chiehn) gawuin (B&D) tiè bossu (B&D) KYAMA aduanéhia (Aub.) annéhia (Aub.; RS) MANDING-MANINKA sula gnaman (A&AA) 'NEKEDIE' unémessini (B&D) **GHANA** VULGAR duabiri (DF) dubiri (DF) ADANGME mlɛto (FRI) AKAN-ASANTE abotokuradua (FRI) botorudua (BD&H)

dubidi (TFC) dubiri (FRI) TWI abotokuradua (DF) otutu-bofunnua (FRI) WASA bulusintim (CV; FRI) dubiri (Enti) dubusintim (BD&H) GA agbanaa (FRI) agbokutʃo = *uncleaned teeth* (FRI) a-ʃamitʃo (FRI) GBE-VHE (Awlan) gbloti (FRI) **NIGERIA** EDO úyẹ̀nghẹ̀n (auctt.) IGBO éhùrù ọ̀hị̄á (the standard transcription); éhùrù ọ̀fị̄á (dialectal variant) = *bush* (*i.e., wild*) *African nutmeg* (auctt.) IGBO (Owerri) érùlù (DRR) YORUBA abo-lakoshin (Ross) efure oguru, oguru: *a squirrel, cf. markings on the plant resemble lines on the back of a squirrel* (anon.) lakesin (KO&S) YORUBA (Ikale) ehinawosin (KO&S) ehinawosin (KO&S)

A small tree to 17 m height of the evergreen and fringing forest, and in secondary and regenerating thickets, occurring from Guinea to W Cameroons and Fernando Po, and from Cameroun and Gabon to Zaïre.

The tree is an attractive ornamental both in foliage and in its flowers which appear before the leaves. It is somewhat demanding as to soil requiring both moisture and depth. It shows some fire-resistance, sprouting again after such injury, and coppices well after felling.

The wood is white, hard, and fairly durable, but usually of comparatively small dimensions. It is tough and does not split well. It does not find much use in local construction-work, but in Ghana the Ga people use poles for cross-pieces in hut-roofs and also in wooden fences (5). In Nigeria the smaller trees furnish material for walking sticks, and the larger ones for hoe and axe-handles. In clearing land for farming, stool-shoots may be left for yam-supports (6). In Zaïre the young shoots are used as ties (2).

The bark has been recorded as used as a medicine for dogs in Sierra Leone but without indication of the purpose (3). The bark and the roots are said to be effective for dysentery and the roots to be used by the Yorubas for toothache (5).

In the Ivory Coast a decoction of the leaves is used in baths to treat body itching of children (1).

The fruits are edible. The seeds are aromatic and widely in the Region they are ground up to a powder and used as a condiment. They may be roasted, ground and rubbed on the skin for skin-diseases (5). They are considered anti-haemorrhagic and cicitrisant in Gabon (7), and to be a source of a good edible oil in S Nigeria (4).

Women in Gabon make necklaces of the seeds (7).

References:

1. Adjanohoun & Aké Assi, 1972: 30. 2. Boutique, 1951: 265. 3. Burbridge 472, K. 4. Carpenter 432, UCI. 5. Irvine, 1961: 14. 6. Unwin, 1920: 264–5. 7. Walker & Sillans, 1961: 69.

### **Neostenanthera gabonensis** (Engl. & Diels) Exell

FWTA, ed. 2, 1: 42–43. UPWTA, ed. 1, 6, as *N. bakuana* Exell.
West African: **LIBERIA** KRU-BASA blahn (C&R) **IVORY COAST** ANYI-AFEMA fravéfu (Aub.)

A shrub or small tree to 6 m high of the dense rain-forest from Liberia to Ghana, and to Gabon and Angola.

In Liberia a snuff is made from the dried leaves which is taken to treat tumour of the nose (1, 2).

References:

1. Cooper 416, K. 2. Cooper & Record, 1931: 17.

### **Neostenanthera hamata** (Benth.) Exell

FWTA, ed. 2, 1: 43. UPWTA, ed. 1, 6, as *N. yalensis* Hutch. & Dalz.
West African: **SIERRA LEONE** KONO k-pamanɛ (S&F) MENDE bɛ lo (FCD) pama-wulo (*def.*-wuli) (FCD; S&F) TEMNE *a*-til (NWT) **LIBERIA** KRU-BASA je-ah-chu (C&R) **IVORY**

**COAST** ABE baué (Aub.) baué fu (Aub.) bawé fu (Aub.) AKYE sainfi (A. Chev.) ANYI surua (A. Chev.) **GHANA** NZEMA osuni-elufoni (auctt.)

A slender tree to 25 m high by 30 cm in diameter, of the evergreen forest from Sierra Leone to Ghana.

The sap-wood is white and the wood is light and almost useless (2, 3). The bark is used in Liberia as a vermifuge (2). The fruits are eaten by wild animals (1).

References:

1. Burtt-Davy & Hoyle, 1937: 6 as *N. yalensis* Hutch. & Dalz. 2. Cooper & Record, 1931: Yale Univ., 17 as *Stenanthera yalensis* Hutch. & Dalz., with timber characters. 3. Irvine, 1961: 14–15.

**Neostenanthera myristicifolia** (Oliv.) Exell

FWTA, ed. 2, 1: 42. UPWTA, ed. 1, 6.
West African: **NIGERIA** EDO úyẹ̀nghẹ́n-ẹ̀zẹ̀ *incl. other Annonaceae;* ẹ̀zẹ̀: *waterside* (JMD; KO&S)

A forest tree to 10 m high recorded only from swampy areas of S Nigeria.

**Pachypodanthium barteri** (Benth.) Hutch. & Dalz.

FWTA, ed. 2, 1: 39.
West African: **NIGERIA** EBIRA (Igara) osoko (KO&S)

A forest tree to 16 m high found only in Kabba Province and near Onitsha in Nigeria.

**Pachypodanthium staudtii** Engl. & Diels

FWTA, ed. 2, 1: 39. UPWTA, ed. 1, 6.

West African: **SIERRA LEONE** KONO mononui (S&F) MENDE malenguli(-i) (DS; S&F) **LIBERIA** DAN gpaladio (GK) gpaladuo (AGV) KRU-GUERE (Krahn) djirrowa-tu (GK) BASA zree-chu (C&R; AGV) **IVORY COAST** ABE aniukéti (auctt.) anokuiti (auctt.) ABURE niangro (A. Chev.; RS) AKYE miedzo (auctt.) ANYI wié-emina (auctt.) AVIKAM niangro (A. Chev.) KRU-BETE diu-diu (auctt.) GUERE vâ (K&B) vahé (Aub.) GUERE (Wobe) yorohutu (A&AA) NZEMA émiengré (auctt.) 'KRU' irobetu (K&B) **GHANA** VULGAR okyiraa (DF) AKAN-ASANTE dankwakyire (CJT) dua-wisa (Enti) okyeraa (auctt.) pae-aduasã = *splits into 30* (*pieces*) (auctt.) TWI dua-wusa (DF) WASA awasa-makyina (CJT) duawisa (CJT) kumdwe, kumdwie = *to kill lice* (CJT) NZEMA fale (FRI; CJT) **NIGERIA** ANAANG ǹtuèn (KW) EFIK ǹtókòn (KW) EFIK ǹtókòn etọ (JMD; KO&S)

A medium-sized tree to about 30 m high of the evergreen forest from Sierra Leone to Nigeria and in Cameroun; trunk straight cylindrical, wood semi-hard and moderately heavy, or soft and light (4), resembling walnut but yellowish-white to greenish-brown, and durable. Because of the tree's good form the volume of timber is high. There is little differentiation between sap- and heart-wood. Uses are for general construction work, planks, house-poles, carpentry and barrel-staves (6, 7, 8, 11, 13).

The bark is used for making hut-walls in W Cameroons (7) and partitions in Gabon (14). The Konos of Sierra Leone use it to make doors in farm houses (1).

The slash is strongly smelling of turpentine, the yellow colour quickly turning brown on exposure (10, 12). The bark has a pungent taste when fresh (7) and is put to a number of medicinal uses. In Ghana it is boiled and used as a chest medicine (4, 8). In the Ivory Coast it is used for bronchitis (3), and in Congo a decoction is used as a cough-medicine (2). The bark is considered to have analgesic properties and so is good for toothache (2). A similar anodynal action is recorded in the Ivory Coast (9) where the bark pulped in a mortar with a kola nut is eaten for gastro-intestinal pains, and a preparation of the plant is used for bronchitis and gastro-intestinal trouble (3). Bark with leaves of *Ficus exasperata* (Moraceae) is pounded with water and clay into a paste which is used in the Ivory Coast topically in friction on oedemas (9).

The bark also has insecticidal and vermifugal properties. In Ghana (8, 12) and in Gabon (14, 15) a bark-decoction is used to kill body-vermin — hence the Wasa name meaning 'to kill lice.' In Liberia a decoction of beaten and boiled bark is considered an excellent vermifuge (5, 6, 9).

The bark is used, along with other ingredients, for tumours (8), and the Guere of the Ivory Coast add the bark to arrow-poison mixtures (9).

Monkeys find the fruit palatable (15).

Traces of alkaloids and of an unstable substance have been reported in the leaves and bark (9), and of tannins in the bark and roots (2).

References:

1. Anon. FHN 13554, SL. 2. Bouquet, 1972: 13. 3. Bouquet & Debray, 1974: 19. 4. Burtt-Davy & Hoyle, 1937: 6. 5. Cooper 296, K. 6. Cooper & Record, 1931: 16 giving timber characters. 7. Dalziel, 1937. 8. Irvine, 1961: 15, 17. 9. Kerharo & Bouquet, 1950: 21. 10. Kunkel, 1965: 156. 11. Savill & Fox, 1967: 42. 12. Taylor, 1960: 87. 13. Voorhoeve, 1965: 54. 14. Walker, 1953, a: 19. 15. Walker & Sillans, 1961: 69.

**Piptostigma fasciculata** (De Wild.) Boutique

FWTA, ed. 2, 1: 39, as *Brieya fasciculata* De Wild.

French: baouéfou à grandes feuilles (Aubréville).

West African: GHANA AKAN-TWI dankwakyere (DF) AKAN-WASA dankwakyere (Enti)

A lower storey tree of the evergreen forest, reaching 25 m high, and occurring in Ivory Coast and Ghana, and also in Gabon, Zaïre and Angola.

**Polyceratocarpus parviflorus** (Bak. f.) Ghesq.

FWTA, ed. 2, 1: 45.

West African: NIGERIA EDO ákọ̀sọ̀ (KO&S)

A tree of the lowland high rain-forest recorded from Guinea to S Nigeria, and to Gabon, to 12 m high with straight bole.

Bark is fibrous and faintly aromatic.

**Uvaria afzelii** Sc. Elliot

FWTA, ed. 2, 1: 38.

UPWTA, ed. 1, 7.

West African: GUINEA-BISSAU CRIOULO banana santcho (JDES) FULA-PULAAR (Guinea-Bissau) guélè-bálè (JDES) MANDING-MANDINKA sambafim (*def.*-ô) (JDES) MANDYAK begundja (JDES) bugunha (JDES) MANKANYA begundja (JDES) bugunha (JDES) PEPEL gundje (JDES) **SIERRA LEONE** KRIO pus-finga = *puss-finger; cf 'finga', U. chamae' P. Beauv.* (FCD) LOKO naagɔnde (NWT) MENDE gɔnɛ-yɔvɔta = *cat's testicles* (FCD; Boboh) TEMNE *an*-mut-*an*-yavi = *cat's paw* (M&H; FCD) **LIBERIA** KRU-BASA gbar-bee-mleh *meaning unknown, but referring to the medicinal properties of the roots* (C&R) **IVORY COAST** ADYUKRU okpap (A&AA) KRU-GUERE (Chiehn) bofo titi (B&D) boko titi (B&D) goffo titi (B&D) KWENI bla (B&D) **GHANA** AKAN-ASANTE abotokuradua (FRI) moto kuradua **NIGERIA** EDO ákósà (JMD)

A scrambling shrub or small tree to 5 m high, of secondary scrub, occurring from Guinea to southern Nigeria.

The fruit is edible (7, 9, 11).

The leaves are used in fevers (12) and, boiled with pepper, are taken in draught, or rubbed on the skin for 'yellow fever' (2) in Nigeria. The plant is held to be good for bronchial troubles and for stomach-ache in the Ivory Coast (4), and in the Gagnoa area pulped leaves are eaten with (? oil-) palm seeds for cough (10). Also in the Ivory Coast, leaf-sap is used in nasal instillation for epileptic fits and fainting, and a decoction is used to wash a person with small-pox or mange (4).

The bark is used in Nigeria for infections of the liver, kidneys and bladder, as a purge and a febrifuge and for coughs (13). The bark or pieces of the stem are sometimes put into palm-wine to add potency, or are added to the distillate to give colour to the spirit (3). The bark is used by the Nzema of Ghana as a cough medicine (9).

The root is used in Nigeria for treating gonorrhoea, and the root-bark itself is taken internally for catarrhal inflammations of mucous membranes and bronchitis as well as for gonorrhoea. In Liberia, a decoction of macerated roots is used for inflammation of the bladder and kidneys, and as a stomachic (5, 6). The Mendes of the Moselelo area of Sierra Leone take a root-decoction as an antidote for poisoning due to eating bad food (8).

Examination of the roots has shown no alkaloid present (1). The bark contains some tannin (13).

References:

1. Adegoke & al., 1968. 2. Ainslie, 1937: sp. no. 353. 3. Boboh, 1974. 4. Bouquet & Debray, 1974: 19. 5. Cooper 471, K. 6. Cooper & Record, 1931: 17. 7. Dalziel, 1937. 8. Deighton 2643, K. 9. Irvine, 1961: 18. 10. Kerharo & Bouquet, 1950: 21. 11. Melville & Hooker 201, K. 12. Oldham, Mus. K. 13. Oliver, 1960: 40.

**Uvaria angolensis** Welw. ex Oliv. **ssp. angolensis.**

FWTA, ed. 2, 1: 38, as *U. angolensis* Welw. ssp. *guineensis* Keay.

A shrub or small straggly tree, occurring from Sierra Leone eastwards in the forested area to N Rhodesia and Angola.

The species has a hard light brown wood (1). No use is recorded.

Reference:

1. Keay, 1953: 71 as *U. angolensis* Welw. ssp. *guineensis* Keay.

**Uvaria chamae** P. Beauv.

FWTA, ed. 2, 1: 38. UPWTA, ed. 1, 7.

English: finger-root (S. Leone).

West African: **SENEGAL** BALANTA sézei (K&A) BANYUN sikaral (K&A) DIOLA bu lèv (JB) bu riay (JB) DIOLA (Bayot) n'taba (K&A) DIOLA (Brin/Seliki) buhal baré (K&A) buléo (K&A) DIOLA (Fogny) fuléafo (K&A) fuléyo-afu (K&A) furay (K&A) DIOLA (Pointe) bananaru, bananiaroli = *monkey's banana* (K&A) DIOLA (Tentouck) boléo (K&A) DIOLA-FLUP buleo (K&A) FULA-PULAAR (Senegal) boélénimbo (K&A) kéleṅ baley (K&A) MANDING-MANDINKA sâbafim(-ô) (K&A) sâbéfin (K&A) sâbifiṅ (K&A) MANKANYA boguna (K&A) SERER mbélam (K&A) yidi (JB) WOLOF hasao = *bad smell* (AS) sédada (JB) **THE GAMBIA** DIOLA (Fogny) bu lev = *spicy* (DF) bu riay = *edible* (DF) fuléafo = *the sound of clapping* (DF) **GUINEA-BISSAU** CRIOULO banana santcho (JDES) FULA-PULAAR (Guinea-Bissau) guélè-bálè (JDES) MANDING-MANDINKA sambafim (*def*.-ô) (JDES) MANINKA furigna (Aub.) MANDYAK begundja (JDES) bugunha (JDES) MANKANYA begundja (JDES) bugunha (JDES) PEPEL gundje (JDES) **GUINEA** FULA-PULAAR (Guinea) boélémimbo (K&A) boïlé **SIERRA LEONE** KISSI kɛmbɔŋyundoŋ (FCD) koi-yondoe (FCD) KRIO finga, finger, fingers = *fingers; after the* *shape of the fruit* (auctt.) finger-root (L-P) LOKO hondowa (FCD) MENDE bilui *the root* (JMD) ndɔgbɔ-jɛlɛ (FCD) ndɔgbɔ-jɛlɛ-j gbɔu = *ripe bush banana;* jɛlɛ: *banana;* gbɔu: *ripe* (FCD) nɛgbɔta (*def.*-tei) *the fruit* (FCD) njɔpɔ-jɛgbɔu njɔpɔ: *youngbush;* jɛ: *probably a shortened form of jɛlɛ* (FCD) MENDE-KPA nɛgbɔta (FCD) SUSU-DYALONKE kan-yaduŋga-na (FCD) TEMNE *an-lanɛ* (NWT; FCD) **MALI** FULA-PULAAR (Mali) boilé (A. Chev.) MANDING-MANINKA (Wasulunka) fiin (A. Chev.) **IVORY COAST** ANYI ado massa (Aub.) **GHANA** AKAN-FANTE akotompo (Enti) akotompotsen = *long akotompo* (FRI, Enti) TWI akotumpetsin (BD&H) anweda (FRI) DAGAARI aura (AEK) GA aŋmɛdãa (FRI) aŋweda (FRI; KD) GBE-VHE agbana-asile (FRI) VHE (Awlan) agbana (FRI; JMD) VHE (Kpando) gbanagbana (FRI; JMD) HAUSA atore (BD&H; FRI) darigaza (BD&H; FRI) LOBI worsalla (AEK) SONGHAI-ZARMA sai (AEK) **TOGO** 'DIFALE' KABRE pereng (Volkens) YORUBA-IFE OF TOGO liasa (Volkens) padiwin (Volkens) **NIGERIA** HAUSA kas kaifi *probably this sp.* (Abrahams) IGBO ḿmị̀mị̀ ọ̀hị̄á = *bush Dennettia* (Singha) YORUBA akisan (JRA) ẹrújù, erú, *Xylopia;* iju: *bush, or desert* (JMD) okó ajá = *dog's penis* (JMD)

The various vernacular names mostly cover more than this one species.

A scandant shrub or small tree to about 4.5 m high, of savanna or secondary forest throughout the Region and eastwards and southwards into Zaïre.

The fruit carpels are in finger-like clusters, the shape giving rise to many vernacular names translated as 'bush banana' or the like implying wildness. The Sierra Leone Krio names 'fingers' and 'finger-root' for the roots are also from the fruit shape. The fruits are yellow when ripe and have a sweet pulp which is widely eaten. In Lagos the fruit is vulgarly called *okó-ajá* ('dog's penis') where it is a common ingredient in *agbo,* chiefly for febrile conditions in children.

All parts of the plant are fragrant. The plant is used to make a pomade in Ghana (8). The root and root-bark have a widespread reputation. The latter yields an oleo-resin and is taken internally for catarrhal inflammation of mucous membranes, bronchitis and gonorrhoea in Nigeria (2), while at one time a fluid extract entered into the composition of a stock hospital prescription in Ghana for dysentery (6). The properties are mainly astringent and styptic, and it is used in native medicine as a specific for piles, useful also for menorrhagia (for which it is taken mixed with Guinea grains and added to food), epistaxis, haematuria, haematemesis and haemoptysis.

In Sierra Leone and Lagos the root is credited with having purgative and febrifugal properties (5, 14). In Sierra Leone the root or the root-bark is boiled with spices and the decoction drunk for fevers classed locally as 'yellow-fever', including almost any indisposition accompanied by jaundice (5, 14), and in the Ivory Coast it enters into a treatment for a form of jaundice (3). In the Casamance of Senegal, leaves and roots are macerated for internal use as a cough mixture (13), and mixed with those of *Annona senegalensis,* dried and pulverized are considered strong medicine for renal and costal pain (9). The roots are used in the Casamance for healing sores, and a concoction called *n'taba* in the Bayot dialect is reputed to cure infantile rickets. In Nigeria a root-decoction is also held to be stomachic and vermifugal, and is used as a lotion; sap from the root and stem is applied to wounds; and the root is made into a drink and a body-wash for oedematous conditions (12). Amongst the Fula peoples of Senegal, the root has a reputation as the 'Medicine of Riches', and is taken for conditions of lassitude and senescence. It is also considered to be a 'woman's medicine' used for amenorrhoea and to prevent miscarriage (10), and in Togo a root-decoction is given for the pains of childbirth. In Ghana severe abdominal pain is treated by a root-infusion with native pepper in gin, and the root with Guinea grains (probably *Piper guinense* Schum. & Thonn. (Piperaceae)) is used in application to the fontanelle for cerebral diseases (8).

The sap of leaves, roots and stems is widely used on wounds and sores (2, 4, 5, 7, 8, 9) and is said to promote rapid healing. A leaf-infusion is used as an eye-wash and a leaf-decoction as a febrifuge (2).

The crushed seeds with those of *Piper guinense* are rubbed on the body. The crushed root, along with *Capsicum* or other rubefacient substances, is rubbed on as a local counter-irritant.

Several workers have reported the presence of alkaloids in small amount in the roots (1, 11, 15). Saponins have been recorded absent (15), and tannins present (11).

References:

1. Adegoke & al., 1968. 2. Ainslie, 1937: sp. no. 354. 3. Bouquet & Debray, 1974: 19. 4. Burtt-Davy & Hoyle, 1937: 7. 5. Dalziel, 1937. 6. Easman, 1891. 7. Irvine, 1930: 428. 8. Irvine, 1961: 19–20. 9. Kerharo & Adam, 1962. 10. Kerharo & Adam, 1964, b: 584. 11. Kerharo & Adam, 1974: 150–1. 12. Millen, s.n. Jan. 1892, K. 13. Portères, s.d. 14. Singha, 1965. 15. Taylor-Smith, 1966: 539–40.

## **Uvaria doeringii** Diels

FWTA, ed. 2, 1: 38.
West African: **GHANA** GA áŋmàdã (BD&H; FRI) GBE-VHE agbana (BD&H; FRI) **TOGO** GBE-VHE agbana (FB)

A tree or scrambling shrub to about 7 m, of coastal thickets of Ghana and Togo.

The whole tree has a strong spicy odour (1). The small green fruits are commonly eaten (1, 2, 3). In Ghana, the roots, mixed with 'mako' pepper (*Capsicum* sp. (Solenaceae)) are rubbed on to sore skin (1, 3), and gin with 'wusa' pepper (sp.?) in which roots have been steeped is drunk for abdominal pains (1). A leaf-concoction is drunk for jaundiced conditions ('yellow fever') (3), to which 'mako' peppers may be added (1). The leaf-concoction is also taken for piles, palpitations and pain, and leaves are made into a soup for newly-delivered women who also rub the seeds with black peppers on their skin (3).

References:

1. Burtt-Davy & Hoyle, 1937: 7. 2. Busson, 1965: 184. 3. Irvine, 1961: 20.

## **Uvaria ovata** (Dunal) A. DC.

FWTA, ed. 2, 1: 36. UPWTA, ed. 1, 7, as *U. globosa* Hook. f.
West African: **GHANA** ADANGME (Krobo) nyɔtʃo *breast tree, from shape of the fruits* (FRI) AKAN-FANTE akotompo (BD&H; FRI) KWAWU nyɛtʃo (FRI) TWI otutumɔfunnua (FRI) GA aanyɛle (FRI) anyɛnyɛ le (FRI) GBE-VHE gbemishi kpemitsi (FRI) VHE (Awlan) agbana (FRI) kpetsi (FRI) VHE (Pecí) gbanagbana (FRI)

A shrub to 2.5 m high, mainly of savanna areas and confined to West Africa from Sierra Leone to Togo. Two subspecies are recognized, *ovata* and *afzeliana* (DC.) Keay, the latter occurring in rocky hill forest to the western side of the range.

The pulp of the globose fruiting carpels is sweet and edible (1, 2), and is sometimes an article sold in Ghanaian markets (2). The stems are used for hurdle-fencing on farms, and also for basket-weaving (2).

References:

1. Busson, 1963: 184. 2. Irvine, 1961: 20.

## **Uvaria scabrida** Oliv.

FWTA, ed. 2, 1: 38.
West African: **IVORY COAST** KYAMA bapo (B&D) niania vabua (B&D)

A climbing shrub or small tree occurring from Liberia eastwards to Zaïre.

In the Ivory Coast, the Kyama use it in a treatment for insanity (1). The bark finds use in Zaïre as fishermen's bow-nets (2).

References:

1. Bouquet & Debray, 1974: 19. 2. Irvine, 1961: 20.

**Uvaria sofa** Sc. Elliot

FWTA, ed. 2, 1: 38.
West African: **GUINEA** FULA-PULAAR (Guinea) boilé (Langdale-Brown) boilé-niaddé (A. Chev.)

A scrambling shrub of thickets and secondary growth from Guinea to Ghana.

**Uvaria thomasii** Sprague & Hutch.

FWTA, ed. 2, 1: 38.
West African: **SENEGAL** DIOLA u lèv iné (JB) DIOLA (Brin/Seleki) buhal bainé (K&A)

A twining shrub recorded from the Casamance of Senegal and Sierra Leone. It is used in the Casamance in the form of a leaf-decoction for catarrh and colic (1).

Reference:

1. Kerharo & Adam, 1974: 151.

**Uvaria tortilis** A. Chev. ex Hutch. & Dalz.

FWTA, ed. 2, 1: 38.
West African: **IVORY COAST** MANDING-MANINKA dazuan (B&D)

Recorded only from the Ivory Coast, the Kyama use it in a treatment of amenorrhoea (1).

Reference:

1. Bouquet & Debray, 1974: 19.

**Uvariastrum pierreanum** Engl.

FWTA, ed. 2, 1: 47.
West African: **GHANA** AKAN-ASANTE ɔtwe-ehi (FRI) BRONG ankumabaka (CV; FRI)

A tree of the lowland high rain-forest, occurring from Ghana to S Nigeria within the Region, and into Cameroun, Gabon and Zaïre eastwards. It attains its maximum development in the western part of its range reaching 30 m height by over 1 m girth, while to the east it remains shrubby or a small tree.

The wood is hard and is used (? in Ghana) for gun-stocks (2).

The flowers after falling have a distinct smell of sulphur dioxide (1).

References:

1. Brenan 8901, K. 2. Irvine, 1961: 21–22.

**Uvariodendron angustifolium** (Engl. & Diels) R. E. Fries

FWTA, ed. 2, 1: 46.
West African: **GHANA** AKAN-ASANTE ɛsono-kwadu (FRI) ɔbɔmmɔfo-kwaadu (CV; FRI) FANTE ɛsung-kodo (CV) **NIGERIA** YORUBA igbere (KO&S)

An understorey tree to about 15 m high, of the high-forest, recorded only from Ghana to W Cameroons.
The fruits smell strongly of lemons (1).

Reference:

1. Keay et al., 1960: 43.

**Uvariodendron calophyllum** R. E. Fries

FWTA, ed. 2, 1: 46.

An understorey tree of the high-forest, to 20 m high by 60 cm girth, occurring from Ghana to W Cameroons, and in E Cameroun.
The leaves are usually over 30 cm long and often attain 1 m by 25 cm across (2).
The seeds are embedded in a colourless jelly (1).

References:

1. Irvine, 1961: 22. 2. Keay & al., 1960: 43–45.

**Uvariodendron occidentalis** Le Thomas

FWTA, ed. 2, 1: 46, as *U. mirabilis* sensu Keay, p.p.
French: misiti à grandes feuilles (Aubréville).
West African: **IVORY COAST** AKYE misiti (Aub.) **GHANA** AKAN-ASANTE ɛsono kwadu = *elephant's banana* (BD&H; FRI) osun-kodu (Andoh)

A small tree to 10 m high with a straight trunk to 15 cm diameter, of the primary forest from Ivory Coast to S Nigeria.
The wood is hard and sweet-smelling (1).

Reference:

1. Le Thomas, 1967.

**Uvariopsis** Engl. & Diels

FWTA, ed. 2, 1: 50.

Little is recorded on the properties of the West Tropical African species. A trace of alkaloid has been reported present in the leaves and in greater quantity in the bark and roots of *U. solheidii* (De Wild.) Rob. & Ghesq. in Congo, and also the presence of tannins (1). West African material should be examined.

Reference:

1. Bouquet, 1972: 13.

**Uvariopsis congoensis** Robyns & Ghesq.

An understorey tree to 8 m high of forest in Kenya, Uganda, Rhodesia and Zaïre, and recently recorded in Ghana (2).
The wood is white, seasoning to pale brown, hard and close-grained of even texture (1).

References:

1. Dale & Greenway, 1961: 41–42. 2. Morton A3394, K.

**Uvariopsis dioica** (Diels) Robyns & Ghesq.

FWTA, ed. 2, 1: 50.
West African: **GHANA** ANYI-SEHWI kyeatoa **NIGERIA** EDO ákósä (Ross; KO&S)

An undergrowth shrub of the dense rain-forest, occurring from Ghana to southern W Cameroons and into E Cameroun.
The wood is soft.

**Uvariopsis guineensis** Keay

FWTA, ed. 2, 1: 50.
West African: **GHANA** ANYI-AOWIN kyeatoa (FRI) SEHWI kyeatoa (FRI)

An understorey forest tree to about 11 m high recorded from Guinea to Ivory Coast.
The wood is reported to be white and soft (Foggie fide 1).

Reference:

1. Irvine, 1961: 23.

**Xylopia** Linn.

FWTA, ed. 2, 1: 41.
West African: **SIERRA LEONE** MENDE hewe (*def*.-i) *all spp., but especially X. aethiopica (Dunal) A. Rich.* (S&F) **IVORY COAST** ABE elo *all spp. with hard usable wood* (Aub.) fondé *all spp. with soft wood* (Aub.)

**Xylopia acutiflora** (Dunal) A. Rich.

FWTA, ed. 2, 1: 42. UPWTA, ed. 1, 8.
English: the fruit — mountain spice (of Sierra Leone).
French: elo à petites feuilles (1).

West African: **SIERRA LEONE** MANDING-MANINKA (Koranko) kenɛ (S&F) MENDE hewe (NWT) kpa-hina (*def*.-hinei) (S&F) ngele-hewe (*def*.-i), ngele: *sky* (S&F) nja hewe = *Xylopia (hewe) growing in water (nja)* (Aylmer; JMD) TEMNE *an*-lane (NWT) *am*-pos; *an*-pos (JMD) *ma*-tsel *the spice* (JMD) **GHANA** ANYI-SEHWI dwombobre (FRI) GBE-VHE tsyo (Akpabla) HAUSA kimbáá (Akpabla) **NIGERIA** EDO úniẹ̀nẹ̀zẹ̀, ẹ̀zẹ̀: *waterside* (JMD) **WEST CAMEROONS** 'BAMILEKE' ki (Johnstone)

A shrub or medium-sized tree to 13–15 m high of the rain-forest from Sierra Leone to W Cameroon, and extending to Zaïre and Rhodesia.
The wood is yellowish. It is used for cross-bows (3) and in Gabon, for paddles (5, 6) and the shafts of harpoons (6).
The bark is thin but is used in Gabon to make walls and partitions of houses (5, 6). In the Ivory Coast it is used to treat bronchio-pneumonial affections and for febrile pains (2).
The fruit is red or blackish, 4.5–5.0 cm long by 1.0–1.5 cm thick, curved, turgid when ripe with a coral-red pulp having a spicy taste like several other *X.* species (4).

References:

1. Aubréville, 1959: 138. 2. Bouquet & Debray, 1974: 19. 3. Dalziel, 1937. 4. Keay & al., 1960: 41. 5. Le Thomas, 1969: 169–72.

**Xylopia aethiopica** (Dunal) A. Rich.

FWTA, ed. 2, 1: 42. UPWTA, ed. 1, 8–9.

English: African pepper, Guinea pepper* (Cooper & Record, Dalziel); Ethiopian pepper (Dalziel, Oliver); spice tree (Sierra Leone, Dalziel).

French: fausse maniguette (Schnell); piment noir de Guinée* (Dalziel); poivre de Guinée* (Irvine, Kerharo & Adam); poivrier de Ethiopie (Aubréville, Kerharo & Adam, Schnell); poivrier de Guinée* (auctt.); poivrier de Sedhiou, poivrier negre (Aubréville, Kerharo & Adam, Schnell).

Portuguese: malaguetta preta, pimenta da Guiné* (Feijao).

*NOTE: not to be confused with *Piper guineense* Schumm. & Thonn., Piperaceae.

West African: **SENEGAL** DIOLA baslèv (JB) uhéu (auctt.) DIOLA (Fogny) buheo (K&A) buleno (K&A) DIOLA-FLUP kaléo (K&A) FULA-PULAAR (Senegal) gilé (K&A) gilé bélé (K&A) MANDING-BAMBARA kani *general for pepper or spice* (JMD) kani fin(-g) = *black pepper* (JMD; JB) MANDINKA kaani (*def.* kaanoo) (Sebia Jatta) 'SOCÈ' kani (K&A) MANDYAK brobleké (K&A) SERER-NON haledé (AS) WOLOF ndiar (auctt.) ndier (K&A) **THE GAMBIA** MANDING-MANDINKA kaanifiŋ (*def.*-o) *probably this sp.* (DRR) **GUINEA-BISSAU** BALANTA sem-unte-pulhe (JDES) BIDYOGO equeché (EPdS) equèche (JDES) CRIOULO malagueta (RS; Aub.) malagueta da Guiné (JDES) malagueta preta (JDES; GeS) pimenta da Guiné (EPdS) pimenta preta (EPdS) FULA-PULAAR (Guinea-Bissau) guilé-bale (JDES; EPdS) guilè-bétè (auctt.) MANDING-MANDINKA djanafi (*def.*-ô) (JDES) MANINKA bolofaran (FB; RS) kani (JMD; RS) MANDYAK irú (JDES) PEPEL djô-djô (EPdS; JDES) iebogôfo (JDES; GES) **GUINEA** FULA-PULAAR (Guinea) guilé (JMD) **SIERRA LEONE** BULOM (Kim) sɔ (FCD) BULOM (Sherbro) son-dɛ (FCD; S&F) GOLA seve (FCD) KISSI siawõ (S&F; FCD) KONO kandi (FCD; S&F) KRIO siminii (S&F) siminji (S&F) spais-tik = *spice tree* (FCD; S&F) LOKO seve (FCD; S&F) MENDE hewe (auctt.) nila (*def.*-nilei) (DS) nja hewe *Xylopia* (*hewe*) *growing in water* (*nja*) (JMD) TEMNE *am*-pos, *an*-pos *the plant* (JMD; Jarr) *ma*-tsel *the spice* (JMD) VAI lɔmbɔɛ (FCD) **LIBERIA** KRU-GREBO deo (JMD; RS) BASA deo (C&R) **UPPER VOLTA** FULA-FULFULDE (Upper Volta) guili (K&B) **IVORY COAST** ABE fondé *incl. other X. spp.* (auctt.) ABURE evafé (B&D) AKAN-BRONG indian (K&B) ANYI efomu (A. Chev.; Aub.) BAULE sindian (K&B) FULA-FULFULDE (Ivory Coast) guili (K&B) KRU-GUERE (Wobe) gbépoo (A&AA) KULANGO sogodio (K&B) MANDING-MANINKA bolo faran (Aub.) guili (K&B) kani (Aub.) 'SOCE' kanifingo (Aub.) **GHANA** ADANGME-KROBO soo (FRI) AKAN-ASANTE hwentia (Enti) TWI hwentia = *slender nose, referring to the shape of the fruits* (FRI) WASA hwentia (Enti) GA soo (FRI) GBE-VHE tsyo HAUSA chimba, kimba (FRI) **TOGO** TEM (Tshaudjo) akatapure sõsi (Volkens) sõsi (Volkens) **NIGERIA** ARABIC kyimba (KO&S) ARABIC-SHUWA kumba (JMD) kyimba (JMD) BEROM lɛ́ŋ (LB) BOKYI kenya (Catterall; KO&S) unie (Vermeer) DEGEMA akada (JMD) EDO únię̣ (auctt.) EFIK átà (auctt.) FULA-FULFULDE (Nigeria) kimbaahre (*pl.* kimbaaje) (auctt.) HAUSA kimbáá (auctt.) IBIBIO átâ (KO&S; Kaufman) IGBO ụ́dà (KO&S) IGBO (Owerri) ọ́dà, ụ̄dà (JMD) NUPE tsunfyányá *the fruit* (Banfield; KO&S) tsunfyányáci *the tree* (Banfield) TERA kìmbílì *? this sp., or others known as 'pepper'* (Newman) YORUBA ẹ̀ẹ̀rù (Lowe) erinje (IFE) ẹ̀rú (auctt.) ẹ̀ru-awonka (JRA) erunje (KO&S) ẹ̃runjẹ̃ (JMD) kharu (Barter) ọlorin (AST; IFE) sẹ̀sẹ̀dó (JMD)

A tree to 20 m high or more, with a clear straight bole, to 75 cm girth, often with short prop roots, smooth grey bark, scented when fresh, in lowland rain-forest, coastal brackish swamps and littoral formations, and deciduous and fringing forests of the Guinean savanna zones, often cultivated near villages, and in the forest often protected, widespread in the Region from the Casamance of Senegal to S Nigeria, and eastwards into East Africa.

The slash is white, sappy and very fragrant (6), and the fresh scented bark is kept in hut rooms in Gabon (25). The thick, fibrous bark peels readily and yields a cordage (9). The bark is used to make doors and partitions, and in Gabon to wrap around torches (14, 26).

A fruit-extract, or decoction of the bark, as of the fruit, is useful in the treatment of bronchitis and dysenteric conditions, and also as a medicine for biliousness (9) and febrile pains (5). In Congo bark is steeped in palm-wine which is given for attacks of asthma, stomach-aches and rheumatism at the rate of one or two glasses per day (4).

The wood is white or pale yellowish brown in the heart, and fairly hard. It is said to be light and brittle in Sierra Leone (21), and to be of no commercial application, but elsewhere to be heavy, strong and elastic, and is used for purposes requiring resilience such as boat-construction, masts, oars, paddles and spars (8, 9, 13). In Togo (9) and Gabon (26) it is used for bows and crossbows for hunters and warriors. The wood is resistant to termite attack and is used in hut-construction for posts, scantlings, roof-ridges and joists (6, 9, 26). It burns with a hot flame and has found use as a steamboat fuel (9, 24, 26).

The root-wood can be used as a cork (9).

The root is strongly aromatic (16) and a concentrated root-decoction is used as a mouthwash for toothache (15) in the Casamance of Senegal. The powdered root is used as a dressing for sores and to rub on gums for pyorrhoea and in local treatment of cancer in Nigeria, and when mixed with salt is a cure for constipation (1). The powdered bark is dusted onto ulcers, and a decoction of leaves and roots is a general tonic in Nigeria for fevers and debility, and enters an *agbo* prescription (1).

The leaves have a pungent smell. A decoction is used in Gabon against rheumatism and as an emetic (25), and as a macerate in palm-wine it makes a popular intoxicating drink (25, 26). In Congo powdered leaves are taken as snuff for headaches, and used in friction on the chest for bronchio-pneumonia. The leaf-sap mixed with kola nut is given at the time of epileptic fits, and the fruit is used to season the patient's food. Concurrently suitable prayers and offerings must be made (4).

The fruit is the most important part of the tree. They are narrow, slightly torulose, dark brown or black, about 5 cm long, borne many (separate carpels) together on a stout peduncle. In the Middle Ages they were exported to Europe as a 'pepper.' They remain an important item of local trade throughout the Region as a spice, and flavouring for food and for medicine. They are sometimes put into jars of water to purify the water. Pulverized they are added to snuff to increase the pungency (9).

Medicinally the principal uses are as a cough-medicine, a carminative, purgative and revulsive to counter pain. The fruit is a common ingredient of the Yoruba *agbo* (9). The fruits are smoked like tobacco in Sierra Leone (22), and the smoke from a mixture of dried pulped fruit and tobacco is inhaled to relieve respiratory ailments in Liberia (8). Fruits are particularly recommended to women who have newly given birth as a tonic in the Ivory Coast where they are also used as an anthelmintic (5, 18). They are often incorporated in preparations for enemas and external uses calling on its revulsive properties for pains in the chest, sides, ribs etc. (8), and generally for any painful area, lumbago, neuralgia and in the treatment of boils and skin eruptions (8, 9, 21, 22). A fruit-decoction, or bark, or both, is useful in the treatment of bronchitis and dysenteric conditions, and also as a medicine for biliousness. As a woman's remedy it is taken to encourage fertility and for ease in childbed. This prescription was employed at one time in the Government Hospital in Ghana, and in combination with *Newbouldia laevis* (Bignoniaceae) was used for inducing or increasing menstrual flow, and was accordingly deemed to have abortifacient properties (10).

The fruit, ground up with Capsicum peppers, is mixed with kola nuts as a repellent for the Kola weevil.

The seeds, as separate from the fruit, are a substitute for pepper, and have cosmetic, revulsive and stimulant uses (9, 15, 25). Crushed they are rubbed on the forehead for headache and neuralgia, or a poultice of leaves and fruit may be applied (9). Mixed with other spices they are rubbed on the body as a cosmetic and scent, and are commonly used as a perfume for clothing. In S Nigeria Igbos grind the seed and give it to lactating mothers (2), and the Edo of Uzalla (Benin) use the seed (probably of this plant) as an ingredient of a medicine to aid conception (24). An extract of the seeds is also taken as a vermifuge for roundworms, and as an emetic for biliousness (9). Some of the medicinal uses are applied also for cattle and domestic animals (9).

The plant is said to contain *anonaceine,* which is an alkaloid resembling morphine in action according to some authorities, and according to others is a glycoside, and the fruit contains a volatile aromatic oil, a fixed oil and *rutin* (17, 27). Recent work has shown the presence of other substances in Nigerian material (11, 20). A pharmacological investigation of fruit-extract used in Nigerian folk-medicine for treatment of skin-infections has shown some action on Gram +ve organisms: *Sarcina lutea* and *Mycobacterium phlei,* and no action on *Staphylococcus aureus,* and no anti-fungal action (19).

References:

1. Ainslie, 1937: sp. no. 4. 2. Anon. FHI 16586, K. 3. Aubréville, 1959: I, 138. 4. Bouquet, 1969: 60. 5. Bouquet & Debray, 1974: 19. 6. Burtt-Davy & Hoyle, 1937: 8. 7. Busson, 1965: 184–5. 8. Cooper & Record, 1931: 18. 9. Dalziel, 1937. 10. Easmon, 1891. 11. Ekong & al., 1969, b. 12. Feijão, 1960–63. 13. Gomez e Sousa, 1930: 39. 14. Irvine, 1961: 23–24. 15. Kerharo & Adam, 1962. 16. Kerharo & Adam, 1963, b. 17. Kerharo & Adam, 1974: 151–2. 18. Kerharo & Bouquet, 1950: 21–22. 19. Malcolm & Sofowora, 1969. 20. Ogan, 1971. 21. Oliver, 1960: 40. 22. Savill & Fox, 1967: 43. 23. Schnell, 1950, b: 211. 24. Vermeer 78, UCI, in litt. 25. Walker, 1953, a: 19. 26. Walker & Sillans, 1961: 72. 27. Watt & Breyer-Brandwijk, 1962: 62.

**Xylopia elliotii** Engl. & Diels

FWTA, ed. 2, 1: 42.

French: fondé à petits feuilles (Aubréville).

West African: **SIERRA LEONE** KONO pasia (S&F) MENDE gbajina (*def.*-jinei) (S&F) gbelo (*def.*-i) = *yellow fever* (S&F)

A small forest tree to 10 m high recorded from Guinea, Sierra Leone and Ivory Coast.

**Xylopia parviflora** (A. Rich.) Benth.

FWTA, ed. 2, 1: 42. UPWTA, ed. 1, 9, as *X. vallotii* Hutch. & Dalz.

French: fondé des rivières (Aubréville); poivres de Sédhiou (the carpels, Dalziel); poivrier de Sédhiou (Kerharo & Adam).

West African: **SENEGAL** DIOLA bulèv bèyné (JB) DIOLA (Brin/Seleki) bosobul biéhu (K&A) **THE GAMBIA** FULA-PULAAR (The Gambia) wile (DAP) MANDING-MANDINKA kaanifiŋ (*def.*-o) (DAP) WOLOF jar (DAP) **SIERRA LEONE** KONO yi-kandi-kɔne (S&F) MENDE nja-hewe (*def.*-i) (auctt.) **GHANA** AKAN-TWI obaa-kokoo (DF) ANYI-SEHWI gyambobre (FRI) **TOGO** TEM (Tshaudjo) tshabola buanda (Volkens) **NIGERIA** EDO àghàkó (AHU; KO&S) HAUSA kimbaa (ZOG) YORUBA ẹrinjẹ (IFE) ẹ̀rù (IFE) sẹ̀sẹ̀dó (JMD; KO&S) vini (IFE)

A shrub or small tree to 17 m high in fringing forest of the lowland rain-forest and savanna zones from Senegal to S Nigeria, and into the Sudan, Uganda and Angola.

The wood is yellow, hard, fine-grained and elastic. Its flexibility and durability render it suitable for oars and masts. It is believed to resist termites, and is used for house-posts, scantlings and general carpentry (3).

The bark is fibrous from which a cordage can be made. In Gabon it is used to make screens for trapping fish (7).

Dalziel notes that the roots have medicinal uses but gives no detail (3). In the Casamance, an aqueous decoction of equal parts of roots and leaves is taken by mouth as a bechic and expectorant (4, 5).

The fruiting carpels are cylindrical, 3.75–4.0 cm long by about 2 cm thick, slightly constricted transversely and with longitudinal ridges. They are spicy. The fruit and the stem-bark are put into a Nigerian remedy for skin infections, but examination for anti-bacterial or anti-fungal action showed no activity (6). Likewise no active principle has been found in Congo material (2).

References:

1. Aubréville, 1959: 1, 138. 2. Bouquet, 1972: 13. 3. Dalziel, 1937. 4. Kerharo & Adam, 1962. 5. Kerharo & Adam, 1974: 152–3. 6. Malcolm & Sofowora, 1969. 7. Walker & Sillans, 1961: 74.

**Xylopia quintasii** Engl. & Diels

FWTA, ed. 2, 1: 42, 757. UPWTA, ed. 1, 9.

English: Negro pepper (Ainslie).

French: elo (from the Abé name, Aubréville).

West African: **SIERRA LEONE** MENDE gbajina (*def.*-jinei) (DS) kpa-hina *i.e., the male Kpa tree;* kpa: *Beilschmiedia mannii* (*Meisn.*) *Benth. & Hook. f.* (FCD) kpaina (*def.* kpainii) (JMD; S&F) **LIBERIA** KRU-BASA gbay (C&R) gbay-dee (C&R) **IVORY COAST** ABE elo (Aub.) fondé (Aub.) KYAMA brala (Aub.) elalé (B&D) **GHANA** AKAN-ASANTE kaba(-h) (BD&H; FRI) obaa (CJT) ɔbaa (auctt.) waba (FRI) TWI obaa (CJT; DF) WASA asima (auctt.) asimba (auctt.) obaa (CJT) ANYI-SEHWI obaa (CJT) NZEMA erale (CJT) **NIGERIA** BOKYI bolonge (Catterall; KO&S) EDO àghàkó (JMD; Kennedy) oviunien (Kennedy) ovunien (KO&S) IGBO ūdà ọ́fị́á = *bush Xylopia* (KO&S) YORUBA ẹ̀ẹ̀rùàwọ̀nká (JRA) ẹ̀rù (JRA) opalifon (KO&S) palufon (JMD)

A slender tree to 30 m with a clean straight bole, of wet evergreen or deciduous lowland forest, from Sierra Leone to S Nigeria, and into Gabon and Zaïre.

The wood is yellowish to brown, heavy and hard, scented (7) or with a fetid odour (4) when fresh. It is reasonably durable against termites (10). People use it for house-posts (3–7, 10), and for timbers when large (5). Its examination for suitability for transmission poles in Sierra Leone has been recommended (9). The grain of the wood is straight and it finishes well. It is harder and finer-textured than the wood of other *Xylopia* species. Its resilience and toughness make it suitable in small sizes for purposes requiring elasticity with strength, such as for tool-handles, pestles, spear-shafts, canoe-paddles, etc. (6, 7). In Gabon it is used to make bows and cross-bows (11).

The bark is fragrant (10). In the Ivory Coast it is used in the treatment of broncho-pneumonic affections and for febrile pains (2). In Liberia, the inner bark is beaten and rubbed on the hands to reduce knot-like swellings, and scrapings of the inner bark soaked in water make a mouthwash for pyorrhoea (6) and powdered bark is dusted on ulcers in Nigeria (1).

The bark strips easily and the inner bark yields a fibre which is used for cordage.

The powdered root mixed with salt is considered a cure for constipation in Nigeria, and the powdered root alone is used in a dressing for sores and for rubbing on gums in cases of pyorrhoea, and in local treatment of cancer (1).

A decoction of leaves and roots is used in Nigeria as a general tonic and in the making of *agbo,* while the fruit is eaten to assist in childbirth, and is taken for mucous discharges such as bronchitis and gonorrhoea. The fruit is used as a stimulant and in treatment of menorrhagia (1). In Ghana the smoke-dried fruit is said to have medicinal use (8).

References:

1. Ainslie, 1937: sp. no. 361. 2. Bouquet & Debray, 1974: 19. 3. Burtt-Davy & Hoyle, 1937: 8. 4. Cooper 222, K. 5. Cooper 337, K. 6. Cooper & Record, 1931: 18 with timber characters. 7. Dalziel, 1937. 8. Irvine, 1961: 25. 9. Savill & Fox, 1967: 44. 10. Taylor, 1960: 89. 11. Walker & Sillans, 1961: 73.

**Xylopia rubescens** Oliv.

FWTA, ed. 2, 1: 41. UPWTA, ed. 1, 9.

French: fondé des marais (Aubréville).

West African: **IVORY COAST** ABE fondé ANYI efomu (A. Chev.) KYAMA dandu (Aub.) **NIGERIA** EDO úniẹ̀n-ẹ̀zẹ̀ (KO&S) EFIK átà rabọ̈n (KO&S) YORUBA ofún òkè (KO&S)

A tree to 30 m high, of evergreen fresh-water swamp forest from Liberia to S Nigeria, and extending to Zaïre and Uganda.

The wood is white, light in weight and believed to be immune to insect attack. In Cameroons it is used for the feet of chairs, etc., made of the local rattans (2). In

Zaïre it is recognized as a 'good' wood, but the reference gives no indication in what way (1).

The bark is used for hut-walls in the Cameroons (2), and in hut-construction in Gabon (3).

References:

1. Boutique, 1951: 322–3. 2. Dalziel, 1937. 3. Le Thomas, 1969: 159–62.

**Xylopia staudtii** Engl. & Diels

FWTA, ed. 2, 1: 41. UPWTA, ed. 1, 8.

English: bush pepper, Guinea pepper tree (Cooper & Record).

French: fondé (from the Abé name, Aubréville).

West African: **SIERRA LEONE** MENDE yengetoma (*def.*-tomei) (auctt.) **LIBERIA** DAN so-pueh (GK) KRU-BASA drehn (C&R) **IVORY COAST** ABE fondé (A. Chev.; Aub.) AKYE fonfi (A. Chev.) ANYI efomu (Aub.) KYAMA aduébê (Aub.) **GHANA** VULGAR duanan (DF) AKAN-WASA duanan = *four legs, i.e., the buttresses* (auctt.) ANYI efomu (FRI) NZEMA alari (auctt.) donga (CJT)

A tree of the evergreen forest, to 30 m high, straight slender bole, buttressed or sometimes stilted, occurring from Sierra Leone to the W Cameroons, and in E Cameroun, Gabon, Zaïre and Angola.

The wood is whitish, darkening to light brown, with a pungent odour when fresh (5), light and soft but tough and strong, of rather coarse texture and straight grain, easily worked and finishing smoothly, but not very durable (6, 7). The wood has no commercial use, but finds use locally. The slender stems are used for house-poles and can readily be sawn into planks (4, 6). In Zaïre the wood is used for carpentry and making chairs (2), and in Gabon for construction work (12).

The bark-slash is sweetly aromatic (9, 10, 11). Powdered bark is taken in the Ivory Coast as a sinus decongestant in the case of colds or for headache (1), and a bark-macerate is used in Gabon for headache (12).

Pieces of bark are used to make hut-walls (3, 6, 7, 8), and a tough cordage is made from the stringy inner layers (6, 7, 11, 12).

The fruits, which are about 5 cm long, fairly thick, black and 2–4-seeded, are a form of 'bush pepper' (7).

References:

1. Bouquet & Debray, 1974: 19. 2. Boutique, 1951: 335–6. 3. Cooper 60, K. 4. Cooper 139, K. 5. Cooper 234, K. 6. Cooper & Record, 1931: 19 with timber characters. 7. Dalziel, 1937. 8. Irvine, 1961: 25–26. 9. Kunkel, 1965: 206. 10. Savill & Fox, 1967: 44–45. 11. Taylor, 1960: 89–91. 12. Walker & Sillans, 1961: 73–74.

**Xylopia villosa** Chipp

FWTA, ed. 2, 1: 42. UPWTA, ed. 1, 9–10.

French: elo pubescent (*elo* from Abé, Aubréville).

West African: **IVORY COAST** ABE elo (A. Chev.) mossohué (A. Chev.) AKYE bello (A. Chev.) KYAMA ahué (Aub.; B&D) van-mê (Aub.) **GHANA** AKAN-ASANTE ɔbaa (BD&H) ɔbaa fufuo = *white Xylopia* (CJT) TWI obaa fufuo = *white Xylopia* (DF) obaa-pete (DF) **NIGERIA** EDO àghàkó (auctt.) IGBO ọ̀dà *a dialectal variant of ụ́dà* (JMD) ụ̀dà (JMD; KO&S) YORUBA palùfòn dúdú = *black palufon* (Punch; KO&S)

A tree to 26 m high of the high-forest occurring from the Ivory Coast to S Nigeria. The bole is slender, and slightly fluted, sap-wood is light brown and hard (7). The wood is elastic, very durable and tough. It is not attacked by termites (3, 4, 5), and finds use as house-posts, axe-handles (6) and like purposes.

The slash is light brown and sweetly-scented (7), and the powdered bark is used in the Ivory Coast as a sinus decongestant in cases of colds or headache (2).

The seeds are pounded to make a poultice for boils in S Nigeria (8).

References:

1. Aubréville, 1959: 1, 140. 2. Bouquet & Debray, 1974: 19. 3. Dalziel, 1937. 4. Irvine, 1961: 26. 5. Keay & al., 1961: 38. 6. Kennedy 415, K. 7. Taylor, 1961: 92. 8. Thomas, N. W., 1951, Nig. Ser., K.

**Xylopia sp. indet.** (Catterall 51, 61, K).

FWTA, ed. 2, 1: 42.
West African: **NIGERIA** IJO-IZON (Kolokuma) ẹngẹ́ (KW)

# APOCYNACEAE

**Adenium obesum** (Forssk.) Roem. & Schult.

FWTA, ed. 2, 2: 76. UPWTA, ed. 1, 365–6, as *A. hongel* A. DC.
English: desert rose.
French: baobab du chacal (*baobab of the jackal*); faux baobab (*false baobab*) (Senegal, Kerharo & Adam).
West African: **MAURITANIA** ARABIC ('Maure') teïduma es seba (Aub.) **SENEGAL** ARABIC (Ferlo) hongel (GR) ARABIC ('Maure') teïduma es seba (Aub.) BEDIK nya-makoelimbar (Ferry) FULA-PULAAR (Senegal) dara bogel (K&A) darabuki (K&A) darbugé (K&A) darbugöl (K&A) darbuki (K&A) lekki pöuri (Aub., ex K&A) TUKULOR dar bogel (K&A) daraboghé (Aub.; Laffitte) MANDING-MANINKA bulu kuruné (Aub.) fukala sitâdi (auctt.) kakalasita (Aub.; K&A) kongosita (Aub.; K&A) kulukuruné (K&A) sita kolokuru (Aub.; K&A) tukala sitâdi (auctt.) SONINKE-SARAKOLE fama (K&A) kidi sarané (Aub.; K&A) WOLOF guy sidéri (JB; K&A) lisugar (auctt.) **GUINEA-BISSAU** FULA-PULAAR (Guinea-Bissau) djindje pétè, djinje: *flower*; pétè: *rock* (*cf. petra, Greek*) (JDES; EPdS) MANDING-MANINKA bulu kuruné (CHOP; Aub.) fukala sitandi (Aub.) kakalsita (Aub.) kongasita (Aub.) sita kolokuru (Aub.) tukala sitandi (Aub.) **MALI** ARABIC (Upper Niger) hongel (GR) MANDING-MANINKA bulu kuruné (Aub.) fukala sitandi (Aub.) kakalasita (Aub.) kongosita (Aub.) sita kolokuru (Aub.) tukala sitandi (Aub.) SONINKE-SARAKOLE kidi sarané (Aub.) **GHANA** BIMOBA saa tesaga = *tornado thunder* (Cardinall) **NIGER** ARABIC shagar-el-sim (Aub.) ARABIC-SHUWA kuka meru (Aub.) HAUSA karya (Aub.) **NIGERIA** ARABIC shagar-el-sim = *poison tree* (JMD) sim-es-samak = *poison the fish* (JMD) ARABIC-SHUWA ɓokki *as for Adansonia digitata* (MM) kuka meru (JMD) FULA-FULFULDE (Nigeria) dar ɓokki = *false baobab* (JMD; MM) lekki peewuri = *medicine against coldness; probably an epithet* (JMD; MM) HAUSA gariya (MM) kariya, ƙaryaà = *a lie; alluding to free-flowering but rarely fruiting* (JMD; ZOG) KANURI kàwàmèrú (C&H)

A shrub to 3 m high or more with a stout swollen stem resembling a miniature baobab, especially when leafless, semi-succulent branches, common and widely dispersed throughout the drier parts of the Region, often planted, and occurring commonly through the drier areas of tropical Africa.

The plant is generally regarded as poisonous. No stock will graze it (1, 10). It is planted fairly commonly in towns and villages as an ornamental for its curious form and its attractive pink, red or white flowers. It can be grown as a decorative hedge (2, 4, 6). In Tanganyika (Usambaras) it is planted to mark the position of graves (5).

The leaves and stem exude when cut a slightly milky sap. This is applied in northern parts of Nigeria to septic wounds, intractable ulcers and to carious teeth. The Hausa use the root, with other roots, e.g. *Landolphia florida,* boiled in guinea-corn pap with blacksmith's slag for a day or two; the mixture is drunk copiously two or three times daily as a cure for venereal disease (4). The latex is dangerous to the eyes, and it is used in arrow-poisons (2). Medicinemen appreciate the plant's toxicity, and nowhere is it recorded used for internal medication, except perhaps as an ordeal or in magic (7). An aqueous macerate of the bark is used in friction and bathes for dermal affections, psoriasis and lice-infection (7). The bark of the stem and young branches is used in Senegal as a rat-poison and compounded into arrow-poison (3). The root is pounded and used to stupefy fish in Adamawa (4) and also in the soudano guinean region and for criminal purposes (2). In Northern Kenya the stems are powdered and mixed with water which is used for killing vermin on camels (8) and domestic stock (9). Preparations are also used for arrow-poisons in Kenya (8) and in Tanganyika (5), and in the latter state as a fish-poison (12). The root, with another unspecified bulbous plant, is added to bait to poison jackals and hyaenas in Somalia (McKinnon fide 6).

Preparations of the plant act as a cardiac poison like *digitalis* with associated affects on the central nervous system, the nerve mechanism of the heart and the heart muscle (4, 7, 11). Cardio-active glycosides are present which collectively have been called *adeniine.* A large number of glycosides have now been isolated and identified. There is evidence of races of differing phytochemistry (7).

In N Togo the plant is held to have magical attributes of keeping lightning away — hence its local name meaning 'tornado thunder.' In N Nigeria women who wish to have no more children wear the plant as a girdle (4).

References:

1. Adam, 1966, a. 2. Aubréville, 1950: 445. 3. Chevalier, 1937, b: 169. 4. Dalziel, 1937. 5. Greenway 4114, K. 6. Irvine, 1961: 609–11. 7. Kerharo & Adam, 1974: 153–5, with phytochemistry. 8. Mwangangi 1424, K. 9. Mwangangi & Gwynne 1102, K. 10. Naegelé, 1958, b: 878. 11. Oliver, 1960: 17, 43–44. 12. Semsei FH. 2162, K.

**Alafia barteri** Oliv.

FWTA, ed. 2, 2: 73. UPWTA, ed. 1, 366.

West African: **SIERRA LEONE** LOKO kpeɛŋgɛ (NWT) MENDE gbɛnge (JMD) ndambi *general term for woody climbers with thin stems* (NWT; FCD) **IVORY COAST** ANYI si-diafua-angbé (PG) **GHANA** AKAN-ASANTE momunimo (BD&H) FANTE edru = *medicine* (Armitage) TWI momonimo (auctt.) momorohemo (FRI) ANYI bodda (Armitage) **NIGERIA** IGBO (Obompa) ọta nza (NWT; JMD) YORUBA agbárí ẹtù = *Guinea fowl's crest* (Millson) ìbò-agba (Yor. Dict.)

A climbing shrub of closed-forest and secondary bush from Sierra Leone to W Cameroons.

The twining stems contain a bast fibre and are used as a binding material for roofs, etc. The plant is used as a fever medicine (2, 3). It is said, without further detail, to be poisonous (4). The Igbo name, *ota,* meaning 'a bow' suggests the stems may be used to make them. The plant contains a latex which has been used as an adulterant of better latices (1, 3).

References:

1. Burtt Davy & Hoyle, 1937: 9. 2. Dalziel, 1937. 3. Irvine, 1961: 611. 4. Jones 1531/FH1 6480, K.

**Alafia benthamii** (Baill.) Stapf

FWTA, ed. 2, 2: 74.

West African: **SIERRA LEONE** MENDE gbɛnge (NWT) wutɛngui (NWT) TEMNE *a*-tak (NWT)

A climber of the forest, recorded from Guinea, Sierre Leone, Liberia and S Nigeria, and extending into Gabon and the Congo basin.

The plant has been recorded used for fever in Sierra Leone (1).

Reference:

1. Thomas, N. W. 15, K (Arch.).

**Alafia lucida** Stapf

FWTA, ed. 2, 2: 73. UPWTA, ed. 1, 366 (excl. vernaculars).
West African: **IVORY COAST** BAULE tiegbanhema folué (B&D) KYAMA ebuliniama (B&D) **GHANA** AKAN-TWI momonimo (JMD)

A shrub climbing to 6 m of the closed-forest from Liberia to S Nigeria and also from E Cameroun to Angola and Uganda.

The plant is used in Ivory Coast in the treatment of jaundice and of swollen glands (2) and in Gabon for eye-troubles (4) and to heal sores (Sillans fide 2). A leaf-decoction is used in Congo to wash sores and is taken by draught for stomach complaints (1). Unnamed alkaloids have been reported in the seeds (2, 5). The latex is said to be abundant, dirty yellow, coagulating with sap of *Costus* (Costaceae) and is used in Congo to poison arrows, a statement in need of confirmation (3).

References:

1. Bouquet, 1969: 62. 2. Bouquet & Debray, 1974: 21. 3. Dalziel, 1937. 4. Walker & Sillans, 1961: 75. 5. Willaman & Li, 1970.

**Alafia multiflora** (Stapf) Stapf

FWTA, ed. 2, 2: 73. UPWTA, ed. 1, 372, as *Holafia multiflora* Stapf.
West African: **GHANA** AKAN-TWI *o*-kum adada = *it kills sores* (K&S; FRI) **NIGERIA** YORUBA aagbá (Punch; JMD) ibo-gidi (Punch; JMD)

A stout climber of the forest, and occurring in Ghana to W Cameroons and Fernando Po, and to Congo, Egypt and Sudan.

The liane contains a brown latex which is non-coagulating (2). It is applied in Ghana to wounds which are not healing properly, and can also be given diluted by draught. The Twi name *kumadada* means 'it kills the sores' (1). The latex is said to contain alkaloids.

References:

1. Irvine s.d. 2. Punch 49, 51, K.

**Alafia scandens** (Thonning) De Wild.

FWTA, ed. 2, 2: 73–74. UPWTA, ed. 1, 366, as *A. landolphioides* K. Schum.
West African: **SENEGAL** DIOLA fu nõng (JB) **THE GAMBIA** DIOLA (Fogny) fu nong = *carelessness* (DF) **GUINEA** FULA-PULAAR (Guinea) délébel (A. Chev.) murtégé **SIERRA LEONE** KONO teinsa *the fruit* (FCD) **IVORY COAST** KRU-GUERE (Chiehn) korodu (K&B) 'SENUFO' g-bande béllèsi yalaba (K&B) **GHANA** AKAN-TWI momonimo (auctt.) **NIGERIA** YORUBA ẹru-sheshon (Millson)

A scandent shrub or stout woody climber to 20 m long, of damp forest from Senegal to N and S Nigeria, and also in the Congo basin.

The flowers are waxy-white with a pink throat in beautiful terminal clusters (2). The plant is worthy of cultivation.

A leaf-decoction is taken in draught and the lees are applied in friction for rheumatism in the Sinematiali District of northern Ivory Coast (4). The latex is said to cause itching (Moor in 1, 3), and to be an ingredient of arrow-poison in Congo (3).

References:

1. Burtt Davy & Hoyle 1937: 9, as *A. landolphioides* K. Schum. 2. Dalziel 1937. 3. Irvine, 1961: 611, 613. 4. Kerharo & Bouquet, 1950: 183, as *A. landolphioides* K. Schum.

**Alafia schumannii** Stapf

FWTA, ed. 2, 2: 74.
West African: **IVORY COAST** AKYE biébum seu (A&AA)

A climbing shrub of the forest, recorded from Sierra Leone and Nigeria, and extending across Africa to Congo, Uganda and Tanganyika.
The plant bears attractive white-flowered densely corymbose inflorescences.

**Alafia sp. indet.** (Farmar 552)
West African: **GHANA** AKAN-TWI homafuntum (Farmar)

An unidentified species of the above Twi name has been recorded as used for coagulating rubber (1), presumably *Funtumia* latex.

Reference:

1. Irvine, 1961: 613.

**Allamanda** spp.

FWTA, ed. 2, 2: 54.
English: (yellow) allamanda.
West African: **GHANA** AKAN-TWI ɔtofaben (Asiedu)

A tropical American genus with some showy species that have become pantropical under cultivation. Most commonly grown are:
1. *A. cathartica* Linn. with bright yellow flowers. An infusion of the leaves is cathartic. It can be grown as an informal hedge but needs trimming which spoils its flowering. It makes a good stock into which less vigorous species of the genus can be grafted.
2. *A. nereiifolia* Hook., a semi-erect shrub with golden-yellow flowers smaller than those of *A. cathartica*.
3. *A. schottii* Pohl with large pale yellow flowers. It makes a good hedge and can be trained over an arch or pergola with good effect.

**Alstonia boonei** De Wild.

FWTA, ed. 2, 2: 68. UPWTA, ed. 1: 366, as *A. congensis* sensu Dalziel.
English: alstonia; timber trade — pattern wood, stoolwood.
French: emien (timber trade, from Ivory Coast vernacular).

West African: **SENEGAL** BANYUN ti keung (K&A) DIOLA bain (K&A) búdafélèk (JB; K&A) butétup (JB; K&A) tigög (K&A) FULA-PULAAR bâtâforo (K&A) PULAAR (Senegal) légéré (K&A) MANDING-MANINKA bâtâ foro (JMD; K&A) MANDYAK bikes (K&A) **THE GAMBIA** DIOLA (Fogny) bain = *winged termite* (DF) **GUINEA-BISSAU** CRIOULO portagaré (D'O) FULA-PULAAR (Guinea-Bissau) bantera-fôrô (D'O) MANDING-MANDINKA bantan-forô (D'O) **GUINEA** FULA-PULAAR (Guinea) leguéré (RS) KISSI tiendo (RS) KONO ioro (RS) KPELLE lâpra (RS) LOMA zolo (RS) MANO ioro (RS) **SIERRA LEONE** MENDE kalo wulo (*def.*) = *wooden plate or basin tree, from* kalo *wooden plate, basin;* wulo: *tree* (auctt.) kalowui (JORU) kalui kalo (*def.*-i) = *wooden plate* (FCD; S&F) kaowuli (DS) kauwi (JMD) kawuli *contraction of 'kalo wuli', and more commonly used* (FCD) **LIBERIA** DAN yung (GK; AGV) KRU-GUERE (Krahn) gona-tu (GK) **UPPER VOLTA** FULA-FULFULDE (Upper Volta) léguéré (K&B) **IVORY COAST** ABE onguié, honguié (auctt.) ABURE ebien (B&D) ADYUKRU idjhièlle (auctt.) AKAN-ASANTE amia (B&D) niamé dua (K&B) sindolo (K&B) sindro (B&D) AKYE kokpè (A&AA) kokué (auctt.) ANYI emien (auctt.) méa (K&B) AVIKAM o-uruzi (A. Chev.; K&B) BAULE emien (auctt.) miéï (K&B) DAN kliméné (RS) FULA-FULFULDE (Ivory Coast) léguéré (K&B; Aub.) KRU-GUERE moh(a)in (Aub.; K&B) muê (RS) GUERE (Chiehn) kié (K&B) tié (K&B; B&D) tien (B&D) GUERE (Wobe) kolatu (RS) KULANGO senuro (K&B) tenulo (K&B) KULANGO (Bondoukou) leroï (K&B) lerué (A. Chev.; RS) lerwé

(K&B) KWENI uro (auctt.) KYAMA aménê (B&D) korogbé (auctt.) korokué (auctt.) leroĩ (A. Chev.; K&B) lerué (RS) lerwé (K&B) NZEMA émien (A.Chev.; K&B) 'PATOKLA' kolaton (RS) **GHANA** VULGAR sinduro (DF) ADANGME adawura (FRI; CJT) sinu (FRI) AHANTA bakunin (auctt.) AKAN ɔnyame-dua *the tree when young* (DF) sinuru, sinduru *the tree when mature* AKAN-AKYEM senuro (FRI) ASANTE nyame-dua (Enti) onyame-dua = *Sky God's tree, when young* (FRI; CJT) senuru (Enti) sindru *applied when the tree is fully grown* (auctt.) sinuro (Enti) FANTE *o*nyame-dua = *sky God's tree; when young* (FRI; BD&H) sindru *applied to a full grown tree* (auctt.) TWI o-nyamedua = *Sky God's tree, when young* (auctt.) osen-nuru (FRI) *o*-sinuru (Enti) sinuru *the tree when mature* WASA sindru, sinuro (CJT, Enti) ANYI emiɛ (FRI) ANYI-AOWIN ɛmee (FRI) SEHWI ɛmee (FRI; CJT) sindru (BD&H) BAULE emiɛ (FRI) GBE-VHE o-nyemidua (FRI) VHE (Pecí) siakɛtɛkrɛ (FRI) NZEMA bakunin (FRI; CJT) emenle (CJT) nyamelebaka = *Sky God's tree* (FRI; CJT) **NIGERIA** BEMBI ofem (AHU) BOKYI bokuk (AHU; KO&S) EDO úkhú (auctt.) EFIK ébó (auctt.) idu (DRR) úkpò (AHU; KO&S) EJAGHAM etiap (auctt.) oguk (auctt.) EJAGHAM (Obang) etiap (AHU) oguk (AHU) EJAGHAM-ETUNG oguk (DRR) ENGENNI uguwa (JMD; DRR) IGALA ano (H-Hansen) IGBO ẹ́gbū (JMD; KO&S) IGBO (Awarra) égbú-ọ̀rà = *égbú of the people* (KW) IJO (Nembe) kìngbòù (KW) IJO-IZON (Egbema) kungbo (Tiemo) IZON (Kolokuma) kígbó (KW) IZON (Mein) èndóúndóú (auctt.) IZON (Oporoma) owéí bálá = *male bala* (KW) ISEKIRI okugbo (KO&S) MUNGAKA bekugbro (DRR) NKEM oguk (DRR) URHOBO ukpukuhu (KO&S) YEKHEE uk(h)u (JMD) YORUBA ahùn (auctt.) ako-ibepo (JRA) àwiń, awùn (auctt.) ogudu(g)bu (Punch; DRR) **WEST CAMEROONS** DUALA bòkùkà (JMD) KOOSI kuge (JMD) KPE wokuka (JMD) KUNDU kanja (JMD) LONG bokuk (JMD)

*NOTE: Names in S. Nigerian languages may also refer to A. congensis.*

A tree to 40 m high by over 3 m in girth, bole cylindrical and long to 27 m, with high narrow buttresses, of the evergreen and deciduous forest in damp situations, throughout the Region from Senegal to W Cameroons and extending across Africa to Egypt, Sudan, Uganda and Zaïre.

Till quite recently most literature concerning this species in the Region has been placed under *A. congensis* Engl. The latter species is however relatively rare and limited only to S Nigeria. Such references more correctly apply to *A. boonei* De Wild. (28).

The wood is yellowish white, soft, light, fine-textured, sap and heart-woods undifferentiated, interspersed with latex vessels and inclined to be gummy but works well. It is permeable and seasons well and is suitable for light carpentry, veneers, boxes, mouldings, match-splints, etc. It is perishable in the ground, and is not resistant to termites and borers, nor to blue stain unless treated with preservative (13, 16). Density is 0·5 fresh, 0·45 dry (24). In Sierra Leone it is considered a timber of secondary importance (9), and, as its Mende name (*kalo-wuli,* wooden plate tree) implies, it is used for making wooden bowls and plates. In Liberia it is used for bowls, toys, masks, canoes, etc. Export prospects are said to be doubtful but there is local industrial potential for domestic articles, wood-wool for packing bananas, matches etc. (28). In Ivory Coast it is used for benches and domestic things (4) and is said to be suitable for joinery and cabinetry (24). In Ghana it is carved into dolls and fetish emblems and made into dishes, platters, basins and stools (7, 8, 26), and it is currently recommended for boxes and crates, handcrafts, general joinery, pattern-making, plywood and veneers (3). The well-known Asante stools are often made of it, as also are stools in S Nigeria and Cameroons where it is made into carved toys, spoons, clogs, plates, images 'devil masks', etc. (10). It is officially recommended in Nigeria for the outer box of matchboxes (2). The Ijo carve small paddles from it for use in dancing (29). The sound box of an Yoruba musical instrument, *asọlogun,* a kind of zither, is made of it (27). In the Cameroons it is made into flasks for holding powder and into canoes (10); in the Sudan into boats and war-drums (10); in Uganda for bowls for holding food (11), and in hut-construction for rafters and window-frames (25).

The bark, and the root, are febrifugal and are said in Nigeria to be very effective in the case of ordinary malaria (1, 10, 22). A bark-decoction is also taken in Ghana for malaria (26) and in Cameroun (20). The bark of an *Alstonia* sp. is used in India for malaria and chronic diarrhoea. It is said to be inferior to cinchona bark but leaves no after-effects, e.g. no buzzing in the ears (22). In decoction it is used in Ivory Coast—Upper Volta to cleanse suppurating sores and exposed fractures (19); in Nigeria for sores and ulcers (1); on snakebite in Liberia (28); and for snakebite and arrow-poison in Cameroons (Mildbraed fide 10). The bark, leaves and roots are all used to relieve rheumatic pain and other pains (1, 6, 10, 28). The bark has a widespread use in Ghana to assuage toothache, and the Akan name *sindru* is a corruption of the words meaning 'tooth medicine' (26). In Sierra Leone, a chicken killed by a male child is cooked with pounded bark; the stomach becomes exceedingly bitter and is taken by those, especially women, suffering from intestinal disorders. The boy who killed the chicken must also partake. This treatment is also followed for curing barrenness in women over 30 years of age, and by women with umbilical suppuration — after eating, some pounded bark is bandaged over the navel (5). The bark is taken in macerate in Ivory Coast for jaundice, and sap for cough and sore throat, and externally for some skin-complaints (6). In Ghana a decoction is given after childbirth to promote expulsion of the afterbirth (16). The bark has anthelmintic use in Sierra Leone: it may be boiled and the liquor strained and taken, especially for children (12), or simply left to stand in a bottle of water (15).

Two indolic alkaloids, *echitamine* and *echitamidine,* have been determined in the bark, which in concentration appears to vary with location: Ghana 0·38–0·56%, Nigeria 0.15–0.31%, and Cameroons 0·18%, total alkaloids, principally echitamine. This is paralysing to the motor nerves similar to the action of curare. A lactone and triterpenes, *amyrine* and *lupeol,* have also been reported. (6, 14, 16, 18, 22.)

The latex is dangerous to the eyes and can cause blindness. It gives an inferior resinous coagulate which has been used to adulterate better rubbers. It has been used as a birdlime (10). The latex is applied to snake-bite after lancing in Ivory Coast, or it may be taken by draught (19). The latex is boiled in Nigeria and the concoction is taken for fever, especially in children (1). In Casamance (Senegal) latex is applied to refractory skin-troubles in children (17, 18). It is also smeared onto 'Calabar Swellings' caused by *Filaria* infection in Cameroons and the area is bandaged with latex and the crushed bark of *Erythrophleum guineense* (Leguminosae: Caesalpinioideae) (Mildbraed fide 10). It is considered galactogenic and is given to Baakpe women of the Cameroons Mountain area at childbirth (21). The latex is supposed to be an antidote for *Strophanthus* poison.

The leaves, pulped to a mash, are applied topically in Ivory Coast to reduce oedemas (19), and leaf-sap is used to cleanse sores in Casamance (18).

The tree has religious association for the Akan races in Ghana as shown by the names meaning 'Sky-God's tree.' This arises from the whorled branches of a young shoot being used to support fetish bowls holding food for spirits at domestic shrines (10, 26).

References:

1. Ainslie, 1937: sp. no. 22, as *A. congensis*. 2. Anon, 1965, b. 3. Anon. s.d.: 1. 4. Aubréville, 1959: 3, 189, as *A. congensis* Engl. 5. Boboh, 1974. 6. Bouquet & Debray, 1974: 21. 7. Burtt Davy & Hoyle, 1937: 9, as *A. congensis* Engl. 8. Chipp, 1922: as *A. congensis* Engl. 9. Cole, 1968: as *A. congensis*. 10. Dalziel, 1937. 11. Dawe 710, K. 12. Deighton 2993, K. 13. Eggeling & Dale, 1952: 24–26. 14. Goodson, 1932. 15. Herbalists, Joru Village, S.L., Nov. 1973. 16. Irvine, 1961: 613–5. 17. Kerharo & Adam, 1963, a. 18. Kerharo & Adam, 1974: 155–7, with phytochemistry and pharmacology. 19. Kerharo & Bouquet, 1950: 184, as *A. congensis* Engl. 20. Lehman, s.n., 18/4/1922, K. 21. Maitland 765, K. 22. Oliver, 1961: 18, 45–46; as *A. congensis*. 23. Savill & Fox, 1967, 46–47. 24. Schnell, 1950, b: as *A. congensis* Engl. 25. Styles 39, K. 26. Taylor, 1960: 93. 27. Unwin, 1920: 391, as *A. congensis* Engl. 28. Voorhoeve, 1965: 59. 29. Williamson 103, UCI.

**Alstonia congensis** Engl.

FWTA, ed. 2, 2: 68.

West African: **NIGERIA** IJO-IZON (Egbema) kùngbò (KW) IZON (Kolokuma) kígbó (KW) IZON (Mein) èndòùndòù (KW) IZON (Oiyakiri) kùgbò (KW) IZON (Olodiama) kùgbò (KW) YORUBA awùn (JMD) **WEST CAMEROONS** DUALA bòkùkà (Ithmann)

*NOTE: Names under A. boonei in S. Nigerian languages may also apply.*

A tree to over 30 m high of the evergreen forest in damp situations, occurring only in S Nigeria in the Region, and extending to Zaïre and Cabinda.

This tree is very similar in general appearance to *A. boonei* and it is probable that it has the same applications. It is distinguished from *A. boonii* in possessing a nearly glabrous inflorescence.

**Ancylobotrys amoena** Hua

FWTA, ed. 2, 2: 60. UPWTA, ed. 1, 373, as *Landolphia amoena* Hua.
West African: **GUINEA-BISSAU** MANDING-BAMBARA tukamalé (Brossart) MANINKA kondané (Paroisse) **MALI** MANDING-BAMBARA kunda ni nombo (A. Chev.; H&C) kurumale (H&C) MANINKA dudu (A. Chev.) kondané (H&C)

A liane to 20 m long by 15 cm girth, commonly in galleried and savanna woodland on the edge of water, and in open savanna, often in montane situations from Guinea across the northern part of the Region to N Nigeria, and extending to E Africa.

The plant contains a white latex which may be sparse, or abundant. It does not coagulate naturally, nor can it apparently be made to do so (3), but it dries to an useless powder (2). There are however references to it being a rubber yielder in the Bassa area of N Nigeria (Elliot fide 2), at Ilorin (1), and in Sudan (6).

In Tanganyika the stems are woven into baskets and fish-traps, and the Sukuma drip the sap straight into the eye for suppuration (4).

The fruits have edible pulp and seeds (1, 3), and they are commonly taken by monkeys in N Nigeria (1). In Tanganyika the fruits are used on sores (5).

The leaves have unspecified fetish uses (3).

References:

1. Ajayi FHI.19294, K. 2. Dalziel, 1937. 3. Pichon, 1953: 274–80. 4. Tanner 610, K. 5. Tanner 4351, K. 6. Turner 131, K.

**Ancylobotrys pyriformis** Pierre

FWTA, ed. 2, 2: 60.

A liane reaching to 30 m long by 32 cm girth, rarely in the forest zone of S Nigeria, and extending commonly into Zaïre.

The latex is abundant, white or bluish-white, or resinous, coagulating spontaneously to a white powder. This has been used to adulterate the rubber of *Clitandra cyumlosa*. It is prepared as an enema for administration to young children who refuse to suckle. The seeds are sometimes eaten in Zaïre.

Reference:

1. Pichon, 1953: 280–3.

**Ancylobotrys scandens** (Schum. & Thonn.) Pichon

FWTA, ed. 2, 2: 60. UPWTA, ed. 1, 375, as *Landolphia scandens* Didr.
West African: **GHANA** AKAN-AKUAPEM abonta (Thonning) TWI ɔbɔmene (FRI) **NIGERIA** EDO ùbó-ámiọ̀ghọ̀n (JMD) IGBO (Asaba) ọtọ poi (JMD) ọtọ-frifedi (JMD) YORUBA ìbó (JMD)

A liane to 20 m long by 50 cm girth, principally of the coastal region, occasionally inland, from Ivory Coast to S Nigeria, and extending into the Congo basin.

The flexible lianes are used in Congo as ties in hut-construction (3).

The stems yield an abundant white latex which is thick and coagulates to an inferior sticky substance that is somewhat plastic (1, 2, 3).

The fruits have an edible pulp, translucent and acid. They are much relished by monkeys (3).

References:

1. Dalziel, 1937. 2. Millen 105, K. 3. Pichon, 1953: 286–90.

**Anthoclitandra nitida** (Stapf) Pichon

FWTA, ed. 2, 2: 58.

A climber of limited distribution in Liberia and Ivory Coast.

The plant yields a rubber, but of unspecified value (1).

Reference:

1. Pichon, 1953: 230–1.

**Anthoclitandra robustior** (K. Schum.) Pichon

FWTA, ed. 2, 2: 58. UPWTA, ed. 1, 369.

A liane reaching 20 m long by 50 cm girth of primary forest and galleried forest in S Nigeria and W Cameroons, and extending to Zaïre and Cabinda.

The lianous stems are rigid and are used for this quality in Zaïre to make traps (2).

Latex may be almost absent or present in abundance and then of variable property, either gummy and useless, or yielding a good quality rubber, which has been exploited in some parts of the Congo basin.

The fruits have an edible, acidic pulp which is much relished. The seeds are eaten by monkeys (2).

References:

1. Dalziel, 1937. 2. Pichon, 1953: 231–5.

**Aphanostylis leptantha** (K. Schum.) Pierre

FWTA, ed. 2, 2: 55.

Liane 3–8 m long of the forest and rare in S Nigeria, and occurring more commonly in Cameroun and Gabon.

The plant yields a white latex of unspecified value. The plant has at one time been under plantation cultivation in Gabon (1).

Reference:

1. Pichon, 1953: 237–8.

**Aphanostylis mannii** (Stapf) Pierre

FWTA, ed. 2, 2: 59. UPWTA, ed. 1, 369.

English: false 'jawe' (the rubber, from Mende, Sierre Leone, Imperial Institute).

West African: **GUINEA-BISSAU** FULA-PULAAR (Guinea-Bissau) djacoram (JDES; EPdS) **SIERRA LEONE** MENDE jawa (Smythe) jiawa (*def.*-i) (JMD)

A liane attaining 35 m long by 50 cm girth of diverse forest types from swampy to dry, and to 800 m altitude, and in open sites, from Guinea to S Nigeria and extending into Zaïre and Angola.

The stems are used to make traps (2).

The plant contains a latex which may be very sparse or abundant, useless or coagulating to a good rubber (2). In Sierra Leone it was the source of a product known as *jawe, false jawe* or *Manoh twist*.

The fruit pulp is edible, astringent and acid (2).

References:

1. Dalziel, 1937. 2. Pichon, 1953: 241–4.

**Baissea axillaris** (Benth.) Hua

FWTA, ed. 2, 2: 80. UPWTA, ed. 1, 367.
West African: **NIGERIA** YORUBA ìmu (Punch)

A climbing shrub to 3 m high of the rain-forest of S Nigeria and W Cameroons, and extending to Angola.

The liane is used in S Nigeria to make palm-wine more intoxicating (2). A decoction of leafy twigs is taken in Congo for kidney disorders (1).

References:

1. Bouquet, 1969: 63. 2. Punch 77, K.

**Baissea breviloba** Stapf

FWTA, ed. 2, 2: 79.
West African: **NIGERIA** EDO èbálẹ̀lẹ̀ (Farquhar)

A climbing shrub to 3 m high of the closed-forest and secondary growth in Guinea to S Nigeria.

**Baissea lane-poolei** Stapf

FWTA, ed. 2, 2: 79.
West African: **SIERRA LEONE** MANDING-MANINKA (Koranko) kisse (D.G.Thomas)

A climbing shrub of secondary vegetation in Sierra Leone and Liberia.

**Baissea laxiflora** Stapf

FWTA, ed. 2, 2: 79.

A stout woody climber to 13 m, of the rain-forest in S Nigeria and W Cameroons, and extending to the Congo basin.

The lianous stems have been used in the Uburubu area of S Nigeria as rope for traps (1).

Reference:

1. Thomas, N. W. 2266 (Nig. Ser.), K.

**Baissea multiflora** A. DC.

FWTA, ed. 2, 2: 79. UPWTA, ed. 1, 367.
French: liane étoilée (Berhaut).
Portuguese: landolfia (Guinea-Bissau, d'Orey).

West African: **SENEGAL** BANYUN kikof kidigen (K&A) BASARI a-ndyakɔrẹnd (Ferry) a-ngwẹtàn (Ferry) ɛ-kutàn (K&A; Ferry) BEDIK ga-nèmbẹ́t (Ferry) DIOLA (Fogny) bufem befol *the liane* (K&A) sifemb *the fruit* (K&A) sikénay (K&A) DIOLA (Tentouck) bufem befol *the liane* (K&A) sifemb *the fruit* (K&A) sikénay (K&A) FULA-PULAAR (Senegal) ńalâporé (K&A) salanôbo (K&A) KONYAGI nọnkọt, ɔ ntak ɔ-rẹnk (Ferry) MANDING-BAMBARA kunda ninôbo (JB; K&A) MANDINKA salanombo (*def.*-o) (Sebia Jatta) MANDYAK bépélam (K&A) SERER ngutin (JB; K&A) WOLOF dam tap (JB; K&A) **THE GAMBIA** DIOLA (Fogny) sifemb = *it cracks* (DF) MANDING-MANDINKA salanombo (*def.*-o) (Fox) **GUINEA-BISSAU** FULA-PULAAR (Guinea-Bissau) poré (D'O) **GUINEA** MANDING-MANINKA kondané nombo (JMD) sala nombolé (JMD) **SIERRA LEONE** MENDE ndabe-gulo (*def.*-guli) (LAK-C; JMD); *perhaps* = ndamba-wulo (FCD) sawa *a group name; see Strophanthus gratus* (FCD) **MALI** MANDING-BAMBARA kunda ni nombo (JMD) m-vugu (GR) MANINKA kondané nombo (JMD) sala nombolé (JMD)

A woody climber to 8 m, of the woody savanna from Senegal to Niger, and also in Cabinda.

The plant is an ornamental climber with copious panicles of white or pinkish sweet-scented flowers (4–6, 12). The stems are very strong and are used for tying house-roofs in Sierra Leone (11). The bark in The Gambia is made into rope (13). The plant contains a white latex which has in the past been collected to produce a good quality rubber (The Gambia, 13; Ghana, 2, 3). Cattle in Senegal will browse the foliage (1).

The plant has a good reputation with Senegalese medicinemen who use the roots and bark in decoction after prolonged boiling for calming colic without causing diarrhoea, for infantile diarrhoea (10) and for treating female sterility (7). An aqueous decoction of the liane is considered diuretic and is taken by draught for rheumatism, arthritis, kidney troubles, lumbago, and for general lassitude while the lees are applied in frictions (7, 9, 10). The bark, and more especially leafy twigs rich in latex are taken in various preparations for oedemas arising through deficiencies (10). Root-powder in water is used for conjunctivitis, and mixed with food or drink is a treatment for appendicitis (10).

The Basari of western Senegal prepare a leaf-macerate in a bath as a stimulant and fetish for the hunt (8).

References:

1. Adam 1966, a. 2. Burtt Davy & Hoyle 1937: 91. 3. Chipp 148, K. 4. Dalziel, 1937. 5. Gomes e Sousa, 1930: 81. 6. Irvine, 1961: 615, as *B. caudiloba* Stapf. 7. Kerharo & Adam, 1962. 8. Kerharo & Adam, 1964, a: 408 9. Kerharo & Adam, 1964, b: 410. 10. Kerharo & Adam, 1974: 157–8. 11. King-Church 12/1922, K. 12. Roberty 3055, IFAN. 13. Williams, F. N., 1907: 379.

**Baissea zygodioides** (K. Schum.) Stapf

FWTA, ed. 2, 2: 79.
West African: **LIBERIA** MANO ba bɛlɛ (Har.)

A woody climber of the forest of Guinea to Ghana.

**Callichilia monopodialis** (K. Schum.) Stapf

FWTA, ed. 2, 2: 64.

A shrub to 2 m high, known only from S Nigeria and W Cameroons, and in E Cameroun.

An unnamed alkaloid is reported present in the roots, leaves, stems and seeds (1).

Reference:

1. Willaman & Li, 1970.

**Callichilia stenopetala** Stapf

FWTA, ed. 2, 2: 64.

A climbing shrub to 1.7 m high, of thickets and secondary jungle in Liberia and S Nigeria.

The alkaloid, *vobtusine,* has been reported present in the aerial parts and in the roots (1).

Reference:

1. Willaman & Li, 1970.

**Callichilia subsessilis** (Benth.) Stapf

FWTA, ed. 2, 2: 64. UPWTA, ed. 1, 367.

West African: **SIERRA LEONE** LOKO (Magbile) vaimi (NWT) TEMNE *a*-kpafoto (NWT; JMD) **GHANA** AKAN-ASANTE aba-nua (CV; FRI) TWI aba-nua (CV)

A climbing shrub to 2.5 m high of the understorey of the evergreen or deciduous forest, from Guinea to S Nigeria, and in Cameroun.

The plant contains white latex (1). The alkaloids, *callichiline* and *vobtusine,* have been reported present in the roots, stems and leaves (2).

References:

1. Burtt Davy & Hoyle, 1937: 9. 2. Willaman & Li, 1970.

**Carissa edulis** Vahl

FWTA, ed. 2, 2: 54. UPWTA, ed. 1, 367.

English: Carrisse; Carandas plum (but correctly this is *C. carandas* Linn., the Karaunda of India and SE Asia).

West African: **SENEGAL** BEDIK ga-kàràbú (Ferry) FULA-PULAAR (Senegal) gubé (K&A) kâboro (CHOP, ex K&A) MANDING-MANINKA kuma kuma (Aub., ex K&A) SERER ndini (JB; K&A) **GUINEA** FULA-PULAAR (Guinea) kamboro (CHOP) MANDING-MANINKA kuma kuma (Aub.) **MALI** MANDING-MANINKA kuma kuma (Aub.) MANINKA (Wasulunka) tomboro ba (A. Chev.) **UPPER VOLTA** 'SENUFO' suruku n'tombolo (Aub.) **IVORY COAST** MANDING-MANINKA kuma kuma (FB) 'SENUFO' suruku n'tombolo (Aub.; FB) **GHANA** ADANGME aflamŋme (FRI) akokɔ-bɛsã = *the fowl will be consumed* (FRI) buɛtʃo (FRI) ŋmeetʃo, ŋmei: *thorn;* tʃo: *trees* (FRI) AKAN-BRONG dakumena (FRI) TWI nkokɔ-bɛsa *from* akokɔ: *fowl*; besa: *will finish,* i.e., = *the fowl will be consumed* (FRI, Enti) nkɔŋkyere (FRI) DAGAARI melimegoa (AEK) nolimagoa (AEK, ex FRI) GA akɔ́kɔbɛsã (FRI; KD) GBE-VHE lande (FRI) VHE (Awlan) botsu(i) (FRI; JMD) **DAHOMEY** GBE-FON ahazo (Aub.; Laffitte) VHE (Awlan) ahozo (Laffitte) **NIGERIA** HAUSA bagozaki (JMD; ZOG) ciìzaáki, gizaki (auctt.) kauci (McElderry) leèmùn tsuntsuu (auctt.) úwaá-bánzaá (JMA; ZOG)

A spiny shrub to about 5 m high, of dry deciduous forest and coastal thickets, extending across the Region from Senegal to W Cameroons, and throughout the drier parts of tropical Africa and across Asia to Indochina. It has been reported as a parasite on henna plants in N Nigeria (14).

The abundant branching habit and the presence of stiff spines to 5 cm long make the plant suitable for planting as a protective hedge (7). Propagation is easy by seed and by cuttings. The white or purple flowers which are carried for much of the year add an attractiveness (2, 10).

The plant is browsed by goats and camels in the dry parts of Sudan (10). In Guinea the boiled leaves are applied as a poultice to relieve toothache (16).

The bark contains a white latex (7) of no recorded property.

The root if crushed emits a strong smell of methyl salicylate and if rubbed onto the fingers produces a prickly sensation (13). The root-bark is mixed with spices and used as an enema for lumbago and other pains in Ghana (7). Root-scrapings are used in the soudano-guinean region for glandular inflammation (adenites) (1). A root-decoction is taken in Senegal to which pimento is added as an anthelminthic, especially against *Taenia* (11, 12). The plant is known in Tanganyika for anthelminthic action in both man and animals (10) and the ground-up root as a remedy for venereal disease (18). In Ghana the root is chewed and the saliva swallowed, the root-sap being considered tonic and restorative of virility (7, 9, 10). They are also used as 'bitters', macerated in rum, gin, etc. (7) and as an expectorant (15). The roots are put into water-gourds to impart an aggreeable taste to the water, and they are sometimes added to soups and stews for the same purpose, or to disguise a strong smell. The Twi name meaning 'the fowl will be consumed' implies an imparted palatibility (7, 10). In E Africa, the root is used for chest-complaints (3), cough remedy, tonic and abortifacient while the Masai take it with goat meat as a treatment for gastric ulcer (18). In Kenya a piece of the root is fixed into a hut-roof as a snake-repellant (18).

The presence of cardiotonic glycosides in the roots of related species in Asia is well documented (5, 17), but in this species there are conflicting reports, and the reported presence of *carissin* is not confirmed by several observers (12).

The fruit, often borne in pairs, are red-black ripening blue-black, and are sweet and are pleasant to eat (1, 2, 4, 6, 7, 10). In Ghana they are often added to the food of a sick person as an appetizer (9, 10). Vinegar can be made from them by fermentation (7, 8, 10). They are made into a jam in the soudan area of the Region (1) and in Kenya (Gardner fide 10).

References:

1. Aubréville, 1950: 445. 2. Aubréville, 1959: 3, 190. 3. Bally, 1937. 4. Berhaut, 1967: 94. 5. Bisset, 1957. 6. Busson, 1965: 407. 7. Dalziel, 1937. 8. Irvine, 1948: 264–5. 9. Irvine, 1952, a: 30, 36. 10. Irvine, 1961: 616–8. 11. Kerharo & Adam, 1964, b: 420. 12. Kerharo & Adam, 1974: 158. 13. Laffitte, s.n. 6 May 1937, IFAN. 14. McElderry FHI.16451, K. 15. Oliver, 1960: 21, 52. 16. Pobéguin, 1912: 41–42. 17. Singh et al., 1963. 18. Watt & Breyer-Brandwijk, 1962: 80–81.

## **Catharanthus roseus** (Linn.) G. Don

FWTA, ed. 2, 2: 68.

English: Madagascar periwinkle, oldmaid (Irvine).

French: pervenche de Madagascar.

West African: **SIERRA LEONE** KRIO joy-sie *the mauve form, for the 'joy' it gives* (FCD) waitie = *white; the white form* (FCD)

A semi-woody perennial (occasionally annual) herb to about 50 cm tall with white or pink flowers, originally from tropical America and now dispersed and more or less naturalized in the tropics.

The plant is distinctly ornamental and is widely grown in gardens. It is said to have entered western horticulture by seeds from Madagascar hence the English and French vernaculars. How or when the plant reached Madagascar is not known.

The plant does not enter into West African local medicine (1, 3) indicating perhaps a relatively recent advent to the Region. It is widely known throughout its dispersal elsewhere in the world as a remedy for diabetes. There is however some doubt about its efficacy for tests have shown no lowering of blood-sugar levels. Such benefit as may arise from medication with the plant may come from its digitalis-like action. In Asia particularly the plant has innumerable applications to almost every common ailment (2).

Clinical interest in the plant dates back to the 1950s when certain alkaloids were isolated that hopefully would give control of Hodgkin's disease and chorio-carcinoma, carcinoma of the breast, and leukaemia. The most promising of the alkaloids were *vinblastine, vincristine, vinleurosine* and *vinrosidine,* the first two finding use in therapy but may be but palliative, not curative. A very extensive series of alkaloids has been extracted and identified from all parts of the plant with an extensive diversity of pharmacological activity (2–8).

References:

1. Bouquet & Debray, 1974: 22. 2. Farnsworth, 1961: with folk-medicine, and many references to pharmacology. 3. Kerharo & Adam, 1974: 158–62, with phytochemistry, pharmacology and bibliography. 4. Quisimbing, 1950: 725–7, with references. 5. Svoboda, 1962. 6. Svoboda, 1964, for pharmacology and phamaceutics. 7. Willaman & Li, 1970, with many references on phytochemistry. 8. Watt & Breyer-Brandwijk, 1962: 23 as *Lochnera rosea* Reichb., with references.

## **Chonemorpha macrophyllum** G. Don

FWTA, ed. 2, 2: 54.

A woody climber to the tops of tall trees in its area of dispersal from the Himalayas to Ceylon and Java. It has been introduced into the West African region as an ornamental for its large, white, sweet-scented flowers.

The bark is a source of fibre which is resistant to both fresh and salt water. In SE Asia it is used to make fishing nets. The stems yield a rubber which flows poorly and contains much resin. It is of little economic value. The bark and leaves contain an unnamed alkaloid (1). *Chonemorphine* has been found in the roots (2).

References:

1. Burkill, 1935: 531. 2. Willaman & Li, 1970.

## **Clitandra cymulosa** Benth.

FWTA, ed. 2, 2: 57. UPWTA, ed. 1, 369.

West African: **GUINEA-BISSAU** KISSI yuruan (A. Chev.) KPELLE yabua (A. Chev.) **SIERRA LEONE** VULGAR jenjé *the rubber* (Conteh) poré *the vine* (Conteh) LOKO ibɛŋ guru (NWT) **IVORY COAST** DAN yuhaba (A.Chev.) **NIGERIA** BASSA (Kwomu) marodi (Elliott) HAUSA baƙin danko = *black rubber the product* (JMD) bassa moradi (ZOG)

A robust vine sometimes attaining 100 m long by 1.25 m girth, of the forest and forest margins of the dry zone from Guinea to N Nigeria and extending to Zaïre, Angola, Uganda and Tanganyika.

The plant contains an abundant latex, watery, somewhat sticky and generally valueless, but occasionally plants yield a black rubber of good quality (3). The plant has been exploited in Guinea and Ivory Coast for its black rubber where along with *Landolphia owariensis* they were the main source of vine-rubber (1). In Sierra Leone its rubber has been used as an adulterant (4). In N Nigeria it has been used to produce a rubber known in Hausa as *bakin dankwo* (= black rubber), and in eastern Nigeria a product described as good black and brown rubber. It is said that the root rubber of Benin called 'red ball' was derived from this species. Uganda vine rubber known as *kapa* or *kapa gambwa* is also from this plant. Coagulation in Nigeria was achieved by dilution with water, boiling and stirring till the coagulum sticks to the stirrer. The latex obtained in Ivory Coast was apparently more difficult to coagulate, prolonged boiling without dilution being necessary. The newly prepared coagulum is white but becomes brown in time and acquires elasticity (1).

The latex is used medicinally in Zaïre, taken by mouth for stomach aches under the premise that by coagulating in the stomach it will enfold the trouble and carry it out, but occasionally such medication results in occlusion of the intestine and death (3).

The fruit is edible, acid and is avidly sought after by both man and monkeys (3). It forms a major source of the chimpanzee's diet in W Cameroons which also consume the seeds (2).

References:

1. Dalziel, 1937. 2. Gartlan 34, K. 3. Pichon, 1953: 205–11. 4. Smythe 320, K.

**Cylindropsis parviflora** Pierre

FWTA, ed. 2, 2: 61.

A liane to 40 m long by 30 cm girth, of the forest and recorded only from Oban, S Nigeria in the Region, but occurring from Gabon to Zaïre and Cabinda.

The plant contains an abundant latex coagulating to give a black rubber of no recorded worth.

The fruit serve as food for monkeys. The seeds are also eaten.

Reference:

1. Pichon, 1953: 329–30.

**Dictyophleba leonensis** (Stapf) Pichon

FWTA, ed. 2, 2: 59. UPWTA, ed. 1, 368 as *Carpodinus macrophyllus* A. Chev. and 374 as *Landolphia leonensis* Stapf.

English: witch vine (Liberia, after Basa, Cooper and Record).

West African: **SIERRA LEONE** MENDE hole (*def.*-i) (JMD; MP) kpuwi (FCD) nali (JMD; MP) **LIBERIA** KRU-BASA way-doo, way-du = *witch vine* (C&R) MENDE nali **IVORY COAST** KYAMA atruan niama (B&D)

A forest liane occurring from Guinea to Ghana.

The plant yields a latex of variable quality, viscid and useless, or coagulating to a rubber of good quality (4, 6). The latex can be used as a bird-lime (5b).

The leaves are used in Liberia with the bark of a plant called *gbu-aye* (Basa) which is probably *Xanthoxylum gilletii* (De Wild) Waterm. (syn. *Fagara macrophylla* Engl., Rutaceae) to make a poultice to apply to the swollen parts in mumps (3, 4, 5b). In Ivory Coast a leaf-decoction is given by the Kyama for rheumatism and arthritis. Examination of the leaves for alkaloids has revealed none — but see under *D. lucida* Pierre.

The fruits have an edible pulp and are occasionally eaten (2, 5a, 6).

References:

1. Bouquet & Debray, 1974: 23. 2. Busson, 1965: 408. 3. Cooper 423, K. 4. Cooper & Record, 1931: 106. 5a. Dalziel, 1937: as *Carpodinus macrophyllus* A. Chev. 5b. Dalziel, 1937: as *Landolphia leonensis* Stapf. 6. Pichon, 1953: 256–9.

**Dictyophleba lucida** (K. Schum.) Pierre

FWTA, ed. 2, 2: 59.

A liane reaching 35 m long by 50 cm girth, much-branched, of primitive forest, rare in Nigeria, and occurring widespread eastwards to Tanganyika, Mozambique and S Rhodesia.

The plant contains a white latex of variable quality, sparse or abundant, fluid or tacky, coagulating only to a poor quality rubber (2).

A root-decoction is given to small children to drink in Tanganyika as a vermifuge (1).

Phytochemical assay of the plant has revealed the prescence of *chonemorphine, dictyodiamine* and three other alkaloids (4).

The fruit is sweet-smelling with an edible pulp of sweet/sour flavour (2). It is common food of the chimpanzee in Tanganyika (3).

References:

1. Haerdi, 1964: 130. 2. Pichon, 1953: 262–5. 3. Toyoshima 271, K. 4. Willaman & Li, 1970.

**Dictyophleba ochracea** (K. Schum.) Pichon

FWTA, ed. 2, 2: 59.

A liane to 40 m long, slender, 16 cm girth, of the forest, and rare in S Nigeria, extending to Zaïre and Cabinda.

The plant contains a white latex which is very variable, sparse or abundant, yielding an inferior or a good quality rubber. The fruit has an edible pulp of a sweet agreeable flavour (1).

Reference:

1. Pichon, 1953: 251–6.

**Dictyophleba rudens** Hepper

FWTA, ed. 2, 2: 59.

A lofty climber of streamside savanna forest, of Adamawa, N Nigeria.

The vine contains a quantity of white latex of undetailed quality (1).

Reference:

1. Hepper, 1963.

**Ervatamia coronaria** (Jacq.) Stapf

FWTA, ed. 2, 2: 54.

English: East Indian rosebay; grape-jasmine.

An ornamental shrub reaching to 3 m tall, native probably of northern India, but widely dispersed by man, and now cultivated as an ornamental in the West African region.

In its native area the wood is burnt as incense and is used to make perfume, and medicinally to provide a refrigerant (cooling) drink. A decoction of the roots is used in Indonesia to stop diarrhoea. Pulp surrounding the seed is used in the Himalayan area to produce a red dye (1).

Alkaloids *coronaridine, dregamine, tabernaemontanine* and others unnamed are reported present in the plant (2).

References:

1. Burkill, 1935: 941. 2. Willaman & Li, 1970.

**Farquharia elliptica** Stapf

FWTA, ed. 2, 2: 74.

A lofty climber of the closed primary forest and secondary forest, recorded in Ivory Coast, Ghana and S Nigeria, and extending to Congo.

The plant has an abundance of pure white sweet-scented flowers in terminal cymes. It also has a white latex.

**Funtumia africana** (Benth.) Stapf

FWTA, ed. 2, 2: 74. UPWTA, ed. 1, 371.

English: rubber tree; false rubber tree: 'male' funtum, 'white' ofrumtum (from Akan dialects, Ghana, Dalziel); trade — the rubber: Lagos rubber (Nigeria, Dalziel); bush rubber (Ghana, Taylor).

French: trade — the rubber: pri (Francophone territories, from Anyi, Ivory Coast).

Portuguese: pao cadeiro (S. Tomé).

West African: **SIERRA LEONE** KISSI tendo (S&F) waŋgolo (S&F) KONO bobo (FCD; S&F) LOKO watia (S&F) MANDING-MANINKA (Koranko) bandapare (S&F) buŋkankoŋ (S&F) poraŋ (S&F) MENDE bobo (*def.*-i) (FCD; S&F) TEMNE *an*-wathia (auctt.) **LIBERIA** KRU-BASA bu-ay-boh (C&R) MENDE bobo (C&R) **IVORY COAST** ABE po, puo, poyu, pwo = *rubber* (auctt.) ABURE botroromi (B&D) ADYUKRU bancero (auctt.) gbossoro (A&AA) AKAN-ASANTE flomondu (B&D) AKYE krokué (A. Chev.) pé-sain (A. Chev.) pri (Banco; A. Chev.) pusso ué (RS; Aub.) sohué (A. Chev.) KULANGO (Bondoukou) manan-wala *the rubber* (A. Chev.) wala (A. Chev.) KYAMA adiakoi, adiakua (A. Chev.) boko (B&D) diakua (RS; Aub.) NZEMA afomnondu (A.Chev.) 'KRU' dodocé (RS; Aub.) **GHANA** ADANGME *o*-sɛsɛ(-o) *the latex* (FRI) AKAN-ASANTE mama, maamae = *sticky, referring to the sticky latex* (auctt.) ɔkae, okan (Enti) ɔkan (Enti) TWI (DF) ɔkan o-sɛsɛ (FRI) WASA okae, ɔkan ɔkan (Enti) ɔkan, ɔkae, maamae (Enti) GA osɛ́sɛ, sɛ̀sɛ (FRI; KD) GBE-VHE kpoli (FRI) kpomi (FRI; CJT) VHE (Awlan) kpomi = *hunch-back* (FRI) VHE (Pecí) kpomi = *hunch-back* (FRI) NZEMA fulmuntu-okai (CJT) **NIGERIA** BOKYI nkwame (JMD; KO&S) EDO áyọ̀n = *rubber*, or *latex* (auctt.) básábású (JMD; JRA) IGBO mba-miri (JMD; KO&S) YORUBA akọ iré̩ = *male ire̩* (auctt.) iré̩

A tree to 25 m high, bole slender cylindrical unbuttressed to 1.60 m girth, of deciduous and evergreen forests from Guinea-Bissau to W Cameroons and Fernando Po, and widely dispersed elsewhere in tropical Africa.

The tree is of rapid growth, and is an evanescent colonizer of abandoned land. The wood is soft, white, uniform in texture and fine-grained. It is easy to work and is ubiquitously used for carving, stools, bowls, clogs, doors, paddles, and miscellaneous household requirements. (4, 7, 10, 11, 12.) It is said to be durable enough in Liberia for sawing into planks and house-timbers though it is subject to stain and to attack by beetle (5, 6).

The bark contains a latex that is gluey and inferior. It has been used to adulterate better rubber. (See *F. elastica* Stapf.) It is suitable for use as a bird-lime.

The dried powdered leaves are used in Liberia to treat fire-burns (5, 6). A number of alkaloids has been reported present in the leaves, and other parts (1, 3, 13). Principally are *funtumine* and *funtumidine*. They are hypotensive and locally anaesthetic. They are easily transformed into steroidal hormones (3). The roots pounded and mixed with palm-wine and water have been used in the Benin area of Nigeria for incontinence of urine (2). A root-decoction is drunk for amoebic dysentery in Tanganyika (9).

The seed pod contains a fine floss surrounding the seeds. This is used for stuffing pillows in Ghana (10) and in Sierra Leone (8). In the latter country it is often preferred over *Ceiba* kapok (11).

References:

1. Adegoke et al. 1967. 2. Ainslie 1937: sp. no. 162. 3. Bouquet & Debray, 1974: 23–24. 4. Burtt Davy & Hoyle 1937: 11. 5. Cooper 373, K. 6. Cooper & Record, 1931: 106. 7. Dalziel, 1937. 8. Glanville 249, K. 9. Haerdi, 1964: 130, as *F. latifolia* (Stapf) Stapf. 10. Irvine 1961: 621. 11. Savill & Fox, 1967: 47–48. 12. Walker & Sillans, 1961: 81, as *F. latifolia* Stapf. 13. Willaman & Li, 1970: as *F. africana* (Benth.) Stapf and *F. latifolia* (Stapf) Stapf.

**Funtumia elastica** (Preuss) Stapf

FWTA, ed. 2, 2: 74, 76. UPWTA, ed. 1. 371–2.

English: West African rubber tree; Lagos silk rubber tree; bush rubber (Ghana, Taylor); 'female' gboi-gboi (Sierra Leone, from Mendi; Smythe, K); 'female' funtum (Ghana, from Akan); 'female' iré (Nigeria from Yoruba).

West African: **SIERRA LEONE** KISSI tendo (S&F) waŋgolo (S&F) KONO bobo (S&F) LOKO watia (S&F) MANDING-MANINKA (Koranko) bandapare (S&F) buŋkankoŋ (S&F) poraŋ (S&F) MENDE bobo (*def*.-i) (JMD; S&F) TEMNE *an*-wathia (auctt.) **LIBERIA** DAN clien (GK) KRU-GUERE (Krahn) dido (GK) dirro (GK) MANO si g(u)illi (JMD) **IVORY COAST** AKAN-FANTE amané-dua = *rubber tree* (A. Chev.; K&B) o-funtum, poyu dua = *rubber tree* (A. Chev.) TWI plapodorosé-populu (A. Chev., ex K&B) AKYE amale(u) (K&B; A&AA) pé (A. Chev.; K&B) pé-chi (A. Chev.; K&B) pri (auctt.) sué (K&B) sué amalé (RS; Aub.) ANYI efurumundu (auctt.) fumundu (auctt.) BAULE potombo (auctt.) potomo (A&AA) GAGU kni (B&D) KRU-BETE trui (K&B) uruba su (A.Chev.; K&B) GREBO (Plapo) dorosé populu (A. Chev.) GUERE diô (RS) gluatu (K&B) ulua (RS) ulua (RS) urobo (RS) GUERE (Chiehn) kricko (B&D) GUERE (Wobe) uluatu (RS) wolobatu, wolowatu (A&AA) NEYO plapodorosé-populu (A. Chev., ex K&B) twi (A. Chev.) KWENI fotobo (auctt.) KYAMA frumudu (B&D) NZEMA amanĕ dua = *rubber tree* (A.Chev.) o-funtum (A.Chev.) o-funtun poyudua (K&B) poyu dua = *rubber tree* (A.Chev.) 'KRU' (Lower Cavally) bébéti (A. Chev.; K&B) 'PATOKLA' têtu (RS) **GHANA** AKAN fruntum (DF) funtum (anon; Armitage) AKAN-AKUAPEM ofruntum, ofuruntum (Enti) ASANTE aman = *rubber* (CJT) *o*-fruntum (auctt.) fruntum (auctt.) o-furuntum (FRI) FANTE o-fruntum (auctt.) TWI *o*-fruntum (auctt.) funtum (CJT) *o*-furuntum (FRI) ofruntum, ofuruntum (Enti) WASA fruntum (BD&H) o-fruntum (Enti) ofuruntum (FRI) ANYI frummundu (FRI) fummundu (FRI) ANYI-AOWIN efunmuntum (CJT) efunumundum (FRI) SEHWI o-frundum (FRI) BAULE potombo (FRI) GBE-VHE funtum (JMD) guni (CV; FRI) ofuruntum (FRI) puni (auctt.) NZEMA efunmuntum (CJT) efunumundum (FRI) fulmuntu (CJT) fulmuntu (CJT) NZIMA fulmuntu **NIGERIA** ABUA ạmạmạ̀ạm k'ákẹ̀ (JMD; DRR) BOKYI nkwame (JMD; KO&S) EDO árábà-nékhùi (Amayo) áyọ̀n = *rubber, latex* (auctt.) básábásá (JMD) EJAGHAM (Keaka) okon-anungi (DRR) EJAGHAM-ETUNG ebuk-etikpa (DRR) okpo-njok (DRR) EKPYE míní-emā (JMD; DRR) ENGENNI abakwa (JMD; DDR) IGBO mba (KO&S) IGBO (Arochukwu) mba (JMD; DRR) IGBO (Awka) àchàrà-mba (DRR) IGBO (Nsokpo) àbá-ọ̀jị́ (JMD; DRR); = *branch of an iroko tree; meaning not clear* (KW) IGBO (Owerri) mba (JMD; DRR) IJO-IZON (Egbema) kpasụ́karaghá (Tiemo) kpàsụ̀kàràgha (KW) MUNGAKA nda (DRR) NUPE 'cigbàn'te (Banfield; KO&S) YORUBA irẹ́ (auctt.) **WEST CAMEROONS** DUALA dìnyɔŋgɔ̀, èbòŋgo (Ithmann) èbòŋgo a ma nyɔ̀ŋɔ̀ (Ithmann) KPE manjongo (JMD) KUNDU dinjongo (JMD)

*NOTE: These names apply more or less to F. africana also. This is the true 'rubber tree' and is regarded as 'female', and F. africana as 'male'.*

A tree to 30 m high, with not quite straight, cylindrical, unbuttressed bole, 2.50 m in girth, of the deciduous forest from Guinea to W Cameroons and in the Congo basin, and along the Nile basin in Egypt, Sudan and Uganda.

Growth is rapid and it is an evanescent species in secondary growth on abandoned cultivated land. It is amenable to cultivation in forest plantations (3, 8). The wood is white and soft, and undifferentiated between sap and heart. It is not durable. It is used in Sierra Leone for carving spoons, bowls and other household utensils (13). At one time it was commonly used in Ghana for making Asante stools (7), and still occasionally is (10). In southern Sudan it is used as a timber for beams and rafters *(miriq)* in buildings (16). It has been found very suitable in match-manufacture for the inner and outer boxes and for match-splints, and is recommended for these purposes in Nigeria. In burning property it is said to be superior to *Gmelina arborea* (Verbenaceae) (3).

The bark contains a white latex which coagulates readily. This is the true West African rubber. Though it was traded under the name of 'Lagos rubber' and the Yoruba name *ire* and variations, the tree appears to have been first exploited for its rubber in Ghana about 1883, and shortly after in Ivory Coast. Not for another 10 years did it attract attention in Nigeria. Plantations were established in several West African territories, and may still even now be exploited to a limited extent in this way in Cameroun. Though the quality of the rubber is comparably good with that of *Hevea* rubber, it can in no way compete in yield and therefore economically except in time of dire necessity. Yield is very low (9). During the 1939–45 war, exploitation was renewed and in 1942 from the western province of Nigeria 1000 tons of rubber were obtained by tapping every available tree. While *Hevea* responds to regular and frequent tapping, commonly alternate-daily, the trees of *F. elastica* were tapped once only in 1942, and then yielded only half as much when tapped once again after 10 months rest, and those tapped a second time in 1942 with less than 10 months rest yielded but a fraction. (2).

The bark is very astringent. It is pounded up and taken in spirit in Nigeria to cure haemorrhoids (4). It is also used for this in Ghana (8). In Congo it is put into prescriptions for troubles associated with blennorrhoea and for painful menstruation. It is also used as a laxative and vermifuge, while the latex is applied to cracked sores of the feet (Fr. 'deshydroses plantaires'), to cutaneous fungal infections and to sores (5). In Ivory Coast it is reported as an ingredient of Guere arrow-poison, but there is some doubt whether its function is as a poison, or more likely as an adhesive (11).

The young leaves are taken by mouth or in enemas for diarrhoea in Ivory Coast (6), or are mixed with kaolin and administered by enema (1). The young leaves are also mixed with those of *Phyllanthus muellerianus* (Euphorbiaceae) and are taken to improve male fertility (1). In Congo they are used for chest-affections and particularly for whooping-cough (5). A number of alkaloids is present in the leaves (6, 17).

The seed-pod contains a fine white floss which is used for stuffing pillows and cushions. It is preferred in Sierra Leone over the floss of *Bombax* and *Ceiba* (13, 15) but it has no commercial interest though spinning trials have indicated a suitability for manufacture (9).

The seeds contain about 26% oil with a bitterness in the cake, making for unfitness for edible purposes (9). An unidentified alkaloid is present in the seed (16). There has been some commercial interest in the seeds as a substitute for *Strophanthus* seed as a source of *strophanthin*. Substitution is however quite a different proposition from adulteration, and while the former proved unavailing, the latter has often been practiced. *Funtumia* seeds are flat, naked, slightly striped with a distinct line along one face. *Strophanthus* seeds are convex or slightly keeled on one face with silky hairs on a reddish-brown background.

The generic name *Funtumia* was created for the African species segregated from the Asian species of *Kickxia,* and was said to be based on the Fante vernacular *funtum* (12, 14). *Fruntumia* or *Ofruntumia* would have been more appropriate.

References:

1. Adjandhoun & Aké Assi, 1972: 37. 2. Anon, 1943, a: 232–3. 3. Anon, 1965, b. 4. Ainslie, 1937: sp. no. 163. 5. Bouquet, 1969: 63. 6. Bouquet & Debray, 1974: 23–24. 7. Burtt Davy & Hoyle 1937: 11. 8. Dalziel, 1937. 9. Holland, 1908–22: 453–61. 10. Irvine, 1961: 621–2. 11. Kerhaor & Bouquet, 1950: 185. 12. Preuss, 1899. 13. Savill & Fox 1967, 48–49. 14. Stapf, 1900: & 1901. 15. Taylor, 1960: 96. 16. Turner 138, K. 17. Willaman & Li, 1970.

## **Hedranthera barteri** (Hook. f.) Pichon

FWTA, ed. 1, 2: 64–65. UPWTA, ed. 1, 367.

West African: **NIGERIA** EDO ékíawā (JMD) IGBO (Asaba) ámụ̀ńkịtā = *dog's penis* (NWT; KW) IGBO (Awka) útùn̲kịtā = *dog's penis* (KW) IGBO (Enugu) útùn̲kịtā = *dog's penis* (KW) IJO-IZON (Kolokuma) òkòtìtákụ̀ = *goat's testicles* (KW) YORUBA àgbo ọmọdé, agbo: *a medicinal infusion;* ọmọdé: *a child* (auctt.) doa (Macgregor) dodo ? (RJN) okó ajá = *dog's penis* (auctt.) oligborogan (RJA) ọmú ajá = *dog's breast* (EWF)

A shrub to 2 m high, in the understorey in damp situations of the closed-forest in Ghana, N and S Nigeria and W Cameroons, and also in Zaïre.

The large white tubular flowers with fragrant scent make the plant decorative and worthy of cultivation.

The plant contains a white latex that does not coagulate (4).

A leaf-decoction is drunk by the Igbo of S Nigeria for dizziness (6), and the leaf is applied to tumours (1).

The fruit is taken in Nigeria in treatment for gonorrhoea and as a vermifuge (1). The Yoruba name *àgbo ọmọde* is not a specific name, but refers to the *àgbo* infusion, of which the plant may be an ingredient, given to children (*ọmọde*) as a laxative (3). The fruit is readily eaten by wild animals (2). The plant (part unspecified, but perhaps the fruit on the Theory of Signatures — see below) is used by the Igbo to prevent miscarriage (5).

A number of alkaloids has been reported present in the plant (8).

The Ijo of S Nigeria give the fruit to teething children to play with and as a dummy (9).

The shape of the free bi-carpelate fruit evokes the bawdy Ijo name meaning 'goat's testicles', and also the Yoruba name 'dog's penis.' The Yoruba in this context also call upon the plant in an Odu incantation to prevent adultery in women on the simile that a dog in coitus finds it difficult to withdraw (7).

References:

1. Ainslie, 1937: sp. no. 66, as *Callichilia barteri*. 2. Akpabla 1105, K. 3. Dalziel, 1937. 4. Punch 48, K. 5. Thomas, N. W. 1640 (Nig. Ser.), K. 6. Thomas, N. W. 2068 (Nig. Ser.), K. 7. Verger, 1967: no. 46, as *Callichilia barteri* Stapf. 8. Willaman & Li, 1970: as *Callichilia barteri* (Hook. f.) Stapf. 9. Williamson 144, UCI.

**Holarrhena floribunda** (G. Don) Dur. & Schinz

FWTA, ed. 2, 2: 68–69. UPWTA, ed. 1, 372–3, as *H. africana* A. DC and *H. wulfsbergii* Stapf.

English: false rubber tree; male of ire, male of false rubber tree (Nigeria); the stem bark — conessi bark, kurchi bark (British Pharmaceutical Codex).

French: holarrhène; holarrhène du Senegal.

West African: **SENEGAL** BALANTA féash (Aub., ex K&A) BANYUN kilon (K&A) DIOLA bumatiap (Aub.) fu kuma (JB) fu mataf (auctt.) kerko (auctt.) DIOLA (Fogny) fukunafu (K&A) fulib (K&A) kalibaku (K&A) kélib (K&A) kölib (K&A) FULA-PULAAR (Senegal) ndama, indama (CHOP; Laffitte, ex K&A) tar(a)ki (K&A) KONYAGI tembo (K&A) MANDING-BAMBARA fufu (JB; K&A) kédan (K&A) nôfô (Aub., ex K&A) MANDINKA tariko (Aub., ex K&A) tarko (K&A) MANINKA kuna sana (auctt.) numoli-soro (Aub.) numuké (Aub.) sulu (Aub.) yate (RS) yété (RS) SERER ken (K&A) kena (auctt.) kôna (RS) ngas a kob (JB, ex K&A) SERER-NON ndôgsay (auctt.) ndôgsay (auctt.) WOLOF salali = *fetish charm* (K&A) séulu (auctt.) **THE GAMBIA** DIOLA (Fogny) bumatiap = *colour changes when ripening* (DF) foulib = *crowd* (DF) fu mataf = *leaf black like match-head* (DF) fukuma = *causes blindness: in reference to the sap* (DF) kalibaku = *sharp leaf* (DF) kélib, kölib = *crowd around a tree: on cultural activity* (DF) **GUINEA-BISSAU** BIDYOGO ete-eri (JDES; EPdS) FULA-PULAAR (Guinea-Bissau) charque (D'O) endame (D'O) MANDING-MANDINKA ticharque (D'O) **GUINEA** FULA-PULAAR (Guinea) gaulen (Aub.; RS) i-ndama (auctt.) lenge *the fruit* (JMD; RS) MANDING-MANINKA kuna sana (Aub.; RS) numoli-soro (Aub.) numuké (Aub.) sulu (Aub.) yate (JMD; RS) yété (JMD; RS) **SIERRA LEONE** KISSI tendo (S&F) waŋgolo-waŋgolo (S&F) KONO beiŋga-kɔne *applied loosely to soft-wooded trees* (S&F) g-basa (S&F) LOKO nuku(ɔ) (FCD; S&F) MANDING-MANDINKA nonokende, nono: *milk;* kende: *fresh* (FCD) MANINKA (Koranko) basa (S&F) boŋgakoŋ (S&F) MENDE nuku (*def.*-i) (auctt.) SUSU-DYALONKE keneyigi-na (KM) kenyekhigi-na, kenye: *mother's milk;* khigi: *water* (FCD) TEMNE *a*-mats (auctt.) *an*-wathia *properly Funtumia spp.* (FCD) **MALI** MANDING-BAMBARA fufu (Laffitte; Aub.) kunafin (GR) nonfon (Laffitte; Aub.) MANINKA kuna sana (Aub.; RS) numoli-soro, numuké, sulu (Aub.) yate, yété (RS) **IVORY COAST** AKAN-ASANTE kuminin (K&B) BRONG cécé (K&B) sésé (K&B) AKYE amaleu (A&AA) sohué (K&B; Aub.) suhè (A&AA) ANYI cécé (K&B; Aub.) sésé (K&B) BAULE sébé (auctt.) DAN kuro (auctt.) FULA-FULFULDE (Ivory Coast) gaulen (Aub.) indama (Aub.) GAGU suabé (B&D) KRU-BETE sagéï (K&B) GUERE (Chiehn) malé (K&B) KULANGO manguibé (K&B) KWENI toro toro (Aub., ex K&B) MANDING-DYULA duadé (K&B) MANINKA kuna sana (K&B; Aub.) séhè (B&D) **GHANA** VULGAR sese (FD) AKAN-ASANTE o-sɛsɛ (Enti) FANTE ɔsɛsɛ (FRI) TWI o-sɛsɛo (JMD) *o*sese, *o*-sɛsɛ (auctt.) WASA sese (CJT) ANYI sese (FRI) ANYI-AOWIN bɔsema (FRI) SEHWI bɛsema (FRI) BAULE sebe (FRI) GBE-VHE gakpoti (auctt.) kpomi (JMD; FRI) kpomli (JMD) NZEMA edrokoza (CJT) **DAHOMEY** BATONNUN hondéhan (Aub.)

YORUBA-NAGO irẹna, irẹ-oju-ọna **NIGERIA** VULGAR ako-ire = *the male 'ire'* (LC) FULA-FULFULDE (Nigeria) nyiiwahi = *elephant tree* (MM) HAUSA baàkín maayuu = *mouth of the sorcerer* (JMD) baakin mutum (auctt.) baƙin itache = *black tree* (RES) gaman sauwa (JMD) sandan maayun = *sorcerers' rod* (JMD) IGBO mba *properly Funtumia* (JMD; BNO) YORUBA akọ irẹ́ = *male iré,* ako: *male;* iré: *rubber tree, properly Funtumia africana* (Benth.) Staff (auctt.) irẹ́ (auctt.) irẹ́ ìbejì = *twin rubber tree* (auctt.) irẹ-ako (auctt.) irẹ-baṣabaṣa (auctt.) irẹna (auctt.) irẹnọ (auctt.) isai (auctt.)

A shrub or tree to 17 m high by 1 m girth of deciduous forest, savanna woodland and in secondary regeneration, common throughout from Senegal to N and S Nigeria, and extending to the Congo basin. The tree does not enter the evergreen forest. In Ghana it may attain 26 m high by near 2 m girth (13) and in savanna scrub it may not surpass shrub-size of 1–2 m height (17). It is frequent around villages. It is not normally planted but if missing farmers in Sierra Leone are said to broadcast seed (21). There is obviously some element of conservation.

The plant is ornamental with shining foliage and white fragrant flowers.

The wood is soft and uniformly white with no distinction between sap and heart-woods. Grain is straight. It works well but is perishable and is not resistant to termites. It is used for carvings, combs, spoons, stirers for the rice-pot and handles for axes and small implements (5, 7, 8, 9, 13, 21). The young stems are flexible and can be bent to make the crook handle of a walking-stick (9). It is suitable for shaving into wood-wool for packing fruit. It takes nails well and is usable for temporary purposes as huts and for packing cases (7). In Ghana it is considered by the stool-carvers to be the best white wood available (22). It is also used in Ivory Coast for carved stools (18).

The bark yields a copious white sticky latex which is resinous, and is inferior. The tree is known as the 'false rubber tree' and its latex has been used to adulterate good latices. The tree may be confused, accidentally or wilfully, with *Funtumia elastica* (Preuss) Stapf which is the 'true rubber tree' of West Africa. *H. floribunda* can be recognized by its bark-slash which is thin, soft, granular and very distinctly and characteristically layered (21). The latex has been used in Ivory Coast in arrow-poisons (4).

The bark is widely known as a dysentery cure. It is taken in decoction or macerated in palm-wine both as a prevention and as a cure for dysentery and fever (1, 9). In Senegal the Fula use either the bark or the roots (16, 17). In Casamance an aqueous macerate of the roots is taken for general stomach-complaints, and a root-decoction is given to women as an antiabortifacient (15, 17). It is diuretic and is taken by men in Senegal with urethral discharge (16, 17) and for gonorrhoea in Nigeria (20).

The tree yields a large number of alkaloids (17, 23). Total alkaloids are greatest in the roots (2%), decreasing in the stem bark (1·0–1·5%) with least in the leaves and flowers (< 1·0%). There is also seasonal variation. Of all the alkaloids present, *conessine* has attracted special attention for its antidysenteric and antiamoebic properties. It is claimed to be as effective as *emetine* and active in cases resistant to the latter and even on the cysts and to be less toxic (5, 14, 17, 18, 20). It is hypotensive and a cardiac depressant (5, 17).

Bark-infusion is commonly taken in substitution for quinine in treatment for malaria (1, 3, 7, 9) but alkaloids extracted from the root-bark have not proved antimalarial (2, 17). A bark-decoction in enema and in baths is used to treat certain skin-affections in Ivory Coast (18).

The alkaloid *holarrhenine,* also found in the plant, has narcotic and local anaesthetic properties, but in contrast is also an irritant (5, 17).

The leaves are used in antimalarial treatment. They are boiled in Ghana and the liquor is added to a bath (5, 7). Similar application is followed in Nigeria, and in addition the liquor is taken in draught (1). In Ivory Coast the leaves, and the bark, are held to be good for treatment of diarrhoea, and the leaves for amenorrhoea (18). Like other lactiferous plants, it is considered galactogenic, and herdsmen of northern Sierra Leone add leaves to salt-lick for their cows to improve the milk-yield (10). Also in Sierra Leone, the leaves are pounded up into a poultice 'for ladies in an interesting condition' (19), a practice perhaps on the same line of thought, though the manner of application is not indicated. Diuretic action of the leaves is used in a mixture with kola nut which is eaten for gonorrhoea in Sierra Leone (10) and in the Dakar area of Senegal they enter a treatment for diabetes (17).

Recent work has shown interesting development in the isolation of steroidal hormones, and in particular of *progesterone,* a so-called 'sex-hormone' and a precursor of hydrocortisone (17).

The fruit husk is used in Guinea-Bissau to prepare a remedy for dropsy (11).

The floss from the fruit is used for filling pillows (9, 21).

The plant is credited with magical attributes. The Hausa of N Nigeria use this, or similar plants, popularly as a charm and secretly in witchcraft. The forked peduncle, with three or four developing carpels, resembles the tripod stove for a cooking pot; the word *sauwa* means 'putting water in the pot on the fire', and the Hausa name, *gaman sauwa,* implies a happy conjunction in the rites of forecasting, etc. The wizard also uses a twig of it to rub his teeth and whomsoever he curses suffers evil or dies. A piece of it carried in the hand when going to the bush, or on business, acts as a favourable charm, both to ensure success or to escape punishment for evil-doing, etc.; for the same purpose it may be burnt in the house. A lover takes a stem and splits it: if the parts are equal he takes courage; if unequal he awaits a more favourable time, the plant thus serving as a fortune-teller *(bakin mayi).* (9). The Arago people of the Benue region are recorded as holding sacred a tree of this species that produced no fruit (12). The Shien of Ivory Coast in the event of a small-pox epidemic seek to protect their villages by sprinkling around a decoction of the leaves (18).

References:

1. Ainslie, 1937: sp. no. 181, as *H. wulfsbergii.* 2. Anon, 1945: as *H. africana.* 3. Aubréville, 1950: 443. 4. Aubréville, 1959: 3, 204, as *H. africana* A. DC. 5. Bouquet & Debray, 1974: 25. 6. Burtt Davy & Hoyle 1937: 12, as *H. wulfsbergii* Stapf. 7. Chalk et al., 1933: 9, as *H. wulfsbergii* Stapf. 8. Dalziel 971, K. 9. Dalziel, 1937. 10. Deighton 4218, K. 11. D'Orey 191, 202, K. 12. Hepburn 49, K. 13. Irvine, 1961: 623–4, as *H. wulfsbergii* Stapf. 14. Kerharo, 1967. 15. Kerharo & Adam, 1962: as *H. africana* A. DC. 16. Kerharo & Adam, 1964, b: 550. 17. Kerharo & Adam, 1974: 162–72, with extensive phytochemistry, pharmacognosy and pharmacology. 18. Kerharo & Bouquet, 1950: 186–7, as *H. africana* A. DC. 19. Lane-Poole 61, K. 20. Oliver, 1960: 7, 67, as *H. africana* and *H. wulfsbergii.* 21. Savill & Fox, 1967, 49–50. 22. Taylor, 1960: 97–98 as *H. wulfsbergii* Stapf. 23. Willaman & Li, 1970: with phytochemistry and references.

## **Hunteria eburnea** Pichon

FWTA, ed. 2, 2: 62. UPWTA, ed. 1, 375–6, as *Picralima elliotii* sensu Dalziel, in part.

A tree to 10 m or more high (sometimes to 40 m) of the closed-forest from Guinea to S Nigeria. Records for this species under *Picralima elliotii* auctt. and *Hunteria elliotii* auctt. have been confused with those of *H. elliotii* (Stapf) Pichon. The latter is distributed from Senegal to Sierra Leone, with an overlap of the two species only in Guinea and Sierra Leone in which territory *H. eburnea* is relatively rare (10).

The wood is yellow, hard and close-grained. It is used in Ghana to make shuttles for weaving cloth and for hair-combs (8), hoe-handles, plane-blocks and spoons (6).

The bark possesses a strong and lasting hypotensive and sympathicostenic action (2, 3, 5, 9). No less than 28 alkaloids have been detected in the trunk-bark. Seeds have given a dozen and the leaves eight (3, 11).

In Ivory Coast—Upper Volta, medicinemen specializing in treatment of leprosy, prepare fresh root-bark into a paste which is applied to the sores. The high toxicity of the medication is known, and fatalities are on record. The fruits are also toxic and are used for criminal purposes (1, 7). Trunk and branch-bark has a bitter taste like quinine (1). The fruit is rich in latex which is an ingredient of Guere arrow-poison in Ivory Coast (Kerharo & Bouquet fide 6).

References:

1. Adam, 1963, b: 26, as *Hunteri elliotii* sensu Adam. 2. Aubréville, 1959: 3, 208. 3. Bouquet & Debray, 1974: 27. 4. Dalziel, 1937, as *Picralima elliotii* sensu Dalziel, in part. 5. Hamet, 1955. 6. Irvine, 1961: 624, as *H. elliotii* sensu Irvine, in part. 7. Kerharo & Bouquet, 1950: 188, as *P. elliotii* sensu Kerharo & Bouquet. 8. McAinsh 890, K. 9. Oliver, 1960: 78. 10. Savill & Fox, 1967, 46. 11. Willaman & Li, 1970: (with references).

**Hunteria elliotii** (Stapf) Pichon

FWTA, ed. 2, 2: 62. UPWTA, ed. 1, 375, as *Picralima elliotii* sensu Dalziel, in part.

A shrub or tree to 10 m high, understorey of the closed-forest in Senegal to Sierra Leone.

The wood is yellow and hard. It is used in Sierra Leone for making combs (4), hoe-handles, police-batons and miscellaneous small wooden articles (1, 3). The bark has a milky sap (2), and a decoction is taken in Sierra Leone as a stomachic (Lane Poole fide 1) and used as a wash for fever (5). The fruit is eaten.

References:

1. Dalziel, 1937. 2. Samai 522, K. 3. Savill & Fox, 1967, 46. 4. Scott Elliot 5690, K. 5. Thomas, N. W. 19, K.

**Hunteria simii** (Stapf) H. Huber

FWTA, ed. 2, 2: 62.

West African: **SIERRA LEONE** MENDE kɔfa (*def.* kɔfei) *applies to several apocynaceous shrubs*(DS)

A shrub of the forest in Sierra Leone, Liberia and Ivory Coast.

The wood is hard and yellow. It is used to make small objects in Ivory Coast (1).

Reference:

1. Aubréville, 1959: 3: 206, as *Tetradoa simii* (Stapf) Pichon.

**Hunteria umbellata** (K. Schum.) Hallier f.

FWTA, ed. 2, 2: 62. UPWTA, ed. 1, 376.

West African: **GHANA** AKAN-ASANTE kanwini (CJT) FANTE akuama (CJT) TWI kanwini (CJT) WASA kanwene (Enti) NZEMA kakooli (CJT) **NIGERIA** EDO ósù (JMD; KO&S) YORUBA àárín (Ross) erin (auctt.)

A tree to 10–12 m high by 55 cm girth, of the understorey in closed-forest in Ghana and S Nigeria, and also in Cameroun and Ubangi-Shari.

The wood is yellow, very hard and fine-grained (4, 5). It is considered one of the best local woods for tool-handles. It is said to be termite-proof and durable. Forked stems are used as hut-posts. It is cut into combs (4).

The bark (4) and the root (3) are made into a bitter tonic in Nigeria, and powdered root and root-decoction are used to prevent miscarriage and in treatment of menorrhoea (2).

A number of alkaloids has been detected in the plant (1, 6).

References:

1. Adegoke & al., 1968. 2. Ainslie, 1937: sp. no. 277, as *Picralima umbellata*. 3. Akpabla 1101, K. 4. Dalziel, 1937. 5. Punch 138, K. 6. Willaman & Li, 1970: as *P. umbellata* (K. Schum.) Stapf, and *H. umbellata* Hallier, with references.

**Isonema buchholzii** Engl.

FWTA, ed. 2, 2: 70.
West African: **NIGERIA** IGBO (Umuahia) mkpọ (AJC)

A woody climber of swamp-forest, recorded only in S Nigeria and W Cameroons, and also in E Cameroun.

**Isonema smeathmannii** Roem. & Schult.

FWTA, ed. 2, 2: 70. UPWTA, ed. 1, 373.
West African: **GUINEA-BISSAU** PEPEL ubimba (JDES, ex EPdS) **SIERRA LEONE** LIMBA bubɔtɛ (NWT) **IVORY COAST** KYAMA n-tua (B&D)

A climbing shrub of the swamp-forest in Guinea-Bissau to Ghana.
The young leaves are eaten as spinach in Sierra Leone (2). The latex is used by the Kyama of lower Ivory Coast to treat old sores (1).

References:

1. Bouquet & Debray 1974: 28. 2. Deighton 662, K.

**Landolphia** P. Beauv.

FWTA, ed. 2, 2: 54. UPWTA, ed. 1, 373–5.
English: rubber vine (the plant); vine rubbers (the product).
French: liane à caoutchouc.
West African: **GHANA** ADANGME-KROBO abɔeka (FRI) fuka (FRI) hevendze (FRI) AKAN-ASANTE apɛya (auctt.) bede-bede = *bundle-bundle, because coagulated latex is carried in bundles* (Enti) gyamaa (FRI) jama (TFC; BD&H) FANTE ɔpaena (FRI) TWI abɔntɔre (FRI) bɔmene (FRI) nkontoma (FRI) ɔbɔwe (FRI) pao (FRI) pempene (FRI) pɔi (FRI) WASA kwantama (auctt.) ANYI-AOWIN abɔntɔle *from Twi* (FRI) faya (auctt.) kiakia (FRI) pumbune (FRI) SEHWI pumbune (FRI) GA abantolí (KD) GBE-VHE (Awlan) enge (TFC; BD&H) VHE (Pecí) enge (TFC; BD&H) NZEMA ahana (FRI) amale (auctt.) pompune = *elephant's heel* (FRI)

Species of the genus all contain rubber in varying degree and quality. Some species were at one time the principal source of wild rubber before the cultivation of *Hevea brasiliensis*. A number of vernacular names are generally applied.

**Landolphia calabarica** (Stapf) E. A. Bruce

FWTA, ed. 2, 2: 55. UPWTA, ed. 1, 373, as *L. bracteata* Dewèvre.
West African: **SIERRA LEONE** MENDE hole (*def.*-i) (FCD) jɛŋjɛŋ (FCD) kpoliyagɛ lɛ *the fruit* (FCD) nali *the stem* (FCD) TEMNE *an*-boi (FCD) *an*-lel-boi (FCD) **GHANA** ADANGME-KROBO kiakia (Ramsey) AKAN-AKUAPEM abontera (Ramsey) hamah (Ramsey) **NIGERIA** IGBO (Asaba) autopoi (AHU)

A liane to 7 m long, of the forest areas of Sierra Leone to S Nigeria.
It yields a white sticky latex (3) that is without worth (2, 4). It has been called 'white rubbervine' which is probably misapplied.
The fruit when ripe is edible (1).

APOCYNACEAE

References:

1. Cole 126, FBC. 2. Dalziel, 1937. 3. Deighton 3947, K. 4. Pichon, 1953: 138–41.

## **Landolphia congolensis** (Stapf) Pichon

FWTA, ed. 2, 2: 57.

A liane to 10 m long, of the primary forest of SE Nigeria and extending to Zaïre and Cabinda.

The plant contains an abundance of a sticky latex, non-elastic and of no recorded usage. The fruit when cut open emits a disagreeable smell. It is said to be both edible and inedible.

Reference:

1. Pichon 1953: 171–6.

## **Landolphia dulcis** (R. Br.) Pichon

FWTA, ed. 2, 2: 56–57. UPWTA, ed. 1, 368, as *Carpodinus barteri* Stapf and *C. dulcis* Sab.

West African: **SENEGAL** BANYUN kibod (K&A) DIOLA bu ñnohol (JB) fu somason (JB) DIOLA (Diembereng) énot (K&A) DIOLA (Fogny) asom asom (K&A) ésébisom (K&A) DIOLA (Tentouck) fubot (K&A) DIOLA-FLUP buńehol (K&A) **THE GAMBIA** DIOLA (Fogny) busom asom = *sticky: of the sugary fruit* (DF) fu somason = *root bad taste* (DF) MANDING-MANINKA sunkutu foolee = *girl's rubber* (DF) **GUINEA-BISSAU** BALANTA impèquéce (JDES) CRIOULO cibode (JDES) MANDING-MANDINKA suncutófóleo (JDES) MANDYAK becute (JDES) blambô (JDES) PEPEL ubimba (JDES) SONINKE-SARAKOLE codudú (JDES) **GUINEA** SUSU uengi (Farmar) wenyi (JMD) **SIERRA LEONE** BULOM (Sherbro) lɔ̣g-lɛ (FCD; JMD) KONO kɔndoo (Pyne) KRIO kushumin, kushumɛnt *probably a corruption of 'pishamin': sweet p. = Diospyros virginica, the American date-plum, and here misapplied* (auctt.) MENDE gɔndi (*def.*-i) (Pyne) gɔndɔ (*def.*-i) (auctt.) hole (*def.*-i) (JMD) kɔndo (*def.* kɔndui) (auctt.) kpoliyagɛlɛi the fruit (FCD) kpuwi (FCD; JMD) lɔgi (FCD; JMD) nali (JMD) ndakpao nyandei = *a handsome youth, an epithet of flattery whereby the hungry traveller will find the plant with edible, not rotten, fruit* (Boboh) powi (JMD) SUSU wei (NWT; JMD) weinyi (NWT; JMD) weyɛnkame (NWT) TEMNE *an*-lɔnk (auctt.) **MALI** FULA-PULAAR (Mali) kodudu (A. Chev.) MANDING-BAMBARA kodudu (A. Chev.) **IVORY COAST** ANYI amalin (K&B) bédé-bédé (K&B) **NIGERIA** EDO ákhẹ̀ (JMD) IGBO ụ̀tụ̀ (BNO) IGBO (Asaba) àkwarị̄ (JMD) àkwárī (AHU) ésó *the latex or rubber* (JMD) ọfọn-àkwárī, ọtọ-àkwárī *perhaps this transcription:* àkwárī: *tough, wild liana – used by hunters in traps* (KW) IGBO (Okpanam) àkwálị *perhaps dialectal variant of àkwárị̄: see L. hirsuta* (NWT; KW) YORUBA ìbò *general for rubber vines* (auctt.)

A climber with stout stem to 10 m long. Two varieties are recognized in the Region: var. *dulcis* occurring in the soudano-guinean savanna region of Senegal to Guinea, becoming infrequent in upper Ivory Coast; and var. *barteri* (Stapf) Pichon more widely dispersed in the dense forest from Guinea to S Nigeria and extending to Congo. The varieties are considered together here.

The plant is held in Senegalese medicine to have healing properties. Serious wounds are treated with a decoction of leafy twigs and powdered bark (5). Plants from Sierra Leone have yielded alkaloid and saponin from the leaves, and saponin from the bark (10). Decoctions of roots and stems are used in Casamance (Senegal) for external massaging and in baths and ointments for arthritis and kidney pains (3, 5), and decoctions of trunk-bark and root are used as a galactogogue by application to the breast (4, 5). Weaning is facilitated for the mother simply by stopping further application. Laboratory investigation has not given consistent galactogenic action, but did reveal some cardio-tonic action (5, 6). The root has unspecified medicinal use in Sierra Leone (8) and in S Nigeria a root-preparation is applied to sores (11).

The bark yields an abundant white latex which does not coagulate but yields a sticky substance used in Sierra Leone as a bird-lime (2, 9). The rubber is inferior and has found no use.

The fruit is edible but of variable palatability, sometimes sweet, sometimes acid, sometimes astringent (7). Perhaps these are varietal differences. The Krio name *kushumin* is likely to be a corruption of *pishamin* (= persimon) brought back from America by returning freed slaves and is applied loosely on account of the similarity of the fruit to persimon. The Mende of Sierra Leone use a cryptic name for the fruit, that is, meaning 'a handsome youth', the story being that if a hungry traveller refers to the plant by its proper name, the fruit will turn rotten and maggoty, but if the cryptic name is used the fruit can be picked sound and palatable (1).

References:

1. Boboh, 1974. 2. Dalziel, 1937. 3. Kerharo & Adam, 1962. 4. Kerharo & Adam, 1963, a. 5. Kerharo & Adam, 1974: 174. 6. Kerharo & Bouquet, 1950: 184. 7. Pichon, 1953: 163–9. 8. Pyne 106, K. 9. Scott Elliot 4294, K. 10. Taylor Smith, 1966: 539–40. 11. Thomas N. W. 1685 (Nig. Ser.), K.

**Landolphia foretiana** (Pierre ex Jumelle) Pichon

FWTA, ed. 2, 2: 57.
West African: **LIBERIA** MANO trodê (JCA) **NIGERIA** EDO ùbònọ̀khùa (AHU)

A robust liane to 80 m long by 90 cm girth, of the rain-forest area, somewhat rare in Liberia to S Nigeria, and occurring more commonly from E Cameroun to Zaïre and Angola.

The wood is used in Zaïre to make canoes (2).

The latex is white and abundant. Some reports of it in Zaïre indicate that it produces a good rubber: others that it is resinous and will not coagulate. It is used as a vermifuge (2).

The fruit is edible, sweet to slightly acid. In parts of Zaïre it is said to be purgative (2). In the Nimba mountain area of Liberia it is much sought after by monkeys (1).

References:

1. Adam, 1971: 375. 2. Pichon, 1953: 179–83.

**Landolphia heudelotii** A. DC.

FWTA, ed. 2, 2: 56. UPWTA, ed. 1, 374.

French: liane gohine (from the Maninka name).

West African: **SENEGAL** BALANTA p-sobé (auctt.) BANYUN dikof (K&A) BASARI a-táléɗy (Ferry) BEDIK ga-ngèɗy (Ferry) CRIOULO fognie (MP) fole (AS, ex K&A) DIOLA barakâ (aucct.) bu fèmb (JB) sôkon (K&A; MP) sokonasu (auctt.) DIOLA (Brin/Seleki) bufemb (K&A) DIOLA (Fogny) bufem (K&A) bufembabu (MP; K&A) fufembabu (K&A) DIOLA (Mampelago) fauli (MP) DIOLA (Sedhiou) buenn (MP) DIOLA ('Kwaatay') né-hemp (K&A) DIOLA-FLUP buhemck = *sour* (MP) buńohol (K&A) FULA-PULAAR (Senegal) poré (JMD, ex K&A) TUKULOR mana (MP) manan (MP) MANDING-BAMBARA goin (auctt.) MANDINKA foolee (*def.* fooleo) (auctt.) nta (A. Chev.) MANINKA goin (A. Chev.; K&A) 'SOCE' folé (A. Chev.; K&A) foré (JMD, ex K&A) MANDYAK nta (AS, ex K&A) SERER folé (JB; K&A) SERER-NON guïnn (MP) ho (K&A) hô (MP) u (K&A) û (MP) uk (auctt.) ûk (MP) WOLOF madd (MP) tol (auctt.) **THE GAMBIA** DIOLA (Fogny) barakâ = *gummy* (DF) bu fèmb *onomatopoeic from the sound of the fruit breaking* (DF) bufem, bufembabu, fufembabu *the sound of a dry stick being broken, i.e. of the fruit broken open by hand* (DF) sôkon *from the habit of growth* (DF) sokonasu *alluding to bunch-fruiting* (DF) MANDING-MANDINKA foolee (*def.*-fooleo) lit. = *rubber; general for any sp. with rubber* kàbaa (DAP) kabaanumba (*def.*-o) (auctt.) kassa-foolee = *rubber from Kassa* (*Casamance, Senegal*) (DF) **GUINEA-BISSAU** BIDYOGO n-batano, umbatano (JDES) CRIOULO fole (JDES) FULA-PULAAR (Guinea-Bissau) debol-poledje (JDES) póre (D'O) porè-lárè (JDES) MANDING-MANDINKA fóleossum (*def.*-ô) (JDES) MANDYAK betá (JDES) PEPEL mutaba (GES) **GUINEA** FULA-PULAAR (Guinea) bahi (MP) bohi-(é) (MP) furé (MP) laré (Bouronville) poré (CHOP; MP) poré-kono (MP) MANDING-BAMBARA goïn (A. Chev.) mana(-n) (MP) MANINKA bahi (MP) baï (JLT) foiro-gué (MP) gohine (CHOP) goïn (A. Chev.) SUSU foré (CHOP; Paroisse) paré (Dubois) SUSU-DYALONKE poré (MP) **SIERRA LEONE** MENDE jɛŋjɛŋ (FCD) TEMNE *an*-boi (FCD) *an*-lel-boi (L-P; FCD) **MALI** BOBO bona (MP) buéna (MP) buna (MP) FULA-PULAAR (Mali) dunda (Seguer) laré (Miguel) MANDING-BAMBARA boré, goé, gohine (MP) goï (A. Chev.) goïne, goue, gouéï, gouenn, gouinn (MP) goïtz (GR) laré (Miguel) MANINKA gohine (A. Chev.) TURUKA fetu (MP) 'SENUFO' daféré (MP) daterele (MP) kontere (MP) **UPPER VOLTA** NYENYEGE niabaho(-n) (MP) SENUFO-TUSIA tan (MP) TURUKA fien (MP) 'LOKI' boma (MP) 'MBOING' nfugurhi (MP) **IVORY COAST** ANYI amané-mbépé (MP) mbépé (MP) MANDING-DYULA k-popo (MP) MANINKA gbégui (A&AA) n-beni, n-bessi (B&D) SENUFO-BAMANA popo (MP) SIANAMANA (Kong) pâfaré (MP) pâfaré (MP) TAFILE beï (Aub.) **GHANA** SISAALA pempen (AEK) **NIGER** FULA-FULFULDE (Niger) poré (AS) MANDING-MANINKA n-beï (MP) n-dambo (MP) n-déï (MP) ngéi (MP) ngoyo (MP)

Bushy or lianous with stems attaining 1.25 m in girth basally by 15 m long, of the soudanian region, in Senegal to N Ghana. This species occurs further north (10–12°N) and west than does *L. owariensis* and extends no further eastwards than Ghana.

The plant is said to withstand bush-fires and grazing (8). The lianes are used in Senegalese medicines. A decoction of them or of the roots, is given for intestinal pains. It is not purgative. Vapour from a boiling concoction of leafy twigs is inhaled orally for tooth troubles (6). In Ivory Coast, the plant (part not stated) is used in draughts and added to squat-baths in treating haemorrhoids (1).

The plant contains an abundance of white latex and the rubber obtained from it is of good quality (3, 7). It has in the past been exploited and it was the most important source of rubber in the westernmost part of the Region (Senegal, 5, 9; The Gambia, 10; Guinea-Bissau, 4). As a result of over-cutting of the lianes, and even of the roots, to strip off the bark in the process of winning the rubber, the plant became rare and authority encouraged cultivation in Guinea, Mali and northern Ivory Coast, an interest long since killed by the development of hevea rubber. There is no use of the latex or rubber in Senegalese medicine (5), but a decoction of the roots, and of the fruit pulp, with some lime-juice is added to baths as a defatiguant (6).

The fruit pulp is edible and refreshing. It is slightly acidulous and mucilaginous and in Senegal is considered eupeptic (5, 6, 7). It is fermented to make an alcoholic drink (3, 7). In Guinea the fruits are an item of the diet of the chimpanzee (2). The seeds have unspecified medicinal use in Sierra Leone (8).

References:

1. Bouquet & Debray, 1974: 28. 2. Bouronville, 1967. 3. Dalziel, 1937. 4. Gomes e Sousa 1930: 81. 5. Kerharo & Adam, 1962. 6. Kerharo & Adam, 1974: 174–5. 7. Pichon, 1953: 129–35. 8. Tindall 71, K. 9. Trochain, 1940: 300. 10. Williams, F. N., 1907: 373.

## **Landolphia hirsuta** (Hua) Pichon

FWTA, ed. 2, 2: 57. UPWTA, ed. 1, 368.

English: 'Ibo tree' of Lagos (Williams).

West African: **SENEGAL** DIOLA bu kélor (JB) bu muk (JB) simonk (H&C) FULA-PULAAR (Senegal) murtégé (H&C) MANDING-MANDINKA bombompale (H&C) kabaa foro (*def.*-o) (H&C) **THE GAMBIA** DIOLA (Fogny) bu kélor = *slice* (DF) bu muk = *the decayer* (DF) **GUINEA** FULA-PULAAR (Guinea) murtégue (Paroisse) SUSU bonkhé (auctt.) **SIERRA LEONE** MENDE hole (*def.*-i) (auctt.) jɛŋjɛŋ (FCD) kpuwi (MP) nali (FCD) powi (MP) TEMNE *an*-boi (FCD) *an*-mar (FCD) **IVORY COAST** ABURE abo (B&D) AKAN-ASANTE diama (A. Chev.) uma pohué (A. Chev.) AKYE m-bafa (Aub.) **GHANA** AKAN-ASANTE bede bede (WTSB, ex FRI) BRONG akɔntɔma (FRI) TWI pumpene (FRI) **NIGERIA** BASSA (Kwomu) alibada (JMD) EDO ùbò *general for rubber vines* (JMD) ùbònọ̀khùa (JMD) ùhònọ̀khùa (Amayo) HAUSA alibida, alubada (ZOG) alubuda (Lamb) IGBO (Asaba) àkwárī = *wild liana* (JMD) ésó *the latex, or the rubber* (JMD) ọfọn àkwárī, ọtọ àkwárī *see L. dulcis* NUPE alibada, alubada (Yates) YORUBA ate *general for any plant yielding bird-lime, but especially this sp.* (Mackay) ìbò (auctt.) ìbò eleki (JMD) ìbò elekiti (JMD)

A liane to 20 m long by 10–15 cm girth, of the forest and galleried forest, often on river-banks, from Senegal to W Cameroons.

This is the *ibo* tree of Lagos. It yields an abundant latex producing occasionally a good quality rubber, but more often resinous and tacky. As much as 66% resin is reported. The rubber was a part of that produced in Ghana and Nigeria at the beginning of this century which was marketed as 'paste rubber', 'Accra paste', 'flake rubber', 'lump rubber', 'root rubber', 'brown cluster', or 'medium brown.' Even into the 1930's there was an export (1933, 188 tonnes; 1934, 120 tonnes) from Ivory Coast of its rubber known as *glu* (1). When coagulated by boiling and stirring a sticky product results which has been widely used as bird-lime, and is an excellent adhesive. Coagulation can also be achieved by addition of lime-juice. The rubber has been used to adulterate *Funtumia* rubber. (4, 5, 7, 8, 9.)

In Ivory Coast fresh bark-sap, after removal of the rubber is used as a cough-mixture, and decoction of the root-bark is taken by draught for blennorrhoea and by enema for haemorrhoids (2).

The fruit is globose, 5–8 cm in diameter, with pulp that is sweet, somewhat acidulous and edible. It is occasionally eaten. (1, 3, 4, 7.)

References:

1. Aubréville 1959: 3, 190. Bouquet Debray, 1974: 28. 3. Busson, 1965: 408. 4. Dalziel, 1937. 5. Jones 309, K. 6. Mackay s.n., 3/4/1940, K. 7. Pichon, 1953: 193–6. 8. Williams, F. N., 1907: 374. 9. Yates 45, K.

**Landolphia landolphioides** (Hall. f.) A. Chev.

FWTA, ed. 2, 2: 56. UPWTA, ed. 1, 368.

English: Cameroons Mountain rubber vine.

Trade: manjonga (the rubber).

West African: **WEST CAMEROONS** KPE maliba ma manjonga *the latex* (JMD) manjonga *the rubber* (MP) njoma, wuoma, yoma *the fruit* (JMD)

A stout liane to 25 m long by 1 m in girth of the closed-forest, lowland in S Nigeria and montane in W Cameroons, and extending to Zaïre and Uganda.

The liane contains an appreciable amount of latex which coagulates immediately on exposure to air. The rubber is of good quality (1, 2).

The fruit is pear-shaped, yellow and about 7.5 cm in diameter. It is edible and is sold in markets in W Cameroons (1).

References:

1. Dalziel, 1937. 2. Pichon, 1953: 53–55.

**Landolphia macrantha** (K. Schum.) Pichon

FWTA, ed. 2, 2: 55.
UPWTA, ed. 1, 368.
West African: **SIERRA LEONE** TEMNE (Kunike) *am*-fenke (FCD) *an*-lonk (FCD)

A liane of the forests of Guinea and Sierra Leone.

The liane yields a white latex (3) containing rubber (5).

The fruit pulp (5) and the seeds (3) are edible (4) and are occasionally eaten (2). In Guinea the fruits form a part of the diet of the chimpanzee (1).

References:

1. Bouronville, 1967. 2. Busson, 1965: 408. 3. Deighton 4087, K. 4. Pichon, 1953: 146–8. 5. Scott Elliott 4924, K.

**Landolphia membranacea** (Stapf) Pichon

FWTA, ed. 2, 2: 57.

West African: **SIERRA LEONE** SUSU fore fafia (NWT) TEMNE *ra-sa-ra-tɔn* = *bitch's udder* (FCD)

A slender liane of the forest zone of Guinea to Ivory Coast.

The plant is said to yield in Liberia a good rubber (2). In Sierra Leone the fruits are considered edible (1).

References:
1. Deighton 2932, K. 2. Pichon, 1953: 176–7.

**Landolphia micrantha** (A. Chev.) Pichon

FWTA, ed. 2, 2: 55.

A liane to 20 m high of the forest zone of Liberia and Ivory Coast.

The rubber is said to be of good quality (1).

Reference:
1. Pichon, 1953: 84–85.

**Landolphia owariensis** P. Beauv., var. **owariensis**

FWTA, ed. 2, 2: 55–56.
UPWTA, ed. 1, 374–5.

English: the plant — white rubber vine; the rubber — white ball rubber; Accra niggers, Addah niggers, Krepi ball (Ghana, Irvine).

West African: **THE GAMBIA** MANDING-MANDINKA cabajugo *the 'caba' tree* (DF) kaba (Fox) **GUINEA-BISSAU** CRIOULO fole de elephante = *elephant's stomach* (GeS; JDES) FULA-PULAAR (Guinea-Bissau) lamúdè (JDES) lamúquè (JDES) MANDING-MANDINKA cabádjuó (JDES) MANDYAK menta (JDES) **GUINEA** FULA-PULAAR (Guinea) poré = *rubber* (JMD; MP) poré-bété (MP) poré-mulutigué (CHOP) KISSI mantamba (MP) MANDING-MANINKA tugue (MP) SONGHAI dindji (JMD; MP) SUSU fore (Bouery) **SIERRA LEONE** VULGAR poré (Conteh) MANDING-MANINKA (Koranko) bɛnɔmbe (NWT) MENDE hole (Boboh) jɛŋjɛŋ *the latex or rubber* (auctt.) SUSU fɔre (JMD) fure (JMD) yuré (GFSE) TEMNE *an*-boi (JMD; MP) *an*-lel-boi (JMD; MP) **IVORY COAST** ADYUKRU bumbari (Jolly) bumboni (Jolly) bumbui (Jolly) léthiéthié (Jolly) muné-muné (Jolly) **GHANA** ADANGME-KROBO kiakia (FRI) AHANTA amanɛ *the rubber* (FRI) AKAN-ASANTE agyaama (Enti) agyaamaa (Enti) aman agyaamaa (Enti) *a*-gyama(a) (auctt.) homa-funtun = *rubber vine, general for such lactiferous plants* (JMD; MP) *o*-pawya (auctt.) TWI abɔntɔre (auctt.) agyamaa (JMD; MP) agyamaa (FRI) akɔntɔma (Bunting) aman = *rubber* (Enti) nkɔntɔma (auctt.) obɔwe (auctt.) pempene (JHH, ex FRI) WASA kwantama (FRI) ANYI-AOWIN faya (FRI) GA alabɔntɔle *from Twi* (FRI) kiakia (auctt.) GBE-VHE abɔeka (FRI) VHE (Awlan) engye (JMD; MP) VHE (Pecí) abne (FRI) engye (JMD; MP) fuka (FRI) hevengye (FRI) NZEMA pompune **TOGO** AVATIME lepapa (Gruner) GBE-AJA (Sikpi) boé-ka (Schumann) VHE abŏeka (Gruner) a-ngeka (Gruner) **NIGERIA** ANAANG énwàñ (KW) okpo-énwàñ *general for several spp.* (JMD) BASSA (Kwomu) arobo *a varietal name* (Elliott) ati fufu (Elliott) BEROM dìnákà the plant (LB) eyέ *the fruit* (LB) DEGEMA ọtuokpo (JMD) EDO ùbó-amióghọ (auctt.) EFIK ọ̀kpộ = *rubber vine* (RFGA, Lowe) ọ̀kpộ-eñwaña (Winston) HAUSA ciiwóò (LB) cìwoó (ZOG) danko (Lamb) hánjín raàkúmií (ZOG) jan danko *a red cluster rubber* (JMD; Yates) opopoi (Yates) oùwoó (ZOG) ubaghi *a variety* (JMD) IBIBIO énwàñà, énwâñ (KW) IGBO ọtọ-frifredi (JHH) ụ̀tụ̀ (JMD) ụ̀tụ̀ ísí ényī (JMD) IGBO (Asaba) ésó = *rubber; a general term* (JMD; MP) ubaghi (JMD) ube *general for several spp.* (JMD; MP) NUPE pakogi (Yates) TIV ḿníngyēm' *general for 'rubber'* (JMD) YORUBA ìbò (JMD) ìbò tabon(g) *the plant* (EWF; JMD)

A shrub in the savanna or a huge liane of secondary deciduous and dense forests, attaining 100 m long by over 1 m girth, occurring from Guinea to W Cameroons, and extending across central Africa to Sudan, Uganda and southern Tanganyika. Three varieties are recognized of which var. *owariensis* only is in the West African region.

This is certainly the commonest of the *Landolphia* spp. It survives bush-fires, and its rhizomes after a burn often put up a coppice of short shoots which even before becoming lignified will bear flowers and fruits (12). Before the days of monopoly of natural rubber sources by hevea plantations, this was a very important source of vine-rubber. It has been exploited as a forest-produce throughout its range, and plantations of it have at times been established in Ivory Coast. During the 1939–45 war it regained an ephemeral interest as a substitute for hevea rubber. In the Nigerian and Cameroons high-forest, yields of 22–180 kg per sq mile of forest (period not stated, but presumably per year) were recorded, and a skilled tapper could obtain $7\frac{1}{2}$ kg per month, at the expense of much dangerous tree-climbing (1). The normal method of extraction is by incision, but excision tapping as for hevea has been practised in Sierra Leone (2). In earlier times wasteful methods of extraction were employed. In Nigeria (5, 10, 13, 16) and in Gabon (15), the bark of both stems and roots was stripped, pounded, soaked and boiled to produce a commercial product known as 'red cluster rubber', or 'root rubber'. Stripping of the roots killed the plant there and then. Stripping of the lianous stems only postponed somewhat inevitable death. In Ghana legislation required the rubber to be made up in string-form which was then wound into balls and traded as 'ball rubber' or according to its place of origin, e.g. 'Adele ball rubber' from the Adeli District of Togo (5). The latex normally is white but may be red, pink or amber, and when obtained by incision or excision is variable. Usually it coagulates immediately, but at some seasons or at certain places it remains fluid enough to run into a collecting vessel when coagulation is done by lime-juice or salt-water (5). The resultant rubber is of good quality, but latex from some lianes will not coagulate and remains tacky. This is commonly used as a bird-lime to catch small birds and other animals (12), especially on rice farms in Sierra Leone to trap birds depredating the crop at ripening time. This latex is also used in the Region in lotions and is taken internally against intestinal worms (12), or it may be used as an enema for the same purpose (14).

The twigs are used in Ghana as chewsticks (7, 9).

The roots and leaves are used in The Gambia in medicines (6) for unspecified purposes. Leaves are boiled in some parts for application to sprains (12). In Congo sap expressed from the leaves is dripped into the eyes and used to wash the patient's face in a treatment for giddiness and epilepsy; sap is rubbed with massage into scarifications over areas of oedema and rheumatism; and decoction of roots or green fruits is drunk as a purgative and for urethral discharge. The liquid of this preparation is used in steam-baths for feverish aches (3).

A trace of flavones is reported in the leaves, and tannin, steroids and terpenes in the roots (4).

The fruits, resembling small oranges, are edible and are esteemed in all areas. The pulp is acid. It is recorded as a source of vitamins (11). In various parts it is fermented to give an alcoholic drink (5, 8, 9).

References:

1. Anon., 1943, b. 2. Boboh, 1974. 3. Bouquet, 1969: 64. 4. Bouquet, 1972: 14. 5. Dalziel, 1937. 6. Fox 68, K. 7. Irvine 765, K. 8. Irvine, 1948: 267. 9, Irvine, 1961: 627–8. 10. Lamb 5, K. 11. Okiy, 1960: 121. 12. Pichon, 1953: 109–28. 13. Unwin s.n., K. 14. Walker, 1953, a: 21. 15. Walker & Sillans, 1961: 83–84. 16. Yates 46, K.

**Landolphia parvifolia** K. Schum., var. **johnstonii** (A. Chev.) Pichon

FWTA, ed. 2, 2: 55.

A lofty liane of the forest zone of Nigeria, and W Cameroons, and extending into Zaïre and Angola, with two other varieties (var. *parvifolia*, and var. *thollonii* (Dew.) Pichon) occurring in central and E Africa.

The latex is white and variably sparse to very abundant. It coagulates quickly and has been found to be without worth. Rhizomatous shoots of var. *thollonii* are said to yield a good rubber which has been called *caoutchouc des herbes* (2).

Var. *parvifolia* (part unknown) is used in Tanganyika in arrow-poison (1) apparently for its cohesive property rather than a toxicity (3).

References:

1. Bally, 1937. 2. Pichon, 1953: 102. 3. Watt & Breyer-Brandwijk, 1962: 67, 85.

**Landolphia subrepanda** (K. Schum.) Pichon

FWTA, ed. 2, 2: 57. UPWTA, ed. 1, 368

West African: **NIGERIA** EFIK onton (McLeod)

A small liane to 6 m high and girth to about 4–5 cm, recorded only from the forested zone of the Cross River of S Nigeria and in W Cameroons, and also in E Cameron to Zaïre.

The plant yields an abundant white latex, which is of no particular value. The fruits are edible.

References:

1. Pichon, 1953: 184–6.

**Landolphia togolana** (Hallier f.) Pichon

FWTA, ed. 2, 2: 55. UPWTA, ed. 1, 370.

West African: **NIGERIA** YORUBA agba (EWF)

A stout liane to 5 m height with stems to 15 cm diameter, of open forest and savanna bush from Ivory Coast to Nigeria.

The latex is of poor quality (1).

Reference:

1. Pichon, 1953: 87–88.

**Landolphia uniflora** (Stapf) Pichon

FWTA, ed. 2, 2: 57. UPWTA, ed. 1, 369.

A slender liane to 9 cm girth, of the forest in S Nigeria and in Cameroun and Gabon.

The plant contains an abundant white latex which is resinous, of unrecorded use, and probably of little value.

References:

1. Dalziel, 1937. 2. Pichon, 1953: 178–9.

**Landolphia utilis** (A. Chev.) Pichon

FWTA, ed. 2, 2: 56.

A liane to 3 m long with stems to 10 cm in girth, occurring only in the coastal area of Ivory Coast.

Latex is said to be in small quantity, but the rubber is of good quality (1, 2).

References:
1. Chevalier, 1920: 401, as *Clitandra laurifolia* A. Chev. (nom. nud.). 2. Pichon, 1953: 200–1.

**Landolphia violacea** (K. Schum.) Pichon

FWTA, ed. 2, 2: 56. UPWTA, ed. 1, 369.

A climber with stems to 16 cm diameter recorded only from S Nigeria in the Region, and occurring in Cameroun to Zaïre.
The latex is white and resinous. The resultant rubber is inferior. The fruits are much sought after by monkeys in Gabon.

References:
1. Pichon, 1953: 157–8. 2. Walker & Sillans, 1961: 78.

**Malouetia heudelotii** A. DC.

FWTA, ed. 2, 2: 76.
West African: **SIERRA LEONE** MENDE bɔnje (*def.*-i) (FCD) SUSU yɛte (NWT) TEMNE *a*-ponthi-*a*-ro-bath (FCD)

A shrub to 1·60 m high on stream-banks and in swamps of Guinea and Sierra Leone.
No usage is recorded in the Region, though by the vernacular names recorded it appears to be recognized. The only other African species of the genus, *M. bequaertiana* Woodson, has been found rich in alkaloids in leaves, bark and roots with steroids and terpenes also present (1).

Reference:
1. Bouquet, 1972: 14.

**Mascarenhasia arborescens** A. DC.

FWTA, ed. 2, 2: 72.

A tree to about 8 m high of lowland forest usually on stream-sides, recorded from Togo, possibly as a subspontaneous escape from cultivation, and occurring widely in E Africa, Mozambique and Madagascar. It is under cultivation in Victoria Botanic Gardens, W Cameroons.
The tree yields a rubber which at one time attracted some interest in south-eastern Africa, but it contains a high amount of gum (68%) and is of no commercial value.

Reference:
1. Watt & Breyer-Brandwijk, 1962: 88.

**Motandra guineensis** (Thonning) A. DC.

FWTA, ed. 2, 2: 80. UPWTA, ed. 1, 375.
West African: **SIERRA LEONE** MENDE yolo (*def.* yolei) *also applied to several other plants* (FCD) **IVORY COAST** AKAN-ASANTE amaleniana (B&D) BAULE agba (B&D) GAGU golia (B&D) yéré-yéré (B&D) yurubini (K&B) KRU-BETE gala kuku (K&B) GUERE (Chiehn) goléa kwoyon (B&D) KWENI goléakoyon (B&D) grakwè (B&D) urobéné (K&B) **GHANA** AKAN-ASANTE *a*-mamfohae (auctt.) **TOGO** TEM (Tshaudjo) dshingia (Gaisser) tshingia (Gaisser) **NIGERIA** YORUBA agba doje (Kennedy) bọ́dẹkádún (AHU)

A scandent shrub with thin stems, of deciduous and secondary forests from Guinea to E Nigeria and W Cameroons and extending to Uganda and Angola.

The stems are hollow, and cut into short lengths they are threaded as beads by Bundu girls in Sierra Leone (3). The latex is abundant but of no use (2). Bete of Ivory Coast express leaf-sap for eye-installation, and to make a mouthwash and to massage into the gums (4). Leaf-sap is also used in nasal-instillation. The initial response is one of irritation of the mucosae followed by sedation. This treatment is used in cases of syncope, on patients who are too somnolent or those with headache; also to quieten those with fits of raving madness. Bark-sap is also reported used in Ivory Coast as an enema to assuage stomach-pains in newly-delivered women.

References:

1. Bouquet & Debray, 1974: 28. 2. Dalziel, 1937. 3. Deighton 3942, K. 4. Kerharo & Bouquet, 1950: 187–8.

**Nerium odorum** Sol.

A shrub similar to *N. oleander,* but with shorter leaves and scented flowers, of Central Asia from Iran to Japan and now widely dispersed by man as an ornamental and present in the West African region.

It has poisonous properties like *N. oleander*.

**Nerium oleander** Linn.

FWTA, ed. 2, 2: 54.

English: oleander.

French: laurier-rose.

Portuguese: loureiro, rosa (Gomes e Sousa).

West African: **SENEGAL** WOLOF tortor (K&A) **MALI** ARABIC addefla (RM) defla (RM) **GHANA** GA fɔfɔitʃo = *flower plant* (FRI)

A bush reaching to 4 m high, native of southern Europe and the Mediterranean from Portugal to Iran and widely dispersed by man to most warm countries. It is commonly grown in the Region for its decorative flowers which range in colour from white to deep rose pink. While growing best under high rainfall, the plant is very tolerant of dry conditions.

All parts of the plant are very toxic. A number of glycosides is present. They have a paralysing action on the heart similar to that of *digitalis*. One of those present, *oleandrin,* has found use in treatment of cardiac deficiencies in persons who cannot tolerate *digitalis* or *ouabain* (1, 3, 6, 9). Material collected from plants growing in seasonal watercourses (*oued*) is said to be particularly good (7). All stock are susceptible to poisoning should they eat the leaves, and there is a long history of poisoning. Alexander in his military campaigns is said to have lost men after eating meat skewered on *Nerium* twigs. In this respect the fresh bark is more poisonous than the leaf. In a garden with *Nerium* plants particular care needs to be taken to prevent children playfully sucking the flower stems, etc. (9).

Wood-ash is added to dates or to butter taken with sour milk by Tidikett women of Central Sahara as a fattening diet (4). The ash mixed with saltpetre is used to make a good gunpowder in Western Sahara (5).

A decoction of powdered leaves, or a leaf-extract is widely used in treatment of skin-complaints and vermin (3, 5, 6, 8). In many countries, the leaf principally, but also the latex, leaf-sap, bark and roots are compounded into medicines applied to tumours, warts, corns, ulcerations and carcinomas (2).

References:

1. a. Bissett, 1953. b. Bissett, 1958. c. Bissett, 1961. 2. Hartwell, 1967. 3. Kerharo & Adam, 1974: 175–8, with phytochemistry & pharmacognosy. 4. Maire, 1933: 169. 5. Monteil, 1953: 118. 6. Oliver, 1960: 32, 74. 7. Schunck de Goldfiem, 1945: 153. 8. Walker & Sillans, 1961: 84. 9. Watt & Breyer-Brandwijk, 1962: 69, 88.

**Oncinotis glabrata** (Baill.) Stapf

FWTA, ed. 2, 2: 80. UPWTA, ed. 1, 375.
West African: **NIGERIA** YORUBA agbaboje (MacGregor)

A woody climber of the closed-forest from Guinea to S Nigeria and W Cameroons, and extending to Angola and (?) Tanganyika.
The plant is decorative. The stem is used in S Nigeria for bowstring (1).

Reference:

1. Thomas, N. W. 1946 (Nig. Ser.), K.

**Oncinotis gracilis** Stapf

FWTA, ed. 2, 2: 80. UPWTA, ed. 1, 375.

A climbing shrub to 13 m high of deciduous forest from Guinea to N and S Nigeria, and also in Cameroun.
The plant is a decorative climber. It contains latex yielding a sticky coagulum which has no recorded usage (2). No active principle is present in the plant and in contrast from *O. nitida* (see there) it has no medicinal application (1).

References:

1. Bouquet & Debray, 1974: 29. 2. Dalziel, 1937.

**Oncinotis nitida** Benth.

FWTA, ed. 2, 2: 80. UPWTA, ed. 1, 375.
West African: **SIERRA LEONE** KRIO quiah (L-P)

A climbing shrub to 8 m high of the closed and galleried-forests from Guinea-Bissau to S Nigeria.
The plant is a decorative climber. In Ivory Coast it is used medicinally to prevent abortion. Two alkaloids are reported present, *oncinotine* and *iso-oncinotine* (1).

Reference:

1. Bouquet & Debray, 1974: 29.

**Orthopichonia schweinfurthii** (Stapf) H. Huber

FWTA, ed. 2, 2: 58. UPWTA, ed. 1, 370, as *Clitandra visciflua* K. Schum.

A climber occurring rarely in S Nigeria, and extending across Africa to Sudan.
It yields a sticky latex. The fruit pulp is strongly acid but edible (1).

Reference:

1. Pichon, 1953: 213–6, as *Orthandra schweinfurthii* (Stapf) Pichon.

**Orthopichonia staudtii** (Stapf) H. Huber

FWTA, ed. 2, 2: 58.

A forest liane occurring rarely in W Cameroons and extending into Zaïre. It contains an abundant latex which is of no value (1).

Reference:

1. Pichon, 1953: 219–20, as *Orthandra staudtii* (Stapf) Pichon.

**Picralima nitida** (Stapf) Th. & H. Dur.

FWTA, ed. 2, 2: 62. UPWTA, ed. 1, 376.

West African: **SIERRA LEONE** SUSU balunyi (NWT) ninge-ɛxunye (NWT) **IVORY COAST** ABURE ebissi (B&D) AKAN-ASANTE aboya (B&D) ANYI k-baba pempé (K&B) KRU-GUERE (Chiehn) krigbé (B&D) **GHANA** VULGAR ekuama (DF) ADANGME kpɛtɛ-kpɛtɛtʃo (FRI) AKAN-ASANTE akuama *refers mainly to the seed* (Enti) akuama, ɛkuama (FRI, Enti) kanwini, ka: *to taste;* nwin: *bitter* (FRI) nwẽma (FRI) FANTE akuama (auctt.) ɔnwẽma (FRI) owema (Enti) KWAWU akuama (Enti) owema (Enti) ɔwẽnma (FRI) TWI akuama, ekuama akuama (Enti) atiridii = *fever* (FRI) o-kanwen, ka: *to taste;* nwen: *bitter* (auctt.) o-nwẽma *not commonly used* (FRI) WASA akuama, ekuama (Enti) akuama (Enti) akuamma (CJT) kanwene (FRI) panwe *not commonly used* (FRI) ANYI-SEHWI kahana (?kakana) (CJT) NZEMA kakooli (CJT) **DAHOMEY** GBE-FON dangné (Laffitte) **NIGERIA** IGBO òsú igwe (auctt.) IGBO (Ogwashi) òsú abwa (NWT; JMD) YORUBA abere (IFE) agègè = *run-run* (Verger) agègè-arin (JRA; FRI) erin (KO&S)

A tree reaching 25 m, usually less (Ghana, 10 m by 60 cm girth, 15; Ivory Coast, 20 m by 1.25 m girth, 2), in understorey of the high deciduous forest from Ivory Coast to W Cameroons, and extending across central Africa to the Congo basin and Uganda.

The wood is pale yellow, hard, elastic, fine-grained and taking a high polish. It is used in Ivory Coast to make incense holders, combs and small objects (2); in Ghana for walking-sticks, weaver's shuttles (19), dolls, plane-blocks and handles for carpenter's tools (9, 15); in S Nigeria for spade-handles (16, 17); W Cameroons for arrows and spoons; in Gabon for carvings, paddles, incense holders, etc. (21) and bows, arrows and so on elsewhere.

All parts of the plant are bitter. The bark is widely used as a febrifuge (2, 5). It is used in Gabon against venereal disease (20, 21), and in Uganda the Banyoro chew it as a vermifuge (6, 7). A bark-decoction is taken in Congo as an anthelmintic and purgative, and also to treat hernia, and with other drug-plants to relieve blennorrhoea (3). The root in Gabon is held to be vermifugal (20, 21). A bark-decoction is taken in draught in Ivory Coast for jaundice and 'yellow fever' (10), and also in Nigeria (1). The roots in various parts are used for fevers, pneumonia (1, 2). In Gabon the Pahoin chew a little of the fruit and bark to allay hunger while on long marches in the bush (14).

The leaves are commonly used as a vermifuge in Dahomey (11). Leaf-sap is dripped into the ears for otitis in Congo (3).

The fruits while still green are used by Wasa in Ghana as a fish-poison (9). The fruit shell after removing the flesh and seeds is filled with palm-wine which is drunk after it has had time to absorb the bitter principle present as a fever treatment (5). The Asante and Twi names meaning 'to taste bitter' arise from this. Small dippers and spoons are made from the shell.

The seeds are extensively used in place of quinine for treatment of fevers in Ghana (13), and in the neighbouring territories of Ivory Coast (2) and Nigeria (1). The powdered seeds are also given in Nigeria for pneumonia and other chest-conditions (1). In Gabon the seeds are recognized as toxic and use appears to be restricted to external treatment for abscesses (2). The Bete of Ivory Coast use the seeds in treatment (method not disclosed) of hernia: two seeds are said to be enough (4). A seed-decoction is given as an enema in Ghana, but crushed seed taken by mouth for chest-complaints, pneumonia, etc. and acute stomach-troubles is not considered purgative (9).

The seeds have entered literature under the name *akuamma,* said to be the name given to them by the Ghanaian woman who first brought them to Accra. This name properly refers to *Pentaclethra,* and the name has doubtless been misapplied (9). Nevertheless, a series of crystalline alkaloids have been identified in the seeds, and have acquired names based on it: *akuammine, akuammidine, akuammiline, akuammigine,* etc. Akuammine is the principle one present. It is a powerful sympathicostenic and has a local anaesthetic action almost equal to that of cocaine. Akuammidine is hypotensive, and weaker but longer lasting in effect than yohimbine. It has strong local anaesthetic action. Akuammigine is identical with yohimbine (8, 10, 12, 14, 22). Only amorphous alkaloids have been detected in the bark and leaves (5). Experimental work has not found extracts of *Picralima* to be active on avian malaria (5, 8, 12).

The plant features in an Yoruba Odu incantation to attract a woman (18).

References:

1. Ainslie, 1937: sp. no. 276. 2. Aubréville, 1959: 3, 206. 3. Bouquet, 1969: 64–65. 4. Bouquet & Debray, 1974: 29–30. 5. Dalziel, 1937. 6. Dawe 707, K. 7. Eggeling & Dale, 1952: 28–29. 8. Henry, 1939: 624–6. 9. Irvine, 1961: 629–30. 10. Kerharo & Bouquet, 1950: 188. 11. Laffitte 70, IFAN. 12. Oliver, 1960: 34, 77–78. 13. Oppenheimer Son & Co. s.n., K. 14. Raymond-Hamet, 1951. 15. Taylor, 1960: 99. 16. Thomas, N. W. 2061 (Nig. Ser.), K. 17. Thomas, N. W. 2101 (Nig. Ser.), K. 18. Verger, 1967: no. 20. 19. Vigne 864, K. 20. Walker, 1953, a: 21. 21. Walker & Sillans, 1961: 85. 22. Willaman & Li, 1970.

**Pleiocarpa bicarpellata** Stapf

FWTA, ed. 2, 2: 63.

West African: **GHANA** AKAN-ASANTE ɔkanwen(e) (FRI) WASA kanwene ANYI-AOWIN kakana (FRI) SEHWI kakana (BD&H; FRI) NZEMA kakali (auctt.)

A shrub or tree to 15 m high, of the undergrowth of high-forest of Ghana (1, 2) and W Cameroons.

The wood is yellow, hard and close-grained, and is used for combs and plane-blocks (1, 2).

References:

1. Burtt Davy & Hoyle, 1937: 14. 2. Irvine, 1961: 630.

**Pleiocarpa mutica** Benth.

FWTA, ed. 2, 2: 63–64. UPWTA, ed. 1, 377.

West African: **SIERRA LEONE** LOKO rafɔi (NWT) MENDE kafa (*def.* kafei) *group name for several shrubs* (auctt.) TEMNE *a*-lap-*a*-ro-bamp (NWT) *a*-poli (NWT) TEMNE (Yoni) *am*-baleng (FCD) **IVORY COAST** ABE efi (Aub.) ABURE kakémé (B&D) taboo (B&D) AKAN-ASANTE kakané (B&D) ANYI kakana (Aub.) DAN dikpa *probably this sp.* (K&B) KYAMA mambeya (B&D) mogba popo (Aub.) 'KRU' frantu *probably this sp.* (K&B) nuliaïé *probably this sp.* (K&B) **GHANA** AKAN-ASANTE kanwene, ka: *to taste;* nwene: *bitter* (auctt.) TWI o-kanwene (auctt.) WASA kanwene (CV) kanwini (CV) ANYI kakana (FRI) ANYI-AOWIN kakana (FRI) SEHWI kakana (FRI)

A shrub or tree to 8 m high in the undergrowth of evergreen and deciduous forests of Sierra Leone to W Cameroons, and into E Cameroun and the Congo basin.

The wood is yellow and hard, dense and tough. In Ghana it is used to make combs, shuttles for weaving and plane-blocks (3, 4, 6, 8); in Sierra Leone for combs (5); and in Ivory Coast canoe paddles and pestles (2) and sundry small objects (1).

In Ivory Coast the plant is considered a general panacea for all ills by the coastal peoples. Of special application is the use by the Anyi of a decoction of the grated bark for stomach-pains, and a similar preparation by the Kyama for oedema of the legs due perhaps to kidney malfunction (2). Towards the Liberian frontier various parts of a *Pleiocarpa sp. indet.*, which is perhaps this, are used as a pulp in frictions and ointments, and internally as a decoction or macerate as an antispasmodic and febrifuge (7).

In Ghana the bitter roots are used for fever, jaundice and 'yellow fever' (Martinson fide 6).

A number of alkaloids has been reported from the bark and roots (9).

The plant has clusters of narrow tubular, sweet-scented flowers. It is ornamental and worthy of cultivation.

References:

1. Aubréville, 1959: 3, 206. 2. Bouquet & Debray, 1974: 31. 3. Burtt Davy & Hoyle, 1937: 14. 4. Dalziel, 1937. 5. Deighton 3270, K. 6. Irvine, 1961: 631. 7. Kerharo & Bouquet, 1950: 188, as *Pleiocarpa sp. indet*. 8. Vigne 269, K. 9. Willaman & Li, 1970.

**Pleiocarpa pycnantha** (K. Schum.) Stapf

FWTA, ed. 2, 2: 63. UPWTA, ed. 1, 376 as *P. flavescens* Stapf, and 377 as *P. micrantha* Stapf.

West African: GHANA ADANGME kpɛtɛkpɛtɛ (Howes) ADANGME-KROBO kpɛtɛ-kpɛtɛ-tʃo (JMD) AKAN-TWI o-kanwen, ka: *to taste;* nwen: *bitter* (JMD; FRI) ANYI-AOWIN kakapembe (FRI) SEHWI kakapembe (FRI) GA kpetekpete (Howes) NZEMA kakapimbe (FRI) NIGERIA YORUBA irokoro (Kennedy)

A shrub or tree attaining 30 m high in secondary jungle, or in the lower-storey of the high-forest from Mali to S Nigeria, and across Africa to Zaïre, Angola, Uganda and Zanzibar.

The wood is hard and yellow, and is used to make combs, plane-blocks and sundry small objects (2, 3, 4).

The roots are added to palm-wine in Ghana to give it potency, and ground and mixed with guinea-grains and palm-wine are taken 'to promote freedom of the bowels' (4). The leaves are bitter (4).

Numerous alkaloids have been reported from all parts of the plant (1, 5).

References:

1. Adegoke & al., 1968. 2. Aubréville, 1959: 3, 206. 3. Dalziel, 1937. 4. Irvine, 1961: 631–2, incl. p. 630–1, *P. micrantha* Stapf. 5. Willaman & Li, 1970: as *P. pycnantha* (K. Schum.) Stapf, *P. flavescens* Stapf and *P. tubicina* Stapf — with references.

**Pleiocarpa talbotii** Wernham.

FWTA, ed. 2, 2: 64.

A shrub or tree, semi-scrambling of the forest understorey, of only S Nigeria and W Cameroons.

The plant yields a white sticky latex (1).

Reference:

1. Keay FHI. 28277, K.

**Pleioceras barteri** Baill.

FWTA, ed. 2, 2: 76. UPWTA, ed. 1, 377.

West African: IVORY COAST ABURE abobo libi (B&D) AKAN-ASANTE adufa (B&D) KRU-BETE suè (K&B) GUERE (Chiehn) suè (K&B) KYAMA ayubu (B&D) GHANA AKAN-TWI kakapenpen *misapplied here* (JMD) ANYI-AOWIN abrafodɔ (CV; FRI) NZEMA bakapembe (auctt.) blɔfodɔ(ɔ) = *European's floss, from* dɔ: *silky floss* (FRI) NIGERIA YORUBA dàgbá, dàgbá ọmọde (JMD) efo

(Millson) irẹ́-ọna-kekeré = *small irẹnọ* (*Holarrhena floribunda*) (JMD) pari ọmọde *from* pa ori ọmọde: = *cause child's head to turn* (JMD)

An erect or climbing shrub or small tree of evergreen forest, secondary growth and scrub in Liberia to W Cameroons, and also in Cameroun.

The plant has gynaeco-medico-magical applications. The bark, and more especially the seeds are used in Ivory Coast as an emmenagogue: strong dosage is abortifacient and careless usage may cause death (4). Nevertheless it is taken by women once or twice a month during pregnancy albeit with care to avoid abortion and because the medication is held to induce movement of the foetus. For this reason it is given shortly before full-term to ensure a head-presentation and explains too the significance, apparently medico-magical, for the Yoruba name meaning *cause child's head to turn*. The prescription requires half a fruit crushed and cooked with a fish called *àrọ* or *ẹjãro,* a Silurid which is known in connexion with fetish worship and which has the habit of turning over on the surface of the water. Sometimes the brown rat is substituted for the fish (2). A root-decoction is used in Ivory Coast as a vaginal douche for female sterility, and a bolus of the crushed plant is placed in the vagina for vaginal malformation (1).

The bark crushed with a fruit of *Ricinodendron africanum* (Euphorbiaceae) is applied in Ivory Coast to suppurating glands in the groin to hasten maturation; a plaster of leaves is applied against rheumatism; and the fruit pericarp prepared as an ointment is used to treat epistaxis (1).

The seeds are the most toxic part of the plant with total alkaloids amounting to 0·3%. The fruit pericarp contains 0·1% and the root-bark 0·01% (5, 6). The bark has been used as an adulterant of *Holarrhena* (Apocynaceae) (5).

The floss is sometimes used to stuff cushions (2, 3, 5).

References:

1. Bouquet & Debray, 1974: 31. 2. Dalziel, 1937. 3. Irvine, 1961: 632. 4. Kerharo & Bouquet, 1950: 189. 5. Oliver, 1960: 34, 78. 6. Willaman & Li, 1970.

## Plumeria spp.

FWTA, ed. 2, 2: 54.

English: (white, yellow, red) frangipani; temple flower.

French: frangipanier (the tree); frangipani (the flower).

Portuguese: franipana.

A genus of shrubs and small trees of tropical America of which several species have been widely distributed by man as ornamentals. *P. rubra* Linn. is the commonest in the West African region. *P. acutifolia* Poir. (*P. rubra* var. *acutifolia*). Ait., *P. alba* Linn. and perhaps others are also present.

All the introduced species are ornamental with white, red or yellow flowers which are sweetly scented, reminiscent to the first Europeans to encounter the plant in the Caribbean of a perfume known as 'frangipani' in France whereby the plant acquired the same name. The name 'temple flower' arises from the practice in Asia of planting it in temple precincts and cemeteries.

All are poisonous. The latex or sap is strongly purgative, and the bark diuretic. Use in the New World as a febrifuge has not been borne out by tests on avian malaria for root-extracts by mouth and subcutaneously have given no antimalarial action (3). The latex contains some rubber but an excess of resin.

A number of active principles have been isolated. Of these *plumierin* is a crystalline bitter, non-toxic, but causing purgation by chemical changes in the intestines. A lactone, *fulvoplumierin,* is bacteristatic to *Mycobacterium tuberculosis*. The reader is referred to extensive phytochemistry, pharmacology and pharmacognosy references given by (1, 2, 4–6).

In the chemicals noted above, the name of Charles Plumier (1646–1704), the French biologist, is honoured. The genus too is named after him but Linneus following an orthographic aberration by Tournefort, who proposed the pre-Linnean name, spelt it *Plumeria*.

The flowers as already stated are sweet-scented. A volatile oil containing *geraniol, citronellol* and other etheric substances can be obtained from them. The flowers are potentially usable for scent-manufacture.

References:

1. a. Bissett, 1953. b. Bissett, 1958. c. Bissett, 1961. 2. Burkill, 1935: 1776–9. 3. Claude et al., 1974. 4. Kerharo & Adam, 1974: 178–9. 5. Quisumbing, 1951: 732–5. 6. Watt & Breyer-Brandwijk, 1962: 94.

**Pycnobotrya nitida** Benth.

FWTA, ed. 2, 2: 68.

A woody climber to 17 m long, of the forest zone of S Nigeria, and extending to Congo.

No usage is recorded in the Region. In Congo the leaves are eaten with other food for chest-affections and the latex is taken for haematuria, diarrhoea and dysenteries. These treatments are completed by wearing a cord made from this and *Haumania* (Marantaceae) together round the neck, stomach or chest. A fibre obtained from the bark is used in Congo to make crossbow cords.

Only steroids and terpenes have been recorded in the plant (2).

References:

1. Bouquet, 1969: 65. 2. Bouquet, 1972: 14.

**Rauvolfia caffra** Sond.

FWTA, ed. 2, 2: 69. UPWTA, ed. 1, 378, as *R. welwitschii* Stapf.
West African: NIGERIA HAUSA wada (JMD; ZOG) YORUBA awa (SOA)

A tree to 23 m high, bole to 1·30 m girth, of the forest especially in swampy areas and of the savanna in riverine situations in N and S Nigeria and W Cameroons, and widespread elsewhere in tropical and S Africa.

The tree is planted as an ornamental shade tree in towns on the Benue River (1, 2). Its timber is sometimes used on the Lower Niger (2). In Kenya it is used for hut-building (3).

The bark contains a fibre which is used in E Cameroun to make bow-strings (7). In Tanganyika it is used for beer-making, and a liquor obtained from boiling bark is given to children affected by 'some red intestinal parasite' (5). A root-decoction is taken by draught in Tanganyika, or the dried bark cooked to a mash is used (?poultice), for maturating hard abscesses (6).

The plant contains a latex that thickens to a sticky mass that is employed in Kenya as a bird-lime (4).

The number of alkaloids, including *reserpine, ajamalicine* and *ajamaline,* also present in *R. vomitoria,* has been recorded present in the roots. The leaves are also with some alkaloid. (6, 8).

References:

1. Dalziel 796, K. 2. Dalziel, 1937. 3. Glover & al. 2416, K. 4. Graham 1742, K. 5. Greenway & Kanuri 12176, K. 6. Haerdi, 1964: 131. 7. Leeuwenberg 5295, K. 8. Willaman & Li, 1970: with references.

**Rauvolfia macrophylla** Stapf

FWTA, ed. 2, 2: 69. UPWTA, ed. 1, 377.

English: false alstonia (Nigeria, Keay & al.).

West African: **WEST CAMEROONS** DUALA enoŋgo (Ithmann) KPE kanja (AHU) KUNDU enonge (AHU)

A tree to 25 m high, of the forest zone of W Cameroons, and perhaps of S Nigeria (Onitsha Province), and occurring in E Cameroun and S Tomé.

The wood is dull yellowish, resembling that of *Alstonia congensis* Engl. The bark contains a fibre that is used in Gabon to make bow-strings (3), and a cordage in W Cameroons (1). It also holds an abundance of white latex that is bitter to the taste. The bark is purgative (1, 3). A root-macerate with fruits of *Brenania brieyi* (De Wild.) Petit (*Randia walkeri*) (Rubiaceae) enters into treatment of intestinal worms and of syphilis in Gabon (3). The presence of alkaloids in the plant is recorded without detail (2).

References:

1. Dalziel, 1937. 2. Oliver, 1960: 81. 3. Walker & Sillans, 1961: 86.

**Rauvolfia mannii** Stapf

FWTA, ed. 2, 2: 69.

West African: **NIGERIA** IJO-IZON (Kolokuma) indóndó (KW) òwéí kórómó (KW) IZON (Oporoma) ẹ̀kpẹ́sịkpẹ̀sị (KW) YORUBA dodo, idodo (N&E)

A shrub or tree to 10 m high of the forest of S Nigeria and W Cameroons, extending to Zaïre.

The Ijo use the plant (part not stated) in gin and take a teaspoonful when tired (2).

The alkaloid *vincamajine* has been reported present in the leaf (1).

References:

1. Willaman & Li, 1970. 2. Williamson KW.34, UCI.

**Rauvolfia serpentina** Benth.

A shrubby plant of open deciduous forest and savanna areas widely dispersed over India and SE Asia, and introduced into the Region. The plant has been used from olden times in India for relief of central nervous system disorders, both psychic and motor, including anxiety conditions, all forms of insanity, epilepsy, etc., intestinal disorders, and many other complaints. The main alkaloid out of many present is *reserpine* with strong hypotensive properties. India used to be the chief world supply centre of this drug widely used for treating hypertension and official in many pharmacopoeas. Over-exploitation in India has resulted in other sources of this drug being sought and *R. vomitoria* Afzel is now a prime source — see there (2).

This drug appears in Sanskritic works as *sarpagandha, sarpa* being a snake (and, of course, the English 'serpent'), and was amongst the Indian 'snake-roots' that were credited with power to cure snake-bite (1). It has been noted under *R. vomitoria* that this latter plant is used in the event of snake-bite in Nigeria and is fetish to the Snake Sect of Ivory Coast.

References:

1. Burkill, 1935: 1583 under *Ophiorrhiza,* and 1885 under *Rauvolfia*. 2. Krishnamurthi, 1969: 377–90.

**Rauvolfia vomitoria** Afzel.

FWTA, ed. 2, 2: 69. UPWTA, ed. 1, 377.

English: swizzle-stick.

West African: **SENEGAL** BASARI a-pínàs (Ferry) BEDIK gi-pilás (Ferry) DIOLA bural (JB, K&A) giupa (JB; K&A) DIOLA (Brin/Seleki) mérianni (K&A) FULA-PULAAR (Senegal) moda tatel (Laffitte, ex K&A) moyatalal (Aub., ex K&A) KONYAGI a-pɛ (Ferry) MANDING-BAMBARA kolidiohi (JB) MANINKA kolidohi (Aub., ex K&A) **THE GAMBIA** DIOLA (Fogny) bural = *better* (DF) giupa (DF) **GUINEA** FULA-PULAAR (Guinea) moyia tialel (RS) KONO uogni-seguélé (RS) uonsé-uru (RS) MANDING-MANINKA kolidiohi (RS) MANO moyala-iri (RS) **SIERRA LEONE** BULOM (Sherbro) bon-dɛ (S&F) KISSI chuŋchuŋka (FCD; S&F) KONO beiŋga-kɔne *loosely applied to soft-wooded trees* (S&F) duŋga (S&F) g-basa (S&F) LIMBA sɛnke (NWT) LOKO koroga (S&F) ndegbe (NWT) MANDING-MANINKA (Koranko) taŋgbɛsowakoloma (S&F) MENDE koga (*def.* kogei) (L-P; S&F) kowogea (*def.* kowogei) (auctt.) koyoga (SKS) SUSU besiwuri (NWT) g-besi (NWT) TEMNE *a*-bobən (FAM; S&F) *a*-fenkre (S&F) **LIBERIA** MANO mõñ a yiddi (JMD) MENDE kawoga (C&R) **IVORY COAST** VULGAR tèrè (A&AA) ABE n-déchavi *a general name* (auctt.) ABURE agbamassan (B&D) gongonkiur (A. Chev.; K&B) ADYUKRU ligbogun (A&AA) AKAN-ASANTE atiablé, bakimbé (B&D) kaka penpe (K&B) kakapimbé (B&D) BRONG kaka pempé (K&B) AKYE n-guessèbi, n-kichèbi (A&AA) ANYI bakaégbi (Aub.; RS) embi-siembi (A. Chev.) kaha paye paye (K&B) BAULE baka pimblé (B&D) kakakwé (B&D) niahui (K&B; B&D) DAN bué mali (K&B) FULA-FULFULDE (Ivory Coast) moyiatialal (Aub.) GAGU bi (B&D) KRU-BETE dobuéï (K&B) dugbéï (K&B) GUERE diablan, dialon, yablan (K&B) yablon (A&AA) GUERE (Chiehn) guéto (K&B) NEYO nia tatté (K&B) KULANGO kaka pempé (K&B) KWENI to (B&D) toto (K&B) KYAMA brokuadiomué (auctt.) MANDING-MANINKA kolidiohi (Aub.) 'KRU' terré (K&B) 'NEKEDIE' to (B&D) **GHANA** ADANGME-KROBO apɔtɔtʃo, apɔtɔ: *to crush, or mash – to obtain the seeds* (FRI) AKAN kakapenpen (Enti) AKAN-AKYEM kaka penpen (Enti) ASANTE kakapenpen *from* kaka: *to bite;* penpen: *to break – resembling snapping of the brittle twigs* (FRI) FANTE kakapenpen (Enti) TWI kakapenpen (auctt.) WASA kakapenpen (Enti) kakapenpen = *brittle* (FRI) nsusuwa-dua (Enti) nsusuwa-dua (auctt.) ANYI baka egbe (FRI) ngbe ngbe (FRI) ANYI-SEHWI aneɛneaa (FRI) aneɛnia (FRI) GBE-VHE dɔɖemakpɔwoe (FRI) NZEMA bakaɛmbe (FRI) **DAHOMEY** GBE-FON leti (Laffitte) **NIGERIA** EDO ákátà (auctt.) EFIK mmọñebà = *milk* (Winston) ḿmọ́ñébà utó ènyìn (RFGA; Lowe) ùtò enyìn = *yellow fever, jaundice* (KO&S; Winston) HAUSA wada (ZOG) IGBO akanta (auctt.) YORUBA asofẹ́yẹjẹ̀ = *fruit for bird to eat* (auctt.) ìra (Millen; JRA) ìra-igbó (JRA; JMD) orà igbó = *lost bush* (Verger; IFE) **WEST CAMEROONS** DUALA bòndoŋgɛ̀ (Ithmann)

A shrub or tree seldom over 10 m tall, of the forest and common in secondary growth throughout the Region from Senegal to W Cameroons, and extending across Africa to Egypt, Sudan, Tanganyika, Uganda and Zaïre.

The tree has cream and white sweet-scented flowers and masses of bright red fruits. It is widely planted as an ornamental (7, 8) and in the Bukoba township of Kenya it is grown as an avenue tree (19). It is suitable for a live fence. In Gabon it is planted as a support for vanilla and as a shade-bearer for young cacao (25, 26). The tetramerous whorled branching is used in Sierra Leone for supporting pieces of papaya as a bait with a horizontal stick as a perch for trapping birds: the tree is thus sometimes known as 'papaw table.' The young twigs with the side branches trimmed short serve as mixers for drinks, hence the English name 'swizzle-stick' (7, 8, 20). Larger branches are similarly used to stir the indigo brew in dyeing (8). The tree is a wild host of the pathogen causing 'collar crack' of cacao (10).

The wood is white and fine-grained, reddening with age and is fairly hard in the heart of older stems. It is of little economic importance (7, 8). It is used as firewood in Sierra Leone (20). It is a substitute for boxwood.

The root has neuro-sedative properties. A decoction can be given for maniacal symptoms inducing several hours sleep (8). In Ivory Coast the roots with the leaves of other drug plants are taken by mouth or the pulp is placed in the nose for epilepsy (2). A 48-hours macerate in palm-wine along with the root of *Cogniauxia podolaena* Baill. (Cucurbitaceae) is taken in Congo to prevent erotic dreams and spermatorrhoea (5). In contrast the root is considered aphrodisiac and genital excitant in Ghana when it is taken macerated in gin with guinea-grains (Anthony fide 11) and in Ivory Coast where the Kweni give in enema water in which bruised roots have been steeped (15). The root-bark is given to children in small dosage as a sedative in Ivory Coast (2), and for convulsion in Nigeria (8). The Yoruba invoke the fruit as well as the bush in separate incantations for the cure of mad people (23). The roots have been effectively used in Nigeria to treat jaundice (1) and in Ghana root-bark preparations with spices are given by enema for jaundice and gastro-intestinal conditions (8). The bark has bitter properties and is a powerful purgative and emetic (8). The Wobe of Ivory Coast swallow some of the pulped-up bark of young twigs against poisoning (2). If not in decoction or macerate, the bark may be given along with food, often a ripe banana (5, 26). Administration requires care. A decoction is reported used in Senegal for blennorrhoea (4), and a decoction in palm-wine in Ivory Coast (2). In Ivory Coast—Upper Volta it is administered for a number of serious illnesses, e.g. leprosy, urethral discharge, and dysentery (in which the disease is held to be caused by wild animals entering the stomach and a vigorous purge being necessary to evict them) (14). At Magbena in Sierra Leone it is compounded with *Mareya micrantha* (Euphorbiaceae), *Xanthoxylum gillettii* (syn. *Fagara macrophylla,* Rutaceae), *Microdesmis puberula* (Euphorbiaceae) and newly-worked black soil from the top of a termite heap to rub into paralyzed limbs (17). In Congo a root-decoction is put into vapour baths and applied by massage for rickets, and is used for rheumatism and fatigue; a mouth-wash is used for gingivitis and thrush (5). Root-bark on the tongue is said to cause numbness (16). The Igbo of S Nigeria use the root as an abortifacient, and the name *akanta* comes into an Igbo song warning young women not to drink the root-extract (12). At Asaba they use the bark for rheumatism (21). The root is also applied powdered by the Yoruba for snake-bite (3). In Liberia a bark-infusion is used for fever (8), and a macerate of bark and leaves for infantile fever in Gabon (26). An extract of the root-bark as eye-drops with pulped-up leaves smeared over the face is a Liberian treatment for vertigo (8). Bark-sap instilled into the eye is a Congo cure for epilepsy (5).

Freshly pulped roots are considered throughout Ivory Coast to be a good antipsoric and are applied with some citron-juice and clay to the affected parts (15). The powdered bark is sprinkled into areas of nettle-rash, measles, herpes, etc. in Nigeria (3) and a decoction is used against head-lice (22). Powdered stem-bark with clay is sprinkled over the eruptions of chickenpox in Ivory Coast (2). Medicinemen in Joru, Sierra Leone, use the bark for curing chronic sores (9). Bark and root-powder, or compounded as ointment with water or palm oil, are used in Gabon to kill fleas and vermin (25, 26). In N Nigeria, Hausa traders use a bark-extract mixed with papaya juice to preserve kola nuts from infestation by the kola weevil (*Balanogastris colae*) (8). In Congo the powdered roots are applied to sores caused by skin-parasites after cleansing with a decoction, a treatment said to be very effective for this and also for eczema and ringworm (5). A decoction of the bark is similarly used by the Fula of Guinea (4).

A yellow dye from the bark is reported used in Sierra Leone (20).

The older woody parts of the plant contain no latex. It is present in the young green leafy twigs, which are powerfully emetic. They are slightly styptic with a strongly bitter taste lingering long on the tongue. Two or three leaves swallowed raw cause vomiting and a handful infused in hot water furnishes a medicine of which three of four teaspoonfuls produce free vomiting, and if continued cause violent purging. Use is indicated as an expectorant in paroxysmal or inflammatory coughs. Externally a compress of the leaves, or root-bark with sap of a plantain stem, sometimes with lime-juice and spices as a counter-irritant, is put on sprains and swellings (8). In Ivory Coast leaves mixed with those of *Nephrolepis biserrata* (Pteridophyta) and *Baphia nitida* (Leguminosae: Papilioniideae) and some melegueta peppers (*Aframomum melegueta,* Zingiberaceae) are applied as a paste for costal pain, and for contusions leaves softened over a fire are rubbed on the area and to inflammation between the toes (2). In Congo leaf-pulp is taken in draught and used in massage for chest-pains, and leaf-sap, or powdered root, is deemed able to arrest loss of hair, and even to restore it (5). The leaves cooked up with djave nut butter (*Baillonella toxisperma* Pierre, Sapotaceae) are applied as an ointment to inflammations, dislocated joints and limbs affected by rheumatism in Gabon, and the leaves in macerate are given to infants with fever (25, 26). The Tiv of the Benue, Nigeria, take leaf-sap in draught for internal disorders (21).

The fruits, like the root of the plant, are also strongly emetic. In Congo the fruit-pulp and crushed seeds are effectively used against filaria and body-lice (5). The seeds are used by Krobo girls as beads, for bracelets and the Krobo name *apoto,* to crush, comes from the way the fruits are squeezed to obtain them (11).

A considerable number of alkaloids is present in the plant. Stem-bark has been found to have 0.59% total alkaloid, while root-bark has 1.7% (14). In the roots the bark contains 7–18 times as much as the wood, and the wood, and total content varies considerably from root to root according to size (27). Early interest in the *Rauvolfia* alkaloids centred on the Indian *R. serpentina* as a source of the alkaloid *reserpine* with sedative and hypotensive properties and for treating high blood pressure. Reserpine has been found in *R. vomitoria* with a yield of 0.1–0·2% from the roots and this has to a large extent replaced *R. serpentina* as the commercial source. *Ajamaline*, active on the sympathetic nervous system, and a long series of other pharmacologically active alkaloids has been identified and isolated (6, 14, 18, 28).

In Ivory Coast, the 'Serpent Sect' of the Man region consider the plant fetish (15).

References:

1. Adegoke et al., 1968. 2. Ajanohoun & Aké Assi, 1972: 40. 3. Ainslie, 1937: sp. no. 298. 4. Anon 7/1939, IFAN. 5. Bouquet, 1969: 65–66. 6. Bouquet & Debray, 1974: 32–33, with clinical uses and phytochemistry. 7. Cooper & Record, 1931: 106–7. 8. Dalziel, 1937. 9. Herbalists. Joru Village, Nov. 1973. 10. Howes 1002, K. 11. Irvine, 1961: 633–4. 12. Irvine, s.d. 13. Kerharo & Adam, 1963, b. 14. Kerharo & Adam, 1974: 179–86, with phytochemistry, pharmacognosy and pharmacology. 15. Kerharo & Bouquet, 1950: 189. 16. Laffitte 91/1937, IFAN. 17. Massaquoi, Nov. 1973. 18. Oliver, 1960: 9, 81. 19. Proctor 891, K. 20. Savill & Fox, 1967, 50. 21. Thomas, N. W. 1673 (Nig. Ser.) K. 22. Thomas, N. W. 2330 (Nig. Ser.) K. 23. Verger, 1967: nos. 49, 150. 24. Vermeer 63, UCI. 25. Walker, 1953, a: 21. 26. Walker & Sillans, 1961: 86–87. 27. Watt & Breyer-Brandwijk, 1962: 96–100. 28. Willaman & Li, 1970.

## **Saba florida** (Benth.) Bullock

FWTA, ed. 2, 2: 61. UPWTA, ed. 1, 373, as *Landolphia florida* Benth.

English: paste rubber (the product, Imperial Institute).

West African: **SENEGAL** DIOLA bundif (H&C) fufufole (MP) MANDING-BAMBARA bili (H&C) MANDINKA kaabaanombo (*def.*-o) WOLOF mad(-a) (MP) **THE GAMBIA** DIOLA (Fogny) bundif = *sour* (DF) WOLOF kaban-dombo = *kaba with tendrils that wrap around* (DF) **GUINEA-BISSAU** MANDING-MANDINKA caba-forô = *pure caba* (JDES) PEPEL mutaba (JDES) **GUINEA** MANDING-BAMBARA saba (JMD) MANINKA saba (JMD) **SIERRA LEONE** MANDING-MANDINKA morulate (FCD) SUSU-DYALONKE birɛ-na (FCD) TEMNE *an*-mar (FCD) **MALI** MANDING-BAMBARA abo(-h), aboli *the fruit* (MP) bili (H&C) saba (MP) **IVORY COAST** MANDING-MANINKA gbèī (B&D) **GHANA** AKAN-ASANTE akontoma (CV) **TOGO** AKAN-ASANTE akarapotu (Seefrieds) BASSARI adjulŏ (Gaisser) GBE-AJA (Sikpi) kadia (JMD; MP) VHE legla (JMD; Kersting) KABRE lo (Kersting) lowu (Kersting) TEM (Tshaudjo) lo (Kersting) loku (Kersting) **NIGERIA** BEROM dìnáká *the plant* (LB) eyɛ́ *the fruit* (LB) BOLE dinza (AST) EDO ùbó (JMD) FULA-FULFULDE (Nigeria) nyanaare gaduuru = *pig's melon* (JMD; MM) GWARI nini (JMD) niniche *from* eche: *rubber* (JMD) HAUSA ciiwóó (LB) cìwoó, cùwoó (ZOG) kuranga, kuringa, kwaranga = *ladder; applied to any rubber-vine* (JMD; ZOG) IGBO ọtọ, utu (auctt.) ụ̀tụ̀ ehi (BNO) utu-isi-enyi (JMD; MP) NUPE bŏnù (auctt.) èbŏ *the fruit* (auctt.) YORUBA ìbò (JMD; MP) ìbò-akitipa (auctt.) ìbò-gidi = *true rubber vine, here misapplied* (JMD; MP) odi-oje (Rowland)

A liane to 40 m long by 1.90 m girth of the closed-forest, fringing forest and savanna woodland, often on stream-banks, commonly throughout the Region from Senegal to W Cameroons, and across Africa to Sudan, Zanzibar and Tanganyika.

The plant contains an abundant white latex which may become pinkish on exposure, sticky and coagulable on prolonged boiling to a product called 'paste rubber.' It is generally worthless as rubber, but occasionally good coagulum may be obtained. It has been used as an adulterant of the better quality rubbers. It is usable as bird-lime (2, 7). In contrast there is often a high presence of resin which results in a brittle product (10, 11) and in N Nigeria the latex boiled with gum copal from *Daniellia thurifera* Benn (Leguminosae: Caesalpinioideae) has been used to prepare a 'flake rubber' (13).

In central Africa the latex is used to heal sores, and is applied to maturate abscesses (7). In Congo it is dripped into the eyes to improve vision (1). In Tanganyika the milky sap together with that of *Diplorhynchus condylocarpon* (Müll.-Arg.) Pichon (Apocynaceae) is drunk as a galactogogue (5), and Fula cattlemen in N Nigeria give the sap to their cattle for the same purpose and as a general food (9). Such practices must surely rest on a knowledge that the local plants do not produce latex which will coagulate in the intestines and cause occlusion.

Bark and root-decoctions are taken by draught in Tanganyika as an aphrodisiac, a root-decoction as a vermifuge and a leaf-decoction as a laxative (5). The root together with other substances is given in Nigeria as a remedy for gonorrhoea (2) and in Tanganyika it enters a snakebite remedy (6). In Congo the plant (part unspecified) is used to treat jaundice (1).

The fruit has an edible, pleasantly acid pulp. It is much relished by nearly all races wherever the plant occurs, and by monkeys (2, 7). To the Acholi of Uganda it is considered a famine-food (3). It is said to be purgative (7, 8). In Zanzibar a fruit drink is taken not only as a pleasant drink but to cure pimples (11).

The seeds are eaten raw in some places of Zaïre (7).

In Togo a blue dye is made from the crushed leaves, flowers and twigs (Volkens fide 2, 4).

References:

1. Bouquet, 1969: 66. 2. Dalziel, 1937. 3. Eggeling E.2343, K. 4. Greenway, 1941. 5. Haerdi, 1964: 132, as *S. comorensis* (Boj.) Pichon. 6. Koritschoner 1422, K. 7. Pichon, 1953: 303–16, as *S. comorensis* (Boj.) Pichon. 8. Tanner 1992, K. 9. Thornewill 101, K. 10. Walker & Sillans, 1961: 82–83. 11. Williams, R. O., 1949: 323. 12. Yates 44, K.

## **Saba senegalensis** (A. DC.) Pichon

FWTA, ed. 2, 2: 61. UPWTA, ed. 1, 375, as *Landolphia senegalensis* Kotsch. & Peyr.

French: liane saba (after the Maninka name).

West African: **SENEGAL** BALANTA beñde (K&A) bengdé (auctt.) BASARI a-ngúŋ (K&A; Ferry) BEDIK gi-wòm (K&A; Ferry) CRIOULO fole-grandi (MP) DIOLA buitipobu (MP) sikokinad (AS) DIOLA (Brin/Seleki) budur (K&A) DIOLA (Fogny) bu hindik (K&A) budimbob *the liane* (K&A) bulanay (K&A) ekenay (K&A) élanay (K&A) sindipasu *the fruit* (K&A) skèanay (K&A) DIOLA-FLUP buindip (K&A) FULA-PULAAR (Senegal) dabadombo (JLT) lamudé, lamuné = *bitter, or salty* (K&A) laré (H&C) pétigué (Segeur) põre laré (H&C) KONYAGI a-kab (Ferry) MANDING-BAMBARA saba (JB; K&A) saba-bili (MP) MANDINKA folé = *rubber* (MP) kaba kàbaa (*indef.* and *def.*) (auctt.) saba (A. Chev.) MANINKA saba (K&A; MP) saba bili (JMD; MP) 'SOCE' kaba (K&A) MANDYAK kaba (FB) SERER mand (auctt.) SERER-NON mat (auctt.) WOLOF foré-fikne (Maclaud) lamagui (Etesse) mad, mada, madd, made (auctt.) tioh, tiojutiopti, tiorh *the roots* (MP) tohl (MP) **THE GAMBIA** DIOLA (Fogny) bugimb = *suck* (DF) buitipobou = *good medicine leaf* (DF) ekenay, skèanay = *tuber* (*onion*) (DF) FULA-PULAAR (The Gambia) laro (DRR) MANDING-MANDINKA kàbaa (*indef.* and *def.*) kabaanumba (*def.*-o) (DAP) WOLOF madd (Leprieur) **GUINEA-BISSAU** FULA-PULAAR (Guinea-Bissau) poré (D'O) **GUINEA** FULA-PULAAR (Guinea) laré (MP) poré (Bouronville) pore-laré (MP) MANDING-BAMBARA saba-bili (CHOP; MP) MANINKA (Guinea) minadiaba (MP) n-zaban (MP) saba-bâ (MP) MANINKA (Guinea)(Koranko) kamu (Ganay) SUSU bonkhé (Paroisse) SUSU-DYALONKE laré (MP) **MALI** MANDING-BAMBARA bili (JMD; Coppins) goe (MP) laré (MP) lingui (A. Chev.) saba (auctt.) saba bili (JMD) saba-laré (Coppins) sagua (JMD; MP) MANINKA bili (JMD) saba (JMD; A. Chev.) SONGHAI lingui (JMD; A. Chev.)

SONINKE-SARAKOLE saba (A. Chev.) **UPPER VOLTA** BISA gulugu (FB) DAGAARI ora (K&B) KIRMA natu (K&B) MANDING-BAMBARA sagua (K&B) sawa (K&B) DYULA zama (K&B) MANINKA mi (auctt.) minadiaba (auctt.) MOORE watega (K&B) wedgha (K&B) SENUFO-TUSIA m'bungo (MP) mi (MP) 'SENUFO' dabirilé (auctt.) dabri (auctt.) dabrumo (K&B) **IVORY COAST** LOBI lwo (A&AA) MANDING-MANINKA saba (K&B; A&AA) SENUFO-TAFILE zama (MP) TAGWANA kobri (K&B) 'SENUFO' dabirile (auctt.) dabri (auctt.) dabrumo (K&B) **GHANA** AKAN-ASANTE ɛsono-nantin = *elephant's heel* (CV; FRI) TWI anoma (FRI) anumba edwiani (CV; BD&H) edwiani (Moor) DAGAARI ora (auctt.) MOORE weda (auctt.)

A liane to over 40 m long by 47 cm girth, of forest fringes, galleried forest and thickets, of the soudanian region and becoming bushy in the drier regions, from Senegal to Togo.

The plant has medico-magical attributes and is often preserved on the outskirts of villages (4, 8, 11).

It contains an abundance of white sticky latex which hardens on exposure. It was exploited at one time for its rubber in Senegal, Guinea, Ivory Coast and Niger (2) and in The Gambia (14) but the rubber was of poor quality and its main use was an adulterant of better quality rubbers (13). Fula of Senegal use the latex medicinally for pulmonary troubles and tuberculosis. The stem on cutting is immediately placed in a gourd with the right amount of water and as the latex expissates the gourd is shaken so that an emulsion of the latex is made which is drunk. It is antitussive and emetic (10, 11). Pallaka hunters in Ivory Coast use the latex as an adhesive for poison preparations onto arrows (12).

The Fula of Senegal prepare the leaves in sauces and condiments as an appetizer with a salty tang — hence their name for the plant *lamudé* or *lamuné,* meaning 'salty' (10, 11). The leaves are eaten in Guinea to stop vomiting (3) and the Fula of Senegal make a steam-bath with them inhaling the vapour for chronic headache (10, 11), while a similar preparation is used in Casamance for the eyes to try to stall off approaching blindness (7, 11). In Ivory Coast—Upper Volta leaf and bark-decoctions are taken for dysenteriform diarrhoea and food-poisoning, and with other plants it is put into leprosy medications (12). In Tanganyika the leaves after crushing and soaking in water are applied as a haemostatic, antiseptic and cicitrisant dressing to wounds (5).

The roots are considered good for urethral discharges by the Fula of Senegal who take a macerate by draught (10). The Dagaari of Upper Volta use the powdered root-bark to heal sores, it being held particularly efficacious on burns on children (12).

The fruits are ovoid, about 7 cm long by 8 cm diameter with a yellowish sweet-sour, soft edible pulp. They are much prized and besides the medico-magical attributes of the plant are a very good reason for the conservation of the liane in the vicinity of villages. The fruit in season may often appear for sale in markets (1, 4, 6). In upper Ivory Coast and Upper Volta Senoufo and Kirma women eat a macerate of the fruits along with other drug plants as a sterility treatment. This may be a belief in the Theory of Signatures hinging on the ovoid 'pregnant' shape of the fruits (12).

References:

1. Busson, 1965: 408. 2. Aubréville, 1959: 3, 189. 3. Chevalier, A., 1920: 408–9, as *Landolphia senegalensis* Kotschy & Peyr. 4. Dalziel, 1937. 5. Haerdi, 1964: 319. 6. Irvine, 1961: 634–5. 7. Kerharo & Adam, 1962. 8. Kerharo & Adam, 1963, b. 9. Kerharo & Adam, 1964, a: 432. 10. Kerharo & Adam, 1964, b: 572–3. 11. Kerharo & Adam, 1974: 186–7. 12. Kerharo & Bouquet, 1950: 187, as *Landolphia senegalensis* Kotschy & Peyr. 13. Pichon, 1953: 316–22. 14. Rosevear, 1961, as *Landolphia senegalensis*.

**Saba thompsonii** (A. Chev.) Pichon

FWTA, ed. 2, 2: 61. UPWTA, ed. 1, 375, as *Landolphia thompsonii* A. Chev.

West African: **DAHOMEY** GBE-AJA komerô-akowa (Le Testu) **NIGERIA** YORUBA ìbò (Punch; JMD) ìbò gidi (EWF; JMD)

A liane to 30 m long to 20 cm girth of closed and fringing forests from Ivory Coast to S Nigeria.

The tree contains a white latex that can be coagulated by heat, but the resultant rubber has no elasticity and is valueless. It may at one time have been a source of 'paste rubber.'

Reference:

1. Dalziel, 1937.

**Strophanthus barteri** Franch.

FWTA, ed. 2, 2: 72. UPWTA, ed. 1, 378.

West African: **GHANA** AKAN-ASANTE manpohan (CV)

A slender climber of closed-forest and scrub recorded from Guinea to W Cameroons.

The roots have unspecified medicinal uses in S Nigeria (1).

Reference:

1. Akpabla 1122, K.

**Strophanthus gracilis** K. Schum. & Pax

FWTA, ed. 2, 2: 72.

A slender climber of the evergreen forest, recorded only from S Nigeria and in E Cameroun and Gabon.

This species is one of the *S. spp.* containing the largest quantity of glycosides. Nine have been isolated. The seeds and wood are used in arrow and fish-poisons (1).

Reference:

1. Oliver, 1960: 38, 85.

**Strophanthus gratus** (Hook.) Franch.

FWTA, ed. 2, 2: 70. UPWTA, ed. 1, 378.

West African: **SIERRA LEONE** MENDE gohɔndo (*def.* gohɔndui) (FCD; JMD) gɔndo *a contraction of gohɔndo* (JMD) sawa (*def.*-i) *a general term for medicine used to cure violations of secret society rules, and also to ensure good crops, made of a decoction of leaves; more specifically applies to Gouania longipetala* (*Rhamnaceae*) **LIBERIA** MANO konen (Adams) **IVORY COAST** AKYE kalanmeni (Planchon) m-gbété (Ivanoff) BAULE m-moropo (K&B) siniabié (K&B) KYAMA salo bego-esé (K&B) **GHANA** AKAN-ASANTE eguro-eguro (anon.; FRI) matwã (Enti) TWI o-mããtwa, o-matwã *general for all S.spp. used in treating leprosy* (Enti) omããtwanini, nini: *male, = male ɔmããtwa* (auctt.) **NIGERIA** IGBO ọta *general* (NWT; JMD) YORUBA ìṣa (Millen; JMD) ìṣa gidi = *true Strophanthus* (JMD; Verger) ìṣagere (JRA) ìṣa-ogbubu (JMD) ṣagere (JRA) YORUBA (Ukwu Nzu) òsú

A vigorous evergreen climber to the canopy of deciduous forest and secondary growths, occurring fairly commonly from Sierra Leone to W Cameroons, and extending to Zaïre.

The plant has shining foliage and showy white, pinkish to purple flowers which are sweetly scented in the evening. It has decorative value.

The leaves are used in Sierra Leone for gonorrhoea (8), and are added to other medicinal prescriptions to render them efficacious (7). Leaf and twig-decoction is taken orally in Ivory Coast for blennorrhoea (4, 11), and leaf-sap is put onto ulcerated sores (11). The leaves are mashed and applied to guinea-worm sores in Ghana (10). They are used as a dressing for sores in Nigeria, leaf and stem decoctions are taken for constipation (2), and the leaves are rubbed on to fever (*eba*) patients (15) or used to prepare a wash in fever treatment (16). The leaves (or those of *Alstonia congensis,* Apocynaceae) are held in Sierra Leone to be an antidote to the poison of the black-headed cobra (*Naja nigricollis*) (10).

A decoction of the crushed stem is taken in Ghana for severe sickness with weakness. It is dangerous to drink water after taking it, and the patient is kept isolated, indicating the risk involved in using this remedy, which is also said to produce remarkable cures (10). Sap from the fresh bark is mixed with that of *Parquetina nigrescens* (Afzel.) Bullock (Periplocaceae) to produce an arrow-poison in Congo (3). The root-bark is used in Ivory Coast near the Liberian border as an antidote in food-poisoning (11). Root and bark-decoctions, and the root taken powdered have an action similar to that of potassium iodine (rapid absorbtion, and excretion by the kidneys, salivary glands, mucosae, etc., and value as an expectorant) and these preparations are used in Nigeria in cases of syphilis (2). The root-tincture is said to be aphrodisiac (2). The plant (part unstated) is used with incantation by the Yoruba to cure craw-craw (17).

The seeds are used in some parts for making an arrow-poison (Liberia, 1; Nigeria, 9; Cameroun, 5; Gabon, 18). A water extract alone is a violent poison, but for the hunt in Gabon the seeds are mixed with the sap of other plants such as *Cococasia* (Araceae), *Aframomum* (Zingiberaceae), *Palisota* (Commelinaceae), etc. (18), and it was affixed in Cameroun and Gabon to arrows-shafts made of *S. gratus* wood (7, 10). An antidote to this poison is said to be external application of the powdered bark of *Erythrophloeum guineense* (Leguminosae: Caesalpinioideae), and internally by draught the sap of *Alstonia congensis* (Apocynaceae) (18). The seeds of *Garcinia cola* (Guttiferae) are also said to be an antidote to this poison (7). In parts of S Nigeria the plant is cultivated by hunters especially for the seeds (14).

The seed is an important pharmaceutical source of *strophanthin-G,* or *ouabain* which acts on the heart like digitalis. It is quick-acting by intravenous injection and is used in cases of heart-failure (6, 13). It raises blood pressure and is also diuretic. In E Cameroun *S. gratus* is cultivated especially for the seeds and supplies the whole of the French market (4). The seeds have no smell but are extremely bitter. Ouabain content is 4%. Plantation practice is to grow support trees of *Spondias cytherea* Sonner (Anacardiaceae) for the plant to climb. *S. cytherea* has a spreading crown furnishing an extended climbing area and has in addition the advantage of producing an edible fruit (7).

The plant enters into superstitions and magical uses, principally to bring good luck. In Sierra Leone, to be fortunate in gambling, a leaf is rubbed between the hands, or when up against an ordeal by fire or other trial by superstition (12). Its strength is recognized in the Mende name *tsa-wa,* big, or powerful (medicine). It is used in the ritual of the women's society 'Humo' (Lane-Poole fide 7), and in Nigeria it has many secret and juju uses (2).

References:

1. Adames 813, K. 2. Ainslie, 1937: sp. no. 327. 3. Bouquet, 1969: 66. 4. Bouquet & Debray, 1974: 35. 5. Brass & Woodward 20854, K. 6. British Pharmaceutical Codex, 1959, 516–7. 7. Dalziel, 1937. 8. Deighton 275, K. 9. Foster 139, K. 10. Irvine 1961: 636–8. 11. Kerharo & Bouquet, 1950: 190. 12. Lane-Poole 147, K. 13. Oliver, 1960: 85. 14. Ross, R.74, K. 15. Thomas, N. W. 1983 (Nig. Ser.) K. 16. Thomas, N. W. 2079 (Nig. Ser.) K. 17. Verger, 1967: no. 123. 18. Walker & Sillans, 1961: 87.

**Strophanthus hispidus** DC.

FWTA, ed. 2, 2: 72. UPWTA, ed. 1, 379–81.

English: 'arrow poison' (Ghana, Irvine); brown strophanthus (the seeds, commercial, Dalziel).

West African: **SENEGAL** BANYUN tifem (K&A) DIOLA (Fogny) funiafu (K&A) funiu (K&A) muriékolo (K&A) DIOLA (Tentouck) fupumben (K&A); = *a gun* (DF, The Gambia)

FULA-PULAAR (Senegal) tokéré = *poison; a general term* (K&A) KONYAGI ato (K&A) MANDING-BAMBARA baga = *poison; a general term* (K&A) baga iri, hunayô, kuna, kunadé (K&A) kunañõ, kunañõ (JB) WOLOF lengé = *supple* (JMD) tio = *bran* (RS; JMD) **THE GAMBIA** DIOLA (Fogny) funiafu, funiu = *liver* (DF) muriékolo = *monkey's sweat* (DF) **GUINEA-BISSAU** BALANTA n-denglè (EPdS) FULA-PULAAR (Guinea-Bissau) quindembode (EPdS) toke (EPdS) MANDING-MANDINKA solanamb (*def.*-ô) (EPdS) MANKANYA biètê (JDES; EPdS) **GUINEA** FULA-PULAAR (Guinea) kindé-toké (RS) toké (RS; Laffitte) KONO klannéné (RS) kola-nélé (RS) **SIERRA LEONE** BULOM (Sherbro) nyamkεε-lε (FCD) LOKO yawai (JMD) MANDING-MANDINKA pole (NWT) MENDE sawa *a general term; see S. gratus* SUSU ninge-firifiri (NWT) TEMNE *a*-kɔlε (FCD) *an*-mar (NWT) **LIBERIA** MANO kuonné (RS) **MALI** BOBO n-vaga (JMD) MANDING-BAMBARA kuna (A. Chev.; RS) kuna dié (Vuillet) kuna ion (A. Chev.) **UPPER VOLTA** BOBO n-vaga (K&B) FULA-FULFULDE (Upper Volta) tokéré = *a poisonous plant* (K&B) GRUSI von (K&B) MANDING-BAMBARA baga (K&B) baga iri (K&B) konna ion (K&B) kuna (K&B) kuna dié (K&B) DYULA baga (K&B) bagairi (K&B) MOORE yobro (K&B) NYENYEGE suru, suro *the drug preparation* (K&B) tchli *the plant* (K&B) tchubia *the seeds* (K&B) SENUFO-KARABORO sula (K&B) 'SENUFO' si-yalma (K&B) **IVORY COAST** AKAN-ASANTE akotom (K&B) BRONG makua(-n) (K&B) mekua (K&B) AKYE bisibidia (K&B) tantsiya (Curasson) ANYI makua(-n) (K&B) BAULE akuéyama (B&D) akuiniama (B&D) KRU-GUERE bidu (K&B) zredubu (K&B) KULANGO makua(-n), mekua (K&B) KYAMA salo bego, salo, *syphilis;* bego, *liane* (Ivanoff) salubé (B&D) MANDING-DYULA baga (K&B) bagairi (K&B) MANINKA hiwenié (K&B) kun-nakla (A&AA) kura (B&D) SENUFO-TAGWANA suwel (K&B) DYIMINI sépéwé (K&B) 'KRU' soouru (K&B) 'SENUFO' si-yalma (K&B) **GHANA** ADANGME ahom (Bunting, ex FRI) totum natʃo (Moore, ex FRI) ADANGME-KROBO ometwa (FRI) AKAN-ASANTE mamfoham (FRI) ɔ-matwa (BD&H) TWI ɔmããtwaa (BD&H; FRI) o-mããtwaa, *o*-matwã DAGAARI bulong (Gaisser) GA áklò (auctt.) omlεtswã (FRI) GBE-VHE matwa (auctt.) KONKOMBA bulong (Gaisser) yabaga (K&S; FRI) MOORE yabaga (AEK; FRI) **TOGO** BASSARI bulõ (Gaisser) KABRE soẽ (Gaisser) sõu (Gaisser) MOORE-NAWDAM raabia (Gaisser) TEM (Tshaudjo) sõdo (Gaisser) sõwé (Gaisser) 'DIFALE' su (Gaisser) **DAHOMEY** FULA-FULFULDE (Dahomey) toké (anon) **NIGERIA** BASSA (Kwomu) ire *either this sp. or S. sarmentosus* (Elliott) BEROM hwaàl ǹdɔ̀m = *poisonous medicine* (LB) FULA-FULFULDE (Nigeria) awdi tooke = *poisonous seeds; an epithet;* (MM) maada (JMD) tantsiyaari *from Hausa* (JMD; MM) tooke, tookere = *poison; the plant, an epithet* (JMD) GOEMAI lauenne (JMD) GWARI obwa (JMD) HAUSA dafi = *poison* (JMD) kwaŋkwánií (LB) tantsiya (JMD; ZOG) yaɓi, zabgai = *to smear poison; for the poison* (JMD) IGBO anu mmii (BNO) NUPE ègwa (Banfield) TIV àgbùlCfl (JMD) YORUBA iṣa (MacGregor; JMD) iṣa fọju fọju (IFE) iṣagere (auctt.) oró = *poison* (JMD)

A climbing shrub reaching to 16 m long of the open savanna woodland, rarely penetrating the more humid forest, occurring widely from Senegal to W Cameroons, and to Congo, Cabinda and Uganda.

The plant is the true arrow-poison plant of much of the Region. It is common in the bush and has in the past been frequently cultivated, and even though cultivation is now proscribed by several West African Governments it is still grown in the more remote areas (6, 9). It is reported planted in Nasarawa, N Nigeria, by pagan tribes (3) and around villages in N Ghana where under pruning it becomes a small tree bearing much fruit from July onwards (10). In Upper Volta it is commonly planted by hunters as a source of poison, and the trees as well as their agricultural land are so important as to be passed down by the same customs of inheritance. In both Ivory Coast and Upper Volta they are often planted in sacred groves and are considered important fetish plants. Their importance for medicinal use is such that medicine-men consider the plant a gift from God and its use is, out of deference, accompanied by particular rites (13).

The slender stems, stripped and stained black after soaking in pond mud are used in N Nigeria for the end pieces of reed-screens called *asabari* in Hausa; they are also used for the cotton carding bow (*masaɓi*) (6).

The translucent sap from crushed leaves or leafy twigs is used in Guinea against body-vermin (14), or that from young shoots which have been torrified and crushed is applied to kill head-lice; the pulped stem is put onto guinea-worm sores and an eye-medicine is made from a decoction of the stem-bark (16). Leaf and stem-decoctions are taken in Nigeria for constipation and to dress sores (2), and a concoction of the plant is used for constipation in Ghana (9, 10) and fever (5). The stems in decoction are taken in Ghana, as for *S.*

*gratus*, for severe illnesses with weakness (9). Preparations of twig or root-bark are given in Ivory Coast—Upper Volta by draught, enema or topically for serious complications of syphilis (13) and a root-decoction is used on skin-eruptions, sores, ulcers, both localized and extensive, and is given to pregnant women for stomachache (4). In Sierra Leone, in the event of snake-bite, leaves are chewed, and with the quid held in the mouth to serve as protection for the operator, the bite is sucked to draw out the poison. The Sherbro name for the plant, *nyamkεε-lε*, is the same as that given to the poison of the black mamba (7).

Root-bark, pounded to a pulp, is applied externally in Senegal to areas of guinea-worm infection and of skin-diseases; a root-decoction taken in draught is good for treating for intestinal parasites, as a diuretic and for urethral discharge; and a macerate of roots and twigs is taken for agalaxy and ascites (11, 12). The root is also taken in Guinea for venereal disease (16). The Kyama of Ivory Coast call it 'syphilis liane' (13). In Ghana the roots are used for rheumatic disease (8) and for many other ailments, and are regarded as specific for venereal disease (6). The Krobo of Odumase, Ghana, use the root for syphilis (10), while a root-decoction or the powdered root is similarly used in Nigeria (2). The roots in spirit are held to be aphrodisiac in Ghana, but an over-indulgence is dangerous (6). An application of chopped-up leaves of *Ficus exasperata* (Moraceae) and latex of *Saba florida* (Apocynaceae) with an over-all dressing of the dried powdered leaves of *S. hispidus* and *Saba florida* is a treatment in Ivory Coast for inflammation of neck-glands (1).

The seeds are highly poisonous and are the source of arrow-poisons prepared commonly throughout the Region. The poison is the sap extracted from them which usually is mixed with other ingredients, often a stem of a cactiform *Euphorbia*. The Konkomba and Yendi of NE Ghana add scorpions and snake-heads (9, 10). The brew is evaporated by boiling to a syrupy consistency, and when smeared on the arrow or spear this dries to a varnish. The poison is called in Hausa *yaɓi,* literally 'to smear poison', or *zabgai,* or simply *dafi,* poison, and in Fula *toke*. References to this attribute are found in several of the languages as may be seen in the vernacular name list above. To ensure the greatest toxicity Fante in Ivory Coast are said to ferment the pounded seed and leaves in contact with banana flowers. Seeds are also sometimes used as a fish-poison. Because of their great toxicity, the seeds do not enter into local medicines, nor into use as ordeals (6).

Various antidotes are in repute, often the secret of the local hunter. The seeds of *Garcinia kola* Heckel (Guttiferae) have been indicated as such. Many of them contain tannin which is known to have precipitant action on glycosides. Acetic acid, or acetates or vinegar have been found to be effective precipitants. In Malawi the flesh of an animal killed by a poisoned arrow is said to be made fit to eat by placing bark of the baobab in the wound (6); this however must refer to arrow-poison made from *S. kombe* Oliv., the Kombe arrow-poison plant. In the Cameroons an infusion of *Alstonia congensis* Engl. is said to be antidotal against both arrow and snake-poisons (6).

The seeds contain 4–8% glycosides under the name of *strophanthin-H,* or *pseudostrophanthin*. This is an amorphous complex of four separate substances (17), and has been equated with the crystalline strophanthin present in *S. gratus* and the E African *S. kombe,* but to have only two-fifths the activity of the latter. Strophanthin acts on striped muscle, especially heart-muscle, and has found use in medicine in treatment of cardiac deficiency. The principle supplies come from Malawi from seeds of *S. kombe* which are official in several pharmacopoeas. Seeds of *S. hispidus* are very similar in appearance to those of *S. kombe* and are often added as adulterants or even entirely to replace those of *S. kombe* in commerce and are known as 'brown strophanthus' (6, 9). Alkaloids *choline* and *trigonelline* are also present and the sugar *rhamnose* (13, 15). The seeds contain 32% of oil consisting of a mixture of a number of acidic glycerides of which the major ones are oleic acid (36%) and linoleic acid (16%) (12).

References:

1. Adjanohoun &Aké Assi, 1972: 43. 2. Ainslie, 1937: sp. no. 307. 3. Binga FHI. 16386, K. 4. Bouquet & Debray, 1974: 36. 5. Burtt Davy & Hoyle, 1937: 14–15. 6. Dalziel, 1937. 7. Deighton 2472, K. 8. Irvine 80, K. 9. Irvine, 1961: 638–40. 10. Irvine s.d. 11. Kerharo & Adam, 1962. 12. Kerharo & Adam, 1974: 187–9, with phytochemistry and pharmacology. 13. Kerharo & Bouquet, 1950: 191–2, with references. 14. Laffitte 59, et coll. alia, IFAN. 15. Oliver, 1960: 85. 16. Pobéquin, 1912: 60.

**Strophanthus preussii** Engl. & Pax

FWTA, ed. 2, 2: 72. UPWTA, ed. 1, 381.

West African: **LIBERIA** MANO gbo yiddi (JMD) konên (JGA) **IVORY COAST** AKYE napiabaté (Ivanoff) KYAMA abepopo (K&B) **GHANA** AKAN-ASANTE dietwa (FRI; JMD) mamfoham (FRI) manpohan (CV) TWI dietwa (FRI; JMD) o-mããtwa-nini (auctt.) **NIGERIA** YORUBA ìṣa-kékéré = *small* (JMD) ìṣa-wẹ́wẹ́ = *small* (JMD)

A robust creeper or climbing shrub to 4 m high, of deciduous and secondary forest and in scrub from Guinea to W Cameroons and Fernando Po, and extending to Angola across central Africa to Tanganyika.

The stems are used in S Nigeria to make bows for hunting (5).

The sap is healing and is used in Zaïre on wounds and sores, and is given to women in childbirth (6). The sap contains tannin. It is used by the Akye of Ivory Coast in treatment of urethral discharge (4). The young leaves are said to be cooked and eaten as a vegetable in Gabon (6). The reported use of the sap for coagulating *Funtumia* rubber in Ghana is perhaps in error for the sap of *S. sarmentosus* (3) — see there.

The Mano of the Nimba mountains of Liberia add the seed to poison mixtures but it is less toxic than the seed of *S. gratus* (1). Its use in the Upper Niger (? Mali) has been recorded, but the Bambara of that area prefer the use of *S. hispidus* (2). The alkaloids present appear not yet to have been worked out. Sarmentogenin has not been found.

References:

1. Adam, 1971: 377. 2. Dalziel, 1937. 3. Irvine, 1961: 640. 4. Kerharo & Bouquet, 1950: 192. 5. Thomas, N. W. 2161 (Nig. Ser.), K. 6. Watt & Breyer-Brandwijk, 1962: 105–6.

**Strophanthus sarmentosus** DC.

FWTA, ed. 2, 2: 70–72. UPWTA, ed. 1, 381.

West African: **SENEGAL** BALANTA biori (A. Chev.) bodi (K&A) ten (JMD, ex K&A) BANYUN ifakum (K&A) BASARI a-mɔnyé (Ferry) BEDIK ga-mondyé (Ferry) DIOLA (Brin/Seleki) butauma (K&A) DIOLA (Fogny) fulâdafo (K&A) fulań (K&A) MANDING-BAMBARA kuna (JB) kunâkalé (JMD, ex K&A) kunamkala (K&A) MANDINKA bidańâ (K&A) kesesoy (K&A) kuna kalo = *arrow poison* (DF) kuna nombo = *poison vine* (DF) kunô = *poison* (K&A) MANINKA kunâkale (K&A) kunamkala (K&A) 'SOCE' bidańâ (K&A) kesesoy (K&A) kunô (K&A) kutandé buna (A. Chev.) pinti nguguet (A. Chev.) MANDYAK brâdtu (K&A) SERER ngabakok (K&A) SERER-NON ngap (AS, ex K&A) ngap (AS, ex K&A) NON (Nyominka) ndolor(-é) (K&A) WOLOF bôdé (JB, ex K&A) tio (A. Chev.; JB) ťox (K&A) **THE GAMBIA** DIOLA (Fogny) fulâdafo = *snake-skin* (DF) FULA-PULAAR (The Gambia) tantsiyari (DAP) tokere (DAP) MANDING-MANDINKA , sulawalado = *monkey's shuttle* (Fox; DF) kunakalo = *arrow poison* (auctt.) kuno nombo = *bird vine* (DF) solanamb(-ô) WOLOF chok = *bran/husk* (DAP) **GUINEA-BISSAU** BALANTA tene (GES) MANDING-MANDINKA kunakalo, sálánambó (*def.*) salanumbá (G&S) PEPEL uhape (GES) **GUINEA** FULA-PULAAR (Guinea) kindambodie (Langdale-Brown) toké (JMD) MANDING-MANDINKA tuman bélé (Brossart) MANINKA (Wasalunka) kunalé (A. Chev.) **SIERRA LEONE** LOKO yawai (NWT) MANDING-MANDINKA gondu (NWT) MENDE (yolei *def.*) *a group term for certain woody climbers* (FCD; SKS) fɛlɛ = *fanner for winnowing; the husks used by children in play as fanners* (FCD; Pyne) sawa *a general term; see S. gratus* SUSU poli (NWT) **MALI** MANDING-BAMBARA kunamkala (A. Chev.) kunankalé (A. Chev.) MANINKA kunankala (A. Chev.) kunankalé (A. Chev.) **UPPER VOLTA** MANDING-BAMBARA kuna (Ivanoff) SENUFO-KARABORO dugua bélé (K&B) 'SENUFO' sayié (K&B)

**IVORY COAST** AKYE atodan (K&B) kpedi (K&B) tzapé (K&B) BAULE niasebaté (K&B) KIRMA bagomo (K&B) KYAMA aberure (K&B) MANDING-MANINKA kun ankala (K&B) sayié (K&B) **GHANA** ADANGME tʃopa yapa (auctt.) AKAN-ASANTE mamfoham (auctt.) matwã (Enti) ɔman (Enti) o-matwa, matwã (Enti) FANTE ɔman (auctt.) TWI o-mããtwãã, o-matwa, matwã (auctt.) omããtwa nini (FRI) ɔman (Enti) GA ɔmããtwãa (auctt.) GBE-VHE akitplale (sodzati) (JMD; Ewe Dict.) NZEMA adwokuma (FRI) **NIGERIA** BERUM iiwaal ndɔm [illegible] *poisonous medicine* (LB) CHAMBA me-ni (FNH) EDO ọvíẹn-órà (Ross) FULA-FULFULDE (Nigeria) awdi tooke = *poisonous seeds; an epithet* (JMD) maada (JMD) tantsiyaari *from Hausa* (JMD) tooke, tookere = *poison; the plant, an epithet* (JMD) GOEMAI lauenne (JMD) GWARI obwa (JMD) HAUSA gama sowa (Ryan) gwasha (FNH) kwaŋkwánií (LB) tantsiya (JMD; ZOG) IGBO ọta = *a bow* (JMD; AJC) ọta nta = *a hunting bow* (NWT; JMD) TIV àgbùlCfl (JMD) YORUBA agan olugbo (RJN) akan (RJN) ako-iṣa (Ross) ilagbà ọmọdé, lãgba-ọmọde = *little child's horse-whip* (Verger; JMD) irẹ (IFE) iṣa (Oluakpata; IFE) iṣakékeré = *lesser* (Ross; JMD) sagere (Millen)

A scandent shrub or lofty climber of transition forest, dry deciduous forest and in thickets of savanna country throughout the Region from Senegal to N Nigeria, and occurring in Camerouns to Congo and Cabinda.

The plant has glabrous foliage; the flowers are white to mauve, turning yellow, or are markedly red, funnel-shaped with long yellow tails to 8 cm long. The Yoruba name, *ilagbà ọmọde,* 'little child's horsewhip', perhaps refers to the tails. The flowers are showy and open mostly while the plant is leafless. It has been brought into cultivation in other countries. It is common throughout the bush in West Africa and in clearing land for habitation it is conserved, and, though not cultivated, is tended.

The stems are used in N Nigeria for reed-screens in houses (3). The Igbo of S Nigeria use the wood to make bows, calling the plant *ọta nta,* 'bow for hunting' (12). The sap has been used for coagulating *Funtumia* latex in Ghana, under which usage the plant has been called *diecha-juice* plant, a case perhaps of misidentity for *dietwa* is the Twi name for *S. preussii* (4, 5). A leaf-decoction is used in Ivory Coast by the Senufo for conjunctivitis and trachoma giving rapid relief and cure. The Baule take a twig-decoction for arthritic rheumatism, and the Akye give leaf-sap by mouth and a bark-macerate in enema for urethral discharge. The action is said to be diuretic and soothing (9). A macerate of pounded roots is taken on an empty stomach in Senegal for treatment of pains in the joints, and also for hernia (6, 8). A leaf-decoction is considered in Ivory Coast to be emetic and antidiarrhoetic (9), and in Senegal a decoction of leaves and roots is used to treat infantile diarrhoea (7, 8). A macerate of the roots is taken by draught and in baths as a vermifuge and invigorator, and to treat sleeping-sickness (7, 8). The powdered roots cooked with [the grain of] *Digitaria exilis* (Graminae) are taken in Senegal for abdominal flatulence with constipation giving rapid relief without painful purging (8). The leaf-sap or latex is held to be cicitrisant in both Ivory Coast (9) and in Senegal (6, 8), and is used on sores and wounds to give rapid healing, and for this reason in some villlages of Casamance it is considered a fetish of the circumcision ceremony.

The plant has a strong reputation amongst Senegalese medicinemen specializing in the treatment of syphilis and associated complications. It is used in conjunction with other drug plants. The roots enter into a treatment for leprosy. It is also used for insanity (8).

The bark is fibrous and a cordage can be made from it (Moor fide 2, 5). Hats and mats are said to be made from it (Dalziel fide 5).

The seeds are widely recognized as highly poisonous and are used in various parts of Africa for making arrow-poison as well as in medicines. There appear to be variations however in potency of the prepared poison. Material from Nasarawa, N Nigeria, is said to be so highly toxic as to be unusable because it causes meat to rot too quickly. While material from Lagos area produced from plants introduced from inland is recorded as being not poisonous if used alone, and having to be mixed with seeds of *S. kombe* to produce an arrow-poison (10).

The seeds have attracted commercial interest as a starting-point for the preparation of *cortisone,* a drug that has raised considerable medical attention in treatment of arthritis, cardiac rheumatism, etc. The seeds contain a cardiotonic glycoside *sarmentocymarin* which in hydrolysis yields *sarmentogenin* of very close physical structure to cortisone and into which it can be converted. Physico-chemical and microbiological manipulation has shown the possibility of producing other corticosteroids and sex-hormones (8). Variation noted in the preceding paragraph has been observed in varietal differences of the species affecting both qualitatively and quantitatively the glycosides present, the plants richest in sarmentocymarin coming from the drier northern region, and those from the more humid south being poorest in glycosides. Medical expectations, however, at one time high, are now abated somewhat and interest in the plant is more historic than practical. (8, 9, 11).

References:

1. Binga FHI.16387, K. 2. Burtt Davy & Hoyle, 1937: 15. 3. Dalziel, 1937. 4. Imperial Institute 1110, K. 5. Irvine, 1961: 640–2. 6. Kerharo & Adam, 1962. 7. Kerharo & Adam, 1964, c: 322–3. 8. Kerharo & Adam, 1974: 189–91, with phytochemistry and pharmacology. 9. Kerharo & Bouquet, 1950: 192, with references. 10. Millen 35, K. 11. Schnell, 1960, c. 12. Thomas, N. W., 2096 (Nig. Ser.) K.

## **Tabernaemontana** Linn.

FWTA, ed. 2, 2: 65.

West African: **GHANA** AKAN-ASANTE amantannua, amantam: *to grasp* (FRI) atwe ada (FRI) *o*-furuma, *o*-fruma (FRI) kakapepe (FRI) o-banawa (FRI) ɔ-bonawa, fruma (FRI; Enti) paapaku, paapae: *to split easily* (FRI) FANTE kakapenpen (Enti) KWAWU bonawa (Enti) kakapenpen (Enti) ɔ-bonawa, fruma (FRI; Enti) TWI kakapepen (Enti) ɔ-bonowa (Enti) fruma (FRI; Enti) o-furuma, o-fruma papaku (FRI) WASA kakapenpen (Enti) kakapepen (FRI) ANYI-AOWIN pekyi-pekyeri (FRI) NZEMA foba (BD&H; FRI) fola (FRI) **NIGERIA** EDO íbù (Lowe)

## **Tabernaemontana brachyantha** Stapf

FWTA, ed. 2, 2: 65–66. UPWTA, ed. 1, 370, as *Conopharyngia brachyantha* Stapf.

West African: **NIGERIA** BOKYI kema-atung (Catterall) **WEST CAMEROONS** KPE etongongon (JMD)

A shrub or tree to 26 m high, in the understorey of the closed high-forest of S Nigeria and W Cameroons, and also in E Cameroun.

In Cameroons the twigs are crushed and mixed with 'feverleaf' (*Ocimum sp.,* Labiatae) as a febrifuge (1). Five alkaloids have been reported present in the bark: *coronaridine, conopharyngine, ibogaine, voacangarine* and *voacangine* (2).

References:

1. Dalziel, 1937. 2. Willaman & Li, 1970.

## **Tabernaemontana chippii** (Stapf) Pichon

FWTA, ed. 2, 2: 66. UPWTA, ed. 1, 370, as *Conopharyngia chippii* Stapf.

West African: **GHANA** AKAN-ASANTE *o*-fruma, *o*-funuma = *navelcord* (Enti) kakapempe (FRI) KWAWU bonawa (Enti) TWI *o*-fruma, *o*-funuma (Enti) kakapenpen = *easily broken* (JMD) o-bonowa (Enti) WASA kakapimpen (BD&H)

A spreading tree to 6 m high in the understorey of the closed wet forest of Liberia and Ghana.

The flowers are large, white and fragrant.

The plant has an abundance of white latex which was permitted at one time as an ingredient of paste rubber called *bede-bede* under the bye-laws of the Asante chiefs of Ghana. (1, 2.)

References:

1. Dalziel, 1937. 2. Irvine, 1961: 662–3.

**Tabernaemontana contorta** Stapf

FWTA, ed. 2, 2: 66.
West African: **NIGERIA** IGBO (Umuahia) pete pete = *very soft* (AJC)

A tree to 12 m high, understorey of the closed-forest in S Nigeria and W Cameroons, and also in E Cameroun.

*Conopharyngine, voacangine* and three other alkaloids have been reported present in the bark (1).

Reference:

1. Willaman & Li, 1970.

**Tabernaemontana crassa** Benth.

FWTA, ed. 2, 2: 66. UPWTA, ed. 1, 370, as *Conopharyngia crassa* Stapf and *C. durissima* Stapf.
West African: **SIERRA LEONE** BULOM (Sherbro) benfukɛ-lɛ (FCD; S&F) KISSI kafayɔlo (S&F) KONO k-poŋgbo (S&F) MENDE kofa (*def.* kofei) (auctt.) loni (*def.*-i) (auctt.) SUSU kunye-ɛfiexe (JMD) ninge-ɛkunyi (NWT) ninge-ɛxunyi (JMD) ninge-wuri (NWT) wuri-ɛfiexe (NWT) TEMNE *ka*-lato (NWT) **LIBERIA** KRU-BASA bo-gar (C&R) weh-boh (C&R) **IVORY COAST** ABE m-piégba (auctt.) ABURE opuko (B&D) ADYUKRU ekre (auctt.) AKAN-ASANTE patié patié (B&D) BRONG napêra (K&B) FANTE atsim (A. Chev.; K&B) AKYE bogbon (A. Chev.; K&B) choha (A. Chev.; K&B) ANYI kwakié-kwakié (A. Chev.; K&B) pakié-pakié (A. Chev.; K&B) BAULE dégdé dégdé (B&D) deguédegué (B&D) KRU-BETE dogbuéï (K&B) GUERE (Chiehn) degué-degué (K&B) tiépéwowo (B&D) KULANGO kutu kwaku logrodo = *testicles of the panther* (K&B) KYAMA foba (B&D) glagla (Aub.; K&B) gragra (Aub.; K&B) **GHANA** AKAN-ASANTE ofuruma (FRI) KWAWU pepae (AEK) TWI obonowa (Enti) WASA kakakie-kwakie (FRI) kakapempen (auctt.) ANYI-AOWIN pekyi-pekyere (JMD) NZEMA ɛzɛnu-foba (FRI) **NIGERIA** IGBO (Umuahia) pete-pete (JMD)

A shrub or tree to 23 m tall, bole to 60 cm in girth sometimes buttressed, or with several trunks, in wet locations of the closed-forest and in forest clearings, from Sierra Leone to W Cameroons, and extending to Zaïre.

The wood is yellowish-white, fine-grained, moderately hard, easy to work and finishes smoothly. It is not resistant to fungal decay. It is occasionally used in Liberia to make rice mortars, and is suitable for general carpentry, plywood, boxboards and miscellaneous common uses where decay and fungal stain is not important (3).

The bark-sap is lactiferous and produces an inferior rubber that remains sticky. It has been used as an adulterant for better rubbers. The sap is extremely caustic. One drip in the eye may cause blindness, and it is an ingredient in the Daola region of Ivory Coast of an arrow-poison. The sap is used in Ivory Coast as a disinfectant and haemostatic and on sores and wounds and on leprous areas (6). One or two drops are instilled into the nose in Ivory Coast to soothe headache, acting as a counter-irritant on the nasal mucosae. It is used also as a seditive in insanity (2). In Liberia the sap is applied to areas of ringworm after scarification (3). In Congo it is commonly applied as a healing dressing to sores, abscesses, furuncles and to anthrax pustules, and to dermal infections such as filaria, ringworm and other fungal troubles, and is taken internally as an anthelmintic (1). Like many lactiferous plants it is considered in Congo to be galactogenic (1).

A bark-decoction is taken in Congo for constipation, ovarian troubles, haematuria, and blennorrhoea (1) and is given as an enema in Ivory Coast for kidney-troubles, and rheumatism and stubborn constipation (2).

A leaf-decoction is considered in Ivory Coast to be strengthening and defatiguant and is applied as a friction; it is massaged onto rachitic children, and onto adults to combat fatigue (6). Large leaves are placed in the roof-thatch of houses in Gabon in the belief that their bitter taste keeps away cockroaches (8, 9).

A considerable number of alkaloids has been reported in the root, bark, seeds, etc. (2, 5, 6, 7, 10).

The flowers are showy, highly scented and ornamental.

References:

1. Bouquet, 1969: 66. 2. Bouquet & Debray, 1974: 36. 3. Cooper & Record, 1931: 105, as *Conopharyngia durissima* Stapf, with timber characteristics. 4. Dalziel, 1937. 5. Hanna, 1964: as *Conopharyngia durissima*. 6. Kerharo & Bouquet, 1950: 105 as *Conopharyngia durissima* Stapf. 7. Oliver, 1960: 23, 51, as *Conopharyngia durissima*. 8. Walker, 1953, a: 20, as *Conopharyngia durissima* Stapf. 9. Walker & Sillan, 1961: 79–80, as *Conopharyngia crassa* Stapf. 10. Willaman & Li, 1970: as *Conopharyngia durissima* (Stapf) Stapf and *C. jollyana* Stapf.

## **Tabernaemontana eglandulosa** Stapf.

FWTA, ed. 2, 2: 66.

A climber of thickets of closed-forest and secondary jungle of S Nigeria and Fernando Po, and in E Cameroun, Gabon and Zaïre.

The plant is lactiferous, yielding sticky rubber which has been used as an adulterant of better rubbers.

A trace of alkaloid has been detected in the leaves of Congo material, more in the bark and an abundance in their roots (1). *Voacangine* and several other alkaloids have been named (2, 3).

References:

1. Bouquet, 1972: 14. 2. Cava & al., 1962: as *Gabunia eglandulosa* Stapf (*G. eglandosa* sphalm). 3. Willaman & Li, 1970: as *T. eglandulosa* Stapf & *Gabunia eglandulosa* Stapf.

## **Tabernaemontana longiflora** Benth.

FWTA, ed. 2, 2: 66. UPWTA, ed. 1, 370.

West African: **SENEGAL** DIOLA bongniendé (A. Chev.) FULA-PULAAR (Senegal) buah (A. Chev.) **GUINEA-BISSAU** PEPEL utá-leite (EPdS) **GUINEA** KISSI kêfa (RS) MANDING-MANINKA bakoroni guenda = *goat's testicles* (RS; Aub.) MANO manga-iri (RS) SUSU ningué khrigni = *cow's milk* (RS; Aub.) **SIERRA LEONE** LOKO (Magbile) iyumbai (NWT) SUSU kunye-fiexe (NWT) ninge-ɛkunyi (JMD) ninge-ɛxunyi (JMD) wuri-ɛfiexi (JMD) TEMNE *a*-boi (NWT) **LIBERIA** MANO lu tŏ kŏlĕh (JMD) tŏ kŏlĕh (JMD)

A shrub or tree to 8 m high occurring from sandy coastal areas to montane forest from Senegal to Ivory Coast.

The plant has white, waxy, long-tubular, strongly scented flowers. It is worthy of cultivation, and has been introduced into horticulture in other regions (2).

The wood is yellowish white, soft and easy to work (1). The wood-ash has been used in soap-making in Guinea, and to provide a mordant for indigo dyeing (3).

An alkaloid, *conoflorine,* has been found in the leaves and bark (4).

References:

1. Aubréville, 1959: 3; 210, 212. 2. Dalziel, 1937. 3. Portères, s.d. 4. Willaman & Li, 1970.

**Tabernaemontana pachysiphon** Stapf

FWTA, ed. 2, 2: 66. UPWTA, ed. 1, 370 as *Conopharyngia pachysiphon* Stapf and *C. cumminsii* Stapf.

West African: **GHANA** AKAN-AKYEM ɔ-banawa (FRI) o bonowa (Enti) ASANTE badawa (BD&H) o-banawa (FRI) KWAWU o-banawa, ɔ-bonowa (Enti) TWI o-bonawa, obonowa (Enti) obonowa (BD&H) **TOGO** GBE-VHE dai (Volkens) **NIGERIA** EDO íbù (auctt.) ìkhúian, iwian (JMD) IGALA oogele (Hendrick) IGBO pete-pete (JMD; KO&S) IGBO (Achi) ivuru (DRR) IGBO (Awka) ngu (FRI) YORUBA abodòdo, abo: *female* (*of 'dodo'*) (JRA; JMD) dòdo (auctt.) kpokpaka (auctt.)

A shrub or tree to 10 m tall in the lower storey of the closed-forest. Two varieties are recognized: var. *pachysiphon* in Togo, Dahomey and S Nigeria; var. *cumminsii* (Stapf) Huber from Ghana, Togo and S Nigeria. They are considered together here.

The wood is yellow and hard, with sap-wood and heart not differentiated (11). It has no economic importance.

The plant has an abundance of white latex. It does not coagulate and has been used to adulterate better (*Funtumia*) rubbers (7). It thickens to a bird-lime and is used to trap birds and to mend broken pots and calabashes (3). It is applied as a cicitrisant on ulcers (1).

The bark contains a fibre which is used by the Igbo of S Nigeria at Uburubu to make *ɛbwa* cloth (9) and as Asaba *ufa* clothes (2), also known as 'dodo-cloth' in Lagos. In Benin small ropes are made from the inner bark. Children use it to trap birds and small animals (8).

A concoction of the root-bark is used in Nigeria in treatment of insanity (1). In this respect the root-bark of an unspecified species of this genus is powerfully seditive (5). The bark and seeds of *T. pachysiphon* have been reported to contain *conopharyngine, voacangine,* and a number of other alkaloids (12). Bark-sap is applied in Benin to inflamed breasts (8).

The leaves are dried and powdered, and the powder is applied to sores, and even to old ulcers (6). The pounded fresh leaves are applied as a pulp to the hair by women in Togo (Volken fide 4).

The fruit is said to be inedible (10).

References:

1. Ainslie, 1937: sp. no. 112, as *Conopharyngia pachysiphon*. 2. Anon. Dept. of Agr. s.n. K. 3. Boston C.16, K. 4. Dalziel, 1937. 5. Irvine, 1961: 621–1, as *Conopharyngia cumminsii* Stapf & incl. *C. sp.* 6. Irvine, s.d. 7. Punch 146, K. 8. Rosevear, 1975. 9. Thomas, N. W. 2100 (Nig. Ser.), K. 10. Unwin 109, K. 11. Unwin, 1920: 391. 12. Willaman Li, 1970.

**Tabernaemontana penduliflora** K. Schum.

FWTA, ed. 2, 2: 65. UPWTA, ed. 1, 371 as *Conpharyngia penduliflora* Stapf.

West African: **NIGERIA** EDO òvìẹn-ìkhúian (JMD) òvìẹn-iwian (Amayo) YORUBA dodo (IFE)

A shrub or tree to 3 m high, occasionally to 8 m, in the understorey of high-forest of S Nigeria and W Cameroons, and in E Cameroun, Congo and Sudan.

Of no recorded usage, but probably with yellowish wood and white latex.

**Tabernaemontana ventricosa** Hochst.

FWTA, ed. 2, 2: 66.

A shrub or tree to 13 m tall, of montane forest in Ghana, S Nigeria and W Cameroons, and widely distributed across Africa to Zaïre, E and South-Central Africa, Natal and Mozambique.

The wood is yellow and very hard. There is white latex which contains 12–20% rubber and 34–50% resin. The Chagga of Tanganyika apply the latex to wounds and it is said that healing, even of old wounds, is rapid. (1).

Reference:

1. Watt & Breyer-Brandwijk, 1962: 81–82, as *Conopharyngia usambarensis* Stapf.

**Thevetia neriifolia** Juss.

FWTA, ed. 2, 2: 54. UPWTA, ed. 1, 382.

English: yellow oleander; exile tree: exile oil tree; milk bush (Irvine).

French: chapeau de Napoléon (Berhaut); laurier jaune des Indes (Kerharo & Adam).

West African: **IVORY COAST** AKYE achiko (A&AA) **GHANA** ADANGME ŋmɔkɔtʃo (FRI) AKAN-TWI nnye me nnyere me = *do not take me and bind me* (FRI) *o*-sibisaba (FRI) *o*-sibi-dua (FRI) GA abɔdɔitʃo (FRI) àbɔdɔitso (KD) kpɔtʃo (FRI) osábisàba (FRI; KD) sibitʃo (FRI) **NIGERIA** YORUBA olómiòjò (JMD; Singha) olómiòjò tilawa oloje *the kernels* (IFE)

A shrub to 6 m high, native of central and tropical S America from Mexico and the West Indies to Brazil, and widely distributed by man, occurring in the West African region as a garden cultivate.

The foliage is not grazed by stock. The plant makes an useful enclosure hedge (1, 5). It stands clipping, endures drought and suppresses plants from its shade (8). In the West Indies half a leaf is known as an emetic and purgative. In Java Indian immigrants are reported to smoke the dried leaves (2). Tincture of bark is also emetic and purgative, and has been used as a febrifuge (2, 10, 14), but tests with bark and fruit extracts have proved inactive against avian malaria (3). In Senegal a macerate of bark and leaves may occasionally be used with precautions for amenorrhoea (9).

The wood is used in Ghana sometimes to make axe-handles (8). In Indonesia it is reported used as a fish-poison (7). The fruits have shown some insecticidal activity (6), and especially the kernels (12).

All parts of the plant are poisonous. It has been used in southern Africa as an arrow or ordeal-poison (14). A number of glycosidal derivatives of *cardenolide* have been detected of which *thevetin* is present throughout except in the leaves and fruit-pulp. The leaves contain several other glycosides including *neriantin*. The former substance is cardio-toxic acting like *digitalis*, but only fractionally as strong. It has been suggested as a substitute for use in digitalis-intolerance. The kernel contains another and more toxic bitter principle which causes tetanic convulsions and speedy death for which it is commonly used in India for suicide and murder. In southern Africa it has acquired the name 'be-still nut.' There is in the kernel a pale yellow fixed oil amounting to 57–62%. Raw, of course, it is poisonous, and is used in external application in India to areas of skin-infections. It can be freed of the toxic principles and is then fit for culinary purposes. It consists mainly of glycerides of oleic acid. (2, 4, 9, 10, 12, 13, 14.)

The fruit pulp is sometimes eaten in Ghana without any apparent ill-effect (8).

Latex is found in all parts of the tree. It sets to a resinous substance rich in ash: 47.9% resins, 20.4% ash (2). The plant also contains a dark coloured gum which is tasteless and mainly insoluble but swelling in water (7).

References:

1. Aubréville, 1959: 3: 190 as *T. peruviana* K. Schum. 2. Burkill, 1935: 2154–5, as *T. peruviana* K. Schum. 3. Claude & al., 1947: as *T. peruviana*. 4. Dalziel, 1937. 5. Kesby 22, K. 6. Heal & al., 1950: 104, as *T. peruviana*. 7. Howes, 1949: 81. 8. Irvine, 1961: 643–4, as *T. peruviana* (Pers.) Merr. 9. Kerharo & Adam, 1974: 192–5, with phytochemistry and pharmacognosy. 10. Oliver, 1960: 39, 88. 11. Quisumbing, 1951: 740–2, with references, as *T. peruviana* (Pers.) Merr. 12. Tattersfield & al., 1948: as *T. peruviana*. 13. Watt, J. M. 1967: as *T. peruviana* Schum. 14. Watt & Breyer-Brandwijk, 1962: 70, 107–9 as *T. peruviana* Schum.

**Vahadenia caillei** (A. Chev.) Stapf

FWTA, ed. 2, 2: 60.
West African: **SIERRA LEONE** MENDE hole (*def.*-i) (Fox)

A liane to 30 m long, of the forest zone from Guinea to Ivory Coast.
The latex is tacky and is used in Sierra Leone as bird-lime. The fruits are edible.

Reference:
Fisher 84, K.

**Vahadenia laurentii** (De Wild.) Stapf

FWTA, ed. 2, 2: 60.

A liane to 50 m long by 65 cm girth, of forest, often riverain, rare in S Nigeria and W Cameroons, and extending widespread to Zäire and Angola.
A plant contains a white thick and abundant latex of no use.
The fruit has an acidulous pulp, edible or not. The seeds are eaten roasted like groundnuts.

Reference:
Pichon, 1953: 266–70.

**Voacanga africana** Stapf

FWTA, ed. 2, 2: 67. UPWTA, ed. 1, 383.
West African: **SENEGAL** BANYUN ksiso rumbel (K&A) DIOLA kagis (JB; K&A) MANDING-MANDINKA sulabérékilo = *monkey's testicles* (K&A) SERER garada (JB, ex K&A) SERER-NON (Nyominka) ibalak (K&A) SOCE narado (K&A) **THE GAMBIA** DIOLA (Fogny) kagil (DF) **SIERRA LEONE** MENDE gboni (*def.*-i) (auctt.) SUSU ninge-ɛkunyi (JMD) ninge-ɛxunyi (NWT) TEMNE *a*-faf (NWT) **IVORY COAST** ABE m-piégba (A&AA) AKAN-ASANTE kwatié kwatié (B&D) AKYE gbon gbon (A&AA) BAULE bunbo (B&D) dégbé dégbé (B&D) déguédégué (B&D) **GHANA** AKAN-AKYEM o-banawa, o-bonowa (Enti) ASANTE kaka pempe (auctt.) o-bonawa, fruma (Enti) ofuruma (BD&H; FRI) paapaku (Enti) KWAWU o-banawa, o-bonowa (Enti) o-bonawa, fruma (Enti) TWI fruma (Enti) o-banawa, o-bonowa (Enti) o-bonawa, fruma (Enti) paapaaku paapae: *to split easily* NZEMA foba (FRI) **TOGO** TEM (Tshaudjo) kongkong (Volkens) **NIGERIA** EDO òvịẹn-íbù (Amayo) IGBO pete pete = *very soft* (auctt.) IGBO (Asaba) akẹte (NWT) YORUBA akọ dòdo = *male dodo, the female being Tabernaemontana pachysiphon* (auctt.) dòdó, igi dòdó (auctt.) giwini (EWF) YORUBA (Ijebu) shẹrẹnkpẹn (JMD)

Tree to 11 m high, low-branching, in understorey of forest, secondary jungle and savanna woodland, throughout the Region from Senegal to W Cameroons and Fernando Po and across Africa to Egypt and Uganda.
The wood is soft and the bole often hollow (5).
The bark contains a fibre which is used in parts of S Nigeria to make a yarn (13, 14), and which may also be admixed with cotton or other fibres for making mats (4). There is an abundant white latex in the bark and other parts. It does not coagulate and has been used to adulterate better rubbers (11). The latex is applied to wounds in Senegal (8, 9) and into a carious tooth in Nigeria (4). Latex from the fruit is used in Nigeria to make a bird-lime (Chesters fide 5).
An extract of the bark is used in S Nigeria for washing sores (3). A root-decoction is given by mouth to women in Senegal to ward off the untoward consequences of premature or precipitant parturition. The same prescription is used for painful hernias (7, 9). A root-decoction is taken in Tanganyika for dysmenorrhoea and a bark or root-decoction for heart-troubles (spasms, angina ?) (6). A decoction is similarly taken in Congo for heart-troubles and

also blennorrhoea (1). The plant (parts unspecified, possibly bark-sap) is used in Congo for treating sores, furuncles, abscesses, fungal infections, filaria and eczema. (1))

In Ivory Coast a leaf-decoction is taken by enema for diarrhoea, in baths for general oedema, by frictions and draughts for leprosy, and in a lotion for convulsions in infants; sap of the leaves is given as nose-drops in insanity (2). A decoction of leafy twigs is prepared in baths and draughts in Senegal for the treatment of some undefined diseases, and decoction of leaves taken by mouth is considered strengthening (9) and is a treatment for fatigue due to shortness of breath (8).

All parts of the plant are rich in alkaloids, especially the bark: root-bark 5–10%, stem-bark 4–5%, seeds 1.5%, leaves 0.3–0.45%. *Voacamine* is the principle alkaloid present followed by *voacangine, voacangarine, voacorine* and *vobtusine.* They are hypotensive, have ventricular cardio-stimulant action and a slight action on the sympathetic and parasympathetic nervous systems (10). Many other alkaloids have been identified, and also the presence of small amounts of tannins and flavonoids. For considerable detail and bibliographic references of phytochemistry and pharmacology the reader is referred especially to (9), but also to (12), (15) and (16).

In S Nigeria the plant has been recorded with magical attributes to 'cleanse' the house (14).

References:

1. Bouquet, 1969: 67. 2. Bouquet & Debray, 1974: 38. 3. Carpenter 307, UCI. 4. Dalziel, 1937. 5. Irvine, 1961: 644–5. 6. Haerdi, 1964: 133. 7. Kerharo & Adam, 1963, b. 8. Kerharo & Adam, 1964, c: 326. 9. Kerharo & Adam, 1974: 195–200, with phytochemistry and pharmacology. 10. Oliver, 1960: 40, 90. 11. Punch 146, K. 12. Thomas, D. W. & Bieman, 1968. 13. Thomas, N. W. 1675 (Nig. Ser.), K. 14. Thomas, N. W. 1809 (Nig. Ser.) K. 15. Watt & Breyer-Brandwijk, 1962: 111. 16. Willaman & Li, 1970.

**Voacanga bracteata** Stapf

FWTA, ed. 2, 2: 67. UPWTA, ed. 1, 383.

West African: **SIERRA LEONE** MENDE gboni, boni (NWT; JMD) TEMNE *a*-kpafoto (NWT; JMD) **LIBERIA** KRU-BASA voo-fohn (C&R) **IVORY COAST** ABURE oflafa (B&D) **GHANA** AKAN-WASA fruma (Enti)

A shrub or tree to 6 m high in the understorey of the closed-forest from Sierra Leone to W Cameroons, and in E Cameroun to Zaïre.

The plant contains an abundant latex. It has been used in Liberia as an adulterant in better rubbers (2).

The root is pulped up and used in Congo in topical frictions for rheumatism (1).

The alkaloids, *voacangarine* and *epivoacangarine* have been recorded in the bark (3).

References:

1. Bouquet, 1969: 67. 2. Cooper & Record, 1931: 107. 3. Willaman & Li, 1970.

**Voacanga thouarsii** Roem. & Schult.

FWTA, ed. 2, 2: 67. UPWTA, ed. 1, 383, as *V. obtusa* K. Schum.

West African: **SENEGAL** DIOLA bu lukuñ (JB) **THE GAMBIA** DIOLA (Fogny) bu lukun = *tasty sap* (DF) MANDING-MANDINKA kuto (JMD; DAP) kuto-jambo *the leaves* (JMD) **SIERRA LEONE** MENDE gboni (auctt.) SUSU ninyge-ɛxunyi (NWT) **LIBERIA** KRU-BASA je-ray-krehn (C&R) **IVORY COAST** FULA-FULFULDE (Ivory Coast) landa édi (Aub.) **NIGERIA** GWARI knubwobwoyi (JMD) HAUSA ƙòƙíyár bírii = *Strychnos of the monkey* (auctt.)

A tree to 13 m high by up to 1 m girth, low-branching, sometimes stilt-rooted in swampy ground, of valley bottoms, ravines, swampy ground and stream-banks of the wooded savanna from Senegal to N Nigeria, and widely dispersed across tropical Africa to as far south as Natal and in Madagascar.

The wood is of little value. It is reddish brown, fine grained, tough, difficult to saw, planing unsmoothly — the grain pulling out — and turning indifferently (3). Poles are used in Liberia for hut-posts (1) and the wood is burnt in Sudan to produce a vegetable salt (4).

The bark is said to furnish a fibre used for making hunting nets in E Africa (8). It is lactiferous. The latex of the branches is recorded as white and of the trunk as clear (5). The latex produces an inferior rubber, usable when mixed with others as a bird-lime and as a glue for fastening handles to knife-blades and to repair baskets (8). The latex is vesicant on the skin and dangerous to the eyes, and is used in Sierra Leone as a remedy for toothache (7).

The ripe seeds are scattered on the ground of rice-farms in Liberia when the grain is ripening to scare off wild pigs which depradate the rice crop. (1, 2)

The roots and bark contain the alkaloids *voacamine, vobtusine* and *voacangine* (8, 9) which are hypotensive, cardiotonic and sympatholytic (6).

References:

1. Cooper 431, K. 2. Cooper & Record, 1931: 107, as *V. obtusa* K. Schum. 3. Eggeling & Dale, 1952: 31 as *V. obtusa* K. Schum. 4. Irvine, 1961: 645, as *V. obtusa* K. Schum. 5. Leeuwenberg 2486, K. 6. Oliver, 1960: 40, 90. 7. Pyne 25, K. 8. Watt & Breyer-Brandwijk, 1962: 112, incl. *V. obtusa* K. Schum. 9. Willaman & Li, 1970.

**Voacanga sp. indet.**

West African: **IVORY COAST** KRU sagukwé (K&B)

An unidentified tree is recognized by the 'Kru' of Ivory Coast as a fetish tree to protect children from illness and evil spirits. Protection is first acquired by the child on reaching the age of unaided walking by being bathed in a leaf-macerate each morning (1).

Reference:

1. Kerharo & Bouquet, 1950: 199.

## APONOGETONACEAE

**Aponogeton subconjugatus** Schum. & Thonn.

FWTA, ed. 2, 3: 15.

An aquatic with floating leaves arising from a rhizomatous rootstock, of fresh water pools in the drier northern part of the Region from Senegal to N Nigeria, and also in E Cameroun.

People in Senegal cook the rhizomes and are said to eat them with relish (1). The starchy tuberous roots of several other species are eaten in their native areas.

Reference:

1. Van Bruggen, 1973: 211.

**Aponogeton vallisnerioides** Bak.

FWTA, ed. 2, 3: 15.

An aquatic herb of rocky pools in Senegal to N Nigeria, and in Sudan, Uganda, Kenya and the Congo basin.

In Katanga, Zaire, a decoction of the leaves and roots is used for washing injuries (2).

This species appears to offer some scope as a water-plant in tropical aquaria. Many *Aponogeton spp.* have been found quite suitable, their only disadvantage being the necessity for introducing a resting period (1).

References:

1. James, 1973. 2. Van Bruggen, 1973: 197.

## AQUIFOLIACEAE

**Ilex mitis** (Linn.) Radik.

FWTA, ed. 2, 1: 623.

West African: NIGERIA FULA-FULFULDE (Nigeria) ndiyamhi = *water plant* (FNH; MM)

A tree to about 13 m tall of montane forest, usually by streams occurring in a scattered distribution in Guinea, Sierra Leone, N Nigeria, W Cameroons and Fernando Po, and extending widespread across Africa to Eritrea, E Africa and South Tropical Africa.

The wood is hard, close-grained, white with a silver grain effect when quarter-sawn, and is said to 'blue' readily (1, 3). It is cut for timber in E Africa for furniture, brake-blocks, railway-sleepers and building construction (7), but is nowhere common. There is no record of the timber being exploited in W Africa. It is burnt as a poor firewood in Lesotho (5).

The bark is purgative and a preparation of pounded bark and leaves is used to wash influenza patients in southern Africa (5, 7). The Masai in Kenya are reported to eat the bark and roots boilded up into a broth (4).

The flowers are white, sweet-scented. Honey from the tree is white and is very highly-rated in Kenya (1). Wild honey-bees in the Cameroon Mountain area frequent the flowers (6).

The plant has a reputation in witchcraft in Lesotho (5, 7) and in Madagascar (2).

References:

1. Dale & Greenway, 1961: 50. 2. Debray & al., 1971: 52. 3. Eggeling & Dale, 1952: 31–32. 4. Glover & al. 1137, K. 5. Guillarmod, 1971: 436. 6. Maitland 212, K. 7. Watt & Breyer-Brandwijk, 1962: 112.

## ARACEAE

**Alocasia macrorhiza** Schott

FWTA, ed. 2, 3: 113.

English: giant taro; elephant's ear.

A large erect herb with trunk to 2 m high or more by 20 cm diameter, bearing leaves which are entire attaining 1 m long and broad; native probably of Ceylon, but dispersed in SE Asia and into the Pacific basin in prehistory and later to tropical Africa. Its climatic requirements have limited its dispersal to areas of evenly distributed rainfall with a temperature constantly over 10°C, and a regular availability of soil-water but not in water-logged sites.

The plant is grown as an ornamental. Forms with coloured leaves are known in Indonesia.

The principal interest in the plant lies in the stem from which a form of *taro* can be prepared. Cropping can be carried out in 1–1$\frac{1}{2}$ years, though sometimes appreciably longer. The stems can be stored in the ground for three months. The raw stem is poisonous and carries raphides of calcium oxalate. The preparation of edible taro to remove these elements is labour-consuming and the return is not high. For these reasons cultivation as a source of food has not caught on in the face of more convenient starchy food-plants. The flour when properly prepared is, however, very white and easily digested. The plant also produces underground cormels which can also be prepared for eating. The plant is recorded as possibly toxic to cattle in Australia, but in Brazil it has been grown as pig food. It has also been investigated as a source of alcohol. (1–4.)

References:

1. Burkill, 1935: 106–8. 2. Coursey, 1968: 25–30. 3. Kay, 1973: 71. 4. Purseglove, 1972: 1, 58–59.

### **Amauriella hastifolia** (Engl.) Hepper

FWTA, ed. 2, 3: 120.

A stemless herb, with leaf-petioles to about 30 cm high, recorded from Ivory Coast, Ghana and S Nigeria, and also in E Cameroun.

A decoction is used in vapour baths in Congo to treat trypanosomiasis (1).

Reference:

1. Bouquet, 1969: 68, as *Anubias hastifolia* Engl.

### **Amorphophallus abyssinicus** (A. Rich.) N.E. Br.

FWTA, ed. 2, 3: 118. UPWTA, ed. 1, 480.

English: Barter's arum (Morton).

West African: **NIGERIA** FULA-FULFULDE (Nigeria) bugulli (JMD) gugulli (JMD) HAUSA bùrár kàreé = *dog's penis; also applies to a phalloid fungus sp.* (JMD) koòdoódòn-kwaàdoó (JMD; ZOG) kunnen jaakii = *donkey's ear* (auctt.) maƙododo (ZOG) mákoòrií maƙododo (JMD) NUPE dùkú dagba (JMD) lúkolúko (RB)

A fleshy stemless plant with leaves arising from a tuber to 30 cm high; spathe dark reddish-purple, slender; of damp places in savanna from Ghana to W Cameroon, and widespread in savanna in eastern Africa from Ethiopia to Rhodesia.

The root is eaten in Sudan (3) and in Tanganyika (2) in time of shortage. It is used in N Nigeria as a 'cure for delayed birth' (1), presumably to induce delivery.

References:

1. Lely 159, K. 2. Michelmore 1153, K. 3. Myers 13852, K.

### **Amorphophallus aphyllus** (Hook.) Hutch.

FWTA, ed. 2, 3: 118. UPWTA, ed. 1, 480.

West African: **SENEGAL** BASARI a-pàty (Ferry) ɛ-kàmb ɓólà (Ferry) BEDIK gi-ngí (Ferry) i-gàmb fali (Ferry) KONYAGI ngamp ɔ-fẹla, a-ngay (Ferry) MANDING-BAMBARA fali foro ba (JB) SERER dul o môn (JB) **GUINEA** KISSI bombole-so (A. Chev.) foko-foko (A. Chev.) massa-polo (A. Chev.) **SIERRA LEONE** BULOM (Sherbro) kɔŋgotigba-lɛ (FCD) KISSI pɔmbiaũ-ma-pianduaũ (FCD) LOKO n-gondi (FCD) SUSU-DYALONKE kalakunde-na (FCD) TEMNE *a*-roth-*a*-kɛrɛ (FCD) **MALI** MANDING-BAMBARA baga, bagani = *poison; a general term* (A. Chev.) soforo = *horse's penis;* (A. Chev.; FB) **UPPER VOLTA** GRUSI leuru (K&B) NYENYEGE tiundaho (K&B)

A fleshy stemless plant with leaves arising from a tuber about 5 cm across; spathe dark purple-red, to 30 cm high, leaves produced after flowering; occurring commonly in the soudano-guinean region from Senegal to Togo.

It is quite a spectacular plant and worthy of cultivation but for its unpleasant smell.

The tuber is eaten in time of dearth, but only after special treatment (1). In Senegal the tuber is dried and then boiled to remove acridity (2, 3). In Upper Volta the tuber is boiled in a vegetable-ash lye, then left in running water till the bitterness has disappeared (4). Its toxicity is such that it is considered in Upper Volta suitable for putting into arrow-poisons (4).

References:

1. Busson, 1965: 517. 2. Dalziel, 1937. 3. Irvine, 1952, a: 29. 4. Kerharo & Bouquet, 1950: 246.

**Amorphophallus dracontioides** (Engl.) N.E. Br.

FWTA, ed. 2, 3: 118. UPWTA, ed. 1, 480.

English: dragon's football (Morton).

West African: **NIGERIA** ARABIC-SHUWA gunta (JRA) BEROM tùtúgéy (LB) HAUSA , kinciya (ZOG) bùrár jaakii = *donkey's penis* (JMD) bùrár kàree = *penis of the dog* (ZOG) gwáázán kwaaɗóó (LB) gwaázar-giiwaá (ZOG) gwazar giwa = *elephant's coco-yam* hántsàr giíwaá = *elephant's udder* (JMD; ZOG) kinchiya kunnan jạki, kunnen jaakii = *ears of the donkey, from the shape of the leaf* (ZOG) IJO èré ótóró = *female otoro* (KW) YORUBA akufọdẹwa (JMD)

A fleshy stemless plant with leaves arising from a stout tuber, appreciably larger than that of *A. aphyllus;* spathe a lurid mixture of purple, brown, grey, olive on the outside, and reddish-purple with white vertical lines inside (7), to about 30 cm long, appearing before the leaves; of seasonally wet land in the savanna, distributed south and to the east of *A. aphyllus* from Ghana to Nigeria.

The sap is intensely acrid (3) and causes skin-irritation (5). Saponins are present. The tuber can be eaten, but is considered only a famine-food. Special treatment is necessary to remove the acridity — slicing, repeated washing, soaking and boiling for one or two day (1, 2, 4–6, 8). In N Nigeria the root is an occasional ingredient of some arrow-poisons (4). It is also considered a valuable remedy in asthmatic infections, and made into a paste it is applied topically to piles (1, a). It is perhaps also this species which is taken in N Nigeria internally and externally to relieve acute rheumatism and sciatica, etc. (1, b) and in S Nigeria to give to a woman during pregnancy to create a feeling of well-being till the time of her delivery: the corm is pealed and cooked with yam or plantain and fish with pepper and salt to taste; the woman will urinate off the *otoro,* and feel at ease (9). The Fula of N Nigeria use the tuber in a medico-magical treatment for snake-bite: to an infusion of the cut-up tuber are added scrapings, and washings of a woman's hair oily with grease, etc., and probably also other ingredients, and this concoction is given to the sufferers to drink (4).

References:

1. Ainslie, 1937: a. sp. no. 27 b. sp. no. 26, as *A. sp.* 2. Busson, 1965: 517. 3. Dalziel 564, K. 4. Dalziel, 1937. 5. Irvine 4527, K. 6. Irvine, 1952, a: 29. 7. Morton, 1961: 43–44. 8. Okiy, 1960: 119. 9. Williamson 76, UCI.

**Amorphophallus flavovirens** N.E. Br.

FWTA, ed. 2, 3: 18.

English: yellow arum (Morton).

West African: **SENEGAL** BASARI a-pač (K&A) a-peùdy ɓólà (Ferry) BEDIK i-gàmb fali, gingí (Ferry) DIOLA éken (JB; K&A) = *sucker* (DF, The Gambia) DIOLA ('Kwaatay') kaiata (K&A)

FULA-PULAAR (Senegal) tubano fauru (K&A) KONYAGI i-pol ɔ-fẹla (Ferry) MANDING-MANDINKA bahôdina (K&A) SERER lar (JB, ex K&A) lar mbind (JB)

A fleshy stemless plant with leaves arising from a circular corm, flower-peduncle 30–60 cm long, spathe yellowish green 7–15 cm long, spadix sometimes to 30 cm long, yellow or flushed dark red, of deep rich soil in soudanian savanna-woodland, from Senegal throughout the Region, and extending into Gabon.

The plant is of attractive appearance, but with a revolting scent. It is recognized in Senegal as toxically dangerous (1).

Reference:

1. Kerharo & Adam, 1974: 200, 202, as *A. consimilis* Blume.

**Amorphophallus johnsonii** N.E. Br.

FWTA, ed. 2, 3: 118. UPWTA, ed. 1, 480.

English: Johnson's arum (Morton).

West African: **GHANA** AKAN-ASANTE anadwoɔ = *snake* (FRI) ɔpɛ (Cox)

A fleshy stemless plant with leaves arising from a corm to about 10 cm across, spathe dark purple-red to 7 cm long, spadix 30 cm, usually less, very occasionally more, peduncle even to over 1 m, of forest shade from Mali to S Nigeria.

It is a striking and attractive plant. The corm, like those of other *Amorphophallus spp.* is taken internally and applied topically in Ghana for snake-bite and acute rheumatism whereby the plant is called 'snake' by Akan people (1).

Reference:

1. Irvine, 1930: 26.

**Anchomanes difformis** (Bl.) Engl.

FWTA, ed. 2, 3: 121–2. UPWTA, ed. 1, 480.

English: forest anchomanes (Morton).

West African: **SENEGAL** DIOLA éken (JB; K&A) = *sucker* (DF, The Gambia) **SIERRA LEONE** KISSI n-dɔndɔ (FCD) MENDE kipɔnɔ (*def.*-i) (FCD) MENDE-KPA kalilugbo (FCD) SUSU-DYALONKE alatala-kunde-na (FCD) TEMNE *a*-thoŋbothigba (NWT; FCD) **UPPER VOLTA** MANDING-DYULA dé (K&B) **IVORY COAST** ABURE kohodié (B&D) AKAN-ASANTE eupé (K&B) niamatimi (B&D) BRONG pê (K&B) AKYE alomé (A&AA) ANYI tupain (K&B) BAULE niamé kwanba (B&D) séréusso kwama (B&D) topi topi (B&D) DAN dina tali (K&B) linna batari (K&B) KRU-BETE bédro-bédro (K&B) GUERE don (K&B) vianakwäï (K&B) GUERE (Chiehn) dobli (K&B) dobré-dobré (K&B) yaa plè (B&D) yaprè (B&D) GUERE (Wobe) méapolodè (A&AA) KULANGO dé (K&B) KWENI diri (K&B) dobli dobli (B&D) KYAMA niamitma (B&D) MANDING-DYULA dé (K&B) 'KRU' blima (K&B) **GHANA** ADANGME-KROBO kwai-aŋma-tʃo (Bunting, ex FRI) AKAN ɔpɛ AKAN-ASANTE ɔpɛ = *dry season, alluding to the plant's renewal of growth at this time* (FRI) FANTE atõe (FRI) nyame kyin = *God's umbrella* (FRI) TWI ɔpɛ = *dry season, alluding to the plant's renewal of growth at this time* (FRI) DAGBANI lukpogu (Ll.W) GA bata foia kani = *bush-pig's cocoyam* (FRI, ex JMD) GBE-VHE deviofɛ-tsini = *children's umbrella* (FRI, ex JMD) dɔli (FRI) **TOGO** TEM nau (FB) TEM (Tshaudjo) nau (Gaisser) **NIGERIA** EDO ólíkhọ̀rọ̀r (Kennedy) EFIK eba enàn̈ = *cow's udder* (FRI; Winston) eba-nne-enàn̈ = *madame cow's udder* (FRI; Winston) FULA-FULFULDE (Nigeria) bugulli, gugulli *the tuber* (JMD) GWARI chakara *the tuber* (JMD) hantsar gada = *duiker's udder; the fruit* (JMD) hantsar giwa = *elephant's udder* (JMD) HAUSA cakara, chakara *the tuber* (auctt.) hántsàr gàdaá = *duiker's udder; the fruiting spadix with red berries* (JMD) hántsàr giawaá = *udder of the elephant* (ZOG) IGBO (Owerri) olumahi (FRI) IGBO (Umuahia) oje (FRI) IJO-IZON (Kolokuma) bòù bèkèódù = *bush new-cocoyam* (KW) IZON (Oporoma) bòù ódù = *bush cocoyam* (KW) KANURI gàsə̀ nàngái (C&H) YORUBA ìgọ (JMD; IFE) lángbòdó (IFE) ọ̀gìrìṣákó *the red berries and ripe spadix* (JMD; Verger)

A large herbaceous plant with stout prickly stem (=leaf-petiole) to 2 m high bearing a huge much-divided leaf, spathe 20–25 cm long, both stem and spathe arising from a horizontal tuber to 80 cm long by 20 cm across, occurring in the forest of Sierra Leone to W Cameroons. Var. *pallidus* with style densely warted occurs only in Fernando Po.

The rhizome is everywhere eaten in time of scarcity but only after special preparation (4–6, 8, 10, 12). Treatment requires prolonged washing and cooking. The Fula add hearth-ashes to the cooking water, then leave the root in water to macerate and ferment for several days. It can then be sundried and stored (5, 6). Irritant properties are due to a saponin and raphides. The rhizome has medicinal uses. In Guinea the rhizomes are used to make rubefacients and vesicants for external application, and alteratives for internal medication, but care has to be exercised on account of the caustic nature of the sap (Pobéguin fide 5). In Ivory Coast the plant is considered to be a powerful purgative and is used to treat oedemas, difficult child-birth, as a poison antidote, and as a strong diuretic for treating urethral discharge, jaundice and kidney-pains: for these the root or the leaves and stems may be used (3, 7). The root pulped with potter's clay is applied to maturate abscesses (9). The rhizome is considered in Gabon to be lactogenic (12) while in Casamance (Senegal) the leaves are used for this (7, 8). Sap from the stem is used in Ghana as an eye-medicine (5) and the liquid obtained after cooking the crushed leaves with other drug-plants is drunk in Ivory Coast as a cough-cure (2).

Rhizomes from Ivory Coast have been reported to contain: carbohydrates 77%; proteins 12%; fats 0·6%; minerals 5%, etc., and a quantity of amino-acids (4). A strong presence of alkaloids is found in Nigerian material (1).

The plant enters into certain superstitious practices. The Hausa consider the fruiting spadix with red berries (*hantsar gada*) a sort of love-charm. The Yoruba invoke the red berries in an incantation for protection against *ṣọpọnna* (smallpox) under the name of *ògìrìṣákọ́, ògìrì* being a Yoruba food or flavouring made from the fermented kernels of *Citrullus colocynthis* (Linn.) Schrad. (Cucurbitaceae). (5, 11). The Ijo of the Niger Delta use the corm in making sacrifice to the dead (13).

References:

1. Adegoke & al., 1968. 2. Adjanohoun & Aké Assi, 1972: 46. 3. Bouquet & Debray, 1974: 47. 4. Busson, 1965: 517, 522 with rhizome analysis. 5. Dalziel, 1937. 6. Irvine, 1952, a: 29. 7. Kerharo & Adam, 1963, a. 8. Kerharo & Adam, 1974: 202. 9. Kerharo & Bouquet, 1950: 246. 10. Morton, 1961: 45–46. 11. Verger, 1967: no. 142. 12. Walker & Sillans, 1961: 92. 13. Williamson, K. 66, UCI.

**Anchomanes giganteus** Engl.

FWTA, ed. 2, 3: 121.

A large forest liane to over 15 m long from a massive corm to 20 cm in diameter, of Zaïre, but now recorded in Ivory Coast.

The corm is held in Zaïre to have purgative and diuretic properties. An aqueous decoction is given for blennorrhoea, stomach-complaints in women, hernia and oedema. A macerate is taken in palm-wine for tachycardia and stomach-pains. The corm pounded with kaolin or palm-oil is applied topically to relieve localized oedemas, rheumatism and other pains. Sap expressed from the corm is added to bath-water for washing persons with fits of insanity, epilepsy and vertigo (1).

Reference:

1. Bouquet, 1969: 67–68, under *A. difformis* Engl.

**Anchomanes welwitschii** Rendle

FWTA, ed. 2, 3: 121.

English: savanna anchomanes (Morton).

West African: IVORY COAST MANDING-MANINKA bakaba (A&AA) san wulu (A&AA) **NIGERIA** EFIK ǹkọ̀kọ̀t = *growing boy; from the unfolding of the leaf* (auctt.) HAUSA tsakara (JMD) IGBO (Onitsha) òpì-olōló = *bottle-horn* (AJC) IGBO (Owerri) uji (AJC) uto (AJC)

Large herbaceous plant with stout prickly stem (leaf-petiole) to about 1.30 m high, spathe greenish-yellow/orange to about 15 cm long on a prickly peduncle, common in savanna from Ghana to S Nigeria, and widespread in tropical African savanna.

The tuberous root is sometimes eaten in time of dearth, but only after careful preparation of washing and cooking to remove poisonous constituents (2–4).

In Ivory Coast pains in various parts of the body are treated by rubbing with the ash of the incinerated root mixed with seed-oil of *Carapa procera* DC. (Meliaceae); leprosy sores are treated by topical application of the pounded rhizome made into a paste with leaves of *Ceratotheca sesamoides* Endl. (Pedaliaceae) (1).

References:

1. Adjanohoun & Aké Assi, 1972: 47. 2. Busson, 1965: 517. 3. Dalziel, 1937. 4. Morton, 1961: 45–46.

**Anchomanes spp. indet.**

West African: **IVORY COAST** MANDING-MANINKA baga (B&D) bagba (B&D) bara (B&D) **NIGERIA** YORUBA olùbọn gaga *a c.var. with large tubers* (JMD)

**Caladium bicolor** Vent.

FWTA, ed. 2, 3: 122.

English: heart of Jesus (U.S.A., Bates).

French: palette de peintre (Berhaut; Bouquet).

West African: **SIERRA LEONE** BULOM (Kim) ɛ-lɛba-sɔnwai (FCD) FULA-PULAAR (Sierra Leone) jabɛrɛ-parinyɛtɛ (FCD) GOLA duu-sisɛ (FCD) KONO g-busi (FCD) KRIO flawa-koko (FCD) MENDE hɔna-gboji = *bewitched cocoyam* (FCD) kpoji (FCD) SUSU galanataŋga (FCD) TEMNE *an*-rɛŋ (FCD) VAI sua-k-posi **NIGERIA** EFIK ḿkpọ̀ñ́ ékpò = *cocoyam of ghost* (RFGA; Lowe)

A perennial stemless herb with a crown of mottled and coloured leaves arising from a tuberous rootstock, native of the tropical S American forest and Caribbean, and dispersed as an ornamental to many tropical and subtropical countries. The plant has been subject to hybridisation over nearly two centuries and it is probable that the material which has run wild and become naturalized is of the *C.* × *hortulanum* group of which *C. bicolor* is the principle parent.

The plant is grown for its decorative coloured leaves which give rise to a number of fanciful names. In its native home the rhizome of *C. bicolor* is locally eaten after boiling, and the fresh rhizome is considered to be emetic and purgative (2, 3). The property of material passing under this name in W Africa is not recorded, but in Congo the rhizome is used as a revulsive and analgesic, the rhizome being applied topically to treat localized pains, oedemas, boils, abscesses and ulcers; the leaves are considered sedative and are taken as a vegetable by insomniacs and those subject to nightmares (1).

References:

1. Bouquet, 1969: 68. 2. Uphof, 1968: 92. 3. Usher, 1974: 110.

**Cercestis afzelii** Schott

FWTA, ed. 2, 3: 126. UPWTA, ed. 1, 481.

West African: **SENEGAL** DIOLA ka lõnk (JB); = *luck* (DF, The Gambia) **SIERRA LEONE** KISSI pambule (FCD) KONO baŋga-yamba (FCD) MANDING-MANINKA (Koranko) firakoyɛsira (NWT) MENDE mbɛmbɛ (*def.*-i) *also applied to epiphytic orchids* (auctt.) TEMNE *a*-toel (JMD) **IVORY COAST** AKAN-ASANTE matakinigma (B&D) matatuoué (B&D) KYAMA nambléblé (B&D) **GHANA** AKAN-ASANTE *m*-batawene (Enti) matatwene = *to bind the drums* (FRI) TWI *m*-batatwene (FRI) WASA *m*-batawene (Enti) NZEMA mpatatwen (FRI)

A creeper, climber or epiphyte to 17 m high, of the forest zone throughout the Region from Senegal to S Nigeria.

The stems are strong. They are used for weaving into mats and baskets in Ghana (3), for making small fishing-nets, and for binding in various purposes. As the Asante name indicates, they are used to bind drums (2).

Preparations from the plant are somewhat toxic and are taken in Ivory Coast as a purgative and to lessen cardiac erethism (1).

References:

1. Bouquet & Debray, 1974: 47. 2. Irvine, 1930: 96. 3. Moor B.31, K.

**Colocasia esculenta** (Linn.) Schott

FWTA, ed. 2, 3: 119. UPWTA, ed. 1, 481.

English: coco yam; dasheen (Caribbean); eddo, or eddoes; scratching edda (an acrid variety); taro (from Polynesia).

French: taro; taro de Chine; taro de Polynésie.

Portuguese: inhame; i. branco, i. da-costa, i. taioba (cultivars).

West African: **SENEGAL** BASARI a-néwùrè (Ferry) ɛ-ɗéwùrè (Ferry) ɔ-léwùrè (Ferry) BEDIK ɛ-ɓɛn, ɛ-mɛn (Ferry) DIOLA usub = *cooler* (DF, The Gambia) KONYAGI va-ngond (Ferry) MANDING-BAMBARA dabéré (JB) **THE GAMBIA** MANDING-MANDINKA oosoo (DA) **GUINEA-BISSAU** BALANTA táem (JDES) BIDYOGO em-átorse (JDES) FULA-PULAAR (Guinea-Bissau) djáberè (JDES) MANKANYA umpinta (JDES) PEPEL umpapa (JDES) **GUINEA** FULA-PULAAR (Guinea) iabéré-koko (FB) jabéré-koko (JMD) KISSI uai (FB) MANDING-MANINKA (Koranko) yabéré (FB) SUSU bari (FB) **SIERRA LEONE** VULGAR koko (FCD) BULOM (Kim) ɛ-lɛba (FCD) FULA-PULAAR (Sierra Leone) yabɛrɛ-ŋanya (FCD) GOLA duu (FCD) KISSI wai (DA) wayelẽ (FCD) waye-pɔmbɔe (FCD) KONO g-busi (FCD) KRIO koko (FCD) krach-koko = *scratch-koko; an acrid var.* (FCD) LIMBA (Tonko) yabɛrɛ (FCD) LOKO k-pohi (FCD) MANDING-MANDINKA fira-koko (FCD) MANINKA (Koranko) fira-koko (FCD) yabɛre (JMD) MENDE koko (*def.*-i) *of Eurasian origin, via English* (JMD; IHB) kpoji (*def.*-i) (FCD; JMD) SUSU bare (FCD) SUSU-DYALONKE yabɛri-gbel-la (FCD) yabɛri-na (JMD) yagbɛri-na (FCD) TEMNE *an*-koko (FCD) *a*-koko-*a*-nɛth (FCD) *a*-rɛŋ (JMD) **LIBERIA** MANO gbĩa (JMD) **MALI** MANDING-BAMBARA jabéré (JMD) **IVORY COAST** KWENI kólúo (Grégoire) **GHANA** VULGAR kooko (FRI) AKAN kooko (auctt.) DAGBANI koúku (Gaisser) GA kolíkò *the fried tuber* (KD) kontómlè *the leaf* (KD) kratse *acrid forms* (IHB) GBE-VHE asɔvle *a c.var.* (JMD) koko (IHB) ulẽ (JMD) **TOGO** BASSARI guge (Gaisser) KABRE gba(e)gbeng (Gaisser; FB) MOORE-NAWDAM gbagbare (Gaisser) gberi-gberin (Gaisser) TEM ku (FB) TEM (Tshaudjo) kóu (Gaisser) **NIGER** HAUSA guaza (FB) **NIGERIA** ANAANG èkà-íkpọ̀ñ (KW) ARABIC kolokass (JMD) BEROM jwɛ (LB) jwɛ pwɛŋ *a white c. var* (LB) jwɛ sunàŋ *a red c. var* (LB) DEGEMA ùkóvúvū (*pl.* à-) (KW) EDO íyòkhó (JMD) EFIK àta ḿkpọ̀ñ = *true cocoyam* (Winston) ḿbọ́ *a general term for lateral suckers from a main tuber* (RFGA; Lowe) ḿkpọ̀ñ (auctt.) ṇkọ́k íkpọ̀ñ *seedling cocoyams* (RFGA) ENGENNI átọ̀ḍị́ạ̀ (KW) EPIE òkílē (*pl.* ì-) (KW) FULA-FULFULDE (Adamawa) bonntoore (*pl.* bonntooje) *probably this sp., edible, known only in Adamawa* (MM) FULFULDE (Nigeria) goojare (*pl.* goojaaje) *only in S Zaria* (MM) meeroore (*pl.* meerooje) = *the vegetable that causes roughness in the throat* (MM) meeroore baaɗi *a wild form, probably this sp.; lit.* = *monkey's cocoyam; inedible by humans; in S Zaria only* (MM) tanndawre (*pl.* tanndawje) (JMD; J&D) GWARI koko (IHB) kwokwo (IHB) HAUSA gwaázaá (ZOG) gwáázáá (LB) gwaàzaà-maì-gudaajií = *lumpy yam; the lateral suckers from the main tuber* (JMD; ZOG) gwaàzà-gwaázaà (JMD; ZOG) gwamba *probably a nickname* (JMD) gwargwaza *archaic* (JMD) kamu *a loan-word used in Kabba for the lateral tubers* (JMD) kunnen jaakii = *donkey's ear; general for aroids* (JMD; ZOG) tagwamba (JMD) IBIBIO íkpọ̀ñ (JMD; Kaufman) ọ̀fọ́b íkpọ̀ñ

= *roasted cocoyam; of this sp. or of X. mafaffa* (Kaufman) IGBO édè nkiti, ogu édè (BNO) ńkàsị IGBO (Awka) édè *general for cocoyam* édè ọ́kụ̄ = *cocoyam of fire* (KW) édē ōnịlị (KW) édè óyìbó = *European cocoyam*, (KW) IGBO (Onitsha) àkàsị (KW) àkàsị ìtè *perhaps = akasi of the pot* (KW) IGBO (Owerri) àkàsị (JMD; IHB) ákàsị ìtè = *cocoyam of the pot* (JMD) édè árọ̀ = *Aro cocoyam* (AJC) èdè nwá jí = *cocoyam child of yam* (AJC; JMD) ìlipotò *the leaves* (JMD) ńkàsị (JMD) IGBO (Umuahia) édè éfù = *ordinary cocoyam*; éfù: *ordinary* (KW) édè ọ́bà = *calabash cocoyam* (KW) IGBO-ABOH àkàshị (KW) akaši (KW) IJO (Nembe) ìkèrèḅùrú, ìkèrèḅú, ḅùrú: *yam* (KW) IJO-IZON òdú (KW) IZON (Arogbo) lòdú (KW) IZON (Egbema) odú (Tiemo) IZON (Kabou) lòdú (KW) IZON (Kolokuma) Ị̀zọ́n ódú = *Ijo cocoyam* (KW) ISEKIRI kóko (KW) ISOKO ódú (KW) KATAB gwaza (Meek) KHANA gẹ́ẹ̀rẹ́ (KW) MAMBILA kuri (Lowe) MARGI tandauji *from Fula?* (MM) TIV mòndò (JMD) URHOBO údù (*pl.*, ídū) (auctt.) údù úrhòbò = *cocoyam of Urhobo* (Faraclas) údùákà *a c.var from Benin* (Faraclas) YORUBA iṣu kókò (JMD) kóko (JMD; IHB) kóko funfun *a white-rooted c.var.* (JMD) kóko pupa *a yellow-rooted c.var.* (JMD) **WEST CAMEROONS** VULGAR dinde (JMD) DUALA dìnde (Ithmann) NGEMBA akù (*pl.*) *c.vars* (Chumbow) mákàb (Chumbow)

*NOTE: Some names probably apply also to Xanthosoma mafaffa (Araceae).*

A stout herbaceous plant with a whorl of leaves erect to 1–2 m high arising from the crown of a corm, variable but usually cylindrical to 30 cm long by 15 cm diameter, native of SE Asia, and dispersed in antiquity throughout the Old World, and later over the Pacific. Its appearance in the New World is relatively recent. The basic form is with a single large corm producing but a few small side corms, or cormels. This is var. *esculenta* and forms of this fall into the *dasheen* group. It is thought that sporting occurred in China or Japan in which the central corm became reduced and an abundance of cormels were produced with also a tendency to greater side suckering. This is var. *antiquorum* (Schott) Hubbard & Rehder, or the *eddoe.* Both varieties have many forms.

The name *Colocasia* is probably of Eurasian origin. It appears in the writing of Virgil 37 B.C., Dioscoides, c. A.D. 78, though for *Nelumbium speciosum* Willd., and of Pliny, c. A.D. 79 for the present species. Philology connects this word with *kulkas* or *qorqas* in Egyptian Arabic and *kalo-kacu* and *kalo-kuchoo* in Bengal and Ceylon. Semantically these words imply a 'food plant that is strongly coloured.' While the Polynesian word *taro* may seem to be remote from this root, it too carries a connotation of colour — deep blue, or indigo (2, 10).

The plant's advent in West Africa has been dated as at least 2000 years ago. It could have come from E Africa, whither it must assuredly have been distributed by Arab traders in the Indian Ocean, and thence across the humid equatorial forest-belt to SE Nigeria. No satisfactory linguistic connexion, however, supports this. It is well-known that under the influence of Islamic arms the plant was carried the length of the Mediterranean and into the Iberian Peninsula. There is no reason to doubt that similar Islamic influences were at work spreading westwards across Africa south of the Sahara, and there is linguistic evidence to support the arrival of the cocoyam by this route. The most common vernacular name in the Region is *koko* or a derivation of this. It has been ascribed to the Fante race of whom there is evidence in their language of an origin, and a subsequent migration from the area of modern Sudan to Ghana. There are acceptable linguistic processes between source languages and receiving languages to equate *koko* with the Arabic *kulkas* or *qorqas,* and from the point of entry in the north central part of the region for *koko* to have spread over the Region. This has probably been the main means of entry and dispersal in the Region. There is linguistic evidence however for another smaller, but equally important line of communication via Lake Chad, NE Nigeria and southwestwards into the Delta area. Shuwa Arabic *kolokass* may have given rise to *akasi* and *ukase* in Igbo at Onitsha. It is curious that this etymological chain from the northeast does not appear to have progressed further for the Igbo to the south at Owerri and Umu Ahia use *e, ede* and *ete,* and this in turn is *odú* in Ijo of the SW Niger Delta. Here again the chain is broken, blocked perhaps by the cocoyam being tabu to some of the western Delta peoples. (2, 4, 5, 8, 16.) More linguistic evidence needs to be collected. To this discussion, however, must be added a consideration of the general term *eddo* which appears to have arisen first amongst the slave community in the Caribbean and has been ascribed to Ghanaian origin (Oxford English Dictionary). *E* is used in Igbo at Umu Ahia and Overri. *Do* in the Benin area implies an edible tuber. What more easy in a polyglot community to combine the two as *e-do* or *eddoe*? Or conceivably to the slaves the plant, so well known, might be quite simply the 'Edo root', i.e., from Edo, or the old Kingdom of Benin. Indeed the plant is so ingrained in Ijo culture as to be their own 'Ijo Cocoyam', or *ìzọ̀n ódú*. (See 16.)

One further derivation needs to be mentioned and perhaps sheds a little light on a subsequent and relatively recent migration. Chinese emigrants must certainly have taken the plant with them to Trinidad and the Caribbean. Indeed, the common term in that area, *dasheen* is but a corruption through French Creole of *taro de Chine*. The word, as well as planting material of the dasheen form have been brought to West Africa by freed slaves on repatriation.

The plant favours damp forest sites. It has become naturalized in many parts of the Region. Numerous varieties occur differing in shape, size, colour of the flesh of the tuber, degree of acridity, food value, etc., and in cultural needs. It is an 8–18 months crop requiring constant humidity and a high rainfall, and it is tolerant of poor drainage. In the Pacific the plant assumes the greatest importance as a staple food. It is of considerable importance in the West Indies. Its position in W Africa is being eroded by another introduced plant the tannia, *Xanthosoma* (Araceae). It has a particular robustness withstanding hard conditions, and encompassing a far greater range of latitude than other important staple food plants such as the greater yam, *Dioscorea alata*. The eddoes form is in general hardier than the dasheen form and can be grown under drier and poorer conditions. Certain cultivars show a marked tolerance to salinity and flooding. Mechanization of cultivation has been successfully applied, and as a food crop it has a valuable potential awaiting yet to be more fully exploited (7).

The tuber contains calcium oxalate rhaphides which cause acridity and much irritation to the mouth and digestive tract. They need to be removed by cooking and repeated washing before the tuber can be eaten. Though some selection work has been done to obtain improved strains, this needs to be taken much further to eliminate acridity, obtain higher yields and better cooking and dietetic properties. The root yields a high quality meal and an easily digested starch, sometimes known as Portland arrowroot. The root is used to make 'foo-foo.' It can be sliced and dried, and stores well (9). The prevalence of certain diseases has, however, been associated with the eating of this plant as a major food-source. A goitrogenic substance is present and also a sapotoxin thought to increase susceptibility to leprosy. Debility and a high incidence of nephritis have been ascribed to it. Toxicity is possibly a seasonal affair being highest at the end of the dry season. On the credit side, the eating of this plant instead of cereals as a carbohydrate source leads to a marked reduction in dental caries (14). In recent years taro flour has been used in speciality foods in preventing certain allergic diseases in susceptible children and as a cereal substitute in coeliac diseases (7).

The tubers assay at: water 63–85%, protein 1.4–3.0%, fat 0.2–0.4%; carbohydrate 13–29%; fibre 0.6–1.2%; and ash 0.6–1.3%. Vitamins B and C are present in appreciable amount (11). The tubers can be a source of alcohol, and mucilage is also present which has some potential in paper manufacture (7). Zulu in S Africa ferment the root with sugar and corn (*Zea*) to make a beer (14).

The tuber has no recorded medicinal use in the Region. In Gabon raspings from it are applied to maturate boils (12, 13). Use of the root as a poultice for snake-bite and rheumatism has been recorded (14). A trace of alkaloid has been reported in Nigerian material (?root) (1). Some races in southern Africa grow the plant near their huts in the belief that it will keep termites away (14).

The young leaves are edible, but the main ribs need removal first, and then cooking with soda. The leaf-petioles are said to be edible after the removal of the cuticle. Acridity after accidental chewing can be relieved by lime-juice (4, 6). The leaves mixed with those of *Tephrosia* (Leguminosae-Papilionoiideae) are used in Gabon as a fish-poison (13). The leaf is less toxic than the root, but a considerable variation in human susceptibility has been found. Cholesterol-containing foodstuffs appear to neutralize toxicity, and the administration of raw adrenal glands is curative (14). The alkaloid *tyramine* has been reported in the leaves at 34 mg/kg fresh weight in eddoes (15).

Igbo of S Nigeria use caps or helmets and a kind of armour woven from a fibre obtained from the petiole. The Temne of Sierra Leone associate the plant with another Aracea, *Caladium bicolor,* whose variegated and coloured leaves are regarded as a protection against thunder and lightning (4). The Ijo of S Nigeria hold a festival to the plant called *odúfi,* or 'eating the cocoyam' (16).

References:

1. Adegoke et al., 1968. 2. Burkill, 1938. 3. Busson, 1965: 520, with root-analysis. 4. Dalziel, 1937. 5. Irvine, 1952, b: 28. 6. Irvine, 1956: 37–38. 7. Kay, 1973: 168. 8. McIntosh, 1978. 9. Okiy, 1960: 118. 10. Portères, 1960. 11. Purseglove, 1972: 1: 61–69. 12. Walker, 1953, a: 22. 13. Walker & Sillans, 1961: 93. 14. Watt & Breyer-Brandwijk, 1962: 113. 15. Wheaton & Stewart, 1970. 16. Williamson, K. 1970.

## **Culcasia** P. Beauv.

FWTA, ed. 2, 3: 122–5.

Some 13 species represent the genus in W Africa. Most are climbing or epiphytic plants. Leaves are commonly given for stomachache, but the sap is probably poisonous: it causes irritation on the skin (1).

Reference:

1. Morton, 1961: 48–49.

## **Culcasia angolensis** Welw.

FWTA, ed. 2, 3: 124. UPWTA, ed. 1, 482.

English: greater climbing arum (Morton).

West African: **IVORY COAST** KYAMA sau niama (B&D) **GHANA** AKAN-ASANTE *n*-kɔnkroahan (FRI) TWI *n*-kɔnkroahan (FRI) GBE-VHE aklãmakpa (FRI) tɔnyã (FRI)

A robust forest climber to over 30 m high with thick tough stem adhering to the host tree by clasping roots, occurring in Guinea to Ghana and in Fernando Po.

The plant is recognized in Ivory Coast as toxic and abortifacient (1).

Reference:

1. Bouquet & Debray, 1974: 48.

## **Culcasia lancifolia** N.E. Br.

FWTA, ed. 2, 3: 124.

A slender climber on tree stems, to 5 m high, occurring in N and S Nigeria, W Cameroons and Fernando Po, and extending to Congo.

The plant has a scent of *coumarin*. In Gabon the leaves and roots are worn around the neck and ears, and sometimes they are powdered and put on the head. Distillation has given good returns of coumarin (1), cf. *C. striolata* Engl.

Reference:

1. Walker & Sillans, 1961: 94.

**Culcasia parviflora** N.E. Br.

FWTA, ed. 2, 3: 124.
West African: **SIERRA LEONE** LIMBA gbilan-gbalan (Glanville) **LIBERIA** KRU-BASA dwehn-doo (C)

A slender climber of the forest in Sierra Leone to W Cameroons and Fernando Po, and also in E Cameroun.
The leaves are used in Liberia to cure boils and 'bumps' on the body: the pulped leaves are rubbed over the body and the 'bumps' are pushed back inside the body, and then a strong purge is taken to chase them out (1). The larger leaves are used in Sierra Leone to wrap kola nuts and are said to prevent damage by weevils (2).

References:
1. Cooper 419, K. 2. Glanville 439, K.

**Culcasia saxatilis** A. Chev.

FWTA, ed. 2, 3: 124, 126.
West African: **SIERRA LEONE** MENDE nɛni (FCD) **NIGERIA** EFIK nkpalaeku kigbadah (Gregory)

An erect profusely-branched plant to about 1.30 m high, with a few aerial roots, of riverain forest in the savanna zone, throughout the Region from Senegal to S Nigeria, and extending to the Congo basin.
The plant sap is reported to be irritant to the skin (1).

Reference:
1. Adames 135, K, FBC.

**Culcasia scandens** P. Beauv.

FWTA, ed. 2, 3: 124. UPWTA, ed. 1, 482.
English: common climbing arum (Morton).
West African: **SIERRA LEONE** KISSI kɛmbɛŋ-hɔnda (FCD) pe-pɛndoa (FCD) KONO baŋga-yamba (FCD) LIMBA gbilan-gbilan (FCD) MENDE kpowa-bɛmbɛ (FCD) nɛni *a general name for certain epiphytes* (JMD) TEMNE *a*-tampene (JMD) **IVORY COAST** AKAN-ASANTE gyamnkawa *from* gyam: *to mourn;* ankaw *or* kawa: *a small fish in the river; alluding to use as a fish-poison* (FRI) *n*-kɔnkroahan (FRI) FANTE ɔtwa tegyrima = *dog's tongue* (JMD) TWI *n*-kɔnkroahan (FRI) GBE-VHE aklãmakpa (FRI) tɔnyã (FRI) KRU-GUERE druhin (K&B) 'KRU' blagéï (K&B) maraka yaya (K&B)

An epiphitic climbing herb with slender wiry stems, of the fringing forest and savanna in Liberia to S Nigeria, and also in E Cameroun.
Sap is irritant on the skin, and is used as a fish-poison (3). In western Ivory Coast the pulped leaves are used in topical applications for headache, inter-costal pain, etc., and a dressing of fresh leaves is put over ulcers (4). In S Nigeria the leaf is sometimes used in medicine for stomachache (5), and in a veterinary preparation for goats suffering from a disease causing inability to stand (6). In Gabon the leaves are made into a douche for blennorrhoea (7, 8). Women in Gabon eat the leaves during pregnancy cooked with groundnuts or with *odika chocolate* (kernel butter of *Irvingia gabonensis* Baill., Ixonanthaceae); they also drink the water in which leaves have been steeped for several days (7, 8). A tisane is held in Congo to be anti-abortive and antemetic, and a decoction is used to bathe rachitic children; sap from the leaves is instilled into the ears for ear-inflammation and deafness, and the ashes of the burnt plant are used to soothe headache (1).

In Malawi the seeds and roots are dried and powdered and mixed with maize seed at sowing. The crop is said to be much increased (2).

Unnamed alkaloid has been detected in the leaf and stem of Nigerian material (9).

References:

1. Bouquet, 1969: 68. 2. Chapman 148, K. 3. Dalziel, 1937. 4. Kerharo & Bouquet, 1950: 247. 5. Thomas, N. W. 2132 (Nig. Ser.) K. 6. Thomas, N. W. 2133 (Nig. Ser.) K. 7. Walker, 1953, a: 22. 8. Walker & Sillans, 1961: 95. 9. Willaman & Li, 1970.

**Culcasia striolata** Engl.

FWTA, ed. 2, 3: 124.

An erect herb to 30 cm high with stout stilt-roots, of forest floor, widely dispersed from Guinea to W Cameroons, and also in E Cameroun.

The plant has a scent of *coumarin*. In Gabon the leaves and roots are worn around the neck and sometimes they are powdered and put on the head. Distillation has given good returns of coumarin (1), cf. *C. lancifolia* N.E. Br.

Reference:

1. Walker & Sillams, 1961: 94.

**Culcasia tenuifolia** Engl.

FWTA, ed. 2, 3: 126.

West African: **SIERRA LEONE** MENDE kpowa-bɛmbɛ (FCD) nɛni (FCD)

A climber or epiphyte on forest trees from Sierra Leone to W Cameroons and in E Cameroun and Cabinda.

**Culcasia sp. indet.**

West African: **IVORY COAST** AKAN-ASANTE gangulan (B&D) BAULE kumalui (B&D) SENUFO-TAGWANA léwo (K&B) DYIMINI léwo (K&B)

An unidentified species is considered in Ivory Coast to be fetish and tonic. Leafy stems are macerated for six days and the filtered liquid is taken to give strength and to preserve one's well-being, and while on a journey as an envigorant (1).

Reference:

1. Kerharo & Bouquet, 1950: 247.

**Cyrtosperma senegalense** (Schott) Engl.

FWTA, ed. 2, 3: 113–4. UPWTA, ed. 1, 482.

English: swamp arum (Morton)

West African: **SENEGAL** DIOLA hubam amata (JB) **GUINEA-BISSAU** FULA-PULAAR (Guinea-Bissau) nopicóbo (JDES) **SIERRA LEONE** BULOM (Sherbro) kin-dɛ (FCD) KISSI yɔ-hũdɔ (FCD) MENDE dina (*def.* dinɛi) (FCD) SUSU baifinɛ (NWT) TEMNE *a*-barɛn (auctt.) **IVORY COAST** AKAN-ASANTE sukoko (B&D) **GHANA** AKAN-FANTE kokoahatew (FRI) NZEMA ɛvuniẽ (FRI) **NIGERIA** EFIK ŋnyọ́rọ́ ọ́tọ́ñ (RFGA; Lowe) IGBO ẹko (AJC; JMD) mgbo ẹko (AJC; JMD) IJO-IZON (Kolokuma) bòù àká = *bush maize* (KW) IZON (Oporoma) okúo (KW) YORUBA ọ̀pẹ̀ igọ̀ = *igo palm* (Millson)

An herbaceous plant with large sagittate leaves borne on prickly petioles 1.70 m long from a rhizome, and with a large spathe on a peduncle to nearly 4 m high, of forest margins, swamps and ravines in savanna country throughout the Region, and dispersed to the Congo basin.

The leaves are eaten as a vegetable in Gabon (8), but consumption in the W African region appears to be more restricted to the young leaf in Sierra Leone in time of dearth (5) and as an ingredient of palaver sauce (4). In Congo the leaves are given to women in childbirth to accelerate delivery: it promotes expurgation (1). In Ivory Coast, the leaf-sap has been used taken by mouth for fits of hiccups (2). In some parts of Gabon the leaves are used to wrap [?for cooking] dumplings of tapioca flour (8).

The rhizome reduced to flour is used by the Ijo of S Nigeria in an unspecified manner in medicine (9). In Gabon the rhizomes are used against ulcers and a maceration is administered to newly-delivered women to drink (7, 8). A decoction is analgesic and sedative. It is taken in Congo as a cough-cure, and in larger dosage for nervousness and agitation; a pulp is applied in frictions for general pains (1).

The plant is burnt to an ash from which a vegetable-salt is obtained in Sierra Leone (3, 6), a use perhaps also known in Gabon (8).

The fruits are added to remedies for gonorrhoea and dysentery in S Nigeria (3).

A decoction of the bark [sic] is dropped in the eyes of a fowl or witch as an ordeal in Sierra Leone (N. W. Thomas fide 3). The Ijo use the berries in making sacrifice to the dead, and the plant's name features in a song in an activity (? game) known as *imbĭoe* or *pịrị* (9).

References:

1. Bouquet, 1969: 69. 2. Bouquet & Debray, 1974: 48. 3. Dalziel, 1937. 4. Deighton 1513, K. 5. Melville & Hooker 261, K. 6. Morton, 1961: 46. 7. Walker, 1953, a: 22. 8. Walker & Sillans, 1961: 95. 9. Williamson, K.49, UCI.

**Dieffenbachia gen. & spp.**

FWTA, ed. 2, 3: 113.

English: dumb cane, dumb plant, mother-in-law's tongue plant, tuft-root (Caribbean, U.S.A., Bates).

French: canna de Imbé (*D. seguina* (Jacq.) Schott., Antilles, Usher).

A genus of some 30 herbaceous, rhizomatous species of tropical America, in forest shade. The leaves are often variegated and undergo natural mutation. Hybridization, particularly using *D. seguina* (Jacq.) Schott and *D. maculata* (Lodd.) G. Don (syn. *D. picta* Schott), has been undertaken resulting in a number of ornamental cultivars which are now widely dispersed throughout the tropics and as house-plants in temperate countries.

The plant is tolerant of neglect and rough conditions. To be kept in prime condition as a foliage pot-plant it should be prevented from flowering by removing the inflorescences. A certain amount of care needs to be exercised in handling the plants as the sap is intensely acrid and irritant to the skin. In the West Indies the pounded plants are used as a revulsive for rheumatism, rashes, prurigo and skin itches. If the plant is chewed the mucosae of the mouth may swell and dumbness lasting several days may ensue, a property recognized in the English names.

**Nephthytis afzelii** Schott

FWTA, ed. 2, 3: 12.

West African: **IVORY COAST** BAULE bono kokó (B&D)

A rhizomatous plant, rhizome about 1 cm thick, with 1–3 hastate leaves to about 30 cm high, in forest in Sierra Leone to Ghana.

Ivorean material has shown reactions for traces of alkaloid and saponin in the leaves. Tests for contraceptive activity have given negative results (1).

Reference:

1. Bouquet & Debray, 1974: 48.

**Pistia stratiotes** Linn.

FWTA, ed. 2, 3: 113. UPWTA, ed. 1, 482–3.

English: water lettuce; great duck-weed (Dalziel); Sudd plant (Ainslie).

French: herbe à la chance (Berhaut).

West African: **SENEGAL** MANDING-BAMBARA ko bua (JB) SERER fas (JB) hubar mbèl (JB) mburu lur (JB) WOLOF tambalay (JB) **GUINEA-BISSAU** MANDING-MANDINKA sarebáfae (EPdS; JDES) **SIERRA LEONE** BULOM (Sherbro) temabɔ-lɛ (FCD) KRIO oj(u)oro (FCD) MENDE mbole (FCD) **LIBERIA** KRU-GUERE (Krahn) lo-ulo (K&B) **MALI** DOGON téne (C-G) SONGHAI soru-fusso (A. Chev.) 'SOURAYE' farka ana (Leclerq) **IVORY COAST** 'KRU' dara (K&B) **GHANA** AKAN-ASANTE ntanowa (FRI, Enti) todia (TFC) FANTE ntãtãaba (FRI) TWI ntaya (FRI) GA ntaya (FRI) tɛtrɛmantɛ (FRI) GBE-VHE (Awlan) aflɔ (FRI) **NIGERIA** FULA-FULFULDE (Nigeria) essodiya (JMD) GWARI mbwosa (JMD) HAUSA dalam, zakankau *the plant ash* (JMD) kaínúwaá (auctt.) IJO-IZON (Kolokuma) èkérékù, èkérégù (KW) NUPE tsamvogi (JMD) YORUBA ojúoró (JMD; IFE)

A floating fresh-water herb, rooting in shallow water, occurring throughout the Region from Senegal to W Cameroons and widely dispersed through the tropics.

This is the *sudd* plant of the upper reaches of the Nile in Sudan. There and in many other African waterways it multiplies into thick mats obstructing navigation and occluding the water to aeration, thus preventing the presence of fish. It is common in swamps and lakes of the forest zone of the Region. It occupies quiet reaches of the Niger River and its tributaries. In the Volta Lake of Ghana it is becoming a multiple nuisance by impeding the movement of boats, and harbouring snails that can carry schistosomiasis and larvae of *Mansonia* mosquito that is the vector of the filaria causing elephantiasis. The mosquito larvae are able to pierce the roots of *Pistia* to obtain adequate air for respiration (6, 7). In the lagoon area of Ivory Coast where the water fluctuates between being sweet and brackish, the plant multiplies exceedingly when the water is fresh and then it harbours the larvae of malaria—carrying *Anopheles gambica,* but when the water turns saline the plant dies, rots and liberates sulphuretted hydrogen poisoning the water and killing the fish in it. It is recorded that a salinity of 1 gr NaCl/litre $H_2O$ is optimal for the plant.

The plant is eaten in Sudan in time of famine (4). It has been so eaten in India, and to this day Chinese eat the young leaves after cooking. The taste is said to be insipid at first, then biting (5). In Nigeria the leaves are taken as a stomachic, but an overdose may cause acute diarrhoea (2). Powdered leaves are taken in water or in palm-wine in Ivory Coast for male-sterility (8), and the dried powdered leaves are used as a healing disinfectant on rebellious sores (3). In Asia they are similarly applied to boils, syphilitic eruptions and to various skin-complaints (5). A leaf-infusion is made into an eyewash in Nigeria (2) and in The Gambia where the plant is called 'eye-pity' (6). Use of the plant as a poultice for piles is recorded in Nigeria (2), Gabon (11), Sudan (4) and Asia (5). A decoction is used in the Region in fumigations for rheumatic pains, fever, etc. (6). The leaves in decoction or infusion are considered demulcent and cooling (Nigeria, 2; Sudan, 4). The plant mixed with rice and coconut-water is taken in some parts against dysentery (11). The roots are laxative and emollient (2, 4).

The plant is able to take up minerals and in parts of the Region it is burnt to produce a vegetable salt. In Hausa the salt ash is called *zakankau,* or *dalam* and is used as a substitute for sodium chloride (6). In Gabon this product assumed such importance before the import of proper salt that when the latter did become available, some races still referred to it by the name of the plant-ash (11). The fresh plant yields about 1% ash, and chemical analyses of Indian material indicate the major component to be potassium chloride (75%) and sulphate (22.6%), (5) but salts of sodium, magnesium, calcium, iron and aluminium are also present (10). A trace of alkaloid has been reported (1).

The ash is used in N Nigeria to apply to ulcerative conditions of the mouth, thrush, etc., due to avitaminosis, and is taken internally in food for intestinal disorders due to worm-infection (6). The ash is applied in ringworm in Nigeria (2), in Sudan, specifically of the scalp (4), and India (5). Fula women in W Africa make a pomade of the ash in fresh cow-butter for the hair, and in some places of the Region it is used for making soap (6).

In N Nigeria the plant is fed to ostriches (6). In Asia Chinese feed it to pigs, and encourage the plant in carp fish-ponds (5). It is used in an undisclosed manner in veterinary medicine in Ivory Coast (8).

There is so much potash in the plant that it has been suggested as a possible source of it. It can serve as an agricultural manure thrown on the land to rot to liberate the potash (5).

References:

1. Adegoke & al., 1968. 2. Ainslie, 1937: sp. no. 281. 3. Bouquet & Debray, 1974: 48. 4. Broun & Massey, 1929: 369. 5. Burkill, 1935: 1756–7. 6. Dalziel, 1937. 7. Hall, J. & al., 1971. 8. Kerharo & Bouquet, 1950: 247. 9. Portères, 1951, b. 10. Quisumbing, 1951: 144–5. 11. Walker & Sillans, 1961: 96.

**Rhaphidophora africana** N.E. Br.

FWTA, ed. 2, 3: 114.

English: banana arum (Morton).

A lofty forest climber occurring from Sierra Leone to S Nigeria and Fernando Po.

The plant is used in Ivory Coast to produce a mouth-disinfectant (1).

Reference:

1. Bouquet & Debray, 1974: 48.

**Rhektophyllum mirabile** N.E. Br.

FWTA, ed. 2, 3: 122.

West African: **NIGERIA** IGBO kpakpari (Lowe) IGBO (Onitsha) ékò (AJC) IJO-IZON (Kolokuma) bòù òdú = *bush cocoyam* (KW)

A stout forest climber to 10 m high with long pendulous roots and also clasping roots, occurring from Dahomey to W Cameroons and Fernando Po, and across the Congo basin to Uganda and Angola.

In Gabon the young leaves and spadices are eaten as a vegetable; the leaves are cooked in *djave* nut butter (*Baillonella toxisperma* Pierre, Sapotaceae) with an unspecified fungus, or with the bark of various trees, the rhizome of *Sarcophrynium sp.* (Marantaceae) and seeds of chili as treatment for insanity, and for stitch and liver-complaints (3, 4). Leaf-sap is taken in draught with kaolin, maleguetta pepper and rock salt in Congo for heart-troubles and as an antemetic (1).

The long pendulous roots are used in S Nigeria as ties for yams (2), and in Gabon the central fibrous core of these roots after the outer layers of tissue have been stripped off serves as fishing line (4).

In Congo it is considered that leaf-sap applied to the soles of the feet and to the legs of one setting out on a journey will confer protection (1).

References:
1. Bouquet, 1969: 69. 2. Lowe, 1973. 3. Walker, 1953, a: 23. 4. Walker & Sillans, 1961: 96

**Stylochiton hypogaeus** Lepr.

FWTA, ed. 2, 3: 114–6.
English: Barter's ground arum (Morton).
West African: **SENEGAL** BASARI a-nyél ɛwùrè (Ferry) KONYAGI ngon tanyankẹli, yafoton (Ferry) SERER ndukul mbap (JB) WOLOF tabal (JB) **GHANA** GRUSI-KASENA worola (CV) MOORE gwadaba (FRI)

A small herbaceous plant with thick rootstock recorded from Senegal, Ivory Coast and Ghana to Nigeria and on to Gabon.
Stock shun grazing it in Senegal but hares are said to eat the roots (1). In Ghana the rhizome is boiled and mashed for application to boils (3). The inflorescence is eaten in the Northern Territory of Ghana (2).

References:
1. Adam, 1966, a. 2. Vigne 3757, 4530, K. 3. Irvine 4601, K.

**Stilochiton lancifolius** Kotschy & Peyr.

FWTA, ed. 2, 3: 114. UPWTA, ed. 1, 383, as *S. warneckei* Engl.
English: Warnecke's ground arum (Morton).
West African: **GHANA** MOORE guadaba (FRI) **TOGO** GURMA (Mango - Jendi) kaka sui (Mellin) **NIGERIA** ARABIC-SHUWA umm burku (JMD) FULA-FULFULDE (Nigeria) guraare (*pl.* guraaje) (JMD; MM) nguraare (*pl.* nguraaje) (JMD; MM) HAUSA guzami *this sp. or an allied plant* (JMD) gwandai *the plant* (JMD) kinciya *the root* (JMD; ZOG) kunnen jaakii = *donkey's ear; the leaf* (JMD) maƙari (JMD) mákoòrií maƙododo *on its first appearance above the soil* (Bargery) HAUSA (West) akwari (JMD) matinga *the root* (JMD) KANURI ngúrà (C&H) TIV imondo kuna kuna

A small herbaceous plant with short rootstock of savanna-woodland from Senegal to S Nigeria, and widespread in the drier savanna of tropical Africa.
The young leaves are eaten as a pot-herb in N Nigeria. They need prolonged boiling to remove their acridity. Cooking is carried out with common salt or Asben salt, a product produced locally and known as *beza*. The rhizome is eaten in time of famine. This requires frequent washing in lye of ashes before cooking. Rhizomes of the genus generally are eaten in famine and require treatment to remove bitterness (1–4). A yellow dye is obtained from the plant which is used in Togo for dyeing cloth (2).

References:
1. Dalziel 237, K. 2. Dalziel, 1937. 3. Irvine, 1952, a: 29, 33. 4. Morton, 1961: 47–48, as *S. warneckei* Engl.

**Typhonium trilobatum** (Linn.) Schott

FWTA, ed. 2, 3: 113.

An herbaceous plant, 30–50 cm high, of waste places in tropical Asia, now recorded in Ivory Coast.
In Malaya the plant has acquired a position as a weed around towns (1). The plant is grown in S India for the rhizome which is eaten (2). It is also eaten in

Indonesia (1). Other species are similarly eaten in the Australasian region. The fresh rhizome is very acrid and is a powerful stimulant. It is pounded into a poultice for use in India on scirrhous tumours. The acrid principle is volatile, and cooking or drying renders the rhizome innocuous. It is said to relax the bowels and to give relief in cases of haemorrhoids, and taken with bananas to cure stomach-complaints (2).

Analysis of Indian material shows the edible portion of the rhizome to consist of: water 69.9 %; protein 1.4%; fat 0.1%; fibre 1.0%; other carbohydrates 26.0%; and minerals 1.6%. Alcoholic extracts have given $\beta$-sitosterol, two undetermined sterols and a crystalline compound (2).

The leaves are also eaten in India. In SE Asia the leaves are fed to fish which greedily eat them (1).

References:

1. Burkill, 1935: 2195–6. 2. Chadha, 1976, a: 401–2.

**Xanthosoma mafaffa** Schott

FWTA, ed. 2, 3: 119. UPWTA, ed. 1, 483, as *X. sagittifolium* sensu Dalziel.

English: coco yam; new coco yam; also Caribbean names used in English: tania, tannia, tannier; taye, tayonne; yautia.

French: chou caraïbe (Berhaut).

West African: **SENEGAL** VULGAR makabo (JB) malanga (JB) **GUINEA** KPELLE g-biné-elin (FB) g-bune (FB) uayé (FB) LOMA gputé (FB) MANDING-MANINKA diabereen puté (FB) gbéna-uku (FB) **SIERRA LEONE** BULOM (Kim) ɛ-lɛba (FCD) BULOM (Sherbro) koku-lɛ (FCD) FULA-PULAAR (Sierra Leone) jabɛrɛ (FCD) jabɛrɛ-koko (FCD) yabɛrɛ (FCD) GOLA duu (FCD) KISSI waye-bɛnde (FCD) wayele (FCD) wayilẽ (FCD) KONO koko (FCD) KRIO koko (FCD) LIMBA (Tonko) kogo (FCD) LOKO koko (FCD) MANDING-MANDINKA yabɛre (FCD) MENDE koko (FCD) SUSU koko (FCD) SUSU-DYALONKE yagbɛri-na (FCD) TEMNE *an*-gbaŋkaŋ (FCD) *an*-koko (FCD) *an*-kol (FCD) TEMNE (Kunike) *am*-bɛroŋ (FCD) TEMNE (Port Loko) *an*-yabɛrɛ (FCD) TEMNE (Sanda) *an*-yabɛrɛ (FCD) TEMNE (Yoni) *am*-bɛroŋ (FCD) VAI k-posi (FCD) **GHANA** VULGAR amankani = *yam of the nation* (FRI) nkontommere *the leaves* (FRI) ADANGME amankani (Lowe) kotomle *the leaves* (FRI) ADANGME-KROBO amãkãni (FRI) amankani hiɔ *a c.var with white-skinned corms* (JMD) amankani tchu *a c.var with red-skinned corms* (JMD) amankanidzẽ *a c.var with red-skinned corms* (JMD) AKAN-ASANTE amankani (FRI) antwibo *a c.var* (JMD) TWI amankani, mankani (auctt.) amankani antwibo *a c.var with small pink corms* (JMD) amankani fita = *white cocoyam; a c.var with small white corms* (JMD, Enti) amankani fufu *a c.var with white corms* (JMD) amankani kokoo = *red cocoyam; a c.var with pink corms* (JMD, Enti) amankani pa = *the real, or edible cocoyam; a c.var with pink corms* (JMD, Enti) antwibo *a c.var* (JMD) GA amankani *from Akan-Twi* (FRI) GBE-VHE amankani (FRI) **TOGO** GBE-VHE mankani (FB) **NIGERIA** EDO íyòkhó àkàrá = *the yam from Accra* (JMD) EFIK àta ḿkpọ̀ñ mbàkárá, idídùòt ḿkpọ̀ñ mbàkárá = *true, or red cocoyam of the European* (*m̀bàkárá*) (Winston) HAUSA gwaàzaà-maì-goòraá, gwaàzaá: *Colocasia esculenta*; gora: *reed, or bamboo — hence straightness, alluding to the tuber being more elongated than that of C. esculenta* (JMD) kan-birii = *monkey's head; a c.var* (JMD; ZOG) kúmaàtún mùzuúruú = *cat's cheek; a c.var* (JMD; ZOG) IBIBIO íkpộñ *correctly Colocasia esculenta* (Kaufman) IGBO édè eko (BNO) IGBO (Awka) édè ékò (KW) IGBO (Owerri) àkàsị óyìbó = *cocoyam of the European* (JMD) édè árọ̀ (JMD) édẹ bẹkẹè = *cocoyam of the European* (JMD) édẹ ọ̄kpọ́rọ́ = *ordinary cocoyam* (JMD) ǹkàsí bèkéè = *cocoyam of the European* (JMD) IGBO (Umuahia) édè ọ̀hị̄á = *bush cocoyam* (JWD; KW) ukurukuru *a nickname* (JMD) IJO-IZON (Kolokuma) bèké òdù = *European cocoyam* (KW) URHOBO údū òyìbó = *cocoyam of the European* (Faraclas) **WEST CAMEROONS** VULGAR makara *the plant* (JMD) mankamo (JMD) BAFOK makao (JMD) DUALA dìkàbò (Ithmann) KOOSI mbanga (JMD) KPE nda *the tuber* (JMD) KUNDU bende (JMD) LONG makabo (JMD) LUNDU mesengu (JMD) MBONGE nda-mukala (JMD) TANGA bamboko (JMD) WOVEA nda *the tuber* (JMD)

*NOTE: Names given under Colocasia esculenta (Araceae), the true cocoyam, may often apply.*

A fleshy herbaceous plant, stemless or briefly caulescent with leaves arising from the crown of a central corm usually surrounded by a mass of cormels, native of the northern part of S America and distributed by man to many tropical countries.

The genus contains a number of species and several have been domesticated in tropical America from an early time. The plant has been confused with *Colocasia esculenta* because of a close likeness in appearance. A ready distinction lies in the junction of the leaf lamina with the petiole: in *Xanthosoma* it is on the margin, and in *Colocasia* towards the centre. Though Spanish and Portuguese explorers were fully cognisant of the plant, its dispersal outside the New World by their agency is undocumented. The plant appears to have reached W Africa from the W Indies only in 1843. The bulk of the material in W Africa has been taxonomically ascribed to *X. mafaffa* though it would be most surprising if *X. sagittifolium* (Linn.) Schott, which has been dispersed throughout the tropics, was not also present. They are considered equivalent here.

The plant is grown for its edible root in the same way as is the coco yam (*Colocasia esculenta*) and for their similarity this is known as the 'new coco yam.' Its exotic origin is also recognized in other vernacular names ascribing it in Nigeria to European or Ghanaian sources. In Ghana however it has acquired such importance as to be known as 'the yam of the nation.'

The main corm is often not entirely free of acridity, but the cormels usually are and are palatable. In the West Indies *Colocasia* corms are preferred, but in West Africa the choice is reversed. The tuber is much in demand for making into *fufu,* and in Ghana the plant is found useful as a nurse-crop for cacao (5). Consequently in Ghana and elsewhere in the Region where soil and climate are suitable it is displacing the true cocoyam in the traditional pattern of food production. Unlike cocoyam it cannot stand waterlogging. It requires a rich deep well-drained soil, pH 5.5–6.5, and a regularly distributed rainfall of at least 1000 mm over its growing period. It is a 9–12 months crop, but a crop can be taken sometimes in 6 months, and a 'cut-and-come-again' husbandry may be possible for up to six years before a decline in yield sets in (4). The composition of the corms has been given as approximately: water 70–77%; protein 1.3–3.7%; fat 0.2–0.4%; carbohydrate 17–26%; fibre 0.6–1.9%; and ash 0.6–1.3% — not far different from that of cocoyam corms (5; also see 1). A number of varieties exists based on yield, the colour of the flesh or skin, size of corm, storage quality, palatibility, etc. Starch grain are larger than those of true cocoyam, a character useful for distinguishing the two apart. Storage in a well-ventilated space is possible up to six months, but usually a loss of quality becomes apparent much earlier. In the Cameroons storage in enclosed pits has been found better than on trays in good ventilation. Bruised corms will not store (4).

The young leaves are eaten as a green vegetable (2; 3; Gabon, 6). The older leaves tend to be bitter and in Gabon are used together with other leaves as a fish-poison (6).

References:

1. Busson, 1965: 521, with corm analysis. 2. Dalziel, 1937. 3. Irvine, 1956: 35: as *X. sagittifolium*. 4. Kay, 1973: 160: as *X. sagittifolium* Schott. 5. Purseglove, 1972: 1: 70–74: as *X. sagittifolium* (Linn.) Schott. 6. Walker & Sillans, 1961: 97: as *X. sagittaefolium* Schott.

**Xanthosoma violaceum** Schott

English: blue taro, blue-stemmed taro (U.S.A., Bates).

Similar to *Xanthosoma mafaffa* but with violet leaf-margins and primary lateral leaf-veins, native of northern S America and dispersed to a number of tropical countries.

It produces corms of excellent eating quality and like *X. mafaffa* is becoming more commonly grown in W Africa in place of *Colocasia esculenta* (1).

Reference:

1. Busson, 1965: 521–2, with corm analysis.

**Zantedeschia angustiloba** (Schott) Engl.

FWTA, ed. 2, 3: 120.

A marsh-herb with leaves and inflorescence to 60 cm or more high, and flower-spathe a conspicuous yellow, occurring on the Mambila Plateau of N Nigeria, and in Angola, Tanganyika and Zambia.

No usage is recorded in Nigeria. Two species in Lesotho yield dyestuffs: *Z. africana* (=? *Z. aethiopica* (Linn.) Spreng.), a good dye for wool, and *Z. albomaculata* (Hook. f.) Baill. (syn. *Z. oculata* (Lindl.) Engl.), a yellow-green dye. Both are eaten as a pot herb. The W African species merits examination (1).

Reference:

1. Guillarmod, 1971: 458.

# ARALIACEAE

**Cussonia arborea** Hochst.

FWTA, ed. 2, 1: 750–1, as *C. barteri* Seeman. UPWTA, ed. 1, 345, as *C. barteri* Seem., *C. djalonensis* A. Chev., *C. nigerica* Hutch. in part and *C. longissima* Hutch. & Dalz.

West African: **SENEGAL** BASARI a-ŋeurɔ̀t (Ferry) BEDIK ga-pɛs, gi-ndam (Ferry) MANDING-BAMBARA bolo koro (JB) **GUINEA-BISSAU** CRIOULO papaia do mato (JDES) **GUINEA** MANDING-MANINKA bulukuntu (Aub.) SUSU bulukuntu (JMD) bulukunyu (JMD) **SIERRA LEONE** MANDING-MANDINKA bolo-kundo (FCD) SUSU-DYALONKE kanya-kanyen-na (FCD) kɛnya-kɛnyɛ-na (FCD) **MALI** MANDING-BAMBARA bolokoloni (Aub.) bolo-koro ni = *stump of an amputated limb* (A. Chev.; Aub.) bolo-kuruni (Aub.) dzinjama (A. Chev.; Aub.) MANINKA bulukuntu (Aub.) SONGHAI karebanga (Aub.) **UPPER VOLTA** MANDING-BAMBARA bolo-kuruni (Aub.) bulu kuruni = *cut hand* (K&B) bulu-kuntu (RS) DYULA bobo (K&B) borukunu (K&B) botuo (K&B) **IVORY COAST** AKAN-BRONG korozingia (K&B) ANYI borokum (K&B) kuao béré félé (Aub.) BAULE akongo (Aub., ex K&B) akuampgo (Aub., ex K&B) bobo (Aub., ex K&B) boglo (Aub., ex K&B) bombo (Aub., ex K&B) essuipugbo (B&D) KRU-GUERE (Chiehn) zakorakwéssu (B&D) KULANGO bogotéï (K&B) KWENI kulédié iri = *tree which kills the tortoise* (K&B) MANDING-DYULA bobo (K&B) botuo (K&B) brukunu (K&B) MANINKA n-gboto (A&AA) SENUFO-TAGWANA fittuwo (K&B) **GHANA** AKAN-ASANTE besimankuma (auctt.) kokobidua (FRI) kurummɔtɔ (FRI) BRONG koronziga (FRI) tufu-tufoi (CV) TWI saa-borɔfere, saa: *plain;* brófre: *papaya; = papaya of the savanna* (FRI) BAULE a-kponkpo (FRI) DAGBANI gangulagu (Saunders) GBE-VHE avlɔ (FRI) awlua (FRI) GUANG-GONJA sámáŋdóró (Rytz) KRACHI kantoboa (auctt.) KONKOMBA ogulongu (Gaisser) **TOGO** BASSARI kigalongo (Gaisser) GBE-FON gothi (Aub.) gotti (Volkens) VHE adiméti (Aub.) fegblo (Volkens) KABRE abu (Gaisser) sondeterĕ (Gaisser) KPOSO obbo (Volkens) MOORE-NAWDAM borogo (Gaisser) SORUBA-KUYOBE triho (Aub.) TEM (Tshaudjo) gongolu (Volkens) kongoli (Volkens) kongolu (Volkens) 'MISAHOHE' bonugu (Volkens) **DAHOMEY** BATONNUN chebulu (Aub.) sainburu (Aub.) sainhunu (Aub.) GBE-GEN hiovoclo (Aub.) SORUBA-KUYOBE triho (Aub.) YOM golur (Aub.) YORUBA-NAGO sigo (Aub.) **NIGERIA** FULA-FULFULDE (Nigeria) bumarlahi (auctt.) HAUSA gwabsa (JMD) hánnúm kútúruú = *leper's hand; a nickname* (JMD) tàkàndár giíwaá = *elephant's sugarcane* (auctt.) TIV ityovor (JMD; KO&S) YORUBA ako-sigo (JRA) ṣigo (auctt.)

A tree to 10 m high, with tortuous trunk, very thick corky bark, of the savanna from Guinea to S Nigeria, and extending into E Africa.

In the dry season the tree becomes completely defoliated, and the thick stumpy branches resembling amputated and deformed limbs sticking up into the sky invoke the Bambara name in Mali 'stump of an amputated limb' and in Upper Volta 'cut hand' (12), and the Hausa name in N Nigeria 'leper's hand'.

The wood is dirty white, soft, brittle and rots easily. As timber it is of little value. Some tribes hollow it out to make quivers (9). The wood of this or another species (vernacular: *indoabaku,* Mangu) in Togo is used to make cases in which to keep gunpowder and cartridges dry (Mellin fide 9). The Madi of Uganda use the wood to make trumpets (7).

A black ink used by Malams is prepared from a wood-decoction (14). The wood-ash is rich in potash. It is used in Togo and in northern Ghana to mordant indigo dyes (Coull fide 9). In Guinea the ash is used to make soap with the oil of *Carapa procera* (Meliaceae) (14).

The bark when slashed exudes a gum which appears to be somewhat variable: that of *C. nigerica* is said to form into 'pencils' (9) and that of *C. longissima* is a sticky ropy juice (18). The gum has a slightly irritant property (Dalziel fide 11).

From the likeness of the defoliated tree to deformed limbs, the plant is used, on the Theory of Signatures, in the treatment of leprosy. In Mali the stem is macerated and taken as a purgative and applied as a lotion (8). It is also used in Upper Volta (12) and in Ivory Coast (5) where the leprous sores may also be dressed with powdered stem-bark (2). Leafy twigs are used in Ivory Coast-Upper Volta with magical rites for 'yellow fever', oedemas, paralysis and sleeping-sickness (12). The plant is an emeto-purgative and diuretic by which actions cure is sought. It is also prescribed as a poison-antidote (5). Leaf-decoction is used as an eye-wash for conjunctivitis in Ivory Coast (2). Root and stem-decoctions are given in Nigeria for painful menstruation, and a root-decoction as an emetic for biliousness (3). The root is prepared in vapour-baths in Tanganyika in the treatment of gonorrhoea (4).

The water in which leaves have been boiled is purgative and is taken in Nigeria for constipation, and a leaf-decoction is used as a massage in cases of epilepsy (3). The latter application is found also in Ghana for treating epileptic children (10). Pulped up young shoots are eaten in Ivory Coast for diarrhoea (2).

The plant has a number of unspecified medicinal uses in Mali (13) and unspecified parts are used to treat urethral discharge in women, and are taken by men as an aphrodisiac (8). The plant has use against fever in Tanganyika (15).

In Zambia the fruits are said to be eaten in time of dearth (6).

The plant does not appear to have any marked physiological activity. It is more or less atoxic, and chemical analysis of Ivorean material has shown no alkaloid, glycoside or saponoside present (12). A trace of alkaloid is reported in Nigerian material (1).

References:

1. Adegoke & al., 1968: as *C. barteri* Seem. 2. Adjanohoun & Aké Assi, 1972: 48, as *C. barteri* Seemann. 3. Ainslie, 1937: sp. no. 121, as *C. nigerica*. 4. Bally, 1937. 5. Bouquet & Debray, 1974: 48, as *C. barteri*. 6. Bradley 1690, K. 7. Brasnett 307, K. 8. Chevalier, 1937, b: 170, as *C. nigerica* Hutch. and *C. longissima* Hutch. & Dalz. 9. Dalziel, 1937. 10. Irvine, 1930: 140, as *C. longissima* Hutch. & Dalz. 11. Irvine, 1961: 576, as *C. barteri* Seem. 12. Kerharo & Bouquet, 1950: 172, as *C. djalonensis* A. Chev. 13. Laferrere 122, K. 14. Portères, s.d. 15. Tanner 5246, K.

**Cussonia bancoensis** Aubrév. & Pellegr.

FWTA, ed. 2, 1: 751.UPWTA, ed. 1, 345, as *Cussonia* sp. under *C. nigerica* Hutch.

West African: **IVORY COAST** ABE ringhalla (A&P; Aub.) BAULE bongo (B&D) GAGU bango (B&D) KRU-BETE kodufleufleu (Aub.) KYAMA n-komi popossi (A&P; Aub.) **GHANA** VULGAR kwaebrofere (DF) AKAN-AKYEM abubu adenkum (FRI) ASANTE kwae-bɔɔfrɛ = *forest papaya* (auctt.) tuntun (BD&H; FRI) TWI kwae-bɔɔfrɛ = *forest tree like papaya* (Enti)

A tree to 30 m high by 3 m girth of the closed-forest of Ivory Coast, Ghana and Nigeria.

The wood is very soft, light, and with a large pith. It is perhaps somewhat elastic rendering it suitable in Akim, Ghana, for making side-drums. It is also used in Ghana for making women's combs, handles of chiefs' swords (*afona*) and the ribs of chiefs' umbrellas (2). The wood-ash is said to be used in Ivory Coast for making soap (3).

The fresh flowers are malodorous and attract many flies (1).

References:

1. Aubréville, 1959: 3: 98. 2. Irvine, 1961: 574. 3. Oldeman 323, K.

**Polyscias fulva** (Hiern) Harms

FWTA, ed. 2, 1: 750. UPWTA, ed. 1, 345.

West African: **WEST CAMEROONS** KPE ekwo (AJC) 'BAMILEKE' akukwai (Johnstone)

A tree to 26 m or more high of montane forest in Guinea, Ghana, Nigeria, W Cameroons and Fernando Po, and widely dispersed across Africa.

The wood is white and very soft and is said to be practically useless (Kenya, 3; Uganda, 4). It is reported to be used in the Kigezi Province of Uganda for making harps (4)! In W Cameroons it has been recorded as used to make doors (2) and burnt as firewood (1).

References:

1. Carpenter 101, K. 2. Dalziel, 1937. 3. Semsei 1195, K. 4. Styles 230, K.

**Polyscias guilfoylei** (Cogn. & March.) Bailey

English: 'wild coffee' (Bailey).

West African: **SIERRA LEONE** KRIO anjɛlika = *angelica* (FCD)

A shrub to about 5 m high of eastern Malesia and Polynesia which has been introduced to many tropical countries. It is ornamental with great variability of the foliage. It can be grown as a screen or hedge. The leaves have a smell like parsley whereby it has acquired by association the Krio name in Sierra Leone *anjɛlika*. The leaves are widely used in Malesia as a flavouring. Leaves and roots are diuretic and a decoction is given in Java for stone and gravel (1).

Reference:

1. Burkill, 1935: 1795.

**Schefflera barteri** (Seem.) Harms

FWTA, ed. 2, 1: 751, incl. *S. hierniana* Harms.

West African: **SIERRA LEONE** MENDE nja-gbamei (GFSE, ex Tennant) **IVORY COAST** DAN tuanli (Aub.)

A tree or epiphytic shrub in swamp-forest or on rocky hills of the forest zone, widely dispersed from Guinea to W Cameroons, and extending to Zaïre and into E Africa.

It is planted in W Cameroon as a hedge (3). The roots possess a toxicity (4), and liquid in which they have been boiled is drunk in Tanganyika by women just before giving birth as an ecbolic (2).

The flowers are very attractive to insects (1).

References:

1. De Wilde 2282, K. 2. Greenway 6030, K. 3. Keay & Lightbody FHI. 28519, K. 4. Koritschoner 1559, K.

# ARAUCARIACEAE

**Araucaria columnaris** (Forst.) Hook.

An exotic species of Polynesia which has been introduced (under the name of *A. cookii* R. Br.) into the highland area of West Cameroons. In its own habitat it attains a height of 60 m in a compact columnar habit as its specific name implies. It is grown in warm temperate and sub-tropical countries for decorative purposes. Its timber does not appear to have any commercial value outside its native country (1).

Reference:

1. Harrison, 1966: 113.

**Araucaria heterophylla** (Salisb.) Franco

English: Norfolk Island pine.

An exotic species of Norfolk Island in the Pacific which has been introduced into the highland area of West Cameroons (under the name of *A. excelsa* R. Br.), and perhaps to other similar plateau areas. In its own habitat it attains 60 m height, and is of extreme grace. It is now a popular ornamental tree in many subtropical countries. A large number of cultivars are recognized but distinctions are not very clear (1). It is being extensively cultivated in S Africa where it may become an important timber tree.

Reference:

1. Harrison, 1966: 114.

# ARISTOLOCHIACEAE

**Aristolochia albida** Duchartre

FWTA, ed. 2, 1: 81. UPWTA, ed. 1, 16.

English: Dutchman's pipe.

West African: **NIGERIA** HAUSA dúmán duútseè = *gourd of the rocks* (auctt.) fiyaka (JMD; ZOG) gaɗahuka, gaɗakuka, gaɗaukuku *from Fula?* (JMD) kadacin kasa (ZOG) mádaàcín ƙásà = *medicine of the earth* (auctt.)

A twining climber of the Sahel zone of the Region from Senegal, Mali, N and S Nigeria, and also Chad and Angola.

An infusion of the dried leaves, sometimes with dried root added, is used in Nigeria by Hausa and Fula as an anthelmintic (1). The leaf is applied in Nigeria to certain (unspecified) painful skin-diseases (2), and crushed and mixed with castor-oil is applied topically on pimples (1). To get rid of guinea-worm, the leaf may be applied, or a poultice composed of powdered root with seeds of *Lepidium sativum* Linn. (Cruciferae), garlic and native natron, and an infusion of the same mixture is drunk (2).

The root is bitter. It is sold in markets in light-coloured pieces 8–10 cm long for taking as a stomachic and tonic for which an infusion is made by pouring water repeatedly on to it through a strainer (2). The root mixed with lime-juice is given in cases of snake-bite, scorpion-stings, etc., against which the flowers are sometimes worn as a juju or charm (1). In local medicine the root of this is probably sometimes confused with that of *Cissampelos* (Menispermaceae) (2), and appears to be interchangeable with *A. bracteolata* Lam.

Root of *A. reticulata* was official in the *British Pharmacopoea 1932* and the *British Pharmaceutical Codices* of 1934 and 1949 for use in the preparations of a compound bitter. The pharmacological activity of *A. albida* merits examination.

References:

1. Ainslie, 1937: sp. no. 39 (as *Aristolochia* spp.). 2. Dalziel, 1937.

**Aristolochia bracteolata** Linn.

FWTA, ed. 2, 1: 81. UPWTA, ed. 1, 16 (as *A. bracteata* Retz.)

English: Dutchman's pipe.

West African: **NIGERIA** HAUSA dúmán duútseè (JMD; ZOG) fiyaka (JMD) gaɗahuka, gaɗakuka, gaɗaukuka *from Fula?* (JMD) kadacin kasa (ZOG) mádaàcín ƙásà (JMD; ZOG)

A small glabrous shrub occurring in the Sahel zone of the Region from Mali to N Nigeria, and in Tropical E Africa, Arabia and India.

An infusion of the dried leaves, sometimes with the dried root added, is used in Nigeria by Hausa and Fula as an anthelmintic (1), a use that is also known in India (4). The freshly bruised leaves are mixed with castor-oil and used in Nigeria topically on pimples (1). In India the plant is used to treat scabies (4), and in the Ogaden of Ethiopia on leg-itch (3).

The root is bitter. Roots mixed with lime-juice are taken for snake-bite, scorpion-stings, etc., in N Nigeria (1). East of Lake Chad also the root is applied to scorpion-sting (2).

The flowers are sometimes worn in N Nigeria as a juju or charm against snake-bite and scorpion-stings (1).

The fresh root yields two acidic crystalline compounds, one with bright yellow needles, m.p. 275–7°C, is identical with *aristolochic acid,* the other has orange yellow needles, m.p. 240–52°C. The seeds also contain the same two substances and also a greenish-brown non-drying fixed oil (4). An unnamed alkaloid is reported present in the root and stem of Indian material (5).

The vernacular names and uses of this species are probably similar to those of *A. albida* with which it is doubtless inter-changeable.

References:

1. Ainslie, 1937: sp. no. 39 (as *Aristolochia* spp.). 2. Dalziel, 1937. 3. Getahun, 1975: (as *A. bracteata* Retz.). 4. Watt & Breyer-Brandwijk, 1962: 118 (as *A. bracteata* Retz.). 5. Willaman & Li, 1970: (as *A. bracteata* Retz.).

**Aristolochia brasiliensis** Mart. & Zucc.

FWTA, ed. 2, 1: 81.

An exotic climber from tropical America, introduced and under cultivation as an ornamental in Sierra Leone (1). Perhaps becoming sub-spontaneous in Zaïre (2).

References:

1. Deighton 3892, K. 2. Hauman, 1948: 382.

**Aristolochia elegans** Mart.

FWTA, ed. 2, 1: 81.

English: Dutchman's pipe (Irvine).

An exotic from tropical America and cultivated as an ornamental at places in Sierra Leone, Ghana and Nigeria, and doubtless elsewhere. Perhaps becoming sub-spontaneous in Zaïre (1).

Reference:
1. Hauman, 1948: 382.

**Aristolochia gibbosa** Duch.

FWTA, ed. 2, 1: 81.

An exotic climber from tropical America introduced as an ornamental and under cultivation in Sierra Leone and Ghana.

**Aristolochia ridicula** N.E. Br.

FWTA, ed. 2, 1: 81.

An exotic from tropical America, introduced and cultivated in the Aburi Botanic Gardens, Ghana.

**Aristolochia ringens** Vahl

FWTA, ed. 2, 1: 81.
French: bel de Calao.

A bushy climber, native of tropical America, introduced to most West African countries as a garden ornamental. It has become naturalized in roadside bush in Sierra Leone, and in many places in Nigeria.

**Paristolochia flos-avis** (A. Chev.) Hutch. & Dalz.

FWTA, ed. 2, 1: 79
West African: **GHANA** AKAN-ASANTE sunai (CV; FRI)

A stout liane to over 30 m high of the dense forest from Sierra Leone to W Cameroon, and in E Cameroun and Equatorial Guinea.

**Paristolochia goldieana** (Hook. f.) Hutch. and Dalz.

FWTA, ed. 2, 1: 79. UPWTA, ed. 1, 16.
West African: **SIERRA LEONE** TEMNE *a*-bare (JMD) **GHANA** AHANTA aheriheri (BD&H) AKAN-ASANTE atchweni-monta (CV) otwene monta (FRI) TWI kotoku saabore (FRI) **NIGERIA** EDO ugbogiẹlimi *name of the black hat of the Mother of Ovia, the founder of a secret society* (NWT) úgbógiórìnmwìn = *grove of the King of the dead from*; úgbó: *a grove or field*; ógie: *a king*; órìní: *a corpse, or the dead* (A.H.Green) EFIK ùbọ́ñ-édòp = *antelope's ubong, or wild ubong, from* ubong: *Telfairia spp., which has gourds of shape similar to its flowers* (Thompson) IGBO ekommili = *fragile bellows from* eko: *bellows*; mmili: *water or anything soft, weak or fragile* (A. H. Green)

A forest climber to 7 m long from a thickened rootstock, occurring in Sierra Leone, N and S Nigeria, Fernando Po and E Cameroun.
The Igbo name, *ekommili*, is derived from ɛ*ko*, bellows; *mmili*, water or anything soft, weak or fragile: hence 'fragile bellows'. The miniature flowers before the perianth splits open resemble primitive bellows used by local black smiths. The Edo name is of more fanciful origin. *Ugbogiorimi* is a compound word from *ugbo*, a grove or field; *ogie*, a king; *orimi*, a corpse or the dead: hence 'The grove of the King of the Dead'. The meaning of this is not known.

Some say it is on account of the smell of decaying flesh of the mature flower. Another version is that the opened flower is the haunt of a small venomous viper that uses the smell to attract its prey (2). *Ugbogielimi,* also an Edo name, is after the name given to the black hat of the mother of Ovia, the founder of a secret society amongst the Edo people (3).

The plant is cultivated in hot houses (1) in temperate countries.

References:

1. Dalziel, 1937. 2. Green, 1951. 3. Thomas, N. W., 1910: 1: 38.

**Paristolochia promissa** (Mast.) Keay

FWTA, ed. 2, 1: 79. UPWTA, ed. 1, 16 (as *A. flagellata* Stapf).

A woody climber of the forest attaining 10 m long, occurring in Ghana, S Nigeria and W Cameroon, and in E Cameroun and Zaïre.

No use has been recorded in the Region. In Zaïre the stems are made into snares and traps for small mammals (1).

Reference:

1. Hauman, 1948: 384–7 (as *Aristolochia congolana* Hauman).

## ASCLEPIADACEAE

**Asclepias curassavica** Linn.

FWTA, ed. 2, 2: 92.

English: blood-flower; red-head; swallow-wort; West Indian ipecacuanha; wild ipecacuanha.

French: asclepias de Curaçao.

West African: GUINEA-BISSAU MANDING-MANDINKA bafurma (JDES)

A herbaceous shrub to nearly 2 m high, native of tropical America, and widely dispersed throughout the tropics by man as an ornamental. Known only under cultivation in the Region throughout from Senegal to Fernando Po.

The plant first came to the knowledge of the Western World during the Spanish Voyages of Discovery when it was seen in Mexico where it was used as an emetic and purgative. It was brought to Europe and dispersed by the Spanish as a substitute for ipecacuanha. The root contains a glycoside, *asclepiadin,* to which those properties are due. In large doses it causes death, and the plant may be a suspect cattle-poison (1, 6).

The plant contains an abundant white latex which is used in the Caribbean and in the Pacific basin on warts and corns (4). The plant is considered cicitrisant in Madagascar (3). Tests of plant extracts on avian malaria have shown no activity (2).

The stems contain a fibre which can be spun. The seed-pods have a floss surrounding the seeds. This is too elastic for spinning unless altered by chemical treatment after which it can be spun admixed with cotton (1). In some places in West Africa the stems are used tied in bundles as brooms (Dalziel fide 5). The floss can be used for filling pillows.

Honey made from the plant in Guyana is reported bitter, dark and thick (1).

References:

1. Burkill, 1935: 261–3. 2. Claude & al., 1947. 3. Debray & al., 1971: 37. 4. Hartwell, 1967. 5. Irvine, 1961: 645–6. 6. Oliver, 1960: 19, 48.

**Aspidoglossum interruptum** (E. Mey.) Bullock

FWTA, ed. 2, 2: 93.

An annual herb to 40 cm high from a woody, rootstock, of savanna grassland of the northern part of the Region from Guinea–Bissau to W and widely dispersed in savanna lands throughout tropical and South Africa.
The root is considered edible in Lesotho (1).

Reference:

1. Watt & Breyer-Brandwijk, 1962: 137.

**Brachystelma bingeri** A. Chev.

FWTA, ed. 2, 2: 199, as *sp. dub.*, perhaps=*Pentagonanthus* (Periplocaceae).
UPWTA, ed. 1, 383.

West African: **SENEGAL** DIOLA hiama (JMD) FULA-PULAAR (Senegal) nda-fegué (JMD) MANDING-BAMBARA fié (JMD) fié-gué (JMD) fié-hié (JMD) figué (JMD) MANDINKA fikongo (JMD) **MALI** FULA-PULAAR (Mali) nda-fegué (JMD) MANDING-BAMBARA fié (JMD) fié-gué (JMD) fié-hié (JMD) figué (JMD)

A plant of doubtful identity recorded from the savanna of Mali and Dahomey.
The tuber is 5–7.5 cm or more in diameter, with white flesh exuding latex when cut. The taste is slightly bitter, due to a resinous substance. The flesh consists largely of carbohydrate matter, and is refreshing rather than nutritious. It is eaten raw in the quite fresh state, after removing the more milky resinous outer layer. Europeans have used it boiled in water as a substitute for turnips. Several other S African and Ethiopian species have edible tubers (1). The W African spp. are more by way of a famine-food (2).

References:

1. Dalziel, 1937. 2. Irvine, 1952, b.

**Brachystelma constrictum** J. B. Hall

A perennial herb arising from a disc-shaped tuber 6 cm diameter by 2 cm thick, branches decumbent to 10 cm long, occurring in open grassy areas of submontane forest in Ghana and N Nigeria. The tuber is reported used by the Fula of N Nigeria to treat stomach-ache (1).

Reference:

1. Daramola FHI. 62731, K.

**Brachystelma togoense** Schltr.

FWTA, ed. 2, 2: 99.

An erect perennial herb, to 30 cm, recorded from Ghana to N Nigeria, in lowlands to montane situations.
The tuber is said to be edible raw (1). The flowers are evil-smelling.

Reference:

1. Keay FHI. 25856, K.

**Calotropis procera** (Ait.) Ait. f.

FWTA, ed. 2, 2: 91. UPWTA, ed. 1, 384–6.

English: auricula tree; Dead Sea apple; Sodom apple; swallow–wort.

French: arbre à soie (tree with floss, *soie* silk); arbre à soie du Sénégal; pomme de Sodome (Sodom's apple, but referring properly to some *Solanum spp.*).

West African: **MAURITANIA** ARABIC (Hassaniya) turja (AN) ARABIC ('Maure') achur (Aub.) kerenka (Aub.) korunka (Aub.) turdja (Aub.) turjé (Aub.) BERBER taurja (Aub.) tourjé (Aub.) **SENEGAL** ARABIC ushar (GR) ARABIC ('Maure') achur (Aub.) kerenka (Aub.) korunka (Aub.) turdja (Aub.) turjé (Aub.) BASARI a-tápúŋ (K&A; Ferry) BEDIK gi-potár (Ferry) CRIOULO bombardeira, bomborderu (JMD) bordéru (Aub.) DIOLA bupumba pumb (JB, ex K&A) FULA-PULAAR (Senegal) babadi (K&A) bamâbi (K&A) bamanbé (Aub.) bamanbi-bauwami (Abdoul Oumar) bandambi (Aub.) bauane (Aub.) bawam bawam (K&A) bawoam (K&A) kupâpâ (K&A) TUKULOR baamwaani (Ba) bawane (JMD) KONYAGI mbontal (Ferry) MANDING-BAMBARA fogofoko (JB) m-pôpôpogolo (K&A) ngeyi (JMD) MANINKA fugofogoiri (Aub.; K&A) mpompompogolo (Aub., ex K&A) ngeyi (K&A) ngoyo (auctt.) 'SOCE' dimpâpaô (K&A) SERER bodafot (K&A) mbadafot (auctt.) SERER-NON hugé (auctt.) ôgu (auctt.) NON (Nyominka) bodafor (K&A) imbudafot (K&A) SONINKE-SARAKOLE tulumpa (K&A) WOLOF faftan (auctt.) fafton (Aub.; K&A) paftan (Abdoul Oumar) **THE GAMBIA** FULA-PULAAR (The Gambia) bawane (DAP) MANDING-MANDINKA kupampaŋ (*def.*-o) *onomatopoeic: children burst the fruits like rubber balloons* (DF) WOLOF faftan (DAP) **GUINEA-BISSAU** BALANTA bagueuóne (JDES) n-olim-nhe, um-olim-nhe (JDES) n-olininhe (EPdS; JDES) CRIOULO bombardeira (EPdS; JDES) DIOLA-FLUP belápse (JDES) FULA-PULAAR (Guinea-Bissau) pama (EPdS; JDES) MANDING-MANDINKA cumpampam (*def.*-ô) (JDES) kupampaŋ (JDES) pampam i.e., *kupampaŋ* (JDES) pópo (*def.*-ô (auctt.) MANKANYA belápse (EPdS; JDES) PEPEL bfó (EPdS) bfô, bió, ufô (JDES) SUSU bussuma (JDES) **GUINEA** MANDING-MANINKA mpompomogolo (Aub.) nguyo (Aub.) **SIERRA LEONE** KRIO inglish-kɔtin = *English cotton* (FCD) MENDE puu-vande (*def.*-i) = *foreign, or English cotton* (FCD) **MALI** ARABIC (Sahara) korunka (Barth) DOGON pòbu (C-G) FULA-PULAAR (Mali) bamanbé (Aub.) bandambi (Aub.) bauane (Aub.) bawane (GR) MANDING-BAMBARA nyeyi (JMD; Aub.) MANINKA mpompompogolo (Aub.) ngoyo (JMD) nguyo (Aub.) SONGHAI turdja (A. Chev.) turia (A. Chev.) TAMACHEK krunka (Barth; RM) tezera (Aub.) tirza (Aub.) toreha (Barth) torha (Barth) toucha (Aub.) tourha (Barth) tourjé (Aub.) tursha (Barth) **UPPER VOLTA** BISA hurégo (Prost, ex K&B) FULA-FULFULDE (Upper Volta) ganganpi (K&B) HAUSA tomfania (K&B) KIRMA diawara (K&B) MANDING-BAMBARA tomo n'déké (K&B) DYULA furo fogo (K&B) tumo tigi (K&B) MOORE potu (K&B) putrempugu (K&B) putru pouga (K&B) puwo (K&B) 'SENUFO' niapi djara (K&B) **IVORY COAST** FULA-FULFULDE (Ivory Coast) ganganpi (K&B) MANDING-DYULA furo fogo (K&B) tumo tigi (K&B) MANINKA togo fogo (K&B) SENUFO-NIAGHAFOLO nopiada (K&B) 'SENUFO' niapi djara (K&B) **GHANA** AKAN-TWI mpatu-asa? (FRI) DAGBANI wolaporhu (Volkens & Gaisser) wolapugo (Volkens & Gaisser) GA blɔfo tɔtɔ (FRI; KD) gbé'kẽbii-awuɔ = *children's fowl; alluding to the floss like feathers* (FRI) owula kofi ba (FRI) GBE-VHE a-gbo-loba (FRI) wutsoe-wutsoe (FRI) GUANG-GONJA pòlípòlí (Rytz) KONKOMBA unablapong (Volkens & Gaisser) **TOGO** ANYI-ANUFO tambutiji (Volkens&Gaisser) BASSARI inawokodu (Volkens; Gaisser) KABRE kudjohĕ (Volkens & Gaisser) tschofu (Volkens & Gaisser) TEM (Tshaudjo) tshawŏu (Volkens; Gaisser) **NIGER** ARABIC (Niger) eshar (Aub.) esshero (Aub.) koruga (Aub.) ochar (Aub.) oshar (Aub.) rhalga (Aub.) FULA-FULFULDE bamanbé (Aub.) bandambi (Aub.) bauane (Aub.) HAUSA tumfafia (Grall; Aub.) SONGHAI sáagéy (*pl.* -à) (D&C) TAMACHEK tezera (Aub.) tirza (Aub.) toreha (Barth) torha (Barth) toucha (Aub.) tourha (Barth) tourjé (Aub.) tursha (Barth) TUBU lifini (Aub.) TUBU (Kaningou) kayio (Aub.) kulunhun (Aub.) **NIGERIA** ARABIC ushar (JMD) FULA-FULFULDE (Nigeria) babambi (J&D; MM) bambambi (J&D; MM) bambami (JMD) bembambi (MM) tumpaapahi *from Hausa* (MM) GWARI pwom pwomohi (JMD) HAUSA bambambele (ZOG) tùmfaáfiyaá (auctt.) JUKUN kupa (JMD) KANURI kayôu (JMD) YORUBA bomubómú (auctt.)

A shrub to about 6 m high, occurring abundantly in arid conditions, often in near pure stands, throughout the northern part of the Region, and dispersed across N Africa to the Middle-East and India.

To some extent the plant is an anthropogene and it occurs commonly around villages, perhaps planted though it is not necessarily tended. Its presence in the bush may mark an abandoned village site and exhausted soil, but also its presence is held to indicate subsoil water, an aspect widely understood in the Hoggar of Central Sahara where water may be expected at 0.5–4.0 m depth (27), in Somalia (28) and Eritrea (7).

The fresh foliage is in general regarded as poisonous (25). It is not grazed by stock (2) except for want of better (Barth fide 10; 20; 30). There is conflicting evidence whether camels will or will not browse the plant. Goats appear to be partial to the flowers (Mauritania 30; Somalia 28; Kenya 36). However, if the leafy branches are cut and left to wilt overnight the foliage is rendered safe to feed to stock.

A sort of manna has been reported appearing on the plant in The Gambia and that this is used as an asthma cure (Dragendorff fide 37). The closely related *C. gigantea* R.Br. is the source of 'shukuri tigha' manna in India. *C. procera* R.Br. is in a list of Iranian manna plants (16).

In the western soudanian region two or three leaves placed in a hole in the sand through which water is percolated are said to clarify the water (10). In N Nigeria the Fula use them in an undisclosed way in pottery-making (21), and in Mauritania larger leaves are used to cover poultices while staining with henna (30).

The leaves and bark contain an abundance of a caustic latex which is vesicant, irritant and rubefacient. It rots clothing. It is used as a depilatory in tanning (4, 5, 10), and it will injure the coats of horses and other animals, even should they only rub against the bark (10, 29). It is commonly compounded into arrow-poisons (5, 8, 25, 26, 31, 37). It is violently purgative, and has widely featured in ordeal-poisons throughout Africa (5, 37). All races in Ivory Coast–Upper Volta recognize the latex as being dangerous to the eyes (26), but in Mauritania it is used for eye-troubles (30), and in N Nigeria and Ghana for treating conjunctivitis (10) inducing tears and a local anaesthesia (31). The leaves bound over the eyes may also be used in Nigeria (4) and The Gambia (33). In Senegal the latex is known for its antiseptic and sedative action (25), and an analgesic effect may be what is sought in the use of the latex for treating points of pain and discomfort in rheumatism, chest-complaints, toothache, etc. (4, 10, 37). In The Gambia the leaves are applied warm to sprains, headache and for other pains (38). Leaf-sap acts on the mucosae and it is squeezed into the nose for headache and catarrh bringing relief by causing sneezing (10). The latex is used on cutaneous affections such as ringworm and aphthous sores in the mouths of children (10). It is applied in Mauritania to open wounds (30); in Senegal to small injuries and the bites of various animals (24); in Nigeria to syphilitic ulcers, yaws, etc. (4), and in the soudanian region to lymphatic ulcers and camel's sores (5). A drop may be applied to a guinea-worm blister to help extract the worm, and in N Nigeria to a scorpion-sting, while someone about to handle a scorpion may rub some latex over his hands as a preventative (10). It is also applied with the latex of certain Euphorbias to 'yaws' in horses (10).

At least seven glycosides have been determined in the latex (8, 25, 31), all cardio-toxic, of which the principal one, *calotropin,* is said to be more potent than *Strophanthus* drugs and causes death by paralysis of the heart-muscles. The latex on standing separates into a resinous coagulum and a clear yellowish serum. Both retain toxicity, and it remains for very long in the resin (10). The latex has been used in Iran as an adulterant of opium (37). The Hausa burn the leafy twigs to make a smoke-inhalation for asthma, cough, etc., or the leaves are smoked in a pipe. The root after soaking in the latex is similarly used in India (10). The resin is smoked by asthmatics in the western Sahara (29). The powdered leaf has been used as a snuff in Nigeria and is said to give relief in cases of pulmonary tuberculosis (4). The leaf has been used in Senegal as an insecticide against poultry-lice (10). In Tanganyika it has been used against bed-bugs (37). The latex has also been used in the soudanian region against poultry-fleas (Curasson fide 5). The plant is held to be repellent to termites, and leafy branches are placed amongst the timber frames of huts in Upper Volta (26) and under floor-mats in Jebal Marra (19). Tests for insecticidal activity have shown only modest action (17). The leaves also enter into anthelmintic preparations (10). An aqueous macerate of three leaves was prescribed in the absence of other drugs in Ivory Coast–Upper Volta during the 1939–45 war as a vermifuge with apparent success (26).

The absence of alkaloids (? from the leaves) has been noted in Nigerian material (3). Alkaloids in the stem of Pakistani material is recorded. Besides the presence of glycosides in West African plants, the enzyme *trypsin* is present (37) and also a very active non-toxic proteolytic enzyme, superior to *papain* has been reported and named as *calotropain* (25). This has the property of curdling milk — more readily if boiled than when fresh. The Fula use the latex in cheese-making. Pagan tribes of N Nigeria add the latex to a corn mash to ferment it to beer (10), while in Jebel Marra leaves are used to cover the germinating grain in wort (19).

Though the plant attains but little height, the stems may reach nearly 1 m in girth (18), and are often thick enough, where other timber in arid areas may be lacking, to serve as joists on *Borassus* (Palmae) supports to carry mud-roofs. They are used for the foundations of large mud-built grain receptacles because the wood is resistant to termite attack (10). The wood is soft and spongy. In western Sahara it is cut into 'slates' for school children (29). The wood, or the pith alone, is used in western Mali (10) and in western Sahara as a fire-tinder when rubbed with a piece of hardwood, usually *Acacia sp.* (Leguminosae: Mimosioideae) (29). Similar use is recorded in Kenya (12). It makes a good charcoal which in central Sahara (37) and in the Chad area (as in India) is used for gunpowder with local nitre and imported sulphur, or else, for occult reasons, the plant is added, along with tamarind, onions and resins, in the preparation of gunpowder (10). In the Hoggar the charcoal is powdered and made into an ointment for treating camel's skin-troubles (27). Pieces of stem serve as floats for fishnets (10), and the Tuareg of central Sahara make the wood into saddles (37). It is said to be suitable for paper-pulp (10). Ash mixed with butter-milk is applied by the Masai to bull's horns as a polish (13).

The bark is fibrous, and in India *madar* fibre is extracted from the inner-bark. It is also produced in Sudan as *ushar* fibre. It is costly to produce as on account of the resin present retting is not possible. In India bark is beaten by hand (9). In Nigeria the bark is macerated for three days and the fibre separated by hand (10). If obtained properly, however, the fibre produces strong cordage which has been reported on fairly well (14). Though the fibre is on average about the same length as that of cotton it is more variable being longer and shorter, and this with an absence of twist might make spinning difficult. If machinery could be adopted a soft fabric of economic quality might be makeable, or the fibre might prove satisfactory admixed with cotton. It is produced here and there in the Region and elsewhere in Africa according to demand. It resists water and has found use in fishnets and lines. It was the standard fibre for fishnets on Lake Chad till displaced by the availability of nylon (35). It has been exported as 'kapok' from India but this is a falsification and a poor substitute. It is commonly used in E Africa for stuffing pillows (37).

The bark contains a bitter principle, *mudarin,* which is emeto-cathartic (5, 25). This substance, with others, is in the resin fraction of the latex. It is very toxic and has been used for criminal purposes (10, 29). In Senegal the emeto-cathartic property is used as a poison-antidote for which the powdered bark (or the leaves or root) is taken in curdled milk. Yet on the other hand the latex, trunk-bark and leaves are fed to cows to increase milk-yield (23). Dosage for internal usages is intentionally restricted because of cardio-toxicity. Powdered dried twigs are added to soup as a stomachic and antidiarrhoetic in northern Ivory Coast and Upper Volta, and the Bobo take these with tamarind juice as a strong diuretic (26). The Fula of N Nigeria take powdered bark for severe stitch (21). In Nigeria the bark, powdered and mixed with oil, is used as a liniment for rheumatism (4). At one time the latex and bark enjoyed a reputation in India as a remedy for syphilis sufficient for it to be given the name 'vegetable mercury' (10). Similarly the root was used in N Nigeria for syphilis after pounding and macerating for a day or two in water in which corn had been soaked, the liquid was drunk copiously, causing profuse vomiting and diarrhoea with discharge, as was believed, of the spawn of syphilis (10).

The root bark has other important medicinal uses. It has been used as a substitute for ipecacuanha and as a bitter tonic. It has been taken with beneficial results in at least mild cases of dysentery, but action on the amoebae of amoebic dysentery is doubtful (31). As in the stem-bark, *mudarin* is present. It is used by various tribes as a stomachic and medicine for colic. It is thought to be galactogenic. The roots are used as tooth-sticks and their use is said to relieve toothache (10). In Senegal the powdered root-bark quaffed in water is taken for snake-bite, and some of the liquid is used to wash the bite (1). Root-bark is used by several races internally and externally in treatment for leprosy (4, 6, 34) and for guinea-worm infection (34). Charcoal made from the roots and root-bark is prepared as an ointment for external application in Senegal to skin-eruptions, syphilitic sores, leprosy, foul ulcers, camels' sores, etc. (10, 25). The pungent, aromatic smell of the root-bark is thought to be snake-repellant and for this the root is crushed into small pieces and is scattered about dwellings in The Gambia (38).

Nectar from the flowers is said to be poisonous (37) so that one should view with caution honey from this source. Leaves and fruit are boiled together in Nigeria (4) and in Ghana (20) to make a decoction in which a guinea-worm affected limb is soaked repeatedly and for several hours to extract the worm. In Jebel Marra the fruit is used for treating eye-troubles (19).

The ripe fruit contains a floss, or 'vegetable silk' known as *madar floss* or *akund,* an Indian commercial term. It can be used for stuffing cushions but is a poor substitute for kapok. It does not withstand rough treatment nor waterlogging (10). It can be spun and in Sierra Leone is known as 'English cotton', i.e. introduced, and is made into a strong cloth (11). A cotton is also made from it in Somalia (28) but industrial attempts to use it have not been entirely successful owing to short staple length, black-spotting under high humidity, and drawbacks attendant on bleaching (9). In Niger (15) and in Mauritania (30) the floss is used as tinder in raising fire by friction.

Though *calotropin* is the primary bitter principle in the vegetative parts of the plant which together with other related glycosides are responsible for the plant's toxicity, it is present in the seeds in only very small amounts. The seeds yield a series of glycosides of a different sort of which the most abundant is *coroglaucigenin,* 0.437%. Examination of Nigerian material showed non-toxic action on the heart (25, 37).

The plant is endowed with special powers in witchcraft, both for and against. In N Nigeria the name of a person to be frightened is written in cock's blood on a leaf. A spell arising from a brew of some other plant may be countered with an antidotal concoction made the stronger and more potent for being cooked over twigs of *bambami* (Hausa). Witches are believed to detest the smell of the plant. (39). The Nyominka of Senegal use the plant as a fetish to confer protection of their huts by fastening branches over the door (24). Perhaps this is in the same line of thought as the use of the plant as a snake-repellent. The plant is also placed over doorways in Senegal as a protection against witchcraft (10).

References:

1. Abdoul Oumar, 1962: 13. 2. Adam, 1966, a. 3. Adegoke & al., 1968. 4. Ainslie, 1937: sp. no. 67. 5. Aubréville, 1950: 447. 6. Ba, 1969. 7. Bally 7050, K. 8. Bouquet & Debray, 1974: 49. 9. Burkill, 1935: 413–15. 10. Dalziel, 1937. 11. Deighton 2493, K. 12. Gardner 3703, K. 13. Glover & Samuel 2944, K. 14. Goulding, 1937: 53. 15. Grall, 1945: 26. 16. Harrison, S. G., 1950. 17. Heal & al., 1950: 147. 18. Holland, 1908–22: 463–4. 19. Hunting Technical Surveys, 1968. 20. Irvine, 1961: 646–9. 21. Jackson, 1973. 22. Kerharo & Adam, 1964, a: 410–12. 23. Kerharo & Adam, 1964, b: 416. 24. Kerharo & Adam, 1964, c: 296. 25. Kerharo & Adam, 1974: 211–14, with phytochemistry and pharmacognosy. 26. Kerharo & Bouquet, 1950: 194–5 with many references. 27. Maire, 1933: 171. 28. McKinnan S/77, K. 29. Monteil, 1953: 119. 30. Naegelé, 1958, b: 877. 31. Oliver, 1960: 5, 21, 51–52. 32. Roberty, 1953: 449. 33. Rosevear, 1961. 34. Schnell, 1953, c: as *C. procera* (Willd.) Ait. 35. Sikes, 1972: 183. 36. Tweedie 420, K. 37. Watt & Breyer-Brandwijk, 1962: 67, 69, 125–7, with extensive review of uses and phytochemistry. 38. Williams, F. N., 1907: 375. 39. McIntosh, 26/1/79: in litt.

**Caralluma dalzielii** N.E. Br.

FWTA, ed. 2, 2: 103. UPWTA, ed. 1, 386.
West African: **SENEGAL** FULA-TUKULOR buri nánéwi (K&A) buri nañey (K&A) burñéñé (K&A) **UPPER VOLTA** BISA dudumosu (K&B) horba ya kalé (K&B) FULA-FULFULDE (Upper Volta) m-bolla (K&B) MOORE myebzoya (Prost, ex K&B) **IVORY COAST** FULA-FULFULDE (Ivory Coast) m-bolla (K&B) **DAHOMEY** YOM nenoch (A. Chev.) **NIGERIA** FULA-FULFULDE (Nigeria) gubehi (JMD) HAUSA káràn-másállaáchíí = *mosque reed* (JMD; ZOG) wútsíyàr dámoó = *monitor's tail* (JMD; ZOG)

A succulent perennial, erect sparsely-branched to 40 cm high, with quadrangular branches, of dry Sahel locations in Senegal to NW Nigeria, and also in the Sahara region.

Cattle do not browse it, and Fula cattle-folk generally regard the latex as toxic (6). However, the stems are reported crushed and eaten raw as a tonic and stimulant for faintness due to fasting, pain in the chest and epigastrium, etc. (3). A decoction is used in the Tenkodogo area of Upper Volta as an antemetic (6). Moorish people in the western Sahara extract a strong poison by maceration in sheep's urine (7).

The plant is easy to grow from cuttings. It is planted in N Nigerian towns (1), and sometimes close by houses (2); also at prayer grounds, etc. as a charm against evil (3). It is ascribed with strong magical attributes by Fula witch-doctors in Senegal who use it for cleansing against illness, and to confer protection against spells, to exorcise spirits, etc. (4, 5).

References:

1. Dalziel 317, K. 2. Dalziel 367, K. 3. Dalziel, 1937. 4. Kerharo & Adam, 1964, b: 418. 5. Kerharo & Adam, 1974: 214–15. 6. Kerharo & Bouquet, 1950: 195. 7. Monteil, 1953: 121.

**Caralluma decaisneana** (Lem.) N.E. Br.

FWTA, ed. 2, 2: 103. UPWTA, ed. 1, 386.
West African: **MAURITANIA** ARABIC ('Maure') abila (A. Chev.) aboïla (A. Chev.) abuaïla (A. Chev.) **SENEGAL** MANDING-BAMBARA dīlissli vili (JB) WOLOF sol (K&A)

A slender perennial succulent not passing 20 cm high, much branched, stems slightly angled, decumbent and rooting, of the dry Sahel regions of Mauritania, Senegal and Mali, and occurring through the Sahara to NE Africa and Arabia.

Stock in the Hoggar of Central Sahara will not graze it, and the plant there is considered toxic to both man and animal (3).

In Senegal the Wolof of Cayor and Cap-Vert apply the latex to a carious tooth to quell toothache (2).

Like *C. dalzielii* N.E. Br., this species has superstitious uses, and is probably planted at mosques, etc., to dispel evil influences (1). In Senegal it enters into a number of medico-magical treatments for mental troubles, epilepsy, and perhaps also for Parkinson's disease (2).

References:

1. Dalziel, 1937. 2. Kerharo & Adam, 1974: 215. 3. Maire, 1933: 173–4, as *C. venenosa* Maire.

**Caralluma mouretii** A. Chev.

FWTA, ed. 2, 2: 103. UPWTA, ed. 1, 386.
West African: **MAURITANIA** ARABIC ('Maure') abila (A. Chev.) aboïla (A. Chev.) abuaïla (A. Chev.)

A succulent with round stems, to 30 cm high, of dry sandy locations in Mauritania, and also in Morocco.

The plant is said to be edible raw (A. Chevalier fide 1).

Reference:

1. Dalziel, 1937.

**Caralluma russelliana** (Courb. ex Brong.) Cufod.

FWTA, ed. 2, 2: 103, as *C. retrospiciens* (Ehrenb.) N.E. Br. UPWTA, ed. 1, 386, as *C. tombuctuensis* N.E. Br.

West African: **SENEGAL** FULA-PULAAR (Senegal) balamadi (K&A) dúklabébé (K&A) kédun war (K&A) téinduwar (K&A) TUKULOR kéidun war (K&A) WOLOF barasan, mbarasan (K&A) mborosan (K&A) **MALI** ARABIC ('Maure') tadenua (A. Chev.) tiduar (A. Chev.) tinedwar (A. Chev.) DOGON tianga (A. Chev.) FULA-PULAAR (Mali) mbola (A. Chev.) SONGHAI mbolla (A. Chev.) TAMACHEK okua (A. Chev.) radjiba (A. Chev.) redjebba (A. Chev.) taiberu (RM; A. Chev.) **NIGER** HAUSA amankett (A. Chev.) ekuwa (A. Chev.) **NIGERIA** HAUSA amankett (ZOG) ekuwa (ZOG)

An erect, sparsely-branched succulent, 1–2 m high, with quadrangular stems to 7 cm across, in the dry Sahel region of Mauritania, Senegal and Mali, and occurring across Africa in semi-arid situations to the Red Sea and northern Kenya.

The plant is toxic and is not grazed by any herbivorous animals (1, 7). The latex is particularly poisonous. Nor does the plant enter into the local medicines in West Africa (3, 4). In northern Kenya the plant sap is applied to wounds (6).

Some tribes in W Africa regard the plant as a fetish (2). Witch-doctors of the Turkana people in N Kenya apply the plant in magic to protect cattle from theft by other tribesmen (5).

The flowers are mal-odorous with a rank meaty smell which attracts many insects.

References:

1. Adam, 1966, a: as *C. retrospiciens* and *C. tombouctouensis*. 2. Dalziel, 1937. 3. Kerharo & Adam, 1964, b: 419, as *C. retrospiciens* (Ehrenb.) N.E. Br. 4. Kerharo & Adam, 1974: 215, 217, as *C. retrospiciens* (Ehrenb.) N.E. Br. 5. Mwangangi 1471, K. 6. Mwangangi & Gwynne 1056, K. 7. Naegelé, 1958, b: 878, as *C. retrospiciens* (Ehrnb.) N.E. Br.

**Ceropegia spp.**

FWTA, ed. 2, 2: 99–102. UPWTA, ed. 1, 386.

West African: **SENEGAL** BASARI a-nyẹkòfẹ́n *C. campanulata* (Ferry) BEDIK gi-ndyòmó *C. campanulata* (Ferry) MANDING-BAMBARA ulu ndïolond missé *C. aristolochioides* (JB) uluku mbiré *C. sankurensis* (JB) SERER falãngoñ *S. sankurensis* (JB) sèmbélèh *C. sankurensis* (JB) WOLOF sambalèh *C. sankurensis* (JB) simbôm *C. sankurensis* (JB) **GUINEA** MANDING-MANINKA fuyé *C. sp. indet.* (H&P) fuyé-diu *C. sp. indet.* (H&P) fuyé-musso *C. sp. indet.* (H&P) fuyé-nké *C. sp. indet.* (H&P) tébiré *C. sp. indet.* (H&P) **SIERRA LEONE** SUSU bɔge C. sankurensis (NWT) **MALI** MANDING-BAMBARA ulu n'dioloko miyé *C. aristolochioides* (GR) **NIGERIA** HAUSA dodoria *C. porphryotricha* (JMD) kafan fakara = *francolin's foot; C. racemosa* (Lely) YORUBA okorun *C. talbotii* (RJN)

Slender, erect herbs or twining climbers arising from a tuberous root-stock, discoid, depressed-globose, long-pointed or spherical, usually of dry savanna, and occurring throughout the drier part of the Region.

The tubers of several species are regularly eaten in NE Africa principally for the liquid contained in them. The leaves are acidulous and are eaten cooked as spinach. The cooked tuber is said to taste like that of the Jerusalem artichoke (*Helianthus tuberosus* Linn. Compositae). The root of *C. fusiformis* N.E. Br. is eaten in Ghana and is reported to be sweet (4). The Maninka of Guinea eat the tubers of several unspecified species (1) but in the Region generally the roots are regarded as a famine-food (2).

In Mali *C. aristolochioides* Dec'ne (part and purpose unstated) is prepared as a wash used for infants (2).

References:

1. Dalziel, 1937. 2. Huber s.n., IFAN. 3. Irvine, 1952. 4. Irvine, s.d.

**Cynanchium adalinea** (K. Schum.) K. Schum.

FWTA, ed. 2, 2: 89.

Slender climber of secondary jungle, present in two subspecies in the Region: ssp. *adalinae* in Mali, Ghana, S Nigeria, W Cameroons and Fernando Po; ssp. *mannii* (Sc. Elliot) Bullock in Sierra Leone and Liberia.

No usage is recorded within the Region. There is unspecified medicinal use in Gabon, and its fruits are eaten there (1).

Reference:

1. Walker & Sillans, 1961: 98, as *C. acuminatum* Schum.

**Dalzielia oblanceolata** Turrill

FWTA, ed. 2, 2: 95.

West African: **SIERRA LEONE** MENDE nyinawe, nyina: *rat; perhaps a general name for some small shrubs* (NWT)

A lactiferous shrub to 2 m high of dry river-beds, reported from Guinea, Sierra Leone and Ivory Coast on the Guinea border.

**Dregea abyssinica** (Hochst.) K. Schum.

FWTA, ed. 2, 2: 97.

West African: **UPPER VOLTA** 'SENUFO' bergu (K&B) **IVORY COAST** 'SENUFO' bergu (K&B)

A shrub or woody climber of secondary scrub and margins of the closed-forest, from Senegal to Nigeria, and widely dispersed in tropical Africa.

The plant with pale grey-brown twigs, bright green young shoots, whitish, scented flowers nearly 1 cm across is not unattractive as an ornamental. The flowers are freely visited by insects. In Sudan the plant yields a fibre (2).

The Senoufo of Ivory Coast-Upper Volta use the latex in ear-instillation for ear-troubles (3).

The leaves are eaten in Uganda as a vegetable (7). The plant has unspecified medicinal uses in Kenya and Tanganyika (6). In the latter country the roots are considered aphrodisiac (4) and the green leaves and twigs are pounded with water and added to dog's food by the Sukuma for treating dogs with a dry cough (5).

The seeds are reported to contain a number of glycosides (1).

References:

1. Bouquet & Debray, 1974: 49. 2. Broun & Massey, 1929: 254, as *Marsdenia spissa* S. Moore. 3. Kerharo & Bouquet, 1950: 196 as *Marsdenia spissa* Moore. 4. Koritschoner 1606, K. 5. Tanner 1079, K. 6. Watt & Breyer-Brandwijk, 1962: 134, as *Marsdenia spissa* S. Moore. 7. Wilson 241, K.

**Dregea crinita** (Oliv.) Bullock

FWTA, ed. 2, 2: 97.
West African: **SIERRA LEONE** MENDE tɛnawulo (*def.*-wuli) (FCD)

A stout woody climber of the closed-forest, recorded from Guinea-Bissau to S Nigeria, and occurring also in Gabon and Angola.
The plant is without latex. The sap is used in Sierra Leone to arrest bleeding (1).

Reference:

1. Deighton 3653, K.

**Dregea schimperi** (Dec'ne) Bullock

FWTA, ed. 2, 2: 97.

A climber of savanna scrub, recorded only from W Cameroons, but occurring widely in E Africa and in Sudan, Somalia and Aden.
The plant is poisonous to stock. Poisoned stock stand stiffly and shake and have diarrhoea sometimes with abdominal pain. Death follows in most cases. A Somali treatment, which is said to be beneficial, is to feed a warm sheep-soup to induce further diarrhoca (2).
The wood in Kenya is used to make panga handles (1).

References:

1. Glover & al., 2440, K. 2. Peck 239, 277, K.

**Glossonema boveanum** (Dec'ne) Dec'ne, **ssp. nubicum** (Dec'ne) Bullock

FWTA, ed. 2, 2: 89. UPWTA, ed. 1, 387, as *G. nubicum* Dec'ne.
West African: **MAURITANIA** ARABIC (Hassaniya) achakan (AN) m'gueïle = *in the shade* (AN) **SENEGAL** MANDING-BAMBARA marka digini (JB) **MALI** MANDING-BAMBARA marku djiguini (GR) **NIGER** FULA-FULFULDE (Niger) gobbel duroobe, cobbel waynaabe = *gobbel, or cobbel (meaning not known) of the herdsmen* (PF) SONGHAI gànd-báa háwrù (*pl.* lò) = *part of the evening meal of the earth* (D&C) **NIGERIA** FULA-FULFULDE (Nigeria) cakum-cako (JMD; MM) gadoil (?) (JMD) gobbel duroobe, cobbel waynaabe (MM) HAUSA taarin-gidaa (auctt.) tafo ka shamamarka, mama: *a woman's breast; – an epithet referring to the use of the plant as a galactogogue* (JMD) tataridʼa (JMD; ZOG)

A herb to 30 cm high, weed of cultivation and in waste places of the Sahel zone, from Mauritania to N Nigeria and W Cameroons, and extending throughout N and NE Africa to Arabia and NW India.
The plant is edible raw, especially the young flowering top and fruits. It provides a good fodder for all stock (1, 3, 6), and serves as a famine-food for man (1, 5). It contains a copious amount of milky sap and is taken by women in N Nigeria to increase lactation, hence the Hausa epithet for the plant.

References:

1. Bullock, 1955, b: 616–18. 2. Dalziel, 1937. 3. Irvine, 1952, a: 34. 4. Irvine, s.n. 5. Moiser, s.n. 5/3/1922, K. 6. Naegelé, 1958, b: 878.

**Gomphocarpus fruticosus** (Linn.) Ait. f.

FWTA, ed. 2, 2: 92–93.

A perennial herb to 2 m high of savanna grassland in Guinea, and occurring widespread in tropical and southern Africa.

The inner bark yields a white bast fibre which has been reported in the same terms as that of *G. physocarpus* (5). In Botswana the fibre is spun into a cotton for sewing clothes and for snaring birds (9). In Somalia it is commonly used as string for snares (7) and to make waistbands (4). The floss from the fruits is also spun in Somalia by women into a white cotton which they use for decoration and to make into belts (2). In Lesotho the floss is used for filling pillows (6).

The dried leaves, flowers and young shoots are ground up and used in Lesotho as a snuff given to seriously ill patients. If the patient does not sneeze the case is considered hopeless! A decoction of the plant is taken by mouth for asthma and for difficulty in breathing. The patient must vomit after taking the dose. The plant (part unspecified) is taken in small doses for excess of bile; it is also used (method not stated) for headache (6). In Tanganyika the plant is used for sores, boils and headache (12), and the root is taken (? in decoction) for abdominal pain (8). In Uganda the plant is said to be used to treat swellings of the neck (10) and in Madagascar swellings generally are treated by massage with a decoction of the plant and *Vernonia appendiculata* (Compositae) and for lumbago the plant is dried and mixed with honey to a paste for application with massage (3). The latex is used against toothache, and a decoction of the seeds as a cough-medicine (3).

All parts of the plant of Madagascan material have shown the presence of alkaloids (3). The presence of cardio-active glycosides, named *gofruside A* and *B,* in African material has been reported and the plant is said to be used in southern and eastern Africa as an arrow or ordeal-poison (13).

The rootstock is eaten in Lesotho as a vegetable (6). Masai of Tanganyika eat the fruit (11).

In Somalia the stems are used by shepherds for driving sheep instead of using big sticks (1).

References:

1. Burne 3, K. 2. Collenette 8, K. 3. Debray & al., 1971: 54. 4. Gillett 3980, K. 5. Goulding, 1937: 53, as *Asclepias fruticosa*. 6. Guillarmod, 1971: 414, as *Asclepias fruticosa* Linn. 7. Keogh 113, K. 8. Koritschoner 1374, K. 9. Miller, 1952: 72, as *Asclepias fruticosa* L. 10. Snowdon 1521, K. 11. Tanner 3795, K. 12. Tanner 5632, K. 13. Watt & Breyer-Brandwijk, 1962: 69, as *Asclepias fruticosa* L.

## **Gomphocarpus physocarpus** E. Mey.

FWTA, ed. 2, 2: 92. UPWTA, ed. 1, 383, as *Asclepias semilunata* N.E. Br.

A perennial herb to 2 m tall of montane seasonal swamp grassland, recorded from Senegal, N Nigeria and W Cameroons, and occurring widespread in tropical and S Africa.

The inner bark yields a strong bast fibre. Examination of material from Uganda has been fairly well reported on. Fibre length is uneven though an average is about that of cotton. There is an absence of twist which may make it difficult to spin, but if it can be properly handled a soft fabric should be producable. It may be suitable for admixture with cotton (1). The fruits also contain a floss.

Reference:

1. Goulding, 1937: 53, as *Asclepias semilunatus*.

## **Gongronema angolense** (N.E. Br.) Bullock

FWTA, ed. 2, 2: 98.

West African: **SIERRA LEONE** LIMBA (Warawara) pəlefɔ (NWT) TEMNE *a*-gbam (NWT)

A slender climber of secondary vegetation, recorded only rarely from Sierra Leone and N Nigeria, but occurring widely elsewhere in tropical Africa.

All parts of the plant are lactiferous. The thick fleshy roots are eaten by children in Ruanda (Bouquet fide 1)

Reference:

1. Bullock, 1961: 199–200.

**Gongronema latifolium** Benth.

FWTA, ed. 2, 2: 98. UPWTA, ed. 1, 388, as *Marsdenia latifolia* K. Schum.

West African: **SENEGAL** SERER gasub (JLT) **SIERRA LEONE** KISSI n-dondo-polole (FCD) KRIO rope quiah (L-P) MENDE buli-yeyakɔ, nyiya yeyakɔ (*def*.-i) *the leaves, from* ngeyakɔ: *a creeper* (FWHM; FCD) tawa-bembe(-i) *the leaves, from* bembe: *to encircle, as of a creeper* (FWHM; FCD) yɔnigbagbɔi *the leaves, from* gbɔle: *drink* (auctt.) TEMNE *ra*-bilong (FCD) **GHANA** AKAN-ASANTE aborode-aborode (FRI) akam? (FRI) kurutu (FRI) nsurogya = *does not fear fire, a general name* (FRI) **NIGERIA** IGBO utazi (BNO) YORUBA arọ̀kẹ̀kẹ̀ (IFE)

A climber from a tuberous base, of deciduous and secondary forests from Guinea-Bissau to W Cameroons, and widely dispersed elsewhere in tropical Africa.

The stems are soft and pliable. They are used in Sierra Leone as chew-sticks (2), and cut up and boiled with lime juice (2), or infused in water over three days (3), the liquor is taken as a purge for colic and stomach-pains, and symptoms connected with worm-infection (1). The infusion is taken as a cleansing purge by Mohammedans during Ramadan (3). It is given to a new-born baby in the Joru area of Sierra Leone to make it grow rapidly (4). In Ghana the leaves are rubbed on the joints of small children to help them to walk (5), and in S Nigeria the leaves serve as a vegetable (7, 8).

The bark contains a quantity of latex and though it has been viewed with potential interest for its rubber (10), it has apparently never been exploited. A closely related species *M. reichenbachii* Triana in Ecuador and Columbia is the source of cundurango bark containing glycosides *cundurangin* and *cundurit* and from which cundurago wine is prepared. This is used as an aromatic bitter and stimulant in treatment of dyspepsia (9, 11, 12). The bark of *G. latifolium* merits examination as an official dyspeptic.

In Ghana the boiled fruits are put into soup as a laxative (6).

References:

1. Dalziel, 1937. 2. Deighton 681, K. 3. Deighton 2428, K. 4. Herbalists, Joru village, Sierra Leone, 1973. 5. Irvine 506, K. 6. Irvine, 1961: 650. 7. Latilo al. FHI 34985, K. 8. Lowe 407, UCI. 9. Oliver, 1960: 30, 42, 70–71. 10. Schlechter, 1900. 11. Uphof, 1968: 333. 12. Usher, 1974: 380.

**Gymnema sylvestre** (Retz.) Schultes

FWTA, ed. 2, 2: 95. UPWTA, ed. 1, 387.

West African: **SENEGAL** SERER gasub a tèk (JB) **MALI** DOGON diì gíru (C-G) **NIGERIA** HAUSA yaryaɗin kura = *hyena's vine; applied loosely* (JMD)

A scandent shrub or many-stemmed climber, of thickets in open savanna, and of deciduous open wooded savanna, from Mauritania to S Nigeria, and widely distributed throughout tropical and southern Africa and across Asia.

The leaves are put into soup by the Igbo of S Nigeria (10). The leaves when chewed have the capacity of destroying for a few hours taste for sweetness and bitterness. Taste for acid, astringent or pungent substances is not affected, and for salt but slightly, if at all. This property is due to the presence of a glycosidal substance, *gymnemin,* which is the potassium salt of gymnemic acid. The plant has been used in India to treat diabetes. The leaves are diuretic and laxative due to the presence of anthraquinone derivatives, and are given in India (6) and in

Indian medicine in Zanzibar (4) to diabetic patients to chew to reduce glycosuria. Tests, however, of the leaf-powder and of alcoholic extract showed no drop in blood-sugar level. Hypoglycaemia has been produced in test animals which has been ascribed, not to any direct action, but indirectly by stimulating increased insulin secretion by the pancreas (6, 8, 11).

The green pounded-up leaves are used in Tanganyika for stitch in the sides for which the pulp is rubbed into small cuts over the pain by the Sukuma (9). These people also use the pounded and cooked root for epilepsy, given twice daily. The root is applied topically and taken internally for snake-bite in Sudan (1), India, Africa and Australia (2), and in Botswana they are made into an ointment for boils (5).

Besides the active principles already mentioned, a number of other substances has been reported (8). Alkaloids, at least three, *betaine, choline* and *trimethylamine,* are present in the leaves (12). In the twigs saponins are abundant (3). A neutral bitter substance has been isolated from the leaves which is sialagogic. Leaf-powder is a stimulant of the heart and circulating system, and activates the uterus (8).

The fruits appear to have similar constituents (8).

The flowers are sweet-scented and attract insects (7).

References:

1. Broun & Massey, 1929: 254. 2. Dalziel, 1937. 3. Debray & al., 1971: 54. 4. Greenway 2658, K. 5. Miller, 1952: 72. 6. Oliver, 1960: 27, 66. 7. Richards 11823, K. 8. Sastri, 1956: 276–7, with chemical analysis. 9. Tanner 1481, K. 10. Thomas, N. W., 1683 (Nig. Ser.), K. 11. Watt & Breyer-Brandwijk, 1962: 133. 12. Willaman & Li, 1970.

**Kanahia laniflora** (Forssk.) R. Br.

FWTA, ed. 2, 2: 91.

A woody shrub, erect to 3 m tall, of mixed deciduous forest along seasonal streams and rivers from Ivory Coast to W Cameroons, and widely distributed across Africa from NE to SW Africa.

In Kenya the plant is reported eaten by cattle, and sometimes by goats (4). The plant is very lactiferous and in Sudan has the Arabic name, *abou lebben,* mother of milk (3). In Uganda the latex is put onto sores (1). In Tanganyika a root-decoction with sap from the leaves is given to small children with malarial rigor (2).

References:

1. Dyson-Hudson 147, K. 2. Haerdi, 1964: 135. 3. Muriel S/14, K. 4. Mwangangi & Gwynne 1037, K.

**Leptadenia arborea** (Forssk.) Schweinf.

FWTA, ed. 2, 2: 98–99.

West African: MALI ARABIC ('Maure') al lendé (GR) FULA-PULAAR (Mali) lahiri (GR)

A slender climber of scrub vegetation, recorded in Niger, N Nigeria and W Cameroons, and extending to Egypt and Arabia.

It provides forage for camels in Mali (4) and for camels, sheep and goats in Somalia (2).

The fruits when ripe, or even nearly ripe, are eaten by children in Chad (1).

The plant is amenable to cultivation and can be trained as a vine over a wall or pergola (3).

References:

1. De Wilde c.s. 5215, K. 2. McKinnon S/73, K. 3. Meyer 8727, K. 4. Roberty 865, IFAN.

**Leptadenia hastata** (Pers.) Dec'ne

FWTA, ed. 2, 2: 98. UPWTA, ed. 1, 387, as *L. lancifolia* Dec'ne.

West African: SENEGAL BALANTI brôgbé (K&A) BANYUN sora (K&A) BASARI a-nyadíwù (Ferry) BEDIK ga-nyádẹ̀m (K&A; Ferry) DIOLA (Fogny) futakadaf (K&A) FULA-PULAAR (Senegal) sapato (K&A) sapatoy (K&A) savato (K&A) sawat (K&A) sawato (K&A) tapatoy (K&A) KONYAGI a-nẹrẹvel (Ferry) MANDING-BAMBARA zoṅé (JB, ex K&A) MANDINKA sapata (*def.* sapatoo) (K&A) 'SOCE' sora (K&A) SERER gasub (K&A) n-gasa, gasu (K&A) sarafat (FB; JLT) SERER-NON (Nyominka) ingasub (K&A) ndis wâdan (K&A) ngazu (K&A) SONINKE-SARAKOLE sarafaté (K&A) sarafato (K&A) WOLOF ḋarhat (K&A) nkarkat (K&A) sahatt = *cough* (K) talal (FB; JLT) ṫarhat (K&A) **THE GAMBIA** DIOLA (Fogny) futakadaf *onomatopoeic resembling the sound made when the plant is being cut* (DF) FULA-PULAAR (The Gambia) safato (DRR) MANDING-MANDINKA so-ora *the newly-sprouted leaf used in soup* (DRR; DF) **GUINEA-BISSAU** PEPEL bissacra (JDES) **GUINEA** MANDING-MANINKA son-niugu (Brossart; FB) **MALI** SONGHAI anu (A. Chev.; FB) gaô *the fruit* (A. Chev.) **UPPER VOLTA** GRUSI benaduru (FRI, ex K&B) HAUSA yadiha (K&B) MANDING-DYULA kosafla (K&B) MOORE lélongo (K&B) **IVORY COAST** MANDING-DYULA kosafla (K&B) SENUFO-TAGWANA iriban (K&B) DYIMINI yéfuké hinzri (K&B) **GHANA** GRUSI benaduru (FRI) HAUSA yaɗiya (FRI) **NIGER** SONGHAI dúlá (*pl.* -à), fàttàgá (*def.* là) (D&C) **NIGERIA** FULA-FULFULDE (Nigeria) sobotoro, sobotorooji (Taylor, ex JMD) yaaɗiyowol *from Hausa* (JMD; MM) HAUSA adizindir *the fruit* (JMD) ɗan zindiri (Bargery, ex JMD) ɗánbaàkúwá (JMD; ZOG) ɗán-bàraáwoò (ZOG) yaádíyaá (auctt.) yaɗiya (MM) HAUSA (East) alizindir (ZOG) KANURI kàlímbó (C&H) YORUBA iran-aji igbó (Millson, ex JMD) isanaje-igbó (JRA)

A many-stemmed climber, becoming bushy at the base, of dry savanna from Senegal to W Cameroons, and extending across Africa to Ethiopia, Kenya and Uganda.

This plant is frequently parasitized by an aphid, *Aphis nerii* Boyer de Fonscolombe (syn. *Siphonophora leptadenieae* Vallot) which is eaten by a coccinellid *Cydonia vicinia*. This latter has been used in attempted biological control of *Aphis craccivora* Koch (syn. *A. laburni* auctt. non Kltb.), *A. gossypii* Glover and *Melanaphis* (*Longiunguis*) *sacchari* Zehntner) (syn. *A. sorghi* Theobald), said to cause transmission of rosette disease of groundnuts, and to attack sorghum and cotton. The cultivation of parasitized *L. hastata* in areas of these crops has therefore been recommended as a measure to generate a population of the predatory coccinellid (7, 21). Work in E Africa, however, on *A. gossypii* has given completely negative results in tests for transmission, and *A. sorghi* is also thought to be an unlikely vector. Another aspect is that the fecundity of coccinellid is considerably affected by the aphids upon which they are reared, and aphids themselves accumulate plant poisons for their own protection (8). In this respect *L. hastata* appears to be benign with a lack of toxicity. The practicality of its planting in areas of cultivation clearly merits further investigation which might lead to interesting findings.

The leaves, young shoots and flowers are eaten throughout the Region and in E Africa, usually cooked and in soups (3, 5, 6, 7, 10, 12, 18, 20). In Cameroun the shoot-tips are eaten fresh (9). In the Kaya area of Upper Volta its consumption by lepers is forbidden on fetish grounds (18).

In Senegal it is said to be relished by horses and donkeys, and to be occasionally grazed by other stock, though in some regions stock will not take it (1, 2). In northern Kenya camels and goats are recorded as being very fond of it, and donkeys but slightly (19).

The leaves of plants grown at Dakar have been analysed as containing: carbohydrates 46%, protein 18%, ash 14%, oil 6% by weight. Calcium was present at 2.06%, potassium 1.74%, with magnesium, sodium, phosphorous and numerous minor elements, the principal one being iron, and amino-acids (5; also 17).

The plant contains a milky sap which suggests, by allusion, usages for ailments symptomized by discharges, or a lack of them. The sap, or the whole leaf-petiole rolled up into a spill, is put in the nose in Senegal for head-colds; the latex is put on wounds; a root macerate is taken for anuria or constipation; a macerate of the whole plant for urethral discharge; a macerate of leaves for lack of breast-milk and impotence. In all these treatments the result is copious diuresis (14, 15, 17). In Upper Volta a decoction of the whole plant is taken for abdominal complaints, and the latex is instilled into the nose for head-pains (18). In Togo and the Northern Territories of Ghana (10), and in N Nigeria (3) the leaves are boiled and the liquor is drunk for treating gonorrhoea. Hausa give this liquor to cure stomach-ache (*chuwan chiki,* Hausa) in children (13). The sap (14), or the root in decoction (16) is used for ophthalmia in Senegal. The powdered roots in water are taken in Nigeria as a stomachic (3).

In association with other plants, it is used in Senegal for suckling babies with green diarrhoea, for all vein troubles, varicose veins, bleeding and painful haemorrhoids, poisonings, anuria, syphilis, leprosy (except by Wolof medicine-men who consider it interdited for this disease), trypanosomiasis, etc. — in short as a general panacea (16, 17). In N Nigeria it is used with the root of *Smilax* (Smilacaceae) for tertiary syphilis (6).

The plant enters into veterinary medicine. A macerate of pounded root, preferably with one or two eggs is given by draught to horses with colic in Senegal (16). Similar usage is recorded in Sudan for flatulence in horses and cattle (4). In Casamance of Senegal the plant is considered to be a contraceptive for mares (15).

A fibre, without recorded use, can be obtained from the stem (6, 11) and the dried fruits serve as a tinder (6).

Magical properties are ascribed to the plant. Fula women of N Nigeria prepare a sweetmeat, apparently as a love-charm, from the buds and flowers called *pindi,* and from the fuits called *chode* (6). The Dyimini of Ivory Coast prepare an ointment of the pulped leaves which spread over the body confers protection to a traveller from evil spirits (18).

References:

1. Adam, 1960: 372. 2. Adam, 1966, a. 3. Ainslie, 1937: sp. no. 210, as *L. lancifolia*. 4. Broun & Massey, 1929: 254–5, as *L. lancifolia* Dec'ne. 5. Busson, 1965: 409, with leaf analysis. 6. Dalziel 95, K. 7. Dalziel, 1937. 8. Eastop, in litt., K. 9. Hepper 3986, K. 10. Irvine 417, K. 11. Irvine, 1930: 262, as *L. lancifolia* Dec'ne. 12. Irvine, 1952, a: 34, as *L. lancifolia*. 13. Irvine, s.d. 14. Kerharo, 1967. 15. Kerharo & Adam, 1963, a. 16. Kerharo & Adam, 1964, b: 555. 17. Kerharo & Adam, 1974: 217–18. 18. Kerharo & Bouquet, 1950: 195–6, as *L. lancifolia* Dec'ne. 19. Mwangangi 1396, K. 20. Rosevear, 1961. 21. Trochain, 1940: 268.

## **Leptadenia pyrotechnica** (Forssk.) Dec'ne

FWTA, ed. 2, 2: 98. UPWTA, ed. 1, 388.

West African: **MAURITANIA** ARABIC (Hassaniya) titarik (auctt.) ARABIC ('Maure') titorekt (Aub.) **SENEGAL** ARABIC ('Maure') titorekt (Aub.) FULA-PULAAR (Senegal) sabaïe (Aub.) **MALI** ARABIC assabaï (Aub.) ARABIC (Hassaniya) titarik (GR) FULA-PULAAR (Mali) sabaïe (Aub.; GR) SONGHAI saabé (A. Chev.) sabeil (JMD) TAMACHEK anah (Aub.) asabai (RM) hana (Aub.) **NIGER** ARABIC (Niger) am dokum al kelb = *dog's kapok* (Aub.) marakh (Aub.) marhaïe (Aub.) r(e)tem (Grall; Aub.) FULA-FULFULDE sabaïe (Aub.) HAUSA kalumbo (Aub.) KANURI kàlímbó (Aub.) TAMACHEK anah (Aub.) hana (Aub.) TUBU kalembo (Grall) kezen (Aub.) kozeum (Grall; Aub.) TUBU (Kaningoù) kalimbo (Aub.) **NIGERIA** ARABIC marakh (JMD) HAUSA námíjìn yaàdíyaá (JMD; ZOG) KANURI kàlímbó (EWF; JMD)

A leafless erect shrub of the northern dry sandy Sahel region from Mauritainia to N Nigeria, and in the semi-desert areas across Africa to western India.

It is a sand-dune plant (3), and possibly has some value against shifting sand. In the Sahel and the Sahara it provides fuel for fires (3). The twiggy branches are made into switches carried by female marabouts in Mauritanian Islamic schools for administering mild punishment (8), and in Somalia the twigs are woven into wickerwork containers for milk and water (6, 10). The twigs are full of a bitter watery sap (12). The plant is browsed by all stock, but especially by camels for which it is considered a good fodder (Senegal, 1; Mali, 14; 'soudano-guinea', 2; Morocco, 11; Sudan, 4), but rather mediocre for other stock (Senegal, 1; Niger, 7; Morocco, 11). The twigs make a tinder and an inextinguishable slow-match. The Latin specific name arises from this attribute.

The bark yields a fibre which is not easy to extract (5) but when obtained produces an excellent non-rotting fibre good for cordage, fishing-lines, etc. (2, 11–14). In Niger, NW of Lake Chad, the fibre is used to make trip-snares for catching small animals and birds (7).

The plant (part not stated, but presumed to be the twigs) is macerated in water in the Hoggar region and the macerate is taken for urine-retention (9). A macerate of the seeds is used in the soudano-guinean region as an eye-lotion (2).

In Niger shepherds eat the flowers raw (2). In Somalia the flowers are frequented by butterflies (6).

References:

1. Adam, 1966, a. 2. Aubréville, 1950: 448, as *L. spartinum* Wight. 3. Bayard, 1947: 10. 4. Broun & Massey, 1929: 255, as *L. spartium* Wight. 5. Dalziel, 1937. 6. Gillett 4438, K. 7. Grall, 1945: 5, 36, as *L. spartium* and *L. spp*. 8. Le Riche, 1952: 980, as *L. spartum* L. 9. Maire, 1933: 173. 10. McKinnon S/153, K. 11. Monteil, 1953: 122. 12. Muriel L/125, K. 13. Portères, s.d.: as *L. spartinum* Wight. 14. Roberty, 1953: 451, as *L. spartinum*.

**Leptadenia sp. indet.**

West African: **SENEGAL** BANYUN ɗarirô (K&A) FULA-PULAAR (Senegal) sabato (K&A)

An undetermined species of the above names is recorded used in the Casamance of Senegal as a macerate of the aerial portion as a laxative for babies less than ten days old, and as a leaf-macerate in draught for kidney-trouble and stiffness (1).

Reference:

1. Kerharo & Adam, 1974: 218.

**Margaretta rosea** Oliv.

FWTA, ed. 2, 2: 91.

An annual herb to about 30 cm high from perennial tuberous rootstock, of upland savanna in N Nigeria and W Cameroons, and widespread in East and South tropical Africa.

In Tanganyika the leaves are boiled and drunk as a male aphrodisiac (3) and similarly also is the dried powdered root added to native beer (1). In the Adamawa area of Nigeria and W Cameroons the Fula dig the roots for consumption (2).

References

1. Haerdi, 1964: 135. 2. Hepper 1563, K, IFAN. 3. Tanner 3890, K.

**Oxystelma bornouense** R. Br.

FWTA, ed. 2, 2: 89–90. UPWTA, ed. 1, 388.

West African: **MALI** MANDING-BAMBARA nomboni, nonfo *probably incl. other twiners* (A. Chev.) **NIGERIA** FULA-FULFULDE (Nigeria) layol gora layol: *a twiner, or vine* (JMD) HAUSA diya iyar kadda (BM) hánjin-raàgoó (JMD; ZOG) harshe rago = *ram's tongue* (MM) harshe tunkiya = *sheep's tongue* (MM)

A climber of the wooded savanna, commonly riverain, occurring from Senegal to W Cameroons, and extending across to NE Africa.

The plant is an attractive climber with fairly large white flowers suffused crimson, and is worthy of cultivation. It has magical use in Kenya in the case of delayed childbirth when the patient ties a stem around her middle (1).

Reference:

1. Adamson 159, K.

**Pachycarpus lineolata** (Dec'ne) Bullock

FWTA, ed. 2, 2: 93. UPWTA, ed. 1, 383, as *Asclepias lineolata* Schltr.

West African: **SENEGAL** MANDING-BAMBARA fiyé (JB) **GUINEA-BISSAU** FULA-PULAAR (Guinea-Bissau) cupunco (EPdS) MANDING-MANDINKA bafurma (EPdS) **NIGERIA** GWARI mullibe (JMD) HAUSA rízgár kùreégeé, rízgá: *the tuber of Plectanthus esculentus* (*Labiatae*); kùreégeé: *a ground squirrel* (JMD; ZOG)

An erect simple-stemmed perennial, recorded from Senegal, Ghana, N Nigeria and W Cameroon in savanna lands, and occurring widely in the savanna of tropical Africa.

The plant (part unstated) is used in Sudan to treat dog-bite (4). Sap which oozes from the wood when heated is used in Zambia on the skin to relieve itch (9, 10).

The roots are thick and spindle-shaped with a white flesh and a milky resinous sap. In Nigeria a root-decoction is taken with local soda for gastro-intestinal troubles, and a weak cold infusion is given to new-born babies (5). In E Africa they are used as a stomachic (8), an effect attributable probably to the resin (1). In Uganda the swollen roots are eaten for sore-throat (7). In Tanganyika the dry, powdered tuber is drunk with local beer as an aphrodisiac, and a root-macerate is similarly used (2). In Zambia the root is used to catch birds by virtue of an alleged narcotic effect (9, 10). Glycoside is reported in the roots (5). Inulin and starch are also present, and the Kyama of Lower Ivory Coast are said to eat the roots like tapioca, pounded and boiled, or even uncooked (1, 3, 6).

Juice extracted from the seeds is mixed with red peppers (*Capsicum,* Solenaceae) in Tanganyika for application to the breasts to encourage milk-flow (8).

References:

1. Dalziel, 1937. 2. Haerdi, 1964: 135. 3. Irvine, 1952, a: 27, as *Asclepias lineolata.* 4. Jackson 2290, K. 5. Oliver, 1960: 19, 48, as *Asclepias lineolata.* 6. Portères, s.n.: as *Asclepias lineolata* Schltr. 7. Thomas, A. S., Th. 3569, K. 8. Wallace 44, K. 9. Watt, 1967: as *Asclepias lineolata* Schltr. 10. Watt & Bryer Brandwijk, 1962: 123, as *Asclepias lineolata* Schltr.

**Pachycarpus schweinfurthii** (N.E. Br.) Bullock

FWTA, ed. 2, 2: 93.

A hispid perennial, similar to *P. lineolatus* (Dec'ne) Bullock, recorded from N Nigeria and W Cameroons, and widely distributed in the savanna of tropical Africa.

The roots have unspecified medicinal use in Sudan (1) and Angola (2).

References:

1. Andrews 1913, K. 2. Dawe 325, K.

**Pentarrhinum insipidum** E. Mey.

FWTA, ed. 2, 2: 90.

A herbaceous climber to about 2 m high, of montane savanna lands on the Cameroon Mountain, but widely dispersed in NE, E, S, and SW Africa.

Tests to confirm alleged reports of toxicity to cattle have proved negative. The young leaves and fruits have in the past served as human food in southern Africa. Vitamin C content of the young leaf has been recorded as 16.1 mg/100 gm (4). In Tanganyika water in which leaves have been heated is used by the Sukuma to wash boils which are then covered with the hot leaves (3), and the water in which roots have been soaked is used to wash sores which are then covered by leaves heated on a fire (2). The plant has undisclosed medicinal use in Botswana (1).

References:

1. Reyneke 209, K. 2. Tanner 1291, K. 3. Tanner 1408, K. 4. Watt & Breyer-Brandwijk, 1962: 134.

**Pentatropis spiralis** (Forssk.) Dec'ne

FWTA, ed. 2, 2: 90–91.
West African: **SENEGAL** SERER fogèl (JB) gasub i ndèn (JB) safsafu (JB)

A woody twiner of dry sandy places and seasonal streams in Senegal, Mali and N Nigeria, and extending in dry savanna across Africa to the Middle-East and into India.

No usage is recorded in the Region. In Kenya the young shoots appear to provide minor browsing for stock (2), though in India the plant is said to be acrid and emetic (1). The tubers are sweet and are eaten in India. Dried roots in decoction are given as an astringent and cooling alterative, and are used in gonorrhoea treatment (1).

References:

1. Deshaprabhu, 1966: 308. 2. Rawlins 125, K.

**Pergularia daemia** (Forssk.) Chiov.

FWTA, ed. 2, 2: 90. UPWTA, ed. 1, 388–9, as *P. extensa* N.E. Br.
West African: **SENEGAL** SERER ndin (JB, ex K&A) WOLOF puni (K&A) tat i ganar = *cloaca of the hen* (K&A; JB) **LIBERIA** MANO nyo yi bɛlɛ (JMD) **MALI** MANDING-BAMBARA fune-fune (A. Chev.) **UPPER VOLTA** MANDING-DYULA kuasuafé (K&B) **IVORY COAST** ADYUKRU mébufu (A&AA) BAULE eflenzué (B&D) lè san nya = *3-day leaves* (PG) néfren zuè (B&D) GAGU bubélé (B&D) wonda (K&B) KRU-GUERE plu airïé (K&B) pulairïé (K&B) GUERE (Chiehn) dipagné (K&B; B&D) dipé (B&D) KWENI drignon, grigno iri (B&D) sienfélé (K&B) solo (B&D) KYAMA abu konbengo (B&D) MANDING-DYULA kuasuafé (K&B) 'KRU' diliwoplu (K&B) libiwopru (K&B) **GHANA** AKAN-TWI nsurogya = *does not fear fire* (FRI) GA kàba (FRI; KD) GBE-VHE kponkehi (FRI, ex JMD) kponkeki (DS; FRI) **NIGERIA** IGBO utaezi (Singha) YORUBA i-joyun (auctt.) kọ́léọrọ́gbà = *building house on top of fence, from* kọ́le ori ọgbà, *after its habit* (auctt.) yanyan (RJN)

A herbaceous or semi-woody climber of littoral scrub (Senegal, 12), damp savanna and forest margins, occurring from Senegal to W Cameroon, and widely dispersed elsewhere in tropical Africa, and into Arabia and Asia.

The flowers are white or greenish and sweet-scented, and the plant has been brought into cultivation as an ornamental in tropical countries. It makes a good pergola plant as may be inferred from its Yoruba name meaning 'building house on top of fence'.

The stems yield a fibre which is used in India as a flax-substitute and to make fishing-lines (20). In Kenya a strong twine and fishing-lines are made from it (7, 20), and the Masai make a rope of the stems (6). The Ghanaian Twi name meaning 'does not fear fire' arises from the capacity of the stems to be slower in being burnt through than other creepers, and for this the stems are used in tying where there may be exposure to fire. One other creeper, *Adenia lobata* Engl. (Passifloraceae), has the same capacity and receives the same Twi name *usurogya,* the former being the 'small *usurogya',* the latter the 'large *usurogya'* (10, 11).

The plant is grazed by all stock in Somalia (18). The leaves have a mousey smell, contain 5.6 mg/100 gm of vitamin C, are anthelmintic, expectorant, emetic and emmenagogue, and in India are used in catarrhal infections and for infantile diarrhoea (3, 17, 20). In Senegal the plant enters into a prescription with others as a dysentery treatment (12). The leaves are boiled and drunk in Ghana to cure stomachache and are taken crushed up with a grey clay for gripe, and with *Capsicum* peppers (Solenaceae) are given in enema for tetanus (11). For treating small children with tetanus only 3–4 peppers are used. The young leaf and shoot are eaten in soup in southern Africa (20), and in soup or as spinach in Ghana where in particular they are given to newly-delivered women (10, 11). Sap expressed from the leaves is held to cure sore eyes and is so used in Ghana (9, 10, 11) and in Tanganyika (8). The leaves chopped up with peppers are given in Ghana to turkey poults with diarrhoea, and sap obtained by chewing a stem after it has been passed through a fire is a good cough-cure (10, 11).

In S Nigeria the plant is used in combination with others for fever, etc. (4). The latex or a poultice of the leaves is appled to boils and abscesses, a usage known in India (3, 17). In Ivory Coast the latex is applied to maturate abscesses, and for this the Adyukru prefer that obtained from the fruit (1). A leaf-dressing is also applied to furuncles allegedly effecting a cure in three days, hence the Baule name meaning '3-day leaves' (5). The milky sap is applied in Ivory Coast as a liniment to areas of rheumatism and oedema and for kidney pains (13), and the plant is particularly used to regulate the menstrual cycle and intestinal functions, and also the heart in attacks of tachycardia arising through over-exertion or fright (2).

The bark is fried in palm-oil in Liberia and the oil is rubbed into areas of 'craw-craw' (Harley fide 3).

The root seems to have no recorded usage in the Region. In Tanganyika it is known as providing a cough-medicine (15), and in decoction with the roots of *Tacazzea apiculata* Oliv. (Periplocaceae) is taken by draught for gonorrhoea (8).

In Turkana, N Kenya, the fruit is eaten (16).

A number of active principles has been recorded in Indian material. The alkaloid *daemine* is present in the leaves and roots. Bitter principles and a glycoside with an action like that of pituitrin on the uterus are present. Reports on Nigerian material have shown the presence of glycosides and absence of alkaloids, of cardiotonic action of extracts of the seeds, and none by extracts from leaves, roots and stems. Further investigation is merited. For phytochemistry and pharmacology see (17, 20).

In parts of Ivory Coast the plant is credited with magical properties. By Guere villages it is often left to grow freely around fetish shelters, and it has more magical than medicinal application in treatment of various illnesses. Pounded with china-clay and palm-wine and left to dry in the sun, a piece tied around the ankle is a protection against snake-bite, and dissolved in water and drunk is a cure for snake-bite (13). In S Nigeria to the Igbo it serves as a protection against 'insult-fever' that is, a fever resulting from insult by a boy of a big man, who, if he suffers such should immediately wash in a preparation of the plant (19). The root has unspecified magical use in Tanganyika (14).

References:

1. Adjanohoun & Aké Assi, 1972: 49. 2. Bouquet & Debray, 1974: 49. 3. Dalziel, 1937. 4. Dawodu 249, K. 5. Garnier PG/UB 11, K. 6. Glover & al., 2400, K. 7. Graham 1646, K. 8. Haerdi, 1964: 136. 9. Irvine 29, K. 10. Irvine, 1930: 330, as *P. extensa* R. Br. 11. Irvine, s.d.: as *P. extensa* N.E. Br. 12. Kerharo & Adam, 1974: 218–20, with phytochemistry and pharmacology. 13. Kerharo & Bouquet, 1950: 197–8, as *P. extensa* N.E. Br. 14. Koritschoner 930, K. 15. Koritschoner 946, 1320, K. 16. Mathew 6736, K. 17. Oliver, 1960: 33, 67, as *P. extensa.* 18. Peck 81, K. 19. Thomas, N. W. 2116 (Nig. Ser.), K. 20. Watt & Breyer-Brandwijk, 1962: 134.

**Pergularia tomentosa** Linn.

FWTA, ed. 2, 2: 90. UPWTA, ed. 1, 389.
West African: **MAURITANIA** ARABIC (Hassaniya) umu éjlud = *mother of skins, hide* (AN) **MALI** DOGON pɔ́liõ pɔpɔ́liõ (C-G) TAMACHEK sellakha (RM) teshilshit (Rodd) **IVORY COAST** GAGU monbula (K&B) SHIEN (Chiehn) sokolu (K&B) **NIGER** FULA-FULFULDE (Niger) bakambi (PF) **NIGERIA** FULA-FULFULDE (Nigeria) bakambi (PF) endumiye, endu: *breast; incl. other galactogogue plants* (JMD) enende (JMD) HAUSA damargu rafi (RES) HAUSA (East) sallenke (JMD) HAUSA (West) fatakka (JMD; ZOG) fatakko (JMD; ZOG) fatako (Lely) patako (BM)

A scandent or climber arising from a woody rot-stock, in dry savanna and wooded savanna, from Mauritania and Mali to N Nigeria, and extending to the Congo basin, N Africa to Arabia, Pakistan and India.

The plant is poisonous, and is dangerous to stock which will not normally browse it (4, 6), though the Fula of N Nigeria, give it, along with *Cissus populnea* (Vitidaceae), to their cows to increase milk-yield (1). In N Nigeria and in the Hoggar of Central Sahara the plant is pulped to a paste which is rubbed over the outer side of the hide and left for some hours (overnight) after which the fur will come away with rubbing. In N Nigeria the latex is also similarly used. In order to prepare a dehaired hide for bating so that it absorbs the tannin readily and evenly, a decoction of the plant is prepared in which the hide is placed, or the whole plant is placed in a pot with water and wood-ash and allowed to ferment for 2–3 days, after which the skin is dipped. This preparation has been an item of market trade in Sokoto, and the term *sari* is given to it and to other plants similarly used, less as a vernacular name than as indicating the usage (1). The Arabic name in Mauritania meaning 'mother of skins' also indicates usage (6). The plant is used in Morocco as a depilatory (? cosmetic) (5).

The latex/sap is irritant. In Ivory Coast the whole plant is crushed with or without pimento and taken in draught or administered by enema for dysenteriform diarrhoea, and the leaf-sap given as an eye-instillation is considered a sovereign remedy for headache (3). In N Nigeria a guinea-worm treatment is by applying a drop or two of the milky juice to an incision made in the blister. In the semi-desert regions, the plant is sometimes used to curdle milk for cheese-making, the stems being soaked in warmed milk, and some of the latter is then added to the milk to curdled (1, 2).

In the Hoggar of Central Sahara the fresh root is pounded and cooked with a piece of goat's meat which is given to a person suffering from bronchitis and haemoptysis. After eating the patient is covered and made to sweat (4).

References:

1. Dalziel, 1937. 2. Irvine, 1952, a: 32. 3. Kerharo & Bouquet, 1950: 198. 4. Maire, 1933: 172–3, 239. 5. Monteil, 1953: 122. 6. Naegelé: 878.

**Sarcostemma viminale** (Linn.) R. Br.

FWTA, ed. 2, 2: 93, 95. UPWTA, ed. 1, 390.
West African: **MALI** MANDING-BAMBARA tumèna (GR)

A rampant climber over vegetation and rocks, stems usually leafless, of dry savanna and Sahel areas, recorded only from Ghana and Nigeria, but widespread elsewhere in tropical and southern Africa.

No usage is recorded of the plant in the Region. The Zulu of southern Africa eat the stems, raw or cooked, and in the Eastern Province of S Africa it is put into salads (17). It is eaten by Somalis and is much liked by pregnant women (6). The Masai of E Africa (2, 17) and the Somalis (5) chew the stems as a thirst-quencher. There are conflicting reports whether the plant is edible to stock (17). Sheep in S Africa after eating the plant have suffered convulsions like those produced by strychnine, and then paralysis (16). The active ingredient is not reported. In Somalia it is said to be unfit for grazing, but to be apparently not a violent poison (3). The stem tastes of oxalic acid, and Somalis say that it is more bitter in the morning than in the afternoon (6), an instance, perhaps, of a diurnal rhythm of a build-up of toxicity during the hours of darkness and dissipation during the passage of the daylight hours.

The whole plant has medicinal uses. The Zulu consider it an emetic used in heartburn. In S Africa and Madagascar it has been used for uterine haemorrhage (17). The stems dried and powdered are mixed with the dried powdered penis of a ram in Tanganyika and taken with milk as an aphrodisiac (14) — surely an example of sympathetic magic.

Latex is abundant. It is used as a fish-poison in Tanganyika (13), and Kenya (11, 12). In Kenya the latex is recognized as being very caustic (8) and poisonous (10). Masai say that it will cause blindness in humans (9), yet the Zulu of S Africa apply the latex to the eyes to relieve the pain caused by entry of the sap of the 'rubber' tree (17). The 'rubber' tree is not identified. The latex is used in the Port Alfred area of S Africa, on veld sores, varicose ulcers, etc., but some persons are allergic and this treatment may produce urticaria and oedema (17), and it is given to both humans and to cows as a galactogogue. In India it is recorded as being actually used as a milk-substitute (17).

The powdered root is used in Tanganyika as an emetic, but there is suspicion of this use causing poisoning (14).

From the variation in character between poisonous and benign, one must suppose that there are edaphic effects at work just as in Somalia there is a diurnal variation in bitterness: these merit further investigation. It has been suggested, albeit with some doubt, that this plant was the Soma, the Divine Plant of the Ancient Ayrians sanctified in the Rigveda (15).

The flowers are sweetly-scented. In Ethiopia (1), Somalia (7) and Tanganyika (4) they are reported much visited by butterflies and bees, and in Ethiopia by sunbirds (1).

The fruit is said to be edible (17).

References:

1. Ash 302, K. 2. Bally, 1937. 3. Collenette 79, K. 4. Faulkner 1435, K. 5. Fawcett 17, K. 6. Gillett 4072, K. 7. Gillett 12673, K. 8. Glover & al., 207, K. 9. Glover & al., 2032, K. 10. Glover & al., 2638, K. 11. Graham V.325/1812, K. 12. Graham AA.334/1813, K. 13. Greenway 3708, K. 14. Koritschoner 1815, K. 15. Tyler, 1966. 16. Watt, J. M., 1967. 17. Watt & Breyer-Brandwijk, 1962: 136.

**Schizoglossum petherickianum** Oliv.

UPWTA, ed. 1, 390.

West African: **NIGERIA** HAUSA tafo ka shamamarka *an epithet applied loosely to milky-juiced asclepiads* (JMD) HAUSA (East) iyah (JMD)

A moderately stout-stemmed herb recorded rarely from N Nigeria, and occurring in the Upper Nile valley of Sudan and Uganda.

**Secamone afzelii** (Schultes) K. Schum.

FWTA, ed. 2, 2: 88. UPWTA, ed. 1, 390, as *S. myrtifolia* Benth.

West African: **SENEGAL** DIOLA (Fogny) bubében mil = *medicine of milk* (K&A) MANDING-BAMBARA koron foin (JB; K&A) **THE GAMBIA** FULA-PULAAR (The Gambia) barankato (DRR) geilogo (DRR) **SIERRA LEONE** KISSI teŋgealapowo (FCD) MENDE ndipagba (*def.*-i) *?= a small brown bird* (FCD) SUSU pulloka (Afzelius) TEMNE *a*-bamp *a loose term covering several plants* (NWT; JMD) *a*-lil-*a*-ro-kant (NWT) *a*-put-*a*-gboya (FCD; M&H) TEMNE (Kunike) *an*-thɔth-*an*-thi (FCD) TEMNE (Port Loko) *an*-thɔth-*an*-thi (FCD) **UPPER VOLTA** MANDING-DYULA mussokoroni-singié (K&B) **IVORY COAST** ABURE abengogo (B&D) AKAN-ASANTE kwantima (K&B) niamablé (B&D) BRONG kointima (K&B) takiwamitié (K&B) AKYE akpeubi-té (A&AA) ANYI buru burua (K&B) n'dessorguin brika (K&B) niablika (K&B) BAULE donniania (B&D) kotébuè

(B&D) niama (K&B) GAGU zara (K&B) KRU-BETE bulekpéï (K&B) GUERE diadubu (K&B) niénedubu (K&B) niénie dugu (K&B) GUERE (Chiehn) korodu = *cord of the tortoise* (K&B) menati, ménatiti, munétiti *when used as medicine* (K&B; B&D) KWENI bassialewin, bisianévuin, ganganovuin, korodu (B&D) mabréka (K&B) KYAMA atianambi (B&D) MANDING-DYULA mussokoroni-singie (K&B) MANINKA donomane (B&D) SENUFO-TAGWANA tri (K&B) 'KRU' mlinima (K&B) **GHANA** GA kotohume (auctt.) yaabrotsiri (FRI) ya'burotire (FRI) **NIGERIA** IGBO (Agukwu) èkètè (NWT) YORUBA arilu (auctt.) ati itakun *the whole plant* (IFE) òlógbọn gbũrú (IFE) re aku *the whole plant* (IFE)

A scandent shrub or climber, of secondary jungle and savanna thickets, common on unkempt farmland and in boundaries, from Senegal to S Nigeria, and also in Cameroun and Gabon.

A Shien name in Ivory Coast meaning 'cord of the tortoise' suggests that the thin flexible stems are used to make some sort of fibre or binding material. A related species, *S. whytei* N.E. Br. *vel aff.*, yields a fibre used for string in Malawi (12).

The Akye of Ivory Coast eat the leaves ground to a paste with palm oil for palpitations, and a drink of the ground-up leaves along with those of several other species and three chilis is taken for pneumonia, while the body is wiped with powdered leaves of *Secamone* in water (2). The whole plant is very lactiferous. In Senegal the Diola name meaning 'medicament of milk' indicates the use of the plant as a galactogogue. For this the leaves are macerated and the macerate is drunk and rubbed on the breasts daily for a month (6, 7). Similar application is made in Ivory Coast by the Guere and Kru who massage the lees onto the kidneys and back as well as onto the breasts (8). The Akye consider the plant can aid towards a normal pregnancy, and a wash is prepared from its leaves admixed with leaves of *Cassia obtusifolia* Linn. (syn. *C. tora* Linn., Leguminosae: Caesalpinioideae) and a small chili; and sap expressed from the leaves of these two species after passing over a fire is given by mouth (2). In The Gambia a plant, probably this species, is reported used as an astringent on cuts (11). In Sierra Leone either the latex (5) or the leaves in poultice (10) are applied to maturate boils, and the leaf-macerate is taken as a purge. Leafy twigs may be boiled with rice and the fluid drunk freely by the Susu of Sierra Leone as a quick-acting, non-griping purge (4). In Ivory Coast the plant is used internally as an antispasmodic and antidiarrhoetic and in particular, the crushed leaves are taken to arrest excessive purging caused by *Anchomanes difformis* (Araceae); it is considered tonic as an aperitive and antianaemic; in enema it is administered to treat female sterility, to facilitate pregnancy and easy delivery (3). It is also used for minor affections, internally or externally, for sore-throat, intercostal pains, colic, oedemas (8). The Dyula of northern Ivory Coast use the whole plant crushed with palm-oil seed into an oily paste which is cooked with food and taken for blennorrhoea (8). The plant is used medicinally in an undisclosed manner for children in SW Nigeria (9).

Though much of the foregoing medical usage is by draught, there has been evidence from Sierra Leone of poisoning ascribed to this plant. Death followed after vomiting and convulsions, and post-mortem examination showed inflammation of the stomach without any abnormality in the bowel (4).

Examination of Nigerian material has given signs of a trace of alkaloid present (? the roots) (1), but has shown hardly any cardiotonic action (7).

Magical attributes are ascribed to the plant in Ivory Coast. The Gagou give sap crushed from the plant to young children to drink as a protection against evil spirits, and the Brong give a decoction by draught or in baths to children in treatment of 'epilepsy' caused by the passage overhead of certain birds, e.g. the hornbill, or crowned crane (8).

References:

1. Adegoke & al., 1968. 2. Adjanohoun & Aké Assi, 1972: 50. 3. Bouquet & Debray, 1974: 47, 50. 4. Dalziel, 1937. 5. Deighton 2136, K. 6. Kerharo & Adam, 1962: as *S. myrtifolia* Benth. 7. Kerharo & Adam, 1974: 220. 8. Kerharo & Bouquet, 1950: 198 & 199, as *S. myrtifolia* Benth. 9. Macgregor 4, K. 10. Melville & Hooker 204, K. 11. Rosevear, 1961. 12. Williamson, J., 1956: 107, as *Secamone sp.*

**Solenostemma oleifolium** (Nect.) Bull. & Bruce

UPWTA, ed. 1, 390, as *S. argel* Hayne.
West African: **MALI** ARABIC argel, hargel (Bullock & Bruce) hargal (JMD) TAMACHEK ghalisum (Rodd)

An herbaceous plant with stems to 60 cm tall from a woody root-stock, of dry sandy semi-desert areas, known from Egypt and Arabia, and recently recorded from Chad, Niger, Central Sahara and Mali on the northern limit of the Region.

Sheep graze the foliage a little at Hoggar, Central Sahara, but other stock scarcely touch it (3). In northern Mali the plant is known for curative properties and is used, as it is in the Air region of Niger, to cleanse sores on camels after which a dressing of the dried powdered leaves is applied (4). The leaves and flowers are also taken by man internally as a blood purifier (1). At Tibesti in Chad the leaves are boiled to make a sort of tea which is taken with sugar (2). At Hoggar a leaf-infusion is taken for rheumatism, blennorrhoea and haemoptysis. For these, the dried powdered leaves are boiled in milk, sweetened with dates or sugar and the infusion is drunk hot (3). This promotes diuresis by making the patient drink a lot of water.

The leaves are said to be used in Egypt like henna (1).

References:

1. Dalziel, 1937. 2. Hinchingbrooke 30, K. 3. Maire, 1933: 170–1. 4. Rodd in litt., K.

**Telosma africanum** (N.E. Br.) Colville

FWTA, ed. 2, 2: 97.
West African: **SIERRA LEONE** MANDING-MANDINKA analuwulikelmɔ (NWT) TEMNE *a*-namkuna (NWT)

A slender climber of riverain and damper deciduous forest, from Guinea to W Cameroons, and widely dispersed in tropical Africa and south to Natal.

No usage is recorded for the Region. In Tanganyika fresh roots are shaken in a bottle of water and the water is drunk as a vermifuge and for venereal disease (1).

Reference:

1. Greenway 6672, K.

**Toxocarpus brevipes** (Benth.) N.E. Br.

FWTA, ed. 2, 2: 88–89.

A slender twining shrub, recorded only from S Nigeria in the Region, and from Cameroun to Zambia.

The plant contains a quantity of white latex. Examination of the plant has shown traces of alkaloids in the leaves and bark, and of steroids and terpenes in the leaves, bark and roots.

References:

1. Bouquet, 1972: 15. 2. Willaman & Li, 1970.

**Trachycalymma pulchellum** (Dec'ne) Bullock

FWTA, ed. 2, 2: 92.

An annual herb to 50 cm high arising from a tuberous rootstock, of savanna woodland in N Nigeria, and widespread in savanna lands throughout tropical Africa.

The flowers are white with purple spots, or white tinged with pink to pinkish mauve, 1–2 cm across, star-shaped, with up to 15 or more in an inflorescence. It is decorative and though rather weak growing is worthy of garden cultivation.

**Tylophora congolana** (Bartl.) Bullock

FWTA, ed. 2, 2: 96.
West African: **SIERRA LEONE** TEMNE *a*-bɔbɔ (NWT) *a*-sul (NWT)

An erect herb to 60 cm high from a woody base, of grassland in Guinea and Sierra Leone, and also in Cameroun and Congo.

**Tylophora conspicua** N.E. Br.

FWTA, ed. 2, 2: 96. UPWTA, ed. 1, 390.
West African: **GHANA** AKAN-ASANTE abɛkammo (FRI)

A large climber of the closed-forest of Liberia and Ghana, and occurring across Africa from Cameroun to E Africa.

The Asante use the leaves, ground and mixed with peppers, then laid on the fire, on the eruptions of yaws; the roots mixed with black pepper (*Piper guineense, n'suro wisa* Asante) are used as an enema to treat pains at the base of the back (? lumbago); and leaves mashed and applied to ulcers and wounds effect rapid healing (1, 2).

References:
1. Irvine 518, K. 2. Irvine, 1961: 654.

**Tylophora glauca** Bullock

FWTA, ed. 2, 2: 96.

A vigorous forest climber, known only from S Nigeria, and the Congo area.

No usage is recorded in the Region. In Congo the leaves are eaten as a vegetable for heart and lower stomach troubles. The Kôyô of Congo bathe their children in a leaf-decoction to make them handsome, big and strong (1).

Reference:
1. Bouquet, 1969: 70–71.

**Tylophora sylvatica** Dec'ne

FWTA, ed. 2, 2: 96.
West African: **SIERRA LEONE** MENDE lɔkɔkɔmi (NWT), kɔmi: *honey, or nasal mucus; lɔkɔ: perhaps the Loko people* (FCD) TEMNE *ra*-biloŋ (FCD) *a*-tap (NWT)

A slender twiner of thickets and secondary closed jungle from Senegal to W Cameroons and Fernando Po, and widely dispersed elsewhere in tropical Africa and Madagascar.

The leaf is applied to sores in S Nigeria (3). Leaf-sap is used in Congo on sores and on skin-affections (1). The plant (? in decoction) is used in S Nigeria in treating long-standing fever (2). In Congo a decoction or the plant-sap is taken by draught to stave off a threatened abortion, and in cases of menstrual troubles; it is also used as a cough medicine (1).

Unnamed alkaloid has been reported present in the root, leaf, stem and seeds (4).

Magical applications in Congo are: to prevent rheumatism, the liane is tied around the leg; to avoid storms and rain when on a journey, the liane is tied around the waist (1).

References:

1. Bouquet, 1969: 71. 2. Thomas, N. W. 1690 (Nig. Ser.), K. 3. Thomas, N. W. 1729 (Nig. Ser.), K. 4. Willaman & Li, 1970.

**Xysmalobium heudelotianum** Dec'ne

FWTA, ed. 2, 2: 93. UPWTA, ed. 1, 391.

West African: **SENEGAL** MANDING-BAMBARA fiyé (JB) WOLOF yahhop (Moloney) **THE GAMBIA** MANDING-MANDINKA bumbaŋ (*def.*-o) (Fox) **NIGERIA** HAUSA lujiya (auctt.) rujiyar-mahalba (JMD)

An erect herb to 60 cm high from a perennial rootstock of grassy savanna from Senegal to N Nigeria, and occurring widely in savanna land of tropical and South Africa.

The small tuberous root is lactiferous. It is cooked and eaten in N Nigeria for stomach troubles (2), and is used as a bitter tonic and stomachic (3). Dried, pounded and mixed with mud, the root makes a plaster used in The Gambia for surfacing the walls of houses (1).

References:

1. Fox 84, K. 2. Hepburn 3, K. 3. Oliver, 1960: 41.

# AVICENNIACEAE

**Avicennia germinans** (Linn.) Linn.

FWTA, ed. 2, 2: 448, as *A. africana* P. Beauv. UPWTA, ed. 1, 453, as *A. nitida* Jacq.

English: white mangrove; black mangrove; olive mangrove (Irvine).

French: palétuvier blanc; mangle blanc; faux palétuvier; mangle boton (Irvine-? in error for *Languncularia racemosa* Gaertn, the white button wood, or false mangrove).

Portuguese: mangue amarelo.

West African: **SENEGAL** BANYUN bukelek (K&A) DIOLA buhek (JB; K&A); = *barrier* (DF, The Gambia) MANDING-MANDINKA jubukuŋ (*def.*-o) (DF) 'SOCE' ɗubukumô (K&A) SERER mbugâd *the fruit* (auctt.) SERER-NON (Nyominka) burhan (K&A) ibuhâd (K&A) mbuhan (K&A) mburhan (K&A) WOLOF mâglé (AS, ex K&A) mbagé (K&A) mbagé ndiar (JB) ndiar (JB, ex K&A) sana(r) (auctt.) **THE GAMBIA** FULA-PULAAR (The Gambia) malanga (DRR) MANDING-MANDINKA jubuŋ (*def.*-o) (DF) WOLOF jekum (DRR) **GUINEA-BISSAU** BALANTA io (JDES) petá (JDES) BIDYOGO cobaca (JDES) CRIOULO tarafe (auctt.) DIOLA-FLUP cabêço (JDES) MANDING-MANDINKA djibicum (JDES) taraf (*def.*-ô) (JDES) MANDYAK pebadje (JDES) púle (JDES) MANKANYA úle (JDES) NALU iófo (JDES) PEPEL búle (JDES) **GUINEA** SUSU ufiri (auctt.) **SIERRA LEONE** BULOM (Sherbro) buɛ-dintɛ (FCD; S&F) buɛ-dintɛ-lɛ (FCD) buɛ-lɛ (FCD; S&F) KRIO mangro (FCD) sɔl-wata-mangro = *salt water mangrove* (FCD; S&F) LOKO makindi (FCD; S&F) MENDE gbɛlɛti (*def.*-i) (auctt.) njaya (*def.* najei) (FCD; S&F) njaya gole (JMD) SUSU wofere (NWT; FCD) TEMNE *an*-bure (auctt.) **IVORY COAST** KRU biaza (RS; Aub.) **GHANA** VULGAR asopro (DF) AKAN-FANTE asudur (FRI) esukuru (FRI, Enti) GA áŋmàa (auctt.) asokpolo (auctt.)

mutukutʃo (FRI) tratʃo? (FRI) GBE-VHE amutsi = *lagoon tree* (auctt.) NZEMA asokoro (FRI) **TOGO** GBE-VHE amu-ati (Volkens) **NIGERIA** EDO ẹ̀dẹ́ (JMD; KO&S) EFIK àfia nnùnùñ = *white mangrove* (auctt.) àta-nnùnùñ = *true mangrove* (Winston) nnùnùn (KW) ọdọnọmịn = *gorilla mangrove* (RFGA) YORUBA ogbun (auctt.) **WEST CAMEROONS** DUALA bwànjò (Ithmann) 'BAMUSO' janju (DRR)

A tree to about 15–17 m high, of brackish lagoons and river estuaries, common at such localities throughout the Atlantic seaboard of the whole Region, and extending to Zaire and also on the tropical western side of the Atlantic Ocean.

The sap-wood is white and the heart-wood pale brown, darkening on exposure, fairly hard and durable, and termite-proof. It lasts in uses underwater, e.g. piles and wharf construction, and is recommended in Ghana for marine work (2). It is used in boat-building on the coast (6, Sierra Leone, 13; Gabon, ribs of boats, 16). It is also used for house-building, furniture, gunstocks, etc., and it makes a good firewood and charcoal. (3, 6, 9, 10, 14, 16.) It has recently been tried in Sierra Leone as telegraph poles after treating with creosote (13).

Tannin is present in the bark at about 12.5% concentration. It is used locally for tanning (6, 8, 9, 14, 16), but probably less so than formerly because of synthetic substitutes. The bark also yields a red dye used in Sierra Leone (6, 7). The powdered bark, dry and with warm water, used in Nigeria as a paste, is effective in treating various kinds of dermatitis (1). In Gabon the powdered bark is made into an ointment with palm-oil base for use against skin-itches, jiggers and fleas (15, 16). Commonly throughout Senegal a decoction of bark and leafy twigs is taken by draught and put into baths to expedite labour in child-birth and the expulsion of the afterbirth (11, 12). The leaves are used in enemas in Liberia for piles (9).

A decoction of the roots is taken in draught in Senegal for troubles of the lower intestines (11, 12). The leaves are said to have quite a salty taste, and they and the roots, are used in the Niger Delta to prepare a vegetable salt (6).

The fruits which consist of a single embryo and two enlarged and fleshy cotyledons have been used in Senegal in the coastal islands as a famine food. They are lethally toxic and require special and careful preparation to remove the poisonous substances (5). The toxic chemicals are not recorded, but the seeds are reported as having 84% carbohydrates and 7% proteins (4).

References:

1. Ainslie, 1937: sp. no. 42, as *A. nitida*. 2. Anon., s.d.: *A. africana*. 3. Aubréville, 1959: 3: 234, as *A. nitida* Jacq. 4. Busson, 1965: 394/6; seed analysis 399–400, as *A. africana* P. Beauv. 5. Chevalier, 1931, b: 1000–1. 6. Dalziel, 1937. 7. Deighton 2362, K. 8. Greenway, 1941, as *A. nitida* Jacq. 9. Irvine, 1961: 750, as *A. africana* Beauv. 10. Keay & al., 1964: 437, as *A. africana* P. Beauv. 11. Kerharo & Adam, 1964, c: 295, as *A. africana* P. Beauv. 12. Kerharo & Adam, 1974: 234–5, as *A. africana* P. Beauv., with phytochemistry and pharmacology. 13. Savill & Fox, 1967, 51, as *A. africana*. 14. Taylor, 1960: 373, as *A. nitida* Jacq. 15. Walker, 1953, b: 317, as *A. alba* P. Beauv. 16. Walker & Sillans, 1961: 421, as *A. nitida* Jacq.

## BALANITACEAE

**Balanites aegyptiaca** (Linn.) Del.

FWTA, ed. 2, 1: 364. UPWTA, ed. 1, 309–11.

English: soapberry tree (Dalziel); thorn tree (Dalziel); desert date (the dried fruit, Burtt Davy & Hoyle, Dalziel, Irvine); Egyptian myrobalan (the unripe fruit, Dalziel); Zachun oil (the seed kernel oil, Dalziel).

French: dattier du désert (Kerharo & Adam); datte du désert (the fruit, Busson); myrobalan d'Egypt (Kerharo & Adam).

West African: **MAURITANIA** ARABIC (Hassaniya) teïchat (Leriche & Mokhtar Ould Hamidoun) ARABIC ('Maure') hadjlidj (Aub.) tchaïchot (Aub.) tichtaya (Aub.) BERBER taïchecht (Aub.) **SENEGAL** ARABIC ('Maure') hadjlidj (Aub.) tchaïchot (Aub.) tichtaya (Aub.) FULA-

PULAAR (Senegal) golétéki (K&A) mur(o)toki (auctt.) mutéki, mutoki (K&A) tani (Aub.; K&A) MANDING-BAMBARA ségéné (JB; K&A) ségiré, séréné (K&A) MANDINKA sumpo (K&A) MANINKA ségéné (K&A) ségiré (K&A) séréné (K&A) 'SOCE' sumpo (K&A) SERER lôl (JB) model (auctt.) SERER-NON lol (AS) lol (AS) SONINKE-SARAKOLE ségéné (K&A) ségiré (K&A) séréné (K&A) WOLOF sump (auctt.) **THE GAMBIA** FULA-PULAAR (The Gambia) murtoki (DAP) MANDING-MANDINKA sumpo (auctt.) WOLOF sumpo (DAP) **MALI** DOGON mɔ̀nɔ (Aub.; C-G) FULA-PULAAR (Mali) murot(a)oki (Aub.; GR) MANDING-BAMBARA séguéné, zéguéné (auctt.) SONGHAI garbaïe (Aub.) TAMACHEK hadjlidj (RM) taborak (JMD; Aub.) taïchot (RM) teboragh (Aub.) **UPPER VOLTA** DAGAARI gongo (K&B) FULA-FULFULDE (Upper Volta) talé (K&B) GRUSI saakuin (K&B) GURMA bupapabu (K&B) kankoabu (K&B) HAUSA adua (Aub.; K&B) MANDING-BAMBARA segainé (K&B) seguiné (K&B) MOORE kia kalaka (K&B) tia galgha (K&B) **IVORY COAST** MANDING-MANINKA séréné (K&B) sereno (K&B) **GHANA** AKAN-BRONG kabawoo (auctt.) DAGAARI gongogua (AEK) gungo (FRI) GUANG-KRACHI gushiocho (Volkens) MOORE chiala (AEK) NANKANNI kyeko (Gaisie) WALA gongo (AEK; FRI) **TOGO** GURMA (Mango) kunja napeule (Volkens) **NIGER** ARABIC (Niger) hadjilidjé (Aub.) FULA-FULFULDE murotauki (Aub.) GURMA bupapabu (Aub.) HAUSA adua (auctt.) KANURI béttò *the fruit* (Aub.) SONGHAI garbaye (FB) garbey (Robin) TAMACHEK taborak (Aub.) teboragh (Aub.) TUBU olo (Aub.) TUBU (Kaningou) tchungo (Aub.) **NIGERIA** ARABIC heglig (JRA; JMD) hejlij (JMD) lalob *the fruit* (JMD) ARABIC-SHUWA angollo *the fruit* (JMD) hajlij (JMD; KO&S) DERA ŋgóɗəm (Newman) FULA-FULFULDE (Nigeria) dubakara *the flowers and leaves for ceremonial use* (JMD) morotodi (Barter) tannere (*pl.* tanni) *the fruit* (JMD; MM) tanni (*pl.* tanne) *the tree* (auctt.) HAUSA aduwa (auctt.) dubagira *a young plant* (Abrahams) furasa *the kernel* (JMD) gallo *the unripe fruit* (JMD) kandambi *the fruit husk* (JMD) kango *a drink made from the fruit-pulp* (JMD) kwaikwaye, kwaikwayo *the kernel* (JMD) walkin dalla = *loin cloth of the (?) wren-warbler; the fruit husk* (JMD) HAUSA (East) annaki *the kernel* (JMD) awgalele *a cereal food with the fruit juice* (JMD) boro *the kernel* (JMD) dabagira, damagira *the flowers and leaves for ceremonial use* (JMD) HAUSA (West) ɓalangade, biribiri *a cereal food with fruit juice* (JMD) ƙaron aduwa *the resin* (JMD) KANURI béttò *the fruit* (JMD) cíngó, cúngó (auctt.) dàwágərá *the leaf* (C&H)

A tree to 12 m high, usually less, with a crown of tangled thorny twigs, drooping at the distal ends, tending to form thickets; on impoverished soils of the drier soudano-sahelian savanna but often on sites liable to inundation, occurring from Mauretania to N Nigeria, and widespread in the dry tropical areas of Africa and into Palestine. The plant hitherto known under this name in India is *B. roxburgii* (Roxb.) Planchon.

The plant is considered very useful and is often respected, if not actually protected. It can be grown from seed but quicker propagation is by stakes (26). It is commonly planted even far south of its natural habitat. Its spiny branches make it a good fencing material for cattle-pens for which it is used by pastoral folk in both West and East Africa (15, 18, 21, 26). The plant's habit of suckering adds thickness to any planted barrier. The foliage, in spite of the thorns, is browsed by all stock (2, 7, 15, 26). Chemical analysis shows high protein, free nitrogen and minerals, and low fibre (26). An oil content of 3·8% is also recorded (42).

The leaves are eaten as a vegetable by the Bagirmi in Chad (8b), by the Fula of N Nigeria (27) and in time of shortage in Jebel Marra of Sudan (22). Leaves containing a bitterness can be made palatable by a preliminary boiling. In Ghana the Nankanni add the leaves to soup (26), and by the Shuwa Arabs of NE Nigeria (15, 26). The young shoots are used as a seasoning in Jebel Marra (22).

The Wolof of Senegal consider the leaves stimulatory and the Fula use them powdered on malignant septic carbuncles (28, 29). In Libya and Eritrea the bark has detersive use on malignant wounds (42) and in Jebel Marra young shoots are chewed to a paste which is applied as a wound-dressing (22). In Niger there is unspecified medicinal use (38).

The wood is pale yellow to yellowish-brown, moderately well figured. It is hard, compact and fine-grained, easily worked, durable and resistant to insects. It is suitable for turnery and cabinetry, and is used for tool-handles. It is widely used for making writing and prayer-boards (15, 22, 26, 27, 34, 41). In the Sahara it is made into saddles for pack-animals, carrying-chairs, spoons, ladles, platters, etc. (32, 35). It provides one of two woods normally used to make measures for grain called *nəfga* and *mudd* north of the Sahara, four of the former being equal to one of the latter. A measure equivalent to the former called a *coppe* (*cup,* English) is used in Ivory Coast (31). In Ethiopia ploughs are made of it (Schimper fide 15), and clubs for threshing *Eleusine corocana* Gaertn. (Gramineae) in Tanganyika (43). In N Nigeria it is considered one of the best firewoods as it gives off almost no smoke and can thus be burnt inside a hut (15). In Jebel Marra it is valued for its hot burning (22). Its charcoal is of high quality and is used to make ink in Ghana (26) and by the Fula of N Nigeria (27). The ashes are used in Mali and the Chad region in indigo dyeing (Portères fide 26). Throwing-sticks are carved from the wood in Sudan. These are pared down to carry a sharp cutting edge. Previously used in warfare, they are now for hunting and ceremonial (6). The twigs are used as chew-sticks in Sudan (22) and in Senegal where they are called *sotio* (41). Chips of wood, known as *hangugu* in N Nigeria are used as a soap-substitute (Bargery fide 15) and thorns from the twigs are used by the Fula of N Nigeria to apply cosmetic eye-black (27).

The bark yields a strong fibre and contains a resin which can be collected in tears of globular pieces of varying size and greenish-yellow to orange-red in colour. The exudation is known as *ƙaron aduwa*, Hausa, is soft and full of fluid pleasant to suck when taken fresh from the tree (15). The gum hardens to a firm cement which is used in the Narok district of Kenya for fixing arrow and spear-heads to the shaft (17, 18). It melts with heat and is mixed, in that area,with maize-meal for chest-complaints or taken ground up into a powder with milk for pneumonia (17, 18). The bark alone may also be chewed (17). In Central Sahara the powdered bark is administered for angina and bronchial troubles (32). Bark on live embers provides an inhalation taken by man and animals for colds in the Richard-Toll area of Senegal (10). The Fula of N Nigeria prepare the bark as a cough-medicine (27), and in Adamawa the Margi tribe take the liquid in which bark has been boiled, together with a strong laxative, as an abortifacient (33). Yedina at Lake Chad burn the bark as a healing fumigation for circumcision wounds (15).

The root-bark is ground to a powder in Senegal, added to water which is drunk for snake-bite and some is used to wash the wound (1, 29). It also has application for treating insanity, jaundice, yellow-fever and syphilitic conditions. The Fula of Senegal use a root macerate as a purgative and to soothe colic (28, 29). In Tanganyika the pounded root is eaten with porridge as a galactogogue, and sliced and boiled with a flour (*mtama*) the liquid is drunk for syphilis (40).

In the guinean and forest zone of W Africa the common fish-poison comes from *Tephrosia vogelii* (Leguminosae: Papilionoiideae) but this does not extend into the drier north where the bark of *B. aegyptiaca* is widely used. Both bark and roots can be used. They lather in water and are used for laundering clothes: cf. above, the wood chips as a soap substitute. Their piscicidal property appears to lie in the contained saponin (7, 12, 15, 29, 30, 36). Some races in Mali add leafy stems of *Cissus quadrangularis* (Vitidiaceae) to *Balanites* bark for synergistic effect (30). Trunk bark is also anthelmintic and is used as such by the Fula of Senegal (28) and elsewhere (42).

The flowers are sucked by children to obtain the nectar (42). They are added to soup as a supplementary food in West Africa (24). Flowers and young leaves may be added to *daudawa* (Hausa), a food prepared from *Parkia filicoidea* (Leguminosae: Mimosoideae) in N Nigeria for ceremonial consumption (15, 24).

The fruit is broadly oblong-ellipsoid, 2·5–3·75 cm long, green, tomentellous when young and ripening to yellow and glabrous. Its resemblance to the date with a sticky pulp surrounding the hard woody nut enclosing the kernel brings it the common name *desert date.* There is some variation in the reported proportions of the different parts of the fruit: pulp 15–45%; nut 40–50%; kernel 9·5–12%; oil 4·14–6·4% of the fruit, 8·21% of the nut, 30–60% of the kernel (5, 11, 15, 30, 39, 42). The fruit, and hence the tree, is of great value to all the desert races so that it enters superstitions and ceremonies. In Bornu a proverb says 'a *bito* tree and a milk-cow are just the same' (8a). In Hausa the numerous uses of the fruit are proverbial in the expression *azabora mai-riba goma* (Bargery fide 15). Superstitious prescriptions (*guba,* Hausa) contain the fruit to ensure immunity in fight, protection against defeat in boxing, etc. In Bornu the fruit is an ingredient, along with tamarind and others, as an antidote to arrow-poison, to be taken internally. In E Africa chips of the wood are placed in elephants' dung as a charm against the ferocity of the elephant. Mistletoe (*kauchin aduwa,* Hausa) growing on *Balanites* forms part of a magical prescription, of which the hoopoe bird and washings of Koran texts are necessary ingredients, which imparts zeal or intelligence to youthful scholars (15). In Upper-Volta bark taken for fish-poisoning may only be removed from the branches by men who must avoid meeting pregnant women. The bark is knocked off with a cadence of blows accompanied by songs and dancing imitating swimming fish, and is then placed in the stream with cries of *ibo! ibo!* (die, die) (30). In Saharan Morocco a fruit hung round the neck is a protective amulet against sorcerers who drink blood (35).

The fruit-pulp is fibrous, dark brown, oily and gummy. It is edible with a bitter-sweet taste and may be laxative. Fruits are a common item of market merchandise. The pulp is used to make sweetmeats, and macerated in water it affords a pleasant drink (*kango,* Hausa) and in many places is fermented to an alcoholic beverage. It has been suggested as a source of alcohol. The juice is added to cereals to make a food known in Hausa as *agwalele, ɓalangade* (Katsina), or *biri biri* (7, 15, 23). The pulp contains 38–40% reducing sugars, 15% acidic substances, 45·6% carbohydrates, 1·4% minerals (11, 39, 42). It is nourishing and particularly palatable to children. A popular sort of paste is made of the pulp mixed with the gum (11). The fruit is also eaten by all stock and is in fact an important forage in season (normally March and November, 4) in the sahelian zone (2).

The unripe fruit is astringent, and more markedly bitter than when ripe (15, 42). It has been called Egyptian myrobalan.

The fruit and the seed-oil are used in Senegal as an embrocation for rheumatism (28, 29), and in spite of edibility the fruit is considered anthelmintic, emetic as well as purgative, and a remedy for liver and spleen conditions (42). The roots and seeds in tests for antimalarial activity have shown no action on avian malaria (13).

The seed is is very hard. It is used as pieces in a *warri* board game. Smaller ones may be pierced for stringing into rosaries and necklaces. In N Nigeria they may sometimes be made into dolls (44). In Ivory Coast the powdered shell is used as a fish-poison, and after 5–6 years contact with it, the fishermen's eyesight may be lost (30). The husk is said to be suitable for making industrial activated charcoal (3), but suitable machinery would have to be devised for removing the pulp and kernel.

The kernels are nutritious. As the husk is hard to break, the kernels are normally obtained by hand-cracking (4). They contain about 40–60% oil, known as *zachun oil,* which is prized in Africa as an inunction and wound-dressing (15, 36). Composition of the oil is: linoleic acid, 38·6%; oleic acid, 33·7%; palmitic acid, 16·4%; stearic acid, 11·3% (11). It is edible and good for making soap. The kernels also contain a high amount of protein — 27·6%, and the plant has been recommended as of interest for arid zone cultivation (20). Commonly the kernels are used for making into a sort of bread and added to soup. In the Shari and Chad areas the seeds constitute a staple for some peoples (23, 24). The fruit can be sun-dried and stored for a long period. The kernels are commonly sold in N Nigerian markets as a medicine for feverish chills and aches. Along with other oil-seeds, it is used to yield an oily extract (*kufi,* Hausa) which is applied to sores on camels and for parasitic skin-conditions, etc. (15).

The meal left after the removal of the oil is rich in proteins, as rich, in fact, as that of ground-nut and soya-bean. Two proteins are present: *balanine* and *aegyptine* (11, 29, 42). A bitter principle, *balanitine,* is also present but such is the value of the kernel, it does not restrict usage. The bitterness can be eliminated by twice cooking (11).

Of particular interest is the presence of steroidal saponins in the plant, present probably all over, but especially in the fruit-pulp and kernel. Action of the bark and roots as piscicides, already mentioned, is due to this, not to rotenone. Fruit-pulp is recorded as containing as much as 7·2% saponin. In the kernel and roots it is about 1%; in stem bark 0·7% (29, 36). It has been found an effective molluscicide, acting on fresh water snails which are the intermediate hosts of the *Schistosoma* trematode and on the free-living larval stages of this parasite, the causal agent of schistosomiasis. It is also lethal to *Cyclops,* the water-flea, which carries guinea-worm disease. As the fruits are freely eaten by man and animals, it appears that the saponins do not confer any toxicity towards them, and wells and water supplies may therefore be safely treated with the fruit (3, 14, 16, 26, 36, 42). The planting of *Balanites* around water-holes and along stream-banks is recommended. Very strong emulsions of the fruit are used in E Africa as a fish-poison (14, 16).

Analysis of the saponins present has shown there to be a mixture of isomers, *diosgenin,* one-third of the whole, and *yamogenin,* two-thirds. Diosgenin is a commercially important substance as a starting-point for the manufacture of cortisone and other valuable cortico-steroid drugs, especially sex-hormones. Large-scale handling of the fruit for the extraction of these chemicals has been hampered by the difficulty of decorticating the seed, but it has been found that disogenin is produced during germination and it is possible that germinating seed rather than dormant seed is more practical to handle (5, 9, 19, 20, 29). Further study is necessary. A method of cutting the hard seeds in half, leaving the embryo end viable and capable of growth, has been devised. The other half of the seed is then submitted to solvent-extraction (20).

References:

1. Abdoul Oumar, 1962: 13. 2. Adam, 1966, a. 3. Ainslie, 1937: sp. no. 44. 4. Anon., 1946, b: 142. 5. Anon., 1961. 6. Arkell, 1939. 7. Aubréville, 1950: 366. 8. Barth, 1857–8: (a) 2: 314; (b) 3: 449. 9. Bouquet & Debray, 1974: 175. 10. Boury, 1962: 15. 11. Busson, 1965: 322 with chemical analyses. 12. Chevalier, 1937, a: 20. 13. Claude & al., 1947. 14. Dale & Greenway, 1961: 533. 15. Dalziel, 1937. 16. Eggeling & Dale, 1952: 405–7. 17. Glover, Gwynne & Samuel, 270, K. 18. Glover, Gwynne & Samuel, 723, K. 19. Hardman, 1969: 202, 207–9. 20. Hardman & Sofowora, 1972. 21. Howes, 1946. 22. Hunting Tech. Survey, 1968. 23. Irvine, 1948: 233, 267. 24. Irvine, 1952, a: 31, 34, 38. 25. Irvine, 1956: 39. 26. Irvine, 1961: 208–9, with leaf analysis. 27. Jackson s.d. 28. Kerharo & Adam, 1964, b: 410–1. 29. Kerharo & Adam, 1974: 790–4, with phytochemistry and pharmacology. 30. Kerharo & Bouquet, 1950: 152. 31. Leriche, 1951: 1228. 32. Maire, 1933: 144. 33. Meek, 1931: 1: 231. 34. Monad, 1950: 70. 35. Monteil, 1953: 87. 36. Oliver, 1960: 20, 48. 37. Roberty, 1953: 449. 38. Robin, 1947: 58. 39. Schunck de Goldfiem, 1942. 40. Tanner, 1222, K. 41. Trochain, 1940: 275. 42. Watt & Breyer Brandwijk, 1962: 1064–5. 43, Wigg 324, K. 44. McIntosh, 26/1/79.

## **Balanites wilsoniana** Dawe & Sprague

FWTA, ed. 2, 1: 364. UPWTA, 311.

West African: **IVORY COAST** ABE béchieta (Aub.) AKYE béchieta (Aub.) kobo, koibo (A&AA) BAULE laule (Aub.) ulélé (Aub.) **GHANA** VULGAR krobodua (DF) kurobow *the gum* (FRI; JMD) ADANGME-KROBO krobodua (Enti) AKAN-AKYEM krobodua (Enti) ASANTE krobodua, krobo: *the gum obtained;* dua: *tree* (auctt.) waka (auctt.) ASANTE (Denkyera) krobodua (Enti) FANTE krobodua (CJT) WASA krobo (JMD; CJT) krobodua (CJT) ANYI-AOWIN krobo-baka (FRI) SEHWI krobo (JMD; Taylor) krobo-baka (FRI) BAULE laule (FRI) ulele (FRI) GA krɔbɔo *the gum; from Akan-Twi* (FRI) **NIGERIA** EDO úbòghò (JMD; KO&S) YORUBA budare (JMD; KO&S) egungun-ekún = *bone of leopard* (Kennedy; KO&S)

A large tree attaining 38 m height by over 3 m girth and becoming a dominant species of the closed forest, recorded from Ivory Coast to S Nigeria, and from Cameroun to Zaïre and Uganda. The species in Kenya and Tanganyika hitherto known as this one is now referred to *B. maughamii* Sprague, a south-eastern African species (Sands, MS in Herb. K, 1981).

The trunk has high buttresses and twisted fluting. The wood is soft and straight-grained, white but yellowing in time, fairly heavy. It works quite easily, polishes well, and is suitable for general construction work (10, 11).

The bark contains numerous horizontal ring-like markings which are the cells containing a copious quantity of scented gum. This is commonly collected in Ivory Coast and in Ghana for making into a cosmetic. The gum is allowed to dry and ground to a powder for use as a dusting-powder or the powder is mixed up with a little water to make a pomade for application to the neck or armpits (1, 2, 4, 6, 10). The ointment is also applied in Ghana to newborn babies (11) and in Ivory Coast to suckling babies to make them grow big (1).

Morphologically the fruit is a drupe. A steroidal saponin, *diosgenin,* is present in the fruit-pulp and in the kernel. This is a substance of interest as a starting point for the preparation of pharmaceutical steroids (3, 8). Whether the fruit-pulp has molluscicidal and arthropodicidal properties as have other *B. spp.* merits investigation.

The seeds are oil-bearing. West African material has been reported as containing 30% oil of the kernel on dry weight, but since the seed-coat amounts to 79% of the whole seed, the net quantity of oil is relatively small (9). The oil is brownish, acidic, and of an unpleasant taste. The seeds are, however, said to be edible, and are used as food by the Baamba people of Uganda who also make an unguent of the oil (7).

The roots of uncertain provenance have been tested for anti-malarial activity and found to be ineffective against avian malaria (5).

References:

1. Adjanohoun & Aké Assi, 1972: 309. 2. Aubréville, 1959: 2: 128. 3. Bouquet & Debray, 1974: 175. 4. Burtt Davy & Hoyle, 1937: 133. 5. Claude & al., 1947. 6. Dalziel, 1937. 7. Eggeling & Dale, 1952: 407–8. 8. Hardman, 1969: 202, 207–9. 9. Hébert, 1914: as *B. tieghemi* A. Chev. 10. Irvine, 1961: 209–10. 11. Taylor, 1960: 325.

# BALANOPHORACEAE

**Thonningia sanguinea** Vahl

FWTA, ed. 2, 1: 667. UPWTA, ed. 1, 298.

English: ground pineapple (Ghana, Irvine).

West African: **SIERRA LEONE** LOKO ngororugu (NWT) **IVORY COAST** ABURE bétianbitibe (B&D) AKAN-BRONG sassiabéréké (K&B) ANYI buro abéle = *pineapple of the bush* (K&B) KRU-GUERE gruhain (K&B) grukoma (K&B) GUERE (Chiehn) niéssagiué (B&D) KULANGO nabianihorongo (K&B) KYAMA duatuigui (B&D) **GHANA** ADANGME-KROBO ablɛfotã (Bunting, ex FRI) kpadei (Bunting) AKAN-ASANTE kwabɛ dwea = *infertile forest oil-palm* (FRI) TWI ananse-abedwaa (Bunting) kwabɛ dwea, kwa(e): *forest;* (a)bɛ: *oil-palm;* dwea: *infertile; i.e.* = *infertile forest oil-palm* (FRI) NZEMA adzilɛ ananse = *ground pineapple* (FRI; JMD) **TOGO** TEM (Tshaudjo) tshitshing (Gaisser) **NIGERIA** EDO èdín-égũi (JMD; Lowe) èdín-òtọ̀ (JRA; JMD) HAUSA kubla (JMD) kúllaá (JMD; ZOG) IGALA obi atu = *duiker's kolanut* (Boston) IGBO (Owerri) akankwanza (JMD) YORUBA adélẹ̀, adé-ilẹ̀ = *crown of the ground* (JMD; Verger) óóyá-ilẹ̀ = *comb of the ground* (JMD; Millen)

The plant is an obligate parasite of trees and perennial woody plants generally producing a stout long rhizome, tuberous at the point of attachment to the host's roots. Only the scaly flower-heads appear above ground up to 7 cm high. Recorded in West Africa from Sierra Leone to W Cameroons, and occurring widely throughout the rest of tropical Africa.

The plant parasitises plantation crops, e.g. Hevea, oil-palm, cacao. Infestation is not normally fatal, but it must cause some loss of vigour in the host, though, heavy infestation has been recorded on *Hevea* in the Southern Province of Sierra Leone as the cause of killed trees (13).

The cryptic habit of the plant has led to it being endowed with magical attributes, and fanciful names. The imbricate flower-heads result in comparison to a pineapple fruit and to the oil-palm (trunk) — see vernacular names above. Similar names are recorded in the equatorial region of Africa (Staner fide 7). The Yoruba, in calling it *Crown of the Earth,* invoke it for longevity and to avoid death (15). The Igala name, *duiker's kola nut,* arises from a belief that antelopes eat the inflorescences (4).

The rhizomes are a common article of market trade in N Nigerian markets. The Hausa use them as a flavouring for soups, etc., and also for medicinal use as a vermifuge (7, 9, 10). In S Nigeria the Yorubas use the flower-heads, along with other medicines, as a vermifuge, and take a decoction for sore-throat (7). The whole plant is pounded up and used as an astringent; it is thought to be an aphrodisiac in Nigeria (3) and in Ivory Coast (2) where, as an anthropomorphic simile, one eats the active meristematic part of the young flower-head just burst out of the ground. In central Nigeria an ointment is prepared for treating swellings on the neck and around the ears (4). In Ghana the rhizome and flowers, without the red bracts, are made into an ointment for application to skin-diseases (7, 8, 11). In Congo the whole plant is prepared as a plaster to maturate abscesses, and crushed and diluted in water is a mouth-wash for dental caries, gingivitis and mouth-infections, and sap is given to a suckling infant with fever as an embrocation applied to the infant's body after the tummy has been pricked with the flower-head scales (5). The plant is used in medicine in N Nigeria for dysentery (Barter fide 7) and in Zaïre for dysentery and blennorrhoea (14) and in Congo the rhizome is made into a tisane taken for rheumatism (5).

The red-coloured flowers, perhaps on the Theory of Signatures, are crushed with a pimento into a paste in eastern Ivory Coast for use as an enema for haemorrhoids. In that territory it is also rubbed onto stiffnecks, and the Guere and Kru incorporate the flowers into arrow-poisons. Along with other plants, the inflorescence is used for leprosy, cutaneous infections, and paralysis (6, 12). The plant is also used in Congo for paralysis (5). Ash from burnt flowers is applied in Gabon to ulceration on the soles of the feet (? yaws) (16). Sap expressed from the flower-heads is used as an eye-instillation for rachitic children and premature babies in Congo (5).

In Ghana (Vigne fide 7, 8) and in Ivory Coast (2) the flower-heads are tied to the ankles of young infants to hasten their learning to walk. The pointed scales prevent sitting down in comfort!

A trace of alkaloid has been recorded in the seeds (1).

References:

1. Adegoke & al., 1968. 2. Ajanohoun & Aké Assi, 1972: 51. 3. Ainslie, 1937: sp. no. 338. 4. Boston C.28, K. 5. Bouquet, 1969: 71. 6. Bouquet & Debray, 1974: 51. 7. Dalziel, 1937. 8. Irvine, 1930: 413–4. 9. Irvine, 1952, b. 10. Irvine, 1952, a: 30. 11. Irvine, s.d. 12. Kerharo & Bouquet, 1950: 141. 13. Robey, 1970–76. 14. Staner, 1948. 15. Verger, 1967: No. 9. 16. Walker, 1953, a: 23.

## BALSAMINACEAE

**Impatiens balsamina** Linn.

English: garden balsam.

A herbaceous annual, native of India, but now widely dispersed as a garden ornamental in many sorts of cultivars. It is present and grown here and there in West Africa.

It is widely known in Asia as a substitute for henna for dyeing the fingernails for which the flower petals are used though it is said the leaves also yield a dye. The leaves are eaten in Indonesia and may be used for poulticing broken and torn nails (1, 2). The seeds are edible and contain 27% of a green viscous oil which can be used for cooking and lighting. An alcoholic extract of the flowers has a marked anti-biotic activity, the active principle being *2-methoxy-1, 4-naphthoquinone* (3).

References:

1. Burkill, 1935: 1227–8, with references. 2. Quisumbing, 1951: 554–5, with references. 3. Sastri, 1959: 168, with oil analysis and references.

## **Impatiens burtonii** Hook. f.

FWTA, ed. 2, 1: 161, incl. *I. deistelii* Gilg.

A herb with prostrate and then ascending stem to about 1 m high, of river-banks and swampy sites in montane forest of W Cameroon, and spreading to Zaïre, Tanganyika, Kenya and Uganda.

No use is recorded in the Region. In Zaïre it is noted that the leaves are used to dress sores and that the flowers are sought out by bees (2). The genus *Impatiens* is known to be a useful bee-plant supplying nectar (1).

References:

1. Howes, 1945. 2. Wilczek & Schulze, 1960: 413–4.

## **Impatiens irvingii** Hook. f.

FWTA, ed. 2, 1: 161–2. UPWTA, ed. 1, 40, as *I. villosa-calcarata* Warb. & Gilg.

West African: **LIBERIA** MANO fa lah (Har.) gbala laa klégẽn (JGA) gbele laa gêkéléi (JGA) gbolo la killi gĕ (Har.) **NIGERIA** IGBO (Obu) ìnìnè ḿmīlí (NWT) IJO-IZON (Kolokuma) òwéí ịlálị = *male ilali* (KW) (Oporoma) ọ̀bọ́ọ̀lọ̀ (KW)

A semi-aquatic herb attaining more than 1 m height, of various forms, occurring from Guinea to N and S Nigeria and W Cameroon, and in central and eastern Africa.

The leaves, but not the fruits, are eaten as a vegetable in the Nimba region of Liberia, though some races of Liberia consider the plant to be toxic perhaps deluded by witch-doctors and sorcerers who make use of the plant (1, 7). The plant is used to prepare a vegetable-salt in Zaïre (9), Gabon (8), and in Sudan (2), and the product is considered to be one of the best vegetable-salts (Portères fide 6).

Dan medicine-men in Ivory Coast use the young leaves for haematuria and schistosomiasis (5). In Congo it is used in the same manner as is *I. niamniamensis* Gilg (3) and in Zaïre to treat burns (9). No active principle has been detected in the plant (4).

References:

1. Adam, 1971: 375. 2. Andrews 1453, K. 3. Bouquet, 1969: 71–72. 4. Bouquet, 1972: 15. 5. Bouquet & Debray, 1974: 51. 6. Busson, 1956: 359. 7. Dalziel, 1937. 8. Hallé, 1962, a: F1. 45–48. 9. Wilczek & Schulze, 1960: 415–6.

## **Impatiens macroptera** Hook. f.

FWTA, ed. 2, 1: 162.

An erect branched herb to about 1 m high, of deep forest shade in S Nigeria and W Cameroon and in montane locations on Fernando Po and extending to Zaïre.

The plant has uses in Congo similar to those of *I. niamniamensis* Gilg (1).

Reference:

1. Bouquet, 1969: 71–72.

**Impatiens niamniamensis** Gilg

FWTA, ed. 2, 1: 161.

A herb to 1 m high, recorded from montane regions of W Cameroon and Fernando Po, and extending into central and East Africa.

No usage is recorded for the West African region. In Congo, where it is the commonest species of the genus, it, like the genus as a whole, is used to make poultices and wet dressings to assuage painful affections such as whitlows, violent migraines and painful joints. The leaves are eaten as a vegetable for heart-troubles and for serious illnesses due to evil spirits (1). In Zaire the leaves are eaten as a vegetable and used to make a vegetable-salt. The plant has medicinal use for treating seizure of the joints and in dressings on wounds and sores (2).

References:

1. Bouquet, 1969: 72. 2. Wilczek & Schulze, 1960: 410.

**Impatiens sakerana** Hook. f.

FWTA, ed. 2, 1: 162, as *I. sakeriana* Hook. f. UPWTA, ed. 1, 40, as *I. sakeriana* Hook. f.

West African: **WEST CAMEROONS** DUALA boloma (Ithmann) KPE boloma (JMD)

A stout erect herb to nearly 3 m high (? epiphytic, 3) of montane forest localities in W Cameroon and Fernando Po, and in Cameroun and Gabon.

The flowers, deep red with yellow standard petal marked with two crimson blotches, are highly ornamental.

The fruits are eaten by people on the Cameroon Mountain (1, 2).

References:

1. Dalziel, 1937. 2. Adam, 1971: 375. 3. Hallé, 1962, a: 22.

## BASELLACEAE

**Basella alba** Linn.

FWTA, ed. 2, 1: 155. UPWTA, ed. 1, 38.

English: Ceylon spinach; Indian spinach; Malabar night-shade.

French: Epinard indien (Busson).

West African: **SIERRA LEONE** VULGAR bolongi (GFSE) KRIO brɔd-bɔlɔgi = *broad bologi* (FCD) **GHANA** AKAN-TWI ɛfan , fan GA ʃwie *general for pot herbs* (FRI; JMD) GBE-VHE ama (FRI) **DAHOMEY** BATONNUN sakonu (A.Chev.; FB) **NIGERIA** YORUBA amúnútutù (CWvE) bọlọgi *correctly refers to Crassocephalum biafrae* (Oliv. & Hiern) S. Moore (JMD) ẹ̀fọ̀ òyìbo = *European herb* (JMD) ṣjẹ̀ sọ̀ọ̀rọ̀ = *causing blood to be yielded freely* (JMD)

A glabrous annual, or short-lived perennial, succulent scrambling twiner, a polymorphic Afro-asian plant now dispersed pan-tropically by man, occurring in West Africa in cultivation and subspontaneously on margins of cultivated land, in thickets, on forest edges, often by water, from sea-level to montane situations. It will tolerate conditions of high humidity but prefers a drier tropical climate. It is easily raised from seed and by cuttings.

There are two main races, one green, the other red in overall colour. The young succulent stems and leaves of both are eaten as a vegetable (2, 3, 5–9). In Ruanda the fruits are eaten in time of dearth (5). The leaves are said to be high in calcium and iron, and to be a good source of vitamins A, B and C (10). On account of the mucilage contained in the leaves and tender stems, poultices are made of them in several Asian countries (1, 9, 10) and in the Antilles (4). The leaves are slightly laxative and have been used in SE Asia for treating constipation in children and pregnant women, and also for cases of urticaria (1, 9, 10).

A dye matter can be obtained from the red form which has been used as a colouring in foods though much use of it is not recommended (1). The dye has found use in China at one time for colouring official seals as well as for a cosmetic rouge (1, 10).

Sap from the red form is considered in Java helpful in cases of conjuctivitis as eye-drops, and the root in the Philippines as a rubefacient (1).

References:

1. Burkill, 1935: 307–8, as *B. rubra* Linn. with references. 2. Busson, 1965: 155–6 with leaf-analysis. 3. Dalziel, 1937. 4. Hartwell, 1967. 5. Hauman, 1951, c. 6. Irvine, 1930: 49–50. 7. Irvine, 1956: 36. 8. Irvine, s.d. 9. Manjunath, 1948: 159, as *B. rubra* Linn., with leaf-analysis. 10. Quisumbing, 1951: 285–6, as *B. rubra* Linn. with references.

## BEGONIACEAE

**Begonia fusicarpa** Irmsch.

FWTA, ed. 2, 1: 761, and at 1: 220 as *B. fissicarpa* Irmsch. UPWTA, ed. 1, 64 as *B. rubro-marginata* sensu Dalziel, non Gilg.

A woody herb reaching 1 m high, epiphytic on trees of the forest areas of Sierra Leone and Ghana. The flowers are pinkish-white. The plant is of possible ornamental value.

The plant-sap is applied to wounds in Sierra Leone (1).

Reference:

1. Deighton 709, K.

**Begonia mannii** Hook.

FWTA, ed. 2, 1: 220. UPWTA, ed. 1, 64.
West African: **NIGERIA** EDO ẹmùnọ̀mùerhán (Kennedy)

A scrambling epiphyte on trees of the forest zone in lowlands to montane elevations in Sierra Leone to W Cameroon and Fernando Po, and also occurring in E Cameroun and Gabon.

No use is recorded of this species in the West African region. In Gabon it is eaten as vegetable, and constitutes a favourite item of food for chimpanzees and gorillas (2). In Congo certain tribes use it exclusively for treating inflammation of the vagina and of the uterus by eating the leaves as a vegetable and by vaginal douches and hip-baths of an aqueous decoction of the leaves with bark of *Cyclodiscus gabonensis* (Leguminosae: Mimosoideae) (1).

References:

1. Bouquet, 1969: 72. 2. Walker & Sillans, 1961: 100.

**Begonia oxyloba** Welw.

FWTA, ed. 2, 1: 218.
English: bush okra (Liberia, Leeuwenberg).
West African: **LIBERIA** EDO (Nimba Mts.) okré (AJML)

A succulent herb to 60 cm high from woody rootstock, occurring in lowlands to over 1,000 m altitude from Guinea to W Cameroon and Fernando Po, and into Uganda, Tanganyika and Angola.

Though *Begonia spp.* are not usually cropped by game, heavy grazing by antelopes of a near-pure stand of this species has been recorded on the Nimba Mountain in Liberia (1).

Reference:

1. Adam, 1971: 373–4.

**Begonia quadrialata** Warb.

FWTA, ed. 2, 1: 218. UPWTA, ed. 1, 64.
West African: **IVORY COAST** ANYI tarié (A. Chev.)

A forest herb, on stream sides, occurring fairly widely in the Region from Guinea to W Cameroon, and in E Cameroun.

**Begonia spp.**

FWTA, ed. 2, 1: 216.
West African: **IVORY COAST** 'KRU' bokia hao (K&B)

An unspecified species with the 'Kru' name *bonia hao* is used by some medicine men in the SW corner of Ivory Coast for bronchial affections and asthma: the powdered dried leaves are taken with a little water or palm-wine (2). The dried powdered leaves of *Begonia* species generally seem to be taken in Ivory Coast for attacks of asthma (1).

Several species are sufficiently ornamental to merit cultivation.

References:

1. Bouquet & Debray, 1974: 51. 2. Kerharo & Bouquet, 1950: 43.

## BIGNONIACEAE

**Bignonia capreolata** Linn.

FWTA, ed. 2, 2: 385.
English: cross vine, quarter vine, trumpet flower (U.S.A., Bates).

A climber reaching 17 m long, native of southeastern N America and introduced to many warm countries as an ornamental for its showy yellow-red flowers 5 cm long. The liane of this and some other related species has a cross-like arrangement of the tissues which accounts for the first English name, and leads to those plants having religious and superstitious uses in the New World.

**Crescentia cujete** Linn.

FWTA, ed. 2, 2: 385.
English: tree calabash.
French: calebassier.
West African: **SIERRA LEONE** BULOM (Kim) g-bas (FCD) BULOM (Sherbro) thɔk-i-pɛpɛ-lɛ (FCD) KISSI n-gulu-kokui (FCD) KONO kɔm-bara (FCD) KRIO kalbas-tik = *calabash tree* (FCD) LIMBA (Tonko) tatogo (FCD) LOKO k-pombi *a calabash cut across the stem end to*

*make a bowl* (FCD) n-guru-g-bulɔ (FCD) tala *a calabash split lengthwise* (FCD) MENDE ngulu-gbula, ngulu gbulɔ (FCD) SUSU leŋge-wuri (FCD) VAI koŋgo-kɔ̃s (FCD) **GHANA** AKAN-ASANTE nsã-kora-dua, nsã: *palm wine;* kora: *calabash;* dua: *tree* (Enti) FANTE dwera aba (FRI)

*NOTE: The vernaculars usually mean 'tree calabash'. The ordinary calabash is Lagenaria. For the calabash itself the names usually apply to either species.*

A handsome dark-foliaged tree to about 13 m high, native of the Caribbean region and widely dispersed by man in tropical countries.

The tree bears flowers with a yellowish-red or purple-veined corolla and fruits to 30 cm in diameter like a gourd for which the tree is grown and as an ornamental.

The wood is light brown with darker veins. It is soft, but tough and flexible. In the West Indies orchid growers use blocks of the wood as a base on which to grow epiphytic orchids (7).

The bark is soft and spongy. It is used in Sumatra in decoction to clean wounds (2), and elsewhere it is said to be good for treating diarrhoea with mucous (4).

The flowers yield nectar for honey (3).

In Brazil juice from the young unripe fruit is taken for colds and lung-ailments and in the Virgin Islands the fruit-pulp boiled to a thicker liquor is medicine for respiratory conditions (6). An alcoholic extract of the not quite ripe fruit is taken in small dosage as a mild aperient: in larger doses it is drastic, but without griping or other ill effect. An alcoholic extract of the ripe fruit is also laxative, and expectorant as well. Fresh pulp boiled in water turns into a black paste which has been used against erysipelas (4). The fresh fruit-pulp is taken in Gabon as a laxative (5). In West Africa and the Caribbean it is macerated in water and is considered depurative, cooling and febrifugal, and good for application in headache and to burns 4, 6) and in West Africa the ash of the roasted fruit as mildly purgative and diuretic (4). In South-east Africa the fruit has been recorded as edible, but at the same time poisonous to small birds and mammals, and also as diuretic (1).

The seed is burnt to an ash which is applied in Transvaal to snake-bite and taken internally (6). Seeds contain 37% content of a fixed oil resembling groundnut and olive oils which consists of oleic acid 59%, linoleic acid 19%, linolenic acid 2% and saturated acids 20%; sugar is also present at 2.6% (6). Kernel analysis is 45% fatty substance and 38.5% protein, and particularly high thiamin (3).

The principal product of the tree is the fruit shell or calabash which is used as a container for all manner of purposes. When ripe they have very hard shells and by curing them with smoke become more durable and water proof. In the native home of the tree, there is a practice of tying the immature fruit whereby the shape may be greatly altered (1).

References:

1. Burkill, 1935: 680. 2. Heyne, 1927: 1372. 3. Irvine, 1961: 535, and seed analysis lxxviii–lxxix, as *C. alata*. 4. Quisumbing, 1951: 874–5. 5. Walker & Sillans, 1961: 100–1. 6. Watt & Breyer-Brandwijk, 1962: 142. 7. Williams, R. O., 1949: 214–5.

**Cybistax donell-smithii** (Rose) Siebert

FWTA, ed. 2, 2: 496.

English: primavera (? from Spanish; U.S.A., Bates).

A tree to 25 m high, native of Mexico and Guatemala, introduced to various tropical countries, and grown in W Africa as an ornamental for its yellow tubular-campanulate flowers. In its native home it is an important timber tree.

**Haplophragma adenophyllum** (Wall.) P. Dop.

FWTA, ed. 2, 2: 385.

A stout tree to 15 m high, native of NE India to the Malay Peninsula, introduced to the Region and cultivated as an ornamental for its yellowish-brown flowers borne in panicles about 20 cm long. The leaves are unusually large attaining 60 cm in length.

**Jacaranda mimosifolia** D. Don

FWTA, ed. 2, 2: 385.

English: jacaranda; blue jacaranda; green ebony (the wood).

French: jacaranda.

Portuguese: caroba (not to be confused with the carob tree, *Ceratonia siliqua,* Leguminosae: Caesalpinioideae); jacaranda.

A deciduous tree to 15 m high or more, native of NW Argentine, and distributed by man to most warm countries.

The tree is principally grown for its attractive violet coloured flowers. It likes a seasonal climate with a marked dry season, and it must have a well-drained situation. Clonal propagation when superior individuals have been come by is easy by cuttings and suckering. The wood is very beautiful and somewhat scented; it is moderately hard and heavy, well-textured and is easy to work. It is one of a group of timbers known as 'green ebony' and is used in carpentry, cabinetry etc.

In S America the bark and leaves are used in medicines for syphilis and urethral discharge. A leaf-infusion is a pectoral, and the powdered leaves a vulnery. Bark-infusion is made into a lotion for ulcers (2). A dye is extracted from the wood which furnishes green and other compound shades (1).

In India the tree is a host of the Indian lac insect (2).

References:

1. Greenway, 1941. 2. Sastri, 1959: 277.

**Kigelia africana** (Lam.) Benth.

FWTA, ed. 2, 2: 385. UPWTA, ed. 1, 443, incl. *K. acutifolia* Engl., *K. aethiopica* var. *bornuensis* Sprague and *K. elliptica* Sprague.

English: sausage tree.

French: saucissonnier; faux baobab (Dalziel).

West African: **SENEGAL** MANDING-BAMBARA sidiamba (auctt.) MANINKA dindon (Aub.) lambâ (Aub., ex K&A) limbi (Aub.; K&A) tuda (Aub.) SERER sayo (auctt.) SERER-NON humbul (auctt.) WOLOF ɗabal, diambal (auctt.) dabolé, dombalé (K&A) **THE GAMBIA** FULA-PULAAR (The Gambia) jirlahi (DAP) MANDING-MANDINKA limbi (DAP) WOLOF jambal (DAP) **GUINEA** MANDING-MANINKA dindon (Aub.) limbi (Aub.) limbi lamban (CHOP) tuda (Aub.) SUSU tuda (CHOP) **SIERRA LEONE** TEMNE *an*-gbonthi (FCD) *an*-tua (GFSE) **MALI** MANDING-BAMBARA sidiamba (Aub.) MANINKA dindon (Aub.) limbi (Aub.) tuda (Aub.) SONGHAI kombolgna (A. Chev.) **UPPER VOLTA** MANDING-DYULA findia(n) (Aub.) findiam (K&B) siai (Aub.) sidia findia (K&B) **IVORY COAST** ABE tombo (auctt.) AKAN-BRONG assongui (K&B) AKYE mia lébé (RS; Aub.) ANYI brimau (RS; Aub.) BAULE blima (B&D) blimo (Aub.; K&B) brimbo (Aub.) KRU-GUERE (Chiehn) blumo (B&D) KULANGO kuruko (K&B) KWENI non (B&D) MANDING-DYULA findia(n) (Aub.; K&B) findiam (K&B) siai (Aub.) sidia (K&B) sidia findia (K&B) MANINKA lamban (Aub.) lemba (Aub.) limbi (Aub.) tuda (Aub.) **GHANA** ADANGME lele (FRI; JMD) AKAN-AKUAPEM nana beretee (FRI) *o*-nufuten(-e) = *hanging breast* (FRI) ASANTE nana beretɛɛ (FRI) nufutene = *hanging breast* (auctt.) FANTE *e*-nufutsen (DF) nufutsen

= *hanging breast* (Moloney, ex JMD; FRI) TWI nufoten (DF) nufuten = *hanging breast* (auctt.) nufutsen (JMD) WASA nufuten (auctt.) ANYI-AOWIN anyafɔtene (FRI) SEHWI anyafɔtene (FRI) BAULE blimmo (FRI) GA fufɔ̃-akplele = *hanging breast* (FRI; KD) tʃɔtɔtʃo (FRI; JMD) GBE-VHE nyakpẽ (FRI; JMD) nyãkpekpẽ (FRI; JMD) NZEMA anyafɔtene (FRI) **TOGO** GBE-GEN nyakpokpo (Volkens) VHE niagpé (Aub.) nyakpekpẽ (Volkens) TEM (Tshaudjo) abilu (Volkens) **DAHOMEY** GBE-FON gwam blipo (Laffitte) niapopo (Aub.) nio (Aub.) **NIGER** ARABIC (Niger) kuk (Aub.) mechtur (Aub.) HAUSA rahunia (Aub.) KANURI bùlóngó (Aub.) **NIGERIA** ARABIC um shutur (JMD) BEROM ràhéyná (LB) EDO úgbòn-gbọn (JMD; KO&S) ùsuọnbọn (auctt.) EJAGHAM-ETUNG ebe-njok (DRR) FULA-FULFULDE (Nigeria) jirlaare *the fruit* (JMD) jirlahi (*pl.* jirlaaje) (auctt.) HAUSA hántsàr giíwaá = *elephant's udder* (JMD; ZOG) kiciiciyaa (JMD; ZOG) noónòn giíwaá = *elephant's milk* (auctt.) ràháínáá (JMD; auctt.) rawuya (auctt.) IGBO alambọrọgoda (DRR) uturubein (DRR, ex JMD; KO&S) uturukpa (BNO) IGBO (Amankalu) itemi (DRR, JMD) iteni (DRR) IGBO (Arochukwu) amọ-ibi (DRR) uturu-bein (DRR) IGBO (Nsokpo) izhi (DRR) IGBO (Nsukka) izhi (DRR; JMD) IGBO (Onitsha) alambọrọgọda (DRR) uturu-bein (DRR) IGBO (Owerri) óké ọgírìsị = *male ogirisi* (*Newbouldia laevis*) IGBO (Umuakpo) umu-aji (DRR; JMD) IJO-IZON (Kolokuma) ògírízì (KW) KANURI bùlóngó (JMD; C&H) NKEM ubung (DRR) NUPE béci (Banfield; K O&S) OLULUMO-OKUNI itiwa-enyi (DRR) itiwa-enyi (DRR) TIV tyembegh (JMD) YORUBA orora (DRR) pandọ̀rò (auctt.) uyan (JMD) **WEST CAMEROONS** DUALA èsembea (Ithmann) èwùŋgè (Ithmann) KOOSI sosong (AHU) KPE bulule (JMD) wulule (Rider)

A tree to 23 m, usually less, extremely variable, of the rain-forest, guinean and soudanian savanna, usually in damp sites, often riverain; throughout the Region and widespread in tropical Africa. A large number of *Kigelia spp.* has been described for Africa, but it is considered that the genus is monospecific and variations are due to ecological influences, affecting also phytochemistry.

The tree has a low-branching trunk, sometimes tortuous (Senegal, 24), with a large spreading crown. It is often planted as a village 'palaver' tree and roadside shade-tree, and for the flowers and fruits which are in places regarded as fetish (Sierra Leone, 13; Ivory Coast 25, 26; Nigeria 11, 12; Fernando Po, usually in hedges, 43; Gabon 37; Uganda 32). Hausa expatriates in Congo often plant it near their dwellings (6). To the Ijo of Kaiama it serves as a boundary tree to mark territorial limits (42), but their use is forbidden to the Yenagoa Ijo (41). In Central Africa the tree is deemed holy and religious meetings may be held in its shade (38).

The wood is light, white with pale brown heart (4, 20, 21, 23). It is not valued commercially. In N Nigeria the Fula use it for fence-posts claiming that it does not rot (22). In Botswana (28), S Rhodesia (38) and Malawi (40) it is considered good for dugout canoes, so good in fact in Malawi as to be protected. It is also used in the last named territory for tool-handles, mortars, drums, boxes, and in E Africa for stools (40). In Sudan the branch-wood is used for bows (8). At Kumba in southern W Cameroons the small branches are hollowed out to make enema tubes for use on children (30).

The leaves are sometimes used by the Baule and Dyula of Ivory Coast to treat dysentery and stomach and kidney complaints (25, 26). The Ijo of S Nigeria take five leaves and seven peppers [? maleguetta], chew them together and 'usi' [? meaning] a person who has fainted and has just been brought round again: this weakens or cures the illness that caused the fainting (42). An infusion of the leaves with other leaves is used for treating an undefined venereal disease (*kaswende,* ciNyanja, Malawi, 40).

The bark has a somewhat bitter taste. In Senegal it is used with the bark of *Mitragyna inermis* (Rubiaceae), the roots of *Xanthoxylum xanthoxyloides* (Rutaceae) and the seeds of *Sterculia setigera* (Sterculiaceae) for epilepsy. It enters into various prescriptions for leprosy and the roots into medico-magical treatments for sterility (24). The Baule and Dyola of Ivory Coast use the bark for dysentery and stomach and kidney complaints, but only in conjunction with other drug-plants. Administration is by enema or draught. They use similar preparations in vapour-baths to treat snake-bite — this softens the wound facilitating the action of medication applied afterwards (25, 26). In Sierra Leone heated bark is applied to women's breasts to hasten their return to normal after a suckling child has been weaned (13). Similarly in Tanganyika the Sambaa use the bark to reduce swelling of the breasts (34). In Nigeria the bark is pounded and taken internally for dysentery; the bark and fruits are powdered and the dust is applied to sores, while of the same an oily ointment is made to rub on rheumatic parts and on malignant tumours (2). Rheumatism and dysentery are treated in Ghana with a bark-preparation (12). The bark is used in N and S Nigeria for syphilitic conditions and for gonorrhea, *e.g.*, boiled with red natron and guinea-corn flour, or in infusion along with other herbs and grains-of-paradise, and in S Nigeria the bark is prepared along with the fruit as a wash and drink given to young children (12). The inner bark in W Cameroons is applied after crushing to chronic wounds and sores: the application is painful. It is also dried and applied as a powder (12). Similar use is found in Tanganyika where the bark-macerate is also used to cleanse the injured place, and a root-decoction is drunk to accompany the external treatment (18). The Baakpe of W Cameroons use the bark in enemas (30) and internally for dysentery, and with *Jateorhiza* (Menispermaceae) for snake-bite (12). A bark-decoction, or a fruit-decoction, is taken in Congo to relieve asthma (6), and in Tanganyika the inner bark is soaked in water which is then used for washing in treating ringworm (35). The latter preparation is also taken internally for constipation.

The bark is reported to contain a bitter principle and tannic acid (38). Alkaloid has been found absent from Nigerian plant material (1).

In E Africa the root is used as a remedy for boils and sorethroat (38), and in Tanganyika a medicine is made of them for infertility in women (33). In Malawi the root along with other roots is soaked in water which is drunk for syphilis (40). The bitter root in Ghana is taken in decoction for constipation and for tapeworm (20).

The flowers in long lax pendulous panicles to nearly 1 m long are maroon coloured, yellow-veined, and attractive. They are eaten by cattle and game in Rhodesia (38) and by antelope in Botswana (28). In the Komba Division of W Cameroons sap expressed from the buds is used for sore eyes (30).

The fruits are the most commonly used part of the tree. Pendulous, to 45 cm long by 15 cm diameter, resembling in fact a 'german' sausage whereby has come both the English and the French names, they are prized as fetish emblems in many parts, and are commonly sold in markets for their fetish and medicinal virtues. The fruit is much used by witch-doctors in Dahomey (27), and in Tanganyika (38). Senoufo of Ivory Coast hang the fruits up in the rafters of their huts as a good fetish for fecundity (25, 26) and the Ghanaian names refer to their shape as a 'hanging breast' (20, 21). In eastern Africa they are hung as a charm outside huts or their ashes are mixed with seed-maize as a magic medicine to increase the crop (12). In Ghana the fruit, painted in various colours, is said to be used by fetish men in divining the cause of a disease (Moloney fide 12). In Senegal village matrons prepare a pulp of the ripe fruit which is given in decoction to take by draught by young girls before puberty and by embrocation and massage onto their breasts to promote an ample development, a practice said to be proven by recorded measurements (24). In Ivory Coast the fruit with pimento is made into a decoction which is taken in draught as a galactogogue, while the breasts are annointed and massaged with an ointment of the fruit pulp in *karite* butter (26). Luvale of S Africa rub a piece of the fruit over the breasts to increase milk-flow, and over the baby to make it fatter but avoiding the head for fear of producing hydrocephalus (38). The powdered fruit is prepared as a poultice in Jebel Marra for treatment of complaints in the breasts (19), possibly mastitis or cancer (39). In Congo in an opposing sense it is forbidden for women to touch the fruit under pain of suffering their breasts or womb (*sein,* French) falling, or if pregnant of having a miscarriage (6). The fruit is thought in places to act as a charm to secure riches and good fortune, but if taken too freely it is said to result in scrotal elephantiasis (12); yet in contrast it is used in Ivory Coast for treating scrotal elephantiasis and oedema of the legs (7). The fruit and roots along with the 'male' tassel of the plantain inflorescence are boiled together to make a women's nostrum in

Ghana and the bark and fruit are used to heal sores and to restore taste (20, 21).

The unripe fruit is not edible, nor is the ripe fruit. It is purgative and toxic (Nigeria, 2; Gabon, 36; Tanganyika, 38; Malawi, 40) and is used in Central Africa in poultices for syphilis and rheumatism (38). The fruit in Tanganyika is said to be intoxicant and a sexual stimulant (5, 38). It is a common additive to ferment in preparing beer to increase potency or to add to the flavour (3, 9, 29, 31). Sometimes the fruit is baked first and the fleshy part is added to the brew to increase strength (Tanganyika, 16; Uganda, 14; Kenya, 10); or sometimes it is the rind (12). Masai and Kipsigis of Kenya boil the fruit and bark as an additive (15). These additions have been postulated as leading to increased fermentation resulting in the formation of amyl-alcohol and explain the severe hang-over after inebriation (38) Squirrels in Malawi are said to be very fond of the fruit; whether it is to eat the pulp and seeds is not certain, but it seems that sap which they get in gnawing open the end of the husk makes them quite tipsy. Rhinoceros also eat the fruit (40), but it is not recorded whether they become affected.

A decoction of the fruits with peppers is taken in Nigeria for constipation and piles, and the powdered fruit is applied as a dressing to ulcers and for rheumatism. The fruit-ash when powdered is said to have disinfective and curative properties after the style of boracic powder (2) and is used in Sudan (8) and in E Africa (12) for sores on animals. The fruit is also used in Ivory Coast for chancres and rheumatic pain (7) and in S Africa for ulcers (38). The fruits, cut into pieces, are boiled with the roots of *Anthocleista sp.* (Loganiaceae) and the liquid is taken by draught or enema in Ghana for piles and lumbago (21). A paste is made from the fruits in Kumba District, W Cameroons, for applying to boils (30). The fruit contains tannin (17, 24).

In parts of Malawi where stones for a fire-hearth are scarce the fruits are used to stand pots on in a fire as they are almost fire-proof (40). In E Africa the husk is hollowed-out, fitted with a noose and bait and used as a mouse-trap. Masai girls make dolls of it. Some people make ladles and cups from the cut husk (12). A black dye can be obtained from the fruit (20, 21).

The seeds in Malawi are eaten after roasting in time of dearth (38).

References:

1. Adegoke & al., 1968. 2. Ainslie, 1937: sp. no. 201, as *K. pinnata*. 3. Akiley 5058, K. 4. Aubréville, 1950: 494. 5. Bally, 1937: as *K. aethiopica* Dec'ne. 6. Bouquet, 1969: 72. 7. Bouquet & Debray, 1974: 51. 8. Broun & Massey, 1929: 325, as *K. aethiopica* Dec'ne. 9. Burtt 5259, K. 10. Dale & Greenway, 1961: 60, as *K. aethiopum* (Fenzl) Dandy. 11. Dalziel 105, K. 12. Dalziel, 1937. 13. Deighton 4134, K. 14. Eggeling & Dale, 1952: 38, as *K. aethiopica* Dec'ne. 15. Glover & al., 1975, K. 16. Greenway 2024, K. 17. Greenway, 1941: as *K. pinnata*. 18. Haerdi, 1964: 148, as *K. aethiopica* Dec'ne. 19. Hunting Technical surveys, 1968. 20. Irvine, 1930: 251, incl. *K. acutifolia* Engl. 21. Irvine, 1961: 736, incl. *K. angolensis* Welw. 22. Jackson, 1973. 23. Keay & al., 1964: 424. 24. Kerharo & Adam, 1974: 236–7, with phytochemistry and pharmacology. 25. Kerharo & Bouquet, 1947. 26. Kerharo & Bouquet, 1950: 226–7. 27. Laffitte 174–1937, IFAN. 28. Miller, 1952: 79, as *K. pinnata* DC. 29. Newbould & Gifford 1653, K. 30. Rosevear, 1975. 31. Semsei FH.2074, K. 32. Styles 259, K. 33. Tanner 326, K. 34. Tanner 3158, K. 35. Tanner 3751, K. 36. Walker, 1953, a: 24. 37. Walker & Sillans, 1961: 101. 38. Watt & Breyer-Brandwijk, 1962: 143, as *K. aethiopica* Dec'ne and *K. pinnata* DC. 39. Wickens 1365, K. 40. Williamson, J. 1955: 72: as *K. aethiopica* Dec'ne. 41. Williamson, K. 4, UCI. 42. Williamson, K. A. 9, UCI. 43. Wrigley & Melville 681, K.

**Markhamia lutea** (Benth.) K. Schum.

FWTA, ed. 2, 2: 387.
West African: **IVORY COAST** AKAN-BRONG tomboro (K&B) DAN blu (K&B) KRU-GUERE poi un dubu (K&B) GUERE (Chiehn) niété brissu (B&D) **GHANA** VULGAR efuo-bese (DF) ADANGME mɔmɔtʃo (FRI) nɔkɔtʃo (FRI) AKAN-ASANTE ɛfuɔ-bese (BD&H; FRI)

A shrub or tree to 15 m high by 1 m girth, of fringing and savanna forest from Ghana to W Cameroons and Fernando Po, and extending to the Congo basin.

The tree is occasionally cultivated in Zaïre. The wood is soft, easy to saw, of an exceptional durability and is held to be an excellent timber for interior carpentry and cabinetry (4). It is also used for paddles on river-craft (2).

The tree is considered in Ivory Coast to be effective on skin-affections, sores and itch: leaves and bark are pounded up with citron juice to a paste, and the liquid is expressed for use as a lotion, while the residual lees may be used with vigorous rubbing. In the case of sores, the lees are applied as a wet dressing under a bandage (3). The plant is also used as a rejuvant and diuretic, and is given [methods not stated] for oedema of the legs and elephantiasis of the scrotum, to treat chancres and rheumatic pain, and is taken for treatment of the respiratory tract and in swamp-fever (1).

This species is very similar in appearance to *M. tomentosa*. They share several vernacular names. The other usages of *M. tomentosa* may very well also apply.

References:

1. Bouquet & Debray, 1974: 51. 2. Dalziel, 1937. 3. Kerharo & Bouquet, 1950: 227. 4. Liben, 1977: 28–30.

## **Markhamia tomentosa** (Benth.) K. Schum.

FWTA, ed. 2, 2: 387. UPWTA, ed. 1, 444.

West African: **SENEGAL** BALANTA blis (Aub., ex K&A) BANYUN kisal (K&A) sibupal sudicam (Aub.) DIOLA kasungkares (JB, ex K&A) DIOLA (Fogny) bunamkarésabu (Aub., ex K&A) FULA-PULAAR (Senegal) kafanâdu (Aub., ex K&A) MANDING-MANDINKA irigańa (JLT, ex K&A) naŋam MANDYAK bétali (Aub., ex K&A) **THE GAMBIA** DIOLA (Fogny) asungkares (DF) bunamkarésabu = *sour sweat – sugary* (DF) **GUINEA-BISSAU** FULA-PULAAR (Guinea-Bissau) n'álè (JDES) um-hálè (JDES) **GUINEA** FULA-PULAAR (Guinea) kafau andu (Aub.; RS) **SIERRA LEONE** MANDING-MANDINKA kodo (FCD) MANINKA (Koranko) sundu kumasore (D.G.Thomas) waralankɔ (NWT) SUSU-DYALONKE filɛtɛ-na (FCD) luguna (NWT) TEMNE *a*-lil-*a*-korankɔ (NWT) **IVORY COAST** AKAN-BRONG tomboro (Aub.; K&B) BAULE kravaka (B&D) DAN blu (K&B) KRU-GUERE poi un dubu (K&B) KWENI vorone (B&D) vuluné (B&D) vuruni (B&D) KYAMA bacombi (B&D) MANDING-MANINKA kokè (B&D) **GHANA** ADANGME nɔkɔtʃo (FRI) AKAN-ASANTE ɛ foɔ-bese (BD&H; FRI) kwaensa (TFC; FRI) BRONG tomboro (FRI) KWAWU obogyanebuɔ = *stone depends upon stone* (FRI; DF) GA nɔkɔtʃo (FRI) **TOGO** YORUBA-IFE OF TOGO tschitschine (Volkens) **NIGERIA** EDO ogie-ikhimwim (KO&S) ògìkhìnrùwìn (Amayo) IGBO echero (DRR) onyiri akikara (auctt.) IGBO (Enugu) og bano (DRR) IGBO (Umuahia) egbo (JMD) ògírìsì (JMD) YORUBA akoko (Phillips; JRA) ìrù àáyá, ãyá: *a species of monkey;* iru: *Parkia filicoidea* (JMD; KO&S) iwe (Dawodu) **WEST CAMEROONS** DUALA bòbèdu (Ithmann) malanga (AHU) KOOSI abbe (AHU) KPE mawelu (AHU)

A shrub or tree to 15 m high, of the relic, fringing, transition and savanna forests, throughout the Region, and extending southward to Angola.

The tree carries large yellow flowers in long terminal racemes and is quite decorative when in flower. It is grown as an ornamental in Guinea-Bissau (6). The timber is pale brown, hard and good for carpentry. It resembles that of *Newbouldia laevis,* and is used in Gabon (13). In S Nigeria it is made into knife-handles (3).

The bark is pulped up and used in Casamance (Senegal) as a poultice in the armpit for localized pain. (7, 8). In Ivory Coast the plant is held to be particularly effective in treating skin-afflictions, sores and scabies: leaves and bark are pulped up with citron-juice to a soft paste; the liquid is squeezed out and used as an embrocation, while the lees may be used in vigorous rubbing over the affected parts, or applied as a wet dressing under a bandage to sores (9, 10). Plant preparations [methods not stated] are also administered as rejuvant and diuretic medicines: for oedema of the legs and elephantiasis of the scrotum, on chancres and for rheumatic pain and in treatment of the respiratory tract and in bouts of swamp-fever (2). In Nigeria a decoction of bark and leaves is given as a mild laxative and in cases of fever; a leaf-decoction and chewed leaves are used for general pains, head, back, etc. (1). In Sierra Leone the bark is used in tanning (5).

The flower is melliferous (6). The buds are used in play by children like those of *Spathodea campanulata* (Bignoniaceae) (4). The juice from the buds is used in S Nigeria for painful eyes (11).

The tree probably has superstitious attributes in S Nigeria (4). The Igbo at Ezi are recorded as using the wood for carving images (12).

References:

1. Ainslie, 1937: sp. no. 224. 2. Bouquet & Debray, 1974: 51. 3. Carpenter 78, UCI. 4. Dalziel, 1937. 5. Deighton 4210, K. 6. Gomes e Sousa, 1930: 42. 7. Kerharo & Adam, 1963, b. 8. Kerharo & Adam, 1974: 237. 9. Kerharo & Bouquet, 1947: 252–3. 10. Kerharo & Bouquet, 1950: 227. 11. Thomas, N. W. 2349, K. 12. Thomas, N. W. 2356, K. 13. Walker & Sillans, 1961: 101.

**Millingtonia hortensis** Linn. f.

English: Indian cork tree
West African: **GHANA** GA oshí'shiu (KD)

A small tree with compound deciduous leaves and white fragrant flowers; native of Burma and Thailand, and now widely distributed as an ornamental in many hot countries including West Africa. There is no other usage recorded for it in the Region. The tree is commonly planted in Tanganyika as a street-tree. The wood is yellowish-white and is said to be usable for making furniture. The bark yields in Asia an inferior cork. The roots have medicinal use in Burma, and in Thailand the flowers are added to tobacco for smoking as treatment for throat ailments.

**Newbouldia laevis** Seem.

FWTA, ed. 2, 2: 388. UPWTA, ed. 1, 444.

West African: **SENEGAL** BALANTA gimgid (K&A) BANYUN kibompor (K&A) sibompol (Aub., ex K&A) DIOLA bu gompa (JB) DIOLA (Efok) égompa = *slippery* (K&A) DIOLA (Fogny) fugompefu (K&A) fugompö (K&A) FULA-PULAAR (Senegal) kôdomburu (K&A) KONYAGI pasal (K&A) MANDING-BAMBARA kolokolo (K&A) MANDINKA kundio buro (Aub.) kundiu mburo, kudu mbioro (K&A) kuṅdo burô (Aub.; K&A) MANINKA kinkin (auctt.) kúdum burô (K&A) moquiquiri (RS) toré (RS) 'SOCE' kúndu buro (K&A) MANDYAK bukob (K&A) MANKANYA bukuf (Aub., ex K&A) SERER gamb (JMD, ex K&A) ngam (JB) WOLOF gam, ngam (auctt.) vosvosor (JB, ex K&A) walakur (JB; K&A) **THE GAMBIA** FULA-PULAAR (The Gambia) kallihi (DRR; DAP) sukunde (DAP) MANDING-MANDINKA kunjumburo = *comes back to life, i.e. coppices well* (DF) SERER gamb (JMD) WOLOF gam, ngam (JMD; DAP) **GUINEA-BISSAU** BIDYOGO canhom (JDES) CRIOULO manduco de feticero (JDES) FULA-PULAAR (Guinea-Bissau) canhómburi (JDES) sucúndè (JDES) MANDYAK becuape (JDES) MANKANYA boukouf (Aub.) NALU singête (JDES) **GUINEA** FULA-PULAAR (Guinea) sukundé (auctt.) KONO tré-tré (RS) LOMA tolé (RS) MANDING-MANDINKA kunjumborong MANINKA kinkin (RS) mofanie (Farmar) moquiquiri (CHOP; RS) toré (RS) MANO déin (RS) dien (RS) SUSU kinki (CHOP; RS) **SIERRA LEONE** BULOM (Sherbro) tisi-lɛ (FCD; S&F) GOLA zɔdɔ (FCD) zɔrɔ (FCD) KISSI teɔ (S&F) teɔ̃ (FCD) tiɔ (S&F) tiɔ̃ (FCD) tuiɔ (S&F) tuiɔ̃ (FCD) KONO bɔidɛ-yamba (FCD) bɔidɛ-yamba (S&F) tɔkɛ (? tɔlɛ) (S&F) tɔlɛ (FCD) tuɛ (S&F; FCD) KRIO snof-lif = *snuff-leaf* (FCD; S&F) LOKO toɛ (FCD; S&F) MANDING-MANDINKA kidiŋkanya (FCD) MENDE pomamagbɛ, poma: *corpse;* magbɛ: *drive on* (auctt.) SUSU kinki (NWT; FCD) SUSU-DYALONKE dantilikofo-na (FCD) TEMNE *an*-yɔl (auctt.) VAI pomamagbe (FCD) **LIBERIA** MANO dĩ(y)a lah (JMD) din-a-lah (RS) **MALI** MANDING-MANINKA kinkin (RS) moquiquiri (RS) toré (RS) **IVORY COAST** ABE g-bâlie (auctt.) k-palié (Aub.; A&AA) ADYUKRU k-poierem (RS; Aub.) AKAN-ASANTE tokonzui (B&D) BAULE tunzué (auctt.) FULA-FULFULDE (Ivory Coast) sukundé (Aub.) KRU-BETE gba buï (K&B) siddo (K&B) GUERE bolu (K&B) gbo-u (RS) tolo-tolo (RS) zotu (K&B) GUERE (Chiehn) sido (B&D) GUERE (Wobe) gbotu (RS; A&AA) KWENI ding on (B&D) dingno dingon (B&D) KYAMA bama (auctt.) borna (A&AA) MANDING-MANINKA kinkin (Aub.) korokoro (B&D) 'NEKEDIE' bagulé (B&D) bomakrokro (B&D) **GHANA** AKAN-ASANTE sesemasa (Enti) BRONG sonsonangsayng (BD&H) FANTE esisimansa (FRI) TWI sesemasa (auctt.) WASA sesemasa (Enti) ANYI-AOWIN atronzuo (FRI) SEHWI atronzuo (FRI) BAULE tonzue (FRI) GA hĩatʃo = *man's tree* (FRI; JMD) hĩiabaatso (FRI) hĩibaatso (FRI) ogbolitso (FRI) GBE-VHE aviãtilifui (FRI) lifui (Ewe Dict.) VHE (Awlan) avĩa (FRI) GUANG-ANUM a-bɔ-anyɛ (FRI) KRACHI bonchu (CV, ex JMD) bonkyu (CV, ex FRI) NZEMA atronzuo (FRI) dupwan (FRI) **TOGO** GBE-VHE lifui (Volkens) TEM (Tshaudjo) akinale (Volkens) YORUBA IFE OF TOGO nboboe (Volkens) **NIGERIA** ANAANG ọniộk (KW) BOKYI nsor (Catterall) EDO íkhímì (JMD; KO&S) ikhimwin (Amayo) EFIK ọ́bọ́tì (auctt.) EJAGHAM isinn (JMD) HAUSA àdùrúkù (auctt.) ba-reshe (JMD) bareshi (ZOG) IBIBIO itömö (Okpon) IGBO egbo (JMD) ogbu (Talbot) ogilisi (auctt.) ogirisi (auctt.) IGBO (Ala) ẹbwo (NWT; JMD) IGBO (Awka) ògílìsì (KW) IGBO (Owerri) ògírìsì = *blacker than head* (AJC; JMD) IJO-IZON (Kolokuma) ògírízì, ùgúrízì (KW) JUKUN agishi (JMD) NUPE dìn bèrè cìn a'milĕ = *it stretches its neck and looks into the compound* (RB) TIV ashishan (Vermeer) chiluali (Vermeer) kontor (JMD; KO&S) YORUBA akòko (auctt.)

A shrub or small tree reaching to 7–8 m high in the west (Senegal, 26) to 20 m in the east (Nigeria, 21) of the Region, and to 2.70 m in girth (S. Leone, 35); shrubby, or erect with vertically ascending branches, of wooded savanna and deciduous forest, occurring across the Region and into the Congo basin.

The plant has shiny dark green leaves and bears large showy terminal purple flowers. It is often grown as an ornamental and is easily propagated by cuttings. It is a familiar live-fence and boundary-tree throughout its distribution (13, 36; Senegal, 26; Sierra Leone, 14, 35; Nimba Mt. Guinea/Liberia, 37; Ivory Coast, 2, 4; Ghana, 16, 18–20; Nigeria 8, 9, 40; Gabon, 42; Zaïre, 29). When planted as a fence, it is often permitted to grow into a stockade. In Ijo custom the tree is used to mark the territorial boundary between non-relatives (45), *Erythrina senegalensis* (Leguminosae — Papilionoiideae) being used between relatives (44). On the Benue Plateau, the Tiv use it commonly as a screen to give privacy around the washing areas (40). To the Igbo, it is more or less a sacred or symbolic tree, often planted in small groves in front of a chief's house (13). To the Efik, Ekoi and Ibibio it is a symbol of the deities: it is found in Efik and Ibibio graveyards and sacred places (32) and when Efik and Ibibio set up a new settlement a cutting or sapling is always brought from the old one (13). In Gabon (42) and in Ivory Coast (36), a tree is planted near to tombs and in villages as a protective talisman. The Mende name meaning 'corpse drive on' derives from the use of leafy branches of the tree being used to fan a corpse to help its spirit on its way, and to keep off flies (13, 35). In both Yorubaland and Hausaland the tree is held in regard: a leaf is placed on the head of a new chief, and cutting the tree with an axe or burning as fuel is avoided (13).

The wood is pale brown (Nigeria, 21) or yellow or yellowish-pink (Ghana, 20), even-textured, moderately hard and durable. The wood when cut remains alive and does not decay for a long time so that it is often used for posts, bridges and out-of-doors woodwork in Ghana (19, 20), fence (38) and house posts (39), and stakes for yams (8) in S Nigeria. The wood is used for knife-handles (13), and in Sierra Leone to make specifically round machete-handles (35). It makes a good firewood.

In Sierra Leone the dried bark and young twigs are pounded up with spices (*Xylopia* (Annonaceae), etc.) are given in decoction or infusion for such complaints as uterine colic, dysmenorrhoea, etc. (13). A decoction of the bark is given to children in Ivory Coast (27, 28) and Nigeria (33) for epilepsy and convulsions. The bark is used in Ghana as a stomachic and in the form of an enema for constipation and piles; the bark is also said to cure septic wounds (19, 20). The bark is used in Guinea to treat snake-bite (36), while in Ghana the chewed-up leaves are applied to the wound which is then sucked to draw out the venom (19). Analgesic properties are said to reside in the bark. One or two sniffs or a snuff made of the sun-dried bark ground up with palm salt ($K_2CO_3$) or 'sel de Taoudenit' and the fruits of *Piper guineense* Schum. & Thonn. (Piperaceae) are taken for headache, sinusitis, head-colds, etc. in Ivory Coast, and will dispel the most obstinate migraine (27, 28). Perhaps similar use as snuff accounts for the Krio name 'snuff-leaf' in Sierra Leone. In Gabon bark heated in a little boiling water is patted on the head for headache (10). Bark pulped up to a paste is used in Casamance (Senegal) on rheumatism, especially painful arthritis in the knee. In some cases a plaster is applied after massage, and in refactory cases where walking has become impossible various parts of *Trichilia prieuriana* (Meliaceae) are added to an aqeous infusion of *N. laevis* roots for internal and external use (23, 26). A bark-decoction is taken in Nigeria (3), and Togo (43) for dysentery. The outer bark is decocted with chilis in Ghana and the liquor drunk for chest-pain (19, 20) and the inner soft bark is put into the ear for earache (13). In Gabon a preparation (? infusion) is used in lotion for headache and in a gargle for toothache (41). A decoction of stem and root barks was in the past used with some success in Sierra Leone for acute malaria with splenic enlargement, by application of the crushed leaves with fruits of *Xylopia aethiopica* in poultice over the spleen, for dysenetry and post-partum and other forms of passive bleeding (15). Bark boiled in water or palm-wine is commonly used in Congo for cough and diarrhoea (5). Bark-preparations are also considered healing. In Ghana they are applied to sore feet and septic scores, and as a poultice to aching limbs (13). In Ivory Coast the bark is used to make washes and hip-baths for chancres, and the liquid obtained by beating the bark in a little water upon a copper coin is used to wash the sore (2). In Casamance (Senegal) preparations are topically applied in dracontiasis and snake-bite, and to abscesses and ulcers (22, 26). It is used on breast-tumour in Ghana (17), and in Nigeria the bark and roots grounded up and mixed with oil and human faeces have been applied as an ointment on wounds (3). The plant (part not stated, but probably the bark) is used in Upper Guinea sometimes against leprosy, and then is said to be some benefit (36). In Congo the chest is rubbed with sap obtained by pounding the bark with leaves of *Kalanchoe sp.* (Crassulaceae) for pulmonary affections (5). As a veterinary medicine, bark is fed to horses in Senegal to improve their appetite (13).

The roots and leaves are often used together. They are a familiar remedy for scrotal elephantiasis, or for any form of orchitis, a decoction being drunk or the materials pounded up together and applied hot. They are also credited with aphrodisiac properties (13). In Nigeria they are boiled together for administration as a febrifuge (3). The roots alone are pounded up with *Lophira* (Ochnaceae) in Senegal for massaging onto areas of oedema arising through dietary deficiency (25, 26), and a macerate or decoction of the roots is taken by mouth as a vermifuge for roundworm in Senegal (23, 24, 26), and in Guinea (13), and to treat hernia in Senegal and syphilis in The Gambia. The treatment is purgative and is regarded as more or less toxic. The roots are used in Senegal (22) and in The Gambia (13) against dysentery and for rheumatic swellings. They are also used in Nigeria as a roundworm vermifuge and stomachic, and for migraine and earache (33). In Liberia root-scrapings mixed with chili are put into a carious tooth (13). A plaster is made from the roots for treating bad feet in The Gambia (34). Stem-bark is used by herbalists in Nigeria for treating skin-infections. Examinations of boiled water extracts showed some activity against Gram +ve *Sarcina lutea,* but no action against *Staphylococcus aureus* and *Mycobacerium phlei,* nor against Gram −ve organisms; nor was there any anti-fungal activity (30).

The leaves are used in decoction in Nigeria as an eye-wash in conjunctivitis, ophthalmia etc. (3; 31); they are cooked in palm-oil soup in Ghana and taken by pregnant women in order to effect easy delivery, and after parturition to promote a rich milk supply (19, 20). Belief in the facilitating of childbirth pertains also in Ivory Coast (7). A decoction of the leaves with those of *Psidium guajava* (Myrtaceae) is taken in Ivory Coast for diarrhoea and dysentery (2). In Sierra Leone leaf-ash mixed with salt is taken as a remedy for pain over the heart (13–? heartburn). Leaves are used in Sierra Leone as a wrapper to hold tobacco snuff (12). It is said that stains on the hands can be removed by the leaves (20).

Tannins are present in the bark of Nigerian material (1) and in plants from Guinea-Bissau and Congo (26). Screening of the roots for antimalarial activity has shown no action against avian malaria (11). Action of extracts of roots and leaves as an antidysenteric and antihaemorrhagic is said to be due not to astringency but possibly due to tonic effect on involuntary muscle and mucous membrane (15). Physiological action on muscle tone of the duodenum and ileum of the rabbit and guinea-pig is reported (26).

References:

1. Adegoke & al., 1968. 2. Adjanohoun & Akè Assi, 1972: 52. 3. Ainslie, 1937: sp. no. 243. 4. Aubréville, 1959: 3: 244. 5. Bouquet, 1969: 73. 6. Bouquet, 1972: 15. 7. Bouquet & Debray, 1974: 51. 8. Carpenter 207, UCI. 9. Carpenter 401, UCI. 10. Chevalier (Fleury) 26,323, K. 11. Claude & al., 1947. 12. Cole 65, FBC. 13. Dalziel, 1937. 14. Deighton 509, K. 15. Easmon, 1891: 54–58. 16. Fishlock 18, K. 17. Hartwell, 1968. 18. Irvine 209, K. 19. Irvine, 1930: 304–5. 20. Irvine, 1961: 738–9. 21. Keay & al., 1964: 428–30. 22. Kerharo & Adam, 1962. 23. Kerharo & Adam, 1963, b. 24. Kerharo & Adam, 1964, b: 561–2. 25. Kerharo & Adam, 1974: 237–9, with phytochemistry & pharmacology. 27. Kerharo & Bouquet, 1947: 253. 28. Kerharo & Bouquet, 1950: 227–8. 29. Liben, 1977: 25. 30. Malcolm & Sofowora, 1969. 31. McLeod s.n. (UIH 3431), UCI. 33. Oliver, 1960: 32, 74. 34. Rosevear, 1961. 35. Savill & Fox, 1967, 52–53. 36. Schnell, 1950: 248. 37. Schnell, 1952: 516. 38. Thomas, N. W., 1949 (Nig. Ser.), K. 39. Thomas, N. W. 2257 (Nig.Ser.), K. 40. Vermeer 13, UCI. 41. Walker, 1953, a: 24. 42. Walker & Sillans, 1961: 101–2. 43. Williams, L1. 42, K. 44. Williamson, K. 146, UCI. 45. Williamson, K. 147, UCI.

**Pandorea pandorana** (Andr.) van Steenis

FWTA, ed. 2, 2: 385.

English: wonga wonga vine (Australia).

A woody vine, native of Malasia and northern Australia, and brought into horticultural cultivation in many warm countries for its decorative flowers. The corolla is yellowish-white, streaked with purple, and is often spotted, to nearly 2 cm long. Cultivar *'rosea'* has pale rose-coloured flowers.

The leaves have given positive tests for alkaloids and the plant is suspect of causing poisoning to stock in Australia (Webb fide 1).

Reference:

1. Deshaprabhu, 1966: 221.

**Parmentiera cereifera** Seem.

FWTA, ed. 2, 2: 385.

English: candle tree (U.S.A., Bates.).

A small spreading tree, native of Mexico and Central America, and dispersed to many warm countries as an ornamental. The flower has a large brownish calyx and a white corolla to 7·5 cm long. The tree gets its English name from the fruit which is pendant to 1.30 m long, yellowish and resembling a candle.

The fruits are fleshy and are said to be eaten in Panama and to be fed to cattle. The leaves contain an unidentified glycoside, and traces of hydrocyanic acid have been detected in the fruits, leaves and roots (1, 2).

References:

1. Deshaprabhu, 1966: 266. 2. Quisumbing, 1951: 1046.

**Pyrostegia venusta** (Ker-Gawl.) Miers

FWTA, ed. 2, 2: 385.

English: flame flower, flame vine, flaming trumpet, golden shower (U.S.A., Bates.).

French: liane aurore (Berhaut).

An evergreen climbing shrub, native of Brazil and Paraguay, and introduced to many warm countries as an ornamental for its brilliant display when in bloom.

**Saritaea magnifica** (Sprague) Dugand

FWTA, ed. 2, 2: 385.

A vine with stems to about 3 m long, native of Colombia and introduced to many warm countries for its showy rose-pink to pale purple flowers to 7.5 cm long.

**Spathodea campanulata** P. Beauv.

FWTA, ed. 2, 2: 386–7. UPWTA, ed. 1, 445.

English: tulip tree; African tulip tree; flame tree; fountain tree (Irvine).

French: tulipier du Gabon; arbre-flamme; 'baton du sorcier'; tulipier (trade, the timber).

Portuguese: tulipeiro do Gabào.

West African: **SENEGAL** BALANTA blalo (Aub.) BANYUN sissal (Aub.) KONYAGI a-tyilil (Ferry) SERER mâm (Aub.; RS) WOLOF étidômô (JGA) fèr, fèhr (auctt.) **THE GAMBIA** FULA-PULAAR (The Gambia) sukunde (DAP) MANDING-MANDINKA sula-selo = *monkey climbs* (DAP; DF) **GUINEA-BISSAU** BALANTA pikeriko (GES) piquério (JDES) FULA-PULAAR (Guinea-Bissau) cafauano (GeS; JDES) culassaque (DF) sekunde (GeS) suncúndè (JDES) MANDING-MANDINKA sula-selô (GeS; JDES) PEPEL teme (GES; JDES) **GUINEA** FULA-PULAAR (Guinea) diapélédé (Aub.; RS) sukundé (RS) MANDING-MANINKA tunda (auctt.) **SIERRA LEONE** MANDING-MANINKA (Koranko) dumɛntili (NWT) dundunturi (NWT) MENDE baine, gobane (Aylmer) gele-gɔ, ngele gɔwɔ (auctt.) tombo-lɛmbi, tombo: *a deserted village;* lɛmbi: *remained long* (S&F) TEMNE *a*-leop-*a*-ro-bath (NWT) **UPPER VOLTA** MANDING-DYULA missiboiri (K&B) tiéré (K&B) **IVORY COAST** ABE boro (auctt.) ABURE kokomayur (A. Chev.) AKAN-ASANTE sinséré (K&B) AKYE kotchu (A&AA) sé (A. Chev.; RS) ANYI asrélé (K&B) BAULE biébié (RS) biébié biébié (K&B; Aub.) biébié sirili (A&AA) diébéserélé (B&D) GAGU vovo (B&D) KRU-BETE zéblé zébré (K&B) zibli (K&B) GUERE (Chiehn) zabré (B&D) zéblé zébré (K&B) GUERE (Wobe) pautu (A.Chev.; RS) KYAMA gbagbihia (auctt.) MANDING-DYULA missiboiri (K&B) tiéré (K&B) MANINKA kokwè (B&D) SENUFO-TAGWANA assien (K&B) **GHANA** ADANGME ɔdumanki (Farmar, ex FRI) votʃo (FRI) AKAN-ASANTE akuakuo ninsuo (Enti) akuakuoninsu = *Akuakuo's tears* (auctt.) akuɔkɔ *the flowers* (FRI; JMD) kɔkɔ-anidua (FRI; CJT) kokonisuo (DF) kɔkɔnsu = *red tears, or = red water* (auctt.) KWAWU osisiri (FRI) TWI akuakuaninsu (CJT) akuakuo-ninsuo (Enti.) aninsu *not commonly used* (FRI) *o*-sisiriw (auctt.) WASA osisiri (BD&H; FRI) BAULE biebie (FRI) GBE-VHE adadase (FRI; JMD) adatsigo = *tear pod* (auctt.) SENUFO-TAGWANA (Chibbi, N. T.) abeni (auctt.) **TOGO** GBE-VHE adadase (Volkens) YORUBA-IFE OF TOGO gbetachi-gbetschi (Volkens) **NIGERIA** BOKYI kenshie (Catterall; KO&S) EDO ókuèkuè, ókwèkwè (auctt.) EFIK èsèǹnīm (auctt.) EJAGHAM ekok (DRR) IGBO ímí éwū = *goat's nose* (KO&S) utu ogbolo ḿmìrì (BNO) IGBO (Amankalu) ọbọ-ọmi (DRR; JMD) IGBO (Awka) akpoti (DRR; JMD) ímí éwū = *goat's nose* (DRR; KW) IGBO (Enugu) ugwogo (DRR) IGBO (Owerri) oruru (DRR) MBEMBE okiníkene (DRR) MUNGAKA e-nko nebang YORUBA mójútòrò = *making eye clear; applies to the flowers* (IFE) orórù (auctt.) **WEST CAMEROONS** DUALA bwèle ba mbonjì (AHU; Ithmann) KPE mbako (JMD) KUNDU etoto (DRR) etutu (JMD)

A tree to about 23 m high by 2.70 m in girth of fluted bole, short branches and a compact crown, in deciduous forest and secondary jungle, transition and wooded savanna, but only a shrub in open savanna (4), from Guinea to S Nigeria and Fernando Po, and extending to Angola and to NE and E Africa. Now widely dispersed by man to many tropical countries.

The tree is quick-growing and bears numerous large trumpet-shaped crimson or flame-red flowers in clusters. It is showy and is commonly planted for decorative purposes. It is however, liable to wind damage and should be kept at a distance from buildings, and its shallow-rooting habit may disturb foundations. The name 'tulip' appearing in European languages arises from the shape of the flower like that of the tulip. A showy buttercup-yellow form has been

found in Uganda and this merits propagation (12, 13). The presence of adventive trees is said to indicate good soil (10, 16, 17). Its presence in the bush may also mark the site of an abandoned settlement, as the Mende name *tombo* (deserted village)
*-lɛmbei* (remained long) implies (24).

The wood is dirty white, very light, soft, fibrous and liable to rot (9, 10, 13, 16–18, 24, 27), though in Zaïre it is recorded as being hard and insect resistant (11). Wood density is recorded as 0.363 (7, 25). It has been suggested for pulping in paper-manufacture (8, 16, 25), and it has at one time been exported from Gabon, under the trade name 'tulipier', to Europe for making plywood (30). In Ghana it is used to make the two sides of blacksmith's bellows (16, 27), in S Nigeria sometimes to make drums (10), and in Gabon to make certain (unspecified) musical instruments (30). It makes such poor firewood that a Ghanaian saying refers to the inability of the wood to heat sufficiently to cure a chronic sore (16). In SE Asia inferior crates are made of the wood which splits and will not hold nails.

A bark macerate is taken by enema in Ivory Coast for kidney and back-pains, and the pulped-up bark is used in frictions on swellings, fungal infections, impetigo, herpes and other skin-affections (19, 20). Bark is used also for stomachache intercostal pain, cough, haematuria, and urethral discharge (6). Sap expressed from the bark is used with kola fruit on guinea-worm sores, the action probably being to reduce inflammation in the lymphatic system (19, 20). The bark is commonly used in Nigeria on ulcers, oedemas and skin-eruptions, either dried and powdered, or the fresh inner bark is applied, or a black decoction, which is sometimes made from the flowers or leaves, is used as a lotion (3, 10, 22). The plant, probably the bark, is used in Tanganyika to treat scabies (26). The inner bark and the flowers are used in Gabon on wounds (10, 29, 30). In Nigeria a bark-infusion is taken for constipation, diarrhoea and dysentery (3, 22), and in Ghana for dysentery (14) and as a stomachic (17). In Congo bark is used to prepare a bath for someone suffering fever and a decoction is taken by draught for scrotal hernia and is used to treat syphilitic sores (5). The bark and flowers in Guinea are principally used in veterinary applications, but preparations are also sometimes given to women for pelvic disorders (23).

Leaves are eaten in soup in Sierra Leone (28). A cold infusion of the leaves is said to be good for urethral inflammation (10, 11, 22). A leaf-decoction is taken in draught in Ivory Coast as a poison-antidote (19) and for urethral discharge (20), and the leaves together with leaves of a number of other plants and clay are made into a paste which is let down in water and drunk in treatment of Pott's disease (spinal tuberculosis) (2).

The unopened flower buds hold a sweet watery red sap to which the Asante and Twi name *kɔkɔnsu*, red tears, and the Ewe name *adatsigo*, tear-pod, refer. It is a common children's game played in Ghana to squeeze this sap into one another's eyes causing red tears to flow (16). This, or an analogous game, is also known in Cameroun in which the sap is squirted as from a water-pistol (21). The central column of the fruits when they have hardened is used by hunters in Ghana: they boil the centre and leave the liquor in the bush where game on drinking it dies soon after. The meat is edible if taken at once (16, 17). The seed is said to be more or less edible (10, 11, 24).

Neither alkaloid nor glycoside has been detected in the bark of Ivorean material, but tannin is present (6). Nigerian material [? leaves] has shown a fairly strong presence of alkaloid (1), and leaves and bark of tannin accounting for their astringency (22). Rather weak insecticidal activity has been found in stem-bark (15).

The plant enters folk-lore in Ghana (16): when the flowers open farmers know it is time to plant the corn. Its indifference as fire-wood has already been noted. The same saying is applied also to a bad debtor who, having acquired a small sum of money, remains unable to wipe off the debt. The ease with which the tree is uprooted is also source of moral comment.

References:

1. Adegoke & al., 1968. 2. Adjanohoun & Aké Assi, 1972: 53. 3. Ainslie, 1937: sp. no. 319. 4. Aubréville, 1959: 3: 248. 5. Bouquet, 1969: 73. 6. Bouquet & Debray, 1974: 52. 1974: 52. 7. Broun & Massey, 1929: 334, as *S. nilotica* Seem. 8. Cole, 1968. 9. Dale & Greenway, 1961: 64, 66, as *S. nilotica* Seem. 10. Dalziel, 1937. 11. De Wildeman, 1906. 12. Eggeling, in litt. K. 13. Eggeling & Dale, 1952: 42. 14. Farmar 470, K. 15. Heal & al., 1950: 106. 16. Irvine, 1930: 390–1. 17. Irvine, 1961: 739–40. 18. Keay & al., 1964: 426. 19. Kerharo & Bouquet, 1947: 253. 20. Kerharo & Bouquet, 1950: 228. 21. Leeuwenberg 7337, K. 22. Oliver, 1960: 37, 83–84. 23. Pobéguin, 1912: 60. 24. Savill & Fox, 1967, 53. 25. Schnell, 1950: 261. 26. Tanner 5890, K. 27. Taylor, 1960: 101. 28. Thomas, N. W. 2002, K. 29. Walker, 1953, a: 24. 30. Walker & Sillans, 1961: 102.

## **Stereospermum acuminatissimum** K. Schum.

FWTA, ed. 2, 2: 386. UPWTA, ed. 1, 445.

West African: **SENEGAL** FULA-PULAAR (Senegal) golumbi (RS) TUKULOR bani (RS) MANDING-MANINKA moro iri (RS) WOLOF bolnac = *don't cheat!* (auctt.) fèhr (RS) **GUINEA** MANDING-MANINKA moro iri (RS) **SIERRA LEONE** MANDING-MANINKA (Koranko) bumbusoi (D.G.Thomas) MENDE lubɛ (*def.*-i) (Aylmer, ex JMD) tombo-lɛmbi (*def.*-i) *see under Spathodea campanulata* (auctt.) SUSU-DYALONKE yale-na (KM) TEMNE *ka*-bɛk (NWT) **MALI** MANDING-MANINKA moro iri (RS) **UPPER VOLTA** GURMA nali limebu (RS) **IVORY COAST** ABE balié (A. Chev.; RS) AKYE fara (K&B; Aub.) tchu(o) (A. Chev.; RS) DAN duo (Aub.; K&B) FULA-FULFULDE (Ivory Coast) urte (Aub.) KRU-BETE zaba (Aub.) GUERE demontué (K&B) KWENI vué-buri iri (K&B; Aub.) **GHANA** VULGAR osontokwakofuo (DF) AKAN-ASANTE ɛsono-tokwa kufu = *elephant's fighter* (Enti) nufutene (LAK-C) tokwakufuo (CV, ex FRI) **NIGER** GURMA nali limebu (RS) HAUSA sansami (RS) KANURI kavogu (RS) **NIGERIA** EDO oshuobon (KO&S) YORUBA ẹru ìyeye = *slave of ìyeye* ẹrú yeyè (auctt.) YORUBA (Ondo) alakiriti (Kennedy, JMD) paripakoje (Kennedy, JMD)

Tree to 40 m high by 3 m girth of savanna-forest and dry closed-forest, occurring commonly from Guinea to W Cameroons and into E Cameroun.

The tree is not unlike *Spathodea campanulata* when not in flower so that to some races the two trees have the same names, e.g., Mende, *tombo lɛmbei*; Edo, *okwekwe*. The tree carries handsome pale pink, purplish or red flowers, a little larger than those of *S. kunthianum*. It is well worth planting as an ornamental tree in the forested zone. Seeds, however, are not easy to germinate and propagation by transplanting wild seedlings or suckers is recommended (2, 3).

The bole is straight and slightly buttressed. The wood is white, soft and not durable (2–4, 9). The wood is said to be used in Sierra Leone (8) but no detail is given. Use is likely to be limited to purposes not requiring strength and durability.

The bark is haemostatic and cicitrisant. It is commonly used on sores and wounds in Ivory Coast — a warm decoction to cleanse, and sap from the fresh bark in topical application (1, 5, 6).

The blossom is much visited by bees (8).

The fruit has unspecified medicinal use at Kumasi, Ghana (7). The seeds are surrounded by a pulp [? edible, cf. *S. kunthianum*].

References:

1. Bouquet & Debray, 1974: 52. 2. Dalziel, 1937. 3. Irvine, 1961: 740–1. 4. Keay & al., 1964: 428. 5. Kerharo & Bouquet, 1947. 6. Kerharo & Bouquet, 1950: 228. 7. King-Church 869, K. 8. Miszewski 2, K. 9. Savill & Fox, 1967, 52.

## **Stereospermum kunthianum** Cham.

FWTA, ed. 2, 2: 386. UPWTA, ed. 1, 445–6.

West African: **SENEGAL** BANYUN kibokok (K&A) BASARI a-màl (K&A; Ferry) BEDIK ga-ngór (Ferry) FULA-PULAAR (Senegal) bănidaney = *white bani* (*Pterocarpus erinaceus*) (K&A) galumbi (K&A) golôbi (K&A) golumbi (K&A) tchingoli (GR) KONYAGI a-kufú dyàx (Ferry)

MANDING-BAMBARA madiri (K&A) moiro = *man's tree* (K&A) mori iri (JB) moro iri (JB; K&A) MANINKA madiri (K&A) mogo kolo (Aub.; K&A) moiro = *man's tree* (K&A) moro yéri (Aub.; K&A) moroiri (K&A) moroiro (K&A) 'SOCE' buapalô (K&A) SERER bol nak (JB) mab (JB, ex K&A) mam (K&A) mamb (JB; JLT) SONINKE-SARAKOLE itôkulé (K&A) tafé (Aub., ex K&A) WOLOF bolinaq (Aub., ex K&A) étidömö (JB; JGA) étudamô = *wand of a wizard* (K&A; DF) étudèmô, dèmô: *sorcerer* (JLT; JGA) fehr; fèr (auctt.) fex (K&A) yetudomo = *wand of the sorcerer* (K&A) **THE GAMBIA** MANDING-MANINKA dafino (Fox) **GUINEA-BISSAU** MANDING-MANDINKA meire (EPdS) móre (JDES) **GUINEA** MANDING-MANINKA mogokolo (Aub.) moro yéri (Aub.) **MALI** DOGON popólo (C-G) MANDING-BAMBARA moro iri (auctt.) soguirini (A. Chev.) MANINKA mogo kolo (Aub.) moro yéri (Aub.) SONINKE-SARAKOLE tafé (Aub.) **UPPER VOLTA** MOORE nihilenga (Aub.) vuiga (Aub.) **GHANA** AKAN-ASANTE ɛsonoetok-wakofoɔ (BD&H; FRI) BRONG kuti-kani-misa (FRI) kuti-kani-misa (FRI) tunturei (BD&H; CV, ex FRI) DAGBANI lengerigongo (auctt.) tepiliga (CV, ex FRI) zugubyetia (CV, ex FRI) GUANG-GONJA jèbòté ɓ kètá (Rytz) KRACHI kuli-kanimisa (BD&H; CV) NANKANNI ylinga (Lynn, ex FRI) SISAALA bisuma (AEK; FRI) **TOGO** TEM (Tshaudjo) essobelia (Volkens) YORUBA-IFE OF TOGO eke-deka = *only one root* (Volkens) **DAHOMEY** BATONNUN benuhebe (Aub.) gurubonbula (Aub.) GBE-FON nsandi (Aub.) YORUBA-NAGO ayada **NIGER** ARABIC (Niger) arad (Aub.) ess (Aub.) khashkhash (Aub.) khess (Aub.) GURMA nali limebu (Aub.) HAUSA sansami (Aub.) KANURI kavogu (Aub.) **NIGERIA** ARABIC-SHUWA arad (JMD) khashkash (JMD) FULA-FULFULDE (Nigeria) buldumhi (Taylor) golombi (*pl.* golombe) *Note: metathesis with gomboli* (MM) gomboli (auctt.) GWARI weknavunihi (JMD) HAUSA ɗán sárkín-ítaátúwàa = *son of the chief of the trees; a superstitious epithet* (JMD) daraɓake *this species?* (Bargery) hachin-tumkia = *goat's corn* (Robinson; RES) jiri (auctt.) jiri dán-sárkín-ítaátuwàa *a superstitious epithet* (JMD) sansami (auctt.) túrkèn-doókii (ZOG) KANURI gòlómbì (auctt.) TIV umana tumba (JMD; K, O&S) YORUBA ajade *an honourable title amongst hunters* (Ross; JMD) ayadá (auctt.)

A slender tree to 15 m high, of the wooded savanna, and less (5–6 m) when in the drier Sahel region, present throughout the Region and widespread across Africa to the Red Sea and southwards to Malawi, the Congo basin and Angola.

The trunk is rarely straight, and the branches too are crooked. The inflorescence is a drooping panicle of pink or purplish flowers which are borne when the tree is wintering. The tree in flower is very showy and well worth cultivation as an ornamental. The seeds however have a very poor germination and propagation by suckers is recommended (10, 11). The scaly bark confers some degree of resistance to fire damage.

The wood is whitish tinged with yellow or pink and is fairly hard (4, 11, 13). Air-dried wood of E African origin is recorded as weighing 60 lb/cu. ft. (3, 5); when big enough it is used to make mortars (1, 4, 11). For superstitious reasons, as well as for its medicinal use, the tree is not cut for fire-wood; in Sokoto, N Nigeria, the smoke is said to conduce one to leprosy (4). In Uganda the wood is known as being useless for making charcoal as in the burning it passes through to ash (8).

The bark on chewing is bitter and nauseous (14), but it is a widespread practice for girls to chew the young bark to stain their lips reddish-brown, perhaps to imitate the effect of chewing kola, or as a substitute for it (4). In Senegal the bark and the roots are commonly prescribed for internal and external use in treatments for a condition called *siti* (primary syphilis) which is manifested by cracks on the joints and fingers that degenerate into ulcers (14, 18), for syphilitic complications, for leprosy, vomiting and fevers, stomach ulcers and phagadenic ulcers (powder used externally) (18). A bark-preparation is used in Malawi on ulcers (28). People in N Nigeria value the bark as a remedy for venereal disease. The Fula take a straight bark-decoction (12). In Sokoto a decoction is taken with natron and a white variety of guinea-corn. It causes sweating and diarrhoea (4). In Tanganyika root and bark decoction together with those of *Tamarindus indicus* (Leguminosae: Caesalpinioiideae) is drunk for leprosy, root and bark-ash with oil is rubbed into scarifications for leprosy, and bark-decoction with the bark of *Dalbergia boehmii* (Leguminosae: Papilionoideae) is given to children in malarial rigor as a drink and for bathing (9). In Senegal the bark with that of *Ozoroa insignis* (Anacardiaceae) is considered a vermifuge, and with *Parkia biglobosa* (Leguminosae: Mimosoideae) a poison-antidote (18). The trunk-bark in decoction is given for refactory cough, bronchitis and pneumonia, (15, 18). Roots and bark are used for stomachache and the bark when crushed is applied to wounds in Sudan. The bark is also used for skin-eruptions in Malawi (11) and ulcers (28). The Hausa and Fula value the bark as a remedy for diarrhoea and dysentery, and as a horse medicine (4).

The roots and leaves are given by Fula in Senegal for venereal diseases, respiratory diseases and gastritis (16, 18). The roots are considered to be strongly diuretic and are used by the Serer of Senegal for anuria, urethral discharge, and schistosomiasis (18). In Nigeria the root, along with other roots is a remedy for a disease called *rana* with symptoms of haematuria (4). Diverse use is made of the roots in Tanganyika: a root decoction with leaf-sap is taken as a cough-mixture (9); for venereal disease (7); roots and leaves are boiled in water which is drunk for stomach-pains by the Zigua (25); the liquor after boiling the roots is drunk for sudden sharp itching rashes (26) [? urticaria]; the Sukuma pound the core of the peeled root which is soaked in water, and the liquid is drunk for constipation and syphilis — diarrhoea is caused within half an hour and lasts for three days (23); the pounded roots are applied by the Sukuma to syphilitic sores after they have been washed (24). In Malawi the roots are used in the treatment of a condition resembling asthma (28).

The leaves are macerated and put into baths in Senegal for general debility (17). They are put into steam-baths in Tanganyika but the purpose is not recorded (20), and burnt, the ash is mixed with oil and applied to scabies (25). In Uganda an infusion is used for washing wounds (5). In The Gambia they are used in dyeing (6). The leaves and small branches are relished by horses in Sudan, and the dried leaves have been analysed as containing 3·8% proteins (11).

The long sinuous fruit pod are considered edible in Guinea (22). They are used in Tanganyika as a cough-cure (2), and the shoot and pod are chewed with salt for the same purpose (27). The pods are also chewed with salt in S Rhodesia (21).

The tree is endowed with a number of superstitious ideas. In Senegal, the Wolof in calling it 'the wand of the sorcerer', bestow upon the tree power to exorcise ghosts (18). The Hausa of Nigeria and the Yoruba give it honorific epithets. The bark is regarded as a preventative against witchcraft, and is either finely powdered and used like snuff or wrapped up and worn as a charm. Pagan tribes lay a stem across the entrance of a hut to prevent thieves. Mistletoe (*kauchin sansami,* parasite of *sansami*) found growing on the tree is considered a lucky find. (4). Bark is used in Tanganyika for devil-worship (20).

References:

1. Aubréville, 1950: 497. 2. Bally, 1937. 3. Dale & Greenway, 1961: 66. 4. Dalziel, 1937. 5. Eggeling & Dale, 1952: 42–44. 6. Fox 75, K. 7. Gane 61, K. 8. Graham 2218, K. 9. Haerdi, 1964: 149. 10. Irvine, 1930: 395–6. 11. Irvine, 1961: 741, lxxxii, leaf-analysis. 12. Jackson 1973. 13. Keay & al., 1964: 427–8. 14. Kerharo, 1967. 15. Kerharo & Adam, 1964, a: 426. 16. Kerharo & Adam, 1964, b: 578. 17. Kerharo & Adam, 1964, c: 322. 18. Kerharo & Adam, 1974: 239–40. 19. Koritschoner 182, K. 20. Koritschoner 1709, K. 21. Pardy, 1952. 22. Pobéguin 180, K. 23. Tanner 1251, K. 24. Tanner 1565, K. 25. Tanner 3164, K. 26. Tanner 3758, k. 27. Watt & Breyer-Brandwijk, 1962: 144. 28. Williamson, J. 1956: 113.

**Tabebuia rosea** (Bertol.) DC.

FWTA, ed. 2, 2: 385.

English: pink poui (Trinidad, Irvine; U.S.A., Bates); rosy trumpet tree (U.S.A., Bates),

West African: GHANA GA óbòntólì (KD)

A tree to 25 m high, (Ghana, 10 m, 2), native of Mexico to Venezuela and Ecuador, and introduced to several tropical countries for its purplish-pink to nearly white flowers to about 8 cm long. This is one of the most common and showy of the flowering trees of the New World tropics and sub-tropics.

Several *Tabebuia spp.* yield excellent timber. One *T. argentea,* has shown some activity against avian malaria, but others tested have not (1). The properties of *T. rosea* are not recorded.

References:

1. Claude & al., 1947. 2. Irvine, 1961: 741–2.

**Tecoma capensis** (Thunb.) Lindley

FWTA, ed. 2, 2: 385, as *Tecomaria capensis* (Thunb.) Spach.

English: Cape honeysuckle (S Africa, Watt & Breyer-Brandwijk; U.S.A. Bates); Cape trumpet flower, Kaffir honeysuckle (Watt & Breyer-Brandwijk); red tecoma (Ghana, Irvine); yellow Cape honeysuckle (c.var. *'aurea',* U.S.A., Bates).

A rambling shrub to 2 m or more, with orange-red to scarlet flowers 5 cm long, native of tropical and S Africa, now widely grown as an ornamental in many tropical and subtropical countries. There are some cultivars, of which c.var. *aurea* has yellow flowers.

The plant lends itself to training against a wall or over a pergola. The long trailing stems are easily rooted. It is grown in southern U.S.A. as a hedge plant (1).

In S Africa the plant is readily browsed by stock. In Transvaal powdered bark is taken in cases of fever and pneumonia, and is said to relieve pain and to induce sleep. The bark is also taken for abdominal troubles, and the powder is rubbed onto bleeding gums. A leaf-decoction is taken for diarrhoea and enteritis. Leaves and flowers contain sterols, but not other active principles, and the plant has given negative antibiotic tests. (2).

References:

1. Bates, 1976: 1099. 2. Watt & Breyer-Brandwijk, 1962: 144, as *Tecomaria capensis* Spach.

**Tecoma stans** (Linn.) H. B. & K.

FWTA, ed. 2, 2: 385.

English: trumpet bush; yellow bells; yellow bignonia (Bates, U.S.A.).

A shrub or tree to 5 m high with bright yellow bell-shaped flowers to 5 cm long, native of Central America, and introduced to many warm countries as an ornamental.

The plant in bloom is very decorative and is well worthy of cultivation. Under equable humid climatic conditions it can be kept more or less continuously in flower by plucking off the flower-heads as soon as the flowers fade. The plant tends to become straggly. It survives heavy cutting back and quickly comes into bloom again. It makes a decorative flowering hedge.

The larger timber is light brown, close-grained, takes a good polish and is durable, but it is seldom large enough for making anything but small objects. In the plant's native Mexico it is considered of little value though it was formerly used by the Indians to make bows. The roots are a powerful diuretic, and the plant has been held to have tonic, antisyphilitic and vermifugal properties. A decoction of the flowers and bark has been used for stomach-pains, and in some parts of Mexico the plant has a reputation for alleviating and even curing diabetes. The root has been recorded as being used at Guadalajara, Mexico, in making a kind of beer (2).

The flowers are slightly fragrant and are said to yield much honey (2).

Phytochemical work has shown the presence of six alkaloids, of which *tecomine (tecomanine)* and *tecostanine* have potential hypoglycaemic properties (1; 3). The seeds have a fatty oil content of 23% which is composed of *octadecatrionoic acid* (41%), *octadecadienoic acid* (24%), *octadecatetraenoic acid* (19%), *octadeconic acid* (7%), *palmitic acid* (6%) and *stearic acid* (3%). Unnamed triterpenes, hydrocarbons, resins and a volatile oil are also reported present in the plant (1).

References:

1. Chadha, 1976, a: 134. 2. Standley, 1920–26: 1318–9. 3. Willaman & Li, 1970.

# BIXACEAE

**Bixa orellana** Linn.

FWTA, ed. 2, 1: 183. UPWTA, ed. 1, 45.

English: anatto; annatto; arnatto; yellow dye (Sierra Leone).

French: rocou; roucouyer.

Portuguese: anato, arnato, rok, uruku (the dye extract, Feijão).

West African: **GUINEA-BISSAU** CRIOULO djanfaraná (JDES) FULA-PULAAR (Guinea-Bissau) djambaraná (JDES) **SIERRA LEONE** BULOM (Kim) kamgo-poto (FCD) KONO bundu (FCD) KRIO rɛd-rɛd (FCD) LIMBA (Tonko) kudonia (FCD) LOKO mbundona (NWT) MENDE mbundɔ (*def.*-i) *properly applies to Camwood, Baphia nitida* (auctt.) pu-bundɔ (*def.*-i), pu: *European* (JMD) SUSU kamonyi (FCD) SUSU-DYALONKE lugbagbel-la (FCD) TEMNE *a*-kam *properly applies to Camwood, Baphia nitida* (auctt.) *a*-kam-*a*-loli (NWT) **IVORY COAST** KRU-GUERE kuiguéhé (Aub.) **GHANA** AKAN-ASANTE daagyene (Enti) konin (Soward, ex FRI) BRONG dagyiri (FRI) TWI brɔfo agyama (auctt.) WASA daagyeni (Enti) GA á'jama (KD) GBE-VHE bernitiku (Volkens, ex FRI) **TOGO** GBE-VHE berniticu (Volkens, ex JMD) TEM (Tshaudjo) kirane (Volkens, ex JMD) **NIGERIA** IGBO úhíé (KW) úhíé árọ̀ *the body paint, cf. Baphia nitida* (JMD) IGBO (Awka) ḿkpụ́lụ́ ọ́fịá = *seed of the bush* (NWT) IGBO (Ibusa) úfié (JMD) uhia nkum (AJC) IGBO (Onitsha) ula (JMD) ula machuku (JMD) IGBO (Umuahia) úhíé n̄kū, úhíé: *camwood* (Singha; KW) YORUBA oṣùn búkẹ *the fruits* (IFE)

An abundantly branched shrub or small tree to about 5 m high, with showy pink or white flowers and brown, bristly bi-valved fruit pods bearing numerous red seeds within; a native of tropical America, now pantropically dispersed by man.

The plant is a decorative shrub and makes a thick, quick-growing hedge of dark green foliage, or it can be grown, though the result is somewhat laxer, to form a screen 4–5 m high. In the West Indies and in E Africa, it is said to be useful as supports for vanilla (4). In Malaya it has been reported able to grow up through *Imperata cylindrica* grass (4) and may thereby be of some assistance in regenerating fertility in degraded land where rainfall is not less than 1,000 mm p.a. The bark will produce a fairly good fibre. The wood is very light and appears to have no uses.

The most interesting aspect of the plant lies in the dye 'anatto', obtainable primarily in the seed coat or in the pulp surrounding the seeds. The dye varies between red, orange and bright yellow depending on the proportions of the red pigment, *bixin*, normally about 70–80%, and an unnamed yellow element. The plant has entered into commercial cultivation for the production of this dye which is used mainly in the food-industry and for colouring dairy products such as butter and cheese, margarine, edible oils, etc. The dye is also used in Brazil in pottery and as an insect-repellent, and in the Philippines in floor, furniture and shoe-polishes, nail-varnish, brass lacquer, hair-oil, etc. Jamaica and S India have been major producers of the top quality product. From elsewhere the dye has lacked the bright colour wanted. Work in Nigeria suggests that a commercially marketable product should be producible in that country (11. 15). Anatto dye is, however, not durable and its use on fabrics, for example, in Europe has entirely gone out (4).

The dye is used like camwood, *Baphia nitida* Lodd. (Leguminosae: Papilionoiideae), by some races in West Africa as a cosmetic on the face in folk-dancing, or for staining utensils, cotton, matting, etc. (2, 3, 5, 7, 8, 9, 12, 13, 16, 17).

The seeds are considered purgative and some races in Gabon eat the fruit-pulp (17), which is taken in West Africa as a febrifuge and an astringent for dysentery and in kidney diseases (7, 13). Screening of the seeds for anti-malarial activity has shown no activity against avian malaria (6). In Congo fruit and seed are ground to a paste and applied with energetic rubbing after washing to areas of itch (3) and seeds and sap have been used on cancerous conditions in the mouth in Mexico and Paraguay (10).

A leaf-decoction is taken in Gabon to arrest vomiting (17), and in Congo as a gargle for sore-throat and quinsy (3). The leaf is used in Sierra Leone as a remedy for pain in the ribs caused by *wanka,* i.e. magic against theft (N.W. Thomas fide 7).

Besides dye substances present in the plant, a resin has been reported (14), and a trace of alkaloid (1).

References:

1. Adegoke & al., 1968. 2. Aubréville, 1959: 3, 5. 3. Bouquet, 1969: 73. 4. Burkill, 1935: 330–2, with references. 5. Burtt-Davy & Hoyle, 1937: 19. 6. Claude & al., 1947. 7. Dalziel, 1937. 8. Deighton 1859, K. 9. Greenway, 1941. 10. Hartwell, 1968. 11. Ingram & Francis, 1967: with many references. 12. Irvine, 1930: 56. 13. Irvine, 1961: 71–72. 14. Oliver, 1960: 20. 15. Raymond & Squires, 1951. 16. Thomas, N. W., 1990 (Nig. Ser.), K. 17. Walker & Sillans, 1961: 103–4.

# BOMBACACEAE

**Adansonia digitata** Linn.

FWTA, ed. 2, 1: 334. UPWTA, ed. 1, 112–5.

English: baobab; monkey-bread tree; sour gourd; cream of tartar tree.

French: baobab; pain de singe (the fruit); arbre aux calebasses; calebassier du Sénégal.

Portuguese: baobab; calabaceira.

West African: **SENEGAL** BALANTA laté (K&A) BASARI a-màk (K&A; Ferry) BEDIK a-mák, ga-mák (K&A; Ferry) DIOLA (Fogny) babaq (JB; K&A) bubakabu (auctt.) FULA-PULAAR (Senegal) boiö (K&A) boki (auctt.) boré (K&A) bóy (K&A) KONYAGI a-mbu (Ferry) MANDING-BAMBARA mŏlōdo *a varietal name* (Vuillet) sira (auctt.) sito (K&A) tedum (Barth) MANDINKA sita (*def.* sitoo) = *to tie* (auctt.) MANINKA sira (auctt.) sito (K&A) 'SOCE' sito (K&A) MANKANYA bedôal (K&A) SERER bak (auctt.) mbak (K&A) o kandalé *the bark-cordage* (N'Diaye) o sag *a drop-net made from bark-cloth* (N'Diaye) SERER-NON ba (auctt.) boh (auctt.) NON (Nyominka) bak (K&A) ibak (K&A) SONINKE-SARAKOLE kide (K&A) WOLOF bui *the fruit pulp or flour* (auctt.) gif *the seeds* (auctt.) gui *the tree* (JMD; K&A) lalo *the leaves, or a mixture of dried powdered leaves of which this is predominant, for a food or medicine* (JMD; K&A) ndaba *the mucilage* (K&A) téga *the bark* (Aub.; K&A) **THE GAMBIA** DIOLA ebakai (DRR) DIOLA (Fogny) bubakabu = *the taller* (DF) FULA-PULAAR (The Gambia) boki (DAP) MANDING-MANDINKA sita (*def*.. sitoo) (auctt.) WOLOF bui *the fruit* (DRR) gui *the tree* (DRR) **GUINEA-BISSAU** BALANTA látè (JDES) BIDYOGO uáto (JDES) CRIOULO cabaceira (JDES; GeS) calabaceira (auctt.) FULA-PULAAR (Guinea-Bissau) bóè (EPdS; JDES) MANDING-MANDINKA citô (JDES) MANDYAK bebaque (JDES) bedom-hal (JDES) brungal (JDES) burungule-burúnque (JDES) PEPEL burungule (JDES) **GUINEA** FULA-PULAAR (Guinea) bŏki (CHOP) MANDING-MANINKA bŏki (CHOP) sira (Aub.) SUSU kiri (CHOP) **SIERRA LEONE** KONO sela (S&F) KRIO baobab (S&F) mɔki-brɛd = *monkey bread* (FCD; S&F) LIMBA kutidi (NWT) LOKO sakwi mbawi (AHU) MANDING-MANDINKA sida (FCD) sira (FCD) MANINKA (Koranko) sirɛ (auctt.) MENDE gbowulo (*def*.-wulii) (S&F) SUSU kiri (NWT) SUSU-DYALONKE kidi-na (NWT) TEMNE *an*-derəbai, *an*-dɛr-*a*-bai = *The chief's body, alluding to the belief that a root decoction taken with food makes a man stout* (FCD; S&F) *a*-kiri (S&F) **MALI** BAGA (Koba) kö-basera, kö; *tree* (Hovis) DOGON gényè *the fruit* (C-G) ɔ́rɔ (auctt.) pepèru *the flower* (C-G) tigɛ̀ *the fruit* (C-G) FULA-PULAAR (Mali) boki (Aub.; GR) bokki (Aub.) MANDING-BAMBARA mŏlōdo *a varietal name* (JMD) sira (GR; FB) MANINKA sira (Aub.) SONGHAI kò (*pl.* kòà) (Aub.; D&C) konian (Aub.) **UPPER VOLTA** BISA mor (Prost, ex K&B) DAGAARI tuo (JMD, ex K&B) GRUSI-LYELA kukulu (Nicholas) HAUSA kuka (Aub.) MANDING-BAMBARA sira (K&B) MOORE toéga (auctt.) toyéga (K&B) 'SENUFO' ngigué (Aub.) **IVORY COAST** BAULE fromdo (auctt.) KRU-GUERE go (*pl.* gwê) (Bertho.) GUERE (Wobe) gblé-tu (Bertho) NGERE go (*pl.* gwê) (Bertho.) KWENI bèlé (Grégoire) MANDING-DYULA sira (K&B) 'SENUFO' ngigué (Aub.) nguigué (FB) **GHANA** ADANGME-KROBO salɛtʃo (FRI) salo (FRI) AKAN-AKUAPEM ɔdadeɛ (FRI) ASANTE odadeɛ (Enti) ɔdadeɛ (Enti) BRONG ala (JMD; FRI) kɛlai (FRI) nilai (FRI) TWI ɔdade(ɛ) (FRI) ɔtɔtɔwaa (FRI) WASA ɔdadeɛ (Enti) BAULE fromodo (FRI) BIMOBA toreg (FRI; JMD) DAGAARI tuo (FRI; JMD) DAGBANI kantong *a food-stuff from the seeds* (JMD) tua (FRI; JMD) tukare *dried powdered leaves admixed with others similarly used* (JMD) GA sàalo,shàaje (KD) GBE-VHE a-dido (FRI) dodo (FRI) VHE (Awlan) alãgba VHE (Pecí) dindo dodo GUANG totɔ (FRI; JMD) GUANG-GONJA kèlárà (Rytz) KRACHI kɛlai (JMD) KONKOMBA nitule (JMD) MOORE toyega (FRI; JMD) NABT tuwa (BD&H) NANKANNI tua (Gaisie) NZEMA ekuba (FRI) SISAALA teliŋ (Blass) **TOGO** BASSARI niturr (JMD) GBE-VHE adido (Volkens) dudo (Aub.; FB) GUANG-KRACHI kelle (Volkens) KABRE taelu (Gaisser) tschodum *the leaves and seeds* (Gaisser) MOORE-NAWDAM kalim (JMD) kekim *the whole fruit* (JMD) telo (JMD) SOMBA turubu (Aub.) TEM (Tshaudjo) taelu, telu (Gaisser) tshodum *the leaves and seeds* (Gaisser) **DAHOMEY** BATONNUN chonbu (Aub.; FB) chonmu (Aub.) conmu (FB) sona (Aub.) BUSA fon (Bertho) GBE-FON kpassa (Aub.) zizon (Aub.) GEN dido (Aub.) SOMBA turubu (Aub.) YOM tolro (Aub.) YORUBA-NAGO ìgì óshè **NIGER** ARABIC (Niger) hamar (Aub.) hamaraya (Aub.) FULA-FULFULDE bokki (Aub.) HAUSA kuka (Aub.) SONGHAI konian (FB) SONGHAI-ZARMA kwo (Robin) **NIGERIA** ARABIC el omarah (JMD) gongoleis *the fruit* (JMD) homar (JMD) humar (JMD) oufa (JMD) tabaldi (JMD) ARABIC (Niger) hamar (JMD) ARABIC-SHUWA hamaraya (JMD) BUSA kuka *from Hausa* (Bertho.) DERA kúrnjé (Newman) EDO ùsì (JMD; KO&S) FULA-FULFULDE (Nigeria) ɓohere (*pl.* ɓohe, ɓoye) *the fruit* (MM) ɓokki (*pl.* ɓoɗe, ɓoge, ɓogeeje, ɓokko) *the leaf* (MM) ɓokko li'o *leaves for soup* (MM) gadiyaare (*pl.* gadiyaaje) *the seeds* (JMD) goromi *the seeds* (JMD) gulumbur *the young leaves* (JMD) mbuja *a food-stuff from the seeds* (JMD) njuulaandi *the fruit pulp* (JMD) nyaande (*pl.* nyaandeeji) *the rough bloom on the fruit, or the meal within the fruit* (Taylor) saaɓeho (*pl.* saɓehe) *fresh leaves* (MM) *or dried powdered leaves* (JMD) saaɓeho *the fresh leaves* (MM) GWARI kwahi (JMD) HAUSA bakko *the leaves, from Fula* (JMD) bambu (JMD) charari *the pounded seeds as a famine-food* (JMD) chusar doki *a horse-food containing mostly powdered leaves of this sp.* (JMD) daddawar ka-tsame, daddawar kuka *food-stuffs from the seeds* (JMD) damsa *the fruit-pulp* (JMD) danana *fruit-pulp and curdled milk* (JMD) dandare *a food-stuff from the fruit-pulp* (JMD) dullu *a decoction of the fruit-pulp* (JMD) dunku *the fruit, seeds, pulp, young leaves; a food-stuff from the seeds; a decoction of the fruit-pulp* (JMD) fanko = *good for nothing; the wood* (JMD) fartako *leaf-fibre left after powdering the leaves* (Bargery) garin kuka *a meal made from the seed kernels and millet* (JMD) gatsika *the young leaves* (JMD) gubdi *the fruit-pulp* (JMD) gujuguju *a decoction of the fruit-pulp* (JMD) gulullutu *the flower* (JMD) gumayi, guntsu, gwargwami *the seeds* (JMD) jar kuka *a variety* (JMD) kalun kuka *dried powdered leaves* (JMD) karkachijeho, katsame *foodstuff from the seeds* (JMD) kata *water filtered through the seed-pod ashes used to make soup* (MM) katambiri *a cosmetic made from the pods reduced to ash* (MM) kirɓbe *a soup of young leaves* (JMD) kirta *inner bark fibre* (JMD) kolo *a foodstuff from the seeds* (JMD) ku(m)bali, kulambali *the flower* (JMD) kuka (auctt.) kulkuli *the fruit-pulp* (JMD) kuukà (Lowe) kwa(i)kwayo *the fruit husk* (JMD) kwalaba *a decoction of the fruit-pulp* (JMD) ƙwame, kwámé, ƙwami kwámii *the fruit* (JMD; ZOG) kwatakwari *fruit-pulp and curdled milk* (JMD) kwatambo *a decoction of the fruit-pulp* (JMD) maiwa *a variety* (JMD) miyar kuka *dried powdered leaves* (JMD) zullu *a decoction of the fruit-pulp* (JMD) HAUSA (West) bumbu (ZOG) kubali, kulambi, kumbali (ZOG) múrnaà (ZOG) IGALA obobo (Odoh) KANURI kàlkúwà *dried powdered leaves* (JMD) kuka *from Hausa* (auctt.) kúwà (C&H) NUPE èmu *the fruit* (JMD) kúka *the leaves when eaten; from Hausa* (JMD) muci *the tree* (Banfield; KO&S) SHANGA hwon (Bertho) YORUBA lũrú *dried powdered leaves* (JMD) oṣè (auctt.) **WEST CAMEROONS** BAFOK njobwih (JMD) KOOSI njobwele (JMD) KOSSI njobwele (JMD) KUNDU ngubwele (JMD) LONG njobwih (JMD) LUNDU njubwele (JMD) MBONGE ngubwele (JMD) TANGA ngubwele (JMD)

A tree to about 15 m high, with an enormous squat trunk to 20 m girth or more (2, 11), occurring in the soudano-sahel savanna, rarely in the guinean wooded savanna, across the Region from Senegal to W Cameroon, and in a discontinuous belt across Africa north and south of the Equator (9, 12, 27). Such is the diverse utility of the tree to man and animals by which agencies its dispersal has been augmented that its original habitat is no longer clear. It is thought not to be native to Sierra Leone (36), nor to The Gambia (35) for it is found only around habitations. Its presence in Senegal is disjuncted and is mainly around villages or old village sites (27). The name baobab arises from the Arabic *bu hibab* meaning *fruit with many seeds*. Its first mention in European literature is in Alpino (Venice, 1592): *De Plantis aegypti liber,* as *ba hobab,* referring to the fruits commonly sold in Egypt for their edible pulp (1, 27). From this it must appear to have entered western European languages as *baobab,* though to Adanson (1727–1806), whose name the genus bears and by whose work so much has been learnt of West African botany, it was known as *arbre aux calebasses* (calabash tree).

The wood is light and spongy. It is not durable and is easily attacked by fungi with the result that local uses are rare. Furthermore the shape of the tree does not make felling easy: an axe is more likely to bounce off the wood than to cut into it and in clearing trees for the Kariba Dam tractor-drawn wire-hawsers were used. A Hausa epithet for the wood, *fanko,* means 'good for nothing.' The wood makes poor firewood unless thoroughly dried out. A large amount of mucilage present, which enables the tree to withstand desiccation, obstructs the drying process. It is not good for cutting into planks, but wide and light canoes can be made, and wooden plates, trays, floats for fishing-nets, etc. The wood can be pulped for paper-making, but quality is doubtful (12, 39). It can be turned into a poor quality charcoal, sometimes used for want of better (1), and burnt to yield a vegetable salt (21, 39). Wild animals are said to chew the wood (39) perhaps to obtain the salt from the sap.

The tree living in very dry situations with its enormous trunk of spongy wood carries a great quantity of water. A good tree may hold as much as 1,000 gallons (39), and girth may vary according to weather conditions. Man has undoubtedly planted trees to be able to tap the aqueous sap as well as for its other multifarious uses. Hollows may be carved out from a small hole which is then corked so that the liquid may collect and be readily drawn off, or even the whole tree hollowed out to form a tank though medical officers may view this with disfavour as furnishing mosquito breeding sites. In E Africa the trunk may be hollowed out to provide shelters and form storage rooms (39). Livingstone in his explorations in Mozambique recorded use as dwellings. The fabled longevity of the tree is not satisfactorily confirmable, but the age of a tree cut down for the Kariba Dam project was determined by $C^{14}$ dating as $1010 \pm 100$ years old, with an inference that really large individuals could, indeed, be several thousands of years old (37). The odd appearance of the tree has resulted in magical and superstitious uses. In Upper Volta it is left standing when clearing the bush as a fetish tree (28). Primitive tribes of N Nigeria reverence it by cutting symbols in the bark (29). The hollowed-out trunk has been recorded as used as tombs (1), and a place where a body denied burial may be suspended between earth and sky for mummification (12). In places it is worshipped as a fertility symbol. Rock-art in the Limpopo Valley depicts women's breasts as baobab pods (39). In Upper Volta children of the Ela born under the sign of this tree (*kukulu,* Lyela) are given the patronymic *kukulu,* boys, or *ekulu,* girls (32).

The bark is fibrous. It is commonly stripped off the lower bole. The tree appears to be able to survive considerable rough treatment and to regenerate the bark. Fibre from the inner bark is particularly strong and durable. It is commonly used to make rope and cordage, harness-straps, string for musical instruments, baskets, nets, fishing-lines, etc. It lacks tenacity and fineness for spinning, and loses strength on drying. The bark can be dried and beaten to yield a crude bark-cloth usable as aprons and loin-cloths, fishing-nets and as sacking and packing material for local trade. (1, 6, 12, 14, 21, 26, 31, 33–35, 38, 39, 41–43.) At one time the bark was exported to Europe to make a strong packing paper (12). In N Nigeria flat pieces serve as the soles of sandals threaded with baobab fibres as toe thongs (43).

Reports that the bark is eaten in Senegal (19, 20, 39) are not corroborated by other authors. At any rate the bark contains a quantity of edible, insoluble, acidic tragacanth-like gum (18). This is used for cleaning sores. The bark has medicinal properties as a febrifuge in West Africa and in other parts of the world. It has been used as a quinine-substitute (5, 12, 21). It has been imported to Europe as *cortex cael cedra*. It is diaphoretic and anti-periodic. Its benefit as a febrifuge has not been detected in experimental malaria (10, 39). It has a bitter taste. An unnamed alkaloid is recorded present (40), and also the presence of the alkaloid, *adansonin,* but examination of Nigerian material gave inconclusive results on the presence of alkaloids (4). Adansonin has a strophanthin-like action, yet in E Africa the bark is said to be antidotal to *Strophanthus* arrow-poison (1, 12, 39). These anomalies clearly merit further examination. In some countries the bark is used for tanning (12, 15). In Congo the bark-decoction is used to bathe ricketty children (7), and in Tanganyika as a mouth-wash for toothache (16). A soap-lye can be made from the bark-ash (1), and the ash of all parts has value as a fertilizer (21).

The leaves, and especially the young leaves, are a popular item of diet as a spinach or to make soups and sauces. In some parts the trees are pollarded so as to produce an abundance of young leaves. In Senegal leaves of a glabrous form are preferred, tomentose forms being considered unsuitable. In Mali the leaves of a variety called *mŏlódo* (Bambara) are not eaten, and in Zaria, N Nigeria, two forms, *jar kuka* and *maiwa* (Hausa), are recognized (12), but the distinctions are not stated. In N Nigeria the first leaves are said to be unfit for use till they have been washed by the rains (43). The most common use of the leaves is to dry and powder them into an article known throughout much of W Africa as *lalo*. The leaves are rich in mucilage containing uronic acids, rhamnose and other sugars (9, 27, 33, 39). Tannins, potassium tartrate, catechins and a flavonic pigment, *Adansonia flavonoside,* are also present (33). On alkaline soils at least, if not on others, the leaves contain a high calcium content. It is said that the Dakaroise consume 35–49 gm of lalo daily providing an adequate calcium intake (1). The fresh leaf is rich in vitamin C but this is lost on drying in producing lalo (1, 9). Lalo is a term applied to other similar ingredients used as food, condiment and seasoning. It is also used to denote medical preparations based on the dried leaf. The leaves have hypotensive and antihistaminic properties (28, 33). They are diaphoritic and promote sweating, but are recorded also as used to treat excessive perspiration (12, 33, 39). General applications of the leaf are for kidney and bladder diseases, asthma, general fatigue, tonic, blood-cleanser, prophylactic and febrifuge, diarrhoea, inflammations, insect bites, expulsion of guinea-worm, internal pains and other affections. (1, 5, 6, 8, 12, 21, 23, 24, 25, 26, 28, 33.) Dysentery is treated by mouth or in hip-baths, and asthma, sedation, colic, fevers, inflammations, diseases of the urinary tract, ear-troubles, backache, ophthalmias, wounds and tumours, respiratory difficulty, etc., are treated by lalo by mouth or in liquid preparations (1).

The leaves are browsed by stock, and are fed to horses (1, 22, 33). In Nigeria the leaves are the most usual ingredient of a horse-food (*chusar doki,* Hausa) which given in large quantities is said to keep a horse in good condition on a journey. Given in smaller amounts it is tonic, blood-making, and good for subcutaneous swellings caused by insect-bites (12).

The roots may be cooked and eaten as a food (1). It is said by the Temne of Sierra Leone that a root-decoction taken with food causes stoutness hence their vernacular name for the plant, *an-derəbai* meaning 'The Chief's Body' (13). The dried powdered root is prepared as a mash which is taken for malaria (16), perhaps as a tonic. The roots are used in E Africa to yield red dye (15).

The earliest European name for the seed pods is that of the French travellers in Senegal, calabasse du Sénégal. The pods, or more correctly the mealy fruit pulp within them, are the Monkey Bread or Pain de Singe (French). They are of variable shape, subspherical to ovoid, 15 to 35 cm long by 7–15 cm in diameter. The husks are good for burning and a potash-rich vegetable salt may be obtained from this ash which is usable for making soap (1). The powdered husk or the powdered peduncle may be smoked like tobacco (1). The whole husk can be used as a dipper or to hold liquids and is sometimes fashioned into snuff-boxes (39, 42). Fibres lining the inner surface of the husk are given in decoction to treat amenorrhoea (1, 27, 33).

The seeds are embedded in the dry acidulous, mealy pulp which is rich in mucilage, pectins, tartrates and free tartaric acid (9). The presence of tartrates gives rise to the name *Cream of Tartar tree.* Calcium and, when eaten raw, vitamins B, and C are abundant (1). It is important in diet as a seasoning and appetiser, and in time of dearth it is eaten as a foodstuff. It can easily be made into a gruel with water, with millet, or milk which it curdles, etc., and provides a refreshing drink (1, 9, 12). It has medicinal uses as a febrifuge and antidysenteric. It appears in small-pox and measles treatments as an eye-instillation (1, 5, 6, 12, 27). At one time the pulp was exported to Europe as an imposture of *terra Lemnia,* or *terra sigillata,* a medicine of the Ancients mined from the Island of Lemnos in the Aegaean (12). It has been used in West Africa as a coagulant for latex resulting in a good quality coagulum (12, 14). It has been used as a coagulant for *Ceara* latex in E Africa (39). The pulp burns with a smoke which can be used for preserving and drying fish. Its acrid smell is also useful to drive away stinging insects troublesome to stock (1, 21).

The seed-pods are burnt as a fuel in N Nigeria and the ash is used to make soap and for curing an illness known as *dankanoma* (? Hausa). Water which has been filtered through the ash is sometimes added to soup in the belief that it will kill germs, e.g., those on unwashed baobab leaves (see above), and will cure stomach troubles (43). The white mealy substance surrounding the seeds is chewed by children and animals. When soaked in water it produces a milky solution which is taken as a milk-substitute (43). The seeds have a relatively thick shell which is not easy to separate from the kernel. This shell is recorded in a Dakar sample as 55·46% of the whole. The kernel is edible but the difficulty of decortication limits its usefulness. It is rich in protein and thiamine, 100 gm daily being adequate to supply an adult's requirement (1). Oil-content by ether-extraction is recorded as 68% of the kernel. The oil is non-drying and consists of stearic, palmitic and oleic acids (27). The kernel is free of starch, alkaloid and cyanogenetic glycosides (39). Parching and crushing can be practised to decorticate the seed. The oil can be obtained by boiling it off (12) and this has uses for gala occasions in Senegal (1). Roasted seeds are crushed to a paste which is applied to diseased teeth and gums. When burnt, a potash-rich salt can be got from the ash suitable for soap-making (1).

The white shoot of the germinating seed and the roots of very young seedlings are edible (12).

References:

1. Adam, 1962. 2. Adam, 1963. 3. Adam, 1966, a. 4. Adegoke & al., 1968. 5. Ainslie, 1937: sp. no. 12. 6. Aubréville, 1950: 165–7. 7. Bouquet, 1969: 74. 8. Boury, 1962: 15. 9. Busson, 1965: 302, with chemical analyses. 10. Claude & al., 1947. 11. Condamin & Lèye, 1964. 12. Dalziel, 1937. 13. Deighton 2759, K. 14. Gomes e Sousa, 1930: 50. 15. Greenway, 1941. 16. Haerdi, 1964: 87. 17. Hartwell, 1968. 18. Howes, 1949: 67. 19. Irvine, 1952, a: 30, 32, 33. 20. Irvine, 1952, b. 21. Irvine, 1961: 185–8, with chemical analyses. 22. Jackson, 1973. 23. Kerharo, 1967. 24. Kerharo & Adam, 1963, a. 25. Kerharo & Adam, 1964, b: 401–2. 26. Kerharo & Adam, 1964, c: 291. 27. Kerharo & Adam, 1974: 241–5, with phytochemistry and pharmacognosy. 28. Kerharo & Bouquet, 1950: 62–63. 29. Meek, 1925: 2, 24. 30. Monteil, 1953: 106. 31. N'Diaye, 1964: 118. 32. Nicholas, 1953: 832. 33. Oliver, 1960: 17, 43. 34. Roberty, 1953: 449. 35. Rosevear, 1961. 36. Savill & Fox, 1967: 55. 37. Swart, 1963. 38. Walker & Sillans, 1961: 104. 39. Watt & Breyer-Brandwijk, 1962: 144–7. 40. Willaman & Li, 1970. 41. Williams, F. N., 1907: 203. 42. Williams, R. O., 1949: 106. 43. McIntosh, 26/1/79.

**Bombacopsis glabra** (Pasq.) A. Robyns

FWTA, ed. 2, 1: 335, as *Bombax sessile* (Benth.) Bakh.

English: English ground-nut; whiteman's ground-nut (Sierra Leone, Deighton; Savill & Fox).

French: noyer d'Amérique (in error).

West African: **SIERRA LEONE** KRIO granat-tri = *groundnut tree* (FCD; S&F) TEMNE (Port Loko) *an*-kantr-*a*-potho (FCD)

A tree to nearly 10 m high introduced from Central America and planted around villages in the West African region. In Zaïre it is said to be able to grow spontaneously on sandy soils in certain localities (3).

The young leaves are said to be eaten in Equatorial Africa (Sillans fide 2). The seeds are oily and are eaten in Sierra Leone (1, 4), Liberia (5) and in Zaïre (3), and doubtless elsewhere. In Sierra Leone the seeds are likened to groundnuts, hence the local English and Krio names, *English,* or *Whiteman's Groundnut*. The common French name in Africa, *noyer d'Amérique,* is misapplied and refers properly to *Pachira aquatica* Aubl. (Bombacaceae) with similar dietetic use (3).

References:

1. Deighton 4148, K. 2. Irvine, 1961: 190, as *Bombax sessilis* (Benth.) Bakh. 3. Robyns, 1963: 199, with reference to the seeds. 4. Savill & Fox, 1967: 55. 5. Voorhoeve, 1965: 65.

**Bombax buonopozense** P. Beauv.

FWTA, ed. 2, 1: 334. UPWTA, ed. 1, 116–8, p.p., incl *B. angulicarpum* Ulbrich and *B. flammeum* Ulbrich.

English: red-flowered silk-cotton tree; red cotton tree; bombax (trade, Liberia Kunkel); West African bombax (trade, Sierra Leone, Savill & Fox).

French: kapokier.

Portuguess: poilão-ferro (Feijão).

West African: **SIERRA LEONE** BULOM (Sherbro) seŋgbeŋ-dɛ (FCD; S&F) FULA-PULAAR (Sierra Leone) jɔhɛ (FCD) GOLA gege (FCD) sona-wowo (FCD) KISSI peiŋgo (FCD; S&F) KONO fua (FCD; S&F) fua-kɔne (S&F) fula (FCD; S&F) fura (FCD) KRIO rɛd-kɔtin-tri = *red cotton tree* (FCD; S&F) LIMBA (Tonko) poŋgi (FCD) LOKO togba (FCD; S&F) MANDING-MANDINKA bumbu (FCD) MANINKA (Koranko) disile (S&F) MENDE kinjulu *old or obsolete* (FCD) titi (*def.*-i) *the flower* (auctt.) yawumbo (*def.*-bui) (FCD; S & F) yi-ndi *the stamens of 'titi'* (Migoed) SUSU lokhe (FCD) SUSU-DYALONKE lokho-na (FCD) TEMNE *an*-folaŋ (FCD; S&F) *an*-poŋk-poŋk (FCD; S&F) VAI tondo-gbanda (FCD) **LIBERIA** DAN gweh-gbonoh (GK) swa-ü (GK) KRU-GUERE (Krahn) sre-tu (GK) **UPPER VOLTA** GRUSI fofo (Houard) GRUSI-LYELA fwo (Nicholas) HAUSA guidjia (K&B) kuria (K&B) MANDING-BAMBARA bumbu (K&B) bumu (K&B) MOORE vaka (Bégué) waka (Houard) SONGHAI-ZARMA forgo (K&B) **IVORY COAST** ABE oba (auctt.) AKAN-ASANTE akongodié (K&B) AKYE mboba (A. Chev.; K&B) ANYI akong'dui (Aub.; K&B) ekuo (K&B; RS) BAULE puka (K&B) DAN gué (auctt.) KRU-BETE gô (auctt.) GUERE koa djo (Aub.) GUERE (Wobe) klimintio (K&B) koikindio (K&B; Aub.) koimintio (RS; Aub.) KYAMA agnébro (B&D) NZEMA adobo (A.Chev.; K&B) ehouen (A.Chev.; K&B) **GHANA** VULGAR akata (DF) ADANGME-KROBO agudesi (FRI) akudoni (FRI) AKAN-ASANTE akata (FRI) akɔnkodeɛ, ŋakokɔdeɛ (auctt.) akonkodie (CJT) okoo (auctt.) ASANTE (Denkyera) akata (Enti) FANTE akɔnkɔre (Enti) eku (BD&H) TWI akata (auctt.) akɔnkodeɛ (auctt.) akɔŋkɔdeɛ (FRI) akonkodie (Enti) WASA akata (auctt.) ANYI akuondwi (FRI) ANYI-AOWIN ekuo (FRI; CJT) ekur (FRI) SEHWI ekuo (FRI) BAULE puka (FRI) DAGAARI vorga (FRI) DAGBANI vabga (FRI) vabɔga (FRI) yabaga (FRI) GA ahéneyeɛŋ tso = *white dried sticks for chiefs*; from TWI (FRI; KD) akrɔŋ-krɔŋ (FRI) kafɔte-yɛŋɛ-tʃo = *white kapok tree* (FRI) GBE-VHE wu (FRI) wudese (FRI) VHE (Pecí) agudese GRUSI kafro (FRI) KONKOMBA bufo (JMD) NANKANNI tua (auctt.) voŋa (auctt.) NZEMA ekuba (FRI) **TOGO** ANYI-ANUFO sambugo (JMD) BASSARI afobil (JMD) GBE-VHE wu (JMD) wudesé (JMD) GURMA (Mango) sambugo (JMD) KABRE hotŏ (JMD) MOORE-NAWDAM kula (JMD) tode (JMD) TEM folo (JMD) TEM (Tshaudjo) ƒolŏ (JMD) fulŏ (JMD) YORUBA-IFE OF TOGO juna (JMD) upolo (von Doering; JMD) **DAHOMEY** BATONNUN muroru (Houard) DENDI forgo (Houard) kponpara (Houard) FULA-FULFULDE (Dahomey) kuluhi (Howard) GBE-FON unséfé (Howard) **NIGERIA** VULGAR johi *from Fulfulde* (JMD) ABUA okpọnumwu (JMD) ARABIC-SHUWA joho (JMD) BOKYI chakum (JMD; KO&S) EDO olikhàtùlòkò (JMD) ùgbọ̀khà (auctt.) EFIK úkím (RFGA) FULA-FULFULDE (Nigeria) joyi (JMD; MM) GWARI knuellehi (JMD) HAUSA gurjiya, gúrjíyaá *the leaves, hence also the tree* (auctt.) kuriya, kúríyaá, kúryaá (auctt.) IBIBIO úkím (JMD) IGBO ákpụ̄ (auctt.) ngārá (JMD) ṉgārá-ákpụ̄ (JMD) IGBO (Awarra) ndukuru (DRR) ubumekpu (DRR) IGBO (Isu) akwotokoro (DRR) IGBO (Onitsha) ákpụ̄ ògìrì (JMD) IGBO (Owerri) ákpụ̄ ọ̄kpọ́rọ́ = *ordinary silk-cotton tree* IGBO (Umuakpu) [illegible] dukpu (DRR) IGBO (Uzuakoli) atụnịaka (FRI, ex JMD) IJO ido-undu (auctt.) IJO-IZON (Egbema) asịsaghá (Tiemo) KANURI gèltà (JMD; C&H) NUPE kútúkpáçi (Yates; Banfield) TIV genger (JMD) YORUBA ẹṣo *a cryptic name used superstitiously to avoid using the proper name* (Millson) lailu (JMD) olu kondo *the flower buds* (IFE) pópòǹlà = *big post* (auctt.)

*NOTE: Many of these vernacular names may apply to B. costatum in those language areas where both species occur.*

A large tree attaining 35–40 m height by over 3 m girth, buttressed, corky bark, warted, spined on young trees. It is a tree of the forested zone and occurs from Sierra Leone to West Cameroon in the Region, and in E Cameroun, Gabon and Rio Muni. It has for long been confused with *B. costatum* Pellgr. & Vuillet which is a savanna species. Where distribution overlaps, uses are perhaps identical, and the species are not separately distinguished (Ghana, 3; Ivory Coast-Upper Volta, 10).

The wood is dirty white, darkening to light brown, and yellow or reddish at the heart, porous, soft, coarse-textured but strong, easy to work, but warping if not seasoned (5, 9, 15, 16). It is not durable and is readily attacked by borers. It is used to make dugout-canoes and water-troughs. Smaller branches are hollowed out for quivers, and the wood is cut or carved to make platters, domestic utensils, doors, stools, drums, shoes, saddles, etc., and into walking sticks called *kwarakwara* in Hausa (5, 7, 9, 15). Currently the wood is proposed for aircraft construction and modelling, boxes and crates, insulation, pulp and paper in Ghana (1).

The bark contains a gum-resin and has emollient properties. In Ivory Coast bark is pulped and applied as a plaster to treat ringworm, and to remove scurf from hairy leather (2). The bark along with the attached spine is an item of trade in markets for pulverizing and mixing with oil to an ointment for application to skin-diseases, craw-craw, etc. Stem-bark, commonly used in Nigeria to treat skin-conditions, has been examined and no bacteriostatic nor fungistatic action has been found (11). In Sierra Leone an infusion of the beaten bark in cold water is drunk or rubbed on the head for dizziness or is also taken in cooked rice (6, 15). The Yoruba regard a bark-decoction as an emmenagogue (5), and in Gabon it is considered febrifugal (18).

Sheets of bark are used as roofing for small huts or temporary shelters, and the fresh root-bark, beaten out and dried, is used as wadding for shot in hunters' guns (5, 9).

A reddish-brown dye is prepared from the bark in N Nigeria (5), and in Ghana where it is used to dye hunters' clothes and is called *kuntunkuri* (Twi) (9).

The warty protuberances and their spines may assume an appreciable size and are used for carving and making into little figures and other objects. In Sierra Leone they are carved into pieces for playing a game like draughts called *ki* (Mende) (6, 15). In Ghana with letters carved on them, they are used as embossers and in printing (5, 9).

The young leaves can be dried and powdered and used like those of the *baobab,* or eaten as a pot-herb. When fresh they are also good for goat-fodder (5).

The flowers have fleshy brown calices, are mucilaginous and are eaten in soup or made into sauces in a similar manner to the use of *Hibiscus sabdariffa* Linn. (Malvaceae) (4, 5, 8). The immature fruits can be sliced and dried for use in cooking. The dried calyx, after the red flower has fallen, is mixed with the spines or bark, or both, and with local natron for chewing, either with a view to oral hygiene or as a mild substitute for kola when the latter is scarce or expensive, or merely to give the appearance of having chewed kola, the procedure being known as *kemu* (Hausa, Fula) and *taune*. The bark and attached spine ground to a powder with tobacco flowers is also rubbed on the teeth or is chewed. The red flowers alone are sometimes rubbed on the teeth by elderly women as young people of both sexes use tobacco flowers (5).

The fruits are normally 10–18 cm long by 4–7 cm broad but may attain 25×8 cm. They contain an abundance of white or greyish floss of excellent quality and of staple length longer and finer than that of *Ceiba* kapok. It is used for stuffing cushions, mattresses, etc., and as tinder. As the pods burst on the trees, climbing for collection, especially in view of the presence of spines, may be difficult. Fruiting does not normally start till the tree is at least 6–7 years old. Propagation by cuttings is easy, thus facilitating multiplication of desirable mother-trees.

The young fruit and calyx are used as tops in Sierra Leone (6).

The seeds are borne in the floss. They are oily containing about 29–34% oil of the whole seed of which palmitic, linoleic and oleic acids are the major fatty acids. The oil is used in cooking (4, 5).

The tree enters somewhat into superstitious uses. The Yoruba, who call it *pòpònlá,* or 'big post', invoke its spirit as a cudgel by which to send evil to one's enemy (17). They also call it *ẹsọ* in a cryptic way believing that its medicinal uses would be ineffective if called by its proper name (5, 13). The tree is held sacred by the Tera of N Nigeria (12), and to be more or less sacred by the Igbo (5). In Upper Volta the Lyela bestow the patronym *Befwo* on boys and *Éfwo* on girls from the name for the tree *fwo* (14).

References:

1. Anon., s.d. 2. Bouquet & Debray, 1974: 53. 3. Burtt Davy & Hoyle, 1937: 19–20, as *B. buonopozense* P. Beauv. and *B. flammeum* Ulbrich. 4. Busson, 1965: 307–8, with chemical analyses of calyx and seed. 5. Dalziel, 1937. 6. Deighton 1584, K. 7. Gomes e Sousa, 1930: 49. 8. Irvine, 1952, a: 34. 9. Irvine, 1961: 188–90. 10. Kerharo & Bouquet, 1950: 63. 11. Malcolm & Sofowora, 1969. 12. Meek, 1931: 1: 176. 13. Millson s.n., Feb. 1890, K. 14. Nicholas, 1953: 829, as *B. angulicarpum.* 15. Savill & Fox, 1967, 55. 16. Taylor, 1960: 105–6. 17. Verger, 1967: no. 167. 18. Walker & Sillans, 1961: 106, as *B. flammeum* Ulbrich.

**Bombax costatum** Pellegr. & Vuillet

FWTA, ed. 2, 1: 334–5. UPTWA, ed. 1, 115–6, as *B. buonopozense* sensu Dalziel, p.p.

English: Kapok.

French: Kapokier; faux kapokier (Berhaut); kapokier rouge (Berhaut); kapokier à fleurs rouges (Kerharo & Adam).

West African: **SENEGAL** BALANTA kikela (Aub., ex K&A) BANYUN siludia (Aub.) BASARI a-ngákhúrè (K&A; Ferry) BEDIK gi-pòwɔ̀l (Ferry) DIOLA bu dimb (JB) DIOLA (Fogny) buńâbu (Aub., ex K&A) FULA-PULAAR (Senegal) bumbuvi (Aub; K&A) diohi (Houard) dioy (K&A) djoi (Aub.) doy (K&A) TUKULOR johi (Houard) kuruhi (Aub.) KONYAGI a-kam (Ferry) MANDING-BAMBARA bumbu (JB; K&A) bumu (K&A) MANDINKA bunkuŋ (*def.*-o) MANINKA bumbu (Aub.) bumu (Aub.) MANDYAK mbetauar (Aub.; K&A) MANKANYA blofo (Aub., ex K&A) SERER bak (Aub.) n-dondul (K&A) SERER-NON ba (Aub.) bâ (Aub.) boh (Aub.) SOCE bumkuô (K&A) SONINKE-SARAKOLE griomé (Houard; Aub., ex K&A) WOLOF dodol (Aub.) dundul (auctt.) garab (Aub.) garab laobé (K&A) garab(i)laobé (auctt.) koyo koyo, kuio kuio (JGA) **THE GAMBIA** FULA-PULAAR (The Gambia) johi (DRR; DAP) MANDING-MANDINKA bunkuŋ (*def.*-o) *from* bungo: *house;* kungo: *head, alluding to the use of the wood in hut-building* (auctt.) WOLOF gerab-laube = *Laube's tree: Laube is the name of a tribal group* (DF) kaltupa (DRR) **GUINEA-BISSAU** BALANTA bumbum (EPdS; JDES) buúforè (EPdS; JDES) CRIOULO fidalgo (EPdS) poilão encarnado (GeS; JMD) poilão foro (EPdS) polóm fidalgo (JDES) polóm fôro (JDES) FULA-PULAAR (Guinea-Bissau) djóè (EPdS; JDES) djóia (JDES) djoóè (EPdS) luncum (DF) MANDING-MANDINKA bantaŋforo (*def.*-o) = *pure bantam* buncum (*def.*-ô) (JDES) bungkungo (*def.*) MANDYAK belofa (JDES) belôfa (EPdS) MANKANYA belofa (JDES) belôfò (EPdS) PEPEL ulôfò (EPdS; JDES) **GUINEA** MANDING-MANINKA bumbu (JMD; Aub.) bumu (Aub.) SUSU lorongui (JMD; Aub.) **MALI** DOGON tógodo (Aub.; C-G) FULA-PULAAR (Mali) bumbui (Houard) diohi (Houard) johi (Houard) MANDING-BAMBARA bumu (Houard; GR) togodo (Aub.) MANINKA bumbu (Aub.) bumu (Aub.) SONGHAI fórgò (Aub.) 'SENUFO' belaguon (Houard) bolognon (Aub.) **UPPER VOLTA** GRUSI fofo (Aub.) HAUSA gurdjia (Aub.) kuria (Aub.) MOORE waka **IVORY COAST** BAULE puka (Aub.) véko (Aub.) MANDING-MANINKA bumu (A&AA) SENUFO-TAGWANA zangoro (Aub.) **GHANA** KONKOMBA bufo (Aub.) SISAALA fufuluŋ (Blass) **TOGO** BASSARI afobil (Aub.) BATONNUN melonlu (Aub.) monoru (Aub.) mulodu (Aub.) muroru (Aub.) MOORE-

NAWDAM kula (Aub.) tode (Aub.) SOMBA fokubu (Aub.) TEM folo (Aub.) **DAHOMEY** BATONNUN melonlu (Aub.) monoru (Aub.) mulodu (Aub.) muroru (Aub.) DENDI fórgò (Aub.) kponpara (Aub.) YORUBA-NAGO póponlà òdàn = *pōpọnlą of the savanna* **NIGER** DENDI fórgò kponpara FULA-FULFULDE (Niger) bumbuyi (Aub.) djoi (Aub.) kuluhi (Aub.) kuruhi (Aub.) HAUSA ab-blugar (Virgo) gurdjia (Aub.) kuria (Aub.) SONGHAI bántàm (*pl.* -ò) (D&C) fórgò (*pl.* -à) (Aub.; D&C) **NIGERIA** DERA ɗáŋəlàŋ (Newman) ťàm (Newman) FULA-FULFULDE (Nigeria) joohi (J&D) kuruhi (KO&S; MM) HAUSA gurjiya (auctt.) kuriya, kúríyaá, kúryaá (auctt.) HAUSA (West) gúrijíyaá (ZOG) NUPE kútúkpáci (K O&S)

*NOTE: Many of these vernacular names apply to B. buonopozence in those language areas where both species occur.*

A tree to 15 m high by 1 m girth, straight trunk slightly thickened basally, thick corky bark, spiny when young. It is a savanna species of the soudanian zone, but disappearing from the sahel and guinean woodland, of Senagal to N Nigeria, and on to Ubangi-Shari. It has been frequently misidentified as *B. buonopozense* P. Beauv.

It is a common tree of the soudanian part of Senegal and the leaves and fallen flowers provide fodder for all stock. Its conservation has been advocated in order to improve pasturage by providing shade and fodder (1). Its thick bark affords protection against fire damage (2). In Sierra Leone stakes are planted to make a live fence (4).

The wood is soft, dirty white, reddish at the heart, and is used for making domestic articles. It is used in The Gambia to make lintels and doors of huts, and this gives rise to the Mandinka name *bungkungo,* derived from *bungo,* a house, and *kungo,* head (Harper fide 3).

The bark of stem and roots is recognized in Senegal to have diuretic properties (5, 6). In Sierra Leone it is said to be more efficient medicinally than the bark of *B. buonopozense* and is used for the same purposes, and in decoctions for 'yellow-fever' and headaches, and 'to make a woman's breasts fine.' For headache a compress is also tied on the head (4).

The leaves are emollient and a warm bath of the decoction may be made for feverish patients, especially children (3). In Senegal they are prescribed with other drug plants for blennorrhoea and diarrhoea (6).

The fruit contains a large amount of white floss usable for stuffing mattresses, pillows, cushions, etc., and which at one time has been an item of export from francophane West Africa. Unfortunately many trees are unproductive owing to fire-damage at the critical time of flowering in the dry season. (2).

References:

1. Adam, 1966, a: 2. Aubréville, 1950: 170. 3. Dalziel, 1937. 4. Deighton, 3622, K. 5. Kerharo & Adam, 1964, b: 413. 6. Kerharo & Adam, 1974: 245.

**Ceiba pentandra** (Linn.) Gaertn.

FWTA, ed. 2, 1: 335. UPWTA, ed. 1, 118–22.

English: cotton tree; cottonwood tree; silk-cotton tree; white silk-cotton tree; white-flowered silk-cotton tree; kapok tree; ceiba.

French: fromager; fromager commun; fromager des Antilles; fromager d'indo-malaisie; fromager inerme du Golf de Guinée; kapokier; faux kapokier; kapokier à fleurs blandes; kapokier du Togo.

Portuguese: mafumeira; poilão; polão (Feijão).

West African: **SENEGAL** BANYUN kidem (K&A) BASARI a-ndín (K&A; Ferry) BEDIK gi-ndìi (K&A; Ferry) DIOLA (Brin/Seleki) busana (K&A) DIOLA (Fogny) busanay (K&A) DIOLA ('Kwaatay') étufay (K&A) FULA-PULAAR (Senegal) bâtigéhi (Aub., ex K&A) bâtinévi (K&A) KONYAGI a-man (Ferry) MANDING-BAMBARA bamâṅ (K&A) batâṅ (K&A) MANDINKA kantaŋ (after K&A) MANINKA bana (auctt.) bana-bâdâ (Aub., ex K&A) bana-bandan (RS) bâtân (K&A) busana (AS; K&A) 'SOCE' bêtanô (K&A) SERER m-buday (auctt.) SERER-NON len (AS, ex K&A) NON (Nyominka) buday (K&A) budey (K&A) WOLOF bêtéñé (auctt.) **THE GAMBIA** DIOLA-FLUP bosanobo = *canoe* (DRR; DF) FULA-PULAAR (The Gambia) bantehi (DAP) MANDING-MANDINKA bantalŋforo (*def.*-o) (JMD) bantaŋ (*def.*-o) (auctt.) WOLOF bentenki (DRR) betenbi (JMD) **GUINEA-BISSAU** BALANTA psáhè (JDES) rumbum (JDES) BIAFADA brêgue (JDES) BIDYOGO cob-bê (JDES) CRIOULO poilão (auctt.) polóm (JDES) FULA-PULAAR (Guinea-Bissau) bantanhe (JDES; EPdS) MANDING-MANDINKA bantaŋ(-ô) bantango (*def.*) bintaforo (*def.*) MANDYAK péntia (JDES) MANKANYA pèntè (EPdS) pentene (JDES) PEPEL metchene (JDES) n'teme (JDES) n'tene (untene) (JDES) **GUINEA** BAGA (Koba) kö-porõ kö: *tree* (Hovis) FULA-PULAAR (Guinea) bantignei (CHOP; RS) bentégniévi (CHOP) KISSI banda (RS) KONO bara (RS) KPELLE uyé (RS) MANDING-BAMBARA banan (CHOP) MANINKA bana (CHOP) bâna (RS) bana-bandan (RS) SUSU kondé (CHOP) TEMNE *am*-polon (Hovis) **SIERRA LEONE** BULOM (Kim) poloŋɛ (FCD) BULOM (Sherbro) polon-dɛ (FCD; S&F) FULA-PULAAR (Sierra Leone) banta (FCD) GOLA sona (FCD) KISSI g-banda (FCD; S&F) KONO g-banda (auctt.) KRIO kɔtin-tri = *cotton tree* (auctt.) LIMBA (Tonko) kutɛnɛ (FCD) LOKO n-gukhɔ(i) (FCD; S&F) MANDING-MANDINKA *g*bandaŋ (FCD) MANINKA (Koranko) banda (S&F) MENDE nguwa (*def.*-wei) (auctt.) SUSU konde (FCD) SUSU-DYALONKE konde-na (FCD) kundi-na (FCD) TEMNE *am*-poloŋ (auctt.) VAI g-banda (FCD) **LIBERIA** DAN gwe (GK) gwèh (AGV) KRU-GUERE (Krahn) dju (GK) MANO geh (JMD) guéh (RS) MENDE nguwa (*def.*-wɛi) (C & R) **MALI** DOGON dámu *the floss* (C-G) jiũ, jũ (C-G) FULA-PULAAR (Mali) bantignei (RS) bantiguehi (RS) MANDING-BAMBARA bana(n) (Houard) MANINKA bâna (RS) bana-bandan (RS) **UPPER VOLTA** BOBO pi (Houard) FULA-FULFULDE (Upper Volta) bantan (RS) bantignei (RS) HAUSA rimi (K&B) KIRMA belon (Houard) MANDING-BAMBARA banan (Bégué, ex K&B) DYULA banda (K&B) MOORE gunga (K&B) SONGHAI-ZARMA bantan (K&B; Aub.) bonetan (K&B; Aub.) TURUKA blo (Houard) **IVORY COAST** ABE gbi (A. Chev.; K&B) ABURE enivé (B&D) AKAN-ASANTE akuondi (B&D) gna (B&D) BRONG guima (K&B) AKYE muong (A. Chev.; K&B) nguéhié (A. Chev.; K&B) won (A&AA) ANYI egniè, egnien (A&AA) enia (auctt.) enya (A. Chev.; K&B) AVIKAM egna (K&B; Aub.) etchui (A. Chev.; K&B) BAULE angbo (B&D) gna (B&D) gniè, gnien (A&AA) nyé (auctt.) DAN guê (RS) FULA-FULFULDE (Ivory Coast) banatan (Aub.) bantan (RS) bantignei (RS) bantiguehi (Aub.) GAGU gué (RS) molongué (K&B) KRU-BETE gô (auctt.) GUERE diô (RS) tiô (RS) tiô (RS) tshyo (*pl.* tshui) (Bertho.) GUERE (Chiehn) go (B&D) GUERE (Wobe) djô (*pl.* dje) (auctt.) NGERE tshyo (*pl.* tshui) (Bertho.) KULANGO ton'go (auctt.) ton'ko (A. Chev.; K&B) toonko (K&B) KWENI dangué (B&D) *n*-gué (auctt.) gwɛ̃ɛ̃ (Grégoire) tyɛ̀nɛ *the kapok* (Grégoire) KYAMA agué (A. Chev.; RS) allotegué (B&D) anié (auctt.) MANDING-MANINKA bana (B&D) bana-bandan (K&B; Aub.) MANO ghê (RS) NZEMA eguina (A.Chev.; K&B) eniémé (A.Chev.; K&B) enyam'gua (K&B) enyan'gua (A.Chev.) SENUFO-TAGWANA sérigné (auctt.) 'ONELE' tiu (Houard; K&B) **GHANA** VULGAR enyena (TFC) onyina (DF) ADANGME lɛno (FRI; CJT) sokpe *a spineless var.* (FRI) AKAN-ASANTE onyina (Enti) ASANTE (Denkyera) onyina (Enti) BRONG danta (FRI) ekile (auctt.) ɔdanta-pu (FRI) FANTE onyãã (auctt.) onyãã (FRI) onyina *anyina: firewood (Fante) cited by authors as the meaning of the name for this species is incorrect; onyina has no connexion* (Enti) TWI onyã-hene = *king silk cotton tree, a spineless var.* (FRI; JMD) onyina (auctt.) WASA onyãã (auctt.) onyina (auctt.) onyina (Enti) ANYI enyaa (FRI) ANYI-AOWIN enya(a) (auctt.) SEHWI enyaa (auctt.) BAULE nye (FRI) BIMOBA gbang (FRI; JMD) DAGAARI gongu (FRI) goni (FRI; JMD) DAGBANI gumbihi *the seed kernel* (Gaisser) guna (FRI) guŋa (FRI) gunga (FRI) gunguma-gumdi *the floss* (Gaisser) gungumli *the fruit* (Gaisser) gung-vale (FRI) kantong *a paste made from the seeds* (Coull; FRI) GA ayigbe ogbedei, Ayigbe: *the Ewe people* (JMD); *a thornless var.* (FRI) onyáá, onyãí (KD) GBE-VHE atepre (FRI) əuti (FRI; CJT) ofu (FRI) wudese (FRI) VHE (Awlan) əuti (FRI; CJT) ofwho (TFC; BD&H) vulẽ *a spineless var.* (FRI) VHE (Kpando) atepré *a spineless var., dehiscent fruit* (Ulbrich) lɔe (FRI) lɔɛ *a spineless var.* (FRI) VHE (Pecí) əuti (FRI) ofua (auctt.) GRUSI gung (auctt.) GUANG-GONJA kàkèlə̀ (Rytz) kàkílíyà (Rytz) kakre (CV; FRI) kìlèntírékpèmbì *the fruit husk* (Rytz) KRACHI kekyafu (auctt.) HAUSA rimi (auctt.) KONKOMBA bufo *the fruit* (Gaisser) bufo-sõgbum (Ulbrich) kpugbum (Ulbrich) tubungbing *the floss* (Gaisser) umfobille *the seed-kernel* (Gaisser) MAMPRULI gunga (JMD) NANKANNI gonga (FRI; JMD) gonga (FRI; JMD) NZEMA enyenna (CJT) enyɛnɔa (FRI) SISAALA kuŋkomo (Blass) kuŋ-kumuŋ *the fruit-pod* (Blass) kuŋkunuŋ *the floss* (Blass) **TOGO** BASSARI bubumbu (Ulbrich) bufu (Ulbrich) tubumbum *the floss* (Gaisser) yigbum *the fruit* (Gaisser) GBE (Bɛ) aloe *a spineless var.* (FRI) eloe (FRI) loe-ti *a spineless var., indehiscent fruit* (FRI) GBE-FON huti (Ulbrich) vuti (Ulbrich) wuti (Ulbrich) GEN lovi *a spineless var., indehiscent fruit* (Ulbrich) VHE aloe, eloe *a spineless var., indehiscent fruit* (Ulbrich) evu (Ulbrich) ewu (Ulbrich) vu (Ulbrich) wu, wudese *a spiny var., dehiscent fruit* (Ulbrich) wuti (Ulbrich) KABRE botu *the floss* (Gaisser) botu-kisemto, botu-kocholemotu botu: *floss, vars with indehiscent fruit* (Houard) kolombolu *the indehiscent fruit* (Houard) komu (Ulbrich) kpong *the fruit* (Gaisser) KPOSO igboa *a spineless var., indehiscent fruit* (Ulbrich) ju *a spiny var., dehiscent fruit* (Ulbrich) juna (Ulbrich) MOORE-NAWDAM bahun (Ulbrich) gomu-dschiade *the floss* (Gaisser) gomu-schiere *the fruit* (Gaisser) ubombĕ (Ulbrich) TEM (Tshaudjo) bagbasse *the fruit* (Gaisser) komu (Ulbrich) YORUBA-IFE OF TOGO huti (Ulbrich) vuti (Ulbrich) wuti (Ulbrich) NAGO agú *a spiny var., dehiscent fruit* (Ulbrich) oguvé *a spineless var., indehiscent fruit* (Ulbrich) **DAHOMEY** BASEDA guénesso (A.Chev.) BATONNUN guma (Houard) BUSA gbê (Bertho) DENDI bantan (Houard) FULA-FULFULDE (Dahomey) linihi

(Houard) rinihi (Houard) GBE-FON adjoro hun (A. Chev.) gpati dêkrun (A. Chev.) hun-ti = *the tree of the canoes* (A. Chev.) FON (Gũ) bentan habu (Houard) VHE (Awlan) hunti (Houard) FON dehon (A.Chev.) gué dehunsu (A.Chev.; Houard) patin dehun *var. with indehiscent fruit* (Houard; A.Chev.) HWEDA hunsufu (Grivot) YORUBA-NAGO igi èégun, igi àràbà (A.Chev.) ogufé (Houard) **NIGER** FULA-FULFULDE (Niger) bantiguélil (RS) SONGHAI-ZARMA forgo (Robin) **NIGERIA** VULGAR araba *from Yoruba* (JMD) okha (JMD) ABUA ukem akabi (Kennedy) ù-mùùm (JMD) ù-mùùm (*pl.* àrù-mùùm) (JMD; KW) ANAANG úkúm (JMD) ARABIC-SHUWA rum (JMD; KO&S) BOKYI bokum (JMD; KO&S) BUSA gbée (Houard) gbiê-li (Bertho.) EDO okha (auctt.) EFIK úkím (KO&S) FULA-FULFULDE (Adamawa) boju *the floss* (MM) *used for tinder* (JMD) FULFULDE (Nigeria) bantahi (*pl.* bantaaje) (auctt.) bantahi (MM) taamu *the floss used for tinder* (JMD) teka *the floss* (Westermann) GWARI gehi (JMD) gyehi (JMD) HAUSA rimi (auctt.) rimin Masar = *Egyptian silk cotton tree; a var. with few or no spines* (JMD) HAUSA (West) abdugar rimi *the floss* (JMD) alhawami *the floss used for tinder* (JMD) ganɗiɗo *the roasted seeds* (JMD) rini (JMD; ZOG) IBIBIO úkím (KO&S) IGALA agwu, agwugu (H-Hansen) IGBO ákpụ̄ (auctt.) IGBO (Arochukwu) ákpụ̄-ugu (Kennedy) IGBO (Onitsha) ákpụ̄ ógwū (Kennedy) IGBO (Owerri) ákpụ̄ ùdèlè = *silk-cotton tree of the vulture* (Kennedy) mbom (Kennedy) IJO afalafase (KO&S) IJO-IZON (Egbema) àsịsàghá (KW) ogungbologhá (Tiemo) IZON (Kolokuma) ịsàgháị (KW) ISEKIRI ẽgungun (Kennedy) ISOKO ahe (Singha) KANURI tôm *the tree* (K O&S; C&H) NUPE kúci (Banfield; KO&S) lembúbúrú *the fruit* (Banfield) SHANGA gbê-siê (Bertho) konngô (Bertho) TIV vàmbè (JMD; KO&S) URHOBO óháhèn (auctt.) YEKHEE ọkho (Kennedy) YORUBA àràbà *the tree* (auctt.) ẹ̀ẹ́gun (JMD) ogungun (KO&S) òwú ẹ̀ẹ́gun *the floss* (JMD) **WEST CAMEROONS** DUALA bǔmà (Ithmann) kabò (Ithmann) KPE buma (JMD) wuma (Reder) LUNDU bum (JMD)

A tree to as much as 65 m high by 10 m or more in girth, with long cylindrical bole and huge buttresses to 8 m high and wide spreading; bole and branches spined when young; main branches horizontal and often bracketted below to the stem; of the secondary forest, seldom if ever in virgin forest, and conspicuous in savanna near habitations.It is said to be the largest tree of the West African region and occurs throughout. The species is thought to be originally from tropical America and it has been postulated that the light seed with the floss could have been wind-carried across the Atlantic (10). The floss, however, is not attached to the seed which is readily shaken out. The plant is now dispersed pan-tropically and a number of varieties and cultivars are recognized: var. *caribaea* is that of America and Africa, and var. *indica* is in Asia from which the bulk of the world's supply of kapok comes.

The English name, *cotton tree,* or *silk-cotton tree,* is derived from the floss of the seed pod. The floss resembles that of the cotton plant, *Gossypium,* but is silkier. The name *kapok* is from Malay for the floss and is a universal trade name. The French name *fromager* originated from French settlers in the Antilles of the Sixteenth century who likened the wood to being as soft as cheese (*fromage*) (10).

The tree is to some extent an anthropogene. It does not occur in virgin forest. It is an evanescent, quickly appearing in abandoned land. Its occurrence in forested areas is held to be a sign of disturbance. Nor does it occur in grass savanna which is subjected to annual burning as the tree does not survive fire, but in savanna land the tree is to be found in the proximity of habitations, albeit perhaps abandoned. The sight of a tree on the horizon to a traveller is a welcome direction post. In towns and villages it is planted as an avenue and shade tree. In built-up areas it will prove to be a troublesome one as the roots effect forceful entry into cracks in buildings, roads, drains, etc., and pass through or under and disturb foundations. In SE Asia the Asian variety has been planted as a shade tree for coffee and support for pepper (8). There may be a use in this way for cacao. In the soudan zone, it is the normal 'palaver' tree. Perhaps the best known is the specimen in the centre of Freetown, Sierra Leone, at the foot of which slaves returning from the Western Hemisphere were symbolically set free. Because of its size, it has become an important sacred and fetish tree, and is held to have magical properties. Over most of the soudanian region it is thought to be inhabited by the divine Python, symbol of maleness. Indeed the very extensive surface spreading lateral root-system suggests long snakes around the base of the tree, and for some races the roots especially are sacred as evoking a giant serpent (20). Other trees are also considered sacred, but the *Ceiba* is the most important. At the commencement of the rains offerings are made to it. In a village the tree is always planted to the southward, the direction from which the 'beneficial forces' come (24). It is also planted at the entrance to sacred groves during initiation ceremonies (29), and at tombs where offerings may be made at its base to the Shades of the Ancestors or to protective genie. In Gabon two trees may be planted before a house where twins have been born (32). In Sierra Leone at certain places it is the centre of a ceremony to pray for long life, wealth, good harvest, prosperity and the well-being of the population. The leader of the celebrants should be an elderly person and a leading citizen of the village (4). When bark is required for medical treatment it is common to take it from either the east or the west side, a mark of sun-worship evinced by all races even those mohammedanized (21). In N Nigeria the leaves, along with other herbs, enter into prescriptions to ensure popularity (Meek fide 11).

The wood is white, sometimes with yellowish or greyish streaks, very soft and light, and brittle when dry. It is liable to insect and fungal attack, and decomposes rapidly if unseasoned. It is resistant to impregnation, but can be used for rough crates though it is liable to split and does not hold nails nor glue well. Paper-pulp and the core of plywood are possible uses but large-scale commercial use seems unlikely. The wood wears tolerably well in domestic and household articles if seasoned, and it is used to make chairs, dishes, boxes, drums, carved figures, idols, stamps and dies, for modelling, musical instruments, etc. It is one of the woods commonly used to make Ashanti stools, and at one time it was the principal timber for making rice-mortars in Guinea. Prisoners-of-war in Japanese hands in Thailand, 1942–45, used the wood to make clogs and wooden soles of sandals for which it served very well. Plane articles such as doors, table-tops, plates, trays, etc., are made from the buttresses. As firewood it is of no value as it only smoulders, but this is put to use to fumigate huts or clothing, etc. (3, 10, 11, 13, 17, 27–31).

The very light density of the wood (0·28–0·35 dry wt.) (29), and the ease with which it can be cut makes the wood valuable for canoes and these are to be found on all West African rivers. The Nzema of Ghana even make seagoing canoes of it (30). In Dahomey the tree is known as 'the tree of the canoe', and trunks and the larger branches are hollowed out, and enormous dugouts fashioned to hold as many as 100–150 people (10). The buoyancy is such that even if swamped the canoe remains afloat.

Wood-ash is widely used in Africa as a kitchen salt and in soap-making. In Casamance, Senegal, ash from the pods is used in making snuff and in Guinea at Fouta Djalon in the indigo industry (17).

The bark contains a blackish mucilaginous gum which swells in water and resembles tragacanth. It is astringent and is used in India and Malaya for bowel-complaints and in West Africa for diarrhoea (6, 8, 16, 20, 33). The bark also contains tannin recorded as 10·82% (20, 33) which is too low a concentration for tanning. There is also a reddish-brown dye used in E Africa on fabrics (14). The bark is used in folk-medicine on skin-infections in Nigeria. Examination of the bark has shown no action on Gram +ve or Gram −ve bacteria, nor on fungi (22). Tests for alkaloids have indicated none present (1). Root and stem-bark has shown the presence of hydrocyanic acid (33) and bark-extracts have given curare-like action on anaesthetized cat-nerves (21, 23).

A bark-decoction is used for tooth-troubles in Senegal (19) and in Liberia as a mouth-wash, and for dysentery, and topically on swollen fingers (23). Stem and root-barks are considered emetic (10, 32, 33) and antispasmodic (6, 33). In Ivory Coast a tisane is taken for diarrhoea and localized oedemas and a decoction is used to wash sores, furuncles and leprous macules (6). In Congo a bark-decoction is taken by mouth to relieve stomach complaints, diarrhoea, hernia, blennorrhoea, heart-trouble and asthma, and in mouth-washes and gargles for gingivitis, aphtes and sometimes toothache (5). In Nigeria a bark-infusion is taken as a febrifuge (2). Homeopathically a bark-decoction is given to ricketty children on the precept of the tree's rapid growth, and bark-sap is given to sterile women to promote conception by reason of the fecundity of the seed in Ivory Coast-Upper Volta (21) and in Congo (5).

In Gabon the bark of young trees, with the spines removed, is used to make hut-walls (32). The bark of the Asian variety contains a reddish fibre (8).

The root-sap in India is said to cure diabetes (23), and the root enters into various remedies for leprosy in Ivory Coast-Upper Volta (21).

The young leaves are sometimes cooked and eaten in West Africa as a soup herb (11, 28). They serve as goat-fodder. The mature leaves contain a mucilage which can be obtained by boiling and is used to remove foreign bodies from the eye in Ivory Coast (6). They are held to be emollient and sedative in Gabon (32). In Congo leaf-sap is given in draught to mental cases, and at the same time the head is washed with a bark-decoction and a circlet of bark is tied round the head (5). In NW Senegal freshly pounded leaves are steeped in water which is drunk for general fatigue (7), and in southern Senegal in Casamance the leaves are used to counter fatigue and lumbago (18). Senegalese medicine-men use them also to prepare a decoction for eye-instillations to treat conjunctivitis (20). The leaves are used in Nigeria as an alterative and laxative, and an infusion is given as a cure for colic in man and in stock (2). In the Kano area of N Nigeria they are pounded to a fine state to apply as a curative dressing on sores (34). A wet poultice of pulped leaves is used to maturate tumours in Guinea (15, 25) and on whitlows in Congo and massage with leaf-pulp and baths in bark-decoction are considered excellent for evening fevers, especially those deemed to arise from evil influence (5). Its use for gonorrhoea in West Africa is also recorded (33).

The flowers are used in Guinea for constipation (25). Flowers and fruit are emollient (33). In W Cameroons the whole flower, or more usually just the calyx, is eaten (12). The flowers may be bat-pollinated (27), but nevertheless are visited by bees and the amber-coloured honey produced has a characteristic taste (17).

The fruit pod is pendulous and normally 10–15 cm long but may be as much as 30–37 cm according to variety. Floss is contained within it and this is the kapok. Typical trees have pods which burst open while still attached. Harvesting is usually undertaken by knocking the pods off before they reach this degree of ripeness. There is variability between trees in this respect. There are some varieties which spontaneously shed their pods before bursting. Both these are important characters. In a large tree some climbing is necessary if pods are to be knocked off, so a third crucial matter is the presence or absence of spines on the trunk and main branches. Some work has been done to propagate trees with white floss, indehiscent pods and spinelessness, but this field requires more attention. Desirable mother-trees can be cloned by cuttings. Trees normally come into bearing at the age of 6–8 years. A considerable amount of selection work has been done in Indonesia on var. *indica,* but on introduction to Africa this variety has not thrived.

The floss is normally greyish. White is a desirable colour and cultivars with white or snow-white floss are known. The fibre is normally up to about 2·7 cm long, fine, silky and too smooth to be spun. Attempts to put a crimp into it have not been economically worthwhile. It is widely used in West Africa and throughout the world to stuff cushions, pillows, mattresses, etc. The fibre retains air and has a resistance to wetting so that it maintains a remarkable buoyancy making it an excellent material for lifebelts and emergency rafts on board ship. In this respect it is superior to cork. It loses only 10% of its buoyancy after 30 days in water and this, on drying, is fully restored. The floss is usable for surgical preparations, replacing cottonwool. It is very inflammable when thoroughly dry and can be used as tinder with flint and steel. It has been used as a base for gun-cotton and in fireworks (8, 11, 17). It is a good acoustic insulant.

The fibre is irritant to the mucosae of the eyes, nose and throat, and during the period of pod-burst, the air filled with floss may set up allergies. In Senegal outbreaks of conjunctivitis are attributed to this (20). For this reason as well as disturbance caused to the foundations of buildings and roads, the tree is really not suitable as a shade or avenue tree in village or town areas. Chemicals identified in the floss are pentosans and uronic anhydrides (20).

The seeds are contained loosely within the floss and separation is usually effected by beating. They are oil-bearing and their oil, known as kapok oil, is similar to ground-nut oil with potential applications in industrial manufacturing as well as for food. Content is 22–25% of which about three-quarters can be obtained by expression. Composition is variable according to region but is of the order: oleic acid 50–53%, linoleic acid 26–29%, palmitic acid 10–16%, stearic acid 2–5%, traces of arachidic and myristic acids and phytosterol (20, 26). A large number of animo-acids are present in the seed of which glutamic acid and arginine are the most abundant (9). Kapok oil is usable in ointments and is official in the *British Pharmaceutical Codex 1959* as a substitute or adulterant of cotton-seed oil and olive oil. It can be used for illumination, soap and paint manufacture and lubrication as well as for cooking (17). The oil is said to be used in Nigeria for rheumatism (2).

The seeds are commonly eaten in West Africa and still containing the oil are pounded and ground to a meal, and cooked in soup. The Hausa of N Nigeria roast the seed to prepare a foodstuff called *gandido*, and in Ghana Dagbani prepare them into a paste called *kantong* which is taken as a food or as a seasoning (11, 17).

The seed cake is a good cattle-feed and has a high protein content, 39% dry weight of the kernels (9). It is also a good agricultural fertilizer.

References:

1. Adegoke & al., 1968. 2. Ainslie, 1937: IFI, sp. no. 82. 3. Anon., s.d. 4. Boboh, 1974. 5. Bouquet, 1969: 74. 6. Bouquet & Debray, 1974: 53. 7. Boury, 1962: 15. 8. Burkill, 1935: 501–5. 9. Busson, 1965: 307–8, with chemical analysis of seeds. 10. Chevalier, 1937, c. 11. Dalziel, 1937. 12. Gartlan 6, K. 13. Gomes e Sousa, 1930: 48. 14. Greenway, 1941. 15. Hartwell, 1968. 16. Howes, 1949: 71. 17. Irvine, 1961: 190–3. 18. Kerharo & Adam, 1962. 19. Kerharo & Adam, 1964, c: 298. 20. Kerharo & Adam, 1974: 245–7. 21. Kerharo & Bouquet, 1950: 64. 22. Malcolm & Sofowora, 1969. 23. Oliver, 1960: 22, 54. 24. Paques, 1953: 1645. 25. Pobéguin, 1912: 9, as *Eriodendron anfractuosum*. 26. Quisumbing, 1951: 595–8. 27. Robyns, 1963: 204. 28. Savill & Fox, 1967: 56. 29. Schnell, 1950, b: 231, 259. 30. Taylor, 1960: 107. 31. Voorhoeve, 1965: 64. 32. Walker & Sillans, 1961: 106. 33. Watt & Breyer Brandwijk, 1962: 148. 34. McIntosh, 26/1/79.

**Ochroma pyramidale** (Cav.) Urb.

FWTA, ed. 2, 1: 334, as *O. lagopus* Sw.

English: balsa, bolsa.

French: patte de lièvre (Berhaut).

A tree of the West Indies which has been introduced to many points in the Region. The wood is extraordinarily light. It can be used as floats and has a greater buoyancy than cork. It has good insulation properties.

**Pachira aquatica** Aubl.

French: châtaignier de Cayenne, cacao sauvage (Berhaut).

A tree of Amazon and American tropical forest between latitudes 4°N. and 8°S. It has been introduced into Senegal.

The fruit resembles that of cacao (*Theobroma,* Sterculiaceae) and the seeds therein contain a pale yellow fat which amounts to 40–50% of the kernel. Many of the physical and chemical characteristics resemble those of palm-oil.

The kernel is edible either raw or cooked and is said to taste like sweet chestnut, and after roasting tastes like cocoa and is sometimes used for the preparation of beverages. The kernel contains about 9% water, 10% starch and 16% protein in addition to the fat (1).

Reference:

1. Bruin al., 1963.

**Pachira insignis** (Sw.) Sav.

A tree introduced to many tropical countries from America, often becoming subspontaneous in India and Africa. Its presence in Ghana is recorded (1). The seeds are edible.

Reference:

1. Burtt Davy & Hoyle, 1937: 20., as *P. affinis* (Mont.) Dec'ne.

**Rhodognaphalon brevicuspe** (Sprague) Roberty

FWTA, ed. 2, 1: 335, as *Bombax brevicuspe* Sprague. UPWTA, ed. 1, 116, as *B. brevicuspe* Sprague.

Trade: alone (Ivory Coast, Voorhoeve).

West African: **SIERRA LEONE** MENDE sangulo (*def.* sanguli) (FCD; S & F) **LIBERIA** DAN swa-uh (AGV) KRU-BASA ju-eh (C&R; AGV) ju-ihn (C&R) **IVORY COAST** ABE kondroti (Aub.; RS) AKYE akogaouan (RS) ANYI kuobéné (RS) **GHANA** VULGAR onyina koben (DF) AKAN-ASANTE kuntunkuni (CJT) kuntunkuri (FRI) onyina kobin (auctt.) TWI enyĩna-kɔbina (FRI; JMD) kuntunkun(-i) (auctt.) kuntunkuri (FRI) kwaseantwa (Thompson) onyaa koben (FD) onyinakoben (CJT) WASA onyina koben (CJT) onyinakoben (Enti) ANYI kuobene (FRI) ANYI-AOWIN kɔbene = *red ekuo, from the red-brown dye* (FRI) SEHWI kɔbene = *red ekuo; from the red-brown dye* (FRI) NZEMA ekuba (CJT) engyakobini (CJT) **NIGERIA** BOKYI nyamenyok (KO&S) EDO ògiúgbòkhà (auctt.) HAUSA kúríyaá, kúryaá (ZOG) HAUSA (West) gúrjíyaá (ZOG) IGBO ákpū ùdèlè = *vulture's akpu* (KO&S) YORUBA awori (JRA; KO&S) **WEST CAMEROONS** KUNDU buma (Schultze)

Tree to 50 m high with long straight clear bole, 5 m in girth and to as much as 30 m long in Sierra Leone (8), buttresses small, bark rugged, thick and fibrous without spines. A species of evergreen and secondary forests from Sierra Leone to W Cameroons, and in E Cameroun and Gabon.

The sap-wood is wide and nearly white, becoming brown. Heart-wood is bright red when fresh turning to brown or pale reddish to dark brown (5, 10) in Liberia, or yellowish (9) in Ghana. Grain is straight, medium soft and readily workable with sharp tools. It does not polish well. In Liberia it is used to make canoes and is sawn into planks (4, 5). In Sierra Leone it is used where durability and strength are not of great importance (8). It is liable to insect-attack. In Ghana smaller branches are sometimes hollowed out to make quivers (7).

The bark-slash is bright red becoming brown on exposure. The bark is easily detached from the wood. A red dye is extracted from it by boiling and this is used in various parts of West Africa for dying cloth (Liberia, 5, 10; Ivory Coast, 2; Ghana, 3, 6, 7). In Ghana it is used to dye funeral cloths called *kuntunkuni* a saffron colour (9). The dye is said to be fast. The dye-decoction is also used in Ivory Coast on sores causing them to heal rapidly (2). In Liberia a liquor prepared from the bark and leaves is used internally to treat venereal diseases and externally in poultices on 'blue boils' (4, 5, 6).

In Nigeria the root is powdered and applied externally to swellings and dislocations. When mixed with water the powder forms a mucilaginous paste. This is taken internally for rheumatism and dysentery (1).

The flowers are pink to reddish and are showy. The fruit pods, in distinction from *Bombax spp.* which burst open on the tree, normally fall unopened. They contain a yellowish brown or reddish, crisp floss of inferior quality which is used locally for stuffing pillows and cusions. It has been suggested as usable for paper-pulp (6, 7).

No use is recorded of the seeds, but one may expect them to be oil-bearing with an oil content similar to that of *Bombax spp.*

References:

1. Ainslie, 1937: sp. no. 56 as *Bombax brevicuspe*. 2. Aubréville, 1959: 2, 264, as *Bombax brevicuspe*. 3. Chipp, 1922: 56, as *Bombax brevicuspe*. 4. Cooper 421, K. 5. Cooper & Record, 1931: 48–49, as *Bombax brevicuspe*, with timber characters. 6. Dalziel, 1937. 7. Irvine, 1961: 188, as *Bombax brevicuspe* Sprague. 8. Savill & Fox, 1967: 58. 9. Taylor, 1960: 103, as *Bombax brevicuspe* Sprague. 10. Voorhoeve, 1965: 70.

# BORAGINACEAE

## **Arnebia hispidissima** (Sieber ex Lehm.) DC.

FWTA, ed. 2, 2:324. UPWTA, ed. 1, 424.
West African: **NIGERIA** HAUSA jinin mutum = *blood of man; alluding to the deep red-coloured roots* (JMD)

A densely hispid much-branched herb from a perennial woody base, of dry situations in N Nigeria and W Cameroons, and occurring over N tropical Africa to N India.

The root is blood-red whereby is the Hausa name, and provides a dye giving red or purple colours (1).

Reference:

1. Dalziel, 1937.

## **Coldenia procumbens** Linn.

FWTA, ed. 2, 2:321. UPWTA, ed. 1, 424.
West African: **THE GAMBIA** MANDING-MANDINKA farakonoɲaama (*def.* f.ɲaamoo) = *swamp grass* (DF) fatta jambo = *stone leaf* (Hayes) **GUINEA-BISSAU** MANDING-MANDINKA farakomanyamo (*def.*) fatta jambo PEPEL onchelma (JDES; EPdS) **MALI** MANDING-BAMBARA tumu tigui (A. Chev.) SONGHAI sané (A. Chev.)

A prostrate spreading herb from a woody rootstock, in sandy sites liable to inundation and desiccation in the drier northern zone from Senegal to N Nigeria, and extending to NE and E Africa and to Angola.

The foliage is recorded as being not grazed by stock in Zambia (3). The fresh leaves are pulped-up in Sudan and applied to areas of rheumatic swelling, and the dried plant with equal parts of fenugreek seeds (*Trigonella foenum-graecum* Linn., Leguminosae: Papilionoiideae) is reduced to a powder and applied to boils which are caused quickly to maturate (1). The leaf added to a bath has been used for tumours (Dragendorff fide 2).

References:

1. Brown & Massey, 1929:304. 2. Hartwell & al., 1968: 3. Vesey-Fitzgerald 4308, K.

## **Cordia africana** Lam.

FWTA, ed. 2, 2:320. UPWTA, ed. 1, 424, as *C. abyssinica* R. Br.
French: sebestier d'Afrique (Portères).

West African: **GUINEA** FULA-PULAAR (Guinea) bamébani (Aub.; FB) **MALI** FULA-PULAAR (Mali) bamébani (Aub.) **NIGERIA** ARABIC inderab (JMD) ARABIC-SHUWA birtjuk, ngirli FULA-FULFULDE (Nigeria) lillibaare (*pl.* lillibaaje) *the fruit* (MM) lillibali (*pl.* lillibaaje) *the tree* (MM) lillibani (KO&S) HAUSA alilliba (auctt.) KANURI álúwá (auctt.)

A tree to about 8 m high, of the savanna woodland in Guinea and N Nigeria, and widely dispersed outside the Region in Africa.

The tree is frequently planted as a shade-tree in N Nigerian villages (9, 10). In Zaïre it is planted in villages and as a coffee shade-tree (11). The wood is pale brown, hard and durable. It is suitable for carpentry, being easy to work, stable and warping little under changing conditions, and taking a good polish (4). It has been traded in Sudan as 'Sudan teak', not so much owing to any resemblance of the wood as of the broad leaves, and has been used there for cabinet-work (1, 7). In Uganda the wood is commonly used to make drums and bee-hives (4), and in Zaïre mortars, 'gongs' [? drums] and canoes (11). Twigs serve as fire-sticks in the Jebel Marra of Sudan (7). Wood-ash mixed with butter is applied to certain skin-troubles in Ethiopia (5).

The leaves are dried and powdered to sprinkle over wounds in Tanganyika, and a root-decoction is drunk for schistosomiasis (6).

The flowers are much sought after by bees.

The fruit-pulp is everywhere eaten (2). In N Nigeria it is added by the Fula to sweeten 'porridge' (3, 10), and it is also made, usually with honey, into a sweetmeat called *alewa* (Arabic, *halaua*) (3). A stimulating tonic for fatigue and exhaustion in man and horse while on a journey is prepared from the bark and fruits along with the stems of *Abelmoschus esculentus* (Linn.) Moench. (syn. *Hibiscus esculentus* Linn., Malvaceae) (3). In Jebel Marra of Sudan a preparation called *dari* is made during Ramadan from the fruits of this, *Balanites aegyptiaca* (Balanitaceae), *Ziziphus spina-christi* (Rhamnaceae), *Tamarindus indicus* (Leguminosae: Caesalpinioideae) and *Grewia tenax* (Tiliaceae). The concoction is made into a paste with flour for eating or into a refreshing drink with honey (7).

The kernels resemble walnut in flavour. In Jebel Marra the nuts are boiled and the liquor is bottled in gourds for two days to produce a drink called *ambila* (7).

References:

1. Broun & Massey, 1929: 302–3, as *C. abyssinica* R. Br. 2. Busson, 1965: 377–9. 3. Dalziel, 1937. 4. Eggeling & Dale, 1952: 46–48, as *C. abyssinica* R. Br. 5. Getahun, 1975. 6. Haerdi, 1964: 150, as *C. abyssinica* R. Br. 7. Hunting Technical Surveys, 1968. 8. Irvine, 1948: 265, as *C. abyssinica*. 9. Lamb 79, K. 10. Peter & Tuley 6, K. 11. Taton, 1971: 4–6.

## **Cordia aurantiaca** Bak.

FWTA, ed. 2, 2:320.

West African: **NIGERIA** EDO úrìghóẹ̀n (Ross; KO&S) IGBO idumuye (KO&S)

A small tree to about 16 m high of the rain-forest in Nigeria, W Cameroons and Fernando Po, and extending to Angola.

In Zaïre it is occasionally planted as a village-tree, and the thin fruit-pulp is used as a gum (1). Like other *Cordia spp.* the fruit-pulp is probably edible.

The Igbo of S Nigeria prepare a wash from the leaves which is used during dance-ceremonies (2).

References:

1. Taton, 1971: 6–7. 2. Thomas, N. W., 2103 (Nig. Ser.), K.

**Cordia millenii** Bak.

FWTA, ed. 2, 2:320–1. UPWTA, ed. 1, 425.

English: drum tree (Ghana, Irvine).

West African: **GHANA** AKAN-AKYEM akaboa (Bunting) ASANTE kyeneboa = *drum tree, from* kyene: *drum* (FRI) kyenedua = *drum tree* (JMD) tweneboa (FRI) tweneboakodua (Enti) tweneduru (FRI) ASANTE (Denkyera) kyeneboa (Enti) tweneboa (Enti) FANTE kyeneboa (FRI; DA) kyenedua = *drum tree, from* kyene: *drum* (FRI) TWI akyaboa (auctt.) kyeneboa, kyenedua = *drum tree, from* kyene: *drums* (auctt.) kyeneduru (FRI) tweneboa (DF) tweneboakodua (FRI) WASA kyeneboa (Enti) tweneboa (Enti) tweneduru (FRI) **NIGERIA** EDO òmà (JMD; KO&S) IJO kiebo eke (KO&S) ISEKIRI egin ogume (KO&S) URHOBO erheigede (KO&S) YORUBA ọmọ̀ (auctt.) **WEST CAMEROONS** DUALA bòòmba (JMD; Ithmann) KPE bomba (JMD) jombomba (JMD) yombomba (JMD) KUNDU bola (AHU; JMD) MUNGAKA jom (JMD) yom (JMD)

A tree reaching 20 m high (Ghana, 10: 13 m in Nigeria, 11), of the forest, in Ivory Coast to W Cameroons, and widely dispersed in tropical Africa.

The tree has a fine spreading crown, and it is often planted in towns and villages as a shade-tree, especially in W Nigeria (10, 11) and Ivory Coast (3). In Ghana it may be grown as a village fetish-tree (10, 13). Sap-wood is white, heart-wood pale brown, close-grained, fairly light and soft, but durable and resistant to termites (6, 10–12). It is used in Zäire for external carpentry (14). It seasons and finishes well, and in E Africa is said to be very suitable for furniture and cabinet-making (5, 8, 9). In several Ghanaian languages the tree is known as the 'drum tree' (cf. Asante *kyene,* drum) and it is used for making drums and is one of those used for the Asante 'talking drums' (10, 15). In Zäire, the wood is used for gongs [? drums] and small bells to hang around the neck of hunting-dogs (14). Similarly in Uganda it has vernacular names denoting any tree, mainly *Cordia spp.,* used for making drums and musical instruments (8, 9, 13). In Ghana the timber is split into shingles which are recorded as lasting four years (4, 6, 10). It is made into bowls and other domestic articles. In Uganda the trunks are hollowed out for dugout canoes being adzed quite easily, and because of its lightness the canoes are unsinkable (9). The wood also has sufficient strength to be the keel-piece of larger boats (7). It is also used in Zäire for canoes and tool-handles (14).

In Nigeria a decoction of leaves is taken for asthma, colds, coughs, etc., and the leaves are dried and smoked for these afflictions. Pulverized seeds mixed with palm-oil are used for ringworm, itch and other epidermal troubles (2). The fruits are added to soap in Zäire to give it a pleasant smell (14).

Tests for alkaloids have not indicated the presence of any (1).

References:

1. Adegoke & al., 1968. 2. Ainslie, 1937: sp. no. 113. 3. Aubréville, 1959: 3: 224. 4. Burtt Davy & Hoyle, 1937: 21. 5. Dale & Greenway, 1961: 70. 6. Dalziel, 1937. 7. Eggeling 96, K. 8. Eggeling 1155, K. 9. Eggeling & Dale, 1952: 48. 10. Irvine, 1961: 727–8. 11. Keay & al., 1964: 421–2. 12. Millen 12, K. 13. Moor 854, K. 14. Taton, 1971: 10–13. 15. Vigne 109, K.

**Cordia myxa** Linn.

FWTA, ed. 2, 2:320. UPWTA, ed. 1, 425.

English: Assyrian plum, sapistan, sebesten plum.

West African: **SENEGAL** BASARI a-mbòn ɓɔnɛ̀r (Ferry) FULA-PULAAR (Senegal) ƭamanohi (Aub., ex K&A) MANDING-BAMBARA ndégé (JB) MANDINKA daramâ tunko (K&A) MANINKA darama (Aub.) dég(u)é daramba (Aub; K&A) ndéké (Aub.; K&A) SERER narr (K&A) sub (JMD, ex K&A) sub djuam (JLT, ex JMD) WOLOF mbey (JLT; K&A) mbey-gilé (JLT; K&A) tampus (JMD, ex K&A) **GUINEA-BISSAU** FULA-PULAAR (Guinea-Bissau) somadjo (JDES; EPdS) MANDING-MANDINKA sanadjô (JDES) **GUINEA** MANDING-MANINKA darama (Aub.) dégué daramba (Aub.) ndéké (Aub.) MANINKA (Wasulunka) ndien (A. Chev.) ntu (A. Chev.) **MALI** FULA-PULAAR (Mali) tiamanohi (Aub.) MANDING-BAMBARA daramba (A. Chev.) tungué (A. Chev.) MANINKA darama (Aub.) dégué daramba (Aub.) ndéké (Aub.) **UPPER VOLTA** DAGAARI tango

(K&B) tungbo (K&B) MANDING-BAMBARA n-déké (K&B) **IVORY COAST** VULGAR dédé (Aub.) BAULE lobotili gbli (Aub.; FB) FULA-FULFULDE (Ivory Coast) tiamanohi (Aub.) MANDING-MANINKA darama (FB) ndédé (Aub.) **GHANA** BAULE lobotili gbli (FRI) DAGAARI tungbo (AEK, ex FRI)

A shrub or tree to 12 m high with a stout stem (tortuous in Senegal, 7), native of the Near and Middle East and now naturalized in the soudanian and guinean zones of the Region, from Senegal to Ghana.

The plant is the *sebesten* of Egypt and of ancient cultivation. It occurs spontaneously in the Region around villages and old abandoned habitations. It may at one time have been cultivated in the western soudanian area. The plant has been valued from olden times for its sticky mucilaginous pulp which is edible and which is the source of the well-known medicine of the Near and Middle East called 'sapistan', useful for coughs and chest-complaints on account of its demulcent property (5). Also it has had widespread use as a bird-lime. In Ivory Coast-Upper Volta it furnishes a glue, and an emollient plaster to maturate abscesses (8). Similar applications to tumours in olden times are recorded (6). The kernel is edible (3, 7). It is oily, containing palmitic, stearic, olic and linoleic acids and $\beta$-sitosterol (2, 7). There is also a substance which will stain linen, but not indelibly (5).

The wood is yellow-brown, polishes well, soft but strong, and is suitable for furniture-making and cabinetry. It has been marketed as 'Khartoum', or 'Sudan teak' (9). The bark is fibrous and can be used to yield a cordage and for caulking boats. In the Near-East the wood has been used as firc-sticks and in tombs of Ancient Egypt (5).

A macerate of the leaves is used in the soudanian zone as a treatment for sleeping-sickness taken internally, and applied externally as a lotion to the fly-bites (1, 4). In Ivory Coast the leaves are put onto sores. Stereols and a gum have been recorded in them (2).

Washing the body first thing in a morning with a leaf-macerate is said by the Serer of Senegal to ensure good fortune (7).

References:

1. Aubréville, 1950: 490. 2. Bouquet & Debray, 1971: 53. 3. Busson, 1965: 379, with kernel analysis. 4. Chevalier, 1937, b: 170. 5. Dalziel, 1937. 6. Hartwell, 1968. 7. Kerharo & Adam, 1974: 247–9, with phytochemistry and pharmacology. 8. Kerharo & Bouquet, 1950: 222. 9. Uphof, 1968: 172.

**Cordia platythyrsa** Bak.

FWTA, ed. 2, 2:321. UPWTA, ed. 1, 425.

Trade: pooli (the timber, Sierra Leone).

West African: **SIERRA LEONE** KONO sao (S&F) MENDE puli (*def.*-i) (auctt.) TEMNE *an*-fundoba (FCD; S&F) *an*-ranko (Burbridge; JMD) **IVORY COAST** AKYE g-bon (Aub.) ANYI ehuno (Aub.) kanédaguru (Aub.) BAULE aundé (Aub.) KRU-GUERE goléhéhiré (Aub.) **GHANA** AKAN-ASANTE tweneboa (CJT; FRI) FANTE kyeneboa (DF) tweneboa (CJT) TWI kyeneboa (FRI) tweneboa (CJT; DF) ANYI ehuno (FRI) kanedaguru (FRI) ANYI-SEHWI achaboa (CJT) BAULE aunde (FRI) NZEMA twenedoleye (CJT) **NIGERIA** YORUBA ako-ledo (AHU, JMD) ọ̀mọ̀ (KO&S) ọ̀mọ̀ wẹ́wẹ̀ (Dawodu)

Tree to over 30 m high by 3 m girth of closed and secondary forest from Sierra Leone to S Nigeria, and also in E Cameroun.

The tree is often planted as a village shade-tree, especially in Ivory Coast (1). The trunk may be low-branching. It changes direction at each persistent whorl when young resulting in an irregular bole in old trees. It is without buttress and old logs may have a pulpy or brittle heart. Sap-wood is white and is liable to turn green when cut with iron or steel implements (4). Heart-wood is brown. The wood is close-grained, works well and takes a good polish. It is valuable for high-class furniture. It is a possible substitute for some grades of mahogany and has been traded as 'pooli.' It is light in weight, and resists both fungal and insect-attack (1–3). The wood is used to make drums, canoes, and ornamental seats in Ivory Coast (1). In Ghana its Akan vernacular names arise from the use of the wood for making drums (*kyene, twene*) like *C. millenii* (4), and it is used also for domestic articles. In Cameroun the wood is taken for the keys of locally made xylophones (Mildbraed fide 2).

The bark can be stripped from the trunk in large pieces and serves to make hut-walls and partitions in Cameroun. It is said that removal of the bark can be effected without damage to the cambium (2). The bark slash is slightly aromatic.

References:

1. Aubréville, 1959: 3:222. 2. Dalziel, 1937. 3. Savill & Fox, 1967: 59–60. 4. Taylor, 1960: 110.

## **Cordia sebestena** Linn.

A shrub or small tree, native of tropical America and fairly widely distributed by man in the tropics, is recorded present in Ghana and Nigeria.

It is easily propagated by seed and cuttings and is useful for decorative purposes in dry regions (2). A trace of alkaloid has been reported in the plant [? leaves] (1).

References:

1. Adegoke et al., 1968. 2. Irvine, 1961: 729–30.

## **Cordia senegalensis** Juss.

FWTA, ed. 2, 2:320.

West African: **SENEGAL** SERER su (auctt.) sub (JB; K&A) suup (K&A) WOLOF béhi (Aub.; K&A) mbey (JB; K&A) mbeygilé (JB; K&A) **THE GAMBIA** FULA-PULAAR (The Gambia) lilibahi (DAP) n-dologa (DRR; DAP) MANDING-MANDINKA tomboroŋ (*def.*-o) = *to start a fire* (DF) WOLOF deke = *spike* (DAP; DF) mbey (DAP) **UPPER VOLTA** MANDING-DYULA yacuma foroto (K&B) **IVORY COAST** AKYE bona *lit.*, bon na = *yellow bon, C. platythyrsa* (Aub.; K&B) MANDING-DYULA yacuma foroto (K&B) **GHANA** VULGAR kyeneboa (DF) tweneboa (DF) AKAN-ASANTE tweneboa (FRI) TWI kyeneboa (FRI)

A shrub or tree of the savanna-forest in Senegal and closed-forest in Ivory Coast to W Cameroons, and also in E Cameroun, Zäire and NE Africa. In the western part of its range in the Region it attains only 7–8 m height with irregular contorted trunk, while in the eastern part it reaches 25 m height with trunk near to 2 m girth.

The wood is yellowish and is distinguished in Ivory Coast as the 'yellow' *bon* from *C. platythyrsa* with greyish-white wood which is *bon* to the Akye (1). It is used by them to make drums and canoes.

The bark contains a little gum. It is also fibrous, and the fibre is extracted in The Gambia for common domestic uses (3, 6).

The leaves are prepared in a macerate taken by draught in Senegal for colic, and put into baths for general fatigue and stiffness (4). A decoction of the crushed leaves is drunk in Ivory Coast for kidney-pains (2, 5). The leaves have unspecified medicinal use in The Gambia (3).

The fruit pulp is sweet and edible.

References:

1. Aubréville, 1959: 3, 220. 2. Bouquet & Debray, 1974: 53. 3. Fox 58, K. 4. Kerharo & Adam, 1974: 249–50. 5. Kerharo & Bouquet, 1950: 222. 6. Rosevear, 1961.

**Cordia sinensis** Lam.

FWTA, ed. 2, 2:320, as *C. rothii* Roem. & Schult. (=*C. gharaf* (Forssk.) Ehrenb.), UPWTA, ed. 1, 425, as *C. gharaf* Ehrenb.

West African: **MAURITANIA** ARABIC ('Maure') akjül (Aub.) **SENEGAL** ARABIC ('Maure') akjül (Aub.) SERER subduam (JB; K&A) suo mâg (JB, ex K&A) WOLOF mbey mbey (K&A) nehneh (JB; K&A) tâpus (JB; K&A) **MALI** SONGHAI barmadangaïe (Aub.) fifrigui (Aub.) TAMACHEK tadanu (Aub.) **GHANA** GA gonyontɛo (FRI) **NIGER** ARABIC (Niger) hen'darabaïe (Aub.) HAUSA tadanat (Aub.) tidâni (Aub.) KANURI kabula (Aub.) kabulu (Aub.) TAMACHEK tadanu (Aub.) TUBU kohul **NIGERIA** ARABIC inderab (JMD) KANURI kabilla (JMD)

A shrub or tree of savanna riverbanks or damp sites, to 8 m high, from Senegal to N Nigeria, and widely dispersed in the drier parts of tropical Africa and into Asia to India.

The plant is well-suited to dry conditions. It has white sweetly-scented flowers and it is planted as an ornamental shrub in N Nigeria. In NE Africa and in India it attains a larger stature. In the township of Merca in Somalia it has been planted as an avenue-tree (2). In Kenya (12) and in India (15) the stems are used in hut-construction and to make agricultural implements. The bark is used for covering huts in northern Kenya (12). The heart-wood is brown and slightly scented. It is used as a substitute for sandalwood in Sudan (4). In the area of Lake Chad, the wood is used for arrow-shafts (6), and by the Masai of Kenya for clubs and spear-handles (16). In Tanganyika walking-sticks are made of it (10) and the dried sticks are used to make fire in Somalia (3). Larger pieces are hollowed out in the South Turkana into pots (13).

The bark contains a gum (4, 6, 9). The bark is astringent and in India is used to prepare a gargle (15). A decoction of root and bark is used by the Masai to wash inflamed eyes in cattle (3, 16). The inner bark is fibrous and a fibre can be extracted to produce a strong cordage (4, 6), and a caulking material for boats which is widely used (E and W Africa, 16; India, 15).

The foliage is grazed a little by cattle in Senegal, but not at all in the area of Lake Chad (1). Stock take it in Kenya (8). The leaves are used in lower Senegal alone or in mixture with other drug-plants against fever (11).

The root is chewed by the Masai and the saliva is swallowed as an abortifacient (3, 16).

Ubiquitously the fruits are eaten. They have a gelatinous, sweet pulp which has a slightly astringent drying after-taste. Besides consumption as a dessert, they are eaten in porridge (5) and with *baobab* meal (17) in Tanganyika, and also fermented into a beer (14). The fruit in E Africa is an important famine-food (16).

References:

1. Adam, 1966, a: as *C. rothii*. 2. Bally, B. 9358, K. 3. Bally, 1937: as *C. gharaf* Ehrenb. 4. Brown & Massey, 1929: 303, as *C. rothii* Roem. & Schult. 5. Bullock 3448, K. 6. Dalziel, 1937. 7. Gillett 4445, K. 8. Gillett 12571, K. 9. Howes, 1949: 72, as *C. rothii*. 10. Irvine, 1961: 729, as *C. rothii* Roem. & Schult. 11. Kerharo & Adam, 1974: 249, as *C. rothii* Roem. & Schult. 12. Mathew 6311, K. 13. Mathew 6600, K. 14. Renvoize & Abadalla 2157, K. 15. Sastri, 1950: 346–7, as *C. rothii* Roem. & Schult. 16. Watt & Breyer-Brandwijk, 1962: 148, as *C. gharaf* Ehrenb. 17. Wigg 984, K.

**Cordia tisserantii** Aubréville

FWTA, ed. 2, 2:321.

A shrub or lianous plant of dry scrub in Senegal and N Nigeria, and also in Ubang-Shari.

The fruit-pulp is sweet and is said to be eaten (1).

Reference:

1. Aubréville, 1950: 491.

**Cordia vignei** Hutch. & Dalz.

FWTA, ed. 2, 2:320.

West African: **SIERRA LEONE** MENDE fule (*def.*-i) (FCD) **IVORY COAST** ABE ko-bona (Aub.)

A shrub of the closed-forest from Sierra Leone to Ivory Coast.

In Sierra Leone, the Mende take a leaf-decoction as a purgative, tie powdered leaves to the body for rheumatism and use a bark-decoction for washing sores, and the Temne use the young leaves ground-up as a dressing on fresh wounds (1).

Reference:

1. Deighton 2213, K.

**Cordia sp. indet.**

West African: **IVORY COAST** KRU-GUERE gulégué (K&B) **NIGERIA** YORUBA ọmọ̀ (Punch)

An unidentified *C. sp.* is recorded used in Ivory Coast for cough — the inner bark is pounded up into a bolus which is sucked as an emollient (1).

References:

1. Kerharo & Bouquet, 1950: 223.

**Cynoglossum amplifolium** Hochst.

FWTA, ed. 2, 2:324.

A perennial herb to 1 m high or more, of montane localities in W Cameroons, and also in Central and E Africa.

In Kenya the plant is grazed by all domestic stock (1).

Reference:

1. Glover & al. 1491, K.

**Cynoglossum lanceolatum** Forssk.

FWTA, ed. 2, 2:324.

A herb up to 1.80 m high, of two subspecies, *lanceolatum* scabrid, *geometricum* (Bak. & Wright) Brand pilose, the former from Liberia to W Cameroons and widespread in the Old World tropics and subtropics, the latter from W Cameroons and in Central and E Africa.

In S Nigeria the plant is recorded used in soup (4). In Kenya ssp. *geometricum* is grazed by all domestic stock (1). In S Africa the plant is considered diaphoretic and expectorant. The Basuto apply the crushed plant to wounds as a plaster (5) and the plant is made into a colic-medicine for children (2, 5). The roots of ssp. *lanceolatum* are used in Zaïre for treatment of eye-troubles, and the plant of ssp. *geometricum* as a vermifuge against *Taenia* (3).

Tests for bitters, alkaloid, volatile oil, hydrocyanic acid, saponin and triterpenoids have given negative results (5).

References:

1. Glover & al. 1493, K. 2. Guillarmod, 1971: 422. 3. Taton, 1971: 50 54. 4. Thomas, N. W. 1745 (Nig. Ser.), K. 5. Watt & Breyer-Brandwijk, 1962: 149.

**Echium horridum** Batt.

West African: **MAURITANIA** ARABIC (Hassaniya) el arche (AN)

A herb to 25 cm high of stream-banks in Mauritania and in N Africa. The plant provides good fodder for all stock (1).

Reference:

1. Naegelé, 1958: 878–9.

**Echium humile** Desf.

West African: **MALI** TAMACHEK uchem (RM)

A hairy, almost bristly, herb to 10–20 cm high, of rocky places in seasonal river-beds, occurring across N Africa and into the Sahara to Hoggar on the northern limit of the Region.

**Ehretia cymosa** Thonning

FWTA, ed. 2, 2:318. UPWTA, ed. 1, 426.

West African: **SIERRA LEONE** TEMNE *a*-ruŋku (NWT) **IVORY COAST** AKYE lauso (Aub.) lusso (Aub.) BAULE alébé (B&D) KRU-GUERE (Chiehn) graku (B&D) GUERE (Wobe) gotué (DF; Aub.) KWENI béléku (B&D) bliku (B&D) **GHANA** ADANGME dutʃo (FRI) muatʃɔtʃo (FRI) tatu huŋwa = *small black ant's pot* (FRI) AKAN-TWI okosua (FRI) WASA kweniŋ (auctt.) GA labaasaatso (FRI) ɔsoŋkoni gbekɛ̃biiaŋmetʃo = *children's oil palm; from the red shining fruits* (FRI) GBE-VHE abovro (auctt.) VHE (Awlan) edɔ (FRI) gbɔmitsi (FRI) GUANG-KRACHI o-koni (Volkens) tatu hungwa chu (BD&H) **NIGERIA** YORUBA jàokè (Phillips; JMD)

A shrub or tree to 7 m high, of savanna and secondary jungle, recorded from Sierra Leone to N and S Nigeria. A variable species with two varieties, var. *cymosa* and var. *zenkeri* (Gürke) Brenan, recognized in the Region, the former extending to Uganda, the latter to E Cameroun and S. Tomé. Other varieties occur in tropical Africa outside the Region.

The wood is described as perishable (var. *sylvatica* Gürke, 6). It is however used in Zäire (14) and in Kenya (8) to make handles for tools, and in Ethiopia yokes (7). The stems are sometimes used in Ghana as chewsticks (12, 13).

Animals are said to browse the foliage in Sudan (2, 3), but in the Makuyuni District the leaves are held to be toxic (9). The leaves are commonly used in the Yoruba *agbo* infusion given in draught and used as a wash for fever, children's convulsions, etc. Sap from the fresh leaves is a mild laxative for children (5, 11–13). The plant (probably the leaves) is used in Gabon as a laxative and a febrifuge (15). The leaf, usually after pounding with that of *Newbouldia laevis* (Bignoniaceae) and a guinea pepper, is tied on the head for headache (5).

The root, like the leaves, is considered toxic in Makuyuni District of Tanganyika (9, 10) but no application is ascribed to it. A decoction of the roots and leaves is taken in Ghana for infantile tetanus (5, 13), and for dysentery (Field fide 13).

In Ivory Coast a bark-decoction is taken for amenorrhoea, and the decoction when left to cool separates to a supernatant layer of oil which is applied to skin-affections (4).

The fruit is edible (3, 13) and is used in Ghana as bait for trapping birds (5, 13).

A trace of alkaloid has been detected in the plant (? leaves) (1).

References:

1. Adegoke & al., 1968. 2. Andrews, A.501, K. 3. Andrews, A.787, K. 4. Bouquet & Debray, 1974: 53. 5. Dalziel, 1937. 6. Eggeling & Dale, 1952: 50, as *E. silvatica* Guerke. 7. Friis 142, K.

8. Glover et al., 2126, K. 9. Koritschoner 1350, K. 10. Koritschoner 1524, K. 16. Irvine 208, K. 12. Irvine, 1930: 172–3. 13. Irvine, 1961: 731, as *E. thonningiana* Exell. 14. Taton, 1971: 22–23. 15. Walker & Sillans, 1961: 107.

**Ehretia trachyphylla** C. H. Wright

FWTA, ed. 2, 2:318.

West African: **IVORY COAST** ABE kombui (Aub.) ANYI assajué (Aub.) **GHANA** AKAN okyine (DF) AKAN-TWI okyine (Enti) WASA okyini (auctt.) ANYI asadwe (FRI) **NIGERIA** YORUBA jasoke (IFE)

A tree to about 16 m high in the understorey of the high-forest of Ivory Coast and Ghana.

The wood is very speckled. It is used in Ivory Coast (1) and in Ghana (3, 4) for tool-handles. In Ivory Coast a bark-decoction is taken for amenorrhoea, and the decoction when left to cool separates to a supernatant layer of oil which is applied to skin-affections (2).

References:

1. Aubréville, 1959: 3, 218. 2. Bouquet & Debray, 1974: 53. 3. Irvine, 1961: 731. 4. Vigne 4781, K.

**Heliotropium bacciferum** Forssk.

FWTA, ed. 2, 2: 322. UPWTA. ed. 1, 427, as *H. undulatum* Vahl.

West African: **MAURITANIA** ARABIC (Hassaniya) hâbaliya (AN) hbaïlia (AN) hebaliye (AN) **NIGERIA** ARABIC ghareir (JMD) FULA-FULFULDE (Nigeria) gerohi (JMD) HAUSA bafilaátànaá, námíjin-roòmaà-faáda (ZOG) ba-filatana *= a Fulani woman; alluding to galactogenic properties* (Abraham fide MM) namijin roma fada *meaning not known* (MM)

A sub-erect or prostrate herb to 50 cm high from a perennial woody rootstock, of dry sahel areas in Mauritania, Senegal, Mali and N Nigeria, and dispersed across N Africa, Arabia and into tropical Asia.

The plant provides good fodder for camels and all stock in Mauritania (6) and the Hoggar (4), though in Senegal stock, except goats, will not take it (1, 3). At Chinguetti in Mauritania the sap is applied direct as a burn-ointment (6). In the Hoggar dried powdered leaves are added to curdled milk or to water for treatment of ringworm (4). The plant is applied (part and method not stated) topically for headache in N Nigeria, and it is used internally for gonorrhoea and to increase lactation (2).

A macerate of the plant is prepared in the western Sahara as an ink (5).

References:

1. Adam, 1966, a. 2. Dalziel, 1937. 3. Kerharo & Adam, 1974: 250. 4. Maire, 1933: 176–7, as *H. undulatum* Vahl. 5. Monteil, 1953: 25, as *H. undulatum* Vahl. 6. Naegelé, 1958, b, 879.

**Heliotropium indicum** Linn.

FWTA, ed. 2, 2: 321. UPWTA, ed. 1, 426.

English: cock's comb (The Gambia), Indian heliotrope, turnsole.

French: herbe à verrues ('plant with warts') (Berhaut).

Portuguese: heliotrópio-indiano (Feijão).

West African: **SENEGAL** MANDING-BAMBARA ńâgiku (JB; K&A) MANDINKA merriño (K&A) MANINKA nasimko (K&A) SERER-NON lamfam (AS) WOLOF nag um der (auctt.) xeteram (K&A) **THE GAMBIA** MANDING-MANDINKA daajulu (*def.* daajuloo) = *sorrel fibre* (DF) merrenjo (DA) **GUINEA-BISSAU** MANDING-MANDINKA dajulo **GUINEA** MANDING-MANINKA nasinko (Brossart) **UPPER VOLTA** MANDING-DYULA nondingko (K&B) **IVORY COAST** BAULE kotokoro kombo (B&D) KPALAGHA berininga (K&B) KRU-GUERE (Chiehn) tapentiti (K&B, B&D) KWENI klauri (B&D) tapérodia (B&D) MANDING-DYULA nondingko (K&B) MANINKA nansifo (B&D) nossiko (B&D) SENUFO-TAGWANA nungro (K&B) **GHANA** AKAN-ASANTE akɔmfɛm-tiko (Enti) akɔmfɛtikoro (FRI) akɔnfem atiko, akɔnfem: *Guinea fowl*; atiko: *back of head* (Enti) akɔnfem kɔn-akyi, akɔnfem: *Guinea fowl;* kɔn: *neck*; akyi: *back ; i.e. in allusion to the back of the bird's neck* (Enti) FANTE akɔkɔtubatuba (FRI) akokɔturbaturba (FRI) TWI akomfem-kon-akyi = *the back of the Guinea-fowl's neck* (FRI) akɔmfɛm-tiko (FRI) ansam-kon-akyi = *the back of the Guinea-fowl's neck* (FRI) asam kon-akyi, asam: *Guinea fowl;* kon: *neck ;* akyi: *back ; i.e., in allusion to the back of the bird's neck* (Enti) asam-atiko (Enti) ANYI-AOWIN apusopusuo (FRI) GA gbé'kɛ̃bìì-awuɔ = *children's fowl; from the shape of the flowering stalk* (FRI) kɔ́kɔdeneba = *frog's leaf* (FRI) GBE-VHE (Awlan) koklotɔtsu = *cock's comb* (FRI) **NIGERIA** EFIK èdísím̀mɔ̄ñ (RFGA; Lowe) HAUSA (West) kalkashin kooramaa (Ryan, ex JMD; ZOG) IGBO (Agukwu) ilolo isi mwa-eku (NWT) IGBO (Ibusa) ụ̀tábā ànị = *tobacco of the ground* (NWT; KW) IGBO (Owerri) azu uzo (AJC; Lowe) IJO-IZON (Kolokuma) ùmbú (kórómọ̀) dírí = *medicine to make navel (cord) fall off* (KW) NUPE vòkpa gùlǔ = *the vulture's knee* (RB) YORUBA àgógó igùn = *beak of the vulture* (Lowe; auctt.) ogbe àkùko = *cock's comb* (Lowe; auctt.) ogbe orí àkùko, àkúko omade (IFE)

A vigorous annual reaching to 1 m high, of inhabited areas throughout the Region except in the driest parts, and pan-tropical in dispersal.

The plant is not eaten as a vegetable by man. In Sudan it is taken by animals (3), but in Senegal it has the reputation of being toxic to certain animals (13). From antiquity, the plant has been used on warts and in poultices for inflammatory tumours (10). In the W African region, the leaves are very widely used in like manner. In Senegal leaf-powder is applied to dermatoses and especially to suppurating eczema and impetigo in children (13). In Nigeria (2) and in Ghana (11) a leaf-infusion is applied topically to sores, stings, pimples, etc., and the sap to gumboils, to cleanse ulcers, to the eyes for ophthalmia, and mixed with castor oil to stings and poisonous bites, etc. Similarly relief in headache is sought in Sudan in the application of the leaves in poultice (12). In Ghana (11), Liberia (6), and Nigeria (15) the plant is applied to erysipelas and in Liberia is known as the 'erysipelas plant' (6). Leaf-ash in oil is applied to scabies in Tanganyika (9). Triturated leaves are applied to the gums (24) and to the genitals (25) for inflammation in Gabon. The leaf in decoction is used in Indonesia for thrush, and in Indochina in poultices for herpes and rheumatism (5). Similar uses are also practiced in the Philippines (17), and elsewhere. Use of a leaf-decoction is recorded in Sierra Leone to wash new-born babies (22). The Ijo of Nigeria use the plant (method not disclosed, but probably by leaf-poultice) to treat umbilical hernia, and for this reason call the plant 'navel medicine' (28). In Ivory Coast the dried powdered leaves are taken up the nose as a decongestant in colds and sinusitis (4).

The plant is a common ingredient of the Yoruba *agbo* infusion for fever in children (6, 7). The Igbo use a leaf-decoction as a wash for a feverish child (20). In Guinea a decoction of the whole plant is taken as a febrifuge, and to arrest diarrhoea (16). In Zaïre a leaf-infusion is a fever medicine (18). In Ivory Coast the plant in decoction or the sap is used for diarrhoea (4, 14), and in Gabon it is considered suitable to administer in enemas during pregnancy (24, 25). The Igbo give a leaf-preparation for worms (21) and for convulsions (19). There is general appreciation of the plant for treating gonorrhoea, recorded from The Gambia (leaf-infusion, 27), Ivory Coast (leaf-sap with citron-juice, 4, 14), Nigeria (leaf-infusion, 2), Ghana (11). The plant has diuretic properties and its action may be in this respect. In Ghana Asante women take a concoction of the leaves with clay as an antiabortive (11), yet in Ivory Coast it has use to facilitate labour (4).

Three alkaloids have been detected in Ghanaian material of which the principal two are *indicine* and *retronecine* (4, 14, 26). Nigerian material has shown traces in the roots and (?) the leaves (1).

Fibres are obtained in The Gambia from the whole plant buried in the mud to rot away the fleshy tissue. The fibres are plaited together to make false hair for wearing by women (8).

The curved arching form of the inflorescence lends the plant to fanciful names as indicated above. The Yoruba say that the plant can help to turn away evil, for in calling it 'beak of vulture' when it bears fruit it turns the wrong way like the vulture's beak when feeding (23).

References:

1. Adegoke & al., 1968. 2. Ainslie, 1937: sp. no. 177. 3. Andrews 771, K. 4. Bouquet & Debray, 1974: 54. 5. Burkill, 1935: 1136. 6. Dalziel, 1937. 7. Dawodu 3, K. 8. Fox 87, K. 9. Haerdi, 1964: 186. 10. Hartwell, 1968. 11. Irvine, s.d. 12. Jackson 4145, K. 13. Kerharo & Adam, 1974: 250–2, with phytochemistry and pharmacology. 14. Kerharo & Bouquet, 1950: 223. 15. Oliver, 1960: 28. 16. Portères, s.d. 17. Quisumbing, 1951: 776–8. 18. Taton, 1970: 29–30. 19. Thomas, N. W., 1774 (Nig. Ser.), K. 20. Thomas, N. W., 1989 (Nig. Ser.), K. 21. Thomas, N. W., 2124 (Nig. Ser.), K. 22. Thomas, N. W., 16, K. 23. Verger, 1967: no. 23. 24. Walker, 1953, a: 25. 25. Walker & Sillans, 1961: 108. 26. Willaman & Li, 1970. 27. Williams, F. N., 1907: 373. 28. Williamson, K., 21, UCI.

**Heliotropium ovalifolium** Forssk.

FWTA, ed. 2, 2:322. UPWTA, ed. 1, 426.

West African: **NIGERIA** HAUSA fárín-kaa-fi-maálàm = *whiter (?brighter) than Teacher* (auctt.) shaá-ní-kà-sán-nì = *drink and know me; referring to its medicinal effects* (JMD; ZOG) tumkiyar rafi = *sheep of the stream* (JMD; RES) KANURI kiska kunkum bel (JMD) levee kiska = *beer plant* (FRI)

A perennial decumbent herb to 60 cm high from a woody rootstock, of savanna from Senegal to N Nigeria, and widely dispersed in tropical Africa, and into Asia.

The plant is grazed by all stock in Senegal (1) and in Kenya (5). It is however not grazed in Zambia (8). In Kenya the leaves are chewed as a subsitute for tobacco (4). The plant has analgesic action. In Tanganyika, the Sukuma prepare the dried plant with ghee into a hot poultice which they apply to areas of pain in cases of severe fever (6). Though one suspects sympathetic medicine because of the scorpioid shape of the flowers, the plant in Ethiopia is rubbed on to scorpion-stings (3). In Bornu, N Nigeria, the plant is said to be poisonous, causing diarrhoea and vomiting, and to give rise to symptoms resembling intoxication by local beer: cf. the Kanuri name meaning 'beer plant.' It is used in N Nigeria for its drastic effect to purge syphilis from the system, but more frequently its use is limited to local application for syphilitic ulcers, etc. (2). In Congo the plant enters a medication for syphilis (7).

References:

1. Adam, 1966, a. 2. Dalziel, 1937. 3. Getahun, 1975. 4. Mathew 6349, K. 5. Mwangangi 1392, K. 6. Tanner 932, K. 7. Taton, 1971: 32–34. 8. Vesey-Fitzgerald 4309, K.

**Heliotropium pterocarpum** (DC. & A. DC.) Hochst. & Steud. ex Bunge

FWTA, ed. 2, 2: 322.

A sub-erect or prostrate herb from a perennial woody rootstock, of the dry sahel in Senegal, Mali and Nigeria, and occurring also in NE Africa and Arabia.

In Eritrea it provides forage for camels. It is the dominant plant in certain sandy places and these become a favourite nursery site for locusts (1).

Reference:

1. Bally B.6938, K.

**Heliotropium strigosum** Willd.

FWTA, ed. 2, 2:322.UPWTA, ed. 1, 426.
West African: **SENEGAL** WOLOF timin-timin *said to be an onomatopoeic name, but the significance is not clear* (DF) **NIGER** SONGHAI àrkúu bòŋ kàaréỳ (*pl.* lá) = *white head of the old man* (D&C)

An erect or spreading perennial herb to 30 cm high from a woody rootstock, of dry sandy waste places from Senegal to N Nigeria, and widely distributed elsewhere in tropical Africa, Egypt, Arabia and Australia.

Stock will not graze it in Senegal (1), but in Sudan it is said to provide good fodder during the rains for camels (3). In Tanganyika (6), and in Zaïre (7) the whole plant is prepared into a paste with butter for application to abscesses of the breast. On the basis of sympathetic medicine because of the scorpioid inflorescences, the plant is deemed in Sudan to be specific against scorpion-stings, and juice rubbed on the skin is a safe protection against getting stung for anyone needing to handle scorpions (4). In India the plant is held to be valuable in cases of snake-bite and poisonous stings (2), to relieve pains in the limbs and sore eyes, and to heal boils, wounds and ulcers (5).

Alkaloids have been reported present in the aerial portion of the plant (8).

References:

1. Adam, 1966, a. 2. Dalziel, 1937. 3. Harrison 978, K. 4. Jackson 3900, K. 5. Sastri, 1959: 30. 6. Tanner 4154, K. 7. Taton, 1971: 34–35. 8. Willaman & Li, 1970.

**Heliotropium subulatum** (Hochst.) Vatke

FWTA, ed. 2, 2:322. UPWTA, ed. 1, 427, as *H. zeylanicum* sensu Dalziel.
West African: **NIGERIA** HAUSA gwanja kusa *an epithet, from Gonja, a district in Ghana exporting kola;* kusa: *to be over-ripe, or tired of waiting, a substitute for kola* (RES) maágànin-kunaámaà = *medicine of the scorpion* (JMD; ZOG) námíjìn roòmaà-faáda = *male Scoparia* (auctt.)

A perennial herb, coarsely branched to 70 cm long from a woody rootstock, and lax scorpioid inflorescences to 20 cm long or more, or dry savanna from Senegal to N Nigeria, and throughout tropical Africa and in India.

The plant is grazed by all stock in Somalia (3).

In N Nigeria the plant is applied to scorpion-stings on the basis of the shape of the tail-tipped petals resembling a scorpion's sting. Along with *Indigofera pulchra* (Leguminosae: Papilionoiideae) it is put on sore breasts. It is also eaten as a bitter tonic and stimulant in the belief that it has some of the properties of the kola nut, making the lips or the saliva red, and hence it is sometimes facetiously referred to as *gwanja kusa* (2). In the Lake Province of Tanganyika the leaves are prepared into poultices alone or with butter and without heating for boils (4). The stems are crushed and soaked, then dried and applied to parts affected by yaws, or the leaves made into a paste with ghee are used (5). The stems and leaves are similarly used in Congo (6). In Kenya a cold root-infusion which has stood overnight is taken as a stomachic and laxative (1).

References:

1. Bally 3682, K. 2. Dalziel, 1937. 3. Peck 73, K. 4. Tanner 656, 4218, K. 5. Tanner 1236, K. 6. Taton, 1971: 31–32.

**Heliotropium supinum** Linn.

FWTA, ed. 2, 2:322.
French: héliotrope couchée (Berhaut).

West African: **SENEGAL** SERER ñamdoh ñamdoh (JB)

A decumbent perennial herb from a woody rootstock, occurring in the drier northern part of the Region from Senegal to Lake Chad, and extending to N and S Africa, the Canary Islands, S Europe and the African tropics.

The plant was known to the Ancients and is mentioned by the celebrated surgeon Paulus Aegineta in his medical treatise for its use on warts (2). A number of alkaloids has been detected in it (4).

The plant is not grazed by cattle in Senegal (1), but it is grazed in Sudan (3).

References:

1. Adam, 1966, a: 2. Hartwell, 1968. 3. Jackson 1904, K. 4. Willaman & Li, 1970.

**Moltkia callosa** (Vahl) Wettst.

West African: **NIGER** ARABIC inchal (Grall) TUBU gonogono (Grall)

A herbaceous plant dispersed in dry sandy localities, and recorded in Niger and occurring in N Africa.

The plant is much relished by camels in Hoggar (1).

Reference:

1. Maire, 1933: 179.

**Trichodesma africanum** (Linn.) Lehm.

FWTA, ed. 2, 2:323. UPWTA. ed. 1, 427.

West African: **MALI** TAMACHEK halka (B&T) talkait (Foureau) **NIGERIA** FULA-FULFULDE (Nigeria) ɗemngal nagge = *cow's tongue* (auctt.) kaawa taba = *uncle of tobacco* (Taylor; MM) limse korɗo = *a female slave's garments* (JMD; MM) HAUSA jinmutu (JRA) walkin tsofo = *old man's loin-cloth* (EWF; JMD) wàlkin waáwaá = *fool's loin-cloth* (JMD; ZOG) wàlkin-tshoóhoó (ZOG) warkin tsoho (ZOG) KANURI bulusoana (Golding)

An annual to 1 m high, widely distributed in the drier parts of the Region from Mauritania to W Cameroons and occurring in the Cape Verde Islands, NE and S Africa.

Though the leaves and stems are harshly scabrid (hence the Fula name), the plant provides an excellent fodder for camels and other stock in the Hoggar (6). Cattle in Senegal appear not to take it (1).

The leaves in infusion and decoction are diuretic and are so used in Nigeria (2) and in Sudan (3). In N Africa the leaves are a remedy for diarrhoea, boiled for a short time and mixed with cereal flour and shea-butter or oil and natron to make a paste which is eaten while the attack lasts (4). A hot poultice also is made of the leaves for application for febrile and inflammatory conditions (4).

The root is infused and mixed with natron by the Fula of N Nigeria and the medicine is drunk for *lekki sawora,* a form of hepatitis (*sawora*) (5).

References:

1. Adam, 1966, a. 2. Ainslie, 1937: sp. no. 346. 3. Broun & Massey, 1929: 306. 4. Dalziel, 1937. 5. Jackson, 1973. 6. Maire, 1933: 178.

**Trichodesma gracile** Battandier & Trabut

West African: **MALI** TAMACHEK bedjig (RM) halka (B&T)

A scrubby annual or perennial herb of arid desert localities in the Hoggar on the northern boundary of the Region.

The plant provides some grazing for stock.

# BROMELIACEAE

**Ananas comosus** (Linn.) Merrill

UPWTA, ed. 1, 467.

English: pineapple; the fibre — pina silk.

French: ananas.

Portuguese: abacaxi; ananás; nanas; ananaseiro.

West African: **GUINEA-BISSAU** FULA-PULAAR (Guinea-Bissau) fúnhe (JDES) **GUINEA** FULA-PULAAR (Guinea) fuunyé *from Susu* (JMD) **SIERRA LEONE** BULOM (Kim) nɛsɛ (FCD) BULOM (Sherbro) nɛsi-ɛ (FCD) FULA-PULAAR (Sierra Leone) nanas (FCD) GOLA keve (FCD) kevie (FCD) KISSI bɛ-kpandeo (*pl.* g-betau-kpandelaũ) (FCD) g-be-bandeo (JMD; FCD) KONO fɛtu (JMD; FCD) KRIO painapul = *pineapple* (FCD) LIMBA ku pampa (JMD; FCD) LIMBA (Tonko) nanas (FCD) LOKO horogu (FCD) MANDING-MANDINKA pampadɛ (FCD) yabibi (FCD) MANINKA (Koranko) pampe (JMD) ya-bibi (JMD) MENDE keve (FCD) mbɛlu (*def.*-i) (FCD) nɛɛsi (FCD) SUSU fonye (FCD) funye (FCD) SUSU-DYALONKE fuŋa-na (FCD) funya-na (FCD) yabibi-na (JMD; FCD) TEMNE *a*-nanas (JMD; FCD) VAI kefe (FCD) **IVORY COAST** ABÈ adodjè (A&AA) AKAN-ASANTE abliblé (K&B) AKYE akodin (A&AA) ANYI ablialé (K&B) BAULE abléblé (B&D) ablèlè (A&AA) abrèlè (A&AA) KWENI bɔlɛ́ (Grégoire) nianganba (B&D) **GHANA** ADANGME-KROBO blɛfota = *European's oil-palm* (FRI) AKAN-ASANTE aborobɛ-hemma = *Queen-pineapple, a smooth-skinned c. var.* (FRI) aborobɛ-toro *a smooth-skinned c. var.* (FRI) abrɔbɛ, aburɔbɛ (Enti) asante aborobɛ = *Asantes' pineapple, a smooth-skinned c. var.* (FRI) ASANTE (Denkyera) abrɔbɛ, aburɔbɛ (Enti) FANTE abuɔbɛ, aburɔbɛ (Enti) TWI aborobɛ *from* abrofo: *European;* abɛ: *oil-palm* (FRI) aborobɛ-fitaa *a c. var.* (FRI) aborobɛ-fufuo = *white pineapple, a c. var.* (FRI) aborobɛ-kɔkɔ = *red pineapple, a c. var.* (FRI) abrɔbɛ, aburɔbɛ (Enti) WASA abrɔbɛ, aburɔbɛ (Enti) GA ánànse *the fruit, or the fibre* (KD) blɔfóŋmè = *European's oil-palm* (FRI) GBE-VHE abable (JMD) ablɛndi (FRI) atɔtɔ (FRI) VHE (Pecí) ablɛndi (FRI) GUANG-CHIRIPON ablɛmmɛ (Bunting; FRI) GONJA flèàyé, lémfià (Rytz) SISAALA abiribɛ (Blass) **NIGERIA** BEROM jàáŋ (LB) EDO ẹ̀dínébō (DA) EFIK éyòp m̀bàkárá = *oil-palm tree of the white man* (auctt.) FULA-FULFULDE (Nigeria) onima (Westermann) HAUSA àbàrbaá *from Akan, Ghana* (JMD; ZOG) hántsàr giíwaá = *elephant's udder* (Bargery; ZOG) moda (JMD) moódaá (ZOG) IGBO nkwuaba (BNO) IGBO (Asaba) ákwú óyìbó (DA) IGBO (Awka) akwú-óyìbó = *European palm-fruit* (DA) ṉkwụ́ólū (DA) IGBO (Onitsha) ákwú ólū (DA) IGBO (Owerri) ákwụ́ ólū *the leaves* (Lowe) IJO-IZON (Egbema) beké kọkọrọ (Tiemo) KANURI àbàràbá *from Hausa* (C&H) NUPE kpàcigbè = *cough medicine* (RB) tíro nukpà = *beak of the tiro-bird* (*red-billed hornbill*) (Banfield) TIV mbuer akporo (DA) mbuer udam (DA) mbuer uke (DA) YORUBA ọ̀gẹ̀dẹ̀-òyìbó = *whiteman's plantain* (JMD) ọ̀pẹ̀-òyìbó = *whiteman's palm* (JRA) ọ̀pọ̀n-òyìbó (DA) **WEST CAMEROONS** BAFOK yijang (DA) KOOSI ajang disambe (DA) KPE liangi (DA) KUNDU diala-mukala (DA) LONG yanga (DA) LUNDU esome yadiangi (DA) MBONGE esome (DA) NGEMBA ànánásə̀ (Chumbow) TANGA diala mukala (DA) WOVEA liangi (DA)

Semi-woody, sprawling to erect stems to about 80 cm high, with a dense rosette of long narrow leaves, often saw-edged, the stem topped by a fleshy syncarpellaceous fruit with a further crown of smaller leaves, occurring from Guinea to W Cameroons, cultivated everywhere and naturalized in the rain-forest zone, native of Brazil, but now dispersed by man throughout the humid tropics and subtropics.

The plant is the source of the well-known pineapple fruit. It is thought to be originally from the Paraná-Paraguay rivers, in an area of Brazil where several other related *Ananas spp.* occur. It is perhaps a budsport or ennobled by generations of selection. By the time of Colombus's second voyage, 1493–96, it had been widely dispersed in the Americas for Columbus met it on the Island of Guadeloupe in November 1493. The Portuguese in their settlement in the Santa Cruz area of Brazil adopted the Tupi Indian name *nana,* and as *nana* they carried it far and wide to other countries. It is thought to have reached Africa in the Sixteenth Century. It was under cultivation by 1605 at the Portuguese fort at Mine in the Gulf of Guinea (13). *Nana* and derivatives of it are found throughout the world and occur in many of the West African vernacular names listed above. The Spaniards saw in the fruit a likeness to a pine-cone and so called it *pinas* from which the English *pineapple* is a derivative.

Several of the Regional vernaculars recognize a similarity to the oil-palm in the criss-cross spiral symmetry of the fruit with that of the leaf-bases of the oil palm trunk. The exotic origin of the pineapple is also noted by the epithet 'European', 'white man' or some other foreign designation in the vernacular.

Innumerable cultivars occur, but a relatively small number is grown commercially. The fruit is canned in Hawaii, Thailand, Brazil, Malaysia, Mexico, etc. for which special clones have been raised for size of fruit, minimal depth of flower-pitting, sugar-acid ratio, etc. Gregarious flowering has also been developed so that all the plants of any one area may be harvested together under mechanization. Production for fresh fruit trade occurs in other areas e.g. Caribbean, the Atlantic Islands, Queensland, to supply markets on the fringes, but ripe fruit does not travel well, 4–5 days being about the limit before overripening and fermentation sets in. Fruit picked short of optimum ripeness while travelling somewhat better loses in quality. Markedly unripe fruit is toxic. Refrigerated transport now enables fresh fruit to reach distant markets in good condition. Fermentation of the fruit has been carried out producing both a wine and vinegar. Production of industrial alcohol is a distinct possibility.

The leaves contain about 3% of a fibre of a fine white silky texture. Different cultivars have been selected for this; the plant may be grown in partial shade to draw it up in quick growth. The Philippines are the principle producing country and there the cloth from it is highly prized (16). Fibre from Ghana has been reported as a suitable substitute for flax (8). Nets, gametraps, hammocks and the like are made of it in Gabon (18), cord for stringing chiefs' jewellery in Ghana (10), and to make chiefs' caps and special capes in Nigeria (9).

The unripe fruit is purgative, diuretic, anthelmintic, expectorant and aborifacient. It is taken in Nigeria as an emmenagogue (2), while the flowers are taken for the same effect in Gabon (18) and in Guinea (14). It is also used in Guinea as a vermifuge especially against ascaris (4). The unripe fruit is regarded in Guinea as good for bladder-troubles, and the root for dropsy, perhaps by causing a diuretic effect. Leaf-sap is administered, with powdered pimento added, as an enema in Ivory Coast for urethral discharge (12), while in Ghana immature fruits are taken for venereal disease (10). Juice from young fruit is taken in Congo for vertigo, and with *Costus afer* Ker-Gawl. (Costaceae) and salt it is used to wash small-pox sores (3). Leaf-sap is considered in Gabon to be good for burns (18) and wounds (17). In Sierra Leone a warm infusion of the leaves has been reported good for fomenting a spider-bite, alternating the treatment with application of pieces of the fruit (7). The leaves, ground with copper or brass filings and palm-oil have been found effective in Ghana in healing ulcers (10). Difficulty in breathing is treated in Congo by taking a root-decoction (3). Juice from a roasted fruit is made in Ghana into a thick gruel called *flaku* which is given to children and sick persons (10).

The fruit is rich in vitamin C, ascorbic acid being present at 24.4 to 96.3 mg per cent: it is present in the fresh young leaf also at 29.4 mg per cent (19). The fruit, ripe and unripe, the leaves and the stalk all contain *bromelin,* a powerful proteolytic enzyme similar to papain found in *Carica papaya* Linn. (Caricaceae). It does not, however, coagulate milk as does papain. It is anthelmintic and is used in tropical America, Brazil and India (19). Alkaloid has also been reported in the fruit (20) and (?) root (1).

There is a superstition in Ghana that a sick person near to death will express a great desire to eat pineapples (10).

References:

1. Adegoke & al. 1968. 2. Ainslie, 1937: sp. no. 29. 3. Bouquet, 1969: 74–75. 4. Bouquet & Debray, 1974: 54. 5. Burkill, 1935: 148–54. 6. Buson, 1965: 112–5. 7. Dalziel, 1937. 8. Goulding, 1937: 53. 9. Holland, 1908–22: 676–9. 10. Irvine, 1930: 28–29. 11. Irvine, 1952: 36. 12. Kerharo & Bouquet, 1950: 240. 13. Mauny, 1953. 14. Pobéguin, 1912: 15. 15. Purseglove, 1972: 1, 76–91. 16. Quisumbing, 1951: 147–8. 17. Walker, 1953, a: 25. 18. Walker & Sillans, 1961: 108. 19. Watt & Breyer Brandwijk, 1962: 150–1. 20. Willaman & Li, 1970.

# BURSERACEAE

**Aucoumea klaineana** Pierre

French: oukoumé (from a Gabonese name; Alba, Aubréville).

A tree reaching 35–40 m height, with bole to 2 m in diameter of the guinean forest of Gabon, Rio Muni and Lower Congo. It has been introduced into Ivory Coast where it is said to be 'doing well' (2).

It is a relatively fast growing tree, faster even than *Guarea cedrata* (A. Chev.) Pellegr. (Meliaceae) (1). In Gabon it is the most important timber species, and provides the major portion of exported timbers for making cigar-boxes and packaging, plywood, furniture and naval construction (3, 4, 5). Locally it is used for canoes (3, 5). The wood is resinous making it somewhat inflammable.

The bark contains a gum-resin smelling of turpentine and usable in torches and for incense. It finds use in Gabonese hospitals for treating abscesses. The bark is considered astringent and anti-diarrhoetic. Scrapings of bark with seed of maleguetta pepper and leaves of a bitter *Solanum sp.* are taken by girls at the commencement of puberty (5).

References:

1. Alba, 1956. 2. Aubréville, 1959: 2, 137. 3. Aubréville, 1962: 57–62. 4. Troupin, 1950: 122. 5. Walker & Sillans, 1961: 109–10.

**Boswellia dalzielii** Hutch.

FWTA, ed. 2, 1: 694. UPWTA, ed. 1, 315.

English: frankincense tree (Dalziel; Irvine).

French: arbre à encens (Kerharo & Bouquet).

West African: **UPPER VOLTA** DAGAARI pianwogu (K&B) pienwogu (K&B) MOORE condrényogho (K&B) gonéniogo (Aub.) komhenyegho (K&B) kubré niango (K&B) **GHANA** DAGBANI pianwogu (FRI; JMD) pienwogu (FRI; JMD) GUANG-NCHUMBULU kabona (FRI; JMD) **NIGER** FULA-FULFULDE (Niger) andakehi gorki (Aub.) HAUSA hano (Aub.) **NIGERIA** FULA-FULFULDE (Adamawa) juguhi (JMD) FULFULDE (Nigeria) anndakehi (JMD; J&D) janawhi (KO&S; MM) shabilabi? (JMD) HAUSA ararraɓi (MM; ZOG) ba-samu = *prosperity* (JMD) cibɗi (ZOG) ɗan magami *the gum-resin, incl. other fragrant resins* (JMD) hánoó, hánuú *from* hano, *to prevent, or to obstruct* (auctt.) hánoó kai *the gum-resin* (JMD) kadayan (EWF) tarmasika *the gum-resin* (JMD) HAUSA (West) chibɗi (Bargery) KANURI káfi dukkán (C&H)

A tree to 13 m high of the wooded savanna, with characteristically pale papery bark, peeling and ragged, locally abundant sometimes to near pure stands, from northern Ivory Coast to N Nigeria and into Cameroun and Ubangi-Shari.

The small white flowers which may appear while the tree is leafless are fragrant. The tree is sometimes planted in northern parts of Ivory Coast and Upper Volta as an ornamental (4). It is planted as a village stockade on the Vogel Peak massif of N Nigeria (3) and often as a live-fence to bring prosperity (*ba-samu*) or to prevent (*hanu*) bad luck. Hence the Hausa names. The name *hanu* may also have significance to certain prejudices regarding the use of the resin (2).

The bark contains a whitish exudate which dries readily and is friable. It is fragrant and is burned alone or with other fragrant resins to fumigate clothing and in rooms to drive out flies, mosquitoes, etc. (2). It is used by Catholic Missions as a substitute for true incense (4). It may be added to the juice of *Acacia* berries used in ritual mummifications practised by various tribes in N Nigeria (6). Along with an extract of bark of *Vitex doniana* Sweet (Verbenaceae), it is sometimes an ingredient of malam's ink (2).

A bark-decoction is used as an antiseptic wash for sores in Ivory Coast, and is an ingredient of a complicated prescription for leprosy (1, 4). In N Nigeria the bark is boiled up in large quantity to make a wash for fever, rheumatism, etc., and the fluid is taken internally for gastro-intestinal troubles (2). The Fula of N Nigeria use a cold infusion for snake-bite (5). The fresh bark (? of the root) is eaten in Adamawa to cause vomiting after a few hours and thus relieve symptoms of giddiness and palpitations (2). Both root and bark are held to be antidotes to arrow-poison, e.g., the root is combined with that of *Daniellia oliveri* (Rolfe) Hutch. & Dalz. (Leguminioideae: Caesalpinioideae) to make a decoction which is drunk by the wounded person, and is said to be effective without causing diarrhoea (2). A root-decoction boiled with *Hibiscus sabdariffa* Linn. (Malvaceae) is taken in copious draughts as a remedy for syphilis (2).

The bark-exudate is an oleo-gum-resin. The gum contains *bassorin,* the resin *boswellinic acids,* and have been used in western medicine in fumigatory preparations, and sometimes in plasters and in urinary antiseptics. It was official in the *British Pharmaceutical Codex 1934* (7).

References:

1. Bouquet & Debray, 1974: 54. 2. Dalziel, 1937. 3. Hepper, 1965: 447. 4. Kerharo & Bouquet, 1950: 154. 5. Jackson, 1973. 6. Meek, 1931: 1, 197. 7. Oliver, 1960: 5, 49–50.

**Boswellia odorata** Hutch.

FWTA, ed. 2, 1: 694.

English: frankincense (N. Nigeria, McIntosh).

West African: **NIGER** FULA-FULFULDE (Niger) andakehi debi (Aub.) anndakehi gorki, anndakehi debi = *male frankincense, female frankincense, respectively, and probably both referring to this sp.,* (PF) HAUSA hano (Aub.) **NIGERIA** FULA-FULFULDE (Nigeria) anndakehi (JMD; MM) janawhi juguhi *applied to the gum-resin* (JMD; MM) shabilabi? *applied to the gum-resin* (JMD) HAUSA ararabi (MM) ararraɓi (MM; ZOG) arrabi (MM) ba-samu = *prosperity,* (auctt.) hanu (JMD) hanu (JMD) HAUSA (West) chibɗi (JMD)

A tree similar to *B. dalzielii,* but of more restricted distribution, occurring only in Niger and N Nigeria, and in E Cameroun.

The bark is more resinous than that of *B. dalzielii* (1). It is fragrant and is burnt in the Yola district of N Nigeria to fumigate clothing (2). The bark is said to be an antidote for arrow-poison (3). There are unspecified medicinal uses which no doubt are the same as of *B. dalzielii.*

An unnamed alkaloid has been reported present in the bark (4).

References:

1. Aubréville, 1950: 371. 2. Dalziel 167, K. 3. Shaw 95, K. 4. Willaman & Li, 1970.

**Canarium schweinfurthii** Engl.

FWTA, ed. 2, 1: 697. UPWTA, ed. 1, 315.

English: African elemi; incense tree: bush candle tree.

French: elèmier d'Afrique; the gum-resin: elèmi d'ouganda, elèmi de Moahum (Ivory Coast, Kerharo & Bouquet).

Trade: the timber — African Canarium, Canarium, white mahogany; francophone territories: elèmier du Gabon (Gabon, Dalziel); aiélé; okumé (Ivory Coast); otua (Nigeria, H. Hansen).

West African: **GUINEA-BISSAU** FULA-PULAAR (Guinea-Bissau) modjetchalè (JDES) PEPEL oclanca (JDES) **GUINEA** KISSI dollo (RS) LOMA sava (RS) MANDING-MANINKA (Konya) ghiémana (RS) **SIERRA LEONE** KISSI dolo (S&F) KONO sawa (S&F) MANDING-MANINKA (Koranko) dollɛ (S&F) MENDE m̲bele (auctt.) TEMNE *a*-menəp, *a*-ŋenəp *the gum-resin* (FCD; S&F)

**LIBERIA** DAN beeng (AGV) bien-g (GK) KRU-GUERE (Krahn) bo-tu (GK) po-tu (AGV) BASA goekwehn (auctt.) MANO bi (Har. ex AGV) MENDE mbele **UPPER VOLTA** VULGAR cien (Vuillet) paja (Vuillet) **IVORY COAST** ABE aiélé (auctt.) labé (auctt.) ABURE muneu (B&D) ADYUKRU aiéré *the gum-resin; in general use* (A. Chev.) AKYE senie (auctt.) senyan (auctt.) ANYI ahié (auctt.) ahiélé (auctt.) kerendja-égué KRU-BETE mosu (Aub; K&B) nosu (RS, Aub.) uléguinali (Aub.) GREBO (Plapo) khiala *the amber-coloured gum-resin of C. khiala of A. Chevalier* (A. Chev.) yatu (A. Chev.; K&B) KYAMA muamohia (auctt.) muemia (B&D) muénohia (auctt.) NZEMA ahié (K&B) ahielé (auctt.) 'KRU' dirutu (K&B) gueritu (K&B) **GHANA** AKAN-ASANTE bedi-wu-nua = *you will have sexual connexion with your sister* (auctt.) TWI bediwonua (FD) bɛdi-wo-nua (FRI) bediwuna (CJT) bɛwɛ-wo-nua *not commonly used* (Enti) kurutwe (FRI) WASA amoukyi (CJT) kurutwe (auctt.) ANYI ahie (FRI) ANYI-AOWIN kandangunuu (FRI) SEHWI kandangunuu (FRI) kantankrui (CJT) NZEMA ɛyɛrɛ (CJT) ɛyɛ lɛ (FRI) **NIGERIA** ABUA ịrẹ́ (DRR; JMD) BEROM pwat (LB) BOKYI boshu-basung (auctt.) ofingot (Kennedy) EDO èkpákpōghò (DRR; JMD) órúnmùnkhìokhìo = *incense tree of the khiokhio bird* (JMD; DRR) EFIK ébèn étìrìdọ̀ñ (auctt.) EJAGHAM njasun (auctt.) EJAGHAM (Keaka) njasung (DRR) EJAGHAM-ETUNG njasin (DRR) njassong (Kennedy) oju-ngon (DRR) ENGENNI abigwa (DRR; JMD) HAUSA atile (auctt.) atilia (MM) atílis (auctt.) IDOMA oda (Odoh) IGALA oda (H-Hansen, Odoh) IGBO ùbé agba, ùbé mkpuru aki (BNO) ùbé-óhịá = *bush pear* (LAKC) IGBO (Amankanu) ùbé = *pear* (DRR; JMD) IGBO (Aro-Chuku) ùbé ōkpókó (DRR; JMD) IGBO (Arochukwu) ùbé wemba (DRR; JMD) IGBO (Awka) ùbé-ọ̀sà (DRR; JMD) IGBO (Bonny) ùbé ọ́hị̄á = *ùbé of the bush, i.e., wild* (KW) IGBO (Nsokpo) ibwaba (DRR; JMD) IGBO (Onitsha) ùbé-ōkpókó (DRR; JMD) ùbé-ọ̀sà (DRR; JMD) IGBO (Owerri) ùbwé aba (DRR; JMD) MUNGAKA tamfre (DRR) NKEM siselung, ute-siselung *the fruit* (DRR) NUPE èshiá (Banfield) itali (Lamb) OLULUMO-OKUNI kọfarẹ (DRR) YEKHEE uda (Kennedy) YORUBA àkó = *gathering* (Verger) anikantuhu (DRR) ibagbo (DRR) órigbó (KO&S) **WEST CAMEROONS** DUALA bòsao b'eyidi (JMD; Ithmann) KPE wotua (JMD) KUNDU hehe (JMD)

A large tree, commonly to 40 m high, but attaining 50 m or more in Sierra Leone (18) and Ghana (19), of the evergreen and deciduous forest throughout the Region from Senegal to W Cameroons, and extending to Ethiopa, Tanganyika and Angola.

The bole is straight and cylindrical attaining 20 m long, usually less, by 4.50 m in girth in Sierra Leone (18) yielding a high volume of timber. The sapwood is whitish to about 10 cm thick. Heart-wood is light brown to pinkish, darkening on exposure to a light brown mahogany colour. It is coarse, softish, even woolly, and is scented. Sawing may be difficult owing to the presence of silica, and it tends to blunt tools, but otherwise it can be worked well. It is used locally for furniture, cabinetry, mortars, planks, canoes and more or less general construction work. Sometimes the grain is spiralled, and distortion may occur in seasoning. It is susceptible to decay, stain, termites and to borers, and it resists impregnation by preservatives which seriously limits its utility. It can be cut for plywood (13), and it has been exported in a small way for use as a core-veneer (22). It makes a good fuel igniting readily and burning with a vigorous hot heat but not long-lasting. The flame is very smoky (17). The tree is sometimes planted and it appears to be amenable to mixed culture plantation husbandry (22), but commercial futures are uncertain.

The bark is thick. On young trees it is fairly smooth and is split off in Gabon to make cylindrical boxes (23). With age it becomes increasingly scaly and fissured. It contains an oleo-resin gum, heavy, sticky, with a smell of turpentine. It is obtained by bark-slashing and allowing the colourless expissation to trickle to the ground where it solidifies into a sulphur-yellow opaque resin. Tribes in SE Nigeria on clearing land for farming will fire ring-girdled trees. This causes resin to flow after a few days which is then collected from the ground (17). There is variation between trees in the quantity produced. The resin has a not-unpleasant taste. It burns readily: as a primitive illuminant it is used as hunters' flares or 'bush candles', sometimes wrapped in leaves, in Nigeria; and as incense it releases a smell like lavender. (1, 8, 10, 15, 19, 24.) The flame is very smokey and soot is collected as carbon-black from the outsides of pots held over it for use in Liberia in tattooing and to make ink (5, 6, 8). The resin melts readily and is compounded with oil or shea butter, clay, etc., to make a fragrant unguent for anointing the body (10, 11, 18). The resin used in Uganda to depilate the whole body of a young bride is perhaps of this

(9). It is also used to fumigate dwellings to drive away mosquitoes (10). It is melted and used to repair broken pottery (10). In Liberia it is a pitch for caulking boats (6), and in southern Africa for fastening an arrowhead to its shaft (24). The resin has been in the past exported to Europe for pharmaceutical use (2, 14). It was found during the 1939–45 World War to be a good substitute for gum-mastic in making wound-dressings.

The resin contains 8–20% of an essential oil, the main constituent of which is *limonene.* It is rich in *phellandrenes,* and contains also resins and a bitter principle (14, 16, 24). Composition is very variable.

A bark-decoction is widely used for dysentery (10). In Nigeria the resin is used for roundworm infection (16) and in Liberia a decoction is taken for intestinal parasites (22). The resin is considered emollient, stimulating and diuretic in Gabon (23). The bark however contains only a small amount of tannin, 0.66% (24). It is taken in small quantity in Liberia for gonorrhoea (22). In Sierra Leone a bark-decoction is drunk for coughs and chest-pains (11), and in Congo the sap is eaten with cassava for pulmonary affections (2) where a decoction is also taken in draught for stomach-complaints, food-poisoning, gynaecological conditions requiring purging and emesis (2). The resin is generally held to have action on skin-affections. It is used for eczema in Congo rubbed into the areas affected. In Liberia the inner bark is pounded with water to rub on the skin for leprosy, and on to ulcers. The patient on this treatment is under taboo not to eat catfish, nor to have sexual intercourse, the bonus for observation being cure within one year (Harley fide 22). In Cameroons the bark is used for chancre (10). The fresh bark is administered as an enema in Ivory Coast for intestinal pains and haemorrhoids and jaundice, and is sometimes given to pregnant women as a tonic (3, 14). In Congo it is put into steam-baths for rheumatism (2). In Ghana the bark is said to provide an aphrodisiac (19). The Yoruba invoke the tree as a cure for small-pox (21).

In collecting *Funtumia* rubber, the bark was used as a coagulant in Liberia (22). Some tribes of SE Nigeria put it into palm-wine (17).

The root is used in Ubangi for treating adenites: the root scrapings are made into a poultice (20).

The fruit is 2.5–3.7 cm long, blue-black with a persistent triangular calyx. It is sold in markets wherever the tree occurs. The slightly greenish outer pulp is of an oily consistency and is edible (10), but acceptance is not general (13, 23). Some people eat the fruit raw, others, Ekoi, for example, steep them in hot water and then bake between leaves in live embers (17). Simply softening them in warm water is said to improve palatibility (13). The pulp-oil is about 71% palmitic acid, 18% oleic acid and small amounts of several other fatty acids (4).

The seed is a hard-shelled, fluted stone. They are often drilled and strung into necklaces. Sometimes they are carved (10, 13). In SE Nigeria the seeds are also strung for attaching to calabashes as musical instruments: the Igbo (Awka) *obago juju,* and the Isibori *jujunam.* The Isibori also use the stringed seeds to decorate their hunting guns, and some Igbo races smooth off the flutings and use the seed as shot (17). The seed-kernel is oily and edible. They are eaten cooked, and in N Nigeria are sometimes prepared into a vegetable-butter and eaten as a substitute for shea-butter (25). Fatty acids present are: oleic 36%, linoleic 28%, palmitic 26%, stearic 7%, and traces of others (4).

Hornbills seek out the fruits in Ghana and appear to be the main seed-dispersal agent (19).

References:

1. Aubréville, 1959: 2, 138. 2. Bouquet, 1969: 75–76. 3. Bouquet & Debray, 1974: 54. 4. Busson, 1965: 336–8, with chemical analysis of the fruit. 5. Cooper 125, K. 6. Cooper 192, K. 7. Cooper 385, K. 8. Cooper & Record, 1931: 89. 9. Cooper, W., 1971: 85. 10. Dalziel, 1937. 11. Deighton 4738, K. 12. Eggeling & Dale, 1952: 52. 13. Irvine, 1961: 508–10. 14. Kerharo & Bouquet, 1950: 154. 15. Kunkel, 1965: 62. 16. Oliver, 1960: 21. 17. Rosevear, 1975. 18. Savill & Fox, 1967: 61–62. 19. Taylor, 1960: 112. 20. Tisserant 581, K. 21. Verger, 1967: no. 30. 22. Voorhoeve, 1965: 75. 23. Walker & Sillans, 1961: 110. 24. Watt & Breyer-Brandwijk, 1962: 152. 25. Yates 21, K.

**Canarium zeylanicum** (Retz.) Blume

FWTA, ed. 2, 1: 697.

A large much-branched tree of Ceylon which has been introduced into the Region.

The wood is light, soft and pale to brownish white. It is used for making packing-cases and coffins in India. The bark yields a clear fragrant gum-resin which is used as an illuminant and fumigant. It contains 10–15% of an essential oil, ± 45% of which is $\alpha$-phellandrene (1).

Reference:

1. Sastri, 1950: 55.

**Commiphora africana** (A. Rich.) Engl., var. **africana**

FWTA, ed. 2, 1: 695. UPWTA, ed. 1, 316.

English: African bdellium; African myrrh.

French: bdellium d'Afrique; myrrh africaine.

West African: **MAURITANIA** ARABIC (Hassaniya) adres (Leriche) ARABIC ('Maure') adras (Aub.) adress (Aub.) adriss (Aub.) **SENEGAL** ARABIC ('Maure') adras (Aub.) adress (Aub.) adriss (Aub.) FULA-PULAAR (Senegal) badadi (Aub.; K&A) badi (AS; K&A) MANDING-BAMBARA barakâti (K&A) darasé (JB, ex K&A) SERER mirdit (JB) ngolotôt (JB) ngoñan (JLT; K&A) sãg'h (JB) SERER-NON bopbop (auctt.) WOLOF hammont *the gum-resin* (DeCordemoy) ngotot (auctt.) niotot, n-mootut = *little ones* (auctt.) **MALI** ARABIC adres (Leriche) adrəs = *bowl made of wood* (Monteill) DOGON ínu bánuma (Aub.; C-G) FULA-PULAAR (Mali) badadi (Aub.) badi (JMD) MANDING-BAMBARA badi (Aub.) barakanté (Aub.) darasé (GR) SONGHAI badâdi (Aub.) barkanté (Aub.) kórómbéy (*pl.* -à) (Aub.; D&C) TAMACHEK adras (JMD; Aub.) taghalbas *the gum-resin* (JMD) **UPPER VOLTA** MANDING-BAMBARA barakanti (K&B) MOORE kodemtabéga (Aub.; K&B) **GHANA** DAGBANI narga (Mellin; FRI) **TOGO** SORUBA-KUYOBE kuénu (Aub.) **DAHOMEY** SORUBA-KUYOBE kuénu (Aub.) **NIGER** ARABIC adress (Grall) ARABIC (Niger) gafal (Aub.) mbarkat (Aub.) FULA-FULFULDE badadi (Aub.) badangereehi (PF) kaalihi (PF) HAUSA dashi (Aub.) ikitchi (Aub.) iskici (Aub.) KANURI kâbi (Aub.) TUBU dachi (Grall) digui (Aub.) **NIGERIA** ARABIC gafal (JMD) um el barka (Vogel) CHAMBA kus-sum (FNH) FULA-FULFULDE (Nigeria) badaaɗi (auctt.) badangereehi (MM) kaalihi (MM) kalihi (MM) watanta jambere (JMD) HAUSA ba-zara (ZOG) bìi-zaáná (ZOG) ɓiskiti (JMD) daàshií (auctt.) daàshií mai-yawan rai = *daàshií with seven lives* (JMD) danbaka sayaba *an epithet for the gum-resin* (JMD; ZOG) jawul *the gum*, (ZOG) HAUSA (West) biskiti (ZOG) iskici (JMD; ZOG) kororo (JMD) KAMWE linga-linga (FNH) KANURI káfí (C&H) KUKA kadige (Elliott)

A shrub or tree to about 6.5 m high, with short bole and straggling branches, usually spined, of dry sahel savanna woodland from Senegal to N Nigeria and widely dispersed elsewhere in drier parts of tropical Africa.

The plant is often grown as a hedge in N Nigeria. It establishes itself quickly from stakes and makes an efficient barrier. It is well suited to dry areas (4, 11). The foliage is readily browsed by cattle, sheep and goats and is often cut by graziers (1, 22). The plant's capacity to withstand cutting leads to the Hausa nickname for it, the *'dashi' with seven lives* (4). To the Tuareg it is a symbol of immortality, and they place it on graves, etc. (4).

The leaves are aromatic and medicinally are taken in West Africa as a stomachic after pounding with bulrush millet and milk (4). A macerate of crushed leaves in oil is drunk in Ivory Coast-Upper Volta as a sedative and soporific (18).

The wood is soft. It serves as a chew-stick in N Nigeria and is carved into rosary beads (4). It burns with a fragrant smoke with which women in N Ghana fumigate their clothing (14). In the Sahara the wood is used to raise fire by friction with a stick of hardwood (22).

In Ethiopia the stems are used as hoe-handles (27) and in Kenya for making cattle stalls (7), and the larger trunks are carved out for water vessels, milk jugs and the like (23). In the Sahara region it is one of two woods (the other is *Balanites aegyptiaca*) used to make two standard market measures known as *nəfga* and *mudel* for selling various fruits, gum, henna leaves and liquids. A measure similar to the former is a *coppe* (English, cup) used in Ivory Coast (21). It is also the wood used to make the Berber violin known as *tidinit,* an instrument of four strings of hair (20).

The bark contains an aromatic oleo-gum-resin which is obtained by making incisions on the trunk. The gum is slow to exude, 6–7 days being necessary to produce a lump about 1 cm in diameter which sets into a tear, first white and clear then yellowing. Under pounding and warming in the sun these soften and are kneaded into larger balls, light brown, dark brown or black according to the amount of extraneous matter contained. This is *African Bdellium,* or *African Myrrh,* aromatic but differing from true myrrh of E Africa in odour, less bitter in taste and slightly acid (4). Between 1900–1910 this product was exported from Senegal and Guinea to Europe for medicinal use in plasters and for addition to varnishes. It contained 70% resin soluble in alcohol, and 29% gum. Under steam dillation it yielded 6–8% of a volatile oil. It has no application now in western medicine (17).

The gum gives an agreeable smell on burning and is used in Ivory Coast-Upper Volta to create a pleasing atmosphere inside dwellings (18). The smoke is used to fumigate clothing, or the dried pulverized bark is likewise burnt or is simply laid between clothing (4). Bark-extracts have been shown to have some insecticidal activity (8), and to be termite-repellant (13). The gum is widely prepared as a sweet-smelling cosmetic unguent for rubbing on to the body. In Senegal the burning gum is used medicinally for antisepsis, and in fatigue and respiratory conditions (16, 17). The gum is widely used to prepare antiseptic washes and baths for skin-infections and sores, and leprosy (3, 17). The bark, chewed with natron, is applied to scorpion stings in Nigeria, and a treatment for inflammation of the eyes is to hold the head over a steaming pot with bark in it (4). In central Africa the bark with salt is applied to snake-bite (29). In Tanganyika the dried powdered bark is eaten as a mash for malaria (9). More fanciful uses are: a decoction is a sovereign remedy in Ivory Coast-Upper Volta for male sterility (18) and for fits of insanity by the Fula of N Nigeria (15).

In Kenya the gum is masticated like chewing-gum (7), and the bark is a source of tannin for curing leather (6). The gum has adhesive property used in Tanganyika for fixing axe-heads to handles (2).

In some parts the roots are dug up to eat: on the Vogel Peak massif they are treated like manihot (10); in Ethiopia they are eaten raw (27); in Jebel Marra they are described as sweet (12). A root-decoction is prepared as a warm poultice in Tanganyika for application to a stiff-neck caused by chilling (9). A root-decoction is used in some parts of Nigeria as a taenicide (24), and in Tanganyika it is taken for stomach-pains and to stop loose stools (26) and to induce delayed childbirth (25).

The fruit, about 1 cm long, consists of a pulp surrounding a single stone. The pulp is edible, and has vulnerary properties. In Kenya it is used as an oral styptic and haemostatic (28), and in Tanganyika it has unspecified veterinary application (19).

The hard coated seeds are drilled and strung into rosaries and scented necklaces in Ivory Coast-Upper Volta (18).

In Nigeria a seed-decoction is held to be a very effective purgative and vermifuge. A dose of 60 gm of powdered seed in a glass of water is certain to expel a tapeworm (24). The unripe seed pounded and steeped in water in Uganda make a warming drink when one is cold (5). The seeds contain a tannin, a dye-stuff and a fixed oil (24).

References:

1. Adam, 1966, a. 2. Bally 7872, K. 3. Bouquet & Debray, 1974: 54. 4. Dalziel, 1937. 5. Dyson-Hudson 2041, K. 6. Gillett 13759, K. 7. Glover & al., 2456, K. 8. Heal & al., 1950: 3: 107. 9. Haerdi, 1964: 119. 10. Hepper, 1965: 489. 11. Howes, 1946. 12. Hunting Technical Surveys,

1968. 13. Irvine, 1955. 14. Irvine, 1961: 510. 15. Jackson, 1973. 16. Kerharo & Adam, 1964, b: 431. 17. Kerharo & Adam, 1974: 252–3, with phytochemistry and pharmacology. 18. Kerharo & Bouquet, 1950: 155. 19. Kornitschoner 1111, K. 20. Leriche, 1950: 745. 21. Leriche, 1951: 1228. 22. Monteil, 1953: 93. 23. Mwanjanji & Gwynne 1193, K. 24. Oliver, 1960: 6, 57. 25. Tanner 3775, K. 26. Tanner 4021, K. 27. Turton DAT 96, K. 28. Van Someren 147, K. 29. Watt & Breyer-Brandwijk, 1962: 152.

## **Commiphora dalzielii** Hutch.

FWTA, ed. 2, 1: 696.

A shrub or tree to 8 m high forming dense thickets of open coastal savanna in Ghana only.

The plant grows well from cuttings and makes a good hedge in dry regions. Its branching habit of short spiny branchlets renders hedges and thickets impenetrable. It smothers out all other vegetation from beneath it, and as it is leafless during the usual period of grass fires, it survives fire damage. Its wood will however burn well, but because it is awkward to handle it is not normally cut for fuel (1).

Leaves and stems are strongly resinous and aromatic. A form of African myrrh is obtained from the plant which when burnt gives an agreeable smell and is used to fumigate clothing in Ghana. It was at one time included with true African myrrh from *C. africana* for shipment to Europe for use in varnishes and medicinal plasters (2).

References:

1. Adams, C. D., 1956. 2. Irvine, 1961: 510.

## **Commiphora kerstingii** Engl.

FWTA, ed. 2, 1: 695. UPWTA, ed. 1, 317.

West African: **NIGERIA** FULA-FULFULDE (Nigeria) chaɓɓuli *properly = Ximenia* (Taylor) daleji (JMD) kaabiwal (*pl.* kaabiije) (auctt.) HAUSA árár(r)ábií (ZOG) bagana (JMD) bar na gada, bar *or* beri: *to allow;* gada: *to inherit, inferring long existence; a nickname* (JMD) bazana (auctt.) dali (Abraham) hànà goòbárá (Abraham; ZOG) HAUSA (East) gurzun dali (JMD) hana gobara, hana: *to prevent;* gobara: *conflagration or the result of one; a nickname* (JMD; KO&S) ka-ƙi-ganin-bula, ƙi: *to hate, or to refuse;* ganinbula (Fula?): *ruins of a town after burning; a nickname* (JMD) HAUSA (West) dale, dali (JMD) garkuwar-wuta = *shield against fire* (JMD) TIV kwaor (Vermeer)

A tree to 10 m high, of the savanna from Togo to Nigeria, and on to Ubangi-Shari.

The wood is soft (1, 5). It is used to make saddles, etc., and is sometimes hollowed out to make quivers in the Yola area of N Nigeria (2, 3). The tree is often planted as a live-fence in towns (3, 4), for its aesthetic value (6) and as a shade-plant (4). The evergreen bark seems to have engendered an idea that the tree is little likely to burn, so that it has acquired vernacular names suggestive of protection against fire, and survival of property and therefore of inheritance (3).

The bark is sometimes used as an antidote to arrow-poison (3).

Fula herdsmen feed the leaves to goats, and in a superstitious token lay sticks of the plant across graves (4), perhaps in the same sense of conferring protection.

References:

1. Aubréville, 1950: 375. 2. Dalziel 165, K. 3. Dalziel, 1937. 4. Jackson, 1973. 5. Keay & al., 1964: 250. 6. Vermeer 10, UCI.

**Commiphora pedunculata** (Kotschy & Peyr.) Engl.

FWTA, ed. 2, 1: 695. UPWTA, ed. 1, 317.

West African: **UPPER VOLTA** MOORE sabnughagha (Aub.) **NIGERIA** ARABIC luban (JMD) ARABIC-SHUWA luban (JMD) HAUSA daashin jeji (ZOG) lubban *the gum-resin* (JMD) námíjìn-daáshií = *the male dashi* (auctt.)

A tree to about 6.5 m high with stout trunk to 1 m in girth, of savanna from Senegal to N Nigeria, and extending across Africa to Sudan.

The wood is soft and is of little use (1).

The bark yields an oleo-gum-resin like myrrh which is used as incense (2).

The small, black plum-like fruit is said to be edible (2).

References:

1. Aubréville, 1950: 376. 2. Dalziel, 1937.

**Dacryodes edulis** (G. Don) H. J. Lam

FWTA, ed. 2, 1: 696. UPWTA, ed. 1, 317, as *Pachylobus edulis* G. Don.

English: African pear (Adams); native pear (Nigeria); bush butter.

French: safoutier (Bouquet).

West African: **IVORY COAST** ABE vi (FB) AKYE tsai (FB) ANYI kerendja **NIGERIA** ABUA ù-bé (*pl.* àrù-bé) (DRR; KW) BOKYI boshu (auctt.) EDO órúnmwun *indicating something edible* (auctt.) EFIK ébèn (auctt.) ébèn m̀bàkárá = *white man's pear; also applied to Persea gratissima Gaertn. f.* (JMD) EJAGHAM aju (DRR) nwan *a small form* (DRR) EJAGHAM (Keaka) aju-obok = *monkey's aju; a small form* EJAGHAM-ETUNG oju (DRR) IGBO ùbé *lit., a pear* (Singha; KO&S) ùbé-óyìbó = *white man's pear, also applied to Persea gratissima Gaertn. f.* (JMD) IGBO (Awka) ùbwé (DRR; JMD) IGBO (Owerri) ùbwé ọyọ (DRR; JMD) IGBO (Umuakpo) ùbé (DRR; JMD) IJO (EAST) - KALABARI ibẹ (AJC) MUNGAKA bakwa (DRR) bakwangok = *bird's bakwa; a small form* (DRR) pakwa (DRR) NKEM uju (DRR) OLULUMO-OKUNI kọfe (DRR) URHOBO orumu (JMD) YORUBA elemi (KO&S) ibagbo (DRR) **WEST CAMEROONS** DUALA bòsao *the fruit* (JMD; Ithmann) KPE bo-sau *the fruit* (JMD) sau *the fruit* (JMD)

A tree to 20 m high of the evergreen forest of S Nigeria and W Cameroons, and extending to Zäire.

The tree is planted for shade and its fruit, and is grown from either seeds or cuttings. The coming of the leaves into bud marks in S Nigeria that the planting season for field crops has arrived (1, 4).

The wood is greyish white to pinkish, and heavy, elastic and used for axe-handles, occasionally for mortars, and suitable for carpentry, etc. (4, 7). The wood contains an oil which on petrol-ether extraction has been found to be composed of fatty acids and their esters (5).

The bark is aromatic and on injury yields a resin which is used as pitch on the inner surfaces of calabashes and for mending earthenware. It can be burnt as a primitive lamp-oil or bush-candle (4, 5). The resin has medicinal use in Nigeria for treatment of parasitic skin-diseases, jiggers, etc. (4). Pulped-up bark is used in Gabon as a wound cicatrisant (7). In Congo a bark-decoction is used for gargles and mouth-washes, for tonsillitis; it is taken powdered with maleguetta pepper as an anti-dysenteric, and for anaemia, spitting blood and as an emmenagogue; with palm-oil it is applied topically to relieve general pains and stiffness and to treat cutaneous conditions (2). Root-bark in decoction is taken for leprosy (2).

The resin under steam distillation has been reported to yield a peppery essential oil rich in *sabinene, β–phellandrene* and *limonene,* and a non-volatile fraction of crystalline *canaric acid,* a keto acid and the corresponding hydroxy acid (5).

The leaves are eaten raw with kola nut as an antemetic in Congo. Leaf-sap is instilled into the ear for ear-trouble, and a leaf-decoction is prepared as a vapour-bath for feverish stiffness with headache (2). The leaves in Gabon yield a dye (7).

The principle value of the tree lies in its fruit which is about 7 cm long by 3 cm in diameter. The leathery shelled stone is surrounded by a pulpy butyraceous pericarp about 5 mm thick which is the portion eaten, either raw or cooked to form a sort of 'butter.' It has a mild smell of turpentine and is oily with palmitic acid 36.5%, oleic acid 33.9 %, linoleic acid 24.0% and stearic acid 5.5% (3). The pulp is also rich in vitamins (3, 6).

The seed kernel is also rich in oil of the same fatty acids and approximately in similar amounts (3).

References:

1. Adams, R. F. G., 1943: as *Pachylobus edulis*. 2. Bouquet, 1969: 76. 3. Busson, 1965: 336–8, with fruit analyses. 4. Dalziel, 1937. 5. Ekong & Okogun, 1969, a. 6. Okiy, 1960: 121. 7. Walker & Sillans, 1961: 113, as *Pachylobus edulis* G. Don.

## **Dacryodes klaineana** (Pierre) H. J. Lam

FWTA, ed. 2, 1: 696–7. UPWTA, ed. 1, 317, as *Pachylobus deliciosa* Pellegr.

English: monkey plum (Liberia, Voorhoeve).

Trade: adjuaba (Ivory Coast).

West African: **SIERRA LEONE** KRIO damzin = *damson* (FCD; S&F) MENDE kafe (*def.*-i) *a group name* (auctt.) **LIBERIA** DAN sion(-g) (GK) zeon (AGV) KRU-GUERE (Krahn) suma-tu (GK) BASA pohn (C&R) **IVORY COAST** ABE n-saniasé (Banco; Aub.) vi (auctt.) ABURE essamé (B&D) essanvi (B&D) ADYUKRU maugun (auctt.) AKAN-ASANTE krindia (B&D) krinja (B&D) AKYE sai (auctt.) sinh (A&AA) tsain (Aub.; RS) ANYI kerendja (auctt.) KRU-GUERE (Wobe) luaintu (JMD) KYAMA agbaya (auctt.) NZEMA adjouaba (auctt.) **GHANA** VULGAR adwea *from Akan - Twi* (DF) AKAN-ASANTE adwea (Enti) FANTE adweaba (FRI) TWI adwea (DF) WASA adwea (CV; FRI) ANYI krandwoa (FRI) GUANG-NKONYA ponsa ponto (BD&H; FRI) NZEMA odwuaba (FRI) **NIGERIA** EDO gólógóló (JMD) gólógózi, ègólógóló (JMD)

An understorey tree to 20 m tall of deciduous and fringing forests from Sierra Leone to W Cameroons, and on into Gabon.

The wood is variously described as grey (Ivory Coast, 2; Nigeria, 6), yellowish grey-brown (Ivory Coast, Bertin fide 4), and pink (Ghana, 5). It is hard with a wide lighter-coloured sap-wood. Marketable logs in Liberia are mostly defective and very large ones often hollow (8). The wood is difficult to work though it is used for building purposes, for rice-mortars and as well for firewood (5). It is presently recommended for telegraph poles in Ghana (1). Though its value as a commercial timber in Liberia is said to be unpromising (8), it is considered to be valuable in Ivory Coast and of good quality (2).

The bark-slash smells strongly of turpentine and exudes a little watery or gummy aromatic liquid (2, 7, 8). No usage is recorded except perhaps that it is the bark that is reported used on Ivory Coast for tachycardia and for cough (3).

The fruit, about 2.5 cm long, has a fleshy pericarp which is edible, raw or cooked to form a sort of 'butter' (5). The appearance of the fruit like a plum gives rise to the name *monkey plum* in Liberia, and to *damzin* (damson) in Sierra Leone. Small mammals are fond of the fruit (8). In Ivory Coast two varieties based on the fruit are recognized: one globular with a taste of turpentine is *saniasé* (Abe); the other is ovoid without a taste of turpentine (2), and seemingly of unrecorded name.

Tannins and saponoside have been recorded in the leaves (3).

References:

1. Anon., s.d. 2. Aubréville, 1959: 2, 140. 3. Bouquet & Debray, 1974: 54. 4. Dalziel, 1937. 5. Irvine, 1961: 511. 6. Keay & al., 1964: 255–6. 7. Kunkel, 1965: 86. 8. Voorhoeve, 1965: 80.

**Santiria trimera** (Oliv.) Aubrév.

FWTA, ed. 2, 1: 696. UPWTA, ed. 1, 318, as *Pachylobus trimera* Engl.

French: adjouaba à racines aériennes (from Nzema: *adjuaba* = *Dacryodes klaineana,* Aubréville).

West African: **SIERRA LEONE** KONO domboɛ (FCD; S&F) KRIO damzin = *damson* (FCD; S&F) MENDE kafe (*def.*-i) *a group name* (auctt.) TEMNE *an*-thaŋka (FCD; S&F) **LIBERIA** KRU-BASA poh (C) MANO néné-iri (RS) **NIGERIA** EDO gólógóló (KO&S) órúnmwún-ẹ́zẹ (Lowe)

A tree to 26 m high with bole to 65 cm in girth, with thin buttresses and stilt-rooted, of the lower storey of the closed-forest from Sierra Leone to S Nigeria, and extending to Zaïre.

The timber is fine-grained and even-textured, greyish to yellowish brown. The tree is occasionally felled in Sierra Leone and the timber used for carving (6). The wood in Gabon is used to make a number of domestic articles and the adventive aerial roots are used to decorate axe-handles and hunting-bows (8).

The bark is aromatic with a smell like balsam, and yields an oleo-resin. Use for therapeutic purposes within the West African region is not recorded. In Gabon the bark is used against chest complaints and is reduced to a powder which is sprinkled on the sores of yaws (8). In Congo powdered bark with salt and palm-oil is given to children for whooping-cough, and is considered to be vermifugal. Bark-decoctions are used in massages and vapour-baths for fever-pains and for eczema (1).

The fruits are black and up to 2.5 cm long, reminiscent of the temperate damson from which the Krio name in Sierra Leone is derived. The tree is cultivated in that country for the fruits which are a usual item of market trade (5), but their taste and smell is of turpentine (4) which is not to everybody's liking. In Gabon the taste is described as vinous and the fruit pulp is eaten there raw and is much appreciated (8). They are eaten also in Liberia (7).

The seed is oily and is eaten in Liberia (3).

Tannins, steroids and terpenes have been recorded in the bark and roots (2).

References:

1. Bouquet, 1969: 77. 2. Bouquet, 1972: 16. 3. Cooper 110, K. 4. Deighton 2628, K. 5. Deighton 4963, K. 6. Savill & Fox, 1967: 63. 7. Voorhoeve, 1965: 75, 105. 8. Walker & Sillans, 1961: 111, as *Pachylobus balsamifera* (Engl.) Guilaum., and 115 as *P. trimera* (Oliv.) Guillaum.

# CACTACEAE

**Cereus peruvianus** (Linn.) Mill.

French: ciege du Pérou (Berhaut).

A South American plant, introduced into horticulture for its ornamental value. It is present in Senegal (2).

The plant has been reported to contain *tyramine* (1). A related specimen *C. grandiflorus* Mill. of Jamaica is the source of a heart stimulant drug.

References:

1. Agurell, 1969. 2. Berhaut, 1967.

**Nopalea (Opuntia) spp.**

French: nopal à cochenille (*N. cochenillifera* (Linn.) Salm. Dyck.).

West African: **SIERRA LEONE** KISSI tandaletando (FCD) MENDE ngele-gonu (*def.*-goni) (FCD)

*N. (Opuntia) cochenillifera,* a shrub-like xerophytic plant from Mexico has been recorded present in Senegal (1). This is the principal food-plant of the cochineal insect *(Dactylopius coccus)*. The red dye, used in the food industry, is obtained from the dried female insect, but cochineal now is made synthetically.

The fruits are edible and the stems of the plant are used as poultices for bruises, etc.

An unidentified *Nopalea* recorded from Sierra Leone with the Kissi and Mende names stated above is planted by Mendes and Temnes to ward off thunder, and has been found planted, perhaps to afford protection, as a memorial over a child's grave at Gbinti (2).

References:

1. Berhaut, 1967. 2. Deighton 2223, K.

**Opuntia spp.**

FWTA, ed. 2, 1: 221.

English: prickly pear.

French: pomme raquette (Berhaut): figuier de Barbarie (Senegal, correctly *O. vulgaris* Mill., Kerharo & Adam).

West African: **SENEGAL** SERER ngala *O. tuna?* (JB; K&A) ngalga (JB; K&A) WOLOF gargâ bos *O. tuna* (JB; K&A) **THE GAMBIA** WOLOF garga mbosé *O. tuna* (DF) **SIERRA LEONE** BULOM (Kim) beakogan (FCD) BULOM (Sherbro) boktampel-lɛ (FCD) FULA-PULAAR (Sierra Leone) malentaŋ (FCD) GOLA g-banda-gbila (FCD) KISSI nyɛŋgeõ (FCD) KONO bandabia (FCD) bandabila (FCD) KRIO kaktos = *cactus* (FCD) MENDE ngele-gba (*def.*-i) (FCD) ngele-gonu (FCD) VAI g-banda-gbila (FCD) **GHANA** AKAN-FANTE nkantonsoɛ (FRI) nketefuw *O. vulgaris Mill.* (Andoh) nkɛtsefuw (FRI) TWI akraate (FRI) GA agbamu *spp. with an edible fruit* (KD) á'klatè (FRI; KD) **NIGERIA** FULA-FULFULDE (Nigeria) paɗe annabiijo = *the Prophet's shoes; O.? tuna* (JMD) YORUBA oro agogo agogo, *a bell* (Burton)

An American genus of succulent plants, perennial, stems flattened and very spiny, to 1.5 m high, occurring commonly in drier areas from Senegal to Nigeria. The specific identity of the common *Opuntia* in W Africa is not clear. *O. tuna* Mill. is recorded for Senegal (1) and *O. vulgaris* Mill. in Ghana. It is probable that the species most widespread and naturalized is *O. tuna,* but others have been introduced and may have become established in limited places. Prickly pears are particularly common along the sandy coastal area of Senegal and Ghana. They are able to invade land and in many parts of the world have been a noxious weed. The usefulness of grazing land can be destroyed and the plant is very difficult to eradicate. The effective curbing of the plant in Australia by the introduction of a moth whose larvae feed on the plant's vascular system is a classical example of biological control.

The plant is able to form dense impenetrable thickets. It is easy to propagate and quick to grow. Further, it is commonly used in West Africa, as in other parts of the world, as a barrier-plant. *O. tuna* is grown in Senegal as a living hedge (1, 6). In Ghana, *Opuntia sp.* is often used for fences (*aklati-afaba,* Ga), especially for cattle enclosures, and pieces of the plant are put outside houses as 'burglar alarms', (5) presumably where a barefooted intruder may carelessly step on one. It is now planted throughout Yorubaland in Nigeria for protection (2). It is said that Tippoo Sahib, 1753–99, Sultan of Mysore in India, during the British occupation of Mysore State in 1799 strengthened the defences of Seringapatam with plantings of *O. dillenii* Haw. in depth around his fortifications (4).

The fleshy stems have been fed to stock in S Africa, Australia and U.S.A. and are considered a useful famine-food for stock, supplying especially a source of water. Nutrient value is low, and a large consumption may cause scouring. Calcium oxalate and calcium malate have both been found in *O. tuna* but the amount seems to be variable between origins of the material and the seasons. 'Sore mouth' may arise in sheep due to the thorns and the presence of minute barbed needles, or glochids, in the leaf (7). Burning has been practised in Australia to remove the thorns prior to turning the plant into silage.

The fleshy stems are pounded and made into a poultice for application on tumours or for rubbing on areas of rheumatism in Senegal (6), and mixed with chewed palm-kernels and applied to maturate whitlows in Ghana (5). In S Africa the 'leaves' are used to poultice ulcers, sores, boils, etc., and in Transvaal a boiled leaf-decoction with sugar is taken as a remedy for whooping-cough. A leaf-decoction administered for diabetes in S Africa and in Australia is said to have given some relief but no cure (7).

In Ethiopia it has been found that stems boiled and then dried are a good medium on which to grow epiphytic orchids (3).

Prickly pear stems have been considered as a source of fibre. One hundred tons of fresh material is estimated to yield 1 ton of paper-pulp (7). The stem yields a sort of gum.

The flowers are yellow and pink and are showy. The fruit is pear-shaped and covered with short spines, hence the English name. They are eaten in many countries, and can be made into relishes, preserves and fermented drinks. The fruit is astringent and eaten in quantity is constipating and in great excess may cause serious bowel obstruction and haematuria. There is no fat present, and the sugars are glucose and fructose, and no sucrose. Ascorbic acid is present (5–7). The fruit-juice contains a natural yeast, *Saccharomyces opuntiae,* and this is responsible for spontaneous fermentation of the ripe fruit. The yeast is unable to ferment sucrose, only glucose and fructose. In S Africa a beer is prepared from the fruit. Production of alcohol is possible but is evidently not economic (7).

The fruit-juice is reported as used as an ingredient of red ink (5).

References:

1. Berhaut, 1967. 2. Burton 22-3-1950, in litt. K. 3. Getahun, 1975. 4. Howes, 1946. 5. Irvine, 1961: 90, 91; fruit analysis, lxxx. 6. Kerharo & Adam, 1974: 254–5, with phytochemistry and pharmacognosy. 7. Watt & Breyer-Brandwijk, 1962: 154–7, with references.

**Rhipsalis baccifera** (J. S. Mill.) Stearn

FWTA, ed. 2, 1: 761, and 1: 221 as *R. cassutha* Gaertn.
West African: **NIGERIA** YORUBA igilligilli (AHU)

A fleshy, articulated shrub, epiphytic, or growing over rocks, recorded from Sierra Leone to W Cameroon and Fernando Po, and widely distributed by man in the tropics from the New World.

## CAMPANULACEAE

**Wahlenbergia lobelioides** (Linn. f.) A. DC. **ssp. riparia** (A. DC.) Thulin

FWTA, ed. 2, 2: 309, as *W. riparia* A. DC.

An erect annual herb to 60 cm high, of moist sites in the savanna of the northern part of the Region from Senegal to N Nigeria and occurring across to NE Africa.

In Sudan, where in places it may become rather abundant, subspecies *nutabunda* (Guss.) Murb. provides fodder for goats and sheep (1).

Reference:

1. Mohd. Eff Ismail A 3460, A 3510, A 3532, K.

**Wahlenbergia perrottetii** (A. DC.) Thulin

FWTA, ed. 2, 2: 311, as *Cephalostigma perrottetii* A. DC. UPWTA, ed. 1, 121, as *C. perrottetii* A. DC.

West African: **SENEGAL** MANDING-BAMBARA fi ni ségé (JB) **NIGERIA** YORUBA ìgbálẹ̀ ọ̀dàn (Dawodu)

An annual herb, variable from a few cm to 60 cm high, of grassland, woodland, waste places and weed of cultivation on sandy soils, recorded commonly across the Region from Senegal to W Cameroons and occurring in Central, NE and E Africa, Comoro Island and also in S America.

The plant is rubbed on the limbs in S Nigeria to relieve pain (3) and has other unspecified medicinal uses (2). A trace of alkaloid has been detected in the plant [? leaves] (1).

References:

1. Adegoke al. 1968: as *C. perrottetii*. 2. Dalziel, 1937. 3. Thomas, N. W. 1843 (Nig. Ser.), K.

# CANNABACEAE

**Cannabis sativa** Linn.

FWTA, ed. 2, 1: 623.

English: hemp; Indian hemp; the resin and preparations: hashish; marihuana.

French: chanvre indien; 'tabac Congo' (Gabon, Walker); the resin and preparations: hachich.

Portuguese: cañhamo (Feijão).

Others: bhang (India, E Africa); dagga (S Africa); ganja (India); hashish (Middle East, Egypt); kif (Morocco, N Africa), etc.

West African: **SENEGAL** WOLOF yâmba (JB; K&A) **SIERRA LEONE** KONO yambɛ (FCD) KRIO diamba (FCD) LOKO yamba (FCD) MENDE jamba (FCD) SUSU-DYALONKE yamba-na (FCD) **MALI** ARABIC el kerneb (RM) **NIGERIA** YORUBA igbó = *bush, forest* igbó *barking* (Verger)

An annual herb attaining as much as 5 m tall, native of temperate central western Asia, and of very ancient cultivation in Asia and the Mediterranean region. Now dispersed by man to very many countries, and cultivated, often illegally, including occasionally the territories of the West African region.

The plant has three economic uses: for fibre, for seed-oil, and for its narcotic resin.

Cultivation of the plant for its fibre is carried out in certain temperate countries and the fibre is extracted by retting. Under the best husbandry the fibre is of high quality and usable for rope, sacking, canvas, etc. There is no record of its production, nor its use in the West African region. A mild humid climate with a growing season temperature of 16–26° C is required.

Cultivation of hemp for its narcotic resin requires dry tropical conditions with higher temperatures than for fibre. The preparations of the plant are: (A) the dried tops of the female plant harvested before the seeds have developed and from which no resin has been removed. This is official 'Cannabis indica' of various pharmacopoeas. Hashish is another name for it. Dross from this carrying more resin and therefore more potency is separated as 'chur.' (B) Resin removed by various means is 'charas', and (C) leaves of either male or female plants dried without any particular care is 'bhang' and is the lowest quality. Official use as a sedative and hypnotic in insomnia and in depressive mental conditions, is now bypassed and the principal use is in veterinary medicine as an analgesic and hypnotic for horses (4).

Unofficial use is usually by smoking 'reefers' of the leaf alone or mixed with tobacco or by use of a special 'hookah' pipe. The effect is to produce a sort of intoxication with hallucinations and a feeling of unreality and complete detachment, sometimes with delirium. Criminal use by dacoits to stupefy their victims is recorded. Yorubas have an incantation to the plant (*igbó,* barking) for the smoke to make their enemies mad (6). The habitual taker is likely eventually to suffer mental and nervous deterioration, and ultimately insanity. The plant has extensive ritual use in many countries (8).

The active principles of cannabis and cannabis resin are *cannabidiol, tetrahydrocannabinol* and *cannabinol.* The relative amounts appear to vary with geographic origin, and the amount of the second named which is the most active is in correlation with climatic factors such as sunshine, temperature and rainfall. Samples from any one country in any one year tend to be alike, but it is not possible to be certain of the country of origin (2).

The seed-oil, known as hemp-oil, is a greenish yellow limpid liquid, with good drying properties (Iodine value 150–166) making it suitable for paint manufacture and as a substitute or adulterant for linseed-oil. It is a good illuminant, and produces a good quality soap. The race of *C. sativa* which gives rise to hemp-oil is not that yielding hashish. The seed however is probably not innocent of narcotic properties. It is said to produce a cake good for cattle-feed, but too frequent use requires some caution. Galen, AD 130–c.200., the Greek physicist, has recorded that it was the done thing in high society to offer seeds to one's guests at the start of a party to promote hilarity.

The Sanskritic names *indrâçana,* food of Indra, and *gadjâçana,* food of the elephant, do not appear to have passed into modern usage. Bhang and gañjá are both present-day Indian and have entered into general use. The Wolof *yâmba* has been postulated as of Brazilian origin (3), but is more likely to be a derivation of ganja as are the other names recorded for Sierra Leone. The plant is Asian, not of the New World. Another point of philological interest is the origin of the English word 'assassin', coming from the olden Arabic *hashshashin,* i.e. a hashish-eater, applied to the official executioner who thus intoxicated himself in preparing to despatch a king or public person, and hence one undertaking death by treacherous violence (*OED*).

References:

1. Burkill, 1935: 437–41. 2. Jenkins & Patterson, 1973. 3. Kerharo & Adam, 1974: 311–2, with phytochemistry and pharmacology. 4. Oliver, 1960: 5, 51. 5. Sastri, 1950: 58–64. 6. Verger, 1967: no.110. 7. Watt & Breyer-Brandwijk, 1962: 759, with clinical physiology and pharmacology. 8. Emboden, 1972.

## CANNACEAE

**Canna edulis** Ker.-Gawl.

English: African arrowroot; edible canna; purple arrowroot; Queensland arrowroot; Sierra Leone arrowroot.

French: tous les mois (the flour, Caribbean).

A herbaceous plant with purple stems in clumps 1–2.50 m high with terminal red or orange-red inflorescences, from a lumpy rhizome, leaves purple on the underside or with a purple border, of S America, and widely distributed by man.

The root is starchy, yielding about 24% starch which is the commercial source of a form of arrowroot that is difficult to distinguish from true arrowroot of *Maranta* (Marantaceae). The starch is easily digested, soluble in water and is an excellent invalid or infant food. It is extracted by rasping, washing and decanting. It has been known in cultivation in S America from prehistory and samples have been identified in archeological sites dated c. 2500 B.C.

Recent taxonomic studies have sunk this species into *C. indica* Linn. (4, 5). It must appear therefore that *C. edulis sensu stricto* is an ennobled strain of very great antiquity selected on the basis of palatability and digestibility of its flour. The modern plant has been commercially exploited in the Caribbean, Queensland, Sierra Leone, Hawaii and elsewhere, yet nowhere does it seem to have become an economically viable crop. It is variable with a wide range of vegetative characters, a feature which should lend itself to breeding and yet further selection, and cloning of superior strains.

The root and tops are a good cattle food which can be given raw, or better still after cooking. Harvesting has been carried out in Hawaii after four months, but in general it is an eight months crop. The rhizome tip is usable as planting material while the older portion goes into the harvest.

References:

1. Burkill, 1935. 435–6, as *C. edulis* Ker. Gawl. 2. Purseglove, 1972: I: 93, as *C. edulis* Ker. Gawl. 3. Richardson & Smith, 1972: as *C. indica* Linn. 4. Sastri, 1950: 158, as C. edulis Ker. Gawl. 5. Segeren & Maas, 1971: as *C. indica* Linn.

## **Canna indica** Linn.

FWTA, ed. 2, 3: 79. UPWTA, ed. 1, 474, as *C. bidentata* Bertolini.

English: Indian shot; wild canna lily (Morton).

French: balisier (from *ballieri,* Carib.); faux sucrier (false sugarcane).

Portuguese: erva-conteira; c.vars: cana-de-jardim, cana-florífera (Feijão).

West African: **SENEGAL** BASARI ɔ-ngànɛ́r (Ferry) BEDIK ma-ngalàl (Ferry) KONYAGI u-ngwar (Ferry) **SIERRA LEONE** BULOM (Kim) k-pɔwe (FCD) yumɛn (FCD) BULOM (Sherbro) k-pɔklɔ-lɛ (FCD) FULA-PULAAR (Sierra Leone) fɔrɔndɔ (FCD) KISSI g-bɔlɔ̃ (FCD) g-bɔlɔ̃kpɔ̃-siaŋa (FCD) KONO g-bɔɛ (FCD) g-bɔ-mɛsɛ (FCD) g-bɔyɛ (FCD) KRIO ɔkpɔlɔ-bid = *okpollo bead,* = *frog's bead; from Yoruba,* ọ̀pọ̀lọ̀: *frog* (FCD; JMD) LOKO fɔrɔ (NWT; FCD) tɔmbega (FCD) MANDING-MANDINKA fɔrɔndɔ-jɔ̃ (FCD) MENDE fɔlo (*def.*-fɔli) fɔlɔ-gbama, giema-vɛlɔ, tomboya-vɛlɔ (FCD) ngiye ma vɔlɔ (FCD) tombo ya vɔlɔ (FCD) SUSU bɔhɔri (FCD) fɔrɔn-dai (FCD) SUSU-DYALONKE g-bɛgbɛ-khɔkhɔri-na (FCD) g-bɛgbɛ-na (FCD) TEMNE *an*-pɔlɔ (JMD; FCD) VAI fɔle (FCD) **IVORY COAST** AKAN-ASANTE akondogu (B&D) BRONG amboron dobia (K&B) AKYE anguè-bun (A&AA) BAULE woko (B&D) GAGU bakoré (K&B) KRU-BETE baore (K&B) GUERE (Chiehn) bauré (K&B; B&D) bobo (B&D) n-goko (B&D) KULANGO kulia gbé (K&B) KWENI torowuin (B&D) KYAMA indiré (B&D) MANDING-MANINKA bambru (B&D) nigoko (B&D) **GHANA** ADANGME blɛfo-tobi = *European's rosary* (FRI) ADANGME-KROBO blɛfo-tobi = *European's rosary* (FRI) AKAN-ASANTE aburobia (FRI) ahabia (FRI) TWI aburobia (FRI) GA ahabía (FRI; KD) GBE-VHE tovi aku = *European's rosary,* tovi: *a rosary;* aku: *European* (FRI) **NIGERIA** EDO èbèsä̀lébō (JMD) EFIK ń̄kwà édọ̀ñ (RFGA; Lowe) HAUSA baƙaleƙale (JMD) baƙalele (ZOG) báƙáreƙáre (JMD; ZOG) gwangwama (auctt.) IGBO (Umuahia) abereka mwọ (JMD) ágịdị *the seed* (KW) YORUBA ìdò (JMD; Verger) ìdòrò *the seeds* (JMD) **WEST CAMEROONS** DUALA èsìmà (Ithmann)

A rhizomatous herb to about 1.30 high, with scarlet or orange-red flowers, native of tropical America, but naturalized throughout tropical Asia and Africa, and found in the W African region mainly in the forest belt near habitations.

The latin specific name belies the plant's origin, but the fact of its original attribution to India indicates that it has been dispersed from its native home for so long that its true origin had been forgotten.

The leaves are washed and used in water as a cure for fever in Nigeria (3) and in Gabon, expecially for children (17, 18). In Ghana they are pounded and put into baths for fever (8, 10). For jaundice the Brong of Ivory Coast take a leaf-macerate in draught and in eye-instillations (12) and the Akye add the pounded leaves to a prescription with other drug-plants for taking by draught and as a wash (2). The tender shoots are applied to bruises and cuts in Nigeria (3). The stems produce an emollient and analgesic action, and this is made use of in Ivory Coast to assuage rheumatic pains, buboes, urethritis and even fractures, and for coughs, fevers and jaundice (6, 12). In Congo a tisane is given to children to sooth paroxysmal coughing in whooping-cough, and the sap is applied to sores and to arrest bleeding (5). The Shien of Ivory Coast cook the stems wrapped in *Maranta* or banana leaf and apply the juice which is expressed as an embrocation for painful breasts (12). In India the stalks are chopped up and boiled in rice-water with pepper and fed to cattle as an antidote after eating poisonous grasses (15).

The leaves serve as wrapping for food in Ghana (9, 10) and doubtless elsewhere in W Africa. In India and SE Asia the leaves are commonly used to wrap parcels (7). A fibre can be extracted from the plant and is of a quality to substitute jute in the making of twine and sacking (15).

The roots are starchy. Starch has been extracted in a small way in Indochina (7). They are eaten in Asia (7), and have been eaten in W Africa in time of dearth (8, 100). In parts of Kenya the root (4) and in Malawi the whole plant (14) is cultivated as a cattle-food. More generally the roots have medicinal applications. The powdered root is taken in Nigeria as a cure for diarrhoea and dysentery (3). In Gabon the rhizome is used in enemas against dysentery and intestinal worms (17, 18), and an aqueous decoction is taken in Congo by women with irregular menses (5). In India the roots are recognized as diaphoretic and diuretic and are administered in fevers and dropsy (15).

The flowers are said in Ghana to be good for curing eye-disease. They contain a little sweet nectar which is used as a bait to trap birds (10).

The seeds are black, hard and the size of a pea. The English name, 'Indian shot', derives from their occasional use in India as shot for guns (19). In Ghana children use them in popguns (9, 10). Throughout Africa and Asia they are used as beads for stringing into necklaces and rosaries, and making into rattles. Several Ghanaian names refer to 'European's rosary' indicating an exotic origin. The seeds are used in S Nigeria as counters in a game of chance called *ido,* the name being taken from the Yoruba name of the plant *ido* or *idora*. The looser of the game acquires the title *ọmọ-odobo,* lit. 'awkward child' (13). No medicinal usage of the seeds is recorded for the Region. In SE Asia they may be pounded to a paste for poulticing headaches (7). A trace of alkaloid has been reported in Nigerian material (1). The seeds also yield an attractive evanescent purple dye (15).

The plant enters into a Yoruba invocation for protection against wizards and witches who are said not to eat *ido,* and to help little children to stand (16). A purple-leafed form is used in ordeal trials in Gabon in cases of alleged adultery (18).

References:

1. Adegoke & al., 1968: as *C. bidentata*. 2. Adjanhoun & Aké Assi, 1972: 78. 3. Ainslie, 1937: sp. no. 69. 4. Bally 7736, K. 5. Bouquet, 1969: 78, as *C. bidentata* Bertolini. 6. Bouquet & Debray, 1974: 62 as *C. bidentata*. 7. Burkill, 1935: 438, as *C. orientalis* Rosc. 8. Dalziel, 1937. 9. Irvine 58, K. 10. Irvine, 1930: 76–77. 11. Irvine, 1952, a: 28 as *C. indica* and *C. bidentata*. 12. Kerharo & Bouquet, 1950: 243 as *C. bidentata* Bertolini. 13. Newberry, 1938, a. 14. Rabson Phiri 150, K. 15. Sastri, 1950: 58, as *C. orientalis* Rosc. 16. Verger, 1967: no. 107, as *C. bidentata* Bertolini. 17. Walker, 1953, a: 26. 18. Walker & Sillans, 1961: 116. 19. Watt, G., 1889–93: 2: 102.

**Canna hybrids**

The common horticultural canna is an adjunct of virtually all tropical gardens, and is to be found also in subtropical, and even in temperate countries. Several species, robust, tall erect, all native of the New World, have long been known in cultivation for their attractive flowers and bronzed foliage. Hybridization was attempted in Europe in the middle of the Nineteenth Century, and in 1862 a dwarfing mutation, to 1.5 m high, or less appeared which has been fixed in all the innumerable hybrids now available with white, lemon, pink, red and purplish flowers and variegated, green and bronzed foliage.

# CAPPARACEAE

**Boscia angustifolia** A. Rich.

FWTA, ed. 2, 1: 93. UPWTA, ed. 1, 18.

West African: **SENEGAL** BASARI a-mìɗ ɔ-ráw̃ (Ferry) BEDIK ga-mẹnɛ-mẹnɛ (Ferry) FULA-PULAAR (Senegal) kiréwi (K&A) tirewi (K&A) tirey (K&A) MANDING-BAMBARA béréde (JB, ex K&A) MANINKA ɗâba ginaɗu (Aub.; K&A) ginégu (Aub.; K&A) somô késébéré (Aub.; K&A) ťékoni kolo (Aub.; K&A) tutiġi (Aub.; K&A) SERER ndeyis (JB; K&A) SERER-NON (Nyominka) dâbâ dâbâ (K&A) isus (K&A) sus (K&A) WOLOF nos (auctt.) **THE GAMBIA** WOLOF nos *pronounced nasally* (DF) **GUINEA** MANDING-MANINKA diaba guinadiu (Aub.) guineguiu (Aub.) somon késébéré (Aub.) tiekoni kolo (Aub.) tutigui (Aub.) **MALI** MANDING-MANINKA diaba guinadiu (Aub.) guineguiu (Aub.) somon késébéré (Aub.) tiékoni kolo (Aub.) tutigui (Aub.) SONGHAI hassu (Aub.) TAMACHEK tadent (JMD) **NIGER** FULA-FULFULDE (Niger) gigile, jigile (PF) HAUSA agahini (Aub.) SONGHAI sáncíligà (*pl.* -à) (D&C) TAMACHEK tadent (JMD) **NIGERIA** FULA-FULFULDE (Nigeria) anzagi *from Hausa* (JMD; KO&S) HAUSA dilo (JMD) 'false' anza *in distinction from 'anza', B. senegalensis* (KO&S) hamza (ZOG) hanza (MM)

A shrub or small tree to 6 m high with a contorted fluted bole, of the dry savanna of the northern Region, often on termite mounds, from Senegal to Niger and N Nigeria, and across Africa to Sudan, Ethiopia, East and South Tropical Africa and Arabia.

The wood is hard and is used for carpentry in Ghana (10) and in Sudan (3). In Kenya it is used to make playing boards for the game of *bao* (Swahili) (8, 14). This game is played in numberless variations throughout Africa and the countries of the Indic basin (9) and is known as *warri* or *ayo* in W. Africa. In Kenya pieces of wood are boiled and used to sweeten milk (7).

The foliage is grazed by all stock but perhaps only in default of grass though wild game appears to relish it. Nevertheless the leaves cut up small are fed in a nose-bag as a strengthening food to horses and camels in poor condition or suffering from mucal diarrhoea, etc.; similarly berries crushed in water are given to cause purging (6). In Senegal, Nyominka use the leaves as a cholagogue (11, 13).

The bark, after the removal of the cuticle, is eaten as a supplementary food pounded up with cereals in the same way as *Cadaba farinosa* (Hausa, *bagayi*), hence the expression *anza rashin bagayi* (i.e. 'for lack of *bagayi*') is applied, or the bark may be pounded up and put into soup (6). Some races of Senegal apply the dried bark burnt over live embers as a fumigation for ophthalmias and sometimes for neuralgia (12, 13). Similarly the tree in Katanga is known as 'tree of the head' in the Basanga dialect for its use in treating headache (4). The bark is also used in Senegal to treat stiffness of the neck and kidney pains and in poultice for swollen feet (13).

The berries are edible but bitterish, and the seeds are eaten cooked (5, 6). The fruit are reported eaten in Kenya (1), and taken by game in Tanganyika (2).

References:

1. Bogdan AB 4099, K. 2. Brooks 81, K. 3. Burnett 3, K. 4. Burtt-Davy 17880, K. 5. Busson, 1965: 193. 6. Dalziel, 1937. 7. Gillett 13442, K. 8. Glover & al., 342, K. 9. Hall, 1953. 10. Irvine, 1961: 42. 11. Kerharo, & Adam, 1964, c: 295. 12. Kerharo & Adam, 1964, b: 414. 13. Kerharo & Adam, 1974: 313. 14. Rammell 3484, K.

**Boscia salicifolia** Oliv.

FWTA, ed. 2, 1: 93. UPWTA, ed. 1, 18–19.

West African: **SENEGAL** FULA-PULAAR (Senegal) tientirgaye (Aub.) tiréï (Aub.) **THE GAMBIA** FULA-PULAAR (The Gambia) ansayi (DAP) WOLOF jendum (DAP) **MALI** FULA-PULAAR (Mali) tientirgaye (Aub.) tiréï (Aub.) SONGHAI sáncíligà TAMACHEK kitshagass (Aub.) **NIGER** FULA-FULFULDE (Niger) tientirgaye (Aub.) tiréï (Aub.) HAUSA zuré (Aub.) TAMACHEK kitshagass (Aub.) **NIGERIA** BOLE zande (Gilman; AST) HAUSA cankas, caskas *the fruit* (JMD; ZOG) legel (Anyadiegwe) mandingiji *the leaves and flowers* (Bargery) zande (AST) zure (auctt.)

A tree with drooping salix-like foliage attaining 13 m height, of the dry savanna zone and commonly in association with termite mounds (4, 8, 9), occurring from Senegal to Niger and N Nigeria, and extending across Africa to the Red Sea and in E Africa southwards to Bechuanaland.

The leaf is sometimes eaten in N Nigeria (4, 14), and it is an ingredient of soups in Mali and Niger (3). In cultivated areas it is conserved as an emergency food (8) and as a forage crop (9). In Tanganyika some peoples use the leaf as a vegetable (2).

In N Rhodesia the leaf is a remedy for a malady called 'chiufa', the symptoms of which include an inflamed and gaping anus, the powdered leaf being blown into the anus of the inverted patient. The treatment is said to be successful (14). In Tanganyika the leaf is given to cattle with fever (1, 14), leaf-sap is dripped into a fresh wound as a dressing and pounded leaves are applied to maturate abscesses (7).

The bark is pounded and put into soup in N Nigeria (4, 12). Pounded bark in Tanganyika is made into an infusion used to irrigate the eyes for conjunctivitis, or a steam-bath of the same can be used, and a root and bark-decoction is drunk as an aphrodisiac (7).

In the Zambesi area the root is said to be soft and starchy and is roasted and eaten by the people (10), though in Tanganyika the root is regarded as toxic (11) and in Northern Rhodesia its consumption is solely as a famine-food (13). In Kenya the root is added to a special brew of beer with other drug-plants which is taken for rheumatism (5).

The seeds are eaten after cooking, not raw, in Uganda (6).

References:

1. Bally, 1937. 2. Bullock 3442, K. 3. Busson, 1965: 193, 202–3 with leaf analysis. 4. Dalziel, 1937. 5. Delap s.n., 25/11/1960, K. 6. Dyson-Hudson 424, K. 7. Haerdi, 1964: 73. 8. Irvine, 1961: 42 with chemical analyses. 9. Keay al. 1960: 77. 10. Kirk, s.n., (Zambesi Expedition) 1858, K. 11. Koritschoner 1456, K. 12. Thornewill 92, K. 13. Trapnell 1658, K. 14. Watt & Breyer-Brandwijk, 1962: 160.

**Boscia senegalensis** (Pers.) Lam.

FWTA, ed. 2, 1: 93–95. UPWTA, ed. 1, 19.

West African: **MAURITANIA** ARABIC (Hassaniya) aizen (AN) ayzin (Kesby) bokkhelli *the fruits* (AN) ARABIC ('Maure') aïzen (Aub.) mandiarha (Aub.) **SENEGAL** VULGAR ńdiandam *from Wolof* ARABIC ('Maure') aïzen (Aub.) mandiarha (Aub.) BASARI a-miɗ ɔ-ráw a-tyán (Ferry) BEDIK ga-mẹnɛ-mẹnɛ (Ferry) FULA-PULAAR (Senegal) gidili (K&A) gigilé (K&A) giǰili (auctt.) gisili (K&A) MANDING-BAMBARA béréfin (JB; K&A) SERER mbańa (JB, ex K&A) SERER-NON bagnan (JMD) bagnan (JMD) bańâ (auctt.) bańâ (Sébire; AS, ex K&A) SONINKE-SARAKOLE gigilé (K&A) WOLOF ɗâda (K&A) ɗâdum (K&A) mbagno (AS) mbum guelem = *mbum* (*Moringa oleifera*) *of honey* (AS; DF) ńâdam (K&A) nâdom (K&A) ńdadam (K&A) n-diandam (auctt.) nos (A. Chev.) **MALI** FULA-PULAAR (Mali) guidjili (Aub.) n-guiguilé (Aub.) MANDING-BAMBARA béré (auctt.) SONGHAI hóoréy (auctt.) orha (Aub.) TAMACHEK sihir (Aub.) tadahant (Aub.) tadent (A. Chev.; Aub.) tadomet (A. Chev.; Aub.) **IVORY COAST** MOORE nabédéga (Aub.) **GHANA** HAUSA dila (FRI) **NIGER** FULA-FULFULDE (Niger) guidjili (Aub.) n-guiguilé (Aub.) HAUSA anza (Aub.) damdam shiya (ZOG) damdamshiya (JMD) dilo *the fruit* (Aub.) KANURI bûltù (Aub.) SONGHAI hóoréy (*pl.* -à) (D&C) TAMACHEK sihir (Aub.) tadahant (Aub.) tadent (A. Chev.; Aub.) tadomet (A. Chev.; Aub.) TUBU modu (Aub.) TUBU (Kaningou) bultu (Aub.) **NIGERIA** ARABIC kursan *the fruit* (JMD) mokheit *the plant* (JMD) ARABIC-SHUWA makheit (JMD) mikheid (JMD) FULA-FULFULDE (Nigeria) anzagi *from Hausa* (JMD) gigile, jigile (PF) HAUSA anza (KO&S; MM) ɗan loma (JMD) danloma (ZOG) dilo (F&M) hamza, lilo (ZOG) hanza (KO&S; ZOG) legel (J) HAUSA (East) bilo (JMD) zeyi (JMD) HAUSA (West) dulunhwa *the fruit* (JMD; ZOG) KANURI bûltù (JMD)

A shrub to 3 m high, often branching from the base, often in thickets, frequently on abandoned termite mounds (18) and on barren and fire-scorched soil of the Sahel and hence more northerly and in drier conditions than *B.*

*angustifolia* A. Rich., distributed from Mauritania to Niger, N Nigeria and NW Cameroons, and across Africa to Sudan and Ethiopia.

The wood when large enough is used for hut-construction in Mauritania (16). As a firewood it burns making much smoke (14).

The leaves have an obnoxious smell but they and the fruit are sold in northern markets for use as human food, cooked and put in soup, or mixed with cereal pap or *couscous,* etc., or for stock (6, 7). Stock will browse the foliage if nothing better is available (1, 16) especially towards the end of the dry season when it appears to assume some importance (8).

The leaves are widely used in Senegal for intestinal troubles, especially colics, the treatment causing no purging (11, 12), and together with other drug-plants are given for serious jaundice and haematuria. A Wolof treatment for stomach complaints is to pick the leaves at dawn, sun-dry them and let them soak with some salt in curdled milk. This is taken first thing in the morning on an empty stomach giving rapid relief without causing diarrhoea or diuresis (12). In a comparable treatment practised at Richard-Toll in NW Senegal, a pap of leaves with millet flour is taken on an empty stomach each morning to treat for worms (5). Leaves in infusion are used in Sudan as an eyewash (19), while in Senegal powdered leaves in decoction are used in instillation for pruritus of the eye arising from syphilitic complications, and also a draught for schistosomiasis and certain mental disorders (12). At Richard-Toll, the leaves with those of *Salvadora persica* Linn. (Salvadoraceae), are eaten as a powder for piles (5). In Mauritania the fresh leaves are cooked in butter and then applied to guinea-worm sores, and fresh uncooked leaves are placed over swellings and left *in situ* for the whole of a day (16) and in N Nigeria dried powdered leaves, or fruit, are applied topically to chronic ulcers, swellings, etc. (7).

In veterinary medicine horses in the Richard-Toll area are treated for colds with an inhalation of the powdered leaf (5).

In Senegal the roots and bark are equally valued as are the leaves for treating intestinal troubles (12).

The fruit is edible, and like the leaves is sold in northern markets for human and stock consumption (6, 7). In the driest sub-Saharan and Sahel zones they may be eaten in season as a regular item of food. Further south the fruit may be a casual supplement, for example, boys in Jebel Marra are reported to eat them (9). But in the main the fruit constitutes a valuable emergency food in time of dearth and famine (3, 7, 10, 17). In the serious drought and famine in Kordofan, Sudan, in 1900, many people are said to have subsisted on them (15). Some reports say that the fruit pulp is sweet (4, 14): another refers to its bitterness (17): others make no observation. At the Pass of Amojjar in Mauritania the fruit is boiled like a vegetable for consumption (13), but other observers state that before it can be eaten a lengthy preparation is required lasting 3–4 days either of boiling or maceration with changes of water (10, 17). This it seems is to remove any bitterness the fruit pulp may have. The fruit is fermented in Sudan to make a beer (2).

The fruit is an ingredient in the dietetic and medicinal treatment of syphilis in N Nigeria and pounded in water it is given to camels as a purge to deplete [sic] the blood at the end of the rains (7).

The seed is edible after steeping in a little water for a week. They are eaten, cooked like haricot beans in Mauritania (16). The roasted seed has been used as a coffee-substitute (7). Oil-content of the seed is less than 1% (12).

References:

1. Adam, 1966, a. 2. Andrews A 758, K. 3. Anon, 1946, b: 153. 4. Berhaut, 1967: 251. 5. Boury, 1962: 15. 6. Busson, 1965: 193. 7. Dalziel, 1937. 8. Foster & Munday, 1961: 316. 9. Hunting Technical Surveys, 1968. 10. Irvine, 1961: 42–44. 11. Kerharo & Adam, 1964, b: 414. 12. Kerharo & Adam, 1974: 314–5 with seed oil analysis. 13. Kesby 43, K. 14. Monteil, 1953: 65. 15. Murid L/170, K. 16. Naegelé, 1958, b: 880. 17. Portères, s.d. 18. Trochain, 1940: 205. 19. Broun & Massey, 1929: 60, as *B. octandra* Hochst.

**Buchholzia coriacea** Engl.

FWTA, ed. 2, 1: 93. UPWTA, ed. 1, 19.

English: musk tree (Unwin 411)

French: kola pimenté, oignon de gorille (Gabon, Walker).

West African: **SIERRA LEONE** MENDE ndo (*def*.-i) (FCD) **LIBERIA** KRU-BASA doe-fiah (C) **IVORY COAST** ABE amo(n) (auctt.) ABURE até (JMD; K&B) AKYE abo (auctt.) mon (A&AA; JMD) ANYI amizi (auctt.) AVIKAM abazi (JMD; K&B) GAGU muin oroko (K&B) KRU-GREBO (Trepo) do (A. Chev., ex K&B; JMD) GUERE saagnan (K&B) saïno (K&B) GUERE (Chiehn) lébé (K&B) GUERE (Wobe) duhutu (A&AA) KWENI grogolégoné (B&D) KYAMA amoulenya (B&D) 'KRU' dantu (Aub.; RS) duétu (K&B) 'SOUBRE' brachi (auctt.) **GHANA** AKAN-ASANTE ɛson-besẽ = *elephant's kola* (BD&H, Enti) ɛsono-bise (FRI) osono-bese (DF) ANYI amigi (FRI) **NIGERIA** EDO ówì (JMD; KO&S) **WEST CAMEROONS** MUNGA banda (JMD)

An evergreen understorey tree of the lowland rain-forest, to 20 m high, occurring from Guinea to West Cameroons, and in E Cameroun and Gabon.

The tree is often planted around Gabon villages (12). The wood is yellowish white, soft and somewhat fibrous (2, 5, 8, 9). It is probably suitable for house-building. The closely related *B. macrophylla* Pax is used in construction work in Gabon (13).

The bark-slash is deep red, and the sap exudes with a violently spicey pungent smell that causes sneezing (2, 10). The bark is made in the Ivory Coast into a pulp for inhalation (1) or into a snuff to relieve headache, sinusitis and nasal congestion in headcolds, also otitis and ophthalmias (3, 10). A bark-decoction is applied externally as a general reconstituent (1) or as a revulsive for pains in the chest, bronchitis, pleurisy and kidney-pains (3, 10). The Gagou of Ivory Coast administer bark-sap as an enema for kidney-pains (10). The fresh bark is used for earache in Ghana (Vigne fide 8). A bark-decoction is used to wash persons with small-pox (3), and in Gabon the crushed bark is used in frictions on skin-itch (12).

The Guere of Ivory Coast incorporate the bark in arrow-poisons and it is highly effective (10).

The leaves, whole are applied to boils in Sierra Leone (6), and for a minute or two to bruised limbs, etc. as a revulsive — this treatment is said to be 'hot' (7). The leaves and fruit are pounded in Sierra Leone with white (*hoji*, Mende) clay and rubbed on the body for fever (6). In Liberia the seeds are used on skin-eruptions and internally for worms and pains (4), while crushed up they are pasted over the stomach in Ivory Coast for difficult childbirth (1). In Gabon the fruit is considered anthelmintic (12). The fruit has a disagreeable smell on cutting open. The seeds have a hot spicy flavour. The seeds are covered in a purple aril which is chewed in Ivory Coast like the kola and has a sharp pungent taste like that of *Capsicum frutescens* (Solenaceae) (11). In S Nigeria the Edo boil and eat the fruit after storage for a few days (8).

The root has no recorded usage in the Region. In Gabon it is considered poisonous (3).

A number of chemical substances has been isolated from the plant and the substances determined (3).

References:

1. Adjanohoun & Aké Assi, 1972: 79. 2. Aubréville, 1959: 1,166. 3. Bouquet & Debray, 1974: 63 (see for phytochemistry). 4. Cooper 326, K. 5. Dalziel, 1937. 6. Deighton 3914, K. 7. Deighton 5367, K. 8. Irvine, 1961: 44. 9. Keay & el. 1960: 79. 10. Kerharo & Bouquet, 1950: 28. 11. Leeuwenberg 2511, K. 12. Walker, 1953, a: 26. 13. Walker & Sillans, 1961: 117.

**Cadaba farinosa** Forssk.

FWTA, ed. 2, 1: 90. UPWTA, ed. 1, 19.

West African: **MAURITANIA** ARABIC ('Maure') azrom (Aub.) szrom (Aub.) zerom (Aub.) zrum (Aub.) **SENEGAL** ARABIC ('Maure') azrom (Aub.) szrom (Aub.) zerom (Aub.) zrum (Aub.)

FULA-PULAAR (Senegal) balamji (Aub.) sińsiń (K&A) sińsińi (K&A) têńsen (AS, ex K&A) tsinsińi (K&A) MANDING-BAMBARA dómadugu (JB, ex K&A) tomani (JMD, ex K&A) [illegible] (K&A) késébéré tamba (K&A) ńogu (Aub., ex K&A) SERER ndébaré (K&A) ndégareg (JMD; FB) ndégarèk (JB) SERER-NON gavargi (AS; K&A) gavargi (AS; K&A) WOLOF ndébarga (K&A) ndébargé (auctt.) n-débarka (auctt.) **GUINEA** FULA-PULAAR (Guinea) quinquemini (Aub.) MANDING-MANINKA bérékunan (Aub.) késébéré tamba niogu (Aub.) **MALI** ARABIC ('Maure') azrom (GR) FULA-PULAAR (Mali) balamji (Aub.) MANDING-BAMBARA to-magny (GR) to-magny (A. Chev.; FB) SONGHAI hassu ueil = *male hassu* (A. Chev.; FB) kwemkwemini (A. Chev.) uggar (FB) TAMACHEK abago (Aub.) **NIGER** FULA-FULFULDE (Niger) balamji (Aub.) HAUSA bagaïe (Aub.) KANURI gursimé (Aub.) marga (Aub.) SONGHAI hassu ueil (Aub.) uggar (Aub.) TAMACHEK abago (Aub.) TUBU harikanelifi (Aub.) harkanelifi (Aub.) **NIGERIA** ARABIC surreih (JMD) BOLE gúno (AST) FULA-FULFULDE (Nigeria) balamji (Taylor) balɗamhi (Taylor) beeliɗamhi = *plant with sweet, pleasant tasting sap* (MM) belɗamhi (MM) HAUSA bagaji *from Fula* (MM) bagayi *from Fula* (JMD; MM) dangarafa (MM) gúno (AST) hadza (ZOG) HAUSA (East) anza, hanza (JMD) handja (JMD) HAUSA (West) anza (ZOG) KANURI bûltù (JMD)

A shrub, usually much branched from the base, reaching 2–3 m high, of the Sahel and northern soudanian zone of the Region from Mauritania and Senegal to Niger and N Nigeria, and widespread across Africa, even to 1,600 m altitude, to the Red Sea and Indian Ocean seaboards and in Arabia and India. It is often found on termite mounds (9).

The leaves and young twigs are edible. In N Nigeria after pounding they are boiled into a gruel for eating (11) and also often after pounding with cereals and drying a dry brown pudding or cake is made for consumption. This appears in N Nigerian markets as dark soapy irregular pieces more or less chocolate-coloured and is called *farsa* and *balambo* in Bornu and *baleno* or *tsawa* in eastern Hausa, also *tigiraganda* (4). In Senegal and Mali the pounded leaves are cooked in *couscous* (3).

The bark alone is eaten with cereals (5) and flowers macerated in water are added as a sweetener to scones of millet flour (10).

The plant is browsed by all stock (2, 10), though horses and cattle in Senegal do not take it (1).

Leaves pounded into a paste are used in Senegal on skin-complaints, more particularly for anthrax (7) and in S Africa a root-decoction is considered an anthrax remedy (12). The plant is credited with analgesic properties. The Wolof of Senegal use the leaves (6, 7) and the Tubu of Niger make an infusion of the shoots (10) for stomach-pains. The Fula of Fouta Toro use the roots with other drug plants for rheumatism (6, 7). Ash of the plant is rubbed into the skin in Tanganyika to relieve general body-pains (12). The plant also has a reputation for the treatment of respiratory and chest-complaints and feverish conditions. A decoction of roots and leaves is used in Senegal (7) or the pounded leaves alone as a cough-medicine (10), to which millet flour may be added to make a sweetened medicine (2). The plant is also used to treat dysentery in Guinea (Pobéguin fide 10) as an antidote in food-poisoning in Egypt, Arabia and India (12), and the leaf and flower buds as a stimulant, antiscorbutic, purgative, emmenagogue, antiphlogistic and anthelmintic (especially for roundworm) in India (12). The pulverised leaves mixed with rust, or with coffee, are taken in Senegal after fasting, as an iron tonic (4). The iron content is however lower then in other Capparaceae at 125 g per 1000 g dry weight in material from Bamako, Mali (3). Copper content is markedly high at 60 g per 1000 g.

In northern Ghana, the bark yields a cordage (8).

A bitter alkaloid and two organic acids resembling cathartic acid have been detected in the leaf (12). The presence of nitrogenous bases, sterols, aliphatic alcohols and heterosides have also been reported (7).

References:

1. Adam, 1966, a. 2. Aubréville, 1950: 51. 3. Busson, 1965: 195–6 (leaf analysis given). 4. Dalziel, 1937. 5. Irvine 1952, a: 30. 6. Kerharo & Adam, 1964, b: 415–6. 7. Kerharo & Adam, 1974: 315. 8. Kitson 504, K. 9. Roberty, 1953: 449. 10. Schnell, R. 1953, d. 11. Thornewill 112, K. 12. Watt & Breyer-Brandwijk, 1962: 160.

**Cadaba glandulosa** Forssk.

FWTA, ed. 2, 1: 90.
West African: **MALI** ARABIC (Hassaniya) teheist (GR) FULA-PULAAR (Mali) hassu (Aub.) uadagoré (Aub.) SONGHAI todi farssa (Aub.) TAMACHEK tabeïbaret (Aub.) tahalist (Aub.) tarbéret (Aub.) téhanizt (Aub.) teïs (Aub.) **NIGER** TUBU doburu (Aub.)

A low much branched shrub occurring in Mali and Niger, and extending across Africa to E Africa, Somalia and Arabia.

**Capparis decidua** (Forssk.) Edgew.

FWTA, ed. 2, 1: 89. UPWTA, ed. 1, 20 as *C. decidua* Pax.
English: capers (the flower buds), salt bush, siwak tree (Lake Chad, Sikes).
French: câprier, câprier sans feuilles.
West African: ARABIC sodäd *the fruit* (JMD) tan zub (Burton) tundub (Burton) **MAURITANIA** ARABIC (Hassaniya) iguîn (Leriche & Mokhtar Ould Hamidoun) ARABIC ('Maure') agani (Aub.) aïguin (Aub.) ignïn (Kesby) iguenine (Aub.) **SENEGAL** ARABIC ('Maure') agani (Aub.) aïguin (Aub.) iguenine (Aub.) FULA-PULAAR (Senegal) bugi (K&A) gumi (K&A) gumi balévi (K&A) gurmel (K&A) gurmel balévi (K&A) WOLOF gumel (K&A) gurmel *from Pulaar* (auctt.) **GUINEA** FULA-PULAAR (Guinea) gumi danévi (Aub.) **MALI** ARABIC sakkul *the fruit* (A. Chev) ARABIC (Hassaniya) iguini (GR) FULA-PULAAR (Mali) gumi danévi (Aub.) TAMACHEK aujumgum (A. Chev.) boghelelli *the fruit* (A. Chev.) igenin (A. Chev.) iguini (A. Chev.) **NIGER** KANURI tchiarbun (Aub.) TUBU kozzom (A. Chev.) TUBU (Kaningou) kussomo (Aub.) maria (Aub.) **NIGERIA** ARABIC-SHUWA syak (Sikes) HAUSA haujarin mutane = *people's haujari* (RES) KANURI kigu (Sikes)

A much-branched bush attaining 3–4 m height, with short recurved spines, often forming dense impenetrable thickets, often leafless except for young shoots and non-flowering branchlets which are leafy only briefly in the rainy season, occurring in the Sahel area of the north of the Region from Senegal to N Nigeria, and in Mauritania and north and south of the Sahara across Africa to Arabia and India.

The wood is very hard and yellow. In Mauritania it is used to make camel-saddles (8), and in Morocco camel-saddles, frames for palanquins and porringers (9). In Sudan the larger stems are hollowed out into water-pipes (3). In the Lake Chad area the younger branches are burnt to obtain a vegetable salt (12), and in Morocco the bark-ash is used as a haemostatic (9).

The young leafless twigs furnish a fodder which is readily taken by camels (3, 8, 11), but less so by cattle (11).

The leaves or the root-bark, are made into a concoction with millet flour and taken for venereal disease discharges in the region of Richard-Toll in Senegal (2), and in the neighbouring Fouta Toro the bark is thought to be diuretic and is therefore used for the same purpose either alone or in conjunction with other drug-plants, principally *Zizyphus mucronata* Willd. (Rhammaceae) (6, 7). The root and root-bark are recognized as pungent and bitter in Iranian and Indian medicine and are given for intermittent fever and rheumatism (4), and for boils, swellings of the joints etc. The centre of the root is macerated and the liquid used as eye-drops for conjunctivitis at Richard-Toll (2).

Commercial capers are the pickled flower buds of *Capparis spinosa* Linn. and of this species, the trade being supplied mainly from Mediterranean sources. The sharp taste of capers is similar to the sharpness of various Cruciferae such as mustards and horse-radish. Their consumption is as a digestive stimulant. *Sinapin* is present. The fruits also are edible and are taken in Morocco (9), Jebel Marra (5) and Sudan (3).

Alcoholic extracts of flowers, seeds and fruits have shown strong antibiotic action on a number of bacteria and no antifungal action (7). A volatile suphuroil from the roots has a potential use in treating rheumatism and fever (10), and has shown both antibacterial and antifungal activity (7).

References:

1. Adam, 1966, a. 2. Boury, 1962: 15. 3. Dalziel, 1937. 4. Hooper, 1931: 307 under *C. spinosa* Linn. 5. Hunting Technical Surveys, 1968. 6. Kerharo & Adam, 1964, b: 417. 7. Kerharo & Adam, 1974: 316–7 with references. 8. Kesby 29, K. 9. Monteil, 1953: 66. 10. Oliver, 1960: 21. 11. Roberty, 1953: 449. 12. Sikes, 1972.

## **Capparis erythrocarpos** Isert

FWTA, ed. 2, 1: 89. UPWTA, ed. 1, 20.

West African: **GUINEA-BISSAU** BALANTA simbus (JDES; EPdS) MANKANYA binherre (JDES) PEPEL brerem-mela-n'sata (JDES) breren mela-sata (EPdS) **SIERRA LEONE** MENDE gele-yeŋgalu = *kite's claw* (FCD) **IVORY COAST** BAULE patafué (B&D) **GHANA** AKAN-ASANTE patabofuo (BD&H; FRI) woresenakyiame = *wait a bit* (FRI, ex JMD); = *when passing salute me* (FRI) FANTE ɔkyerabran = *catch giants, referring to reflexed thorns* (FRI) TWI apana (FRI) okyere-abrane (Enti) GA aŋmaŋma (FRI) kpitikpiti *a name for spiny shrubs* (FRI) GBE-VHE aŋŋɔ = *thorn* (FRI)

A very thorny, much-branched shrub with recurved hooks, or small tree reaching 6 m height, usually scandent, common in grassy savanna, occurring from Guinea to S Nigeria, and extending across Africa to Sudan, Uganda and Angola.

In Zaïre the root-bark is ground up and applied to the head for headache (5) and pounded roots are applied to the breasts over scarifications to stimulate milk-production (3). In Tanganyika the root-bark after pounding is bound on to hard pusy abscesses to hasten maturation, and it is put into a steam-vapour bath for treating conjunctivitis (2).

The fruit contains a sweetly insipid pulp (1, 4, 5) which is eaten by children in Ghana (5), and is much sought after by children and birds in Zaïre (3). The pulp attracts a species of black ant which often colonizes the fruit (4, 5).

References:

1. Burtt-Davy & Hoyle, 1937, 30. 2. Haerdi, 1964: 73. 3. Hauman & Wilczek, 1951: 468–9. 4. Irvine, 1930: 79. 5. Irvine, 1961: 46–47.

## **Capparis fascicularis** DC., **var. fascicularis**

FWTA, ed. 2, 1: 90, as *C. rothii* Oliv.

West African: **NIGERIA** HAUSA haujari (JMD) haujarin mutane *i.e., of men* (JMD)

A scrambling shrub similar to *C. sepiaria* var. *fischeri* attaining about 6–7 m height, of the savanna woodland from The Gambia to Niger and N Nigeria and extending across Africa to Ethiopia and E. Africa.

The leaves are sold as food in N Nigerian markets (1).

Reference:

1. Dalziel 139, K.

## **Capparis sepiaria** Linn. **var. fischeri** (Pax) De Wolf

FWTA, ed. 2, 1: 90, as *C. corymbosa* Lam. UPWTA, ed. 1, 19, as *C. corymbosa* Lam.

West African: **MAURITANIA** ARABIC ('Maure') bauier (Aub.) bulgui (Aub.) **SENEGAL** ARABIC ('Maure') bauier (Aub.) bulgui (Aub.) FULA-PULAAR (Senegal) gumba (Aub.) gumi (Aub.) gumi balévi (Aub.) MANDING-BAMBARA m-bukari (JB) n-bukari (Aub.) tabuti (Aub.; JB) **MALI**

FULA-PULAAR (Mali) gumba (Aub.) gumi (Aub.) gumi balévi (Aub.) MANDING-BAMBARA tabuti (Aub.) SONGHAI cobigna (Aub.) **UPPER VOLTA** MANDING-BAMBARA n-bukari (Aub.) tabuti (Aub.) MOORE gaongo (Aub.) **NIGERIA** FULA-FULFULDE (Nigeria) gorko nyangudoohi = *tree of the sarcastic man* (J&D; MM) HAUSA haujari (JMD) haujarin mutane *i.e., of men* (JMD; MM)

A scrambling prickly shrub or liane of the Sahel savanna from Senegal to Niger and N Nigeria and extending across Africa to Sudan and E Africa.

The leaves are said to be edible to man, being put into soups (3, 4), but graziers consider the plant toxic to stock (2) which will not graze it (1).

The root is said to be poisonous (Sébire fide 5) and a root-flour is prepared by the Fula of N Nigeria to use as a hunting poison (6).

The spherical yellowish-orange fruit is edible and is eaten by hunters when out on the chase (4). The flesh has the flavour of ether and is considered to be aphrodisiac. The skin is very bitter (2).

References (as *C. corymbosa* Lam.):

1. Adam, 1966, a. 2. Aubréville, 1950: 50. 3. Busson, 1962: 196, 202–3 with leaf analysis. 4. Dalziel, 1937. 5. Irvine, 1961: 46. 6. Jackson, 1973.

**Capparis spinosa** Linn.

English: capers.

French: câprier.

Portuguese: alcaparra (Feijão).

West African: **MALI** TAMACHEK qabbar (RM)

A bush essentially of the Mediterranean but extending south of the northernmost limit of the West African Region. This is the principal source of commercial capers. (See also *C. decidua* (Forssk.) Edgew.)

In the Hoggar people use the plant for treating rheumatism. Dried and powdered leaves with curdled milk or butter are used to rub on to skin-infections on camels. Camels will not graze the bush. (1).)

Reference:

1. Maire, H., 1933: 100–1.

**Capparis thonningii** Schum.

FWTA, ed. 2, 1: 90. UPWTA, ed. 1, 20.

English: 'humbug' (after Arabic, *tundub*).

West African: **GHANA** GBE-VHE aŋŋɔ-ka = *thorny liane* (JMD) kpitipiti (FRI) VHE (Awlan) aŋɔka = *thorny creeper* (FRI) **NIGERIA** YORUBA awara (JRA) èékán-àwòdì = *hawk's claw* (JMD) ẹ̀wọ̀n ẹ̀kiri = *wild goat's bramble* (JMD; EWF)

A scrambling or climbing prickly shrub of savanna woodland, occurring from Sierra Leone to S Nigeria.

The only recorded uses for this species all come from S Nigeria where the leaf is eaten as an aphrodisiac; a leaf-decoction and crushed leaves are applied to tumours; leaf-sap is used for snake-bite, swollen tonsils, sores and ulcers; the root is used as a cure for coughs and spitting of blood (1) and the fruit is a Lagos remedy for tapeworm (2).

References:

1. Ainslie, 1937: sp. no. 70. 2. Dalziel, 1937.

**Capparis tomentosa** Lam.

FWTA, ed. 2, 1: 89–90, incl. *C. biloba* Hutch. & Dalz., and *C. polymorpha* Guill. & Perr. UPWTA, ed. 1, 20.

French: câprier d'Afrique; câprier de brousse.

West African: **MAURITANIA** ARABIC ('Maure') diemar (Aub.) **SENEGAL** ARABIC ('Maure') diemar (Aub.) BALANTA purage (Aub.) DIOLA fungok (JB; K&A) kânog (Laffitte; K&A) FULA-PULAAR (Senegal) bugi baley (K&A) gubi (K&A) gubi balédé = *black gubi* (K&A) gumibalewi (K&A) gumibaley = *black gubi* (K&A) MANDING-BAMBARA donkori (JB) dukari (Laffitte; K&A) MANINKA ɗatabeli kilifara (Aub.; K&A) SERER ngubor (auctt.) ngufor (auctt.) SERER-NON bufa (AS; K&A) nguo (AS, ex K&A) nguo (AS; K&A) NON (Nyominka) gufor (K&A) igufor (K&A) ingufor (K&A) SONINKE-SARAKOLE teko (K&A) WOLOF kareń, kéreń (auctt.) xareń = *clever* (K&A; DF) **THE GAMBIA** MANDING-MANDINKA bambaɲiŋ (*def.*-o) = *crocodile's tooth* (DF) belibelo (JMD) **GUINEA** MANDING-MANINKA diatabeli kilifara (Aub.) **MALI** FULA-PULAAR (Mali) dalévi (Aub.) gumi (Aub.) **UPPER VOLTA** DAGAARI wagua (K&B) GRUSI galo (K&B) galu (K&B) MOORE kalengoré (K&B) lemboitéka (K&B) silikoré (K&B) **IVORY COAST** FULA-FULFULDE (Ivory Coast) dalévi (Aub.) gumi (Aub.) MANDING-MANINKA diatabeli kilifara (Aub.) **GHANA** DAGBANI sansangwa (AEK; FRI) **NIGER** HAUSA haujeri (Aub.) KANURI záji (Aub.) **NIGERIA** ARABIC-SHUWA mardo (JMD) BEROM fwí (LB) FULA-FULFULDE (Nigeria) ngumi daleewi (MM) HAUSA cii zááƙíí ƙábdóódo (LB) haujari (MM) haujarin raƙumi *i.e., of the camel* (JMD; MM) ja ni baibai (*lit.*) *pull me back a little* (JMD) ƙabdarai (JMD; ZOG) kabdodo (auctt.) ƙábdóódo (auctt.) ƙàbdoódò, ƙadabebe, kaùdoódò (ZOG) ƙadabebe (JMD) ƙaudodo (JMD; MM) KANURI záji (JMD; C&H)

A thorny bush to 1.5 m high or climber to 5 m, occurring throughout the dry grassy savanna of the northern part of the Region, often on termite mounds, from Senegal to Niger and N Nigeria, and widespread in Africa generally from Ethiopia to S Africa.

In West Africa the leaves are considered good forage for camels (1, 4, Chevalier fide 4) but the flowers are toxic (1). In the Sahel de Nioro of Mali the leaves are reported to be edible [? to man] (11). In Tanganyika it is regarded as a browse-plant for camels (13) but in Sudan (14) and in E Africa generally it is said to be fatal to the camel and to horned stock except the goat (Schweinfurth fide 4, 13). Whereas in Ethiopia the leaves are said to be appreciated by goats (Schimper fide 4), or on the other hand, to cause their immediate death (5). The plant is toxic to camels in Ethiopia causing death two weeks after eating if diarrhoea does not intervene to cause voiding of the intestines — treatment is to give a purgative (5).

A similar ambivalence is recorded regarding the fruit. In Bornu (Vogel fide 4) and in other parts of Nigeria (4) the fruit is said to be edible [? to man]. In Ethiopia goats appreciate the fruit but it is held to be dangerous to man (Schimper fide 4) and an infallible poison to camels and horned stock except the goat (Schweinfurth fide 4). In the Sahel de Nioro the fruit is said to be edible (11). In Senegal their toxicity is recognized (Sébire fide 4, 7, 9), but snakes and birds freely eat the fruit (Sébire fide 4). In N Rhodesia the fruit is thought to be poisonous but feeding tests have shown no ill-effect (13).

Clearly such contradictory evidence merits careful examination of all the factors affecting the growing conditions of the plant.

As regards the roots there is no ambiguity: all reports agree they are very poisonous.

The plant enters widely into local pharmacopoeas throughout Africa, often with magico-medical attributes. A decoction of leafy-stems in draught is a common treatment for venereal disease discharges (3, 10) in the Ivory Coast. The roots producing diuretic and purgative effects are used internally in Senegal for treating urethral discharges and syphilis, and externally powdered roots are applied to chancres, and orchites (serious cases with internal dosing with other drugs as well) (7, 9). In Upper Volta the powdered root is taken in a little water for snake-bite and a root-decoction with *Gardenia sp.* (Rubiaceae) for a poison-antidote (10). A root macerate with *Waltheria americana* (Sterculiaceae) is used in the Ivory Coast—Upper Volta on the eyelids for ophthalmia (10) and a steam-fumigation from an infusion of the pounded root is used for conjunctivitis in Tanganyika (6). The pounded leaf is an Ethiopian eye-remedy (13). Sap (? root) is instilled into the ear for inflammation in the Ivory Coast (3), and the fruit, or the root, are used in Senegal for diseases calling for clearing of the bowels, e.g. hernia and sterility (7, 9). East of Lake Chad a powder of stem and leaf is a wound-dressing (4). In Tanganyika pounded root-bark is used to maturate abscesses (6), and as a cough cure (2). The plant has applications in southern Africa for treating colds, coughs and bronchial troubles, malaria, jaundice, etc. and also veterinary uses by the Zulu for stomach-troubles in cattle and sore teats in the cow (13).

The panaceal quality of the plant has attracted a number of medico-magical uses for treating madness, poisoning, spells, protection, leprosy, invocations of rainfall or to ward off storms etc. (7, 8, 9).

The alkaloid, *l-stachydrine,* has been isolated from the pericarp, endocarp and seed-husk of the fruit (12, 13), and also it is in the roots of Senegalese material (9). An alkaloid of the pyrrolidine group has been found in Tanganyikan material (6). A sulphur-oil is present in the root (13). The seed is moderately rich in oil with a 29·1% content consisting, in a Sudan sample, of oleic acid 29%, linoleic acid 26%, palmitic acid 23% and stearic acid 21%. The seed-kernel contains 38% protein with very low fibre, and the pressed cake can make a valuable cattle-fodder (14).

References:

1. Aubréville, A. 1950: 50. 2. Bally, 1937: as *C. persicifolia* A. Rich. 3. Bouquet & Debray, 1974: 63. 4. Dalziel, 1937. 5. Getahun, 1975. 6. Haerdi, 1964: 74 as *C. tomentosa* Lam. & *C. persicifolia* A. Rich. 7. Kerharo & Adam, 1964, b: 418. 8. Kerharo & Adam, 1964, c. 296–7 as *C. polymorpha* Guill. & Perr. 9. Kerharo & Adam, 1974: 318–9. 10. Kerharo & Bouquet, 1950: 28–29. 11. Roberty, 1953: 449. 12 Watt, J. M. 1967. 13. Watt & Breyer-Brandwijk, 1962, 161–3. 14. Grindley, D. N., 1950.

## **Capparis viminea** Hook. f. & Thoms.

FWTA, ed. 2, 1: 89.

A scrambling spiny bush with branches attaining up to 40 cm length, of the forest, and rare in Dahomey and N and S Nigeria, and in Zaïre, Angola and Tanganyika.

No use is recorded in the Region. In Tanganyika leaf-sap is drunk for epilepsy, and the bark is pounded, cooked and then cooled for applying as a poultice to maturate hard pusy abscesses (1).

Reference:

1. Haerdi, 1964: 74.

## **Cleome** Linn.

FWTA, ed. 2, 1: 86, including *Gynandropsis* DC, 1: 87–88.
English: spider flower, bastard mustard.

West African: **SENEGAL** SERER-NON (Nyominka) nakayorel (A. Chev.) **GUINEA** MANDING-MANINKA nonselé (Brossart) MANINKA (Waslunka) passo-ni-kuna (A. Chev.) SERER korgona (JLT) **MALI** SERER-NON (Nyominka) nakayorel (A. Chev.) **GHANA** AKAN-ASANTE tɛtɛ (JMD) TWI tɛtɛ (JMD) GBE-VHE amãgã (Westermann) awumɛ (Westermann) sɔbui (JMD) sɔgbee (JMD) sɔlwi (FRI) VHE (Pecí) afumoe (JMD) NANKANNI nanjinda (JMD) **NIGERIA** HAUSA gasaya (JMD) HAUSA (West) tabadamashe, tubadai, y'ár ùngúwaá = *a denizen of the suburbs*

*NOTE: The foregoing names appear in literature as 'Cleome spinosa'. This is a neo-tropical species, not recorded in West Africa. The species referred to is perhaps C. gynandra or the widely-cultivated C. hassleriana, but anyhow the vernacular names are loosely applied to all species of the genus wherever they occur in the region.*

A genus of 12 or perhaps 14 species recorded for the Region. Their leaves are used as spinach (2). Some species of the genus are useful as bee-plants (1).

References:

1. Howes, 1945. 2. Irvine, 1956: 39.

**Cleome afrospinosa** Iltis

FWTA, ed. 2, 1: 87, as *C. spinosa* Jacq.

West African: **NIGERIA** FULA-FULFULDE (Nigeria) gasiyaahu (JMD) GWARI amaddo (JMD) HAUSA gasaya (JMD) kinaski HAUSA (West) tabadai, tabadamashe, y'ár ùngúwaá = *a denizen of the suburbs* YORUBA ekùyá (Phillips; JMD) èkùyáko *from* oko: *farm* (JMD) èkùyalé *from* ilé: *house, when cultivated* (JMD)

A much-branched erect herb to 1.5 m occurring in N and S Nigeria and W Cameroons and extending from the Cameroun Republic to the Congo region. *FWTA*, ed. 2, equates the species in Africa with the neotropical S American *C. spinosa* Linn. which is now considered to be separate (2), but which has been recorded as a garden cultivate at Douala and Edea in E. Cameroun outside the West African Region (1).

The leaves are said to be edible (3).

The showy and widely cultivated garden plant known as the 'spider flower' has for long been referred to as 'C. spinosa' is correctly *C. hassleriana* Chod. (1).

References:

1. Kers, 1969. 2. Iltis, 1967. 3. Walker & Sillans, 1961: 117–8.

**Cleome arabica** Linn.

West African: **MAURITANIA** ARABIC (Hassaniya) mraïnzè = *the stinker* (AN) **MALI** TAMACHEK mkhenza (RM)

A desert plant recorded as far south as 20°N and which may be found just within the northern limit of the Region, 18°N.

Moorish women of Mauritania sun-dry the whole plant, reduce it to powder which is added to milk and drunk. This is considered a good fattening food (2). It is also considered to give good pasturage (2). However in the Hoggar camels refuse the plant, and goats and sheep rarely take it (1).

In the Hoggar dried or powdered leaves are added to food as a diuretic for treating rheumatism, or to cause sweating (1). In Mauritania grilled leaves are cooked into food for kidney and back ailments and to give virility to men (2).

References:

1. Maire, 1933: 3: 99. 2. Naegelé, 1958, b: 880.

**Cleome gynandra** Linn.

FWTA, ed. 2, 1: 87–88, as *Gynandropsis gynandra* (Linn.) Briq. UPWTA, ed. 1, 21–22, as *Gynandropsis pentaphylla* DC.

English: cat's whiskers, bastard mustard.

West African: **SENEGAL** MANDING-BAMBARA naségé (JB, ex K&A) SERER korgona(b) (auctt.) safoybidum (JB; K&A) SERER-NON (Nyominka) nakayorel (A. Chev.) WOLOF gor bu di daw = *old man* (K&A); = *male runner* (DF) nakayorel (K&A) n-gor si bidaw = *old man* (JB; K&A) **GUINEA** MANDING-MANINKA nonselé (Brossart) MANINKA (Waslunka) passo-ni-kuna (A. Chev.) **MALI** DOGON ɔ́gɔ lubɔ̀ (C-G) SONGHAI húbéy (*pl.* -à) (D&C) **IVORY COAST** AKYE kpawun (A&AA) ANYI uataï (K&B) BAULE wètè (A&AA) KRU-GUERE (Chiehn) niamé (B&D) GUERE (Wobe) sèhè (A&AA) KWENI sango(n) (K&B; B&D) **GHANA** ADANGME kɛtɛ (FRI) tɛtɛ (FRI) ADANGME-KROBO tɛtɛ (Bunting) AKAN-ASANTE tɛtɛ (FRI) tɛtɛtɛtɛ *from the noise of the bursting fruits* (FRI) TWI tɛtɛ (FRI) GA kɛtɛ (FRI) tɛtɛ (FRI) GBE-VHE amãgã (Westermann) awumɛ (Westermann) sɔbui (JMD) sɔgbee (JMD) solui (FRI) sɔlwi (FRI) VHE (Awlan) sɔbui (FRI) VHE (Pecí) afumoe (FRI) HAUSA garsar (FRI) KUSAL kaa (FRI) MOORE tiɛlɛbdo (FRI) NANKANNI nanjinda (JMD) **NIGERIA** CHAMBA tee-bu(-noh) noh: *any medicinal herb* (FNH) FULA-FULFULDE (Nigeria) gasiyaahu (JMD) kinaski (auctt.) GWARI amaddo (JMD) HAUSA gaasayaa (auctt.) HAUSA (West) diyar ungwa (BM) tabadai (JMD; ZOG) tabadamashe (JMD; ZOG) y'ár ùngúwaá (auctt.) NUPE lanti bókùn (RB) TIV gashia (Vermeer) YORUBA ekesi-masun (RJN) èkùyá (auctt.) èkùyáko *from* ako: *a farm, when cultivated* (JMD) èkùyalé *from* ilé: *a house, when cultivated* (JMD)

A herb to about 60 cm high, a common plant of waste places and weed of cultivated land, and often itself cultivated, occurring throughout all territories of the Region, and pan-tropics generally.

The plant is eaten in all parts of the Region and in other countries of its worldwide distribution, often as a pot-herb, taken cooked, fresh or dried (20, 21) or as spinach and in soups (2, 3). The leaves are made into sauces in N Nigeria, called *míya* in Hausa (24), and in Gabon (33). In Senegal they are added to *cuscus* (31). In N Rhodesia they are made into a relish (30), and in the Sudan Arabs eat them as a salad (9), though in Tanganyika it is recorded that the leaves require soaking in several changes of water to remove the bitterness (25), an action which cooking also effects (10, 36, 37).

The leaves are said to provide a piquant taste or a sourness, perhaps due to the conditions under which the plant has grown. They probably contain *sinapin,* the substance which gives mustard its 'biting' taste (21). They are considered very good for the stomach and to have antiscorbutic properties (3). Their vitamin C content has been recorded at 6.0 mg per 100 gm dry weight (37). They are rich in minerals and content of aluminium (1,390 ppm) and of iron (470 ppm) is unusually high (11). Though the plant is everywhere considered edible it is curious that it has been used as a fish-poison (36, 37). There are opposing reports on its acceptance by cattle: in Senegal it is said to be ungrazed or only infrequently by cattle (1) whereas in Somalia (12), Kenya (14) and Tanganyika (17) all stock is said to take it readily.

The leaves have medical uses everywhere the plant occurs, most commonly as a counter-irritant for local pain, the leaves being merely rubbed on the part affected or applied as a poultice (13, 22). The leaves are considered disinfectant and a good remedy for rheumatism in Nigeria (26), and pounded are applied externally for rheumatism, lumbago, etc., but if left too long blisters are produced — indeed, this preparation is used in Nigeria as a vesicant (3). A leaf-macerate with pimento is given in the Ivory Coast as an enema for rheumatism (8, 23), and an infusion is taken for bronchitis in Zaïre and used as an enema (15). A leaf-mash is warmed and laid on swollen armpit bubos in the Ivory Coast (2) and over the kidneys and poulticing is known and practised in SE Asia (10). The leaves are rubbed on the hands and inhaled like smelling salts for headache in West Africa (13). A headache cure in Tanganyika which is said to be effective in about five minutes is to rub the leaves into small cuts on the temples (28), while leaf-sap is given in nasal instillation in the Ivory Coast (8). Widespread use is made of the leaves for ear-trouble. The juice expressed after heating, alone or mixed with oil, is dripped into the ear for earache (13) or a leaf-infusion is an effective irrigation for the outer ear (27). Leaf-sap is instilled into the ears for otitis in the Ivory Coast — Upper Volta (2, 8, 23), and with oil added is commonly used in Ghana (19). This treatment is said to be painful to both ears and eyes, the administration requiring some caution. Elswhere in Africa, Congo (6) and East and South Africa (37) and in

India and SE Asia the same application is known (10). Leaf-sap in minute quantities is used in Nigeria as an eyewash (26), and in Tanganyika for inflamed eyes (16). The Dan people of Ivory Coast use a leaf-infusion for bathing infants (27), while in general it is recognized as a revigorant and defatigant especially in treating debilitation of old men, hence the Wolof name *gor bu di dâw,* meaning 'old man' (22).

The seed capsule is eaten as a condiment in Gabon (33), and in S Africa a capsule may be inserted into the outer ear as an effective method of softening and removing wax (37).

The seed, and the plant as a whole, contains a volatile oil rich in *senevol* with properties resembling sulphur derivatives found in garlic and mustard oils (22). An acrid fixed oil and a brown resin are also present (37). The concentration of the oil is about 17·6% (22) and it is regarded as edible and suitable for soap-making. A thick greenish drying oil has also been reported in the seeds (36). The seeds have anthelmintic properties (32, 33) and their oil is used in India to expel roundworm (37). They contain a substance called *cleomin* which is vermifugal (26, citing *Chem. Abstr.* 1938, 2137). The oil is used in the Upper Nile area and in India as a hairdressing to kill lice. The oil is furthermore used as a fish-poison (10, 35). In time of dearth the seeds are eaten in Tanganyika (4), and in India they serve as a substitute for mustard (19). The roots are cooked and eaten in Uganda (18) and in N Rhodesia (34). In Tanganyika they are used to facilitate childbirth and for internal disorders (5) in undisclosed ways, and the sap from pounded roots is instilled into the ear for earache (29). They are recorded as cyanogenetic (7) with no other active principles present in the plant.

For many of these uses the whole plant is also used. It is like enough to *Cleome viscosa* Linn., *C. monophylla* Linn. and *C. afrospinosa* in appearance to be confused with them and doubtless all are to a greater or lesser degree interchangeable in usages.

References:

1. Adam, 1966, a: as *G. gynandra* (Linn.) Briq. 2. Adjanohoun & Aké Assi, 1972: 81, as *G. gynandra* (Linn.) Briq. 3. Ainslie, 1937: sp. no. 172, as *G. pentaphylla* DC. 4. Allnutt 39, K. 5. Bally, 1937: as *G. gynandra* Briq. 6. Bouquet, 1969: 78–79, as *G. pentaphylla* DC. 7. Bouquet, 1972: 16, as *G. gynandra* (Linn.) Briq. 8. Bouquet & Debray, 1974: 63, as *C. gynandra* (Linn.) Briq. 9. Broun 7, K. 10. Burkill, 1935: 1119, as *G. gynandra* (Linn.) Briq. 11. Busson, 1965: 200, as *G. gynandra* (L) Briq., with leaf analysis. 12. Collinette 280, K. 13. Dalziel, 1937. 14. Glover et al. 689, K. 15. Hauman & Wilczek, 1951: 519–21, as *G. gynandra* (Linn.) Briq. 16. Haerdi, 1964: 75, as *G. gynandra* (Linn.) Briq. 17. Hornby 85, K. 18. Imperial Institute 46731, K. 19. Irvine, s.d. 20. Irvine, 1952, b: as *G. gynandra* (Linn.) Briq. 21. Irvine, 1956: 39, as *G. gynandra* (Linn.) Briq. 22. Kerharo & Adam, 1974: 320–1 as *G. gynandra* (Linn.) Briq. 23. Kerharo & Bouquet, 1950: 24, as *G. pentaphylla* Linn. 24. Lely 158, K. 25. Michelmore 1402, K. 26. Oliver, 1960: 28, 66, as *G. gynandra*. 27. Portères, s.d. 28. Tanner 586, 974, K. 29. Tanner 3782, K. 30. Trapnell 1768, K. 31. Trochain, 1940: 270–1, as *G. gynandra* (Linn.) Briq. 32 Walker, 1953, a: 26 as *G. gynandra* Merrill. 33. Walker & Sillans, 1961: 118, as *G. gynandra* (Linn.) Briq. 34. Walters 3, K. 35. Watt, G., 1889–93: 4: 190–2, as *G. gynandra* (Linn.) Briq. 36. Watt & Breyer-Brandwijk, 1962: 164, as as *G. gynandra* Briq. 37. Watt & Breyer-Brandwijk, 1962: 165, as *G. pentaphylla* DC.

**Cleome hirta** (Klotsch) Oliv.

A bushy herbaceous plant to over 1 m high of the grassy savanna, swamps, roadsides, village areas, etc. occurring in NE, E, Central and SW Africa. *FWTA* ed. 2 does not record it present in West Africa. It is recorded however as used as a leaf vegetable in the Region, (1) perhaps as an introduction.

Reference:

1. Irvine F.R. 1952, b.

**Cleome monophylla** Linn.

FWTA, ed. 2, 1: 87. UPWTA, ed. 1, 20–21.

West African: **SENEGAL** MANDING-BAMBARA ngélu (JB) **NIGERIA** HAUSA a'a kai ka fito = *out you come* (JMD; RES) HAUSA (East) rimin samara (Grove)

An erect stocky annual herb to 60 cm height, a weed of cultivated land, common around villages, and especially near to cattle-pounds, recorded from the Sahel areas of the Region from Senegal to N Nigeria, and occurring throughout tropical and subtropical Africa.

The plant has an unpleasant smell and acrid taste. Cattle will not graze it (1). Curiously in contrast the plant is widely used in food by man in Africa. The leaves are taken as a pot-herb in West Africa (7), as indeed they are in all other countries of tropical Africa: the fruit is sometimes eaten in the Congo region and Sudan (2); the leaves are made into a relish in Malawi (8), and used as a flavouring in Tanganyika (4); the whole plant, except the roots, is cooked with porridge in Zambia (3); young shoots, flowers and older leaves are cooked along with groundnuts or tomatoes to make a side-dish in Malawi, or the leaves are sometimes mixed with those of *Cleome gynandra* and *Amaranthus lividus* (Amaranthaceae) and this when prepared in the rainy season is slightly bitter and more bitter in dry season (14).

The vitamin C content of the leaves is given as 4.3 mg/100 gm on a dry basis (13).

In Nigeria crushed leaves are rubbed on the head for headache (12), and the leaf, finely ground, is sometimes put in the eye to remove irritating particles — hence perhaps the expression sometimes applied to the plant, *a'a kai ka fito,* a sort of incantation meaning 'out you come' (see also under *Sida linifolia*) (5). In Tanganyika the pounded leaves, dried and ground are put on sores, while the roots are chewed morning and evening for cough and the whole plant is used externally for swellings (11).

The seeds and leaves of *Cleome monophylla* of the violet-flowered form are used with or without those of *Gynandropsis pentaphylla* (= *C. gynandra* Linn.) to make the Ayurvedic drug called 'hurhur' in India. This drug is used on ulcers, boils and wounds and is said to prevent the formation of pus. The sap expressed and added to water is a common remedy for treatment of ear discharges. The pounded leaves tied on to swellings caused by plague hasten maturation. The leaf-sap is a sudorific in fever, and the seeds are anthelmintic, rubefacient and vesicant (10).

The plant is a weed of cultivated land. In Malawi it is known to be the host of the tobacco aphid (9).

The seed has been used as a source of vegetable oil in Tanganyika (6).

References:

1. Adam, 1966, a. 2. Andrews A 525, K. 3. Bush 12, K. 4. Chancellor 19, K. 5. Dalziel, 1937. 6. Davies 606, K. 7. Irvine, 1952. 8. Jackson 1810, k. 9. Lawrence 545, K. 10. Puri, 1971. 11. Tanner 1257, 4216, K. 12. Thornewill 174, K. 13. Watt & Breyer-Brandwijk, 1962: 163. 14 Williamson, J., 1955: 36.

**Cleome rutidosperma** DC.

FWTA, ed. 2, 1: 87, as *C. ciliata* Schum. & Thonn. UPWTA, ed. 1, 21, as *C. ciliata* Schum. & Thonn.

West African: **GHANA** AKAN-AKYEM tɛtɛ (FRI) ASANTE tɛtɛ (FRI) NANKANNI nanjinda (JMD; FRI) **NIGERIA** FULA-FULFULDE (Nigeria) kinaski ciile (J&D) HAUSA garseya (Yates) IGBO (Owerri) àkídìmmọ̀ó = *beans of the dead* (AJC) IJO-IZON (Kolokuma) àgbàlàlà (KW) IZON (Oporoma) àgbàlàlá (KW) kàlá àwọ̀ụ̀ ẹ̀gïnà = *small children's pepper* (KW) NUPE èyà kapangi (RB) YORUBA ẹtarẹ (auctt.) **WEST CAMEROONS** KPE lovanga (Waldau)

Several species of *Cleome* are included in the loosely applied Hausa name *namijin gasaya* and the Yoruba *ekiye*.

A common annual weed of waste places attaining about 90 cm height, occurring from Guinea to N and S Nigeria, West Cameroons and Fernando Po, and into E Cameroun, Zaïre and Angola.

The leaves are edible, have a taste like mustard and are occasionally taken as a potherb for which also the plant may sometimes be grown (2, 4, 5, 7).

It has medicinal uses similar to those of *Cleome gynandra* Linn. with which it is confused (4). The leaf-sap is used in ear-instillations in Ghana for earache (6), in Gabon for ear-inflammation (9, 10) and in Zaïre for deafness (5). A leafy-extract is made into a lotion for irritable skin conditions, prickly heat, etc, in Ghana (7) and by the Igbos of S Nigeria (3). In S Nigeria it is used for convulsions (8).

There are opposing reports regarding the presence (11) and absence (1) of alkaloid in the plant.

References:

1. Adegoke & al. 1968: as *C. ciliata* Schum. & Thonn. 2. Busson, 1965: 196, as *C. ciliata* Schum. & Thonn., with leaf analysis. 3. Carpenter 2 (UIH 1105), UCI. 4. Dalziel, 1937. 5. Hauman, & Wilczek, 1951: 514–5, as *C. ciliata* Schum. & Thonn. 6. Irvine 562, K. 7. Irvine, s.d. 8. Thomas, N. W., 2006 (Nig. ser.), K. 9. Walker, 1953, a: 26, as *C. ciliata* Schum. & Thonn. 10. Walker & Sillans, 1961: 117, as *C. ciliata* Schum. & Thonn. 11. Willaman & Li, 1970: as *C. ciliata* Schum. & Thonn.

**Cleome scaposa** DC.

FWTA, ed. 2, 1: 87

An erect slender and aromatic herb of the Sahel in Mali and N Nigeria, also in arid territory across Africa to Arabia.

In good rainy seasons it provides grazing for camels and sheep in Somalia (1).

Reference:
1. Harrison 947, K.

**Cleome speciosa** H.B. & K.

FWTA, ed. 2, 1: 88, as *Gynandropsis speciosa* DC.

A herbaceous plant resembling *C. gynandra* Linn., but with larger more showy flowers. It has been introduced to the Region and is grown as an ornamental. No other use is recorded in West Africa, but it is likely that it will find the same application as *C. gynandra*. In Ethiopa it is used to make a tea infusion which is drunk for two weeks as a cure for kidney stone (1).

Reference:
1. Getahun, 1975.

**Cleome tenella** Linn. f.

FWTA, ed. 2, 1: 87.

An erect abundantly forked shrub to 60 cm high of the Sahel area from Senegal and Mali to Sudan, Zanzibar, Madagascar, Sokotra and India.

The plant is not grazed by cattle (1).

Reference:
1. Adam, 1966, 1.

**Cleome viscosa** Linn.

FWTA, ed. 2, 1: 87, 758. UPWTA, ed. 1, 21.

English: wild mustard

French: cléome visqueuse

West African: **NIGER** SONGHAI jéerí-lèmtì (*pl.* -ò) = *gazelle's sesame* (D&C) **NIGERIA** NUPE èyà'zo = *relative of the bean* (RB)

An erect herb of waste places to 60 cm high, with viscous foliage, occurring widespread over the whole Region from Senegal to Nigeria, and commonly across Africa and throughout the tropics.

The plant has an acrid taste rather like that of mustard, and is sometimes used in tropical Africa as a potherb. It is possibly grown for this purpose in Africa as it certainly is in Asia where the pungent seeds are used in curries or eaten as a condiment like mustard or as horse-radish may be in Europe (2).

It is not grazed by cattle (1).

The leaves are used as a counter-irritant for headache or local pain, the leaves being rubbed on the parts affected, or applied as a poultice, or rubbed in the hands and inhaled like smelling-salts. These stimulating properties are the same as for *Cleome gynandra* Linn. and are said to be probably stronger than in the latter (3). The plant is a well-known drug-plant in India and Southeast Asia. The leaves and seeds are used alone or with those of *Cleome gynandra* and/or *C. monophylla* to make the Ayurvedic drug called 'hurhur' — see *C. monophylla* for Indian uses. The use of 'hurhur' for intestinal troubles is known also in Malaya where a decoction of the plant is used for colic and even advanced dysentery. In Guam the seeds are taken as a vermifuge, and in support of effectiveness in this property African material has been examined and volatile oil distilled from the seeds is found to be vermifugal (2). The seeds also contain a yellow fatty fixed oil which has a possible use as an edible oil. From it a flavone has been isolated. Water extracts of the seeds have shown strong presence of saponins, but not of alkaloid nor tannin. The leaves have given positive tests for sterols, alkaloids and tannins (5).

The plant is shallow rooted. Its ploughing-in is easy and in areas where it occurs in abundance, for example, at Pong Tamale in northern Ghana, it presents a possible use as a cover-plant and greenmanure (4).

References:

1. Adam, 1966. 2. Burkill, 1935: 581, as *C. icosandra* Linn., with references. 3. Dalziel, 1937. 4. Irvine, s.d. 5. Puri, 1971.

**Crateva adansonii** DC.

FWTA, ed. 2, 1: 90, as *Crateva religiosa* sensu Keay. UPWTA, ed. 1, 21, as *Crataeva adansonii* DC.

West African: **SENEGAL** BASARI a-ŋàtɛdyíy (Ferry) FULA-PULAAR (Senegal) dâta kulagé (Aub., ex K&A) naiki (K&A) naiko (Aub.; K&A) nayibi (K&A) MANDING-BAMBARA balasirani (K&A) bani d́ugu (K&A) gâdolo (K&A) mogo iri (K&A) mongo kula (K&A) muńê (K&A) sunamê (K&A) MANINKA balasirani (Aub.; K&A) banid́ugu (Aub.; K&A) gâdolo (Aub.; K&A) mogo iri (Aub.; K&A) mongo kula (Aub.; K&A) muńê (Aub.; K&A) sunamê (Aub.; K&A) SERER ngoral (K&A) ngorèl (JB) ngorol (JMD; K&A) safoy (K&A) SERER-NON sek (K&A) NON (Nyominka) ingorol (K&A) ngorol (K&A) WOLOF kred (FB; K&A) kred kred (Aub.; K&A) kulel (K&A) kurel (K&A) kurit (K&A) orel, horel (auctt.) red(d) (auctt.) també = *whip* (auctt.) xulel (K&A) xurel (K&A) **THE GAMBIA** FULA-PULAAR (The Gambia) n-gududi (DAP) MANDING-MANDINKA bani-sirali *this is the Koranic name for what is now Israel, and is perhaps applied here in confusion with the Asian C. religiosa* (DF) WOLOF horel (DAP) **GUINEA-BISSAU** CRIOULO pau de bola (EPdS) pó de bola (JDES) **GUINEA** MANDING-MANINKA balasirani (A. Chev.; Aub.) banidiugu (Aub.) gandolo (Aub.) mogo iri (Aub.) mongo kulu (Aub.) mugnien (Aub.) sunamin

(Aub.) **MALI** DOGON ánabe, anave (C-G)) FULA-PULAAR (Mali) danta kulague (Aub.) naïko (Aub.) MANDING-BAMBARA bulusirani (Aub.) banidiugu (Aub.) gandolo (Aub.) mogo iri (Aub.) mongo kulu (Aub.) mugnien (Aub.) sunamin (Aub.) MANINKA bakasirani (Aub.) banidiugu (Aub.) gandolo (Aub.) mogo iri (Aub.) mongo kulu (Aub.) mugnien (Aub.) sunamin (Aub.) SONGHAI adjétef (Aub.; FB) lélé (*pl.* -à) (auctt.) **UPPER VOLTA** DAGAARI dumko (K&B) HAUSA guvé (K&B) MANDING-DYULA kodra iri (K&B) MOORE kalegaintohiga (auctt.) NYENYEGE koyani (K&B) **IVORY COAST** MANDING-DYULA kodra iri (K&B) **GHANA** NANKANNI chelum punga (Lynn; FRI) kuliguŋa (Gaisie) SISAALA chie (AEK; FRI) kohier (FRI) **TOGO** AHLO camu (Aub.) GURMA (Mango) dengma (Volkens) MOORE-NAWDAM tschengunga (Volkens) TEM (Tshaudjo) anomolum (Volkens) **NIGER** FULA-FULFULDE (Niger) danta kulague (Aub.) naïko (Aub.) HAUSA gude (Aub.) **NIGERIA** ARABIC-SHUWA dabkar (Aub.) FULA-FULFULDE (Nigeria) lamɗam baali = *sheep's salt* (JMD) HAUSA bududu (EWF) gude (Lowe) ingidido (Abraham) ingidudu (EWF) kalu (JMD; ZOG) ùnguduudù (Lowe; KO&S) HAUSA (East) engedidi (JMD) HAUSA (West) gudai (auctt.) ìngúdiídì (JMD; ZOG) ƙoƙirmo (JMD; ZOG) ungududu (JMD) IGBO amakarode (Kennedy; KO&S) KANURI ngálídò (C&H) YORUBA ẹgun-ọ̀run (auctt.) taniyá (JMD)

A small handsome tree of the galleried forest and savanna woodland, often on river-banks, from Senegal to N Nigeria and across Africa to Zaïre, Tanganyika and Madagascar. As understood here this species is confined to Africa but bears very close affinity to the Asian *C. religiosa* Forst. f. with which it has been equated by some authorities.

The tree attains 7 m height or more. The trunk is irregular, seldom straight, but it is worthy of cultivation as an ornamental for its dense masses of white flowers borne at the ends of all the shoots. In the bush, owing to grass-burning, which it survives, and repeated stripping of its leaves, the tree is often stunted. The wood is soft and yellow (2, 5) and strong-smelling when cut (9). It is of no practical use though it is reported in Togo to be usable like boxwood (5) — presumably for small items of joinery.

The leaves have a disagreeable smell when crushed (2, 4). They are however eaten in soups or mixed with cereals. They are boiled and added to a mixture, called in Hausa *kwaɗo*, containing a paste of locust beans, *Parkia spp.* (Leguminosae: Mimosoideae), with salt, pepper, etc. (cf. *Hibiscus cannabinus* Linn. Malvaceae). The Yoruba consume the leaves as a potherb. In Upper Volta they are an ingredient of sauces (13). To some peoples the leaves are taken only in time of dearth and then they are sold in northern markets (4, 5, 14). In the Northern Territory of Ghana the tree has been used as a fodder tree though the leaves are found to be not very palatable (8). The use of the plant as a dye source has been recorded (Dalziel ms. fide 8).

The Yoruba of S Nigeria (5) and races in Zaïre (6) apply the leaf to the head as a mild counter-irritant for headache. Powdered leaves and bark are considered by the Fula of Nguénar in Senegal to be rubefacient especially for use on cysts, and the bark removed from the base of the trunk has been recorded as used at Mutam internally and externally for treating sterility (10, 12). Also in Senegal the roots figure in several treatments for syphilis, and the leaves are used in fumigations for jaundice and yellow fevers, perhaps on the Theory of Signatures for the yellow colour of the wood (11, 12). For all troubles due to poor vision the leaves are used in Senegal in steam-fumigation over the face (11). The bark is widely used for stomach-troubles: in Upper Volta (13), in Ivory Coast (3), in Nigeria (1, 15), and in Jebal Marra (7). In Upper Volta it is used in association with *Flacourtia flavescens* Willd. (Flacourtiaceae) as a treatment for leprosy (13). It is held to have tonic properties and to be a counter-irritant for headache in Nigeria (15) where it is also used as an application for rheumatic conditions after powdering and boiling in oil (1). In Jebel Marra a bark-paste is used as a poultice on swellings (7). The root is used as a febrifuge boiled with natron and eaten with guinea-corn pap (5, 8). The Nankanni of Ghana apply the the dried ground roots to swollen parts of the body (Lynn fide 8).

The fruit contains a white mealy pulp and is occasionally eaten, usually roasted (4, 5). It is commonly taken by birds (5). The seeds have (unspecified) medicinal use by the Nankanni in Ghana (Lynn fide 8).

References:

1. Ainslie, 1937: sp. no. 116. 2. Aubréville, 1950: 45, as *C. religiosa* Forst. 3. Bouquet & Debray, 1974: 63. 4. Busson, 1965: 196, 202–3, with leaf analysis, as *C. religiosa* Forst. 5. Dalziel, 1937. 6. Hauman & Wiczek, 1951: 478–9, as *C. religiosa* Forst. 7. Hunting Technical Surveys, 1967. 8. Irvine, 1961: 48. 9. Keay et al. 1960: 74. 10. Kerharo & Adam, 1964, b: 432, as *C. religiosa* Forst. f. 11. Kerharo & Adam, 1964, c: 301, as *C. religiosa* Forst. f. 12. Kerharo & Adam, 1974: 319–20, as *C. religiosa* Forst. f. 13. Kerharo & Bouquet, 1950: 29. 14. Lely 806, K. 15. Oliver, 1960: 23, as *C. religiosa.*

## **Euadenia eminens** *Hook. f.*

FWTA, ed. 2, 1: 93, incl. *E. pulcherrima* Gilg. & Benedict.

West African: **SIERRA LEONE** KISSI kilonden (FCD) MENDE blɔloni (FCD) nyanɛku (*def.*-i) (FCD) MENDE-KPA ngasawa (FCD) SUSU hontɛgitɛge (NWT) TEMNE (Yoni) *an*-thonk-nin (Pyne) **LIBERIA** MANO tŏ a lah (Har., ex JMD) **IVORY COAST** BAULE bu (B&D) kbu (B&D) GAGU garela (B&D) KRU-GUERE (Chiehn) geto kwé (B&D) gudo kwé (B&D) KWENI dogo (B&D) vunadabla (B&D) zundabla (B&D) **GHANA** AKAN-ASANTE dinsinkoro (FRI)

A soft-wooded shrub of the undergrowth of deciduous forest, attaining 1.5 m height, occurring from Sierra Leone to Ghana. The flowers are greenish-yellow borne in a terminal panicle. They are very attractive though with a peculiar pungent smell. The plant is worthy of cultivation as an ornamental (6, 7).

The plant is said to be poisonous (Chevalier fide 2). Nevertheless a decoction of leaves, which after standing for 24 hours becomes a strongly red gelatinous mass, is eaten in the Ivory Coast as a tonic and antianaemic, and to combat nausea (1). In Sierra Leone the plant is said to make an eye-medicine and in Ghana to cure swellings on the back of the hand (7). The seeds are eaten in Sierra Leone and taste like pepper. The pulp surrounding them is also eaten and is said to be the best of aphrodisiacs (3).

The sap is used in Sierra Leone as an eye-medicine (9), and in Ivory Coast for headache and ear-inflammation by nasal drops, and in frictions for chest-troubles, kidney-pains and general pains (1). The root is used in Ghana for earache (5, 6, 7), and a root-decoction is administered in the Ivory Coast for anuria, painful micturation and as an aphrodisiac (1). The root is piercingly aromatic. Cut up into small pieces and enclosed in a bottle with a little water and exposed to sunlight the resultant vapour in the bottle is sniffed like smelling salts as a cure for headache in Sierra Leone (4, 8).

In Gabon the root is used as an arrow-poison or an antidote (10).

References:

1. Bouquet & Debray, 1974: 63. 2. Dalziel, 1937. 3. Deighton, 2406, K. 4. Gledhill, s.n., April 1962, FBC. 5. Irvine, 534, K. 6. Irvine, 1930: 188. 7. Irvine, 1961: 48–49. 8. Pyne 112, K, SL. 9. Scott-Elliot 5447, K. 10. Walker & Sillans, 1961: 118.

## **Euadenia trifoliolata** (Schum. & Thonn.) Oliv.

FWTA, ed. 2, 1: 93. UPWTA, ed. 1, 21.

West African: **IVORY COAST** AKYE dzotinkin (A&AA) n-zo téké (auctt.) uhon (auctt.) BAULE pu (B&D) pwo (B&D) KRU-BETE g-bababa (K&B) gugokwayé (K&B) GUERE dowé (K&B) dubué (K&B) granié bahu (K&B) KYAMA mobo (auctt.) 'KRU' witaro (K&B) **NIGERIA** EDO ólíkà (JMD) ólíkàn-lérhān (Kennedy) YORUBA lógbònkíyàn (auctt.) ológbe-kuyan, ológbe: *crested or having a cock's comb* (JMD) ológbomodu (JRA) sangi (Lamborn)

An undershrub of the dense forest attaining 4 m height, occurring from Ivory Coast to West Cameroons, in the lowlands and up to 1000 m altitude.

The leaves are used as a potherb in S Nigeria (5), and in the Region they are put into soups or used in cereal dishes (4).

As for *E. eminens* Hook. f., a decoction of leaves is eaten in the Ivory Coast as a tonic and antianaemic, and to combat nausea (3).

A decoction of the whole plant is used by Ivorean midwives to wash newly-delivered young women (6). This is perhaps in an antiseptic role which is certainly attributed to the roots in Ivory Coast medicine. Gratings of the root bark are placed in the outer ear passage by the Akye for earache. This causes a burning sensation (1). A root-decoction is used as a gargle for gingivitis and sores in the mouth, and as an instillation for otitis and purulent ophthalmias, or sap is given in nasal instillation for headache and ear-inflammation. The pulped root with citron, or the sap, are used in frictions for chest-troubles, kidney-pains or general pains (3, 6). A root-decoction is given for anuria, painful micturation or as an aphrodisiac (3). A Yoruba prescription in S Nigeria includes the roots with several other plants as a tonic and aphrodisiac (2).

The roots emit a strong smell.

In Gabon the roots, like those of *E. eminens,* are used in arrow-poisons or as an antidote (7).

References:

1. Adjanohoun & Aké Assi, 1972: 80. 2. Ainslie, 1937: sp. no. 152. 3. Bouquet & Debray, 1974: 63. 4. Busson, 1965: 200 with leaf analysis. 5. Dawodu 135, K. 6. Kerharo & Bouquet, 1950: 29. 7. Walker & Sillans, 1961: 118.

## Maerua angolensis DC.

FWTA, ed. 2, 1: 88–89. UPWTA, ed. 1, 22.

West African: **MAURITANIA** ARABIC ('Maure') léguinaye (GR) **SENEGAL** ARABIC ('Maure') aneb *the fruit* (Aub.) athil (Aub.) iatil (Aub.) libti (Aub.) FULA-PULAAR (Senegal) bagu (K&A) buguhi, logul bahi, sogui (Aub.) MANDING-BAMBARA bélébélé (auctt.) lébo lébo (Aub.; K&A) lébu lébu (K&A) MANINKA bré bré (Aub.) kokali (Aub.; K&A) SERER safoy (JB) WOLOF toȷ̈ (K&A) **THE GAMBIA** MANDING-MANDINKA belibelo (JMD) **GUINEA** FULA-PULAAR (Guinea) buguhi (Aub.) logul bahi (Aub.) sogui (Aub.) MANDING-MANINKA bré bré (Aub.) kokali (Aub.) **MALI** DOGON belaya (Aub.) FULA-PULAAR (Mali) buguhi (Aub.) logul bahi (Aub.) MANDING-BAMBARA bélé bélé (auctt.) lébu lébu (Aub.) **UPPER VOLTA** FULA-FULFULDE (Upper Volta) leguilnaye (Aub.) KIRMA kudia lampéré (K&B) MANDING-BAMBARA bélé bélé (Bégué, ex K&A) MOORE kessiga (Aub.) lambatagha (K&B) zélogo (K&B) **IVORY COAST** FULA-FULFULDE (Ivory Coast) loguilnaye (Aub.) **GHANA** NABT pugodigo (BD&H) **NIGERIA** ARABIC shegara el zeraf = *giraffe tree* (JMD; KO&S) FULA-FULFULDE (Nigeria) baguhi, bugi, buguhi (auctt.) leggal baali = *tree of sheep* (JMD; KO&S) HAUSA babbajuji (AST) ciciwa (KO&S) gazare (ZOG) jìgá (JMD; ZOG) manɗewa (Abraham) zúmùwaá (JMD; ZOG) HAUSA (East) tsua (JMD) HAUSA (West) kiyafa (ZOG) manɗewa (Bargery; ZOG) miyafa (auctt.) KANURI àpcí (C&H)

A small tree to 10 m high of the savanna and galleried forest areas of the Region from Senegal to Niger and N Nigeria, and widespread in the savanna areas of tropical Africa, and extending southwards to Angola and Transvaal.

The tree is an attractive ornamental for garden planting especially in the drier parts (8, 9). The straggling branches droop at the ends which carry the abundant, conspicuous white flowers. It is often planted on graves in the Nupe area of Nigeria (5). The wood is hard and heavy, yellowish, fine-grained and capable of taking a fine polish: it is suitable for small cabinet work, but is brittle. In Tanganyika it has been used for hut-posts but is said to be not particularly good (2), and for charcoal and canoes (23).

The leaves are eaten by races in Senegal (1) and elsewhere in soups (7) but perhaps in some areas only in time of dearth, for example, NW Nigeria (13). The Mossi of Upper Volta use them as a condiment and in sauces (16). In contrast from this, in The Gambia the leaves are considered to be poisonous to cattle, and though thought to be non-poisonous to man are nevertheless not eaten there (9). Stock in Senegal do not take them (1). In Ghana the tree is said to afford good fodder for sheep and goats (12). In Tanganyika the powdered leaves are used as fish-poison (10).

Similar ambivalence appears to exist regarding the fruit. In Guinea it is said to be edible (Pobéguin fide 3). In parts of NE Nigeria it is considered poisonous but equally is recorded as eaten by boys (9). In Tanganyika it is also said to be toxic (19). In Kenya the raw fruit crushed in water is taken to cleanse out the stomach (4). In Senegal opinion on the poisonous or non-poisonous nature of the fruit to stock is divided (15).

The seeds are eaten in Tanganyika (2).

The leaves are used medicinally either alone or with other drug-plants for stomach-troubles in Senegal (14, 15), and powdered leaves taken with food are prescribed for asthenia and anorrhexia, and a snuff made of a mixture of leaves of this species with leaves of *Ximenia americana* (Olacaceae) is taken for headaches (15). The leaves are considered to have an analgesic action. In The Gambia they are laid on a painful area to give relief (9), in Upper Volta a leaf-decoction is taken in draught for rheumatism while the lees are used in friction on the area of pain (16). In the Ivory Coast a decoction is given in enema for rheumatism (6), while in Tanganyika the whole plant is put into a medicine for epilepsy (22) and pounded leaves are laid over hard pusy abscesses to maturate them and a root-decoction is drunk (10). In that country too, leaf-sap is dropped into fresh wounds as an antiseptic dressing (10).

The roots find use in Tanganyika for hydrocoele (21), for influenza (17) and for toothache (18).

In N Nigeria the fruit is used by youths as a love-charm, mixed with *tozali* (galena, or so-called antimony) and rubbed on the eyelids to render themselves irresistible to girls (9). Root and bark-decoctions are drunk in Tanganyika as aphrodisiacs. (10).

Material from Tanganyika showed the presence of alkaloid and saponoside (10). Examination of the roots for insecticidal activity gave inconclusive results (11).

References:

1. Adam, 1966, a. 2. Allnutt 20, K. 3. Aubréville, 1950: 55. 4. Bally 2088, K. 5. Baxter 1703, K. 6. Bouquet & Debray, 1974: 63. 7. Busson, 1965: 200, with leaf analysis. 8. Dalziel 8122, K. 9. Dalziel, 1937. 10. Haerdi, 1964: 75. 11. Heal & al. 1950: 108. 12. Irvine, 1961: 50. 13. Jibirin FHI 5153, K. 14. Kerharo & Adam, 1964, b: 557. 15. Kerharo & Adam, 1974: 321–2. 16. Kerharo & Bouquet, 1950: 30. 17. Koritscher 1139, K. 18. Koritscher 1484, K. 19. Koritscher 1576, K. 20. Richards 10144, K. 21. Tanner 2665, K. 22. Tanner 4406, K. 23. Watkins 39 (FH 2326), K.

## **Maerua crassifolia** Forssk.

FWTA, ed. 2, 1: 88. UPWTA, ed. 1, 22.

West African: **MAURITANIA** ARABIC (Hassaniya) atîl (Leriche & Mokhtar Ould Hamidoun) ARABIC ('Maure') aneb (Aub.) athil (Aub.) iatil (Aub.) libti (Aub.) **SENEGAL** ARABIC ('Maure') aneb (Aub.) athil (Aub.) iatil (Aub.) libti (Aub.) FULA-PULAAR (Senegal) sogui (Aub.) MANDING-BAMBARA bérédiu (Aub.) **MALI** ARABIC atil (A. Chev.; GR) FULA-PULAAR (Mali) sogui (Aub.) TAMACHEK adiar (Aub.) adjar (A. Chev.) agar (A. Chev.; Aub.) atil (A. Chev.) tagart (Aub.) teghert (B&T) **NIGER** HAUSA jiga (Aub.) ziga (Aub.) SONGHAI hàsì (*pl.* -ò) (D&C) TUBU arkenn (Aub.) harikane (Aub.) TUBU (Kaningou) ngusuri (Aub.) **NIGERIA** ARABIC sarah (JMD; KO&S) HAUSA gazari (MM) jiga (Lely; KO&S) HAUSA (West) jirga (ZOG) KANURI ngə́zə̀rì (JMD; C&H)

A small tree to 10 m high of the dry savanna and desert, occurring in Mauritania and Senegal and eastwards across the northern Sahel zone of the Region to E Africa, Egypt and Arabia.

The wood is whitish, very hard and is used to make handles for weapons, implements, ploughs and water-troughs in the northern part of the Region (2, 4), chew-sticks in Ghana (4) and staves and toothpicks in Morocco (6). It burns with a nauseating smell (2, 4). The ash furnishes a black dye used by the Masai of E Africa to colour their shields (8).

The leaves enter into human diet in Senegal and the leafy twigs yield a good forage for stock but horses will not browse it (1). It is especially of value in the dry season. The flowering shoots are much appreciated by camels (2). Crushed leaves are used in the Central Sahara as a febrifuge and an infusion of dried leaves for arresting vomiting (5) and for stomach-disorders (2). Leaves in decoction are considered in the western Sahara to be a specific against skin-affections of the head and leaves pounded with the bark and taken in draught in hot milk constitute a cure for fever and toothache (6). They are said to be usable like those of *Cadaba farinosa* Forssk. (3). The calcium content of leaves from Sudan are reported to be very high (4).

The fruit is edible and is reported eaten in the northern part of the Region (2) in Mauritania, where it is known as *eb nembe* (Chevalier fide 3), and in Hoggar (5).

The bark is used in the Kordofan and Darfur area of Sudan for purification of water (7).

References:

1. Adam, 1966, a: 517. 2. Aubréville, 1950: 55. 3. Dalziel, 1937. 4. Irvine, 1961: 50. 5. Maire, 1933: 100, 237. 6. Monteil, 1953: 68. 7. Samia al Azharia Jahn, 1976. 8. Watt & Breyer-Brandwijk, 1962: 165.

**Maerua duchesnei** (De Wild.) F. White

FWTA, ed. 2, 1: 92, as *Ritchiea duchesnei* (De Wild.) Keay.
West African: **IVORY COAST** VULGAR chidongho (Aub.) **NIGERIA** EDO sabere (KO&S) YORUBA isuru (KO&S) logbòkihàn (Chappel)

A small evergreen tree to about 8 m high, of the wooded savanna and drier parts of the lowland rain-forest, occurring from Sierra Leone to West Cameroons, and across Africa to Sudan, Uganda and Zaïre.

Wood is soft, white, fibrous, hard and with an unpleasant odour (2, 3). A tree identified as *lógbòkihàn*, Yoruba, which furnishes timber used for carving in the Egba-Egbada area of Nigeria (4) is thought on its wood morphology to be this species.

No use is recorded within the Region. In Congo sap extracted from the roots by passing them through a flame is used in ear-instillation for inflammation of the outer ear, 1 drop × 3 daily (1).

References (as *R. duchesnei* (De Wild) Keay:

1. Bouquet, 1969: 79. 2. Irvine, 1961: 51. 3. Keay et al. 1960: 76. 4. Chappel, 1976/77.

**Maerua oblongifolia** (Forssk.) A. Rich.

FWTA, ed. 2, 1: 89. UPWTA, ed. 1, 22 (as *M. rogeoni* A. Chev.).
West African: **SENEGAL** SERER gul (JB) WOLOF n-gen (JB; K&A) n-gensé (K&A)

A low woody bushy under-shrub, sometimes scandent, to 2–3 m high, with thick rootstock and thick leaves, flowers strongly scented, occurring in savanna woodland from Senegal to N Nigeria, and in Sudan to the Red Sea and Arabia.

The plant survives annual burning by throwing up shoots from its thick rootstock (1).

It is recorded as used by a Wolof medicineman in Senegal along with eight other drug-plants to treat syphilis (2).

References:

1. Dalziel, 1937. 2. Kerharo & Adam, 1974: 322.

**Maerua pseudopetalosa** (Gilg & Bened.) De Wolf

FWTA, ed. 2, 1: 95, 758, as *Courbonia virgata* Brongn.
West African: **NIGERIA** KANURI kùmkúm, kə̀mkə̀m (C&H)

A glaucous, glabrous shrub of the dry savanna in flood-land, occurring from Senegal to N Nigeria and to Sudan, Ethiopia, Somalia and southwards to Uganda.

The foliage provides much-relished browsing for goats in Somalia (2). In parts of southern Sudan the plant is eaten but only as a famine-food after careful preparation to remove the toxic principle (3). *Tetramethylammonium iodide,* known as tetramine for short, is reported present in the tuberous root, shoot and leaf. This substance has proved fatal to humans within a quarter hour. *Di-, tri-* and *tetra-methylamine hydroxide* have also been found (7). The fruit is eaten in Sudan 'to make one strong' (1). The root when chewed is at first bitter, then a sweetness follows, and the Sudanese use the root to make sweet drinks (6), a practice that surely needs viewing with caution. Fruits and roots are used in topical application to the chest for coughs in Nigeria (4). The bark is used by Masai medicinemen, and the plant's toxic properties appear to be well known in E Africa. The vegetative parts contain 14·5% fats. The seed-husk and seed-kernel are also oil-bearing, and the following composition as percentage of total fats has been reported present in Sudan material:

| Glycerides of | Root | Seed-husk | Seed-kernel |
|---|---|---|---|
| Linolenic acid | 3·5 | — | — |
| Linoleic acid | 57·1 | 10·7 | 21·5 |
| Oleic acid | 37·7 | 57·9 | 27·0 |
| Saturated acids | 1·7 | — | — |
| Palmitic acid | — | 13·2 | 12·6 |
| Stearic acid | — | 18·2 | 38·9 |

The root is said to be an efficient precipitant of suspensions in water and is used in Sudan in water purification and storage in rural areas (6). Related plants, *Maerua subcordata* and *M. glauca* are said to be used in E Africa for clearing water muddied by cattle. The roots are chopped up and thrown into the water (5). The chemicals in them carry a heavy charge of ions.

References:

1. Aylmer 51, K. 2. Glover & Gilliland 911, 962, K. 3. Grindley, 1950: as *C. virgata* Brongn. 4. Oliver, 1960: 23, as *C. virgata*. 5. Polhill, 1971. 6. Samia al Azharia Jahn, 1976. 7. Watt & Breyer-Brandwijk, 1962: 163, as *C. virgata* Brongn.

**Ritchiea albersii** Gilg

FWTA, ed. 2, 1: 92.

A small tree with short thick trunk to 11 m high, of the montane forest at 1000–1600 m altitude, from S Nigeria and W Cameroons, and widespread in E Africa.

The wood is soft and white (1).

Reference:

1. Keay & al., 1960: 76.

**Ritchiea capparoides** (Andr.) Britten var. **capparoides**

FWTA, ed. 2, 1: 91–92, as *R. longipedicellata* Gilg, *C. capparoides* sensu Keay, p.p. and *R. fragariodora* Gilg. UPWTA, ed. 1, 22, as *R. fragrans* R. Br. and *R. fragariodora* Gilg.

West African: SENEGAL BANYUN kalal (K&A) kaluf (K&A) xalal (K&A) xaluf (K&A) DIOLA kunana amata (JB); = *shepherd's banana* (DF, The Gambia) DIOLA (Fogny) bubében barasit = *medicament of the ganglions* (K&A) burane (K&A) fu somsom (K&A) kabuful (K&A) kabulafuko (K&A) somsom (K&A) DIOLA (Tentouck) busamay (K&A); = *leopard coloured* (DF, The Gambia) DIOLA-FLUP banana hamata = *shepherd's banana* (K&A) MANDYAK bukokökes (K&A) **THE GAMBIA** DIOLA (Fogny) bouran (DF) bubében barasit, barasit *a species of snake medicine* (DF) — cf. under Senegal fou somsom = *it's sweat* (DF) kabuful, kabulafuko = *flushed* (DF) somsom = *sweat* (DF) **GUINEA-BISSAU** BIDYOGO nucunoduco (JDES; EPdS) DIOLA-FLUP bussámáéba (D'O) **SIERRA LEONE** MENDE nga sawei = *3-leaved* (Boboh) nyanɛku (NWT) **NIGERIA** EDO ólíkàn-líri (Kennedy) YORUBA lógbònkíyàn (JMD) olögbe-kuyan (JMD)

A lianescent shrub to 6 m high of the guinean forest zone from the Casamance of Senegal to W Cameroon, on to Angola, Zaïre, Zambia, Mozambique and Zanzibar. The species is represented by two varieties, var. *capparoides* with flowers of 4 petals and var. *longipedicellata* (Gilg) De Wolf with flowers of 8 petals or more. Var. *capparoides* is generally widespread. Var. *longipedicellata* is restricted to Togo, Dahomey and S Nigeria and Zambia.

The plant has a high reputation amongst Casamance medicinemen. The leaves are reputed to be effective in cases of poisoning. They and the roots are used externally and internally for snake-bite, and externally on the guinea-worm (3, 5). Leafy twigs made into a poultice are applied to the ganglia of the neck as part of a treatment for sleeping-sickness (3, 5). Powder or decoction of roots are taken in draught for stomach-complaints and a decoction for refactory cough (3, 5). Leaf-sap or macerate or decoction is used in instillation or as an eye-bath for conjunctivitis and other eye-troubles (3, 5). A more complex method is for the patient with his eyes open to breathe and inhale heavily through his nose in a confined box containing finely powdered leaves which have been colllected in the dry season (4, 5). In the Uburubu area of S Nigeria the root of var. *longipedicellata* (Gilg) De Wolf is used 'for pain in the eye' (6).

In Gabon the leaves are used as a vulnary (7, 8). In Tanganyika leaf-sap and root-decoction are taken for gonorrhaea, and a root-decoction is taken in draught to maturate abscesses. Powdered bark in a wash is taken for stomach-pains, and root-scrapings steeped in water are put in the ear for earache (2). Similar to this last usage, sap extracted from a root softened by passing through a flame is dripped into the ear passage for inflammation in Congo (1).

The Tanganyikan material has been found to give reactions to the presence of alkaloids and saponosides the same as for *Maerua angolensis* DC. but stronger (2, 5).

The plant enters into circumcision rites in The Gambia (9).

References:

1. Bouquet, 1969: 79 as *R. fragrans* R. Br. 2. Haerdi, 1964: 75. 3. Kerharo & Adam, 1962: as *R. fragrans* R. Br. 4. Kerharo & Adam, 1963, b. 5. Kerharo & Adam, 1974: 322–3. 6. Thomas, N. W. 2322 (Nig. ser.), K. 7. Walker, 1953, a: 27, as *R. fragrans* R. Br. 8. Walker & Sillans, 1961: 118–9. 9. Dept. Forests.

**Ritchiea reflexa** (Thonn.) Gilg & Benedict

FWTA, ed. 2, 1: 92. UPWTA, ed. 1, 22.

West African: **GHANA** ADANGME-KROBO *g*-bowe tʃo (auctt.) nyɔtʃo (FRI) GA áàyélèbí (KD) GBE-VHE alẽvo (FRI) VHE (Awlan) alẽvo (FRI)

A scrambling or erect shrub of the grassy and wooded savanna, often in scrub of abandoned farms occurring from Guinea to Togo.

The plant has decorative greenish white flowers. It has been taken into cultivation in Florida, U.S.A., where it can be trained into a dense trellis vine (1).

In Ghana the leaves are steeped in water which is inhaled through the nose, or is drunk as a cough cure. The pounded leaves are used on abscesses and to treat guinea-worm. Root-bark is placed in the ear to cure earache. The fruit contains a sweet [? edible] pulp (2).

References:

1. Dalziel, 1937. 2. Irvine, 1961: 51.

## CAPRIFOLIACEAE

**Lonicera spp.**

English: honeysuckle.
French: chèvrefeuille.
Portuguese: madre (s) silva.

Climbers with sweet-scented flowers, mostly are temperate or subtropical plants. A few, e.g. *L. japonica* Thunb. and *L. longiflora* DC. can be cultivated under tropical conditions and are to be found in the Region as ornamentals.

## CARICACEAE

**Carica papaya** Linn.

FWTA, ed. 2, 1:201–1. UPWTA, ed. 1, 52–53.
English: papaya, pawpaw, papaw, melon tree (Holland), mummy apple (Holland).
French: papayer.
Portuguese: mamoeiro; mamao (the fruit).
West African: **SENEGAL** VULGAR papaya, papaye BALANTA paɓa (K&A) BASARI a-ndáɓáná (Ferry) BEDIK ga-ndyέntὲn (Ferry) DIOLA bu mpapa (JB) DIOLA (Fogny) bupâpa (K&A) FULA-PULAAR (Senegal) papayi (K&A) papayo (K&A) KONYAGI papiya (Ferry) MANDING-BAMBARA mãndé (JB) manguié (Vuillet) papia (K&A) papiu (K&A) MANDINKA paapiyaa (*indef.* and *def.*) (Sebia Jatta) 'SOCE' papia (K&A) papiu (K&A) SERER papayo (JB; K&A) SERER-NON (Nyominka) impapa key (K&A) WOLOF papayo (JB; K&A) **THE GAMBIA** VULGAR pawpaw FULA-PULAAR (The Gambia) budi (DAP) MANDING-MANDINKA papakayaa WOLOF popokaiyo (DRR) popokaye (DAP) **GUINEA-BISSAU** BIDYOGO umpandá (JDES) FULA-PULAAR (Guinea-Bissau) budibaga (JDES) MANDYAK pupá (JDES) MANKANYA pedum-hal (JDES) **GUINEA** BAGA (Koba) a-papari (Hovis) BAGA (Tumue) kö-papay, kö: *tree* (Hovis) FULA-PULAAR (Guinea) budi baga, budi beli, budi: *pumpkin* (JMD) **SIERRA LEONE** BULOM (Kim) nas (FCD) paga (FCD) BULOM (Sherbro) pakai-lɛ (FCD) FULA-PULAAR (Sierra Leone) budi-lɛdɛ (FCD) mandɛmbeli (FCD) GOLA diakwi (FCD) KISSI chalɛ̃-kɔŋguluma (FCD) KONO kɔŋgbi (FCD) kɔŋgbili (FCD) KRIO pɔpɔ (FCD) LIMBA wugbondi (JMD) LIMBA (Tonko) papala (FCD) LOKO baibai (NWT) paivai (FCD) MANDING-MANDINKA yiriye (FCD) MENDE fakai, fakali (*def.*-i) (auctt.) nyini-fakali = *woman's breast, i.e., pear-shaped* (JMD) SUSU fɔfia(i) (FCD) SUSU-DYALONKE bɔŋgo-na (FCD) wudiŋa-na(?) (FCD) TEMNE *a*-bap-*a*-ro-kant (NWT) *am*-papai (FCD) VAI pakai (FCD) **LIBERIA** MANO gbã gah (JMD) **MALI** MANDING-BAMBARA manguié (FB) **IVORY COAST** ABE olôkô (A&AA) ABURE eplé (B&D) AKAN-ASANTE boflé (K&B) bofré (K&B) BRONG boflé (K&B) bofré (K&B) AKYE m-bomu (A&AA) BAULE boflè (B&D) offlè (A&AA) GAGU baké (K&B) KRU-GUERE fakwau (K&B) vatré (K&B) vatu (K&B) GUERE (Chiehn) badié (K&B) GUERE (Wobe) gnoèyu-kporota (A&AA) KULANGO boflé (K&B) bofré (K&B) KWENI vadien (B&D) KYAMA kpakpa (B&D) MANDING-MANINKA badien (B&D) **GHANA** ADANGME-KROBO gɔ (auctt.) AKAN-ASANTE bɔɔfe-nini *the male tree* (Enti) bɔɔfrɛ (Enti) borɔfere, bɔɔfre (auctt.) brɔfe nini *the male tree* (FRI; JMD) ASANTE (Denkyera) bɔɔfe-nini *the male tree* (Enti) borɔsor, bɔɔfrɛ (Enti) DENKYERA borɔsor, bɔɔfrɛ (Enti) FANTE borɔfere (Enti)

borɔsow-nyin, b. nini *the male tree* (Enti) brɔsow TWI bɔɔfe-nini *the male tree* (Enti) bɔɔfrɛ (Enti) borɔfere (FRI) borɔsor, bɔɔfrɛ (Enti) brɔfre (FRI) WASA bɔɔfe-nini *the male tree* (Enti) bɔɔfrɛ (Enti) borɔsor ANYI-AOWIN berɛferɛ (FRI) SEHWI berɛferɛ (FRI) DAGBANI gondelɛ (Gaisser) GA adiba *from Ewe* (FRI) akpakpá (FRI; KD) a-kũnyãnyela (FRI) GBE-VHE aɖiba (FRI) aɖiba nyutɔ *a varietal name* (JMD) akɔdu-ɖiba *a varietal name* (JMD) ɔpɔpɔ *usually with the equivalent word for 'European'* (FRI) yevuɖiba = *European's pawpaw; a long-fruited variety* (FRI) GUANG-GONJA gɔ́ndɔ̀ (Rytz) KONKOMBA sagbir (Gaisser) NZEMA papa (FRI) SISAALA bɔrifira (Blass) **TOGO** BASSARI bŏrfule (Gaisser) KABRE samasse summelae *a var. with 6 longitudinal furrows* (Gaisser) sumolu (auctt.) KONKOMBA sagbir (FB; Froelich) MOORE-NAWDAM budebalod (Gaisser) filefille (Gaisser) TEM poripori (FB) TEM (Tshaudjo) pŏripŏri (Gaisser) **NIGER** SONGHAI dèndì-múféy (*pl.* -á) (D&C) **NIGERIA** ANAANG ókpòt (JMD) ARABIC-SHUWA bambus al beit = *pumpkin of the house* (JMD) kabbus (JMD) umm takalak (JMD) DERA lə̀bèt *wild papaya* (Newman) EDO úhòrò (JMD) EFIK étígí ḿbàkárá = *white man's okra* (auctt.) pọ̀pọ, ọpọ̀pọ *usually with the equivalent word for 'European'* (Winston) FULA-FULFULDE (Nigeria) dukkuhi, (*pl.* dukkuuje), rukkuhi (*pl.* rukkuuje) *the fruit* (JMD) kaabuse (MM) rukuure (*pl.* dukuuje) (auctt.) GWARI gwanda (JMD) kabu (JMD) kabushe (JMD) HAUSA gonda (JMD) gwadda, gwándàr daji (ZOG) gwandaa (auctt.) gwándàr Másàr *i.e., of Egypt; cf. gw. daji, Annona senegalensis* (JMD) gwándàr rebeji *a varietal name* (DA, ex JMD) IDOMA ape (Odoh) IGALA echibakpa (Odoh) IGBO ḿgbíḿgbí (Singha) okwuru bèkéè (BNO) ọ̀pọ̀pọ́ *from English* (JMD) IGBO (Asaba) egummọ (JMD) IGBO (Awka) ọ́kwụ̀lụ̀ IGBO (Onitsha) ọ̀gèdè-óyìbó = *European's plantain* (JMD) ọ̀kwụ̀lụ̀ ézì = *pig's okra* ọ̀kwụ̀lụ̀ óyìbó = *European's okra* pọpọ nlí ézì, pọpọn̠ní ézì IGBO (Owerri) ḿgbím̱gbí (AJC) ndidi (AJC) ọgẹde-ojo (JMD) ojo (JMD) ọ́kwọ̀rọ̀bèkéè = *European's okra* (AJC; JMD) IGBO (Umuahia) ḿbí-ḿgbí (JMD) KANURI gúndà *from Hausa* (C&H) káw ùsè (C&H) káwùsè (C&H) LEGBO akolo (AJC) TIV mbuer (JMD) URHOBO étò (Singha; Faraclas) YORUBA ìbẹpẹ ìbẹ́pẹ́ (JRA; JMD) **WEST CAMEROONS** VULGAR popo *from English* (JMD) NGEMBA pɔ̀pɔ̃ (Chumbow)

A fast-growing semi-woody tree to about 6 m tall, stem normally not branched, and becoming hollow with age; male and female flowers usually on separate plants, reaching maturity within one year and living for 5–6 years only. It is widely cultivated throughout West Africa around habitations in the forest zone and may occasionally become subspontaneous where conditions of adequate rainfall and free-draining fertile soil pertain. Full sunlight is necessary. The plant is native of central tropical America and is now spread by man to all warm countries.

The Spanish appear to have first met the plant on the mainland of central tropical America. They assisted its dispersal in the New World. *Ababai* is recorded as a Carib name in Breton's *Dictionaire Caraibe,* and Oviedo writing in 1535 gives *papaya* as the common name in Hispaniola. It was under this name that the Spanish took it westwards to the Philippines, and still under this name or derivation of it that it travelled across Asia and into Africa where it arrived in the Sixteenth century (24). How *papaw* or *pawpaw* entered the English language is not clear. At the best it can only be a corruption of *papaya,* which must remain the preferred name. The papaya travelled to Brazil where the Portuguese colonists, seeing a resemblance in the shape of the fruit to a woman's breast, called it *mamão,* and perhaps the name *mummy apple* arises from this line of thinking. An interesting parallel lies in the Mende *uying-fakali,* that is *pear-shaped* like a woman's breast. Being an exotic plant several West African vernaculars show an *ad hoc* reference to a well-known plant, e.g. pumpkin or okra, or a reference to a foreign origin, e.g. European.

The leaves are eaten by some [? West] African tribes as a vegetable (13). They contain, as does the whole plant, a watery-white sap in which is a proteolytic ferment of two distinct enzymes, one digesting protein to peptone, the other peptone to amino-acids. Meat, preferably moistened with vinegar, is sometimes wrapped in leaves for some hours before cooking in order to tenderize it by the action of these enzymes. The same effect can be achieved by adding a piece of unripe fruit to the water when cooking.

The leaf-pulp is haemostatic and is put on sores to promote healing (4, 17, 26). The Nyominka of Senegal apply a pad of dried leaves to the head for persistent headache, and young girls suffering from ailments of fatigue connected with puberty are thought to benefit by lying on a bed of fresh leaves of papaya for the duration of the illness (17, 18). At Sedhiou in southern

Senegal, the leaves, as also the bark of the trunk, are deemed to be antidotal for venoms and rabies (15). A leaf-decoction is given in Ivory Coast as a vigorous purgative in cases of difficult child-birth, and also as an abortifacient, in treatment of hernia and affections of the urino-genital system, blennorrhoea, orchitis and chancres (19). In Nigeria water in which young leaves have been crushed and squeezed is drunk three times daily for similar complaints; leaves are also used as a febrifuge and are said to be beneficial in diabetes (1). The action is diuretic.

The leaves enter into veterinary medicine. The Fula of Senegal dose horses with them for colic either alone or mixed with pimento and fresh eggs (15, 17).

In Gabon the leaves may sometimes be mixed with those of *Tephrosia* (Leguminosae: Papilionoiideae) in the preparation of fish-poison (25). In Guinea at Fouta Djallon tanners wrap the hide of a newly-killed animal, after it has been rubbed over with woodash, in papaya leaves. Left overnight it is easily de-haired (21).

In Gabon, the leaves are used as a soap-substitute for laundering cloth (25), and in Sierra Leone the leaves are known to form an effective bleach when clothes are soaked in a leaf-decoction for some hours (2). Similar usage is recorded in Ghana (10) and in E Africa (26).

The leaf-petiole is made in Gabon into a child's toy trumpet (25), and in Sierra Leone the petiole is used in certain witchcraft to simulate a gun (2).

The roots are purgative. In Ivory Coast they are used in treating intestinal troubles, oedemas and venereal diseases (4). In Ghana, the root mixed with salt and pounded into a paste is used as an enema, and together with the seeds is taken internally as an abortifacient (10), while in Nigeria the root of the male tree is powdered and taken for headache (1). In Gabon, the roots of young plants are administered for dysentery (25), and they are considered an excellent remedy for this in Congo where they are also given for blennorrhoea (3). A root-decoction is taken in E Africa as an anthelmintic, and an infusion as a remedy for syphilis (26). The roots have also been used in Ghana to cure yaws and piles (10). The enzyme *myrosin* and *potassium myronate* and a glycoside similar to *sinigrin* have been found in the roots (18, 20).

Root-ashes are used in francophone Africa to produce a vegetable-salt (21) and ash of the whole plant to make soap in Sierra Leone (2).

The flowers are eaten as a sweetmeat in SE Asia (5). Tests for antimalarial activity of the flowers showed no action against avian malaria (7).

The fruit is eaten raw when ripe, or cooked as a vegetable while still green. There is considerable variation in shape and size of fruit, and in sweetness of the pulp. There are yellow-fleshed and pink-fleshed forms, but in free-standing, self-regenerating village populations of short-lived trees desirable strains cannot be maintained. Propagating by hand-pollination and grafting is possible and is necessary if selected races are to be kept. Water content of the pulp amounts to about 90%. Sugars and carbohydrates account for most of the balance, and of the former a half is glucose with somewhat less fructose and a little saccharose (5, 6). The fruit does not lend itself to transportation, but storage by drying, with therefore the possibility of carriage, is practised in SE Asia (5).

The fruit is relatively rich in *papain*. It is obtained by scarifying the fully-grown but unripe, still green fruit. Many countries have undertaken production but Ceylon and E Africa are the principal commercial sources. Its main industrial uses are in 'chill-proofing' beer, in medicine and pharmacy, the food industry and textile manufacture, especially shrink-proofing and other treatments of wool (23). In medicine papain is official in several pharmacopoeas for its proteolytic action and digestive ferment. It is also anthelmintic but its medical favour is now superseded by synthetic drugs. In the Casamance of Senegal (15, 18) and in Ivorean medicine (4) the latex is used as a galactogogue. In Casamance it is used for blennorrhoea (15, 18), and in Ivory Coast (4, 19) and in Gabon (25) for intestinal worms. Besides papain, alkaloids *carpine* and *pseudo-carpine* have been found in the latex. They are very bitter and the former has anthelmintic action on *Ascaris, Trichinia* and

*Enterobius vermicularis*. Both alkaloids are heart-depressants, and may have some value in treatment of heart-disease (19, 20).

The milky juice is used in various countries to draw boils (5, 22, 26) and is widely used to 'cauterize' warts, corns, tumours, etc. (1, 9, 26). In Sierra Leone juice from the leaf-petiole is applied to burns (2). The green fruits are used in Casamance of Senegal in a treatment for jaundice (15) and E Africa young fruit split in half is applied over scarifications over the spleen to reduce its size (26). Unripe fruit is widely held in Asia to be dangerous to pregnant women who in eating may suffer abortion.

Insect-repellent properties are ascribed to the juice by Hausa traders in Sierra Leone who use it along with a cold infusion of the bark of *Rauvolfia vomitoria* (Apocynaceae) to preserve kola nuts against weevils, etc. It is said that a 'hedge' of papaya trees around a dwelling site will keep away mosquitoes; on the other hand, the soft hollow stems of old plants have been reported to become breeding sites for mosquitoes (8).

The seeds have a sharp spicy flavour and contain a substance similar to sinigrin that characterizes the seeds of *Brassica nigra* (Cruciferae). The germinating seeds are eaten like asparagus in Guinea. They are vermifugal and are said to be able to cause abortion (8), a belief strongly held in India (26). The seeds contain somewhat more than 25% of a fixed oil and about as much protein, but seemingly of no application. The presence of *carpasemine,* an alkaloid with amoebicidal action, has been recorded (20, 26).

Though a normal population of papaya trees will be made up of male and female plants, it is known for some male plants to bear fertile female flowers so that one gets fruit on male trees. These fruit are usually small but produce fertile seed. Such a phenomena must surely have given rise to ethnological attributes which should make an interesting study. The pulp of such half-ripe fruit is made into a crystallized preserve in SE Asia (5).

The stem, with bark removed, has been used as a famine-food in Indonesia. The central pith can be eaten raw (5). A fibre suitable for cordage can be made from the bark (8).

References:

1. Ainslie 1937: sp. no. 76. 2. Boboh, 1974. 3. Bouquet, 1969: 79. 4. Bouquet & Debray, 1974: 64. 5. Burkill, 1935: 459–64, with many references. 6. Busson, 1965: 207–8, fruit analysis given. 7. Claude & al., 1947. 8. Dalziel, 1937. 9. Hartwell, 1968. 10. Irvine, 1930: 82–83. 11. Irvine, 1952, b: 36. 12. Irvine, 1952, a: 32. 13. Irvine, 1956: 37–38. 14. Irvine, 1961: 86–88. 15. Kerharo & Adam, 1962. 16. Kerharo & Adam, 1964, b: 419. 17. Kerharo & Adam, 1964, c: 297. 18. Kerharo & Adam, 1974: 323–9, with much phytochemistry. 19. Kerharo & Bouquet, 1950: 41–42. 20. Oliver, 1960: 5, 52. 21. Portères s.d. 22. Quisumbing, 1951: 632–5, with many references. 23. Smith, 1952: with many references. 24. Trochain, 1940: 265. 25. Walker & Sillans, 1961: 19–20. 26. Watt & Breyer-Brandwijk, 1962: 167–73, with many references.

**Cyclomorpha solmsii** (Urb.) Urb.

FWTA, ed. 2, 1:221.

Tree with a single straight stem bearing sharp conical prickles, to about 25 m high, of forest edges, secondary growths and farmland of the savanna woodland areas of S Nigeria and W Cameroon.

The wood is soft; sapwood white; bark-slash produces some white latex (1). The hollowed-out stem is used in W Cameroon for bee-keeping (3).

A closely related species, *C. parviflora* Urb., has medicinal uses in the treatment of mental illnesses, schistosomiasis, cardiac oedema and toothache (2). In view of this and its affinity to *Carica papaya,* an examination of its phytochemistry and pharmacognosy seems indicated.

References:

1. Ejiofor FHI 30082, K. 2. Haerdi, 1964: 83. 3. Keay FHI 37365, K.

# CARYOPHYLLACEAE

**Drymaria cordata** (Linn.) Willd.

FWTA, ed. 2, 1:131.

A straggling procumbent herb with stems to nearly 1 m long, occupying grassland, forest margins, roadsides and cultivated areas, often under shade, at mid to higher elevations, and recorded from Guinea to Fernando Po, and widely dispersed in the tropics and subtropics.

The plant appears spontaneously as a weed of cultivation. It is common under coffee in Ethiopia and East Africa. It has been tried as a cover under tea in Ceylon with non-beneficial results of reduced yield and an inferior appearance of the manufactured tea. It has been found useful in India, however, as a ground cover to prevent erosion, especially on steep slopes. It is also a useful fodder-plant (8).

The sap has an aromatic pungency (9) whereby the plant is much used in many countries for respiratory chest-ailments: in Zaïre for colds and bronchitis (1); in Tanganyika as an inhalation for colds (5), as an inhalation of the roasting leaf for headache and head-colds (2, 12), and the dried leaf smoked like a cigarette for chest-complaints (7); and in Ruanda for bronchitis (10). The plant is scalded and the steam is used as an eye-fumigation for eye-troubles in Tanganyika (5), while in Madagascar it is an ingredient of a decoction administered as a cerebral stimulant, especially for children (4). The sap is said to be laxative and antifebrile (8).

The plant also has vesicant properties leading to its use in application to oedemas of the feet and to leprosy in Congo (3); as a poultice on injuries and yaws eruptions in Tanganyika (5); and on leprosy in Gabon (11). There are records of its use in the West Indies and S America on sores (12) and on tumours in Mexico (6). Topical application must be done with caution as prolonged treatment causes burning.

The flower, fruit, seed and root have given very weak positive reactions for the presence of haemolytic saponin, and the leaf and stem a negative response (12).

References:

1. Balle, 1951: 139–40. 2. Bally, 1937. 3. Bouquet, 79. 4. Debray & al., 1971: 57. 5. Haerdi, 1964: 159. 6. Hartwell, 1968. 7. McHardy 3, K. 8. Sastri, 1952: 113–4. 9. Tanner 2494, 2990, 5547, K. 10. Van der Ben 1112, K. 11. Walker & Sillans, 1961: 120. 12. Watt & Breyer-Brandwijk, 1962: 175.

**Drymaria villosa** Chamb. & Schlect.

A straggling prostrate or ascending herb with stems to about 45 cm long, a native of Central America and recorded also in Malesia. It has recently been reported in W Cameroon as a common adventive on palm and pepper plantations near Victoria (2), and as a medicinal herb at Bambui Experiment Station, but its medicinal use is not stated (1). It is likely to be similar to *D. cordata*.

References:

1. Brunt 496, K. 2. Chuml 96B, K.

**Polycarpaea** Lam.

FWTA, ed. 2, 1: 131.
West African: **NIGER** SONGHAI àrkús-bòŋ káaréỳ (*pl.* -á) = *white head of the old man* (D&C)

**Polycarpaea billei** Le Brun

FWTA, ed. 2, 1:132, as *P. eriantha* Hochst., quoad Vigne 4660.

See *P. eriantha* Hochst.

**Polycarpaea corymbosa** (Linn.) Lam.

FWTA, ed. 2, 1: 132, 759. UPWTA, ed. 1, 29, as *P. glabrifolia* DC.
West African: **SENEGAL** SERER-NON (Nyominka) bautaguerlet (Kaichinger, ex A. Chev) ndoggiar (JLT) wajo (JLT) **GUINEA-BISSAU** BADYARA mama-cúncoe (JDES) **GUINEA** MANDING-MANINKA dugumayetele (Brossart, ex A. Chev.) gumania (Brossart, ex A. Chev.) **GHANA** KUSAL pukumpɛlung (FRI) MOORE pogalum (FRI) **NIGERIA** HAUSA baàkin-suùdáa = *mouth, or speech of suùdáa (a bird)* (JMD: ZOG) fuùlár tsoóhoó, huùlar tsoóhoó = *old man's cap* (JMD; ZOG) máguúdìyaá (JMD; ZOG) mai-nasara = *luck-bringer* (JMD)

A small annual tufted herb of sandy savanna sites on waste places, and a weed of cultivated ground, occurring throughout the northern part of the Region, and widespread in Africa, Asia and America.

The plant provides forage for stock (1). An infusion is used as a wash for fever in Guinea (A. Chevalier fide 1). In India the leaves are made into a poultice for boils and inflamed swellings (3), and the powdered leaf is applied to the bites of venomous reptiles and animals. The leaf is also taken internally in the treatment of jaundice and snake-bite (4). The flowering head with a part of the leafy stem is used as a demulcent and astringent in Malaya (4).

As for *P. linearifolia,* this species is used in Nigeria as a 'good luck' plant (1), and to make a defatiguant drink in Ghana (2).

References:

1. Dalziel, 1937. 2. Irvine, s.d. 3. Krishnamurthi, 1969: 189. 4. Watt & Breyer-Brandwijk, 1962: 175.

**Polycarpaea eriantha** Hochst.

FWTA, ed. 2, 1:132, incl. *P. corymbosa* (Linn.) Lam. var. *effusa* (Oliv.) Turrill, pp. 1: 132, 759; excl. Vigne 4660. UPWTA, ed. 1, 29.
English: Tamale heather (Ghana, Williams).
West African: **GUINEA-BISSAU** BADYARA mama-cúncoe (EPdS)

A much-branched, tufted annual herb of dry ground on roadsides and waste places, and a weed of cultivation, reaching to about 20 cm high, often less. Recorded from Senegal to N Nigeria, and generally widespread in tropical Africa. The species, *sensu* FWTA, ed. 2, has been now recognized as containing a second species separated as *P. billei* Le Brun (2), which is restricted to Upper Volta, northern Ghana and Togo. The usages reported here may be taken to refer to both.

The plants are readily grazed by stock (1, 3). In Ghana an infusion is drunk when on the march against fatigue. The reaction may possibly be only psychological (1).

References:

1. Irvine, s.d. 2. Le Brun, 1970. 3. Williams 835, K.

**Polycarpaea linearifolia** (DC.) DC.

FWTA, ed. 2, 1:132. UPWTA, ed. 1, 29.
West African: **SENEGAL** FULA-PULAAR (Senegal) arinayel (K&A) mamadâdâdi (K&A) SERER vokat (JB) wokat (K&A) WOLOF n-didöbop = *big head* (K&A; JB) ndogġar (JLT; K&A) wajo (JLT; JGA) **MALI** DOGON gòmudu (C-G) **GHANA** HAUSA kaiduba (Williams) **NIGERIA** BOLE rowoni (AST) HAUSA baákin suùdáa = *mouth, or speech of a bird 'suùudáa'* (JMD; ZOG) fuùlár tsoóhoó, huùlár tsoóhoó = *old man's cap* (JMD; ZOG) máguúdìyaá (auctt.) mai-nasara = *luck-bringer* (JMD) YORUBA amórí taná = *put flowers in a head* (JMD)

An annual herb, occasionally perennial, of grassland savanna, to about 60 cm high, occurring throughout the Region from Senegal to S Nigeria, and widespread in tropical Africa.

Its principal applications are almost entirely superstitious. In Nigeria it is considered a charm to assure success in love or war — hence the Hausa name *mai-nasara,* luck-bringer (2). Traders and travellers in the North carry some as a protection (5). The Hausa name, *bakin suda,* or speech of *suda,* refers to a species of shrike whose song only the initiated can interpret (2). In Senegal the powder of dried fruits is held to have great magical properties and is much used by sorcerers. It is an ingredient of numerous potions beneficial to mental health and longevity of old people (3,4). In Nigeria, too, the plant is used to make an infusion which is drunk to combat fatigue while on the march, probably for a psychological effect (2).

The plant is grazed in Senegal by all stock, though horses and cattle may sometimes shun it (1), and in Nigeria (2).

References:

1. Adam, 1966, a. 2. Dalziel, 1937. 3. Kerharo & Adam, 1964, b: 568. 4. Kerharo & Adam, 1974: 330. 5. Thornewill 124, K.

**Polycarpaea tenuifolia** (Willd.) DC.

FWTA, ed. 2, 1:132.
West African: **SENEGAL** MANDING-BAMBARA nambla dé (JB)

An annual herb recorded from Senegal to Dahomey, and widespread in tropical Africa.

**Polycarpon prostratum** (Forssk.) Asch. & Schweinf.

FWTA, ed. 2, 1:131.
West African: **SENEGAL** MANDING-BAMBARA séré (JB) **GUINEA-BISSAU** BALANTA pôlu-úh (JDES) **SIERRA LEONE** MANDING-MANDINKA funtunčai (NWT)

An annual herb of sand-banks and river-beds occurring throughout the Region, and widely dispersed in the tropics.

It provides grazing for all stock in Mauritania (2) and Senegal (1).

References:

1. Adam, 1966, a: as *P. depressum*. 2. Kesby 54, K.

**Stellaria media** (Linn.) Vill.

FWTA, ed. 2, 1:129.
English: chickweed.
French: mouron des oiseaux (Adam).
Portuguese: morugem (Feijão).

An annual herb with diffuse leafy stems, recorded in montane situations in Guinea, W Cameroon and Fernando Po. A native of north temperate regions of the Old World but now a cosmopolitan weed.

The plant is very quick-growing, and flowers almost continuously. The seed is relished by cage-birds and is readily taken by poultry. The plant is grazed by stock though lambs are reported to suffer gastro-intestinal upsets by fermentation of lumps of the plant in the stomach. The plant has been used for haemorrhoids, eye-inflammations, blood-diseases and eczema, and in medicinal baths (2); also on warts and cancer (1). The seed contains 4.8% of a fixed oil and no saponin. The plant yields 150–550 mg/100 gm of ascorbic acid (2) and an unnamed alkaloid is reported in the aerial portion (3).

References:

1. Hartwell, 1968. 2. Watt & Breyer-Brandwijk, 1962: 177, with references. 3. Willaman & Li, 1970: with reference.

## CASUARINACEAE

**Casuarina equisitifolia** Forst.

English: casuarina; beefwood; ironwood; Queensland swamp oak; whistling pine; she-oak.

French: filao.

Portuguese: casuarina; árvore-da-tristeza; salgueiro-chorão (Feijão).

A rapid-growing tree to 25 m high or more, slender when young but developing a bole to 2 m in girth, straight, somewhat buttressed and fluted at the base. Native of Australasia; a seashore plant of essentially sandy soil but able to grow well on other soils except under waterlogging; dispersed by man pantropically and occurring commonly in the West African region.

The tree is very graceful and is often planted for decoration. A number of forms exist, some with drooping, some with with horizontal to erect branches. Crown is small and casts little shade. It can be planted as a roadside tree producing an attractive avenue. It is resistant to wind-damage and in monsoon areas of SE Asia is often planted as a windbreak. It grows particularly well on sandy beachheads to high-tide level and is useful for resisting sandwash and erosion. It stands clipping and will give a good quick-growing hedge. Topiary is possible.

Sap-wood is pale brown, heart-wood reddish-brown, hard but of limited durability in open air. In Asia however the poles are used in house-construction, for mine-props, electricity transmission lines, etc. (2). It provides masts for local sailing craft in the Indic basin (2, 3). The wood has a degree of resistance to termites (1, 2) and is durable in salt water so that it is used in E Africa as wharf-fenders (3). The wood is difficult to season and not easy to work. Its toughness and elasticity lends itself to use in wheels, rollers and wheelwright's work, hammer-handles, oars, yokes, etc. but it splits readily, and does not hold nails. It provides a firstclass firewood, burning even when green (1, 2). With destruction of forests in SE Asia as sources of firewood, casuarina is commonly planted near towns and villages as a substitute. In S Thailand it is a very remunerative cash-crop interplanted with avenue-planted young *Hevea brasiliensis*.

The roots contain bacterial nodules able to fix nitrogen.

The bark of Indian material contains 6–18% tannin. It is said to penetrate hide quickly and to result in a pliant pale reddish-brown leather. It can be used to toughen fishing lines and to arrest diarrhoea and dysentery (1, 2).

References:

1. Burkill, 1935: 491–3. 2. Sastri, 1950: 101–3. 3. Williams, R. O., 1949: 182–3.

## CECROPIACEAE

**Musanga cecropioides** R. Br.

FWTA, ed. 2, 1:616. UPWTA, ed. 1, 284, as *M. smithii* R. Br.

English: umbrella tree (from its shape); corkwood.

French: parasolier (for its resemblance to a parasol): bois bouchon (corkwood).

West African: **GUINEA** KISSI pédo (FB) peindo (RS) KONO uê (RS) KPELLE ué (RS) uéna (FB) uyo (RS) LOMA gozogui (FB) MANDING-MANINKA uonjo (RS) MANO ulo (RS) **SIERRA LEONE** BULOM (Sherbro) herka (JMD) GOLA didi (FCD) KISSI peindo (FCD; S&F) KONO wunsonɛ (FCD; S&F) LOKO n-gogho (FCD; S&F) MANDING-MANINKA (Koranko) wunsoŋ (S&F) MENDE ngovo (*def.*-i) (auctt.) SUSU feŋka (NWT) TEMNE *an*-fekaŋ = *food of the python* (auctt.) VAI wonje (FCD) **LIBERIA** DAN glu (GK) KRU-BASA doe (C; C&R) MANO wolo (JMD) MENDE ngovo **IVORY COAST** ABE loho (auctt.) AKAN-ASANTE edui (B&D) kwavréfé (B&D) AKYE agumi (A. Chev.; FB) moin (A. Chev.; A&AA) ANYI abomé (auctt.) agbomé (A. Chev.; K&B) egui (auctt.) BAULE adjuin (A&AA) DAN ulo (RS) KRU-BETE kodé (auctt.) GUERE deulo (K&B) doé (RS) do-ué (RS) magnéon (K&B) GUERE (Chiehn) go(d)dé (K&B; B&D) GUERE (Wobe) dué (auctt.) téré (RS) KULANGO (Bondoukou) djauna (A. Chev.; K&B) djima (A. Chev.; K&B) guima (A. Chev.; K&B) KWENI vogoba (auctt.) vokoba (B&D) KYAMA amohia (auctt.) egoni (B&D) 'KRU' tutué (K&B) **GHANA** VULGAR dwuma *from Akan* (DF) ADANGME odzuma (FRI) AKAN-ASANTE ajama (TFC) odwuma (Enti) odwumafufuo *the stipules* (Bowdich) ojamba (TFC) ojuma, odwuma (auctt.) ASANTE (Denkyera) odwuma (Enti) FANTE odwuma (CJT) TWI *o*dwuma (auctt.) odwuma (Enti) WASA agyemkama *the fruit, not commonly used* (FRI; Enti) odwuma (auctt.) ojuma (BD&H; TFC) ANYI eguĩ (FRI) ANYI-AOWIN eguĩ (FRI; CJT) eguin (auctt.) SEHWI eguĩ (FRI; CJT) GA odzuma (FRI) GBE-VHE (Pecí) *a*-juma (TFC; BD&H) NZEMA egunli (FRI; CJT) **NIGERIA** ABUA ọbonia (DRR; JMD) BOKYI bok(u)obe (auctt.) EDO óghóhẹ̀n (Lowe; auctt.) EFIK òtótò únọ̀ *the inflorescence, or fruit* (JMD) únộ (auctt.) únọ̀ ìdìm, idim: *stream; alluding to sap from the stem and roots* (JMD; KW) EJAGHAM egimamfuk (Talbot; KO&S) egum-amfuk (DRR) EJAGHAM (Keaka) egumamfuk (DRR) EJAGHAM-ETUNG egum-amfuk (DRR) ENGENNI uboniboni (JMD; DRR) IGBO ụ́rụ̀ (KW) ụ́rụ̀ (KO&S) IGBO (Arochukwu) ọru (Kennedy; DRR) IGBO (Awarra) égbū (JMD) IGBO (Awka) ụ́lụ̀ (Kennedy) IGBO (Onitsha) ọ́nrụ̀ (Kennedy) IGBO (Owerri) ọ́nrù (DRR) ụjụ̀jụ̀ (AJC) ụ̄rụ̀ (Kennedy) IGBO (Umuakpo) ụ́rụ̀ (Kennedy; DRR); *may be heard as* únrù, *a nasalisation not shown in the standard orthography* (KW) IJO (Nembe) óbòónyà (KW) ọ́bọ̀ọ́nyà (KW) IJO-IZON (Kolokuma) àfánfán (KW) IZON (Oporoma) àgbáwèì (KW) ISEKIRI egbesu (Kennedy; KO&S) MUNGAKA taku (DRR) OLULUMO-OKUNI mkpenga (DRR) URHOBO úhọ̄vbè (auctt.) YEKHEE ufogho (Kennedy; KO&S) YORUBA àga (auctt.) agbàwọ̀ = *grab the hook* (auctt.) **WEST CAMEROONS** DUALA bɔsengé (JMD) KOOSI ekombo (JMD) KPE eseng (JMD) lisengi (JMD) KUNDU bokombo (JMD)

A very rapidly growing tree to 20 m tall with an umbrella-shaped crown — hence the English and French names. Trunk straight cylindrical and up to 2 m in girth with stilt roots to 3 m above ground level and spreading branches. It is sun-loving and a pioneer species of regeneration occupying old clearings, first singly, then gregariously by coppicing from the stilt-roots. An unmistakable species, short-lived, of the mixed deciduous forest zone from Guinea to West Cameroons and Fernando Po, and on to Zäire and Uganda.

The tree canopy produces a dense leaf-litter which creates a heavy layer of humus. This serves as a nursery for other hardwood species which take over in succession. Its rapid growth may be an adverse factor in forest management if left unchecked, but with suitable control practices it could no doubt be used to good effect in regeneration after felling (20, 22). In Liberia it is sometimes used as a shade for coffee plantings (15).

The wood is a whitish grey to slightly pink. It is exceptionally light (density 0·20–0·25 when dry), and is one of the lightest African woods (4, 21). It is soft, coarse-grained, easily worked though not planing, nor finishing well; strength is poor but it is said that with proper seasoning this can be improved. Sap-wood is white and with a slight sheen; heart-wood is white (22). It is easily split and is used for making palings for enclosing compounds and fields, rough partitions in temporary huts, shingles for interior lining of roofs having an insulating effect. As roof-rafters it is said to last two years. It can be worked into a variety of domestic articles such as stools, musical instruments, walking-sticks, trays, baskets, toy popguns, etc. It is used as a cork-substitute whereby are the English and French names 'corkwood'. Its extreme lightness lends itself to make fishing-net floats and rafts. It has been used to float heavy bridging timbers to inaccessible river-bank sites, and it is recorded that on Lake Bosumtwi in Ghana where canoes or constructed craft are taboo rafts of it are used instead. Larger trunks are used to fashion out canoes and dugouts, long drums and blacksmith's bellows. (9, 10, 13, 19, 20, 26.)

The wood is currently recommended in Ghana for use in industrial insulation and in aircraft construction and for models (3). It yields a strong paper and has found recommendation as pulp (3, 4, 8, 10, 26). It is used in SE Nigeria as firewood (19) but there is no record of this usage elsewhere in the Region. In Gabon in olden times it provided the fuel for anyone condemned to death by burning (26). Wood-ash from freshly felled trees provides a vegetable salt for use in cooking in some parts (7; Guinea, 13), and a lye for soap-making (Guinea, 13).

The bark is intricately layered: light grey outer layer, then green, white, pink, white, all becoming brown on exposure. The outer part exudes a red-brown juice (22). In Ivory Coast this exudate is mixed with maize pap and is taken in the belief that it is a galactogogue (14). This belief is recorded again in Cameroun where women who have been taking it over a period of several days experience an increased milk-flow, and even those without child-at-breast (16). A number of substances are used in Zäire to increase milk-flow, and some are known even to induce lactation in virgin animals and before puberty. The sap of this plant has been investigated and found to contain an oestrogen and a galactogen with these properties (12). The treatment of dysmenorrhoea in Ivory Coast may perhaps be explained by this presence (6).

The plant is held to have some analgesic properties in Ivory Coast, and is administered for asthenia and loss of appetite; fumigation of bark and leaves mixed with leaves of *Adenia lobata* (Passifloraceae) is given to relieve asthenia in infants in which it acts as an expectorant and dehydrant (6). In Gabon a bark-macerate is used as a gargle for toothache and as a decoction for pulmonary troubles, or to the same end a piece of bark is chewed (25, 26). Also in Gabon a strip of the heated bark placed over the lumbar region relieves lumbago (25, 26). Root-bark is chewed in Sierra Leone with kola nuts as a cough-cure, and bark from callouses is tied onto wounds where it is supposed to effect a rapid healing (20) — sympathetic, may be, but perhaps with justification. Bark-scrapings in Zäire are added to fermenting sugar-cane sap to increase the potency (11).

The sap which tends to be tacky turns black on exposure and is used in Cameroun as ink (16).

The bark is fibrous. Long strong fibres can be extracted amounting to 25–30% by weight. They can be bleached and turned into a resistant paper, or made into twine (10, 13).

The aerial stilt-roots and also the younger branches are noted for their capacity of yielding a large amount of potable sap. 'Half a bucketful' is said to be obtainable from a single tree overnight (13). It is colourless, odourless and of an insipid sweetish taste (5). This source of drinking water is of great importance to several tribes in SE Nigeria and W Cameroons for sometimes whole villages will depend upon it in dry seasons (19), and similarly in Zäire (11). Hunters and others break off stems to draw an impromptu drink, and even monkeys have learnt to do this. As a regular supply a renewed flow is obtained by recutting the cut surface and beating the severed limb. This evokes the Efik name *uno idim, idim* being 'a stream' (10). The Itung of SE Nigeria use it for filming their earthenware pots. They and the Efik consider it medicinal for women (19). In Congo sap from the larger roots is drunk as a blood-purifier, to clean the stomach, for blennorrhoea, cough and chest affections, as a galactogogue, and commonly as a wash for persons with sleeping sickness, leprosy and fevers to relieve aches and pains, asthenia and rheumatism (5). The Kyama of Ivory Coast use the root-sap in topical embrocation for pulmonary congestion (18). Also in Ivory Coast ash from powdered roots mixed with palm oil into a paste is applied as a healing dressing to circumcision wounds (14).

The leaf and inflorescence buds are enclosed in a red stipular sheath which may be 20 cm long. This attracts attention, in part, at least, on the Theory of Signatures, for treatment of gynaecological conditions. To hasten childbirth, the whole sheath boiled in soup is recorded used by the Asante as a powerful emmenagogue (10, 13). The leaf is used by the Igbo of Nigeria (23), while the whole shoot is similarly used in Gabon (25, 26), and in Zaïre (5). The leaves enter a prescription in Ivory Coast to prepare a vaginal douch for painful menstruation (1) and in Congo pulped buds are given to women for leucorrhoea and other vaginal affections (5). The buds in Ivory Coast are crushed and boiled in water which, after filtering, is taken by draught and by enema for abdominal troubles (14). They are compounded in Sierra Leone with the bark of *Uapaca guineensis* (Euphorbiaceae), pepper (either a red chili or other pungent seed), salt and newly deposited soil from the *top* of a termite mound, and boiled up; the liquid after filtering is taken by mouth for swollen stomach and swellings in other parts of the body (17). The action is presumed to be diuretic. In Congo, the terminal bud is crushed whole and taken, often with sap added, to calm attacks of epilepsy and insanity, to treat blennorrhoea and heart-pains; sap expressed from the bud is an eardrop for earache and is applied topically for localized swellings (5). Hairs from the inside of the stipule are considered good in Congo for burns and healing sores (5).

The catkins are cooked with groundnuts and are taken to facilitate childbirth in Gabon (25, 26).

The fruit is 10–13 cm long by 5–6 cm wide, yellowish green, succulent flesh with embedded seeds. The flesh is edible, though seemingly not much relished in the Region. They are sought after by birds and bats which ensures distribution of the small seeds (20).

The plant enters into magic. In Liberia pieces of bark are placed over a doorway as a protection against lightning (9). It is imprecated in an Yoruba incantation against heart-disease (24). In Congo sap is used to wash new-born twins as a protection from evil spirits (5).

References:

1. Adjanohoun & Aké Assi, 1972: 198. 2. Ainslie, 1937: sp. no. 239 (as *M. smithii*). 3. Anon., For. Prod. Res. Inst., Ghana, s.d.: Inf. Bull. No. 1. 4. Aubréville, 1959: 1: 64/66. 5. Bouquet, 1969: 171 (as *M. smithii* R. Br.). 6. Bouquet & Debray, 1974: 124. 7. Busson, 1965: 111 (as *M. smithii* R. Br.). 8. Cole, 1968: 9. Cooper & Record, 1931: 79 (as *M. Smithii* R. Br.). 10. Dalziel, 1937. 11. Hauman, 1948: 88–89 (as *M. smithii* R. Br.). 12. Herman, 1956: 1345–68. 13. Irvine, 1961: 446–7. 14. Kerharo & Bouquet, 1950: 134 (as *M. smithii* R. Br.). 15. Kunkel, 1965: 144. 16. Leeuwenberg 5961, K. 17. Massaquoi, 1973. 18. Portères, s.d. 19. Rosevear, 1975 (as *M. smithii* R. Br.). 20. Savill & Fox, 1967: 191–2. 21. Schnell, 1950: I, 257 (as *M. smithii*). 22. Taylor, C. J., 1960: 253. 23. Thomas, N. W. 2360 (Nig. Ser.) K. 24. Verger, 1967: No. 15. 25. Walker, 1953: 6: 296 (as *M. Smithii* R. Br.). 26. Walker & Sillans, 1961: 300.

## **Myrianthus arboreus** P. Beauv.

FWTA, ed. 2, 1:614. UPWTA, ed. 1, 285.

French; arbre à pain indigène; grand 'wounian' (from Abe, Ivory Coast, Aubréville).

Portuguese: pernambuc (Feijão).

West African: **GUINEA** KONO bâ (RS) LOMA gbalué (RS) **SIERRA LEONE** GOLA fɔvɔ (FCD) KISSI g-bando (FCD; S&F) k-pando (FCD; S&F) KONO kaamba (FCD; S&F) LOKO fɔfɔ (S&F) MANDING-MANDINKA wakawaka (FCD) MANINKA (Koranko) wagale (S&F) MENDE fɔfɔ (*def.*-i) (auctt.) SUSU feŋkai (FCD) SUSU-DYALONKE khambu-na (FCD) mulukho-na? (FCD) TEMNE *a*-waka (FCD; S&F) VAI g-bã (FCD) **IVORY COAST** ABE wunian (auctt.) ABURE atolahié (A. Chev.; K&B) atoraé (B&D) ADYUKRU kenu-ikun (auctt.) AKAN-ASANTE niankoma (K&B) AKYE djélété (auctt.) djienkonguie (A. Chev.; K&B) jin (A&AA) ANYI kumaniangama (Aub.; RS) niangama (auctt.) BAULE agama (B&D) DAN gboâ (RS) KRU-GUERE tébo (RS) tobo-ué (RS) GUERE (Chiehn) tikriti (K&B; B&D) GUERE (Wobe) taratu (RS) tonohue (A.Chev.; FB) KWENI doba (auctt.) KYAMA anianahia (auctt.) aniéré (A. Chev.; K&B) yanguma (B&D) 'NEKEDIE' tikiritisu (B&D) **GHANA** ADANGME nfohwe (auctt.) AKAN nyankom *the seeds of the genus generally* (auctt.)

AKAN-ASANTE kokua-adua ba *a children's name;* kokuo: *a kind of monkey* (FRI) *o*-nyankomaa = *God's heart* (auctt.) yankoma (TFC) FANTE nyankoma-bere, bere: *female* (Enti) KWAWU kwesi popuro *not commonly used* (FRI, Enti) TWI anyaŋkãmãã (Twi Dict., ex FRI) nyankomaa = *God's heart* (auctt.) WASA nyankomaa (Enti) ANYI nyangama (FRI) ANYI-AOWIN ɛ-nyangama (FRI; BD&H) SEHWI ɛ-nyangama (FRI) GBE-VHE avagolo *the name for a woven fibre bag: connexion not known* (FRI) NZEMA nyangama (FRI) **TOGO** GBE-VHE avogalo (Volkens; FB) **NIGERIA** BOKYI kekeku (auctt.) EDO íhièghè (auctt.) EFIK n̠dìsọ̀k (auctt.) EJAGHAM echimbuk (DRR) EJAGHAM (Keaka) echimbuk EJAGHAM-ETUNG echimbuk (DRR) EKAJUK ebakan (DRR) IGALA apulu (Boston) IGBO ụ́jụ̀jụ̀ (auctt.) MUNGAKA betak-inok (DRR) NKEM igbebere (DRR) NUPE tsàkpàci *? this sp., or Burkea africana* (*Leguminosae: Caesalpinoideae* (auctt.) OLULUMO-OKUNI ereturuni (DRR) YEKHEE oseghe (Kennedy; KO&S) YORUBA ìbíṣèrè = *soup tree* (auctt.) ṣapo-obibere (DRR) **WEST CAMEROONS** DUALA bokɛku (JMD) KOOSI bokukulende (AHU) KPE bokĕre (JMD) wokĕku (JMD)

A tree to about 20 m high, short bole to 1 m in girth, often divided near the base, much-branched and with stilt-roots, the roots forming a network structure above ground; of secondary jungle often in damp situations and on stream-banks of the forest zone from Guinea and Sierra Leone to West Cameroons, and extending across Africa to Sudan, Tanganyika and Angola.

The wood is yellowish-white, soft, fibrous and difficult to work. Though perishable it is used for fencing (7, 8). It is burnt as firewood in SE Nigeria (13). Lye can be extracted from the ash and this is used in Guinea in making soap (Portères in 8).

The bark is said to be variable in appearance: in parts of SE Nigeria, at least, it may be greenish white and slightly flaky, or almost white and smooth (13). The slash is slightly tinted but rapidly darkens to brown; there is little or no exudation (13). The bark is used in Ghana for chest-complaints (8) and in Gabon, scrapings cooked in palm-oil are taken to relieve sore-throat (15). In Congo a bark-tisane is said to be cholagogic and antidysenteric (4). The bark is taken in Nigeria as a taenifuge (12).

The leaves are enormous, to 70 cm across, digitately compound of 7–9 leaflets, the largest attaining 50 cm length by 25 cm breadth. When they fall and lie on the ground they form a good groundcover retaining moisture and rotting down to form a thick humus. In SE Nigeria the young leaves are commonly eaten in vegetable-soup, hence the tree being known in Yoruba as the 'soup tree' (11, 13). The soup is so highly considered by the Egba people as to evoke the saying that 'one will kill his child for the sake of *ibishere* soup (or perhaps at the foot of the *ibishere* tree)' (7). Some races in Ghana also take the leaves in their diet (7, 8).

An extract of the leaves is made in Nigeria with *Alchornea* (Euphorbiaceae) for taking in cases of dysentry (7, 12), and leafy shoots are chewed by peoples on the Cameroon Mountain for this purpose. In Sierra Leone the liquid in which young leaf-flushes and a peeled green banana have been boiled is a medicine taken little and often to stop diarrhoea and vomiting (10). In the Igala area of Nigeria the leaves are an ingredient of a febrifuge given to small children (3). The Boki of SE Nigeria beat the leaf-petiole into a plaster for application to boils (13) while the bruised leaf is similarly used in Gabon (14). In Congo the leaves chopped up small are eaten raw with salt for heart-troubles, pregnancy complications, dysmenorrhoea and incipient hernia, and sap from young leaves, or the terminal buds is applied topically for toothache, or applied to the chest for bronchitis, or as a throat-paint for laryngitis or sore throat (4). Of the three *Myrianthus spp.* present in Ivory Coast, this one is the most commonly used as an analgesic for muscular pains and to 'reduce' fractures and to put into enemas for haemorrhoids. However, sap or powdered leaves when added to soup or to palm wine are said to induce 'madness', but other information is that cooking renders the leaves harmless and safe to eat with impunity (5). The Akye of Ivory Coast pound the leaves with those of *Holarrhena floribunda* (Apocynaceae) and a chili to a paste which is diluted with warm water for administration as an enema for pain in the back and loins (1).

The aerial roots yield an abundant amount of sap when cut up. This is drunk in Congo as an anti-tussive and anti-diarrhoeic, and as a remedy for haematuria and blennorrhoea (4). Agni-Asante people in eastern Ivory Coast prepare a vapour-bath from the diced roots with maleguetta pepper for headache (9).

The fruit is heart-shaped and may attain 10–15 cm in diameter. It is sufficiently cherished in Ghana to be called 'God's heart.' It is very hard and green when unripe, but turns yellow and soft when mature. In Congo the whole fruit is boiled in sap from the tree or in palm wine or other fruit ferments to take as an emeto-purgative, and is preferred to the less active bark or leaves which may be used for the same purpose (4). The fruit contains 5–15 seeds, each surrounded by a sweet or acidulous pulp which is commonly eaten. The seeds are enclosed in a woody pericarp amounting to 60–65% in weight of the whole. The kernel is about 1 cm long by 5–7 mm across. It is eaten in various territories after cooking (Ivory Coast, 2; Ghana, Nigeria, Zaïre, 8; Gabon, 15). It is rich in oil containing a 45% content, of which linoleic acid amounts to 93% in a sample of Ivorean material. Sugars are present at 19%, and proteins 30%, and also a large number of amino acids of which the most important feature is the unusual amount of cystine of potential value to a population suffering from a chronic deficiency of sulphur-bearing amino acids (6).

References:

1. Adjanohoun & Aké Assi, 1972: 199. 2. Aubréville, 1959: 1: 62/64. 3. Boston C.1, K. 4. Bouquet, 1969: 172. 5. Bouquet & Debray, 1974: 124. 6. Busson, 1965: 118 (with kernel analysis). 7. Dalziel, 1937. 8. Irvine, 1961: 447–8. 9. Kerharo & Bouquet, 1950: 135. 10. Massaquoi, 1973, 11. Oke, 1966: 128–32. 12. Oliver, 1960: 32. 13. Rosevear, 1975 [ined.]: 14. Walker, 1953: 297. 15. Walker & Sillans, 1961: 301.

## **Myrianthus libericus** Rendle

FWTA, ed. 2, 1:616. UPWTP, ed. 1, 286.

West African: **GUINEA** KISSI g-bando (RS) KONO bâ (RS) KPELLE g-balo (RS) LOMA gbâlé (RS) MANO balé (RS) **SIERRA LEONE** GOLA fɔvɔ (FCD) KISSI g-bando (FCD; S&F) k-pando (FCD; S&F) KONO kaanɛ (FCD; S&F) LIMBA kuroko (JMD) LOKO fɔfɔ (S&F) MANDING-MANDINKA wakawaka (FCD) MENDE fɔfɔ (*def.*-i) (auctt.) SUSU feŋkai (FCD) SUSU-DYALONKE khambu-na (FCD) mulukho-na? (FCD) TEMNE *a*-waka (auctt.) VAI g-bã (FCD) **LIBERIA** KRU-BASA vahn, vahn-vehn (C) MANO gban (Har.) gbõng (JMD) **IVORY COAST** ABE wunian (Aub.; FB) AKYE djien (FB) KRU-GUERE tabatué (RS) tébo (RS) têbo (RS) GUERE (Wobe) tabatué (RS) taratu (RS) KYAMA anianakia (Aub.; RS) **GHANA** AKAN-ASANTE nyankom-nini, nyankoma-nini (FRI, Enti) FANTE o-nyankõma, nynankoma nini (Enti) TWI o-nyankõma, nynankoma nini (Enti)

A shrub or small tree to 10 m high, rarely more, unbuttressed, of secondary jungle in the closed forest zone from Guinea to Ghana.

The wood is yellowish-brown, light and strong (2, 4, 7) and varies between being soft (Liberia, 4) to hard (somewhat more than *M. serratus,* Ghana, 2, 8). It is used in Liberia to make domestic articles such as wooden spoons, combs, stools, etc. (6).

The young leaves are eaten in Liberia as a pot-herb (4) and are used for 'palaver' sauce (5). A small quantity of tannin has been detected in Ivorean material (1).

The fruit is edible and is sometimes eaten, e.g. in Liberia (5). The seed has an oil-rich kernel with 39.0% oil, 18.0% sugars and 35.7% proteins on dry weight in an Ivorean sample. The oil is 89.2% linoleic acid and 7.7% oleic acid. A large number of amino acids are also present with an unusually high amount of sulphur-bearing cystine (3).

References:

1. Bouquet & Debray, 1974: 125. 2. Burtt Davy & Hoyle, 1937: 80. 3. Busson, 1965: 113–8 (with kernel analysis). 4. Cooper 128, K. 5. Cooper 285, K. 6. Cooper & Record, 1931: 79. 7. Dalziel, 1937. 8. Irvine, 1961: 448.

**Myrianthus serratus** (Trécul.) Benth.

FWTA, ed. 2, 1:614, 616. UPWTA, ed. 1, 286.

French: 'wounian' des rivières (from Abe Ivory Coast, Aubrèville).

West African: **SENEGAL** MANDING-BAMBARA takola (JB) MANDINKA bailliri (A. Chev.) **SIERRA LEONE** GOLA fɔvɔ (FCD) KISSI g-bando (FCD; S&F) k-pando (FCD; S&F) KONO kaanɛ (FCD; S&F) LIMBA kuroko (JMD) LOKO fɔfɔ (S&F) MANDING-MANDINKA wakawaka (FCD) MENDE fɔfɔ (*def.*-i) (auctt.) nja fɔfɔ (S&F) SUSU feŋkai (FCD) SUSU-DYALONKE khambu-na (FCD) mulukho-na? (FCD) TEMNE *a*-waka (auctt.) VAI g-bã (FCD) **LIBERIA** KRU-BASA vahn, vahn vehn (C; C&R) MANO gbŏng (JMD) MENDE fɔfnɔ (C&R) **MALI** FULA-PULAAR (Mali) podi (RS) MANDING-MANINKA kangaba (RS) takala (A. Chev.; FB) tangaba (FB) **IVORY COAST** ABE wunian (auctt.) AKYE djien (auctt.) AVIKAM godé (A. Chev.; K&B) niangama (A. Chev.; K&B) KRU-GUERE teutieu win (K&B) **GHANA** AKAN-AKYEM nyankõm (FRI) nyankõm-nini (Enti) ASANTE onyankoma (FRI) onyankoma-nini (BD&H; FRI) FANTE nyankom-nyin, nyin: *male* (Enti) KWAWU *o*-nyankõma (FRI) TWI *o*-nyankoma-nini (DF) nyankuma-nini, nini: *male* (Enti) WASA *o*-nyankoma, *o*-nyankoma-nini (Enti) NZEMA baŋgama (FRI) **NIGERIA** HAUSA farin ganye = *white leaf; referring to the leaves* (JMD) IJO-IZON (Kolokuma) òfólò (KW) IZON (Oporoma) ọ̀fọ́fọ̀ (KW) JUKUN (Abinsi) apulu (JMD)

A shrub or tree, exceptionally to 20 m high, of the closed-forest and forest outliers in savanna in stream banks from Senegal to S Nigeria.

The trunk is cylindrical and erect, but branching low down and with stilt-roots. The timber is yellowish-white without differentiation between sap and heart-wood, light and firm. It is used in Liberia to make domestic articles such as wooden spoons, combs, stools, etc. (3).

The leaves are used in N Nigeria for wrapping kola seeds and are given the Hausa name meaning 'white leaf' (4). The young leaves are said to be eaten as a vegetable in Liberia (3), but a fairly high amount of saponside, together with mucilage, is reported in Ivorean material (1). Also in Ivory Coast, the use of the plant in arrow-poison is recorded (6). Though the part of the plant is not specified and the authors express doubt on the validity of this information, the presence of saponosides may be an explanation.

Elephants are said to eat the flowers (4). The fruits are 2.5–3.75 cm across composed of a cluster of 7 or so heads united basally. They contain a little flesh which is edible. The kernels are also edible and are sometimes eaten after cooking (4, 5). They are rich in oil. Ivorean material is reported to contain 31·7% oil, 30% sugars and 30·8% proteins, dry weight. The oil contains 74% linoleic acid and 21% oleic acid, and, like *M. arboreus,* the kernel holds a high number of amino acids with an unusually high amount of sulphur-bearing cystine and three peptidic alkaloids (2).

References:

1. Bouquet & Debray, 1974: 124–5. 2. Busson, 1965: 113–8 (with kernel analysis). 3. Cooper & Record, 1931: 79–80. 4. Dalziel, 1937. 5. Irvine, 1961: 448–9. 6. Kerharo & Bouquet, 1950: 135.

## CELASTRACEAE

**Elaeodendron buchananii** (Loes.) Loes.

FWTA, ed. 2, 1:626, as *Cassine buchananii* Loes.

A shrub or tree to as much as 30 m high of savanna and forest, recorded from Sierra Leone and Togo only in the Region, but occurring widespread across Africa to NE, E and South Central tropical Africa.

The timber is pale brown, fine-textured, rather hard and heavy, but works well when dry. It is fairly durable in contact with the ground at higher altitudes (3, 8).

The tree yields a gum (2), but of unrecorded utility. A close relative, *E. roxburgii* Wight & Arn., is the source of a high quality gum in India (6, 7).

The plant is known in Kenya to be extremely poisonous. If children have been eating the fruits or the leaves and then take a drink of milk, they will die. Sheep are also fatally poisoned by browsing (1, 3, 4, 5).

Young trees are often covered by web-spinning gregarious caterpillars (3), presumably one of the African silk-moths.

References:

1. Bally B.4670, K. 2. Battiscombe 642, K. 3. Dale & Greenway, 1961: 133, as *Cassine buchananii* Loes. 4. Glover & al. 214, K. 5. Glover & al. 2583, K. 6. Howes, 1949: 57. 7. Sastri, 1952: 141–2, as *E. glaucum* Pers. 8. Wye 640, K.

## **Hippocrataea** Linn.

West African: **IVORY COAST** ABURE olofé (B&D) orofé (B&D) BAULE komédada (B&D) krélé (B&D) KRU-GUERE (Chiehn) den titi (B&D) didali kpokpo (B&D) didali poto (B&D) kolétiti (B&D) koré titi (B&D) uamku (B&D) uenikru (B&D) wéléwéléku (B&D) KYAMA denderiniama (B&D) tiamalabébo (B&D) 'NEKEDIE' onaniku (B&D)

## **Hippocratea africana** (Willd.) Loes.

FWTA, ed. 2, 1:628. UPWTA, ed. 1, 289, as *H. richardiana* Cambess.

West African: **SENEGAL** FULA-PULAAR delbi (K&A) tálel wâdu MANDING-BAMBARA mãngana (JB) SERER ndèl (JB) tèl (JB) WOLOF taf = *to seal* (JB; DF) **THE GAMBIA** MANDING-MANDINKA kesayso (Fox) **GUINEA-BISSAU** MANDYAK onchom (JDES) **SIERRA LEONE** LOKO njabo(li) (JMD) MENDE mbalu (*def.*-i) TEMNE *an*-tonke (GFSE) **GHANA** AKAN-ASANTE nnoto (BD&H; FRI) **NIGERIA** FULA-FULFULDE (Adamawa) biwa (JMD) FULFULDE (Nigeria) balandibi (JMD) HAUSA godayi (JMD; ZOG) gwaɗayi (AST; JMD) TIV ipungwa (JMD) YORUBA pọnju òwiwí (IFE)

A woody liane with tough wiry stems, of savanna woodland and riverain fringes, commonly throughout the Region, and widespread in African, Asian and Australasian tropics.

The toughness of the stems is reflected in the Hausa saying *gwadayi saranka ba ja* ('it cannot be broken by pulling but has to be cut with a matchet'). They are not attacked by termites (2), and are favoured for this and their durability as binding-material, after splitting, in N Nigeria (3), and in The Gambia (4) for hut-roof thatching. In Kenya the stems are used to make ropes (5). In Tanganyika the bark is removed to make a fibre (1), and in Ethiopia the stems are used to bind the timbers in constructing granaries, and to weave into baskets (7). In a superstitious sense of the bond of tribalism, the stems enter into prescriptions for decoctions drunk by Fula youths preparatory to the ordeal of entering manhood (*sharo*) (3).

The foliage is said (in Tanganyika) to be rarely or never eaten by stock (6). In Senegal the leaves are made into a tisane for colds (Sébire fide 11) and occasionally a decoction of the liane is used in baths and by mouth to treat oedemas (8, 9).

In Sierra Leone the root is used to prepare a beverage called *djendjeng* (12). The root is used in Nigeria in treatment for skin-infections. Examination has shown no activity against fungi, nor Gram +ve, or Gram −ve bacteria (10).

References:

1. Bally 420, K. 2. Dalziel 217, K. 3. Dalziel, 1937. 4. Fox 57, K. 5. Gillett 13658, K. 6. Hornby 26, K. 7. Hundessa 67, K. 8. Kerharo & Adam, 1964, b: 549. 9. Kerharo & Adam, 1974: 332. 10. Malcolm & Sofowora, 1969: as *Hippocratea richardiana* Cambess. 11. Schnell, 1953, b: as *H. richardiana* Cambess. 12. Scott Elliot 4965, K.

**Hippocratea apocynoides** Welw., **ssp. guineensis** (Hutch. & M. B. Moss) Robson

FWTA, ed. 2, 1:627, as *H. guineensis* Hutch. & M. B. Moss. UPWTA, ed. 1, 289, as *H. guineensis* Hutch. & M. B. Moss.

West African: **NIGERIA** HAUSA godayi (ZOG) gwaɗayi (JMD; ZOG)

A climber to 13 m or more of secondary growths and in open gaps of developed secondary deciduous forest, occurring from Guinea to W Cameroons, and on across Africa to Uganda and Tanganyika.

The woody stems are used as a tie in N Nigeria because of their resistance to termites (1).

Reference:

1. Dalziel 204, K.

**Hippocratea indica** Willd.

FWTA, ed. 2, 1:627, as *H. indica* sensu Blakelock, p.p. UPWTA, ed. 1, 289, as *H. loesneriana* Hutch. & M. B. Moss.

West African: **SENEGAL** BANYUN dabufobénal (K&A) sińas sińas (K&A) DIOLA ('Kwaatay') noro uy (K&A) MANDING-MANDINKA kubeh-jara = *cures everything* (DA) **GUINEA** FULA-PULAAR (Guinea) gniako (A. Chev.) **SIERRA LEONE** MENDE mbalu (NWT, JMD) wawa (*def.*-i) (L-P), *probably a group name for certain woody climbers with slender stems* (FCD) **LIBERIA** KRU-BASA wehn-flay (C&R) MANO gie gbinni (JMD) popio lo (JMD) **NIGERIA** YORUBA pọnju-òwiwí (Dawodu)

A shrub or slender climber to 4 m long of the closed-forest and dry deciduous forest, occurring from Guinea and Mali to S Nigeria and Fernando Po, and widely distributed across tropical Africa and Asia.

The lianes are strong and are used as binding material in constructing huts (5).

The plant (part unstated) is prepared as a pulp for application to guinea-worm sores, and as a draught for respiratory troubles by the Baule of Ivory Coast (2). The leaves are steeped in water in Casamance (Senegal) for washing — a common practice for women and children (8, 9), and in decoction along with the roots for washing infants with various illnesses and fevers, and as a strengthening tonic for adults (7, 9). A decoction of macerated leaves is used in Liberia for cleansing sores and wounds (3, 4). The leaf-sap, and also a root-decoction, are drunk in Tanganyika for malaria (6). A trace of alkaloid has been detected in the leaves (1).

The macerated root is applied as a poultice to sores in Liberia (4). The root chewed with hot oil is blown on the skin of a patient as a revulsive (5). A root-decoction is taken in draught in Casamance to relieve colic, and in treatment for female sterility. A secondary action is to cause diuresis (8, 9).

Root-extracts show an antibiotic action against Gram +ve bacteria. *Pristimerin,* a strong Gram +ve antibiotic substance, is present. Root-extracts have no action on Gram −ve organisms (9).

References:

1. Adegoke & al., 1968. 2. Bouquet & Debray, 1974: 65. 3. Cooper 409, K. 4. Cooper & Record, 1931: 80, as *Hippocratea loesneriana* Hutch. & M. B. Moss. 5. Dalziel, 1937. 6. Haerdi, 1964: 107. 7. Kerharo & Adam, 1962. 8. Kerharo & Adam, 1963, b: 9. Kerharo & Adam, 1974: 334–6, with phytochemistry and pharmacology.

**Hippocratea iotricha** Loes.

FWTA, ed. 2, 1:628. UPWTA, ed. 1, 289.

West African: **SIERRA LEONE** MENDE ndambi *for woody climbers generally* (FCD) tɛli (NWT)

A woody climber of the closed forest, recorded from Sierra Leone, Liberia and S Nigeria, and in E Cameroun.

**Hippocratea macrophylla** Vahl.

FWTA, ed. 2, 1:629.

A shrub, small tree or climber of fringing forest from Guinea to W Cameroons, and extending to Zäire and Angola.

An oil from the seeds is rubbed on the body in Ghana (1).

Reference:

1. Irvine, 1961: 455.

**Hippocratea myriantha** Oliv.

FWTA, ed. 2, 1:627.

A climbing shrub of the closed forest from Sierra Leone to W Cameroons, and on to central Africa, Zäire and Angola.

The wood is very tough. The lianes are used in Zäire for lashing together palm-oil presses (3).

The plant is sometimes used in Ivory Coast to treat diarrhoea in suckling babies. It has emetic action for which it is used to treat coughs and as a poison-antidote (2). The bark of this along with other drug-plants is macerated for two days in water and used for diarrhoeas and dysenteries in Congo (1).

The roots burnt and powdered with the bark of *Heisteria parvifolia* (Olacaceae) and rock-salt, are held to be anodynal for headache and costal pains in Congo, the prescription being applied onto epidermal scarification at the area of pain (1).

References:

1. Bouquet, 1969: 135. 2. Bouquet & Debray, 1974: 64. 3. Wilczek, 1960, b: 170.

**Hippocratea pallens** Planch.

FWTA, ed. 2, 1:627.

West African: **SENEGAL** BASARI a-peutẹiɛ́wɔ́ (Ferry) BEDIK gi-puwɛɗ (Ferry) KONYAGI a-nkaryankank (Ferry) **GUINEA** MANDING-MANINKA sungala-lé (CHOP) **SIERRA LEONE** SUSU furi (NWT) **IVORY COAST** MANDING-MANINKA bongu (B&D)

A climber to 13 m long, of the closed-forest on riversides, from Senegal to S Nigeria, and on to Zäire, Kenya and Angola.

The stems are cut into batons called *soungala* in Guinea used by women in dancing (2).

A bark-decoction is used in baths and steam-fumigations in Ivory Coast to relieve rheumatic pains and headache, and in draught for dysentery (1).

References:

1. Bouquet & Debray, 1974: 64. 2. Pobéguin 815, K.

**Hippocratea paniculata** Vahl

FWTA, ed. 2, 1: 627.
West African: **SENEGAL** DANYUN damoré guren (K&A) DIOLA (Efok) bufembõ (K&A) DIOLA (Fogny) bufembe (K&A) **GUINEA-BISSAU** FULA-PULAAR (Guinea-Bissau) quelissá (D'O)

A climbing shrub of the galleried forest occurring from Senegal to S Nigeria, and also in Zaïre, Angola and E Africa.

The strong flexible lianes are used by Kipsigis and Masai in Kenya as ties in hut-building (1). In Casamance (Senegal) the leaves are crushed and macerated to put into baths and to take by draught for lumbago and general pains (2, 3).

References:

1. Glover al., 236, 378, K. 2. Kerharo & Adam, 1962. 3. Kerharo & Adam, 1974: 332–3.

**Hippocratea rowlandii** Loes.

FWTA, ed. 2, 1: 628.
West African: **GHANA** AKAN-ASANTE ntwea (BD&H; FRI) ogogo (Williams)

A bush or woody climber of drier closed-forest from Guinea to W Cameroons, and also in E Cameroun.

The flexible stems are used in Ghana as ties for rafters in hut-building (1). The fruits are laxative (2).

References:

1. Irvine, 1961: 455–6. 2. Ll. Williams 271, K.

**Hippocratea velutina** Afzel.

FWTA, ed. 2, 1: 629. UPWTA, ed. 1, 289.
West African: **SIERRA LEONE** MENDE tombe (Dawe)

A woody climber to 16 m long of the secondary closed-forest in damp situations from Guinea to W Cameroons, and extending widespread across tropical Africa.

The young branches are useful as binding material. In Sierra Leone the leaves are dried, pounded and boiled to be applied as a poultice to the temples for headache (Afzelius fide 1). The Igbo of S Nigeria use the plant as magic to keep termites away from yams (2). In Senegal the seeds are used for headaches and fever (Lanessan fide 1).

References:

1. Dalziel, 1937. 2. Thomas, N. W., 1852 (Nig. Ser.), K.

**Hippocratea welwitschii** Oliv.

FWTA, ed. 2, 1: 628. UPWTA, ed 1, 289.
West African: **LIBERIA** MANO gie gbini (Har.) **GHANA** ADANGME akladefi (?akradefi) (Johnson, ex FRI) akladekpa (?akradekpa) (Johnson, ex FRI) **NIGERIA** YORUBA ijan (Kennedy)

A shrub or climber of closed primary or mature secondary forest, or in thickets of secondary scrub, from Guinea to W Cameroons, and widespread across Africa to Angola, Uganda and Tanganyika.

The roots of Ghanaian material have been found to contain a sort of gutta, but only in small amounts, 2.58% crude, or 1.22% pure (2), of no recorded utility.

The plant (part unstated) is considered in the lagoon area of Ivory Coast to be helpful in easing labour and delivery at childbirth (1).

References:

1. Bouquet & Debray, 1974: 65. 2. Dalziel, 1937.

**Maytenus acuminata** (Linn. f.) Loes.

FWTA, ed. 2, 1:625, as *M. acuminatus* (Linn. f.) Loes. [NOTE: *Maytenus* Molina is in feminine gender.]

An understorey shrub or small tree of closed montane forest, recorded only from W Cameroons in the Region, but occurring extensively across Africa to E and S Africa.

The wood possesses great strength and is prized in Kenya (1), and in Uganda (2) for walking-sticks, and in Lesotho for fighting-sticks (3). A fibre is obtained from the bark in Lesotho which is used for tying the wooden frames of huts (3).

References:

1. Brasnett 1507, K. 2. Byabainzi 111, K. 3. Guillarmod, 1971: 440.

**Maytenus senegalensis** (Lam.) Exell

FWTA, ed. 2, 1:624–5. UPWTA, ed. 1, 287–8, as *Gymnosporia senegalensis* Loes.

West African: **MAURITANIA** ARABIC ('Maure') eich (Aub.) **SENEGAL** ARABIC ('Maure') eich (Aub.) BASARI a-ngùyì, a-ngɔ̀ỹ (K&A; Ferry) BEDIK ga-ngɔndy, ga-ngoyindy (Ferry) DIOLA bu fimbok (JB, ex K&A) FULA-PULAAR (Senegal) giagut (K&A) gialgoti (K&A) gielgotel (K&A) TUKULOR dialgoti (JLT) KONYAGI aŋaty, a-ngoy (Ferry) MANDING-BAMBARA ntogoyo (JB, ex K&A) MANINKA gégé (K&A) gogé (K&A) gogué (Aub.) guégué (Aub.) ngéké (K&A) ngigé (K&A) n-guéké (Aub.) n-guigué (Aub.) tolé (Aub., ex K&A) 'SOCE' kasabaro (K&A) SERER ndafar (JB) n-tafar (JB; K&A) n-tafara (K&A) SERER-NON ndukut (auctt.) pori (auctt.) NON (Nyominka) dafar (K&A) indafar (K&A) SONINKE-SARAKOLE ńalé (K&A) WOLOF dori (auctt.) énidek = *head load of thorns* (K&A; DF) gédek, génedek = *spiked fish* (K&A; DF) génidek (K&A) génôdek (K&A) n-gidek = *of spike/thorn* (K&A; DF) **THE GAMBIA** FULA-PULAAR kelele (DRR; DAP) keleli (DAP) PULAAR (The Gambia) jalkoti (DAP) jalugotet (DRR) tiapule-ranehi (DRR) WOLOF kada (DRR; DAP) ngendi (DAP) ngenidik (DRR) weke (DAP) **GUINEA** FULA-PULAAR (Guinea) dialgoti (Aub.) MANDING-MANDINKA kada MANINKA gogué (Aub.) guégué (Aub.) n-guéké (Aub.) n-guigue (Aub.) tolé (Aub.) **MALI** MANDING-MANINKA gogué (Aub.) guégué (Aub.) n-guéké (Aub.) n-guigue (Aub.) tolé (Aub.) SONGHAI hassana (A. Chev.; Aub.) TAMACHEK hasahanna (Aub.) 'SENUFO' kafukoinan (Aub.) **UPPER VOLTA** DAGAARI koktripa (K&B) GRUSI cesiu (K&B) cessio (K&B) KIRMA niembélé (K&B) wésam (K&B) MANDING-BAMBARA kussié (K&B) DYULA guegué (K&B) guéké (K&B) MOORE tokobuguri (K&B) tokovuguri (Aub.; K&B) SENUFO-KARABORO naplafantien (K&B) 'OUROUDOUGOU' sukon (Aub.) 'SENUFO' nanienga (K&B) **IVORY COAST** MANDING-DYULA guegué (K&B) guéké (K&B) MANINKA guegué (K&B) guéké (K&B) n-gbékémon (A&AA) tonson-uni (B&D) SENUFO-TAGWANA mone (K&B) 'SENUFO' nanienga (K&B) **GHANA** AKAN-ASANTE kumakuafo = *kills farmers, because of its very sharp spines* (auctt.) BRONG asandurowa (BD&H; FRI) kiliummulu (FRI) kulummulu (FRI) primbiri (FRI) DAGAARI zaigoli (JMD) DAGBANI zaigoli (FRI) GBE-VHE nowe (FRI) wɔtsinotsi (FRI) GUANG-GONJA bàndɔ̀ŋbìrà (Rytz) KRACHI primbiri (CV; BD&H) NCHUMBULU kulumbila (Coull; FRI) MOORE tokobuguri (AEK; FRI) tokoyogwri (AEK; FRI) SISAALA bademi (AEK; FRI) **TOGO** GBE-'KPEVE' nowɛ (FRI) SORUBA-KUYOBE atiéklie (Aub.) TEM (Tshaudjo) mlimlisaure (Volkens) YORUBA-IFE OF TOGO nowoe (Volkens) **DAHOMEY** BATONNUN sakisakiné (Aub.) GBE-FON djaduma (Aub.) SORUBA-

KUYOBE atiéklie (Aub.) **NIGER** GURMA lidiamali (Aub.) SONGHAI hassana (Aub.) TAMACHEK hasahanna (Aub.) **NIGERIA** FULA-FULFULDE (Nigeria) bakororo (JMD) cabbi-taniki (JMD) tultulhi, tultulki (*pl.* tultulɗe) (JMD; KO&S) yaree lesdi (MM) HAUSA kumban-shafu (EWF) ķunķushewa, ķyalbuwa, mangaladi (ZOG) ƙunƙushewa (MM) námíjìn tsaádaà = *male tsaádar* (*Ximenia americana, Olacaceae, wild olive*) (ZOG) HAUSA ([illegible]) [illegible] (Lely) ƙunƙushewa (JRA; JMD) ƙurunƙushewa, ƙurunƙushiya (JMD) kyalbuwa (JMD) mangaladi (Lely; JMD) namijin tsada = *male Ximenia* (JMD; KO&S) IGALA ucholom (Odoh) NUPE momfofoji (K O&S) YORUBA ìṣẹ̀polóhùn (JMD) ṣépólóhùn (JRA; KO&S)

A shrub or tree to 8 m high by 25 cm (sometimes to 70 cm, 21) in girth, of savanna from sea-level to montane situations, throughout the Region from Senegal to Nigeria, and widespread in the savanna regions across Africa.

The wood is whitish to red-brown, hard, fine-grained and durable (6, 9, 19, 21). A vegetable salt is obtained from the wood and the leaves reduced to an ash in N Nigeria (9) and in Sudan (5, 9, 35), and from the leaves in Zaïre (36). The stems are commonly used in N Kenya to make Turkana stools, wooden platters and spoons (28). The Kipsigis use the wood for axe-handles, and the Masai carve ear-trinkets out of it (11). The branches with their spiny armature are favoured by cattle-folk for making intrusion-proof cattle-enclosures (12, 13, 14). Cattle-bells are made of the wood with a characteristic sound resulting in the Turkana name for the tree: *koro-koro-koro* (28). The thorns are used by butcher-birds (shrikes) in S Rhodesia for impaling insects they catch (Pardy fide 19).

The leaves and twigs are grazed by cattle, sheep and goats in Senegal (1) and by camels and goats in N Kenya (28). Kenyan Masai add the leaves to soup and broth as a flavouring (14). Though the leaves are widely used in medicines they are free enough of toxic substances not to kill laboratory test mice, unlike the root-bark which can cause a 40% mortality (26). A leaf-infusion does however intoxicate fish. A decoction of leafy twigs is used to bathe new-born infants at Lokoja in N Nigeria (9). A leaf-decoction is widely used for toothache, tooth-abscesses and mouth-infections (Senegal, 23, 24, 25; Soudano-guinea, 6; Ivory Coast-Upper Volta, 26; Ivory Coast, 2; Ghana, 18, 19; Togo, 9; Nigeria, 29).

The leaves have a slight laxative action and are taken in many parts for gastro-intestinal troubles and dysentery as a vermifuge (Senegal, 24, 25; Ivory Coast-Upper Volta, 26; Nigeria, 29; Tanganyika, 17). In Senegal the powdered dried leaves are mixed with milk as a vermifuge for children (Trochain fide 9). In Zaïre the young shoots are employed to relieve blennorrhoea (36). Leaf-sap is recommended for use in eye-trouble in Senegal (25). Similar use is recorded in Uganda (20). Sap from pounded leaves with sugar is taken in Tanganyika for schistosomiasis and with a root-decoction of *Cyperus papyrus* (Cyperaceae) for female sterility (24). Green leaves pounded up are commonly used in Tanganyika as a plaster for sores (33). In Ivory Coast-Upper Volta a decoction of leaves and twig-bark, sometimes seasoned with bull's pancreas, is prepared in baths, steam fumigation and in draughts for generalized oedemas, while the leaves with those of *Crossopteryx febrifuga* (Rubiaceae) are decocted for giving as an enema for strengthening debilitated children (26).

The leaves and branches contain some dulcite and tannins. The leaves also contain a wax which is mainly esters of ceryl alcohol, and also a sterol, flavenol, flavonic glycoside, a holoside, and a substance which appears to be a rubber (26, 29). Saponosides, flavone derivatives and tannins are present in the bark (7). The closely related *M. undata* (Thunb.) Blakel. is recorded as yielding a manna (see there). The present species should be examined for this character.

The bark is commonly used in Senegal for infants with fevers, loss of appetite and general ill-health, and for adults with jaundice and costal pains (23). It is also considered excellent for treating gastric ulcers, obviating surgical intervention (23). In Upper Volta the Moore use a bark-decoction for washing sores after which they apply a dressing of the powdered bark along with the bark of *Terminalia macroptera* Guill. & Perr. (Combretaceae) (26). In India an insecticidal action is claimed for the powdered bark mixed with mustard oil in dusting on the head for *Pediculus capitis* (34), but inconclusive insecticidal activity for the root and bark is also reported (16).

The root is slightly bitter in taste and has a mild laxative action. It is widely used in the soudanian region for all gastro-intestinal troubles (6, 9, 26) and is a common market commodity in Dakar markets for it is considered one of the more active drugs in the Senegalese pharmacopoea (22, 23, 25). The root is used against tertiary syphilis, female complaints, leprosy, dysentery, blennorrhoea, etc. The root-bark is used in Nigeria in infusions for long-standing dysentery, and a decoction is used to relieve pain especially at childbirth (4). A root-infusion is used in Tanganyika for violent stomach-ache (15, 32). In Zäire a root-infusion is put onto sores (36).

Examination of Nigerian material for alkaloids showed none present (3).

The flowers are much visited by bees (30). A yeast culture is made from the fruit in Sudan (31), and in Kenya both flowers and young fruit are sought after by birds to eat (28). A decoction of seeds is taken in Senegal for catarrh (24).

References:

1. Adam, 1966, a. 2. Adjanohoun & Aké Assi, 1972: 83. 3. Adegoke & al., 1968. 4. Ainslie, 1937: sp. no. 83, as *Celastrus senegalensis*. 5. Andrews A.1525, K. 6. Aubréville, 1950: 349, as *Gymnosporia senegalensis*. 7. Bouquet & Debray, 1974: 65. 8. Dale & Greenway, 1961: 136. 9. Dalziel, 1937. 10. Eggeling & Dale, 1952: 82, as *Gymnosporia senegalensis* (Lam.) Loes. 11. Glover & al., 771, K. 12. Glover & al., 2330, K. 13. Glover & al. 2357, K. 14. Glover & Samuel 3083, K. 15. Harley 9169, K. 16. Heal & al., 1950: 109, as *Gymnosporia senegalensis*. 17. Haerdi, 1964: 107. 18. Irvine 1744, K. 19. Irvine, 1961: 456–8. 20. Jarrett 195, K. 21. Keay & al., 1964: 199–200. 22. Kerharo, 1967. 23. Kerharo & Adam, 1964, b. 24. Kerharo & Adam, 1964, c: 314. 25. Kerharo & Adam, 1974: 333–4, with phytochemistry and pharmacology. 26. Kerharo & Bouquet, 1950: 136–7. 27. Koritschoner 1594, K. 28. Mwangangi 1384, K. 29. Oliver, 1960: 30, 71. 30. Proctor 2630, K. 31. S. [N.D. . . .] 7505, K. 32. Tanner 365, K. 33. Tanner 2072, 2968, 3583, 3638, K. 34. Watt & Breyer-Brandwijk, 1962: 183, as *Gymnosporia senegalensis* Loes. 35. Whitehead 7, K. 36. Wilczek, 1960, a: 122.

## **Maytenus undata** (Thunb.) Blakelock

FWTA, ed. 2, 1:624.

A shrub or tree to 8 m high by 70 cm girth, of both lowland and montane savanna areas from Guinea to W Cameroons, and widespread in E Africa.

The wood is red, hard, dense and heavy (3), with a certain amount of flexibility making it a favourite wood for fighting-sticks in Lesotho (4), and for axe-handles by the Masai and Kipsigis in Kenya (6). The stems are also used for hut-building and for fuel in E Africa (6, 11).

The foliage is eaten by camels in Sudan (1), and cattle-folk of Kenya feed the leafy branches to cattle in time of drought (7). The bark is eaten in Kenya as a relish (13), and boiled and the liquid mixed with milk is given as a tonic to babies and young children (6). It is also boiled in broth as a vegetable (7) and as a general tonic (8). Pounded and steeped in cold water, a red liquor results which is taken as a beverage (7). A manna containing 54% dulcite, 6.4% dextrose sugar and 6.6% sucrose is reported obtained as a white incrustation on the plant in NW Rhodesia (10).

The root is boiled with milk and given to babies with stomach-trouble in Kenya (2). In Tanganyika the root is considered beneficial for diseases of the urethra (12), and root and bark are boiled together for vapour-baths over which fever patients squat (15). The plant (? bark) is considered in Tanganyika to provide a cough-cure (14), and to be a good chest-medicine.

In Mozambique the fruit is recorded as being edible (9).

References:

1. Andrews A.3587, K. 2. Bally 3616, K. 3. Dale & Greenway, 1961: 136–8. 4. Guillarmod, 1971: 441. 5. Gane 69, K. 6. Glover & al., 644, K. 7. Glover & al., 1181, K. 8. Glover & al., 2612, K. 9. Gomes e Sousa 4351, K. 10. Harrison, 1950: as *Gymnosporia deflexa*. 11. Kanya 12, K. 12. Koritschoner 1237, K. 13. Rammell 1053, K. 14. Semsei 1273, K. 15. Tanner 1293, .

**Mystroxylon aethiopicum** (Thunb.) Loes.

FWTA, ed. 2, 1:625, as *Cassine aethiopica* Thunb.

A bush or tree to about 7 m high, occasionally more, of montane forest in Guinea and W Cameroons and probably elsewhere in the Region; dispersed widely over tropical Africa in savanna vegetation from sea level to over 2000 m in altitude (1, 3).

No usage is recorded in the Region.

In S Africa the wood is used by the Xhosa to make knob-kerries and as fuel (13). Cattle-folk in Kenya also use it as firewood (8) and for hut-construction and furniture (6).

The green leaves are used pulped up on sores in Tanganyika (10). Small traces of alkaloid in the leaves and twigs of Madagascan material have been reported (2).

The bark yields a brownish dye (1, 9). Crumbled and infused, the bark is much used in Kenya to make a tea (4, 5, 7, 8). It is considered to be a good stomach-medicine (6) and boiled in water with milk added it is given to young babies as a stomachic (5). The plant is amongst the most commonly used drug-plants in Madagascar. The bark-decoction is given for stomach and pulmonary complaints as a stimulant and tonic, for hypertension, antiabortifacient, diuretic, etc. (2). The root and bark are administered, presumably for diuretic action, for urinary infections (11) and gonorrhoea (12) in Tanganyika. A bark-infusion in milk or whey is a drink given by the Zulu to worm-infected cattle in S Africa (13). The plant is said to be free of alkaloids but to contain abundant leucanthocyanins and tannins, some saponins (not in the root bark) and a trace of sterols (2).

The fruit is a drupe about 1 cm long with a sweet edible pulp.

References:

1. Dale & Greenway, 1961: 132, as *Cassine aethiopica* Thunb. 2. Debray & al., 1971: 17–18, 57–58. 3. Eggeling & Dale, 1952: 84. 4. Glover & al., 77, K. 5. Glover & al., 202, K. 6. Glover & al., 1971, K. 7. Glover & al., 2522, K. 8. Glover & Samuel 3163, K. 9. Graham 1566, K. 10. Tanner 3677, K. 11. Tanner 4404, K. 12. Tanner 4356, K. 13. Watt & Breyer-Brandwijk, 1962: 177, as *Cassine aethiopica* Thunb.

**Salacia camerunensis** Loes.

FWTA, ed. 2, 1:633, incl. p. 634, *S. sp. D.*

West African: **GHANA** AKAN-ASANTE nse dua-nse huma (S. T. Jackson)

A lianescent shrub of the closed forest from Ghana to W Cameroons, and to Zäire.

**Salacia chlorantha** Oliv. **ssp. dalzielii** (Hutch. & M. B. Moss) Hallé

FWTA, ed. 2, 1:632, sp. b. under *S. senegalensis* (Lam.) DC.

West African: **GHANA** AKAN-ASANTE eplatokɔkɔe-wiriwiria (FRI)

A shrubby climber recorded from Guinea to Nigeria. The petiole contains a latex (1).

Reference:

1. Hallé, 1962, b: 223–6.

**Salacia columna** Hallé

FWTA, ed. 2, 1:631, as *S. senegalensis* sensu Blakelock p.p.

A liane attaining 10 m long, occurring in Sierra Leone, Liberia and Ivory Coast.

The fruit is sub-spherical up to 4–4.5 cm in diameter. The fruit of var. *akearsi* Hallé has an elongated fruit to 7.5 cm long. The flesh is sweet and edible (1).

Reference:

1. Hallé, 1962, b: 226–9.

**Salacia cornifolia** Hook. f.

FWTA, ed. 2, 1:633, as *S. erecta* sensu Blakelock p.p. UPWTA, ed. 1, 290.
West African: GHANA GA ayamensa (Imp. Inst; FRI) TOGO GA echenjongo (von Doering)

A liane, recorded from Sierra Leone to W. Cameroons, and to Ubangi-Shari.
The fruit pulp is sweet and edible (1–3).

References:

1. Dalziel, 1937. 2. Hallé, 1962, b: 191–4. 3. Tuley 720, K.

**Salacia debilis** (G. Don) Walp.

FWTA, ed. 2, 1:633. UPWTA, ed. 1, 290.
West African: GHANA AKAN-ASANTE hama-kyerebeŋ = *green mamba climber* (BD&H; FRI) GA kpleŋ (FRI)

A forest climber to 7 m high, often riverain, of the closed-forest, from Guinea-Bissau to Fernando Po, and on into Zaïre.
The pulped leaves in a plaster, or the leaf-sap as an embrocation, are put into areas of skin-itch and ringworm in Congo (1).

Reference:

1. Bouquet, 1969: 136.

**Salacia erecta** (G. Don) Walp.

FWTA, ed. 2, 1:633, as *S. erecta* sensu Blakelock, p.p.
West African: SIERRA LEONE GOLA kɔtigikpoŋ (Pyne) MENDE-KPA gi-kɔtigibuwe (Pyne) SUSU woni (NWT) TEMNE *an*-kəŋkərəs (FCD) IVORY COAST KRU-GUERE (Chiehn) bago titi (B&D) kpokpo (B&D) 'NEKEDIE' urizanabagri (B&D) GHANA GA ayamensa (FRI)
*NOTE: S. cornifolia Hook. f. may perhaps also refer, and S. cerasifera Welw. to the Sierra Leone vernaculars*

A shrub or liane to about 5 m high, of the closed-forest, dispersed throughout the Region from Guinea to Nigeria, and on across Africa to Angola, Uganda and Tanganyika.
The plant is considered to be a good soothing medicine for children fearful of darkness. A glassful of tisane taken on going to bed leads to peaceful and deep sleep (1).

Reference:

1. Bouquet & Debray, 1974: 65.

**Salacia lateritia** Hallé

FWTA, ed. 2, 1:632, as *S. pyriformis* sensu Blakelock, p.p., and 1:634, as *S. sp. E.*
West African: LIBERIA MANO giligbon (Adames)

A liane occurring in Liberia, Ivory Coast and Nigeria.
The petiole contains latex (1).

Reference:

1. Hallé, 1962, b: 216–8.

**Salacia lehmbachii** Loes.

FWTA, ed. 2, 1:633.

A bush or small tree to 3 m high, in understorey of closed-forest, galleried-forest and perhaps savanna. Recorded only in W Cameroon in the Region, and occurring also to Zäire and in E Africa.

The wood is yellow and very hard (2). It is used in Tanganyika for tool handles (1).

The grated root is made up into a wet pad for placing over umbilical hernias and haemorrhoids in Tanganyika (1).

References:

1. Greenway 6052, K. 2. Wilczek, 1960, b: 214–5.

**Salacia leptoclada** Tul.

FWTA, ed. 2, 1:633, as *S. erecta* sensu Blakelock, p.p. (syn. *S. baumannii* Loes.).

A shrub or liane of the coastal region of Ivory Coast, Ghana and Togo, also in East and South Central Africa.

The fruit has sweet edible pulp (1, 2).

References:

1. Hallé, 1962, b: 194–6. 2. Irvine 894, K.

**Salacia nitida** (Benth.) N.E.Br.

FWTA, ed. 2, 1:633, p.p., incl. p. 634. *S. sp. C.*

West African: **GHANA** NZEMA mumue (BD&H; FRI) **NIGERIA** IGBO (Owerri) otuwerehi (Tuley)

A liane to 10 m long, of primary and secondary forest and savanna, from Sierra Leone to W Cameroons, and extending to Zäire.

The bark is used for constipation and syphilitic conditions in Congo (1). An unnamed alkaloid has been detected in the leaves and stems (5).

The fruit which is 6–10 cm long by 2.5–4.0 cm in diameter has a sweet edible pulp (2–4).

References:

1. Bouquet, 1969: 136. 2. Busson, 1965: 355. 3. Hallé, 1962, b: 168–70. 4. Irvine, 1961: 460. 5. Willaman & Li, 1970: with reference.

**Salacia oliveriana** Loes.

FWTA, ed. 2, 1:631, as *S. senegalensis* sensu Blakelock, p.p.

A species from Ivory Coast, previously held under *S. senegalensis* (Lam.) DC. The petiole contains latex. The subcylindrical fruit 10 cm long by 3·5 cm in diameter has a sweet edible pulp (1).

Reference:

1. Hallé, 1962, b: 218–20.

**Salacia pallescens** Oliv.

FWTA, ed. 2, 1:632.

A forest understorey shrub to 3 m tall, recorded from Guinea, Sierra Leone and N and S Nigeria, and in Gabon and Zäire.

The plant enters into treatment for skin-infections in Nigeria. Examination of leaf-extracts has shown no antibacterial nor antifungal activity (1).

Reference:

1. Malcolm & Sofowora, 1969.

**Salacia pyriformis** (Sab.) Steud.

FWTA, ed. 2, 1:632, in part.
West African: **SIERRA LEONE** MENDE bowu-gigbɔ, bugigbɔ (FCD) TEMNE *an*-kəŋkərəs (FCD) *an*-kɛp *a general term, not strictly correct* (FCD)

A forest climber recorded from Senegal to Nigeria, and dispersed across Africa to Zäire and to NE and E Africa.

The leaf petioles contain latex (2). In Tanganyika the leaves are used as a 'stomach' medicine and also a 'small children's remedy' (4), of unspecified significance.

The fruit-pulp is eaten in Sierra Leone (1), and in Tanganyika (3). It is larger than the fruit of *S. senegalensis* and is said to be the largest of the *gigboi* fruits in Sierra Leone (1).

References:

1. Deighton 2772, K. 2. Hallé, 1962, b: 211–3. 3. Musk 90, K. 4. Yusufu Mohamedi 9099, K.

**Salacia senegalensis** (Lam.) DC.

FWTA, ed. 2, 1:631, in part. UPWTA, ed. 1, 1:290.

English: beacon bush; beacon fire (Dalziel).
West African: **SENEGAL** DIOLA bu kãndikãng (JB) DIOLA (Fogny) bukâgir (K&A) épumbay (K&A) DIOLA-FLUP bulel (K&A); = *fetching firewood* (DF, The Gambia) bulil (K&A); = *private parts* (DF, The Gambia) MANDING-MANDINKA karra (K&A) SERER-NON (Nyominka) n-doloré (K&A) WOLOF hebet (K&A) **THE GAMBIA** DIOLA (Fogny) butagir (DF) épumbay *onomatopoeic, as the sound of a gun when the fruit is broken open by hand* (DF) DIOLA-FLUP bu layo (DF) FULA-PULAAR (The Gambia) n-dama (DRR; DAP) MANDING-MANDINKA karro = *dye* (Hayes) sinjang-koyo = *white sinjang* (*Cassia sieberana; Leguminosae: Caesalpinoideae*) (DAP) WOLOF kebett (Williams) kibil (DAP; Barter) **GUINEA-BISSAU** BALANTA blende (JDES) BIDYOGO ngeuêdja (JDES) FULA-PULAAR (Guinea-Bissau) suncurô-fólé (EPdS) MANDING-MANDINKA sinjang-koyo suncurò-fóleò (JDES) MANDYAK becubar (JDES) **SIERRA LEONE** KRIO beacon bush (JMD) beacon fire (JMD) MENDE dogbojelegbo (SKS) gigbɔ (NWT, SKS) TEMNE *an*-kɛp (FCD)

*NOTE: S. chlorantha Oliv. is very similar, and the Sierra Leone vernaculars may perhaps also apply to it.*

A shrub to 2 m high or climbing to the tree canopy of primary and secondary forest, of damp situations from Senegal to S Nigeria, and widespread to Zäire and E Africa.

The stems are used in Zäire to make axe-handles and fishing-rods (12). The leaf-petioles, as also the seeds, contain a latex (4). A leaf-infusion is used in West Africa as a lotion for sick children (2) and a preparation (part unstated) is used in S Nigeria for 'painful skin' (10).

A bark-decoction is taken in Nigeria (1) and in The Gambia (8) for stomach pains. The roots in decoctions are used as a mild aperient in Nigeria (1) and are given to children with diarrhoea in Senegal (6, 7). In Tanganyika water in which the roots have been boiled is drunk for sharp stomach-pains (9), and for malaria (3). The powdered root is also dusted on old infected wounds (3). A root-macerate is taken in Senegal for blennorrhoea and as a cough-cure (5, 7).

The fruit has been reported used in S Nigeria for washing cotton thread (11). The fruit-pulp is edible.

References:

1. Ainslie, 1937: sp.no. 307. 2. Dalziel, 1937. 3. Haerdi, 1964: 108. 4. Hallé, 1962, b: 221–3. 5. Kerharo & Adam, 1962. 6. Kerharo & Adam, 1964, c: 319. 7. Kerharo & Adam, 1974: 336. 8. Rosevear, 1961. 9. Tanner 3717, K. 10. Thomas, N. W., 2328 (Nig.Ser.), K. 11. Thomas, N. W., 2358 (Nig.Ser.), K. 12. Wilczek, 1960, b: 200–1.

**Salacia staudtiana** Loes.

FWTA, ed. 2, 1:633, incl. *S. caillei* A. Chev. UPWTA, ed. 1, 290, as *S. caillei* A. Chev.

West African: **SIERRA LEONE** KISSI yilio (FCD) **LIBERIA** KRU-BASA du-kpay (C) **GHANA** ADANGME kpleŋ (FRI) GA kpleŋ (Johnson; FRI)

A shrub or liane to 7 m high, of forest understorey, sometimes riverain, from Guinea to W Cameroons and extending to Zäire.

The root contains a deep yellow gutta, 5.02% crude, or 3–4% when purified of resin and colouring matter. Stems yield 2.85% and 1.03% respectively. Quantities are too low to be of economic interest (1).

The fruit pulp is edible (1, 2).

References:

1. Dalziel, 1937. 2. Deighton 2407, K.

**Salacia stuhlmanniana** Loes.

FWTA, ed. 2, 1:632, as *S. lomensis* Loes, and p. 634, *S. gilgiana* Loes. UPWTA, ed. 1, 290, as *S. lomensis* Loes.

West African: **GHANA** AKAN-TWI abɔntɔre (FRI) GA kpleŋ (auctt.)

A scrambling shrub to 2 m high, of savanna areas from Ivory Coast to Togo.

The leaves contain a latex (3), reported present also in the fruit-pulp and seed-coat (1, Loesener fide 2, 4, 5). The fruit-pulp is orange-red when ripe. It is sweet and edible (1, 4, 5).

References:

1. Busson, 1965: 355, as *S. lomensis* Loes. 2. Dalziel, 1937. 3. Hallé, 1962, b: 209–10. 4. Irvine 1648, K. 5. Irvine, 1961: 459, as *S. lomensis* Loes.

**Salacia togoica** Loes.

FWTA, ed. 2, 1:632. UPWTA, ed. 1, 290.

West African: **GHANA** ADANGME kpleŋ (FRI) AKAN-TWI abɔntɔre (JMD) GA kpleŋ (FRI)

A climbing shrub to 10 m high recorded from the savanna zone from Sierra Leone to Nigeria.

The root has been recorded as containing 13.10% of a crude yellow gutta, which, when purified of colouring matter and resin, amounted to 11.35% rubber. The stems contained 4.47% and 2.10% respectively. Commercial value is doubtful (2).

The subglobose fruit is 2.5–3.0 cm in diameter. The orange-red pulp when ripe is edible (1, 2, 3). The roots (3) and the fruit (2) are said to have unspecified medicinal use.

References:

1. Busson, 1965: 355. 2. Dalziel, 1937. 3. Irvine, 1961: 461.

**Salacia tuberculata** Blakelock

FWTA, ed. 2, 1:633.

West African: **SIERRA LEONE** MANDING-MANINKA (Koranko) kɛnde (NWT) MENDE gigbɔ

A scandent shrub of the closed-forest in Sierra Leone and Nigeria, and extending to Zäire.

**Salacia whytei** Loes.

FWTA, ed. 2, 1:633, as *S. nitida* (Benth.) N.E.Br., p.p.

A liane recorded from Guinea to W Cameroons, and on into Zäire.

The fruit, up to 4 cm in diameter, has a sweet, edible mucilaginous pulp (1).

Reference:

1. Hallé, 1962, b: 186–8.

# CERATOPHYLLACEAE

**Ceratophyllum demersum** Linn.

FWTA, ed. 2, 1: 65.

A fresh water aquatic herb forming underwater meadows in shallow muddy sheltered places, or becoming detached floating in masses, recorded from various localities from Senegal to S Nigeria and quite recently in the Volta Lake of Ghana (1). A cosmopolitan.

Floating masses of this plant afford harbour for snails of the fresh-water genera *Bulinus* and *Physopsis* (3) which are the alternate hosts for *Schistosoma haematobium,* the causal parasite of schistosomiasis in West Africa (4). In Central Africa the causal parasite is *S. mansonii* and the host *Planorbis spp.* (2), and these may occur in West Africa, but are rare.

References:

1. Hall, J. & al., 1971. 2. Hauman, 1951, d: 116. 3. Lock, 1976. 4. Woodruff, 1975.

# CHENOPODIACEAE

**Agathophora alopecuroides** (Del.) Bunge

A perennial bushy plant to 25 cm high of desert and sandy waste places of N Africa and the Near East, occurring southwards across the Sahara to the northern limit of the West African Region.

The plant is readily eaten by camels but not by other stock. At Hoggar the green plant is crushed and an infusion is taken for liver complaints. It produces an emeto-cathartic effect (1).

Reference:

1. Maire, 1933: 85, as *Halogeton alopecuroides* (Del.) Moq.

**Bassia muricata** (Linn.) Aschers

West African: **MAURITANIA** ARABIC (Hassaniya) dkhaïné (AN) **MALI** TAMACHEK rebbir (RM)

A herb of waste sandy places, to 20 cm high, recorded in Mauritania, and occurring in the Sahara to the northern limit of the Region and in N Africa.

The plant provides a good fodder for camels and other stock (1).

Reference:

1. Maire, 1933: 81.

**Beta patellaris** Moq.

West African: **MAURITANIA** ARABIC (Hassaniya) bu izurgan (AN)

A herbaceous plant becoming woody at the base, of desert mountainous places in the Sahara and Atlantic Islands, and recorded from Mauritania (1).

Reference:

1. Naegelé 1958, b: 881.

**Chenolea lanata** (Masson) Moq.

UPWTA, ed. 1, 33 (as *C. canariensis* Moq.).

West African: **MAURITANIA** ARABIC damrâne (A. Chev.)

A dwarf herbaceous shrub to 20 cm high, branched basally and becoming woody, of NW Africa and the Atlantic Islands, and in Mauritania where it is used for fuel (1).

Reference:

1. Dalziel, 1937.

CHENOPODIACEAE

**Chenopodium ambrosioides** Linn.

FWTA, ed. 2, 1: 144, 759. UPWTA, ed. 1, 33–34.

English: Indian wormseed; sweet pigweed; Mexican tea; Jesuit's tea.

French: ansérine; herbe à vers (Berhaut).

Portuguese: ambrósia-do-México; erva-formigueira; quenopódio (Feijão).

West African: **SIERRA LEONE** KRIO makru-leaf = *pile leaf* (FCD) **GHANA** GBE-VHE agbali soko (FRI) **NIGERIA** YORUBA ewe imí = *leaf of excreter* (Opayemi & Adewole) manturusi (auctt.)

A very polymorphic annual, occasionally perennial, herb to over 1 m high, covered with aromatic glandular hairs and with a strong rank smell when bruised. Originally a native of Mexico, it has been spread by man to all countries but the coldest. It is established in West Africa in Senegal, Ghana, S Nigeria and West Cameroon, and probably occurs throughout the southern zone. It is very closely related to *C. anthelminticum* Linn., the true American wormseed and official source of Chenopodium oil, that some authorities unite them as varieties of the same species.

The plant is sometimes cultivated principally for medicinal uses in West Africa becoming naturalized, or subspontaneous around villages (Senegal, 14; Sierra Leone, 9; Ghana, 7, 13; Dahomey, 7; S Nigeria, 7, 15) also in Gabon (6) and Zaïre (5). The leaves have a pungent taste and are added as a flavouring to soup in Ghana (13), and pounded leaves are applied to sores (12), or to swellings on the body and to areas of pain, and are taken as a purgative (13). In Congo leaf-sap is applied to oedemas and areas of local pain, and the aromatic smell is inhaled for headache (4). Kipsigis of Kenya make an infusion of the pounded leaf which is given to calves with swelling of the neck-glands (10). In S Nigeria the whole plant is pounded and eaten as a laxative, and an infusion is used as a febrifuge for indigestion, as a laxative and for coughs and tuberculosis (2). In Gabon a hot infusion is used for fever, and to wash sores and abscesses (6). In Sierra Leone the young leaves are used by the Krio for haemorrhoids, a use alluded to in their name *makru leaf,* and for fevers and headache (8). The Banso people of Bamenda, W Cameroon, boil the leaves and drink the water for various ailments apparently stomach pains (11). In Senegal the plant is used with allegedly good results for diabetes (14). In Tanganyika boiled leaves are used to treat rash (16) and in S Africa for eczema and erysipelas (19). In connexion with the latter, a plant extract is reported to be slightly bacteristatic and fungistatic.

All parts of the plant contain saponins, the highest concentration being in the roots, ± 2.5%, and increases with age (19). Examination of Nigerian material gave doubtful results (1).

The principal interest of the plant lies in the essential oil which is in the glandular hairs covering the plant, but more especially on the fruit pericarp. The oil has important medicinal uses for treating intestinal parasites, and is effective against hookworm, ascaris, dysenteric amoeba, etc., in humans and in animals. Administration requires care as dosage above tolerance leads to serious toxic symptoms. The oil is obtained by steam-distillation, is pale yellow to orange-yellow, has an unpleasant smell and a bitter burning taste. The anthelmintic component is *ascaridol.* In American wormseed this may amount to 65% or more of the oil, but oil from *C. ambrosioides* is usually less, and, indeed, the composition of oil varies with geographical locality (3, 14, 15, 18, 19). Oil from *C. ambrosioides* is potentially a source of Chenopodium oil, and West African material merits investigation.

In Nigeria the seeds and buds are used as an anthelmintic and vermifuge (2, 15).

The plant is said in Congo to repel snakes and to cure snake bite (4).

References:

1. Adegoke & al., 1968. 2. Ainslie, 1937: sp. no. 87. 3. Ashby, 1941: 11. 4. Bouquet, 1969: 87. 5. Carlier 34, K. 6. Cavaco, 1963, a: 20. 7. Dalziel, 1937. 8. Deighton 2702, K. 9. Deighton 4053, K. 10. Glover & al., 179, K. 11. Hepper 2022, K. 12. Irvine 1725, K. 13. Irvine, s.d. 14. Kerharo & Adam, 1974: 337–8, with pharmacological references. 15. Oliver, 1960: 5, 54. 16. Pirozynski P.347, K. 17. Quisumbing, 1951: 261–2, 1019. 18. Sastri, 1950: 127–8, with oil analysis. 19. Watt & Breyer-Brandwijk, 1962: 187, with references.

## **Chenopodium murale** Linn.

FWTA, ed. 2, 1: 144 UPWTA, ed. 1, 34.

French: anserine.

West African: **MAURITANIA** ARABIC (Hassaniya) tal koda (AN) tarwal (Kesby; AN) **MALI** ARABIC lessig (A. Chev.) TAMACHEK tekauit (RM) tibbi (Foureau) TUBU kono quechi (A. Chev.)

An annual herb attaining 1 m high, widespread in tropics and temperate regions and occurring in W Africa across the northern part from Mauritania eastwards into Senegal and Mali. The plant has anthropogenic tendencies and is spread by man.

The whole plant is used in NE Senegal and in the Saharan oases to make sauces (2, 3, 8). It is cultivated in the Chinguetti area of Mauritania (7). The Bantu of S Africa commonly eat it as a spinach (9). Stock in many territories browse it: Mauritania (5), Sudan (1), Somaliland (4), and in Morocco it is said to provide good pasturage (6), though in Australia there is inconclusive evidence of it causing stock-poisoning (9).

The seed is eaten by North American Indians as a cereal (9), and in Morocco it is taken in time of dearth (6).

References:

1. Andrews 3534, K. 2. Busson, 1965: 133. 3. Dalziel, 1937. 4. Keogh 121, K. 5. Kesby 38, K. 6. Monteil, 1953: 47. 7. Naegelé, 1958, a: 293–305. 8. Trochain, 1940: 265. 9. Watt & Breyer-Brandwijk, 1962: 192.

## **Cornulaca monacantha** Del.

UPWTA, ed. 1, 34.

West African: **MALI** ARABIC had (JMD) hamad (JMD) TAMACHEK had (RM) tahara (JMD) tasra (JMD) tazera (JMD) TUBU sri (A. Chev.) **NIGER** ARABIC had (Grall)

A woody shrub of sandy waste places, much-branched basally, of N Africa and the Near East and southwards across the Sahara into Mali and Niger. According to A. Chevalier the true limit of the Sahara and the Sahel is indicated by the southward limit of this plant and the northern limit of *Euphorbia balsamifera,* these limits coinciding (fide 1).

The plant is salt-tolerant. It is grazed by camels and is used as fuel (1).

Reference:

1. Dalziel, 1937.

## **Haloxylon articulatum** (Moq.) Bunge

UPWTA, ed. 1, 34.

West African: **MALI** TAMACHEK ichafa (JMD) remt(z) (JMD; RM) rummef (JMD) uan ihedan (RM)

A woody salt-bush of sandy waste places of North Africa and the Near East, and southwards across the Sahara to the northern limit of the Region.

The wood is sold for fuel, and the ash is mixed with tobacco-snuff (A. Chevalier fide 1).

Reference:

1. Dalziel, 1937.

**Nucularia perrini** Batt.

French: askaf du sud (Naegelé).

West African: **MAURITANIA** ARABIC (Hassaniya) arjem (AN) **MALI** TAMACHEK askaf (RM)

A low bush of desert rocky sites, endemic to the Sahara, recorded in Mauritania and along the northern limit of the Region.

The plant is relished by camels, but less so by other stock (1). It is a halophyte and supplies salt required in a camel's diet (2).

References:

1. Maire, 1933: 82. 2. Naegelé, 1958, b: 882, as *Arthrocnemum fruticosum* (Linn.) Moq.

**Salsola baryosma** (Schult.) Dandy

FWTA, ed. 2, 1: 144, 759. UPWTA, ed. 1, 34 (under *S. tetragona* Del. as *S. foetida* Del.).

West African: **MAURITANIA** ARABIC (Hassaniya) aghacel (AN) gil *applied to larger plants* (AN) tasra (AN) **MALI** TAMACHEK ressal (RM) **NIGER** TAMACHEK ecchi (A. Chev.) TUBU zri-che (Hinchingbrooke)

A stout much-branched shrub to about 1 m high of saline and waste sandy places, recorded across the north of the Region from Mauritania to Niger, and occurring across N Africa to India.

The plant is very salty. Camels will graze it but it is taken only reluctantly by other stock (4, 5, 6). It is commonly used as camel fodder in the Middle East.

The plant is used in Mauritania for laundering cloth (6). It is used as a fuel in Mauritania and Mali (Chevalier fide 2). In NW India it is burnt to obtain a crude sodium carbonate. The ash is applied to itch, and the plant is considered vermifugal (1). Alkaloids *betaine* and *piperidine* and two others unnamed have been found in the leaf and stem (8).

A form of manna is obtained from the leaves in the Middle East. It is a hardened exudation which is scraped off for eating (1, 3, 7).

References:

1. Chadha, 1972: 184. 2. Dalziel, 1937. 3. Harrison, 1950. 4. Kesby 35, K. 5. Maire, 1933: 83–84, as *S. foetida* 6. Naegelé, 1958, b: 882–3. 7. Watt, G., 1889–93: 6 (2): 392, as *S. foetida*. 8. Willaman & Li, 1970.

**Salsola tetrandra** Forssk.

FWTA, ed. 2, 1: 144, 759. UPWTA, ed. 1, 34, as *S. tetragona* Del.

West African: **MALI** ARABIC belbel (A. Chev.) rassel (A. Chev.) TAMACHEK lharad (A. Chev.)

A bush of desert sandy places recorded doubtfully from Senegal and occurring in the Sahara and N Africa.
In Central Sahara the plant is burnt as a fuel (1).

Reference:

1. Dalziel, 1937.

**Suaeda fruticosa** Forssk.

FWTA, ed. 2, 1: 144.

An erect or ascending perennial herb to over 1 m high, very widespread in Europe, Africa, Asia, Australia and N America, and occcurring in W Africa on the coast of Senegal.
The plant is poisonous to stock (1) and ingestion is said to produce persistent black diarrhoea in sheep followed by death. Sheep not familiar with the plant will eat it (2).
Medicinal uses in India have been as a leaf-poultice for ophthalmia and sores and in infusion as an enema. The woolly excrescence on the branch tips is mixed with an empyreumatic oil for application to sores on camels backs (2).

References:

1. Hardy 619, K. 2. Watt & Breyer-Brandwijk, 1962: 192, with references.

**Suaeda monoica** Forssk.

FWTA, ed. 2, 1: 759.

A shrub to 6 m high with succulent leaves of xerophytic and brackish sites, recorded only in Mali, but occurring in E Africa and in Asia.
The plant is halophytic, a saline soil-indicator and able to tolerate frequent sea-water flooding. In Somalia it is often creeping and thus acts as a sand-dune binder (2). The foliage provides browsing in Somalia (2, 3) and in Kenya (4) for camels and goats. A root-decoction is drunk for sore-throat in Kenya (1).

References:

1. Adamson 324, K. 2. Glover & Gilliland 583, K. 3. McKinnon S/207, K. 4. Mwangangi & Gwynne 1230, 1259, 1265, K.

**Suaeda vermiculata** Forssk.

FWTA, ed. 2, 1: 759.
West African: **MAURITANIA** ARABIC (Hassaniya) solid (AN) **MALI** ARABIC suid (RM) zoggid (RM)

A small shrub to 1 m high with succulent leaves occurring in sandy places of Mauritania and Senegal and across northern Africa to Arabia and Iraq.
In Somalia it is grazed by camels, sheep, goats and horses (2), but in central Sahara it is not readily taken, and if camels do so on an empty stomach it causes colic (1).

References:
1. Maire, 1933: 82. 2. McKinnon S/151, K.

**Traganum nudatum** Del.

UPWTA, ed. 1, 34.

West African: **MALI** ARABIC askaf (A. Chev.) belbala, belbela (A. Chev.) damrâne (A. Chev.) TAMACHEK abelbal (A. Chev.) demran (RM) eshium (Rodd) tahara, tasra, terchit (A. Chev.) terhit, tirehit (RM)

A straggling bush to 60 cm high of rocky waste places and desert pasturage in North Africa and the Near East and southwards across the Sahara to the northern limit of the Region.

The foliage is readily browsed by camels but less so by other stock. It is collected in winter, dried and the frass obtained by crushing between the hands serves as a tinder in the Hoggar (1).

Reference:

1. Maire, H., 1933: 82.

# CHRYSOBALANACEAE

**Acioa barteri** (Hook. f.) Engl.

FWTA, ed. 2, 1: 431.

English: monkey fruit (Liberia, Cooper & Record).

West African: **SIERRA LEONE** MENDE nye-galei (SKS) **NIGERIA** IGBO àhàbà (Siguade) àhàwà *dialectal variant* (anon) àràbà (BNO) IGBO (Umuahia) íchékū (AJC)

A shrub or small tree to 12 m high, slender branches more or less scandent, fluted bole, of the rain-forest and in particular on river-banks from Sierra Leone to SE Nigeria, and in Cameroun to Zaïre.

The tree is sometimes planted. It coppices readily and some races in S Nigeria engaged on subsistence tillage plant it in rows about 2 m (6ft) apart. Food crops are grown between for about two years by which time the coppice growth will have begun to close overhead. The branches are then cut and used for firewood, and cropping is begun again (4). The timber is also used as fuel in Liberia (2). The wood is hard and dark red (3) and appears to have no other usage.

A liquor made from the bark is used in Liberia as a purge (2).

The flowers have an unpleasant scent (1) but the fruit is sweet-smelling and flavoured, and is consumed by wild animals (2).

References:

1. Carpenter AJC.304 (UIH.1948), UCI. 2. Cooper & Record, 1931: 58. 3. Letouzey & White, 1978: 10–13. 4. Nye, 1957.

**Acioa dewevrei** De Wild. & Th. Dur.

FWTA, ed. 2, 1: 433, as *A. unwinii,* sensu Keay, quoad S Nigeria.

A lianous plant to 5 m long, recorded only in S Nigeria in the Region, but extending over the Congo basin.

In Congo powdered bark mixed with palm-oil is used in frictions for fever pains (1).

Reference:

1. Bouquet, 1969: 203, as *A. brazzae* De Wild.

**Acioa dinklagei** Engl.

FWTA, ed. 2, 1: 433. UPWTA, ed. 1, 167.
West African: **GHANA** AKAN-TWI atwere (DF) WASA atwiri (auctt.)

A tall shrub or small tree of river-banks from Liberia and Ghana.

**Acioa hirsuta** A. Chev.

FWTA, ed. 2, 1: 433.
West African: **IVORY COAST** KRU-GUERE (Wobe) uaraha (Aub.)

An understorey tree of the forest zone, known only from Ivory Coast. The wood is dirty white and hard (1).

Reference:

1. Aubréville, 1959: 1: 192.

**Acioa johnstonei** Hoyle

FWTA, ed. 2, 1: 433. UPWTA, ed. 1, 167.
West African: **WEST CAMEROONS** 'BAMILEKE' tileelee (Johnstone)

A shrub or tree to 3 m high, trunk to 25 cm diameter, of stream-banks in forest or savanna, and known only from the mountains of southern W Cameroons.

**Acioa lehmbachii** Engl.

FWTA, ed. 2, 1: 433, as *A. rudatisii* De Wild.
West African: **NIGERIA** EFIK úkáñ = *charcoal* (Olu Akpata)

A tree to 60 m high, of secondary jungle and known only from Calabar, Nigeria, and W Cameroons.
The wood is hard and is used in S Nigeria to make charcoal, hence the Efik name meaning 'charcoal' (1).

Reference:

1. Olu Akpata FHI.3939, K.

**Acioa scabrifolia** Hua

FWTA, ed. 2, 1: 432, excl. W Cameroons. UPWTA, ed. 1, 167.
West African: **GUINEA** FULA-PULAAR (Guinea) boillé démon (Langdale-Brown) **SIERRA LEONE** KISSI yɔmiswendɛ (S&F) KONO nyɛ-ku (S&F) MENDE nyɛgai (auctt.) nyɛgala (*def.*-lei) *correctly for this sp., but often applied to other shrubs* (auctt.) **LIBERIA** KRU-BASA see (C&R) **IVORY COAST** FULA-FULFULDE (Ivory Coast) guilinti (Aub.) kébé-fitoba (Aub.) SUSU kébé (Aub.)

A small to medium-sized tree to 20 m height (but to as much as 27 m by 1.70 m in girth in Sierra Leone, 6), an understorey tree of the high closed rainforest, and as a relic in farmlands, but sometimes on swamp edges (Fouta Djalon, Guinea, 1), and no more than a shrub, from Guinea to Liberia, and possibly in Ghana (5).

The tree is often locally abundant, and in managed forestry is considered a weed species (6). The wood is coarse, heavy, hard and durable provided it is not exposed to decaying conditions. In Liberia it is made into rice-pestles and mortars, and is considered suitable for implement frames and for heavy construction work if not exposed to decay (2, 3). It converts into excellent charcoal, and in Sierra Leone it has been favoured by Mende blacksmiths for smelting iron (4).

The boiled leaves provide a decoction taken in Sierra Leone for dysentery (6). The fruit contains an edible seed (6).

References:

1. Aubréville, 1959: 1: 190. 2. Cooper 98, K. 3. Cooper & Record, 1931: 58, with timber characteristics. 4. Deighton 3779, K. 5. Irvine, 1961: 260. 6. Savill & Fox, 1967: 215–6.

**Acioa whytei** Stapf

FWTA, ed. 2, 1: 431.

West African: **SIERRA LEONE** BULOM (Sherbro) tikonko (Garrett) MENDE nyɛgala (*def.*-lei) (SKS)

A shrub or tree to 10 m high common over all of Sierra Leone, and in Liberia.

Of no recorded usage, but bearing the Mende name *nyɛ galai* that is applied to trees of this genus, and of similar appearance and not readily distinguished apart; they probably all have common uses.

**Acioa sp.**

West African: **LIBERIA** KRU-BASA dee-waye (C&R)

A Liberian tree of the above name is recorded used for rice-mortars because of its strength and durability, and occasionally for small timber (1); cf *A. scabrifolia*.

Reference:

1. Cooper & Record, 1931: 58, as *Acioa* sp. (Cooper 99).

**Bafodeya benna** (Sc. Elliot) Prance

FWTA, ed. 2, 1: 429, as *Parinari benna* Sc. Elliot, excl. Senegal. UPWTA, ed. 1, 168, as *P. benna* Sc. Ell.

West African: **GUINEA** FULA-PULAAR (Guinea) kura (Bouronville) sigo(n) (JMD; Aub.) MANDING-MANINKA sigonaïa (Aub.) SUSU sigon (Aub.) siko (Aub.) **SIERRA LEONE** SUSU-DYALONKE siko-na (FCD; KM)

A tree to about 7 m high of the savanna, known only from Mali, Guinea and Sierra Leone at 900–1,000 m elevation.

The fruit of most *Parinari* (*sens. lat.*) species has a tawny layer of a soft cottony substance lining the inside of the endocarp which is used over tropical Africa as a fire-tinder, cf. *Neocarya macrophylla* (Sab.) Prance. In *B. benna* this layer has a harsher texture. The hairs are perfectly straight, apparently hollow, needle-shaped spines which project stiffly into the loculus cavity. These are used in Guinea as a vermifuge corresponding in a way to the mechanical irritation caused by the hairs of, for example, *Mucuna pruriens* (Leguminosae: Papilionioideae) (1). A record of the fruit used in northern Sierra Leone as a vermifuge (2) is evidently of this also.

References:

1. Dalziel, 1931. 2. Miszewski 5, K.

**Chrysobalanus icaco** Linn.

FWTA, ed. 2, 1: 426, as *C. orbicularis* Schum. and *C. ellipticus* Soland. UPWTA, ed. 1, 168, as *C. orbicularis* Schum. & Thonn., and 167, as *C. ellipticus* Soland.

English: coco plum; icaco.

French: prune icaque; prunier d'Amerique; prunier des anses.

West African: **SENEGAL** SERER banara (JB) vanara (auctt.) SERER-NON (Nyominka) ibanara (K; K&A) WOLOF rad (JB; K&A) vorač (auctt.) **THE GAMBIA** FULA-PULAAR uoratch (DAP) WOLOF vorach (JMD; DAP) **GUINEA-BISSAU** BIDYOGO ebenga (JDES; EPdS) MANDYAK bôpace (JDES; EPdS) **GUINEA** BOKE grubé (Aub.) moholo (Aub.) **SIERRA LEONE** BULOM (Sherbro) bɛmbɛ-lɛ (FCD; S&F) KRIO bimbi (FCD; S&F) MENDE gogo, gɔgivi (*def.*-i) (FCD; S&F) mapo (*def.*-i) (FCD; S&F) SUSU sufe-wuri (NWT) VAI mapɔi (FCD) **LIBERIA** KRU-BASA pahn-doh (C&R) tah (C&R) **IVORY COAST** ADYUKRU mikni (Aub.) ANYI hanfuru (A. Chev.) KYAMA aké (Aub.) grubé (Aub.) **GHANA** GBE-VHE (Awlan) fɔtigba (FRI) NZEMA abɛblɛ (auctt.) abrɛbɛrɛ (FRI) ɛdu (FRI) **DAHOMEY** SOMBA yuhabu (Aub.) **NIGERIA** YORUBA amukan (Ross) awọ̀nrinwán (KO&S) ikate (auctt.) YORUBA (Ijebu) elewu (JRA)

A tree of 25 m or more height, widely dispersed in the guinean forest zone of the Region, recognized as of two subspecies, ssp. *icaco* occurring principally in coastal thickets from Senegal to W Cameroons, and in central, east and south central Africa, and ssp. *atacorensis* (A. Chev.) F. White in more localized sites, coastal and inland, from Sierra Leone to Nigeria, and in Central Africa. The African plant is held to be the same as that of the New World (17, 20). It has become naturalized in the Seychelles, Vietnam and Fiji. The name icaco has entered European languages from *ekakes* of Curaçao.

It forms an attractive glossy, deep-green-leaved bush or small tree, and is often cultivated in tropical and subtropical America as an ornamental. It is vigorous growing. In the Seychelles it has been found to colonize foot-hill slopes following denudation of forest becoming co-dominant with *Cinnamomum zeylanicum* (Lauraceae), thereby giving protection against erosion. In the Congo River, ssp. *atacorensis* is the first plant to colonize newly formed land (10). The wood is reddish-brown, very hard and dense, but workable. In Liberia it is used where strength and durability count, and, if well-grown, it provides heavy timbers (7, 8: also 9, 11, 19). It is used as fencing and firewood in Ghana (11), and valued for the latter purpose in Nigeria (18).

The bark is astringent and contains tannins (19). A trace of alkaloid has also been reported (1). Tannin from the fruits has been tested in the U.S.A., and is reported to produce a soft and porous leather; it has been used admixed with other tanning materials (6).

The fruits, leaves and roots are used in Senegal (12, 14) and in Guinea (16) against refactory diarrhoea, and the fruits are taken by the Nyominka of Senegal as a laxative (12, 13). In Congo a bark-decoction of ssp. *atacorensis* is used to wash persons affected by itch or dermatitis (5).

The fruit is in general considered edible, and the plant is widely cultivated in the tropics for them, but reports vary very much on their palatability. Fruit in Senegal is said to be somewhat sweet (12) and the pulp of fruit sold in Dakar has been found to contain 13.7% sugars (14). Elsewhere the fruit is said to be vinous (4), or bitter, unpleasant and astringent (9). Such variability suggests scope for selection, but in merit for attention it can hardly rank high. In Angola the fruit pulp is eaten after drying (3), a factor which suggests that the pulp can be stored dry. In the W Indies it is candied (6).

The seed is oil-bearing. The kernel of American material is recorded as containing 20.25% oil of undisclosed constituents (9). In The Gambia the oiliness of the seed is such that the seeds have been used to burn as an illuminant in the manner of a candle (9). They have unspecified medicinal use in Gabon/Cameroun (15, 19). The oil is used in Nigeria in treatment of dysentery (7). The kernel of material from Dakar is recorded containing 37% sugars and 23% lipids (14).

References:

1. Adegoke & al., 1968: 13–33, as *C. ellipticus*. 2. Ainslie, 1937: sp. no. 90, as *C. orbicularis*. 3. Anon. 29, K. 4. Aubréville, 1959: 1, 172. a. as *C. ellipticus* Soland., b. as *C. orbicularis* Schum. & Thonn. 5. Bouquet, 1969: 204, as *C. atacorensis* A. Chev. 6. Burkill, 1935: 533. 7. Cooper 191, K. 8. Cooper & Record, 1931: 59, as *C. ellipticus* Soland. 9. Dalziel, 1937. 10. Dawe 44, K. 11. Irvine, 1961: 262, as *C. ellipticus* Soland. and *C. orbicularis* Schum. 12. Kerharo, 1966: as *C. orbicularis* Schum. 13. Kerharo & Adam, 1964: 299, as *C. orbicularis* Sch. & Th. 14. Kerharo & Adam, 1974: 677–8, as *C. orbicularis* Schum. 15. Letouzey & White, 1978: 60–67. 16. Pobéguin, 1912: 24. 17. Prance, 1972. 18. Sankey 12, K. 19. Walker & Sillans, 1961: 385, as *C. ellipticus* Soland. and *C. icaco* Linn. 20. White, 1976.

**Licania elaeosperma** (Mildbr.) Prance & White

FWTA, ed. 2, 1: 427, as *Afrolicania elaeosperma* Mildbr. UPWTA, ed. 1, 167, as *A. elaeosperma* Mildbr.

English: the fruits — mahogany nuts; nico, niko, nikko; the fruit oil — poyak, from Sherbro, Sierra Leone.

West African: **SIERRA LEONE** BULOM (Kim) ma-g-biti (FCD) ma-k-povɛ (FCD) BULOM (Sherbro) poyok-ɛ (FCD; S&F) poyo-lɛ (FCD; S&F) MENDE nja-gbɔlɔ (*def.*-lei) (FCD; S&F) **GHANA** AKAN-TWI takorowa (DF) WASA kajibiri (DF) takorowa (DF) NZEMA kokorobe (auctt.) kokrobe (DF) **NIGERIA** EDO ógúi-àhá (KO&S) YORUBA ikate oro igbó (KO&S) YORUBA (Ondo) aghen funfun (Kennedy) **WEST CAMEROONS** DUALA ndonda (Mildbraed)

A tree to 17 m high, with crooked, irregular, trunk, buttressed and to 1.30 m in girth, of coastal bush and riverain forest, in Sierra Leone to S Nigeria, and extending into Gabon.

The warty, ovoid fruit are the *nico, nikko* or *mahogany nuts* of trade, source of *po-yok* oil, often incorrectly rendered as *po-yoak,* which is potentially usable for making paints, varnishes, polishes, etc. How nico or nikko has entered the trade vocabulary is not clear for these are derived from S Nigerian names referring to *Maranthes robusta* (Oliv.) Prance (syn. *Parinarium robustum*) and *Chrysobalanus sp.* (7), in no way connected with po-yok oil. Nor are they mentioned amongst the *Useful Plants of Nigeria* (5). The correct and unambiguous term for the nuts must be *po-yok nuts*.

The nut is up to 3 cm long and contains a single seed of two thick, fleshy, concave cotyledons enclosing a large cavity. They are oil-bearing. The percentage of kernel to shell is high, 61% to 39% respectively, and the kernels yield on extraction 52% of a yellow oil which dries fairly rapidly to a varnish-like mass. Its composition is reported as: licanic acid 44%, eleostearic acid 34%, saturated fatty acids (not itemized) 13%; oleic and linoleic acids 9%, and calculated iodine value 207–8 which is unusually high (4). The oil is thus potentially valuable as a drying oil in paint manufacture. It is reported inferior to tung oil and superior to linseed oil which it can substitute. The fruit shell is brittle and the cotyledons lie loose within, factors making for ease of decortication (2). Trees are slow to mature and 20–30 years old trees under observation were still not bearing nuts. A 'good' tree is said to produce 'half a bushel' (40 lbs, or c. 18 kg) of nuts a year (3). Trees normally grow near coastal water, and fruits shed into it are carried away by the tide to be collected between March and June along the coastal strand at high water mark (6). Po-yok nuts have been shipped to Europe from Liberia and Sierra Leone, but the trade is an insignificant amount, and as a crop it may not be amenable of expansion.

There is local use of the nuts in Sierra Leone. They are pounded, either fresh or dry, and the oil is boiled out in water and skimmed off. The oil is used mixed with clay by secret society (*Bundu*) women as a body-scent. Women use the oil alone as a hair-dressing which is said to stimulate hair-growth (2).

The residual cake after commercial oil extraction is reported unsuitable as a cattle-feed (1).

Though the tree is well-known in the coastal area of SE Sierra Leone, no usage is made of the wood (3).

References:

1. Anon., 1942: as *Afrolicania elaeosperma*. 2. Dalziel, 1937. 3. Deighton, s.d., n.s., K. 4. Eckey, 1954: 472, as *Parinarium sherbroense*. 5. Holland, 1908–22. 6. Savill & Fox, 1967, 215, as *Afrolicania elaeosperma* Mildbr. 7. Unwin, 1920: 267 as *Parinarium robustum* & 269, as *Chrysobalanus sp.*

**Magnistipula conrauana** Engler

FWTA, ed. 2, 1: 430, as *Hirtella conrauana* (Engl.) A. Chev.

A tree reaching 10–12 m high, or a shrub to 5 m, of rain-forest between 900–1,000 m altitude from W Cameroons only (2).

It is occasionally planted as a hedge for enclosures (1).

References:

1. Letouzey & White, 1978: 79–81. 2. White, 1976: 289.

**Magnistipula cupheiflora** Mildbraed, **ssp. leonensis** F. White

FWTA, ed. 2, 1: 430, as *Hirtella cupheiflora* (Mildbr.) Mildbr.
West African: **SIERRA LEONE** MENDE kafi (D.G.Thomas)

A tree to 10 m high of the evergreen rain-forest in Sierra Leone only. Ssp. *cupheiflora* occurs in E Cameroun and Gabon.

The trunk is fluted and crooked. It is said, however, to provide timber in Sierra Leone (1).

The fruit is recorded as being edible (1).

Reference:

1. Thomas, D. G. 118, K.

**Magnistipula tessmannii** (Engl.) Prance

A tree attaining 40 m high, with a fluted irregular trunk to 1 m diameter or more, of primary, evergreen rain-forest in SE Nigeria, rare, at 600–700 m altitude, and occurring in Cameroun to Congo.

The fruit is edible (1).

Reference:

1. Letouzey & White, 1978: 88–93.

**Magnistipula zenkeri** Engl.

FWTA, ed. 2, 1: 430, as *Hirtella fleuryana* A. Chev.

A forest tree to 15 m high, of evergreen rain-forest especially on river-banks and in swampy places from Liberia, Ivory Coast, and in W Cameroons, and extending from E Cameroun to Gabon.

The tree is to some extent myrmecophilous. The flowers have glands which secrete honey that attracts ants.

References:

1. Letouzey & White, 1978: 93–96. 2. White, 1976: 293–4.

**Maranthes aubrevillei** (Pellegr.) Prance

FWTA, ed. 2, 1: 428, as *Parinari aubrevillei* Pellegr. UPWTA, ed. 1, 168, as *P. aubrevillei* Pellegrin.

French: aramon (from Abe) à feuilles dentées (Aubréville).

West African: **SIERRA LEONE** MENDE ndondelole (S&F) **LIBERIA** DAN su-ti (GK) zu-ti = *black zu* (AGV) **IVORY COAST** ABE aramon (Aub.) AKYE aramon (Aub.)

A tree to 30 m high, trunk straight, cylindrical to 80 cm diameter, of evergreen rain-forest, from Sierra Leone to Ghana and W Cameroons, and also in E Cameroun.

Sap-wood and heart-wood are yellowish white, and extremely hard making the wood unfit for usage (1–3). The Dan name derives from the blackish cast of the bark (3).

References:

1. King 286, K. 2. Savill & Fox, 1967: 220, as *Parinari aubrevillei* Pellegr. 3. Voorhoeve, 1965: 317, as *P. aubrevillei* Pellegr.

**Maranthes chrysophylla** (Oliv.) Prance, **ssp. chrysophylla**

FWTA, ed. 2, 1: 428, as *Parinari chrysophylla* Oliv. UPWTA, ed. 1, 170, as *P. sp.* near *P. chrysophylla* Oliv.

West African: **SIERRA LEONE** MENDE ndawa-hina = *male adawa, as opposed to Parinari exelsa: female ndawa* (S&F) **LIBERIA** DAN su-nasa (GK) KRU-BASA ve-ay-du-ah, ve-ay: *to hide oneself;* du-ah: *a species of red monkey with the habit of hiding itself in the foliage* (C&R) **IVORY COAST** KRU-GUERE (Wobe) kioro (Aub.) **GHANA** AKAN-WASA afam (Enti) kajibiri (DF)

A tree to 30–40 m high, trunk straight, up to 1.20 m in diameter, of the evergreen rain-forest from Sierra Leone to S Nigeria, and on to Gabon.

Sap-wood is not particularly thick (to 5 cm); wood is reddish becoming darker towards the centre, dense and hard (5, 6). The wood is used in Liberia for planks and timbers, and though strong and heavy is reckoned to be not durable on exposure (2–4). In Gabon it is used for larger construction works (8).

The bark is poisonous (1, 3, 4).

This species, which has recently been recorded in SE Sierra Leone, is easily confused with *M. excelsa* and the former is distinguished by the Mende from the latter on the edibility of their fruits. *M. chrysophylla* has inedible fruit and is thus called 'male *ndawa*', while the fruit of *M. excelsa* is deemed edible and so is 'female *ndawa*' (7).

References:

1. Cooper 232, K. 2. Cooper 297, K. 3. Cooper & Record, 1931: 60, as *Parinarium sp.* 4. Dalziel, 1937. 5. Letouzey 11138, K. 6. Letouzey & White, 1978: 100–4. 7. Savill & Fox, 1967: 217, as *Parinari chrysophylla.* 8. Walker & Sillans, 1961: 359, as *P. chrysophylla* Oliv.

**Maranthes gabunensis** (Engl.) Prance

FWTA, ed. 2, 1: 428, as *Parinari gabunensis* Engl.

A shrub 6–8 m high, or a tree reaching to 20 m, trunk straight 60–80 cm in diameter, tapering, of evergreen forest in S Nigeria and W Cameroons, and in E Cameroun, Gabon and Zaïre.

A decoction of plant (part not stated) is used in Congo as a draught and in steam-fumigations for treating paralysis (1).

Reference:

1. Bouquet, 1969: 204, as *Parinari gabunensis* Engl.

**Maranthes glabra** (Oliv.) Prance

FWTA, ed. 2, 1: 428, as *Parinari glabra* Oliv. UPWTA, ed. 1, 169, as *P. glabra* Oliv.

Trade: aramon (the timber in Ivory Coast, but in confusion with *M. robusta* Aubréville).

West African: **SIERRA LEONE** BULOM (Sherbro) sas-lɛ (FCD; S&F) MENDE ndondelole (Fox; S&F) sasi, sakasi (*def.*-i) (FCD; S&F) **LIBERIA** KRU-BASA zwahn (C&R) MANO kpo kala (Har.) **IVORY COAST** ABE aramon (Aub.) AKYE aramon (Aub.) ANYI amalarué (Aub.) bukuma (Aub.) KWENI zéri zéri (Aub.) KYAMA alobo (Aub.) **GHANA** AKAN-WASA kagyibiri (auctt.) takrowa (Beveridge, ex FRI) ANYI-AOWIN pinini (FRI) SEHWI pinini (FRI) NZEMA punini (auctt.) **NIGERIA** EDO óghòyè (Kennedy; KO&S) YORUBA dabadogun (Ross)

A large tree to over 30 m high, bole cylindrical and straight (Ghana, 13) or low-branched (1), up to 4 m in girth, sometimes slightly buttressed; crown dense and spreading; of evergreen and deciduous forests, from Sierra Leone to S Nigeria, and extending across central Africa to Zaïre and Angola.

The wood is hard and heavy, sinking when fresh (6, 14). Sap-wood is white to yellow-brown with attractive longitudinal veining; heart-wood pink or reddish (5, 6, 10, 13). The wood is throughout its range considered of little value for carpentry. In Sierra Leone it is said to be difficult to work and has no use (12) except for making canoes (8). In Ghana it is used for beams and stakes and provides an excellent, though expensive fuel (8, 9).

The bark-slash exudes a little colourless or pinkish liquid (10, 12, 13). The inner bark is soaked for 24 hours in Liberia to treat dysentery (4, 5), and sap from the cambial area is instilled into the eye as an ordeal-poison in Zaïre (11). An aqueous decoction of the bark is used in Congo to wash a person suffering from chronic itch, after which the patient is covered with a prescription consisting of palm-oil, sap of the leaves of *Tephrosia vogelii* (Leguminosae: Papilionniideae) and powdered root of *Maranthes glabra*. A similar prescription in which the *Tephrosia* is replaced by bark of *Croton haumanianus* (Euphorbiaceae, but not recorded for W Africa) is applied to domestic animals for skin-troubles (2, 11).

The flowers have a strong smell of honey when fresh which turns to that of cows [!] when old (7). The fruit is used in Zaïre as a bait in traps for river-hog (*Potamochoerus*) and antelope (11). The seeds are edible and oily.

*Maranthes kerstingii* (Engl.) Prance is not recorded for Ivory Coast and a reference to *Parinari kerstingii* sensu Bouquet & Debray (3) used in that territory is probably to *P. glabra* Oliv. (=*M. glabra* (Oliv.) Prance): a bark decoction is drunk three times daily for anaemia, and as a tonic for pregnant women.

References:

1. Aubréville, 1959: 1, 175–82 as *Parinari glabra* Oliv. 2. Bouquet, 1969: 204, as *P. glabra* Oliv. 3. Bouquet & Debray, 1974: 146, as *P. kerstingii* sensu Bouq. & Deb. 4. Cooper 345, K. 5. Cooper & Record, 1931: 59–60, as *Parinarium kerstingii* sensu Cooper & Record with timber characteristics. 6. Dalziel, 1937. 7. Deighton 2482, K. 8. Deighton 3716, K. 9. Irvine, 1961: 265, as *P. glabra* Oliv. 10. King 68, K. 11. Letouzey & White, 1978: 108–12. 12. Savill & Fox, 1967, 220, as *P. glabra* Oliv. 13. Taylor, 1960: 285, as *P. glabra* Oliv. 14. Vigne 216, K.

**Maranthes kerstingii** (Engl.) Prance

FWTA, ed. 2, 1: 428, as *Parinari kerstingii* Engl. UPWTA, ed. 1, 169, as *P. kerstingii* Engl. in part.

West African: **IVORY COAST** ABE aramon (Aub.) KYAMA alobo (B&D) bokwo (B&D) **GHANA** ATWODE (Kromase) kakyiki (Jenik & Hall) **TOGO** TEM (Tshaudjo) ningelia (Gaisser) YORUBA-IFE OF TOGO okpe (von Doering) **NIGERIA** HAUSA ƙaiƙayi (auctt.) kyarka (ZOG)

A tree to 20 m high, or more, by 1.70 m in girth, in fringing forest of the wetter part of the soudanian region from Ghana to Nigeria, and on to E Cameroun and the Central African Republic.

The wood is brown, hard and heavy, and appears to have no recorded usage (2, 3). A decoction of the plant (part not stated) is used in Congo with the addition of rock-salt in draughts, baths and steam-fumigations for broncho-pneumonic affections and for feverish pains (1, 4). The plant is also considered to be emetic and purgative.

References:

1. Bouquet, 1969: 204, as *Parinari kerstingii* Engl. 2. Dalziel, 1937. 3. Keay & al., 1960: 314 as *P. kerstingii* Engl. 4. Letouzey & White, 1978: 112–5.

**Maranthes polyandra** (Benth.) Prance

FWTA, ed. 2, 1: 428, as *Parinari polyandra* Benth. var. *polyandra* and var. *cinerea* Engl. UPWTA, ed. 1, 170, as *P. polyandra* Benth.

West African: **SENEGAL** MANDING-BAMBARA tudu (A. Chev.) **GUINEA** MANDING-MANINKA Wasalunka tudu (A. Chev.) **UPPER VOLTA** GURMA bumansabu (Aub.; K&B) HAUSA gongea kussa (K&B) KIRMA goyoma (K&B) LOBI pwapatula oro (K&B) MANDING-BAMBARA tutu (K&B) DYULA tutu (K&B) 'SENUFO' subokoro (K&B) **IVORY COAST** BAULE hanvien (K&B) MANDING-DYULA tutu (K&B) SENUFO-NIAGHAFOLO tubozofi (K&B) DYIMINI komologuo (K&B) **GHANA** AKAN-ASANTE abrabesi (auctt.) apodwo (TFC; FRI) DAGBANI lara(-k) pirri (AEK; FRI) GUANG-GONJA pɔ́té pɔ́té (Rytz) SISAALA poti poti (Andoh; FRI) pumfumbile (AEK; FRI) **TOGO** TEM (Tshaudjo) bende-moso (Kersting) bende-noso (Kersting) **DAHOMEY** BATONNUN bakuku (Aub.) GBE-FON huantu hové hové (Aub.) huantu ui ui (Aub.) YORUBA-NAGO keroko djikassi (Aub.) **NIGERIA** FULA-FULFULDE (Nigeria) ciiɓooli, *pl.* ciiboole (Taylor) ciiɓooli debbe (*pl.* ciiɓoole debbe) = *female c.* (Taylor) ciiɓooli gorɗe (*pl.* ciiɓoole gorɗe) = *male c.* (Taylor) HAUSA gwànjaá kúsá *cf. Heliotropium subulatum* (*Boraginaceae*) (JMD; ZOG) ƙaiƙani (JMD) kaikayi (KO&S) TIV ibyua kuna (JMD; KO&S) YORUBA abere (JRA) akọ ìdòfún (JRA; JMD) ìdòfún (KO&S) ìdòfún-abo (Ross)

A tree to 8 m high of the wooded savanna from Ghana to S Nigeria, and on to Gabon and Sudan.

The bole is gnarled and twisted (5), but the stems are usable for building huts and sheds (3, 4, 8). The wood is used in the Region for charcoal (3, 4) and also in Cameroun/Gabon (7). Nupe blacksmiths claim it makes the best charcoal they use (3). The wood reduced to ash yields a vegetable salt (7).

The bark is rich in tannin and is used in Togo for tanning (3). A bark-decoction is used in Nigeria for fevers, and is said to have aphrodisiac properties (2). In Cameroun/Gabon a bark-preparation is used to bathe and to rub over fractures (7). An examination of Nigerian material has shown no alkaloid in the bark (1).

The leaves are sometimes chewed, with or without some of the bark, and like kola are said to redden the mouth, and they may sometimes be used as a red dye (3). The crushed leaves are used in Cameroun/Gabon on wounds and fractures, and are taken in draught and baths for fevers there (7), and in Ivory Coast/Upper Volta (6).

The roots are dried and powdered for use as a cure for syphilis both internally and on the ulcers (2).

The fruits are ellipsoid, dark red or blackish-purple to 2.5 cm long or less. They are hardly edible (3, 4, 7).

References:

1. Adegoke & al., 1968: as *Parinari polyandra*. 2. Ainslie, 1937: sp. no. 261, *Parinarium polyandrum*. 3. Dalziel, 1937. 4. Irvine, 1961: 266, as *Parinari polyandra* Benth. 5. Keay et al., 1960: 314–6, as *P. polyandra* Benth. 6. Kerharo & Bouquet, 1950: 90, as *P. polyandrum* Benth. 7. Letouzey & White, 1978: 116–8. 8. Unwin, 1920: 269, as *P. polyandrum* Benth.

**Maranthes robusta** (Oliv.) Prance

FWTA, ed. 2, 1: 428, as *Parinari robusta* Oliv. UPWTA, ed. 1, 170, as *P. robusta* Oliv.

English: 'mahogany nut' (the fruit, Dalziel).

Trade: aramon (the timber in Ivory Coast from Abe, but confused with *M. glabra,* Aubréville).

West African: **IVORY COAST** ABE aramon (A. Chev.) ABURE araba (A. Chev.) oroba (A. Chev.) AKYE boamué (A. Chev.) ko-aramon, ko: *tree* (Aub.) ANYI bokoma (A. Chev.) bukuma (A. Chev.) DAN ponéné (A. Chev.) punini (A. Chev.) KYAMA alobo (Aub.) aroba (Aub.) **GHANA** VULGAR kuokuo dua (DF) AKAN-TWI kukuodua (auctt.) WASA kagyibiri (auctt.) GBE-VHE kiẹ̃ (FRI) klae (CJT) MOORE kwinabuku (FRI) **NIGERIA** EDO dabadogun (Farquhar; KO&S) HAUSA kàshè-kaàjí, kaskawani (ZOG) IGBO àhàbà-újì (KO&S) YORUBA aghaghe (Millen) aiye (AHU) aiyena (AHU) awewe (KO&S) ìdòfún (Ross; KO&S)

A deciduous tree of swamp-forest to 13 m high and low-branching in coastal areas, or to 35 m or more inland with a cylindrical bole to 1.70 m girth (5, 6), in Ivory Coast to Nigeria.

This species is confused with *M. glabra*. The two species share the same name for the timber in Ivory Coast (1). The sap-wood is thick, white, yellowish or pinkish-white: heart-wood is hard, dark red, fibrous, resembling mahogany, and darkening on exposure. It is said to be termite- and borer-proof and durable. It finds local use in house-building and is favoured in northern Ghana, but it is too hard and heavy for ordinary construction purposes, though suitable for piles and possibly for sleepers (3, 5, 6).

The bark-slash exudes a red watery gum.

The fruit are flattened-obovoid, reddish-brown when dry, up to 5 cm long. They have been called 'mahogany nut', a term better applied to this species than to *Licania elaeosperma,* though the significance of the name is not recorded. The kernel is oily and has a characteristic smell (3, 4). The endocarp of this species is lined with a velvety layer of hairs in distinction from the loose cottony wool filling the loculus cavity in *Parinari spp.* Whether this is usable as tinder in the same way, or has medicinal usage does not seem to be recorded.

References:

1. Aubréville, 1959: 1: 175, 182–4. 2. Dalziel, 1931: 99. 3. Dalziel, 1937. 4. Irvine, 1961: 266. 5. Keay et al., 1960: 317. 6. Taylor, 1960: 286.

**Neocarya macrophylla** (Sab.) Prance

FWTA, ed. 2, 1: 429, as *Parinari macrophylla* Sab. UPWTA, ed. 1, 169, as *P. macrophylla* Sab.

English: gingerbread plum; neou oil tree; the fruit: ginger plum (Sierra Leone, Deighton); 'rotten plum' (locally, Dalziel).

French: pomme (pommier) du Cayor (apple, or apple tree of Cayor).

Portuguese: matapaz-grande (from Crioulo, G. Bissau, Feijão).

West African: **SENEGAL** BALANTA fé (Aub.; FB) BANYUN kifokum (K&A) BASARI a-núnú (Ferry) BEDIK gi-núnú (Ferry) DIOLA bahab (AS; FB) bel (auctt.) bu ba (JB) bumafaye (Aub.); = *sucking* (DF, The Gambia) DIOLA (Efok) beul (K&A) buel (K&A) DIOLA (Fogny) ba (K; K&A) baa (DF) baabu = *over-grown fruit* (auctt.) ahab (DF) DIOLA ('Kwaatay') nuhuntan (K&A) DIOLA-FLUP bu ngafay (JB); = *hand-climber* (DF, The Gambia) FULA-PULAAR (Senegal) naodé (K; K&A) neudi (K; K&A) MANDING-BAMBARA danga (Aub.) tétu (Vuillet) vo (JB) wo (K&A) MANDINKA tamba tambakumbaa (*indef.* and *def.*) MANDYAK bunou (auctt.) SERER daf (auctt.) idaf (K; K&A) SERER-NON ferah (JMD; Aub.) néva (AS; Aub.) nif (AS; Aub.) WOLOF néo (auctt.) néu (K; K&A) **THE GAMBIA** DIOLA (Fogny) baa, bahab, baabou *over-grown fruit* (DF) bahab (AS) bel (AS) FULA-PULAAR (The Gambia) kura-bansuma(-i) (DRR; DAP) naude (JMD; DAP) MANDING-MANDINKA tambaa (*indef.* and *def.*) (auctt.) tamba-kumbaa (*indef.* and *def.*) = *big headed 'tamba'* (DF) WOLOF neong néu (JMD; FD) **GUINEA-BISSAU** BALANTA n-bute (JDES) n-djapô (JDES) n-japo (EPdS) téhé (JDES) umbatú (JDES) undjapô (JDES) BIDYOGO nórònóròdó (JDES) nororodo (EPdS) orodjô (JDES) CRIOULO mampataz (GeS; JMD) mampataz grande (EPdS) FULA-PULAAR (Guinea-Bissau) cura-bussuma (JDES) curanaco (EPdS) nando (EPdS) naudo (JDES) tambacumba (GeS; RdoF) MANDING-MANDINKA néo tambacumba (JDES) MANDYAK bénô bénô (JDES) bitiague (JDES) menau (JDES) MANKANYA bénô bénô (JDES) bitiague (JDES) menau (JDES) **GUINEA** FULA-PULAAR (Guinea) kura-bansuma (CHOP) néudi (Aub.) niamui (Aub.) SUSU bansuma (CHOP) **SIERRA LEONE** BULOM (Sherbro) kisin-dɛ (FCD; S&F) KRIO bɛnjamin (FCD; S&F) LIMBA makalonkalɔ (Glanville; FCD) MANDING-MANDINKA tamba (JMD) MENDE foni-lawa (*def.*-lawei) (FCD; S&F) gise, gisini (*def.*-i) (auctt.) ndawa (JMD) SUSU kobira-fire (FCD) tamui (FCD) VAI gisa (FCD) jɔ (FCD) **LIBERIA** KRU-BASA tifi (Barker) **MALI** MANDING-BAMBARA danga (FB) **IVORY COAST** VULGAR néu (K&B) **GHANA** NANKANNI nya (Lynn; FRI) **NIGER** HAUSA gaosa (Aub.) SONGHAI gámsà (*pl.* -à) (D&C) **NIGERIA** FULA-FULFULDE (Nigeria) naawdi (*pl.* naawde) *the fruit* (MM) nawarre *the tree* (JMD; KO&S) node(l) *the fruit* (JMD) HAUSA bakar rura (RES) gàwàsaá (auctt.) NUPE kóbenci *the fruit* (Banfield) pútú (Banfield; KO&S) pútú yiwó *'female'* (Banfield) pútú'bá *'male'* (Banfield)

A tree to 10 m high, often less, of disjuncted distribution, along the coastal strip from Senegal to Liberia to 300 km inland in sandy localities, and secondly 700–1,000 km inland Mali to Niger and N Nigeria on banks of sandy seasonal watercourses and on sandstone cliffs, conjecturally a relic vegetation from an earlier wetter climate (22).

The tree appears to survive annual firing in savanna and in northern Sierra Leone it is often the only tree of any stature attaining double the normal height to 20 m, with girth of 1.30 m (21).

The wood is light brown and fairly hard. It works and polishes well and is suitable for planks and building timber (5, 13). The Nankanni of Ghana use the wood for making canoes (Lynn fide 12). In Liberia it serves to produce charcoal (2), and firewood in The Gambia (20).

A decoction of the bark, leaves or dried fruit pulp is much used over the Region as a gargle-cum-mouthwash for toothache (5, 12; Senegal, 14–17). The leaves may also be chewed (5), or applied topically (15) for the same end. A decoction of bark or leaves is taken internally for respiratory troubles, and a lotion made from the macerated bark is instilled into the eyes for inflammation (5). In The Gambia powdered bark is used in application over deep-seated pain (23).

In Senegal (Casamance) the powdered root in a water decoction is used as a gargle against toothache and the root-bark is considered haemostatic and cicitrisant on wounds, especially by the Banyun and Efok in circumcision (14–17). In Cayor (Senegal) the powdered root is taken with *fuf* (Wolof, *Securidaca longepedunculata* Fres., Polygalaceae) as a poison-antidote (14, 17). The root has unspecified medicinal use in The Gambia (10).

The fruit is ellipsoid about 5 cm long. Before maturity fishermen in Senegal obtain a sticky substance from it for smearing on their tackle and to stop cracks in leaking pots (Sébire fide 5). The endocarp is surrounded by a soft mealy pulp that is edible and evokes the name 'rotten plum' (5). The pulp is everywhere eaten but it is not ubiquitously favoured, perhaps on account of flavour, perhaps because other choicer fruits are in season. In The Gambia it is not so much appreciated as is the closely related *mampato* fruit (Mandinka, *Parinari excelsa*) (6). Nor is it liked by the Limba of northern Sierra Leone (7). Yet in Guinea-Bissau it is said to be much liked (11), and farmers in Casamance (Senegal) clearing the bush retain the tree for the sake of its fruit: thus the tree enters into a form of semi-cultivation (1). In Sokoto market in N

Nigeria the fruit is said to be available all the year round (19). In parts of Senegal a fruit-decoction is taken against diarrhoea (14, 17).

The endocarp bears within it a layer of hairs which can be used for tinder (5). A similar tinder occurs in other African species, and also in an Indo-Chinese species (3). The hairs are also held to be anthelmintic in children. In Ivory Coast, because they tend to stick in the gullet, the hairs are administered with a ripe banana taken first thing in the morning on an empty stomach (18). In Guinea the endocarp is sold in markets as *sikuni* (Susu) as a vermifuge for taking with milk or better still with a banana. In Konakry Government hospital tests conducted showed evidence of affectiveness against helminths (14, 18). In this treatment the endocarpic hairs were placed in capsules for ease of swallowing. The hairs have been shown to contain *ceryl palmitate.* See also *Bafodeya benna.*

The kernel is edible. It is the source of *neou oil* which takes its name from the Wolof vernacular. It is an excellent drying oil and is composed of oleic acid 40%, eleostearic acid 31 %, linoleic acid 15%, palmitic acid 12% and stearic acid 2% (4, 9). The kernel also contains some protein and two phytosterols: *parincerium sterol A* and *B* (18). Neou oil finds local use in Sierra Leone mixed with palm-oil as a pleasant-smelling body-unguent (7). The fruit-rind is also similarly used (5). The ratio of the endocarp to the kernel is about 85: 15%. The kernel has been recorded as containing 62% oil, while 9% has been found in the endocarp (14). The nut is burnt to ash in The Gambia (23), and in Sierra Leone (8) for making soap.

The plant has superstitious usage in Senegal. In parts the root enters into a charm against bad luck (14), and branches are placed before new huts about to be occupied for the first time as a fetish to bring good luck (16). The Wolof endow the roots with power to relieve one possessed or tormented by Shades of the Departed (17).

References:

1. Aubréville, 1950: 205, as *Parinari macrophylla* Sab. 2. Barker 1242, K. 3. Burkill, 1935: 1666, *P. annanense.* 4. Busson, 1965: 220, as *P. macrophylla* Sab. 5. Dalziel, 1937. 6. Dawe 21, K. 7. Deighton 2492, K. 8. Deighton 5302, K. 9. Eckey, 1954: 472, as *Parinarium macrophyllum.* 10. Fox 82, K. 11. Gomes e Sousa, 1930: 69, as *Parinarium macrophyllum* Sabine. 12. Irvine, 1961: 265, as *P. macrophylla* Sab. 13. Keay et al., 1960: 317, as *P. macrophylla* Sabine. 14. Kerharo, 1966: 78–79, as *P. macrophylla.* 15. Kerharo & Adam, 1962: as *P. macrophylla* Sab. 16. Kerharo & Adam, 1963, b: as *P. macrophylla* Sab. 17. Kerharo & Adam, 1974: 680–2 as *P. macrophylla* Sab., with phytochemistry and pharmacology. 18. Kerharo & Bouquet, 1950: 90, as *P. macrophyllum* Sab. with references. 19. Lely 804, K. 20. Rosevear, 1961, as *P. macrophylla.* 21. Savill & Fox, 1967, 216, as *P. macrophylla* Sabine. 22. White, 1976: 308–10. 23. Williams, F. N., 1907: 96, as *P. macrophyllum* Sabine.

## **Parinari congensis** F. Didr.

FWTA, ed. 2, 1: 429. UPWTA, ed. 1, 170, as *P. subcordata* Oliv.

French: sougué des rivières (*sugué* (Susu for *P. excelsa* Sab.) *of the rivers,* Aubréville).

West African: **SIERRA LEONE** MENDE nja-lawa (*def.*-lawei) (FCD; S&F) TEMNE *am*-bis-*an*-ro-bath (FCD; S&F) **IVORY COAST** AKYE kotosoma (Aub.; RS) **GHANA** ADANGME adabutʃo (FRI) AKAN-BRONG tulingi (CV; FRI) GUANG-GONJA koto trampɔ (CV; BD&H) NCHUMBULU kototrampɔ (FRI) **TOGO** GURMA (Mango) insuo-pangi (Volkens) MOORE-NAWDAM pekire (Volkens) TEM (Tshaudjo) bende-noso (Volkens) **NIGERIA** EJAGHAM ekamufit (auctt.) IGBO àhàbà (KO&S) IJO unga (auctt.)

A tree reaching to about 30 m high, usually less, by 60 m in diameter, trunk irregular, twisted, sometimes forked or low-branching, of river-banks in fringing forest of soudanian savanna; occurring in Guinea and Mali to Nigeria, from Cameroun to Zaïre (1, 8–10).

The wood is orange-brown. It is said to resemble that of *P. curatellifolia*. It is hard and is reputed to be termite-proof (4, 6, 7). It is used in Nigeria (3) for hut-posts, and in Ghana (6) for poles and rafters.

The bark is purgative. It is used in Zäire in treatment for leprosy (8), and in Congo a tisane is administered for dysentery (1).

The fruit, which is an ellipsoidal drupe up to 3.75 cm long, has an edible pulp (5), and is used as a bait in fish-traps in Zäire (8).

References:

1. Aubréville, 1959: 1: 180–2. 2. Bouquet, 1969: 204. 3. Corbin 50, K. 4. Dalziel, 1937. 5. Irvine 2876, K. 6. Irvine, 1961: 263. 7. Keay & al., 1960: 318. 8. Letouzey & White, 1978: 124–6. 9. Savill & Fox, 1967: 216. 10. White, 1976: 321–3.

## **Parinari curatellifolia** Planch.

FWTA, ed. 2, 1: 429. UPWTA, ed. 1, 168.

West African: **SENEGAL** FULA-PULAAR (Senegal) mâpata baley = *black mapata* (K; K&A) TUKULOR naode (FB) MANDING-BAMBARA ntama (JB) tâmba (K&A) tudu (JMD) turu (JB) tutu (JMD; K&A) MANINKA tâmba (auctt.) tutu (auctt.) SOCE tâmba (K&A) tutu (K&A) **GUINEA-BISSAU** FULA-PULAAR (Guinea-Bissau) mámpara-djom-áe (JDES) mámpara-djon-áe (EPdS) **GUINEA** FULA-PULAAR (Guinea) benna (Langdale Brown) kura dombi (Aub.) MANDING-MANINKA tamba (Aub.) tutu (Aub.; FB) tutu kuma (Aub.) **SIERRA LEONE** KRIO rof skin plɔm (Thompson-Clewry) **MALI** MANDING-BAMBARA tamba (Aub.; FB) tutu (Aub.) tutu kuma (Aub.) MANINKA tamba (Aub.) tutu (Aub.) tutu kuma (Aub.) 'SENUFO' suboroko (Aub.) **UPPER VOLTA** GURMA bumansabu (Aub.; K&B) HAUSA gongea kussa (K&B) KIRMA goyoma (K&B) LOBI pwapatula oro (K&B) MANDING-BAMBARA tutu (K&B) DYULA tutu (K&B) MOORE uamtanga (Aub.; FB) 'SENUFO' subokoro (K&B) **IVORY COAST** BAULE hanvi(a)en (auctt.) mvia (Aub.) LOBI mlo-mlo (A&AA) MANDING-DYULA tutu (K&B) MANINKA sièra ughwé (B&D) tutu (Aub.; A&AA) SENUFO-NIAGHAFOLO tubozofi (K&B) DYIMINI komologuo (K&B) 'SENUFO' suboroko (Aub.; FB) **GHANA** AKAN-ASANTE atena (FRI) BRONG apodwono (BD&H) apogyono (BD&H) atina (BD&H) BAULE hamvia (FRI) mvia (FRI) DAGAARI ɛri (AEK) DAGBANI bong kapalla ?*this language or other northern language as spoken at Kulmasa/Sagalo* (AEK) papalatutubu (AEK) GBE-VHE (Pecí) potu-poti (BD&H) GRUSI kaleason (BD&H; FRI) GUANG-KRACHI pɔ́té pɔ́té (auctt.) SISAALA fun(g)funga (AEK; FRI) WALA eri (AEK; FRI) **TOGO** GURMA (Mango) insofani-woche (Volkens) TEM molemole (FB) TEM (Tshaudjo) melĕmĕlo (Volkens) molemole (Volkens) 'KPEDSI' yafo (Volkens) **DAHOMEY** BATONNUN pakuiku (Aub.; FB) pâschon (Aub.) GBE-FON huantu (Aub.) YORUBA-NAGO idofun (Aub.) **NIGER** HAUSA gauhassa (Aub.) **NIGERIA** FULA-FULFULDE (Nigeria) nawarre-baaɗi = *monkey's plum* (JMD; KO&S) GWARI bobwohi (JMD) HAUSA fara rura (ZOG) gwànjáa kúsá (auctt.) ƙaiƙayi (JMD) rura (auctt.) IDOMA odaubi (Odoh) IGALA ijakẹrẹ (Odoh) KANURI mándǝ̀ (C&H) NUPE kóbenci *the fruit* (Banfield) pútú *the tree* (auctt.) TIV ibyua (KO&S) ibyua i uter (JMD) YORUBA abo ìdòfún (JMD) ìdòfún (auctt.) ìdòfún akọ (Ross)

A tree to 8 m high, bole to 1 m girth, often stunted, and bole and branches twisted, of savanna forest from Senegal throughout the Region to W Cameroons, and widely dispersed across tropical Africa.

The tree shows some measure of fire-resistance and is widespread and locally frequent in secondary grassland of the transition zone between soudanian and guinean savanna which is subject to annual burning (22).

The wood is light brown to yellowish-red. In Nigeria it is said to be soft, easily worked, planing with a dull, smooth finish (Lely fide 8; 12). But in Togo it has been described as oak-like, very hard and heavy, remarkably fine and uniform, too hard for easy working, but especially suited for objects requiring great durability, and suggested for pegs (Volkens fide 8). In Ghana it is said to be hard, heavy, durable and not easily worked, but is used for rough timber for hut-poles and building (11). In Cameroun/Gabon it is hard and heavy (17). In Rhodesia it is reported as hard and heavy, and very hard on saws because of its silica content (21), and in Tanganyika to be 'useful' (5). The timber characters especially of Nigerian material seem to merit further examination. The timber is used in the Region for making mortars, burning as firewood and for conversion into charcoal (8, 11, 24). The young twigs, stripped of bark, are used [? in Nigeria] as a toothbrush (8). Young shoots are similarly used in Zanzibar and Pemba (23).

The bark is an ingredient of Yoruba *agbo* infusions (8, 18). A bark-decoction is taken by draught in Senegal by Fula (Fouladou) for nasal catarrh and bronchial affections (13–15). In Ivory Coast a bark-decoction is used for batheing a fracture after which it is dressed with a compress of crushed leaves and ground seed of maleguetta pepper and then bound in splints (16). Bark is part of a prescription for treating hookworm and insanity in Tanganyika (10), and is considered a malaria-remedy, blood-tonic and cardiac stimulant (21). Alkaloids have been reported absent from Nigerian material (1). The root-bark enters into Nigerian folk-medicine in treatment of skin infections, and water-extracts have been found to act on Gram +ve organism: *Sarcina lutea, Staphylococcus aureus* and *Mycobacterium phlei.* It had no action on Gram —ve organism, nor was it fungistatic (19).

The young leaves are often pinkish in colour. They are sometimes chewed instead of kola in the soudanian-guinean region (3, 8) and have in N Nigeria the Hausa name *Gwanja kusa,* or 'Gwanja is near', Gwanja, or more properly Gonja being a district in northern Ghana from which kola nuts are exported. Chewing the leaves, with which sometimes bark is added, is said to redden the mouth in the same manner as does kola (8, 24), but in Central Africa the result is said to be black (21). The young leaf in W Africa yields a red dye (8). The process by which it is obtained is not stated, but similar reference to the production of a black dye in central Africa is as the result of chewing the leaves (21). A leaf-decoction is used in Ivory Coast-Upper Volta in draughts and in baths as a febrifuge (16). In Ivory Coast sap from young leaves is swallowed as a cough-medicine, and young leaves dried for a day are powdered with those of *Piliostigma thonningii* (Leguminosae: Caesalpinioideae) for the preparation of a sporting powder ('poudre de chasse', Fr.) (2).

The fruit is sub-globose, 2.5–3.3 cm long. It has a pale yellow to reddish mealy pulp which is edible and is said to be very palatable. The kernel is also edible and yields on expression 17.7% of fatty substances, reddish, pleasant smell, viscous and thickening rapidly in air to give a skin. The kernel by solvent extraction may yield 39% of its weight as oil. It could be a source of a drying oil for use in the manufacture of paint, varnish, soap, etc., but the seed is difficult to decorticate and the return is low: in a Sudan sample 150 gm of whole nuts produced only 8·79 gm of kernel. (4, 6, 9, 20.) The endocarp is filled with a cottony wool surrounding the kernel. This is used over tropical Africa as a fire-tinder (7).

The plant enters into magic in the soudanian zone. Banda (Ligbi) hunters in northern Ivory Coast tie a piece of the root to the stocks of their guns for success in the chase, and the root has other unspecified magical attributes (3). In Ivory Coast protection against evil spirits is sought by drinking and washing with a leaf-decoction, and a leaf decoction made with leaves from the edge of a pathway, i.e., those brushed against by passers-by, is effective against asthma if one washes with it at night while all the world sleeps (2).

References:

1. Adegoke & al., 1968. 2. Adjanohoun & Akè Assi, 1972: 247. 3. Aubréville, 1950: 207. 4. Aubréville, 1959: 1: 175. 5. Brenan & Greenway, 1949: 476. 6. Busson, 1965: 216. 7. Dalziel, 1931: 99. 8. Dalziel, 1937. 9. Grindley, 1950: 152, with analysis of the oil. 10. Haerdi, 1964: 41. 11. Irvine, 1961: 263. 12. Keay & al., 1960: 319. 13. Kerharo, 1966: 78. 14. Kerharo & Adam, 1964, b: 564. 15. Kerharo & Adam, 1974: 678. 16. Kerharo & Bouquet, 1950: 90, as *Parinarium curataefolium* Planch. 17. Letouzey & White, 1978: 126–130. 18. MacGregor 149, K. 19. Malcolm & Sofowora, 1969. 20. Portères, s.d. 21. Watt & Breyer-Brandwijk, 1962: 890. 22. White, 1976: 323–33. 23. Williams, R. O., 1949: 400, as *Parinarium curatellifolium.* 24. Yates 23, K.

**Parinari excelsa** Sabine

FWTA, ed. 2, 1: 429 UPWTA, ed. 1, 168–9, incl. *P. tenuifolia* A. Chev., p. 170.

English: grey plum; Guinea plum; rough skin plum; rough-skinned plum; 'dawi'; plum (Sierra Leone, Sc. Elliot).

French: prunier de Guinée.

Portuguese: matapaz (from Crioulo, Guinea-Bissau, Feijão).

Trade: parinari (Liberia); sougué sugué (a foresters' term in francophone territories, and for the timber in Ivory Coast; kokoti (foresters' and saw-millers' term in Ghana and Sierra Leone).

West African: **SENEGAL** BALANTA fé (K&A) féfé (K; K&A) ndjano (FB) BANYUN inare (K&A) BASARI a-ngúd (Ferry) BEDIK gi-wùd (Ferry) 'CASAMANCE' mampata (Aub.) DIOLA bussuah (RS) DIOLA (Fogny) bu songay (JB, ex K&A) buyel (K&A) DIOLA (Tentouck) buyel (K; K&A) DIOLA ('Kwaatay') niñia (K&A) DIOLA-FLUP gulih = *throat* (auctt.) FULA-PULAAR (Senegal) mâpata (K; K&A) mâpatadé (K; K&A) MANDING-MANDINKA mampata (*def.* mampatoo) (auctt.) tambaa (*indef.* and *def.*) (K; K&A) tambakumbaa (*indef.* and *def.*) (auctt.) MANDYAK mampata (FB) SERER lo (auctt.) SOCE tâmba (K; K&A) tâmba kunda (K; K&A) WOLOF lay (JB; K&A) mâpata (JB; K&A) **THE GAMBIA** DIOLA-FLUP koelako (DRR; DF) FULA-PULAAR (The Gambia) kura-nako (DAP) MANDING-MANDINKA mampata (*def.* mampatoo) = *human complexion* (auctt.) WOLOF mampato (DAP) **GUINEA-BISSAU** BALANTA meile (JDES) meli meli (D'O) n-djano (auctt.) n-djápo (RdoF) undjano (JDES) BIAFADA mantchôl (RdoF) BIDYOGO nhêg-cuneme (JDES; RdoF) nhêg-ugene (RdoF) ugnene (JDES) CRIOULO mampataz (auctt.) FULA-PULAAR (Guinea-Bissau) cura (EPdS; RdoF) cura (JDES) curanaco (auctt.) kuranako (GeS) MANDING-MANDINKA mampatá (*def.* mampató) (JDES) MANDYAK betchalam (RdoF) bitchala(m) (D'O; JDES) n-tchalame (EPdS) MANKANYA minquela (JDES; RdoF) PEPEL minquelma (JDES; RdoF) **GUINEA** FULA-PULAAR (Guinea) kura (auctt.) KISSI gballo (RS) LOMA dava (RS) davagui (RS; FB) MANDING-MANINKA kura (JMD; FB) MANO kouin (RS) SUSU sugue (auctt.) **SIERRA LEONE** BULOM (Sherbro) bal-lɛ (FCD; S&F) FULA-PULAAR (Sierra Leone) kura (FCD) GOLA kuẽ (FCD) KISSI kwalo (FCD; S&F) KONO foni-lawei (Fox) koa (FCD; S&F) kola (FCD; S&F) kurɛ (S&F) KRIO roffin-plɔm (S&F) rof-skin-plɔm = *rough-skinned plum* (FCD; S&F) LOKO n-dawa (FCD; S&F) MANDING-MANDINKA kura (NWT; FCD) MENDE ndawa (*def.* ndawei) (auctt.) ndawa-hei, hei: *a woman, hence 'female ndawa' as opposed to Maranthes chrysophylla, 'male ndawa'* (S&F) SUSU suge (auctt.) SUSU-DYALONKE suge-na (FCD) TEMNE *am*-bis (auctt.) VAI kula-kɔ̃ (FCD) **LIBERIA** DAN g-boh (AGV; GK) KRU-GUERE (Krahn) djirro (GK) ko-tu (GK) BASA kpar (auctt.) umbah (C) MANO koine (JGA) MENDE ndawi-baji (C&R) **IVORY COAST** ABE so (auctt.) ABURE kotessima (A. Chev.; B&D) AKAN-ASANTE uallé (B&D) AKYE mosé (A. Chev.) mussé (A. Chev.) ANYI kotosuma (auctt.) piolo (RS; Aub.) DAN faulé-kokolé (A.Chev.) FULA-FULFULDE (Ivory Coast) kura (Aub.) KRU-GUERE (Wobe) nitu (A.Chev.) KYAMA assain (RS; Aub.) patobi (A. Chev.) 'KRU' tabotu (RS; Aub.) **GHANA** VULGAR afam (DF) kotosema (DF) ofam (DF) AKAN-ASANTE afam (DF) kwanedua (auctt.) ɔfam (FRI) TWI afamfufuo (CJT) ofam (DF) ɔpam (FRI) WASA afam (auctt.) afamfufuo (CJT) kotɔsɛma (Enti) kɔtɔsima (auctt.) ɔfam (FRI) ANYI kɔtɔsima (FRI) piolo (FRI) ANYI-AOWIN kotosima (FRI) SEHWI kotosima (auctt.) NZEMA faanle (CJT) kotosima (CJT) **NIGERIA** EDO ésàghò (JMD; KO&S) IJO dee (KO&S) YORUBA yínrínyìnrín YORUBA (Ikale) ohehe (auctt.)

An evergreen tree, to 40 m high by over 4 m in girth in humid rain-forest, but less in guinean forest and reaching only 8 m high in limiting conditions (29); in lowlands and into montane situations of 1,000 m elevation; common throughout, locally abundant and often forming nearly pure stands (2), from Senegal to W Cameroons, and widespread across tropical Africa.

The tree is used as a shade-bearer in *Cinchona* plantations (20). It is a rapid grower. In areas of managed forestry, it appears difficult to control and it tends to suppress desirable tree species present with it (24). It often occurs in open farmland and in Sierra Leone in rows which may indicate farm boundaries (24).

Sap-wood is creamy-white. Heart-wood is brown. There is no sharp demarcation between the two. The wood is hard and heavy. It sinks when fresh. It is a timber of secondary importance and logging is now not commonly carried out, though it has been traded as *sugué* from Ivory Coast for shipment to Europe as a substitute for oak. The timber is difficult to work. It is so full of silica that saws blunt rapidly. It is best sawn whilst still green and with special saws. Yet it finds use locally and has potential. It is cut into planks and timbers and made into furniture. If exposed to weathering it is not durable and may be attacked by borers, though if protected it is good for house-timbers. If pressure-treated with preservative, it is excellent for railway sleepers (6–8, 12, 13, 23–27). The wood makes good firewood and charcoal (21). The wood-ash contains tannin, which with tannin from the bark-ash, is used for tanning hides (8, 10, 12, 28).

The bark-slash has the smell of sour milk (2), and exudes a watery sap. The bark alone or with other drug plants is taken in decoction in Senegal against diarrhoea and other stomach-disorders (14, 17, 19). The bark pounded, macerated and chewed is applied as a cicitrisant to fresh wounds, especially in circumcision (22). In Sierra Leone a bark-decoction is taken to relieve stomachache, and roasted bark is added to palm-wine to improve its flavour (24). In Ivory Coast a bark-decoction is taken for anaemia and by women during pregnancy as a tonic (3), while in Liberia the inner bark soaked in rum or gin is considered good medicine (7). The bark, together with earth from the fungus-nursery in a termite nest, is taken in Tanganyika for hookworm, and it enters into a treatment of insanity (11). Analysis of Congo material showed a trace of flavones in the stem-bark, and an abundance in the roots, and plenty of tannin in both but no other active principal (19).

A root-macerate is taken internally in Senegal for migraine and stomach-pains, and for female sterility. Externally it is used as a haemostatic and cicitrisant. Root-decoctions are prepared for washes, baths, massages and fumigations in tiredness, fevers and chest-pains and rheumatism (14, 15, 19).

Steam from a boiling decoction of leafy twigs is considered in Senegal to soothe gingivitis, stomatitis and toothache (14, 17, 19).

The flowers are sweet-scented and meliferous. The fruit is ellipsoid 3.25 cm long with a warty surface. The pulp is yellowish when fresh, soft, edible with a flavour not unlike that of avocado pear (*Persea americana* Mill., Lauraceae). It has been reported to contain 38% sugars (19). There are occasions when it has served as an useful emergency food. Palatability and/or acceptability varies. In Guinea-Bissau they are not much appreciated (9). In Liberia they are highly esteemed (7, 26). In some parts of W Africa the pulp is fermented to an alcoholic drink known as *dhiaou* and *kounangui;* the fermented pulp is extracted by water and the sweet liquid is filtered off, concentrated by boiling and then fermented (5; De Wildemian fide 8). An infusion of the fruit is taken as a drink for diarrhoea and dysentery (8). The fruit is sometimes used as forage (25) and on the Nimba Mountain is sought out by wild animals for food (1). The endocarp is filled with a loose cottony wool surrounding the kernel. This is used over tropical Africa as a fire-tinder (30).

The kernel is also edible, and is oil-bearing. It is usually eaten after roasting and then mixed with other foods. A fat can be extracted from it, but quickly turns rancid (5, 8). The oil component is a drying-oil (25).

The seed shell and pulp of the fruit yield a dye (28).

References:

1. Adam, 1971: 376. 2. Aubréville, 1959: 1: 180, and as *P. holstii* Engl., 179. 3. Bouquet & Debray, 1974: 146. 4. Burtt Davy & Hoyle, 1937: 103, as *P. tenuifolia* A. Chev. 5. Busson, 1965: 216. 6. Cooper 93, 270, K. 7. Cooper & Record, 1931: 59, as *Parinarium excelsum* Sab., with timber characteristics. 8. Dalziel, 1937. 9. Gomes e Sousa, 1930: 68, as *Parinarium excelsum* Sab. 10. Greenway, 1941: sp. no. 189. 11. Haerdi, 1964: 41. 12. Irvine, 1961: 264–5. 13. Keay & al., 1960: 318–9. 14. Kerharo, 1966: 78. 15. Kerharo & Adam, 1962. 16. Kerharo & Adam, 1963, b. 17. Kerharo & Adam, 1964, b: 564–5. 18. Kerharo & Adam, 1964, c: 316. 19. Kerharo & Adam, 1974: 680. 20. Killian, 1953: 910. 21. Letouzey & White, 1978: 130–4. 22. Pobéguin, 1912: 51. 23. Sankey 19, K. 24. Savill & Fox, 1967: 217–9. 25. Schnell, 1950: 213, & as *Parinarium excelsum,* p. 257. 26. Voorhoeve, 1965: 314. 27. Walker & Sillans, 1961: 359. 28. Watt & Breyer-Brandwijk, 1962: 891. 29. White, 1976: 333–47. 30. Dalziel, 1930–31: 99.

## 'Parinari' spp indet.

West African: **SENEGAL** BASARI a-nóỹ (Ferry) KONYAGI a-né, u-sa (Ferry) **IVORY COAST** ABURE fuanté (B&D) **GHANA** AKAN-KWAWU sun gusubiri (Brent) TWI afrane (FRI)

A plant of the above Abure name is recorded as having medicinal use in Ivory Coast. *Parinari* sensu *FWTA*, ed. 2, is now recognized as falling into four genera, *Parinari* Aubl., *Maranthes* Blume, *Neocarya* (DC.) Prance and *Bafodeya* Prance (3). Plants of this group are used in traditional Ivorean medicine as analgesics for local pain, administered as a decoction of the bark in draught and as a plaster of the freshly crushed leaves or bark (1).

The Ghanaian names refer to a small tree of the savanna forest (2).

References:

1. Bouquet & Debray, 1974: 146–7. 2. Irvine, 1961: 266. 3. White, 1976: 265–350.

## CISTACEAE

**Helianthemum spp.**

English: rock-rose.

French: ciste; hélianthème.

Portuguese: cisto.

West African: **MALI** TAMACHEK rega *H. kahiricum* (RM) regig *H. lipii* (RM)

A genus of low woody herbs, essentially occurring in Africa north of the Sahara, but three species *H. ellipticum* (Desf.) Pers., *H. lippii* (Linn.) Pers. and *H. kahiricum* Del. occur across the Sahara to the Hoggar and the northern border of the Region. The first two named (1), and probably the last named also, provide a palatable fodder for camels and stock.

Reference:

1. Maire, H., 1933: 160.

## COCHLOSPERMACEAE

**Cochlospermum planchonii** Hook. f.

FWTA, ed. 2, 1: 185.

West African: **SENEGAL** BASARI a-nyẹ̀kɛ̀n-kɛ̀n (K&A; Ferry) **SIERRA LEONE** SUSU filɛrai onyi (NWT) SUSU-DYALONKE khɔleegɛsɛ-na (FCD; KM) TEMNE *a*-sin-*a*-ro-lal (NWT) **IVORY COAST** BAULE bro-guiessé (A&AA) MANDING-MANINKA trigba (A&AA) **GHANA** AKAN-ASANTE kokrosabia (FRI) DAGBANI biberetugu (FRI) biberetutugu (FRI) GBE-VHE kakalito (FRI) GUANG-NCHUMBULU bole (FRI) **NIGERIA** CHAMBA toh (FNH) FULA-FULFULDE (Nigeria) ambulolooji gaaduru (*pl.*) (J&D) ambuloloowol (*pl.* ambulolooji) (Taylor) jurɓaango (*pl.* jurɓaaɗi) (J&D; Taylor) yarudi (MM) HAUSA belge, zúnzùnaá (ZOG) NUPE gómbara (RB) YORUBA àwọ̀ òwú, òwú: *cotton;* àwọ: *likewise* (Phillips; JMD) fé rú *perhaps from* fé: *repeatedly;* rú: *sprouting up* (Dawodu; JMD) ṣẹwutu (Dawodu, ex JMD)

A shrub to about 2–2.5 m high, widespread in the Region from Senegal to W Cameroons, and into E Cameroun; with bright yellow flowers borne basally as in *C. tinctorium* but the two species are distinguishable on time of flowering, *C. planchoni* during the rainy season, *C. tinctorium* in the dry season.

The stem-bark yields a fibre which is used in northern Sierra Leone (8) and in N Nigeria (3, 4, 6) for making string and rope. It is cultivated spasmodically by the Chambas of the Vogel Peak area of N Nigeria who use the threaded seeds as beads (3, 4).

The root is a source of a yellow dye which is used in Sudan (9) and in Nupe and elsewhere (5). Hausas in N Nigeria add indigo to obtain green shades (9).

The root is used in Lagos in cooking soup when oil is not available (1). A root-decoction is drunk in northern Sierra Leone as a gonorrhoea treatment

(2). An extract of the plant (part not stated) is said to control menstruation (9), and the plant has unspecified medicinal uses in east Senegal (7).

The Fula of N Nigeria claim that a leaf-infusion bestows magical protection (6).

References:

1. Dawodu 202, K. 2. Deighton 4231, K. 3. Hepper 1298, K. 4. Hepper, 1965: 449. 5. Irvine, 1961: 73–74. 6. Jackson, 1973. 7. Kerharo & Adam, 1964, a: 434. 8. Miszewski 16, K. 9. Usher, G., 1974: 163.

## **Cochlospermum religiosum** (Linn.) Alston

FWTA, ed. 2, 1: 185, as *C. gossypium* (Linn.) DC. and 1: 760.

A sparsely branched shrub, or 'naked' tree of the drier parts of India that has been introduced into the Region and is cultivated in several areas.

The tree yields a gum, *katira gum,* which is insoluble in water but swells in it, and mixed with gum-arabic gives a water-borne adhesive paste. The gum has some value in cigar and ice-cream manufacture, and can be used as a substitute for gum tragacanth in various industrial processes. It is sweetish, cooling and sedative and helpful in cough medicine. The dried leaves and flowers are said to be stimulant. The floss surrounding the seeds is an inferior substitute for kapok. The seeds contain a non-drying oil reported in Indian material to amount to 14–15% and to be usable in soap-manufacture. The residual seed-cake is a suitable cattle concentrate, or can be used as a manure. The wood is soft, light and of little value. The bark contains a cordage fibre.

Reference:

1. Sastri, 1950: 261–2.

## **Cochlospermum tinctorium** A. Rich.

FWTA, ed. 2, 1: 185. UPWTA, ed. 1, 45.

West African: **SENEGAL** BASARI a-tyeùkúrbís (Ferry) BEDIK gi-ndyès, ga-tɔ̀fásyɛ (Ferry) DIOLA buɗâk (JB, ex K&A) FULA-PULAAR (Senegal) ɗâdéré (K&A) ɗâduré (K&A) darundé (K&A) faɗu râdé (K&A) nɗadéré (K&A) KONYAGI mbirlob, a-sukurbis (Ferry) MANDING-BAMBARA bu dãnk (JB) bu lulumay (JB) n-dilibara (JB) MANINKA tibibara, tribo, turugba (K&A) tiriba (JMD; K&A) uruba (JMD) SERER fayar (JB; K&A) per (JB; K&A) WOLOF fayar (auctt.) forray *the dye* (A. Chev.) **THE GAMBIA** FULA-PULAAR (The Gambia) dafe (DRR) MANDING-MANDINKA foosea (Ozanne) kunturro (DRR) tiribe (*def.* tiriboo) (Fox) **GUINEA-BISSAU** FULA-PULAAR (Guinea-Bissau) djánderè (JDES) djarúndjè (JDES) djaundéré (JDES, ex EPdS) tirbom (JDES; EPdS) **GUINEA** FULA-PULAAR (Guinea) diarundé (CHOP) MANDING-MANINKA tiriba (JMD) uruba (JMD) SUSU filira gĕsé (CHOP) **MALI** DOGON sólo anyu (C-G) MANDING-BAMBARA n-dli-bara (GR) **UPPER VOLTA** BISA lugur (Prost; K&B) DAGAARI beluma (K&B) GRUSI tampo (K&B) LOBI sory (K&B) MANDING-DYULA bédiéra korandi (K&B) konlo koroni (K&B) sandé koroni (K&B) MOORE sansa (K&B) sanséghé (K&B) soasgha (K&B) 'SENUFO' tikwélégué (K&B) **IVORY COAST** BAULE auniguéssébé (K&B) babigna (K&B) broguessé (Bégué) diéssé (K&B) kadiendi (K&B) krédé diéssé (K&B) siripopo (K&B) KPALAGHA gapoli (K&B) KULANGO pugutipu (K&B) MANDING-DYULA bédiera korandi (K&B) konlo koroni (K&B) sandé koroni (K&B) MANINKA tiriba (K&B) tiribara (K&B) tirigba (K&B) tribga tiama (K&B) turugba (K&B) SENUFO-TAGWANA kikwu (K&B) kukuo (K&B) DYIMINI sinbellébé (K&B) 'SENUFO' tikwélégué (K&B) **GHANA** DAGBANI biberetugu (FRI) biberetutugu (FRI) GBE-VHE kakalito (BD&H; FRI) GUANG-NCHUMBULU bole (FRI) MOORE tɔngvóle = *ground ochro, i.e., the red fruits* (FRI) **TOGO** ANYI-ANUFO uanyise (Volkens) TEM (Tshaudjo) lombo (Volkens) **NIGERIA** ARABIC-SHUWA maghr (JMD) BEROM tútok sunàŋ, hwaàl jey (LB) CHAMBA toh (FNH) FULA-FULFULDE (Nigeria) jaarundal (*pl.* jaarunde) (JMD) njaareendahi = *plant that grows on sandy soil* (J&D) GWARI gwagwa = *yellow; i.e., the dye* (JMD) HAUSA balagande (JMD) balge (JMD) kukur (ZOG) kyamba (JMD; ZOG) raawáya, raàwáyaà (auctt.) turri (Hill) ƴan kyamba (JMD; ZOG) zabiibii (JMD; ZOG) zúnzùnaá (Bargery; ZOG) HAUSA (West) balagande, balge (ZOG) IGBO (Nkalike) abangi (DRB) IGBO (Nsukka) abazi (DRB) KANURI masauwe; massowai (C&H) TIV kpáyāndē (JMD) YORUBA rawaye *from Hausa* (JMD; IFE) ṣẹwutu (JRA) ṣwẹwutu (JRA)

A bushy plant attaining about 30 cm tall with annual shoots arising from a perennial woody stock, and handsome golden yellow flowers which render it worthy of horticultural cultivation. Of widespread occurrence in savanna and scrub land throughout the drier parts of the West African Region, and, indeed, seemingly to prefer devastated, rocky and annually burnt areas. It occurs also in Cameroun, Sudan and Uganda.

Pulped leaves are used in Ivory Coast in a wet dressing to maturate abscesses and furuncles, and a decoction of twigs, or roots, is taken in draught or in baths for urino-genital disorders, and kidney and intercostal pain. There may well be some toxic substance present for related *Cochlospermum* species in Cameroun are added to arrow-poison formulae (4). Cattle will not graze the plant (1) even in time of shortage (12), though it appears to have been considered at one time as a possible fodder plant in northern Ghana (8). Unripe fruit capsules are however eaten by hunters to allay thirst, and porcupines will eat the roots (5).

The fruit contains a floss which can be used to stuff cushions. In Togo it is spun into necklace cords and the plant is regarded as the 'father of cotton' (5). The young stem-bark yields a useful fibre (5, 7, 8, 16). The fruit is mixed with that of *Tamarindus* and made into a concoction which is drunk by the Chambas of N Nigeria for snake-bite (6), while the powdered root is applied topically in Ivory Coast-Upper Volta for the same purpose (13). The seeds yield an oil which has found use in treatment of leprosy (3, 8), and the root along with other drug-plants is similarly used in Ivory Coast-Upper Volta (13).

The conspicuous flowers appearing from the perennating rootstock before the plant has put out any annual shoots attract for the plant numerous attributes by medicine-men though in the main it is the root of the plant which receives most attention. A yellow or brownish-yellow dye is obtained from it with the result that the root, on the Theory of Signatures, is used, with or without other drug-plants, in Senegal for jaundice and liverish fevers (10, 11), and amenorrhoea (9). The roots are used against ascites, beriberi and various oedemas (11), the diuretic action probably being helpful. Similarly in Ivory Coast-Upper Volta the roots are used for oedematous conditions, for orchites, schistosomiasis, jaundice, fevers, epilepsy, pneumonia, intercostal pains, and bronchial affections, in eye-instillations for conjunctivitis, and for indigestion and stomach-pains (13). Water in which roots have been boiled is given to women in childbirth in The Gambia (17), and the root is chewed in Nigeria as a tonic (3). In Senegal (11, 12) and in Ivory Coast (13) the root has a reputation as an efficient decongestant, and as a venal vaso-constrictor it is said to be effective in reducing haemorrhoids where surgery would normally be indicated.

A root-infusion is used by the Fula cattlemen in Guinea (15) to arrest diarrhoea in calves.

The yellow or brownish-yellow dye is used to dye cloth, thread, mats, leather, etc. The root may be rubbed while fresh onto the article to be dyed, or dried, pulverized and pounded into a paste and rubbed on (5); or the roots may be crushed up with potassium salts obtained from vegetable ash and boiled up, and with the addition of indigo a wider range of colour is achieved (16). The dye is used to colour shea butter and cooking oil, and may perhaps also impart some flavour. It is said to stain the mouth. Shea butter and other oils so coloured are used in applications to burns (5). Besides the yellow dye-stuff, much mucilage is present, and also sugar, tannin and some alkaloid (2, 5).

The root is used in Nigerian folk medicine to treat skin infections and water extracts have shown activity against Gram +ve organisms, but no activity on Gram −ve organisms, nor on fungae (14).

References:

1. Adam, 1966, a. 2. Adegoke & al., 1968. 3. Ainslie, 1937: sp. no. 102a. 4. Bouquet & Debray, 1974: 66. 5. Dalziel, 1937. 6. Hepper, 1965: 449–50. 7. Irvine, 1930: 114. 8. Irvine, 1961: 74. 9.

Kerharo, 1967. 10. Kerharo & Adam, 1963, a. 11. Kerharo & Adam, 1964, b: 426–7. 12. Kerharo & Adam, 1974: 339–41, with references. 13. Kerharo & Bouquet, 1950: 38, as *C. niloticum* Oliv. & p. 39. 14. Malcolm & Sofowora, 1969. 15. Pobéguin, 1912: 26. 16. Portères, R., s.d. 17. Williams, F. N., 1907: 200.

**Cochlospermum vitifolium** (Willd.) Spreng.

A small tree with attractive yellow flowers of Mexico and the Caribbean now introduced into the Region and under cultivation in Ghana at the Cacao Research Institute Station.

In Belize the wood is used to make fish-net floats (2). The bark yields fibre used in Mexico to make rope (3, 4) and a bark concoction is taken for jaundice (1).

References:

1. Hinton 3596, K. 2. Schipp 45, K. 3. Uphof, 1968: 141. 4. Usher, 1974: 163.

# COMBRETACEAE

**Anogeissus leiocarpus** (DC.) Guill. & Perr.

FWTA, ed. 2, 1: 280. UPWTA, ed. 1, 73–74, as *A. leiocarpus* Guill. & Perr. and *L. schimperi* Hochst.

English: anogeissus (trade, Nigeria, Lucas); "chew-stick" (forestry, Kano, N Nigeria).

French: bouleau d'Afrique, = *African birch* because of the silvery cast of the foliage like the temperate birch (auctt.).

West African: **MAURITANIA** ARABIC ('Maure') arueidja (Aub.) awithegé (Aub.) awithegi (Aub.) gerk (Aub.) **SENEGAL** BASARI a-ngàn (K&A; Ferry) BEDIK ga-ngál (K&A; Ferry) FULA-PULAAR (Senegal) godoli (K&A) kodol (K&A) kodoli (K&A) KONYAGI a-kar-gatè (Ferry) MANDING-BAMBARA kalama (Vuillet; K&A) kérékéto, krékété, krékri, krétété (K&A) n-galama (JB) MANDINKA kérékéto (K&A) krékété (K&A) krékri (K&A) krétété (K&A) MANINKA kérékéto (K&A) krékété (K&A) krékri (K&A) krétété (K&A) 'SOCE' kérékéto (K&A) krékété (K&A) krékri (K&A) krétété (K&A) SERER ngogil (K&A) ngoǰil (auctt.) SERER-NON godal (auctt.) godal (AS, ex K&A; Aub.) WOLOF geyd (K&A) geyt (K&A) mara nor (G&P) ngégan (K&A) n-geǰ = *pointed* (K&A; DF) **THE GAMBIA** FULA-PULAAR kojoli (DAP) MANDING-MANDINKA kalamaa (*indef.* and *def.*) = *calabash spoon* (DAP; DF) kerketto (DAP) MANINKA kalaama (DAP) WOLOF mara (DAP) wej (DAP) **GUINEA** MANDING-MANDINKA kalamaa MANINKA kréké (CHOP) krékété (CHOP) krékré (CHOP) **MALI** DOGON sìgilu (C-G) uan FULA-PULAAR (Mali) kodioli (GR) MANDING-BAMBARA n-galama (GR) KHASONKE kerkété (Aub.) MANINKA krékété (Aub.) SONINKE-SARAKOLE uaye (Aub.) **UPPER VOLTA** DAGAARI sinki (K&B) GURMA buhiébu (Aub.) HAUSA maréké (K&B) mariki (K&B) MANDING-BAMBARA kalama (K&B) krékété (K&B) DYULA guiméni (K&B) MOORE piega (K&B) sigha (K&B) SONGHAI-ZARMA gonga (K&B) 'SENUFO' niulepiaï (K&B) **IVORY COAST** VULGAR kerkété *from Maninka* (A. Chev.) ANYI kakaleka (A. Chev.) kakalema (K&B) BAULE kalima (A. Chev., ex K&B) MANDING-DYULA guiméni (K&B) MANINKA kerkété (Martineau) kré kré (CHOP) SENUFO-TAGWANA gla (K&B) 'SENUFO' niulepiaï (K&B) **GHANA** VULGAR kane (DF) ADANGME *o*-sãkane(-ɛ) (auctt.) AKAN-ASANTE kane (auctt.) BRONG akula (FRI) akuta (JMD) KWAWU osakaneɛ = *desert lamp* (FRI) DAGBANI sia (auctt.) ʃia (FRI) GA sankane (auctt.) GBE-VHE hehɛ (BD&H; FRI) hihe (FRI) tsetse (auctt.) VHE (Pecí) ɣuɣɔe (FRI) GUANG-KRACHI kakanla (Volkens) NCHUMBULU kakanli (JMD; FRI) NANKANNI shia (Gaisie) tcheean (Lynn; FRI) **TOGO** ANYI-ANUFO kalékété (Aub.) BATONNUN kakala (Aub.) kakara (Aub.) GBE-VHE echeche (Volkens) réré (Aub.) tsetse (Volkens) HAUSA binaselimebu (Aub.) KABRE kõlu (Gaisser) KPOSO oga (Volkens) MOBA nasiéga (Aub.) MOORE-NAWDAM hiega (Aub.) SOMBA karosufa (Aub.) TEM (Tshaudjo) kodelia (Volkens) YORUBA-NAGO àyìń **DAHOMEY** BATONNUN kakala (Aub.) kakara (Aub.) GBE-FON chleho (Volkens) hilihaye (Aub.) VHE réré (Aub.) HAUSA binaselimebu (Aub.) YOM sira (Aub.) **NIGER** FULA-FULFULDE (Niger) kadioli (Aub.) GURMA hovsiébu (Aub.) HAUSA maréké (Aub.) SONGHAI-ZARMA gonga (Aub.) TAMACHEK akuku (Aub.) **NIGERIA** ARABIC-SHUWA sahab (JMD) BEROM kasi (LB) FULA-FULFULDE (Nigeria) galaldi *a form* *of this sp.* (MM) kojoli (auctt.) HAUSA farin gamji *a form of this sp.* (MM) gàngàmaá (ZOG) ganuwa *ash of the burnt wood* (JMD) maariƙéé (auctt.) máárƙéé (auctt.) HAUSA (West) ƙwanƙila *ash of the burnt wood* (JMD) IGBO abakaliki atara (DRR) atara (auctt.) KANURI ànəm (auctt.) YORUBA àyìn (KO&S) àyìn *pronounced 'anyi'* (JMD) orín òdan (Lucas)

A graceful tree of the Sahel to the forest zone from Senegal to W Cameroons, to 15–18 m high in Senegal (20), 22 m in Ghana (14) and to nearly 30 m in Nigeria (16), straight tapering bole, branching from low down, common in places, often gregarious and effectively killing out grass (9, 11). Its range extends across Africa to Sudan and Ethiopa, and southwards to Zaïre.

In drier regions it grows into a good shade tree, and prefers acid but good soil in proximity to water (26). It is potentially a reafforestation species for the soudanian savanna (6, 9).

The sap-wood is rather wide, dirty white or yellowish-grey and is susceptible to borers. Heart-wood is dark dull brown, streaky, becoming sometimes almost ebony-black, and is very hard, dense and fine-textured. It is insect and termite-proof but not durable in contact with damp ground. The wood is resistant to treatment with preservatives, seasons rapidly with some distortion, is difficult to work but has good turning properties especially when there is black heart-wood. It finishes well and takes a good polish (5, 6, 8, 9, 14, 23).

The wood compares favourably with that of *Tamarindus indica* (Leguminosae: Caesalpinioideae), *Diospyros mespiliformis* (Ebenaceae) or *Terminalia glaucescens* (Combretaceae). It is used for the building of rural and temporary buildings as posts and roofing, and is good for tool-handles and sports goods (9, 23). As firewood it is excellent giving out great heat, and it makes good charcoal (14). The wood is commonly used in the upper framework of wells (*ahola,* Hausa) in N Nigeria.

Ash of the burnt wood is used in N Nigeria as a dehairing agent in the preparation of skins for the tanning bath (9). The Yoruba of S Nigeria and Fula of the north use it as a mordant for *Lonchocarpus* (Leguminosae: Papilionoideae) indigo dye (9, 15). It is used as a lye in washing clothes and a basis for making soap (5, 9). The ash is in some parts an item of trade (9).

The bark is generally grey and scaly. It is fire-resistant (3). Slash is yellow and exudes a not very soluble gum, (*marike,* Hausa), which is eaten in N Nigeria and is regarded by some as the best gum for chewing (9, 10, 12, 13). It contains 22% uronic acid (20). It is of good viscosity and usabie as a substitute for or adulterant of gum-arabic. It is used in beating cloth, and to make ink. It has adhesive properties (5, 6, 21). A cold infusion is considered a palatable beverage and a decoction is given to a new-born baby to drink. It is laxative (4). It can be used in pharmacy as an emulsifying agent (9, 24). The bark itself is chewed to obtain the gum (9, 13). Powdered bark is applied to wounds and ulcers and a lotion made from the bark is similarly used (9). In Senegal the Fula of the Ferlo region use the bark to treat diarrhoea, and the Tandanke in a medico-magical application suggestive of sun-worship take the bark in the morning from the east and west sides of the trunk for infantile diarrhoea (17, 20). In Ivory Coast – Upper Volta the powdered bark with that of a *Terminalia sp.* (Combretaceae) is applied to the gums for toothache and a decoction of the bark from the twigs is used to clean sores and syphilitic chancres (21). Root-bark is considered by the Soce to be stimulant and aphrodisiac (19). In some regions of Upper Guinea the bark is used as a febrifuge in hot lotions and infusions (25), and in Ivory Coast – Upper Volta prescriptions of leaves and bark with other drug-plants are taken for leprosy (21). In Ghana a bark-decoction with red peppers is recorded as taken internally for body and chest-pains, and with *Aframomum* (Zingiberaceae) added for external application (14). In N Nigeria some parts of the tree are boiled in water together with potash and the liquid is taken as a cure for stomach pains and for schistosomiasis (27).

In veterinary medicine, the bark, or more often the fruit or seeds, is used as a taenicide for horses and donkeys (9, 21, 24).

The bark has been recorded having 17% tannin, not a high enough concentration for profitable export, but usable locally for tanning (9, 11). The resultant leather is yellowish-brown, stiff and rather harsh.

In N Nigeria the tree (part unspecified) is used as a chew-stick so that in forestry circles in Kano it is known in English as "chew-stick" (27), but more specifically the roots are widely used in Nigeria and in Ghana for this purpose, even far from their source of origin (4; 8; 9). An examination of chew-stick materials at Ibadan, Nigeria, detected no antibiotic activity (22). Pulped roots are applied to sores in Ivory Coast – Upper Volta to promote healing (21). In the Sahel de Nioro of Mali a good yellow dye is extracted from them (26).

The leaves furnish a yellow dye (*gangamu,* Hausa), used in N Nigeria for cloth and leather (9), and in Dahomey the tree is said to be specially planted close by villages for obtaining the dye. In the soudanian region beside the dye the liquor is used for washes and feverish colds (5, 6). A decoction of leaves is applied to the skin in Ivory Coast to alter the pigmentation, and is used as an eye-wash for certain complaints. In the Jebel Marra of Sudan leafy twigs are infused to make a drink (11). In Senegal the foliage is browsed a little by all stock (1). It serves as fodder in N Nigeria (27).

The leaves are rich in tannin and have been found to contain saponin. The twig and root-barks are rich in tannin, saponin and sterols (21).

A trace of alkaloid may also be present in the bark (2).

References:

1. Adam, 1966, a. 2. Adegoke & al., 1967. 3. Alasoadura 82, UCI. 4. Ainslie, 1937: sp. no. 33. as *A. schimperi.* 5. Aubréville, 1950: 135–9. 6. Aubréville, 1959: 3, 70. 7. Bouquet & Debray, 1974: 66. 8. Burtt Davy & Hoyle, 1937: sp. no. 33. as *A. schimperi* Hochst. 9. Dalziel, 1937. 10. Howes, 1949: 67, 69. 11. Hunting Technical Surveys, 1968. 12. Irvine, 1952, a: 30–31, as *A. schimperi.* 13. Irvine, 1952, b. 14. Irvine, 1961: 112–4. 15. Jackson, 1973. 16. Keay & al., 1960: 159–61. 17. Kerharo & Adam, 1964, a: 406–8, 434. 18. Kerharo & Adam, 1964, b: 407. 19. Kerharo & Adam, 1964, c: 294. 20. Kerharo & Adam, 1974: 341–2. 21. Kerharo & Bouquet, 1950: 47–48, as *A. schimperi* Hochst. and *A. leiocarpus* Guill. & Perr. 22. Lowe, 29/9/1976. 23. Lucas, 1967. 24. Oliver, 1960: 19, 43. 25. Pobéguin, 1912: 15. 26. Roberty, 1953: 449. 27. McIntosh, 26/1/1979.

## **Combretum** Loefl.

West African: **THE GAMBIA** MANDING-MANDINKA kataŋkoyo = *white katang* (DRR; DF) kulungkalaŋ (*def.*-o) (DRR) kunundindoo = *small birds intoxicating drink, applied loosely to all C. spp.* (JMD) susujango (DRR) WOLOF alom (DRR) rhatbuye (DRR) **SIERRA LEONE** KISSI hoya (S&F) tomɛso (S&F) LOKO wundindi (S&F) MENDE hakpa-namu (*def.*-i) (FCD) TEMNE *a*-kati (S&F) **NIGERIA** FULA-FULFULDE (Nigeria) ɓuski (*pl.* ɓusɗe) *often with qualifying colour:* daneehi: *white;* baleehi, boɗeehi: *black* (JMD) doki HAUSA googà-jìkí (ZOG) mumuye *the gum* (Lowe) taramniya, tarauniya *general for gum-yielding spp.* (KO&S); *often with qualifying colour:* fari: *white;* ja: *red;* baki: *black* (JMD) KANURI kàtáàr, káwúlá (C&H)

## **Combretum aculeatum** Vent.

FWTA, ed. 2, 1: 273. UPWTA, ed. 1, 75.

West African: **MAURITANIA** ARABIC (Hassaniya) ikik (Leriche & Mokhtar Ould Hamidoun) ARABIC ('Maure') ikik (Aub.) savât (GR) **SENEGAL** VULGAR saut *from Wolof* (Bouery) FULA-PULAAR (Senegal) bulapâl (Aub.) laońâdi (Aub.; K&A) MANDING-BAMBARA kabana (K&A) kôti (K&A) wolo (K&A) wolo koli (JB; K&A) MANINKA kôti (K&A) wolo (K&A) 'SOCE' kunu ndin dolo (K&A) SERER ńalafum (K&A) ńélafum (K&A) ńelafun (Aub.) ñélafund (JB) sambé (auctt.) SERER-NON sabe (AS; K&A) sambé (AS; Aub.) WOLOF saut, savat (auctt.) sowat (JMD; K&A) **GUINEA** FULA-PULAAR (Guinea) bulapâl (Aub.) lahon niandi (Aub.) laugni (Aub.) MANDING-MANINKA konti (Aub.) uolo (Aub.) **MALI** DOGON pɛ̀ɛ̀lie nínyilu (C-G) FULA-PULAAR (Mali) bulapâl (Aub.) lahon niandi (A. Chev.; Aub.) laugni (Aub.) MANDING-BAMBARA kolobé (A. Chev.) konti (Aub.) uolo (Aub.) wolo-konti (A. Chev.) MANINKA konti (Aub.) uolo (Aub.) SONGHAI buburé (A. Chev.) **UPPER VOLTA** BISA gulugu (K&B) laré (K&B) MOORE kodem tabaga (K&B) kodentigla (K&B) koditambiga (auctt.) kotintaabora = *causing pain when stepped on* (FRI) kudugulungu (K&B) **NIGER** FULA-FULFULDE (Niger) bulapâl (Aub.) lahon niandi (Aub.) laugni (Aub.) HAUSA bubukia (Aub.) SONGHAI búrbúrá (*pl.* -á) (D&C) SONGHAI-ZARMA buburé (Aub.) **NIGERIA** FULA-FULFULDE (Nigeria) lawnyi (MM) sakaata-saare (*pl.* saakata-saareeji) (auctt.) HAUSA farar geza (Taylor; JMD)

A scandent or rambling shrub of dry savanna, sometimes riverain, recorded from Mauritania to N Nigeria and across Africa to NE Africa, northern Kenya, Tanganyika and Uganda.

The leaf-petioles become spines with age, and evoke the Moore name in Upper Volta meaning 'causing pain when stepped on' (7). Nevertheless the plant provides browsing for all stock in Senegal (1), Jebel Marra (6) and Sudan (4), and in northern Kenya where buck will also take it (13).

The lianous branches are supple and are used in Kenya to make donkey panniers (12) and wicker-baskets for holding milk-vessels (5). In Senegal they are a part of the construction of a fish-lure used in the Senegal River (8, 10).

The plant has diuretic properties. The water in which leaves have been boiled is drunk in NW Senegal to promote micturition in cases of venereal disease obstruction of the urethra (3). The plant is also purgative. It is prescribed in Senegal as a purging medicine, and by causing expurgation, to treat colitis, blenorrhoea, helminthiasis, leprosy, loss of appetite and wasting (10). It is also used in Upper Volta for leprosy (11). In Senegal root-powder is rubbed over the body in a leprosy treatment (8). The Soce of Senegal claim that a root-decoction has a well-established reputation in the treatment of catarrh (9), and Serer use sap from the centre of the stem for eye-troubles (10). The boiled roots are taken in Kenya for stomach complaints (2).

The seeds are eaten by people of N Kenya (13).

References:

1. Adam, 1966, a. 2. Bally B 3691, K. 3. Boury, 1962: 15. 4. Carr 816, K. 5. Gillett 13345, K. 6. Hunting Technical Surveys, 1968. 7. Irvine 4679, K. 8. Kerharo & Adam, 1964, b: 427. 9. Kerharo & Adam, 1964, c: 300. 10. Kerharo & Adam, 1974: 342. 11. Kerharo & Bouquet, 1950: 48. 12. Mathew 6734, K. 13. Mwangangi & Gwynne 1255, K.

## **Combretum bracteatum** (Laws.) Engl. & Diels

FWTA, ed. 2, 1: 274. UPWTA, ed. 1, 75.

West African: **NIGERIA** EDO ọ̀kọ́sọ̀ (Amayo) IGBO ṁmányá ṉzā, ṁ.ńzīzá = *palm-wine of the sun-bird* (JMD) IGBO (Awka) àchịchà ṉzā (NWT) YORUBA ọ̀gàn (JMD) ọ̀gàn dúdu = *black ogan* (Dawodu)

A scandent shrub of wooded savanna, recorded from S Nigeria, W Cameroons and Fernando Po extending to Zaïre.

The flowers are showy, yellowish and red. The plant has been brought into cultivation. The flowers are sucked by children and visited by sun-birds to obtain the nectar, and the hollow stems are used in S Nigeria in tapping palms for wine (1).

Reference:

1. Dalziel, 1937.

## **Combretum cinereipetalum** Engl. & Diels

FWTA, ed. 2, 1: 273.

A scandent shrub or forest liane recorded only in W Cameroons in the Region, but extending to Zaïre and Uganda.

The plant is used in Gabon as a cicitrisant (1).

Reference:

1. Walker & Sillans, 1961: 121.

**Combretum collinum** Fresen.

**ssp. binderianum** (Kotschy) Okafor

FWTA, ed. 2, 1: 271 as *C. binderianum* Kotschy.

West African: **GHANA** AKAN-BRONG domapowa (FRI) MOORE kwegenga (AEK) **NIGERIA** HAUSA farar geza (KO&S)

**ssp. geitonophyllum** (Diels) Okafor

FWTA, ed. 2, 1: 271 as *C. crotonoides* Hutch. & Dalz., *C. lamprocarpum* Diels and *C. geitonophyllum* Diels UPWTA, ed. 1, 77 as *C. kerstingii* Engl. & Diels, and 79 as *c. verticillatum* Engl.

West African: **SENEGAL** BANYUN kifati fati (K&A) BASARI a-noxɔ̀x (K&A; Ferry) BEDIK ga-nɔ́hɔ́ (Ferry) DIOLA (Fogny) bagen déma (K&A) FULA-PULAAR (Senegal) dókidéwi = *female doki* (K&A) MANDING-BAMBARA dirinimble (JB; K&A) tâgaré dé (JB, ex K&A) MANDINKA mad́éfo (K&A) madifo (K&A) 'SOCE' mad́éfo (K&A) madifo (K&A) **THE GAMBIA** DIOLA (Fogny) bagen déma = *stout herb* (DF) FULA-PULAAR (The Gambia) buki-danehi (DAP) doki (DAP) MANDING-MANDINKA jambakataŋ (*def.*-o) = *leaf-scabies* (DAP) WOLOF kunaji (DAP) rat *onomatopoeic, after the snapping of a dry twig* (DF) **GUINEA** MANDING-MANINKA dirinіblé (Aub.) djirinimblé (Aub.) tiangara dié (Aub.) tiemankaran dié (Aub.) **MALI** MANDING-BAMBARA dirinіblé (Aub.) tiangara dié (Aub.) tiemankaran dié (Aub.) MANINKA dirinіblé (Aub.) djirinimblé (Aub.) tiahgara dié (Aub.) tiemankaran dié (Aub.) 'SENUFO' nanfolorho (Aub.) **UPPER VOLTA** GURMA okanguiamonnu (Aub.) HAUSA farar taramnia (Aub.) taramnia (Aub.) **IVORY COAST** MANDING-MANINKA bogbo (B&D) **GHANA** AKAN-ASANTE mwansanta (FRI) nnuan (FRI) BRONG mwansanta (BD&H) DAGBANI ulinga (BD&H; FRI) urinpielega (JMD) NANKANNI kɔnkɔka (Gaisie) kunkunka (Lynn; FRI) **TOGO** BATONNUN banguboguru (Aub.) sinabidékuku (Aub.) SORUBA-KUYOBE fetanpuara (Aub.) mufanpahida (Aub.) TEM (Tshaudjo) alembole (Kersting) **DAHOMEY** BATONNUN banguboguru (Aub.) sinabidékuru (Aub.) GBE-FON dosso (Aub.) SORUBA-KUYOBE fetanpuara (Aub.) mufanpahida (Aub.) YORUBA-NAGO àńràgba **NIGER** GURMA okanguiamonnu (Aub.) HAUSA farar taramnia (Aub.) taramnia (Aub.) **NIGERIA** FULA-FULFULDE (Nigeria) ɓuski (Aub.) ɓuski daneehi = *white danehi* (JMD) dookehi (JMD) dooki (JMD) dooki daneehi (JMD) HAUSA bauli, dagara, dageera (ZOG) farar tàrámnı́yaá = *white taramniya* (JMD; ZOG) kantakara *from Kanuri* (JMD) tàrámnı́yaá (JMD; KO&S) KANURI kàtáàr (JMD) KILBA shafa *a sacred symbol for taking oaths made from the leaves of this sp. tied up with other objects* (Meek) YORUBA ajantiro (Kennedy; KO&S) arọ (KO&S; IFE)

**ssp. hypopilinum** (Diels) Okafor

FWTA, ed. 2, 1: 271 as *C. hypopilinum* Diels. UPWTA, ed. 1, 76 as *C. hypopilinum* Diels.

West African: **THE GAMBIA** FULA-PULAAR (The Gambia) buki-danehi (DAP) MANDING-MANDINKA kataŋ (*def.*-o) kataŋ-koyo (DAP) WOLOF rat (DAP) **GUINEA** FULA-PULAAR (Guinea) dooki (Aub.) MANDING-MANINKA uahia uahia (Aub.) **MALI** FULA-PULAAR (Mali) dooki (Aub.) MANDING-BAMBARA uahia uahia (Aub.) MANINKA uahia uahia (Aub.) **UPPER VOLTA** HAUSA jar ganyi (Aub.) jar tàrámnı́yaá (Aub.) **GHANA** DAGBANI urinli (JMD; FRI) GRUSI-KASENA lampone (Gaisie) NABT porpiliga (BD&H; FRI) NANKANNI da-koonga (Lynn; FRI) SISAALA chinchɛpula (AEK; FRI) **TOGO** BATONNUN buhangossa (Aub.) denguéborumé (Aub.) **DAHOMEY** BATONNUN buhangossa (Aub.) denguéborumé (Aub.) **NIGER** FULA-FULFULDE (Niger) dooki (Aub.) HAUSA jar ganyi (Aub.) jar tàrámnı́yaá (Aub.) **NIGERIA** FULA-FULFULDE (Nigeria) ɓuski daneehi = *white danehi* (JMD) HAUSA farar tàrámnı́yaá = *white taramniya* (JMD) ján ganyeé (ZOG) ján tàrámnı́yaá (Lely; KO&S) jár tàrámnı́yaá = *red taramniya* (JMD) kantakara *from Kanuri* (JMD; ZOG) tàrámnı́yaá (JMD) tàraúnı́yá, jár-tàraúnı́yá (JMD; ZOG) YORUBA arọ (KO&S; IFE)

A shrub or tree, variable, with 11 subspecies recognized in Africa of which 3 are in West Africa (15). Subspecies *geitonophyllum* is the most widespread occurring in the savanna from Senegal to S Nigeria, but not beyond. Subspecies *binderianum* occurs from Ghana to Nigeria and is widespread in E Africa, and ssp. *hypopilinum* is in Guinea to S Nigeria and across to Sudan and Uganda. In this survey of the species, the three West African subspecies are taken together.

Height is variable with ssp. *binderianum* 3–8 m high, ssp. *geitonophyllum* to 10 m, and ssp. *hypopilinum* 6–13 m. The wood (ssp. *geitonophyllum*) is recorded as brownish-grey, hard and heavy (6b, 8c) and greenish-yellow, tough and difficult to work (6c, 9). Subspecies *binderianum* is said to be used in Sudan for hut-poles (8a), and the wood of ssp. *geitonophyllum* is cut for firewood in Togo (Kersting fide 6b, 8c).

The bark yields a gum (ssp. *binderianum,* 19; ssp. *geitonophyllum,* 6c; ssp. *hypopilinum,* 6a, 7), loosely called *taramniya* (Hausa) in N Nigeria. It is used to cure toothache (6a) or to plug a carious tooth (6c). It is eaten in Sudan (3).

The leaves and leafy twigs (ssp. *geitonophyllum*) are used in Senegal in decoction drunk as a cough-medicine and for pneumo-bronchial troubles with the lees applied in poultice (12, 13), in aqueous decoction for bathing the body as a defatigant (10, 11), and the dried baked leaves in infusion as a cholagogue (11). In Guinea a leaf-macerate is considered effective in treating diarrhoea, and as a depurative and cholagogue. Its action in diarrhoea is perhaps to hasten evacuation of the cause for in The Gambia (6a) and elsewhere in Africa (16b) a leaf-macerate, as also of the bark in Tanganyika (17), is given as a purge. In Ivory Coast the leaves are held to be strongly diuretic and of benefit in cases of general oedemas (5). Similar use is recorded in Tanganyika to reduce swollen tummies of children (18). The taste of a leaf-decoction is said to be unpleasant (16a), but stock, except cattle, will browse the foliage in Senegal (1). In Sudan it is taken by all animals (3).

A root-decoction is taken for stomach-ache in The Gambia (6a), and the roots are said to have unspecified medicinal use in the soudano-guinean region of francophone Africa (4). It is held in Tanganyika that if the roots grow into a well the water tastes bitter (20). An examination of the roots of Nigerian material of ssp. *geitonophyllum* revealed no alkaloid present (2).

The plant enters superstitious uses. The Kilba of N Nigeria prepare a bundle of leaves tied up with a piece of lion's tail or hog's hair, a bug, a louse, an arrowshaft, porcupine quill, stalk of grass and a stick used by grave diggers. These are supposed to symbolize the evil things that guilty persons stand for. The bundle is called *shafa* upon which the Kilba take oaths (14). In Ghana ssp. *hypopilinum* is common in Tengani sacred groves (Lynn fide 8b). In Senegal the plant is considered to be the female of *Combretum glutinosum* and is called *dókidéwi,* or female *doki,* by the Fula.

References:

1. Adam, 1966, a: as *C. lamprocarpum* Diels. 2. Adegoke & al., 1968: as *C. lamprocarpum* Diels. 3. Andrews A700, A702, A760, K. 4. Aubréville, 1950: 103 as *C. hypopilinum* Diels. 5. Bouquet & Debray, 1974: 66 as *C. lamprocarpum* Diels. 6. Dalziel, 1937: a. as *C. hypopilinum* Diels, b. as *C. kerstingii* Engl. & Diels, c. as *C. verticillatum* Engl. 7. Howes, 1949: 72. 8. Irvine, 1961: a. 114–5 as *C. binderianum* Kotschy, b. 117 as *C. hypopilinum* Diels, c. 117–8 as *C. lamprocarpum* Diels. 9. Keay & al., 1960: 150–1 as *C. binderianum* Kotschy, *C. lamprocarpum* Diels, *C. hypopilinum* Diels. 10. Kerharo & Adam, 1962: as *C. crotonoides* Hutch. & Dalz. 11. Kerharo & Adam, 1963, a: as *C. crotonoides* Hutch. & Dalz. 12. Kerharo & Adam, 1964, b: 421 as *C. crotonoides* Hutch. & Dalz. 13. Kerharo & Adam, 1974: 342 as *C. crotonoides* Hutch. & Dalz. 14. Meek, 1931: 1, 225 as *C. verticillatum.* 15. Okafor, 1967. 16a. Portères s.d., as *C. crotonoides.* b. as *C. kerstingii.* 17. Tanner 2966, K. 18. Tanner 4734, K. 19. Watt & Breyer Brandwijk, 1962: 192, as *C. binderianum* Kotschy. 20. Wigg s.n., K.

## **Combretum comosum** G. Don

FWTA, ed. 2, 1: 273–4. UPWTA, ed. 1, 75.

West African: **SENEGAL** WOLOF kindindolo (JMD) **SIERRA LEONE** KISSI chimbule (FCD) KONO bwandi (FCD) LIMBA babe, bufɛ (NWT) LOKO wundindi (FCD) MENDE kpuandi *the water-shoots after leaf-fall but with persistent woody spike-like leaf-stalks* (FCD) tomba-mɛnyɛ (FCD) MENDE (Komboia) tomba-mɛnyɛ (*def.*-i) (FCD) MENDE-KPA hakpa-namu (*def.*-i) (FCD) UP MENDE ngenge-tumba (*def.*-tumbei) (FCD) TEMNE *a*-kati (NWT) **IVORY COAST** KRU-GUERE (Chiehn) degré nelé (B&D) KYAMA apokomondi (B&D) **GHANA** NZEMA eveleni (FRI)

A scandent shrub or liane of the evergreen forest occurring from The Gambia to Ghana, and in Cameroun and Sudan.

The flowers are pink, red or crimson, and conspicuously showy. The plant is worthy of cultivation. Children suck the flowers to get the nectar (1,2).

The plant (? root) is used in Nigerian folk-medicine as a remedy for skin-infections and examination of a water extract of the roots shows it to be active against Gram +ve *Sarcina lutea, Staphylococcus aureus,* and *Mycobacterium phlei,* and against Gram –ve *Pseudomonas aeruginosa*. The extract showed no anti-fungal action.

References:

1. Dalziel, 1937. 2. Irvine, 1961: 115. 3. Malcolm & Sofowora, 1969.

**Combretum constrictum** (Benth.) Laws.

FWTA. ed. 2, 1: 273. UPWTA, ed. 1, 75.
West African: **NIGERIA** YORUBA ọ̀gàn ìbúlẹ̀ = *finger lies on the ground* (Verger)

A shrub or climber, usually low, of riverain forest, often in damp sites, and of tidal forest, of N and S Nigeria and in E Africa.

The flowers are bright red and the plant has ornamental value (1). The root is said to be an excellent remedy for worms in children (3).

The Yoruba name, *ògan ìbúlè*, meaning 'finger lies on the ground', features in an Odu incantation to achieve victory over an enemy (2).

References:

1. Dalziel, 1937. 2. Verger, 1967: No. 155. 3. Watt & Breyer Brandwijk, 1962: 193.

**Combretum dolichopetalum** Engl. & Diels

FWTA, ed. 2, 1: 273. UPWTA, ed. 1, 75.
West African: **SIERRA LEONE** MENDE hakpa-nomu (FCD) **GHANA** AKAN-TWI ɔhwirem (FRI) DAGBANI garizegu (JMD; FRI) **NIGERIA** EDO ọ́kọ́sọ́ (Farquhar; JMD) IGBO àchịchà ńzā, ḿ.ṉzā *lit., = food of the small bird, nza:* achịcha: *dried plantain* (JMD) ḿmányá ṉzī̩zá = *sun-bird's wine* (auctt.) IGBO (Azumini) ubim (AJC) IGBO (Umuahia) ḿmị́ị ńzī̩zá = *sun-bird's wine* (AJC)

A scandent shrub or forest liane of deciduous forest and in secondary re-growth in areas receiving at least 1250 mm rainfall, usually near rivers (6) occurring from Sierra Leone to W Cameroons.

The leaves are used by Igbo for burns (7), and in decoction as a purgative (1). A root-extract is also taken for dysentery (4), and indigestion (2).

In northern Ghana an infusion of roots, leaves and stems is administered to cattle suffering from *garli* (Fula) (5), a condition of 'stomach staggers', cf. *C. paniculatum* Vent.

The Igbo names referring to 'the wine of birds' suggest that honey-suckers frequent the flowers, and perhaps the flowers are bird-pollinated (3).

References:

1. Carpenter 3, UCI. 2. Carpenter 303, UCI. 3. Carpenter 355, UCI. 4. Carpenter 667, UCI. 5. Dalziel, 1937. 6. Irvine, 1961: 115–6. 7. Thomas, N. W. 2148 (Nig. Ser.), K.

**Combretum etessi** Aubréville

FWTA, ed. 2, 1: 274.

West African: **SENEGAL** BASARI a-kaxeúr ɔ-xèɗy (Ferry) BEDIK ga-ndyátyàr (Ferry) **GUINEA-BISSAU** FULA-PULAAR (Guinea-Bissau) djambacatam (JDES; EPdS) MANDING-MANDINKA djambacatam (*def*.-ô) (JDES)

A tree recorded only from the dry savanna of Casamance, Senegal, and Guinea-Bissau.

**Combretum fragrans** F. Hoffman

FWTA, ed. 2, 1: 271 as *C. ghasalense* Engl. & Diels. UPWTA, ed. 1, 75 as *C. dalzielii* Hutch., and 76 as *C. ghasalense* Engl. & Diels.

West African: **SENEGAL** FULA-PULAAR (Senegal) buski (Aub.) MANDING-MANINKA sama m'bali (Aub.) **GUINEA-BISSAU** CRIOULO jambacatá (D'O) jambacatam (D'O) FULA-PULAAR (Guinea-Bissau) djambacatam (D'O) MANDING-MANDINKA djambacatam (*def*.-ô) = *leaf-scabies* (JDES) **GUINEA** FULA-PULAAR (Guinea) buski (Aub.) dooki (Aub.) MANDING-MANINKA sama m'bali (Aub.) SUSU limbi (auctt.) lindi (GFSE; Aub.) **SIERRA LEONE** MANDING-MANDINKA kolokolo-yiri (FCD) MENDE djendjeng *a preparation including the roots of this tree* (GFSE) SUSU lindi (GFSE) SUSU-DYALONKE samataga-na *from* samana: *elephant; the elephant cannot break this tree as the branches bend but do not break* (FCD) **MALI** FULA-PULAAR (Mali) buski (Aub.) dooki (A. Chev.) MANDING-BAMBARA diangara (A. Chev.) sama m'bali (Aub.) tiangara (GR) MANINKA kangara (A. Chev.) sama m'bali (Aub.) 'SENUFO' diadiuna (Aub.) **UPPER VOLTA** HAUSA dalo (Aub.) **IVORY COAST** 'SENUFO' diadiuna (Aub.) **GHANA** AKAN-BRONG atɛna (CV; FRI) DAGAARI pamara (AEK; FRI) DAGBANI tachale *incl. other spp.* (Dudgeon) tachayi (Dudgeon) GBE-VHE (Pecí) wɛtʃo *probably incl. other spp.* (FRI) GRUSI-KASENA kamagiri, kamagali (Gaisie) MOORE kwaginyanga (JMD; FRI) kwanganga (auctt.) kwegenga (JMD; FRI) **TOGO** BASSARI diginegri (Aub.) BATONNUN takunta (Aub.) MOBA tantabili (Aub.) SORUBA-KUYOBE tamtamtuéni (Aub.) TEM alimebolé (Aub.) **DAHOMEY** BATONNUN takunta (Aub.) SORUBA-KUYOBE tamtamtuéni (Aub.) **NIGER** HAUSA dalo (Aub.) **NIGERIA** FULA-FULFULDE (Nigeria) ɓuski (*pl.* ɓusɗe) (JMD) ɓuski boɗeehi (J&D) HAUSA bákár tàrámníyaá = *black taramniya* (JMD; ZOG) bakin tàrámníyaá (KO&S; RES) ciiriiri, tsiri (JMD; ZOG) dalo (JMD; ZOG) mumie *the gum* (EWF) námíjìn-tàrámníyaá = *male taramniya* (JMD; ZOG) tsiriri (JMD) wiyandamo (EWF) KANURI zə̀ndí (auctt.)

A tree to 8 m high by 70 cm or more in girth; bole twisted and low-branching; of savanna areas from Guinea-Bissau to S Nigeria and on across Africa to Sudan and E Africa.

The wood is said to be unbreakable by the elephant: it just bends — hence the Sierra Leone Dyalonke name (5). It is used for hut-poles in Sudan (9) and for firewood generally (8, 13). Smoke from the wood is used as a perfume by Sudanese women (3).

The bark exudes a soluble, red-coloured gum which is more or less edible. The tree is said to be the source of much of the gum collected in the Niger basin south of 14° N (4, 8).

The leaves yield a yellow dye on boiling (13). Leaves and leafy twigs are sometimes mixed with thatching grass in Bornu (4). The foliage is not browsed by animals in Sudan (1). In Guinea-Bissau the leaves enter a medicine for fevers (6). Branches, free of fruit, are used to prepare an infusion in Liberia for washing the body for pain (12).

The Fula of N Nigeria prepare an infusion of the bark which is taken with natron for *lekki beernde* (pains in chest), and the bark together with a mistletoe which commonly parasitizes the tree is made into an infusion for washing the body (10). The bark and roots provide a decoction used in W Africa (13) and in Tanganyika (16) for abdominal pains and for low backache, and in Tanganyika the fresh roots ground up and dried are put on sores (15) or prepared as a decoction for treating primary sores of syphilis (11). In Sierra Leone the root is used to prepare *djendjeng,* Mende, a beverage (14) — see also *Dissotis grandiflora* Benth. (Melastomataceae).

In the dry season when the trees are bare of leaves there is usually an intense burst of yellow flowers, sweet-scented, which attract many insects to the nectar (2, 7). The trees may have some value for honey-production.

References:

1. Andrews A658, K. 2. Aubréville, 1950: 102, as *C. ghasalense* Engl. & Diels. 3. Broun & Massey, 1929: 113, as *C. glasalense* Engl. & Diels. 4. Dalziel, 1937. 5. Deighton 4193, K. 6. D'Orey 177, K. 7. Eggeling 1153, K. 8. Howes, 1949: 72, as *C. dalzielii*. 9. Irvine, 1961: 116, as *C. ghasalense* Engl. & Diels. 10. Jackson, 1973. 11. Koritschoner 1675, K. 12. Oldeman 318, K. 13. Porteres s.d. as *C. ghasalense* Engl. & Diels. 14. Scott Elliot 4966, K. 15. Tanner 391, K. 16. Tanner 4428, K.

## **Combretum glutinosum** Perr.

FWTA, ed. 2, 1: 271. UPWTA, ed. 1, 76; and 78 as *C. passargei* Engl. & Diels.

French: khât (colloquial, West Africa, from Wolof, Roberty); bois d'éléphant (Pobèguin fide Dalziel).

West African: **MAURITANIA** ARABIC ('Maure') tikfit (Aub.) **SENEGAL** ARABIC ('Maure') tikfit (Aub.) BANYUN sidiabakatan (Aub.) sikinn (Aub.) BASARI a-ndẹ̀kẹ́ɗár (K&A; Ferry) BEDIK ga-ndyabàkátánà (K&A; Ferry) DIOLA kalâkudun (Aub., ex K&A) FULA-PULAAR uski (GR) PULAAR (Senegal) dóki (K&A) dóki déwi = *female dóki; leaves without fasciation* (K&A) dóki gori = *male dóki; leaves with fasciation, 'gori' from* gorko: *male* (K&A) doko woro *form with fasciated leaves* (K&A) dóóki (Aub.; K&A) dooko (K&A) ndukoworo (K&A) nokégoré (K&A) nooko (K&A) TUKULOR dodjié kéwudé (JLT) KONYAGI a-mbusẹla (Ferry) MANDING-BAMBARA ďâbakatâ (K&A) diangara (JMD) tâgara (K&A) ťâgara (JB; K&A) MANDINKA jambakataŋ (*def.*-o) MANINKA ďâbakatâ (K&A) demba (Aub.) djamba (Aub.) khattan (Aub.) tâgara (K&A) ťâgara (K&A) tiangara (Aub.) 'SOCE' ďâbakatâ (K&A) ďâbakataô SERER yay, yaye = *mamma; i.e., the mother of medicines* (auctt.) SERER-NON pemben (AS, ex K&A) pemben (AS, ex K&A) NON (Nyominka) yay, yaye = *mamma* (K&A) SONINKE-SARAKOLE dooko (K&A) tafé (K&A) téfé (K&A) WOLOF khat (GR) rat(t) (auctt.) **THE GAMBIA** FULA-PULAAR (The Gambia) buki (DAP) doki (DAP) dotihi (DRR) MANDING-MANDINKA jambakataŋ (*def.*-o) = *leaf-scabies* (auctt.) jambakatango (Fox) jangara (DAP) WOLOF rat (DAP) rat-buye (DAP) **GUINEA** FULA-PULAAR (Guinea) dooki (Aub.) MANDING-MANINKA demba (Aub.) diamba (CHOP) djamba (Aub.) khattan (Aub.) simba beli (CHOP) tiangara (Aub.) **MALI** BOBO intianon (Aub.) FULA-PULAAR (Mali) dooki (GR) MANDING-BAMBARA demba (Aub.) djamba (Aub.) khattan (Aub.) tiangara (GR; Aub.) MANINKA demba (Aub.) djamba (Aub.) khattan (Aub.) tiangara (Aub.) SONINKE-MARKA banamba (Aub.) **UPPER VOLTA** HAUSA dalo (Aub.) farin taramnid (Aub.) MOORE kugunga (Aub.) kwegenga (Ouedraogo) SONGHAI-ZARMA kokolobé (Aub.) kokorbé (Aub.) **IVORY COAST** MANDING-MANINKA naniaragbwé (B&D) **GHANA** GRUSI-KASENA vakogu (FRI) vọkoŋ (Gaisie) NABT nkunga (BD&H; FRI) NANKANNI urinperiga (Saunders) **TOGO** GBE-VHE akpiti (Aub.) **DAHOMEY** GBE-FON dosso (Aub.) GEN atissainsain (Aub.) VHE akpiti (Aub.) YORUBA-NAGO bodomi (Aub.) **NIGER** ARABIC (Niger) hébil (Aub.) FULA-FULFULDE dooki (Aub.) GURMA lifapélu (Aub.) HAUSA dalo (Aub.) farin taramnid (Aub.) SONGHAI dèeli-ñá' (*pl.* -ñóŋó) (D&C) kókórbéy (*pl.* -à) (D&C) TUBU (Kaningou) kadagar (Aub.) **NIGERIA** FULA-FULFULDE (Nigeria) ɓoodi (*pl.* ɓoode) (Taylor) ɓuski (*pl.* ɓusɗe) (JMD; KO&S) ɓuski daneehi (J&D) HAUSA dageera, dangeera (ZOG) dalo (JRA; JMD) faran wiski, faran: *white* (FNH) farin taramniya (JMD; Lely) talonia (EWF) tàrámníyaá *a general term for gum-yielding plants* (auctt.) turaar turauniya (ZOG) HAUSA (West) wúyàn-dámoó (ZOG) YORUBA daguro (JRA)

A small tree to 11 m high by 60 cm girth, bole rarely straight, bark warty; leaves thick, sticky above, near white below, of the Sahel region from Senegal to S Nigeria and across Africa to Sudan.

The wood is yellowish, hard and very durable. It is used for hut-posts (7) and tool-handles (4) and makes a good firewood. The wood-smoke is used in Nigeria as an incense and for fumigation (2). The alkaline ash is used in Senegal in indigo-dyeing, and the root and bark yield a yellow dye (7).

The tree is particularly resistant to arid conditions, surviving where grass will not, and recovering into leaf quickly after a burn. It occurs often in degraded soils (17). The leaves furnish a useful forage in the Sahel zone which is relished by all stock, sometimes even when dry (1). In clearing land for cultivation this tree is commonly left standing (19).

The leaves play an important role in the Senegalese pharmacopoea and are commonly sold in markets: as a diuretic in treating urinary troubles, for liver and kidney complaints, fevers, oedemas, etc. The leaves have a high reputation for treating chest-ailments, colics and stomach-trouble (6, 9, 11, 13, 14). A leaf-preparation is used as an expectorant in bronchitis and pneumonia (9), as a decoction in draught for bronchial troubles, nasal catarrh and constipation (10), and as a macerate for persistent headaches (13). In The Gambia (18) and Nigeria (2) the liquid in which the leaves have been boiled is taken as a purge. In Ivory Coast the Maninka take a leaf-decoction in baths and by draught against general fatigue (5). The leaves are generally used in decoctions for cleansing sores, and crushed as a dressing for sores and wounds (4, 15). The dried powdered leaves are also used as a wound-dressing (7). Pounded leaves are said to stem post-circumcision heamorrhage (4, 7) and a decoction is taken internally for menorrhagia (2). Flavonic heterosides are reported present in the leaves (15).

The plant in Senegal is often affected by a fasciation of the leaves. To this is attached great importance and only fasciated leaves are used by some races. Fasciated plants are designated 'male' and normal plants 'female' (12, 14).

The bark yields a gum, very variable in composition, but basically consisting of uronic acid materials which hydrolyse to a number of sugars (3). The gum is recorded as used in Senegal to fill the cavity of a carious tooth (6). The pounded bark produces a sort of lint which is used with success on wounds (4). The Fula of Nigeria use a bark-infusion to bathe cases of influenza, rheumatism and temporary insanity (8).

The roots are used in Senegal to treat blennorrhoea (9). The root-bark is thought to be anthelmintic (12). Young shoots and roots are held to be aphrodisiac by the Soce (10).

In Senegal the dried green fruits are powdered and applied to syphilitic chancres and to sores. The fruit is also used in veterinary medicine (12, 14).

Scientific observation has shown proven effect in promoting diuresis, reduction of hypertension, discharge of stone from the bladder, and cure of hepatitis (14).

In Upper Volta the plant, to the exclusion of all others, has funerary use: two items fastened in a cross are placed in a hole in which a newly dead corpse is laid for shaving and washing (16). The plant is very subject to infestation by mistletoe and the Fula of N Nigeria use the parasite in infusions to poultice swellings (8) — perhaps a case of sympathetic medico-magic.

References:

1. Adam, 1966, a: 577. 2. Ainslie, 1937: spp. nos 107, 109. 3. Aspinall & Bhavandan, 1965: as *C. leonense*. 4. Aubrèville, 1950: 102. 5. Bouquet & Debray, 1974: 66. 6. Boury, 1962: 15. 7. Dalziel, 1937. 8. Jackson, 1973. 9. Kerharo, 1967. 10. Kerharo & Adam, 1963, a. 11. Kerharo & Adam, 1964, a: 412. 12. Kerharo & Adam, 1964, b: 427–8. 13. Kerharo & Adam, 1964, c: 300. 14. Kerharo & Adam, 1974: 345–5, with pharmacology. 15. Oliver, 1960: 23, 57. 16. Ouedraogo, 1950: 442. 17. Roberty, 1953: 450. 18. Rosevear, 1961. 19. Trochain, 1940: 243.

## **Combretum grandiflorum** G. Don

FWTA, ed. 2, 1: 273. UPWTA, ed. 1, 76.

West African: **SENEGAL** DIOLA fu yayos (JB) **THE GAMBIA** DIOLA (Fogny) fu yayos = *scratchy when touched* (DF) **GUINEA-BISSAU** DIOLA-FLUP pelae-djacumáe (JDES) pláè-djacumáè (JDES, ex EPdS) **SIERRA LEONE** KISSI g-bekulɔpio (FCD, ex JMD) hoya (FCD) MANDING-MANDINKA banapare (NWT) waiya (NWT) MENDE hakpa-namu (*def.*-i) *incl. several spp. esp. red-flowered climbers* (auctt.) kpuandi (NWT) ngenge-tumba (*def.*-tumbei) (JMD) tomba-mɛnyɛ *incl. several spp. esp. red-flowered climbers* (JMD) MENDE (Daru) bɔhu (FCD) MENDE (East) ngengele-ha (*def.*-hei) *incl. several spp. esp. red-flowered climbers* (JMD) MENDE-KPA hakpa-namu (*def.*-i) (FCD) UP MENDE ngenge-tumba *incl. several spp. esp. red-flowered climbers* (FCD; JMD) TEMNE *a*-kati *incl. several spp. esp. red-flowered climbers* (NWT) **GHANA** AKAN-ASANTE ohwiremnini (FRI) WASA ohwiremnini (FRI)

A climbing shrub, over-topping small trees to about 6–7 m high in old clearings of the evergreen forest, occurring from The Gambia to Ghana.

The flowers are scarlet in dense spikes and make the plant worthy of cultivation.

The flowers are pulled off and sucked by children to obtain the nectar. The plant has superstitious and ceremonial uses by the Temne of Sierra Leone (1).

Reference:

1. Dalziel, 1937.

### **Combretum hispidum** Laws.

FWTA, ed. 2, 1: 274.

West African: **SIERRA LEONE** MANDING-MANDINKA gbaŋgiyaũ (FCD) SUSU-DYALONKE g-baŋgiŋɛ-na (FCD) **NIGERIA** IGBO (Uburubu) ákwụ́kwọ̀ ósò = *leaf of pepper* (NWT; KW)

A straggly scandent shrub or liane to about 6 m high, of transition-forest between closed-forest and savanna-woodland, from Guinea to S Nigeria, and on into Zaïre and Angola.

Igbo of S Nigeria are recorded as using the leaf to draw out thorns (2), and in a superstitious way to place a leaf upon a child's head so that the child may grow (1).

References:

1. Thomas, N. W. 2302 (Nig. Ser.), K. 2. Thomas, N. W. 2314 (Nig. Ser.), K.

### **Combretum lecardii Engl. & Diels**

FWTA, ed. 2, 1: 273. UPWTA, ed. 1, 77.

West African: **SENEGAL** BASARI a-mbệx ɔ-rembàl (K&A; Ferry) BEDIK gi-ndeúgẹlɛmbàr (Ferry) DIOLA busamontaf (JB; K&A) buseytaf (JB; K&A) FULA-PULAAR (Senegal) kôtân ́tolen̈ (K&A) MANDING-BAMBARA demba fura (auctt.) MANDINKA kunundindolo (*def.*-o) (K&A) MANINKA demba fura (Aub.; K&A) tiambéré (Aub.) SERER lumel (JB; K&A) ndadel (JLT; K&A) WOLOF kérindolo (JB; K&A) kindindolo *incl. other red-flowered spp.* (auctt.) **THE GAMBIA** FULA-PULAAR (The Gambia) dati (DRR) MANDING-MANDINKA kunundindolo (*def.*-o) = *small bird drink* (auctt.) naka burayo = *sticks forever* (DAP; JMD) **GUINEA-BISSAU** MANDING-MANDINKA cundundim (*def.*-ô) (JDES) MANINKA demba fura (Aub.) tiambéré (Aub.) **MALI** MANDING-BAMBARA demba fura (Aub.) tiambéré (Aub.) MANDINKA demba fura (Aub.) tiambéré (Aub.) **UPPER VOLTA** MOORE kudugulugu (Aub.)

A scandent shrub, 4–5 m high, of the soudanian region of Senegal to Sierra Leone.

The plant is recorded with medico-magical properties in Senegal for stomach-complaints. A *Tapinanthus sp.* (Loranthaceae) which parasitizes it is used in a prescription for female sterility or to prevent miscarriage (1).

Reference:

1. Kerharo & Adam, 1974: 345–6.

### **Combretum micranthum** G. Don

FWTA, ed. 2, 1: 271–2. UPWTA, ed. 1, 77–78.

French: kinkéliba (colloquial in West Africa after various West African vernaculars).

West African: **MAURITANIA** ARABIC ('Maure') bufum el eid (Aub.) dafo (Aub.) **SENEGAL** VULGAR kinkeliba *from Wolof* ARABIC ('Maure') bufum el eid (Aub.) dafo (Aub.) BALANTA psâgala (K&A) BANYUN bussankeul (K&A) kusanköl (K&A) BEDIK gi-mbeúd́ (Ferry) DIOLA (Diatok) bulusor (K&A) DIOLA (Efok) buté kabo (K&A) DIOLA (Fogny) butek (K&A) butik (K&A) futik (K&A) DIOLA-FLUP butik (K&A); = *pink* (DF, The Gambia) FULA-PULAAR (Senegal) gumuni (Aub.) talli (Aub.; K&A) tallika (K&A) MANDING-BAMBARA baraulé (Aub.; K&A) kofina (K&A) kolobe (K&A) n-golobe (auctt.) singolobe (K&A) MANDINKA bara (*def.*

baroo) = *fork* kulunkalaŋ (*def*.-o) = *pestle handle* (DF) pakwiakaro (JMD; Aub.) MANINKA baraulé (Aub., ex K&A) kofina (Aub.; K&A) kolobé (Aub.; K&A) n-golobé (Aub.; K&A) singolobé (Aub.; K&A) 'SOCE' bara iro (K&A) baro (K&A) MANDINKA barairo (K&A) MANDYAK bòk (K&A) SERER daj (K&A) lakak (auctt.) ndag (auctt.) sésed (JB; K&A) SONINKE-SARAKOLE kâde (K&A) WOLOF késeu (auctt.) kinkéliba (Aub.; K&A) séreu (Perrot&Lefèvre, Aub.) [illegible] (K&A) sexéo (K&A) xaseo = *nasty smell* (K&A; DF) **THE GAMBIA** FULA-PULAAR (The Gambia) m-bara (DRR; DAP) tali (DAP) MANDING-MANDINKA bara (*def*. baroo) = *a fork* (auctt.) barajamba (*def*. barajamboo) = *Combretum leaves* (DRR; DAP) kinkilibaa (*indef*. and *def*.) (auctt.) kunundindolo (*def*.-o) = *small bird's wine, a loose term for Combretum spp. & possibly for some other plants* (JMD) WOLOF keseu (DAP) kinkiliba *from Diola* (auctt.) seweyo (DAP) **GUINEA-BISSAU** BIDYOGO upatocuma (JDES) CRIOULO buco (auctt.) FULA-PULAAR (Guinea-Bissau) cancalibá (JDES) canquelibá (EPdS) tade (JDES) MANDING-MANDINKA baracólómó = *river stick* cancalibá (JDES) PEPEL buéco (JDES) **GUINEA** FULA-PULAAR (Guinea) gugomi (Aub.) gumuni (Aub.) talli (Aub.) MANDING-MANINKA bara ulé (CHOP) SUSU kinkéliba (CHOP; Aub.) **SIERRA LEONE** SUSU kankeliba (GFSE) kankilibanyi (KM) **MALI** DOGON kuyó (C-G) FULA-PULAAR (Mali) gumuni (Aub.) talli (A. Chev.) MANDING-BAMBARA baraulé (Aub.) golobé (JMD; Aub.) kofina (Aub.) kolobé (JMD; Aub.) nyolobé (A. Chev.) simba beli (JMD) singolobé (Aub.) MANINKA barulé (Aub.) golobé (Aub.) kofina (Aub.) kolobé (Aub.) singolobé (Aub.) SONINKE-MARKA tuba (Aub.) 'SENUFO' kolobé (Aub.) **UPPER VOLTA** VULGAR kinkeliba (K&B) HAUSA faraguéza (Aub.) guiéza (Aub.) MANDING-BAMBARA singolobé (Dubois; K&B) MOORE dandegha (K&B) kuégna (AJML) langaga (Aub.) randega (auctt.) towu (Aub.) SONGHAI-ZARMA kubu (Aub.) **IVORY COAST** VULGAR kinkeliba (K&B) MANDING-MANINKA bara mussoma (K&B) 'SENUFO' kolobé (Aub.; K&B) **GHANA** HAUSA siesa (jiesa) (AEK) MOORE landaga (AEK; FRI) NANKANNI gee-an (Lynn; FRI) SISAALA pamamenga (AEK; FRI) **DAHOMEY** YORUBA-NAGO okan (Aub.) **NIGER** GURMA etiani (Aub.) HAUSA faraguéza (Aub.) guiéza (Aub.) SONGHAI tíŋgílá (*pl*. -á) (D&C) SONGHAI-ZARMA kubu (Aub.; Robin) **NIGERIA** VULGAR kinkeliba FULA-FULFULDE (Nigeria) gumumi (auctt.) HAUSA farar geézaà (ZOG) gaiza (Ryan) gaza (EWF) geézaà (auctt.) YORUBA ọ̀gàn bulẹ̀ (Dawodu; JMD) ọ̀gàn ibulẹ̀ (IFE) ọkán (JRA; JMD)

A plant of the Sahel, Soudan and guinean savanna forest. On rocky hills it is a shrub, but in woodlands it becomes a tree with a stem to 10 cm in diameter, or a liane reaching 15–20 m long. In some places near the coast it forms thickets, and stands of it are recorded on the tableland of Thiès in Senegal (5, 14). Its distribution is from Senegal to Nigeria in the Region and on into Gabon. In The Gambia it is said to be always on termite heaps (20).

The plant is very drought and fire-resistant (5). It quite commonly invades cultivated land (10). It is a browse-plant for cattle and is of some importance as forage in the dry season in N Nigeria (7). It has been recommended for conservation in the Sahel zone to improve pasturage for the provision of shade and forage, though in Senegal it appears to be readily browsed only by donkeys, sheep and goats, cattle taking it but occasionally (1).

The stems are tough and pliable when young and are used split in N Nigeria to make large wicker-baskets *(kantalma, kwando,* Hausa) which are often smeared inside and out with cow-dung and used to store grain or to hold fowls. The unsplit stems are used for the interior centre part of a thatch-roof. The wood is very hard and in Sokoto it is used for charcoal. Occasionally walking-sticks are made from the stems (5).

The plant is the source of general panacea *kinkeliba*. The leaves are ubiquitously so well-known for their diuretic, febrifugal and digestive properties that medicine-men make no mystery of their use (13, 14), and to be of such diverse application that the name *kinkeliba* has become synonymous with the word *medicine* (15). The leaves are official in the *French Pharmacopoea, 1937,* and are also in the *Spanish Pharmacopoea* as cholagogues and antipyretics. They are also used as diuretics and astringents (18). The leaves are given for coughs, bronchitis, malaria, bilious haematuric fevers and all ailments of the liver and gall-bladder (15); for colic and nausea to prevent vomiting, in vapour-baths or washes for fevers and lumbago (5); in association with other drug-plants for beriberi, infantile and adult diarrhoea, haemorrhages, leprosy, enuresis, blennorrhagia, etc. (13, 14, 15). Its action in diuresis may double or treble the excretion in part by vasodilation of the kidneys, and in part by action on the renal epithelium (15). The leaves contain a number of mineral salts, organic acids, tannin, flavones, amines, etc. (15). The presence of maleic, glycolic and glyceric acids, potassium nitrates and a catechin may account for the diuretic action, and tannins for the stimulation of the gall-bladder (18).

At one time a leaf-infusion was drunk as a daily matter-of-course by Europeans in francophone West Africa (4, 6), and an infusion known as 'bush tea' was taken in The Gambia (8, 20) and in Nigeria (3), seemingly in part at least for a prophylaxis against illness. In Sierra Leone the leaves are made into an infusion taken by Muslims especially during Ramadan (17).

The bark is also used medicinally. It contains tannins (5). and alkaloids (15). In Casamance of Senegal powdered bark mixed with palm-oil or karite-butter is applied in massage to contusions and sprains. The Soce massage in this preparation when they wish to cover long distances (11). The bark yields a gum which is marketed in the Dosso Province of Niger (19).

A root-decoction is vermifugal and is used as a wash for sores (5). In Senegal the roots are used against syphilis (13) and in the Region in draughts and lotion for fever (4). It is recorded that in Casamance a root-decoction has been an essential ingredient of diet for treating female sterility (12). A trace of alkaloid has been detected in the roots (2) and a water-extract has shown action on Gram +ve bacteria, *Sarcina lutea* and *Staphylococcus aureus,* but no action on Gram −ve bacteria, nor on fungi (16).

The dried powdered fruits are made into an ointment with oil which is applied to suppurating swellings, abscesses, etc. whether of syphilitic or other origin (5).

The seeds have a flavour of ground-nuts and are commonly eaten by Moore boys in the Ouagadougou area of Upper Volta (9).

References:

1. Adam, 1966, a. 2. Adegoke & al., 1968. 3. Ainslie, 1937: sp. no. 108. 4. Aubréville, 1950: 114. 5. Dalziel, 1937. 6. Deighton 4037, K. 7. Foster & Munday, 1961: 316. 8. Fox 66, K. 9. Irvine 4682, K. 10. Irvine, 1961: 118. 11. Kerharo & Adam, 1962. 12. Kerharo & Adam, 1963, a. 13. Kerharo & Adam, 1964, b: 429. 14. Kerharo & Adam, 1974: 346–9 with phytochemistry and pharmacology. 15. Kerharo & Bouquet, 1951: 48. 16. Malcolm & Sofowora, 1969. 17. Miszewski 30, K, SL. 18. Oliver, 1960: 6, 57. 19. Robin, 1947: 58. 20. Rosevear, 1961.

## **Combretum molle** R. Br.

FWTA, ed. 2, 1: 270. UPWTA, ed. 1, 79, as *Combretum sokodense* Engl.

West African: **SENEGAL** BASARI a-nyéwɔrɛ̀nd (K&A; Ferry) BEDIK ga-nyómàr (Ferry) FULA-PULAAR (Senegal) ńańakawi (K&A) ńańakey (K&A) ńańatavi (K&A) MANDING-BAMBARA mańaka (JB; K&A) wanaka (K&A) MANINKA mańaka (K&A) maniaka (Aub.) uaniaka (Aub.) wanaka (K&A) **GUINEA** MANDING-MANINKA maniaka (Aub.) uanika (Aub.) **MALI** MANDING-BAMBARA maniaka (Aub.) uaniaka (Aub.) MANINKA maniaka (Aub.) uaniaka (Aub.) 'SENUFO' kandiara (Aub.) **UPPER VOLTA** DAGAARI bal ora (K&B) GRUSI deyugu (K&B) HAUSA uyane damo (Aub.) wuindamo (K&B) KIRMA somonto (K&B) LOBI harbara (K&B) MANDING-DYULA kélété gnéma (K&B) kélété kaïri (K&B) MOORE kwegenga (Aub.) lugulé (K&B) **IVORY COAST** BAULE biassuabaka (Aub.) bliassubaka (K&B) utréruaka (K&B) yassuabaka (K&B) MANDING-DYULA kélété gnéma (K&B) kélété kaïri (K&B) MANINKA naniabaka (Bégué; K&B) woingnankan, wolobatu (A&AA) SENUFO-TAGWANA kakiélé (K&B) DYIMINI iania ulé (K&B) 'SENUFO' kandiara (Aub.) wagniara (K&B) **GHANA** BEDU kuli (CV; FRI) DAGBANI gburega (Coull; FRI) tachale (Dudgeon, ex FRI) tachayi (Ll.W; FRI) GUANG-GONJA techidamma (AEK) KONKOMBA digingel (Gaisser) KUSAL dapinga (FRI) MOORE kwaginyanga, kwanganga (FRI) kwegenga (AEK; FRI) NABT takiringa (BD&H; FRI) 'KAGONGA' tante bulense (Harper) **TOGO** ANYI-ANUFO bensékomi (Aub.) BASSARI digingire (Gaisser) fao (Aub.) kubughon (Aub.) BATONNUN hinubadu (Aub.) sunudunguéra (Aub.) MOBA dionbibinega (Aub.) MOORE-NAWDAM limnorarager (Gaisser) TEM (Tshaudjo) sissiku (Volkens) **DAHOMEY** BATONNUN hinubadu (Aub.) sunudunguéra (Aub.) **NIGER** GURMA otifebigu (Aub.) HAUSA uyane damo (Aub.) wuindamo (K&B) **NIGERIA** FULA-FULFULDE (Nigeria) boodi (Taylor) ɓuski (J&D) damoruhi (JMD; KO&S) HAUSA dinyar gata (RES) joogà-jikí (ZOG) wuyan damo (MM) HAUSA (East) wiyan damo (JMD) wiyan demmu (JMD) HAUSA (West) gogen damo = *scrape the lizard* (JMD; RES) gogen jiki, goge: *to scrape*; jiki: *the body* (JMD) wúyàn dámoó = *neck of the monitor lizard* (auctt.) YORUBA ànràgba

A shrub or small tree to 10 m high, rarely to 16 m, with a straight regular bole to 1 m girth, of savanna forest, from Senegal to W Cameroons, and widespread in tropical Africa.

The stems are durable underground and are valuable for house-posts. The wood is brownish or yellowish-green, very hard and compact, strong and durable, but difficult to work (3, 6). In Jebel Marra of Sudan the wood is used in making drums (5).

The bark is dark-coloured, fissured, corky and scaly on the smaller branches, evocative of the common Hausa name, *wuyan damo,* 'neck of the monitor lizard', and of the Hausa (Katsina) name *gogo jiki,* 'to scrape the body', as in the chafing of the skin when a bundle of branches is carried (3). The bark-slash exudes a gum, known as *mumuye* gum, which has been a minor source of trade from Bornu and Adamawa in N Nigeria. It has been examined for commercial potential and found inferior as a substitute for gum-arabic (3, 4).

The bark has medicinal used. In Senegal it is held to be cholagogic, but inferior to *C. crotonoides* and *C. micranthum,* and an aqueous suspension of powdered bark together with the mumuye gum is used as a gargle and in draught for sore-throat (7, 8). The Dagaari of Upper Volta apply powdered bark to sores (9). The bark along with cereal foods is taken (? in N Nigeria) for dysentery, and is used in ceremonial preparation for young children to prevent sickness and other troubles (3).

Leaves are prepared as a decoction for baths and draughts, or powdered and added to food in treatment of dropsy, ascites and oedemas in Senegal, (7, 8). In Ivory Coast (9) and Nigeria (10) this plant is used in the absence of that popular panacea, *kinkeliba (Combretum micranthum* G. Don) For treating jaundice and yellow fever, and in Upper Volta it is given for abdominal complaints, diarrhoea etc., blennorrhoea, anuria, etc., and sometimes to women in childbirth to hasten the expulsion of the after-birth (9). Lobi of Upper Volta often take a decoction of leafy twigs in draughts and baths for bronchial affections, and Bobo of Ivory Coast consider the plant a poison-antidote (9). The Maninka also treat whitlows by steeping the affected part in a leaf-decoction (2). A black dye is obtained from the leaves in Zaïre (6). Alcoholic extracts of leaves with water extracts of twigs have shown capacity to reduce sarcoma tumours in animals (8).

Cattle in Senegal do not browse the plant (1).

In the southern part of Senegal the plant is ascribed with magical properties by the Fula and Fouladou to promote courage in battle by inhaling smoke emitted from a fire on which bark and branches have been placed. The wood is never used to stoke domestic hearths (8, 9).

References:

1. Adam, 1966, a. 2. Adjanohoun & Aké Assi, 1974: 86. 3. Dalziel, 1937. 4. Howes, 1949: 72. 5. Hunting Technical Surveys, 1968. 6. Irvine, 1961: 119. 7. Kerharo & Adam, 1964, b: 429–30. 8. Kerharo & Adam, 1974: 349–50. 9. Kerharo & Bouquet, 1950: 49–50. 10. Oliver, 1960: 23, 57.

## **Combretum mooreanum** Exell

FWTA, ed. 2, 1: 274. UPWTA, ed. 1, 78.

West African: **SIERRA LEONE** KISSI g-bekulɔpio (FCD) MENDE tomba (*def.*-i) (FCD) MENDE (Komboia) tomboa-ɛnyɛ (*def.*-i) (FCD) MENDE-KPA (Kpa) hakpa-namu (*def.*-i) (FCD) UP MENDE ngenge-tumba (FCD)

A straggling shrub to 2 m high, of marshy localities in Sierra Leone.

The young leaves are used in soup (1, 2, 3).

References:

1. Busson, 1965: 284. 2. Dalziel, 1937. 3. Deighton 2380, K.

**Combretum nigricans** Lepr.

FWTA, ed. 2, 1: 270–1. UPWTA, ed. 1, 75–76, as *C. elliotii* Engl. & Diels; 77 as *C. lecanthum* Engl. & Diels.

West African: **SENEGAL** BALANTA funtzi (K&A) mpôzé (K&A) BANYUN kikuda und (K&A) ťurum da hum (K&A) BASARI a-tyɛ̀l-khɔnd (Ferry) BEDIK ga-tyẹr-wànd (Ferry) DIOLA funt (JB; K&A) FULA-PULAAR (Senegal) buiki (Aub.; K&A) buiti (K&A) busdé (K&A) busdi (K&A) dokigori (Aub.) MANDING-BAMBARA dâgara (A. Chev.; K&A) iribéléni (K&A) samambali (auctt.) simbabali (K&A) ťâkara (K&A) MANDINKA kulunkalaŋ (*def.*-o) = *pestle handle* (DF) MANINKA dâgara (Aub.; K&A) iribéléni (Aub.; K&A) kulôkalaâ (K&A) kulukalaô (K&A) kulum kalâṅ (K&A) samambali (Aub.; K&A) simbabali (Aub.; K&A) ťâkara (Aub.; K&A) 'SOCE' kulôkalaâ (K&A) kulukalaô (K&A) kulum kalân (K&A) SERER bes (JB; K&A) WOLOF damrat (JB; K&A) taab = *a boil* (JLT; DF) tap (auctt.) toč (Aub; K&A) **THE GAMBIA** DIOLA (Fogny) funt = *mortar* (DF) FULA-PULAAR (The Gambia) buki (DAP) MANDING-MANDINKA jambakataŋ (*def.*-o) = *rough leaf* (DAP; DF) jangara (DAP) WOLOF tab (DAP) **GUINEA-BISSAU** FULA-PULAAR (Guinea-Bissau) dodje-górè (JDES) MANDING-MANDINKA janbakataŋ (*def.*-ô) **GUINEA** FULA-PULAAR (Guinea) buhiki (Aub.) dokigori (Aub.) dooki (Aub.) MANDING-MANINKA diangara (Aub.) iribéléni (Aub.) sama m'bali (Aub.) simbabali (Aub.) tiankara (Aub.) SUSU fubécine (Farmar, ex JMD) **MALI** BOBO nihinpuin (Aub.) FULA-PULAAR (Mali) buhiki (Aub.) dokigori (Aub.) MANDING-BAMBARA diangara (Aub.) diangara ké (A. Chev.) iribéléni (Aub.) sama m'bali (Aub.) simbabali (Aub.) tiankara (Aub.) MANDINKA diamakata kéma = *male janbakatango* (A.Chev.) diangara (Aub.) iribéléni (Aub.) sama *m*bali (Aub.) simba bali (Aub.) tiankara (Aub.) SONINKE-MARKA kékébe (Aub.) **UPPER VOLTA** HAUSA chiriri (Aub.) dagéra (Aub.) MOORE kuaremtuaga (Aub.) **GHANA** DAGBANI gburega (Coull) tachayi (Ll.W) MOORE kwaginyanga (FRI) kwanganga (FRI) kwegenga (FRI) **TOGO** BASSARI diokundio (Aub.) BATONNUN kulogutémini (Aub.) MOBA tantabili (Aub.) SORUBA-KUYOBE alemebé (Aub.) mufopaïe (Aub.) **DAHOMEY** BATONNUN kulogutémini (Aub.) GBE-FON hégnimale (Aub.) SORUBA-KUYOBE alemebé (Aub.) mufopaïe (Aub.) **NIGER** FULA-FULFULDE (Niger) buhiki (Aub.) dokigori (Aub.) GURMA okuamonnu (Aub.) HAUSA chiriri (Aub.) dagéra (Aub.) SONGHAI-ZARMA deligna (Aub.) delinia (Robin) **NIGERIA** FULA-FULFULDE (Nigeria) ɓuski (*pl.* ɓusɗe) (JMD) buyki (MM) HAUSA cìiriirì (auctt.) dagara, dageera, ɗàgeera (auctt.) HAUSA (West) tsiriri (ZOG) TIV alo (KO&S)

A small tree to 10 m high by 1 m girth with smooth bark and bole often twisted, of savanna and fringing forest of dry regions. The species is represented by two varieties: var. *nigricans* is recorded from Senegal and The Gambia, and var. *elliotii* (Engl. & Diels) Aubrév. from Senegal across the Region to Nigeria, and on into Sudan. Var. *nigricans* has the leafy stems puberulous, and var. *elliotii* is distinguishable by their near glabrousness.

The wood is dirty white to yellowish-green at the heart. It is very hard and is included in the general designation of 'breakaxe' (Hausa, *karya gatari)* (14). When dry it is reported to be susceptible to borer attack. It has no recorded use except for firewood (6, 11).

The bark in the hot season freely yields a gum known as *chiriri* in Hausa, which is an item of market-trade in the sudano-guinean region (1, 4, 5, 12). It is white, yellow or red-brown, water-soluble and of good viscosity. It might substitute *Acacia* gums but it does not appear to have been accepted for industrial use. It is edible. It has good adhesive properties. In N Nigeria it finds use in leatherwork and in making ink (4). The tree is the most abundant gum-yielding tree of NW Nigeria.

The bark and leaves of var. *nigricans* is used in Senegal for enteralgia, in cough-medicine and as an expectorant (9, 10), and an aqueous macerate is taken for colic and various intestinal complaints (7). The bark of var. *elliotii* is prepared into a draught for rheumatism treatment and the residual lees are applied in frictions. The bark also has a reputation for action against neuralgia (8, 9, 10).

The leafy twigs and roots are held in the Casamance of Senegal to be, if not toxic, at least repellent to certain animals (7, 10). Branches are placed in water in Senegal and Mali to incapacitate fish and crocodile and force them to the surface (1, 3, 10).

The roots of var. *nigricans* together with other plants are used by Fula in the Fouta Tora of Senegal in treatment of insanity (9, 10).

Leaf-powder macerated in water is given to horses and donkeys to combat fatigue in the Richard-Toll area of Senegal (2).

References:

1. Aubréville, 1950: 102. 2. Boury, 1962: 15. 3. Chevalier, 1937, b: 170. 4. Dalziel, 1937. 5. Howes, 1949: 72. 6. Irvine, 1961: 119. 7. Kerharo & Adam, 1962. 8. Kerharo & Adam, 1963, a. 9. Kerharo & Adam, 1964, b: 430–1. 10. Kerharo & Adam, 1974: 350–1. 11. Portères, s.d. 12. Robin, 1947: 58.

## **Combretum paniculatum** Vent., **ssp. paniculatum**

FWTA, ed. 2, 1: 273. UPWTA, ed. 1, 78.

West African: **SENEGAL** BEDIK gi-ndeúgẹlembar (Ferry) DIOLA (Fogny) busuémet (K&A) SERER bindil (JB; K&A) lumel (auctt.) ndadel (JB; K&A) WOLOF kindindolo *incl. other red flower spp.* (auctt.) tundal (AS; K&A) **THE GAMBIA** DIOLA (Fogny) busuémet = *old thread: like spider's gossamer falling from the sky after the rains* (DF) **GUINEA** FULA-PULAAR (Guinea) yaré safele (Langdale-Brown) SUSU funfunsaré (Farmar) **SIERRA LEONE** MENDE (Dia) jɛmɛlɔ (FCD) MENDE (Komboia) tomba-mɛnyɛ (*def.*-i), tombo, *a deserted village;* mɛnyɛ: *nearly ripe* (FCD; JMD) MENDE-KPA (Kpa) hakpa-namu, hakpa: *pot-herbs in general;* namu: *slimy or sticky* (JMD; FCD) UP MENDE ngenge-tumba, ngenge: *Corchorus olitorius L.;* tomba: *to boil vigorously* (FCD; JMD) **IVORY COAST** VULGAR dawahuru (A&AA) AKYE yatandza (A&AA) KYAMA evéléné belé (B&D) **GHANA** AKAN-ASANTE kyeramoa (BD&H) o-hwirɛmo (Enti) omeha (Green, ex FRI) WASA omeha (Green, ex FRI) DAGBANI garizegu (Ll.W) **NIGER** HAUSA goon-aby (Virgo) kooboo (Virgo) **NIGERIA** HAUSA geza (Lely) kariya (ZOG) YORUBA enyiro? (RJN)

A scandent shrub or robust liane attaining 15 m length, of soudanian humid forest and into savanna fringing forest, occurring throughout the Region and widespread in Tropical Africa.

The plant has vivid scarlet flowers, very spectacular and is worthy of cultivation. The stems are used to make a rope in Tanganika (11), to weave into winnowing baskets in Kenya (16) and stems and roots to tie up beehives (15). The wood is said to be good for implement-handles (4).

The leaves are used in soups and are often cooked with other vegetables. In Sierra Leone the Mende name *hakpanamoi* implies that when put in with spinach it makes the whole pleasantly 'slippery' (5) due to their viscid nature. Chemical analysis has recorded 50% carbohydrates, 20.2% protein and 1.3% fatty substances (3).

An infusion of roots, leaves and stems is used in northern Ghana as a *garli* medicine (12). *Garli* (Fula) is a condition of 'stomach staggers' affecting all cattle (7). Leaf-sap is used externally for gonorrhoea in Tanganyika (14), while in Ivory Coast the tomentum off the leaves is taken in draught to prevent vomiting (2). In Ivory Coast leaves which have been galled are ground up with salt and the paste is applied to the tongue and inside the mouth of babies with stomatitis, and a decoction of galled leaves is used in hipbaths and vaginal douches for haemorrhoids (1).

The roots crushed with a pimento make an enema for haemorrhoids in Ivory Coast (2). In Senegal an aqueous decoction of the roots is recorded as used for diarrhoea (8), and in Tanganyiea for stomach-ache and for gonorrhoea (17) and for fever (10).

It is recorded that children suck the flowers (4) to get the nectar. In Ethiopia the sap expressed from freshly ground-up flowers is used to treat conjunctivitis and other eye-troubles (6). The aqueous extract of the inflorescence has shown some action against carcinomous tumours (9).

Fruits eaten by children at Ikire, S Nigeria, made them very ill and were found to have a paralysing action on test animals, an effect analogous to that of acetycholine (13).

In Gabon the plant is a fetish plant for muscular strength (18).

References:

1. Adjanohoun & Aké Assi, 1972: 87. 2. Bouquet & Debray, 1974: 66. 3. Busson, 1965: 284–7, with leaf-analysis. 4. Dalziel, 1937. 5. Deighton, 1930, K. 6. Getahun, 1975. 7. Irvine, 1961: 120, 807. 8. Kerharo & Adam, 1962. 9. Kerharo & Adam, 1974: 351. 10. Koritschoner 1601, K. 11. Lewis 3183, K. 12. Lloyd Williams 719, K. 13. Lowe 30/6/1975. 14. Markham 328, K. 15. Mwangi 1959, K. 16. Rodgers s.n., K. 17. Tanner 3183, K. 18. Walker & Sillans, 1961: 121.

**Combretum platypterum** (Welw.) Hutch. & Dalz.

FWTA, ed. 2, 1: 274. UPWTA, ed. 1, 78.

West African: **SIERRA LEONE** KISSI yekpandeochuãboã (FCD) MENDE hakpa-namu (Dawe; JMD) tomba-mɛnyɛ (*def.*-i) (JMD) TEMNE *a*-kati (NWT) **LIBERIA** MANO kpã dah (Har.) kpã lah (Har.) **GHANA** AKAN-ASANTE kyeramoa (FRI) o-hwirɛmo ɔpaka (auctt.) TWI aserewa gyama = *sunbird's gyama* (*Alchornea*) (auctt.) pepea (FRI) **NIGERIA** IGALA itado dudu itado: *dark* (Boston) IGBO m̱mányá ṉzā, m̱ńzīzá = *palm-wine of the sunbird* (JMD) IGBO (Awka) àchịchà ṉzā (JMD) okólò (NWT; JMD) YORUBA ọ̀gàn (JMD) ọ̀gàn dúdu (Dawodu) ọ̀gàn ìbulẹ̀ (IFE)

A straggling scandent shrub or liane of secondary deciduous forest from Guinea to W Cameroons, and extending to Sudan, Zaïre and Angola.

The flowers are profuse and showy, yellow to red, followed by red fruits nearly 5 cm across. The plant is worthy of cultivation.

The flowers are sucked by children and visited by sun-birds to obtain the nectar (2).

The wood is hard (3). The hollow stems are used in tapping palms for wine (2; Chesters fide 3, 5) by the Igbo.

The young leaf is sometimes put into soups in Sierra Leone, and a leaf-decoction is used in Lagos as a tonic and febrifuge (2). The leaf is used in Central Africa as a cough-remedy (6) and in Congo leaf-sap furnishes a collyrium for conjunctivitis and diluted with hot water is put into hip-baths for post-partum haemorrhage (1).

The plant has magical application in Congo to afford protection to a man against afflictions that can arise through sleeping with a woman in broad daylight (Sandberg fide 1).

References:

1. Bouquet, 1969: 88. 2. Dalziel, 1939. 3. Irvine 314, K. 4. Irvine, 1961: 120. 5. Thomas, N. W., 1706 (Nig. Ser.), K. 6. Watt & Breyer-Brandwijk, 1962: 194.

**Combretum racemosum** P. Beauv.

FWTA, ed,. 2, 1: 272. UPWTA, ed. 1, 78.

English: Cristmas rose (Port Harcourt, Nigeria, Stubbings): false bougainvillea (Kenya, Gardner, Tweedie).

West African: **SENEGAL** BALANTA kindé kindé (K&A) DIOLA (Fogny) kalakarẽg (JB) kalakarẽng (JB) **THE GAMBIA** MANDING-MANDINKA kotura (Fox) WOLOF topp (Williams) **SIERRA LEONE** LIMBA bede (NWT) LOKO bondi (NWT; JMD) k-pondi (JMD) MENDE hakpa-namu (L-P; JMD) kpuandi (auctt.) SUSU fumunsare (NWT) TEMNE *a*-kati (NWT; JMD) **MALI** MANDING-BAMBARA kotri (A. Chev.) **IVORY COAST** AKYE bétzo (A&AA) KRU-GUERE (Chiehn) ligué degre (B&D) 'NEKEDIE' bélégwé (B&D) **GHANA** AKAN-AKYEM ɔ-hwiremo = *piercing through* (auctt.) ASANTE o-hwirɛmo (BD&H; FRI) wotã (BD&H; FRI) TWI ɔhwirem (BD&H; FRI) **NIGERIA** EDO ọ́kọ́sọ́ (JMD) IGBO (Umuahia) alagame = *thorns near the ground* (AJC) IJO-IZON (Kolokuma) ìgbálị́ = *a large thorn* (KW) ịgbálị (KW) òwéí ịgbálị = *male thorn* (KW) YORUBA ọ̀gàn (RJN; JMD) ọ̀gàn pupa, pupa: *red* (JMD)

A straggling shrub, scandent or liane to 15 m long, of mixed deciduous forest from Senegal to S Nigeria and Fernando Po, and widespread across Africa to Sudan, Kenya, Zaïre and Angola.

The plant bears a mass of crimson flowers, and is very spectacular and worthy of cultivation. It flowers in December/January in S Nigeria, gaining the local English name of 'Christmas rose' (9). The flowers attract bees (5) and may prove useful for honey-production. In Gabon it is frequently planted at village entrances, or a branch is hung over a house-door to keep away spells. Its great floriferousness is a fecundity talisman (10).

The leaves are said to be used in Sierra Leone to season soup (Lane-Poole fide 4). The young leaves are said to be anthelmintic (4). They are used in The Gambia to kill roundworm in children (11). The leaves find local use in Nigeria

(8) and leafy twigs are the basis for various preparations used in Casamance of Senegal against internal parasites (6, 7). The Akye of Ivory Coast prepare a draught from the young leaves with pimento and salt for taking in the morning, or eat them cooked with pimento and salt wrapped in a leaf of *Thaumatococcus danielli* (Marantaceae) as a vermifuge (1).

In Congo the plant is used for all genito-urinary and gastro-intestinal affections accompanied by bleeding: root-macerate or decoction in draught for dysenteries; leaf-sap for haemorrhoids; bark-pulp for bleeding during pregnancy; powdered leaves or roots for haematuria; for convulsive coughing and tuberculosis; the sap as a haemostatic and cicitrisant; powdered bark or leaves to circumcision wounds. These applications may be on the Theory of Signatures because of the red colour of the flowers (2).

The leaf-sap when mixed with water produces a greenish liquid which gels in time. The jelly is taken in Ivory Coast for male sterility (3).

The bark produces a gum which is reported used in The Gambia for toothache (11).

References:

1. Adjanohoun & Aké Assi, 1972: 88. 2. Bouquet, 1969: 88. 3. Bouquet & Debray, 1974: 66. 4. Dalziel, 1973. 5. Frith 280, K. 6. Kerharo & Adam, 1962. 7. Kerharo & Adam, 1974: 353. 8. Oliver, 1960: 23, 57. 9. Stubbings 132, K. 10. Walker & Sillans, 1961: 121. 11. Williams, 1907: 95.

## **Combretum rhodanthum** Engl. & Diels

FWTA, ed. 2, 1: 274.

West African: **SIERRA LEONE** MENDE kpuandi (E. Macdonald) tomba-mɛnyɛ (NWT)

A scandent shrub or liane of the rain-forest and secondary growths, occurring from Guinea to Ghana, and in Cameroun, Zäire, Uganda and Sudan.

The plant is used in Sierra Leone for pains in the tummy (2). The wood contains a jelly-like sticky sap (1).

References:

1. Leeuwenberg 2401, K. 2. Thomas, N. W. 77, K (Arch.)

## **Combretum sericeum** G. Don

FWTA, ed. 2, 1: 270. UPWTA. ed. 1, 78–79.

West African: **GHANA** KUSAL peytuba (FRI) **NIGERIA** GWARI fwainko tnubwa = *hare's ear* (JMD) HAUSA karya garma = *break hoe; referring to the hard root-stock troublesome to farmers* (JMD; ZOG) taro (JMD) HAUSA (West) tabara (Bunny & Ryan) tasagi (Bunny & Ryan)

A shrub of annual erect shoots to about 1 m high from a perennial woody creeping rootstock, of savanna woodland from Guinea to S Nigeria, and in Ubangi-Sheri and Sudan.

The hard rootstock creates difficulty for farmers in clearing land, hence the Hausa name meaning 'breakhoe.' The plant yields a gum but of inferior quality to gum-arabic (2). Hausas of N Nigeria prepare a decoction of the plant for giving to a newly born baby, and for use as a wash for it. The fruit has superstitious use in prescriptions to secure immunity against cutting weapons. This effect is called *kaskaifi,* Hausa, 'edge destroyer' (1).

References:

1. Dalziel, 1937. 2. Howes, 1949: 72.

**Combretum smeathmannii** G. Don

FWTA, ed. 2, 1: 272. UPWTA, ed. 1, 78 as *C. mucronatum* Schum. & Thonn.

West African: **SENEGAL** DIOLA (Fogny) ba nélé (JB) MANDING-BAMBARA koko toro (JB) **THE GAMBIA** DIOLA (Fogny) fufunukafu (DF) **GUINEA-BISSAU** FULA-PULAAR (Guinea-Bissau) iári-sáfi-bátè (EPdS) iári-sap-bátè (JDES) **SIERRA LEONE** LOKO bondi (JMD) k-pondi (NWT; JMD) MENDE hakpa-namu (FCD) tomba (*def.*-i) (JMD; FCD) tomba-mɛnyɛ (L-P; FCD) TEMNE *a*-kati (JMD; FCD) **GHANA** AKAN-AKYEM ɔ-hwiremo (FRI) ASANTE o-hwirɛmo (FRI; Enti) TWI ohwirem (JMD; FRI) **NIGERIA** YORUBA agbọ̀n igbó, agbọ̀n: *basket* (Millson; JMD) agbọ̀n ọ̀dan, agbọ̀n: *basket* (Millson; JMD) efirin? (RJN) lawo (Dawodu) ọ̀kàn, ọ̀gàn

A scandent shrub of secondary forest or forest liane to 13 m long, occurring from The Gambia to S Nigeria and on to Zaïre.

The leaves are boiled and eaten in Lagos as a preventative against illness (6), and have unspecified medicinal use in Sierra Leone (2). The Kweni (Gouro) of Ivory Coast consider the leaf-sap haemostatic and healing in topical applications (1) and pounded leaves are used to cure wounds in Cameroun (4). The leaf-sap is used in Zaïre to produce a black dye (Staner fide 3).

The roots, cut small, are boiled with *Capsicum* peppers (Solenaceae) or wood-ash and the concoction is drunk for chest-pains and gonorrhoea, and an infusion of young leaves with natron is used as a vermifuge in Ghana (Saunders fide 3).

The flowers are sweet-scented and with a strong smell of honey (5). They may be of value for production of honey.

References:

1. Bouquet & Debray, 1974: 66. 2. Dalziel, 1937. 3. Irvine, 1961: 123. 4. Leeuwenberg 7370, K. 5. Letouzey & Villiers 10398, K. 6. Millen 52, K.

**Combretum tarquense** J. J. Clark

FWTA, ed. 2, 1: 273.

A scandent shrub or climber to 3 m of evergreen and secondary forests of Ivory Coast and Ghana.

The inflorescences are borne in terminal racemes of red flowers followed by crimson fruit. The plant is strikingly showy and worthy of cultivation.

**Combretum tomentosum** G. Don

FWTA, ed. 2, 1: 270.

West African: **SENEGAL** BANYUN kignat bagid (K&A) BASARI a-ndèg (Ferry) DIOLA (Fogny) fu ruk (JB); = *big spread* (DF, The Gambia) MANDING-BAMBARA ñañaka (JB) MANDINKA salanombo (K&A) SERER sérédéd (JB) **THE GAMBIA** MANDING-MANDINKA kabaa (*indef.* and *def.*) (Frith) **GUINEA-BISSAU** FULA-PULAAR (Guinea-Bissau) cancalibá-déo (JDES) canquelibá-deo (EPdS) iári-sáfi (JDES; EPdS) PEPEL ulô (JDES; EPdS) **GUINEA** FULA-PULAAR (Guinea) safiri (Langdale-Brown) **UPPER VOLTA** GRUSI lokoan (K&B) KIRMA somania (K&B)

A straggly lianous shrub with stems 12–15 m long forming dense thickets, of the soudanian savanna, recorded from Senegal to Upper Volta.

A decoction of leaves is prescribed in Upper Volta in lotions and baths to strengthen infants and old men (4). In the Casamance of Senegal a root-macerate is taken against coughs (2, 3).

The fruits are edible and are reported eaten in The Gambia (1).

References:

1. Frith 51, K. 2. Kerharo & Adam, 1963, a. 3. Kerharo & Adam, 1974: 353. 4. Kerharo & Bouquet, 1950: 50.

**Combretum zenkeri** Engl. & Diels

FWTA, ed. 2, 1: 273. UPWTA, ed. 1, 79.

West African: **IVORY COAST** ANYI foro (K&B) BAULE fono (B&D) **GHANA** ADANGME adatʃo? (FRI) ADANGME-KROBO tadatʃo (auctt.) **NIGERIA** YORUBA ọ̀gàn (Macgregor; JMD)

A climbing shrub or scandent forest liane to 27 m high, of savanna and secondary forests; common in places and widely distributed from Guinea to S Nigeria and in Cameroun.

The plant is recorded having unspecified medicinal use in Lagos (3). Igbo use the leaves in a worm-treatment (4). In Ivory Coast a piece of twig is chewed by women to relieve menstrual pain (2) and Anyis prepare the roots with stem of *Aframomum* (Zingiberaceae) and a small pimento into a suppository for use in dysentery (2). The fermented leaf-sap with that of *Struchium sparganophora* (Linn.) O. Ktze (Compositae) is used by the Baule for male sterility. The plant is also used internally and externally for certain oedemas (1).

References:

1. Bouquet & Debray, 1974: 66. 2. Kerharo & Bouquet, 1950: 50. 3. Macgregor 209, K. 4. Thomas, N. W., 1879 (Nig. Ser.), K.

**Combretum spp. indet.**

West African: **SENEGAL** BASARI a-kúlángán, a-ngálásy (Ferry) BEDIK ga-kangàl (Ferry) **THE GAMBIA** FULA-PULAAR (The Gambia) kondyam-cholli = *bird's wine; a red-flowered sp.* (DAP) MANDING-MANDINKA jambakataŋ (*def.* j. musoo) = *female jambakato: C. collinum prox.* (DRR) WOLOF do-oki *C. collinum prox.* (DRR) kindingdolo = *bird's wine, a red-flowered sp.* (DAP) kunaji *C. collinum prox.* (DRR) naka burayo = *bird's wine, a red-flowered sp.* (DAP) **GUINEA-BISSAU** BALANTA messancala (DF) FULA-PULAAR (Guinea-Bissau) boilamauba (DF) MANDING-MANDINKA djambacatam-quéó (JDES) hiremonssôlo (DF) **UPPER VOLTA** GRUSI kamar péru (K&B) KIRMA térétan (K&B) LOBI felsi (K&B) MOORE zanpitaga (K&B) **IVORY COAST** AKAN-ASANTE flan (B&D) BAULE borolié néniku (B&D) KRU-GUERE (Chiehn) borohonliku (B&D) KWENI tirika tirika (B&D) trika (B&D) KYAMA afuyaniama (B&D) 'NEKEDIE' boroleniléu (B&D) **GHANA** AKAN-ASANTE hõmaben (FRI) **NIGERIA** IGBO ana (BNO) ọtugọ *probably applies to several spp.* (JMD)

An unspecified species is recorded used in Upper Volta in root-decoctions for mouthwashes and gargles for toothache. The fruit is eaten in time of dearth but with circumspection as it causes irritation to the mouth and throat. The fruit is used in treatment for leprosy, and a root decoction is used by midwives to hasten child-delivery.

The Ijo of S Nigeria used an unidentified species for curing *ukú*, an undefined condition, and leaves squeezed in water are applied to ease chest and liver pains (3).

In The Gambia the leaves of a small tree *Combretum* species, *C. collinum prox.*, are used for boils (2).

References:

1. Kerharo & Bouquet, 1950: 51. 2. Rosevear, 1961. 3. Williamson 19, UCI.

**Conocarpus erectus** Linn.

FWTA, ed. 2, 1: 279–80, 762. UPWTA, ed. 1, 79.

English: button wood; West Indian alder.

French: manglé-boton (Irvine); manglier gris (Dalziel); palétuvier gris (Berhaut); palétuvier zaragosa (Dalziel); petit manglier (Aubréville).

West African: **SENEGAL** DIOLA bu dakãnd (JB) DIOLA (Fogny) fu lensal (JB) SERER gnara djuam (JLT) ñara (JB) ndãmb (JB) WOLOF ñarañara (JB) **THE GAMBIA** DIOLA (Fogny) bubangalatab (DF) fu lensal = *salty* (DF) **GUINEA-BISSAU** BIDYOGO nedêg-dêg-ká (EPdS) MANDING-MANDINKA tarafô (EPdS) **SIERRA LEONE** TEMNE *an*-saa (FCD) **IVORY COAST** SUSU kinsi kundji = *captive of the mangrove* (Aub.) 'KRU' nja (Aub.) **GHANA** ADANGME koka (FRI) AKAN-FANTE asukuru, asu: *water* (Enti) GA asokpolo (auctt.) kaba (BD&H; FRI) **NIGERIA** YORUBA abododo (JRA)

A shrub to about 5 m high, of brackish swamps and mangrove communities along the coast from Senegal to W Cameroons and along the Atlantic coast to Angola and in S America on both Atlantic and Pacific coasts, and in Samoa.

The wood is suitable for small posts being durable in the ground and in salt water. It can be used for piles and is regarded as producing a good firewood (3, 5).

The bark from a Belize sample contains 18% tannin which produces good leather (3, 4).

In Nigeria the leaves are reported eaten and a decoction drunk for fever (1). Tests for anti-malarial activity proved negative on avian malaria (2). Also in Nigeria, the ground-up roots are boiled and the liquid is taken for gonorrhoea and catarrhal conditions of the respiratory channels, and latex is used as a styptic (1).

References:

1. Ainslie, 1937: sp. no. 111. 2. Claude & al., 1947. 3. Dalziel, 1937. 4. Holland, 1908–22: 308. 5. Irvine, 1961: 124.

## **Guiera senegalensis** J. F. Gmel.

FWTA, ed. 2, 1: 275, 762. UPWTA, ed. 1, 79–80.

English: Moshi medicine (Ghana, Kerharo & Bouquet).

French: guier du Sénégal.

West African: **MAURITANIA** ARABIC ('Maure') agu (Aub.) liené (Aub.) liyina (Aub.) **SENEGAL** VULGAR nger, n'guer *from Wolof* (K&A) n'guère (JB) BALANTA buisi (K&A) ḋosé (Aub.; K&A) BANYUN kifudo (K&A) sifôdum (Aub.; K&A) BASARI a-pɛsy ɛ-syiw̃ (Ferry) BEDIK gi-nyím-tyingeus (Ferry) DIOLA (Fogny) bu fatikay (JB) épérum (K&A) fufanikay, fufunikay, fufunuk, fufunukafu, fusabel (K&A) fupirum (Aub.) DIOLA (Tentouck) bufuluk (K&A) DIOLA-FLUP buhunuk (K&A); = *chocolate* (DF, The Gambia) FULA-PULAAR (Senegal) elloko, eloko, gelodé (K&A) géloki (JMD; K&A) géram, ngéram *from Wolof?* (K&A) iéloko (Aub.) kormégélok *the galls* (K&A) ndiéloki (Aub.) TUKULOR hud (AS) KONYAGI a-nganant (Ferry) MANDING-BAMBARA kuḋêmbé (K&A) kundié (JMD) kuñé (JB) MANDINKA maamakunkoyi (*def.* maamakunkoyoo) MANINKA kunḋé (auctt.) kuńé (Aub.; K&A) 'SOCE' kâkatanô (K&A) MANDYAK bentzins (Aub.; K&A) binsilič, mêté (K&A) MANKANYA binzéhon (Aub.) binzéô (K&A) SERER hud (Aub.; K&A) ngud (JMD; JB) ngut (K&A) SERER-NON hud (AS) hud (AS) SONINKE-SARAKOLE kamé (K&A) xamé (K&A) WOLOF guier, n-ger (auctt.) n-domonger *the galls* (K&A) nger (JB; K&A) **THE GAMBIA** DIOLA (Fogny) fufanikay, etc. (as under Senegal) *in allusion to dust trapped on the tomentum of the leaves and twigs* (DF) FULA-PULAAR (The Gambia) geloki (DAP) MANDING-MANDINKA kankanaŋ (*def.*-o) (auctt.) kunjee (*def.* kunjeo) *from Wolof* (DAP) maamakunkoyi (*def.* maamakunkuyoo) = *old woman's head* (DAP; DF) WOLOF n-ger **GUINEA-BISSAU** BALANTA biôcê (JDES) iuci (JDES) CRIOULO badosdoce (JDES) badossosso (JDES) FULA-PULAAR (Guinea-Bissau) elócò (EPdS) guêlodi (DF) helócò (JDES) MANDYAK bissem antchom (JDES) bissilintche (JDES) bitchiante (JDES) MANKANYA bissem antchom (JDES) bitchiante (JDES) **GUINEA** FULA-PULAAR (Guinea) bali niama (CHOP) MANDING-MANINKA ko-nguélé (CHOP) kugnie, kundié, kunié (Aub.) **MALI** DOGON góbu pílu (C-G) FULA-PULAAR (Mali) geloki (JMD) iéloko (Aub.) n-diéloki (Aub.) MANDING-BAMBARA kudiengbé (Aub.) kundié (JMD) MANINKA kugnie (Aub.) kundié (Aub.) kunie (Aub.) SONGHAI sàabàrà (Aub.) TAMACHEK tuhila (Aub.) **UPPER VOLTA** GRUSI sobara (K&B) HAUSA sabara (Aub.; K&B) KIRMA tupo (K&B) MOORE wilinwiga (FRI; K&B) wilinwissi (K&B) NYENYEGE sumu inga (K&B) SONGHAI-ZARMA pabré (Aub.; K&B) 'SENUFO' kubélégelman (K&B) **IVORY COAST** MANDING-MANINKA kudiengbé (Bégué, ex K&B) **NIGER** ARABIC (Niger) abesh (Aub.) FULA-FULFULDE iéloko (Aub.) ndiéloki (Aub.) HAUSA sabara (Aub.; K&B) KANURI kàáshì (Aub.) SONGHAI kánfúló (*pl.* -á) (D&C) sàabàrà (D&C) SONGHAI-ZARMA sabré (Robin) TAMACHEK tuhila (Aub.) **NIGERIA** ARABIC obeish, ghobeish (JMD) ARABIC-SHUWA kabeish (JMD) FULA-FULFULDE (Nigeria) geelooki (*pl.* geeloɗe); gelooki (*pl.* gelooɗo); gellooki (*pl.* gelloudɗe) (auctt.) jelloki, jolooki (MM) puri (Taylor) HAUSA bàrbáttaá *the leaves* (JMD; ZOG) ƙululu *the galls* (Bargery; ZOG) ƙuruƙuru, ƙurƙure *the roots as firewood* (JMD) sàabarà, saàbáraà (auctt.) KANURI kàáshì (JMD)

A shrub reaching 3 m high, of low rainfall areas and light dry soils occurring throughout the Sahel region from Mauritania to N Nigeria, and across Africa to Sudan.

The wood is whitish or tinged red, coarse-grained, knotted and short, but very hard. It used in Sahel de Nioro of Mali for the framework of wells, bed-posts, etc. (22). The shrub is commonly cut to fence farms (5). The plant is capable of colonizing tracts of land, which might otherwise be bare, to form pure stands. Though restriction of its spread is recommended in Senegal to improve the quality of Sahel pasturage (1), it does provide a forage for stock including camels, and may assume some importance in N Nigeria and the drier regions during the dry season (6).

The leaves have a bitter taste and a widespread acknowledgement in African medicine as a cure-all, in places being equated with *kinkeliba (Combretum micranthum* G. Don). They are commonly sold in markets as the dry leaf and may be transported far from where the plant grows. The usual form of preparation for internal use is in decoctions or mixed with food preparations. The leaf is widely administered for pulmonary and respiratory complaints, for coughs, as a febrifuge, colic and diarrhoea, syphilis, beriberi, leprosy and impotence, for rheumatism, diuresis and expurgation (2–4, 11–18). In Guinea the leaves are commonly made into an infusion with those of *Combretum micranthum* for fever, chest and rheumatic conditions, and as a nasal douche for cold in the head; mixed with tamarind pulp they are held to be good laxative and appetizer (5). In Ivory Coast-Upper Volta annual rainy season outbreaks of choleriform diarrhoea and dysentery, known in Upper Volta as *sada wubre* (Moore, lit. 'diarrhoea-vomiting'), and particularly the serious epidemic of 1945, are treated with marked success by leaf-infusions (18). Similarly in official hospitals in Guinea leaf infusions are used as an anti-dysentric and for acute enteritis (19). In Northern Nigeria powdered leaves are mixed with food as a general tonic and blood restorative after any exhausting condition, and especially to women after childbirth as a galactogogue. The fruit and leaves are common ingredients in more or less ceremonial prescriptions for strengthening and preventing disease (*dauri,* Hausa) in young children. In Sokoto the plant has special reputation as a preventive of leprosy and many people drink a cold decoction every morning and evening; in particular it is given to the newborn child, and to the child of a leper parent, or when there is the least suspicion of hereditary taint or early symptom (5).

In external application the leaves are considered vulnerary, antiseptic and healing. In Senegal they are applied to wounds, sores in the mouth, syphilitic chancres and phagadenic ulcers (17). In Nigeria the leaves are applied to skin infections and in poultices on inflammatory swellings and for guinea-worm (5). In Guinea leaves have been applied to tumours to maturate them (21) and in the soudanian region for skin-diseases (2). The Fula of N Nigeria will chew the twigs for symptomatic relief of scorpion stings (10).

A leaf-decoction is used in N Nigeria as a body-wash (5). The Moore of Upper Volta use a leaf-decoction for bathing new-born babies (8). In The Gambia the dried leaves are smoked in a pipe and the smoke is blown through the nose for colds (24). In Casamance of Senegal the dried leaves are mixed with tobacco for smoking for coughs and respiratory trouble (12, 17). The 'Senoufo' of Ivory Coast-Upper Volta add the powdered leaves to a snuff taken for headache and sinusitis (18).

In a superstitious sense the Bedik of Senegal administer a bath of a macerate of leafy twigs to persons possessed by evil spirits, a treatment considered able also to further affairs of the heart and of fortune (14, 17). In The Gambia it is said that when people of the Mandinka race are buried some leaves of the plant are placed in the grave with them (7).

The leaves have veterinary use. They are fed to stock in Senegal to ensure increase in weight, fertility and milk-yield (15, 17). Calabashes or other utensils used in milking in Guinea are washed with an infusion of leaves to ensure abundance of yield or richness in cream (20). In N Nigeria leaves are fed to cattle as a tonic and digestive, and especially to cows to increase the milk. The leaves are considered a remedy for internal troubles to horses: leaves crushed with other substances, e.g. capsicum peppers, leaves of *Hyptis pectinata* (Labiatae) and dye-pit indigo, with warm water added, are boiled and placed before the horse to inhale. The dried leaves and twigs are commonly burnt in cattle-pens and horse-stables and around camps of domestic stock as a fumigation against biting and other flies and to prevent chills (5).

The plant becomes commonly infested with galls which enter into local pharmacopoeas, and are sold in northern markets and have vernacular names. Charcoal made from the galls *(kormé gélok,* Wolof) is recorded as being very diuretic and is given in Senegal for oliguria and anuria and for attacks of pernicious fevers. Another application with twigs of *Combretum aculeatum* and salt is for colics and vomiting (17). In N Nigeria the galls are said to be effective in curing sores on children's heads (9).

The bark yields a gum which is marketed in the Dosso region of Niger (23).

The root is considered a good firewood in N Nigeria (5). It is also used in medicines. The Moore of Upper Volta put the roots into a medico-magical prescription for leprosy (18). The Fula of N Nigeria wash in water in which roots have been soaked to ensure good hunting (10), and in Bornu the roots powdered and boiled are considered a remedy for diarrhoea and dysentery (von Beurman fide 5). The roots are commonly split and used as chew-sticks and tooth-picks (5, 20).

The fruits baked and reduced to a powder with salt added to mask the bitter taste are considered a sovereign cure for hiccups in Ivory Coast-Upper Volta (18).

Leafy stems have been found to contain traces of alkaloids and tannins. Ash of the roots and leaves appear to be particularly rich in magnesium, calcium, strontium, titanium, iron and aluminium. (See also 9 for leaf-analyses.) Pharmacologically the plant has positive action on coughs, is hypotensive, antidiarrhetic and anti-inflammatory (17).

References:

1. Adam, 1966, a: 517. 2. Aubréville, 1950: 90. 3. Bouquet & Debray, 1974: 32: 67, with references. 4. Boury, 1962: 15. 5. Dalziel, 1937. 6. Foster & Munday, 1961: 316. 7. Fox 44, K. 8. Irvine 4684, K. 9. Irvine, 1961: 124–6, with leaf-analyses. 10. Jackson, 1973. 11. Kerharo, 1967. 12. Kerharo & Adam, 1962. 13. Kerharo & Adam, 1963, a. 14. Kerharo & Adam, 1964, a: 420. 15. Kerharo & Adam, 1964, b: 547. 16. Kerharo & Adam, 1964, c: 308. 17. Kerharo & Adam, 1974: 353–6, with detail and references on phytochemistry and pharmacology. 18. Kerharo & Bouquet, 1950: 51–52. 19. Oliver, 1960: 27, 66. 20. Pobéguin 817, K. 21. Pobéguin, 1912: 27. 22. Roberty, 1953: 451. 23. Robin, 1947: 58. 24. Rosevear, 1961.

**Laguncularia racemosa** Gaertn.

FWTA, ed. 2, 1: 280–1. UPWTA, ed. 1, 80.

English: white mangrove (Burtt Davy & Hoyle; Dalziel; Taylor) – not to be confused with *Avicennia africana;* white button wood (Dalziel; Irvine).

French: paléturvier noir [!] (Berhaut).

West African: **SENEGAL** DIOLA flẽs (JB) SERER bak (JB) ndas bal (JB) **THE GAMBIA** DIOLA (Fogny) fles = *broom* (DF) FULA-PULAAR (The Gambia) kajo (DRR; DAP) MANDING-MANDINKA baatomanki (*def.* baatomankoo) = *swamp mangrove* (auctt.) **GUINEA-BISSAU** CRIOULO tarafe (JDES; EPdS) tarafe preto (D'O) DIOLA-FLUP cahaguela (JDES) MANDYAK pfẽque (JDES) NALU n'concom (JDES) unconcom (JDES) PEPEL btèque (JDES) oellha (D'O) **SIERRA LEONE** BULOM (Sherbro) chechem-dɛ (FCD) TEMNE *an*-kənt-*a*-bi (FCD; S&F) **GHANA** AKAN-FANTE abin (auctt.) asopru (auctt.) NZEMA asokru (CJT) **NIGERIA** IJO orke (KO&S) IJO-IZON (Egbema) ọ̀kẹ̀ (KW)

A shrub or small tree to 6 m high of the mangrove community in brackish and tidal swamps, occurring throughout the coastline of the Region from Senegal to W Cameroons and also in tropical America. In The Gambia it is said to occur mainly on the landward side of mangrove swamps where the land-building is hardening to dry land (5). It is often gregarious.

The wood is heavy, hard, strong and close-grained (3), but use appears limited to firewood which is of good quality (6). The bark contains tannin. Caribbean samples have varied from 12–24% content, hardly of economic proportions, but leather made from Belize material has been reported to be of excellent quality (3). The leaves also contain tannin recorded as 10–20% content (4). Dyeing material and unspecified medicaments are also obtained from the tree (1, 4, 7).

The foliage is browsed by camels (4). The fruit are recognized as edible in Guinea-Bissau (2).

References:

1. Dalziel, 1937. 2. D'Orey 36, K. 3. Holland, 1908–22: 309. 4. Irvine, 1961: 126–7. 5. Rosevear, 1961. 6. Taylor, 1960: 152. 7. Unwin, 1920: 56, 374.

**Pteleopsis habeensis** Aubrév.

FWTA, ed. 2, 1: 275.
West African: **MALI** DOGON guan (Aub.) gum (Aub.) MANDING-MANINKA kolobé (Aub.) **NIGERIA** HAUSA lállèn giíwaá (Lowe; ZOG)

A shrub forming thickets with *Combretum micranthum* with which it shares the Maninka vernacular name in Mali, and extending into N Nigeria.

In the Bandiagara area of Mali the branches are used to weave into baskets (1).

Reference:

1. Aubréville, 1950: 141.

**Pteleopsis hylodendron** Mildbr.

FWTA, ed. 2, 1: 275.
West African: **IVORY COAST** VULGAR ko-framiré (DF; Aub.)

A large tree reaching to over 30 m high by 6 m girth, of the closed rain-forest from Ivory Coast to S Nigeria and on into Zaïre.

The tree resembles *Terminalia ivorensis* with which it is confused, but the wood appears to be little used.

**Pteleopsis suberosa** Engl. & Diels

FWTA, ed. 2, 1: 275. UPWTA, ed. 1, 80.
West African: **SENEGAL** BASARI a-nẹ̀kúɗy (K&A; Ferry) BEDIK ga-ndẹkwéɗy (Ferry) MANDING-BAMBARA dana (JB; K&A) téréni (Aub.; K&A) **MALI** MANDING-BAMBARA téréni (Aub.) **IVORY COAST** LOBI kpossinkpo (A&AA) **GHANA** KONKOMBA legbarembal (Gaisser) **TOGO** BASSARI digbar (Gaisser) digbaré (Aub.) KABRE kessissinõ (Gaisser) MOORE-NAWDAM assissinõ (Gaisser) TEM sisinan (Aub.) TEM (Tshaudjo) sissina (Volkens) **DAHOMEY** BATONNUN kulu kula (Aub.) BULBA (Tanguieta) nuaka (Aub.) GBE-FON kulu kuli (Aub.) SOMBA di kointonné (Aub.) **NIGERIA** HAUSA wúyàn giíwaá (ZOG)

A shrub or tree to over 10 m high, of open and wooded savanna of Senegal to Nigeria.

Restriction on its spread is recommended for the improvement of Sahel pasturage (1).

Lobi of Ivory Coast prepare an extract from the chopped up roots and smaller shoots as a cough-medicine (2).

Basari medicine-men in Senegal endow the plant with fetish power of protection of millet, a few fragments being added to the grain. Equally the tree is protected in the neighbourhood of millet fields (3, 4).

References:

1. Adam, 1966, a. 2. Adjanohoun & Aké Assi, 1972: 89. 3. Kerharo & Adam, 1964, a: 423–4. 4. Kerharo & Adam, 1974: 356–7.

**Quisqualis indica** Linn.

FWTA, ed. 2, 1: 275. UPWTA, ed. 1, 80–81.

English: Rangoon creeper; 'Love and Innocence' (Geitlinger).

West African: **TOGO** ANYI-ANUFO gargu (Volkens) **NIGERIA** YORUBA ọ̀gàn fúnfún (auctt.) ọ̀gàn-igbó (JMD)

A woody climber of savanna forest occurring from Mali to S Nigeria and W Cameroons, widespread in tropical Africa and Asia, and cultivated throughout the Tropics.

The species has been under cultivation as an ornamental and a drug-plant for so long its native home is by no means certain. The plant occurs throughout India and SE Asia as a cultivate, perhaps arising from the east side of the Bay of Bengal and confirming the appropriateness of the English name *Rangoon creeper*. The *Flora of Tropical East Africa* suggests that it may also be indigenous in the Iringa District of Tanganyika (8).

The flexible stems are used in Togo for basket weaving, and in making fish-weirs and fish-traps (4).

The fruits (5) and the seeds (1) are used in Nigeria and in Ivory Coast (2) for their anthelmintic properties and to cure diarrhoea, as also in Gabon (7). These applications are widely used in Asian medicine. The fruit is best picked half-ripe. A root-extract is also vermifugal. The seeds are oil-bearing and the vermifugal property is said to lie in the oil (3, 6). Dosage requires some caution. A large dose of the fruit-extract produces hiccough and an overdose unconsciousness, the active principle being particularly concentrated in the ovary-wall and seed-coat. Ripe seeds are sweet and if the ovary-wall and seed-coat are removed are pleasant to eat but there are cases of people becoming ill on eating only two or three. Excess causes drowsiness (3).

References:

1. Ainslie, 1937: sp. no. 296. 2. Bouquet & Debray, 1974: 67. 3. Burkill, 1935: 1860–1. 4. Dalziel, 1937. 5. Singha, 1965. 6. Quisumbing, 1951: 654–6. 7. Walker & Sillans, 1961: 122. 8. Wickens, 1973: 68.

**Quisqualis latialata** (Engl.) Exell

FWTA, ed. 2, 1: 275.

A scandent shrub to 20 m long of forest clearings, secondary forest of the rain-forest area of SE Nigeria and extending to Zaïre and Angola into savanna habitats.

No use is recorded in the Region. In Gabon the plant is known as an antidiarrhoetic (2). In Congo it is a general drug-plant for genito-urinary and gastro-intestinal troubles, and is used also for diarrhoea, dysentery, haemorrhoids, stomach-troubles, female sterility and costal pains (1).

References:

1. Bouquet, 1969: 81. 2. Walker & Sillans, 1961: 121.

**Strephonema mannii** Hook. f.

FWTA, ed. 2, 1: 264.

A shrub to about 6 m high, by rivers, recorded from S Nigeria and W Cameroons, and occurring also in Gabon.

The wood is yellowish-brown, very dense and hard. The bark has (unspecified) medicinal uses in Gabon (1).

Reference:

1. Walker & Sillans, 1961: 122.

**Strephonema pseudocola** A. Chev.

FWTA, ed. 2, 1: 264. UPWTA. ed. 1, 81.

French: faux colatier (Dalziel); poto-poto (Ivory Coast, from Abe, A. Chevalier).

West African: **SIERRA LEONE** KONO demgbeyawi (SKS) MENDE degbeme-wulo, degbeme-yawi (S&F) demgbeyani *a variant pronunciation* (Fox; SKS) kovo (*def.* kovui) (FCD; S&F) **LIBERIA** KRU-BASA tentout (C&R) **IVORY COAST** ABE boto (Aub., ex K&B) poto-poto = *swamp* (Aub., ex K&B) ABURE ploplo (B&D) potopoto (B&D) AKAN-BRONG dum (K&B) DAN kéré (K&B) kialé (K&B) KULANGO pilé (K&B) 'KRU' kuguetu (K&B) **GHANA** VULGAR abutusimu (DF) AKAN-WASA awuruku (auctt.) tutuabu (BD&H; FRI)

An understorey tree of the evergreen rain-forest, to over 20 m high and trunk girth to 2–3 m, recorded from Sierra Leone to Ghana. The tree is fairly common in the high-forest, and in Sierra Leone it is considered an important forest-regeneration weed-species (8).

The wood is yellowish-white or pale olive, hard and heavy but not durable (4, 5, 8). It is occasionally logged in Sierra Leone, but the quantity is limited and the timber is used in general construction work in mixture with other woods (8). Log-ends become gummy and the wood is not easy to work though it gives a smooth finish and is usable locally for carpentry.

The bark-slash yields a copious quantity of clear watery gum which thickens to a translucent gelatinous glue. This substance is the *poto-poto* of Ivory Coast which the Abe liken to a swamp, hence their vernacular name for the species (1, 7). In that country the gum, heated over stones, is used topically for yaws on the soles of the feet (2), and in Sierra Leone it is applied to rheumatic areas to alleviate the pain (8). A bark-decoction is astringent and is taken in Liberia for diarrhoea (3, 4, 5).

The Kru of Ivory Coast use a leaf-decoction in draught and baths for dropsy and oedemas (7).

A root-decoction is recorded as used in fumigations and frictions for localized oedemas in Ivory Coast, eye-wash for ophthalmias and in draught as a diuretic (2). Also in Ivory Coast a root-bark decoction prepared together with other drug-plants is given in cases of very difficult childbirth (7).

The seed contains two cotyledons, and in external shape and colour is so alike the kola nut as to be used sometimes as an adulterant. The seed however is not edible and is very bitter. The seed-oil is used in Sierra Leone to rub on the body for craw-craw and on the head for lice (6).

References:

1. Aubréville, 1959: 3, 72. 2. Bouquet & Debray, 1974: 67. 3. Cooper 121, K. 4. Cooper & Record, 1931: 35, with timber characters. 5. Dalziel, 1937. 6. Deighton 3055, K. 7. Kerharo & Bouquet, 1950: 53. 8. Savill & Fox, 1967: 104.

**Terminalia** Linn.

FWTA, ed. 2, 1: 277.
West African: **SENEGAL** BASARI a-peùlá mẹ-nàngàr (Ferry) **SIERRA LEONE** MENDE foni-baji (*def.*-i) *savanna spp. from* foni: *grassfield, and* baji: *T. ivorensis* (S&F) **NIGERIA** BEROM pyòso (LB) HAUSA báúshe (LB) YORUBA ìdí *the closing* (Verger)

A genus represented by 11 species in West Africa occurring in the closed forest and in savanna, and counting amongst them important timber-trees. The genus is important in India as sources of tanning and dyeing 'myrobalan' materials (2) of which a number of species have been introduced.

Removal of *Terminalia spp.* from cattle land in the Sahel of Senegal is recommended for the improvement of pasturage (1).

A Yoruba incantation invokes the genus under the general name of *ìdí*, 'the closing' of the paths of evil and death, for the safe conduct of an *abiku* child during its life on this earth (3).

References:

1. Adam, 1966, a: 517. 2. Greenway, 1941. 3. Verger, 1967: no. 106.

**Terminalia albida** Sc. Elliot

FWTA, ed. 2, 1: 279.
West African: **SENEGAL** BALANTA kakufufa (Aub.) kénacu (Aub.) DIOLA kéo (JB; Aub.) FULA-PULAAR (Senegal) bori billel (Aub.) MANDING-BAMBARA volo ni dié (JB) MANDINKA woloforo (*def.*-o) (after Aub.) MANDYAK bugundio (Aub.) **THE GAMBIA** FULA-PULAAR (The Gambia) bodehi (DAP) bori-bila (DRR) boro-bila (DAP) MANDING-MANDINKA wollo-koio (Fox) wolo (*def.*-o) (auctt.) wolo-iro, iro: *tree* (DRR) **GUINEA-BISSAU** MANDING-MANDINKA sirafitom (JDES) MANKANYA bidianali (Aub.) **GUINEA** FULA-PULAAR (Guinea) bori billel (Aub.) SUSU copera figue (Aub.) **SIERRA LEONE** MANDING-MANINKA (Koranko) wasɛ (S&F) wɔlsayamba (NWT) MENDE foni-baji (S&F) SUSU kuberefiexai (NWT) **MALI** FULA-PULAAR (Mali) bori billel (Aub.) MANDING-BAMBARA uolonidié (Aub.)

A shrub or tree to about 13 m high, with corky fissured bark, of the savanna woodland from Senegal to Mali. It is one of the commonest small trees of The Gambia (2). Its thick bark confers on it some protection against fire (3).

Its chief use in The Gambia is for firewood but when found in large sizes it is used for house-building (2).

An aqueous extract of the bark is used in The Gambia as an eye-lotion (1).

References:

1. Fox 83, 110, K. 2. Rosevear, 1961. 3. Savill & Fox, 1967, 103.

**Terminalia arjuna** Bedd.

A large tree, native of India, Ceylon and Burma, and introduced into the Region in Ivory Coast (1).

The bark of Indian material yields about 15% tannin which is extensively used for producing leather of good quality. The bark also has medicinal uses. The timber is brown and very hard, and is used for carts, agricultural implements, house and boat-building (2).

References:

1. Aubréville, 1959: 3, 69. 2. Burkill, 1935: 2136.

**Terminalia avicennioides** Guill. & Perr.

FWTA, ed. 2, 1: 279. UPWTA, ed. 1, 81.

West African: **SENEGAL** BANYUN kiminkel gunpan (K&A) BASARI a-nòkɔ́rɔ́ɓ (Ferry) BEDIK ga-pẹɗà kɔ́lɛ́té (Ferry) DIOLA (Fogny) ku furok (JB) DIOLA-FLUP kauôq (Aub.; K&A); = *cemetary* (DF, The Gambia) FULA-PULAAR (Senegal) pulémi (Aub.; K&A) KONYAGI a-pasal (Ferry) MANDING-BAMBARA horo (JMD) wolo (JMD) wolodé (auctt.) wolofi (Aub.; K&A) woloténi (Aub.; K&A) MANDINKA woloforo (*def.*-o) (Aub.; K&A) MANINKA wolodé (Aub., ex K&A) wolofi (Aub., ex K&A) woloténi (Aub., ex K&A) 'SOCE' wolokoyo (Aub.; K&A) SERER mbulem (JB; K&A) WOLOF bobröb (K&A) robrob (K&A) röbröb (auctt.) **THE GAMBIA** FULA-PULAAR (The Gambia) bodi (DAP) MANDING-MANDINKA wolo (*def.*-o) (DAP) wolo-koyi (*def.* w. koyoo) = *white wolo* (DAP; DF) **GUINEA-BISSAU** FULA-PULAAR (Guinea-Bissau) culune (JDES) MANDING-MANDINKA sirafitom (JDES) **MALI** MANDING-BAMBARA bodèyi (GR) uolo (A. Chev.) uolodié, uolofi, uolotiéni (Aub.) wolo-dyè (GR) **UPPER VOLTA** FULA-FULFULDE (Upper Volta) loko (K&B) GRUSI ko (K&B) HAUSA baotché (Aub.) baotchi (Aub.) MANDING-BAMBARA wolo (K&B) MOORE kwɔtɔndé (FRI) wayaba (K&B) **IVORY COAST** BAULE koma (K&B; B&D) MANDING-DYULA waa iri (K&B) SENUFO-TAGWANA koô (K&B) **GHANA** AKAN-BRONG ɔngo (FRI) DAGAARI napéla (K&B) DAGBANI korli (JMD; FRI) GUANG-GONJA tepiala (AEK) MOORE kwodanga (JMD; FRI) kwodianga (JMD; FRI) NANKANNI petni (Lynn; FRI) pɔteri (Gaisie) SISAALA chinchɛpula (Blass) pagha (AEK) sugulugu (AEK; FRI) **TOGO** ANYI-ANUFO koma (Aub.) BASSARI kõbre (Gaisser) BATONNUN bélo (Aub.) MOBA potili (Aub.) TEM (Tshaudjo) sua (Volkens) **DAHOMEY** BATONNUN bélo (Aub.) GBE-FON alotu diésama (Aub.) **NIGER** GURMA lesakuala (Aub.) HAUSA baotché (Aub.) baotchi (Aub.) KANURI kumâda (Aub.) SONGHAI kàarèy-háŋ (*pl.* -á) = *crocodile's ear* (D&C) SONGHAI-ZARMA fàrkà-háŋ (*pl.* -á) = *donkey's ear* (auctt.) **NIGERIA** BEROM pyòso (LB) FULA-FULFULDE (Nigeria) boɗeyi, boɗehi (auctt.) boɗi (JMD) kuulahi (*pl.* kuulaji) (J&D) GOEMAI nkeng (JMD) HAUSA báúshe, baúsheè (auctt.) KANURI bàrbár (C&H) NUPE kpace, kpaci (Yates; Banfield) TIV kuegh YORUBA idi (auctt.)

A small tree with short bole to 10 m high, sometimes bushy, branching from the base (Senegal), of savanna woodland from Senegal to N and S Nigeria, and on into Cameroun and Ubangi-Shari.

The rough grey corky bark renders the tree somewhat fire-resistant (18). It occurs on the fixed sand-dunes of the Djolof area of Senegal (12), and may perhaps contribute to stabilization.

The wood is yellowish-brown, hard, tough and durable. The Hausa of N Nigeria make bows and walking-sticks or cudgels from the roots, which are first wrapped in leaves and scorched to remove the bark, and then soaked in oil. Any staff or cudgel called *gwalmi* or *kwaleru* (Hausa) is usually said to be made of wood of this species (5). In Niger the wood is used in house-construction and in revetting wells (17). For carpentry however it is inferior and the twisted grain of the sap-wood makes it difficult to finish (7, 16). It is valued for charcoal in the Sahel de Nioro of Mali (16). Fallen or felled trees rot to a light spongy reddish material with a strong smell which is sold in markets in Mali under the name of *uolo mogo* (Bambara) for burning as incense (4).

The root-bark is made into a decoction along with other drug-plants by the Baule of Ivory Coast for use by draught and enema for severe jaundice, and the lees left over are used in friction over the body (13), and the Maninka apply a compress of root-decoction to old sores which will not heal (3). In Casamance of Senegal the root-bark is considered cleansing and healing on refractory sores (9, 12). Fula hunters of N Nigeria wash in a bark-infusion especially when going to hunt the roan antelope (8). Bark-powder is emetic and charlatans in Senegal administer it in milk resulting in their 'patients' vomiting up a red liquid which the credulous believe is taking away the cause of their illness (12).

In Ivory Coast a root-decoction is taken by draught and by enema for diarrhoea and dysenteries. Children with these affections are given powdered root in boiled milk sweetened with honey. The powdered root is applied topically to sores and ulcers and is rubbed on the gums for toothache. The roots also enter into a prescription for leprosy (13), and in NE Nigeria the Jukun use the roots in a ceremonial treatment of syphilis (Meek fide 5). In Senegal a root-decoction is considered by the Fula active in the treatment of ascites, oedemas and of *diangara cayor* arising from tertiary syphilis. They also give an aqueous root-macerate to women with babies-at-breast suffering from eye-troubles (10, 12).

The roots are used as chew-sticks in the Ibadan area of Nigeria for their alleged effect against dental caries, but an examination of chew-stick material revealed no antibiotic activity (14). The roots, however, enter into treatments for skin-infections and a separate examination for anti-microbial activity showed action against *Sarcina lutea, Staphylococcus aureus* and *Mycobacterium phlei*, all Gram +ve organisms. No action was found on Gram −ve bacteria; nor on fungi (15).

A yellow dye is extracted from the beaten roots in Upper Volta and is used for dyeing cloth (6).

The pulverized leaves are used in N Nigeria on burns and bruises (5), and a general fetish of the Soce of Senegal is that no medicine is any good unless seven leaves of *T. avicennioides* have been added to it (11, 12). The leaves are often galled, called in N Nigeria *galulai-baushi* (Hausa), and these have medicinal use: reduced to ashes and mixed with roasted bulbs of *Crinum* (Amaryllidaceae), an ointment is prepared with fresh cow's butter for rubbing on rheumatic and swollen joints (5).

The leaves provides forage for sheep and a little for cattle in Senegal (1).

References:

1. Adam, 1966, a. 2. Aubréville, 1950: 127. 3. Bouquet & Debray, 1974: 67. 4. Chavalier, 1937, b: 170–1. 5. Dalziel, 1937. 6. Irvine 4697, K. 7. Irvine, 1961: 129–30. 8. Jackson, 1973. 9. Kerharo & Adam, 1963, b. 10. Kerharo & Adam, 1964, b: 581. 11. Kerharo & Adam, 1964, c: 324. 12. Kerharo & Adam, 1974: 357–8, with pharmacology. 13. Kerharo & Bouquet, 1950: 53. 14. Lowe, 29/9/76. 15. Malcolm & Sofowora, 1969. 16. Roberty, 1953: 452. 17. Robin, 1947: 58. 18. Savill & Fox, 1967, 103.

**Terminalia belerica** Roxb.

A tall tree found throughout India and in Malesia, introduced to Ivory Coast (1) and doubtless to other parts of the Region.

The fruit is one of the myrobalans of commerce but has a lower tannin content, 25–35%, than the fruit of *T. cheluba*. The tanning is used for tanning and dyeing, the resultant colour being black. The fruits are also used to make an ink in India. The bark also contains tannin and a quantity of an insoluble gum. The wood is apparently not durable though steeping in water evidently prolongs its life (2, 3, 4).

References:

1. Aubréville, 1959: 3, 69. 2. Burkill, 1935: 2136–7. 3. Greenway, 1941. 4. Hathway, 1959.

**Terminalia brownii** Fres.

FWTA, ed. 2, 1: 279.

A tree to 13 m high confined to rocky hill sites of N Nigeria but widespread in E Africa from Ethiopia to Tanganyika and Uganda.

The wood is yellow-brown and fairly hard (11). It is said to be resistant to boring insects (6) and termites (3). In Zäire the wood is used to make canoes (13), and for house-building, piles for grain-stores, beds, tool-handles, etc., in Uganda (3, 5). Axe-handles are made of it in Ethiopia (15), and in Kenya it is used for hut-pillars, pestles, and axe-handles (8), and in Tanganyika for grain-mortars (12) and to make a charcoal used in iron-smelting (3).

The bark is used for tanning skins for water-bottles in Somalia, and a bark-infusion is taken for tuberculosis (9). The powdered bark is infused and taken for chest-complaints – perhaps pneumonia (4). In Kenya a bark-decoction is taken for colds (1), or the bark is chewed for colds and an infusion drunk for fever (14). Similarly in Uganda the inner bark is beaten and infused in water for colds (7).

The inner bark contains a small amount of a sticky exudate (10) and the wood burns with a scented smoke with which Boran women perfume their hair (2).

References:

1. Bally 3656, K. 2. Bally 6391, K. 3. Brasnett 326, K. 4. Dowson 3, K. 5. Dyson-Hudson 125B, K. 6. Dyson-Hudson 195, K. 7. Dyson-Hudson 414, K. 8. Gardner 3608, K. 9. Gillett 4424, K. 10. Hughes 200, K. 11. Keay & al., 1960: 158. 12. Lewis 240, K. 13. Liben, 1968: 100. 14. Spjut 2860, K. 15. Turton 84, K.

**Terminalia catappa** Linn.

FWTA, ed. 2, 1: 277.

English: Singapore almond (Burkill); Indian almond; white bombwe.

French: badamier; amandier; amandier de Gambie (Berhaut).

Portuguese: amendoeira-da-Índia (Feijão).

West African: **SENEGAL** WOLOF gerté tubab = *nut of the European* (auctt.) **NIGERIA** EFIK ḿbàńsāñ m̀bàkárá = *groundnut (of) whiteman* (RFGA; Lowe)

A tree to 20–25 m high with whorled horizontal branching, a seashore plant native of Malesia and the western Pacific, introduced and present in the West African Region in higher rainfall areas from Senegal to W Cameroons.

The tree is an ideal shade and avenue tree with an unbuttressed trunk, open spreading crown and a rooting system that does not unduly disturb the foundations of buildings, sidewalks, roadside drains, etc. The foliage is decorative on the approach of wintering turning yellow and deep red, a feature unusual in tropical trees. It is widely planted as a shade tree in Asia, and is now so used in many parts of West Africa (Sierra Leone; Liberia, 11; Ivory Coast, 1; Ghana, 8; Nigeria, 6, 9; Gabon, 12).

The timber is reddish, with a crossed, curled, twisted grain, and is elastic. It is easy to work and seasons well. In the western Pacific it is used for purposes where toughness and durability are wanted, and in Malaya for boat and house-construction (2). The wood is said to have a pleasant smell (12). Chips of the wood soaked in water give a yellow solution (2) and the bark yields a black dye (13), and also contains tannins recorded at 11–23% used for tanning (5). The bark is used in Asian medicine as an astringent for dysentery, etc. (2). It has also been recommended for use as a decoction for gonorrhoea and leucorrhoea, bilious fever and stomach-cramp (10). The tree yields an insoluble gum of the bassoin type (4, 7, 13).

The leaves and flowers also contain tannin, and the presence of a sterol is reported (13). In the Philippines and S India sap from young leaves is made into an ointment for scabies, leprosy and other cutaneous diseases (10). The leaves when applied externally are refreshing and sudorific, and appear to exercise an anodynal effect on pain for they are used for headache, on rheumatic joints, or in an oily ointment for breast-pain (2, 10). In Nigeria the leaves macerated in palm-oil have been used as a remedy for tonsilitis (6). Leaf in Java has shown some antibiotic activity (13).

The fruit contains 6–20% tannin. The fleshy pericarp is eaten in Gabon (12). The kernel is edible, and has a very subtle flavour making it a delicacy. The husk however is tough, not amenable to separation, so that the kernel is difficult to obtain. The kernel contains 51–63% of a fixed oil known as Indian Almond oil, oil of Badamier, or from the Philippines as Talisay oil. Material from Ivory Coast has been analysed to contain glycerides of palmitic acid 34.4%, oleic acid 32.1%, linoleic acid 27.5% and stearic acid 6% (3). It closely resembles sweet almond, cotton seed, kapok and ground nut oils, which it might substitute for dietetic and other industrial uses.

References:

1. Aubréville, 1959: 3, 68. 2. Burkill, 1935: 2137–9. 3. Busson, 1965: 287–8, with chemical analysis. 4. Decary, 1946: 49. 5. Greenway, 1941. 6. Holland, 1908–22: 306–7. 7. Howes, 1949:

51, 79. 8. Irvine, 1961: 131. 9. Keay & al., 1960: 159. 10. Quisumbing, 1951: 657–9, with references and chemical analysis. 11. Voorhoeve, 1965: 85. 12. Walker & Sillans, 1961: 122. 13. Watt & Breyer Brandwijk, 1962: 195 with references.

**Terminalia cheluba** Retz.

A deciduous tree reaching to about 12 m high, of India, Ceylon and Burma, and introduced to Ivory Coast (1) and doubtless elsewhere in the Region.

The dried fruit of this species is the *myrobalans* of commerce though other species may be involved. Principal production is in India where the fruit is picked just yellow to quite yellow when it is completely ripe. It is sun-dried. Tannin content varies 28–46% in Indian material, apparently on place of origin. There it is a forest product. In E Africa, where it is grown under cultivation, fruits yield 19–51% (3). Leather produced from this tanning material is a soft yellow. In the British tanning trade it is mixed with wattle tannin to produce red-coloured leather (2, 3).

In Indian production, a small proportion of the fruit crumble internally to a powder which is made into ink (2).

The tree is grown in Zaïre as an ornamental (5). The timber is durable and is used in India for furniture making (2).

References:

1. Aubréville, 1959: 3, 69. 2. Burkill, 1935: 2139–40. 3. Greenway, 1941. 4. Hathway, 1959: with many references. 5. Liben, 1968: 102.

**Terminalia glaucescens** Planch.

FWTA, ed. 2, 1: 279. UPWTA, ed. 1, 81.

West African: **SENEGAL** MANDING-BAMBARA vara sa (JB) **GUINEA** FULA-PULAAR (Guinea) bori (Aub.) bosi (Langdale-Brown) daroth (Aub.) **SIERRA LEONE** MANDING-MANDINKA walisa (FCD) MANINKA (Koranko) wo (S&F) wɔlsa (NWT; KM) MENDE foni-baji (SKS) SUSU-DYALONKE korɛtɛ-na (FCD; KM) kɔrətɛ-na (FCD) **MALI** FULA-PULAAR (Mali) bori (Aub.) daroth (Aub.) MANDING-BAMBARA horo (JMD) wolo (JMD) **UPPER VOLTA** HAUSA kandaré (Aub.) kandari (Aub.) **IVORY COAST** BAULE koma (A. Chev.; Aub.) MANDING-DYULA en'ga (A. Chev.) MANINKA koma kièni (A&AA) woagna (B&D) **GHANA** AKAN-ASANTE ɔngo (auctt.) wangu (auctt.) BRONG no (BD&H) noakyi (auctt.) wongwong (TFC) wonwon (JMD) TWI ɔngo (auctt.) BAULE koma (FRI) DAGBANI korli (FRI) kworli (FRI) GBE-VHE dzogbedodo (auctt.) **TOGO** GBE-VHE epi (Aub.) YORUBA-IFE OF TOGO opeti (Volkens) **DAHOMEY** GBE-FON alotu diésama (Aub.) VHE epi (Aub.) YORUBA-NAGO idi, idi ọ̀dàn (Aub.) **NIGER** HAUSA kandaré (Aub.) kandari (Aub.) **NIGERIA** BEROM pyòso (LB) FULA-FULFULDE (Adamawa) kuulahi (*pl.* kuulaje) FULFULDE (Nigeria) baushishi (RES) boɗi (JMD) ɓooɗeyi (JMD) GOEMAI nkeng (JMD) HAUSA báúshe (auctt.) IGBO èdò (auctt.) KANURI bàrbár (C&H) NUPE kpace, kpaci (Yates) TIV kuegh (JMD) YORUBA idi (auctt.) idi-apata (JRA) idi-ọ̀dàn, idi: *Terminalia gen.;* ọ̀dàn: *open country* (auctt.)

A tree to 20 m high with short gnarled bole, dark grey, thick, deeply fissured bark, of the savanna woodland from Guinea to W Cameroons and on across Africa to Sudan and Ethiopia.

The corky bark confers some degree of fire-resistance to the tree (19). The wood is light yellowish-brown, hard, coarse and not easy to work (8, 12, 13). The wood is regarded in Nigeria as being of little value. It is not resistant to termites. It is used for canoes and for interior house-construction: rafters, forked branches for posts, and for these it is considered fairly durable (7, 8). In Sierra Leone it is used for its toughness for making pounding mortars (12). Charcoal made from this wood is used in Sudan for smelting iron (4).

The bark-decoction is taken as a purgative in Sierra Leone (12). Bark is chewed in S Nigeria as a laxative and alterative (2). The root-bark enters a treatment for burns and for sores in general in Ivory Coast (6). In Ubangi-Shari root-bark is used on wounds and is said to have an effect like tincture of iodine (5). The bark provides a tanning material in Ethiopia (18).

The roots are chewed in S Nigeria as a laxative and alterative (2). A root-decoction is taken in Ivory Coast-Upper Volta by draught and enema for diarrhoeas and dysenteries and the powdered root is added to milk for treating babies with these affections. Powdered root is applied to sores and ulcers, and to gums for toothache, and the roots are put into prescriptions for leprosy (14) and syphilis (5). The Baule of Ivory Coast take by draught and enema a root-decoction to which other drug-plants have been added for jaundice, and friction the lees over the body (14). The roots are used in S Nigeria as chew-sticks and are held to have aphrodisiac properties (2). A trace of alkaloid has been detected in them (1). Examination for alleged effect against dental caries of 13 plant materials used as chew-sticks in the Ibadan area of Nigeria showed the roots of this plant to be the only one having an antibiotic activity, and this against *Staphylococcus aureus* (15).

Root-wood is tough and is used in S Nigeria to make bows and walking-sticks. A dye is obtained from the roots in Sudan which is used to darken cloth (3).

Young leaves in macerate are used in the soudanian region as a cough-remedy (5) and a leaf and bark-decoction is similarly used in Sierra Leone (10). A leaf-decoction is used in Ivory Coast to wash the head for headache, and is taken internally for stomach-ache (6).

In Ubangi-Shari the detached bark is used to make beehives (5). On human values the flowers in Sierra Leone have an unpleasant smell (10, 11) and an odour resembling fermenting *Parkia* (Leguminosae: Mimosioideae) seeds which nevertheless is attractive to bees for it is a favourite bee plant (16), and is visited between dawn and noon (17). The honey produced is said to be particularly good (9).

The fruit is used in Nigeria as a vermifuge (2).

The tree in Guinea is the host-plant of a lepidopterous larva which spins a silk cocoon (8).

References:

1. Adegoke & al., 1968. 2. Ainslie, 1937: sp. no. 332. 3. Andrews A 1455, K. 4. Andrews A 1522, K. 5. Aubréville, 1950: 127. 6. Bouquet & Debray, 1974: 67. 7. Chalk & al., 1933: 39. 8. Dalziel, 1937. 9. Deighton 4223, K. 10. Deighton 4817, K. 11. Deighton 4821, K. 12. Irvine, 1961: 131–3. 13. Keay & al., 1960: 157–8. 14. Kerharo & Bouquet, 1950: 53–54. 15. Lowe, 29/9/76. 16. Miszewski 1, K. 17. Miszewski 59, SL. 18. Mooney 5460, K. 19. Savill & Fox, 1967, 103.

## **Terminalia ivorensis** A. Chev.

FWTA, ed. 2, 1: 279. UPWTA, ed. 1, 81–82.

English: black afara (Chalk & al., Kunkel); black bark (Dalziel, Kunkel, Irvine); black barked terminalia (Chalk & al.,); brimstone wood (Chalk & al.); satinwood (Chalk & al., Dalziel, Kunkel); shingle wood (Dalziel, Irvine, Kunkel); yellow terminalia (Chalk & al., Dalziel, Irvine); yellow pine (Kunkel).

French: bois satiné (Kerharo & Boquet); framiré (auctt).

West African: **GUINEA-BISSAU** KPELLE bassi (RS) **SIERRA LEONE** GOLA badi (FCD) bari (FCD) KISSI basio (FCD; S&F) KONO g-basi (auctt.) KRIO ronko (FCD; S&F) LIMBA (Tonko) koroŋgo (FCD) LOKO bahi (FCD; S&F) MANDING-MANINKA (Koranko) fira-wasɛ (S&F) MENDE baji (*def.*-i) = *yellow* (auctt.) SUSU woli (FCD) TEMNE *an*-roŋko (FCD; S&F) VAI basi (FCD) **LIBERIA** DAN bai-ti = *black bai* (AGV) KRU-GUERE (Krahn) blie (GK; AGV) BASA baye (C; AGV) MANO bai (AGV) MENDE baji (AGV) **IVORY COAST** ABE m-boti (auctt.) ABURE kahuri (A. Chev.) kauri (A. Chev.; K&B) yapi (A. Chev.; K&B) AKYE buma (LC) buna (A. Chev.; K&B) mboti (auctt.) yabi (A. Chev.) ANYI anhidja (A. Chev.; K&B) framiré (auctt.) BAULE diba bulu (B&D) flaméné (B&D) DAN banidi (auctt.) GAGU bulu (B&D) KRU-BETE blié (A. Chev.) dibebrui (A.Chev.; K&B) GUERE blié (K&B) GUERE (Wobe) blié (auctt.) KULANGO (Bondoukou) anhidja (A. Chev.; K&B) mbonoi-cauri (LC) onidjo (A. Chev.) tuhidja (LC) uhidji (LC) KWENI brogba (Aub.; RS) brogha (K&B) gbeï (B&D) KYAMA degonbroya (auctt.) MANDING-BAMBARA féla (auctt.) NZEMA framiré (A. Chev.) 'SOUBRE' buri (auctt.) **GHANA** VULGAR emeri (LC; DF) idigbo (auctt.) ADANGME emiri (CV) ADANGME-KROBO emere (auctt.) AHANTA amire (auctt.) AKAN-ASANTE amire (FRI) emeri (auctt.) emil (auctt.) ASANTE

(Denkyera) amire, emeri (Enti) BRONG noakyi (CV; FRI) FANTE amire (FRI) emeri (auctt.) TWI amire (FRI) emire (auctt.) WASA amire (FRI) emeri (auctt.) emri (LC) ANYI ɛ-frammire = *black T. superba* (FRI) ANYI-AOWIN ɛ-frameri (auctt.) SEHWI ɛ-frammire *black T. superba* (FRI) efremeli (CJT) farayemile *black afaraa, T. superba* (auctt.) BAULE koma (FRI) DAGBANI korli (FRI) kurli (FRI) kworli (FRI) GA emeri (auctt.) GBE-VHE dzogbedodo (auctt.) VHE (Pecí) emere (FRI) NZEMA ɛ-frammire = *black T. superba* (FRI) emere (auctt.) ɛsiɛ (FRI) faraeneri (CJT) **NIGERIA** BOKYI kekange (auctt.) EDO ẹ̀ghọ̀n-nébì (auctt.) ẹ̀ghọ̀n-nékhũi, nékhũí: *black* (JMD; Lowe) EFIK àfiã (JMD; KO&S) EJAGHAM (Keaka) nkombe (Kennedy) nkondi (Kennedy) okpoha (Kennedy) IGALA uji-oko (H-Hansen) IGBO awunshin (JMD; KO&S) awunshin-oji (JMD) IJO-IZON (Kolokuma) dị́ị̀ (KW) YORUBA afàrà dúdú, afàrà: *Terminalia*; dúdú: *black*; i.e. *black Terminalia* (LC; JMD) idi (Sankey) idígbó, idi-igbó, igbó: *forest* (auctt.) opepe (Punch) YORUBA (Ijebu) epepe, oweive (auctt.) YORUBA (Ikale) epepe (LC) oweive (LC)

A massive forest tree to 50 m high, with a strong clean bole, to 35 m long by 5 m girth, broad, blunt buttresses, occurring principally in the partly deciduous forest zone between the true evergreen rain-forest and savanna woodland, from Guinea to W Cameroons.

It is one of the principle timber trees of West Africa with numerous expatriate names as indicated above, and traded in anglophone territories as *idigbo* (Yoruba) and in francophone territories as *framiré* (Anyi). The timber is yellow to light brown, moderately hard, heavy, coarse-textured, grain straight or wavy, fairly strong, easy to work. It seasons easily, shrinking a little. The timber is fairly resistant to fungal and insect attack except by pinhole borers and termite in the sap-wood. The wood is said to be resistant to penetration by preservatives (4, 7, 9, 12, 14).

The timber is used for primary construction work and general utility purposes for which it is available in large baulks; for house-building, planking, door and window-frames, etc. It saws well, nails and glues well, and its decorative graining makes it a desirable wood for high-class joinery, cabinetry, furniture and domestic uses. It is suitable for plywoods and veneers. The wood splits easily and is much used in Ghana for roof-shingles which are said to have a life of 15–20 years. (1, 2, 4, 5, 7, 9, 12–15.)

Natural forest stands have, where accessible, been much depleted. The tree is sun-loving, fast growing and has a vigorous natural regeneration in secondary forest. Furthermore the tree appears amenable to plantation husbandry. There have been problems of insect-damage in monoculture, but the 'taungya' system with intercropping or avenue planting in forest or mixed plantations offer prospect for successful cultivation (2, 8, 15). The tree is naturally not longlived, and in old age the heart-wood becomes inferior, hollow or brittle (9, 14). Cultivation should ensure felling at the optimum age.

Other local uses of the wood are to make frames for mud-huts, and the logs for dug-out canoes and rice mortars (7). In Ghana its use for canoes is restricted to the SW of the country in the absence of the preferred *Triplochiton scleroxylon* K. Schum. (14), and in Sierra Leone the life-span of canoes made of it is given as three years (12). Large boles in Sierra Leone are hollowed out to make long drums and the timber is made into knee-joints for seagoing boats (12). The open verticillate nature of the branching makes the living tree useful as a shade tree for cocoa in Ivory Coast (2), and avenue plantation culture in Sierre Leone lends itself to intercropping with shade-requiring species (12).

A yellow dye can be obtained from the wood (9) and more especially from the bark and is used for dyeing cloth and fibres for basket-work, hammocks, etc. This dye is known as *kuntunkuri* (Asante) in Ghana (4, 7, 10–13). A bark decoction or macerate yields a red liquor which is rich in tannins and is used for treating sores (3, 6, 7). The powdered bark in Ivory Coast is dusted over ulcers (10) and pulped it is rubbed over areas of muscular and rheumatic pain (3). In Ghana the bark is used for a condition called *ŋnewee* (Twi) which may be rheumatism (9).

Sap expressed from young leaves is applied to cuts in Sierra Leone (12) and is taken in draught with a bark-decoction by enema for blennorrhoea and kidney disorders, and as an aphrodisiac (3).

The wood has been found to contain a saponin which may induce allergy in persons working with it (3).

References:

1. Anon, s.d. 2. Aubréville, 1959: 3, 69. 3. Bouquet & Debray, 1974: 67. 4. Chalk & al., 1933: 35. 5. Cole, 1968. 6. Cooper 370, K. 7. Dalziel, 1937. 8. Foggie, 1957: 142. 9. Irvine, 1961: 133–4. 10. Kerharo & Bouquet, 1950: 54. 11. Lane Poole 276, K. 12. Savill & Fox, 1967, 99–101. 13. Schnell, 1950: 238, 260. 14. Taylor, 1960: 153. 15. Voorhoeve, 1965: 85.

**Terminalia laxiflora** Engl.

FWTA, ed. 2, 1: 279. UPWTA, ed. 1, 83, as *T. sokodensis* Engl.

West African: **SENEGAL** BASARI kobri (Aub., ex K&A) DIOLA (Fogny) kavâga (K&A) FULA-PULAAR (Senegal) kulémi (K&A) MANDING-'SOCE' wolokoyo (K&A) SERER balak (K&A) SERER-NON (Nyominka) balak (K&A) **THE GAMBIA** DIOLA (Fogny) kavâga = *tall grower* (DF) **MALI** 'SENUFO' koma ba (Aub.) koma messin (Aub.) **UPPER VOLTA** 'OUROUDOUGOU' zango pas (Aub.) zango pélé (Aub.) **IVORY COAST** AKAN-BRONG won (K&B) KULANGO honguigon (K&B) 'SENUFO' koma ba (Aub.) koma messin (Aub.) **GHANA** AKAN-ASANTE ɔngo (FRI) BRONG kuku(k)we (BD&H; FRI) ngu (CV) no (BD&H) ɔŋŋo (FRI) **TOGO** BASSARI kobri (Aub.) TEM suha (Aub.) **NIGER** GURMA tipanpansanfati (Aub.) **NIGERIA** BEROM pyòso (LB) HAUSA fárín báúshe, fárín baúsheè (auctt.) zindi (Lowe; KO&S) YORUBA idi inu ọ̀dan (IFE) idi-ọ̀dan (auctt.) orin idi (IFE)

A tree to 12 m high by nearly 1 m girth, short usually crooked bole, with dark grey, deeply fissured scaly bark, of the savanna woodland and in much the same sites as *T. macroptera* Guill. & Perr., with which it is frequently confused, occurring from Senegal to N and S Nigeria and W Cameroons, and across Africa to Sudan.

The tree is fire-resistant because of its thick corky bark (10). The wood is hard, heavy and coarse: sap-wood yellow with greenish tinge, about 5 cm wide; heart-wood pale brown (1). The tree is reported in Sudan to yield a gum (9).

The bark is used at Bondoukou, east Ivory Coast, to treat yaws, the pustules being washed with a hot bark-decoction to which citron is added, and then they are covered with powdered bark (8).

The roots give a brown dye used for clothing (2, 3).

The leaves enter a culinary preparation in Casamance, Senegal, against dysentery: fresh leaves from which the sap has been expressed are mixed up in cow's butter and the mixture cooked with rice or millet (4). The Fula of Senegal consider the plant anti-dysenteric. The leaves are used, again only after the sap has been expressed, or alternatively the roots can be used after the outer layer of bark has been removed (5, 7). In the Isle of Saloum, Senegal, various parts of the plant are used for tubercular coughing with vomiting and bloody sputum (6, 7).

Various substances have been identified in the plant and aqueous extract of stem bark has shown some antibiotic action against *Sarcina lutea* and *Staphylococcus aureus* (7).

References:

1. Chalk & al., 1933: 40. 2. Dalziel, 1937. 3. Irvine, 1961: 134. 4. Kerharo & Adam, 1963, b. 5. Kerharo & Adam, 1964, b: 582. 6. Kerharo & Adam, 1964, c: 324. 7. Kerharo & Adam, 1974: 358–9, with phytochemistry and pharmacology. 8. Kerharo & Bouquet, 1950: 54. 9. Murid L/150, K. 10. Savill & Fox, 1967, 103.

**Terminalia macroptera** Guill. & Perr.

FWTA, ed. 2, 1: 279. UPWTA, ed. 1, 82.

French: badamier du Sénégal (Berhaut; Kerharo & Adam).

West African: **SENEGAL** BANYUN kifudal (K&A) sisal ɗafasô (Aub.; K&A) BASARI a-peùlá (Ferry) BEDIK ga-pedã (K&A; Ferry) CRIOULO masiti (K&A) DIOLA bu âga (JB; K&A) buhinkabu (Aub.) kauangakon (Aub.) masiti (K&A) DIOLA (Brin/Seliki) bubum émâdé (K&A) bufumay sibé

(K&A) bulibam (K&A) DIOLA-FLUP busalaba (K&A) buson uyag = *fickle wood: because it does not burn properly* (K&A); *or* = *firewood hole* (DF, The Gambia) FULA-PULAAR (Senegal) bodévi (Aub.; K&A) bodey (K&A) bodi (K&A) KONYAGI a-ntud (Ferry) apela (K&A) MANDING-BAMBARA olofira (JB) wolo (K&A) wolo ba (GR) wolo muso (Aub.; K&A) MANDINKA uolossa (Aub.) wolo (*def*.-o) (K&A) wolo ɗambo (K&A) woloforo (*def*.-o) (K&A) MANINKA kasaulé (Aub.; K&A) 'SOCE' wolokoyo (K&A) MANDYAK biɗabar (Aub.; K&A) MANKANYA blopo (K&A) bolobo (K&A) SERER balat (JMD) m-balak (auctt.) SONINKE-SARAKOLE furbé (K&A) kamba (K&A) WOLOF guy dema (JB; K&A) wolo *from Mandinka* (auctt.) **THE GAMBIA** DIOLA (Fogny) bu âga (DF) bujingkabo (DF) FULA-PULAAR (The Gambia) bodehi (DAP) MANDING-MANDINKA wolo (*def*.-o) (auctt.) WOLOF wolo *from Mandinka* (DAP) **GUINEA-BISSAU** BALANTA fadi(h) (JDES; GES) CRIOULO macête (GeS; JDES) FULA-PULAAR (Guinea-Bissau) bodé (JDES; EPdS) bói (auctt.) MANDING-MANDINKA hóló (GeS) hóló-fóro (JDES) MANDYAK betáli (JDES) betcháli (JDES) betêlêdji (JDES) braqui (JDES) MANKANYA blopo (Aub.) bolóbô (untulam) (JDES) PEPEL n'tula (GES) n'tulam (untulam) (JDES) **GUINEA** FULA-PULAAR (Guinea) bodévi (Aub.) MANDING-MANINKA woro-ba (JMD) **MALI** FULA-PULAAR kulahi (GR) PULAAR (Mali) bodévi (Aub.) bodeyi (JMD) MANDING-BAMBARA horo (JMD) uolo musso (Aub.) wolo (JMD) 'SENUFO' mango figué (Aub.) **UPPER VOLTA** DAGAARI dagnéla (K&B) kotéli (K&B) kwatri (K&B) HAUSA bauché bochy (K&B) MANDING-BAMBARA uolo (Bégué; K&B) MOORE kondré (K&B) kwadianga (K&B) kwodaga (K&B) **IVORY COAST** 'SENUFO' mango figué (Aub.) **GHANA** DAGAARI kawtri (AEK; FRI) kwatiri (FRI) DAGBANI korli (FRI) korli-nyao (FRI) kurli-langban (FRI) kworli (FRI) kworli-ngbandi = *kworli with skin, because of the large leaves* (FRI) GUANG-GONJA pàkpáká (Rytz) MOORE kwodaga (AEK; FRI) kwodianga (AEK; FRI) NABT potiri (BD&H; FRI) **TOGO** BATONNUN béro (Aub.) SOMBA mukindimu (Aub.) TEM (Tshaudjo) soria-dau (Volkens) sua-dau (Volkens) **DAHOMEY** BATONNUN béro (Aub.) GBE-FON pavu (Aub.) SOMBA mukindimu (Aub.) **NIGER** FULA-FULFULDE (Niger) bodévi (Aub.) HAUSA bauché bochy (K&B) **NIGERIA** BEROM pyòso (LB) FULA-FULFULDE (Adamawa) kuulahi (*pl.* kuulaje) FULFULDE (Nigeria) ɓoodi (*pl.* ɓoode); ɓoodi (MM) kamdare (Lely) kandarahi (JMD) HAUSA ɓaushe (MM) kandare, kandari, kwandare, kwandari, kalangon daji (ZOG) kwandare (JMD; KO&S) HAUSA kandare (JMD) YORUBA orin idi ọ̀dan (IFE)

A tree to about 13 m high, short bole, thick black, deeply fissured bark, of the savanna woodland usually in moist or occasionally flooded sites, often gregarious, sometimes dominant, e.g. Nasia Swamp in Ghana where it was recorded as 70% of the stand (5); occurring from Senegal to N and S Nigeria and across Africa to Sudan and in Uganda.

The tree is fire-resistant because of its thick corky bark (24). It has been recommended as a desirable species for reafforestation especially where it was formerly abundant (6, 11).

The wood is yellowish or light brown, hard, coarse uneven grained, rarely straight and difficult to work (6, 11, 12)), though it is said to be valued for carpentry in Guinea-Bissau (19), and to have an attractive appearance. Its timber in Sudan has been traded as 'teak' (4). It has a reputation for resistance to borers and termites (2), and is used for house-posts, frames of boats (2, 6, 11) and is said to produce a very good charcoal (2, 6, 11, 23). The heart-wood is scented and is used in Senegal in a perfume called *amulguéné* (Wolof), *amu* (Fula–Fouta Toro), and *suma-diala* (Maninka) (2). The wood as fuel is said to be avoided by the Mandinka of The Gambia as the fumes are alleged to affect young children, the antidote to which is a decoction made from the same tree's roots (7). The thicker branches are hollowed out by the Acholi of Uganda to use as beehives and storage boxes (18).

The stem and root bark contains tannin. It is haemostatic and cicitrisant. Powdered bark is used in Nigeria to treat piles, diarrhoea and dysentery (21). In Ivory Coast-Upper Volta a decoction is used to cleanse sores, and the Lobi of Upper Volta administer an enema for piles (20). The root-bark yields a yellow-brownish dye used for dyeing cloth in Guinea (22), Ivory Coast-Upper Volta (20) and Ghana (11). The bark of many Indian species is said to contain cardiac-stimulant substances (21).

A root-decoction is considered a strengthening tonic in Senegal and is given in treating many illnesses causing debilitation and depression; also for fevers, jaundice, syphilis and as an aphrodisiac. The roots are diuretic, helpful in treating urethral discharge and urinary trouble especially in women during pregnancy (14–19). A root-decoction is considered haemostatic and is used topically on wounds (16), and to be purgative (13). Roots chopped up into small pieces and macerated to a lotion are used in the soudanian region for sprains, and a decoction is given to fractious cattle and to epileptic persons (2).

The tree is said to contain a gum (4, 10).

The leaves are little grazed by cattle (1). Like the bark, they contain tannin. A hot decoction is used as a wash and fumigation in fevers, and can be used for ringworm and other skin diseases (6). The plant (? part) is recorded as used to treat *culfetin*, a sort of leprosy, in Senegal (13). In Guinea-Bissau a leaf-infusion is used as a depurative (9). The leaves provide a deep-black dye used for cloth (2), and to make ink (6).

Fruits which have been galled are used as an astringent for dysentery (6).

Flavones and steroids and other substances are reported present in the plant (3, 19), Chlorogenic acid in the leaves is perhaps responsible for cholagogic action.

References:

1. Adam, 1966, a. 2. Aubréville, 1950: 127. 3. Bouquet & Debray, 1974: 67. 4. Broun & Massey, 1974: 109–10. 5. Burtt Davy & Hoyle, 1937: 36. 6. Dalziel, 1937. 7. Dawe 15, K. 8. Eggeling 1339, K. 9. Gomes e Sousa, 1930: 56. 10. Howes, 1949: 79. 11. Irvine, 1961: 134–5. 12. Keay & al., 1960: 155. 13. Kerharo, 1967. 14. Kerharo & Adam, 1962. 15. Kerharo & Adam, 1963, b. 16. Kerharo & Adam, 1964, a: 428. 17. Kerharo & Adam, 1964, b: 582. 18. Kerharo & Adam, 1964, c: 324–5. 19. Kerharo & Adam, 1974: 359–60. 20. Kerharo & Bouquet, 1950: 55. 21. Oliver, 1960: 39, 87. 22. Pobéguin, 1912: 27. 23. Roberty, 1953: 452. 24. Savill & Fox, 1967, 103.

**Terminalia mantaly** H. Perrier

French: mantaly (from a Madagascan name).

The tree 10–20 m high, endemic of Madagascar, introduced into Senegal (1).

The bark and wood are used in Madagascar for dyeing and in treatment of dysentery (2).

References:

1. Berhaut, 1967: 233. 2. Perrier de la Bâthie, 1954: 50–51.

**Terminalia mollis** Laws.

FWTA, ed. 2, 1: 279. UPWTA, ed. 1, 83 as *T. reticulata* Engl.

West African: **GHANA** AKAN-ASANTE ɔngo (auctt.) BRONG kuku(k)we (BD&H; FRI) DAGBANI kworli (CV; FRI) **NIGERIA** HAUSA baúshin giíwaá (ZOG)

A small tree to 10 m high by 60 cm girth, low branching, crooked bole with rough corky bark, of the savanna woodland and recorded only from Guinea, Sierra Leone, Ivory Coast, Ghana, Togo and Nigeria, but widespread elswhere in tropical Africa.

The wood is hard, and in southern Zaïre is used to make axle-blocks of wheels and a good quality charcoal (3). The root, beaten in water, yields a crude red-brown dye used for cloth in Ghana. The colour darkens in time (1, 2).

In Tanganyika the plant (? bark) is used to treat diarrhoea and gonorrhoea (4).

References:

1. Dalziel, 1937. 2. Irvine, 1961: 135. 3. Liben, 1968: 98–99. 4. Tanner 5111, K.

**Terminalia scutifera** Planch.

FWTA, ed. 2, 1: 277. UPWTA, ed. 1, 83.

English: beach oak (Sierra Leone, Savill & Fox).

West African: **GUINEA-BISSAU** BIDYOGO cabor (JDES) CRIOULO salangue (JDES; EPdS) TEMNE *an*-taraba (CHOP) **SIERRA LEONE** BULOM (Sherbro) rak-lɛ (FCD; S&F) KRIO bich-ok = *beach oak* (FCD; S&F) ok = *oak* (S&F) MENDE baji (L-P)

A tree to 15 m high with a gnarled bole branching from low down, of the foreshore and upper littoral of Guinea-Bissau, Guinea and Sierra Leone, and withstanding inundation at spring high-tides.

The wood is used in Sierra Leone to make knee-pieces and curved parts of the hulls of seagoing fishing-boats (3, 4).

A yellow dye is obtained from the bark (1, 2, 3) in Sierra Leone and a wash for sore feet (3).

References:

1. Dalziel, 1937. 2. Lane Poole 87, K. 3. Savill & Fox, 1967; 101, 103. 4. Unwin, 1920: 56.

**Terminalia superba** Engl. & Diels

FWTA, ed. 2, 1: 277. UPWTA, ed. 1, 83.

English: afara (from Yoruba); white afara; ofram (from Asanti, Twi); shingle wood; yellow pine; white mukonja (W Cameroons); Congo walnut.

French: franké (from Anyi); limba (ex Congo); limbo (ex Gabon); limbo blanc, or limbo clair (with light-coloured wood); limbo noir (with dark-coloured heart-wood); noyer du moyambe (i.e. Congo walnut, with dark-coloured heart-wood).

West African: **GUINEA-BISSAU** KISSI mamno (RS) KONO g-bê (RS) KPELLE bassi (RS) LOMA bazi (RS) MANO gbeï (RS) **SIERRA LEONE** GOLA jiekũ (FCD) KISSI kongo (S&F) KONO kone (S&F) MANDING-MANINKA (Koranko) bese (S&F) kumburibe (S&F) MENDE baji (LC) gbili (L-P; LC) kojaga (*def.* kojagei) (auctt.) TEMNE *an*-rɛn (FCD; S&F) VAI salafo (FCD) salafuro (FCD) **LIBERIA** DAN going (AGV) guwing (AGV) g(u)weng (GK) KRU-GUERE (Krahn) blie (GK) BASA baye (C&R) MENDE kojaga (C&R) **IVORY COAST** ABE pai (auctt.) AKYE fé (Aub.; RS) té (A. Chev.) ANYI franké (auctt.) BAULE balé (A. Chev.) fra (A. Chev.) tra (Aub.) DAN kobaté (Aub.; RS) KRU-BETE kom'brihi (A.Chev.) sanuhé (A.Chev.) solo (Aub.; RS) GUERE (Wobe) saha(i)n (auctt.) KULANGO (Bondoukou) fram (A. Chev.; LC) KWENI gbeï MANDING-MANINKA guè (B&D) 'SOUBRE' saro (Aub.) **GHANA** VULGAR afara (LC) ofram *from Akan* (LC; DF) ADANGME-KROBO ofram (LC; FRI) AKAN-ASANTE framo (BD&H; FRI) framoo (auctt.) óframfram (auctt.) FANTE ɔ-fram (auctt.) TWI ɔ-fram (auctt.) WASA fram (auctt.) framoo (Enti) ɔfram(o) (auctt.) ANYI franko (FRI) ANYI-AOWIN farayen (auctt.) ɛ-frameri (auctt.) ɛ-frany (FRI) SEHWI afraa (auctt.) ɛ-frany (FRI) BAULE tra (FRI) GA ofram (FRI) GBE-VHE kɛgblale (auctt.) VHE (Pecí) afɔdɔnkɔ (auctt.) dɔnkɔ (JMD) frango (auctt.) NZEMA frane (auctt.) **NIGERIA** BOKYI kèpurudotkekange (Kennedy) EDO ẹ̀ghọ̀n-nófũá, nófũá: *white; referring to the flaking bark* (auctt.) ẹ̀ghọ̀nọ́fũa (Amayo) EFIK àfia ètò = *white tree; as the wood is whiter than that of T. ivorensis* (auctt.) IGALA uji-oko (H-Hansen) IGBO èdò (auctt.) èdò ọ́chá = *white edo* (LC) IGBO (Amufu) ojiloko (DRR) IGBO (Nkalagu) ojiroko (DRR, JMD) IGBO (Owerri) èdò ọ́chá = *white edo* (DRR) IJO-IZON (Egbema) apaụpaụ tịín (Tiemo) IZON (Kolokuma) ámbị̀lárị̀ tìn = *umbrella tree* (KW) ISEKIRI egonni (KO&S) egonni (auctt.) NUPE eji (auctt.) URHOBO unwon ron (auctt.) YEKHEE aghoin (Kennedy; LC) YORUBA afaa (LC; JMD) áfàrà (auctt.) **WEST CAMEROONS** DUALA bòkɔ̀mɛ̀, mùkonya (Ithmann) KPE djombe (JMD) KUNDU bokombe (JMD)

A magnificant tree with long straight bole, with heavy 'plank' buttress to about 2.5–3.0 m from the ground, whorled branches, spreading open crown, of high-forest, savanna woodland and farmlands wherever rainfall exceeds about 1500 mm and soil is not waterlogged, emergent in the closed forest to 30 m, even to 45 m in height by 4–5 m in girth, common throughout, sometimes gregarious from Guinea to W Cameroon and on to Zäire and Angola.

The timber is light yellowish-brown, of lighter colour than that of *T. ivorensis* A. Chev., hence various names referring to its whiteness. Sap-wood and heart-wood are not normally distinct but there are forms with black streaks and, especially from Zäire, trees with a walnut-like heart. The wood is light medium density, rather soft, saws well and is easy to work and takes a good polish. It is not resistant to decay nor to termites and pinhole borers, thus its value for outdoor work is limited. Logs must be removed quickly after felling and timbers given a protective treatment after sawing. The wood is suitable for numerous indoor applications: for furniture where strength is not required, match-boxes and match-splints, door-panels, office-fittings, shingles, doorposts, packing crates, planking, veneers, plywood, etc. (3–6, 9, 12–14, 17, 19). It is a valuable substitute for softwoods imported into the Region, and for many years especially from the Camerouns, Congo and Angola there has been an export trade to Europe. The tree has shown itself very amenable to plantation management. It is now extensively planted in a number of West African states (6, 12, 15, 17, 19).

In addition to the variation in the timber already indicated above, in the heartwood of some trees there is a tendency to brittleness, and though in general the wood is reported liable to termite attack, some degree of immunity is recorded in material from the Ivory Coast (8), and the active chemical is thought to be the cause of allergy manifest by persons working on the wood. These variations suggest that there is scope for selection work, even for breeding, since some defects in meeting commercial standards seem to be partly genetic. It is suggested that seed be collected for planting from selected mother-trees after felling and examination of their timber (20). Light-coloured wood is the most valuable.

Not all the usual applications of wood are possible with the timber of this species. It is not recommended for turnery in Nigeria (4) though it is so used and recommended in Ghana (5, 19). In paper-making it requires a higher than normal quantity of soda, and the bleached paper is dark (9), though the rough paper is fairly strong (10).

Other uses of the wood are as firewood, canoes, paddles, framework of mud houses, and occasionally for coffins, boxes, stools and domestic utensils (12).

The bark contains a yellow (12, 14, 18) or black (21) dye which finds local use, as, for example in Congo to blacken *Pandanus* (Pandanaceae) leaf-matting (7) and in the Camerouns to stain vegetable fibres.

A bark-infusion or decoction is astringent (2) and is administered in Ivory Coast (8) and in Congo as an anti-dysenteric. It is given in Ivory Coast as an antemetic (8). In Congo it is held to have emetic and expectorant properties and is administered alone or in mixtures for pneumo-bronchial affections, and is taken by women for sterility or threatened miscarriage or other ovarian troubles (7). The inner-bark in aqueous macerate is prepared as a mouth-wash for gingivitis and thrush (7). A bark-macerate is used in Liberia as an antiseptic on sores and wounds (11) and in Congo on swellings and areas of general pain (7).

A trace of alkaloid is reported present in the bark (1).

The powdered leaf is used in Nigeria to assist in childbirth (2).

The roots contain resin and tannins and are said to be used in Nigeria as a laxative (16).

References:

1. Adegoke & al., 1968. 2. Ainslie, 1937: sp. no. 333. 3. Anon, 1965, a: 1. 4. Anon, 1965, b. 5. Anon s.d. 6. Aubréville, 1959: 3, 70. 7. Bouquet, 1969: 89. 8. Bouquet & Debray, 1974: 67. 9. Chalk & al., 1933: 30. 10. Coomber, 1952. 11. Cooper & Record, 1931: 35–36., with timber characters. 12. Dalziel, 1937. 13. Foggie, 1957. 14. Irvine, 1961: 135–6. 15. Kunkel, 1965: 192. 16. Oliver, 1960: 39. 17. Savill & Fox, 1967, 102–3. 18. Schnell, 1950: 238. 19. Taylor, 1960: 155. 20. Voorhoeve, 1965: 89. 21. Walker & Sillans, 1961: 122–3.

## COMMELINACEAE

**Aneilema aequinoctiale** (P. Beauv.) Kunth

FWTA, ed. 2, 3: 30.

West African: **NIGERIA** YORUBA (Ondo) ẹfiajija (AJC)

A scrambling or erect robust herb to 2 m high, of forest and sometimes farmland from Liberia to S Nigeria, and widely dispersed in forest regions of tropical and S Africa.

The plant is grazed by ungulates in Kenya (2). In the Ondo area of Nigeria the roots are used to make a wash for babies (1). There is medicinal use of the plant for colds in Zanzibar (3).

References:

1. Carpenter (Faqubemi) 1077, UCI. 2. Magogo & Glover 487, K. 3. Williams, R. O., 1949: 122.

**Aneilema beninense** (P. Beauv.) Kunth

FWTA, ed. 2, 3: 31. UPWTA, ed. 1, 464.

West African: **SIERRA LEONE** LOKO unkɛngi (NWT) wukɛŋgi (NWT) SUSU kurukɔre (NWT) **GHANA** AKAN-ASANTE akotia (JMD) Nyame bewu na mawu (Enti) onyamenwuna mɛwu (FRI) TWI akotia (JMD) O-nyame nwu na mɛwu (JMD) **NIGERIA** EDO óhīovbù (JMD) IGBO (Agukwu) ọ̀bọ́-ọ̀gụ̀ úkú (NWT; JMD) óké òbá ògụ̀ ójīī (NWT; JMD) YORUBA gòdọ̀bọ̀-fúnfún (RJN) gòdọ̀bọ̀-odò, odò: *stream* (Millson) itọ́-ìpére (Macgregor)

A robust straggling herb to 1 m high, of the rain-forest throughout the Region from Senegal to W Cameroons and Fernando Po, and widespread in the rain-forest regions of tropical Africa.

A decoction of the leaves is used in S Nigeria as a mild aperient, especially for children (1). The leaves are pounded up for use as a stronger laxative in enemas in Nigeria (1) and Ghana (6, 7). The sap is recorded as used in Ivory Coast in treatment for amenorrhoea, and the plant is prescribed in steam-baths to induce sleep in febrile conditions (4). A preparation of the plant is used in S Nigeria for a skin-trouble symptomized by black spots (8). To promote walking in rachitic or retarded children, the legs in Congo are rubbed with the leaf-sap in palm-oil (2). No active principle has been found in the plant (3).

The Ghanaian names refer to other species and genera also, and in general apply to commelinaceous plants with deep, complicated roots. Their presence in forests indicates good cacao soil. The plant, however, may become a pest in areas of forest under regeneration (5).

References:

1. Ainslie, 1937: IFI, Oxford, Paper No. 7, sp. no. 32. 2. Bouquet, 1969: Mém. ORSTOM 36: 90. 3. Bouquet, 1972: 19. 4. Bouquet & Debray, 1974: 69. 5. Dalziel, 1937. 6. Irvine 458, K. 7. Irvine, 1930: 30. 8. Thomas, N. W. 1868 (Nig. Ser.), K.

**Aneilema lanceolatum** Benth.

FWTA, ed. 2, 3: 31. UPWTA, ed. 1, 464.

West African: **GHANA** DAGBANI nahi nyamere *from Fula* (JMD) FULA-FULFULDE (Ghana) nai hinere = *cattle nose* (JMD) SISAALA fufura (Blass) **NIGERIA** BOLE raban zabuwa (AST) CHAMBA il-en-goy-yah = *plant for a thief to boil yams in* (FNH) HAUSA burabaya (AST) karya garma = *break-hoe from* garma: *sort of hoe* (auctt.) tsiìdan-kàreé = *devil's thorn* (*Tribulus terrestris, Zygophyllaceae*) *of the dog* (JMD; ZOG) IGALA ẹbọnọgu (Boston)

A perennial herb with tough, tuberous rootstock and prostrate to ascending stems to about 60 cm, of savanna woodland and grassland, in Guinea and Mali to N and S Nigeria, and also across Africa to NE and E Africa.

The Hausa name meaning 'break-hoe' is applied to this and to other plants with fibrous deep-penetrating rootsystems which cause damage to a type of fenestrated hoe called *garma*. The root is bulbous or thickened, with rather stout almost tuberous long rootlets (2). The Chamba of the Vogel Peak area call the plant 'a plant for a thief to boil yams in', meaning that if no water is available, a cooking pot may be lined with the plant and it will provide enough sap for boiling yams (3, 4). The plant is grazed by stock (1).

The Hausa of N Nigeria use the seeds for eye-diseases: crushed and placed in the eye they cause tears to flow, the mucilaginous property probably helping to effect removal of foreign matter (2).

References:

1. Andrews A. 735, K. 2. Dalziel, 1937. 3. Hepper 1307, K. 4. Hepper, 1965: 496.

**Aneilema pomeridianum** Stanfield & Brenan

FWTA, ed. 2, 3: 31. UPWTA, ed. 1, 461, as *A. lanceolatum* Benth., in minor part.

West African: **GHANA** DAGBANI nahi nyamere (auctt.) **NIGERIA** HAUSA gada machiji = *duiker's snake* (Grove; RES)

A perennial herb to 1 m high resembling *A. lanceolatum,* of grassland, woodland and farmland, in Ghana to Niger and N and S Nigeria.

The plant is a weed of cultivation. In northern Ghana the plant is used for healing sores on the feet (1).

Reference:

1. Williams, L1. 163, K.

**Aneilema setiferum** A. Chev.

FWTA, ed. 2, 3: 32.

West African: **IVORY COAST** MANDING-MANINKA tigbé (B&D)

A perennial herb to 60 cm high from a tuberous rootstock, of savanna grassland, from Mali to N Nigeria.

The plant in Ivory Coast is incorporated in a drink taken for leprosy (2). Var. *pallidicilliatum* J. K. Morton, restricted to Ghana, is reported to be very mucilaginous (1, 3, 4).

References:

1. Andoh 5053, K. 2. Bouquet & Debray, 1974: 69. 3. Hall 1979, K. 4. Morton GC. 7502, K.

**Aneilema silvaticum** Brenan

FWTA, ed. 2, 3: 31.

West African: **NIGERIA** EDO óhīovbù (Kennedy)

A slender straggling herb to 30 cm high of the forest area of S Nigeria, and also in Cameroun and the Congo basin.

**Aneilema umbrosum** (Vahl) Kunth

FWTA, ed. 2, 3: 30–31.

West African: **GHANA** AKAN-ASANTE onyame bewu na mewu (FRI) TWI akotia (Dade; FRI) **NIGERIA** IJO-IZON (Kolokuma) dìinàá (KW)

A straggling or decumbent herb with erect branches to 1 m high, of the forest or damp shaded places in savanna from Sierra Leone to W Cameroons and

Fernando Po, and extending across the Congo basin to E Africa.
The leaves are mashed in Ghana for use in enemas for constipation (1).

Reference:

1. Irvine, 1930: 31, as *A. ovato-oblongum* P. Beauv.

**Coleotrype laurentii** K. Schum.

FWTA, ed. 2, 3: 36.

A robust straggling herb, stems often stoloniferous to 2 m long by 1·30 m high, of rain-forest in Ivory Coast to S Nigeria, and in the Congo basin and Uganda.
A decoction of the whole plant is made in Congo into a gargle for sore throat, quinsy and tonsilitis (1).

Reference:
1. Bouquet, 1969: 90.

**Commelina africana** Linn.

FWTA, ed. 2, 3: 45.

A slender prostrate herb, of five varieties recognized in the Region, stem brief to 1 m long, from sea-level to montane situations over 2000 m altitude, throughout the Region, and widespread elsewhere in tropical and S Africa.
No particular usage is recorded in W Africa. In Kenya the leaves are cooked and eaten as a vegetable (4), and the plant is grazed by all stock (1). In the Kwale District an infusion of the plant is used as a wash to reduce fever (3), while the Zulu of S Africa bathe the body, especially of a child, with a cold infusion in cases of restless sleeping; a leaf from the infusion may also be rubbed over the sleeping-place for the same end (5). In Lesotho a decoction of the plant with *Tephrosia capensis* (Leguminosae: Papilionioideae) is taken for a weak-heart and nervousness (2, 5), and in the Congo region the root is used for the former (5). The plant cooked with *Haplocarpa scaposa* Harv. (Compositae) and another unspecified edible root is given in Lesotho as medicine to a young woman supposedly barren, while an infusion of the plant is drunk and its ash is rubbed over the loins as a fertility charm (5). In S Africa a root-decoction is taken as a treatment for venereal disease and by women with unduly troublesome menstruation; this preparation is also used for pains around the hips and for bladder complaints (5).
The Masai of Kenya put the plant into a milk-container with water, milk and honey, and then sprinkle the liquid around the cattle enclosure as a charm to drive away disease (1).

References:

1. Glover & al. 241, K. 2. Guillarmod, 1971: 419. 3. Magogo & Glover 905, K. 4. Mwangangi 1937, K. 5. Watt & Breyer-Brandwijk, 1962: 196–7.

**Commelina benghalensis** Linn.

FWTA, ed. 2, 3: 48. UPWTA, ed. 1, 465.

West African: **SENEGAL** BASARI ɛ-bokó kòlòr (Ferry) MANDING-BAMBARA ténébra (JMD; JB) SERER dãgasisa (JB) mbap gôr (JB) WOLOF hép a hép (JB) véréyâ, véréyâ bumak, bumak: *large, i.e., the large vereyan* (JMD) **THE GAMBIA** MANDING-MANDINKA muso kafo jio, jio: *woman's group; a troublesome weed left to women to work: hard to eradicate* (DF)

**GHANA** AKAN-TWI O-nyame bewu na mawu = *God will die before I die* (FRI) **NIGER** SONGHAI tòktòkò (*pl.* -à) (D&C) **NIGERIA** CHAMBA pong-sah (FNH) EDO óhīovbù (JMD) FULA-FULFULDE (Nigeria) uppurwa (J&D) HAUSA bàlaàsáanaá (auctt.) bulabula (RES) IGBO ọbọgu (JMD; BNO) IGBO (Ogwashi) óbó ògù (NWT) **WEST CAMEROONS** 'BAMILEKE' nkwaa (AJC)

Prostrate or scrambling ascending herb to about 1 m, of two varieties: var. *benghalensis* in open cultivated and waste ground, and savanna; var. *hirsuta* C.B. Cl. of the forest and in montane grassland; occurring from The Gambia to W Cameroons, and widely dispersed elsewhere in tropical Africa, and in Asia.

The plant is succulent and mucilaginous, and resists desiccation resulting in the fanciful Twi name meaning 'God will die before I die!' (cf. *C. diffusa*). It is difficult to eradicate and becomes a troublesome weed of cultivation. In parts of the Region the leaves are cooked and eaten as a vegetable (5; Gurunshi, Ghana, 10). In India the leaves are said to be eaten only as a famine-food (4, 12). In Sudan (1) and E Africa (all collectors) the plant is grazed by domestic stock, an advantage being that it provides a significant amount of an animal's water requirement (6). In N Ghana it is a favourite for pigs and poultry (9; 15), though in southern Africa its use as pigfeed is restricted to time of dearth as it is thought to cause the animals to come out in a sort of 'measles' (14).

The plant is astringent and both vegetative and flowering parts yield hydrocyanic acid (14). In S Nigeria it is made into a poultice for sore feet (13). No other medicinal usage is recorded in the Region. Plant-sap is used in E Africa for ophthalmia, sore-throat and burns. More specifically liquid from the flower-spathe is used in Zanzibar for eye-complaints (16) and in Tanganyika in topical application to thrush in infants (2; 14). In Lesotho the plant is made into a medicine to counter barrenness in women (8; 14). In India it is said to be beneficial in leprosy (2), and in the Philippines it is used as an emollient collyrium and taken for strangury (11).

The rhizomes are starchy and mucilaginous. In India (12) and Sudan (3) they are commonly cooked and eaten, and are said to be a wholesome food.

A dye is obtained in India and China from the sap of the flowers (7).

References:

1. Andrews A. 3463, K. 2. Bally, 1937. 3. Broun & Massey, 1929: 387. 4. Burkill, 1935: 645–6. 5. Busson, 1965: 439, with leaf-analysis. 6. Glover & al. 447, K. 7. Greenway, 1941. 8. Guillarmod, 1971: 419. 9. Irvine 4598, K. 10. Irvine, 1930: 126. 11. Quisumbing, 1951: 149. 12. Sastri, 1950: 312–3. 13. Thomas, N. W. 2057 (Nig. Ser.), K. 14. Watt & Breyer-Brandwijk, 1962: 197. 15. Williams, Ll. 850, K. 16. Williams, R. O. 1949: 208.

## **Commelina bracteosa** Hassk.

FWTA, ed. 2, 3: 48.

West African: **THE GAMBIA** MANDING-MANDINKA muso-kafo-jilo (Fox)

A perennial straggling herb of Senegal to N Nigeria, and occurring in E Cameroun and in eastern Africa from Sudan to Mozambique and Malawi.

No usage is recorded in the Region. In Kenya it is browsed by all domestic stock and by ungulates (1). The sap is used in Tanga Province of Tanganyika by instillation for sore eyes; it is also put into magic nostrums for coping with bad luck (2).

References:

1. Magogo & Glover 409, K. 2. Tanner 3162, K.

## **Commelina capitata** Benth.

FWTA, ed. 2, 3: 47.

West African: **SENEGAL** MANDING-BAMBARA ba folo (JB) **SIERRA LEONE** VAI k-piye (FCD)

A straggling robust herb to 1.30 m high of high rain-forest throughout the Region, and extending to Angola and Uganda.

**Commelina congesta** C. B. Cl.

FWTA, ed. 2, 3: 49.
West African: **UPPER VOLTA** MANDING-DYULA korongbé (K&B) **IVORY COAST** MANDING-DYULA korongbé (K&B) **NIGERIA** IJO-IZON (Kolokuma ịkpịrịbụ́ịkpị(rị)bụ (KW)

Creeping perennial herb, of the forest, occasionally in the open, from Guinea to S Nigeria and Fernando Po, and into central Africa.

In Ivory Coast it is an ingredient of a complicated treatment for jaundice and 'yellow' fever: compounded with other drug-plants it is taken by draught. Also a decoction of the whole plant with citron is taken internally for gastro-intestinal pains, and a plaster of the plant is applied for seven days to reduce swelling after a fracture (1). The Ijo of S Nigeria use the plant as a cure for general debility: the leaves and stems are mixed with locally-made soap to wash the body when the sun is hot. The treatment is very irritating (2).

References:

1. Kerharo & Bouquet, 1950: 239. 2. Williamson, K. 41, UCI.

**Commelina diffusa** Burn. f.

FWTA, ed. 2, 3: 47. UPWTA, ed. 1, 465, as *C. nudiflora* Linn.
West African: **IVORY COAST** AKYE sa baté (A&AA) **GHANA** AKAN-ASANTE nyame bewu ansã na mewu = *God will die before I die - in allusion to its tenacity to life* (Enti) onyame bewu na mawu (FRI) TWI Onyame bewu na mawu = *God will die before I die* (FRI) nyame bewu ansã na mewu (FRI) GA too-lílɛ̀í = *sheep's tongue* (FRI; KD) GBE-VHE agbenokui nokui (FRI) agbɔ maku maku = *a grass that will not die* (FRI) NZEMA nyamɛnlẽ wua me ngɛ wu (FRI) **NIGERIA** EFIK ákpáfrí-íkáñ (auctt.) FULA-FULFULDE (Nigeria) waalwaalnde (*pl.* baalbaalɗe) (Taylor) HAUSA bàlaàsaá (JMD) bàlaàsánaá (JMD) bàlaàsáyaá (JMD) kununguru (JRA) HAUSA (West) baàlaàsáa (ZOG) IGBO òbògù (JMD) IGBO (Umuahia) áhịhịá èbisango, áhịhịá: *grass*; èbisango: *a bird with a bright blue tail-feather* (AJC) IJO-IZON (Kolokuma) òwéí ịkpịrịbụ́ị kpịbụ (KW) YORUBA gòdọ̀gbọ̀ (JRA) gòdọ̀gbọ̀-odò (JMD; SOA) itõpére (JMD)

An annual or perennial herb, prostrate and rooting nodally, variable with two subspecies recognized in the Region, of open wet places from the lowlands to 2000 m elevation, dispersed throughout the Region, and widespread in the tropics.

The leaves are very mucilaginous, and the plant resists desiccation, a feature which results in Ghana in fanciful names: 'grass that will not die'; 'God will die before I die.' The Efik name is the principle of a saying — *akpafri ikaŋ ikpaha ndaeyo ikpaha ukwɔ*, 'the herb *akpafri ikaŋ* does not die in the dry season, does not die in the time of flood'; thus a live-wire person is liable to be called *nte akpafri ikaŋ*, i.e. like the *akpafri ikaŋ* (1). The Igbo name likening the plant to the *ɛbisango* bird with a bright blue tail feather refers to the blue flowers.

In SE Asia the leaves are used as a vegetable and cattle freely graze the plant which in Indochina has been used for stall-feeding. In Malaya cattle are said to improve on it (5). It has been recorded as cattle-fodder in Hausaland (8) and in Sudan (6), but on the other hand it is said to be harmful to sheep causing foaming at the mouth and death if eaten in excess (9). It appears to be a plant of various races which may result in such differences; perhaps also involved are edaphic and phenologic factors which need to be investigated.

The leaves have medicinal uses. They are taken in Nigeria as an aperient and a decoction is used in fever (3, 10). A leaf-infusion is a Nigerian eye-wash, and the leaves pounded and mixed with the seeds of *Leea guineensis* G. Don (Vitaceae) are applied topically to buboes and swollen glands, and as a rubefacient in rheumatism (3). Similar use is reported in Ghana where it is said that the swellings in the groin of a malady called *okwaha* can be maturated and burst with consequent relief after three days treatment (9). Temne of Sierra Leone use the plant as a wound-dressing after circumcision (N. W. Thomas fide 7). Akye of Ivory Coast crush the leaves with those of several other drug-plants and kaolin to make a paste which is let down with water and drunk for Potts disease (spinal tuberculosis) (2). In Congo leaf-sap is used on abscesses, buboes, etc. to the same end, and for headache (4), while in Gabon a macerate of the stems is used for the latter purpose (11, 12). Leaf-sap is also instilled into the ear for inflammation of the ear-passage in Congo, and the leaves are held to be aphrodisiac (4).

A root-decoction finds use in Nigeria in treatments for gonnorrhoea and for women suffering severe menstrual pains (3).

As a play-thing, Hausa roll the leaves into pellets for discharge from popguns (Bargery fide 7). In Sierra Leone the plant features in a ceremony to protect rice on the farm (N. W. Thomas fide 7).

The claim that the plant is able to supress lalang grass (*Imperata cylindrica* (Linn.) P. Beauv.) is not tenable.

References:

1. Adams, 1943: as *C. nudiflora*. 2. Adjanohoun & Aké Assi, 1972: 91. 3. Ainslie, 1937: sp. no. 110, as *C. nudiflora*. 4. Bouquet, 1969: 90, as *C. nudifera* Linn. (Sphalm: *C. nudiflora* Linn.) 5. Burkill, 1935: 646, as *C. nudiflora* Linn. 6. Broun & Massey, 1929: 387, as *C. nudiflora* Linn. 7. Dalziel, 1937. 8. Holland, 1908–22: 711–2, as *C. nudiflora*. 9. Irvine, 1930: 216, as *C. nudiflora* Linn. 10. Oliver, 1960: 23, as *C. nudiflora*. 11. Walker, 1953, a: 27 as *C. nudiflora*. 12. Walker & Sillans, 1961: 123, as *C. nudiflora*.

## **Commelina erecta** Linn.

FWTA, ed. 2, 3: 49. UPWTA, ed. 1, 465.

West African: **SENEGAL** FULA-PULAAR (Senegal) bura-bura ba (JMD) wal-wal dé (JMD) MANDING-BAMBARA ténébra (JMD) SERER dâgasisa (JLT; JB) WOLOF véréyâ (JMD) véréyâ bu gôr (JB) **MALI** FULA-PULAAR (Mali) bura-bura ba (JMD) wal-waldé (JMD) MANDING-BAMBARA ténébra (JMD) **GHANA** AKAN-TWI nyame bewu na mawu (FRI) GA too-líĺɛ̀i = *sheep's tongue* (FRI) GBE-VHE agbenokui nokui (FRI) agbɔmaku maku = *grass that will not die* (FRI) NZEMA nyamɛnlẽ wua me ngɛ wu (FRI) **NIGERIA** FULA-FULFULDE (Nigeria) waalwaalnde (*pl.* baalbalɗe) (JMD; J&D) HAUSA ba-kiskis (JMD; ZOG) bàlaàsánaà (JMD) bùrábayà (ZOG) hánjín kúdaá, kunun curu, tubanin dawaki (ZOG) HAUSA (East) balasar dawaki , burabaya, hanjin kuda, kununguru, tubanin dawaki HAUSA (West) baàlaàsáa, baàlaàsár-dáwaákìi (ZOG) balasaya IGBO áhị̀hịá èbisango, áhị̀hịá: *grass;* èbisango: *a bird with a bright blue tail feather* (JMD) YORUBA ilẹ̀kẹ̀ ọ̀pọ̀lọ̀ = *frog's spawn, alluding to mucilage in the spathe* (auctt.) itọ̀ (i-) pére *from* itó ìpére: *snail's spittle, or* itọ̀paire: *creeper that kills the farm crops* (Dennett; JMD)

A prostrate to weakly erect herb up to 2 m long, variable, of three subspecies in the Region, widespread, and elsewhere in tropical Africa.

The plant is a weed of cultivation, resists desiccation; hence the Ewe name meaning 'grass that will not die', and is difficult to eradicate. In Kenya the Digo know it as 'father-in-law' because it is always around (3), cf. *C. imberbis*. It is grazed by all domestic stock including poultry, and in markets in Mali it is sold as camel-food and is regarded as a tonic for horses (1). Fula of N Nigeria feed it to horses (2) but with some discretion (see *C. forskalaei*) (1). In Kenya it is grazed by bush-buck and duiker (3). The Yoruba invoke the plant in an incantation 'to give an itch' to someone (4).

References:

1. Dalziel, 1937. 2. Jackson, 1973. 3. Magogo & Glover 540, K. 4. Verger, 1967: no. 116.

## **Commelina forskalaei** Vahl

FWTA, ed. 2, 3: 48. UPWTA, ed. 1, 465.

West African: **MAURITANIA** ARABIC ('Maure') agherf (JMD) **SENEGAL** FULA-PULAAR (Senegal) bura-bura-ba (JMD) wal-wal-dé (JMD) werkin (K&A) MANDING-BAMBARA gomblé ni (JB; K&A) ténébra (auctt.) SERER dâgasisa (auctt.) mbap (JB; K&A) wéhéhan (JB; K&A) WOLOF véréyâ (AS; K&A) **THE GAMBIA** MANDING-MANDINKA yamo = *grass* (Frith) **MALI** DOGON burú gɛgɛ̀ɛru (C-G) FULA-PULAAR (Mali) bura-bura-ba (JMD) wal-waldé (JMD) MANDING-BAMBARA ténébra (JMD) **GHANA** AKAN-TWI nyame bewu na mawu (FRI) GA too-lìlɛ̀í = *sheep's tongue* (FRI) GBE-VHE agbenokui nokui (FRI) agbɔmaku maku = *grass that will not die* (FRI) NZEMA nyamɛnlẽ wua me ngɛwu (FRI) **NIGERIA** BOLE gajuruwi (AST) FULA-FULFULDE (Nigeria) waalwaalnde (*pl.* baalbaalɗe) (JMD) HAUSA ba-kiskis (JMD; ZOG) bàlaàsánaá (JMD) balasa ta dawaki (BM) balasan dawaki (BM; JMD) burabaya (AST; JMD) bùrábayà, daban dari (RES) hánjín kúdáa, kunan curu, tubanin dawaki (ZOG) HAUSA (East) hanjin kuda (JMD) kununguru (JMD) tubanin dawaki (JMD) HAUSA (West) baàlaàsáa, baàlaàsár-dáwaákii (ZOG) balasaya (JMD) IGALA ẹbọnọgu *this is female in distinction from Cyanotis longifolia* (Boston) IGBO (Owerri) ńtī ọké = *rat's ear* (AJC)

*NOTE: These vernaculars appear interchangeable with C. erecta Linn.*

An annual prostrate, straggling weed, of grassland, open cultivated ground and waste spaces, from Senegal to N Nigeria, and widespread in tropical Africa outside of the forested zones, and in Madagascar, Arabia and India.

The plant is grazed with relish by all stock including poultry (6; Senegal, 1; Nigeria, 8, 12, 13). In Mali it is traded in markets as a camel food and is regarded as a tonic for horses (6). However in Igala, Nigeria, it has been reported that horses eating it may die (3). The Fula of N Nigeria give the name *walwalnde* to more than one species, but they recognize only one with small leaves as suitable for fodder. Likewise the Igala give the same name to *Cyanotis longifolia,* though distinguish the latter as 'male.' There appears to be room for confusion. In Sudan the plant is grazed by goats and sheep but not by cattle (5), while in Somalia the plant is considered harmful to sheep and causes symptoms (anaemia and diarrhoea) similar to those of infection by the widespread nematode *Haemonchus contortus* (Peck). The 'army-worm' (*Spodoptera exempta* (Walk.), Lepidoptera, Noctuidae) is a pest dispersed over tropical Africa and beyond. It destroys cereals and grass pasture, and also Cyperaceae but not Commelinaceae. In E Africa *C. forskalaei* is particularly abundant on the black cotton soils and is reported to take over pasture where army-worm has destroyed the grass (2). The various reports of good grazing and toxic grazing provided by this plant suggest the need for further examination. Edaphic and phenological factors may be important, but if the conditions can be understood, there may be a place of some benefit in agriculture where army-worm is a nuisance.

The plant has medicinal uses. The sap is used by the Fula of Senegal in treatment for small sores and for minor injury such as is caused by splinters and thorns (9, 10). In N Nigeria the plant is used to prepare a soothing medicine for sores (8), and in Igala it is an ingredient of medicines for small children (3).

In Sudan the stem is cut up and serves as a material to smoke when tobacco is not available (7).

A substance has been obtained from the flower-spathe which has a physiological action on pre-oestral rate of development (4, 10).

References:

1. Adam, 1966, a. 2. Bally 631, K. 3. Boston C.14, K. 4. Bouquet & Debray, 1974: 69. 5. Carr 829, K. 6. Dalziel, 1937. 7. Evans Pritchard 79, K. 8. Grove 10, K. 9. Kerharo & Adam, 1964, b: 431. 10. Kerharo & Adam, 1974: 360–1. 11. Peck 2, K. 12. Thornewill 138, K. 13. Thornewill 183, K.

**Commelina imberbis** Ehrenb.

FWTA, ed. 2, 3: 48.

Prostrate or straggling stems to near 2 m long, of cultivated land in S Nigeria and perhaps exotic from eastern Africa, Sudan to Rhodesia.

No W African usage is recorded. When the plant invades cultivation it is difficult to eradicate. In Kenya the Digo call it 'father-in-law' as it is always there (2), and they use it in a wash to reduce fever (3). In the Narok District goats and sheep readily browse it, and to the Masai it is held to have magical powers: they drape it around a pot containing milk from which milk and water is drunk on ceremonial occasions (1).

References:

1. Glover & Samuel 2918, K. 2. Magogo & Glover 677, K. 3. Magogo & Glover 904, K.

**Commelina lagosensis** C. B. Cl.

FWTA, ed. 2, 3: 48–49.
West African: **SIERRA LEONE** SUSU balɛkɔre (NWT) **NIGERIA** HAUSA (West) baàlaàsár dáwaákíí (ZOG) IGBO (Agolo) ọ̀bọ́-ọ̀gụ́

Prostrate herb, stems to 1 m long, recorded from Senegal to S Nigeria, and occurring also in E Africa.

**Commelina nigritana** Benth.

FWTA, ed. 2, 3: 50.
West African: **SENEGAL** BASARI i-royɛ́ké, a-yɛ̀royɛ́kè (Ferry) **NIGERIA** YORUBA gbògọ̀bọ̀ (N&E) godogbògọ̀dọ̀ (N&E)

A slender decumbent or erect herb to 30 cm high of savanna and grassland, throughout the Region, and in E and S Central tropical Africa.

**Commelina subulata** Roth

FWTA, ed. 2, 3: 47.
West African: **NIGER** HAUSA balassa (Bartha)

A slender annual herb of wet places in Senegal, Ghana and Niger, and widespread in E and S Africa, and also in S India.

**Commelina thomasii** Hutch.

FWTA, ed. 2, 3: 47.
West African: **SIERRA LEONE** SUSU malkɔre (NWT) sunyugi (NWT)

A straggling sprawling perennial herb with stems to 1 m long, of lowland rain-forest, in Sierra Leone to S Nigeria.
The plant is a persistent weed of cultivation.

**Commelina zambesica** C. B. Cl.

FWTA, ed. 2, 3: 48.

A sprawling robust herb with weak upright stems to over 1 m high, of woodland and grassland, in N Nigeria, and occurring in E and south-central tropical Africa.
The plant is recorded as used as a vegetable on Pemba Island, and also as an andidote [? for the bites] for *maji moto* ('hot water') ants (1).

Reference:

1. Williams R. O., 1949: 208.

**Commelina spp. indet.**

West African: **THE GAMBIA** FULA-PULAAR (The Gambia) jedda-unki (DRR) **IVORY COAST** AKAN-ASANTE niaméwowokohu (B&D) KWENI okublio (B&D) MANDING-MANINKA tébélé (B&D) tigbé (B&D) 'NEKEDIE' okublio (B&D) **NIGERIA** NUPE lḱokúnkún (RB)

**Cyanotis arachnoidea** C. B. Cl.

FWTA, ed. 2, 3: 38, 40.

A succulent, prostrate and matting perennial herb, stems to 0.5 m long, from Liberia to N and S Nigeria, and in E Cameroun.

The plant is recorded as having unspecified medicinal use at Ijebu in Nigeria (1).

Reference:

1. Latilo FHI. 20267, K.

**Cyanotis caespitosa** Kotschy & Peyr

FWTA, ed. 2, 3: 38. UPWTA, ed. 1, 466.
West African: **NIGERIA** HAUSA tamrano (JMD; ZOG)

A perennial herb with stout villous base but glabrous above, of grassland from Ivory Coast to W Cameroons, and widespread across central Africa to Angola and E Africa.

In Sudan it is grazed by animals (1).

The root is put into a magical potion by hunters of the cane-rat (Hausa, *gyazbi*) in N Nigeria to enable them to come upon it unawares (2).

References:

1. Andrews A.655, K. 2. Dalziel, 1937.

**Cyanotis lanata** Benth.

FWTA, ed. 2, 3: 40. UPWTA, ed. 1, 466.
West African: **MALI** DOGON sá nìì tòmolo ána, sá nìì tòmolo yà (C-G) **IVORY COAST** AKAN-BRONG séssèro adiaïra (K&B) KULANGO boro-boro (K&B) **GHANA** MOORE tɛnkwitɛnkwi = *the ground will be dry before it dries* (FRI) **NIGERIA** HAUSA raábaá (JMD)

An annual herb, usually woolly when young, of seasonally wet places and weed of cultivation, from lowlands to 1500 m altitude, throughout the Region, and widely distributed throughout tropical Africa to as far south as Transvaal.

The leaves and stems are very fleshy and are resistant to desiccation — hence the Moore vernacular name. It is put in footbaths in Ivory Coast for treating soggy conditions of the skin (Fr. 'hyperhydroses'), and also a decoction is taken by draught to ease childbirth (2).

In Ethiopia it is said to provide grazing (3), but in Hausaland to cause a sort of eczema on the muzzle of horses browsing amongst it; thus the Hausa word *raɓa* means dew, which is supposed to have the same effect, also a disease of goats and sheep (1).

A plant which is perhaps this species is grown in masses on thatched roofs in N Nigeria as a juju or sacred object (1).

References:

1. Dalziel, 1937. 2. Kerharo & Bouquet, 1950: 239. 3. Taddesse Ebba 206, K.

**Cyanotis longifolia** Benth.

FWTA, ed. 2, 3: 37–38.
West African: **THE GAMBIA** MANDING-MANDINKA muso kafo jilo ba, ba· *large* (Fox) **SIERRA LEONE** SUSU buli (NWT) **NIGERIA** IGALA ẹbọnọgu (Boston) ọkọ ẹbọnọgu = *male ẹbọnọgu* (Boston)

A perennial herb, variable, in three varieties in the Region, slender to somewhat robust, 1 m high, of savanna grassland, widely dispersed from The Gambia to W Cameroons, and throughout the savanna regions of tropical Africa.

In N Nigeria Igala refer to it as the 'male' *ẹbọnọgu* to distinguish it from *Commelina forskalei* Vahl which is *ẹbọnọgu* without gender.

Cattle graze it a little, or not, in Senegal (1). In Kenya it is said to remain undepradated in farm paddocks after 'armyworm' have taken all the grass (3). In Tanganyika it has unspecified medicinal use (2).

References:

1. Adam, 1966, a. 2. Musk 98, K. 3. Symes 765, K.

**Dichorisandra thyrsiflora** Mikan. f.

FWTA, ed. 2, 3: 23.

A semi-woody herb to 1 m or more, with glossy green not variegated leaves and deep blue-violet flowers, native of Brazil, and distributed by man to many warm countries including the West African Region as an ornamental.

**Floscopa africana** (P. Beauv.) C. B. Cl.

FWTA, ed, 2, 3: 28. UPWTA, ed. 1, 466.
West African: **THE GAMBIA** MANDING-MANDINKA muso kafo jilo (DA) **IVORY COAST** SHIEN (Chiehn) niakamako lukuyié (K&B) **NIGERIA** IGBO ọ́dụ̄ ñgwèlè = *lizard's tail, a general term* (JMD) IGBO (Onitsha) opungwi (AJC) YORUBA gòdògbọ̀ (N&E)

An erect or straggling variable herb to 1 m high, of three subspecies, recorded from The Gambia to W Cameroons, and over the Congo basin to Uganda.

The plant is endowed with magical properties of protection from all illnesses in Ivory Coast. It is pounded with a leafy twig of *Microglossa volubilis* DC (= *M. pyrifolia* (Lam.) O. Ktze, Compositae) and a little water and the sloughed skin of a house-spider, rolled into a pellet and dried and stored in an antelope's horn or in a fold of leather about the house (1).

Reference:

1. Kerharo & Bouquet, 1950: 239–40.

**Floscopa aquatica** Hua

FWTA, ed. 2, 3: 28.

A creeping or floating herb of swamps, rivers and rice fields, from Senegal S Nigeria, and in the Congo basin.

The plant is a weed of rice paddies and may become troublesome.

**Floscopa glomerata** (Willd.) Hassk.

FWTA, ed. 2, 3: 28–29.
West African: **SIERRA LEONE** SUSU-DYALONKE fangnyerasakhana (Haswell)

An annual or perennial herb, erect or straggling to about 1.70 m long, of marshes and swamps from Senegal to Nigeria, and generally widespread in tropical Africa.

**Murdannia simplex** (Vahl) Brenan

FWTA, ed. 2, 3: 24–26.
West African: **NIGERIA** HAUSA baàlaàsár dáwaákií (ZOG) HAUSA (West) baàlaàsáa (ZOG)

*NOTE: Both names given as M. nudiflora, an Asian species recorded in Sierra Leone, probably refer here.*

A robust, erect or sprawling herb to 1.30 m high recorded throughout the Region, and widespread in the rest of tropical Africa, and also in Madagascar and Asia.

It is sometimes grown in gardens in Ghana as an ornamental (2).

In Kenya the plant is grazed by all domestic stock (1). There are unspecified medicinal uses in Tanganyika (4), and the root is recognized as possessing some toxic properties (3).

References:

1. Glover & al. 1782, K. 2. Irvine, 1930: 31, as *Aneilema sinicum* Linn. 3. Koritschoner 566, K. 4. Musk 94, K.

**Palisota ambigua** (P. Beauv.) C. B. Cl.

FWTA, ed. 2, 3: 35.
West African: **NIGERIA** EDO ìghíguẹ̀wé (Kennedy) iguewe-nowẹẹ (Kennedy) IGBO (Umuahia) íkpèrè nwá átụ́rụ̄ = *knee of little sheep* (AJC) **WEST CAMEROONS** KUNDU kimbimba (Dundas) metumba (Dundas)

A robust herb to over 2 m high, of lowland rain-forest in S Nigeria and W Cameroons, and extending to the Congo basin.

The plant provides good goat food (2).

Traces of steroids and terpenes have been reported in the roots of Congo material, but no other active principle (1).

References:

1. Bouquet, 1972: 19. 2. Carpenter 354, UCI.

**Palisota barteri** Hook.

FWTA, ed. 2, 3: 35–36. UPWTA, ed. 1, 466.
West African: **NIGERIA** EDO ighíguẹ̀wé (Kennedy) **WEST CAMEROONS** KPE ugbodo (Gregory)

A stemless herb reduced to a rosette of leaves at ground level, of rain-forest, from Sierra Leone to W Cameroons and Fernando Po, and into the Congo basin.

A decorative horticultural form is known with an inflorescence of very numerous small white flowers longer and looser than the wild form (1).

The plant is used in W Camerouns as a fish-trap: its leaves are placed on the surface of a river, and fish come to take their rest immediately beneath where they are caught (2).

References:

1. Dalziel, 1937. 2. Gregory 83, K.

**Palisota bracteosa** C. B. Cl.

FWTA, ed. 2, 3: 35. UPWTA, ed. 1, 466.

West African: **SIERRA LEONE** LOKO toawombi (JMD) MENDE ka silo luwu gbɔlu (?) (FCD) ndomu (*def.*-i) (JMD) **GHANA** AKAN-ASANTE akwabe (FRI)

A stemless herb reduced to a rosette of leaves at ground level, of rain-forest from Guinea to W Cameroons, and in E Cameroun and S. Tomé.

The leaves have medicinal use in Sierra Leone and Ghana similar to *P. hirsuta* (1). The fruit is covered with urticating hairs (2), and the fruit sap is said to be also irritant (3).

References:

1. Dalziel, 1937. 2. Deighton 3452, K. 3. Pyne 77, K.

**Palisota hirsuta** (Thunb.) K. Schum.

FWTA, ed. 2, 3: 35. UPWTA, ed. 1, 466.

West African: **SENEGAL** BANYUN tigugalé kidika = *female tigual* (*Costus afer*) (K&A) **SIERRA LEONE** VULGAR tologbelo *understood by all races at Blama* (FCD) KISSI tologbelo (FCD) tundui (FCD) LOKO toagumbe (NWT; JMD) toawombi (JMD) MANDING-MANDINKA kumbe (JMD) tumbe (JMD) MENDE ndomu (auctt.) SUSU kosakumbegine (NWT) kosaxumbe (JMD) kɔsɛgbi (NWT; JMD) TEMNE *an*-sita (auctt.) **LIBERIA** KRU-BASSA kor-dru-boe = *swollen knee, alluding to the thickened nodes* (C&R) MANO kpuokẽ (JMD) **IVORY COAST** ABURE branbran (B&D) AKAN-ASANTE diéssanara (B&D) guéssan clan (B&D) BRONG zoméhini (K&B) AKYE wuchakpè (A&AA) BAULE gonkobiessoa (B&D) guéssanuhama (B&D) niéssanounama (B&D) nissanganama (B&D) wokuesua (B&D) DAN genkulu (K&B) GAGU zokwo (K&B) KRU-BETE kokurum boké (K&B) GUERE dondré (K&B) kogboago (A&AA) kogloago (A&AA) GUERE (Chiehn) blékolopiti (B&D) brikro féfé (B&D) luébo (B&D) KULANGO zomésagan (K&B) zumésan (K&B) KWENI tonton (B&D) KYAMA woza wona (B&D) MANDING-MANINKA duébo (B&D) senzédu (B&D) 'KRU' néléwa (K&B) 'NEKEDIE' lébo (B&D) **GHANA** AKAN-ASANTE somɛ-nini, sommɛ: *Costus* (FRI; JMD) FANTE sombenyin (JMD) TWI akwabe (auctt.) mpentemi (FRI; JMD) somɛ-nini (FRI) GBE-VHE adutsyrɔʃfoti (FRI; JMD) sumbe (Ewe Dict.; FRI) NZEMA nzahuara (JMD) 'NEKEDIE' nakutchorpor (A.S.Thomas) **NIGERIA** EDO ìghíguẹ̀wé (Vermeer) IGBO íkpèlè átụ̀lụ̄, ìkpèrè átụ̀rụ̄ = *sheep's knee* ikpele oku (NWT) IGBO (Awka) ikpele oku (NWT; JMD) IGBO (Onitsha) íkpèlè átụ̀lụ̄ = *sheep's knee* (AJC; JMD) IGBO (Umuahia) akpọrọ onye (AJC) íkpèrè átụ́rụ̄, íkpèlè átụ́rụ̄ = *sheep's knee* (AJC; JMD) YORUBA àkéréjùpọ̀n (IFE) jàngbórókún *kneecap* (Dennett; JMD) ọjọ (auctt.)

*NOTE: This species may be confused in the vernacular with Costus afer Ker. (Costaceae).*

A robust, perennial herb to 3 m high or more, of lowland rain-forest throughout the Region, and extending over the Congo basin.

The internodes reach up to 30 cm long with swollen nodes made larger by short ragged sheaths of the nodal whorl of leaves. This character is noted in the Liberian Bassa name meaning 'swollen knee', and Nigerian Igbo 'sheep's knee' and Yoruba 'knee-cap'. The plant is grown as an ornamental and is sometimes a component of hedges (6).

This species is the most commonly used of the Commelinaceae. It appears to have analgesic and antiseptic properties. In Ivory Coast a plaster of pulped stems or the sap in a compress is applied as a dressing to furuncles, whitlow, craw-craw sores, contusion, fractures, adenites and arthritic pains; a draught of plant-sap is taken for cough, bronchitis and chest-pains; a decoction of the leafy stems is used in frictions and in baths for oedema and a decoction of the whole plant for urethral discharge, or a decoction of the stem with root of *Alchornea* (Euphorbiaceae) in draught and in enema for haematuria and urethral discharge; the sap is haemostatic and is also applied to yaws and guineaworm sores; a preparation is taken internally for difficult childbirth, female sterility and as an antemetic and antidysenteric; the crushed root is made into a suppository as an aphrodisiac (4, 12). The Guere use the stem-pith made into a paste with pimento and water in enemas for general fatigue, claiming that it is an excellent 'pick-me-up', while the Akye use the dried powdered leaves in water as an antidysentery enema (1). In Ghana (9) and Nigeria (2) the stem is chewed as a sedative for cough; the dried leaves are smoked for toothache; the roots are added to soup taken by women in pregnancy and pounded up with peppercorns of *Piper guineense* (Piperaceae) are applied externally to sprains, and put in warm water constitute an enema for constipation. The Tiv people believe that the leaves and roots are an aid to conception (16), and the Igbo of Obompa prepare an ointment of the plant for gun-shot wounds and swellings (15). In Sierra Leone the roots washed, cut up and boiled with lime are held to cure gonorrhoea in three days if taken immediately (13); sore feet are treated by fumigation, the roots being made to smoulder in a hole in the ground covered by wicker (6), and are put into medicine for stomachache (8). The plant is used in Ghana for stomach-pains (9), and a leaf-infusion is taken for piles and given to babies to heal the navel (14). Shavings of the stem are used in Gabon to promote healing of wounds particularly of the umbilicus; heated leaves are applied over the lumbar region for kidney-pains; cooked with groundnuts, the leaves are taken by suckling mothers to cleanse their milk; and pieces of the stem, after exposure to the sun are made into a draught for urethral discharge (17, 18). In Liberia the plant is held to be good for treating deafness (5), and sap from the roasted leaf is instilled in the ear for earache (6).

In Casamance (Senegal) the plant is used in conjunction with *Adenia lobata* (Jacq.) Engl. (Passifloraceae) as a fish-poison (10, 11). In Gabon arrow-poison is made from the stems mixed with *Tephrosia* (Leguminosae: Papilionioideae), or the sap is added to the powdered seeds of *Strophanthus* (Apocynaceae) (18). The dried plant is used sometimes in Sierra Leone for soap-making (7).

The flowers are fragrant and are frequented by bees (3).

In superstitious practice amongst the Temne the plant is carried in the hands in witch-divination (N. W. Thomas fide 6), and in Ghana it is regarded as a deterrent to the spiritual influences caused by plague and is placed on paths, etc., during epidemics which the plague cannot pass (9).

References:

1. Adjanohoun & Akė Assi, 1972: 92. 2. Ainslie, 1937: sp. no. 258. 3. Bates 239, K. 4. Bouquet & Debray, 1974: 70. 5. Cooper 359, K. 6. Dalziel, 1937. 7. Deighton 310, K. 8. Deighton 5922, K. 9. Irvine, 1930: 319, as *P. thyrisflora* Benth. 10. Kerharo & Adam, 1963, b. 11. Kerharo & Adam, 1974: 361–2, with phyto-chemistry. 12. Kerharo & Bouquet, 1950: 240. 13. Lane-Poole 171, K. 14. Thomas, A. S. D.29, K. 15 Thomas, N. W. 1651 (Nig. Ser.), K. 16. Vermeer 77, UCI. 17. Walker, 1953, a: 27. 18. Walker & Sillans, 1961: 124.

## Palisota spp. indet.

The leaves of an unidentified plant, probably *P. hirsuta*, are crushed by hand in a little water at Magbena, Sierra Leone, and the preparation is drunk for gonorrhoea (1).

Reference:

1. Massaquoi, 1937.

**Pollia condensata** C. B. Cl.

FWTA, ed. 2, 3: 33. UPWTA, ed. 1, 467.
West African: **GHANA** AKAN-ASANTE ewi-ani, ewi-aniwa, ɛwi: *a sp. of antelope*; ani(-wa): *eyes* (auctt.) ɔtwe-ani-wa, ɔtwe: *a sp. of small antelope* (FRI)

A stout herb, stems decumbent or stoloniferous, or erect to over 1 m high, in forest shade, occurring in Sierra Leone to W Cameroons, and generally throughout tropical Africa.

The Asante names refer to the blue bead-like fruits likening them to the eyes of a species of small black antelope (1).

The plant is very mucilaginous. In Cameroun the stems are used as a wound-dressing (2), and in Gabon they are crushed and made into an infusion which is drunk for urethral discharge (3, 4). The leaves are put in water used in Ghana as a wash during pregnancy to ensure speedy delivery (1).

In Gabon the plant is a fetish of elephant hunters (4).

References:
1. Irvine, 1930: 348. 2. Leeuwenberg 7074, K. 3. Walker, 1953, a: 27. 4. Walker & Sillans, 1961: 125.

**Polyspatha paniculata** Benth.

FWTA, ed. 2, 3: 42.

A decumbent or stoloniferous herb with erect flowering stem to 60 cm high, of lowland rain-forest from Guinea to W Cameroons and Fernando Po, and extending across the Congo basin to Uganda.

Its presence is said in Ghana to indicate good soil (1).

Reference:
1. Thomas, A. S. D.34, K.

**Rhoeo spathacea** (Sw.) Stearn

FWTA, ed. 2, 3: 23.
English: purple-leaved spiderwort, oyster plant, boat lily, Moses on a raft (M. in a boat, cradle, the bulrushes), man (two, three men) in a boat (U.S.A., Bates).

An attractive foliage plant with some cultivars, native of W Indies and Mexico dispersed by man to warm countries including the W African region as an ornamental.

**Stansfieldiella imperforata** (C. B. Cl.) Brenan

FWTA, ed. 2, 3: 23.
West African: **GHANA** AKAN-ASANTE Onyame nwu namen wu *this and variants apply generally to commelinaceous herbs* (FRI)

A herb, decumbent, stoloniferous, from a few to 60 cm high, in lowland rain-forest of Sierra Leone to W Cameroons.

In Ghana the Asante of Amentia crush the leaves for use as an enema for constipation (1).

Reference:
1. Irvine 442, K.

**Zebrina pendula** Schnizl.

FWTA, ed. 2, 3: 23.

English: wandering Jew, inch plant (Bates).

A decumbent herb, with leaves striped white above, purple below, and several cultivars with variations of leaf-colour, native of Mexico and widely dispersed by man to warm countries and as a house plant in temperate countries, as an ornamental, especially as a trailing plant.

## COMPOSITAE

**Acanthospermum hispidum** DC.

FWTA, ed. 2, 2: 241–2. UPWTA, ed. 1, 414.

English: star-bur (S. Africa).

West African: **SENEGAL** DIOLA (Fogny) ńamusigüroÿ éḱeďḱda = *piquant* (K&A) DIOLA (Tentouck) nabati kalimô (K&A) ńabati kalimu (K&A) FULA-PULAAR (Senegal) dagasalum (K&A) MANDING-BAMBARA nanengu (auctt.) suraka wôni (JB; K&A) SERER nob *properly Tribulus terrestris Linn. and applied here for the similarity of the burred achenes* (auctt.) sakarkasâg (JB; K&A) WOLOF dag i Ganar = *Tribulus* (*dag*) *of the Mauritanians* (auctt.) gasconi *a nickname after a deputy of that name* (JLT) **GUINEA-BISSAU** BALANTA misquito (JDES) singuir (JDES; EPdS) CRIOULO nhara-siguido (EPdS) FULA-PULAAR (Guinea-Bissau) búlè-n'baba (JDES) n'arè-sáquè (JDES) umbaba (JDES) um-nhárè-sáquè (JDES) MANDING-MANDINKA berentam (*def.*-ô) (JDES) berentão (EPdS) PEPEL buchigado (JDES) SUSU manguera-górè (JDES) **SIERRA LEONE** MANDING-MANDINKA ŋwɔni (FCD) sahiligbin (FCD) sofaligbie-na (FCD) TEMNE *an*-gbəntmər *general for small plants with hooked fruits* (NWT; FCD) **MALI** DOGON deguru kúú, kɛ̀bɛ (C-G) FULA-PULAAR (Mali) gié (FNH) MANDING-BAMBARA nioninju (FNH) **IVORY COAST** VULGAR cram-cram *properly Cenchrus biflorus Roxb., and applied here for the similarity of their burred fruits* (K&B) GAGU tovenlé (K&B) KRU-GUERE (Chiehn) sukawuïa (K&B) KWENI béna iri (K&B) bohuédèri (B&D) MANDING-DYULA iukubassa moni (K&B) SENUFO-TAGWANA koakuru (K&B) **GHANA** ADANGME-KROBO awusagbe = *Hausa herb* (FRI) AKAN-ASANTE mpũpuã (Enti) mpupuaa (Enti) FANTE ɔkyer mfanti (Deakin) petekunsɔe = *hyaena's thorns* (Deakin) TWI mpupuaa (Enti) sraha nsɔe (FRI) GA awusaŋme = *Hausa thorns* GBE-VHE awusagbe (FRI) awusatsoe (FRI) sɛnutsoe (FRI) SISAALA sɔɔ = *thorn; a general term* (Blass) VHE (Awlan) dugba VHE (Pecí) awusagbe = *Hausa herb* (FRI) awusatsoe (FRI) senutsoe = *Hausa thorns* sɛnutsoe (FRI) **NIGERIA** FULA-FULFULDE (Nigeria) kassiyaawo *probably from Hausa* (J&D) yaawol (RES) HAUSA kaashin yaawoo (auctt.) YORUBA dágunró (Abraham; JMD)

A bushy annual to about 50 cm high, of open waste places, cultivated land, and an undershrub in shade, widely dispersed throughout the Region, a native of tropical America and spreading into many tropical countries.

The plant has spiny achenes like *Tribulus terrestris* Linn. (Zygophyllaceae) with which it may be locally equated. Its thorny nature evokes several of the vernacular names. The Wolof *gasconi* is taken from a Deputy of that name whose election coincided with the first appearance of the plant in Senegal (Trochain fide 6). No stock will graze it (1). The thorns obviously are liable to injure the mouth, but in the case of sheep in Ghana it is said to be because of the bitter taste of the leaf; hence the Twi name *duankyene* (9). The bur is generally troublesome to stock-farmers (7) and it is said in S Africa that if the skin of the foot is punctured by stepping on one, the lesion may take as long as six weeks to heal (13). The Fante name meaning 'hyaena's thorns' is because hyaenas cannot pass through masses of the plant (9). Though stock avoid it, its dispersal is primarily by the adherence of the achenes to the coats of animals and it is always present where stock are collected (11). The plant is also a pest of arable land, invading cultivation.

The leaves are squeezed by boys in Ghana to put the sap into lime-juice to add to the flavour (9). In Ivory Coast a decoction is drunk as a purgative and counter-poison, and an aqueous macerate is drunk and put into baths for arthritis (10) and rheumatism (5). In Congo the plant is used to treat stomach complaints, wounds and migraine applied as either the undiluted sap of the whole fresh plant or this diluted with water as nose-drops, or the sun-dried leaf is powdered for topical application with or without scarification (3). The Ga of Ghana are said to use the leaves to cure *kpiti* (leprosy) (6, 9).

In northern Ivory Coast in Boundiali Prefecture, the Dyula are recorded as using the ash to make soap (12).

Traces of alkaloid have been reported present in the whole plant and in the leaves (2, 4, 5, 13).

References:

1. Adam, 1966, a. 2. Adegoke & al., 1968. 3. Bouquet, 1967: 91. 4. Bouquet, 1972: 19. 5. Bouquet & Debray, 1974: 70. 6. Dalziel, 1937. 7. Egunjobi, 1969. 8. Irvine, 1930: 5. 9. Irvine, s.d. 10. Kerharo & Bouquet, 1950: 212. 11. Leeuwenberg 4423, K. 12. Portères, s.d. 13. Watt & Breyer-Brandwijk, 1962: 197.

**Adenostemma caffrum** DC.

FWTA, ed. 2, 2: 286.

A weak herb of damp and swampy sites from Mali to W Cameroons and extending across to E Africa and southern Africa.

In Uganda the leaves are used as a cough-medicine (2), and in Kenya an inhalation is made of the boiling roots for chest-complaints (1).

References:

1. Broadhurst-Hill 151, K. 2. Shillito 112, K.

**Adenostemma perrottetii** DC.

FWTA, ed. 2, 2: 286. UPWTA, ed. 1, 414.

West African: **GUINEA** FULA-PULAAR (Guinea) farmatuli (A. Chev.) kumbé tiangol (A. Chev.) **SIERRA LEONE** KONO yandigboinei (FCD) **IVORY COAST** SHIEN (Chiehn) bopiti (B&D) **GHANA** AKAN-AKYEM akoramfidie (FRI) ASANTE akoramfidie (FRI) nkokɔnkatie (FRI) **NIGERIA** YORUBA imí-ẹ̀ṣù (JMD)

A weak annual herb of damp shaded sites, occurring throughout the Region, and widespread across tropical Africa.

The sap from the crushed stems together with that of *Carpolobea lutea* (Polygalaceae) is used in Ivory Coast for rheumatism (1). In the area of Lakes Edward and Albert in eastern Zäire the plant is considered a remedy for syphilis (2).

References:

1. Bouquet & Debray, 1974: 70. 2. Van der Ben 638, K.

**Aedesia baumannii** O. Hoffm.

FWTA, ed. 2, 2: 271.

West African: **NIGERIA** YORUBA ope-kaua kaua (Macgregor)

An erect perennial herb to 30 cm high, of savanna grassland from Mali to S Nigeria, and also in Cameroun.

**Aedesia glabra** (Klatt) O. Hoffm.

FWTA, ed. 2, 2: 271.

West African: **SENEGAL** BASARI bandji tati (K&A) **SIERRA LEONE** MANDING-MANINKA (Koranko) wuseyambe (NWT) MENDE ninge (Macdonald) **IVORY COAST** MANDING-MANINKA konon nimila (A&AA) kwoniagbé (B&D)

A perennial herb to 60 cm high from a stout knotted rootstock, of damp grassland, from Senegal to N Nigeria, and also in Cameroun and Congo.

The plant is held to have pain-relieving properties in Ivory Coast. It is used for rheumatism, and inhalations are given for sore-throat (1).

Reference:

1. Bouquet & Debray, 1972: 70.

**Ageratum conyzoides** Linn.

FWTA, ed. 2, 2: 287. UPWTA, ed. 1, 414–5.

English: goat weed, billy goat weed (Australia).

French: herbe aux sorciers (Berhaut).

Portuguese: cachacim (from Crioulo, Cape Verde); folha-male (S. Tomé) (Feijão).

West African: **SENEGAL** DIOLA é kerkéda (JB; K&A) MANDING-BAMBARA nungu (JB; K&A) WOLOF gobu (JB; K&A) **THE GAMBIA** FULA-PULAAR (The Gambia) chikara-pre (DRR) MANDING-MANDINKA hatayajambo = *green tea leaf* (Fox; DF) jambo-serila = *aromatic leaf* (DRR; DF) **GUINEA-BISSAU** CRIOULO balquiama (JDES) balquiana (EPdS) FULA-PULAAR (Guinea-Bissau) laboel (JDES) luboel (EPdS) qúiçala-púrè (JDES) MANDING-MANDINKA bóròbórò-menchena (JDES) **GUINEA** FULA-PULAAR (Guinea) kumba-dongul (Langdale-Brown) **SIERRA LEONE** KISSI ŋyanikpiɔ (FCD) yaniyo (FCD) KONO yandigbɛne yani (FCD) KRIO wet-ed-lif = *white head leaf* (FCD) MENDE ngulu-gbɛ *a general term for several annual weeds of farmland* (FCD) yani-gbɛ *a general term for some annual farm weeds* (FCD) SUSU-DYALONKE khampu-na (FCD) khempo-no (Haswell) TEMNE *an*-bal-*an*-yan (FCD) **LIBERIA** BASA omalu-ana (Okeke) MANO dah vŏ (Har.; JMD) **IVORY COAST** AKAN-ASANTE gua koro (K&B) gua kubo, vanvan (B&D) ANYI gua koro (K&B) BAULE kondrè (B&D) kudrè (B&D) kundré (K&B) DAN dussuo (K&B) duzi (K&B) GAGU maingué, manigué (K&B) taoné (B&D) KRU-BETE bonwo (K&B) GREBO (Nyabo) pono (K&B) ponopan (K&B) GUERE zanuin (K&B) GUERE (Chiehn) nébuï piti (K&B) néuripiti, nirimuri titi, urifapiti (B&D) pruli titi (K&B) NEYO logoniokui (K&B) KULANGO boiokro (K&B) KWENI leplaurauni (B&D) trubié (B&D) SENUFO-TAGWANA nufon (K&B) 'KRU' poni (K&B) 'NEKEDIE' uritapiti (B&D) **GHANA** AKAN-AKYEM adwowakuro (FRI) ASANTE guakuro (Rattray; JMD) FANTE efũmɔmoe = *turtle salt fish* (FRI) TWI guakro (JMD) gu-ɛkura = *fail village* (FRI) GBE-VHE mimã (auctt.) **NIGERIA** ANGAS yima RES EDO èbéghó-ẹ̀dòré (JMD; Lowe) EFIK íkọ̀ñ ífuộ èyèn = *leaf of the excreta of a child* (auctt.) IGBO (Asaba) ágádī-ísí-awa = *old person with grey hair* (NWT) ágádī-ísí-áwō-ọ̀chá = *old person with white hair* (NWT) anwuliriwani (NWT) IGBO (Onitsha) oso angweri ngwa (AJC) IGBO (Owerri) áhị̀hị́á-nwá-òshì n'ákà = *grass that smells in the hand* (auctt.) ákwụ̄kwọ́-nwá-òshì-n'ákà = *leaf that smells in the hand* (auctt.) IGBO (Umuahia) osu angweri ngwa (AJC; JMD) IJO (Nembe) fụ̀rụ̀ ítùkà = *smelling herb* (KW) IJO-IZON (Kolokuma) fụ́rụ́ túá = *smelling grass* (KW) fụ́rụ́ túó = *smelling herb* (KW) IZON (Oporoma) òbórífị̀ghá = *goat doesn't eat it* (KW) TIV hùhù (Vermeer) YORUBA ákọ̀ yúnyún (JMD) arùnsánànsánàn = *smell around* (Verger) imí-eṣú = *locust's excreta* (auctt.) **WEST CAMEROONS** DUALA èwùdu a njɔ̃ (Ithmann)

*NOTE: Many Sierra Leonean names refer to 'the plant which drives away Axonopus grass'* (FCD): yani *and variants = Axonopus spp.*

An annual herb to 1 m high, common throughout the Region, except in the driest situations, in open spaces, disturbed sites, etc., distributed pantropically and subtropically.

The plant is to some extent an anthropogene being common in the proximity of habitation. In Ghana its presence in too common abundance is deemed a bad omen as is implied by the Twi name (18, 19). The plant has a peculiar, even a rank, smell, likened in Australia to that of a male goat, by the Fante to the strong smell of salt fish, and by the Yoruba with more graphic expletives. Leopards in Kenya are said to be attracted by the smell (14). The Efik name arises not from the smell but from the use of the leaf to clean the bottoms of children (1).

The plant is not toxic, but it appears to be scarcely eaten by man as an item of diet. It is recorded only as added to soup in S Nigeria (27, 28). Notwithstanding the Ijo (Oporoma) name, all stock including goats browse it. The Yoruba consider it a delicacy for domestic guinea-pigs, and in Java it has been fed to horses and cattle (10).

A strong concoction of the leaves is considered in Nigeria to be a good tonic (3). The leaves are taken in Benin (Nigeria) with certain foodstuffs as an aid to conception (33). But conversely in Malaya it has been recorded that it is taken to prevent the birth of a quick succession of children (8). The leaves are considered to be antiseptic. Preparations are commonly applied to craw-craw in the Region (3, 10, 17), and to itch in SE Asia (8, 24). In Congo the sap is put onto prurient affections of the skin (6). The leaves are cicitrisant. They are applied to chronic ulcers (10, 23), to bruises, cuts and sores (3), and circumcision wounds (28) in Nigeria; to cuts and sores in Gabon (34, 35), Tanganyika (26) and in Ethiopia (13); as a haemostatic topically on wounds and haemorrhoids and intravaginally for uterine bleeding in Ivory Coast (7). The sap or the plant, dried and powdered, is a wound-dressing in Tanganyika (16), and is valued especially for burns (36); similar uses are recorded in SE Asia (8, 24). The leaves may have some analgesic action: powdered leaves are applied to the forehead for headache in The Gambia (25); the whole green leaf is so used in Nigeria (3); the sap in Congo (6), and mixed with clay in Ivory Coast-Upper Volta for headache and chest-pains (21). Leaves baked in palm-oil are used for rheumatism in Gabon (34, 35). A decoction makes a mouthwash in Congo for toothache (6). Similar anodynal applications are recorded from many other countries (36).

The leaves are taken internally in an aqueous macerate in Ivory Coast for all gastro-intestinal pains (21) and in Benin (Nigeria) (32). In Gabon they are given as a sedative and painkiller during pregnancy (34, 35), and in Ivory Coast in the event of prolonged and painful childbirth (7). Sap from the leaves is widely used for eye-troubles in Senegal (20), Ivory Coast and Upper Volta (7, 21), Ghana (18, 19), Nigeria (4, 23) and in Congo (6). Cattle-folk in eastern Uganda are said to use the plant as a prophylactic and as a cure for trachoma in their cattle (9). The leaves crushed in water are taken in Sierra Leone as an emetic (10), and a decoction in Togo (18, 19) and in Nigeria (10) as a febrifuge and for gonorrhoea (29). An infusion is taken in Gabon for blennorrhoea (34). In Congo leaf-sap is used as ear-drops for deafness, and is taken by mouth for cough, stomachache and tachycardia, and in decoction as a wash for fever in children (6). It is taken in draught in Ivory Coast for heart-pains and palpitations, as a vermifuge and to arrest diarrhoea and hiccups, and in snake-bite; sap is instilled in to the eye for jaundice (7). In Cabinda the plant is used for treating sleeping-sickness (11) and in Tanganyika for cough (32).

The roots are chewed in Tanganyika for indigestion (15) and abdominal pains (5, 36). A decoction of the flowers is taken in Madagascar for abdominal complaints (12) and the Ijo of S Nigeria prepare a decoction of the heads in fruit with alligator peppers (*Aframomum meleguetta* K. Schum., Zingiberaceae) for hernia (38). In Ghana young stems are poked into burnt charcoal and then rubbed on children's eyebrows to make the hair grow (19).

The whole plant in distillation yields a volatile oil, strong but pleasant smelling. Concentration is 0.02% in fresh material, 0.16% in dried. Constituents are phenolic esters similar to ethyleugenol which oxidize to release the strong smell of *vanillin* (20, 36). *Coumarin* is also present. Alkaloids have been reported (2, 8, 12, 20, 36, 37). Vegetative and reproductive parts yield hydrocyanic acid (8, 20, 36).

The plant has a number of magical and superstitious attributes. In Senegal, perhaps due to the strong smell released when bruised, it finds use in the treatment of mental disorders amongst other medico-magic (20). In Liberia pneumonia in children is treated by rubbing an extract of the leaves on the chest and then 'transferring to a stick' (Harley fide 10). In Ivory Coast it has protective fetish properties for the followers of the Snake Sect against snake-bite, and along with other plants it confers protection against ghosts and evil spirits (21). It enters into Yoruba incantations on the strength of its smell 'to placate the witches' (31a) and 'to kill bad medicine' (31b). In Gabon it is used in sorcery affecting a 'possessed' house (35). In Congo, leaf-sap on the hands of card players 'improves' their luck, and it is applied in a mild form of trial by ordeal — sap is spread on the accused's hand which is then pricked with a needle; only if guilty will any pain be felt (6).

The plant has been reported to be readily attacked by the nematode, *Heterodera radicola* (30), and as such might be useful as a trap to protect valuable but susceptible plants.

References:

1. Adams, R. F. G., 1943. 2. Adegoke & al., 1968. 3. Ainslie, 1937: sp. no. 20. 4. Alasoadura 3, UCI. 5. Bally, 1937. 6. Bouquet, 1969: 91. 7. Bouquet & Debray, 1974: 71. 8. Burkill, 1935: 71–72. 9. Cadbury s.n., Dec. 1956, K. 10. Dalziel, 1937. 11. Dawe 59, K. 12. Debray & al., 1971: 59. 13. Getahun, 1975. 14. Graham B.374, K. 15. Haarer 768, K. 16. Haerdi, 1964: 164. 17. Imperial Institute s.n., K. 18. Irvine, 1930: 12–13. 19. Irvine, s.d. 20. Kerharo & Adam, 1974: 222–3, with phytochemistry. 21. Kerharo & Bouquet, 1950: 212–3. 22. Koritschoner 872, K. 23. Oliver, 1960: 18. 24. Quisumbing, 1951: 958–60. 25. Rosevear, 1961. 26. Tanner 5744, K. 27. Thomas, N. W. 1604 (Nig. Ser.), K. 28. Thomas, N. W. 1616a. (Nig. Ser.), K. 29. Thomas, N. W. 1633 (Nig. Ser.), K. 30. Trochain, 1940: 267. 31. Verger, 1967. a. no. 48. b. no. 117. 32. Vermeer 72, UCI. 33. Vermeer 73, UCI. 34. Walker, 1953, a: 27. 35. Walker & Sillans, 1961: 125. 36. Watt & Breyer-Brandwijk, 1962: 197–8. 37. Willaman & Li, 1970. 38. Williamson, K. 28, UCI.

## **Ageratum houstonianum** Mill.

FWTA, ed. 2, 2: 287.

A herbaceous plant, more robust than *A. conyzoides,* native of Mexico, and introduced into cultivation. The plant is showy and makes an attractive annual bedding plant. An unnamed alkaloid has been detected in the seed (1).

Reference:

1. Willaman & Li, 1970.

## **Ambrosia maritima** Linn.

FWTA, ed. 2, 2: 268. UPWTA, ed. 1, 415.

West African: SENEGAL SERER nit ńiti (JB; K&A) nonâ ambel (JB; K&A) WOLOF ngâdal nak (JB; K&A) nginé (JB; K&A) **NIGERIA** HAUSA baàba, bàabaà, babbaba *loosely applied to plants like indigo or associated with it* (JMD; ZOG) baba more (JMD) ƙaiƙayi, makarfo, tutubidi (ZOG) HAUSA (East) matsermama (JMD) HAUSA (West) tutu bidi (JMD)

An annual herb or shortlived perennial from woody rootstock to 1 m high, of open waste spaces and riverain grassland, sometimes forming monospecific stands (Senegal, 5) from Senegal to N Nigeria, and widely distributed in the Mediterranean, NE, E and S Africa and Mascarene Islands.

The plant is aromatic. It is not touched by stock (1). It is sometimes added to soup in N Nigeria to flavour it (3). The plant is used with others by the Wolof of Senegal in the treatment of syphilis and as a stimulant (5), while in Ivory Coast the medicinemen appear not to use it (2). The leaves have been used in N Nigeria to draw inflammation from a whitlow (3). The plant has unspecified medicinal use in Uganda (6), and in Tanganyika the pounded-up root is steeped in cold water which is drunk to cause vomiting in the treatment of snake-bite (7). There is record of use of the plant on tumours in southern Europe (4).

The plant contains sesquiterpenic lactones, and *ambrosin* and *damsin* have been isolated. These substances act mildly on the heart, blood-pressure, uterus and intestines (2, 5).

References:

1. Adam, 1966, a. 2. Bouquet & Debray, 1974: 70. 3. Dalziel, 1937. 4. Hartwell, 1968. 5. Kerharo & Adam, 1974: 223. 6. Maitland 621, K. 7. Tanner 1531, K.

**Anisopappus africanus** (Hook. f.) Oliv. & Arn.

FWTA, ed. 2, 2: 258, as *A. dalzielii* Hutch. p.p., and *A. suborbicularis* Hutch. & Burtt, excl. *A. africanus* sensu FWTA, ed. 2.

A herb to 1.30 m high, of wooded savanna and montane grassland from Ghana to W Cameroons, and dispersed over the remainder of tropical Africa.
The dried powdered leaves are sternutatory and are used as a snuff in Uganda for headcolds (1).

Reference:

1. Purseglove P.464, K.

**Anisopappus chinensis** (Linn.) Hook. & Arn.

FWTA, ed. 2, 2: 258, as *A. dalzielii* Hutch. p.p.
West African: **NIGERIA** YORUBA (Ilorin) arójokú (Clarke)

A perennial herb to 1 m high, of wooded savanna and rough open grassland, from Guinea and Mali to Nigeria, and in central and south central tropical Africa and across Asia to Japan.

**Artemisia spp.**

UPWTA, ed. 1, 415.
West African: **NIGERIA** HAUSA (East) tagargade (JMD) tezaragade *perhaps a loan-word from Tamachek* (JMD) HAUSA (West) tagyalgyade (JMD) tazargade (JMD)

*NOTE: These names refer to the powder prepared from the leaves and shoots of A. judaica Linn., A. maciverae Hutch. & Dalz. and perhaps others.*

*A. judaica* Linn. ssp. *sahariensis* (Chevalier) Maire, which occurs in N Africa, is exported in some quantity to Mali where it is used as an aromatic condiment. Small bags filled with the dried plant are hung by the Tuareg in fig plantations in the Hoggar along the northern limit of the Region to ensure development of the fruit (2). This species, or *A. maciverae* Hutch. & Dalz., or a mixture of the two, is sold in N Nigerian markets as a fragrant powder with bitter aromatic properties under the names given above. Other species occurring in the Central Sahara are used as a vermifuge (1).

References:

1. Dalziel, 1937. 2. Maire, 1933: 212.

**Aspilia africana** (Pers.) C.D. Adams

FWTA, ed. 2, 2: 238–9. UPWTA, ed. 1, 415, as *A. latifolia* Oliv. & Hiern.
English: haemorrhage plant; wild sunflower (Liberia, Okeke).
West African: **SIERRA LEONE** KISSI nyana (FCD) KONO woyonɛ (FCD) MANDING-MANDINKA fidaŋwanyane (FCD) MANINKA (Koranko) woya (NWT; JMD) MENDE ngoyo (FCD) nguyu (auctt.) tɔnyɛ (FWHM; FCD) SUSU-DYALONKE sɔyɔ-na (Haswell; FCD) TEMNE *a*-surikbɛne (NWT) *a*-suru-*a*-ro-kant (NWT) **LIBERIA** MANO winnih (Har.; JMD) **GHANA** ADANGME-KROBO nasa (FRI) AKAN-AKYEM fofo (Ll.W) ASANTE mfũfũ (Enti) FANTE mfũfũ (auctt.) TWI nfõfõ, mfũfũ (Enti) GA nasa (FRI) NZEMA afole (FRI) afɔle (JMD) **NIGERIA** EFIK édémè èdọ̀ñ = *sheep's tongue* (auctt.) FULA-FULFULDE (Nigeria) nyarki (RES) HAUSA jamajina = *to draw up mucus* (JMD; ZOG) kalankuwa = *headband* (JMD; Singha) nanake (auctt.) toozalin-ƴan maataa (JMD; ZOG) IGBO ọ̀ráǹjìlá (Singha) ụ̀ráǹjìlà IGBO (Owerri) azuzo (Lowe) ọ̀rámejìlà

(Lowe) IGBO (Umuahia) ọ̀rámējìnà (JMD) IJO-IZON (Kolokuma) ịyọ́ụ́nkọ́rị́, òwéí ịyọ́ụ́nkọ́rị́ = *male iyounkori* (KW) odedé túá = *flowering grass* (KW) NUPE èyà yaká = *friend of pepper* (RB) YORUBA ákọ̀ yúnyún, ako : *male* (Egunjobi) yinrin-yinrin (JRA) yúnriyun (auctt.) yúnyún (auctt.) **WEST CAMEROONS** KPE mbnaso (AJC; Lowe)

*NOTE: These names may apply also to other A. spp., as well as to other genera with similar marigold-like flowers, e.g., Melanthera.*

A semi-woody herb from a perennial woody root-stock to 2 m high, very polymorphic with at least four varieties recognized in the Region, occurring throughout the Region on waste land of the savanna and forested zones, and widely distributed across tropical Africa.

The plant is a weed of cultivated land and fallows. It is of very rapid growth. It has a somewhat aromatic carroty smell. In Nigeria it is grazed by cattle and sheep, and is much used in the Western State as a food for rabbits and hares (2, 6). It has a high crude protein content.

The plant has a wide reputation and use as a haemostatic (Sierra Leone, 4; Liberia, 10; Nigeria, 1, 3, 11, 15) and in Liberia it is even credited with the capacity of arresting bleeding of a severed artery (3). The fresh bruised leaves and flowers are used. Their application is said to draw up exudations (hence the Hausa vernacular name meaning 'draw up mucus') and to promote rapid healing. Besides using the fresh leaf on cuts the Ijo of S Nigeria will also apply leaf-ash onto wounds and sores (15). A decoction has been recommended for use in treating pulmonary haemorrhage, and haemostasis is thought to be due to vaso-constriction (3, 11). In Tanganyika a root-decoction is taken for tuberculosis (7) and in Ghana the leaves are made into a cough-medicine for children (8). In Uganda a leaf-concoction is taken for gonorrhoea (5).

Water in which leaves have been squeezed with a little salt and lime-juice added is much used in Ghana as an eye-medicine to remove corneal opacities (8, 9). In Nigeria such water with only salt added is used as an eye-lotion for sun-blindness (1). The leaf-sap is also an eye-medicine in Tanganyika, dripped into the eyes for eye-pains of no apparent origin (7). The Igbo instill a leaf-decoction into the eyes for headache (12), and a decoction is used to wash the face and eyes to relieve feverish headache in Ghana (8, 9) and in S Nigeria (3).

A leaf-infusion is taken in Ghana to assist in childbirth (8, 9), and also in Nigeria (1). The leaf-sap and a leaf-decoction are rubbed onto the breasts and made into an inhalation in Tanganyika to promote milk-flow (7). An infusion mixed with white clay is taken by the Akan people of Ghana for stomach-troubles (14). Ijo of S Nigeria squeeze out the leaf-sap which is given with salt to revive a fainting person or to someone suffering from an attack of nerves (16). The Yoruba use the plant to cure crawcraw, and it features in an incantation to this end: the Yoruba name *yun yun* means 'scratch' (13). In francophone territories it is used in fumigation to remove guinea-worm (11)—see also *A. rudis* Oliv. & Hiern.

The plant is said to be used in decoction for washing horses, and the leaves to be admixed with other materials for plastering mud floors (3).

Amongst the Hausa, superstitious uses are prominent as a love-philtre, etc. A charm prepared from the plant and tied around the forehead attracts the 'glad eye' (*kalankuwa* means a headband); or a youth hides the plant in a maiden's house (3). For the proverb referring to jealousy between *Aspilia* and *Bidens* see *B. pilosa*. To the Ijo this species is male, its female counterpart being *Melanthera scandens* (Compositae).

References:

1. Ainslie, 1937: sp. no. 41, as *A. latifolia*. 2. Alasoadura 125, UCI. 3. Dalziel, 1937. 4. Deighton 4553, K. 5. Eggeling 23, K. 6. Egunjobi, 1969. 7. Haerdi, 1964: 165. 8. Irvine 478, K. 9. Irvine, 1930: 42–43, as *A. latifolia* Oliv. & Hiern. 10. Okeke 9, K. 11. Oliver, 1960: 19, 48, as *A. latifolia*. 12. Thomas, N. W. 64 (Nig. Ser.), K. 13. Verger, 1967: no. 180, as *A. latifolia*. 14. Williams, Ll. 7, K. 15. Williamson, K. 25B, UCI. 16. Williamson, K. 26, UCI. 17. Williamson, K. 27, UCI.

**Aspilia angustifolia** Oliv. & Hiern.

FWTA, ed. 2, 2: 239, incl. *A. linearifolia* Oliv. & Hiern and *A. paludosa* Berhaut.

An annual herb to 2 m high, of wet places in savanna country, in Senegal, Mali, Ghana and N Nigeria.

The golden-yellow or orange flowers to 2.5 cm across are very showy. The plant may have some value as an ornamental.

**Aspilia helianthoides** (Schum. & Thonn.) Oliv. & Hiern.

FWTA, ed. 2, 2: 239–40. UPWTA, ed. 1, 415, and 416 as *Blainvillea prieuriana* DC.

West African: **THE GAMBIA** MANDING-MANDINKA niambi borango (DF) **GHANA** AKAN-ASANTE sareso fõfõ (FRI) **NIGERIA** HAUSA kálánkúwaá, kalanwuka, jamajina, nanake, toozalin-y'an-maataa (ZOG) YORUBA abo yúnyún, abo: *female* (Egunjobi) yúnriyun fúnfún (JMD) yúnyún (JMD)

A herb to 1 m high, polymorphic with four subspecies recognized in the Region, of exposed places in grass-savanna, occurring throughout the Region, and extending to Congo and Angola.

The plant has an aromatic smell. It enters into herbal medicine in an unspecified manner in Nigeria, and it is also used as food for rabbits (1).

It is recognized by the Yoruba as the female (*abo yun yun*) of *Aspilia africana* (*ako yun yun*).

Reference:

1. Egunjobi, 1969.

**Aspilia kotschyi** (Sch. Bip) Oliv.

FWTA, ed. 2, 2: 239. UPWTA, ed. 1, 415.

West African: **NIGERIA** HAUSA jamajina = *to draw up mucus* (JMD; ZOG) jinin barewa (RES) nanake (JMD; ZOG) toozalin-y'an-maataa (ZOG) yankar mala (RES)

An erect herb to 1.3 m high with purple or brownish-red flowers in var. *kotschyi*, and white flowers in var. *alba* Berhaut, occurring in the soudanian region of Senegal to northern W Cameroons (Adamawa), and extending across Africa to E Africa and Angola.

The plant has the same Hausa names in N Nigeria as *A. africana* and will have the same applications to clean up exudations and promote healing. In Congo (1) and in Tanganyika (4) the leaf-sap is used as eye-drops for eye complaints. In Congo it is also used as ear-drops, and is rubbed on the chest and made into a tisane to drink for chest-affections (1). In Gabon the plant is used to soothe headache (5) and in Uganda a decoction of leaves is taken for gonorrhoea (3). The powdered root is applied to cuts by the Sukuma of Tanganyika causing any suppuration quickly to dry up (4).

Some tannin has been recorded in the plant but no other active principle (2).

Like *A. africana* it has superstitious applications. It is put into a number of philtres in Gabon (5). In Congo, canoemen claim the plant can protect them from crocodiles, and the plant-sap in nasal-instillations is said to revive the dead (1).

References:

1. Bouquet, 1969: 91–92. 2. Bouquet, 1972: 19. 3. Maitland 7, K. 4. Tanner 1424, K. 5. Walker & Sillans, 1961: 126.

**Aspilia rudis** Oliv. & Hiern.

FWTA, ed. 2, 2: 238, incl. *A. spencerana* Muschl.
West African: IVORY COAST BAULE lalobi (K&B) MANDING-MANINKA takan (B&D)

A coarse herb, often perennial, to 1 m high, of savanna grassland and woodland, from Guinea and Mali to Niger and N Nigeria.

The Baule of Ivory Coast prepare a decoction of this plant with the bark of *Spathodea campanulata* P. Beauv. (Bignoniaceae) for use in a vapour-bath to treat dracontiasis caused by the guinea-worm, *Dracunculus medinensis*. Treatment is given as soon as the swelling appears, and the swelling is lanced to facilitate removal of the worm (2). The plant also has antiphlogistic and antitussive properties (1).

References:

1. Bouquet & Debray, 1974: 71, as *A. spencerana* Muschl. 2. Kerharo & Bouquet, 1950: 213.

**Atractylis aristata** Batt.

FWTA, ed. 2, 2: 291.
West African: MAURITANIA ARABIC (Hassaniya) cherb djemel = *(inside of the) camel's cheek* (AN)

A prickly herb of dry areas, in Mauritania and Niger, and dispersed in the general Saharan region.

It produces good forage for all domestic stock (1, 2) and in Mauritania it is said to 'provide the camel with its hump' (2).

References:

1. Monod, 1950: 65. 2. Naegelé, 1958, b: 883.

**Ayapana triplinervis** (Vahl) King & Robinson

FWTA, ed. 2, 2: 285, as *Eupatorium triplinerve* Vahl.

A perennial herb, semi-woody at base, partly decumbent, to 60 cm high, recorded only from Dahomey, native of tropical America and widely dispersed by man in the tropics. The generic name is taken from a Brazilian vernacular (4).

The plant has at some time been official in the French and Indian Pharmacopoeias. It has long been held in esteem as a medical plant, but its virtues are overrated. It has no recorded usage in the Region. In Gabon it is used as a digestive, stimulant and sudorific in infusion (3). In India it is taken in infusion in small doses as a stimulant and tonic, in larger doses as a laxative, and in a hot infusion as an emetic and sudorific (2). In the Philippines it has use as a sudorific and tonic, especially in fevers, and in several other countries also (1). A decoction of the plant and leaf-sap are said to be detergent and are used on foul ulcers. A leaf-decoction is haemostatic, and an aqueous extract of the dried leaves and shoots are a cardiac stimulant (3).

The leaves yield on steam distillation an essential oil present at 1.0–1.4% concentration. The principal constituent is *thymohydroquinone dimethyl ether*. A *sesquiterpene* and *coumarin* are also present. A substance called *ayapanin,* which has haemostatic properties, has been isolated from the leaves (2).

The plant is often grown in Indian gardens as an ornamental aromatic shrub.

References:

1. Quisumbing, 1951: 983–4, as *Eupatorium triplinerve* Vahl. 2. Sastri, 1952: 233, as *E. triplinerve* Vahl, with pharmacological information. 3. Walker & Sillans, 1961: 128, as *E. ayapana* Vent. 4. Watt, G., 1889–93: 3 (DAC-GOR): 293.

**Berkheya spekeana** Oliv.

FWTA, ed. 2, 2: 288.

A herb to 1.30–2.00 m high, with spiny leaves, of montane grassland in W Cameroons, and occurring in NE, E and central Africa.

No usage is recorded for the Region. The plant is grazed by donkeys in Kenya (1). In Tanganyika a preparation is given for diarrhoea (3), and to children for debility (2).

References:

1. Glover & al., 2304, K. 2. Tanner 4505, K. 3. Tanner 5842, K.

**Bidens** Linn.

General names for the genus: English: beggar's ticks; burmarigold; stick-tight; sweethearts; tickseed.

The generic name meaning 'twice-toothed' refers to the teeth on what is commonly called the seed, but is correctly a dry single-seeded fruit, or achene. These teeth cause them to adhere firmly to clothing or the fur of passing animals, hence the vernacular names.

**Bidens bipinnata** Linn.

FWTA, ed. 2, 2: 234. UPWTA, ed. 1, 416.

English: hemlock beggar's ticks; Spanish needles.

French: herbe à aiguilles (Berhaut).

West African: **SIERRA LEONE** MANDING-MANDINKA sanyi (FCD) SUSU-DYALONKE saŋyi-na (FCD)

An erect, sub-bushy, annual to 1.3 m high, recorded only from Guinea, Sierra Leone, Ghana and N Nigeria, but widely dispersed pantropically.

The leaf has a bitter astringent taste. It is eaten in the Region as a pot-herb and in other countries of the plant's occurrence. The whole shoot is said to have a piquant flavour and to be eaten fresh or dried and stored in southern Africa. Cattle relish the young plant which has a high food value. The young leaves have a vitamin C content of 40 mg/100 gm. There is also present in comparative abundance a volatile oil of unpleasant smell which, however, may taint milk (1, 3).

The plant is able to set up localized irritation. The flowers and stem have given positive antibiotic action on *Staphylococcus aureus,* but none on *Escherichia coli* (3). In Sierra Leone the leaves are squeezed over boils and are also eaten to cure them (2).

Uses of *B. pilosa* are probably in common with this species.

References:

1. Busson, 1965: 417–8. 2. Deighton 4508, K. 3. Watt & Breyer-Brandwijk, 1962: 205.

**Bidens biternata** (Lour.) Merrill & Sherff

FWTA, ed. 2, 2: 234.

An annual weedy herb recorded briefly from Ivory Coast, Ghana and N Nigeria, and occurring from Ethiopia and Sudan to Angola.

A trace of alkaloid has been reported present in the leaves (1).

Reference:

1. Adegoke & al., 1968.

**Bidens pilosa** Linn.

FWTA, ed. 2, 2: 234. UPWTA, ed. 1, 416.

English: black jack; black fellows (Australia).

French: sornet.

West African: **SIERRA LEONE** KISSI dada *a general name* (FCD) KONO kandanɛ *a general name* (FCD) KRIO agbɛri-oku (FCD) nidul-lif = *needle leaf* (FCD) MANDING-MANDINKA sanyi (FCD) MENDE nana *a general name* (FCD) tombo-maga (*def.*-gei) (NWT) SUSU-DYALONKE saŋyi-na (FCD) **LIBERIA** KRU-GUERE (Krahn) niani (K&B) MANO zikilli wissi (Har.; JMD) **UPPER VOLTA** MANDING-DYULA nanguadian (K&B) **IVORY COAST** ABURE amonoablanfè (B&D) AKAN-ASANTE anasipagné (K&B) gilingui (K&B) BRONG abissawa (K&B) dinenkui (K&B) BAULE alongoa (B&D) alongoï (K&B; B&D) zagaï zagagbé (K&B) DAN sosolé (K&B) GAGU diandu (K&B) légué (K&B) nangua (K&B) KRU-GREBO (Nyabo) zagoi ini (K&B) GUERE lebason (K&B) tabason (K&B) tebasson (A&AA) GUERE (Chiehn) iréné (K&B; B&D) iuna (B&D) kukwe kwo (K&B) zebeyuzébogue (B&D) zegbei zegbagwè (B&D) GUERE (Wobe) kiradalé (A&AA) NEYO klakuo (K&B) KULANGO kokosa (K&B) manamendigo (K&B) KWENI gonoretti (B&D) pétéoré (B&D) MANDING-DYULA nanguadian (K&B) MANINKA passoklo (B&D) 'KRU' diaani (K&B) tagiaani (K&B) **GHANA** ADANGME-KROBO dsetʃi (FRI) AKAN-AKYEM kurofidie (FRI) ASANTE aseduro (Enti) TWI anansee mpaane = *spider's needles, referring to the narrow pointed achenes* (FRI) dwirantwi (FRI) gyinantwi (FRI) GBE-VHE adzrɔkpi (FRI) VHE (Pecí) dzani pipi (FRI) **NIGERIA** YORUBA abẹ̀rẹ (IFE) abẹ̀rẹ olóko = *farmer's needles'* (auctt.) agánran-mọyan (JMD) akẹsan (JRA) akẹsin-másọ, kẹshin: *to spur a horse* (JMD; IFE) ẹlẹshin-má-sọ = *has a spear but does not throw it* (JMD) kete kete (AJC) YORUBA (Ago-Are) eyinata (IFE)

An erect annual herb to 1.5 m high, of open disturbed ground, native of S America, but now distributed pantropically, and occurring commonly throughout all parts of the Region.

The leaves are eaten in West Africa as a pot-herb and in soups (7, 9, 13, 14, 16). They have a bitter astringent taste, and the young shoots 2.5–5.0 cm long are a favourite dish throughout southern Africa eaten fresh, or dried and stored for later use (19). The plant is readily browsed by all domestic stock including poultry and has a high nutritive value. The aromatic oil present with an objectionable smell may taint milk. It has been much used as fodder in S Africa especially for pigs (Howes fide 13, 14). But despite its use as food and fodder the plant is regarded as a topical irritant (19), yet withal it has soothing anodynal and analgesic uses. The plant-sap with or without the addition of pepper is used in the ear for earache and alone in the eye for conjunctivitis and as a styptic to arrest bleeding (2, 9, 10, 13, 14). In Ivory Coast eye-instillations are a treatment for jaundice (15). In Tanganyika the sap is applied to burns (11). In Ivory Coast (15) and in Nigeria (2) the leaf-sap or infusion is soothing in colitis and diarrhoea. The young leaves are also used in Ivory Coast for dysentery: softened over fire the leaves are admixed with the ash of banana skin, and taken by mouth. For babies the leaf only is given, or is administered by enema reduced to a paste in water (1). In Gabon sap from the stem is used for filaria in the eye (17). In Ivory Coast the plant is considered to be a nematicide and to relieve muscular pain; it is advocated as a poison-antidote, for snake-bite and small-pox (6, 15). Sap or a leaf-infusion is widely used as a cough-medicine in which tannin present may play the beneficial part. In Congo whole plant decoction is taken as a poison-antidote, to ease child-birth and to relieve hernia; a decoction is taken in draught, bath or vapour-bath, with the lees of freshly pulped leaves massaged topically onto the body for headache, fevers, and intercostal pain, and as a sovereign treatment for whooping-cough; the sap is used for headache, earache, ophthalmia and dental caries, and the leaves are rolled up into a rectal suppository for piles (4). The Zulu of S Africa give the powdered leaf in water as an enema for abdominal trouble and people of European descent take strong decoctions of the leaf in large doses and at frequent intervals for any inflammation (19). An infusion of the plant is given as a tea to babies inLiberia (3) (perhaps superstitious — see below), but an infusion is given in Ivory Coast as a sedative to agitated children (6) and in Tanganyika to children to remedy fits (18). In Mexico it is used as a tea-substitute, tonic and stimulant (19). A macerate in palm-wine alone or with *Melanthera scandens* (Compositae) is a counter-poison commonly used in the Lower Ivory Coast (15).

The flowers are considered a remedy for diarrhoea in Nigeria (2), and they are used in Ivory Coast-Upper Volta for bronchial and intestinal affections (15). No antimalarial activity has been found (8). The seeds, burnt and powdered, show a local anaesthesia, and are rubbed into cuts in Nigeria (2), and into scarifications on the body in S Africa (19) to relieve pain.

Traces of alkaloids and saponins have been recorded throughout Madagascan material (10), traces of steroids and terpene in the roots of Congo material and saponin in fair quantity in the leaves (5), and the presence of an unnamed alkaloid in the seeds (20). Other work shows the presence of flavones and antibiotic substances (6, 19).

The plant, as with the genus generally, is a good honey-plant (12).

In Ivory Coast-Upper Volta it is considered a fetish plant *par excellence* for children: a bath containing it gives strength and protection from illness, and afflictions resulting from a crane flying overhead can be averted by rubbing the limbs with its sap and scattering some of the lees on the ground so that passers-by take the ill-effects away with them (15). The plant features in Ghanaian proverbs. Asantes say that 'the black seeds of *Bidens* pray that the yellow *Aspilia* may become equally black' (9). *Bidens pilosa* often grows in competition with *Aspilia latifolia,* and an Akyem proverb runs '*Fofo* [*A. latifolia*] only lives to see *Gyinantwi* [*P. pilosa*] withered' (13).

The seeds are used by children in Gabon to play a game suspending them from the eyelashes, whereby comes the Fang vernacular name meaning 'tears of beauty' (17).

References:

1. Adjanohoun & Aké Assi, 1972: 94. 2. Ainslie, 1937: sp. no. 50. 3. Baldwin, Jr. 6781, K. 4. Bouquet, 1969: 92. 5. Bouquet, 1972: 19. 6. Bouquet & Debray, 1974: 71. 7. Busson, 1965: 418, 424–5 (with leaf analysis). 8. Claude & al., 1947. 9. Dalziel, 1937. 10. Debray & al., 1971: 59. 11. Haerdi, 1964: 165. 12. Howes, 1945. 13. Irvine, 1930: 55. 14. Irvine, s.d. 15. Kerharo & Bouquet, 1950: 213. 16. Schnell, 1960, a. 17. Walker & Sillans, 1961: 126. 18. Watt, 1967. 19. Watt & Breyer-Brandwijk, 1962: 205. 20. Willaman & Li, 1970.

**Blainvillea gayana** Cass.

FWTA, ed. 2, 2: 237.

West African: SENEGAL FULA-PULAAR (Senegal) légilégirdé (K&A)

An annual herb to nearly 1 m high, of open ground in the sahel/soudanian regions from Senegal to N Nigeria, and occurring in the Cape Verde Islands and in NE Africa and Arabia.

In Senegal the plant is not grazed by any stock (1) but in Zambia it is said to be heavily taken by wild game (? elephant) (2). In Senegal the Fula hold the plant to be toxic, but use the leafy stems in decoction and the powdered seeds as antiseptics in eye-treatment (3, 4).

References:

1. Adam, 1966, a. 2. Astle 5611, K. 3. Kerharo & Adam, 1964, b: 412. 4. Kerharo & Adam, 1974: 233–4.

**Blumea aurita** (Linn. f.) DC.

FWTA, ed. 2, 2: 261. UPWTA, ed. 1, 416.

West African: SENEGAL MANDING-BAMBARA boylé buti (JB, ex K&A) 'SOCE' frasâbukonaô (K&A) frâsesâkukunau (K&A) lubuńô (K&A) SERER bubuńgor (K&A) datotâ

(Merlier, ex K&A) lubuñ gor (JB) WOLOF ngun, ngun (JB, ex K&A) **GUINEA-BISSAU** PEPEL ompempene (JDES) umpempene (JDES) **GHANA** AKAN-FANTE afũn nena = *hunchback's mother* (FRI) TWI plaaduru (FRI) GA plaaduru (FRI; KD) **NIGERIA** IGBO (Awka) abanaadẹne = *vulture's excrement* (NWT, JMD) YORUBA (Ilorin) eru-taba = *slave of tobacco; meaning not known* (Ajayi)

An annual or biennial herb to 1.50 m high, of dry waste places from Senegal to S Nigeria. Var. *foliosa* (DC.) C. D. Adams of Senegal to Guinea occurs in damper, or even swampy localities. The species is widely dispersed throughout tropical Africa.

The plant is strongly aromatic and its use has been suggested for driving away insects and as a source of insect-powder (6). Its strong smell, even described as rank, evokes the Igbo vernacular meaning 'vulture's excrement.' The plant is not, or very rarely, grazed by stock (2), but var. *foliosa* occurring in the rice padis of Senegal is a good green-manure (1). The leaves are held to have cicatrisant properties. They are applied in Ghana to heal cuts especially those made by bush-knives (7, 8) and in Nigeria to bruises (3). In Senegal the leaves are bound as a last resort to phagedenic and chronic ulcers (9, 10). The pulped-up plant is applied to whitlows in Tanganyika (5). Relief of rheumatic pain is sought in Angola (11) and in Tanganyika (4) by topical application of an infusion. The leaves prepared as an enema are used in Ghana to cure constipation and dysentery (8), and the powdered plant or infusions are given in Nigeria for dyspepsia and indigestion (3).

References:

1. Adam, 1960: 366. 2. Adam, 1966, a. 3. Ainslie, 1937: sp. no. 52. 4. Bally 8298, K. 5. Haerdi, 1964: 165. 6. Holland, 1908–22: 383–4, under *B. lacera* DC. 7. Irvine 732, K. 8. Irvine, 1930: 58. 9. Kerharo & Adam, 1964, c: 295. 10. Kerharo & Adam, 1974: 224. 11. Watt & Breyer-Brandwijk, 1962: 207.

**Blumea gariepina** DC.

FWTA, ed. 2, 2: 261.

A herb or undershrub to 60 cm high, recorded in Niger, Sudan and south tropical Africa.

The plant has unspecified medicinal use in southern Africa. An aromatic oil has been obtained by steam-distillation of the leaf which consisted of 66% of *cineol*. The oil concentration was 0.5% of the leaf (1).

Reference:

1. Watt & Breyer-Brandwijk, 1962: 207.

**Blumea perrottetiana** DC.

FWTA, ed. 2, 2: 261.

A herb, erect to 1.30 m high, of open scrub and grass savanna from Senegal to N Nigeria and W Cameroons, and dispersed throughout the African tropics.

The plant is a common weed of savanna farms. It is strongly aromatic. Goats browse it in the Narok District of Kenya (2). The leaves are used in enemas in Nigeria and are said to have styptic properties. The powdered bark taken in milk or any bland drink is held to cure piles. The powdered root is stomachic and the whole plant is pounded and used as a febrifuge and to treat diarrhoea. The leaf-sap is a useful anthelmintic and anti-scorbutic, and eye-wash is made of it (1). In Tanganyika the plant-sap is taken for hookworm (ankylostomiasis), and is given to children for malaria (3).

References:

1. Ainslie, 1937: sp. no. 53, as *B. lacera*. 2. Glover & al., 1859, K. 3. Haerdi, 1964: 165.

**Bothriocline schimperi** Oliv. & Hiern

FWTA, ed. 2, 2: 283, as *Erlangea schimperi* (Oliv. & Hiern) S. Moore.

A shrub to about 1.30 m high, recorded in montane situations in Sierra Leone, S Nigeria and W Cameroons, and occurring also in NE and E Africa.

The leaves are strongly aromatic. The flowers are worked by hive-bees in Ethiopia (1).

Reference:

1. Ash 1469, K.

**Carthamnus tinctorius** Linn.

English: safflower.

An annual herb to 50 cm high, of the Mediterranean-Asia Minor region, occurring southwards across the Sahara, and cultivated in Mali (1). The plant is the source of safflower dyes, red and yellow (2), and are used for colouring butter, liqueurs, cosmetics, etc. The fruit yields a dry oil of an iodine value 140–150, and is the source of roghum and afridi waxes used for waterproofing oilcloth, tent-canvas, tarpaulins, etc. The fruits are edible. They are said to be good poultry food, and when dried to put into chutneys. The pressed seedcake is suitable for cattle food (3).

References:

1. Adams, C. D., 1956, a. 2. Greenway, 1941. 3. Uphof, 1968: 109.

**Centaurea nigerica** Hutch.

FWTA, ed. 2, 2: 291.
West African: **NIGERIA** HAUSA dáyii (JRA)

A perennial herb, erect to 1 m high, in Guinea and N Nigeria.

The roots have a bitter tonic property, and in large doses are emetic (1).

Reference:

1. Ainslie, 1937: sp. no. 86.

**Centaurea perrottetii** DC.

FWTA, ed. 2, 2: 292. UPWTA, ed. 1, as *C. alexandrina* sensu Dalziel.
English: star thistle (with other *C. spp.*).
West African: **SENEGAL** WOLOF gargam bosé (Merlier, ex K&A) homhom *general for spiny weeds* (JMD; K&A) xomxom (K&A) **MALI** SONGHAI alladjer kordji (A. Chev.) TAMACHEK akekchakar (JMD) **NIGER** SONGHAI yò-kérjí (*pl.* -ià) = *camel's thorn* (D&C) **NIGERIA** FULA-FULFULDE (Adamawa) suruuduyel (*pl.* suraaduhoy) (Taylor) FULFULDE (Nigeria) chaile (JMD) gi'el geeloobi = *camel's thorn* (MM) HAUSA ba-susa (JMD) ɗandar mahalba, danji, danyi, fárár kábàa (ZOG) daúdàr maáguzaáwaá = *dirt of the heathen* (*Hausa*) (JMD) ɗaudar maguzawo (ZOG) daúdàr-mahalbáa = *dirt of the hunters; alluding to the involucral scales which may pierce the feet* (JMD; ZOG) dáyii (JMD; ZOG) suraddu (ZOG) suraddu (JMD; ZOG) HAUSA (East) dai (Grove) deyi (JMD) HAUSA (West) danyi (JMD) farar kaba (Lely) kùbuúbúwar màkaàfií = *blind man's cobra* (JMD) surarad, surardu, surandu (JMD; ZOG)

A perennial herb, prostrate or erect to 50 cm high, leaves and flower-heads spiny, of dry sandy places from Mauritania to Niger, N Nigeria and W Cameroons, and through the Sahara.

The plant furnishes a common cattle fodder in N Nigeria (2, 3). It is also taken by horses (3) and other stock (4). It has use in N Nigeria in soup, but the whole plant is bitter and is a stomachic and purgative (3). There are closely related species official in several European pharmacopoeas for these actions (6). The Wolof of Senegal consider the plant valuable taken internally with other drug-plants for blennorrhoea, syphilis and fevers. They use the whole plant pounded to a paste for external application to the neck for ear-troubles, and dressings of leaves on ulcers and on dermal infections and parasites (5). The aerial part of the plant has been tested for action on avian malaria with but slight result (1).

A bitter principle, *calcitrapin,* has been isolated. This is probably the same as *centaurin.* The flowers contain pectins and a blue dyestuff, and perhaps glycosides (5).

Boys undergoing the circumcision ceremony are said to carry the spiny flower-heads to keep away others who might venture too near (3).

References:

1. Claude & al., 1947. 2. Dalziel 176, K. 3. Dalziel, 1937. 4. Grove 32, K. 5. Kerharo & Adam, 1974: 224–5, with phyto-chemistry. 6. Oliver, 1960: 22.

## **Centaurea praecox** Oliv. & Hiern

FWTA, ed. 2, 2: 291. UPWTA, ed. 1, 416, incl. *C. rhizocephala* Oliv. & Hiern.
English: star thistle (with other *C. spp.*).
West African: **MALI** SONGHAI alladjer (A. Chev.) **GHANA** DAGBANI balinyiri (auctt.) **NIGERIA** FULA-FULFULDE (Adamawa) suruuduyel (*pl.* suraaduhoy) (Taylor) FULFULDE (Nigeria) chaile (JMD) gi'el geeloobi = *camel's thorn* (JMD; MM) HAUSA ba-susa (JMD) daúdàr maáguzaáwaá, d.maguzawo (JMD; ZOG) dauɗar mahalba = *dirt of the hunters* (Hepburn; JMD) daúdàr-máhalbáa (ZOG) dáyií, danji, danyi (JMD; ZOG) farar kaba (ZOG) suraddu (JMD) HAUSA (West) danyi (JMD) farar kaba (Lely; JMD) kùbuúbúwar màkaàfií = *blind man's cobra* (JMD) suraradu, surandu, surardu (JMD; ZOG)

An annual herb to 60 cm high from a perennial woody rootstock, flowers in prickly heads at ground-level appearing before the vegetative shoots, from Mali to N and S Nigeria, and extending to Sudan.

Because the spiny-heads are at ground-level they are apt to cause injury by piercing bare feet: hence the vituperative Hausa name for the plant (3). The plant is used in northern Ghana to ease pains of child-birth (4). The roots have a bitter tonic property, and in large doses are emetic (1). The stems are used as fodder for camels and horses (2).

References:

1. Ainslie, 1937: sp. no. 86. 2. Dalziel, 1937. 3. Hepburn 15, K. 4. Williams 164, K.

## **Centaurea pungens** Pomel

West African: **MAURITANIA** ARABIC (Hassaniya) bu neger (AN) hachalilah (AN)

A herbaceous plant with spined bracts on the flowerhead, of rocky and sandy pasturage in Mauritania, and occurring throughout the Sahara.

It provides good grazing for all stock. The leaves are used by the Hassaniya in Mauritania for poulticing the nipples (1), but the purpose is not stated.

Reference:

1. Naegelé, 1958, b: 883–4.

**Centaurea senegalensis** DC.

FWTA, ed. 2, 291–2.
French: centaurée du Sénégal.
West African: SENEGAL FULA-PULAAR (Senegal) kodebâbi (K&A) kodem baba (K&A) SERER dakar ńig (K&A) dakarńé (K&A) ndakar ñik (JB) nogo ńtasé (K&A) WOLOF homhom (JB; K&A) xomxom (K&A)

An annual or perennial herb, prostrate or to 60 cm high, of the sahel from Mauritania to Nigeria and W Cameroons, and extending on to Sudan.

The Fula of Senegal take in draught a root-decoction or macerate of the whole plant crushed and steeped for 12 hours for stomachache due to constipation, and for dysuria (1, 2). The Serer prepare a paste of the crushed roots with millet flour which is plastered around the scrotum for orchitis (2). They also use the plant as protection against evil spirits and malign illnesses (2).

References:

1. Kerharo & Adam, 1964: 424. 2. Kerharo & Adam, 1974: 225–6.

**Centipeda minima** (Linn.) A. Br. & Aschers

FWTA, ed. 2, 2: 250.

A small weedy herb of damp situations, common throughout tropical Asia and the Indic basin, Australasia and the Pacific Islands, and now reported from Liberia.

The powdered leaves and minute seeds induce sneezing. They are used as a snuff for this effect in India for head-colds. Also taken as snuff the eyes are made to water and this is a Chinese treatment for ophthalmia in China and Malaya. In India an infusion is used for ophthalmia, and a thick paste of the leaves is applied to toothache (1, 3). The plant appears in the Chinese *materia medica* for polypus of the nose (2).

The plant contains an alkaloid, a glycoside, traces of saponin, an essential oil *myriogynic acid,* and a bitter principle *myriogynin* (3, 4).

The plant is said to cause poisoning of stock in Queensland (1).

References:

1. Burkill, 1935: 507–8, as *C. orbicularis* Lour. 2. Hartwell, 1968. 3. Sastri, 1950: 118. 4. Uphof, 1968: 118, as *C. orbicularis* Lour.

**Centratherum angustifolium** (Benth.) C. D. Adams

FWTA, ed. 2, 2: 283. UPWTA, ed. 1, 417.
West African: SIERRA LEONE MENDE kikpɔ-hina = *male kikpoi* (FCD) TEMNE *an*-balkayan-*a*-ro-bath, balkayan: *Ageratum sp.* (NWT; JMD) *an*-fola (FCD) *an*-lotho *perhaps a general name; correctly Crassocephalum crepidioides* (FCD)

A simple annual herb recorded from Guinea, Sierra Leone, Ivory Coast and N Nigeria.

The leaves are said to be eaten by fish at Njala, Sierra Leone (1).

Reference:

1. Deighton 3443, K.

**Ceruana pratensis** Forssk.

FWTA, ed. 2, 2: 255.
West African: SENEGAL WOLOF lilié (JB) muladĩ (JB)

An erect semi-woody annual to 60 cm high, of seasonally dry marshy ground and river-banks, across the northern part of the Region from Senegal to N Nigeria, and extending to NE Africa.

**Chromolaena odorata** (Linn.) King and Robinson

FWTA, ed. 2, 2: 285, as *Eupatorium odoratum* Linn.

English: Siam weed (Malaysia); Malay weed (Thailand,!); Enugu plantation weed (Nigeria, Jones).

West African: **NIGERIA** IJO-IZON (Kolokuma) fụ̄rụ̄tuo (KW) ògìdì dìrí = *medicine for matchet wounds* (KW) sèì tùó = *bad herb* (KW)

A semi-woody perennial shrub to 2.50 m high, of open places in Ivory Coast to Nigeria; of S American origin, it has been in the Asian tropics for a long time and has aggressively occupied vast areas of disturbed land in India and SE Asia. It first was observed in Nigeria at Enugu about 40 years ago occupying road-sides, waste and fallow lands (1). It is spreading in the Region and its distribution is certain to be much greater than that recorded.

In Asia it is an obnoxious weed invading plantations, forestry regeneration and agricultural land competing and interfering with crop-growth. Successful control has been achieved in Nigeria (1) and in SE Asia by spraying with 2, 4-dichlorophenoxyacetic acid (2, 4-D) or this in mixture with 2, 4, 5-trichlorophenoxyacetic acid (2, 4, 5-T). The plant is a strong light demander and can be shaded out if permanent crops can be closed up to form a canopy. The plant is also shallow-rooted and in India grubbing up and burning before flowering is advocated (2).

An unnamed alkaloid has been recorded present in the seed (3).

References (as *E. odoratum* Linn.):

1. Egunjobi, 1969. 2. Sastri, 1952: 223–4. 3. Willaman & Li, 1970.

**Chrysanthellum indicum** (Linn.) Vatke, var. **afroamericanum** Turner

FWTA, ed. 2, 2: 234–5. UPWTA, ed. 1, 417, as *C. procumbens* Pers.

West African: **SENEGAL** MANDING-BAMBARA fura kuna (JB) **GHANA** MOORE niba = *heart-leaf* (FRI) **NIGERIA** FULA-FULFULDE (Nigeria) giito gortogalhi (J&D) HAUSA goshin baana (Grove) HAUSA (East) ganshin gona (Grove) HAUSA raáriyár kása = *underground drain !* (auctt) IGBO (Asaba) ágádī-isí-awO = *old person with grey hair* (NWT) YORUBA abilẹ̀rẹ̀ (Philipps) ayigi (Macgregor) oyigi (Dawodu)

A faintly aromatic herb to 30 cm high, of open spaces, and dry rocky places of the forest zone from Senegal to S Nigeria, and widely distributed in the tropics. Old and New World plants hitherto considered as one species, *C. americanum* (Linn.) Vatke, are now separated.

The leaves in some parts of N Nigeria are mixed with henna for tinting the nails (3). The plant (part not stated, presumed to be the aerial portion) is used in Ondo province to make a poultice for maturating boils (2), and near Kaduma a medicine for fever in small children (4). In Igboland the plant is reported used for gonorrhoea (7). An infusion of the whole plant without the roots together with *Tamarindus indicus* (Leguminosae: Caesalpinioideae) is taken by the Fula of N Nigeria for *saworo* (hepatitis) (6) and similarly a decoction of the entire plant is taken in Congo for jaundice with urinary complications (1). Leaves mashed and mixed in shea butter, or in any other oil are taken in Ghana to cure heart-troubles in the northern territory; hence the Moore vernacular meaning 'heart-leaf' (5).

References:

1. Bouquet, 1969: 92. 2. Carpenter 1125, UCI. 3. Dalziel, 1937. 4. Grove 36, K. 5. Irvine 4911, K. 6. Jackson, 1973, as *C. procumbens*. 7. Thomas, N. W. 1621 (Nig. Ser.), K.

**Conyza aegyptiaca** (Linn.) Ait.

FWTA, ed. 2, 2: 254.

West African: **SENEGAL** MANDING-BAMBARA bolo mundi (JB) **SIERRA LEONE** TEMNE *an*-bɔ-*a*-ro-petr (NWT)

An annual or biennial herb reaching to 1 m high, of open waste places and cultivated land outside the forest but of the forest zone, from Senegal to W Cameroons, and throughout tropical Africa and Asia.

No usage is recorded for the Region. A decoction of the leafy stems is used by the Masai of Kenya as a body-wash for convalescents and for persons with skin-diseases for which it is said to be very soothing (1). The plant-sap is taken in Tanganyika as an anthelmintic for oxyurides (2).

References:

1. Glover & Samuel 2972, K. 2. Haerdi, 1964: 166.

**Conyza attenuata** DC.

FWTA, ed. 2, 2: 254, as *C. persicifolia* (Benth.) Oliv. & Hiern.

An erect annual herb to nearly 2 m high, of open waste places from Guinea to W Cameroons and Fernando Po, and widespread elsewhere in tropical Africa.

The leaves yield a black dye which is used in E Africa and in Zäire (1). They are used in Tanganyika for snake-bite (2).

References:

1. Greenway, 1941: as *C. persicifolia* Oliv. & Hiern. 2. Koritschoner 941, K.

**Conyza bonariensis** (Linn.) Cronq.

FWTA, ed. 2, 2: 253, as *Erigeron bonariensis* Linn.

An erect herb to 45 cm high, of open waste spaces in Ivory Coast and Ghana, and occurring in S. Tomé, Principe, Sudan and E and southern Africa.

No usage is recorded for the Region. The plant is browsed by goats in Sudan (1). In Lesotho, a leaf-decoction is taken for sore-throat, and is used in a preparation for ringworm and in lotion for washing sick children (2).

References:

1. Andrews 3613, K. 2. Guillarmod, 1971: 426.

**Conyza canadensis** (Linn.) Cronq.

English: Canadian flea-bane.

West African: **IVORY COAST** ABURE santika (B&D) KRU-GUERE (Chiehn) bliro (B&D) KYAMA adika (B&D) MANDING-MANINKA nuguban (B&D)

A herbaceous plant, native of eastern north America, now to be found in several tropical and subtropical countries, and present, perhaps introduced, in Ivory Coast.

The plant at some time has had a place in orthodox medicine in South Africa. In the United States it has been used as a haemostatic and remedy for diarrhoea and dropsy, for haemoptysis, cystitis, gonorrhoea and other urino-genital diseases. Its use against diarrhoea and dysenteries and as a wound-dressing is recorded, and a leaf-infusion has been successfully administered to counter diarrhoea in children allergic to milk who have been fed on soy-bean milk (4). It is used in Ivory Coast to relieve headache, fainting fits and eyesight troubles, and as a gum-friction for teething children (1). In Lesotho a leaf-decoction is drunk for sore-throat and as a wash for sick children. In northern United States American Indians add the plant to sweat-baths. The sap is however irritant to skin as is also the powdered leaf. The plant makes the eyes smart and grazing cattle suffer irritation of the nostrils. In parts of Africa the plant has become a troublesome weed (4).

The plant is aromatic with an essential oil that is principally *limonene,* the proportion of which varies with the stage of flowering. Tannins and gallic acids are also present (1). The powdered seed is sharply aromatic and a small but rather indefinite insecticidal activity has been recorded (2, 3, 4). There is some anti-bacterial activity in plant extracts, and leaf and flower extracts have given positive action against *Mycobacterium tuberculosis* (4).

References (as *Erigeron canadensis* Linn.):

1. Bouquet & Debray, 1974: 72. 2. Heal & al., 1950: 112. 3. Tattersfield & al., 1948. 4. Watt & Breyer-Brandwijk, 1962: 227–8.

**Conyza pyrrhopappa** Sch. Bip.

FWTA, ed. 2, 2: 251, as *Microglossa angolensis* Oliv. and *M. caudata* O. Hoffm. & Muschl.

An erect shrub to 1.30 m high, of stream-sides and forest-margins in hilly situations of N Nigeria and W Cameroons, and widely dispersed elsewhere in tropical Africa.

No usage is recorded for the Region. In the Narok District of Kenya it furnishes browsing for goats (1). The leaves pounded up and soaked in luke-warm water produce a very bitter infusion which is taken for malaria. The action is purgative and emetic (2). Masai of this district use the dried stems as firesticks (2). In Tanganyika the plant is recorded as used to treat boils (4), and the root for influenza (3).

References:

1. Glover & al., 467, K. 2. Glover & al., 637, K. 3. Semsei FH. 2912, K. 4. Tanner 5890, K.

**Conyza stricta** Willd.

FWTA, ed. 2, 2: 255.

An annual herb reaching to near 1 m high, of montane situations in Guinea and N Nigeria and occurring in the mountains of tropical Africa.

In Kenya the flowering tops are readily taken by cattle where the plant occurs in good grazing (1).

Reference:

1. Fox 3269, K.

**Conyza subscaposa** O. Hoffm.

FWTA, ed. 2, 2: 254.

A perennial herb to about 30 cm high of montane grassland in W Cameroons, and distributed across central north-eastern and eastern Africa.

The rootstock is recorded containing latex (1).

Reference:
1. Keay, FHI, 28020, K.

**Conyza sumatrensis** (Retz.) E. H. Walker

FWTA, ed. 2, 2: 263, as *Erigeron floribundus* (H. B. & K.) Sch. Bip.
West African: **SIERRA LEONE** KISSI nɛwa (FCD) **LIBERIA** MANO zo a gili (Har.) **IVORY COAST** KRU-GUERE gotuba (A&AA) **NIGERIA** IGBO (Umuahia) ázụ́ghụ̀zụ́ (AJC) IJO-IZON (Kolokuma) bịbị́ímbẹ́lẹ̀mọ́ (KW)

A herb to nearly 2 m high, of open waste spaces, roadsides, weed of cultivation in lowlands of transition zones between forest and savanna, and in montane situations, in Guinea to W Cameroons and Fernando Po, and widely dispersed in all warm countries.

The Guere of Ivory Coast pulp the leaves in a mortar with a little water and run off the green liquid which is then mixed with clay and rubbed on to the body of a suckling infant for fever (1). The plant has a reputation in Congo for treating ophthalmia, feverish conditions with cramp and stiffness for which the leaf-sap is instilled into the nose or eyes and rubbed on the patient's thorax (2). Nasal drops are also given for vertigo, epilepsy and attacks of insanity, and the leaves are made into cigarettes for tuberculosis and asthma (2). The plant is recorded as a 'fever medicine' in Tanganyika (3). In that country too, dried leaf-powder is spread over burns (5).

Goats graze the plant in the Narok District of Kenya (4).

References:
1. Adjanohoun & Aké Assi, 1972: 96, as *Erigeron floribundus* (H. B. & K.) Sch. Bip. 2. Bouquet, 1969: 93, as *E. floribundus* (H. B. & K.) Sch. Bip. 3. Ford 543, K. 4. Glover & al., 998, K. 5. Haerdi, 1964: 168, as *E. floribundus* (H. B. & K.) Sch. Bip.

**Coreopsis barteri** Oliv. & Hiern

FWTA, ed. 2, 2: 232.
West African: **NIGERIA** YORUBA yorungobi (Phillips)

An annual herb to nearly 1 m high of grassy savanna from Ghana to S Nigeria.

No usage is recorded for the Region. It is a weed of cultivation, fields and grassland (1).

Reference:
1. Irvine, s.d.

**Coreopsis borianiana** Sch. Bip.

FWTA, ed. 2, 2: 232.
French: coreopsis de Guinée (Berhaut).
West African: **SENEGAL** MANDING-BAMBARA kanalé (JB) **NIGERIA** YORUBA (Ilorin) eru-apata-aparo (Clarke)

An annual herb to nearly 2 m high, of open waste spaces, common throughout the Region from Senegal to W Cameroons, and across Africa to Sudan.

The golden-yellow flower heads 5.0–7.5 cm across are showy. No usage is recorded for the Region. In Sudan the plant (? leaves) is eaten with the addition of sesame, and it is browsed by goats (1).

Reference:

1. Andrews 37, K.

**Cotula anthemoides** Linn.

FWTA, ed. 2, 2: 250.

A spreading prostrate annual herb of open spaces, recorded only from Senegal in the Region, but widely dispersed in warmer countries of the Old World.

The plant is weakly aromatic resembling camomile (*Anthemis sp*). The leaves and roots are taken in Lesotho for colic (1, 2). In S Africa it is a remedy for head and chest-colds and is said to have been much used in the worldwide flu epidemic of 1919. Sometimes the leaf is crushed and stuffed into the nose (2). The plant has occasionally been used in southern Europe as an aromatic lawn in substitution for camomile (2), but its annual habit must be a drawback unless, as with camomile, non-flowering perennial cultivars can be found.

References:

1. Guillarmod, 1971: 420. 2. Watt & Breyer-Brandwijk, 1962: 221.

**Cotula cinerea** Del.

West African: **MALI** TAMACHEK gertufa (RM)

A herb to about 15 cm high, of sandy damp places occurring in Mali and along the northern border of the Region, and throughout N and NE Africa.

The Tuareg of the Hoggar use the plant as a condiment especially on meat that has begun to go off (1).

Reference:

1. Maire, 1933: 209.

**Crassocephalum crepidioides** (Benth.) S. Moore

FWTA, ed. 2, 2: 246. UPWTA, ed. 1, 418.

West African: **SIERRA LEONE** BULOM (Sherbro) kikpɔ-lɛ (FCD) kumeny-ɛ (NWT; JMD) GOLA gimbo (FCD) KONO kumbɔnɛ (FCD) LOKO gigboi (JMD) gipoi (JMD) jigboi (JMD) kigboi (JMD) kipoi (NWT) MANDING-MANDINKA kumbɛ (FCD) MANINKA (Koranko) kumba (NWT) MENDE kikpɔ, kipɔ (FCD; JMD) SUSU-DYALONKE kumɛn-na (FCD) TEMNE *a*-bwɔ (NWT) *an*-lotho (Glanville; FCD) VAI gɛnɛ (FCD) kimbɔ (FCD) **LIBERIA** MANO gũe (Har.; JMD) **GHANA** HAUSA babahun (FRI) babaku (FRI) **NIGERIA** HAUSA babohoh (DA) IGBO alapolo (NWT) ṉtị-ènē = *antelope's ear*, cf. éné: *harnessed antelope* YORUBA ebòlò (JMD; CWvE) ẹ̀fọ ébòlò, ẹ̀fọ́-ebúre

A stout erect herb to 1 m high, a weed of cultivation, disturbed land and waste places, occurring across the Region from Guinea to W Cameroons and Fernando Po, and generally distributed over tropical Africa.

The whole young plant and the semi-succulent leaves are mucilaginous and are eaten as a vegetable and in soups and sauces in the same manner as *C. rubens* (Juss.) S. Moore with which it shares many of the same vernacular names (4, 7, 14). The wild plant is collected and sold in northern Ghana markets (10), while in S Nigeria it is said to be cultivated for sale at Lagos (6). In Sierra Leone the leaves are eaten with especial relish with groundnuts (8). The plant is also eaten while young as a vegetable in Kenya (3). The leaves are mildly stomachic (7). They are used in S Nigeria for indigestion (5), and in Congo the leaf-sap is taken for upset tummy with colic and flatulence (2). The leaves prepared as a lotion (7) or a decoction (12) are also used in Nigeria as an analgesic for headache, and in Tanganyika the plant-sap, together with sap from *Cymbopogon giganteus* (Hochst.) Chiov. (Gramineae), is considered soothing when rubbed over the body in epilepsy, while at the same time the sap is taken internally (9). Also in Tanganyika the powdered leaves are used as a snuff to stop nose-bleeding (1, 14) and smoked in a pipe for sleeping-sickness (11).

The root is reported to contain tannin (14). It is cut up and boiled in water which is taken in Tanganyika internally twice daily for swollen lips (11).

The plant has magical attributes as a 'medicine for wrestling' at Ibuza, S Nigeria (13).

References:

1. Bally, 1937. 2. Bouquet, 1969: 93. 3. Broadhurst-Hill 157, K. 4. Busson, 1965: 418. 5. Carpenter 280, UCI. 6. Dalziel 1167, K. 7. Dalziel, 1937. 8. Deighton 4513, K. 9. Haerdi, 1964: 167. 10. Irvine 4781, K. 11. Tanner 1487, K. 12. Thomas, N. W., 787 (Nig. Ser.), K. 13. Thomas, N. W., 2019 (Nig. Ser.), K. 14. Watt & Breyer-Brandwijk, 1962: 221.

## **Crassocephalum montuosum** (S. Moore) Milne-Redhead

FWTA, ed. 2, 2: 248.

A much-branched herb, erect to 1.30 m high from a fibrous root, of open clearings in forested hilly districts of W Cameroons and Fernando Po, and dispersed in Congo, Sudan and E Africa.

It is a common weed of cultivation. In Kenya donkeys are reported to browse it (1) and in Tanganyika the plant is used for infected eyes (2).

References:

1. Glover & al., 1153, K. 2. Tanner 5918, K.

## **Crassocephalum picridifolium** (DC.) S. Moore

FWTA, ed. 2, 2: 248.
West African: **SIERRA LEONE** MANDING-MANINKA (Koranko) fugumba (NWT) SUSU-DYALONKE burukumena (Haswell)

A scrambling or erect herb to 60 cm high of open damp sites from Mauritania across the northern part of the Region to N Nigeria and in Fernando Po, and distributed into Sudan, E Africa, Congo and Angola. In Zambia it is reported to occur on lake water as floating mats.

No usage is recorded for the Region. In Kenya it is grazed by sheep and goats (1) and is fed to rabbits (3). In Tanganyika the leaves are used on wounds (2).

References:

1. Glover & al., 2036, K. 2. Koritschoner 1367, K. 3. Napier 2120, K.

## **Crassocephalum rubens** (Juss.) S. Moore

FWTA, ed. 2, 2: 248–9. UPWTA, ed. 1, 418, as *Gynura cernua* Benth.
West African: **GUINEA** FULA-PULAAR (Guinea) kumbé (JMD) **SIERRA LEONE** BULOM (Sherbro) kikpɔ-lɛ (FCD) LOKO gigboi (JMD) gipoi (JMD) jigboi (JMD) kigboi (JMD) kipoi (NWT) MANDING-MANDINKA kumbɛ (FCD) MANINKA (Koranko) kumba (JMD) MENDE kikpɔ, kipɔ (*def.*-i) *general names for this sort of plant* (FCD) SUSU kumenyi (JMD) SUSU-DYALONKE

kumɛn-na (FCD) TEMNE *a*-bwɔ (JMD) *an*-lotho (Glanville) **LIBERIA** MANO goên (JGA) goîn (JGA) **GHANA** AKAN-FANTE agyegyen-suani, agyegyen: *clear;* su: *water;* ani: *surface; i.e., clear water-surface when put in* (Enti) GBE-VHE (Pecí) banfa-banfa (Easmon) **NIGERIA** HAUSA baba hun (JMD) baba huun (JMD) IGBO n̠tì-ènē *see C. crepidioides* (JMD) IGBO (Afikpo) ewa isi (AJC) IGBO (Umuahia) n̠tì élē = *deer's ears* (AJC) JUKUN anigge (JMD) YORUBA ebòlò (JMD) èfó-ebúre, ẹfọ ébòlò (auctt.)

An erect herb to 1 m high, of open disturbed land in lowlands and montane situations from Guinea and Mali to W Cameroons, and widespread in the tropics.

The whole young plant and the semi-succulent leaves are mucilaginous and are a pot-herb eaten in soups and sauces (1, 5, 7, 9, 10, 15). In places the plant is cultivated and is an item of market trade (5, 7, 12).

In northern Sierra Leone two varieties are recognized and the leaves of both are eaten and are especially relished with groundnuts (8). The leaves are also eaten in Malawi with groundnuts and tomatoes but the product there is said to cause stinging in the mouth and to be not well-liked (16).

The leaves are slightly laxative. In the Nimba Mountain area of Liberia they are given to women after childbirth for this effect (1). In Sierra Leone the leaves are used as a medicine for 'belly palava' (stomachache) (11), and in Nigeria the whole plant may be so used (3). In Ubangi-Shari the leaves are eaten in quantity for liver-complaints (13), and in Gabon a leaf-infusion is taken against colds and leaf-poultices are applied to burns (14).

Leaf-sap is applied to sore eyes in S Nigeria (6, 7). In Congo the sap is instilled into the eye to remove filaria parasites (4), and in Malawi the leaves crushed in water are rubbed into the ear for earache (16).

A trace of alkaloid has been reported present in the leaves (2).

The powdered root has been used prepared as a paste for external application to breast-cancer in Nigeria (3).

Like garlic, the whole plant has repellant properties to crocodiles (7).

References:

1. Adam, 1971: 374. 2. Adegoke & al., 1968. 3. Ainslie, 1937: sp. no. 173, as *Gynura cernua*. 4. Bouquet, 1969: 93. 5. Busson, 1965: 421: leaf-analysis, 424–5. 6. Carpenter 297, UCI. 7. Dalziel, 1937. 8. Deighton 4513, K. 9. Irvine, 1948: 258–9, as *Gynura cernua*. 10. Irvine, 1956: 39–41, as *Gynura cernua*. 11. Jordan 333, K, FBC. 12. Lamb 37, K. 13. Portères, s.d.: as *Gynura cernua* Benth. 14. Walker, 1953, a: 28, as *Gynura cernua* Benth. 15. Walker & Sillans, 1961: 129, as *Gynura cernua* Benth. 16. Williamson, J., 1956: 41.

**Crassocephalum sarcobasis** (Boj.) S. Moore

FWTA, ed. 2, 2: 249.

An annual herb to about 30 cm high, of open disturbed land in montane situations of N Nigeria and W Cameroons, and occurring widespread throughout tropical Africa and Madagascar.

It is a weed of cultivated land. No usage is recorded for the Region. As it is similar to *C. crepidioides* it may possibly have the same applications. In Madagascar the dried haulm is powdered and put on the skin showing symptoms of leprosy. Traces of alkaloid have been recorded in the stem and roots, and an abundance of tannin in the stem (1).

Reference:

1. Debray & al., 1971: 60.

**Crassocephalum vitellinum** (Benth.) S. Moore

FWTA, ed. 2, 2: 248.

A weak straggling herb to 1 m high, of grassy clearing in montane situations in S Nigeria, W Cameroons and Fernando Po, and dispersed over the mountains of tropical Africa.

No usage is recorded for the Region. In Kenya it is recorded as browsed by sheep and goats (1–3), and in Tanganyika as used for infected eyes (4). The Haya of E Africa use the plant as a gonorrhoea remedy, for suppurations generally, and to improve the quality of a woman's milk (5).

References:

1. Bogdan AB.4168, K. 2. Glover & Samuel 3138, K. 3. Glover & al., 1930, 2318, 2578, K. 4. Tanner 5577A, K. 5. Watt & Breyer-Brandwijk, 1962: 222.

**Crassocephalum sp. indet.**

West African: **NIGERIA** FULA-FULFULDE (Nigeria) manda mbaala *cf. Nelsonia canescens* (*Acanthaceae*); = *sheep's salt* (J&D)

A plant, undetermined as to species, is used by the Fula of N Nigeria for hepatitis: broken fragments are taken in fresh milk (1).

Reference:

1. Jackson, 1973.

**Dichrocephala chrysanthemifolia** (Blume) DC.

FWTA, ed. 2, 2: 256.

An aromatic straggling herb to 1 m high, of open grassland and semi-shaded stream-sides in montane situations of W Cameroons and Fernando Po, and occurring throughout tropical Africa on high ground, and in Asia and Madagascar.

The plant is grazed by all stock in the Narok District of Kenya (2), and the reports that the plant is highly toxic with a reputation for producing abortions in domestic stock in a fashion similar to ergot have not been proven (3). The Masai of Kenya boil the roots, add milk and give the liquid to women after childbirth to relieve pain (1).

References:

1. Glover & al., 1114, K. 2. Glover & Samuel 1712, K. 3. Watt & Breyer-Brandwijk, 1962: 222.

**Dichrocephala integrifolia** (Linn. f.) O. Ktze

FWTA, ed. 2, 2: 256.

An annual herb to nearly 1 m high, of grassland and semi-shaded areas on high ground from Guinea to W Cameroons and Fernando Po, and widespread in tropical and subtropical Africa, Asia and Europe.

In the Narok District of Kenya the plant is reported grazed by all stock (2). One or two drops of the plant-sap are instilled into the eye in Congo to kill filaria parasites in the mucosae (1), and the plant dried and reduced to a powder is sprinkled over old injuries as a wound-dressing in Tanganyika (3).

References:

1. Bouquet, 1969: 93. 2. Glover & al., 1177, K. 3. Haerdi, 1964: 167.

**Dicoma sessiliflora** Harv.

FWTA, ed. 2, 2: 287.

West African: **SENEGAL** DIOLA (Fogny) sâbum elilit (K&A) FULA-PULAAR (Senegal) wordénâdé (K&A) MANDING-BAMBARA ńamé ngoin sina (JB; K&A) **THE GAMBIA** DIOLA (Fogny) sâbum elilit = *over-loaded sâbum* (DF) **NIGERIA** HAUSA doda (JRA) duda (JRA) HAUSA (East) alkamman kwaɗi (JMD)

A perennial herb to 1.30 m high from a woody rootstock, of wooded savanna in rocky and stony places in Senegal, Ghana and N Nigeria, and extending to Sudan, E Africa and Angola.

The whole plant is strongly bitter and is used in Nigeria as a febrifuge, particularly for children (1). In Casamance (Senegal) a decoction of the roots is taken as a cough-medicine and for stomach-pains (5). The root is used in Tanganyika for stomach-complaints (6), and a decoction is taken by draught to maturate hard abscesses (3). The plant has analgesic properties. Plant ash is rubbed into scarifications on the forehead and temples in Tanganyika for headache (3), and similarly root-ash is rubbed onto the sides and ribs to relieve stiffness due to unwonted effort (2). Ash is rubbed into scarifications on the chest and a root-decoction drunk for pneumonia, and leaf-sap and root-decoction for dysentery (3). In the Fouladou region of Senegal the root is considered aphrodisiac (4).

References:

1. Ainslie, 1937: sp. no. 134. 2. Allnutt 2, K. 3. Haerdi, 1964: 167. 4. Kerharo & Adam, 1964, b: 438. 5. Kerharo & Adam, 1974: 226. 6. Semsei F. H. 2137, K.

**Dicoma tomentosa** Cass.

FWTA, ed. 2, 2: 287. UPWTA, ed. 1, 417.

West African: **SENEGAL** MANDING-BAMBARA sali sali (JB) **NIGERIA** HAUSA dauɗa (JRA; JMD) farin dàyií (JMD) kwardauɗa (JMD; ZOG) surandu (JMD; ZOG)

An annual herb to about 60 cm high, of dry fields and sandy arid places in Senegal to N Nigeria, and extending to NE Africa and into Asia, and to Angola and Mozambique.

The whole plant is strongly bitter and is used in Nigeria as a febrifuge, particularly for children (2). It has febrifugal use in India, especially after childbirth (6). Cattle do not graze it (1). The plant is applied as a dressing to septic wounds in N Nigeria (3, 4). In Sudan the plant is used in a fumigation for skin-itch (5). In Tanganyika for cold in the head the Sukuma prepare a roll of bruised leaves which is inserted into the nostrils and left there for about an hour, and for pain in the testicles, leaves and fruit are burnt in a hole over which the patient squats (7).

References:

1. Adam, 1966, a. 2. Ainslie, 1937: sp. no. 134. 3. Dalziel, 1937. 4. Oliver, 1960: 25. 5. Patel & El Kheir 73, K. 6. Sastri, 1952: 57. 7. Tanner 1302, K.

**Echinops amplexicaulis** Oliv.

FWTA, ed. 2, 2: 291.

A stout herb to 2.50 m high with large spherical flower-heads to 12 cm across, of montane situations in W Cameroons, and also in NE and E Africa.

No usage is recorded for the Region. In Tanganyika the plant is used for stomach-pains (1).

Reference:

1. Tanner 5619, K.

**Echinops giganteus** A. Rich., var. **lelyi** (C. D. Adams) C. D. Adams

FWTA, ed. 2, 2: 291

An erect herb to 1.70 m high with large spherical flower-heads, of rough open country and on cultivated land, of N Nigeria and W Cameroons, and is limited to the Region. Var. *giganteus* is recorded from NE and E Africa.

No usage is recorded for var. *lelyi*. The flower-heads and leaves of var. *giganteus* provide grazing for oxen in Ethiopia (1).

Reference:

1. Pifford 94, K.

**Echinops longifolius** A. Rich.

FWTA, ed. 2, 2: 290. UPWTA, ed. 1, 417.

West African: GHANA AKAN-BRONG lunawefö (CV) GUANG-GONJA kasawuli bunku (Andoh) NIGERIA CHAMBA zé-ganó (FNH) FULA-FULFULDE (Adamawa) nyaalelhi = (?) *plant of the cattle heron* (MM) FULFULDE (Nigeria) boginaahi (Taylor) HAUSA guùluúlùn zoómoó (JMD; ZOG) HAUSA (West) boginahi (ZOG) YORUBA àgbẹ̀ (Dawodu)

A perennial herb to 1 m high, of savanna land, from Mali to N and S Nigeria and W Cameroons, and extending across Africa to NE and E Tropical Africa.

A decoction of the spherical flower-heads is used in N Nigeria as an eye-medicine (1, 3). In S Nigeria (6) and in Kenya (2) the plant has unspecified medicinal uses. In the Vogel Peak area the roots are boiled and taken by the Chamba for stomachache. They also boil up the plant and give the liquor to children who are shedding their milk-teeth: this is said to help loosen teeth to be shed (4, 5).

The plant has magical uses. To the Chamba a decoction added to bathwater confers protection against suspected witchcraft (4, 5). To the Fula of N Nigeria, the leaves cut up and sown with seed-corn ensures an increased yield (3). The Yoruba invoke the plant in an incantation to obtain money (7).

References:

1. Adegoke & al., 1968. 2. Bally 2469, K. 3. Dalziel, 1937. 4. Hepper 1286, K. 5. Hepper, 1965: 452. 6. Macgregor 152, K. 7. Verger, 1967: no. 16.

**Echinops spinosus** Linn. ssp. **bovei** (Boiss.) Maire

West African: MALI TAMACHEK tasegra (RM)

A perennial herb to 60–80 cm tall, very polymorphic, of the Mediterranean region, and ssp. *bovei* extending southwards across the Sahara to the northern limit of the Region, and now recorded in Mali.

The plant provides good forage for camels and other domestic stock (1, 2). In the Hoggar the dried stems are used as touchwood (1).

References:

1. Maire, 1933: 215. 2. Monod, 1950: 65.

**Eclipta alba** (Linn.) Hassk.

FWTA, ed. 2, 2: 241, as *E. prostrata* (Linn.) Linn. UPWTA, ed. 1, 417.

French: herbe à l'encre ('ink plant', Berhaut).

West African: **SENEGAL** WOLOF élektag (K&A) **SIERRA LEONE** KISSI bihûdo? (FCD) MENDE kpawu (FCD) **IVORY COAST** BAỤLE n-dalo-blé (A&AA) KRU-BETE grobidia (AJML) GUERE (Chiehn) kleiri iwonné (B&D) klériuémé (B&D) 'NEKEDIE' blignablé (B&D) **GHANA** AKAN-ASANTE ntum (FRI; JMD) **NIGERIA** HAUSA rimin sauro (Lely) IGBO (Onitsha) àgbịrịgbá ọzàrà, ọzàrà: *of the bush* (KW) IGBO (Uburubu) àgbịlịgbá ọzàlà (NWT) IJO-IZON (Kolokuma) òbírímá, pọ́ụ̀, óbírímá = *waterside obirima* (KW) TIV ichigh-ki-usugh (Vermeer) YORUBA abikolo (Dawodu; JMD; Macgregor) arójòkú (JRA; Millson; JMD)

A straggling herb to 50 cm high, of waste damp sites occurring throughout the Region and the whole of the tropics and part of the subtropics.

The plant is to some extent an anthropogene occurring frequently around houses, open spaces in towns and villages and on river-bank clearings, a weed of cultivation and in disturbed ground, along roadsides and railway lines, i.e. in man-made habitats. In Gabon it may even be planted (20). It is eaten as a pot-herb here and there in the Region (7, 8) and in Indonesia (6). It is sometimes put into chutneys in India (16). In Senegal cattle are reported not to graze it, but sheep do (1), while in N Kenya it is taken readily by all stock (15). The leaves are mildly laxative and in large doses emetic. An extract is given for constipation in Ghana (11, 12), and an infusion for diarrhoea in Nigeria (3). They are used in Ivory Coast for infantile diarrhoea after reduction to a paste in warm water and given as an enema (2). The plant's purgative and emetic properties especially in the roots are well-known in Asia (6, 16). The leaves are used in Ivory Coast to treat intestinal bleeding and for jaundice and convulsions in young children (5). The healing and cleansing property of the leaves is recognized in application of preparations to cuts and sores in Sierra Leone (14) and in Ivory Coast (5) and by the Tiv in Nigeria (18). In Tanganyika dried powdered leaves are sprinkled over injuries as a wound-dressing (10). A burn-treatment in Gabon is to compound the leaves with djavé nut butter (*Baillonella toxisperma,* Sapotaceae) into an ointment, and a macerate of leaves with citron juice is applied to circumcision wounds and in embrocations for many different ailments (19, 20). The sap from crushed leaves is taken for bronchial affections, and mixed with that of a pineapple and a pimento for massage over areas of oedema and kidney pains in Congo (4). With honey it makes a popular medicine for catarrh in infants (16).

The roots are used internally by the Wolof for liver complaints (13) and sap from root and stem is taken in Nigeria for disorders of the liver and spleen and for dropsy (3). A root-decoction is used in Tanganyika for pains in the abdomen (10).

The leaves contain a black dye furnishing an indelible stain, which explains the French name. For this either the sap can be used fresh or the leaves can be boiled and the liquor used (22). It has the capacity of blackening scar-tissue, so that when the leaves are used as a wound-dressing the scar heals a bluish-black. The sap is thus also used in tattooing, a practice known in Nigeria (3) and in E Africa (9) and India (16). It has been used in Asia from olden times as a hair-dye, and by extension put into preparations to promote hair-growth (6, 16).

An extract of the whole plant has given promising action against the virus causing Raniket disease. Inconclusive results have been obtained for action on cancers (13). Positive antibiotic action against *Staphylococcus aureus* and *Escherichia coli* is reported (16). An unnamed alkaloid has been found present in the leaf and stem (21), and *nicotine* at 0.078% dry weight (16).

The eel-worm, *Heterodera radicola,* is reported to be very partial to attacking this plant (17) which therefore offers the possibility of attracting and so removing this pest from cultivated plots.

The plant may be ascribed with magical properties. In Gabon it is put into a special beverage for people 'possessed', and into a lotion with other plants to anoint a supplicant for special favours from influential persons (20).

References:

1. Adam, 1966, a. 2. Adjanohoun & Aké Assi, 1972: 95, as *E. prostrata* (Linn.) Linn. 3. Ainslie, 1937: sp. no. 141. 4. Bouquet, 1969: 93, as *E. prostrata* (Linn.) Linn. 5. Bouquet & Debray, 1974: 71, as *E. prostrata*. 6. Burkill, 1935: 889–90. 7. Busson, 1965: 421, as *E. prostrata* Linn. 8.

Dalziel, 1937. 9. Greenway, 1941. 10. Haerdi, 1964: 168 as *E. prostrata* (Linn.) Linn. 11. Irvine 467, K. 12. Irvine, 1930: 172. 13. Kerharo & Adam, 1974: 226–7 as *E. prostrata* (L.) L. 14. Leeuwenberg 4539, FBC. 15. Mwangangi & Gwynne 1169, K. 16. Sastri, 1952: 127–8. 17. Trochain, 1940: 272. 18. Vermeer 42, UCI. 19. Walker, 1953, a: 5: 28. 20. Walker & Sillans, 1961: 127. 21. Willaman & Li, 1970. 22. Williamson, K. 35, UCI.

**Elephantopus mollis** Kunth

FWTA, ed. 2, 2: 269.
French: herbe à vache (Senegal, Berhaut).
West African: **NIGERIA** YORUBA arójòkú = *die in the rainy season*

A perennial herb, usually about 1 m high but occasionally attaining 2.60 m, of open grassy places in savanna woodland, fringing forest and cultivated land throughout the Region from Senegal to W Cameroons and Fernando Po; a tropical American plant now widely distributed in the tropics.

The French name suggests association with cows but there is no record of how this is. The plant contains a bitter principle and is used in Nigeria for inflammation of the uterus and ovaries, in anaemia and for coughs and fever (2). Its diuretic and febrifugal properties are made use of in Gabon (3). The leaf-sap is part of a Tanganyikan treatment for cardiac oedema (1).

References:

1. Haerdi, 1961: 168. 2. Oliver, 1960: 25, as *E. scaber*. 3. Walker & Sillans, 1961: 127, as *E. scaber*.

**Elephantopus spicatus** B. Juss.

FWTA, ed. 2, 2: 268–9.
West African: **SIERRA LEONE** KISSI bɛndiaiyondo? (FCD) MENDE kpagbanyo (FCD)

An erect herb to 60 cm high, native of tropical America and now reported from Guinea to Liberia, and spreading into Asia.

**Eleutheranthera ruderalis** (Sw.) Sch. Bip.

FWTA, ed. 2, 2: 236.
West African: **SIERRA LEONE** MENDE ndaima (NWT) TEMNE *a*-pethe-pethe (FCD)

An annual herb to 60 cm high, of wet sites near the sea from Senegal to W Cameroons and Fernando Po and pan-tropical in general distribution.

The plant is a common weed of cultivation.

**Emilia coccinea** (Sims) G. Don

FWTA, ed. 2, 2: 244. UPWTA, ed. 1, 417, as *E. sagittata* DC.
West African: **SIERRA LEONE** LOKO koya-gipoi (JMD) ndogbɔ-jigboi (JMD) MANDING-MANDINKA kumbɛ (NWT) MENDE ndɔgbɔ-gipɔ (*def.*-i), ndɔgbɔ: *regrowth of the bush or forest* (FCD) SUSU kumenyi (NWT) TEMNE *a*-bwɔ (NWT) **LIBERIA** MANO dein blih su (Har.; JMD) **MALI** MANDING-MANINKA kononi-abien (Brossart) **IVORY COAST** 'KRU' kluapo (K&B) **NIGERIA** IGBO ṇtị-ènē *see Crassocephalum crepidioides* (NWT; JMD) IJO-IZON (Kolokuma) kàlámà tọ̀rụ̀ ẹ̀dẹ̀dẹ̀ = *Kalama River flower* (KW) ọ̀kọ̀lọ̀màtọ̀rụ́ ẹdẹdẹ = *Okolohoma River flower* (KW) YORUBA ọ̀dúndún-odò (auctt.)

An annual herb, weak-stemmed, to 1 m high, of roadsides and clearings of forest country, dispersed from Guinea to N and S Nigeria, W Cameroons and Fernando Po, and occurring through E Africa from Sudan to Mozambique, and into tropical Asia.

The leaves are eaten fresh in salad or cooked as spinach in the Region (4, 5), in Kenya (6) and Tanganyika (15, 20). The plant is fed to rabbits and guinea-pigs in Gabon, and they eat it with relish (19).

The leaf is used by Igbo to cover sores (17). A leaf-poultice mixed with copper filings is used in Gabon to dress ulcers (18, 19). The sap is applied in Congo to ulcers, craw-craw, abscesses of the breast, leprous maculae and the ulcers of yaws, and also for mange, lice and ringworm (3). The dried powdered leaves are used in Tanganyika on sores after cleaning, and the fresh leaf pulped with ghee is put onto swollen legs (13). The plant is considered in Lagos to be a children's medicine, and in Port Harcourt has been reported used in poultice to heal the navel of a new-born baby (7). A leaf-decoction is put into a bath for the newly-born in Sierra Leone (5). A baby in Congo which cries too much, has convulsions, or has gross enlargement of the spleen is rubbed over with the leaf-sap (3), which is also used in Ivory Coast (9) and in Nigeria (12) for epilepsy and vertigo. A root-preparation is given to babies in Tanganyika for colic (2). For the eyes, fresh sap is instilled in Nigeria (1, 5, 12) and Gabon (18, 19) for soreness, and in Tanganyika for pink-eye, and a compress of leaves is applied for inflammation (2, 21).

A decoction of the leaves is used as a febrifuge in Nigeria, and has mild laxative properties (1). The leaves are pounded up with guinea-grains and are used for sore-throat and backache in Nigeria (1) and similarly in Tanganyika sap is squeezed out and taken with vegetables for sore-throat (16). The water in which leaves have been boiled is drunk as a treatment for syphilis (14), while the leaves enter a treatment for 'nose disease' (? syphilitic) (11). The root is also used for syphilis (10). In Congo leaf-sap is given for gonococcal infection, hernia and blood in the sputum and as an antiabortifacient and for menstrual troubles (3). In Kenya the root is said to be a chest-medicine (8) and in Gabon a leaf-macerate is taken for heart-affections (18, 19).

The Ijo of S Nigeria put the leaves with fish and cut-up plantain into a medico-magical preparation for treating painful nerves in the legs: the concoction must be prepared in an earthenware (not metal) pot used by a woman. A further superstitious use is as a love-charm: a man urinates on the leaves and calls the woman's name four times, and he must then contrive to get the leaves into her bath-water (22).

References:

1. Ainslie, 1937: sp. no. 144, as *E. sagittata.* 2. Bally, 1937. 3. Bouquet, 1969: 93. 4. Busson, 1965: 421. 5. Dalziel, 1937. 6. Graham N.402, 408, K. 7. Gregory 11, K. 8. Jeffrey K.528, K. 9. Kerharo & Bouquet, 1950: 214, as *E. sagittata* DC. 10. Koritschoner 583, K. 11. Koritschoner 1079, K. 12. Oliver, 1960: 28, as *E. sagittata* (Sims) G. Don. 13. Tanner 506, K. 14. Tanner 2918, K. 15. Tanner 3011, 3662, K. 16. Tanner 3592, K. 17. Thomas, N.W. 781 (Nig. Ser.) K. 18. Walker, 1953, a: 28, as *E. sagittata* DC. 19. Walker & Sillans, 1961: 128 as *E. sagittata* DC. 20. Wallace 127, K. 21. Watt & Bryer-Brandwijk, 1962: 226, as *E. sagittata* DC. 22. Williamson 24, UCI.

**Emilia praetermissa** Milne-Redhead

FWTA, ed. 2, 2: 244.
West African: **SIERRA LEONE** MENDE kipɔ (FCD) koyagipɔ (FCD) ndɔgbɔ gipɔ (*def.*-i) (FCD)

An erect herb to 1 m high, of open ground in the forest zone from Sierra Leone to S Nigeria, and a common plant in brackish *Avicennia* swamp.

Till recently this plant has not been distinguished from *E. coccinea.* It is eaten as a spinach in Sierra Leone (1) and liquid in which the plant has been boiled is used to wash new-born babies (2). It is certain to be used in the Region in other ways like *E. coccinea,* and it carries in Sierra Leone the general Mende name that is applied to all similar plants.

References:

1. Deighton 489, K. 2. Deighton 1188, K.

**Emilia sonchifolia** (Linn.) DC.

FWTA, ed. 2, 2: 244. UPWTA, ed. 1, 417.

West African: **GUINEA-BISSAU** KISSI kumba (FB) MANDING-MANINKA kumbi (FB) MANINKA (Koranko) kumbi (FB) MANO déblisou (FB) **SIERRA LEONE** MENDE kipɔ (JMD) **GHANA** AKAN-ASANTE guakuro (Ll.W; FRI) **NIGERIA** HAUSA hurahuran boka (RES) YORUBA ọ̀dúndún odò (SOA)

An annual weed to about 30 cm high, of open waste spaces, and weed of cultivated land from Senegal to S Nigeria and Fernando Po, and widespread throughout the tropics. The other two *E. spp.* of the Region are tetraploids. *E. sonchifolia* is diploid, 2n = 10, and is said to prefer a drier habitat and has a less extensive north-south range (1, 2, 7). Its presence in the drier northern states, which is not recorded in *The Flora of West Tropical Africa,* ed. 2, should be expected.

The plant is edible and is commonly used as a salad plant in SE Asia. It is said to be useful in areas with a pronounced dry season. The young leaves can be eaten raw if plucked before the flowering stalk lengthens; if after that the whole stalk and leaves can be eaten cooked. The taste is a delicate one, slightly acid with a touch of bitterness (3). It has a normal dietetic value (4).

The leaves are mixed with guinea-grains and limes as a remedy in Ghana for sore-throat (9). In Malaya the plant is administered to the form of a decoction for coughs and phthisis (3) and in India as a febrifuge and sudorific in infantile tympanites and bowel complaints and the root is taken for diarrhoea (8). The plant is used internally in Indo-China and externally in Java against fever (3). Tests for action against avian malaria have however proved negative (5). Leaf-sap is considered healing and is applied in India to cuts and wounds and is instilled into eyes and ears for soreness and for night-blindness (8). It is put in the eyes made dim by the sun and into the ears for soreness in Indonesia (3), and on to tumours (6).

The plant has a reputation as a bad omen in Ghana if present in a village for it is frequently found growing on ruins (9).

References:

1. Baldwin, 1946. 2. Baldwin & Speese, 1951. 3. Burkill, 1935: 921–2. 4. Busson, 1965: 421, 424–5. 5. Claude & al., 1947. 6. Hartwell, 1968. 7. Milne-Redhead, 1951. 8. Sastri, 1952: 172. 9. Williams 3, K.

**Enhydra fluctuans** Lour.
**E. radicans** (Willd.) Lack

FWTA, ed. 2, 2: 242, as *E. fluctuans* sensu C. D. Adams.

Perennial herb of swampy ground in coastal areas, till recently considered as a single species under the first name, but now recognized (2) to be two: *E. fluctuans* only in the Niger Delta, but widespread in the tropics, and *E. radicans* from Senegal to Dahomey and Fernando Po.

No usage of either species is recorded for the Region. The leaves of *E. fluctuans* are somewhat bitter and are eaten as a salad or vegetable in several tropical countries. They are said to be a laxative, antibilious and demulcent (1, 4). They are used in India in skin and nervous affections (4), and in the Philippines are applied to certain herpetic eruptions (3). A concentration of 0.21% dry weight of essential oil is present (4).

In Zaïre *E. fluctuans* has been reported a favourite food of the hippopotamus (5).

References:

1. Burkill, 1935: 924. 2. Lack, 1980: 6. 3. Quisumbing, 1951: 980–1. 4. Sastri, 1952: 173–4. 5. Van der Ben 706, K.

**Epaltes gariepina** (DC.) Steetz

FWTA, ed. 2, 2: 259, as *E. alata* (Sond.) Steetz.

A herb to 30 cm high, of the sahel zone of Mali and Niger, and widely dispersed across Africa to Sudan and southern Africa.

The plant is heavily grazed in Kenya (1). In S Africa poisoning of sheep has been traced to this plant when taken in the late flowering and seeding stage, but if given younger than this no ill-effect resulted in test animals. It is suspect of being toxic to other stock as well (5).

In Lake Province of Tanganyika the roots and leaves are pounded up and applied to sores after cleaning (4). In the Handeni District the presence of the plant is held to be an indicator of salt (2) and at Chifipa it itself is a source of salt (3).

References:

1. Anderson 1190, K. 2. Burtt 4914, K. 3. Michelmore 1484, K. 4. Tanner 1395, K. 5. Watt & Breyer-Brandwijk, 1962: 227.

**Ethulia conyzoides** Linn. f.

FWTA, ed. 2, 2: 284. UPWTA, ed. 1, 417.

West African: **SENEGAL** MANDING-BAMBARA varani do (JB) **GUINEA-BISSAU** FULA-PULAAR (Guinea-Bissau) socòra-bátè (JDES; EPdS) **LIBERIA** MANO suo lŏngo lah (JMD) swo hongo la (Har.) **NIGERIA** HAUSA mashenkuturu (Lely)

A herb to 1.30 m high, of wet grassland or in riverain situations, widely dispersed over the Region from Senegal to W Cameroons and elsewhere in the African tropics and into Asia.

The plant may become a troublesome weed of cultivation. In Zanzibar and Pemba it invades rice-padis (11). Some races in Gabon use the leaves for yaws, and administer a leaf-macerate to children to calm palpitations (8, 9). In central Tanganyika the plant is used as a birth-control for which the leaves are pounded and mixed with water which is drunk 2–3 days after conception (1). A decoction of the fine roots is also taken in Tanganyika as an abortifacient (3). By contrast the leaves are taken with food in Liberia to prevent abortion (2). In Gabon the cooked leaves are given at childbirth as an ecbolic (8, 9). Leaf-sap is taken in Tanganyika for diarrhoea (3) and in Gabon to relieve distension (9). It is a Zulu medicine for intestinal parasites, abdominal disorders and colic (10). The root with red pepper is given in an enema form in Liberia for constipation (2). The plant is considered anodynal: in Liberia the sap is squeezed in the eyes for headache, and a decoction of the leaves with those of *Baphia* (Leguminosae: Papilionoiideae) is made into a 'hot squat' for a patient with pelvic pain (2). Leaf-sap is squeezed into the eyes in Gabon for sore eyes (8, 9) and in Tanganyika for painful blood-shot eyes (6). The leaves are pressed into fresh cuts in the latter country (7) and a root-decoction is considered aphrodisiac (3). In Madagascar pounded leaves are applied to tumours (10), and over sprains and fractures, and the boiled leaves are used for wound and traumatic haemorrhages (5). The plant is also used for scabies (5).

The seed is recorded as containing saponin (10).

References:

1. Bally (Wilhelmi s.n.) B.11638, K. 2. Dalziel, 1937. 3. Haerdi, 1964: 169. 4. Hartwell, 1968. 5. Quisumbing, 1951: 982–3. 6. Tanner 2101, 3482, K. 7. Tanner 3037, K. 8. Walker, 1953, a: 28. 9. Walker & Sillans, 1961: 128. 10. Watt & Breyer-Brandwijk, 1962: 228. 11. Williams, R. O., 1949: 247.

**Fleishmannia microstemon** (Cass.) King & Robinson

FWTA, ed. 2, 2: 205, as *Eupatorium microstemon* Cass.

An annual herb to 1 m high, native of central America and the Caribbean, now occurring in Ivory Coast and Ghana.

It is a common weed of farms, roadsides and waste places of the forest zone. It first came to notice in Ghana in 1953, but it is likely to have been present for a long time before this as it has entered folk-lore as the 'male' counterpart of the more robust 'female' *Ageratum conyzoides* (1).

Reference:

1. Adams, C. D., 1964, a: 131.

**Gaillardia spp.**

Herbaceous plants with showy flowers, several species have been brought into cultivation as garden ornamentals and may be found grown in the Region. Some species have hairy leaves which may cause urticaria.

**Galinsoga parviflora** Cav.

FWTA, ed. 2, 2: 230

An erect annual herb to about 45 cm high, of open waste spaces in montane situations in W Cameroons. The plant is native of S America and has become dispersed to many upland situations in tropical countries. It must be expected to occur in other montane locations in the Region.

The plant is swift to colonize fallow and waste ground. It invades cultivation and is a serious weed-pest. In Kenya it is reported to smother plantings of young maize and millet (3).

The leaves are eaten as a vegetable in Tanganyika (5), and in Indonesia (1). Stock readily browse it in Kenya (3, 4) and in India (6). It is also taken by small antelope ('dikdik') in Kenya (3). Material in India has been analysed as containing: protein, 11%; fat, 0.8%; carbohydrate, 34%; fibre, 38%; ash, 16% (6).

Extract of fresh leaves and inflorescences is used in Ethiopia to dress new wounds and cuts, and also rawness and saddle sores (? on horses and camels) resulting in rapid healing (2). In Indonesia the leaves are said to be rubbed on the skin as an anodyne for nettle-stings (1). In Lake Province of Tanganyika the pulped foliage is applied externally for low backache (7).

The plant has strong cardio-vascular effects on laboratory test animals, and increases respiration rate in the dog. Tests for anti-bacterial activity have given negative results (8). Tests for insecticidal activity have also proved negative (1).

References:

1. Burkill, 1935: 1043. 2. Getahun, 1975. 3. Glover & al., 688A, K. 4. Glover & al., 1298, 1731, K. 5. Mtali 7, K. 6. Sastri, 1956: 98. 7. Tanner 4683, K. 8. Watt & Breyer-Brandwijk, 1962: 224.

**Galinsoga urticifolia** (Kunth) Benth.

FWTA, ed. 2, 2: 230, as *G. ciliata* (Raf.) Blake.
West African: **WEST CAMEROONS** 'BAMILEKE' chu mambọ (AJC)

An annual herb to about 30 cm high, native of S America, and now recorded from montane locations in W Cameroons.

The plant is alike to *G. parviflora,* but appears to be less aggressive as a weed of cultivation, and occupies more shaded positions.

**Geigeria alata** (DC.) Oliv. & Hiern

FWTA, ed. 2, 2: 257.

An aromatic herb to about 30 cm high, of sahel locations in Mauritania, Senegal, Mali and Niger, and a common weed of the drier parts of central and south-tropical Africa and on into Arabia.

It does not appear to be grazed by stock in Senegal except camels (1), and but lightly by wild game in Kenya (4). Various *Geigeria spp.* have been shown to cause disease and death in stock in south and central Africa. The active principle is *geigerin* and related substances (6). The W African species merits investigation.

It is added to spiced sauces by Arabs in Eritrea (2). In Swaziland a leaf and root-decoction is taken for giddiness (5).

In Tibesti it enters superstition and is a remedy against the 'evil eye' (3).

References:

1. Adam, 1966, a. 2. Bally B.6901, K. 3. Monod, 1950: 68. 4. Stewart 497, K. 5. Watt, J. M., 1967. 6. Watt & Breyer-Brandwijk, 1962: 230–4, with much detail.

**Gnaphalium luteo-album** Linn.

FWTA, ed. 2, 2: 266.

A bush, usually annual, of waste wet sandy places, wet pastures, cultivations, etc., throughout the Region from Senegal to W Cameroons and Fernando Po, and widely dispersed in warm countries.

In S Africa the cooked leaf is eaten (3). It is suspected of causing stock poisoning but tests have been inconclusive (3). Masai of Kenya burn the plant to ash which is mixed with ghee and given on the finger to babies as a tonic (1). In southern Africa the plant is burnt in the hut of a feverish child to drive away the illness (2, 3) and Australian aborigines consider the plant a general remedy for sickness (3).

Positive tests have been obtained for the presence of alkaloids (3).

The cut haulm is said to make a good improvized bed to lie on for want of anything better (1). In Lesotho it is laid in bundles on the ground to make a mattress on which skins are worked to make them supple (2).

References:

1. Glover & al., 984, 1125, K. 2. Guillarmod, 1971: 430. 3. Watt & Breyer-Brandwijk, 1962: 234.

**Gnaphalium polycaulon** Pers.

FWTA, ed. 2, 2: 266 as *G. indicum* sensu FWTA, ed. 2, non Linn.
West African: MALI DOGON dìì dóó (C-G)

A small prostrate herb covered with silvery hairs, to about 10 cm high, of waste places in the drier northern belt of the Region, and widely distributed in tropical Africa.

**Grangea maderaspatana** (Linn.) Poir.

FWTA, ed. 2, 2: 256.

West African: **SENEGAL** MANDING-BAMBARA bama kuo (JB) é fulimful (JB) **MALI** MANDING-BAMBARA fila-fila (Lean) **NIGERIA** HAUSA manjarafi (Ujor) YORUBA ponponla-tobi (JRA)

An annual or perennial herb or shrub of damp places, river-banks and seasonal river-beds from Senegal to N Nigeria and W Cameroons, and widely spread elsewhere in tropical Africa and in Asia.

The leaves are eaten raw in Nigeria, and are taken in infusion or in soup as a stomachic. The leaf is said to have a somewhat narcotic effect (1). Stock take it as a fodder in Mali (3), but in Tanganyika on the Momba River, though apparently the only green plant over large tracts, it does not seem to be eaten by game which abounds thereabouts (2).

Leaf-sap is used in Nigeria for earache and the powdered dry leaves or wet-leaf compress are applied to contusions (1).

References:

1. Ainslie, 1937: sp. no. 170. 2. Bullock, ad not., K. 3. Lean 23, K.

**Guizotia abyssinica** (Linn. f.) Cass.

FWTA, ed. 2, 2: 230.

English: Niger seed (the seeds, trade).

An erect annual herb to 1.30 m high, of open waste places in montane situations in W Cameroons, and widely dispersed, but sporadically from Ethiopia to Malawi, and occurring also in India and as a casual in Europe.

The plant is the source of commercial 'niger seed' for which it is cultivated in E Africa and in India. The black seed has an adherent thick seed coat, and can be stored for a year or so without deterioration, though the oil when extracted has the disability of rather rapid deterioration. Standard quality Indian seed assays at: moisture 7.8%, protein 19.40%, oil 31.3%, carbohydrate 39.7% and ash 1.8%. Oil content varies from 30 to 50% reaching a maximum at 45 days after flower-opening. The oil is semi-drying and the iodine value increases from 90 to 126 during the later stages of ripening. Fatty acids are: linoleic acid 51.6–54.3%, oleic acid 31.1–38.9%, palmitic acid 5.0–8.4%, and smaller quantities of a number of others. African oil contains as much as 70% linoleic acid which improves its drying and makes it better than Indian oil for paint manufacture. Niger-seed oil in comparison with linseed oil is considered inferior by British paint manufacturers and its use is mainly as an extender. Best quality oil is good for eating purposes. It is free of odour and has a pleasant nutty taste. It readily absorbs smells and appears to offer some scope in cosmetics.

The residual seed-cake can be fed to cattle, though it appears to be less palatable than sunflower-seed cake. It has a high manurial value.

The seed can be cooked and eaten whole. It is put into chutneys and condiments and serves as a cage-bird food in India.

Variation is known in the seed, and though little selection work has been done, this should be worth undertaking. Edaphic and climatic factors also need examination.

Cattle in India will not browse the plant and it is often planted around other crops to discourage trespass by cattle.

References:

1. Pearman & al., 1951. 2. Purseglove, 1968: 65–67. 3. Sastri, 1956: 270–5.

**Guizotia scabra** (Vis.) Chiov.

FWTA, ed. 2, 2: 230.

An erect semi-woody herb to 1.8 m high, of open waste places in montane situations in N and S Nigeria and W Cameroons, and extending widely from NE to south-central Africa.

The plant is cultivated on the Jos plateau of Nigeria (4), and in Sudan (5) and Ethiopia (2). The leaves are eaten in soup in Nigeria and the seeds are pounded and eaten raw (4). A leaf-decoction is taken in Uganda for stomachache and for gonorrhoea (6), and the plant (part not stated) is used in Tanganyika for syphilis (7). The plant is grazed by stock in Sudan (1), but apparently is not so in the North Kavirondo District of Kenya (3).

References:

1. Andrews 78, K. 2. Bally (Singer) 6561, K. 3. Gillett 19350, K. 4. Hepper 1160, K. 5. Lea 114, K. 6. Maitland 38, 182, K. 7. Tanner 4681, K.

**Gutenbergia nigritana** (Benth.) Oliv. & Hiern

FWTA, ed. 2, 2: 284.
West African: **SIERRA LEONE** MANDING-MANDINKA tondo (FCD) SUSU-DYALONKE burekume-na (FCD) kundindigli-na (KM) **NIGERIA** IGALA ẹbeju gbigbili (Boston)

A herb to 1.30 m high, of open grassland in Guinea, Sierra Leone, Ghana and N and S Nigeria.
The plant is used as an ingredient in the Igala Division of Nigeria of a medicine given to small children (1).

Reference:

1. Boston C.12, K.

**Gutenbergia rueppellii** Sch. Bip.

FWTA, ed. 2, 2: 284.

An annual herb to 60 cm high, of rough waste places and open land in N Nigeria, and occurring in NE and E Africa.
The plant is a weed of cultivation in Ethiopia and E Africa.

**Gynura miniata** Welw.

FWTA, ed. 2, 2: 243–4. UPWTA, ed. 1, 417, as *G. amplexicaulis* Oliv. & Hiern.
West African: **SIERRA LEONE** MANDING-MANINKA (Koranko) fugumba (NWT) **NIGERIA** HAUSA sitta (Singha) IGBO ṉtị = *ear* (Singha) TIV sètá (JMD)

A perennial, semi-succulent herb to 1.30 m high from hard woody tubers, of rocky hill savanna, from Guinea to N Nigeria and W Cameroons. This plant is perhaps the same as *G. pseudochina* Linn. of India, and if so the latter name takes precedence.
The plant is commonly grown in E Nigeria for use in soups and sauces. The leaves are mucilaginous. Protein and mineral contents are good, but oxalic acid may be relatively abundant though the soluble fraction may be removed in the preparation for eating and so render it safe. Selection of low oxalic acid strains is recommended (4). In the Benue area of N Nigeria it is planted to provide medicine for use in fever (2, 3). The fresh leaves are used in Nigeria for their demulcent property and the fresh sap is instilled into the eyes for soreness (5). The plant has a strong smell of musk (1), and it is said to have the same property as garlic in protecting against crocodiles (3).

References:

1. Adams, C. D., 1957: 114–5. 2. Dalziel 656, K. 3. Dalziel 1937. 4. Oke, 1966, b: 128–32, as *G. amplexicaulis*. 5. Singha, 1965.

**Gynura procumbens** (Lour.) Merrill

FWTA, ed. 2, 2: 243 (as *G. sarmentosa* (Blume) DC.). UPWTA, ed. 1, 418 (as *G. sarmentosa* DC.).

West African: **SIERRA LEONE** LOKO gigboi (JMD) gipoi (JMD) jigboi (JMD) kigboi (JMD) MANDING-MANINKA (Koranko) kumba (JMD) MENDE gimbo *a general term for some herbs* (FCD) kipɔ (JMD) SUSU kumenyi (JMD) TEMNE *a*-bwɔ (NWT) *an*-lotho (Glanville) **NIGERIA** IGBO n̠tị-ènē *see Crassocephalum crepidioides* (JMD) YORUBA eburẹ̀ (JMD) ẹ̀fọ́-eburẹ̀ (JMD) eru ebolo = *slave of ebolo*

A vigorous herbaceous climber of forest margins and thickets, native of SE Asia, and occurring, probably introduced, from Sierra Leone to S Nigeria.

The plant is grown in the rice area of Mindanao in the Philippine Islands as a 'remedy for rice aphids' (3), but the manner of its action is not clear. Florific horticultural cultivars of the species have been developed and are popular as house-plants in temperate countries.

The leaves serve in Malaya as a flavouring for food, and in SE Asia the plant is used for kidney-troubles, dysentery and as a febrifuge (1, 5). In Nigeria the semi-succulent leaves are applied to relieve rheumatic pains (2), while in Sierra Leone it is the boiled leaves that are applied for general body-pains (4).

References:

1. Burkill, 1935: 1122. 2. Dalziel, 1937. 3. Davies, 1980. 4. Deighton 240, K. 5. Quisumbing, 1951: 986.

**Helianthus annuus** Linn.

English: sunflower.

French: hélianthe, tournesol.

Portuguese: helianto, gira(s)sol.

A robust annual, usually about 2 m high, but sometimes attaining 4–5 m, of temperate countries, widely dispersed by man to most countries as an ornamental and as a commercial crop.

The plant is principally grown in Russia, Argentina, Roumania and eastern Europe for its oil-bearing seeds. The oil is valuable for cooking, margarine and other food products. The meal is rich in protein, and if kept free of the husk in milling is a valuable food-stuff. Testing in Nigeria produced normal amounts of oil at 22–25% of the whole seed and 40–52% of the kernel. This and samples from Ghana gave commercially satisfactory results. Decorticating machinery as that used for groundnuts has been found suitable.

References:

1. Bray & Major, 1945. 2. Major & al., 1948.

**Helichrysum** Mill.

FWTA, ed. 2, 2: 263.

English: everlastings.

French: immortelles.

The genus is one of a group of plants which as dried cut flowers retain form and colour and give rise to the English and French names.

An essential oil is extracted from some temperate species for use in a small way in perfumery. The E African species *H. kilimanjari* has been examined and reported on as producing a potentially useful aromatic oil (1). None of the W African species appear to have been tested, and examination seems warranted.

Reference:

1. Brown, 1950.

**Helichrysum cymosum** (Linn.) D. Don

FWTA, ed. 2, 2: 264, as *H. cymosum* (Linn.) Less.

An erect perennial herb or shrub to 1.30 m high, of montane grassland in W Cameroons and Fernando Po, and widely dispersed elsewhere in tropical Africa.
The roots are used in Tanganyika as an emetic and purgative (1).

Reference:

1. Koritschoner 921, K.

**Helichrysum foetidum** (Linn.) Moench

FWTA, ed. 2, 2: 264.

An aromatic herb to about 1.70 m high, of montane situations in S Nigeria, W Cameroons and Fernando Po, and widely dispersed elsewhere in tropical Africa.
The plant is a weed of cultivation with a strong aromatic smell of chrysanthemum. It furnishes browsing for goats in Kenya (1). The plant is used in Tanganyika for bad colds (4), and more specifically the roots for influenza (3) and for eye-troubles (2). The leaf is said to make an excellent dressing for a festering sore and in the Cape the plant has been used as an aromatic and astringent (5).
The involucral leaves contain *helichrysin* (5).

References:

1. Glover & al., 1159, K. 2. Koritschoner 784, K. 3. Koritschoner 1371, K. 4. Maber s.n., June 1924, K. 5. Watt & Breyer-Brandwijk, 1962: 238.

**Helichrysum glumaceum** DC.

FWTA, ed. 2, 2: 264.

An annual or perennial bushy herb, of littoral sand-dunes in Mauritania and Senegal, and in eastern Africa.
Around Dakar, Senegal, it is a codominant species on fixed dunes forming clumps to nearly a metre across by 30 cm high (2), and so contributes to stabilization of the sand.
The plant provides grazing for goats and sheep in Sudan (1). The Masai of Kenya burn the plant to ash which is mixed with ghee and fat and given on the finger-tip to small babies as a tonic (3).

References:

1. Andrews 3609, 3623, K. 2. Broadbent 32, K. 3. Glover & Samuel 3152, K.

**Helichrysum mechowianum** Klatt.

FWTA, ed. 2, 2: 266.

A perennial herb, reaching barely 20 cm high, in montane grassland subject to burning, in Togo to W Cameroons, and occurring across central Africa to Tanganyika, Rhodesia and Angola.

The powdered dry leaves are put over wounds in Congo, and the pulped leaves are rubbed on to pains in the joints and for lumbago. Leaf-sap is taken in palm-wine for anaemia, liver malfunction and stomach upsets. The plant is added to the drinking water of poultry and is said to be effective against 'chicken cholera' [? raniket disease] (1).

A trace of alkaloid has been detected in the whole plant (2).

References:

1. Bouquet, 1969: 94. 2. Bouquet, 1972: 19.

**Helichrysum nudifolium** (Linn.) Less.

FWTA, ed. 2, 2: 264.
West African: **WEST CAMEROONS** 'BAMILEKE' mba kokeka (Richards)

A perennial herb of montane grassland in Sierra Leone and W Cameroons, and widely dispersed elsewhere in tropical Africa. The Regional material is var. *leiopodium* (DC.) Moeser.

The plant is aromatic. Var. *nudifolium* and var. *leiopodium* are both used to make an infusion as a tea-substitute in southern Africa. The young leaves are preferred and the flavour is said to be not unpleasant. The plant is a South African cold-remedy: a decoction of the leaves is taken or the leaf itself may be eaten. In olden days it was regarded as demulcent and an infusion was taken for catarrh, phthisis and other pulmonary troubles. The root-decoction is also used as a remedy for coughs and colds, and as an emetic. A steam-bath of an infusion is taken for fever and bad dreams. The leaves are a South African wound-dressing, and in poultices are applied to swellings. A leaf-decoction is given in enema form to children with colic (1, 2).

Both the leaves and the roots are said to contain *helichrysin* (2).

References:

1. Guillarmod, 1971: 433. 2. Watt & Breyer-Brandwijk, 1962: 238–9.

**Helichrysum odoratissimum** (Linn.) Less.

FWTA, ed. 2, 2: 264.

A perennial herb, erect or straggling, to 1.70 m high, of montane grassland in S Nigeria, W Cameroons and Fernando Po, and widely distributed elsewhere in tropical Africa.

The plant is strongly aromatic. It produces an attractive, but non-persistent head of mustard-yellow flowers. In Lesotho it is burnt inside huts as a fumigation, and it is compounded with fat to make a pleasant smelling ointment used by women (2, 4). A leaf-decoction is taken in E Africa to relieve stitch and abdominal pains; the leaf-sap is given for heart-burn (4). The root is used for coughs and colds, and the leaf as a wound-dressing (4). The plant is also used in Tanganyika for syphilis (3), and ground up with fat is given in Kenya to children with stomach-troubles (1).

The roots and the leaves are said to contain *helichrysin*. Flowering tops of Kenyan material yield 0·2% of a clear pale yellow limpid volatile oil with a slightly camphoraceous smell, and consists largely of terpenes (4).

References:

1. Glover & al., 2138, K. 2. Guillarmod, 1971: 433. 3. Tanner 4689, K. 4. Watt & Breyer-Brandwijk, 1962: 239.

**Inula** Linn.

FWTA, ed. 2, 2: 259

*Inula helenium* Linn. of central and southern Europe is the Elecampane root, traditionally used as a flavouring for sweetmeats. The oil is put in vermouth and absinthe, and in medicine for chest-complaints and said to kill tuberculosis bacterium. The three W African species, *mannii* (Hook. f.) Oliv. & Hiern and *subscaposa* S. Moore occurring in the mountains of W Cameroons, and *klingii* O. Hoffm. of Guinea, Togo, N Nigeria and W Cameroons, need examining.

**Lactuca capensis** Thunb.

FWTA, ed. 2, 2: 293. UPWTA, ed. 1, 417.

West African: **GUINEA** FULA-PULAAR (Guinea) punta-bowali (Caille) **NIGERIA** FULA-FULFULDE (Nigeria) kaaɗe-kaɗɗe (MM) kaɗkaɗɗe (J&D) HAUSA nonokwarai (DA) YORUBA yánrin-oko (Macgregor; JMD)

A variable herb, from a few cm to over 1.70 m high from a woody rootstock, commonly dispersed from Guinea and Mali to W Cameroons and Fernando Po, and widespread in tropical and subtropical Africa and in the Mascarenes.

The young plant is eaten in Lesotho as a pot-herb (3). Mature leaves are made into an infusion and drunk by the Fula of N Nigeria for venereal disease, and pounded with natron are given to horses as a vermifuge (4). The powdered or pulped root is applied in Congo to sores, ulcers, leprosy and eczema (1). A trace of alkaloid has been detected in the roots, none in the leaves, nor any other active principle (2).

Reference:

1. Bouquet, 1969: 94. 2. Bouquet, 1972: 19. 3. Guillarmod, 1971: 437. 4. Jackson, 1973.

**Lactuca schulzeana** Büttner

FWTA, ed. 2, 2: 293.

An erect herb to 2 m high, of montane situations in W Cameroons, and occurring also in Congo, Angola and Uganda.

The sap ('one coffeespoonful') from the inflorescence-stalk is given to children as a vermifuge in Congo (1).

Reference:

1. Bouquet, 1969: 94.

**Laggera alata** (D. Don) Sch. Bip.

FWTA, ed. 2, 2: 262. UPWTA, ed. 1, 418.

French: menthe des Pygmées (Gabon, Walker).

West African: **MALI** ARABIC ('Maure') aura (A. Chev.) FULA-PULAAR (Mali) furu-furu (Caille) guilɗi (A. Chev.) MANDING-BAMBARA kundindan (A. Chev.) nitoro (A. Chev.) **UPPER VOLTA** FULA-FULFULDE (Upper Volta) mérigné (K&B) **IVORY COAST** AKAN-ASANTE saman'muto (K&B) FULA-FULFULDE (Ivory Coast) mérigné (K&B) KRU-BETE blaï kuri (K&B) GUERE bladeï = *sheep's ear* (K&B) GUERE (Chiehn) légé tamme = *false tobacco* (K&B) **GHANA** KRU-GUERE (Bappaw) agbaflu *a general name for composits* (A.S.Thomas) **NIGERIA** IGBO (Uburubu) ɟi ulili (NWT; JMD) YORUBA agẹmọ-kògùn, agẹmọ: *chameleon*, = *chameleon does not climb it* (Dawodu; Verger)

A stout herb to 3 m high, occurring throughout the Region from Guinea and Mali to W Cameroons, with var. *montana* C.D. Adams, somewhat smaller than var. *alata* and reaching only 2 m high, confined to the mountains of W Cameroons. The species is widespread elsewhere in tropical Africa and on into Asia.

The plant is strongly aromatic. The tender young leaves yield an ethereal oil (7) which might have some application in perfumery (5, 12). The leaf is smoked in parts of Gabon in place of tobacco. It has a narcotic effect (10, 11). Crushed leaves with ashes of the plant in frictions, and leaf-sap in draughts are given in Ivory Coast-Upper Volta for chest and intercostal pains, and a leaf-infusion is used as a fumigation in Gabon against rheumatic pains and fever (10, 11). Similar applications are found in Tanganyika (4): leaf-sap and root-decoctions are taken by draught for pneumonia, and root scrapings are rubbed into scarifications on the chest; and a plant-decoction is applied hot for muscular rheumatism and ashes of the plant are rubbed into scarifications. Inhalations and washes for the head are taken in Madagascar for headache, and the crushed leaves are massaged on the face and forehead for vertigo (2), and are used as a general disinfectant (7). The liquid resulting from prolonged boiling of leafy stems is used as a wash for sick persons by the Masai of Kenya (3). For enlargement of the spleen root-scrapings are rubbed over the area in Tanganyika (4), and the leaves pounded to a paste are applied to sore eyes (8). The plant-sap is taken in Ivory Coast-Upper Volta as an emmenagogue (5). In both Mali (1, 6) and Upper Volta (5) the plant is used as a taenicide: leaves, fruits and twigs are dried and reduced to powder. A small spoonful is taken in curdled milk after a day of fasting, and the tapeworm will be expelled with neither griping nor diarrhoea.

Saponins have been reported present in the plant (2).

Magical attributes are ascribed to it. The chameleon, a common prop of a wizard's paraphernalia, is said in S Nigeria never to climb the plant, and so the plant is invoked by the Yoruba to confer protection against witchcraft (9) — hence the significance of the Yoruba vernacular name. The Anyi of Ivory Coast put the sap into nasal instillations as a rapid cure for attacks of madness caused by evil influence (5). In Gabon the plant is planted around houses to counteract malign influences of sorcerers (11).

References:

1. Chevalier, 1937, b: 171. 2. Debray & al., 1971: 62. 3. Glover & al., 601, K. 4. Haerdi, 1964: 169–70. 5. Kerharo & Bouquet, 1950: 214, with references. 6. Portères, s.d. 7. Sastri, 1962: 25. 8. Tanner 3481, K. 9. Verger, 1967: no. 21. 10. Walker, 1953, a: 29. 11. Walker & Sillans, 1961: 129. 12. Watt & Breyer-Brandwijk, 1962: 244, with references.

**Laggera heudelotii** C. D. Adams

FWTA, ed. 2, 2: 262.

West African: **NIGERIA** YORUBA agẹmọ-kògùn (Macgregor)

An erect herb to nearly 2 m high, widely dispersed from Senegal to W Cameroons, and occurring also in the Congo basin and Angola.

The plant is used in Lagos, Nigeria, as a children's medicine (2). In Congo the sap from the leaves is used externally on localized pains and sprains. It is said to have analgesic properties. It is also a remedy for snake-bite (1).

References:

1. Bouquet, 1969: 94. 2. Macgregor 178, K.

**Laggera oloptera** (DC.) C. D. Adams

FWTA, ed. 2 2: 262.

A herb to 60 m high from a woody rootstock, of grassy savanna from Senegal to N Nigeria, and extending to Sudan, and Angola.

No usage is recorded for the Region. In Gabon the leaves are smoked in some parts as a substitute for tobacco, and produce a narcotic effect. The plant is also used medicinally in infusions, and in fumigations against fever and rheumatism (1).

Reference:

1. Walker & Sillans, 1960: 129, as *L. oblonga* Oliv. & Hiern.

**Laggera pterodonta** (DC.) Sch. Bip.

FWTA, ed. 2, 2: 262.

West African: **SENEGAL** MANDING-BAMBARA ko kuna sigi (JB) **NIGERIA** YORUBA taba agbe (AJC)

A robust herb to 1.70 m high, of open waste spaces and partially shaded galleried forest, recorded from Senegal, Sierra Leone, Nigeria and W Cameroons, and probably occurring elsewhere in the Region. It is widespread in other parts of tropical Africa and into Asia.

The plant is viscid and strongly aromatic. It is a weed of cultivation. In Lake Province of Tanganyika it is reported to kill tobacco plants if growing nearby (1).

Reference:

1. Tanner 973, K.

**Launaea chevalieri** O. Hoffm. & Muschl.

FWTA, ed. 2, 2: 296, as *Sonchus chevalieri* (O. Hoffm. & Muschl.) Dandy. UPWTA, ed. 1, 420, as *S. prenanthoides* Oliv. & Hiern.

West African: **NIGERIA** HAUSA námíjin dàyií (JMD; ZOG) námíjin surendi (JMD)

An annual or perennial herb to 60 cm high, in the drier northern part of the Region from Mauritania to Niger and N Nigeria, and in NE and E Africa.

The foliage is bitter.

**Launaea cornuta** (Oliv. & Hiern) C. Jeffrey

FWTA, ed. 2, 2: 296, as *Sonchus exauriculatus* (Oliv. & Hiern) O. Hoffm.

A perennial herb to 1.30 m tall, of the Vogel Peak area of W Cameroons and widely dispersed from NE Africa to South Central Africa.

The foliage is commonly eaten as a vegetable in Kenya (4, 5, 8, 11) and in Tanganyika (1–3, 15, 17, 19). It is bitter like quinine, and serves also to flavour food in Kenya (10). It is grazed by sheep and rabbits (14), and is fed to cattle in Tanganyika to increase milk-yield (6).

Water in which the leaves have been cooked to eat as spinach is used in Tanganyika as a hair-wash to kill lice (2).

The plant is held to have an analgesic property. The whole plant is taken in Tanganyika for pain in the spleen (18), and sap is instilled into the ear for earache (7). A cold water root-infusion is drunk in Malawi for stomachache (9) and a root-decoction with leaf-sap is taken for the same trouble in Tanganyika (7). Chopped-up into small pieces and cooked in a little water, the liquor is taken in Kenya for sore-throat (13). The root is used in Tanganyika for gonorrhoea (7, 15), syphilis (16) and cough (17), and as a lotion for eye-troubles (12). It is used in Tanganyika (7, 16) and in Malawi (9) for hookworm.

References:

1. Archbold 2, 565, K. 2. Bally 8064, K. 3. Faulkner 4101, K. 4. Graham 324, 1614, K. 5. Graham 1797, K. 6. Haarer 31 B, K. 7. Haerdi, 1964: 171, as *Sonchus exauriculatus* (Oliv. &

Hiern) O. Hoffm. 8. Jeffrey K13, K. 9. Lawrence 329, K. 10. MacNaughton 28, K. 11. Magogo & Glover 813, K. 12. Michelmore 1129, 1405, K. 13. Mwangangi 277, K. 14. Mwangangi 1285, K. 15. Tanner 926, K. 16. Tanner 1339, K. 17. Tanner 2945, K. 18. Tanner 4271, K. 19. Watt & Breyer-Brandwijk, 1962: 291, as *Sonchus bipotini* Aschers.

**Launaea intybacea** (Jacq.) Beauverd

FWTA, ed. 2, 2: 294, as *Lactuca intybacea* Jacq.

An annual herb, erect to nearly 2 m high, in the sahel zone of Senegal, Ghana, Niger and N Nigeria, and distributed throughout the Old and New World tropics.

The plant is grazed by goats in Eritraea (2), and by all stock in Sudan (1) and Somalia (4). It is said in Somalia to be unfit (February) for lambs to eat it or they will die (3), cf. *Sonchus oleraceus*.

References:

1. [? Andrews] A. 3477, K. 2. Bally B.6625, K. 3. Burne 17, K. 4. Peek 12, K.

**Launaea nana** (Bak.) Chiov.

FWTA, ed. 2, 2: 296, as *Sonchus elliotianus* Hiern.
West African: **IVORY COAST** ABURE apuifa (B&D)

A perennial herb with leafless flowering shoots to a few cm high from a thickened lactiferous rhizome, occurring widely dispersed over the Region from Guinea to W Cameroons, and throughout tropical Africa.

The plant is thought to be toxic and to cause violent vomiting (1).

Reference:

1. Watt & Breyer-Brandwijk, 1962, 291.

**Launaea taraxacifolia** (Willd.) Amin

FWTA, ed. 2, 2: 293–4, as *Lactuca taraxacifolia* (Willd.) Schum. UPWTA, ed. 1, 418, as *L. taraxacifolia* Schum. & Thonn.

English: wild lettuce.

French: langue de vache (=*tongue of the cow,* Dalziel).

West African: **SENEGAL** WOLOF valovalo *i.e. of Walo, a district of Senegal* (JB; DF) **SIERRA LEONE** KISSI bekuhoa-pɔmboɛ̃? (FCD) KRIO ɛfɔ-nyɔri (FCD) MENDE kipɔ (JMD) TEMNE *a*-mɔthsəra (FCD) **GHANA** ADANGME-KROBO kusu (A.S. Thomas; FRI) ʃwie (FRI) AKAN-AKUAPEM nne-noa = *boil today* (FRI) TWI dadedru (FRI) mmrobo (FRI) nne-noa = *boil today* (Enti) GA àgblɔke (KD) fie (Bunting, ex FRI) GBE-VHE bɛblɛ tama = *frog's tobacco* (FRI) bɛlɛ tama (FRI) VHE (Awlan) aŋɔto (FRI) VHE (Pecí) bɛblɛ tama, bɛblɛ: *frog;* tama: *tobacco* (FRI) bɛlɛ tama = *frog's tobacco* (FRI) **DAHOMEY** GBE-VHE latotué (A. Chev.) niontoto (A. Chev.) **NIGERIA** HAUSA námíjìn dàyií *applied loosely* (JMD; ZOG) nomen barewa (Bargery; ZOG) nonanɓarya (Bargery; ZOG) YORUBA ẹ̀fọ́ yánrin = *wild lettuce* (Verger) ọ̀dúndún odò (SOA) yánrin (auctt.)

A herb with basal rosette of leaves and erect stems to 1.3 m high from a woody rhizome, from Senegal to S Nigeria, and dispersed to Sudan and Ethiopia.

The leaves are eaten fresh as a salad or cooked in soups and sauces. The plant is often grown for cropping the leaves which are sold in markets as cooked or rolled-up balls prepared for use. (1, 2, 5–8.) The leaves are fed to cows-in-milk in N Nigeria to increase the yield, and to sheep and goats mixed with natron to produce multiple births (2). Mineral-content of the leaves is relatively high

(21·8%–1), and the plant is burnt to ash in N (4), and S Nigeria (3) to prepare a vegetable-salt. The leaves mixed with fine ash are rubbed onto the sores of yaws in Ghana (9) and the boiled leaves are applied to the head of a newly-born baby if the bones have not knit together properly (8). The limbs of backward children are rubbed with the leaves in Ghana to induce them to walk (Bunting fide 2, 5).

The plant is so ubiquitously appreciated that it features in an Yoruba invocation for someone to become well-known in the community (10).

References:

1. Busson, 1965: 421: leaf-analysis, 424–5, as *Lactuca taraxacifolia* (Willd.) Schum. 2. Dalziel, 1937. 3. Gwynn 86, K. 4. Gwynn 136, K. 5. Irvine, 1930: 252–3, as *L. taraxacifolia* Schum. & Thonn. 6. Irvine, 1948: 259, as *L. taraxacifolia*. 7. Irvine, 1956: 38, as *L. taraxacifolia*. 8. Irvine, s.d. 9. Thomas, A. S. D. 5, K. 10. Verger, 1967: no. 87, as *L. taraxacifolia* Schum. & Thonn.

## **Melanthera elliptica** O. Hoffm.

FWTA, ed. 2, 2: 241. UPWTA, ed. 1, 418.

West African: **GHANA** GBE-VHE adzrɔkpi (Ewe Dict.) NZEMA anamawuramfo (WTSB) **NIGERIA** HAUSA ja majina = *to draw out sickness* (JMD) kalankuwa (JMD) nanake (JMD) tozalinyam mata (JMD)

*NOTE: Ghanaian names under M. scandens may perhaps apply.*

A perennial bushy herb to 1.3 m high, of the savanna, occurring from Mali to W Cameroons, and extending to Ubangi-shari.

The plant shares vernaculars with *M. scandens* and *Aspilia africana*, and it probably has similar applications.

## **Melanthera gambica** Hutch. & Dalz.

FWTA, ed. 2, 2: 240.

West African: **SENEGAL** FULA-PULAAR (Senegal) surkémé (K&A)

A herb to 60 cm high, of the wooded soudanian and guinean savanna in Mauritania, Senegal, The Gambia and Guinea-Bissau.

The Fula in Senegal compound the roots of this species with those of *Cochlospermum tinctorium* (Cochlospermaceae) and the leaves of *Combretum glutinosum* (Combretaceae) in a prescription for jaundice and 'yellow fever'. The preparation is purgative and diuretic (1).

Reference:

1. Kerharo & Adam, 1974: 227.

## **Melanthera scandens** (Schum. & Thonn.) Roberty

FWTA, ed. 2, 2: 240–1. UPWTA, ed. 1, 418, as *M. brownei* Sch. Bip.

West African: **SIERRA LEONE** KISSI nyanya (FCD) KONO woyonɛ (FCD) MANDING-MANDINKA fidaŋwanyane (FCD) MENDE ngoyo (FCD) tɔnyɛ (FCD) SUSU-DYALONKE sɔyɔ-na (FCD) **LIBERIA** MANO wẽnẽ (Har.; JMD) **UPPER VOLTA** MANDING-DYULA missifabimbii (K&B) SENOUFO nanalékalé (K&B) **IVORY COAST** VULGAR himbien-na (A&AA) ABURE natuaté (B&D) nétuan até (B&D) AKAN-BRONG waga waga (K&B) BAULE aofuin (K&B; B&D) lalobè (B&D) GAGU niania mo (B&D) niblida (K&B) KRU-BETE zégnon (K&B) GUERE (Chiehn) wénokomé (K&B; B&D) zagnon, zanion (B&D) KULANGO waga waga (K&B) KWENI niania (K&B) zuruné tabunia (B&D) MANDING-DYULA missifabimbii (K&B) SENUFO-TAGWANA sindikamba (K&B) 'SENUFO' nanalékalé (K&B) **GHANA** ADANGME-KROBO nasa (FRI) ŋŋasa (FRI) AKAN-ASANTE nfõfõ (auctt.) FANTE mfũfũ (JMD) TWI mfũfũ (Enti) nfõfõ (FRI; JMD) GA nasa (JMD) ŋŋasa (FRI) GBE-VHE alawalawatsɛ (Ewe Dict.; FRI) NZEMA afɔle (JMD) **NIGERIA**

IGBO (Asaba) anwuliriwani (NWT) IGBO (Umuahia) azuzọ (AJC) IJO-IZON (Kolokuma) èré ịyọ́ụ̀nkọ̀rị = *female breath-catcher*, (KW) ịyọ́ụ̀nkọ̀rị (KW) YORUBA abo yúnyún, abo : *female* (EWF; JMD) ewé-agbugbo (JMD) yunri yungbodo (EWF; JMD)

*NOTE: Ghanaian names may perhaps also apply to M. elliptica.*

A scrambling or scandent herb of waste thickets, cultivation edges and forest margins, commonly dispersed in the forested areas of the Region, and extending widely across tropical Africa.

The plant is often confused with *Aspilia africana*. It may be given the same names and have the same uses.

It provides forage for all stock in thicket edges around villages. It is a fast-growing weed of cultivation becoming troublesome in S Nigeria (6) and in W Cameroons where it is deemed a pest on oil-palm plantations (3). The leaves are a favourite food of hares in S Nigeria (4), and are commonly put into soups (11, 12). The whole leaves are very scabrid and can cause irritation to the skin if rubbed against. On the other hand, pulped, decocted or macerated leaves are cicatrisant. Preparations are haemostatic and are used on cuts and wounds. Like *A. africana* these are said to draw up exudations from open sores, to curb inflammation, and to promote healing (2, 4, 5, 8, 13). The Ijo use either leaf-sap or the leaf reduced to ash on wounds (14) and apply the ash to burns (15). The Igbo tie on a leaf as a wound-dressing after circumcision (11). Leaf-sap or decoction with citron-juice added is used in Ivory Coast-Upper Volta for eye-troubles, even for trachoma, but the treatment is irritant and is given only with some hesitation (5). It is also used in Ivory Coast as eyedrops in syncope (2), and Akye use it for chickenpox internally and externally (1).

Infusions made from leaves, stems and roots are emetic (13). The leaves are a drastic purgative, and leaf-sap is used in Ivory Coast to accelerate childbirth and to treat for poisoning (5). The Ijo rub the leaves with salt, then mix them (? chopped up) with gin to take as an antidote against poisoning (15). In Ivory Coast a leaf-decoction is used as a soothing cough-mixture and for sore-throat (2) and in Tanganyika a 24-hour infusion of powdered leaves is taken for hiccups (9).

The twigs serve in Tanganyika for teeth-cleaning (10), and in the Acholi-West district of Uganda bundles of sticks are burnt to make acrid smoke to drive off bees from a nest in raiding it for honey (7).

The plant probably has superstitious attributes similar to *Aspilia africana*. In Ivory Coast-Upper Volta, a preparation of leaves is rubbed on the patient's head for epilepsy (5). To the Ijo and the Yoruba the plant is female, the male counterpart being *Aspilia africana*.

References:

1. Adjanohoun & Akė Assi, 1972: 97. 2. Bouquet & Debray, 1974: 72. 3. Chuml 80, K. 4. Dalziel, 1937. 5. Kerharo & Bouquet, 1950: 214–5. 6. Obaseki FHI. 23837, K. 7. Okello-Degaouchi 15, K. 8. Portères, s.d. 9. Tanner 555, K. 10. Tanner 1047, K. 11. Thomas N. W. 1616B, 1616C (Nig. Ser.), K. 12. Thomas, N. W. 1677 (Nig. Ser.), K. 13. Watt & Breyer-Brandwijk, 1962: 250. 14. Williamson, K. 25.A, UCI. 15. Williamson, K. 39, UCI.

**Microglossa afzelii** O. Hoffm.

FWTA, ed. 2, 2: 251. UPWTA, ed. 1, 418.

West African: **SIERRA LEONE** MENDE gimbo (FCD) gimbo yufe (auctt.) TEMNE *an*-kai (FCD) *an*-kai-*a*-bɛra (FCD) **LIBERIA** MANO fai bulu lah (JMD) gai bulu da (Har.) **IVORY COAST** AKYE peu chi dzakoè (A&AA) BAULE donienia donien (B&D) DAN klékélé (K&B) mlébulé (K&B) KRU-GUERE kokuiribaï (K&B) zagu (K&B) GUERE (Chiehn) limré (K&B; B&D) 'KRU' sofré (K&B) **GHANA** AKAN-FANTE ɛsono-mbabe (FRI) TWI asomerewa (Enti) ɛsono-mbabaa = *elephant's sticks* (FRI) pofiri (FRI)

A scrambling shrub to 5 m high of brakes and forest margins and secondary jungle, in Senegal to W Cameroons and on into E Cameroun: also in Sudan and Uganda.

The leaves are eaten by the Temne of Sierra Leone in 'palaver sauce', and are taken medicinally with rice for headache and dizziness (3). The Mende prepare a leaf-decoction as eye-drops for ophthalmia, and bathe with a decoction for rheumatism, pains in the body and fever (3). Leafy stems in decoction are taken in draught in Nigeria (7) and in Ivory Coast (2, 6) for bronchial troubles, coughs and sore-throats. In serious cases in Ivory Coast the fresh leaves are frictioned on the chest and lumbar region. Sap is squeezed into the ears in Ghana for earache (4), and the leaves with guinea-grains and rum are taken for chest-complaints (5).

Diuretic properties are recognized in Ivory Coast for which the plant is used in the treatment of blennorrhoea, and oedema in pregnant women. The Dan of Ivory Coast put it into an antivenin mixture (6).

The liquid squeezed out of the roots, ground up to a paste with those of *Setaria chevalieri* (Gramineae) and a little water is used in nasal instillation in Ivory Coast for migraine (1).

This species is alike to and confused with *M. pyrifolia*. Their uses are probably the same.

References:

1. Adjanohoun & Aké Assi, 1972: 98. 2. Bouquet & Debray, 1974: 72. 3. Deighton 1921, K. 4. Irvine 157, K. 5. Irvine, 1961: 718. 6. Kerharo & Bouquet, 1950: 215. 7. Oliver, 1960: 31, 71.

**Microglossa pyrifolia** (Lam.) O. Ktze

FWTA, ed. 2, 2: 251. UPWTA, ed. 1, 419.

West African: **SENEGAL** DIOLA (Fogny) bubum éñab = *medicine of the elephant* (K&A) **THE GAMBIA** DIOLA (Fogny) bubum éñab = *suckable bubum* (DF) — cf. under Senegal **SIERRA LEONE** BULOM (Sherbro) baŋgbawapɛ (FCD) KISSI yeniyɔlɔ (FCD) yenyɔlɔ (FCD) KONO yamba-duknɛ (FCD) MENDE gimbo (*def.* gimbui) (FCD) gimbo-yufe (JMD; FCD) TEMNE *an*-kai (FCD) *an*-kai-*a*-bɛra (NWT) *an*-kai-*a*-runi (FCD; FAM) **LIBERIA** KRU-BASA gban-gbah (C&R) **UPPER VOLTA** MANDING-DYULA simblé sama (K&B) **IVORY COAST** ABURE kénini (B&D) AKAN-BRONG monfragué-neda (K&B) AKYE logbu baté (A&AA) BAULE assuebo (B&D) essubo (K&B; B&D) DAN buzé dundi (K&B) KRU-BETE koagni sélébéï (K&B) GUERE bagigo (K&B) bau kutué (K&B) blongbé (K&B) GUERE (Chiehn) lititi (B&D) pitibokobé (K&B; B&D) KULANGO kazongula (K&B) KWENI furu (K&B) kofélorotru (B&D) KYAMA lonbongbué (B&D) MANDING-DYULA simblé sama (K&B) 'KRU' pukuyobaïé (K&B) 'NEKEDIE' dièméu (B&D) **GHANA** ADANGME plɔke = *secondary bush* (A.S.Thomas) AKAN-FANTE ɛsono-mbabe = *elephant's sticks* (Ll.W., ex FRI) TWI asõmerewa (JMD) asɔmmerewa (FRI) ɛsono-mbabaa = *elephant's sticks* (Ll. W., ex FRI) **WEST CAMEROONS** KPE bendem-bende (Waldau)

An erect or scandent shrub, to 5 m high, of wooded savanna, secondary jungle and regrowth in abandoned farm-land occurring throughout the Region and widespread in tropical Africa and Asia.

It is sometimes cultivated in gardens, probably not as an ornamental, but for its medicinal virtues. The leaves are not eaten, and their use in medicines taken by mouth is limited on account of their toxicity (16, 17, 20). They are said to be poisonous to goats in Ghana (14), but in Kenya they browse the plant (9, 10). The leaf is aromatic, and has an unpleasant smell. A leaf-infusion is taken for fever with headache, and also as a lotion or fumigation, or inhalation to cause perspiration (5, 7, 20, 22). The liquid of a leaf-decoction is commonly taken in Kenya for malaria: it is said to be very bitter (8), and to be emetic and purgative (10, 11). In Sierra Leone a decoction has been considered specific for 'yellow fever' and for dropsy, and to be reliable for blackwater fever (5). In Ghana it is said to be suitable to give (method not disclosed) to infants with fever (14, 26) while in Uganda children with fever are bathed in a leaf-decoction, and for fever generally a root-decoction is drunk (19). Bathing has a sedative action. In Ivory Coast it is used in treating insanity (2), and in E Africa for epilepsy and fits in children (25, 26). A leaf-decoction is taken in Sierra Leone by women when in labour (18). In Liberia an infusion is used as a nasal inhalant (3). The powdered root is taken as a snuff in the Region to relieve colds (5); in Uganda the root-sap is sniffed up the nose for headache (21), and in Tanganyika the leaf is placed inside the nostril of both man and cattle for head-colds (25). The sap is said to be irritant to the mucosae and is applied as nose-drops in Ivory Coast as a restorative in fainting and as an anodyne in headache, and in enema form as an aphrodisiac and to treat female sterility (2). Sap from young crushed roots is applied in Ghana to the eyes for eye-troubles and as a specific for cataract causing a burning sensation for 10–15 minutes. The lees from the crushed roots are mixed with shea butter and smeared over the lids at bedtime to reduce the swelling. Treatment is repeated weekly (5). In Congo leaf-sap is instilled into the eye for ophthalmia and to kill filaria (1). The plant is anodynal to aches, pains and swellings. It is used in Ivory Coast for various stomach complaints, jaundice and oedemas (2). In Casamance the root-powder, usually taken as snuff via the nose, is commonly used for toothache (15). In Congo pulped leaves are prepared as a wet dressing for inflammations, and for pulmonary troubles when the material is massaged over the chest (1). It is a decongestant causing diuresis and thus relief in urethral obstruction and is given in Ivory Coast (2, 17) and Congo (1) for gonorrhoea. Purgative action is used in treatment of abdominal troubles, leprosy, and as an abortifacient in Ivory Coast. In the case of leprosy the leaves pulped with clay are also applied topically and cause a 'burning' of the skin (17). In Liberia and Nigeria a leaf-infusion is taken as a vermifuge (4, 6, 20), while in E Africa the plant is considered a remedy for hookworm (25). The sap from warmed leaves is put into sore eyes in Sierra Leone, and on to the scalp to cure ringworm whereby the preparation is called *ta-hale,* 'skull medicine' (5). A root-decoction is applied as a compress to draw lanced abscesses in Tanganyika (12).

The Dan of Ivory Coast compound the plant with *Piper umbellatum* Linn. (Piperaceae) and *Mansonia altissmia* A. Chev. (Sterculiaceae) into an arrow-poison which is used for hunting larger game such as antelope. For smaller animals, especially monkeys, they compound it with *Pentaclethra macrophylla* Benth. (Leguminosae: Mimosoideae) and *Agelaea sp.* (Connaraceae) (17). The use of the plant in arrow-poisons does not apparently occur in Senegal (16).

Phyto-chemical analysis of leafy material from Ivory Coast showed the presence of traces of alkaloids, steroids and a strongly haemolytic substance, propably a saponin (16, 20). Extracts of roots, stems, leaves and flowers have shown a weak insecticidal activity (13).

The Igbo of Uburubu rub the plant on the skin to bring good luck (23), and in Ivory Coast it is incorporated into a protective talisman against evil spirits and illnesses (17).

The Masai of Kenya use the stems as firesticks (9).

References:

1. Bouquet, 1969: 95. 2. Bouquet & Debray, 1974: 72. 3. Cooper 449, K. 4. Cooper & Record, 1931: 116, as *M. volubilis* DC. 5. Dalziel, 1937. 6. Daniel 85, K. 7. Deighton 685, K. 8. Glover & al., 236, K. 9. Glover & al., 975, K. 10. Glover & al., 1489, K. 11. Glover & al., 2346, K. 12. Haerdi, 1964: 170. 13. Heal & al., 1950: 114, as *M. volubilis* DC. 14. Irvine, 1961: 718–9. 15. Kerharo & Adam, 1963, b: as *M. volubilis* DC. 16. Kerharo & Adam, 1974: 227–8. 17. Kerharo & Bouquet, 1950: 215, as *M. volubilis* DC. 18. Lane-Poole 142, K. 19. Maitland 39, K. 20. Oliver, 1960: 31, 71, as *M. volubilis* DC. 21. Roscoe s.n., 5 Aug. 1920, K. 22. Thomas, N. W., 1694 (Nig. Ser.), K. 23. Thomas, N. W., 2073 (Nig. Ser.), K. 24. Watt, J. M., 1967. 25. Watt & Breyer-Brandwijk, 1962: 250. 26. Williams 32, K.

**Microglossa sp. indet.** (Ross 261, K).

West African: **NIGERIA** EDO ásivbógò (Ross)

A shrub to 4 m high of abandoned clearings.

**Mikania cordata** (Burm. f.) B. L. Robinson

FWTA, ed. 2, 2: 286. UPWTA, ed. 1, 419, as *M. scandens* Willd.

English: climbing hemp-weed (Dalziel).
West African: **SENEGAL** WOLOF kumbańul (JB; K&A) **THE GAMBIA** MANDING-MANDINKA niankokono singo = *bushy leg* (DF) **GUINEA-BISSAU** MANDING-MANDINKA fricoiô (JDES) **GUINEA** FULA-PULAAR (Guinea) noré (Farmar) **SIERRA LEONE** MENDE gimbo (FCD) gimbo-yufe (FCD) ndondokɔ (FCD) **IVORY COAST** AKAN-ASANTE banakumanu (B&D) GAGU gazienla (K&B) KRU-GUERE don (K&B) GUERE (Chiehn) bazéru (K&B; B&D) KYAMA kokotobangui (B&D) 'MAHO' kuagbo (K&B) **NIGERIA** YORUBA ejon (RJN)

A climber to 8 m or more, of regeneration on old farms and in secondary jungle and waste places generally, widespread throughout the Region; probably of S American origin, but now pantropical. Two varieties are recognized in West Africa, var. *cordata* and var. *chevalieri* C. D. Adams, the latter being confined to Sierra Leone to S Nigeria in the Region, but occurring also in Congo and Angola.

The plant has been tried as a soil-cover to prevent erosion (6). In the East it has been planted as a smother-crop to keep down weeds in coconut plantations. The plant is vigorous and takes up a great amount of potash which must be returned to the soil (green manuring), or the soil will be impoverished (4).

In some places the cooked leaves are added to soup (5, 6; Sierra Leone, 7). In Congo there is said to be a variety with rather fleshy flowers which is cultivated and eaten like spinach (19). Cattle graze the plant (6). In Uganda the leaves are a part of the gorilla's diet (13).

In Nigeria a decoction of the plant is taken for cough (6, 14, 16), and the leaf-sap is a remedy for sore eyes (6, 14, 17). In Ivory Coast sap from the whole plant is crushed with pimento and taken by draught, repeated daily, for cough and bronchitis and as a vermifuge (3, 11). It is used as a vermifuge in Congo and for stomach-complaints (2), and as a vermifuge in Tanganyika (15). A leaf-decoction is taken in Sierra Leone (8) and in Ivory Coast (3) as a sedative for abdominal and intercostal pains. In Congo vapour-baths of a decoction of the whole plant, followed by frictions with the lees, are taken as an analgesic for lumbago and rheumatic pains, and as a febrifuge (2). Frictions are given in Senegal for fever aches and pains (10). The plant crushed with a pimento is rubbed on to the sides for stitch in Gabon (18, 19). The sap mixed with that of *Desmodium canum* (J. E. Gmel.) Schinz & Thell. (syn. *D. mauritianum* (Willd.) DC., Leguminosae: Papilionoiidae) is instilled into the ear for earaches in Ivory Coast (11); the leaves are used for headache in Tanganyika (12) and drops of the leaf-sap are instilled into the eyes or nose in Congo for migraines, conjunctivitis and to prevent vertigo (2). Sap is also taken in Congo for urethral discharges and to treat senile impotence (2), and the leaves macerated in citron-juice are taken in Gabon for the former trouble (18, 19). It is sometimes used for small-pox (also see below) and for jaundice in Ivory Coast (3), and in Tanganyika the leaf-sap is drunk as an anti-malarial and diuretic and also for schistosomiasis (9). The sap from the entire plant is held in Congo to be antiseptic and is used with vigorous rubbing for psoriasis (2). Similar usage is reported in Malaya (4). A sulphuric acid extract has been found to inhibit growth of *Staphylococcus aureus* (20). The plant is held to be effective for snake-bite in Senegal (10) and for snake-bite and scorpion-sting in Mozambique: similar uses of *Mikania spp.* are recorded from Brazil (6, 20). The leaves are applied to cuts and wounds in Gabon (18, 19), and similar use occurs in southern Africa and Indonesia (4, 20).

In order to prevent goats eating tapioca plants, in some parts of Congo, sap of *M. cordata* is rubbed over them (2).

Active principles have not been detected in African material. A saponin has been detected in S American plants (10, 11).

Magical attributes have been accorded the plant. In Sierra Leone, a paste of the plant in goat-fat is rubbed on the feet of infants slow to learn to walk. It must be goat-fat and no other as the goat is an active animal (1). In Ivory Coast to prevent the passage of small-pox from village to village, the plant is macerated in palm-wine with *Cassia occidentalis* (Leguminosae:

Caesalpinioideae) and pimentos and the preparation is spread along tracks between villages. It is forbidden to pass the points so treated (11).

References:

1. Boboh, 1974. 2. Bouquet, 1969: 95. 3. Bouquet & Debray, 1974: 72. 4. Burkill, 1935: 1470, as *M. scandens* Willd. 5. Busson, 1965: 421, 424. 6. Dalziel, 1937. 7. Deighton 255, K. 8. Fisher 96, K. 9. Haerdi, 1964: Afr. Heilpfl. 170. 10. Kerharo & Adam, 1974: 228. 11. Kerharo & Bouquet, 1950: 216, as *M. scandens* Willd., with references. 12. Koritschoner 853, K. 13. Schaller 369, K. 14. Oliver, 1960: sp. no. 221, as *M. scandens*. 15. Tanner 5727, K. 16. Thomas, N. W. 2026 (Nig. Ser.), K. 17. Thomas, N. W. 2188 (Nig. Ser.), K. 18. Walker, 1953, a: 29, as *M. scandens* (Linn.) Willd. 19. Walker & Sillans, 1961: 131, as *M. scandens* Willd. 20. Watt & Breyer-Brandwijk, 1962: 251.

## **Nidorella spartioides** (O. Hoffm.) Cronq.

FWTA, ed. 2, 2: 255, as *Conyza spartioides* O. Hoffm.

A perennial herb to about 60 cm high, of montane situations in W Cameroons, and also in E and South-tropical Africa.

The dried plant has a strong smell of coumarin (1). The Bemba of eastern Africa reduce the flower to a powder for use as a snuff in treatment of head-colds. It causes irritation (2).

References (as *Conyza spartioides* O. Hoffm.):

1. Adams, C. D., 1956, a: 62. 2. Watt & Breyer-Brandwijk, 1962: 221.

## **Odontospermum graveolens** Forssk.

FWTA, ed. 2, 2: 257, as *Bubonium graveolens* (Forssk.) Maire.

A tough bushy herb to 50 cm high, occurring from Mauritania to Niger in the sub-desert region, and in N Africa and across to NE Africa and Arabia.

The plant provides grazing for camels and other domestic stock (1). In the Tibesti region it is known as 'food of the donkeys' (Teda, *arma budoï*) (2).

References:

1. Maire, 1933: 208, 227. 2. Monod, 1950: 63.

## **Piloselloides hirsuta** (Forssk.) C. Jeffrey

FWTA, ed. 2, 2: 288, as *Gerbera piloselloides* (Linn.) Cass.

A perennial stemless herb with flower peduncles to 40 cm high, of hill grassland in Guinea, Ghana and W Cameroons, and widespread through NE, E, central and southern Africa.

No usage is recorded for the Region. In Kenya the plant is grazed by all stock (1), and the Masai burn the plant to ash which mixed with ghee is given on the fingertip to babies as a tonic (2). In Lesotho the plant is burnt inside the hut of a person suffering from a headcold to fumigate it (3, 4), and is used as a purgative for stomach troubles (3); also a root decoction is given to treat tuberculosis and miners' phthisis (3); or in milk decoction or infusion for chest complaints generally (4). A root-infusion in human urine is a Zulu remedy for earache (4).

References:

1. Glover & al., 1233, K. 2. Glover & al., 1645, K. 3. Guillarmod, 1971: 429, as *Gerbera piloselloides* (Linn.) Cass. 4. Watt & Breyer-Brandwijk, 1962: 231, as *G. piloselloides* Cass.

**Pleiotaxis chlorolepis** C. Jeffrey

FWTA, ed. 2, 2: 287 as *P. newtonii* sensu FWTA, ed. 2. UPWTA, ed. 1, 419, as *P. newtonii* sensu Dalziel.
West African: **SENEGAL** BASARI a-mbɛrí-syambàt (Ferry) **GUINEA** FULA-PULAAR (Guinea) dononturu (Brossart)

A perennial herb to 1.50 m high, of savanna in Senegal, Guinea-Bissau and Guinea, and occurring also in Angola.
The plant is used as a purge in Guinea (1).

Reference:
1. Dalziel, 1937.

**Pluchea ovalis** (Pers.) DC.

FWTA, ed. 2, 2: 260.
West African: **SENEGAL** WOLOF sãnda-sãnda (JB)

A shrub to 1.70 m high, of open scrub, alluvial flats, etc., usually by water, from Mauritania to Togo, Cape Verde Islands and widespread in NE and E Africa.
The plant is strongly aromatic. It is much relished for grazing by goats and sheep in eastern Africa (Ethiopia, 1; Kenya, 2).

References:
1. Carr 689, K. 2. Glover & al., 2393, K.

**Porphryostemma chevalieri** (O. Hoffm.) Hutch. & Dalz.

FWTA, ed. 2, 2: 260.

A dwarf herb to 60 cm high, of dry sandy localities in Senegal, Guinea-Bissau, Ghana and N Nigeria, and also in E Africa.
No usage is recorded for this species, but it is scarcely distinguishable from *P. grantii* Benth. which is widely distributed in E Africa and Angola. The whole plant of this latter is reported used for stomach pains (1) in the Lake Province of Tanganyika.

Reference:
1. Tanner 4248, K.

**Pulicaria crispa** (Forssk.) Oliv.

FWTA, ed. 2, 2: 258. UPWTA, ed. 1, 419.
West African: **MAURITANIA** ARABIC (Hassaniya) llaïna (AN) **SENEGAL** 'MUSSO' koroni dé (JB) **MALI** SONGHAI tarkundé dierfendu (A. Chev.) TAMACHEK attasa (RM) **DAHOMEY** YOM paria zupen (A. Chev.) **NIGERIA** HAUSA bafuri, bálbeélàa *general for a hoary plant* (JMD; ZOG) farar saura = *white gleanings* (JMD; ZOG) ƙurar shanu = *cow's dust* (MM; ZOG)

An annual decumbent or erect herb to 60 cm high, of the sahel zone from Mauritania to N Nigeria and across Africa to Arabia.
The plant is grazed by domestic animals and camels in the Hoggar (3). It is aromatic and in the Sahara is made into an infusion for drinking. It has a slight gingery taste and is sometimes sold as a market merchandise in Dahomey and Nigeria for medicinal use. In Nigeria it is applied topically to swellings and

bruises, is rubbed on the temples for headache, and a decoction taken for febrile conditions (1).

The related *P. dysenterica* (Linn.) Gaertn. is the traditional British insecticide, 'fleabane'. Tests on *P. crispa* have shown insignificant insecticidal action (2)

References:

1. Dalziel, 1937. 2. Heal & al., 1950: 114. 3. Maire, 1933: 206–7, 242.

## **Pulicaria undulata** (Linn.) C. A. Mey.

FWTA, ed. 2, 2: 258.

A herb to about 30–40 cm high, of the sahel zone from Mauritania to N Nigeria and across Africa to Arabia.

The plant is strongly aromatic with a spicy smell suggestive of caraway oil. In Eritrea, the smell, and therefore presumably the oil-content, is reported to be stronger at high altitudes than lower ones (1). Tests on the plant tops and flowers for insecticidal activity showed insignificance (2).

References:

1. Bally B.6645, B7037, K. 2. Heal & al., 1950: 114.

## **Sclerocarpus africanus** Jacq.

FWTA, ed. 2, 2: 235. UPWTA, ed. 1, 420.

West African: **SENEGAL** MANDING-BAMBARA gôni (JB) **NIGERIA** IGBO n̠lí-átúlụ̄ = *food of the sheep* (NWT; JMD)

An annual herb to 1.30 m high, of open waste places, and widely dispersed in tropical Africa and in India.

An extract of the leaves in the water in which a sheep has been cooked is taken by the Igbo as a remedy for gonorrhoea (1).

Reference:

1. Thomas, N. W., s.n. (Nig. Ser.) K.

## **Senecio abyssinicus** Sch. Bip.

FWTA, ed. 2, 249–50. UPWTA, ed. 1, 420.

English: 'ragwort' (Nigeria, Ainslie).

West African: **NIGERIA** YORUBA amùnimúyè (auctt.)

An annual herb to about 50 cm high, of open places and disturbed land of lowlands and montane elevations in N and S Nigeria and W Cameroons, and widely distributed in central and east tropical Africa.

The powdered root or the root in decoction is taken in Nigeria both internally and externally in treatment of syphilis (2), and in Tanganyika the plant is a remedy for syphilis and yaws (4). The plant is considered stomachic and a blood-purifier in Nigeria, and the bruised leaves are applied topically to painful areas of rheumatism, and to bruises and cuts (2). It is held to have a cicatrizing effect in skin diseases and is used in Zaïre to promote healing of cuts and sores, and to treat diseases of the eyes (4). The plant is poisonous if taken internally in large doses. However no alkaloid has been found in the leaves or roots (1).

The Yoruba names meaning 'bringing' or 'catching life', or 'catching a person's thoughts' to prevent someone knowing about a matter suggest magical uses (3).

References:

1. Adegoke & al., 1968. 2. Ainslie, 1937: sp. no. 314. 3. Dalziel, 1937. 4. Watt & Breyer-Brandwijk, 1962: 283.

**Senecio baberka** Hutch.

FWTA, ed. 2, 2: 250.
English: 'ragwort' (Nigeria, Ainslie).
West African: **NIGERIA** HAUSA (West) baberka (JMD)

Annual herbaceous shoots to about 30 cm high arising from a perennial wood rootstock in grassy savanna of N Nigeria, and also in Cameroun and Sudan.
The plant produces a bitter medicine (2) and has the same medicinal uses as *S. abyssinicus* Sch. Bip. (1).

References:

1. Ainslie, 1937: sp. no. 314. 2. Dalziel 390, K.

**Senecio biafrae** Oliv. & Hiern

FWTA, ed. 2, 2: 246, as *Crassocephalum biafrae* (Oliv. & Hiern) S. Moore. UPWTA, ed. 1, 420.
West African: **SIERRA LEONE** KISSI bɔlɔgi (FCD) lambe pundo = *English spinach* (FCD) **IVORY COAST** KRU-GUERE (Chiehn) kokolé titi (B&D) 'NEKEDIE' balo dédé (B&D) **GHANA** AKAN-FANTE yankonfeh (Easmon) **NIGERIA** IGBO (Owerri) ota eke (AJC) YORUBA akọ amùnimúyè = *male of amùnimúyè* g-bọlọgi *also applied loosely to some other herbaceous plants* (Miller) worowo (CWvE) YORUBA (Ilorin) rọrọwọ (JMD)

A climbing herb of secondary jungle, roadsides, waste places and disturbed land of hilly country from Guinea to W Cameroons, and extending into central tropical Africa.
The plant has succulent leaves and stems which are eaten as a spinach, but there are several cultivars of superior quality grown as a leaf-vegetable in the Region and in the adjoining territories of its occurrence. A leaf-infusion is also taken as a drink (3, 4, 6–10, 12). In northern Sierra Leone the leaves are cooked in palaver sauce, the first cooking-water being discarded (5, 10). The use by the Krio of Sierra Leone of the Yoruba name for the plant (*bologi*) must raise a query whether the plant is truly native of as far west as Sierra Leone and Guinea for there seems to be a break in its distribution between Sierra Leone and Ivory Coast. Furthermore the Kissi of Sierra Leone infer an exotic origin in calling it 'English spinach.' It is the true *bọlọgi* of the Yorubas, *Basella alba* being 'broad *bọlọgi*' and *Talinum* 'Lagos *bọlọgi*.'
The leaves are applied in Nigeria as a wound-dressing (4). In Ivory Coast the plant is pulped into a paste for application to the breasts as a galactogene and the sap is taken by draught for cough in children (2). A preparation is taken by the Igbo of S Nigeria for 'hot belly' (? indigestion) (11). In Congo the plant has a reputation as a cough-cure, for heart-troubles, and to be aperitive and tonic, and for these it is eaten as a vegetable with meat or fish. The sap is also rubbed on the body to relieve rheumatic pain, prurigenic allergies and localized oedemas (1).
It has superstitious uses in Congo: to wash initiates at sect ceremonies, to assure rest to the spirit of a departed sect member, and by eating some of the leaves someone pleading before a tribunal will put forward a good case (1).

References:

1. Bouquet, 1969: 92. 2. Bouquet & Debray, 1974: 71. 3. Carpenter 182, UCI. 4. Dalziel, 1937. 5. Deighton 3508, K. 6. Irvine, 1948: 258–9. 7. Irvine, 1952, b. 8. Irvine, 1956: 41. 9. Millen 96, K. 10. Morton & Jorr SL 1694, FBC. 11. Thomas, N. W., 2234 (Nig. Ser.), K. 12. Walker & Sillans, 1961: 130.

**Senecio mannii** Hook. f.

FWTA, ed. 2, 2: 246, as *Crassocephalum mannii* (Hook. f.) Milne-Redhead.

A shrub, or tree to 8 m high, of montane situations, usually in clearings in Nigeria, W Cameroons and Fernando Po, and widely distributed from Ethiopia to south tropical Africa.

The wood is soft and the plant is of very rapid growth. It is planted on the Mambila Plateau as a village stockade plant (5). Cultivation as a living hedge is practised in Ethiopia (7) and in Tanganyika (3, 8). In Kenya it is planted as boundary markers for agricultural holdings and as a support for fruiting bananas (2), and the lower stems are cut to make stands for gourds (1). In Tanganyika where the stems may attain 10 cm in diameter they are used as rafters in Swahili huts (11).

The young shoots are said to be eaten by the rock hyrax (10).

Leaf-sap or a root-decoction is taken as a constituent of a mixture used in Tanganyika as a sedative in treatment of insanity (4). The plant is also used for gonorrhoea (9). A decoction of leaves is taken with local salt in Uganda for acid stomach (6).

References:

1. Broadhurst-Hill 296, K. 2. Dale & Greenway, 1961: 157. 3. Haaver 503, K. 4. Haerdi, 1964: 167. 5. Hepper 1900, K. 6. Maitland 9, K. 7. Meyer 7992, K. 8. Tanner 551, K. 9. Tanner 4680, K. 10. van Someren 113, K. 11. Wigg 316, K.

**Senecio ruwenzoriensis** S. Moore

FWTA, ed. 2, 2: 250.

A perennial herb with fleshy stems and leaves to 60 cm high from a tuberous rootstock, of montane grassland in W Cameroons, and distributed to the Congo basin and E Africa.

This plant is a farm-weed in E Africa and is believed to be the cause of seneciosis in cattle in Kenya, the symptoms being a persistent diarrhoea, staring coat, anorexia, frenzy, coma and death in two to three days. Alkaloids *ruwenine* (0.11%) and *ruzorine* (0.34%) have been detected. The former is the more toxic. The plant has a cicatrizing effect on skin-diseases, and in Zaïre is used to reduce inflammation of the limbs (1).

Reference:

1. Watt & Breyer-Brandwijk, 1962: 262–3, 287.

**Sigesbeckia orientalis** Linn.

FWTA, ed. 2, 2: 242.

An erect wiry herbaceous annual to nearly 1 m high of damp sites in montane situations of W Cameroons, and occurring widespread in other regions of tropical Africa, and the Old World tropics.

No usage is recorded for the Region. The plant is bitter to taste. It is considered cardiotoxic, diaphoretic, antiscorbutic, sialagogic and anthelmintic.

It is widely used in Asian medicine to treat gangrenous ulcers and sores, and fresh sap applied over sores leaves a varnish-like covering when it dries. A fluid extract has been used on the skin lesions of leprosy, syphilis and venereal diseases, and as a remedy for ringworm and other parasitic infections. An extract is a renal tonic and stimulates urine secretion. It has been used for rheumatism. Extracts show some hypoglycaemic activity, and action against the virus of Raniket disease of poultry; also some insecticidal action against some cockroach species.

The root contains an essential oil that has been recommended for use in perfumery and essential oil trades. There is also a substance resembling salicylic acid and a bitter glycoside, *daratoside,* or *darutine.*

References:

1. Chadha, 1972: 326. 2. Quisumbing, 1951: 994–5. 3. Watt & Breyer-Brandwijk, 1962: 289.

**Sonchus asper** (Linn.) Hill

FWTA, ed. 2, 2: 296.

A coarse erect or decumbent herb to 1 m high, a temperate ruderal of cosmopolital distribution, now present in the mountains of W Cameroons.

The latex has been used on warts (1).

Tests on North American material has shown absence of alkaloids and tannins and an inability to cause haemolysis, and the presence of flavonoids and sterols (2).

References:

1. Hartwell, 1968. 2. Watt & Breyer-Brandwijk, 1962: 291, as *S. oleraceus* var. *asper.*

**Sonchus oleraceus** Linn.

FWTA, ed. 2, 2: 296.

English: sow-thistle.

French: laiteron (Adam).

West African: **MAURITANIA** ARABIC (Hassaniya) tadkernit (AN) **SENEGAL** WOLOF luguJ u valo (JB) ngésu (JB)

An erect annual herb to 1 m high, a temperate ruderal, now a cosmopolitan weed of cool situations and present in the Region in suitable locations from Senegal to W Cameroons and Fernando Po.

In olden times it was cultivated in Europe as a pot-herb. It appears in the Region as a weed of cultivation, and the aerial parts are taken to eat as a vegetable. It is readily grazed by stock and is good for poultry. There has however been suspicion of it poisoning horses in Australia, and sheep have refused to eat it (5). In Somalia lambs are reported to die if they eat it (March, cf. *Launaea intybacea*) (1). It is fed to domestic rabbits in Gabon (4). The young leaf contains 4.1 mg per 100 gm of vitamin C, and the latex is 1.44% rubber (5). The latex has been used in many countries in treatment of warts and other cancers (2).

The root has been used in Tanganyika as an abortifacient (3), and as a vermifuge, particularly for roundworm (5).

A brownish gum obtained by evaporating the plant to dryness is said to be a powerful cathartic, and has been used in the treatment of the opium habit (5). The plant has also been used for liver-troubles and jaundice and as a blood-cleanser, and the sap as eyedrops (5).

COMPOSITAE

References:

1. Burne, 42, K. 2. Hartwell, 1968. 3. Koritschoner, 820, K. 4. Walker & Sillans, 1961: 130–1.5. Watt & Breyer-Brandwijk, 1962: 291.

**Sonchus schweinfurthii** Oliv. & Hiern

FWTA, ed. 2, 2: 296.

An erect herb to nearly 2 m high, occurring in montane situations in W Cameroons, and across central Africa to E and South-central Africa.

The leaves are browsed by stock in Kenya (3). They are however emetic (4) and used on infected wounds (5) in Tanganyika, and a decoction is given to children suffering from fits, and to treat chickenpox and chronic abortion (6, 7). It is also used for sore eyes (7).

In Uganda a vegetable salt is obtained from the plant (2).

References:

1. Broadhurst-Hill 159, K. 2. Chancellor 200, K. 3. Glover & al., 2022B, K. 4. Koritschoner 894, K. 5. Tanner 4685, K. 6. Watt, 1967. 7. Watt & Breyer-Brandwijk, 1962: 291.

**Sphaeranthus angustifolius** DC.

FWTA, ed. 2, 2: 267. UPWTA, ed. 1, 420, as *S. nubicus* Sch. Bip.

West African: **NIGERIA** HAUSA hurahura (JMD; ZOG) kaúdeè (Bargery; ZOG) roógòn-mákìyaàyaá (Bargery; ZOG) HAUSA (West) fura fura (auctt.)

An erect or diffuse annual herb to about 60 cm high, of the sahel zone from Senegal to Niger and N Nigeria, and also in Sudan.

The plant is aromatic.

**Sphaeranthus flexuosus** O. Hoffm.

FWTA, ed. 2, 2: 266–7.

An annual herb of grassland, open spaces and river banks in N Nigeria, and in Sudan, Congo and southern Africa.

The plant has a powerfully spicy scent. It is grazed by cattle in Botswana (1).

Reference:

1. Biegal & al., 4098, K.

**Sphaeranthus senegalensis** DC.

FWTA, ed. 2, 2: 267. UPWTA, ed. 1, 420.

West African: **MAURITANIA** ARABIC ('Maure') cheuguet (A. Chev.) chinguet (A. Chev.) **SENEGAL** DIOLA é fulimful (JB, ex K&A) é kolinkol (JB, ex K&A) FULA-PULAAR (Senegal) depé (K&A) téduwawa (K&A) tédwawa (K&A) MANDING-BAMBARA boylé buti téni (JB, ex K&A) MANDINKA kubédaro (K&A) 'SOCE' kubédaro (K&A) SERER lubuń (K&A) lugut (JB) lulubuń (JLT; K&A) ndulubuń (JB; K&A) SERER-NON (Nyominka) ayalor (K&A) lubuń (K&A) WOLOF karbobaba (K&A) **THE GAMBIA** FULA-PULAAR (The Gambia) landaniari (DRR) MANDING-MANDINKA cummu-cummu = *sour-sour* (William; DF) julifio (DRR) lookidge (JMD) lookrij (Williams) ninsi-kumbaliŋ (*def.*-o) = *cow's knee* (auctt.) **GUINEA-BISSAU** FULA-PULAAR (Guinea-Bissau) dépè (JDES) MANDING-MANDINKA potororô (JDES) **NIGERIA** HAUSA dodoya (JRA)

An annual herb sprawling or erect to near 1 m high, of damp or swampy, even brackish, localities from Mauritania to N Nigeria, and also in Zambia and Mozambique, and in tropical Asia.

The plant is strongly aromatic with an agreeable spicy smell. All stock is said to graze it in Senegal, but there is also contradictory evidence (1). To the Soce of Senegal it is a cure-all (6). The powdered flowers and leaves are applied to skin-diseases of various sorts in Nigeria, and in decoctions are taken as a diuretic and blood-purifier (2). In Mauritania (3) and in Senegal (4, 6) the crushed leaves are used to maturate boils and furuncles. In The Gambia it is said to relieve rheumatic pains, and after treatment to induce sleep if the leaves are thickly strewn on the bed on which the patient is laid (3, 7). The Fula of Senegal prepare a leaf-decoction which is taken with food for rheumatism and the roots reduced to ash are also used in a prescription with other plants (4, 6). The powdered root is considered in Nigeria to be stomachic and digestive. When mixed with groundnut or other vegetable oil the powdered root is a treatment for piles (2). The roots and seeds are held to have anthelmintic properties (2). In N Nigeria the plant is an ingredient of a decoction of herbs which is given to young infants to give them strength and to make them walk quickly (3).

References:

1. Adam, 1966, a. 2. Ainslie, 1937: sp. no. 317. 3. Dalziel, 1937. 4. Kerharo & Adam, 1964, b: 577. 5. Kerharo & Adam, 1964, c: 321. 6. Kerharo & Adam, 1974: 228–9. 7. Williams, F. N., 1907: 381, as S. hirtus.

**Spilanthes** Jacq.

FWTA, ed. 2, 2: 235.

*S. acmella* auctt. of ethnobotanical and phytochemical literature is now split into *S. costata* Benth., *S. uliginosa* Sw. and *S. filicaulis* (Schum. & Thonn.) C. D. Adams in *The Flora of West Tropical Africa,* ed. 2. The attributes of *S. acmella sens. lato* are here put under the last named which is the commonest and most widely dispersed species of the three. It is possible that the first two have been used in like manner.

**Spilanthes costata** Benth.

FWTA, ed. 2, 2: 236.
West African: **IVORY COAST** 'NEKEDIE' gaïbé (B&D)

A fleshy decumbent herb occurring near the sea in Liberia, Ivory Coast and Ghana.

The leaf sap has a piquant taste and is used in Ivory Coast for toothache (1).

Reference:

1. Bouquet & Debray, 1974: 72.

**Spilanthes filicaulis** (Schum. & Thonn.) C. D. Adams

FWTA, ed. 2, 2: 236. UPWTA, ed. 1, 420, as *S. acmella* sensu Dalziel.
English: Brazil cress; Para cress.
West African: **UPPER VOLTA** MANDING-DYULA saraktro (K&B) **IVORY COAST** KRU-GUERE gahé (K&B) MANDING-DYULA saraktro (K&B) **GHANA** AKAN-ASANTE nyame nyen? (FRI) **NIGERIA** HAUSA parpehi (RES) IGBO ilulonoica (NWT) IGBO (Awka) osana (NWT; JMD) IGBO (Okpanam) ósē ànì = *pepper of the ground* (NWT; KW) IJO-IZON (Kolokuma) kírí ẹ̀bẹ̀dẹ̀ = *ground new foliage* (KW) kírí ịgịná = *ground pepper* (KW) kírí ịsànị = *ground pepper* (KW) YORUBA awere pépé (auctt.) **WEST CAMEROONS** KPE sekke (Waldau)

A creeping herb with flowers on ascending peduncles, often forming near-pure stands, and in water, or very damp or shaded localities, from lowland to montane situations, and occurring commonly through the Region from Guinea to W Cameroons and Fernando Po; an American plant that has become pan-tropical.

It has a pungent taste which explains the English name of cress. The plant is regarded as antiscorbutic, diuretic and a digestive tonic. It induces salivation. It is added to salads in West Africa (7) and in Gabon (15) to add to the flavour. In Indonesia it is eaten both raw and cooked (6) and is given as fodder to cattle and horses (7). Ijo girls of S Nigeria use it as a pepper-substitute when cooking food in play (16): hence the Ijo name meaning 'groundpepper.' Wherever the plant grows it is chewed, leaves, or more especially the flower heads, for toothache, both within the Region (7; Liberia, 3; Nigeria, 2, 12) and beyond (Gabon, 15; Tanganyika, 4; S Africa, 7) and into Asia (India, 7, 11; Malaya, 6). The Ijo of S Nigeria grind the leaves with guinea-grains and gin to a paste which is applied to a cleaned aching tooth: the patient is enjoined not to drink water left by others (17). The Igbo chew the seed (13), and they pulp the flower or seed to rub on the head for headache (14). In Cameroun the pulped leaves with water and salt are eaten or placed as a bolus in the nose for headache (9). In Gabon the plant is used for earache (15). The flowering tops are made into topical application in Ivory Coast as a haemostatic and to promote healing of cuts caused by sharp objects and in circumcision (8). The Ijo consider the plant an antidote against poison (16) and ground with guinea-grain and a little chalk they use it as a vaginal suppository for uterine troubles (17). The plant, and especially the flowers in large quantity are purgative and emetic. Herbalists in Sierra Leone give a preparation to 'clean' the tummy of someone who fancies he has been given bad food while in a dream (5). In Cameroun the plant is eaten in a preparation with water and salt for dysentery (9). In W Cameroon the plant with others is chewed in a snake-bite treatment (7). Sore lips and gums in children in S Africa are treated with the powdered leaf (7), and in Gabon a quid of leaves with one or two maleguetta peppers, masticated by a nursing mother is slipped into the mouth of her suckling baby as cough treatment (15). The sore mouth of sprue is treated in Indonesia by a preparation of the dried flower heads (6). The plant with its diuretic effect is said to dissolve out stones in the bladder (6, 11).

The plant and especially the flowers, contain a local anaesthetic substance, *spilanthol,* which is iso-butylamide of decadenic acid. Analgesic action manifest in the foregoing applications is doubtless due to it. It also has a little insecticidal action (8, 10). *Pyrethrin* is present. This acts like piperin of pepper, and is irritant and stimulates the digestion (6). A trace of alkaloid has been detected in the leaves of Nigerian plants (1).

References:

1. Adegoke & al., 1968. 2. Ainslie, 1937: sp. no. 321, as *S. acmella.* 3. Baldwin, Jr. 6767, K. 4. Bally, 1937: as *S. acmella.* 5. Boboh, 1974. 6. Burkill, 1935: 2065–6, as *S. acmella.* 7. Dalziel, 1937. 8. Kerharo & Bouquet, 1950: 217, as *S. acmella.* 9. Leeuwenberg 6217, K. 10. Oliver, 1960: 37, 84, as *S. acmella.* 11. Quisumbing, 1951: 999–1000, as *S. acmella.* 12. Sampson s.n., Oct. 1900, K. 13. Thomas, N. W. 1699 (Nig. Ser.), K. 14. Thomas, N. W. s.n. (Nig. Ser.), 1911, K. 15. Walker, 1953, a: 5: 29, as *S. acmella.* 16. Williamson, K.37, UCI. 17. Williamson, K. 63, UCI.

**Spilanthes uliginosa** Sw.

FWTA, ed. 2, 2: 236.

French: herbe mal aux dents (Senegal, Berhaut).
West African: **IVORY COAST** AKYE n-talé (A&AA)

A herbaceous plant to 30 cm high, of swampy and damp sites, and roadsides, and a weed of cultivation, of the forest zone from Senegal to Ghana, and also in Cameroun and Tanganyika and the Caribbean.

The French name suggests that in Senegal, at least, it is known as a toothache cure. The plant has use in Cameroun to cure (? relieve the itching of) fly bites (4).

It has a piquant taste and promotes the flow of saliva when chewed. Lightly pulped it is considered in Senegal to promote rapid healing of sores (3). In Ivory Coast an extract expressed from the leaves mixed with a little tobacco and salt is taken to stave off a threatened abortion: this is amplified by placing a few drops on the head, and a repetition of the treatment, if necessary (1).

Tests for alkaloids, flavones, saponins, quinones, hydrocyanic acid, steroids and terpenes have been negative. *Spilanthol* has been recorded in the flowering shoots. This substance has local anaesthetic action (3), and its presence no doubt accounts for the action of the plant in quelling toothache. Plant extracts have shown a little insecticidal activity (2, 3), and some action on bloodpressure of cats and dogs (3).

References:

1. Adjanohoun & Aké Assi, 1972: 100. 2. Heal & al., 1950: 115, 150. 3. Kerharo & Adam, 1974: 229–30. 4. Leeuwenberg 6384, K.

**Stomatanthes africanus** (Oliv. & Hiern) King and Robinson

FWTA, ed. 2, 2: 285, as *Eupatorium africanum* Oliv. & Hiern.

West African: **NIGERIA** HAUSA (West) hakkorin fara (JMD)

A herb or shrub to 1 m high, of hilly grassland throughout the Region from Guinea to W Cameroons and widespread in tropical Africa.

The plant has unspecified medicinal use in Gabon (4). In Congo a leaf-decoction is made as a mouthwash for aphthera, and a tisane of the boiled root as a cough-medicine and an antidiarrhoetic (1). Traces of alkaloid and flavonins have been detected in leaves and roots (2).

In Kenya the plant may be used to make the heads of brooms and brushes (3).

References (as *E. africanum* Oliv. & Hiern):

1. Bouquet, 1969: 94. 2. Bouquet, 1972: 19. 3. Carrall H.5, K. 4. Walker & Sillans, 1961: 128.

**Struchium sparganophora** (Linn.) O. Ktze

FWTA, ed. 2, 2: 269. UPWTA, ed. 1, 424, as *Sparganophorus vaillantii* Gaertn.

West African: **SIERRA LEONE** KRIO wata-bitas = *water-bitters* (FCD) **IVORY COAST** AKAN-ASANTE sufien (B&D) AKYE bé vê (A&AA) BAULE kilifien (B&D) muzivien (B&D) sufien (B&D) KRU-GUERE (Chiehn) bogogo (B&D) bolowé (K&B) KWENI futruden (B&D) KYAMA aniebron (B&D) 'NEKEDIE' botiti (B&D) **NIGERIA** IGBO ṉtị̄ mg̀bàdà = *ear of Maxwell's duiker* (NWT) IJO-IZON (Kolokuma) bòù kìrìólògbò = *bush bitter leaf* (KW) kírí-kòròmọ̀ní̩-i̩-kọ̀ri̩ (KW) YORUBA ewúrò (JRA) ewúrò odò = *water bitter leaf* (JMD) ewúrodò (CWvE)

A semi-succulent annual herb, decumbent or erect to 1.30 m high, of damp sites across the Region from Senegal to W Cameroons and Fernando Po, and widely distributed pan-tropically.

It is eaten by the Yoruba in Nigeria as a pot-herb (6, 7). Cattle in Java are said to graze it (5, 6). Decoctions of stems and leaves are used for headaches in Nigeria (2). The plant (methods not disclosed) is a headache cure in Ghana (6, 8) and Ivory Coast (4), and in Gabon the leaves are mixed with the powdered seed of *Monodora myristica* (Annonaceae) (10, 11). The leaf-sap is instilled into the eye in Ivory Coast for vertigo and fainting (4, 9), and a preparation is taken for dysentery and guinea-worm infection, at childbirth to ease delivery, and is prescribed for sterility (4). A decoction is drunk in Nigeria for gonorrhoea (2). As an antidote against serious poison, 'pick four leaves and chew them with seven "lark" [? small] peppers and some salt and you will soon feel well' say the Ijo (12). The leaves are used against itch in Gabon (10, 11).

Traces of alkaloid in the roots and an abundance in the leaves of Nigerian material have been reported (1). Material from Congo has assayed at alkaloids 0.1–0.3% of the whole plant (3).

The plant invades rice-padis in Malaya and forms a useful part of the woods turned in as green manure (5).

References:

1. Adegoke & al., 1968. 2. Ainslie, 1937: sp. no. 318, as *Sparganophorus vaillantii* Gaertn. 3. Bouquet, 1972: 19. 4. Bouquet & Debray, 1974: 72. 5. Burkill, 1935: 2060, as *Sparganophorus vaillantii* Gaertn. 6. Dalziel, 1937. 7. Holland, 1908–22: 378–9, as *Sparganophora vaillantii* Gaertn. 8. Irvine, 1930: 389–90, as *Sparganophorus vaillantii* Gaertn. 9. Kerharo & Bouquet, 1950: 217, as *Sparganophora vaillantii* Gaertn. 10. Walker, 1953, a: 5: 29, as *Sparganophorus vaillantii* Gaertn. 11. Walker & Sillans, 1961: 131, as *Sparganophorus vaillantii* Gaertn. 12. Williamson 43, UCI.

## **Synedrella nodiflora** (Linn.) Gaertn.

FWTA, ed. 2, 2: 229. UPWTA, ed. 1, 421.

French: herbe à feu (Berhaut).

West African: **SIERRA LEONE** MENDE ngulu-gbɛ, yani-gbɛ *general for some herbaceous weeds* (FCD) TEMNE *an*-bal-*ka*-yan *general for some herbaceous weeds* (FCD) *an*-bal-*ka*-yan-*a*-runi *from* bal-ka-yan: *Ageratum* (FCD) **IVORY COAST** KRU-GUERE (Chiehn) waka (K&B) **GHANA** AKAN-ASANTE *n*-kwadupɔ (auctt.) tutummerika kohwɛ ɛpɔ = *run quickly, go look at the sea, significance unknown* (FRI; JMD) TWI mampoŋfo-apɔw = *Mampong people are civilized, significance unknown* (FRI) **NIGERIA** YORUBA akara ajẹ (IFE) àlùgànnbí (DF; JMD) aworo ona (Egunjobi) ewé pòpò (IFE) kanju-kanmu (N&E) zannapọso (DF; JMD)

A straggling annual herb to nearly 2 m high, from a fibrous woody rootstock, native of tropical America and now dispersed pan-tropically, and occurring throughout the Region in waste places. Superficially it may be confused with *Ageratum conyzoides,* but it is not aromatic if crushed, while the latter is strongly pungent. The Temne of Sierra Leone apply to both plants the same name. In Ghana, Akan people give *S. nodiflora* fanciful names, but of unrecorded significance.

The plant is a weed of cultivation. Its eradication is difficult. It is so tough that slashing knives are quickly blunted (6, 7). The French name for it in Senegal suggests that it can survive fire, or very soon regenerates. The seeds are quick to germinate, five days of warm humid conditions being adequate (4). The foliage is readily eaten by stock in Cameroons (7) and in Ghana (3). In Indonesia the young foliage is eaten as a vegetable (2) and in Ghana an infusion of leaves is drunk as a laxative (10). Leaf-sap is used in Congo for mouth-affections and is rubbed on the gums for shrinkage (1). The plant is prepared as a paste for use in Ivory Coast in topical embrocation for oedemas and the leonine form of leprosy, and the liquid obtained from the leaves crushed to a paste with those of *Cyathula prostrata* (Amaranthaceae), seeds of *Aframomum melegueta* K. Schum. (Zingiberaceae), salt and kaolin is drunk as a sovereign remedy for heart-troubles (8). In Malaya it is used for poulticing sore legs and the head for headache after confinement, and the sap is instilled into the ear for earache, while in Indonesia the leaf-sap together with other materials, is applied for stomachache, and the plant is used in embrocation for rheumatism (2).

The roots are pounded and cooked in Tanganyika, and the decoction is drunk as a cough-mixture (5).

Unnamed alkaloid is reported present in the leaf and seed (9).

In Ivory Coast it enters into a magical rite for isolating villages suffering from small-pox in the same way as *Mikania scandens* is used (see there) (8).

References:

1. Bouquet, 1969: 95. 2. Burkill, 1935: 2116–7. 3. Dalziel, 1937. 4. Egunjobi, 1969. 5. Haerdi, 1964: 172. 6. Irvine, 1930: 401. 7. Irvine, s.d. 8. Kerharo & Bouquet, 1950: 217–8. 9. Willaman & Li, 1970: with references. 10. Williams, L1. 10, K.

**Tagetes spp.**

English: African marigold.

Herbaceous plants with showy flowerheads, a number of species have been brought into cultivation as ornamental garden plants, and may be found grown in the Region.

Orange, yellow, brown and black dyes can be obtained from the flowers (1). *T. patulus* Linn. is used in Gabon as a seasoning (2).

References:

1. Greenway, 1941. 2. Walker & Sillans, 1961: 131.

**Tanacetum cinerariifolium** (Trév.) Sch. Bip. (syn. *Pyrethrum cinerariaefolium* (Trév. Vis.)

English: pyrethrum.
West African: MALI TAMACHEK nogud (RM)

A herbaceous plant of southern Europe and the Mediterranean region, now cultivated in many countries, especially E Africa, as the source of insecticidal pyrethrum powder. The powder is made from the dried flowerheads. It is official in many pharmacopoeas, the content being not less than 1% pyrethrins and cinerins of which the former should be not less than a half. The plant has been successfully grown in the Bamenda Highlands of W Cameroons, and assayed at 1.04% total pyrethrins. It is considered a potentially quite satisfactory crop.

The flowers also contain an aromatic oil, a resin, a glycoside and an alkaloid.

References:

1. Major, 1945. 2. Oliver, 1960: 9.

**Tridax procumbens** Linn.

FWTA, ed. 2, 2: 230.
West African: NIGERIA CHAMBA nuke-noh, nuke: *grasshopper;* noh: *a medicinal plant, indicating it had grown up from grasshopper's dung* (FNH) HAUSA harantama (RES) YORUBA sábárúmá (Macgregor) YORUBA (Ilorin) kodelẹ yiri (Clarke)

A creeping annual herb of open waste places, native of central tropical America, now widely occurring in the tropics and present throughout the Region except in the drier parts.

It is a weed of roadsides and cultivation. The rootstock is brittle making it troublesome to eradicate (3). The plant invades lawns (S. Leone, 3; Nigeria 4; Kenya 6), and under mowing becomes perennial (4). It has great drought-resistance, outliving all grasses and most broad-leaved plants, a capacity due perhaps to dew-absorption by its leaf-surface hairs (9).

The plant is used in Nigeria as a green feed for poultry (4). Cattle in Malaya graze it; the plant is recorded having a high potash content (1).

In Kenya the leaves are chewed followed by a drink of water by the Masai for stomachache and malaria (5), and in Tanganyika a decoction of the whole plant in water is taken for persistent low backache (11), and a root decoction for infantile diarrhoea (10). In Madagascar the plant is used to prevent or to staunch bleeding (2).

Small harvester ants have been observed in the Vogel Peak area collecting the seed in large quantity (7, 8).

References:

1. Burkill, 1935: 2181–2. 2. Debray & al., 1971: 18. 3. Deighton 6071, K. 4. Egunjobi, 1969. 5. Glover & Samuel 3255, K. 6. Greenway 1822, K. 7. Hepper 1228, K. 8. Hepper, 1965: 455. 9. Jenik & Longman, 1965. 10. Tanner 4145, K. 11. Tanner 4250, K.

**Triplotaxis stellutifera** (Benth.) Hutch.

FWTA, ed. 2, 2: 269. UPWTA, ed. 1, 421.

West African: **SIERRA LEONE** KISSI pe-chialachale *possibly general for some small herbs* (FCD) **LIBERIA** MANO pilli pilli (Har.; JMD) **IVORY COAST** DAN guièbokengla (A&AA) KRU-GREBO (Nyabo) kupaniénié (K&B) GUERE bohon (K&B) païseo (K&B) GUERE (Chiehn) kategnini (B&D) MANDING-MANINKA ibéhua (B&D) 'KRU' diblikė (K&B) 'NEKEDIE' guzére (B&D) **GHANA** AKAN-ASANTE kokɔɔ (FRI; JMD) **NIGERIA** YORUBA (Ijesha) afuku (AJC) **WEST CAMEROONS** 'BAMILEKE' abui abui (AJC; JMD)

An herbaceous annual, decumbent or erect to 60 cm high, of forest paths, clearings and undergrowth from Guinea to W Cameroons and Fernando Po, and on to Uganda and Angola.

A leaf-decoction is given to Ghanaian children as a cough-cure (5, 6). A tisane is considered febrifugal in Congo (2). In Ivory Coast it is held to produce an excellent medication for soft chancres: leaf preparations are taken by draught, baths, frictions, enemas and urethral douches. The liquid [? decoction] is very bitter and results when absorbed in copious stools (7). It is also given by enema for dysmenorrhoea (3). In Liberia the dried pulverized leaves mixed with palm-oil are a remedy for head-lice (Harley fide 4), or the whole plant may be used pounded up to a pulp to apply to the head (1). The plant is also active against fleas (1). Convulsions in infants are treated in Congo by eye-instillation of the leaf-sap, and intercostal pain by massage with the pulped-up plant (2).

References:

1. Adam, 1971: 377. 2. Bouquet, 1969: 95. 3. Bouquet & Debray, 1974: 72. 4. Dalziel, 1937. 5. Irvine 533, K. 6. Irvine, 1930: 421–2. 7. Kerharo & Bouquet, 1950: 218.

**Verbesina encelioides** (Cav.) A. Gray

FWTA, ed. 2, 2: 241.

An annual herb to 1 m high, of open waste spaces, native of tropical and subtropical America, and now reported from Senegal, Sudan and Southern Africa.

The leaf, flower and the whole plant have been used in Argentina and New Mexico on warts, cancers, malignant tumors and hardness of the stomach (1). Some insecticidal action, generally low, has been reported (2).

References:

1. Hartwell, 1968. 2. Heal & al., 1950.

**Vernonia adoensis** Sch. Bip.

FWTA, ed. 2, 2: 276, as *V. kotschyana* Sch. Bip., *V. stenostegia* (Stapf) Hutch. & Dalz., and *V. tenoreana* Oliv. UPWTA, ed. 1, 423, as *V. kotschyana* Sch. Bip.

West African: **NIGERIA** EDO óríwò (Amayo) HAUSA cikà fágeé *probably applying to several plants* (JMD; ZOG) domashi (JMD; ZOG) kùmbùrà fágeé (JMD; ZOG) YORUBA ewúrò-ọ̀dàn = *ewúro of the plain or grassy field* (JMD) ewúrò-oko = *ewúrò of the farm* (auctt.)

A shrub to 2 m high, of savanna, from Senegal to N and S Nigeria, and extending across Africa to Ethiopia.

The leaves are used in Kenya by the Masai crushed in cold water and applied to cattle sores caused by ticks (3).

The root is prepared in N Nigeria into a bitter medicine and is used as a digestive and appetizer (1, 2). In S Nigeria the root is a chew-stick (2). In

Tanganyika a root-infusion is taken for stomach-pains (5), and for tuberculosis (6), and fresh roots sliced and cooked with milk and flour for gonorrhoea (4). The Sukuma cut up and soak the fresh roots for a short time in water which is used to wash white blotches, which feel a little 'soapy', on the skin of children (4).

References

1. Dalziel, 173, K. 2. Dalziel, 1937. 3. Glover & al., 2336, K. 4. Tanner 1217, K. 5. Tanner 4029, 4803, K. 6. Tanner 4238, K.

**Vernonia ambigua** Kotschy & Peyr.

FWTA, ed. 2, 2: 281. UPWTA, ed. 1, 421.
West African: NIGERIA YORUBA orungọ (auctt.)

An annual herb to 60 cm high, occurring throughout the drier parts of the Region, and widely dispersed in similar parts of tropical Africa.

The herb has a bitter juice (1). The roots are chewed in Tanganyika raw, or a decoction is taken by draught as an expectorant for coughs and colds (2).

References:

1. Dalziel, 1937. 2. Haerdi, 1964: 172.

**Vernonia amygdalina** Del.

FWTA, ed. 2, 2: 277–9. UPWTA, ed. 1, 421.
English: bitter leaf.
West African: **GUINEA** FULA-PULAAR (Guinea) bantara bururé (CHOP) dakuna (CHOP) MANDING-MANINKA kossa fina (CHOP; FB) **SIERRA LEONE** KRIO bita-lif = *bitter leaf* (FCD) MENDE nje nyani = *spoil goat, from the bitter quality* (JMD) TEMNE *a*-bita-lif *from the Krio* (FCD) *a*-dif-wir = *kill goats* (FCD) *an*-gbɔnthɔ (FCD) **GHANA** ADANGME agba (A.S.Thomas) AKAN-ASANTE bowin (Sampeney) mponasere = *concubine laughs* (FRI) FANTE o-bɔwen(e) (auctt.) TWI awonwẽnẽ *from* nwẽnẽ: *sour* (BD&H; FRI) awonywẽnẽ *o*-bɔwen (Enti) GA agɔaflu (FRI) akpa (Field; FRI) tà (KD) GBE-VHE avenya (FRI, ex JMD) gbɔ (FRI, ex JMD) gbɔti (FRI, ex JMD) gbɔtsi = *goat tree; for its poisonous property* (FRI, ex JMD) VHE (Pecí) gbɔ (BD&H; FRI) gbɔti (FRI) gbɔtsi = *goat tree; from its poisonous property* (FRI, ex JMD) GUANG-GONJA sàŋká (Rytz) **TOGO** GBE-VHE avenya (Volkens; FB) TEM tusima (FB) TEM (Tshaudjo) tingma (Volkens) tusima (Volkens) **DAHOMEY** TEM aloma (A. Chev.) aloma kulu (A. Chev.) **NIGERIA** BEROM etulùp (LB) DERA wólóm (Newman) EDO óríwò (JMD) FULA-FULFULDE (Nigeria) kaɗkaɗɗe (JMD) siwaakewal (*pl.* siwaakeeje) = *tooth-stick* (JMD) HAUSA chusar doki *a horse tonic food containing the leaves* (JMD) fatefate, mayemaye *a food prepared from the leaves* (JMD) shiwáákáá, shìwaákaá (auctt.) IBIBIO àtídòt = *bitter leaf* (SOA; Kaufman) IGBO òlúgbù (*standard transcription*); òlúbì, òlúbú, ònúgbù, ònúbú (*variants*) (BNO) IGBO (Umuahia) òlúbì, òlúgbì = *bitter leaf* (AJC) IJO-IZON (Kolokuma) kíríòlògbò = *ground cat* (KW) TIV ityuna (Vermeer) YORUBA ewúrò (auctt.) ewúrò jijẹ = *edible ewúrò* (Ross; JMD) ewúrò oko = *farm ewúro* (Ross; JMD) orín, pákò *general for chew-stick roots* (JMD) **WEST CAMEROONS** NGEMBA yiŋŋá (Chumbow)

*NOTE: These names are generally interchangeable with V. colorata.*

A shrub or tree to 5 m high, of savanna and forest-margins, often forming thickets, from Guinea to W Cameroons and Fernando Po, and widely distributed throughout tropical Africa.

This species and *V. colorata* are very similar in appearance and are probably not distinguished apart. They share many of the same vernacular names and uses. This species is often planted to have on hand materials, which are frequently items of market-merchandise, for its many uses.

The leaves are very bitter. Bitterness can be abated by boiling, or in the young leaves by soaking in several changes of water. They are held to be antiscorbutic and are added to soups or eaten as spinach (5; Sierra Leone, 9;

Nigeria, 8, 24; Edo, 26; Igbo, 6, 31; Tiv, 32; Yoruba, 3, 23; W Cameroons, 18, 21). The Krio of Sierra Leone, and others who have become creolized, cultivate the plant for the leaves which are put into "palavar sauce" and eaten with *fufu* (cooked tapioca) (10, 27). The leaves are taken in Nigeria as an appetizer and digestive tonic (28). The Hausa of N Nigeria prepare a special food called *fatefate* or *mayemaye* from the leaves with butter and condiments, etc. This is taken by men as a food but by women in a belief that it renders themselves sexually more attractive (8). The leaves are added to horsefeed with bran, natron, etc., in N Nigeria to provide a strengthening or fattening tonic called in Hausa *chusar doki* (8). The leaves are used in Ethiopia as hops in preparing *tela* beer (4, 13).

The leaves are widely used for fevers and are known as a quinine-substitute (Guinea, 25; Nigeria, 2, 24, 28; Ethiopia, 13; Tanganyika, 29). A leaf-decoction is taken as a laxative in Nigeria (24), and in Tanganyika (12) and the sap in Ethiopia (13). A purgative enema is made in Ghana by macerating the leaves through a cloth and adding peppers and spices (16). A cough-medicine is made of them in Ghana (16, 17), Nigeria (as an expectorant, 24) and in Tanganyika (Fraser fide 17). The leaves are rubbed directly onto the body for itch, parasitic affections, ringworm, etc., or a cold infusion is applied as a wash in Nigeria (8, 24, 28). During the puerperium a mother may take a decoction of the leaves to affect her milk so as to act as a prophylactic against worms in the baby. The leaves are added to horse-feed as a vermifuge and to treat internal disorders symptomized by mucal discharge from the nose. Leaves are rubbed on the breasts for weaning infants (8). They are rubbed on the inside of honey jars in Ethiopia to add bitterness or a laxative property (8). In Zaïre they are put into fish-traps as a lure (19). Cattle will more or less readily browse the foliage in Tanganyika (14), and this is qualified by another report of browsing but specifically in the evenings (Burtt fide 17). See reference below to flower fragrance. Are cattle here showing a preference for flowering material sweetened by nectar secretion?

The wood, most usually from the root after the bark has been removed by scorching, provides one of the common chewsticks in Nigeria and the plant may be specially grown for the purpose. It is valued as a tooth-cleaner and more especially as a stomachic and appetizer (8). It has an alleged beneficial effect on dental caries, but no antibiotic activity has been found in material from Ibadan, S Nigeria (20). In N Nigeria the Hausa use these chewsticks with natron for gastro-intestinal troubles (8). The root has been described as a substitute for ipecacuanha (24). A root-infusion is taken in Nigeria as an anthelmintic as well as for enteritis (2). The twigs also are used as chew-sticks, and as well are pounded for application to cuts as a haemostatic and cicatrizant. The wood is used for cleaning teeth in Uganda (30). The plant is grown as a hedge or compound fence (Ghana, 16, 17; Nigeria, 8), and as a shade tree for coffee in Kenya (Battiscombe fide 17). In E Africa the branches are used as lining-stakes for plantations as they resist termite-attack (7, 11) and for the same reason the more slender stems are used as withies (35). Drinking 'straws' are made of the smaller twigs in Tanganyika after the central pith has been removed (33).

The bark of stem and roots is particularly bitter and is more or less astringent. Infusions are commonly taken for fever and for diarrhoea. In Nigeria an infusion is used for rheumatism (2). A cold infusion of root-bark with *Vigna unguiculata* (Leguminosae: Papilionoiideae) is taken in southern Africa for schistosomiasis. The fruit is also eaten for the same purpose (34).

The flowers are fragrant in the evening. They are much visited by bees, and the plant in Ethiopia is a well known bee-plant (22). The honey is said to have a strong flavour (4). Dried flowers are used in Ethiopia to treat stomach-disorders (13). The fruit is used in Nigeria as an aphrodisiac, and the powdered seeds as an anthelmintic (2).

A cardio-tonic glycoside, *vernonin,* has been isolated from the root with an action comparable with that of digitalin (24). No alkaloid has been detected (1).

References:

1. Adegoke & al., 1968. 2. Ainslie, 1937: sp. no. 255. 3. Alasoadura 114, UCI. 4. Archer 9368, 9625, K. 5. Busson, 1965: 424–5, with leaf-analysis. 6. Carpenter 416, UCI. 7. Dale & Greenway, 1961: 161–2. 8. Dalziel, 1937. 9. Dawe 409, K. 10. Deighton 2448, K. 11. Eggeling & Dale, 1952: 99. 12. Gane 67, K. 13. Getahun, 1975. 14. Hornby 5, K. 15. Irvine 46A, K. 16. Irvine, 1930: 431. 17. Irvine, 1961: 720–1. 18. Latilo & Daramola FHI. 28786, K. 19. Liben 3063, K. 20. Lowe, 1976. 21. Maitland 316, K. 22. Meyer 8094, K. 23. Okpon s.n. (UIH. 3209) UCI. 24. Oliver, 1960: 40, 90. 25. Pobéguin, 1912: 65, as *V. senegalensis*. 26. Ross R.149, K. 27. Scott-Elliot 5753, K. 28. Singha, 1965. 29. Tanner 4876A, K. 30. Thomas, A. S. 2278, K. 31. Thomas, N. W. 2308 (Nig. Ser.), K. 32. Vermeer 19, 21, UCI. 33. Watkins 330, K. 34. Watt & Breyer-Brandwijk, 1962: 296. 35. Wigg FH.3046, K.

**Vernonia andohii** C. D. Adams

FWTA, ed. 2, 2: 277.

West African: **GHANA** AKAN-ASANTE npunasiri = *concubine laughs* (Andoh) TWI npunasiri = *concubine laughs* (Adams)

A scrambling shrub to 3 m or more high, of open forest in Ivory Coast and Ghana.

In Ghana it shares the same Twi name as *V. amygdalina* Del. and *V. conferta* Drake (*n*(*m*)*punasiri*) with which it is confused and perhaps it has the same applications.

**Vernonia auriculifera** Hiern

FWTA, ed. 2, 2: 277.

A shrub or tree to 5 m high, of montane grassland or forest margins in W Cameroons, and extending over E Africa.

No usage is recorded for the Region. In Kenya Masai use the stems and leaves in hut-construction and Kipsigis use these materials to make platforms in hut-roofs for storing grain (3). The root soaked in water is a physic for children in Tanganyika (1).

It is said that when the plant flowers it is time to sow the millet (1).

References:

1. Broadhurst-Hill 683A, K. 2. Bullock 3443, K. 3. Glover & al., 843, K.

**Vernonia biafrae** Oliv. & Hiern

FWTA, ed. 2, 2: 276. UPWTA, ed. 1, 422.

West African: **SIERRA LEONE** TEMNE *an*-kai-*a*-bɛra (FCD) **GHANA** ADANGME hu (FRI) ADANGME-KROBO hutʃo (FRI) AKAN-TWI asɔmmerewa (Bunting, ex FRI) GA hutʃo (FRI)

A scandent scrambling shrub of deciduous and closed secondary jungle and openings in closed-forest, from Guinea to W Cameroons and Fernando Po, and also in Cameroun.

The crushed leaves are used as a snuff in Ghana for headache (3, 4), and a decoction of leaves as a bath for fever (Bunting fide 2, 3). It is also used for unspecified women's ailments. Leaf-sap is used for eye troubles in Congo and to kill filaria in the conjunctive tissue of the eye. This medicament is very strong and only one drop at a time should be used (1). In Ghana the pounded roots are prepared as a poultice for the eyes to treat iritis. This causes painful swelling, but cure follows (Bunting fide 2–4). Cooked leaves with palm-oil are used in repeated applications in Congo to treat abscesses (1).

References:

1. Bouquet, 1969: 96. 2. Dalziel, 1937. 3. Irvine, 1930: 431–2. 4. Irvine, 1961: 721.

**Vernonia camporum** A. Chev.

FWTA, ed. 2, 2: 280.
West African: **SIERRA LEONE** SUSU-DYALONKE kundiditu-na (FCD)

A woody herb to 2 m high, of the savanna across the Region from Senegal to W Cameroons.

**Vernonia cinerea** (Linn.) Less.

FWTA, ed. 2, 2: 283. UPWTA, ed. 1, 422.
French: bouton violet (Berhaut).
West African: **SENEGAL** MANDING-MANINKA kunguéni (K&A) SERER kébelgamu (AS; K&A) SERER-NON kébelgamu (AS; K&A) WOLOF Jibi Jêba (K&A) **GUINEA** MANDING-MANINKA kunguéni (Brossart) **SIERRA LEONE** MENDE kipɔ *a general term* (NWT) ngulu-gbɛ *general for several annual weeds* (NWT) TEMNE *a*-bwɔ (NWT) *a*-bwɔ-*a*-ro-kant (NWT) **IVORY COAST** KRU-GUERE grabi (K&B) GUERE (Chiehn) bri yro (K&B) **GHANA** GBE-VHE osikonu (FRI) **NIGERIA** EDO óríwò (Amayo) YORUBA bójúrẹ́ (auctt.) ẹlẹgbẹ-oju (Dodd; JMD) ewé jẹdíjẹdí = *pilewort* (JMD) orungọ (JMD)

An annual herb reaching 1 m high, widespread in waste places, roadsides, farmland, etc., usually in damp sites of the soudanian zone and southwards, and widely dispersed throughout the tropics.

The leaves are sometimes put into soup in S Nigeria (12) and are eaten as a vegetable in Tanga Province of Tanganyika (11) and in S Africa (14). It is eaten in Java as a quite young plant, and animals will browse it (4). With age it becomes bitter and traces of hydrocyanic acid have been found in all parts (10, 14).

In Senegal the root is used as a bitter febrifuge and vermifuge, and the leaves are applied in frictions during fevers (7). In Indian and Ceylonese medicine the plant has a reputation as sudorific inducing sweating in febrile conditions and taken in conjunction with quinine has been helpful in malaria (10). In Ivory Coast the plant is used in topical frictions to relieve intercostal and pleural pain: with other plants it is taken in dracontiasis (8). The roots are used as a vermifuge in Gabon (13) and for roundworm in East Africa (14). In Senegal and Guinea an infusion of the plant is used to wash a new-born infant, and is given also to children with incontinence of urine. In India it is used for spasms of the bladder and strangury (10). In Lagos it is an ingredient of *agbo* infusions (5). Though there is no precise record of the use of the plant in S Nigeria for piles, the Yoruba name meaning 'pilewort' infers this. The sap is a pile-remedy in India (10, 14). A leaf-macerate is instilled into the nose in Gabon for headache (13), and for this a poultice of leaves may be applied to the head in Malaya (4). The plant has common use for chest-conditions: for cough in Gabon (13), Indonesia (4) and the Philippines (10); leaf-sap for pneumonia in Tanganyika (6); for asthma, bronchitis and tuberculosis in India (10). In Tanganyika leaves and flowers are taken as a stomachic, and the leaves are used for snake-bite (9). Sap from the whole plant is given as an ecbolic in Indonesia (4).

Negative and positive results have been obtained for alkaloids and triterpenes (3, 14). Nigerian material has shown a trace of alkaloid (1). Taiwanese material has also shown some (15). The seeds contain a relatively low amount of oil (3.8%), but that oil has an unusually high unsaponifiable component — 78% (7).

References:

1. Adegoke & al., 1968. 2. Bally, 1937. 3. Bouquet, 1972: 20. 4. Burkill, 1935: 2226–7. 5. Dalziel, 1937. 6. Haerdi, 1964: 173. 7. Kerharo & Adam, 1974: 230–1, with seed-oil analysis. 8. Kerharo & Bouquet, 1950: 218. 9. Koritschoner 940, K. 10. Quisumbing, 1951: 1003–4, 1048. 11. Tanner 2950, K. 12. Unwin, 1920: 410. 13. Walker, 1953, a: 5: 30. 14. Watt & Breyer-Brandwijk, 1962: 296. 15. Willaman & Li, 1970.

**Vernonia cinerascens** Sch. Bip.

FWTA, ed. 2, 2: 283.

A shrub to 1 m high, of dry localities in Senegal and Niger, and extending to NE Africa and into India, and also in Angola and SW Africa.

In Somalia the plant provides fodder for camels and goats (2). The Masai of Kenya char the tips of the stems and use the burnt ends to decorate the skin (1).

References:

1. Glover & Samuel 2954, K. 2. McKinnon S/278, K.

**Vernonia cistifolia** O. Hoffm.

FWTA, ed. 2, 2: 282.

An undergrowth shrub to 2 m high, recorded from N Nigeria, Congo and E Africa.

No usage is reported for the Region. In Tanganyika a root-decoction is drunk for coughs and colds (1).

Reference:

1. Haerdi, 1964: 173.

**Vernonia colorata** (Willd.) Drake

FWTA, ed. 2, 2: 277. UPWTA, ed. 1, 421–2.

English: bitter leaf, bitters tree (The Gambia, Williams).

West African: **SENEGAL** BASARI a-tyátyá lùf-lùf (K&A; Ferry) BEDIK bantárá tyangɔl (Ferry) DIOLA butahat (auctt.) kasipa (JB; K&A) FULA-PULAAR (Senegal) kofesafuna (K&A) kosafana (K&A) KONYAGI ndara (K&A) MANDING-MANDINKA bâtara burué (K&A) MANINKA kosafina (Aub.) kosafuné (Aub.) SERER mam (auctt.) ndu(m)barkat (auctt.) SERER-NON tudutj (AS) NON (Nyominka) indubarkat (K&A) indubarkz (K&A) ndubarkat (K&A) WOLOF luguJ *said to be onomatopoeic: significance not known* (K&A; DF) luguJ ndöbarkat (K&A) ndumburgat (auctt.) zidor (auctt.) **THE GAMBIA** DIOLA (Fogny) butakaga (DF) kasipa (DF) FULA-PULAAR (The Gambia) banta-buruda (DRR) MANDING-MANDINKA kunakalo (DRR) WOLOF n-dumburgat (JMD) **GUINEA-BISSAU** CRIOULO pó de sabom (JDES) sucudera (JDES) FULA-PULAAR (Guinea-Bissau) bantara-burúrè (DF; JDES) nabi (JDES) MANDING-MANDINKA bantara-burúrè (JDES) nabicôssô (JDES) MANDYAK benitaha (JDES) umpimpia (JDES) PEPEL pampáe-gôfe (JDES) **GUINEA** FULA-PULAAR (Guinea) bantara bururé (CHOP) dakuna (CHOP) MANDING-MANDINKA kosafina (CHOP; Aub.) **SIERRA LEONE** MANDING-MANDINKA dakuna (FCD) MENDE nje nyani = *spoil goat, from the bitter quality* (JMD) SUSU-DYALONKE daku-na (FCD) TEMNE *a*-dif-wir = *kills goats* (FCD) *an*-gbɔnthɔ (FCD) **MALI** MANDING-MANINKA kosafina (Aub.) kosafune (Aub.) **UPPER VOLTA** MANDING-BAMBARA ko safuna (K&B) DYULA brahia (K&B) **IVORY COAST** VULGAR todzo (A&AA) AKAN-ASANTE aohué (B&D) AKYE gbinto (A&AA) iaonvi (Aub.; K&B) BAULE abovi (A&AA) aboyu (B&D) abué (K&B) KRU-BETE kugopo (K&B) GUERE (Chiehn) ugopo (B&D) ukopo (K&B) GUERE (Wobe) gbaninron (A&AA) KWENI niénié (B&D) MANDING-DYULA brahia (K&B) MANINKA dakuin (B&D) kosafina = *quinine of the Blacks* (Aub.) kosaflan (A&AA) séguindi (B&D) SENUFO-TAGWANA anna (K&B) DYIMINI kpo (K&B) 'KRU' bitali (K&B) 'MAHO' dakuma (K&B) **GHANA** ADANGME agba (A.S.Thomas; FRI) AKAN-ASANTE mpena-sere = *concubine laughs* (FRI) mponasere = *concubine laughs* (FRI) FANTE awonyõn (FRI) awonyõn (Enti) o-bɔwen(e) (auctt.) TWI awodwene (BD&H) awonwẽnẽ (FRI) *o*-bɔwen (Enti) DAGBANI biebingira (AEK; FRI) GA agɔaflu (A.S.Thomas; FRI) akpa (Field; FRI) tà (KD) GBE-VHE avenya (FRI) gbɔ (FRI) gbɔti (FRI) gbɔtsi = *goat tree; for its poisonous property* (FRI) GUANG-GONJA sàŋká (CV; FRI) SISAALA fuodabia (AEK; FRI) **TOGO** GBE-VHE avenya (Volkens) TEM (Tshaudjo) tingma (Volkens) tusima (Volkens) **DAHOMEY** TEM aloma (A. Chev.) aloma kulu (A. Chev.) **NIGERIA** EDO óríwò (JMD) FULA-FULFULDE (Nigeria) kaɗkaɗɗe (JMD) siwaakewal (*pl.*

siwaakeeje) = *tooth-stick* (JMD) HAUSA chusar doki *a horse tonic food containing the leaves* (JMD) fate fate, may emaye *foods prepared from the leaves* (JMD) shiwáákáá, shiwaákaá (JMD; ZOG) shwaka (Shaw) IGBO òlúgbù *see V. amygdalina* òlúgbù ọhīá (BNO) IJO-IZON (Kolokuma) kíríòlògbò = *ground cat* (KW) YORUBA eéro oko, ewúrò oko – *bitter leaf of the farm* (JMD; Verger) ewúrò (JMD; IFE) ewúrò jíjẹ = *edible ewúro* (JMD) orin, pákò *general for chew-stick roots* (JMD)

*NOTE: These names are generally interchangeable with V. amygdalina.*

A variable plant, undershrub or tree to 8 m high (Ghana), usually 3.5 m, of deciduous soudanian savanna across the Region from Senegal to W Cameroons, and throughout central and south tropical Africa.

This species and *V. amygdalina* are very similar in appearance. They share many of the vernacular names and uses are the same. This plant is a savanna species but its dispersal has been amplified by man who has planted it in the vicinity of villages and has taken it into the forest zone for its manifest medicinal uses (2). Leaves are often sold in markets. They are extremely bitter and are taken for fevers, and are known as a quinine-substitute. In Ivory Coast a leaf-decoction is administered in small doses throughout the day with honey or sugar to mask the bitterness. Besides producing a febrifugal effect, it promotes diuresis, expurgation and vomiting (16). The leaves are taken in decoction in Sierra Leone for fever and anaemia (5), while in Uganda a cold infusion of crushed leaves is taken for fever (19). In The Gambia the leaves are chewed (24) or put into soup (18) for their astringent and stomachic properties. In Senegal leaves or roots are macerated for 48 hours for all stomach-complaints (12). A decoction is used as a mouth-wash for tonsillitis and stomatitis in Tanganyika (7), and a cough-medicine in Ghana (8) and for stomachache (20). In Ivory Coast a preparation is taken as an anthelmintic especially for ascaris, in the treatment of jaundice (1, 4), with *Piper guineensis* (Piperaceae) for chickenpox (1), and in decoction with *Argemone mexicana* (Papaveraceae) in draughts and baths for pneumonia (16). In Congo the sap is given for gastro-intestinal complaints and for urethral discharge (3). The leaves are said to have action on skin-conditions: in Senegal they are rubbed on the body for all sorts of eruptions (11, 14, 15). An antipsoric lotion is made of the leaves in Congo (3) and in Ivory Coast the crushed leaves softened over a fire are applied to heal sores (1). The Yoruba prescribe the leaf and invoke it in an incantation against *kọ́nú* disease (2). A leaf and flower-decoction is given in inhalations for heart-failure and epilepsy in Senegal (14). A hypotensive decoction is prepared in Ivory Coast with the bark of *Blighia unijugata* (Sapindaceae) for taking by draught and by nasal and eye instillations (1).

The Hausa of N Nigeria mix the leaves with butter, condiments, etc., to make special foods eaten by both sexes, by men principally as a food, and by women in a belief that they improve their appearance and sexual attraction. Bitterness can be reduced by soaking in several changes of water or boiling (9).

The bark of branches, trunk and roots is highly regarded in parts of Senegal for the treatment of schistosomiasis, impotence and female sterility. For added erogenic effect the bracts surrounding the cobs of a red variety of maize, which resemble a phallus, may be added along with other drug-plants (14, 15). Twig-bark is pulped up with ginger and water and taken in a draught in Ivory Coast for stomach-complaints (1). Root-bark scrapings mixed with castor-oil are rubbed on to scabies in Tanganyika (7).

The roots in decoction are taken for gonorrhoea in Ivory Coast-Upper Volta (16) and in Tanganyika (7). In the latter country the roots of *Rauvolfia vomitoria* (Apocynaceae) are added. The root is considered a poison-antidote in Tanganyika (17).

The wood is soft. Twigs are cut for chew-sticks in Senegal (15), Ghana (10, 23), and Nigeria. Wood-ash is used in Zaïre and Ubangi-Shari as a vegetable-salt (10).

The flowers and seeds, in contrast from the bitterness of other parts of the plant, are eaten in Uganda for their sweetness (19). The flowers are heavily worked for honey by bees (Howes fide 10).

Alkaloids are probably absent from the plant, but bitter principles are present. Toxicity of the plant in Sierra Leone is reflected in the Temne name meaning 'kills goats.' Extracts have killed laboratory mice, but bitters isolated have shown little toxicity (15). Twig-bark is rich in saponosides (4). A cardiotonic glycoside has been reported in Nigerian material under the name of *vernonin*. Two terpenic lactones have been isolated, *vernolide* and *hydroxyvernolide* (15, 21). The seeds contain 3.1% oil containing glycerides of oleic acid 12%, linoleic acid 15% and 12.13-epoxyoleic acid 38% (6, 15).

The Bete of Ivory Coast ascribe fetish powers to the plant. The leaves crushed with white sand are spread along tracks and at certain points in a village to confer protection from small-pox (16).

References:

1. Adjanohoun & Aké Assi, 1972: 103. 2. Aubréville, 1959: 3: 313. 3. Bouquet, 1969: 96. 4. Bouquet & Debray, 1974: 73. 5. Deighton 837, K. 6. Ewing & Hopkins, 1967. 7. Haerdi, 1964: 173. 8. Irvine 46, K. 9. Irvine, 1956: 41. 10. Irvine, 1961: 722. 11. Kerharo, 1967. 12. Kerharo & Adam, 1963, b. 13. Kerharo & Adam, 1964, b: 584–5. 14. Kerharo & Adam, 1964, c: 325. 15. Kerharo & Adam, 1974: 231–2. 16. Kerharo & Bouquet, 1950: 218. 17. Koritschoner 1216, K. 18. Rosevear, 1967. 19. Tanner 6038, K. 20. Thomas, A. S. D.98, K. 21. Toubiana & Gaudemer, 1967. 22, Verger, 1967: no. 70. 23. Vigne 947, K. 24. Williams, F. W., 1907: 381.

**Vernonia conferta** Benth.

FWTA, ed. 2, 2: 277. UPWTA, ed. 1, 422.

English: cabbage tree, soap tree.

West African: **GUINEA** MANO lulu (RS) **SIERRA LEONE** MENDE hɛgɛ-wulu = *soap tree* (JMD) nyina-nyini *a general name* (auctt.) **LIBERIA** KRU-GREBO glabe (auctt.) **IVORY COAST** ABE pupuia (RS; Aub.) ABURE obuboé (B&D) AKAN-ASANTE oflan (B&D) ANYI ohomo (K&B) KRU-BETE boko bokoaëï (K&B) poko pokowëï (K&B) GUERE blotué (K&B) guétué (K&B) GUERE (Chiehn) boko boko (K&B; B&D) KWENI pokopokowéi (B&D) KYAMA amambé (Aub.) amandé (RS) bugulé (B&D) manbenbi (B&D) 'KRU' ufuo (K&B) **GHANA** ADANGME-KROBO flakwa (A. S. Thomas; FRI) AKAN-ASANTE awudefo kete = *murderer's mat, alluding to the large leaves* (BD&H; FRI) ɔfana (BD&H) ofena (auctt.) ɔfena (Enti) TWI ofena (BD&H; FRI) WASA ɔfena (Enti) NZEMA kululɛtɛnde (FRI) kululɛtɛnde (FRI) **NIGERIA** EFIK òkpón íkọ̀ñ = *big leaf* (RFGA; Lowe) oriweni (JMD; KO&S) IGBO (Owerri) ujuju (FRI) IGBO (Ukwuani) ụ́bọ̀ọ̄ (JMD; KW) IGBO (Umuahia) òlúbì, òlúgbì (AJC; JMD) YORUBA ṣapo (JMD; KO&S) **WEST CAMEROONS** DUALA bòpòlò-pòlò (JMD)

Shrub or tree to 9 m, of secondary regeneration in clearings and old farmland from Guinea to S Nigeria and Fernando Po, and extending through central Africa to Uganda and Angola.

The trunk is slender and the wood very soft (9). It is sometimes planted in Ghana as a fence (8). The Mende name meaning 'soap tree' indicates one use, for the branches are burnt to ash to make soap in Sierra Leone (11), Liberia (5) and Ivory Coast (2). A liquid of the extracted ashes is evaporated and the dried residue pulverized to mix with tobacco-snuff in Sierra Leone (Lane Poole fide 6). The ashes are used as a vegetable salt in Zaïre (Staner fide 8).

The leaves which may attain a metre in length by 20 cm in breadth evoke the Asante name meaning 'murderer's mat' and the Efik name, both cited above.

The young leafy shoots are boiled into a soup given to newly-delivered women in Ivory Coast to induce an abundant milk-supply causing the baby to thrive and grow quickly (10). A leaf-macerate is considered aphrodisiac in Ivory Coast (10). A leaf and bark-decoction is taken in Sierra Leone for stomachache (7), especially as a laxative for children (6). A tisane of leaves is considered good for whooping-cough, convulsive coughing, bronchitis and asthma in Congo (3). The young leaves are crushed and put on cuts in Congo as a wound-dressing; they are haemostatic and cicatrizant (3). In Gabon the leaves are put into the sores of yaws (12, 13), and the plant has much use in

Ivory Coast for skin-complaints, abdominal pains and jaundice (4). A leaf-decoction along with the bark of *Blighia sapida* (Sapindaceae) is taken by mouth as a poison-antidote (10). The action of the plant is said to be diuretic (4). Leaves are put into steam baths in Ivory Coast for treating gonococcal orchitis (10).

In Ghana the bark is given for diarrhoea and constipation (8). In Sierra Leone the bark is boiled with a plantain and fed as a tonic to invalids (Lane Poole fide 6). Also the dried powdered bark is boiled with beniseed and the liquor is given by draught to children as an anthelmintic (7). In Congo a bark-decoction is used as a vermifuge, and also to treat intestinal and urino-genital ailments: administration is by draught and enema (3). Women in Gabon take a bark-decoction as a galactogogue; the leaf is also considered a remedy for ophthalmias and to maturate abscesses (12, 13).

The roots of Nigerian material are reported to contain a little alkaloid (1), while none has been detected in plants from Ivory Coast, but bitter principles are present showing a lethal toxicity to test mice (10).

References:

1. Adegoke & al., 1968. 2. Aubréville, 1959: 3: 314. 3. Bouquet, 1969: 96. 4. Bouquet & Debray, 1974: 73. 5. Cooper & Record, 1931: 116. 6. Dalziel, 1937. 7. Deighton 1070, K. 8. Irvine, 1961: 722–3. 9. Keay & al., 1964: 420. 10. Kerharo & Bouquet, 1950: 219. 11. Lane Poole 25A, K. 12. Walker, 1953, a: 30. 13. Walker & Sillans, 1961: 132.

## **Vernonia doniana** DC.

FWTA, ed. 2, 2: 277.

West African: **SIERRA LEONE** MENDE gbɔnɛtɛ (Burbridge)

A shrub to 2 m high, recorded from only Sierra Leone and Liberia.

At Freetown, the leaves are given to children as a laxative (1).

Reference:

1. Burbridge 522, K.

## **Vernonia frondosa** Oliv. & Hiern

FWTA, ed. 2, 2: 276. UPWTA, ed. 1, 422.

West African: **NIGERIA** YORUBA akisa (Sankey; JMD) **WEST CAMEROONS** DUALA bòpòlò-pòlò (JMD)

An erect shrub or tree to 6 m high by 30 cm girth, of forest clearings in Ivory Coast, S Nigeria and W Cameroons, and also in E Cameroun.

The leaves are remarkably large at nearly 1 m long by 50 cm width. The flower panicle may also attain nearly 1 m across. The wood is cream-coloured and very soft (1).

Reference:

1. Keay & al., 1964: 420.

## **Vernonia galamensis** (Cass.) Less.

FWTA, ed. 2, 2: 280, as *V. pauciflora* (Willd.) Less. UPWTA, ed. 1, 423, as *V. pauciflora* Less.

West African: **SENEGAL** MANDING-BAMBARA siginiba (JB; K&A) MANDINKA faransambo kunango (K&A) 'SOCE' frâsâ bukanaô (K&A) kunturô (K&A) MANDINKA faransambo kunaŋ (*def.*-o) (after K&A) **THE GAMBIA** MANDING-MANDINKA faransambo kunaŋ (*def.*-o) = *bitter* *faransambo* (Hayes; DF) nyama kunturula = *sharp grass* (Fox; DF) **NIGERIA** HAUSA toozalin bareewaa = *eye-shadow of the gazelle* (JMD; RES) NUPE etikó yíwógi = *wife of the big-head* (RB)

An annual herb to 1.30 m high, occurring throughout the Region and across Africa to NE and E Africa.

The plant is considered toxic in Senegal and particularly so against termites. It is used (method not stated) to protect palisades, construction timber, etc. (4).

The leaves are smoked in cigarettes as a tobacco-substitute in Ethiopia (2, 5). In Tanganyika the leaves are cooked in porridge, or drunk in tea for chest-pains (6); stock browse it (3); and the roots have undisclosed medicinal use (7).

A preparation of the flowers crushed along with scent is used by hunters in N Nigeria either smeared on the body or worn as a charm to attract 'beef' (*barewa,* gazelle, or other antelopes) (1).

References:

1. Dalziel, 1937. 2. Getahun, 1975. 3. Hornby 128, K. 4. Kerharo & Adam, 1964, c: 326. 5. Taddesse Ebba 719, K. 6. Tanner 1332, K. 7. Williams 59, K.

**Vernonia gerberiformis** Oliv. & Hiern

FWTA, ed. 2, 2: 279–80.

English: veld corn flower (Kenya, suggested name, Harvey).

West African: **NIGERIA** HAUSA tozalin kura = *eye-shadow of the hyena* (JMD; RES)

A stemless herb arising from a perennial woody rootstock, to about 12 cm high, recorded from the Jos Plateau of N Nigeria, and occurring widely in NE, E and South Central Africa.

The purplish blue flower is decorative.

**Vernonia glaberrima** Welw.

FWTA, ed. 2, 2: 283. UPWTA, ed. 1, 422.

West African: **SIERRA LEONE** MANDING-MANDINKA setun (FCD) **NIGERIA** HAUSA shiwáákáár (shiwaákár)-ján-gágárii = *Vernonia of the red clay soil* (auctt.)

An erect undershrub to 2 m high, of open hillside grassland in Guinea to N Nigeria and W Cameroons, and central Africa to Angola.

No usage is recorded for the Region. In Congo leaf-sap is sniffed up the nose for chronic and very painful migraine (1). A leaf-decoction is used in central Africa as an antipsoric (2).

References:

1. Bouquet, 1969: 96. 2. Watt & Breyer-Brandwijk, 1962: 297.

**Vernonia glabra** (Steetz) Vatke

FWTA, ed. 2, 2: 280–1.

A herb to 1.30 m (var. *occidentalis* C. D. Adams) and to 1.65 m (var. *hillii* (Hutch. & Dalz.) C. D. Adams) of disturbed ground in montane situations of N Nigeria and W Cameroons and occurring also in E Africa.

No usage is recorded for the Region. The plant is browsed by stock in Tanganyika (3). A crushed leaf is applied to a burn. It adheres closely to the skin excluding the air (1). Sap is dripped into a fresh wound as a wound-dressing (2). Leaf-sap or a root-decoction together with *Harrisonia abyssinica*

Oliv. (Simaroubaceae) (large-leafed form) is taken for intestinal obstruction, and leaf and root-decoctions for gonorrhoea (2). A root-decoction is also taken as a diuretic and in treatment of amoebic dysentery (2).

References:

1. Allnutt 21, K. 2. Haerdi, 1964: 174. 3. Hornby 132, K.

**Vernonia guineensis** Benth., vars.

FWTA, ed. 2, 2: 282. UPWTA, ed. 1, 422–3.

West African: **GUINEA** MANDING-BAMBARA sula guéku (A. Chev.) MANINKA kunguéni (Brossart) sumantoro (Brossart) **SIERRA LEONE** KISSI yufɛho (FCD) MANDING-MANDINKA funyembe (NWT) SUSU kubɛfiɛxe (NWT) kumorufiɛxe (NWT) **UPPER VOLTA** MANDING-DYULA léiguéde (K&B) **IVORY COAST** BAULE awon nugba (K&B) katinuaba = *savanna tapioca* (B&D) KWENI buibu (B&D) MANDING-DYULA léiguédé (K&B) MANINKA bubémėssi (B&D) darama-ku (A&AA) tion (B&D) SENUFO-TAGWANA kolab (K&B) kolabu (K&B) DYIMINI lawon (K&B) nawa (K&B) **GHANA** DAGBANI nyeri-kobiga (Ll.W.) **NIGERIA** YORUBA ewúrò (IFE)

A variable species with three varieties (var. *guineensis,* var. *cameroonica* C. D. Adams, var. *procera* (O. Hoffm.) C. D. Adams) recognized in the Region; herbaceous with strong erect stems from a perennial woody rootstock, to 1.70 m high, distributed across the Region from Mali to W Cameroons, and across central Africa from Cameroun to Sudan.

Though the Baule name means 'savanna tapioca', there is no record of the plant being eaten in an alimentary sense. The plant has medicinal uses. The leaves are held to have analgesic properties. Pounded leaves are applied to the face to relieve toothache in Guinea (3). A paste of the leaves is applied to the tooth for toothache and on to sores in Congo, where also a tisane is given for kidney-pains and incipient hernia with massage with the lees (1).

The whole plant, or the leaves or roots alone, are administered in decoction in Ivory Coast-Upper Volta as a purgative, anti-dysenteric and counter-poison (5), and a root-decoction is taken for urethral discharge (2). In Guinea an antemetic is made of the roots (3). In Ivory Coast-Upper Volta the plant is administered with other drug-plants for feverish conditions, jaundice and as an anti-malarial (5). In Gabon the plant is given as a snake bite antidote (6, 7). The roots are chewed raw in Sierra Leone as an aphrodisiac (4), while in Ivory Coast Tagwana compound the leaves with other drug-plants to increase vigour and to promote spermatogenesis in male sterility.

The plant in northern Ghana is one of a group used in infusion to treat a disease of cattle known as *garli* (8).

No alkaloid is present, but there are bitter principles (5).

References:

1. Bouquet, 1969: 97. 2. Bouquet & Debray, 1974: 73. 3. Dalziel, 1937. 4. Deighton 979, K. 5. Kerharo & Bouquet, 1950: 219–20. 6. Walker, 1953, a: 30. 7. Walker & Sillans, 1961: 132. 8. Williams 717, K.

**Vernonia hymenolepis** A. Rich.

FWTA, ed. 2, 2: 276, as *V. calvoana* (Hook. f.) Hook. f., *V. insignis* (Hook. f.) Oliv. & Hiern, *V. mokaensis* Mildbr. & Mattf. and *V. leucocalyx* O. Hoffm. UPWTA, ed. 1, 422, as *V. calvoana* Hook. f.

A shrub or tree to 5 m high, of grassland and openings in montane forest of S Nigeria, W Cameroons and Fernando Po.

The leaves are said to be eaten by people on the Cameroon Mountain (1).

Reference:

1. Dalziel, 1937.

**Vernonia lasiopus** O. Hoffm.

FWTA, ed. 2, 2: 276, as *V. iodocalyx* O. Hoffm.

An erect shrub of montane locations in Guinea, and also in E Africa.
No usage is recorded for the Region. In Tanganyika an extremely bitter decoction is made from the whole plant which is used for epilepsy, indigestion and in childbirth. The root is also used to facilitate parturition.

References (as *V. iodocalyx* O. Hoffm.):
1. Bally, 1937. 2. Watt, J. M., 1967. 3. Watt & Breyer-Brandwijk, 1962: 297.

**Vernonia macrocyanus** O. Hoffm.

FWTA, ed. 2, 2: 280. UPWTA, ed. 1, 423, as *V. primulina* sensu Dalziel.
West African: NIGERIA HAUSA saàbúlùn yán-maátaá (JMD; ZOG) saàbúlùn-maátaá = *women's soap* (JMD) toozalin-kuuraa (ZOG)

A perennial herb to 60 cm high from a fleshy rhizome, recorded from Mali, N Nigeria and W Cameroons, and occurring in E Africa and Angola.
The leaves are used in N Nigeria for scrubbing and the Hausa name implies that the plant is a soap-substitute (1). In Angola the plant is used as an antispasmodic for the relief of colic and as a fish-poison (2).

References:
1. Dalziel, 1937. 2. Watt & Breyer-Brandwijk, 1962: 297.

**Vernonia migeodii** S. Moore

FWTA, ed. 2, 2: 282–3. UPWTA, ed. 1, 423.
West African: NIGERIA YORUBA ẹru ewuru = *slave of the bitter leaf* (Phillips; JMD)

A perennial herb to 1.30 m high, recorded from Ghana to N Nigeria, and W Cameroons, and into Ubangi-Shari.
The sap is bitter (1). No usage is recorded.

Reference:
1. Dalziel 32, K.

**Vernonia myriantha** Hook. f.

FWTA, ed. 2, 2: 277.

A shrub or tree to 6 m high, of montane forest in W Cameroons and Fernando Po.
The wood is soft.
It is a common plant around gardens near Buea (? subspontaneous), and the leaves are eaten there as spinach (1).

Reference:
1. Maitland 256, K.

**Vernonia nestor S. Moore**

FWTA, ed. 2, 2: 282.

A perennial herb to 60 cm high, recorded from Guinea, N and S Nigeria and W Cameroons, and extending to Malawi.

The latex is red and is recorded as used at Calabar, S. Nigeria, to cure ringworm (1).

Reference:

1. Daramola FHI.55262, K, FBC.

**Vernonia nigritiana** Oliv. & Hiern

FWTA, ed. 2, 2: 280. UPWTA, ed. 1, 423.

West African: **SENEGAL** DIOLA-FLUP giénom = *erect yourself* (JB; K&A) FULA-PULAAR (Senegal) dononturu (Brossart) sâban (K&A) MANDING-MANDINKA jubu jamba (K&A) kame kumo = *head of the guinea-fowl* (K&A) 'SOCE' kumareturo (K&A) kurukey (K&A) SERER yetãngol (JB) yetngol (JLT; K&A) WOLOF baťaťor (auctt.) **THE GAMBIA** MANDING-MANDINKA juba jamba, jamba: *leaf* (Brown-Lester; Williams) kumare-two = *crested-crane bird's hat* (Frith) **GUINEA-BISSAU** FULA-PULAAR (Guinea-Bissau) birre-djom (JDES) MANDING-MANDINKA cúmarô-túrô (JDES) **SIERRA LEONE** MANDING-MANDINKA nonomorandakofin (FCD) SUSU-DYALONKE kundiditu-na (FCD) **UPPER VOLTA** MANDING-DYULA duodru (K&B) **IVORY COAST** AKAN-BRONG tiatowihan-goso (K&B) BAULE saminia (K&B) MANDING-DYULA duodru (K&B) SENUFO-TAGWANA gopité (K&B) **GHANA** AKAN-ASANTE saanunum (FRI) BRONG gyakuruwa (Brown; FRI) GBE-VHE avegbɔ? (Ewe Dict.)

An annual herb or shrub to 60 cm high from a somewhat tuberous, perennial rootstock, of savanna grass and woodland, in Senegal to W Cameroons.

The flower-heads attain 4–5 cm across with bright red involucral bracts making the plant showy and worthy of cultivation as an ornamental. The Adamawa Fula call the plant 'husband of pumpkin' and plant or retain it in plantings of cucurbitaceous crops to increase their yield (3).

A leaf-decoction is used in Sierra Leone to wash out calabashes for holding milk. This causes curdling (4). Leaves are put into soups in Nigeria for colic and generally as a stomachic (1).

The small tuberous roots are fascicled and woolly on the surface. They are commonly sold in markets in Senegal, The Gambia and Guinea and are of ancient application as an emetic, vermifuge, diuretic, antidysenteric, febrifuge, rheumatic medicine (8, 9), for coughs (7) and piles (6). In The Gambia the pounded roots are boiled and taken as a purge (11). They are considered in Senegal to be an aphrodisiac *par excellence*, as the Flup name implies (7, 9). In Ivory Coast the roots are considered mainly as febrifugal and purgative. They are used with *Lippia adoensis* (Verbenaceae) and *Cassia occidentalis* (Leguminosae: Caesalpinioideae) for 'yellow fever' and serious jaundice (10). In Ghana the boiled roots are given as a diuretic, purgative and emetic, for constipation and colic, to relieve piles, and to purify the blood (5). A root-decoction is used in lotion form for eye-troubles in Senegal (3).

In Ivory Coast the flowers, perhaps on the Theory of Signatures for their red colour, are used as an emmenagogue (2, 10).

A bitter glycoside, *vernonin*, has been obtained from the roots. This has an action like *digitalin* but weaker (3, 9, 10). While alkaloid has been reported absent from the root (3), recent work has shown reaction from the presence of alkaloids in the root-bark (2). The seeds have an oil-content of 7.6%, which contains glycerides of oleic acid 19%, linoleic acid 43% and 12.13-epoxyoleic acid 8% (9).

References:

1. Ainslie, 1937: sp. no. 356. 2. Bouquet & Debray, 1974: 32: 73–74. 3. Dalziel, 1937. 4. Deighton 4470, K. 5. Irvine, 1961: 723–4. 6. Kerharo, 1967. 7. Kerharo & Adam, 1963, b: 70. 8. Kerharo & Adam, 1964, c: 325. 9. Kerharo & Adam, 1974: 231–3, with pharmacology. 10. Kerharo & Bouquet, 1950: 220. 11. Williams, F. W., 1907: 381.

**Vernonia oocephala** Bak.

FWTA, ed. 2, 2: 281.
West African: **NIGERIA** HAUSA (West) zagar demmu (JMD)

A shrub to 1.30 m high, of undergrowth in montane situations in N Nigeria and W Cameroons, and occurring in central and south tropical Africa.
No usage is recorded.

**Vernonia perrottetii** Sch. Bip.

FWTA, ed. 2, 2: 281. UPWTA, ed. 1, 423.
West African: **SENEGAL** MANDING-BAMBARA ségini (JB) **THE GAMBIA** FULA-PULAAR (The Gambia) dattal (DRR) MANDING-MANDINKA eata-silo (DRR) silli (*def.* silloo) (Fox) **NIGERIA** GWARI bachigehi (JMD) HAUSA burzu (JMD; ZOG)

An annual herb to 60 cm high, a common plant of fields and waste places, occurring throughout the Region, and widely dispersed across tropical Africa.
In The Gambia it is regarded as a serious weed of cultivation (2). The leaves are said to be used as a vegetable in Sierra Leone, and in Nigeria are regarded as being slightly purgative (1). The plant is used in The Gambia to prepare a sort of ink (2). The stems can be used for spindles for thread and in the upper Nile region the plant is burnt with *Hygrophila auriculata* (Acanthaceae) to obtain a vegetable salt from the ash (1).

References:
1. Dalziel, 1937. 2. Fox 26, K.

**Vernonia poskeana** Vatke & Hildebrandt var. **elegantissima** (Hutch. & Dalz.) C. D. Adams

FWTA, ed. 2, 2: 281.
West African: **NIGERIA** NUPE tsùla bishe = *hen's bitterleaf* (RB)

An annual herb to 30 cm high, recorded from Senegal, Ghana and Nigeria. Other varieties occur elsewhere in tropical Africa.
No usage is recorded for the West African variety. That in Tanganyika is used as a decoction of the whole plant for cough (1), to reduce fever (2) and to prevent miscarriage (3).

References:
1. Tanner 4206, K. 2. Tanner 4228, K. 3. Tanner 4255, K.

**Vernonia pumila** Kotschy & Peyr.

FWTA, ed. 2, 2: 282. UPWTA, ed. 1, 423.
West African: **GHANA** AKAN-BRONG foforo-bankye = *new cassava* (FRI) **NIGERIA** HAUSA ba-gashi, sányár ƙasà (JMD; ZOG) sheèkár-zoómoó (JMD; ZOG) sheƙani (JMD; ZOG)

An annual herb of variable height to 60 cm, arising from a perennial woody rootstock with small spindle-shaped roots, in Senegal to N and S Nigeria, and extending to Sudan and Uganda.
The roots are bitter, but may be eaten raw or boiled with cereal foods as a stomachic, and for gonorrhoea, etc., acting without purging (1).

Reference:
1. Dalziel, 1937.

**Vernonia smithiana** Less.

FWTA, ed. 2, 2: 282.

A perennial herb to 1 m high, of hill grassland in Guinea, Mali and W Cameroons, and extending to Congo, Angola and E Africa.

No usage is recorded for the Region. In Congo, an aqueous decoction of the whole plant is taken in a vapour-bath to relieve rheumatic pains; for heart-troubles the leaves are chopped up with maleguetta peppers and chewed, the saliva being swallowed; and a tisane of roots is said to be good for venereal discharge, and stomach-complaints and to be antiabortifacient (1).

Reference:

1. Bouquet, 1969: 97.

**Vernonia subuligera** O. Hoffm.

FWTA, ed. 2, 2: 279.

A shrub to 4 m high, of montane situations in W Cameroons, and occurring also in E Africa.

No usage is recorded for the Region. In Kenya it is planted as a hedge and to stop erosion (1). In Tanganyika the root is taken as a galactogogue (2).

References:

1. Dale & Greenway, 1961: 163. 2. Koritschoner 1208, K.

**Vernonia theophrastifolia** Schweinf.

FWTA, ed. 2, 2: 279.

A shrub or tree to 5 m high of uplands in N Nigeria and W Cameroons, and occurring also in Sudan.

It is recorded as being used as a compound fencing tree on the Mambila Plateau (1).

Reference:

1. Hepper, 1966: 115.

**Vernonia thomsoniana** Oliv. & Hiern

FWTA, ed. 2, 277.

French: quinine des noirs (Gabon, Walker).

A shrub to 3 m high, of upland savanna in Guinea and W Cameroons, and extending through Cameroun and the Congo basin to Sudan, E Africa and Angola.

The leaves are very bitter, hence the French name given to the plant in Gabon where a leaf-infusion substitutes for quinine and is taken for fever. The leaves are eaten in Gabon as a vegetable but only after boiling in several changes of water. The leaves are said to be piscicidal, and are used against itch. In maceration, the liquor is soapy and is used for bathing and for laundering clothes.

The plant is grown in Gabon as a hedge.

Reference:

1. Walker & Sillans, 1961: 132–3.

**Vernonia undulata** Oliv. & Hiern

FWTA, ed. 2, 2: 283.

A perennial herb to 1.30 m high from a woody rootstock, of grassland in Guinea to W Cameroons and to Sudan, Congo, Uganda and Angola.
A leaf-decoction is used in Gabon in enemas for constipation and piles (1, 2).

References:

1. Walker, 1953, a: 30. 2. Walker & Sillans, 1961: 133.

**Vernonia spp. indet.**
West African: **IVORY COAST** DAN blante (Laffitte) 'KRU' bidali (K&B) **NIGERIA** NUPE tsùla

Plants of the above vernaculars are reported used in Ivory Coast as a febrifuge in decoction, and as an analgesic by friction with the pulped up plant (1).

Reference:

1. Kerharo & Bouquet, 1950: 220.

**Vicoa leptoclada** (Webb) Dandy

FWTA, ed. 2, 2: 259.
West African: **SENEGAL** SERER a tomdofan (JB) **THE GAMBIA** MANDING-MANDINKA silli foro (*def.*-o) = *pure sillo* (Fox) **IVORY COAST** SENUFO-TAGWANA piéragulu (K&B)

A herb to 1.30 m high, of open places outside the high-forest from Senegal to W Cameroons, and widespread elsewhere in tropical Africa.
The plant is a weed of cultivation (2). It is not grazed by stock (1). In Ivory Coast it is held to have magical properties and is thus used in a prescription for treating hysteria and epilepsy (3).

References:

1. Andrews 30, K. 2. Fox 45, K. 3. Kerharo & Bouquet, 1950: 221.

**Wedelia trilobata** (Linn.) Hitchc.

FWTA, ed. 2, 2: 236–7.
English: creeping ox-eye, marigold (Jamaica, Adams).

A creeping prostrate herb with flowering stems to 30 cm high, of open damp sites, a tropical American plant now occurring in Guinea and Sierra Leone.
The yellow flower which is up to 2.5 cm in diameter is attractive. The plant may have some value as an ornamental. The plant is boiled and taken as a tonic in Belize by the Creole (1).

Reference:

1. Schipp 689, K.

# CONNARACEAE

**Agelaea dewevrei** De Wild. & Th. Dur.

FWTA, ed. 2, 1: 746, 763.

A scrambling lianescent shrub to 15 m high of the closed-forest in W Cameroons, and extending to Zäire and Angola.

The white flowers are fragrant.

In Congo pains in the sides and chest are treated by scarification and dusting with a powder prepared from the twigs with roots of *Rauvolfia sp.* (Apocynaceae) after heating, and for urethral discharge a leaf-tisane is taken (1). In Zäire a bark-decoction is used for earache (2).

References:

1. Bouquet, 1969: 97. 2. Troupin, 1952: 104–6.

**Agelaea hirsuta** De Wild.

FWTA, ed. 2, 1: 745.

A scandent shrub to about 1.5 m high, of the closed-forest and secondary jungle in S Nigeria, and extending to Zäire.

No use is recorded in the Region. In Zäire it yields a certain amount of [? potable] sap from the stem (1).

Reference:

1. Troupin, 1952: 100.

**Agelaea mildbraedii** Gilg

West African: **IVORY COAST** AKAN-ASANTE abokoponko (K&B) KRU-GUERE bagnimoney (K&B) 'KRU' twi woplu (K&B)

A shrubby liane of thickets in secondary jungle, occurring in Gabon to Zäire, and reported in Ivory Coast where red sap exuded from crushed leaves is taken by draught and applied as a lotion for anaemia, and a root-bark decoction with a pimento added is taken for urethral discharge and female sterility (1).

Reference:

1. Kerharo & Bouquet, 1950: 170.

**Agelaea nitida** Soland.

FWTA, ed. 2, 1: 746.

A straggling bush to 3 m high, occurring in Guinea, Sierra Leone and Liberia.

The white fragrant flowers and the red fruits make this an attractive ornamental.

**Agelaea obliqua** (P. Beauv.) Baill.

FWTA, ed. 2, 1: 745–6. UPWTA, ed. 1, 342.

West African: **IVORY COAST** KRU-GUERE (Chiehn) kapussé (B&D) 'KRU' niawri kluabu (K&B) 'NEKEDIE' digripayuassu (B&D) **GHANA** AKAN-ASANTE homabiri (BD&H) TWI apɔsẽ (auctt.) **TOGO** TEM (Tshaudjo) akulagbe-tiko (Gaisser) **NIGERIA** IGBO (Onitsha) alanhita nta (NWT) IGBO (Umudike) ehu (Ariwaodo) YORUBA egu (Millen) esura (JMD)

A scrambling shrub or climber to 6 m high of the forest and secondary jungle, widespread from Guinea-Bissau to West Cameroons.

The white fragrant flowers and the scarlet fruits make this an attractive ornamental.

A leaf-decoction is taken by draught by newly-delivered women in Ivory Coast (3), and should a suckling baby fall ill while its mother is observing certain taboos, it is said to be cured by the mother taking a dose of the leaf-sap (1).

The root-bark chopped up with *Costus sp.* (Costaceae) is said in Ivory Coast to be aphrodisiac (1).

The fruits are used to rub the teeth (2).

References:

1. Bouquet & Debray, 1974: 75. 2. Dalziel, 1937. 3. Kerharo & Bouquet, 1950: 170.

**Agelaea pilosa** Schellenb.

FWTA, ed. 2, 1: 745.

A climbing shrub of the forest zone of S Nigeria.

The wood is used as chew-sticks (1).

Reference:

1. Thomas, N. W. 2168 (Nig. Ser.), K.

**Agelaea trifolia** (Lam.) Gilg

FWTA, ed. 2, 1: 746. UPWTA, ed. 1, 342–3.

West African: **SIERRA LEONE** MENDE gongonga (*def.*-i) (DS) TEMNE *an*-bis-*a*-kɔnth (M&H) **LIBERIA** MANO bono takpa (Har.) kpa (Har.) wuẽ lĕh (Har.) **IVORY COAST** KYAMA niamablé (B&D)

A scrambling shrub or climber to 8 m high, or rarely to 30 m, of the forest, secondary jungle and scrub from Guinea-Bissau to S Nigeria.

The creamy-white flowers are scented. The fruits are scarlet. The plant is an attractive ornamental.

The wood is used as chew-sticks in S Nigeria, and the stems are used in Liberia in the construction of thatch-roofing (2).

In Ivory Coast a bark-decoction is given by enema and by draught for fever and jaundice (1). A bark-decoction is reported as used as a witch ordeal in Sierra Leone (N. W. Thomas fide 2).

The fruit is used unsectioned for brushing the teeth in Sierra Leone (3).

References:

1. Bouquet & Debray, 1974: 75. 2. Dalziel, 1937. 3. Melville & Hooker 202, K.

**Agelaea spp. indet.**

West African: **IVORY COAST** DAN vezagu (K&B) KRU-GUERE dien dubu (K&B) looku (K&B)

Three unidentified species are recorded as used in Ivory Coast (1):

1. *looku* (Guere) is considered toxic and is put into arrow-poison;
2. *dien dubu* (Guere) yields a clear liquid from the cut liane which is used in nasal instillation as a febrifuge, and to wash sores and ulcers, especially lepromas;
3. *vezagu* (Dan) is incorporated in arrow-poisons.

Reference:

1. Kerharo & Bouquet, 1950: 170.

**Byrsocarpus coccineus** Schum. & Thonn.

FWTA, ed. 2, 1: 741. UPWTA, ed. 1, 343.

West African: **SENEGAL** DIOLA (Fogny) bu sèngèt (JB) **GUINEA-BISSAU** BIDYOGO endúrè *as B. viridis (Gilg) Schellenb., but probably B. coccineus* (JDES) **SIERRA LEONE** MANDING-MANDINKA manawala (NWT) SUSU ningɛrɛwutinyi (NWT) TEMNE *an*-tempəs (FCD) **IVORY COAST** BAULE botan (B&D) wotan (B&D) **GHANA** ADANGME awenade (FRI) AKAN-FANTE awennade, awinadze (Enti) KWAWU awenade (Enti) samantini (AEK; FRI) TWI awennade = *to eat iron* (Hartwell; FRI) awinadze (BD&H) GA awen-dade (JMD) GBE-VHE ayitsi (FRI) ʋumalia (FRI) VHE (Pecí) hlese (JMD) xlese (FRI) **TOGO** GBE-VHE hesre (Volkens) TEM (Tshaudjo) samala (Volkens) **NIGERIA** HAUSA kasa, tsaamiyar-kasa (ZOG) kímbá (Singha) kímbár máhàlbaá = *huntsman's pepper* (JMD; ZOG) tsaamiyar kurmi *tamarind of the valley* (MM) HAUSA (West) hallilua (LAKC) IGBO (Ila) oke abolo (NWT; Singha) YORUBA ado (JRA) kanti-kanti, amùjẹ wẹ́wẹ = *small amùjẹ* (JMD; Singha) orikotẹni (auctt.) yẹri eti ọmọde = *children's ear-ring*

A scandent shrub or liane of savanna thickets and secondary jungle, widely dispersed from Guinea to West Cameroons, and in other parts of tropical Africa.

The plant with its delicate, pink-tinged foliage, and white, sweet-scented flowers has ornamental value.

A decoction of leaves is drunk in Ghana as a remedy for piles (8), and in Nigeria for piles and flatulence. In the soudanian region the leaves are used for venereal diseases (de Gironcourt fide 9), and in S Nigeria the Yoruba take a cold water macerate of pounded leaves for gonorrhoea and dysentery (2). The plant in general is held to have urinary sedative action, and to be healing. The Yoruba name *amùjẹ* implies haemostasis: *mú*, to arrest; *eje*, blood. The leaves are used in Nigeria on swellings and tumours (6). In Ghana a decoction is used for sores in the mouth and on the skin (9), and in Sierra Leone the leaves are prepared into a general embrocation (Pelly fide 9).

The roots ground and mixed with water are taken by Yoruba to treat jaundice and worms. The powdered root is also used for application to cuts, etc. (2). The Ghanaian Twi name *awennade,* 'to eat iron', arises from the use of the root-bark, after grinding up with black peppers (*Piper guineense,* Piperaceae), to treat cuts and bruises caused by iron implements (7, 8). The roots enter into prescriptions along with other drug-plants in Ivory Coast for sexual stimulants (5) and in Akwapem, Ghana, for impotence (9). The scraped roots, macerated in water or with guinea-grains in palm-wine, are taken in Ghana for jaundice (9), while in the soudanian region root-sap is used for earache (3). Anodynal action is also sought of the plant by its use (part unspecified) in Ivory Coast for sore-throat and muscular and rheumatic pains (5).

The plant has been shown to be poisonous in Togo (Lewin fide 6, 9). Examination of material from Congo has given traces of flavonins, saponin and tannins in the bark, but no active principle in the leaves (4). Nigerian material has shown a trace of alkaloid in the leaves and roots (1).

References:

1. Adegoke & al., 1968. 2. Ainslie, 1937: sp. no. 64. 3. Aubréville, 1950: 416. 4. Bouquet, 1972: 20. 5. Bouquet & Debray, 1974: 75. 6. Dalziel, 1937. 7. Irvine 7, K. 8. Irvine, 1930: 69. 9. Irvine, 1961: 568–70.

**Byrsocarpus dinklagei** (Gilg) Schellenb.

FWTA, ed. 2, 1: 741.

A scandent shrub to 4 m high of the dense rain-forest of S Nigeria, and in Cameroun to Zäire.

No usage is recorded for the Region. In Zäire the plant is used to treat blennorrhoea (1).

Reference:

1. Troupin, 1952: 94, 96.

**Byrsocarpus poggeanus** (Gilg) Schellenb.

FWTA, ed. 2, 1: 741.

A laxly lianous shrub to about 2 m high of damp and swampy localities of the rain-forest of S Nigeria, and in Cameroun to Zäire and Angola.

No usage is recorded for the Region. In Congo the leaves are given to convalescents as a tonic and appetizer. They are also considered good for tachycardia (1). Tannins have been detected in the leaves, bark and roots, but no other active principle (2).

References:

1. Bouquet, 1969: 98. 2. Bouquet, 1972: 20.

**Byrsocarpus viridis** (Gilg) Schellenb.

FWTA, ed. 2, 1: 741.

A shrub or liane to 20 m long, of the closed rain-forest of S Nigeria, and also in Cameroun to Zäire and Angola.

No usage is recorded in the Region. Leaves cooked and steeped in oil, or served cooked as a vegetable, are taken for cough, asthma and generally for all gastro-intestinal pains in Congo (1). Similar use occurs in Zäire (2). The leaves are considered in Congo to be beneficial in tachycardia, and leaf-sap diluted in water with a little laundry blue is given for insanity (1). A macerate of young leaves is taken in draught for pneumonia and put into baths for children with fever (1). A leaf-infusion is emetic (2).

References:

1. Bouquet, 1969: 98. 2. Troupin, 1952: 96–97.

**Castanola paradoxa** (Gilg) Schellenb.

FWTA, ed. 2, 1: 746. UPWTA, ed. 1, 343.

West African: **GHANA** AKAN-ASANTE nsɛ dua = *not like a tree* (Enti) nsɛ homa = *not like a liana* (Enti) TWI abokodidua (auctt.) borɔdedwo, duaborɔdedwo (BD&H; FRI) **NIGERIA** EDO ogbe mavben (Kennedy; JMD) YORUBA ọsọdu (Ross)

A woody climber or small tree with a fluted trunk to 9 cm in diameter occurring widely in deciduous forest and secondary jungle from Sierra Leone to S Nigeria, and on to Zäire.

The stems are used in Ghana as chew-sticks (1, 2).

The leaves are put into soup by the Edo of S Nigeria (3). In Ghana the leaves have unspecified medicinal use (Moor fide 2), and similarly the plant in Gabon (4).

The flowers are sweet-scented.

References:

1. Irvine 127, K. 2. Irvine, 1961: 570. 3. Kennedy 1744, K. 4. Walker & Sillans, 1961: 133.

**Cnestis corniculata** Lam.

FWTA, ed. 2, 1: 743. UPWTA, ed. 1, 343.

West African: **GUINEA-BISSAU** FULA-PULAAR (Guinea-Bissau) talquidáua (DF) **SIERRA LEONE** MENDE yɛle (NWT) TEMNE *an*-totwokər (Lindall; Pyne) **IVORY COAST** BAULE diérétian (B&D) KRU-GUERE (Chiehn) digbé (B&D) kagnon (B&D) kossokablé (B&D) KWENI diéléka (B&D) **NIGERIA** YORUBA àkàrà àjẹ́ = *witch's bean cake* (Verger) esise *cf. esinsin, general for plants with stinging hairs* (JMD)

A woody climber of the humid-forest zone occurring in Guinea, Sierra Leone and Liberia.

The leaves and fruits are covered with pungent urticating hairs. The effect on the skin may be purely mechanical, but is possibly due partly at least to a brownish fluid in the cavity of the hair. The leaves are said to be strongly astringent, and are used as an internal remedy for gonorrhoea, being taken along with food (1).

Though *C. corniculata* is not recorded from Nigeria a plant said to be this and likely to be *C. longiflora* or perhaps *C. grisea,* both Nigerian and both possessing the same pungent irritant hairs, is invoked in an Yoruba incantation under the name of 'witches bean cake' to make the vagina swell and so to harm a hostile woman, or to expel an intruder from one's home (2).

References:

1. Dalziel, 1937. 2. Verger, 1967: no. 26.

## **Cnestis ferruginea** DC.

FWTA, ed. 2, 1: 743. UPWTA, ed. 1, 343.

West African: **SENEGAL** BANYUN diakanare diré (K&A) DIOLA (Diembereng) énoté kuruk (K&A) DIOLA (Efok) ufel édâgi = *vulva of the bitch* (K&A) DIOLA (Fogny) budanâd (K&A) fupéléen = *vulva of the bitch* (K&A) DIOLA-FLUP épéléhèn = *bitches vagina* (JB) ufélâdâgi = *bitch's vagina* (K&A) MANDING-MANDINKA dûdamo bɛsi (JMD, ex K&A) MANKANYA kadéra, kadira = *anus, or buttocks* (K&A) **THE GAMBIA** DIOLA (Fogny) fupéléen = *bitch's vagina* (DF) MANDING-MANDINKA dandamo-besi (JMD) **GUINEA-BISSAU** BALANTA treventi-ito (DF; JDES) BIDYOGO naporó (JDES) MANDING-MANDINKA manterim-ô (JDES) MANDYAK peduto-ubusse (JDES) utonque-ubule (JDES) utonque-ubusse (JDES) MANKANYA beduto-ubule (JDES) utonque-ubusse (JDES) PEPEL barniate (JDES) **SIERRA LEONE** BULOM (Sherbro) bɛltampel-lɛ (FCD) KISSI selialiwɔ (FCD) KONO wulukɛsɛ (FCD) MENDE bɔgɔjɛli (NWT; JMD) nyamawa (auctt.) SUSU kulenyi-mabari (NWT) kulewuri (NWT) TEMNE *an*-totwokər (FCD) **LIBERIA** KRU-GUERE sanhanguin (K&B) tilihizoto (K&B) MANO wuẽ-lĕh (JMD) **UPPER VOLTA** MANDING-DYULA tangolo sébé (K&B) **IVORY COAST** VULGAR djèhuon (A&AA) ADYUKRU alabalu (Aub.) AKAN-ASANTE afossé (K&B) ahuénadi, apoissien, plakassé, wansien blakassa (B&D) BRONG apossien (K&B) ANYI wusien blakassa = *wood which kills flies* (K&B) BAULE wonsien blakassa (B&D) wusien blakassa = *wood which kills flies* (K&B) KRU-BETE béblédiduku (K&B) kossoblé = *kola of the margouillat* (*a sp. of grey lizard*) (K&B) GUERE diazré (K&B) diézé (K&B) GUERE (Chiehn) diédiésoko (B&D) yéyésoko (K&B; B&D) zirakpagwè (B&D) zirikugèrè (B&D) GUERE (Wobe) siapiyi (A&AA) KULANGO karanwin (K&B) KWENI diéléka (B&D) diérika (B&D) yonoléssé (B&D) yunétia (B&D) KYAMA ahuénani (B&D) lobate (B&D) subuin (B&D) MANDING-DYULA tangolo sébé (K&B) SENUFO-TAGWANA koiméni (K&B) kumani (K&B) 'KRU' giakari (K&B) guienkarie (K&B) **GHANA** AKAN-ASANTE apɔsẽ (FRI) FANTE akitasẽ (auctt.) TWI apɔsẽ (auctt.) GA àhũidade (KD) GUANG-ANUM adonwaɛbɛ (FRI) NZEMA pudaegye (FRI; JMD) **NIGERIA** ANAANG útín éwà (KW) EDO úkpò-ìbìeká, ìbìeká: *boy* (JMD) EFIK ùsièrè ébuà, útín ébuà, ébuá: *dog* (auctt.) HAUSA fura amarya (RES) otito (Ross) IGBO àmụ̀ ṇkịtā = *dog's penis* (Singha; KW) ọ̀kpụ̀ṇkịtā = *dog's bone* (Lowe; KW) IGBO (Owerri) ?ọ̀kpụ́ ńkịtā = *leaf of the dog* (KW) ámụ̀nkịtā (JMD) íkè ńkịtā = *buttocks of the dog* ọ́kpụ̄kpụ̄ ńkịtā = *dog's bone* (Lowe; KW) ọ̀kpụ̀ńkịtā = *buttocks of the dog* (JMD) IGBO (Umuahia) ọ̀kpụ̀ọ́chá ṇkítā (AJC) URHOBO agwọla (?) (JMD) YORUBA àkarà-ojẹ́ = *witch's bread* (JMD) bónyinbónyin (JRA) ekoro (auctt.) esise (auctt.) gbóyingbóyin (JMD; Lowe) ọyàn-àjẹ́ = *witch's breast* (JMD)

A shrub or tree to 6 m high of deciduous forest and secondary scrub, common throughout the Region from Senegal to West Cameroons and in other parts of tropical Africa.

The plant with its scarlet fruits has value as an ornamental. The stems are used by Igbo in S Nigeria for bows (20) and in Zäire they are cut to yield a potable sap (22). The bark yields a red dye which is used by the Mende of

Sierra Leone for dyeing clothing (8). The powdered bark is rubbed into gums in S Nigeria for pyorrhoea (2). In Ivory Coast a paste of root-bark is rubbed on the forehead for headache, and with the addition of ash of burnt bark of *Calpocalyx aubrevillei* (Leguminosae: Mimosoideae) (as a vegetable salt) is given as an appetite stimulant in cases of illness (1).

The leaves are laxative and are taken by the Yoruba in decoction (2, 7). A leaf-decoction is given in Zaïre to treat bronchitis and as an abortifacient (22). Leaf-sap is placed on the eyelids and instilled into the eyes in Ivory Coast for eye-troubles (5, 14, 16), and the leaves, or the roots, are used for dysmenorrhoea (17). Also in Ivory Coast sap expressed from leafy twigs is taken by draught for fevers (14), and a paste of young leaves compounded with a little wood-ash is let down with warm water and administered as an enema for heartburn (1). The leaf is rubbed on the body in S Nigeria for *eba* (fever) (21), and the pulped plant is similarly used in Ivory Coast-Upper Volta for all manner of pains (14), mange, asthenia and as a sedative in insanity (5).

The roots are recognized in Zaïre as a purgative (22). They enter into remedies for treating skin-infections in Nigeria, and examination has shown action against *Sarcina lutea* and *Staphylococcus aureus,* but no action against Gram -ve organisms, nor fungi (15). The roots are used by women in Benin Province as a skin-ointment (6). Root-decoction is taken by draught in Ivory Coast-Upper Volta as an aphrodisiac, and by enema for gynaecological troubles, and for dysentery and urethral discharge (14). The roots are held to be a remedy against snake-bite in Senegal (11, 13) though doubt has been expressed on their efficacy in Ghana (10). Sap from the roots, or root-powder, is applied to the nostrils for migraine and sinusitis in Ghana (10) and in Ivory Coast-Upper Volta (14). Powdered roots enter into magical treatments in Ivory Coast-Upper Volta for anyone suffering madness due to misfortune, and into a philtre sprinkled on roads to prevent the spread of small-pox during an epidemic (14). Igbo of S Nigeria use the roots against toothache (10).

The fruit contains a soft, juicy, somewhat bitter and acid pulp. This is widely used to rub on the teeth to clean and whiten them. It leaves a refreshing taste in the mouth (7; 16; Soudano-guinea, 3; Ivory Coast-Upper Volta, 14; Sierra Leone, 18; Ghana, 9, 10, 23, 24; Nigeria, 2, 17; Congo, 4). In this respect the Ghanaian Akan names mean 'clean the teeth' (7, 10). Fruit-pulp is also rubbed on the skin and is used as a medicine for the throat (7). It is taken as a tonic, and Efik give a medicine of this and allied species to weakly children to encourage them to walk — an instance of sympathetic magic held to impart the agility of the dog, both Efik and Igbo names referring to 'dog' (7). Other names also apply which remain in need of explanation, e.g. the Yoruba 'witch's bread' which infers magical attributes, but see also under *C. corniculata.*

Fruit-juice is an eye-instillation in Senegal for various eye-complaints, principally conjunctivitis (11, 12, 13). Throughout the Congo the fruits are used to treat bronchial affections, especially whooping-cough and tuberculosis (4). In Zaïre fruit-juice is applied to wounds (22). The fruit with seeds are ground up with spirit (2) or boiled in wine (19) to produce a remedy for snake-bite in Nigeria.

Leaves, bark and roots have been recorded as containing tannins, but no other active principle (4, 13). The fruits are considered toxic in Senegal (13). The presence of glycoside may be suspected (17).

References:

1. Adjanohoun & Aké Assi, 1972: 105. 2. Ainslie, 1937: sp. no. 99. 3. Aubréville, 1959: 1, 193. 4. Bouquet, 1969: 98. 4a. Bouquet, 1972: 20. 5. Bouquet & Debray, 1974: 75. 6. Carpenter 26, UCI. 7. Dalziel, 1937. 8. Deighton 2333, K. 9. Howes 999, K. 10. Irvine, 1961: 570–2. 11. Kerharo & Adam, 1962. 12. Kerharo & Adam, 1963, a. 13. Kerharo & Adam, 1974: 362. 14. Kerharo & Bouquet, 171. 15. Malcolm & Sofowora, 1969: 512–17. 16. Mangenot, 1957, a. 17. Oliver, 1960: 22, 55–56. 18. Scott Elliot 3882, K. 19. Singha, 1965. 20. Thomas, N. W. 38 (Nig. Ser.), K. 21. Thomas, N. W. 1807 (Nig. Ser.), K. 22. Troupin, 1952: 118–20. 23. Vigne 110, K. 24. Williams 34, K.

**Cnestis longiflora** *Schellenb.*

FWTA, ed. 2, 1: 743. UPWTA, ed. 1, 344.
West African: NIGERIA YORUBA amùjẹ = *arrester of blood* (JMD) amùjẹ wẹwẹ (JMD)

A woody climber to 4 m high of secondary jungle and scrub from Ivory Coast to Nigeria.

The scarlet fruits are covered with pungent urticating hairs. The plant is said to be poisonous to goats (1). See also under *C. corniculata*.

Reference:

1. Ibafidon 396, K.

**Cnestis racemosa** G. Don

FWTA, ed. 2, 1: 743.
West African: SIERRA LEONE KISSI kawa-kawa (FCD)

A climbing shrub to 3 m high of the humid forest zone of Sierra Leone and Liberia.

**Connarus africanus** Lam.

FWTA, ed. 2, 1: 748. UPWTA, 344.
West African: **SENEGAL** DIOLA (Fogny) bu mem (JB) ka dahund (JB) MANDING-BAMBARA musso sana (JB) **THE GAMBIA** DIOLA (Fogny) ka dahund = *where are you going* (DF) **GUINEA-BISSAU** BIDYOGO cadjime (JDES) endurè (JDES) **GUINEA** FULA-PULAAR (Guinea) sira-wonuon (Caille) SUSU seri-gbeli = *red medicine* (JMD) **SIERRA LEONE** KISSI kawa-kawa (FCD) MANDING-MANDINKA woɛnčari (NWT) TEMNE *an*-bis-*a*-kənth (FCD) *an*-lilipənth (auctt.) **IVORY COAST** KWENI balo iri (B&D) bla (B&D)

A shrub or scandent to 7 m high or more of the evergreen wet forest from Senegal to S Nigeria.

The plant is ornamental, often climbing, and can be grown from stakes to make a hedge (2, 3).

The bark contains some resinous matter giving it tonic and astringent properties (Lannessan fide 2). A hot decoction of bark is used in Guinea to wash ulcers, and as an astringent for bleeding wounds the bark is applied finely powdered or mixed with palm-oil (Afzelius fide 2). The root-bark in some districts of Sierra Leone is used as a taenicide.

Leaf-sap is irritant to mucosae, and is used in Ivory Coast in nasal instillation together with bark-sap of *Chrysophyllum perpulchrum* Mildbr. (Sapotaceae) in cases of syncope (1). Leaf-sap with salt and pimento is considered aphrodisiac in Ivory Coast (1).

The seeds, after removal of the aril and sun-drying, are ground up in Sierra Leone and taken as a purge and vermifuge which is said to be particularly effective against tapeworm (4). The preparation is normally taken with boiled rice, or as a decoction or infusion. Sixty gm of the powder is said to act with certainty (2).

The seeds are used by Sherbro boys to bait hooks for fishing (3).

There is colouring matter in the bark and seeds but only a fatty substance and some tannin have been recorded. The taenicidal action may be due to the latter (2).

References:

1. Bouquet & Debray, 1974: 75. 2. Dalziel, 1937. 3. Lane-Poole 181, K. 4. Scott-Elliot 5603, K.

**Connaris griffonianus** Baill.

FWTA, ed. 2, 1: 748.

A shrub to 8 m high or lofty liane with stems to 15 cm diameter, of the closed-forest in wet sites in S Nigeria, West Cameroon and Fernando Po, and extending to Zaïre and Angola.

The plant has unspecified medicinal use in Gabon (2). The bark-slash exudes a colourless sour slime which turns dark red on exposure (1).

References:

1. Breteler 1850, K. 2. Walker & Sillans, 1961: 134.

**Hemandradenia mannii** Stapf

FWTA, ed. 2, 1: 749.

A tree to 6.5 m high of forest undergrowth in S Nigeria, and extending to Congo.

The bark is diuretic and purgative, and is used for these actions in Congo to treat oedemas and food-poisoning (1).

References:

1. Bouquet, 1969: 98.

**Jaundea pinnata** (P. Beauv.) Schellenb.

FWTA, ed. 2, 1: 742. UPWTA, ed. 1, 344.
West African: **GHANA** GBE-VHE ipose (FRI) **NIGERIA** IGBO (Ala) nninzā (NWT) IGBO (Onitsha) ọwa (NWT; JMD)

A shrub to 6–7 m tall, or lianescent to the forest canopy, widely dispersed from Guinea to S Nigeria, and onto Angola and NE and E Africa.

The stems beaten and dried are burnt as a torch in S Nigeria (2).

The fruits are said to be lethal to sheep and goats in Kenya if eaten (1).

References:

1. Leone 7, K. 2. Thomas, N. W. 1870 (Nig. Ser.), K.

**Jaundea pubescens** (Bak.) Schellenb.

FWTA, ed. 2, 1: 742.

A shrub or climber to 5 m high of the closed rain-forest of S Nigeria and in Cameroun to Zaïre.

No usage is recorded for the Region. In Zaïre the bark is used for stomach-complaints and the leaf-sap for eye-troubles; the wood is made into arrow-shafts; and the fruits serve as bait in fish-traps. Large specimens will yield an abundant potable sap (1).

Reference:

1. Troupin, 1952: 84, 86.

**Manotes expansa** Soland.

FWTA, ed. 2, 1: 747. UPWTA, ed. 1, 344.
West African: **SIERRA LEONE** BULOM (Sherbro) sokobalin le (FCD) LIMBA komonori (FCD) MENDE yamawa (FCD) TEMNE *an*-bis-*a*-kənth (Pyne) **LIBERIA** MANO zolo bɛlɛ (JMD) **GHANA** NZEMA awɔhanya

A scandent shrub to 6.5 m high of the rain-forest in Guinea to Liberia.

The tough resilient branches are made in Sierra Leone into oval-frame fishing-nets much used by women (1).

Reference:

1. Pyne 3, K.

**Manotes longiflora** Bak.

FWTA, ed. 2, 1: 747. UPWTA, ed. 1, 344.
West African: **SIERRA LEONE** LIMBA komonori (FCD; JMD) **LIBERIA** MANO zolo bɛlɛ (JMD) **IVORY COAST** ABURE aturufa (B&D) ADYUKRU aur-kpoeti (auctt.) orkpèti (A&AA) AKAN-ASANTE augna (B&D) awagna (B&D) wansien blakassa (B&D) KYAMA kotié brédué (B&D) **GHANA** NZEMA awɔhanya (FRI)

A scandent shrub to 6.5 m high of secondary jungle and in cultivated fields of the forest-savanna transition zone from Ivory Coast to Nigeria.

Sap expressed from young leaves is used by many tribes in Ivory Coast to relieve eye-troubles (1; Jolly fide 2, 4, 5), even in the new-born (3). The sap is also used for headaches, and the plant is considered a specific antidote for poisoning by *Solanum torvum* (Solenaceae) which is sometimes used criminally in Ivory Coast to induce insanity (3).

References:

1. Adjanohoun & Aké Assi, 1972: 106. 2. Aubréville, 1959: 1: 193. 3. Bouquet & Debray, 1974: 76. 4. Kerharo & Bouquet, 1950: 172. 5. Mangenot, 1957, b.

**Manotes zenkeri** Gilg

FWTA, ed. 2, 1: 747.
West African: **NIGERIA** IGBO ámụ̄-èbùlù = *ram's penis* (auctt.)

A scrambling shrub to 5 m high in thickets of the forest zone of S Nigeria and West Cameroons, and on into Gabon.

Though known to the Igbos, no usage is recorded. Its wood is very pale yellow turning light brown (1).

Reference:

1. Breteler 1276, 1708, K.

**Roureopsis obliquifoliolata** (Gilg) Schellenb.

FWTA, ed. 2, 1: 740.

A scandent shrub of the rain-forest of S Nigeria and extending to Zäire and Angola.

The pulped leaves are used in Congo on wounds as a haemostatic and cicitrisant. The leaf-sap is said to be able to calm bees to enable one to take honey from their nest. Leaf-sap and root-sap are also held to cure snake-bite (2).

The root is made in Nigeria into a bitter tonic given in treatment of rheumatism and diabetes. It is said to have antiscorbutic properties (1).

Tannin, and no other active principle, has been recorded in the leaves, bark and roots (3).

References:

1. Ainslie, 1937: sp. no. 304. 3. Bouquet, 1969: 99. 3. Bouquet, 1972: 20.

**Santaloides afzelii** (R. Br.) Schellenb.

FWTA, ed. 2, 1: 746–7. UPWTA, ed. 1, 344, as *S. gudjuanum* Schellenb.
West African: **GUINEA** MANDING-MANINKA kononi soro (CHOP) **IVORY COAST** ABURE ekapa (B&D) BAULE niama boboahue (B&D) niania (B&D) GAGU botrobatra tiama (B&D) MANDING-MANINKA déni m'bro (A&AA) kononé soro (Aub.) **GHANA** AKAN-BRONG huma-tarakwa (CV; FRI) huma-tarakwa (CV; FRI) **DAHOMEY** BATONNUN marisoma (Aub.) sessenkergu (Aub.) GBE-FON ganganrissé (Aub.) **NIGER** FULA-FULFULDE (Niger) yangara buhili (Aub.) HAUSA kímbár máhàlbaá (Aub.) **NIGERIA** EGGON mbaching (Hepburn; JMD)

A shrub to 2 m high or climber reaching to 20 m high, on river-banks of the closed-forest and guinean savanna from Guinea-Bissau to N Nigeria, and extending to Angola and across Africa to Sudan and Zambia.

The Maninka of Ivory Coast prepare a macerate of the leaves and use the liquid as a wash in order to have a stable matrimonial home (1).

The fruit has a scanty yellowish flesh which is sweet and edible. Though eaten by man (Guinea, 2, 6; Ghana, 5; N Nigeria, 4), it is more often taken by birds (3). The taste is like that of a cherry.

References:

1. Adjanohoun & Aké Assi, 1972: 107. 2. Aubréville, 1950: 416. 3. Dalziel, 1937. 4. Hepburn 73, K. 5. Irvine, 1961: 573. 6. Pobéguin 820, K.

**Spiropetalum heterophyllum** (Bak.) Gilg

FWTA, ed. 2, 1: 748. UPWTA, ed. 1, 344.
West African: **SIERRA LEONE** MENDE kumajɛjɛ (FCD) **GHANA** AKAN-WASA homa-kyem (auctt.)

A shrub or lofty climber of the evergreen forest, often on river-sides, from Sierra Leone to W Cameroons, and extending into Zaïre.

Though recognized in both Sierra Leone and Ghana, there appears to be no recorded usage, but as it has blood-red latex it is held in Ghana to be a god. Local people will not cut it without carrying out rituals such as sacrificing fowls or providing eggs (1).

Reference:

1. Enti, 1981-82.

**Spiropetalum reynoldsii** (Stapf) Schellenb.

FWTA, ed. 2, 1: 748.

A straggling shrub or climber of the forest of Liberia and Ivory Coast.

The plant is held to have analgesic properties. Small amounts of saponosides and tannins are present in the leaves (1).

Reference:

1. Bouquet & Debray, 1974: 76.

CONNARACEAE

**Spiropetalum solanderi** (Bak.) Gilg

FWTA, ed. 2, 1: 748.
West African: **SIERRA LEONE** MENDE kumajɛjɛ *presumed to apply* (FCD)

A shrub or climber of the forest zone of Sierra Leone and Liberia.

## CONVOLVULACEAE

**Anisera martinicensis** (Jacq.) Choisy

FWTA, ed. 2, 2: 343.
West African: **LIBERIA** MANO koo-su (Har.)

A slender perennial twining herb, occurring commonly throughout the Region and widespread over the tropics.
No use is recorded in the Region. In Malaya the leaves are said to be eaten as a vegetable (1).

Reference:
1. Burkill, 1935: 160.

**Argyreia nervosa** (Burm. f.) Bojer

FWTA, ed. 2, 2: 496.
English: elephant creeper (India, Manjunath); woolly morning glory (U.S.A., Bates).

A strong creeper with densely pubescent lower surface of the leaves, native of India, and brought into cultivation as an ornamental in several countries. It is recorded from Upper Volta in the Region.
The lower leaf-surface hairs are irritant. In India they are used as a rubefacient in skin-disease and to promote maturation of boils. The root is considered alterative and tonic, and to be useful in rheumatism and diseases of the nervous system (2). The seeds have been found to contain large amounts of hallucinogenic ergoline alkaloids (1).

References:
1. Der Marderosian, 1967. 2. Manjunath, 1948: 116, as *A. speciosa* Sweet.

**Argyreia tiliaefolia** Wight

French: liane argent (Berhaut).

A scandent or twiner with hoary or glabrescent silvery leaves, native of India and often cultivated for its large showy rose-purple flowers. It is reported under cultivation in Senegal (1).

Reference:
1. Berhaut, 1967: 268 (as *A. tiliifolia*).

**Astripomoea malvacea** (Klotzsch) Meeuse

FWTA, ed. 2, 2: 344.

A subshrubby perennial of several ill-defined varieties: in West Africa var. *involuta* (Rendle) Verdc. prostrate; var. *floccosa* (Vatke) Verdc. subprostrate or erect to 1 m high, both occurring only in N Nigeria in open grass savanna and waste spaces; these and other varieties occurring generally in NE, E, central and southern Africa.

No usage is recorded in the Region. Cattle browse it (var. *volkensii* (Dammer) Verdc.) in Uganda and it can evidently assume some importance as fodder ('It is their grass', 1).

A root-infusion (var. not stated) is used in topical application in Malawi for ophthalmia, and a poultice of crushed roots for swellings and inflammation; also the sap of leaves and flowers, or from the leaves alone is applied for inflammation of the eyeball: the treatment is said to sting (6). Of var. *floccosa,* leaf-sap is added in Tanganyika to a root-decoction which is taken by draught for hard abscesses, and a root-decoction alone for hookworm (2), for abdominal pain (3), and put into cough-medicine (4). The root of var. *malvacea* is pounded up by the Sukuma who take it thrice daily with food from which salt has been excluded for hookworm (5).

References:

1. Dyson-Hudson 170, K. 2. Haerdi, 1964: 177. 3. Koritschoner 1192, K. 4. Lazarus-Thomas 11, K. 5. Tanner 1505, K. 6. Watt & Breyer-Brandwijk, 1962: 306.

**Astripomoea sp. indet.**

West African: **NIGERIA** FULA-FULFULDE (Nigeria) nofru mbe'el (J&D)

An unidentified *Astripomoea sp.* of this Fula name is eaten in N Nigeria as spinach (1).

Reference:

1. Jackson, 1973.

**Bonamia thunbergiana** (Roem. & Schult.) F. N. Williams

FWTA, ed. 2, 2: 339. UPWTA, ed. 1, 435, as *B. cymosa* Hall. f.
West African: **SIERRA LEONE** TEMNE *ra*-kil *a general term* (auctt.) **LIBERIA** KRU-BASA doo (C&R) MANO lo-lo (Har.)

A woody twiner recorded from The Gambia to S Nigeria and in E Cameroun.

It is reported that pounded and moistened leaves are inserted into the nostrils of hunting dogs to improve their ability to scent (1).

Reference:

1. Portères, s.d.

**Calycobolus africanus** (G. Don) Heine

FWTA, ed. 2, 2: 338.
West African: **IVORY COAST** KRU-GUERE (Chiehn) gurulu (B&D) KYAMA edianbego (B&D) **GHANA** AKAN-ASANTE mutuo (CV; BD&H) TWI mutuo (CV; BD&H)

A woody climber of the forest, from S Leone to S Nigeria, and extending to the Congo basin.

The leaves are eaten or taken in enemas for fever in Ivory Coast (1).

Reference:

1. Bouquet & Debray, 1974: 77.

**Calycobolus heudelotii** (Bak. ex Oliv.) Heine

FWTA, ed. 2, 2: 337–8.

West African: **SENEGAL** DIOLA bu kunger (JB) ukoken u mbir (JB; K&A) DIOLA (Tentouck) furkunger (K&A) **THE GAMBIA** DIOLA (Fogny) bu kunger = *sharp eye* (DF) ukoken u mbir = *tied that one* (DF) **GUINEA-BISSAU** BALANTA másfi (JDES; EPdS) **SIERRA LEONE** MENDE koleiveiye (*def*.-i) (Macdonald) **IVORY COAST** ABURE apumu (B&D)

A lianous shrub to 15 m high of the forest, from Senegal (Casamance) to W Cameroons, and also in E Cameroun.

In Casamance the stems are used as ties for roof-thatching, and a decoction of the leaves is taken as a popular medicine for stiffness and fatigue by draught and in steam-baths (2). In Ivory Coast the bark is pulped up into an embrocation massaged onto the stomach for intestinal pain (1).

References:

1. Bouquet & Debray, 1974: 77. 2. Kerharo & Adam, 1974: 363.

**Calycobolus parviflorus** (Mangenot) Heine

FWTA, ed. 2, 2: 338.

A climber to 10 m high, of swamp-forest in Liberia and Ivory Coast.

The powdered bark is taken with food in Ivory Coast to soothe abdominal pains during pregnancy (1).

Reference:

1. Bouquet & Debray, 1974: 77.

**Convolvulus arvensis** Linn.

English: bindweed.

French: liseron (des champs).

Portuguese: trepadeira.

West African: **MAURITANIA** ARABIC (Hassaniya) anasfal (AN)

A trailing, twining herb from a perennial semi-woody rootstock, native of Europe and northern Africa and occurring as far south as Mauritania (1). It is a troublesome weed of cultivation and because of its very deep-penetrating root difficult to eradicate.

Reference:

1. Naegelé, 1958, b: 884–5.

**Convolvulus microphyllus** Sieb. ex Spreng.

FWTA, ed. 2, 2: 340.

West African: **MAURITANIA** ARABIC (Hassaniya) kraa el ghazel (AN)

A prostrate twining perennial herb from a woody rootstock, of grassy savanna in Mauritania and Senegal, and widely dispersed in NE Africa and western Asia.

The plant is not browsed by cattle in Senegal (1), but in Mauritania it is said to provide good pasturage for all stock and is especially sought after by gazelle, as is indicated by the Hassaniya name (2).

References:

1. Adam, 1966, a. 2. Naegelé, 1958, b: 885.

**Cressa cretica** Linn.

FWTA, ed. 2, 2: 339.
West African: **MAURITANIA** ARABIC (Hassaniya) tarmoyia (AN) **SENEGAL** SERER lupuń (K&A) SERER-NON (Nyominka) lupuń (K&A) WOLOF turmbel (JB; K&A)

A small herb from a few cm to 30 cm high from a woody perennial rootstock, halophytic in salt pans or near the sea, in Senegal to Guinea, and probably along all the Region's Atlantic coast, and occurring almost worldwide on all coasts.

It occurs in the rice-padis of the Senegal River at Richard Toll, not particularly as a weed, but its presence in increasing or decreasing amount is an indication of brackishness of the water (1). In Mauritania only camels graze it, and then but occasionally (7). In Iraq (4) and in Saudi Arabia (2) it is said to provide good fodder for sheep and goats. The plant has medicinal applications in Senegal in maceration with the barks of *Vitex cuneata* (Verbenaceae) and *Acacia albida* (Leguminosae: Mimosoideae) for treating bronchitis (6), in Sudan as a tonic (3), and in India as an alterative, stomachic, tonic and aphrodisiac (8). It has a sour unpleasant taste.

The plant was used by the Greek physician, Dioscorides, as a suppository for uterine tumours (5), but examination for anti-tumoral activity has not given significant results (6).

References:

1. Adam, 1960: 368. 2. Baker s.n., 30/12/1953, K. 3. Broun & Massey, 1929: 323. 4. Guest & al. 16145, K. 5. Hartwell, 1968. 6. Kerharo & Adam, 1974: 364. 7. Naegelé, 1958, b. 8. Sastri, 1952: 367.

**Cuscuta australis** R. Br.

FWTA, ed. 2, 2: 336. UPWTA, ed. 1, 436, as *C. chinensis* sensu Dalziel, non Lam.
West African: **GHANA** ADANGME-KROBO mprakɛ (FRI) AKAN-TWI dɔme atrɛ = *if you love me, spread* (FRI) mpenãbegu, mprabegu = *the lover will be dropped, i.e., if it does not spread* (FRI) ntentene = *spread out a long way* (FRI) GBE-VHE vuvudranyi (FRI) VHE (Pecí) mprakɛ (FRI) HAUSA soyanyaa (FRI) NZEMA brɛzinlẽ nyɛmã (FRI) mɛrɛhyiazo (FRI) **NIGERIA** HAUSA soyayya = *mutual affection* (JMD) YORUBA ganagana (JMD) gingin *referring to the slender stems* (JMD) ọmọnígèdègédé = *lonely child* (Verger; JMD) ọmọnígínigíní (JMD) YORUBA (Ijebu) ọmọnigèlègélé, ọmọ: *child;* oni: *possessing;* gèlè: *a woman's kerchief* (JMD)

A leafless parasitic slender-stemmed twiner forming tangled masses on marsh and forest vegetation, from Sierra Leone to W Cameroons and widely dispersed in E Africa and the Old World tropics.

Because of the tangled manner of growth, the plant is used for superstitious purposes. As is implied in several of the Ghanaian vernacular names it is used as a love-charm, or to test affection (1, 2). The Yoruba in calling it 'lonely child' imply that such a child keeps secrets to itself and thus they use it in an invocation to conceal a secret (4).

No medicinal, nor work-a-day use appears to be made of the plant. The seed of a related species, *C. reflexa* Roxb. in India, contains a wax composed of esters of higher aliphatic alcohols and saturated fatty acids, and also a semi-drying oil with an iodine value of 97 (3). Seeds of West African *Cuscuta spp.* ought to be examined.

References:

1. Dalziel, 1937. 2. Irvine, 1930: 139, as *C. chinensis*. 3. Sastri, 1950: 403–4, under *C. reflexa* Roxb. 4. Verger, 1967: no. 158, as *C. chinensis*.

**Evolvulus alsinoides** (Linn.) Linn.

FWTA, ed. 2, 2: 339. UPWTA, ed. 1, 436

West African: **SENEGAL** MANDING-BAMBARA koni ka koa (JB) **GUINEA** MANDING-MANINKA dubryémetré (Brossart) **GHANA** HAUSA ka-fi malam (FRI) **NIGERIA** FULA-FULFULDE (Nigeria) ndottihon (PF fide MM) ndottiyel = *little old man* (JMD; MM) HAUSA kaà-fi-mallam = *better than teacher* (JMD; ZOG) matakin kurcia = (?) *stair/step of the Senegal blue-winged dove* (MM) YORUBA ẹ̀fúnlè (JMD) itanna-dudu (JRA)

A bushy herb to 20–30 cm high, of open dry waste places, mainly in the northern part of the Region from Senegal to W Cameroons, and widely dispersed throughout the tropics.

The plant bears numerous small sky-blue flowers and is cultivated in Gabon as a herbaceous ornamental (15). It is grazed by all stock (Senegal, 1; Sudan, 3; Kenya, 7). The leaves, however, are bitter and are used in Nigeria (2), Ethiopia (6), Sudan (4), Philippines (11) and in India (14) to prepare a bitter tonic and febrifuge for taking in fever. This preparation is said in India to be indicated in fever accompanied by indigestion or diarrhoea (14). It is used in the Philippines for certain bowel irregularities (11). It is also vermifugal (6, 11, 14). Infusions of roots, stalks and leaves are all used in Nigeria as stomachics (2).

In Kenya (Kwale Province) sores are treated by application of the powdered leaves (8), and in Tanganyika (Lake Province) the pounded leaves are put onto enlarged glands in the neck (13). The Sukuma burn the dried leaves in a pipe as a leprosy-cure (12). Leaves are also smoked in Nigeria (2) and in India (11) in cases of asthma and chronic bronchitis. The leaves produce a somewhat fragrant smoke that is used in N Nigeria to perfume houses (Barter fide 10). In the old Sudanese Kingdom it was an ingredient along with other herbs as a charm to exorcise an evil spirit causing disease and women would burn the plant to fumigate the hut during the puerperium, and use a warm infusion as a wash during the forty days purification (Meek fide 5). The plant is sold in N Nigeria principally as a charm worn as a girdle or circlet on the arm, etc., to procure love or a favour. The Hausa name means 'better than a malam' for such purposes. The Fula add the plant to milk to bring success (5). The Hausa of Ghana similarly use the plant in love-potions and in religious practices (9, 10).

The plant with oil has been used in India in the belief of promoting growth of the hair (11).

References:

1. Adam, 1966, a. 2. Ainslie, 1937: sp. no. 155. 3. Andrews A.722, K. 4. Broun & Massey, 1929: 322–3. 5. Dalziel, 1937. 6. Getahun, 1975. 7. Glover & al. 1528, 1808, K. 8. Graham Q.574, K. 9. Irvine 283, K. 10. Irvine, 1930: 192–3. 11. Quisumbing, 1951: 756–7. 12. Tanner 1420, K. 13. Tanner 4078, K. 14. Sastri, 1952: 233–4. 15. Walker & Sillans, 1961: 135.

**Evolvulus nummularius** (Linn.) Linn.

FWTA, ed. 2, 2: 339.

A perennial prostrate herb of open waste places, in Ivory Coast and Ghana, and occurring elsewhere throughout tropical Africa, and in tropical America and India.

In India in some parts it invades the turf of grass-lawns and becomes an established weed (2).

The plant has been recorded in India as containing sedative and anticonvulsive substances (1).

References:

1. Bouquet & Debray, 1974: 76. 2. Sastri, 1952: 234.

**Hewittia sublobata** (Linn. f.) O. Ktze

FWTA, ed. 2, 2: 342–3.
West African: **SIERRA LEONE** MANDING-MANINKA (Koranko) kɔkbe (NWT) MENDE ndondokɔ *general name for some herbaceous climbers, esp. Ipomoea involucrata* (NWT) **NIGERIA** IGBO (Onitsha) áfị̣fị̣á (KW) urunsi = *die quickly* (AJC) IGBO (Owerri) áhị̣hị̣á = *grass, weeds; a general term* (KW)

A slender, prostrate or twining herbaceous perennial, but of rampant growth over bush and grass, widely dispersed from The Gambia to W Cameroons, and throughout tropical Africa, Asia and Polynesia.

The leaves are said to be eaten in the Mombasa area of Kenya (2). Onitsha Igbo rub the leaves onto sores (1). A root-decoction is drunk in Tanganyika for *Oxyuris* threadworm (3).

References:

1. Carpenter 17, UCI. 2. MacNaughton 88, K. 3. Haerdi, 1964: 178.

**Ipomoea** Linn.

*Ipomoea spp.* are known honey-yielders. *I. triloba* is particularly useful to bee-keepers in Cuba as 'campanilla' yielding honey equal in flavour and colour to that from lucerne and sage, and also producing a very white wax (2).

The seeds of 'morning glory', *I. violacea* and related species have been used for hallucinatory and divinatory purposes in central America. Ergoline alkaloids previously known from certain fungi have recently been found in *Ipomoea*. The W African species merit careful examination. The absence of hallucinogenic indole alkaloids in W African material of certain species is reported under each relevant species (1).

References:

1. Der Marderosian & Youngken, 1966. 2. Howes, 1945.

**Ipomea acuminata** (Vahl) Roem.

FWTA, ed. 2, 2: 352 incl. *I. learii* Lindl.
English: morning glory.

A prostrate or twining herbaceous plant, native of central America, but now pantropical, is cultivated in the Region as an ornamental with a blue or bluish-purple flower. *I. learii* is considered a cultivated form.

**Ipomoea aitonii** Lindl.

FWTA, ed. 2, 2: 352. UPWTA, ed. 1, 439, as *I. pilosa* Sweet.
West African: **SENEGAL** SERER lul *general for convolvulus* (JMD) WOLOF laulau = *runner; general for convolvulus* (JMD) **NIGER** SONGHAI dàar-dàarà (*pl.* -à) (D&C) gènjì-hàabù (*pl.* -ò) = *bush-cotton* (D&C) hàabù-báasé (*pl.* -ó) = *cotton's cousin* (D&C) jéerí-hàabù (*pl.* -ò) = *gazelle's cotton* (D&C) **NIGERIA** FULA-FULFULDE (Nigeria) buta wuta *this sp?* (JMD) dale *a general name* (JMD) pampaali (PF fide MM) yako (JMD) HAUSA adbugar barewa (PF fide MM) dambuntsofi (PF fide MM) dankwon kuyangi *a medicine made from dried leaves* (JMD) hantsar gada *a love-charm made from the thickened roots* kùnnén-kuúsuù, námíjin baá-maà-táboo, táboo: *scar; perhaps alluding to cicatrizing property* (JMD; ZOG) yako (ZOG)

A perennial twiner recorded from Senegal, Ghana and N Nigeria.

In Senegal the plant is browsed by all domestic stock (1). The dried leaves are used in N Nigeria to apply to burns (2). The seeds along with those of *Hibiscus sabdariffa* (Malvaceae) are used as a purgative, and the thickened root is considered in Damagaram to be a love charm (2).

References:

1. Adam, 1966, a. 2. Dalziel, 1937.

**Ipomoea alba** Linn.

FWTA, ed. 2, 2: 346. UPWTA, ed. 1, 435, as *Calonyction aculeatum* House.

English: moon-flower.

West African: **SIERRA LEONE** KISSI k-panja-humdõŋ (FCD) kulukɛnyɔ? (FCD) KONO bayugɛnɛ (FCD) LIMBA bukbui *a general name* (JMD) MENDE bukbɔ (*def.*-i) (GFSE; JMD) kpokpo *general for Convolvulus* (JMD) kpokpo-hina, hina: *male* (FCD) SUSU liti *general for Ipomoea* (NWT; JMD) **NIGERIA** EFIK ndiami (JMD)

A prostrate or twining annual or perennial herb of secondary vegetation, native of tropical America, but now widely dispersed by man in the tropics, and recorded from Ghana to Nigeria in the Region.

The flowers are large, white, tubular, opening at night, sweet-scented. The plant is commonly grown as an ornamental. The seeds are hard-coated and scarification gives better germination.

The leaves are said to be sometimes eaten in Sierra Leone as a vegetable (Scott Elliot fide 2), a practice widely found in Asia (1). In Ghana the leaves give a soap-substitute and are used for bathing (3), and in S Nigeria they are used as a wash for headache (7). The plant is recognized as toxic in Gabon (8). Unnamed alkaloids have been detected in the leaves, stems, fruits and seeds (9), and a resin (6), but West African material is reported free of hallucinogenic indole alkaloids (4). Most parts of the plant serve in snake-bite medicines in India (1). The leaves are prepared into poultices in the Caribbean region for application to tumours (5).

References:

1. Burkill, 1935: 405, as *Calonyction aculeatum* House. 2. Dalziel, 1937. 3. Deakin 115, K. 4. Der Marderosian & Youngken, 1966. 5. Hartwell, 1968: as *C. aculeatum* (Linn.) House. 6. Sastri, 1950: 17, as *C. aculeatum* House. 7. Thomas, N. W. 1672, 2053 (Nig. Ser.), K. 8. Walker & Sillans, 1961: 135, as *C. aculeatum* House. 9. Willaman & Li, 1970.

**Ipomoea aquatica** Forssk.

FWTA, ed. 2, 2: 349. UPWTA, ed. 1, 439, as *I. reptans* Poir.

French: patate aquatique (Dalziel).

West African: **SENEGAL** MANDING-BAMBARA bafaraka (JB, ex K&A) SERER n-dapul (JB, ex K&A) nof mbal (JB, ex K&A) **SIERRA LEONE** MENDE kogidi (FCD; JMD) kpokpo (*def.*-i) *common for vines and creepers* (FCD; JMD) **MALI** DOGON dìì bibíle (C-G) **GHANA** GA kɔkɔle biakɔ (FRI) **NIGER** SONGHAI tàlhánà (*pl.* -à) (D&C) **NIGERIA** FULA-FULFULDE (Nigeria) delbol (PF fide MM) laylayduuji = *creeping things; perhaps a general term* (MM) HAUSA awarwaro (JMD; ZOG) fùrén gàdú = *warthog's flower* (JMD; ZOG) tarfo, taufau (Wedderburn) terfai (RES) IJO-IZON (Kolokuma) èméin fị̀ tùò = *herb eaten by manatees* (KW) tànàïn túó = *creeping herb* (KW) IZON (Oporoma) tọ́rụ́bẹ́ịn = *crosses river* (KW)

An aquatic, trailing or floating herb, usually perennial in suitably damp and wet sites throughout the Region and into montane situations, and of pantropical distribution.

The young leafy shoots are sometimes eaten in W Africa after cooking (3, 4, 6), but it appears to be considered more as an item of diet for time of dearth. The plant assumes a far greater importance in SE Asia where it is a common market commodity. Cultivars are recognized and it is said that plants grown on somewhat drier land have a more pleasant taste (2). It grows vigorously. Propagation by slips is easy. It rapidly colonizes any suitable piece of water or damp ground. Requirements to the Region seem to be met almost entirely by collection from the wild, but cultivation may occasionally be practised. It is common on the edges of the flood plain of the Sokoto River in N Nigeria where it creates a suitable breeding site for Anophelene mosquitoes (11). In Sudan it is an important factor in bonding together the *sudd* vegetation (1). It is valuable as feed for domestic animals. Cattle and pigs readily take it. In SE Asia it is the principal vegetable fed to pigs by Chinese market-gardeners. The plant contains about 90% water. Dry weight analysis shows about 48% carbohydrates, 24% proteins, 13% ash. It is rich in minerals and vitamins (3, 9, 10).

Though the plant is so extensively eaten in Asia, it appears occasionally to have toxic properties. Taken in large amounts it has been found in India to be a drastic purgative and toxic irritant. In Australia it is thought to have caused mortality to horses and it is listed as suspect of poisoning stock (8). Certainly in India it has medicinal uses. The plant-sap is given as an emetic in cases of opium and arsenical poisoning and the dried sap is purgative. In NE India it is given for nervous and general debility, while in Indo-china the plant is applied as a poultice in febrile delirium. The flower buds are used on ringworm (2, 9, 10). In Tanganyika the leaf-sap is compounded with a root-decoction of *Nymphaea spp.* to take by draught as a sedative in insanity (5).

The roots are sometimes eaten (7). They are sought after by the wart-hog both in the W African soudan (4, 6, 7) and in India (10), and the plant is noted by the Hausa as 'The Wart-hog's flower.'

References:

1. Broun & Massey, 1929: 316–17. 2. Burkill, 1935: 1250, as *I. reptans* Poir. 3. Busson, 1965: 375–6, with leaf analysis. 4. Dalziel, 1937. 5. Haerdi, 1964: 178. 6. Irvine, 1952, a: 29. 7. Irvine, 1952, b: 46. 8. Kerharo & Adam, 1974: 364–5. 9. Quisumbing, 1951: 757–8. 10. Sastri, 1959: 237–8 with phytochemistry. 11. Wedderburn FHI 42780, K.

## **Ipomoea arborescens** (Humb. & Bonpl.) G. Don

FWTA, ed. 2, 2: 352.

A small tree to 6 m high, native of Mexico, and recorded as cultivated in Senegal as an ornamental for its white star-like flower to 5 cm long.

Hallucinogenic indole alkaloids are reported absent from W African material (1).

In Mexico, deer eat the fallen flowers (2).

References:

1. Der Marderosian & Youngken, 1966. 2. Standley 1277, K.

## **Ipomoea argentaurata** Hallier f.

FWTA, ed. 2, 2: 347. UPWTA, ed. 1, 436.

West African: **MALI** DOGON ómɔnɔ kómɔ kulɔgu (C-G) **GHANA** DAGBANI ukpali (JMD) **NIGERIA** BOLE kofoitila (AST) GWARI gammu (JMD) HAUSA fárín gámó = *good luck* (auctt.) kaa-fi-boókaá = *better than a medicineman* (JMD; ZOG) YORUBA ado (Rowland) inuwo-elepe (Kennedy; JMD)

Prostrate or ascending herbaceous stems from a woody rootstock, of the savanna from Sierra Leone to Nigeria, and into Cameroun.

A decoction of the plant is taken in Ivory Coast with kola nut in the belief that it promotes spermatogenesis (1). Perhaps this, but at least its other uses are probably mainly superstitious, as medicine for witch-craft worn as an amulet, armlet or girdle, etc. Clothing is fumigated with it, not as a scent, but as a charm for the same purpose, or for luck. The flowers open in the morning, and the Hausa name *farin gamo,* means 'luck in trading early in the day' (2).

References:

1. Bouquet & Debray, 1974: 76. 2. Dalziel, 1937.

**Ipomoea asarifolia** (Desr.) Roem. & Schult.

FWTA, ed. 2, 2: 348. UPWTA, ed. 1, 439, as *I. repens* Lam.

West African: **MAURITANIA** ARABIC (Hassaniya) benaman (Kesby; AN) **SENEGAL** DIOLA (Fogny) éralak (JB; K&A) FULA-PULAAR (Senegal) ndénat (K&A) suparnao (K&A) MANDING-BAMBARA baforoko (JB; K&A) 'SOCE' barakora (K&A) SERER furtut (JB) sufar nak (JB; K&A) SERER-NON (Nyominka) éfurtut (K&A) furtut (K&A) opurtut (K&A) WOLOF n-dénat *general for convolvulus* (auctt.) **THE GAMBIA** DIOLA (Fogny) éralak = *the spreader* (DF) MANDING-MANDINKA samajewo = *elephant pumpkin* (Fox; DF) **GUINEA-BISSAU** BALANTA n-tome, untome (JDES) CRIOULO lacacom (JDES) **MALI** DOGON lolóriõ (C-G) SONGHAI talala (A. Chev.) **UPPER VOLTA** MOORE kokwaka (AJML) **IVORY COAST** BAULE alédan bliassu (K&B) **GHANA** BULISHA buntunti (CV) **NIGERIA** DERA lóŋgúlóorí (Newman) FULA-FULFULDE (Nigeria) layre ngabbu = *creeper of the hippopotamus* (Taylor; MM) wababoo (PF fide MM) woba boje, boje: *hares* (JMD) HAUSA dúmán kadaá = *gourd of the crocodile* (auctt.) dúmán raàfíi = *gourd of the river* (auctt.) NUPE dumma *from Hausa,* (JMD) YORUBA gbọọrọ ayaba, gbọọrọ: *long and stately;* ayaba = aya ọba: *queen, for the plant itself* (auctt.) òdòdó = *scarlet, but refers in general to bright flowers* (JMD) òdòdó oko *incl. other common I. spp.* (JMD)

Long-trailing herbaceous perennial, sometimes twining, of sandy areas and waste places throughout the Region, and in C Verde Islands, tropical Asia and America.

The plant is not eaten by man, nor normally by stock (1, 10). A purgative toxic resin is probably present. It is said to cause in N Nigeria diarrhoea in horses if accidentally grazed, and madness and death in camels (2), but camels are recorded taking it in Senegal (1) and a little by sheep in Mauritania (9). The plant has medicinal uses, in particular in Senegal for various gynaecological purposes: urinary problems during pregnancy, haemorrhage, as an ecbolic and abortifacient; also in general as a wound-dressing, and for treating ophthalmias, neuralgia, headaches, arthritic pain and stomach-ache (5, 6, 7). The pulped-up leafy stems are mixed with citron and water and taken in Ivory Coast as an ecbolic (8). In N Nigeria a leaf-decoction is taken internally and as a wash for feverish chills and rheumatic pains, or the face is steamed over a hot decoction of the plant along with husks of bulrush millet (2). A leaf-poultice is applied to guinea-worm sores in Nigeria (2) and Mauritania (10), and the flowers boiled with beans are eaten as a remedy for syphilis in Nigeria (2). The leaves have undefined medicinal use in The Gambia (3).

In the Northern Territory of Ghana cattle affected with *garli* disease are treated by concoctions of stems and roots of various plants of which this is usually the most important ingredient. An infusion of the materials is given to the animal internally, and a charcoal of the burnt plants is pulverized, mixed with shea butter and rubbed on the joints (2).

In Senegal a decoction of the plant is used to stain cloths and the hair black (Sébire fide 2), while in Mauritania the ashes of the plant are mixed with indigo to provide a blue dye for cloth (10), or ashes of the leaves alone are used (9). The dried stems are used as a tinder, and the leaves are sometimes used to wrap the feet or hands in applying henna (2).

The plant trailing over sand-dunes is a good sandbinder (4). The long stems are used as ties for tying up produce (2). In Senegal certain agriculturists growing groundnuts stretch out the long runners over the top end of stakes in their fields as 'a good example' to the groundnuts to encourage the plants to emulate the long stems and many pods of the *Ipomoea* (11).

References:

1. Adam, 1966, a. 2. Dalziel, 1937. 3. Fox 31, K. 4. Irvine, 1930: 240. 5. Kerharo & Adam, 1963, a. 6. Kerharo & Adam, 1964, c: 311. 7. Kerharo & Adam, 1974: 365. 8. Kerharo & Bouquet, 1950: 226. 9. Kesby 47, K. 10. Naegelé, 1958, b: 885. 11. Trochain, 1940: 269.

**Ipomoea batatas** (Linn.) Lam.

FWTA, ed. 2, 2: 350–1. UPWTA, ed. 1, 436–8.

English: sweet potato; Spanish potato; yam (U.S.A., erroneously).

French: batate; patate; patate douce.

Portuguese: batata; batata doce.

West African: **SENEGAL** BASARI fútutè, ɔ-pútutè (Ferry) BEDIK peutètè (Ferry) KONYAGI tata (Ferry) MANDING-BAMBARA konduba (auctt.) patato (K&A) toma ulé (JB; K&A) WOLOF patas (auctt.) patat (auctt.) **THE GAMBIA** MANDING-MANDINKA pataati (*def.* pataatoo) WOLOF linene (JMD) patat (JMD) **GUINEA-BISSAU** BALANTA patate (JDES) FULA-PULAAR (Guinea-Bissau) pute (JDES) MANDING-MANDINKA patatô (JDES) PEPEL uatata (JDES) **GUINEA** FULA-PULAAR (Guinea) wusé (CHOP) MANDING-MANINKA kinkio (CHOP; FB) ussu (CHOP) ussu-gué *the common white c. var.* (CHOP) ussu-ulé *c. var. with small red tubers* (CHOP) SUSU torekélé-uré *c.var. with small yellow tubers, considered the best* (CHOP) uré (CHOP; FB) uré firke *the common white c.var.* (CHOP) uré gbuéli *the common red c.var.* (CHOP) **SIERRA LEONE** BULOM (Kim) g-bamɛ (FCD) g-bami (FCD) BULOM (Sherbro) gbam-dɛ (FCD) FULA-PULAAR (Sierra Leone) kunduwana *c. var. with white tubers* (Glanville) wusɛ (FCD) wusen boɗe, boɗe: *red; c. var. with red tubers* (Glanville) GOLA n-joule (FCD) KISSI g-banja, k-panja (FCD) k-pande-wendẽ *the root* (FCD; JMD) KONO bundɛ *c.var with lobed leaves* (FCD; JMD) yukɛnɛ *c. var. with unlobed leaves* (FCD) KRIO pɛtɛtɛ (FCD) LIMBA ampete (FCD; JMD) bɔkié (JMD) potekai kai *a c.var.* (Glanville) tokpohun *c.var. with dirty white tubers* (Glanville) yenkisa *a c.var.* (Glanville) LIMBA (Tonko) ntada (FCD) LOKO g-bɔgu (FCD) MANDING-MANDINKA mandoroka *c.var. with reddish-purple tuber* (Glanville) tokpohunka *c.var. with dirty white tuber and pure white flesh* (Glanville) wuse (FCD) wusengbe *c.var. with short vines, white tubers* (Glanville) yenkisa ka *a c.var.* (Glanville) MANINKA (Koranko) balawura (NWT; JMD) bɔkui (FCD) gbanka yanka *c.var. with reddish-purple tuber and yellowish flesh* (Glanville) lofolafan *c.var. with reddish-purple tuber and white flesh* (Glanville) wusi (FCD; JMD) MENDE njowo (*def.*-i) (NWT; JMD) SUSU wurɛ (auctt.) SUSU-DYALONKE bongo-lakhabi-na *c.var. with dirty white tuber and salmon pink flesh* (Glanville) lahhabi-na (FCD; JMD) TEMNE *a*-bɔko *c.var. with red tuber* (JMD) *a*-muna *c.var. with white tuber* (FCD; JMD) VAI n-jowe (FCD) **LIBERIA** MANO kwisõ (JMD) **MALI** DOGON banakúú *from Bambara* (C-G) MANDING-BAMBARA konduba (FB) **IVORY COAST** BAULE tingo (K&B) KRU-GUERE bassi (K&B) deguibo (K&B) teguibo (K&B) KWENI baàtɛ́n (Grégoire) KYAMA gonembi (B&D) **GHANA** ADANGME-KROBO anãgo (FRI) atomo (FRI) AKAN-AKYEM abrɔdwobãa *a climber* (FRI; JMD) asikuma (JMD) ASANTE abrɔdwobãa (FRI) asikuma (FRI) ntɔmmɔ (FRI; JMD) FANTE saantom (FRI; JMD) TWI abrɔdwemaa (Enti) ntɔmmɔ (FRI; JMD) BIMOBA kampidiem (JMD) DAGBANI wulejɔ *the tubers* (JMD) FRA-FRA nubienko (Ll.W.) GA atómò (FRI; KD) blɔfo-atómò (FRI; JMD) GBE-VHE anagote (FRI) dzete (FRI) VHE (Awlan) anago VHE (Pecí) anago-te GRUSI nenure (JMD) GUANG-GONJA jìbélbí (Rytz) KONKOMBA wuridschɔ *c.var. with red tubers* (Gaisser) MAMPRULI nanyuya (JMD) SISAALA naanuuruŋ (Blass) **TOGO** BASSARI uridschŏ *c.var. with white tubers* (Gaisser) GBE-VHE anago (FB) anago-te = *yam of the Lagos* (*Anago*) *people; i.e. from the Coast* (FRI) KABRE auihae (FB) MOORE-NAWDAM jarenio-ronde (Gaisser) sama (Gaisser) TEM anago (FB) TEM (Tshaudjo) toonini (Gaisser) 'DIFALE' agundae (Gaisser) agundae aelalo, aelalo: *female* (Gaisser) agundae aloba, aloba: *male* (Gaisser) **DAHOMEY** BATONNUN dantin (FB) **NIGER** SONGHAI kúudékà (*pl.* -à) (D&C) SONGHAI-ZARMA kudaku (Robin) **NIGERIA** ANAANG edia-makara (JMD) ARABIC bombe *a general name* (JMD) ARABIC-SHUWA aya (JMD) BEROM dànkál (LB) EDO iyan-ebo (JMD) EFIK biâ ḿbàkárá = *yam of the European* (JMD; Lowe) FULA-FULFULDE (Nigeria) dudepurre (J&D) kudaku (*pl.* kadakuuji) (JMD; Taylor) kuudaku (PF fide MM) laire = *creeper; a general term* (MM) laire ngabbu *the plant* (JMD) nyamduuji = *things to eat; i.e., food* (MM) peembere (*pl.* peembereeji) (JMD; Taylor) HAUSA bà-fádámeè *c.var. grown in marshy land* (JMD; ZOG) ba-lawure *c.var. with white tubers* (JMD) ba-nkuli (Singha) ciì-mázà *c.var. with red tubers* (JMD; ZOG) daakataá *c.var. with red tubers* (JMD; ZOG) daŋkálii, dànkálìi *a general term, but originally to red-tubered c.var.* (auctt.) dukuma *generally for red-tubered c.var.* (JMD) gàmàgàrí *c.var. with white tuber, a name applied to other favoured products-kola, sorghum, groundnut, etc., vars.* (JMD; ZOG) iyaya *for c.var. with large tubers* (JMD) kudaku *generally for c.var. with red* *tubers* (JMD) kugundugu *from Yoruba* (JMD) laáwúr (JMD; ZOG) shaà-kuùshé *an epithet* (JMD; ZOG) turkai *also an okra var.* (JMD) warina (JMD) yáryaàdií *c.var. with a white, forked tuber* (JMD; ZOG) IBIBIO èdiàm; èdiàm-ùmánī *a wild sweet potato* (*? sp.*) (AJC, ex JMD, Kaufman) ùdlá-mákàrà (JMD) IGBO ji nwa nnu (BNO) ji-bèkéè = *European yam* (auctt.) IGBO (Awka) ọ̀tọ́lị̀ (JMD) IGBO (Onitsha) jí-óyìbó = *white man's yam* (JMD) kúkúǹdùkú (JMD) IGBO (Owerri) jí-bèkéè = *European yam* (KW) oi-bèkéè (JMD) IJO-IZON (Kolokuma) bèkè búrù = *European yam* (KW) KANURI dàngálì, kúndùwú, kuwunduwu (C&H) MAMBILA dangura (Lowe) NUPE dùkú (Banfield) TIV atsaka (JMD) URHOBO ímìtátà (Faraclas) ọ̀lẹ́-òyìnbó (JMD) YORUBA ànàmọ́ (JMD) ẹdumṣi *c.var. with red tubers* (JMD) kúkúndùnkún, dùn: *sweet;* kú: *full* (JMD; Singha) òdukún, òdunkún (JMD) òdukún fúnfún *c.var. with white tubers* (JMD) òdukún pupa *c.var. with red tubers* (Dodd; JMD) **WEST CAMEROONS** BAFOK mabongo (JMD) KOOSI ndoko (JMD) KPE ndoko (JMD) KUNDU metika (JMD) LONG mabongo (JMD) MBONGE metika (JMD) TANGA metika (JMD) WOVEA ndoko (JMD)

Prostrate or ascending, rarely climbing annual (or ever-growing in equable climatic conditions) stems from underground tubers, native of tropical America, introduced and cultivated throughout the Region, and almost ubiquitously through the tropics and subtropics.

The plant is principally grown for its sweet starchy tuber. There are innumerable varieties which tend to fall into three categories: (1) dry and mealy when cooked, (2) soft and watery because of a tendency to convert their starch to sugar (and erroneously called 'yams' in the United States of America) and (3) coarse-fleshed, suitable for animal feed or for industrial use. Classification may be by skin-colour — red or white, or by root-flesh colour — white or yellow. Shape of tuber, leaf and stem are also diagnostic cultivar characters. The plant is in brief very polymorphic. There is ample scope for breeding and selection of improved varieties.

The tuber is rich in starch but cultivars show a wide range, 8 to 29%, with a mean about 20–25%. Sugar is about 4–5% and protein 2–3.5%. Cooking breaks down the starch so that sweetness is accentuated by hydrolysis. Little use has been made industrially of the sweet potato as a source of starch. Some extraction has been undertaken in Japan. In Nigeria tubers have been used for distillation to 'tumbo spirit.' Vitamins are well represented, particularly the yellow-fleshed cultivars.

The plant is usually a 3–4 months crop, but some cultivars run to 6 months. A rainfall of 500 mm during the growing season is necessary, but in low rainfall areas irrigation is satisfactory. Primitive husbandry may be by grouting at the base of the plant to pull out any tubers found, leaving the plant *in situ*. More conventionally the plant is lifted in due season and tubers not required immediately put into store. This is done in a sand-filled pit (if to be kept viable), or packed with grass in baskets placed in the rafters of a house, or they are sliced and dried and stored in the rafters where the hearth-smoke helps to keep them free from insects. The tubers however do not store well and are very liable to insect-damage and fungal rot in which they become bitter, toxic and inedible. Because of poor storing quality, other root-crops tend to find preference in West Africa. The plant is not attacked by locusts, and timely planting in E Africa has done much to alleviate famine (11, 12).

Little medicinal use is made of the tubers. They are used in frictions on the skin in Ivory Coast to prevent loss of pigmentation (3). A purgative tisane is made from the root with leaves of *Cassia occidentalis* (Leguminosae: Caesalpinioideae) and the bark of *Bridelia ferruginea* (Euphorbiaceae) in Congo (2). Bactericidal and fungicidal substances have been isolated from the tuber and the haulm.

The young leaves are commonly eaten by man and leafy stems are fed to stock. They are a good source of vitamins and minerals especially calcium and especially in the purple-leafed forms. They are antidiabetic and antiscorbutic. In Senegal poultices for abscesses are made of the leaves (15). The leaf-sap is used on burns in Ivory Coast, the pounded leaves are made into an enema given to avert miscarriage (3), and leaves are applied in topical frictions to relieve intercostal pain and in mouth-wash and gum-massage for toothache (16). Toxic substances have been reported and excessive ingestion is known to cause diarrhoea, and even death. Some alkaloid has been recorded in the stems and leaves (3) and in the roots (1).

In Zanzibar the leaves are pounded with water added to produce a green dye (10). Similarly in Gabon dye from the leaves is used to colour fishing lines (21).

*I. batatas* is not known in the wild. Columbus met the plant in extensive cultivation in Cuba in 1492, on his first Voyage of Discovery. The two vessels of his expedition must surely have been reprovisioned with tubers, and Columbus's return journey to Spain of 51 days was just within the storage life of the tuber but whether it was on this occasion or from a subsequent voyage that material was established in Spain and Portugal is unrecorded. Nevertheless then or soon after the sweet potato passed into cultivation in SW Europe and the word *batata* from one of the Caribbean languages entered European languages. The Portuguese carried the plant and the name down the coast of Africa where the name has persisted pure or in clearly recognizable form in several of the languages of the Senegambian area, and even in Urhobo (*matata*) as far removed as S Nigeria. The exoticness of the plant is also reflected in some other languages, such as Igbo and Efik as 'European's Yam' and in Ewe in Togo as 'yam of the Lagos People.' *Solanum tuberosum* did not get carried to Europe till nearly 80 years later and when it did it became a staple root-crop in more northern Europe outside the range of climatic tolerance of *I. batatas*. The English language with no better name available took *batata* by transference to become *potato* for the newer plant.

References:
1. Adegoke & al., 1968. 2. Bouquet, 1969: 99. 3. Bouquet & Debray, 1974: 76. 4. Burkill, 1935: 1246–8. 5. Burkill, 1954: historical. 6. Busson, 1965: 374 — with tuber analysis. 7. Chadha & Dakshinamurthy, 1965. 8. Dalziel, 1937. 9. Debray & al., 1971: 67. 10. Greenway, 1941. 11. Irvine, 1952, a: 24. 12. Irvine, 1952, b: 34. 13. Irvine, 1956: 37. 14. Kay, 1973: 144. 15. Kerharo & Adam, 1974: 366–7, with phytochemistry and pharmacology. 16. Kerharo & Bouquet, 1950: 225. 17. Oliver, 1960: 8, 68. 18. Purseglove, 1968: 79–88. 19. Quisumbing, 1951: 758–9. 20. Sastri, 1959: 238–47. 21. Walker & Sillan, 1961: 135. 22. Watt & Breyer-Brandwijk, 1962: 307.

**Ipomoea cairica** (Linn.) Sweet

FWTA, ed. 2, 2: 351. UPWTA, ed. 1, 438.

English: railway creeper (India, Sastri).

French: ipomée du caire (Berhaut).

West African: **SENEGAL** FULA-TUKULOR botiré (auctt.) SERER-NON (Nyominka) nburbop (K&A) WOLOF laulau (auctt.) **THE GAMBIA** FULA-PULAAR (The Gambia) gwangwangi-daniejo (DRR) MANDING-MANDINKA gulundingo (DRR) **SIERRA LEONE** MENDE ndɔgbɔ-yuwo (JB) **IVORY COAST** AKYE n-kpokrobè (A&AA) **GHANA** AKAN-FANTE supripi (Deakin) **NIGERIA** HAUSA yako (JRA)

A slender perennial twiner from tuberous root-stock from seashore, forest-clearings, grassland and damp sites, throughout the Region and tropical Africa and Asia to the Far East.

The violet-purple to white flowers are attractive and the plant is grown as an ornamental. It can be trained up a trellis and has been much used as a screen in this manner on Indian stations, thus acquiring the name 'railway creeper' (8).

The foliage is said to be eaten by the giraffe in Kenya (7) and by goats (4). It is fed to rabbits, guinea-pigs, goats and pigs in Gabon (10), but in Australia the plant is suspect of poisoning horses (9).

Fibres from the stems are made into sponges in the Cape Coast district of Ghana (3). This absorbent-cum-medicinal effect is used in Senegal for treating eye-troubles: the whole plant is firmly tied in a bundle, immersed in water, boiled and then withdrawn and while still hot used as a sponge to wash the eyes (5, 6). Crushed leaves are taken in a draught in S Africa for body-rashes, especially if accompanied by fever (9). The vines are used in India as cordage (8), and in Gabon are believed to be a lucky charm for those who wish to catch big fish (10).

The tuber and the stems, though bitter are used in Hawaii as food, but both are slightly cyanogenetic (8). Steroids and terpenes have been recorded in the whole plant in Congo, without other active principles (2).

The seeds are used in Nigeria (1) and in India (8) as a strong purgative, an action probably due to the presence of a yellow glycosidic compound resembling *muricatin A*, and perhaps other substances. The seeds also contain a non-purgative fixed oil containing glycerides of palmitic (8%), stearic (11%), arachidic (3%), behenic (1%), oleic (24%), linoleic (33%) and linolenic (5%) acids. Unsaponifiable material contains β-sitosterol (8, 9). Considerable antibiotic action has been found in the plant, but its presence is not consistent (9).

References:

1. Ainslie, 1937: sp. no. 193. 2. Bouquet, 1972: 20. 3. Deakin 23, K. 4. Glover & al. 1847, K. 5. Kerharo & Adam, 1964, c: 310–11. 6. Kerharo & Adam, 1974: 367–8, with phytochemistry and pharmacology. 7. Nesbitt-Evans 9, K. 8. Sastri, 1959: 247–8, with phytochemistry. 9. Watt & Breyer-Brandwijk, 1962: 308. 10. Walker & Sillans, 1961: 136.

**Ipomoea coptica** (Linn.) Roth

FWTA, ed. 2, 2: 350. UPWTA, ed. 1, 438, as *I. dissecta* Willd.

West African: **SENEGAL** SERER ndaf a tèk (JB) ndégèt fa mbé (JB) WOLOF sagar i surga = *little sugar* (DF) **GHANA** GA kléŋme (FRI; KD) **NIGERIA** HAUSA saawuu-dubuu = *a thousand foot-prints; from the very deeply dissected pattern of the leaves* (JMD; ZOG)

A slender prostrate or twining annual to about 1.5 m long, of grassland, waste places and woodland, and a weed of cultivation, occurring in Senegal, Mali, Ghana and Nigeria, and dispersed throughout tropical Africa, South Africa and Australasia.

The leaves are used in Ghana to treat chest-complaints in children (3, 4). In N Nigeria a cold infusion of the plant is taken as a remedy for giddiness or intoxication (2). A decoction is said in Egypt to provide a cooling lotion (5). All domestic stock relish it as grazing in Senegal (1).

The plant has been reported to contain a cyanogenetic glycoside (5).

References:

1. Adam, 1966, a: as *I. dissecta*. 2. Dalziel, 1937. 3. Irvine 710, K. 4. Irvine, 1930: 241, as *I. dissecta* Willd. 5. Watt & Breyer-Brandwijk, 1962: 308.

**Ipomoea coscinosperma** Hochst.

FWTA, ed. 2, 2: 350.

West African: **SENEGAL** MANDING-BAMBARA ñiñéni (JB)

A trailing annual herb, of grassland and weed of cultivation in the drier part of the Region from Senegal to N Nigeria, and extending across Africa to Ethiopia, and also in S Africa.

The plant is relished by cattle in Senegal, and provides a little grazing in the Chad area (1).

Reference:

1. Adam, 1966, a.

**Ipomoea eriocarpa** R. Br.

FWTA, ed. 2, 2: 350. UPWTA, ed. 1, 438, as *I. hispida* Roem. & Schult.
West African: **SENEGAL** MANDING-BAMBARA gabi (JB) SERER butkundo (JB) fohos o gay (JB) **THE GAMBIA** MANDING-MANDINKA julindiŋ (*def*.-o) = *little fibre* (Fox; DF) sanding sulo = *small rabbit root* (Ingram; DF) **GUINEA-BISSAU** MANDING-MANDINKA djulinding **SIERRA LEONE** MENDE kpokpo (*def*.-i) *general for twiners* (FCD; JMD) **NIGERIA** ARABIC-SHUWA liwai (Musa Daggash) CHAMBA tala *general for climbers* (FNH) GWARI amugini (JMD) HAUSA yaryaɗi, yimɓururu, yinɓururu *common for twiners, esp. convolvulus* (auctt.) YORUBA okorowu (N&E)

A slender twining annual herb of grassland, waste spaces and a weed of cultivation, occurring throughout the Region, and widespread in the Old World tropics and N Australia.

The plant is sometimes grown around village huts in N Nigeria (1) and is occasionally used in soup or mixed with other food (probably including other *Ipomoea* species) (2). The leaves are boiled and eaten in Tanganyika (4). It is commonly eaten in India as a vegetable as are also the seeds which are nutritious with 22% proteins, 10% fat/oil, 44% carbohydrate, etc. The seed is also reported to contain an irritant purgative resin (5).

The plant has been successfully cultivated as an arable crop in India to provide green fodder for cattle. It is hardy and drought-resistant. As fodder it is equated with sunnhemp (*Crotalaria juncea* Linn., Leguminosae: Papilionioideae) and harvested at the flowering stage gave an analysis of: moisture 16%; proteins 16%; ether extract 2%; carbohydrates 35%; fibre 21%, etc. It is useful for milch-cattle. The plant is an effective soil-binder and smotherer of weeds (5).

The plant has unspecified medicinal use in The Gambia (2). An oil extract of the plant is used in India for external application in headache, rheumatism, leprosy, epilepsy, ulcers and fevers. It is also applied to the neck-sores of [? draft] bulls (5).

References:

1. Dalziel 177, K. 2. Dalziel, 1937. 3. Fox 23, K. 4. Newbould & Harley 4298, K. 5. Sastri, 1959: 248–9.

**Ipomoea hederifolia** Linn.

FWTA, ed. 2, 2: 347.
French: liseron rouge (Berhaut).

An annual twiner, native of tropical America, but widely dispersed by man to tropical and subtropical countries, and naturalized here and there in the Region.

The plant is commonly grown as an ornamental with scarlet flowers.

**Ipomoea intrapilosa** Rose

FWTA, ed. 2, 2: 352.

A small tree to about 4 m high with branches disposed to twine, native of Mexico, and cultivated at Ibadan, Nigeria, at least, and probably elsewhere in the Region for its large white flowers.

**Ipomoea involucrata** P. Beauv.

FWTA, ed. 2, 2: 347. UPWTA, ed. 1, 438–9.

West African: **THE GAMBIA** MANDING-MANDINKA makumbu (Brown-Lester) **GUINEA** KISSI g-pandia polo (FB) LOMA lovolovohi (FB) MANDING-MANINKA (Koranko) lofolofo (FB) **SIERRA LEONE** KISSI n-goŋgbo-pɔmbɔ (FCD) KONO tofo-mɛsɛ (FCD) MANDING-MANDINKA sawane (NWT) MENDE kpokpo ndondokɔ (*def.*-i) (FCD; JMD) SUSU fotal ɛliti (NWT, JMD) liti *a general term* (NWT; JMD) TEMNE *a*-niban-*a*-ro-bath (NWT) *a*-thotho (JMD; FCD) *a*-thotho-*a*-runi *incl. other I. spp* (NWT; JMD) **LIBERIA** MANO ko su (JMD) lo-lo-kŏ (JMD) **IVORY COAST** KYAMA aïmé (B&D) **NIGERIA** ARABIC-SHUWA hantu'd (JMD) HAUSA dúmán-kwaàdii = *frog's gourd; a loose term for this sort of plant* (JMD; ZOG) IGBO fifi lori = *bearing no fruit* (auctt.) m̀gbá-n'àlà = *close to the ground* (auctt.) YORUBA àlùkọrẹṣẹ (auctt.) apiiti (IFE) awarwaroo (Lowe) inuwo elepe (IFE) òdodó (RJN) òdodó oko (JMD; Lowe)

A slender but vigorous, sprawling or twining annual or perennial herb, of grassland, secondary scrub and forest, widespread throughout the Region, and very common throughout tropical Africa.

The plant by virtue of its active growth has been found suitable as a natural forestry cover for plantations in S Nigeria (8), and in W Cameroons (6), usage reflected in one of the Igbo names. It can be satisfactorily grown over a frame as a screen. Its invasive habit is considered to be a good talisman in Gabon for fecundity so that pregnant women sometimes wear a liane around the waist (11). In Congo a length of the stem is sometimes tied around a baby's loins to promote walking (2).

In Guinea, the Lélé cook and eat the leaves as a spinach with rice or with *fonio* (*Digitaria exilis,* Graminae) (5). In Uganda it frequently invades pasture and is grazed with relish by stock (10).

No alkaloid (1), nor active principle (3) has been recorded in the plant. It, however, enters into various medicinal uses. In Lagos the plant is made into an infusion drunk as a stimulant, or preventative of fever, and in Sierra Leone a decoction of the fresh sap is taken as a remedy for gonorrhoea (7). The leaves are used (? method) in Nigeria for asthma (9). In Ivory Coast, a plant preparation is added to baths or made into a lotion for treating jaundice (4). In Congo leaf-sap is applied and rubbed into areas of localized oedema and is instilled into the eyes for filarial infection; an aqueous decotion is taken by women for dysmenorrhea, and at child-birth to hasten expulsion of the after-birth; and a compress of pounded up stems is used for headache (2).

References:

1. Adegoke & al., 1968. 2. Bouquet, 1969: 99–100. 3. Bouquet, 1972: 20. 4. Bouquet & Debray, 1974: 76. 5. Busson, 1965: 376–7. 6. Dalziel, 1937. 7. Deighton 1026, K. 8. Obpon s.n. (UIH. 3347), UCI. 9. Oliver, 1960: 68. 10. Purseglove P.2727, K. 11. Walker & Sillans, 1961: 136.

**Ipomoea mauritiana** Jacq.

FWTA, ed. 2, 2: 351. UPWTA, ed. 1, 438, as *I. digitata* sensu Dalziel, non Linn.

West African: **SENEGAL** MANDING-MANDINKA bulobulu (K&A) **THE GAMBIA** FULA-PULAAR (The Gambia) gwelling-gwelling (DRR) MANDING-MANDINKA jongmuso jongo julo = *slave woman's bonds,* julo: *rope* (Hayes) julindi-messaŋ (*def.*-o) = *tiny fibre robe* (DRR; DF) **SIERRA LEONE** BULOM (Sherbro) yɛke-kɔl-lɛ = *monkey's cassava* (FCD) KISSI yambale-kɛɛnda (FCD) yambe-kɛɛnda (FCD) KONO ba-bundɛ (FCD) LOKO ndɔbobɔgo (NWT) ndɔbohande (NWT) MENDE hele gbokpo = *elephant's convolvulus* (JMD) hoke-yowo (*def.*-i) (auctt.) SUSU foriwuri (NWT; JMD) kɔse-wure (FCD) SUSU-DYALONKE futa-na (FCD) khabi-na (FCD) **LIBERIA** MANO gia gba (Har.; JMD) **IVORY COAST** ABURE aba abasanité (B&D) AKAN-ASANTE baloa (B&D) niaméolui (B&D) taduéné (B&D) **GHANA** AKAN-BRONG nzansea (Plumtre) TWI amanin (Plumtre) dinsinkɔro (FRI) GA loloa (FRI) **NIGERIA** FULA-FULFULDE (Nigeria) ɓoore (JMD) IGBO m̀gbá-n'àlà (JMD) YORUBA àtẹ̀wọ́ (JRA) àtẹ̀wọ́ ẹdun = *palm of monkey's hand* (auctt.)

A large perennial liane with a tuberous root, of lowland rain-forest, riverain forest, savanna and secondary scrub, occurring throughout the Region, and pan-tropical in dispersal.

The plant has rose-coloured flowers and is grown as a decorative ornamental vine.

The leaves are not eaten by man as a food, but are taken in Ivory Coast in soup as a purgative and diuretic (4). Cattle will browse it a little in Senegal (1). The leafy stems are fed to cattle in India (11).

The tuberous root is used in dried and powdered form in Senegal as an abortifacient. It is also used as a purge in Senegal, but not for pregnant women (7, 8), and in Ghana (6) and Nigeria (9). Yet in contrast it is prepared as a decoction and administered as an enema in Ivory Coast for kidney-pains, female sterility, and to ensure a good pregnancy and to avoid miscarriages (4). The root is said in Nigeria to be a tonic, alterative and aphrodisiac; the rootstock mixed with palm-wine is given to nursing mothers as a galactagogue (3, 9). In India the root has been used from ancient Sanskritic times. It is considered tonic, alterative, aphrodisiac, demulcent, galactogogic and cholagogic. It is recommended for emaciation in children, and is put into a compound decoction which is nutritive, diuretic, expectorant and useful in fevers and bronchitis, in disease of the spleen and menorrhagia (5, 11).

The root of Indian material has been reported to contain 1.3% content of a fixed oil consisting of glycerides of oleic (60%), linoleic (19%), palmitic (8%) and linolenic (1%) acids, $\beta$-sitosterol, glycoside, reducing sugars and mucilage (8). A resin resembling jalap resins is present (10). Nigerian material is said to be free of alkaloid (2). Ether soluble substances which are hypotensive and relaxant have been obtained; also a non-ether soluble substance which raises blood pressure, increases respiration and stimulates smooth muscle and the uterus. A heterosidal substance, *paniculatine,* which acts in the manner of the latter has been isolated (8).

The seeds have been used in India to coagulate milk (11).

References:

1. Adam, 1966, a: as *I. digitata.* 2. Adegoke & al., 1968. 3. Ainslie, 1937: sp. no. 192, as *I. digitata.* 4. Bouquet & Debray, 1974: 76. 5. Burkill, 1935: 1248–9, as *I. digitata* Linn. 6. Irvine, s.d. 7. Kerharo & Adam, 1963, a: as *I. digitata* Linn. 8. Kerharo & Adam, 1974: 368–9, with phytochemistry and pharmacology. 9. Oliver, 1960: 28, 68, as *I. digitata.* 10. Quisumbing, 1951: 759–60, as *I. digitata* Linn. 11. Sastri, 1959: 248, as *I. digitata* Linn.

## **Ipomoea muricata** (Linn.) Jacq.

FWTA, ed. 2, 2: 347. UPWTA, ed. 1, 436, as *Calonyction muricatum* G. Don.

West African: **SENEGAL** SERER ɗorombolan (JB) lulan (JLT) SERER-NON diégnediguelit (AS) WOLOF lautan (auctt.) léméléme (JB; AS) n-dénat *a general term* (AS) **SIERRA LEONE** LIMBA am-pɛte-kofufe (Glanville) LOKO pɛte-yeobo (Glanville) SUSU-DYALONKE liti kungbe-na (Glanville) **GHANA** MOORE kanyɔɣe (FRI) **NIGERIA** FULA-FULFULDE (Nigeria) dale baali, baali: *sheep* (JMD; MM) HAUSA yáryaádií *general for twining herbs* (JMD; ZOG) YORUBA eksale (RJN)

A strongly climbing annual, native of tropical America, distributed pantropically by man, and recorded here and there in the Region.

In Senegal and Guinea the roasted seeds have been used as a substitute for coffee and are laxative (1). In the Philippines the seeds are used as a vulnery and are considered a very efficacious antidotal remedy for poisoning, and the plant sap is used as an insecticide (3). Unnamed alkaloid has been reported from the seed (4), but W African plant material is said to be free from hallucinogenic indole alkaloids (2).

References:

1. Dalziel, 1937. 2. Der Marderosian & Youngken, 1966. 3. Quisumbing, 1951: 755–6, as *Calonyction muricatum* (Linn.) G. Don. 4. Willaman & Li., 1970.

**Ipomoea nil** (Linn.) Roth

FWTA, ed. 2, 2: 351–2. UPWTA, ed. 1, 138, as *I. hederacea* sensu Dalziel.
English: morning glory.
French: liseron bleu (Berhaut).
West African: **SIERRA LEONE** MENDE ndɔgbɔ-yuwo (*def.*-i) (JMD) **NIGERIA** HAUSA yako (JMD; ZOG) NUPE edzògi (Banfield) YORUBA ejirin-ọ̀dàn *applied loosely to any trailing plant* (JMD)

A prostrate or twining annual or perennial, found in Sierra Leone, Ghana, and Nigeria in waste ground, and occurring throughout the tropics.

The plant has blue flowers opening in the early morning and turning pink during the day, the process being slower in hilly situations and cooler temperatures. It is an attractive plant and is often grown over fences or a trellis.

The plant has medicinal and superstitious uses in N Nigeria the same as has *I. aitonii* Lindl. The dried leaves are applied to burns (2). A fairly strong presence of alkaloids has been found in Nigerian material (1), but specifically hallucinogenic indole alkaloids are absent (3). The root is said in Gabon to be poisonous (6). It is a love charm in N Nigeria (2).

The principal interest lies in the seeds. In N Nigeria they are used along with *Hibiscus sabdariffa* (Malvaceae) as a purgative (2). The seeds are official in the *Indian Pharmacopoea* as a purgative and are a substitute for true jalap (*Ipomoea purga*). Their taste is at first sweetish, then acrid and disagreeable. Purgative action was at one time ascribed to the presence of a glycosidal resin, *pharbisitin,* which can be extracted from the seeds, along with inert resinous material by alcohol, but recent work has shown the non-glycosidal resin to be responsible (5). The seeds also contain a fixed oil, saponin, mucilage and tannin. The oil, of unpleasant taste, contains glycerides of palmitic (6%), stearic (20%), arachidic (8%), behenic (1%), linolenic (6%), linoleic (15%) and oleic (44%) acids (5). Unnamed alkaloids have also been reported (7). The seeds are regarded in China as diuretic, anthelminthic and deobstruant and are prescribed for dropsy and constipation, and to promote menstruation and cause abortion (4).

The flowers contain anthocyanim pigments (5).

References:

1. Adegoke & al., 1968. 2. Dalziel, 1937. 3. Der Marderosian & Youngken, 1966. 4. Quisumbing, 1951: 760, as *I. hederacea.* 5. Sastri, 1959: 249–51. 6. Walker & Sillans, 1961: 136, as *I. hederacea.* 7. Willaman & Li, 1970.

**Ipomoea obscura** (Linn.) Ker-Gawl. aggregate

FWTA, ed. 2, 2: 349, incl. *I. ochracea* (Lindl.) G. Don, *I. trichocalyx* Schumm. & Thonn. and *I. acanthocarpa* (Choisy) Ascherson & Schweinf.
French: ipomée jaune (*I. ochracea,* Berhaut).
West African: **SENEGAL** SERER ɗorombolan ala *'I. acanthocarpa'* (JB) pis lulãnd *'I. acanthocarpa'* (JB) **NIGERIA** IGBO (Uburuku) ògbanànị *'I. ochracea'* (NWT) YORUBA ododo oko ododo owuro

Slender, prostrate or twining perennial herbs of the savanna; a diffuse aggregate of doubtfully separate taxa occurring throughout the Region, and at least in part commonly widespread in tropical Africa.

The leaves are eaten in soup in S Nigeria (6) and as a vegetable in Kenya (5). All stock graze it in Senegal (1a), Kenya (4) and Uganda (3), but stock at Chad will not take it (1b). Leaf-sap has medicinal use in Congo for treating fits of insanity (2).

References:

1. Adam, 1966, a. a. as *I. ochracea*. b. as *I. kentrocarpa* (=*I. ochracea*). 2. Bouquet, 1969: 100, as *I. obscura*. 3. Dyson-Hudson 441, K, as *I. obscura*. 4. Glover & al. 1563, K, as *I. obscura*. 5. Graham 2075, K, as *I. obscura*. 6. Thomas, N. W. 2076 (Nig. Ser.), K, as *I. ochracea*.

**Ipomoea pes-caprae** (Linn.) Sweet, ssp. **brasiliensis** (Linn.) van Ooststr.

FWTA, ed. 2, 2: 347–8. UPWTA, ed. 1, 439.

English: goat's foot convolvulus.

French: ipomée pied de chèvre (Berhaut); liseron pied de chèvre (Kerharo & Adam).

West African: **SENEGAL** MANDING-'SOCE' bababarakora (K&A) SERER duam (JB; K&A) furtut duam (JB) purtut (JLT; K&A) WOLOF n-dénat (K&A) n-dénat i gèti (JB) purtut (K&A) **SIERRA LEONE** BULOM (Sherbro) hontolon-dɛ (FCD) MENDE hondolo (FCD) hondolo-hina, hina: *male* (FCD) **GHANA** AHANTA aupa nziŋa (FRI) AKAN-FANTE papan (FRI) NZEMA akpa **NIGERIA** YORUBA npiiti (IFE) olinuwo ẹlopo *the roots* (IFE)

A long-trailing creeping, or scrambling, vine, perennial, rooting at the nodes, of sandy foreshores down to the limits of the high tide level, occurring throughout the Region's seaboard, and pantropical in distribution. The other subspecies, ssp. *pes-caprae,* is confined to the Indic basin.

The plant is a pioneer colonizer of sand-dunes and is a sand-binder *par excellence*. The leaves are eaten as a vegetable in Zanzibar (10). Most animals will eat the foliage. Leafy stems are commonly fed by Chinese pig-farmers to their stock in SE Asia. Cows fed on the leaves are said to yield tainted milk (1, 6). Pulped leaves are rubbed on fishing nets in Malawi as a lure to entice the fish to enter (11).

An unnamed alkaloid has been reported present in the leaves (9) but other reports indicated the absence of toxic substances such as alkaloid, glycoside, saponin, hydrocyanic acid, etc. (8). Material from Florida has been reported to contain mucilage, a complex resin, volatile oils, fats, a bitter pigment, etc. The leaves are inactive against *Staphylococcus aureus,* but are said to have beneficial effect on bed-sores. Strong anti-tumour action has been shown by extracts of the stems (3, 4). In the Philippines the leaf is used to extirpate fungoid growth of ulcers (5), and in the Philippines (5), Australia and India (8), a leaf-preparation is anodynal in rheumatism. Leaf-sap is applied to fish-stings in Malaya (1). In Senegal the leaves are used as an emollient, applied in a hot poultice on ulcerous and other sores (4). Leaf-preparations are commonly applied to boils, carbuncles, swelling, ulcers, piles, and other surface ailments in Asia (1, 5, 6). The leaf is considered diuretic and laxative: it is used in India for dropsy (6, 8) and for urethral discharge in Madagascar (2) and Indo-china (6).

The long stoloniferous stems are relatively strong. They are made into ropes in Malawi for hauling in fish-nets (11) and in Gabon have been used for tug-of-war and singly as skipping ropes (7).

The root is starchy. A saponin has been reported present (6). They are diuretic and are taken boiled to bring relief in bladder-diseases (1, 8). They are also purgative, and have been used as an adulterant or falsification of Indian jalap (4).

The seed is said in Indonesia to be a good remedy for stomach-ache and cramp (1).

References:

1. Burkill, 1935: 1249–50. 2. Debray & al., 1971: 38. 3. Hartwell, 1968. 4. Kerharo & Adam, 1974: 369–70, with phytochemistry and pharmacology. 5. Quisumbing, 1951: 761–2. 6. Sastri, 1959: 251. 7. Walker & Sillans, 1961: 137. 8. Watt & Breyer-Brandwijk, 1962: 309. 9. Willaman & Li, 1970. 10. Williams, R. O., 1949: 310. 11. Williamson, J., 1956: 70.

**Ipomoea pes-tigridis** Linn.

FWTA, ed. 2, 2: 347.

West African: SENEGAL MANDING-BAMBARA bugu mugu (JB) SERER ñahan (JB) WOLOF lahal lur (JB)

A herbaceous annual twiner of the sahel zone from Senegal to Niger and N Nigeria, and dispersed across tropical Africa and into Asia and Australasia.

The plant provides good grazing in Senegal for all domestic stock (1). In northern India it is an important fodder, growing profusely during the rainy season and remaining green and succulent for 3–4 months after. In nutritive value it compares favourably with legumes, being rich in protein, calcium and phosphorus. Indian material assayed at: crude protein 12%, fat 3%, fibre 26%, N-free balance 48%. Digestibility was good (6).

The leaves are used in Indonesia (2, 5) and the Philippines (5) for poulticing sores and pimples. In Tanganyika they are used for whitlow (4). The plant is applied as an emollient on tumours in India, Ceylon and SE Asia (3).

The root is purgative and is said to contain a resin (6).

References:

1. Adam, 1966, a. 2. Burkill, 1935: 1250. 3. Hartwell, 1968. 4. Koritschoner 1143, K. 5. Quisumbing, 1951: 762–3. 6. Sastri, 1959: 251–2.

**Ipomoea purpurea** (Linn.) Roth

FWTA, ed. 2, 2: 352.

English: morning glory.

A herbaceous twining annual, native of tropical America, but now dispersed to most warm countries as an ornamental with white, pink or magenta flowers. Cultivars with double flowers are known.

The leaves are eaten in Nigeria and are purgative (1). They are also considered in Nigeria (1) and in S Africa (4) to be antisyphilitic, but are ineffective as such. Both root and stem are purgative. Investigation has shown the presence of a resin in the stem at 4.8% concentration which is the active purging principle and to be the same as *convolvulin*. Aerial portions of the plant have given some positive, some negative, antibiotic results (3, 4). The presence of a number of alkaloids is also reported in the leaf, stem, root and principally the seeds (5), but hallucinogenic indole alkaloids appear to be absent from West African material (2).

References:

1. Ainslie, 1937: sp. no. 194. 2. Der Marderosian & Youngken, 1966. 3. Sastri, 1959: 252. 4. Watt & Breyer-Brandwijk, 1962: 309. 5. Willaman & Li, 1970.

**Ipomoea quamoclit** Linn.

FWTA, ed. 2, 2: 347.

English: cypress vine; Cupid's flower; 'Johnny Walker' (Ainslie).

French: cheveux de Venus (Berhaut).

A graceful, slender twining annual, native of tropical America but distributed by man to many tropical and subtropical countries, and found naturalized here and there throughout the Region.

The delicate deeply dissected leaves and the crimson, scarlet or white flowers make the plant an attractive garden ornamental. The leaves are eaten in India as a pot-herb (4), though in Congo they are said to be taken as a somniferic (2). The leafy stems have been found to contain unnamed alkaloid (6) and to be cyanogenetic (4), but W African material is reported free of hallucinogenic indole alkaloids (3). The leaves are pounded up in Nigeria to apply as a plaster to boils, ulcers, etc. (1). Use as poultices for bleeding piles and carbuncles is also recorded (4).

In Gabon the plant is a fetish for eloquence (5).

References:

1. Ainslie, 1937: sp. no. 295, as *Quamoclit pennata*. 2. Bouquet, 1969: 100. 3. Der Marderosian & Youngken, 1966. 4. Sastri, 1959: 252–3. 5. Walker & Sillans, 1961: 138, as *Q. pennata* Bojer. 6. Willaman & Li, 1970: as *Q. pennata* (Desr.) Bojer.

**Ipomoea sepiaria** Roxb.

FWTA, ed. 2, 2: 349, as *I. hellebarda* Schweinf.

West African: **GHANA** ADANGME-KROBO odumace (FRI)

A slender trailing herb of open places.

**Ipomoea setifera** Poir.

FWTA, ed. 2, 2: 349.

French: liane à faux (Berhaut).

West African: **SENEGAL** DIOLA (Fogny) élalak yéta bu yit (JB); = *elalak of the swamp* (DF, The Gambia) MANDING-BAMBARA ulonido (JB)

A robust twiner and trailer, native of tropical America, and introduced and naturalized in Senegal, The Gambia, Sierra Leone and Ivory Coast.

The plant has purple-red flowers up to 10 cm long, and is grown in the Region as an ornamental.

The seed is reported to contain at least three hallucinogenic indole alkaloids (1).

Reference:

1. Der Marderosian, 1967.

**Ipomoea setosa** Ker-Gawl.

FWTA, ed. 2, 2: 352.

A perennial twiner, native of Brazil, and brought into cultivation in many warm countries for its showy rose-purple flowers to 7.5 cm long.

**Ipomoea sinensis** (Desr.) Choisy

FWTA, ed. 2, 2: 349–50.

A slender trailing or twining annual of scrub, throughout the Region and in the Old World tropics.

In Kenya it is grazed by all stock, and cattle-folk consider that it results in good milk-yield (1, 2). Feeding tests have shown it to be non-toxic to sheep (4). It is grazed also by wild game (3).

References:

1. Glover & al. 828, K. 2. Glover & Samuel 3209, K. 3. Stewart 499, K. 4. Watt & Breyer-Brandwijk, 1962: 310.

**Ipomoea stolonifera** (Cyrill.) J. F. Gmel.

FWTA, ed. 2, 2: 350. UPWTA, ed. 1, 440.

West African: **SENEGAL** WOLOF mbory (JLT) **GUINEA-BISSAU** BALANTA n-tome, untome (JDES) BIAFADA bubafgale (JDES)

A long-trailing stoloniferous perennial plant occurring along all the Region's seaboard at suitably sandy sites, and distributed pan-tropically.

The plant trails over beach-head sand-dunes and serves as a sand-binder. It grows into wetter conditions than are tolerated by *I. pes-caprae,* and also better withstands burying — to at least 15 cm.

**Ipomoea tricolor** Cav.

FWTA, ed. 2, 2: 352.

A stout perennial twiner, native of Mexico, and distributed by man to many warm countries as an ornamental. Several very decorative cultivars are known.

The plant contains a number of hallucinogenic indole compounds (2). It was used by the Aztecs as a hallucinogen for their religious ceremonies and in medicine.

References:

1. Der Marderosian, 1967. 2. Willaman & Li, 1970: as *I. tricolor* Cav. and *I. rubro-caerulea* Hook.

**Ipomoea triloba** Linn.

FWTA, ed. 2, 2: 350.

A slender trailing and twining herb, native of tropical America, but found throughout the Region and pan-tropically generally.

The use in Malaya of a poultice for the head in headache may refer to this plant (1). W African material is reported to be free of hallucinogenic indole alkaloids (2).

References:

1. Burkill, 1935: 1250. 2. Der Marderosian & Youngken, 1966.

**Ipomoea tuba** (Schlechtend.) G. Don

FWTA, ed. 2, 2: 352.

French: grande sultane (Berhaut).

A perennial twiner of the American tropics, widely distributed as an ornamental for its white flowers to 10 cm long.

**Ipomoea vagans** Bak.

FWTA, ed. 2, 2: 349.

West African: **SENEGAL** SERER nof ndol (JB) nof o mbam (JB) WOLOF na bu digen (JB) **GUINEA-BISSAU** FULA-PULAAR (Guinea-Bissau) tirde (JDES; EPdS) **MALI** DOGON ɔ́mɔnɔ kɔ́mɔ pílu (C-G) **NIGERIA** HAUSA walkin machiji (BM)

A trailing herb of Senegal Ghana, Niger and N Nigeria and also in Sudan.

In Senegal, all stock will graze it, though at Chad it is not taken (1).

Reference:

1. Adam, 1966, a.

**Ipomoea spp. indet.**

West African: **THE GAMBIA** MANDING-MANDINKA ilundimba (DRR) juluwaila = *spreading fibre* (DRR; DF) kurubhobo (DRR) **GHANA** AKAN-FANTE supripi (Deakin) **NIGERIA** FULA-FULFULDE (Nigeria) ɓoore (J&D)

The following are recorded without detail:

sp. a. *ilundimba, juluwaila* (Mandinka, The Gambia, 2). A climbing herb.
sp. b. *kurubobo* (Mandinka, The Gambia, 2). A climbing herb.
sp. c. *ɓoore* (Fula, N Nigeria, 1). A weed, not eaten.

References:

1. Jackson, 1973. 2. Rosevear, 1961.

**Jacquemontia ovalifolia** (Vahl) Hallier f.

FWTA, ed. 2, 2: 340.

An annual or biennial prostrate or creeping herb to 3 m long, rooting at the nodes, from a semi-woody rootstock over seasonally exposed mud and sandy wastes behind beach-heads in Ghana and Togo, and also in Somalia, Angola and SW Africa.

At Accra it occupies sites behind the shoreline and in dry sandy places where no other plants will grow (1, 2). It must help to stabilize sand.

References:

1. Irvine 668, K. 2. Irvine, 1930: 243.

**Jacquemontia pentantha** (Jacq.) G. Don

FWTA, ed. 2, 2: 340.

A climber with showy light blue to violet flowers, native of the Caribbean, and introduced by man as an attractive ornamental to many tropical and subtropical countries including the W African Region.

**Jacquemontia tamnifolia** (Linn.) Griseb.

FWTA, ed. 2, 2: 340. UPWTA, ed. 1, 440, as *J. capitata* G. Don.

West African: **SENEGAL** FULA-PULAAR (Senegal) tirdé (K&A) MANDING-BAMBARA batu (JB; K&A) SERER fafay (JB; K&A) lul *general for convolvulus* (auctt.) WOLOF laulau (JMD; K&A) mbifir (K&A) möför (K&A) neför (JB; K&A) nopilbam (auctt.) **GHANA** GBE-VHE boeboe (FRI) **NIGERIA** HAUSA dúmán gwaázaá (JMD; ZOG) yaɗon gwaázaá (JMD) yaraɗi (RES) yaraɗin tudu *general for herbaceous twiners* (Lely; JMD)

A slender annual twiner to about 1 m long, of dry sandy places, grassland, scrub, etc., native of tropical America, but now naturalized in tropical Africa and recorded in the Region from Senegal to N Nigeria and W Cameroons.

The plant is grazed by stock (Senegal, 1; Sudan, 2). The leaves are mashed and added to soup in Ghana by the Vhe (8, 9), and eaten as a vegetable in Kenya (5) and Tanganyika (11). In N Nigeria the leaves are crushed with scent and applied for headache (3), and in Senegal are dried and powdered with those of *Nauclea latifolia* (Rubiaceae) to take as snuff for neuralgia (10). Sap is dripped into the eyes in Tanganyika for conjunctivitis and the plant reduced to ash is rubbed with caster-oil into scarifications for leprosy (6). Traces of hallucinogenic indole alkaloids have been reported present in the seeds (4, 12). Extracts of the whole plant and of the fruits have shown a slight insecticidal activity (7).

The plant bears heads of bright blue flowers and can be trained over a trellis as an ornamental climber (3).

References:

1. Adam, 1964. 2. Andrews 2, K. 3. Dalziel, 1937. 4. Der Marderosian, 1967: as *Ipomoea tamnifolia*. 5. Graham B.521, K. 6. Haerdi, 1964: 178. 7. Heal & al., 1950: 116. 8. Irvine 691, K. 9. Irvine, 1930: 243, as *J. capitata* G. Don. 10. Kerharo & Adam, 1974: 370. 11. Tanner 3688, K. 12. Willaman & Li, 1970: as *Ipomoea tamnifolia* Linn.

**Lepistemon owariense** (P. Beauv.) Hallier f.

FWTA, ed. 2, 2: 343.

West African: **GHANA** ADANGME dɔkɔ (FRI) **NIGERIA** YORUBA ewe jenoko = *let me sit down*

A robust twining perennial of rain-forest and waste scrub from Guinea-Bissau to W Cameroons and in central, NE and E Africa and Angola.

The plant is frequently hairy, urticant or pungent. The leaves are boiled and eaten in the Shai Plains of Ghana as a vegetable (1).

Reference:

1. Irvine 664, K.

**Merremia aegyptia** (Linn.) Urban

FWTA, ed. 2, 2: 342. UPWTA, ed. 1, 440, as *M. pentaphylla* Hall. f.

West African: **SENEGAL** DIOLA-FLUP é gonnoray (JB) MANDING-BAMBARA sunzan ka tamugu (JB) SERER hayéhay (JB) maf i nogoy (JB) WOLOF laulau (JB) **GUINEA-BISSAU** FULA-PULAAR (Guinea-Bissau) tirde (JDES; EPdS) **SIERRA LEONE** SUSU bagifiri (NWT) **NIGERIA** HAUSA baa-maa-taboo, bàr-mà-tàbó (JMD; ZOG) dankwon kuyangi (auctt.) yako (JMD; ZOG)

A robust annual twiner occurring throughout the Region, and widespread in the Tropics.

Horses will not graze it; other stock in Senegal may or may not take it (1). The dried leaves are applied in Nigeria as a dressing for burns (2). The stems are used in Senegal (3), and probably elsewhere as ties (2).

References:

1. Adam, 1966, a. 2. Dalziel, 1937. 3. Farmar 36, K.

**Merremia cissoides** (Lam.) Hallier f.

FWTA, ed. 2, 2: 342.

A slender climber of central America is occasionally cultivated as an ornamental.

**Merremia dissecta** (Jacq.) Hallier f.

FWTA, ed. 2, 2: 342.
West African: **NIGERIA** IGALA ọlubọ (Boston)

A perennial twiner, native of tropical America, introduced to parts of the Region, and to many countries of the tropics.

The white/purple flowers are ornamental and the plant is often grown around houses. The flowers open in the evening and unlike most ornamental 'ipomoeas' remain open till the following late afternoon (2). In Igala Province, Nigeria, the plant is reported to keep snakes away from houses where it is grown, and the leaves in infusion are a major ingredient of snake-bite medicines for taking by draught and by topical application (1). The leaves are said to produce hydrocyanic acid and to be poisonous to cattle. They have the smell of oil of bitter almonds, and have been used [? country] in the preparation of liqueur (3).

References:

1. Boston C.27, K. 2. Deighton 2818, K. 3. Sastri, 1962: 347.

**Merremia hederacea** (Burm. f.) Hallier f.

FWTA, ed. 2, 2: 341–2.
West African: **UPPER VOLTA** SENUFO-TUSIA sâpu pora = *herb of the sooth-sayers* (Hébert)

A slender prostrate or twining herb of thickets, grassland and open sandy places, in the northern drier part of the Region from Senegal to W Cameroons and dispersed across Africa and Asia to Australasia and the Pacific.

Stock will graze it, and American work has shown that given nothing else, animals will thrive on it (1).

In Malaya leaves are made into a poultice with turmeric and broken rice to apply to chapped hands and feet (1).

Known as 'herb of the soothsayers' in Upper Volta, the plant is used there along with *Sclerocarya birrea* (Anacerdiaceae) in fortune-telling (2).

References:

1. Burkill, 1935: 1456. 2. Guilhem & Hébert, 1965.

**Merremia kentrocaulos** (C.B. Cl.) Rendle

FWTA, ed. 2, 2: 342.
West African: **SENEGAL** SERER sanabra (JB; K&A) **NIGERIA** BOLE zovigewandu (AST)

A perennial twiner of wooded savanna, grassland and sandy places in Senegal and Nigeria only, but occurring widely elsewhere in tropical Africa and tropical Asia to N Australia and the Pacific Islands.

It is used in Senegal in association with *Pericopsis laxiflora* (syn. *Afrormosia laxiflora,* Leguminosae: Papilionoiideae) for the treatment of jaundice and liver-malfunction (1). The root is used in Tanganyika as a fever-remedy (2).

References:

1. Kerharo & Adam, 1974: 370–1. 2. Watt & Breyer-Brandwijk, 1962: 310.

**Merremia pinnata** (Hochst. ex Choisy) Hallier f.

FWTA, ed. 2, 2: 341

West African: **SENEGAL** WOLOF sagar i surga = *little sugar* (DF) **NIGERIA** FULA-FULFULDE (Nigeria) leyleydi (PF fide MM)

An annual trailing or twining herb of grassland and open waste places of the drier northern part of the Region from Senegal to N Nigeria, and widespread elsewhere in tropical Africa.

It is relished for grazing by all domestic stock (1).

Reference:

1. Adam, 1966, a.

**Merremia pterygocaulos** (Steud. ex Choisy) Hallier f.

FWTA, ed. 2, 2: 342. UPWTA, ed. 1, 440.

West African: **SIERRA LEONE** LOKO n-dangeha (NWT) SUSU wori-wuri (NWT) **NIGERIA** HAUSA ƙáfàr kaàzaá = *fowl's foot; prob. refers to several twiners with digitate leaves* (JMD; ZOG)

A perennial climber of savanna woodland and grassland by stream and swamps, from The Gambia to Nigeria, and widespread throughout tropical Africa.

**Merremia tridentata** (Linn.) Hallier f., ssp. **angustifolia** (Jacq.) Ooststr.

FWTA, ed. 2, 2: 341. UPWTA, ed. 1, 440, as *M. angustifolia* Hall. f.

West African: **SENEGAL** MANDING-'SOCE' duludigô (K&A) SERER law mbambé (JB; K&A) lébel (K&A) lébèl pul (JB) nof ndol (JB; K&A) pul (K&A) yuran (JB; K&A) WOLOF salaulit (auctt.) **THE GAMBIA** MANDING-MANDINKA muso jong julo = *slave woman's ropes* (Fox; DF) **SIERRA LEONE** LOKO n-dangeha (NWT) **GHANA** GBE-VHE vudrai (FRI) HAUSA yamburu (FRI) **NIGERIA** FULA-FULFULDE (Nigeria) leeɓol pullo, leebol: *a pat of butter, or a* (*single*) *hair,* thus, *hair*, or *butter of a Fulani* (JMD; MM) leyleydi (PF fide MM) HAUSA gadon machiji = *snake's bed* (JMD) gámmòn baawaa = *slave's head-pad* (JMD; ZOG) koòrén-hàwaíniyàá = *ringworm of the chameleon* (auctt.) maganin kunama = *medicine for scorpion* (JMD) yamɓururu, yámɓururu (auctt.) yimɓururu (JMD) KANURI tattir (JMD)

A prostrate or twining annual of wooded savanna, grassy savanna and open sandy places, polymorphic with only *ssp. angustifolia* in the West African region throughout, and the species generally occurring widespread elsewhere in tropical and southern Africa, Mascarenes, and tropical Asia.

The plant is relished by all domestic stock in Senegal (1), and by cattle in India (9). The plant obviously has some claim to anodynal, if not analgesic, effect. This is coupled with superstitious magic in N Nigeria where the root eaten with bran is said to confer immunity for a year or more from scorpion-sting provided one does not deliberately, apart from in food, eat salt during that period. The plant is thus loosely known as 'medicine for scorpion' (5). In E Africa the leaves are taken as an antidote for snake-bite (2). In Senegal a decoction of the whole plant gives rapid relief and cure from various ophthalmias (7, 8), and in Congo sap from grated roots is similarly used (2). In India (*ssp.* not stated) the whole plant or the roots are used for rheumatism and the roots of *ssp. hastata* (Desr.) Ooststr. are prepared as mouthwash for toothache (9). The plant is bitter, astringent, calefacient and tonic. In India the whole plant or the roots are used for hemiplegia, piles, swellings and urinary disorders (9). A decoction of the whole plant is taken with natron in Sokoto, N Nigeria, for gonorrhoea. In Tanganyika, *ssp. angustifolia* is pounded and left to macerate for a week to prepare a wash for small children with malaria (6). In Malaya the leaves of *ssp. hastata* are made into a poultice for the head in cases of fever (1).

References:

1. Adam, 1966, a. 2. Bally, 1937. 3. Bouquet, 1969: 100. 4. Burkill, 1935: 1455–6, as *M. hastata* Hall. f. 5. Dalziel, 1937. 6. Haerdi, 1964: 179. 7. Kerharo & Adam, 1964, c: 314. 8. Kerharo & Adam, 1974: 371. 9. Sastri, 1962: 347.

**Merremia tuberosa** (Linn.) Rendle

FWTA, ed. 2, 2: 342.
English: woodrose; yellow morning glory; Brazilian jalap (the root).

A robust twiner, native of tropical America and cultivated in various tropical countries including occasionally in West Africa.
The plant is an attractive ornamental, grown for its flowers and woody fruit pods which find a place in dry flower arrangements in Western countries. The tuber is a drastic purgative and contains a resin present at 12–15% which has been used as an adulterant for true jalap (*Exogonium purga,* Convolvulaceae).

**Merremia umbellata** (Linn.) Hallier f. ssp. **umbellata.**

FWTA, ed. 2, 2: 342. UPWTA, ed. 1, 440.
West African: **SIERRA LEONE** MANDING-MANDINKA soriondibi (NWT) MENDE kpokpo SUSU liti *general for convolvulus group* (NWT; JMD) TEMNE *a*-gbungabo (FCD) **MALI** MANDING-BAMBARA ulou nin tulu (A. Chev.) **NIGERIA** IJO-IZON (Oporoma) èpírí kọ̀rị (KW)

A slender perennial twiner or climber of thickets in swampy places by streams and sandy sites behind beach-heads, from The Gambia to W Cameroons and in tropical America. *Ssp. orientalis* (Hallier f.) Ooststr. is in E Africa and Asia.
The plant is reported to act as a sand-binder, on the Ghana coast (2).
*Ssp. orientalis* has a number of uses in Asia which may be found relevant to the W African subspecies. The young leaves are eaten as a pot-herb, and the leaves are applied in poultices on burns, scalds and sores in Malesia (1). It is considered useful in India for fistulae, pustules and tumours; the seeds yield a mucilage which is laxative and alterative in cutaneous diseases, and contains a fatty oil and a resin (4). A plant decoction is also used as a diuretic and alterative in rheumatism, neuralgia, headache, etc., and is instilled into the ear for auricular ulcers and abscesses (3). A leaf-powder is taken as snuff in epilepsy (3).

References:

1. Burkill, 1935: 1456–7. 2. Chipp 59, K. 3. Quisumbing, 1951: 764–5. 4. Sastri, 1962: 348.

**Neuropeltis acuminata** (P. Beauv.) Benth.

FWTA, ed. 2, 2: 238. UPWTA, ed. 1, 440.
West African: **GHANA** AKAN-ASANTE mutuo (CV; FRI) ANYI-SEHWI mutuo (auctt.)

A climbing shrub occurring in forest areas of Sierra Leone to S Nigeria and W Cameroons, and extending to Angola.
The leaves are eaten in Gabon, and the wiry stems are used as lashings in building huts in Ghana (1, 2).

References:

1. Dalziel, 1937. 2. Irvine, 1961: 734.

**Neuropeltis velutina** Hallier f.

FWTA, ed. 2, 2: 338
West African: **SIERRA LEONE** MENDE ndàvà (*def.-vui*) *general for certain woody climbers which make string rope* (FCD) **LIBERIA** MANO dei wã bele (Har.) gene zolo yıdi (Har.) pini la (Har.) tue sao goh (Har.)

A forest climber from Sierra Leone, Liberia and S Nigeria.
It is valued as a source of good rope in Sierra Leone (1).

Reference:

1. Deighton 3772, K.

**Operculina macrocarpa** (Linn.) Urban

FWTA, ed. 2, 2: 340. UPWTA, ed. 1, 440, as *Merremia alata* Rendle.
French: liane tonnelle, rose-en-bois (Berhaut).
Portuguese: batata de purga (Brazil, Heine).
West African: **GHANA** AKAN-AKUAPEM abia (FRI) FANTE ebia (Hall) TWI abia (FRI)

A stout climber, native of Brazil and the Caribbean, and naturalized in Ivory Coast, Ghana and Togo.

The plant is grown, in French parlance, to make an arbour or screen ('tonnelle') (1).

The ripe seeds are black, and statements (2, 5) that white seeds are used as beads must refer to unripe seeds (4). It is not clear whether the superstitious use of the seeds to make an *abam* bracelet in Ghana requires white or either-coloured seeds. An *abam* is a string of a few seeds worn as a bracelet by Fante women who have borne twins. A fetish priest can converse with the spirit of the *abam,* and by a trick of ventriloquism the spirit will tell him the future of the twins (3, 4). The use of the seeds as beads may have come to Ghana through the agency of Portuguese Roman Catholic priests from Brazil, or by slaves returning to Ghana who may have used them for exchange (4). In Brazil the plant is well known as a purgative, as the Portuguese name recognizes.

References:

1. Berhaut, 1967: 264. 2. Dalziel, 1937. 3. Hall 17/9/1959: ad not., K. 4. Heine, 1960: 398. 5. Irvine, 1930: 282–3, as *Merremia alata* Rendle.

**Seddera latifolia** Hochst. & Steud.

An undershrub to 30 cm high, of dry bush, in Mauritania and Niger, and extending over NE and E Africa, and into India.

It is browsed by animals in Mauritania, and it is used to make an infusion for drinking (1).

Reference:
1. Kesby 34, K.

**Stictocardia beraviensis** (Vatke) Hallier f.

FWTA, ed. 2, 2: 352. UPWTA, ed. 1, 440.
West African: **LIBERIA** MANO kpua-to (Har.) **NIGERIA** YORUBA abẹsundigbaro (EWF; JMD)

A strong woody climber of open wooded grassland and riverain forest, from Mali to S Nigeria, and extending across the Congo basin to E Africa and Madagascar.

The flowers are bright crimson and decorative.

In S Nigeria leaf-sap is dropped into wounds (2). A related species, *S. tiliafolia* (Desr.) Hall., has been reported containing at least six hallucinogenic indole alkaloids (1). *S. beraviensis* should be screened.

References:

1. Der Marderosian, 1967. 2. Thomas, N. W. 1904 (Nig. Ser.), K.

# COSTACEAE

**Costus** Linn.

West African: **IVORY COAST** ABURE auinfa (B&D) AKAN-ASANTE agniaï (B&D) agnian (B&D) BAULE agnan (B&D) agnié (B&D) alloso (B&D) KRU-GUERE (Chiehn) zazaboto KWENI yoga (B&D) zolu (B&D) KYAMA emprobego (B&D) toetoeya (B&D) MANDING-MANINKA koagbè (B&D) kwabgé (B&D)

*NOTE: See also C. afer.*

Without distinguishing between species in Ivory Coast the genus is recorded with the following medicinal and superstitious usages: the inflorescence is a good remedy for tachycardia, cough and stomach complaints; leaf-sap is used in eye-instillations for eye-troubles and headache, and in frictions for oedemas and fever; the stem-sap is used against urethral discharges, to prevent miscarriage and to treat jaundice; root-pulp will maturate buboes and abscesses; the plant is put into many prescriptions for constipation, nausea, haemorrhage in pregnancy, 'yellow' fever, etc., and in medico-magical formulations to protect people and villages from evil spirits and illness. In these applications *Aframomum spp.* (Zingiberaceae) seem to be used also more or less indeterminately (1). *Costus spp.* recorded for Ivory Coast are *afer* Ker-Gawl; *deistelii* K. Schum.; *dubius* (Afzel.) K. Schum.; *englerianus* K. Schum.; *lucanusianus* J. Braun & K. Schum.; *schlechteri* Winkler; and *spectabilis* (Fenzl) K. Schum.

Reference:

1. Bouquet & Debray, 1974: 175.

**Costus afer** Ker-Gawl.

FWTA, ed. 2, 3: 78. UPWTA, ed. 1, 472.

English: ginger lily; common ginger lily (Morton); 'bush-cane'.

West African: **SENEGAL** BALANTA belêgôfódô (K&A) BANYUN gugali (K&A) tigugal (K&A) DIOLA (Fogny) bomay (K&A) bumay (K&A) DIOLA-FLUP yumay karêg (JB; K&A); = *yumay of the forest reserve* (DF, The Gambia) FULA-PULAAR (Senegal) timbiyâba (K&A) MANDING-MANDINKA belêgôfodô (K&A) MANINKA bira kurubafira (CHOP; K&A) **THE GAMBIA** DIOLA (Fogny) bumay = *easily snapped* (DF) **GUINEA-BISSAU** FULA-PULAAR (Guinea-Bissau) gògódje-súto (EPdS) PEPEL rum-rum (EPdS) **GUINEA** MANDING-MANINKA bira kurubafira (CHOP) **SIERRA LEONE** BULOM (Sherbro) sayina-lɛ (FCD) GOLA sawa (FCD) KISSI siaŋdẽ (FCD) KONO tofa (FCD) LOKO hɔvai (auctt.) MANDING-MANDINKA kumbɛ (NWT) timba (FCD) tumba (FCD) MENDE hɔwa (auctt.) SUSU sinkoinye (NWT; JMD) timbanyi (FCD) SUSU-DYALONKE khɛmu-na (FCD) TEMNE *a*-sul *the cane* (JMD) *an*-tap (NWT; FCD) VAI tɔfa (FCD) **IVORY COAST** AKAN-ASANTE sumé (K&B) AKYE leussin (A&AA) DAN sungho (K&B) KRU-BETE doï (K&B) GUERE do (K&B) dodré (K&B) don (K&B) GUERE (Chiehn) zazaboto (K&B) MAHO yaya (K&B) MANDING-MANINKA kogbèhun (A&AA) 'KRU' tanton (K&B) 'MAHO' koyéyé (K&B) loko yaya (K&B) **GHANA** AKAN-ASANTE sommɛ (Rattray; FRI) FANTE bɛ (FRI) TWI o-sommɛ (FRI) GBE-VHE asumbeɛ (FRI) NZEMA ɛnyanɛ (FRI) 'MAHO' tchone (A.S.Thomas) **TOGO** TEM (Tshaudjo) bomire (Gaisser) **NIGERIA** EDO úkhúeréohã (JRA; Dennett) EFIK m̀bríté̩m̀ (auctt.) HAUSA kákìi-zuwaà (Singha) kákìi-zúwaà-Háusá = *thou refusest to come to Hausa* (JMD; ZOG) tùmfaáfiyár kádaá = *crocodile's* [illegible] (JMD; ZOG) [illegible] [illegible] (Lowe; Kaufman) IGBO ókpètè (Singha) ópètè *a variant form* (Lowe) IGBO (Agolo) ókpètè (JMD) IGBO (Onitsha) ókpètè, ọkètè *other dialectal forms:* ókpòtò, ókpìtì (KW) IGBO (Owerri) ópètè (KW) IGBO (Umuahia) ópòtò (JMD) IJO-IZON (Kolokuma) ere ogbodo = *female ogbodo* (KW) JUKUN andura (JMD) TIV achiklu (JMD) YORUBA atare tẹ̀tẹ̀-ẹ̀gún (JMD; Singha)

*NOTE: Many names also apply to other C. spp., especially C. lucanusianum.*

A tall perennial semi-woody herb with leafy canes to 3 m high bearing terminal inflorescences of white and yellow flowers, of the forest zone in moist places from Senegal to Nigeria and Fernando Po and widespread throughout the forest region of tropical Africa.

The plant while occurring fairly commonly in the wild, particularly in the higher rainfall areas, is often planted in gardens. The significance of the Hausa name 'Thou refusest to come to Hausa' is not clear for the plant is common in Hausaland.

The sap is somewhat rubefacient, and on open wounds is burning, yet it is also anodynal and healing. Sap from a fresh plant or a decoction of the dried plant is used in Ivory Coast–Upper Volta in eye-instillation for various eye-affections and for headache, and in frictions for oedemas and fever (6, 14). The Akye of Ivory Coast mix the leaves with those of *Aframomum sceptrum* (Oliv. & Hanb.) K. Schum. (Zingiberaceae) and use the expressed sap in nasal instillation for headache with vertigo: the mixed leaves are also applied to the face (3). A stem-decoction is widely used for treatment of rheumatism (3, 7, 15) and in Casamance (Senegal) (12, 13) and in Gabon (17) crushed stems rubbed on to wounds immediately after their infliction give good results for cleansing and healing. The stem in decoction or the pounded fruit are commonly used in Nigeria as a cough-medicine (7, 15). In Sierra Leone the stem is mashed (16) or chewed (8) to relieve cough. The pulped-up stems taken in a little water are strongly diuretic. This is used in Ivory Coast–Upper Volta to relieve urethral discharge (14). The Ijo of S Nigeria express the sap which is taken for malaria and to clear urine (18). The deleafed and debarked stem is used in Nigeria against attacks of nausea (4) and young stems are sucked by the Efik to quench thirst (1). Leaf-sap in a root-decoction is drunk in Tanganyika for malaria (9).

The outer layer of the stems is used in Lower Dahomey to weave into small table mats, kola baskets, etc., and the plant is sometimes cultivated for this purpose (5, 7). The stem has some value as a paper material, and good quality brown and white papers have been produced (5). Children in S Nigeria use the stems for cleaning school slates (1). The sap can be used to coagulate rubber (1).

Nigerian material has been found free of alkaloids (2).

The root is used in the Benue region for cough, and a decoction of roots and epiphytic orchids has been used as a remedy for sleeping-sickness (7). In Casamance (Senegal) the roots mashed to a thick paste are applied topically to abscesses and ulcers (11, 13) and they are put into a popular medicine for constipation (13). In Nigeria a root-infusion is taken for stomach-ache and is considered to be a stimulant and an aphrodisiac (4). In Gabon the root is boiled to a pap and taken for syphilis (16), and a decoction is taken in Tanganyika for leprosy and gonorrhoea (9).

In Sierra Leone the canes are laid under bodies of the dead, and when twins are born are laid by an anthill. In Ghana the plant has many superstitious uses. It is placed on a cultivated field, or path, or entrance to a house for protection, and planted in sacred groves. It is used in religious ceremonies. Its smell is regarded as inimical to ghosts and evil influences, and it is offered to a spirit occupying a tree (7, 10).

References:

1. Adams, R. F. G., 1947. 2. Adegoke & al., 1968. 3. Adjanohoun & Aké Assi, 1972: 307. 4. Ainslie, 1937: sp. no. 115. 5. Anon., 1924: 424. 6. Bouquet & Debray, 1974: 175. 7. Dalziel,

1937. 8. Deighton 4570, K. 9. Haerdi, 1964: 194. 10. Irvine, 1930: 131–2. 11. Kerharo & Adam, 1962. 12. Kerharo & Adam, 1963, a. 13. Kerharo & Adam, 1974: 787–8. 14. Kerharo & Bouquet, 1950: 242. 15. Oliver, 1960: 23, 58. 16. Thomas, N. W. 80, K. 17. Walker, 1953, b: 319. 18. Williamson, K. 8, UCI.

**Costus dubius** (Afzel.) K. Schum.

FWTA, ed. 2, 3: 78.
West African: **SIERRA LEONE** SUSU sinkoinyi (NWT) TEMNE *a*-sul (NWT) **LIBERIA** KRU-BASA dɔ̃ (Barker) **IVORY COAST** AKYE leussin (A&AA) MANDING-MANINKA kogbing (A&AA)

A perennial semiwoody herb to 2 m high with terminal inflorescences or inflorescences on short leafless stems at the base, of rainforest, in Sierra Leone to Nigeria and extending to Gabon.

In Liberia the sap from inflorescences is used for sore eyes (2). In Liberia (2) and Ivory Coast (1) the stem is chewed for cough, and in Ivory Coast a decoction of the inflorescence with citron juice and powdered guinea-grain (*Xylopia aethiopica,* Annonaceae) is drunk when miscarriage is threatening.

References:

1. Adjanohoun & Aké Assi 1972: 308 as *C. albus* A. Chev. 2. Barker 1085, K.

**Costus englerianus** K. Schum.

FWTA, ed. 2, 3: 78.
English: dwarf forest ginger lily (Morton).
West African: **IVORY COAST** AKAN-ASANTE kmamegan (B&D)

A creeping fleshy herb forming dense ground cover in rainforest, in Ivory Coast to S Nigeria and Fernando Po, and in E Cameroun.

**Costus lucanusianus** J. Braun & K. Schum.

FWTA, ed. 2, 3: 78–79. UPWTA, ed. 1, 473.
West African: **SIERRA LEONE** MENDE howa (FWHM) **IVORY COAST** ANYI aniakoinia (K&B) zaza boto (K&B) BAULE juboma alosso (K&B) DAN nion (K&B) GAGU bikwo (K&B) KRU-GUERE zobié (K&B) zoé (K&B) GUERE (Chiehn) zazaboto (K&B) SENUFO-TAGWANA tekuato (K&B) **NIGERIA** EDO úkhúeréohā (Kennedy) IJO-IZON (Kolokuma) ògbódó, òwéi ógbódó = *male ogbodo* (KW)

*NOTE: See also C. afer.*

A semiwoody herb with stems to 3 m high bearing a terminal inflorescence, of the forest, from Guinea to W Cameroons and Fernando Po, and on to Gabon.

Taxonomic distinction, if tenable, between this and *C. afer* is slight. Medicinal usages, at least, of both tend to be more or less identical (Ivory Coast–Upper Volta, 2; Nigeria 3). The Ijo of S Nigeria while using them alike recognize this one as 'male' (*owěi ogbodo*) and *C. afer* as 'female' (*erě ogbodo*).

All parts have an acidulous taste, and the young shoots are cooked and eaten in Gabon (4, 5). A decoction of the stem is used in Ivory Coast as an eye-wash (2), while in Gabon the expressed sap is instilled into the eye for filarial infection (4, 5). The sap alone or with copper filings is used on wounds and cuts in Gabon, and raspings of the stem are applied hot to ulcers (4, 5). The plant is held to have febrifugal and analgesic properties. It is administered internally for coughs and bronchitis (Ivory Coast–Upper Volta, 2; Nigeria, 3; Gabon, 4, 5). It is considered antemetic in Ivory Coast–Upper Volta (2). It is applied in topical frictions for feverish stiffness and rheumatism (Ivory Coast, 2; Nigeria, 3). The Ijo of S Nigeria use sap expressed from the central pith of the stem for malaria and to clear urine (6).

Sap of this species and of allied species is said to have been used in S Nigeria with lime-juice to coagulate *Landolphia* latex, and in the equatorial region that of *Funtumia* (1), (both Apocynaceae).

An essential oil is present in the plant (2).

The plant is fetish and has superstitious uses to confer protection in Ivory Coast–Upper Volta to villages against epidemics and to repell evil spirits (2). The sap is used in Gabon in benediction rituals (5).

References:

1. Dalziel, 1937. 2. Kerharo & Bouquet, 1950: 242, with references. 3. Oliver, 1960: 23, 58. 4. Walker 1953, b: 319–20. 5. Walker & Sillans, 1961: 430. 6. Williamson, K. 7, UCI.

**Costus schlechteri** Winkler

FWTA, ed. 2, 3: 78.

English: hairy ginger lily (Morton).

Semi-woody with terminal inflorescence on stems to 2 m long, in Liberia to W Cameroons.

**Costus spectabilis** (Fenzl) K. Schum.

FWTA, ed. 2, 3: 78. UPWTA, ed. 1, 473.

English: dwarf savanna ginger lily (Morton).

West African: **SENEGAL** BASARI a-ngwẹ̀ ɛ-ndáng (Ferry) BEDIK go-mbẹrɛndón (Ferry) **NIGERIA** FULA-FULFULDE (Nigeria) dengere (JMD) HAUSA kunnan zoómoó = *hare's ear* (JMD; Lowe) tàabármár zoómoó = *hare's mat* (auctt.) taákálmàn zoómoó (*plur.*) = *hares' sandal* (JMD; ZOG)

Herb with rosette of a few fleshy leaves on the ground and sessile inflorescence with bright orange or yellow flowers, of rocky savanna, in Senegal to W Cameroons, and to Sudan, E Africa, Rhodesia and Angola.

The plant is ornamental but not particularly easy to cultivate. In N Nigeria the leaves are rubbed on the feet to relieve an itching condition experienced in the rainy period (? *Tinea* infection) (2). In Congo sap is instilled into the eyes to remove foreign matter, and is used to treat injury caused in, for example, epilepsy and convulsions (1).

References:

1. Bouquet, 1969: 246. 2. Dalziel, 1937.

## CRASSULACEAE

**Bryophyllum pinnatum** (Lam.) Oken

FWTA, ed. 2, 1: 116. UPWTA, ed. 1, 28 (as *B. pinnatum* S. Kurz).

English: never-die (Dalziel), resurrection plant (Dalziel), air plant (Deighton), life plant (Deighton)—from the viviparous and tenacious properties of the plant.

West African: **SIERRA LEONE** BULOM (Kim) pɔm-ta-bolɛ? (FCD) BULOM (Sherbro) bɔbɔ-lɛ (FCD) KISSI sosala (FCD) KONO g-ba-yama (FCD) k-pa-yamba (FCD) KRIO nobadai = *never die* (FCD) MENDE gbola, kpola, kpulu-la (*def.*-i) (JMD) kpoto (FCD) TEMNE (Kunike) *an*-kil-*an*-arɔŋ (FCD) TEMNE (Port Loko) *an*-ləbin-ləbin TEMNE (Sanda) *an*-ləbin-ləbin TEMNE (Yoni) *an*-kil-*an*-arɔŋ (FCD) VAI k-pokpui (FCD) **IVORY COAST** ABURE kokolé (B&D) ANYI akwolé (K&B) kpolembli (K&B) BAULE akpoléblè (B&D) KRU-BETE kpokorokpo (K&B) GUERE kpoapo (K&B) GUERE (Chiehn) kpakolé (K&B) kpotroko (B&D) kwotrotro (K&B) pakolo (B&D) KWENI lala (B&D) **GHANA** ADANGME gbɔtue nwa = *gbɔ with big ears* (FRI) ADANGME-KROBO kokonadu (Bunting; FRI) AKAN-ASANTE bosompra *when associated with the fetish grove* (Enti) bosompra-egorɔ (JMD) ɛgorɔ (FRI; JMD) FANTE egporɔ (FRI) TWI bosompra (FRI) bosompra-ɛgorɔ (FRI) ɛgorɔ (FRI) GA gbɔ̀ (FRI; KD) tamiawa (FRI) GBE-VHE aflaa (JMD) a-flaatoga (FRI) kokonadu (FRI) NZEMA aporokɛ (FRI) ɛkplokɛ (FRI) **NIGERIA** EDO danweshin (auctt.) èkpókpö (Amayo) EFIK àfiáíyọ̀ (auctt.) IJO-IZON (Kolokuma) ùmbú (kòròmọ̀) dìrì = *medicine for the navel-cord (to fall off)* (KW) YORUBA àbámodá = *what you want to do, you do* (JMD) àbámọdá = *I make a proposition* (Verger) ẹrú-ọ̀dúndún = *slaves' ọ̀dúndún* (*Kalanchoe integra*) (JMD) ọ̀dúndún (IFE) rẹ̀ẹ̀rẹ̀ (IFE)

A succulent perennating herb 0.60–1.20 m high, of Madagascan origin, but spread as an anthropogene pantropically. It is naturalized in the territories of the Region from Sierra Leone to S Nigeria, but has not penetrated into the drier states.

The plant has remarkable tenacity of life with an ability to root freely from buds lying in the serrations of the leaf-margins should they be in contact with the ground. A collection (10) in Kew Herbarium from Sierra Leone put out an axillary bud sideshoot 2 cm long while in the plant press over two months, while another (2) is annotated as having continued to grow for 2–3 weeks while in the press. This capacity has naturally attracted descriptive English names. The generic name appears also to allude to the leaf vivipary: *brúo,* from Greek, to be full of, or to burst forth, and *phúllon,* a leaf.

The plant is widely grown in the Region as an ornamental and for medicinal use. In Sierra Leone a cough-medicine is made from the roots (6, 10) and in Ivory Coast root-sap is taken by draught for attacks of epilepsy (9). In Congo leaf-sap is given to children with convulsions and epilepsy (4). The Shien of Ivory Coast take a decoction of the whole plant as a febrifuge and tranquillizer, and for refractory cough (9). In Ivory Coast too (9) and in Ghana (8) the leaves are rubbed on feverish children, and in Gabon on small children with colds (14, 15). Commonly leaves are rubbed or tied on the head for headache (6) and in Ghana they are mashed and inserted in the nostrils for this (8). Curative and analgesic action is also claimed for the sap in the eyes for eye-complaints in Ivory Coast (9), for ear-ache and ophthalmia in emergencies (11) and in Ghana (8). Instillations into nose, ears and eyes of sap, prepared preferably from leaves passed over a fire, for ear, naso-pharyngeal and eye-troubles are practised in Congo (4). Leaf-sap is used on dermal conditions in Ghana (8), for itch in Gabon (15), for inflammation of an allergic nature or fungal or eczematous infection in Congo (4). For these uses the leaf-mash alone or compounded with palm oil or djave or shea butter may be used, as also for wounds, burns, abscesses, ulcers, sores, swellings, rheumatic and intercostal pain (3, 4, 5, 6, 8, 14, 15).

Similarly abdominal poultices are used for more deep-seated intestinal pain in Ivory Coast (5) and Nigeria (3). In Ivory Coast the sap is used on cuts to stop bleeding (5).

Leaf-sap is used as a diuretic and antemetic in Ivory Coast (5, 9) and Nigeria (3). The diuretic action is said to be due to the presence of a high concentration of potassium mallate (9, 11). Concentration varies on a diurnal rhythm reaching a low point in the afternoon when the concentration of potassium iso-citrate is at a peak (13). Ascorbic and other organic acids (9) and traces of alkaloid (in Nigerian material, 1) have been detected. A common use is to dose by draught a newly delivered woman, and to squeeze some of the sap into the baby's mouth (6, 8), and in Congo it is an ingredient of a prescription to hasten expulsion of the after-birth (4). A second prescription in Congo is given to regularize the ovarian cycle (4).

Two drops of leaf-sap up the nose before going to sleep is considered in Congo to prevent snoring (4).

Leaf-sap will raise a foam suggesting the presence of saponins which however seems not to have been reported. The sap is used as a lather for shaving the head, a practice recorded in Sierra Leone (10, also 6, 11).

In Ghana the leaves are sometimes used to make spoons for feeding babies with water or other liquid, and in certain villages the leaves are specially fed to sheep (7, 8).

The peculiar vegetative properties of the plant naturally lead to it being used in superstitions and fetishism. It features in a Yoruba Odu incantation for the acquisition of money (12). In Gabon because of its ability to regenerate from a single leaf it has fetish properties to render soil fertile (15). In Congo the presence of a plant beside a house makes the intentions of all persons approaching it friendly (4).

References:

1. Adegoke & al., 1968. 2. AEK 1029, K. 3. Ainslie, 1937: sp. no. 61 as *B. calycinum*. 4. Bouquet, 1969: 100. 5. Bouquet & Debray, 1974: 77. 6. Dalziel, 1937. 7. Irvine, 1930: p. 65, as *B. pinnatum* Kurz. 8. Irvine, s.d. 9. Kerharo & Bouquet, 1950: 33, as *B. pinnatum* S. Kurz, with references. 10. Lane-Poole 159, K. 11. Oliver, 1960: 50. 12. Verger, 1967: no. 3, as *B. pinnatum* S. Kurz. 13. Vickery & Wilson, 1958: as *B. calycinum* Salisb. 14. Walker, 1953, a: 31 as *B. calycinum* Salisb. 15. Walker & Sillans, 1961: 138 as *B. pinnatum* (Lamk.) S. Kurz.

**Kalanchoe crenata** (Andr.) Haw.
**Kalanchoe integra** (Medic) O. Ktze

FWTA, ed. 2, 1: 117 (as *K. crenata* sensu Keay). UPWTA, ed. 1, 29 (as *K. crenata* sensu Dalziel).

English: never-die (*K. integra,* Rowlands).

West African: **SIERRA LEONE** MENDE gbola, kpola, kpulu-la (JMD) **IVORY COAST** VULGAR togbô (A&AA) ANYI akwolé (K&B) kpolembli (K&B) KRU-BETE kpokorokpo (K&B) GUERE kpoapo (K&B) GUERE (Chiehn) kpakolé (K&B) kwotrotro (K&B) **GHANA** ADANGME-KROBO gbɔ (FRI; JMD) AKAN-FANTE aporo (FRI; JMD) ɛgoro (JMD) TWI ɛgoro (FRI) ɛgorɔ (JMD) GA gbɔ̀ (FRI; JMD) GBE-VHE (Pecí?) aflaa (JMD) aflaa-toga (FRI; JMD) **NIGERIA** HAUSA harfifi (JRA) IGBO (Owerri) unwa (JMD) YORUBA eletí, etí: *an ear, K. integra* (JMD) ọ̀dúndún = *K. integra* (JMD; Verger)

Fleshy perennating plants to 1.5 m high, recorded from Mali to W Cameroon, and occurring widespread in tropical Africa. Two species, as named above (4), appear to be involved in *K. crenata* sensu FWTA, ed. 2, with the bulk of West African material being referred to *K. integra*. It is not possible with certainty however to separate the available information to each relevant species. It is here taken together with the species indicated if known. Probably there is no practical difference in application.

Exterior applications are the same as for *Bryophyllum pinnatum* (5, 10, 13). Warmed leaf-sap is a common treatment for ear-ache. In Ghanaian medicine the leaves are deemed to have both purifying and mitigating properties. Boiled leaves are used for asthma (*K. integra,* 9), and one leaf soaked in water when taken internally for palpitations acts as a sedative (7, 8). In Nigeria a leaf-infusion and decoction is taken as a sedative, often in asthma (2) and *ọ́dúndún* (Yoruba) features in an Odu incantation to produce calmness (12). The whole plant in decoction is administered as an anthelmintic enema; a root-decoction is taken by women in pregnancy as a tonic, and the fresh root is used as a snuff (or is chewed) to treat colds (2). In Ivory Coast the leaf-sap is used as an antemetic and to calm intercostal and intestinal pain. Leaves are put into cuts to staunch bleeding and are rubbed on feverish infants, while a leaf-decoction is given to pregnant women by enema as a tonic (3). In Tanganyika leaf-sap is drunk as a galactogogue, and root and leaf-sap as an antimalarial (6). In Zaïre, leaf-sap is used to cicatrize wounds, cure colds and eye and ear troubles (11).

A trace of alkaloid has been reported in the leaves of Nigerian material and a strong presence in the roots (1).

The plant is cultivated sometimes in villages.

References:

1. Adegoke & al., 1968. 2. Ainslie, 1937: sp. no. 198. 3. Bouquet & Debray, 1974: 77. 4. Cufodontis, 1969. 5. Dalziel, 1937. 6. Haerdi, 1964: 162. 7. Irvine, 1930: 249. 8. Irvine, s.d. 9. Johnson s.n., K. 10. Kerharo & Bouquet, 1950: 33. 11. Toussant, 1951: 565. 12. Verger, 1967: no. 153. 13. Walker & Sillans, 1961: 139.

**Kalanchoe laciniata** (Linn.) DC.

FWTA, ed. 2, 1: 117.

An erect simple herb to nearly 2 m high recorded from S Nigeria and the W Cameroons mountains. Indigenous to tropical Africa but now widespread throughout the Tropics.

The plant is decorative and worthy of horticultural cultivation.

In Central Africa the leaf is applied in frictions for the relief of itch (5). In India the leaves are considered styptic, astringent and antiseptic and to be good for cleaning ulcers and allaying inflammation, to stem bleeding on fresh cuts and abrasions, and in poultices for application to venomous insect-bites. The leaves are taken internally for diarrhoea, dysentery, lithiasis and phthisis (3, 4). In Malaya they are made into a chest-poultice for coughs and colds and into a lotion for small-pox (1). In the Philippines and in the West Indies they are said to be used for headache and in Amboina for poulticing fevered heads. The leaf is emollient and has been reported to contain a fat, a yellow organic acid, cream of tartar, calcium sulphate, tartaric acid, calcium oxalate and malic acid (2).

References:

1. Burkill, 1935: 1276–7. 2. Quisumbing, 1951: 351–2, with references. 3. Sastri, 1959: 315–6. 4. Watt, G., 1889–93: 4 (GOS-LIN): 562. 5. Watt & Breyer-Brandwijk, 1962: 324, with references.

**Kalanchoe lanceolata** (Forssk.) Pers.

FWTA, ed. 2, 1: 118.

A succulent herb to about 1.20 m height with yellow to salmon-pink flowers. Recorded only from Mali, Ghana and N Nigeria but occurring widespread in tropical Africa.

The plant is decorative and worthy of horticultural cultivation.

## CRUCIFERAE

**Anastatica hierochuntica** Linn.

English: rose of Jericho.

French: rose de Jéricho, main de Fathma (Algeria).

West African: **MAURITANIA** ARABIC (Hassaniya) kamché (AN) **MALI** TAMACHEK komecht-en-nebi (RM)

A rough squat annual with indurated incurving branches, hygroscopic and uncurling when wet, of the N and Central Sahara to Hoggar and Mauritania on the northern limit of the Region. It is extremely resistant to desiccation. It is grazed by donkeys (1).

Reference:

1. Naegelé, 1958, b: 886.

CRUCIFERAE

**Brassica campestris** Linn.

FWTA, ed. 2, 1: 97.
English: field mustard.

An annual herb to 1 m or more high, originally of the Afghanistan/NW Indian area but now a world-wide cultivated plant or weed of cultivation. It is grown here and there in the West African Region.

The leaf is eaten in many countries, though it may be reserved sometimes as an emergency food in time of shortage. It is often fed to cattle, but under certain circumstances it may be toxic due to the presence of the glycoside *sinigrin* which hydrolyses to a highly irritant volatile mustard oil. This property is used in India for treating skin-disease and snake-bite.

The plant is a very important seed-oil crop in India, and varieties are recognized. Two principal sorts of oil are produced, *sarsan* and *toria* which are used for cooking (1, 2). The oil is one of the chief sources of erucic acid, a fatty acid of the oleic acid series and which has important applications in food and industry (1, 3).

References:

1. Manjunath, 1948: 215–7, with oil analyses. 2. Watt & Breyer-Brandwijk, 1962: 325–6. 3. Zuckerman & Grace, 1949.

**Brassica integrifolia** (West) O. E. Schulz

FWTA, ed. 2: 1: 97.

A glabrous herb to 1 m high, recorded only in Ivory Coast. An enigmatic species that is perhaps but a form of the polymorphic *B. juncea,* and to which the remarks there equally apply.

**Brassica juncea** (Linn.) Coss

FWTA, ed. 2, 1: 97. UPWTA, ed. 1, 23.
English: Indian mustard, Chinese mustard, wild lettuce (Liberia, Okeke 18, K), leaf mustard (Sierra Leone, Deighton.)
West African: **GUINEA-BISSAU** MANDING-BAMBARA karadali (A. Chev.) MANINKA timbo (A. Chev.) MANINKA (Wasulunka) karadali (A. Chev.) **SIERRA LEONE** KRIO bush-kabej = *bush cabbage* (FCD) kabej-pla-sas = *cabbage palavar sauce* (FCD) TEMNE *a*-mɔthsəra (auctt.)

An erect annual herb to 1 m high, recorded from Guinea, Sierra Leone and Liberia, but it would be surprising if it was not actually more widespread in the Region. The plant is probably of African origin, but the main cultivation centres lie between eastern Europe, India and China. It is widespread in the tropics, grown for use as a vegetable or for its seed (1, 8).

It is grown in Sierra Leone and Guinea as a pot-herb and peppery condiment (2, 3, 4, 5, 6). It is cooked as a side-dish in Malawi, and for storage the leaves are sun-dried (14). In Zäire it is used as a flavouring (11), and in Livingston's Zambesi Expedition of 1867 it was found to be an ingredient of salads near Murchison Falls. Similar uses are found throughout Asia (1, 10). Twice-cooking of the leaves is often practised to rid them of the glycoside *sinigrin* (9). It is said that excessive consumption will give the skin a smell that is repellent to mosquitoes (12).

The root has been fed to cattle at Amani Research Station to promote milk-production (7), and at other places in Tanganyika (12).

In India the plant is considered to have bitter, aperient and tonic properties, and the volatile oil to be stimulant and counter-irritant (12). The sun-dried leaf and flower is smoked in Tanganyika like hemp (*Cannabis*) 'to get in touch with the spirits.' The effect is said to be weaker than that of true cannabis (12).

The seeds yield a fixed oil of 30–47% content and a volatile mustard oil of 2.89%. The former is widely used in India for cooking and to anoint the body. The volatile oil is allyl, not crotonyl, mustard oil (12). An unnamed alkaloid has been detected in the seed (13).

References:

1. Burkill, 1935:361–2. 2. Dalziel, 1937. 3. Irvine, 1948: 259–60. 4. Irvine, 1952. 5. Irvine, 1956: 36. 6. Jordan 989, K. 7. Koritschoner 811, K. 8. Manjunath, 1948: 217–8, 222–7 (oil and cake analyses). 9. Purseglove, 1962: 91–92, with leaf analysis. 10. Quisumbing, 1951: 332–4, as *B. integrifolia* (West) O. E. Schutz. 11. Robyns & Boutique, 1951: 530. 12. Watt & Breyer-Brandwijk, 1962: 326. 13. Willaman & Li, 1970. 14. Williamson, J. 1956: 26.

**Brassica napus** Linn. var. **sahariensis** A. Chev.

FWTA, ed. 2, 1: 97. UPWTA, ed. 1: 23 (under *B. juncea*).

English: rape, swede.

West African: MALI ARABIC (Upper Niger) left (A. Chev.) left wafran (A. Chev.) SONGHAI lifiti (A. Chev.) TAMACHEK afran (A. Chev.)

A herbaceous plant, cultivated in gardens in Mali and in oases of the Sahara. The species is of Mediterranean origin and has been in cultivation from very early times. Varieties are numerous. In Europe forms with a caulescent root, the annual is summer rape, the biennial is winter rape. A form where the root swells into a tuberous shape is the swede. It is the swede form that is recorded in Mali and the Sahara (Chevalier fide 1) where it is grown as a vegetable.

Rape is not much grown in the tropics but it is grown in the plains of India in the cold season to provide a highly-prized vegetable (2), which is rich in ascorbic acid (3). The plant is grown in Japan as a valuable edible oil seed crop and is the source of Colza oil, or rape oil and rape cake in South Africa. Though the cake is normally fed freely to stock, it does contain some sinigrin-like substance which on hydrolysis yields an irritant sulphur-containing oil. Poisoning of stock can occur, the symptoms being acute or haemorrhagic gastroenteritis leading to death. The risk is reduced if the ferment responsible is killed by treating the broken cake in boiling water. Rape oil is a fixed oil and the content of the seeds is 30–45%. It is edible if cold-pressed. Mustard oil which is a volatile oil can also be obtained from the seed. In Eastern Europe the early sprouts are used dietetically and used as a seasoning. The plant has shown antibiotic activity. Isothiocyanates, known to be biotically active, have been identified in the volatile seed oil. Many other chemical substances are present (4).

The entries under *B. campestris* Linn. in general apply to *B. napus* Linn.

References:

1. Dalziel, 1937. 2. Manjunath, 1948: 218, 222–7, with oil and cake analyses and many references. 3. Purseglove, 1968: 92, 98, with leaf analysis. 4. Watt & Breyer-Brandwijk, 1962: 325–7.

**Brassica oleracea** Linn.

FWTA, ed. 2, 1: 97. UPWTA, ed. 1, 23.

English: sea cabbage, wild cabbage, and many cultivar names.

West African: MALI ARABIC (Upper Niger) korub (A. Chev.) SONGHAI kurumbu (A. Chev.) NIGERIA BEROM kábɛ̀j *from English* (LB) EFIK éféré ḿbàkárá = *soup of the European; cabbage sp. indet. but correctly Portulaca oleracea* (*Portulacaceae*) (Lowe)

A herbaceous caulescent of many wild forms in the Mediterranean region and SW Europe where it was brought into cultivation 4–5 millenia ago and which have given rise to the garden cabbage, kale, broccoli, cauliflower and brussels sprouts. The cabbage form, var. *capitata* Linn., is cultivated in gardens and oases of Mali and the Sahara and here and there throughout the Region. Cultivars are available for successful cultivation at higher elevations or in cold seasons of the tropics, and a few with careful husbandry can succeed in the lowlands.

The leaf contains a fair amount of vitamins A and the B complex. A vitamin D-like substance is produced in the leaf-extract after irradiation. Seed-extracts have given positive antibiotic tests but variably so between cultivars (1, 2).

A volatile oil which is not mustard oil has been found in the seeds on maceration in water. This oil has antibacterial and antifungal properties (2).

References:

1. Purseglove, 1968: 94–95, 98, with leaf analysis. 2. Watt & Breyer-Brandwijk, 1962: 328.

**Brassica rapa** Linn.

FWTA, ed. 2, 1: 97.

English: turnip.

French: newet.

A biennial with a swollen tuberous taproot. Native of south and central Europe and in cultivation by man for over 4 millenia resulting in worldwide dispersal and the selection of numerous variants. The plant can be successfully grown in the tropics and is found cultivated in the Region for its edible root and foliage (1).

The plant contains *allylthiocarbimide* and the enzyme *myrosin,* but no sinigrin. Excessive consumption may cause gastro-intestinal irritation in humans and in stock. The seed yields 30–40% of a fixed oil which is commercially classified as Rape oil or Colza oil. If the seed is pressed cold the crude oil is edible. Aqueous leaf and root extracts are biotically active inhibiting *Escherischia coli,* but not against *Staphylococcus aureus.* Root peelings yield an oily yellow substance, called *rapine,* which at 1: 100,000 concentration inhibits growth of bacteria, fungi, yeasts and other parasites of man and animals (2).

References:

1. Purseglove, 1968: 95–96, with leaf analysis. 2. Watt & Breyer-Brandwijk, 1962: 329.

**Capsella bursa-pastoris** (Linn.) Medik.

English: shepherd's purse.

French: bourse à pasteur.

Portuguese: bolsa-de-pastor.

An annual or biennial herb to about 20–25 cm high, ruderal, very polymorphic of Europe and N Africa. It has been introduced to Dakar, Senegal and is established in montane areas in tropical Africa. A cosmopolitan.

The plant has a number of medical uses. In Europe it has been used for menorrhagia and malaria; mixed with normal diet to inhibit the oestrous cycle and in a hot infusion as an emmenagogue, and a diaphoretic. The plant has a bitter taste and has been used as a tonic, antiscorbutic, astringent and diuretic. In India the plant is considered antiblennorrhagic and a remedy for atrophy of the limbs.

Experimental work has shown the plant to have a stimulating action on the uterine muscles and to lower blood pressure, this latter being ascribed to *acetylcholine* found present. A weak anti-pyretic action is also recorded (2).

The literature shows many records of use in cancerous conditions of the uterus and stomach and for undefined cancer (1).

The alkaloid, *choline,* is reported in the aerial part of the plant, and also other alkaloids (3).

References:

1. Hartwell, 1969: 83. 2. Watt & Breyer-Brandwijk, 1962: 329, with many references. 3. Willaman & Li, 1970: with references.

**Cardamine hirsuta** Linn.

FWTA, ed. 2, 1: 98.

English: hairy bittercress.

West African: **WEST CAMEROONS** VULGAR m'p' fogo (JRA)

A small, much branched or basally tufted annual attaining 10–30 cm height. The specific name is misleading for the plant is glabrous or nearly so. It has been recorded in West Cameroons and Fernando Po above 2500 m altitude on humid newly exposed ground. This is a plant of temperate conditions, common in Europe and occurring in Africa in montane localities.

The leaves are boiled up into a 'soup' which is said to have stomachic properties in West Cameroons (1).

Reference:

1. Ainslie, J. R. 1937: sp. no. 74.

**Crambe abyssinica** Hochst.

A herbaceous annual of E and NE Africa, which like other Cruciferae has oily seed. The first commercial exploitation was in Russia in 1932. Trials are now being conducted in N Nigeria. Oil content of the seeds is 25–40%, protein 20–29%. The oil is semidrying, non-viscous, transparent, and agreeable to taste. It is very similar to rape oil and contains erucic acid 57–60%, oleic acid 20–21%, linoleic acid 6–9%, sicosenoic acid 3.5%, linolenic acid 2.6–5.6%, palmitic acid 2.0–2.8%, and others in trace. It is suitable for soap and margarine manufacture and in other foods, as a mineral lubricant additive, in manufacture of greases and as a mould lubricant in steel castings. (1).

The meal contains 46–48% crude protein and is of good nutritional quality and a potential animal foodstuff. There is however 8–10% of thioglycosides containing certain growth inhibitors and this may present some objections (1).

An unnamed alkaloid has been reported present in the seed (2).

References:

1. Cornelius & Simmons, 1969: with many references. 2. Willaman & Li, 1970.

**Crambe kilimandscharica** O. E. Schulz

FWTA, ed. 2, 1: 758.

An annual herb reaching to 1 m high, recorded at Aburi, Ghana—perhaps introduced, and occurring in Zaïre, and E Africa.

In Ruanda, children are reported to eat the fruits (1).

Reference:

1. Robyns & Boutique, 1951: 527–8.

**Crambe kralikii** Coss

A vigorous shrubby plant, endemic in the Sahara, and recorded from Hoggar just north of the Region.

The plant is edible to man and provides an excellent fodder for camels and other stock (1).

Reference:

1. Maire, 1933: 112–3.

**Diplotaxis acris** (Forssk.) Boiss.

An annual herb of sandy and rocky places of the central Sahara southwards to the Hoggar where it provides good fodder for camels and other stock (1), and along the northernmost border of the Region.

Reference:

1. Maire, 1933: 107.

**Diplotaxis virgata** (Cav.) DC.

West African: MAURITANIA ARABIC (Hassaniya) awinar (AN)

A leafy annual occurring throughout Algeria and Mauritania along the northernmost limit of the Region (1) and probably provides grazing for animals.

Reference:

1. Naegelé, 1958, b: 886.

**Eremobium aegyptiacum** (Spreng.) Asch.

FWTA, ed. 2, 1: 96.

West African: **MAURITANIA** ARABIC (Hassaniya) lehmé (AN) **MALI** TAMACHEK elmarujet, telmarujet *general* (A. Chev.) hema (RM)

An annual or perennial herb with woody base, prostrate or to 30 cm high, of sandy places, recorded from Mauritania to Niger and across the Sahara and North Africa to the Near East and Arabia.

The plant provides at Hoggar a fodder that is relished by camels and stock (1).

Reference:

1. Maire, 1933: 105–6.

**Eruca sativa** Mill.

English: rocket; taramira (India).

French: roquette.

A herbaceous annual to about 1 m height, native of Central and Southern Europe, northern Africa and eastwards into Asia. There are subspecies and selected cultivars taken to many countries where escapes have become naturalized. The north African distribution spreads southward to the southern limit of the Sahara and the northern border of the Region.

The plant was originally cultivated for its edible young leaves which were considered a palatable addition to salads, and to be diuretic and stomachic. The Touareg of Hoggar eat the leaves either fresh or cooked, and there the plant furnishes good palatable fodder for camels and other stock (1). The leaves also have pharmaceutical use as a rubefacient on skin.

The seeds are of commercial importance being oil-bearing, yielding on expression about 32% of a pungent semi-drying oil known as Jamba oil (3) or Taramira oil (2) which is used in India for making pickles. The oil initially is acrid but after six months loses its acridity when it can be used as an ordinary cooking oil along with other Brassica oils. It may appear as an adulterant in rape oil. It is usable for lubrication having in some respects a superiority over castor-oil or just as an additive. It is also an illuminant oil in India burning, it is said, with a more luminous and a less sooty flame than Brassica oils.

Taramira oil contains about 1% of a volatile oil which can be separated by steam-distillation, or by steam-distillation of the seeds crushed in water. In the latter instance the yield can be increased slightly by leaving the crushed seeds to stand for a while to allow the action of the enzyme myrosin on the glycosides to proceed further.

The residual cake is a good cattle feed, and it is said that cattle fed on it are immune from tick attack.

The plant is quite extensively grown in northern India as a cold weather crop, even in areas under drought conditions. It appears to be a potentially promising crop for the drier parts of the West African Region and so merits investigation.

References:

1. Maire, 1933: 107–8. 2. Sastri, 1952: 190–2, with many references. 3. Usher, 1974: 234.

**Farsetia aegyptiaca** Turr.

UPWTA, ed. 1, 26.

West African: **MALI** TAMACHEK uar ames (RM) urtemess (B&T)

A herb to 60 cm high with woody rootstock throughout the desert region. It has been recorded at Hoggar 20°N just outside the W African Region.

It is relished as fodder by camels and other stock (1).

Reference:

1. Maire, 1933: 103.

**Farsetia hamiltonii** Royle

A bushy perennial with upright branches of the Sahara southwards to the northern limit of the Region.

At Hoggar it is relished as fodder by camels and other stock (1).

Reference:

1. Maire, 1933: 105.

**Farsetia ramosissima** Hochst.

FWTA, ed. 2, 1: 97. UPWTA, ed. 1, 26.

West African: **MALI** TAMACHEK elmarujet (A. Chev.) telmarujet (A. Chev.) tozokamit (A. Chev.)

A twiggy annual herb recorded in seasonally dry stream beds (*oueds*) and hillsides of the sahel zone of Senegal to Niger and at Lake Chad in NE Nigeria, and extending across Africa to Sudan.

The plant is recorded as providing good fodder for cattle to Senegal (1) and for camels and other stock in Hoggar (3).

The Moors in the Sahara make an infusion of it in milk as an effective treatment for colds (Chevalier fide 2).

References:

1. Adam, 1966, a. 2. Dalziel, 1937. 3. Maire, 1933: 103–4.

**Farsetia stenoptera** Hochst.

FWTA, ed. 2, 1: 97.

An erect herb of the Sahel recorded from Senegal, Mali and Niger and extending across Africa to Sudan.

The plant provides very good grazing for camels in Sudan (2). Cattle in Senegal do not take it (1).

References:

1. Adam, 1966, a. 2. Harrison 491, K.

**Lepidium sativum** Linn.

FWTA, ed. 2, 1: 98. UPWTA, ed. 1, 24.

English: common cress, garden cress.

French: cresson alénois.

Portuguese: mastruço.

West African: MALI ARABIC hab-el-rechad (A. Chev.) **NIGERIA** BEROM làbsúr (LB) FULA-FULFULDE (Nigeria) takere foondu (?), foondu: *dove* (JMD; MM) HAUSA algarif, algarup *the seeds* (JMD) algarup ja *the seeds of a red variety* (JMD) labsûr, làbsûr, lafsur, lausur, laùsúr, laùsúrù (auctt.) HAUSA (West) làfusûr (ZOG) zamantarori, zumantarori *the seeds of a red variety* (JMD) KANURI algarabbu *the seeds* làpsûr (JMD)

An erect annual to about 20–30 cm high, very variable. The plant is originally of Levantine origin. It has been cultivated from early times and is frequently mentioned in Greek and Latin works. It was grown in England as early as 1548 and is mentioned by the herbalist John Gerard as being eaten with bread and butter or with other salad plants. The wild form, var. *silvestre* Thell. occurs from Sudan to the Himalayas. The cultivated form, var. *vulgare* Spenn. is the one occurring in West Africa where it is commonly cultivated and in some places has escaped to become naturalized in the north of the Region in Mali, Guinea, Niger and N Nigeria.

Three varieties of seed are commonly sold in northern markets. These are red, black and white. The seed becomes mucilaginous when soaked in water and the seed is sold in small cohesive ball-like masses. It is commonly cultivated in Ethiopia and is generally on sale in markets there (3). The mucilage can be used as a substitute for tragacanth and gum-arabic and consists of a mixture of cellulose (18.3% in Indian material) and uronic acid-containing polysaccharides (5).

The plant is grown in N Nigeria, e.g. Abinsi and vicinity (Dalziel fide 4)—along with onions under irrigation and is sold as *laussur* in Kano market (1). The plant is pungent becoming too hot for eating with age, but it can be taken while the shoots are still young and tender. The shoots are used medicinally as an irritant or counter-irritant, and they contain relatively high amounts of iodine, iron, phosphates and potash. The bruised leaves are strongly inhibiting to growth of *Bacillus subtilis* and *Staphylococcus aureus,* and less markedly towards *Escherichia coli* (6). The plant has been used as a fish-poison (6).

The seeds are slightly pungent. They are given for diarrhoea and dysentery, and are considered tonic, aphrodisiac and diuretic. Crushed and mixed with water they are taken in draught for rheumatic pains, swellings, etc., or they are made into liniments, and also as a dressing for sores on camels and horses. In N Nigeria the red variety is favoured for this (2). In Ethiopia the seeds have several medical uses: as a livestock drench for stomach-disorders, on human skin-disorders, chapped lips and sunburn, for amoebic infection, as an insect-repellant, pasted on the skin soldiers use it to engender a feeling of warmth at night, for stomach cramps, and in a preparation called *fetto fitfit* to increase the appetite—but what is sauce for the gourmet is sauce for the glutton too!—it is taken to relieve pains caused by over-eating. A further use is to prepare a solution [part of the plant is not specified] to massage infected cows' udders (3).

The seeds contain a fixed fatty oil of 50–60% content, used in India as an illuminant and in soap manufacture (5). The enzyme *myrosin* is also present, and in contact with water a volatile oil, cress oil, is produced in small quantity. This contains *phenylacetonitrile* and *benzylisothiocyanate*. The latter has long been known for its fungistatic and bacteristatic properties and at a concentration of 0.001–0.002% can pass unchanged through the human body (5, 6).

References:

1. Dalziel 333, K. 2. Dalziel, 1937. 3. Getahun, 1975. 4. Irvine, s.d. 5. Sastri, 1962: 71–73. 6. Watt & Breyer-Brandwijk, 1962: 332.

**Morettia canescens** Boiss.

UPWTA, ed. 1, 24.

West African: **MALI** TAMACHEK aslar (RM) asselar (B&T) jelghum (RM)

A prostrate perennial of the Sahara extending southwards to the northernmost limit of the West African Region. It is very common in clay areas of Mauritania during the rainy season and provides good grazing for all stock in Mauritania (1) and Hoggar (2). Its nutritive value is not high (1).

References:

1. Adam, 1966, b: 338. 2. Maire, 1933: 102.

**Moricandia arvensis** (Linn.) DC.

West African: **MALI** TAMACHEK krenb (RM)

An annual, biennial, or longer-lived plant of the Sahara extending southwards to the northern limit of the West African Region.

In Hoggar it provides excellent fodder for camels, but it is less attractive to other stock (1).

Reference:

1. Maire, 1933: 109.

**Raphanus sativus** Linn.

UPWTA, ed. 1, 24.

English: radish, Chinese radish, Japanese radish, oriental radish.

French: radis.

West African: **MALI** ARABIC (Sahara) bukir (A. Chev.) bu-tum (A. Chev.) **NIGERIA** ARABIC figl (JMD) HAUSA fijil (JMD)

An annual or biennial herb with swollen rootstock and leafy crown attaining about 20 cm height, of very many forms. It is thought to be of European origin but perhaps also of Asia Minor. Though a cool climate of about 15°C is optimal, extensive selection has now made it possible for the radish to be cultivated under much higher temperatures so that it is grown throughout the sub-tropics and tropics. Its cultivation has for long been known in the oases of the Sahara and in Mali and it should be possible at most places in the Region.

The plant is grown mainly for its root which is eaten raw. Some Asian forms are cooked. The leaves may also be eaten and are said to be highly nutritious. Sometimes the plant is fed to stock.

The plant has the characteristic pungency of the Cruciferae and Capparaceae due to the presence of a glycoside and an enzyme interacting to produce isothiocyanates.

The seed produces an oil similar to colza oil and is 30–50% of the seed weight. Oleic acid accounts for 60% of the oil, erucic acid 22% with smaller quantities of other fatty acids. The seeds are used medicinally in India and China. Their action is diuretic, laxative stomachic and expectorant.

References:

1. Burkill, 1935: 1866–7. 2. Kay, 1973: 127.

**Rorippa humifusa** (Guill. & Perr.) Hiern.

FWTA, ed. 2, 1: 97.

English: watercress, African cress, Nigerian watercress.

French: cresson du Sénégal (Berhaut), cresson sauvage (Gabon, Walker & Sillans).

West African: NIGERIA YORUBA omisuru (JRA)

A low annual herb with much reduced stem, in damp positions on stream-banks and sand-bars, often ruderal, recorded in Senegal, Ivory Coast, Ghana and N and S Nigeria, and also in Zäire, Angola, Uganda and Madagascar.

The plant is eaten as a salad in Nigeria (1), Gabon (3) and Zäire (2). It is said to be cultivated in Gabon, and also to make a good grazing for sheep (3). The leaves have antiscorbutic properties and are considered in Nigeria to purify the blood (1).

References:

1. Ainslie, 1937: sp. no. 242, as *Nasturtium humifusum*. 2. Robyns & Boutique, 1951: 534. 3. Walker & Sillans, 1961: 139, as *Nasturtium humifusum* Guill. & Perr.

**Rorippa nasturtium-aquaticum** (Linn.) Hayek

FWTA, ed. 2, 1: 97.

English: watercress.

French: cresson.

West African: SIERRA LEONE KRIO wata-krɛs = *water-cress* (FCD)

A much branched perennial herb, procumbent with stoloniferous stems to 30–60 cm long rooting at the nodes forming extended colonies usually in running water. This is the common watercress of Europe and has been introduced to West Africa where it has become naturalized in a few localities in Mali, Niger and on the Cameroon Mountain of W Cameroons. It has been naturalized at a number of other localities in Africa, usually at the altitudes ± 2000 m.

No use is recorded for it in the Region, but it must surely be eaten. It is recorded as eaten raw as a relish and cooked as a pot-herb in Lesotho (1).

Reference:

1. Guillarmod, 1971: 449.

**Savignya parviflora** (Del.) Webb

An annual occurring thoughout the northern Sahara and southwards to the northern limit of the West African Region, providing in season desert pasturage. In Hoggar it is much appreciated by camels and all domestic stock (1).

Reference:

1. Maire, 1933: 108–9.

**Schouwia purpurea** (Forssk.) Schweinf.

West African: **MALI** TAMACHEK djirdjir (RM)

An annual herb of the Sahel towards the boundary of the Sahara, common, appearing after the rains in Mauritania and the Central Sahara.

The Touareg eat the leaves either cooked or dried without salt. The young leaves add an agreeable flavour to a salad similar to that of rocket, *Eruca sativa* Mill. (Cruciferae) (2). The plant is relished by camels either green or dry (1) but less so by other stock (2). It has no outstanding nutritional properties (1).

References:

1. Adam, 1968, b: 342. 2. Maire, 1933: 111.

**Schouwia schimperi** Jaub. & Spach

FWTA, ed. 2, 1: 98. UPWTA, ed. 1, 24 (as *S. arabica* DC.)
West African: **MALI** TAMACHEK alwas (Foureau) alwat (auctt.) djirdjir (B&T)

A woody annual to 1 m high, common in the Sahel and Saharan oases, recorded from Mauritania and Central Sahara, and extending to Sudan, Egypt, Ethiopia and Arabia.

The plant is much appreciated by camels when fresh and appears to reduce the camel's need for water. The dry plant serves as fuel (1).

Reference:

1. Dalziel, 1937.

**Sisymbrium reboudianum** Verlot

A herbaceous plant of North Africa spreading southwards through the Sahara to the northern limit of the West African Region. In Hoggar it often forms pure stands in the valleys after the rains when it provides grazing for camels, sheep and goats (1).

Reference:

1. Maire, 1933: 106.

CRUCIFERAE

**Zilla spinosa** (Turra) Prantl

West African: **MAURITANIA** ARABIC Hassaniya achenfarach *probably this sp.* (AN) **MALI** TAMACHEK chebreg (RM)

A spiny shrub of N Africa occurring over the Sahara southwards into Mauritania and the Hoggar area to the northern limit of the West African Region. It provides excellent fodder for camels and other stock (1).


Reference:

1. Maire, 1933: 112, as *Z. spinosa* and *Z. macroptera.*


## CTENOLOPHONACEAE

**Ctenolophon englerianus** Mildbr.

FWTA, ed. 2, 1: 357. UPWTA, ed. 1. 38.

West African: **NIGERIA** IJO me (KO&S)

A tree to 26 m high of freshwater swamp and riparian forest occurring in S Nigeria and on into Cameroun, Gabon and Angola.

Sapwood is dirty white. Sap and heart-wood is very hard and durable. The Ijo (2) and Eket (1) peoples use the wood to make canoes and to provide durable, termite-proof uprights for house-building and other construction work.


References:

1. Amachi FHI 24306, K. 2. Kennedy, 1574, K.


## CUCURBITACEAE

**Bambekea racemosa** Cogn.

FWTA, ed. 2, 1: 208.

A climbing herb, of forest clearings, recorded from S Nigeria and W Cameroons, and from E Cameroun to the Congo.

Pulp from the leafy twigs crushed up with sap is applied as a poultice to the throat for throat and ear-affections in Congo (1).


Reference:

1. Bouquet, 1969: 101.


**Citrullus colocynthis** (Linn.) Schrad.

FWTA, ed. 2, 1: 213, as *Colocynthis vulgaris* Schrad. UPWTA, ed. 1. 53.

English: colocynth, bitter gourd, wild gourd.

French: coloquinte.

West African: **MAURITANIA** ARABIC (Hassaniya) aferziz (AN) hadje *a single fruit* (AN) ihadej lehmar *the fruit* (AN) ilif lehmar *the plant* (AN) **SENEGAL** BASARI ɔ-syẹlis ɔr ɛ-peúkèl, a-tyẹlis and ɛ-peúkèl (Ferry) DIOLA (Fogny) sisigi *var. with bitter flesh* (JB; AS, ex K&A); = *sheep* (DF, The Gambia) MANDING-BAMBARA sara (JB; K&A) sora (K&A) MANDINKA sara (K&A) SERER béref a koy (JB; K&A) ségel ngoy (JB; K&A) SERER-NON (Nyominka) idêg (K&A) SONINKE-SARAKOLE ɗaridôdônané (K&A) WOLOF hal u buki = *melon of the wolf* (auctt.) yomba mbot, yombebut (K&A) **GUINEA** FULA-PULAAR (Guinea) koron mboddi

= *snake's gourds* (JMD) SUSU séréré (CHOP) **MALI** ARABIC hadj (RM; JMD) DOGON náá gabá jɔ̀gɔ (C-G) TAMACHEK alkat (JMD) alked (JMD) halkat (JMD) tagalel (A. Chev.) tagellit (RM) tedjellet (Foureau) **NIGERIA** ARABIC handal (JMD) hanzal (JMD) HAUSA kwartowa, kwattowa *the pulp* (JMD; ZOG) IGBO egwusi, elili (BNO)

A perennial trailing or climbing herb of desert and semi-desert areas of the northern limit of the Region in Senegal and Mali and cultivated here and there. The plant is distributed across the Sahara from the Atlantic Islands to Kenya and into western Asia. It favours drier conditions than *C. lanatus* (Thunb.) Mansf.

The vegetative parts are browsed by donkeys and goats in Senegal (1), and are said to be taken by wild game, but the fruit only by donkeys (3), gazelles and ostriches (7). The whole of the plant is very bitter, but especially the fruit due to the presence of a bitter glycosidal principle, *colocynthin*. In the past the purgative action has been ascribed to this, but another intensely bitter weakly basic amorphous alkaloid has been detected and this is drastically purgative. Another less powerful alkaloid has also been found and also other substances (5, 9, 12). Colocynth of commerce is official in most pharmacopoeias as a purgative and has been used for dropsical and other conditions, and sometimes as a vermifuge. It is the dried pulp of the peeled fruit freed of seeds. The powder is irritating to mucous membranes, strongly hydragogue, cathartic and is usually administered mixed with other drugs on account of its griping action (9). In the British Army of World War I, 'No. 9', given for so many complaints on medical inspection parades, was based on this substance. In small dosage it is violent. In larger doses it is lethal. Its use as a drug was first recorded in Rome during the reign of Emperor Claudius, A.D. 41–54, whereupon, in a craze of hypochondria by the well-to-do Roman citizen, it attained much popularity—even for political murder, and perhaps of Claudius himself. Bulk to line the stomach wall is evidently able to reduce its action, and it is possible that the dish of pottage which was inedible to the multitude in the Bible, II Kings 4: 38–41 because, perhaps unwittingly, of the presence of colocynth fruit was miraculously made edible by Elisha who added meal to it (11).

It is used as a purge for man and animals in Mauritania (8), but recognition of its violence and toxicity in Senegal (4), Nigeria (2), Gabon (10), and in general appears to have very greatly restricted its use in this way. In Mauritania the baked fruit is rubbed on camels affected by itch, and an unripe fruit cooked in hot sand is used to treat blennorrhagia (? gonorrhoea) in man: the cooked fruit is bored centrally and the glans of the penis is inserted and kept there for an hour in which time a cure is said to be effected (8). Arabs use the ripe fruit, charred in a fire and pulverized, to prepare a gunpowder, tinder and fuses, and in Egypt, the green fruit is crushed on a piece of flannel fabric which absorbs the juice and when dried it acts like tinder. Fruits are supposed to have the property or keeping moths, etc. away from clothing, and pieces of the dried fruit (sometimes along with black pepper) pulverised and wrapped in paper are used for this purpose (3). In Mauritania the root is held to be effective in killing headlice: the root cooked in hot sand is powdered and mixed with butter from ewe's or camel's milk for application to the infested head of women (8).

A bitter black extract prepared from the rind is sometimes smeared on water-bags by the Arabs to keep camels, etc. away from them (3). The fruit pulp is also used (12).

Though the fruit is poisonous, the seeds are sometimes eaten. In the Hoggar area the Tebbou boil them for a whole day with a change of water. The seeds are then dried and eaten. The Tuareg of the same area steam the seeds to drive off a black oil after which the seeds can be eaten or dried for storage (6). In Sudan the seed is eaten as an emergency food (3, 12). The seed-oil which resembles pumpkin seed-oil is used in India as a remedy for snake-bite, scorpion-sting, epilepsy and to promote hair-growth. Content is about 21% and it is semi-drying. It is usable for illumination and as a dye to darken grey hair (12). The Tuareg of Hoggar treat skin-troubles on camels with it (6).

References:

1. Adam, 1966, a: as *Colocynthis vulgaris*. 2. Ainslie, 1937: sp. no. 95, as *Citrullus vulgaris*. 3. Dalziel, 1937. 4. Kerharo, 1967: as *Cucumis colocynthis*. 5. Kerharo & Adam, 1974: 375–7, as *Colocynthis vulgaris* Schrad., with references and phytochemistry. 6. Maire, 1933: 199–200, as *Colocynthis vulgaris* Schrad. 7. Monteil, 1953: 135, as *Colocynthis vulgaris*. 8. Naegelé, 1958, b: 888, as *Colocynthis vulgaris* Schrad. 9. Oliver, 1960: 6, 56, as *Colocynthis vulgaris*. 10. Walker & Sillans, 1961: 139. 11. Wasson & Wasson, 1957: 61–3, as *Colocynthis*. 12. Watt & Breyer-Brandwijk, 1962: 349–50, as *Colocynthis vulgaris* Schrad.

## **Citrullus lanatus** (Thunb.) Mansf.

FWTA, ed. 2, 1: 213, as *Colocynthis citrullus* (Linn.) O. Ktze. UPWTA, ed. 1, 54–55, as *Citrullus vulgaris* Schrad.

English: water melon.

French: pastèque; melon d'eau.

West African: **MAURITANIA** ARABIC (Hassaniya) fundi (Kesby) **SENEGAL** SERER béref (JB; K&A) ségal (JB; K&A) WOLOF béref *sweet melon* (auctt.) hal *melon* (JB) hatar *the plant* (FB; JMD) xal *sweet-fruited forms* (K&A) **THE GAMBIA** MANDING-MANDINKA saura (Ingram; JMD) **GUINEA** MANDING-MANINKA sara (FB; CHOP) séré (CHOP) **SIERRA LEONE** BULOM (Kim) bɔhrɔ (FCD) yɛntɛ (FCD) BULOM (Sherbro) cham-dɛ (FCD) sãhũ-bombom-dɛ (FCD) sãŋga-lɛ (FCD) FULA-PULAAR (Sierra Leone) budi (FCD) GOLA legiɛ-kwi (FCD) KISSI leŋguõ (FCD) KONO kɔ-saa (FCD) saa (FCD) KRIO bara-ɛgusi (FCD) e-gbara (FCD) ɛgusi (FCD) MANDING-MANDINKA sara (FCD) sarɛ (FCD) MANINKA (Koranko) sara-kumba (FCD) MENDE koja (FCD) pu-goja (FCD) SUSU-DYALONKE sara-na (FCD) TEMNE *an*-tent *the plant* (FCD; JMD) *ma*-tent *the fruit* (JMD; FCD) *ma*-tent-*ma*-potho *the fruit* (FCD) VAI sa (FCD) sara (FCD) **MALI** ARABIC hadj en-nas (A. Chev.) MANDING-BAMBARA missi tsara *the wild form:* missi: *ox* (Vuillet) tsara *the cultivated form* (FB; JMD) SONGHAI kànéỳ molli *the wild form* TAMACHEK fundi *the fruit* (A. Chev.) ilif *the plant* (A. Chev.) takinkeint, talejest, tiledjest (A. Chev.) **GHANA** AKAN-FANTE akate (FRI; JMD) akatsewa *the seeds* (FRI; JMD) akyɛkyẽa (FRI; JMD) anamuna, anemuna *var. with black inedible seeds* (FRI; JMD) TWI akat(s)ewa *the seeds* (FRI; JMD) akyẽkyẽa (FRI; JMD) anemuna *the fruit* (FRI, JMD) ɛfɛrɛ (FRI; JMD) DAGBANI neri (FRI) GA agúshi (*pl*.-i) *the seed* (KD) wátrɛ̀ (FRI) GBE-VHE atsẽtsẽa (FRI) VHE (Awlan) dzamatre VHE (Pecí) aŋyaŋye HAUSA àgúshií (FRI) KONKOMBA inabe (Gaisser) NZEMA akitsiwa = *niri oil* (auctt.) NZEMA (Atwabo) nfatee nzima (Chakravarty) **TOGO** BASSARI inam (Gaisser) KABRE kamie (Gaisser; FB) KONKOMBA inabe (FB) MOORE-NAWDAM fude (Gaisser) TEM kanjinga (FB) TEM (Tshaudjo) kanjinga (Gaisser) **NIGER** SONGHAI kànéỳ (*def*. -à) (FB; D&C) **NIGERIA** ANAANG íkpán (KW) ARABIC-SHUWA batteikh al hamdal *sour form* (JMD) batteikh al masak *sweet form* (JMD) BEROM cɛ́s (LB) EDO ikpogi *the seeds* (JMD) ògì (JRA; JMD) EFIK íkòn *the plant* (auctt.) íkpán *the gourd with its seeds* (KW) FULA-FULFULDE (Adamawa) cikilje pahe *a varietal name* FULFULDE (Nigeria) cikirre (*pl*. chiklje) (JMD; Taylor) dende (JMD) dene (*pl*. denaje) (JMD; Taylor) GWARI mimihi (Edgar) HAUSA àgúshií *from Yoruba* (JMD: ZOG) bambus *a varietal name* (JMD: ZOG) gúna (JMD: LB) IBIBIO íkôn (KW) íkpán *the dried and crushed seeds* (KW) IGBO ègúsí (AJC) élìlì (AJC) nwanru (BNO) ogili-irele (Onochie) IGBO (Asaba) ògìlì *the seeds* (JMD) ògìlì-élìlì = *seeds of melon* (JMD) IGBO (Awka) nkbuluko (JMD) IGBO (Owerri) ègúsí (KW) élìlè, élìlì (KW) KANURI bàmbúsə̀ (auctt.) fálí, gúnà (C&H) NUPE epín, epíngi *the seeds* (Banfield) pàràgi (Banfield) TIV icegher (JMD) URHOBO elebo (DA) ikpógrì *the seeds* (Faraclas) YORUBA bàrà *the plant, or the fruit* (JRA; JMD) ẹ̀gúsí (auctt.) ẹ̀gúsí agbè = *water melon* (Verger) ẹ̀gúsí bora (Johnston) YORUBA (Ijesha) egúnsi (JMD) ẹ̀gúsí maga *a large variety* (JMD) YORUBA (Oshogbo) ṣọ̀fin (JMD) **WEST CAMEROONS** BAFOK esaka bawu (DA) KPE ngondo (DA) KUNDU esaka (DA) LONG esaka (DA) LUNDU esaka (DA) MBONGE osaka (DA) TANGA esaka (DA) WOVEA ngondo (DA)

An annual climbing or trailing herb to 3 m long, of grassy savanna and bush-savanna, occurring as an introduced cultivated plant throughout the West African region. The wild plant must originally have come from NE Africa and passed into cultivation in the Near East, Middle East and Western Asia in prehistorical time. It favours a dry climate and is mainly a dry season crop in monsoon areas, requiring only limited rainfall.

In some sources of information there is confusion between this species and *C. colocynthis* (Linn.) Schrad., compounded perhaps by the similarity in synonymy. Vernacular names may be mixed. *C. lanatus* is the edible water-melon and is an annual plant. *C. colocynthis* is the poisonous bitter colocynth and is a perennial.

The fruit is variable in size from about 7 cm in diameter to over 20 cm, in shape from round to long-marrow, in outside coloration, the usual pattern being a variegation of green stripes, and inside the flesh may be red or white and the seeds black, red or pale coloured. The flesh amounts to about 65% of the whole fruit, and of this 95% is water. Indeed the fruits in near-desert areas where the plant has run wild constitute a valuable source of water. The plant has become naturalized in many drier parts of West Africa. The wild unselected fruit tend to be slightly bitter, due to the presence of *cucurbitacin,* the bitter principle common to the Cucurbitaceae, and in places, e.g., Senegal (17), Mali (6, 10), N Nigeria (6), are fed only to cattle. Throughout the Region sweet-fleshed forms arc cultivated, especially in Mali and Niger (12). Each territory seems to have its own special cultivars, and dry conditions are conducive to raising the sweetest fruit, high atmospheric humidity apparently suppressing the formation of sugars. The flesh is eaten raw or cooked, the choice usually depending on the degree of sweetness or bitterness. In these cultivars the content of cucurbitacin has in varying measure been selected out.

The fruit has but few medicinal uses in West Africa. Bitter forms are used in Senegal as a drastic purge, and are considered poisonous. Of more usual forms, the pulp is taken as a diuretic and purgative (11). Fruit-pulp in NE Hausaland is a remedy for a urinary condition suggestive of gravel and stone in the bladder. The Yoruba of Nigeria take the whole fruit of a variety called *bàrà,* cut it up and boil it with onions and the root of an unidentified plant (*epatun*) to produce a decoction taken for gonorrhoea, and for leucorrhoea in women (6). The fruit is invoked by the Yoruba in an incantation for the good delivery of a pregnant woman, doubtless on an anthropomorphic basis because the watermelon normally fruits abundantly and is the 'owner of many children' (18). In SW Africa the fruit-juice is used to stick the powdered dried entrails of the pupal larvae of *Diamphidia simplex* and *D. locusta*, which contain a toxic albumin, to arrowheads as an arrow-poison (19).

Hausas in Ghana use the hollowed-out fruit in which a henna preparation has been placed to insert their hands to dye the nails red (9).

The plants of both bitter- and sweet-fleshed forms yield seeds sold in W African markets under the Yoruba name *ègúsí.* This is the name for *Cucumeropsis mannii* Naud., an indigenous species, and when *Citrullus lanatus* arrived in West Africa in prehistory it acquired, by its similarities, the same name. That the watermelon is known by its Yoruba name throughout the Region may perhaps suggest its coming from NE Africa via S Nigeria. The seeds are chiefly used as a masticatory, and for medicine, food and oil. They may be extracted from the pulp when this is cut up for food, or the fruits are stacked in heaps or buried during the rains and the seed later recovered after the flesh has rotted off. The seeds are generally sun-dried for marketing. Besides being taken as a masticatory, they are roasted and ground up to a pulp which is added to soup or made into a sauce or a porridge, or the oil may be extracted for use in cooking. A Yoruba food or flavourer called *ògìrì* is made from the fermented kernels; *ìgbãlò* is another food made from the seeds roasted, pounded, wrapped in a leaf and then boiled. The seeds are oil-bearing but of variable oil content from 15–45%. The oil consists of glycerides of linoleic, oleic, palmitic and stearic acids (4), and of an inedible variety from Sudan the relative amounts were 68%, 6%, 22% and 4% respectively (20). It is semi-drying with an iodine value of 122, and has a number of potential uses: culinary; soap-making; medicinal vehicle; illuminant. The quantity and quality of oil varies between cultivars and there appears to be a field for breeding and selection. Oil from the best seeds is called in Yoruba *òróro-ègúsi* and this is used in cooking (6). Protein content of the seeds is about 34% (5), carbohydrates about 5% and 12% crude fibre and ash (15). The cake left after expression of the oil is a good cattle-feed (4).

The seed can be roasted and used as a coffee-substitute (6).

The seeds are used as a vermifuge in Senegal (11), and juice squeezed from pulp roasted in fire-ash is drunk in S Nigeria as a worm-medicine (6). They are used as a diuretic and for their strengthening properties in India (14, 15) and their beneficial use in acute cystitis and capacity to lower the blood pressure is recorded (14).

The leaves, as well as the fruit, provide grazing for all stock (1). In Lesotho and Tanganyika the leaf is eaten as a vegetable (19). In Congo sap from the leaves is instilled into the eye in cases of fainting (3) and in Tanganyika it is taken for malaria and the pounded leaf applied as a dressing to whitlows (8).

Traces of alkaloid have been detected in the fruit and leaves of Nigerian-grown material (2).

The root is believed by the Sukuma of Tanganyika to prevent hoarseness and singers eat the root before a long session to prevent it (16).

References:

1. Adam, 1966, a. 2. Adegoke & al., 1968: as *Colocynthis citrullus*. 3. Bouquet, 1969: 101, as *Citrullus vulgaris* Schrad. 4. Burkill, 1935: 560–1, as *Citrullus vulgaris* Linn. with references. 5. Busson, 1965: 413, 416 with seed analysis of amino acids. 6. Dalziel, 1937. 7. Trochain, 1940: 267, as *Citrullus vulgaris*. 8. Haerdi, 1964: 79. 9. Irvine, 1930: 104–5, as *Citrullus vulgaris* Schrad. 10. Irvine s.d. 11. Kerharo & Adam, 1974: 374–5, as *Colocynthis citrullus* (L.) O. Ktze, with references and phytochemistry. 12. Mauny, 1953: 715, as *Citrullus vulgaris*. 14. Quisumbing, 1951: 931–3, as *Citrullus vulgaris* Schrad. with references and phytochemistry. 15. Sastri, 1950: 186–8, as *Citrullus vulgaris* Schrad. with analyses. 16. Tanner, 828, K. 17. Trochain, 1940: 267, as *Citrullus vulgaris* Schrad. 18. Verger, 1967: no. 89, as *Citrullus vulgaris* Schrad. 19. Watt & Breyer-Brandwijk, 1962: 346–9, as *Colocynthis citrullus* O. Ktze, with references. 20. Grindley, 1950.

**Coccinia adoensis** (A. Rich.) Cogn.

FWTA, ed. 2, 1: 216, as *C. sp. C.* of Keay.

A climbing or trailing annual herb to 3 m long arising from a swollen perennial rootstock, recorded in N Nigeria, and occurring widespread in eastern Africa from Ethiopia to the Transvaal.

No use is recorded in West Africa. In Sudan the tuberous roots are commonly eaten (1), and the leaves are also boiled and eaten as a vegetable (2). In Zambia the fruits are eaten (4).

A root-decoction with leaf-sap is used in Tanganyika as a mouth-wash for tooth-abscesses (3).

References:

1. Andrews 434, 908, 1310, 1610, K. 2. Andrews 498, K. 3. Haerdi, 1964: 80. 4. Macrae 1689, K.

**Coccinia barteri** (Hook. f.) Keay

FWTA, ed. 2, 1: 125. UPWTA, ed. 1, 63, as *Physedra barteri* Cogn.

West African: **SIERRA LEONE** KISSI sɛŋgũ-yondoŋ (FCD) MENDE ndɔgbɔ-goja (*def.*-i) (FCD) tɛlɛgɛ (NWT) TEMNE *an*-tin-*a*-ro-lal (NWT) **GHANA** AKAN-AKYEM asaman katewa (FRI) TWI asaman kyekewa = *spirit's water melon* (FRI) **WEST CAMEROONS** KPE efoto (Waldau)

A herbaceous climber or trailer to 6 m long, recorded from lowland rain-forest of Sierra Leone to W Cameroons and Fernando Po, and occurring in Cameroun, Central and East Africa.

A cold infusion of the plant is used in parts of Nigeria and W Cameroons for venereal diseases, taken internally; the powdered dried leaves are applied as a dressing for chancre, etc. In Cameroons, also (*fide* Santesson) the juice of the boiled leaves, mixed with a few drops of water is an ear-instillation for ear-ache (1).

The Akyem of Ghana take the contents of a fruit mixed with fresh lime juice and apply it with a feather to the umbilical cord, tied near the body with a strand of pineapple or plantain fibre, of a new-born baby. The application is repeated till the cord falls off (2).

References:

1. Dalziel, 1937. 2. Irvine s.n.

**Coccinia grandis** (Linn.) J. O. Voight

FWTA, ed. 2, 1: 215. UPWTA, ed. 1, 53, as *Coccinia cordifolia* Cogn.

West African: **SENEGAL** MANDING-BAMBARA bagéna ka dégé (JB) SERER safu gaynak (JB) WOLOF yomb u mbot = *gourd of the toad* (JB) **MALI** DOGON gɛ̃ũ (C-G) SONGHAI lombaria (A. Chev.) **NIGERIA** HAUSA gùrjín-dájìi (JMD; ZOG) gwanduwa (JMD; ZOG)

An annual climbing or trailing herb to about 6 m long arising from a perennial swollen rootstock, of wooded or grassy savanna in Senegal and N Nigeria, and occurring in Cameroun and widespread in NE and E Africa, Arabia, India and SE Asia.

The leaves are eaten as a vegetable in Ethiopia (14), Kenya (5) and Tanganyika (12). They are also eaten in India (10) and in SE Asia (3). The plant is grazed by all stock except cattle in Senegal (1), and by all stock in Jebel Marra (6). The Zigua of Tanganyika prepare a boiling of leaves and give the liquid to a woman to drink for delayed childbirth (13). In SE Asia the leaves are used on skin-complaints and a decoction is taken internally for gonorrhoea (3).

The fruits are red when ripe with a creamy pulp of an agreeable slightly acid flavour (9). They are eaten in Senegal (2), in Ethiopia (11) and Kenya (8) where they are also food for birds (15) and game (7). They are recorded as having been an item of market sale in Timbuktu (Chevalier fide 4).

The fruit in Nigeria has sometimes been used as an eye-medicine (4). The roots and stems have a number of medical applications in India (10) and SE Asia (3).

References:

1. Adam, 1966, a: as *C. cordifolia*. 2. Berhaut, 1967: 261. 3. Burkill, 1935: 593, as *C. indica* Wight & Arn. 4. Dalziel, 1937. 5. Graham 2073, K. 6. Hunting Technical Surveys, 1968. 7. Leuthold 134, K. 8. Mathew 6262, 6348, K. 9. Roberty, 1953: 450, as *C. cordifolia*. 10. Sastri, 1950: 257, as *C. indica*. Wight & Arn. 11. Taddesse Ebba 624, K. 12. Tanner 2030, K. 13. Tanner 2691, K. 14. Turton 36, K. 15. Tweedie 2285, K.

**Coccinia keayana** R. Fernandes

FWTA, ed. 2, 1: 216, as *C. sp. A* of Keay.

West African: **SIERRA LEONE** MANDING-MANINKA (Koranko) nala (NWT) MENDE ndɔgbɔ-goja = *bush blow-blow* (*Ruthalacia eglandulosa*) (FCD)

A herbaceous climber to 3 m long, of forest areas from Sierra Leone to S Nigeria, and also in E. Cameroun.

In Sierra Leone the plant is pounded and tied warm to the body for side-pains (1).

Reference:

1. Deighton 2217, K.

**Cognìauxia podolaena** Baillon

FWTA, ed. 2, 1: 200.

A herbaceous climber to 6 m high or trailer forming clumps on the ground, tender when young, but becoming robust with age, occurring in the bush or in clearings; common in E Cameroun to Zaïre and into Angola, it has been recorded in W Cameroons.

No usage is recorded for the West African Region. In Gabon dried leaves are powdered and put on burns and the fruits are recognized as toxic (3, 4).

The root is used as an extremely vigorous emeto-cathartic in Congo, with also diuretic and revulsive properties. Care in administration is necessary especially to pregnant women owing to the risk of inducing abortion. It is used as a root-decoction in palm-wine for obstinate constipation, to treat ascites, oedema and elephantiasis of the scrotum; by enema for hernias, urethral discharges and haematuria; by sap from the roots crushed with sugar cane for costal and stomach pains; by root-pulp applied by friction over scarifications as a revulsive for intercostal pain, rheumatism and headache; as plasters to maturate furuncles, buboes and breast abscesses, care being taken to protect the skin from scarring if the plaster has to be kept on for long by an application of palm-oil. Root-sap is insecticidal and parasiticidal. It is given in draught or enema as an anthelmintic, and in eye-instillation for filaria in the conjunctiva of the eye. It is painted on the wood of beds and cupboards as a repellant for bugs and cockroaches. To calm fits of insanity, a small spoonful of root-decoction is taken morning and evening, and to stop nightmares and erotic dreams a macerate of palm-wine with roots of *Cogniauxia sp.* and *Rauwolfia vomitoria* (Apocynaceae) is drunk before retiring (1).

Analyses of material from Congo showed the presence of alkaloids—leaves a trace, bark 0.1–0.3%, and roots up to 1%; saponins—leaves none, bark and roots abundant; steroids and terpenes present in the roots (2).

References:

1. Bouquet, 1969: 101. 2. Bouquet, 1972: 21. 3. Walker, 1953, a: 31. 4. Walker & Sillans, 1961: 1940.

**Corallocarpus boehmi** (Cogn.) C. Jeffrey.

FWTA, ed. 2, 1: 213 as *Cucumis prophetarum* sensu Keay, non Linn., in part.

West African: **NIGERIA** HAUSA (East) kam-faraka = *francolin's head; cf. Cucumis prophetarum Linn.* (JMD)

A climbing or trailing herb from a perennial rootstock, recorded only from N Nigeria in the Region but occurring widespread in open savanna elsewhere in tropical Africa. The Nigerian record is a *mixtum compositum* with *Cucumis prophetarum* Linn. and bears the Hausa vernacular cited above. It is possible that besides sharing the same vernaculars, *C. boehmi* may have the same uses in N Nigeria as *C. prophetarum.*

**Ctenolepis cerasiformis** (Stocks) Naud.

FWTA, ed. 2, 1: 208. UPWTA, ed. 1, 53 as *Blastania fimbristipula* Kotschy & Peyr.

West African: **SENEGAL** MANDING-BAMBARA tiga sina (JB) SERER madar o yar (JB) ndunngal o rif (JB) **NIGERIA** HAUSA bàrkoònon-hàwaínìyàa = *chili of the chameleon* (auctt.) ɗiyar hàwaínìyàá = *offspring, or fruit of the chameleon* (JMD) námíjin gàraàfúnii = *male garafuni* (*Momordica balsamina*) (JMD)

A slender climber with creamy-white flowers and scarlet cherry-like fruit, occurring in Senegal, Mali and N Nigeria, and extending to the eastern side of Africa from NE to South, and into India.

The plant is shunned by all stock for grazing (1). The fruit is an ingredient in some prescriptions for tertiary syphilis (2).

References:

1. Adam, 1966, a. 2. Dalziel, 1937.

**Cucumeropsis mannii** Naud.

FWTA, ed. 2, 1: 214, incl. *C. edulis* (Hook. f.) Cogn. UPWTA, ed. 1, 56.

West African: **GUINEA-BISSAU** CRIOULO tomatom (EPdS) FULA-PULAAR (Guinea-Bissau) mantem (EPdS) MANDING-MANDINKA mantéo (EPdS) **SIERRA LEONE** BULOM (Kim) yɛntɛ? (FCD) BULOM (Sherbro) cham-dɛ (FCD) sãhũ-tuntun-dɛ *small var. with elongated white fruit and white edible seeds* (FCD) sãŋga-lɛ (FCD) FULA-PULAAR (Sierra Leone) budi (FCD) GOLA legiɛ (FCD) KISSI heŋguo (FCD) seŋgo (FCD) seŋgo-pɔmbɔ (FCD) seŋgu-chielachalo *a small-fruited var.?* (FCD) KONO kõ-sa? *small var. with elongated white fruit and white edible seeds* (FCD) sa (FCD) sa-mɛsɛ (FCD) KRIO ɛgusi (FCD) wet-ɛgusi = *white egusi; a small var. with elongated white fruit and white edible seeds* (FCD) LIMBA (Tonko) mayentɛrɛba (FCD) LOKO koeja (FCD) MANDING-MANDINKA sara (FCD) MANINKA (Koranko) sara-mɛsɛ (FCD) sarɛ (FCD) MENDE fɛŋɛ-goja *a short (3 months) season var.* (FCD) koja (*def.*-i) (FCD) koja-mumu (*def.*-i) *small var. with elongated white fruit and white edible seeds* (FCD) koja-wawa *a long (6–10 months) season var.* (FCD) SUSU kɔbɛ (FCD) SUSU-DYALONKE sara-na (FCD) TEMNE *an*-tent (FCD) *ma*-teint-*ma*-temne *small var. with elongated white fruit and edible white seeds* (FCD) VAI sa (FCD) sara (FCD) **LIBERIA** KISSI hiŋgo (FCD) MANO pinni guo (JMD) **GHANA** AKAN-ASANTE akatewa (FRI) FANTE akate (FRI) akatsewa (FRI) TWI akatewa *with creamy white fruits* (FRI) akyẽyẽa *with green white-spotted fruits* (FRI) GA agúshi (FRI) GBE-VHE abgesi (FRI) gusi (FRI) susu mase (Howes; FRI) VHE (Awlan) agushi (FRI) VHE (Pecí) atsẽtsẽa (FRI) HAUSA àgúshií (FRI) NZEMA akatoba *from* toba: *bottle* (FRI) ngatia-fufule *a white climbing var.* (FRI) ngatia-kokole *a yellow-red climbing var.* (FRI) **NIGERIA** EFIK òkôkòn (auctt.) IGBO àhụ̀ (AJC; JMD) àhụ̀ elu, aki (BNO) IGBO (Onitsha) ògìlì *the seed* (FRI) IGBO (Owerri) àhụ̀ = *melon* (KW) YORUBA ẹ̀gúsí-itóò (auctt.) itóò, ẹgúsí itóò (JMD) ìtóò (JMD) itórò (JMD) YORUBA (Ilorin) itoro (Clarke)

A herbaceous climber reaching 4 m height, of forest areas from Guinea-Bissau to W Cameroons and in E Cameroun to Angola and Uganda.

This is the true indigenous *egusi* of West Africa—see *Citrullus lanatus* (Thunb.) Mansf. It is commonly cultivated on farms and in gardens throughout its area of distribution. The flesh of the fruit is edible but as an item of diet it appears to be less important than the oily seeds for which the plant is mainly grown. The seeds are removed from the flesh usually by stacking the fruit to allow decomposition to take place. The seeds are washed out after 10–15 days, and are then dried and can be stored. In N Ghana storage is done in pits in the ground. The seeds are a common article of trade in markets under the Hausa name *agushi* or the Yoruba *egusi*. They are prepared for consumption by parching and pounding to free the seed-coat from the kernel which can be eaten either raw or cooked, or more usually when ground to a powder it is added to soups and stews. In flavour they resemble groundnuts. They are rich in oil and contain more protein than the latter. Comparative figures are:

| | Egusi | Groundnuts |
|---|---|---|
| Fatty matter | 45% | 36–47% |
| Protein | 34% | 23–30% |

Amino acids are particularly well represented and are appreciably more abundant than in groundnuts (2). They are thus of high food value.

The oil is semi-drying. It is a good substitute for cotton-seed oil and is suitable for soap-manufacture and for illumination. It is used in cooking and can readily be refined into a superior product for table use (4). Ivorean material is reported to consist of: linolenic acid, 64.9%; linoleic acid, 12.4%; stearic acid, 11.8% and palmitic acid, 10.9% (2).

Juice from the fruit mixed with other ingredients is applied in Ghana to the navel of a new-born baby for five days till the cord-relic drops off (4).

The dried fruit-shell of a form with small elongated fruit is pierced and used in Sierra Leone as a warning horn by cattle boys (3).

The leaves are used in Gabon to make a macerate for purging a suckling baby either by administration direct to the baby or by putting some on the mother's breasts before she nurses it (5).

In E Cameroun the root has been recorded as being poisonous (1).

References:

1. Bates 261, K. 2. Busson, 1965: 416–7 (with seed analysis, as *C. edulis*). 3. Deighton 2379, K. 4. Irvine, s.d. 5. Walker & Sillans, 1961: 140.

**Cucumis anguria** Linn.

English: West Indian gherkin, wild cucumber (Irvine).

West African: **SIERRA LEONE** SUSU fɔtɛnalinyi (NWT)

An annual climbing or trailing herb to 2.5 m long, of deciduous woodland and bushland and grassland of mid-elevations of Tanganyika, S and SW Africa. It is a well naturalized alien in the West Indies whence it may have reached West Africa to be recorded in Senegal and Sierra Leone.

The wild plant of Africa is var. *longipes* (Hook. f.) A. Meeuse. The cultigen, gherkin, is var. *anguria* seems to have arisen in the West Indies. It is in abundant cultivation around Thiès in Senegal where the immature fruits are pickled green. The ripe fruit is also edible (1).

Reference:

1. Irvine, 1952, a: 36.

**Cucumis figarei** Naud.

FWTA, ed. 2, 1: 213, as *C. pustulatus* Hook. f. UPWTA, ed. 1, 57, as *C. pustulatus* Hook. f.

West African: **SENEGAL** WOLOF beref u buki = *sweet melon of the wolf* (JMD) **NIGERIA** HAUSA bakin duniya = *all the world;* (JMD) goólón zaakii (ZOG) gúnàr kuúraá = *hyena's watermelon* (JMD; ZOG) kaashin gwankii = *dung of the roan antelope* (JMD; ZOG) kuƙaimi maƙaimi = *spiny one; from* ƙaimi: *a spur* (JMD; ZOG) HAUSA (East) golon ƶaki (JMD) noónòn kàreé = *hyena's milk* (JMD; ZOG) HAUSA (West) tsúwaàwún-zaákiì (ZOG; JMD)

An annual trailing herb to 2 m long from perennial woody rootstock, recorded from Senegal and N Nigeria and occurring doubtless in other dry parts in the north of the West African region. It is distributed across Africa to NE, E and SW Africa.

The plant is said to provide grazing for all stock during the rains in Sudan (3).

The fruit is mixed with bran and used as a horse-medicine in N Nigeria. It is also a medicine for fowls, placed in their drinking water to help growth and prevent disease, render them immune to predatory hawks, increase egg-laying, etc.—a usage which is probably wholly superstitious, but see *C. prophetarum* (1). In the Hoggar of Central Sahara, scorpion-stings and snake-bites, after scarifying, are rubbed with the flesh of the green fruit, or they are dusted with powder from the dried pulverized fruit (2).

The taproot, dried and finely pulverized, is used in N Nigeria like snuff in the belief that it relieves toothache (1).

References:

1. Dalziel, 1937. 2. Maire, 1933: 200–1. 3. Peers BM 3, K.

**Cucumis melo** Linn.

FWTA, ed. 2, 1: 213, incl. *C. dipsaceus* sensu Keay. UPWTA, ed. 1, 56.

English: melon.

French: melon.

West African: **SENEGAL** MANDING-BAMBARA kongo dïé (JB) SERER khad (JB) nak nãngoy (JB) WOLOF hal u mbot = *melon of the toad* (JB) yombebute, yomba mböt, yomba: *gourd;* bute, mböt: *toad, loosely applied to small cucurbits* (JMD) **THE GAMBIA** MANDING-MANDINKA kurubombonmesengo = *tiny kurubombong* (Fox) **MALI** DOGON íne gabá jógɔ (C-G) SONGHAI al bata (JMD) TAMACHEK itekel (JMD) **GHANA** DAGBANI neri *the seeds, but properly of Citrullus* (JMD) GA wátɛ̀rɛ, watrɛ (KD) GRUSI yengani (Ll.W) KONKOMBA inabe (Gaisser) NANKANNI sarma (Ll.W) **TOGO** BASSARI inabe (Gaisser) KABRE kaniiga (FB) kanjinga (Gaisser) KONKOMBA inabe (FB) MOORE-NAWDAM kanjinga (Gaisser) TEM kaenia (FB) TEM (Tshaudjo) kaenja (Gaisser) **NIGER** SONGHAI múné (*pl.* -à) (D&C) TAMACHEK ermana (JMD) **NIGERIA** ARABIC-SHUWA fagus (JMD) shamman (JMD) FULA-FULFULDE (Nigeria) cikiire (*pl.* chikilje) (JMD) denaare (*pl.* denaage) (JMD) dene bamɗi = *donkey's melon* (MM) nyanaare (*pl.* nyanaaɗe) (JMD) HAUSA batanya (JMD) gulli (JMD) gunaà, gurjií (ZOG) gurji (JMD) gurji kwantal (JMD) gurli (JMD) gurli ƙwantal (ZOG) gwandar beri (MM) kuliri (JMD; ZOG) kwantal (JMD) shamman (JMD) IGBO isikiripum *this and other C. spp.* (BNO) TIV icegher *probably this sp.* (Vermeer) **WEST CAMEROONS** KUNDU esaka (JMD) LONG esaka (JMD) LUNDU esaka (JMD) MBONGE osaka (JMD) TANGA esaka (JMD)

A herbaceous trailing plant, often rampant in dry exposed sandy situations of savanna land, and occurring throughout the West African region, often under cultivation around villages, and widespread in the tropics and sub-tropics. The plant has been in domestication for so long that its area of origin cannot be determined, but probably it lies in Africa. Wild or subspontaneous forms are numerous. Selection seems to have begun in western Asia and has given rise to the cultivated melon, eaten for its sweet flesh, and to other races with bitter flesh but edible seeds.

The fruit is very varied in size. Sweet-fleshed cultivars vary from the size of a hen's egg to 15 cm diameter or so. Shape is also variable: round, elongated, ribbed or smooth. Those with edible flesh may be eaten raw, or cooked in a soup, and some are eaten before ripening. The flesh can be preserved by drying after removal of the seeds (10).

The flesh contains little nutriment with over 90% of water. The rind contains more than double vitamin A at 39·3 mg/100 gm than is in the flesh (9). The juice is demulcent, diuretic and cooling (8). The rind and other residues have a high percentage of potash which has manurial value (4).

The seeds are commonly eaten after drying and crushing. They are nutritive and diuretic (7). The Hausa of N Nigeria prepare from them fermented cakes called *daddawar gulli* or *d. gurji* (4). Globulin, gluten and a number of proteins are present (8). The kernels contain about 40% of an edible, semi-drying fixed oil consisting of glycerides of linoleic acid, 56–59%; oleic acid, 27–29%; palmitic acid, 10%; and smaller amounts of stearic and myristic acids (8, 9). At some time there has been an export trade in the seed from Sudan under the Arabic name *senat* (4).

Both the leafy trailers and the fruit provide good forage for all stock (1, 4).

The leaves are eaten as a vegetable and relish in S Africa and in soups in Tanganyika. Plant extracts have been shown to inhibit fungal activity (9).

The root has been found to contain an emetic principle (8).

References:

1. Adam, 1966, a. 2. Burkill, 1935: 696–7. 3. Busson, 1965: 414. 4. Dalziel, 1937. 5. Irvine, 1948: 255. 6. Irvine, 1952, a: 23–40. 7. Irvine s.d. 8. Quisumbing, 1951: 934–5, with references. 9. Watt & Breyer-Brandwijk, 1962: 353, with references. 10. Williamson, J. 1956: 43.

**Cucumis metuliferus** E. Mey.

FWTA, ed. 2, 1: 213. UPWTA, 56–57.

English: horned cucumber (Dalziel); English tomato (Sierra Leone, Morton).

French: concombre metulifere (Holland).

West African: **SENEGAL** DIOLA sidèldéla (JB) SERER dandalub (JB) **THE GAMBIA** DIOLA (Fogny) sidèldéla = *fowl* (DF) **NIGERIA** HAUSA buùrár zaákiì (JMD; ZOG) noónòn-kuúraá = *hyena's milk* (JMD; ZOG)

A herbaceous annual climbing or trailing herb, recorded from Senegal to N Nigeria in drier parts of the West African region, and occurring widespread across eastern, central and south tropical Africa.

All parts of the wild form are bitter. The foliage gives a strongly frothing extract which contains saponin. Thus the plant is considered poisonous. Cultivars have been selected and cultivation is found in some of the drier parts of the Region (2). The leaves are cooked and eaten in parts of Malawi at certain times of the year (5) but whether these are of a cultivar or the season permits use of the wild form is not indicated.

The fruits of both wild and cultivar forms when immature are like a green cucumber and become red on ripening and are covered with sharp spines which readily injure the skin in handling. Wild forms are very bitter and are considered inedible, though in the Kalahari area of S Africa game-animals eat them, and in time of scarcity are fed to cattle and are even eaten by the Bushmen (1). Edible cultivars are very similar in appearance and can only be distinguished from the wild fruit by trial (4). Cultivation is carried on in northern Sierra Leone and the fruit is a common item of market produce in Kabala market under the name *English tomato* (3).

References:

1. Dalziel, 1937. 2. Jeffrey, 1964. 3. Morton 3340, FBC, K. 4. Watt & Breyer-Brandwijk, 1962: 353. 5. Williamson, J., 1956: 43.

**Cucumis prophetarum** Linn.

FWTA, ed. 2, 1: 213, excl. Dalziel 123, and incl. *C. ficifolius* sensu FWTA, ed. 2. UPWTA, ed. 1, 57, incl. *C. ficifolius* sensu Dalziel, p. 56.

West African: **MAURITANIA** ARABIC (Hassaniya) agtastaf (AN) tagasrarit (Kesby; AN) **NIGER** SONGHAI céŋ-kànì (*pl.* -ìà) = *rat's melon* (D&C) **NIGERIA** FULA-FULFULDE (Nigeria) gerlal = *bush fowl* (MM) HAUSA cicudu (ZOG) gunar zaki = *lion's melon* (PF fide MM) kaƙaimi, maƙaimi = *spiny one* (JMD; ZOG) kam-makwarwaa (ZOG) kàncíkúlkùl, kan-fakaraa (ZOG) HAUSA (West) cicidu (ZOG) hamanya (JMD; ZOG) kam fakara, kam makwarwa = *francolin's head; from the striped fruit* (JMD) yamanya (JMD; ZOG) KANURI gùnò gùnó = *white melon, from L.Chad area: perhaps this sp.* (C&H)

An annual climber or trailer to 2 m long from a perennial woody rootstock, of semi-desert scrub in Senegal, Mali and N Nigeria, and extending across Africa to NE and E Africa and Angola, and on into Arabia and India.

The plant provides grazing for all stock (Senegal, 1; Mauritania, 9; Sudan, 8; Somalia, 7; Uganda, 3: Kenya, 5). It is said in Mauritania to be good for milk-production (9). The plant, however, part unspecified but probably the fruit, is used as an abortifacient for women and to hasten expulsion of the placenta for cows in Ethiopia (4). Stock at Lake Chad will not eat the fruit (1). The fruits are recognized in Ethiopia as highly poisonous, and are reported as used to treat rabies (4). In N Nigeria the fresh fruit with an end cut off is applied thimble-like as a dressing for an inflamed finger. The fruit has veterinary use as a vermifuge with the addition of natron for horses by the Hausa. It is also used as a medicine for fowls (cf. *C. figarei* Naud.). In some places it is an ingredient of medicines for syphilis (2), and as an emetic and in small doses with honey as a stomachic for children (6).

The seeds are oil-bearing (10).

References:

1. Adam, 1966, a. 2. Dalziel, 1937. 3. Dyson-Hudson 209, K. 4. Getahun, 1975. 5. Glover & al. 670, K. 6. Holland, 1908–22: 335. 7. Keogh 122, K. 8. Mohamed Ismail A 3553, K. 9. Naegelé, 1958, b: 888–9 (incl. *C. ficifolius*). 10. Roberty, 1953: 450.

**Cucumis sativus** Linn.

English: cucumber.

French: concombre.

West African: **GUINEA-BISSAU** BALANTA betbinho (JDES) CRIOULO pepino (JDES) **SIERRA LEONE** GOLA kipɔsi (FCD) KISSI kumba (FCD) KONO kɔkumba (FCD) KRIO kɔkumba (FCD) LIMBA (Tonko) kɔkumba (FCD) MENDE kikpɔsi (FCD) kɔkumba *ex Krio* (FCD) TEMNE (Kunike) *ma*-kɔnkəbos *a combination of* bos: *Lagenaria, with* kɔkumba (FCD) TEMNE (Port Loko) *ma*-kɔkumba *the fruit* (FCD) TEMNE (Sanda) *ma*-kɔkumba *the fruit* (FCD) TEMNE (Yoni) *ma*-kɔnkəbos *a combination of* bos: *Lagenaria, with* kɔkumba (FCD) VAI kipɔsɛ (FCD) **MALI** ARABIC faggus (RM) **GHANA** GA kokómbà (KD) **NIGERIA** KANURI ngùrlí (C&H)

A herbaceous annual, introduced and widely cultivated in West Africa and in all warm countries. A cultigen, the common Cucumber, of lost botanical origin in western Asia or in Africa, now of innumerable cultivars; its advent to West Africa must be relatively recent judging by the paucity of true African vernacular names.

The plant is primarily grown for its fruit which is eaten while still immature either fresh or cooked. The very young fruit can be pickled as gherkins. The best fruit for eating is parthenocarpic, some modern cultivars being only female-flowered. If fertilization takes place the fruit tends to become bitter. On bisexually flowered plants the male flower buds are better removed before anthesis to prevent pollination. The fruit juice is slightly purgative and diuretic. A small amount of ascorbic acid is present (7). It is a good source of vitamins B, C and G and of iron and calcium. It is low in vitamin A and nutritional calorific value (5). The juice has commercial cosmetic uses to impart a smell and perhaps for an emollient effect (7). The fruit rubbed over the skin is said to keep it soft, and is cooling, healing and soothing. It has been incorporated into a soap (5). The juice is said to banish fish-moth and woodlice, and peel left on the floor at night where cockroaches can reach it and eat it will kill them after 3–4 nights (7).

The seed, which is, of course, produced only in fruit from fertilized flowers, is edible. The kernels are much used in confectionery in India (6). Reports (1, 3, 5) on the presence of alkaloid, *hypoxanthine,* in the seed appear to be based on assays of seed that had undergone deterioration (? poor storage) for its presence is not reported in Henry: *The Plant Alkaloids,* ed. 3, 1939. There is evidence however of the presence of a saponin. The seeds have been used by Europeans in the Transvaal and elsewhere as an anthelmintic which action is ascribed to contained saponin (1, 5, 7).

The kernels contain 42% protein and 42.5% fatty matter. The protein is rich in phosphates and the seed cake after expression of the oil has a $P_2\,O_5$ content of 11.17% (6). The oil is pale yellow and semi-drying, and consists of glycerides of the following fatty acids: oleic, 58.5%; linoleic, 22.3%; palmitic, 6.8%, and stearic, 3.7%, and also some phytine and licithine (5). It is suitable for cooking.

The leaves are used as a vegetable in Malaya (1) and they have medicinal use in India for throat affections and as a diuretic (2, 6).

References:

1. Burkill, 1935: 697–8. 2. Chakravarty, 1968: 439. 3. Henry, 1924: 336. 4. Henry, 1939. 5. Quisumbing, 1951: 935–6 with references. 6. Sastri, 1950: 391–2, with analyses and references. 7. Watt & Breyer-Brandwijk, 1962: 355–6, with references.

**Cucurbita maxima** Duch.

FWTA, ed. 2, 1: 215.

English: squash; squash-gourd; melon pumpkin; also (Whitaker & Davis; Purseglove) pumpkin, winter squash.

French: (Kerharo & Adam)—courge; courge-potiron; potiron.

West African: **SENEGAL** FULA-PULAAR (Senegal) budi *a general term* (JMD) MANDING-BAMBARA ié (Vuillet) MANINKA guié (JMD) SERER a dyeng (JLT) WOLOF bâg (K&A) banga (AS) dombos (AS) nadé (AS) naḓé (K&A) yomba (AS) **THE GAMBIA** MANDING-MANDINKA miraŋ (*def.*-o) *= something one drinks with* (DF) WOLOF laket (JMD) **GUINEA** FULA-PULAAR (Guinea) budi *a general term for 'pumpkin'* (JMD) MANDING-MANINKA guié (JMD) **SIERRA LEONE** KISSI njalin (JMD) KONO g-bi (JMD) MENDE towa (*def.* towei) (JMD) TEMNE *ma*-ali *the seeds* (JMD) a-ŋali (*pl.* ɛ-yali) (FCD) *a*-kali (*pl.* a-tali) *the plant, or the fruit* (FCD) **LIBERIA** MANO bili gah (JMD) **MALI** TAMACHEK akesaïm (A. Chev.) takasaïm (A. Chev.) **GHANA** AKAN-ASANTE ɛfere, ɛfre (Enti) FANTE ɛfir (JMD) TWI ɛfrɛ (Enti) DAGBANI ɣoɣli (FRI) kaere (Gaisser) kilbundo (Gaisser) GA sakribonte (JMD) GUANG-GONJA kàwúrènchú (Rytz) KONKOMBA dedschaelbɔne (Gaisser) **TOGO** BASSARI kaere (-tukurn) (Gaisser) katekatego (Gaisser) TEM (Tshaudjo) kaere (Gaisser) kodshŏdo (Gaisser) **NIGERIA** ARABIC arä (JMD) garä (JMD) ARABIC-SHUWA murr BEROM bátìrí *a small c. var= Hausa, goojíí* (LB) ebɛ́ (LB) DERA yáaŋgú (Newman) EDO éyèn (JMD) EFIK ḿfrì ǹdìsè, ḿfrì: *fruit; thus the pumpkin itself* (JMD; KW) ǹdìsè (JMD) ǹnàñí (RFGA) ọkpọ ǹnàñí *the plant* (JMD; Winston) FULA-FULFULDE (Nigeria) feraare (MM) fol'yere (*pl.* pol'ye) *a young pumpkin* (JMD; Taylor) mborho *the leaves* (Taylor) waygoore (*pl.* baygooje) (JMD) GWARI knuba (JMD) knugba (JMD) HAUSA akwato, bàkánùwaá, gámmòn fátaàkeé, ganwon fatake, gesuma, rugudu, ruguguwa, waáwán-goónaá, ƴar garii (ZOG) goojíí *a general name for this type of fruit* (JMD; LB) kàbaíwaà, kàbeéwaà, kúbeèwaá (ZOG) kankana (Abraham) laki(-t) (JRA) HAUSA (West) balayal (ZOG) goòjií (ZOG) kabus, kabushi (ZOG) IBIBIO ǹdìsì (JMD) IGBO ukọrọ *the fruit* (JMD) IGBO (Onitsha) ụ́gbọ̀gụ̀lụ̀ (KW) IGBO (Owerri) ányū (JMD) IGBO (Umuahia) ákwṳ̄kwọ̀ *= leafy; a general term* (KW) ụ́gbọ̀ghọ̀rọ̀ (AJC; KW) ụ́gbọghụ̀rụ̀ (KW) KANURI kàfétò *from Hausa* (JMD) sààdə (C&H) NUPE ébè (Banfield) TIV àgbàdù (JMD) furum (JMD) URHOBO ètò-úrhòbò (DA) koko-eyen (DA) YORUBA apala *a general term* (JMD) élégédé (JMD; CWvE) **WEST CAMEROONS** BAFOK dibok (JMD) DUALA dìbɔ̌ (Ithmann) KOOSI abok (JMD) KPE diboke (JMD) KUNDU dibuke (JMD) LONG dibok (JMD) LUNDU dibuke (JMD) MBONGE dibuke (JMD) NGEMBA nə̀bɔ̀´ (Chumbow) TANGA diboke (JMD) WOVEA diboke (JMD)

A herbaceous annual, usually trailing, rarely bushy, not harsh, polymorphic; native of Peru and now dispersed as a cultigen throughout most of the world. A number of varieties or groups and cultivars are recognized (11, 16).

The plant is grown in all West African territories. It may be found in village cultivated areas and around houses often trained against house-walls flowering and fruiting on the roof. The young leaves and shoots, and even the flowers, are used as a pot-herb, and, as for *C. pepo,* the end-of-season leafy shoots and undeveloped fruits are eaten by the Hausa of N Nigeria (4, 6, 7). The young boiled shoots are commonly pickled in vinegar in Gabon (14). In Malawi the leaves are the most commonly eaten and most universally liked of all edible leaves, and if not eaten fresh are dried for storage (17).

The fruit flesh is generally boiled and eaten in pieces (8) or put into stews and soups (14, 15). In the drier climate of the north of the Region e.g. in N Nigeria and northern Ghana, the pulp is also sliced and dried for storage (4, 8).

The fruit-pulp is used as a poultice for boils and carbuncles. In Tamale, Ghana, the fruit cut into pieces, mixed with ashes and left to stand overnight is used for dehairing hides (5), and in this the fruit-pulp appears to be a vehicle for the depilatory substance of the ash. (See also *C. pepo.*)

Chemical composition of the fruit varies greatly with the cultivar: water may be 87–97%, protein about 1.4–2.0%, fat 0.1–0.5%, total sugar 3.0–5.0%, starch 0.3–1.3%. Calcium, iron, phosphorus, vitamins A, B and $B_2$, and C, and niacin, etc. are present (2, 9, 12, 13, 16). The fruit is considered to be good fattening fodder for stock (8, 15).

The seeds are eaten in parts of West Africa and are much appreciated by the Asante (8). In India (13) the seed is considered edible. The seed however has medicinal uses. Freshly powdered it is taken to expell tapeworm in Nigeria (1).

In India the seed is considered taenicide, diuretic and tonic (2, 13). As a taenicide a dose of 10–15 gm is said to be neither irritant nor toxic, but is effective because of its saponin content. The active principle is thought to lie in the cotyledons. The pericarp contains a bitter resin but this is not anthelmintic (15).

The seeds yield a reddish fixed oil of 30–42% content consisting of glycerides of: linoleic acid 41–45%, oleic acid 35–37%, palmitic acid 12–13%, stearic acid 6% and a trace of arachidic acid (13, 15).

The fruit is perhaps the largest of the known gourds and weights of over 100 kilograms have been recorded (13). The husk serves as a container, and to make domestic utensils. In West Africa they are often carved. In N Nigeria they are used as floats (3), and to make single membrane drums (10—probably this sp.).

In Senegal sap from the roots is used in ear-instillations for otitis (9).

References:

1. Ainslie, 1937: sp. no. 120. 2. Chakravarty, 1968: 459–60. 3. Dalziel 119, K. 4. Dalziel, 1937. 5. Irvine 4596, K. 6. Irvine, 1952, a: 34. 7. Irvine, 1956: 36. 8. Irvine s.d. 9. Kerharo & Adam, 1974: 377–9, with phyto-chemistry. 10. Meek, 1925: 2: 156. 11. Purseglove, 1968: 119. 12. Quisumbing, 1951: 937–8, with references. 13. Sastri, 1950: 394, with analyses. 14. Walker & Sillans, 1961: 141. 15. Watt & Breyer-Brandwijk, 1962: 356–7 with references. 16. Whitaker & Davis, 1962: 48, 194–210, with analyses. 17. Williamson, J. 1956: 44–45.

## **Cucurbita moschata** (Duch.) Duch.

English: pumpkin, winter squash, musk melon (Burkill), soft-leaved autumn, or winter pumpkin (Chakravarty).

West African: **SIERRA LEONE** TEMNE *ma*-ali *the seeds* (NWT) **MALI** SONGHAI mara kachia (A. Chev.) **GHANA** DAGAARI yər (FRI) GUANG-GONJA bùlji, kíngámà (Rytz) KUSAL yua (FRI) MOORE yoɣole (FRI) SISAALA kaamiŋ (Blass) WALA yəgəre (FRI)

Annual trailing herb, of Central American origin, domesticated in Mexico at least c. 5000 B.C. and in Peru c. 3000 B.C., now widely distributed as a cultigen throughout the world. It is said to be the most commonly cultivated *Cucurbita* in the American tropics. Three general groups of cultivars based on fruit form are recognized in America (3). It stands hotter conditions than other *Cucurbita spp.* and requires 4–5 months for its growth. It is surprising that it is not more commonly grown in West Africa where it occurs only in a small way, e.g., northern Sierra Leone and Ghana.

Its attributes are similar to the other *Cucurbita spp.* The leaves are eaten in northern Ghana (2). The fruits are sweet-fleshed and are eaten cooked. The fruits are said to be storable for a long time and to stand transport (1).

References:

1. Burkill, 1935: 699. 2. Irvine 4641, K. 3. Whitaker & Davis, 1962: 52.

## **Cucurbita pepo** Linn.

FWTA, ed. 2, 1: 215.

English: pumpkin; marrow, vegetable marrow—(long fruited forms); ornamental gourd.

French: courge, citrouille, courge citrouelle, giraumon, potiron.

West African: **SENEGAL** BASARI ɔ-syẹlis, a-tyẹlis, ɛ-tyẹlìs (Ferry) BEDIK i-dìm, mi-rìm (Ferry) FULA-PULAAR (Senegal) budi *a general term for 'pumpkin'* (JMD) KONYAGI i-tyirir (Ferry) MANDING-BAMBARA ié (Vuillet) MANINKA guié (JMD) SERER a dyeng (JLT) WOLOF banga (AS) dôbos (K&A) dombos (AS) nadé yomba (AS) **THE GAMBIA** MANDING-MANDINKA miraŋ (*def.*-o) = *something one drinks with* (DF) WOLOF laket (JMD) **GUINEA-BISSAU**

BALANTA bleessim (JDES) BIDYOGO cartbáe (JDES) FULA-PULAAR (Guinea-Bissau) búdi (JDES) MANDYAK ussanufo (JDES) MANKANYA umbôgre (JDES) **GUINEA** FULA-PULAAR (Guinea) budi *a general term for 'pumpkin'* (JMD) MANDING-BAMBARA ié (Vuillet) MANINKA guié (JMD) guié-ni *elongated green fruit, white blotches* (JMD) kabu-guié *a var. grown by the Gabou* (JMD) malinka-guié *large white-fruited var.* (JMD) **SIERRA LEONE** BULOM (Kim) lam (FCD) lamgo? (FCD) BULOM (Sherbro) lam-dɛ (FCD) FULA-PULAAR (Sierra Leone) bodi (FCD) budi (FCD) GOLA die (FCD) KISSI chalẽ (FCD) njalin (JMD) KONO g-bi (FCD) g-bili (FCD) g-bi-yawa *fruit yellowish, elongated, curved* (FCD) korankɔ-g-bi *fruit green, watery* (FCD) KRIO põkin (FCD) LIMBA (Tonko) kutu (FCD) LOKO kuwe (NWT) tea (FCD) yeŋgelesi *fruit bottle-shaped, mottled* (FCD) MANDING-MANDINKA gbili (FCD) jee (FCD) ye (FCD) MANINKA (Koranko) g-bili (FCD) gbulɛ ye yee (FCD) ye-gbɛ *fruit green, watery* (FCD) ye-kumba *fruit large, dark, not mottled* (FCD) ye-nyɛ̃ *fruit spherical, mottled* (FCD) MENDE blamange *fruit spherical, mottled* (FCD) koja-lowa *fruit spherical, mottled* (FCD) sɛgbula-lowa *fruit bottle-shaped, mottled* (FCD) towa (*def.* towei) (FCD) towa-wawa *fruit green, watery* (FCD) MENDE-KPA (Kpa) kɔŋgɔ-lowa *fruit bottle-shaped, mottled* (FCD) pama-dowa *fruit green, watery* (FCD) SUSU-DYALONKE gabu-na *fruit green, watery* (FCD) kɔndi-na *fruit bottle-shaped, mottled* (FCD) nali-na (FCD) ŋali-na (FCD) nyɛli-na (FCD) TEMNE *a*-kali-*a*-bana *fruit green, watery* (FCD) *a*-ŋali (FCD) *a*-yiŋkɔres *fruit bottle-shaped, mottled* (FCD) VAI k-po (FCD) pori (FCD) **LIBERIA** MANO bili gah (JMD) **MALI** SONGHAI labtenda (A. Chev.) TAMACHEK akesaïm (A. Chev.) **GHANA** AKAN-ASANTE ɛfrɛ (Enti) FANTE akamfir *of variable shape, with or without bulged end, warted, ribbed or not, grown by Akans in Cape Coast* (FRI) ɛfir (FRI) ɛfrɛ (Enti) TWI akomfem-fere = *guinea-fowl's pumpkin, referring to its speckled skin* (FRI) ampesi *fleshy pulp sometimes dried in strips* (FRI) ɛfere-pa = *good pumpkin, a form* (FRI) ɛfrɛ (Enti) fere-fita *large white fruit* (FRI) DAGBANI kaere (Gaisser) kilbundo (Gaisser) GA sakribonte (FRI) GUANG-GONJA bùlji, kíngámà (Rytz) KONKOMBA dedschaelbɔne (Gaisser) SISAALA kaamiŋ (Blass) **TOGO** BASSARI kaere (-tukurn) (Gaisser) katekatego (Gaisser) **NIGER** SONGHAI léptándà (*pl.* -à) (D&C) **NIGERIA** ARABIC arä (JMD) garä (JMD) garä kossa *a marrow* (JMD) garä stambuli *a pumpkin* (JMD) ARABIC-SHUWA murr BEROM yɛ̀ɛ́ (LB) DERA yáaŋgú (Newman) EDO éyèn (JMD) EFIK ḿfrì ǹdìsè, ḿfrì: *fruit; thus the pumpkin itself* (JMD) ǹdìsè (JMD; RFGA) ǹnàñí (JMD) ọkpọ ǹnàñí *the plant* (JMD) FULA-FULFULDE (Nigeria) foɗɗere (*pl.* poɗɗe, po'e) *the fruit pulp* (JMD) fol'yere (*pl.* pol'ye) *small young fruit* (JMD) huccere (*pl.* kutje) *the fruit-pulp* (JMD) kucceere *a whole fruit* (JMD) mbọrho *young leaves and shoots* (JMD) waygoore (*pl.* baygooje) (auctt.) GWARI knuba (JMD) knugba (JMD) HAUSA akwato *a marrow var.* (JMD) bàkánùwaà *a white-fruited var.* (JMD; ZOG) balaya *a marrow var.* (JMD) gamon fatake *a marrow var.* (JMD) ganwon fatake *a marrow var.* (JMD) gesuma (ZOG) goòjíí *a general term* (auctt.) gundar kabewa *small immature fruits* (JMD) gwaɗaɗɗasai *end of season leafy shoots with undeveloped fruit* (JMD) harsa *the fruit pulp* (JMD) harza *the fruit pulp* (JMD) kàbaíwaà, kàbeéwaà, kúbeèwaá (ZOG) kabééwa (JMD; LB) kabus, kabushi, rugudu, ruguguwa *a long-fruited marrow var.* (JMD) kalaye *the fruit pulp dried in strips* (JMD) katirga *end of season leafy shoots with undeveloped fruit* (JMD) wawan gona *a marrow var.* (JMD) y'ar-garii *a marrow var.* (JMD; ZOG) zazzaɓe *end of season leafy shoots with undeveloped fruit* (JMD) IBIBIO ǹdìsè (auctt.) IGBO ukọrọ *the fruit* (JMD) IGBO (Onitsha) ụ́gbọ̀gụ̀lụ̀ (KW) IGBO (Owerri) ányụ̃ IGBO (Umuahia) ụ́gbọghụ̀rụ̀, ụ́gbọ̀ghọ̀rọ̀ (KW) IJO-IZON (Egbema) udẹmụdẹ́ (Tiemo) utí (Tiemo) KANURI fámfam (C&H) fànnâ díwì (C&H) kàfétò *from Hausa* (JMD) sàádə̀ (C&H) NUPE bĕshé *after quartering and boiling* (RB) èbĕ (Banfield) TIV àgbàdù (JMD) furum (JMD) icegher-nyian *red fruited var.* (JMD) icegher-pupuu *white-fruited var.* (JMD) URHOBO ètò-úrhòbò (DA) koko-eyen (DA) YORUBA apala *general for pumpkin, etc.* (JMD) élégédé (JMD) gbọ̀rọ̀, gbọ̀rọ̀ esi *young leaves and shoots* (JMD) langbade (IFE) légedẹ (IFE) **WEST CAMEROONS** BAFOK dibok (JMD) KOOSI abok (JMD) KPE diboke (JMD) KUNDU dibuke (JMD) LONG dibok (JMD) LUNDU dibuke (JMD) MBONGE dibuke (JMD) NGEMBA nə̀bɔ̀´ (Chumbow) TANGA diboke (JMD) WOVEA diboke (JMD)

An annual, coarsely herbaceous, climbing, trailing or bushy, polymorphic plant; a native of Central America where it was domesticated in Mexico at least as early as 7000–5500 B.C. (18). It is now cultivated throughout the world except in arctic regions in numerous cultivars.

The polymorphic nature of the plant lends confusion to the vernacular names and some of the names given above may also refer to *C. maxima* Duch. Three general varieties may be recognized (18):

(1) var. *pepo*—field pumpkins

(2) var. *medullosa* Alef.—vegetable marrows

(3) var. *melopepo* Alef.—bush squash and pumpkins; summer squashes.

Other divisions into vegetative and fruit-forms have been attempted (22). Three kinds of *ɛfere* (pumpkin) are recognized in Ghana (14):

(1) *ɛfere pa*, or 'good pumpkin'—large, white fruit.

(2) *Akomfem-fere,* or 'guinea-fowl's pumpkin'—spotted or speckled fruit, spherical and smaller than (1).

(3) *Akan-fir,* or 'The Akan's pumpkin', commonly grown in the Cape Coast area. The shape varies: a large curved fruit bulged at one end, or not bulged; sometimes with warty or ribbed skin.

Elsewhere in West Africa two or three varieties are recognized, e.g., large, white; smaller, elongated, green with white blotches; green with reddish-yellow flesh, etc., and in Sierra Leone some half dozen cultivars are commonly grown.

The plant is, of course, grown principally for the fruit. The pulp is eaten as a vegetable or in soup. It is also cut into strips and dried for storage. The young leaves and shoots, and even the flowers, are used as a pot-herb (3, 5, 6, 11–13, 20, 21), and the end-of-season leafy shoots with undeveloped fruits are taken as a vegetable in N Nigeria (5, 13). The leaves are dried in southern Africa for winter use (21). The fruit-pulp contains 87–94% water, 4–8% carbohydrate, 0.5–1.8% protein and small amounts of fibre, oil and minerals, especially iron, phosphorus and calcium (4, 14, 15, 22). Fresh fruit-pulp contains vitamin C and niacin, riboflavin and thiamine, and the level of these declines on maturity (15, 21), and some bitter forms have a high amount of cucurbitacin (15).

The raw fruit-pulp makes a good poultice for minor burns (20), boils and inflamed swellings, or applied as a cooling compress for headache and neuralgia (5). It has been recorded as used for tumours of the eye, liver and corns on the feet (7). The pulp is used in some localities to dehair hides, and as a bate in tanning.

The fresh seeds are said to be edible (4, 10, 17), and the fruits to be excellent food for pigs and cattle in Ghana, either raw or cooked (14), and are fed to cattle in the U.S.A. (21). The general applicability of this information needs viewing with caution and may arise through local circumstances and judicious selection of cultivar. The seeds are commonly regarded as anthelmintic and useful as a taenicide in very many countries (2, 5, 16, 17, 19, 20). They have been official in the *British Pharmaceutical Codex* 1934 and *Pharmacopoeia Helvetica.* If freely eaten by poultry and ostrich in S Africa a sort of inebriation results with a loss of the use of the limbs and of the birds becoming 'hooked' on the seeds, a habit which seems difficult to break. Stock owners have also repeatedly suspected the seed of causing 'craziness' and symptoms of paralysis in cattle (21). *C. pepo, C. maxima* and *C. moschata* grown in Lebanon have been reported to contain a water soluble amino acid, 3-amino-3-carboxypyrrolidine, or *cucurbitin,* which is taenifugal but only in large dosage (16). A resin, which may be taenicidal, and a toxalbumin have been reported (17, 19). A fairly strong presence of alkaloids in Nigerian material is recorded (1). The seeds lightly torrefied and crushed in water are given in Congo to a woman in labour to promote delivery (2).

The chemical composition of the whole seed varies: 36–40% oil, 30–41% crude protein, 15–18% fibre, 3–4% ash and 2% carbohydrates (3, 14, 19). The oil consists of glycerides of the following acids: linoleic 43%, oleic 34%, palmitic 16% and stearic 8% (3). Cold press extraction which produces about 30–35% of the oil content is said to be suitable for edible purposes, and the lower grade for burning-oil (19).

The flowers are recorded as used cosmetically in Iran to improve the complexion, and medically for chest-troubles (8). The pollen grains are large and sticky and are much sought after by honey bees who also visit the flowers for nectar (9).

In N Nigeria children make toy trumpets called *bututu* (Hausa) from the hollow stem (5).

References:

1. Adegoke & al., 1968. 2. Bouquet, 1969: 102. 3. Busson, 1965: 414, 416–7 with seed analysis. 4. Chakravarty, 1968: 457–9. 5. Dalziel, 1937. 6. Guillarmod, 1971: 421. 7. Hartwell, 1969: 95. 8. Hooper, 1931: 312. 9. Howes, 1945. 10. Irvine, 1930: 138. 11. Irvine, 1952, a: 34. 12. Irvine, 1952, b: 47. 13. Irvine, 1956: 36. 14. Irvine s.n. 15. Kerharo & Adam, 1974: 379–81, with

phytochemistry and references. 16. Mihranian & Abou-chaar, 1968. 17. Oliver, 1960: 6, 59. 18. Purseglove, 1968: 122–4. 19. Sastri, 1950: 394–5. 20. Walker & Sillans, 1961: 141–2. 21. Watt & Breyer-Brandwijk, 1962: 357, with references. 22. Whitaker & Davis, 1962: 44–59, 194–210 with analyses.

**Diplocyclos palmatus** (Linn.) C. Jeffrey

FWTA, ed. 2, 1: 214, as *Bryonopsis laciniosa* sensu Keay.
English: lollipop climber (Rhodesia, Pole Evans).

A slender much-branched climber with perennial fleshy rootstock of forest on Fernando Po only in the Region, but recorded elsewhere in tropical Africa, Asia and Australasia.
The plant has been grown as an ornamental in gardens in Kenya (1) and in Rhodesia (2) for its decorative fruits which when ripe are bright carmine with pure white stripes.

References:
1. Bally 7788, K. 2. Pole Evans 6800, 6801, K.

**Kedrostis foetidissima** (Jacq.) Cogn.

FWTA, ed. 2, 1: 210.
West African: SENEGAL SERER mãndafura (JB)

A slender climbing or trailing herb to about 2.5 m long of forest margins, woodland and wooded savanna recorded from the northern part of the Region from Senegal to Niger and N Nigeria, and occurring widespread across Africa to E and S Africa, and into tropical Asia to Burma.
The plant has a rank foetid smell.

**Kedrostis hirtella** (Naud.) Cogn.

FWTA, ed. 2, 1: 210, quoad Dalziel 128 as *Toxanthera sp.* vel aff., under *K. foetidissima* (Jacq.) Cogn. UPWTA, ed. 1, 58 as *K. foetidissima* sensu Dalziel.
West African: NIGERIA HAUSA bàrkònón bírii = *monkey's chili* (JMD; ZOG)

A climber or trailer to 2 m long from a perennial tuberous rootstock, of deciduous bushland recorded from Senegal and N Nigeria, and occurring eastward and across Africa to Ethiopia, E, S and SW Africa.
The Hausa name refers to the scarlet beaked fruit and applies perhaps to other species of Cucurbitaceae (1).

Reference:
1. Dalziel, 1937.

**Lagenaria breviflora** (Benth.) Roberty

FWTA, ed. 2, 1: 206, as *Adenopus breviflorus* Benth. and *A. ledermannii* Harms. UPWTA, ed. 1, 53, as *Adenopus breviflorus* Benth.
English: wild colocynth (Ainslie).
French: calebasse tigrée (Berhaut).

West African: **SENEGAL** BASARI a-kása-kása (Ferry) BEDIK gi-nyudùŋ (Ferry) DIOLA é sigir (JB) ku batak (JB) FULA-PULAAR (Senegal) denni biram dau, denni: *water melon* (A. Chev.) MANDING-BAMBARA ka bara ni (JB) SERER mbomb (JB) **THE GAMBIA** DIOLA-FLUP é sigir = *heart* (DF) kufatak = *lamb* (DF) **GHANA** AKAN-TWI asãmãn-akyẽkyẽa = *spirit's water melon* (FRI) ANYI-AOWIN aboa ngateɛ (FRI) SEHWI aboa-ngateɛ (FRI) GA ánúwátrɛ̀ (FRI; KD) NZEMA aboa-ngatseɛ = *leopard's groundnut, from the blotched fruit* (FRI) **NIGERIA** HAUSA gojin jima, gúnàr jiímaà, jiímaà: *to tan* (JMD; ZOG) IGBO ányụ́ḿmụ̄ọ́ = *anyu of the spirits* (Singha); *any plant qualified by* ḿmụ̄ọ́: *spirits, dead, is inedible, as opposed to another which is eaten* (KW) uriem (BNO) YORUBA eso gbegbe, eso: *fruit* (IFE) eso gbo ayaba *the seed* (IFE) ito (JRA) tagiiri (auctt.)

A perennial climber ascending to the forest canopy, occurring from Senegal to W Cameroons, and generally widespread in tropical Africa.

The leaves very scabrid and sandpapery. The stem when broken has an unpleasant smell (4), and a decoction from it is said to be used in Nigeria for headache (2). The root is used in Tanganyika as a purgative (7) and in Nigeria as a vermifuge (2).

The fruits are dark green with creamy blotches, and are ovoid to 9 cm long. They are commonly used in Nigeria for depilating hides. The fruits are cut up, put in water with lye of wood-ashes and in this hides are left to soak for one or two days (6, 8). Alternatively the hides are stretched and the inner surface scraped clean, and then the fruit pulp is rubbed in followed by a free application of dry wood-ash. Depilating is done after the folded hide has been steeped for a further day in the lye of wood-ash (6). The fruit is perhaps also used in bating-bath to prepare skin to receive the tanning material (5, 6).

The fruit is extremely bitter and contains a strong amount of alkaloids (1). The fruit is used in Nigeria as a cathartic (2).

The seeds are used in some places to stupefy fish (6). In Sudan the seeds are said to be chewed while smoking tobacco to induce a sort of intoxication (Schweinfurth fide 6). In S Nigeria a seed-decoction is reported given to pregnant women (9) but the purpose is not stated.

References:

1. Adegoke & al., 1968: as *A. breviflorus*. 2. Ainslie, 1937: sp. no. 13, as *A. breviflorus*. 3. Chakravarty, 1968: 406–7, as *A. breviflorus* Benth. 4. Chandler, 1673, K. 5. Dalziel 735, K. 6. Dalziel, 1937. 7. Koritschoner 893, K. 8. Singha, 1965: as *A. breviflorus*. 9. Thomas, N. W., 2105 (Nig. Ser.), K.

## **Lagenaria guineensis** (G. Don) C. Jeffrey

FWTA, ed. 1, 1: 206 as *Adenopus guineensis* (G. Don) Exell.

West African: **IVORY COAST** KWENI biéla (B&D)

A herbaceous climber of secondary bush occurring in Sierra Leone and Ivory Coast, and in Cameroun and Central Africa.

The sap is used in Ivory Coast as a collyrium for ophthalmias (1).

Reference:

1. Bouquet & Debray, 1974: 78 as *A. guineensis*.

## **Lagenaria rufa** (Gilg) C. Jeffrey

FWTA, ed. 2, 1: 206, as *Adenopus rufus* Gilg.

West African: **SIERRA LEONE** KRIO tagiri (FCD) MANDING-MANDINKA kani-ŋjɔ̃ (FCD) MENDE kɔli-goko (FCD)

A herbaceous climber of the secondary bush in Sierra Leone, and in E Cameroun.

Though known to races in Sierra Leone, it is apparently of no recorded usage.

**Lagenaria siceraria** (Molina) Standl.

FWTA, ed. 2, 1: 206. UPWTA, ed. 1, 58–60, as *L. vulgaris* Seringe.

English: calabash; calabash gourd; gourd calabash; and forms: white pumpkin; bottle gourd; club gourd; etc.

French: calebasse; calebasse cultivée; courge; cougourde; and forms: gourde massue (club gourd); gourde trompette (trumpet gourd); gourde de pèlerins (traveller's gourd); etc.

West African: **SENEGAL** BASARI a-ngɔ̀w̃ (Ferry) BEDIK gi-ngóm (Ferry) DIOLA ka tuk (JB) KONYAGI i-ntyen, i-yawù (Ferry) MANDING-BAMBARA bara (JB; K&A) fié (K&A) filé (JB; K&A) MANDINKA kalamaa (*indef.* and *def.*) (K&A) SERER fed (JB; K&A) limb (JB; K&A) mbed (auctt.) WOLOF gamba (auctt.) kok *small gourds, split open* (AS) leket *large, round gourds* (AS) mbag *gourds for drawing water* (AS) mbatu *medium sized gourds, split for drinking* (AS; DF) nanu *small gourds* (AS) patu *gourds for making butter* (AS) som (AS) taglu *gourds for palm wine* (AS) tah *small gourds* (AS) teletj *long gourds* (AS) yad (AS) yomba = *cheap, easy* (auctt.) **THE GAMBIA** DIOLA (Fogny) ka tuk = *bamboo* (DF) MANDING-MANDINKA miraŋ (*def.*-o) = *something one drinks with* (Williams) **GUINEA-BISSAU** BALANTA fôóti (JDES) BIDYOGO eparrá (JDES) omparsa (JDES) CRIOULO cabaça (JDES) FULA-PULAAR (Guinea-Bissau) córè (JDES) còré (EPdS) fahándu (JDES) lami-córè (JDES) ordè (JDES) MANDING-MANDINKA mirandjô-lô (JDES; EPdS) MANDYAK pucúo (JDES) MANKANYA udungue (JDES) PEPEL ecanda (EPdS) **GUINEA** FULA-PULAAR (Guinea) horde (CHOP) MANDING-MANINKA bolin-bara *bottle-gourd* (CHOP) fé (FB) fé lemba *large, round gourd* (CHOP) fé lémé nkuru *small gourd, cut for saucers* (CHOP) kalama léga *long gourd, cut for spoons* (CHOP) kulu félé *a c. var. with the largest gourds* (CHOP) **SIERRA LEONE** BULOM (Kim) sasa *var. used as a rattle* (FCD) taga (FCD) BULOM (Sherbro) chãsa-lɛ (FCD) gbos-lɛ *for a gourd used to collect palm-wine* (FCD) pɛpɛ-lɛ *split gourd used as a bowl* (FCD) FULA-PULAAR (Sierra Leone) hɔrədɛ *the plant, and the gourd of common var.* (FCD) pɛlɛtɛ *fruit of var. used as a rattle* (FCD) GOLA gɛrɛ *variety used as a rattle* (FCD) k-pai, k-pali *the plant, and whole gourd* (FCD) k-puo, k-puɔ *split calabash used as a bowl* (FCD) KISSI bala *var. with medium sized fruit* (FCD) hemdo *var. used as a rattle* (FCD) tala *a large-fruited var.* (FCD) KONO ba (FCD) g-bala, g-bara *the plant, and the gourd* (FCD) ta, tala *a split gourd used as a bowl* (FCD) KRIO kalbas = *calabash* (FCD) LIMBA (Tonko) datogo *the plant* (FCD) kaləmã *var. used as a rattle* (FCD) kəbulo *fruit of ordinary var.* (FCD) LOKO kua (FCD) MENDE kpula,kpulɔ *the plant* (auctt.), or *spherical fruit, long neck used for palm-wine* (FCD) sɛgbula *small fruit, long neck used as a rattle* (FCD) tawa *large spherical fruit, very little neck* (FCD) tawa-wawa *large spherical fruit, short neck* (FCD) TEMNE *an*-bos (JMD) *a*-kalma *var. used as a rattle* (FCD) ɛ-pɛpɛ *a gourd split in 2 halves* (FCD) VAI koŋgoe (FCD) **LIBERIA** MANO gah (JMD) **MALI** DOGON gabá *general for a gourd-bearing plant* (C-G) gabá íí, kɛ̀mɛ *general for a gourd* (C-G) MANDING-BAMBARA fié (A. Chev.) filé (Vuillet; FB) **GHANA** ADANGME-KROBO akpe, kpatu, tɔ *gourds of the largest size* (JMD) daka *a symmetrical bottle-gourd for holding liquids* (FRI; JMD) sasagné (JMD) tɔtɔtʃo = *ladle, a bottle-shaped gourd with handle that can be cut and used as a large ladle or spoon* (FRI; JMD) tʃimi *a round calabash* (FRI) AKAN-ASANTE adidi-pakyie *a large calabash for storing meat* (Enti) bɛntowa *bottle-shaped calabash with long, straight neck* (Enti) FANTE adenkum-mfoaa *symmetrical bottle-gourd for holding liquids* (FRI) adidi apakyi *a large calabash for storing meat* (FRI) apakyi(e) *gourd of the largest size, for storing clothes, food, etc.* (FRI; JMD) apakyiwa *small calabash for holding beads, etc.* (FRI) bɛntua *bottle-shaped calabash with long straight neck* (FRI; JMD) ɔ-danka *bottle-gourd with a constriction, for holding liquids* (FRI; JMD) dwereba *a round calabash* (FRI) ngotoa *a bottle-gourd for holding oil* (FRI) nsutoa *a bottle-gourd for holding water* (FRI) TWI adenkum *a bottle-gourd* (FRI; JMD) adidi apakyi *large calabash for storing meat* (FRI) apakyi(e) *gourd of the largest size, for storing clothes, food etc.* (FRI; JMD) apakyiwa *small calabash for holding beads, etc.* (FRI; JMD) apebentutu *small tuberculed gourd* (JMD) apɛbɛntutu *small tuberculed gourd used only by men for holding their fetishes* (FRI) bentoa, bɛntoa, bentowa *bottle or club-gourd with long straight neck used in local medicine* (FRI; JMD) kora *a round calabash giving a bowl used for measuring palm wine* (FRI) korawa *a small form of 'kora'*, wa: *small* (FRI) krokuma (korokuma) *large calabash in which 'kora' calabashes are kept* (FRI) mfoaa *a bottle-gourd* (FRI; JMD) ngotoa *bottle-gourd for holding oil* (FRI) nkaŋ-kruwa *a small bottle-shaped gourd with neck cut in 2, used for spoons* (FRI) nsɔase *a round calabash for collecting palm-wine* (FRI) nsutoa *bottle-gourd for holding water* (FRI) ɔdanka *bottle-gourd for holding gunpowder* (FRI) sã-kora *a round calabash for holding palm-wine, and cut in two to make drinking cups, from 'kora', a calabash cup measure for* nsã: *palm-wine* (FRI) toa *bottle-gourd with a constriction for holding liquids* (FRI; JMD) DAGBANI garle *bottle-shaped* (Gaisser) kjirge *club-shaped, halved for ladles* (Gaisser) tschocho *large calabash* (Gaisser) tumbe *bottle-shaped* (Gaisser) GA adéŋkũ *a calabash rattle* (KD) akpaki *a large calabash* (FRI; JMD) bén'toa *the plant, or its fibre* (KD) kyene (JMD) tɔ *a symmetrical bottle-gourd for holding liquids* (FRI; JMD) tʃene *a round calabash* (FRI) GBE-VHE adaŋga *small club-gourd for enemas, cf. 'adagogoe'* (Ewe Dict; FRI) aɖaŋga *bottle gourd* (FRI) adangagoe *large gourd for liquids* (JMD) aha go = *palmwine gourd; from* aha: *palmwine;* go: *bottle* (FRI) ako *large gourd for liquids or bottle-gourd for drinking* (FRI; JMD) akpẽ, akpaku *general term for calabash* (JMD) akugoe *long-necked bottle-gourd for liquids* (FRI) asɔe *bottle-gourd with thick spout used for 'sieving' palm pericarp in preparing palm soup* (JMD) dugoe = *gunpowder container; a small bottle-gourd* (FRI) ekplẽ *the fruit* (JMD) go *large gourd for liquids* (JMD) sasa goe *small bottle-gourd with long straight neck for enemas* (FRI) tigoe *'black' (medicinal) powder container; a very small gourd* (FRI) tigwe *bottle or club-gourd* (JMD) trɛ *a round calabash used as a cup for palmwine on special occasions* (FRI; JMD) trɛkpe *the unopened fruit* tsi go = *water gourd; from* tsi: *water;* go: *bottle, a symmetrical bottle gourd for water* (FRI) tsi gui *a bottle-gourd; or small bottle-gourd with long straight spout for enemas* (FRI; JMD) VHE (Awlan) akatse goe *gourd used as cymbals for shaking with beads* (FRI) akpaku *a large calabash* (FRI) VHE (Pecí) adanga *bottle or club-gourd* (JMD) akayɛ *gourd used as cymbals for shaking with beads* (FRI) akogo (JMD) akpɛ, akwɛ *a large calabash* (FRI; JMD) dizebome (JMD) tigwe, tsigui *bottle or club-gourd* (JMD) GRUSI sunga (Ll.W.) GUANG-GONJA chɛ̀fɔ̀l *var. with a necked-gourd* (Rytz) dèŋkèŋ, kàtùrbí (Rytz) jɔ̀tá (Rytz) kàwíẽ *the plant* (Rytz) lɔ̀ŋkɔ́ŋ *the fruit* (Rytz) MOORE gole (FRI) wamde (FRI) **TOGO** MOORE-NAWDAM filinga *club-shaped, split for ladles* (Gaisser) TEM langa (FB) sununga (FB) TEM (Tshaudjo) dschola *bottle-shaped* (Gaisser) dshendsha *small tuberculed gourd* (JMD) langa *bottle-shaped* (Gaisser) sununga *club-shaped, split for ladles* (Gaisser) tjikare *a large calabash* (Gaisser) **DAHOMEY** BUSA kpê-on (Bertho) **NIGER** SONGHAI gáasí (*pl.* -ó), tàndà (*pl.* -à) (D&C) **NIGERIA** ABUA éekó (KW) òte (KW) AGWAGWUNE (Abini) urok (KW) AKPET ekori (KW) AKPET-EHOM aboh (KW) ANAANG ìkìm *the plant* (KW) ìkpɔ̀ *a shaped calabash* (KW) ARABIC-SHUWA bukhsa langa *small-mouthed bottle-gourd* (Letham) gumbul *bottle-gourd* (Letham) gumbul alme *bottle-gourd* (Letham) ATTE u-bene (*pl.* i-bene) (KW) AYU ice (KW) BA'BAN otoh (KW) BADE-NGIZIM fùnà (Kraft ex KW) kurtu (Meek) BEROM cɔ̀ *a general term for all shapes* (LB) gòfùs *general for large calabashes carried in a pannier* (LB) kácɔ̀ŋ *general for a bottle-gourd* (LB) kwɛ̀t *general for elongated calabashes* (LB) lété *general for club-shaped calabashes used as a spoon or ladle* (LB) syìí *general for a calabash, not elongated* (LB) BOKYI diba (KW) BOLE gewi *bottle-gourd* (Benton) kula *calabash* (Benton) liggide *spoon* (Benton) BUSA mkpini (Bertho.) CHAWAI ajingvwo (Meek) CHAWAI-KURAMA laura (Meek) ne-jááró (*pl.*: á- or n̩-) *calabash* (Herrman Jungraithmayr ex KW) DEGEMA ụ̀ḅàḅá (*pl.* ị) (KW) DERA gə̀là *gourd bottle* (Newman; KW) gila (Meek) gwákə̀rák (*pl.* gwákə̀rnjén) *large calabash* (Newman; KW) líɓè (*pl.* lípén) *calabash* (Newman; KW) EBIRA (Etuno) mgbana (KW) EDO ẹ̀kpérẹ̀, uko ẹ̀kpérẹ (Amayo) okpan (JMD) EFIK àyàrá *small flask-shaped gourd* (JMD) ékíkóp *small flask-shaped gourd* (JMD) ekpat *long narrow gourd* (JMD) ìkìm *small flask-shaped gourd without opening: used as floats or by wine tappers* (RFGA; Winston) ìkó *general term for gourd or calabash* (auctt.) ìkpáñ = *spoon* (JMD; RFGA) m̀bọ̀nì *mark made by blood-letting; hence a small gourd used for cupping* (JMD; KW) òkpóñ *a calabash used as a pot* (RFGA) ENGENNI ạ̀gbị̀nạ̀ (KW) EPIE egbele (KW) ègbèlè (*pl.* ị-) (KW) EVANT u-kwlobu (*pl.* vi-kwlobu) (KW) FULA-FULFULDE (Nigeria) ɓiiloonde *large bottle-gourd curved neck, cut open on one side for use as a hand-dipper* (JMD) ɓirdude *wide-mouthed bottle-gourd to hold milk* (JMD) bunga *horn made from a gourd:* shantu, *Hausa;* janturu, *Fula* (JMD) dawɗere (*pl.* dawdeeje) = *gourd for milking; lit. a gourd held between the legs* (MM) dumbaare (*pl.* dumbaaje) *a large c. var.* (Taylor) eegirde (*pl.* eegirɗe) *a large c. var. for water* (Taylor) faandu (*pl.* paali) *water-bottle gourd* (JMD) faandu dawa, faandu dawara = *ink-pot gourd* (JMD) faandu hecceru (JMD; Taylor) faandu heccuuru (*pl.* paali kecci) *a large white-fleshed pumpkin, bottle- or club-gourd; used for floats, fishing keep-net or courage, drums, e.g.,* talleeru: *hunting drum* (MM) faandu layru (*pl.* layruuji) *small gourd for pomade etc.* (JMD) faandu urdi *small gourd for scent* (JMD; MM) fawruɗe (*pl.* pawruɗe) *a c. var with largest calabash, used in markets to hold smaller ones, from* fawa: *to pack one's goods and chattels* (JMD; MM) fenndirɗe (*pl.* penndirɗe) *wide-mouthed bottle-gourd for milk* (JMD) futeere (*pl.* puteeje) *ovoid, tuberculed gourd* (JMD) gonogono (*pl.* gonogonooji) *cucumber shaped fruit, edible* (JMD; Taylor) goraru (*pl.* goraji) (JMD) gulum butu *cucumber shaped fruit, edible* (JMD) gummbal (*pl.* gummbe) *calabash with small opening for honey, water, etc.* (JMD) gummbal nalle *club-shaped or cylindrical for immersing hands for henna staining, nalle:* henna (JMD; MM) gurmusal (*pl.* gurmuse) *a small calabash* (Taylor) horde (*pl.* kore) *a calabash spoon* (Taylor) janndorde *calabash for souring milk* (JMD) jantuuru (*pl.* jantuuji) *long narrow gourd* (JMD) jollooru (*pl.* jallooji) *water-bottle gourd* (JMD) junguru *club-shaped or cylindrical gourd for immersing hands for henna-staining* (JMD) kuccere *whole calabash with pulp: a general term* (JMD; MM) litaare *small calabash with hole in it for carrying by string* (JMD) loonde (*pl.* looɗe) *a large pot such as used for a cooler* (Taylor) mabakachi *c. var.* (JMD) masaki *c. var.* (JMD) moɗaare *large bottle-gourd, curved neck, cut open on one side for* *a hand-dipper* (JMD) namarde (*pl.* namarɗe) *c. var. with largest calabash, used in markets, for floats* (JMD) putteputre *ovoid, tuburculed gourd; perhaps variant of futeere* (MM) sanndorɗe (*pl.* caandorɗe) *calabash for souring milk* (JMD; Taylor) tummbude (*pl.* tummbuɗe), tummude (*pl.* tummuɗe) *a generic term for calabash* (auctt.) tunduwol (PF fide MM) yonkirde (*pl.* yonkirɗe) *gourd for making butter by shaking* (JMD; MM) yoogirde (*pl.* yoogirɗe) *large calabash, esp. for water* (Taylor) GA'ANDA kurupta (Meek) tíɓé (Kraft ex KW) GURE-KAHUGU wara (Meek) GWANTO kucha (KW) GWARI bvokun *long-necked bottle gourd* (JMD) bwebihi *c.var.* (JMD) esshi *small gourd* (JMD) gurra *c.var.* (JMD) gwabwi *long-necked bottle gourd* (JMD) obwe *c.var.* (JMD) zungkwa *bottle-gourd with long narrow neck, split for spoons, ladles, etc.* (JMD) HAUSA agofata *large, shallow gourd for carrying loads* (JMD) akwato *small gourd for a soap-dish* bango *large neckless gourd* (JMD) borin danki *the smallest gourd* (JMD) bututu *bottle or club-gourd – the neck used as a funnel* (JMD) búúta *general for a gourd, not elongated* (LB) ɗan jallo *small pear-shaped gourd for water for ceremonial ablution, or a journey* (JMD) ɗan kwakwangi *small gourd for a soap-dish* (JMD) ɗan muda *small gourd* (JMD) dasa *water-bottle gourd* (JMD) dúmaá, dúmáá *general term for all shapes of fruit* (ZOG; LB) dúmán gaùraákaa *large round gourd of irrigated farms* (Bargery; ZOG) dúmán kwáryaá *c.var.* (Bargery; ZOG) dúmán luúdàyii *c.var.* (Bargery; ZOG) gago, gako *bottle-gourd with solid neck* (JMD) godo *small bottle-gourd for tobacco* (JMD) gooráá *general for bottle gourds used for water-bottles and floats* (auctt.) gumbali *calabash with small opening for honey, water, etc.* (JMD) guraka *large gourd used as floats* (JMD) gyanɗama *water-bottle gourd* (JMD) jallo *small pear-shaped gourd for water for ceremonial ablution, or a journey* (JMD) kabewa (JRA) ka-fi-ɗa-wuya *a very small gourd* (JMD) kan doki *elliptical gourd* (JMD) kan kare *a long narrow gourd* (JMD) karuguna (ZOG) kata *small gourd to hold butter* (JMD) kololo *a large dipper* (JMD) kúlùbuútuù *cucumber-like gourd, edible* (JMD; ZOG) kundumasa (ZOG) kurtun lalle, kurtutu *bottle or club-gourd; wide end for immersing hands for henna* (lalle) *staining; neck end for a horn* (JMD) kurtun tadawa, kurtun tawada *small gourd for ink-pot* (JMD) kurtúú *general for large gourds carried on a pannier* (LB) ƙurzunu *cucumber-like gourd, edible* (JMD) kwachare *small gourd* (JMD) kwáryaá *general term for an ordinary calabash cut vertically* (JMD; ZOG) kwoton tadawa *ovoid, tuberculed gourd for ink-pot, etc.* (JMD) leɓan uwar miji *small calabash with a lip* (JMD) lúúdayíí *general for club-shaped gourds with long narrow neck spllt for spoons, ladles, etc.* (JMD; LB) makamfachi *a large dipper* (JMD) másákíí *general for elongated gourds* (LB) matar dakare *water-bottle gourd* (JMD) moɗa *large bottle-gourd, curved neck, cut open on one side for a hand-dipper* (JMD) murgwi *large gourds as floats* (JMD) rudu *gourd for beer* (JMD) san doki *spherical gourd* (JMD) shantu *long narrow gourd used as a horn* (JMD) tsana *cucumber-like gourd, edible* (JMD) zunguru, zuru *club-shaped or cylindrical gourd for immersing the hands for henna-staining* (JMD) HAUSA (West) ba-yammi *c.var* (DA) bumbu *whole gourd with pulp* (JMD; ZOG) ɗan kwakwangi *small gourd for a soap-dish* (JMD) jemo *calabash with a lip* (JMD) kanho *a large dipper* (JMD) karuguna *whole fruit with pulp* (Bargery) ki ta ɓewa *c.var* (JMD; ZOG) ƙoƙo *small calabash cup* (JMD) ƙoƙuwa *small calabash cup* (JMD) ƙululu *large water-gourd* (JMD) kumbo *snuff-box gourd, or small cup with cover* (JMD) ƙundumasa *whole fruit with pulp* (Bargery) kurumbo *calabash cut horizontally above the middle with a narrow mouth* (JMD) kwachiya *small calabash cup* (JMD) kwalalo *large dipper* (JMD) kwargo *small calabash with a hole in it for carrying by string* (JMD) lita *small calabash with a hole in it for carrying by string* (JMD) makurkura *small calabash cup* (JMD) tamaula *small calabash with a hole in it for carrying by string* (JMD) zomodo, zumudi *calabash with a lip* (JMD) HWANA ɗɛ́ŋdà (Kraft ex KW) kuɓana (Meek) HYAM hep (Meek) IBIE (North) ukoko (KW) ICEN mwetsa (Meek) ICHEVE e-tsəndəgə (*pl.* ne-tsəndəgə) (KW) IDOMA ɔbàtu (KW) ógò (KW) IGBO abwẹ *large gourd for floats* (JMD) ágbè (KW) agbo (BNO) ágbụ̀gbà (KW) akbẹlẹ, akpẹlẹ *long narrow gourd for flutes, etc. with mouth-hole in the centre* (JMD) ebele *bottle or pot-shaped gourds* (BNO) èbèlè (KW) iko *cup, or tumbler-shaped gourds* (BNO) ḿbùbò (KW) m̀gbọ̀lọ̀ (KW) nkpọ *bottle gourd split for spoons* (JMD) ǹkúkú (KW) ñtìkpó (KW) ọba *bowl-shaped gourds* (BNO) ọ́bà (KW) òbò (KW) okbili *long narrow gourd used as flutes, etc.* (JMD) okutu *long gourd* (JMD) ọ̀nụ̀nụ̀ *small bottle-shaped gourd used as a powder-flask* (KW) òpì *horn, from the shape* (KW) ọyọ *bottle gourd* (JMD) ụgba *gourds with wide mouths like pans* (BNO) ụ́gbā (KW) ugbugba (BNO) ùkó *small cup-like gourd* (JMD) IGBO (Awka) mbùbù (JMD) ọ́bà (JMD) IGBO (Onitsha) ágbè *a long, narrow gourd for carrying water or wine* (KW) m̀kpá *general for a small calabash, or gourd* (KW) ọ́bà *general for calabash* ọbwọlọ (JMD) ọ́gbụ̄gbá *a gourd prepared for fetching water* (KW) úgbā *general for calabash* (KW) IGBO (Owerri) èbèlè *the plant* (JMD) mgbam (AJC) IJO-IZON àgbá (KW) igògò (KW) pìghàn (KW) ụ̀gbá (KW) ụ̀gbàgbá (KW) ùgbàlá (KW) ùgbégbé (KW) IJO EAST (Kalabari) àḅụ̀rọ̀ (Williamson) ḅàḅà (Williamson) IKWERE àgbọ̀rọ̀ (KW) ọ́bèlè (KW) ISEKIRI ugbá (KW) JANJO la (Meek) JARA gila (Meek) gila (Meek) JUKUN (Gwana) akũri (Shimizu after Meek) JUKUN (Kona) aku (Meek) JUKUN (Takum) kusã̀ (KW) JUKUN (Wukari) kwi (Meek) JUNKUN (Donga) bùnà (Meek) KAJE kashiom (Meek) KAKA ŋgap (KW) KANUFI-KANINGKON kergba (KW) KARSHI igwa (ikan) (KW) KANURI dəmbá *c. var. with largest calabash*

(JMD) gùwá (JMD) jeni *bottle-gourd with long narrow neck, split for spoons, ladles, etc.* (JMD) jìwí *water-bottle gourd* (JMD) kumo (JMD) ngùwú *the largest calabash used in markets and for separating grain from husk, and as a float* (JMD) KAREKARE ɗàyí = *calabash* (Kraft ex KW) jewi (Meek) kare (Meek) KATAB kurum (Meek) KHANA ékób (KW) KPAN nyásó (KW) u-kpán (*sing.*) (KW) ù-tikà (*sing.*) (KW) LONGUDA gwaraki (Meek) NDORO tongi (Meek) NGAMO gõbo (Meek) shókó (Kraft ex KW) NUPE bàbò *a gourd net-rattle made from zùngùrù* (RB) bàbò *bottle-necked form* (Banfield) bingi *small bottle-gourd* (Banfield) bingi *smaller than bàbò, and exclusively associated with charms* (RB) evo *general* (Banfield) gbàla *large gourded form* (Banfield) kókpa *long tubular gourded form* (Banfield) kondò *a spherical calabash used as a fisherman's creel, and also as the body of a large single-headed drum* (RB) kondò *large gourds* (Banfield) kpasà *large ladle-shaped form* (Banfield) vàtà *a flat circular gourd used for carrying things on the head* (RB) zùngùrù *a thin tubular gourd used to make a musical instrument open at both ends known as 'santo' (from Hausa: shantu), said to have now disappeared from Nupeland. This type is used for applying henna to the hands* (RB) OBOLO ògbòkót (KW) OGORI-MAGONGO ọbọ-ukwa (KW) OKPAMHERI ugo (KW) uko (KW) OKPE apele (KW) ORING (Ufia) ù-gà (*pl.* ì-gà) (KW) OTANK i-jendir (*pl.* í-jendir) (KW) PITI ribo (Meek) SHALL nba (KW) SHANGA fiê fiê küe TERA bungda (Meek) dəɓá *? this sp., or others yielding 'calabashes'* (Newman) TERA-PIDLIMNDI bungdi (Meek) dı̃rbí (Kraft ex KW) TIV aluku *c.var.* (JMD) icegher *c.var.* (JMD) ijọndogh *a general term* (JMD) íjọ́ndúgh (KW) ikpete *c.var.* (JMD) ikpokpo *c.var.* (JMD) iyongu *c.var.* (JMD) kapu *c.var.* (JMD) kwèsé *c.var.* (JMD) mkem *c.var.* (JMD) tsogh (*sing.*), icogh (*pl.*) *c.var.* (JMD) UBAGHARA (Biakpan) ngbana (KW) UBAGHARA (Ikun) ikim (KW) itut (KW) UHAMI-IYAYU ugban (KW) UKPE eki (KW) URHOBO okpan (JMD) òwàrá *bottle-gourd* (Faraclas) UZEKWE lì-ggó (*pl.* là-ggó) (KW) YEDINA hebi *water-bottle gourd* (JMD) maruadde *bottle-gourd with long narrow neck, split for spoons, ladles, etc.* (JMD) YORUBA àdó *small gourd* (auctt.) agbè *bottle-gourd* (JMD) ahá *small gourd* (JMD) akèngbè *bottle-gourd* (JMD) akèrengbe *bottle-gourd* (JMD) akoto *a deep calabash* (JMD) arọ̀ *gourd used as a quiver* (JMD) atọ́ *long-necked gourd, for spoons* (JMD) igbá (JMD) igbá-(a)jẹ *large gourd* (JMD) irere atiowoeyo (IFE) itakun igbá *the plant* (JMD) koto *a deep calabash* (JMD) pánsá *a dry unopened gourd containing the seeds* (JMD) pọọkọ́ = *a small ladle* (JMD) YUNGUR dagumra (Meek) **WEST CAMEROONS** BAFOK eyengo (DA) DUALA èkàŋgà *a gourd* (Ithmann) mbàmbe *a large calabash* (Ithmann) KUNDU ekondokia (DA) LONG eyengo (DA) LUNDU ekodokon (DA) MBONGE efimbiriki (DA) NGEMBA atɛ̃Cfl (Chumbow) TANGA ekodokori (DA)

*NOTE: These vernacular names in general refer to the gourd rather than to the plant, or occasionally to some article made from the gourd as indicated above. Some names may also apply to other genera with similar fruits, e.g., Cucurbita, Citrullus, Cucumeropsis, etc.*

A herbaceous annual, climber or trailer to 4–5 m long, subspontaneous in savanna and bushland and cultivated throughout the West African region and in all countries of the tropics throughout the world. The plant is certainly one of the most ancient to have been taken into cultivation by man. Its original home is lost in the antiquity of prehistory, but the region of the Afro-Asian conjunction is the most probable. If not actually indigenous in West Africa, it must have reached there at an extremely early age, for it has been found in deposits in Mexico dated at least 5500 B.C. and its advent to the Western Hemisphere is thought to have been by floating on Atlantic currents from West Africa. One should note that the calabash of the Caribbean is the fruit of quite a different plant, *Crescentia cujete* Linn. (Bignoniaceae).

The fruits display a remarkable diversity in size, shape, colour and patterning. Large spherical ones may be as much as 60 cm diameter. Bottle-shaped gourds may attain 50 cm in length of varying diameter at the wider distal end and of varying length, shape and diameter of the neck. In general the fruit is too bitter for use as food. If not actually poisonous, it is purgative. It has been for the use of the dried gourds as containers that man has been principally interested in their cultivation. The shell of the ripe fruit is hard, woody and impermeable to water and when the pulp and seeds have been removed serve as containers for all manner of purposes. For lightness and strength and relative ease of production, they are admirable, and invaluable for transportation and storage of liquids such as water and palm-wine, before the availability of pottery and glassware which anyhow are but expensive, heavy and fragile alternatives. The diversity of usage may be seen in the list of vernacular names given above. Gourds tend to be given cultivar names according to shape and use and to be used only for specific purposes. Gourds are however plastic in the younger stages of growth and may within limits be moulded by cultural practices to desired shapes. Once gourds have been used for a certain purpose, e.g. to contain liquids, they may acquire the taint of the contents and so be unusable for different substances.

In some races bitterness of the fruit has been selected out or greatly reduced and so for the pulp to be edible. Nevertheless they are eaten while still young and tender, and some are held to be as good as a pumpkin (6). The pulp is used in Asian medicine as a diuretic and antemetic, and to soothe coughs and as an antidote against certain poisons (5).

The seeds are relatively large and are surrounded by a tough outer skin. In a sample from Sudan the kernel amounted to 68% of the seedweight (14). They are a masticatory and are commonly eaten. Some cultivars are grown for the kernel-oil which commonly amounts to 40–50% of the seed in W African material (7, 9), or to 67% of the kernel (13). The Sudan sample cited above gave 52% of a pale yellow oil consisting of linoleic acid 42%, oleic acid 38%, stearic acid 13% and palmitic acid 7%. Linolenic acid is absent and the oil has poor drying qualities making it of little use in paint manufacture as the low iodine value, 105, indicates (14). The oil is used in cooking. Protein in W African seed is recorded as 5% (7, 9), and as high as 30.7% of the kernel-weight in Sudan seed in which also the crude fibre was very low at 1.6%. The cake after expression of the oil is suitable for cattle-fodder. The seed is taken internally, or the seed-oil is applied externally in India for headache (5). The seed-oil is considered anthelmintic (11). In Gabon, the seeds are prepared into a sauce, or crushed to make a paste for consumption (12). Saponin has been detected in the seeds (13), and no alkaloid (1).

The vegetative parts of most varieties become bitter. They are thought to contain amygdalin and thus to be cyanogenetic under certain conditions (13). However, in many countries the young shoots and tender stems are eaten as a vegetable (6, 7, 13). They are a rich source of minerals and vitamins (4, 10, 11). Leaves are used in Indian medicine as a purgative and a soup of young shoots is tonal against constipation (5). In India (5) and in Nigeria (2) a leaf-decoction is given for jaundice. A dressing of crushed leaves and palm-oil is applied for urticaria caused by caterpillars in Congo and the use of leaves as baby's nappies is recorded (3).

References:

1. Adegoke & al., 1968. 2. Ainslie, 1937: sp. no. 202. 3. Bouquet, 1969: 102. 4. Busson, 1965: 414–6 with leaf analysis. 5. Chakravarty, 1958: 403, 411–2. 6. Irvine, 1948: 254. 7. Irvine, 1952, a: 36, 38. 8. Irvine, 1956: 37. 9. Irvine s.d. 10. Okiy, 1960: 121. 11. Quisumbing, 1951: 938–40, with references. 12. Walker & Sillans, 1961: 142. 13. Watt & Breyer-Brandwijk, 1962: 359. 14. Grindley, 1950.

## **Luffa acutangula** Roxb.

FWTA, ed. 2, 1: 207. UPWTA: ed. 1, 61.

English: loofah; angular loofah; angular sponge loofah; fluted loofah; vegetable sponge; luffa sponge; dish cloth; dishrag gourd; snake gourd.

French: liane torchon; papangaie (Réunion); pipangaie (Réunion).

West African: **SENEGAL** MANDING-BAMBARA ko barani (K&A) zinzan tigi (JB; K&A) MANINKA foro foro (Laffitte) SERER nemnem sérer (JB; K&A) WOLOF poṫok (K&A) **SIERRA LEONE** BULOM (Kim) bɔndɔ-ma-poto (FCD) GOLA bɔndɔ-ku (FCD) KISSI chiula-hɛŋgɔ (FCD) g-basama-puluã (FCD) KONO pu-bɔnduɛ (FCD) LOKO pu-hakpatoro (FCD) MENDE pu-bɔndɔ (FCD) SUSU forəto-sɛ lonyi (FCD) TEMNE *an*-rɔnthma-*a*-potho (FCD) VAI po-boŋgbo (FCD) **GHANA** AKAN-AKYEM akatong (FRI) **NIGERIA** HAUSA soson wanka, soso: *the fibre;* wanka: *to wash* (JMD) soson yama, soso: *the fibre;* yama: *the West* (JMD) YORUBA kàànkan-aiya *the fruit* (JMD) orira *the plant* (JMD)

A herbaceous annual climber or trailer, very similar to *L. cylindrica* (Linn.) M. J. Roem., but distinguished by less deeply lobed leaves, an angled stem and

10 raised longitudinal ridges along the fruit. The plant is found wild in India where it is probably indigenous but it has been taken by man throughout the world and is found widely in the West African region.

The species is a source of commercial loofahs but is less exploited than *L. cylindrica*. At Sokoto, the fibrous network is said to be less harsh and therefore more suitable as a scrubber, hence the Hausa name *soson wanka,* a term applied to the sponge of the European from *wanka,* to wash, and *soson yama,* from *yama,* west (4). [See also remarks under *L. cylindrica*.] The lesser harshness lends itself also as a strainer and its cultivation as a source of loofahs for oil-filters was considered as a war measure (11). The fibre of the fruit is used in Senegal, Ghana and Nigeria for making hats (11).

The fruit is very variable in bitterness. In some wild forms in Asia this is intense to the extent of being toxic. Sweet edible races have been selected and these are widely cultivated in India (10) and SE Asia (2, 9) as a vegetable for which the fruits are picked young. They are also eaten in West Africa (3, 4 ,6). The Akyem of Ghana cultivate it by training the plant up trees (7). The fruit is said to be a fair source of vitamin B, and has a vitamin A potency of 7.9–8.9 IU/gm in Indian material (11) and is a good source of calcium, iron and phosphorus (9). The fruit, presumably of a non-edible strain, is considered anthelmintic in Mauritius, and has some potential as a fish-poison (11).

The seeds are very purgative. Bitter principles are present of which the major one is *cucurbitacin B* as in *L. cylindrica* (8, 10). In Mauritius 7–10 seeds powdered up are considered an opening dose; 15–20 cause vomiting and 30–40 to be fatal (3, 8). Oil is also present to a concentration of about 48%. It is semi-drying and of the same constituents as the oil from *L. cylindrica* seeds, and as such one may expect the oil to be edible provided the extraction is carried out so as to leave the bitter purgative substances in the cake. The seed contains a glycosidal saponin and an enzyme capable of hydrolysing it rendering the cake poisonous and unfit for feeding to cattle as a concentrate, but since it is rich in nitrogen and phosphorous it is suitable as a fertilizer.

The leaves and expressed sap are applied to sores in West Africa (4). Made into a poultice they are put on to cutaneous eruptions and on to guinea-worm sores to kill the worm in Senegal: also the leaf-sap is sometimes used as an eyewash (8). The leaves are used in India as a poultice for piles, leprosy and splenitis, and leaf-sap for granular conjunctivitis in children. Leaf-decoction has use in Java for uraemia and amenorrhoea (2, 10). Leaves may also be put on the bites of poisonous animals and for itch (2).

The root is commonly used in India and Asian Russia as a drastic purge (4, 11), and this treatment is used in India for dropsy (3, 11). A little alkaloid has been detected in Nigerian material (1).

The entire plant, and seed are insecticidal (5, 11). The plant contains no rotenone (11).

References:

1. Adegoke & al., 1968. 2. Burkill, 1935: 1370–1. 3. Chakravarty, 1968. 4. Dalziel, 1937. 5. Heal & al., 1950: 116. 6. Irvine, 1952, a: 36. 7. Irvine s.d. 8. Kerharo & Adam, 1974: 382–3. 9. Quisumbing, 1951: 940. 10. Sastri, 1962: 177–9. 11. Watt & Breyer-Brandwijk, 1962: 360.

**Luffa cylindrica** (Linn.) M. J. Roem.

FWTA, ed. 2, as *L. aegyptiaca* Mill. UPWTA, ed. 1, 61.

English: loofah; loofah gourd; smooth loofah; vegetable sponge; luffa sponge; dishcloth; dishrag gourd; snake gourd.

French: éponge végétable (vegetable sponge); liane torchon (dishcloth liane); courge torchon (dishcloth gourd); gourde serviette (serviette gourd).

West African: **SENEGAL** BASARI ɓẹ-xúl (Ferry) BEDIK ɓẹ-feúɗ (Ferry) DIOLA é forafora (JB; K&A) MANDING-BAMBARA kofu (JB; K&A) ńabésé (auctt.) MANINKA saradô (JMD; K&A) SERER nemnem durubab (JB; K&A) SERER-NON (Nyominka) dapanoyaye (A. Chev.) WOLOF dapanoyay (JMD; K&A) ńapé (K&A) **THE GAMBIA** FULA-PULAAR (The Gambia) chochore

(Frith) MANDING-MANDINKA suusaaraŋ (*def.*-o) = *scrubber* (Frith; DF) **GUINEA-BISSAU** BALANTA fuáski (JDES) BIDYOGO empenche (JDES) CRIOULO djadar (JDES) FULA-PULAAR (Guinea Bissau) [illegible] (JDES) lotórcò (JDES) MANKANYA noéntè (JDES) SUSU fúti (JDES) **GUINEA** MANDING-MANINKA sara dion (CHOP) **SIERRA LEONE** BULOM (Kim) saloɛ (FCD) FULA-PULAAR (Sierra Leone) lauyirikɔ (FCD) GOLA guagei (FCD) guagɛrɛ (FCD) KISSI n-dapalẽ (FCD) ţuila-hɛnge-chiu (FCD) KONO ma-sasa (FCD) sasaa (FCD) KRIO sapo (FCD; JMD) LIMBA (Tonko) malega (FCD) LOKO n-gondi (NWT) n-gongo (NWT) yaraba (FCD) yaumba (NWT) MANDING-MANDINKA mansara (FCD) yisera (FCD) MANINKA (Koranko) yɛsira (NWT) MENDE kalohuwua (FCD) safo (*def.*-i) *from Krio* (FCD; JMD) SUSU nalinyi (NWT) SUSU-DYALONKE leka-na (FCD) TEMNE *a*-bos-*a*-rakai (NWT) *an*-rɔnthma (FCD; JMD) ta-giri (GFSE) VAI bũlɔko (FCD) gbundɔku (FCD) **LIBERIA** MANO koi pělě (JMD) kwa yi pělě (JMD) **MALI** DOGON gabá nà (C-G) **GHANA** AKAN-TWI *a*-borɔfo sapɔw = *European's sponge* (FRI) DAGBANI lɔcha (Gaisser) (FRI) GA blɔfo kotʃa = *European's sponge* (FRI) GBE-VHE gbeklɔ (FRI) treklɔnu (FRI) yakutsa (FRI) **TOGO** BASSARI bindumpo (Gaisser) gudscha (Gaisser) TEM (Tshaudjo) genaenjau (Gaisser) gnaejau (Gaisser) **NIGERIA** BIROM sò-só (LB) EDO íhíon *sponge* (JMD) íhíon-òsà *sponge-soap* (JMD) íhíon-oyibo *sponge of the white man* (JMD) FULA-FULFULDE (Nigeria) giggirɗum = *a thing for rubbing with* (MM) loonirde = *thing to wash with* (MM) HAUSA baska (JMD) sòòsóó *the fruit, or the sponge* (auctt.) IGBO nza (BNO) IGBO (Asaba) akb'an'ude (NWT) IGBO (Awka) ásịsá *the loofah fibre* (JMD) IGBO (Onitsha) àgbọ̀ *a growing calabash* agwo *a c.var producing an elongated calabash used for cups* IGBO (Owerri) ahia mmaia, ọsa mme (AJC) IGBO (Umuahia) ahia mme (JMD) IJO-IZON (Kolokuma) ẹ̀lẹ̀lẹ̀pán (KW) NUPE bongi (Banfield) rŭmakà (Banfield) TIV kileiyongo (JMD) kileyongo (Vermeer) YORUBA ẹrún (JMD) kàànkan, kànkàn (JMD) kàànkan òyibo = *white man's sponge* (JMD) kànrìnkàn = *sponge* (auctt.) kànrìnkàn-ayaba = *queen's sponge* (Macgregor; IFE) **WEST CAMEROONS** DUALA dinyɛŋgɛ (Ithmann)

A herbaceous annual climber or trailer to 6 m or more long, a cultivated plant but naturalized in all kinds of vegetation, common throughout the Region, and outside it is dispersed pantropically and subtropically to all countries where rainfall is high enough, but not excessive, or adequate watering can be given. The plant is native of the Old World tropics and is of such ancient cultivation that its original home, whether in Africa or in Asia, cannot now be determined.

The fruit is the source of the commercial loofah or vegetable sponge which is the hard fibro-vascular network found within the ripe fruit after the intervening tissue has been rotted away. Japan, by a careful grading system to ensure uniformity, was the primary world producer before 1942. Loofahs have since been satisfactorily grown in Malawi and there seems no reason why improvement on the village market product in West Africa could not be achieved. Commercial grade loofahs are light uniform in colour, clean and free of seeds and extraneous matter, and preferably over 35 cm long. Besides use as a washing sponge, the fibre has commercial uses in hat-manufacture (18); as in-soles of shoes, car-wipers, marine engine filters, etc. (8); in the manufacture of pot-holders, table-mats, door and bath-mats, gloves, etc., and for their shock and sound absorbing properties in military steel helmets and armoured vehicles (13). The fibre may also have some potential as a paper-material (4). In Ghana loofahs are used for filtering water and palm-wine (10); in Gabon to brush clothes (17) and as cleaning squabs in Jebel Marra (7) as well as scrubbers and sponges for washing. The frequency of the reference to 'European' or 'white-man' in the vernaculars suggests that this use in West Africa as a sponge is a relatively recent innovation. Nevertheless the Yoruba of S Nigeria address the plant in an incantation *to wash the badness out of the body* alluding to the way the plant creeps in the forest like a liane (16). But may not this refer to its purgative attributes?

The wild fruits are bitter and poisonous. In Liberia the plant is regarded as so poisonous that if one eats with hands soiled by it violent catharsis results (Harley fide 4). A bitter substance and a saponin have been isolated. The fruit is used in Guinea on tumours (Pobéguin in 6). The fruit-juice is reported used on the inside of the nose to treat apoplexy and in Queensland, Australia, unripe fruit have been used as a fish-poison (18). The pulp of young fruit may also be applied as a poultice to swellings, etc. (4). The fruit however is eaten in West

Africa and is often cultivated for this purpose (4, 9, 10, 17, 18), but they have to be picked young before the fibrous vascular bundles harden and before the purging substance develops towards ripening. In various countries edible cultivars have been selected. The Susu of Guinea are said to have an edible variety (Pobéguin fide 4) and edible races have been selected in India (15) and in the Philippines (14), where they are commonly cultivated. In Togo the whole fruit is steeped in guinea-corn beer for 12–15 hours to strengthen the fermentation (Gaisser fide 4). A seedless variety is known in West Africa (4), but the benefit of this character is not clear.

The entire seeds are emetic and cathartic. An aqueous or alcoholic emulsion is reported usefully anthelmintic (14), and the seeds are eaten for this purpose with meat in Guinea (12). An oil is present at about 45–51% weight of the kernel. It is variously described as colourless (8), green (3), brownish green (4) and dark red (15), perhaps dependent on the method of extraction, of a faint odour, pleasant taste and semi-drying. It is edible and consists of a mixture of mainly linoleic, 43%, and oleic acids, 40%, with smaller amounts of palmitic and stearic acids (15, 18). It is proposed as a good substitute for culinary olive oil (18) and has been used in the U.S.A. in soap-manufacture (8). The seed-cake is bitter and toxic and is unfit for cattle-feed, but by virtue of being rich in nitrogenous matter, 41%, and phosphorus ($P_2O_5$), 1.8% (15) could be used in agricultural fertilizers. The seed, or oil, is said in various countries to have medicinal uses, but this is more likely due to active principles of the kernel rather than the oil itself. Glycosides are present (11) and a bitter probably identical with cucurbitacin B. (15).

The leaves have medicinal uses: in Gabon to promote healing of wounds (17); in Congo in a plaster to maturate abscesses and to kill filaria (2); in S Africa a leaf-infusion is taken by the Zulu for stomach-ache (18); in Tanganyika leaf-sap is added to a a root-decoction to prevent abortion (5), a surprising application considering that the root is a drastic purge even in minute quantity (11, 14). A haemolytic saponin, as in the fruit of wild forms, is present (11) and also a trace of alkaloid (1). A root-preparation is said in Gabon to be an effective remedy for cancer of the nose (17).

If the fruits are scarified before normal harvesting and the plant has an incision in the stem about 25 cm above the ground, a clear liquid is expressed which in Japan is held to have medicinal value for respiratory diseases (8).

References:

1. Adegoke & al., 1968. 2. Bouquet, 1969: 102. 3. Burkill, 1935: 1371–2. 4. Dalziel, 1937. 5. Haerdi, 1964: 81. 6. Hartwell, 1969: 96. 7. Hunting Technical Services, 1968. 8. Ingram, 1952–3. 9. Irvine, 1952, a: 36. 10. Irvine s.d. 11. Oliver, 1960: 30, 70. 12. Portères, s.d. 13. Purseglove, 1960: 129–30. 14. Quisumbing, 1951: 941–3. 15. Sastri, 1962: 179–81, with analyses. 16. Verger, 1967: no. 130. 17. Walker & Sillans, 1961: 143. 18. Watt & Breyer-Brandwijk, 1962: 360–1.

## **Momordica angustisepala** Harms

FWTA, ed. 2, 1: 212. UPWTA, ed. 1. 62, as *M. bracteata* Hutch. & Dalz.

West African: **GHANA** AKAN-ASANTE sapɔ (Enti) FANTE ɛsaw (Enti) essewu (Andoh) TWI ahensaw (FRI) sapɔ = *sponge* (auctt.) GA ansao (FRI) GBE-VHE akutsa (FRI; JMD) VHE (Pecí) adoka? (FRI) **NIGERIA** IGBO ògbo (Chizea)

A large forest climber of the closed-forest zone known only from Ghana, S Nigeria and Cameroun.

The thick stems are pounded and impurities are washed out leaving only the white fibers which the Asante of Ghana use as a washing sponge. The English word 'sponge' is derived from the Asante name *sapow* (2, 3)). A similar use is recorded from Benin, S Nigeria (1). The plant is sometimes cultivated in Ghana (3).

References:

1. Chizea FHI 8286, K. 2. Irvine, 1930: 289, as *M. bracteata* Hutch. & Dalz. 3. Irvine, 1961: 89.

**Momordica balsamina** Linn.

FWTA, ed. 2, 1: 212. UPWTA, ed. 1, 62.

English: balsam apple.

French: margose (Kerharo & Adam).

West African: **SENEGAL** DIOLA édé hindel (JB; K&A) tébétébed (JB; K&A) FULA-PULAAR (Senegal) burbog (K&A) mburbok (K&A) MANDING-BAMBARA zara (JLT) SERER birbof (K&A) mbirbop (JB, K&A) WOLOF m-barböf (JMD; K&A) yombebute (JMD) **THE GAMBIA** DIOLA (Fogny) édé hindel = *honey* (DF) tébétébed = *it has been twisted* (DF) **SIERRA LEONE** SUSU kuru-kurinyi (NWT) **LIBERIA** MANO gã gẽ su lu (JMD) **MALI** SONGHAI lumba-lumba (A. Chev.) TAMACHEK manamat (A. Chev.) **GHANA** AKAN-ASANTE nya-nya (JMD) TWI nya-nya (JMD) nyinya (FRI; JMD) GA nyanyra (FRI) GBE-VHE kaklẽ (FRI) **NIGER** HAUSA garauni (Bartha) SONGHAI bàdóomà (*pl.* -à) (D&C) **NIGERIA** DERA ndákɗì (Newman) FULA-FULFULDE (Nigeria) dagdaggi (JMD) habiiru (JMD; J&D) lele duji (JMD) GOEMAI hashinashiap (JMD) HAUSA daddagu *from Fula* (JMD; ZOG) gàraàfúnií (JMD; ZOG) garahuni (JMD; ZOG) IGBO (Ibusa) akban ndene (NWT) KANURI dàgdágə (JMD) MBULA garahanu (Meek) YORUBA ejìnrin (JMD) ejirin (JMD)

A climber or trailer with annual stems attaining 4–5 m length, a plant of dry savanna and clearings in secondary bush of Senegal, Mali, Niger and N Nigeria, and outside the Region it is a common plant in drier parts of the tropics.

This species is closely related to *M. charantia* Linn. which occurs in areas of greater rainfall. They are probably not clearly distinguished apart by the countryman. Many vernaculars appear to be common to both.

The whole plant is used as a bitter stomachic, an emetic and a purgative (4, 12). The Fula of Senegal use it as a vermifuge (10). Juice expressed from the leaves is taken by Yoruba for roundworm (*Ascaris*) and given to children for threadworm. It is an ingredient of the Yoruba *agbo* pot. A macerate of the whole plant, to which salt is added, is used in Senegal as a galactogogue by draught, and by massage to the chest. This latter application serves also to treat intercostal pains. Fula herdsmen in Senegal also use this preparation to increase milk-yield of cows (9, 10). An infusion is used in the Region as a wash for fever and for yaws, and for these affections a decoction with natron added is taken internally. This preparation is used for horses. The Fula of Senegal ascribe tranquillizing properties to the plant which are of benefit in cases of mental illness (10). Zulu of South Africa make an infusion or decoction as a sedative for an irritable stomach.

Notwithstanding the emetic and purgative properties, leaves, and sometimes the fruit, are eaten in sauces and soups in the Region (4, 7, 8, 10). In Jebel Marra of Sudan (14) the leaves serve as a vegetable, and the Pedi of S Africa eat the young leaves as a pot-herb though they recognize the fruit as being deadly poisonous (13). The Kanuri of N Nigeria are said to relish the bitter taste (6). Consumption though is thought to be less as a foodstuff than as of a vehicle for its medicinal properties, as, for example, its inclusion together with other drug-plants in a Hausa food called *fatefate* (4). A trace of alkaloid has been detected in the leaves (2).

Donkeys, cattle, sheep and goats are recorded as grazing the plant in Senegal, but not horses (1). Feeding trials on sheep in N Rhodesia produced no ill effects (13).

The leaves can be used to clean metals, and leaves and fruit give a lather in water and are used as soap in N Nigeria (3). It can be used for washing the hands and body but not clothes (4).

The fruit mixed with any bland oil can be made into a drawing ointment for festers, inflammations, swellings, yaws, burns, etc. (4, 12). A bitter principle, *momordicin,* is present (10, 12). In U.S.A. compounded with olive or almond oil it has been used for chapped hands and for piles (13), and as a salve on open sores of long standing (5).

The fruit is emetic and cathartic. The seed soaked in water and then inserted in the neck of the womb is a method of producing abortion practiced by the

Mbula tribe of N Nigeria (11a). The plant is also added to *Stropanthus* arrow-poisons by Benin tribes. The fruit is thought to have caused poisoning of pigs in Queensland, but nevertheless the seed is said in Australia to be edible after steeping in salt water and cooking (13, 7).

The root is sometimes an ingredient of aphrodisiac prescriptions, and, as are the fruit seeds, is used as an abortifacient (4). The leaves are put in water for ceremonial washing after digging a grave amongst the Ngizim in Bornu (11b).

References:

1. Adam, 1966, a. 2. Adegoke & al., 1968. 3. Dalziel 131, K. 4. Dalziel, 1937. 5. Hartwell, 1969: 96. 6. Irvine, 1952, a: 33. 7. Irvine s.d. 8. Jackson, 1973. 9. Kerharo & Adam, 1964, b: 560. 10. Kerharo & Adam, 1974: 386–7. 11a. Meek, 1931: 1: 58. 11b. Meek, 1931: 2: 259. 12. Oliver, 1960: 72. 13. Watt & Breyer-Brandwijk, 1962: 362–3. 14. Wickens 1377, K.

**Momordica cabraei** (Cogn.) C. Jeffrey

FWTA, ed. 2, 1: 211, as *Dimorphochlamys mannii* Hook. f. UPWTA, ed. 1, 58, as *D. mannii* Hook. f.

West African: **NIGERIA** YORUBA ahara (Dawodu; JMD)

A twining herbaceous climber to 5 m high, of secondary bush and forest clearings, recorded from Ivory Coast, S Nigeria, W Cameroon and Fernando Po, and distributed to Zäire.

The black seeds are oily (1).

Reference:

1. Walker & Sillans, 1961: 142.

**Momordica charantia** Linn.

FWTA, ed. 2, 1: 212. UPWTA, ed. 1, 62–63.

English: African cucumber; balsam pear.

French: concombre africain (Kerharo & Adam); liane merveille (Berhaut); margose (Kerharo & Adam).

West African: **SENEGAL** FULA-TUKULOR beurbŏh (AS) MANDING-BAMBARA zara (JLT) WOLOF m-barböf (JMD; K&A) m-burböf (K&A) yombebute = *gourd of the toad* (JMD) **GUINEA-BISSAU** CRIOULO sancaetano (JDES) FULA-PULAAR (Guinea-Bissau) burbóqui (JDES) buroki (EPdS) JAKHANKHE cassêlaha (JDES; EPdS) cosselaha (JDES) MANDING-MANDINKA isróbódô (JDES; EPdS) **SIERRA LEONE** VULGAR agini (Jarr) KRIO sapodila (FCD) SUSU kuru-kurinyi (NWT; JMD) **LIBERIA** MANO gã gẽ su lu (JMD) **MALI** SONGHAI lumba-lumba (A. Chev.) TAMACHEK manamat (A. Chev.) **IVORY COAST** ABURE acoatiango (B&D) ADYUKRU sing biep (A&AA) AKAN-ASANTE nia-nia (K&B) AKYE ato m-bomu (A&AA) ANYI nia-nia (K&B) KRU-BETE nienbélé (K&B) GUERE n'guéné boué (K&B) n'guéré (K&B) GUERE (Chiehn) bobobo (K&B; B&D) NEYO nania-nania (K&B) 'KRU' zagué zru (K&B) **GHANA** AKAN-ASANTE nya-nya (JMD) FANTE nya-nya (Williams) TWI nya-nya (Williams; JMD) nyinya (FRI; JMD) GA nyanylã *this is the female of okú nyanylã, Cardiospermum grandiflorum* (*Sapindaceae*) (KD) GBE-VHE kaklẽ (FRI; JMD) VHE (Awlan) kaklẽ (FRI) VHE (Pecí) kaklẽ (FRI) **NIGERIA** DERA ndákɗì (Newman) FULA-FULFULDE (Nigeria) dagdaggi (JMD) habiiru (JMD) lele duji (JMD) GOEMAI hashinashiap (JMD) HAUSA daddagu *from Fula* (JMD; ZOG) gàraàfùnií (JMD; ZOG) garahuni (JMD; ZOG) IGALA iliahia (Boston) ilialihia (Boston) IGBO kakayi (Singha) IGBO (Ibusa) akban ndene (JMD) KANURI dàgdágə (JMD) YORUBA akara aję (IFE) ejìnrìn (JMD) ejìnrìn nla (IFE) ejìnrìn wẹẹri (IFE) ejirin (auctt.) ejirin-wẹ́wẹ́ (IFE) igbólé ajá (IFE)

A climber or trailer with annual stems to about 5 m long, very similar to *M. balsamina* Linn., but of lowland rain-forest and wooded areas with a higher rainfall than is required by the latter, occurring commonly from Senegal to W Cameroons, and widespread pantropically.

The plant appears to be almost entirely wild in Africa and is in fact one of the commonest cucurbits of the continent (6). In Asia it is plentifully cultivated for the edible fruits and tender shoots. Both parts are bitter and in its long history of domestication in Asia there has been selection which has eliminated some, but not all, of the bitterness. The presence of two wild forms in francophone West Africa is reported and both are said to be cultivated (19). The plant reached the New World during the Slave Trade.

The young fruits and shoots are sometimes eaten in West Africa as a supplementary or emergency food (12). In India and SE Asia the plant is widely cultivated and is eaten in curries and pickles. The bitterness of the fruit can be reduced by peeling and steeping it in water before cooking. The fruit is rich in iron, calcium and phosphorus, and in vitamins (18, 24).

The principal uses of the plant in West Africa are as a laxative, for stomachache, a taenifuge and anthelmintic, and to treat fevers (4, 8, 17, 20). The fruits are used as a purgative and vermifuge in Senegal (14, 15); in Ghana and Nigeria the leaves are steeped in water for taking internally for diarrhoea and dysentery, and also by enema and are said to have powerful astringent properties (1, 11, 13). In Ivory Coast–Upper Volta the application of the plant as both purgative and antidysenteric, while seemingly in opposition, is based on the precept that dysentery is a sign of "beasts in the abdomen" and the best treatment is to charge them out with a good purge (16). In Ivory Coast–Upper Volta (16) and in Congo (3) the plant is considered to provide a taenifuge especially good for children. The seeds are used in Zaïre for roundworm, and in Brazil a dose of 2–3 seeds is anthelmintic. The seed contains 32% of an oil but the effective part is said to lie in the embryo (24).

The plant is commonly used in Ghana and in Nigeria as a febrifuge, either as a wash in bathwater, or by draught in drinking water or more often in palmwine (1, 11, 13). The Igbo of Tjele prepare a wash and a drink from the plant for *ɛba*, a feverous condition (22). The leaves are used as a febrifuge in Senegal (15). In Ivory Coast–Upper Volta 'yellow fevers' and jaundice are treated by enema of the entire plant in water and eye-instillation of the leaf-sap (16). Tests for anti-malarial activity using the 'upper part' of the plant showed no action on avian malaria (17).

The plant is used in Ghana (11, 13) and Nigeria (1) as an aphrodisiac and administered in larger doses for gonorrhoea. Use as an aphrodisiac is practised in Ivory Coast–Upper Volta where the leaf-sap is added to a calabash of palmwine which is drunk during the course of the day! (16).

Gbe (Peki) people of Ghana rub a pad made from the plant stems and leaves with some soap added over areas affected by yaws (13). In S Nigeria Igbo at Ibuzo prepare a wash from the plant for yaws (23), and in Senegal it is treated by a root-preparation for taking internally and for external application (15). Similarly use on cutaneous affections is known in India where the whole plant along with other drug-plants is made into an ointment for psoriasis, scabies and other diseases, and also in Japan (24). In Igala, S Nigeria, the liquid in which the plant has been cooked is used externally on boils, ulcers, septic swellings and infected feet (2). A plaster of the pulverized plant is also used in Nigeria on malignant ulcers and of the pounded leaf on cancer of the breast (1, 17). Use on malignant ulcers in Guam and on the mouth in Brazil has been recorded (9). In Senegal a plaster of the crushed leaves is used on cutaneous parasites such as filaria and guinea-worm (15).

The plant has been used as an insecticide in Haiti (24) and leaves and stems have shown in tests to have some insecticidal properties (10).

In Nigeria a fruit-infusion in oil is applied to burns, ulcers, etc., and fruit cut open or mashed into a poultice is applied as a vulnerary (1). Such healing treatment is found in Malaya (5). The leaves are used in Senegal for menstrual troubles, and the roots for syphilis and rheumatism, and the whole plant together with other drug-plants is prescribed for snake-bites (15). In Congo the leaves crushed in the hands are massaged into the body for fever aches and pains, and may be also used in vapour-baths, and with other drug-plants is used for heart conditions expecially tachycardia (3).

Because of its bitter taste, Anyi of the Ivory Coast apply a leaf-macerate in palm-wine to a mother's teats to wean a suckling child (16).

Leaf-material grown in Sierra Leone showed the presence of no alkaloid, nor saponin (21), but material of the whole plant from Congo has shown traces of both (3). The seed yields a highly aromatic volatile oil, a fixed oil, carotene, a resin, two alkaloids and a saponin. The fixed oil is recorded as 32–35% of the seed and contains 17% stearic acid, with also oleic and linoleic acids. One of the alkaloids is the bitter principle *momordecin,* probably an active constituent of an arrow-poison mixture made in the Philippines using material of this plant (24). Momordecin, oil and a resin are also present in the leaves.

References:

1. Ainslie, 1937: sp. no. 232. 2. Boston C5, K. 3. Bouquet, 1969: 102. 4. Bouquet & Debray, 1974: 78. 5. Burkill, 1935: 1485–6. 6. Chakravarty, 1968: 433–4. 7. Claude & al., 1947. 8. Dalziel, 1937. 9. Hartwell, 1969: 97. 10. Heal & al., 1950: 117, 151. 11. Irvine, 1930: 289–90. 12. Irvine, 1952, a: 36. 13. Irvine s.d. 14. Kerharo, 1967. 15. Kerharo & Adam, 1974: 387–90 with phytochemistry and references. 16. Kerharo & Bouquet, 1950: 42–43. 17. Oliver, 1960: 31, 72. 18. Quisumbing, 1951: 944–8, with references. 19. Roberty, 1954: 794. 20. Singha, 1965. 21. Taylor-Smith, 1966: 539. 22. Thomas, N. W. 1805 (Nig. Ser.), K. 23. Thomas, N. W. 2012 (Nig. Ser.), K. 24. Watt & Breyer-Brandwijk, 1962: 363–4, with references.

## **Momordica cissoides** Planch.

FWTA, ed. 2, 1: 211. UPWTA, ed. 1, 63.

West African: **SIERRA LEONE** MENDE tɛlɛge (NWT; JMD) **GHANA** AKAN-ASANTE sopropo (FRI) TWI *n*-tonto (JMD; FRI) GA tonto (JMD) **NIGERIA** IGBO (Ukwuani) is-ugu (NWT; JMD) IJO-IZON (Kolokuma) béínmó = *fill it to the brim* (KW) IZON (Oporoma) béínmọ̀ (KW) YORUBA akọ ejìrin (Millson; JMD) ogbomorhan (Rowland)

A climbing or trailing herb of bush and secondary forest, common throughout the Region from Guinea-Bissau eastwards to Sudan, E Africa, Zäire and Angola.

The plant is used in Nigeria for fever. The Igbo prepare it as a wash for *ɛba* (5), and the Ijo put the leaves into a special prescription for malaria (6). In Cameroun a preparation is put on the heads of children suffering from fever (3). It is said to be used as a laxative for children (1). The Ijo also give the leaves squeezed up in gin and a little chalk to women to promote growth of a foetus and use the leafy stem in bath preparations to cure a child suffering from *endeé* during a pregnancy of its mother (6). Igbo at Tjele in S Nigeria consider the plant a charm against ill-luck (4).

In W Cameroon the Bakundu eat the seed kernels (2).

References:

1. Dalziel, 1937. 2. Gartlan 29, K. 3. Leeuwenberg 6172, K. 4. Thomas, N. W. 1800 (Nig. Ser.), K. 5. Thomas, N. W. 2207 (Nig. Ser.), K. 6. Williamson, K. 64, UCI.

## **Momordica foetida** Schum. & Thonn.

FWTA, ed. 2, 1: 212, incl. *M. cordata* Cogn. UPWTA, ed. 1, 63.

French: concombre sauvage (Walker).

West African: **SIERRA LEONE** TEMNE *a*-bos-*a*-wir (NWT; JMD) **IVORY COAST** AKAN-ASANTE aoasongo (B&D) BAULE gaayama (B&D) nanïa (B&D) n-gessannia (B&D) KRU-GUERE (Chiehn) boobo (B&D) KWENI bobonowron (B&D) vovolé (B&D) vovoné vono (B&D) KYAMA hepa (B&D) **GHANA** AKAN-ASANTE nyanya-nua (Enti) sɔprɔpɔ (FRI) TWI kakle (FRI) ɔwɔduan, ɔwɔ: *snake;* aduan: *food;* i.e., *snake's food* (FRI) sɔprɔpɔ (FRI) **NIGERIA** EDO ìsûgũ (Vermeer) IGBO alu-osi (Singha) IGBO (Asaba) akb'an'udẹne (NWT) YORUBA ejìnrìn (JMD; RJN) ejìrìn (auctt.) ejìrin-nia (IFE) isugu (Kennedy) tsekiri (IFE)

A herbaceous creeper from stout perennial rootstock, of open clearings in the forest zone, abandoned farmland, roadsides, etc., recorded from Guinea to W Cameroons and Fernando Po, ascending to montane elevations, and generally widespread in tropical Africa.

The leaves have a rank smell and a bitter taste. Nevertheless some peoples in Gabon eat them as a vegetable after suitable preparation, and after soaking and drying in the sun also use them to stuff cushions (23). The leaves are cooked for food in Sudan (2), and in Malawi cooked with pumpkin leaves as a side-dish and eaten as a women's food, men refusing their bitterness (26). The plant is occasionally grazed by stock in Sudan (1), but in Kenya it is held to be poisonous and stock do not graze it (8). The leaves' unpleasant smell more readily suggests medicinal attributes for the plant. In S Nigeria the Edo take leaf-sap by draught for intestinal disorders (21) and Igbo at Asaba take the plant for *iba ozi* (? bilious fever) (18). In Ivory Coast a preparation is taken by women as an emmenagogue and to facilitate childbirth, and also as an aphrodisiac (5). Similarly by the Igbo of Tjele, S Nigeria, it is given to women in cases of difficult delivery (19), and at Idumuje the parched leaves are given to pregnant women (20). In Gabon the leaves are regularly used as an emetic and in enemas (22, 23). A leaf-decoction is used in Ivory Coast to wash those suffering from small-pox (5).

In Tanganyika the young leaves are eaten for stomach-ache (9) and are used for dropsical conditions (17). Leaves crushed in the hands and rubbed on parts affected by the spitting cobra, *Naja nigrocollis,* will prevent inflammation, especially if applied within a few minutes of the attack: the leaves may also be chewed to augment this treatment (6). The leaf-sap is also drunk for snake-bite, and for nose-bleeding and severe headache (10). The leaf is used for earache (4). In Malawi the head is bound up with the plant stem for headache (14). The leaf is recorded as used in tropical Africa against round-worm and in Uganda a leaf-infusion, or a root-infusion, are used as an abortifacient and an ecbolic (25).

The orange fruits are attractive in appearance especially when they have burst open exposing the red pulp studded with the black seeds. The fruit is edible. They are eaten in Ghana by some tribes (12), and occasionally in Gabon (23). Consumption in E Africa is also recorded: in Sudan (3), and in Tanganyika (11, 24). In Malawi the fruit fixed with hairs of a cow's tail is used as bait to trap birds (16). There is, however, record that the fruit-pulp is regarded in Tanganyika as poisonous to weevils, moths and ants and is used as a repellant (25).

The root is tuberous and is considered edible in Sudan (2, 3, 15), but in Tanganyika it is recorded as being purgative (13). It is also used in Tanganyika to wash small children and their mother's breasts as a tonic in malaria (10). In S Africa a root-decoction containing other drug plants is taken for boils (25).

The plant is said to contain alkaloids (25).

Its presence in vegetation is considered to indicate good cacao soil (7).

References:

1. Andrews 699, K. 2. Andrews 845, K. 3. Andrews 1720, K. 4. Bally, 1937. 5. Bouquet & Debray, 1974: 78. 6. Culwick 4, K. 7. Dalziel, 1937. 8. Glover & al. 1161, K. 9. Haarer 44B, K. 10. Haerdi, 1964: 81. 11. Hukui 50, K. 12. Irvine s.d. 13. Koritschoner 705, K. 14. Lawrence 614, K. 15. Mohamed Ismail Sherif A 3992, K. 16. Pawek 5518, K. 17. Tanner 4991, K. 18. Thomas, N. W. 1658 (Nig. Ser.), K. 19. Thomas, N. W. 1803 (Nig. Ser.), K. 20. Thomas, N. W. 2143 (Nig. Ser.), K. 21. Vermeer 65, UCI. 22. Walker 1953, a: 32. 23. Walker & Sillans, 1961: 143. 24. Wallace 536, K. 25. Watt & Breyer-Brandwijk, 1962: 364. 26. Williamson, J. 1956: 82.

**Mukia maderaspatana** (Linn.) M. J. Roem.

FWTA, ed. 2, 1: 209 as *Melothria maderaspatana* (Linn.) Cogn. UPWTA, ed. 1, 62, as *Melothria maderaspatana* Cogn.

West African: SENEGAL FULA-PULAAR (Senegal) pomey (K&A) MANDING-BAMBARA basa furãda (JB) SERER pôm (JB) WOLOF hal u mbot = *melon of the toad* (JMD; K&A) ngon soré na = *evening is far* (JB; DF) yombebute = *gourd of the toad* (JMD) **THE GAMBIA** WOLOF ngon jogena = *evening is near* (DF) **SIERRA LEONE** MANDING-MANINKA (Koranko) yɛsira (NWT) TEMNE *an*-sisɔ (NWT) **NIGERIA** BEROM egìrí (LB) HAUSA gautan kaji (RES) gáútan zóómóó = *hare's tomato* (JMD; LB) maàlàmií, máálami, maálamií-ná-maátaá, maalamìn máátáá *indicates superstitious use of the plant as a love-charm* (auctt.) YORUBA erinkanyaba (N&E) okí-ọ̀ka, ọ̀ka: *a children's disease of the head for which the plant is held to be a cure or preventative* (JMD) oré-ọ̀ka (Macgregor)

An annual scandant or trailing herb to about 4 m long, of open, not forested, localities, throughout the West African region, and widespread in tropical Africa, Asia and into Australia.

A decoction of young shoots and leaves is used in Nigeria as an aperient, especially for children (2). The tender shoots and bitter leaves are also used as an aperient in India and are taken for vertigo and biliousness (8). In Tanganyika the leaf-sap is used as a wound-dressing, leaves in poultice for burns, and the sap given to small children for amoebiasis; dried powdered leaves are dusted over scabies, and plant-ash in castor oil is rubbed over scarifications and the temples for headache (4). The plant (part not stated) is regarded by the Yoruba of S Nigeria as a preventive or a cure for *òkà* a disease of children's heads (7). The plant is said to have expectorant properties but tests conducted in India gave unsatisfactory results (8).

Seeds when chewed, or in decoction, are taken in Nigeria to cause perspiration (2). In India they are considered sudorific (8).

Stock animals, except horses, are recorded as browsing the plant in Senegal (1), but information from the Department of Animal Health, Zambia, suggests that the fruit is deadly poisonous to cattle (9). In India cattle are said to like the plant but that fruits destroy their eyes (8). The fruits are however commonly eaten by birds in India (8) and are greedily taken by the francolin in Tanganyika (3). The fruit in Senegal is used as a vermifuge (6).

The root is chewed to relieve facial neuralgia, toothache, etc, in Nigeria (2), and for toothache and, in decoction, for flatulence in India (8).

The Fula of Dianguel in Senegal ascribe magical properties to the fruit and in this sense use them as poison-antidotes (5, 6). In Cayor, Senegal, the leaves enter prescriptions for treating mental troubles (6).

References:

1. Adam, 1966, a: as *Melothria maderaspatana.* 2. Ainslie, 1937: sp. no. 62, as *Bryonia scabrella.* 3. Bullock 3608, K. 4. Haerdi, 1964: 82. 5. Kerharo & Adam, 1964, b: 538–9 as *Melothria maderaspatana* (L.) Cogn. 6. Kerharo & Adam, 1974: 385–6, as *Melothria maderaspatana* (L.) Cogn. 7. Macgregor 56, K. 8. Sastri, 1962: 336, as *Melothria maderaspatana* (Linn.) Cogn. 9. Trapnell CRS 400, K.

**Oreosyce africana** Hook. f.

FWTA, ed. 2, 1: 210.

A slender climber to 3–4 m high of montane localities in W Cameroons and Fernando Po, and widespread in upland grassland and swampy forest across to E Africa, South Tropical Africa and Madagascar.

The plant is cooked together with *Justicia heterocarpa* T. Anders (Acanthaceae) and the soup is taken to accelerate childbirth in Tanganyika, and the leaf is used to rub on areas infected by ringworm. (1).

Reference:

1. Haerdi, 1964: 82.

**Peponium vogelii** (Hook. f.) Engl.

FWTA, ed. 2, 1: 215.

West African: **GHANA** AKAN-FANTE antomona (Chakravarty) GBE-VHE kaktesui (Chakravarty)

A climber or trailer to 8 m long of lowland and montane rain-forest occurring in Ghana, Dahomey, S Nigeria and Fernando Po, and distributed to Zaïre, NE and E Africa and Natal.

The pulped-up leaves are used to maturate abscesses and furuncles in Congo (1). In Tanganyika the dried powdered leaves are rubbed into scarifications for leprosy (3).

The fruits are very bitter while unripe but when ripe are sweet (6) and are much appreciated by Kipsigi and Masai children in the Narok area of Kenya (2). In Ethiopia the fruits are eaten and are said to be good for stomach-ache (5). The fruit provides food for bush-babies in Tanganyika (7).

The root and the fruits (? unripe) are said to be toxic (4) in Tanganyika.

References:

1. Bouquet, 1969: 102. 2. Glover & al. 2268, 2339, K. 3. Haerdi, 1964: 82. 4. Koritschoner 610, K. 5. Taddesse Ebba 589, K. 6. Tanner 1180, K. 7. Willan 364, K.

**Raphidiocystis chrysocoma** (Schumach.) C. Jeffrey

FWTA, ed. 2, 1: 215, as *R. caillei* Hutch. & Dalz.
West African: **SIERRA LEONE** KISSI mufo (FCD)

A perennial climber to 6 m long of lowland rain-forest from Guinea to Togo, and occurring in Congo, Angola, Uganda and Tanganyika.

The fruit has a strong smell of cucumber when cut.

**Ruthalicia eglandulosa** (Hook. f.) C. Jeffrey

FWTA, ed. 2, 1: 214, as *Physedra eglandulosa* (Hook. f.) Hutch. & Dalz.
UPWTA, ed. 1, 63, as *P. eglandulosa* Hutch. & Dalz.
West African: **SIERRA LEONE** KRIO tagiri (FCD) MENDE kojo *applied to certain spherical hard-shelled fruits, but properly this is Strychuos spinosa* (*Loganiaceae*) (FCD) kɔli-gojo (*def.*-i) = *leopard's blow-blow* (*trumpet*) (Fisher; FCD) TEMNE (Port Loko) *an*-gbende (FCD) TEMNE (Sanda) *an*-gbende (FCD) **LIBERIA** MANO gã gẽ su (JMD) **IVORY COAST** BAULE floméné (B&D) KRU-GUERE zapinkii (K&B) zapuki (K&B) KWENI tiètièpalo (B&D) KYAMA ahobégu (B&D)

A climber of bush and forest areas, to about 6 m long, widespread in the Region from Mali to S Nigeria. The fruit is bright scarlet with yellow markings.

The plant is very poisonous (3), though the sap is used as a poison-antidote in Ivory Coast and to treat burns (1). The Guere of Ivory Coast use the leaf-sap in enemas and in eye-instillations in treatment of jaundice (6). In Sierra Leone the fruit of a large form is sold in Freetown market for crushing and cooking to make a liquid with which Mendes wash babies with fever (5). Nevertheless the ripe fruit is eaten by animals (2). A decoction of the whole plant is sometimes used in Ivory Coast–Upper Volta as a parasiticidal wash (6).

The fruit shell is hollowed out, pierced with holes and used as a horn by boys in Sierra Leone, one of several fruits bearing the Mende name *kokoi* (5). This is *kɔlo-gokoi,* the leopard's blow-blow. *Strychnos sp.* (Loganiaceae) is *ngolo-gokoi,* the baboon's blow-blow.

The plant has medico-magical application in Liberia to treat shortness of breath: the leaves are rubbed on the chest and then on a stick whereby the malady is 'transferred' to the stick (Harley fide 4). Superstitious use is also recorded by the Temne of Port Loko area of Sierra Leone: a section of leafy stem bearing a fruit is cut and attached with suitable 'swear' to plants to prevent thieving. The hands of a thief will swell up (5).

References:

1. Bouquet & Debray, 1974: 78, as *Physedra eglandulosa*. 2. Chakravarty, 1968: 447–51. 3. Chevalier, 1920: 295, as *Cephalandra sylvatica* A. Chev. 4. Dalziel, 1937. 5. Deighton 2801, K. 6. Kerharo & Bouquet, 1950: 43, as *Physedra eglandulosa* Hutch. & Dalz.

**Ruthalicia longipes** (Hook. f.) C. Jeffrey

FWTA, ed. 2, 1: 214, as *Physedra longipes* Hook. f.
West African: **GHANA** AKAN-TWI asaman akyekyea akim (Chakravarty)

A robust herbaceous climber, of semi-deciduous forest and secondary scrub, with brilliantly scarlet conspicuous fruit when ripe, recorded from Liberia to S Nigeria and Fernando Po, and occurring also in Cameroun, Gabon and Rio Muni.

A decoction of the plant is used in Ivory Coast to relieve stomach-ache, scrotal elephantiasis and jaundice (1).

Reference:
1. Bouquet & Debray, 1974: 78, as *Physedra longipes*.

**Sechium edule** (Jacq.) Sw.

English: chayota, chayote, choko (the fruit), vegetable pear.
French: chaiote, chayote, chouchou, christophine.
West African: **SIERRA LEONE** KRIO cho-cho (FCD)

A robust climbing annual herb from a large perennial rootstock; native of tropical South America and now spread through the tropics, and preferring a cool climate is more readily cultivated at an elevation. It is cultivated in a limited way in Ghana and perhaps elsewhere.

The young stem-shoots can be eaten like asparagus, and the stems when older yield a fibre which finds use in East Africa. The fruit is fleshy and edible. It is an important item of food in some tropical countries, especially in the Western Hemisphere. The large swollen rootstock contains 20–25% starch and is also an important foodstuff. The starch is easily digestible and is deemed to be a suitable substitute for arrowroot (1).

Peelings of the fruit have been found to cause such numbness to the skin as to suggest its possible use as a local anaesthetic (2).

The plant merits wider cultivation in West Africa.

References:
1. Irvine s.n. 2. Watt & Breyer-Brandwijk, 1962: 365–6.

**Telfairea occidentalis** Hook. f.

FWTA, ed. 2, 1: 211.
English: oyster nut, fluted pumpkin, oil nut (trade, Irvine).
West African: **SIERRA LEONE** GOLA gɔniŋgbe (FCD) gɔnuŋgbe (FCD) KISSI lambulambo (FCD) KONO pɔndo-koko (FCD) KRIO ɔroko (FCD) MENDE gɔnugbe (FCD) **GHANA** ADANGME krobonko *from Twi* (FRI) ADANGME-KROBO krobonko (Bunting, ex FRI) AKAN-AKYEM krobonko (FRI) ASANTE bomonfradaa (Enti) bɔmɔnfradaa (Enti) krobonko (Enti) TWI bɔmmɔfrana (Bunting; FCD) krobonko (FRI; JMD) GA ansao *probably from Twi* sapɔw: *sponge* (JMD) krobonko *from Twi* (FRI) GBE-VHE ahinsaw *from Twi* (FRI) VHE (Pecí) ahinsaw *probably from Twi* sapɔw: *sponge* (auctt.) **NIGERIA** EFIK ùbọ̀ñ (JMD) IBIBIO íkọ̄ñ úbọ̀ñ (Lowe) IGBO (Owerri) óhí *the seed* (KW) ụ́gū̩ *the plant* (JMD) IGBO (Umuahia) óhí *the fruit or seed* (AJC) ụ́gū̩ (AJC) YORUBA apiroko (IFE) ẹgúsí (IFE) ìrókò (JMD; CWvE)

A liane with herbaceous stems or bushy in the older parts, climbing to 20 m or more, along the fringes of the closed forest and may be a relic of previous cultivation, recorded from Sierra Leone to S Nigeria and Fernando Po, and to the Congo area and Angola.

The leaves and young shoots are frequently eaten as a pot-herb (2, 5, 6, 9). The plant is cultivated in some places especially in S Nigeria and by some tribes in Ghana. It is grown on stakes or trained up trees and thrives best in closed-forest country. The leaves and young shoots are picked continuously as the plant grows (7, 8). They are recorded as being rich in minerals (9).

The stems are macerated in preparations for the fibre to be used as a sponge (4, 7).

The fruits may attain as much as 60 cm in length by 25 cm diameter. The bright yellow fibrous flesh contains 30–40 red seeds over 2.5 cm in diameter. The cotyledons have curious lumpy swellings. The seeds are a popular item of diet and are cooked whole or ground up and put into soups. They have an agreeable almond-like flavour. They are very nutritious and rich in oil which consists of oleic acid 37%, stearic and palmitic acids 21% each, linoleic acid 15% and smaller quantities of others, and of minerals and vitamins (2, 7). The oil is extensively used in African cooking (2, 3, 5). It is non-drying and is considered suitable for soap-manufacture. In the past the seeds have been shipped to Britain in small quantities as 'oil nuts' (8) and bear a close similarity to the nuts of *T. pedata* (Sims) Hook., the commercial source of true oyster nut oil (7, 10).

A trace of alkaloid has been detected in the seeds and none in the roots (1).

The seeds are used for polishing locally-made earthenware pots, and the dry shell of the fruit is sometimes used for utensils (3, 7).

References:

1. Adegoke & al. 1968. 2. Busson, 1965: 416–7, with cotyledon analysis. 3. Dalziel, 1937. 4. Irvine, 1930: 406. 5. Irvine, 1948: 255–6. 6. Irvine, 1956: 35–36. 6. Irvine, 1961: 89–90, with seed and kernel analyses. 8. Irvine s.d. 9. Okiy, 1910: 119, 121. 10. Watt & Breyer-Brandwijk, 1962: 367–8.

**Trichosanthes cucumerina** Linn.

English: snake-gourd; English tomato (Sierra Leone, Melville & Hooker)
West African: **SENEGAL** SERER mété durubab (JB) **GUINEA-BISSAU** CRIOULO camatom (JDES) **SIERRA LEONE** KRIO snɛk-tamatis = *snake tomato* (FCD) TEMNE *an*-rɔthma (M&H) **NIGERIA** YORUBA tòmátò eléjò (CWvE)

A herbaceous annual climber with perennial rootstock, a native of the SE Asian-Australasian region, and now cultivated throughout the tropics.

The botanical identity of this species seems probably to be that of a cultigen selected in antiquity from the species of the genus which in general have bitter inedible fruits and medicinal attributes. The fruit is eaten young as a vegetable for with age it accumulates a glycoside and a purgative substance, *elaterin,* and becomes very fibrous. Its taste is like that of a tomato—hence the Sierra Leone English market name.

An infusion of the young shoots is mildly aperient. The leaf sap is emetic, and the seeds are anthelmintic and antiperiodic.

**Trochomeria macrocarpa** (Sond.) Hook. f.

FWTA, ed. 2, 1: 206–7 as *T. macroura* Hook. f., *T. dalzielii* Bak. f. and *T. atacorensis* A. Chev. UPWTA, ed. 1, 64 as *T. dalzielii* Bak. f., and *T. macroura* Hook. f.
West African: **THE GAMBIA** MANDING-MANDINKA alalata (Fox) **NIGERIA** HAUSA akwalu (JMD; ZOG) HAUSA (West) basgo (JMD; ZOG)

A herbaceous annual liane to about 2·5 m long from perennial fleshy, tuberous rootstock, of deciduous woodland, bushland and savanna from Senegal eastwards to N Nigeria, and across Africa to Sudan and Ethiopia, Tanganyika and SW Africa.

The large yam-like root, often deep in the ground, is dug up when mature after the rains; it is boiled and eaten, and is sometimes also used medicinally in N Nigeria. The Fula prize it as a superstitious medicine for cattle, e.g. to promote fertility (1, 2). The fruit is non-toxic (4). But in The Gambia the plant (part unstated) is recognized as poisonous (3).

References:

1. Dalziel 736, K. 2. Dalziel, 1937. 3. Fox 163, K. 4. Watt & Breyer-Brandwijk, 1962: 368.

**Zehneria hallii** C. Jeffrey

FWTA, ed. 2, 1: 209, as *Melothria deltoidea* Benth., p.p.

A slender climber of the savanna, recorded from Senegal to N Nigeria.

The powdered leaves are given in Nigeria in treatment for tapeworm, and are said also to be a sedative (1).

Reference:

1. Ainslie, 1937: sp. no. 226 (as *Melothria deltoidea*).

**Zehneria scabra** (Linn. f.) Sond.

FWTA, ed. 2, 1: 209 as *Melothria mannii* Cogn., *M. fernandensis* Hutch. & Dalz. and *M. punctata* (Thunb.) Cogn.

A perennial herb, climber or trailing to 6 m, in montane forest localities of W Cameroons and Fernando Po, and widespread in tropical and S Africa, Madagascar and in Java.

No usage is recorded for the plant in West Africa. In Kenya the Masai recognize it as poisonous, but goats in the Narok District readily browse it (1, 2). The Masai pound up the fresh leaves into a paste with water to wash calves to rid them of fleas (1). In Tanganyika at Mamba, S. Pare, the plant is similarly used to treat scabies, especially on babies, and fever (4). The Sukuma also use the plant for fever: fresh leaves are heated in a pot over which the patient sits covered by a blanket (5). It is also used on sores (6). In Ethiopia it is used against alopecia (7).

The stems make a useful twine used in Kenya (3).

*Bryonia scabra,* stated to be emetic, and as an infusion in wine and brandy to be purgative (8), is probably this plant.

References:

1. Glover & al., 859, 954, K. 2. Glover & al., 1741, 2287, K. 3. Mabberley 375, K. 4. Mshigeni 1014, K. 5. Tanner 1370, K. 6. Tanner 5001, K. 7. Tedla Bairu s.n., 1972, K. 8. Watt & Breyer-Brandwijk, 1962: 345.

**Zehneria thwaitesii** (Schweinf.) C. Jeffrey

FWTA, ed. 2, 1: 209, as *Melothria tridactyla* Hook. f.

West African: SENEGAL WOLOF barbo (K&A) barböf (K&A)

A slender climber or trailer to 1.2 m long from fibrous rootstock in damp or swampy localities from Senegal to S Nigeria, and widespread in E Africa and into Asia.

# CYPERACEAE

West African: **SENEGAL** MANDING-MANDINKA kuntiampo (JMD) kwoto-murridié (JMD) SERER khein (JMD) rŏl (JMD) WOLOF kheter (JMD) ndidan (JMD) ndiran = *mob/crowd* (DF) ndupentan (JMD) niaje-payas (JMD) niakala-khat (JMD) tiéhomtioli (JMD) tiohamtuile (JMD) tiokom-tiokom *onomatopoeic, as when chewing* (DF) **THE GAMBIA** FULA-PULAAR (The Gambia) fungul (JMD) **GUINEA** MANDING-MANINKA n-togon (JMD) **SIERRA LEONE** BULOM (Sherbro) santil-lɛ (JMD) KISSI kayundu (JMD) LOKO koibɛre (JMD) koibwɛre (JMD) m-bembe-ambawa (JMD) m-bembe-ange (JMD) MANDING-MANDINKA deran (JMD) MANINKA (Koranko) babwie (JMD) mɛle (JMD) togon (JMD) MENDE njewɔ *a general term, esp. for Scleria spp.* (FCD) njewɔ-ha (*def.*-hei) ha: *female-? small, soft spp.* njewɔ-hina: *male-? tough, scabrid spp.* njewɔ-mumu *sharp-leaved sedges* (JMD) njewɔ-wa *large sedges* (JMD) tugbɛ, tugbɛlɛ SUSU fainye-saxei (JMD) fidera-saxei (JMD) fili-saxei (JMD) lela (JMD) mela (JMD) melaxame (JMD) nela (JMD) nela-fimbe (JMD) nela-gine (JMD) nela-xame (JMD) nela-xuxuri (JMD) siaxei (JMD) tigerin-wuri (JMD) tigerinyi (JMD) TEMNE *an*-kolma (JMD) *an*-kolma-*an*-tar (JMD) *a*-ro-gban (JMD) *an*-siri (JMD) *an*-siri-*an*-rini (JMD) *an*-siri-*a*-ro-beth (JMD) *an*-siri-*a*-ro-kant (JMD) *a*-soi (JMD) *an*-soi-*a*-rungi (JMD) *a*-wor-woraŋ for larger sedges (FCD) *a*-wor-woraŋ-*a*-ro-bath (JMD) *a*-wor-woraŋ-*a*-ro-bunko (JMD) *an*-yetɔ (JMD) *an*-yetɔ-*a*-ro-bath, ro-bath: *in a fresh-water swamp, i.e., sedges of the swamp* (FCD) *an*-yetɔ-*a*-ro-kai (JMD) **LIBERIA** MANO pehpeh (JMD) pi pi (JMD) **MALI** SONGHAI tara (JMD) taré gué (JMD) **GHANA** AKAN-ASANTE aberewa sekan = *old woman's knife* (Enti) ɛtene (JMD) TWI tene, ɛtene (JMD, Enti) WASA ɛtene (FRI) GBE-VHE ahɔ (JMD) NZEMA tekɛ (JMD) **NIGERIA** ARABIC-SHUWA jamsindi (JMD) kajije nguru *fragrant-rooted sedges* (JMD) suweji (JMD) EFIK ásái *sharp-leaved sedges* (JMD) HAUSA alwanzan *sharp-leaved sedges* (JMD) ayáá-ayáá (JMD) badayi *sharp-leaved sedges* (JMD) geémùn kwaàdoó *grass-like sedges* (JMD) kaàjiíjìn daájiì (JMD) kajijin shanu (JMD) kudunduriniya *sharp-leaved sedges* (JMD) IGBO ashiayi (JMD) ńné íkútē = *female, or mother grass, i.e., softer than the male* (JMD) óké íkútē = *male grass* YORUBA dogbodogobo (JMD) hujẹhujẹ (JMD) imeremere (JMD) làbẹlàbẹ = *it cuts, it cuts: for sharp-leaved sedges, cf.* ọbẹ: *knife* (JMD; Verger) matisan *sharp-leaved sedges* (JMD)

*NOTE: All these vernaculars generally cover several genera, and sometimes include grasses.*

## **Afrotrilepis pilosa** (Boeck.) J. Raynal

FWTA, ed. 2, 3:374. UPWTA, ed. 1, 515, as *Catagyna pilosa* Beauv.

English: 'devil grass' (Sierra Leone, Dawe).

West African: **GUINEA** KISSI nongura (A. Chev.) **SIERRA LEONE** KISSI nuŋgo (FCD) KONO nuŋbu (FCD) nuŋgu (FCD) MENDE munyɛ (auctt.) tinyi (auctt.) SUSU tigerinyi (NWT; JMD) SUSU-DYALONKE tentɛri-na (FCD) tintiriŋhu-na (FCD) TEMNE (Kunike) *an*-fulu (FCD) TEMNE (Port Loko) *a*-kek-*a*-suaŋla (FCD) TEMNE (Sanda) *a*-kek-*a*-suaŋla (FCD) TEMNE (Yoni) *an*-fulu (FCD) **IVORY COAST** DAN mu (RP) muta (RP) MANDING-MANINKA foa (RP)

A stout perennial, rhizome often erect or ascending, matting, on rock, from Guinea and Mali to W Cameroons and into Gabon.

The curious manner of growth results in a sort of turf pioneering a cover over bare rocks and where felling, fire and erosion have left devastation. Into this cover grit and humus accumulate in which other plants eventually grow (Chevalier fide 2; 3, 5, 6). At Binkolo in northern Sierra Leone the mat covering on granite outcrops is recorded as 1.25–2.50 cm thick (4). The leaves vary from about 10 to 40 cm long and are used in Guinea (1) and in northern Sierra Leone (3, 4, 5) for thatching hut-roofs. Also in northern Sierra Leone in the Mabonto-Bumban area, a plant of this species is put on the top of the main post of each house (4). This is presumably in a superstitious sense similar to the practice in Upper Cavally of Ivory Coast where the turf is planted on the top of huts to ward off lightning (Chevalier fide 2). The English name 'devil grass' from Sierra Leone implies magical attributes.

Deer are said to graze the foliage (7).

References:

1. Baldwin 9759, K. 2. Dalziel, 1937. 3. Dawe 509, K. 4. Deighton 1964, K. 5. Fisher 22, K. 6. Schnell, 1950, a: 910. 7. Scott-Elliott 5644, K.

**Ascolepis capensis** (Kunth) Ridley

FWTA, ed. 2, 3:327.

An annual of wet grassland, from Mali to S Nigeria, and throughout tropical and S Africa.

The plant probably provides grazing for cattle in Ethiopia (1), but it is untouched in Tanganyika (2).

References:

1. Gilbert 11, K. 2. McGregor 1, K.

**Bulbostylis abortiva** C.B. Cl.

FWTA, ed. 2, 3:318.
West African: **SENEGAL** SERER-NON (Nyominka) n-dipiandapa (A. Chev.)

A tufted annual of grassland cultivated and waste places, throughout from Senegal to S Nigeria, and distributed over tropical Africa and Madagascar.

**Bulbostylis barbata** (Rottb.) C.B. Cl.

FWTA, ed. 2, 3:316.
West African: **THE GAMBIA** MANDING-MANDINKA n'chongo mesongo = *small n'chongo* (Fox) **GHANA** GBE-VHE havi (FRI) **NIGERIA** YORUBA huję huję (Macgregor)

A tufted annual to 20 cm tall, of waste spaces especially on sandy soil, occurring throughout the Region, and widespread in the Old World tropics.

It is a weed of cultivated land.

**Bulbostylis congolensis** De Wild.

FWTA, ed. 2, 3:318.
West African: **SIERRA LEONE** LOKO koibɛre (NWT) MANDING-MANINKA (Koranko) babwie (NWT) MENDE nyina-voni, nyina: *rat;* voni: *grass* (Fisher) SUSU fainye-saxei (NWT)

A tufted annual of lateritic flats and wet sites, throughout the Region from Guinea and Mali to S Nigeria, and dispersed throughout tropical Africa.

**Bulbostylis densa** (Wall.) Hand.-Mazz.

FWTA, ed. 2, 3:318.
West African: **SIERRA LEONE** MENDE nyina-voni *applied to several spp.* (FCD)

A tufted variable sedge, of grassland in the lowlands and in montane (W Cameroons) situations throughout the Region, and dispersed through the Old World tropics.

**Bulbostylis laniceps** C.B. Cl.

FWTA, ed. 2, 3:316.
West African: **SIERRA LEONE** MENDE nyina voni (NWT)

A densely tufted plant of grassland in Guinea to W Cameroons, and extending across central Africa to Angola and Zambia.

**Bulbostylis lanifera** (Boeck.) Kük.

FWTA, ed. 2, 3:317.
West African: **UPPER VOLTA** MOORE daua njiaré kobodo (Scholz)

A tufted plant of savanna woodland across the Region from Senegal to S Nigeria, and extending also to Chad and Congo.

**Bulbostylis metralis** Cherm.

FWTA, ed. 2, 3:317.
West African: **SIERRA LEONE** SUSU fidera-sanoi (NWT) TEMNE *an*-putobaŋ (NWT)

A tufted perennial, with brief rhizome, stems 30–40 cm tall, of montane grassland or savanna, from Guinea and Mali to N and S Nigeria, and in central and east tropical Africa.

**Bulbostylis pilosa** (Willd.) Cherm.

FWTA, ed. 2, 3:316. UPWTA, ed. 1, 515.
West African: **GHANA** GBE-VHE (Kpando) angoxabe (FRI; JMD)

A tufted perennial, stems to 30–40 cm long from a woody rhizome, of savanna woodland and grassland throughout the Region, and widespread over tropical Africa.
The stems are used in Ghana to make brooms (1).

Reference:

1. Dalziel, 1937.

**Bulbostylis pusilla** (A. Rich.) C.B. Cl.

FWTA, ed. 2, 3:318.
West African: **SIERRA LEONE** LOKO fawu (NWT)

A slender annual of moist grassy places in Mali and Guinea to N Nigeria, and also across Africa to Ethiopia.

**Bulbostylis scabricaulis** Cherm.

FWTA, ed. 2, 3:316.
West African: **UPPER VOLTA** MOORE sourdjiongo (Scholz)

A tufted perennial, briefly rhizomed, stems 30–35 cm high, of moist places throughout the Region, and dispersed over central, east and southeast Africa and Madagascar.
Cattle are said to avoid grazing it in Guinea (1).

Reference:

1. Scott-Elliott 5467, K.

**Bulbostylis sp. indet.**

West African: **SENEGAL** BASARI a-làp ɔ-dẹ̀naw̃ (Ferry)

**Carex chlorosaccus** C.B. Cl.

FWTA, ed. 2, 3:349.

A tufted perennial to 1 m high, of montane forest on Fernando Po, and in NE and E Africa.
In Kenya it provides grazing for domestic stock and buffaloes (1).

Reference:

1. Glover & al. 1772, K.

**Carex echinochloë** G. Kunze

FWTA, ed. 2, 3:349.

A tufted perennial, to about 1 m high of shaded montane forest in W Cameroons, and also in NE and E Africa.

In both Ethiopia (2) and in Tanganyika (1) it is recorded as being not grazed by cattle. The foliage is not scabrid, so the reason must lie in another direction.

References:

1. McGregor 2, K. 2. Reading University 45, K.

**Cladium mariscus** (Linn.) Pohl, ssp. **jamaicense** (Crantz) Kük.

FWTA, ed. 2, 3:333.

English: saw grass.

A robust, thick-stemmed sedge, culms to 1·60 m high, of marshy woodland and dunes at Cape Verde, Senegal, and widely dispersed in E and S Africa, Congo basin, Sokotra, India and tropical and subtropical America.

The leaves are tough and very sharply serrated: hence the English name. The plant has been used in the Americas for producing a cheap paper (2, a). It is considered an excellent thatching material in Uganda (1). Ssp. *mariscus* which occurs in more temperate areas is known as 'thatching sedge', or 'fen sedge', and is a first-class thatching material (2, b).

An unnamed alkaloid is reported present in the seeds (3).

References:

1. Thomas A. S. 4246, K. 2. Uphof, 1968: 135: a. as *C. effusum* (Sw.) Torr. (Syn. *M. jamaicense* (Crantz) Britt.) b. *C. mariscus* Pohl. 3. Willaman & Li, 1970.

**Cyperus alopecuroides** Rottb.

FWTA, ed. 2, 3: 285.

West African: **NIGER** SONGHAI dúgú (*pl.* -ò) *also to closely related spp.* (D&C)

A perennial sedge, of short rhizome and tufted, short stems to 1.25 m high, in swamps and on the edge of water.

**Cyperus alternifolius** Linn.

FWTA, ed. 2, 3:289.

French: faux papyrus (Berhaut).

A robust, rhizomatous perennial, culms to 2 m high, native of E and S Africa and introduced to W Africa and to other areas of the tropics.

It is grown as an ornamental in the Region (Ghana, 2; W Cameroons, 1) forming thick clumps. In Kenya the Masai say that it is grazed by all domestic stock (3). In Tanganyika the root is administered as medicine to children for stomach-ache (6), and the dried plant is reduced to an ash for extraction by water which is evaporated and the residual salt is applied to fresh wounds as a disinfectant (4).

An unnamed alkaloid is reported present in the root and leaf (7).

The plant enters superstitious ritual for the cure of sickness in Ethiopia (5). Culms are laid across the kraal to rid people and cattle of illness. Its green and cool appearance is held to cool the malady. If a cow behaves strangely and climbs into a bush like a goat, a switch of the grass is used to splash a mixture of milk and water over it, and then some of the grass is laid down at the side of the cow.

References:

1. Brunt 1041, K. 2. Deighton 3785, K. 3. Glover & al. 2008, K. 4. Haerdi, 1964: 206. 5. Strecker s.n., s.d., K. 6. Tanner 2299, K. 7. Willaman & Li, 1970.

**Cyperus amabilis** Vahl

FWTA, ed. 2, 3:291.

West African: **NIGERIA** HAUSA gìrìgìrì (BM)

An annual to 20 cm high, of open savanna cultivated on waste ground, often on sandy soil, throughout the Region, and distributed pan-tropically and subtropically.

In Senegal cattle graze it, but not in Chad (1).

Reference:

1. Adam, 1966, a.

**Cyperus articulatus** Linn.

FWTA, ed. 2, 3:285. UPWTA, ed. 1, 516.

French: souchet articulé, souchet odorant (Berhaut, Kerharo & Adam); grand jonc (Berhaut).

West African: **SENEGAL** BASARI a-ngeul ɛ-gàrè (Ferry) BEDIK ga-nẹkabírín (Ferry) FULA-PULAAR (Senegal) gowé (K&A) SERER yig (JB; K&A) yih (JB) SERER-NON (Nyominka) oyé = *mat; incl. all plants used for matting* (K&A) WOLOF gawé, gowé *the fragrant tuber* (K&A) ndegit (auctt.) **THE GAMBIA** MANDING-MANDINKA kuntama (Fox) **GUINEA-BISSAU** BALANTA bum-ane (JDES) mussumárrè (JDES) BIAFADA n-pôpa, umpôpa (JDES) BIDYOGO ussóè (JDES) CRIOULO mampufa (JDES) MANDING-MANDINKA coutumô (JDES) NALU n'tende (JDES) untende (JDES) PEPEL modjotè (JDES) SUSU guleme (JDES) **GUINEA** FULA-PULAAR (Guinea) govhé (CHOP) MANDING-MANINKA madia (JMD) maya (JMD) SUSU kŏlumé (A.Chev.) turunyi (JMD) **SIERRA LEONE** BULOM (Kim) sɔ (FCD) FULA-PULAAR (Sierra Leone) gɔvɛ (FCD) GOLA ge (FCD) KONO sikoɛ (FCD) LIMBA (Tonko) kutagana (FCD) MENDE yiwɔ (FCD) MENDE-KPA hegɔ, higɔ (FCD) SUSU kura-kɔləme (FCD) SUSU-DYALONKE semu-na (FCD) TEMNE *an*-kolma, *an*-koɔlma *the mat made from it; name general for 'Cyperus' and allied genera* (FCD) TEMNE (Kunike) *an*-siri (FCD) TEMNE (Port Loko) *an*-koɔlma (FCD) TEMNE (Yoni) *an*-siri (FCD) VAI ze, zie *for the c.var grown for its fragrant roots* (FCO) (FCD) **MALI** MANDING-BAMBARA maya (JMD) **UPPER VOLTA** MANDING-DYULA njoro (JMD) **GHANA** AKAN-FANTE kokyi (FRI) TWI kyakya (FRI) kyerekyerewa (FRI) peprɛ (FRI) DAGBANI nasagti-pura (Ll.W., FRI) GA ŋɔi (FRI) GBE VHE ketsi (FRI) VHE (Awlan) ketsi (FRI) **TOGO** TEM (Tshaudjo) sang (Gaisser) **NIGERIA** FULA-FULFULDE (Nigeria) goye gulbi = *'goye' of the stream* (MM) goye maayo = *'goye' of the river* (MM) woyre (*pl.* go'e, goye) *incl. several sedges with fragrant roots* (JMD; Taylor) HAUSA gajiji (JMD) jiji *incl. several other spp.* (JMD) kaàjiíjiì *from a Bornu name, the root* (JMD; ZOG) kaàjiíjii-nágulbií, tùraàrén gulbií *incl. other spp. with fragrant roots* (JMD; ZOG) IGBO óké ubiri (BNO) óké íkútē = *male grass, loosely applied* KANURI kogo (Golding; JMD) NUPE efákó (JMD) TIV chagu (JMD) ilyogh (Vermeer) ishọhọ i toho (JMD) YORUBA ifeie (Dawodu) ifin *the plant* (JMD) òré *properly the mattress made from the plant* (JMD)

A robust perennial, culms to 1·75 m high by 8 mm in diameter arising from a rhizome, of sites in or near water, and occurring throughout the Region, and widespread in tropical and subtropical Africa and America, and into India.

The plant occurs wild in the Region. It is common in coastal salt-marshes, and forms extensive stands in the Senegal River estuary (11) and in Ghana (10). It is reported in the Volta Lake (19). In Lake Chad it occupies the seasonal flooded zone (7). The culms are commonly harvested for making high-class mats (Guinea, 2; Sierra Leone, 5; Ghana, 10; Nigeria 4); also for mattresses and sacking. In the 1920s and 30s it was made into mats for packing bananas for export from Guinea (2). In Ghana the leaves soaked in damp ashes, sundried and beaten flat are woven into square-bottomed fish-baskets (10).

The plant is commonly found under cultivation in villages throughout W Africa grown primarily for its aromatic rhizome. The wild races are said, in Sierra Leone at least, to be not aromatic (6). The smell of rhizomes of cultivars is pungent and musk-like (11). They are sold in markets everywhere. In N Nigeria Hausa traders offer two varieties, one black, the other red (3) but no difference in property is recorded. When dried and powdered, they are used as a fumigant, commonly mixed with scented resins, for the clothing, and to sweeten the air of rooms in the rainy season. In the Lake Chad area Shuwa women uproot the rhizomes from the flood-zone of the lake after the waters have receded to burn them over fires as a mosquito-repellant (7). Some tribes make necklaces and waist-girdles to keep insects away (3). In Gabon the rhizome is made into an application to the head for migraine (14, 15). Similar use is recorded in Senegal, or in the form of inhalations (11). In Nigeria the powdered root is admixed with grains-of-Paradise for this purpose, and with clay for putting on areas of craw-craw, etc. (3). The Tiv of Nigeria apply the root in poultice to swollen areas (13). Other analgesic relief is sought in Congo where the pulped roots are rubbed into epidermal scarifications for oedema and rheumatism and onto the bodies of babies with fever (1). In a medico-magical application in N Nigeria, the rhizome is added to water which has been used to wipe Koran texts from a writing-board, and after evaporation, the residue is used to fumigate the body in sickness. As a body-perfume, as well as a cleanser for the skin, it is mixed with oil or soap (3). Cultivation for commercial soap-making has been suggested (11). The Fula of Inner Senegal consider the scent aphrodisiac (11).

The rhizomes are used internally in Nigeria as a cough-medicine, or, if sleepless on account of cough a piece is placed in the mouth (3). A decoction is drunk in Congo for all respiratory troubles (1). The rhizomes are eaten raw in Tanganyika as a vermifuge and a decoction is drunk as an antimalarial (8). They are held to be aphrodisiac, vermicidal and beneficial in menstrual affections in Congo (1). The taste is bitter and it is antiemetic and sedative. It has been used to arrest vomiting in cases of 'yellow fever.' The root is an E African toothache remedy (16).

The plant enters into superstitious practices. In N Nigeria the root is chewed by a defendant in court as a charm to secure acquittal. It enters into prescriptions to ensure success in trade and to exorcise an evil spirit causing disease (Meek fide 3). In Gabon a mother who has given birth to twins will wear on her body some of the culm as protection for herself and the babies against evil influences (15). Similar faith is shown by a Sukuma mother after child-birth in Tanganyika by wearing a necklace of the rhizome. If the child gives cause for concern the mother chews the root and the child is quietened (12). The plant is commonly grown in sacred groves, and the scented rhizomes are offered to the Shades of the Departed. Witch-doctors sprinkle plant debris over the body of their clients to give them strength (15).

References:

1. Bouquet, 1969: 103. 2. Chevalier, 1931, a: 443. 3. Dalziel, 1937. 4. Dawodu 258, K. 5. Deighton 929, K. 6. Deighton 1417, K. 7. Golding & Gwynn, 1939. 8. Haerdi, 1969: 206. 9. Hall, J. & al, 1971. 10. Irvine, s.d. 11. Kerharo & Adam, 1974: 390–1. 12. Tanner 818, K. 13. Vermeer 26, UCI. 14. Walker, 1953, a: 32. 15. Walker & Sillans, 1961: 145. 16. Watt & Breyer Brandwijk, 1962: 373.

## **Cyperus bulbosus** Vahl

FWTA, ed. 2, 3:286.

West African: **SENEGAL** SERER lãgama (JB) **NIGER** SONGHAI dúg-ñá' (*pl.* -ñóŋó) (D&C) **NIGERIA** HAUSA jiji (Golding)

A slender rhizomatous perennial with leaves arising from a swollen base, of damp sandy places throughout the northern part of the Region from Mauritania to Niger and N Nigeria, and in N Africa through to Arabia to Australia.

It is a troublesome weed of cultivation in Sudan (3), and in parts of Tanganyika (5). It provides grazing for stock in Sudan (3) and in Somalia growing even on salt areas (4).

The swollen base is fed to infants in Upper Volta (1). Somalis eat the base, and guinea fowl and other wild animals also eat it (2).

The plant is similar to *C. rotundus* some of whose attributes may apply here.

References:

1. Chevalier, 1920: 689. 2. Godfrey-Fausset 70, K. 3. Harrison 67, K. 4. McKinnon S/228, K. 5. Terry, s.d.

**Cyperus compressus** Linn.

FWTA, ed. 2, 3: 288.
West African: **SIERRA LEONE** MENDE tugbɛlɛ (FCD)

An annual, tufted, culm to 60 cm high, of waste, often sandy, places, throughout the Region, and of pantropical dispersal.

It is an anthropogene and is commonly found near habitations as well as being a weed of cultivated and disturbed land (2). In SE Asia it provides grazing for buffaloes and cattle (1).

References:

1. Burkill, 1935: 733. 2. Deighton 2082, K.

**Cyperus congensis** C.B. Cl.

FWTA, ed. 2, 3: 288.
West African: **SIERRA LEONE** SUSU lela-fimbe (NWT)

A perennial with culms to about 1 m high, of river-banks and damp places, known only from Senegal and Sierra Leone in the Region, and occurring also in Gabon and Congo.

**Cyperus conglomeratus** Rottb.

FWTA, ed. 2, 3: 292.
West African: **MALI** TAMACHEK saad (RM)

A tufted plant with deep-penetrating wiry roots, of sand-dunes, in Mauritania, Senegal and Mali, into NE Africa and to India.

In Sudan (3) and in Somalia (1) it grows over sand-dunes often in extensive dominant stretches. Because of its deep-penetrating roots it tends to remain green for a while in the first part of the dry season and then it provides useful fair grazing for camels and stock (1, 2). It may also contribute to stabilization of sand.

References:

1. Gillett 4434, K. 2. Harrison 966, 1321, K. 3. Jackson 2435, K.

**Cyperus cuspidatus** Kunth

FWTA, ed. 2, 3: 291.
West African: **SIERRA LEONE** LOKO yɛngu (NWT)

A slender annual to 15 cm high, of sandy soil, on rocks, in turf and damp places generally.

Cyperus **dichroöstachyus** Hochst.

FWTA, ed. 2, 3: 290.

A leafy rhizomatous perennial to 80–100 cm high, of wet ground in W Cameroons and occurring throughout central and east tropical Africa and into Asia.

In Kenya it is grazed by all domestic stock (1, 2), and is used for hut-thatch (1).

References:

1. Glover & al. 1334, K. 2. Glover & al. 2001, K.

**Cyperus difformis** Linn.

FWTA, ed. 2, 3:290.
West African: **SIERRA LEONE** TEMNE *a*-kek-*a*-pot (Glanville) **NIGERIA** HAUSA geémùn kwaàdoó *a general term* (RES) YORUBA imeremere (Dawodu)

An annual, to 90 cm high, of wet sites, rice-fields, ditches, etc., throughout the Region, and distributed pantropically and subtropically. The plant is a common weed of rice-fields (2). In Kenya it is reported as grazed by all domestic stock (1).

References:

1. Glover & al. 1823, K. 2. Terry, s.d.

**Cyperus digitatus** Roxb., ssp. **auricomus** (Spreng.) Kük.

FWTA, ed. 2, 3:284. UPWTA, ed. 1, 516, as *C. auricomus* Sieber.
West African: **THE GAMBIA** MANDING-MANDINKA bankaŋ (*def.*-o) (Macluskie) n'chongo (Macluskie) **GUINEA-BISSAU** MANDING-MANINKA mela dion (JMD) **MALI** MANDING-BAMBARA dukon (Lean; JMD) labipobi (Lean) **NIGERIA** HAUSA aya aya (AST) gizgiri (JMD) kaajiijii-na-fadamaa *probably incl. other spp.* (JMD; ZOG)

A perennial herb with culms to 3 m high, of aquatic places throughout the northern part of the Region from Mauritius to Niger and W Cameroons, and dispersed across tropical Africa to central and east Asia and into S America.

The culm is suitable for pulping. It is not good for thatching (3). In The Gambia it is invasive of rice-padis and may be a nuisance (2). The tuberous rhizome is slightly fragrant. It is sometimes used as a substitute for genuine *kajiji* (Hausa, *C. maculatus* Boeck.) which is commonly sold in markets of the Soudanian region (1).

References:

1. Dalziel, 1937. 2. Macluskie 3, K. 3. Watt & Breyer Brandwijk, 1962: 493.

**Cyperus dilatatus** Schum. & Thonn.

FWTA, ed. 2, 3:286.
West African: **SENEGAL** WOLOF fiojamtiulé (JLT) **NIGERIA** IJO-IZON (Kolokuma) òwéí ángị = *male angi* (KW) NUPE efángi (RB)

A slender perennial herb arising from many slender curved rhizomes, of damp ground and cultivated land, throughout from Senegal to N and S Nigeria, and widespread in tropical Africa.

The plant becomes a pest of lawns in Ghana and is difficult to eradicate (1). The Ijo of the Niger Delta have medicinal uses. For cuts and wounds, the stem is chewed into a quid which is bandaged over the place and left for 3 days; this is replaced by a new quid till the place has healed. For gonorrhoea (*tu bẹịmọ ọgị*) the rhizome is washed and pounded in a mortar with some honey; a part of the preparation is drunk, and a part is rubbed around the waist (2).

References:

1. Irvine, s.n., s.d., K. 2. Williamson, K. 12.C, UCI.

**Cyperus distans** Linn. f.

FWTA, ed. 2, 3:287. UPWTA, ed. 1, 516.

West African: **GUINEA** MANDING-MANINKA togon (JMD) toyo (A. Chev.) **SIERRA LEONE** MENDE njewɔ-hina = *male sedge* (JMD) SUSU nela-xuxuri (JMD) **NIGERIA** IJO-IZON (Kolokuma) òwéí ángí = *male angi* (KW)

A perennial to about 1 m high, culms from a rhizome, in cultivated land, waste ground and damp grassland, occurring throughout the Region, and generally pantropical.

The plant is a common weed of many crops, but rarely, if ever, in E Africa, at least, does it become a serious problem (2).

In Guinea the rhizomes are used in sauces (1). The Ijo use this in the same manner for cuts and wounds, and for gonorrhoea as *C. dilatatus* (3).

References:

1. Chevalier, 1920: 690–1. 2. Terry, s.d. 3. Williamson, K, 12B, UCI.

**Cyperus dives** Del.

FWTA, ed. 2, 3:284.

West African: **SENEGAL** WOLOF hérèntan (JB)

A robust perennial similar to *C. exaltatus,* of marshes and recorded only in Senegal in W Africa, but otherwise pantropical African, in Madagascar, NE Africa and the Near East.

No usage is recorded in the Region. In Uganda the culms are used in basket-making (1).

Reference:

1. Thomas, A. S. 2775, K.

**Cyperus esculentus** Linn.

FWTA, ed. 2, 3:286. UPWTA ed. 1, 516–7.

English: tiger nut; earth almond, earth nut; chufa (U.S.A. from Spanish settlers in the southern states); rush nut; yellow nut-grass; Zulu nut.

French: souchet comestible; amande de terre.

Portuguese: chufa.

West African: **SENEGAL** DIOLA épaympay (JB, ex K&A) MANDING-BAMBARA ntô (JB; K&A) ntogô (JMD, ex K&A) togô (K&A) SERER ról (auctt.) WOLOF nâgaro *a c.var.* (AS; K&A) nder (auctt.) **THE GAMBIA** MANDING-MANDINKA n'chongo (Macluskie) n'chongo ba = *big n'chongo, a c.var.* (Fox) **GUINEA** MANDING-MANINKA toki (JMD; FB) **SIERRA LEONE** KRIO gramanti (FCD) kramanti (FCD) TEMNE *an*-kolma-*a*-ro-gban *an indefinite term, incl. other genera* (JMD) *an*-soi (KM) **MALI** MANDING-BAMBARA njoro, nton (A. Chev.) n-togon *general for sedges* (auctt.) togon-kayogo (A. Chev.) **UPPER VOLTA** MANDING-BAMBARA njoro (K&B) nton (K&B) togon (K&B) DYULA tchoro (K&B) toro (K&B) **IVORY COAST** MANDING-DYULA tchoro (K&B) toro (K&B) MANINKA toki (K&B) **GHANA** ADANGME fie (FRI) ADANGME-KROBO fɛ (FRI; JMD) AKAN-FANTE atadwe (FRI) atadwe awehea, awe, *to chew;* awhẽa, *sand, i.e., 'to chew sand', referring to the absence of nourishment* (FRI) atadwe fufuu *a white c.var.* (FRI) atadwe hene, hene: *king, a white c.var. with large nuts* (FRI) atadwe tsintsin *a white* c.var. *with slender roots* (FRI) atadwe tuntum *a black* c.var. (FRI) TWI atadwe (FRI) atadwe awehea, awe, *to chew;* awhẽa, *sand, i.e., 'to chew sand', referring to the absence of nourishment* (FRI) atadwe fufuu *a white* c.var. (FRI) atadwe hene, hene: *king, a white* c.var. *with large nuts* (FRI) atadwe tsintsin *a white* c.var. *with slender roots* (FRI) atadwe tuntum *a black* c.var. (FRI) DAGBANI nansaga (JMD) GA ataŋme (FRI) GBE-VHE fie (FRI; JMD) fie ge *a c.var.* (JMD) fie yibɔ *a c.var.* (JMD) fio (FRI; JMD) VHE (Awlan) fie (FRI) VHE (Pecí) fio (FRI) KONKOMBA idabongmale (Gaisser) **TOGO** BASSARI dimori (Gaisser) DAGBANI nansacha (Gaisser) KONKOMBA idabongmale (FB) idabunyal amo (Froelich) indabongmale (FB) indabunyal amo (Froelich) TEM amo (FB) TEM (Tshaudjo) amo (Gaisser) mõẽ (Gaisser) **NIGER** SONGHAI hántí (*pl.* -ó) (D&C) **NIGERIA** ARABIC saad (JMD) ARABIC-SHUWA nab al oziz *also for the kernel of Bambara groundnut* (JMD) suget (JMD) BEROM pyò (LB) EFIK ísíp akȓâ = *Accra Kernel; i.e., introduced from Accra* (auctt.) FULA-FULFULDE (Nigeria) ayaare (*pl.* ayaaje) (auctt.) waccuure (*pl.* waccuuje) (auctt.) HAUSA árìgízà (JMD; ZOG) ayáá, áyaá, àyaà-áyaà (auctt.) áyaá rigiza (JMD) ba-dusaya *a c.var.* (JMD; ZOG) bákár-áyaá *a c.var.* (JMD; ZOG) chiso (JMD) cizo (ZOG) daƙuwa *a sweetmeat from the nuts* (JMD) guzu-guzu *a c.var.* (JMD; ZOG) jar áyaá (JMD) jiji (Golding) rìgízà (JMD; ZOG) tamangarusa, tsinakuwa *the roasted nuts* (JMD) HAUSA (East) haya (ZOG) HAUSA (West) bakar (ZOG) KANURI nebu (JMD) nufu (JMD) NUPE efá (JMD) TIV ìshọ̀họ̀ (JMD) YORUBA imùmú òfio õmu YORUBA (Ijesha) ẹrũnshà (JMD)

A perennial herb to about 30–40 cm high from a slender rhizome and tubers at the base of each group of leafy stems, of cultivated and waste land, common throughout the whole Region and distributed into India to Indochina, and into Madagascar, the Mediterranean and S. Europe.

The plant originated from the Mediterranean and SW Asian area. It is of ancient cultivation. Tubers have been identified in the funery of tombs of the Twelfth Dynasty of Egypt, *c.* 2000 B.C. (4). The plant has been dispersed by man under cultivation, and it has in turn become a serious and common weed of many crops especially in highland areas (18). Its advent in W Africa is lost in prehistory and a study of the vernacular names does not reveal any suggestion of exotic origin; only the Efik name meaning 'kernel of Accra', i.e., of introduction from Accra, acknowledges a very limited movement (2).

The tuber is the part of the plant of principal interest. To some tribes it is an important food-stuff. For example, it is much grown on the coastal strip of Ghana and it enters into many Ghanaian aphorisms (8). The Hausa of N Nigeria manage to obtain two crops in the rainy season. Heavy consumption is said in Lesotho to cause constipation (7). To other races it is but a famine food. It can be grown under a wide range of conditions. Light sandy soil of pH 5.5–6.5 is considered best. It will tolerate salt and will grow on reclaimed coastal soil. Under optimum conditions it is a 3–4 months crop. The tuber is cooked as a vegetable or roasted on a tray of sand. Average size is 1–2 cm long by half that in width. There has been some ennoblement and strains with larger tubers are known and also some strains that can be eaten raw. The flavour is nutty, not unlike almonds (5, 8–12, 13). Composition is of the order: water 7–14%; fatty substances 20–36%; carbohydrates 37–50% of which as much as a half may be sucrose or other sugars; protein 1–5%; fibre 6–14%; ash 2% (4, 5, 12, 14, 17, 21).

A number of preparations is made from the tubers. Hausa of N Nigeria make a sweetmeat. Very commonly the fresh tubers are ground finely and the sappy liquid strained off for boiling with wheat flour and sugar. Constant stirring is necessary to prevent lumping and cooking is continued till the required sticky pap-like consistency is reached. This is 'tiger-nut milk' or in Ghana '*atadwe* milk.' It must be eaten at once as fermentation sets in quickly rendering the preparation unfit to eat. In Ghana white or yellow-tubered cultivars are preferred (8, 12). This white jelly-like substance is considered in the Kong area of Ivory Coast, perhaps on the Theory of Signatures, to be lactogenic (3, 15). In S Europe (Portugal, Spain and Italy) the plant is cultivated for the tubers for consumption and for the preparation of a frozen or chilled drink known as *horchata de chufas* in Spain obtained by expressing the sap and emulsified oils (19, 20). In the Keta area of Ghana the sundried tubers are ground to a very fine powder which with sugar added can be stored till required. Roasted tubers may be similarly ground to a powder known in Vhe (Awlan) as *fie-dzowe*, from *fie*, tiger-nut: *dzo*, fire: *we* powder. These meals may be eaten alone or with water added to make a beverage (12). *Fie-dzowe* is reputed to be aphrodisiac.

The tubers are made into a kind of 'chocolate' in Sierra Leone, and can be used as a substitute for or adulterant of coffee or cocoa, and in lieu of almonds in confectionery. No caffein is present (5, 8, 12, 13, 21). Because of the rich oil-content flour prepared from the tubers is of high calorie content. In India the tubers have been satisfactorily fed to cattle and to pigs (17). In S Africa the tubers are chewed to relieve indigestion especially when accompanied by halitosis. The tubers are added to food taken by young Zulu girls to hasten the inception of menstruation, and in China the tuber is considered stimulant, stomachic, sedative and tonic (21).

The fatty substance of the tuber contains a non-drying pleasantly-flavoured oil similar to olive or sweet almond oil the composition of which is mainly oleic acid 73%, palmitic acid 12–13%, linoleic acid 6–8%, stearic acid 5–6% with traces of others (4, 14, 17, 21). It is extracted by cold expression and is used commercially in India and in Europe for cooking and soap-manufacture (13, 17). There is a commercial potential in the preparation of starch from the cake after extraction of the oil, and of alcohol by fermentation of the tubers (13).

The haulm is readily grazed by all stock (1). In Lesotho rough ropes are plaited from it (7). It is suitable also for making a paper-pulp (13). In Sierra Leone bees are reported to visit the flowers as a source of pollen (16). The roots are burnt in N Nigeria to produce a scented smoke (6). In Senegal the rhizomes (including the tubers) are taken in decoction as a draught for stomach-troubles and as a refreshing drink, while the leaves are applied in a poultice to the forehead for migraine (14).

References:

1. Adam, 1966, a. 2. Adams, R. F. G., 1943. 3. Bouquet & Debray, 1974: 79. 4. Busson, 1965: 441. 5. Dalziel, 1937. 6. Golding 14/14A, K. 7. Guillarmod, 1971: 422. 8. Irvine, 1930: 143–4. 9. Irvine, 1948: 266. 10. Irvine, 1952, a: 30. 11. Irvine, 1952, b: 32. 12. Irvine, s.d. 13. Kay, 1973: 50. 14. Kerharo & Adam, 1974: 391–2, with phytochemistry and pharmacology. 15. Kerharo & Bouquet, 1950: 251. 16. Miszewski 41, K. 17. Sastri, 1950: 423–4. 18. Terry s.d. 19. Uphof, 1968: 169. 20. Usher, 1974: 195. 21. Watt & Breyer-Brandwijk, 1962: 373.

## **Cyperus exaltatus** Retz.

FWTA, ed. 2, 3:284. UPWTA, ed. 1, as *C. dives* sensu Dalziel.

West African: **NIGERIA** ARABIC-SHUWA kajiji nguru (JMD) FULA-FULFULDE (Nigeria) gurguhọ (JMD) wurguhọ (JMD) HAUSA gwaigwaya (auctt.) karan masallachin kogi (JMD) mashin kogi = *spears of the stream* (JMD)

*NOTE: The foregoing vernaculars probably refer here.*

A robust perennial, culms to 1·5 m high, in aquatic sites, occurring throughout the Region from Senegal to S Nigeria, and dispersed pantropically.

The culms are strong. They are used in play by boys in N Nigeria — hence the Hausa name meaning 'Spear of the stream.' They are used in hut-building, and split to weave into reed-mats. In the Lake Chad this and other large sedges are burnt to produce a vegetable salt (1). In Kenya the culms are used as thatching (3). The rhizome has medicinal applications in Tanganyika: grated in poultice to whitlow and to swollen buboes in cases of blood poisoning to draw and maturate the pus; grated and eaten, and in dressings to scarifications over the spleen in cases of chronic malaria; and together with the cane of *Saccharum officinarum* Linn. (sugar cane, Graminae) in application to swollen breasts to promote milk-flow (2).

References:

1. Dalziel, 1937. 2. Haerdi, 1964: 206. 3. Lamprey 382, K.

## **Cyperus fenzelianus** Steud.

FWTA, ed. 2, 3:285.

A perennial herb, culms to 1 m high from slender rhizomes, of marshes and at the edge of open water, across the Sahel zone of the Region and extending into India.

The distinction between this and *C. longus* Linn. ssp. *longus* is not well defined. The latter as var. *pallidus* (Boeck.) Kük. occurs in E Africa where it provides grazing for all stock (3) and game (2); in Somalia it has been used to prepare poison-bait for locust-control (1). Var. *tenuiflorus* (Rottb.) Boeck. is a weed of irrigation in Kenya (4). True *C. longus* of Europe is the 'galingale', sweet scented and used in perfumery. W African material requires examination.

References:

1. Bally s.n., July 1947, K. 2. Glover & al. 2968, K. 3. Mwangangi & Gwynne 1224, K. 4. Terry s.d.

**Cyperus haspan** Linn.

FWTA, ed. 2, 3:291. UPWTA, ed. 1, 517.
West African: **SIERRA LEONE** MENDE njewɔ (JMD) njewɔ mumu *usually applied to sharp-leaved sedges* (JMD) njewɔ wa, wa: *large* (NWT; JMD) tugbɛlɛ SUSU lela (JMD) melai (JMD) nelai (JMD) TEMNE *a*-kek-*a*-pot (Glanville) *an*-siri, *an*-roi-*a*-rungi (JMD) **NIGERIA** HAUSA àyaà-áyaá (Taylor)

A tufted perennial, soft almost succulent, of open, damp sites, occurring throughout the Region, and distributed pantropically and subtropically.

In E Africa it is burned to provide a vegetable salt (2). Potassium salts can be obtained by lixiviation of the plant-ash (1). In SE Asia it is common in rice-padi fallows and cattle are let loose to graze it. The pith is said in Malaya to furnish lamp-wicks (1).

References:

1. Burkill, 1935: 734. 2. Dalziel, 1937.

**Cyperus imbricatus** Retz.

FWTA, ed. 2, 3:284.
West African: **MALI** MANDING-BAMBARA dukon (Lean)

A perennial herb, culms to 1·2 m high, of aquatic sites, throughout the Region from Senegal to W Cameroons, and dispersed pantropically.

The plant provides grazing for animals in Sudan (1).

Reference:

1. Andrews A.646, A.647, K.

**Cyperus iria** Linn.

FWTA, ed. 2, 3:288.

An annual herb, culms to 75 cm high, of damp sites from Senegal to N Nigeria, and distributed pantropically.

In Tanganyika it is considered a minor weed of rice-padis (2). In Malaya it is used for matting, and cattle eat it (1).

References:

1. Burkill, 1935: 734. 2. Parker R, K.

**Cyperus jeminicus** Rottb.

FWTA, ed. 2, 3:292.
West African: **NIGER** HAUSA guémé n'dari (Bartha)

A tufted plant with wiry roots, of sandy places, in Mauritania, Senegal and Niger, and across NE Africa to Arabia.

This is very closely related to *C. conglomeratus* Rottb. and has been known as var. *multiculmis* (Boeck.) Kük. of that species.

In Senegal it provides grazing which is relished by cattle. It is also grazed a little at Chad (1).

Reference:

1. Adam, 1966, a: as *C. cruentus*.

**Cyperus latifolius** Poir.

FWTA, ed. 2, 3:285.

Robust perennial plant, culms to 1·2 m high, of swampy ground, recorded only from W Cameroons in the Region, but widespread elsewhere in tropical and S Africa and in Madagascar.

The culms are used for thatching in Tanganyika (1), and in both Kenya (2) and Tanganyika (3, 4) are used to make spirit-houses.

References:

1. Archbold 56, K. 2. Magogo & Glover 576, K. 3. Tanner 819, K. 4. Tanner 1010, K.

**Cyperus ledermannii** (Kük.) Hooper

FWTA, ed. 2, 3:293.
West African: **NIGERIA** HAUSA (West) farin kai = *white head* (Taylor)

A perennial, up to 90 cm high, of savanna and open woodland in Dahomey and N Nigeria, and in E Cameroun and E Africa.

The N Nigerian vernacular name refers to the whitish appearance of the inflorescence.

**Cyperus maculatus** Boeck.

FWTA, ed. 2, 3:285.
West African: **SENEGAL** MANDING-BAMBARA gonéni (JB) WOLOF ralélé n-dau (RP) **GUINEA** FULA-PULAAR (Guinea) tiacktal (RP) **SIERRA LEONE** SUSU lela-fimbe (JMD) nela-fimbe (JMD) **MALI** MANDING-BAMBARA guenun (anon) n-tioko (GR) SONGHAI haagui (A. Chev.; RP) **GHANA** DAGBANI kulisaa (Ll.W.) **NIGERIA** ARABIC-SHUWA si'id *incl. several spp. with fragrant roots* (JMD) FULA-FULFULDE (Nigeria) goye ɓaleeje = *black 'goye'* (JMD) HAUSA àyaà-áyaá *the fragrant dried roots, incl. several spp.* (JMD) dan Tunuga, kajiji ɗan Tunuga *from Tunuga, a place in SW Niger* (JMD) turare = *perfume; the fragrant dried roots* (JMD) NUPE efági keni = *Hausa tiger-nut* (JMD)

A rhizomatous perennial with slender culms arising from hard round tubers, of sandy places near to water and in flood-plains. Throughout the Region from Mauritania to S Nigeria, and generally dispersed over tropical Africa and into Madagascar.

The foliage is edible to stock and provides much-relished grazing (Senegal, 1; Sudan, 2). In Ghana an infusion of the leaves and roots is put into a preparation for treating *garli* cattle disease (6). The tubers are edible and are aromatic. They are sold in northern markets for the making of fragrant sachets (5) and perfume (4), the Hausa name *turare* for the dried roots meaning 'perfume.' *Tunugu* in the Hausa names is said to refer to the place in Niger Republic from where much of the marketed produce comes. A number of fragrant-tubered species may be involved. The tubers are also burnt in hut-fires in the Nupe area to create a pleasant smell (3).

References:

1. Adam, 1966, a. 2. Andrews A. 645, K. 3. Barter 1571, K. 4. Dalziel, 1937. 5. Portères, s.d. 6. Williams, Ll. 723, K.

**Cyperus margaritaceus** Vahl

FWTA, ed. 2, 3:292.

A tufted plant, compactly rhizomatous with contiguous swollen stem-bases, of sandy places throughout the Region, and through tropical Africa and into S Africa.

No attribute is recorded for the plant in W Africa. In Malawi it provides while still young, grazing for stock (1).

Reference:

1. Fenner 203.

**Cyperus maritimus** Poir.

FWTA, ed. 2, 3:291–2.

West African: **SENEGAL** SERER tol ɗuam (JB) WOLOF gowé (JB)

A perennial to 20–30 cm high from tough long rhizomes of the coastal strand throughout the length of the Region's coast, and in E and S Africa and Madagascar.

The plant grows down to high-tide level. It is a good sand-binder and is frequently a pioneer in this respect. The rhizomes are slightly aromatic (1).

Reference:

1. Tanner 1928, K.

**Cyperus papyrus** Linn.

FWTA, ed. 2, 3:284. UPWTA, ed. 1, 517.

English: papyrus; 'bulrush' (of the Bible).

French: papyrus.

Portuguese: papiro.

West African: **NIGERIA** ARABIC birdi (JMD) burdi (JMD) ARABIC-SHUWA umm ganagan (JMD) KANURI fole (Golding) kotolo (Golding; JMD)

A perennial herb, stems to 5 m long, of lakes and waterways, often forming floating mats, but usually anchored by a massive wooden rhizome in shallow water, occurring here and there throughout the Region, and elsewhere in tropical Africa, S Africa and Madagascar. It has been widely dispersed by man under cultivation.

The plant was first described by Theophrastus (*c.*372–287 B.C.) from material under cultivation in the Nile delta. He referred to the 'wood' [rhizome] being used for utensils and fuel and the stem for boats, sails, mats, cloth, cord and writing-material. The pith was commonly eaten either raw or cooked, and used for caulking seams in boats. Of the inflorescence he commented that it could be used for nothing but garlands for the shrines of the Gods. The reference in the Bible in Isaiah 18:2 to 'vessels of bulrushes upon the waters' was certainly to boats made of papyrus which were generally used to paddle about in the shallow waters. The principal use of papyrus in the Ancient World was the manufacture of parchment for writing upon. The stems were split lengthwise and laid side by side to the required width, other strips were laid at right angles across them and after soaking the 'woven' strips were bonded together by beating with mallets and drying in the sun. The best quality parchment came from the widest strips taken at the centre of the culm. The very extensive cultivation of papyrus that was carried on in Egypt for this purpose died in the 9th and 10th Centuries A.D. by competition from new sorts of paper (5). The culms are suitable for pulping and could be used in paper-making.

In Lake Chad, Yedina make canoes and rafts of the stems, and weave water-tight baskets (1). The plant forms a fringe around the lake and where it is rooted to the substrate it is said to mark the lake's lower water-level. But it also forms floating islands growing out of a mass of rotting vegetation (2). The plant has been recorded in Dahomey but has not so far (1971) appeared in the Volta Lake where it could become a noxious weed forming large areas of floating swamp (4).

Though use of the culms in hut-building may seem attractive, it is not a good thatching material. In Gabon it is plaited into large mats and made into house-partitions (6). With the outer cortex removed, cut into sections and sun-dried, it is used to stuff mattresses and cushions.

The alkaloids *tyramine* and *octopamine* have been recorded present in the leaves (7). A root-decoction together with the leaf-sap of *Maytenus senegalensis* (Lam.) Exell (Celastraceae) is taken by women in Tanganyika for sterility (3).

The Hausa of N Nigeria pull apart the leaf and the nature of the tear indicates either the rupture (*sa ani*) or the cementing of friendship (1). In Gabon the chewed dried rhizomes are used to ward off evil spirits either by using them to beat the limbs, or by steeping them in water which is used for ablution (6).

References:

1. Dalziel, 1937. 2. Golding 23, K. 3. Haerdi, 1964: 207. 4. Hall, J. & al., 1971. 5. Thompson, 1929. 6. Walker & Sillans, 1961: 146. 7. Wheaton & Stewart, 1970.

**Cyperus podocarpus** Boeck.

FWTA, ed. 2, 3:289.
West African: NIGERIA HAUSA (East) wuchar sharindo (Grove)

A slender annual of variable size, of marshes and in rocky outcrops, from Senegal to N Nigeria, and in central and NE Africa.

**Cyperus procerus** Rottb.

FWTA, ed. 2, 3:285.

A stout perennial herb, rhizomatous with basal tubers, of wet sites in the northern part of the Region from Mali to N Nigeria, and dispersed across to NE Africa to India, Malesia and Australia.

No attribute is recorded of the plant in W Africa. In India it is a troublesome weed of rice-padis and is difficult to eradicate. A robust variety (var. *lasiorrhacis* Clarke) occurring in Madras rice-fields is used to make mats (2). In Java the stems are used on a considerable scale as string (1, 3).

References:

1. Burkill, 1935: 735. 2. Haines, 1921–25: 902 [1961 reprint. p.945]. 3. Sastri, 1950: 424.

**Cyperus pustulatus** Vahl

FWTA, ed. 2, 3:288.
West African: **SIERRA LEONE** TEMNE *an*-soi (NWT) **NIGERIA** HAUSA àyaà-áyaà (Taylor)

A tufted annual reaching 1 m tall, of marshy and open-water sites, over the Region from Senegal to S Nigeria, and distributed throughout tropical Africa.

**Cyperus reduncus** Hochst.

FWTA, ed. 2, 3:290.
West African: **DAHOMEY** BATONNUN gbassoki (DA) **NIGERIA** HAUSA aska (RES)

A weak annual, to 30 cm high, of wet localities, throughout the Region, and in central, NE, E and south-central Africa.
The plant is reported as a weed of rice-fields in Zambia (1).

Reference:

1. Van der Veben, s.n. Sept. 1962, K.

**Cyperus renschii** Boeck.

FWTA, ed. 2, 3:289.
West African: **SIERRA LEONE** MANDING-MANINKA (Koranko) fudinyɛ (NWT)

A robust perennial, of secondary and gallery forest by streams, from Guinea to W Cameroons, and widely distributed elsewhere in tropical Africa.
Sap expressed from the leaf and rhizome is taken with the frond of *Cyclosorus gongylodes* (Schkukr) Link (Pteridophyta) in Tanganyika for cardiac hyrocele (cardiac oedema) (1).

Reference:

1. Haerdi, 1964: 207.

**Cyperus rotundus** Linn.

FWTA, ed. 2, 3:285. UPWTA, ed. 1, 517.
English: nut-grass.
West African: **SENEGAL** FULA-PULAAR (Senegal) gowé (A. Chev.) hissel (JMD) hisser (JMD) MANDING-BAMBARA n-togon (JMD) SERER rôl (JMD) WOLOF ndidan (JMD) ndiran *= crowd, as applied to a rushing crowd* (DF) tiehomtioli (JMD) tiohamtiule (JMD) **SIERRA LEONE** MENDE njewɔ (JMD) njewɔ mumu *for sharp-leaved sedges* (JMD) njewɔ wa, wa: *large, general for large sedges* (JMD) tugbɛlɛ SUSU lela (JMD) melai (JMD) nelai (JMD) TEMNE *an*-roi-*an*-rungi (JMD) *an*-siri (JMD) **MALI** DOGON numii sami (C-G) FULA-PULAAR (Mali) gué (A. Chev.) hissel (JMD) hisser (JMD) MANDING-BAMBARA digityeh (FNH) n-tioko (GR) n-togon *general* (JMD) SONGHAI digai sa (anon) **GHANA** DAGBANI kulisaa (FRI; JMD) GA ngɔi

(FRI) NZEMA mbubule (FRI; JMD) **NIGER** HAUSA guiraguri (Bartha) SONGHAI dúgú ɓì (*pl.* -bíò) *also to closely related spp.* (D&C) **NIGERIA** FULA-FULFULDE (Nigeria) ayaare (JMD) goye *general* (JMD) HAUSA àyaà áyaà (JMD; RES) giragiri (JMD) girigiri (BM) gwaigwaya (JMD; ZOG) jigi, jiji (JMD; ZOG) KANURI nû (C&H) TIV ishọhọ i toho (JMD)

*NOTE: These vernacular names often apply to other sedges, and in some cases to sedges in general.*

A herb to about 50 cm high from a slender rhizome with small nut-like tubers at the base of each group of stems, of cultivated land and damp sites, common throughout the Region and dispersed pantropically and subtropically.

The plant is a noxious weed of cultivation and waste places. In E Africa it has become particularly serious in some irrigation schemes and in coffee grown under minimal tillage (12). Dispersal is principally by abundant seeding. The plant provides some grazing for stock, though not for horses (1), and it is said that the seed will pass through animals' intestines undigested and ready to germinate (2). Active rhizome growth promotes further local spread. The plant is difficult to eradicate.

The tubers are edible, and are occasionally eaten. They may be regarded as a famine food. In Mali they are taken as an aphrodisiac (11). They form some attraction for wild animals and are recorded as being dug up in Kenya by warthogs (8), guinea-fowl (4) and francolin (5). In Tanganyika they serve as bait for catching rats (6). They are somewhat fragrant. An essential oil is present, and they are made into a cough-medicine for children, and in northern Ghana are put into a medicine for treating *garli* cattle disease (3). In Jebel Marra they are crushed and made into a scented ointment with ghee (7), and in W Africa they are used to perfume clothing and to repel insects (3). Besides the volatile essential oil there are a fixed oil, a wax and a substance capable of dissolving several times its own weight of lecithin and other amino-acid substances associated with the formation of urinary stones. The tubers are regarded as stimulant, stomachic, diuretic, astringent and beneficial in diarrhoea. They are widely used in Africa and Asia for urinary troubles, indigestion, childbirth, jaundice, malaria and many other conditions (2, 9, 10, 13). The tubers are official in various pharmacopoeias.

This species is similar to and confused with other tuberous sedges, e.g. *C. tuberosus* Rottb. which may be, in fact, but a subspecies, *C. esculentus* Linn., *C. bulbosus* Vahl, *C. fenzelianus* Steud, etc.

References:

1. Adam, 1966, a. 2. Burkill, 1935: 735–7. 3. Dalziel, 1937. 4. Glover & al. 434, K. 5. Glover & al. 2946, K. 6. Harwood 60, K. 7. Hunting Technical Surveys, 1968. 8. Mwangangi & Gwynne 1030, K. 9. Oliver, 1960: 24. 10. Quisumbing, 1951: 113–4. 11. Roberty 2131, IFAN. 12. Terry s.d. 13. Watt & Breyer-Brandwijk, 1962: 374.

## **Cyperus sphacelatus** Rottb.

FWTA, ed. 2, 3:286. UPWTA, ed. 1, 518.

West African: **SENEGAL** FULA-PULAAR (Senegal) gowé (JMD) hissel (JMD) hisser (JMD) MANDING-BAMBARA n-togon (JMD) SERER rôl (JMD) WOLOF ndidan (JMD) ndiran (JMD) tiehomtioli (JMD) tiohamtiule (JMD) **SIERRA LEONE** MENDE njawa-wa, wa: *large* (JMD) njewɔ (JMD) njewɔ mumu (JMD) tugbɛlɛ SUSU lela (JMD) melai (JMD) nela(-i) (JMD) tigerinyi (NWT; JMD) TEMNE *a*-kɔlma-*a*-ro-bath (FCD) *an*-roi-*an*-rungi (JMD) *an*-siri (JMD) **MALI** FULA-PULAAR (Mali) gué (JMD) hissel (JMD) hisser (JMD) MANDING-BAMBARA n-togon (JMD) **GHANA** DAGBANI kulisaa (JMD) GA ngɔi (JMD) NZEMA mbubule (JMD) **NIGERIA** FULA-FULFULDE (Nigeria) ayaare (JMD) goye (JMD) HAUSA àyaà-áyaà (JMD) giragiri (JMD) gwaigwaya (JMD) jiji (JMD) IGBO (Umuahia) óké íkútē = *strong grass; hence considered male* (AJC) IJO-IZON (Kolokuma) àngị́ (KW) TIV ishọhọ i toho (JMD) YORUBA ewa (West)

*NOTE: Most of the foregoing names are taken from C. rotundus and are probably applicable here.*

A rhizomatous perennial, without tubers, of disturbed ground and damp grassy places, common throughout the Region and pantropical.

The plant provides somewhat indifferent fodder, but it is said to be relished in Ghana by sheep and goats till it reaches seeding time (1). The rhizome is scented ('slightly', 2; 'strongly', 3). As it shares the same vernacular names in Ghana with *C. rotundus*, it perhaps has the same attributes.

References:

1. Beal 12, K. 2. Dalziel, 1937. 3. Irvine 4843, K.

**Cyperus tenuiculmis** Boeck.

FWTA, ed. 2, 3:287.
West African: **SIERRA LEONE** MENDE ngete-wulo-hina, hina: *male* (FCD)

A slender perennial with a short, swollen stem-base, of grassland, swamp and sandy waste places, common throughout the Region, and dispersed pantropically.

In Nigeria it is recorded as a weed of cultivation, pastures and lawns especially in wetter places (2). It has unspecified use as a children's medicine in Tanganyika (1).

References:

1. Boaler 472, K. 2. Egunjobi, 1969.

**Cyperus tonkinensis** C.B. Cl. var **baikiei** (C.B. Cl.) Hooper

FWTA, ed. 2, 3:288.
West African: **SIERRA LEONE** SUSU nela (NWT) **MALI** MANDING-BAMBARA n-dougan (anon.) n-douin (anon.) **NIGERIA** HAUSA dan Tunuga, kajijiɗan Tunuga (ZOG)

A tufted plant with pseudobulbous stem-bases on river sand-banks from Senegal to S Nigeria, and from Chad to Congo. The typical form is from SE Asia.

**Cyperus tuberosus** Rottb.

FWTA, ed. 2, 3:286.
West African: **MAURITANIA** ARABIC (Hassaniya) lekmir (AN)

A herb to about 50 cm high similar to *C. rotundus* but with laxer spikes, of damp grassy and waste places from Mauritania to S Nigeria and dispersed throughout the Old World tropics.

The plant is grazed by cattle in Senegal (1) and is said to provide good pasturage for all stock in Mauritania (2). It is a common weed of lowland cultivation and in E Africa it has become a serious nuisance in the coastal area (3).

References:

1. Adam, 1966, a. 2. Naegelé, 1958, b: 889. 3. Terry, s.d.

**Cyperus zollingeri** Steud.

FWTA, ed. 2, 3:287

An annual herb to about 20 cm high of grassland, rare in Senegal, Ghana and Togo, but widely distributed elsewhere in tropical Africa and in Madagascar, Malesia and N Australia.

It provides a little grazing for cattle in Senegal (1).

Reference:

1. Adam, 1966, a.

**Cyperus spp. indet.**

West African: **SIERRA LEONE** MANDING-MANDINKA tiri (NWT) **NIGERIA** FULA-FULFULDE (Nigeria) kukuuri (J&D) woyre (J&D)

Unidentified *Cyperus spp.* of the foregoing vernacular names have recorded uses: (a) *tiri* in Sierra Leone for swollen feet (2); (b) *woyre* in N Nigeria the rhizomes are burnt in huts for colds, flu and rheumatism (1).

References:

1. Jackson, 1973. 2. Thomas, N. W. 173, K.

**Eleocharis acutangula** (Roxb.) Schult.

FWTA, ed. 2, 3:314.
West African: **SIERRA LEONE** SUSU koline (NWT)

A tufted stoloniferous plant of swamps, ponds and streams, occurring throughout the W African Region, and pantropical.

It is similar to *E. mutata* and probably is not fit for grazing.

**Eleocharis dulcis** (Burm. f.) Henschel

FWTA, ed. 2, 3:313.
English: Chinese water chestnut.
West African: **SIERRA LEONE** TEMNE *a*-kɔlma-*a*-tar (FCD)

A stoutly tufted annual with photo-synthesising leafless stems arising from a corm or tuber, with radiating stolons each ending in a tuber, aquatic and occurring throughout the Region, and across the Old World tropics to Japan.

The stems are silicaceous and are not grazed by stock (1). They are used in Sumatra to make mats which are of limited durability (3).

The corms are 1·0–1·5 cm in diameter. They are edible cooked, are palatable and nutritious and are widely eaten in the Far East. On dry weight, starch-content is nearly 60% and protein 7%. Some sugars are present. *E. tuberosus* Schult. is held by some authorities to be a cultigen derived in China from *E. dulcis*. The selection has resulted in fewer but larger corms which measure about 2·5–4·0 cm in diameter. It is commonly grown in China and Japan, and more recently in the Philippines, Hawaii, India and U.S.A. Cultivation is in the same manner as for wet-rice, the fields being allowed to dry off before harvest. A flour made from the corms is traded in N China. Commercial starch can be prepared, and there is an export trade of the corms, either fresh or canned. There is some variability in the chemical composition of the corms. Those of Chinese origin have been recorded as containing 77% carbohydrates (starch and sugars in equal proportions) and 8% albuminoids (2–7).

Sap expressed from the corms has been found to possess an antibiotic action against *Staphylococcus aureus, Escherichia coli* and *Aerobacter aerogenes.* The active principle has been named *puchiin* (5).

References:

1. Adam, 1966, a: as *Heleocharis plantaginea.* 2. Bates, 1976: 420. 3. Burkill, 1935: 906–7. 4. Kay, 1973: 43. 5. Sastri, 1952: 142–3. 6. Uphof, 1968: 197. 7. Usher, 1974: 226.

**Eleocharis mutata** (Linn.) Roem. & Schult.

FWTA, ed. 2, 3:314.
West African: **SENEGAL** BALANTA cunduman amésima (Dubois) DIOLA fu pepe (FB) MANDING-MANDINKA sorto (Dubois) SERER ́délah (JB) ́dèmtépetèk (JB) **SIERRA LEONE** TEMNE *a*-kɔlma-*a*-kuisisi (Glanville)

A tufted stoloniferous plant of brackish and coastal swamps from Senegal to S Nigeria and widely dispersed on the western Atlantic seaboard.

The plant withstands a high degree of brackishness becoming a weed in marginal coastal rice-padis and in cleared mangrove areas (2). The stems are silicaceous and are not grazed by stock (1).

References:

1. Adam, 1966, a: 519 as *Heleocharis mutata* and *H. fistulosa.* 2. Glanville 211, K.

**Fimbristylis aphylla** Steud.

FWTA, ed. 2, 3:323.
West African: **SIERRA LEONE** TEMNE *a*-kek-*a*-pot (Glanville)

A stout, tufted perennial to 75 cm high, of swamps and damp ground in savanna and near rivers, etc., in Sierra Leone, Ghana and Nigeria, and extending into the Congo basin.

**Fimbristylis cioniana** Savi

FWTA, ed. 2, 3:324.
West African: **SIERRA LEONE** MANDING-MANDINKA šime (NWT) SUSU fili-saxei (NWT) saxei (NWT)

A small tufted annual to 15 cm high, of sandy grassland and river-banks from Senegal to Nigeria, and widely dispersed in tropics and subtropics.

**Fimbristylis debilis** Steud.

FWTA, ed. 2, 3:323.
West African: **NIGERIA** HAUSA (West) bunsulu fadama (Grove)

A slender annual to 30 cm high, of savanna and scrub, especially near water in laterite out-crops.

**Fimbristylis dichotoma** (Linn.) Vahl

FWTA, ed. 2, 3:320–1.
West African: **SENEGAL** DIOLA é fok (JB) **SIERRA LEONE** LOKO dondoi (NWT) TEMNE *an*-soi (NWT) *an*-soi-*a*-bɛra (NWT) **NIGERIA** CHAMBA ka-san (FNH) ka-zan = *fowl; alluding to pigeons eating the fruit* (FNH) HAUSA áyaá (FNH) geémùn kwaàdoó *a general term* (RES)

A tufted annual or perennial, culms to 1 m high, variable with three recognised varieties in the Region, var. *dichotoma,* var. *laxa* (Vahl) Napper and var. *pluristria* (C.B. Cl.) Napper; of stream-banks, swamps, damp grassland and cultivation, common throughout the Region, and generally pan-tropical.

Cattle graze it especially when still young. Seeds are said to pass through the intestines undigested, thus causing distribution (1, 2). Other species of *Fimbristylis* also provide fodder and an analysis of some Indonesian material has given: protein 4·8–6·8%, fat 0·6–1·0%, carbohydrates 43–48%, and fibre 30–36% (5). In some parts of India the roots are aromatic and are collected for this property (5). Pigeon on the Adamawa massif eat the seeds (3, 4). An inferior matting is made of the culms in India (5) and the Philippines (1). The plant is a weed of rice-fields, and in ploughing the fallows in Malaya it serves as a green manure (1).

References:

1. Burkill, 1935: 1017. 2. Fenner 305, K. 3. Hepper 1385, K. 4. Hepper, 1965: 498. 5. Sastri, 1956: 41–42.

## **Fimbristylis ferruginea** (Linn.) Vahl

FWTA, ed. 2, 3:321.

West African: **SENEGAL** SERER dalah (JB) **THE GAMBIA** MANDING-MANDINKA n'chongo mesongo (Fox)

A stout tufted perennial, culm erect to 75 cm high, of brackish swamps and damp grassland, from Senegal to Niger and S Nigeria, and widespread in pantropics.

The plant produces a strong network of roots. It is tolerant of seawater and colonises sandy areas of cleared mangrove land (3), thus contributing to stabilisation. It is a weed of brackish tidal rice-fields (The Gambia, 2). In Somalia the culm are beaten to soften the fibres and then plaited into screens for use in huts (1).

References:

1. Collenette 16, K. 2. Fox 121, K. 3. Small 114, K.

## **Fimbristylis hispidula** (Vahl) Kunth

FWTA, ed. 2, 3:324. UPWTA, ed. 1, 518, as *F. exilis* Roem. & Schult.

West African: **SENEGAL** SERER hasukar (JB) WOLOF tokal pul (JB) **THE GAMBIA** MANDING-MANDINKA n'chongo mesengo (Fox) **SIERRA LEONE** MENDE nyina vɔni = *ratgrass, general for Panicum, etc.* (JMD) **MALI** SONGHAI taré-gué (A. Chev.) **IVORY COAST** MANDING-DYULA barkorni buzi (RP) **NIGER** HAUSA guemé n'kussu (Bartha) **NIGERIA** FULA-FULFULDE (Nigeria) dutiel (JMD) duttir (JMD) HAUSA geémún beéraá = *whisker of the rat* (auctt.) geémùn kwaàdoó = *whisker of the frog* (auctt.) geémùn-kuúsuù = *rat's whisker* (auctt.) liìdíin-tuújií (JMD; ZOG) riídín tuújií = *bustard's sesame* (JMD; ZOG) YORUBA irugbọn arugbo

A densely tufted annual, or perennial, variable and represented in the Region by three subspecies, *hispidula, brachyphylla* (Cherm.) Napper and *senegalensis* (Cherm.) Napper; of grassland and savanna on sandy soils, on seasonally damp sites and on cultivated land; occurring throughout the Region and the second-named subspecies extending over tropical and SW Africa.

The plant is a common weed of cultivated areas. It is grazed a little by cattle in Senegal (1), probably for want of better fodder as in Tanganyika it is said to be 'not liked by stock' (4) and to be ungrazed by cattle and goats in Uganda (3). The rhizome is fragrant as is perhaps also the whole plant which is burnt along with the leaves of *Cordia africana* Lam. (Boraginaceae) to fumigate fingers suffering from the effects of handling cotton thread (2). This species along with other robust species may sometimes be used in mat-making.

The Hausa name meaning 'rat's whisker' has a parallel in Uganda with a name meaning 'beard of the lion' (5).

References:

1. Adam, 1966, a: as *F. exilis*. 2. Dalziel, 1937. 3. Dyson-Hudson 17, K. 4. Marshall 39, K. 5. Purseglove P.1574, K.

**Fimbristylis littoralis** Gaud.

FWTA, ed. 2, 3:323.

A tufted leafy annual to about 60 cm tall, of rice-padis, marshes and standing water; native of tropical Asia; rare in Senegal, Ivory Coast, Ghana and Nigeria, and probably introduced with cultivated rice.

The plant is a weed of cultivated rice in Asia. It becomes abundant in the fallows and its seed survive ploughing-in, germinating as the new rice crop is planted. It provides cattle-fodder and the seed appears to pass undigested through the intestines (1).

Reference:

1. Burkill, 1935: 1018, as *F. miliacea* sensu Burkill.

**Fimbristylis ovata** (Burm. f.) Kern

FWTA, ed. 2, 3:324.

A slender tufted rhizomatous perennial of grassland and savanna in Ivory Coast, Ghana and N Nigeria, and generally widespread in the tropics.

No usage is recorded in the Region. In Kenya it provides grazing for all stock and is favoured by horses (1, 2). The culm-bases are swollen into bulbils and these are dug up by guinea-fowl, francolins and rodents (2). The Digo plait the flowering culms to wear as bangles around the wrists for rheumatism (3).

References:

1. Glover & al. 133, 1303, 1666, K. 2. Glover & al. 718, K. 3. Magogo & Glover 384, K.

**Fimbristylis pilosa** Vahl

FWTA, ed. 2, 3:321

West African: **NIGERIA** FULA-FULFULDE (Nigeria) alkamaari, alkama (*properly wheat, from Arabic via Hausa* (MM) label kurasa (LKS?) HAUSA alkamari (LHS) àyaà-áyaà (LHS)

A perennial, rarely annual, with culms to 75 cm high, of damp and swampy places in grassland and savanna, from Senegal to S Nigeria, and extending to the Congo basin and Uganda.

The plant is said in N Nigeria to produce an underground 'nut' [? corm] but it is not recorded whether the 'nut' is used in any way (1).

Reference:

1. Saunders 17, K.

**Fimbristylis scabrida** Schumach.

FWTA, ed. 2, 3:323.
West African: **NIGERIA** HAUSA (West) kajijin daji (Taylor)

Forming a coarse matted turf in damp grassland and savanna from Sierra Leone to Nigeria, and extending to the Congo basin.

**Fimbristylis schoenoides** (Retz.) Vahl

FWTA, ed. 2, 3:321.

An annual, or perennial, of swampy places in grassland in Senegal, Sierra Leone and Ghana; widely dispersed in India and SE Asia to Australia.

In Malaya it forms a fair proportion of the weeds of somewhat dry rice fallows where it gets turned-in in ploughing to make green manure (1).

Reference:
1. Burkill, 1935: 1019.

**Fimbristylis triflora** (Linn.) K. Schum.

FWTA, ed. 2, 3:324.

A stout rhizomatous perennial, culms crowded, to 60 cm high, of grassland, especially near the coast, in Ghana and Togo, and widespread elsewhere in the tropics.

No utility is recorded in the Region. In E Africa the plant forms large tufted colonies on beachheads down the beach to high tide level, and even tolerating occasional inundation (Kenya, 1; Tanganyika, 2, 3). It must thus contribute to stabilisation of sand.

References:
1. Bogdan 2631, K. 2. Faulkner 1818, K. 3. Greenway 1859, K.

**Fimbristylis spp. indet.**

West African: **GHANA** GBE-VHE (Awlan) ketsi (FRI) **NIGERIA** FULA-FULFULDE (Nigeria) mordiho (J&D)

The Ghanaian *ketsi*, which forms pure communities in up to 1 m depth of brackish water in coastal lagoons, is probably *F. triflora*.

**Fuirena ciliaris** (Linn.) Roxb.

FWTA, ed. 2, 3:326.
West African: **SENEGAL** DIOLA sit (JB)

A lax tufted plant, of damp places throughout the Region and the Old World tropics.

The plant is a weed of rice-fields in Tanganyika (2). In Kenya it is said to be readily eaten by elephants (1, 3).

References:
1. Dougall Ab.5481, K. 2. Harris & Gardiner BJH.6577, K. 3. Napier Bax M/E/2, K.

**Fuirena leptostachya** Oliv.

FWTA, ed. 2, 3:326.
West African: **MALI** MANDING-BAMBARA m-bi (Néa)

A hairy annual, laxly tufted, of rice-fields and similar damp places in Mali, Upper-Volta and N Nigeria, and scattered throughout tropical Africa.

**Fuirena stricta** Steud.

FWTA, ed. 2, 3:325–6.
West African: **SIERRA LEONE** TEMNE *an*-yaba (FCD)

A slender, tufted plant of swamps, rice-fields and wet savanna, found all over the Region and occurring throughout tropical Africa and in Madagascar. Var. *chlorocarpa* (Ridl.) Kük. occurs in wet montane situations of N Nigeria and W Cameroons.

The plant is a weed of rice cultivation.

**Fuirena umbellata** Rottb.

FWTA, ed. 2, 3:325. UPWTA, ed. 1, 518.
West African: **SENEGAL** MANDING-'SOCE' komuruni (Monod) WOLOF benteu diojum (JLT) **GUINEA-BISSAU** BALANTA mangatchaca (JDES) **GUINEA** FULA-PULAAR (Guinea) dandandi (A. Chev.) MANDING-MANINKA soti (Langdale-Brown) **SIERRA LEONE** LOKO n-jabai (NWT) MENDE njewɔ *general for sedges* (NWT; JMD) TEMNE *an*-worworaŋ (NWT; JMD) *ka*-yaba *as for shallot, because of the similarity of the swollen stem bases* (FCD; JMD) *an*-yaba-*an*-ro-bath (Glanville; JMD) **LIBERIA** MANO la bele (JMD) la bele dike (Har.) **IVORY COAST** MANDING-BAMBARA kumuru (Bégué) **NIGERIA** IJO-IZON (Ikibiri) bòù àdẹ́ịn = *bush knife* (KW) IZON (Oporoma) ụ̀gbịádẹ́ịn (KW) YORUBA abo làbẹ̀làbẹ̀ = *female làbẹ̀làbẹ̀* (Macgregor; JMD) làbẹ̀làbẹ̀ (Macgregor)

A generally robust rhizomatous sedge, culms to 40–60 cm high, of river-banks and damp places, even in brackish sites, common throughout the Region from Mauritania to W Cameroons, and dispersed pantropically.

The plant will grow in brackish sites and on tidal mud-banks (3). It may serve as a mud-binder to resist tidal scouring. It is an universal weed of rice-padis and it contributes to green-manure on ploughing in (Senegal, 1; SE Asia, 5). The basal node of the flowering culm is swollen into a bulbil so that the tufted plant resembles a shallot, and in Sierra Leone the two plants bear the same Temne name. These bulbils render the plant as a rice-farm weed difficult to eradicate (7). The plant has been recorded as providing good grazing for stock at all times in Ghana (4, 6). Yet in Senegal the plant is reported to be silicaceous and is not eaten by stock (2). There is material in *Herb. Kew* from Mali, northern Sierra Leone and N Nigeria which is plainly scabrid and harsh to the touch. Other collections are of soft, hairy stems and leaves. Similar differences occur in collections from E Africa. The existence of two races of the plant seems probable and this should merit examination. The Ijo name meaning 'bush knife' (9) perhaps also refers to a harsh race.

In Liberia the plant is burnt to obtain salt from the ashes (Harley fide 6).

New-born babies in The Gambia are washed in an infusion of the leaves (8).

References:

1. Adam, 1960: 369–70. 2. Adam, 1966, a: 519, incl. *F. glomerata*. 3. Anderson 1412, K. 4. Beal 29, K. 5. Burkill, 1935: 1038. 6. Dalziel, 1937. 7. Deighton 4049, K. 8. Williams, F. N., 1907: 90. 9. Williamson, K. 152, UCI.

**Hypolytrum heteromorphum** Nelmes

FWTA, ed. 2, 3:336.
West African: **SIERRA LEONE** SUSU wuri-kale (NWT)

A moderately robust, wiry-rooted sedge, briefly stemmed, leaves up to 30–40 cm long, on wet rocks and damp places in the forest, from Guinea to S Nigeria, and extending across the Congo basin to Uganda and Tanganyika.
In Congo a decoction, as also of other plants of close affinity, is used as a face-wash in cases of insanity (1).

Reference:
1. Bouquet, 1969: 103.

**Hypolytrum poecilolepis** Nelmes

FWTA, ed. 2, 3:336. UPWTA, ed. 1, 518, as *H. heterophyllum* sensu Dalziel.
West African: **SIERRA LEONE** MENDE njewɔ-hina = *male njewɔ* (JMD) njewɔ-wa (JMD) papa (L-P; JMD) TEMNE *an*-worworaŋ-bera (JMD) *an*-worworaŋ-*a*-ro-gbɔ (JMD)

*NOTE: H. africanum, H. heteromorphum and H. purpurascens may also apply.*

A robust, stoutly rhizomatous plant with leaves to 1 m long by 3–5 cm wide, of damp sites in high-forest, from Sierra Leone to Ghana.
In Sierra Leone when a Mende woman bears twins, they are placed on a mat specially made of the leaves of this plant before they are washed. This is to bring good luck (1).

Reference:
1. Lane-Poole 143, K.

**Hypolytrum purpurascens** Cherm.

FWTA, ed. 2, 3:336.
West African: **SIERRA LEONE** KONO dɛɛ *a general name for sedges with saw-like leaf-margins* (FCD) MENDE fia ya (*def.*-i) fia: *twins*; ya: *mat* (FCD) tugbɛlɛ *a general term for several small grasses and sedges* (FCD) TEMNE *a*-nyeta-*a*-bana (Glanville) *an*-worworaŋ-*a*-ro-gbɔ (FCD) **LIBERIA** MANO la bele gon (Har.)

*NOTE: See under H. poecilolepis.*

A robust, rhizomatous plant with leaves up to 80 cm long, of stream-banks and damp places in the forest, from Sierra Leone to W Cameroons and Fernando Po, and extending over the Congo basin to Angola.
In Sierra Leone a leaf-decoction is used as a cough-medicine (1).

Reference:
1. Glanville 413, K.

**Kyllinga brevifolia** Rottb.

FWTA, ed. 2, 3:307.

A spreading sedge with long slender rhizomes and culms to about 30–40 cm high, of damp places in Liberia and W Cameroons, and dispersed over the tropics and subtropics of the Old and New Worlds.

In E Cameroun it may become a troublesome weed on palm plantations (1).

Reference:

1. Chuml CDC.374, K.

**Kyllinga bulbosa** P. Beauv.

FWTA, ed. 2, 3:304.
West African: **SIERRA LEONE** MANDING-MANINKA (Koranko) tɔhɔ (NWT) MENDE tugbɛlɛ (FCD)

A slender rhizomatous and tuberiferous perennial, to 10–30 cm tall, of grassy areas and disturbed habitats, from Mali to W Cameroons, and dispersed throughout tropical Africa.

In E Africa it is a bad and persistent weed of lawns and open grassland, often becoming dominant (1, 3). In the tea-growing area of Kenya it is a serious weed of young tea (2).

References:

1. Greenway 3925, K. 2. Terry, s.d. 3. Verdcourt 966, K.

**Kyllinga elatior** Kunth

FWTA, ed. 2, 3:305.

A rhizomatous sedge, culms to 30 cm high, of montane grassland and forest pathsides in W Cameroons and Fernando Po, and in E Cameroun and NE to S Africa.

In Ethiopia it is a common and sometimes troublesome spreading weed of cultivation, the Ari vernacular name *s'as'i* meaning 'making land waste' (1).

Reference:

1. Fukui 775, K.

**Kyllinga erecta** Schumach.

FWTA, ed. 2, 3:307. UPWTA, ed. 1, 518.
West African: **SENEGAL** FULA-TUKULOR todiugolo (JLT) WOLOF todiugolo (JLT) **THE GAMBIA** MANDING-MANDINKA n'chongo mesongo (Fox) **GUINEA** MANDING-MANINKA doron (Brossart) tonku (A. Chev.) **SIERRA LEONE** MENDE njawa, njawa-wa *incl. several Mariscus spp.* (JMD) tugbɛlɛ SUSU tigerin-wuri (JMD) tigerinyi (JMD) TEMNE *an*-siri *more especially applies to Cyperus articulatus* (JMD; FCD) *an*-siri-*an*-rini (JMD) *an*-siri-*a*-ro-kant (JMD) *ka*-soi *the stem or foliage; incl. other sedges* (JMD) *an*-soi *the plant* (JMD) **MALI** MANDING-BAMBARA togon *applied generally* (A. Chev.) **NIGERIA** HAUSA àyaà-áyaà *applied generally* (auctt.) geémùn kwaàdoó *general for grass-like sedges* (JMD) turare *= scent; the roots* (Afzelius) IJO-IZON (Kolokuma) àngị (KW) YORUBA dogbodogbo *= K. peruviana, but probably applies* (JMD)

*NOTE: These vernacular names may apply also to K. pumila, K. squamulata and K. tenuifolia.*

A robust perennial sedge with contiguous culms arising from a rhizome, culms sometimes swollen at the base; a variable species present in the Region in three recognised varieties; distributed throughout the Region in damp grassy places, and elsewhere in tropical and S Africa, Madagascar and the Mascarenes.

The plant is a weed of cultivation. In E Africa it invades lawns developing a dense mat to the exclusion of desired grass species (4). The rhizome is aromatic with a rather bitter taste. It and other *Kyllinga* species which also have aromatic roots are sold in N Nigerian markets under the Hausa name *turare* meaning 'scent' for use as fumigants and to put in food and medicine as a flavour (2). In Congo a tisane is prepared by boiling the plant in palm-wine which is taken as a depurative for treating certain prurient skin-affections (1). In Lesotho bundles of the plant are placed beneath skins being worked to make them supple (3).

References:

1. Bouquet, 1969: 103. 2. Dalziel, 1937. 3. Guillarmod, 1971: 437. 4. Terry, s.d.

**Kyllinga nemoralis** (Forst.) Dandy

FWTA, ed. 2, 3:307.
West African: **SENEGAL** BASARI a-syurúrú (Ferry) **NIGERIA** IGBO (Umuahia) óké íkútē (AJC)

A rhizomatous sedge with spaced culms to 30–40 cm tall, of secondary forest and disturbed habitats, from Guinea to W Cameroons, and pantropics.
The plant is a weed of cultivation.

**Kyllinga nigritana** C.B. Cl.

FWTA, ed. 2, 3:304.

A tufted perennial, of the savanna in Ivory Coast, Ghana and N Nigeria, and in central Africa. This or the closely-related *K. alba* Nees group is widespread in tropical Africa.
The roots are aromatic. The scent is likened to that of citronella (1).

Reference:
1. Gillett 12758, K.

**Kyllinga odorata** Vahl

FWTA, ed. 2, 3:304.
West African: **SIERRA LEONE** TEMNE *an*-siri-*an*-rini (NWT)

A tufted perennial with slender rhizome, culms to 30–50 cm tall, of general damp sites from Guinea to W Cameroons and Fernando Po, and over tropical Africa, in Madagascar and America.
In the Aberdare area of Kenya horses and mules are said to dislike it, but that it can be grazed without apparent ill-effect (1). It is a common weed of crops in Ethiopia (3) and throughout the highlands of E Africa invading coffee, tea and pyrethrum cultivation (4), and also lawns to become the principal species of the sward (2).

References:

1. Grant 1250, K. 2. Huxley 13, K. 3. Parker E. 50, K. 4. Terry, s.d.

**Kyllinga peruviana** Lam.

FWTA, ed. 2, 3:305.
West African: **SIERRA LEONE** TEMNE *a*-kek-*a*-pot (Glanville) *a*-kɔlma *a general term, but more especially refers to Cyperus articulatus* (NWT) **NIGERIA** YORUBA dogbodogbo (Macgregor)

A rhizomatous sedge with closely-spaced culms rising to about 30 cm high, of dunes, tidal swamps and coastal strand throughout the coast-line of the Region, extending to Congo, and in central and tropical S America.
The plant is used in Ghana for stuffing mattresses (1).

Reference:
1. Irvine s.d.

**Kyllinga pumila** Michx.

FWTA, ed. 2, 3:305. UPWTA, ed. 1, 518.
West African: **SIERRA LEONE** MENDE tugbɛlɛ (FCD) SUSU tigerinyi (NWT)

*NOTE: See under K. erecta.*

A tufted plant to about 40 cm tall, of damp or wet places throughout the Region, and the African tropics and in temperate and tropical America.
The roots are fragrant and are used as a fumigant. They may be chewed and added as a flavourer to food or medicine. This species together with others of similar attributes, of which *K. erecta* is the principal one, are loosely known in N Nigerian markets as *turare* (Hausa, 'scent') (1).

Reference:
1. Dalziel, 1937.

**Kyllinga squamulata** Thonn.

FWTA, ed. 2, 3:304. UPWTA, ed. 1, 518.

A weak leafy annual, stems up to 15 cm long, of open sandy areas, common throughout the Region, and pantropics.
It grows forming low prostrate mats on the ground. It is not grazed by domestic stock in Senegal (1). The culms are swollen at the base and the bases along with those of others of like nature are fragrant and are sold in N Nigerian markets under the general name *turare* (Hausa: 'scent'). They are used as a fumigant and are chewed or added to food and medicine as a flavour (2). In Tanganyika and Uganda at middle elevations it is a weed of annual crops (3).

References:
1. Adam, 1966, a. 2. Dalziel, 1937. 3. Terry, s.d.

**Kyllinga tenuifolia** Steud.

FWTA, ed. 2, 3:305. UPWTA, ed. 1, 518, as *K. triceps* Rottb.
West African: **NIGERIA** HAUSA ayáá-ayáá (Taylor) HAUSA (West) burzu (DA)

*NOTE: See under K. erecta.*

A tufted sedge, briefly-rhizomed with closely-spaced culms, swollen at their base, rising to 30–40 cm high, of damp sites in savanna throughout the Region, and widespread in the Old World tropics.
The rhizome with the swollen culm-bases, along with other similar *Kyllinga* species (*K. erecta, K. pumila, K. squamulata*), are sold in N Nigerian markets under the Hausa name *turare* meaning 'scent' for use in fumigation, to chew

and for addition to food and medicine as a flavour. The taste is aromatic and slightly bitter (1).

Reference:

1. Dalziel, 1937.

**Kyllinga welwitschii** Ridley

FWTA, ed. 2, 3:305.
West African: NIGERIA HAUSA (West) gemin kusu = *tail of the rat* (Grove)

A briefly-rhizomed and closely-culmed tufted sedge, with culms to 30–40 cm tall, in sandy ground of the savanna, in Mauritania to N Nigeria, and throughout tropical Africa.

It is grazed by all domestic stock in Senegal (1).

Reference:

1. Adam, 1966, a: as *K. blepharinota.*

**Lipocarpa albiceps** Ridl.

FWTA, ed. 2, 3: 328.
West African: UPPER VOLTA MANDING-BAMBARA torongoyogo (Scholz) NIGERIA HAUSA ayáá-ayáá (Taylor) HAUSA (West) kai-barowa (Taylor)

A laxly-tufted sedge, sometimes with short rhizomes, of damp positions in grassland and open woodland.

**Lipocarpa chinensis** (Osb.) Kern

FWTA, ed. 2, 3:328.
West African: GUINEA FULA-PULAAR (Guinea) doyarama djangol (Langdale-Brown) SIERRA LEONE MENDE njawa *a general name* (NWT) SUSU tubinyi (NWT)

A tufted perennial with stout culms to about 50 cm high, of stream-sides and other damp places, throughout the Region, and widely dispersed over tropical and S Africa and across tropical Asia to Australia.

The plant offers a little grazing to stock (1). In Tanganyika the plant-ash is rubbed with sap from a lime, *Citrus aurantifolia* (Christen.) Swingle (Rutaceae), into scarifications on the temple and forehead as an analgesic for headache (2).

In Gabon the plant is ascribed with magical properties to overcome crises in madness. Some of the plant is carried in the hair or on the wrists (3).

References, as *L. senegalensis* (Lam.) Th. & Dur.:

1. Adam, 1966, a. 2. Haerdi, 1964: 207. 3. Walker & Sillans, 1961: 147.

**Lipocarpa gracilis** (L. C. Rich.) Pers.

FWTA, ed. 2, 3:328 (as *L. sphacelata* (Vahl) Kunth).
West African: NIGERIA HAUSA geémùn kwaàdoó (RES)

An annual, tufted to 30 cm high, thriving on poor soil of waste places and in brackish sites. Common throughout the Region, and occurring in Congo, Madagascar, central America, India and Thailand.

**Mapania linderi** Hutch.

FWTA, ed. 2, 3:335.
West African: **LIBERIA** GOLA pa (AJML) MANO suru la (Har.)

A robust, fibrous-rooted, almost stemless plant with a crown of leaves to nearly 1 m long by up to 5 cm across, of shaded situations in riverain and montane forest, from Guinea to Ivory Coast.

In the Yoma area of Liberia the leaves are reported used for roof-thatching (1).

Reference:

1. Leeuwenberg 4838, K.

**Mariscus alternifolius** Vahl

FWTA, ed. 2, 3:296. UPWTA, ed. 1, 519, as *M. umbellatus* Vahl.
West African: **SENEGAL** WOLOF îl (JB) tiokom tiokom (JLT; FB) **THE GAMBIA** MANDING-MANDINKA n'chongo mesongo (Fox) **SIERRA LEONE** KISSI sande-sieyo (FCD) sisɛmbɛlo-piando (FCD) KONO dipoi (Fisher) MANDING-MANINKA (Koranko) tɔgɔŋ (NWT) MENDE ngete-wulo-hina (FCD) njawa, njawa-wa *general for several spp. of sedge* (JMD) tugbɛlɛ (JMD; Fisher) SUSU tigerinyi *general for sedges* (JMD) TEMNE *a*-nyɛtɔ *here applied loosely, but more especially refers to Scleria boivinii* (JMD; FCD) *an*-siri (NWT; JMD) *an*-siri-*a*-ro-bath *for sedges in marshy places* (JMD) **NIGERIA** HAUSA áyaá *for the tuberous base* (JMD) IGBO (Onitsha) ataku mainya (NWT; JMD) IGBO (Umuahia) n̄né íkútē = *female, or mother grass* (JMD) IJO-IZON (Kolokuma) órú ángị́ = *Gods' angi* (KW) òwéí ángị́ = *male angi* (KW) NUPE efo'aba (RB) YORUBA ṣamikoko (West) YORUBA (Ilorin) alubosa eranko (Ward)

A tufted plant, rhizomatous with closely-packed swollen culm-bases, common in damp grassy places throughout the Region, and pantropical.

The plant is a weed of cultivation, roadsides, lawns, waste places, especially in wet sites (3). The rhizome is aromatic and a food-flavouring (4). The swollen culm-bases are edible after cooking (1, 2) and the plant along with others similarly edible is called *aya* in Hausa, a name strictly referring to *Cyperus esculentus* Linn. (2). The haulm is relished by domestic stock (5). This plant like other large sedges is included in the Mende term *njawa-wai* applied to those used for thatching and making brushes (2).

The Ijo of the Niger Delta use the plant medicinally. The chewed stem is bandaged on a cut or wound for at least three days, being replaced thereafter as necessary till healing has been effected. The swollen stem-bases after washing are ground in a mortar and then mixed into some honey. A dessertspoonful is taken by mouth and some more is rubbed around the waist. This is a treatment for gonorrhoea (*tu bẹịmọ ọgị*) (6).

References:

1. Busson, 1965: 442 as *M. umbellatus* Vahl. 2. Dalziel, 1937. 3. Egunjobi, 1969: as *M. umbellatus* Vahl. 4. Irvine, 1952, a: 30, as *M. umbellatus*. 5. Ward 0044, K. 6. Williamson, K. 12.A, UCI.

**Mariscus flabelliformis** Kunth

FWTA, ed. 2, 3:296.
West African: **SIERRA LEONE** MENDE ndɔgbɔ niwi, ndɔgbɔ: *secondary bush* (FWHM) **NIGERIA** IGBO (Agukwu) ntọ̄i ẹgẹ (NWT)

A tufted plant of open damp ground, throughout the Region and in Chad and tropical America.

In W Cameroons it is said to be a troublesome weed on estates, especially in nurseries, and survives herbicide spraying (1). The root is aromatic (2).

References:

1. Chuml 193, K. 2. Migoed 273, K.

**Mariscus ligularis** (Linn.) Urb.

FWTA, ed. 2, 3:295.
West African: **SENEGAL** SERER gut (JB) **SIERRA LEONE** BULOM (Sherbro) santil-lɛ (FCD) SUSU wurəwurai (FCD) TEMNE *a*-kek-*a*-pot (Glanville) *an*-wor-woraŋ (FCD) **GHANA** AKAN-FANTE mbew (FRI)

A coarse perennial to 1·70 m high of swamps and marshes, usually near the sea, occurring throughout the regional coast-line, and extending to Congo, Mascarenes and in tropical America.

The culms are used in Ghana for making brushes which are used for applying whitewash to houses (1).

Reference:

1. Irvine, 1930: 279, as *M. rufus* H.B. & K.

**Mariscus longibracteatus** Cherm.

FWTA, ed. 2, 3:295.
West African: **SIERRA LEONE** MANDING-MANDINKA dɛrari (NWT) MENDE njewɔ-hina = *male njewɔ* (FCD) TEMNE *a*-kek-*a*-pot (Glanville) *an*-tet-*a*-wir (NWT)

A tufted plant to 1·20 m high, of swampy ground, distributed throughout the Region from Senegal to W Cameroons and Fernando Po and all over tropical Africa.

No usage is recorded in the Region. In the Narok area of Kenya it provides grazing for all domestic stock (1–3), and the culms are used for thatching (1) and weaving into baskets (3). The bulbous culm-bases are said to be eaten by rats and francolins (4).

References:

1. Glover & al. 185, K. 2. Glover & al. 2316, K. 3. Glover & al. 2561, K. 4. Magogo & Glover 549, K.

**Mariscus soyauxii** (Boeck.) C.B. Cl.

FWTA, ed. 2, 3:295.
West African: **NIGERIA** YORUBA lamarin (West)

A small annual herb occurring here and there on disturbed land over the Region and extending into Gabon.

**Mariscus squarrosus** (Linn.) C.B. Cl.

FWTA, ed. 2, 3:294.
West African: **SENEGAL** SERER dug o gay (JB) ndol o ñan (JB) WOLOF ndupèntan (JB) **THE GAMBIA** MANDING-MANDINKA ńchongo (Fox) **NIGER** HAUSA yan tafké (RP)

An annual herb, of open ground, occurring throughout the Region and widespread pantropically and subtropically.

The plant is a weed of cultivated land (2).

In Senegal (1) and in Niger (3) it provides forage for cattle.

References:

1. Adam, 1966, a: as *M. aristatus*. 2. Fox 169, K. 3. Portères, s.d.: as *Cyperus aristatus* Roth.

**Nemium spadiceum** (Lam.) Desv.

FWTA, ed. 2, 3:309, as *Scirpus angolensis* C.B. Cl., var. *brizaeformis* (Hutch.) Hooper.

A slender tufted annual with stems to 35 cm long, on damp rocks, in lateritic places and damp grassy places throughout the Region, and across central Africa to Uganda and Zambia.

The sedge provides a little grazing for cattle in Senegal (1).

Reference:

1. Adam, 1966, a: as *Scirpus briziformis*.

**Pycreus acuticarinatus** (Kük.) Cherm.

FWTA, ed. 2, 3:302.
West African: **GHANA** AKAN-ASANTE anansi-atadwi (Andoh)

A tufted plant from a woody base, of wet grassland in the savanna region from Senegal to S Nigeria, and into E Cameroun.

**Pycreus capillifolius** (A. Rich.) C.B. Cl.

FWTA, ed. 2, 3:301.
West African: **SIERRA LEONE** TEMNE *a*-kek-*a*-pot (Glanville)

A small tufted plant of wet situations throughout the Region and tropical Africa, Madagascar and Brazil.

**Pycreus elegantulus** (Steud.) C.B. Cl.

FWTA, ed. 2, 3:300.

A plant with culms to about 85 cm high, of montane swamps in N Nigeria, W Cameroons and Fernando Po, and throughout other parts of tropical and S Africa.

In Kenya it is grazed by all domestic stock (1).

Reference:

1. Glover & al. 2004, 2317, K.

**Pycreus lanceolatus** (Poir.) C.B. Cl.

FWTA, ed. 2, 3:300.
West African: **SIERRA LEONE** TEMNE *an*-siri-*a*-ro-bath (NWT) **NIGERIA** IGBO íkútē (SOA)

A loosely-tufted perennial of river-banks, marshes and other wet places, occurring throughout the Region, and over tropical Africa, Madagascar and in tropical America.

**Pycreus macrostachyos** (Lam.) J. Raynal

FWTA, ed. 2, 3:301.
West African: **SENEGAL** DIOLA èndèm (JB) èt (JB) SERER deb (JB) gélèt (JB) **THE GAMBIA** MANDING-MANDINKA n'chongo (Macluskie) nyama foro = *pure grass* (Macluskie)

A robust annual, in or near open water or other damp places, throughout the Region, and pantropical.

The roots are said to be eaten in Bornu (1).

Reference:

1. Kennedy 8008, K.

**Pycreus mundtii** Nees

FWTA, ed. 2, 3:302.

A semi-aquatic rhizomatous plant from Mauritania to N Nigeria, and dispersed throughout Africa, and in Madagascar and the Mediterranean.

The haulm is said to be suitable for pulping and not for thatching (1).

Reference:

1. Watt & Breyer-Brandwijk, 1962: 493.

**Pycreus nitidus** (Lam.) J. Raynal

FWTA, ed. 2, 3:300.

A robust perennial plant with culms to 60 cm high, in or near permanent water in Senegal, Dahomey, Niger, N Nigeria and W Cameroons, and distributed across Africa south of the Sahara into E Africa and Madagascar, and in the Levant.

Plants from swamps in Uganda were, before commercial salt became available, burnt and the ash lixiviated to obtain a cooking salt (1). In Tanganyika the culms are used for making mats (2). The rhizome is scented and in Lesotho it is put amongst clothing to freshen them, and it is made into a medicine for chest-colds (3). The plant is one of those suspected of causing 'vlei' disease of sheep in S Africa (3).

References:

1. Eggeling 237, K. 2. McGregor 4, K. 3. Watt & Breyer-Brandwijk, 1962: 374, as *P. umbrosa* Nees.

**Pycreus polystachyos** (Rottb.) P. Beauv.

FWTA, ed. 2, 3:301.
West African: **SIERRA LEONE** MENDE tugbɛ *loosely applied* (FCD) TEMNE *a*-kek-*a*-pot (Glanville) *an*-soi (Glanville)

A neat, tufted plant of coastal sand-dunes, rice-swamps and other wet sites, coastal or inland throughout the Region, and the Old World tropics.

The plant (var. *laxiflorus* (Benth.) C.B. Cl.) is a troublesome weed in rice-padis on the fringe of the tidal area in the coastal rice-farms of Sierra Leone (2). It appears to tolerate a higher degree of salinity than rice will (3).

Domestic stock will not graze it in Senegal (1). Cattle in Tanganyika graze it (4), and in Vietnam it is commonly fed to stock.

References:

1. Adam, 1966, a. 2. Glanville 216, K. 3. Jordan 84, K. 4. Richards 19948, K.

**Remirea maritima** Aubl.

FWTA, ed. 2, 3:297. UPWTA, ed. 1, 519.
West African: **SIERRA LEONE** SUSU baratigɛrinyi (NWT; JMD) **NIGERIA** YORUBA ikotio olokun (Ward)

A strand plant to 10–20 cm high, rhizomatous, of the sandy beaches to high-tide level, from Sierra Leone to S Nigeria and widely dispersed throughout tropical coastal areas.

The plant colonises open sandy areas of the beach-head, down to the high-tide mark, surviving occasional inundation and growing on its own or in association with other strand-plants such as *Ipomoea pes-caprae* (Linn.) Sweet (Convolvulaceae). It is rhizomatous and/or stoloniferous in habit with trailers reaching several metres in length. It appears to survive in driven sand even when buried to unusual depths. On all coasts it acts as a valuable sand-binder.

Though the leaves are pungent, goats are recorded as browsing them (3).

The rhizome is aromatic (1), astringent and diuretic. An infusion is said to be used in Brazil and Guyana as a sudorific and diuretic (2).

References:

1. Burkill, 1935: 1888. 2. Krishnamurthi 1969: 393. 3. Ward 0005, K.

**Rhynchospora corymbosa** (Linn.) Britt.

FWTA, ed. 2, 3:331.

French: herbe à couteau (Berhaut; Kerharo & Adam).

West African: **THE GAMBIA** FULA-PULAAR (The Gambia) gwobe (DRR) MANDING-MANDINKA kuntu-mango = *aromatic grass* (DRR) **SIERRA LEONE** BULOM (Sherbro) santil-lɛ (FCD) KISSI yawa (*pl.* wawa) (FCD) KONO dɛ-wa (FCD) dɛ-wawa (FCD) LOKO m-bɛambɛa (FCD) MANDING-MANDINKA deran (NWT) MENDE njawa wa (Fisher) njewɔ SUSU melai (NWT) TEMNE *an*-wor-woraŋ (FCD) **LIBERIA** MANO la bele (Har.) **GHANA** AKAN-ASANTE aberewa sekan = *old woman's cutlass* (FRI) **NIGERIA** HAUSA kudunduru iya (AST) IJO-IZON (Kolokuma) bòù ẹ̀dẹ́ịn = *bush knife; also applies to Scleria spp.* (KW) IZON (Oporoma) ụ̀gbịádẹ́ịn (KW) YORUBA làbẹlàbẹ(-nla) (auctt.)

A robust sedge with culms to 1 m or more in the wild state, usually less in disturbed places, of swamps, rice-farms and in or near standing water generally; distributed throughout the Region, and pantropics.

The leaves are silicaceous and are not grazed by stock (1). Leaf-edges are sharp and cutting, hence the French name meaning 'knife grass', and the Asante 'old woman's cutlass' (5, 6). The plant colonises badly drained and brackish land, and swamp clearings where it may be put to good use in tsetse fly control. Its invasiveness prevents, as do all similar 'grasses', the deposition of tsetse puparia and breeding (10). It is a weed of rice-nurseries in Sierra Leone (7), rice-padis and rice-fallows, where in Asia it achieves gregarious stands, and is ploughed in as green manure (3, 4). The culms are used in the Philippines, either split or whole, for making mats, sandals, baskets, screens, etc. (3, 4). In Tanganyika they are made into a rough string (8).

In The Gambia a decoction of the seed is given to children for abdominal pains (9), and in Nigeria for colic (2).

References:

1. Adam, 1966, a: 519. 2. Ainslie, 1937: sp. no. 301. 3. Burkill, 1935: 1906. 4. Chadha, 1972: 24. 5. Irvine 954, K. 6. Irvine, 1930: 370, as *R. aurea* Vahl. 7. Jordan 114, K. 8. McGregor 88, K. 9. Rosevear, 1961. 10. Thornewill 87, K.

**Rhynchospora holoshoenoides** (L. C. Rich.) Herter

FWTA, ed. 2, 3:329.

West African: **SIERRA LEONE** TEMNE *an*-binthi (FCD)

A stout plant, culms to 1 m high but often dwarfed, of marshes and damp sites generally, sometimes brackish, occurring from Senegal to Ghana and Niger, and elsewhere throughout tropical and S Africa, Madagascar and tropical America.

In Sierra Leone the plant is a weed of empoldered rice-padis (1).

Reference:

1. Deighton 4383, K.

**Rikliella kernii** (Raym.) J. Raynal

FWTA, ed. 2, 3:310, as *Scirpus kernii* Raymond.

A small herbaceous sedge with stems to 25 cm long on seasonally wet ground, from Senegal to S Nigeria, and across Africa to Ethiopia, and into India.

The plant is grazed by cattle in Senegal (1).

Reference:

1. Adam, 1966, a: as *Scirpus squarrosus*.

**Scirpus aureiglumis** Hooper

FWTA, ed. 2, 3:310. UPWTA, ed. 1, 519 as *S. supinus* Linn.
West African: **SENEGAL** MANDING-MANDINKA kunguleu (JLT) **NIGERIA** HAUSA kúbeèwár kwaàdoó (Golding; ZOG)

A tufted annual to 60 cm high, on the margins of ponds and swamps, from Senegal to S Nigeria, and in E Africa.

The plant is reported used in W Africa after pounding up in water as an antidote for scorpion-sting (1).

Reference:

1. Dalziel, 1937.

**Scirpus brachyceras** Hochst.

FWTA, ed. 2, 3:311.
West African: **MALI** MANDING-BAMBARA marjasina (Lean)

A robust tufted rush-like plant with stems to 2 m high, in or near standing water, in Mali to W Cameroons and Fernando Po, and extending to E Africa.

In Kenya the stems are grazed by all domestic stock, especially in time of drought (1, 2). Kipsigi children chew the white stem tips (2).

References:

1. Glover & al. 894, K. 2. Glover & al. 1070, K.

**Scirpus cubensis** Poeppig & Kunth

FWTA, ed. 2, 3:311.

Stems to 50 cm high from slender rhizomes, in or near permanent marsh, often floating, occurring throughout the Region, and widespread over tropical Africa and America.

The plant occurs in parts of the Volta Lake of Ghana, forming a floating community with *Pistia* (Araceae) in thick mats which impede navigation (1).

Reference:

1. Hall, J. & al. 1971.

**Scirpus holoschoenus** Linn. var. **australis** (Linn.) Koch

West African: MALI TAMACHEK semmar (BM)

With stems to 2 m high, of seasonal watercourses in Mali (Hoggar) on the northern boundary of the Region, and in N Africa.
The stems are not hollow. They are used in Libya for weaving mats (1).

Reference:

1. Keith 241, K.

**Scirpus jacobii** C. E. C. Fischer

FWTA, ed. 2, 3:310.
West African: SENEGAL SERER yig (JLT)

A slender rush-like sedge, on edges of standing water, marshes and temporary pools on laterite, throughout the Region from Mauritania to N Nigeria, and dispersed across tropical and S Africa, and into India.

**Scirpus litoralis** Schrad.

FWTA, ed. 2, 3:311.
West African: GHANA GBE-VHE (Awlan) ketsi (FRI)

A perennial rush-like sedge with pithy stems to 1·5 m long, in or near saline water, from the coastal area of Senegal to Ghana, and widely dispersed in temperate and tropical parts of the Old World.
In Ghana it grows in pure communities in brackish water near the coast and the Vhe (Awlan) cut the stems to stuff mattresses (1). Other pithy tall-growing species, e.g. *S. maritimus*, appear to offer the same possibility.

Reference:

1. Irvine 2782, K.

**Scirpus maritimus** Linn.

FWTA, ed. 2, 3:309.
West African: MAURITANIA ARABIC (Hassaniya) ssaád (AN)

A creeping perennial sedge with leafy stems to 1 m high, of saline or brackish marshes in Mauritania to Mali in the Region, and occurring wide-spread in the tropical and subtropical regions.
No usage is recorded in the Region. In the Harar District of Ethiopia the stems are used for thatching (2), and in Somalia they have been used in locust-bait (1). An unnamed alkaloid has been recorded in the aerial parts of the plant (3).

References:

1. Bally s.n., July 1947, K. 2. IECAMA 1–68, K. 3. Willaman & Li, 1970.

**Scirpus mucronatus** Linn.

FWTA, ed. 2, 3:310.
West African: **GUINEA** FULA-PULAAR (Guinea) cosda (Langdale-Brown)

A leafless, rhizomatous perennial of swamps and open waters, from Guinea, N Nigeria and W Cameroons, and widely distributed in the Old World temperate and tropical regions.

**Scirpus pterolepis** (Nees) Kunth

FWTA, ed. 2, 3:311. UPWTA, ed. 1, 519, as *S. lacustris* sensu Dalziel.

Stems tufted, pithy, to 1·20 m high, from a woody rhizome, near open water, in Senegal, and from Chad to E and S Africa.
The pithy stems retain a hygroscopicity even when dry. They have been used in the past for caulking palm-oil casks (1), a reference which must appear to refer to material from the Gabon — Congo area as the plant is not recorded from the oil-palm growing area of W Africa.

Reference:
1. Dalziel, 1937.

**Scirpus roylei** (Nees) Parker

FWTA, ed. 2, 3:310.

A slender sedge with clustered besom-like stems to about 30 cm long, of shallow water and swampy grassland, from Mauritania to N Nigeria, and in Chad, Congo, Angola, E and SW Africa, and India.
In Kenya it is recorded as a weed of rice-padis and irrigated land (1).

Reference:
1. Bogdan AB.4444, K.

**Scleria boivinii** Steud.

FWTA, ed. 2, 3:340. UPWTA, ed. 1, 519, as *S. barteri* Boeck.
English: sword grass
French: liane rasoir
West African: **SIERRA LEONE** BULOM (Sherbro) santil-lɛ (FCD) KISSI kayɛndẽ (FCD) kayendo (FCD) kayundũ (FCD) KONO dɛɛ (FCD) dɛ-mɛsɛ (FCD) **MANDING-MANDINKA** komuru (FCD) MENDE njewɔ (auctt.) **MENDE-KPA** ngewɔ (FCD) SUSU nɛlai (NWT) wuriwuri (FCD) SUSU-DYALONKE dɛrɛ-na (FCD) TEMNE *a*-nyɛtɔ (FCD) VAI n-juli (FCD) **LIBERIA** MANO pipi (Har.) **IVORY COAST** AKYE chiètin ayéa (A&AA) DAN pi (K&B) KRU-GUERE niandri (K&B) niaré (K&B) GUERE (Wobe) gnatèrè (A&AA) 'KRU' dirimé (K&B) n-nia (K&B) **GHANA** AKAN-ASANTE ɛtene (CV) **NIGERIA** IGBO (Onitsha) asay (AJC) obẹ (NWT) YORUBA làbẹ̀làbẹ

*NOTE: Also general names for larger and sharp-leaved sedges apply.*

A perennial scandent to 12 m high, leaves with razor-sharp edges, forming impenetrable thickets in scrub, bush and secondary forest, from Senegal to W Cameroons, and spreading across the Congo basin to Uganda and Angola, and in Pemba and Madagascar.
The sharp-edged leaves are used as 'razors' in Congo (2). If the scabrid parts are eaten they cause damage to the gut. Criminal use in this way is recorded

from the Liberia/Ivory Coast border region (5). A leaf-decoction, from which all scabrid material is removed or avoided, is taken in Ivory Coast for coughs (3, 5) and used as a wash on snake and other venomous animal bites (5). A warm decoction is said to soothe toothache when used as a mouthwash (4). Aerial parts of the plant are used in Ivory Coast to relieve white patches of the cornea: stems, leaves and inflorescences are crushed with salt and the expressed liquid is administered as an eye-instillation (1). The aerial parts are also compounded into a formulation with several other plants, and the liquid after cooking is taken as a drink for cough (1). A leaf-macerate is said in Gabon to ease or hasten childbirth (6).

A root-macerate is taken in draught in Ivory Coast–Upper Volta for hiccups, especially when refactory (5). A root-decoction is used against blennorrhoea in Gabon (6) while the whole plant is similarly used in Congo and is also ascribed with aphrodisiac properties (2). A root-decoction is used in Congo to treat irregular menses or too abundant menstruation, and also for haematuria; the dried powdered roots are applied topically over epidermal scarifications for headache and leprous sores (2).

The nut-like seeds are used as beads (4).

In Gabon the young shoots of this plant together with other herbs are cooked and given to secret society initiates to eat when the intoxicating effects of taking *Tabernanthe iboga* Baill. (Apocynaceae) are wearing off. The latter plant, which is not recorded from W Africa, is said to be strongly stimulating, defatiguant, inebriant and aphrodisiac (6). *S. boivinii* is also deemed to be a good talisman when the backs of the hands are beaten with the leaves together with leaves of *Staudtia stipitata* Warb. (Myristicaceae) for catching naked-handed crustaceans and shellfish (6).

References, as *Scleria barteri* Boeck.:

1. Adjanohoun & Aké Assi, 1972: 111. 2. Bouquet, 1969: 104. 3. Bouquet & Debray, 1974: 79. 4. Dalziel, 1937. 5. Kerharo & Bouquet, 1950: 251. 6. Walker & Sillans, 1961: 147.

### **Scleria depressa** (C.B. Cl.) Nelmes

FWTA, ed. 2, 3:340. UPWTA, ed. 1, 519, as *S. racemosa* Poir.

English: sword grass (Irvine.)

West African: **SENEGAL** DIOLA é liba (JB) solahir (AS, ex RP) **THE GAMBIA** DIOLA (Fogny) é liba = *knife* (DF) **GUINEA** FULA-PULAAR (Guinea) mala (Langdale-Brown) MANDING-MANINKA kômum (Langdale-Brown) **SIERRA LEONE** MENDE njewɔ (NWT) SUSU mela(i) (NWT) TEMNE *a*-nyɛtɔ-*a*-ro-bat (NWT) *a*-nyɛtɔ-*a*-ro-kai (NWT) **LIBERIA** MANO pipi (Har.) **GHANA** GBE-VHE ahõ (FRI) **NIGERIA** IJO-IZON (Kolokuma) bòù ẹ̀dẹ́ịn = *bush knife* (KW)

A robust perennial with culms 1–2 m tall, of swamps in savanna and forest areas throughout the Region from Mauritania to W Cameroons.

The leaves are very sharp-edged, and have a name meaning 'bush knife' to the Ijo of the Niger delta who use the leaves to make small cuts on the body into which to rub ointment, e.g. over the breasts for administration of alligator pepper (5) as a stimulant. Sap from the base of the mucilaginous young shoots is applied in E Cameroun to heal cuts and wounds, and a root-decoction is taken for blennorrhoea (4).

The plant has till recently been considered a variety of *S. racemosa* Poir., var. *depressa* C.B. Cl., the present segregation being based only on the minutae of the achene structure. The gross morphological characters are similar. The use by Ugandan blacksmiths of burning the culms of the latter in their braziers to increase the heat of the charcoal (3) may well be possible by the W African plant. In Tanganyika a root-decoction of *S. racemosa* is taken with that of *Mimosa pigra* Linn. (Leguminosae–Mimosoideae) for dysmenorrhoea, and a macerate of the pounded root for gonorrhoea (1).

The seeds of *S. depressa* are used in Ghana as beads (2).

CYPERACEAE

References:

1. Haerdi, 1964: 209, under *S. racemosa* Poir. 2. Irvine 1585, K. 3. Maitland 87, K, under *S. racemosa* Poir. 4. Portères, s.d., under *S. racemosa* Poir. 5. Williamson, K. 20, UCI.

**Scleria foliosa** Hochst.

FWTA, ed. 2, 3:343.
West African: **NIGERIA** HAUSA kudunduriniya (Kennedy)

A tufted annual, culms to 1 m high, in seasonally wet grassland, in Senegal, Ivory Coast, Ghana and N Nigeria, and widespread over the rest of tropical Africa and into Madagascar.

In Tanganyika the rhizome is pounded in water and the macerate is taken for gonorrhoea, and a decoction of the roots with those of *Mimosa pigra* Linn. (Leguminosae–Mimosoideae) is taken for dysmenorrhoea (1).

Reference:

1. Haerdi, 1964, 208.

**Scleria iostephana** Nelmes

FWTA, ed. 2, 3:342.

A stout perennial sedge, with culms erect to 2 m high, of stream-sides and damp places in rain-forest in Ivory Coast, Ghana and N Nigeria, and distributed over the Congo basin, Uganda and to Zambia.

Water in which the roots have been pulped-up is taken by draught in Congo by sufferers of piles (1).

Reference:

1. Bouquet, 1969: 103.

**Scleria lagoënsis** Boeck.

FWTA, ed. 2, 3:342.
West African: **SENEGAL** BASARI a-nyinoseùng, ỹinoseùng (Ferry) **SIERRA LEONE** MENDE njewɔ (FCD)

A tufted perennial, culms swollen at the base, and rising to 1·50 m high, of damp usually seasonally waterlogged grassland, across the Region from Senegal to S Nigeria, and throughout tropical Africa and south tropical America.

**Scleria lithosperma** (Linn.) Swartz

FWTA, ed. 2, 3:343.

A slender perennial sedge with wiry rhizome and culms up to 1 m high, usually in dense shade of forest or woodland, in Ivory Coast and Ghana, and pantropically dispersed.

In Tanganyika a decoction of the plant is taken by draught for dysmenorrhoea and to stave off a threatened miscarriage (2). In Malaya a root-decoction is administered after childbirth (1).

References:

1. Burkill, 1935: 1982. 2. Haerdi, 1964, 208.

**Scleria melanomphala** Kunth

FWTA, ed. 2, 3:340.
West African: **SIERRA LEONE** LOKO m-bɛambɛa-mbawa (NWT) m-bɛambɛa-nge (NWT) nuŋwa (NWT) MENDE njewɔ (Dawe) SUSU mela (NWT) TEMNE *a*-nyɛtɔ (NWT) *an*-wor-woraŋ (NWT) **LIBERIA** MANO la bele (Har.)

A robust perennial to 2 m high of swamps and very wet sites in grassland and forest, from Guinea to W Cameroons, and throughout tropical and S Africa, Madagascar and in S America.
In Tanganyika the whole plant with its rootstock is decocted and the liquor is drunk for dysmenorrhoea (1).

Reference:

1. Haerdi, 1964: 208.

**Scleria naumanniana** Boeck.

FWTA, ed. 2, 3:342. UPWTA, ed. 1, 519.
West African: **SIERRA LEONE** MANDING-MANINKA (Koranko) kɔkmie (NWT) mɛle (NWT) MENDE njewɔ (auctt.) SUSU nelaxame (NWT) TEMNE *a*-nyɛtɔ (NWT)

A perennial sedge, culms to 1·20 m high of damp sites in forest, savanna and marshes, on sand-dunes and near mangrove swamps throughout the Region from Senegal to S Nigeria, and extending to Angola.
A warm decoction of the leaves is used as a mouthwash to relieve toothache, and the nut-like seeds are used as beads (1).

Reference:

1. Dalziel, 1937.

**Scleria pterota** Presl

FWTA, ed. 2, 3:342.
West African: **SIERRA LEONE** MENDE njewɔ (FCD) njewɔ-wa (FCD)

A perennial sedge, culms to 1·20 m high, of river-banks and damp shady places, in Senegal to S Nigeria, and distributed over tropical Africa, Madagascar and south tropical America.
In Tanganyika a root-decoction is drunk for dysmenorrhoea, and plant-ash is given in small doses to infants for colds. The plant is also used in veterinary medicine: two or three plants are cut up fine and cooked, and the resultant brew is given to cattle for rinderpest (*nagana,* bovine trypanosomiasis) (1).

Reference:

1. Haerdi, 1964: 208.

**Scleria spiciformis** Benth.

FWTA, ed. 2, 3:340.
West African: **SIERRA LEONE** BULOM (Sherbro) santil-lɛ (FCD) MENDE njawa-wa (FCD) sawawa (FCD)

A tufted perennial to 75 cm high, of damp places on rocky outcrops and on seasonally wet grassland.

**Scleria verrucosa** Willd.

FWTA, ed. 2, 3:340.
West African: **LIBERIA** MANO pipi (Har.) **NIGERIA** IGBO (Umuahia) àsháyì = *a sharp grass* (AJC) IJO-IZON (Kolokuma) bòù ẹ̀dẹ́ịn = *bush knife* (KW) YORUBA gewinin (N&E) làbẹlàbẹ (Millen; West)

A robust perennial with culms arising to 3 m high, of river-banks and marshes, from Senegal to W Cameroons, and extending to E Cameroun and western Uganda and Tanganyika.

**Scleria spp. indet.**

West African: **GHANA** AKAN-FANTE ɛtsin = *creeping* (FRI) **NIGERIA** IJO (Nembe) ịnwànwàn (KW) IJO-IZON (Kolokuma) gbànràịn ẹ̀dẹ́ịn *'Gabrain Taylor Creek knife'* (KW)

Of the foregoing Sclerias unidentified as to species that of Ghana is a scandent with sharp-edged leaves capable of lacerating the flesh (1).

Reference:

1. Irvine, 1930: 379.

**Torulinum odoratum** (Linn.) Hooper

FWTA, ed. 2, 3:297.
West African: **NIGERIA** YORUBA làbẹlàbẹ (Millen)

Robust perennial to 2·5 m high, occurring near open-water in Senegal, Ivory Coast, Ghana, Dahomey and S Nigeria; pantropical.
Tests of extracts from the roots for activity against avian malaria have given negative action (1).

Reference:

1. Claude et al., 1947.

## DICHAPETALACEAE

**Dichapetalum albidum** A. Chev.

FWTA, ed. 2, 1: 436, as *D. pallidum* sensu Keay, in part.
West African: **SIERRA LEONE** LOKO tɛhɛ (NWT) MANDING-MANINKA (Koranko) kɔgbe (NWT) MENDE magbɛvi (FCD) MENDE (Gaura) kponetei (Pyne) TEMNE *a*-bonk (JMD) *ma*-nunk *the fruit* (JMD) *a*-nunk (JMD)

A shrub or liane to 20 m long, stems by 5 cm diameter, of the closed rain-forest or semi-deciduous forest, of Guinea to Ivory Coast (3).
A plant, under the name *D. pallidum,* but is perhaps this species, is prescribed in Ivory Coast for apparently analgesic effects: the leaves in plasters and poultices are applied to chronic sores and on old painful urethrites (2), and the leaves along with other drug plants are decocted and used in washes and steam fumigations for rheumatism (1). It is also taken as an emmenagogue (2).

References:

1. Adjanohoun & Aké Assi, 1972: 84, as *D. pallidum* (Oliv.) Engl. 2. Bouquet & Debray, 1974: 79, as *D. pallidum*. 3. Breteler, 1973: 48–52.

**Dichapetalum barteri** Engl.

FWTA, ed. 2, 1: 436. UPWTA, ed. 1, 171.
West African: **NIGERIA** IGBO m̀gbú ēwú = *goat killer* (auctt.)

A shrub or small tree to 13 m high, usually less, of secondary scrub, and undergrowth of galleried forest or drier parts of rain-forest of Ivory Coast, Ghana and Nigeria.

The Igbo name, 'goat-killer', suggesting that the plant has poisonous properties, is likely to be correct as there are very closely related species in E Africa known to be toxic (1).

Reference:

1. Breteler, 1973: 82–86.

**Dichapetalum heudelotii** (Planch.) Baill.

FWTA, ed. 2, 1: 438, incl. *D. acutisepalum* Engl., *D. linderi* Hutch. & Dalz., *D. johnstonii* Engl., *D. subauriculatum* (Oliv.) Engl., *D. kumasiense* Hoyle and *D. ferrugineum* Engl.

Shrub of the forest area from Guinea to W Cameroons.

*Mono, di* and *tri-fluoroacetic acids* have been detected in this species (1).

Reference:

1. Vickery, B. & al., 1973: 145–7.

**Dichapetalum madagascariense** Poir.

FWTA, ed. 2, 1: 436–7, as *D. guineense* (DC.) Keay incl. *D. subcordatum* (Hook. f.) Engl., *D. floribundum* (Planch.) Engl., *D. thomsonii* (Oliv.) Engl., and *D. chrysobalanoides* Hutch. & Dalz. UPWTA, ed. 1, 171, as *D. flexuosum* Engl.
West African: **SIERRA LEONE** MENDE makpavi (FCD) TEMNE *a*-bonk, *a*-nunk (JMD) **IVORY COAST** BAULE sumolié (Aub.; K&B) DAN guahiélu (Aub.; K&B) **GHANA** VULGAR nkronua (DF) AKAN-ASANTE akwakoraa gyihenim (FRI) denkyera hwerewa (FRI) kyekyereantena (FRI) TWI nkorodua (FRI) nkron-nua (FRI) ofenwa-biri (DF) ɔfẽwa-biri (FRI) ofoabiri (BḌ&H; FRI) ANYI-AOWIN sumolie (FRI) SEHWI asunwi(n)dia (auctt.) ɛson-windia (CV) GA antro (FRI) GBE-VHE folie, fɔlie *the genus generally?* (FRI) tsrokpati (FRI) VHE (Awlan) kletsi **NIGERIA** YORUBA alo (JRA; KO&S) ekusan (Ross) ikun imú àgbò (IFE) kukumarugbo (Macgregor; Ross)

A shrub, or tree to over 20 m high by 1·70 m girth, of savanna and forest, from Sierra Leone to Nigeria, and perhaps on to the Congo basin.

The bark exudes a little brownish gum when slashed (1, 9). The fresh cut wood is white and turns brownish. It is hard and contains black veins (1, 4, 8, 9). The smaller stems are used as chew-sticks (5, 8) and larger pieces find unspecified domestic uses (6, 8).

The plant is said to be poisonous to stock (5, 6, 8). There is strong but so far inconclusive evidence of the presence of monofluoroacetic acid (12). This is the toxic substance present in *gifblaar, D. cymosum* Engl., the cause of very considerable loss of stock in S Africa (3, 13). It is present in the leaves, more especially the younger ones. The substance of itself is not poisonous, but toxicity follows from enzymatic action in digestion. The addition of monosodium acid phosphate to stock drinking water conveys complete protection against dichapetalosis (13). See also *D. toxicaricum* below.

Sap expressed from the pulped leaves, together with the leaves of other drug-plants, is used by the Baule of Ivory Coast in nasal instillations for jaundice (10), and the leaves are made into plasters and poultices for treating sores and old painful urethrites (2). In S Nigeria, the Igbo use the leaf with soap for washing (11).

The fruit contains an edible pulp. The seed is also edible (5, 6, 7, 8).

References:

1. Aubréville, 1959: 2: 10, as *D. guineense* (DC.) Keay. 2. Bouquet & Debray, 1974: 79. 3. Breteler, 1973: 36. 4. Burtt Davy & Hoyle, 1937: 32. 5. Dalziel, 1937. 6. Irvine 337, K. 7. Irvine 1964, K. 8. Irvine, 1961: 267, as *D. guineense* (DC.) Keay. 9. Keay & al., 1960: 325–5, as *D. guineense* (DC.) Keay. 10. Kerharo & Bouquet, 1950: 91, as *D. flexuosum* Engl. 11. Thomas, N. W., 2098 (Nig. Ser.), K. 12. Vickery, B. & al., 1973. 13. Watt & Breyer-Brandwijk, 1962: 375–83, under *D. cymosum* Engl.

**Dichapetalum pallidum** (Oliv.) Engl.

FWTA, ed. 2, 1: 436 as *D. pallidum* sensu Keay, in part. UPWTA, ed. 1, 171.
West African: **IVORY COAST** ABURE abea (B&D) AKAN-ASANTE mazania (B&D) **TOGO** GBE-VHE follye (Busse) **NIGERIA** YORUBA kukumarugbo (JMD)

A shrub, erect or scandent of the closed-forest zone of swampy sites and in secondary growths from Ghana to S Nigeria, and extending into Gabon. This species has previously been confused with *D. albidum* Chev. (1).

The leaves are used in Togo for diarrhoea (Busse fide 2, 3) and the seeds (? fruits) are said to be edible and are eaten there (2, 3).

References:

1. Breteler, 1973: 48–52, under *B. albidum* A. Chev. 2. Dalziel, 1937. 3. Irvine, 1961. 268.

**Dichapetalum tomentosum** Engl.

FWTA, ed. 2, 1: 438. UPWTA, ed. 1, 171, incl. *D. acutifolium* Engl.

A scandent shrub or woody climber of the forest zone of S Nigeria and W Cameroons.

The fruit is said to be very poisonous and is known in W Cameroons as 'broke-back' like *D. toxicarium* (Engler fide 1).

Reference:

1. Dalziel, 1937.

**Dichapetalum toxicarium** (G. Don) Baill.

FWTA, ed. 2, 1: 438. UPWTA, ed. 1, 171–2.

English: rat's bane (Deighton); West African rat's bane (Dalziel; Irvine); broke-back (Sierra Leone, auctt.).

West African: **SIERRA LEONE** KISSI koli-tɔmda? (FCD; S&F) makpafi (FCD; S&F) KONO magbavi (FCD; S&F) KRIO broko-bak = *broke-back* (FCD; S&F) LIMBA meme (GFSE) LIMBA (Tonko) makpalaba (FCD) MENDE makpavi (*def.*-i) (auctt.) SUSU mɛmɛ (FCD) TEMNE *a*-bonk (JMD) *a*-nunk (auctt.) *ma*-nunk *the fruit* (auctt.) **IVORY COAST** KRU-GUERE suin (K&B) **GHANA** AKAN-FANTE ekum-nkura = *kill mice* (FRI; JMD) TWI kum-nkura = *kill mice* (Enti)

A shrub or tree to about 16 m tall as a forest understorey species or in secondary bush from Guinea to Ghana.

The leaves are prescribed in Ivory Coast for use in poultices on chronic sores and on old painful urethrites (1). A herbalist at Magbena, Sierra Leone, prescribes the leaves, cooked with a chicken, for heart-palpitations (6).

Bark under the name of *ekum-nkura* ('killer of mice', Fante) in Ghana is probably of this plant. It is sold as a poison for mice, and crushed to a powder it is rubbed over swellings or taken as a sneezing powder to restore one in a faint (Onacoe Amnah fide 2, 3).

The fruit-pulp surrounding the seeds is edible (2, 3, 7).

The kernels are well-known as a rodenticide. They are an item of market trade in Sierra Leone under the name of 'broke-back', a name arising from the symptoms of poisoning: two to five hours after taking, there is a sudden seizure with convulsions, difficulty of breathing and shortage of oxygen in the blood (cyanosis), followed by paralysis of the hind-limbs, or occasionally of the whole body. In fatal cases death follows in about 12 hours, sometimes preceded by partial recovery and recurring seizures. In non-fatal cases convulsions occur and recur once or twice before recovery. The poison is cumulative, being but slowly excreted from the body, so that repeated small doses have ultimately the same effect as a single dose of the same total amount. The poisonous property is not destroyed by heat. Susceptibility to the poison varies within wide limits (2). Man is known to be affected and an antidote is said to be to drink a pint of water (5). Broke-back is used to kill other animals and it is recorded to have been used to poison water-supplies (2). The Guere of SW Ivory Coast add it to their arrow-poisons (1).

The plant contains traces of alkaloids and a non-haemolytic saponin (4). Toxicity, however, is due to monofluoroacetate and fluoride ions present in varying concentrations throughout the plant. These substances are synthesized in the young leaves and the ion is stored in the small leaves adjacent to the flowers till it is drawn into the young seeds to be converted into long chain fluoro-fatty acids. Plants in general contain 0·1–10 ppm dry weight of fluorine, the concentration being highest in the hot dry season. At no time is the plant not toxic but it is least so when the seeds are developing and young leaf formation is halted (8).

References:

1. Bouquet & Debray, 1974: 79. 2. Dalziel, 1937. 3. Irvine, 1961: 268–9. 4. Kerharo & Bouquet, 1950: 91. 5. Kirk, 20, K. 6. Massaquoi, 1973. 7. Savill & Fox, 1967, 98. 8. Vickery, B. & Vickery, M. L., 1972.

**Tapura fischeri** Engl.

FWTA, ed. 2, 1: 439.
West African: NIGERIA YORUBA aṣaṣa-igbó (KO&S)

A shrubby much-branched tree of forest understorey to 13 m high, often in fresh-water swamp, recorded from Ivory Coast, Ghana and Nigeria, and extending to Zaïre, Sudan and E Africa.

The wood is hard and tough. It is used in Uganda for digging implements (1), and in Tanganyika for hut-poles (3).

The bark of a related species in Guyana, *T. guianensis* Aubl. is weakly narcotic and is used as a fish-poison (2).

References:

1. Eggeling & Dale, 1952: 84–85. 2. Fanshawe, 1953. 3. Sensei, FH.2919, K.

## DILLENIACEAE

**Dillenia indica** Linn.

FWTA, ed. 2, 1: 180.
English: elephant apple.

A small tree to 10 m high of India and Western Malesia which has been introduced into West Africa and is under cultivation in Sierra Leone, Ghana,

Nigeria and West Cameroons and probably elsewhere with an adequate rainfall. It is recorded as able to do well in W Cameroons (3).

The timber in India is said to be red, close-grained and moderately hard but seldom yielding useable timber because the trunks are so crooked, though small timber finds miscellaneous uses (1). A red dye is obtainable from the bark (2).

The fruits are fleshy with a sour apple taste. They are used in India as a flavouring for curry and to make jam (1). The fruit pulp is used in peninsular Thailand to wash the hair (1) and in Zanzibar as a hair-dressing (4).

References:

1. Burkill, 1935: 809–10. 2. Quisumbing, 1951: 612–3. 3. Unwin, 1920: 421. 4. Williams, R. O., 1949: 231.

**Tetracera affinis** Hutch.

FWTA, ed. 2, 1: 181.

West African: **GHANA** AKAN-ASANTE efwiema (BD&H; FRI) efwirinema (FRI) TWI twẽhama = *a fibre; from* ɔtwẽ: *a kind of antelope;* hama: *fit to bind* (FRI)

A scrambling or twining shrub or small tree to 13 m high with strong cutting hairs.

The roots are used for yaws in Ghana (Vigne fide 1), and the flowers are much visited by bees (1). The Akan-Twi name suggests the use of the stem as a cordage material, but such a use is not recorded.

Reference:

1. Irvine, 1961: 69–70.

**Tetracera alnifolia** Willd.

FWTA, ed. 2, 1: 180–1. UPWTA, ed. 1, 44.

French: liane à eau (auctt.)

West African: **SENEGAL** DIOLA hu hut (JB; K&A) humoka (JB; K&A) DIOLA (Fogny) fufut (K&A) WOLOF lala (JB; K&A) **THE GAMBIA** DIOLA (Fogny) fufut = *rotten* (DF) DIOLA-FLUP hu hut = *rotten* (DF) hu moka = *decayer* (DF) **GUINEA-BISSAU** BIDYOGO eberigom (JDES) FULA-PULAAR (Guinea-Bissau) goróluga (JDES) **SIERRA LEONE** BULOM (Sherbro) katata-lɛ *probably refers* (FCD) KISSI diamɔle-kolẽŋ *probably refers* (FCD) KONO kuin-nɛnɛnɛ *probably refers* (FCD) LOKO n-dɔpararo *probably refers* (FCD) MENDE katata (FCD) katata-wa (*def.*-i) = *big katata* (JMD) ndɔpa-nɛɛ (auctt.) SUSU nintɛ *probably refers* (FCD) SUSU-DYALONKE firi-na *probably refers* (FCD) TEMNE *ra*-nəth (FCD) VAI kwenɛ, juu *probably refers* (FCD) **MALI** MANDING-BAMBARA trélégué (A. Chev.) 'SENUFO' sensenrré (A. Chev.) **UPPER VOLTA** MANDING-BAMBARA trélégue (K&B) DYULA noma (K&B) 'SENUFO' sensenrré (A. Chev.) **IVORY COAST** AKAN-BRONG assassinian-luo (K&B) DAN buzadulé (K&B) GAGU muniziba (K&B) nima (K&B) KRU-BETE bimien (K&B) GUERE séérbu (K&B) zérébu (K&B) GUERE (Chiehn) g-bimien (K&B; B&D) KULANGO bogoro (K&B) bonanga (K&B) KWENI tirili (B&D) trilidè (B&D) MANDING-BAMBARA trélégue (K&B) DYULA noma (K&B) MANINKA bassakwo (B&D) bassoko (B&D) kanangwo (B&D) 'SENUFO' sensenrré (A. Chev.) **GHANA** AKAN-TWI akotopa (FRI) ANYI-AOWIN ɛhweɛ-nyamaa (FRI) enwuma (FRI) SEHWI ɛhweɛ-nyamaa (FRI) enwuma (FRI) NZEMA ehwenyema (BD&H; FRI) nyaminle (BD&H; FRI)

A liane or multi-stemmed climber to 20 m high, or shrubby tree to about 8 m, or trailing in grassland; of savanna, thickets, forest margins, mangrove communities by coastal swamps, recorded from Senegal to W Cameroons and Fernando Po, and also extending to Angola.

The stem yields an abundant limpid sap which is potable. At one time the Bapanu of Gabon specially planted the plant in savanna regions against time of water-shortage (12). In Senegal the sap is dripped from a cut stem direct into the eye for 'clouding' (? cataract) and eye-troubles (5, 6). The same practice is followed in Ivory Coast for conjunctivitis (1). In Gabon the sap is added to water and drunk for colic, and lactating mothers take the sap in which sweet tapioca has been macerated as a galactogogue (11, 12). In Congo the sap is similarly given to mothers-in-milk. It is used to 'purify' both mother and child immediately after birth and is given to a baby with its first suckle and regularly to twins to strengthen them (2).

The plant is held in Ivory Coast to have high therapeutic value in treatment of pain. Leafy twigs are ground up and mixed into a paste with palm-oil for application in cases of headache, intercostal and abdominal pain, rheumatism, etc., and a leaf-powder is added to food (7). Leaves are said to relieve stomach-ache, hernia, haematuria and food-poisoning (3). In Gabon a leaf-decoction is taken for dysentery and, as a strengthening food, powdered, dried stems cooked with groundnuts are given to women in pregnancy (11, 12). A length of leafy stem screwed up into a ball is boiled in water which is drunk for dysentery in Sierra Leone (8). In Congo it is used as a vermifuge and purgative and with other plants for gastro-intestinal troubles (2).

A snake-bite remedy which is held in great repute in Senegal is prepared from the leaves and roots together with other drug-plants. As preparation requires 48 hours this is normally made in anticipation and held in stock by Casamance medicine-men (5, 6). The preparation is taken both internally and externally. An alcoholic macerate of leafy-twigs in palm-wine is taken in Ivory Coast for asthma, and is considered to be also febrifugal. A root macerate is used for urethral discharges and is given as an enema to strengthen children with rickets (7). Leaves crushed with salt and pimento are taken in Ivory Coast as an aphrodisiac (3, 7).

Reports that the plant is piscicidal are probably erroneous (4). The plant is considered non-toxic by Ivorean medicine-men (7). In Zäire the young leaves are eaten as a vegetable (10). The leaves however have a quantity of flavones and mucilage and flavones are present in the root-bark (3). *Dilleniaceae* in general are rich in tannin. Investigation of the use of the stem in Nigerian folk-medicine for dermal infections has shown no action on Gram −ve organisms, and no anti-fungal action (8).

References:

1. Adjanohoun & Akė Assi, 1972: 112. 2. Bouquet, 1969: 105. 3. Bouquet & Debray, 1974: 79. 4. Dalziel, 1937. 5. Kerharo & Adam, 1962. 6. Kerharo & Adam, 1974: 393. 7. Kerharo & Bouquet, 1950: 37. 8. Malcolm & Sofowora, 1969. 9. Massaquoi, 1974. 10. Sillans, 1953, b: 82–99. 11. Walker, 1953, a: 32–33. 12. Walker & Sillans, 1961: 149.

**Tetracera leiocarpa** Stapf

FWTA, ed. 2, 1: 181.
English: water tree (Liberia, Cooper & Record).

A scandent shrub to 8 m high occurring in the forested areas of Guinea, Liberia and Ivory Coast.

A potable sap is obtained from the freshly cut stems in Liberia (1).

Reference:

1. Cooper & Record, 1931: 25.

**Tetracera podotricha** Gilg

FWTA, ed. 2, 1: 181.
French: liane à eau.

A forest liane of S Nigeria and W Cameroons; also in E Cameroun, Zäire and Angola.

Leaves and stems bear urticant hairs (2).

The stem yields a clear potable sap which can be drunk in time of necessity. No usage is recorded for the plant in the Region. In Gabon (3, 4), and Congo (1) it is put to similar uses as are *T. alnifolia* Willd. and *T. potatoria* Afzel.

References:

1. Bouquet, 1969: 105. 2. Okafor FHI 35862, K. 3. Walker, 1953, a: 32–33. 4. Walker & Sillans, 1961: 150.

**Tetracera potatoria** Afzel.

FWTA, ed. 2, 1: 180. UPWTA, ed. 1, 45.

English: Sierra Leone water tree.

French: liane à eau.

West African: **SENEGAL** BANYUM sigump kidigen = *female sigump* (K&A) DIOLA fufut (JB) DIOLA (Efok) kokotau (K&A) DIOLA (Fogny) fu fut (K&A) DIOLA ('Kwaatay') esseńata (K&A) **SIERRA LEONE** BULOM (Sherbro) katata-lɛ (FCD) KISSI diamɔle-kolẽŋ (FCD) diamɔle-nawele = *cow's tongue* (JMD) KONO kuin-nɛnɛnɛ (FCD) LIMBA y'ebe (NWT) LOKO *n*dɔpalaro (NWT) *n*dɔpararo (FCD) MANDING-MANDINKA sagbɛ (NWT) MANINKA (Koranko) budanagara (NWT) molɛnakara (NWT) MENDE *n*dɔpa-nɛɛ (NWT; Dawe) katata (*def.* -tei) (auctt.) MENDE-KPA *n*dɔpa-nɛ *from* ndɔpa: *the lesser bush-buck, or harnessed antelope*; nei: *tongue — referring to the scabrid leaf* (FCD; JMD) SUSU nintɛ (auctt.) SUSU-DYALONKE firi-na (FCD) TEMNE *ra*-nəth (auctt.) VAI juu(?) (FCD) kwenɛ(?) (FCD) **LIBERIA** MANO zok péi bété (JGA) zokpei bélé (Adames) **IVORY COAST** GAGU moulia bien (K&A) **GHANA** AKAN-TWI akotopa (FRI) twihama (FRI) twihõma (FRI)

A scandent or lianous shrub attaining 10 m height, of scrub, wooded and forested areas from Senegal to S Nigeria, and in central and eastern Africa.

The stems hold a clear watery sap which is potable and is obtained by cutting a length of stem and letting the liquid run out into a container. The quantity is variable according to the climate, dry conditions resulting in a limited return whereas plants from humid high-forest may yield several litres of potable liquid. Such a phenomenon has naturally led to medicinal applications, and to a certain amount of magic. Curiously the use of the sap to relieve thirst is barely noted (8) though it must be of value to anyone on the march.

In Nigeria the sap is used for toothache and cough (9), and in Ivory Coast for tachycardia, ophthalmias and to remove foreign objects from the eyes (3). In Congo the sap is used to 'purify' mother and child immediately after birth. It must be given, as if it is a sort of colostrum, to a baby for its first suckle and is given regularly to twins to strengthen them, and to lactating mothers as a galactogogue. The sap is drunk by those with serious pulmonary affections, and the excitable and mentally ill. The lotion is used to relieve purulent ophthalmias, sores and oedemas (2).

The sap is used as a vehicle for the preparation of a macerate of the plant itself or of other drug-plants. The plant's lianous stems are macerated in its own sap in Senegal and taken in draught, baths or washes for leprosy (6, 7). A leaf-decoction is a remedy for toothache and a root-decoction for gonorrhoea in Sierra Leone (5). The powdered leaf may be used or a decoction of it for cough or held in the mouth to relieve toothache, or as a medicinal bath; also a decoction of leaf and root for venereal diseases (4). In Congo the plant's leaves or a portion of liane boiled in its own sap is held to be a powerful diuretic and used for urethral discharges and generalized oedemas, and to be vermifugal and purgative for stomach complaints. The sap is also concocted with other drug-plants in treatment for gastro-intestinal affections and haemoptysis (2).

No active principle apart from a trace of saponin has been recorded in the plant (4, 7).

The lianous stems serve as binding material (5).

The flowers are visited by bees (1).

References:

1. Adam, 1971: 377. 2. Bouquet, 1969: 105. 3. Bouquet & Debray, 1974: 79. 4. Dalziel, 1937. 5. Glanville 120, K. 6. Kerharo & Adam, 1962. 7. Kerharo & Adam, 1974: 394. 8. Kerharo & Bouquet, 1950: 37. 9. Oliver, 1960: 88.

# DIONCOPHYLLACEAE

**Habropetalum dawei** (Hutch. & Dalz.) Airy Shaw

FWTA, ed. 2, 1: 191. UPWTA, ed. 1, 47, as *Dioncophyllum dawei* Hutch. & Dalz.

West African: **SIERRA LEONE** BULOM (Kim) saŋgi (auctt.) BULOM (Sherbro) sangei-lɛ (TLG) MENDE tɔma (*def.*-i) *loosely used for fish-poisons* (auctt.)

A shrub or scandent of open places, usually on sandy coastal soil, of Pleistocene origin, recorded only in Sierra Leone.

The plant is very floriferous and showy in flower. The stems are used for tying house-posts. Split they are regarded as specially good rope. Kwako people who are noted for producing specially high quality piassava (*Raphia spp.,* Palmae) tie up their bundles of piassava with this material as a sort of trade-mark. The young leaves, pounded, yield a fish-poison, and when mixed with palm-oil and applied to affected feet to be effective in killing jiggers (Deighton fide 1). A phenol, present in the plant in unusually high concentration, has been isolated. This is found to kill fish at dilutions down to 10 ppm. (2).

References:

1. Airy Shaw, 1952: 336–9. 2. Hanson, 1976.

**Triphyophyllum peltatum** (Hutch. & Dalz.) Airy Shaw

FWTA, ed. 2, 1: 194. UPWTA, ed. 1, 47, as *Dioncophyllum peltatum* Hutch. & Dalz.

West African: **SIERRA LEONE** MENDE sangi (JMD) tɔma (*def.*-i) (auctt.) **LIBERIA** KRU-BASA goe-doo (C&R) goe-du (C&R) MANO– ma-bɛlɛ (Har.)

A scandent shrub or liane of the high forest occurring commonly in SE Sierra Leone, and in Liberia and western Ivory Coast.

The stems are used in Sierra Leone (Deighton fide 1; 5, 8) and in Liberia (5, 7) for tying house-posts and rafters. The root is said to be purgative, and very poisonous in overdose, a death being recorded at Bo, Sierra Leone, in 1936 within two hours (Deighton fide 1). In Liberia the inner-bark and leaves pounded finely are rubbed on parts of the body affected with elephantiasis, and made into poultices for abdominal pains, particularly for women (3, 4, 5).

A trace of alkaloid has been recorded in the leaves and bark, and 0.1–0.3% in the roots together with a strong presence of saponin and quinine and some tannin (2).

The stem contains much 'water' (6).

References:

1. Airy Shaw, 1952: 327–47. 2. Bouquet, 1972: 21. 3. Cooper 303, K. 4. Cooper & Record, 1931: 27. 5. Dalziel, 1937. 6. De Wilde & Leeuwenberg 3577, K. 7. Harley 1133, K. 8. Lane-Poole 186, K.

# DIOSCOREACEAE

## **Dioscorea** Linn.

FWTA, ed. 2, 3: 148. UPWTA, ed. 1, 488.

English: yam.

French: igname.

Portuguese: inhame.

West African: VULGAR g-niambi *from Bambara and throughout the Region* (A. Chev.) **SENEGAL** FULA-TUKULOR kappé (JMD) WOLOF ɲâbi (K&A) ɲamis (K&A) **THE GAMBIA** FULA-PULAAR (The Gambia) kape (DRR) MANDING-MANDINKA kalantaŋ (*def.*-o) (DRR) WOLOF ɲyambi (JMD) **GUINEA** BAGA da-tsak (Hovis) FULA-PULAAR (Guinea) kapé (*pl.*) (CHOP) MANDING-MANINKA ku (CHOP) SUSU khabi (CHOP) **SIERRA LEONE** BULOM (Kim) yamis (FCD) BULOM (Sherbro) yams-ɛ (FCD) FULA-PULAAR (Sierra Leone) kapɛ (FCD; JMD) GOLA siɛ (FCD) KISSI kulẽ (FCD; JMD) KONO ku *the tuber* (FCD; JMD) ra-ku *the plant* (JMD) KRIO yams (FCD) LIMBA (Tonko) yamis (FCD) LOKO m-boe (FCD) MANDING-MANDINKA ku *the tuber* (FCD; JMD) ra-ku *the plant* (JMD) MANINKA (Koranko) kui (FCD) MENDE mbole, mbui, mbuli, kpuli, etc. *general names* (FCD; JMD) SUSU khabi (FCD; JMD) SUSU-DYALONKE ku-na (FCD) nyal-la (JMD) TEMNE *ra*-boŋk *the plant* (JMD) *a*-yams (FCD) VAI sinabe (FCD) **LIBERIA** MANO dini *various c. vars.* (Har., ex JMD) kambo *various c. vars.* (Har., ex JMD) neng *various c. vars.* (Har., ex JMD) sõ *various c. vars.* (Har., ex JMD) soũ *various c. vars.* (Har., ex JMD) **MALI** MANDING-BAMBARA ku *c. var.* (JMD) nyambi *general* (JMD) uombo (Dumas) MANINKA ku *c. vars.* (JMD) nyambi *general* (JMD) **IVORY COAST** BAULE duo (JM) KWENI sɛi *a wild yam, in general* (Grégoire) yá *yam in general* (Grégoire) KYAMA nepu (JMD) MANDING-MANINKA ku (JM) **GHANA** AKAN bayerɛ (FRI) AKAN-FANTE ɛdwew (Enti) TWI bayirɛ (Enti) BIMOBA nuga (JMD) DAGBANI nyuya (JMD) GBE-VHE ehlo *wild spp.* (JMD) klo (JMD) tɛ (JMD) SISAALA piiŋ *general for 'yam'* (Blass) **TOGO** BASSARI dunore (Gaisser) DAGBANI nyule (Gaisser) KABRE hae (Gaisser) KONKOMBA lenul (Gaisser) MOORE-NAWDAM gbina, gebina *yams grown in moist places* (Gaisser) haere (Gaisser) ronde (*pl.* rona) (Gaisser) TEM (Tshaudjo) fudu **NIGERIA** ANAANG edia (JMD) èdiá (KW) m̀bọ̀kọ̀ ékpọ̄ (KW) DERA gújáanò (Newman) DOKO-UYANGA etun (*pl.* batun) (Talbot) EBIRA (Igara) ẹ̄nụ̄ (Moomo) EDO ìyán (JMD) EFIK biâ (auctt.) ǹdọ̀ñ ùbọk *yams requiring no staking* (auctt.) EJAGHAM (Obang) eyu (*pl.* ayu) (Talbot) EJAGHAM-EKIN eo (*pl.* ao) (Talbot) FULA-FULFULDE (Nigeria) arasje (*pl.*) *the aerial tubers* (JMD) ɓulumwol (*pl.* ɓulumji) *from Adamawa dialect and borrowed from neighbouring tribes* (JMD; MM) doya *from Hausa and understood by all* (MM) dunnduure (*pl.* dunnduuje) *from Sokoto dialect and perhaps a borrowed word* (MM) happere (*pl.* kappe) (MM) GWARI shama (JMD) shema (JMD) HAUSA doóyaà, dóóya (auctt.) dóóyar daji *wild yams* (JMD) dóóyar danga (RES) IBIBIO ùdiá (JMD; KW) IGBO akbọmana, asẹlẹ, asilẹ, ẹdo *red, or yellow yams* (JMD) jí *the general term: this may then be qualified but some qualifiying words may stand alone* (auctt.) IGBO (Onitsha) nboẹke *red, or yellow yams* (NWT) nchala, obiashe, obute oka, oku, onoku *red, or yellow yams* (NWT) umẹfu, umiakbo *red, or yellow yams* (NWT) IJO bùrú (KW) IJO-IZON (Kolokuma) òwéí bùrú *Guinea yams, in general* (KW) JUKUN atswi (JMD) tsi (JMD) KANURI bə̀rmá (JMD) KOROP karaia (*pl.* buria) (Talbot) NKUKOLI hellong (*pl.* olong lebamo) (Talbot) NUPE bàje (Banfield) banfyági (Banfield) báyìdzà (Banfield) bondzúrúgi *red yams* (Banfield) dagbaci (Banfield) ebù *seed bearing yams* (Banfield) eci (JMD) eci-kóyà *wild yams* (JMD) TERA cax (Newman) TIV agatu (Vermeer) agbo (JMD) ájíē (JMD) íjíē *aerial tubers* (JMD) íngyáwā = *finger yams* (DA) ínímbē (JMD) ìyógh (JMD) URHOBO ọ̀lẹ́ (Faraclas) YORUBA ègbodò *a new, fresh yam* (JMD) iṣu (JMD; IFE) **WEST CAMEROONS** BAFOK bekui (DA) BALONG bekui (DA) DUALA dibaŋga *a poisonous yam* (Ithmann) KOOSI beku (DA) KPE masawa (DA) KUNDU bekue (DA) LUNDU beye (DA) MBONGE bekue (DA) NGEMBA àzúɔ̀ (Chumbow) mə̀nəlȳìnə *c.vars* (Chumbow) ndɔ̀ŋbuɔ̀ (Chumbow) TANGA bekue (DA) WOVEA masuwa (DA)

Products – **NIGERIA** HAUSA alibo, alubo, èlubó *a flour* (JMD) sokwara *a food preparation of mashed yams* (JMD) NUPE cingini *the pounded yam* (RB)

Annual or perennial climbers with annual or perennial underground tuber, of some score species in the Region including two valuable Asian exotics; of numerous cultivars.

The word 'yam' in English and its equivalents in European languages have been derived from the common vulgar *niam* or *niambi* used in West Africa (2, 5). Yams have held from antiquity a position of great importance in West African food economy. Even now with introduced food-plants (e.g. tapioca, new coco-yam, sweet potatoes, etc.) which yield more for less effort and whose flour and starch is produced more easily, yams remain an extremely important crop. The main area of cultivation straddles the southern guinean savanna and forest areas to about 10°N lying in a zone with marked ethnic and linguistic frontiers that has been called 'the civilization of the yam' (12). Ninety per cent of the yam production of the African continent is said to be raised here. Rainfall is critical during growth and also good drainage. Intercropping is the traditional method of husbandry, but monoculture is now becoming more widely practiced. Yams are mostly eaten as *fufu* or *fou-fou* (in Francophone territories), a glutinous dough made from the peeled and boiled root. It does not keep and consumption is necessary soon after preparation. Production of *fufu* provides an extensive cottage industry. Some species are preferred to others for this purpose. A preparation which can be stored is yam flour, or *kokonte,* made from the tubers, sliced, sun-dried and coarsely ground. This is a very variable product and refinements in preparation can result in an improved product. Nutritionally the tuber is equivalent to the common potato providing 80–90% carbohydrates, about 5–8% protein and 3·5% minerals. With a normally high daily consumption protein and mineral intake is said to be adequate. Certain species afford promising commercial exploitation as sources of starch, and there is scope for the application of food technology to diversify utilization. (3, 4, 6, 7, 8, 10, 13.)

The alkaloid *dioscorine* is characteristic of the genus. It is bitter and toxic. In the species used for food any toxicity is either very weak or has been eliminated by ennoblement. Certain wild species (e.g. *D. dumetorum* (Kunth) Pax) may be eaten after suitable preparation as famine-food. Other important pharmacological substances occur in several wild species, steroidal sapogenins related to sex hormones and corticosteroids have been isolated. These provide a starting point for making progesterone and cortisone, oral contraceptives and sex-hormones. (1, 9, 11, 14.)

References:

1. Bampton, 1961. 2. Burkill, 1938. 3. Burkill, 1951. 4. Busson, 1965: 426–35, with tuber analyses. 5. Chevalier, 1946. 6. Coursey, 1966. 7. Ingram & Greenwood-Barton, 1962. 8. J.S.I. & L.H.G-B [Ingram & Greenwood-Barton], 1962. 9. Kaul & al., 1969. 10. Kay, 1973: 190. 11. Kerharo & Adam, 1974: 394–5. 12. Miège, 1952. 13. Purseglove, 1972: 1, 97–99. 14. Watt & Breyer-Brandwijk, 1962: 383–4.

**Dioscorea abyssinica** Hochst.

FWTA, ed. 2, 3: 153, incl. *D. sagittifolia* sensu Miège, p.p.
West African: **DAHOMEY** BATONNUN dika (A.Chev.) GURMA diabongua (A. Chev.) **NIGERIA** BEROM edɔ̀ŋ (LB) HAUSA dóòya (LB)

A non-spiny climber to 4 m high, twining right-handed, from the northern part of the Region from Mali to N Nigeria, and extending along the sahel zone to Ethiopia and E Africa.

The tuber is edible and palatable. It is thrust deep into the ground and is thus hard to get (1). It is cultivated in E Africa.

It is conjecturally one of the parents of *D. cayenensis.*

Reference:

1. Burkill, s.d.

**Dioscorea alata** Linn.

FWTA, ed. 2, 3: 152–3. UPWTA, ed. 1, 489, incl. *D. colocasiifolia* Pax, p. 491.

English: winged yam; water yam; white yam; ten-months yam; greater yam; greater asiatic yam; large leaf yam (Gambia, Dawe); red yam (form with red beneath the skin of the tubers, S. Leone, Deighton); Lisbon yam (from Portuguese, Burkill); Reuter yam (? a corruption of water yam, West Indies).

French: igname blanc; igname d'eau.

Portuguese: inhame bravo; inhame cicorero (*cicorero*, calabash, Burkill); inhame-da-china; inhame-da-índia; inhame-de-cariolá; inhame-da-Lisboa; inhame-da-S. Thomé.

West African: **SENEGAL** BALANTA bosétogué (A. Chev.) DIOLA e kama (JB; K&A) MANDING-BAMBARA danda ba, ba: *large* (JMD) gua, guagara, guara (A. Chev.) massa ku *a c. var.* (JB) MANDINKA balantaɲambi (after A.Chev.) massaku (K&A) ɲambi-ba = *large yam* (A.Chev.) WOLOF ɲam-ba = *large yam* (A. Chev.) **THE GAMBIA** DIOLA-FLUP kakama = *yam* (DF) WOLOF ku-ɲambo *a general term* (JMD) ɲamba ba (Dawe; JMD) **GUINEA-BISSAU** BALANTA-MANE umbóce (JDES) **GUINEA** KPELLE gua (A. Chev.) guangana (A. Chev.) MANDING-MANINKA gbara-gué *a white var.* (CHOP) gbara-ulé *a var., red under the skin* (CHOP) SUSU khabi-gbueli (CHOP; FB) **SIERRA LEONE** KRIO agbana *general for all yams not white* (JMD; FCD) MENDE nja-mbole = *water yam* (FCD; JMD) TEMNE *a*-wata-yams *from English and Krio* (FCD) **MALI** KWENI-TURA dô (A. Chev.) MANDING-BAMBARA danda ba (FB) gua, guagara, guara (A. Chev.) uagara (A. Chev.) **IVORY COAST** BAULE aken zaya *a form of bété bété, = beard, from the adherent roots* (JM) alua ualé *a form of bété bété,* alua, *dog;* uale, *thigh = dog's leg* (JM) bété bété, wété wété *general terms for a c. var.* (JM) bété bété alengbe *a form of bété bété* (JM) bété bété kpa *a form of bété bété,* kpa, *good* (JM) blu blu *a form of bété bété* (JM) douoblé *a form of n-za, 'black duo'* (JM) duo ofué *a form of bété bété,* duo, *yam:* ofué, *white* (JM) lengbé nziua *a form of n-za* (JM) n-ti kpri *a form of bété bété,* n-ti, *head:* kpri, *large* (JM) n-za *a general term for a c. var., = good* (JM) n-za blé *a form of n-za, = black n-za* (JM) n-za kpa *a form of n-za* (JM) n-za n'dré *a form of n-za,* ńdre, *hair* (JM) n-za tendéké *a form of n-za,* tendé, *long;* ke, *base, referring to the elongated root* (JM) nziua *a form of n-za* (A. Chev.; JM) plan *a form of bété bété* (JM) sépié *a form of bété bété,* se, *to cut up:* pié, *small, in allusion to only small pieces of the tuber being required for propagation* (JM) suiguié *a form of bété bété,* sui, *elephant;* guié, *tooth, from the shape of the tuber curved like a tusk* (JM) toro gua (A. Chev.) uoduo *a form of n-za,* uo, *snake, from the form of the tuber* (JM) uoduo blé *a form of n-za* (JM) KYAMA bera-bera (A. Chev.) bra-bra (A. Chev.) **GHANA** ADANGME-KROBO alamoa (auctt.) alamoa gaga = *long water-yam* alamoa gu = *ordinary water-yam* alamoa kani, alamoa ku = *male water-yam* alamoa puka, alamoa sale (A.S.Thomas) alamoa senya = *backward water-yam, from the manner in which the root is bent back* (JMD; A.S.Thomas) alamoa tun, alamoa tun gaga = *dark water-yam* asete alamoa = *Asante water-yam* gbani yebletchi = *seed producer* yorbli (A.S.Thomas) AKAN-AKYEM nuhõ *a food perparation from afaseɛ* (FRI) ASANTE afaseɛ (Enti) ko-ase-kohwe = *go look underneath, a c. var. with large tubers* (FRI) FANTE ɛdwew (Enti) TWI ad-ammã-wo-ba = *you eat it without giving some to your children; a c.var of afasew of good flavour* (FRI; Enti) adjugo somadrewe *a c. var.* (FRI) afasew (auctt.) afasew aba = *seed water-yam* afasew anamasu, afasew apuka, afasew kani, afasew pa = *real water-yam* afasew biri *a c. var. with reddish-purple flesh beneath the tuber skin* (FRI) afasew ɔdepa, pa: *good;* ɔde: *white, i.e.= good white water-yam* (FRI) afasew tinlin = *long water-yam* afasew tuntum = *black water-yam — a c.var. with reddish-purple flesh beneath the tuber skin* (FRI) apuka *a c.var. with round tubers* (FRI) butuabaso = *sitting on eggs, referring to the shape of the tuber like a sitting hen* (A. S. Thomas) 'dwaa *a c.var. of smooth skin and very white flesh* (FRI) namonsi *a c.var. of afasew with long tubers* (FRI) nwoma *a food preparation* (FRI) GA afasɛ́o, afasó (*pl.*-i) (KD) dã̀na (Addo) GBE-VHE adzigo, adzugo (A.S.Thomas) adzugo bolobolo *soft water-yam* (A.S.Thomas) adzugo dzedze = *branched water-yam* (A.S.Thomas) adzugo eto *fruit water-yam* (A.S.Thomas) adzugo kpuka (A.S.Thomas) adzugo nkani (A.S.Thomas) adzugo toko *water-yam with knobs on* (A.S.Thomas) adzugo tsetse *bearing water-yam* (A.S.Thomas) VHE (Pecí) adzigo, adzugo (JMD; FRI) ahamante (FRI, ex JMD) dzɔbɔli (FRI, ex JMD) GUANG-GONJA kùbòrú (Rytz) **DAHOMEY** BATONNUN sakuru (auctt.) DENDI sakata (A. Chev.) GBE-FON fuho, fuo (A. Chev.) anugan (A.Chev.) SORUBA-KUYOBE sinnoré (A.Chev.) YORUBA (Idacha) kiamfa (A. Chev.) YORUBA-NAGO iṣu ewura = *water yam* **NIGERIA** BEROM dém pwɛŋ = *white yam* (LB) dém sunàŋ = *red yam* (LB) kit jey = *outside yam; because the tubers are emergent from the ground* (LB) kit kácìk = *straight yam; for the long, large tubers, bigger than dém, smooth* (LB) EDO ígiorùa (JMD) udin (JMD) EFIK èbìgè (auctt.) èbìgè ǹtán, è. òsúkpà, ǹpoyo èìgè, ọsoñ-ikpọk èbìgè = *tough-skinned water yam* etc. *varieties* (auctt.) HAUSA baƙar dóóya *a wild form, bitter or acrid, with rough skin* (JMD) jikin mutun = *man's body; a wild form* (JMD) sakataa, sakatáá (auctt.) IBIBIO àbìrè (Kaufman) IGBO jí àbàlà (BNO) jí m̀bàlà àdàkà *a c.var* (JMD) jí ḿbàlà ḿme *a c.var* (JMD) jí ḿbàlà ọ́chá *a c.var* (JMD) jí ḿbàlà ubute oka *a c.var* (JMD) m̀bàlà, jí-m̀bálá m̀bàlà: *general for water-yam* ọ́màrìmà a wild water-yam (KW) IGBO (Onitsha) jí-àbànà (JMD; Lawton) mbwẹde (JMD) ngbẹde (JMD) onoko (JMD) IJO-IZON (Kolokuma) ịyọ́rọ́bùrú = *female yam; cf., male yam, D. cayenensis* (KW) NUPE wùrà *from Yoruba* (JMD) TIV àgbò *general, with many vars.* (JMD; Vermeer) àgbò tsábàgú *a c.var.* (JMD) URHOBO akẹnẹdo (JMD) edialukpakọn (JMD) ega (JMD) eni (JMD) igyolowa (JMD) ijorua (JMD) kediavo (JMD) ọkorhọ (JMD) ọniya (JMD) udin (JMD) YORUBA arunfanfan *a var.* (DA) atti *a var.* (DA) duduku *a white-tubered var.* (JMD) ègbódo (JMD) ewùrà *sometimes known as 'female yam'* (auctt.) ewùrà funfun *a white-tubered var.*

(JMD) ewùrà pupa *a yellow-tubered var.* (DA) ịṣu ewùrà lanṣẹjẹ *a var.* (DA) onẹdo *a var.* (DA)
**WEST CAMEROONS** VULGAR joma (JMD)

A vigorous climber to 15 m high, stem quadrangular, winged, twining right-handed, a very polymorphic cultigen originating in SE Asia and not found wild, and dispersed by man over the Pacific and Indian Oceans in prehistory, and now widespread in the Region and pantropics.

The plant produces usually one enormous tuber. Giant yam production is associated with religious cults in New Guinea and tubers 64 kg in weight and 3·5 m long have been reported. By special cultural methods tubers of 20 kg were commonly raised, and one of 110 kg has been recorded (6). They are deep-penetrating and require under cultivation a large mound of soil, but are tolerant of poorer soil than other yam species. Crop period is 9–10 months and the tubers are very watery, characteristics accounting for the names '10-months yam' and 'water yam.' It has a dormancy period of at least 2 months and for this reason undoubtedly man has used it for stocking canoes for his Pacific Ocean island-to-island migrations. Portuguese brought the plant to West Africa probably in the 16th Century and used the tubers for stocking slave-trade ships crossing the Atlantic. At this time there was a slave-market in Lisbon whither yams were sent from San Thomé so that in Portuguese the yams became known as 'Lisbon yams' or 'S. Thomé yams' (2, 3).

The plant has numerous forms which fall into two main groups: (a) those whose foliage is always green, and (b) whose foliage is more or less red. These are respectively the *bété-bété* and the *n'za* groups of the Baule yam culture in Ivory Coast (8). Within these groups numerous cultivars exist on tuber forms (4, 8). A curious race from SE Asia produces a downward-growing tuber which then turns upwards till it extrudes from the ground. This arrests growth, but growth will continue if it is earthed up. Lateral tubers arising from the main tuber also grow upwards. This form appears to be represented in certain cultivars in W Africa; see especially Ghanaian vernaculars above (5).

It is widely grown for consumption after peeling and boiling or baking. It is not mashed nor made into *fufu*. The uncooked tuber is toxic, and is said to produce narcosis. Saponin is present and cooking renders the tubers safe to eat (9, 10). Though it is an exotic plant it has gained and maintains a position of considerable importance in the W African food economy (8), and it has been pointed out that in the Delta area of Nigeria where traditionally subsistence hinged on fishing, agriculture has been made entirely and only possible on the arrival of exotic food plants amongst them the water-yam. The other important species are the coco-yam and the plantain (11). The tuber is high in carbohydrates, 88%. Protein is about 7% in Ivorean material (4). In various countries of the world attempts are being made to develop processed products (7). Notwithstanding the tuber's capacity for easy transportation, processing of W African material merits examination. It should be a source of starch. It can be converted into a meal that stores well for upwards of a year.

Material raised in Nigeria has been found free of alkaloid (1).

References:

1. Adegoke & al., 1968. 2. Burkill, 1935: 814–6. 3. Burkill, s.d. 4. Busson, 1965: 429–30, with tuber-analysis, 434–5. 5. Dalziel, 1937. 6. Haynes & Coursey, 1969. 7. Kay, 1973: 213. 8. Miège, 1952: 147–52. 9. Watt, 1967. 10. Watt & Breyer-Brandwijk, 1962: 384. 11. Williamson, K., 1970.

**Dioscorea bulbifera** Linn.

FWTA, ed. 2, 3: 152. UPWTA, ed. 1, 490.

English: potato yam; aerial yam; air potato; bulbil yam; bulbil-bearing yam; turkey liver yam, top yam (Ghana, Sampson); otaheite yam (an Indian race in Ghana, Burkill).

West African: **SENEGAL** VULGAR dana (A. Chev.; IHB) danda *cf. 'danda yams', West Indies* (A. Chev.; IHB) BALANTA busu bulé (A. Chev.) BASARI ɔ-fɔngwón, a-pɔngwón (Ferry) BEDIK ga-pángol (Ferry) DIOLA diɓan (A. Chev.) éborimbor (JB; K&A) kanum (A. Chev.; IHB); *= a sharpener, cf. a hard dead wood used for sharpening machets* (DF, The Gambia) karamba (A. Chev.) DIOLA-FLUP kamako (A. Chev.; IHB) KONYAGI ɓẹ-ngwaf u-naɗ, tẹmp-tẹmp (Ferry) MANDING-BAMBARA dâda (JMD; K&A) dana (K&A) MANDINKA bayulo (A.Chev.) kama (*def.* kàmoo) (A.Chev.) kamao (A.Chev.; IHB) tubaabuɲambi (*def.* t. ɲamboo) MANINKA dâda (JMD; K&A) dana (K&A) dan-dan (A. Chev.) SERER ań (JB; K&A) sipa (JB; K&A) WOLOF ań (JB; K&A) ngolgol (IHB) **THE GAMBIA** MANDING-MANDINKA tubaabuɲambi (*def.* t. ɲamboo) *= whiteman's cassava* (DF) **GUINEA-BISSAU** BALANTA n-dome (JDES; EPdS) BALANTA-MANE undome (JDES) CRIOULO genebra (JDES; EPdS) FULA-PULAAR (Guinea-Bissau) púri (JDES; EPdS) MANDING-MANDINKA camó dandandim (*def.*-ô) (JDES) dandaõ (JDES) MANDYAK timbom (JDES; EPdS) MANKANYA catoco (JDES) SUSU dané (auctt.) **GUINEA** FULA-PULAAR (Guinea) puri-balé (CHOP) MANDING-MANINKA dana (CHOP) danda (auctt.) dan-dan *the wild form* (CHOP) MANINKA (Guinea) guinfiné (CHOP) **SIERRA LEONE** KISSI kule-yondolẽ *var. anthropophagorum* (FCD) KONO kũyawa *var. sativa* (FCD) KRIO ɛmina *var. sativa* (IHB; FCD) LIMBA kəyoro (JMD) LOKO m-bo boni (NWT) n-gengange (NWT) MANDING-MANDINKA danda *var. sativa* (FCD) MANINKA (Koranko) digegbe (JMD; IHB) MENDE fɔli *var. sativa* (auctt.) mbole (auctt.) ngawu (FCD; IHB) puli (auctt.) TEMNE *am*-boŋk (auctt.) *ra*-boŋk *the vine* (auctt.) *an*-wun (FCD; IHB) **MALI** VULGAR dana (A. Chev.; IHB) danda *cf. 'danda yams', West Indies* (A. Chev.; IHB) MANDING-BAMBARA dana (JMD) danda (JMD; FB) kamu *the aerial tubers* (JMD, ex IHB) KHASONKE murugo (A. Chev.) **IVORY COAST** AKAN-ASANTE akaï (B&D) AKYE achi a hu (A&AA) achi a ko fonfon (A&AA) BAULE bronu kaha (A&AA) ka (A. Chev.) DAN beidé (A.Chev.) mi (A.Chev.) **GHANA** ADANGME-KROBO akam, akom (IHB) AKAN-ASANTE akam *'acom' of auctt.* (auctt.) FANTE man *cf. 'man yam', Jamaica* (auctt.) TWI akam *'acom' of auctt.* (auctt.) akam bonwoma, bonwoma: *bile, a wild form bitter like bile* (FRI) akam moto (auctt.) ananse akam *= spider's yam, a wild form* (FRI) DAGBANI friguma (Gaisser; IHB) GBE-VHE gbe-tɔ *= bush yam* (auctt.) tɔ (auctt.) **TOGO** BASSARI nubulentschor (auctt.) KABRE nbanioke (auctt.) TEM agbanio (FB) TEM (Tshaudjo) agbanio (auctt.) **DAHOMEY** BATONNUN mokuru (A.Chev.) DENDI dundu mbissa, dundu = D. dumetorum (A. Chev.) GBE-FON agbabli (A. Chev.) guite sindé (A. Chev.) YORUBA (Idacha) dɟite(s)odé (A. Chev.) guite sindé (A. Chev.) guite sodé (A. Chev.) **NIGERIA** VULGAR akam (IHB) BEROM edɔ̀ŋ (LB) etɔt *tuber protected, bristly* (LB) etɔt goy *= yam of the sun; not edible* (LB) kwáŋ *var. with very large tubers* (LB) yɔ̀ŋ *var. with smaller tubers than kwáŋ* (LB) EFIK édómō (auctt.) HAUSA dóóya (LB) dóóyan ítaacéé (LB) doóyar-bísà, dóóyar-bisa, doóya: *yam*; bísà: *topside; i.e. 'the top-side yam' in allusion to the aerial tubers* (auctt.) kamu *from Bambara/Maninka (Mali), the aerial tubers* (auctt.) túwón bírii *= monkey's food* (auctt.) IGBO (Agukwu) àbànà ófịá *= water-yam of the bush* (NWT; JMD) àbànà ófịá baba, à.ó.gbara, à.ọ.mbaba *vars.* (JMD; IHB) ádụ̀ (Lawton) NUPE kàndù (JMD) TIV ájíē (JMD) íjémbēàóndó *a var.* (DA; JMD) íjíē (JMD) íjiē pupuu *a white fleshed var.* (JMD) íjiē shìbèrá *a var.* (JMD) YORUBA aparia iṣu *the tubers of an edible c.var.* (IFE) ẹmina (auctt.) ẹmina-esi (Dawodu; IHB) èsúrú (IFE) ẹwura-esi (auctt.) iṣu (IFE) iṣu ahum *the aerial bulbils* (IFE) iṣu eleso *the plant* (IFE) ori iṣu *the tubers of a wild form* (IFE) **WEST CAMEROONS** DUALA mbă bèdimò *perhaps this sp., but cited as 'D. sativa'* (Ithmann)

A glabrous-leafed, non-spiny climber twining left-handed to about 6 m high from a small woody tuber and producing aerial axillary bulbils, widely distributed over the Indic basin and extending into W Africa throughout along the coast and penetrating inland along migration and trade and travel routes.

It is a species of many races. The wild ones have globose, dark brown to liver-coloured, non-angular bulbils which serve as a famine-food as do the tubers. To be made edible prolonged preparation is necessary without which consumption may cause death. Wild strains are often planted intermixed with or on the perimater of plantings of improved races as a protection against thieving. Cattle eating them accidentally may be fatally poisoned showing frothing at the mouth and bloating (12, 13). The species is in the process of ennoblement and selected cultivars show in varying degree bitterness and poisonousness. Of some races even after prolonged preparation the bulbils remain bitter. Superior races are said to be very palatable and sweet, and to be entirely free from toxic substances so that consumption, even raw, is safe. The skin is grey, lighter coloured than the wild forms, and the flesh is pale yellow to near white. Nevertheless the best are still considered inferior to most of the common yams (5, 6, 11).

The bulbils of selected cultivars tend to be angular with a flattened shape and a skin-colour which evokes the name 'turkey liver yam.' They may attain as much as 2 kg in weight but an average weight is about 0·5 kg. Races with increased bulbil production tend to show a reduction of the tuber, and in those with the highest bulbil return the tuber is but a woody rootstock. Crop production is usually 24 months. Bulbils are ready for harvesting when they fall off the plant at a slight touch. The flesh is mucilaginous. Ivorean material has shown 80% carbohydrates and the production of starch and flour is possible. They are not converted into *fufu* (5, 6, 9). In Gabon a small piece of the bulbil may be added to palm-wine to promote fermentation. This results in a frothy head and a stronger brew, but excessive treatment causes vomiting and strong diarrhoea (13).

Both the tuber and the bulbil of wild races have medicinal uses. In Gabon crushed bulbils are massaged onto areas of rheumatism and for troubles of the breasts and for jiggers (13). In Congo both parts are used in dressings for dermal parasitic and fungal infections, and an ointment of crushed bulbils in palm-oil is used to relieve the pain of rheumatism (4). In Madagascar the dried and rasped bulbils are used on wounds, sores and inflammations (8). In India the tuber is considered to be diuretic and to be a remedy for diarrhoea and haemorrhoids and to be an excellent dressing for boils (1, 14). The fruits are used in S Nigeria for boils and for fever (7).

Sap expressed from the vine stems is applied in Congo for purulent ophthalmias, and for snake-bite (4). In Tanganyika the leaves are used, often by steam-distillation, against pink-eye (3, 14). The Akye of Ivory Coast instil sap expressed from the leaves into the eyes to keep one awake (2).

*Dioscorine* has been detected in the tuber (15) though certain Nigerian material has been reported free of alkaloid (1). Alkaloids have been reported from the leaves and stems and particularly in the fruits (8). In Mexico *diosgenin* has been detected at 0·45% concentration (14). Saponin is present and a number of other pharmacologically active substances (10).

References:

1. Adegoke et al., 1968. 2. Adjanohoun & Akė Assi, 1972: 113. 3. Bally, 1937. 4. Bouquet, 1969: 106. 5. Burkill, s.d. 6. Busson, 1965: 430, with tuber and bulbil analysis, 434–5, incl. *D. latifolia* Benth., 432. 7. Dawodu 145, K. 8. Debray, 1971: 68. 9. Kay, 1973: 229. 10. Kerharo & Adam, 1974: 395–6, with phytochemistry and pharmacology. 11. Purseglove, 1972: 1, 103–4. 12. Walker, 1953, a: 33. 13. Walker & Sillans, 1961: 152. 14. Watt & Breyer-Brandwijk, 1962; 384. 15. Willaman & Li, 1970.

## **Dioscorea burkilliana** Miège

FWTA, ed. 2, 3: 153.

West African: **SIERRA LEONE** MENDE mbole SUSU g-beli-gbeli (NWT) TEMNE *an*-buk (NWT) *an*-tankali (NWT) *an*-tantali (NWT) *ra*-won *the vine* (NWT) **IVORY COAST** ANYI kokua (JM) BAULE n-guku (JM)

A climber with spiny annual stems twining right-handed to 8 m high from a perennial fibrous tuber protected by a woody rootstock, occurring only in the bush and forest zone of Sierra Leone and Ivory Coast. It may occur in Nigeria (1).

The plant is very similar to *D. minutiflora* and *D. smilacifolia,* but it is recognized in Ivory Coast as different (2). Information is lacking regarding its uses. As there is some protection of the tuber it is perhaps edible, but probably as of the two related species only after special preparation.

References:

1. Lawton, 1967: 9. 2. Miège, 1958.

**Dioscorea cayenensis** Lam.

FWTA, ed. 2, 3: 153, as *D. cayenensis* Lam. ssp. *cayenensis*. UPWTA, ed. 1, 490.

English: yellow Guinea yam; twelve-month yam; Lagos yam (Sierra Leone, Deighton); 'cut-and-come-again' (West Indies).

French: igname jaune; igname-Guinée.

Portuguese: inhame.

West African: **SENEGAL** BASARI kħáɓ, ɔ-ŋáɓ (Ferry) BEDIK ɛ-gɛ̀b (Ferry) KONYAGI u-nkwav (Ferry) MANDING-BAMBARA dãnda (JB) fasaka (JB; K&A) MANDINKA ɲambi (*def.* ɲamboo) *a general term* (A.Chev.; Ayensu & Coursey) **SIERRA LEONE** KRIO agbani, agbani-yams *general for any yam not white* (JMD) MENDE mbo *wild form* (FCD) mbole-gbɔlu *cultivated form* (FCD) TEMNE *a*-mas (FCD) *a*-nyame (Ayensu & Coursey) *a*-nyams-*a*-yin *yellow yam* (FCD) TEMNE (Kunike) *an*-nai *cultivated form* (FCD) *an*-nəi *cultivated form* (FCD) TEMNE (Yoni) *an*-ku *cultivated form* (FCD) *an*-nai *cultivated form* (FCD) *an*-nəi *cultivated form* (FCD) **MALI** MANDING-BAMBARA fasaka (JMD; FB) **IVORY COAST** BAULE akra n'dufon *a form of krengle* (JM) assaua *a form of lokpa* (JM) assobare *a form of lokpa* (JM) fru *a form of krengle* (JM) kangba *a form of krengle* (JM) kangba blé *a form of krengle* (JM) kangba kokolé *a form of krengle* (JM) kangba ofué *a form of krengle* (JM) kpona *a form of lokpa* (JM) krenglé *general term for a c. var.* (JM) lokpa *a general term for a c. var.* (JM) n-détré *a form of lokpa* (JM) sépélé *a form of lokpa* (JM) zrézrou *a form of lokpa* (JM) **GHANA** ADANGME-KROBO kani (auctt.) kani-gu *a c.var., = ordinary yellow yam* (auctt.) kani-hini *a c.var., = king of yellow yams* (auctt.) kani-kuku *a c.var.* (auctt.) kani-tʃu *a c.var., = red yellow yam* (auctt.) otri-kani *a c.var = white yellow yam* (auctt.) tchomatchom *a c.var. of poor quality* (auctt.) AKAN-ASANTE afũũ *a c.var., much cultivated large tuber* (aucct.) afũũ-ba *a c.var., large tubers* (auctt.) nkontia *a c.var., large tubers* (auctt.) TWI akwakwɔ-bedina *a c. var.* (auctt.) aniwa-aniwa *a c. var. = many eyes* (auctt.) *n*-kani (auctt.) n-kani-fufu *a c. var., = white yellow yam* (auctt.) n-kani-hene (or -hini) *a c. var., = King yellow yam* (auctt.) n-kani-kuku *a c. var.* (auctt.) n-kani-pa *a c. var., = proper yellow yam* (auctt.) nkuku *a c. var.* (auctt.) GA nkani (FRI; A.S.Thomas; JMD) GBE-VHE dzafutu (auctt.) dzoboli *general for all yellow yams* (auctt.) kani gu *a c.var.* (auctt.) kani hene *a c.var.* (auctt.) kani kuku *a c.var.* (auctt.) kani tchu *a c.var.* (auctt.) kani yie *a c.var. = white yellow yam* (auctt.) kpasa (auctt.) n-kani (auctt.) otre kani *a c.var.* (auctt.) **TOGO** GBE-VHE avete *a c. var.* (Metzer) boto *a c. var.* (Metzer) dza (Metzer) kani *a c. var.* (Metzer) loboko *a c. var.* (Metzer) **NIGERIA** DEGEMA ògbı́rō̃ (*pl.* ì-) (KW) EDO íkpèn (JMD) òkpòdọ̀ghòn (JMD) EFIK àkpàná (JMD; RFGA) àkpàná-ntàntà *a variety* (JMD) èfiáñ (auctt.) òtúk-ọ̀kpọ̄ *a variety* (JMD; RFGA) ENGENNI ẹ̀dị̀à (KW) EPIE àdị̀à (*pl.* ị̀-) (KW) HAUSA doóyàr kúdù = *yam of the South* (JMD; ZOG) IBIBIO àkpàànà (JMD; Kaufman) àkpàànà-ńtãńtá *a wild form* (JMD; KW) èfiààñ (DA) èfiáàñ údiã *root crooked, conseq. v. hard to dig out in one piece* (JMD; Kaufman) IGBO jí ogbagada (BNO) jí òkù (KW) jí-òkò (NWT) jí-òkò-ọ̀chá, ọ̀chá: *white* (Lawton) IJO bùrú (KW) IJO-IZON (Kolokuma) òwéí bùrú = *male yam; cf. D. alata, female yam* (KW) ISOKO ọ̀lẹ́ (KW) NUPE eci dzuru (RB) TIV íyogh (Vermeer) nwángē *with numerous varietal names* (JMD) URHOBO okpen (JMD) YORUBA àgándán (JMD) aginipa (JMD) àlọ̀ = *riddle* (auctt.) apèpe (JMD) ìgángán (JMD; Lawton) igángán alo *a c.var.* (IHB) iṣu = *tuber* (IFE) olo (Lawton) oparaga (JMD) panṣágà = *prostitute* (Verger; JMD)

*NOTE: Some Ivorean names may refer to D. rotundata as Miège (JM) places this species under D. cayenensis.*

A cultigen, not occurring wild, with annual stems twining right-handed, tuber surface-rooting, common throughout the southward part of the Region, and dispersed by man to areas with high and regular rainfall. FWTA, ed. 2 has reduced this to a subspecies of a larger species including *D. rotundata* (Poir.) Miège. This is regarded as unacceptable lumping (1, 2), and the two are considered here as separate species (*sensu stricto*).

This species is the Yellow Guinea yam. In spite of its specific latin name it is a West African plant and a cultigen developed entirely by the African himself, the history of which is lost in antiquity. Both Guinea yams are thought to be amongst the first plants cultivated in W Africa. To the Ijo of the Niger Delta these plants are synonymous with 'eating' (7). The Yoruba in dubbing the Yellow Guinea yam 'riddle' invoke the plant in a repetitive incantation against death, and again as 'prostitute' or 'beautiful woman' imply a similar desire for biological continuity (5). The epicentre of cultivation is the Gulf of Guinea from Ivory Coast to S Nigeria. It has a 10–12 months growing cycle and requires a fairly regular wetness throughout the year; hence it is restricted to the region south towards the coast while *D. rotundata* with a 7–8 months crop period to complete its growing cycle occurs to the northward.

The tuber is formed near the surface of the ground. It has yellow flesh though it is white in some races and some vernacular names imply whiteness. It can be dug up as required and a common practice is to expose one side of the tuber and to cut away the distal part for consumption leaving the upper part to regenerate; hence the name 'cut-and-come-again.' This practice may be repeated for upward of three years. The tuber has a very brief resting period and does not store well. It is in fact best stored, if necessary, in the earth. There are many races and some are reputed to be pleasant to eat though it is usually said to be inferior to the White Guinea yam (*D. rotundata*). It is of lesser commercial importance, but it has a higher yield and can be harvested over a longer period (4).

The root contains the highest amount of carbohydrate of the yams. Ivorean material is reported to have 91%, but protein is low, 2·5% (3). The root is normally eaten boiled. It can be made into *fufu* but the product is not of the best, and if required for this purpose they are added to better grade yams.

An unspecified alkaloid has been reported present in the leaves and stem (6).

References:

1. Ayensu & Coursey, 1972. 2. Burkill, s.d. 3. Busson, 1965: 431, with tuber analysis 434–5. 4. Kay, 1973: 237. 5. Verger, 1967: nos. 31, 164. 6. Willaman & Li, 1970. 7. Williamson, K. 1970.

**Dioscorea dumetorum** (Kunth) Pax

FWTA, ed. 2, 3: 151. UPWTA, ed. 1, 491.

English: bitter yam; cluster yam (when tubers are bunched); three-leaved yam; trifoliate yam.

West African: **SENEGAL** BEDIK ɛ-ɗyíɗy, gi-nyíɗy (Ferry) MANDING-MANINKA bodu (auctt.) budé *wild form* (auctt.) **GUINEA-BISSAU** FULA-PULAAR (Guinea-Bissau) mabaia (JDES; EPdS) **GUINEA** MANDING-MANINKA bodu *cultivated form* (auctt.) budé *wild form* (auctt.) **SIERRA LEONE** KRIO ɛsuru-yams (FCD) LOKO n-gengange (NWT) **MALI** MANDING-BAMBARA kuba *wild form* (A. Chev.; IHB) MANINKA budé *wild form* (auctt.) 'SONINKE' laliman (A. Chev.) **IVORY COAST** AKYE ado (A&AA) **GHANA** ADANGME-KROBO kamfo (auctt.) nkampo (auctt.) nkanfo = *praise* (auctt.) nya = *mouth* (auctt.) AKAN-ASANTE nkanfoɔ (Enti) TWI akori ekyi *a var. of nkamfo, yellow flesh, said to be the best* (auctt.) nkamfo (auctt.) nkanfo = *remembrance; praise* (auctt.) oworo-woro *a var. of nkamfo, long sweet tubers* (auctt.) yaw serewin, serewin: *hairy thigh, a var. of nkamfo, tuber covered with hairs* (auctt.) GA akori ekyi *a var. of nkamfo, yellow flesh* (auctt.) nkampo (auctt.) ŋkanfo = *praise* (auctt.) oworo-woro *var. of nkamfo, long sweet tubers* (auctt.) yaw serewin, serewin: *hairy thigh; a var. of nkamfo, tuber covered with hairs* (auctt.) GBE-VHE kamfo (auctt.) kanfu (auctt.) kangfo (auctt.) **TOGO** GBE-VHE adekute (Metzer) **DAHOMEY** BATONNUN yéséku *cultivated form* (A.Chev.) DENDI dundu kiré *cultivated form* (A. Chev.) GBE-FON e-léfé *cultivated form* (A. Chev.) YORUBA (Idacha) pansuréru *cultivated form* (A. Chev.) **NIGERIA** VULGAR esuri yam (Corkill) ANAANG ìfòmò (KW) EDO ùfúa, èmówè *applies to all white yams* (auctt.) EFIK afia edidia = *think of soup; i.e. eatable, sufficient as staple* (Winston; KW) àfia èdidìà iwá, àfia iwá = *white cassava* (auctt.) àkpànà (auctt.) eba edi = *sow's teats* (auctt.) èdèm ìnàẏ = *four-sides* (auctt.) èdiá anañ (JMD; IHB) èdidìà ìwá = *eatable cassava* (Winston) ìwá, ìwá ekoi, ìwá ḿfìm, ìwá: *cassava, Manihot esculenta* (*Euphorbiaceae*); Ekoi: *Efik name for the Ejagham, a tribe of the Cross River state;* ḿfìm: *wild; a wild and poisonous form* (auctt.) ǹdìsime ìwá = *foolish cassava* (auctt.) òbubit ìwá = *black cassava* (auctt.) HAUSA gursami *tubers short in radiate cluster* (auctt.) kisra *the flour* (FRI) koósán-roógoò *a cultivated form* (auctt.) roogon biri, rogwain biri = *monkey's tapioca; a wild form* (auctt.) IBIBIO àkpààǹà (Corkill) ánêm, énêm (Kaufman) IGBO ádụ̀ (JMD; IHB) akpana (NWT) atọka (auctt.) eba edi, obubit iwa *c.vars.* (NWT) IGBO (Owerri) ọ̀nà IGBO (Umuahia) ọ̀nụ̀ TIV ínímbē (*sing.*), anembe (*pl.*) (auctt.) URHOBO olimẹhi *a yellow yam* (auctt.) owabọ *a yellow yam* (auctt.) ufua *a yellow yam* (auctt.) YORUBA èsúru (auctt.) ewu eleso *a wild form* (IFE) fẹ́lẹ *a var.* (auctt.) ganhun-ganhun *a var. with yellow or white tubers* (auctt.) gùdùgúdú (auctt.) iṣu, iṣu-èsúrú (IFE) iṣu-ira (JRA) itakun iṣu *a cultivated form* (IFE) **WEST CAMEROONS** DUALA mbǎ (Ithmann) NGEMBA enkoa *a wild form* (A.Chev., ex IHB)

Climber twining left-handed from a shallow-seated annual single tuber or tubers in a cluster, spiny stem to about 3 m high, common throughout the savanna region from Senegal to W Cameroons, and extending across the Congo basin to Angola.

Wild forms are in general poisonous due to the presence of the alkaloids *dioscorine* and *dihydrocortisone* (1, 9, 14, 15, 21–23). Both are convulsants. The tuber is an important famine-food (5, 8, 11, 20). It has been so used in Ghana (11), and in Sudan in 1938 it was the principal amongst several roots and tubers (7). The plant is often purposely grown as an insurance at spare points and in hedgerows around a farm and especially on the edges of yam fields as a deterrent to human and animal marauders. It grows easily and is a heavy cropper. If it is to be eaten there must be a detoxification which is carried out by slicing and steeping the root for 3–5 days in running water. A test for completion of the cleansing process is said to be to see if shrimps or fish are feeding around the basket in which the cut-up roots are suspended. Another test is to feed some to a chicken and watch for it showing signs of giddiness, or if the expressed juice causes smarting if dripped into an eye. If running fresh water is not available, salt water can be used. The roots can also be cleaned by burying them for three days in black cotton soil, or steeping in cotton soil and water, or with tamarind pods (*Tamarindus indica* Linn. – Leguminosae – Caesalpinioidae) or desert dates (*Balanites aegyptiaca* (Linn.) Del. – Balanitaceae) (11).

There appears to be some variation in toxicity perhaps due to edaphic, phenologic or other cryptic factors. Material from the Congo has been reported free of active principles (4). On the Benue Plateau the Tiv recognize it as a food-plant but do not greatly appreciate it and do not eat it if other yams are available (19). Genetic variation there certainly is: the plant has undergone a degree of ennoblement by man's selection so that there is recognized a number of edible cultivars. In Nigeria there are yellow, white and pale yellow forms (8). In N Nigeria two races, one poisonous, one not, can be distinguished by hairiness, the poisonous one being the more tomentose (5). Cultivars are numerous in Dahomey (6). The meal is coarse and starchy, and in Gabon is said still to retain a bitter taste even after cooking (20). It is particularly rich in scorbutic acid especially when prepared with tamarind pods. Detoxified tubers form the basis for brewing a beer in Sudan (13). In Ivory Coast the tuber is boiled and eaten in treatment of jaundice and malaria: the liquid is also drunk for these troubles (2).

Physiological effect of the poisonous tubers is to paralyse the respiratory system. In S Africa mealies treated with the root are prepared for catching monkeys that become stupified after eating the bait. In Malawi the root is mixed with *Strophanthus* arrow-poison and some races in S Africa use it topically as an anodyne to relieve pain. In Tanganyika the root is considered a cure for schistosomiasis (7, 14, 22). In Sudan the root is used on rheumatic arthritis (16).

In Ghana it is planted along a farm roadside as a protection of the farm from remarks of passers-by: if the farm is praised too much it might become weedy or the crops die, the inference being that the presence of inferior yams will not tempt providence (17); hence the Twi name meaning 'remembrance' or 'praise'! The Yoruba of S Nigeria evoke the plant in a sinister incantation to bestow virile power (18). In Congo the plant is acknowledged to be toxic not only to humans but to spirits and able to exorcize devils. It is planted on graves to drive away spirits and ghosts (3).

References:

1. Adegoke & al., 1968. 2. Adjanohoun & Aké Assi, 1972: 114. 3. Bouquet, 1969: 106. 4. Bouquet, 1972: 21. 5. Burkill, s.d. 6. Busson, 1965: 431; root analysis, 434–5. 7. Corkill, 1948. 8. Dalziel, 1937. 9. Haerdi, 1964: 202. 10. Irvine, 1948: 260. 11. Irvine, 1952, a: 25–26. 12. Irvine, 1952, b. 13. Kay, 1973: 203. 14. Kerharo & Adam, 1974: 396–8, with pharmacology. 15. Oliver, 1960: 61. 16. Patel & El Kheir 55, K. 17. Thomas, A. S. 36, K. 18. Verger, 1967: no. 102. 19. Vermeer 20, 50, UCI. 20. Walker & Sillans, 1961: 151. 21. Watt, J. M., 1967. 22. Watt & Breyer-Brandwijk, 1962: 385. 23. Willaman & Li, 1970.

**Dioscorea esculenta** (Lour.) Burkill

FWTA, ed. 2, 3: 152. UPWTA, ed. 1, 493.

English: lesser yam; asiatic yam; lesser asiatic yam; 'Chinese yam' (general and Departments of Agriculture, in error); Hausa potato (Ghana, Irvine).

French: igname des blancs (Busson).

West African: **IVORY COAST** ABE brofié mbu (FB) AKYE brofué shie (FB) BAULE brofue duo (FB) **GHANA** ADANGME-KROBO biɛfo hiɛ (FRI) AKAN-TWI ɔde-duanan, ɔdebɔ duanan *from* ɔde: *white yam;* duanan: *forty* (FRI) GA ant-wakɔle = *does not cross the Korle lagoon, i.e., is limited to Accra* (FRI) GBE-VHE anago-tɛ = *Lagos yam* (FRI) yevu-tɛ (FRI) **TOGO** GBE-VHE anago té (FB) **NIGERIA** YORUBA iṣu (IFE) iṣu àlùbọ́sà (FRI; JMD) iṣu ànàmọ̀ (IFE) odunkun (IFE)

A spiny climber to 12 m high twining left-handed, with numerous shallow-rooted tubers, native of SE Indochina, and widespread in the East, recently introduced to W Africa and found under cultivation around the coast, particularly from Ivory Coast to Nigeria.

The plant requires a somewhat seasonal climate. It is a 6–10 months crop with short dormancy period. The tubers are small and are found in clusters of some 5 to 20 slightly below the soil surface. Unselected forms produce spiny roots lying above the tubers as a protection. Selection has eliminated some or all of the spininess in certain cultivars. Yield is high and the tubers are palatable and nutritious. Tubers grown in Ivory Coast have been recorded producing 83% starch and 12% protein. Many races in Asia are slightly sweet. The tubers do not store well. They are quick to sprout if left in the ground and are easily damaged in harvesting. Six months storage is said to be possible of sound roots in a dry well-ventilated store. They are not suitable for transport to distant markets, nor for turning into *fufu.* There appears to be scope for growing the plant under mechanical cultivation.

References:

1. Burkill, 1935: 818. 2. Busson, 1965: 431, with tuber analysis, 434–5. 3. Kay, 1973: 224.

**Dioscorea hirtiflora** Benth.

FWTA, ed. 2, 3: 152. UPWTA, ed. 1, 491.

French: igname étoilé (Berhaut).

West African: **SENEGAL** DIOLA kãngund (JB) **GUINEA** MANDING-MANINKA danda dion (CHOP) denda fara, denda fore (IHB) **SIERRA LEONE** KISSI lumula-kuwe-yundo (FCD) MANDING-MANDINKA boroborome (FCD) MANINKA (Koranko) mbotiu (JMD) MENDE mbole (JMD) MENDE-KPA mbote (NWT) SUSU g-beli-gbeli (NWT) TEMNE *a*-bent-*a*-bala (auctt.) *a*-bɔti (auctt.) *am*-buŋk (JMD) TEMNE (Port Loko) *an*-chak (FCD) TEMNE (Sanda) *an*-chak (FCD) **NIGERIA** YORUBA eṣeṣu wẹ́wẹ́ (RJN)

A climber to 6 m high, stems twining right-handed, tubers lobed, annual, occurring throughout the guinean forest area of the Region, and extending across Africa to Tanganyika and Zambia.

The tuber is caustic and is not normally eaten. It is however said to be grown in Sierra Leone and N Nigeria, a practice that must surely be but a form of insurance against dearth. The tuber is eatable only after detoxification (2, 3, 5, 6). In Rhodesia the tuber is considered to be a famine-food. Though scarcity is likely to come during their resting stage, food-gatherers are led to them by the dried vine haulm. Tubers can also be taken from under living vines in time of dire need (1).

In Tanganyika the mashed up tuber is applied to freshly washed areas of scabies, and the leaf-sap is taken by mouth for hard pusy abscesses (4).

References:

1. Burkill, s.d. 2. Busson, 1965: 432. 3. Dalziel, 1937. 4. Haerdi, 1964: 202. 5. Irvine, 1952, a: 26. 6. Okiy, 1960: 118.

**Dioscorea lecardii** De Wild.

FWTA, ed. 2, 3: 154, incl. *D. sagittifolia* sensu Miège p.p.
West African: **SENEGAL** BASARI kh̃áɓ i-faràs (Ferry) BEDIK ɛ-gɛ̀ɓ i-ɗyák, mɛ̀ɓ (Ferry) FULA-TUKULOR kappé (A. Chev.) MANDING-BAMBARA g-niambi (A. Chev.) **SIERRA LEONE** TEMNE *an*-won (Glanville) **GHANA** GBE-VHE yɔzi (Akpabla)

An unarmed climber to 8 m high, twining right-handed from a deeply-rooted tuber protected by spiny roots in the Sahel zone from Senegal to Upper Volta, and into Central Africa and Gabon.
The tuber is edible (1), and in Gabon is said to be cultivated (2).

References:
1. Akpabla 54, K. 2. Walker & Sillans, 1961: 153, as *D. sagittifolia.*

**Dioscorea mangenotiana** Miège

FWTA, ed. 2, 3: 153.
West African: **IVORY COAST** BAULE suiduo = *elephant's yam* (JM) **NIGERIA** YORUBA ẹgẹu ẹsụsu (Dawodu)

A vigorous annual climber to 30 m long, twining right-handed from a massive perennial vertical tuber, of the forest in Liberia, Ivory Coast and S Nigeria.
The root is long-lived, increasing year by year with extra lobes meriting from the likeness to an elephant's foot the Baule name meaning 'elephant's yam.' In time the tuber may attain as much as 60 kg in weight (1, 2).
The main part of the root is too woody to eat, but the current year's lobes are fleshy and edible. They are thus obtained only by dint of laborious digging. Nor is the tuber amenable to multiplication and replanting for the plant will grow only from the whole or most of the tuber.

References:
1. Haynes & Coursey, 1969. 2. Miège, 1958.

**Dioscorea minutiflora** Engl.

FWTA, ed. 2, 3: 153.
West African: **SENEGAL** MANDING-MANDINKA kubara (A.Chev.) **SIERRA LEONE** KONO ko (Pyne) LIMBA ndama (Pyne) LOKO n-degange (NWT) MENDE didi (auctt.) hehe-mbo (FCD) mbo (FCD) ngawu (FCD) vai-mbo (FCD) MENDE-KPA (Kpa) majaaja (FCD) UP MENDE didi (FCD) TEMNE *am*-buŋk (JMD) *an*-gbuk (FCD) *an*-gpuk (Pyne) *an*-lik (NWT) *an*-won (FCD) TEMNE (Port Loko) *ra*-kubaŋ (FCD) TEMNE (Sanda) *an*-gbuk (NWT) *a*-won-*a*-raŋk = *elephant yam; for the root* (FCD) **IVORY COAST** BAULE akponi(-n) (JM) DAN yogo (RP; A.Chev.) KRU-BETE tiri buru (A.Chev.) tiri rikpo (A.Chev.) tiri rikwé (A.Chev.) GUERE (Chiehn) tiribi (K&B) 'KRU' tikaru (K&B) **GHANA** AKAN-TWI aha-bayerɛ *from* ɛha: *forest, bush = bush yam* (FRI) GA yɛlɛ (Addo) **NIGERIA** IJO-IZON (Kolokuma) ògbóródìì (KW) IZON (Oporoma) ògbóródìgi (KW)

A perennial climber with spiny stems twining right-handed to 10 m high in forest margins throughout the Region from Senegal to W Cameroons, and extending to the Congo basin into Angola and Uganda.
The tuber, usually only the lower portion and while still young, is eaten in time of dearth, the upper part being left for regeneration. Before consumption prolonged soaking is required, not so much because of any traces of toxicity as because of the woodiness of the tissue (1–3, 5, 6). It may even be cultivated in a small way.
In Ivory Coast the leafy stems are pulped up for topical application to skin-affections (4). The Ijo of the Niger Delta squeeze the leaves in water which is then added to gin for taking as a treatment for jaundice (7).

References:

1. Busson, 1965: 432. 2. Irvine, 1948: 260. 3. Irvine, 1952, a: 6: 27. 4. Kerharo & Bouquet, 1950: 248. 5. Pyne 107, K. 6. Walker & Sillans, 1961: 153. 7. Williamson, K. 5, UCI.

## **Dioscorea odoratissima** Pax

FWTA, ed. 2, 3: 154 as *D. liebrechtsiana* De Wild. & *D. praehensilis* sensu Miège in minor part.

West African: **SIERRA LEONE** MANDING-MANDINKA dia (FCD) MENDE mbo (NWT) ngawu (FCD) TEMNE *an*-lik (NWT) *an*-won (FCD) **IVORY COAST** BAULE broduo (JM) **NIGERIA** HAUSA doóyàr daájiì (Lawton) IGBO jí-abai (Lawton) YORUBA iṣu awún (EWF)

A climber to 8 m high, twining right-handed from a surface-rooting tuber protected by spiny roots, very similar to *D. praehensilis* Benth., occurring in Sierra Leone to S Nigeria, and in E Africa.

The tuber tends to be elongated. The flesh is white and edible (1, 2). The plant is sometimes cultivated.

References:

1. Busson, 1965: 432, as *D. liebrechtsiana* De Wild. 2. Deighton 1352, K.

## **Dioscorea praehensilis** Benth.

FWTA, ed. 2, 3: 153–4, *p. majore parte*. UPWTA, ed. 1, 492.

English: bush yam; forest yam; 'white yam' (in cultivation, at least in part).

French: igname de brousse.

West African: **SENEGAL** BASARI kh̃áɓ a-liyán (Ferry) BEDIK ɛ-gɛ̀b (Ferry) DIOLA yendoli (JB; K&A) FULA-PULAAR (Senegal) kapé (K&A) MANDING-BAMBARA ńâbi (JB; K&A) ńana (JB; K&A) MANDINKA ɲambi (*def.* ɲamboo) (after K&A) MANINKA ɲambi kéo *probably = 'male'* (AS) SERER hab (JB; K&A) sipa (AS) SERER-NON gap (AS) tât (AS) WOLOF dolohom (AS) ńâbal (K&A) ɲâbi (K&A) ngolgol (auctt.) nyambi ala = *tapioca of the bush* (AS) **THE GAMBIA** DIOLA (Fogny) yendilay (DF) MANDING-MANINKA ku-ɲyambi (JMD) ɲambi-mesongo = *small-leaved yam* (JMD) wulakono ɲambi (JMD) **GUINEA-BISSAU** MANDING-MANDINKA canhambô (JDES) PEPEL n'pabe, umpabe (JDES) titè (JDES) **SIERRA LEONE** KISSI lumula-kuwe (FCD) LIMBA kəyoro (NWT; JMD) LOKO n-gange (NWT) MANDING-MANINKA (Koranko) digɛgbe (NWT; JMD) kɔnkɔdige (NWT; JMD) MENDE didi *applied to several D. spp.* (JMD; FCD) fɔli (JMD) mbo (FCD; JMD) mbui, mbole , mbo, ngeya *the tuber* (FCD) mbo-gbama (FCD) mbo-mboto (FCD) ngawu (NWT; JMD) MENDE-UP MENDE majaaja (FCD) SUSU xabi (NWT) SUSU-DYALONKE bude mana (Haswell) TEMNE *ma*-gbuk *the tuber* (FCD; JMD) *ra*-gbuk *the vine* (FCD; JMD) *ra*-kubang *the vine* (FCD; JMD) *an*-lekaŋ (FCD) *an*-lik (JMD) *an*-lilik (NWT) *an*-won (JMD; FCD) TEMNE (Port Loko) *an*-chak (FCD) TEMNE (Sanda) *an*-chak (FCD) **MALI** MANDING-BAMBARA fasaka (JMD) MANINKA dian-fasaka (CHOP) nyambi (CHOP) **IVORY COAST** BAULE sopéré *a c. var.* (A. Chev.) KRU-GUERE simmien (K&B) GUERE (Chiehn) semen (K&B) **GHANA** ADANGME-KROBO akwalikwa (JMD) bale (JMD) kati *the commonest form of forest yam* (JMD) AKAN-ASANTE eurum bayiera (Moor) FANTE batafo-nto = *traveller does not collect* (JMD) TWI bayerɛ *general for spiny-rooted edible wild yams* (JMD) ɛwuram bayerɛ, ɛwuram: *bush;* bayerɛ: *yam,* i.e., *wild* (*bush*) *yam* mensã *a man's name, with c. vars.* (FRI) ose kye (Glover) DAGBANI kyenkyetoe *a c.var.* (JMD) GA bayerɛ *general for spiny-rooted edible wild yams* (JMD) GBE-VHE e-hlo *from* hlo: *a thorn; general for forest yams* (JMD; Kyei) hlo-matofui, hlo-tofui *vars. of* hlo (JMD) kɔkɔli-makɔe (JMD) **DAHOMEY** BATONNUN susu *a c.var.* (JMD) **NIGERIA** BEROM edɔ̀ŋ (LB) EDO ògígbàn (Kennedy) ùghó *general for bush yam* (JMD) EFIK biâ íkọ̀t = *yam of the bush* (JMD; KW) HAUSA dóóya (LB) doóyàr daájiì *a wild form* (JMD; ZOG) doóyàr duútseè *a wild form* (auctt.) doóyàr giíwaá = *elephant's yam* (JMD; ZOG) káyàr-giíwaá = *elephant's thorn* (JMD; ZOG) maágùraázaà *a wild form* (JMD; ZOG) IGBO jí àbị (BNO) IGBO (Agukwu) mbọ-utulu = *duiker's yam* (JMD) TIV nwángē *general for the sp., with vars.* (JMD) YORUBA akọ iṣu = *male yam* (JMD) ẹ̀gún èsúrú (JMD) iṣu odẹ = *hunter's yam, an edible wild form* (JMD)

*NOTE: Some names are loosely applied between this species, D. rotundata, D. cayenensis, D. minutiflora, D. togoensis, etc.*

A sturdy climber to 20 m high twining right-handed from a massive tuber protruding above ground and armed with curving protective roots, occurring from Sierra Leone to N and S Nigeria

The vine is very vigorous. In Sierra Leone it is said to ascend and smother the canopy of trees nearly 20 m high (2). The tuber flesh is bitter. It is used as a famine-food after careful preparation (2, 6). The tuber has a number of downward growing processes. It is only these that are eaten in Sierra Leone and at certain stages of their growth (4). In Gabon the tuber is eaten only when young and after long cooking (7). In Ivory Coast it is considered fetish and is eaten by women wishing to have children (5); cf. the Yoruba incantation invoking *D. rotundata*. In Congo the boiled root is taken to accelerate childbirth, and sometimes to relieve rheumatism (1).

The young growing vine shoots are eaten in Congo like asparagus, and the older stem is cut up and used to make an infusion which is taken for stomach complaints, urethral discharge and oedemas (1).

The wild form has been to some extent ennobled. In some cultivars the protective thorny armature of roots has been reduced to slender short spiny roots (3).

References:

1. Bouquet, 1969: 106. 2. Burkill, s.d. 3. Dalziel, 1937. 4. Deighton 2547, 2872, K. 5. Kerharo & Bouquet, 1950: 248. 6. Okiy, 1960: 118. 7. Walker & Sillans, 1961: 153.

**Dioscorea preussii** Pax

FWTA, ed. 2, 3: 152. UPWTA, ed. 1, 492.

West African: **SENEGAL** DIOLA kankurẽg (JB) SERER puloh a kob (JB) **THE GAMBIA** DIOLA (Fogny) kangulay = *daily activity* (DF) MANDING-MANDINKA wulakonoɲambi (*def.* w. ɲamboo) = *bush cassava* (DF) **GUINEA-BISSAU** BALANTA n-paba (JDES) n-pabe (EPdS) umpaba (JDES) MANDING-MANDINKA dandam(-ô) (JDES) malá (JDES) MANDYAK bombôpale (JDES; EPdS) MANKANYA etóco-n'sanha (JDES) etóco-unsanha (JDES) PEPEL etêtê (JDES) etoe (JDES; EPdS) **GUINEA** MANDING-MANINKA dena fare (CHOP) **SIERRA LEONE** MENDE foli (JMD) puli (NWT) TEMNE *am*-boŋk (FCD) **NIGERIA** EDO ígìorùaèsì, ígìorùèsì (Amayo) IGBO ainyelo (auctt.)

A robust climber to 30 m high, stems twining left-handed, from deeply-buried tubers, in the guinean forest zone from Senegal to W Cameroons and extending to Central Africa.

The tubers are very caustic and are eaten only as a famine-food after suitable prolonged soaking and washing (2–5). Treatment is said to require 15 days (1). In some districts it may be eaten by hunters while out on the chase.

References:

1. Busson, 1965: 432–3, with tuber analysis, 434–5. 2. Dalziel, 1937. 3. Irvine, 1948: 260. 4. Irvine, 1952, a: 27. 5. Okiy, 1960: 118.

**Dioscorea quartiniana** A. Rich.

FWTA, ed. 2, 3: 151.

West African: **SENEGAL** BASARI a-tabe tyelóngẹ́t (Ferry) BEDIK nyi-ɗyín (Ferry) **GHANA** AKAN-TWI ŋkamfo-bɔŋwoma (Akpabla)

A climber to about 6 m high, non-spiny stems twining left-handed from annual tubers, not common from Senegal to N Nigeria in hills above 300 m altitude, and extending across Africa south of the Sahara and common in numerous varieties from Ethiopia to S Africa.

The tuber is edible only after detoxification treatment, and then it is eaten as a famine-food (1).

Reference:

1. Irvine, 1948: 260.

**Dioscorea rotundata** Poir.

FWTA, ed. 2, 3: 153, as *D. cayenensis* Lam., ssp. *rotundata* (Poir.) Miège. UPWTA, ed. 1, 492.

English: white Guinea yam; 'half-a-yam' (West Indies from *afa,* French Antilles ex *afasew,* Akan-Twi, *D. alata* Linn.); connie yam (West Indies, from *nkam* Ghana).

French: igname; igname blanc; afa (French Antilles, from *afasew,* Akan, Ghana, Burkill).

West African: **SENEGAL** FULA-PULAAR (Senegal) kapé (IHB) **GUINEA** MANDING-MANINKA dian fasaka (CHOP) ku-gue (CHOP) uraka (CHOP) SUSU khabi **SIERRA LEONE** MENDE mbole-gowe = *white yam* (FCD) TEMNE *an*-won (JMD) *a*-yams-*a*-fɛra (FCD) **IVORY COAST** BAULE kiri kiri, kuana, sopéré *good c. vars.* (A. Chev.) **GHANA** ADANGME yɛ (JMD) ADANGME-KROBO ade, akim bali, akwakor, akwakor pechiwa, akwakwaleekwa (A.S.Thomas) atief *a forest yam* (A.S.Thomas; IHB) deeka *a corruption of the name of the man who introduced the c.var.* (A.S.Thomas) hiɛ (JMD) jeow, kate (A.S.Thomas) kentem (A.S.Thomas) kumiyo *from the name of the man who introduced the c.var.* (A.S.Thomas) mãade (JMD) mali, obobi (A.S.Thomas) odjam *the name of the man who introduced the c.var.* (A.S.Thomas) odornor = *sorrowful, the sweetest forest yam - sorrow if it is unavailable* (A.S.Thomas) oho hiɛ *with round tubers* (JMD) okoryeownye *name of the woman who introduced the c.var.* (A.S.Thomas) osiecutchie *name of the man who introduced the c.var.* (A.S.Thomas) osuban *the commonest form of c.var.* hiẽ (A.S.Thomas; JMD) AKAN-AKYEM bayerɛ *incl. other D. spp., not the yellow nor winged yams* (JMD) ɛ-dwo (JMD) ɔdeɛ (JMD) ASANTE bayerɛ *a general term for yams incl. other D. spp., not the yellow nor winged yams* (JMD) FANTE ayebir, pona *c.vars.* (JMD) TWI ade *a c. var.* (A. S. Thomas) adipa *a corruption of the name of the man who introduced the c. var.* (A. S. Thomas) akimbayeray, akuku, akwakor *c. vars.* (A. S. Thomas) aso bayerɛ (Patterson) ayebir *c.var.* (Enti) bayerɛ *a general term for wild yams* jeow, kokolimakoe, krukupã *c. vars.* (A. S. Thomas) kumiyow *from the name of the man who introduced the c. var.* (A. S. Thomas) mãade *a form of* ɔde: *strong-smelling* (A. S. Thomas) nana-ntɔ *a c. var., = too good to sell to stranger* (A. S. Thomas) obobi *from the name of the man who introduced the c. var.* (A. S. Thomas) odannã *a c. var.* (A. S. Thomas) ode (Enti) ode bayerɛ = yam and forest yam, i.e., a c. var. between domestic and wild forms (A. S. Thomas) ɔde kwasea kwasea: *a fool = foolish yam; very large tuber, rapid growth, soft flesh* (A. S. Thomas) ɔde pa = *proper yam* (A. S. Thomas) osibã (Patterson) pona *c. var.* (Enti) tonto (Glover) DAGBANI bamegu (JMD) dakpan (JMD) gungunsale (JMD) gunuukple *with round tubers* (JMD) kpasajɔ (JMD) kpeney (JMD) larbakɔ (JMD) larebakɔ (JMD) lili(a) (JMD) perenga, kprenga *from gbelenga, Konkomba* (JMD) sagalanga (JMD) sanaraje (JMD) ziglanbo (JMD) zon (JMD) zugulangbɔ (JMD) zulanbo (JMD) GA dãna (JMD) male (JMD) tonto (JMD) yɛlɛ (JMD) GBE-VHE ade, akuku (A.S.Thomas) dreeka *corruption of the name of the man who introduced the c.var.* (A.S.Thomas) dze-tɛ, ge-tɛ (JMD) hlo = *thorn; referring to the thorns at the top of the tuber; a general term for a forest yam* (A.S.Thomas; IHB) klewu, kokolimakoe (A.S.Thomas) nkam (IHB) sonka (JMD) sunka *a c.var. with one large tuber and several small ones* (JMD) tɛ, tɛ-dze (JMD) te-gba = *real yam* (A.S.Thomas; JMD) te-la = *foolish yam; it grows too big, too quickly, becoming soft* (A.S.Thomas; K; JMD) VHE (Pecí) ade (Kyei) djafuto (Kyei) dzɔgɔli (JMD) klewu (Kyei) kpasa (Kyei) tɛkpuri = *short and stout yam* (JMD) GUANG-GONJA kújɔ́ (Rytz) KONKOMBA gbelenga (JMD) **TOGO** BASSARI lili (JMD) **NIGERIA** ANAANG èkà èdiá = *mother of yams* (KW) èkárá édiã (KW) BEROM kit ryáŋ *perhaps this sp.* (LB) EDO èmọ̀wè, ígièrù, ómí (DA) ómí funfun *from Yoruba* (DA) òrí (DA) EFIK afia oko (Johnson) àfia òkò, òkò: *white* (auctt.) èkà biâ = *mother of yams* (JMD) ekefià *the commonest c.var.* (JMD) èkò (JMD; RFGA) n̄diahà *I don't eat* (*it*) (Winston) oko (Johnson) okpo (Johnson) ọ̀kpọ́ (JMD; RFGA) okpo uman (Johnson) ọ̀kpọ́ úmàn íwã (JMD; KW) okpuru *a c. var. grown with D. cayenensis for early cropping* (IHB) okpuru (Johnson) otuk okpo (Johnson) EPIE àdị̀à (*pl.* ị̀-) (KW) HAUSA dóòya *general for yam* (JMD) IBIBIO èkò (JMD; Kaufman) ǹdiágà *a c.var planted late, giving highest yield* (JMD; Kaufman) ọ̀kpọ́ *a small white yam, keeping well for up to two years* (JMD; Kaufman) IGBO àbí, jí àbí (NWT) àgà, jí àgà (JMD) ala (NWT) alafolo, jí alafolo alẹ̀fulu (JMD) asokolo, awudo, jí awudo (NWT) asuku, ẹkpẹ, (JMD) jí àyọ̀bè (NWT) jí íkē = *strong yam* jí ọ́chá = *white yam* jí òkò ji-akero (accro) (Lawton; JMD) ólũ (NWT; JMD) omi (JMD) IJO bùrú (KW) ISOKO ọ̀lẹ́ (KW) TIV nwángē (JMD) URHOBO àbẹ́ (JMD) okpuru (JMD) ọ̀lẹ́ *general, not specific* (JMD) olumuda (JMD) ómí (JMD) ùjèrú (Faraclas) ukpokoro (JMD) urevbose (JMD) YORUBA ààlàoko (Johnson; JMD) àbàjẹ (JMD) aga (Johnson) agake (Johnson; JMD) agẹmọkun, aginni, agogo, aladẹ (JMD) alo (Johnson) àlọ̀lọ, ayin, bọki (JMD) apepe (DA) awure (Johnson) bùnbun = *lazy* (Verger) dodoro (JMD) efùrù = *dry* (auctt.) ehuru (JMD) ẹlẹyintu (JMD; DA) esinméérin, fodu (JMD) gàmù gámù = *bad conduct* (Verger) gbẹ̀hinrà (JMD) gùdùgbú = *short* (Verger) igun (Johnson; JMD) ihobia (Johnson) ilọlọ (JMD) iṣu (JMD; IFE) iṣu funfun, iyawo, iyawo alaaji, iyawo ọlọ́run, janyin-janyin, jọyin-jọyin (JMD) kangi = *strong* (Verger; DA) kangì-ojúnlaja, kọ̀tisàn, kukundu, lásanrin (JMD) layinbo (Johnson; JMD) ọdọ (JMD; DA) olóféèré = *hazy* (Verger; JMD) olonko (Johnson; JMD) olotùn, olotùn iyaagbà (JMD) omi funfun (DA) omifun (Lawton) oparaga, oputu (JMD) ṣaja, sofini, wawajı (JMD)

A climber with stems often prickly, twining right-handed, tuber surface-rooting, of numerous cultivars grown throughout the Region and in many tropical countries. See also notes under *D. cayenensis,* para. 1.

This species is the White Guinea yam. It is a cultigen developed entirely by the African himself. Material was carried by slaves to the Western Hemisphere and the plant has become a staple on both sides of the Atlantic. In W Africa it is of greater importance than is the yellow Guinea yam and it has been subject to more selection. It is of increasing importance in the Region from west to east reaching its greatest in Nigeria where large storage racks are built for it in populous villages to store it after harvest. It has a growing cycle of 6–8 months followed by a marked dormancy period. It is tolerant of drier conditions than the yellow Guinea yam. Its cultivation is therefore of a far greater penetration inland into the drier Sahel zone. The tuber is very palatable and it makes good *fufu.* It is normally found close to the surface of the soil, is about 30 cm or more long and sometimes may reach a great weight, as much as 26 kg being recorded in Nigeria (1, 3). In Ghana the production of large yams is a part of the religious cult of the Vhe people's New Yam Festival. These are preserved at the abode of the yam spirit from one season to the next (2). The plant is invoked by the Yoruba in a number of incantations: to make someone generous, to save people from witches, to give bad medicine to an enemy, to get a child, to arouse dead people, and to make a policeman forget about a case (4).

References:

1. Burkill, s.d. 2. Haynes & Coursey, 1969. 3. Kay, 1973: 233. 4. Verger, 1967: 62, 66, 99, 101, 129, 147.

## **Dioscorea sansibarensis** Pax

FWTA, ed. 2, 3: 152. UPWTA, ed. 1, 491, as *D. macroura* Harms.

West African: **DAHOMEY** GBE-FON gudu-gudu (A. Chev.) YORUBA (Idacha) gué-gué (A. Chev.)

A vigorous climber to 6 m high, stems non-spiny, square in cross-section, twining left-handed from a massive perennial tuber and producing bulbils along the stem; an East African species that is found at a few points from Ivory Coast to Nigeria, and dispersed across Central Africa.

The leaves contain glands filled with a fluid holding nitrifying bacteria (9). Leaf-sap and root-decoction are taken by draught for epilepsy in Tanganyika (4).

The tuber is sometimes of great weight. As much as 30 kg has been recorded (1, 5). The wild form is toxic but can be eaten if the poison is removed, usually by cutting up, boiling and washing. In Madagascar the tuber has been used as an ordeal-poison, whereby is a Malagassy name *haranara,* meaning 'search' (2). The plant is often grown on the edge of yam-fields in the hope that if there be thieves their roots would be stolen instead of the valuable yams. It is also sometimes planted in Dahomey around sorghum granaries in, apparently, an act of superstition. The tuber, and the bulbil, are used for poisoning wild animals, pigs and other destructive kinds (2), and as a fish-poison (10). Though the tuber is primarily considered a famine-food, it is spasmodically grown in many parts. There is evidence of its wider cultivation in olden times (6, 8). Some selection to reduce toxicity has been practised to the extent at least that in some races toxicity is confined to the upper part of the tuber (3).

Bulbils are produced at the leaf-axils, dark purplish brown and usually larger than those of *D. bulbifera.* The bulbils are adapted for water dispersal. At first they sink in water, then float and on being left by receding water will germinate. They are toxic and inedible.

The tuber contains *dioscorine, dihydrodioscorine* and other unnamed alkaloids (10, 11).

References:

1. Burkill, 1960. 2. Burkill, s.d. 3. Busson, 1965: 433. 4. Haerdi, 1964: 203. 5. Haynes & Coursey, 1969. 6. Irvine, 1948: 260. 7. Irvine, 1952, a: 25. 8. Irvine, 1952, b. 9. Orr, 1923. 10. Watt & Breyer-Brandwijk, 1962: 387. 11. Willaman & Li, 1970.

## **Dioscorea schimperana** Hochst.

FWTA, ed. 2, 3: 152.

A climber to about 6 m high, vines twining right-handed from a tuber and bearing axillary bulbils, of open upland savanna country and recorded only from N Nigeria and W Cameroons in the Region, and occurring over central and eastern Africa. In Malawi it appears to favour termite mounds and to sprawl over rocks on river-margins.

The bulbils are abundant and are eaten in eastern Africa in time of scarcity; then too the roots are sought but it is laborious to obtain them and the return is small (1).

Reference:

1. Burkill, s.d.

## **Dioscorea smilacifolia** De Wild.

FWTA, ed. 2, 3: 153. UPWTA, ed. 1, 493.

West African: **SIERRA LEONE** LOKO n-dange (NWT) n-degange (JMD) TEMNE *am*-bunk (JMD) *an*-gbuk (FCD; Glanville) *an*-lik (NWT; JMD) **IVORY COAST** BAULE akponi(-n) (JM) **GHANA** AKAN-TWI kokora (Plumtre) **NIGERIA** IGBO (Izi) ikwolo-ji-oku (NWT; JMD) YORUBA ẹwo (Millen; JMD)

A perennial climber with wiry, often spiny stems to about 4 m high, twining right-handed and sometimes bearing axillary bulbils, arising from a ground-tuber, of the forest zone from Sierra Leone to W Cameroons and Fernando Po, and extending over the Congo basin.

The species is sometimes grown in a small way, but the root is only eaten in time of scarcity, and then after careful preparation (3–5, 7, 9, 10). Even then a bitterness remains after cooking (11). Some alkaloid has been detected in Nigerian material (1).

In Ghana the leaves are rubbed onto cuts (8), and in Ivory Coast three leaves are rubbed on the temples in three applications to cure epilepsy (6). In Congo leaves passed through a fire are pulped to a paste for rubbing on the head to relieve headache, and a decoction of the stem is drunk for vertigo and hernia (2).

References:

1. Adegoke et al., 1968. 2. Bouquet, 1969: 107. 3. Dalziel, 1937. 4. Irvine, 1948: 260. 5. Irvine, 1952, a: 25. 6. Kerharo & Bouquet, 1950: 248. 7. Okiy, 1960: 118. 8. Plumtre 87, K. 9. Thomas, N. W. 2081 (Nig. Ser.), K. 10. Thomas, N. W. 2343 (Nig. Ser.), K. 11. Walker & Sillans, 1961: 153.

**Dioscorea togoensis** Knuth

FWTA, ed. 2, 3: 153.

West African: **THE GAMBIA** MANDING-MANDINKA kuuɲambi (*def.* kuuɲamboo) = *cassava* (Dawe; DF) ɲambi mesongo = *small-leaved yam* (Dawe) wulakonoɲambi (*def.* w. ɲamboo) *wild yam* (Dawe) **SIERRA LEONE** LIMBA kəyoro (NWT) LOKO n-degange (NWT) MANDING-MANINKA (Koranko) digɛgbe, digɛgbwe (NWT) kainya (NWT) yigɔile (NWT) MENDE didi (NWT) fɔli (NWT) mbole (NWT) ngawu (NWT) TEMNE *a*-tak, *an*-buŋk (NWT) *a*-tak-*a*-runi (NWT)

An annual rather weak climber with stems twining right-handed to about 2 m high, bearing axillary bulbils, from a slender deep-rooted tuber, of savanna and dry guinean forest from Senegal to N Nigeria.

It is a wild species but colonizes well-cultivated land reverting to scrub (1). In Ghana the small bulbils are eaten in the Kita district in time of shortage (2).

References:

1. Burkill, s.d. 2. Irvine 2554, K.

**Dioscorea** spp. indet.

West African: **SENEGAL** MANDING-BAMBARA gniambi alla *wild sp. of Soudanian region* (anon.) **GUINEA-BISSAU** BALANTA bgá (JDES) FULA-PULAAR (Guinea-Bissau) cápè-caladje (JDES) MANDYAK barafe (JDES) **GUINEA** SUSU tabe-khabi *a yellow-fleshed c.var.* (CHOP) **SIERRA LEONE** MENDE kanya-bo *a wild yam, sp. indet.* (FCD) **LIBERIA** MANO yẽ, yeng *wild edible yams* (Har.) **MALI** MANDING-BAMBARA boru *c. var., black skin* (Dumas; JMD) guara lo *c. var., flat tuber* (Dumas; JMD) guara simbessi = *elephant's foot; c. var., tuber flat sinuate margin* (Dumas; JMD) saruko *c. var., white skin* (Dumas; JMD) uokuni *c. var., tuber lobed, palmate* (Dumas; JMD) **IVORY COAST** ABURE oafé (B&D) **GHANA** ADANGME-KROBO akwakwatikwa *probably a c.var. of* ɔde (FRI) AKAN-AKYEM nana-ntɔ, nana: *grandparent;* ntɔ, *does not buy, i.e., a very good yam, therefore not sold to aged people* (Enti) ɔdeɛ (FRI) ɔde-kwasea, kwasea: *a fool; 'foolish yam', very large tubers, rapid growth resulting in soft flesh; perhaps the water-yam* (FRI) ASANTE asɔbayerɛ *a form of bayerɛ* (FRI) *a*-tasiɛ *a yam with a very large tuber of rapid growth resulting in soft flesh; perhaps a water-yam* (Enti) TWI akwakɔ *a good form of* ɔ*de* (FRI) akwakɔ-mpakyiwa *a form of* ɔ*de: akwak*ɔ*, a kind of* ɔde; mpakyiwa = apakyiwa, *a small calabash, referring to the round tubers* (FRI) amanyãkun *a form of* ɔde, *large tubers* (FRI) *a*-tasiɛ *a yam with a very large tuber of rapid growth resulting in soft flesh; perhaps a water-yam* (Enti) bayerɛ biri *a dark-coloured form of 'bayerɛ'* (FRI) dika *a form of* ɔ*de; large tuber* (FRI) dokowa (*pl.* nnokowa) dɔkɔdɔkɔ: *sweet; a sweet form of* ɔ*de* (FRI) krukrupa *a form of* ɔ*de, tubers white, round, good food value* (FRI) nkuku (FRI) ntonto *a form of bayerɛ* (FRI) odanna(-n) *a form of* ɔ*de; pure white flesh* (FRI) onyame bayerɛ = *God's bayerɛ* (FRI) opokuase bayerɛ *a form of bayerɛ originally grown at Opokuasa: cf.* opotuase *Akan* (*Asante*)*: fence inside which yams are stored* (FRI) osampam *a form of mɛnsã* (FRI) osu *a form of* ɔ*de* (FRI) GA ŋkani (FRI) GBE-VHE kɔkɔli-makɔe *probably a form of* ɔ*de* (FRI) **NIGERIA** ANAANG akpana (DA) ebirẹ (DA) efiang (DA) eko (DA) enem (DA) ndiaha (DA) nkorito (DA) nkpuk (DA) BEROM edɔ̀ŋ cwà, cwà: *an unidentified tree up which the yam is grown; a yam within the dóóya* (*Hausa*) *group* (LB) kit *a yam within the dóóya* (*Hausa*) *group* (LB) kit tàl = *curved yam; within the dóóya* (*Hausa*) *group* (LB) EFIK èmìnè àbìà *yam buds and shoots* (Lowe) ényìn-énìn = *elephant's eye; small yam* (JMD) ǹkọ̀rọ̀tọ̀ (JMD; RFGA) HAUSA aduru *c.var.* (JMD) amaru *c.vars.* (JMD) anige *c.vars.* (JMD) arukaruka *c.var.* (JMD) ashake-shake *c.vars.* (JMD) ba-gwandara *c.vars.* (JMD) baƙin ganye *c.vars.* HAUSA (West) dúndúu *c.vars.* fárín ganyee *c.vars.* gumji *c.vars.* hákoórín-giíwaá *c.vars.* inoga *c.vars.* kwato *c.vars.* mà-ámfá *c.vars.* tuùlún-moowaà *c.vars.* IBIBIO àfiá énēm = *white sweet yam* (DA; KW) àfiá údiā = *white yam* (DA; KW) ánêm, énêm *all sweet yams* (KW) èkà údiā = *mother of yams; the small tubers used for replanting* (KW; JMD) òbúbít énêm = *black sweet yam* (KW) IGBO abanane (JMD) abanẹke (JMD) àdàkā (JMD) ádụ̀ *a yam bearing seeds* (JMD) akbalaji (JMD) ákụ̀rụ̀ a white, long yam (KW) alitu *a waterside yam* (JMD) alọji (JMD) NUPE eci-kokana, kana: *a small monkey; a c.var.* (JMD) URHOBO ógwà *digitate-leaved* (Faraclas) òkpéyìn *digitate-leaved; yellow yam harvested in the dry season* (Faraclas) ọ̀lẹ́ éjà = *yam of woman digitate-leaved, of waterlogged areas, thornless, harvested in wet season* (Faraclas) **WEST CAMEROONS** DUALA lô *?an aerial yam* (Ithmann) mùkɔndɛ *a large red-tubered c. var* (Ithmann)

The foregoing vernaculars are recorded for a miscellany of non-specified dioscoreas.

# DIPSACACEAE

**Succisa trichotocephala** Baksay

FWTA, ed. 2, 2: 223. UPWTA, ed. 1. 414, as *Scabiosa succisa* Linn.
West African: **WEST CAMEROONS** KPE beyembe (JMD)

A sparingly-branched herb, to 1.25 m high, of montane grassland in W Cameroons.

Unnamed alkaloid has been reported in the aerial portion of the plant (1)

Reference;
1. Williaman & Li, 1970.

# DIPTEROCARPACEAE

**Monotes kerstingii** Gilg

FWTA, ed. 2, 1: 235. UPWTA, ed. 1, 68.
West African: **SENEGAL** MANDING-MANINKA béré béré (Aub.) gbrégbré (Aub.) kukuru (Aub.) **GUINEA** MANDING-MANINKA béré béré (Aub.) gbrégbré (Aub.) kukuru (Aub.) **MALI** MANDING-BAMBARA béré béré (Aub.) gbrégbré (Aub.) guéré-guéré (Aub.) kuru ru (Aub.) MANINKA béré béré (Aub.) gbrégbré (Aub.) kukuru (Aub.) **UPPER VOLTA** 'SENUFO' gandama (Aub.) kadongnuon (Aub.) **IVORY COAST** 'SENUFO' gandama (Aub.) kadongnuon (Aub.) **GHANA** KONKOMBA bilengbile (Gaisser) **TOGO** BASSARI birangbiram (Gaisser) TEM (Tshaudjo) kesang (Volkens) **NIGERIA** HAUSA farar rura, rura: *Parinari curatellifolia Planch.* (Lely; ZOG) gasa kura (Lely; ZOG) hantso (KO&S) wasani (JMD; ZOG)

A shrub or small tree to 16 m high, of savanna woodland and locally abundant throughout the Guinean zone from Mali to N Nigeria, and on across Africa to Sudan.

The plant has been confused with the shrub, *Parinari curatellifolia* Planch. (Chrysobalanaceae), hence one of the Hausa names.

Sap-wood is pinkish-brown, heart-wood darker, close-grained with reflections, hard and heavy, and difficult to work (1, 2, 4, 5). The species is the far-western outflier of the *Dipterocarpaceae,* a large family centred on SE Asia and an extremely important source of timber. The wood of *M. kerstingii* appears to be of use only as firewood (3).

The bark is boiled and the decoction is taken for dysentery (Aubréville in 4).

References:
1. Bancroft 1934: with timber characters. 2. Dalziel 1937. 3. Holland 1920: 139 4. Irvine 1961: 90. 5. Keay & al. 1960: 128.

# DROSERACEAE

**Drosera indica** Linn.

FWTA, ed. 2, 1: 122.
English: sun-dew, dew plant
French: piège à papillon (*butterfly trap,* Berhaut).

A caulescent annual to 50 cm long, fairly widespread in West Africa from Senegal to N Nigeria in boggy sites, and is recorded from tropical Africa, Asia and Australia. It is an indicator of acid soil.

All *Drosera spp.* are bitter and caustic to taste. Cattle do not graze them and in some hydrocyanic acid has been detected (1). *D. indica* is commonly used in Indian medicine as a powerful rubefacient (1, 5) and a macerate is applied topically on corns and warts in Indochina (2, 3).

As one would expect of an insectivorous plant, proteolytic enzymes have been found in leaf-extract (6).

References:

1. Burkill, 1935: 861. 2. Hartwell, 1967: 106. 3. Sastri, 1952: 218. 4. Usher, 1974: 218. 5. Uphof, 1968: 186. 6. Quisumbing, 1951: 349, with references.

## **Drosera madagascariensis** DC.

FWTA, ed. 2, 1: 121.

A caulescent perennial to 25 cm long of boggy places in grassland and open places of higher elevations, recorded only in Guinea, N Nigeria and W Cameroons, but occurring widespread in tropical and southern Africa and Madagascar.

In Madagascar an infusion of the whole plant is taken for urine incontinence, and as a diuretic and is used for conjunctivitis. An absence of alkaloid and an abundance of saponins is reported (1). It is also used to preserve teeth and in a remedy for dyspepsia and cough (2).

References:

1. Debray & al., 1971: 68. 2. Uphof, 1968: 186, as *D. ramentacea* Burch.

## REFERENCES

ABDOUL OUMAR, FALL, 1962: Traitement des morsures de serpents avec des plantes du Djolof (Sénégal), in J. G. ADAM: Les plantes utiles en Afrique occidentale, *Notes Afr.* 93: 13–14.

ADAM, J. G., 1960: Quelques plantes adventrices des rizières de Richard-Toll, *Bull. Inst. Franç. Afr. Noire,* A, 22: 361–84.

— 1961: Adaptation de trois plantes cultivées des pays tempérés à Dakar (Tétragone, Capucine, Céleri), *Notes Afr.* 89: 13–14.

— 1962: Le Baobab (Adansonia digitata L.) *Notes Afr.* 94: 33–44.

— 1963,a: Le plus gros Baobab du Sénégal n'est plus celui de Dakar, *Notes Afr.* 98: 50–53.

— 1963,b: Les forêts de Symmeria-Hunteria des berges de la Gambie-Koulountou (Sénégal sud-oriental), *Bull. Inst. Franç. Afr. Noire,* A, 25: 24–37.

— 1966,a: Les pâturages naturels et postculturaux de Sénégal, *Bull. Inst. Franç. Afr. Noire,* A, 28: 450–537.

— 1966,b: Composition chimique de quelques herbes mauritanniennes pour dromadaires, *J. Agr. trop. Bot appl.* 13: 337–42.

— 1971: Quelques utilisations de plantes par les Manon du Libéria (Monts Nimba), *J. Agric. trop Bot. appl.* 18: 372–8.

ADAMS, C. D., 1956,a: New records of flowering plants in West Africa, II, Compositae, *J. W. Afr. Sci. Ass.* 2: 61–66.

— 1956,b: Commiphora dalzielii Hutch., *Kew Bull.* 11: 541–4.

— 1957: New records of flowering plants in West Africa III, Compositae, *op. cit.* 3: 111–22.

— 1960: New records of flowering plants in West Africa, IV, Compositae, *op. cit.* 6: 149–55.

— 1964,a: New records of flowering plants in West Africa, V, Compositae, *op. cit.* 8: 127–33.

— 1964,b: New records of flowering plants in West Africa, VI, Compositae, *op. cit.* 8: 134–40.

ADAMS, R. F. G., 1943: Efik vocabulary of living things, *Nigerian Field* 11.

— 1947: II. *op. cit.,* 12: 23–24.

ADANDÉ, A., 1953: Le maïs et ses usages dans le Bas-Dahomey, *Bull. Inst. Franç. Noire,* 15: 220–82.

ADEGOKE, E. A., A. AKISANYA & S. H. Z. NAQVI, 1968: Studies of Nigerian medicinal plants: I. A preliminary survey of plant alkaloids, *J. W. Afr. Sci. Ass.* 13: 13–33.

ADJANOHOUN, E. & L. AKÉ ASSI, 1972[?]: *Plantes pharmaceutiques de Côte d'Ivoire;* Abidjan, Ivory Coast (mimeographed).

AGURELL, S., 1969: Cactaceae alkaloids, I: *Lloydia* 32: 206–16.

AINSLIE, J. R., 1937: A list of plants used in native medicine in Nigeria, *Imp. Forest. Inst. Oxford,* Inst. Paper 7 (mimeographed).

AIRY SHAW, H. K., 1952: On the Dioncophyllaceae, a remarkable new family of flowering plants, *Kew Bull.* 6: 327–47.

— 1972: The Euphorbiaceae of Siam, *Kew Bull.* 26: 191–363.

— 1975: The Euphorbiaceae of Borneo, *Kew Bull.,* Add. Ser. IV.

ALBA, P., 1956: Le développment de la foresterie en Afrique occidentale Française, *J. W. Afr. Sci. Ass.* 2: 158–71.

ALBRIGHT, J. D., J. C. van METER & L. GOLDMAN., 1965: Alkaloid studies, IV: Isolation of Cephaeline and Tubulosine from Alangium lamarckii, *Lloydia* 28: 212–7.

AMIN, M. A., A. A. DAFFALA & O. A. EL MONEIM, 1972: Preliminary report on the molluscicidal properties of habat el-mollok, Jatropha sp., *Trans. Roy. Soc. Trop. Med. & Hygiene* 66: 805.

ANON., 1904: *Bull. Imp. Inst.* 227–9.

— 1924: Investigations of paper-making materials: Costus afer from Uganda, *Bull. Imp. Inst.* 22: 424–5.

— 1930: The feeding value of Para rubber seed meal, *Bull. Imp. Inst.* 28: 459–60.

— 1938: Water lily seed from the Sudan, *Bull. Imp. Inst.* 36: 470–2.

— 1939: Cassava, *Bull. Imp. Inst.* 37: 205.

---

When the manuscript for this volume was being prepared it was intended that the volume would cover the useful plants of Families A to E. At a very late state of preparation mounting printing costs necessitated a reduction of coverage to Families A to D, and species descriptions of the E Families have been held over to appear in the next volume. It has, however, been impracticable to cull from the list of references those exclusively relevant to the E Families, so the original list stands here at the risk of some slight superfluity.

---

— 1941: Sur l'extraction des alcaloïdes de l'Holarrhena africanna et leur utilisation possible. Rapport sur le fonctionement technique de l'Institut Pasteur de l'Afrique Occidentale Française en 1940, Dakar, 1941, pp. 102–9. Abstract in *Bull. Inst. Franç. Afr Noire* 5, 1945: 210.
— 1942: Po-yok fruits from Sierra Leone, *Bull. Imp. Inst.* 40: 99–103.
— 1943,a: Castor oil from Nigeria as a lubricant, *Bull. Imp. Inst.* 41: 223–6.
— 1943,b: Rubber in Nigeria, *Bull. Imp. Inst.* 41: 232–3.
— 1944: Euphoria tirucalli resin from South Africa, *Bull. Imp. Inst.* 42: 1–13.
— 1945: *Bull. Inst. Franç. Afr. Noire,* 7: 210.
— 1946,a: Cashew Nut Shell Oil, *Bull. Imp. Inst.* 44: 17–20.
— 1946,b: Calendrier agricole pour le Sénégal, *Bull. Inst. Franç. Afr. Noire* 8: 138–63.
— 1961: Balanites as a source of diosgenin, *Trop. Sci.* 3: 132–3.
— 1965,a: Terminalia superba (Afara), *Fed. Dept. For. Res., Ibadan: For. Prod. Repts.* No. F.P.R. L/1.
— 1965,b: Nigerian Timbers for match-making, *Fed. Dep. For. Res. Ibadan, For. Prod. Res. Repts,* No. F.P.R., L/5.
— s.d.: Use-guide for Ghanian timbers, *Inf. Bull. For.* Prod. Res. Inst., Ghana, No. 1. (Forest Products Research Institute, Kumasi, Ghana.).
APPIA, B., 1940: Superstitions Guinéennes et Sénégalaises, *Bull. Inst. Franç. Afr. Noire.* 2: 358–95.
ARKELL, A. J., 1939: Throwing-sticks and Throwing-knives in Darfur, *Sudan Notes and Records* 23: 251–267.
ASHBY, M., 1941: Wartime drug supplies and Empire production, I. *Bull. Imp. Inst.* 39: 1–17.
ASPINALL, G. O. & V. P. BHAVANANDAN, 1965: Combretum leonense gum, *J. Chem. Soc.* 1965 (April): 2693–700.
AUBRÉVILLE, A., 1950: *Flore Forestière Soudano–guinéenne, A.O.F.–Cameroun–A.E.F.,* Paris Société d'Editions Geographiques Maritimes et Coloniales.
— 1959: *La Flore forestière de la Côte d'Ivoire* Ed. 2., 3 vols; Centre Technique Forestier tropicale, Nogent-sur-Marne.
— 1962: *Flora du Gabon,* Vol. 3, *Irvingiacées, Simaroubacées, Burseracées;* Muséum National d'Histoire Naturelle, Paris.
AUBRÉVILLE, A. & F. PELLEGRIN, 1938: Sapindacées et Euphorbiacées nouvelles d'Afrique Occidentale, *Bull. Soc. Bot. France* 85: 290–3.
AYENSU, E. S. & D. G. COURSEY, 1972: Guinea Yams. The Botany, Ethnobotany, Use and possible future of Yams in West Africa, *Econ. Bot.* 26: 301–18.
B[ray], G. T., 1947: Oil of Tetracarpidium conophorum, *Bull. Imp. Inst. 45:* 131–3.
BA, OUMAR, 1969: Traitement de la Lèpre chez les Toucouleurs, *Notes Afr.* 124: 126.
BAILEY, L. H., 1901: *Cyclopedia of American Horticulture,* London, Macmillan & Co., Ltd.
BALDWIN, J. T. Jr., 1946: Cytogeography of Emilia Cass. in the Americas, *Bull. Torrey Bot. Club.* 73: 18–23.
BALDWIN, J. T. Jr., & B. M. SPEESE, 1951: Cytogeography of Emilia in West Africa, *Bull. Torrey Bot. Club* 76: 346–51.
BALLE, S., 1951: Fam. 30, *Caryophyllaceae,* in R. BOUTIQUE, *Flore du Congo-Belge et du Ruanda-Urundi,* Spermatophytes 2, I.N.É.A.C., Brussels.
BALLY, P. R. O., 1937: Native Medicinal and Poisonous Plants of East Africa, *Bull. Misc. Inf.,* 1937: 10–26.
BAMPTON, S. S., 1961: Yams and Diosgenin, *Trop. Sci.* 3: 150–3.
BANCROFT, H., 1934: New material of Monotes kerstingii from the Gold Coast, *Bull. Misc. Inf.,* 1934: 233–7.
BARTH, H., 1857–8: *Travels and Discoveries in North and Central Africa, being a Journal of an expedition undertaken under the auspices of H.B.M.'s Government in the years 1849–1855,* 5 vols, Ed. 2. Longman, Brown, Green, Longmans & Roberts, London.
BATES, D. M., (Ed.), 1976: *Hortus Third. A concise Dictionary of the Plants cultivated in the United States and Canada,* Macmillan, New York.
BAYARD, Lieut., 1947: Aspects principaux et conistance des Dunes (Mauritanie), *Bull. Inst. Franç. Afr. Noire* 9: 1–17.
BEILLE, H., 1927: *Phyllanthus,* pp. 571–608, In M. H. LECOMTE: *Flore générale de L'Indo-Chine,* Masson & Cie., Paris.
BENNET, H., 1950: Alchornea cordifolia leaves and bark from Nigeria, *Colon. Pl. Anim. Prod.* 1: 132–61.
BERHAUT, J., 1967: *Flore de Sénégal,* Ed. 2, Clairafrique, Dakar.
BERTHO, J., 1951,a: Quatre dialectes Mandé du Nord-Dahomey et de la Nigeria anglaise, *Bull. Inst. Franç. Afr. Noire* 13: 126–71.
— 1951,b: La place des dialectes Géré et Wobê par rapport aux autres dialectes de la Côte d'Ivoire, *op. cit.:* 1271–80.

— 1953: La place des dialectes Dogon (dogõ) de la falaise de Bandiagara parmi les autres groupes linguistiques de la zone soudanaise, *op. cit.* 15: 405–41.

BÉZANGER-BEAUQUESNE, L., 1955: Contribution des plantes à la défence de leurs semblages, *Bull. Soc. bot. Franç.* 102: 548–75.

BISSET, N. G., 1933: The steroid glycosides of the Apocynaceae, I, *Indonesian J. for Nat. Sci.* 109: 173–211.

— 1957: Cardiac glycosides, I, Apocynaceae: the genera Carissa L., Vallaris Burm. f., Beaumontia Wall., Anodendron A. DC. and others, *Ann. Bogor.* 2: 193–210.

— 1958: The occurrence of alkaloids in the Apocynaceae, *Ann. Bogor.* 3: 105–236.

— 1961: The occurrence of alkaloids in the Apocynaceae, II. A Review of recent developments. *op. cit.* 4: 65–144.

BLAKELOCK, R. A., 1956: Notes on African Celastraceae: I, *Kew Bull.* 11: 237–47.

BOBOH, J. L., 1974: in litt., 24 April 1974.

BOUQUET, A., 1969: Féticheurs et Médecines traditionnelles du Congo (Brazzaville), *Mém. O.R.S.T.O.M.* 36.

— 1972: Plantes médicinales du Congo-Brazzaville: Uvariopsis, Pauridiantha, Diospyros, *Trav. Doc. O.R.S.T.O.M.* 13.

BOUQUET, A. & M DEBRAY, 1974: Plantes médicinales de la Côte d'Ivoire, *Trav. Doc. O.R.S.T.O.M.* 32.

BOURNONVILLE, D de., 1967: Contribution à l'étude du Chimpanzé en République de Guinée, *Bull. Inst. Fond. Afr. Noire, A,* 29: 188–1269.

BOURY, N'DIAYE JABSA, 1962: Végétaux utilisés dans la Médecine africaine, dans la région de Richard-Toll (Sénégal). In J. ADAM: Les Plantes utiles en Afrique Occidentale, *Notes Afr.* 93: 14–16.

BOUTIQUE, R., 1951: Fam. 35. *Annonaceae* in R. BOUTIQUE, *Flore du Congo-Belge et du Ruanda-Urundi,* Spermatophytes 2, I.N.É.A.C. Brussels.

BRAY, G. T. & F. MAJOR, 1945: Sunflower seed from Nigeria, *Bull. Imp. Inst.* 43: 83–86.

BRENAN, J. P. M., 1952: Notes on African Commelinaceae, *Kew Bull.* 7: 179–208.

— 1960: Notes on African Commelinaceae, II. The genus Buforrestia C.B.Cl. and a new related genus Stanfieldiella Brenan, *op. cit.* 14: 280–6.

— 1961: Notes on African Commelinaceae III, *op. cit.* 15: 207–28.

BRENAN, J. P. M. & P. J. GREENWAY, 1949: *Imp. For. Inst. Checklist 5, Tanganyika Territory,* Part II, Imperial Forest Institute, Oxford.

BRETELER, F. J., 1973: The African Dichapetalaceae, *Meded. Landbouwhogesch. Wageningen* 73–113.

BRETSCHNEIDER, E., 1895: Botanicum Sinicum, III: Botanical investigations into the materia medica of the Ancient Chinese, *J. China Br. Roy. As. Soc.* 29: 1–623.

BRITISH PHARMACEUTICAL CODEX, 1959: The Pharmaceutical Press, London.

BROUN, A. F. & R. E. MASSEY, 1929: *Flora of the Sudan,* Sudan Govt. Office, London.

BROWN, E., 1950: Oil of Helichrysum kilimanjari, *Colon. Pl. Anim. Prod.* 1: 117–8.

BRUIN, A. de, J. E. HEESTERMAN & M. R. MILLS, 1963: A preliminary examination of the fat from Pachira aquatica, *J. Sci. Fd. Agric.* 14: 758–60.

BULLOCK, A. A., 1952: Notes on African Asclepiadaceae I. *Kew Bull.* 7: 405–26.

— 1953,a: II *op. cit.* 8: 51–67.

— 1953,b: III, *op. cit.* 8: 329–62.

— 1954: IV, *op. cit.* 9: 349–73.

— 1955,a: V. *op. cit.* 9: 579–94.

— 1955,b: VII, *op. cit.* 10: 611–26.

— 1956: VIII, *op. cit.* 11: 503–22.

— 1961: IX, *op. cit.* 15: 193–206.

— 1963: X, *op. cit.* 17: 183–96.

BURKILL, I. H., 1935: *A dictionary of the Economic Products of the Malay Peninsula,* Crown Agents for the Colonies, London.

— 1938: The contact of Portuguese with African food-plants which gave words such as 'yam' to European languages, *Proc. Linn. Soc. Lond.* Sess. 150 (1937–38): 84–95.

— 1951: The rise and decline to the greater yam in the service of man, *Advancement Sci.* 7(28).

— 1954: Aji and Batata as group names within the species Ipomoea batatas, *Ceiba* 4: 227–40.

— 1960: The organography and the evolution of the Dioscoreaceae, *J. Linn. Soc. Lond. (Bot.)* 56: 319–412.

— s.d. Manuscript notes on Dioseorea; Dioscorea vernacular names – MS list; Reliquae (in Kew Herb.).

BURTT DAVY, J. & A. C. HOYLE [Ed.], 1937: *Check-lists of the Forest Trees and Shrubs of the British Empire.* No 3. *Draft of First Descriptive Check-list of the Gold Coast,* Imperial Forest Institute, Oxford.

BUSNEL, R. G., 1959: Étude d'un appeau acoustique pour la pêche, utilisé au Sénégal et au Niger, *Bull. Inst. Franç. Afr. Noire,* A. 21: 346–60.

# REFERENCES


BUSSON, F., 1965: *Plantes alimentaires de l'ouest african*, Leconte, Marseilles.

ÇAVA, M. P., S. K. TALAPATRA, J. A. WEISBACH, R. F. RAFFAUF & B. DOUGLAS, 1962: Triterpenoid constituents of Gabunia odoratissima, *Lloydia* 25: 222–4.

CAVACO, A., 1963,a: *Chenopodiacées,* in A. AUBRÉVILLE, *Flore du Gabon* 7: 16–20, Muséum National d'Histoire Naturelle, Paris.

— 1963,b: *Amarantacées,* in *op. cit.* 7: 21–48.

CHADHA, Y. R., [Ed.], 1972: *The Wealth of India – Raw Materials* 9(Rh-So), C.S.I.R., India, New Delhi.

— 1976,a: *op. cit.* 10(Sp-W).

— 1976,b: *op. cit.* 11(X-Z, indices).

CHADHA, Y. R. & J. DAKSHINAMURTHY, 1965: Sources of starch in Commonwealth Territories, Part V: Sweet potato, *Trop. Sci.* 7: 56–66.

CHAKRAVARTY, H. L., 1968: Cucurbitaceae of Ghana, *Bull. Inst. Fond. Afr. Noire,* A, 30: 400–68.

CHALK, L., J. BURTT DAVY, H. E. DESCH & A. C. HOYLE, 1933: *Twenty West African Timber Trees,* Clarendon Press, Oxford.

CHAMPAULT, A., 1970: Étude caryosystématique et écologique de quelques Euphorbiacées herbacées et arbustives africaines, *Bull. Soc. bot. France,* 117: 137–68.

CHAPPEL, T. J. H., 1976/77: Personal communications.

CHEVALIER, A. 1920: *Exploration botanique de l'Afrique occidentale française, I: Énumération des Plantes récoltées,* Paul Lechevallier, Paris.

— 1931,a: Progrès de la culture du Bananier en Guinée française. *Rev. Bot. appl. Agr. trop.* 11: 435–47.

— 1931,b: Les graines d'Avicennia comme aliment de famine, *Rev. Bot. appl. Agr. trop.* 11: 1000–1.

— 1935: Les Iles du Cap Vert. Flore de l'Archipel, *Rev. Bot. appl. Agr. trop.* 15: 733–1090.

— 1937,a: Plantes ichyotoxiques des Genres Tephrosia et Mundulea, *Rev. Bot. appl. Agr. trop.* 185: 9–37.

— 1937,b: Une Equête sur les Plantes médicinales de l'Afrique occidentales, *Rev. Bot. app. Agr. trop.* 187: 165–75.

— 1937,c: Abres à Kapok et Fromagers, *Rev. Bot. appl. Agr. trop.* 188: 245–68

— 1946: Nouvelles recherches sur les Ignames cultivées. *Rev. Bot. appl. Agr. trop.* 26: 26–31.

— 1950: Sur quelques Crinum d'Afrique tropicale, *Rev. Bot. appl. Agr. trop.* 30: 610–25.

CHIPP, T. F., 1922: *The Forest Officers' Handbook of the Gold Coast, Ashanti and the Northern Territories,* London.

CHITTENDEN, A. E. & H. E. COOMBER, 1948: Castor stems from Ceylon, *Bull. Imp. Inst.* 46: 223–7.

CHITTENDEN, A. E., C. G. JARMAN, D. MORTON & G. B. PICKERING, 1954: Three timbers from Kenya, Neoboutonia macrocalyx Pax., Macaranga kilimandscharica Pax, and Croton macrostachys Hochst, as paper-making materials, *Colon. Pl. Anim. Prod.* 4: 46–52.

CHOPRA, R. N., 1933: *Indigenous Drugs of India: Their medical and economic aspects,* Calcutta.

CLAUDE, F. et al., 1947: Survey of plants for antimalarial activity. *Lloydia,* 10: 145–74.

CLIFFORD, H. T., 1958: On hybridisation between Amaranthus dubius and A. spinosus in the vicinity of Ibadan, Nigeria. *J. W. Afr. Sci. Ass.* 4: 112–116.

COLE, N. H. AYODELE, 1968: *The vegetation of Sierra Leone,* Njala University College Press, Sierra Leone.

CONDAMIN, M. & T. LÈYE, 1964: À la recherche du plus gros Baobabs du Sénégal, *Notes. Afr.* 101: 29–30.

COOMBER, H. E., 1952/53: Pulping studies with Colonial tropical hardwoods as paper-making materials, *Colon. Pl. Anim. Prod.* 3: 13–27.

COOPER, G. P. & S. J. RECORD, 1931: The Evergreen Forests, of Liberia, *Yale University School of Forestry, Bull. No. 31.*

COOPER, W., 1971: *Hair. Sex, Society, Symbolism,* Aldus Books. London.

CORKILL, N. L., 1948: The poisonous wild cluster yam, D. dumetorum Pax, as a famine food in the Anglo-Egyptian Sudan, *Ann. Trop. Med. Parasit.* 42: 278–87.

CORNELIUS, J. A., 1966: Cashew Nut Shell Liquid and Related Materials, *Trop. Sci.* 8: 79–84.

CORNELIUS, J. A. & E. A. SIMMONS, 1969: Crambe abyssinica – a new commercial oil seed, *Trop. Sci.* 11: 17–22.

COURSEY, D. G., 1966: Food Technology and the Yam in West Africa, *Trop. Sci.* 8: 152–9.

— 1968: The Edible Aroids, *World Crops* 20: 25–30.

CROIZAT, LEON, 1938: Euphorbia (Diacanthium) deightonii. A new succulent from West Africa, with brief notes on some allied species, *Bull. Misc. Inf.* 1938: 53–59.

CUFODONTIS, G., 1969: Über Kalanchoe integra (Med.) O. Kuntze und ihre Beziehung zu K. crenata (Andr.) Haworth, *Österr. Bot. Zeitschr.* 116: 312–20.

DALE, I. R. & P. J. GREENWAY, 1961: *Kenya Trees and Shrubs,* Buchanan's Kenya Estates, Ltd & Hatchards, London.

DALZIEL, J. M., 1931: The hairs lining the loculi of fruits of species of Parinarium, *Proc. Linn. Soc. London* 1930–31: 99.

— 1937: *The useful plants of West Tropical Africa,* Crown Agents for the Colonies, London.

DAVIES, F., 1980: verb. comm.

DEBRAY, M., H. JACQUEMIN & R. RAZAFINDRAMBOA, 1971: Contribution à l'inventaire des plantes médicinales de Madagascar, *Trav. Doc. O.R.S.T.O.M.,* No. 8, Paris.

DECARY, H., 1946: Plantes et animaux utiles de Madagascar, *An. Mus. Colon. Marseille,* 6e. sér., 4: 6–234.

DEIGHTON, F. C., 1957: *Vernacular botanical vocabulary for Sierra Leone,* Crown Agents for Overseas Governments and Administrations, London.

DER MARDEROSIAN, A., 1967: Hallucinogenic Indole Compounds from higher plants, *Lloydia* 30: 23–38.

DER MARDEROSIAN, A. & H. W. YOUNGKEN, 1966: The distribution of Indole alkaloids among certain species and varieties of Ipomoea, Rivea and Convolvulus (Convolvulaceae), *Lloydia* 29: 35–42.

DESHAPRABHU, S. B., [Ed.], 1966: *The Wealth of India – Raw Materials* Vol. 7 (N-Pe), C.S.I.R., India, New Delhi.

DE WILDEMAN, E., 1906: *Notices sur les plantes utiles ou intéressantes de la Flore du Congo.* 2, 1: 145–6, Brussels.

DUCKWORTH, E. H., 1947: re Jatropha curcas *Nigerian Field* 12: 58.

EASMON, J. F., 1891: in *Colonial Reports – Miscellaneous No. 1 – Gold Coast,* H.M.S.O., London.

ECKEY, E., 1954: *Vegetable Fats and Oils,* Reinbold, New York.

EGGELING, W. J., & I. R. DALE, 1952: *The Indigenous Trees of the Uganda Protectorate,* Government Printer, Uganda.

EGUNJOBI, J. K., 1969: Some common weeds of Western Nigeria. *Bull. Res. Div. Min. Agr. Nat. Resources, Western State, Nigeria.*

EKONG, D. E. U. & J. I. OKOGUN, 1969, a: Terpenoids of Dacryodes edulis, *Phytochemistry* 8: 669–71.

EKONG, D. E. U., E. U. OLAGBEMI & F. A. ODUTOLA, 1969, b: Further diterpenes from Xylopia aethiopica (Annonaceae), *Phytochemistry* 8: 1053.

EMBODEN, W. A., 1972: *Ritual use of Cannabis sativa L.: a historical- ethnographic survey,* pp. 214–36, in P. T. FURST, *Flesh of the Gods. The ritual use of hallucinogens.* Praeger, New York.

ENTI, A. A., 1981–82: Pers. comm.

EPENHUIJSEN, C. W van., 1971: *A collection of market vegetables in SW Nigeria,* (MS. in Kew Herb.)

ESPÍRITO SANTO, J. Do., 1963: Nomes vernáculos de algumas plantas da Guiné Portuguesa, *Estudos, Ensaios e Documentos,* No. 104, Junta de Investigacões do ultramar, Lisbon.

EWING D. F. & C. Y. HOPKINS, 1967: *Can. J. Chem.* 45: 1259–65 [*Biol. Abstr.* 1967 113712]

FANSHAWE, D. B., 1953: Fish Poisons in British Guiana, *Kew Bull.* 8: 239–40.

FARNSWORTH, N. R., 1961: The pharmacognosy of the periwinkles: Vinca and Catharanthus, *Lloydia* 24: 105–38.

FARNSWORTH, N. R., R. N. BLOMSTER, W. M. MESSMER, J. C. KING, G. J. PERSINOS & J. D. WILKES, 1969: A phytochemical and biological review of the genus Croton, *Lloydia* 32: 1–28.

FEIJÃO, R. d'O., 1960–63: *Elucidário Fitológico Plantas Vulgares de Portugal continental, insular e ultramarino (Classificação, nomes vernáculos e aplicações).* Instituto Botanico de Lisbon. 3 vols.

FERNANDEZ, J. W., 1972: *Tabernanthe iboga. Narcotic ecstasis and the work of the Ancestors,* pp. 237–60, in P. T. FURST, *Flesh of the Gods. The ritual use of hallucinogens,* Praeger, New York.

*Flora of West Tropical Africa,* ed. 2. 1954–72:

— Keay (Ed.], 1954: 1 (1)

— Keay [Ed.], 1958: 1 (2)

— Hepper [Ed.], 1963: 2

— Hepper [Ed.], 1968: 3 (1)

— Hepper [Ed.], 1972: 3 (2)

— Alston, 1959: *Fern and Fern-allies,* Supplement. Crown Agents for Overseas Government and Administrations, London.

## REFERENCES


FOGGIE, A., 1957: Forestry problems in the Closed Forest Zone of Ghana, *J. W. Afr. Sci. Ass.* 3: 131–47.

FOSTER, W. H. & E. J MUNDY, 1961: Forage species of Northern Nigeria, *Trop. Agr.* 38: 311–8. [Reprinted in *Samaru Res. Bull.* No. 14. 1961.]

FRISON, [Ed.], 1942: De la présence de corpuscules siliceux dans le bois tropicaux en général et en particulier dans le bois du Parinari glabra Oliv. et du Dialium klainei Pierre. Utilisation de ces bois en construction maritime, *Bull. Agricole du Congo Belge.* 33: 91–105.

FRITZ, F. & G. GAZET du CHATELIER, 1967: Sur le Tragia benthami Baker, Euphorbiacées africaines; Etude botanique et pharmacodynamique, *J. Agr. trop. Bot. appl.* 14: 339–58.

FURLONG, J. R. 1942: Nigerian Cassava Starch, *Bull. Imp. Inst.* 40: 257–68.

GETAHUN, A. 1975 (ined.) *Some common medicinal and poisonous plants used in Ethiopian folk-medicine,* (MS. in Kew Herb.).

GOLDING, F. D. & A. M. GWYNN, 1939: Notes on the vegetation of the Nigerian shore of Lake Chad, *Bull. Misc. Inf.* 1939: 631–43.

GOMES E SOUSA, A. de F., 1930: Subsidios para o conhecimento da Flora da Guiné Portugesa, *Mem. Soc. Brot.* 1.

GOODSON, A., 1932: Echitamine in Alstonia Barks. *J. Chem. Soc.* 1932: 2626–30.

GOULDING, E., 1937: Textile fibres of vegetable origin. Forty years of investigation at the Imperial Institute, *Bull. Imp. Inst.* 35: 27–56.

GRALL, Lieut., 1945: Le Secteur nord du Cercle de Goure, *Bull. Inst. Franç. Afr. Noire* 7: 1–46.

GREEN, A. H., 1951: Pararistolochia goldieana (Hook. f.) Hutch. & Dalz., *Kew Bull.* 6: 132.

GREENWAY, P. J., 1941: Dyeing and Tanning Plants in East Africa, *Bull. Imp. Inst.* 39: 222–45.

GRINDLEY, D. N., 1950: The component fatty acids of various vegetable oils, *J. Sci. Fd. Agr.* 1: 152–5.

GRIVOT, R., 1949: La pêche chez les Pedah du lac Ahémé, *Bull. Inst. Franç. Noire* 11: 106–28.

GUILHEM, M. & R. P. G. HÉBERT, 1965: Notes additive sur "Les Divins en pays Toussain", *Notes Afr.*, 107: 92–95.

GUILLARMOD, A. J., 1971: *Flora of Lesotho (Basutoland),* J. Cramer.

HAERDI, F. in HAERDI, F., F. J. KERHARO & J. G. ADAM, 1964: *Afrikanische Heilpflanzen,* Basel.

HAINES, H. H. 1921–25: *The Botany of Bihar and Orissa,* 6, Vols. Govt. Bihar & Orissa [Reprint, 1961].

HALL, F. J. & L. BANKS, 1965: Cashew Nut Processing — Part, 1, *Trop. Sci.* 7: 12–26.

HALL, J., P. PIERIE, & G. LAWSON, 1971: *Common Plants of the Volta Lake,* University of Legon, Ghana.

HALL, R. de Z., 1953: Bao, *Tanganyika Not. Rec.* 34: 57.

HALLÉ, N., 1962,a: *Melianthacées, Balsaminacées, Rhamacées,* in A. AUBRÉVILLE, *Flore du Gabon,* 4, Mus. Nation. Hist. Nat., Paris.

— 1962,b: Monographie des Hippocratéacées d'Afrique occidentale, *Mém. Inst. Franç. Afr. Noire* 64.

HAMET, R., 1955: Sur quelques propriétés physiologiques d'une apocynacée africane, *Comptes rendus hebdomadaire des Séances de l'Académie des Sciences,* 240: 1470–2.

HANNA, R., 1964: Neutral constituents of Conopharyngia durissima, *Lloydia* 27: 40–46.

HANSON, S. W., 1976: Comm. 5th June, 1976.

HARDMAN, R., 1969: Pharmacutical products from plants steroids, *Trop. Sci.* 11: 196–228.

HARDMAN, R. & E. A. SOFOWORA, 1972: A reinvestigation of Balanitas aegyptiaca as a source of Steroidal Sapogenins, *Econ. Bot.* 26: 169–73.

HARGREAVES, B J., 1978: Killing and Curing; *The Society of Malawi Journal,* 31: 21–30.

HARRISON, S. G., 1950: Manna and its Sources, *Kew Bull.* 5: 407–17.

— 1966: *Dallimore and Jackson's Handbook of Coniferae and Ginkoaceae,* ed. 4, Arnold, London.

HARTWELL, J. L., 1967: Plants used against Cancer. A survey. *Lloydia* 30: 379–436.

— 1968: *op. cit.* 31: 71–170.

— 1969: *op. cit.* 32: 79–107, 153–205, 247–96.

— 1970: *op. cit.* 33: 97–194, 208–392.

HATHWAY, D. E., 1959: Myrobalans: an important tanning material, *Trop. Sci.* 1: 85–106.

HAUMAN, L., 1948: Fam. 18, *Aristolochiaceae,* in R. BOUTIQUE, *Flore du Congo-Belge et du Ruanda-Urundi, Spermatophytes* 1, I.N.É.A.C., Brussels.

— 1951,a: Fam. 24, *Amaranthaceae, op. cit.,* 2: 12–81.

— 1951,b: Fam. 27, *Aizoaceae, op. cit.,* 2: 27.

— 1951,c: Fam. 29, *Basellaceae, op. cit,* 2: 128–9.

— 1951,d: Fam. 32, *Ceratophyllaceae, op. cit.,* 2.

HAUMAN, L. & R. WILCZEK, 1951: Fam. 41, *Capparidaceae,* in R. BOUTIQUE, *Flore du Congo-Belge et du Ruanda-Urundi, Spermatophytes,* 2, I.N.É.A.C., Brussels.
HAYNES, P. H. & D. G. COURSEY, 1969: Gigantism in the Yam, *Trop. Sci.* 11: 93–96.
HEAL, R. E., E. F. ROGERS, R. T. WALLACE & O. STARNES, 1950: A survey of plants for insecticidal activity, *Lloydia* 13: 89–162.
HÉBERT, M. A., 1914: Les "Balanites" et leur utilisation possible, *J. Agr. trop.* 156: 171–2.
HEINE, H., 1960: Operculina macrocarpa (L.) Urban (Convolvulaceae) in West Tropical Africa, *Kew Bull.* 14: 397–9.
— 1966: Révision du Genre Thomandersia Baill. (Acanthaceae), *Bull. Jard. Bot. Nation. Belg.* 36: 207–48.
HENRY, T. A., 1924: *The Plant Alkaloids,* ed. 2; 1939, ed. 3. Churchill, London.
HEPPER, F. N., 1963,a: Plants of the 1957–58 West African Expedition, III, *Kew Bull.* 16: 451–9.
— 1963,b: *Flora of West Tropical Africa,* ed. 2, Vol. 2, Crown Agents for Overseas Governments and Administrations, London.
— 1965: The vegetation and flora of the Vogel Peak Massif, Northern Nigeria, *Bull. Inst. Franç. Afr. Noire,* A, 27: 413–513.
— 1966: Outline of the vegetation and flora of Mambila Plateau, Northern Nigeria, *Bull. Inst. Franç. Afr. Noire,* A. 28: 91–127.
— 1967: Culcasia scandens P. Beauv. (Araceae) and allied species in West Africa, *Kew Bull.* 21: 315–26.
— 1968: *Flora of West Africa,* ed. 2. Vol. 3, pt. 1, Crown Agents for Overseas Governments and Administrations, London.
— 1972: *Flora of West Tropical Africa,* ed. 2. Vol. 3. Pt. 2, Crown Agents for Overseas Governments and Administrations, London.
HEPPER, F. N. & F. NEATE, 1971: Plant collectors in West Africa, *Reg. Veg.* 74.
Herbalists, Joru Village, Sierra Leone: Meeting of herbalists, Nov. 1973.
Herbaria, BM: British Museum (Natural History).
— FBC: Fourah Bay College, Sierra Leone.
— IFE: University of Ife, Nigeria.
— IFAN: University of Dakar, Senegal.
— K: Royal Botanic Gardens, Kew.
— P: Muséum d'Histoire naturelle, Paris.
— SL: Njala University College, Sierra Leone.
— UCI: Ibadan University, Nigeria.
HERMAN, J. P., 1956: Etude des propriétés galactogènes et oestrogènes de la sève du Parasolier (*Musanga cecropioides* R. Brown), *Bull. Agr. Congo-belge* 47: 1345–68.
HEYNE, K., 1927: *De nuttige planten van Nederlandsch-Indië,* ed. 2. Buitenzorg.
HOLDSWORTH, M., 1961: The flowering of Rain Flowers, *J. W. Afr. Sci. Ass.* 7: 28–36.
HOLLAND, J. H., 1908–22: The useful Plants in Nigeria, *Bull. Misc. Inf. Add. Ser.* IX.
— 1937: *Overseas Plant Products,* John Bale, Sons & Curnow, London.
HOOKER, J. D., 1878: *Curtis's Botanical Magazine,* Reeve, London.
HOOPER, D., 1931: Some Persian Drugs, *Bull. Misc. Inf.* 1931: 299–344.
HOVIS, M., 1953: Le système promominal et les classes dans les dialectes baga, *Bull. Inst. Franç. Noire.* 15: 381–404.
HOWES, F. N., 1930: Fish Poison Plants. *Kew. Bull.* 1930: 129–153.
— 1945: *Plants and Bee Keeping,* Faber & Faber, London.
— 1946: Fence and Barrier Plants in Warm Climates. *Kew Bull,* 1: 51–87.
— 1949: *Vegetable gums and resins,* Chronica Botanica, Waltham, Mass., USA.
HUNTING TECHNICAL SURVEYS, LTD., 1968: *Land and Water Resources Survey of the Jebel Marra Area, Republic of the Sudan: Reconnaissance vegetation Survey,* ref. LA: SF/SUD/17, F.A.O., Rome.
HUTCHISON, J. & J. M. DALZIEL, 1937: Tropical African Plants, XV, *Bull. Misc. Inf.* 1937: 54–63.
ILTIS, H. H., 1967: Studies in the Capparidaceae, XI. Cleome afrospina, a tropical African endemic with neo-tropical affinities, *Amer. J. Bot.* 54: 953–62.
INGRAM, J. S., 1952/3: The Luffa Plant and its Uses, *Colon. Pl. Anim. Prod.* 3: 165–73.
INGRAM, J. S. & B. J. FRANCIS, 1969: The Annatto Tree (Bixa orellana L.). A guide to its occurrence, cultivation, preparation and uses, *Trop. Sci.* 11: 97–102.
INGRAM, J. S. & L. H. GREENWOOD-BARTON, 1962: The Cultivation of Yams for Food, *Trop. Sci.* 4: 82–86.
I......., J. S. & L. H. G-B. [Ingram & Greenwood-Barton], 1962: Experimentally produced Nigerian Yam Starches, *Trop. Sci.* 4: 103–4.
IRVINE, F. R., 1930: *Plants of the Gold Coast,* O.U.P.
— 1948: The indigenous food plants of West African Peoples, *J. New. York Bot. Gard.* 49: 225–36, 254–67.

— 1952,a: Supplementary and Emergency Food Plants of West Africa, *Econ. Bot (New York Bot. Gard.)* 6: 23–40.
— 1952,b: Food Plants of West Africa, *Lejeunia* 16: 27–51. [Report of A.E.T.F.A.T. Brussels Conferences, 1952.]
— 1955: West African insecticides, *Colon. Pl. Anim. Prod.* 5: 34–8.
— 1956: Cultivated and semi-cultivated leafy vegetables of West Africa, *Materiae Vegetabiles* 2.
— 1961: *Woody Plants of Ghana,* O.U.P., London.
— s.d. Reliquae, Library, R.B.G., Edinburgh.
IRVINE, F. R. & R. S. TRICKETT, 1953: Water lilies as Food, *Kew Bull.* 8: 363–70.
JACKS, T. J., T. P. HENSARLING, & L. Y. YATSU, 1972: Curcurbit seeds: I. Characterizations and uses of Oils and Proteins. A Review, *Econ. Bot.* 26: 135–41.
JACKSON, G., 1973, ined.: MS re *Fulani in N. Nigeria,* Botany Department, Ibadan University.
JAEGER, P., 1965: Espèces végètales de l'étage altitudinal des Monts Loma (Sierra Leone), *Bull. Inst. Franç. Afr. Noire,* A. 27: 34–120.
JAMES, B., 1975: Aponogetons, *Aquarist and Pond-Keeper.* 39: 472.
JEFFREY, C., 1964: Key to the Cucurbitaceae of West Tropical Africa, with a guide to localities of rare and little-known species, *J. W. Afr. Sci. Ass.* 9: 79–97.
JENIK, J. & K. A. LONGMAN, 1965: A new type of drought resistance, *J. W. Afr.* Sci. Ass. 10: 61–80. [Abstract in *Conference Report of the 5th Biennial Conference of the West African Science Association, Freetown, Sierra Leone,* 1965.]
JENKINS, R. W. & D. A. PATTERSON, 1973: The relationship between chemical composition and geographical origin of Cannabis, *Forensic Sci.* 2: 59–66.
JENNINGS, D. L., 1957: Further studies in breeding cassava for virus resistance, *E. Afr. Agric. J.* 22: 213–9.
JOHNSON, R. M. & W. P. RAYMOND, 1965: The chemical composition of some Tropical Food Plants, IV: Manioc, *Trop. Sci.* 7: 109–15.
KAUL, B., S. J. STOHS & E. J. STABA, 1969: Dioscorea tissue cultures, III. Influence of various factors on Diosgenin production by Dioscorea deltoidea callus and suspension cultures, *Lloydia* 32: 347–59.
KAY, D. E., *T.P.I. Crop and Product Digest, No. 2: Root Crops,* The Tropical Products Institute. H.M.S.O. London.
KEAY, R. W. J., 1953: Revision of the "Flora of West Tropical Africa," III, *Kew Bull.* 8: 69–82.
— 1954: *Flora of West Tropical Africa,* ed. 2. Vol. 1 (1), Crown Agents for Overseas Governments and Administrations, London.
— 1958: *op. cit.* 1 (2).
KEAY, R. W. J., C. F. A. ONOCHIE & D. P. STANFIELD, 1960: *Nigerian Trees,* Vol 1, Government Printer, Lagos.
— 1964: *op. cit.* Vol. 2, Nigerian National Press, Apapa.
KERHARO, J., 1966: La Pharmacopée Sénégalaise: Note sur les Rosacées utilisées en médecin traditionelle, *J. W. Afr. Sci. Ass.* 11: 77–80.
— 1967: A propos de la pharmacopée Sénégalaise: aperçu historique concernant les recherches sur la flore et des plantes médicinales du Sénégal, *Inst. Fond. Afr. Noire,* A. 29: 1391–1434.
KERHARO, J. & J. G. ADAM, 1962: Premier inventaire des plantes médicinales et toxiques de la Casamance (Sénégal), *Ann. Pharm. Franç.* 20: 726–44, 823–41.
— 1963,a: Deuxième inventaire des plantes médicinales et toxiques de la Casamance (Sénégal), *Ann. Pharm. Franç* 21: 773–92.
— 1963,b: *op. cit.* 21: 853–70.
— 1964,a: Note sur quelques plantes médicinales des Bassari et des Tandanké du Sénégal oriental, *Inst. Franç. Afr. Noire.* A. 26: 403–437.
— 1964,b: Plantes médicinales et toxiques des Peul et des Toucouleur du Sénégal, *J. Agr. trop. Bot. appl.* 11: 384–444, 543–99.
— 1964,c: Les plantes médicinales, toxiques et magiques des Niominka et des Socé des Iles du Saloum (Sénégal), in HAERDI, KERHARO & ADAM, *Afrikanisches Heilpflanzen: Plantes médicinales africaines,* Basel.
— 1974: *La Pharmacopée Sénégalaise traditionelle. Plantes médicinales et toxiques.* Vigot Frères, Paris.
KERHARO, J. & A. BOUQUET, 1947: Note sur l'utilisation de quelques Bignoniacées dans la thérapeutique indigène de la Côte d'Ivoire, *Bull. Soc. bot. Franç* 94: 251–3.
— 1950: *Plantes médicinales et toxiques de la Côte-d'Ivoire – Haute-Volta,* Vigot Frères, Paris.
KERS, L. E. 1969: Cleome spinosa Jacq. new to Africa, *Bot. Not.* 122: 294–5.
KILLIAN, C., 1953: Observations sur l'écologie et les besoins édaphiques du Quinquina, *Bull. Inst. Franç Afr. Noire,* 15: 901–71.

KIRBY, R. H., 1963: *Vegetable Fibres,* Leonard Hill, London.

KRISHNAMURTHI, A., [Ed.], 1969: *The Wealth of India. Raw Materials.* Vol. 8 (Ph-Re), C.S.I.R., New Delhi.

KULLENBERG, B., 1955: Quelques observations sur les Apides en Côte-d'Ivoire faites en août 1954, *Bull. Inst. Franç Afr. Noire* 17: 1125–31.

KUNKEL, G., 1965: *The Trees of Liberia.* German Forestry Mission to Liberia, Report No. 3, Munich.

LACK, H. W., 1980: The genus Enhydra (Asteraceae, Heliantheae) in West Tropical Africa, *Willdenowia* 10.

LAPERRINE, Général, 1919: Notice sur la cendre d'Aferegak, *Bull. Soc. Hist. Nat. Afr. Nord.* 10: 31.

LAWTON, J. R. S., 1967: A key to the Dioscorea species in Nigeria, *J. W. Afr. Sci. Ass.* 12: 1–9.

LEANDRI, J., 1958: Le problème du Casearia bridelioides Mildbr. ex Hutch. et Dalz., *Bull. Soc. bot. Franç.* 105: 512–7.

LE-BRUN, J. P., 1970: Un Polycarpaea nouveau d'Afrique tropicale, *Adansonia* 2, 10: 135–7.

LÉONARD, J., 1955: Notulae systematicae, XIX: observations sur divers Bridelia africains (Euphorbiacées), *Bull. Jard. bot. Nation. Belg.* 25: 359–74.

— 1956: Notulae systematicae, XXIII: contribution à l'étude des Croton africains (Euphorbiacées), *op. cit.* 26: 383–97.

— 1958: Notulae systematicae, XXIII: notes sur divers Euphorbiacées africaines des Genres Croton, Crotonogyne, Dalechampia, Grossera et Thecacoris, *op. cit.* 28: 111–21.

— 1960: Notulae systematicae, XXIX: révision des Cleistanthus d'Afrique continentale (Euphorbiacées), *op. cit.* 30: 421–61.

— 1961: Notulae systematicae, XXXII: observations sur des espèces africaines de Clutia, Ricinodendron et Sapium (Euphorbiacées), *op. cit.* 31: 391–406.

— 1962,a: Le cas amusant de Tetrorchidium minus (Euphorbiacée africaine), *Bull. Soc. Roy. bot. Belg.* 94: 29–34.

— 1962,b: Fam. 71, *Euphorbiaceae* (Part), in R. BOUTIQUE, *Flore du Congo et du Rwanda-Burundi,* Spermatophytes 8 (1) I.N.É.A.C., Brussels.

LERICHE, A., 1950: Instruments de musique maure et Griots, *Bull. Inst. Franç. Afr. Noire* 12: 744–50.

— 1951: Mesures maures: note préliminaire, *op. cit.* 13: 1227–56.

— 1952: De l'enseignement arabe féminin en Mauritanie, *op. cit.* 14: 975–83.

LEROUX, H., 1948: Animisme et Islam dans la subdivision de Maradi, *Bull. Inst. Franç. Afr. Noire* 10: 595–697.

LE THOMAS, A., 1967: A propos de L'Uvariodendron mirabile R. E. Fries., *Adansonia* n.s. 7: 251–3.

— 1969: Fam. 16, *Annonacées* in A. AUBRÉVILLE, *Flore du Gabon,* Muséum National d'Historie Naturelle, Paris.

LETOUZEY, R. & F. WHITE, 1970: *Ebénacées,* In A. AUBRÉVILLE & J. F. LEROY, *Flore du Cameroun* (Fam. 11) & *Flore du Gabon* (Fam. 18), Muséum National d'Histoire Naturelle, Paris.

— 1978: *Chrysobalanaceae,* opp. cit., (Fams. 20 & 24 respectively).

LIBEN, L., 1968: *Combretaceae,* in *Flore du Congo, du Rwanda et du Burundi, Spermatophytes,* I.N.É.A.C., Brussels.

— 1977: *Bignoniaceae,* in P. BAMPS, *Flore d'Afrique centrale (Zaire-Rwanda-Burundi), Spermatophytes,* I.N.É.A.C., Brussels.

LOCK, J. M., 1976: in litt., 20/1/76.

LOWE, J., [Ed.], 1973: *Index to Plants in the Nigerian Field, Vols 1–30 (1931–1965),* Dept. of Botany, University of Ibadan (mimeographed).

— 1975: in litt. 30/6/75.

— 1976: in litt. 29/6/76 – *List of chew-stick plants for the Ibadan area, Nigeria* – by B. I. Asuquo, 1975/76.

LUCAS, E. B., 1967: The properties of some Savanna Timber Trees, *Fed. Dept. For. Res., Ibadan, For. Res. Repts.* No. FCJ.R.L/11, 1967.

MCINTOSH, M., 1978: in litt., 23/2/78.

— in litt., 26/1/79.

MAIRE, H., 1933: Études sur la Flore et la Végétation du Sahara Central, *Mem. Soc. Hist. Nat. Afr. Nord. Algiers,* 1933.

MAIRE, R., 1925: Sur le Chrozophora brocchiana Schweinf., *Bull. Soc. Hist. Nat. Afr. Nord.* 16: 42.

MAJOR, F., Pyrethrum Flowers from Nigeria, *Bull. Imp. Inst.* 43: 7–8.

MAJOR, F., W. S. A. MATHEWS & G. SMITH, 1948: Sunflower seeds from Nigeria and the Gold Coast, *Bull. Imp. Inst.* 46: 197–209.

## REFERENCES


MALCOLM, S. A. & E. A. SOFOWORA, 1969: Antimicrobial activity of selected Nigerian folk-remedies and their constituent plants, *Lloydia* 32: 512–7.

MANGENOT, G., 1957,a: *Icones Plantarum Africanarum, IV. fasc. 81, Cnestis ferruginaea DC., Connaracée,* I.F.A.N., Dakar.

— 1957,b: *op.cit., IV. fasc. 89 Manotes longiflora Baker, Connaracée.*

MANJUNATH, B. L., [Ed.] 1948: *The Wealth of India, Raw Materials,* Vol. 1 (A–B), C.S.I.R., New Delhi.

MASSAQUOI, JOHN (Herbalist, Magbena, S. Leone), Nov. 1973: Herbal prescriptions, verb. comm.

MAUNY, R., 1953: Notes historiques autour des principales plantes cultivées d'Afrique occidentale, *Bull. Inst. Franç. Afr. Noire* 15: 684–730.

MEEK, C. K., 1925: *The Northern Tribes of Nigeria,* 2 vols., OUP.

— 1931: *Tribal studies in Northern Nigeria,* 2 vols. Kegan Paul, Trench Trubner.

MIÉGE, J., 1952: L'importance économique des Ignames en Côte d'Ivoire. Répartition des cultures et principales variétés, *Rev. Bot. app. Agr. trop.* 32: 144–55.

— 1958: Deux Ignames ouest-africaines à tubercules vivaces, *Bull. Inst. Franç. Afr. Noire.* A, 20: 39–59.

MIHRANIAN, V. H. & C. I. ABOU-CHAAR, 1968: Extraction, detection and estimation of cucurbitin in Cucurbita seeds, *Lloydia* 31: 23–9.

MILLER, O. B., 1952: *The Woody Plants of the Bechuanaland Protectorate,* Kirstenbosch, S. Africa. [Reprinted from *J. S. Afr. Bot.* 18, 1952.]

MILLSON, A., 1891: Indigenous Plants of Yorubaland, *Bull. Misc. Inf.* 1891: 206–19.

MILNE-REDHEAD, E., 1949: Euphorbia geniculata Orteg (Euphoriaceae), *Kew Bull.* 1948: 457–8.

— 1951: Emilia praetermissa Milne-Redhead, *Kew Bull.* 1950: 375–6.

MODEBE, A. N. A., 1963: Preliminary trial on the value of dried cassava (Manihot utilissima Pohl) for pig feeding, *J. W. Afr. Sci. Ass.* 7: 127–33.

MONOD, TH., 1950: Vocabulaire Botanique Teda in R. MAIRE & TH. MONOD, Études sur la flore et la végétation du Tibesti, *Mém. Inst. Franç. Afr. Noire* 8.

MONTEIL, V., 1953: *Institut des Hautes Études Marocaines. Notes & Documents, VI: Contribution à l'étude de la Flora du Sahara Occidental,* II, Larose, Paris.

MORTIMER, W. G., 1901. *Peru. History of Coco. "The Divine Plant of the Incas",* Vail & Co., New York.

MORTON, J. F., 1961: The cashew's brighter future, *Econ. Bot.* 15: 57–78.

MORTON, J. K., 1961: *West African Lilies and Orchids,* in H. J. SAVORY [Ed.], West African Nature Handbooks, Longmans, Green & Co. Ltd., London.

— 1979: An overlooked species of Nelsonia (Acanthaceae) from Africa, *Kew Bull.* 33: 399–402.

MUNAKATA, K., SHINGO MARUMO & KEIICHI OHTA, 1965: Justicidin A and B, the fish-killing component of Justicia hayatai var. decumbens, *Tetrahedron Let.* 47: 4167–70.

MURRAY, H. J. R., 1952: *A history of Board Games other than Chess,* Clarendon Press, Oxford.

NAEGELÉ, A., 1958,a: Contribution à l'étude de la flore et des groupements végétaux de la Mauritanie. I: note sur quelques plantes récoltées à Chinguetti (Adrar Tmar), *Bull. Inst. Franç. Noire* A, 20: 293–305.

— 1958,b: *ibid,* II: plantes recueillées par Mlle Odette de Puigandeau en 1950, *op. cit.* 876–908.

N'DIAYE, S., 1964: Notes sur le engins de pêcher chez les Sérèr, *Notes Afr.* 104: 116–120.

NELMES, E., & J. T. BALDWIN, Jr., 1952: Cyperaceae in Liberia, *Amer. J. Bot.* 39: 368–93.

NEWBERRY, R. J., 1938,a: Some games and pastimes of Southern Nigeria, I: Some Yoruba games, *Nigerian Field* 7: 85–90.

— 1938,b: II: Okoto, *op. cit.,* 7: 131–2.

— 1939: III: Ayo, *op. cit.,* 8: 75–80.

— 1940: IV: Sundry Yoruba games *op. cit.,* 9: 40–43.

NICHOLAS, F. J., 1953: Onomastique personnelle des l'Éla de la Haute-Volta, *Bull. Inst. Franç Afr. Noire* 15: 818–47.

NICHOLS, R. W. F., 1947: Breeding cassava for virus resistance, *E. Afr. Agric. J.* 12: 184–94.

NORMAND, D., 1937: Le bois de Landa, Erythroxylum du Cameroun, *Rev. Bot. appl. et Agr. trop.* 196: 883–9.

NYE, P. H., 1957: Some prospects for subsistence agriculture in West Africa, *J. W. Afr. Sci. Ass.* 3: 91–95.

OGAN, A. V., 1971: Isolation of cuminal from Xylopia aethiopica, *Phytochemistry* 10: 2823–4.

OKAFOR, J. C., 1967: A taxonomic study of the Combretum collinum group of species, *Bol. Soc. Brot.* Ser. 2, 41: 137–50.

OKE, O. L., 1965: Chemical studies on some Nigerian vegetables, *J. W. Afr. Sci. Ass.* 10: 61–80. [Abstract in *Report of Conference of the 5th biennial conference of the West African Science Association, Freetown, Sierra Leone* 1965.]

— 1966,a: Chemical Studies on some Nigeria Foodstuffs – Kpokpogari (Processed Cassava), *Trop. Sci.* 8: 23–27.
— 1966,b: Chemical Studies on some Nigerian vegetables, *Trop. Sci.* 8: 128–32.
OKIY, G. E. O., 1960: Indigenous Nigerian Food Plants, *J. W. Afr. Sci. Ass.* 6: 117–21.
OLIVER, Bep, 1960: *Medicinal Plants in Nigeria,* Nigerian College of Arts Science and Technology.
ORR, Y., 1923: The Leaf-glands of Dioscorea macroura Harms., *Not. Roy. Bot. Gard. Edin.* 14: 57–72.
OUEDRAOGO, J., 1950: Les funérailles en pays Mossi, *Bull. Inst. Franç. Afr. Noire.* 12: 441–55.
PÂQUES, V., 1953: L'estrade royale des Niaré, *Bull. Inst. Franç. Afr. Noire.* 15: 1642–54.
PARDY, A. A., 1952: Notes on Indigenous Shrubs and trees of S. Rhodesia, *Rhod. Agric. J.* 49: 80.
PEARMAN, R. W., W. D. RAYMOND & J. A. SQUIRES, 1951: Niger seed from Tanganyika, *Colon. Pl. Anim. Prod.* 2: 101–5.
PERRIER DE LA BÂTHIE, H., 1954: Fam. 151, *Combretacées,* in H. HUMBERT, *Flore de Madagascar et des Comores,* Firmin-Didot & Cie, Paris.
PICHON, M., 1953: Monographie des Landolphiées (Classification des Apocynacées XXXV), *Mém. Inst. Franç. Afr. Noire* 35.
POBÉGUIN, H., 1912: *Plantes médicinales de la Guinée, Paris.*
POLHILL, R., 1971: Verb. Comm.
PORTÈRES, R., 1935: Plantes toxiques utilisées par les peuplades Dan et Guéré de la Côte d'Ivoire, *Bull. Comité d'Études Historiques et Scientifiques de l'AOF* 18.
— 1951,a: *Comptes rendus. Première Conférence internationale des Africanistes de l'Ouest,* I.F.A.N., Dakar.
— 1951,b: Les variations des Ceintures hydrophytiques et gramino – helophytiques des eaux vives du système, *Bull. Inst. Franç Afr. Noire* 13: 1011–28.
— 1960: La Sombre Aroidée cultivée: Colocasia antiquorum Schott ou Taro de Polynésie: Essai d'Etymologie sémantique, *J. Agric. trop. Bot. appl.* 7: 169–92.
— s.d. Reliquae, Lab d'Ethnobotanique, Paris.
POWELL, J. W. & D. A. H. TAYLOR, 1967: Alkaloids from West African Crinum species, *J. W. Afr. Sci. Ass.* 12: 50–52
PRANCE, G. T., 1972: *Flora Neotropica. Monograph No. 9: Chrysobalanaceae,* Organisation for Flora Neotropica, Haffner Publishing Coy., New York.
PREUSS, P., 1899: Über Westafrikanische Kickxia- Arten. *Notizblatt des Königl. botanischen Gartens und Museums zu Berlin* 2: 353–360.
PURI, H. S., 1971: Macro and micro-morphology of leaf and seed of Cleoma viscosa L., *J. Agr. trop. Bot. appl.* 18: 566–71.
PURSEGLOVE, J. W., 1968: *Tropical Crops. Dicotyledons,* 2 vols., Longmans Green, London.
— 1972: *Tropical Crops. Monocotyledons,* 2 vols., Longmans Green London.
QUISUMBING, E., 1951: Medicinal Plants of the Philippines, *Dept. Agric. Nat. Resources, Tech. Bull. 16.*
RAYMOND, W. D., 1961: Castor Beans as Food and Fodder, *Trop Sci.* 3: 19–27.
RAYMOND, W. D. & J. A. SQUIRES, 1951: Annatto seed from Nigeria, *Colon. Pl. Anim. Prod.* 2: 114–7.
RAYMOND-HAMET, 1951: Sur une drogue remarquable de l'Afrique tropicale, le 'Picralima nitida' (Stapf) Th. & H. Durant, *Rev. Bot. app. Agric. trop.* 31: 465–85.
RICHARDSON, J. W. & L. B. SMITH, 1972: *Canáceas* in P. R. REITZ, *Flora ilustrada Catarinense.*
RIDLEY, H. N., 1906: Malay Drugs, *Agr. Bull. Straits & Fed. Malay States.* 5: 193–206, 245–254, 269–282.
ROBERTY, G., 1953: Notes de botanique Ouest-africaine, VI: Plantes banales dans le Sahel de Nioro, *Bull. Inst. Franç. Afr. Noire* 15: 442–52.
— 1954: Notes sur la flore de l'Ouest-Africain, IV, *Bull. Inst. Franç. Afr. Noire* 16: 774–96.
— 1955: VI, *op. cit.* 17: 12–79.
ROBEY, M. E. L., 1970–76: Personal communications.
ROBIN, J., 1947: Description de la Province de Dosso, *Bull. Inst. Franç. Afr. Noire* 9: 56–98.
ROBINSON, C. H., 1913: *Dictionary of the Hausa Language: I, Hausa-English; II, English-Hausa.* 2 vols. Cambs. Univ. Press.
ROBYNS, A., 1963: *Bombacaceae* in R. BOUTIQUE, *Flore du Congo, du Rwanda et du Burundi, Spermatophytes X,* I.N.É.A.C., Brussels.
ROBYNS, W. & R. BOUTIQUE, 1951: Fam. 42, *Cruciferae,* in R. BOUTIQUE, *Flore du Congo-Belge et du Ruanda-Urundi, Spermatophytes* 2, I.N.É.A.C. Brussels.
ROBYNS, W. & J. GHESQUIÈRE, 1933: Revision du Genre Enantia Oliv. (Annonacées), *Bull. Jard. bot. Nation. Belg.* 9: 303–16.

ROSEVEAR, D. R., 1961, ined.: *Gambia Trees and Shrubs.* (MS. in Kew Herb.) [Notes to accompany the authors "Forestry conditions in The Gambia" *Emp. Forest J.* 16: 1937.]
— 1975, ined.: *Notes on Nigerian Trees.* (MS. in Kew Herb.)
SAMIA AL AZHARIA JAHN, 1976: Personal communications.
SANDWITH, N. Y., 1946: Gomphrena celosioides Mart., a weed spreading in the Old World Tropics, *Kew Bull.* 1946: 29–30.
SASTRI, B. N. [Ed.], 1950: *The Wealth of India. Raw materials* II(c), C.S.IR., New Delhi.
— 1952: op. cit. 3 (D–E).
— 1956: op. cit. 4 (F–G).
— 1959: op. cit. 5 (H–K).
— 1962: op. cit. 6 (L–M).
SAVILL, P. S. & J. E. D. FOX, 1967: ined. *Trees of Sierra Leone.* (MS in Forest Department, Freetown.)
SCHLECHTER, R., 1900 [1901] *Westafrikanische Kautschuk – Expedition, 1899–1900,* Berlin.
SCHNELL, R., 1950,a: Études préliminaires sur la végétation et la flore des hauts plateaux de Mali, *Bull. Inst. Franç. Afr. Noire* 12: 905–26.
— 1950,b: *Manuels Ouest-africains,* I. *La forêt dense. Introduction à l'étude botanique de la région forestière d'Afrique occidentale,* Paris.
— 1952: Végétation et Flore de la Région Montagneuse du Nimba, *Mém. Inst. Franç. Afr. Noire* 22.
— 1953,a: *Icones Plantorum Africanarum,* I, fasc. 3, *Alternanthera maritima* St. Hilaire, var. *africana* Hauman, I.F.A.N., Dakar.
— 1953,b: *op. cit.* I, fasc. 12, *Hippocratea richardiana* Cambess.
— 1953,c: *op. cit.* I, fasc. 14, *Calotropis procera* (Willd.) Aiton.
— 1953,d: *op. cit.* II, fasc. 25, *Cadaba farinosa* Forsskal (Capparidacées).
— 1953,e: *op. cit.* II, fasc. 28, *Euphorbia balsamifera* Aiton, 1789, (Euphorbiacées).
— 1960,a: *op. cit.* V, fasc. 99, *Bidens pilosa* Linn., Composées.
— 1960,b: *op. cit.* V, fasc. 110, *Jatropha curcas* Linn., 1753, Euphorbiacées.
— 1960,c: *op. cit.* V, fasc. 119, *Sarmentosus* A.P.DC, Apocynées.
SCHUNCK DE GOLDFIEM, J., 1942: Étude chimique de Balanites aegyptiaca (Del.), *Bull. Soc. bot. Franç.* 89: 236–7.
— 1945: Flore utilitaire de la Thyrrénéide, *Bull. Soc. bot. Franç.* 92: 152–3.
SEGEREN, W. & P. J. M. MAAS, 1971: The Genus Canna in northern South America, *Acta Bot. Neerl.* 20: 663–80.
SIDIBÉ, M., 1939: Famille, vie sociale et vie religieuse chez les Birifer et les Oulé (region de Diébougou, Côte d'Ivoire), *Bull. Inst. Franç. Afr. Noire* 1: 697–742.
SIKES, S. K., 1972: *Lake Chad,* Eyre Methuen, London.
SILLANS, R., 1953,a: *Ann. Pharm. Franç.* 11.
— 1953,b: Plantes Alimentaires spontanées d'Afrique centrale, *Bull. Inst. Études Centrafr.* 5: 77–99.
SINGH, P., C. L. MADAN & B. C. KUNDU, 1963: Histological study of the Carissa carandas and Carissa spinarum, *Lloydia* 26: 49–56.
SINGHA, S. C., 1965: *Medicinal Plants of Nigeria,* Nigerian National Press, Ltd., Apapa.
SMITH, E. H. C., 1952: Papain: its production and market, *Colon. Pl. Anim. Prod.* 3: 1–12.
SPICKETT, R. G. W., J. A. SQUIRES & J. B. WARD, 1955: Gari from Nigeria, *Colon. Pl. Anim. Prod.* 5: 230–5.
STANDLEY, P. C., 1920–26: Trees and Shrubs of Mexico, *Contr. U.S. Nation. Herb.* 23 (5 parts).
STANER, P., 1932: Une plante toxique pour le bétail, *Rev. Zool. Bot. Afr.* 23.
STANER, R. P., 1948: *Balanophoraceae,* in R. BOUTIQUE [Ed.]: Flore du *Congo-belge et du Ruanda-Urundi,* I: 394–5, I.N.É.AC., Brussels.
STAPF, O., 1900: Ad not. in *Proc. Linn. Soc. Lond. 1900,* p.2.
— 1901: Funtumia elastica Stapf in *Hooker's Icon. Pl.* ser. 4, 7: t.2694/5.
SVOBODA, G. H., 1962: The current status of research on the alkaloids of Vinca rosea (Catharanthus roseus), *Lloydia* 25: 334–5.
— 1964: The current status of Catharanthus roseus (Vinca rosea) research, *Lloydia* 27: 203–19, 275–6, 361–3.
SVOBODA, G. H., M. GORMAN & M. A. ROOT, 1964: Alkaloids of Vinca rosea (Catharanthus roseus), XXVIII: A preliminary report on hypoglycemic activity, *Lloydia* 27: 361–3.
SVOBODA, G. H., M. GORMAN & R. H. TUST, 1964: Alkaloids of Vinca rosea (Catharanthus roseus), XXV: Lochrovine, Perimivine, Vincoline, Lochrovidine, Lochrovicine, and Vincolidine, *Lloydia* 27: 203–19.
SWART, E. R., 1963: Age of the Baobab tree, *Nature* 198: 708–9.

TATON, A., 1971: *Boraginaceae,* in P. BAMPS, *Flore du Congo, Rwanda et du Burundi, Spermatophytes,* I.N.É.A.C., Brussels.
TATTERSFIELD, F., C. POTTER, K. A. LORD, E. M. GILLHAM, M. J. WAY, & R. I. STOKER, 1948: Insecticides derived from Plants, *Kew. Bull.* 3: 329–49.
TAYLOR, C. J., 1960: *Synecology and Silviculture in Ghana,* Thos. Nelson & Sons, Ltd. for Univ. Coll. of Ghana.
TAYLOR-SMITH, R., 1966: Investigations on plants of West Africa, III: Phytochemical studies of some plants of Sierra Leone, *Bull. Inst. Franç. Afr. Noire* A, 28: 538–41.
TAYLOR-SMITH, R. & D. E. B. CHAYTOR, 1966: Investigations on West African plants, III: Studies of three West African medicinal plants, *Bull. Inst. Franç. Noire* A, 28: 895–8.
TENNANT, J. R., 1961: Notes on African Araliaceae, III: Schefflera barteri now in Eastern Africa, *Kew Bull.* 15: 331–5.
TERRY, P. J., s.d.: *Tropical Pesticides Research Institute, Miscellaneous Report No 924,* East African Community Tropical Pesticides Research Institute, Arusha, Tanzania.
THOMAS, D. W. & K. BIEMANN, 1968: The Alkaloids of Voacanga africana, *Lloydia* 31: 1–8.
THOMAS, N. W., 1910: *Report on Edo-spreading people of Nigeria,* London.
THOMPSON, Sir EDWARD MAUNDE, 1929: Papyrus in *Encyclopaedia Britannica,* ed. 14. 17: 246–8.
TOUBIANA, R. & A. GAUDEMER, 1967: *Tetrahedron Lett.* 14: 1333–6 [in *Biol. Abstr.,* 1967: 71764].
TOUSSANT, L., 1951: Fam. 45, *Crassulaceae,* in R. BOUTIQUE, *Flore du Congo-Belge et du Ruanda-Urundi, Spermatophytes* 2, I.N.É.A.C., Brussels.
TOWNSEND, C. C., 1974: *Flora of Pakistan, No. 71, Amaranthaceae,* Karachi.
— 1975: The Genus Celosia (subgenus Celosia) in Tropical Africa, *Hooker's Icon. Pl.* 38, 1975.
TROCHAIN, J., 1940: La végétation du Sénégal, *Mém. Inst. Afr. Noire* 2.
TROUPIN, G., 1950: Les Burseraceae du Congo- Belge et du Ruanda-Urundi, *Bull. Soc. Roy. Bot. Belg.* 83: 111–28.
— 1952: Fam. 49, *Connaraceae,* in R. BOUTIQUE, *Flore du Congo-Belge et du Ruanda-Urundi, Spermatophytes* 3: 70–136, I.N.É.A.C., Brussels.
TYLER, V. E., Jn., 1966: The physiological properties and chemical constitutents of some habit-forming plants, *Lloydia* 29: 275–92.
UNWIN, A. H., 1920: *West African Forests and Forestry,* Unwin, London.
UPHOF, J. C. Th., 1968: *Dictionary of Economic Plants,* ed. 2., Cramer.
USHER, G., 1974: *A Dictionary of Plants used by Man,* Constable, London.
VAN BRUGGEN, H. E. W., 1973: Revision of the Genus *Aponogeton (Aponogetonaceae),* VI: The Species of Africa, *Bull. Jard. bot. Nation. Belg.* 43: 193–233.
VAN DER VEKEN, P., 1960: *Anacardiaceae,* in R. BOUTIQUE, *Flore du Congo-Belge et du Ruanda-Urundi, Spermatophytes* 9: 5–108, I.N.É.A.C., Brussels.
VAUGHAN, J. G., 1963: The testa structure of Crambe hispanica L. (Cruciferae), *Kew. Bull.* 16: 393–4.
VERGER, P. F., 1967: *Awọn ewe ọsanyin* (Yoruba medicinal leaves), Univ. of Ife.
VERMEER, D. E., 1976: in litt., 28/1/76.
VICKERY, B. & M.L. VICKERY, 1972: Fluoride metabolism in Dichapetalum toxicarium, *Phytochemistry* 11: 1905–9.
VICKERY, B., M. L. VICKERY & J. T. ASHU, 1973: Analysis of Plants for Fluoracetic Acids, *Phytochemistry* 12: 145–7.
VICKERY, H. B. & D. C. WILSON, 1958: Preparation of potassium dihydrogen $L_s(+)$– isocitrate from *Bryophyllum calycinum* leaves, *J. Biol. Chem. (Baltimore)* 233: 14–17.
VOORHOEVE, A. G., 1965: *Liberian high forest trees,* Wageningen.
WALKER, A. R., 1952: Usages pharmaceutiques des plantes spontanées du Gabon, I. *Bull. Inst. Études Centrafr.* n.s.4: 181–6.
— 1953,a: II, *op. cit.* n.s.5: 19–40.
— 1953,b: III, *op. cit.* n.s. 6: 275–329.
WALKER, A. R. & R. SILLANS, 1961: *Les plantes utiles du Gabon,* Paul Lechevalier, Paris.
WASSON V. P. & R. G. WASSON, 1957: *Mushrooms, Russia and History,* Pantheon Books, New York.
WATT, G., 1889–93: *A Dictionary of the Economic Products of India,* 6 vols. (Vol. 6 in 4 parts), Calcutta.
WATT, J. M., 1967: African Plants potentially useful in mental health, *Lloydia* 30: 1–22.
WATT, J. M. & M. G. BREYER-BRANDWIJK, 1962: *The Medicinal and Poisonous Plants of Southern and Eastern Africa,* 2 ed., Livingstone: Edinburgh and London.
WEALTH OF INDIA, THE-RAW MATERIALS, 1948–76, C.S.I.R., New Delhi.
— Manjunath [Ed.]. 1948: 1 (A–B).
— Sastri [Ed.] 1950: 2 (C).

—— 1952: 3 (D–E).
—— 1956: 4 (F–G).
—— 1959: 5 (H–K).
—— 1962: 6: (L–M).
— Deshaprahhu [Ed.] 1966: 7 (N–Pe).
— Krishnamurthi [Ed.] 1969: 8 (Ph–Re).
— Chadha [Ed.] 1972: 9 (Rh–So).
—— 1976: 10 (Sp–W).
—— 1976: 11 (X–Z, indices)

WEBSTER, G. L., 1957: A monographic study of the West Indian species Phyllanthus, II, *J. Arnold Arbor.* 38: 51–79.

WESTERMANN, D. & M. A. BRYAN, 1970: *Handbook of African Languages,* II: *Languages of West Africa,* International African Institute, London.

WHEATON, T. A. & I. STEWART, 1970: The distribution of tyramine, etc. in Higher Plants, *Lloydia* 33: 244–54.

WHITAKER, T. W. & G. N. DAVIS, 1962: *Cucurbits: botany, cultivation and utilisation,* Leonard Hill, London.

WHITE, F., 1956: Notes on Ebenaceae, II: Diospyros piscatoria and allies, *Bull. Jard. bot. Nation. Belg.* 26: 277–307.

— 1957: III, *op. cit.* 27: 515–31.

— 1976: The taxonomy, ecology and chorology of African *Chrysobalanaceae* (excluding Acioa), *Bull. Jard. bot. Nation. Belg.* 46: 265–350.

WICKENS, G. E., 1973: *Combretaceae,* in R. M. POLHILL, *Flora of Tropical East Africa,* Crown Agents, London.

WILCZEK, R., 1960,a: Fam. 75, *Celastraceae,* in R. BOUTIQUE, *Flore du Congo Belge et du Ruanda-Urundi, Spermatophytes* 9: 113–32. I.N.É.A.C., Brussels.

WILCZEK, R., 1960,b: Fam. 76, *Hippocrataceae,* in *op. cit.* 9: 133–232.

WILCZEK, R. & G. M. SCHULZE, 1960: Fam. 81, *Balsaminaceae,* in R. BOUTIQUE, *Flore du Congo-Belge et du Ruanda-Urundi, Spermatophytes 9,* I.N.É.A.C., Brussels.

WILLAMAN, J. J. & HUI-LIN LI, 1970: Alkaloid-bearing plants and their contained alkaloid, *Lloydia.* 33(3A).

WILLIAMS, F. N., 1907: Florula Gambica. Une contribution à la flore de la colonie britannique, *Bull. Herb. Boissier* sér. 2., 7: 81–96, 193–208, 369–86.

WILLIAMS, R. O., 1949: *The useful and ornamental plants in Zanzibar and Pemba,* Zanzibar.

WILLIAMSON, J., 1955, (?1956) *Use Plants of Nyasaland,* Govt. Printer, Zomba.

WILLIAMSON, K., 1970: Some food-plant names in the Niger Delta, *Int. J. Amer. Linguistics* 36: 156–67.

— s.d. (ined.): Field-notes re Niger Delta plant collections, [MS in Ibadan University].

WOODRUFF, A. W., 1975: in litt., 16/10/75.

ZUCKERMAN, A. & H. N. GRACE, 1949: Utilization of erucic acid oils, *Canadian Chem. & Process Indust.* 33: 588–93, 607 [*Biol. Abstr.* 1950, 24: 6054].

## LIST OF AUTHORITIES CITED FOR VERNACULAR NAMES

Authorities are given in this list either by the full name, or names in cases of joint publication, or by cypher which may be identified as indicated, and then taken up in the general bibliographic references.

A&AA – Adjanohoun & Aké Assi.
A&P – Aubréville & Pellegrin.
A.Chev. – A. Chevalier.
AEK – Kitson.
AGV – Voorhoeve.
AHU – Unwin.
AJC – Carpenter.
AJML – Leeuwenberg.
AN – Naegelé.
APDJ – Jones.
AS – Sébire.
AST – Thornewill.
Aub. – Aubréville.
auctt. – citation by 3 or more authorities.
Abdoul Omar: 1962.
Adam, = JGA: 1960 to 1971, and with Kerharo.
Adames: Herb. K.
Adams, C. D.: 1956 to 1964; Herb. K.
Adams, R. F. G.: 1943, 1947.
Adande: 1953.
Addo: ex Irvine.
Adjanohoun & Aké Assi = A&AA: 1972.
Afzelius: ex Dalziel.
Ainslie = JRA: 1937.
Airy Shaw: 1952.
Ajayi: Herb. K.
Akpabla: ex Irvine; Herb, K.
Akpata, or Olu Akpata: Herb. K.
Akwa: Herb. K.
Alasoadura = SOA: Herb. UCI.
Amanu: ex Burkill.
Amayo: ex K. Williamson, pers. comm.
Andoh: ex Irvine; Herb. K.
anon. = anonymous: ex herbaria.
Anthony: ex Irvine.
Anyadiegwe: Herb. K.
Armitage: Herb. K.
Ariwaodo: Herb. K.
Asiedu: ex Irvine.
Asuquo: ex Lowe, 1975.
Aubréville = Aub.: 1950, 1959; ex auctt.
Aubréville & Pellegrin = A&P: 1938.
Ayensu: with Coursey, 1972.
Aylmer: ex Irvine; Herb. K.
B&D – Bouquet & Debray.
B&T – Battandier & Trabut.
BM – Moiser.
BNO –Okigbo, 1980.
Ba: 1969.
Bakshi: Herb. SL.
Banco: Herb. K.
Banfield: ex Dalziel.
Bargery: ex Dalziel.
Barker: Herb. K.
Barter: ex Dalziel; Herb. K.
Barth: ex Dalziel.
Bartha: Herb. K.
Bates: Herb. K.
Battandier & Trabut = B&T: ex Dalziel.
Bayard: 1947.
Bégué: ex Kerharo & Bouquet; Herb. IFAN.
Benton: ex Dalziel.
Berhaut = JB: 1967.
Bertho: 1951; 1953.
Beveridge: ex Irvine.
Blakelock: 1956.
Blench: 1981–82, pers. comm. = RB.
Boboh: 1974.
Boston: Herb. K.
Bouquet & Debray = B&D: 1974.
Bouery: ex Pichon.
Bouquiaux: 1971–72 = LB.
Bouronville: 1967.
Boury: 1962.
Bowdich: ex Irvine, 1961.
Brand: ex Irvine.
Brenan: 1952; 1960; 1961; Herb. K.
Brent: ex Dalziel; ex Irvine; Herb. K.
Breteler: 1973.
Brossart: ex A. Chevalier; ex Dalziel.
Brown = WTSB: ex Irvine; Herb. K.
Brown-Lester: ex Dalziel; Herb. K.
Bullock: 1952–63; Herb. K.
Bunny & Ryan: Herb. K.
Bunting: ex Dalziel; ex Irvine.
Burbridge: Herb. K.
Burkill: 1938; Reliquae.
Burton: Herb. K.
Busse: ex Dalziel.
Busson = FB: 1965.
C – Cooper.
C&H – Cyffer & Hutchinson.
C&R – Cooper & Record.
CG – Calme-Griaule.
CHOP – Pobéguin.
CJT = C. J. Taylor, Herb. K; 1960.
CWvE – Epenhuijsen.
Caille: ex A. Chevalier; ex Dalziel.
Calme-Griaule, 1968 = CG.
Cardinall: ex Dalziel.
Carpenter = AJC: ex Dalziel; ex Rosevear; Herb. UCI.
Catterall: ex Dalziel.
Chakravarty: 1968.
Chalk = LC: 1933.
Chesters: ex Dalziel; Herb. K.
Chevalier, A. = A. Chev.: 1920; ex auctt.
Chipp = TFC: 1922; Herb. K.
Chizea: Herb. K.
Chumbow: 1982.
Clarke: Herb. K.
Clusters: Herb. K.
Cole: 1968.
Cons. For. = DF: of Herbaria.
Conteh: ex Imperial Institute.
Cooper =C: in Cooper & Record; ex auctt.
Cooper & Record = C&R: 1931.
Coppins: ex Pichon.
Corbin: ex Dalziel; Herb. K.
Corkill: 1948.
Corre: Herb. K.
Coull: ex Dalziel; ex Irvine.
Coursey: 1966; 1968; with Ayensu.
Cox: Herb. K.
Croix: ex Dalziel. [? = De la Croix.]
Curasson: ex Kerharo & Bouquet.
Cyffer & Hutchinson, s.d., ms = C&H.
DA = Department of Africulture: of Herbaria.
D&C – Ducroz & Charles, 1978.
DF = Department of Forests, Conservator of Forests, Service Forestier, etc.: of Herbaria.
D'O = D'Orey.
DRR = Rosevear.
DS = Small.
Dade: ex Irvine.

Dalziel = JMD: 1937; Herb. K.
Darko: Herb. K.
Dawe: Herb. K.
Dawodu: Herb. K.
Deakin: ex Irvine.
Decordemoy: ex Dalziel.
Deighton = FCD: 1957; Herb. K, SL.
De la Croix: ex Kerharo.
Dennett: ex Dalziel; Herb. K.
Dept. of Agriculture = DA; ex Dalziel (all territories).
Dept. of Forests (Forestry) = DF, auctt. (all territories).
De Sousa = EPdS: ex Espirito Santo.
Dieterlen: Herb. IFAN.
Dodd: Herb. K.
D'Orey = D'O: Herb. K.
Droit: ex Dalziel.
Dudgeon: ex Dalziel; ex Irvine.
Dumas: ex Dalziel.
Dundas: Herb. K.
Dunlap: ex Dalziel; Herb. K.
Dunnett: ex. Dalziel; ex Pichon; Herb. K.
E – Engler.
E&D – Engler & Diels.
EK – unknown: Herb. K.
EPdS – De Sousa.
EWF – Foster.
Eady: Herb. K.
Easmon: ex Dalziel; ex Irvine.
Edgar: ex Dalziel.
Edy [? Eady]: Herb. K.
Eggeling: Herb. K.
Egunjobi: 1969.
Ejiofor: Herb. K.
Elliott: ex Dalziel; Herb. K.
Engler = E: ex Dalziel.
Engler & Diels = E&D: ex Dalziel.
Epenhuisjen = CWvE: Herb. K.
Esp. [Espirito] Santo = JDES: 1963; Herb. K.
Etesse: ex Pichon; Herb. K.
Evans: Herb. K.
Ewe Dictionary = Ewe Dict.: ex Irvine.
F [? = RF]: unknown, (Sierra Leone).
F&M – Foster & Munday.
FB – Busson.
FAM – Melville.
FCD – Deighton.
FNH – Hepper.
FP – Pellegrin.
FRI – Irvine.
FWHM – Migoed.
Faraclas: 1982.
Farmar: ex auctt.; Herb. K.
Farquhar: ex Dalziel; Herb. K.
Feijão = RdOF: 1960–63.
Fernandes: Herb. K.
Field: Herb. K.
Fisher: Herb. K.
Foggie: ex Irvine.
Foster = EWF: Herb. K.
Foster & Munday = F&M: 1961.
Foureau: ex A. Chevalier; ex Dalziel.
Fox: Herb. K.
Frith: Herb. K.
Froelich: 1954.
G&H – Guilhem & Hébert.
G&P – Guillemin & Perottet.
GeS – Gomes e Sousa.
GFS-E – Scott-Elliot.
GK – Kunkel.
GR – Roberty.
Gaisie: ex Irvine.
Gaisser: ex Dalziel.
Gamar: ex Trochain.
Garnier = PG: Herb. K.
Garrett: Herb. K.
Gbile: 1980 = ZOO.
Gent: Herb. K.
Gilman: ex Dalziel; Herb. K.
Giwa: Herb. UCI.
Glanville: ex Dalziel; ex Deighton; Herb. K.
Gledhill: Herb. FBC.
Glover: Herb. K.
Golding: with Gwynne, 1939; Herb. K.
Gomes e Sousa = GeS; 1930.
Graham: ex Dalziel; Herb. K.
Grall: 1945.
Gray: Herb. K.
Green, A. H.: 1951.
Green (Nigeria, ?J. Green): Herb. K.
Green, T. L. = TLG: Dioncophyllaceae, Herb. K.
Green[e], probably G S Green in Hepper & Neate (1971). (Ghana – Greene in Herb. K.; Green in Irvine).
Grégoire: 1975.
Gregory: Herb. K.
Grivot: 1949.
Grove: ex Brenan; Herb. K.
Gruner: ex Dalziel; ex Pichon.
Guilhem & Hébert = G&H: 1965.
Guillemin & Perottet = G&P; ex Dalziel.
Gurney: Herb. K.
Gwynne: with Golding, 1939.
H&C – Hua & Chevalier.
H&D – Hutchinson & Dalziel.
H&J – Hoyle & Jones.
Har. – G. W. Harley and W. J. Harley.
HS – Hunting Surveys.
Hall – Herb. K.
Hallé: 1962, b.
Harley = Har. (G. W. Harley and W. J. Harley): ex Dalziel; Herb. K.
Harper: Herb. K.
Hartwell: ex Irvine.
Haswell: Herb. FBC.
Hayes: ex Dalziel.
Hébert: with Guilhem; 1965.
Hédin: ex Dalziel.
Hendrick: 1979.[?]
Henry: with Amanu; (Dioscoreaceae, Togo) ex Burkill.
Hepburn: ex Dalziel; Herb. K.
Hepper = FNH: 1965; Herb. K.
Heudelot: ex Dalziel; Herb. K.
Hill, probably A. G. Hill: (Hausa, Nigeria) Herb. K.
Hinchingbrooke: Herb. K.
Holland = JHH: 1908–22; ex Dalziel, ex Irvine.
Hopkinson: ex Dalziel.
Houard: ex Dalziel.
Hovis: 1953.
Howes: ex Dalziel; ex Irvine; Herb. K.
Hoyle & Jones = H&J: Herb. K.
Hua & Chevalier = H&C.
Hunting Serveys = HS: 1968.
Huntting: ex Dalziel; Herb. K.
Hutchinson & Dalziel = H&D: 1937.
IHB – Burkill.
Imp. Inst. – Imperial Institute.
IFE – Departments of Pharmaceutics and Botany, University of Ife.
Imperial Institute = Imp. Inst.: Herb. K.
Ingram: ex Dalziel; Herb. K.
Irvine: 1930 to 1961; s.d.; Herb. K.
Ithmann: s.d.[?]

Ivanoff: ex Herharo & Bouquet.
J – G. Jackson.
J&D – Jackson & David.
JB – Berhaut.
JCA – Adam.
JDES – Espirito Santo.
JHH – Holland.
JLT – Trochain.
JM – Miège.
JMD – Dalziel.
JORU – Herbalists of Joru Village, Sierra Leone.
JRA – Ainslie.
Jackson, G. = J: (Nigeria) Herb. K.
Jackson, S. T.: (Ghana) Herb. K.
Jackson & David = J&D: in Jackson, G. 1973.
Jarr: Herb. FBC, K, SL.
Jenik & Hall: Herb. K.
Jibirin: Herb. K.
Johnson = W. H. Johnson (Ghana; S. Nigeria): Herb. K; ex Dalziel; ex Irvine.
Johnstone: ex Dalziel, (Nigeria, W Cameroons).
Jolly: ex Kerharo & Bouquet; ex Pichon; Herb. K, IFAN.
Jones, A. P. D., = APDJ: Herb. K.
Jordan: Herb. K, FBC.
K – Kerharo.
K&A – Kerharo & Adam.
K&B – Kerharo & Bouquet.
K&S – Katz & Schmutz.
KD – Kropp-Dabuku.
KM – Misczewski.
KO&S – Keay, Onochie & Stanfield.
KW – K. Williamson.
Kaichinger: ex A. Chevalier ex Dalziel.
Kasamany: Herb. K.
Katz & Schmutz = K&S.
Kaufmann: 1972.
Keay, Onochie & Stanfield = KO&S: 1960; 1964.
Kennedy: ex Dalziel; Herb. K.
Kerharo = K: 1966; 1967.
Kerharo & Adam = K&A: 1962 to 1974.
Kerharo & Bouquet = K&B: 1947; 1950.
Kersting: ex Dalziel; ex Pichon; Herb. K.
Kesby: Herb. K.
King, E. L.: Herb. K.
King-Church = LAK-C: Herb. K.
Kitson =AEK: ex Dalziel; ex Irvine; Herb. K.
Krause: ex Irvine.
Kropp-Dabuku: 1973 = KD.
Kucera: Herb. K.
Kunkel = GK: 1965.
Kyei: Herb. K.
LAK-C – King-Church.
LB – Bouquiaux: 1971–72.
LC – Chalk et al.: 1933.
LHS – L. H. Saunders.
Ll-W – Lloyd Williams.
L-P – Lane-Poole.
Laffitte: ex Kerharo & Bouquet; Herb. IFAN.
Lamb: Herb. K.
Lamborn: Herb. K.
Lancaster: Herb. K.
Lane-Poole = L-P: Herb. K.
Langdale-Brown: Herb. K.
Lawton: 1967.
Lean: Herb. K.
Lecerf: ex Pichon.
Leclerq: ex Dalziel.
Leeuwenberg = AJML: Herb. K.
Lefèvre: with Perrot; ex Dalziel.
Lely: Herb. K.
Léonard: 1955 to 1962.
Leprieur: ex Pichon.
Leriche: 1950 to 1952, (Mauritania).
Leroux: 1948.
Lester-Brown = Brown-Lester; Herb. K.
Letestu: ex Pichon.
Letham: ex Dalziel.
Lindall: Herb. K.
Lloyd-Williams: ex Dalziel; ex Irvine; Herb. K.
Lowe: 1973.
Lucas: 1967.
Lynn: ex Dalziel; ex Irvine.
Lyon: ex Irvine.
M&H – Melville & Hooker.
MM – M. McIntosh.
MP – Pichon.
McAinsh: Herb. K.
MacDonald: Herb. K.
McClintock: Herb. K.
McElderry: Herb. K.
MacGregor: Herb. K.
McIntosh: 1978.
MacKay: Herb.K.
Maclaud: ex Pichon.
McLeod: ex Dalziel, ex Pichon.
Macluskie: Herb. K.
Maire = RM: ex Dalziel.
Maitland: ex Dalziel; Herb. K.
Martineau: Herb. IFAN, K.
Martinson: ex Irvine.
Meek: ex Dalziel.
Mellin: ex Dalziel; ex Irvine; ex Thoms.
Melville = FAM: Herb. FBC, K.
Melville & Hooker = M&H: Herb. K.
Merlier: ex Kerharo & Adam, 1974.
Metzer: Herb. K (Dioscoreaceae, Togo).
Miège = JM: 1952.
Migoed = FWHM: ex Dalziel
Miguel: ex Pichon.
Mildbraed: ex Dalziel.
Millen: ex Dalziel; Herb. K.
Millson: ex Dalziel; Herb. K.
Misczewski = KM: Herb. K.
Moiser = BM: Herb. K.
Moktar Ould Hamidou: with Leriche, (Mauritania).
Moloney: ex Dalziel.
Monod: 1950; Herb. IFAN.
Monteil: 1953.
Moomo: 1982.
Moor: ex Irvine; Herb. K.
Musa Daggash: Herb. K.
N&E – Newberry & Etim.
NWT – N. W. Thomas.
Naegelé = AN: 1958a, b.
N'Diaye: 1964.
Néa: Herb. IFAN.
Newberry = RJN: Herb. K.
Newberry & Etim = N&E: Herb. K.
Newman: 1964, 1974.
Nicholas: 1953.
Obpon: Herb. UCI.
Odoh: 1978.[?]
Odukwe: Herb. UCI.
Oke: 1966.
Okeke: Herb. K.
Okpon: Herb. UCI.
Okusi: Herb. UCI.
Oldeman: Herb. K.
Oldfield: ex Dalziel; Herb. K.
Olu Akpata (Akpata, or Oluakputu): Herb. K.
Onochie: in Keay, Onochie & Stanfield, 1960, 1964; Herb. K.

Osnan: ex Irvine.
Ouedrago: 1950.
Ozanne: ex Dalziel; ex F. N. Williams.
PF –Peyre de Fauberge.
PG – P. Garnier.
Paroisse: ex Pichon.
Parsons: Herb. K.
Patterson: Herb. K.
Pax: ex Dalziel.
Pellegrin = FP: with Aubréville.
Pelly: Herb. SL.
Perrot: in Perrot and Lefèvre; ex Dalziel.
Perrottet: ex Pichon.
Peyre de Fauberge = PF.
Phillips: ex Dalziel; Herb. K.
Pichon = MP: 1953.
Planchon: ex Kerharo & Bouquet.
Plumtre: ex Dalziel; ex Irvine; Herb. K.
Pobéguin = CHOP: ex Dalziel; 1912.
Poisson: ex Pichon.
Port Development Syndicate = Port Develop. Synd.; Herb. K.
Portères = RP: s.d.; ex A. Chevalier; ex Kerharo & Bouquet.
Prost: ex Kerharo & Bouquet.
Punch: Herb. K.
Pyne: Herb. K.
RB – Blench.
RF: unidentified: ? = F; Herb. K.
RJN = Newberry.
RM = Maire.
RP = Portères.
RS = Schnell.
Ramsey: Herb. K.
Rattray: ex Dalziel; ex Irvine; Herb. K.
Reder: ex Dalziel.
Remondiére: ex Pichon.
Richards: ex White; Herb. BM, UCI.
Rider: ex Dalziel.
Roberty = GR: Herb. IFAN.
Robin: 1947.
Rodd: Herb. K.
Rogeon: ex Dalziel.
Rosevear = DRR: 1961; 1975.
Ross: ex Dalziel; Herb. K.
Rousselot: (Hauser, Niger).
Rowland: ex Pichon; Herb. K.
Ryan: Herb. K.
Rytz: sd
S&F – Savill & Fox.
SKS – Samai.
SOA – Alasoadura.
Sam: Herb. K.
Samai: Herb. K.
Sampenay: ex Dalziel; Herb. K.
Samuels: Herb. FBC.
Sankey: ex Dalziel; Herb. K.
Saunders: probably H. N. Saunders, ex Irvine; Herb. K. (Ghana).
Saunders: probably L. H. Saunders, Herb. K. (Cyperaceae, Nigeria).
Savill & Fox: 1967.
Sawyerr: Herb. K.
Schnell = RS: 1950–1960.
Scholz: Herb. K.
Schultze: ex Dalziel.
Schumann: ex Dalziel.
Scott-Elliot = GFS-E: ex Dalziel.
Sébire = AS: ex Dalziel; ex Kerharo & Adam; ex Portères.
Seefrieds: ex Dalziel; ex Pichon.
Segeur: ex Dalziel; ex Pichon.
Service Forestier – DF: (Ivory Coast).
Shaw: Herb. K.
Sidibé: 1939.
Siguade: Herb. IFE.
Sikes: 1972.
Singha: 1965.
Small = DS; Herb. K.
Smythe: Herb. K.
Soward: ex Dalziel; ex Irvine; Herb. K.
Symington: Herb. K.
TFC – Chipp.
TLG – T. L. Green.
Taiwo: Herb. K.
Talbot: ex Dalziel.
Taylor, C. J. = CJT, Herb. K, 1960.
Taylor: ex Dalziel; several persons, not individualised.
Tennant: 1961.
Thomas, A. S.: ex Irvine; Herb, K.
Thomas, D. G: ex Dalziel; Herb. K.
Thomas, N. W.: Herb. K.
Thoms: ex Dalziel
Thompson: Herb. K; ?H. M. Thompson – Ghana; ?C. P. Thompson – Nigeria.
Thompson-Clewry: Herb. K; FBC.
Thonning: ex Pichon.
Thornewill = AST.
Tiemo: 1968, ex K. Williamson.
Tindall: Herb. K.
Trochain = JLT: 1940; ex Dalziel.
Tudhope: ex Irvine.
Tuley: Herb. K.
Twi Dictionary – Twi Dict.; ex Irvine.
Ujor: Herb. K.
Ukpon: Herb. UCI.
Ulbrich: ex Dalziel.
Unwin = AHU; 1920; ex Dalziel; ex Pichon.
Vaillant: Herb. K.
Verger: 1968.
Vermeer: Herb. UCI.
Vigne = CV: ex Irvine; Herb. K.
Virgo: Herb. K.
Vogel: ex Dalziel.
Volkens: ex Dalziel.
Von Doering: ex Dalziel; Herb. K.
Voorhoeve = AGV: 1965.
Vuillet: ex Dalziel.
WTSB – Brown (Ghana).
Waldau: ex Dalziel.
Wallace: Herb. K.
Ward: Herb. K.
Warnecke: ex Volkens ex Dalziel.
Wedderburn: Herb. K.
West: Herb. K.
Westerman: ex Dalziel.
White, F.: 1956, and with Letouzey, 1970, 1978.
Winston: 1974–82, pers. comm.
Willey: Herb. K; ex Pichon.
Williams, F. N.: 1907; Herb. K.
Williamson, K.
Wulfsberg: ex Irvine.
Yates: ex Dalziel; Herb. K.
Yoruba Dictionary = Yor. Dict.: ex Dalziel.
Zenker: ex Dalziel.

# INDEX OF PLANT SPECIES

(The names of species treated in the text are given in roman print and can be located in the section of the family cited. Names used in *The Useful Plants of West Tropical Africa*, ed. 1, or in *The Flora of West Tropical Africa*, ed. 2, which have been reduced to synonymy, are given in italics with the present name alongside.)

C. lecardii Engl. & Diels
micranthum G. Don
molle R. Br.
mooreanum Exell
*mucronatum* Schum. & Thonn. = C. smeathmannii G. Don
nigricans Lepr.
paniculatum Vent., ssp. paniculatum.
*passargei* Engl. & Diels = C. glutinosum Perr.
platypterum (Welw.) Hutch. & Dalz.
racemosum P. Beauv.
rhodanthum Engl. & Diels
sericeum G. Don
smeathmannii G. Don
*sokodense* Engl. = C. molle R. Br.
tarquense J. J. Clark
tomentosum G. Don
*verticillatum* Engl. = C. collinum Fresen., ssp. geitonophyllum (Diels) Okafor
zenkeri Engl. & Diels
*sp. B.* of Keay = C. blepharopetala Wickens
spp. indet.
Commelina africana Linn. – **Commelinaceae**
benghalensis Linn.
bracteosa Hassk.
capitata Benth.
congesta C.B.Cl.
diffusa Burm. f.
erecta Linn.
forskalaei Vahl
imberbis Ehrenb.
lagosensis C.B.Cl.
nigritana Benth.
*nudiflora* Linn. = C. diffusa Burm. f.
subulata Roth
thomasii Hutch.
zambesica C.B.Cl.
spp. indet.
Commiphora africana (A. Rich.) Engl. – **Burseraceae**
dalzielii Hutch.
kerstingii Engl.
pedunculata (Kotschy & Peyr) Engl.
Connarus africanus Lam. – **Connaraceae**
griffonianus Baill.
Conocarpus erectus Linn. – **Combretaceae**
*Conopharyngia brachyantha* Stapf = Tabernaemontana brachyantha Stapf – **Apocynaceae**
*chippi* Stapf = Tabernaemontana chippii (Stapf) Pichon
*crassa* Stapf = Tabernaemontana crassa Benth.
*cumminsii* Stapf = Tabernaemontana pachysiphon Stapf
*durissima* Stapf = Tabernaemontana crassa Benth.
*pachysiphon* Stapf = Tabernaemontana pachysiphon Stapf
*penduliflora* Stapf = Tabernaemontana penduliflora Stapf
Convolvulus arvensis Linn. – **Convolvulaceae**
microphyllus Sieb.
Conyza aegyptiaca (Linn.) Ait.– **Compositae**
C. attenuata DC.
bonariensis (Linn.) Cronq.
canadensis (Linn.) Cronq.
*persicifolia* (Benth.) Oliv. & Hiern = C. attenuata DC.
pyrrhopappa Sch. Bip.
*spartioidea* O. Hoffm. = Nidorella spartioides (O. Hoffm.) Cronq.
stricta Willd.
subscaposa O. Hoffm.
sumatrensis (Retz.) E. H. Walker
Corallocarpus boehmi (Cogn.) C. Jeffrey – **Cucurbitaceae**
Cordia *abyssinica* R. Br. = C. africana Lam. – **Boraginaceae**
africana Lam.
aurantiaca Bak.
*gharaf* (Forssk.) Ehrenb. = C. sinensis Lam.
millenii Bak.
myxa Linn.
platythyrsa Bak.
*rothii* Roem. = C. sinensis Lam.
sebestena Linn.
senegalensis Juss.
sinensis Lam.
tisserantii Aubrév.
vignei Hutch. & Dalz.
sp. indet.
Coreopsis barteri Oliv. & Hiern – **Compositae**
boriania Sch. Bip.
Cornulaca monacantha Del. – **Chenopodiaceae**
Costus Linn. – **Costaceae**
afer Ker-Gawl.
dubius (Afzel.) K. Schum.
englerianus K. Schum.
lucanusianus J. Braun & K. Schum.
schlecteri Winkler
spectabilis (Fenzl) K. Schum.
Cotula anthemoides Linn. – **Compositae**
cinerea Del.
*Courbonia virgata* Brogn. = Maerua pseudopetalosa (Gilg & Bened.) De Wolf – **Capparaceae**
Crambe abyssinica Hochst. – **Cruciferae**
kilimandscharica O. E. Schulz
kralikii Coss.
Crassocephalum *biafrae* (Oliv. & Hiern ) S. Moore = Senecio biafrae Oliv. & Hiern – **Compositae**
crepidioides (Benth.) S. Moore
*mannii* (Hook. f.) Milne-Redhead = Senecio mannii Hook. f.
montuosum (S. Moore) Milne-Redhead
picridifolium (DC.) S. Moore
rubens (Juss.) S. Moore
sarcobasis (Boj.) S. Moore
vitellinum (Benth.) S. Moore
sp. indet.
*Crataeva adansonii* DC. (orth. aberr.) = Crateva adansonii DC. – **Capparaceae**
Crateva adansonii DC. – **Capparaceae**
*religiosa* Forst. f. = C. adansonii DC.

C. reduncus Hochst.
renschii Boeck
rotundus Linn.
sphacelatus Rottb.
tenuiculmis Boeck.
tonkinensis C.B.Cl., var. baikkiei (C.B.Cl.) Hooper
tuberosus Rottb.
zollingeri Steud.
spp. indet.
Cyrtosperma senegalensis (Schott) Engl. – **Araceae**
Dacryodes edulis (G. Don) H. J. Lam – **Burseraceae**
klaineana (Pierre) H. J. Lam
Dalzielia oblanceolata Turrill – **Asclepiadaceae**
Dennettia tripetala Bak. f. – **Annonaceae**
Dichapetalum *acutifolium* Engl. = D. tomentosum Engl. – **Dichapetalaceae**
*acutisepalum* Engl. = D. heudelotii (Planch.) Baill.
albidum A. Chev.
barteri Engl.
*chrysobalanoides* Hutch. & Dalz. = D. madagascariense Poir.
*ferrugineum* Engl. = D. heudelotii (Planch.) Baill.
*flexuosum* Engl. = D. madagascariense Poir.
*floribundum* (Planch.) Engl. = D. madagascariense Poir.
*guineense* (DC.) Keay = D. madagascariense Poir.
heudelotii (Planch.) Baill.
*johnstonii* Engl. = D. heudelotii (Planch.) Baill.
*kumasiense* Hoyle = D. heudelotii (Planch.) Baill.
*linderi* Hutch. & Dalz. = D. heudelotii (Planch.) Baill.
madagascariense Poir.
pallidum (Oliv.) Engl.
*pallidum* sensu Keay, p.p. = D. albidum A. Chev., p.p. = D. pallidum Engl.
*subauriculatum* (Oliv.) Engl. = D. heudelotii (Planch.) Baill.
*subcordatum* (Hook. f.) Engl. = D. madagascariense Poir.
*thomsonii* (Oliv.) Engl. = D. madagascariense Poir.
tomentosum Engl.
toxicarium (G. Don) Baill.
Dichorisandra thyrsiflora Mikan. f. – **Commelinaceae**
Dichrocephala chrysanthemifolia (Blume) DC. – **Compositae**
integrifolia (Linn. f.) O. Ktze
Dicliptera elliotii C.B.Cl. – **Acanthaceae**
laxata C.B.Cl.
verticillata (Forssk.) C. Christens.
Dicoma sessiliflora Harv. – **Compositae**
tomentosa Cass.
Dictyophleba leonensis (Stapf) Pichon – **Apocynaceae**
lucida (K. Schum.) Pierre

D. ochracea (K. Schum.) Pichon
rudens Hepper
Dieffenbachia Schott – **Araceae**
maculata (Lodd.) G. Don
*picta* Schott = D. maculata (Lodd.) G. Don
seguina (Jacq.) Schott
*Digera alternifolia* sensu FWTA, ed. 2 = Cyathula prostrata (Linn.) Blume – **Amaranthaceae**
Dillenia indica Linn. – **Dilleniaceae**
*Dimorphochlamys mannii* Hook. f. = Momordica cabraei (Cogn.) C. Jeffrey – **Cucurbitaceae**
*Dioncophyllum dawei* Hutch. & Dalz. = Habropetalum dawei (Hutch. & Dalz.) Airy Shaw – **Dioncophyllaceae**
*peltatum* Hutch. & Dalz. = Triphyophyllum peltatum (Hutch. & Dalz.) Airy Shaw
Dioscorea Linn. – **Dioscoraceae**
abyssinica Hochst.
alata Linn.
bulbifera Linn.
burkilliana Miège
cayenensis Lam.
*cayenensis* Lam. ssp. *rotundata* (Poir.) Miège. = D. rotundata Poir.
*colocasiifolia* Pax. = D. alata Linn.
dumetorum (Kunth) Pax
esculenta (Lour.) Burkill
hirtiflora Benth.
lecardii De Wild.
*liebrechtsiana* De Wild. = D. odoratissima Pax
*macroura* Harms. = D. sansibarensis Pax
mangenotiana Miège
minutiflora Engl.
odoratissima Pax
praehensilis Benth.
*praehensilis* sensu Miège p.p. = D. odoratissima Pax
preussii Pax
quartiniana A. Rich.
rotundata Poir.
*sagittifolia* sensu Miège p.p. = D. abyssinica p.p. = D. lecardii De Wild.
sansibarensis Pax
schimperana Hochst.
smilacifolia De Wild.
togoensis Knuth
spp. indet.
Diplocylos palmatus (Linn.) C. Jeffrey – **Cucurbitaceae**
Diplotaxis acris (Forssk.) Boiss. – **Cruciferae**
virgata (Cav.) DC.
Dischistocalyx T. Anders. – **Acanthaceae**
obanenis S. Moore
thunbergiiflorus (T. Anders.) Benth.
Dregea abyssinica (Hochst.) K. Schum. – **Asclepiadaceae**
crinita (Oliv.) Bullock
schimperi (Dec'ne) Bullock

L. owariensis P. Beauv.
parvifolia K. Schum.
*scandens* Didr. = Ancyclobotrys scandens (Schum. & Thonn.) Pichon
*senegalensis* Kotsch. & Peyr = Saba senegalensis (A. DC.) Pichon
subrepanda (K. Schum.) Pichon
*thompsonii* A. Chev. = Saba thompsonii (A. Chev.) Pichon
togolana (Hall. f.) Pichon
uniflora (Stapf) Pichon
utilis (A. Chev.) Pichon
violacea (K. Schum.) Pichon
Lankesteria brevior C.B.Cl. – **Acanthaceae**
elegans (P. Beauv.) T. Anders.
hispida (Willd.) T. Anders.
thyrsoidea S. Moore
Lannea acida A. Rich. – **Anacardiaceae**
*acidissima* A. Chev. = L. welwitschii (Hiern) Engl.
*afzelii* Engl. = L. nigritana (Sc. Elliot) Keay
barteri (Oliv.) Engl.
egregia Engl. & K. Krause
fruticosa (Hochst.) Engl.
humilis (Oliv.) Engl.
*kerstingii* Engl. & K. Krause = L. barteri (Oliv.) Engl.
microcarpa Engl. & K. Krause
nigritana (Sc. Elliot) Keay
schimperi (Hochst.) Engl.
velutina A. Rich.
welwitschii (Hiern) Engl.
spp.
Launaea chevalieri O. Hoffm. & Munschl – **Compositae**
cornuta (Oliv. & Hiern) O. Jeffrey
intybacea (Jacq.) Beauverd
nana (Bak.) Chiov.
taraxacifolia (Willd.) Amin
Lepidagathis alopecuroides (Vahl) R. Br. – **Acanthaceae**
collina (Endl.) Milne-Redhead
heudelotiana Nees
serica Benoist
Lepidium sativum Linn. – **Cruciferae**
Lepistemon owariense (P. Beauv.) Hall. f. – **Convolvulaceae**
Leptadenia arborea (Forssk.) Schweinf. – **Asclepiadaceae**
hastata (Pers.) Dec'ne
*lancifolia* Dec'ne = L. hastata (Pers.) Dec'ne
pyrotechnica (Forssk.) Dec'ne
sp. indet.
Licania elaeosperma (Mildbr.) Prance & White – **Chrysobalanaceae**
Limeum diffusum (Gay) Schinz – **Aizoaceae**
indicum Stocks
*linifolium* Fenzl = L. diffusum (Gay) Schinz
pterocarpum (Gay) Heimerl
viscosum (Gay) Fenzl
Limnophyton angolense Buchen. – **Alismataceae**
obtusifolium (Linn.) Miq.
L. *obtusifolium* sensu Dalziel = L. angolense Buchen.
Lipocarpa albiceps Ridl. – **Cyperaceae**
chinensis (Osb.) Kern
gracilis (L. C. Rich.) Pers.
*sphacelata* (Vahl) Kunth = L. gracilis (L. C. Rich.) Pers.
Lonicera Linn. – **Caprifoliaceae**
japonica Thumb.
longiflora DC.
Luffa acutangula Roxb. – **Cucurbitaceae**
*aegyptiaca* Mill. = L. cylindrica (Linn.) M. J. Roem.
cylindrica (Linn.) M. J. Roem.
Maerua angolensis DC. – **Capparaceae**
crassifolia Forssk.
duchesnei (De Wild.) F. White
oblongifolia (Forssk.) A. Rich.
pseudopetalosa (Gilg & Bened.) De Wolf
*rogeoni* A. Chev. = M. oblongifolia (Forssk.) A. Rich.
sp.
Magnistipula conrauana Engl. – **Chrysobalanaceae**
cupheiflora Mildbr. ssp. leonensis F. White
tessmannii (Engl.) Prance
zenkeri Engl.
Malouetia heudelotii A. DC. – **Apocynaceae**
Mangifera indica Linn.– **Anacardiaceae**
Manotes expansa Soland. – **Connaraceae**
longiflora Bak.
zenkeri Gilg
Mapania linderi Hutch. – **Cyperaceae**
Maranthes aubrevillei (Pellegr.) Prance – **Chrysobalanaceae**
chrysophylla (Oliv.) Prance
gabunensis (Engl.) Prance
glabra (Oliv.) Prance
kerstingii (Engl.) Prance
polyandra (Benth.) Prance
robusta (Oliv.) Prance
Margaretta rosea Oliv. – **Asclepiadaceae**
Mariscus alternifolius Vahl – **Cyperaceae**
flabelliformis Kunth
ligularis (Linn.) Urb.
longibracteatus Cherm.
soyauxii (Boeck.) C.B.Cl.
squarrosus (Linn.) C.B.Cl.
*umbellatus* Vahl = M. alternifolius Vahl
Markhamia lutea (Benth.) K. Schum. – **Bignoniaceae**
tomentosa (Benth.) K. Schum.
*Marsdenia latifolia* K. Schum = Gongronema latifolium Benth. – **Asclepiadaceae**
Mascarenhasia arborescens A. DC. – **Apocynaceae**
Maytenus acuminata (Linn. f.) Loes. – **Celastraceae**
senegalensis (Lam.) Exell
undata (Thunb.) Blakelock
Melanthera *brownei* Sch. Bip. = M. scandens (Schum. & Thonn.) Roberty – **Compositae**

M. elliptica O. Hoffm.
gambica Hutch. & Dalz.
scandens (Schum. & Thonn.) Roberty
*Melothria deltoidea* Benth. p.p. = Zehneria hallii O. Jeffrey – **Cucurbitaceae**
*fernandensis* Hutch. & Dalz. = Zehneria scabra (Linn. f.) Sond.
*maderaspatana* (Linn.) Cogn. = Mukia maderaspatana (Linn.) M. J. Roem.
*mannii* Cogn. = Zehneria scabra (Linn. f.) Sond.
*punctata* (Thunb.) Cogn. = Zehneria scabra (Linn. f.) Sond.
*tridactyla* Hook. f. = Zehneria thwaitesii (Schweinf.) C. Jeffrey
Merremia aegyptia (Linn.) Urban – **Convolvulaceae**
*alata* Rendle = Operculina macrocarpa (Linn.) Urban
*angustifolia* Hall. f. = M. tridentata (Linn.) Hall. f., ssp. angustifolia (Jacq.) Ooststr.
cissoides (Lam.) Hall. f.
dissecta (Jacq.) Hall. f.
hederacea (Burm. f.) Hall. f.
kentrocaulos (C.B.Cl.) Rendle
*pentaphylla* Hall. f. = M. aegyptia (Linn.) Urban
pinnata (Hochst.) Hall. f.
pterygocaulos (Steud.) Hall. f.
tridentata (Linn.) Hall. f. ssp. angustifolia (Jacq.) Ooststr.
tuberosa (Linn.) Rendle
umbellata (Linn.) Hall. f. ssp. umbellata
Microglossa afzelii O. Hoffm. – **Compositae**
*angolensis* Oliv. = Conyza pyrrhopappa Sch. Bip.
*caudata* O. Hoffm. & Muschl = Conyza pyrrhopappa Sch. Bip.
pyrifolia (Lam.) O. Ktze
sp. indet.
Mikania cordata (Burm. f.) B. L. Robinson – **Compositae**
*scandens* Willd. = M. cordata (Burm. f.) B. L. Robinson
Millingtonia hortensis Linn. f. – **Bignoniaceae**
Mimulopsis solmsii Schweinf. – **Acanthaceae**
*violacea* Lindau = M. solmsii Schwienf.
Mischogyne elliotianum (Engl. & Diels) R. E. Fries – **Annonaceae**
Mollugo cerviana (Linn.) Seringe – **Aizoaceae**
nudicaulis Lam.
pentaphylla Linn.
Moltkia callosa (Vahl) Wettst. – **Boraginaceae**
Momordica angustisepala Harms – **Cucurbitaceae**
balsamina Linn.
*bracteata* Hutch. & Dalz. = M. angustisepala Harms
cabraei (Cogn.) C. Jeffrey
charantia Linn.
M. cissoides Planch.
*cordata* Cogn. = M. foetida Schum. & Thonn.
foetida Schum. & Thonn.
Monanthotaxis barteri (Baill.) Verdc. – **Annonaceae**
diclina (Sprague) Verdc.
foliosa (Engl. & Diels) Verdc.
laurentii (De Wild.) Verdc.
stenopetala (Engl. & Diels) Verdc.
vogelii (Hook. f.) Verdc.
whytei (Stapf) Verdc.
Monechma ciliatum (Jacq.) Milne-Redhead – **Acanthaceae**
depauperatum (T. Anders.) C.B.Cl.
Monodora Dunal – **Annonaceae**
brevipes Benth.
crispata Engl. & Diels
myristica (Gaertn.) Dunal
tenuifolia Benth.
Monotes kerstingii Gilg – **Dipterocarpaceae**
Morettia canescens Boiss. – **Cruciferae**
Moricandia arvensis (Linn.) DC. – **Cruciferae**
Motandra guineensis (Thonn.) A. DC. – **Apocynaceae**
Mukia maderaspatana (Linn.) M. J. Roem. – **Cucurbitaceae**
Murdannia simplex (Vahl) Brenan – **Commelinaceae**
Musanga cecropioides R. Br. – **Cecropiaceae**
*smithii* R. Br. = M. cecropioides R. Br.
Myrianthus arboreus P. Beauv. – **Cecropiaceae**
libericus Rendle
serratus (Trécul) Benth.
Mystroxylon aethiopicum (Thunb.) Loes. – **Celastraceae**
Nelsonia *campestris* R. Br. = N. canescens (Lam.) Spreng. – **Acanthaceae**
canescens (Lam.) Spreng.
smithii Orsted
Nemium spadiceum (Lam.) Desv. – **Cyperaceae**
Neocarya macrophylla (Sab.) Prance – **Chrysobalanaceae**
Neostenanthera *bakuana* Exell = N. gabonensis (Engl. & Diels) Exell – **Annonaceae**
gabonensis (Engl. & Diels) Exell
hamata (Benth.) Exell
myristicifolia (Oliv.) Exell
*yalensis* Hutch. & Dalz. = N. hamata (Benth.) Exell
Nephthytis afzelii Schott – **Araceae**
Nerium odorum Sol. – **Apocynaceae**
oleander Linn.
Neuropeltis acuminata (P. Beauv.) Benth. – **Convolvulaceae**
velutina Hall. f.
Newbouldia laevis Seem. – **Bignoniaceae**
Nidorella spartioides (O. Hoffm.) Cronq. – **Compositae**
Nopalea cochenillifera (Linn.) Salm. Dyck. – **Cactaceae**
spp.

Nothosaerva brachiata (Linn.) Wight. – **Amaranthaceae**
Nucularia perrini Batt. – **Chenopodiaceae**
Ochroma *lagopus* Sw. = O. pyramidale (Cav.) Urb. – **Bombacaceae**
pyramidale (Cav.) Urb.
Odontonema cuspidatum (Nees) O. Ktze – **Acanthaceae**
Odontospermum graveolens Forssk. – **Compositae**
Oncinotis glabrata (Baill.) Stapf – **Apocynaceae**
gracilis Stapf
nitida Benth.
Operculina macrocarpa (Linn.) Urban – **Convolvulaceae**
Opuntia *cochenillifera* Linn. = Nopalea cochenillifera (Linn.) Salm. Dyck. – **Cactaceae**
tuna Mill.
vulgaris Mill.
spp.
Oreosyce africana Hook. f. – **Cucurbitaceae**
Orthopichonia schweinfurthii (Stapf) H. Huber – **Apocynaceae**
staudtii (Stapf) H. Huber
*Oxymitra gracilis* (Hook. f.) Sprague & Hutch. = Friesodielsia gracilis (Hook. f.) v. Steenis – **Annonaceae**
Oxystelma bornouense R. Br. – **Asclepiadaceae**
Ozoroa insignis Del. – **Anacardiaceae**
pulcherrima (Schweinf.) R. & A. Fernandes
Pachira aquatica Aubl. – **Bombacaceae**
insignis (SW.) Sav.
Pachycarpus lineolata (Dec'ne) Bullock – **Asclepiadaceae**
schweinfurthii (N. E. Br.) Bullock
Pachypodanthium barteri (Benth.) Hutch. & Dalziel – **Annonaceae**
staudtii Engl. & Diels
*Pachylobus deliciosa* Pellegr. = Dacryodes klaineana (Pierre) H. J. Lam – **Burseraceae**
*edulis* Don. = Dacryodes edulis (Don) H. J. Lam
*trimera* Engl. = Santiria trimera (Oliv.) Aubrév.
Pachystachys coccinea (Aubl.) Nees – **Acanthaceae**
Palisota ambigua (P. Beauv.) C.B.Cl. – **Commelinaceae**
barteri Hook.
bracteosa C.B.Cl.
hirsuta (Thunb.) K. Schum.
spp. indet.
Pancratium *hirtum* A. Chev. = P. tenuifolium Hochst. – **Amaryllidaceae**
tenuifolium Hochst.
trianthum Herb.
Pandiaka heudelotii (Moq.) Hook. f. – **Amaranthaceae**
involucrata (Moq.) Hook. f.
Pandorea pandorana (Andr.) van Steenis – **Bignoniaceae**
Parinari *aubrevillei* Pellegr. = Maranthes aubrevillei (Pellegr.) Prance – **Chrysobalanaceae**
*benna* Sc. Ell. = Bafodeya benna (Sc. Ell.) Prance
*chrysophylla* Oliv. = Maranthes chrysophylla (Oliv.) Prance
congensis F. Didr.
curatellifolia Planch
excelsa Sab.
*gabunensis* Engl. = Maranthes gabunensis (Engl.) Prance
*glabra* Oliv. = Maranthes glabra (Oliv.) Prance
*kerstingii* Engl. = Maranthes kerstingii (Engl.) Prance
*macrophylla* Sab. = Neocarya macrophylla (Sab.) Prance
*polyandra* Benth. = Maranthes polyandra (Benth.) Prance
*robusta* Oliv. = Maranthes robusta (Oliv.) Prance
*subcordata* Oliv. = P. congensis F. Didr.
*tenuifolia* A. Chev. = P. excelsa Sab.
*sp.* nr. *P. chrysophylla* Oliv. = Maranthes chrysophylla (Oliv.) Prance
'Parinari' sp. indet.
Paristolochia flos-avis (A. Chev.) Hutch. & Dalz. – **Aristolochiaceae**
goldieana (Hook. f.) Hutch. & Dalz.
promissa (Mast.) Keay
Parmentiera cereifera Seem. – **Bignoniaceae**
*Paulowilhelmia polysperma* Benth. = Eremomastax speciosa (Hochst.) Cufod. – **Acanthaceae**
*Pentagonanthus sp. dub.* (Periplocaceae) of FWTA. ed. 2 = Brachystelma bingeri A. Chev. – **Asclepiadaceae**
Pentarrhinum insipidum E. Mey – **Asclepiadaceae**
Pentatropis spiralis (Forssk.) Dec'ne – **Asclepiadaceae**
Peponium vogelii (Hook. f.) Engl. – **Cucurbitaceae**
Pergularis daemia (Forssk.) Chiov. – **Asclepiadaceae**
*extensa* N. E. Br. = P. daemia (Forssk.) Chiov.
tomentosa Linn.
Peristrophe bicalyculata (Retz) Nees – **Acanthaceae**
Phaulopsis barteri (T. Anders.) Lindau – **Acanthaceae**
falcisepala C.B.Cl.
imbricata (Forssk.) Sweet
*Phaylopsis falcisepala* C.B.Cl. = Phaulopsis falcisepala C.B.Cl. – **Acanthaceae**
*parviflora* Willd. = Phaulopsis imbricata (Forssk.) Sweet
Philoxerus vermicularis (Linn.) P. Beauv. – **Amaranthaceae**
*Physedra barteri* Cogn. = Coccinia barteri (Hook. f.) Keay – **Cucurbitaceae**

Rhoeo spathacea (Sw.) Stearn – **Commelinaceae**
Rhus longipes Engl. – **Anacardiaceae**
natalensis Bernh.
tripartita (Ucria) Grande
Rhynchospora corymbosa (Linn.) Britt. – **Cyperaceae**
holoschoenoides (L. C. Rich.) Herter
Rikliella kernii (Rayon.) J. Raynal – **Cyperaceae**
Ritchiea albersii Gilg – **Capparaceae**
capparoides (Andr.) Britten var. capparoides, & var. longipedicellata (Gilg) De Wolf
*duchesnei* (De Wild.) Keay = Maerua duchesnei (De Wild.) F. White
*fragiodora* Gilg = R. capparoides (Andr.) Britten var. capparoides
*fragrans* R. Br. = R. capparoides (Andr.) Britten var. capparoides
*longipedicellata* Gilg = R. capparoides (Andr.) Britten var. longipedicellata (Gilg) De Wolf
reflexa (Thonn.) Gilg & Benedict
Rorippa humifusa (Guill. & Perr.) Hiern – **Cruciferae**
nasturtium-aquaticum (Linn.) Hayek
Roureopsis obliquifoliolata (Gilg) Schellenb. – **Connaraceae**
Ruellia praetermissa Schweinf. – **Acanthaceae**
primuloides (T. Anders.) H. Heine
tuberosa Linn.
Rungia eriostachya Hua – **Acanthaceae**
grandis T. Anders.
*grandis* sensu Heine, p.p. = R. guineensis Heine
*grandis* sensu Dalziel, p.p. = R. guineensis Heine
guineensis Heine
Ruspolia hypocrateriformis (Vahl) Milne-Redhead – **Acanthaceae**
Ruthalicia eglandulosa (Hook. f.) C. Jeffrey – **Cucurbitaceae**
longipes (Hook. f.) C. Jeffrey
Saba florida (Benth.) Bullock – **Apocynaceae**
senegalensis (A. DC.) Pichon
thompsonii (A. Chev.) Pichon
Sagittaria guayanensis Kunth – **Alismataceae**
Salacia *baumannii* Loes. = S. leptoclada Tul. – **Celastraceae**
*caillei* A. Chev. = S. staudtiana Loes.
camerunensis Loes.
chlorantha Oliv. ssp. dalzielii (Hutch. & M. B. Moss) Hallé
columna Hallé
cornifolia Hook. f.
debilis (G. Don) Walp.
erecta (G. Don) Walp.
*erecta* sensu Blakelock p.p. = E. cornifolia Hook. f., p.p. = S. erecta (G. Don) Walp., p.p. = S. leptoclada Tul.
*gilgiana* Loes. = S. stuhlmanniana Loes.
lateritia Hallé
lehmbachii Loes.
leptoclada Tul.
S. *lomensis* Loes. = S. stuhlmanniana Loes.
nitida (Benth.) N. E. Br.
*nitida* sensu Blakelock p.p. = S. whytei Loes.
oliveriana Loes.
pallescens Oliv.
pyriformis (Sab.) Steud.
*pyriformis* sensu Blakelock p.p. = S. lateritia Hallé
senegalensis (Lam.) DC.
*senegalensis* sensu Blakelock p.p. = S. columna Hallé
staudtiana Loes.
stuhlmanniana Loes.
togoica Loes.
tuberculata Blakelock
whytei Loes.
*sp. B* of Blakelock = S. chlorantha Oliv. ssp. dalzielii (Hutch. & M. B. Moss) Hallé
*sp. C* of Blakelock = S. nitida (Benth.) N. E. Br.
*sp. E* of Blakelock = S. lateritia Hallé
Salsola baryosma (Schult.) Dandy – **Chenopodiaceae**
*foetida* Del. = S. baryosma (Schult.) Dandy
*tetragona* Del. = S. tetrandra Forssk.
tetrandra Forssk.
Santaloides afzelii (R. Br.) Schellenb. – **Connaraceae**
*gudjuanum* Schellenb. = S. afzelii (R. Br.) Schellenb.
Santiria trimera (Oliv.) Aubrév. – **Burseraceae**
Sarcostemma viminale (Linn.) R. Br. – **Asclepiadaceae**
Saritaea magnifica (Sprague) Dugand – **Bignoniaceae**
Satanocrater berhautii Benoist – **Acanthaceae**
Savignya parviflora (Del.) Webb – **Cruciferae**
*Scabiosa succisa* Linn. = Succisa trichotocephala Baksay – **Dipsacaceae**
Scadoxus cinnabarinus (Dec'ne) Friis & Nordal – **Amaryllidaceae**
multiflorus (Martyn) Raf.
sp. indet.
Schefflera barteri (Seem.) Harms – **Araliaceae**
*hierniana* Harms = S. barteri (Seem.) Harms
Schinus molle Linn. – **Anacardiaceae**
terebinthifolius Raddi
Schizoglossum petherickianum Oliv. – **Asclepiadaceae**
Schouwia *arabica* DC. = S. schimperi Jaub. & Spach – **Cruciferae**
purpurea (Forssk.) Schweinf.
schimperi Jaub. & Spach
Scirpus *angolensis* C.B.Cl. var. *brizaeformis* (Hutch.) Hooper = Nemium spadiceum (Lam.) Desv. – **Cyperaceae**

## INDEX OF PLANT AND PRODUCT NAMES

*in English, French, Portuguese and some other languages, and Trade names.*


A

Abacaxi – *Ananas comosus,* **Bromeliaceae**

Abeokuta bark – *Enantia polycarpa,* **Annonaceae**

Acajou – *Anacardium occidentale,* **Anacardiaceae**

Accra Niggers – *Landolphia owariensis,* **Apocynaceae**

Addah Niggers – *Landolphia owariensis,* **Apocynaceae**

Adjuaba – *Dacryodes klaineana,* **Burseraceae**

Adouaba à racines aériennes – *Santira trimera,* **Burseraceae**

Aerial yam – *Dioscorea bulbifera,* **Dioscoreaceae**

Afa – *Dioscorea rotundata,* **Dioscoreaceae**

Afara – *Terminalia superba,* **Combretaceae**

Afara, black – *Terminalia ivoriensis,* **Combretaceae**

Afara, white – *Terminalia superba,* **Combretaceae**

African arrowroot – *Canna edulis,* **Cannaceae**

African bdellium – *Commiphora africana, C. dalzielii,* **Burseraceae**

African canarium – *Canarium schweinfurthii,* **Burseraceae**

African cress – *Rorippa humifusa,* **Cruciferae**

African elemi – *Canarium schweinfurthii,* **Burseraceae**

African marigold – *Tagetes spp.,* **Compositae**

African myrrh – *Commiphora africana, C. dalzielii,* **Burseraceae**

African nutmeg – *Monodora myristica, M. tenuifolia,* **Annonaceae**

African pear – *Dacryodes edulis,* **Burseraceae**

African pepper – *Xylopia aethiopica,* **Annonaceae**

African spinach – *Amaranthus hybridus,* **Amaranthaceae**

African tulip tree – *Spathodea campanulata,* **Bignoniaceae**

African whitewood – *Enantia polycarpa,* **Annonaceae**

African yellowwood – *Enantia chlorantha, E. polycarpa,* **Annonaceae**

Agave – *Agave sisalana,* **Agavaceae**

Aiélé – *Canarium schweinfurthii,* **Burseraceae**

Air plant – *Bryophyllum pinnatum,* **Crassulaceae**

Air potato – *Dioscorea bulbifera,* **Dioscoreaceae**

Akoua – *Antrocaryon micraster,* **Anacardiaceae**

Alcaparra – *Capparis spinosa,* **Capparaceae**

Alder, West Indian – *Conocarpus erectus,* **Combretaceae**

Allamanda – *Allamanda spp.,* **Apocynaceae**

Alligator apple – *Annona glabra,* **Annonaceae**

Almond, Indian – *Terminalia catappa,* **Combretaceae**

Almond, Singapore – *Terminalia catappa,* **Combretaceae**

Alone – *Rhodognaphalon brevicuspe,* **Bombacaceae**

Alstonia – *Alstonia boonei,* **Apocynaceae**

Alstonia, false – *Rauvolfia macrophylla,* **Apocynaceae**

Amande de terre – *Cyperus esculentus,* **Cyperaceae**

Amandier (de Gambie) – *Terminalia catappa,* **Combretaceae**

Amaranth, globe – *Gomphrena globosa,* **Amaranthaceae**

Amaranth, green – *Amaranthus lividus, A. viridis,* **Amaranthaceae**

Amaranth, prickly – *Amaranthus spinosus,* **Amaranthaceae**

Amaranth, spiny – *Amaranthus spinosus,* **Amaranthaceae**

Amaranth, wild – *Amaranthus graecizans, A. lividus,* **Amaranthaceae**

Amaranthe bord de mer – *Philoxerus vermicularis,* **Amaranthaceae**

Amaranthe épineuse – *Amaranthus spinosus,* **Amaranthaceae**

Amaranthe du Soudan – *Amaranthus hybridus,* **Amaranthaceae**

Amazon lily – *Eucharis grandiflora,* **Amaryllidaceae**

Ambrósia-do-México – *Chenopodium ambrosioides,* **Chenopodiaceae**

Amendoeira-da-Índia – *Terminalia catappa,* **Combretaceae**

American mastic – *Schinus molle,* **Anacardiaceae**

Ananas – *Ananas comosus,* **Bromeliaceae**

Ananás – *Ananas comosus,* **Bromeliaceae**

Ananaseiro – *Ananas comosus,* **Bromeliaceae**

Anato – *Bixa orellana,* **Bixaceae**

Anatto – *Bixa orellana,* **Bixaceae**

Anchomanes, forest – *Anchomanes difformis,* **Araceae**

Anchomanes, savanna – *Anchomanes welwitschii,* **Araceae**

Angular (sponge) loofah – *Luffa acutangula,* **Cucurbitaceae**

Annatto – *Bixa orellana,* **Bixaceae**

Annone – *Annona senegalensis,* **Annonaceae**

Anogeissus – *Anogeissus leiocarpus,* **Combretaceae**

Anserine – *Chenopodium murale,* **Chenopodiaceae**

Ansérine – *Chenopodium ambrosioides,* **Chenopodiaceae**

Apple, alligator – *Annona glabra,* **Annonaceae**

Apple, balsam – *Momordica balsamina,* **Cucurbitaceae**

## INDEX OF PLANT AND PRODUCT NAMES

Batata doce – *Ipomoea batatas*, **Convolvulaceae**
Batata de purga – *Operculina macrocarpa*, **Convolvulaceae**
Batate – *Ipomoea batatas*, **Convolvulaceae**
Baton du sorcier – *Spathodea campanulata*, **Bignoniaceae**
Bdellium, African – *Commiphora africana*, *C. dalzielii*, **Burseraceae**
Bdellium d'Afrique – *Commiphora africana*, **Burseraceae**
Beach oak – *Terminalia scutifera*, **Combretaceae**
Beacon bush – *Salacia senegalenis*, **Celastraceae**
Beacon fire – *Salacia senegalensis*, **Celastraceae**
Bean, lizard – *Celosia argentea*, **Amaranthaceae**
Beautiful crinum – *Crinum zeylanicum*, **Amaryllidaceae**
Beefwood – *Casuarina equisitifolia*, **Casuarinaceae**
Beggar's ticks – *Bidens gen.*, **Compositae**
Bengal clock vine – *Thunbergia grandiflora*, **Acanthaceae**
Bhang – Cannabis sativa, **Cannabaceae**
Bignonia, yellow – *Tecoma stans*, **Bignoniaceae**
Billygoat weed – *Ageratum conyzoides*, **Compositae**
Bindweed – *Convolvulus arvensis*, **Convolvulaceae**
Bittercress, hairy – *Cardamine hirsuta*, **Cruciferae**
Bitter gourd – *Citrullus colocynthis*, **Cucurbitaceae**
Bitterleaf – Vernonia amygdalina, **Compositae**
Bitter yam – *Dioscorea dumetorum*, **Dioscoreaceae**
Black afara – *Terminalia ivorensis*, **Combretaceae**
Black bark (-ed) terminalia – *Terminalia ivorensis*, **Combretaceae**
Black-eyed Susan – *Thunbergia alata*, **Acanthaceae**
Black fellows – *Bidens pilosa*, **Compositae**
Black Jack – *Bidens pilosa*, **Compositae**
Black mangrove – *Avicennia germinans*, **Avicenniaceae**
Blood-flower – *Scadoxus cinnabarinus*, *S. multiflorus*, **Amaryllidaceae**; *Asclepias curassavica*, **Asclepiadaceae**
Blood plum – *Haematostaphis barteri*, **Anacardiaceae**
Blue bells – *Barleria cristata*, **Acanthaceae**
Blue jacaranda – *Jacaranda mimosifolia*, **Bignoniaceae**
Blue (-stemmed) taro – *Xanthosoma violaceum*, **Araceae**
Blue trumpet vine – *Thunbergia grandiflora*, **Acanthaceae**
Boat lily – *Rhoeo spathacea*, **Commelinaceae**
Bois bouchon – *Musanga cecropioides*, **Cecropiaceae**
Biois d'éléphant – *Combretum glutinosum*, **Combretaceae**
Bios satiné – *Terminalia ivorensis*, **Combretaceae**
Bolsa – *Ochroma pyramidale*, **Bombacaceae**
Bottle gourd – *Lagenaria siceraria*, **Cucurbitaceae**
Bombax – *Bombax buonopozense*, **Bombacaceae**
Bombwe, white – *Terminalia catappa*, **Combretaceae**
Bougainvillea, false – *Combretum racemosum*, **Combretaceae**
Boule de feu – *Scadoxus multiflorus*, **Amaryllidaceae**
Bouleau d'Afrique – *Anogeissus leiocarpus*, **Combretaceae**
Bouton violet – *Vernonia cinerea*, **Compositae**
Brazil cress – *Spilanthes filicaulis*, **Compositae**
Brazilian jalap – Merremia tuberosa, **Convolvulaceae**
Brazilian pepper (tree) – *Schinus molle*, *S. terebinthifolius*, **Anacardiaceae**
Brimstone wood – *Terminalia ivorensis*, **Combretaceae**
Broke-back – *Dichapetalum toxicarium*, **Dichapetalaceae**
Brown strophanthus – *Strophanthus hispidus*, **Apocynaceae**
Bulbil (-bearing) yam – *Dioscorea bulbifera*, **Dioscoreaceae**
Bullock's heart – *Annona reticulata*, **Annonaceae**
Bulrush – *Cyperus papyrus*, **Cyperaceae**
Bur, khaki – *Alternanthera pungens*, **Amaranthaceae**
Burmarigold – *Bidens gen.*, **Compositae**
Bush butter – *Dacryodes edulis*, **Burseraceae**
Bush candle tree – *Canarium schweinfurthii*, **Burseraceae**
Bush-cane – *Costus afer*, **Costaceae**
Bush clock vine – *Thunbergia erecta*, **Acanthaceae**
Bush greens – *Amaranthus hybridus*, **Amaranthaceae**
Bush pepper – *Xylopia staudtii*, **Annonaceae**
Bush rubber – *Funtumia africana*, *F. elastica*, **Apocynaceae**
Bush yam – *Dioscorea praehensilis*, **Dioscoreaceae**
Butter, bush – *Dacryodes edulis*, **Burseraceae**
Button-wood – *Conocarpus erectus*, **Combretaceae**

C

Cabbage, sea – *Brassica oleracea*, **Cruciferae**
Cabbage tree – *Vernonia conferta*, **Compositae**
Cabbage, wild – *Brassica oleracea*, **Cruciferae**
Cacao sauvage – *Pachira aquatica*, **Bombacaceae**

Herbe à verrues – *Heliotropium indicum*, **Boraginaceae**
Hog plum – *Spondias mombin*, **Anacardiaceae**
Holarrhène – *Holarrhena floribunda*, **Apocynaceae**
Holarrhène du Sénégal – *Holarrhena floribunda*, **Apocynaceae**
Honeysuckel – *Lonicera spp.*, **Caprifoliaceae**
Honeysuckle, Cape – *Tecoma capensis*, **Bignoniaceae**
Honeysuckle, Kaffir – *Tecoma capensis*, **Bignoniaceae**
Honeysuckle, yellow Cape – *Tecoma capensis*, **Bignoniaceae**
Horned cucumber – *Cucumis metulifera*, **Cucurbitaceae**
Horse purslane – *Trianthemum portulacastrum*, *Zaleya pentandra*, **Aizoaceae**
Humbug – *Capparis thonningii*, **Capparaceae**

I
Ibo tree – *Landolphia hirsuta*, **Apocynaceae**
Icaco – *Chrysobalanus icaco*, **Chrysobalanaceae**
Igname – *Dioscorea gen. & spp.*, **Dioscoreaceae**
Igname blanc – *Dioscorea alata*, *D. rotundata*, **Dioscoreaceae**
Igname des blancs – *Dioscorea esculenta*, **Dioscoreaceae**
Igname de brousse – *Dioscorea praehensilis*, **Dioscoreaceae**
Igname d'eau – *Dioscorea alata*, **Dioscoreaceae**
Igname étoilé – *Dioscorea hirtiflora*, **Dioscoreaceae**
Igname-Guinée – *Dioscorea cayenensis*, **Dioscoreaceae**
Igname jaune – *Dioscorea cayenensis*, **Dioscoreaceae**
Immortelles – *Helichrysum gen.*, **Compositae**
Immortelle à bouton – *Gomphrena globosa*, **Amaranthaceae**
Inca wheat – *Amaranthus hybridus*, **Amaranthaceae**
Incense tree – *Canarium schweinfurthii*, **Burseraceae**
Inch plant – *Zebrina pendula*, **Commelinaceae**
Indian almond – *Terminalia catappa*, **Combretaceae**
Indian heliotrope – *Heliotropium indicum*, **Boraginaceae**
Indian hemp – *Cannabis sativa*, **Cannabaceae**
Indian mustard – *Brassica juncea*, **Cruciferae**
Indian shot – *Canna edulis*, **Cannaceae**
Indian spinach – *Basella alba*, **Basellaceae**
Indian wormseed – *Chenopodium ambrosioides*, **Chenopodiaceae**
Inhame – *Colocasia esculenta*, **Araceae;** *Dioscorea gen. & spp.*, **Dioscoreaceae**
Inhame bravo – *Dioscorea alata*, **Dioscoreaceae**
Inhame-de-cariolá – *Dioscorea alata*, **Dioscoreaceae**
Inhame-da-China – *Dioscorea alata*, **Dioscoreaceae**
Inhame cicorero – *Dioscorea alata*, **Dioscoreaceae**
Inhame-da-Índia – *Dioscorea alata*, **Dioscoreaceae**
Inhame-da-Lisboa – *Dioscorea alata*, **Dioscoreaceae**
Inhame-da-S. Thomé – *Dioscorea alata*, **Dioscoreaceae**
Ipecacuanha, West Indian – *Asclepias curassavica*, **Asclepiadaceae**
Ipecacuanha, wild – *Asclepias curassavica*, **Asclepiadaceae**
Ipomée du caire – *Ipomoea cairica*, **Convolvulaceae**
Ipomée jaune – *Ipomoea obscura*, **Convolvulaceae**
Ipomée pied de chèvre – *Ipomoea pes-caprae*, **Convolvulaceae**
Ire, female – *Funtumia elastica*, **Apocynaceae**
Ire, male of – *Holarrhena floribunda*, **Apocynaceae**

J
Jacaranda (blue) – *Jacaranda mimosifolia*, **Bignoniaceae**
Jalap, Brasilian – *Merrimia tuberosa*, **Convolvulaceae**
Japanese radish – *Raphanus sativus*, **Cruciferae**
Jawe, false – *Aphanostylis mannii*, **Apocynaceae**
Jesuit's tea – *Chenopodium ambrosioides*, **Chenopodiaceae**
Johnny Walker – *Ipomoea quamoclit*, **Convolvulaceae**
Johnson's arum – *Amorphophallus johnsonii*, **Araceae**
Joseph's coat – *Graptophyllum pictum*, **Acanthaceae**

K
Kaffir honeysuckle – *Tecoma capensis*, **Bignoniaceae**
Kandia bark – *Enantia polycarpa*, **Annonaceae**
Kapok tree – *Ceiba pentandra*, **Bombacaceae**
Kapokier – *Bombax buonopozense*, *Ceiba pentandra*, **Bombacaceae**
Kapokier à fleurs blanches – *Ceiba pentandra*, **Bombacaceae**
Kapokier du Togo – *Ceiba pentandra*, **Bombacaceae**
Khaki bur – *Alternanthera pungens*, **Amaranthaceae**
Khaki weed – *Alternanthera pungens*, **Amaranthaceae**
Khât – *Combretum glutinosum*, **Combretaceae**
Kif – *Cannabis sativa*, **Cannabaceae**
King's mantle – *Thunbergia erecta*, **Acanthaceae**

Kinkéliba – *Combretum micranthum*, **Combretaceae**

Kokoti – *Parinari excelsa*, **Chrysobalanaceae**

Kola pimenté – *Buchholzia coriacea*, **Capparaceae**

Krepi ball – *Landolphia owariensis*, **Apocynaceae**

Kurchi bark – *Holarrhena floribunda*, **Apocynaceae**

L

Lady's tears – *Ozoroa pulcherrima*, **Anacardiaceae**

Lagos spinach – *Celosia argentea*, **Amaranthaceae**

Lagos yam – *Dioscorea cayenensis*, **Dioscoreaceae**

Lagos rubber – *Funtumia africana*, **Apocynaceae**

Lagos silk rubber tree – *Funtumia elastica*, **Apocynaceae**

Laiteron – *Sonchus oleraceus*, **Compositae**

Landolfia – *Baissea multiflora*, **Apocynaceae**

Langue de vache – *Launaea taraxacifolia*, **Compositae**

Large leaf yam – *Dioscorea alata*, **Dioscoreaceae**

Laurier jaune des Indies – *Thevetia neriifolia*, **Apocynaceae**

Laurier-rose – *Nerium oleander*, **Apocynaceae**

Leaf mustard – *Brassica juncea*, **Cruciferae**

Lesser (Asiatic) yam – *Dioscorea esculenta*, **Dioscoreaceae**

Lettuce, water – *Pistia stratiotes*, **Araceae**

Lettuce, wild – *Launaea taraxacifolia*, **Compositae**; *Brassica juncea*, **Cruciferae**

Liane aurore – *Pyrostegia venusta*, **Bignoniaceae**

Liane à caoutchouc – *Landolphia gen.*, **Apocynaceae**

Liane à l'eau – *Tetracera alnifolia*, *T. podotricha*, *T. potatoria*, **Dilleniaceae**

Liane étoilée – *Baissea multiflora*, **Apocynaceae**

Laine à faux – *Ipomoea setifera*, **Convolvulaceae**

Liane gohine – *Landolphia heudelotii*, **Apocynaceae**

Liane merveille – *Momordica charantia*, **Cucurbitaceae**

Liane rasoir – *Scleria boivinii*, **Cyperaceae**

Liane saba – *Saba senegalensis*, **Apocynaceae**

Liane tonnelle – *Operculina macrocarpa*, **Convolvulaceae**

Liane torchon – *Luffa acutangula*, *Luffa cylindrica*, **Cucurbitaceae**

Life plant – *Bryophyllum pinnatum*, **Crassulaceae**

Lily, Amazon – *Eucharis grandiflora*, **Amaryllidaceae**

Lily, boat – *Rhoeo spathacea*, **Commelinaceae**

Lily, Cape Coast – *Crinum gen.*, **Amaryllidaceae**

Lily, common ginger- – *Costus after*, **Costaceae**

Lily, crinum – *Crinum gen.*, **Amaryllidaceae**

Lily, dwarf forest ginger- – *Costus englerianus*, **Costaceae**

Lily, dwarf savanna ginger- – *Costus spectabilis*, **Costaceae**

Lily, eucharis – *Eucharis grandiflora*, **Amaryllidaceae**

Lily, fire-ball – *Scadoxus cinnabarinus*, *S multiflorus*, **Amaryllidaceae**

Lily, forest fire-ball – *Scadoxus cinnabarinus*, **Amaryllidaceae**

Lily, hairy ginger- – *Costus schlechteri*, **Costaceae**

Lily, Hanga – *Crinum gen.*, **Amaryllidaceae**

Lily, Mabanta – *Crinum gen.*, **Amaryllidaceae**

Lily, Madonna – *Eucharis grandiflora*, **Amaryllidaceae**

Lily, pancratium – *Pancratium trianthum*, **Amaryllidaceae**

Lily, small fire-ball – *Scadoxus multiflorus*, **Amaryllidaceae**

Lily, spider – *Hymenocallis littoralis*, **Amaryllidaceae**

Lily, wild canna – *Canna indica*, **Cannaceae**

Limba – *Terminalia superba*, **Combretaceae**

Limbo blanc – *Terminalia superba*, **Combretaceae**

Limbo clair – *Terminalia superba*, **Combretaceae**

Limbo noir – *Terminalia superba*, **Combretaceae**

Lis de brousse – *Crinum gen.*, **Amaryllidaceae**

Lisbon yam – *Dioscorea alata*, **Dioscoreaceae**

Liseron blue – *Ipomoea nil*, **Convolvulaceaee**

Liseron des champs – *Convolvulus arvensis*, **Convolvulaceae**

Liseron pied de chèvre – *Ipomoea pes-caprae*, **Convolvulaceae**

Liseron rouge – *Ipomoea hederifolia*, **Convolvulaceae**

Lizard bean – *Celosia argentea*, **Amaranthaceae**

Local tete – *Amaranthus viridis*, **Amaranthaceae**

Lollipop climber – *Diplocyclos palmatus*, **Cucurbitaceae**

Loloti des savannes – *Lannea nigritana*, **Anacardiaceae**

Loofah – *Luffa acutangula*, *L. cylindrica*, **Cucurbitaceae**

Loofah, angular (-sponge) – *Luffa acutangula*, **Cucurbitaceae**

Loofah, fluted – *Luffa acutangula*, **Cucurbitaceae**

Loofah, smooth – *Luffa cylindrica*, **Cucurbitaceae**

Loofah gourd – *Luffa cylindrica*, **Cucurbitaceae**

Loureiro – *Nerium oleander*, **Apocynaceae**

Love-lies-ableeding – *Amaranthus hybridus*, **Amaranthaceae**

Moses in a boat (cradle, the bulrushes, on a raft) – *Rhoeo spathacea*, **Commelinaceae**
Moshi medicine – *Guiera senegalensis*, **Combretaceae**
Mother-in-law's tongue plant – *Dieffenbachia gen.*, **Araceae**
Mountain spice – *Xylopia acutiflora*, **Annonaceae**
Mouron des oiseaux – *Stellaria media*, **Caryophyllaceae**
Mukonja, white – *Terminalia superba*, **Combretaceae**
Mummy apple – *Carica papaya*, **Caricaceae**
Muscadier de calabash – *Monodora myristica*, **Annonaceae**
Musk melon – *Cucurbita moschata*, **Cucurbitaceae**
Musk tree – *Buchholzia coriacea*, **Capparaceae**
Mustard, bastard – *Cleome gen.*, *C. gynandra*, **Capparaceae**
Mustard, Chinese – *Brassica juncea*, **Cruciferae**
Mustard, field – *Brassica campestris*, **Cruciferae**
Mustard, Indian – *Brassica juncea*, **Cruciferae**
Mustard, leaf – *Brassica juncea*, **Cruciferae**
Mustard, wild – *Cleome viscosa*, **Capparaceae**
Myrobalan d'Egypte – *Balanites aegyptiaca*, **Balanitaceae**
Myrobalan, Egyptian – *Balanites aegyptiaca*, **Balanitaceae**
Myrrh, African – *Commiphora africana*, *C. dalzielii*, **Burseraceae**
Myrrh africaine – *Commiphora africana*, **Burseraceae**

N

Nanas – *Ananas comosus*, **Bromeliaceae**
Native pear – *Dacryodes edulis*, **Burseraceae**
Negro pepper – *Xylopia quintasii*, **Annonaceae**
Neou oil tree – *Neocarya macrophylla*, **Chrysobalanaceae**
Nerve plant – *Fittonia verschaffeltii*, **Acanthaceae**
Never-die – *Bryophyllum pinnatum*, *Kalanchoe crenata*, **Crassulaceae**
New cocoyam – *Xanthosoma mafaffa*, **Araceae**
New Zealand spinach – *Tetragona tetragonioides*, **Aizoaceae**
Newet – *Brassica rapa*, **Cruciferae**
Nico – *Licania elaeosperma*, **Chrysobalanaceae**
Night-shade, Malabar – *Basella alba*, **Basellaceae**
Nik(k)o – *Licania elaeosperma*, **Chrysobalanaceae**
Niger seed – *Guizotia abyssinica*, **Compositae**
Nigerian watercress – *Rorippa humifusa*, **Cruciferae**
Noix d'acajou – *Anacardium occidentale*, **Anacardiaceae**
Noix de gajou – *Anacardium occidentale*, **Anacardiaceae**
Noix muscade, fausse – *Monodora myristica*, **Annonaceae**
Nopal à cochenille – *Nopalea spp.*, **Cactaceae**
Norfolk Island pine – *Araucaria heterophylla*, **Araucariaceae**
Noyer d'Amerique – *Bombacopsis glabra*, **Bombacaceae**
Noyer du Moyambe – *Terminalia superba*, **Combretaceae**
Nut-grass – *Cyperus rotundus*, **Cyperaceae**
Nut-grass, yellow – *Cyperus esculentus*, **Cyperaceae**
Nutmeg, African – *Monodora myristica*, *M. tenuifolia*, **Annonaceae**
Nutmeg, calabash – *Monodora myristica*, **Annonaceae**
Nutmeg, Calabar – *Monodora myristica*, **Annonaceae**
Nutmeg, false – *Monodora myristica*, **Annonaceae**

O

Oak, beach – *Terminalia scutifera*, **Combretaceae**
Oak, she- *Casuarina equisitifolia*, **Casuarinaceae**
Oak, Queensland swamp – *Casuarina equisitifolia*, **Casuarinaceae**
Ofram – *Terminalis superba*, **Combretaceae**
Ofruntum, white – *Funtumia africana*, **Apocynaceae**
Oignon de gorille – *Buchholzia coriacea*, **Capparaceae**
Oil-nut – *Telfairea occidentalis*, **Cucurbitaceae**
Okumé – *Canarium schweinfurthii*, **Burseraceae**
Oleander – *Nerium oleander*, **Apocynaceae**
Oleander, yellow – *Thevetia neriifolia*, **Apocynaceae**
Olive mangrove – *Avicennia germinans*, **Avicenniaceae**
Oriental radish – *Raphanus sativus*, **Cruciferae**
Ornamental gourd – *Cucurbita pepo*, **Cucurbitaceae**
Otaheite apple – *Spondias cytherea*, **Anacardiaceae**
Otaheite yam – *Dioscorea bulbifera*, **Dioscoreaceae**
Otu – *Cleistopohlis patens*, **Annonaceae**
Oukoumé – *Aucoumea klaineana*, **Burseraceae**
Ox-eye, creeping – *Wedelia trilobata*, **Compositae**
Oyster nut – *Telfairea occidentalis*, **Cucurbitaceae**
Oyster plant – *Rhoeo spathacea*, **Commelinaceae**

P

Pain de singe – *Adansonia digitata*, **Bombacaceae**
Palette de peinture – *Caladium bicolor*, **Araceae**
Palétuvier blanc – *Avicennia germinans*, **Avicenniaceae**
Palétuvier, faux – *Avicennia germinans*, **Avicenniaceae**
Palétuvier gris – *Conocarpus erectus*, **Combretaceae**
Palétuvier noir – *Laguncularia racemosa*, **Combretaceae**
Palétuvier zaragosa – *Conocarpus erectus*, **Combretaceae**
Pancratium lily – *Pancratium trianthum*, **Amaryllidaceae**
Pao cadeiro – *Funtumia africana*, **Apocynaceae**
Papangaie – *Luffa acutangula*, **Cucurbitaceae**
Papaw – *Carica papaya*, **Caricaceae**
Papaya – *Carica papaya*, **Caricaceae**
Papayer – *Carica papaya*, **Caricaceae**
Papiro – *Cyperus papyrus*, **Cyperaceae**
Pap-leaf – *Phaulopsis imbricata*, **Acanthaceae**
Papyrus – *Cyperus papyrus*, **Cyperaceae**
Papyrus, faux – *Cyperus alternifolius*, **Cyperaceae**
Para-cress – *Spilanthes filicaulis*, **Compositae**
Parasolier – *Musanga cecropioides*, **Cecropiaceae**
Parinari – *Parinari excelsa*, **Chrysobalanaceae**
Paste rubber – *Saba florida*, **Apocyaceae**
Pastèque – *Citrullus lanatus*, **Cucurbitaceae**
Patate – *Ipomoea batatas*, **Convolvulaceae**
Patate aquatique – *Ipomoea aquatica*, **Convolvulaceae**
Patate douce – *Impomoea batatas*, **Convolvulaceae**
Patte de lièvre – *Ochroma pyramidale*, **Bombacaceae**
Pattern wood – *Alstonia boonei*, **Apocynaceae**
Pawpaw – *Carica papaya*, **Caricaceae**
Pear, African – *Dacryodes edulis*, **Burseraceae**
Pear, balsam – *Momordica charantia*, **Cucurbitaceae**
Pear, native – *Dacryodes edulis*, **Burseraceae**
Pear, vegetable – *Sechium edule*, **Cucurbitaceae**
Pepper, African – *Xylopia aethiopica*, **Annonaceae**
Pepper, Brazilian – *Schinus terebinthifolius*, **Anacardiaceae**
Pepper, bush – *Xylopia staudtii*, **Annonaceae**
Pepper, Ethiopian – *Xylopia aethiopica*, **Annonaceae**
Pepper, Guinea – *Xylopia aethiopica*, *X. staudtii*, **Annonaceae**
Pepper, negro – *Xylopia quintasii*, **Annonaceae**
Pepper fruit – *Dennettia tripetala*, **Annonaceae**
Peppertree, California – *Schinus molle*, **Anacardiaceae**
Peppertree, Brazilian – *Schinus molle*, **Anacardiaceae**
Perfume tree – *Cananga odorata*, **Annonaceae**
Pernambuc – *Myrianthus arboreus*, **Cecropiaceae**
Periwinkle, Madagascar – *Catharanthus roseus*, **Apocynaceae**
Pervenche de Madagascar – *Catharanthus roseus*, **Apocynaceae**
Petit manglier – *Conocarpus erectus*, **Combretaceae**
Philippine violet – *Barleria cristata*, **Acanthaceae**
Pigweed, sweet – *Chenopodium ambrosioides*, **Chenopodiaceae**
Piège à papillon – *Drosera indica*, **Droseraceae**
Piment noir de Guinée – *Xylopia aethiopica*, **Annonaceae**
Pimenta da Guiné – *Xylopia aethiopica*, **Annonaceae**
Pina silk – *Ananas comosus*, **Bromeliaceae**
Pine, Norfolk Island – *Araucaria heterophylla*, **Araucariaceae**
Pine, whistling – *Casuarina equisitifolia*, **Casuarinaceae**
Pine, yellow – *Terminalia ivoriensis*, *T. superba*, **Combretaceae**
Pineapple – *Ananas comosus*, **Bromeliaceae**
Pineapple, ground – *Thonningia sanguinea*, **Balanophoraceae**
Pink poui – *Tabebuia rosea*, **Bignoniaceae**
Pipangaie – *Luffa acutangula*, **Cucurbitaceae**
Plantation weed, Enugu – *Chromolaena odorata*, **Compositae**
Plum, Ashanti – *Spondias mombin*, **Anacardiaceae**
Plum, Assyrian – *Cordia myxa*, **Boraginaceae**
Plum, blood – *Haematostaphis barteri*, **Anacardiaceae**
Plum, carandas – *Carissa edulis*, **Apocynaceae**
Plum, coco – *Chrysobalanus icaco*, **Chrysobalanaceae**
Plum, dawi – *Parinari excelsa*, **Chrysobalanaceae**
Plum, English – *Spondias cytherea*, **Anacardiaceae**
Plum, Gambia – *Spondias purpurea*, **Anacardiaceae**
Plum, ginger (-bread) – *Neocarya macrophylla*, **Chrysobalanaceae**
Plum, grey – *Parinari excelsa*, **Chrysobalanaceae**
Plum, Guinea – *Parinari excelsa*, **Chrysobalanaceae**
Plum, hog – *Spondias mombin*, **Anacardiaceae**
Plum, monkey – *Dacryodes klaineana*, **Burseraceae**
Plum, rotten – *Neocarya macrophylla*, **Chrysobalanaceae**

## PLANT SPECIES BY USAGES

Plant species are listed here by recorded usage. Many usages are well-established and are indisputable. Others are not so well known and readers are enjoined to make their own subjective judgements on the evidence offered and their own additional experience, particularly in instances where risk of human damage may result. There is much yet to be learnt regarding the harvesting of, say, famine-foods, and their preparation for consumption, and in the phytochemical/pharmacological field on edaphic and phenological factors and on methods by which medicinal materials are collected and prepared, as, indeed, on safe and certain identification of the plants themselves.

A: Food; 1, general
2, special diets
3, sauces, condiments, spices, flavourings
4, sweets, sweetmeats
5, masticatory

B: Drink; 1, water/sap
2, sweet, milk-substitutes
3, infusions
4, water-purifiers
5, alcoholic, stimulant

C: Medicines; 1, generally healing
2, skin, mucosae
3, cutaneous, subcutaneous parasitic infection
4, skeletal structure
5, paralysis, epilepsy, convulsions, spasm
6, insanity
7, brain, nervous system
8, heart
9, arteries, veins
10, blood disorders
11, pain-killers
12, sedatives, etc.
13, arthritis, rheumatism, etc.
14, eye treatments
15, ear treatments
16, oral treatments
17, naso-pharyngeal affections
18, pulmonary troubles
19, stomach troubles
20, 'intestines'
21, emetics
22, antemetics
23, laxatives, etc.
24, diarrhoea, dysentery
25, cholera
26, vermifuges
27, liver, etc.
28, kidneys, diuretics
29, anus, haemorrhoids
30, genital stimulants/depressants
31, menstrual cycle
32, pregnancy, antiabortifacients
33, lactation stimulants (incl. veterinary)
34, abortifacients, ecbolics
35, venereal diseases
36, febrifuges
37, small-pox, chickenpox, measles, etc.
38, leprosy
39, yaws
40, dropsy, swellings, oedema, gout
41, tumours, cancers
42, malnutrition, debility
43, food-poisoning
44, antidotes (venomous stings, bites, etc.)
45, homeopathic

D: Phytochemistry; 1, alkali salts (excl. common salt)
2, soap and substitutes
3, salt and substitutes
4, mineral salts
5, fatty acids, etc.
6, aromatic substances
7, starch, sugar
8, mucilage
9, alkaloids
10, glycosides, saponins, steroids
11, tannins, astringents
12, flavones
13, resins
14, hydrogen cyanide
15, fish-poisons
16, insecticides, arachnicides
17, rodenticides, mammal and bird poisons
18, reptile repellants
19, molluscicides
20, arrow-poisons
21, ordeal-poisons
22, depilatories
23, antibiotic, bacteristatic, fungistatic
24, uricant
25, miscellaneously poisonous or repellant

E: Agri-horticulture; 1, ornamental, cultivated or partially tended
2, fodder
3. veterinary medicine
4, bee/honey plants, insect plants
5, land conservation
6, composting, manuring
7, indicators (soil, water)
8, indicators (weather, season, time)
9, shade trees
10, hedges, markers
11, fence-posts, poles, sticks
12, weeds, parasites
13, biotically active

F: Products; 1, building materials
2, carpentry and related applications
3, farming, forestry, hunting, fishing apparatus
4, fuel and lighting
5, household, domestic and personal items
6, pastimes — carving, musical instruments, games, toys, etc.
7, containers, food-wrappers
8, abrasives, cleaners, etc.
9, chew-sticks, etc.
10, fibre
11, floss, stuffing, caulking
12, withies and twigs
13, pulp and paper
14, pottery
15, beehives
16, dyes, stains, ink, tattoos and mordants
17, tobacco, snuff
18, exudations – gums, resins, etc.
19, manna and other exudations

G: Social; 1, religion, superstition, magic
2, ceremonial
3, sayings and aphorisms

**A: Food; 1, cooked or uncooked; staple, supplementary or famine foods (including those eaten only in limited quantity or requiring special treatment before eating to reduce or eliminate toxic substances).**

ACANTHACEAE: *Asystasia calycina* (leaf), *A. gangetica* (leaf), *Barleria brownii* (young leaf), *B. opaca* (leaf), *Blepharis linariifolia* (seed), *Eremomastax speciosa* (leaf), *Hygrophila auriculata, Hypoestes verticillata, Justicia schimperi/striata* (leafy stem), *J. flava, Phaulopsis imbricata, Ruellia praetermissa, Pseuderanthemum ludovicianum, P. tunicatum, Thunbergia alata, T. grandiflora.* AIZOACEAE: *Glinus lotoides, G. oppositifolius, Gisekia pharnaceoides, Limeum indicum* (seed), *L. viscosum* (seed), *Mollugo pentaphylla, Sesuvium portulacastrum, Tetragona tetragonioides, Trianthema portulacastrum, Zaleya pentandra.* ALISMATACEAE: *Caldesia reniformis* (leaf, stem), *Burnatia enneandra* (tuber), *Sagittaria guayanensis* (tuber). AMARANTHACEAE: *Achyranthus aspera, Aerva javanica, A. lanata, Alternanthera maritima, A. nodiflora, A. sessilis, Amaranthus dubius, A. graecizans, A. hybridus, A. lividus, A. spinosus, A. tricolor, A. viridis, Celosia argentea, C. globosa, C. isertii, C. leptostachya, C. pseudovirgata, C. trigyna, Cyathula prostrata, Gomphrena globosa, Philoxerus vermicularis, Sericostachys scandens.* ANACARDIACEAE: *Anacardium occidentale* (young leaf, fruit, nut), *Antrocaryon klaineanum* (fruit, seed), *A. microaster* (fruit), *Fegimanra afzelii* (seed), *Haematostaphis barteri* (fruit, kernel), *Lanna acida* (young leaf, fruit pulp), *L. fruticosa* (root), *L. humilis* (root), *L. microcarpa* (leaf, gum, fruit), *L. nigritana* (fruit), *L. schimperi* (fruit), *L. welwitschii* (fruit), *Mangifera indica* (young leaf, fruit, kernel), *Rhus longipes* (fruit), *R. natalensis* (fruit pulp, seed), *Sclerocarya birrea* (fruit pulp, kernel), *Sorindeia juglandifolia* (fruit), *S. warneckei* (fruit), *Spondias cythera* (young shoot, fruit), *S. mombin* (fruit), *S. purpurea* (fruit), *Trichoscypha acuminata* (fruit), *T. arborea* (fruit), *T. chevalieri* (fruit), *T. longifolia* (kernel), *T. yapoensis* (fruit). ANNONACEAE: *Annona arenaria* (fruit, seed), *A. chrysophylla* (fruit), *A. glabra* (fruit), *A. muricata* (fruit), *A. reticulata* (fruit), *A. senegalensis* (leaf, fruit), *A. squamosa* (fruit), *Annonidium mannii* (fruit), *Hexalobus monopetalus* (fruit, seed), *Isolona deightonii* (fruit), *Monanthotaxis laurentii* (fruit), *M. vogelii* (fruit), *Monodora tenuifolia* (fruit), *Uvaria chamae* (fruit), *U. doeringii* (fruit), *U. ovata* (fruit). APOCYNACEAE: *Ancylobtrys amoena* (fruit, seed), *A. pyriformis* (seed), *A. scandens* (fruit pulp), *Aphanostylis mannii* (fruit pulp), *Carissa edulis* (fruit), *Clitandra cymulosa* (fruit, seed), *Cylindropsis parviflora* (fruit, seed), *Dictyophleba leonensis* (fruit), *D. lucida* (fruit), *D. ochracea* (fruit), *Hunteria elliotii* (fruit), *Isonema smeathmannii* (leaf), *Landolphia calabarica* (fruit), *L. congolensis* (fruit), *L. dulcis* (fruit), *L. foretiana* (fruit), *L. heudelotii* (fruit), *L. hirsuta* (fruit), *L. landolphioides* (fruit), *L. macrantha* (fruit, seed), *L. membranacea* (fruit), *L. owariensis* (fruit), *L. subrepanda* (fruit), *L. violacea* (fruit), *Orthopichonia schweinfurthii* (fruit), *Saba florida* (fruit, seed), *S. senegalensis* (fruit), *Strophanthus preussii* (young leaf), *Thevetia neriifolia* (fruit pulp), *Vahadenia caillei* (fruit), *V. laurentii* (fruit, seed). APONOGETONACEAE: *Aponogeton subconjugatus* (rhizome). AQUIFOLIACEAE: *Ilex mitis* (bark, root). ARACEAE: *Alocasia macrorhiza* (stem), *Amorphophallus abyssinicus* (root), *A. aphyllus* (root), *Anchomanes difformis* (root), *A. welwitschii* (root), *Cyrtosperma senegalense* (leaf), *Pistia stratiotes* (plant), *Rhektophyllum mirabile* (leaf, spadix), *Stylochiton hypogaeus* (inflorescence), *S. lancifolius* (leaf, root), *Typhonium trilobatum* (leaf, root), *Xanthosoma mafaffa* (leaf, root), *X. violaceum* (root). ARALIACEAE: *Cussonia arborea* (fruit). ASCLEPIADACEAE: *Aspidoglossum interruptum* (root), *Brachystelma bingeri* (tuber), *B. togoense* (tuber), *Caralluma mouretii* (plant), *Ceropegia spp.* (leaf, tuber), *Cynanchium adalinae* (fruit), *Dregea abyssinica* (leaf), *Glossonema boveanum* (plant), *Gomphocarpus fruticosus* (root, fruit), *Gongronema angolense* (root), *G. latifolium* (leaf), *Gymnema sylvestre* (leaf), *Leptadenia arborea* (fruit), *L. hastata* (leaf, shoot, flower), *L. pyrotechnica* (flower), *Pachycarpus lineolata* (root), *Pentarrhinum insipidum* (leaf, fruit), *Pentatropis spiralis* (tuber), *Pergularia daemia* (leaf, fruit), *Sarcostemma viminale* (stem, fruit). AVICENNIACEAE: *Avicennia germinans* (fruit). BALANITACEAE: *Balanites aegyptiaca* (leaf, flower, kernel), *B. wilsoniana* (seed). BALSAMINACEAE: *Impatiens balsamina* (leaf, seed), *I. irvingii* (leaf), *I. sakerana* (fruit). BASELLACEAE: *Basella alba* (leaf, stem, fruit). BEGONIACEAE: *Begonia mannii* (leafy stem). BIGNONIACEAE: *Kigelia africana* (seed), *Parmentieria cereifera* (fruit), *Spathodia campanulata* (leaf, seed), *Stereospermum kunthianum* (fruit pod). BOMBACACEAE: *Adansonia digitata* (leaf, root, fruit pulp, kernel, seed plumule), *Bombacopsis glabra* (leaf, seed), *Bombax buonopozense* (leaf, flower), *Ceiba pentandra* (leaf, flowers, seed), *Pachira aquatica* (kernel), *P. insignis* (seed). BORAGINACEAE: *Ehretia cymosa* (fruit), *Cordia africana* (bark, fruit), *C. aurantiaca* (fruit pulp), *C. myxa* (fruit pulp, kernel), *C. senegalensis* (fruit), *C. sinensis* (fruit), *C. tisserantii* (fruit), *Cynoglossum lanceolatum* (plant). BROMELIACEAE: *Ananas comosus* (fruit). BURSERACEAE: *Canarium schweinfurthii* (fruit, kernel), *Commiphora africana* (root, fruit), *C. pedunculata* (*fruit*), *Dacryodes edulis* (fruit pulp), *D. klaineana* (fruit pulp), *Santira trimera* (fruit pulp, seed). CACTACEAE: *Nopalea spp.* (fruit), *Opuntia spp.* (flower, fruit, seed). CANNACEAE: *Canna indica* (root). CAPPARACEAE: *Boscia angustifolia* (bark, fruit, seed), *B. salicifolia* (leaf, bark, root, seed). *B. senegalensis*(leaf, fruit, seed), *Buchholzia coriacea* (seed), *Cadaba farinosa* (leaf, twig, bark, flower), *Capparis decidua* (fruit), *C. erythrocarpos* (fruit pulp), *C. fascicularis* (leaf), *C. sepiaria* (leaf, fruit), *C. tomentosa* (leaf, fruit), *Cleome afrospinosa* (leaf), *C. gynandra* (leaf, root, seed, oil), *C. hirta* (leaf), *C. monophylla* (leaf, fruit), *C. rutidosperma* (leaf), *C. viscosa* (plant), *Crataeva adansonii* (leaf, fruit), *Euadenia eminens* (seed), *E. trifoliolata* (leaf), *Maerua angolensis* (leaf, fruit, seed), *M. crassifolia* (leaf, fruit), *M. pseudopetalosa* (leaf). CARICACEAE: *Carica papaya* (leaf, stem, pith, fruit, seed-shoot). CECROPIACEAE: *Musanga cecropioides* (fruit pulp), *Myrianthus arboreus* (leaf, fruit pulp, seed), *M. libericus* (leaf, fruit), *M. serratus* (leaf, flower, kernel). CELASTRACEAE: *Maytenus undata* (bark, fruit), *Mystroxylon aethiopicum* (fruit), *Salacia columna* (fruit), *S. cornifolia* (fruit), *S. leptoclada* (fruit), *S. nitida* (fruit), *S. oliveriana* (fruit), *S. pyriformis* (fruit), *S. senegalensis* (fruit), *S. staudtiana* (fruit), *S. stuhlmanniana* (fruit), *S. togoica* (fruit), *S. whytei.* CHENOPODIACEAE: *Chenopodium murale* (plant, seed). CHRYSOBALANACEAE: *Acioa scabrifolia* (seed), *Chrysobalanus icaco* (fruit), *Magnistipula cupheiflora* (fruit), *M. tessmannii* (fruit), *Maranthes glabra* (seed), *Neocarya macrophylla* (fruit, kernel), *Parinari congensis* (fruit pulp), *P. curatellifolia* (fruit pulp, kernel), *P. excelsa* (fruit pulp, kernel). COMBRETACEAE: *Anogeissus leiocarpus* (bark gum), *Combretum aculeatum* (seed), *C. collinum* (gum), *C. micranthum* (fruit), *C. mooreanum* (leaf), *C. nigricans* (gum), *C. paniculatum* (leaf), *C. platypterum* (leaf), *C. tomentosum* (fruit), *Laguncularia racemosa* (fruit), *Terminalia catappa* (fruit flesh, kernel). COMMELINACEAE: *Commelina africana* (leaf), *C. benghalensis* (leaf, rhizome), *C. diffusa* (leaf), *C. zambesica* (plant). COMPOSITAE: *Ageratum conyzoides* (leaf), *Bidens bipinnata* (leaf), *B. pilosa* (leaf), *Carthamnus tinctorius* (fruit), *Centaurea perrottetii* (plant), *Coreopsis borianiana* (plant), *Crassocephalum crepidioides* (leaf), *C. rubens* (leaf), *Eclipta alba* (plant), *Emilia coccinea* (leaf), *E. praetermissa* (leaf), *E. sonchifolia* (plant), *Enhydra fluctuans* (leaf), *Galinsoga parviflora* (leaf), *Gnaphalium luteo-album* (leaf), *Grangea maderaspatana* (leaf), *Guizotia abyssinica* (seed, oil), *G. scabra, Gynura miniata* (plant), *Helianthus annuus* (oil), *Lactuca capensis* (plant), *Launaea cornuta* (leaf), *L. taraxacifolia* (leaf), *Melanthera scandens* (leaf), *Mikania cordata* (leaf), *Senecio biafrae* (leaf), *Sonchus oleraceus* (plant), *Spilanthes filicaulis* (leaf), *Struchium sparganophora* (plant), *Synedrella nodiflora* (leaf), *Vernonia amygdalina* (leaf), *V. cinerea* (leaf), *V. colorata* (flower, seed), *V. hymenolepis* (leaf), *V. myriantha* (leaf), *V. perrottetii* (leaf), *V. thomsoniana* (leaf). CONNARACEAE: *Castanola paradoxa* (leaf), *Santaloides afzelii* (fruit pulp). CONVOLVULACEAE: *Anisera martinicensis* (leaf), *Astripomoea sp.* (leaf), *Hewittia sublobata* (leaf), *Ipomoea alba* (leaf), *I. aquatica* (leaf, root), *I. batatas* (leaf, tuber), *I. eriocarpa* (leaf, seed), *I. involucrata* (leaf), *I. obscura* (leaf), *I. pes-caprae* (leaf), *I. purpurea* (leaf), *I. quamoclit* (leaf), *Jacquemontia tamnifolia* (leaf), *Lepistemon owariense* (leaf), *Merremia umbellata* (leaf), *Neuropeltis acuminata* (leaf). COSTACEAE: *Costus lucanusianus* (young shoot). CRUCIFERAE: *Brassica campestris* (leaf, seed-oil), *B. juncea* (leaf), *B. napus* (root, seed-oil, seed-sprout), *B. oleracea* (foliage, stem, flower), *B. rapa* (plant, seed-oil), *Crambe kilimandscharica* (fruit), *C. kralikii* (plant), *Eruca sativa* (leaf, oil), *Lepidium sativum* (leaf), *Raphanus sativus* (leaf, root), *Rorippa humifusca* (plant), *R. nasturtium-aquaticum* (plant), *Schouwia purpurea* (leaf). CUCURBITACEAE: *Citrullus colocynthis* (seed), *C. lanatus* (leaf, fruit, seed, oil), *Coccinia adoensis* (leaf, root, fruit), *C. grandis* (leaf, fruit), *Cucumeropsis mannii* (fruit, seed), *Cucumis anguria* (fruit), *C. melo* (leaf, fruit, seed), *C. metuliferus* (selected c.vars, leaf, fruit), *C. sativus* (leaf, fruit, seed), *Cucurbita maxima* (leafy stem, fruit, seed), *C. moschata* (leaf, fruit), *C. pepo* (leaf, shoots, fruit, seed-oil), *Lagenaria siceraria* (selected c.vars, stem, fruit, seed-oil), *Luffa acutangula* (selected c.vars, fruit), *L. cylindrica* (selected c.vars, fruit, seed-oil), *Momordica balsamina* (leaf, fruit), *M. charantia* (shoots, fruit), *M. cissoides* (kernel), *M. foetida* (leaf, root, fruit), *Peponium vogelii* (ripe fruit), *Sechium edule* (leafy stem, root, fruit), *Telfairea occidentalis* (leaf, young shoot, seed, oil), *Trichosanthes cucumerina* (young fruit), *Trochomeria macrocarpa* (root). CYPERACEAE: *Cyperus bulbosus* (swollen culm-base), *C. esculentus* (tuber, oil), *C. maculatus* (tuber), *C. papyrus* (pith), *C. rotundus* (tuber), *Eleocharis dulcis* (tuber), *Mariscus alternifolius* (culm-base), *Pycreus macrostachyos* (root), *Scirpus brachyceras* (stem-tip). DICHAPETALACEAE: *Dichapetalum madagascariense* (fruit-pulp, seed), *D. pallidum* (seed), *D. toxicarium* (fruit pulp). DILLENIACEAE: *Tetracera alnifolia* (young leaf). DIOSCOREACEAE: *Dioscorea abyssinica* (tuber), *D. alata* (tuber), *D. bulbifera* (c.vars, bulbil), *D. cayenensis* (tuber), *D. dumetorum* (tuber), *D. esculenta* (tuber), *D. hirtiflora* (tuber), *D. lecardii* (tuber), *D. mangenotiana* (young tuber), *D. minutiflora* (young tuber), *D. odoratissima* (tuber), *D. praehensilis* (young shoots, tuber), *D. preusii* (tuber), *D. quartiniana* (tuber), *D. rotundata* (tuber), *D. sansibarensis* (tuber), *D. schimperana* (bulbil, tuber), *D. smilacifolia* (tuber), *D. togoensis* (bulbil).

**A: Food: 2, special diets**

APOCYNACEAE: *Nerium oleander* (wood-ash). ARACEAE: *Culcasia scandens* (leaf). BALANITACEAE: *Balanites aegyptiaca* (fruit). BOMBACAEAE: *Adansonia digitata* (root). BROMELIACEAE: *Annas comosus* (fruit). CAPPARACEAE: *Boscia senegalensis* (fruit), *Cleome arabica* (plant), *Euadenia eminens* (plant). COMPOSITAE: *Vernonia amygdalina* (leaf), *V. colorata, V. conferta* (leafy shoot).

**A: Food: 3, sauces, condiments, spices, flavourings**
AIZOACEAE: *Gisekia pharnaceoides*. AMARANTHACEAE: *Amaranthus dubius, A. lividus, Celosia isertii, C. trigyna*. ANACARDIACEAE: *Schinus molle* (berry), *S. terebinthifolius* (fruit). ANNONACEAE: *Annona senegalensis* (flower), *Dennettia tripetala* (young leaf, seed), *Monodora brevipes* (seed), *M. myristica* (seed), *M. tenuifolia* (seed), *Xylopia acutiflora* (fruit), *X. aethiopica* (fruit, seed), *X. parviflora* (fruit), *X. staudtii* (fruit). APOCYNACEAE: *Carissa edulis* (root), *Saba senegalensis* (leaf). ARACEAE: *Cyrtosperma senegalense* (leaf). ARALIACEAE: *Polyscias guilfoylei* (leaf). BALANITACEAE: *Balanites aegyptiaca* (shoot). BALANOPHORACEAE: *Thonningia sanguinea* (rhizome). BOMBACACEAE: *Adansonia digitata* (calyx, fruit). CACTACEAE: *Opuntia spp.* (flower). CAPPARACEAE: *Capparis decidua* (flower-bud), *C. spinosa* (flower-bud), *Cleome gynandra* (leaf, seed-capsule, seed), *C. monophylla* (leaf), *Crataeva adansonii* (leaf), *Maerua angolensis* (leaf). CECROPIACEAE: *Myrianthus libericus* (leaf). CELASTRACEAE: *Maytenus senegalensis* (leaf). CHENOPODIACEAE: *Chenopodium ambrosioides* (leaf), *C. murale* (plant). COMBRETACEAE: *Combretum racemosum* (leaf). COMPOSITAE: *Acanthospermum hispidum* (leaf-sap), *Ambrosia maritima* (plant), *Artemisia spp.* (plant), *Carthamnus tinctorius* (seed), *Cotula cinerea* (plant), *Crassocephalum crepidioides* (leaf), *C. rubens* (leaf), *Eclipta alba* (plant), *Geigeria alata* (plant), *Guizotia abyssinica* (seed), *Gynura miniata* (plant), *G. procumbens* (leaf), *Launaea cornuta* (leaf), *Microglossa afzelii* (leaf), *Senecio biafrae* (leaf), *Tagetes patulus, Vernonia amygdalina* (leaf). CRUCIFERAE: *Brassica juncea* (leaf). CUCURBITACEAE: *Citrullus lanatus* (seed-meal), *Lagenaria siceraria* (seed), *Momordica balsamina* (fruit). CYPERACEAE: *Cyperus distans* (rhizome), *Kyllinga erecta* (rhizome), *K. pumila* (root), *K. squamulata* (culm-base), *K. tenuifolia* (culm-base), *Mariscus alternifolius* (rhizome). DILLENIACEAE: *Dillenia indica* (fruit-pulp).

**A: Food; 4, sweets, sweetmeats, suckers**
BALANITACEAE: *Balanites aegyptiaca* (gum, flower, fruit). BORAGINACEAE: *Cordia africana* (fruit). CARICACEAE: *Carica papaya* (flower). CUCURBITACEAE: *Citrullus lanatus* (seed-meal), *Cucumis melo* (seed-meal). CYPERACEAE: *Cyperus esculentus* (tuber). DIOSCOREACEAE: *Dioscorea cayenensis* (tuber), *D. rotundata* (tuber).

**A: Food; 5, masticatory**
APOCYNACEAE: *Picralima nitida* (bark, fruit). BIGNONIACEAE: *Stereospermum kunthianum* (bark). BOMBACACEAE: *Bombax buonopozense* (calyx). BORAGINACEAE: *Heliotropium ovalifolium* (leaf), *H. subulatum. BURSERACEAE: Commiphora africana* (gum). CAPPARACEAE: *Buchholzia coriacea* (aril). CHRYSOBALANACEAE: *Maranthes polyandra* (leaf), *Parinari curatellifolia* (leaf, bark). COMBRETACEAE: *Anogeissus leiocarpus* (gum). CUCURBITACEAE: *Citrullus lanatus* (seed), *Lagenaria siceraria* (seed). CYPERACEAE: *Kyllinga pumila* (root), *K. squamulata* (culm-base), *K. tenuifolia* (culm-base).

**B: Drink; 1, water/sap**
ANNONACEAE: *Artabotrys thomsonii* (liane). ASCLEPIADACEAE: *Ceropegia spp.* (tuber), *Leptadenia pyrotechnica* (twig), *Sarcostemma viminale* (stem). BOMBACACEAE: *Adansonia digitata* (wood). CACTACEAE: *Opuntia spp.* (stem). CECROPIACEAE: *Musanga cecropioides* (stilt-root), *Myrianthus arboreus* (aerial root). COCHLOSPERMACEAE: *Cochlospermum tinctorium* (unripe fruit). COMMELINACEAE: *Aneilema lanceolatum*. CONNARACEAE: *Cnestis ferruginea* (stem), *Jaundea pubescens* (stem). COSTACEAE: *Costus afer* (stem). CRUCIFERAE: *Schouwia schimperi* (plant). CUCURBITACEAE: *Citrullus lanatus* (fruit). DILLENIACEAE: *Tetracera alnifolia* (stem), *T. leiocarpa* (stem), *T. podotricha* (stem), *T. potatoria* (stem).

**B: Drink; 2, sweet, milk substitutes**
ANACARDIACEAE: *Sclerocarya birrea* (fruit), *Spondias mombin* (fruit), *Trichoscypha acuminata* (fruit). ANNONACEAE: *Annona reticulata* (fruit). APOCYNACEAE: *Saba florida* (fruit). ASCLEPIADACEAE: *Sarcostemma viminale* (stem). BALANITACEAE: *Balanites aegytiaca* (fruit). BOMBACACEAE: *Adansonia digitata* (fruit-meal). BORAGINACEAE: *Cordia africana* (fruit). CAPPARACEAE: *Maerua pseudopetalosa* (root). CELASTRACEAE: *Hippocratea africana* (root). CYPERACEAE: *Cyperus esculentus* (tuber).

**B: Drink; 3, infusions, tisanes, etc., including substitutes, adulterants**
ANACARDIACEAE: *Rhus natalensis* (bark), *Schinus molle* (fruit), *S. terebinthifolius* (fruit). APOCYNACEAE: *Ervatamia coronaria* (wood). ARACEAE: *Culcasia scandens* (leaf). ASCLEPIADACEAE: *Solenostemma oleifolium* (leaf). BOMBACACEAE: *Pachira aquatica* (kernel). CAPPARACEAE: *Boscia senegalensis* (seed). CELASTRACEAE: *Mystroxylon*

*aethiopicum* (bark). COMBRETACEAE: *Anogeissus leiocarpus* (gum, leafy twig), *Combretum fragrans* (root), *C. micranthum* (leaf). COMPOSITAE: *Bidens pilosa* (plant), *Helichrysum nudifolium* (leaf), *Pulicaria crispa* (plant), *Senecio biafrae* (leaf), *Spilanthes filicaulis* (leaf). CONVOLVULACEAE: *Ipomoea muricata* (seed), *Seddera latifolia* (plant). CUCURBITACEAE: *Citrullus lanatus* (seed). CYPERACEAE: *Cyperus esculentus* (tuber).

**B: Drink; 4, water purifiers**

ANNONACEAE: *Xylopia aethiopica* (fruit). APOCYNACEAE: *Carissa edulis* (root). ASCLEPIADACEAE: *Calotropis procera* (leaf). CAPPARACEAE: *Maerua crassifolia* (bark), *M. pseudopetalosa* (root). CUCURBITACEAE: *Luffa cylindrica* (fruit-fibre).

**B: Drink; 5, alcoholic, stimulant, including additives in fermentation**

AGAVACEAE: *Agave sisalana*. AMARANTHACEAE: *Amaranthus hybridus* (seed). ANACARDIACEAE: *Anacardium occidentale* (fruit), *Antrocaryon micraster* (fruit), *Lannea acida* (fuit-pulp), *L. microcarpa* (fruit), *Sclerocarya birrea* (fruit), *Spondias mombin* (fruit). ANNONACEAE: *Annona muricata* (fruit), *A. squamosa* (fruit), *Xylopia aethiopica* (leaf). APOCYNACEAE: *Baissea axillaris* (liane), *Carissa edulis* (fruit), *Landolphia heudelotii* (fruit), *L. owariensis* (fruit), *Pleiocarpa pycnantha* (root), *Rauvolfia caffra* (bark). ASCLEPIADACEAE: *Calotropis procera* (latex, leaf). BALANITACEAE: *Balanites aegyptiaca* (fruit). BIGNONIACEAE: *Kigelia africana* (fruit), *Tecoma stans* (root). BORAGINACEAE: *Cordia africana* (kernel). BROMELIACEAE: *Ananas comosus* (fruit). BURSERACEAE: *Canarium schweinfurthii* (bark). CACTACEAE: *Opuntia spp.* (fruit). CAPPARACEAE: *Boscia senegalensis* (fruit). CECROPIACEAE: *Musanga cecropioides* (bark). CHRYSOBALANACEAE: *Parinari excelsa* (fruit-pulp). COMPOSITAE: *Vernonia amygdalina* (leaf). CONVOLVULACEAE: *Ipomoea batatas* (tuber). DIOSCOREACEAE: *Dioscorea dumetorum* (tuber).

**C: Medicines; 1, generally healing (cicitrisant, haemostatic, antiseptic, stimulant, tonic, drawing, etc.)**

ACANTHACEAE: *Acanthus montanus* (leaf), *Asystasia gangetica* (whole plant, sap), *Barleria cristata* (leaf, root), *Brillantaisia nitens* (leaf), *B. patula* (leaf), *Elytraria marginata* (leaf sap), *Graptophyllum pictum* (leaf), *Hypoestes verticillata* (sap), *Justicia betonica* (leaf), *J. flava* (whole plant), *J. glabra* (leaf), *J. schimperi/striata* (leaf), *Phaulopsis barteri, P. falcisepala* (dried plant), *Pseuderanthemum ludovicianum* (leaf), *P. tunicatum, Ruellia praetermissa* (leaf, root), *Thomandersia hensii* (twig, root), *Thunbergia chrysops* (leaf sap), *T. grandiflora* (leaf), *T. laurifolia* (leaf). AIZOACEAE: *Gisekia pharnaceoides, Glinus oppositifolius* (leaf), *Limeum indicum, Mollugo nudicaulis* (sap), *M. pentaphylla, Sesuvium portulacastrum, Trianthema portulacastrum* (leaf). AMARANTHACEAE: *Achyranthes aspera* (leaf), *Alternanthera maritima, A. nodiflora* (whole plant, sap), *A. sessilis, Amaranthus viridis* (leaf), *Celosia argentea, C. trigyna* (leaf), *Cyathula achyranthoides, C. prostrata, C. uncinulata, Gomphrena globosa, Pandiaka involucrata, Pupalia lappacea, Seriocostachys scandens*. AMARYLLIDACEAE: *Crinum zeylanicum, Pancratium trianthum, Scadoxus multiflorus*. ANACARDIACEAE: *Anacardium occidentale* (fruit), *Lannea barteri* (bark, root), *L. humilis* (root), *L. velutina* (bark), *Rhus natalensis* (root), *Schinus molle* (bark), *S. terebinthifolius* (leaf, bark), *Sorindeia juglandifolia* (leaf), *S. warneckei* (plant), *Spondias mombin* (leaf, bark), *Trichoscypha acuminta* (bark), *T. longifolia* (bark). ANNONACEAE: *Annona arenaria* (plant), *A. chrysophylla* (leaf, bark, root), *A. muricata* (leaf, fruit), *A. reticulata* (leaf), *A. senegalensis* (leaf), *Annonidium mannii* (bark), *Enantia chlorantha* (bark), *E. polycarpa* (bark), *Isolona campanulata* (plant), *I. cooperi* (leaf), *Monodora myristica* (bark, seed), *M. tenuifolia* (seed), *Uvaria chamae* (leaf, stem, bark, root), *U. doeringii* (leaf), *Xylopia aethiopica* (root, fruit, seed), *X. quintasii* (leaf, bark), *X. villosa* (seed). APOCYNACEAE: *Alafia lucida* (plant), *A. multiflora* (latex), *Alstonia boonei* (leaf, bark), *Ancylobothrys amoena* (fruit), *Baissea multiflora* (liane), *Funtumia africana* (leaf), *F. elastica* (bark), *Hunteria umbellata* (bark, root), *Isonema smeathmannii* (latex), *Landolphia dulcis* (plant), *L. heudelotii* (root), *L. owariensis* (leaf), *Pleioceras barteri* (bark, fruit), *Rauvolfia mannii* (plant), *R. vomitoria* (leaf), *Saba florida* (latex), *S. senegalensis* (leaf, root-bark), *Strophanthus preussii* (sap), *S. sarmentosus* (leaf-sap, root), *Tabernaemontana crassa* (leaf, sap), *T. ventricosa* (latex), *Voacanga africana* (bark, latex). APONOGETONACEAE: *Aponogeton vallisnerioides* (leaf). ARACEAE: *Anchomanes difformis* (tuber), *A. welwitschii* (tuber ash), *Colocasia esculenta* (tuber), *Stylochiton hypogaeus* (tuber). ARISTOLOCHIACEAE: *Aristolochia albida* (root). ASCLEPIADACEAE: *Asclepias curassavica* (latex), *Calotropis procera* (latex), *Caralluma russelliana* (sap), *Dregea crinita* (sap), *Gomphocarpus fruticosus* (plant), *Gymnema sylvestre* (root), *Kanahia lanifolia* (latex), *Pentarrhinum inispidum* (leaf), *Pentatropis spiralis* (root), *Pergularia daemia* (leaf, latex, fruit), *Tylophora sylvatica* (leaf), *Sarcostemma viminale* (plant), *Secamone afzelii* (leaf), *Solenostemma oleifolium* (leaf). BALANITACEAE: *Balanites aegyptiaca* (leaf, bark, seed-oil). BALSAMINACEAE: *Impatiens balsamina* (leaf), *I. burtonii* (leaf), *I. irvingii* (leaf), *I. niamniamensis* (plant). BASELLACEAE: *Basella alba* (leafy stems). BEGONIACEAE: *Begonia fusicarpa* (sap), *B. mannii*. BIGNONIACEAE: *Kigelia africana* (bark, root, fruit), *Markhamia lutea* (plant), *M. tomentosa* (plant), *Newbouldia laevis* (leaf, bark), *Spathodea campanulata* (bark), *Stereospermum acuminatissimum* (bark), *S. kunthianum* (leaf), *Tecoma stans* (plant). BOMBACACEAE: *Adansonia digitata* (leaf, bark), *Ceiba pentandra* (leaf, bark), *Rhodognaphalon brevicuspe* (bark). BORAGINACEAE: *Coldenia procumbens* (plant), *Cordia africana* (leaf), *C. myxa* (fruit pulp), *C. senegalensis* (leaf), *C. vignei* (leaf, bark), *Cynoglossum lanceolatum* (plant), *Heliotropium bacciferum* (sap), *H. indicum* (leaf), *H. subulatum* (leaf), *Trichodesma africanum* (leaf). BROMELIACEAE: *Ananas comosus* (leaf sap). BURSERACEAE: *Aucoumea klaineana* (bark), *Boswellia dalzielii* (bark), *Canarium schweinfurthii* (gum, root), *Commiphora africana* (gum, root, fruit), *Dacryodes edulis* (bark). CACTACEAE: *Nopalea spp.* (fruit), *Opuntia spp.* (stem). CANNACEAE: *Canna indica* (shoot). CAPPARACEAE: *Boscia salicifolia* (leaf), *Cadaba farinosa* (plant), *Capparis erythrocarpos* (root-bark), *C. picridifolium* (leaf), *C. tomentosa* (leaf, stem, root-bark),*C. viminea* (bark), *Cleome gynandra* (plant), *C. monophylla* (leaf), *Crataeva adansonii* (bark), *Euadenia trifoliolata* (leaf), *Maerua angolensis* (leaf), *M. pseudopetalosa* (fruit), *Ritchiea capparoides* (leaf), *R. reflexa* (leaf). CARICACEAE: *Carica papaya* (leaf, latex). CARYOPHYLLACEAE: *Drymaria cordata* (plant), *Polycarpaea corymbosa* (plant), *P. eriantha* (plant). CECROPIACEAE: *Myrianthus arboreus* (leaf, petiole). CELASTRACEAE: *Hippocratea indica* (leaf, root), *H. macrophylla* (seed oil), *Maytenus senegalensis* (leaf, bark, root), *M. undata* (bark), *Mystroxylon aethiopicum* (green leaf, bark), *Salacia lehmbachii* (root), *S. senegalensis* (leaf, root). CHENOPODIACEAE: *Chenopodium ambrosioides* (plant, leaf). CHRYSOBALANACEAE: *Maranthes polyandra* (leaf), *Neocarya macrophylla* (root-bark), *Parinari curatellifolia* (leaf, bark), *P. excelsa* (bark, root). COCHLOSPERMACEAE: *Cochlospermum tinctorium* (leaf, root), *C. religiosum* (flower). COMBRETACEAE: *Anogeissus leiocarpus* (bark), *Combretum cinereipetalum, C. collinum* (leaf), *C. dolichopetalum* (leaf), *C. glutinosum* (leaf, bark, fruit), *C. hispidum* (leaf), *C. micranthum* (root, fruit), *C. platypterum* (leaf), *C. racemosum* (leaf, sap, bark), *C. sericeum* (gum), *C. smeathmannii* (leaf), *C. tomentosum* (leaf), *Guiera senegalensis* (leaf), *Terminalia avicennioides* (leaf, root, root-bark), *T. glaucescens* (stem-bark, root-bark), *T. ivorensis* (sap), *T. macroptera* (leaf, stem-bark, root, root-bark), *T. scutifera* (bark), *T. superba* (bark). COMMELINACEAE: *Aneileme aequinoctiale* (root), *A. pomeridianum* (plant), *Commelina benghalensis* (plant), *C. diffusa* (leaf), *C. forskalaei* (plant), *Palisota bracteosa* (plant), *P. hirsuta* (plant), *Pollia condensata* (stem). COMPOSITAE: *Ageratum conyzoides* (leaf, sap), *Ambrosia maritima* (plant), *Aspilia africana* (plant), *A. kotschyi* (plant), *A. rudis* (plant), *Ayapana triplinervis* (plant), *Bidens bipinnata* (leaf), *Blainvillea gayana* (leaf), *Blumea aurita* (leaf), *B. perrottetiana* (leaf), *Chrysanthellum americanum* (plant), *Conyza aegyptiaca* (leafy stem), *C. canadensis* (plant), *C. pyrrhopappa* (plant), *C. sumatrensis* (leaf), *Crassocephalum crepidioides* (leaf), *C. picridifolium* (leaf), *C. rubens* (leaf), *Dichrocephala integrifolia* (plant), *Dicoma tomentosa* (plant), *Eclipta alba* (leaf), *Emilia coccinea* (leaf), *E. praetermissa* (plant), *E. sonchifolia* (leaf-sap), *Epaltes gariepina* (leaf, root), *Ethulia conyzoides* (leaf), *Galinsoga parviflora* (leaf), *Gnaphalium luteo-album* (ash), *Grangea maderaspatana* (leaf), *Gutenbergia nigritana* (plant), *Helichrysum foetidum* (leaf), *H. glumaceum* (ash), *H. mechowianum* (leaf), *H. nudifolium* (leaf), *H. odoratissimum* (leaf), *Lactuca capensis* (root), *Laggera alata* (leaf), *L. heudelotii* (plant), *Melanthera scandens* (leaf), *Microglossa pyrifolia* (leaf, root), *Mikania cordata* (leaf), *Pilloselloides hirsuta* (ash), *Senecio abyssinicus* (leaf), *S. baberka* (leaf), *S. biafrae* (plant, leaf), *S. ruwenzoriensis* (plant), *Sigesbeckia orientalis* (sap), *Sonchus schweinfurthii* (leaf), *Sphaeranthus senegalensis* (plant), *Spilanthes filicaulis* (plant), *S. uliginosa* (plant), *Tridax procumbens* (plant), *Vernonia amygdalina* (leafy stem), *V. auriculifera* (root), *V. biafrae* (leaf), *V. cinerea* (plant), *V. colorata* (leaf), *V. conferta* (leaf, bark), *V. glabra* (leaf), *Wedelia trilobata* (plant). CONNARACEAE: *Agelaea spp.* (sap), *A. obliqua* (leaf), *Byrsocarpus coccineus* (leaf, root), *B. poggeanus* (leaf), *Cnestis ferruginea* (root-bark, fruit), *Connarus africanus* (leaf, bark), *Roureopsis obliquifoliolata* (leaf). CONVOLVULACEAE: *Argyreia nervosa* (leaf), *Calycobolus heudelotii* (leaf), *Cressa cretica* (plant), *Evolvulus alsinoides* (leaf), *Hewittia sublobata* (leaf), *Ipomoea aquatica* (plant), *I. asarifolia* (plant), *I. batatas* (leaf), *I. involucrata* (plant), *I. mauritiana* (tuber), *I. muricata* (seed), *I. nil* (leaf), *I. pes-tigridis* (leaf), *I. quamoclit* (leaf), *Merremia aegyptia* (leaf), *M. hederacea* (leaf), *M. tridentata* (plant), *M. umbellata* (leaf), *Stictocardia beraviensis* (sap). COSTACEAE: *Costus afer* (sap, root), *C. lucanusianus* (sap), *C. spectabilis* (sap). CRASSULACEAE: *Kalanchoe crenata/integra* (leaf, root), *K. laciniata* (leaf). CRUCIFERAE: *Brassica juncea* (plant), *Capsella bursa-pastoris* (plant), *Lepidium sativum* (seed). CUCURBITACEAE: *Citrullus lanatus* (seed), *Coccinia adoensis* (leaf-sap, root), *Cogniauxia podolaena* (leaf), *Cucumis prophetarum* (fruit), *C. sativus* (fruit), *Cucurbita maxima* (fruit, seed), *C. pepo* (flower, fruit pulp), *Luffa acutangula* (leaf), *L. cylindrica* (leaf), *Momordica balsamina* (fruit), *M. charantia* (fruit), *Mukia maderaspatana* (leaf-sap), *Peponium vogelii* (leaf), *Ruthalicia eglandulosa* (sap). CYPERACEAE: *Cyperus alternifolius* (plant), *C. dilatatus* (stem), *C. distans* (rhizome), *C. esculentus* (tuber), *C. exaltatus* (rhizome), *C. rotundus* (tuber), *Fuirena umbellata* (leaf), *Mariscus alternifolius* (stem), *Scleria boivinii* (leaf, leaf, aerial part). DICHAPETALACEAE: *Dichapetalum madagascariense* (leaf), *D. toxicarium* (leaf). DILLENIACEAE: *Tetracera alnifolia* (sap). DIOSCOREACEAE: *Dioscorea bulbifera* (bulbil, tuber), *D. smilacifolia* (leaf). DROSERACEAE: *Drosera indica* (plant).

**C: Medicines; 2, skin, mucosae (dermal eruptions, inflammation, ulcers, pruritus and skin diseases generally)**

ACANTHACEAE: *Acanthus montanus* (leaf), *Asystasia gangetica* (whole plant), *Barleria prionitis, Dicliptera elliotii* (leaf), *D. verticillata* (leaf), *Graptophyllum pictum* (plant, leaf), *Hygrophila auriculata* (plant), *Justicia betonica* (leaf, ash), *J. extensa* (leaf, root), *J. glabra* (leaf), *Phaulopsis barteri, P. falcisepala* (plant, sap), *P. imbricata* (root), *Rhinacanthus virens* (leaf, root), *Ruellia praetermissa* (leaf, root), *Thomandersia hensii* (sap). AIZOACEAE: *Glinus lotoides, G. oppositifolius.* AMARANTHACEAE: *Achyranthus aspera* (leaf), *A. spinosus, Celosia argentea, C. trigyna* (leaf), *Cyathula achyranthoides* (leaf-sap), *C. prostrata* (leaf, ash), *Pupalia lappacea.* AMARYLLIDACEAE: *Crinum zeylanicum, Scadoxus multiflorus.* ANACARDIACEAE: *Lannea acida* (root bark), *L. barteri* (bark), *L. nigritana* (bark, root), *L. velutina* (bark), *L. welwitschii* (stem bark), *Mangifera indica* (leaf, bark-resin), *Rhus natalensis* (leaf), *Sclerocarya birrea* (bark, root), *Trichoscypha longifolia* (bark). ANNONACEAE: *Annona arenaria* (plant), *A. muricata* (leaf), *A. reticulata* (bark), *A. squamosa* (leaf), *Cananga odorata* (leaf), *Enantia chlorantha* (bark), *E. polycarpa* (bark), *Isolona campanulata* (plant), *Monodora myristica* (bark), *M. tenuifolia* (leaf, seed), *Uvaria afzelii* (leaf), *U. chamae* (seed), *U. doeringii* (fruit), *Xylopia aethiopica* (fruit), *X. parviflora* (stem-bark, fruit), *X. quintasii* (root). APOCYNACEAE: *Adenium obesum* (sap, bark), *Alstonia boonei* (bark, latex), *Funtumia elastica* (bark), *Holarrhena floribunda* (bark), *Nerium oleander* (leaf, most parts), *Picralima nitida* (seed), *Rauvolfia caffra* (bark, root), *R. vomitoria* (bark, root), *Saba florida* (fruit), *Strophanthus gratus* (leaf-sap, root), *S. hispidus* (leaf, stem, root), *Tabernaemontana crassa* (sap), *T. pachysiphon* (leaf, latex), *Voacanga africana* (plant). ARACEAE: *Caladium bicolor* (root), *Culcasia parviflora* (leaf), *C. scandens* (leaf), *Cyrtosperma senegalense* (root), *Dieffenbachia spp., Pistia stratiotes* (leaf, plant, ash). ARISTOLOCHIACEAE: *Aristolochia albida* (leaf), *A. bracteolata* (plant, leaf). ASCLEPIADACEAE: *Calotropis procera* (latex, root-bark), *Gomphocarpus fruticosus* (plant), *Pachycarpus lineolata* (sap), *Pergularia daemia* (bark), *Sarcostemma viminale* (sap), *Tylophora conspicua* (leaf), *T. sylvatica* (leaf). AVICENNIACEAE: *Avicennia germinans* (bark). BALANOPHORACEAE: *Thonningia sanguinea* (rhizome, flowerhead). BASELLACEAE: *Basella alba* (leaf). BIGNONIACEAE: *Jacaranda mimosifolia* (bark), *Markhamia lutea* (leaf, bark), *M. tomentosa* (bark), *Newbouldia laevis* (bark), *Spathodia campanulata* (bark), *Stereospermum kunthianum* (bark, root). BIXACEAE: *Bixa orellana* (fruit, seed). BOMBACACEAE: *Bombax buonopozense* (bark-gum), *Ceiba pentandra* (bark). BORAGINACEAE: *Cordia africana* (wood-ash), *C. millenii* (seed), *Ehretia cymosa* (bark-oil), *E. trachyphylla* (bark-oil), *Heliotropium bacciferum* (leaf), *H. indicum* (leaf). BROMELIACEAE: *Ananas comosus* (leaf). BURSERACEAE: *Canarium schweinfurthii* (bark), *Commiphora africana* (gum), *Dacryodes edulis* (resin), *Santira trimera* (bark). CAPPARACEAE: *Buchholzia coriacea* (bark, seed), *Cadaba farinosa* (leaf), *Capparis spinosa* (leaf), *C. thonningii* (leaf-sap), *Cleome rutidosperma* (leaf-sap), *Maerua crassifolia* (leaf). CARYOPHYLLACEAE: *Stellaria media* (plant). CELASTRACEAE: *Salacia debilis* (leaf, sap), *S. pallescens* (plant), *S. senegalensis* (plant), *Hippocratea africana* (root), *H. indica* (root). CHENOPODIACEAE: *Chenopodium ambrosioides* (leaf), *Salsola baryosma* (ash). CHRYSOBALANACEAE: *Chrysobalanus icaco* (bark), *Maranthes glabra* (bark, root), *M. polyandra* (root), *Parinari curatellifolia* (bark). COCHLOSPERMACEAE: *Cochlospermum tinctorium* (root). COMBRETACEAE: *Combretum comosum* (plant), *Guiera senegalensis* (leaf), *Strephonema pseudocola* (seed-oil), *Terminalia avicennioides* (root), *T. catappa* (leaf-sap), *T. glaucescens* (root), *T. ivorensis* (bark), *T. macroptera* (leaf). COMMELINACEAE: *Aneilema beninensis* (plant), *Cyanotis lanata* (plant), *Palisota bracteosa* (plant), *P. hirsuta* (plant). COMPOSITAE: *Ageratum conyzoides* (leaf, sap), *Aspilia africana* (leaf), *Blumea perrottetiana* (leaf-sap), *Centaurea perrottetii* (plant), *Conyza aegyptiaca* (leafy stem), *C. bonariensis* (leaf), *Dicoma tomentosa* (plant), *Emilia coccinea* (leaf), *Enhydra fluctuans* (leaf), *Ethulia conyzoides* (plant), *Lactuca capensis* (root), *Microglossa pyrifolia* (leaf-sap), *Mikania cordata* (sap), *Senecio abyssinicus* (leaf), *S. biafrae* (plant), *S. ruwenzoriensis* (plant), *Sigesbeckia orientalis* (plant), *Sphaeranthus senegalensis* (leaf, flower), *Vernonia adoensis* (root), *V. amygdalina* (leaf), *V. colorata* (leaf, root-bark), *V. conferta* (plant), *V. glaberrima* (leaf), *V. nestor* (sap), *V. thomsoniana* (leaf). CONNARACEAE: *Cnestis ferruginea* (root). CONVOLVULACEAE: *Argyreia nervosa* (leaf), *Ipomoea aquatica* (flower bud), *I. cairica* (leaf), *I. eriocarpa* (seed oil), *I. pes-caprae* (leaf), *Merremia umbellata* (leaf, mucilage). CRASSULACEAE: *Bryophyllum pinnatum* (sap), *Kalanchoe laciniata* (leaf). CRUCIFERAE: *Brassica campestris* (seed-oil).

CUCURBITACEAE: *Citrullus colocynthis* (fruit), *Coccinia grandis* (leaf), *Lagenaria siceraria* (leaf), *Luffa acutangula* (leaf), *Momordica charantia* (plant), *Mukia maderaspatana* (leaf), *Oreosyce africana* (leaf). CYPERACEAE: *Cyperus articulatus* (rhizome), *Kyllinga erecta* (plant). DIOSCOREACEAE: *Dioscorea bulbifera* (bulbil), *D. hirtiflora* (tuber), *D. minutiflora* (leaf).

**C: Medicines; 3, cutaneous/subcutaneous parasites (Guinea-worm, filaria, jiggers, scabies, etc.)**
ACANTHACEAE: *Eremomastax speciosa* (leaf), *Justicia extensa* (leaf-sap), *Nelsonia canescens* (leaf-sap). AMARANTHACEAE: *Aerva lanata, Amaranthus viridis* (leaf-sap). AMARYLLIDACEAE: *Crinum zeylanicum* (fruit). ANNONACEAE: *Monodora myristica* (bark). APOCYNACEAE: *Alstonia boonei* (latex), *Rauvolfia vomitoria* (fruit-pulp, seed), *Strophanthus gratus* (leaf), *S. hispidus* (leaf, stem, root), *Tabernaemontana crassa* (sap), *Voacanga africana* (plant). ARISTOLOCHIACEAE: *Aristolochia albida* (leaf, root). ASCLEPIADACEAE: *Calotropis procera* (latex, root-bark, flower), *Pergularia tomentosa* (sap). AVICENNIACEAE: *Avicennia germinans* (bark). BIGNONIACEAE: *Crescentia cujete* (fruit), *Kigelia africana* (bark), *Spathodea campanulata* (bark-sap), *Stereospermum kunthianum* (leaf). BOMBACACEAE: *Adansonia digitata* (leaf). BURSERACEAE: *Dacryodes edulis* (resin). CAPPARACEAE: *Boscia senegalensis* (leaf), *Ritchiea capparoides* (leaf, root), *R. reflexa* (leaf). CELASTRACEAE: *Hippocratea indica* (plant). COMBRETACEAE: *Guiera senegalensis* (leaf). COMMELINACEAE: *Palisota bracteosa* (sap), *P. hirsuta* (sap). COMPOSITAE: *Aspilia africana* (leaf), *A. rudis* (plant), *Bidens pilosa* (sap), *Centaurea perrottetii* (plant), *Crassocephalum rubens* (sap), *Dichrocephala integrifolia* (sap), *Emilia coccinea* (leaf), *Microglossa pyrifolia* (leaf-sap), *Struchium sparganophora* (leaf), CONVOLVULACEAE: *Ipomoea asarifolia* (leaf), *I. involucrata* (leaf-sap). COSTACEAE: *Costus lucanusianus* (sap). CUCURBITACEAE: *Cogniauxia podolaena* (root), *Luffa acutangula* (leaf), *Momordica charantia* (plant), *Ruthalicia eglandulosa* (whole plant), *Zehneria scabra* (leaf). DIONCOPHYLLACEAE: *Habropetalum dawei* (young leaf). DIOSCOREACEAE: *Dioscorea bulbifera* (bulbil).

**C: Medicines; 4, skeletal structure, bones, limbs, deformity, rickets**
AMARANTHACEAE: *Cyathula achyranthoides* (sap), *C. prostrata*. AMARYLLIDACEAE: *Crinum zeylanicum*. ANACARDIACEAE: *Lannea acida* (root bark), *L. velutina* (bark, root). ANNONACEAE: *Annona arenaria* (plant), *Cleistopholis patens* (bark), *Uvaria chamae* (root). APOCYNACEAE: *Alstonia boonei* (bark), *Rauvolfia vomitoria* (leaf, bark), *Tabernaemontana crassa* (leaf). ARACEAE: *Culcasia scandens* (leaf). BALANOPHORACEAE: *Thonningia sanguinea* (sap). BOMBACACEAE: *Adansonia digitata* (bark). CHRYSOBALANACEAE: *Maranthes polyandra* (bark). COMMELINACEAE: *Aneilema beninensis* (leaf-sap). COMPOSITAE: *Launaea taraxacifolia* (leaf). DILLENIACEAE: *Tetracera alnifolia* (root).

**C: Medicines; 5, paralysis, epilepsy, convulsions, spasm**
ACANTHACEAE: *Asystasia gangetica* (leaf), *Barleria opaca* (whole plant), *B. prionitis, Brillantaisia patula* (leaf), *Justicia extensa* (leaf, root), *J. flava* (leaf), *Stenandriopsis buntingii* (plant). AMARANTHACEAE: *Amaranthus spinosus, A. viridis* (leaf-sap). ANACARDIACEAE: *Lannea acida* (bark-sap), *L. barteri* (bark), *L. nigritana* (bark-sap), *L. velutina* (bark), *L. welwitschii* (bark), *Rhus natalensis* (root). ANNONACEAE: *Annona arenaria* (plant), *Annonidium mannii* (bark), *Uvaria afzelii* (leaf), *Xylopia aethiopica* (leaf). APOCYNACEAE; *Hedranthera barteri* (leaf), *Landolphia owariensis* (leaf), *Rauvolfia vomitoria* (leaf, bark, root), *Voacanga africana* (leaf, root). ARACEAE: *Anchomanes giganteus* (sap). ARALIACEAE: *Cussonia arborea* (leaf). ASCLEPIADACEAE: *Caralluma dalzielii* (stem), *Gymnema sylvestre* (leaf). BALANOPHORACEAE: *Thonningia sanguinea* (flower). BIGNONIACEAE: *Kigelia africana* (bark), *Newbouldia laevis* (bark). BORAGINACEAE: *Ehretia cymosa* (leaf), *Heliotropium indicum* (leaf). CAPPARACEAE: *Capparis viminea* (leaf), *Cleome rutidosperma* (leaf sap), *Maerua angolensis* (leaf). CECROPIACEAE: *Musanga cecropioides* (leaf bud). CHRYSOBALANACEAE: *Maranthes gabunensis* (plant). COCHLOSPERMACEAE: *Cochlospermum tinctorium* (root). COMBRETACEAE: *Terminalia macroptera* (root). COMPOSITAE: *Conyza sumatrensis* (sap), *Crassocephalum crepidiodes* (leaf), *Eclipta alba* (leaf), *Emilia coccinea* (leaf), *Enhydra fluctuans* (leaf), *Geigeria alata* (leaf, root), *Laggera alata* (leaf), *Microglossa pyrifolia* (plant), *Sonchus schweinfurthii* (leaf), *Triplotaxis stellutifera* (leaf-sap), *Vernonia cinerea* (plant), *V. colorata* (leaf, flower), *V. lasiopus* (plant). CONVOLVULACEAE: *Evolvulus nummularius* (plant), *Ipomoea eriocarpa* (seed-oil), *Merremia umbellata* (leaf). CRASSULACEAE: *Bryophyllum pinnatum* (leaf-sap, root-sap). CUCURBITACEAE: *Citrullus colocynthis* (seed-oil). DIOSCOREACEAE: *Dioscorea sansibarensis* (leaf, root), *D. smilacifolia* (leaf).

**C: Medicines; 6, insanity**
ANACARDIACEAE: *Lannea barteri* (bark), *L. velutina* (bark). ANNONACEAE: *Annona senegalensis* (leaf, root), *Uvaria scabrida* (plant). APOCYNACEAE: *Motandra guineensis* (leaf), *Strophanthus sarmentosus* (root), *Tabernaemontana pachysiphon* (root-bark). ARACEAE: *Rhektophyllum mirabile* (leaf). BALANITACEAE: *Balanites aegytiaca* (root-bark). BURSERACEAE: *Commiphora africana* (bark). CAPPARACEAE: *Boscia senegalensis* (leaf). CECROPIACEAE: *Musanga cecropioides* (leaf-bud). CHRYSOBALANACEAE: *Parinari curatellifolia* (bark), *P. excelsa* (bark). COMBRETACEAE: *Combretum nigricans* (root). COMPOSITAE: *Conyza sumatrensis* (sap), *Microglossa pyrifolia* (plant), *Senecio mannii* (leaf, root). CONNARACEAE: *Byrsocarpus viridis* (leaf-sap). CONVOLVULACEAE: *Ipomoea aquatica* (leaf-sap), *I. obscura* (leaf-sap). CYPERACEAE: *Hypolytrum heteromorphum* (plant).

**C: Medicines; 7, brain/nervous diseases, encephalitis, meningitis**
ACANTHACEAE: *Hypoestes aristata* (leaf, root). ANNONACEAE: *Uvaria chamae* (root). CONVOLVULACEAE: *Argyreia nervosa* (root)

**C: Medicines; 8, heart (tonal, disease, tremor, palpitations, etc.)**
ACANTHACEAE: *Acanthus montanus* (young shoot), *Asystasia gangetica* (leaf), *Brillantaisia patula* (leaf), *Elytraria marginata* (leaf), *Justicia insularis* (leaf). AMARANTHACEAE: *Amaranthus viridis* (leaf-sap), *Celosia trigyna* (leaf), *Cyathula prostrata*. AMARYLLIDACEAE: *Hippeastrum puniceum*. ANACARDIACEAE: *Antrocaryon micraster* (fruit). ANNONACEAE: *Uvaria doeringii* (leaf). APOCYNACEAE: *Rauvolfia serpentina*. ARACEAE: *Anchomanes giganteus* (tuber), *Cercestis afzelii* (plant), *Rhektophyllum mirabile* (leaf-sap). ASCLEPIADACEAE: *Gymnema sylvestre* (leaf), *Pergularia daemia* (plant), *Secamone afzelii* (leaf), *Tylophora glauca* (leaf). BALANITACEAE: *Balanites aegyptiaca* (bark). BALSAMINACEAE: *Impatiens niamniamensis* (leaf). BOMBACACEAE: *Ceiba pentandra* (bark). BURSERACEAE: *Boswellia dalzielii* (bark), *Dacryodes klaineana* (bark). CECROPIACEAE: *Musanga cecropioides* (leaf-bud), *Myrianthus arboreus* (leaf). COMMELINACEAE: *Commelina africana* (plant). COMPOSITAE: *Ageratum conyzoides* (leaf), *Ayapana triplinervis* (plant), *Chrysanthellum americanum* (leaf), *Elephantopus mollis* (leaf), *Emilia coccinea* (root), *Senecio biafrae* (plant), *Sigesbeckia orientalis* (plant), *Synedrella nodiflora* (leaf), *Vernonia colorata* (leaf, flower), *V. smithiana* (plant). CONNARACEAE: *Byrsocarpus poggeanus* (leaf), *B. viridis* (leaf). CUCURBITACEAE: *Momordica charantia* (leaf). DICHAPETALACEAE: *Dichapetalum toxicarium* (leaf). DILLENIACEAE: *Tetracera potatoria* (sap).

**C: Medicines; 9, arteries, veins**
ANNONACEAE: *Enantia polycarpa* (bark). ASCLEPIADACEAE: *Gymnema sylvestre* (leaf), *Leptadenia hastata* (plant). COCHLOSPERMACEAE: *Cochlospermum tinctorium* (root). COMPOSITAE: *Galinsoga parviflora* (plant).

**C: Medicines; 10, blood disorders, anaemia**
ACANTHACEAE: *Brillantaisia patula* (leaf). ASCLEPIADACEAE: *Solenostemma oleifolium* (leaf, flower). BOMBACACEAE: *Adansonia digitata* (leaf). CAPPARACEAE: *Euadenia trifoliolata* (leaf). CARYOPHYLLACEAE: *Stellaria media* (plant). CECROPIACEAE: *Musanga cecropioides* (root-sap). CHRYSOBALANACEAE: *Parinari excelsa* (bark). COMBRETACEAE: *Guiera senegalensis* (leaf). COMPOSITAE: *Elephantopus mollis* (plant), *Helichrysum mechowianum* (leaf-sap), *Senecio abyssinicus* (plant), *Sonchus oleraceus* (plant), *Sphaeranthus senegalensis* (leaf, flower), *Vernonia colorata* (leaf). CONNARACEAE: *Agelaea mildbraedii* (sap). CRUCIFERAE: *Rorippa humifusca* (leaf).

**C: Medicines; 11, pain-killers (revulsive, anodynal, analgesic, anaesthetic)**
ACANTHACEAE: *Asystasia calycina* (whole plant), *A. gangetica* (whole plant), *Brillantaisia nitens* (root), *B. patula* (plant), *Elytraria marginata, Eremomastax speciosa* (leaf), *Graptophyllum pictum* (sap), *Hygrophila auriculata* (whole plant), *Justicia extensa* (leaf), *J. flava* (leaf), *Monechma depauperatum* (plant), *Phaulopsis barteri, P. falcisepala* (sap), *P. imbricata Pseuderanthemum dispersum* (root), *P. ludovicianum* (leaf), *P. tunicatum* (root), *Thunbergia alata* (leaf), *T. chrysops* (leaf). AIZOACEAE: *Gisekia pharnaceoides, Glinus oppositifolius* (leaf). ALISMATACEAE: *Limnophyton angolense* (ash). AMARANTHACEAE: *Achyranthes aspera* (whole plant, leaf), *Aerva lanata, Alternanthera nodiflora* (whole plant, sap), *A. pungens, A. sessilis, Amaranthus hybridus, Celosia trigyna* (leaf), *Cyathula achyranthoides, C. prostrata*. ANACARDIACEAE: *Anacardium occidentale, Lannea acida* (leaf), *L. nigritana* (bark), *L. schimperi* (root), *L. velutina* (bark, root), *Mangifera indica* (leaf, bark), *Ozoroa insignis* (bark,

root), *O. pulcherrima* (bark), *Pseudospondias microcarpa* (bark), *Sclerocarya birrea* (bark), *Spondias mombin* (leak, bark, root), *Trichoscypha acuminata* (bark), *T. chevalieri* (plant), *T. patens* (plant). ANNONACEAE: *Annona arenaria* (plant), *A. senegalensis* (bark, root), *A. squamosa* (root-bark), *Annonidium mannii* (bark), *Cleistopholis patens* (bark-sap), *Enantia chlorantha* (bark), *Greenwayodendron suaveolens* (leaf, leaf-sap, bark), *Monanthotaxis vogelii* (root), *Monodora myristica* (bark, seed), *Pachypodanthium staudtii* (bark), *Uvaria afzelii* (leaf), *U. chamae* (leaf, root), *U. doeringii* (leaf, fruit), *Xylopia acutiflora* (bark), *X. aethiopica* (leaf, fruit, seed), *X. staudtii* (bark). APOCYNACEAE: *Alstonia boonei* (leaf, bark, root), *Carissa edulis* (leaf), *Clitandra cymulosa* (latex), *Holarrhena floribunda* (plant), *Landolphia heudelotii* (twig, root), *Motandra guineensis* (bark), *Pleiocarpa mutica* (bark), *Rauvolfia vomitoria* (leaf, root-bark), *Saba senegalensis* (leaf), *Strophanthus hispidus* (root), *S. sarmentosus* (root), *Tabernaemontana crassa* (sap), *Voacanga thouarsii* (latex). ARACEAE: *Anchomanes giganteus* (tuber), *Caladium bicolor* (root), *Culcasia scandens* (leaf, plant ash), *Cyrtosperma senegalense* (root), *Dieffenbachia spp., Pistia stratiotes* (leaf). ASCLEPIADACEAE: *Brachystelma constrictum* (tuber), *Calotropis procera* (latex, root), *Caralluma decaisneana* (latex), *Gomphocarpus fruticosus* (latex, root), *Gongronema latifolium* (stem), *Gymnema sylvestre* (leaf), *Leptadenia hastata* (latex), *Pergularia daemia* (leaf, latex), *Secamone afzelii* (leaf), *Tylophora conspicua* (root). BALANITACEAE: *Balanites aegyptiaca* (seed-husk). BALSAMINACEAE: *Impatiens niamniamensis (plant)*. BIGNONIACEAE: *Kigelia africana* (fruit), *Markhamia lutea* (plant), *M. tomentosa* (leaf, bark), *Newbouldia laevis* (bark, root), *Spathodea campanulata* (bark), *Stereospermum kunthianum* (root), *Tecoma capensis* (bark), *T. stans* (bark, flower). BOMBACACEAE: *Adansonia digitata* (leaf), *Bombax costatum* (bark), *Ceiba pentandra* (bark). BORAGINACEAE: *Ehretia cymosa* (leaf), *Heliotropium bacciferum* (plant), *H. indicum* (leaf), *H. ovalifolium* (plant), *H. strigosum* (plant). BURSERACEAE: *Dacryodes edulis* (leaf). CAMPANULACEAE: *Wahlenbergia perrottetii* (plant). CANNACEAE: *Canna indica* (shoot, seed). CAPPARACEAE: *Boscia angustifolia* (bark), *Buchholzia coriacea* (leaf, bark), *Cadaba farinosa* (plant), *Capparis erythrocarpos* (root-bark), *Cleome gynandra* (plant), *C. monophylla* (leaf), *C. rutidosperma* (leaf-sap), *C. viscosa* (leaf), *Crataeva adansonii* (leaf, bark), *Euadenia eminens* (sap), *E. trifoliolata* (root, root-bark), *Maerua angolensis* (leaf, root), *M. crassifolia* (leaf, bark), *Ritchiea capparoides* (bark). CARICACEAE: *Carica papaya* (leaf, root). CARYOPHYLLACEAE: *Drymaria cordata* (leaf). CECROPIACEAE: *Musanga cecropioides* (leaf, bark, root), *Myrianthus arboreus* (leaf sap). CELASTRACEAE: *Hippocratea myriantha* (root-ash), *H. pallens* (bark), *H. paniculata* (leaf), *H. velutina* (leaf, seed), *Maytenus senegalensis* (leaf, bark, root), *Salacia senegalensis* (bark, root). CHENOPODIACEAE: *Chenopodium ambrosioides* (leaf). CHRYSOBALANACEAE: *Acioa dewevrei* (bark), *Neocarya macrophylla* (leaf, bark, root, fruit), *Parinari excelsa* (twig, bark, root, endocarp hairs). COCHLOSPERMACEAE: *Cochlospermum tinctorium* (twig, root). COMBRETACEAE: *Anogeissus leiocarpus* (bark), *Combretum collinum* (gum, root), *C. fragrans* (whole plant), *C. glutinosum* (leaf), *C. micranthum* (leaf), *C. molle* (leaf), *C. nigricans* (leaf, bark), *C. paniculatum* (root), *C. racemosum* (gum), *C. rhodanthum* (plant), *C. smeathmannii* (root), *Quisqualis latialata* (plant), *Strephonema pseudocola* (gum), *Terminalia avicennioides* (root), *T. catappa* (leaf-sap), *T. glaucescens* (root), *T. ivorensis* (bark), *T. superba* (bark). COMMELINACEAE: *Commelina africana* (root), *C. congesta* (plant), *C. diffusa* (leaf, root), *Palisota bracteosa* (plant), *P. hirsuta* (plant). COMPOSITAE: *Acanthospermum hispidum* (leaf), *Adenostemma perrottetii* (plant), *Aedesia glabra* (plant), *Ageratum conyzoides* (leaf, root, flower), *Aspilia africana* (leaf), *Bidens pilosa* (plant, seed), *Centaurea praecox* (plant), *C. senegalensis* (plant), *Centipeda minima* (plant), *Conyza bonariensis* (leaf), *Dichrocephala chrysanthemifolia* (root), *Dicoma sessiliflora* (root, ash), *Echinops amplexicaulis* (plant), *E. longifolius* (root), *Emilia sonchifolia* (leaf), *Ethulia conyzoides* (plant), *Grangea maderaspatana* (leaf), *Guizotia scabra* (leaf), *Gynura procumbens* (leaf), *Laggera alata* (leaf, root), *L. oloptera* (plant), *Launaea cornuta* (plant), *Microglossa afzelii* (leaf, root), *M. pyrifolia* (plant, leaf, root), *Porphyrostemma chevalieri* (plant), *Pulicaria crispa* (plant), *Sphaeranthus senegalensis* (plant), *Spilanthes costata* (leaf-sap), *Struchium sparganophora* (leafy stem), *Synedrella nodiflora* (plant), *Triplotaxis stellutifera* (plant), *Vernonia adoensis* (root), *V. galamensis* (leaf), *V. glaberrima* (sap), *V. guineensis* (leaf), *V. smithiana* (plant). CONNARACEAE: *Byrsocarpus viridis* (leaf), *Cnestis ferruginea* (leaf, sap, root-bark), *Manotes longiflora* (sap), *Spiropetalum reynoldsii* (plant). CONVOLVULACEAE: *Astripomoea malvacea* (root), *Calycobolus parviflorus* (bark), *Ipomoea eriocarpa* (seed oil), *I. pes-caprae* (leaf, seed), *I. triloba* (plant), *Jacquemontia tamnifolia* (leaf). CRASSULACEAE: *Bryophyllum pinnatum* (plant, leaf, sap), *Kalanchoe crenata/integra* (sap). CRUCIFERAE: *Eruca sativa* (leaf). CUCURBITACEAE: *Coccinia keayana* (plant), *Cogniauxia podolaena* (root), *Cucumis figarei* (tap-root), *Lagenaria breviflora* (plant), *Momordica balsamina* (plant), *M. foetida* (leaf), *Sechium edule* (fruit peel). CYPERACEAE: *Cyperus articulatus* (rhizome), *Lipocarpa chinensis* (plant ash), *Scleria naumanniana* (leaf). DICHAPETALACEAE: *Dichapetalum albidum* (leaf). DILLENIACEAE: *Tetracera potatoria* (leaf, sap). DIONCOPHYLLACEAE: *Triphyophyllum peltatum* (leaf, bark).

**C: Medicines; 12, sedatives, tranquillisers, hallucinogens, narcotics**
ACANTHACEAE: *Brillantaisia patula* (leaf), *Lepidagathis alopecuroides, Stenandriopsis buntingii* (plant). ANACARDIACEAE: *Pseudospondias microcarpa.* ANNONACEAE: *Annona muricata* (leaf). APOCYNACEAE: *Motandra guineensis* (leaf), *Rauvolfia vomitoria* (root), *Tabernaemontana crassa* (sap). ARACEAE: *Caladium bicolor* (leaf), *Cyrtosperma senegalense* (root). ASCLEPIADACEAE: *Calotropis procera* (latex), *Pachycarpus lineolata* (root). BIGNONIACEAE: *Tecoma capensis* (bark). BOMBACACEAE: *Adansonia digitata* (leaf), *Ceiba pentandra* (leaf). BURSERACEAE: *Commiphora africana* (leaf), *Cleome monophylla* (leaf-sap). CELASTRACEAE: *Salacia erecta* (plant). COCHLOSPERMACEAE: *Cochlospermum religiosum* (gum). COMBRETACEAE: *Combretum glutinosum* (plant), *Terminalia catappa* (leaf-sap). COMMELINACEAE: *Aneilema beninensis* (plant), *Commelina africana* (plant). COMPOSITAE: *Ageratum conyzoides* (leaf), *Aspilia africana* (leaf), *Ayapana triplinervis* (plant), *Bidens pilosa* (plant), *Emilia sonchifolia* (plant), *Ethulia conyzoides* (leaf), *Grangea maderaspatana* (leaf), *Helichrysum nudifolium* (plant), *Laggera alata* (leaf), *L. oloptera* (leaf), *Microglossa pyrifolia* (plant), *Senecio mannii* (leaf, root), *Sphaeranthus senegalensis* (plant). CONNARACEAE: *Agelaea dewevrei.* CONVOLVULACEAE: *Argyreia nervosa* (seed), *Evolvulus nummularius* (plant), *Ipomoea aquatica* (leaf-sap), *I. quamoclit* (leaf). CRASSULACEAE: *Bryophyllum pinnatum* (whole plant), *Kalanchoe crenata/integra* (leaf). CRUCIFERAE: *Brassica juncea* (leaf). CUCURBITACEAE: *Cogniauxia podolaena* (root), *Lagenaria breviflora* (seed), *Momordica balsamina* (plant), *Zehneria hallii* (leaf). CYPERACEAE: *Cyperus articulatus* (rhizome). DILLENIACEAE: *Tetracera potatoria* (sap). DIOSCOREACEAE: *Dioscorea alata* (raw tuber).

**C: Medicines; 13, arthritis, rheumatism, lumbago, muscular and body-pains, stiffness (see also C:11)**
ACANTHACEAE: *Acanthus montanus* (leaf, stem), *Adhatoda buchholzii* (leaf), *A. robusta* (leaf), *A. tristis* (leaf), *Barleria opaca* (whole plant), *B. prionitis* (leafy twigs, leaf-sap), *Brillantaisia patula* (leaf), *Hygrophila auriculata* (leaf, root), *Justicia extensa* (leaf), *Phaulopsis falcisepala* (sap), *P. imbricata* (plant-ash), *Thomandersia hensii* (root). AIZOACEAE: *Trianthema portulacastrum* (leaf). AMARANTHACEAE: *Celosia isertii* (leaf), *C. trigyna* (leaf). ANACARDIACEAE: *Pseudospondias microcarpa* (bark), *Schinus terebinthifolius* (bark), *Trichoscypha acuminata* (bark). ANNONACEAE: *Annonidium mannii* (bark), *Enantia chlorantha* (bark), *Greenwayodendron suaveolens* (leaf), *Isolona campanulata* (root), *Monodora myristica* (bark), *Xylopia aethiopica* (leaf, bark). APOCYNACEAE: *Alafia scandens* (leaf), *Baissea multiflora* (liane), *Carissa edulis* (root-bark), *Dictyophleba leonensis* (leaf), *Landolphia dulcis* (stem, root), *L. owariensis* (leaf), *Pleioceras barteri* (leaf), *Rauvolfia vomitoria* (leaf, root), *Strophanthus hispidus* (root), *S. sarmentosus* (twig), *Tabernaemontana crassa* (bark), *Voacanga bracteata* (root). ARACEAE: *Amorphophallus johnsonii* (tuber), *Anchomanes giganteus* (tuber), *Colocasia esculenta* (tuber), *Dieffenbachia spp.* ASCLEPIADACEAE: *Calotropis procera* (bark), *Solenostemma oleifolium* (leaf). BALANITACEAE: *Balanites aegyptiaca* (fruit, seed). BIGNONIACEAE: *Kigelia africana* (bark, fruit), *Markhamia lutea* (plant), *Newbouldia laevis* (bark, root). BOMBACACEAE: *Rhodognaphalon brevicuspe* (root). BORAGINACEAE: *Coldenia procumbens* (leaf), *Cordia vignei* (leaf), *Heliotropium indicum* (leaf). BURSERACEAE: *Boswellia dalzielii* (bark), *Canarium schweinfurthii* (bark). CACTACEAE: *Opuntia spp.* (stem). CAPPARACEAE: *Boscia salicifolia* (root), *Cadaba farinosa* (root), *Capparis decidua* (root), *C. spinosa* (plant), *Cleome arabica* (leaf), *C. gynandra* (plant), *Maerua angolensis* (leaf). CECROPIACEAE: *Musanga cecropioides* (root-sap). COMBRETACEAE: *Combretum glutinosum* (bark), *C. nigricans* (bark), *Guiera sengalensis* (leaf). COMPOSITAE: *Acanthospermum hispidum* (leaf), *Adenostemma perrottetii* (stem-sap), *Aedesia glabra* (plant), *Ageratum conyzoides* (leaf), *Blumea aurita* (plant), *Gynura procumbens* (leaf), *Laggera alata* (plant), *L. oloptera* (plant), *Microglossa afzelii* (leaf), *Mikania cordata* (plant), *Senecio abyssinicus* (leaf), *Sigesbeckia orientalis* (plant), *Sphaeranthus senegalensis* (plant). *Synedrella nodiflora* (plant), *Vernonia amygdalina* (bark), *V. nigritiana* (root), *V. smithiana* (plant). CONNARACEAE: *Roureopsis obliquifoliolata* (root). CONVOLVULACEAE: *Argyreia nervosa* (root), *Ipomoea asarifolia* (plant), *I. eriocarpa* (seed-oil), *I. pes-caprae* (leaf), *Merremia umbellata* (plant). COSTACEAE: *Costus afer* (stem). CRUCIFERAE: *Lepidium sativum* (seed). CUCURBITACEAE: *Momordica charantia* (root), *Zehneria scabra* (leaf). CYPERACEAE: *Fimbristylis ovata* (culm). DILLENIACEAE: *Tetracera alnifolia* (sap). DIOSCOREACEAE: *Dioscorea bulbifera* (bulbil), *D. dumetorum* (tuber), *D. praehensilis* (tuber).

**C: Medicines; 14, eye treatments (see also C:11)**
ACANTHACEAE: *Dyschoriste perrottetii* (seed), *Elytraria marginata* (leaf-sap), *Hygrophila senegalensis* (seed), *Lepidagathis heudelotiana* (seed), *Peristrophe bicalyculata* (mucilage), *Thomandersia hensii* (root), *Thunbergia alata* (leaf-sap). AMARANTHACEAE: *Achyranthes*

*aspera* (leaf-sap), *Aerva lanata, Amaranthus spinosus, A. viridis* (leaf-sap), *Celosia leptostachya* (fruit, seed), *C. trigyna* (leaf-sap, seed). ANACARDIACEAE: *Lannea acida, Ozoroa insignis* (bark), *Sclerocarya birrea* (bark), *Spondias mombin* (leaf), *Pseudospondias microcarpa* (bark). ANNONACEAE: *Enantia polycarpa* (bark), *Monodora myristica* (bark), *Uvaria chamae* (leaf). APOCYNACEAE: *Alafia lucida* (plant), *Ancylobotrys amoena* (latex), *Baissea multiflora* (root), *Motandra guineensis* (leaf), *Saba florida* (latex), *S. senegalensis* (leaf), *Strophanthus sarmentosus* (leaf). ARACEAE: *Anchomanes difformis* (sap), *Pistia stratiotes* (leaf). ARALIACEAE: *Cussonia arborea* (leaf). ASCLEPIADACEAE: *Calotropis procera* (leaf, latex, flower), *Leptadenia hastata* (root), *L. pyrotechnia* (seed), *Pergularia daemia* (sap), *P. tomentosa* (leaf-sap), *Sarcostemma viminale* (sap). BASELLACEAE: *Basella alba* (sap). BIGNONIACEAE: *Kigelia africana* (flower-sap), *Markhamia tomentosa* (bud-sap), *Newbouldia laevis* (leaf). BOMBACACEAE: *Adansonia digitata* (leaf, fruit-pulp), *Ceiba pentandra* (leaf, mucilage). BORAGINACEAE: *Cynoglossum lanceolatum* (root), *Heliotropium indicum* (leaf). BURSERACEAE: *Commiphora africana* (gum). CANNACEAE: *Canna indica* (leaf, flower). CAPPARACEAE: *Bossia angustifolia* (bark), *B. salicifolia* (bark), *B. senegalensis* (leaf), *Capparis decidua* (root), *C. erythrocarpos* (root-bark), *C. tomentosa* (leaf, root), *Cleome gynandra* (leaf-sap), *C. monophylla* (leaf), *Crataeva adansonii* (leaf), *Euadenia eminens* (plant, sap), *E. trifoliolata* (root), *Ritchiea capparoides* (leaf). CARYOPHYLLACEAE: *Drymaria cordata* (plant), *Stellaria media* (plant). CELASTRACEAE: *Maytenus senegalensis* (leaf-sap). CHENOPODIACEAE: *Suaeda fruticosa* (leaf). CHRYSOBALANACEAE: *Neocarya macrophylla* (bark). COCHLOSPERMACEAE: *Cochlospermum tinctorium* (root). COMBRETACEAE: *Anogeissus leiocarpus* (leaf), *Combretum aculeatum* (stem-sap), *C. paniculatum* (flower-sap), *C. platypterum* (leaf-sap), *Strephonema pseudocola* (root), *Terminalia albida* (bark). COMMELINACEAE: *Aneilema lanceolatum* (seed), *Commelina benghalensis* (plant-sap, flower-spathe), *C. bracteosa* (sap), *C. diffusa* (leaf). COMPOSITAE: *Ageratum conyzoides* (sap), *Aspilia africana* (leaf-sap), *A. kotschyi* (leaf-sap), *Bidens pilosa* (sap), *Blainvillea gayana* (leaf), *Blumea perrottetiana* (leaf-sap), *Centipeda minima* (plant), *Conyza canadensis* (plant), *C. sumatrensis* (plant), *Crassocephalum montuosum* (plant), *C. rubens* (leaf-sap), *C. vitellianum* (plant), *Echinops longifolius* (flower head), *Emilia coccinea* (leaf), *E. sonchifolia* (leaf-sap), *Ethulia conyzoides* (leaf-sap), *Gynura miniata* (leaf), *Helichrysum foetidum* (root). *Melanthera scandens* (leaf-sap), *Microglossa afzelii* (leaf), *M. pyrifolia* (leaf-sap, root-sap), *C. vitellianum* (plant), *Echinops longifolius* (flower-head), *Emilia coccinea* (leaf), *E. abyssinicus* (leaf), *Sonchus oleraceus* (sap), *S. schweinfurthii* (leaf), *Vernonia biafrae* (leaf, root), *V. conferta* (leaf), *V. nigritiana* (root). CONNARACEAE: *Cnestis ferruginea* (leaf-sap, fruit-juice), *Jaundea pubescens* (leaf-sap). CONVOLVULACEAE: *Astripomoea malvacea* (leaf, root, flower), *Ipomoea asarifolia* (plant), *I. cairica* (fibre), *Jacquemontia tamnifolia* (leaf-sap), *Merremia tridentata* (leaf). COSTACEAE: *Costus afer* (sap), *C. dubius* (inflorescence), *C. lucanusianus* (stem), *C. spectabilis* (sap). CRASSULACEAE: *Bryophyllum pinnatum* (sap), *Kalanchoe crenata/integra* (leaf-sap). CONNARACEAE: *Manotes longiflora* (leaf). CUCURBITACEAE: *Coccinia grandis* (fruit), *Lagenaria guineensis* (sap), *Luffa acutangula* (leaf-sap), *Momordica charantia* (plant), *Ruthalicia eglandulosa* (leaf-sap). CYPERACEAE: *Scleria boivinii* (leaf-sap). DILLENIACEAE: *Tetracera alnifolia* (sap), *T. potatoria* (sap). DIOSCOREACEAE: *Dioscorea bulbifera* (stem-sap). DROSERACEAE: *Drosera madagascariensis* (plant).

**C: Medicines, 15, ear treatments (see also C:11)**

ACANTHACEAE: *Barleria prionitis, Brillantaisia nitens* (leaf-sap), *Graptophyllum pictum* (sap), *Peristrophe bicalyculata* (mucilage), *Thomandersia hensii* (root), *Thunbergia laurifolia* (leaf-sap). AMARANTHACEAE: *Cyathula achyranthoides* (leaf-sap), *C. prostreta* (sap, seed), *Amaranthus spinosus*. AMARYLLIDACEAE: *Scadoxus multiflorus* (sap). APOCYNACEAE: *Picralima nitida* (leaf-sap). ARACEAE: *Culcasia scandens* (leaf-sap). ASCLEPIADACEAE: *Dregea abyssinica* (latex). BIGNONIACEAE: *Newbouldia laevis* (bark). BOMBACACEAE: *Adansonia digitata* (leaf). BURSERACEAE: *Dacryodes edulis* (leaf). CAPPARACEAE: *Buchholzia coriacea* (bark), *Cleome gynandra* (leaf-sap, seed-capsule), *C. rutidosperma* (leaf-sap), *Euadenia eminens* (root), *E. trifoliolata* (root), *Maerua duchesnei* (root-sap), *Ritchiea capparoides* (bark), *R. reflexa* (bark). COMMELINACEAE: *Commelina diffusa* (leaf), *Palisota bracteosa* (plant), *P. hirsuta* (plant). COMPOSITAE: *Ageratum conyzoides* (leaf), *Aspilia kotschyi* (leaf-sap), *Bidens pilosa* (sap), *Centaurea perrottetii* (plant), *Crassocephalum rubens* (sap), *Emilia sonchifolia* (leaf-sap), *Grangea maderaspatana* (leaf), *Launaea cornuta* (sap), *Microglossa afzelii* (leaf-sap), *Mikania cordata* (plant-sap), *Piloselloides hirsuta* (root), *Spilanthes filicaulis* (plant), *Synedrella nodiflora* (leaf-sap), *Vernonia biafrae* (leaf). CONNARACEAE: *Agelaea dewevrei* (bark), *Byrsocarpus coccineus* (root-sap). CONVOLVULACEAE: *Merremia umbellata* (plant). CRASSULACEAE: *Bryophyllum pinnatum* (sap), *Kalanchoe crenata/integra* (leaf-sap). CUCURBITACEAE: *Bambekea racemosa* (leaf), *Coccinia barteri* (leaf-sap), *Cucurbita maxima* (root-sap).

**C: Medicines; 16, oral (see also C:11 for toothache treatment)**

ACANTHACEAE: *Barleria prionitis*. ANACARDIACEAE: *Lannea acida* (bark), *L. schimperi* (root), *L. welwitschii* (root), *Sorindeia juglandifolia* (leaf). ANNONACEAE. *Annona arenaria* (plant), *Xylopia aethiopica* (root), *X. quintasii* (bark, root). APOCYNACEAE: *Motandra guineensis* (leaf), *Rauvolfia vomitoria* (bark). ARACEAE: *Pistia stratiotes* (plant ash), *Rhaphidophora africana*. ASCLEPIADACEAE: *Calotropis procera* (latex). BOMBACACEAE: *Ceiba pentandra* (bark). BURSERACEAE: *Dacryodes edulis* (bark). CAPPARACEAE: *Euadenia trifoliolata* (root). COMBRETACEAE: *Combretum paniculatum* (leaf), *Terminalia superba* (bark). COMPOSITAE: *Spilanthes filicaulis* (plant), *Stomatanthes africanus* (leaf), *Synedrella nodiflora* (leaf), *Vernonia colorata* (leaf). CONNARACEAE: *Byrsocarpus coccineus* (leaf).

**C: Medicine; 17, naso-pharyngeal (catarrh, stuffiness, sneezing, sore-throat, cough, phlegm, etc.)**

ACANTHACEAE: *Acanthus guineensis* (fruit), *A. montanus* (leaf), *Adhatoda camerunensis* (whole plant), *Barleria cristata* (leaf, root), *B. opaca* (leaf-sap), *B. prionitis* (leaf, sap), *Dicliptera verticillata* (leaf-sap), *Elytraria marginata* (plant, leaf), *Hygrophila auriculata* (plant), *Hypoestes verticillata* (root), *Justicia flava* (root), *Monechma ciliatum* (leaf), *Thunbergia chrysops* (leaf). AIZOACEAE: *Mollugo nudicaulis*. AMARANTHACEAE: *Achyranthes aspera* (leaf), *Aerva lanata, Alternanthera pungens, Gomphrena globosa, Pupalia lappacea*. AMARYLLIDACEAE: *Pancratium trianthum*. ANACARDIACEAE: *Antrocaryon micraster* (fruit), *Lannea welwitschii* (root), *Pseudospondias microcarpa* (bark), *Spondias mombin* (leaf, bark), *S. cytherea* (leaf). ANNONACEAE: *Annona arenaria* (plant), *A. chrysophylla* (root), *A. glabra* (leaf, shoot, flower), *A. muricata* (flower, fruit), *A. senegalensis* (plant), *Hexalobus monopetalus* (leaf), *Monodora myristica* (seed), *Uvaria afzelii* (leaf, stem-bark, root-bark), *U. chamae* (gum), *U. thomasii* (leaf), *Xylopia aethiopica* (fruit), *X. parviflora* (leaf, root), *X. staudtii* (bark), *X. villosa* (bark). APOCYNACEAE: *Carissa edulis* (root), *Funtumia elastica* (leaf), *Landolphia hirsuta* (bark), *Motandra guineensis* (leaf), *Rauvolfia vomitoria* (leaf), *Saba senegalensis* (latex). ARACEAE: *Anchomanes difformis* (sap), *Cyrtosperma senegalense* (root). ASCLEPIADACEAE: *Calotropis procera* (leaf-sap), *Gomphocarpus fruticosus* (seed), *Leptadenia hastata* (sap), *Pachycarpus lineolata* (root), *Pergularia daemia* (leaf, stem, root), *Secamone afzelii* (leaf). BALANITACEAE: *Balanites aegyptiaca* (bark). BIGNONIACEAE: *Kigelia africana* (root), *Spathodea campanulata* (bark), *Stereospermum kunthianum* (bark, fruit pod). BIXACEAE: *Bixa orellana* (leaf). BORAGINACEAE: *Cordia myxa* (fruit-pulp), *C. sinensis* (bark), *Cynoglossum lanceolatum* (plant), *Heliotropium indicum* (leaf). BROMELIACEAE: *Ananas comosus* (unripe fruit). BURSERACEAE: *Canarium schweinfurthii* (bark), *Dacryodes klaineana* (bark), *Santiria trimera* (bark). CACTACEAE: *Opuntia spp.* (leaf). CANNACEAE: *Canna indica* (shoots). CAPPARACEAE: *Buchholzia coriacea* (bark). *Cadaba farinosa* (plant), *Capparis thonningii* (leaf-sap, root), *Cleome monophylla* (root), *Ritchiea reflexa* (leaf). CECROPIACEAE: *Musanga cecropioides* (bark, root-sap), *Myrianthus arboreus* (bark, leaf-sap, root-sap). CELASTRACEAE: *Hippocratea africana* (leaf), *Maytenus senegalensis* (seed), *M. undata* (plant). CHENOPODIACEAE: *Suaeda monoica* (root). CHRYSOBALANACEAE: *Parinari curatellifolia* (leaf, bark). COCHLOSPERMACEAE: *Cochlospermum religiosum* (gum). COMBRETACEAE: *Combretum aculeatum* (root), *C. collinum* (leafy twig), *C. glutinosum* (leaf), *C. micranthum* (leaf), *C. molle* (bark), *C. platypterum* (leaf), *C. racemosum* (plant), *C. tomentosum* (root), *Conocarpus erectus* (root), *Guiera senegalensis* (leaf), *Pteleopsis suberosa* (shoot, root), *Terminalia brownei* (bark), *T. catappa* (leaf), *T. glaucescens* (leaf, bark). COMMELINACEAE: *Aneilema aequinoctiale* (plant), *Coleotrype laurentii* (plant), *Commelina benghalensis* (plant-sap), *Palisota bracteosa* (plant), *P. hirsuta* (plant). COMPOSITAE: *Adenostemma caffrum* (leaf), *Aedesia glabra* (plant), *Ageratum conyzoides* (leaf), *Anisopappus africanus* (leaf), *Aspilia africana* (leaf), *Aspilia rudis* (plant), *Bidens pilosa* (leaf), *Conyza bonariensis* (leaf), *Dicoma sessiliflora* (root), *D. tomentosa* (leaf), *Elephantopus mollis* (plant), *Emilia sonchifolia* (plant), *Helichrysum foetidum* (plant, root), *H. nudifolium* (plant), *H. odoratissimum* (root), *Launaea cornuta* (plant), *Melanthera scandens* (leaf), *Microglossa pyrifolia* (root), *Mikania cordata* (leaf), *Nidorella spartioides* (flower), *Piloselloides hirsuta* (plant), *Senecio biafrae* (leaf), *Sigesbeckia orientalis* (plant), *Spilanthes filicaulis* (plant), *Stomatanthes africanus* (root), *Synedrella nodiflora* (root), *Triplotaxis stellutifera* (leaf), *Vernonia ambigua* (plant), *V. amygdalina* (leaf), *V. cinerea* (plant), *V. cistifolia* (root), *V. conferta* (leaf), *V. nigritiana* (root), *V. poskeana* (plant). CONNARACEAE: *Byrsocarpus viridis* (leaf), *Cnestis ferruginea* (leaf). COSTACEAE: *Costus afer* (stem), *C. dubius* (stem), *C. lucanusianus* (plant). CRASSULACEAE: *Bryophyllum pinnatum* (plant, root), *Kalanchoe laciniata* (leaf). CRUCIFERAE: *Farsetia ramosissima* (plant), *Raphanus sativus* (seed). CUCURBITACEAE: *Bambekea racemosa* (leaf), *Cucurbita pepo* (flower), *Lagenaria siceraria* (fruit). CYPERACEAE: *Cyperus articulatus* (rhizome), *Hypolytrum purpurascens* (leaf), *Scleria boivinii* (leaf less scabrid part). DILLENIACEAE: *Tetracera potatoria* (sap). DROSERACEAE: *Drosera madagascariensis* (plant).

**C: Medicines; 18, pulmonary (chest, lungs, pneumonia, bronchitis, pleurisy, tuberculosis, asthma, etc.)**

ACANTHACEAE: *Acanthus montanus* (leaf), *Brillantaisia patula* (leaf), *Elytraria marginata* (plant, leaf), *Hypoestes verticillata* (plant), *Justicia extensa* (leaf, root), *J. flava* (plant), *Pseuderanthemum tunicatum* (sap, root), *Ruellia tuberosa* (leaf), *Whitfieldia elongata* (leaf). AIZOACEAE: *Gisekia pharnaceoides* (whole plant), *Mollugo cerviana*. AMARANTHACEAE: *Achyranthes aspera* (leaf), *Amaranthus hybridus, A. lividus, Celosia trigyna* (leaf). AMARYLLIDACEAE: *Crinum purpurascens*. ANACARDIACEAE: *Lannea nigritana* (bark, root), *L. welwitschii* (root), *Mangifera indica* (leaf), *Rhus natalensis* (root), *Spondias mombin* (bark), *Trichoscypha acuminata* (bark). ASCLEPIADACEAE: *Calotropis procera* (leafy twigs, manna, root), *Gomphocarpus fruticosus* (plant), *Margaretta rosea* (root), *Pergularia tomentosa* (root), *Secamone afzelii* (leaf). ANNONACEAE: *Annona arenaria* (plant), *A. senegalensis* (leaf), *Cleistopholis patens* (bark-sap), *Enantia chlorantha* (bark), *Hexalobus monopetalus* (leaf), *Isolona campanulata* (plant, twig-bark), *Pachypodanthium staudtii* (bark), *Uvaria afzelii* (leaf, root-bark), *Xylopia acutiflora* (bark), *X. aethiopica* (leaf, bark, fruit), *X. quintasii* (fruit). APOCYNACEAE: *Carissa edulis* (root), *Picralima nitida* (root, seed), *Pycnobotrys nitida* (leaf), *Saba senegalensis* (latex). AQUIFOLIACEAE: *Ilex mitis* (bark). ARACEAE: *Amorphophallus dracontioides* (root). BALANITACEAE: *Balanites aegyptiaca* (gum, bark). BIGNONIACEAE: *Crescentia cujete* (fruit), *Markhamia lutea* (plant), *M. tomentosa* (plant), *Stereospermum kunthianum* (leaf, bark, root), *Tecoma capensis* (bark). BOMBACACEAE: *Adansonia digitata* (leaf), *Ceiba pentandra* (bark). BORAGINACEAE: *Cordia millenii* (leaf), *C. myxa* (fruit-pulp). BURSERACEAE: *Canarium schweinfurthii* (sap), *Dacryodes edulis* (bark), *Santiria trimera* (bark). CAPPARACEAE: *Cleome gynandra* (plant), *Euadenia trifoliolata* (root), *Maerua pseudopetalosa* (root, fruit), *Ritchiea capparoides* (root). CARYOPHYLLACEAE: *Drymaria cordata* (sap). CECROPIACEAE: *Myrianthus arboreus* (leaf-sap, bark), CELASTRACEAE: *Hippocratea indica* (plant), *Mystoxylon aethiopicum* (bark). CHENOPODIACEAE: *Chenopodium ambrosioides* (plant). CHRYSOBALANACEAE: *Maranthes kerstingii* (plant), *Neocarya macrophylla* (leaf, bark). COCHLOSPERMACEAE: *Cochlospermum tinctorium* (root). COMBRETACEAE: *Combretum collinum* (leafy twig), *C. glutinosum* (leaf), *C. micranthum* (leaf), *C. molle* (leaf), *Terminalia brownii* (bark), *T. laxiflora* (plant), *T. superba* (bark). COMPOSITAE: *Adenostemma caffrum* (root), *Aspilia africana* (root), *A. kotschyi* (leaf), *Bidens pilosa* (flower), *Conyza sumatrensis* (leaf), *Cotula anthemoides* (leaf, root), *Eclipta alba* (leaf), *Emilia coccinea* (root), *Helichrysum nudifolium* (plant), *Laggera alata* (leaf, root), *Microglossa afzelii* (leafy stem), *M. pyrifolia* (leaf), *Mikania cordata* (plant), *Piloselloides hirsuta* (root), *Vernonia adoensis* (root), *V. cinerea* (plant), *V. colorata* (leaf), *V. conferta* (leaf). CONNARACEAE: *Byrsocarpus viridis* (leaf), *Cnestis ferruginea* (fruit). CONVOLVULACEAE: *Cressa cretica* (plant), *Evolvulus alsinoides* (leaf), *Ipomoea coptica* (leaf), *I. involucrata* (leaf). COSTACEAE: *Costus lucanusianus* (plant). CRASSULACEAE: *Kalanchoe crenata/integra* (leaf), *K. laciniata* (leaf). CUCURBITACEAE: *Luffa cylindrica* (fruit exudation). CYPERACEAE: *Cyperus articulatus* (rhizome), *Pycreus nitidus* (rhizome), *Scleria pterota* (plant-ash). DILLENIACEAE: *Tetracera alnifolia* (leafy twig).

**C: Medicine; 19, stomachic (see also C:11 for stomachache and other pains)**

ACANTHACEAE: *Asystasia gangetica* (leaf), *Brillantaisia patula* (plant), *Elytraria marginata, Hygrophila auriculata* (whole plant), *Pseuderanthemum dispersum* (root), *Rhinacanthus virens* (leaf, root), *Thunbergia chrysops* (leaf), *T. grandiflora* (leaf), *Whitfieldia elongata* (leaf). AIZOACEAE: *Glinus oppositifolius* (leaf). AMARANTHACEAE: *Alternanthera pungens, A. sessilis, Amaranthus dubius, A. spinosus, A. viridis* (leaf-sap), *Celosia argentea, C. trigyna* (leaf), *Cyathula uncinulata. AMARYLLIDACEAE: Crinum purpurascens*. ANACARDIACEAE: *Antrocaryon micraster* (fruit), *Lannea barteri* (bark), *L. nigritana* (twig), *Ozoroa insignis* (root), *O. pulcherrima* (root), *Rhus natalensis* (root), *Sclerocarya birrea* (bark), *Spondias mombin* (root). ANNONACEAE: *Annona squamosa* (leaf), *Annonidium mannii* (bark), *Monanthotaxis vogelii* (root), *Monodora myristica* (bark), *Uvaria thomasii* (leaf), *Xylopia aethiopica* (bark, fruit). APOCYNACEAE: *Baissea multiflora* (bark, root), *Holarrhena floribunda* (root), *Hunteria elliotii* (bark). ARACEAE: *Anchomanes giganteus* (tuber), *Culcasia scandens* (leaf), *Pistia stratiotes* (leaf), *Typhonium trilobatum* (root). ARISTOLOCHIACEAE: *Aristolochia albida* (root). ASCLEPIADACEAE: *Leptadenia hastata* (root), *Pachycarpus lineolata* (root), *Tylophora glauca* (leaf), *Xysmalobium heudelotianum* (root). BIGNONIACEAE: *Kigelia africana* (leaf). BOMBACACEAE: *Ceiba pentandra* (leaf, bark). BORAGINACEAE: *Cordia senegalensis* (leaf), *Cynoglossum lanceolatum* (plant), *Heliotropium subulatum* (root). BURSERACEAE: *Canarium schweinfurthii* (bark), *Commiphora africana* (leaf, gum). CAPPARACEAE: *Boscia senegalensis* (leaf, bark, root), *Cadaba farinosa* (shoot), *Cleome viscosa* (leaf, seed), *Crataeva adansonii* (bark), *Maerua angolensis* (leaf), *Ritchiea capparoides* (root). CECROPIACEAE: *Musanga cecropioides* (leaf, bark, root-sap). CELASTRACEAE:

*Hippocratea indica* (root), *Maytenus undata* (root), *Mystroxylon aethiopicum* (bark), *Salacia pyriformis* (plant). COMBRETACEAE: *Combretum aculeatum* (root), *C. dolichopetalum* (root), *C. glutinosum* (leaf), *C. micranthum* (leaf). COMMELINACEAE: *Commelina congesta* (plant). COMPOSITAE: *Ageratum conyzoides* (leaf), *Ayapana triplinervis* (plant), *Blumea aurita* (plant), *B. perrottetiana* (root), *Centaurea perrottetii* (plant), *C. senegalensis* (plant), *Cotula anthemoides* (leaf, root), *Crassocephalum crepidioides* (leaf), *Echinops amplexicaulis* (plant), *Emilia coccinea* (leaf), *Grangea maderaspatana* (leaf), *Helichrysum mechowianum* (leaf-sap), *H. nudifolium* (leaf), *H. odoratissimum* (leaf), *Launaea cornuta* (leaf, root), *Senecio abyssinicus* (plant), *S. mannii* (leaf), *Sphaeranthus senegalensis* (root), *Tridax procumbens* (plant), *Vernonia amygdalina* (wood, leaf, flower), *V. cinerea* (leaf, flower), *V. colorata* (leaf, root), *V. conferta* (leaf, bark), *V. lasiopus* (plant), *V. macrocyanus* (plant), *V. nigritiana* (leaf), *V. pumila* (root), *V. smithiana* (root). CONNARACEAE: *Byrsocarpus viridis* (leaf), *Cnestis ferruginea* (leaf), *Jaundea pubescens* (bark). CONVOLVULACEAE: *Evolvulus alsinoides* (leaf), *Ipomoea pes-caprae* (seed). COSTACEAE: *Costus afer* (root). CRUCIFERAE: *Cardamine hirsuta* (leaf), *Eruca sativa* (leaf), *Raphanus sativus* (seed). CUCURBITACEAE: *Momordica balsamina* (plant), *M. foetida* (leaf), *Ruthalicia longipes* (plant). CYPERACEAE: *Cyperus alternifolius* (root), *C. esculentus* (tuber), *C. rotundus* (tuber), *Rhynchospora corymbosa* (seed). DILLENIACEAE: *Tetracera alnifolia* (sap). DIOSCOREACEAE: *Dioscorea praehensilis* (stem). DROSERACEAE: *Drosera madagascariensis* (plant).

**C: Medicines; 20, 'intestines' (unspecified)**

AIZOACEAE: *Tetragona tetragonioides*. ANACARDIACEAE: *Ozoroa pulcherrima* (root). ANNONACEAE: *Annonidium mannii* (bark). APOCYNACEAE: *Alstonia boonei* (bark), *Strophanthus sarmentosus* (root), *Tabernaemontana crassa* (bark). ARACEAE: *Rhektophyllum mirabile* (leaf). ASCLEPIADACEAE: *Pachycarpus lineolata* (root). AVICENNIACEAE: *Avicennia germinans* (root). BOMBACACEAE: *Ceiba pentandra* (bark). CAPPARACEAE: *Boscia senegalensis* (leaf, bark, root). COMBRETACEAE: *Combretum molle* (leaf). *C. nigricans* (leaf, bark). *Quisqualis latialata* (plant). COMPOSITAE: *Ageratum conyzoides* (leaf).

**C: Medicines; 21, emetic**

ACANTHACEAE: *Acanthus montanus* (leaf), *Justicia flava* (leaf), *Rhinacanthus virens* (leaf, stem), *Ruellia tuberosa* (plant). AMARYLLIDACEAE: *Crinum purpurascens, Cyathula uncinulata*. ANACARDIACEAE: *Lannea welwitschii* (root), *Spondias mombin* (bark). ANNONACEAE: *Annona arenaria* (plant), *A. chrysophylla* (bark), *A. glabra* (seed), *A. muricata* (seed), *A. senegalensis* (bark), *Xylopia aethiopica* (leaf, seed). APOCYNACEAE: *Rauvolfia vomitoria* (leaf, latex, root, fruit), *Saba senegalensis* (leaf, latex), *Strophanthus sarmentosus* (leaf), *Thevetia neriifolia* (leaf, bark). ARACEAE: *Caladium bicolor* (root). ARALIACEAE: *Cussonia arborea* (plant). ASCLEPIADACEAE: *Asclepias curassavica* (plant), *Calotropis procera* (root-bark), *Sarcostemma viminale* (plant). BOMBACACEAE: *Ceiba pentandra* (bark). BURSERACEAE: *Boswellia dalzielii* (bark), *Canarium schweinfurthii* (bark). CECROPIACEAE: *Myrianthus arboreus* (fruit). CELASTRACEAE: *Hippocratea myriantha* (plant). CHENOPODIACEAE: *Agathophora alopecuroides* (plant). CHRYSOBALANACEAE: *Maranthes kerstingii* (plant). COMBRETACEAE: *Terminalia avicennioides* (bark). COMPOSITAE: *Ageratum conyzoides* (leaf), *Ambrosia maritima* (root), *Ayapana triplinervis* (plant), *Centaurea nigerica* (root), *C. praecox* (plant), *Conyza pyrrhopappa* (leaf), *Eclipta alba* (leaf, root), *Helichrysum cymosum* (root), *Launaea nana* (plant), *Melanthera scandens* (leaf, stem, root), *Microglossa pyrifolia* (*leaf*), *Sonchus schweinfurthii* (leaf), *Vernonia colorata* (leaf), *V. nigritiana* (root). CONNARACEAE: *Byrsocarpus viridis* (leaf). CONVOLVULACEAE: *Ipomoea aquatica* (plant, seed). CUCURBITACEAE: *Cucumis melo* (root), *C. prophetarum* (fruit), *Luffa cylindrica* (seed), *Momordica balsamina* (plant), *M. foetida* (leaf), *Trichosanthes cucumerina* (leaf-sap).

**C: Medicines; 22, antemetic**

ANACARDIACEAE: *Lannea humilis* (root). ANNONACEAE: *Monodora myristica* (seed). ARACEAE: *Culcasia scandens* (leaf), *Rhektophyllum mirabile* (leaf-sap). ASCLEPIADACEAE: *Caralluma dalzielii* (stem). BIGNONIACEAE: *Stereospermum kunthianum* (bark, root). BIXACEAE: *Bixa orellana* (leaf). BURSERACEAE: *Dacryodes edulis* (leaf). CAPPARACEAE: *Euadenia eminens* (plant), *E. trifoliolata* (leaf), *Maerua crassifolia* (leaf). COMBRETACEAE: *Combretum micranthum* (leaf), *C. paniculatum* (leaf tomentum), *Terminalia superba* (bark). COMPOSITAE: *Vernonia guineensis* (root). COSTACEAE: *Costus lucanusianus* (plant). CRASSULACEAE: *Bryophyllum pinnatum* (leaf-sap), *Kalanchoe crenata/integra* (leaf-sap). CUCURBITACEAE: *Lagenaria siceraria* (fruit c.vars). CYPERACEAE: *Cyperus articulatus* (rhizome).

**C: Medicines; 23, laxatives, purgatives, drastics, enemas**

ACANTHACEAE: *Acanthus montanus* (leafy twig), *Graptophyllum pictum* (plant), *Justicia schimperi/striata* (leaf), *Lepidagathis heudelotiana* (stem, root), *Phaulopsis falcisepala* (plant). AIZOACEAE: *Gisekia pharnaceoides, Glinus lotoides, Mollugo pentaphylla, Trianthema portulacastrum* (root), *Zaleya pentandra*. AMARANTHACEAE: *Achyranthes aspera* (leaf), *Amaranthus hybridus* (root). AMARYLLIDACEAE: *Crinum purpurascens*. ANACARDIACEAE: *Lannea acida* (kernel), *L. nigritana* (seed), *L. schimperi* (bark), *L. welwitschii* (seed), *Ozoroa insignis* (leaf, root), *Pseudospondias microcarpa* (bark), *Rhus longpipes* (leaf, root), *Schinus molle* (bark), *Sclerocarya birrea* (bark, fruit), *Sorindeia juglandifolia, Spondias mombin* (leaf, bark), *Trichoscypha acuminata* (bark). ANNONACEAE: *Annona Ancistrocladus barteri* (root). ANNONACEAE: *Annona arenaria* (plant), *A. squamosa* (root), *Hexalobus monopetalus* (root), *Greenwayodendron suaveolens* (bark), *Monodora brevipes* (bark), *Uvaria afzelii* (bark), *U. chamae* (root), *Xylopia aethiopica* (root, fruit), *X. quintasii* (root). APOCYNACEAE: *Allamanda cathartica* (leaf), *Ancylobotrys pyriformis* (latex), *Landolphia foretiana* (fruit), *L. owariensis* (root), *Picralima nitida* (bark), *Plumeria spp.* (sap/latex), *Rauvolfia macrophylla* (bark), *Saba florida* (leaf, fruit), *Strophanthus hispidus* (leaf, stem), *Thevetia neriifolia* (leaf, bark). AQUIFOLIACEAE: *Ilex mitis* (bark). ARACEAE: *Anchomanes difformis* (tuber), *A. giganteus* (tuber), *Caladium bicolor* (root), *Cercestis afzelii* (plant), *Pistia stratiotes* (root). ARALIACEAE: *Cussonia arborea* (leaf, stem). ASCLEPIADACEAE: *Asclepias curassavica* (plant), *Calotropis procera* (latex), *Gongronema latifolium* (stem, fruit), *Gymnema sylvestre* (plant), *Secamone afzelii* (leaf). BALANITACEAE: *Balanites aegyptiaca* (bark, root, fruit). BASELLACEAE: *Basella alba* (leaf). BIGNONIACEAE: *Crescentia cujete* (fruit), *Kigelia africana* (fruit), *Markhamia tomentosa* (leaf, bark), *Newbouldia laevis* (bark), *Stereospermum kunthianum* (root). BIXACEAE: *Bixa orellana* (seed). BOMBACACEAE: *Ceiba pentandra* (flower). BORAGINACEAE: *Cordia vignei* (leaf), *Ehretia cymosa* (leaf), *Heliotropium subulatum* (root). BROMELIACEAE: *Ananas comosus* (unripe fruit). CAPPARACEAE: *Capparis tomentosa* (root). CARICACEAE: *Carica papaya* (root). CARYOPHYLLACEAE: *Drymaria cordata* (sap). CECROPIACEAE: *Myrianthus arboreus* (fruit). CELASTRACEAE: *Hippocratea rowlandii* (fruit), *Maytenus senegalensis* (leaf, root), *Salacia nitida* (bark), *S. senegalensis* (root). CHENOPODIACEAE: *Agathophora alopecuroides* (plant), *Chenopodium ambrosioides* (whole plant, leaf). CHRYSOBALANACEAE: *Acioa barteri* (bark), *Maranthes kerstingii* (plant), *Parinari congesta* (bark). COMBRETACEAE: *Anogeissus leiocarpus* (gum), *Combretum aculeatum* (leaf), *C. collinum* (leaf), *C. dolichopetalum* (leaf), *C. glutinosum* (leaf), *Guiera senegalensis* (leaf), *Terminalia glaucescens* (bark, root), *T. macroptera* (root), *T. superba* (root). COMMELINACEAE: *Aneilema beninensis* (leaf), *A. umbrosum* (leaf), *Commelina diffusa* (leaf), *Stansfieldiella imperforata* (leaf). COMPOSITAE: *Acanthospermum hispidum* (leaf), *Ayapana triplinervis* (plant), *Centaurea perrottetii* (plant), *Conyza pyrrhopappa* (leaf), *Crassocephalum rubens* (leaf), *Eclipta alba* (leaf-root), *Emilia coccinea* (leaf), *Enhydra fluctuans* (leaf), *Ethulia conyzoides* (leaf), *Helichrysum cymosum* (root), *Melanthera gambica* (root), *M. scandens* (leaf), *Microglossa pyrifolia* (leaf), *Piloselloides hirsuta* (plant), *Pleiotaxis chlorolepis* (plant), *Sonchus oleraceus* (gum), *Synedrella nodiflora* (leaf), *Vernonia amygdalina* (*leaf*), *V. colorata* (leaf), *V. conferta* (leaf, bark), *V. doniana* (leaf), *V. glabra* (leaf, root), *V. guineensis* (plant), *V. nigritiana* (root), *V. perrottetii* (leaf), *V. undulata* (leaf). CONNARACEAE: *Cnestis ferruginea* (leaf, root), *Connarus africanus* (seed), *Hemandradenia mannii* (bark). CONVOLVULACEAE: *Ipomoea aitonii* (seed), *I. aquatica* (plant), *I. cairica* (seed), *I. mauritiana, I. muricata* (seed), *I. nil* (seed), *I. pes-caprae* (leaf, root), *I. pes-tigridis* (root), *I. purpurea* (leaf, stem, root), *Merremia tuberosa* (tuber), *M. umbellata* (mucilage), *Operculina macrocarpa* (plant). COSTACEAE: *Costus afer* (root). CRUCIFERAE: *Brassica juncea* (plant), *Raphanus sativus* (seed). CUCURBITACEAE: *Citrullus colocynthis* (fruit), *C. lanatus* (fruit-bitter form), *Cogniauxia podolaena* (root), *Cucumeropsis mannii* (leaf), *Lagenaria breviflora* (root, fruit), *L. siceraria* (fruit), *Luffa acutangula* (root, seed), *L. cylindrica* (seed), *Momordica balsamina* (plant), *M. charantia* (plant, fruit), *M. cissoides* (plant), *Mukia maderaspatana* (leafy shoot), *Ruthalicia eglandulosa* (leaf-sap), *Trichosanthes cucumerina* (young shoot). DILLENIACEAE: *Tetracera alnifolia* (leafy stem). DIONCOPHYLLACEAE: *Triphyophyllum peltatum* (root).

**C: Medicines; 24, diarrhoea, dysentery**

ACANTHACEAE: *Justicia betonica* (leaf), *J. extensa* (leaf, flower), *J. flava* (plant, root, flower), *Nelsonia canescens, Phaulopsis imbricata* (leaf-sap), *Rungia grandis* (bark), *Stenandriopsis guineensis* (leaf), *Thomandersia hensii* (leaf). AIZOACEAE: *Gisekia pharnaceoides*. AMARANTHACEAE: *Achyranthes aspera* (leaf), *Aerva lanata, Alternanthera pungens, A. sessilis, Amaranthus spinosus, A. viridis* (root), *Celosia argentea* (flower, seed), *C. trigyna* (leaf), *Cyathula achyranthoides* (sap), *C. prostrata, Pupalia lappacea*. ANACARDIACEAE: *Anacardium occidentale, Lannea acida* (bark), *L. nigritana* (bark), *L. velutina* (bark), *L. welwitschii* (bark), *Mangifera indica* (leaf, bark, kernel), *Ozoroa insignis* (leaf, bark, root), *O. pulcherrima* (bark), *Sclerocarya birrea* (bark), *Spondias mombin* (leaf, bark), *Trichoscypha arborea* (bark). ANNONACEAE: *Annona arenaria* (plant), *A. chrysophylla* (leaf), *A. muricata* (bark, root, fruit), *A. reticulata* (fruit), *A. senegalensis* (young twig, stem-bark, root-bark), *A. squamosa* (root), *Annonidium mannii* (bark), *Enantia chlorantha* (bark-sap), *Hexalobus monopetalus* (leaf), *Monodora tenuifolia* (bark, root), *Uvaria chamae* (root-bark), *Xylopia aethiopica* (bark, fruit). APOCYNACEAE: *Alstonia boonei* (bark), *Baissea multiflora* (bark, root), *Ervatamia coronaria* (root), *Funtumia africana* (root), *F. elastica* (leaf), *Holarrhena floribunda* (leaf, bark, root), *Pycnobotrys nitida* (latex), *Saba senegalensis* (leaf, bark), *Strophanthus gratus* (leafy stem), *Voacanga africana* (leaf). ARACEAE: *Cyrtosperma senegalense* (fruit); *Pistia stratiotes* (leaf). ARALIACEAE: *Cussonia arborea* (shoot). ASCLEPIADACEAE: *Pergularia daemia* (leaf), *P. tomentosa* (whole plant). BALANOPHORACEAE: *Thonningia sanguinea* (flower-head). BIGNONIACEAE: *Kigelia africana* (leaf, bark), *Newbouldia laevis* (leaf, bark, root), *Spathodea campanulata* (bark), *Stereospermum kunthianum* (root), *Tecoma capensis* (leaf). BIXACEAE: *Bixa orellana* (fruit pulp). BOMBACACEAE: *Adansonia digitata* (leaf, fruit-pulp), *Bombax costatum* (leaf), *Ceiba pentandra* (bark), *Rhodognaphalon brevicuspe* (root). BORAGINACEAE: *Heliotropium indicum* (plant), *Trichodesma africanum* (leaf). BURSERACEAE: *Aucoumea klaineana* (bark), *Canarium schweinfurthii* (bark), *Dacryodes edulis* (bark). CANNACEAE: *Canna indica* (rhizome). CAPPARACEAE: *Cadaba farinosa* (plant), *Cleome viscosa* (leaf, seed). CASUARINACEAE: *Casuarina equisitifolia* (bark). CECROPIACEAE: *Myrianthus arboreus* (leaf, bark). CELASTRACEAE: *Hippocratea myriantha* (plant, bark). *H. pallens* (bark), *Maytenus senegalensis* (leaf, root). CHRYSOBALANACEAE: *Acioa scabrifolia* (leaf), *Chrysobalanus icaco* (leaf, root, fruit), *Maranthes glabra* (bark), *Neocarya macrophylla* (fruit), *Parinari congensis* (bark), *P. excelsa* (bark, fruit). COMBRETACEAE: *Anogeissus leiocarpus* (bark), *Combretum collinum* (leaf), *C. dolichopetalum* (root), *C. micranthum* (leaf), *C. molle* (leaf, bark), *C. paniculatum* (root), *C. racemosum* (root), *C. zenkeri* (root), *Guiera senegalensis* (leaf, root), *Quisqualis indica* (fruit, seed), *Q. latialata* (plant), *Strephonema pseudocola* (bark), *Terminalia avicennioides* (root), *T. glaucescens* (root), *T. laxiflora* (leaf, root), *T. macroptera* (bark), *T. mantaly* (bark, wood), *T. mollis* (plant), *T. superba* (bark). COMMELINACEAE: *Palisota hirsuta* (plant), *P. bracteosa* (plant). COMPOSITAE: *Berkheya spekeana* (plant), *Bidens pilosa* (leaf, flower), *Blumea aurita* (plant), *B. perrottetiana* (plant), *Conyza canadensis* (plant), *Dicoma sessiliflora* (sap, root), *Eclipta alba* (leaf), *Emilia sonchifolia* (plant), *Ethulia conyzoides* (leaf-sap), *Gynura procumbens* (plant), *Spilanthes filicaulis* (plant), *Stomatanthes africanus* (root), *Struchium sparganophora* (leaf), *Tridax procumbens* (root), *Vernonia amygdalina* (bark), *V. conferta* (bark), *V. guineensis* (plant), *V. nigritiana* (root). CONNARACEAE: *Byrsocarpus coccineus* (leaf). CRASSULACEAE: *Kalanchoe laciniata* (leaf). CRUCIFERAE: *Lepidium sativum* (seed). CUCURBITACEAE: *Momordica charantia* (plant). CYPERACEAE: *Cyperus rotundus* (tuber). DICHAPETALACEAE: *Dichapetalum pallidum* (leaf). DILLENIACEAE: *Tetracera alnifolia* (leaf). DIOSCOREACEAE: *Dioscorea bulbifera* (tuber). DIPTEROCARPACEAE: *Monotes kerstingii* (bark).

**C: Medicines; 25, cholera**
ANACARDIACEAE: *Ozoroa insignis* (leaf). ANNONACEAE: *Annona senegalensis* (bark).

**C: Medicines; 26, vermifuges, parasite expellants**
ACANTHACEAE: *Justicia flava* (leaf-sap), *Nelsonia canescens* (root), *Rungia grandis* (leaf), *Thomandersia hensii* (twig, root). AIZOACEAE: *Gisekia pharnaceoides, Mollugo nudicaulis.* AMARANTHACEAE: *Aerva lanta, Alternanthera pungens, Amaranthus hybridus, A. viridis* (leaf-sap), *Celosia argentea* (seed), *C. trigyna* (leaf, sap), *Cyathula cylindrica* (leaf), *C. uncinulata.* AMARYLLIDACEAE: *Crinum zeylanicum, Scadoxus multiflorus* (bulb). ANACARDIACEAE: *Lannea acida* (bark), *L. barteri* (bark), *L. schimperi* (fruit), *Mangifera indica* (kernel), *Ozoroa insignis* (leaf, root), *Rhus natalensis* (plant), *Sclerocarya birrea* (root), *Spondia mombin* (leaf, bark), *Trichoscypha acuminata* (bark). ANNONACEAE: *Annona muricata* (leaf, bark, root), *A. reticulata* (leaf, fruit), *A. senegalensis* (root), *Cleistopholis patens* (leaf, bark), *Greenwayodendron suaveolens* (root), *Isolona campanulata* (plant), *Neostenanthera hamata* (bark), *Pachypodanthium staudtii* (bark), *Uvaria chamae* (root), *Xylopia aethiopica* (seed). APOCYNACEAE: *Alstonia boonei* (bark), *Carissa edulis* (root), *Dictyophleba lucida* (root), *Funtumia elastica* (bark), *Hederantha barteri* (fruit), *Landolphia foretiana* (latex), *L. owariensis* (latex), *Picralima nitida* (leaf, bark, root), *Rauvolfia caffra* (bark), *R. macrophylla* (root), *R. vomitoria* (root), *Strophanthus hispidus* (root), *S. sarmentosus* (root). ARACEAE: *Amauriella hastifolia* (plant), *Pistia stratiotes* (plant-ash). ARISTOLOCHIACEAE: *Aristolochia albida* (leaf, root), *A. bracteolata* (leaf, root), *Telosma africanum* (root). ASCLEPIADACEAE: *Calotropis procera* (leaf), *Gongronema latifolium* (stem), *Leptadenia hastata* (whole plant). BALANITACEAE: *Balanites aegyptiaca* (bark, fruit). BALANOPHORACEAE: *Thonningia sanguinea* (rhizome, flower-head).

BALSAMINACEAE: *Impatiens irvingii* (leaf). BIGNONIACEAE: *Kigelia africana* (root), *Newbouldia laevis* (root), *Stereospermum kunthianum* (bark, root), *Tecoma stans* (plant). BORAGINACEAE: *Cordia africana* (root), *Cynoglossum lanceolatum* (plant), *Heliotropium indicum* (leaf). BROMELIACEAE: *Ananas comosus* (unripe fruit). BURSERACEAE: *Canarium schweinfurthii* (bark), *Commiphora africana* (root, seed). CANNACEAE: *Canna indica* (rhizome). CAPPARACEAE: *Boscia senegalensis* (leaf), *Buchholzia coriacea* (seed), *Cadaba farinosa* (plant), *Capparis thonningii* (fruit), *Cleome gynandra* (seed), *C. monophylla* seed), *C. viscosa* (seed). CARICACEAE: *Carica papaya* (root, fruit, seed-shoot). CECROPIACEAE: *Myrianthus arboreus* (bark). CELASTRACEAE: *Maytenus senegalensis* (leaf, leaf-sap). CHENOPODIACEAE: *Chenopodium ambrosioides* (bud, seed, seed-oil), *Salsola baryosma* (plant). CHRYSOBALANACEAE: *Bafodeya benna* (endocarp hairs), *Chrysobalanus icaco* (seed-oil), *Neocarya macrophylla* (endocarp hairs), *Parinari curatellifolia* (bark), *P. excelsa* (bark). COCHLOSPERMACEAE: *Cochlospermum tinctorium* (root). COMBRETACEAE: *Anogeissus leiocarpus* (bark), *Combretum aculeatum* (leaf), *C. constrictum* (root), *C. glutinosum* (root), *C. micranthum* (root), *C. molle* (leaf), *C. smeathmannii* (leaf), *C. zenkeri* (leaf), *Quisqualis indica* (root, fruit, seed), *Terminalia avicennioides* (root), *T. glaucescens* (fruit). COMPOSITAE: *Ageratum conyzoides* (leaf), *Artemisia spp.* (plant), *Bidens pilosa* (plant), *Blumea perrottetiana* (sap), *Conyza aegyptiaca* (sap), *Ethulia conyzoides* (leaf), *Lactuca schulzeana* (leaf, stem sap, flower), *Laggera alata* (plant), *Launaea cornuta* (root), *Microglossa pyrifolia* (leaf), *Mikania cordata* (plant), *Sigesbeckia orientalis* (plant), *Sonchus oleraceus* (root), *Sphaeranthus senegalensis* (root), *Vernonia amygdalina* (whole plant), *V. cinerea* (root), *V. colorata* (leaf, bark), *V. conferta* (bark), *V. glabra* (root), *V. nigritiana* (root). CONNARACEAE: *Byrsocarpus coccineus* (root), *Connarus africanus* (root, bark, seed). CONVOLVULACEAE: *Astripomoea malvacea* (root), *Evolvulus alsinoides* (leaf), *Hewittia sublobata* (root), *Ipomoea nil* (seed). CRASSULACEAE: *Kalanchoe crenata/integra* (plant). CUCURBITACEAE: *Citrullus colocynthis* (fruit), *C. lanatus* (seed), *Cucumis sativus* (fruit-juice, seed), *Cucurbita maxima* (seed), *C. pepo* (seed), *Lagenaria breviflora* (root), *L. siceraria* (seed-oil), *Luffa cylindrica* (seed), *Momordica balsamina* (plant), *M. charantia* (fruit, seed), *M. foetida* (leaf), *Mukia maderaspatana* (fruit), *Trichosanthes cucumerina* (seed). CYPERACEAE: *Cyperus articulatus* (rhizome). DILLENIACEAE: *Tetracera alnifolia* (leafy stem), *T. potatoria* (sap). DIOSCOREACEAE: *Dioscorea dumetorum* (tuber).

**C: Medicines; 27, liver, gall bladder, spleen (biliousness, jaundice, 'yellow fever')**

ACANTHACEAE: *Acanthus montanus* (leafy twig), *Asystasia gangetica* (root), *Barleria opaca* (whole plant), *B. prionitis*, *Brillantaisia patula* (leaf), *Hygrophila auriculata* (root), *Phaulopsis falcisepala* (sap), *P. imbricata* (leaf), *Pseuderanthemum ludovicianum* (whole plant), *Thunbergia cynanchifolia* (leaf). AIZOACEAE: *Mollugo nudicaulis*, *Trianthema portulacastrum* (leaf). AMARANTHACEAE: *Alternanthera sessilis*, *Celosia trigyna* (leaf), *Cyathula achyranthoides* (sap), *C. uncinulata*. AMARYLLIDACEAE: *Crinum purpurascens* (sap). ANACARDIACEAE: *Antrocaryon klaineanum* (gum), *Haematostaphis barteri* (bark), *Ozoroa insignis* (bark), *Pseudospondias microcarpa* (bark), *Sorindeia juglandifolia* (leaf), *Trichoscypha patens*. ANNONACEAE: *Artabotrys stenopetalus* (leaf), *A. thomsonii* (leaf), *Cleistopholis patens* (leaf), *Enantia chlorantha* (bark-sap), *Hexalobus monopetalus* (bark), *Uvaria afzelii* (leaf, bark), *U. chamae* (root), *U. doeringii* (fruit). APOCYNACEAE: *Alafia lucida* (plant), *Alstonia boonei* (bark), *Picralima nitida* (root), *Pleiocarpa mutica* (root), *Rauvolfia vomitoria* (root), *Saba florida* (plant). ARACEAE: *Anchomanes difformis* (tuber), *Rhektophyllum mirabile* (leaf). BALANITACEAE: *Balanites aegyptiaca* (root bark, fruit). BOMBACACEAE: *Bombax costatum* (bark). BORAGINACEAE: *Trichodesma africanum* (root). BURSERACEAE: *Canarium schweinfurthii* (bark). CANNACEAE: *Canna indica* (leaf, shoot). CAPPARACEAE: *Boscia angustifolia* (leaf), *B. senegalensis* (leaf), *Crataeva adansonii* (leaf). CARICACEAE: *Carica papaya* (green fruit). CARYOPHYLLACEAE: *Polycarpaea corymbosa* (leaf). CHENOPODIACEAE: *Agathophora alopecuroides* (plant). CELASTRACEAE: *Maytenus senegalensis* (bark). CECROPIACEAE: *Myrianthus arboreus* (bark). COCHLOSPERMACEAE: *Cochlospermum tinctorium* (root). COMBRETACEAE: *Combretum collinum* (leaf), *C. glutinosum* (leaf), *C. micranthum* (leaf), *C. molle* (leaf, bark), *Terminalia avicennioides* (root-bark), *T. glaucescens* (root). COMMELINACEAE: *Commelina congesta* (plant). COMPOSITAE: *Ageratum conyzoides* (leaf), *Bidens pilosa* (leaf), *Chrysanthellum americanum* (plant), *Eclipta alba* (leaf, root), *Helichrysum mechowianum* (leaf-sap), *Laggera alata* (root), *Launaea cornuta* (plant), *Melanthera gambica* (root), *Microglossa pyrifolia* (leaf), *Mikania cordata* (sap), *Sonchus oleraceus* (plant, gum), *Vernonia colorata* (leaf), *V. conferta* (plant), *V. guineensis* (plant), *V. nigritiana* (root). CONNARACEAE: *Agelaea trifolia* (bark), *Byrsocarpus coccineus* (root). CONVOLVULACEAE: *Ipomoea involucrata* (plant), *Merremia kentrocaulos* (plant). CUCURBITACEAE: *Lagenaria siceraria* (leaf), *Momordica charantia* (plant), *M. foetida* (leaf), *Ruthalicia eglandulosa* (leaf-sap), *R. longipes* (plant). DICHAPETALACEAE: *Dichapetalum madagascariense* (leaf-sap). DIOSCOREACEAE: *Dioscorea dumetorum* (tuber), *D. minutiflora* (leaf).

**C: Medicines; 28, kidneys, micturition, diuresis**

ACANTHACEAE: *Acanthus montanus* (whole plant), *Barleria prionitis, Hygrophila auiculata* (whole plant), *Ruellia tuberosa* (plant). AMARANTHACEAE: *Achyranthes aspera* (leaf), *Aerva lanata, Amaranthus hybridus, A. spinosus, A. viridis* (leaf), *Celosia trigyna* (leaf). ANACARDIACEAE: *Mangifera indica* (bark), *Pseudospondias microcarpa* (bark), *Sorindeia juglandifolia* (leaf), *Spondias mombin* (fruit). ANNONACEAE: *Annona glauca* (root), *A. senegalensis* (bark, root), *Uvaria afzelii* (bark, root). APOCYNACEAE: *Baissea axillaris* (leaf), *B. multiflora* (liane), *Funtumia africana* (root), *Holarrhena floribunda* (leaf, root), *Landolphia dulcis* (stem, root), *Plumeria spp.* (bark), *Pyconobotrys nitida* (latex), *Saba senegalensis* (root), *Strophanthus hispidus* (root), *Tabernaemontana crassa* (bark). ARACEAE: *Anchomanes difformis* (tuber), *A. giganteus* (tuber). ARALIACEAE: *Polyscias guilfoylei* (leaf, root). ASCLEPIADACEAE: *Calotropis procera* (twig), *Gymnema sylvestre* (plant), *Leptadenia hastata* (leaf, root), *L. pyrotechnica* (plant), *Solenostemma oleifolium* (leaf). BALSAMINACEAE: *Impatiens irvingii* (leaf). BIGNONIACEAE: *Jacaranda mimosifolia* (leaf, bark), *Kigelia africana* (leaf), *Markhamia lutea* (plant), *M. tomentosa* (plant), *Spathodea campanulata* (leaf, bark), *Stereospermum kunthianum* (root), *Tecoma stans* (root). BIXACEAE: *Bixa orellana* (fruit-pulp). BOMBACACEAE: *Adansonia digitata* (leaf), *Bombax costatum* (bark). BORAGINACEAE: *Cordia senegalensis* (leaf), *Trichodesma africanum* (leaf). BROMELIACEAE: *Ananas comosus* (unripe fruit). BURSERACEAE: *Boswellia dalzielii* (resin), *Canarium schweinfurthii* (resin). CANNACEAE: *Canna indica* (rhizome). CAPPARACEAE: *Boscia senegalensis* (leaf), *Capparis decidua* (bark), *C. tomentosa* (root), *Cleome arabica* (leaf), *C. speciosa, Euadenia eminens* (root), *E. trifoliolata* (root). CARIACEAE: *Carica papaya* (leaf). CELASTRACEAE: *Hippocratea indica* (root), *Maytenus undata* (root), *Mystroxylon aethiopicum* (bark, root). COCHLOSPERMACEAE: *Cochlospermum tinctorium* (twig, root). COMBRETACEAE: *Combretum aculeatum* (leaf), *C. collinum* (leaf), *C. glutinosum* (leaf), *C. micranthum* (leaf), *C. racemosum* (plant, leaf), *Guiera senegalensis* (leaf), *Quisqualis latialata* (plant), *Strephonema pseudocola* (root), *Terminalia ivorensis* (sap), *T. macroptera* (root). COMMELINACEAE: *Palisota bracteosa* (plant), *P. hirsuta* (plant), *Pollia condensata* (stem). COMPOSITAE: *Centaurea senegalensis* (plant), *Chrysanthellum americanum* (plant), *Gynura procumbens* (plant), *Melanthera gambica* (root), *Microglossa afzelii* (plant), *M. pyrifolia* (leaf), *Sigesbeckia orientalis* (plant), *Sphaeranthus senegalensis* (leaf, flower), *Spilanthes filicaulis* (plant), *Vernonia cinerea* (plant), *V. colorata* (leaf), *V. conferta* (plant), *V. glabra* (root), *V. nigritiana* (root). CONNARACEAE: *Byrsocarpus coccineus* (leaf), *Hemandradenia mannii* (bark). CONVOLVULACEAE: *Ipomoea asarifolia* (plant), *I. mauritiana* (leaf), *I. nil* (seed), *I. pes-caprae* (leaf, root), *Merremia tridentata* (plant, root), *M. umbellata* (plant). COSTACEAE: *Costus lucanusianus* (sap). CRASSULACEAE: *Bryophyllum pinnatum* (leaf-sap), *Kalanchoe laciniata* (leaf). CRUCIFERAE: *Capsella bursa-pastoris* (plant), *Eruca sativa* (leaf), *Raphanus sativus* (seed). CUCURBITACEAE: *Citrullus lanatus* (fruit, seed), *Cogniauxia podolaena* (root), *Cucumis melo* (fruit, seed), *C. sativus* (leaf, fruit-juice), *Cucurbita maxima* (seed), *Lagenaria siceraria* (fruit - c.vars). CYPERACEAE: *Cyperus rotundus* (tuber), *Remirea maritima* (rhizome). DILLENIACEAE: *Tetracera potatoria* (leaf-sap, stem-sap). DIOSCOREACEAE: *Dioscorea bulbifera* (tuber). DROSERACEAE: *Drosera madagascariensis* (plant).

**C: Medicines; 29, anus, haemorrhoids**

ACANTHACEAE: *Asystasia gangetica* (leaf), *Barleria opaca* (leaf), *Thunbergia alata* (leaf-sap). AMARANTHACEAE: *Achyranthes aspera* (leaf), *Amaranthus spinosus, A. viridis* (leaf). ANACARDIACEAE: *Anacardium occidentale* (young leaf), *Lannea acida* (bark), *Lannea welwitschii* (bark), *Mangifera indica* (bark). ANNONACEAE: *Monodora myristica* bark), *Uvaria doeringii* (leaf). APOCYNACEAE: *Funtumia elastica* (bark), *Landolphia heudelotii, L. hirsuta* (root-bark). ARACEAE: *Amorphophallus dracontioides* (root), *Pistia stratiotes* (leaf), *Typhonium trilobatum* (root). ASCLEPIADACEAE: *Leptadenia hastata* (whole plant). AVICENNIACEAE: *Avicennia germinans* (leaf). BALANOPHORACEAE: *Thonningia sanguinea* (flower-head). BIGNONIACEAE: *Kigelia africana* (fruit), *Newbouldia laevis* (bark). BURSERACEAE: *Canarium schweinfurthii* (bark). CAPPARACEAE: *Boscia senegalensis* (leaf). CARICACEAE: *Carica papaya* (root). CARYOPHYLLACEAE: *Stellaria media* (plant). CELASTRACEAE: *Salacia lehmbachii* (root). CHENOPODIACEAE: *Chenopodium ambrosioides* (plant). COCHLOSPERMACEAE: *Cochlospermum tinctorium* (root). COMBRETACEAE: *Combretum paniculatum* (leaf, root), *C. racemosum* (leaf-sap), *Quisqualis latialata* (plant), *Terminalia macroptera* (bark). COMMELINACEAE: *Palisota bracteosa* (plant), *P. hirsuta* (plant). COMPOSITAE: *Ageratum conyzoides* (leaf), *Bidens pilosa* (leaf), *Blumea perrottetiana* (bark), *Sphaeranthus senegalensis* (root), *Vernonia amygdalina* (root), *V.*

*cinerea* (leaf), *V. nigritiana* (root), *V. undulata* (leaf). CONNARACEAE: *Byrsocarpus coccineus* (leaf). CONVOLVULACEAE: *Ipomoea pes-caprae* (leaf), *I. quamoclit* (leaf), *Merremia tridentata* (plant, root). CUCURBITACEAE: *Luffa acutangula* (leaf). DIOSCOREACEAE: *Dioscorea bulbifera* (tuber).

**C: Medicines; 30, genital stimulants/depressants**
ACANTHACEAE: *Asystasia calycina* (twig), *Elytraria marginata* (leaf), *Phaulopsis barteri, P. falcisepala* (sap), *P. imbricata*. AMARYLLIDACEAE: *Crinum purpurascens* (leaf). ANACARDIACEAE: *Ozoroa insignis* (root), *Pseudospondias microcarpa* (plant), *Spondias mombin* (leaf), *Trichosypha acuminata* (bark). ANNONACEAE: *Annona senegalensis* (twig, bark), *Artabotrys stenopetalus* (sap), *A. thomsonii* (sap), *Greenwayodendron suaveolens* (root), *Isolona campanulata* (plant, bark). APOCYNACEAE: *Carissa edulis* (root), *Funtumia elastica* (leaf), *Rauvolfia vomitoria* (root), *Saba florida* (bark, root), *Strophanthus gratus* (root), *S. hispidus* (root). ARALIACEAE: *Cussonia arborea* (plant). ASCLEPIADACEAE: *Dregea abyssinica* (root), *Leptadenia hastata* (leaf), *Margaretta rosea* (leaf, root), *Pachycarpus lineolata* (root). BALANOPHORACEAE: *Thonningia sanguinea* (flower-head). BIGNONIACEAE: *Newbouldia laevis* (leaf, root). BURSERACEAE: *Canarium schweinfurthii* (bark). CAPPARACEAE: *Boscia salicifolia* (bark, root), *Capparis sepiaria* (fruit), *C. thonningii* (leaf), *Cleome arabica* (leaf), *Euadenia eminens* (root, fruit-pulp), *E. trifoliolata* (root), *Maerua angolensis* (bark, root). CHRYSOBALANACEAE: *Maranthes polyandra* (bark). COMBRETACEAE: *Combretum glutinosum* (leafy twig, root), *C. racemosum* (leaf-sap/gel), *Guiera senegalensis* (leaf), *Quisqualis latialata* (plant), *Terminalia ivorensis* (sap), *T. macroptera* (root), *T. glaucescens* (root). COMMELINACEAE: *Commelina diffusa* (leaf), *Palisota bracteosa, P. hirsuta* (root). COMPOSITAE: *Dicoma sessiliflora* (root), *Ethulia conyzoides* (root), *Microglossa pyrifolia* (plant), *Vernonia amygdalina* (flower), *V. colorata* (bark), *V. conferta* (leaf), *V. guinensis* (root), *V. nigritiana* (root). CONNARACEAE: *Agelaea obliqua* (root-bark), *Byrsocarpus coccineus* (root). *Cnestis ferruginea* (root), *Connarus africanus* (leaf-sap). CONVOLVULACEAE: *Ipomoea argentaurata* (plant), *I. mauritiana* (tuber). COSTACEAE: *Costus afer* (root). CUCURBITACEAE: *Momordica charantia* (plant), *M. foetida* (leaf). CYPERACEAE: *Cyperus articulatus* (rhizome), *C. rotundus* (tuber), *Scleria boivinii* (root). DILLENIACEAE: *Tetracera alnifolia* (leaf).

**C: Medicines; 31, menstrual cycle**
ACANTHACEAE: *Brillantaisia patula* (plant), *Hygrophila auriculata* (whole plant), *Justicia flava* (plant), *Pseuderanthemum ludovicianum* (leaf), *Thunbergia laurifolia* (leaf-sap). AMARANTHACEAE: *Achyranthes aspera* (root), *Amaranthus spinosus, Celosia trigyna* (leaf). AMARYLLIDACEAE: *Crinum purpurascens*. ANACARDIACEAE: *Lannea welwitschii* (bark), *Trichoscypha acuminata* (bark), *T. arborea* (bark). ANNONACEAE: *Annonidium manii* (bark), *Annona arenaria* (plant), *Greenwayodendron suaveolens* (leaf), *Uvaria chamae* (root), *U. tortilis* (plant), *Xylopia aethiopica* (fruit), *X. quintasii* (fruit). APOCYNACEAE: *Adenium obesum* (latex), *Funtumia elastica* (bark), *Holarrhena floribunda* (leaf), *Tabernaemontana crassa* (bark), *Thevetia neriifolia* (bark), *Voacanga africana* (root). ARALIACEAE: *Cussonia arborea* (stem, root). ASCLEPIADACEAE: *Gymnema sylvestre* (leaf), *Pergularia daemia* (plant). BIGNONIACEAE: *Kigelia africana* (root), *Newbouldia laevis* (twig, bark). BOMBACACEAE: *Adansonia digitata* (fruit-husk), *Bombax buonopozense* (bark), BORAGINACEAE: *Ehretia cymosa* (bark), *E. trachyphylla* (bark). BURSERACEAE: *Aucoumea klaineana* (bark), *Dacryodes edulis* (bark). CANNACEAE: *Canna indica* (rhizome). CECROPIACEAE: *Musanga cecropioides* (leaf-bract, bark-sap), *Myrianthus arboreus* (leaf), CELASTRACEAE: *Maytenus senegalensis* (root). COCHLOSPERMACEAE: *Cochlospermum planchonii* (plant), *C. tinctorium* (root). COMBRETACEAE: *Anogeissus leiocarpus* (bark), *Combretum glutinosum* (leaf), *C. zenkeri* (twig). COMMELINACEAE: *Aneilema beninensis* (sap), *Commelina africana* (root), *C. diffusa* (root). COMPOSITAE: *Elephantopus mollis* (plant), *Emilia coccinea* (leaf), *Laggera alata* (plant-sap), *spilanthes filicaulis* (plant), *Triplotaxis stellutifera* (leaf), *Vernonia nigritiana* (flower). CONNARACEAE: *Cnestis ferruginea* (leaf, root). CONVOLVULACEAE: *Ipomoea involucrata* (leaf-sap), *I. nil* (seed). CRASSULACEAE: *Bryophyllum pinnatum* (leaf-sap). CRUCIFERAE: *Capsella bursa-pastoris* (plant), CUCURBITACEAE: *Luffa acutangula* (leaf), *Momordica charantia* (leaf), *Trichosanthes cucumerina* (seed). CYPERACEAE: *Cyperus articulatus* (rhizome), *C. esculentus* (tuber), *Scleria boivinii* (root), *S. depressa* (root), *S. foliosa* (root), *S. lithosperma* (plant), *S. melanophala* (whole plant), *S. pterota* (root). DICHAPETALACEAE: *Dichapetalum albidum* (leaf).

**C: Medicines; 32, conception, pregnancy promotion, antiabortifacient**
ACANTHACEAE: *Acanthus montanus* (leafy twigs), *Brillantiasia nitens* (root), *B. patula* (plant), *Elytraria marginata* (leaf-sap), *Pseuderanthemum ludovicianum* (whole plant), *Whifieldia elongata* (leaf), AIZOACEAE: *Gisekia pharnaceodides*. AMARANTHACEAE: *Aerva lanata,*

*Cyathula prostrata, C. uncinulata* (root). ANACARDIACEAE: *Lannea acida* (bark), *L. welwitschii* (bark), *Rhus natalensis* (root), *Spondias mombin* (leaf), *Trichosypha acuminata* (bark), *T. arborea* (bark). ANNONACEAE: *Annona arenaria* (plant), *A. squamosa* (seed), *Artabotrys stenopetalus* (twig), *A. thomsonii* (sap), *Enantia chlorantha* (bark), *Greenwayodendron suaveolens* (bark), *Isolona campanulata* (plant), *Monanthotaxis whytei* (plant), *Uvaria chamae* (root), *Xylopia aethiopica* (fruit, seed). APOCYNACEAE: *Alstonia boonei* (bark), *Baissea multiflora* (bark, root), *Holarrhena floribunda* (root), *Hunteria umbellata* (root), *Oncinotis nitida, Pleioceras barteri* (root), *Saba florida* (root). ARACEAE: *Culcasia scandens* (leaf). ASCLEPIADACEAE: *Secamone afzelii* (leaf), *Tylophora sylvatica* (plant). BEGONIACEAE: *Begonia mannii.* BIGNONIACEAE: *Newbouldia laevis* (leaf). BORAGINACEAE: *Heliotropium indicum* (leaf). CECROPIACEAE: *Myrianthus arboreus* (leaf). CELASTRACEAE: *Hippocratea indica* (root), *Maytenus senegalensis* (leaf-sap), *Mystoxylon aethiopicum* (bark). COMBRETACEAE: *Combretum micranthum* (root), *C. racemosum* (bark), *Quisqualis latialata* (plant), *Terminalia superba* (bark). COMMELINACEAE: *Commelina benghalensis* (plant), *Palisota bracteosa* (leaf, root), *P. hirsuta* (leaf, root). COMPOSITAE: *Ageratum conyzoides* (leaf), *Emilia coccinea* (leaf), *Ethulia conyzoides* (leaf), *Microglossa pyrifolia* (plant), *Sonchus schweinfurthii* (leaf), *Spilanthes uliginosa* (plant), *Vernonia colorata* (bark), *V. poskeana* (plant), *V. smithiana* (root). CONNARACEAE: *Agelaea mildbraedii* (root-bark). CONVOLVULACEAE: *Ipomoea mauritiana* (tuber). COSTACEAE: *Costus dubius* (inflorescence). CUCURBITACEAE: *Momordica cissoides* (leaf). CYPERACEAE: *Cyperus papyrus* (rhizome), *Scleria lithosperma* (plant). DILLENIACEAE: *Tetracera alnifolia* (leaf).

**C: Medicines; 33, lactation stimulants (incl. veterinary)**
ACANTHACEAE: *Justicia betonica* (leaf). AIZOACEAE: *Trianthema portulacastrum* (root). AMARANTHACEAE: *Alternanthera pungens, A. sessilis, Amaranthus spinosus, A. viridis* (leaf-sap). AMARYLLIDACEAE: *Scadoxus multiflorus.* ANACARDIACEAE: *Ozoroa insignis* (leaf, root). ANNONACEAE: *Annona senegalensis* (bark), *Xylopia aethiopica* (sap). APOCYNACEAE: *Alstonia boonei* (latex), *Holarrhena floribunda* (leaf), *Landolphia dulcis* (stem bark), *Pleioceras barteri* (bark, seed), *Saba florida* (latex), *Strophanthus hispidus* (root), *Tabernaemontana crassa* (plant). ARACEAE: *Anchomanes difformis* (leaf, tuber). ASCLEPIADACEAE: *Calotropis procera* (latex, root-bark), *Glossonema boveanum* (latex), *Leptadenia hastata* (leaf), *Pachycarpus lineolata* (seed-sap), *Pergularia tomentosa* (plant), *Sarcostemma viminale* (sap), *Secamone afzelii* (leaf). BALANITACEAE: *Balanites aegyptica* (root). BIGNONIACEAE: *Kigelia africana* (fruit). BORAGINACEAE: *Heliotropium bacciferum* (plant). CAPPARACEAE: *Capparis erythrocarpos* (root). CARICACEAE: *Carica papaya* (fruit). CECROPIACEAE: *Musanga cecropioides* (bark-sap, root-sap). COMBRETACEAE: *Guiera senegalensis* (leaf). COMMELINACEAE: *Palisota bracteosa* (leaf), *P. hirsuta* (leaf). COMPOSITAE: *Aspilia africana* (leaf), *Crassocephalum vitellianum* (plant), *Launaea cornuta* (leaf), *L. taraxacifolia* (leaf), *Senecio biafrae* (leaf), *Vernonia conferta* (leafy shoot, bark), *V. subuligera* (root). CONVOLVULACEAE: *Ipomoea mauritiana* (root-stock), *I. sinensis* (plant). CRASSULACEAE: *Kalanchoe crenata/integra* (leaf-sap). CRUCIFERAE: *Brassica juncea* (plant). CUCURBITACEAE: *Cucumis prophetarum* (plant), *Momordica balsamina* (plant). CYPERACEAE: *Cyperus exaltatus* (rhizome). DILLENIACEAE: *Tetracera alnifolia* (sap).

**C: Medicines; 34, abortifacients, ecbolics, parturition stimulants**
ACANTHACEAE: *Asystasia gangetica* (whole plant), *Brillantaisia patula* (plant). AIZOACEAE: *Trianthema portulacastrum* (root), AMARANTHACEAE: *Alternanthera pungens, Celosia trigyna* (leaf), *Cyathula uncinulata.* ANACARDIACEAE: *Rhus longipes* (leaf, root), *Sclerocarya birrea* (bark), *Spondias mombin* (bark). ANNONACEAE: *Uvaria chamae* (root), *Xylopia aethiopica* (fruit), *X. quintasii* (fruit). APOCYNACEAE: *Alstonia boonei* (bark), *Pleioceras barteri* (bark, seed), *Rauvolfia vomitoria* (root), *Voacanga africana* (root). ARACEAE: *Amorphophallus abyssinicus* (root), *Anchomanes difformis* (tuber), *Culcasia angolensis* (plant), *Cyrtosperma senegalensis* (leaf). ARALIACEAE: *Schefflera barteri* (root). AVICENNIACEAE: *Avicennia germinans* (twig, bark). BALANITACEAE: *Balanites aegyptiaca* (bark). BORAGINACEAE: *Cordia sinensis* (root), *Heliotropium indicum* (leaf). BROMELIACEAE: *Ananas comosus* (unripe fruit). BURSERACEAE: *Commiphora africana* (gum). CAPPARACEAE: *Buchholzia coriacea* (seed), *Cleome gynandra* (root). CARICACEAE: *Carica papaya* (leaf, root, seed). CECROPIACEAE: *Musanga cecropioides* (leaf-bract). CELASTRACEAE: *Hippocratea welwitschii* (plant). COCHLOSPERMACEAE: *Cochlospermum tinctorium* (root). COMBRETACEAE: *Combretum molle* (leaf), *Strephonema pseudocola* (root), *Terminalia superba* (leaf). COMMELINACEAE: *Palisota bracteata* (sap), *P. hirsuta* (sap), *Pollia condensata* (leaf). COMPOSITAE: *Aspilia africana* (leaf), *Bidens pilosa* (plant), *Centaurea praecox* (plant), *Ethulia conyzoides* (leaf, root), *Melanthera scandens* (leaf), *Microglossa pyrifolia* (leaf), *Sonchus oleraceus* (root), *Struchium sparganophora* (leaf), *Vernonia*

*cinerea* (plant), *V. lasiopus* (root). CONNARACEAE: *Cnestis ferruginea* (leaf). CONVOLVULACEAE: *Ipomoea asarifolia* (plant), *I. involucrata* (leaf-sap), *I. mauritiana* (tuber), *I. nil* (seed). CRASSULACEAE: *Bryophyllum pinnatum* (leaf-sap). CUCURBITACEAE: *Coccinia grandis* (leaf), *Cucumis prophetarum* (? fruit), *Cucurbita pepo* (seed), *Momordica balsamina* (seed), *M. foetida* (leaf). CYPERACEAE: *Scleria boivinii* (leaf). DIOSCOREACEAE: *Dioscorea praehensilis* (tuber).

**C: Medicines; 35, venereal diseases, treatment, prophylaxis (See also C: 28 for diuretics.)**
ACANTHACEAE: *Acanthus montanus* (whole plant), *Asystasia gangetica* (leaf), *Blepharis linariifolia* (whole plant), *Elytraria marginata* (leaf), *Hygrophila auriculata* (root), *Lankesteria brevior* (leaf), *L. elegans* (leaf), *Mimulopsis solmsii* (leaf), *Pseuderanthemum tunicatum* (plant, leaf), *Thomandersia hensii* (twig, sap, root), AIZOACEAE: *Trianthema portulacastrum* (leaf, root), *Zaleya pentandra*. AMARANTHACEAE: *Aerva lanata, Alternanthera pungens, Amaranthus spinosus, A. viridis* (leaf), *Celosia argentea, C. trigyna* (leaf), *Centrostachys aquatica* (root), *Cyathula achyranthoides* (sap), *C. prostrata, C. uncinulata* (root), *Pupalia lappacea, Sericostachys scandens* (bark). ANACARDIACEAE: *Anacardium occicentale, Lannea acida* (root-bark), *L. welwitschii* (bark), *Mangifera indica* (bark-sap), *Pseudospondias microcarpa* (bark), *Rhus natalensis* (leaf, root), *Schinus molle* (leaf, fruit-oil), *S. terebinthifolius* (bark), *Sclerocarya birrea* (bark), *Spondias mombin* (leaf). ANNONACEAE: *Annona arenaria* (plant), *A. senegalensis* (root), *Monodora brevipes* (root), *Uvaria afzelii* (root), *U. chamae* (resin), *Xylopia quintasii* (fruit). APOCYNACEAE: *Adenium obesum* (root), *Carissa edulis* (root), *Funtumia elastica* (bark), *Hedranthera barteri* (fruit), *Holarrhena floribunda* (root), *Landolphia hirsuta* (root-bark), *L. owariensis* (root), *Picralima nitida* (bark), *Rauvolfia macrophylla* (root), *R. vomitoria* (bark), *Saba florida* (root), *Strophanthes gratus* (leaf, twig, bark, root), *S. hispidus* (twig-bark, root-bark, root), *S. preussii* (sap), *S. sarmentosus* (plant, leaf), *Tabernaemontana crassa* (bark). ARACEAE: *Anchomanes difformis* (tuber), *A. giganteus* (tuber), *Culcasia scandens* (leaf), *Cyrtosperma senegalensis* (fruit), *Pistia stratiotes* (leaf). ARALIACEAE: *Cussonia arborea* (plant, root). ASCLEPIADACEAE: *Calotropis procera* (latex, bark), *Leptadenia hastata* (whole plant), *Pentatrophis spiralis* (root), *Pergularia daemia* (root), *Secamone afzelii* (leaf), *Solenostemma oleifolium* (leaf), *Telosma africanum* (root). BALANITACEAE: *Balanites aegyptiaca* (root-bark). BIGNONIACEAE: *Jacaranda mimosifolia* (leaf, bark), *Markhamia lutea* (plant), *Newbouldia laevis* (root), *Spathodea campanulata* (bark), *Stereospermum kunthianum* (leaf, bark, root), *Tecoma stans* (plant). BOMBACACEAE: *Bombax costatum* (leaf), *Ceiba pentandra* (leaf, bark), *Rhodographalon brevicuspe* (bark). BORAGINACEAE: *Heliotropium bacciferum* (plant), *H. indicum* (plant), *H. ovalifolium* (plant). BROMELIACEAE: *Ananas comosus* (unripe fruit). BURSERACEAE: *Boswellia dalzielii* (root), *Canarium schweinfurthii* (bark). CAPPARACEAE: *Capparis decidua* (leaf, root-bark), *C. tomentosa* (leafy stem, root), *Crataeva adansonii* (root), *Maerua oblongifolia* (plant), *Ritchiea capparoides* (leaf, root). CARICACEAE: *Carica papaya* (leaf, root, fruit). CECROPIACEAE: *Musanga cecropioides* (leaf-bud, root-sap). CELASTRACEAE: *Maytenus senegalensis* (shoot, root), *Mystroxylon aethiopicum* (bark, root), *Salacia nitida* (bark), *S. senegalensis* (root). CHRYSOBALANACEAE: *Maranthes polyandra* (root). COCHLOSPERMACEAE: *Cochlospermum planchonii* (root). COMBRETACEAE: *Anogeissus leiocarpus* (bark), *Combretum fragrans* (root), *C. glutinosum* (root, fruit), *C. micranthum* (leaf, root, fruit), *C. molle* (leaf), *C. paniculatum* (leaf, root), *C. smeathmannii* (root), *Conocarpus erectus* (root), *Guiera senegalensis* (leaf), *Terminalia glaucescens* (root), *T. ivorensis* (sap), *T. mollis* (plant). COMMELINACEAE: *Commelina africana* (root), *C. diffusa* (root), *Palisota bracteosa* (plant), *P. hirsuta* (plant). COMPOSITAE: *Adenostemma perrottettii* (plant), *Ageratum conyzoides* (leaf), *Ambrosia maritima* (plant), *Aspilia africana* (leaf), *A. kotschyii* (leaf), *Centaurea perrottetii* (plant), *Chrysanthellum americanum* (plant), *Conyza canadensis* (plant), *Crassocephalum vitellianum* (plant), *Emilia coccinea* (leaf, root), *Guizotia scabra* (leaf), *Helichrysum odoratissimum* (plant), *Lactuca capensis* (leaf), *Launaea cornuta* (root), *Microglossa afzelii* (plant), *M. pyrifolia* (leaf), *Mikania cordata* (sap), *Sclerocarpus africanus* (leaf), *Senecio abyssinicus* (root), *S. baberka* (root), *S. mannii* (plant), *Sigesbeckia orientalis* (plant), *Struchium sparganophora* (leaf), *Triplotaxis stellutifera* (leaf), *Vernonia adoensis* (root), *V. colorata* (leaf, root), *V. conferta* (leaf) *V. glabra* (leaf, root), *V. guineensis* (root), *V. pumila* (root), *V. smithiana* (root). CONNARACEAE: *Agelaea dewevrei* (leaf), *A. mildbraedii* (root-bark), *Byrsocarpus coccineus* (leaf), *B. dinklagei* (plant), *Cnestis corniculata* (leaf). CONVOLVULACEAE: *Ipomoea asarifolia* (flower), *I. involucrata* (sap), *I. pes-caprae* (leaf), *Merremia tridentata* (plant). COSTACEAE: *Costus afer* (stem, root). CUCURBITACEAE: *Citrullus colocynthis* (unripe fruit), *C. lanatus* (fruit), *Coccinia barteri* (plant), *Ctenolepis cerasiformis* (plant), *Cucumis prophetarum* (fruit), *Momordica charantia* (plant). CYPERACEAE: *Cyperus dilatatus* (rhizome), *C. distans* (rhizome), *Mariscus alternifolius* (culm base), *Scleria boivinii* (root), *S. depressa* (root), *S. foliosa* (rhizome). DILLENIACEAE: *Tetracera alnifolia* (root), *T. potatoria* (leaf, stem, sap, root). DIOSCOREACEAE: *Dioscorea praehensilis* (stem).

**C: Medicines; 36, febrifuges, sudorifics, temperature control, rigor control, etc.**

ACANTHACEAE: *Asystasia gangetica* (leaf), *Barleria prionitis* (leafy twig, leaf-sap), *Brillantaisia lamium* (whole plant), *Pyschoriste pedicellata* (leaf), *Hygrophila auriculata* (leaf, root), *Hypoestes verticillata* (whole plant), *Justicia extensa* (root), *J. flava* (plant), *Nelsonia smithii* (leaf-sap), *Phaulopsis falcisepala* (sap). AMARANTHACEAE: *Alternanthera pungens, Amaranthus viridis* (leaf), *Pupalia lappacea*. ANACARDIACEAE: *Mangifera indica* (leaf), *Ozoroa insignis* (root, bark), *Rhus longipes* (root), *Spondias mombin* (root, fruit). ANNONACEAE: *Annona arenaria* (plant), *A. glabra* (leaf), *Cleistopholis patens* (leaf), *Enantia chlorantha* (bark), *Greenwayodendron oliveri* (bark), *G. suaveolens* (leaf), *Isolona campanulata* (plant, twig-bark), *Monanthotaxis laurentii* (leaf), *Monodora myristica* (bark), *Uvaria afzelii* (leaf, bark), *U. chamae* (leaf, root, fruit), *Xylopia acutiflora* (bark). APOCYNACEAE: *Alafia barteri* (plant), *A. benthamii* (plant), *Alstonia boonei* (latex, bark, root), *Holarrhena floribunda* (leaf, bark, root), *Hunteria elliotii* (bark), *Landolphia owariensis* (root), *Picralima nitida* (bark, root, fruit-husk), *Pleiocarpa mutica* (root), *Rauvolfia vomitoria* (leaf, bark), *Strophanthus gratus* (leaf), *Tabernaemontana brachyantha* (twig), *Thevetia nereiifolia* (bark). ARALIACEAE: *Cussonia arborea* (plant). ASCLEPIADACEAE: *Kanahia lanifolia* (leaf-sap), *Pergularia daemia* (plant), *Tylophora sylvatica* (plant). BIGNONIACEAE: *Crescentia cujete* (fruit), *Markhamia lutea* (plant), *M. tomentosa* (leaf, bark), *Newbouldia laevis* (leaf, bark, root), *Spathodea campanulata* (bark), *Stereospermum kunthianum* (bark, root), *Tecoma capensis* (bark). BIXACEAE: *Bixa orellana* (fruit-pulp). BOMBACACEAE: *Adansonia digitata* (leaf, bark, fruit-meal), *Bombax buonopozense* (bark), *B. costatum* (leaf), *Ceiba pentandra* (leaf, bark). BORAGINACEAE: *Cordia sinensis* (leaf), *Cynoglossum lanceolatum* (plant), *Ehretia cymosa* (leaf), *Heliotropium indicum* (leaf), *Trichodesma africanum* (leaf). BURSERACEAE: *Boswellia dalzielii* (bark), *Commiphora africana* (bark), *Santira trimera* (bark). CANNACEAE: *Canna indica* (leaf, shoot, rhizome). CAPPARACEAE: *Buchholzia coriacea* (leaf), *Capparis decidua* (root), *Cleome arabica* (leaf), *Crataeva adansonii* (bark), *Maerua crassifolia* (leaf, bark). CARICACEAE: *Carica papaya* (leaf). CARYOPHYLLACEAE: *Drymaria cordata* (sap), *Polycarpaea corymbosa* (plant). CECROPIACEAE: *Myrianthus arboreus* (leaf). CELASTRACEAE: *Hippocratea indica* (leaf-root), *H. velutina* (seed), *Maytenus senegalensis* (bark), *M. undata* (root), *Salacia senegalensis* (root). CHENOPODIACEAE: *Chenopodium ambrosioides* (whole plant). CHRYSOBALANACEAE: *Maranthes polyandra* (leaf, bark), *Parinari curatellifolia* (leaf, bark). COCHLOSPERMACEAE: *Cochlospermum tinctorium* (root). COMBRETACEAE: *Anogeissus leiocarpus* (leaf, bark), *Combretum fragrans* (leaf), *C. glutinosum* (bark), *C. micranthum* (leaf, root), *C. paniculatum* (root), *C. platypterum* (leaf), *Conocarpus erectus* (leaf), *Guiera senegalensis* (leaf), *Terminalia brownii* (bark), *T. macroptera* (leaf). COMMELINACEAE: *Commelina africana* (plant), *C. diffusa* (leaf), *C. imberbis* (plant). COMPOSITAE: *Ageratum conyzoides* (leaf), *Blumea perrottetiana* (plant), *Centaurea perrottetii* (plant), *Chrysanthellum americanum* (plant), *Conyza canadensis* (plant), *C. pyrrhopappa* (leaf), *C. sumatrensis* (leaf-sap), *Dicoma sessiliflora* (plant), *D. tomentosa* (plant), *Elephantopus mollis* (plant), *Emilia coccinea* (leaf), *E. sonchifolia* (plant), *Gynura miniata* (plant), *G. procumbens* (plant), *Helichrysum nudifolium* (plant), *Laggera alata* (plant), *L. oloptera* (plant), *Microglossa afzelii* (leaf), *M. pyrifolia* (leaf), *Mikania cordata* (plant), *Pulicaria crispa* (plant), *Sigesbeckia orientalis* (plant), *Tridax procumbens* (plant), *Triplotaxis stellutifera* (leaf), *Vernonia amygdalina* (bark), *V. biafrae* (leaf), *V. cinerea* (plant, root, leaf), *V. colorata* (leaf), *V. guineensis* (plant), *V. nigritiana* (root), *V. poskeana* (plant), *V. thomsoniana* (leaf). CONNARACEAE: *Agelaea trifolia* (bark), *A. spp.* (sap). CONVOLVULACEAE: *Calycobolus africana* (leaf), *Evolvulus alsinoides* (leaf), *Ipomoea aquatica* (plant), *I. asarifolia* (leaf), *I. eriocarpa* (seed-oil), *Merremia kentrocaulos* (root). *M. tridentata* (plant). COSTACEAE: *Costus afer* (sap, root), *C. lucanusianus* (plant). CRASSULACEAE: *Bryophyllum pinnatum* (whole plant), *Kalanchoe crenata/integra* (leaf-sap, root-sap). CRUCIFERAE: *Capsella bursa-pastoris* (plant). CUCURBITACEAE: *Momordica balsamina* (plant), *M. charantia* (plant, fruit), *M. cissoides* (plant), *Mukia maderaspatana* (seed), *Ruthalicia eglandulosa* (fruit), *Zehneria scabra* (leaf). CYPERACEAE: *Cyperus articulatus* (rhizome), *C. exaltatus* (rhizome). DILLENIACEAE: *Tetracera alnifolia* (leafy twig).

**C: Medicines; 37, small-pox, chicken-pox, measles, etc.**

ACANTHACEAE: *Justicia extensa* (leaf), *Nelsonia canescens* (plant), *Thomandersia hensii* (sap), *Thunbergia chrysops* (leaf), *T. cynanchifolia* (leaf). ANACARDIACEAE: *Trichoscypha acuminata* (bark). ANNONACEAE: *Uvaria afzelii* (leaf). APOCYNACEAE: *Rauvolfia vomitoria* (bark). BOMBACACEAE: *Adansonia digitata* (fruit-pulp). BROMELIACEAE: *Ananas comosus* (unripe fruit). CAPPARACEAE: *Buchholzia coriacea* (bark). COMPOSITAE: *Melanthera scandens* (leaf-sap), *Mikania cordata* (sap), *Sonchus schweinfurthii* (leaf). CRASSULACEAE: *Kalanchoe laciniata* (leaf). CUCURBITACEAE: *Momordica foetida* (leaf).

**C: Medicines; 38, leprosy treatment**

ACANTHACEAE: *Thomandersia hensii* (sap). AMARANTHACEAE: *Amaranthus spinosus, A. viridis* (leaf), *Cyathula achyranthoides* (leaf), *C. cylindrica* (root). ANACARDIACEAE: *Anacardium occidentale* (sap), *Lannea barteri* (bark), *L. velutina* (bark), *Sclerocarya birrea* (bark), *Spondias mombin* (leaf, bark). ANNONACEAE: *Annona arenaria* (plant), *A. senegalensis* (bark), *Enantia polycarpa* (bark). APOCYNACEAE: *Hunteria eburnea* (root-bark), *Rauvolfia vomitoria* (*bark*), *Saba senegalensis* (leaf, bark), *Strophanthus sarmentosus* (root), *Tabernaemontana crassa* (sap), *Voacanga africana* (leaf). ARACEAE: *Anchomanes welwitschii* (root). ARALIACEAE: *Cussonia arborea* (stem, stem-bark). ASCLEPIADACEAE: *Leptadenia hastata* (whole plant). BALANOPHORACEAE: *Thonningia sanguinea* (flower). BIGNONIACEAE: *Kigelia africana* (bark), *Newbouldia laevis* (plant), *Sterospermum kunthianum* (bark, root). BOMBACACEAE: *Ceiba pentandra* (bark, root). BURSERACEAE: *Boswellia dalzielii* (bark), *Canarium schweinfurthii* (bark), *Commiphora africana* (gum), *Dacryodes edulis* (bark). CAPPARACEAE: *Crataeva adansonii* (bark). CARYOPHYLLACEAE: *Drymaria cordata* (plant). CECROPIACEAE: *Musanga cecropioides* (root-sap). CELASTRACEAE: *Maytenus senegalensis* (root). CHRYSOBALANACEAE: *Parinari congensis* (bark). COCHLOSPERMACEAE: *Cochlospermum tinctorium* (root, seed-oil). COMBRETACEAE: *Anogeissus leiocarpus* (bark), *Combretum aculeatum* (leaf, root), *Guiera senegalensis* (leaf), *Terminalia avicennioides* (root), *T. catappa* (leaf-sap), *T. glaucescens* (root), *T. macroptera* (leaf). COMMELINACEAE: *Aneilema setiferum* (plant), *Commelina benghalensis* (plant). COMPOSITAE: *Acanthospermum hispidum* (leaf), *Crassocephalum sarcobasis* (plant), *Emilia coccinea* (leaf), *Lactuca capensis* (root), *Sigesbekia orientalis* (plant), *Synedrella nodiflora* (plant). CONVOLVULACEAE: *Evolvulus alsinoides* (leaf), *Ipomoea eriocarpa* (seed-oil), *Jacquemontia tamnifolia* (leaf-ash). COSTACEAE: *Costus afer* (root). CUCURBITACEAE: *Luffa acutangula* (leaf), *Peponium vogelii* (leaf). CYPERACEAE: *Scleria boivinii* (root). DILLENIACEAE: *Tetracera potatoria* (stem).

**C: Medicines; 39, yaws**

ACANTHACEAE: *Acanthus montanus* (leaf-spines). *Asystasia calycina* (whole plant), *Brillantaisia lamium* (leaf), *B. patula* (leaf), *Justicia flava* (plant), *Thomandersia hensii* (sap). AMARANTHACEAE: *Achyranthes aspera* (leaf), *Amaranthus viridis* (leaf). ANACARDIACEAE: *Anacardium occidentale* (root). ASCLEPIADACEAE: *Calotropis procera* (latex, root-bark), *Tylophora conspicua* (leaf). BORAGINACEAE: *Heliotropium subulatum* (leaf, stem). BURSERACEAE: *Santira trimera* (bark). CARICACEAE: *Carica papaya* (root). COMBRETACEAE: *Strephonema pseudocola* (gum), *Terminalia laxiflora* (bark). COMPOSITAE: *Emilia coccinea* (leaf), *Ethulia conyzoides* (leaf), *Launaea taraxacifolia* (leaf), *Senecio abyssinicus* (root), *Vernonia conferta* (leaf). CUCURBITACEAE: *Momordica balsamina* (plant), *M. charantia* (leafy stem, root). DILLENIACEAE: *Tetracera affinis* (root).

**C: Medicines; 40, dropsy, swellings, oedemas, gout**

ACANTHACEAE: *Barleria prionitis, Blepharis maderaspatana* (plant-ash), *Graptophyllum pictum* (leaf), *Hygrophila auriculata* (root), *Hypoestes verticillata* (leaf), *Justicia betonica* (leaf), *J. flava* (leaf-sap), *Ruellia praetermissa* (leaf, root), *Thomandersia hensii* (root). AIZOACEAE: *Gisekia pharnaceoides, Glinus oppositifolius* (leaf). AMARANTHACEAE: *Aerva lanata, Alternanthera maritima, A. pungens, Cyathula cylindrica* (leaf), *Pupalia lappacea.* ANACARDIACEAE: *Lannea barteri* (bark), *L. velutina* (bark), *L. welwitschii* (bark), *Schinus molle* (bark), *S. terebinthifolius* (bark). ANNONACEAE: *Annona arenaria* (plant), *Greenwayodendron suaveolens* (root-sap), *Monanthotaxis whytei* (plant), *Pachypodanthium staudtii* (bark), *Uvaria chamae* (root). APOCYNACEAE: *Alstonia boonei* (leaf), *Carissa edulis* (root-bark), *Dictyophleba leonensis* (leaf), *Holarrhena floribunda* (fruit-bark), *Landolphia owariensis* (leaf), *Pleiocarpa mutica* (bark), *Rauvolfia vomitoria* (leaf), *Voacanga africana* (leaf). ARACEAE: *Anchomanes difformis* (tuber), *A. giganteus* (tuber), *Caladium bicolor* (tuber), *Culcasia parviflora* (leaf). ASCLEPIADACEAE: *Gomphocarpus fruticosus* (plant), *Secamone afzelii* (leaf). BALANOPHORACEAE: *Thonningia sanguinea* (flower-head). BIGNONIACEAE: *Kigelia africana* (bark), *Markhamia lutea* (plant), *M. tomentosa* (plant), *Newbouldia laevis* (leaf, root), *Spathodea campanulata* (bark). BOMBACACEAE: *Ceiba pentandra* (bark), *Rhodognaphalon brevicuspe* (root). CANNACEAE: *Canna indica* (rhizome). CAPPARACEAE: *Boscia angustifolia* (bark), *B. senegalensis* (leaf), *Capparis decidua* (root), *Crataeva adansonii* (root), *Maerua angolensis* (leaf, root). CARICACEAE: *Carica papaya* (root). CARYOPHYLLACEAE: *Drymaria cordata* (plant). CECROPIACEAE: *Musanga cecropioides* (leaf, bark). CELASTRACEAE: *Hippocratea africana* (liane), *Maytenus senegalensis* (leaf). CHENOPODIACEAE: *Chenopodium ambrosioides* (leaf). COCHLOSPERMACEAE: *Cochlospermum tinctorium* (root). COMBRETACEAE: *Combretum collinum* (leaf), *C. glutinosum* (leaf), *C. micranthum* (bark, fruit), *C. molle* (leaf), *C.*

*zenkeri* (plant), *Strephonema pseudocola* (leaf, root), *Teminalia avicennioides* (root), *T. superba* (bark). COMMELINACEAE: *Commelina congesta* (plant), *Palisota bracteosa* (plant), *P. hirsuta* (plant). COMPOSITAE: *Centaurea senegalensis* (root), *Helichrysum nudifolium* (leaf), *Microglossa afzelii* (plant), *M. pyrifolia* (leaf), *Pulicaria crispa* (plant), *Senecio biafrae* (plant), *Synedrella nodiflora* (plant). CONNARACEAE: *Byrsocarpus coccineus* (leaf), *Hemandradenia mannii* (bark). CONVOLVULACEAE: *Astripomoea malvacea* (root), *Ipomoea involucrata* (leaf-sap), *I. pes-caprae* (leaf), *Merremia tridentata* (plant, root). COSTACEAE: *Costus afer* (sap). CUCURBITACEAE: *Citrullus colocynthis* (fruit), *Luffa acutangula* (root), *Momordica foetida* (leaf), *Ruthalicia longipes* (plant). CYPERACEAE: *Cyperus articulatus* (rhizome). *C. renchii* (leaf, rhizome). DICHAPETALACEAE: *Dichapetalum toxicarium* (bark). DILLENIACEAE: *Tetracera potatoria* (sap). DIONCOPHYLLACEAE: *Triphyophyllum peltatum* (leaf, bark). DIOSCOREACEAE: *Dioscorea praehensilis* (stem).

**C: Medicines; 41, tumours, cancers**

ACANTHACEAE: *Barleria cristata* (leafy twig, root), *Thunbergia alata* (leaf-sap). AIZOACEAE: *Aizoon canariensis, Gisekia pharnaceoides*. AMARANTHACEAE: *Amaranthus hybridus*. ANACARDIACEAE: *Rhus natalensis* (leaf), *Schinus terebinthifolius* (leaf), *Spondias mombin* (leaf), ANNONACEAE: *Annona reticulata* (leaf), *A. squamosa* (leaf), *Neostenanthera gabonensis* (leaf), *Pachypodanthium staudtii* (bark), *Xylopia quintasii* (root). APOCYNACEAE: *Hedranthera barteri* (leaf), *Nerium oleander* (most parts). ARACEAE: *Typhonium trilobatum* (root). ASCLEPIADACEAE: *Asclepias curassavica* (latex). BIGNONIACEAE: *Newbouldia laevis* (bark). BIXACEAE: *Bixa orellana* (sap, seed). BOMBACACEAE: *Adansonia digitata* (leaf), *Ceiba pentandra* (leaf). BORAGINACEAE: *Coldenia procumbens* (leaf), *Heliotropium indicum* (plant), *H. strigosum* (whole plant), *H. supinum* (plant). CACTACEAE: *Opuntia spp.* (stem). CAPPARACEAE: *Capparis thonningii* (leaf). CARICACEAE: *Carica papaya* (latex). CARYOPHYLLACEAE: *Drymaria cordata* (plant), *Stellaria media* (plant). COMBRETACEAE: *Combretum molle* (leaf), *C. paniculatum* (flower). COMPOSITAE: *Centipeda minima* (plant), *Crassocephalum rubens* (root), *Emilia sonchifolia* (leaf-sap), *Ethulia conyzoides* (leaf), *Sonchus asper* (latex), *S. oleraceus* (latex), *Verbesina encelioides* (plant). CONNARACEAE: *Byrsocarpus coccineus* (leaf). CONVOLVULACEAE: *Astripomoea malvacea* (root), *Ipomoea alba* (leaf), *I. pes-caprae* (stem). CUCURBITACEAE: *Cucurbita pepo* (fruit-pulp), *Luffa cylindrica* (root, fruit), *Momordica charantia* (plant). DROSERACEAE: *Drosera indica* (plant).

**C: Medicines; 42, malnutrition, debility**

ACANTHACEAE: *Brillantaisia patula* (leaf), *Dicliptera laxata* (whole plant). AIZOACEAE: *Tetragona tetragonioides*. ALISMATACEAE: *Sagittaria guayanensis* (leaf). AMARANTHACEAE: *Celosia argentea*. ANACARDIACEAE: *Lannea acida* (bark). APOCYNACEAE: *Baissea multiflora* (leafy twig, bark), *Strophanthus gratus* (leafy stem). ARACEAE: *Pistia stratiotes* (plant-ash). BIGNONIACEAE: *Newbouldia laevis* (root). CAPPARACEAE: *Euadenia eminens* (plant). COCHLOSPERMACEAE: *Cochlospermum tinctorium* (root). COMBRETACEAE: *Combretum aculeatum* (leaf), *C. micranthum* (leaf). COMMELINACEAE: *Commelina congesta* (plant). COMBRETACEAE: *Guiera senegalensis* (leaf). COMPOSITAE: *Berkheya spekeana* (plant).

**C: Medicines; 43, food poisoning**

ACANTHACEAE: *Whitfieldia elongata* (leaf). AMARYLLIDACEAE: *Crinum purpurascens* (leaf). ANACARDIACEAE: *Lannea welwitschii* (root). ANNONACEAE: *Annona arenaria* (plant), *Uvaria afzelii* (root). APOCYNACEAE: *Rauvolfia vomitoria* (bark), *Strophanthus gratus* (root-bark). BURSERACEAE: *Canarium schweinfurthii* (bark). CAPPARACEAE: *Cadaba farinosa* (plant), *Ritchiea capparoides* (leaf). COMPOSITAE: *Acanthospermum hispidum* (leaf), *Bidens pilosa* (plant), *Spilanthes filicaulis* (plant), *Vernonia colorata* (root), *V. conferta* (plant), *V. guineensis* (plant). CONNARACEAE: *Hemandradenia mannii* (bark). CONVOLVULACEAE: *Ipomoea muricata* (seed). DILLENIACEAE: *Tetracera alnifolia* (leaf).

**C: Medicines; 44, poison antidotes (venomous stings, bites, etc.)**

ACANTHACEAE: *Asystasia gangetica* (root), *Barleria cristata* (seed), *Crossandra nilotica* (leaf), *Eremomastax speciosa* (leaf), *Hygrophila auriculata* (whole plant), *Justicia betonica* (leaf), *Peristrophe bicalyculata*. AIZOACEAE: *Sesuvium portulacastrum*. AMARANTHACEAE: *Achyranthes aspera* (root), *Aerva lanata* (root), *Alternanthera sessilis, Celosia argentea, Pupalia lappacea*. AMARYLLIDACEAE: *Crinum purpurascens* (leaf). ANACARDIACEAE: *Lannea schimperi* (bark), *Sclerocarya birrea* (leaf, root-bark), *Spondias mombin* (bark). ANNONACEAE: *Annona arenaria* (plant), *A. chrysophylla* (bark), *A. glabra* (root), *A. muricata* (root), *A. senegalensis* (bark), *Annonidium mannii* (bark). APOCYNACEAE: *Alstonia boonei* (latex, bark), *Rauvolfia serpentina, R. vomitoria* (root), *Saba florida* (root), *Strophanthus gratus* (leaf), *S. hispidus* (leaf). ARACEAE: *Amorphophallus dracontioides* (root), *A. johnsonii* (tuber), *Anchomanes difformis* (tuber), *Colocasia esculenta* (tuber). ARALIACEAE: *Cussonia arborea.* ARISTOLOCHIACEAE: *Aristolochia albida* (root), *A. bracteolata* (root). ASCLEPIADACEAE: *Collotropis procera* (latex, bark, root-bark), *Gymnema sylvestre* (root). BALANITACEAE: *Balanites aegyptiaca* (root-bark). BIGNONIACEAE: *Crescentia cujete* (seed-ash), *Kigelia africana* (bark), *Newbouldia laevis* (bark), *Spathodea campanulata* (leaf). BORAGINACEAE: *Heliotropium indicum* (leaf), *H. strigosum* (plant). BROMELIACEAE: *Ananas comosus* (leaf, fruit). BURSERACEAE: *Boswellia dalzielii* (bark, root), *B. odorata* (bark), *Commiphora africana* (gum), *C. kerstingii* (bark). CAPPARACEAE: *Capparis thonningii* (leaf-sap), *C. tomentosa* (root), *Euadenia eminens* (root), *E. trifoliolata* (root), *Ritchiea capparoides* (leaf, root). CARICACEAE: *Carica papaya* (leaf, bark). CARYOPHYLLACEAE: *Polycarpaea corymbosa* (plant). CELASTRACEAE: *Hippocratea myriantha* (plant). CHRYSOBALANACEAE: *Neocarya macrophylla* (root). COCHLOSPERMACEAE: *Cochlospermum tinctorium* (root, fruit). COMBRETACEAE: *Combretum molle* (leaf), *Guiera senegalense* (twig). COMMELINACEAE: *Commelina zambesica* (plant). COMPOSITAE: *Ageratum conyzoides* (leaf), *Ambrosia maritima* (root), *Bidens pilosa* (plant), *Conyza attenuata* (leaf), *Laggera heudelotii* (plant), *Melanthera scandens* (leaf), *Microglossa afzelii* (plant), *Mikania cordata* (plant), *Spilanthes filicaulis* (plant), *Vernonia cinerea* (leaf), *V. guineensis* (plant). CONNARACEAE: *Cnestis ferruginea* (root, fruit), *Roureopsis obliquifoliolata* (leaf-sap, root-sap). CONVOLVULACEAE: *Ipomoea alba* (plant), *I. pes-caprae* (leaf-sap), *Merremia tridentata* (leaf). CRASSULACEAE: *Kalanchoe laciniata* (leaf). CRUCIFERAE: *Brassica campestris* (seed-oil). CUCURBITACEAE: *Citrullus colocynthis* (seed-oil), *Cucumis figarei* (green fruit), *Luffa acutangula* (leaf), *Momordica charantia* (plant), *M. foetida,* (leaf), *Ruthalicia eglandulosa* (sap). CYPERACEAE: *Scirpus aureiglumis* (plant). DILLENIACEAE: *Tetracera alnifolia* (leaf, root). DIOSCOREACEAE: *Dioscorea bulbifera* (stem-sap).

**C: Medicines; 45, homeopathic, Theory of Signatures**

AMARYLLIDACEAE: *Crinum zeylanicum.* ANNONACEAE: *Annona senegalensis* (root). APOCYNACEAE: *Hedranthera barteri* (fruit), *Saba senegalensis* (fruit). ARALIACEAE: *Cussonia arborea* (bark). ASCLEPIADACEAE: *Leptadenia hastata, Sarcostemma viminale.* BALANOPHORACEAE: *Thonningia sanguinea.* BOMBACACEAE: *Adansonia digitata, Ceiba pentandra.* BORAGINACEAE: *Heliotropium ovalifolium, H. strigosum, H. subulatum.* CAPPARACEAE: *Crataeva adansonii.* CECROPIACEAE: *Musanga cecropioides.* COCHLOSPERMACEAE: *Cochlospermum tinctorium.* COMBRETACEAE: *Combretum racemosum.* COMPOSITAE: *Vernonia nigritiana.*

**D: Phytochemistry; 1, alkali salts—excl. common salt. (See also D:2.)**

ACANTHACEAE: *Asystasia gangetica* (leaf, flower), *Barleria opaca* (leaf), *B. prionitis* (whole plant), *Hygrophila auriculata, Hypoestes verticillata* (plant), *Justicia extensa* (leaf, bark), *Rungia grandis* (leaf), *Thunbergia grandiflora* (leaf). AIZOACEAE: *Mollugo pentaphylla* (plant), *Zaleya pentandra.* AMARANTHACEAE: *Amaranthus hybridus, A. spinosus, A. viridis, Celosia trigyna.* ANACARDIACEAE: *Spondias mombin* (bark). ARACEAE: *Pistia stratiotes* (plant-ash). BOMBACACEAE: *Adansonia digitata* (fruit-husk). CHENOPODIACEAE: *Salsola baryosma* (ash). COMBRETACEAE: *Combretum glutinosum* (wood-ash). CUCURBITACEAE: *Cucumis melo* (fruit-husk). CYPERACEAE: *Cyperus haspan* (ash).

**D: Phytochemistry; 2, soap, soap-substitutes**

ACANTHACEAE: *Asystasia gangetica* (whole plant), *Brillantaisia nitens* (leaf), *Eremomastax speciosa* (leaf). AMARANTHACEAE: *Amaranthus viridis, Cyathula cylindrica* (root), *C. uncinulata* (root). ANACARDIACEAE: *Lannea acida* (kernel-oil). *L. nigritana* (wood-ash), *Spondias mombin* (wood-ash). ANNONACEAE: *Annona senegalensis* (wood-ash). APOCYNACEAE: *Tabernaemontana longiflora* (wood-ash). ARACEAE: *Pistia stratiotes* (plant-ash). ARALIACEAE: *Cussonia arborea* (wood), *C. bancoensis* (wood-ash). BALANITACEAE: *Balanites aegytiaca* (wood, bark, root, seed-oil). BOMBACACEAE: *Adansonia digitata* (bark-ash, fruit-husk, seed-ash), *Ceiba pentandra* (wood-ash, oil). CANNABACEAE: *Cannabis sativa* (oil). CAPPARACEAE: *Cleome gynandra* (oil). CARICACEAE: *Carica papaya* (whole plant, leaf). CECROPIACEAE: *Musanga cecropioides* (wood-ash), *Myrianthus arboreus* (wood-ash). CHENOPODIACEAE: *Salsola baryosma* (plant). CHRYSOBALANACEAE: *Neocarya macrophylla* (seed-endocarp). COMBRETACEAE: *Anogeissus leiocarpus* (wood-ash). COMMELINACEAE: *Palisota hirsuta* (plant). COMPOSITAE: *Acanthospermum hispidum* (ash), *Vernonia thomsoniana* (leaf). CONVOLVULACEAE: *Ipomoea alba* (leaf). CUCURBITACEAE: *Momordica balsamina* (leaf). DICHAPETALACEAE: *Dichapetalum madagascariense* (leaf). DILLENIACEAE: *Dillenia indica* (fruit-pulp).

**D: Phytochemistry; 3, salt, salt-substitutes**

ACANTHACEAE: *Brillantaisia nitens* (stem), *Hygrophila auriculata* (whole plant ash), *Justicia flava* (leaf-ash), *J. schimperi/striata* (leaf), *Nelsonia canescens*, *Peristrophe bicalyculata*, *Phaulopsis imbricata*. AIZOACEAE: *Zaleya pentandra*. ALISMATACEAE: *Limnophyton angolense*. AMARANTHACEAE: *Amaranthus spinosus*, *Cyathula achyranthoides*. APOCYNACEAE: *Voacanga thouarsii* (wood-ash). ARACEAE: *Cyrtosperma senegalensis* (plant-ash), *Pistia stratiotes* (plant-ash). AVICENNIACEAE: *Avicennia germinans* (leaf, root). BALSAMINACEAE: *Impatiens irvingii*, *I. niamniamensis* (leaf). BOMBACACEAE: *Adansonia digitata* (wood). CAPPARACEAE: *Capparis decidua* (branch). CARICACEAE: *Carica papaya* (root-ash). CECROPIACEAE: *Musanga cecropioides* (wood-ash). CELASTRACEAE: *Maytenus senegalensis* (leaf, wood). CHRYSOBALANACEAE: *Maranthes polyandra* (wood-ash). COMPOSITAE: *Epaltes gariepina*. *Launaea taraxacifolia* (plant), *Sonchus schweinfurthii* (plant), *Vernonia colorata* (wood-ash), *V. perrottetii* (plant-ash). CYPERACEAE: *Cyperus exaltatus* (culm), *C. haspan* (culm), *Fuirena umbellata* (plant), *Pycreus nitidus* (plant-ash).

**D: Phytochemistry; 4, mineral salts**

AIZOACEAE: *Glinus oppositifolius*, *Mollugo pentaphylla*, *Trianthema portulacastrum*. AMARANTHACEAE: *Achyranthes aspera*, *Alternanthera maritima*, *Amaranthus hybridus*. ANACARDIACEAE: *Mangifera indica* (fruit). ARACEAE: *Pistia stratiotes* (plant-ash). BOMBACACEAE: *Adansonia digitata* (fruit). CAPPARACEAE: *Maerua crassifolia* (leaf). COMBRETACEAE: *Combretum micranthus* (leaf), *Guiera senegalensis* (leaf, root). COMPOSITAE: *Launaea taraxacifolia* (leaf). CONVOLVULACEAE: *Ipomoea aquatica*, *I. batatas*. CRASSULACEAE: *Kalanchoe laciniata* (leaf). CRUCIFERAE: *Lepidium sativum*. CUCURBITACEAE: *Momordica charantia* (fruit), *Telfairea occidentalis* (leaf).

**D: Phytochemistry; 5, fatty acids, oils, waxes**

ACANTHACEAE: *Brillantaisia nitens* (leaf), *Hygrophila auriculata* (plant, seed), *Peristrophe bicalyculata*. AGAVACEAE: *Agave sisalana* (leaf). AMARANTHACEAE: *Amaranthus spinosus* (seed), *Celosia argentea* (seed), *C. trigyna* (seed). ANACARDIACEAE: *Anacardium occidentale* (nut shell, kernel), *Antrocaryon klaineanum* (seed), *A. micraster* (seed), *Haematostaphis barteri* (kernel), *Lannea acida* (kernel), *Mangifera indica* (bark), *Schinus molle* (leaf, fruit), *S. terebinthifolius* (fruit), *Sclerocarya birrea* (seed), *Trichoscypha longifolia* (kernel). ANNONACEAE: *Annona muricata* (seed), *A. reticulata* (seed), *A. senegalalensis* (root), *A. squamosa* (seed), *Cananga odorata* (flower), *Monodora myristica* (seed), *M. tenuifolia* (seed), *Xylopia aethiopica* (fruit). APOCYNACEAE: *Funtumia elastica* (seed), *Strophanthus hispidus* (seed), *Thevetia neriifolia* (seed). ARISTOLOCHIACEAE: *Aristolochia bracteolata* (seed). BALANITACEAE: *Balanites aegyptiaca* (fruit), *B. wilsoniana* (seed). BALSAMINACEAE: *Impatiens balsamina* (seed). BIGNONIACEAE: *Crescentia cujete* (seed), *Tecoma stans* (seed). BOMBACACEAE: *Adansonia digitata* (seed), *Bombacopsis glabra* (seed), *Bombax buonopozense* (seed), *Ceiba pentandra* (seed), *Pachira aquatica* (seed), *Rhodognaphalon brevicuspe* (seed). BORAGINACEAE: *Cordia myxa* (kernel), *Ehretia cymosa* (bark), *E. trachyphylla* (bark). BURSERACEAE: *Canarium schweinfurthii* (resin, fruit-pulp, kernel), *Commiphora africana* (resin, seed), *Dacryodes edulis* (wood, fruit-pulp, kernel), *Santiria trimera* (seed). CANNABACEAE: *Cannabis sativa* (seed). CAPPARACEAE: *Capparis tomentosa* (seed), *Cleome viscosa* (seed), *Maerua pseudopetalosa* (seed). CARICACEAE: *Carica papaya* (seed). CELASTRACEAE: *Maytenus senegalensis* (leaf). CECROPIACEAE: *Myrianthus arboreus* (kernel), *M. libericus* (kernel), *M. serratus* (kernel). CELASTRACEAE: *Hippocrates macrophylla* (seed). CHENOPODIACEAE: *Chenopodium ambrosioides* (glandular hairs). CHRYSOBALANACEAE: *Licania elaeosperma* (kernel), *Maranthes glabra* (seed), *M. robusta* (kernel), *Neocarya macrophylla* (kernel), *Parinari curatellifolia* (kernel), *P. excelsa* (kernel). COCHLOSPERMACEAE: *Cochlospermum religiosum* (seed), *C. tinctorium* (seed). COMBRETACEAE: *Quisqualis indica* (seed), *Strephonema pseudocola* (seed), *Terminalia catappa* (kernel). COMPOSITAE: *Carthamnus tinctorius* (seed), *Guizotia abyssinica* (seed), *Helianthus annuus* (seed), *Helichrysum odoratissimum* (flowering tops), *Veronia cinerea* (seed), *V. colorata* (seed), *V. nigritiana* (seed). CONVOLVULACEAE: *Evolvulus alsinoides* (plant), *Ipomoea nil* (seed), *I. pes-caprae* (plant), *Merrermia umbellata* (seed). CRASSULACEAE: *Kalanchoe laciniata*. CRUCIFERAE: *Brassica campestris* (seed), *B. juncea* (seed), *B. napus* (seed), *B. oleracea* (seed), *B. rapa* (seed), *Crambe abyssinica* (seed), *Eruca sativa* (seed), *Lepidium sativum* (seed), *Raphanus sativus* (seed). CUCURBITACEAE: *Citrullus lanatus* (seed), *Cucumeropsis mannii* (seed), *Cucumis melo* (seed), *C. prophetarum* (seed), *C. sativus* (seed), *Cucurbita maxima* (seed), *C. pepo* (seed), *Lagenaria siceraria* (seed), *Luffa acutangula* (seed), *L. cylindrica* (seed), *Momordica cabraei* (seed), *M. charantia* (seed), *Telfairea occidentalis* (seed). CYPERACEAE: *Cyperus esculentus* (tuber).

**D: Phytochemistry; 6, aromatic substances (scent, cosmetics, coumarin, musk, incense, etc.)**
ACANTHACEAE: *Graptophyllum pictum* (plant), *Hygrophila auriculata* (seed), *H. odora* (flower), *Monechma ciliatum* (seed). ANNONACEAE: *Cananga odorata* (flower), *Greenwayodendron oliveri* (bark), *G. suaveolens* (bark), *Monanthotaxis diclina* (leaf), *Monodora myristica* (seed), *Pachypodanthium staudtii* (bark), *Polyceratocarpus parviflorus* (bark), *Uvaria chamae* (whole plant), *U. doeringii, Xylopia aethiopica* (bark, root, fruit, seed), *X. quintasii* (bark), *X. staudtii* (bark). APOCYNACEAE: *Ervatamia coronaria* (wood), *Plumeria spp.* (flower). ARACEAE: *Culcasia lancifolia* (plant), *C. striolata* (plant). BALANITACEAE: *Balanites wilsoniana* (bark). BORAGINACEAE: *Cordia millenii* (fruit), *C. sinensis* (wood). BURSERACEAE: *Boswellia dalzielii* (bark), *B. odorata* (bark), *Canarium schweinfurthii* (wood, bark), *C. zeylanicum* (bark), *Commiphora africana* (leaf, wood, bark, gum), *C. dalzielii* (leaf), *C. pedunculata* (bark), *Dacryodes edulis* (bark), *D. klaineana* (bark), *Santiria trimera* (bark). CAPPARACEAE: *Buchholzia coriacea* (bark), *Cleome gynandra* (plant). CHRYSOBALANACEAE: *Licania elaeosperma* (oil), *Neocarya macrophylla* (kernel-oil). COMBRETACEAE: *Combretum fragrans* (wood-smoke), *C. glutinosum* (wood-smoke), *Terminalia avicennioides* (rotted wood), *T. macroptera* (heart-wood). COMPOSITAE: *Ageratum conyzoides* (whole plant), *Aspilia helianthoides* (plant), *Ayapana triplinervis* (leaf), *Bidens pilosa* (plant), *Blumea aurita* (plant), *B. gariepina* (plant), *B. perrottetiana* (plant), *Bothriocline schimperi* (leaf), *Conyza canadensis* (plant), *Cotula anthemoides* (plant), *Enydra fluctuans* (plant), *Gynura miniata* (plant), *Helichrysum foetidum* (plant), *H. nudifolium* (plant), *H. odoratissimum* (plant), *Laggera alata* (plant, leaf), *L. pterodonta* (plant), *Microglossa pyrifolia* (leaf), *Nidorella spartioides* (plant), *Pluchea ovalis* (plant), *Pulicaria undulata* (plant), *Sigesbeckia orientalis* (root), *Sphaeranthus angustifolius* (plant), *S. flexuosus* (plant), *S. senegalensis* (plant), *Tanacetum cinerarifolium* (flower). COSTACEAE: *Costus lucanusianus* (plant). CUCURBITACEAE: *Momordica charantia* (seed). CYPERACEAE: *Cyperus articulatus* (rhizome), *C. digitatus* (culm), *C. esculentus* (tuber), *C. maculatus* (tuber), *C. maritimus* (rhizome), *C. rotundus* (tuber), *C. sphacelatus* (rhizome), *Kyllinga erecta* (rhizome), *K. nigritana* (root), *K. pumila* (root), *K. squamulata* (culm base), *K. tenuifolia* (culm base), *Mariscus alternifolius* (rhizome), *M. flabelliformis* (root), *Pycreus nitidus* (rhizome).

**D: Phytochemistry; 7, starch, sugar**
ACANTHACEAE: *Stenandriopsis guineensis* (leaf), *Thunbergia grandiflora* (flower). AGAVACEAE: *Agave sisalana* (bagasse). AIZOACEAE: *Mollugo nudicaulis.* AMARANTHACEAE: *Amaranthus viridis* (leaf). ANACARDIACEAE: *Anacardium occidentale* (fruit, nut), *Sclerocarya birrea* (fruit). ANNONACEAE: *Annona muricata* (fruit). ASCLEPIADACEAE: *Pachycarpus lineolata* (root). BALANITACEAE: *Balanites aegyptiaca* (fruit-pulp). CANNACEAE: *Canna edulis* (root), *C. indica* (root). CAPPARACEAE: *Boscia angustifolia* (wood), *B. salicifolia* (root). CECROPIACEAE: *Myrianthus arboreus* (kernel), *M. serratus* (kernel). CHRYSOBALANACEAE: *Chrysobalanus icaco* (kernel), *Parinari excelsa* (fruit-pulp). COCHLOSPERMACEAE: *Cochlospermum tinctorium* (root). COMMELINACEAE: *Commelina benghalensis* (rhizome). CONVOLVULACEAE: *Ipomoea batatas* (tuber), *I. pes-caprae* (root). CYPERACEAE: *Cyperus esculentus* (tuber). DIOSCOREACEAE: *Dioscorea cayenensis* (tuber), *D. esculenta* (tuber).

**D: Phytochemistry; 8, mucilage**
ACANTHACEAE: *Dyschoriste perrottetii* (seed), *Hygrophila auriculata* (plant), *Peristrophe bicalyculata* (seed), *Stenandriopsis guineensis* (leaf). ANNONACEAE: *Annona senegalensis* (root). ARACEAE: *Colocasia esculenta* (tuber). BASELLACEAE: *Basella alba.* BOMBACACEAE: *Adansonia digitata* (leaf, wood, bark, fruit), *Bombax buonopozense* (flower), *Ceiba pentandra* (leaf, bark). *Rhodognaphalon brevicuspe* (root). BORAGINACEAE: *Cordia myxa* (fruit-pulp). COCHLOSPERMACEAE: *Cochlospermum tinctorium* (root). COMMELINACEAE: *Aneilema lanceolatum* (seed), *A. setiferum* (plant), *Commelina benghalensis* (plant), *C. diffusa* (leaf), *Pollia condensata* (plant). COMPOSITAE: *Crassocephalum crepidioides* (plant), *C. rubens* (plant), *Gynura mintata* (plant). CRUCIFERAE: *Lepidium sativum* (seed). CONVOLVULACEAE: *Ipomoea nil* (seed), *I. pes-caprae* (plant), *Merremia umbellata* (seed). DILLENIACEAE: *Tetracera alnifolia* (leaf).

**D: Phytochemistry; 9, alkaloids**
ACANTHACEAE: *Acanthus montanus* (leaf), *Asystasia gangetica* (leaf, flower), *Brillantaisia patula* (leaf), *Elytraria marginata* (leaf), *Eremomastax speciosa, Graptophyllum pictum, Hygrophila auriculata. Justicia extensa, J. laxa, Lankesteria brevior* (leaf, bark, root), *Peristrophe bicalculata* (leaf, stem), *Thunbergia alata* (leaf). AIZOACEAE: *Mollugo nudicaulis, Tetragona tetragonioides.* AMARANTHACEAE: *Aerva lanata, Alternanthera pungens, A. nodiflora* (leaf), *Pupalia lappacea* (seed). AMARYLLIDACEAE: *Crinum glaucum, C. jagus, C. zeylanicum, Eucharis grandiflora, Scadoxus multiflorus* (bulb), *Zephyranthes spp.* (bulb).

ANACARDIACEAE: *Anacardium occidentale, Schinus molle* (fruit). ANNONACEAE: *Annona chrysophylla* (bark), *A. muricata* (bark), *A. reticulata* (leaf, bark, root), *A. senegalensis* (leaf, bark, seed), *A. squamosa* (leaf, root), *Annonidium mannii* (leaf), *Artabotrys velutinus* (leaf), *Enantia chlorantha* (bark), *E. polycarpa* (bark, root), *Greenwayodendron oliveri* (bark), *G. suaveolens* (bark, root), *Monanthotaxis diclina* (bark, root), *Monodora brevipes* (bark), *M. crispata* (leaf), *M. myristica* (seed), *Pachypodanthium staudtii* (leaf, bark), *Uvaria chamae* (root), *Xylopia aethiopica*. APOCYNACEAE: *Alafia lucida* (seed), *Callichilia monopodialis* (leaf, stem, root, seed), *C. stenopetala* (plant), *C. subsessilis* (leaf, stem, root), *Catharanthus roseus* (plant), *Chonemorpha macrophyllum* (leaf, bark, root), *Dictyophleba lucida* (plant), *Ervatamia coronaria* (plant), *Funtumia africana* (plant), *F. elastica* (leaf, seed), *Hedranthera barteri* (plant), *Holarrhena floribunda* (plant), *Hunteria eburnea* (leaf, bark, seed), *H. umbellata* (plant), *Landolphia dulcis* (leaf), *Oncinotis nitida* (plant), *Pleiocarpa mutica* (bark, root), *Plumeria spp.* (whole plant), *Picralima nitida* (seed), *Rauvolfia caffra* (leaf, root), *R. macrophylla* (plant), *R. vomitoria* (plant), *Strophanthus hispidus* (seed), *S. preussii* (seed), *Tabernaemontana brachyantha* (bark), *T. contorta* (bark), *T. crassa* (whole plant), *T. eglandulosa* (whole plant), *T. longiflora* (leaf, bark), *T. pachysiphon* (bark), *Voacanga bracteata* (bark). ARACEAE: *Anchomanes difformis* (root), *Colocasia esculenta* (leaf, tuber), *Culcasia scandens* (leaf, stem), *Nephthytis afzelii* (leaf), *Pistia stratiotes* (plant). ARALIACEAE: *Cussonia arborea* (plant). ARISTOLOCHIACEAE: *Aristolochia braeteolata* (stem, root). ASCLEPIADACEAE: *Gomphocarpus fruticosus* (whole plant), *Gymnema sylvestre* (plant), *Pergularia daemia* (leaf, root), *Secamone afzelii* (plant), *Toxocarpus brevipes* (leaf, bark), *Tylophora sylvatica* (leaf, stem, root, seed). BALANOPHORACEAE: *Thonningia sanguinea* (seed). BIGNONACEAE: *Newbouldia laevis* (plant), *Pandorea pandorana* (leaf), *Spathodea campanulata* (plant), *Tecoma stans* (plant). BIXACEAE: *Bixa orellana* (plant). BOMBACACEAE: *Adansonia digitata* (bark). BORAGINACEAE: *Cordia sebestena* (plant), *Heliotropium indicum* (plant), *H. strigosum* (aerial parts), *Ehretia cymosa* (plant). BROMELIACEAE: *Ananas comosus* (?root, fruit). BURSERACEAE: *Boswellia odorata* (bark). CAMPANULACEAE: *Wahlenbergia perrottetii* (plant). CANNACEAE: *Canna indica* (plant). CAPPARACEAE: *Cadaba farinosa* (leaf), *Cleome viscosa* (leaf), *Maerua angolensis* (plant), *Ritchiea capparoides* (plant). CARYOPHYLLACEAE: *Stellaria media* (plant). CECROPIACEAE: *Myrianthus serratus* (kernel). CELASTRACEAE: *Hippocratea indica* (leaf), *Salacia nitida* (leaf, stem). CHENOPODIACEAE: *Salsola baryosma* (leaf, stem). CHRYSOBALANACEAE: *Chrysobalanus icaco*. COCHLOSPERMACEAE: *Cochlospermum tinctorium* (root). COMBRETACEAE: *Guiera senegalensis* (leafy stem), *Terminalia superba* (bark). COMPOSITAE: *Ageratum houstonianum* (seed), *Bidens biternata* (leaf), *B. pilosa* (plant), *Centipeda minima* (plant), *Chromolaena odoratum* (seed), *Crassocephalum sarcobasis* (stem, root), *Eclipta alba* (leafy stem), *Gnaphalium lutea-album* (plant), *Helichrysum mechowianum* (whole plant), *Lactuca capensis* (root), *Microglossa pyrifolia* (plant), *Struchium sparganophora* (leaf, root), *Synedrella nodiflora* (leaf, seed), *Tanacetum cinerarifolium* (flower). CONNARACEAE: *Byrsocarpus coccineus* (leaf, root). CONVOLVULACEAE: *Argyreia nervosa* (seed), *Ipomoea spp.* (seed), *I. alba* (leaf, stem, fruit) *I. batatas* (leaf, stem, root), *I. nil* (plant), *I. pes-caprae* (leaf), *I. purpurea* (plant), *I. quamoclit* (leafy stem), *I. tricolor* (plant), *Jacquemontia tamnifolia* (leaf, root). CRUCIFERAE: *Brassica juncea* (seed), *Capsella bursa-pastoris* (plant), *Crambe abyssinica* (seed). CUCURBITACEAE: *Citrullus colocynthis* (fruit), *C. lanatus* (leaf, fruit), *Cogniauxia podolaena* (bark, root), *Cucumis sativus* (seed), *Cucurbita pepo* (seed), *Lagenaria breviflora* (fruit), *Luffa acutangula* (root), *L. cylindrica* (fruit), *Momordica balsamina* (leaf), *M. charantia* (leaf, seed), *M. foetida* (plant), *Telfairea occidentalis* (seed). CYPERACEAE: *Cladium mariscus* (seed), *Cyperus alternifolius* (leaf, root), *C. papyrus* (leaf), *Scirpus maritimus* (aerial part). DICHAPETALACEAE: *Dichapetalum toxicarium* (plant). DIONCOPHYLLACEAE: *Triphyophyllum peltatum* (leaf, bark, root). DIOSCOREACEAE: *Dioscorea bulbifera* (plant), *D. cayenensis* (leafy stem), *D. dumetorum* (plant), *D. sansibarensis* (tuber), *D. smilacifolia* (tuber). DIPSACACEAE: *Succsia trichotocephala* (aerial part).

**D: Phytochemistry; 10, glycosides, saponins, steroids**

ACANTHACEAE: *Asystasia gangetica* (whole plant), *Eremomastax speciosa* (plant). Agavaceae: *Agave sisalana* (bagasse), *Furcraea foetida* (leaf). AIZOACEAE: *Mollugo nudicaulis* (plant), *M. pentaphylla* (plant), *Trianthema portulacastrum* (plant), *Zaleya pentandra* (plant). AMARANTHACEAE: *Amaranthus spinosus* (plant), *Celosia trigyna* (plant), *Pupalia lappacea* (plant). ANACARDIACEAE: *Lannea welwitschii* (bark), *Mangifera indica* (leaf). ANNONACEAE: *Annona chrysophylla* (bark), *A. senegalensis* (root), *Greenwayodendron suaveolens* (root), *Hexalobus crispiflorus* (bark, root), *Xylopia aethiopica* (plant). APOCYNACEAE: *Adenium obesum* (plant), *Landolphia dulcis* (leaf, bark), *Nerium oleander* (whole plant), *Pycnobotrys nitida* (plant), *Strophanthus gracilis* (plant), *S. gratus* (seed), *S. hispidus* (seed), *S. sarmentosus* (seed), *Thevetia neriifolia* (whole plant). ARACEAE: *Amorphophallus dracontioides* (sap), *Anchomanes difformis* (root), *Colocasia esculenta* (leaf, tuber), *Nephthytis afzelii* (leaf). ASCLEPIADACEAE: *Asclepias curassavica* (root), *Calotropis procera* (latex, seed), *Dregea abyssinica* (seed), *Gomphocarpus fruticosus* (whole plant), *Gymnema sylvestre* (leaf, twig), *Pachycarpus lineolata* (root), *Pergularia daemia* (plant). BALANTACEAE: *Balanites aegyptiaca* (plant, seed, kernel) *B. wilsoniana* (fruit-pulp, kernel). BIGNONIACEAE: *Parmentiera cereifera* (leaf). BOMBACACEAE: *Ceiba pentandra* (seed). BORAGINACEAE: *Cordia myxa* (leaf). BURSERACEAE: *Dacryodes klaineana* (leaf). CAPPARACEAE: *Cadaba farinosa* (plant), *Cleome viscosa* (leaf, seed), *Maerua angolensis* (plant), *Ritchiea capparoides* (plant). CARYOPHYLLACEAE: *Drymaria cordata* (root, flower, fruit, seed). CELASTRACEAE: *Mystroxylon aethiopicum* (plant). CHENOPODIACEAE: *Chenopodium ambrosioides* (plant). COMBRETACEAE: *Terminalia catappa* (leaf, flower), *T. ivorensis* (wood), *T. macroptera* (plant). COMMELINACEAE: *Palisota ambigua* (root). COMPOSITAE: *Bidens pilosa* (plant), *Centaurea perrottetii* (flower), *Centipeda minima* (plant), *Helichrysum foetidum* (plant), *H. nudifolium* (leaf, root), *H. odoratissimum* (leaf, root), *Microglossa pyrifolia* (plant), *Sigesbeckia orientalis* (root), *Spilanthes filicaulis* (root), *Vernonia cinerea* (root), *V. colorata* (plant, bark). CONNARACEAE: *Byrsocarpus coccineus* (bark), *Spiropetalum reynoldsii* (leaf). CONVOLVULACEAE: *Ipomoea cairica* (plant), *I. coptica* (plant), *I. mauritiana* (plant), *I. nil* (seed), *I. pes-caprae* (root). CRUCIFERAE: *Brassica campestris* (leaf), *B. juncea* (leaf), *Raphanus sativus* (plant). CUCURBITACEAE: *Citrullus colocynthis* (fruit), *Cogniauxia podolaena* (root), *Cucumis metuliferus* (leaf), *C. sativus* (seed), *Cucurbita maxima* (seed), *Lagenaria siceraria* (seed), *Luffa acutangula* (seed), *L. cylindrica* (fruit, seed), *Momordica charantia* (seed), *Trichosanthes cucumerina* (old fruit). DICHAPETALACEAE: *Dichapetalum toxicarium* (plant). DIONCOPHYLLACEAE: *Triphyophyllum peltatum* (root). DIOSCOREACEAE: *Dioscorea alata* (tuber), *D. bulbifera* (plant). DROSERACEAE: *Drosera madagascariensis* (plant).

**D: Phytochemistry; 11, tannins, astringents, bitters**

ACANTHACEAE: *Justicia extensa, J. flava* (root), *Stenandriopsis guineensis* (leaf), *Thomandersia hensii* (bark, leaf). AIZOACEAE: *Gisekia pharnaceoides* (plant). AMARANTHACEAE: *Amaranthus viridis* (leaf). ANACARDIACEAE: *Anacardium occidentale* (leaf, bark), *Lannea welwitschii* (bark), *Mangifera indica* (leaf, bark-sap), *Schinus molle* (leaf, bark, fruit), *S. terebinthifolius* (plant, fruit), *Spondias mombin* (bark). ANNONACEAE: *Annona reticulata* (bark, seed), *A. senegalensis* (whole plant), *A. squamosa* (root, seed), *Monanthotaxis diclina* (bark, root), *Pachypodanthium staudtii* (bark, root), *Uvaria afzelii* (bark), *U. chamae* (plant). APOCYNACEAE: *Funtumia elastica* (bark, seed), *Picralima nitida* (whole plant), *Rauvolfia vomitoria* (bark), *Strophanthus gratus* (seed), *S. preussii* (sap), *Voacanga africana* (plant). ASCLEPIADACEAE: *Calotropis procera* (whole plant), *Gymnema sylvestre* (leaf). AVICENNIACEAE: *Avicennia germinans* (bark). BIGNONIACEAE: *Kigelia africana* (bark), *Markhamia tomentosa* (bark), *Newbouldia laevis* (bark), *Spathodea campanulata* (bark). BOMBACACEAE: *Adansonia digitata* (bark, seed), *Ceiba pentandra* (bark, seed). BURSERACEAE: *Canarium schweinfurthii* (bark, resin, fruit-pulp, kernel). *Commiphora africana* (resin, seed), *Dacryodes klaineana* (leaf). CAPPARACEAE: *Cleone viscosa* (leaf, seed). CASUARINACEAE: *Casuarina equisitifolia* (bark). CELASTRACEAE: *Maytenus senegalensis* (leaf, root), *Mystroxylon aethiopicum* (plant). CHRYSOBALANACEAE: *Chrysobalanus icaco* (bark, seed), *Maranthes polyandra* (bark), *Parinari excelsa* (wood, stem bark, root, kernel). COCHLOSPERMACEAE: *Cochlospermum tinctorium* (root, seed). COMBRETACEAE: *Anogeissus leiocarpus* (wood, bark), *Combretum micranthum* (leaf, bark), *Conocarpus erectus* (bark), *Guiera senegalensis* (leaf, leafy stem, fruit), *Laguncularia racemosa* (leaf), *Terminalia arjuna* (bark), *T. belerica* (bark, fruit), *T. brownii* (bark), *T. catappa* (leaf, bark, flower, fruit, kernel), *T. cheluba* (fruit), *T. glaucescens* (bark), *T. ivorensis* (bark), *T. macroptera* (stem bark, root bark), *T. superba* (bark, root). COMPOSITAE: *Bidens pilosa* (leaf). *Centaurea perrottetii* (plant), *Centipeda minima* (plant), *Conyza canadensis* (plant), *Crassocephalum crepidioides* (root), *C. sarcobasis* (stem), *Launaea cornuta* (foliage), *Sigesbeckia orientalis* (plant), *Triplotaxis stellutifera* (leaf), *Vernonia adoensis* (root), *V. ambigua* (plant), *V. amygdalina* (leaf, bark), *V. cinerea* (seed), *V. colorata* (vegetative parts, seed), *V. guineensis, V. lasiopus* (plant), *V. migoedii* (sap), *V. pumila* (root), *V. thomsoniana* (leaf). CONNARACEAE: *Byrsocarpus coccineus* (bark), *B. poggeanus* (leaf, bark, root), *Cnestis corniculatus* (leaf), *C. ferruginea* (leaf, bark, root), *Connarus africanus* (bark, seed), *Roureopsis obliquifoliolata* (leaf, bark, root), *Spiropetalum reynoldsii* (leaf), CONVOLVULACEAE: *Evolvulus alsinoides* (plant), *Ipomoea nil* (seed), *I. pes-caprae* (plant), *Merremia tridentata* (plant). CRUCIFERAE: *Brassica juncea* (plant, seed). CUCURBITACEAE: *Citrullus colocynthis* (plant, fruit), *Cucumis metuliferus* (wild plant), *Lagenaria breviflora* (fruit), *L. siceraria* (fruit, seed), *Luffa acutangula* (fruit, seed), *Momordica balsamina* (leaf, fruit). CYPERACEAE: *Remirea maritima* (rhizome). DILLENIACEAE: *Tetracera alnifolia* (leaf, root-bark). DIONCOPHYLLACEAE: *Triphyophyllum peltatum* (root). DIOSCOREACEAE: *Dioscorea bulbifera* (wild bulbils). DROSERACEAE: *Drosera indica* (plant).

**D: Phytochemistry; 12, flavones**

AIZOACEAE: *Mollugo nudicaulis* (plant). AMARANTHACEAE: *Celosia trigyna* (plant). APOCYNACEAE: *Voacanga africana* (plant). CAPPARACEAE: *Cleome viscosa* (seed) COMBRETACEAE: *Combretum micranthum* (leaf), *Terminalia macroptera* (plant). COMPOSITAE: *Bidens pilosa* (plant). DILLENIACEAE: *Tetracera alnifolia* (leaf, root, bark).

**D: Phytochemistry; 13, resin. (cf. also F:18)**

AMARANTHACEAE: *Amaranthus viridis*. ANNONACEAE: *Annona chrysophylla* (bark), *A. squamosa* (seed). BIGNONIACEAE: *Tecoma stans*. BIXACEAE: *Bixa orellana*. BURSERACEAE: *Canarium schweinfurthii* (bark). COMBRETACEAE: *Terminalia superba* (root). COMPOSITAE: *Tanacetum cinererifolium* (flower). CONVOLVULACEAE: *Ipomoea alba* (plant), *I. asarifolia* (plant), *I. mauritiana* (root), *I. nil* (seed), *I. pes-caprae* (plant), *I. pes-tigridis* (root), *I. purpurea* (stem), *Merremia tuberosa* (tuber), *M. umbellata* (seed).

**D: Phytochemistry; 14, hydrogen cyanide**

AMARANTHACEAE: *Amaranthus spinosus* (plant). ANNONACEAE: *Annona muricata* (leaf, bark, root), *A. reticulata* (whole plant), *A. squamosa* (leaf, bark, root). BIGNONIACEAE: *Parmentiera cereifera* (leaf, root, fruit). BOMBACACEAE: *Ceiba pentandra* (stem-bark, root-bark). CAPPARACEAE: *Cleome gynandra* (root). CONVOLVULACEAE: *Ipomoea cairica* (stem, tuber), *I. quamoclit* (leafy stem), *Merremia dissecta* (leaf). CUCURBITACEAE: *Lagenaria siceraria* (vegetative part). DROSERACEAE: *Drosera indica* (plant).

**D: Phytochemistry; 15, fish-poisons**

ACANTHACEAE: *Adhatoda buchholzii* (whole plant), *Eremomastax speciosa* (plant), *Justicia extensa* (plant), *J. laxa* (plant), *Rhinacanthus virens* (plant). AIZOACEAE: *Aizoon canariensis* (plant). AMARYLLIDACEAE: *Crinum zeylanicum* (seed), *Scadoxus multiflorus* (bulb). ANACARDIACEAE: *Anacardium occidentale* (bark). ANNONACEAE: *Annona glauca* (seed), *A. muricata* (seed), *A. squamosa* (plant). APOCYNACEAE: *Adenium obesum* (root), *Picralima nitida* (fruit), *Strophanthus gracilis* (wood, seed), *S. hispidus* (seed), *Thevetia neriifolia* (wood). ARACEAE: *Culcasia scandens* (sap), *Xanthosoma mafaffa* (old leaf). ASCLEPIADACEAE: *Sarcostemma viminale* (latex). BALANITACEAE: *Balanites aegyptiaca* (bark, root, fruit). CAPPARACEAE: *Cleome gynandra* (plant, oil). CARICACEAE: *Carica papaya* (leaf). COMBRETACEAE: *Combretum nigricans* (leafy twig). COMMELINACEAE: *Palisota hirsuta* (plant). COMPOSITAE: *Vernonia macrocyanus* (plant). CRUCIFERAE: *Lepidium sativum* (plant). CUCURBITACEAE: *Lagenaria breviflora* (seed), *Luffa acutangula* (fruit), *L. cylindrica* (unripe fruit). DIONCOPHYLLACEAE: *Habropetalum dawii* (young leaf). DIOSCOREACEAE: *Dioscorea sansibarensis* (tuber).

**D: Phytochemistry; 16, insecticides, arachnicides**

ACANTHACEAE: *Hypoestes verticillata* (plant), *Lepidagathis alopecuroides* (plant). ANACARDIACEAE: *Anacardium occidentale* (bark-gum). ANNONACEAE: *Annona cherimola* (seed), *A. glauca* (seed), *A. muricata* (leaf), *A. reticulata* (leaf, fruit, seed), *A. senegalensis* (root), *A. squamosa* (plant), *Monodora myristica* (seed), *Pachypodanthium staudtii* (bark), *Xylopia aethiopica* (fruit). APOCYNACEAE: *Adenium obesum* (bark), *Nerium oleander* (leaf), *Rauvolfia vomitoria* (bark), *Strophanthus hispidus* (sap), *Tabernaemontana crassa* (leaf), *Thevetia neriifolia* (fruit). ARACEAE: *Colocasia esculenta* (plant), *Culcasia parviflora* (leaf). BALANITACEAE: *Balanites wilsoniana* (fruit pulp, kernel). BIGNONIACEAE: *Spathodea campanulata* (stem bark). BIXACEAE: *Bixa orellana* (dye). BURSERACEAE: *Commiphora africana* (bark). CAPPARACEAE: *Cleome gynandra* (oil). CARICACEAE: *Carica papaya* (plant, sap). CELASTRACEAE: *Maytenus senegalensis* (bark). COMBRETACEAE: *Guiera senegalensis* (smoke of leafy twig), *Strephonema pseudocola* (seed-oil). COMPOSITAE: *Conyza canadensis* (seed), *Launaea cornuta* (water in which leaf has been boiled), *Melanthera scandens* (smoke from sticks), *Microglossa pyrifolia* (leaf, stem, root, flower), *Sigesbeckia orientalis* (plant), *Spilanthes filicaulis* (plant), *Triplotaxis stellutifera* (plant, leaf), *Vernonia galamensis* (plant). CONVOLVULACEAE: *Ipomoea muricata* (sap), *Jacquemontia tamnifolia* (whole plant). CRUCIFERAE: *Brassica juncea* (leaf), *Eruca sativa* (seed-cake). CUCURBITACEAE: *Citrullus colocynthis* (root, fruit), *Cogniauxia podolaena* (root), *Cucumis sativus* (fruit-skin, fruit-sap), *Momordica charantia* (plant), *M. foetida* (fruit), *Zehneria scabra* (leaf). CYPERACEAE: *Cyperus articulatus* (rhizome), *C. rotundus* (tuber), *Scirpus maritimus* (culm as bait).

**D: Phytochemistry; 17, rodenticides, mammal and bird poisons**

APOCYNACEAE: *Adenium obesum* (stem, bark). DICHAPETALACEAE: *Dichapetalum toxicarium* (bark, kernel). DIOSCOREACEAE: *Dioscorca dumetorum* (tuber).

**D: Phytochemistry; 18, reptile-repellents**
APOCYNACEAE: *Carissa edulis* (root). CHENOPODIACEAE: *Chenopodium ambrosioides* (plant). COMBRETACEAE: *Combretum nigricans* (leafy twig). CONVOLVULACEAE: *Merremia dissecta* (plant).

**D: Phytochemistry; 19, molluscicide**
BALANITACEAE: *Balanites aegyptiaca* (plant).

**D: Phytochemical; 20, arrow-poisons (incl. adhesives)**
AMARYLLIDACEAE: *Crinum spp.* (bulb). ANACARDIACEAE: *Anacardium occidentale* (bark). ANNONACEAE: *Annona senegalensis* (root), *Enantia polycarpa* (bark), *Pachypodanthium staudtii* (bark). APOCYNACEAE: *Adenium obesum* (bark, latex), *Alafia lucida* (latex), *A. scandens* (latex), *Alstonia boonei* (bark), *Funtumia elastica* (bark), *Holarrhena floribunda* (latex), *Hunteria eburnea* (latex), *Landolphia parvifolia* (latex), *Saba senegalensis* (latex), *Strophanthus gracilis* (wood, seed), *S. gratus* (bark, seed), *S. hispidus* (seed), *S. sarmentosus* (seed), *Tabernaemontana crassa* (sap), *Thevetia neriifolia* (plant). ARACEAE: *Amorphophallus aphyllus* (tuber), *A. dracontioides* (tuber). ASCLEPIADACEAE: *Calotropis procera* (latex), *Gomphocarpus fruticosus* (plant). BALANOPHORACEAE: *Thonningia sanguinea* (flower-head). CAPPARACEAE: *Buchholzia coriacea* (bark), *Euadenia eminens* (root), *E. trifoliolata* (root). CECROPIACEAE: *Myrianthus serratus* (leaf). COMMELINACEAE: *Palisota hirsuta* (stem). COMPOSITAE: *Microglossa pyrifolia* (plant). CUCURBITACEAE: *Momordica balsamina* (plant). DICHAPETALACEAE: *Dichapetalum toxicarium* (kernel).

**D: Phytochemical; 21, ordeal-poisons**
APOCYNACEAE: *Adenium obesum* (plant), *Thevetia neriifolia* (plant). ASCLEPIADACEAE: *Calotropis procera* (latex), *Gomphocarpus fruticosus* (plant). CANNACEAE: *Canna indica* (plant). CHRYSOBALANACEAE: *Maranthes glabra* (sap). COMPOSITAE: *Ageratum conyzoides* (sap). CONNARACEAE: *Agelaea trifolia* (bark). DIOSCOREACEAE: *Dioscorea sansibarensis* (tuber).

**D: Phytochemistry; 22, depilatory**
ANACARDIACEAE: *Sclerocarya birrea* (wood-ash), ASCLEPIADACEAE: *Calotropis procera* (latex), *Pergularia tomentosa* (plant, latex). BURSERACEAE: *Canarium schweinfurthii* (resin). CARICACEAE: *Carica papaya* (leaf). COMBRETACEAE: *Anogeissus leiocarpus* (wood-ash). CUCURBITACEAE: *Cucurbita pepo* (fruit-pulp), *Lagenaria breviflora* (fruit).

**D: Phytochemistry; 23, antibiotic, bacteristatic, fungistatic**
ACANTHACEAE: *Hygrophila auriculata* (root), *Peristrophe bicalyculata*. ANACARDIACEAE: *Mangifera indica* (leaf, bark, fruit), *Spondias mombin* (bark). ANNONACEAE: *Annona glabra* (wood, bark), *A. squamosa* (plant), *Xylopia aethiopica* (fruit). APOCYNACEAE: *Voacanga africana* (plant). BALANITACEAE: *Balanites aegyptiaca* (fruit). BALSAMINACEAE: *Impatiens balsamina* (flower). BIGNONIACEAE: *Newbouldia laevis* (bark). CAPPARACEAE: *Capparis decidua* (flower, fruit, seed). CELASTRACEAE: *Hippocratea indica* (root). CHENOPODIACEAE: *Chenopodium ambrosioides* (plant). CHRYSOBALANACEAE: *Parinari curatellifolia* (bark). COCHLOSPERMACEAE: *Cochlospermum tinctorium* (root). COMBRETACEAE: *Combretum comosum* (plant), *C. micranthum* (root), *Terminalia avicennioides* (root), *T. catappa* (leaf), *T. glaucescens* (root), *T. laxiflora* (bark). COMPOSITAE: *Bidens bipinnata* (stem, flower), *Conyza canadensis* (plant), *Eclipta alba* (plant), *Mikania cordata* (plant), *Vernonia nigritiana* (leaf). CONNARACEAE: *Cnestis ferruginea* (root). CONVOLVULACEAE: *Ipomoea batatas* (haulm, tuber), *I. cairica* (plant). COSTACEAE: *Costus spectabilis* (leaf). CRASSULACEAE: *Bryophyllum pinnatum* (sap). CRUCIFERAE: *Brassica napus* (plant), *B. oleracea* (seed-oil), *B. rapa* (leaf, root), *Crambe abyssinica* (seed-meal), *Lepidium sativum* (plant, seed-oil). CUCURBITACEAE: *Cucumis melo* (plant). CYPERACEAE: *Eleocharis dulcis* (tuber-sap).

**D: Phytochemistry; 24, urticant**
AMARANTHACEAE: *Aerva javanica* (?).

**D: Phytochemistry; 25, miscellaneously poisonous or repellent. (cf. D: 15 to 21)**
ACANTHACEAE: *Crossandra nilotica*. AIZOACEAE: *Gisekia pharnaceoides* (fruit), *Zaleya pentandra.* (fruit). APOCYNACEAE: *Adenium obesum, Plumeria spp. Thevetia neriifolia.* ARACEAE: *Dieffenbachia spp.* (sap). ASCLEPIADACEAE: *Asclepias curassavica* (whole plant), *Calotropis procera* (whole plant), *Sarcostemma viminale* (latex). BALANITACEAE:

*Balanites aegyptiaca*. BIGNONIACEAE: *Pandorea pandorana* (plant), *Spathodea campanulata* (fruit placenta). CAPPARACEAE: *Capparis sepiaria* (root), *Maerua angolensis* (leaf). COMPOSITAE: *Centipeda minima* (plant), *Crassocephalum rubens* (plant), *Helichrysum odoratissimum*. CONNARACEAE: *Byrsocarpus coccineus, Cnestis longiflora* (plant), *Jaundea pinnata* (fruit). CONVOLVULACEAE: *Ipomoea alba* (plant), *I. aquatica* (plant), *I. asarifolia* (plant), *I. nil* (plant), *Merremia dissecta* (leaf). CRUCIFERAE: *Brassica napus* (seed-cake). CUCURBITACEAE: *Cogniauxia podolaena* (fruit), *Cucumeropsis mannii* (root), *Cucumis metuliferus* (whole plant), *C. prophetarum* (fruit), *Luffa acutangula* (seed), *L. cylindrica* (fruit), *Mukia maderaspatana* (fruit), *Peponium vogelii* (root, unripe fruit), *Ruthalicia eglandulosa* (plant), *Zehneria scabra*. CYPERACEAE: *Cyperus fenzelianus, Pycreus nitidus*. DICHAPETALACEAE: *Dichapetalum barteri* (plant), *D. madagascariensis* (plant), *D. tomentosum* (fruit). DIOSCOREACEAE: *Dioscorea bulbifera* (wild bulbil), *D. dumetorum* (wild plant), *D. sansibarensis* (bulbil).

**E: Agri-horticulture; 1, ornamental, cultivated or partially tended**

ACANTHACEAE: *Acanthus guineensis* (flower), *Asystasia scandens* (flower), *Barleria cristata, B. oenotheroides* (flower), *B. prionitis* (flower), *Brillantaisia spp.* (flower), *Crossandra flava* (flower), *Fittonia verschaffeltii, Graptophyllum pictum, Hygrophila senegalensis* (flower), *Lankesteria hispida, Lepidagathis heudelotiana, Odontonema cuspidatum, Pachystachys coccinea, Pseuderanthemum atropurpureum, P. ludovicianum, Ruellia tuberosa, Ruspolia hypocrateriformis, Satanocrater berhautii, Stenandriopsis guineensis, Thunbergia alata, T. chrysops, T. erecta, T. grandiflora, T. laevis, T. laurifolia* (whole plant, flower), *T. vogeliana, Whitfieldia colorata, W. lateritia*. AGAVACEAE: *Furcuraea foetida*. AMARANTHACEAE: *Alternanthera bettzickiana, Amaranthus hybridus, Celosia argentea, Gomphrena globosa*. AMARYLLIDACEAE: *Crinum spp, Hymenocallis littoralis, Pancratium trianthum, Scadoxus cinnabarinus, S multiflorus, Zephyranthes spp*. ANACARDIACEAE: *Schinus molle* (foliage, fruit). ANNONACEAE: *Artabotrys stenopetalus* (flower), *Monodora crispata* (flower), *M. tenuifolia* (tree), APOCYNACEAE: *Adenium obesum* (plant), *Alafia scandens* (plant), *Allamanda spp.* (flower). *Baissea multiflora* (plant), *Carissa edulis* (plant), *Catharanthus roseus* (plant), *Chonemorpha macrophyllum, Ervatamia coronaria, Farquharia elliptica, Hedranthera barteri, Holarrhena floribunda, Nerium odorum, N. oleander, Oncinotis glabrata, O. gracilis, O. nitida, Pleiocarpa mutica, Plumeria spp.* (flower), *Rauvolfia caffra, R. vomitoria* (tree), *Saba senegalensis* (tree), *Strophanthus gratus, S. hispidus, S. sarmentosus, Tabernaemontana crassa* (flower), *Thevetia neriifolia* (plant). ARACEAE: *Alocasia macrorhiza, Amorphophallus aphyllus, Caladium bicolor, Dieffenbachia spp*. ARALIACEAE: *Polyscias guilfoylei*. ARAUCARIACEAE: *Acaucaria columnaris, A. heterophylla*. ARISTOLOCHIACEAE: *Aristolochia brasiliensis, A. elegans, A. gibbosa, A. ridicula, A. ringens, Paristolochia goldieana*. ASCLEPIADACEAE: *Asclepias curassavica, Dregea abyssinica, Leptadenia arborea, Oxystelma bornouense, Pergularia daemia, Trachycalymma pulchellum*. BALSAMINACEAE: *Impatiens sakerana*, BEGONIACEAE: *Begonia fusicarpa*. BIGNONIACEAE: *Bignonia capreolata, Crescentia cujete, Haplophragma adenophyllum, Jacaranda mimosifolia, Kigelia africana, Markhamia tomentosa, Newbouldia laevis, Pandorea pandorana, Pyrostegia venusta, Saritaea magnifica, Stereospermum acuminatissimum, S. kunthianum, Spathodea campanulata, Tabebuia rosea, Tecoma capensis, T. stans*. BIXACEAE: *Bixa orellana*. BORAGINACEAE: *Cordia sebestena, C. sinensis*. BURSERACEAE: *Boswellia dalzielii, Commiphora kerstingii*. CACTACEAE: *Opuntia spp*. CANNACEAE: *Canna hybrids*. CAPPARACEAE: *Cleome afrospinosa, C. speciosa, Crataeva adansonii, Euadenia eminens, Maerua angolensis, Ritchiea reflexa*. CAPRIFOLIACEAE: *Lonicera spp*. CASUARINACEAE: *Casuarina equisitifolia*. CHRYSOBALANACEAE: *Chrysobalanus icaco*. COMBRETACEAE: *Combretum bracteatum, C. comosum, C. constrictum, C. fragrans* (bark-gum), *C. grandiflorum, C. paniculatum, C. platypterum, C. racemosum, C. tarquense, Quisqualis indica, Terminalia catappa, T. cheluba*. COMMELINACEAE: *Dichorisandra thyrsiflora, Murdannia simplex, Palisota barteri, P. hirsuta, Rhoeo spathacea, Zebrina pendula*. COMPOSITAE: *Ageratum houstianum, Aspilia angustifolia, Ayapana triplinervis, Coreopsis borianiana, Cotula anthemoides, Gynura procumbens, Helianthus annuus, Helichrysum odoratissimum, Tagetes spp., Vernonia gerberiformis, V. nigritiana, Wedelia trilobata*. CONNARACEAE: *Agelaea nitida* (flower, fruit), *A. obliqua* (flower, fruit), *A. trifolia* (flower, fruit), *Byrsocarpus coccineus* (foliage, flower), *Connarus africanus, Cnestis ferruginea*. CONVOLVULACEAE: *Argyreia tiliaefolia* (flower), *Evolvulus alsinoides* (flower), *Ipomoea acuminata, I. alba* (flower), *I. arborescens* (flower), *I. cairica* (flower), *I. hederifolia* (flower), *I. intrapilosa* (flower). *I. mauritiana* (flower), *I. nil* (flower), *I. quamoclit* (flower), *I. setifera, I. tricolor, I. tuba, Jacquemontia pentantha, J. tamnifolia, Merremia cissoides, M. dissecta, M. tuberosa* (flower, fruit-pod), *Stictocardia beraviensis*. COSTACEAE: *Costus spectabilis*. CRASSULACEAE: *Bryophyllum pinnatum*,

*Kalanchoe crenata/integra, K. laciniata, K. lanceolata.* CUCURBITACEAE: *Diplocyclos palmatus.* CYPERACEAE: *Cyperus alternifolius.*
Addendum—BIGNONIACEAE: *Millingtonia hortensis*

**E: Agri-horticulture; 2, fodder (grazing, browsing, or eaten for lack of better)**
ACANTHACEAE: *Asystasia gangetica, Barleria eranthemoides, Blepharis linariifolia, Crossandra nilotica, Hygrophila auriculata, H. senegalensis, Hypoestes verticillata, Justicia betonica, J. flava, Nelsonia canescens, Peristrophe bicalyculata, Phaulopsis imbricata, Ruellia praetermissa, Thunbergia alata, T. grandiflora.* AIZOACEAE: *Gisekia pharnaceoides, Glinus lotoides, G. oppositifolius, Limeum diffusum, L. viscosum, Mollugo cerviana, Sesuvium portulacastrum, Trianthema triquetra, Zaleya pentandra.* AMARANTHACEAE: *Achyranthes aspera, Aerva javanica, A. lanata, Alternanthera nodiflora, A. sessilis, Amaranthus graecizans, A. hybridus, A. spinosus, A. viridis, Celosia argentea, C. trigyna, Cyathula cylindrica, Nothosaerva brachiata, Philoxerus vermicularis, Pupalia lappacea.* ANACARDIACEAE: *Lannea velutina, Ozoroa insignis, Sclerocarya birrea.* APOCYNACEAE: *Baissea multiflora, Carissa edulis.* ARACEAE: *Pistia stratiotes, Stylochiton hypogaeus* (tuber), *Typhonium trilobatum.* ASCLEPIADACEAE: *Calotropis procera, Glossonema boveanum, Kanahia lanifolia, Leptadenia arborea, L. hastata, L. pyrotechnica, Pentatropsis spiralis, Pergularia daemia, Solenostemma oleifolium.* BALANITACEAE: *Balanites aegyptiaca.* BEGONIACEAE: *Begonia oxyloba.* BIGNONIACEAE: *Parmentiera cereifera, Stereospermum kunthianum, Tecoma capensis.* BOMBACACEAE: *Adansonia digitata, Bombax buonopozense, B. costatum, Ceiba pentandra.* BORAGINACEAE: *Cordia sinensis, Cynoglossum amplifolium, C. lanceolatum, Echium horridum, Ehretia cymosa, Heliotropium bacciferum, H. indicum, H. ovalifolium, H. pterocarpum, H. strigosum, H. subulatum, Molktia callosa, Trichodesma africana, T. gracile.* BURSERACEAE: *Commiphora africana, C. kerstingii.* CACTACEAE: *Opuntia spp.* CAMPANULACEAE: *Wahlenbergia lobelioides.* CANNABACEAE: *Cannabis sativa* (seed-cake). CANNACEAE: *Canna edulis, C. indica.* CAPPARACEAE: *Boscia augustifolia, B. salicifolia, B. senegalensis, Cadaba farinosa, Capparis decidua, C. tomentosa, Cleome arabica, C. gynandra, C. scaposa, Crataeva adansonii, Maerua angolensis, M. crassifolia, M. pseudopetalosa.* CARYOPHYLLACEAE: *Drymaria cordata, Polycarpaea corymbosa, P. eriantha, P. linearifolia, Polycarpon prostratum, Stellaria media.* CELASTRACEAE: *Maytenus senegalensis, M. undata.* CHENOPODIACEAE: *Agathophora alopecuroides, Bassia muricata, Chenopodium murale, Cornulaca monacantha, Nucularia perrini, Salsola baryosma, Suaeda monoica, S. vermiculata, Traganum nudatum.* CHRYSOBALANACEAE: *Parinari excelsa* (fruit). CISTACEAE: *Helianthemum spp.* COMBRETACEAE: *Anogeissus leiocarpus, Combretum aculeatum, C. collinum, C. glutinosum, C. micranthum, Laguncularia racemosa, Terminalia avicennioides.* COMMELINACEAE: *Aneilema aequinoctiale, A. lanceolatum, Commelina africana, C. benghalensis, C. bracteosa, C. diffusa, C. erecta, C. forskalaei, Cyanotis caespitosa, C. lanata, Murdannia simplex, Palisota ambigua.* COMPOSITAE: *Ageratum conyzoides, Aspilia africana, A. helianthoides, Atractylis aristata, Berkheya spekeana, Bidens bipinnata, B. pilosa, Blumea perrottetiana, Carthamnus tinctorius, Centaurea perrottetii, C. praecox, C. pungens, Conyza bonariensus, C. pyrrhopappa, C. stricta* (flowering top), *C. sumatrensis, Coreopsis borianiana, Crassocephalum montuosum, C. picridifolium, C. vitellianum, Dichrocephala chrysanthemifolia, D. integrifolia, Echinops giganteus, E. spinosus, Eclipta alba, Emilia coccinea, Epaltes gariepina, Galinsoga parviflora, Grangea maderaspatana, Guizotia abyssinica, G. scabra, Helianthus annuus* (seed-meal), *Helichrysum foetidum, H. glumacceum, Launaea cornuta, L. intybacea, Melanthera scandens, Mikania cordata, Odontospermum graveolens, Piloselloides hirsuta, Pluchea ovalis, Pulicaria crispa, Sonchus oleraceus, S. schweinfurthii, Sphaeranthus flexuosus, S. senegalensis, Spilanthes filicaulis, Struchium sparganophora, Synedrella nodiflora, Tridax procumbens, Vernonia amyggdalina, V. cinerascens, V. cinerea, V. glabra.* CONVOLVULACEAE: *Astripomoea malvacea, Convolvulus microphyllus, Cressa cretica, Evolvulus alsinoides, Ipomoea aitonii, I. aquatica, I. batatas, I. cairica, I. coptica, I. coscinosperma, I. eriocarpa, I. mauritiana, I. involucrata, I. obscura, I. pes-caprae, I. pes-tigridis, I. sinensis, I. vagans, Jacquemontia tamnifolia, Merremia aegyptia, M. hederacea, M. pinnata, M. tridentata, Seddera latifolia.* CRASSULACEAE: *Bryophyllum pinnatum.* CRUCIFERAE: *Anastatica hierochuntica, Brassica campestris, B. juncea, B. napus* (seed-cake), *Crambe abyssinica* (seed-meal), *C. kralikii, Diplotaxis acris, D. virgata, Eremobium aegyptiacum, Eruca sativa, Farsetia aegyptiaca, F. hamiltonii, F. ramosissima, F. stenoptera, Morettia canescens, Moricanda arvensis, Raphanus sativus, Rorippa humifusa, Savignya parviflora, Schouwia purpurea, S. schimperi, Sisymbrium reboudianum, Zilla spinosa.* CUCURBITACEAE: *Citrullus colocynthis, C. lanatus* (plant, seed-cake), *Coccinia grandis, Cucumis figarei, C. melo, C. prophetarum, Cucurbita maxima* (fruit), *C. pepo* (fruit), *Lagenaria siceraria* (seed-cake), *Momordica balsamina, M. foetida, Mukia maderaspatana.* CYPERACEAE: *Afrotrilepis pilosa, Ascolepis capensis, Carex chlorosaccus, Cyperus alternifolius, C. amabilis, C. bulbosus, C. compressus, C. conglomeratus, C. dichroöstachys, C. difformis, C. esculentus, C. fenzelianus, C. haspan, C.*

*imbricatus, C. iria, C. jeminicus, C. maculatus, C. margaritaceus, C. rotundus, C. sphacelatus, C. tuberosus, C. zollingeri, Fimbristylis dichotoma, F. littoralis, F. ovata, Fuirena umbellata, Kyllinga odorata, K. welwitschii, Lipocarpa chinensis, Mariscus alternifolius, M. longibracteatus, M. squarrosus, Nemium spadiceum, Pycreus elegantulus, P. polystachyos, Remirea maritima, Rikliella kernii, Scirpus brachyceras.*

**E: Agri-horticulture; 3. veterinary medicine**

ACANTHACEAE: *Crossandra nilotica ssp. massaica.* AIZOACEAE: *Zaleya pentandra.* AMARANTHACEAE: *Aerva javanica, Alternanthera sessilis.* AMARYLLIDACEAE: *Scadoxus multiflorus.* ANACARDIACEAE: *Ozoroa pulcherrima, Schinus molle, Sclerocarya birrea.* ANNONACEAE: *Annona senegalensis, Monodora tenuifolia, Xylopia aethiopica.* APOCYNACEAE: *Adenium obesum, Carissa edulis, Holarrhena floribunda.* ARACEAE: *Culcasia scandens, Pistia stratioles.* ASCLEPIADACEAE: *Calotropis procera, Dregea abyssinica, Leptadenia hastata, Pergularia daemia, Solenostemma oleifolium.* BALANITACEAE: *Balanites aegyptiaca.* BIGNONIACEAE: *Kigelia africana, Newbouldia laevis, Spathodea campanulata, Sterospermum kunthianum.* BOMBACACEAE: *Ceiba pentandra.* BORAGINACEAE: *Cordia sinensis.* BURSERACEAE: *Commiphora africana.* CANNABACEAE: *Cannabis sativa.* CANNACEAE: *Canna indica.* CAPPARACEAE: *Boscia angustifolia, B. senegalensis, B. salicifolia, Capparis spinosa, C. tomentosa.* CARICACEAE: *Carica papaya.* CELASTRACEAE: *Mystroxylon aethiopicum.* CHENOPODIACEAE: *Chenopodium ambrosioides, Suaeda fruticosa.* CHRYSOBALANACEAE: *Maranthes glabra.* COCHLOSPERMACEAE: *Cochlospermum tinctorium.* COMBRETACEAE: *Anogeissus leiocarpus, Combretum dolichopetalum, C. glutinosum, C. nigricans, C. paniculatum, Guiera senegalensis, Terminalia macroptera.* COMMELINACEAE: *Commelina erecta, C. forskalaei.* COMPOSITAE: *Ageratum conyzoides, Eclipta alba, Galinsoga parviflora, Helichrysum mechowianum, Lactuca capensis, Microglossa pyrifolia, Sigesbeckia orientalis, Vernonia adoensis, V. amygdalina, V. guineensis.* CONVOLVULACEAE: *Ipomoea asarifolia, I. eriocarpa.* CRUCIFERAE: *Lepidium sativum.* CUCURBITACEAE: *Citrullus colocynthis, Cucumis figarei, C. prophetarum, Momordica balsamina, Trochomeria macrocarpa.* CYPERACEAE: *Cyperus maculatus, C. rotundus, Scleria pterotus.*

**E: Agri-horticulture; 4, bee/honey plants, insect plants**

ACANTHACEAE: *Barleria eranthemoides, Hypoestes verticillata, Justicia betonica, J. flava, Phaulopsis imbricata, Thomandersia hensii.* ANACARDIACEAE: *Anacardium occidentale, Lannea barteri, L. schimperi, Ozoroa insignis.* AIZOACEAE: *Glinus lotoides.* AQUIFOLIACEAE: *Ilex mitis.* ASCLEPIADACEAE: *Asclepias curassavica, Dregea abyssinica, Sarcostemma viminale.* BALSAMINACEAE: *Impatiens burtonii.* BIGNONIACEAE: *Crescentia cujete, Markhamia tomentosa, Stereospermum acuminatissimum, Tecoma stans.* BOMBACACEAE: *Ceiba pentandra.* CAPPARACEAE: *Cleome spp.* CELASTRACEAE: *Maytenus senegalensis.* CHRYSOBALANACEAE: *Parinari excelsa.* COMBRETACEAE: *Combretum fragrans, C. racemosum, C. smeathmannii, Guiera senegalensis, Terminalia glaucescens.* COMMELINACEAE: *Palisota hirsuta.* COMPOSITAE: *Bidens pilosa, Bothriocline schimperi, Vernonia amygdalina, V. colorata.* CONVOLVULACEAE: *Ipomoea spp.* CUCURBITACEAE: *Cucurbita pepo.* CYPERACEAE: *Cyperus esculentus.* DILLENIACEAE: *Tetracera affinis, T. potatoria.*

**E: Agri-horticulture; 5, land conservation, cover-plants, sand-binders, erosion prevention, pioneer species, fire-resistant, etc.**

ACANTHACEAE: *Justicia flava, Sclerochiton vogelii.* AGAVACEAE: *Agave sisalana.* AIZOACEAE: *Zaleya pentandra,* ALANGIACEAE: *Alangium chinense.* AMARANTHACEAE: *Alternanthera maritima, Gomphrena celosioides, Philoxerus vermicularis.* ANACARDIACEAE: *Anacardium occidentale.* ANNONACEAE: *Annona glabra, A. glauca, A. senegalensis, Cleistopholis patens, Monodora tenuifolia,* APOCYNACEAE: *Funtumia africana, F. elastica, Landolphia heudelotii, L. owariensis.* ASCLEPIADACEAE: *Leptadenia pyrotechnica.* BIGNONIACEAE: *Stereopermum kunthianum.* BIXACEAE: *Bixa orellana.* BOMBACACEAE: *Bombax costatum, Ceiba pentandra.* BUSERACEAE: *Commiphora dalzielii.* CAPPARACEAE: *Cleome viscosa, Maerua oblongifolia.* CARYOPHYLLACEAE: *Drymaria cordata.* CASUARINACEAE: *Casuarina equisitifolia.* CECROPIACEAE: *Musanga cecropioides.* CHENOPODIACEAE: *Suaeda monoica.* CHRYSOBALANACEAE: *Chrysobalanus icaco, Parinari curatellifolia.* COMBRETACEAE: *Combretum micranthum, Guiera senegalensis, Laguncularia racemosa, Pteleopsis suberosa, Terminalia avicennioides.* COMPOSITAE: *Galinsoga parviflora, Helichrysum glumaceum, Mikania cordata, Vernonia subuligera.* CONVOLVULACEAE: *Ipomoea asarifolia, I. eriocarpa, I. involucrata, I. pes-caprae, I. stolonifera, Jacquemontia ovalifolia, Merremia umbellata.* CYPERACEAE: *Afrotrilepis pilosa, Cyperus conglomeretus, C. maritimus, Fimbristylis ferruginea, F. triflora, Fuirena umbellata, Remirea maritima.*

**E: Agri-horticulture; 6, composting, green-manuring, fertilizer, etc.**
ACANTHACEAE: *Peristrophe bicalyculata.* ALISMATACEAE: *Sagittaria guayanensis.* AMARANTHACEAE: *Achyranthes aspera.* ANNONACEAE: *Monodora myristica* (seed-cake). ARACEAE: *Pistia stratiotes.* BOMBACACEAE: *Adansonia digitata* (bark-ash), *Ceiba pentandra* (seed-cake). CAPPARACEAE: *Cleome viscosa.* CASUARINACEAE: *Casuarina equisitifolia.* CECROPIACEAE: *Myrianthus arboreus* (leaf). COCHLOSPERMACEAE: *Cochlospermum religiosum* (seed-cake). COMPOSITAE: *Blumea aurita, Guizotia abyssinica* (seed-cake), *Struchium sparganophora.* CUCURBITACEAE: *Cucumis melo* (fruit-husk), *Luffa acutangula* (seed-cake), *L. cylindrica* (seed-cake). CYPERACEAE: *Fimbristylis dichotoma, F. schoenoides, Fuirena umbellata, Rhynchospora corymbosa.*

**E: Agri-horticulture; 7, indicators (soil, water)**
ASCLEPIADACEAE: *Calotropis procera.* BIGNONIACEAE: *Spathodea campanulata.* CHENOPODIACEAE: *Suaeda monoica.* COMMELINACEAE: *Aneilema beninensis, Polyspatha paniculata.* COMPOSITAE: *Epaltes gariepina.* CONVOLVULACEAE: *Cressa cretica, Momordica foetida.* DROSERACEAE: *Drosera indica.*

**E: Agri-horticulture; 8, indieators (weather, season, time)**
ACANTHACEAE: *Brillantaisia nitens, Justicia flava, Whitfieldia elongata.* AMARYLLIDACEAE: *Pancratium tenuifolium.* ANACARDIACEAE: *Pseudospondias microcarpa.* BIGNONIACEAE: *Spathodea campanulata.* BURSERACEAE: *Dacryodes edulis.* COMPOSITAE: *Vernonia auriculifera.*

**E: Agri-horticulture; 9, shade-trees, avenue-trees, nursery-plants**
ANACARDIACEAE: *Mangifera indica, Sclerocarya birrea.* ANNONACEAE: *Cananga odorata.* APOCYNACEAE: *Rauvolfia vomitoria, Thevetia neriifolia.* ARACEAE: *Xanthosoma mafaffa.* BIGNONIACEAE: *Kigelia africana.* BOMBACACEAE: *Ceiba pentandra.* BORAGINACEAE: *Cordia aurantiaca, C. millenii, C. platythyrsa, C. sinensis.* BURSERACEAE: *Commiphora kerstingii, Dacryodes edulis.* CASUARINACEAE: *Casuarina equisitifolia.* CECROPIACEAE: *Musanga cecropioides.* CHRYSOBALANACEAE: *Parinari excelsa.* COMBRETACEAE: *Anogeissus leiocarpus, Combretum micranthum, Terminalia catappa, T. ivorensis.* COMPOSITAE: *Vernonia amygdalina.*

**E: Agri-hortculture; 10, living hedges, markers**
ACANTHACEAE: *Barleria cristata, B. prionitis, Graptophyllum pictum, Ruspolia hypocrateriformis, Thunbergia erecta.* AGAVACEAE: *Agave sisalana, Furcraea foetida.* AMARYLLIDACEAE: *Hymenocallis littoralis, Scadoxus multiflorus.* ANACARDIACEAE: *Spondias mombin.* APOCYNACEAE: *Adenium obesum, Allamanda cathartica, A. schottii, Carissa edulis, Rauvolfia vomitoria, Thevetia neriifolia.* ARALIACEAE: *Polyscias guilfoylei, Schefflera barteri.* BALANITACEAE: *Balanites aegyptiaca.* BIGNONIACEAE: *Kigelia africana, Newbouldia laevis, Tecoma capensis, T. stans.* BIXACEAE: *Bixa orellana.* BOMBACACEAE: *Bombax costatum.* BURSERACEAE: *Boswellia dalzielii, Commiphora africana, C. dalzielii, C. kerstingii.* CACTACEAE: *Opuntia spp.* CAPPARACEAE: *Maerua angolensis.* CASUARINACEAE: *Casuarina equistifolia.* CHRYSOBALANACEAE: *Magnistipula conrauana, Parinari excelsa.* COMMELINACEAE: *Palisota hirsuta.* COMPOSITAE: *Guizotia abyssinica, Senecio mannii, Vernonia amygdalina, V. conferta, V. subuligera, V. theophrastifolia, V. thomsoniana.* CONNARACEAE: *Connarus africanus.* CONVOLVULACEAE: *Ipomoea cairica, I. involucrata, I. nil, Jacquemontia tamnifolia, Operculina macrocarpa.*

**E: Agri-horticulture; 11, fence-posts, yam-poles, bean-sticks etc.**
ANACARDIACEAE: *Spondias mombin.* ANNONACEAE: *Monodora tenuifolia, Uvaria ovata.* BIGNONIACEAE: *Kigelia africana, Newbouldia laevis.* BIXACEAE: *Bixa orellana.* CECROPIACEAE: *Musanga cecropioides, Myrianthus arboreus.* CELASTRACEAE: *Maytenus senegalensis.* CHRYSOBALANACEAE: *Chrysobalanus icaco.* COMBRETACEAE: *Guiera senegalensis.* COMPOSITAE: *Senecio mannii.*

**E: Agri-horticulture; 12, weeds, parasites**
ACANTHACEAE: *Hygrophila auriculata, Lankesteria brevior, Nelsonia canescens.* AMARANTHACEAE: *Alternanthera pungens, Amaranthus lividus, Gomphrena celosioides, Pandiaka heudelotii, Philoxerus vermicularis.* ARACEAE: *Pistia stratiotes, Typhonium trilobatum.* BALANOPHORACEAE: *Thonningia sanguinea.* CACTACEAE: *Opuntia spp.* CARYOPHYLLACEAE: *Drymaria cordata, Polycarpaea eriantha.* COMMELINACEAE: *Aneilema lanceolatum, A. pomeridianum, Commelina benghalensis, C. erecta, C. imberbis, C.*

*thomasii, Floscopa aquatica.* COMPOSITAE: *Acanthospermum hispidum, Aspilia africana, Blumea perrottetiana, Chromolaena odoratum, Conyza canadensis, Coreopsis barteri, Crassocephalum montuosum, C. sarcobasis, Eclipta alba, Eleutheranthera ruderalis, Ethulia conyzoides, Fleishmannia microstemon, Galinsoga parviflora, G. urticifolia, Gutenbergia rueppellii, Helichrysum foetidum, Laggera pterodonta, Melanthera scandens, Senecio ruwenzoriensis, Sonchus oleraceus, Synedrella nodiflora, Tridax procumbens, Vernonia perrottetii, Vicoa leptoclada.* CONVOLVULACEAE: *Convolvulus arvensis, Evolvulus nummularius, Ipomoea coptica.* CYPERACEAE: *Bulbostylis barbata, Cyperus bulbosus, C. compressus, C. difformis, C. digitatus, C. dilatatus, C. distans, C. esculentus, C. fenzelianus, C. iria, C. procerus, C. reduncus, C. rotundus, C. tenuiculmis, C. tuberosus, Fimbristylis dichotoma, F. littoralis, F. schoenoides, Fuirena ciliaris, F. stricta, F. umbellata, Kyllinga brevifolia, K. bulbosa, K. elatior, K. erecta, K. nemoralis, K. odorata, K. squamulata, Mariscus alternifolius, M. flabelliformis, M. squarrosus, Pycreus polystachyos, Rhynchospora corymbosa, R. holoshoenoides, Scirpus roylei.*

**E: Agri-horticulture; 13, biotically active**

ARACEAE: *Culcasia scandens.* ASCLEPIADACEAE: *Leptadenia hastata.* COMPOSITAE: *Ageratum conyzoides, Eclipta alba, Gynura procumbens, Laggera pterodonta, Mikania cordata, Vernonia nigritiana.* CYPERACEAE: *Rhynchospora corymbosa.*

**F: Products; 1, building materials**

ACANTHACEAE: *Thomandersia hensii.* ALANGIACEAE: *Alangium chinense.* ANACARDIACEAE: *Anacardium occidentale, Antrocaryon klaineanum, A. micraster, Mangifera indica, Pseudospondias microcarpa* (wood), *Spondias mombin* (pole), *Trichoscypha arborea* (wood), *T. longifolia* (wood), *T. yapoensis* (wood). ANNONACEAE: *Annona chrysophylla* (wood), *A. senegalensis* (stem), *Cleistopholis patens* (wood, bark), *C. staudtii* (bark), *Ennantia chlorantha* (wood, bark), *E. polycarpa* (wood), *Greenwayodendron oliveri* (stem), *G. suaveolens* (wood), *Hexalobus monopetalus* (pole), *Isolona thonneri* (wood), *Monodora crispata* (wood), *M. tenuifolia* (pole), *Pachypodanthium staudtii* (wood, bark), *Xylopia acutiflora* (bark), *X. aethiopica* (wood, bark), *X. parviflora* (wood), *X. quintasii* (wood), *X. rubescens* (bark), *X. staudtii* (wood, bark), *X. villosa* (wood). APOCYNACEAE: *Alstonia boonei* (wood), *Baissea multiflora* (stem), *Funtumia africana* (wood), *F. elastica, Hunteria umbellata* (wood), *Landolphia foretiana* (wood), *Rauvolfia caffra* (wood), *Voacanga thouarsii* (pole). AQUIFOLIACEAE: *Ilex mitis* (wood). ARAUCARIACEAE: *Araucaria heterophylla.* ASCLEPIADACEAE: *Calotropis procera* (stem), *Xysmalobium heudelotianum* (root). AVICENNIACEAE: *Avicennia germinans* (wood). BALANITACEAE: *Balanites wilsoniana* (wood). BIGNONIACEAE: *Cybistax donell-smithii, Kigelia africana, Markhamia lutea* (wood), *Newbouldia laevis.* BOMBACACEAE: *Adansonia digitata, Bombax buonopozense* (wood, bark), *B. costatum* (wood), *Ceiba pentandra, Rhodognaphalon brevicuspe* (wood). BORAGINACEAE: *Cordia africana* (wood), *C. platythyrsa* (bark), *C. millenii* (wood), *C. myxa* (leaf), *C. sinensis* (stem, bark). BURSERACEAE: *Aucoumea klaineana* (wood), *Canarium schweinfurthii* (wood), *Dacryodes klaineana* (wood). CAPPARACEAE: *Boscia senegalensis* (wood), *Buchholzia coriacea* (wood), *Capparis decidua* (wood) *Maerua angolensis* (wood). CASUARINACEAE: *Casuarina equisitifolia.* CECROPIACEAE: *Musanga cecropioides* (wood). CELASTRACEAE: *Elaeodendron buchananii* (wood), *Hippocratea africana* (stem), *Maytenus undata* (wood), *Mystroxylon aethiopicum* (wood). CHRYSOBALANACEAE: *Acioa scabrifolia* (wood), *Chrysobalanus icaco* (wood), *Magnistipula cupheiflora, Maranthes chrysophylla* (wood), *M. glabra* (wood), *M. polyandra* (wood), *M. robusta* (wood), *Neocarya macrophylla* (wood), *Parinari congensis* (wood), *P. curatellifolia* (wood), *P. excelsa* (wood). COMBRETACEAE: *Anogeissus leiocarpus* (wood), *Combretum fragrans* (wood), *C. glutinosum* (wood), *C. micranthum* (stem), *C. molle* (stem), *Conocarpus erectus* (wood), *Guiera senegalensis* (wood), *Strephonema pseudocola* (wood), *Terminalia spp.* (wood). COMPOSITAE: *Aspilia africana* (plant), *Senecio mannii* (stem), *Vernonia auriculifera* (stem). CONVOLVULACEAE: *Ipomoea aitonii* (leaf), CTENOLOPHONACEAE: *Ctenolophon englerianus* (wood). CYPERACEAE: *Afrotrilepis pilosa* (leaf), *Cladium mariscus* (leaf), *Cyperus dichroöstachys* (leaf), *C. exaltatus* (culm), *C. latifolius* (culm), *C. papyrus* (culm), *Mapania linderi* (leaf), *Mariscus alternifolius* (leaf), *M. longibracteatus* (culm), *Scirpus maritimus* (culm). DICHAPETALACEAE: *Tapura fischeri* (pole).

**F: Products; 2. carpentry and related applications**

ANACARDIACEAE: *Anacardium occidentala, Antrocaryon klaineanum, A. micraster, Lannea humilis, L. welwitschii, Ozoroa insignis.* ANNONACEAE: *Cleistopholis patens, Enantia chlorantha, Greenwayodendron spp. Monodora myristica, Pachypodanthium staudtii, Xylopia parviflora, X. quintasii.* APOCYNACEAE: *Alstonia boonei, Funtumia elastica, Picralima nitida, Pleiocarpa bicarpellata, P. mutica, P. pycnantha, Tabernaemontana crassa, T. longiflora.* AQUIFOLIACEAE: *Ilex mitis.* BALANITACEAE: *Balanites aegyptiaca.* BIGNONIACEAE:

*Jacaranda mimosifolia, Markhamia lutea, M. tomentosa.* BORAGINACEAE: *Cordia africana, C. millenii, C. myxa, C. platythyrsa.* BURSERACEAE: *Aucoumea klaineana, Canarium schweinfurthii.* CAPPARACEAE: *Boscia angustifolia, Crataeva adansonii, Maerua angolensis.* COMBRETACEAE: *Anogeissus leiocarpus, Strephonema pseudocola, Terminalia cheluba, T. ivorensis, T. macroptera, T. superba.*

**F: Products; 3, farming, forestry, hunting and fishing apparatus**

ALISMATACEAE: *Limnophyton angolense* (stem, petiole). AMARANTHACEAE: *Pupalia lappacea.* ANACARDIACEAE: *Lannea acida, L. velutina* (wood), *Sclerocarya birrea* (wood), *Spondias mombin* (wood, root). ANNONACEAE: *Annona chrysophylla* (wood), *A. glabra* (root, wood), *A. senegalensis* (stem), *Cleistopholis patens* (trunk), *Greenwayodendron suaveolens* (wood), *Monodora myristica* (seed), *M. tenuifolia* (wood), *Uvariastrum pierreanum* (wood), *Xylopia acutiflora* (wood), *X. aethiopica* (wood), *X. parviflora* (bark), *X. quintasii* (wood), *X. villosa* (wood). APOCYNACEAE: *Alstonia boonei* (latex), *Ancyclobotrys amoena* (stem), *Anthoclitandra robustior* (liane), *Aphanostylis mannii* (stem), *Baissea laxiflora* (liane), *Dictyophleba leonensis* (latex), *Funtumia africana* (latex), *Holarrhena floribunda* (wood), *Hunteria elliotii* (wood), *H. umbellata* (wood), *Landolphia hirsuta* (latex), *L. owariensis* (latex), *Oncinotis glabrata* (stem), *Picralima nitida* (wood), *Pleiocarpa mutica* (wood), *Rauvolfia caffra* (latex), *Saba floribunda* (latex), *Strophanthus preussii* (stem), *S. sarmentosus* (wood), *Tabernaemontana pachysiphon* (latex), *Vahadenia caillei* (latex), *Voacanga africana* (latex from fruit), *V. thouarsii* (fibre, latex). ARACEAE: *Cercestis afzelii* (stem), *Rhektophyllum mirabile* (aerial root). ARALIACEAE: *Cussonia arborea* (wood). ASCLEPIADACEAE: *Gomphocarpus fruticosus* (fibre), *Leptadenia pyrotechnica* (fibre), *Pergularia daemia* (fibre). AVICENNIACEAE: *Avicennia germinans* (wood). BALANITACEAE: *Balanites aegyptiaca* (wood, resin). BIGNONIACEAE: *Kigelia africana* (wood, fruit-pod), *Tecoma stans* (wood), BOMBACACEAE: *Adsonia digitata* (bark), *Bombax buonopozense* (wood, root-bark), *Ochroma pyramidale, Rhodognaphalon brevicuspe* (wood). BORAGINACEAE: *Cordia myxa* (fruit-pulp), *C. senegalensis* (wood), *C. sinensis* (stem, wood), *Ehretica cymosa* (wood, fruit). BROMELIACEAE: *Ananas comosus* (fibre). BURSERACEAE: *Aucoumea klaineana* (wood), *Canarium schweinfurthii* (resin, seed), *Commiphora africana* (wood), *C. kerstingii* (wood), *Dacryodes edulis.* CANNACEAE: *Canna indica* (flower, seed). CAPPARACEAE: *Capparis decidua* (wood), *Maerua crassifolia* (wood). CECROPIACEAE: *Musanga cecropioides* (wood). CELASTRACEAE: *Salacia senegalensis* (stem). CHRYSOBALANACEAE: *Parinari curatellifolia* (leaf). COCHLOSPERMACEAE: *Cochlospermum vitifolium* (wood). COMBRETACEAE: *Combretum aculeatum* (stem), *C. paniculatum* (wood), *Quisqualis indica* (stem), *Terminalia arjuna, T. mollis* (wood). COMMELINACEAE: *Palisota barteri.* CONNARACEAE: *Cnestis ferruginea* (stem), *Connarus africanus* (seed), *Jaundea pubescens* (wood, fruit), *Manotes expansa* (stem). CYPERACEAE: *Cyperus articulatus.* DICHAPETALACEAE: *Tapura fischeri* (wood).

**F: Products; 4, fuel, illuminant, tinder, smoke-maker, etc.**

ALANGIACEAE: *Alangium chinense.* AMARANTHACEAE: *Aerva javanica* (inflorescence). ANACARDIACEAE: *Anacardium occidentale, Lannea welwitschii* (wood), *Mangifera indica, Ozoroa insignis* (wood), *O. pulcherrima* (split stem), *Schinus molle* (wood). ANNONACEAE: *Annona senegalensis* (wood), *Xylopia aethiopica* (wood, bark). APOCYNACEAE: *Rauvolfia vomitoria* (wood). AQUIFOLIACEAE: *Ilex mitis* (wood). ARALIACEAE: *Polyscias fulva* (wood). ASCLEPIADACEAE: *Calotropis procera* (wood, pith, floss), *Leptademia hastata* (fruit), *L. pyrotechnica.* AVICENNIACEAE: *Avicennia germinans* (wood). BALANITACEAE: *Balanites aegyptiaca* (wood). BIGNONIACEAE: *Newbouldia laevis.* BOMBACACEAE: *Adansonia digitata* (wood, seed pod), *Ceiba pentandra* (floss). BORAGINACEAE: *Cordia africana* (wood), *C. myxa* (wood), *C. sinensis* (wood). BURSERACEAE: *Aucoumea klaineana* (bark), *Canarium schweinfurthii* (wood, resin), *C. zeylanicum* (resin), *Commiphora africana* (wood, resin), *C. dalzielii* (wood), *Dacryodes edulis* (bark), *D. klaineana* (wood). CANNABACEAE: *Cannabis sativa* (seed-oil). CAPPARACEAE: *Boscia senegalensis* (wood), *Maerua angolensis* (wood). CASUARINACEAE: *Casuarina equisitifolia* (wood). CECROPIACEAE: *Musanga cecropioides* (wood), *Myrianthus arboreus* (wood). CELASTRACEAE: *Maytenus undata* (wood), *Mystroxylon aethiopicum* (wood). CHENOPODIACEAE: *Chenolea lanata, Cornulaca monacantha, Haloxylon articulatum, Salsola baryosma, S. tetrandra, Traganum nudatum.* CHRYSOBALANACEAE: *Acioa barteri, A. lehmbachii, A. scabrifolia* (wood), *Chrysobalanus icaco* (wood, seed), *Maranthes glabra* (wood), *M. polyandra* (wood), *Neocarya macrophylla* (wood, endocarp hair), *Parinari curatellifolia* (wood, endocarp hair), *P. excelsa* (wood). COMBRETACEAE: *Anogeissus leiocarpus* (wood), *Combretum glutinosum* (wood), *C. micranthum* (wood), *C. nigricans* (wood), *Conocarpus erectus* (wood), *Guiera senegalensis* (root), *Laguncularia racemosa, Terminalia albida, T. avicennioides* (wood), *T. brownii* (wood), *T. glaucescens* (wood), *T. macroptera* (wood), *T. mollis* (wood), *T. superba* (wood). COMPOSITAE: *Conyza pyrrhopappa* (stem), *Echinops spinosus* (stem), *Microglossa pyrifolia* (stem), *Jaundea pinnata* (stem). CONVOLVULACEAE: *Ipomoea asarifolia* (stem). CRUCIFERAE: *Eruca sativa* (seed-oil), *Lepidium sativum* (seed-oil). CUCURBITACEAE: *Citrullus colocynthis* (fruit, seed-oil), *C. lanatus* (seed-oil), *Cucumeropsis mannii* (seed-oil), *Cucurbita pepo* (seed-oil). CYPERACEAE: *Cyperus esculentus* (leaf), *C. haspan* (pith), *C. papyrus* (rhizome), *Scleria depressa* (culm). DIPTEROCARPACEAE: *Monotes kerstingii* (wood).

**F: Products; 5, household, domestic and personal items**

ACANTHACEAE: *Whitfieldia elongata.* ALANGIACEAE: *Alangium chinense.* AMARANTHACEAE: *Aerva javanica* (root). AMARYLLIDACEAE: *Crinum jagus* (bulb), *C. zeylanicum.* ANACARDIACEAE: *Lannea acida, L. nigritana* (bark), *L. velutina* (wood), *L. welwitschii* (wood), *Sclerocarya birrea* (wood), *Sorindeia juglandifolia* (stem), *Trichoscypha cavalliensis* (wood). ANNONACEAE: *Cananga odorata* (wood), *Cleistopholis patens* (bark, seed), *Enantia chlorantha* (wood), *E. polycarpa* (wood), *Hexalobus monopetalus* (wood), *Monodora myristica* (wood, seed), *M. tenuifolia* (stick, pole, seed), *Xylopia quintasii* (wood), *X. rubescens* (wood), *X. staudtii* (wood), *X. villosa* (wood). APOCYNACEAE: *Alstonia boonei* (wood), *Funtumia africana* (wood), *F. elastica, Holarrhena floribunda* (wood), *Hunteria eburnea* (wood), *H. elliotii* (wood), *H. simsii* (wood), *H. umbellata* (wood), *Landolphia owariensis* (twig), *Motandra guineensis* (stem), *Picralima nitida* (wood), *Pleiocarpa mutica* (wood), *P. pycnantha* (wood), *Rauvolfia vomitoria* (seed), *Strophanthus hispidus* (stem), *S. sarmentosus* (stem), *Tabernaemontana crassa* (wood), *T. pachysiphon* (leaf, latex), *Thevetia neriifolia* (wood). AQUIFOLIACEAE: *Ilex mitis* (wood). ARALIACEAE: *Cussonia bancoensis* (wood). ASCLEPIADACEAE: *Gongronema latifolium* (stem). AVICENNIACEAE: *Avicennia germinans* (wood). BALANITACEAE: *Balanites aegyptiaca* (wood, seed). BIGNONIACEAE: *Kigelia africana* (wood, fruit-pod), *Newbouldia laevis* (wood), *Stereospermum kunthianum* (wood). BIXACEAE: *Bixa orellana* (dye). BOMBACACEAE: *Bombax buonopozense* (wood), *B. costatum* (wood), *Ceiba pentandra.* BORAGINACEAE: *Cordia millenii* (wood), *C. platythyrsa* (wood), *Ehretia cymosa* (wood, fruit), *E. trachyphylla* (wood), *Heliotropium indicum* (fibre). BURSERACEAE: *Aucoumea klaineana* (wood), *Canarium schweinfurthii* (seed), *C. zeylanicum* (wood), *Commiphora africana* (wood), *Dacryodes edulis, D. klaineana* (wood), *Santiria trimera* (wood). CANNACEAE: *Canna indica* (seed). CAPPARACEAE: *Capparis decidua* (wood), *Maerua crassifolia* (wood). CECROPIACEAE: *Musanga cecropioides* (wood), *Myrianthus libericus* (wood), *M. serratus* (wood). CELASTRACEAE: *Maytenus acuminata* (wood), *M. undata* (wood), *M. senegalensis* (wood), *Mystoxylon aethiopicum* (wood), *Salacia lehmbachii* (wood), *S. senegalensis* (stem). CHRYSOBALANACEAE: *Acioa scabrifolia* (wood), *Parinari curatellifolia* (wood). COCHLOSPERMACEAE: *Cochlospermum planchonii* (seed). COMBRETACEAE: *Anogeissus leiocarpus* (wood), *C. micranthus* (wood), *Guiera senegalensis* (wood), *Terminalia avicennioides* (root), *T. brownii* (wood), *T. glaucescens* (wood, root), *T. ivorensis* (wood), *T. superba* (wood). COMPOSITAE: *Ageratum conyzoides* (leaf), *Helichrysum odoratissimum* (plant). CONVOLVULACEAE: *Operculina macrocarpa* (seed). COSTACEAE: *Costus afer* (stem). CYPERACEAE: *Cyperus articulatus, C. papyrus* (culm, rhizome), *Kyllinga erecta* (rhizome), *Scleria boivinii* (seed), *S. depressa* (seed), *S. naumanniana* (seed). DICHAPETALACEAE: *Dichapetalum madagascariense* (stem). DILLENIACEAE: *Dillenia indica* (fruit-pulp).

**F: Products; 6, pastimes (carving, musical instruments, games, toys, etc.)**

ACANTHACEAE: *Phaulopsis imbricata* (fruit-capsule), *Thunbergia alata* (fruit). AMARANTHACEAE: *Pupalia lappacea.* ANACARDIACEAE: *Trichoscypha arborea* (wood), *T. yapoensis* (wood). ANNONACEAE: *Cananga odorata* (wood), *Cleistopholis patens* (wood), *Isolona thonneri* (wood). APOCYNACEAE: *Alstonia boonei* (wood), *Funtumia africana* (wood), *Holarrhena floribunda* (wood), *Picralima nitida* (wood). ARALIACEAE: *Cussonia arborea* (wood), *C. bancoensis* (wood), *Polyscias fulva* (wood). BALANITACEAE: *Balanites aegyptiaca* (wood, seed). BIGNONIACEAE: *Kigelia africana* (fruit-pod), *Markhamia tomentosa* (bud, wood), *Spathodea campanulata* (flower-bud, wood). BOMBACACEAE: *Bombax buonopozense* (wood, spine, fruit), *Ceiba pentandra* (wood). BORAGINACEAE: *Cordia africana* (wood), *C. millenii* (wood), *C. platythyrsa* (wood), *C. senegalensis* (wood). BURSERACEAE: *Canarium schweinfurthii* (seed), *Commiphora africana* (wood), *Santiria trimera* (wood). CANNACEAE: *Canna indica* (seed). CAPPARACEAE: *Boscia angustifolia* (wood), CARICACEAE: *Carica papaya* (leaf petiole). CECROPIACEAE: *Musanga cecropioides* (wood). CELASTRACEAE: *Hippocratea pallens* (stem). COMBRETACEAE: *Combretum molle* (wood), *Terminalia ivorensis* (wood). COMMELINACEAE: *Commelina diffusa* (leaf). COMPOSITAE: *Bidens pilosa* (seed), *Spilanthes filicaulis* (plant). CUCURBITACEAE: *Cucumeropsis mannii* (fruit-shell), *Cucurbita maxima* (fruit-shell), *C. pepo* (stem), *Ruthalicia eglandulosa* (fruit-shell), CYPERACEAE: *Cyperus exaltatus* (culm).

**F: Products; 7, containers, food-wrappers**
ARACEAE: *Culcasia parviflora* (leaf), *Cyrtosperma senegalense* (leaf). ASCLEPIADACEAE: *Calotropis procera* (wood). BIGNONIACEAE: *Crescentia cujete* (fruit-shell), *Newbouldia laevis* (leaf). BOMBACACEAE: *Adansonia digitata* (husk). BURSERACEAE: *Commiphora africana* (wood). CANNACEAE: *Canna indica* (leaf). CARICACEAE: *Carica papaya* (leaf). CECROPIACEAE: *Myrianthus serratus* (leaf). COMBRETACEAE: *Terminalia macroptera* (branch). CUCURBITACEAE: *Cucurbita maxima* (fruit-shell), *Lagenaria siceraria* (fruit-shell), *Telfairea occidentalis* (fruit-shell).

**F: Products; 8, abrasives, polishers, cleaners, sponges**
ACANTHACEAE: *Brillantaisia lamium* (leaf), *Hypoestes verticillata* (plant). ANNONACEAE: *Mischogyne elliotianum* (leaf). COMPOSITAE: *Vernonia macrocyanus* (leaf). CONVOLVULACEAE: *Ipomoea cairica* (stem). CUCURBITACEAE: *Lagenaria breviflora* (leaf), *Luffa acutangula* (fruit), *L. cylindrica* (fruit), *Momordica angustisepala* (stem), *M. balsamina* (leaf), *Telfairea occidentalis* (stem-fibre).

**F: Products; 9, chew-sticks, tooth-cleaners**
CHRYSOBALANACEAE: *Parinari curatellifolia* (twig). COMBRETACEAE: *Anogeissus leiocarpus, Guiera senegalensis* (root), *Terminalia avicennioides* (root), *T. glaucescens* (root). COMPOSITAE: *Melanthera scandens* (twig), *Vernonia adoensis* (root), *V. amygdalina* (wood, root), *V. colorata* (twig). CONNARACEAE: *Agelaea obliqua* (fruit), *A. pilosa* (wood), *A. trifolia* (wood), *Castanola paradoxa* (stem), *Cnestis ferruginea* (fruit-pulp).

**F: Products; 10, fibre for ties, cordage and bark-cloth**
ANACARDIACEAE: *Lannea acida, L. barteri, L. humilis, L. schimperi, Sclerocarya birrea.* AGAVACEAE: *Agave sisalana* (leaf), *Furcraea foetida, F. selloa* (leaf). ANNONACEAE: *Annona chrysophylla, A. senegalensis, Cananga odorata, Cleistopholis patens, Dennettia tripetala, Enantia chlorantha, Greenwayodendron oliveri, G. suaveolens, Hexalobus monopetalus, Polyceratocarpus parviflorus, Uvaria scabrida, Xylopia aethiopica, X. parviflora, X. quintasii, X. staudtii.* APOCYNACEAE: *Alafia barteri, Baissea multiflora, Chonemorpha macrophyllum, Pychobotrys nitida, Rauvolfia caffra, R. macrophylla, Strophanthus sarmentosus, Tabernaemontana pachysiphon, Voacanga africana, V. thouarsii.* ARACEAE: *Cercestis afzellii, Colocasia esculenta* (petiole), *Rhektophyllum mirabile* (aerial root). ASCLEPIADACEAE: *Asclepias curassavica* (stem), *Calotropis procera* (bark, fruit-pod), *Dregea abyssinica, Gomphocarpus fruticosus, G. physocarpus, Leptadenia hastata, L. pyrotechnica, Pergularia daemia, Secamone afzelii.* BALANITACEAE: *Balanites aegyptiaca.* BIXACEAE: *Bixa orellana.* BOMBACACEAE: *Adansonia digitata.* BORAGINACEAE: *Cordia myxa, C. senegalensis, C. sinensis, Heliotropium indicum.* BROMELIACEAE: *Ananas comosus* (leaf). CACTACEAE: *Opuntia spp.* CANNABACEAE: *Cannabis sativa.* CANNACEAE: *Canna indica,* CAPPARACEAE: *Cadaba farinosa.* CARICACEAE: *Carica papaya.* CECROPIACEAE: *Musanga cecropioides.* CELASTRIACEAE: *Hippocratea africana, Maytenus acuminata.* COCHLOSPERMACEAE: *Cochlospermum planchonii, C. religiosum, C. tinctorium, C. vitifolium.* CONVOLVULACEAE: *Ipomoea cairica, I. pes-caprae, Merremia aegyptia* (stem), *Neuropeltis acuminata* (stem), *N. velutina.* CUCURBITACEAE: *Sechium edule.* CYPERACEAE: *Cyperus esculentus* (culm) *C. papyrus* (culm), *C. procerus* (culm), *Rhynchospora corymbosa* (culm). DILLENIACEAE: *Tetracera potatoria* (stem). DIONCOPHYLLACEAE: *Habropetalum dawei* (stem), *Triphyophyllum peltatum* (stem).

**F: Products; 11, floss, stuffing and caulking**
AMARANTHACEAE: *Aerva javanica* (inflorescence). ANACARDIACEAE: *Lannea microcarpa.* ANNONACEAE: *Hexalobus monopetalus* (bark). APOCYNACEAE: *Funtumia africana* (seed-pod), *F. elastica* (seed-pod), *Holarrhena floribunda* (wood, fruit), *Pleioceras barteri* (seed-pod). ASCLEPIADACEAE: *Asclepias curassavica* (fruit-pod), *Calotropis procera* (bark, fruit-pod), *Gomphocarpus fruticosus* (fruit-pod), *G. physocarpus* (fruit-pod). BOMBACACEAE: *Bombax buonopozense* (fruit-pod), *B. costatum* (fruit-pod), *Ceiba pentandra* (fruit-pod), *Rhodognaphalon brevicuspe* (fruit-pod). COCHLOSPERMACEAE: *Cochlospermum religiosum* (fruit-pod), *C. tinctorium* (fruit-pod). COMPOSITAE: *Gnaphalium luteo-album.* CUCURBITACEAE: *Momordica foetida* (leaf). CYPERACEAE: *Cyperus articulatus, Kyllinga peruviana* (plant), *Scirpus litoralis* (culm), *S. pterolepis* (culm).

**F: Products; 12, withies (basketry), twigs (brooms), matting, etc.**
ANNONACEAE: *Uvaria ovata.* APOCYNACEAE: *Ancylobotrys amoena, A. scandens, Anthoclitandra robustior.* ARACEAE: *Cercestis afzelii.* ASCLEPIADACEAE: *Asclepias curassavica, Leptadenia pyrotechnica.* BIXACEAE: *Bixa orellana* (dye). BOMBACACEAE: *Adansonia digitata* (bark). CELASTRACEAE: *Hippocratea africana, H. apocynoides, H. indica,*

*H. myriantha, H. paniculata, H. rowlandii, H. velutina* (branch). COMBRETACEAE: *Combretum aculeatum, C. micranthum, C. paniculatum, Pteleopsis habeensis, Quisqualis indica.* COMPOSITAE: *Stomatanthes africanus, Vernonia amygdalina.* CONNARACEAE: *Agelaea trifolia.* CONVOLVULACEAE: *Calycobolus heudelotii, Ipomoea asarifolia.* CUCURBITACEAE: *Zehneria scabra.* CYPERACEAE: *Bulbostylis pilosa, Cyperus dives* (culm), *C. iria, C. papyrus* (culm), *C. procerus* (culm). *Eleocharis dulcis* (culm), *Fimbristylis ferruginea* (culm), *Marisus alternifolius* (culm), *M. ligularis* (culm), *M. longibracteatus* (culm), *Pycreus nitidus, Rhynchospora corymbosa* (culm), *Scirpus holoschoenus* (culm).

**F: Products; 13, pulp and paper**

ANNONACEAE: *Annona glabra* (wood), *Cleistopholis patens* (wood). ASCLEPIADACEAE: *Calotropis procera* (wood). BIGNONIACEAE: *Spathodea campanulata* (wood). BOMBACACEAE: *Adansonia digitata*(wood, bark), *Bombax buonopozense* (wood), *Ceiba pentandra.* CECROPIACEAE: *Musanga cecropioides* (wood). COMBRETACEAE: *Terminalia superba* (wood). COSTACEAE: *Costus afer* (stem). CUCURBITACEAE: *Luffa cylindrica* (fruit). CYPERACEAE: *Cladium mariscus, Cyperus digitatus* (culm), *C. papyrus* (culm), *Pycreus mundtii* (haulm).

**F: Products; 14, pottery**

ACANTHACEAE: *Blepharis linariifolia* (spiny bracts). ASCLEPIADACEAE: *Calotropis procera* (leaf). CUCURBITACEAE: *Telfairea occidentalis* (seed).

**F: Products; 15, beehives**

BORAGINACEAE: *Cordia africana* (wood). CARICACEAE: *Cyclomorpha solmsii* (stem). COMBRETACEAE: *Terminalia glaucescens* (bark), *T. macroptera* (branch).

**F: Products; 16, dyes, stains, inks, tattoos and mordants**

ACANTHACEAE: *Hypoestes verticillata* (root), *Whitfieldia elongata* (leaf). AMARANTHACEAE: *Amaranthus hybridus, Celosia trigyna, Gomphrena globosa* (plant, flower), *Pandiaka heudelotii.* ANACARDIACEAE: *Anacardium occidentale* (leaf), *Lannea barteri* (bark), *L. velutina* (bark), *L. welwitschii* (bark), *Mangifera indica* (bark), *Sclerocarya birrea* (bark), *Sorindeia warneckei* (sap), *Spondias mombin* (wood-ash), *Trichoscypha longifolia* (bark-gum). ANNONACEAE: *Annona chrysophylla* (bark), *A. reticulata* (leaf), *Enantia chlorantha* (bark), *E. polycarpa* (bark, wood), *Uvaria afzelii* (bark). APOCYNACEAE: *Ancylobotrys pyriformis, Ervatamia coronaria* (fruit-pulp), *Rauvolfia vomitoria* (bark), *Saba florida* (leaf, twig, flower), *Tabernaemontana florida* (wood-ash). ARACEAE: *Stilochiton lancifolius* (plant). ARALIACEAE: *Cussonia arborea* (wood). ASCLEPIADACEAE: *Solenostemon oleifolium* (leaf). AVICENNIACEAE: *Avicennia germinans* (bark). BALANITACEAE: *Balanites aegyptiaca* (charcoal, ash). BALSAMINACEAE: *Impatiens balsamina* (leaf, flower). BASELLACEAE: *Basella alba.* BIGNONIACEAE: *Jacaranda mimosifolia* (wood), *Kigelia africana* (fruit-pod), *Stereospermum kunthianum* (leaf). BIXACEAE: *Bixa orellana* (fruit-pulp, seed-coat). BOMBACACEAE: *Adansonia digitata* (root), *Bombax buonopozense* (bark), *Ceiba pentandra* (bark), *Rhodognaphalon brevicuspe* (bark). BORAGINACEAE: *Arnebia hispidissima* (root), *Cordia myxa* (kernel). BURSERACEAE: *Boswellia dalzielii* (bark), *Canarium schweinfurthii* (resin-soot). CACTACEAE: *Opuntia spp.* (fruit-sap). CANNACEAE: *Canna indica* (seed). CAPPARACEAE: *Crataeva adansonii* (plant), *Maerua crassifolia* (wood-ash). CECROPIACEAE: *Musanga cecropioides* (sap). CELASTRACEAE: *Mystroxylon aethiopicum* (bark). CHRYSOBALANACEAE: *Parinari curatellifolia* (leaf), *P. excelsa* (fruit-pulp, seed-coat). COCHLOSPERMACEAE: *Cochlospermum planchonii* (root), *C. tinctorium* (root). COMBRETACEAE: *Anogeissus leiocarpus* (gum), *Combretum fragrans* (leaf), *C. glutinosum* (bark, root), *C. molle* (leaf), *C. nigricans* (gum), *C. smeathmannii* (leaf-sap), *Laguncularia racemosa, Terminalia avicennioides* (root), *T. belerica* (fruit), *T. catappa* (wood, bark), *T. glaucescens* (root), *T. ivorensis* (wood, bark), *T. laxiflora* (root), *T. macroptera* (leaf, root-bark), *T. mantaly* (wood, bark), *T. mollis* (wood), *T. scutifera* (bark), *T. superba* (bark). COMMELINACEAE: *Commelina benghalensis* (flower-sap). COMPOSITAE: *Carthamnus tinctorius* (plant), *Centaurea perrottetii* (flower), *Chrysanthellum americanum* (leaf), *Conyza attenuata* (leaf), *Eclipta alba* (leaf), *Tagetes spp.* (flower), *Vernonia cinerascens* (stem), *V. perrottetii* (plant). CONNARACEAE: *Cnestis ferruginea* (bark), *Connarus africanus* (bark, seed). CONVOLVULACEAE: *Ipomoea asarifolia* (plant), *I. batatas* (leaf), *I. nil* (flower), *I. pes-caprae* (plant), *Citrullus colocynthis* (seed-oil). DILLENIACEAE: *Dillenia indica* (bark).

**F: Products; 17, tobacco, snuff**

ANACARDIACEAE: *Spondias mombin* (wood-ash). ANNONACEAE: *Annona senegalensis* (wood, wood-ash), *Neostenanthera gabonensis* (leaf), *Xylopia aethiopica* (fruit).

APOCYNACEAE: *Thevetia neriifolia* (leaf). ASCLEPIADACEAE: *Calotropis procera* (leaf). BIGNONIACEAE: *Newbouldia laevis* (bark). BOMBACACEAE: *Adansonia digitata* (fruit), *Ceiba pentandra* (pod-ash). CANNABACEAE: *Cannabis sativa* (leaf). CAPPARACEAE: *Buchholzia coriacea* (bark). CHENOPODIACEAE: *Haloxylon articulatum* (ash). COMMELINACEAE: *Commelina forskalaei* (stem). COMPOSITAE: *Centipeda minima* (leaf, seed), *Laggera alata* (leaf), *L. oloptera* (leaf), *Vernonia galamensis* (leaf).

**F: Products; 18, exudations (gum, resin, etc.)**

ANACARDIACEAE: *Anacardium occidentale* (bark), *Antrocaryon klaineanum* (bark), *A. micraster, Haematostaphis barteri* (bark), *Lannea acida* (bark), *L. barteri* (bark), *L. fruticosa* (bark), *L. microcarpa* (bark), *L. schimperi* (bark), *L. welwitschii* (wood, bark), *Mangifera indica* (bark), *Ozoroa insignis* (sap), *Pseudospondias microcarpa* (bark), *Schinus molle* (leaf, trunk) *S. terebinthifolius* (bark), *Sclerocarya birrea* (bark), *Sorindeia grandifolia* (bark), *Spondias cythera* (bark), *S. mombin* (bark), *Trichoscypha arborea* (bark), *T. longifolia* (bark, fruit), *T. preussii* (bark), *T. yapoensis* (bark). ANNONACEAE: *Cleistopholis patens* (bark), *Uvaria chamae* (root-bark). APOCYNACEAE: *Alafia barteri, A. multiflora, Alstonia boonei, Ancylobotrys amoena, A. scandens* (stem), *Anthoclitandra nitida, A. robustior, Aphanostylis leptantha, A. mannii, Baissea multiflora, Callichilia subsessilis, Carissa edulis* (bark), *Chonemorpha macrophyllum* (stem), *Clitandra cymulosa, Cylindropsis parviflora, Dictyophleba leonensis, D. lucida, D. ochracea, D. rudens, Farquharia elliptica, Funtumia africana* (bark), *F. elastica* (bark), *Hedranthera barteri, Holarrhena floribunda* (bark), *Hunteria eburnea* (fruit), *H. elliotii* (bark), *Landolphia calabarica, L. congolensis, L. dulcis, L. foretiana, L. heudelotii, L. hirsuta, L. landolphioides, L. macrantha, L. membranacea, L. micrantha, L. owariensis, L. parvifolia, L. subrepanda, L. togolana, L. uniflora, L. utilis, L. violacea, Mascarenhasia arborescens, Motandra guineensis, Oncinotis gracilis, Orthopichonia schweinfurthii, O. staudtii, Pleiocarpa talbotii, Plumeria spp., Rauvolfia caffra, R. macrophylla* (bark), *R. vomitoria* (young twig), *Saba florida, S. senegalensis, S. thompsonii, Tabernaemontana chippi, T. crassa* (bark), *T. eglandulosa, T. pachysiphon, T. ventricosa, Thevetia neriifolia* (all parts), *Vahadenia caillei, V. laurentii, Voacanga africana* (all parts), *V. bracteata, V. thouarsii.* ARALIACEAE: *Cussonia arborea* (bark). ASCLEPIADACEAE: *Brachystelma bingeri* (tuber), *Dalzielia oblanceolata, Gongronema angolense* (whole plant), *G. latifolium* (bark), *Kanahia lanifolia, Pachycarpus lineolata* (root), *Sarcostemma viminale, Secamone afzelii* (whole plant), *Toxocarpus brevipes, Xysmalobium heudelotianum* (root). BALANITACEAE: *Balanites aegyptiaca* (bark). BOMBACACEAE: *Bombax buonopozense* (bark). BORAGINACEAE: *Cordia myxa* (leaf), *C. senegalensis* (bark), *C. sinensis* (bark). BURSERACEAE: *Aucoumea klaineana* (wood, bark), *Boswellia dalzielii* (bark), *B. odorata* (bark), *Canarium schweinfurthii* (bark), *C. zeylanicum* (bark), *Commiphora africana* (bark), *C. dalzielii* (plant), *C. pedunculata* (bark), *Dacryodes edulis* (bark), *D. klaineana* (bark), *Santiria trimera* (bark). CACTACEAE: *Opuntia spp.* (stem). CELASTRACEAE: *Elaeodendron buchanii, Hippocratea welwitschii* (root), *Salacia chlorantha* (petiole), *S. lateritia* (petiole), *S. oliveriana* (petiole), *S. pyriformis* (petiole), *S. senegalensis* (petiole, seed), *S. staudtiana* (stem, root), *S. stuhlmaniana* (leaf, fruit-pulp, seed-coat), *S. togoica* (stem, root). CHRYSOBALANACEAE: *Maranthes glabra* (bark), *M. robusta* (bark), *Neocarya macrophylla* (fruit-endocarp). COCHLOSPERMACEAE: *Cochlospermum religiosum.* COMBRETACEAE: *Anogeissus leiocarpus* (leaf, bark, root), *Combretum fragrans* (bark), *C. glutinosum* (bark), *C. micranthum* (bark), *C. molle* (bark), *C. nigricans* (bark), *C. rhodanthum* (wood), *C. sericeum, Guiera senegalensis* (bark), *Strephonema pseudocola* (bark), *Terminalia belerica* (bark), *T. brownii* (bark), *T. catappa, T. laxiflora, T. macroptera.* COMPOSITAE: *Conyza subscaposa* (root), *Sonchus asper, Vernonia nestor.* DICHAPETALACEAE: *Dichapetalum madagascariense* (bark).

**F: Products; 19, manna and other exudations**

ASCLEPIADACEAE: *Calotropis procera.* CELASTRACEAE: *Maytenus undata.* CHENOPODIACEAE: *Salsola baryosma.*

**G: Social; 1, religion, superstition, magic**

ACANTHACEAE: *Acanthus montanus, Asystasia gangetica, Brillantaisia patula, Elytraria marginata, Eremomastax speciosa* (leaf), *Justicia flava, Thomandersia hensii.* AMARANTHACEAE: *Aerva lanata, Amarathus spinosus, Celosia argentea, C. trigyna Gomphrena globosa, Nothosaerva brachiata, Philoxerus vermicularis, Pupalia lappacea.* AMARYLLIDACEAE: *Crinum spp., Pancratium trianthum, Scadoxus cinnabarinus, S. multiflorus.* ANACARDIACEAE: *Lannea welwitschii, Pseudospondias microcarpa* (bark), *Spondias mombin, Trichoscypha patens* (bark). ANNONACEAE: *Annona arenaria, A. glauca* (root), *A. senegalensis, Annonidium mannii, Celeistopholis patens, Isolona cooperi* (bark), *Monodora myristica* (seed). APOCYNACEAE: *Adenium obesum, Alstonia boonei, Ancylobotrys amoena* (leaf), *Baissea multiflora* (leaf), *Hedranthera barteri, Holarrhena floribunda, Picralima nitida, Pleioceras barteri, Rauvolfia serpentina, R. vomitoria, Saba senegalensis, Strophanthus gratus, S. hispidus, S. sarmentosus* (leaf-sap), *Voacanga africana, V. sp. indet.* AQUIFOLIACEAE: *Ilex mitis.* ARACEAE: *Amorphophallus dracontioides* (root), *Anchomanes difformis* (tuber, spadix, berry), *Colocasia esculenta, Culcasia sp. indet, Cyrtosperma senegalense, Rhektophyllum mirabile.* ARALIACEAE: *Cussonia arborea.* ARISTOLOCHIACEAE: *Aristolochia albida* (flower), *A. bracteolata* (flower), *Paristolochia goldieana.* ASCLEPIADACEAE: *Calotropis procera, Caralluma dalzielii, C. decaisneana, C. russelliana, Gongronema latifolium* (leaf, stem), *Leptadenia hastata, Oxystelma bornouense, Pergularia daemia, Secamone afzelii, Tylophora glauca, T. sylvatica.* BALANITACEAE: *Balanites aegyptiaca, B. wilsoniana.* BALANOPHORACEAE: *Thonningia sanguinea.* BALSAMINACEAE: *Impatiens irvingii.* BIGNONIACEAE: *Kigelia africana, Markamia tomentosa, Newbouldia laevis, Spathodea campanulata, Stereospermum kunthianum.* BIXACEAE: *Bixa orellana.* BOMBACACEAE: *Adansonia digitata, Bombax buonopozense, Ceiba pentandra.* BORAGINACEAE: *Cordia millenii, C. myxa, Heliotropium indicum.* BROMELIACEAE: *Ananas comosus.* BURSERACEAE: *Boswellia dalzielii, Canarium schweinfurthii, Commiphora africana, C. kerstingii.* CACTACEAE: *Nopalea* (*Opuntia*) *spp.* CANNABACEAE: *Cannabis sativa.* CAPPARACEAE: *Capparis tomentosa, Maerua angolensis.* CANNACEAE: *Canna indica.* CARICACEAE: *Carica papaya.* CARYOPHYLLACEAE: *Polycarpaea corymbosa, P. linearifolia.* CECROPIACEAE: *Musanga cecropioides.* CELASTRACEAE: *Hippocratea velutina.* CHRYSOBALANACEAE: *Neocarya macrophylla, Parinari curatellifolia.* COCHLOSPERMACEAE: *Cochlospermum planchonii.* COMBRETACEAE: *Anogeissus leiocarpus, Combretum collinum, C. constrictum, C. glutinosum, C. grandiflorum, C. hispidum* (leaf), *C. lecardii, C. molle, C. paniculatum, C. platypterum, C. racemosum, C. sericeum* (fruit), *Guiera senegalensis, Pteleopsis suberosa, Terminalia spp.* COMMELINACEAE: *Commelina africana, C. bracteosa, C. erecta, C. imberbis, Cyanotis caespitosa, C. lanata, Floscopa africana, Palisota hirsuta, Pollia condensata.* COMPOSITAE: *Ageratum conyzoides, Aspilia africana, A. kotschyi, Bidens pilosa, Centaurea senegalensis, Crassocephalum crepidioides, Echinops longifolius, Eclipta alba, Emilia coccinea, E. sonchifolia, Fleishmannia microstemon, Geigeria alata, Laggera alata, Melanthera scandens, Microglossa pyrifolia, Mikania cordata, Senecio abyssinicus, S. biafrae, Synedrella nodiflora, Vernonia colorata, V. galamensis, Vicoa leptoclada.* CONNARACEAE: *Agelaea trifolia, Cnestis corniculata, C. ferruginea, Santaloides afzelii.* CONVOLVULACEAE: *Cuscuta australis, Evolvulus alsinoides, Ipomoea aitonii* (root), *I. argentaurata, I. asarifolia, I. nil, I. quamoclit, Merremia hederacea, M. tridentata, Operculina macrocarpa.* COSTACEAE: *Costus afer, C. lucanusianus.* CRASSULACEAE: *Bryophyllum pinnatum, Kalanchoe crenata/integra.* CUCURBITACEAE: *Citrullus lanatus, Coccinea barteri, Cucumeropsis mannii, Cucumis figarei, Momordica cissoides, Mukia maderaspatana, Ruthalicia eglandulosa.* CYPERACEAE: *Afrotrilepis pilosa, Cyperus alternifolius, C. articulatus, C. latifolius, C. papyrus, Hypolytrum poecilolepis, Lipocarpa chinensis, Scleria boivinii.* DILLENIACEAE: *Tetracera potatoria.* DIOSCOREACEAE: *Dioscorea dumetorum, D. praehensilis, D. rotunda, D. sansibarensis.*

**G: Social; 2, ceremonial**

BALANITACEAE: *Balanites aegyptiaca.* BIGNONIACEAE: *Kigelia africana.* BOMBACACEAE: *Adansonia digitata, Ceiba pentandra.* BORAGINACEAE: *Cordia aurantiaca.* BROMELIACEAE: *Ananas comosus.* BURSERACEAE: *Boswellia dalzielii.* CELASTRACEAE: *Hippocratea africana.* COMBRETACEAE: *Combretum molle* (bark), *Guiera senegalensis* (leaf, fruit). COMMELINACEAE: *Commelina diffusa.* COSTACEAE: *Costus lucanusianus.* CUCURBITACEAE: *Momordia balsamina.*

**G: Social; 3, sayings, aphorisms**

ACANTHACEAE: *Brillantaisia nitens.* AMARANTHACEAE: *Amaranthus hybridus, Celosia argentea, Cyathula cylindrica.* ANNONACEAE: *Cleistopholis patens, Monodora myristica* (seed). BIGNONIACEAE: *Spathodea campanulata.* BOMBACACEAE: *Adansonia digitata.* BURSERACEAE: *Commiphora kerstingii.* CECROPIACEAE: *Myrianthus arboreus.* COMPOSITAE: *Bidens pilosa, Launaea taraxacifolia, Struchium sparganophora.* CONVOLVULACEAE: *Cuscuta australis.* CUCURBITACEAE: *Luffa cylindrica.* CYPERACEAE: *Cyperus esculentus, C. papyrus.*

# INDEX OF WEST AFRICAN PLANT NAMES

a dyeng SERER CUCURBITACEAE *Cucurbita maxima, C. pepo*

a tomdofan SERER COMPOSITAE *Vicoa leptoclada*

a'a kai ka fito HAUSA CAPPARACEAE *Cleome monophylla*

aadɔntʃo ADANGME ANACARDIACEAE *Spondias mombin*

aadɔŋtʃo ADANGME ANACARDIACEAE *Lannea nigritana*

aagbá YORUBA APOCYNACEAE *Alafia multiflora*

ààlàoko YORUBA DIOSCOREACEAE *Dioscorea rotundata*

aanyɛle GA ANNONACEAE *Uvaria ovata*

ààrín YORUBA APOCYNACEAE *Hunteria umbellata*

áàyélèbí GA CAPPARACEAE *Ritchiea reflexa*

aba abasanité ABURE CONVOLVULACEAE *Ipomoea mauritiana*

abable GBE - VHE BROMELIACEAE *Ananas comosus*

abadan awurebe YORUBA AMARANTHACEAE *Pandiaka involucrata*

abago TAMACHEK CAPPARACEAE *Cadaba farinosa*

àbàjẹ YORUBA DIOSCOREACEAE *Dioscorea rotundata*

abakaliki atara IGBO COMBRETACEAE *Anogeissus leiocarpus*

abakwa ENGENNI APOCYNACEAE *Funtumia elastica*

abalapuli NZEMA ANACARDIACEAE *Lannea welwitschii*

àbámodá YORUBA CRASSULACEAE *Bryophyllum pinnatum*

àbámọdá YORUBA CRASSULACEAE *Bryophyllum pinnatum*

àbànà ófịá IGBO DIOSCOREACEAE *Dioscorea bulbifera*

àbànà ófịá baba, à.ó.gbara, à.ọ́.mbaba IGBO DIOSCOREACEAE *Dioscorea bulbifera*

abanaadẹne IGBO COMPOSITAE *Blumea aurita*

abanane IGBO DIOSCOREACEAE *Dioscorea misc. & spp. unknown*

abanase-abanase AKAN - TWI AMARANTHACEAE *Alternanthera nodiflora*

abanẹke IGBO DIOSCOREACEAE *Dioscorea misc. & spp. unknown*

abantolí GA APOCYNACEAE *Landolphia*

aba-nua AKAN - ASANTE APOCYNACEAE *Callichilia subsessilis* TWI *C. subsessilis*

abanzi IGBO COCHLOSPERMACEAE *Cochlospermum tinctorium*

àbá-ọ́jị IGBO APOCYNACEAE *Funtumia elastica*

àbàràbá KANURI BROMELIACEAE *Ananas comosus*

àbàrbaá HAUSA BROMELIACEAE *Ananas comosus*

abazi AVIKAM CAPPARACEAE *Buchholzia coriacea*

abbe KOOSI BIGNONIACEAE *Markhamia tomentosa*

ab-blugar HAUSA BOMBACACEAE *Bombax costatum*

abdugar rimi HAUSA BOMBACACEAE *Ceiba pentandra*

àbẹ́ URHOBO DIOSCOREACEAE *Dioscorea rotundata*

abea ABURE DICHAPETALACEAE *Dichapetalum pallidum*

abɛblɛ NZEMA CHRYSOBALANACEAE *Chrysobalanus icaco*

abɛkammo AKAN - ASANTE ASCLEPIADACEAE *Tylophora conspicua*

abelbal TAMACHEK CHENOPODIACEAE *Traganum nudatum*

abéné mulo AKAN - ASANTE AMARANTHACEAE *Alternanthera pungens*

abéné muro AKAN - ASANTE AMARANTHACEAE *Alternanthera pungens*

abeneburo AKAN - ASANTE AMARANTHACEAE *Alternanthera pungens*

abengogo ABURE ASCLEPIADACEAE *Secamone afzelii*

abeni SENUFO - TAGWANA BIGNONIACEAE *Spathodea campanulata*

abepopo KYAMA APOCYNACEAE *Strophanthus preussii*

abere YORUBA APOCYNACEAE *Picralima nitida* CHRYSOBALANACEAE *Maranthes polyandra*

abẹ́rẹ YORUBA COMPOSITAE *Bidens pilosa*

abẹ́rẹ olóko YORUBA COMPOSITAE *Bidens pilosa*

abereka mwọ IGBO CANNACEAE *Canna indica*

aberewa nyansiŋ AKAN - TWI ANACARDIACEAE *Lannea welwitschii*

aberewa sekan AKAN - ASANTE CYPERACEAE *Cyperaceae, Rhynchospora corymbosa*

abéròdéfẹ YORUBA ACANTHACEAE *Blepharis maderaspatensis*

aberure KYAMA APOCYNACEAE *Strophanthus sarmentosus*

abesh ARABIC COMBRETACEAE *Guiera senegalensis*

abẹsundigbaro YORUBA CONVOLVULACEAE *Stictocardia beraviensis*

abgesi GBE - VHE CUCURBITACEAE *Cucumeropsis mannii*

abi JUKUN AMARYLLIDACEAE *Crinum Gen.*

àbị, jí àbí IGBO DIOSCOREACEAE *Dioscorea rotundata*

abia AKAN - AKUAPEM CONVOLVULACEAE *Operculina macrocarpa* TWI *O. macrocarpa*

abigwa ENGENNI BURSERACEAE *Canarium schweinfurthii*

abikolo YORUBA COMPOSITAE *Eclipta alba*

abila ARABIC ASCLEPIADACEAE *Caralluma decaisneana, C. mouretii*

abilẹrẹ̀ YORUBA COMPOSITAE *Chrysanthellum indicum var. afroamericanum*

abilokun YORUBA ACANTHACEAE *Lankesteria elegans*

abilu TEM BIGNONIACEAE *Kigelia africana*

abin AKAN - FANTE COMBRETACEAE *Laguncularia racemosa*

àbìrè IBIBIO DIOSCOREACEAE *Dioscorea alata*

abiribɛ SISAALA BROMELIACEAE *Ananas comosus*

abirimuro AKAN - TWI AMARANTHACEAE *Alternanthera pungens*

abissawa AKAN - BRONG COMPOSITAE *Bidens pilosa*

abisuru YORUBA ACANTHACEAE *Lankesteria elegans*

## INDEX

abléblé **BAULE** BROMELIACEAE *Ananas comosus*

ablɛfotã **ADANGME - KROBO** BALANOPHORACEAE *Thonningia sanguinea*

ablèlè **BAULE** BROMELIACEAE *Ananas comosus*

ablɛmmɛ **GUANG - CHIRIPON** BROMELIACEAE *Ananas comosus*

ablɛndi **GBE - VHE** BROMELIACEAE *Ananas comosus*

ablialé **ANYI** BROMELIACEAE *Ananas comosus*

abliblé **AKAN - ASANTE** BROMELIACEAE *Ananas comosus*

abne **GBE - VHE** APOCYNACEAE *Landolphia owariensis*

abo **ABURE** APOCYNACEAE *Landolphia hirsuta* **AKYE** CAPPARACEAE *Buchholzia coriacea*

abo(-h), aboli **MANDING - BAMBARA** APOCYNACEAE *Saba florida*

àbo **YORUBA** ANNONACEAE *Annona senegalensis*

abo idòfún **YORUBA** CHRYSOBALANACEAE *Parinari curatellifolia*

abo làbẹ̀làbẹ̀ **YORUBA** CYPERACEAE *Fuirena umbellata*

abo làkọ̀ṣẹ **YORUBA** ANNONACEAE *Monodora myristica*

abo yúnyún **YORUBA** COMPOSITAE *Aspilia helianthoides, Melanthera scandens*

aboa ngateɛ **ANYI - AOWIN** CUCURBITACEAE *Lagenaria breviflora*

aboa-ngateɛ **ANYI - SEHWI** CUCURBITACEAE *Lagenaria breviflora*

aboa-ngatseɛ **NZEMA** CUCURBITACEAE *Lagenaria breviflora*

a-bɔ-anyɛ **GUANG - ANUM** BIGNONIACEAE *Newbouldia laevis*

abobo libi **ABURE** APOCYNACEAE *Pleioceras barteri*

aboboe **YORUBA - IFE OF TOGO** BIGNONIACEAE *Newbouldia laevis*

aboboma **AKAN - ASANTE** ANNONACEAE *Annona arenaria* **BRONG** *A. senegalensis* **TWI** *A. arenaria, A. senegalensis*

abododo **YORUBA** COMBRETACEAE *Conocarpus erectus*

abodòdo **YORUBA** APOCYNACEAE *Tabernaemontana pachysiphon*

abɔdɔitʃo **GA** APOCYNACEAE *Thevetia neriifolia*

àbɔdɔitso **GA** APOCYNACEAE *Thevetia neriifolia*

abŏeka **GBE - VHE** APOCYNACEAE *Landolphia owariensis*

abɔeka **ADANGME - KROBO** APOCYNACEAE *Landolphia* **GBE - VHE** *Landolphia owariensis*

aboh **AKPET - EHOM** CUCURBITACEAE *Lagenaria siceraria*

aboïla **ARABIC** ASCLEPIADACEAE *Caralluma decaisneana, C. mouretii*

abok **KOOSI** CUCURBITACEAE *Cucurbita maxima, C. pepo*

abokodidua **AKAN - TWI** CONNARACEAE *Castanola paradoxa*

abokoponko **AKAN - ASANTE** CONNARACEAE *Agelaea mildbraedii*

abo-lakoshin **YORUBA** ANNONACEAE *Monodora tenuifolia*

abɔma **AKAN - BRONG** ANNONACEAE *Annona senegalensis*

abomé **ANYI** CECROPIACEAE *Musanga cecropioides*

abonta **AKAN - AKUAPEM** APOCYNACEAE *Ancylobothrys scandens*

abontera **AKAN - AKUAPEM** APOCYNACEAE *Landolphia calabarica*

abɔntɔle **ANYI - AOWIN** APOCYNACEAE *Landolphia*

abɔntɔre **AKAN - TWI** APOCYNACEAE *Landolphia, Landolphia owariensis* CELASTRACEAE *Salacia stuhlmanniana, S. togoica*

aboro **YORUBA** AMARANTHACEAE *Achyranthes aspera*

aborobɛ **AKAN - TWI** BROMELIACEAE *Ananas comosus*

aborobɛ-fitaa **AKAN - TWI** BROMELIACEAE *Ananas comosus*

aborobɛ-fufuo **AKAN - TWI** BROMELIACEAE *Ananas comosus*

aborobɛ-hemma **AKAN - ASANTE** BROMELIACEAE *Ananas comosus*

aborobɛ-kɔkɔ **AKAN - TWI** BROMELIACEAE *Ananas comosus*

aborobɛ-toro **AKAN - ASANTE** BROMELIACEAE *Ananas comosus*

aborode-aborode **AKAN - ASANTE** ASCLEPIADACEAE *Gongronema latifolium*

àbórrà **EDO** AMARANTHACEAE *Celosia argentea*

abotokuradua **AKAN - ASANTE** ANNONACEAE *Monodora myristica, M. tenuifolia, Uvaria afzelii* **TWI** *Monodora tenuifolia* **WASA** *M. brevipes*

abovi **BAULE** COMPOSITAE *Vernonia colorata*

abovro **GBE - VHE** BORAGINACEAE *Ehretia cymosa*

aboya **AKAN - ASANTE** APOCYNACEAE *Picralima nitida*

aboyu **BAULE** COMPOSITAE *Vernonia colorata*

abrabesi **AKAN - ASANTE** CHRYSOBALANACEAE *Maranthes polyandra*

abrafodɔ **ANYI - AOWIN** APOCYNACEAE *Pleioceras barteri*

abrɛbɛrɛ **NZEMA** CHRYSOBALANACEAE *Chrysobalanus icaco*

abrèlè **BAULE** BROMELIACEAE *Ananas comosus*

abrɔbɛ, aburɔbɛ **AKAN - ASANTE** BROMELIACEAE *Ananas comosus* **TWI** *A. comosus* **WASA** *A. comosus*

abrɔdwemaa **AKAN - TWI** CONVOLVULACEAE *Ipomoea batatas*

abrɔdwobãa **AKAN - AKYEM** CONVOLVULACEAE *Ipomoea batatas* **ASANTE** *I. batatas*

abrɔfontunkum **AKAN - AKYEM** ANNONACEAE *Annona muricata*

abu **KABRE** ARALIACEAE *Cussonia arborea*

abu konbengo **KYAMA** ASCLEPIADACEAE *Pergularia daemia*

abuaïla **ARABIC** ASCLEPIADACEAE *Caralluma decaisneana, C. mouretii*

abubu adenkum **AKAN - AKYEM** ARALIACEAE *Cussonia bancoensis*

abué **BAULE** COMPOSITAE *Vernonia colorata*

abui abui **'BAMILEKE'** COMPOSITAE *Triplotaxis stellulifera*

abuɔbɛ, aburɔbɛ **AKAN - FANTE** BROMELIACEAE *Ananas comosus*

àbụ̀rọ̀ **IJO EAST** CUCURBITACEAE *Lagenaria siceraria*

aburobia **AKAN - ASANTE** CANNACEAE *Canna indica* **TWI** *C. indica*

abutusimu **VULGAR** COMBRETACEAE *Strephonema pseudocola*

abwẹ **IGBO** CUCURBITACEAE *Lagenaria siceraria*

abwoluku **IGBO** ACANTHACEAE *Brillantaisia lamium, B. nitens*

achaboa **ANYI - SEHWI** BORAGINACEAE *Cordia platythyrsa*

achakan **ARABIC** ASCLEPIADACEAE *Glossonema boveanum*

| | |
|---|---|
| aduru | HAUSA DIOSCOREACEAE *Dioscorea misc. & spp. unknown* |
| àdùrúkù | HAUSA BIGNONIACEAE *Newbouldia laevis* |
| adutsyiɔʃfoti | GBE - VHE COMMELINACEAE *Palisota hirsuta* |
| aduwa | HAUSA BALANITACEAE *Balanites aegyptiaca* |
| adwea | VULGAR BURSERACEAE *Dacryodes klaineana* AKAN - ASANTE *D. klaineana* TWI *D. klaineana* WASA *D. klaineana* |
| adweaba | AKAN - FANTE BURSERACEAE *Dacryodes klaineana* |
| adwera-akoa | AKAN - TWI AIZOACEAE *Trianthema portulacastrum, Zaleya pentandra* |
| adwokuma | NZEMA APOCYNACEAE *Strophanthus sarmentosus* |
| adwowakuro | AKAN - AKYEM COMPOSITAE *Ageratum conyzoides* |
| adzigo, adzugo | GBE - VHE DIOSCOREACEAE *Dioscorea alata* |
| adzilɛ ananse | NZEMA BALANOPHORACEAE *Thonningia sanguinea* |
| adzrɔkpi | GBE - VHE COMPOSITAE *Bidens pilosa, Melanthera elliptica* |
| adzugo bolobolo | GBE - VHE DIOSCOREACEAE *Dioscorea alata* |
| adzugo dzedze | GBE - VHE DIOSCOREACEAE *Dioscorea alata* |
| adzugo eto | GBE - VHE DIOSCOREACEAE *Dioscorea alata* |
| adzugo kpuka | GBE - VHE DIOSCOREACEAE *Dioscorea alata* |
| adzugo nkani | GBE - VHE DIOSCOREACEAE *Dioscorea alata* |
| adzugo toko | GBE - VHE DIOSCOREACEAE *Dioscorea alata* |
| adzugo tsetse | GBE - VHE DIOSCOREACEAE *Dioscorea alata* |
| afaa | YORUBA COMBRETACEAE *Terminalia superba* |
| afalafase | IJO BOMBACACEAE *Ceiba pentandra* |
| afam | VULGAR CHRYSOBALANACEAE *Parinari excelsa* AKAN - ASANTE *P. excelsa* WASA *Maranthes chrysophylla, Parinari excelsa* |
| afamfufuo | AKAN - TWI CHRYSOBALANACEAE *Parinari excelsa* WASA *P. excelsa* |
| àfánfán | IJO - IZON CECROPIACEAE *Musanga cecropioides* |
| afara | VULGAR COMBRETACEAE *Terminalia superba* |
| áfàrà | YORUBA COMBRETACEAE *Terminalia superba* |
| afàrà dúdú | YORUBA COMBRETACEAE *Terminalia ivorensis* |
| afaseɛ | AKAN - ASANTE DIOSCOREACEAE *Dioscorea alata* |
| afaséo, afasó (*pl.*-i) | GA DIOSCOREACEAE *Dioscorea alata* |
| afasew | AKAN - TWI DIOSCOREACEAE *Dioscorea alata* |
| afasew aba | AKAN - TWI DIOSCOREACEAE *Dioscorea alata* |
| afasew anamasu, afasew apuka, afasew kani, afasew pa | AKAN - TWI DIOSCOREACEAE *Dioscorea alata* |
| afasew biri | AKAN - TWI DIOSCOREACEAE *Dioscorea alata* |
| afasew ɔdepa | AKAN - TWI DIOSCOREACEAE *Dioscorea alata* |
| afasew tinlin | AKAN - TWI DIOSCOREACEAE *Dioscorea alata* |
| afasew tuntum | AKAN - TWI DIOSCOREACEAE *Dioscorea alata* |
| àfe | YORUBA ANNONACEAE *Annona glabra* |
| afema | AKAN - ASANTE ACANTHACEAE *Justicia flava* |
| aféma | AKAN - ASANTE ACANTHACEAE *Justicia flava* |
| aferziz | ARABIC CUCURBITACEAE *Citrullus colocynthis* |
| àfiā | EFIK COMBRETACEAE *Terminalia ivorensis* |
| afia edidia | EFIK DIOSCOREACEAE *Dioscorea dumetorum* |
| àfia èdidià iwá, àfia iwá | EFIK DIOSCOREACEAE *Dioscorea dumetorum* |
| àfiá énēm | IBIBIO DIOSCOREACEAE *Dioscorea misc. & spp. unknown* |
| àfia étò | EFIK COMBRETACEAE *Terminalia superba* |
| àfia nnùnùñ | EFIK AVICENNIACEAE *Avicennia germinans* |
| afia oko | EFIK DIOSCOREACEAE *Dioscorea rotundata* |
| àfia òkò | EFIK DIOSCOREACEAE *Dioscorea rotundata* |
| àfiá údiā | IBIBIO DIOSCOREACEAE *Dioscorea misc. & spp. unknown* |
| àfiáíyọ̀ | EFIK CRASSULACEAE *Bryophyllum pinnatum* |
| áfifiá | IGBO CONVOLVULACEAE *Hewittia sublobata* |
| áfifiá àchàlà | IGBO ACANTHACEAE *Rhinacanthus virens* |
| afirifiriwa | AKAN - ASANTE ANNONACEAE *Cleistopholis patens* |
| aflaa | GBE - VHE CRASSULACEAE *Bryophyllum pinnatum, Kalanchoe integra: Kalanchoe crenata* |
| a-flaatoga | GBE - VHE CRASSULACEAE *Bryophyllum pinnatum* |
| aflaa-toga | GBE - VHE CRASSULACEAE *Kalanchoe integra: Kalanchoe crenata* |
| aflamŋme | ADANGME APOCYNACEAE *Carissa edulis* |
| aflɔ | GBE - VHE ARACEAE *Pistia stratiotes* |
| afobil | BASSARI BOMBACACEAE *Bombax buonopozense, B. costatum* |
| afɔdɔnkɔ | GBE - VHE COMBRETACEAE *Terminalia superba* |
| afole | NZEMA COMPOSITAE *Aspilia africana* |
| afɔle | NZEMA COMPOSITAE *Aspilia africana, Melanthera scandens* |
| afomnondu | NZEMA APOCYNACEAE *Funtumia africana* |
| afossé | AKAN - ASANTE CONNARACEAE *Cnestis ferruginea* |
| afraa | ANYI - SEHWI COMBRETACEAE *Terminalia superba* |
| afran | TAMACHEK CRUCIFERAE *Brassica napus* |
| afrane | AKAN - TWI CHRYSOBALANACEAE *'Parinari' spp. indet.* |
| afruenba | KYAMA ANNONACEAE *Monodora myristica* |
| afuku | YORUBA COMPOSITAE *Triplotaxis stellulifera* |
| afumoe | GBE - VHE CAPPARACEAE *Cleome Linn.* |
| afumoe | GBE - VHE CAPPARACEAE *Cleome gynandra* |
| afũn nena | AKAN - FANTE COMPOSITAE *Blumea aurita* |
| afũũ | AKAN - ASANTE DIOSCOREACEAE *Dioscorea cayenensis* |
| afũũ-ba | AKAN - ASANTE DIOSCOREACEAE *Dioscorea cayenensis* |
| afuyaniama | KYAMA COMBRETACEAE *Combretum spp. indet.* |
| aga | YORUBA DIOSCOREACEAE *Dioscorea rotundata* |
| àga | YORUBA CECROPIACEAE *Musanga cecropioides* |
| àgà, jí àgà | IGBO DIOSCOREACEAE *Dioscorea rotundata* |

agishi JUKUN BIGNONIACEAE *Newbouldia laevis*
agliko YORUBA - IFE OF TOGO ANACARDIACEAE *Spondias mombin*
agnan BAULE COSTACEAE *Costus Linn.*
agnébro KYAMA BOMBACACEAE *Bombax buonopozense*
agniaï AKAN - ASANTE COSTACEAE *Costus Linn.*
agnian AKAN - ASANTE COSTACEAE *Costus Linn.*
agnié BAULE COSTACEAE *Costus Linn.*
agɔaflu GA COMPOSITAE *Vernonia amygdalina, V. colorata*
agofata HAUSA CUCURBITACEAE *Lagenaria siceraria*
àgógó igùn YORUBA BORAGINACEAE *Heliotropium indicum*
agrada AKAN - ASANTE ANNONACEAE *Hexalobus crispiflorus*
agtastaf ARABIC CUCURBITACEAE *Cucumis prophetarum*
agu ARABIC COMBRETACEAE *Guiera senegalensis*
agú YORUBA - NAGO BOMBACACEAE *Ceiba pentandra*
a-guare-(a)nsra AKAN - ASANTE ACANTHACEAE *Brillantaisia lamium, B. nitens, Eremomastax speciosa* FANTE *Brillantaisia lamium, Eremomastax speciosa* TWI *E. speciosa*
agudese GBE - VHE BOMBACACEAE *Bombax buonopozense*
agudesi ADANGME - KROBO BOMBACACEAE *Bombax buonopozense*
agudugbu YORUBA ANNONACEAE *Greenwayodendron suaveolens*
agué KYAMA BOMBACACEAE *Ceiba pentandra*
agumi AKYE CECROPIACEAE *Musanga cecropioides*
agundae 'DIFALE' CONVOLVULACEAE *Ipomoea batatas*
agundae aelalo 'DIFALE' CONVOLVULACEAE *Ipomoea batatas*
agundae aloba 'DIFALE' CONVOLVULACEAE *Ipomoea batatas*
agushi GBE - VHE CUCURBITACEAE *Cucumeropsis mannii*
agúshi (*pl.*-i) GA CUCURBITACEAE *Citrullus lanatus*
agúshi GA CUCURBITACEAE *Cucumeropsis mannii*
àgúshií HAUSA CUCURBITACEAE *Citrullus lanatus, Cucumeropsis mannii*
aguto KYAMA ANNONACEAE *Cleistopholis patens*
agwo IGBO CUCURBITACEAE *Luffa cylindrica*
agwọla (?) URHOBO CONNARACEAE *Cnestis ferruginea*
agwu, agwugu IGALA BOMBACACEAE *Ceiba pentandra*
agyaama AKAN - ASANTE APOCYNACEAE *Landolphia owariensis*
agyaamaa AKAN - ASANTE APOCYNACEAE *Landolphia owariensis*
agyamaa AKAN - TWI APOCYNACEAE *Landolphia owariensis*
agyegyen-suani AKAN - FANTE COMPOSITAE *Crassocephalum rubens*
agyemkama AKAN - WASA CECROPIACEAE *Musanga cecropioides*
ahá YORUBA CUCURBITACEAE *Lagenaria siceraria*
aha go GBE - VHE CUCURBITACEAE *Lagenaria siceraria*
àhàbà IGBO CHRYSOBALANACEAE *Acioa barteri, Parinari congensis*
àhàbà-úji IGBO CHRYSOBALANACEAE *Maranthes robusta*
aha-bayerɛ AKAN - TWI DIOSCOREACEAE *Dioscorea minutiflora*
ahabia AKAN - ASANTE CANNACEAE *Canna indica*
ahabía GA CANNACEAE *Canna indica*
ahaji HAUSA AMARANTHACEAE *Aerva javanica*
ahamante GBE - VHE DIOSCOREACEAE *Dioscorea alata*
ahana NZEMA APOCYNACEAE *Landolphia*
ahara YORUBA CUCURBITACEAE *Momordica cabraei*
àhàwà IGBO CHRYSOBALANACEAE *Acioa barteri*
ahazo GBE - FON APOCYNACEAE *Carissa edulis*
ahe ISOKO BOMBACACEAE *Ceiba pentandra*
ahéneyeɛŋ tso GA BOMBACACEAE *Bombax buonopozense*
ahensaw AKAN - TWI CUCURBITACEAE *Momordica angustisepala*
aheri NZEMA ANNONACEAE *Cleistopholis patens*
aheriheri AHANTA ARISTOLOCHIACEAE *Paristolochia goldieana*
ahia mmaia, ọsa mme IGBO CUCURBITACEAE *Luffa cylindrica*
ahia mme IGBO CUCURBITACEAE *Luffa cylindrica*
ahie ANYI BURSERACEAE *Canarium schweinfurthii*
ahié ANYI BURSERACEAE *Canarium schweinfurthii* NZEMA *C. schweinfurthii*
ahielé NZEMA BURSERACEAE *Canarium schweinfurthii*
ahiélé ANYI BURSERACEAE *Canarium schweinfurthii*
áhịhịá IGBO CONVOLVULACEAE *Hewittia sublobata*
áhịhịá èbisango IGBO COMMELINACEAE *Commelina diffusa, C. erecta*
áhịhịá-nwá-òshì n'áká IGBO COMPOSITAE *Ageratum conyzoides*
ahinsaw GBE - VHE CUCURBITACEAE *Telfairea occidentalis*
ahõ GBE - VHE CYPERACEAE *Scleria depressa*
ahɔ GBE - VHE CYPERACEAE *Cyperaceae*
ahobégu KYAMA CUCURBITACEAE *Ruthalicia eglandulosa*
ahom ADANGME APOCYNACEAE *Strophanthus hispidus*
ahọn ẹ́kùn YORUBA ACANTHACEAE *Acanthus montanus*
ahọ́n ẹ́kùn YORUBA ACANTHACEAE *Justicia extensa*
ahọ́n ẹ́kùn dúdú YORUBA ACANTHACEAE *Acanthus montanus*
ahozo GBE - VHE APOCYNACEAE *Carissa edulis*
àhụ́ IGBO CUCURBITACEAE *Cucumeropsis mannii*
àhụ́ elu, aki IGBO CUCURBITACEAE *Cucumeropsis mannii*
ahué ABURE AMARANTHACEAE *Cyathula prostrata* KYAMA ANNONACEAE *Xylopia villosa*
ahuénadi, apoissien, plakassé, wansien blakassa AKAN - ASANTE CONNARACEAE *Cnestis ferruginea*
ahuénani KYAMA CONNARACEAE *Cnestis ferruginea*
àhũidade GA CONNARACEAE *Cnestis ferruginea*

# INDEX

akèrengbe YORUBA CUCURBITACEAE *Lagenaria siceraria*
akesaïm TAMACHEK CUCURBITACEAE *Cucurbita maxima, C. pepo*
akẹsan YORUBA COMPOSITAE *Bidens pilosa*
akẹṣin-másọ YORUBA COMPOSITAE *Bidens pilosa*
akẹte IGBO APOCYNACEAE *Voacanga africana*
ákhẹ̀ EDO APOCYNACEAE *Landolphia dulcis*
àkídìmmọ̄ó IGBO CAPPARACEAE *Cleome rutidosperma*
akika, okika YORUBA ANACARDIACEAE *Spondias mombin*
akikan ISEKIRI ANACARDIACEAE *Spondias mombin*
akiko YORUBA - IFE OF TOGO ANACARDIACEAE *Spondias mombin*
akimbayeray, akuku, akwakor AKAN - TWI DIOSCOREACEAE *Dioscorea rotundata*
akinale TEM BIGNONIACEAE *Newbouldia laevis*
akisa YORUBA COMPOSITAE *Vernonia frondosa*
akisan YORUBA ANNONACEAE *Uvaria chamae*
akitasẽ AKAN - FANTE CONNARACEAE *Cnestis ferruginea*
akitplale (sodzati) GBE - VHE APOCYNACEAE *Strophanthus sarmentosus*
akitsiwa NZEMA CUCURBITACEAE *Citrullus lanatus*
akjül ARABIC BORAGINACEAE *Cordia sinensis*
akladefi (?akradefi) ADANGME CELASTRACEAE *Hippocratea welwitschii*
akladekpa (?akradekpa) ADANGME CELASTRACEAE *Hippocratea welwitschii*
aklãmakpa GBE - VHE ARACEAE *Culcasia angolensis, C. scandens*
á'klatè GA CACTACEAE *Opuntia spp.*
áklò GA APOCYNACEAE *Strophanthus hispidus*
akmé AKYE ANACARDIACEAE *Antrocaryon micraster*
ako GBE - VHE CUCURBITACEAE *Lagenaria siceraria*
ákò EDO ANNONACEAE *Dennettia tripetala*
àkó YORUBA BURSERACEAE *Canarium schweinfurthii*
akọ amùnimúyè YORUBA COMPOSITAE *Senecio biafrae*
ako dòdó YORUBA APOCYNACEAE *Tabernaemontana eglandulosa*
akọ dòdo YORUBA APOCYNACEAE *Voacanga africana*
akọ ejìrin YORUBA CUCURBITACEAE *Momordica cissoides*
akọ idòfún YORUBA CHRYSOBALANACEAE *Maranthes polyandra*
akọ irẹ́ YORUBA APOCYNACEAE *Funtumia africana, Holarrhena floribunda*
akọ iṣu YORUBA DIOSCOREACEAE *Dioscorea praehensilis*
ákọ̀ yúnyún YORUBA COMPOSITAE *Ageratum conyzoides, Aspilia africana*
akodin AKYE BROMELIACEAE *Ananas comosus*
akɔdu-ɖiba GBE - VHE CARICACEAE *Carica papaya*
akogaouan AKYE BOMBACACEAE *Rhodognaphalon brevicuspe*
akogo GBE - VHE CUCURBITACEAE *Lagenaria siceraria*
ako-ibepo YORUBA APOCYNACEAE *Alstonia boonei*
ako-ire VULGAR APOCYNACEAE *Holarrhena floribunda*
ako-iṣa YORUBA APOCYNACEAE *Strophanthus sarmentosus*
akoko YORUBA BIGNONIACEAE *Markhamia tomentosa*
akòko YORUBA BIGNONIACEAE *Newbouldia laevis*
akokɔ-bɛsã ADANGME APOCYNACEAE *Carissa edulis*
akɔ́kɔbɛsã GA APOCYNACEAE *Carissa edulis*
akɔkɔtubatuba AKAN - FANTE BORAGINACEAE *Heliotropium indicum*
akokɔturbaturba AKAN - FANTE BORAGINACEAE *Heliotropium indicum*
akɔle ADANGME ANACARDIACEAE *Spondias mombin* GA *S. mombin*
ako-ledo YORUBA BORAGINACEAE *Cordia platythyrsa*
akolo LEGBO CARICACEAE *Carica papaya*
akomfem-fere AKAN - TWI CUCURBITACEAE *Cucurbita pepo*
akomfem-kon-akyi AKAN - TWI BORAGINACEAE *Heliotropium indicum*
akɔmfɛm-tiko AKAN - ASANTE BORAGINACEAE *Heliotropium indicum* TWI *H. indicum*
akɔmfɛtikoro AKAN - ASANTE BORAGINACEAE *Heliotropium indicum*
akondogu AKAN - ASANTE CANNACEAE *Canna indica*
akɔnfem atiko AKAN - ASANTE BORAGINACEAE *Heliotropium indicum*
akɔnfem kɔn-akyi AKAN - ASANTE BORAGINACEAE *Heliotropium indicum*
akong'dui ANYI BOMBACACEAE *Bombax buonopozense*
akongo BAULE ARALIACEAE *Cussonia arborea*
akongodiė AKAN - ASANTE BOMBACACEAE *Bombax buonopozense*
akɔnkodeɛ, ŋakokɔdeɛ AKAN - ASANTE BOMBACACEAE *Bombax buonopozense*
akɔnkodeɛ AKAN - TWI BOMBACACEAE *Bombax buonopozense*
akɔŋkɔdeɛ AKAN - TWI BOMBACACEAE *Bombax buonopozense*
akonkodie AKAN - ASANTE BOMBACACEAE *Bombax buonopozense* TWI *B. buonopozense*
akɔnkɔre AKAN - FANTE BOMBACACEAE *Bombax buonopozense*
akontoma AKAN - ASANTE APOCYNACEAE *Saba florida*
akɔntɔma AKAN - BRONG APOCYNACEAE *Landolphia hirsuta* TWI *L. owariensis*
akoo-ano AKAN - ASANTE ANNONACEAE *Artabotrys insignis, A. stenopetalus, A. velutinus*
akopinolé BAULE AMARANTHACEAE *Aerva lanata*
akorabahia KYAMA ANACARDIACEAE *Antrocaryon micraster*
akoramfidie AKAN - AKYEM COMPOSITAE *Adenostemma perrottetii* ASANTE *A. perrottetii*
akori ekyi AKAN - TWI DIOSCOREACEAE *Dioscorea dumetorum* GA *D. dumetorum*
ákósä EDO ANNONACEAE *Uvaria afzelii, Uvariopsis dioica*
ako-sigo YORUBA ARALIACEAE *Cussonia arborea*
ákọ́sọ̄ EDO ANNONACEAE *Polyceratocarpus parviflorus*
akotia AKAN - ASANTE COMMELINACEAE *Aneilema beninense* TWI *A. beninense, A. umbrosum*
akoto YORUBA CUCURBITACEAE *Lagenaria siceraria*

akwakwatikwa ADANGME - KROBO DIOSCOREACEAE *Dioscorea misc. & spp. unknown*
akwakwɔ-bedina AKAN - TWI DIOSCOREACEAE *Dioscorea cayenensis*
àkwálị IGBO APOCYNACEAE *Landolphia dulcis*
akwalikwa ADANGME - KROBO DIOSCOREACEAE *Dioscorea praehensilis*
akwalu HAUSA CUCURBITACEAE *Trochomeria macrocarpa*
akwari HAUSA ARACEAE *Stylochiton lancifolius*
àkwarị IGBO APOCYNACEAE *Landolphia dulcis*
àkwárī IGBO APOCYNACEAE *Landolphia dulcis*
àkwárị IGBO APOCYNACEAE *Landolphia hirsuta*
akwato, bàkánùwaá, gámmòn fátaàkeé, ganwon fatake, gesuma, rugudu, ruguguwa, waáwán-goónaá, ƴar garii HAUSA CUCURBITACEAE *Cucurbita maxima*
akwato HAUSA CUCURBITACEAE *Cucurbita pepo, Lagenaria siceraria*
akwolé ANYI CRASSULACEAE *Bryophyllum pinnatum, Kalanchoe integra: Kalanchoe crenata*
akwotokoro IGBO BOMBACACEAE *Bombax buonopozense*
ákwú ólū IGBO BROMELIACEAE *Ananas comosus*
ákwụ́ ólū IGBO BROMELIACEAE *Ananas comosus*
ákwú óyìbó IGBO BROMELIACEAE *Ananas comosus*
ákwụ̄kwọ́ IGBO CUCURBITACEAE *Cucurbita maxima*
ákwụ́kwọ́ ịjīji IGBO ACANTHACEAE *Hypoestes verticillaris*
ákwụ́kwọ́ ijiji uku IGBO ACANTHACEAE *Eremomastax speciosa*
ákwụ́kwọ́ ósò IGBO COMBRETACEAE *Combretum hispidum*
ákwụ̄kwọ́-nwá-òshi-n'ákà IGBO COMPOSITAE *Ageratum conyzoides*
akwú-óyìbó IGBO BROMELIACEAE *Ananas comosus*
akyaboa AKAN - TWI BORAGINACEAE *Cordia millenii*
akyẽkyẽa AKAN - TWI CUCURBITACEAE *Citrullus lanatus*
akyɛkyẽa AKAN - FANTE CUCURBITACEAE *Citrullus lanatus*
akyere-nkura AKAN - ASANTE AMARANTHACEAE *Pupalia lappacea* TWI *P. lappacea*
akyẽyẽa AKAN - TWI CUCURBITACEAE *Cucumeropsis mannii*
al bata SONGHAI CUCURBITACEAE *Cucumis melo*
al lendé ARABIC ASCLEPIADACEAE *Leptadenia arborea*
ala AKAN - BRONG BOMBACACEAE *Adansonia digitata* IGBO DIOSCOREACEAE *Dioscorea rotundata*
ala fango GBE - VHE ACANTHACEAE *Barleria opaca*
alabalu ADYUKRU CONNARACEAE *Cnestis ferruginea*
alabɔntɔle GA APOCYNACEAE *Landolphia owariensis*
aládé oko YORUBA ACANTHACEAE *Rungia grandis*
alafolo, jí alafolo alẹfulu IGBO DIOSCOREACEAE *Dioscorea rotundata*
alagame IGBO COMBRETACEAE *Combretum racemosum*
alãgba GBE - VHE BOMBACACEAE *Adansonia digitata*
alakiriti YORUBA BIGNONIACEAE *Stereospermum acuminatissimum*
alakui ANYI ANACARDIACEAE *Trichoscypha arborea*
alalata MANDING - MANDINKA CUCURBITACEAE *Trochomeria macrocarpa*
alambọrọgoda IGBO BIGNONIACEAE *Kigelia africana*
alambọrọgọda IGBO BIGNONIACEAE *Kigelia africana*
alamoa ADANGME - KROBO DIOSCOREACEAE *Dioscorea alata*
alamoa gaga ADANGME - KROBO DIOSCOREACEAE *Dioscorea alata*
alamoa gu ADANGME - KROBO DIOSCOREACEAE *Dioscorea alata*
alamoa kani, alamoa ku ADANGME - KROBO DIOSCOREACEAE *Dioscorea alata*
alamoa puka, alamoa sale ADANGME - KROBO DIOSCOREACEAE *Dioscorea alata*
alamoa senya ADANGME - KROBO DIOSCOREACEAE *Dioscorea alata*
alamoa tun, alamoa tun gaga ADANGME - KROBO DIOSCOREACEAE *Dioscorea alata*
alanhita nta IGBO CONNARACEAE *Agelaea obliqua*
a-làp ɔ-dẹ́naŵ BASARI CYPERACEAE *Bulbostylis sp. indet.*
alapolo IGBO COMPOSITAE *Crassocephalum crepidioides*
alari NZEMA ANNONACEAE *Xylopia staudtii*
alatala-kunde-na SUSU - DYALONKE ARACEAE *Anchomanes difformis*
alawalawatsɛ GBE - VHE COMPOSITAE *Melanthera scandens*
áláyyafóo, àláyyàfuú HAUSA AMARANTHACEAE *Amaranthus hybridus*
albacce buru FULA - FULFULDE AMARYLLIDACEAE *Crinum zeylanicum*
albacce dawaaɗi FULA - FULFULDE AMARYLLIDACEAE *Crinum zeylanicum*
albásar kwaadíi HAUSA AMARYLLIDACEAE *Scadoxus multiflorus*
albásar kwaaɗíi HAUSA AMARYLLIDACEAE *Crinum zeylanicum*
alébé BAULE BORAGINACEAE *Ehretia cymosa*
alédan bliassu BAULE CONVOLVULACEAE *Ipomoea asarifolia*
aleifu buter TIV AMARANTHACEAE *Amaranthus tricolor*
alembole TEM COMBRETACEAE *Combretum collinum ssp. geitonophyllum*
alemebé SORUBA - KUYOBE COMBRETACEAE *Combretum nigricans*
alẽvo GBE - VHE CAPPARACEAE *Ritchiea reflexa*
algarabbu KANURI CRUCIFERAE *Lepidium sativum*
algarif, algarup HAUSA CRUCIFERAE *Lepidium sativum*
algarup ja HAUSA CRUCIFERAE *Lepidium sativum*
alhaji HAUSA AMARANTHACEAE *Aerva javanica*
alhawami HAUSA BOMBACACEAE *Ceiba pentandra*
*ma*-ali TEMNE CUCURBITACEAE *Cucurbita maxima, C. moschata*
alibada BASSA APOCYNACEAE *Landolphia hirsuta*
alibada, alubada NUPE APOCYNACEAE *Landolphia hirsuta*
alibida, alubada HAUSA APOCYNACEAE *Landolphia hirsuta*
alibo, alubo, èlubó HAUSA DIOSCOREACEAE *Dioscorea gen. (products)*
alilliba HAUSA BORAGINACEAE *Cordia africana*

INDEX

atruan niama KYAMA APOCYNACEAE *Dictyophleba leonensis*

atsaka TIV CONVOLVULACEAE *Ipomoea batatas*

atsõtsẽŋ GBE - VHE CUCURBITACEAE *Citrullus lanatus, Cucumeropsis mannii*

atshuro-tshatsho TEM AMARANTHACEAE *Gomphrema globosa*

atsia GBE - VHE ANACARDIACEAE *Anacardium occidentale*

atʃia ADANGME ANACARDIACEAE *Anacardium occidentale*

atsim AKAN - FANTE APOCYNACEAE *Tabernaemontana crassa*

atswi JUKUN DIOSCOREACEAE *Dioscorea Linn.*

attasa TAMACHEK COMPOSITAE *Pulicaria crispa*

atti YORUBA DIOSCOREACEAE *Dioscorea alata*

atukruhia KYAMA ANACARDIACEAE *Lannea welwitschii*

atunjaka IGBO BOMBACACEAE *Bombax buonopozense*

aturufa ABURE CONNARACEAE *Manotes longiflora*

atwẽ AKAN - FANTE ACANTHACEAE *Hygrophila auriculata*

atwe ada AKAN - ASANTE APOCYNACEAE *Tabernaemontana Gen.*

atwere AKAN - TWI CHRYSOBALANACEAE *Acioa dinklagei*

atwiri AKAN - WASA CHRYSOBALANACEAE *Acioa dinklagei*

a-tyàláx KONYAGI ANACARDIACEAE *Spondias mombin*

a-tyàndíw̃ù lokwẹ́ta BASARI AMARYLLIDACEAE *Scadoxus multiflorus*

a-tyátyá lùf-lùf BASARI COMPOSITAE *Vernonia colorata*

a-tyékẹnú KONYAGI ANACARDIACEAE *Lannea acida*

a-tyɛ̀l-k̇hɔnd BASARI COMBRETACEAE *Combretum nigricans*

a-tyeùkúrbís BASARI COCHLOSPERMACEAE *Cochlospermum tinctorium*

a-tyilil KONYAGI BIGNONIACEAE *Spathodea campanulata*

aubé KYAMA ANACARDIACEAE *Spondias mombin*

augna AKAN - ASANTE CONNARACEAE *Manotes longiflora*

auihae KABRE CONVOLVULACEAE *Ipomoea batatas*

auinfa ABURE COSTACEAE *Costus Linn.*

aujumgum TAMACHEK CAPPARACEAE *Capparis decidua*

aunde BAULE BORAGINACEAE *Cordia platythyrsa*

aundé BAULE BORAGINACEAE *Cordia platythyrsa*

auniguéssébé BAULE COCHLOSPERMACEAE *Cochlospermum tinctorium*

aupa nziŋa AHANTA CONVOLVULACEAE *Ipomoea pes-caprae*

aura ARABIC COMPOSITAE *Laggera alata* DAGAARI ANNONACEAE *Uvaria chamae*

áuré KANURI ANACARDIACEAE *Mangifera indica*

aur-kpoeti ADYUKRU CONNARACEAE *Manotes longiflora*

ause YORUBA ANNONACEAE *Monodora brevipes*

autié AKAN - ASANTE ANNONACEAE *Cleistopholis patens*

autopoi IGBO APOCYNACEAE *Landolphia calabarica*

avagolo GBE - VHE CECROPIACEAE *Myrianthus arboreus*

avegbɔ? GBE - VHE COMPOSITAE *Vernonia nigritiana*

avenya GBE - VHE COMPOSITAE *Vernonia amygdalina, V. colorata*

avete GBE - VHE DIOSCOREACEAE *Dioscorea cayenensis*

avĩa GBE - VHE BIGNONIACEAE *Newbouldia laevis*

aviãtilifui GBE - VHE BIGNONIACEAE *Newbouldia laevis*

àviń YORUBA COMBRETACEAE *Anogeissus leiocarpus*

avlɔ GBE - VHE ARALIACEAE *Cussonia arborea*

avogalo GBE - VHE CECROPIACEAE *Myrianthus arboreus*

avonoba NZEMA ANNONACEAE *Monodora myristica*

awa YORUBA APOCYNACEAE *Rauvolfia caffra*

awagna AKAN - ASANTE CONNARACEAE *Manotes longiflora*

awara YORUBA CAPPARACEAE *Capparis thonningii*

awarwaro HAUSA CONVOLVULACEAE *Ipomoea aquatica*

awarwaroo YORUBA CONVOLVULACEAE *Ipomoea involucrata*

awasa-makyina AKAN - WASA ANNONACEAE *Pachypodanthium staudtii*

awdi tooke FULA - FULFULDE APOCYNACEAE *Strophanthus hispidus, S. sarmentosus*

awenade ADANGME CONNARACEAE *Byrsocarpus coccineus* AKAN - KWAWU *B. coccineus*

awen-dade GA CONNARACEAE *Byrsocarpus coccineus*

awennade, awinadze AKAN - FANTE CONNARACEAE *Byrsocarpus coccineus*

awennade AKAN - TWI CONNARACEAE *Byrsocarpus coccineus*

awere pépé YORUBA COMPOSITAE *Spilanthes filicaulis*

awere-aba AKAN - TWI ANNONACEAE *Monodora myristica*

awerewa AKAN - ASANTE ANNONACEAE *Monodora myristica* TWI *M. myristica* WASA *M. myristica*

awewe YORUBA CHRYSOBALANACEAE *Maranthes robusta*

awgalele HAUSA BALANITACEAE *Balanites aegyptiaca*

awiara AKAN - WASA ANNONACEAE *Monodora myristica*

àwiń, awùn YORUBA APOCYNACEAE *Alstonia boonei*

awinadze AKAN - TWI CONNARACEAE *Byrsocarpus coccineus*

awinar ARABIC CRUCIFERAE *Diplotaxis virgata*

awire-aba AKAN - WASA ANNONACEAE *Monodora myristica*

awithegé ARABIC COMBRETACEAE *Anogeissus leiocarpus*

awithegi ARABIC COMBRETACEAE *Anogeissus leiocarpus*

awlua GBE - VHE ARALIACEAE *Cussonia arborea*

àwọ̀ òwú YORUBA COCHLOSPERMACEAE *Cochlospermum planchonii*

awodwene AKAN - TWI COMPOSITAE *Vernonia colorata*

awɔhanya NZEMA CONNARACEAE *Manotes expansa, M. longiflora*

awon nugba BAULE COMPOSITAE *Vernonia guineensis*

awọ́nrinwán YORUBA CHRYSOBALANACEAE *Chrysobalanus icaco*

awonwẽnẽ AKAN - TWI COMPOSITAE *Vernonia amygdalina, V. colorata*

# INDEX

bagnan SERER - NON CAPPARACEAE *Boscia senegalensis*
bagnimoney KRU - GUERE CONNARACEAE *Agelaea mildbraedii*
bago titi KRU - GUERE CELASTRACEAE *Salacia erecta*
bagogwé GAGU ACANTHACEAE *Phaulopsis imbricata*
bâgok BANYUN ANACARDIACEAE *Lannea velutina*
bagomo KIRMA APOCYNACEAE *Strophanthus sarmentosus*
bagozaki HAUSA APOCYNACEAE *Carissa edulis*
bagu FULA - PULAAR CAPPARACEAE *Maerua angolensis*
bagueuóne BALANTA ASCLEPIADACEAE *Calotropis procera*
baguhi, bugi, buguhi FULA - FULFULDE CAPPARACEAE *Maerua angolensis*
bagulé 'NEKEDIE' BIGNONIACEAE *Newbouldia laevis*
ba-gwandara HAUSA DIOSCOREACEAE *Dioscorea misc. & spp. unknown*
bahá MANDING - MANDINKA AMARYLLIDACEAE *Scadoxus multiflorus*
bahab DIOLA CHRYSOBALANACEAE *Neocarya macrophylla*
bahéma BATONNUN ANACARDIACEAE *Lannea microcarpa*
bahi FULA - PULAAR APOCYNACEAE *Landolphia heudelotii* LOKO COMBRETACEAE *Terminalia ivorensis* MANDING - MANINKA APOCYNACEAE *Landolphia heudelotii*
bahô MANDING - MANINKA AMARYLLIDACEAE *Pancratium trianthum*
bahôdina MANDING - MANDINKA ARACEAE *Amorphophallus flavovirens*
bahun MOORE - NAWDAM BOMBACACEAE *Ceiba pentandra*
bai MANO COMBRETACEAE *Terminalia ivorensis*
baï MANDING - MANINKA APOCYNACEAE *Landolphia heudelotii*
baibai LOKO CARICACEAE *Carica papaya*
baifinɛ SUSU ARACEAE *Cyrtosperma senegalense*
bai-kafa MENDE - KPA ANACARDIACEAE *Sorindeia juglandifolia*
bailliri MANDING - MANDINKA CECROPIACEAE *Myrianthus serratus*
bain DIOLA APOCYNACEAE *Alstonia boonei*
baine, gobane MENDE BIGNONIACEAE *Spathodea campanulata*
baiséguma ANYI ANACARDIACEAE *Lannea welwitschii*
bai-ti DAN COMBRETACEAE *Terminalia ivorensis*
bàje NUPE DIOSCOREACEAE *Dioscorea Linn.*
baji (*def.*-i) MENDE COMBRETACEAE *Terminalia ivorensis, T. scutifera, T. superba*
bak SERER BOMBACACEAE *Adansonia digitata, Bombax costatum* COMBRETACEAE *Laguncularia racemosa* SERER - NON BOMBACACEAE *Adansonia digitata*
baka egbe ANYI APOCYNACEAE *Rauvolfia vomitoria*
baka pimblé BAULE APOCYNACEAE *Rauvolfia vomitoria*
bakaba MANDING - MANINKA ARACEAE *Anchomanes welwitschii*
bakaégbi ANYI APOCYNACEAE *Rauvolfia vomitoria*
bakaɛmbe NZEMA APOCYNACEAE *Rauvolfia vomitoria*
baƙaleƙale HAUSA CANNACEAE *Canna indica*
baƙalele HAUSA CANNACEAE *Canna indica*
bakambi FULA - FULFULDE ASCLEPIADACEAE *Pergularia tomentosa*
bàkánùwaà HAUSA CUCURBITACEAE *Cucurbita pepo*
bakapembe NZEMA APOCYNACEAE *Pleioceras barteri*
bakar HAUSA CYPERACEAE *Cyperus esculentus*
baƙar dóóya HAUSA DIOSCOREACEAE *Dioscorea alata*
bakar rura HAUSA CHRYSOBALANACEAE *Neocarya macrophylla*
bákár tàrámnîyaá HAUSA COMBRETACEAE *Combretum fragrans*
bákár-áyaá HAUSA CYPERACEAE *Cyperus esculentus*
báƙáreƙáre HAUSA CANNACEAE *Canna indica*
bakaroro HAUSA CELASTRACEAE *Maytenus senegalensis*
bakasirani MANDING - MANINKA CAPPARACEAE *Crateva adansonii*
baké GAGU CARICACEAE *Carica papaya*
bakha MANDING - MANINKA AMARYLLIDACEAE *Crinum zeylanicum*
baƙin danko HAUSA APOCYNACEAE *Clitandra cymulosa*
bakin duniya HAUSA CUCURBITACEAE *Cucumis figarei*
baƙin ganye HAUSA DIOSCOREACEAE *Dioscorea misc. & spp. unknown*
baƙin itache HAUSA APOCYNACEAE *Holarrhena floribunda*
baƙin kukkuki HAUSA ANACARDIACEAE *Lannea barteri*
bakin tàrámnîyaá HAUSA COMBRETACEAE *Combretum fragrans*
ba-kiskis HAUSA COMMELINACEAE *Commelina erecta, C. forskalaei*
bakko HAUSA BOMBACACEAE *Adansonia digitata*
bakoré GAGU CANNACEAE *Canna indica*
bakorompéku MANDING - BAMBARA ANACARDIACEAE *Lannea velutina*
bakoroni guenda MANDING - MANINKA APOCYNACEAE *Tabernaemontana longiflora*
bakororo FULA - FULFULDE CELASTRACEAE *Maytenus senegalensis*
bakuku BATONNUN CHRYSOBALANACEAE *Maranthes polyandra*
bakunin AHANTA APOCYNACEAE *Alstonia boonei* NZEMA *A. boonei*
bakwa MUNGAKA BURSERACEAE *Dacryodes edulis*
bakwangok MUNGAKA BURSERACEAE *Dacryodes edulis*
bal ora DAGAARI COMBRETACEAE *Combretum molle*
bala KISSI CUCURBITACEAE *Lagenaria siceraria*
balâ balâgan BANYUN ACANTHACEAE *Hygrophila auriculata*
bàlaàsaá HAUSA COMMELINACEAE *Commelina diffusa*
bàlaàsáanaá HAUSA COMMELINACEAE *Commelina benghalensis*
bàlaàsánaá HAUSA COMMELINACEAE *Commelina diffusa, C. forskalaei*
bàlaàsánaà HAUSA COMMELINACEAE *Commelina erecta*
bàlaàsáyaá HAUSA COMMELINACEAE *Commelina diffusa*
balagande, balge HAUSA COCHLOSPERMACEAE *Cochlospermum tinctorium*
balak SERER COMBRETACEAE *Terminalia laxiflora* SERER - NON *T. laxiflora*

INDEX

| | |
|---|---|
| barakanti | MANDING - BAMBARA BURSERACEAE *Commiphora africana* |
| barakâti | MANDING - BAMBARA BURSERACEAE *Commiphora africana* |
| barakola | MANDING - 'SOCE' CONVOLVULACEAE *Ipomoea asarifolia* |
| barankato | FULA - PULAAR ASCLEPIADACEAE *Secamone afzelii* |
| barasan, mbarasan | WOLOF ASCLEPIADACEAE *Caralluma russelliana* |
| baratigɛrinyi | SUSU CYPERACEAE *Remirea maritima* |
| baraulé | MANDING - BAMBARA COMBRETACEAE *Combretum micranthum* MANINKA *C. micranthum* |
| bàrbár | KANURI COMBRETACEAE *Terminalia avicennioides, T. glaucescens* |
| bàrbáttaá | HAUSA COMBRETACEAE *Guiera senegalensis* |
| barbo | WOLOF CUCURBITACEAE *Zehneria thwaitsii* |
| barböf | WOLOF CUCURBITACEAE *Zehneria thwaitsii* |
| bare | SUSU ARACEAE *Colocasia esculenta* |
| *a*-bare | TEMNE ARISTOLOCHIACEAE *Paristolochia goldieana* |
| baré | MANDING - BAMBARA ACANTHACEAE *Peristrophe bicalyculata* |
| *a*-barɛn | TEMNE ARACEAE *Cyrtosperma senegalense* |
| ba-reshe | HAUSA BIGNONIACEAE *Newbouldia laevis* |
| bareshi | HAUSA BIGNONIACEAE *Newbouldia laevis* |
| barhô firila | MANDING - MANINKA AMARYLLIDACEAE *Crinum distichum* |
| bari | GOLA COMBRETACEAE *Terminalia ivorensis* SUSU ARACEAE *Colocasia esculenta* |
| barimbini | DAGBANI AMARANTHACEAE *Pandiaka heudelotii* |
| barkanté | SONGHAI BURSERACEAE *Commiphora africana* |
| bàrkònón bírii | HAUSA CUCURBITACEAE *Kedrostis hirtella* |
| bàrkoònon-hàwaínìyàa | HAUSA CUCURBITACEAE *Ctenolepis cerasiformis* |
| barkorni buzi | MANDING - DYULA CYPERACEAE *Fimbristylis hispidula* |
| barmadangaïe | SONGHAI BORAGINACEAE *Cordia sinensis* |
| barniate | PEPEL CONNARACEAE *Cnestis ferruginea* |
| baro | MANDING - 'SOCE' COMBRETACEAE *Combretum micranthum* |
| barulé | MANDING - MANINKA COMBRETACEAE *Combretum micranthum* |
| basa | MANDING - MANINKA APOCYNACEAE *Holarrhena floribunda* |
| basa furãda | MANDING - BAMBARA CUCURBITACEAE *Mukia maderaspatana* |
| básábásá | EDO APOCYNACEAE *Funtumia africana, F. elastica* |
| ba-samu | HAUSA BURSERACEAE *Boswellia dalzielii, B. odorata* |
| basgo | HAUSA CUCURBITACEAE *Trochomeria macrocarpa* |
| basi | VAI COMBRETACEAE *Terminalia ivorensis* |
| basio | KISSI COMBRETACEAE *Terminalia ivorensis* |
| baska | HAUSA CUCURBITACEAE *Luffa cylindrica* |
| baslèv | DIOLA ANNONACEAE *Xylopia aethiopica* |
| bassa moradi | HAUSA APOCYNACEAE *Clitandra cymulosa* |
| bassakwo | MANDING - MANINKA DILLENIACEAE *Tetracera alnifolia* |
| bassi | KPELLE COMBRETACEAE *Terminalia ivorensis, T. superba* KRU - GUERE CONVOLVULACEAE *Ipomoea batatas* |
| bassialewin, bisianévuin, ganganovuin, korodu | KWENI ASCLEPIADACEAE *Secamone afzelii* |
| bassoko | MANDING - MANINKA DILLENIACEAE *Tetracera alnifolia* |
| ba-susa | HAUSA COMPOSITAE *Centaurea perrottetii, C. praecox* |
| bata foia kani | GA ARACEAE *Anchomanes difformis* |
| bâtâ foro | MANDING - MANINKA APOCYNACEAE *Alstonia boonei* |
| batafo-nto | AKAN - FANTE DIOSCOREACEAE *Dioscorea praehensilis* |
| bâtâforo | FULA - PULAAR APOCYNACEAE *Alstonia boonei* |
| batâǹ | MANDING - BAMBARA BOMBACACEAE *Ceiba pentandra* |
| bâtân | MANDING - MANINKA BOMBACACEAE *Ceiba pentandra* |
| batanga | DAGAARI ANNONACEAE *Annona senegalensis* |
| batani | WALA ANNONACEAE *Annona senegalensis* |
| batanya | HAUSA CUCURBITACEAE *Cucumis melo* |
| bâtara burué | MANDING - MANDINKA COMPOSITAE *Vernonia colorata* |
| bataťor | WOLOF COMPOSITAE *Vernonia nigritiana* |
| *m*-batatwene | AKAN - TWI ARACEAE *Cercestis afzelii* |
| *m*-batawene | AKAN - ASANTE ARACEAE *Cercestis afzelii* WASA *C. afzelii* |
| bâtigêhi | FULA - PULAAR BOMBACACEAE *Ceiba pentandra* |
| bâtinévi | FULA - PULAAR BOMBACACEAE *Ceiba pentandra* |
| bátìrí | BEROM CUCURBITACEAE *Cucurbita maxima* |
| batteikh | ARABIC - SHUWA CONVOLVULACEAE *Citrullus lanatus* |
| batteikh al hamdal | ARABIC - SHUWA CUCURBITACEAE *Citrullus lanatus* |
| batteikh al masak | ARABIC - SHUWA CUCURBITACEAE *Citrullus lanatus* |
| batu | MANDING - BAMBARA CONVOLVULACEAE *Jacquemontia tamnifolia* |
| bâtu bahoń | WOLOF AMARYLLIDACEAE *Scadoxus multiflorus* |
| bau kutué | KRU - GUERE COMPOSITAE *Microglossa pyrifolia* |
| bauane | FULA - FULFULDE ASCLEPIADACEAE *Calotropis procera* PULAAR *C. procera* |
| bauché bochy | HAUSA COMBRETACEAE *Terminalia macroptera* |
| baué | ABE ANNONACEAE *Enantia polycarpa, Neostenanthera hamata* |
| baué fu | ABE ANNONACEAE *Greenwayodendron oliveri, Neostenanthera hamata* |
| bauier | ARABIC CAPPARACEAE *Capparis sepiaria* |
| bauli, dagara, dageera | HAUSA COMBRETACEAE *Combretum collinum ssp. geitonophyllum* |
| bauré | KRU - GUERE CANNACEAE *Canna indica* |
| báúshe | HAUSA COMBRETACEAE *Terminalia Linn.* |
| báúshe, baúsheè | HAUSA COMBRETACEAE *Terminalia avicennioides* |
| báúshe | HAUSA COMBRETACEAE *Terminalia glaucescens* |
| ɓaushe | HAUSA COMBRETACEAE *Terminalia macroptera* |

baúshin giíwaá HAUSA COMBRETACEAE *Terminalia mollis*
baushishi FULA - FULFULDE COMBRETACEAE *Terminalia glaucescens*
bautaguerlet SERER - NON CARYOPHYLLACEAE *Polycarpaea corymbosa*
bawam bawam FULA - PULAAR ASCLEPIADACEAE *Calotropis procera*
bawane FULA - PULAAR ASCLEPIADACEAE *Calotropis procera* TUKULOR *C. procera*
bawé fu ABE ANNONACEAE *Neostenanthera hamata*
bawoam FULA - PULAAR ASCLEPIADACEAE *Calotropis procera*
baxa WOLOF AMARYLLIDACEAE *Scadoxus multiflorus*
ba-yammi HAUSA CUCURBITACEAE *Lagenaria siceraria*
baye KRU - BASA COMBRETACEAE *Terminalia ivorensis, T. superba*
bayerɛ AKAN DIOSCOREACEAE *Dioscorea Linn.* AKAN - AKYEM *D. rotundata* ASANTE *D. rotundata* TWI *D. praehensilis, D. rotundata* GA *D. praehensilis*
bayerɛ biri AKAN - TWI DIOSCOREACEAE *Dioscorea misc. & spp. unknown*
báyìdzà NUPE DIOSCOREACEAE *Dioscorea Linn.*
bayirɛ AKAN - TWI DIOSCOREACEAE *Dioscorea Linn.*
bayugɛnɛ KONO CONVOLVULACEAE *Ipomoea alba*
bayulo MANDING - MANDINKA DIOSCOREACEAE *Dioscorea bulbifera*
ba-zana HAUSA BURSERACEAE *Commiphora kerstingii*
ba-zara HAUSA BURSERACEAE *Commiphora africana*
bazéru KRU - GUERE COMPOSITAE *Mikania cordata*
bazi LOMA COMBRETACEAE *Terminalia superba*
bé AKYE ANACARDIACEAE *Pseudospondias microcarpa*
bɛ AKAN - FANTE COSTACEAE *Costus afer*
bé vê AKYE COMPOSITAE *Struchium sparganophora*
beacon bush KRIO CELASTRACEAE *Salacia senegalensis*
beacon fire KRIO CELASTRACEAE *Salacia senegalensis*
beakogan BULOM CACTACEAE *Opuntia spp.*
bebaque MANDYAK BOMBACACEAE *Adansonia digitata*
bébéti 'KRU' APOCYNACEAE *Funtumia elastica*
bɛblɛ tama GBE - VHE COMPOSITAE *Launaea taraxacifolia*
béblédiduku KRU - BETE CONNARACEAE *Cnestis ferruginea*
bechete PEPEL ACANTHACEAE *Hygrophila auriculata*
béchieta ABE BALANITACEAE *Balanites wilsoniana* AKYE *B. wilsoniana*
béci NUPE BIGNONIACEAE *Kigelia africana*
becuape MANDYAK BIGNONIACEAE *Newbouldia laevis*
becubar MANDYAK CELASTRACEAE *Salacia senegalensis*
becute MANDYAK APOCYNACEAE *Landolphia dulcis*
bede LIMBA COMBRETACEAE *Combretum racemosum*
bede bede AKAN - ASANTE APOCYNACEAE *Landolphia hirsuta*
bede-bede AKAN - ASANTE APOCYNACEAE *Landolphia*
bédé-bédé ANYI APOCYNACEAE *Landolphia dulcis*
bedié KRU - GREBO AMARANTHACEAE *Amaranthus spinosus*
bédiera korandi MANDING - DYULA COCHLOSPERMACEAE *Cochlospermum tinctorium*
bédiéra korandi MANDING - DYULA COCHLOSPERMACEAE *Cochlospermum tinctorium*
bediko MANDING - MANINKA ANACARDIACEAE *Ozoroa pulcherrima*
bediwonua AKAN - TWI BURSERACEAE *Canarium schweinfurthii*
bɛdi-wo-nua AKAN - TWI BURSERACEAE *Canarium schweinfurthii*
bediwuna AKAN - TWI BURSERACEAE *Canarium schweinfurthii*
bedi-wu-nua AKAN - ASANTE BURSERACEAE *Canarium schweinfurthii*
bedjig TAMACHEK BORAGINACEAE *Trichodesma gracile*
bedôal MANKANYA BOMBACACEAE *Adansonia digitata*
bedom-hal MANDYAK BOMBACACEAE *Adansonia digitata*
bédro-bédro KRU - BETE ARACEAE *Anchomanes difformis*
beduto-ubule MANKANYA CONNARACEAE *Cnestis ferruginea*
beeliɗamhi FULA - FULFULDE CAPPARACEAE *Cadaba farinosa*
beeng DAN BURSERACEAE *Canarium schweinfurthii*
ɓẹ-feuɗ BEDIK CUCURBITACEAE *Luffa cylindrica*
begafe MENDE ACANTHACEAE *Rungia guineensis*
begundja MANDYAK ANNONACEAE *Uvaria afzelii, U. chamae* MANKANYA *U. afzelii, U. chamae*
béhi WOLOF BORAGINACEAE *Cordia senegalensis*
beï SENUFO - TAFILE APOCYNACEAE *Landolphia heudelotii*
beidé DAN DIOSCOREACEAE *Dioscorea bulbifera*
beigérelên DIOLA ANACARDIACEAE *Lannea nigritana*
beiŋga-kɔne KONO APOCYNACEAE *Holarrhena floribunda, Rauvolfia vomitoria*
béínmó IJO - IZON CUCURBITACEAE *Momordica cissoides*
béínmọ̀ IJO - IZON CUCURBITACEAE *Momordica cissoides*
*ka*-bɛk TEMNE BIGNONIACEAE *Stereospermum acuminatissimum*
bèkè búrù IJO - IZON CONVOLVULACEAE *Ipomoea batatas*
beké kọkọrọ IJO - IZON BROMELIACEAE *Ananas comosus*
bèké òdù IJO - IZON ARACEAE *Xanthosoma mafaffa*
békiblé KRU - GUERE AMARANTHACEAE *Amaranthus spinosus*
bɛ-kpandeo (*pl.* g-betau-kpandelaũ) KISSI BROMELIACEAE *Ananas comosus*
beku KOOSI DIOSCOREACEAE *Dioscorea Linn.*
bekue KUNDU DIOSCOREACEAE *Dioscorea Linn.* MBONGE *D. Linn.* TANGA *D. Linn.*
bekugbro MUNGAKA APOCYNACEAE *Alstonia boonei*
bekuhoa-pɔmboɛ̃? KISSI COMPOSITAE *Launaea taraxacifolia*
bekui BAFOK DIOSCOREACEAE *Dioscorea Linn.* BALONG *D. Linn.*

bel DIOLA CHRYSOBALANACEAE *Neocarya macrophylla*
belaguon 'SENUFO' BOMBACACEAE *Bombax costatum*
belápse DIOLA - FLUP ASCLEPIADACEAE *Calotropis procera* MANKANYA *C. procera*
belaya DOGON CAPPARACEAE *Maerua angolensis*
belbala, belbela ARABIC CHENOPODIACEAE *Traganum nudatum*
belbel ARABIC CHENOPODIACEAE *Salsola tetrandra*
belɗamhi FULA - FULFULDE CAPPARACEAE *Cadaba farinosa*
*m*bele MENDE BURSERACEAE *Canarium schweinfurthii*
bèlé KWENI BOMBACACEAE *Adansonia digitata*
bélé bélé MANDING - BAMBARA CAPPARACEAE *Maerua angolensis*
bɛlɛ tama GBE - VHE COMPOSITAE *Launaea taraxacifolia*
bélébélé MANDING - BAMBARA CAPPARACEAE *Maerua angolensis*
belêgôfoɗô MANDING - MANDINKA COSTACEAE *Costus afer*
belêgôfódô BALANTA COSTACEAE *Costus afer*
bélégwé 'NEKEDIE' COMBRETACEAE *Combretum racemosum*
béléku KWENI BORAGINACEAE *Ehretia cymosa*
belge, zúnzùnaá HAUSA COCHLOSPERMACEAE *Cochlospermum planchonii*
belibelo MANDING - MANDINKA CAPPARACEAE *Capparis tomentosa, Maerua angolensis*
bello AKYE ANNONACEAE *Xylopia villosa*
bélo BATONNUN COMBRETACEAE *Terminalia avicennioides*
belofa MANDYAK BOMBACACEAE *Bombax costatum* MANKANYA *B. costatum*
belôfa MANDYAK BOMBACACEAE *Bombax costatum*
belôfò MANKANYA BOMBACACEAE *Bombax costatum*
bɛlo-hina (*def*.-hinei) MENDE ANNONACEAE *Enantia polycarpa*
belon KIRMA BOMBACACEAE *Ceiba pentandra*
bɛltampel-lɛ BULOM CONNARACEAE *Cnestis ferruginea*
béluki FULA - TUKULOR ANACARDIACEAE *Lannea humilis*
beluma DAGAARI COCHLOSPERMACEAE *Cochlospermum tinctorium*
bembambi FULA - FULFULDE ASCLEPIADACEAE *Calotropis procera*
bembe MANDING - BAMBARA ANACARDIACEAE *Lannea barteri*
bembe (*def*. bemboo) MANDING - MANDINKA ANACARDIACEAE *Lannea acida, L. velutina*
bembé MANDING - BAMBARA ANACARDIACEAE *Lannea acida* MANINKA *L. barteri, L. microcarpa, L. nigritana, L. velutina*
bembé fing MANDING - BAMBARA ANACARDIACEAE *Lannea acida*
bembé gua gua MANDING - MANINKA ANACARDIACEAE *Lannea velutina*
bembedje (*pl.*) FULA - PULAAR ANACARDIACEAE *Lannea acida, L. nigritana*
bembedje FULA - PULAAR ANACARDIACEAE *Lannea velutina*
bembefiŋ (*def*.-o) MANDING - MANDINKA ANACARDIACEAE *Lannea acida*
bembégaga MANDING - MANINKA ANACARDIACEAE *Lannea velutina*
bɛmbɛ-lɛ BULOM CHRYSOBALANACEAE *Chrysobalanus icaco*
bembem-hei FULA - PULAAR ANACARDIACEAE *Lannea acida, L. nigritana, L. velutina*
bembemusu (*def*. b. musoo) MANDING - MANDINKA ANACARDIACEAE *Lannea velutina*
bembéńańa MANDING - MANDINKA ANACARDIACEAE *Lannea acida* 'SOCE' *L. acida*
bembénugu MANDING - MANINKA ANACARDIACEAE *Lannea acida*
bembey FULA - PULAAR ANACARDIACEAE *Lannea acida*
bembo MANDING - 'SOCE' ANACARDIACEAE *Lannea acida*
bembô (*def*.) MANDING - MANDINKA ANACARDIACEAE *Lannea acida, L. nigritana, L. velutina, Pseudospondias microcarpa*
bembô-fingo MANDING - MANDINKA ANACARDIACEAE *Lannea acida*
bembo-keo, bembo-muso MANDING - MANDINKA ANACARDIACEAE *Lannea acida*
béna iri KWENI COMPOSITAE *Acanthospermum hispidum*
benaduru GRUSI ASCLEPIADACEAE *Leptadenia hastata*
benaman ARABIC CONVOLVULACEAE *Ipomoea asarifolia*
benature GURMA ANACARDIACEAE *Lannea barteri*
bende KUNDU ARACEAE *Xanthosoma mafaffa*
beńde BALANTA APOCYNACEAE *Saba senegalensis*
bendem-bende KPE COMPOSITAE *Microglossa pyrifolia*
bende-moso TEM CHRYSOBALANACEAE *Maranthes polyandra*
bende-noso TEM CHRYSOBALANACEAE *Maranthes polyandra, Parinari congensis*
bɛndiaiyondo? KISSI COMPOSITAE *Elephantopus spicatus*
bendu KISSI ACANTHACEAE *Justicia schimperi striata*
bene-fi MANDING - MANINKA ACANTHACEAE *Lepidagathis heudelotiana*
benempe(-le) MANDYAK ANNONACEAE *Annona senegalensis*
benempé MANDYAK ANNONACEAE *Annona senegalensis*
benfukɛ-lɛ BULOM APOCYNACEAE *Tabernaemontana crassa*
bengdé BALANTA APOCYNACEAE *Saba senegalensis*
ɓẹ-ngwaf u-naɗ, tẹmp-tẹmp KONYAGI DIOSCOREACEAE *Dioscorea bulbifera*
benitaha MANDYAK COMPOSITAE *Vernonia colorata*
bɛnjamin KRIO CHRYSOBALANACEAE *Neocarya macrophylla*
benna FULA - PULAAR CHRYSOBALANACEAE *Parinari curatellifolia*
bénô bénô MANDYAK CHRYSOBALANACEAE *Neocarya macrophylla* MANKANYA *N. macrophylla*
bɛnɔmbe MANDING - MANINKA APOCYNACEAE *Landolphia owariensis*
benotaro MANDYAK ANNONACEAE *Annona senegalensis*
bensékomi ANYI - ANUFO COMBRETACEAE *Combretum molle*
bentan habu GBE - FON BOMBACACEAE *Ceiba pentandra*
*a*-bent-*a*-bala TEMNE DIOSCOREACEAE *Dioscorea hirtiflora*
bentégniévi FULA - PULAAR BOMBACACEAE *Ceiba pentandra*
bentenki WOLOF BOMBACACEAE *Ceiba pentandra*

bitali 'KRU' COMPOSITAE *Vernonia colorata*
bita-lif KRIO COMPOSITAE *Vernonia amygdalina*
*a*-bita-lif TEMNE COMPOSITAE *Vernonia amygdalina*
bitchala(m) MANDYAK CHRYSOBALANACEAE *Parinari excelsa*
bitchiante MANDYAK COMBRETACEAE *Guiera senegalensis* MANKANYA *G. senegalensis*
bitiague MANDYAK CHRYSOBALANACEAE *Neocarya macrophylla* MANKANYA *N. macrophylla*
biwa FULA - FULFULDE CELASTRACEAE *Hippocratea africana*
bla KWENI ANNONACEAE *Uvaria afzelii* CONNARACEAE *Connarus africanus*
bladeï KRU - GUERE COMPOSITAE *Laggera alata*
blagéï 'KRU' ARACEAE *Culcasia scandens*
blahn KRU - BASA ANNONACEAE *Neostenanthera gabonensis*
blaï kuri KRU - BETE COMPOSITAE *Laggera alata*
blalo BALANTA BIGNONIACEAE *Spathodea campanulata*
blamange MENDE CUCURBITACEAE *Cucurbita pepo*
blambô MANDYAK APOCYNACEAE *Landolphia dulcis*
blante DAN COMPOSITAE *Vernonia spp. indet.*
blatiki BAULE ACANTHACEAE *Elytraria marginata*
blé DAN AMARANTHACEAE *Cyathula prostrata*
bleessim BALANTA CUCURBITACEAE *Cucurbita pepo*
blɛfota ADANGME - KROBO BROMELIACEAE *Ananas comosus*
blɛfo-tobi ADANGME CANNACEAE *Canna indica* ADANGME - KROBO *C. indica*
blékolopiti KRU - GUERE COMMELINACEAE *Palisota hirsuta*
blékuré ABE ANACARDIACEAE *Pseudospondias microcarpa*
blende BALANTA CELASTRACEAE *Salacia senegalensis*
bleu DAN ANNONACEAE *Isolona campanulata*
bleuleuf DIOLA - FLUP ANNONACEAE *Annona senegalensis*
bleulof DIOLA - FLUP ANNONACEAE *Annona senegalensis*
bliassubaka BAULE COMBRETACEAE *Combretum molle*
blie KRU - GUERE COMBRETACEAE *Terminalia ivorensis, T. superba*
blié KRU - BETE COMBRETACEAE *Terminalia ivorensis* GUERE *T. ivorensis*
blignablé 'NEKEDIE' COMPOSITAE *Eclipta alba*
bliku KWENI BORAGINACEAE *Ehretia cymosa*
blima BAULE BIGNONIACEAE *Kigelia africana* 'KRU' ARACEAE *Anchomanes*
blimah KRU - BASA ANACARDIACEAE *Trichoscypha arborea, T. yapoensis*
blimah(-pu) KRU - BASA ANACARDIACEAE *Trichoscypha longifolia*
blimmo BAULE BIGNONIACEAE *Kigelia africana*
blimo BAULE BIGNONIACEAE *Kigelia africana*
bliro KRU - GUERE COMPOSITAE *Conyza canadensis*
blis BALANTA BIGNONIACEAE *Markhamia tomentosa*
blo TURUKA BOMBACACEAE *Ceiba pentandra*
blofo MANKANYA BOMBACACEAE *Bombax costatum*
blɔfo kotʃa GA CUCURBITACEAE *Luffa cylindrica*
blɔfo tɔtɔ GA ASCLEPIADACEAE *Calotropis procera*
blɔfo-atómò GA CONVOLVULACEAE *Ipomoea batatas*
blɔfodɔ(ɔ) NZEMA APOCYNACEAE *Pleioceras barteri*
blɔfóŋmè GA BROMELIACEAE *Ananas comosus*
blölöf DIOLA ANNONACEAE *Annona senegalensis*
blɔloni MENDE CAPPARACEAE *Euadenia eminens*
blongbé KRU - GUERE COMPOSITAE *Microglossa pyrifolia*
blopo MANKANYA COMBRETACEAE *Terminalia macroptera*
blotué KRU - GUERE COMPOSITAE *Vernonia conferta*
blu DAN BIGNONIACEAE *Markhamia lutea, M. tomentosa*
blu blu BAULE DIOSCOREACEAE *Dioscorea alata*
blumo KRU - GUERE BIGNONIACEAE *Kigelia africana*
boa MANDING - MANINKA AMARYLLIDACEAE *Pancratium trianthum*
boamué AKYE CHRYSOBALANACEAE *Maranthes robusta*
bobèbé KRU - GUERE AMARYLLIDACEAE *Crinum Gen.*
bòbèdu DUALA BIGNONIACEAE *Markhamia tomentosa*
*am*-bobəl TEMNE ANNONACEAE *Cleistopholis patens*
*a*-bobən TEMNE APOCYNACEAE *Rauvolfia vomitoria*
bobo BAULE ARALIACEAE *Cussonia arborea* KONO APOCYNACEAE *Funtumia africana, F. elastica* KRU - GUERE CANNACEAE *Canna indica* MANDING - DYULA ARALIACEAE *Cussonia arborea, C. arborea* MENDE APOCYNACEAE *Funtumia africana*
bobo (*def.*-i) MENDE APOCYNACEAE *Funtumia africana, F. elastica*
*a*-bɔbɔ TEMNE ASCLEPIADACEAE *Tylophora congolana*
bobobo KRU - GUERE CUCURBITACEAE *Momordica charantia*
bɔbɔɛ KONO AMARANTHACEAE *Amaranthus hybridus*
*am*-bobɔl TEMNE ANNONACEAE *Cleistopholis patens*
bɔbɔ-lɛ BULOM CRASSULACEAE *Bryophyllum pinnatum*
boboma AKAN - BRONG ANNONACEAE *Annona arenaria, A. senegalensis*
bobonowron KWENI CUCURBITACEAE *Momordica foetida*
boboroa DAGBANI AMARANTHACEAE *Amaranthus viridis* GURMA *A. viridis*
bɔbɔ-wa KONO AMARANTHACEAE *Amaranthus hybridus*
bobröb WOLOF COMBRETACEAE *Terminalia avicennioides*
bobwohi GWARI CHRYSOBALANACEAE *Parinari curatellifolia*
bodafor SERER - NON ASCLEPIADACEAE *Calotropis procera*
bodafot SERER ASCLEPIADACEAE *Calotropis procera*

bodbada ARABIC AMARANTHACEAE *Amaranthus graecizans*

bodda ANYI APOCYNACEAE *Alafia barteri*

bodé FULA - PULAAR COMBRETACEAE *Terminalia macroptera*

bôɗé WOLOF APOCYNACEAE *Strophanthus sarmentosus*

bodehi FULA - PULAAR COMBRETACEAE *Terminalia albida, T. macroptera*

bọ́dẹkádún YORUBA APOCYNACEAE *Motandra guineensis*

bodévi FULA - FULFULDE COMBRETACEAE *Terminalia macroptera* PULAAR *T. macroptera*

bodey FULA - PULAAR COMBRETACEAE *Terminalia macroptera*

bodeyi FULA - PULAAR COMBRETACEAE *Terminalia macroptera*

bodèyi MANDING - BAMBARA COMBRETACEAE *Terminalia avicennioides*

boɗeyi, boɗehi FULA - FULFULDE COMBRETACEAE *Terminalia avicennioides*

bodi BALANTA APOCYNACEAE *Strophanthus sarmentosus* FULA - PULAAR COMBRETACEAE *Terminalia avicennioides, T. macroptera* CUCURBITACEAE *Cucurbita pepo*

boɗi FULA - FULFULDE COMBRETACEAE *Terminalia avicennioides, T. glaucescens*

bodo porto MANDING - MANINKA ANACARDIACEAE *Mangifera indica*

bodomi YORUBA - NAGO COMBRETACEAE *Combretum glutinosum*

bodu MANDING - MANINKA DIOSCOREACEAE *Dioscorea dumetorum*

bóè FULA - PULAAR BOMBACACEAE *Adansonia digitata*

boeboe GBE - VHE CONVOLVULACEAE *Jacquemontia tamnifolia*

boé-ka GBE - AJA APOCYNACEAE *Landolphia owariensis*

boélémimbo FULA - PULAAR ANNONACEAE *Uvaria chamae*

boélénimbo FULA - PULAAR ANNONACEAE *Uvaria chamae*

boflé AKAN - ASANTE CARICACEAE *Carica papaya* BRONG *C. papaya* KULANGO *C. papaya*

boflè BAULE CARICACEAE *Carica papaya*

bofo titi KRU - GUERE ANNONACEAE *Uvaria afzelii*

bofré AKAN - ASANTE CARICACEAE *Carica papaya* BRONG *C. papaya* KULANGO *C. papaya*

bofu AKAN - FANTE ANNONACEAE *Cleistopholis patens*

bo-gar KRU - BASA APOCYNACEAE *Tabernaemontana crassa*

bogbo MANDING - MANINKA COMBRETACEAE *Combretum collinum ssp. geitonophyllum*

bogbon AKYE APOCYNACEAE *Tabernaemontana crassa*

bɔge SUSU ASCLEPIADACEAE *Ceropegia spp.*

boghelelli TAMACHEK CAPPARACEAE *Capparis decidua*

boginaahi FULA - FULFULDE COMPOSITAE *Echinops longifolius*

boginahi HAUSA COMPOSITAE *Echinops longifolius*

boglo BAULE ARALIACEAE *Cussonia arborea*

bɔgɔ MANDING - MANINKA ANACARDIACEAE *Pseudospondias microcarpa*

bogogo KRU - GUERE COMPOSITAE *Struchium sparganophora*

bɔgɔjɛli MENDE CONNARACEAE *Cnestis ferruginea*

bogon koreil SONGHAI AMARANTHACEAE *Aerva javanica*

bogoro KULANGO DILLENIACEAE *Tetracera alnifolia*

bogotéï KULANGO ARALIACEAE *Cussonia arborea*

boguna MANKANYA ANNONACEAE *Uvaria chamae*

boh SERER - NON BOMBACACEAE *Adansonia digitata, Bombax costatum*

bõh MANO AMARANTHACEAE *Celosia argentea*

ɓohere (*pl.* ɓohe, ɓoye) FULA - FULFULDE BOMBACACEAE *Adansonia digitata*

bohi-(é) FULA - PULAAR APOCYNACEAE *Landolphia heudelotii*

bohon KRU - GUERE COMPOSITAE *Triplotaxis stellulifera*

bɔhɔri SUSU CANNACEAE *Canna indica*

bɔhrɔ BULOM CUCURBITACEAE *Citrullus lanatus*

bɔhu MENDE COMBRETACEAE *Combretum grandiflorum*

bohuédèri KWENI COMPOSITAE *Acanthospermum hispidum*

*an*-boi TEMNE APOCYNACEAE *Landolphia calabarica, L. heudelotii, L. hirsuta, L. owariensis*

*a*-boi TEMNE APOCYNACEAE *Tabernaemontana longiflora*

bói FULA - PULAAR COMBRETACEAE *Terminalia macroptera*

bɔidɛ-yamba KONO BIGNONIACEAE *Newbouldia laevis*

boilamauba FULA - PULAAR COMBRETACEAE *Combretum spp. indet.*

boile FULA - FULFULDE ANNONACEAE *Hexalobus monopetalus* PULAAR *H. monopetalus*

boilé FULA - PULAAR ANNONACEAE *Artabotrys velutinus, Hexalobus monopetalus, Uvaria chamae, U. sofa*

boïlé FULA - PULAAR ANNONACEAE *Uvaria chamae*

boilé bonno FULA - PULAAR ANNONACEAE *Hexalobus crispiflorus, Monodora tenuifolia*

boilé-niaddé FULA - PULAAR ANNONACEAE *Uvaria sofa*

boili FULA - FULFULDE ANNONACEAE *Annona senegalensis, Hexalobus monopetalus*

boillé démon FULA - PULAAR CHRYSOBALANACEAE *Acioa scabrifolia*

bo-in-dah KRU - BASA ANACARDIACEAE *Trichoscypha yapoensis*

boiö FULA - PULAAR BOMBACACEAE *Adansonia digitata*

boiokro KULANGO COMPOSITAE *Ageratum conyzoides*

boju FULA - FULFULDE BOMBACACEAE *Ceiba pentandra*

bójúrẹ̀ YORUBA COMPOSITAE *Vernonia cinerea*

bok MANDYAK COMBRETACEAE *Combretum micranthum*

*am*-bók TEMNE ANNONACEAE *Cleistopholis patens*

bokan gida HAUSA AMARANTHACEAE *Celosia isertii, C. trigyna*

bona — AKYE BORAGINACEAE *Cordia senegalensis* BOBO APOCYNACEAE *Landolphia heudelotii*
bonanga — KULANGO DILLENIACEAE *Tetracera alnifolia*
ɔ-bonawa, fruma — AKAN - ASANTE APOCYNACEAE *Tabernaemontana Linn.* KWAWU *Tabernaemontana Linn.* TWI *Tabernaemontana Linn.*
bonawa — AKAN - KWAWU APOCYNACEAE *Tabernaemontana chippii*
bonchu — GUANG - KRACHI BIGNONIACEAE *Newbouldia laevis*
bon-dɛ — BULOM APOCYNACEAE *Rauvolfia vomitoria*
bondi — LOKO COMBRETACEAE *Combretum racemosum, C. smeathmannii*
bɔndɔ-ku — GOLA CUCURBITACEAE *Luffa acutangula*
bondolo — NYENYEGE ANNONACEAE *Annona arenaria, A. senegalensis*
bɔndɔ-ma-poto — BULOM CUCURBITACEAE *Luffa acutangula*
bòndoŋgè — DUALA APOCYNACEAE *Rauvolfia vomitoria*
bondzúrúgi — NUPE DIOSCOREACEAE *Dioscorea Linn.*
bonetan — SONGHAI - ZARMA BOMBACACEAE *Ceiba pentandra*
bonfu — MANDING - MANINKA AMARANTHACEAE *Alternanthera pungens*
bong kapalla — DAGBANI CHRYSOBALANACEAE *Parinari curatellifolia*
boŋgakoŋ — MANDING - MANINKA APOCYNACEAE *Holarrhena floribunda*
bongi — NUPE CUCURBITACEAE *Luffa cylindrica*
bongniendé — DIOLA APOCYNACEAE *Tabernaemontana longiflora*
bongo — BAULE ARALIACEAE *Cussonia bancoensis*
bongo-lakhabi-na — SUSU - DYALONKE CONVOLVULACEAE *Ipomoea batatas*
bɔŋgo-na — SUSU - DYALONKE CARICACEAE *Carica papaya*
bongu — MANDING - MANINKA CELASTRACEAE *Hippocratea pallens*
bɔnje (*def.*-i) — MENDE APOCYNACEAE *Malouetia heudelotii*
*a*-bonk — TEMNE DICHAPETALACEAE *Dichapetalum albidum, D. toxicarium*
*a*-bonk, *a*-nunk — TEMNE DICHAPETALACEAE *Dichapetalum madagascariense*
*ra*-boŋk — TEMNE DIOSCOREACEAE *Dioscorea Linn., D. bulbifera*
*am*-boŋk — TEMNE DIOSCOREACEAE *Dioscorea bulbifera, D. preussii*
bonkhé — SUSU APOCYNACEAE *Landolphia hirsuta, Saba senegalensis*
bonkyu — GUANG - KRACHI BIGNONIACEAE *Newbouldia laevis*
bonntoore (*pl.* bonntooje) — FULA - FULFULDE ARACEAE *Colocasia esculenta*
bono kokó — BAULE ARACEAE *Nephthytis afzelii*
bono takpa — MANO CONNARACEAE *Agelaea trifolia*
*a*-bonthila — TEMNE AMARANTHACEAE *Amaranthus dubius, A. hybridus, A. spinosus*
bǒnù — NUPE APOCYNACEAE *Saba florida*
bonugu — 'MISAHOHE' ARALIACEAE *Cussonia arborea*
bonwo — KRU - BETE COMPOSITAE *Ageratum conyzoides*
bónyinbónyin — YORUBA CONNARACEAE *Cnestis ferruginea*
boobo — KRU - GUERE CUCURBITACEAE *Momordica foetida*
ɓooɗeyi — FULA - FULFULDE COMBRETACEAE *Terminalia glaucescens*
boodi — FULA - FULFULDE COMBRETACEAE *Combretum molle*
ɓoodi (*pl.* ɓoode) — FULA - FULFULDE COMBRETACEAE *Combretum glutinosum*
ɓoodi (*pl.* ɓoode); ɓooɗi — FULA - FULFULDE COMBRETACEAE *Terminalia macroptera*
bɔɔfe-nini — AKAN - ASANTE CARICACEAE *Carica papaya* TWI *C. papaya* WASA *C. papaya*
bɔɔfrɛ — AKAN - ASANTE CARICACEAE *Carica papaya* TWI *C. papaya* WASA *C. papaya*
bòòmba — DUALA BORAGINACEAE *Cordia millenii*
ɓoore — FULA - FULFULDE CONVOLVULACEAE *Ipomoea mauritiana, I. spp. indet.*
bôpace — MANDYAK CHRYSOBALANACEAE *Chrysobalanus icaco*
bopbop — SERER - NON BURSERACEAE *Commiphora africana*
bopire — ANYI - SEHWI ANACARDIACEAE *Lannea welwitschii*
bopiti — SHIEN COMPOSITAE *Adenostemma perrottetii*
bòpòlò-pòlò — DUALA COMPOSITAE *Vernonia conferta, V. frondosa*
bör — WOLOF ANACARDIACEAE *Sclerocarya birrea*
bórbórò-déo — FULA - PULAAR AMARANTHACEAE *Celosia trigyna*
bórborò-dórò — FULA - PULAAR AMARANTHACEAE *Amaranthus viridis*
bordéru — CRIOULO ASCLEPIADACEAE *Calotropis procera*
boré — BALANTA ANNONACEAE *Annona senegalensis* FULA - PULAAR BOMBACACEAE *Adansonia digitata*
boré, goé, gohine — MANDING - BAMBARA APOCYNACEAE *Landolphia heudelotii*
bórè — BALANTA ANNONACEAE *Annona senegalensis*
boré poré — ANYI ANACARDIACEAE *Lannea welwitschii*
bɔrəbɔrɛ — MANDING - MANDINKA AMARANTHACEAE *Amaranthus hybridus* MANINKA *A. hybridus*
bɔrebɔre-ba — LIMBA AMARANTHACEAE *Amaranthus hybridus*
bŏrfule — BASSARI CARICACEAE *Carica papaya*
bori — FULA - PULAAR COMBRETACEAE *Terminalia glaucescens*
bori billel — FULA - PULAAR COMBRETACEAE *Terminalia albida*
bori-bila — FULA - PULAAR COMBRETACEAE *Terminalia albida*
bɔrifira — SISAALA CARICACEAE *Carica papaya*
borin danki — HAUSA CUCURBITACEAE *Lagenaria siceraria*
borna — KYAMA BIGNONIACEAE *Newbouldia laevis*
boro — ABE BIGNONIACEAE *Spathodea campanulata* HAUSA BALANITACEAE *Balanites aegyptiaca*
bóró — MANDING - MANDINKA AMARANTHACEAE *Amaranthus hybridus*
bɔrɔ̃ — MANDING - MANDINKA AMARANTHACEAE *Amaranthus hybridus* MANINKA *A. hybridus*
boro-bila — FULA - PULAAR COMBRETACEAE *Terminalia albida*
boroboro — FULA - PULAAR AMARANTHACEAE *Amaranthus hybridus*

bubakabu DIOLA BOMBACACEAE *Adansonia digitata*
bubangalatab DIOLA COMBRETACEAE *Conocarpus erectus*
bubében barasit DIOLA CAPPARACEAE *Ritchiea capparoides*
bubében mil DIOLA ASCLEPIADACEAE *Secamone afzelii*
bubélé GAGU ASCLEPIADACEAE *Pergularia daemia*
bubéméssi MANDING - MANINKA COMPOSITAE *Vernonia guineensis*
buben MANDING - 'SOCE' ACANTHACEAE *Hygrophila senegalensis* SERER *Peristrophe bicalyculata*
bubɔtɛ LIMBA APOCYNACEAE *Isonema smeathmannii*
bubuka DIOLA ANACARDIACEAE *Lannea acida, L. velutina*
bubukia HAUSA COMBRETACEAE *Combretum aculeatum*
bubum émâdé DIOLA COMBRETACEAE *Terminalia macroptera*
bubum éńab DIOLA COMPOSITAE *Microglossa pyrifolia*
bubumbu BASSARI BOMBACACEAE *Ceiba pentandra*
bubuńgor SERER COMPOSITAE *Blumea aurita*
buburé SONGHAI COMBRETACEAE *Combretum aculeatum* SONGHAI - ZARMA *C. aculeatum*
buchigado PEPEL COMPOSITAE *Acanthospermum hispidum*
buco CRIOULO COMBRETACEAE *Combretum micranthum*
búdafélèk DIOLA APOCYNACEAE *Alstonia boonei*
budâk DIOLA COCHLOSPERMACEAE *Cochlospermum tinctorium*
budanâd DIOLA CONNARACEAE *Cnestis ferruginea*
budanagara MANDING - MANINKA DILLENIACEAE *Tetracera potatoria*
budare YORUBA BALANITACEAE *Balanites wilsoniana*
buday SERER - NON BOMBACACEAE *Ceiba pentandra*
budé MANDING - MANINKA DIOSCOREACEAE *Dioscorea dumetorum*
bude mana SUSU - DYALONKE DIOSCOREACEAE *Dioscorea praehensilis*
budebalod MOORE - NAWDAM CARICACEAE *Carica papaya*
budébeli LOMA ACANTHACEAE *Justicia flava*
budek DIOLA ANACARDIACEAE *Pseudospondias microcarpa*
budey SERER - NON BOMBACACEAE *Ceiba pentandra*
budi FULA - PULAAR CARICACEAE *Carica papaya* CUCURBITACEAE *Citrullus lanatus, Cucumeropsis mannii, Cucurbita maxima, C. maxima, C. pepo*
búdi FULA - PULAAR CUCURBITACEAE *Cucurbita pepo*
budi baga, budi beli FULA - PULAAR CARICACEAE *Carica papaya*
budibaga FULA - PULAAR CARICACEAE *Carica papaya*
budidiyo HAUSA ACANTHACEAE *Justicia schimperi striata*
budigi élen DIOLA ANACARDIACEAE *Lannea nigritana*
budi-lɛdɛ FULA - PULAAR CARICACEAE *Carica papaya*
budimbob DIOLA APOCYNACEAE *Saba senegalensis*
budoy DIOLA ANACARDIACEAE *Spondias mombin*
bududu HAUSA CAPPARACEAE *Crateva adansonii*
budur DIOLA APOCYNACEAE *Saba senegalensis*
bué mali DAN APOCYNACEAE *Rauvolfia vomitoria*
buéco PEPEL COMBRETACEAE *Combretum micranthum*
buɛ-dintɛ BULOM AVICENNIACEAE *Avicennia germinans*
buɛ-dintɛ-lɛ BULOM AVICENNIACEAE *Avicennia germinans*
buel DIOLA CHRYSOBALANACEAE *Neocarya macrophylla*
buɛ-lɛ BULOM AVICENNIACEAE *Avicennia germinans*
buéna BOBO APOCYNACEAE *Landolphia heudelotii*
buenn DIOLA APOCYNACEAE *Landolphia heudelotii*
bueta DIOLA ANACARDIACEAE *Spondias mombin*
buétia DIOLA ANACARDIACEAE *Spondias mombin*
buɛtʃo ADANGME APOCYNACEAE *Carissa edulis*
bufa SERER - NON CAPPARACEAE *Capparis tomentosa*
bufem, bufembabu, fufembabu DIOLA APOCYNACEAE *Landolphia heudelotii*
bufem befol DIOLA APOCYNACEAE *Baissea multiflora*
bufemb DIOLA APOCYNACEAE *Landolphia heudelotii*
bufembabu DIOLA APOCYNACEAE *Landolphia heudelotii*
bufembe DIOLA CELASTRACEAE *Hippocratea paniculata*
bufembõ DIOLA CELASTRACEAE *Hippocratea paniculata*
bufira DIOLA ANACARDIACEAE *Lannea acida*
bufo KONKOMBA BOMBACACEAE *Bombax buonopozense, B. costatum, Ceiba pentandra*
bufo-sõgbum KONKOMBA BOMBACACEAE *Ceiba pentandra*
bufu BASSARI BOMBACACEAE *Ceiba pentandra*
bufuluk DIOLA COMBRETACEAE *Guiera senegalensis*
bufum el eid ARABIC COMBRETACEAE *Combretum micranthum*
bufumay sibé DIOLA COMBRETACEAE *Terminalia macroptera*
bugi FULA - PULAAR CAPPARACEAE *Capparis decidua*
bugi baley FULA - PULAAR CAPPARACEAE *Capparis tomentosa*
bugimb DIOLA APOCYNACEAE *Saba senegalensis*
bugu mugu MANDING - BAMBARA CONVOLVULACEAE *Ipomoea pes-tigridis*
buguhi, logul bahi, sogui FULA - PULAAR CAPPARACEAE *Maerua angolensis*
buguhi FULA - PULAAR CAPPARACEAE *Maerua angolensis*
bugulé KYAMA COMPOSITAE *Vernonia conferta*
bugulli FULA - FULFULDE ARACEAE *Amorphophallus abyssinicus*
bugulli, gugulli FULA - FULFULDE ARACEAE *Anchomanes difformis*
bugundio MANDYAK COMBRETACEAE *Terminalia albida*
bugunha MANDYAK ANNONACEAE *Uvaria afzelii, U. chamae* MANKANYA *U. afzelii, U. chamae*

| | |
|---|---|
| buhal bainé | DIOLA ANNONACEAE *Uvaria thomasii* |
| buhal baré | DIOLA ANNONACEAE *Uvaria chamae* |
| buhangossa | BATONNUN COMBRETACEAE *Combretum collinum ssp. hypopilinum* |
| buhek | DIOLA AVICENNIACEAE *Avicennia germinans* |
| buhemck | DIOLA - FLUP APOCYNACEAE *Landolphia heudelotii* |
| buheo | DIOLA ANNONACEAE *Xylopia aethiopica* |
| buhiébu | GURMA COMBRETACEAE *Anogeissus leiocarpus* |
| buhiki | FULA - FULFULDE COMBRETACEAE *Combretum nigricans* PULAAR *C. nigricans* |
| buhinkabu | DIOLA COMBRETACEAE *Terminalia macroptera* |
| buhunuk | DIOLA - FLUP COMBRETACEAE *Guiera senegalensis* |
| bui | WOLOF BOMBACACEAE *Adansonia digitata* |
| buibu | KWENI COMPOSITAE *Vernonia guineensis* |
| buiki | FULA - PULAAR COMBRETACEAE *Combretum nigricans* |
| buindip | DIOLA - FLUP APOCYNACEAE *Saba senegalensis* |
| buisi | BALANTA COMBRETACEAE *Guiera senegalensis* |
| buiti | FULA - PULAAR COMBRETACEAE *Combretum nigricans* |
| buitipobou | DIOLA APOCYNACEAE *Saba senegalensis* |
| buitipobu | DIOLA APOCYNACEAE *Saba senegalensis* |
| bùjé wẹẹrẹ | YORUBA ANACARDIACEAE *Sorindeia warneckei* |
| bujingkabo | DIOLA COMBRETACEAE *Terminalia macroptera* |
| *an*-buk | TEMNE DIOSCOREACEAE *Dioscorea burkilliana* |
| bukâgir | DIOLA CELASTRACEAE *Salacia senegalensis* |
| bukayu | DIOLA ANACARDIACEAE *Anacardium occidentale* |
| bukbɔ (*def*.-i) | MENDE CONVOLVULACEAE *Ipomoea alba* |
| bukbui | LIMBA CONVOLVULACEAE *Ipomoea alba* |
| bukelek | BANYUN AVICENNIACEAE *Avicennia germinans* |
| bukhsa langa | ARABIC - SHUWA CUCURBITACEAE *Lagenaria siceraria* |
| buki | FULA - PULAAR COMBRETACEAE *Combretum glutinosum, C. nigricans* |
| buki-danehi | FULA - PULAAR COMBRETACEAE *Combretum collinum ssp. hypopilinum, C. collinum ssp. geitonophyllum* |
| bukinebélége | DIOLA ANACARDIACEAE *Lannea acida* |
| bukir | ARABIC CRUCIFERAE *Raphanus sativus* |
| bukob | MANDYAK BIGNONIACEAE *Newbouldia laevis* |
| bukobatané | MANDYAK ACANTHACEAE *Nelsonia canescens* |
| bukokökes | MANDYAK CAPPARACEAE *Ritchiea capparoides* |
| bukuf | MANKANYA BIGNONIACEAE *Newbouldia laevis* |
| bukuma | ANYI CHRYSOBALANACEAE *Maranthes glabra, M. robusta* |
| bulabula | HAUSA COMMELINACEAE *Commelina benghalensis* |
| bulanay | DIOLA APOCYNACEAE *Saba senegalensis* |
| bulapâl | FULA - FULFULDE COMBRETACEAE *Combretum aculeatum* PULAAR *C. aculeatum* |
| buldumhi | FULA - FULFULDE BIGNONIACEAE *Stereospermum kunthianum* |
| búle | PEPEL AVICENNIACEAE *Avicennia germinans* |
| buledia | TEM ANNONACEAE *Cleistopholis patens* |
| bulekpéï | KRU - BETE ASCLEPIADACEAE *Secamone afzelii* |
| bulel | DIOLA - FLUP CELASTRACEAE *Salacia senegalensis* |
| búlè-n'baba | FULA - PULAAR COMPOSITAE *Acanthospermum hispidum* |
| buleno | DIOLA ANNONACEAE *Xylopia aethiopica* |
| buleo | DIOLA - FLUP ANNONACEAE *Uvaria chamae* |
| buléo | DIOLA ANNONACEAE *Uvaria chamae* |
| bulèv bèyné | DIOLA ANNONACEAE *Xylopia parviflora* |
| bulgui | ARABIC CAPPARACEAE *Capparis sepiaria* |
| buli | SUSU COMMELINACEAE *Cyanotis longifolia* |
| bulibam | DIOLA COMBRETACEAE *Terminalia macroptera* |
| bulil | DIOLA - FLUP CELASTRACEAE *Salacia senegalensis* |
| buli-yeyakɔ, nyiya yeyakɔ (*def*.-i) | MENDE ASCLEPIADACEAE *Gongronema latifolium* |
| bùlji, kíngámà | GUANG - GONJA CUCURBITACEAE *Cucurbita moschata, C. pepo* |
| bulõ | BASSARI APOCYNACEAE *Strophanthus hispidus* |
| bulobulu | MANDING - MANDINKA CONVOLVULACEAE *Ipomoea mauritiana* |
| bũlɔko | VAI CUCURBITACEAE *Luffa cylindrica* |
| bulɔmbuxo | DAGAARI ANNONACEAE *Annona senegalensis* |
| bulong | DAGAARI APOCYNACEAE *Strophanthus hispidus* KONKOMBA *S. hispidus* |
| bùlóngó | KANURI BIGNONIACEAE *Kigelia africana* |
| bultu | TUBU CAPPARACEAE *Boscia senegalensis* |
| bûltù | KANURI CAPPARACEAE *Boscia senegalensis, Cadaba farinosa* |
| bulu | GAGU COMBRETACEAE *Terminalia ivorensis* |
| bulu kuruné | MANDING - MANINKA APOCYNACEAE *Adenium obesum* |
| bulu kuruni | MANDING - BAMBARA ARALIACEAE *Cussonia arborea* |
| bulu-kuntu | MANDING - BAMBARA ARALIACEAE *Cussonia arborea* |
| bulukuntu | MANDING - MANINKA ARALIACEAE *Cussonia arborea* SUSU *C. arborea* |
| bulukunyu | SUSU ARALIACEAE *Cussonia arborea* |
| bulule | KPE BIGNONIACEAE *Kigelia africana* |
| bululuk | DIOLA ANNONACEAE *Annona glabra* |
| bululumay | DIOLA ANACARDIACEAE *Anacardium occidentale* |
| bulumbogo | DAGAARI ANNONACEAE *Annona senegalensis* DAGBANI *A. squamosa* |
| ɓulumwol (*pl.* ɓulumji) | FULA - FULFULDE DIOSCOREACEAE *Dioscorea Linn.* |
| bulunonima | MANDING - MANINKA AMARANTHACEAE *Amaranthus spinosus* |
| bulusintim | AKAN - WASA ANNONACEAE *Monodora tenuifolia* |
| bulusoana | KANURI BORAGINACEAE *Trichodesma africanum* |
| bulusor | DIOLA COMBRETACEAE *Combretum micranthum* |

bum LUNDU BOMBACACEAE *Ceiba pentandra*
buma AKYE COMBRETACEAE *Terminalia ivorensis* KPE BOMBACACEAE *Ceiba pentandra* KUNDU *Rhodognaphalon brevicuspe*
bŭmà DUALA BOMBACACEAE *Ceiba pentandra*
bumafaye DIOLA CHRYSOBALANACEAE *Neocarya macrophylla*
bumakurin DIOLA ANACARDIACEAE *Lannea nigritana*
bum-ane BALANTA CYPERACEAE *Cyperus articulatus*
bumansabu GURMA CHRYSOBALANACEAE *Maranthes polyandra, Parinari curatellifolia*
bumarlahi FULA - FULFULDE ARALIACEAE *Cussonia arborea*
bumatiap DIOLA APOCYNACEAE *Holarrhena floribunda*
bumay DIOLA COSTACEAE *Costus afer*
bumbaŋ (*def.*-o) MANDING - MANDINKA ASCLEPIADACEAE *Xysmalobium heudelotianum*
bumbari ADYUKRU APOCYNACEAE *Landolphia owariensis*
bumboni ADYUKRU APOCYNACEAE *Landolphia owariensis*
bumbop WOLOF ACANTHACEAE *Hypoestes verticillaris*
bumbu HAUSA BOMBACACEAE *Adansonia digitata* CUCURBITACEAE *Lagenaria siceraria* MANDING - BAMBARA BOMBACACEAE *Bombax buonopozense, B. costatum* MANDINKA *B. buonopozense* MANINKA *B. costatum*
bumbui ADYUKRU APOCYNACEAE *Landolphia owariensis* FULA - PULAAR BOMBACACEAE *Bombax costatum*
bumbum BALANTA BOMBACACEAE *Bombax costatum*
bumbusoi MANDING - MANINKA BIGNONIACEAE *Stereospermum acuminatissimum*
bumbuvi FULA - FULFULDE BOMBACACEAE *Bombax costatum* PULAAR *B. costatum*
bumkumo MANDING - 'SOCE' ANACARDIACEAE *Spondias mombin*
bumkuô SOCE BOMBACACEAE *Bombax costatum*
bumu MANDING - BAMBARA BOMBACACEAE *Bombax buonopozense, B. costatum* MANINKA *B. costatum*
buna AKYE COMBRETACEAE *Terminalia ivorensis* BOBO APOCYNACEAE *Landolphia heudelotii* MANO ANACARDIACEAE *Spondias mombin*
bùnà JUNKUN CUCURBITACEAE *Lagenaria siceraria*
buńâbu DIOLA BOMBACACEAE *Bombax costatum*
bunamabu GURMA ANACARDIACEAE *Sclerocarya birrea*
bunamkarésabu DIOLA BIGNONIACEAE *Markhamia tomentosa*
bunbo BAULE APOCYNACEAE *Voacanga africana*
bùnbun YORUBA DIOSCOREACEAE *Dioscorea rotundata*
buncum (*def.*-ô) MANDING - MANDINKA BOMBACACEAE *Bombax costatum*
bundɛ KONO CONVOLVULACEAE *Ipomoea batatas*
bundif DIOLA APOCYNACEAE *Saba florida*
bundu KONO BIXACEAE *Bixa orellana*
buńehol DIOLA - FLUP APOCYNACEAE *Landolphia dulcis*
bunga FULA - FULFULDE CUCURBITACEAE *Lagenaria siceraria*
bungda TERA CUCURBITACEAE *Lagenaria siceraria*
bungdi TERA - PIDLIMNDI CUCURBITACEAE *Lagenaria siceraria*
bungkungo (*def.*) MANDING - MANDINKA BOMBACACEAE *Bombax costatum*
*am*-bunk TEMNE DIOSCOREACEAE *Dioscorea smilacifolia*
*am*-buŋk TEMNE DIOSCOREACEAE *Dioscorea hirtiflora, D. minutiflora*
buŋkankoŋ MANDING - MANINKA APOCYNACEAE *Funtumia africana, F. elastica*
bunkuŋ (*def.*-o) MANDING - MANDINKA BOMBACACEAE *Bombax costatum*
buńohol DIOLA - FLUP APOCYNACEAE *Landolphia heudelotii*
bunou MANDYAK CHRYSOBALANACEAE *Neocarya macrophylla*
bunsulu fadama HAUSA CYPERACEAE *Fimbristylis debilis*
buntunti BULISHA CONVOLVULACEAE *Ipomoea asarifolia*
bupâpa DIOLA CARICACEAE *Carica papaya*
bupapabu GURMA BALANITACEAE *Balanites aegyptiaca*
bupumba pumb DIOLA ASCLEPIADACEAE *Calotropis procera*
bura bura ba MANDING - BAMBARA AMARANTHACEAE *Amaranthus hybridus*
burabaya HAUSA COMMELINACEAE *Aneilema lanceolatum, Commelina erecta, C. forskalaei*
bùrábayà HAUSA COMMELINACEAE *Commelina erecta*
bùrábayà, hánjín kúdáa, kunan curu, tubanin dawaki HAUSA COMMELINACEAE *Commelina forskalaei*
bura-bura ba FULA - PULAAR COMMELINACEAE *Commelina erecta*
bura-bura-ba FULA - PULAAR COMMELINACEAE *Commelina forskalaei*
bural DIOLA APOCYNACEAE *Rauvolfia vomitoria*
burane DIOLA CAPPARACEAE *Ritchiea capparoides*
bùrár jaakii HAUSA ARACEAE *Amorphophallus dracontioides*
bùrár kàree HAUSA ARACEAE *Amorphophallus dracontioides*
bùrár kàreé HAUSA ARACEAE *Amorphophallus abyssinicus*
burbog FULA - PULAAR CUCURBITACEAE *Momordica balsamina*
burbóqui FULA - PULAAR CUCURBITACEAE *Momordica charantia*
búrbúrá (*pl.* -á) SONGHAI COMBRETACEAE *Combretum aculeatum*
burdi ARABIC CYPERACEAE *Cyperus papyrus*
*an*-bure TEMNE AVICENNIACEAE *Avicennia germinans*
burekume-na SUSU - DYALONKE COMPOSITAE *Gutenbergia nigritana*
burhan SERER - NON AVICENNIACEAE *Avicennia germinans*
buri 'SOUBRE' COMBRETACEAE *Terminalia ivorensis*
buri nánéwi FULA - TUKULOR ASCLEPIADACEAE *Caralluma dalzielii*
buri nańey FULA - TUKULOR ASCLEPIADACEAE *Caralluma dalzielii*
burńéńé FULA - TUKULOR ASCLEPIADACEAE *Caralluma dalzielii*
buro abéle ANYI BALANOPHORACEAE *Thonningia sanguinea*
buroki FULA - PULAAR CUCURBITACEAE *Momordica charantia*

INDEX


dale baali — FULA - FULFULDE CONVOLVULACEAE *Ipomoea muricata*
daleji — FULA - FULFULDE BURSERACEAE *Commiphora kerstingii*
dalémulɪɔr — BANYUN AIZOACEAE *Sesuvium portulacastrum*
dalévi — FULA - FULFULDE CAPPARACEAE *Capparis tomentosa* FULAAR *C. tomentosa*
dali — HAUSA BURSERACEAE *Commiphora kerstingii*
dalkané — WOLOF AMARYLLIDACEAE *Crinum zeylanicum*
dalo — HAUSA COMBRETACEAE *Combretum fragrans, C. glutinosum*
dalôga — SERER AIZOACEAE *Sesuvium portulacastrum*
dalubu — KANURI ANACARDIACEAE *Lannea barteri*
dam tap — WOLOF APOCYNACEAE *Baissea multiflora*
dama — HAUSA ANACARDIACEAE *Sclerocarya birrea* KANURI *Lannea barteri*
damargu rafi — HAUSA ASCLEPIADACEAE *Pergularia tomentosa*
dambrohia — KYAMA ANNONACEAE *Greenwayodendron oliveri*
dambu — ARABIC ANACARDIACEAE *Sclerocarya birrea*
dambun makiyaayaa — HAUSA ACANTHACEAE *Nelsonia canescens*
dambuntsofi — HAUSA CONVOLVULACEAE *Ipomoea aitonii*
damdam shiya — HAUSA CAPPARACEAE *Boscia senegalensis*
damdamshiya — HAUSA CAPPARACEAE *Boscia senegalensis*
damdun makiyaya — HAUSA ACANTHACEAE *Nelsonia canescens*
damfarƙami — HAUSA ACANTHACEAE *Monechma ciliatum* AMARANTHACEAE *Pandiaka heudelotii, P. involucrata*
damoré guren — BANYUN CELASTRACEAE *Hippocratea paniculata*
damoruhi — FULA - FULFULDE COMBRETACEAE *Combretum molle*
damrâne — ARABIC CHENOPODIACEAE *Chenolea lanata, Traganum nudatum*
damrat — WOLOF COMBRETACEAE *Combretum nigricans*
damsa — HAUSA BOMBACACEAE *Adansonia digitata*
dámu — DOGON BOMBACACEAE *Ceiba pentandra*
damugurumo — 'SENUFO' ANNONACEAE *Annona arenaria, A. senegalensis*
damurana — 'SENUFO' ANNONACEAE *Annona arenaria, A. senegalensis*
damzin — KRIO BURSERACEAE *Dacryodes klaineana, Santiria trimera*
dan — MANDING - BAMBARA ANNONACEAE *Annona arenaria, A. senegalensis*
ɗan danya — HAUSA ANACARDIACEAE *Haematostaphis barteri*
ɗan jallo — HAUSA CUCURBITACEAE *Lagenaria siceraria*
ɗan kwakwangi — HAUSA CUCURBITACEAE *Lagenaria siceraria*
ɗan loma — HAUSA CAPPARACEAE *Boscia senegalensis*
ɗan magami — HAUSA BURSERACEAE *Boswellia dalzielii*
ɗan muda — HAUSA CUCURBITACEAE *Lagenaria siceraria*
ɗán sárkin-ìtaátúwàa — HAUSA BIGNONIACEAE *Stereospermum kunthianum*
dan Tunuga, kajiji ɗan Tunuga — HAUSA CYPERACEAE *Cyperus maculatus*
dan Tunuga, kajijiɗan Tunuga — HAUSA CYPERACEAE *Cyperus tonkinensis*
ɗan zindiri — HAUSA ASCLEPIADACEAE *Leptadenia hastata*
dana — VULGAR DIOSCOREACEAE *Dioscorea bulbifera* MANDING - BAMBARA COMBRETACEAE *Pteleopsis suberosa* DIOSCOREACEAE *Dioscorea bulbifera, D. bulbifera* MANINKA *D. bulbifera*
dȁna — GA DIOSCOREACEAE *Dioscorea alata, D. rotundata*
danana — HAUSA BOMBACACEAE *Adansonia digitata*
ɗánbaàkúwá — HAUSA ASCLEPIADACEAE *Leptadenia hastata*
danbaka sayaba — HAUSA BURSERACEAE *Commiphora africana*
ɗán-bàraáwoò — HAUSA ASCLEPIADACEAE *Leptadenia hastata*
danda — VULGAR DIOSCOREACEAE *Dioscorea bulbifera* MANDING - BAMBARA *D. bulbifera* MANDINKA *D. bulbifera* MANINKA *D. bulbifera*
dãnda — MANDING - BAMBARA DIOSCOREACEAE *Dioscorea cayenensis*
danda ba — MANDING - BAMBARA DIOSCOREACEAE *Dioscorea alata*
danda dion — MANDING - MANINKA DIOSCOREACEAE *Dioscorea hirtiflora*
dandalub — SERER CUCURBITACEAE *Cucumis metuliferus*
dandam(-ô) — MANDING - MANDINKA DIOSCOREACEAE *Dioscorea preussii*
dandami — HAUSA AIZOACEAE *Gisekia pharnaceoides*
dandamo-besi — MANDING - MANDINKA CONNARACEAE *Cnestis ferruginea*
dan-dan — MANDING - MANINKA DIOSCOREACEAE *Dioscorea bulbifera*
dandandi — FULA - PULAAR CYPERACEAE *Fuirena umbellata*
dandandim (*def.*-ô) — MANDING - MANDINKA DIOSCOREACEAE *Dioscorea bulbifera*
dandaõ — MANDING - MANDINKA DIOSCOREACEAE *Dioscorea bulbifera*
ɗandar mahalba, danji, danyi, fárár kábàa — HAUSA COMPOSITAE *Centaurea perrottetii*
dandare — HAUSA BOMBACACEAE *Adansonia digitata*
dandegha — MOORE COMBRETACEAE *Combretum micranthum*
dandu — KYAMA ANNONACEAE *Xylopia rubescens*
dané — SUSU DIOSCOREACEAE *Dioscorea bulbifera*
ɗáŋəlàŋ — DERA BOMBACACEAE *Bombax costatum*
danga — MANDING - BAMBARA CHRYSOBALANACEAE *Neocarya macrophylla*
dàngálì, kúndùwú, kuwunduwu — KANURI CONVOLVULACEAE *Ipomoea batatas*
dangan — MANDING - BAMBARA ANNONACEAE *Annona glauca, A. senegalensis*
dangarafa — HAUSA CAPPARACEAE *Cadaba farinosa*
dang-dang — CHAMBA ACANTHACEAE *Justicia schimperi striata*
dangné — GBE - FON APOCYNACEAE *Picralima nitida*
danguan — MANDING - BAMBARA ANNONACEAE *Annona arenaria, A. senegalensis*
dangué — KWENI BOMBACACEAE *Ceiba pentandra*
dangura — MAMBILA CONVOLVULACEAE *Ipomoea batatas*
dânha — MANDING - BAMBARA ANNONACEAE *Annona senegalensis*
daŋinyɛ-fuɛ̃ — MENDE ACANTHACEAE *Brillantaisia nitens*

dietwa AKAN - ASANTE APOCYNACEAE *Strophanthus preussii* TWI *S. preussii*
diézé KRU - GUERE CONNARACEAE *Cnestis ferruginea*
[illegible] dif wir TEMNE COMPOSITAE *Vernonia amygdalina, V. colorata*
digai sa SONGHAI CYPERACEAE *Cyperus rotundus*
digbar BASSARI COMBRETACEAE *Pteleopsis suberosa*
digbaré BASSARI COMBRETACEAE *Pteleopsis suberosa*
digbé KRU - GUERE CONNARACEAE *Cnestis corniculata*
digegbe MANDING - MANINKA DIOSCOREACEAE *Dioscorea bulbifera*
digɛgbe MANDING - MANINKA DIOSCOREACEAE *Dioscorea praehensilis*
digɛgbe, digɛgbwe MANDING - MANINKA DIOSCOREACEAE *Dioscorea togoensis*
diginegri BASSARI COMBRETACEAE *Combretum fragrans*
digingel KONKOMBA COMBRETACEAE *Combretum molle*
digingire BASSARI COMBRETACEAE *Combretum molle*
diginshayib ARABIC - SHUWA AMARANTHACEAE *Aerva javanica*
digityeh MANDING - BAMBARA CYPERACEAE *Cyperus rotundus*
digor WOLOF ANNONACEAE *Annona senegalensis*
digor yéner WOLOF ANNONACEAE *Annona glabra*
digor yénèr WOLOF ANNONACEAE *Annona glauca*
digripayuassu 'NEKEDIE' CONNARACEAE *Agelaea obliqua*
digui TUBU BURSERACEAE *Commiphora africana*
dií IJO - IZON COMBRETACEAE *Terminalia ivorensis*
dìì bibíle DOGON CONVOLVULACEAE *Ipomoea aquatica*
dìì dóó DOGON COMPOSITAE *Gnaphalium polycaulon*
dìì gíru DOGON ASCLEPIADACEAE *Gymnema sylvestre*
diìnàá IJO - IZON COMMELINACEAE *Aneilema umbrosum*
diìnéy (*pl.* -à) SONGHAI ANACARDIACEAE *Sclerocarya birrea*
dika AKAN - TWI DIOSCOREACEAE *Dioscorea misc. & spp. unknown* BATONNUN *D. abyssinica*
dikàbò DUALA ARACEAE *Xanthosoma mafaffa*
dikof BANYUN APOCYNACEAE *Landolphia heudelotii*
dikpa DAN APOCYNACEAE *Pleiocarpa mutica*
dila HAUSA CAPPARACEAE *Boscia senegalensis*
dīlissli vili MANDING - BAMBARA ASCLEPIADACEAE *Caralluma decaisneana*
diliwoplu 'KRU' ASCLEPIADACEAE *Pergularia daemia*
dilo HAUSA CAPPARACEAE *Boscia angustifolia, B. senegalensis*
dimori BASSARI CYPERACEAE *Cyperus esculentus*
dimpâpaô MANDING - 'SOCE' ASCLEPIADACEAE *Calotropis procera*
dìn bèrè cìn a'mìlě NUPE BIGNONIACEAE *Newbouldia laevis*
dina (*def.* dinɛi) MENDE ARACEAE *Cyrtosperma senegalense*
dina tali DAN ARACEAE *Anchomanes difformis*
dìnáká BEROM APOCYNACEAE *Landolphia owariensis, Saba florida*
din-a-lah MANO BIGNONIACEAE *Newbouldia laevis*
dinde VULGAR ARACEAE *Colocasia esculenta*
dìnde DUALA ARACEAE *Colocasia esculenta*
dindji SONGHAI APOCYNACEAE *Landolphia owariensis*
dindo GBE - VHE BOMBACACEAE *Adansonia digitata*
dindon MANDING - MANINKA BIGNONIACEAE *Kigelia africana*
dinégna SONGHAI ANACARDIACEAE *Sclerocarya birrea*
dinenkui AKAN - BRONG COMPOSITAE *Bidens pilosa*
dineygna SONGHAI ANACARDIACEAE *Sclerocarya birrea*
ding on KWENI BIGNONIACEAE *Newbouldia laevis*
dingno dingon KWENI BIGNONIACEAE *Newbouldia laevis*
dini MANO DIOSCOREACEAE *Dioscorea Linn.*
dinjongo KUNDU APOCYNACEAE *Funtumia elastica*
dinkòrò'bá NUPE AMARYLLIDACEAE *Crinum zeylanicum*
dinsinkoro AKAN - ASANTE CAPPARACEAE *Euadenia eminens*
dinsinkɔro AKAN - TWI CONVOLVULACEAE *Ipomoea mauritiana*
dinyar gata HAUSA COMBRETACEAE *Combretum molle*
dinyɛŋgɛ DUALA CUCURBITACEAE *Luffa cylindrica*
dinyɔŋgɔ̀, èbòŋgo DUALA APOCYNACEAE *Funtumia elastica*
dinza BOLE APOCYNACEAE *Saba florida*
diô KRU - GUERE APOCYNACEAE *Funtumia elastica* BOMBACACEAE *Ceiba pentandra*
diohi FULA - PULAAR BOMBACACEAE *Bombax costatum*
diokundio BASSARI COMBRETACEAE *Combretum nigricans*
dionbibinega MOBA COMBRETACEAE *Combretum molle*
diorgud WOLOF ANNONACEAE *Annona senegalensis*
dioy FULA - PULAAR BOMBACACEAE *Bombax costatum*
dipagné KRU - GUERE ASCLEPIADACEAE *Pergularia daemia*
dipé KRU - GUERE ASCLEPIADACEAE *Pergularia daemia*
dipoi KONO CYPERACEAE *Mariscus alternifolius*
dirbi TERA - PIDLIMNDI CUCURBITACEAE *Lagenaria siceraria*
diri KWENI ARACEAE *Anchomanes difformis*
dirimé 'KRU' CYPERACEAE *Scleria boivinii*
diriniblé MANDING - BAMBARA COMBRETACEAE *Combretum collinum ssp. geitonophyllum* MANINKA *C. collinum ssp. geitonophyllum*
dirinimble MANDING - BAMBARA COMBRETACEAE *Combretum collinum ssp. geitonophyllum*
dirro KRU - GUERE APOCYNACEAE *Funtumia elastica*
dirutu 'KRU' BURSERACEAE *Canarium schweinfurthii*
disile MANDING - MANINKA BOMBACACEAE *Bombax buonopozense*
ditopédamen DIOLA ANACARDIACEAE *Lannea acida*

dogbodogobo **YORUBA** CYPERACEAE *Cyperaceae*

dogbojelegbo **MENDE** CELASTRACEAE *Salacia senegalensis*

dogbuéï **KRU - BETE** APOCYNACEAE *Tabernaemontana crassa*

dogo **KWENI** CAPPARACEAE *Euadenia eminens*

dogut **WOLOF** ANNONACEAE *Annona senegalensis*

doï **KRU - BETE** COSTACEAE *Costus afer*

doki **FULA - FULFULDE** COMBRETACEAE *Combretum Loefl.* **PULAAR** *C. collinum ssp. geitonophyllum, C. glutinosum*

dóki **FULA - PULAAR** COMBRETACEAE *Combretum glutinosum*

dóki déwi **FULA - PULAAR** COMBRETACEAE *Combretum glutinosum*

dóki gori **FULA - PULAAR** COMBRETACEAE *Combretum glutinosum*

dókidéwi **FULA - PULAAR** COMBRETACEAE *Combretum collinum ssp. geitonophyllum*

dokigori **FULA - FULFULDE** COMBRETACEAE *Combretum nigricans* **PULAAR** *C. nigricans*

dɔkɔ **ADANGME** CONVOLVULACEAE *Lepistemon owariense*

doko woro **FULA - PULAAR** COMBRETACEAE *Combretum glutinosum*

dokowa (*pl.* nnokowa) **AKAN - TWI** DIOSCOREACEAE *Dioscorea misc. & spp. unknown*

dɔkpɔ̃ **KISSI** AMARANTHACEAE *Cyathula prostrata*

doleke **MANDING - MANINKA** ANACARDIACEAE *Pseudospondias microcarpa*

dɔli **GBE - VHE** ARACEAE *Anchomanes difformis*

doliségi **MANDING - BAMBARA** ANACARDIACEAE *Ozoroa insignis*

dollɛ **MANDING - MANINKA** BURSERACEAE *Canarium schweinfurthii*

dollo **KISSI** BURSERACEAE *Canarium schweinfurthii*

dolo **KISSI** BURSERACEAE *Canarium schweinfurthii*

dologa **FULA - FULFULDE** ANACARDIACEAE *Pseudospondias microcarpa* **PULAAR** *P. microcarpa*

dologa-chango **FULA - PULAAR** ANACARDIACEAE *Trichoscypha longifolia*

dolohom **WOLOF** DIOSCOREACEAE *Dioscorea praehensilis*

domapowa **AKAN - BRONG** COMBRETACEAE *Combretum collinum ssp. binderianum*

domashi **HAUSA** COMPOSITAE *Vernonia adoensis*

domboɛ **KONO** BURSERACEAE *Santiria trimera*

dombos **WOLOF** CUCURBITACEAE *Cucurbita maxima, C. pepo*

dɔme atrɛ **AKAN - TWI** CONVOLVULACEAE *Cuscuta australis*

don **KRU - GUERE** ARACEAE *Anchomanes difformis* COMPOSITAE *Mikania cordata* COSTACEAE *Costus afer*

dondoi **LOKO** CYPERACEAE *Fimbristylis dichotoma*

dondré **KRU - GUERE** COMMELINACEAE *Palisota hirsuta*

donga **NZEMA** ANNONACEAE *Xylopia staudtii*

donienia donien **BAULE** COMPOSITAE *Microglossa afzelii*

dɔnkɔ **GBE - VHE** COMBRETACEAE *Terminalia superba*

donkori **MANDING - BAMBARA** CAPPARACEAE *Capparis tomentosa*

donniania **BAULE** ASCLEPIADACEAE *Secamone afzelii*

donomane **MANDING - MANINKA** ASCLEPIADACEAE *Secamone afzelii*

dononturu **FULA - PULAAR** COMPOSITAE *Pleiotaxis chlorolepis, Vernonia nigritiana*

doo **KRU - BASA** CONVOLVULACEAE *Bonamia thunbergiana*

dookehi **FULA - FULFULDE** COMBRETACEAE *Combretum collinum ssp. geitonophyllum*

dooki **FULA - FULFULDE** COMBRETACEAE *Combretum collinum ssp. geitonophyllum, C. collinum ssp. hypopilinum, C. glutinosum* **PULAAR** *C. collinum ssp. hypopilinum, C. fragrans, C. glutinosum, C. nigricans*

do-oki **WOLOF** COMBRETACEAE *Combretum spp. indet.*

dóóki **FULA - PULAAR** COMBRETACEAE *Combretum glutinosum*

dooki daneehi **FULA - FULFULDE** COMBRETACEAE *Combretum collinum ssp. geitonophyllum*

dooko **FULA - PULAAR** COMBRETACEAE *Combretum glutinosum* **SONINKE - SARAKOLE** *C. glutinosum*

dóóya **HAUSA** DIOSCOREACEAE *Dioscorea abyssinica, D. bulbifera, D. praehensilis, D. rotundata*

doóyaà, dóóya **HAUSA** DIOSCOREACEAE *Dioscorea Linn.*

dóóyan ítaacéé **HAUSA** DIOSCOREACEAE *Dioscorea bulbifera*

doóyàr daájii **HAUSA** DIOSCOREACEAE *Dioscorea odoratissima, D. praehensilis*

dóóyar daji **HAUSA** DIOSCOREACEAE *Dioscorea Linn.*

dóóyar danga **HAUSA** DIOSCOREACEAE *Dioscorea Linn.*

doóyàr duútseè **HAUSA** DIOSCOREACEAE *Dioscorea praehensilis*

doóyàr giíwaá **HAUSA** DIOSCOREACEAE *Dioscorea praehensilis*

doóyàr kúdù **HAUSA** DIOSCOREACEAE *Dioscorea cayenensis*

doóyar-bisà, dóóyar-bisa **HAUSA** DIOSCOREACEAE *Dioscorea bulbifera*

*n*dɔpalaro **LOKO** DILLENIACEAE *Tetracera potatoria*

*n*dɔpa-nɛ, *n*dɔpa-nɛɛ **MENDE - KPA** DILLENIACEAE *Tetracera potatoria*

*n*dɔpararo **LOKO** DILLENIACEAE *Tetracera potatoria*

doréké **MANDING - MANINKA** ANACARDIACEAE *Pseudospondias microcarpa*

dorgot **WOLOF** ANNONACEAE *Annona senegalensis*

ḍori **WOLOF** CELASTRACEAE *Maytenus senegalensis*

ḍorombolan **SERER** CONVOLVULACEAE *Ipomoea muricata*

ḍorombolan ala **SERER** CONVOLVULACEAE *Ipomoea obscura aggregate*

doron **MANDING - MANINKA** CYPERACEAE *Kyllinga erecta*

ḍorosé populu **KRU - GREBO** APOCYNACEAE *Funtumia elastica*

ḍosé **BALANTA** COMBRETACEAE *Guiera senegalensis*

dosso **GBE - FON** COMBRETACEAE *Combretum collinum ssp. geitonophyllum, C. glutinosum*

dotihi **FULA - PULAAR** COMBRETACEAE *Combretum glutinosum*

do-ué KRU - GUERE CECROPIACEAE *Musanga cecropioides*
douoblé BAULE DIOSCOREACEAE *Dioscorea alata*
dowé KRU - GUERE CAPPARACEAE *Euadenia trifoliolata*
dɔwɔ (*def.* dɔwei) MANDING - MANINKA ANACARDIACEAE *Pseudospondias microcarpa*
doxonyá BASARI ACANTHACEAE *Rungia eriostachya*
doy FULA - PULAAR BOMBACACEAE *Bombax costatum*
doya FULA - FULFULDE DIOSCOREACEAE *Dioscorea Linn.*
doyarama djangol FULA - PULAAR CYPERACEAE *Lipocarpa chinensis*
dreeka GBE - VHE DIOSCOREACEAE *Dioscorea rotundata*
drehn KRU - BASA ANNONACEAE *Xylopia staudtii*
drignon, grigno iri KWENI ASCLEPIADACEAE *Pergularia daemia*
druhin KRU - GUERE ARACEAE *Culcasia scandens*
dschola TEM CUCURBITACEAE *Lagenaria siceraria*
dsetʃi ADANGME - KROBO COMPOSITAE *Bidens pilosa*
dshendsha TEM CUCURBITACEAE *Lagenaria siceraria*
dshingia TEM APOCYNACEAE *Motandra guineensis*
dua sika AKAN - ASANTE ANNONACEAE *Enantia polycarpa* TWI *E. polycarpa*
dua sika ɔkoduben AKAN - ASANTE ANNONACEAE *Enantia polycarpa*
duabiri VULGAR ANNONACEAE *Greenwayodendron oliveri, Monodora tenuifolia* AKAN - ASANTE *Greenwayodendron oliveri*
duadé MANDING - DYULA APOCYNACEAE *Holarrhena floribunda*
duam SERER CONVOLVULACEAE *Ipomoea pes-caprae*
duanan VULGAR ANNONACEAE *Xylopia staudtii* AKAN - WASA *X. staudtii*
duantũnkũm AKAN - TWI ANNONACEAE *Annona muricata*
dua-sika VULGAR ANNONACEAE *Enantia polycarpa*
duatuigui KYAMA BALANOPHORACEAE *Thonningia sanguinea*
dua-wisa AKAN - ASANTE ANNONACEAE *Pachypodanthium staudtii*
duawisa AKAN - WASA ANNONACEAE *Pachypodanthium staudtii*
dua-wusa AKAN - TWI ANNONACEAE *Pachypodanthium staudtii*
dubagira HAUSA BALANITACEAE *Balanites aegyptiaca*
dubakara FULA - FULFULDE BALANITACEAE *Balanites aegyptiaca*
dubidi AKAN - ASANTE ANNONACEAE *Monodora myristica, M. tenuifolia*
dubima AKAN - TWI ANNONACEAE *Enantia polycarpa* WASA *E. polycarpa*
dubiri VULGAR ANNONACEAE *Greenwayodendron oliveri, Monodora tenuifolia* AKAN - ASANTE *M. tenuifolia* WASA *M. tenuifolia*
dubryémetré MANDING - MANINKA CONVOLVULACEAE *Evolvulus alsinoides*
dubué KRU - GUERE CAPPARACEAE *Euadenia trifoliolata*
dubukumô MANDING - 'SOCE' AVICENNIACEAE *Avicennia germinans*
dubuma AKAN - WASA ANNONACEAE *Enantia polycarpa*
dubusintim AKAN - WASA ANNONACEAE *Monodora tenuifolia*
ducume FULA - PULAAR ANNONACEAE *Annona senegalensis*
duda HAUSA COMPOSITAE *Dicoma sessiliflora*
dudepurre FULA - FULFULDE CONVOLVULACEAE *Ipomoea batatas*
dudo GBE - VHE BOMBACACEAE *Adansonia digitata*
dudu MANDING - MANINKA APOCYNACEAE *Ancylobothrys amoena*
duduku YORUBA DIOSCOREACEAE *Dioscorea alata*
dudumosu BISA ASCLEPIADACEAE *Caralluma dalzielii*
dué KRU - GUERE CECROPIACEAE *Musanga cecropioides*
duébo MANDING - MANINKA COMMELINACEAE *Palisota hirsuta*
duekɛ KONO ANACARDIACEAE *Pseudospondias microcarpa*
duétu 'KRU' CAPPARACEAE *Buchholzia coriacea*
dug o gay SERER CYPERACEAE *Mariscus squarrosus*
dugba GBE - VHE COMPOSITAE *Acanthospermum hispidum*
dugbéï KRU - BETE APOCYNACEAE *Rauvolfia vomitoria*
dugbruhia KYAMA ANACARDIACEAE *Lannea welwitschii*
dúg-ñá' (*pl.* -ñóŋó) SONGHAI CYPERACEAE *Cyperus bulbosus*
dugoe GBE - VHE CUCURBITACEAE *Lagenaria siceraria*
dugor WOLOF ANNONACEAE *Annona senegalensis*
dugor ianuri WOLOF ANNONACEAE *Annona senegalensis*
dúgú (*pl.* -ò) SONGHAI CYPERACEAE *Cyperus alopecuroides*
dúgú bì (*pl.* -bíò) SONGHAI CYPERACEAE *Cyperus rotundus*
dugua bélé SENUFO - KARABORO APOCYNACEAE *Strophanthus sarmentosus*
dugumayetele MANDING - MANINKA CARYOPHYLLACEAE *Polycarpaea corymbosa*
dugumé FULA - PULAAR ANNONACEAE *Annona senegalensis*
dugur, dugor mèr WOLOF ANNONACEAE *Annona glabra, A. glauca*
duguy SERER ANACARDIACEAE *Lannea acida*
duhutu KRU - GUERE CAPPARACEAE *Buchholzia coriacea*
dukari MANDING - BAMBARA CAPPARACEAE *Capparis tomentosa*
dukkuhi (*pl.* dukkuuje) FULA - FULFULDE ANNONACEAE *Annona senegalensis*
dukkuhi, (*pl.* dukkuuje), rukkuhi (*pl.* rukkuuje) FULA - FULFULDE CARICACEAE *Carica papaya*
dukkuhi ladde (*pl.* dukkuuje ladde) FULA - FULFULDE ANNONACEAE *Annona senegalensis*
dukkumbe ladde FULA - FULFULDE ANNONACEAE *Annona senegalensis*
duklabébé FULA - PULAAR ASCLEPIADACEAE *Caralluma russelliana*
duko KULANGO ANACARDIACEAE *Lannea welwitschii*
dukon MANDING - BAMBARA CYPERACEAE *Cyperus digitatus, C. imbricatus*
du-kpay KRU - BASA CELASTRACEAE *Salacia staudtiana*
dùkú NUPE CONVOLVULACEAE *Ipomoea batatas*
dùkú dagba NUPE ARACEAE *Amorphophallus abyssinicus*
dukuma HAUSA CONVOLVULACEAE *Ipomoea batatas*
dukume FULA - PULAAR ANNONACEAE *Annona senegalensis*

dukumé FULA - PULAAR ANNONACEAE *Annona senegalensis*

dukume porto FULA - PULAAR ANNONACEAE *Annona muricata*

dukumé porto FULA - PULAAR ANNONACEAE *Annona muricata*

dukumi FULA - PULAAR ANNONACEAE *Annona senegalensis*

dul o môn SERER ARACEAE *Amorphophallus aphyllus*

dúlá (*pl.* -à), fàttàgá (*def.* là) SONGHAI ASCLEPIADACEAE *Leptadenia hastata*

dullu HAUSA BOMBACACEAE *Adansonia digitata*

duludigô MANDING - 'SOCE' CONVOLVULACEAE *Merremia tridentata*

dulunhwa HAUSA CAPPARACEAE *Boscia senegalensis*

dum AKAN - BRONG COMBRETACEAE *Strephonema pseudocola*

dúmaá, dúmáá HAUSA CUCURBITACEAE *Lagenaria siceraria*

dúmán duútseè HAUSA ARISTOLOCHIACEAE *Aristolochia albida, A. bracteolata*

dúmán gaùraákaa HAUSA CUCURBITACEAE *Lagenaria siceraria*

dúmán gwaázaá HAUSA CONVOLVULACEAE *Jacquemontia tamnifolia*

dúmán kadaá HAUSA CONVOLVULACEAE *Ipomoea asarifolia*

dúmán kwáryaá HAUSA CUCURBITACEAE *Lagenaria siceraria*

dúmán luúdàyií HAUSA CUCURBITACEAE *Lagenaria siceraria*

dúmán raàfií HAUSA CONVOLVULACEAE *Ipomoea asarifolia*

dúmán-kwaàdií HAUSA CONVOLVULACEAE *Ipomoea involucrata*

dumbaare (*pl.* dumbaaje) FULA - FULFULDE CUCURBITACEAE *Lagenaria siceraria*

dumɛntili MANDING - MANINKA BIGNONIACEAE *Spathodea campanulata*

dumko DAGAARI CAPPARACEAE *Crateva adansonii*

dumma NUPE CONVOLVULACEAE *Ipomoea asarifolia*

dunda FULA - PULAAR APOCYNACEAE *Landolphia heudelotii*

dundu kiré DENDI DIOSCOREACEAE *Dioscorea dumetorum*

dundu mbissa DENDI DIOSCOREACEAE *Dioscorea bulbifera*

dundul WOLOF BOMBACACEAE *Bombax costatum*

dundunturi MANDING - MANINKA BIGNONIACEAE *Spathodea campanulata*

dúndúu HAUSA DIOSCOREACEAE *Dioscorea misc. & spp. unknown*

duŋga KONO APOCYNACEAE *Rauvolfia vomitoria*

dunku HAUSA BOMBACACEAE *Adansonia digitata*

dunnduure (*pl.* dunnduuje) FULA - FULFULDE DIOSCOREACEAE *Dioscorea Linn.*

dunore BASSARI DIOSCOREACEAE *Dioscorea Linn.*

duo BAULE DIOSCOREACEAE *Dioscorea Linn.* DAN BIGNONIACEAE *Stereospermum acuminatissimum*

duo ofué BAULE DIOSCOREACEAE *Dioscorea alata*

duodru MANDING - DYULA COMPOSITAE *Vernonia nigritiana*

dupwan NZEMA BIGNONIACEAE *Newbouldia laevis*

durétche BIAFADA ANNONACEAE *Monodora myristica*

durgo KULANGO ANACARDIACEAE *Lannea welwitschii*

duruku KULANGO ANACARDIACEAE *Lannea welwitschii*

dussuo DAN COMPOSITAE *Ageratum conyzoides*

dutiel FULA - FULFULDE CYPERACEAE *Fimbristylis hispidula*

dutʃo ADANGME BORAGINACEAE *Ehretia cymosa*

duttir FULA - FULFULDE CYPERACEAE *Fimbristylis hispidula*

duu GOLA ARACEAE *Colocasia esculenta, Xanthosoma mafaffa*

duu-sisɛ GOLA ARACEAE *Caladium bicolor*

duutoo MANDING - MANDINKA ANACARDIACEAE *Mangifera indica*

duzi DAN COMPOSITAE *Ageratum conyzoides*

'dwaa AKAN - TWI DIOSCOREACEAE *Dioscorea alata*

dwehn-doo KRU - BASA ARACEAE *Culcasia parviflora*

dwera aba AKAN - FANTE BIGNONIACEAE *Crescentia cujete*

dwereba AKAN - FANTE CUCURBITACEAE *Lagenaria siceraria*

dwirantwi AKAN - TWI COMPOSITAE *Bidens pilosa*

dwɔ ɛmine brɔfo ngme GA AMARYLLIDACEAE *Crinum zeylanicum*

dwombobre ANYI - SEHWI ANNONACEAE *Xylopia acutiflora*

dwuma VULGAR CECROPIACEAE *Musanga cecropioides*

*o*dwuma AKAN - TWI CECROPIACEAE *Musanga cecropioides*

dza GBE - VHE DIOSCOREACEAE *Dioscorea cayenensis*

dzafutu GBE - VHE DIOSCOREACEAE *Dioscorea cayenensis*

dzamatre GBE - VHE CUCURBITACEAE *Citrullus lanatus*

dzani pipi GBE - VHE COMPOSITAE *Bidens pilosa*

dzete GBE - VHE CONVOLVULACEAE *Ipomoea batatas*

dze-tɛ, ge-tɛ GBE - VHE DIOSCOREACEAE *Dioscorea rotundata*

dzinjama MANDING - BAMBARA ARALIACEAE *Cussonia arborea*

dzoboli GBE - VHE DIOSCOREACEAE *Dioscorea cayenensis*

dzɔbɔli GBE - VHE DIOSCOREACEAE *Dioscorea alata*

dzogbe solĩ GBE - VHE AMARANTHACEAE *Philoxerus vermicularis*

dzogbedodo GBE - VHE COMBRETACEAE *Terminalia glaucescens, T. ivorensis*

dzɔgɔli GBE - VHE DIOSCOREACEAE *Dioscorea rotundata*

dzotinkin AKYE CAPPARACEAE *Euadenia trifoliolata*

é fok DIOLA CYPERACEAE *Fimbristylis dichotoma*

é forafora DIOLA CUCURBITACEAE *Luffa cylindrica*

é fulimful DIOLA COMPOSITAE *Sphaeranthus senegalensis* MANDING - BAMBARA *Grangea maderaspatana*

é gonnoray DIOLA - FLUP CONVOLVULACEAE *Merremia aegyptia*

e kama DIOLA DIOSCOREACEAE *Dioscorea alata*

é kerkéda DIOLA COMPOSITAE *Ageratum conyzoides*

e kobolora DIOLA ANNONACEAE *Hexalobus monopetalus*

é kobolora DIOLA ANNONACEAE *Hexalobus monopetalus*

é kolinkol DIOLA COMPOSITAE *Sphaeranthus senegalensis*

é liba DIOLA CYPERACEAE *Scleria depressa*

INDEX

ekoro YORUBA CONNARACEAE *Cnestis ferruginea*
èkpákpōghò EDO BURSERACEAE *Canarium schweinfurthii*
ekpat EFIK CUCURBITACEAE *Lagenaria siceraria*
ẹ̀kpérẹ̀, uko ẹ̀kpérẹ EDO CUCURBITACEAE *Lagenaria siceraria*
ẹ̀kpẹ́sịkpẹ̀sị IJO - IZON APOCYNACEAE *Rauvolfia mannii*
ekpi EJAGHAM ANACARDIACEAE *Spondias mombin*
ekplẽ GBE - VHE CUCURBITACEAE *Lagenaria siceraria*
ɛkplokɛ NZEMA CRASSULACEAE *Bryophyllum pinnatum*
èkpókpö EDO CRASSULACEAE *Bryophyllum pinnatum*
ekre ADYUKRU APOCYNACEAE *Tabernaemontana crassa*
eksale YORUBA CONVOLVULACEAE *Ipomoea muricata*
eku AKAN - FANTE BOMBACACEAE *Bombax buonopozense*
ekualokpoe GBE - VHE ANACARDIACEAE *Lannea acida*
ekuama VULGAR APOCYNACEAE *Picralima nitida*
ekuba NZEMA BOMBACACEAE *Adansonia digitata, Bombax buonopozense, Rhodognaphalon brevicuspe*
ekudan YORUBA ANACARDIACEAE *Lannea egregia*
éküé ANYI ANACARDIACEAE *Antrocaryon micraster*
ekukuaba AKAN - FANTE AMARANTHACEAE *Pupalia lappacea*
ekum-nkura AKAN - FANTE DICHAPETALACEAE *Dichapetalum toxicarium*
ekun YORUBA ACANTHACEAE *Brillantaisia lamium*
ekuo ANYI BOMBACACEAE *Bombax buonopozense* ANYI - AOWIN *B. buonopozense* SEHWI *B. buonopozense*
ekur ANYI - AOWIN BOMBACACEAE *Bombax buonopozense*
ekusan YORUBA DICHAPETALACEAE *Dichapetalum madagascariense*
ɛ-kutàn BASARI APOCYNACEAE *Baissea multiflora*
ekuwa HAUSA ASCLEPIADACEAE *Caralluma russelliana*
ekùyá YORUBA CAPPARACEAE *Cleome afrospinosa*
èkùyá YORUBA CAPPARACEAE *Cleome gynandra*
èkùyáko YORUBA CAPPARACEAE *Cleome afrospinosa, C. gynandra*
èkùyalé YORUBA CAPPARACEAE *Cleome afrospinosa, C. gynandra*
ekwo KPE ARALIACEAE *Polyscias fulva*
el arche ARABIC BORAGINACEAE *Echium horridum*
el kerneb ARABIC CANNABACEAE *Cannabis sativa*
el omarah ARABIC BOMBACACEAE *Adansonia digitata*
élalak yéta bu yit DIOLA CONVOLVULACEAE *Ipomoea setifera*
elalé KYAMA ANNONACEAE *Xylopia quintasii*
élanay DIOLA APOCYNACEAE *Saba senegalensis*
ɛ-lɛba BULOM ARACEAE *Colocasia esculenta, Xanthosoma mafaffa*
ɛ-lɛba-sɔnwai BULOM ARACEAE *Caladium bicolor*
elebo URHOBO CUCURBITACEAE *Citrullus lanatus*
e-léfé GBE - FON DIOSCOREACEAE *Dioscorea dumetorum*
ẹlẹgbẹ-oju YORUBA COMPOSITAE *Vernonia cinerea*
élégédé YORUBA CUCURBITACEAE *Cucurbita maxima, C. pepo*
élektag WOLOF COMPOSITAE *Eclipta alba*
eleku ISEKIRI ANNONACEAE *Greenwayodendron suaveolens*
ẹ̀lẹ̀lẹ̀pán IJO - IZON CUCURBITACEAE *Luffa cylindrica*
elemi YORUBA BURSERACEAE *Dacryodes edulis*
ẹlẹshin-má-sọ YORUBA COMPOSITAE *Bidens pilosa*
eletí YORUBA CRASSULACEAE *Kalanchoe integra: Kalanchoe crenata*
elewu YORUBA CHRYSOBALANACEAE *Chrysobalanus icaco*
ẹlẹyintu YORUBA DIOSCOREACEAE *Dioscorea rotundata*
èlí-émì-ọ̀nụ̄ IGBO AMARANTHACEAE *Celosia trigyna*
élílè, élìlì IGBO CUCURBITACEAE *Citrullus lanatus*
élìlì IGBO CUCURBITACEAE *Citrullus lanatus*
elloko, eloko, gelodé FULA - PULAAR COMBRETACEAE *Guiera senegalensis*
elmarujet, telmarujet TAMACHEK CRUCIFERAE *Eremobium aegyptiacum*
elmarujet TAMACHEK CRUCIFERAE *Farsetia ramosissima*
elo ABE ANNONACEAE *Xylopia Linn., X. quintasii, X. villosa*
elócò FULA - PULAAR COMBRETACEAE *Guiera senegalensis*
eloe GBE BOMBACACEAE *Ceiba pentandra*
elolaè DIOLA - FLUP ACANTHACEAE *Hygrophila auriculata*
em-átorse BIDYOGO ARACEAE *Colocasia esculenta*
embi-siembi ANYI APOCYNACEAE *Rauvolfia vomitoria*
ɛmee ANYI - AOWIN APOCYNACEAE *Alstonia boonei* SEHWI *A. boonei*
èméìn fị tùò IJO - IZON CONVOLVULACEAE *Ipomoea aquatica*
emenle NZEMA APOCYNACEAE *Alstonia boonei*
emere ADANGME - KROBO COMBRETACEAE *Terminalia ivorensis* GBE - VHE *T. ivorensis* NZEMA *T. ivorensis*
emeri VULGAR COMBRETACEAE *Terminalia ivorensis* AKAN - ASANTE *T. ivorensis* FANTE *T. ivorensis* WASA *T. ivorensis* GA *T. ivorensis*
emiɛ ANYI APOCYNACEAE *Alstonia boonei* BAULE *A. boonei*
emien ANYI APOCYNACEAE *Alstonia boonei* BAULE *A. boonei*
émien NZEMA APOCYNACEAE *Alstonia boonei*
émiengré NZEMA ANNONACEAE *Pachypodanthium staudtii*
emil AKAN - ASANTE COMBRETACEAE *Terminalia ivorensis*
ẹmina YORUBA DIOSCOREACEAE *Dioscorea bulbifera*
ɛmina KRIO DIOSCOREACEAE *Dioscorea bulbifera*
ẹmina-esi YORUBA DIOSCOREACEAE *Dioscorea bulbifera*
èmìnè àbìà EFIK DIOSCOREACEAE *Dioscorea misc. & spp. unknown*
emire AKAN - TWI COMBRETACEAE *Terminalia ivorensis*

emiri ADANGME COMBRETACEAE *Terminalia ivorensis*
ẽmọ́ àgbò YORUBA AMARANTHACEAE *Pupalia lappacea*
[illegible] ígierù ómí EDO DIOSCOREACEAE *Dioscorea rotundata*
empenche BIDYOGO CUCURBITACEAE *Luffa cylindrica*
emprobego KYAMA COSTACEAE *Costus Linn.*
emri AKAN - WASA COMBRETACEAE *Terminalia ivorensis*
èmu NUPE BOMBACACEAE *Adansonia digitata*
ẹ̀mùnọ̀mùerhán EDO BEGONIACEAE *Begonia mannii*
ɛ̀-nám BASARI ACANTHACEAE *Nelsonia canescens*
enayele KONKOMBA ANACARDIACEAE *Spondias mombin*
endame FULA - PULAAR APOCYNACEAE *Holarrhena floribunda*
èndèm DIOLA CYPERACEAE *Pycreus macrostachyos*
èndóúndóú IJO - IZON APOCYNACEAE *Alstonia boonei*
èndòùndòù IJO - IZON APOCYNACEAE *Alstonia congensis*
endumiye FULA - FULFULDE ASCLEPIADACEAE *Pergularia tomentosa*
endurè BIDYOGO CONNARACEAE *Connarus africanus*
endúrè BIDYOGO CONNARACEAE *Byrsocarpus coccineus*
endwa AKAN - ASANTE ANNONACEAE *Hexalobus crispiflorus*
eñe tara MANDING - BAMBARA ACANTHACEAE *Blepharis maderaspatensis*
enem ANAANG DIOSCOREACEAE *Dioscorea misc. & spp. unknown*
enende FULA - FULFULDE ASCLEPIADACEAE *Pergularia tomentosa*
enɛ̀p BEROM AMARANTHACEAE *Amaranthus hybridus*
en'ga MANDING - DYULA COMBRETACEAE *Terminalia glaucescens*
enge GBE - VHE APOCYNACEAE *Landolphia*
ẹ̀ngẹ́ IJO - IZON ANNONACEAE *Xylopia sp. indet.* (*Catterall 51, 61,K*)
engedidi HAUSA CAPPARACEAE *Crateva adansonii*
engyakobini NZEMA BOMBACACEAE *Rhodognaphalon brevicuspe*
engye GBE - VHE APOCYNACEAE *Landolphia owariensis*
eni URHOBO DIOSCOREACEAE *Dioscorea alata*
enia ANYI BOMBACACEAE *Ceiba pentandra*
enial DIOLA AMARANTHACEAE *Philoxerus vermicularis*
énidek WOLOF CELASTRACEAE *Maytenus senegalensis*
eniémé NZEMA BOMBACACEAE *Ceiba pentandra*
enivé ABURE BOMBACACEAE *Ceiba pentandra*
enkem DIOLA AMARYLLIDACEAE *Scadoxus multiflorus*
e-nko MUNGAKA BIGNONIACEAE *Spathodea campanulata*
enkoa NGEMBA DIOSCOREACEAE *Dioscorea dumetorum*
eŋmebi GA AIZOACEAE *Sesuvium portulacastrum*
enonge KUNDU APOCYNACEAE *Rauvolfia macrophylla*
enoŋgo DUALA APOCYNACEAE *Rauvolfia macrophylla*
énot DIOLA APOCYNACEAE *Landolphia dulcis*
énoté kuruk DIOLA CONNARACEAE *Cnestis ferruginea*
ẹ̄nū̃ EBIRA DIOSCOREACEAE *Dioscorea Linn.*
énwañ ANAANG APOCYNACEAE *Landolphia owariensis*
énwàñà, énwâñ IBIBIO APOCYNACEAE *Landolphia owariensis*
enwuma ANYI - AOWIN DILLENIACEAE *Tetracera alnifolia* SEHWI *T. alnifolia*
enya ANYI BOMBACACEAE *Ceiba pentandra*
enya(a) ANYI - AOWIN BOMBACACEAE *Ceiba pentandra*
enyaa ANYI BOMBACACEAE *Ceiba pentandra* ANYI - SEHWI *C. pentandra*
enyam'gua NZEMA BOMBACACEAE *Ceiba pentandra*
ɛnyanɛ NZEMA COSTACEAE *Costus afer*
ɛ-nyangama ANYI - SEHWI CECROPIACEAE *Myrianthus arboreus*
enyan'gua NZEMA BOMBACACEAE *Ceiba pentandra*
énye taràà DOGON ACANTHACEAE *Blepharis maderaspatensis*
énye vɛvɛy DOGON ANACARDIACEAE *Spondias mombin*
enyena VULGAR BOMBACACEAE *Ceiba pentandra*
enyenna NZEMA BOMBACACEAE *Ceiba pentandra*
enyɛnɔa NZEMA BOMBACACEAE *Ceiba pentandra*
enyĩna-kɔbina AKAN - TWI BOMBACACEAE *Rhodognaphalon brevicuspe*
ényìn-énìn EFIK DIOSCOREACEAE *Dioscorea misc. & spp. unknown*
enyiro? YORUBA COMBRETACEAE *Combretum paniculatum ssp. paniculatum*
eo (*pl.* ao) EJAGHAM - EKIN DIOSCOREACEAE *Dioscorea Linn.*
eparrá BIDYOGO CUCURBITACEAE *Lagenaria siceraria*
épaympay DIOLA CYPERACEAE *Cyperus esculentus*
épéléhèn DIOLA - FLUP CONNARACEAE *Cnestis ferruginea*
epepe YORUBA COMBRETACEAE *Terminalia ivorensis*
epepe, oweive YORUBA COMBRETACEAE *Terminalia ivorensis*
épérum DIOLA COMBRETACEAE *Guiera senegalensis*
epi GBE - VHE COMBRETACEAE *Terminalia glaucescens*
epín, epíngi NUPE CUCURBITACEAE *Citrullus lanatus*
ẹ̀pírí kọ̀rì IJO - IZON CONVOLVULACEAE *Merremia umbellata*
eplatokɔkɔe-wiriwiria AKAN - ASANTE CELASTRACEAE *Salacia chlorantha*
eplé ABURE CARICACEAE *Carica papaya*
épumbay DIOLA CELASTRACEAE *Salacia senegalensis*
equeché BIDYOGO ANNONACEAE *Xylopia aethiopica*
equèche BIDYOGO ANNONACEAE *Xylopia aethiopica*
éralak DIOLA CONVOLVULACEAE *Ipomoea asarifolia*
erale NZEMA ANNONACEAE *Xylopia quintasii*
èré ínínáịn IJO AMARANTHACEAE *Amaranthus spinosus*
èré ịyọ́ụ́nkọ́rị́ IJO - IZON COMPOSITAE *Melanthera scandens*
èré ógbódó IJO - IZON COSTACEAE *Costus afer*
èré ótóró IJO ARACEAE *Amorphophallus dracontioides*

| | |
|---|---|
| fola | NZEMA APOCYNACEAE *Tabernaemontana Gen.* |
| *an*-fola | TEMNE COMPOSITAE *Centratherum angustifolium* |
| *an*-folaŋ | TEMNE BOMBACACEAE *Bombax buonopozense* |
| fole | CRIOULO APOCYNACEAE *Landolphia heudelotii* KANURI CYPERACEAE *Cyperus papyrus* |
| folé | MANDING - MANDINKA APOCYNACEAE *Saba senegalensis* 'SOCE' *Landolphia heudelotii* SERER *L. heudelotii* |
| fɔle | VAI CANNACEAE *Canna indica* |
| fole de elephante | CRIOULO APOCYNACEAE *Landolphia owariensis* |
| fole-grandi | CRIOULO APOCYNACEAE *Saba senegalensis* |
| fóleossum (*def.*-ô) | MANDING - MANDINKA APOCYNACEAE *Landolphia heudelotii* |
| foli | MENDE DIOSCOREACEAE *Dioscorea preussii* |
| fɔli | MENDE DIOSCOREACEAE *Dioscorea bulbifera, D. praehensilis, D. togoensis* |
| folie, fɔlie | GBE - VHE DICHAPETALACEAE *Dichapetalum madagascariense* |
| follye | GBE - VHE DICHAPETALACEAE *Dichapetalum pallidum* |
| folo | TEM BOMBACACEAE *Bombax buonopozense, B. costatum* |
| folŏ | TEM BOMBACACEAE *Bombax buonopozense* |
| fɔlo (*def.*-fɔli) | MENDE CANNACEAE *Canna indica* |
| fol'yere (*pl.* pol'ye) | FULA - FULFULDE CUCURBITACEAE *Cucurbita maxima, C. pepo* |
| fon | BUSA BOMBACACEAE *Adansonia digitata* |
| fondé | ABE ANNONACEAE *Xylopia Linn., X. aethiopica, X. quintasii, X. rubescens, X. staudtii* |
| fondolo | KISSI AMARANTHACEAE *Amaranthus hybridus* |
| fondolo-bɛndoŋ | KISSI AMARANTHACEAE *Amaranthus hybridus* |
| fondolo-ungu | KISSI AMARANTHACEAE *Amaranthus spinosus* |
| fondulo | KISSI AMARANTHACEAE *Amaranthus hybridus, A. lividus* |
| fondulo saman | KISSI AMARANTHACEAE *Amaranthus hybridus* |
| fondulo sankura | KISSI AMARANTHACEAE *Amaranthus hybridus* |
| fonfi | AKYE ANNONACEAE *Xylopia staudtii* |
| foni-baji (*def.*-i) | MENDE COMBRETACEAE *Terminalia Linn.* |
| foni-baji | MENDE COMBRETACEAE *Terminalia albida, T. glaucescens* |
| foni-lawa (*def.*-lawei) | MENDE CHRYSOBALANACEAE *Neocarya macrophylla* |
| foni-lawei | KONO CHRYSOBALANACEAE *Parinari excelsa* |
| fono | BAULE COMBRETACEAE *Combretum zenkeri* |
| fonye | SUSU BROMELIACEAE *Ananas comosus* |
| foolee (*def.* fooleo) | MANDING - MANDINKA APOCYNACEAE *Landolphia heudelotii* |
| foolee (*def.*-fooleo) | MANDING - MANDINKA APOCYNACEAE *Landolphia heudelotii* |
| foosea | MANDING - MANDINKA COCHLOSPERMACEAE *Cochlospermum tinctorium* |
| fôóti | BALANTA CUCURBITACEAE *Lagenaria siceraria* |
| foranoh | SERER AMARANTHACEAE *Philoxerus vermicularis* |
| fore | SUSU APOCYNACEAE *Landolphia owariensis* |
| foré | MANDING - 'SOCE' APOCYNACEAE *Landolphia heudelotii* SUSU *L. heudelotii* |
| fɔre | SUSU APOCYNACEAE *Landolphia owariensis* |
| fore fafia | SUSU APOCYNACEAE *Landolphia membranacea* |
| foré-fikne | WOLOF APOCYNACEAE *Saba senegalensis* |
| forəto-sɛ lonyi | SUSU CUCURBITACEAE *Luffa acutangula* |
| forgo | DENDI BOMBACACEAE *Bombax buonopozense* SONGHAI - ZARMA *B. buonopozense, Ceiba pentandra* |
| fórgò | DENDI BOMBACACEAE *Bombax costatum* |
| fórgò (*pl.* -à) | SONGHAI BOMBACACEAE *Bombax costatum* |
| fori-wuri | SUSU CONVOLVULACEAE *Ipomoea mauritiana* |
| foro | ANYI COMBRETACEAE *Combretum zenkeri* |
| fɔrɔ | LOKO CANNACEAE *Canna indica* |
| foro foro | MANDING - MANINKA CUCURBITACEAE *Luffa acutangula* |
| fɔrɔn-dai | SUSU CANNACEAE *Canna indica* |
| fɔrɔndɔ | FULA - PULAAR CANNACEAE *Canna indica* |
| fɔrɔndɔ-jɔ̃ | MANDING - MANDINKA CANNACEAE *Canna indica* |
| forray | WOLOF COCHLOSPERMACEAE *Cochlospermum tinctorium* |
| fotai ɛliti | SUSU CONVOLVULACEAE *Ipomoea involucrata* |
| fɔtɛnalinyi | SUSU CUCURBITACEAE *Cucumis anguria* |
| fotie | ANYI - AOWIN ANNONACEAE *Cleistopholis patens* NZEMA - SEHWI *C. patens* |
| fɔtigba | GBE - VHE CHRYSOBALANACEAE *Chrysobalanus icaco* |
| fotobo | KWENI APOCYNACEAE *Funtumia elastica* |
| fou somsom | DIOLA CAPPARACEAE *Ritchiea capparoides* |
| foulib | DIOLA APOCYNACEAE *Holarrhena floribunda* |
| fɔvɔ | GOLA CECROPIACEAE *Myrianthus arboreus, M. libericus, M. serratus* |
| fra | BAULE COMBRETACEAE *Terminalia superba* |
| ɔ-fram | AKAN - FANTE COMBRETACEAE *Terminalia superba* |
| fram | AKAN - WASA COMBRETACEAE *Terminalia superba* KULANGO *T. superba* |
| ɛ-frameri | ANYI - AOWIN COMBRETACEAE *Terminalia ivorensis, T. superba* |
| framiré | ANYI COMBRETACEAE *Terminalia ivorensis* NZEMA *T. ivorensis* |
| framo | AKAN - ASANTE COMBRETACEAE *Terminalia superba* |
| framoo | AKAN - ASANTE COMBRETACEAE *Terminalia superba* WASA *T. superba* |
| frane | NZEMA COMBRETACEAE *Terminalia superba* |
| frango | GBE - VHE COMBRETACEAE *Terminalia superba* |
| franké | ANYI COMBRETACEAE *Terminalia superba* |
| franko | ANYI COMBRETACEAE *Terminalia superba* |
| frantu | 'KRU' APOCYNACEAE *Pleiocarpa mutica* |

INDEX

gã gĕ su lu — MANO CUCURBITACEAE *Momordica balsamina, M. charantia*
gaadal — FULA - FULFULDE AMARYLLIDACEAE *Crinum Gen.*
gaadal (*pl.* gaade) — FULA - FULFULDE AMARYLLIDACEAE *Crinum jagus, C. natans, C. zeylanicum*
gaadal faaɓru (*pl.* gaade paaɓi) — FULA - FULFULDE AMARYLLIDACEAE *Crinum zeylanicum*
gaasayaa — HAUSA CAPPARACEAE *Cleome gynandra*
gáasí (*pl.* -ó), tàndà (*pl.* -à) — SONGHAI CUCURBITACEAE *Lagenaria siceraria*
gaatarin kureegee — HAUSA AMARYLLIDACEAE *Scadoxus multiflorus*
gaayama — BAULE CUCURBITACEAE *Momordica foetida*
gabá — DOGON CUCURBITACEAE *Lagenaria siceraria*
gabá íí, kɛ̀mɛ — DOGON CUCURBITACEAE *Lagenaria siceraria*
gabá nà — DOGON CUCURBITACEAE *Luffa cylindrica*
gabi — MANDING - BAMBARA CONVOLVULACEAE *Ipomoea eriocarpa*
gabué — KRU - GUERE ACANTHACEAE *Justicia laxa*
gabu-na — SUSU - DYALONKE CUCURBITACEAE *Cucurbita pepo*
gada machiji — HAUSA COMMELINACEAE *Aneilema pomeridianum*
gaɗahuka, gaɗakuka, gaɗaukuku — HAUSA ARISTOLOCHIACEAE *Aristolochia albida*
gaɗahuka, gaɗakuka, gaɗaukuka — HAUSA ARISTOLOCHIACEAE *Aristolochia bracteolata*
gadali — HAUSA AMARYLLIDACEAE *Crinum Gen.*
gadiyaare (*pl.* gadiyaaje) — FULA - FULFULDE BOMBACACEAE *Adansonia digitata*
gadoil (?) — FULA - FULFULDE ASCLEPIADACEAE *Glossonema boveanum*
gâdolo — MANDING - BAMBARA CAPPARACEAE *Crateva adansonii* MANINKA *C. adansonii*
gadon machiji — HAUSA AIZOACEAE *Trianthema portulacastrum, Zaleya pentandra* CONVOLVULACEAE *Merremia tridentata*
gádón maciijíí — HAUSA AIZOACEAE *Trianthema portulacastrum, Zaleya pentandra*
gafaa toh — SERER AMARYLLIDACEAE *Scadoxus multiflorus*
gafal — ARABIC BURSERACEAE *Commiphora africana*
gagi, gigi — HAUSA ACANTHACEAE *Blepharis linariifolia*
gago, gako — HAUSA CUCURBITACEAE *Lagenaria siceraria*
gah — MANO CUCURBITACEAE *Lagenaria siceraria*
gahé — KRU - GUERE COMPOSITAE *Spilanthes filicaulis*
gai bulu da — MANO COMPOSITAE *Microglossa afzelii*
gaïbé — 'NEKEDIE' COMPOSITAE *Spilanthes costata*
gaiza — HAUSA COMBRETACEAE *Combretum micranthum*
gajiji — HAUSA CYPERACEAE *Cyperus articulatus*
gajuruwi — BOLE COMMELINACEAE *Commelina forskalaei*
ga-kangàl — BEDIK COMBRETACEAE *Combretum spp. indet.*
ga-kàràbú — BEDIK APOCYNACEAE *Carissa edulis*
gakpoti — GBE - VHE APOCYNACEAE *Holarrhena floribunda*
ga-kurámb — BEDIK ANACARDIACEAE *Ozoroa insignis*
gala — KPALAGHA ANNONACEAE *Annona arenaria, A. senegalensis*
gala kuku — KRU - BETE APOCYNACEAE *Motandra guineensis*
galaldi — FULA - FULFULDE COMBRETACEAE *Anogeissus leiocarpus*
galanataŋga — SUSU ARACEAE *Caladium bicolor*
gallo — HAUSA BALANITACEAE *Balanites aegyptiaca*
galo — GRUSI CAPPARACEAE *Capparis tomentosa*
galu — GRUSI CAPPARACEAE *Capparis tomentosa*
galumbi — FULA - PULAAR BIGNONIACEAE *Stereospermum kunthianum*
gam, ngam — WOLOF BIGNONIACEAE *Newbouldia laevis*
gama sowa — HAUSA APOCYNACEAE *Strophanthus sarmentosus*
gàmàgàrí — HAUSA CONVOLVULACEAE *Ipomoea batatas*
gaman sauwa — HAUSA APOCYNACEAE *Holarrhena floribunda*
ga-mángɔ̀ — BEDIK ANACARDIACEAE *Mangifera indica*
gamb — SERER BIGNONIACEAE *Newbouldia laevis*
gamba — WOLOF CUCURBITACEAE *Lagenaria siceraria*
gambarin marmoru — HAUSA AMARANTHACEAE *Alternanthera nodiflora*
gambe-plɔm — KRIO ANACARDIACEAE *Spondias purpurea*
ga-mẹnɛ-mẹnɛ — BEDIK CAPPARACEAE *Boscia angustifolia, B. senegalensis*
gámmòn baawaa — HAUSA CONVOLVULACEAE *Merremia tridentata*
gammu — GWARI CONVOLVULACEAE *Ipomoea argentaurata*
gamon fatake — HAUSA CUCURBITACEAE *Cucurbita pepo*
ga-mondyé — BEDIK APOCYNACEAE *Strophanthus sarmentosus*
gámsà (*pl.* -à) — SONGHAI CHRYSOBALANACEAE *Neocarya macrophylla*
gàmù gámù — YORUBA DIOSCOREACEAE *Dioscorea rotundata*
ganagana — YORUBA CONVOLVULACEAE *Cuscuta australis*
ga-nány — BEDIK ACANTHACEAE *Nelsonia canescens*
gandama — 'SENUFO' DIPTEROCARPACEAE *Monotes kerstingii*
gandarin damo — HAUSA ACANTHACEAE *Blepharis linariifolia*
gànd-báa háwrù (*pl.* lò) — SONGHAI ASCLEPIADACEAE *Glossonema boveanum*
ga-ndẹkwéɗy — BEDIK COMBRETACEAE *Pteleopsis suberosa*
ganɗiɗo — HAUSA BOMBACACEAE *Ceiba pentandra*
gandolo — MANDING - BAMBARA CAPPARACEAE *Crateva adansonii* MANINKA *C. adansonii*
ga-ndyabàkátánà — BEDIK COMBRETACEAE *Combretum glutinosum*
ga-ndyátyàr — BEDIK COMBRETACEAE *Combretum etessei*
ga-ndyɛ́ntɛ̀n — BEDIK CARICACEAE *Carica papaya*
ga-nẹkabírín — BEDIK CYPERACEAE *Cyperus articulatus*
ga-nẹ̀mbẹ́t — BEDIK APOCYNACEAE *Baissea multiflora*
ga-ngál — BEDIK COMBRETACEAE *Anogeissus leiocarpus*
gàngàmaá — HAUSA COMBRETACEAE *Anogeissus leiocarpus*
ganganpi — FULA - FULFULDE ASCLEPIADACEAE *Calotropis procera*

INDEX

gongolu **TEM** ARALIACEAE *Cussonia arborea*
gongonga (*def*.-i) **MENDE** CONNARACEAE *Agelaea trifolia*
gongonklui **ABURE** APOCYNACEAE *Rauvolfia vomitoria*
gongu **DAGAARI** BOMBACACEAE *Ceiba pentandra*
goni **DAGAARI** BOMBACACEAE *Ceiba pentandra*
gôni **MANDING - BAMBARA** COMPOSITAE *Sclerocarpus africanus*
gɔniŋgbe **GOLA** CUCURBITACEAE *Telfairea occidentalis*
gonkobiessoa **BAULE** COMMELINACEAE *Palisota hirsuta*
gonogono (*pl*. gonogonooji) **FULA - FULFULDE** CUCURBITACEAE *Lagenaria siceraria*
gonogono **TUBU** BORAGINACEAE *Moltkia callosa*
gonoretti **KWENI** COMPOSITAE *Bidens pilosa*
gɔnugbe **MENDE** CUCURBITACEAE *Telfairea occidentalis*
gɔnuŋgbe **GOLA** CUCURBITACEAE *Telfairea occidentalis*
gɔ̀nyɔ **DOGON** ANNONACEAE *Hexalobus monopetalus*
gonyontɛo **GA** BORAGINACEAE *Cordia sinensis*
googà-jìkí **HAUSA** COMBRETACEAE *Combretum Loefl., C. molle*
goojare (*pl*. goojaaje) **FULA - FULFULDE** ARACEAE *Colocasia esculenta*
goojíí **HAUSA** CUCURBITACEAE *Cucurbita maxima*
goòjíí **HAUSA** CUCURBITACEAE *Cucurbita maxima*
goòjíí **HAUSA** CUCURBITACEAE *Cucurbita pepo*
goólón zaakii **HAUSA** CUCURBITACEAE *Cucumis figarei*
goon-aby **HAUSA** COMBRETACEAE *Combretum paniculatum ssp. paniculatum*
gooráá **HAUSA** CUCURBITACEAE *Lagenaria siceraria*
gopité **SENUFO - TAGWANA** COMPOSITAE *Vernonia nigritiana*
gor bu di ɗaw **WOLOF** CAPPARACEAE *Cleome gynandra*
goraru (*pl*. goraji) **FULA - FULFULDE** CUCURBITACEAE *Lagenaria siceraria*
gorko nyangudoohi **FULA - FULFULDE** CAPPARACEAE *Capparis sepiaria*
goróluga **FULA - PULAAR** DILLENIACEAE *Tetracera alnifolia*
goromi **FULA - FULFULDE** BOMBACACEAE *Adansonia digitata*
goshin baana **HAUSA** COMPOSITAE *Chrysanthellum indicum var. afroamericanum*
gothi **GBE - FON** ARALIACEAE *Cussonia arborea*
gotti **GBE - FON** ARALIACEAE *Cussonia arborea*
gotuba **KRU - GUERE** COMPOSITAE *Conyza sumatrensis*
gotué **KRU - GUERE** BORAGINACEAE *Ehretia cymosa*
gɔvɛ **FULA - PULAAR** CYPERACEAE *Cyperus articulatus*
govhé **FULA - PULAAR** CYPERACEAE *Cyperus articulatus*
gowé **FULA - PULAAR** CYPERACEAE *Cyperus articulatus, C. rotundus, C. sphacelatus* **WOLOF** *C. maritimus*
goye **FULA - FULFULDE** CYPERACEAE *Cyperus rotundus, C. sphacelatus*
goye ɓaleeje **FULA - FULFULDE** CYPERACEAE *Cyperus maculatus*
goye gulbi **FULA - FULFULDE** CYPERACEAE *Cyperus articulatus*
goye maayo **FULA - FULFULDE** CYPERACEAE *Cyperus articulatus*
goyoma **KIRMA** CHRYSOBALANACEAE *Maranthes polyandra, Parinari curatellifolia*
gozogui **LOMA** CECROPIACEAE *Musanga cecropioides*
gpaladio **DAN** ANNONACEAE *Pachypodanthium staudtii*
gpaladuo **DAN** ANNONACEAE *Pachypodanthium staudtii*
g-pandia polo **KISSI** CONVOLVULACEAE *Ipomoea involucrata*
gpati dêkrun **GBE - FON** BOMBACACEAE *Ceiba pentandra*
*an*-gpuk **TEMNE** DIOSCOREACEAE *Dioscorea minutiflora*
gputé **LOMA** ARACEAE *Xanthosoma mafaffa*
grabi **KRU - GUERE** COMPOSITAE *Vernonia cinerea*
gragra **KYAMA** APOCYNACEAE *Tabernaemontana crassa*
graku **KRU - GUERE** BORAGINACEAE *Ehretia cymosa*
grakwè **KWENI** APOCYNACEAE *Motandra guineensis*
gramanti **KRIO** CYPERACEAE *Cyperus esculentus*
granat-tri **KRIO** BOMBACACEAE *Bombacopsis glabra*
granié bahu **KRU - GUERE** CAPPARACEAE *Euadenia trifoliolata*
grins **KRIO** AMARANTHACEAE *Amaranthus hybridus*
grins-ɛ **BULOM** AMARANTHACEAE *Amaranthus hybridus*
griomé **SONINKE - SARAKOLE** BOMBACACEAE *Bombax costatum*
grobidia **KRU - BETE** COMPOSITAE *Eclipta alba*
grogolégoné **KWENI** CAPPARACEAE *Buchholzia coriacea*
grubé **BOKE** CHRYSOBALANACEAE *Chrysobalanus icaco* **KYAMA** *C. icaco*
gruhain **KRU - GUERE** BALANOPHORACEAE *Thonningia sanguinea*
grukoma **KRU - GUERE** BALANOPHORACEAE *Thonningia sanguinea*
gua **KPELLE** DIOSCOREACEAE *Dioscorea alata*
gua, guagara, guara **MANDING - BAMBARA** DIOSCOREACEAE *Dioscorea alata*
gua koro **AKAN - ASANTE** COMPOSITAE *Ageratum conyzoides* **ANYI** *A. conyzoides*
gua kubo, vanvan **AKAN - ASANTE** COMPOSITAE *Ageratum conyzoides*
guadaba **MOORE** ARACEAE *Stylochiton lancifolius*
guagei **GOLA** CUCURBITACEAE *Luffa cylindrica*
guagɛrɛ **GOLA** CUCURBITACEAE *Luffa cylindrica*
guahiélu **DAN** DICHAPETALACEAE *Dichapetalum madagascariense*
guakro **AKAN - TWI** COMPOSITAE *Ageratum conyzoides*
guakuro **AKAN - ASANTE** COMPOSITAE *Ageratum conyzoides, Emilia sonchifolia*
guan **DOGON** COMBRETACEAE *Pteleopsis habeensis*
guangana **KPELLE** DIOSCOREACEAE *Dioscorea alata*
guara lo **MANDING - BAMBARA** DIOSCOREACEAE *Dioscorea misc. & spp. unknown*
guara simbessi **MANDING - BAMBARA** DIOSCOREACEAE *Dioscorea misc. & spp. unknown*
*a*-guare-ansra **AKAN - TWI** ACANTHACEAE *Brillantaisia lamium, B. nitens*
guaza **HAUSA** ARACEAE *Colocasia esculenta*
gubdi **HAUSA** BOMBACACEAE *Adansonia digitata*

gulugu BISA APOCYNACEAE *Saba senegalensis* COMBRETACEAE *Combretum aculeatum*
gulullutu HAUSA BOMBACACEAE *Adansonia digitata*
gulum butu FULA - FULFULDE CUCURBITACEAE *Lagenaria siceraria*
gulumbur FULA - FULFULDE BOMBACACEAE *Adansonia digitata*
gulundingo MANDING - MANDINKA CONVOLVULACEAE *Ipomoea cairica*
gum DOGON COMBRETACEAE *Pteleopsis habeensis*
guma BATONNUN BOMBACACEAE *Ceiba pentandra*
gumania MANDING - MANINKA CARYOPHYLLACEAE *Polycarpaea corymbosa*
gumayi, guntsu, gwargwami HAUSA BOMBACACEAE *Adansonia digitata*
gumba FULA - PULAAR CAPPARACEAE *Capparis sepiaria*
gumbali HAUSA CUCURBITACEAE *Lagenaria siceraria*
gumbihi DAGBANI BOMBACACEAE *Ceiba pentandra*
gumbul ARABIC - SHUWA CUCURBITACEAE *Lagenaria siceraria*
gumbul alme ARABIC - SHUWA CUCURBITACEAE *Lagenaria siceraria*
gumel WOLOF CAPPARACEAE *Capparis decidua*
gumi FULA - FULFULDE CAPPARACEAE *Capparis tomentosa* PULAAR *C. decidua, C. sepiaria, C. tomentosa*
gumi balévi FULA - PULAAR CAPPARACEAE *Capparis decidua, C. sepiaria*
gumi danévi FULA - PULAAR CAPPARACEAE *Capparis decidua*
gumibalewi FULA - PULAAR CAPPARACEAE *Capparis tomentosa*
gumibaley FULA - PULAAR CAPPARACEAE *Capparis tomentosa*
gumji HAUSA DIOSCOREACEAE *Dioscorea misc. & spp. unknown*
gummbal (*pl.* gummbe) FULA - FULFULDE CUCURBITACEAE *Lagenaria siceraria*
gummbal nalle FULA - FULFULDE CUCURBITACEAE *Lagenaria siceraria*
gumumi FULA - FULFULDE COMBRETACEAE *Combretum micranthum*
gumuni FULA - PULAAR COMBRETACEAE *Combretum micranthum*
guna DAGBANI BOMBACACEAE *Ceiba pentandra*
guŋa DAGBANI BOMBACACEAE *Ceiba pentandra*
gúna HAUSA CUCURBITACEAE *Citrullus lanatus*
gunaà, gurjií HAUSA CUCURBITACEAE *Cucumis melo*
gúnàr kuúraá HAUSA CUCURBITACEAE *Cucumis figarei*
gunar zaki HAUSA CUCURBITACEAE *Cucumis prophetarum*
gúndà KANURI CARICACEAE *Carica papaya*
gundar kabewa HAUSA CUCURBITACEAE *Cucurbita pepo*
gundje PEPEL ANNONACEAE *Uvaria afzelii, U. chamae*
gung GRUSI BOMBACACEAE *Ceiba pentandra*
gunga DAGBANI BOMBACACEAE *Ceiba pentandra* MAMPRULI *C. pentandra* MOORE *C. pentandra*
gungo DAGAARI BALANITACEAE *Balanites aegyptiaca*
gunguma-gumdi DAGBANI BOMBACACEAE *Ceiba pentandra*
gungumli DAGBANI BOMBACACEAE *Ceiba pentandra*
gungunsale DAGBANI DIOSCOREACEAE *Dioscorea rotundata*
gung-vale DAGBANI BOMBACACEAE *Ceiba pentandra*
guni GBE - VHE APOCYNACEAE *Funtumia elastica*
gúno BOLE CAPPARACEAE *Cadaba farinosa* HAUSA *C. farinosa*
gùnò gùnó KANURI CUCURBITACEAE *Cucumis prophetarum*
gunta ARABIC - SHUWA ARACEAE *Amorphophallus dracontioides*
gunuukple DAGBANI DIOSCOREACEAE *Dioscorea rotundata*
guopatréké 'SENUFO' AMARYLLIDACEAE *Scadoxus multiflorus*
guraare (*pl.* guraaje) FULA - FULFULDE ARACEAE *Stylochiton lancifolius*
guraka HAUSA CUCURBITACEAE *Lagenaria siceraria*
gurdjia HAUSA BOMBACACEAE *Bombax costatum*
gurguhọ FULA - FULFULDE CYPERACEAE *Cyperus exaltatus*
gurguli FULA - FULFULDE AMARYLLIDACEAE *Crinum zeylanicum*
gúrjìyaá HAUSA BOMBACACEAE *Bombax costatum*
gurji HAUSA CUCURBITACEAE *Cucumis melo*
gurji kwantal HAUSA CUCURBITACEAE *Cucumis melo*
gùrjin-dájìi HAUSA CUCURBITACEAE *Coccinia grandis*
gurjiya, gúrjíyaá HAUSA BOMBACACEAE *Bombax buonopozense*
gurjiya HAUSA BOMBACACEAE *Bombax costatum*
gúrjìyaá HAUSA BOMBACACEAE *Rhodognaphalon brevicuspe*
gurli HAUSA CUCURBITACEAE *Cucumis melo*
gurli ƙwantal HAUSA CUCURBITACEAE *Cucumis melo*
gurmel FULA - PULAAR CAPPARACEAE *Capparis decidua* WOLOF *C. decidua*
gurmel balévi FULA - PULAAR CAPPARACEAE *Capparis decidua*
gurmusal (*pl.* gurmuse) FULA - FULFULDE CUCURBITACEAE *Lagenaria siceraria*
gurra GWARI CUCURBITACEAE *Lagenaria siceraria*
gursami HAUSA DIOSCOREACEAE *Dioscorea dumetorum*
gursimé KANURI CAPPARACEAE *Cadaba farinosa*
gurubonbula BATONNUN BIGNONIACEAE *Stereospermum kunthianum*
gurugâhi FULA - FULFULDE ANACARDIACEAE *Ozoroa insignis* PULAAR *O. insignis*
gurulu KRU - GUERE CONVOLVULACEAE *Calycobolus africanus*
gurzun dali HAUSA BURSERACEAE *Commiphora kerstingii*
gushiocho GUANG - KRACHI BALANITACEAE *Balanites aegyptiaca*
gusi GBE - VHE CUCURBITACEAE *Cucumeropsis mannii*
gut SERER CYPERACEAE *Mariscus ligularis*
guùluúlùn zoómoó HAUSA COMPOSITAE *Echinops longifolius*
guvé HAUSA CAPPARACEAE *Crateva adansonii*
gùwá KANURI CUCURBITACEAE *Lagenaria siceraria*
guwing DAN COMBRETACEAE *Terminalia superba*
guy dema WOLOF COMBRETACEAE *Terminalia macroptera*

| | |
|---|---|
| guy sidéri | WOLOF APOCYNACEAE *Adenium obesum* |
| guzami | HAUSA ARACEAE *Stylochiton lancifolius* |
| guzére | 'NEKEDIE' COMPOSITAE *Triplotaxis stellulifera* |
| guzu-guzu | HAUSA CYPERACEAE *Cyperus esculentus* |
| gwaázaá | HAUSA ARACEAE *Colocasia esculenta* |
| gwáázáá | HAUSA ARACEAE *Colocasia esculenta* |
| gwaàzaà-maì-goòraá | HAUSA ARACEAE *Xanthosoma mafaffa* |
| gwaàzaà-maì-gudaajií | HAUSA ARACEAE *Colocasia esculenta* |
| gwaàzà-gwaázaà | HAUSA ARACEAE *Colocasia esculenta* |
| gwáázán kwaaɗóó | HAUSA ARACEAE *Amorphophallus dracontioides* |
| gwaázar-giiwaá | HAUSA ARACEAE *Amorphophallus dracontioides* |
| gwabsa | HAUSA ARALIACEAE *Cussonia arborea* |
| gwabwi | GWARI CUCURBITACEAE *Lagenaria siceraria* |
| gwadaba | MOORE ARACEAE *Stylochiton hypogaeus* |
| gwaɗaɗɗasai | HAUSA CUCURBITACEAE *Cucurbita pepo* |
| gwaɗayi | HAUSA CELASTRACEAE *Hippocratea africana, H. apocynoides* |
| gwadda, gwándàr daji | HAUSA CARICACEAE *Carica papaya* |
| gwagwa | GWARI COCHLOSPERMACEAE *Cochlospermum tinctorium* |
| gwaigwaya | HAUSA CYPERACEAE *Cyperus exaltatus, C. rotundus, C. sphacelatus* |
| gwákərák (*pl.* gwákərnjén) | DERA CUCURBITACEAE *Lagenaria siceraria* |
| gwam blipo | GBE - FON BIGNONIACEAE *Kigelia africana* |
| gwamba | HAUSA ARACEAE *Colocasia esculenta* |
| gwanda | GWARI CARICACEAE *Carica papaya* |
| gwandaa | HAUSA CARICACEAE *Carica papaya* |
| gwandai | HAUSA ARACEAE *Stylochiton lancifolius* |
| gwandar beri | HAUSA CUCURBITACEAE *Cucumis melo* |
| gwándàr daájií | HAUSA ANNONACEAE *Annona senegalensis* |
| gwándar jééji | HAUSA ANNONACEAE *Annona senegalensis* |
| gwándàr kúnneé | HAUSA AIZOACEAE *Zaleya pentandra* |
| gwàndàr kúnneé | HAUSA AIZOACEAE *Trianthema portulacastrum* |
| gwándàr Másàr | HAUSA CARICACEAE *Carica papaya* |
| gwándàr rebeji | HAUSA CARICACEAE *Carica papaya* |
| gwanduwa | HAUSA CUCURBITACEAE *Coccinia grandis* |
| gwangwale | HAUSA ANNONACEAE *Annona senegalensis* |
| gwangwama | HAUSA CANNACEAE *Canna indica* |
| gwangwangi-daniejo | FULA - PULAAR CONVOLVULACEAE *Ipomoea cairica* |
| gwanja kusa | HAUSA BORAGINACEAE *Heliotropium subulatum* |
| gwànjaá kúsá | HAUSA CHRYSOBALANACEAE *Maranthes polyandra* |
| gwànjáa kúsá | HAUSA CHRYSOBALANACEAE *Parinari curatellifolia* |
| gwaraki | LONGUDA CUCURBITACEAE *Lagenaria siceraria* |
| gwargwaza | HAUSA ARACEAE *Colocasia esculenta* |
| gwasha | HAUSA APOCYNACEAE *Strophanthus sarmentosus* |
| gwaza | KATAB ARACEAE *Colocasia esculenta* |
| gwazar giwa | HAUSA ARACEAE *Amorphophallus dracontioides* |
| gwe | DAN BOMBACACEAE *Ceiba pentandra* |
| gwɛ̃ɛ̃ | KWENI BOMBACACEAE *Ceiba pentandra* |
| gwèh | DAN BOMBACACEAE *Ceiba pentandra* |
| gweh-gbonoh | DAN BOMBACACEAE *Bombax buonopozense* |
| gwelling-gwelling | FULA - PULAAR CONVOLVULACEAE *Ipomoea mauritiana* |
| g(u)weng | DAN COMBRETACEAE *Terminalia superba* |
| gwobe | FULA - PULAAR CYPERACEAE *Rhynchospora corymbosa* |
| gyakuruwa | AKAN - BRONG COMPOSITAE *Vernonia nigritiana* |
| *a*-gyama(a) | AKAN - ASANTE APOCYNACEAE *Landolphia owariensis* |
| gyamaa | AKAN - ASANTE APOCYNACEAE *Landolphia* |
| gyambobre | ANYI - SEHWI ANNONACEAE *Xylopia parviflora* |
| gyamnkawa | AKAN - ASANTE ARACEAE *Culcasia scandens* |
| gyanɗama | HAUSA CUCURBITACEAE *Lagenaria siceraria* |
| gyehi | GWARI BOMBACACEAE *Ceiba pentandra* |
| gyinantwi | AKAN - TWI COMPOSITAE *Bidens pilosa* |
| gyolo-kidime | FULA - PULAAR ANACARDIACEAE *Ozoroa insignis* |
| hàabù-báasé (*pl.* -ó) | SONGHAI CONVOLVULACEAE *Ipomoea aitonii* |
| haagui | SONGHAI CYPERACEAE *Cyperus maculatus* |
| haako ndiyam | FULA - FULFULDE AMARANTHACEAE *Amaranthus hybridus, A. spinosus* |
| haako-ndiyam | FULA - FULFULDE AMARANTHACEAE *Amaranthus viridis* |
| hab | SERER DIOSCOREACEAE *Dioscorea praehensilis* |
| hâbaliya | ARABIC BORAGINACEAE *Heliotropium bacciferum* |
| hab-el-rechad | ARABIC CRUCIFERAE *Lepidium sativum* |
| habiiru | FULA - FULFULDE CUCURBITACEAE *Momordica balsamina, M. charantia* |
| hãbue | ADANGME ANNONACEAE *Annona squamosa* |
| habugan | WOLOF ANACARDIACEAE *Lannea humilis* |
| hachalilah | ARABIC COMPOSITAE *Centaurea pungens* |
| hachin-tumkia | HAUSA BIGNONIACEAE *Stereospermum kunthianum* |
| had | ARABIC CHENOPODIACEAE *Cornulaca monacantha* TAMACHEK *C. monacantha* |
| haddo | AKYE ANACARDIACEAE *Antrocaryon micraster* |
| hadj | ARABIC CUCURBITACEAE *Citrullus colocynthis* |
| hadj en-nas | ARABIC CUCURBITACEAE *Citrullus lanatus* |
| hadje | ARABIC CUCURBITACEAE *Citrullus colocynthis* |
| hadjilidjé | ARABIC BALANITACEAE *Balanites aegyptiaca* |
| hadjlidj | ARABIC BALANITACEAE *Balanites aegyptiaca* TAMACHEK *B. aegyptiaca* |
| hadza | HAUSA CAPPARACEAE *Cadaba farinosa* |
| hae | KABRE DIOSCOREACEAE *Dioscorea Linn.* |

ìgángán YORUBA DIOSCOREACEAE *Dioscorea cayenensis*
ìgángán alo YORUBA DIOSCOREACEAE *Dioscorea cayenensis*
igarulal ARABIC AIZOACEAE *Zaleya pentandra*
igbá YORUBA CUCURBITACEAE *Lagenaria siceraria*
igbá-(a)jẹ YORUBA CUCURBITACEAE *Lagenaria siceraria*
igbálẹ̀-ọ̀dàn YORUBA CAMPANULACEAE *Wahlenbergia perrottetii*
ìgbálį IJO - IZON COMBRETACEAE *Combretum racemosum*
ìgbálį IJO - IZON COMBRETACEAE *Combretum racemosum*
igbebere NKEM CECROPIACEAE *Myrianthus arboreus*
igbere YORUBA ANNONACEAE *Uvariodendron angustifolium*
igberi YORUBA ANNONACEAE *Dennettia tripetala*
igbó YORUBA CANNABACEAE *Cannabis sativa*
ìgbó YORUBA CANNABACEAE *Cannabis sativa*
igboa KPOSO BOMBACACEAE *Ceiba pentandra*
ìgbólé ajá YORUBA CUCURBITACEAE *Momordica charantia*
igenin TAMACHEK CAPPARACEAE *Capparis decidua*
ighíguẹ̀wé EDO COMMELINACEAE *Palisota ambigua, P. barteri, P. hirsuta*
igi èègun, igi àràbà YORUBA - NAGO BOMBACACEAE *Ceiba pentandra*
igi óshè YORUBA - NAGO BOMBACACEAE *Adansonia digitata*
igilligilli YORUBA CACTACEAE *Rhipsalis baccifera*
ìgìnàịn, ìgìnèín IJO - IZON ANACARDIACEAE *Spondias mombin*
ìgìnéìn IJO - IZON ANACARDIACEAE *Spondias mombin*
igiorùa EDO DIOSCOREACEAE *Dioscorea alata*
igiorùaèsì, ígìorùèsì EDO DIOSCOREACEAE *Dioscorea preussii*
ignïn ARABIC CAPPARACEAE *Capparis decidua*
ìgọ YORUBA ARACEAE *Anchomanes difformis*
ìgọ̀gò IJO - IZON CUCURBITACEAE *Lagenaria siceraria*
iguenine ARABIC CAPPARACEAE *Capparis decidua*
iguewe-nowẹẹ EDO COMMELINACEAE *Palisota ambigua*
igufor SERER - NON CAPPARACEAE *Capparis tomentosa*
iguîn ARABIC CAPPARACEAE *Capparis decidua*
iguini ARABIC CAPPARACEAE *Capparis decidua* TAMACHEK *C. decidua*
igun YORUBA DIOSCOREACEAE *Dioscorea rotundata*
igwa (ikan) KANUFI - KARSHI CUCURBITACEAE *Lagenaria siceraria*
ígyár´ TIV AMARANTHACEAE *Celosia argentea*
igyolowa URHOBO DIOSCOREACEAE *Dioscorea alata*
ihadej lehmar ARABIC CUCURBITACEAE *Citrullus colocynthis*
iheiffif TAMACHEK AIZOACEAE *Aizoon canariensis*
íhièghè EDO CECROPIACEAE *Myrianthus arboreus*
íhíon EDO CUCURBITACEAE *Luffa cylindrica*
íhíon-òsà EDO CUCURBITACEAE *Luffa cylindrica*
íhíon-oyibo EDO CUCURBITACEAE *Luffa cylindrica*
ihobia YORUBA DIOSCOREACEAE *Dioscorea rotundata*
ijakẹrẹ IGALA CHRYSOBALANACEAE *Parinari curatellifolia*
ijan YORUBA CELASTRACEAE *Hippocratea welwitschii*
íjémbēàóndó TIV DIOSCOREACEAE *Dioscorea bulbifera*
i-jendir (*pl.* í-jendir) OTANK CUCURBITACEAE *Lagenaria siceraria*
ijiē TIV DIOSCOREACEAE *Dioscorea Linn., D. bulbifera*
ijiē pupuu TIV DIOSCOREACEAE *Dioscorea bulbifera*
ijiē shìbèrá TIV DIOSCOREACEAE *Dioscorea bulbifera*
íjíkàrà IGBO ANACARDIACEAE *Spondias mombin*
ijọndogh TIV CUCURBITACEAE *Lagenaria siceraria*
íjọ́ndúgh TIV CUCURBITACEAE *Lagenaria siceraria*
ijorua URHOBO DIOSCOREACEAE *Dioscorea alata*
i-joyun YORUBA ASCLEPIADACEAE *Pergularia daemia*
ikate YORUBA CHRYSOBALANACEAE *Chrysobalanus icaco, Licania elaeosperma*
íkè ńkịtā IGBO CONNARACEAE *Cnestis ferruginea*
ìkéré IGBO ACANTHACEAE *Asystasia gangetica, Hypoestes rosea*
ìkèrèḅùrú, ìkèrèḅú IJO ARACEAE *Colocasia esculenta*
íkhímì EDO BIGNONIACEAE *Newbouldia laevis*
ikhimwin EDO BIGNONIACEAE *Newbouldia laevis*
ìkhúian, iwian EDO APOCYNACEAE *Tabernaemontana pachysiphon*
ikik ARABIC COMBRETACEAE *Combretum aculeatum*
ikim UBAGHARA CUCURBITACEAE *Lagenaria siceraria*
ikim ANAANG CUCURBITACEAE *Lagenaria siceraria* EFIK *L. siceraria*
ikitchi HAUSA BURSERACEAE *Commiphora africana*
iko IGBO CUCURBITACEAE *Lagenaria siceraria*
ìkó EFIK CUCURBITACEAE *Lagenaria siceraria*
íkòn EFIK CUCURBITACEAE *Citrullus lanatus*
íkôn IBIBIO CUCURBITACEAE *Citrullus lanatus*
ìkọ̀ñ EFIK AMARANTHACEAE *Pupalia lappacea*
íkọ̀ñ ífuộ èyèn EFIK COMPOSITAE *Ageratum conyzoides*
íkọ̃ñ úbọ̀ñ IBIBIO CUCURBITACEAE *Telfairea occidentalis*
ikotio olokun YORUBA CYPERACEAE *Remirea maritima*
íkpán ANAANG CUCURBITACEAE *Citrullus lanatus* EFIK *C. lanatus* IBIBIO *C. lanatus*
ìkpáñ EFIK CUCURBITACEAE *Lagenaria siceraria*
íkpèlè átụ̄lụ̄ IGBO COMMELINACEAE *Palisota hirsuta*
íkpèlè átụ̄lụ̄, ìkpèrè átụ̄rụ̄ IGBO COMMELINACEAE *Palisota hirsuta*
ikpele oku IGBO COMMELINACEAE *Palisota hirsuta*
íkpèn EDO DIOSCOREACEAE *Dioscorea cayenensis*
íkpèrè átụ́rụ̄, íkpèlè átụ́rụ̄ IGBO COMMELINACEAE *Palisota hirsuta*

kani-gu ADANGME - KROBO DIOSCOREACEAE *Dioscorea cayenensis*
kani-hini ADANGME - KROBO DIOSCOREACEAE *Dioscorea cayenensis*
kanıga KABRE CUCURBITACEAE *Cucumis melo*
kani-kuku ADANGME - KROBO DIOSCOREACEAE *Dioscorea cayenensis*
kani-ŋjɔ̃ MANDING - MANDINKA CUCURBITACEAE *Lagenaria rufa*
kani-tʃu ADANGME - KROBO DIOSCOREACEAE *Dioscorea cayenensis*
kanja KPE APOCYNACEAE *Rauvolfia macrophylla* KUNDU *Alstonia boonei*
kanjinga KABRE CUCURBITACEAE *Cucumis melo* MOORE - NAWDAM *C. melo* TEM CONVOLVULACEAE *Citrullus lanatus* CUCURBITACEAE *Citrullus lanatus*
kanju KANURI ANACARDIACEAE *Anacardium occidentale*
kanju-kanmu YORUBA COMPOSITAE *Synedrella nodiflora*
kànjuú HAUSA ANACARDIACEAE *Anacardium occidentale*
kankana HAUSA CUCURBITACEAE *Cucurbita maxima*
kankanaŋ (*def.*-o) MANDING - MANDINKA COMBRETACEAE *Guiera senegalensis*
*a*-kan-*a*-kei TEMNE ACANTHACEAE *Lankesteria brevior*
kankeliba SUSU COMBRETACEAE *Combretum micranthum*
kankilibanyi SUSU COMBRETACEAE *Combretum micranthum*
kankoabu GURMA BALANITACEAE *Balanites aegyptiaca*
kankurẽg DIOLA DIOSCOREACEAE *Dioscorea preussii*
kânog DIOLA CAPPARACEAE *Capparis tomentosa*
kànrìnkàn YORUBA CUCURBITACEAE *Luffa cylindrica*
kànrìnkàn-ayaba YORUBA CUCURBITACEAE *Luffa cylindrica*
kantakara HAUSA COMBRETACEAE *Combretum collinum ssp. hypopilinum, C. collinum ssp. geitonophyllum*
kantaŋ MANDING - MANDINKA BOMBACACEAE *Ceiba pentandra*
kantankrui ANYI - SEHWI BURSERACEAE *Canarium schweinfurthii*
kanti-kanti YORUBA CONNARACEAE *Byrsocarpus coccineus*
kantoboa GUANG - KRACHI ARALIACEAE *Cussonia arborea*
kantong DAGBANI BOMBACACEAE *Adansonia digitata, Ceiba pentandra*
kantonọyọ YORUBA ANACARDIACEAE *Anacardium occidentale*
*an*-kantr-*a*-potho TEMNE BOMBACACEAE *Bombacopsis glabra*
kanum DIOLA DIOSCOREACEAE *Dioscorea bulbifera*
kanwene AKAN - ASANTE APOCYNACEAE *Pleiocarpa mutica* WASA *Hunteria umbellata, Picralima nitida, Pleiocarpa bicarpellata*
kanwini AKAN - ASANTE APOCYNACEAE *Hunteria umbellata, Picralima nitida* TWI *Hunteria umbellata* WASA *Pleiocarpa mutica*
kanya-bo MENDE DIOSCOREACEAE *Dioscorea misc. & spp. unknown*
kan-yaduŋga-na SUSU - DYALONKE ANNONACEAE *Uvaria chamae*
kanya-kanyen-na SUSU - DYALONKE ARALIACEAE *Cussonia arborea*
kanyɔɣe MOORE CONVOLVULACEAE *Ipomoea muricata*
kaowuli MENDE APOCYNACEAE *Alstonia boonei*
kape FULA - PULAAR DIOSCOREACEAE *Dioscorea Linn.*
kapé (*pl.*) FULA - PULAAR DIOSCOREACEAE *Dioscorea Linn.*
kapé FULA - PULAAR DIOSCOREACEAE *Dioscorea praehensilis, D. rotundata*
kapɛ FULA - PULAAR DIOSCOREACEAE *Dioscorea Linn.*
kàpííbí GUANG - GONJA ANACARDIACEAE *Lannea nigritana*
kappé FULA - TUKULOR DIOSCOREACEAE *Dioscorea Linn., D. lecardii*
kapu TIV CUCURBITACEAE *Lagenaria siceraria*
kapussé KRU - GUERE CONNARACEAE *Agelaea obliqua*
karadali MANDING - BAMBARA CRUCIFERAE *Brassica juncea* MANINKA *B. juncea*
karagbéi KRU - GUERE ACANTHACEAE *Eremomastax speciosa*
karaia (*pl.* buria) KOROP DIOSCOREACEAE *Dioscorea Linn.*
karakil-kɛnɛ MANDING - MANINKA ANNONACEAE *Cleistopholis patens*
karamba DIOLA DIOSCOREACEAE *Dioscorea bulbifera*
karamoko sunzun MANDING - MANINKA ANNONACEAE *Annona arenaria, A. senegalensis*
karan masallachin kogi HAUSA CYPERACEAE *Cyperus exaltatus*
kàràngíyár beéraá HAUSA AMARANTHACEAE *Cyathula prostrata*
kàràngíyár kuúsuú HAUSA AMARANTHACEAE *Cyathula prostrata*
kárán-másállaáchii HAUSA ASCLEPIADACEAE *Caralluma dalzielii*
karanwin KULANGO CONNARACEAE *Cnestis ferruginea*
karbobaba WOLOF COMPOSITAE *Sphaeranthus senegalensis*
kare KAREKARE CUCURBITACEAE *Lagenaria siceraria*
karebanga SONGHAI ARALIACEAE *Cussonia arborea*
kareń, kéreń WOLOF CAPPARACEAE *Capparis tomentosa*
karfaillé FULA - PULAAR ANACARDIACEAE *Lannea microcarpa*
kariya, ƙaryaà HAUSA APOCYNACEAE *Adenium obesum*
kariya HAUSA COMBRETACEAE *Combretum paniculatum ssp. paniculatum*
karkachijeho, katsame HAUSA BOMBACACEAE *Adansonia digitata*
karoduidiga MOORE ANNONACEAE *Annona arenaria, A. senegalensis*
ƙaron aduwa HAUSA BALANITACEAE *Balanites aegyptiaca*
karosufa SOMBA COMBRETACEAE *Anogeissus leiocarpus*
karra MANDING - MANDINKA CELASTRACEAE *Salacia senegalensis*
karro MANDING - MANDINKA CELASTRACEAE *Salacia senegalensis*
karuguna HAUSA CUCURBITACEAE *Lagenaria siceraria*
karya HAUSA APOCYNACEAE *Adenium obesum*
karya garma HAUSA COMBRETACEAE *Combretum sericeum* COMMELINACEAE *Aneilema lanceolatum*
kas kaifi HAUSA ANNONACEAE *Uvaria chamae*
kasa, tsaamiyar-kasa HAUSA CONNARACEAE *Byrsocarpus coccineus*
kasabaro MANDING - 'SOCE' CELASTRACEAE *Maytenus senegalensis*
ka-san CHAMBA CYPERACEAE *Fimbristylis dichotoma*
kasaulé MANDING - MANINKA COMBRETACEAE *Terminalia macroptera*

kit kácik BEROM DIOSCOREACEAE *Dioscorea alata*
kit ryáŋ BEROM DIOSCOREACEAE *Dioscorea rotundata*
kit tàl BEROM DIOSCOREACEAE *Dioscorea spp. indet.*
kitshagass TAMACHEK CAPPARACEAE *Boscia salicifolia*
kiyafa HAUSA CAPPARACEAE *Maerua angolensis*
kjirge DAGBANI CUCURBITACEAE *Lagenaria siceraria*
klae GBE - VHE CHRYSOBALANACEAE *Maranthes robusta*
klãgbã ADANGME - KROBO AMARANTHACEAE *Alternanthera pungens*
klakuo KRU - NEYO COMPOSITAE *Bidens pilosa*
klannéné KONO APOCYNACEAE *Strophanthus hispidus*
klauri KWENI BORAGINACEAE *Heliotropium indicum*
kleiri iwonné KRU - GUERE COMPOSITAE *Eclipta alba*
klékélé DAN COMPOSITAE *Microglossa afzelii*
kléŋme GA CONVOLVULACEAE *Ipomoea coptica*
kleŋme? GA AIZOACEAE *Sesuvium portulacastrum*
klériuémé KRU - GUERE COMPOSITAE *Eclipta alba*
kletsi GBE - VHE DICHAPETALACEAE *Dichapetalum madagascariense*
klewu GBE - VHE DIOSCOREACEAE *Dioscorea rotundata*
klewu, kokolimakoe GBE - VHE DIOSCOREACEAE *Dioscorea rotundata*
kliméné DAN APOCYNACEAE *Alstonia boonei*
klimintio KRU - GUERE BOMBACACEAE *Bombax buonopozense*
klo GBE - VHE DIOSCOREACEAE *Dioscorea Linn.*
kloadjémiéhuon KRU - GUERE ANNONACEAE *Dennettia tripetala*
kluapo 'KRU' COMPOSITAE *Emilia coccinea*
kmamegan AKAN - ASANTE COSTACEAE *Costus englerianus*
kmijirihi GWARI ANNONACEAE *Annona senegalensis*
kni GAGU APOCYNACEAE *Funtumia elastica*
knuba GWARI CUCURBITACEAE *Cucurbita maxima, C. pepo*
knubwobwoyi GWARI APOCYNACEAE *Voacanga thouarsii*
knuellehi GWARI BOMBACACEAE *Bombax buonopozense*
knugba GWARI CUCURBITACEAE *Cucurbita maxima, C. pepo*
ko GRUSI COMBRETACEAE *Terminalia avicennioides* KONO DIOSCOREACEAE *Dioscorea minutiflora*
kò (*pl.* kòà) SONGHAI BOMBACACEAE *Adansonia digitata*
ko barani MANDING - BAMBARA CUCURBITACEAE *Luffa acutangula*
ko bua MANDING - BAMBARA ARACEAE *Pistia stratiotes*
ko kuna sigi MANDING - BAMBARA COMPOSITAE *Laggera pterodonta*
ko safuna MANDING - BAMBARA COMPOSITAE *Vernonia colorata*
ko su MANO CONVOLVULACEAE *Ipomoea involucrata*
koa KONO CHRYSOBALANACEAE *Parinari excelsa*
koa djo KRU - GUERE BOMBACACEAE *Bombax buonopozense*
koagbè MANDING - MANINKA COSTACEAE *Costus Linn.*
koagni sélébéï KRU - BETE COMPOSITAE *Microglossa pyrifolia*
koakuru SENUFO - TAGWANA COMPOSITAE *Acanthospermum hispidum*
ko-aramon AKYE CHRYSOBALANACEAE *Maranthes robusta*
ko-ase-kohwe AKAN - ASANTE DIOSCOREACEAE *Dioscorea alata*
ko-ba MANDING - MANINKA AMARYLLIDACEAE *Scadoxus cinnabarinus, S. multiflorus*
kö-basera BAGA BOMBACACEAE *Adansonia digitata*
kobaté DAN COMBRETACEAE *Terminalia superba*
kɔbɛ SUSU CUCURBITACEAE *Cucumeropsis mannii*
kóbenci NUPE CHRYSOBALANACEAE *Neocarya macrophylla, Parinari curatellifolia*
kɔbene ANYI - AOWIN BOMBACACEAE *Rhodognaphalon brevicuspe* SEHWI *R. brevicuspe*
kɔbɛwu GBE - VHE ANACARDIACEAE *Lannea barteri*
kobira-fire SUSU CHRYSOBALANACEAE *Neocarya macrophylla*
kobo, koibo AKYE BALANITACEAE *Balanites wilsoniana*
ko-bona ABE BORAGINACEAE *Cordia vignei*
kõbre BASSARI COMBRETACEAE *Terminalia avicennioides*
kobri BASARI COMBRETACEAE *Terminalia laxiflora* BASSARI *T. laxiflora* SENUFO - TAGWANA APOCYNACEAE *Saba senegalensis*
kodé KRU - BETE CECROPIACEAE *Musanga cecropioides*
kodebâbi FULA - PULAAR COMPOSITAE *Centaurea senegalensis*
kodelẹ yiri YORUBA COMPOSITAE *Tridax procumbens*
kodelia TEM COMBRETACEAE *Anogeissus leiocarpus*
kodem baba FULA - PULAAR COMPOSITAE *Centaurea senegalensis*
kodem tabaga MOORE COMBRETACEAE *Combretum aculeatum*
kodemtabéga MOORE BURSERACEAE *Commiphora africana*
kodentigla MOORE COMBRETACEAE *Combretum aculeatum*
kodioli FULA - PULAAR COMBRETACEAE *Anogeissus leiocarpus*
koditambiga MOORE COMBRETACEAE *Combretum aculeatum*
kodo MANDING - MANDINKA BIGNONIACEAE *Markhamia tomentosa*
kodol FULA - PULAAR COMBRETACEAE *Anogeissus leiocarpus*
kodoli FULA - PULAAR COMBRETACEAE *Anogeissus leiocarpus*
kɔdolɔka MANDING - MANDINKA ANACARDIACEAE *Pseudospondias microcarpa*
kôdomburu FULA - PULAAR BIGNONIACEAE *Newbouldia laevis*
kodoyagi-na SUSU - DYALONKE ANNONACEAE *Hexalobus crispiflorus*
kodra iri MANDING - DYULA CAPPARACEAE *Crateva adansonii*
kodshõdo TEM CUCURBITACEAE *Cucurbita maxima*
kodudu FULA - PULAAR APOCYNACEAE *Landolphia dulcis* MANDING - BAMBARA *L. dulcis*
kodufleufleu KRU - BETE ARALIACEAE *Cussonia bancoensis*

kokonadu ADANGME - KROBO CRASSULACEAE *Bryophyllum pinnatum* GBE - VHE *B. pinnatum*
*a*-koko-*a*-nɛth TEMNE ARACEAE *Colocasia esculenta*
kokonisuo AKAN - ASANTE BIGNONIACEAE *Spathodea campanulata*
kɔkɔnsu AKAN - ASANTE BIGNONIACEAE *Spathodea campanulata*
kokoɔ AKAN - ASANTE COMPOSITAE *Triplotaxis stellulifera*
kokora AKAN - TWI DIOSCOREACEAE *Dioscorea smilacifolia*
kokorbé SONGHAI - ZARMA COMBRETACEAE *Combretum glutinosum*
kókórbéy (*pl.* -à) SONGHAI COMBRETACEAE *Combretum glutinosum*
*an*-kɔkɔrɔ TEMNE ACANTHACEAE *Justicia schimperi striata* AMARANTHACEAE *Celosia argentea, C. argentea*
kokorobe NZEMA CHRYSOBALANACEAE *Licania elaeosperma*
*a*-kɔkɔrɔ-*a*-ro-pet TEMNE AMARANTHACEAE *Celosia isertii, C. trigyna*
kokosa KULANGO COMPOSITAE *Bidens pilosa*
kokotau DIOLA DILLENIACEAE *Tetracera potatoria*
kokotobangui KYAMA COMPOSITAE *Mikania cordata*
kókpa NUPE CUCURBITACEAE *Lagenaria siceraria*
kokpè AKYE APOCYNACEAE *Alstonia boonei*
kokrobe NZEMA CHRYSOBALANACEAE *Licania elaeosperma*
kokrosabia AKAN - ASANTE COCHLOSPERMACEAE *Cochlospermum planchonii*
koktripa DAGAARI CELASTRACEAE *Maytenus senegalensis*
kokua ANYI DIOSCOREACEAE *Dioscorea burkilliana*
kokua-adua ba AKAN - ASANTE CECROPIACEAE *Myrianthus arboreus*
kokué AKYE APOCYNACEAE *Alstonia boonei*
kokuiribaï KRU - GUERE COMPOSITAE *Microglossa afzelii*
koku-lɛ BULOM ARACEAE *Xanthosoma mafaffa*
kɔkumba KONO CUCURBITACEAE *Cucumis sativus* KRIO *C. sativus* LIMBA *C. sativus* MENDE *C. sativus*
*ma*-kɔkumba TEMNE CUCURBITACEAE *Cucumis sativus*
kokurum boké KRU - BETE COMMELINACEAE *Palisota hirsuta*
ƙoƙuwa HAUSA CUCURBITACEAE *Lagenaria siceraria*
kokwaka MOORE CONVOLVULACEAE *Ipomoea asarifolia*
kokwè MANDING - MANINKA BIGNONIACEAE *Spathodea campanulata*
kokyi AKAN - FANTE CYPERACEAE *Cyperus articulatus*
*an*-kol TEMNE ARACEAE *Xanthosoma mafaffa*
kola KONO CHRYSOBALANACEAE *Parinari excelsa*
kolab SENUFO - TAGWANA COMPOSITAE *Vernonia guineensis*
kolabu SENUFO - TAGWANA COMPOSITAE *Vernonia guineensis*
kola-nélé KONO APOCYNACEAE *Strophanthus hispidus*
kolaton 'PATOKLA' APOCYNACEAE *Alstonia boonei*
kolatu KRU - GUERE APOCYNACEAE *Alstonia boonei*
*a*-kɔlɛ TEMNE APOCYNACEAE *Strophanthus hispidus*
koleiveiye (*def.*-i) MENDE CONVOLVULACEAE *Calycobolus heudelotii*
kọ̀léọ̀rọ́gbà YORUBA ASCLEPIADACEAE *Pergularia daemia*
kolétiti KRU - GUERE CELASTRACEAE *Hippocratea spp. sens. lat.*
kölib DIOLA APOCYNACEAE *Holarrhena floribunda*
kolidiohi MANDING - BAMBARA APOCYNACEAE *Rauvolfia vomitoria* MANINKA *R. vomitoria*
kolidohi MANDING - MANINKA APOCYNACEAE *Rauvolfia vomitoria*
kɔli-gojo (*def.*-i) MENDE CUCURBITACEAE *Ruthalicia eglandulosa*
kɔli-goko MENDE CUCURBITACEAE *Lagenaria rufa*
kolíkò GA ARACEAE *Colocasia esculenta*
kɔli-la MENDE ACANTHACEAE *Brillantaisia lamium, B. nitens*
koline SUSU CYPERACEAE *Eleocharis acutangula*
koli-tɔmda? KISSI DICHAPETALACEAE *Dichapetalum toxicarium*
*an*-kolma TEMNE CYPERACEAE *Cyperaceae*
*an*-kolma, *an*-koɔlma TEMNE CYPERACEAE *Cyperus articulatus*
*a*-kɔlma TEMNE CYPERACEAE *Kyllinga peruviana*
*a*-kɔlma-*a*-kuisisi TEMNE CYPERACEAE *Eleocharis mutata*
*a*-kɔlma-*a*-ro-bath TEMNE CYPERACEAE *Cyperus sphacelatus*
*an*-kolma-*a*-ro-gban TEMNE CYPERACEAE *Cyperus esculentus*
*an*-kolma-*an*-tar TEMNE CYPERACEAE *Cyperaceae*
*a*-kɔlma-*a*-tar TEMNE CYPERACEAE *Eleocharis dulcis*
kolo HAUSA BOMBACACEAE *Adansonia digitata*
kɔlɔ KISSI ACANTHACEAE *Asystasia gangetica*
kolobe MANDING - BAMBARA COMBRETACEAE *Combretum micranthum*
kolobé MANDING - BAMBARA COMBRETACEAE *Combretum aculeatum, C. micranthum* MANINKA *C. micranthum, Pteleopsis habeensis* 'SENUFO' *Combretum micranthum*
kolokass ARABIC ARACEAE *Colocasia esculenta*
kolokolo MANDING - BAMBARA BIGNONIACEAE *Newbouldia laevis*
kolokolo-yiri MANDING - MANDINKA COMBRETACEAE *Combretum fragrans*
kololo HAUSA CUCURBITACEAE *Lagenaria siceraria*
kolombolu KABRE BOMBACACEAE *Ceiba pentandra*
kŏlu KABRE COMBRETACEAE *Anogeissus leiocarpus*
kŏlumé SUSU CYPERACEAE *Cyperus articulatus*
kólúo KWENI ARACEAE *Colocasia esculenta*
koma ANYI - ANUFO COMBRETACEAE *Terminalia avicennioides* BAULE *T. avicennioides, T. glaucescens, T. ivorensis*
koma ba 'SENUFO' COMBRETACEAE *Terminalia laxiflora*
koma kièni MANDING - MANINKA COMBRETACEAE *Terminalia glaucescens*

kɔmafai MENDE ACANTHACEAE *Stenandriopsis buntingii*

kulummulu AKAN - BRONG CELASTRACEAE *Maytenus senegalensis*
kulungkalaŋ (*def.*-o) MANDING - MANDINKA COMBRETACEAE *Combretum Loefl.*
kulunhun TUBU ASCLEPIADACEAE *Calotropis procera*
kulunkalaŋ (*def.*-o) MANDING - MANDINKA COMBRETACEAE *Combretum micranthum, C. nigricans*
*o*-kum adada AKAN - TWI APOCYNACEAE *Alafia multiflora*
kuma KRU - GUERE AMARANTHACEAE *Cyathula prostrata*
kumâ KRU - GREBO AMARANTHACEAE *Cyathula prostrata*
kuma kuma MANDING - MANINKA APOCYNACEAE *Carissa edulis*
kúmaàtún mùzuúruú HAUSA ARACEAE *Xanthosoma mafaffa*
kumâda KANURI COMBRETACEAE *Terminalia avicennioides*
kumajɛjɛ MENDE CONNARACEAE *Spiropetalum heterophyllum, S. solanderi*
kumakuafo AKAN - ASANTE CELASTRACEAE *Maytenus senegalensis*
kumalui BAULE ARACEAE *Culcasia spp. indet.*
kumani SENUFO - TAGWANA CONNARACEAE *Cnestis ferruginea*
kumaniangama ANYI CECROPIACEAE *Myrianthus arboreus*
kum-anini AKAN - ASANTE ANACARDIACEAE *Lannea welwitschii* TWI *L. welwitschii*
kumareturo MANDING - 'SOCE' COMPOSITAE *Vernonia nigritiana*
kumare-two MANDING - MANDINKA COMPOSITAE *Vernonia nigritiana*
kumba ARABIC - SHUWA ANNONACEAE *Xylopia aethiopica* KISSI COMPOSITAE *Emilia sonchifolia* CUCURBITACEAE *Cucumis sativus* MANDING - MANINKA COMPOSITAE *Crassocephalum crepidioides, C. rubens, Gynura sarmentosa*
kumba-dongul FULA - PULAAR COMPOSITAE *Ageratum conyzoides*
kumban-shafu HAUSA CELASTRACEAE *Maytenus senegalensis*
kumbańul WOLOF COMPOSITAE *Mikania cordata*
kumbe MANDING - MANDINKA COMMELINACEAE *Palisota hirsuta*
kumbé FULA - PULAAR COMPOSITAE *Crassocephalum rubens*
kumbɛ MANDING - MANDINKA COMPOSITAE *Crassocephalum crepidioides, C. rubens, Emilia coccinea* COSTACEAE *Costus afer*
kumbé tiangol FULA - PULAAR COMPOSITAE *Adenostemma perrottetii*
kumbi MANDING - MANINKA COMPOSITAE *Emilia sonchifolia*
kumbo HAUSA CUCURBITACEAE *Lagenaria siceraria*
kumbɔnɛ KONO COMPOSITAE *Crassocephalum crepidioides*
kùmbùrà fágeé HAUSA COMPOSITAE *Vernonia adoensis*
kumburibe MANDING - MANINKA COMBRETACEAE *Terminalia superba*
kumdwe, kumdwie AKAN - WASA ANNONACEAE *Pachypodanthium staudtii*
kumenini VULGAR ANACARDIACEAE *Lannea welwitschii* AKAN - ASANTE *L. welwitschii* WASA *L. welwitschii*
kumɛn-na SUSU - DYALONKE COMPOSITAE *Crassocephalum crepidioides, C. rubens*
kumeny-ɛ BULOM COMPOSITAE *Crassocephalum crepidioides*
kumenyi SUSU COMPOSITAE *Crassocephalum rubens, Emilia coccinea, Gynura sarmentosa*
kuminimba AKAN - WASA ANACARDIACEAE *Antrocaryon micraster*
kuminin AKAN - ASANTE APOCYNACEAE *Holarrhena floribunda*
kumiyo ADANGME - KROBO DIOSCOREACEAE *Dioscorea rotundata*
kumiyow AKAN - TWI DIOSCOREACEAE *Dioscorea rotundata*
kùmkúm, kǝ̀mkǝ́m KANURI CAPPARACEAE *Maerua pseudopetalosa*
kumkumo MANDING - 'SOCE' ANACARDIACEAE *Spondias mombin*
kum-nkura AKAN - TWI DICHAPETALACEAE *Dichapetalum toxicarium*
kumo KANURI CUCURBITACEAE *Lagenaria siceraria* KULANGO ANNONACEAE *Annona arenaria, A. senegalensis*
kum-onini AKAN - ASANTE ANACARDIACEAE *Lannea welwitschii*
kumorufiɛxe SUSU COMPOSITAE *Vernonia guineensis*
kumudoi HAUSA ACANTHACEAE *Monechma ciliatum*
kumuru MANDING - BAMBARA CYPERACEAE *Fuirena umbellata*
kun ankala MANDING - MANINKA APOCYNACEAE *Strophanthus sarmentosus*
kuna MANDING - BAMBARA APOCYNACEAE *Strophanthus hispidus, S. sarmentosus*
ku-na SUSU - DYALONKE DIOSCOREACEAE *Dioscorea Linn.*
kuna dié MANDING - BAMBARA APOCYNACEAE *Strophanthus hispidus*
kuna ion MANDING - BAMBARA APOCYNACEAE *Strophanthus hispidus*
kuna kalo MANDING - MANDINKA APOCYNACEAE *Strophanthus sarmentosus*
kuna nombo MANDING - MANDINKA APOCYNACEAE *Strophanthus sarmentosus*
kuna sana MANDING - MANINKA APOCYNACEAE *Holarrhena floribunda*
kunafin MANDING - BAMBARA APOCYNACEAE *Holarrhena floribunda*
kunaji WOLOF COMBRETACEAE *Combretum collinum ssp. geitonophyllum, C. spp. indet.*
kunâkale MANDING - MANINKA APOCYNACEAE *Strophanthus sarmentosus*
kunâkalé MANDING - BAMBARA APOCYNACEAE *Strophanthus sarmentosus*
kunakalo MANDING - MANDINKA APOCYNACEAE *Strophanthus sarmentosus* COMPOSITAE *Vernonia colorata*
kunalé MANDING - MANINKA APOCYNACEAE *Strophanthus sarmentosus*
kunam 'SENUFO' ANACARDIACEAE *Lannea nigritana*
ku-ɲambo WOLOF DIOSCOREACEAE *Dioscorea alata*
kunamkala MANDING - BAMBARA APOCYNACEAE *Strophanthus sarmentosus* MANINKA *S. sarmentosus*
kunan MANDING - BAMBARA ANACARDIACEAE *Sclerocarya birrea* MANINKA *S. birrea*
kunana amata DIOLA CAPPARACEAE *Ritchiea capparoides*
kunangué 'SENUFO' ANACARDIACEAE *Lannea nigritana*
kunankala MANDING - MANINKA APOCYNACEAE *Strophanthus sarmentosus*

kuobene ANYI BOMBACACEAE *Rhodognaphalon brevicuspe*
kuobéné ANYI BOMBACACEAE *Rhodognaphalon brevicuspe*
kuokuo dua VULGAR CHRYSOBALANACEAE *Maranthes robusta*
kuonné MANO APOCYNACEAE *Strophanthus hispidus*
kupa JUKUN ASCLEPIADACEAE *Calotropis procera*
kupampaŋ MANDING - MANDINKA ASCLEPIADACEAE *Calotropis procera*
kupampaŋ (*def.*-o) MANDING - MANDINKA ASCLEPIADACEAE *Calotropis procera*
kupaniénié KRU - GREBO COMPOSITAE *Triplotaxis stellulifera*
kupâpâ FULA - PULAAR ASCLEPIADACEAE *Calotropis procera*
kura FULA - FULFULDE CHRYSOBALANACEAE *Parinari excelsa* PULAAR *Bafodeya benna, Parinari excelsa* MANDING - MANDINKA *P. excelsa*
kura MANDING - MANINKA APOCYNACEAE *Strophanthus hispidus*
kura MANDING - MANINKA CHRYSOBALANACEAE *Parinari excelsa*
kura dombi FULA - PULAAR CHRYSOBALANACEAE *Parinari curatellifolia*
kura-bansuma FULA - PULAAR CHRYSOBALANACEAE *Neocarya macrophylla*
kura-bansuma(-i) FULA - PULAAR CHRYSOBALANACEAE *Neocarya macrophylla*
kura-kɔləme SUSU CYPERACEAE *Cyperus articulatus*
kuranako FULA - PULAAR CHRYSOBALANACEAE *Parinari excelsa*
kura-nako FULA - PULAAR CHRYSOBALANACEAE *Parinari excelsa*
kuranga, kuringa, kwaranga HAUSA APOCYNACEAE *Saba florida*
ƙurar shanu HAUSA COMPOSITAE *Pulicaria crispa*
kuré DOGON ANACARDIACEAE *Lannea microcarpa*
kurɛ KONO CHRYSOBALANACEAE *Parinari excelsa*
kure pallaandi FULA - FULFULDE AMARANTHACEAE *Achyranthes aspera, Celosia trigyna, C. trigyna*
kurel WOLOF CAPPARACEAE *Crateva adansonii*
kuri MAMBILA ARACEAE *Colocasia esculenta*
kuria HAUSA BOMBACACEAE *Bombax buonopozense, B. costatum*
kuridêdê MANDYAK ANACARDIACEAE *Sorindeia juglandifolia*
kurit WOLOF CAPPARACEAE *Crateva adansonii*
kuriya, kúríyaá, kúryaá HAUSA BOMBACACEAE *Bombax buonopozense, B. costatum*
kúríyaá, kúryaá HAUSA BOMBACACEAE *Rhodognaphalon brevicuspe*
kurli DAGBANI COMBRETACEAE *Terminalia ivorensis*
kurli-langban DAGBANI COMBRETACEAE *Terminalia macroptera*
kúrnjé DERA BOMBACACEAE *Adamsonia digitata*
kuro DAN APOCYNACEAE *Holarrhena floribunda* LIMBA ANACARDIACEAE *Pseudospondias microcarpa*
kurobow VULGAR BALANITACEAE *Balanites wilsoniana*
kurofidie AKAN - AKYEM COMPOSITAE *Bidens pilosa*
kuroko LIMBA CECROPIACEAE *Myrianthus libericus, M. serratus*
kursan ARABIC CAPPARACEAE *Boscia senegalensis*
kurtu BADE - NGIZIM CUCURBITACEAE *Lagenaria siceraria*
kurtun lalle, kurtutu HAUSA CUCURBITACEAE *Lagenaria siceraria*
kurtun tadawa, kurtun tawada HAUSA CUCURBITACEAE *Lagenaria siceraria*
kurtúú HAUSA CUCURBITACEAE *Lagenaria siceraria*
kuru ru MANDING - BAMBARA DIPTEROCARPACEAE *Monotes kerstingii*
kuru ulu MANDING - BAMBARA ACANTHACEAE *Blepharis linariifolia, Hygrophila senegalensis*
kurubhobo MANDING - MANDINKA CONVOLVULACEAE *Ipomoea spp. indet.*
kurubomboŋmesengo MANDING - MANDINKA CUCURBITACEAE *Cucumis melo*
kuruhi FULA - FULFULDE BOMBACACEAE *Bombax costatum* TUKULOR *B. costatum*
kurukey MANDING - 'SOCE' COMPOSITAE *Vernonia nigritiana*
kuruko KULANGO BIGNONIACEAE *Kigelia africana*
kurukɔre SUSU COMMELINACEAE *Aneilema beninense*
kuru-kurinyi SUSU CUCURBITACEAE *Momordica balsamina, M. charantia*
ƙuruƙuru, ƙurƙure HAUSA COMBRETACEAE *Guiera senegalensis*
kurum KATAB CUCURBITACEAE *Lagenaria siceraria*
kurumale MANDING - BAMBARA APOCYNACEAE *Ancylobothrys amoena*
kurumbo HAUSA CUCURBITACEAE *Lagenaria siceraria*
kurumbu SONGHAI CRUCIFERAE *Brassica oleracea*
kurummɔtɔ AKAN - ASANTE ARALIACEAE *Cussonia arborea*
ƙurunƙushewa, ƙurunƙushiya HAUSA CELASTRACEAE *Maytenus senegalensis*
kurupta GA'ANDA CUCURBITACEAE *Lagenaria siceraria*
kurutu AKAN - ASANTE ASCLEPIADACEAE *Gongronema latifolium*
kurutwe AKAN - TWI BURSERACEAE *Canarium schweinfurthii* WASA *C. schweinfurthii*
ƙurẓunu HAUSA CUCURBITACEAE *Lagenaria siceraria*
kusã JUKUN CUCURBITACEAE *Lagenaria siceraria*
kusanköl BANYUN COMBRETACEAE *Combretum micranthum*
kushu KRIO ANACARDIACEAE *Anacardium occidentale*
kushumin, kushumɛnt KRIO APOCYNACEAE *Landolphia dulcis*
kussié MANDING - BAMBARA CELASTRACEAE *Maytenus senegalensis*
kussomo TUBU CAPPARACEAE *Capparis decidua*
kus-sum CHAMBA BURSERACEAE *Commiphora africana*
kusu ADANGME - KROBO COMPOSITAE *Launaea taraxacifolia*
kutagana LIMBA CYPERACEAE *Cyperus articulatus*
kutan MANDING - BAMBARA ANACARDIACEAE *Sclerocarya birrea*
kutaŋ (*def.* kuntaŋo) MANDING - MANDINKA ANACARDIACEAE *Sclerocarya birrea*
kutan dao MANDING - 'SOCE' ANACARDIACEAE *Sclerocarya birrea*
kutandé buna MANDING - 'SOCE' APOCYNACEAE *Strophanthus sarmentosus*
kutchuichivi GUANG - NCHUMBULU ANACARDIACEAE *Lannea barteri*

kuten dao MANDING - 'SOCE' ANACARDIACEAE *Sclerocarya birrea*
kutɛnɛ LIMBA BOMBACACEAE *Ceiba pentandra*
kutidi LIMBA BOMBACACEAE *Adansonia digitata*
kuti-kani-misa AKAN - BRONG BIGNONIACEAE *Stereospermum kunthianum*
kuto MANDING - MANDINKA APOCYNACEAE *Voacanga thouarsii*
kuto blamien BAULE AMARANTHACEAE *Alternanthera pungens*
kuto-jambo MANDING - MANDINKA APOCYNACEAE *Voacanga thouarsii*
kutoro LIMBA ANACARDIACEAE *Spondias mombin*
kutu LIMBA CUCURBITACEAE *Cucurbita pepo*
kutu kwaku logrodo KULANGO APOCYNACEAE *Tabernaemontana crassa*
kútúkpáci NUPE BOMBACACEAE *Bombax buonopozense, B. costatum*
kuudaku FULA - FULFULDE CONVOLVULACEAE *Ipomoea batatas*
kúudékà (*pl.* -à) SONGHAI CONVOLVULACEAE *Ipomoea batatas*
kuukà HAUSA BOMBACACEAE *Adansonia digitata*
kuulahi (*pl.* kuulaji) FULA - FULFULDE COMBRETACEAE *Terminalia avicennioides*
kuulahi (*pl.* kuulaje) FULA - FULFULDE COMBRETACEAE *Terminalia glaucescens*
kuulahi (*pl.* kuulaje) FULA - FULFULDE COMBRETACEAE *Terminalia macroptera*
kuuɲambi (*def.* kuuɲamboo) MANDING - MANDINKA DIOSCOREACEAE *Dioscorea togoensis* MANINKA AMARANTHACEAE *Philoxerus vermicularis*
kúwà KANURI BOMBACACEAE *Adansonia digitata*
kuwe LOKO CUCURBITACEAE *Cucurbita pepo*
kuyawa KONO DIOSCOREACEAE *Dioscorea bulbifera*
kuyó DOGON COMBRETACEAE *Combretum micranthum*
kwa yi pělě MANO CUCURBITACEAE *Luffa cylindrica*
kwaaduko AKAN - ASANTE ACANTHACEAE *Hypoestes verticillaris, Phaulopsis falcisepala, Rhinacanthus virens*
kwabɛ dwea AKAN - ASANTE BALANOPHORACEAE *Thonningia sanguinea* TWI *T. sanguinea*
kwabgé MANDING - MANINKA COSTACEAE *Costus Linn.*
kwachare HAUSA CUCURBITACEAE *Lagenaria siceraria*
kwachiya HAUSA CUCURBITACEAE *Lagenaria siceraria*
kwadianga MOORE COMBRETACEAE *Terminalia macroptera*
kwaduko AKAN - TWI ACANTHACEAE *Hypoestes verticillaris, Rhinacanthus virens*
*n*-kwadupɔ AKAN - ASANTE COMPOSITAE *Synedrella nodiflora*
kwae-bɔɔfrɛ AKAN - ASANTE ARALIACEAE *Cussonia bancoensis* TWI *C. bancoensis*
kwaebrofere VULGAR ARALIACEAE *Cussonia bancoensis*
kwaensa AKAN - ASANTE BIGNONIACEAE *Markhamia tomentosa*
kwaginyanga MOORE COMBRETACEAE *Combretum fragrans*
kwaginyanga, kwanganga MOORE COMBRETACEAE *Combretum molle*
kwaginyanga MOORE COMBRETACEAE *Combretum nigricans*
kwagnan KRU - GUERE AMARANTHACEAE *Alternanthera maritima*
kwahi GWARI BOMBACACEAE *Adansonia digitata*
kwai-aŋma-tʃo ADANGME - KROBO ARACEAE *Anchomanes difformis*
kwaikwaye, kwaikwayo HAUSA BALANITACEAE *Balanites aegyptiaca*
kwakié-kwakié ANYI APOCYNACEAE *Tabernaemontana crassa*
kwa(i)kwayo HAUSA BOMBACACEAE *Adansonia digitata*
kwalaba HAUSA BOMBACACEAE *Adansonia digitata*
kwalalo HAUSA CUCURBITACEAE *Lagenaria siceraria*
kwalo KISSI CHRYSOBALANACEAE *Parinari excelsa*
kwamamusu MENDE ANACARDIACEAE *Lannea nigritana*
kwamamusu (*def.*-i) MENDE - KPA ANACARDIACEAE *Lannea nigritana*
ƙwame, kwámé, ƙwami kwámii HAUSA BOMBACACEAE *Adansonia digitata*
kwáŋ BEROM DIOSCOREACEAE *Dioscorea bulbifera*
kwandare HAUSA COMBRETACEAE *Terminalia macroptera*
kwanedua AKAN - ASANTE CHRYSOBALANACEAE *Parinari excelsa*
kwanganga MOORE COMBRETACEAE *Combretum fragrans, C. nigricans*
ƙwanƙila HAUSA COMBRETACEAE *Anogeissus leiocarpus*
kwaŋkwánìí HAUSA APOCYNACEAE *Strophanthus hispidus, S. sarmentosus*
kwantal HAUSA CUCURBITACEAE *Cucumis melo*
kwantama AKAN - WASA APOCYNACEAE *Landolphia, Landolphia owariensis*
kwantima AKAN - ASANTE ASCLEPIADACEAE *Secamone afzelii*
kwaor TIV BURSERACEAE *Commiphora kerstingii*
kwardauɗa HAUSA COMPOSITAE *Dicoma tomentosa*
kwargo HAUSA CUCURBITACEAE *Lagenaria siceraria*
kwartowa, kwattowa HAUSA CUCURBITACEAE *Citrullus colocynthis*
kwáryaá HAUSA CUCURBITACEAE *Lagenaria siceraria*
kwaseantwa AKAN - TWI BOMBACACEAE *Rhodognaphalon brevicuspe*
kwasi pɛtɛprɛ AKAN - TWI ACANTHACEAE *Eremomastax speciosa, Hypoestes verticillaris*
kwatakwari HAUSA BOMBACACEAE *Adansonia digitata*
kwatambo HAUSA BOMBACACEAE *Adansonia digitata*
kwatié kwatié AKAN - ASANTE APOCYNACEAE *Voacanga africana*
kwatiri DAGAARI COMBRETACEAE *Terminalia macroptera*
kwato HAUSA DIOSCOREACEAE *Dioscorea misc. & spp. unknown*
kwatri DAGAARI COMBRETACEAE *Terminalia macroptera*
kwavréfé AKAN - ASANTE CECROPIACEAE *Musanga cecropioides*
kwegenga MOORE COMBRETACEAE *Combretum collinum ssp. binderianum, C. fragrans, C. glutinosum, C. molle, C. nigricans*
kwemkwemini SONGHAI CAPPARACEAE *Cadaba farinosa*
kwenɛ, juu VAI DILLENIACEAE *Tetracera alnifolia*
kwenɛ(?) VAI DILLENIACEAE *Tetracera potatoria*
kweniŋ AKAN - WASA BORAGINACEAE *Ehretia cymosa*

lállèn giíwaá HAUSA COMBRETACEAE *Pteleopsis habeensis*
lallen shaamuwaa HAUSA AIZOACEAE *Gisekia pharnaceoides*
lalo WOLOF BOMBACACEAE *Adansonia digitata*
lalob ARABIC BALANITACEAE *Balanites aegyptiaca*
lalobè BAULE COMPOSITAE *Melanthera scandens*
lalobi BAULE COMPOSITAE *Aspilia rudis*
lam BULOM CUCURBITACEAE *Cucurbita pepo*
lamagui WOLOF APOCYNACEAE *Saba senegalensis*
lamarin YORUBA CYPERACEAE *Mariscus soyauxii*
lambâ MANDING - MANINKA BIGNONIACEAE *Kigelia africana*
lamban MANDING - MANINKA BIGNONIACEAE *Kigelia africana*
lambatagha MOORE CAPPARACEAE *Maerua angolensis*
lambe pundo KISSI COMPOSITAE *Senecio biafrae*
lambulambo KISSI CUCURBITACEAE *Telfairea occidentalis*
lamɗam baali FULA - FULFULDE CAPPARACEAE *Crateva adansonii*
lam-dɛ BULOM CUCURBITACEAE *Cucurbita pepo*
lamfam SERER - NON BORAGINACEAE *Heliotropium indicum*
lamgo? BULOM CUCURBITACEAE *Cucurbita pepo*
*an*-lami TEMNE ACANTHACEAE *Justicia schimperi striata* AMARANTHACEAE *Celosia argentea, C. argentea*
lami-córè FULA - PULAAR CUCURBITACEAE *Lagenaria siceraria*
lampone GRUSI - KASENA COMBRETACEAE *Combretum collinum* (*c*) *ssp. hypopilinum*
lamudé, lamuné FULA - PULAAR APOCYNACEAE *Saba senegalensis*
lamúdè FULA - PULAAR APOCYNACEAE *Landolphia owariensis*
lamúquè FULA - PULAAR APOCYNACEAE *Landolphia owariensis*
laŋbaŋgunɛ MANDING - MANINKA ACANTHACEAE *Asystasia calycina*
landa édi FULA - FULFULDE APOCYNACEAE *Voacanga thouarsii*
landaga MOORE COMBRETACEAE *Combretum micranthum*
landaniari FULA - PULAAR COMPOSITAE *Sphaeranthus senegalensis*
landan-niari FULA - PULAAR ACANTHACEAE *Nelsonia canescens*
lande GBE - VHE APOCYNACEAE *Carissa edulis*
landjirco FULA - PULAAR CUCURBITACEAE *Luffa cylindrica*
*an*-lane TEMNE ANNONACEAE *Xylopia acutiflora*
*an*-lanɛ TEMNE ANNONACEAE *Uvaria chamae*
langa TEM CUCURBITACEAE *Lagenaria siceraria*
langaga MOORE COMBRETACEAE *Combretum micranthum*
langbade YORUBA CUCURBITACEAE *Cucurbita pepo*
lángbòdó YORUBA ARACEAE *Anchomanes difformis*
lanṣẹjẹ YORUBA DIOSCOREACEAE *Dioscorea alata*
lanti bókùn NUPE CAPPARACEAE *Cleome gynandra*
laoñâdi FULA - PULAAR COMBRETACEAE *Combretum aculeatum*
lapawe YORUBA ANNONACEAE *Hexalobus crispiflorus*
lâpra KPELLE APOCYNACEAE *Alstonia boonei*
*a*-lap-*a*-ro-bamp TEMNE APOCYNACEAE *Pleiocarpa mutica*
làpsûr KANURI CRUCIFERAE *Lepidium sativum*
lar SERER AMARYLLIDACEAE *Crinum zeylanicum* ARACEAE *Amorphophallus flavovirens*
lar mbind SERER ARACEAE *Amorphophallus flavovirens*
lara(-k) pirri DAGBANI CHRYSOBALANACEAE *Maranthes polyandra*
larbakɔ DAGBANI DIOSCOREACEAE *Dioscorea rotundata*
laré BISA COMBRETACEAE *Combretum aculeatum* FULA - PULAAR APOCYNACEAE *Landolphia heudelotii, L. heudelotii, Saba senegalensis, S. senegalensis* MANDING - BAMBARA *Landolphia heudelotii, Saba senegalensis* SUSU - DYALONKE *S. senegalensis*
larebakɔ DAGBANI DIOSCOREACEAE *Dioscorea rotundata*
laro FULA - PULAAR APOCYNACEAE *Saba senegalensis*
laté BALANTA BOMBACACEAE *Adansonia digitata*
látè BALANTA BOMBACACEAE *Adansonia digitata*
*ka*-lato TEMNE APOCYNACEAE *Tabernaemontana crassa*
latotué GBE - VHE COMPOSITAE *Launaea taraxacifolia*
lauenne GOEMAI APOCYNACEAE *Strophanthus hispidus, S. sarmentosus*
laugni FULA - FULFULDE COMBRETACEAE *Combretum aculeatum* PULAAR *C. aculeatum*
laulau WOLOF CONVOLVULACEAE *Ipomoea aitonii, I. cairica, Jacquemontia tamnifolia, Merremia aegyptia*
laule BAULE BALANITACEAE *Balanites wilsoniana*
laura CHAWAI - KURAMA CUCURBITACEAE *Lagenaria siceraria*
lauro YORUBA BOMBACACEAE *Bombax buonopozense*
lauso AKYE BORAGINACEAE *Ehretia cymosa*
lautan WOLOF CONVOLVULACEAE *Ipomoea muricata*
lauyirikɔ FULA - PULAAR CUCURBITACEAE *Luffa cylindrica*
law mbambé SERER CONVOLVULACEAE *Merremia tridentata*
lawnyi FULA - FULFULDE COMBRETACEAE *Combretum aculeatum*
lawo YORUBA COMBRETACEAE *Combretum smeathmannii*
lawon SENUFO - DYIMINI COMPOSITAE *Vernonia guineensis*
lay WOLOF CHRYSOBALANACEAE *Parinari excelsa*
layinbo YORUBA DIOSCOREACEAE *Dioscorea rotundata*
laylayduuji FULA - FULFULDE CONVOLVULACEAE *Ipomoea aquatica*
layol gora FULA - FULFULDE ASCLEPIADACEAE *Oxystelma bornouense*
layre ngabbu FULA - FULFULDE CONVOLVULACEAE *Ipomoea asarifolia*
lè san nya BAULE ASCLEPIADACEAE *Pergularia daemia*

INDEX


leɓan uwar miji HAUSA CUCURBITACEAE *Lagenaria siceraria*
lebason KRU - GUERE COMPOSITAE *Bidens pilosa*
lèbè KRU - GUERE CAPPARACEAE *Buchholzia coriacea*
lébel SERER CONVOLVULACEAE *Merremia tridentata*
lébèl pul SERER CONVOLVULACEAE *Merremia tridentata*
lə́bèt DERA CARICACEAE *Carica papaya*
*an*-ləbin-ləbin TEMNE CRASSULACEAE *Bryophyllum pinnatum*
lebleban FULA - TUKULOR AMARANTHACEAE *Alternanthera sessilis*
lébo 'NEKEDIE' COMMELINACEAE *Palisota hirsuta*
lébo lébo MANDING - BAMBARA CAPPARACEAE *Maerua angolensis*
lébu lébu MANDING - BAMBARA CAPPARACEAE *Maerua angolensis*
leeɓol pullo FULA - FULFULDE CONVOLVULACEAE *Merremia tridentata*
leèmùn tsuntsuu HAUSA APOCYNACEAE *Carissa edulis*
left ARABIC CRUCIFERAE *Brassica napus*
left wafran ARABIC CRUCIFERAE *Brassica napus*
legbarembal KONKOMBA COMBRETACEAE *Pteleopsis suberosa*
legbule (*def*.-i) MENDE ACANTHACEAE *Asystasia calycina, A. gangetica*
legbule MENDE ACANTHACEAE *Asystasia scandens*
légé tamme KRU - GUERE COMPOSITAE *Laggera alata*
légedẹ YORUBA CUCURBITACEAE *Cucurbita pepo*
legel HAUSA CAPPARACEAE *Boscia salicifolia, B. senegalensis*
légéré FULA - PULAAR APOCYNACEAE *Alstonia boonei*
leggal baali FULA - FULFULDE CAPPARACEAE *Maerua angolensis*
legiɛ GOLA CUCURBITACEAE *Cucumeropsis mannii*
legiɛ-kwi GOLA CUCURBITACEAE *Citrullus lanatus*
légilégirdé FULA - PULAAR COMPOSITAE *Blainvillea gayana*
legla GBE - VHE APOCYNACEAE *Saba florida*
légué GAGU COMPOSITAE *Bidens pilosa*
leguéré FULA - PULAAR APOCYNACEAE *Alstonia boonei*
léguéré FULA - FULFULDE APOCYNACEAE *Alstonia boonei*
leguilnaye FULA - FULFULDE CAPPARACEAE *Maerua angolensis*
léguinaye ARABIC CAPPARACEAE *Maerua angolensis*
lehmé ARABIC CRUCIFERAE *Eremobium aegyptiacum*
léiguéde MANDING - DYULA COMPOSITAE *Vernonia guineensis*
léiguédé MANDING - DYULA COMPOSITAE *Vernonia guineensis*
*an*-lekaŋ TEMNE DIOSCOREACEAE *Dioscorea praehensilis*
leka-na SUSU - DYALONKE CUCURBITACEAE *Luffa cylindrica*
leket WOLOF CUCURBITACEAE *Lagenaria siceraria*
lekir KANURI AMARANTHACEAE *Amaranthus viridis*
lekki peewuri FULA - FULFULDE APOCYNACEAE *Adenium obesum*
lekki pöuri FULA - PULAAR APOCYNACEAE *Adenium obesum*
lekmir ARABIC CYPERACEAE *Cyperus tuberosus*
lela SUSU CYPERACEAE *Cyperaceae, Cyperus haspan, C. rotundus, C. sphacelatus*
lela-fimbe SUSU CYPERACEAE *Cyperus congensis, C. maculatus*
*an*-lel-boi TEMNE APOCYNACEAE *Landolphia calabarica, L. heudelotii, L. owariensis*
lele ADANGME BIGNONIACEAE *Kigelia africana*
le-lɛ BULOM ANACARDIACEAE *Spondias mombin*
lélé (*pl.* -à) SONGHAI CAPPARACEAE *Crateva adansonii*
lele duji FULA - FULFULDE CUCURBITACEAE *Momordica balsamina, M. charantia*
lelegbule MENDE ACANTHACEAE *Asystasia gangetica*
lélongo MOORE ASCLEPIADACEAE *Leptadenia hastata*
lemba MANDING - MANINKA BIGNONIACEAE *Kigelia africana*
lemboitéka MOORE CAPPARACEAE *Capparis tomentosa*
lembúbúrú NUPE BOMBACACEAE *Ceiba pentandra*
léméléme WOLOF CONVOLVULACEAE *Ipomoea muricata*
len SERER - NON BOMBACACEAE *Ceiba pentandra*
lɛ́ŋ BEROM ANNONACEAE *Xylopia aethiopica*
lengbé nziua BAULE DIOSCOREACEAE *Dioscorea alata*
lenge FULA - PULAAR APOCYNACEAE *Holarrhena floribunda*
lengé WOLOF APOCYNACEAE *Strophanthus hispidus*
lengerigongo DAGBANI BIGNONIACEAE *Stereospermum kunthianum*
leŋge-wuri SUSU BIGNONIACEAE *Crescentia cujete*
leŋguõ KISSI CUCURBITACEAE *Citrullus lanatus*
lenje HAUSA AIZOACEAE *Gisekia pharnaceoides*
lɛno ADANGME BOMBACACEAE *Ceiba pentandra*
lenul KONKOMBA DIOSCOREACEAE *Dioscorea Linn.*
*a*-leop-*a*-ro-bath TEMNE BIGNONIACEAE *Spathodea campanulata*
lep BULOM ANACARDIACEAE *Spondias mombin*
*an*-ləp TEMNE ANACARDIACEAE *Spondias mombin*
lepapa AVATIME APOCYNACEAE *Landolphia owariensis*
leplaurauni KWENI COMPOSITAE *Ageratum conyzoides*
lep-pogɛ BULOM ANACARDIACEAE *Lannea nigritana*
léptándà (*pl.* -à) SONGHAI CUCURBITACEAE *Cucurbita pepo*
leroï KULANGO APOCYNACEAE *Alstonia boonei*
leroĩ KYAMA APOCYNACEAE *Alstonia boonei*
lerué KULANGO APOCYNACEAE *Alstonia boonei* KYAMA *A. boonei*
lerwé KULANGO APOCYNACEAE *Alstonia boonei* KYAMA *A. boonei*
lesakuala GURMA COMBRETACEAE *Terminalia avicennioides*
lessig ARABIC CHENOPODIACEAE *Chenopodium murale*
létè BEROM CUCURBITACEAE *Lagenaria siceraria*

| | |
|---|---|
| lɔɛ | GBE - VHE BOMBACACEAE *Ceiba pentandra* |
| loe-ti | GBE BOMBACACEAE *Ceiba pentandra* |
| lofolafan | MANDING - MANINKA CONVOLVULACEAE *Ipomoea batatas* |
| lofolofo | MANDING - MANINKA CONVOLVULACEAE *Ipomoea involucrata* |
| lógbònkíyàn | YORUBA CAPPARACEAE *Euadenia trifoliolata, Ritchiea capparoides* |
| logbu baté | AKYE COMPOSITAE *Microglossa pyrifolia* |
| lɔgi | MENDE APOCYNACEAE *Landolphia dulcis* |
| lɔg-lɛ | BULOM APOCYNACEAE *Landolphia dulcis* |
| logoniokui | KRU - NEYO COMPOSITAE *Ageratum conyzoides* |
| loguilnaye | FULA - FULFULDE CAPPARACEAE *Maerua angolensis* |
| logul bahi | FULA - PULAAR CAPPARACEAE *Maerua angolensis* |
| loho | ABE CECROPIACEAE *Musanga cecropioides* |
| lɔkhdɛ-na | SUSU - DYALONKE ANACARDIACEAE *Spondias mombin* |
| lokhe | SUSU BOMBACACEAE *Bombax buonopozense* |
| lokhodi-na | SUSU - DYALONKE ANACARDIACEAE *Spondias mombin* |
| lokho-na | SUSU - DYALONKE BOMBACACEAE *Bombax buonopozense* |
| loko | FULA - FULFULDE COMBRETACEAE *Terminalia avicennioides* |
| loko yaya | 'MAHO' COSTACEAE *Costus afer* |
| lokoan | GRUSI COMBRETACEAE *Combretum tomentosum* |
| lɔkɔkɔmi | MENDE ASCLEPIADACEAE *Tylophora sylvatica* |
| lokpa | BAULE DIOSCOREACEAE *Dioscorea cayenensis* |
| loku | TEM APOCYNACEAE *Saba florida* |
| lokure | SUSU ANACARDIACEAE *Lannea nigritana* |
| lol | SERER - NON BALANITACEAE *Balanites aegyptiaca* |
| lôl | SERER BALANITACEAE *Balanites aegyptiaca* |
| lo-lo | MANO CONVOLVULACEAE *Bonamia thunbergiana* |
| lolo | VAI AMARANTHACEAE *Amaranthus hybridus* |
| loloa | GA CONVOLVULACEAE *Ipomoea mauritiana* |
| lo-lo-kŏ | MANO CONVOLVULACEAE *Ipomoea involucrata* |
| lolóriõ | DOGON CONVOLVULACEAE *Ipomoea asarifolia* |
| loloti | VULGAR ANACARDIACEAE *Lannea welwitschii* ABE *L. welwitschii* |
| lombaria | SONGHAI CUCURBITACEAE *Coccinia grandis* |
| lombo | TEM COCHLOSPERMACEAE *Cochlospermum tinctorium* |
| lɔmbɔɛ | VAI ANNONACEAE *Xylopia aethiopica* |
| lommo | MANDING - DYULA ANNONACEAE *Annona arenaria, A. senegalensis* |
| lonbongbué | KYAMA COMPOSITAE *Microglossa pyrifolia* |
| lóŋgúlóorí | DERA CONVOLVULACEAE *Ipomoea asarifolia* |
| loni (*def.*-i) | MENDE APOCYNACEAE *Tabernaemontana crassa* |
| *an*-lonk | TEMNE APOCYNACEAE *Landolphia macrantha* |
| *an*-lɔnk | TEMNE APOCYNACEAE *Landolphia dulcis* |
| lɔ̀ŋkɔ́ŋ | GUANG - GONJA CUCURBITACEAE *Lagenaria siceraria* |
| lookidge | MANDING - MANDINKA COMPOSITAE *Sphaeranthus senegalensis* |
| lookrij | MANDING - MANDINKA COMPOSITAE *Sphaeranthus senegalensis* |
| looku | KRU - GUERE CONNARACEAE *Agelaea spp. indet.* |
| loonde (*pl.* looɗe) | FULA - FULFULDE CUCURBITACEAE *Lagenaria siceraria* |
| loonirde | FULA - FULFULDE CUCURBITACEAE *Luffa cylindrica* |
| lorongui | SUSU BOMBACACEAE *Bombax costatum* |
| lossion | SENUFO - TAGWANA ANNONACEAE *Cleistopholis patens* |
| *an*-lotho | TEMNE COMPOSITAE *Centratherum angustifolium, Crassocephalum crepidioides, C. rubens, Gynura sarmentosa* |
| lotórcò | FULA - PULAAR CUCURBITACEAE *Luffa cylindrica* |
| lo-ulo | KRU - GUERE ARACEAE *Pistia stratiotes* |
| lovanga | KPE CAPPARACEAE *Cleome rutidosperma* |
| lovi | GBE - GEN BOMBACACEAE *Ceiba pentandra* |
| lovolovohi | LOMA CONVOLVULACEAE *Ipomoea involucrata* |
| lowu | KABRE APOCYNACEAE *Saba florida* |
| ltsúwaàwún-zaákìi | HAUSA CUCURBITACEAE *Cucumis figarei* |
| lu tŏ kŏlĕh | MANO APOCYNACEAE *Tabernaemontana longiflora* |
| luaintu | KRU - GUERE BURSERACEAE *Dacryodes klaineana* |
| luban | ARABIC BURSERACEAE *Commiphora pedunculata* ARABIC - SHUWA *C. pedunculata* |
| lubban | HAUSA BURSERACEAE *Commiphora pedunculata* |
| lubɛ (*def.*-i) | MENDE BIGNONIACEAE *Stereospermum acuminatissimum* |
| luboel | FULA - PULAAR COMPOSITAE *Ageratum conyzoides* |
| lubualansanlu | GURMA ANNONACEAE *Annona arenaria, A. senegalensis* |
| lubuń | SERER COMPOSITAE *Sphaeranthus senegalensis* SERER - NON *S. senegalensis* |
| lubuñ gor | SERER COMPOSITAE *Blumea aurita* |
| lubuńô | MANDING - 'SOCE' COMPOSITAE *Blumea aurita* |
| ludu, lule, nunu | HAUSA ANACARDIACEAE *Sclerocarya birrea* |
| luébo | KRU - GUERE COMMELINACEAE *Palisota hirsuta* |
| lugbagbel-la | SUSU - DYALONKE BIXACEAE *Bixa orellana* |
| luguʃ | WOLOF COMPOSITAE *Vernonia colorata* |
| luguʃ u valo | WOLOF COMPOSITAE *Sonchus oleraceus* |
| lugulé | MOORE COMBRETACEAE *Combretum molle* |
| luguna | SUSU - DYALONKE BIGNONIACEAE *Markhamia tomentosa* |
| lugur | BISA COCHLOSPERMACEAE *Cochlospermum tinctorium* |
| luguṣque | WOLOF ACANTHACEAE *Hypoestes verticillaris* |
| lugut | SERER COMPOSITAE *Sphaeranthus senegalensis* |
| lujiya | HAUSA ASCLEPIADACEAE *Xysmalobium heudelotianum* |
| lúkolúko | NUPE ARACEAE *Amorphophallus abyssinicus* |
| lukpogu | DAGBANI ARACEAE *Anchomanes difformis* |

mákàb NGEMBA ARACEAE *Colocasia esculenta*
makalonkalɔ LIMBA CHRYSOBALANACEAE *Neocarya macrophylla*
[illegible] HAUSA CUCURBITACEAE *Lagenaria siceraria*
makao BAFOK ARACEAE *Xanthosoma mafaffa*
makara VULGAR ARACEAE *Xanthosoma mafaffa*
makargis, makersas TAMACHEK AMARANTHACEAE *Aerva javanica*
maƙari HAUSA ARACEAE *Stylochiton lancifolius*
makheit ARABIC - SHUWA CAPPARACEAE *Boscia senegalensis*
makhmila TAMACHEK AMARANTHACEAE *Aerva javanica*
makindi LOKO AVICENNIACEAE *Avicennia germinans*
maƙododo HAUSA ARACEAE *Amorphophallus abyssinicus*
mákoòrií maƙododo HAUSA ARACEAE *Amorphophallus abyssinicus, Stylochiton lancifolius*
makpafi KISSI DICHAPETALACEAE *Dichapetalum toxicarium*
makpalaba LIMBA DICHAPETALACEAE *Dichapetalum toxicarium*
makpavi (*def.*-i) MENDE DICHAPETALACEAE *Dichapetalum madagascariense, D. toxicarium*
ma-k-povɛ BULOM CHRYSOBALANACEAE *Licania elaeosperma*
makru-leaf KRIO CHENOPODIACEAE *Chenopodium ambrosioides*
makua(-n) AKAN - BRONG APOCYNACEAE *Strophanthus hispidus* ANYI *S. hispidus*
makua(-n), mekua KULANGO APOCYNACEAE *Strophanthus hispidus*
makumbu MANDING - MANDINKA CONVOLVULACEAE *Ipomoea involucrata*
makurkura HAUSA CUCURBITACEAE *Lagenaria siceraria*
mala FULA - PULAAR CYPERACEAE *Scleria depressa*
malá MANDING - MANDINKA DIOSCOREACEAE *Dioscorea preussii*
malagueta CRIOULO ANNONACEAE *Xylopia aethiopica*
malagueta da Guiné CRIOULO ANNONACEAE *Xylopia aethiopica*
malagueta preta CRIOULO ANNONACEAE *Xylopia aethiopica*
malan kochi HAUSA AMARANTHACEAE *Amaranthus viridis*
malanga VULGAR ARACEAE *Xanthosoma mafaffa* DUALA BIGNONIACEAE *Markhamia tomentosa* FULA - PULAAR AVICENNIACEAE *Avicennia germinans*
malan-kotshi HAUSA AMARANTHACEAE *Amaranthus viridis*
male GA DIOSCOREACEAE *Dioscorea rotundata*
malé KRU - GUERE APOCYNACEAE *Holarrhena floribunda*
malega LIMBA CUCURBITACEAE *Luffa cylindrica*
malenguli(-i) MENDE ANNONACEAE *Pachypodanthium staudtii*
malentaŋ FULA - PULAAR CACTACEAE *Opuntia spp.*
malga HAUSA ANACARDIACEAE *Lannea microcarpa*
mali, obobi ADANGME - KROBO DIOSCOREACEAE *Dioscorea rotundata*
maliba ma manjonga KPE APOCYNACEAE *Landolphia landolphioides*
malinguer SONGHAI AÏZOACEAE *Trianthema portulacastrum, Zaleya pentandra*
malinka-guié MANDING - MANINKA CUCURBITACEAE *Cucurbita pepo*
malkɔre SUSU COMMELINACEAE *Commelina thomasii*
mam SERER BIGNONIACEAE *Stereospermum kunthianum* COMPOSITAE *Vernonia colorata*
mâm SERER BIGNONIACEAE *Spathodea campanulata*
mama, maamae AKAN - ASANTE APOCYNACEAE *Funtumia africana*
mama-cúncoe BADYARA CARYOPHYLLACEAE *Polycarpaea corymbosa, P. eriantha*
mamadâdâdi FULA - PULAAR CARYOPHYLLACEAE *Polycarpaea linearifolia*
mamb SERER BIGNONIACEAE *Stereospermum kunthianum*
mambeya KYAMA APOCYNACEAE *Pleiocarpa mutica*
mambumba
*a*-mamfohae AKAN - ASANTE APOCYNACEAE *Motandra guineensis*
mamfoham AKAN - ASANTE APOCYNACEAE *Strophanthus hispidus, S. preussii, S. sarmentosus*
mamno KISSI COMBRETACEAE *Terminalia superba*
mámpara-djom-áe FULA - PULAAR CHRYSOBALANACEAE *Parinari curatellifolia*
mámpara-djon-áe FULA - PULAAR CHRYSOBALANACEAE *Parinari curatellifolia*
mampata (*def.* mampatoo) MANDING - MANDINKA CHRYSOBALANACEAE *Parinari excelsa*
mampata MANDYAK CHRYSOBALANACEAE *Parinari excelsa* 'CASAMANCE' *P. excelsa*
mampatá (*def.* mampató) MANDING - MANDINKA CHRYSOBALANACEAE *Parinari excelsa*
mampataz CRIOULO CHRYSOBALANACEAE *Neocarya macrophylla, Parinari excelsa*
mampataz grande CRIOULO CHRYSOBALANACEAE *Neocarya macrophylla*
mampato WOLOF CHRYSOBALANACEAE *Parinari excelsa*
mampihege DAGBANI ANNONACEAE *Annona glauca*
mampoŋfo-apɔw AKAN - TWI COMPOSITAE *Synedrella nodiflora*
mampufa CRIOULO CYPERACEAE *Cyperus articulatus*
man AKAN - FANTE DIOSCOREACEAE *Dioscorea bulbifera*
mana FULA - TUKULOR APOCYNACEAE *Landolphia heudelotii*
mana(-n) MANDING - BAMBARA APOCYNACEAE *Landolphia heudelotii*
mańaka MANDING - BAMBARA COMBRETACEAE *Combretum molle* MANINKA *C. molle*
manamat TAMACHEK CUCURBITACEAE *Momordica balsamina, M. charantia*
manamendigo KULANGO COMPOSITAE *Bidens pilosa*
manan FULA - TUKULOR APOCYNACEAE *Landolphia heudelotii*
manan-wala KULANGO APOCYNACEAE *Funtumia africana*
manawala MANDING - MANDINKA CONNARACEAE *Byrsocarpus coccineus*
manbenbi KYAMA COMPOSITAE *Vernonia conferta*
mand SERER APOCYNACEAE *Saba senegalensis*
manda mbaala FULA - FULFULDE COMPOSITAE *Crassocephalum sp. indet.*
mãndafura SERER CUCURBITACEAE *Kedrostis foetidissima*

INDEX

mpéku ba MANDING - BAMBARA ANACARDIACEAE *Lannea acida*
mpenãbegu, mprabegu AKAN - TWI CONVOLVULACEAE *Cuscuta australis*
mpena-sere FULA - PULAAR COMPOSITAE *Acanthospermum hispidum*
mpentemi AKAN - TWI COMMELINACEAE *Palisota hirsuta*
m-piégba ABE APOCYNACEAE *Tabernaemontana crassa, Voacanga africana*
mpompomogolo MANDING - MANINKA ASCLEPIADACEAE *Calotropis procera*
mpompompogolo MANDING - MANINKA ASCLEPIADACEAE *Calotropis procera*
mponasere AKAN - ASANTE COMPOSITAE *Vernonia amygdalina, V. colorata*
m-pôpôpogolo MANDING - BAMBARA ASCLEPIADACEAE *Calotropis procera*
ḿpōtó IGBO ARACEAE *Colocasia esculenta*
mpôzé BALANTA COMBRETACEAE *Combretum nigricans*
mprakɛ ADANGME - KROBO CONVOLVULACEAE *Cuscuta australis* GBE - VHE *C. australis*
mpupua AKAN - TWI AMARANTHACEAE *Pupalia lappacea*
mpupuã AKAN - ASANTE AMARANTHACEAE *Cyathula prostrata, Pupalia lappacea*
mpũpuã AKAN - ASANTE COMPOSITAE *Acanthospermum hispidum*
mpupuaa AKAN - ASANTE COMPOSITAE *Acanthospermum hispidum* TWI *A. hispidum*
mraïnzè ARABIC CAPPARACEAE *Cleome arabica*
m'ru SENUFO - TAGWANA ANNONACEAE *Annona arenaria, A. senegalensis*
m-sal BALANTA ANACARDIACEAE *Spondias mombin*
mu ADANGME ACANTHACEAE *Barleria opaca* DAN CYPERACEAE *Afrotrilepis pilosa*
muamohia KYAMA BURSERACEAE *Canarium schweinfurthii*
muatʃɔtʃo ADANGME BORAGINACEAE *Ehretia cymosa*
mubulighra NZEMA ACANTHACEAE *Justicia flava*
mucéhuda SOMBA ANACARDIACEAE *Lannea microcarpa*
muci NUPE BOMBACACEAE *Adansonia digitata*
mué ABE ANNONACEAE *Monodora myristica*
muê KRU - GUERE APOCYNACEAE *Alstonia boonei*
muemia KYAMA BURSERACEAE *Canarium schweinfurthii*
muenla KWENI ANACARDIACEAE *Spondias mombin*
muénohia KYAMA BURSERACEAE *Canarium schweinfurthii*
muésué BATONNUN ANNONACEAE *Hexalobus monopetalus*
mufania SONGHAI - ZARMA ANNONACEAE *Annona arenaria, A. senegalensis*
mufanpahida SORUBA - KUYOBE COMBRETACEAE *Combretum collinum ssp. geitonophyllum*
mufo KISSI CUCURBITACEAE *Raphidiocystis chrysocoma*
mufopaïe SORUBA - KUYOBE COMBRETACEAE *Combretum nigricans*
mugaŋgà DUALA ANACARDIACEAE *Spondias mombin*
mugnien MANDING - BAMBARA CAPPARACEAE *Crateva adansonii* MANINKA *C. adansonii*
muin oroko GAGU CAPPARACEAE *Buchholzia coriacea*
mukindimu SOMBA COMBRETACEAE *Terminalia macroptera*
mùkɔndɛ DUALA DIOSCOREACEAE *Dioscorea misc. & spp. unknown*
mukopé KRU - GUERE ANACARDIACEAE *Trichoscypha patens*
muladĩ WOLOF COMPOSITAE *Ceruana pratensis*
mullibe GWARI ASCLEPIADACEAE *Pachycarpus lineolatus*
mulodu BATONNUN BOMBACACEAE *Bombax costatum*
mulukho-na? SUSU - DYALONKE CECROPIACEAE *Myrianthus arboreus, M. libericus, M. serratus*
mumie HAUSA COMBRETACEAE *Combretum fragrans*
mumue NZEMA CELASTRACEAE *Salacia nitida*
mu-mugga DAGBANI ANACARDIACEAE *Sclerocarya birrea*
mumuye HAUSA COMBRETACEAE *Combretum Loefl.*
*a*-muna TEMNE CONVOLVULACEAE *Ipomoea batatas*
munayem KONKOMBA ANACARDIACEAE *Spondias mombin*
muńê MANDING - BAMBARA CAPPARACEAE *Crateva adansonii* MANINKA *C. adansonii*
múné (*pl.* -à) SONGHAI CUCURBITACEAE *Cucumis melo*
mune(i)ko SENUFO - BAMANA ANACARDIACEAE *Spondias mombin*
muné-muné ADYUKRU APOCYNACEAE *Landolphia owariensis*
muneu ABURE BURSERACEAE *Canarium schweinfurthii*
mungo KONO ANACARDIACEAE *Spondias mombin*
muniziba GAGU DILLENIACEAE *Tetracera alnifolia*
munongbe AKYE AMARANTHACEAE *Aerva lanata*
munyɛ MENDE CYPERACEAE *Afrotrilepis pilosa*
muong AKYE BOMBACACEAE *Ceiba pentandra*
muɔtsu ADANGME AMARANTHACEAE *Amaranthus hybridus*
mupila PEPEL ANACARDIACEAE *Spondias mombin*
murgwi HAUSA CUCURBITACEAE *Lagenaria siceraria*
muriékolo DIOLA APOCYNACEAE *Strophanthus hispidus*
múrnaà HAUSA BOMBACACEAE *Adansonia digitata*
muro SENUFO - TAGWANA ANNONACEAE *Annona arenaria, A. senegalensis*
muroru BATONNUN BOMBACACEAE *Bombax buonopozense, B. costatum*
murotauki FULA - FULFULDE BALANITACEAE *Balanites aegyptiaca*
murot(a)oki FULA - PULAAR BALANITACEAE *Balanites aegyptiaca*
murr ARABIC - SHUWA CUCURBITACEAE *Cucurbita maxima, C. pepo*
murtégé FULA - PULAAR APOCYNACEAE *Alafia scandens, Landolphia hirsuta*
murtégue FULA - PULAAR APOCYNACEAE *Landolphia hirsuta*
murtoki FULA - PULAAR BALANITACEAE *Balanites aegyptiaca*
mur(o)toki FULA - PULAAR BALANITACEAE *Balanites aegyptiaca*
murugo MANDING - KHASONKE DIOSCOREACEAE *Dioscorea bulbifera*

INDEX

## INDEX

ndipagba (*def*.-i) **MENDE** ASCLEPIADACEAE *Secamone afzelii*
n-ḍipiandaṗa **SERER - NON** CYPERACEAE *Bulbostylis abortiva*
ndiran **WOLOF** CYPERACEAE *Cyperaceae*, *Cyperus rotundus*, *C. sphacelatus*
nḍis wâdan **SERER - NON** ASCLEPIADACEAE *Leptadenia hastata*
ndisan **SERER** AIZOACEAE *Trianthema portulacastrum*
ǹdìsè **EFIK** CUCURBITACEAE *Cucurbita maxima*
ǹdìsè **EFIK** CUCURBITACEAE *Cucurbita pepo* **IBIBIO** *C. pepo*
ǹdìsì **IBIBIO** CUCURBITACEAE *Cucurbita maxima*
ǹdìsime ìwá **EFIK** DIOSCOREACEAE *Dioscorea dumetorum*
n̠dìsọ̀k **EFIK** CECROPIACEAE *Myrianthus arboreus*
ndiyamhi **FULA - FULFULDE** AQUIFOLIACEAE *Ilex mitis*
n-djano **BALANTA** CHRYSOBALANACEAE *Parinari excelsa*
ndjano **BALANTA** CHRYSOBALANACEAE *Parinari excelsa*
n-djapô **BALANTA** CHRYSOBALANACEAE *Neocarya macrophylla*
n-djápo **BALANTA** CHRYSOBALANACEAE *Parinari excelsa*
n-dli-bara **MANDING - BAMBARA** COCHLOSPERMACEAE *Cochlospermum tinctorium*
ndo (*def*.-i) **MENDE** CAPPARACEAE *Buchholzia coriacea*
ndoa **KUNDU** ANACARDIACEAE *Spondias mombin* **LUNDU** *S. mombin* **MBONGE** *S. mombin* **TANGA** *S. mombin*
ndöbarkat **WOLOF** COMPOSITAE *Vernonia colorata*
ndɔbobɔgo **LOKO** CONVOLVULACEAE *Ipomoea mauritiana*
ndɔbohande **LOKO** CONVOLVULACEAE *Ipomoea mauritiana*
ndôg **SERER** ANACARDIACEAE *Lannea velutina* **WOLOF** ANNONACEAE *Annona senegalensis*
ndɔgbɔ gipɔ (*def*.-i) **MENDE** COMPOSITAE *Emilia praetermissa*
ndɔgbɔ niwi **MENDE** CYPERACEAE *Mariscus flabelliformis*
ndɔgbɔ-gipɔ (*def*.-i) **MENDE** COMPOSITAE *Emilia coccinea*
ndɔgbɔ-goja (*def*.-i) **MENDE** CUCURBITACEAE *Coccinea barteri*
ndɔgbɔ-goja **MENDE** CUCURBITACEAE *Coccinea keayana*
ndɔgbɔ-jɛlɛ **MENDE** ANNONACEAE *Uvaria chamae*
ndɔgbɔ-jɛlɛ kafa **MENDE - UP MENDE** ANACARDIACEAE *Sorindeia juglandifolia*
ndɔgbɔ-jɛlɛ-gbɔu **MENDE** ANNONACEAE *Uvaria chamae*
ndogbɔ-jigboi **LOKO** COMPOSITAE *Emilia coccinea*
ndɔgbɔ-yuwo **MENDE** CONVOLVULACEAE *Ipomoea cairica*
ndɔgbɔ-yuwo (*def*.-i) **MENDE** CONVOLVULACEAE *Ipomoea nil*
ndoggar **WOLOF** CARYOPHYLLACEAE *Polycarpaea linearifolia*
ndoggiar **SERER - NON** CARYOPHYLLACEAE *Polycarpaea corymbosa*
ndogot **WOLOF** ANACARDIACEAE *Lannea velutina*
n-dogot **WOLOF** ANACARDIACEAE *Lannea velutina*
ndôgsay **SERER - NON** APOCYNACEAE *Holarrhena floribunda*
ndokarok **SERER** ACANTHACEAE *Hygrophila auriculata*
ndoko **KOOSI** CONVOLVULACEAE *Ipomoea batatas* **KPE** *I. batatas* **WOVEA** *I. batatas*
ndol o ñan **SERER** CYPERACEAE *Mariscus squarrosus*
n-dolo **GOLA** AMARANTHACEAE *Amaranthus hybridus*
n-dologa **FULA - PULAAR** BORAGINACEAE *Cordia senegalensis*
n-dõloga **FULA - PULAAR** ANACARDIACEAE *Pseudospondias microcarpa*
ndolor(-é) **SERER - NON** APOCYNACEAE *Strophanthus sarmentosus*
n-doloré **SERER - NON** CELASTRACEAE *Salacia senegalensis*
n-dome **BALANTA** DIOSCOREACEAE *Dioscorea bulbifera*
n-domonger **WOLOF** COMBRETACEAE *Guiera senegalensis*
ndomu (*def*.-i) **MENDE** COMMELINACEAE *Palisota bracteosa*, *P. hirsuta*
ǹdọ̀ñ ùbọk **EFIK** DIOSCOREACEAE *Dioscorea Linn.*
ndɔ̀ŋbuɔ̀ **NGEMBA** DIOSCOREACEAE *Dioscorea Linn.*
ndonda **DUALA** CHRYSOBALANACEAE *Licania elaeosperma*
ndondelole **MENDE** CHRYSOBALANACEAE *Maranthes aubrevillii*, *M. glabra*
n-dɔndɔ **KISSI** ARACEAE *Anchomanes difformis*
ndondokɔ (*def*.-i) **MENDE** ACANTHACEAE *Thunbergia chrysops* COMPOSITAE *Mikania cordata* CONVOLVULACEAE *Hewittia sublobata*, *Ipomoea involucrata*
n-dondo-polole **KISSI** ASCLEPIADACEAE *Gongronema latifolium*
n-dondul **SERER** BOMBACACEAE *Bombax costatum*
ndong **SERER** ANNONACEAE *Annona senegalensis*
ndɔpa-nɛɛ **MENDE** DILLENIACEAE *Tetracera alnifolia*
n-dɔpararo **LOKO** DILLENIACEAE *Tetracera alnifolia*
ndottihon **FULA - FULFULDE** CONVOLVULACEAE *Evolvulus alsinoides*
ndottiyel **FULA - FULFULDE** CONVOLVULACEAE *Evolvulus alsinoides*
n-dougan **MANDING - BAMBARA** CYPERACEAE *Cyperus tonkinensis*
n-douin **MANDING - BAMBARA** CYPERACEAE *Cyperus tonkinensis*
ndɔwɔ **LOKO** ACANTHACEAE *Rhinacanthus virens*
ndu(m)barkat **SERER** COMPOSITAE *Vernonia colorata*
ndubarkat **SERER - NON** COMPOSITAE *Vernonia colorata*
n-duɛge(-ŋ) **LOKO** ANACARDIACEAE *Pseudospondias microcarpa*
n-duguń **SERER** ANACARDIACEAE *Lannea acida*
ndugutj **SERER** ANACARDIACEAE *Lannea acida*
ndukoworo **FULA - PULAAR** COMBRETACEAE *Combretum glutinosum*
ndukul mbap **SERER** ARACEAE *Stylochiton hypogaeus*
ndukuru **IGBO** BOMBACACEAE *Bombax buonopozense*
ndukut **SERER - NON** CELASTRACEAE *Maytenus senegalensis*
ndulubuń **SERER** COMPOSITAE *Sphaeranthus senegalensis*
n-dumburgat **WOLOF** COMPOSITAE *Vernonia colorata*
ndumburgat **WOLOF** COMPOSITAE *Vernonia colorata*

ngo-na-ŋkyene — AKAN - WASA ANNONACEAE *Cleistopholis patens*
ngonar did — SERER ANACARDIACEAE *Lannea humilis*
ngonaro — SERER ANACARDIACEAE *Lannea humilis*
n-gondi — LOKO ARACEAE *Amorphophallus aphyllus* CUCURBITACEAE *Luffa cylindrica*
ngondo — KPE CUCURBITACEAE *Citrullus lanatus* WOVEA *C. lanatus*
ngo-ne-nkyene — AKAN - TWI ANNONACEAE *Cleistopholis patens*
n-goŋgbo-pɔmbɔ — KISSI CONVOLVULACEAE *Ipomoea involucrata*
n-gongo — LOKO CUCURBITACEAE *Luffa cylindrica*
ngónówù — KANURI ANNONACEAE *Annona senegalensis*
n-gor si bidaw — WOLOF CAPPARACEAE *Cleome gynandra*
ngoral — SERER CAPPARACEAE *Crateva adansonii*
ngorèl — SERER CAPPARACEAE *Crateva adansonii*
ngoroba blé — MANDING - BAMBARA AMARANTHACEAE *Amaranthus spinosus*
ngorol — SERER CAPPARACEAE *Crateva adansonii* SERER - NON *C. adansonii*
ngororugu — LOKO BALANOPHORACEAE *Thonningia sanguinea*
ngotoa — AKAN - FANTE CUCURBITACEAE *Lagenaria siceraria* TWI *L. siceraria*
ngotot — WOLOF BURSERACEAE *Commiphora africana*
ngovo (*def.*-i) — MENDE CECROPIACEAE *Musanga cecropioides*
ngoyo — MANDING - MANINKA APOCYNACEAE *Landolphia heudelotii* ASCLEPIADACEAE *Calotropis procera, C. procera* MENDE COMPOSITAE *Aspilia africana, Melanthera scandens*
ngu — AKAN - BRONG COMBRETACEAE *Terminalia laxiflora* IGBO APOCYNACEAE *Tabernaemontana pachysiphon*
ngua — ABE ANACARDIACEAE *Spondias mombin*
n-gua — AKYE ANACARDIACEAE *Spondias mombin*
nguan — BALANTA ACANTHACEAE *Hypoestes verticillaris*
ngubor — SERER CAPPARACEAE *Capparis tomentosa*
ngubwele — KUNDU BOMBACACEAE *Adansonia digitata* MBONGE *A. digitata* TANGA *A. digitata*
ngud — SERER COMBRETACEAE *Guiera senegalensis*
n-gududi — FULA - PULAAR CAPPARACEAE *Crateva adansonii*
nguéhié — AKYE BOMBACACEAE *Ceiba pentandra*
n-guéké — MANDING - MANINKA CELASTRACEAE *Maytenus senegalensis*
n'guéné boué — KRU - GUERE CUCURBITACEAE *Momordica charantia*
n'guéré — KRU - GUERE CUCURBITACEAE *Momordica charantia*
n'guère — VULGAR COMBRETACEAE *Guiera senegalensis*
n-guessèbi, n-kichèbi — AKYE APOCYNACEAE *Rauvolfia vomitoria*
ngufor — SERER CAPPARACEAE *Capparis tomentosa*
n-guigue — MANDING - MANINKA CELASTRACEAE *Maytenus senegalensis*
n-guigué — MANDING - MANINKA CELASTRACEAE *Maytenus senegalensis*
nguigué — 'SENUFO' BOMBACACEAE *Adansonia digitata*
n-guiguilé — FULA - FULFULDE CAPPARACEAE *Boscia senegalensis* PULAAR *B. senegalensis*
n-gukhɔ(i) — LOKO BOMBACACEAE *Ceiba pentandra*
n-guku — BAULE DIOSCOREACEAE *Dioscorea burkilliana*
ngulu-gbɛ — MENDE COMPOSITAE *Ageratum conyzoides*
ngulu-gbɛ, yani-gbɛ — MENDE COMPOSITAE *Synedrella nodiflora*
ngulu-gbɛ — MENDE COMPOSITAE *Vernonia cinerea*
ngulu-gbula, ngulu gbulɔ — MENDE BIGNONIACEAE *Crescentia cujete*
n-gulu-kokui — KISSI BIGNONIACEAE *Crescentia cujete*
ngumi daleewi — FULA - FULFULDE CAPPARACEAE *Capparis tomentosa*
ngun, gun — WOLOF COMPOSITAE *Blumea aurita*
nguo — SERER - NON CAPPARACEAE *Capparis tomentosa*
ngúrà — KANURI ARACEAE *Stylochiton lancifolius*
nguraare (*pl.* nguraaje) — FULA - FULFULDE ARACEAE *Stylochiton lancifolius*
ngùrlí — KANURI CUCURBITACEAE *Cucumis sativus*
n-guru-g-bulɔ — LOKO BIGNONIACEAE *Crescentia cujete*
ngusuri — TUBU CAPPARACEAE *Maerua crassifolia*
ngut — SERER COMBRETACEAE *Guiera senegalensis*
ngutin — SERER APOCYNACEAE *Baissea multiflora*
nguwa (*def.*-wei) — MENDE BOMBACACEAE *Ceiba pentandra*
nguwa (*def.*-wɛi) — MENDE BOMBACACEAE *Ceiba pentandra*
ngùwú — KANURI CUCURBITACEAE *Lagenaria siceraria*
nguyo — MANDING - MANINKA ASCLEPIADACEAE *Calotropis procera*
nguyu — MENDE COMPOSITAE *Aspilia africana*
nhara-siguido — CRIOULO COMPOSITAE *Acanthospermum hispidum*
nhêg-cuneme — BIDYOGO CHRYSOBALANACEAE *Parinari excelsa*
nhêg-ugene — BIDYOGO CHRYSOBALANACEAE *Parinari excelsa*
nia tatté — KRU - NEYO APOCYNACEAE *Rauvolfia vomitoria*
niabaho(-n) — NYENYEGE APOCYNACEAE *Landolphia heudelotii*
niablika — ANYI ASCLEPIADACEAE *Secamone afzelii*
niadukuko — GBE - VHE ANACARDIACEAE *Lannea microcarpa*
niagpé — GBE - VHE BIGNONIACEAE *Kigelia africana*
niahui — BAULE APOCYNACEAE *Rauvolfia vomitoria*
niaje-payas — WOLOF CYPERACEAE *Cyperaceae*
niakala-khat — WOLOF CYPERACEAE *Cyperaceae*
niakamako lukuyié — SHIEN COMMELINACEAE *Floscopa africana*
niama — BAULE ASCLEPIADACEAE *Secamone afzelii*
niama boboahue — BAULE CONNARACEAE *Santaloides afzelii*
niamablé — AKAN - ASANTE ASCLEPIADACEAE *Secamone afzelii* KYAMA CONNARACEAE *Agelaea trifolia*

ɔdumanki ADANGME BIGNONIACEAE *Spathodea campanulata*

ọ̀dúndún YORUBA CRASSULACEAE *Bryophyllum pinnatum, Kalanchoe integra: Kalanchoe crenata*

ọ̀dúndún odò YORUBA COMPOSITAE *Emilia sonchifolia, Launaea taraxacifolia*

ọ̀dúndún-odò YORUBA COMPOSITAE *Emilia coccinea*

odunkun YORUBA DIOSCOREACEAE *Dioscorea esculenta*

odwuaba NZEMA BURSERACEAE *Dacryodes klaineana*

odwuma AKAN - ASANTE CECROPIACEAE *Musanga cecropioides* ASANTE (DENKYERA) *M. cecropioides* FANTE *M. cecropioides* TWI *M. cecropioides* WASA *M. cecropioides*

odwumafufuo AKAN - ASANTE CECROPIACEAE *Musanga cecropioides*

odzuma ADANGME CECROPIACEAE *Musanga cecropioides* GA *M. cecropioides*

oellha PEPEL COMBRETACEAE *Laguncularia racemosa*

ofam VULGAR CHRYSOBALANACEAE *Parinari excelsa* AKAN - TWI *P. excelsa*

ɔfam AKAN - ASANTE CHRYSOBALANACEAE *Parinari excelsa* WASA *P. excelsa*

ɔfana AKAN - ASANTE COMPOSITAE *Vernonia conferta*

ofem BEMBI APOCYNACEAE *Alstonia boonei*

ofena AKAN - ASANTE COMPOSITAE *Vernonia conferta* TWI *V. conferta*

ɔfena AKAN - ASANTE COMPOSITAE *Vernonia conferta* WASA *V. conferta*

ofenwa-biri AKAN - TWI DICHAPETALACEAE *Dichapetalum madagascariense*

ɔfẽwa-biri AKAN - TWI DICHAPETALACEAE *Dichapetalum madagascariense*

offlè BAULE CARICACEAE *Carica papaya*

ofingot BOKYI BURSERACEAE *Canarium schweinfurthii*

òfio YORUBA CYPERACEAE *Cyperus esculentus*

oflafa ABURE APOCYNACEAE *Voacanga bracteata*

oflan AKAN - ASANTE COMPOSITAE *Vernonia conferta*

ofoabiri AKAN - TWI DICHAPETALACEAE *Dichapetalum madagascariense*

ọ̀fọ́b íkpộñ IBIBIO ARACEAE *Colocasia esculenta*

ọ̀fọ́fọ̀ IJO - IZON CECROPIACEAE *Myrianthus serratus*

ọ̀fóló IJO - IZON CECROPIACEAE *Myrianthus serratus*

ọfọn àkwárī, ọtọ àkwárī IGBO APOCYNACEAE *Landolphia hirsuta*

ọfọn-àkwárī, ọtọ-àkwárī IGBO APOCYNACEAE *Landolphia dulcis*

ɔ-fɔngwón, a-pɔngwón BASARI DIOSCOREACEAE *Dioscorea bulbifera*

ofram VULGAR COMBRETACEAE *Terminalia superba* ADANGME - KROBO *T. superba* GA *T. superba*

ófram AKAN - ASANTE COMBRETACEAE *Terminalia superba*

ɔ-fram AKAN - TWI COMBRETACEAE *Terminalia superba*

ɔfram(o) AKAN - WASA COMBRETACEAE *Terminalia superba*

o-frundum ANYI - SEHWI APOCYNACEAE *Funtumia elastica*

ofruntum, ofuruntum AKAN - AKUAPEM APOCYNACEAE *Funtumia elastica*

o-fruntum AKAN - FANTE APOCYNACEAE *Funtumia elastica*

ofruntum, ofuruntum AKAN - TWI APOCYNACEAE *Funtumia elastica*

o-fruntum AKAN - WASA APOCYNACEAE *Funtumia elastica*

ofu GBE - VHE BOMBACACEAE *Ceiba pentandra*

ofua GBE - VHE BOMBACACEAE *Ceiba pentandra*

ofún òkè YORUBA ANNONACEAE *Xylopia rubescens*

o-funtum, poyu dua AKAN - FANTE APOCYNACEAE *Funtumia elastica*

o-funtum NZEMA APOCYNACEAE *Funtumia elastica*

o-funtun poyudua NZEMA APOCYNACEAE *Funtumia elastica*

ofuruma AKAN - ASANTE APOCYNACEAE *Tabernaemontana crassa, Voacanga africana*

o-furuma, o-fruma AKAN - TWI APOCYNACEAE *Tabernaemontana Gen.*

o-furuntum AKAN - ASANTE APOCYNACEAE *Funtumia elastica*

ofuruntum AKAN - WASA APOCYNACEAE *Funtumia elastica* GBE - VHE *F. elastica*

ofwho GBE - VHE BOMBACACEAE *Ceiba pentandra*

og bano IGBO BIGNONIACEAE *Markhamia tomentosa*

oga KPOSO COMBRETACEAE *Anogeissus leiocarpus*

ogáe BIDYOGO ANACARDIACEAE *Spondias mombin*

ọ̀gàn YORUBA COMBRETACEAE *Combretum bracteatum, C. platypterum, C. racemosum, C. zenkeri*

ọ̀gàn bulẹ YORUBA COMBRETACEAE *Combretum micranthum*

ọ̀gàn dúdu YORUBA COMBRETACEAE *Combretum bracteatum, C. platypterum*

ọ̀gàn fúnfún YORUBA COMBRETACEAE *Quisqualis indica*

ọ̀gàn ibulẹ YORUBA COMBRETACEAE *Combretum micranthum, C. platypterum*

ọ̀gàn ìbúlẹ̀ YORUBA COMBRETACEAE *Combretum constrictum*

ọ̀gàn pupa YORUBA COMBRETACEAE *Combretum racemosum*

ọ̀gàn-igbó YORUBA COMBRETACEAE *Quisqualis indica*

ògbanànị IGBO CONVOLVULACEAE *Ipomoea obscura aggregate*

ogbe àkùko YORUBA BORAGINACEAE *Heliotropium indicum*

ogbe mavben EDO CONNARACEAE *Castanola paradoxa*

ogbe orí àkùko, àkúko omade YORUBA BORAGINACEAE *Heliotropium indicum*

ògbo IGBO CUCURBITACEAE *Momordica angustisepala*

ògbódó, òwéí ógbódó IJO - IZON COSTACEAE *Costus lucanusianus*

ògbòkót OBOLO CUCURBITACEAE *Lagenaria siceraria*

ogbolitso GA BIGNONIACEAE *Newbouldia laevis*

ogbomorhan YORUBA CUCURBITACEAE *Momordica cissoides*

ògbóródìgì IJO - IZON DIOSCOREACEAE *Dioscorea minutiflora*

ògbóródìì IJO - IZON DIOSCOREACEAE *Dioscorea minutiflora*

ògbɾ́ō (*pl.* ì-) DEGEMA DIOSCOREACEAE *Dioscorea cayenensis*

ogbu IGBO BIGNONIACEAE *Newbouldia laevis*

ọ́gbụ̄gbá IGBO CUCURBITACEAE *Lagenaria siceraria*

ogbun YORUBA AVICENNIACEAE *Avicennia germinans*

ọgẹde-ojo IGBO CARICACEAE *Carica papaya*
ọ̀gẹ̀dẹ̀-óyìbó IGBO CARICACEAE *Carica papaya*
ọ̀gẹ̀dẹ̀-òyìbó YORUBA BROMELIACEAE *Ananas comosus*
ọ̀ghẹ́dẹ́gbó EDO ANNONACEAE *Anonidium mannii*
óghẹ̀éghẹ̀ EDO ANACARDIACEAE *Spondias mombin*
óghìghẹ̀n URHOBO ANACARDIACEAE *Spondias mombin*
óghóhẹ̀n EDO CECROPIACEAE *Musanga cecropioides*
óghòyẹ̀ EDO CHRYSOBALANACEAE *Maranthes glabra*
ògì EDO CUCURBITACEAE *Citrullus lanatus*
ògìdì dìrí IJO - IZON COMPOSITAE *Chromolaena odoratam*
ogie-ikhimwim EDO BIGNONIACEAE *Markhamia tomentosa*
ògígbàn EDO DIOSCOREACEAE *Dioscorea praehensilis*
ògìkhìnrùwìn EDO BIGNONIACEAE *Markhamia tomentosa*
ògìlì IGBO CUCURBITACEAE *Citrullus lanatus, Cucumeropsis mannii*
ògìlì-élìlì IGBO CUCURBITACEAE *Citrullus lanatus*
ogili-irele IGBO CUCURBITACEAE *Citrullus lanatus*
ogilisi IGBO BIGNONIACEAE *Newbouldia laevis*
ògílìsì IGBO BIGNONIACEAE *Newbouldia laevis*
ọ̀gìrìṣákó YORUBA ARACEAE *Anchomanes difformis*
ogirisi IGBO BIGNONIACEAE *Newbouldia laevis*
ògírìsì IGBO BIGNONIACEAE *Markhamia tomentosa, Newbouldia laevis*
ògírízì IJO - IZON BIGNONIACEAE *Kigelia africana*
ògírízì, ùgúrízì IJO - IZON BIGNONIACEAE *Newbouldia laevis*
ògiúgbòkhà EDO BOMBACACEAE *Rhodognaphalon brevicuspe*
ógò IDOMA CUCURBITACEAE *Lagenaria siceraria*
ɔ́gɔ lubɔ̀ DOGON CAPPARACEAE *Cleome gynandra*
ogogo AKAN - ASANTE CELASTRACEAE *Hippocratea rowlandii*
ôgu SERER - NON ASCLEPIADACEAE *Calotropis procera*
ogudu(g)bu YORUBA APOCYNACEAE *Alstonia boonei*
ogufé YORUBA - NAGO BOMBACACEAE *Ceiba pentandra*
ógúi, ógúiébō EDO ANACARDIACEAE *Mangifera indica*
ógúi-àhá EDO CHRYSOBALANACEAE *Licania elaeosperma*
oguk EJAGHAM APOCYNACEAE *Alstonia boonei* EJAGHAM - ETUNG *A. boonei* NKEM *A. boonei*
ogulongu KONKOMBA ARALIACEAE *Cussonia arborea*
ogungbologhá IJO - IZON BOMBACACEAE *Ceiba pentandra*
ogungun YORUBA BOMBACACEAE *Ceiba pentandra*
oguoto YEKHEE ANNONACEAE *Annona senegalensis*
oguvé YORUBA - NAGO BOMBACACEAE *Ceiba pentandra*
ógwà URHOBO DIOSCOREACEAE *Dioscorea spp. indet.*
óháhèn URHOBO BOMBACACEAE *Ceiba pentandra*
ohehe YORUBA CHRYSOBALANACEAE *Parinari excelsa*
óhí IGBO CUCURBITACEAE *Telfairea occidentalis*
óhīovbù EDO COMMELINACEAE *Aneilema beninense, A. sylvaticum, Commelina benghalensis*
oho hiɛ ADANGME - KROBO DIOSCOREACEAE *Dioscorea rotundata*
ohohiro EDO ACANTHACEAE *Thunbergia vogeliana*
ohomo ANYI COMPOSITAE *Vernonia conferta*
òhúnẹ́gbó EDO ANNONACEAE *Hexalobus monopetalus*
ohwirem AKAN - TWI COMBRETACEAE *Combretum smeathmannii*
ɔhwirem AKAN - TWI COMBRETACEAE *Combretum dolichopetalum, C. racemosum*
ohwiremnini AKAN - ASANTE COMBRETACEAE *Combretum grandiflorum* WASA *C. grandiflorum*
ɔ-hwiremo AKAN - AKYEM COMBRETACEAE *Combretum smeathmannii*
o-hwirɛmo AKAN - ASANTE COMBRETACEAE *Combretum paniculatum ssp. paniculatum, C. platypterum, C. racemosum, C. smeathmannii*
oi-bèkéè IGBO CONVOLVULACEAE *Ipomoea batatas*
ojamba AKAN - ASANTE CECROPIACEAE *Musanga cecropioides*
oje IGBO ARACEAE *Anchomanes difformis*
oji ogoda IGBO ANNONACEAE *Hexalobus crispiflorus*
ojifo BOKYI ANACARDIACEAE *Antrocaryon micraster*
ojiloko IGBO COMBRETACEAE *Terminalia superba*
ojiroko IGBO COMBRETACEAE *Terminalia superba*
ojo IGBO CARICACEAE *Carica papaya*
ójō IGBO ANNONACEAE *Cleistopholis patens*
ọjọ YORUBA COMMELINACEAE *Palisota hirsuta*
oj(u)oro KRIO ARACEAE *Pistia stratiotes*
oju EJAGHAM - ETUNG BURSERACEAE *Dacryodes edulis*
ojuma, odwuma AKAN - ASANTE CECROPIACEAE *Musanga cecropioides*
ojuma AKAN - WASA CECROPIACEAE *Musanga cecropioides*
oju-ngon EJAGHAM - ETUNG BURSERACEAE *Canarium schweinfurthii*
ojúoró YORUBA ARACEAE *Pistia stratiotes*
ok KRIO COMBRETACEAE *Terminalia scutifera*
okae, ɔkan AKAN - WASA APOCYNACEAE *Funtumia africana*
ɔkae, okan AKAN - ASANTE APOCYNACEAE *Funtumia africana*
okan YORUBA - NAGO COMBRETACEAE *Combretum micranthum*
ọkán YORUBA COMBRETACEAE *Combretum micranthum*
ɔkan AKAN - ASANTE APOCYNACEAE *Funtumia africana* TWI *F. africana* WASA *F. africana*
ɔkan, ɔkae, maamae AKAN - WASA APOCYNACEAE *Funtumia africana*
ọ̀kàn, ọ̀gàn YORUBA COMBRETACEAE *Combretum smeathmannii*
okanguiamonnu GURMA COMBRETACEAE *Combretum collinum ssp. geitonophyllum*

| | |
|---|---|
| osisi-izi | IGBO ACANTHACEAE *Lankesteria elegans* |
| osisiri | AKAN - KWAWU BIGNONIACEAE *Spathodea campanulata* WASA *S. campanulata* |
| oso angweri ngwa | IGBO COMPOSITAE *Ageratum conyzoides* |
| ọṣọdu | YORUBA CONNARACEAE *Castanola paradoxa* |
| osoko | EBIRA ANNONACEAE *Pachypodanthium barteri* |
| o-sommɛ | AKAN - TWI COSTACEAE *Costus afer* |
| osomolu | YORUBA ANNONACEAE *Enantia chlorantha* |
| ɔsoŋkoni gbekɛ̃biiaŋmetʃo | GA BORAGINACEAE *Ehretia cymosa* |
| osono-bese | AKAN - ASANTE CAPPARACEAE *Buchholzia coriacea* |
| osontokwakofuo | VULGAR BIGNONIACEAE *Stereospermum acuminatissimum* |
| osopupa | YORUBA ANNONACEAE *Enantia chlorantha, E. polycarpa* |
| osu | AKAN - TWI DIOSCOREACEAE *Dioscorea misc. & spp. unknown* |
| ósù | EDO APOCYNACEAE *Hunteria umbellata* |
| òsú | YORUBA APOCYNACEAE *Strophanthus gratus* |
| òsú abwa | IGBO APOCYNACEAE *Picralima nitida* |
| osu angweri ngwa | IGBO COMPOSITAE *Ageratum conyzoides* |
| òsú igwe | IGBO APOCYNACEAE *Picralima nitida* |
| osuban | ADANGME - KROBO DIOSCOREACEAE *Dioscorea rotundata* |
| osun | YORUBA AMARANTHACEAE *Celosia argentea* |
| oṣùn búkẹ | YORUBA BIXACEAE *Bixa orellana* |
| osuni-elufoni | NZEMA ANNONACEAE *Neostenanthera hamata* |
| osun-kodu | AKAN - ASANTE ANNONACEAE *Uvariodendron occidentalis* |
| osunyane | AKAN - AKYEM ANACARDIACEAE *Pseudospondias microcarpa* |
| ɔ-syẹlis, a-tyẹlis, ɛ-tyẹlis | BASARI CUCURBITACEAE *Cucurbita pepo* |
| ɔ-syẹlis ɔr ɛ-peúkèl, a-tyẹlis and ɛ-peúkèl | BASARI CUCURBITACEAE *Citrullus colocynthis* |
| ọta | IGBO APOCYNACEAE *Strophanthus gratus, S. sarmentosus* |
| ota eke | IGBO COMPOSITAE *Senecio biafrae* |
| ọta nta | IGBO APOCYNACEAE *Strophanthus sarmentosus* |
| ọta nza | IGBO APOCYNACEAE *Alafia barteri* |
| òte | ABUA CUCURBITACEAE *Lagenaria siceraria* |
| ọtẹtẹ | URHOBO AMARANTHACEAE *Amaranthus hybridus* |
| otiemasa | AKAN - ASANTE ACANTHACEAE *Elytraria marginata, Lankesteria brevior* |
| otifebigu | GURMA COMBRETACEAE *Combretum molle* |
| otito | HAUSA CONNARACEAE *Cnestis ferruginea* |
| ọtọ, utu | IGBO APOCYNACEAE *Saba florida* |
| ọtọ poi | IGBO APOCYNACEAE *Ancylobothrys scandens* |
| ɔtofaben | AKAN - TWI APOCYNACEAE *Allamanda spp.* |
| ɔtofammẽn | AKAN - ASANTE AMARANTHACEAE *Amaranthus hybridus* |
| ọtọ-frifedi | IGBO APOCYNACEAE *Ancylobothrys scandens* |
| ọtọ-frifredi | IGBO APOCYNACEAE *Landolphia owariensis* |
| otoh | BA'BAN CUCURBITACEAE *Lagenaria siceraria* |
| ọ̀tọ́lị | IGBO CONVOLVULACEAE *Ipomoea batatas* |
| òtóró, òwéí ótóró | IJO - IZON AMARYLLIDACEAE *Scadoxus spp.* |
| òtótò únọ̀ | EFIK CECROPIACEAE *Musanga cecropioides* |
| ɔtɔtɔwaa | AKAN - TWI BOMBACACEAE *Adansonia digitata* |
| otre kani | GBE - VHE DIOSCOREACEAE *Dioscorea cayenensis* |
| otri-kani | ADANGME - KROBO DIOSCOREACEAE *Dioscorea cayenensis* |
| otta, otta ifo | YORUBA ACANTHACEAE *Lankesteria elegans* |
| ótù | EDO ANNONACEAE *Cleistopholis patens* |
| ọtugọ | IGBO COMBRETACEAE *Combretum spp. indet.* |
| otuk okpo | EFIK DIOSCOREACEAE *Dioscorea rotundata* |
| òtúk-ọ̀kpọ̄ | EFIK DIOSCOREACEAE *Dioscorea cayenensis* |
| ọtuokpo | DEGEMA APOCYNACEAE *Landolphia owariensis* |
| otutu-bofunnua | AKAN - TWI ANNONACEAE *Monodora tenuifolia* |
| otutumɔfunnua | AKAN - TWI ANNONACEAE *Uvaria ovata* |
| otuwerehi | IGBO CELASTRACEAE *Salacia nitida* |
| ɔtwa tegyrima | AKAN - FANTE ARACEAE *Culcasia scandens* |
| ɔtwe-ani-wa | AKAN - ASANTE COMMELINACEAE *Pollia condensata* |
| ɔtwe-ehi | AKAN - ASANTE ANNONACEAE *Uvariastrum pierreanum* |
| otwene monta | AKAN - ASANTE ARISTOLOCHIACEAE *Paristolochia goldieana* |
| oufa | ARABIC BOMBACACEAE *Adansonia digitata* |
| ououd | GOEMAI ANNONACEAE *Annona senegalensis* |
| o-uruzi | AVIKAM APOCYNACEAE *Alstonia boonei* |
| oùwoó | HAUSA APOCYNACEAE *Landolphia owariensis* |
| òvbiàkpẹ̀ | EDO ACANTHACEAE *Asystasia gangetica* |
| ọ̀viẹ́n-èrhẹ̀nbàvbógò | EDO ANNONACEAE *Enantia polycarpa* |
| òvìẹn-íbù | EDO APOCYNACEAE *Voacanga africana* |
| òvìẹn-ìkhúian | EDO APOCYNACEAE *Tabernaemontana pedulifiora* |
| òvìẹn-iwian | EDO APOCYNACEAE *Tabernaemontana peduliflora* |
| ọ̀víẹn-órà | EDO APOCYNACEAE *Strophanthus sarmentosus* |
| oviunien | EDO ANNONACEAE *Xylopia quintasii* |
| ovunien | EDO ANNONACEAE *Xylopia quintasii* |
| ọwa | IGBO CONNARACEAE *Jaundea pinnata* |
| owabọ | URHOBO DIOSCOREACEAE *Dioscorea dumetorum* |
| òwàrá | URHOBO CUCURBITACEAE *Lagenaria siceraria* |
| o-wedgeɛ-aba, moto kura dua | AKAN - ASANTE ANNONACEAE *Monodora myristica* |
| owedia-aba | AKAN - WASA ANNONACEAE *Monodora myristica* |
| o-wediɛ-aba | AKAN - ASANTE ANNONACEAE *Monodora myristica* TWI *M. myristica* WASA *M. myristica* |
| òwéí ángí | IJO - IZON CYPERACEAE *Cyperus distans* |

papaku AKAN - TWI APOCYNACEAE *Tabernaemontana Linn.*
papala LIMBA CARICACEAE *Carica papaya*
papulatutubu DAGBANI CHRYSOBALANACEAE *Parinari curatellifolia*
papan AKAN - FANTE CONVOLVULACEAE *Ipomoea pes caprae*
papaya VULGAR CARICACEAE *Carica papaya*
papaye VULGAR CARICACEAE *Carica papaya*
papayi FULA - PULAAR CARICACEAE *Carica papaya*
papayo FULA - PULAAR CARICACEAE *Carica papaya* SERER *C. papaya* WOLOF *C. papaya*
papia MANDING - BAMBARA CARICACEAE *Carica papaya* 'SOCE' *C. papaya*
papiu MANDING - BAMBARA CARICACEAE *Carica papaya* 'SOCE' *C. papaya*
papiya KONYAGI CARICACEAE *Carica papaya*
pára IJO - IZON ANNONACEAE *Cleistopholis patens*
pàràgi NUPE CUCURBITACEAE *Citrullus lanatus*
paré SUSU APOCYNACEAE *Landolphia heudelotii*
pari ọmọde YORUBA APOCYNACEAE *Pleioceras barteri*
paria zupen YOM COMPOSITAE *Pulicaria crispa*
paripakoje YORUBA BIGNONIACEAE *Stereospermum acuminatissimum*
parpehi HAUSA COMPOSITAE *Spilanthes filicaulis*
pãru kanuri FULA - FULFULDE ANACARDIACEAE *Lannea fruticosa*
paruhi FULA - FULFULDE ANACARDIACEAE *Lannea acida, L. barteri, L. microcarpa*
pasal KONYAGI BIGNONIACEAE *Newbouldia laevis*
pâschon BATONNUN CHRYSOBALANACEAE *Parinari curatellifolia*
pasia KONO ANNONACEAE *Xylopia elliotii*
passoklo MANDING - MANINKA COMPOSITAE *Bidens pilosa*
passo-ni-kuna MANDING - MANINKA CAPPARACEAE *Cleome Linn., C. gynandra*
pataati (*def.* pataatoo) MANDING - MANDINKA CONVOLVULACEAE *Ipomoea batatas*
patabofuo AKAN - ASANTE CAPPARACEAE *Capparis erythrocarpos*
patafué BAULE CAPPARACEAE *Capparis erythrocarpos*
patako HAUSA ASCLEPIADACEAE *Pergularia tomentosa*
patandĕu TEM ANACARDIACEAE *Lannea barteri*
patas WOLOF CONVOLVULACEAE *Ipomoea batatas*
patat WOLOF CONVOLVULACEAE *Ipomoea batatas*
patate BALANTA CONVOLVULACEAE *Ipomoea batatas*
patato MANDING - BAMBARA CONVOLVULACEAE *Ipomoea batatas*
patatô MANDING - MANDINKA CONVOLVULACEAE *Ipomoea batatas*
patié patié AKAN - ASANTE APOCYNACEAE *Tabernaemontana crassa*
patin dehun GBE - FON BOMBACACEAE *Ceiba pentandra*
patobi KYAMA CHRYSOBALANACEAE *Parinari excelsa*
patu WOLOF CUCURBITACEAE *Lagenaria siceraria*
pau de bola CRIOULO CAPPARACEAE *Crateva adansonii*
pautu KRU - GUERE BIGNONIACEAE *Spathodea campanulata*
pavu GBE - FON COMBRETACEAE *Terminalia macroptera*
pawpaw VULGAR CARICACEAE *Carica papaya*
*o*-pawya AKAN - ASANTE APOCYNACEAE *Landolphia owariensis*
pé AKYE APOCYNACEAE *Funtumia elastica*
pê AKAN - BRONG ARACEAE *Anchomanes difformis*
pebadje MANDYAK AVICENNIACEAE *Avicennia germinans*
pébouillé FULA - PULAAR ANACARDIACEAE *Lannea microcarpa*
pé-chi AKYE APOCYNACEAE *Funtumia elastica*
pe-chialachale KISSI COMPOSITAE *Triplotaxis stellulifera*
pedjindo KISSI ACANTHACEAE *Brillantaisia nitens*
pédo KISSI CECROPIACEAE *Musanga cecropioides*
pedum-hal MANKANYA CARICACEAE *Carica papaya*
peduto-ubusse MANDYAK CONNARACEAE *Cnestis ferruginea*
pɛ̀ɛ̀lie nínyilu DOGON COMBRETACEAE *Combretum aculeatum*
peembere (*pl.* peembereeji) FULA - FULFULDE CONVOLVULACEAE *Ipomoea batatas*
pe-folã KISSI AMARYLLIDACEAE *Crinum natans*
pɛgɛre LOKO ACANTHACEAE *Phaulopsis falcisepala*
*p*-pegu DAGBANI ANACARDIACEAE *Lannea velutina*
pegu (?) FULA - PULAAR ANACARDIACEAE *Lannea acida*
pehpeh MANO CYPERACEAE *Cyperaceae*
péhuni MANDING - BAMBARA ANACARDIACEAE *Lannea microcarpa*
peindo KISSI CECROPIACEAE *Musanga cecropioides*
peiŋgo KISSI BOMBACACEAE *Bombax buonopozense*
pekire MOORE - NAWDAM CHRYSOBALANACEAE *Parinari congensis*
peku MANDING - MANINKA ANACARDIACEAE *Lannea barteri, L. velutina*
pékuba MANDING - MANINKA ANACARDIACEAE *Lannea microcarpa*
pekuni MANDING - BAMBARA ANACARDIACEAE *Lannea acida*
pékuni MANDING - BAMBARA ANACARDIACEAE *Lannea acida* MANINKA *L. acida*
pekyi-pekyere ANYI - AOWIN APOCYNACEAE *Tabernaemontana crassa*
pekyi-pekyeri ANYI - AOWIN APOCYNACEAE *Tabernaemontana Gen.*
pelae-djacumáe DIOLA - FLUP COMBRETACEAE *Combretum grandiflorum*
pélé uluzedigpoie LOMA AMARANTHACEAE *Amaranthus lividus*
pélé uluzédikolé LOMA AMARANTHACEAE *Amaranthus spinosus*
pəlefɔ LIMBA ASCLEPIADACEAE *Gongronema angolense*
pɛlɛtɛ FULA - PULAAR CUCURBITACEAE *Lagenaria siceraria*
pélévélé LOMA ACANTHACEAE *Justicia schimperi striata*
pemben SERER - NON COMBRETACEAE *Combretum glutinosum*
pempen SISAALA APOCYNACEAE *Landolphia heudelotii*
pempene AKAN - TWI APOCYNACEAE *Landolphia, Landolphia owariensis*

pɔi AKAN - TWI APOCYNACEAE *Landolphia*
poi un dubu KRU - GUERE BIGNONIACEAE *Markhamia lutea, M. tomentosa*
poilão CRIOULO BOMBACACEAE *Ceiba pentandra*
poilão encarnado CRIOULO BOMBACACEAE *Bombax costatum*
poilão foro CRIOULO BOMBACACEAE *Bombax costatum*
põkin KRIO CUCURBITACEAE *Cucurbita pepo*
poko pokowéï KRU - BETE COMPOSITAE *Vernonia conferta*
pokopokowéi KWENI COMPOSITAE *Vernonia conferta*
pole MANDING - MANDINKA APOCYNACEAE *Strophanthus hispidus*
poli SUSU APOCYNACEAE *Strophanthus sarmentosus*
*a*-poli TEMNE APOCYNACEAE *Pleiocarpa mutica*
pɔ́liõ pɔpɔ́liõ DOGON ASCLEPIADACEAE *Pergularia tomentosa*
pòlípòlí GUANG - GONJA ASCLEPIADACEAE *Calotropis procera*
*an*-pɔlɔ TEMNE CANNACEAE *Canna indica*
polo polo IGBO ACANTHACEAE *Brillantaisia owariensis*
polóm CRIOULO BOMBACACEAE *Ceiba pentandra*
polóm fidalgo CRIOULO BOMBACACEAE *Bombax costatum*
polóm fôro CRIOULO BOMBACACEAE *Bombax costatum*
*am*-polon TEMNE BOMBACACEAE *Ceiba pentandra*
*am*-poloŋ TEMNE BOMBACACEAE *Ceiba pentandra*
polon-dɛ BULOM BOMBACACEAE *Ceiba pentandra*
poloŋɛ BULOM BOMBACACEAE *Ceiba pentandra*
polo-polo IGBO ACANTHACEAE *Brillantaisia lamium, B. nitens*
pôlu-úh BALANTA CARYOPHYLLACEAE *Polycarpon prostratum*
pôm SERER CUCURBITACEAE *Mukia maderaspatana*
pomamagbe VAI BIGNONIACEAE *Newbouldia laevis*
pomamagbɛ MENDE BIGNONIACEAE *Newbouldia laevis*
pɔmbiaũ-ma-pianduaũ KISSI ARACEAE *Amorphophallus aphyllus*
pomey FULA - PULAAR CUCURBITACEAE *Mukia maderaspatana*
pompune NZEMA APOCYNACEAE *Landolphia, Landolphia owariensis*
pɔm-ta-bolɛ? BULOM CRASSULACEAE *Bryophyllum pinnatum*
pona AKAN - TWI DIOSCOREACEAE *Dioscorea rotundata*
pɔndo-koko KONO CUCURBITACEAE *Telfairea occidentalis*
ponéné DAN CHRYSOBALANACEAE *Maranthes robusta*
poŋgi LIMBA BOMBACACEAE *Bombax buonopozense*
pong-sah CHAMBA COMMELINACEAE *Commelina benghalensis*
poni 'KRU' COMPOSITAE *Ageratum conyzoides*
pọnju òwiwí YORUBA CELASTRACEAE *Hippocratea africana*
pọnju-òwiwí YORUBA CELASTRACEAE *Hippocratea indica*
*an*-poŋk-poŋk TEMNE BOMBACACEAE *Bombax buonopozense*
pono KRU - GREBO COMPOSITAE *Ageratum conyzoides*
ponopan KRU - GREBO COMPOSITAE *Ageratum conyzoides*
ponponla-tobi YORUBA COMPOSITAE *Grangea maderaspatana*
ponsa ponto GUANG - NKONYA BURSERACEAE *Dacryodes klaineana*
*a*-ponthi-*a*-ro-bath TEMNE APOCYNACEAE *Malouetia heudelotii*
pọọkọ́ YORUBA CUCURBITACEAE *Lagenaria siceraria*
popio lo MANO CELASTRACEAE *Hippocratea indica*
popo VULGAR CARICACEAE *Carica papaya* SENUFO - BAMANA APOCYNACEAE *Landolphia heudelotii*
pɔpɔ KRIO CARICACEAE *Carica papaya*
pɔ̀pɔ̌ NGEMBA CARICACEAE *Carica papaya*
pọ̀pọ, ọpọ̀pọ EFIK CARICACEAE *Carica papaya*
pópo (*def*.-ô) MANDING - MANDINKA ASCLEPIADACEAE *Calotropis procera*
pọpọ nlí ézì IGBO CARICACEAE *Carica papaya*
popokaiyo WOLOF CARICACEAE *Carica papaya*
popokaye WOLOF CARICACEAE *Carica papaya*
popólo DOGON BIGNONIACEAE *Stereospermum kunthianum*
pópòǹlà YORUBA BOMBACACEAE *Bombax buonopozense*
póponlà òdàn YORUBA - NAGO BOMBACACEAE *Bombax costatum*
pọpọnní ézì IGBO CARICACEAE *Carica papaya*
poposané ABURE ANACARDIACEAE *Spondias mombin*
poraŋ MANDING - MANINKA APOCYNACEAE *Funtumia africana, F. elastica*
poré VULGAR APOCYNACEAE *Clitandra cymulosa, Landolphia owariensis* FULA - FULFULDE *L. heudelotii* PULAAR *Baissea multiflora, Landolphia heudelotii, L. owariensis, Saba senegalensis, S. senegalensis* SUSU - DYALONKE *Landolphia heudelotii*
póre FULA - PULAAR APOCYNACEAE *Landolphia heudelotii*
põre laré FULA - PULAAR APOCYNACEAE *Saba senegalensis*
poré-bété FULA - PULAAR APOCYNACEAE *Landolphia owariensis*
poré-kono FULA - PULAAR APOCYNACEAE *Landolphia heudelotii*
pore-laré FULA - PULAAR APOCYNACEAE *Saba senegalensis*
porè-lárè FULA - PULAAR APOCYNACEAE *Landolphia heudelotii*
poré-mulutigué FULA - PULAAR APOCYNACEAE *Landolphia owariensis*
pori SERER - NON CELASTRACEAE *Maytenus senegalensis* VAI CUCURBITACEAE *Cucurbita pepo*
poripori TEM CARICACEAE *Carica papaya*
pŏripŏri TEM CARICACEAE *Carica papaya*
porpiliga NABT COMBRETACEAE *Combretum collinum ssp. hypopilinum*
portagaré CRIOULO APOCYNACEAE *Alstonia boonei*
*am*-pos, *an*-pos TEMNE ANNONACEAE *Xylopia acutiflora, X. aethiopica*
pɔ́té pɔ́té GUANG - GONJA CHRYSOBALANACEAE *Maranthes polyandra* KRACHI *Parinari curatellifolia*

röbröb WOLOF COMBRETACEAE *Terminalia avicennioides*
rof skin plɔm KRIO CHRYSOBALANACEAE *Parinari curatellifolia*
roffin-plɔm KRIO CHRYSOBALANACEAE *Parinari excelsa*
rof-skin-plɔm KRIO CHRYSOBALANACEAE *Parinari excelsa*
*a*-ro-gban TEMNE CYPERACEAE *Cyperaceae*
*an*-roi-*an*-rungi TEMNE CYPERACEAE *Cyperus rotundus, C. sphacelatus*
rokãndok SERER ACANTHACEAE *Hygrophila auriculata*
rokarot SERER ACANTHACEAE *Hygrophila auriculata* SERER - NON *H. auriculata*
ról SERER CYPERACEAE *Cyperus esculentus*
rôl SERER CYPERACEAE *Cyperus rotundus, C. sphacelatus*
rŏl SERER CYPERACEAE *Cyperaceae*
roltakar SERER AIZOACEAE *Sesuvium portulacastrum*
ronde (*pl.* rona) MOORE - NAWDAM DIOSCOREACEAE *Dioscorea Linn.*
ronko KRIO COMBRETACEAE *Terminalia ivorensis*
*an*-roŋko TEMNE COMBRETACEAE *Terminalia ivorensis*
*an*-rɔnthma TEMNE CUCURBITACEAE *Luffa cylindrica*
*an*-rɔnthma-*a*-potho TEMNE CUCURBITACEAE *Luffa acutangula*
roogon biri, rogwain biri HAUSA DIOSCOREACEAE *Dioscorea dumetorum*
roógòn yáăraá HAUSA AIZOACEAE *Trianthema portulacastrum*
roógòn-mákiyaàyaá HAUSA COMPOSITAE *Sphaeranthus angustifolius*
roógòn-yaàraá HAUSA AIZOACEAE *Zaleya pentandra*
rope quiah KRIO ASCLEPIADACEAE *Gongronema latifolium*
ọrọwọ YORUBA COMPOSITAE *Senecio biafrae*
*a*-rotə-*a*-kɛrɛ TEMNE AMARYLLIDACEAE *Crinum jagus, Hymenocallis littoralis, Scadoxus cinnabarinus*
*a*-roth-*a*-kɛrɛ TEMNE ARACEAE *Amorphophallus aphyllus*
*an*-rɔthma TEMNE CUCURBITACEAE *Trichosanthes cucumerina*
rowoni BOLE CARYOPHYLLACEAE *Polycarpaea linearifolia*
r(e)tem ARABIC ASCLEPIADACEAE *Leptadenia pyrotechnica*
rubbundehi FULA - FULFULDE AIZOACEAE *Glinus lotoides*
rùb-dà-túkúnyá HAUSA AMARANTHACEAE *Amaranthus viridis*
rudu HAUSA CUCURBITACEAE *Lagenaria siceraria*
rujiyar-mahalba HAUSA ASCLEPIADACEAE *Xysmalobium heudelotianum*
rukuɓu HAUSA AMARANTHACEAE *Amaranthus spinosus, A. viridis*
rukuɓuho FULA - FULFULDE AMARANTHACEAE *Amaranthus viridis*
rukuure (*pl.* dukuuje) FULA - FULFULDE CARICACEAE *Carica papaya*
rum ARABIC - SHUWA BOMBACACEAE *Ceiba pentandra*
rŭmakà NUPE CUCURBITACEAE *Luffa cylindrica*
rumbum BALANTA BOMBACACEAE *Ceiba pentandra*
rummef TAMACHEK CHENOPODIACEAE *Haloxylon articulatum*
rum-rum PEPEL COSTACEAE *Costus afer*
*a*-ruŋku TEMNE BORAGINACEAE *Ehretia cymosa*
rura HAUSA CHRYSOBALANACEAE *Parinari curatellifolia*
sa KONO CUCURBITACEAE *Cucumeropsis mannii* VAI *Citrullus lanatus, Cucumeropsis mannii*
sá DOGON ANACARDIACEAE *Lannea acida*
sa baté AKYE COMMELINACEAE *Commelina diffusa*
sá nìì tòmolo ána, sá nìì tòmolo yà DOGON COMMELINACEAE *Cyanotis lanata*
saa KONO CUCURBITACEAE *Citrullus lanatus*
*an*-saa TEMNE COMBRETACEAE *Conocarpus erectus*
saa tesaga BIMOBA APOCYNACEAE *Adenium obesum*
sàabarà, saàbáraà HAUSA COMBRETACEAE *Guiera senegalensis*
sàabàrà SONGHAI COMBRETACEAE *Guiera senegalensis*
saabé SONGHAI ASCLEPIADACEAE *Leptadenia pyrotechnica*
saaɓeho (*pl.* saɓehe) FULA - FULFULDE BOMBACACEAE *Adansonia digitata*
saaɓeho FULA - FULFULDE BOMBACACEAE *Adansonia digitata*
saa-borɔfere AKAN - TWI ANNONACEAE *Annona senegalensis* ARALIACEAE *Cussonia arborea*
saàbúlùn y'án-maátaá HAUSA COMPOSITAE *Vernonia macrocyanus*
saàbúlùn-maátaá HAUSA COMPOSITAE *Vernonia macrocyanus*
saad ARABIC CYPERACEAE *Cyperus esculentus* TAMACHEK *C. conglomeratus*
sàádə̀ KANURI CUCURBITACEAE *Cucurbita maxima, C. pepo*
sáagéy (*pl.* -à) SONGHAI ASCLEPIADACEAE *Calotropis procera*
saagnan KRU - GUERE CAPPARACEAE *Buchholzia coriacea*
saajamba (*def.* saajamboo) MANDING - MANDINKA AIZOACEAE *Zaleya pentandra*
saakuin GRUSI BALANITACEAE *Balanites aegyptiaca*
sàalo,shàaje GA BOMBACACEAE *Adansonia digitata*
saantom AKAN - FANTE CONVOLVULACEAE *Ipomoea batatas*
saanunum AKAN - ASANTE COMPOSITAE *Vernonia nigritiana*
saàre gwiáwà HAUSA ACANTHACEAE *Hygrophila auriculata*
saawuu-dubuu HAUSA CONVOLVULACEAE *Ipomoea coptica*
saba MANDING - BAMBARA APOCYNACEAE *Saba florida, S. senegalensis* MANDINKA *S. senegalensis* MANINKA *S. florida, S. senegalensis* SONINKE - SARAKOLE *S. senegalensis*
saba bili MANDING - BAMBARA APOCYNACEAE *Saba senegalensis* MANINKA *S. senegalensis*
saba-bâ MANDING - MANINKA APOCYNACEAE *Saba senegalensis*
saba-bili MANDING - BAMBARA APOCYNACEAE *Saba senegalensis*
sâbafim(-ô) MANDING - MANDINKA ANNONACEAE *Uvaria chamae*
sabagha MOORE ANACARDIACEAE *Lannea barteri, L. velutina*

sali sali MANDING - BAMBARA COMPOSITAE *Dicoma tomentosa*

sallenke HAUSA ASCLEPIADACEAE *Pergularia tomentosa*

salo ADANGME - KROBO BOMBACACEAE *Adansonia digitata*

salo bego KYAMA APOCYNACEAE *Strophanthus hispidus*

salo bego-esé KYAMA APOCYNACEAE *Strophanthus gratus*

salubé KYAMA APOCYNACEAE *Strophanthus hispidus*

sama MOORE - NAWDAM CONVOLVULACEAE *Ipomoea batatas*

sama *m*bali MANDING - MANDINKA COMBRETACEAE *Combretum nigricans*

sama m'bali MANDING - BAMBARA COMBRETACEAE *Combretum fragrans, C. nigricans* MANINKA *C. fragrans, C. nigricans*

samajewo MANDING - MANDINKA CONVOLVULACEAE *Ipomoea asarifolia*

samala TEM CONNARACEAE *Byrsocarpus coccineus*

samambali MANDING - BAMBARA COMBRETACEAE *Combretum nigricans* MANINKA *C. nigricans*

sámáŋdóró GUANG - GONJA ARALIACEAE *Cussonia arborea*

saman'muto AKAN - ASANTE COMPOSITAE *Laggera alata*

samantini AKAN - KWAWU CONNARACEAE *Brysocarpus coccineus*

samasse summelae KABRE CARICACEAE *Carica papaya*

samataga-na SUSU - DYALONKE COMBRETACEAE *Combretum fragrans*

samba gnagna SERER AIZOACEAE *Sesuvium portulacastrum*

sambafim (*def.*-ô) MANDING - MANDINKA ANNONACEAE *Uvaria afzelii, U. chamae*

sambagha MOORE ANACARDIACEAE *Lannea acida*

sambalèh WOLOF ASCLEPIADACEAE *Ceropegia spp.*

sambaniani SERER AIZOACEAE *Sesuvium portulacastrum*

sambé SERER COMBRETACEAE *Combretum aculeatum* SERER - NON *C. aculeatum*

sambè BALANTA ANNONACEAE *Monodora myristica*

sambiga MOORE ANACARDIACEAE *Lannea microcarpa*

sambituliga MOORE ANACARDIACEAE *Lannea barteri, L. egregia, L. velutina*

sambugo ANYI - ANUFO BOMBACACEAE *Bombax buonopozense* GURMA *B. buonopozense*

same BALANTA ANACARDIACEAE *Spondias mombin*

sa-mɛsɛ KONO CUCURBITACEAE *Cucumeropsis mannii*

ṣamikoko YORUBA CYPERACEAE *Mariscus alternifolius*

saminia BAULE COMPOSITAE *Vernonia nigritiana*

sampane PEPEL ANNONACEAE *Annona senegalensis*

samtuluga MOORE ANACARDIACEAE *Lannea acida*

san doki HAUSA CUCURBITACEAE *Lagenaria siceraria*

san wulu MANDING - MANINKA ARACEAE *Anchomanes welwitschii*

sana(r) WOLOF AVICENNIACEAE *Avicennia germinans*

sanabra SERER CONVOLVULACEAE *Merremia kentrocaulos*

sanadjô MANDING - MANDINKA BORAGINACEAE *Cordia myxa*

sanaraje DAGBANI DIOSCOREACEAE *Dioscorea rotundata*

sancaetano CRIOULO CUCURBITACEAE *Momordica charantia*

sáncíligà (*pl.* -à) SONGHAI CAPPARACEAE *Boscia angustifolia*

sáncíligà SONGHAI CAPPARACEAE *Boscia salicifolia*

sandan maayun HAUSA APOCYNACEAE *Holarrhena floribunda*

sandan maym YORUBA AMARYLLIDACEAE *Scadoxus multiflorus*

sãnda-sãnda WOLOF COMPOSITAE *Pluchea ovalis*

sandé koroni MANDING - DYULA COCHLOSPERMACEAE *Cochlospermum tinctorium*

sande-sieyo KISSI CYPERACEAE *Mariscus alternifolius*

sanding sulo MANDING - MANDINKA CONVOLVULACEAE *Ipomoea eriocarpa*

sandje-bombo FULA - PULAAR ANACARDIACEAE *Sorindeia juglandifolia*

sandji-bombro FULA - PULAAR ANACARDIACEAE *Sorindeia juglandifolia*

sané SONGHAI BORAGINACEAE *Coldenia procumbens*

sané sané SERER ANACARDIACEAE *Ozoroa insignis*

sang TEM CYPERACEAE *Cyperus articulatus*

sãŋga-lɛ BULOM CUCURBITACEAE *Citrullus lanatus, Cucumeropsis mannii*

sangei-lɛ BULOM DIONCOPHYLLACEAE *Habropetalum dawei*

sangi MENDE DIONCOPHYLLACEAE *Triphyophyllum peltatum* YORUBA CAPPARACEAE *Euadenia trifoliolata*

saŋgi BULOM DIONCOPHYLLACEAE *Habropetalum dawei*

sango(n) KWENI CAPPARACEAE *Cleome gynandra*

sangulo (*def.* sanguli) MENDE BOMBACACEAE *Rhodognaphalon brevicuspe*

sanhanguin KRU - GUERE CONNARACEAE *Cnestis ferruginea*

sàŋká GUANG - GONJA COMPOSITAE *Vernonia amygdalina, V. colorata*

sankane GA COMBRETACEAE *Anogeissus leiocarpus*

sanndorɗe (*pl.* caandorɗe) FULA - FULFULDE CUCURBITACEAE *Lagenaria siceraria*

sansa MOORE COCHLOSPERMACEAE *Cochlospermum tinctorium*

sansami HAUSA BIGNONIACEAE *Stereospermum acuminatissimum, S. kunthianum*

sansangwa DAGBANI CAPPARACEAE *Capparis tomentosa*

sanséghé MOORE COCHLOSPERMACEAE *Cochlospermum tinctorium*

santaŋ maloro MANDING - MANDINKA ANACARDIACEAE *Lannea acida*

santika ABURE COMPOSITAE *Conyza canadensis*

santil-lɛ BULOM CYPERACEAE *Cyperaceae, Mariscus ligularis, Rhynchospora corymbosa, Scleria boivinii, S. spiciformis*

santuluga MOORE ANACARDIACEAE *Lannea acida*

sanuguélé MANDING - MANINKA ACANTHACEAE *Monechma depauperatum*

sanuhé KRU - BETE COMBRETACEAE *Terminalia superba*

sányár ƙasà HAUSA COMPOSITAE *Vernonia pumila*

sanyi MANDING - MANDINKA COMPOSITAE *Bidens bipinnata, B. pilosa*

saŋyi-na SUSU - DYALONKE COMPOSITAE *Bidens bipinnata, B. pilosa*

sao KONO BORAGINACEAE *Cordia platythyrsa*

ségini MANDING - BAMBARA COMPOSITAE *Vernonia perrottetii*
ségiré, séréné MANDING - BAMBARA BALANITACEAE *Balanites aegyptiaca*
ségiré MANDING - MANINKA BALANITACEAE *Balanites aegyptiaca* SONINKE - SARAKOLE *B. aegyptiaca*
séguéné, zéguéné MANDING - BAMBARA BALANITACEAE *Balanites aegyptiaca*
séguindi MANDING - MANINKA COMPOSITAE *Vernonia colorata*
seguiné MANDING - BAMBARA BALANITACEAE *Balanites aegyptiaca*
séhè MANDING - MANINKA APOCYNACEAE *Holarrhena floribunda*
sèhè KRU - GUERE CAPPARACEAE *Cleome gynandra*
sɛi KWENI DIOSCOREACEAE *Dioscorea Linn.*
sèì tùó IJO - IZON COMPOSITAE *Chromolaena odoratam*
sek SERER - NON CAPPARACEAE *Crateva adansonii*
sekke KPE COMPOSITAE *Spilanthes filicaulis*
sékör SERER ANNONACEAE *Hexalobus monopetalus*
sekunde FULA - PULAAR BIGNONIACEAE *Spathodea campanulata*
sela KONO BOMBACACEAE *Adansonia digitata*
selialiwɔ KISSI CONNARACEAE *Cnestis ferruginea*
sellakha TAMACHEK ASCLEPIADACEAE *Pergularia tomentosa*
sèmbélèh SERER ASCLEPIADACEAE *Ceropegia spp.*
semen KRU - GUERE DIOSCOREACEAE *Dioscorea praehensilis*
semmar TAMACHEK CYPERACEAE *Scirpus holoschoenus*
semu-na SUSU - DYALONKE CYPERACEAE *Cyperus articulatus*
sem-unte-pulhe BALANTA ANNONACEAE *Xylopia aethiopica*
*a*-senduku TEMNE AMARANTHACEAE *Amaranthus hybridus, A. spinosus*
*an*-senduku TEMNE AMARANTHACEAE *Amaranthus viridis*
sene pelye DOGON ANACARDIACEAE *Ozoroa insignis*
seŋgbeŋ-dɛ BULOM BOMBACACEAE *Bombax buonopozense*
seŋgo KISSI CUCURBITACEAE *Cucumeropsis mannii*
seŋgo-pɔmbɔ KISSI CUCURBITACEAE *Cucumeropsis mannii*
seŋgu-chielachalo KISSI CUCURBITACEAE *Cucumeropsis mannii*
sɛŋgũ-yondoŋ KISSI CUCURBITACEAE *Coccinea barteri*
senie AKYE BURSERACEAE *Canarium schweinfurthii*
sɛnke LIMBA APOCYNACEAE *Rauvolfia vomitoria*
*an*-seŋkowe TEMNE ANNONACEAE *Monanthotaxis stenosepala*
sensenrré 'SENUFO' DILLENIACEAE *Tetracera alnifolia*
senuro AKAN - AKYEM APOCYNACEAE *Alstonia boonei* KULANGO *A. boonei*
senuru AKAN - ASANTE APOCYNACEAE *Alstonia boonei*
senutsoe GBE - VHE COMPOSITAE *Acanthospermum hispidum*
sɛnutsoe GBE - VHE AMARANTHACEAE *Alternanthera pungens* COMPOSITAE *Acanthospermum hispidum, A. hispidum*
sɛnutsõe GBE - VHE AMARANTHACEAE *Amaranthus spinosus*
senyan AKYE BURSERACEAE *Canarium schweinfurthii*
senzédu MANDING - MANINKA COMMELINACEAE *Palisota hirsuta*
sépélé BAULE DIOSCOREACEAE *Dioscorea cayenensis*
sépéwé SENUFO - DYIMINI APOCYNACEAE *Strophanthus hispidus*
sépié BAULE DIOSCOREACEAE *Dioscorea alata*
ṣépólóhùn YORUBA CELASTRACEAE *Maytenus senegalensis*
séré MANDING - BAMBARA CARYOPHYLLACEAE *Polycarpon prostratum* MANINKA CUCURBITACEAE *Citrullus lanatus*
sérédéd SERER COMBRETACEAE *Combretum tomentosum*
séréné MANDING - MANINKA BALANITACEAE *Balanites aegyptiaca* SONINKE - SARAKOLE *B. aegyptiaca*
sereno MANDING - MANINKA BALANITACEAE *Balanites aegyptiaca*
sérére SUSU CUCURBITACEAE *Citrullus colocynthis*
séréu WOLOF COMBRETACEAE *Combretum micranthum*
séréusso kwama BAULE ARACEAE *Anchomanes difformis*
seri-gbeli SUSU CONNARACEAE *Connarus africanus*
sérigné SENUFO - TAGWANA BOMBACACEAE *Ceiba pentandra*
sese VULGAR APOCYNACEAE *Holarrhena floribunda*
*o*sese, *o*-sɛsɛ AKAN - TWI APOCYNACEAE *Holarrhena floribunda*
sese AKAN - WASA APOCYNACEAE *Holarrhena floribunda* ANYI *H. floribunda*
sésé AKAN - BRONG APOCYNACEAE *Holarrhena floribunda* ANYI *H. floribunda*
*o*-sɛsɛ(-o) ADANGME APOCYNACEAE *Funtumia africana*
sésed SERER COMBRETACEAE *Combretum micranthum*
sẹ̀sẹ̀dó YORUBA ANNONACEAE *Xylopia aethiopica, X. parviflora*
sesemasa AKAN - ASANTE BIGNONIACEAE *Newbouldia laevis* TWI *N. laevis* WASA *N. laevis*
sessenkergu BATONNUN CONNARACEAE *Santaloides afzelii*
sésséro adiaïra AKAN - BRONG COMMELINACEAE *Cyanotis lanata*
sètá TIV COMPOSITAE *Gynura miniata*
setane BALANTA ANNONACEAE *Monodora tenuifolia*
setun MANDING - MANDINKA COMPOSITAE *Vernonia glaberrima*
séulu WOLOF APOCYNACEAE *Holarrhena floribunda*
seve GOLA ANNONACEAE *Xylopia aethiopica* LOKO *X. aethiopica*
seweyo WOLOF COMBRETACEAE *Combretum micranthum*
ṣẹwutu YORUBA COCHLOSPERMACEAE *Cochlospermum planchonii, C. tinctorium*
sexen WOLOF COMBRETACEAE *Combretum micranthum*
sexéo WOLOF COMBRETACEAE *Combretum micranthum*
sézei BALANTA ANNONACEAE *Uvaria chamae*
shaà-kuùshé HAUSA CONVOLVULACEAE *Ipomoea batatas*
shaá-ní-kà-sán-nì HAUSA BORAGINACEAE *Heliotropium ovalifolium*
shabilabi? FULA - FULFULDE BURSERACEAE *Boswellia dalzielii, B. odorata*
shafa KILBA COMBRETACEAE *Combretum collinum ssp. geitonophyllum*

shagar-el-sim ARABIC APOCYNACEAE *Adenium obesum*
shama GWARI DIOSCOREACEAE *Dioscorea Linn.*
shamman ARABIC - SHUWA CUCURBITACEAE [illegible] *C. melo*
shantu HAUSA CUCURBITACEAE *Lagenaria siceraria*
sheèkár-zoómoó HAUSA COMPOSITAE *Vernonia pumila*
shegara el zeraf ARABIC CAPPARACEAE *Maerua angolensis*
sheiba ARABIC AMARANTHACEAE *Aerva javanica, Celosia argentea*
sheƙani HAUSA COMPOSITAE *Vernonia pumila*
shema GWARI DIOSCOREACEAE *Dioscorea Linn.*
shẹrẹnkpẹn YORUBA APOCYNACEAE *Voacanga africana*
shia NANKANNI COMBRETACEAE *Anogeissus leiocarpus*
shigel kirgegu KANURI ACANTHACEAE *Justicia schimperi striata*
shiwáákáá, shìwaákaá HAUSA COMPOSITAE *Vernonia amygdalina, V. colorata*
shiwáákáár (shìwaákár)-ján-gágárií HAUSA COMPOSITAE *Vernonia glaberrima*
shiwaákár-ján-gárgárií HAUSA ANACARDIACEAE *Ozoroa pulcherrima*
shókó NGAMO CUCURBITACEAE *Lagenaria siceraria*
shɔkɔtɔ-yɔkɔtɔ KRIO AMARANTHACEAE *Celosia argentea*
shwaka HAUSA COMPOSITAE *Vernonia colorata*
si g(u)illi MANO APOCYNACEAE *Funtumia elastica*
sia DAGBANI COMBRETACEAE *Anogeissus leiocarpus*
ʃia DAGBANI COMBRETACEAE *Anogeissus leiocarpus*
siai MANDING - DYULA BIGNONIACEAE *Kigelia africana*
siakɛtɛkrɛ GBE - VHE APOCYNACEAE *Alstonia boonei*
siaŋdẽ KISSI COSTACEAE *Costus afer*
siapiyi KRU - GUERE CONNARACEAE *Cnestis ferruginea*
siawõ KISSI ANNONACEAE *Xylopia aethiopica*
siaxei SUSU CYPERACEAE *Cyperaceae*
*o*-sibi-dua AKAN - TWI APOCYNACEAE *Thevetia neriifolia*
sibigwi (?) FULA - PULAAR ANACARDIACEAE *Lannea acida*
sibitʃo GA APOCYNACEAE *Thevetia neriifolia*
sibompol BANYUN BIGNONIACEAE *Newbouldia laevis*
sibupal sudicam BANYUN BIGNONIACEAE *Markhamia tomentosa*
sida MANDING - MANDINKA BOMBACACEAE *Adansonia digitata*
siddo KRU - BETE BIGNONIACEAE *Newbouldia laevis*
sidèldéla DIOLA CUCURBITACEAE *Cucumis metuliferus*
sidia MANDING - DYULA BIGNONIACEAE *Kigelia africana*
sidia findia MANDING - DYULA BIGNONIACEAE *Kigelia africana*
sidiabakatan BANYUN COMBRETACEAE *Combretum glutinosum*
si-diafua-angbé ANYI APOCYNACEAE *Alafia barteri*
sidiamba MANDING - BAMBARA BIGNONIACEAE *Kigelia africana*
sido KRU - GUERE BIGNONIACEAE *Newbouldia laevis*
siɛ GOLA DIOSCOREACEAE *Dioscorea Linn.*
siélébé AKYE ANNONACEAE *Hexalobus crispiflorus*
sien doro nani MANDING - BAMBARA AMARANTHACEAE *Achyranthes aspera*
sièn goni MANDING - BAMBARA AMARANTHACEAE *Alternanthera pungens*
sienfélé KWENI ASCLEPIADACEAE *Pergularia daemia*
sièra ughwé MANDING - MANINKA CHRYSOBALANACEAE *Parinari curatellifolia*
siesa (jiesa) HAUSA COMBRETACEAE *Combretum micranthum*
sifemb DIOLA APOCYNACEAE *Baissea multiflora*
sifôdum BANYUN COMBRETACEAE *Guiera senegalensis*
sigha MOORE COMBRETACEAE *Anogeissus leiocarpus*
sìgilu DOGON COMBRETACEAE *Anogeissus leiocarpus*
siginiba MANDING - BAMBARA COMPOSITAE *Vernonia galamensis*
sigo(n) FULA - PULAAR CHRYSOBALANACEAE *Bafodeya benna*
sigo YORUBA - NAGO ARALIACEAE *Cussonia arborea*
ṣigo YORUBA ARALIACEAE *Cussonia arborea*
sigon SUSU CHRYSOBALANACEAE *Bafodeya benna*
sigonaïa MANDING - MANINKA CHRYSOBALANACEAE *Bafodeya benna*
sigump kidigen BANYUM DILLENIACEAE *Tetracera potatoria*
sihir TAMACHEK CAPPARACEAE *Boscia senegalensis*
si'id ARABIC - SHUWA CYPERACEAE *Cyperus maculatus*
sikaral BANYUN ANNONACEAE *Uvaria chamae*
sikénay DIOLA APOCYNACEAE *Baissea multiflora*
sikinn BANYUN COMBRETACEAE *Combretum glutinosum*
siko SUSU CHRYSOBALANACEAE *Bafodeya benna*
sikoɛ KONO CYPERACEAE *Cyperus articulatus*
sikokinad DIOLA APOCYNACEAE *Saba senegalensis*
siko-na SUSU - DYALONKE CHRYSOBALANACEAE *Bafodeya benna*
silikokoré AKAN - ASANTE ANNONACEAE *Enantia polycarpa*
silikoré MOORE CAPPARACEAE *Capparis tomentosa*
silli (*def.* silloo) MANDING - MANDINKA COMPOSITAE *Vernonia perrottetii*
silli foro (*def.*-o) MANDING - MANDINKA COMPOSITAE *Vicoa leptoclada*
siludia BANYUN BOMBACACEAE *Bombax costatum*
simba bali MANDING - MANDINKA COMBRETACEAE *Combretum nigricans*
simba beli MANDING - BAMBARA COMBRETACEAE *Combretum micranthum* MANINKA *C. glutinosum*
simbabali MANDING - BAMBARA COMBRETACEAE *Combretum nigricans* MANINKA *C. nigricans*
simblé sama MANDING - DYULA COMPOSITAE *Microglossa pyrifolia*
simbôm WOLOF ASCLEPIADACEAE *Ceropegia spp.*
simbus BALANTA CAPPARACEAE *Capparis erythrocarpos*

sisibigolo DAGAARI ANACARDIACEAE *Lannea barteri, L. velutina*
sisibu DAGAARI ANACARDIACEAE *Lannea acida*
sisigi DIOLA CUCURBITACEAE *Citrullus colocynthis*
sisinan TEM COMBRETACEAE *Pteleopsis suberosa*
*o*-sisiriw AKAN - TWI BIGNONIACEAE *Spathodea campanulata*
*an*-sisɔ TEMNE CUCURBITACEAE *Mukia maderaspatana*
sissal BANYUN BIGNONIACEAE *Spathodea campanulata*
sissiku TEM COMBRETACEAE *Combretum molle*
sissina TEM COMBRETACEAE *Pteleopsis suberosa*
sissulu DAGAARI ANACARDIACEAE *Lannea acida*
sisubu DAGAARI ANACARDIACEAE *Lannea acida*
sit DIOLA CYPERACEAE *Fuirena ciliaris*
sita (*def.* sitoo) MANDING - MANDINKA BOMBACACEAE *Adansonia digitata*
sita (*def.*. sitoo) MANDING - MANDINKA BOMBACACEAE *Adansonia digitata*
*an*-sita TEMNE COMMELINACEAE *Palisota hirsuta*
sita kolokuru MANDING - MANINKA APOCYNACEAE *Adenium obesum*
sito MANDING - BAMBARA BOMBACACEAE *Adansonia digitata* MANINKA *A. digitata* 'SOCE' *A. digitata*
sitta HAUSA COMPOSITAE *Gynura miniata*
siwaakewal (*pl.* siwaakeeje) FULA - FULFULDE COMPOSITAE *Vernonia amygdalina, V. colorata*
si-yalma 'SENUFO' APOCYNACEAE *Strophanthus hispidus*
ṣḛ̀ sọ́ọ̀rọ́ YORUBA BASELLACEAE *Basella alba*
skèanay DIOLA APOCYNACEAE *Saba senegalensis*
slaha-ŋmei GA AMARANTHACEAE *Amaranthus spinosus*
snɛk-tamatis KRIO CUCURBITACEAE *Trichosanthes cucumerina*
snof-lif KRIO BIGNONIACEAE *Newbouldia laevis*
so ABE CHRYSOBALANACEAE *Parinari excelsa*
sô KRU - GUERE ANNONACEAE *Enantia polycarpa*
sŏ MANO DIOSCOREACEAE *Dioscorea Linn.*
sɔ BULOM ANNONACEAE *Xylopia aethiopica* CYPERACEAE *Cyperus articulatus*
so messin 'SENUFO' ANNONACEAE *Hexalobus monopetalus*
soasgha MOORE COCHLOSPERMACEAE *Cochlospermum tinctorium*
sob WOLOF ANACARDIACEAE *Spondias mombin*
sob tubab WOLOF ANACARDIACEAE *Spondias purpurea*
sobara GRUSI COMBRETACEAE *Guiera senegalensis*
sobotoro, sobotorooji FULA - FULFULDE ASCLEPIADACEAE *Leptadenia hastata*
sobu ABE ANNONACEAE *Cleistopholis patens*
sɔbu ADANGME AMARANTHACEAE *Amaranthus hybridus*
sɔbuɛ ADANGME - KROBO AMARANTHACEAE *Amaranthus hybridus*
sɔbui GBE - VHE CAPPARACEAE *Cleome Linn., C. gynandra*
socòra-bátè FULA - PULAAR COMPOSITAE *Ethulia conyzoides*
sodäd ARABIC CAPPARACEAE *Capparis decidua*
sõdo TEM APOCYNACEAE *Strophanthus hispidus*
soẽ KABRE APOCYNACEAE *Strophanthus hispidus*
sofaligbie-na MANDING - MANDINKA COMPOSITAE *Acanthospermum hispidum*
ṣọ̀fin YORUBA CUCURBITACEAE *Citrullus lanatus*
soforo MANDING - BAMBARA ARACEAE *Amorphophallus aphyllus*
sofré 'KRU' COMPOSITAE *Microglossa afzelii*
sɔgbee GBE - VHE CAPPARACEAE *Cleome Linn., C. gynandra*
sogodio KULANGO ANNONACEAE *Xylopia aethiopica*
sogui FULA - PULAAR CAPPARACEAE *Maerua angolensis, M. crassifolia*
soguirini MANDING - BAMBARA BIGNONIACEAE *Stereospermum kunthianum*
sohn KRU - BASA ANNONACEAE *Enantia polycarpa*
sohué AKYE APOCYNACEAE *Funtumia africana, Holarrhena floribunda*
*a*-soi TEMNE CYPERACEAE *Cyperaceae*
*an*-soi TEMNE CYPERACEAE *Cyperus esculentus, C. pustulatus, Fimbristylis dichotoma, Kyllinga erecta, Pycreus polystachyos*
*an*-soi-*a*-bɛra TEMNE CYPERACEAE *Fimbristylis dichotoma*
*an*-soi-*a*-rungi TEMNE CYPERACEAE *Cyperaceae*
ṣọkọ YORUBA AMARANTHACEAE *Celosia argentea*
sọkọ yọ̀kọ̀tọ̀ YORUBA AMARANTHACEAE *Celosia argentea*
sọkọ yọ̀kọ̀tọ̀ pupa YORUBA AMARANTHACEAE *Celosia argentea*
sokolu SHIEN ASCLEPIADACEAE *Pergularia tomentosa*
sôkon DIOLA APOCYNACEAE *Landolphia heudelotii*
sokonasu DIOLA APOCYNACEAE *Landolphia heudelotii*
sokosaka-lɛ BULOM CONNARACEAE *Manotes expansa*
ṣọkọtọ YORUBA AMARANTHACEAE *Celosia argentea*
ʃokotɔ MANO AMARANTHACEAE *Celosia argentea*
sokoyokotaw HAUSA AMARANTHACEAE *Celosia leptostachya*
sokpe ADANGME BOMBACACEAE *Ceiba pentandra*
sokwara HAUSA DIOSCOREACEAE *Dioscorea gen. (products)*
sol WOLOF ASCLEPIADACEAE *Caralluma decaisneana*
solahir DIOLA CYPERACEAE *Scleria depressa*
solanamb (*def.*-ô) MANDING - MANDINKA APOCYNACEAE *Strophanthus hispidus, S. sarmentosus*
solĩ GBE - VHE AIZOACEAE *Sesuvium portulacastrum*
solid ARABIC CHENOPODIACEAE *Suaeda vermiculata*
solo KRU - BETE COMBRETACEAE *Terminalia superba* KWENI ASCLEPIADACEAE *Pergularia daemia*
sólo anyu DOGON COCHLOSPERMACEAE *Cochlospermum tinctorium*
solui GBE - VHE CAPPARACEAE *Cleome gynandra*

sungul SERER - NON ANACARDIACEAE *Sclerocarya birrea*
sunka GBE - VHE DIOSCOREACEAE *Dioscorea rotundata*
sunku MANDING - 'SOCE' ANNONACEAE *Annona senegalensis*
sunkuŋ (*def.*-o) MANDING - MANDINKA ANNONACEAE *Annona senegalensis*
sunkung (*def.*-ô) MANDING - MANDINKA ANNONACEAE *Annona senegalensis*
sunkuno MANDING - 'SOCE' ANNONACEAE *Annona senegalensis*
sunkutu foolee MANDING - MANINKA APOCYNACEAE *Landolphia dulcis*
sunsu MANDING - MANINKA ANNONACEAE *Annona arenaria, A. senegalensis*
sunsun MANDING - BAMBARA ANNONACEAE *Annona glauca, A. senegalensis*
sunsuŋ MANDING - MANDINKA ANNONACEAE *Annona arenaria*
sunsun sunzu MANDING - MANINKA ANNONACEAE *Annona arenaria, A. senegalensis*
sunudunguéra BATONNUN COMBRETACEAE *Combretum molle*
sununga TEM CUCURBITACEAE *Lagenaria siceraria*
sunyãn AKAN - TWI ANACARDIACEAE *Pseudospondias microcarpa*
sunyugi SUSU COMMELINACEAE *Commelina thomasii*
sunzan ka tamugu MANDING - BAMBARA CONVOLVULACEAE *Merremia aegyptia*
sunzun BAULE ANNONACEAE *Annona muricata*
suo lõngo lah MANO COMPOSITAE *Ethulia conyzoides*
suo mâg SERER BORAGINACEAE *Cordia sinensis*
supa-jambo MANDING - MANDINKA ANACARDIACEAE *Spondias mombin*
suparnao FULA - PULAAR CONVOLVULACEAE *Ipomoea asarifolia*
supripi AKAN - FANTE CONVOLVULACEAE *Ipomoea cairica, I. spp. indet.*
supusapui MENDE ANNONACEAE *Annona muricata*
suraddu HAUSA COMPOSITAE *Centaurea perrottetii, C. praecox*
suraka wôni MANDING - BAMBARA COMPOSITAE *Acanthospermum hispidum*
surandu HAUSA COMPOSITAE *Centaurea perrottetii, Dicoma tomentosa*
surarad u, surardu, surandu HAUSA COMPOSITAE *Centaurea perrottetii*
suraradu, surandu, surardu HAUSA COMPOSITAE *Centaurea praecox*
*a*-surikbɛne TEMNE COMPOSITAE *Aspilia africana*
surkémé FULA - PULAAR COMPOSITAE *Melanthera gambica*
surreih ARABIC CAPPARACEAE *Cadaba farinosa*
suru, suro NYENYEGE APOCYNACEAE *Strophanthus hispidus*
suru la MANO CYPERACEAE *Mapania linderi*
surua ANYI ANNONACEAE *Neostenanthera hamata*
suruku n'tombolo 'SENUFO' APOCYNACEAE *Carissa edulis*
*a*-suru-*a*-ro-kant TEMNE COMPOSITAE *Aspilia africana*
suruuduyel (*pl.* suraaduhoy) FULA - FULFULDE COMPOSITAE *Centaurea perrottetii, C. praecox*
sus SERER - NON CAPPARACEAE *Boscia angustifolia*
sussuguté DAGAARI ANACARDIACEAE *Lannea barteri, L. velutina*
susu BATONNUN DIOSCOREACEAE *Dioscorea praehensilis* PEPEL ANNONACEAE *Annona senegalensis*
susu mase GBE - VHE CUCURBITACEAE *Cucumeropsis mannii*
susujango MANDING - MANDINKA COMBRETACEAE *Combretum Loefl.*
su-ti DAN CHRYSOBALANACEAE *Maranthes aubrevillii*
suup SERER BORAGINACEAE *Cordia senegalensis*
suusaaraŋ (*def.*-o) MANDING - MANDINKA CUCURBITACEAE *Luffa cylindrica*
suweji ARABIC - SHUWA CYPERACEAE *Cyperaceae*
suwel SENUFO - TAGWANA APOCYNACEAE *Strophanthus hispidus*
swa-ü DAN BOMBACACEAE *Bombax buonopozense*
swa-uh DAN BOMBACACEAE *Rhodognaphalon brevicuspe*
ṣwẹwutu YORUBA COCHLOSPERMACEAE *Cochlospermum tinctorium*
ʃwie ADANGME - KROBO COMPOSITAE *Launaea taraxacifolia* GA BASELLACEAE *Basella alba*
switi-sap KRIO ANNONACEAE *Annona squamosa*
swo hongo la MANO COMPOSITAE *Ethulia conyzoides*
syak ARABIC - SHUWA CAPPARACEAE *Capparis decidua*
syiï BEROM CUCURBITACEAE *Lagenaria siceraria*
szrom ARABIC CAPPARACEAE *Cadaba farinosa*
ta, tala KONO CUCURBITACEAE *Lagenaria siceraria*
tà GA COMPOSITAE *Vernonia amygdalina, V. colorata*
taa WOLOF ANACARDIACEAE *Rhus longipes*
taab WOLOF COMBRETACEAE *Combretum nigricans*
tàabármár zoómoó HAUSA COSTACEAE *Costus spectabilis*
taákálmàn zoómoó (*plur.*) HAUSA COSTACEAE *Costus spectabilis*
taamu FULA - FULFULDE BOMBACACEAE *Ceiba pentandra*
taarin-gidaa HAUSA ASCLEPIADACEAE *Glossonema boveanum*
taatsunyaa HAUSA AMARANTHACEAE *Amaranthus viridis*
tab WOLOF COMBRETACEAE *Combretum nigricans*
taba agbe YORUBA COMPOSITAE *Laggera pterodonta*
tabadai HAUSA CAPPARACEAE *Cleome afrospinosa, C. gynandra*
tabadamashe HAUSA CAPPARACEAE *Cleome Linn., C. afrospinosa, C. gynandra*
tabal WOLOF ARACEAE *Stylochiton hypogaeus*
tabaldi ARABIC BOMBACACEAE *Adansonia digitata*
tabara HAUSA COMBRETACEAE *Combretum sericeum*
tabason KRU - GUERE COMPOSITAE *Bidens pilosa*
tabatué KRU - GUERE CECROPIACEAE *Myrianthus libericus*
tabeïbaret TAMACHEK CAPPARACEAE *Cadaba glandulosa*
tabe-khabi SUSU DIOSCOREACEAE *Dioscorea misc. & spp. unknown*
table WOLOF AMARYLLIDACEAE *Crinum zeylanicum*
taboo ABURE APOCYNACEAE *Pleiocarpa mutica*
taborak TAMACHEK BALANITACEAE *Balanites aegyptiaca*
tabotu 'KRU' CHRYSOBALANACEAE *Parinari excelsa*

tabuti MANDING - BAMBARA CAPPARACEAE *Capparis sepiaria*
tachale DAGBANI COMBRETACEAE *Combretum fragrans, C. molle*
tachayi DAGBANI COMBRETACEAE *Combretum fragrans, C. molle, C. nigricans*
tadahant TAMACHEK CAPPARACEAE *Boscia senegalensis*
tadanat HAUSA BORAGINACEAE *Cordia sinensis*
tadanu TAMACHEK BORAGINACEAE *Cordia sinensis*
tadatʃo ADANGME - KROBO COMBRETACEAE *Combretum zenkeri*
tade FULA - PULAAR COMBRETACEAE *Combretum micranthum*
tadent TAMACHEK CAPPARACEAE *Boscia angustifolia, B. senegalensis*
tadenua ARABIC ASCLEPIADACEAE *Caralluma russelliana*
tadkernit ARABIC COMPOSITAE *Sonchus oleraceus*
tadomet TAMACHEK CAPPARACEAE *Boscia senegalensis*
taduéné AKAN - ASANTE CONVOLVULACEAE *Ipomoea mauritiana*
taelu KABRE BOMBACACEAE *Adansonia digitata*
taelu, telu TEM BOMBACACEAE *Adansonia digitata*
táem BALANTA ARACEAE *Colocasia esculenta*
taf WOLOF CELASTRACEAE *Hippocratea africana*
tafé SONINKE - SARAKOLE BIGNONIACEAE *Stereospermum kunthianum* COMBRETACEAE *Combretum glutinosum*
tafo ka shamamarka HAUSA ASCLEPIADACEAE *Glossonema boveanum, Schizoglossum petherickianum*
taga BULOM CUCURBITACEAE *Lagenaria siceraria*
tagalel TAMACHEK CUCURBITACEAE *Citrullus colocynthis*
tâgara MANDING - BAMBARA COMBRETACEAE *Combretum glutinosum* MANINKA *C. glutinosum*
ťâgara MANDING - BAMBARA COMBRETACEAE *Combretum glutinosum* MANINKA *C. glutinosum*
tâgaré dé MANDING - BAMBARA COMBRETACEAE *Combretum collinum ssp. geitonophyllum*
tagargade HAUSA COMPOSITAE *Artemisia spp.*
tagart TAMACHEK CAPPARACEAE *Maerua crassifolia*
tagasrarit ARABIC CUCURBITACEAE *Cucumis prophetarum*
tâgba SERER - NON ANACARDIACEAE *Lannea velutina*
tagbo-mɛsɛ KONO ALISMATACEAE *Limnophyton obtusifolium*
tagellit TAMACHEK CUCURBITACEAE *Citrullus colocynthis*
taghalbas TAMACHEK BURSERACEAE *Commiphora africana*
tagiaani 'KRU' COMPOSITAE *Bidens pilosa*
tagiiri YORUBA CUCURBITACEAE *Lagenaria breviflora*
tagiri KRIO CUCURBITACEAE *Lagenaria rufa, Ruthalicia eglandulosa*
ta-giri TEMNE CUCURBITACEAE *Luffa cylindrica*
taglu WOLOF CUCURBITACEAE *Lagenaria siceraria*
tagwamba HAUSA ARACEAE *Colocasia esculenta*
tagyalgyade HAUSA COMPOSITAE *Artemisia spp.*
tah KRU - BASA CHRYSOBALANACEAE *Chrysobalanus icaco* WOLOF CUCURBITACEAE *Lagenaria siceraria*
tahalist TAMACHEK CAPPARACEAE *Cadaba glandulosa*
tahara TAMACHEK CHENOPODIACEAE *Cornulaca monacantha*
tahara, tasra, terchit TAMACHEK CHENOPODIACEAE *Traganum nudatum*
tahonak TAMACHEK ANACARDIACEAE *Rhus tripartita*
ta-hɔndi MENDE AMARANTHACEAE *Amaranthus spinosus, A. viridis*
taiberu TAMACHEK ASCLEPIADACEAE *Caralluma russelliana*
taïchecht BERBER BALANITACEAE *Balanites aegyptiaca*
taïchot TAMACHEK BALANITACEAE *Balanites aegyptiaca*
*a*-tak TEMNE APOCYNACEAE *Alafia benthamii*
*a*-tak, *an*-buŋk TEMNE DIOSCOREACEAE *Dioscorea togoensis*
tak ara kuléhi FULA - PULAAR ANACARDIACEAE *Ozoroa insignis*
takala MANDING - MANINKA CECROPIACEAE *Myrianthus serratus*
takan MANDING - MANINKA COMPOSITAE *Aspilia rudis*
tàkàndár giiwaá HAUSA ARALIACEAE *Cussonia arborea*
ťâkara MANDING - BAMBARA COMBRETACEAE *Combretum nigricans* MANINKA *C. nigricans*
takara kuléhi FULA - FULFULDE ANACARDIACEAE *Ozoroa insignis* PULAAR *O. insignis*
takasaïm TAMACHEK CUCURBITACEAE *Cucurbita maxima*
takbo KONO AMARYLLIDACEAE *Crinum jagus, C. natans*
takere foondu (?) FULA - FULFULDE CRUCIFERAE *Lepidium sativum*
takinkeint, talejest, tiledjest TAMACHEK CUCURBITACEAE *Citrullus lanatus*
takiringa NABT COMBRETACEAE *Combretum molle*
takiwamitié AKAN - BRONG ASCLEPIADACEAE *Secamone afzelii*
takka cijla FULA - FULFULDE AIZOACEAE *Gisekia pharnaceoides*
takola MANDING - BAMBARA CECROPIACEAE *Myrianthus serratus*
takorowa AKAN - TWI CHRYSOBALANACEAE *Licania elaeosperma* WASA *L. elaeosperma*
takrowa AKAN - WASA CHRYSOBALANACEAE *Maranthes glabra*
*a*-tak-*a*-runi TEMNE DIOSCOREACEAE *Dioscorea togoensis*
taku MUNGAKA CECROPIACEAE *Musanga cecropioides*
takunta BATONNUN COMBRETACEAE *Combretum fragrans*
tal koda ARABIC CHENOPODIACEAE *Chenopodium murale*
tala CHAMBA CONVOLVULACEAE *Ipomoea eriocarpa* KISSI CUCURBITACEAE *Lagenaria siceraria* LOKO BIGNONIACEAE *Crescentia cujete*
talal WOLOF ASCLEPIADACEAE *Leptadenia hastata*
talala SONGHAI CONVOLVULACEAE *Ipomoea asarifolia*

| | |
|---|---|
| talé | FULA - FULFULDE BALANITACEAE *Balanites aegyptiaca* PULAAR ANACARDIACEAE *Spondias mombin, S. mombin* |
| tálel wâdu | FULA - PULAAR CELASTRACEAE *Hippocratea africana* |
| tàlhánà (*pl.* -à) | SONGHAI CONVOLVULACEAE *Ipomoea aquatica* |
| tali | FULA - PULAAR ANACARDIACEAE *Spondias mombin* COMBRETACEAE *Combretum micranthum* |
| talkait | TAMACHEK BORAGINACEAE *Trichodesma africanum* |
| tàllâfà màraàyú | HAUSA ANNONACEAE *Annona senegalensis* |
| talli | FULA - PULAAR COMBRETACEAE *Combretum micranthum* |
| tallika | FULA - PULAAR COMBRETACEAE *Combretum micranthum* |
| talonia | HAUSA COMBRETACEAE *Combretum glutinosum* |
| ṭalquidáua | FULA - PULAAR CONNARACEAE *Cnestis corniculata* |
| tàm | DERA BOMBACACEAE *Bombax costatum* |
| tamakerkait, tamakerzist, timekerkest | TAMACHEK AMARANTHACEAE *Aerva javanica* |
| tamalege | ARABIC - SHUWA AIZOACEAE *Trianthema portulacastrum, Zaleya pentandra* |
| ṭamangarusa, tsinakuwa | HAUSA CYPERACEAE *Cyperus esculentus* |
| tamanohi | FULA - PULAAR BORAGINACEAE *Cordia myxa* |
| tamarza | SONGHAI - ZARMA ANACARDIACEAE *Lannea microcarpa* |
| tamaula | HAUSA CUCURBITACEAE *Lagenaria siceraria* |
| tamba | MANDING - BAMBARA CHRYSOBALANACEAE *Parinari curatellifolia* MANDINKA *Neocarya macrophylla, N. macrophylla* MANINKA *Parinari curatellifolia, P. curatellifolia* |
| tâmba | MANDING - BAMBARA CHRYSOBALANACEAE *Parinari curatellifolia* MANINKA *P. curatellifolia* SOCE *P. curatellifolia, P. excelsa* |
| tâmba kunda | SOCE CHRYSOBALANACEAE *Parinari excelsa* |
| tambaa (*indef.* and *def.*) | MANDING - MANDINKA CHRYSOBALANACEAE *Neocarya macrophylla, Parinari excelsa* |
| tambacumba | FULA - PULAAR CHRYSOBALANACEAE *Neocarya macrophylla* MANDING - MANDINKA *N. macrophylla* |
| tambakumbaa (*indef.* and *def.*) | MANDING - MANDINKA CHRYSOBALANACEAE *Neocarya macrophylla, Parinari excelsa* |
| tambalay | WOLOF ARACEAE *Pistia stratiotes* |
| tambė | WOLOF CAPPARACEAE *Crateva adansonii* |
| tambutiji | ANYI - ANUFO ASCLEPIADACEAE *Calotropis procera* |
| tamé | FULA - PULAAR ALISMATACEAE *Sagittaria guayanensis* |
| tamfre | MUNGAKA BURSERACEAE *Canarium schweinfurthii* |
| tamia | AKAN - TWI ANACARDIACEAE *Pseudospondias microcarpa* |
| tamiawa | GA CRASSULACEAE *Bryophyllum pinnatum* |
| tamiye | ARABIC AMARANTHACEAE *Aerva javanica* |
| *a*-tampene | TEMNE ARACEAE *Culcasia scandens* |
| tampo | GRUSI COCHLOSPERMACEAE *Cochlospermum tinctorium* |
| tampus | WOLOF BORAGINACEAE *Cordia myxa* |
| tamrano | HAUSA COMMELINACEAE *Cyanotis caespitosa* |
| tamtamtuéni | SORUBA - KUYOBE COMBRETACEAE *Combretum fragrans* |
| tamui | SUSU CHRYSOBALANACEAE *Neocarya macrophylla* |
| tan | SENUFO - TUSIA APOCYNACEAE *Landolphia heudelotii* |
| tan zub | ARABIC CAPPARACEAE *Capparis decidua* |
| tana | 'SENUFO' ANACARDIACEAE *Spondias mombin* |
| tànàïn túó | IJO - IZON CONVOLVULACEAE *Ipomoea aquatica* |
| tandaletando | KISSI CACTACEAE *Nopalea* (*Opuntia*) |
| tandauji | MARGI ARACEAE *Colocasia esculenta* |
| tangaba | MANDING - MANINKA CECROPIACEAE *Myrianthus serratus* |
| tangasu | MANDING - BAMBARA ANNONACEAE *Annona glauca, A. senegalensis* |
| taŋgbɛsowakoloma | MANDING - MANINKA APOCYNACEAE *Rauvolfia vomitoria* |
| tango | DAGAARI BORAGINACEAE *Cordia myxa* |
| tangolo sébé | MANDING - DYULA CONNARACEAE *Cnestis ferruginea* |
| tani | FULA - PULAAR BALANITACEAE *Balanites aegyptiaca* |
| taniyá | YORUBA CAPPARACEAE *Crateva adansonii* |
| *an*-tankali | TEMNE DIOSCOREACEAE *Dioscorea burkilliana* |
| tanma | 'SENUFO' ANACARDIACEAE *Spondias mombin* |
| tanndawre (*pl.* tanndawje) | FULA - FULFULDE ARACEAE *Colocasia esculenta* |
| tannere (*pl.* tanni) | FULA - FULFULDE BALANITACEAE *Balanites aegyptiaca* |
| tanni (*pl.* tanne) | FULA - FULFULDE BALANITACEAE *Balanites aegyptiaca* |
| tantabili | MOBA COMBRETACEAE *Combretum fragrans, C. nigricans* |
| *an*-tantali | TEMNE DIOSCOREACEAE *Dioscorea burkilliana* |
| tante bulense | 'KAGONGA' COMBRETACEAE *Combretum molle* |
| tanton | 'KRU' COSTACEAE *Costus afer* |
| tantsiya | AKYE APOCYNACEAE *Strophanthus hispidus* HAUSA *S. hispidus, S. sarmentosus* |
| tantsiyaari | FULA - FULFULDE APOCYNACEAE *Strophanthus hispidus, S. sarmentosus* |
| tantsiyari | FULA - PULAAR APOCYNACEAE *Strophanthus sarmentosus* |
| taoné | GAGU COMPOSITAE *Ageratum conyzoides* |
| *a*-tap | TEMNE ASCLEPIADACEAE *Tylophora sylvatica* |
| *an*-tap | TEMNE COSTACEAE *Costus afer* |
| tap | WOLOF COMBRETACEAE *Combretum nigricans* |
| tapatoy | FULA - PULAAR ASCLEPIADACEAE *Leptadenia hastata* |
| tapentiti | KRU - GUERE BORAGINACEAE *Heliotropium indicum* |
| tapérodia | KWENI BORAGINACEAE *Heliotropium indicum* |
| tâpus | WOLOF BORAGINACEAE *Cordia sinensis* |
| tara | SONGHAI CYPERACEAE *Cyperaceae* |

| | |
| --- | --- |
| tẹ̀tẹ̀ | **YORUBA** AMARANTHACEAE *Amaranthus dubius, A. hybridus, A. viridis* |
| tẹ̀tẹ̀ atẹlẹ danji | **YORUBA** AMARANTHACEAE *Amaranthus viridis* |
| tẹ̀tẹ̀ àtètèdayé | **YORUBA** AMARANTHACEAE *Amaranthus viridis* |
| tẹ̀tẹ̀ ẹ̀gún, tẹ̀tẹ̀ eléẹ̀gún | **YORUBA** AMARANTHACEAE *Amaranthus spinosus* |
| tẹ̀tẹ̀ kékeré | **YORUBA** AMARANTHACEAE *Amaranthus viridis* |
| tẹ̀tẹ̀ ńlá | **YORUBA** AMARANTHACEAE *Amaranthus hybridus* |
| tẹ̀tẹ̀ òyìbó | **YORUBA** AMARANTHACEAE *Amaranthus hybridus* |
| tẹ̀tẹ̀ pòpó | **YORUBA** AMARANTHACEAE *Amaranthus hybridus* |
| tẹ̀tẹ̀-ẹ̀gún | **YORUBA** COSTACEAE *Costus afer* |
| tétégné | **KRU - GUERE** ANACARDIACEAE *Lannea welwitschii* |
| tɛtɛtɛtɛ | **AKAN - ASANTE** CAPPARACEAE *Cleome gynandra* |
| tɛtrɛmantɛ | **GA** ARACEAE *Pistia stratiotes* |
| tétu | **MANDING - BAMBARA** CHRYSOBALANACEAE *Neocarya macrophylla* |
| têtu | **'PATOKLA'** APOCYNACEAE *Funtumia elastica* |
| *an*-tet-*a*-wir | **TEMNE** CYPERACEAE *Mariscus longibracteatus* |
| teutieu win | **KRU - GUERE** CECROPIACEAE *Myrianthus serratus* |
| tɛvɛ | **MENDE** ACANTHACEAE *Justicia schimperi striata* |
| tɛyɛ | **MENDE** ACANTHACEAE *Justicia schimperi striata* |
| tezaragade | **HAUSA** COMPOSITAE *Artemisia spp.* |
| tezera | **TAMACHEK** ASCLEPIADACEAE *Calotropis procera* |
| *an*-thaŋka | **TEMNE** ANACARDIACEAE *Sorindeia collina, S. juglandifolia, S. mildbraedii, Trichoscypha arborea* BURSERACEAE *Santiria trimera* |
| *an*-thaŋka bombar | **TEMNE** ANACARDIACEAE *Sorindeia juglandifolia* |
| *an*-thaŋth-*an*-ləp | **TEMNE** ANACARDIACEAE *Lannea nigritana* |
| thɔk-i-pɛpɛ-lɛ | **BULOM** BIGNONIACEAE *Crescentia cujete* |
| *a*-thoŋbothigba | **TEMNE** ARACEAE *Anchomanes difformis* |
| *an*-thonk-nin | **TEMNE** CAPPARACEAE *Euadenia eminens* |
| *a*-thotho | **TEMNE** CONVOLVULACEAE *Ipomoea involucrata* |
| *a*-thotho-*a*-runi | **TEMNE** CONVOLVULACEAE *Ipomoea involucrata* |
| *an*-thɔth-*an*-thi | **TEMNE** ASCLEPIADACEAE *Secamone afzelii* |
| ti keung | **BANYUN** APOCYNACEAE *Alstonia boonei* |
| tia galgha | **MOORE** BALANITACEAE *Balanites aegyptiaca* |
| tiacktal | **FULA - PULAAR** CYPERACEAE *Cyperus maculatus* |
| tiahgara dié | **MANDING - MANINKA** COMBRETACEAE *Combretum collinum ssp. geitonophyllum* |
| tialé | **FULA - PULAAR** ANACARDIACEAE *Spondias mombin* |
| tiali porto | **FULA - FULFULDE** ANACARDIACEAE *Spondias mombin* |
| tiamalabébo | **KYAMA** CELASTRACEAE *Hippocratea spp. sens. lat.* |
| tiamanohi | **FULA - FULFULDE** BORAGINACEAE *Cordia myxa* **PULAAR** *C. myxa* |
| tiambéré | **MANDING - BAMBARA** COMBRETACEAE *Combretum lecardii* **MANINKA** *C. lecardii* **MANDINKA** *C. lecardii* |
| tianga | **DOGON** ASCLEPIADACEAE *Caralluma russelliana* |
| tiangara | **MANDING - BAMBARA** COMBRETACEAE *Combretum fragrans, C. glutinosum* **MANINKA** *C. glutinosum* |
| tiangara dié | **MANDING - BAMBARA** COMBRETACEAE *Combretum collinum ssp. geitonophyllum* **MANINKA** *C. collinum ssp. geitonophyllum* |
| tiankara | **MANDING - BAMBARA** COMBRETACEAE *Combretum nigricans* **MANDINKA** *C. nigricans* **MANINKA** *C. nigricans* |
| tiapule-ranehi | **FULA - PULAAR** CELASTRACEAE *Maytenus senegalensis* |
| tiatowihan-goso | **AKAN - BRONG** COMPOSITAE *Vernonia nigritiana* |
| tibbi | **TAMACHEK** CHENOPODIACEAE *Chenopodium murale* |
| tiɓé | **GA'ANDA** CUCURBITACEAE *Lagenaria siceraria* |
| tibibara, tribo, turugba | **MANDING - MANINKA** COCHLOSPERMACEAE *Cochlospermum tinctorium* |
| ticharque | **MANDING - MANDINKA** APOCYNACEAE *Holarrhena floribunda* |
| tichtaya | **ARABIC** BALANITACEAE *Balanites aegyptiaca* |
| tidâni | **HAUSA** BORAGINACEAE *Cordia sinensis* |
| tidé | **KRU - GUERE** ANACARDIACEAE *Pseudospondias microcarpa* |
| tiduar | **ARABIC** ASCLEPIADACEAE *Caralluma russelliana* |
| tié | **KRU - GUERE** APOCYNACEAE *Alstonia boonei* |
| tiè bossu | **KRU - GUERE** ANNONACEAE *Monodora tenuifolia* |
| tiébé zago | **KRU - GUERE** ACANTHACEAE *Elytraria marginata* |
| tiegbanhema folué | **BAULE** APOCYNACEAE *Alafia lucida* |
| tiehomtioli | **WOLOF** CYPERACEAE *Cyperus rotundus, C. sphacelatus* |
| tiéhomtioli | **WOLOF** CYPERACEAE *Cyperaceae* |
| tiekoni kolo | **MANDING - MANINKA** CAPPARACEAE *Boscia angustifolia* |
| tiékoni kolo | **MANDING - MANINKA** CAPPARACEAE *Boscia angustifolia* |
| tiɛlɛbdo | **MOORE** CAPPARACEAE *Cleome gynandra* |
| tiemankaran dié | **MANDING - BAMBARA** COMBRETACEAE *Combretum collinum ssp. geitonophyllum* **MANINKA** *C. collinum ssp. geitonophyllum* |
| tien | **AKAN - BRONG** ANNONACEAE *Annona arenaria, A. senegalensis* **KRU - GUERE** APOCYNACEAE *Alstonia boonei* |
| tiendo | **KISSI** APOCYNACEAE *Alstonia boonei* |
| tientirgaye | **FULA - FULFULDE** CAPPARACEAE *Boscia salicifolia* **PULAAR** *B. salicifolia* |
| tiépéwowo | **KRU - GUERE** APOCYNACEAE *Tabernaemontana crassa* |
| tiéré | **MANDING - DYULA** BIGNONIACEAE *Spathodea campanulata* |
| tiètièpalo | **KWENI** CUCURBITACEAE *Ruthalicia eglandulosa* |
| tifahama | **'SENUFO'** ANACARDIACEAE *Ozoroa insignis* |
| tifem | **BANYUN** APOCYNACEAE *Strophanthus hispidus* |
| tifi | **KRU - BASA** CHRYSOBALANACEAE *Neocarya macrophylla* |
| tiga sina | **MANDING - BAMBARA** CUCURBITACEAE *Ctenolepis cerasiformis* |
| tigbé | **MANDING - MANINKA** COMMELINACEAE *Aneilema setiferum, Commelina spp. indet.* |

tigè DOGON BOMBACACEAE *Adansonia digitata*
tigerin-wuri SUSU CYPERACEAE *Cyperaceae, Kyllinga erecta*
tigerinyi SUSU CYPERACEAE *Afrotrilepis pilosa, Cyperaceae, Cyperus sphacelatus, Kyllinga erecta, K. pumila, Mariscus alternifolius*
tigoe GBE - VHE CUCURBITACEAE *Lagenaria siceraria*
tigög DIOLA APOCYNACEAE *Alstonia boonei*
tigugal BANYUN COSTACEAE *Costus afer*
tigugalé kidika BANYUN COMMELINACEAE *Palisota hirsuta*
tigwe GBE - VHE CUCURBITACEAE *Lagenaria siceraria*
tigwe, tsigui GBE - VHE CUCURBITACEAE *Lagenaria siceraria*
tiin diri IJO - IZON AMARANTHACEAE *Alternanthera sessilis*
tikaru 'KRU' DIOSCOREACEAE *Dioscorea minutiflora*
tikfit ARABIC COMBRETACEAE *Combretum glutinosum*
tikiritisu 'NEKEDIE' CECROPIACEAE *Myrianthus arboreus*
tikonko BULOM CHRYSOBALANACEAE *Acioa whytei*
tikriti KRU - GUERE CECROPIACEAE *Myrianthus arboreus*
tikwélégué 'SENUFO' COCHLOSPERMACEAE *Cochlospermum tinctorium*
*a*-til TEMNE ANNONACEAE *Neostenanthera hamata*
*a*-tila TEMNE ACANTHACEAE *Asystasia calycina*
tileelee 'BAMILEKE' CHRYSOBALANACEAE *Acioa johnstonei*
tilihizoto KRU - GUERE CONNARACEAE *Cnestis ferruginea*
timba MANDING - MANDINKA COSTACEAE *Costus afer*
timbá PEPEL ANACARDIACEAE *Sorindeia juglandifolia*
timbanyi SUSU COSTACEAE *Costus afer*
timbindingo DIOLA AMARANTHACEAE *Philoxerus vermicularis*
timbiyâba FULA - PULAAR COSTACEAE *Costus afer*
timbo MANDING - MANINKA CRUCIFERAE *Brassica juncea*
timbom MANDYAK DIOSCOREACEAE *Dioscorea bulbifera*
timin-timin WOLOF BORAGINACEAE *Heliotropium strigosum*
timintiminjoŋ (*def.*-o) MANDING - MANDINKA ACANTHACEAE *Dicliptera verticillata*
tinedwar ARABIC ASCLEPIADACEAE *Caralluma russelliana*
tingbatau TEM ANACARDIACEAE *Lannea barteri*
tíŋgílá (*pl.* -á) SONGHAI COMBRETACEAE *Combretum micranthum*
tingma TEM COMPOSITAE *Vernonia amygdalina, V. colorata*
tingo BAULE CONVOLVULACEAE *Ipomoea batatas*
tingoli FULA - PULAAR ANACARDIACEAE *Lannea acida, L. velutina*
tiṅoli FULA - PULAAR ANACARDIACEAE *Lannea acida*
tiṅolipoley FULA - PULAAR ANACARDIACEAE *Lannea velutina*
*an*-tin-*a*-ro-lal TEMNE CUCURBITACEAE *Coccinea barteri*
tintiriŋhu-na SUSU - DYALONKE CYPERACEAE *Afrotrilepis pilosa*
tinyi MENDE CYPERACEAE *Afrotrilepis pilosa*
tio WOLOF APOCYNACEAE *Strophanthus hispidus, S. sarmentosus*
tiô KRU - GUERE BOMBACACEAE *Ceiba pentandra*
tiɔ KISSI BIGNONIACEAE *Newbouldia laevis*
tiɔ̃ KISSI BIGNONIACEAE *Newbouldia laevis*
tioh, tiojutiopti, tiorh WOLOF APOCYNACEAE *Saba senegalensis*
tiohamtiule WOLOF CYPERACEAE *Cyperus rotundus, C. sphacelatus*
tiohamtuile WOLOF CYPERACEAE *Cyperaceae*
tiokom tiokom WOLOF CYPERACEAE *Mariscus alternifolius*
tiokom-tiokom WOLOF CYPERACEAE *Cyperaceae*
tion MANDING - MANINKA COMPOSITAE *Vernonia guineensis*
tiou 'ONELE' BOMBACACEAE *Ceiba pentandra*
tipanpansanfati GURMA COMBRETACEAE *Terminalia laxiflora*
tirbom FULA - PULAAR COCHLOSPERMACEAE *Cochlospermum tinctorium*
tirde FULA - PULAAR CONVOLVULACEAE *Ipomoea vagans, Merremia aegyptia*
tirdé FULA - PULAAR CONVOLVULACEAE *Jacquemontia tamnifolia*
tiréï FULA - FULFULDE CAPPARACEAE *Boscia salicifolia* PULAAR *B. salicifolia*
tirewi FULA - PULAAR CAPPARACEAE *Boscia angustifolia*
tirey FULA - PULAAR CAPPARACEAE *Boscia angustifolia*
tiri MANDING - MANDINKA CYPERACEAE *Cyperus spp. indet.*
tiri buru KRU - BETE DIOSCOREACEAE *Dioscorea minutiflora*
tiri rikpo KRU - BETE DIOSCOREACEAE *Dioscorea minutiflora*
tiri rikwé KRU - BETE DIOSCOREACEAE *Dioscorea minutiflora*
tiriba MANDING - MANINKA COCHLOSPERMACEAE *Cochlospermum tinctorium*
tiribara MANDING - MANINKA COCHLOSPERMACEAE *Cochlospermum tinctorium*
tiribe (*def.* tiriboo) MANDING - MANDINKA COCHLOSPERMACEAE *Cochlospermum tinctorium*
tiribi KRU - GUERE DIOSCOREACEAE *Dioscorea minutiflora*
tirigba MANDING - MANINKA COCHLOSPERMACEAE *Cochlospermum tinctorium*
tirika tirika KWENI COMBRETACEAE *Combretum spp. indet.*
tirili KWENI DILLENIACEAE *Tetracera alnifolia*
tíro nukpà NUPE BROMELIACEAE *Ananas comosus*
tirza TAMACHEK ASCLEPIADACEAE *Calotropis procera*
tisi-lɛ BULOM BIGNONIACEAE *Newbouldia laevis*
titarik ARABIC ASCLEPIADACEAE *Leptadenia pyrotechnica*
titè PEPEL DIOSCOREACEAE *Dioscorea praehensilis*
titi KRU - BETE ANACARDIACEAE *Spondias mombin*
titi (*def.*-i) MENDE BOMBACACEAE *Bombax buonopozense*
titorekt ARABIC ASCLEPIADACEAE *Leptadenia pyrotechnica*
tiuko FULA - PULAAR ANACARDIACEAE *Lannea acida, L. barteri, L. egregia*
tiuko niabé FULA - PULAAR ANACARDIACEAE *Lannea velutina*
tiundaho NYENYEGE ARACEAE *Amorphophallus aphyllus*
tjikare TEM CUCURBITACEAE *Lagenaria siceraria*

to KWENI APOCYNACEAE *Rauvolfia vomitoria* 'NEKEDIE' *R. vomitoria*
tɔ GA CUCURBITACEAE *Lagenaria siceraria* GBE - VHE DIOSCOREACEAE *Dioscorea bulbifera*
tŏ a lah MANO CAPPARACEAE *Euadenia eminens*
tŏ kŏlĕh MANO APOCYNACEAE *Tabernaemontana longiflora*
toa AKAN - TWI CUCURBITACEAE *Lagenaria siceraria*
tõaa AKAN - ASANTE ANACARDIACEAE *Spondias mombin*
toagumbe LOKO COMMELINACEAE *Palisota hirsuta*
toawombi LOKO COMMELINACEAE *Palisota bracteosa, P. hirsuta*
tobéré KIRMA ANNONACEAE *Annona arenaria, A. senegalensis*
tobero GAGU ANACARDIACEAE *Lannea welwitschii*
tɔbɔ ADANGME - KROBO AMARANTHACEAE *Celosia argentea*
tobo-ué KRU - GUERE CECROPIACEAE *Myrianthus arboreus*
toč WOLOF COMBRETACEAE *Combretum nigricans*
tode MOORE - NAWDAM BOMBACACEAE *Bombax buonopozense, B. costatum*
todi farssa SONGHAI CAPPARACEAE *Cadaba glandulosa*
todia AKAN - ASANTE ARACEAE *Pistia stratiotes*
todiugolo FULA - TUKULOR CYPERACEAE *Kyllinga erecta* WOLOF *K. erecta*
todzo VULGAR COMPOSITAE *Vernonia colorata*
toɛ LOKO BIGNONIACEAE *Newbouldia laevis*
toéga MOORE BOMBACACEAE *Adansonia digitata*
*a*-toel TEMNE ARACEAE *Cercestis afzelii*
toetoeya KYAMA COSTACEAE *Costus Linn.*
tofa KONO COSTACEAE *Costus afer*
tɔfa VAI COSTACEAE *Costus afer*
tofo-mɛsɛ KONO CONVOLVULACEAE *Ipomoea involucrata*
togba LOKO BOMBACACEAE *Bombax buonopozense*
togbô VULGAR CRASSULACEAE *Kalanchoe integra: Kalanchoe crenata*
togbòkihà YORUBA CAPPARACEAE *Maerua duchesnei*
togô MANDING - BAMBARA CYPERACEAE *Cyperus esculentus*
togo fogo MANDING - MANINKA ASCLEPIADACEAE *Calotropis procera*
togodo MANDING - BAMBARA BOMBACACEAE *Bombax costatum*
tógodo DOGON BOMBACACEAE *Bombax costatum*
togon MANDING - BAMBARA CYPERACEAE *Cyperus esculentus, Kyllinga erecta* MANINKA *Cyperaceae, Cyperus distans*
tɔgɔŋ MANDING - MANINKA CYPERACEAE *Mariscus alternifolius*
togon-kayogo MANDING - BAMBARA CYPERACEAE *Cyperus esculentus*
toh CHAMBA COCHLOSPERMACEAE *Cochlospermum planchonii, C. tinctorium*
tohl WOLOF APOCYNACEAE *Saba senegalensis*
tɔhɔ MANDING - MANINKA CYPERACEAE *Kyllinga bulbosa*
toǰ WOLOF CAPPARACEAE *Maerua angolensis*
tokal pul WOLOF CYPERACEAE *Fimbristylis hispidula*
toke FULA - PULAAR APOCYNACEAE *Strophanthus hispidus*
toké FULA - FULFULDE APOCYNACEAE *Strophanthus hispidus* PULAAR *S. hispidus, S. sarmentosus*
tɔkɛ (? tɔlɛ) KONO BIGNONIACEAE *Newbouldia laevis*
tokere FULA - PULAAR APOCYNACEAE *Strophanthus sarmentosus*
tokéré FULA - FULFULDE APOCYNACEAE *Strophanthus hispidus* PULAAR *S. hispidus*
toki MANDING - MANINKA CYPERACEAE *Cyperus esculentus*
tokobuguri MOORE CELASTRACEAE *Maytenus senegalensis*
tokonzui AKAN - ASANTE BIGNONIACEAE *Newbouldia laevis*
tokovuguri MOORE CELASTRACEAE *Maytenus senegalensis*
tokoyogwri MOORE CELASTRACEAE *Maytenus senegalensis*
tokpohun LIMBA CONVOLVULACEAE *Ipomoea batatas*
tokpohunka MANDING - MANDINKA CONVOLVULACEAE *Ipomoea batatas*
tòktòkò (*pl.* -à) SONGHAI COMMELINACEAE *Commelina benghalensis*
tokwakufuo AKAN - ASANTE BIGNONIACEAE *Stereospermum acuminatissimum*
tol WOLOF APOCYNACEAE *Landolphia heudelotii*
tol ɗuam SERER CYPERACEAE *Cyperus maritimus*
tola GAGU ANACARDIACEAE *Spondias mombin*
tolé LOMA BIGNONIACEAE *Newbouldia laevis* MANDING - MANINKA CELASTRACEAE *Maytenus senegalensis, M. senegalensis*
tɔlɛ KONO BIGNONIACEAE *Newbouldia laevis*
tologbelo VULGAR COMMELINACEAE *Palisota hirsuta* KISSI *P. hirsuta*
tolo-tolo KRU - GUERE BIGNONIACEAE *Newbouldia laevis*
tolro YOM BOMBACACEAE *Adansonia digitata*
tôm KANURI BOMBACACEAE *Ceiba pentandra*
tɔma (*def.*-i) MENDE DIONCOPHYLLACEAE *Habropetalum dawei, Triphyophyllum peltatum*
toma ulé MANDING - BAMBARA CONVOLVULACEAE *Ipomoea batatas*
to-magny MANDING - BAMBARA CAPPARACEAE *Cadaba farinosa*
tomani MANDING - BAMBARA CAPPARACEAE *Cadaba farinosa*
tòmátò eléjò YORUBA CUCURBITACEAE *Trichosanthes cucumerina*
tomatom CRIOULO CUCURBITACEAE *Cucumeropsis mannii*
tomba (*def.*-i) MENDE COMBRETACEAE *Combretum mooreanum, C. smeathmannii*
tomba-mɛnyɛ (*def.*-i) MENDE COMBRETACEAE *Combretum comosum, C. grandiflorum, C. paniculatum ssp. paniculatum, C. platypterum, C. rhodanthum, C. smeathmannii*
tombe MENDE CELASTRACEAE *Hippocratea velutina*
tɔmbega LOKO CANNACEAE *Canna indica*
tombo ABE BIGNONIACEAE *Kigelia africana*

| | |
|---|---|
| tubano fauru | FULA - PULAAR ARACEAE *Amorphophallus flavovirens* |
| tubinyi | SUSU CYPERACEAE *Lipocarpa chinensis* |
| tubozofi | SENUFO - NIAGHAFOLO CHRYSOBALANACEAE *Maranthes polyandra, Parinari curatellifolia* |
| tubumbum | BASSARI BOMBACACEAE *Ceiba pentandra* |
| tubungbing | KONKOMBA BOMBACACEAE *Ceiba pentandra* |
| tuda | MANDING - MANINKA BIGNONIACEAE *Kigelia africana* SUSU *K. africana* |
| tudi | HAUSA ANACARDIACEAE *Lannea barteri* |
| tudu | MANDING - BAMBARA CHRYSOBALANACEAE *Maranthes polyandra, Parinari curatellifolia* MANINKA ACANTHACEAE *Phaulopsis falcisepala* CHRYSOBALANACEAE *Maranthes polyandra* |
| tudutj | SERER - NON COMPOSITAE *Vernonia colorata* |
| tuɛ | KONO BIGNONIACEAE *Newbouldia laevis* |
| tue sao goh | MANO CONVOLVULACEAE *Neuropeltis velutina* |
| tufu-tufoi | AKAN - BRONG ARALIACEAE *Cussonia arborea* |
| tugbɛ, tugbɛlɛ | MENDE CYPERACEAE *Cyperaceae* |
| tugbɛ | MENDE CYPERACEAE *Pycreus polystachyos* |
| tugbɛlɛ | MENDE CYPERACEAE *Cyperus compressus, C. haspan, C. rotundus, C. sphacelatus, Hypolytrum purpurascens, Kyllinga bulbosa, K. erecta, Mariscus alternifolius* |
| tughbɛlɛ | MENDE CYPERACEAE *Kyllinga pumila* |
| tugue | MANDING - MANINKA APOCYNACEAE *Landolphia owariensis* |
| tuhidja | KULANGO COMBRETACEAE *Terminalia ivorensis* |
| tuhila | TAMACHEK COMBRETACEAE *Guiera senegalensis* |
| tuhmeya | ARABIC AMARANTHACEAE *Aerva javanica* |
| tuila | TAMACHEK ANACARDIACEAE *Sclerocarya birrea* |
| tuila-hɛnge-chiu | KISSI CUCURBITACEAE *Luffa cylindrica* |
| tuiɔ | KISSI BIGNONIACEAE *Newbouldia laevis* |
| tuiɔ̃ | KISSI BIGNONIACEAE *Newbouldia laevis* |
| tukala sitâdi | MANDING - MANINKA APOCYNACEAE *Adenium obesum* |
| tukala sitandi | MANDING - MANINKA APOCYNACEAE *Adenium obesum* |
| tukamalé | MANDING - BAMBARA APOCYNACEAE *Ancylobothrys amoena* |
| tukare | DAGBANI BOMBACACEAE *Adansonia digitata* |
| ṭuko | FULA - PULAAR ANACARDIACEAE *Lannea acida, L. microcarpa* |
| ṭukoṅabé | FULA - PULAAR ANACARDIACEAE *Lannea velutina* |
| tukoneudu | FULA - PULAAR ANACARDIACEAE *Lannea microcarpa* |
| tulingi | AKAN - BRONG CHRYSOBALANACEAE *Parinari congensis* |
| tultulhi, tultulki (*pl.* tultulɗe) | FULA - FULFULDE CELASTRACEAE *Maytenus senegalensis* |
| tulumpa | SONINKE - SARAKOLE ASCLEPIADACEAE *Calotropis procera* |
| tuman bélé | MANDING - MANDINKA APOCYNACEAE *Strophanthus sarmentosus* |
| tumba | MANDING - MANDINKA COSTACEAE *Costus afer* |
| tumbalaka | YORUBA - IFE OF TOGO ANNONACEAE *Hexalobus monopetalus* |
| tumbe | DAGBANI CUCURBITACEAE *Lagenaria siceraria* MANDING - MANDINKA COMMELINACEAE *Palisota hirsuta* |
| tumèna | MANDING - BAMBARA ASCLEPIADACEAE *Sarcostemma viminale* |
| tùmfaáfìyaá | HAUSA ASCLEPIADACEAE *Calotropis procera* |
| tùmfaáfìyár kádaá | HAUSA COSTACEAE *Costus afer* |
| tumfafia | HAUSA ASCLEPIADACEAE *Calotropis procera* |
| tumkiyar rafi | HAUSA BORAGINACEAE *Heliotropium ovalifolium* |
| tummbude (*pl.* tummbuɗe), tummude (*pl.* tummuɗe) | FULA - FULFULDE CUCURBITACEAE *Lagenaria siceraria* |
| tumo tigi | MANDING - DYULA ASCLEPIADACEAE *Calotropis procera* |
| tumpaapahi | FULA - FULFULDE ASCLEPIADACEAE *Calotropis procera* |
| tumu tigui | MANDING - BAMBARA BORAGINACEAE *Coldenia procumbens* |
| tunda | MANDING - MANINKA BIGNONIACEAE *Spathodea campanulata* |
| tundal | WOLOF COMBRETACEAE *Combretum paniculatum ssp. paniculatum* |
| tundub | ARABIC CAPPARACEAE *Capparis decidua* |
| tundui | KISSI COMMELINACEAE *Palisota hirsuta* |
| tunduwol | FULA - FULFULDE CUCURBITACEAE *Lagenaria siceraria* |
| tungbo | DAGAARI BORAGINACEAE *Cordia myxa* |
| tungué | MANDING - BAMBARA BORAGINACEAE *Cordia myxa* |
| tuntun | AKAN - ASANTE ARALIACEAE *Cussonia bancoensis* |
| tunturei | AKAN - BRONG BIGNONIACEAE *Stereospermum kunthianum* |
| tunzué | BAULE BIGNONIACEAE *Newbouldia laevis* |
| tuo | DAGAARI BOMBACACEAE *Adansonia digitata* |
| tupain | ANYI ARACEAE *Anchomanes difformis* |
| tupo | KIRMA COMBRETACEAE *Guiera senegalensis* |
| turaar turauniya | HAUSA COMBRETACEAE *Combretum glutinosum* |
| turare | HAUSA CYPERACEAE *Cyperus maculatus, Kyllinga erecta* |
| turdja | ARABIC ASCLEPIADACEAE *Calotropis procera* SONGHAI *C. procera* |
| turia | SONGHAI ASCLEPIADACEAE *Calotropis procera* |
| turja | ARABIC ASCLEPIADACEAE *Calotropis procera* |
| turjé | ARABIC ASCLEPIADACEAE *Calotropis procera* |
| turkai | HAUSA CONVOLVULACEAE *Ipomoea batatas* |
| túrkèn-doókìi | HAUSA BIGNONIACEAE *Stereospermum kunthianum* |
| turmbel | WOLOF CONVOLVULACEAE *Cressa cretica* |
| turri | HAUSA COCHLOSPERMACEAE *Cochlospermum tinctorium* |
| tursha | TAMACHEK ASCLEPIADACEAE *Calotropis procera* |
| tursuhi, tursahi (*pl.* tursuuje) | FULA - FULFULDE ANACARDIACEAE *Haematostaphis barteri* |
| turu | MANDING - BAMBARA CHRYSOBALANACEAE *Parinari curatellifolia* |
| turubu | SOMBA BOMBACACEAE *Adansonia digitata* |
| turugba | MANDING - MANINKA COCHLOSPERMACEAE *Cochlospermum tinctorium* |

ťurum da hum BANYUN COMBRETACEAE *Combretum nigricans*
turuŋyi SUSU CYPERACEAE *Cyperus articulatus*
tusima TEM COMPOSITAE *Vernonia amygdalina, V. colorata*
tutiǵi MANDING - MANINKA CAPPARACEAE *Boscia angustifolia*
tutigui MANDING - MANINKA CAPPARACEAE *Boscia angustifolia*
tútok sunàŋ, hwaàl jey BEROM COCHLOSPERMACEAE *Cochlospermum tinctorium*
tutu MANDING - BAMBARA CHRYSOBALANACEAE *Maranthes polyandra, Parinari curatellifolia, P. curatellifolia* DYULA *Maranthes polyandra, M. polyandra, Parinari curatellifolia, P. curatellifolia* MANINKA *P. curatellifolia* SOCE *P. curatellifolia*
tutu bidi HAUSA COMPOSITAE *Ambrosia maritima*
tutu kuma MANDING - BAMBARA CHRYSOBALANACEAE *Parinari curatellifolia* MANINKA *P. curatellifolia*
tutuabu AKAN - WASA COMBRETACEAE *Strephonema pseudocola*
tutubi KYAMA AMARANTHACEAE *Cyathula prostrata*
tutué 'KRU' CECROPIACEAE *Musanga cecropioides*
tùtúgéy BEROM ARACEAE *Amorphophallus dracontioides*
tutummerika kohwɛ ɛpɔ AKAN - ASANTE COMPOSITAE *Synedrella nodiflora*
tuùlún-moowaà HAUSA DIOSCOREACEAE *Dioscorea misc. & spp. unknown*
tuwa NABT BOMBACACEAE *Adansonia digitata*
túwón bíriì HAUSA DIOSCOREACEAE *Dioscorea bulbifera*
twẽhama AKAN - TWI DILLENIACEAE *Tetracera affinis*
tweneboa VULGAR BORAGINACEAE *Cordia senegalensis* AKAN - ASANTE *C. millenii, C. platythyrsa, C. senegalensis* FANTE *C. platythyrsa* TWI *C. millenii, C. platythyrsa* WASA *C. millenii*
tweneboakodua AKAN - ASANTE BORAGINACEAE *Cordia millenii* TWI *C. millenii*
twenedoleye NZEMA BORAGINACEAE *Cordia platythyrsa*
tweneduru AKAN - ASANTE BORAGINACEAE *Cordia millenii* WASA *C. millenii*
twi KRU - NEYO APOCYNACEAE *Funtumia elastica*
twi woplu 'KRU' CONNARACEAE *Agelaea mildbraedii*
twihama AKAN - TWI DILLENIACEAE *Tetracera potatoria*
twihõma AKAN - TWI DILLENIACEAE *Tetracera potatoria*
tyembegh TIV BIGNONIACEAE *Kigelia africana*
tyɛ̀nɛ KWENI BOMBACACEAE *Ceiba pentandra*
tyetyebu GURMA ANACARDIACEAE *Lannea acida*
tzapé AKYE APOCYNACEAE *Strophanthus sarmentosus*
u SERER - NON APOCYNACEAE *Landolphia heudelotii*
û SERER - NON APOCYNACEAE *Landolphia heudelotii*
u lèv iné DIOLA ANNONACEAE *Uvaria thomasii*
uadagoré FULA - PULAAR CAPPARACEAE *Cadaba glandulosa*
uaga guitel FULA - PULAAR ANACARDIACEAE *Rhus longipes, R. natalensis*
uagara MANDING - BAMBARA DIOSCOREACEAE *Dioscorea alata*
uahia uahia MANDING - BAMBARA COMBRETACEAE *Combretum collinum ssp. hypopilinum* MANINKA *C. collinum ssp. hypopilinum*
uai KISSI ARACEAE *Colocasia esculenta*
uallé AKAN - ASANTE CHRYSOBALANACEAE *Parinari excelsa*
uamku KRU - GUERE CELASTRACEAE *Hippocratea spp. sens. lat.*
uamtanga MOORE CHRYSOBALANACEAE *Parinari curatellifolia*
uan DOGON COMBRETACEAE *Anogeissus leiocarpus*
uan ihedan TAMACHEK CHENOPODIACEAE *Haloxylon articulatum*
uaniaka MANDING - BAMBARA COMBRETACEAE *Combretum molle* MANINKA *C. molle*
uanika MANDING - MANINKA COMBRETACEAE *Combretum molle*
uanyise ANYI - ANUFO COCHLOSPERMACEAE *Cochlospermum tinctorium*
uar ames TAMACHEK CRUCIFERAE *Farsetia aegyptiaca*
uaraha KRU - GUERE CHRYSOBALANACEAE *Acioa hirsuta*
uataï ANYI CAPPARACEAE *Cleome gynandra*
uatata PEPEL CONVOLVULACEAE *Ipomoea batatas*
uáto BIDYOGO BOMBACACEAE *Adansonia digitata*
uaye SONINKE - SARAKOLE COMBRETACEAE *Anogeissus leiocarpus*
uayé KPELLE ARACEAE *Xanthosoma mafaffa*
ụ̀bạ̀bạ́ (*pl.* ị̀) DEGEMA CUCURBITACEAE *Lagenaria siceraria*
ubaghi HAUSA APOCYNACEAE *Landolphia owariensis* IGBO *L. owariensis*
ube IGBO APOCYNACEAE *Landolphia owariensis*
ù-bé (*pl.* àrù-bé) ABUA BURSERACEAE *Dacryodes edulis*
ùbé IGBO BURSERACEAE *Canarium schweinfurthii, Dacryodes edulis, D. edulis*
ùbé agba, ùbé mkpuru aki IGBO BURSERACEAE *Canarium schweinfurthii*
ùbé ọ́hị̄á IGBO BURSERACEAE *Canarium schweinfurthii*
ùbé ōkpókó IGBO BURSERACEAE *Canarium schweinfurthii*
ùbé wemba IGBO BURSERACEAE *Canarium schweinfurthii*
u-bene (*pl.* i-bene) ATTE CUCURBITACEAE *Lagenaria siceraria*
ùbé-óhị̄á IGBO BURSERACEAE *Canarium schweinfurthii*
ùbé-ōkpókó IGBO BURSERACEAE *Canarium schweinfurthii*
ùbé-ọ̀sà IGBO BURSERACEAE *Canarium schweinfurthii*
ùbé-óyìbó IGBO BURSERACEAE *Dacryodes edulis*
ubim IGBO COMBRETACEAE *Combretum dolichopetalum*
ubimba PEPEL APOCYNACEAE *Isonema smeathmannii, Landolphia dulcis*
ùbó EDO APOCYNACEAE *Saba florida*
ùbò EDO APOCYNACEAE *Landolphia hirsuta*
ùbó-amiọ́ghọ EDO APOCYNACEAE *Landolphia owariensis*
ùbó-ámiọ́ghọ̀n EDO APOCYNACEAE *Ancylobothrys scandens*
úbòghò EDO BALANITACEAE *Balanites wilsoniana*

úhōvbè **URHOBO** CECROPIACEAE *Musanga cecropioides*
uịnda **KWENI** ANACARDIACEAE *Spondias mombin*
ùjẹ̀rú **URHOBO** DIOSCOREACEAE *Dioscorea rotundata*
uji **IGBO** ARACEAE *Anchomanes welwitschii*
uji-oko **IGALA** COMBRETACEAE *Terminalia ivorensis, T. superba*
uju **NKEM** BURSERACEAE *Dacryodes edulis*
ujuju **IGBO** COMPOSITAE *Vernonia conferta*
ụjụ̀jụ̀ **IGBO** CECROPIACEAE *Musanga cecropioides*
ụ́jụ̀jụ̀ **IGBO** CECROPIACEAE *Myrianthus arboreus*
uk **SERER - NON** APOCYNACEAE *Landolphia heudelotii*
ûk **SERER - NON** APOCYNACEAE *Landolphia heudelotii*
úkáñ **EFIK** CHRYSOBALANACEAE *Acioa lehmbachii*
ukem akabi **ABUA** BOMBACACEAE *Ceiba pentandra*
úkhú **EDO** APOCYNACEAE *Alstonia boonei*
úkhúeréohā **EDO** COSTACEAE *Costus afer, C. lucanusianus*
úkím **EFIK** BOMBACACEAE *Bombax buonopozense, Ceiba pentandra* **IBIBIO** *Bombax buonopozense, Ceiba pentandra*
uko **OKPAMHERI** CUCURBITACEAE *Lagenaria siceraria*
ùkó **IGBO** CUCURBITACEAE *Lagenaria siceraria*
ukoken u mbir **DIOLA** CONVOLVULACEAE *Calycobolus heudelotii*
ukoko **IBIE** CUCURBITACEAE *Lagenaria siceraria*
ukopo **KRU - GUERE** COMPOSITAE *Vernonia colorata*
ukọrọ **IGBO** CUCURBITACEAE *Cucurbita maxima, C. pepo*
ùkóvúvū (*pl.* à-) **DEGEMA** ARACEAE *Colocasia esculenta*
ukpali **DAGBANI** CONVOLVULACEAE *Ipomoea argentaurata*
u-kpán (*sing.*) **KPAN** CUCURBITACEAE *Lagenaria siceraria*
úkpò **EFIK** APOCYNACEAE *Alstonia boonei*
úkpò-ìbìeká **EDO** CONNARACEAE *Cnestis ferruginea*
ukpokoro **URHOBO** DIOSCOREACEAE *Dioscorea rotundata*
ukpokpo **IGALA** ANNONACEAE *Annona senegalensis*
úkpósā **EDO** ANNONACEAE *Monodora brevipes*
ukpukuhu **URHOBO** APOCYNACEAE *Alstonia boonei*
uk(h)u **YEKHEE** APOCYNACEAE *Alstonia boonei*
úkúm **ANAANG** BOMBACACEAE *Ceiba pentandra*
ukurukuru **IGBO** ARACEAE *Xanthosoma mafaffa*
u-kwlobu (*pl.* vi-kwlobu) **EVANT** CUCURBITACEAE *Lagenaria siceraria*
ula **IGBO** BIXACEAE *Bixa orellana* **KRU - GUERE** AMARANTHACEAE *Aerva lanata*
ula machuku **IGBO** BIXACEAE *Bixa orellana*
ulẽ **GBE - VHE** ARACEAE *Colocasia esculenta*
úle **MANKANYA** AVICENNIACEAE *Avicennia germinans*
ulele **BAULE** BALANITACEAE *Balanites wilsoniana*
ulélé **BAULE** BALANITACEAE *Balanites wilsoniana*
ùlì ókō **IGBO** ACANTHACEAE *Asystasia gangetica*
ulinga **DAGBANI** COMBRETACEAE *Combretum collinum ssp. geitonophyllum*
ulo **DAN** CECROPIACEAE *Musanga cecropioides* **MANO** *M. cecropioides*
ulô **PEPEL** COMBRETACEAE *Combretum tomentosum*
ulôfò **PEPEL** BOMBACACEAE *Bombax costatum*
ulonido **MANDING - BAMBARA** CONVOLVULACEAE *Ipomoea setifera*
ulou nin tulu **MANDING - BAMBARA** CONVOLVULACEAE *Merremia umbellata*
ụ́lụ̀ **IGBO** CECROPIACEAE *Musanga cecropioides*
ulu n'dioloko miyé **MANDING - BAMBARA** ASCLEPIADACEAE *Ceropegia spp.*
ulu ndïolond missé **MANDING - BAMBARA** ASCLEPIADACEAE *Ceropegia spp.*
ulua **KRU - GUERE** APOCYNACEAE *Funtumia elastica*
uluatu **KRU - GUERE** APOCYNACEAE *Funtumia elastica*
uluku mbiré **MANDING - BAMBARA** ASCLEPIADACEAE *Ceropegia spp.*
ulungkwana **DAGAARI** ANNONACEAE *Annona senegalensis*
um el barka **ARABIC** BURSERACEAE *Commiphora africana*
um shutur **ARABIC** BIGNONIACEAE *Kigelia africana*
uma pohué **AKAN - ASANTE** APOCYNACEAE *Landolphia hirsuta*
umana tumba **TIV** BIGNONIACEAE *Stereospermum kunthianum*
umangohi **IDOMA** ANACARDIACEAE *Mangifera indica*
umbaba **FULA - PULAAR** COMPOSITAE *Acanthospermum hispidum*
umbah **KRU - BASA** CHRYSOBALANACEAE *Parinari excelsa*
umbatú **BALANTA** CHRYSOBALANACEAE *Neocarya macrophylla*
umbóce **BALANTA - MANE** DIOSCOREACEAE *Dioscorea alata*
umbôgre **MANKANYA** CUCURBITACEAE *Cucurbita pepo*
ùmbú (kóróm̀ọ̀) dírí **IJO - IZON** BORAGINACEAE *Heliotropium indicum*
ùmbú (kòròm̀ọ̀) dìrì **IJO - IZON** CRASSULACEAE *Bryophyllum pinnatum*
umẹfu, umiakbo **IGBO** DIOSCOREACEAE *Dioscorea Linn.*
umegarin **WOLOF** ACANTHACEAE *Asystasia gangetica*
umfobille **KONKOMBA** BOMBACACEAE *Ceiba pentandra*
um-hálè **FULA - PULAAR** BIGNONIACEAE *Markhamia tomentosa*
umm boro **ARABIC - SHUWA** ANNONACEAE *Annona senegalensis*
umm burku **ARABIC - SHUWA** ARACEAE *Stylochiton lancifolius*
umm ganagan **ARABIC - SHUWA** CYPERACEAE *Cyperus papyrus*
umm takalak **ARABIC - SHUWA** CARICACEAE *Carica papaya*
um-nhárè-sáquè **FULA - PULAAR** COMPOSITAE *Acanthospermum hispidum*
umpaba **BALANTA** DIOSCOREACEAE *Dioscorea preussii*
umpandá **BIDYOGO** CARICACEAE *Carica papaya*
umpapa **PEPEL** ARACEAE *Colocasia esculenta*
umpela **MANKANYA** ANACARDIACEAE *Spondias mombin*

vorač — WOLOF CHRYSOBALANACEAE *Chrysobalanus icaco*
vorach — WOLOF CHRYSOBALANACEAE *Chrysobalanus icaco*
vorga — DAGAARI BOMBACACEAE *Bombax buonopozense*
vorone — KWENI BIGNONIACEAE *Markhamia tomentosa*
vosvosor — WOLOF ANACARDIACEAE *Ozoroa insignis* BIGNONIACEAE *Newbouldia laevis*
voti — GBE - VHE ANNONACEAE *Annona muricata*
votsi — GBE - VHE ANNONACEAE *Annona muricata*
votʃo — ADANGME BIGNONIACEAE *Spathodea campanulata*
vovo — GAGU BIGNONIACEAE *Spathodea campanulata*
vovolé — KWENI CUCURBITACEAE *Momordica foetida*
vovoné vono — KWENI CUCURBITACEAE *Momordica foetida*
vu — GBE - VHE BOMBACACEAE *Ceiba pentandra*
vudrai — GBE - VHE CONVOLVULACEAE *Merremia tridentata*
vué-buri iri — KWENI BIGNONIACEAE *Stereospermum acuminatissimum*
vuiga — MOORE BIGNONIACEAE *Stereospermum kunthianum*
vulẽ — GBE - VHE BOMBACACEAE *Ceiba pentandra*
vuluné — KWENI BIGNONIACEAE *Markhamia tomentosa*
ʋumalia — GBE - VHE CONNARACEAE *Byrsocarpus coccineus*
vunadabla — KWENI CAPPARACEAE *Euadenia eminens*
vuruni — KWENI BIGNONIACEAE *Markhamia tomentosa*
vuti — GBE - FON BOMBACACEAE *Ceiba pentandra* YORUBA - IFE OF TOGO *C. pentandra*
vuvudranyi — GBE - VHE CONVOLVULACEAE *Cuscuta australis*
waa iri — MANDING - DYULA COMBRETACEAE *Terminalia avicennioides*
waalwaalnde (*pl.* baalbaalɗe) — FULA - FULFULDE COMMELINACEAE *Commelina diffusa*
waalwaalnde (*pl.* baalbalɗe) — FULA - FULFULDE COMMELINACEAE *Commelina erecta*
waalwaalnde (*pl.* baalbaalɗe) — FULA - FULFULDE COMMELINACEAE *Commelina forskalaei*
waba — AKAN - ASANTE ANNONACEAE *Xylopia quintasii*
wababoo — FULA - FULFULDE CONVOLVULACEAE *Ipomoea asarifolia*
waccuure (*pl.* waccuuje) — FULA - FULFULDE CYPERACEAE *Cyperus esculentus*
wada — HAUSA APOCYNACEAE *Rauvolfia caffra, R. vomitoria*
wadiɛ-aba — AKAN - ASANTE ANNONACEAE *Monodora myristica*
waga waga — AKAN - BRONG COMPOSITAE *Melanthera scandens* KULANGO *M. scandens*
wagale — MANDING - MANINKA CECROPIACEAE *Myrianthus arboreus*
wagniara — 'SENUFO' COMBRETACEAE *Combretum molle*
wagua — DAGAARI CAPPARACEAE *Capparis tomentosa*
wai — KISSI ARACEAE *Colocasia esculenta*
waitie — KRIO APOCYNACEAE *Catharanthus roseus*
waiya — MANDING - MANDINKA COMBRETACEAE *Combretum grandiflorum*
wajo — SERER - NON CARYOPHYLLACEAE *Polycarpaea corymbosa* WOLOF *P. linearifolia*
waka — AKAN - ASANTE BALANITACEAE *Balanites wilsoniana* KRU - GUERE ACANTHACEAE *Phaulopsis imbricata* COMPOSITAE *Synedrella nodiflora* MOORE BOMBACACEAE *Bombax buonopozense, B. costatum*
*a*-waka — TEMNE CECROPIACEAE *Myrianthus arboreus, M. libericus, M. serratus*
wakawaka — MANDING - MANDINKA CECROPIACEAE *Myrianthus arboreus, M. libericus, M. serratus*
wala — KULANGO APOCYNACEAE *Funtumia africana*
walakur — WOLOF BIGNONIACEAE *Newbouldia laevis*
walia bocio — SONGHAI AMARANTHACEAE *Amaranthus graecizans, A. viridis* TAMACHEK *A. graecizans*
walisa — MANDING - MANDINKA ANNONACEAE *Annona senegalensis* COMBRETACEAE *Terminalia glaucescens*
walkin dalla — HAUSA BALANITACEAE *Balanites aegyptiaca*
walkin machiji — HAUSA CONVOLVULACEAE *Ipomoea vagans*
walkin tsofo — HAUSA BORAGINACEAE *Trichodesma africanum*
wàlkin waáwaá — HAUSA BORAGINACEAE *Trichodesma africanum*
wàlkin-tshoóhoó — HAUSA BORAGINACEAE *Trichodesma africanum*
wal-wal dé — FULA - PULAAR COMMELINACEAE *Commelina erecta*
wal-waldé — FULA - PULAAR COMMELINACEAE *Commelina erecta*
wal-wal-dé — FULA - PULAAR COMMELINACEAE *Commelina forskalaei*
wal-waldé — FULA - PULAAR COMMELINACEAE *Commelina forskalaei*
wamde — MOORE CUCURBITACEAE *Lagenaria siceraria*
wanaka — MANDING - BAMBARA COMBRETACEAE *Combretum molle* MANINKA *C. molle*
waŋgolo — KISSI APOCYNACEAE *Funtumia africana, F. elastica*
waŋgolo-waŋgolo — KISSI APOCYNACEAE *Holarrhena floribunda*
wangu — AKAN - ASANTE COMBRETACEAE *Terminalia glaucescens*
wansien blakassa — AKAN - ASANTE CONNARACEAE *Manotes longiflora*
wara — GURE - KAHUGU CUCURBITACEAE *Lagenaria siceraria*
waralankɔ — MANDING - MANINKA BIGNONIACEAE *Markhamia tomentosa*
warina — HAUSA CONVOLVULACEAE *Ipomoea batatas*
warkin tsoho — HAUSA BORAGINACEAE *Trichodesma africanum*
warko — WOLOF AMARANTHACEAE *Celosia trigyna*
wasani — HAUSA DIPTEROCARPACEAE *Monotes kerstingii*
wasɛ — MANDING - MANINKA COMBRETACEAE *Terminalia albida*
wasswassur — FULA - TUKULOR ANACARDIACEAE *Ozoroa insignis*
waswasor, waswasür — WOLOF ANACARDIACEAE *Ozoroa insignis*
wata-bitas — KRIO COMPOSITAE *Struchium sparganophora*
wata-krɛs — KRIO CRUCIFERAE *Rorippa nasturtium-aquaticum*
watanta jambere — FULA - FULFULDE BURSERACEAE *Commiphora africana*

yako FULA - FULFULDE CONVOLVULACEAE *Ipomoea aitonii* HAUSA *I. aitonii, I. cairica, I. nil, Merremia aegyptia*
yakutsa GBE - VHE CUCURBITACEAE *Luffa cylindrica*
yálágèrè BEDIK ANACARDIACEAE *Anacardium occidentale*
yale-na SUSU - DYALONKE BIGNONIACEAE *Stereospermum acuminatissimum*
yamanya HAUSA CUCURBITACEAE *Cucumis prophetarum*
yamawa MENDE CONNARACEAE *Manotes expansa*
yamba LOKO CANNABACEAE *Cannabis sativa*
yâmba WOLOF CANNABACEAE *Cannabis sativa*
yamba-duknɛ KONO COMPOSITAE *Microglossa pyrifolia*
yambale-kɛɛnda KISSI CONVOLVULACEAE *Ipomoea mauritiana*
yamba-na SUSU - DYALONKE CANNABACEAE *Cannabis sativa*
yambɛ KONO CANNABACEAE *Cannabis sativa*
yambe-kɛɛnda KISSI CONVOLVULACEAE *Ipomoea mauritiana*
yambɛli-na SUSU - DYALONKE AMARANTHACEAE *Amaranthus hybridus*
yamburu HAUSA CONVOLVULACEAE *Merremia tridentata*
yamɓururu, yámɓururu HAUSA CONVOLVULACEAE *Merremia tridentata*
yamis BULOM DIOSCOREACEAE *Dioscorea Linn.* LIMBA *D. Linn.*
yamo MANDING - MANDINKA COMMELINACEAE *Commelina forskalaei*
yams KRIO DIOSCOREACEAE *Dioscorea Linn.*
*a*-yams TEMNE DIOSCOREACEAE *Dioscorea Linn.*
yams-ɛ BULOM DIOSCOREACEAE *Dioscorea Linn.*
*a*-yams-*a*-fɛra TEMNE DIOSCOREACEAE *Dioscorea rotundata*
y'an kyamba HAUSA COCHLOSPERMACEAE *Cochlospermum tinctorium*
yan tafké HAUSA CYPERACEAE *Mariscus squarrosus*
yandigbɛne KONO COMPOSITAE *Ageratum conyzoides*
yandigboinei KONO COMPOSITAE *Adenostemma perrottetii*
yanga LONG BROMELIACEAE *Ananas comosus*
yangara buhili FULA - FULFULDE CONNARACEAE *Santaloides afzelii*
yanguma KYAMA CECROPIACEAE *Myrianthus arboreus*
yani KONO COMPOSITAE *Ageratum conyzoides*
yani-gbɛ MENDE COMPOSITAE *Ageratum conyzoides*
yaniyo KISSI COMPOSITAE *Ageratum conyzoides*
yankar mala HAUSA COMPOSITAE *Aspilia kotschyi*
yankoma AKAN - ASANTE CECROPIACEAE *Myrianthus arboreus*
yankonfeh AKAN - FANTE COMPOSITAE *Senecio biafrae*
yánrin YORUBA COMPOSITAE *Launaea taraxacifolia*
yánrin-oko YORUBA COMPOSITAE *Lactuca capensis*
yanyan YORUBA ASCLEPIADACEAE *Pergularia daemia*
yapi ABURE COMBRETACEAE *Terminalia ivorensis*
yaprè KRU - GUERE ARACEAE *Anchomanes difformis*
y'ár ùngúwaá HAUSA CAPPARACEAE *Cleome Linn., C. afrospinosa, C. gynandra*
yaraba LOKO CUCURBITACEAE *Luffa cylindrica*
yaraɗi HAUSA CONVOLVULACEAE *Jacquemontia tamnifolia*
yaraɗin tudu HAUSA CONVOLVULACEAE *Jacquemontia tamnifolia*
yaré safele FULA - PULAAR COMBRETACEAE *Combretum paniculatum ssp. paniculatum*
yaree lesdi FULA - FULFULDE CELASTRACEAE *Maytenus senegalensis*
y'ar-garii HAUSA CUCURBITACEAE *Cucurbita pepo*
yaru YORUBA ANNONACEAE *Enantia polycarpa*
yarudi FULA - FULFULDE COCHLOSPERMACEAE *Cochlospermum planchonii*
yáryaádii HAUSA CONVOLVULACEAE *Ipomoea muricata*
yáryaàdii HAUSA CONVOLVULACEAE *Ipomoea batatas*
yaryaɗi, yimɓururu, yinɓururu HAUSA CONVOLVULACEAE *Ipomoea eriocarpa*
yaryaɗin kura HAUSA ASCLEPIADACEAE *Gymnema sylvestre*
yassuabaka BAULE COMBRETACEAE *Combretum molle*
yatandza AKYE COMBRETACEAE *Combretum paniculatum ssp. paniculatum*
yate MANDING - MANINKA APOCYNACEAE *Holarrhena floribunda*
yate, yété MANDING - MANINKA APOCYNACEAE *Holarrhena floribunda*
yate MANDING - MANINKA APOCYNACEAE *Holarrhena floribunda*
yatu KRU - GREBO BURSERACEAE *Canarium schweinfurthii*
yaumba LOKO CUCURBITACEAE *Luffa cylindrica*
yaw serewin AKAN - TWI DIOSCOREACEAE *Dioscorea dumetorum* GA *D. dumetorum*
yawa (*pl.* wawa) KISSI CYPERACEAE *Rhynchospora corymbosa*
yawai LOKO APOCYNACEAE *Strophanthus hispidus, S. sarmentosus*
yawumbo (*def.*-bui) MENDE BOMBACACEAE *Bombax buonopozense*
yay, yaye SERER COMBRETACEAE *Combretum glutinosum* SERER - NON *C. glutinosum*
yaya MAHO COSTACEAE *Costus afer*
ye MANDING - MANDINKA CUCURBITACEAE *Cucurbita pepo* MANINKA *C. pepo*
yẽ, yeng MANO DIOSCOREACEAE *Dioscorea misc. & spp. unknown*
yɛ ADANGME DIOSCOREACEAE *Dioscorea rotundata*
y'ebe LIMBA DILLENIACEAE *Tetracera potatoria*
yee MANDING - MANINKA CUCURBITACEAE *Cucurbita pepo*
yɛ̀ɛ́ BEROM CUCURBITACEAE *Cucurbita pepo*
yéfuké hinzri SENUFO - DYIMINI ASCLEPIADACEAE *Leptadenia hastata*
ye-gbɛ MANDING - MANINKA CUCURBITACEAE *Cucurbita pepo*
yegbeetsoe GBE - VHE AMARANTHACEAE *Alternanthera pungens*
yəgərе WALA CUCURBITACEAE *Cucurbita moschata*
yɛke-kəl-lɛ BULOM CONVOLVULACEAE *Ipomoea mauritiana*
yekpandeochuãboã KISSI COMBRETACEAE *Combretum platypterum*